SOME FUNDAMENTAL PHYSICAL CONSTANTS

Constant	*Symbol*	*Value*
speed of light in a vacuum	c	2.99×10^8 m/s
gravitational constant	G	6.67×10^{-11} m^3/kg·s^2
Avogadro's number	N_A	6.02×10^{23} mol^{-1}
universal gas constant	R	8.31 J/mol·K
Boltzmann's constant	k	1.38×10^{-23} J/K
elementary charge	e	1.60×10^{-19} C
permittivity of free space	ϵ_0	8.85×10^{-12} C^2/N·m^2
	$1/4\pi\epsilon_0$	8.99×10^9 kg·m^3·s^{-2}·C^{-2}
permeability of free space	μ_0	$4\pi \times 10^{-7}$ T·m/A
electron mass	m_e	9.11×10^{-31} kg
proton mass	m_p	1.67×10^{-27} kg
neutron mass	m_n	1.67×10^{-27} kg
Planck's constant	h	6.63×10^{-34} J·s
$h/2\pi$	$\hbar$	1.05×10^{-34} J·s
		$= 6.58 \times 10^{-22}$ MeV·s
	$\hbar c$	197 Mev·fm
electron charge-to-mass ratio	$-e/m_e$	-1.76×10^{11} C/kg
proton-electron mass ratio	m_p/m_e	1840
molar volume of ideal gas at STP		22400 cm^3/mol
Bohr magneton	μ_B	9.27×10^{-24} J/T
magnetic flux quantum	$\Phi_0 = h/2e$	2.07×10^{-15} Wb
Bohr radius	a_0	0.529×10^{-10} m
Rydberg constant	R_∞	1.10×10^7 m^{-1}

SOME ASTRONOMICAL CONSTANTS

Constant	*Symbol*	*Value*
standard gravity at Earth's surface	g	9.80665 m/s^2
equatorial radius of Earth	R_e	6.374×10^6 m
mass of Earth	M_e	5.976×10^{24} kg
mass of Moon		7.350×10^{22} kg $= 0.0123\ M_e$
mean radius of Moon's orbit around Earth		3.844×10^8 m
mass of Sun	$M_\odot$	1.989×10^{30} kg
mean radius of Earth's orbit around Sun	AU	1.496×10^{11} m
period of Earth's orbit around Sun	yr	3.156×10^7 s
diameter of our galaxy		7.5×10^{20} m
mass of our galaxy		2.7×10^{41} kg $= (1.4 \times 10^{11})\ M_\odot$
Hubble parameter	H	2.5×10^{-18} s^{-1}

SOME SI BASE UNITS

Physical Quantity	Name of Unit	Symbol
length	meter	m
mass	kilogram	kg
time	second	s
electric current	ampere	A
thermodynamic temperature	kelvin	K
amount of substance	mole	mol

SOME SI DERIVED UNITS

Physical Quantity	Name of Unit	Symbol	SI Unit
frequency	hertz	Hz	s^{-1}
energy	joule	J	$kg{\cdot}m^2/s^2$
force	newton	N	$kg{\cdot}m/s^2$
pressure	pascal	Pa	$kg/m{\cdot}s^2$
power	watt	W	$kg{\cdot}m^2/s^3$
electric charge	coulomb	C	$A{\cdot}s$
electric potential	volt	V	$kg{\cdot}m^2/A{\cdot}s^3$
electric resistance	ohm	Ω	$kg{\cdot}m^2/A^2{\cdot}s^3$
capacitance	farad	F	$A^2{\cdot}s^4/kg{\cdot}m^2$
inductance	henry	H	$kg{\cdot}m^2/A^2{\cdot}s^2$
magnetic flux	weber	Wb	$kg{\cdot}m^2/A{\cdot}s^2$
magnetic flux density	tesla	T	$kg/A{\cdot}s^2$

SI UNITS OF SOME OTHER PHYSICAL QUANTITIES

Physical Quantity	SI Unit
speed	m/s
acceleration	m/s^2
angular speed	rad/s
angular acceleration	rad/s^2
torque	$kg{\cdot}m^2/s^2$, or $N{\cdot}m$
heat flow	J, or $kg{\cdot}m^2/s^2$, or $N{\cdot}m$
entropy	J/K, or $kg{\cdot}m^2/K{\cdot}s^2$, or $N{\cdot}m/K$
thermal conductivity	$W/m{\cdot}K$

SOME CONVERSIONS OF NON-SI UNITS TO SI UNITS

Energy:

1 electron-volt (eV) = 1.6022×10^{-19} J
1 erg = 10^{-7} J
1 British thermal unit (BTU) = 1055 J
1 calorie (cal) = 4.186 J
1 kilowatt-hour (kWh) = 3.6×10^6 J

Mass:

1 gram (g) = 10^{-3} kg
1 atomic mass unit (u) = 931.5 MeV/c^2
= 1.661×10^{-27} kg
1 MeV/c^2 = 1.783×10^{-30} kg

Force:

1 dyne = 10^{-5} N
1 pound (lb or #) = 4.448 N

Length:

1 centimeter (cm) = 10^{-2} m
1 kilometer (km) = 10^3 m
1 fermi = 10^{-15} m
1 Angstrom (Å) = 10^{-10} m
1 inch (in or ″) = 0.0254 m
1 foot (ft) = 0.3048 m
1 mile (mi) = 1609.3 m
1 astronomical unit (AU) = 1.496×10^{11} m
1 light-year (ly) = 9.46×10^{15} m
1 parsec (ps) = 3.09×10^{16} m

Angle:

1 degree (°) = 1.745×10^{-2} rad
1 min (′) = 2.909×10^{-4} rad
1 second (″) = 4.848×10^{-6} rad

Volume:

1 liter (L) = 10^{-3} m^3

Power:

1 kilowatt (kW) = 10^3 W
1 horsepower (hp) = 745.7 W

Pressure:

1 bar = 10^5 Pa
1 atmosphere (atm) = 1.013×10^5 Pa
1 pound per square inch (lb/in^2) = 6.895×10^3 Pa

Time:

1 year (yr) = 3.156×10^7 s
1 day (d) = 8.640×10^4 s
1 hour (h) = 3600 s
1 minute (min) = 60 s

Speed:

1 mile per hour (mi/h) = 0.447 m/s

Magnetic field:

1 gauss = 10^{-4} T

PHYSICS

FOR SCIENTISTS AND ENGINEERS

SECOND EDITION

PHYSICS

FOR SCIENTISTS AND ENGINEERS

SECOND EDITION

EXTENDED

PAUL M. FISHBANE
University of Virginia

STEPHEN GASIOROWICZ
University of Minnesota

STEPHEN T. THORNTON
University of Virginia

Prentice Hall International, Inc.

© 1996, 1993 by Prentice-Hall, Inc.
Simon & Schuster/A Viacom Company
Upper Saddle River, New Jersey 07458

Printed in the United States of America
10 9 8 7 6 5 4 3 2 1

ISBN 0-13-268954-5

Prentice-Hall International (UK) Limited, London
Prentice-Hall of Australia Pty. Limited, Sydney
Prentice-Hall Canada Inc., Toronto
Prentice-Hall Hispanoamericana, S. A., Mexico
Prentice-Hall of Japan, Inc., Tokyo
Simon & Schuster Asia Pte. Ltd., Singapore
Editora Prentice-Hall do Brasil, Ltda., Rio de Janeiro
Prentice-Hall, Inc., Upper Saddle River, New Jersey

Dedication

To our students, the most important element in the making of this book

Brief Contents

Contents

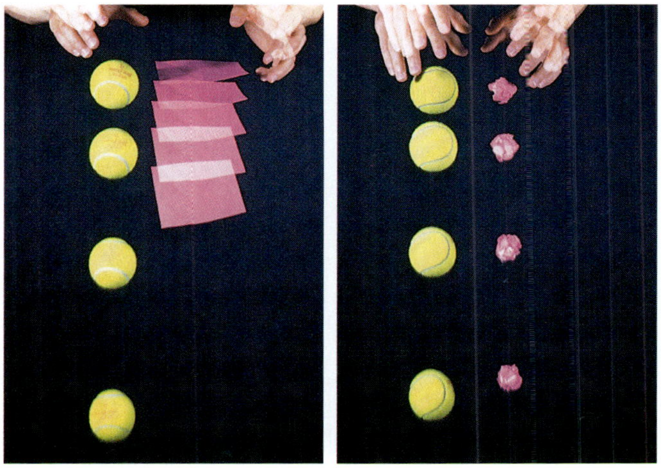

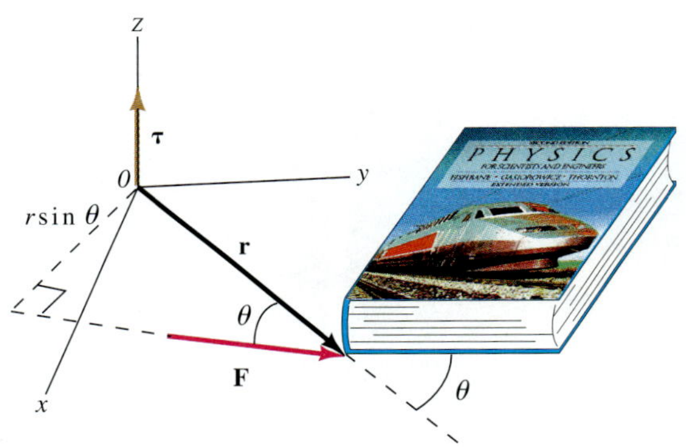

z
τ
O
y
r sin θ
r
θ
F
x
θ
PHYSICS

16 PROPERTIES OF FLUIDS 433

17 TEMPERATURE AND IDEAL GASES 463

18 HEAT FLOW AND THE FIRST LAW OF THERMODYNAMICS 483

19 MOLECULES AND GASES 511

20 THE SECOND LAW OF THERMODYNAMICS 535

33 INDUCTANCE AND CIRCUIT OSCILLATIONS 885

34 ALTERNATING CURRENTS 909

35 MAXWELL'S EQUATIONS AND ELECTROMAGNETIC WAVES 935

36 LIGHT 965

37 MIRRORS AND LENSES AND THEIR USES 989

38 INTERFERENCE 1025

39 DIFFRACTION 1049

40 SPECIAL RELATIVITY 1077

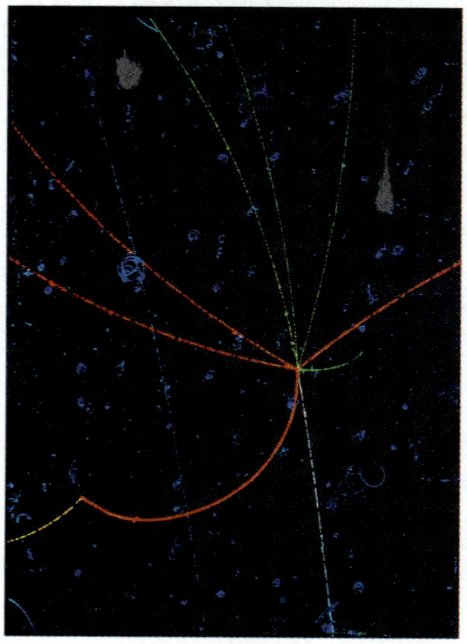

41 QUANTUM PHYSICS 1117

42 QUANTIZATION OF ANGULAR MOMENTUM AND OF ENERGY VALUES 1141

43 QUANTUM EFFECTS IN LARGE SYSTEMS OF FERMIONS AND BOSONS 1167

44 QUANTUM ENGINEERING 1189

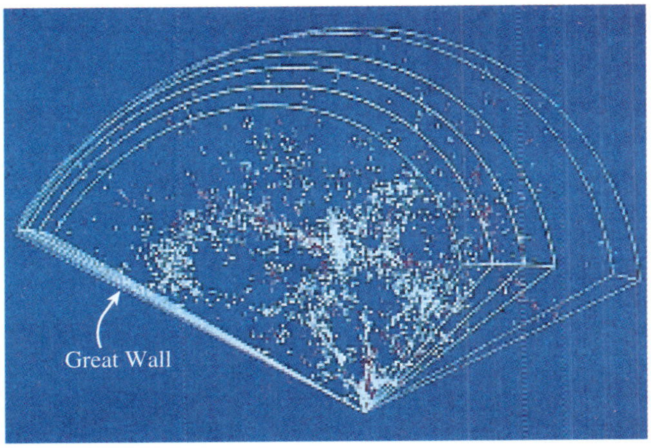

Great Wall

Preface

Our goals for the second edition of this book are primarily to increase its accessibility, relevance, and friendliness for the student, who needs a strong background in physics now more than ever. At the same time, we feel that it is important to maintain the standards that, if anything, have been raised higher in an increasingly technological world. Students must have a conceptual understanding as well as an understanding at the level of problem solving; we have worked hard to maintain and improve the balance between these two crucial aspects of learning physics.

This text is designed for a calculus-based physics course; such a course gives students a solid foundation in understanding the meaning of the fundamental physical laws and teaches them how these laws can be applied to solve problems. The students taking this course are typically in their first or second year and may be prospective engineers, scientists, physicians, or indeed any students who will use technology in their careers. They either have taken or are taking a course in calculus. We believe that it makes little difference which of the above profiles describe the student; it is the nature of physics that the few principles that unify the subject hold whatever the discipline in which they are applied.

Here, we take a look at some of the special features that make this text an effective teaching tool. Whatever these features are, we continue to hope that the most important "feature" of the text is the clarity and correctness with which the principles of physics, their relations, and their applications are set forth. Such an exposition must balance rigor with a more intuitive understanding of the subject, and it must balance explanation and a brevity that too often teaches physics as a set of rules to be memorized. We have tried to maintain these balances throughout.

WHAT'S NEW IN THIS EDITION?

- We have rewritten throughout for focus, clarity, and simplicity in language without loss of depth. In addition, we have made cuts, both large and small, for length control. Sometimes this means more compact writing; sometimes it means removing entire paragraphs of material deemed so far from the essential that the student will not miss it.
- We have rearranged some material that was not, on our own reflection or at the suggestion of users, properly placed or ordered. For example, conservative forces are now introduced in Chapter 6 rather than in Chapter 7. In addition, we have provided new and simpler derivations for certain subjects; for example, the derivation of the wave equation is a new one.
- Conceptual matters play a more prominent role. From the beginning, we have tried to take the tack that physics is more a matter of understanding certain ideas and methods than of being able to plug into a series of formulas. As part of an increased effort to help the student understand rather than simply parrot the solution of stereotypical problems, the questions at the end of each chapter have been strengthened.

- We have added more engineering-oriented application material. This is most noticeably realized in the new application boxes that now appear in each chapter and in new realistic problems. The application boxes are listed in the Table of Contents.
- An increased number of problem-solving boxes are presented throughout the text.
- Margin notes and short set-off remarks and signals have now been divided into four different categories: (i) ordinary side notes, which contain the important results, those that will "be on the test"; (ii) problem-solving notes, containing references to useful techniques for problems; (iii) application notes, signaling device or technology relevance; and (iv) cross references, indicated by the icon , which remind the student that he or she has already seen the referenced material and just *where* he or she has seen it.
- In the context of the problem-solving techniques, a set of new figures take the form of "student sketches," which are drawn in a style that the student can realistically be expected to emulate. To carry through on this idea, all the figures in the end-of-chapter problems are also drawn in this style. In the sense that a sketch represents a first step in problem solving, the sketches in the end-of-chapter problems sometimes provide a crucial hint to the student; more generally, they help the student develop the habit of using sketches to help in problem solving.
- We have eliminated derivation boxes. A few of the derivations themselves have been eliminated entirely, and all the original derivations from these boxes have been simplified; where we feel the derivation can be skipped without great loss, the derivations simply form an optional portion of the running text.
- We have simplified our notation for multi-dimensional integrals with the consistent use of a single integral sign. To avoid possible misunderstanding, we have consistently added labeling on the integral sign that guards against the possibility that the student might confuse a multiple integral with a single integral.
- We have made our examples more useful to students by employing student sketches and changing the solutions to more clearly show students *how* and *why* we reach an answer. The text contains an average of more than 8 examples per chapter, at least 10% of the examples are new, and we have changed up to 30% of the remaining examples.
- Visualization is an important part of learning physics. In addition to redrafting all end-of-chapter illustrations as student sketches, we have added approximately 75 new in-text illustrations. Further, almost every figure has been redrafted for clarity, including the addition of grid lines to many graphs and, often, the presentation of a three-dimensional view instead of a two-dimensional view.
- We have made a major revision to the end-of-chapter problem sets. First, we have added more than ten new level I problems in each chapter to increase the possibility of student drill. Working toward our goal of providing a greater applications orientation, we have also added close to 20% more problems that are entirely new and changed almost 30% of existing problems. Finally, problems that have not proven useful have been eliminated.

Emphasis on Problem Solving. A student's ability to solve qualitative and quantitative problems is the best measure of whether the physics is truly understood. With this in mind, we have included certain features that will strengthen the student's problem-solving ability. Central to these is the end-of-chapter material. This material includes about 700 qualitative, conceptual **questions** under the heading "Understanding Key Concepts." These test the student's understanding of the ideas that underlie a situation. Such questions may not have simple or direct answers. They challenge students to consider concepts and to recognize the meaning of the principles taught in the text.

In addition to the questions, there are some 3000 **problems.** Some problems are keyed to text sections, and are presented in increasing order of difficulty within each section. Others are **general problems,** which bring together material from the entire chapter as well as from previous chapters. The general problems, with a lesser degree of guidance as to the subject being studied, resemble more closely the situations that are met in real-life science and engineering, and they also help to develop the student's appreciation of the links that exist throughout physics. All problems are labeled I, II, or III. Level I problems are "easy," not going much beyond plug-ins. These problems develop student recognition of particular physics concepts and build confidence. Level II problems are typically multistep and require an increased understanding of the interconnectedness of physics; the general problems carry this requirement a step further. Level III problems are especially challenging, in some cases demanding significant synthesis of concepts in the text. We know that you cannot assign every problem, even over many years. But the gradations in problem range and difficulty allow you to tailor the problems you assign to the capabilities of your class and to the subjects that interest you the most.

The text contains several significant aids to the development of the student's problem-solving capacities. First, there are numerous **examples,** an average of 8 or 9 per chapter (some chapters contain more than a dozen examples). Examples represent typical problems, generally at level II. Their solutions illustrate sound approaches to problem solving. We "think out loud" in the example solutions to reveal *how* and *why* we reach our answers rather than providing just the answers themselves. Many of the examples employ a technique that we refer to as **"student sketches,"** which reflect our advice that the student sketch the situation. These drawings are meant to be something that a student could reasonably emulate, not the sort of elaborate drawing that the student—not to mention the instructor—could not reproduce. Second, we have included **problem-solving techniques,** separately boxed throughout the text. This material appears in the key chapters that introduce new ways of looking at problems in physics. Third, we include brief **problem-solving margin notes** adjacent to text discussion of problem-solving methods. Finally, we have included answers to odd-numbered problems at the back of the book, and the new student solution manual contains solutions to a significant portion of all problems.

Mathematics. Three remarks are worth making: First, we introduce the mathematics that students need to know the first time they need to know it, in the context of the physics being presented. We try to make that material self-contained, so that the student can understand the material without having to go elsewhere for mathematical help. In this way, the mathematics appears in progressive degrees of difficulty, and we believe that this approach fosters better understanding and less reliance on formula memorization. Second, we teach the correct usage of significant figures and vary the number of significant figures in examples and in problems—in keeping with the way problems arise in the real world. In this way, students must maintain an awareness of significant figures and are not lulled into thinking that all problems involve the same number of significant figures. Third, the ability to make quantitative estimates is one of the most important skills that a scientist or engineer can have. We have made the development of that ability an important part of our approach, both in the text and in problems.

Applications. Students are interested in knowing how the principles that govern the physical world are associated with today's technology, or with the workings of devices that they see every day around them. With this in mind, we have included in every chapter at least one separate box that describes an application relevant to that chapter's subject. In addition, we have added margin notes that signal applications as they appear in the text.

Modern Physics. Classical topics have lost none of their importance and must form the basis of any first course. However, what is traditionally called "modern" physics—the topics centered on relativity and quantum physics—began about a century ago. It hardly seems possible to ignore these topics in view of their importance for understanding today's world. Many of the ideas of modern physics are not mathematically difficult. However, they can be nonintuitive, and we think that it is important that students begin to develop intuition about this material as early as possible. Without sacrificing the essential aspects of classical physics, we have introduced modern notions from the beginning. Although much of this material appears in optional sections (marked with an asterisk), in many cases it is intertwined with the classical material. We conclude the text traditionally, with chapters on modern physics. We think that the preparation we have laid down for this material will make it more easily assimilated.

Other Study Aids. Effective keys and reviewing aids appear in several forms. A **summary** at the end of each chapter reminds students of the most important results of the chapter. For chapters that introduce numerous or especially difficult concepts, we offer **interim summaries** to give students extra assistance. In addition, two types of **margin notes** that appear adjacent to the corresponding text form study aids. These include margin notes that key important concepts or results, as well as cross references that recall where previously developed concepts are first used when they appear in a later chapter. **Important equations** are boxed, although we have resisted the temptation to box too many of them, a temptation which encourages a plug-in mentality.

Figures and Photographs. A number of features distinguish the figures and photographs in this text from those in other texts. Labeled photos clearly illustrate the physical principles present at each point in an object's movement. Sequence photos help "bring the physics to life" for the student. Student sketches are introduced in this edition; these sketches are used in the end-of-chapter problems to provide hints and are presented within the text to help students learn to draw sketches during problem solving. For clarity, a significant number of figures have been changed from a two-dimensional to a three-dimensional rendition. Finally, color is used in this text as a pedagogical aid, primarily where the consistent use of color to represent physical quantities and analytic elements in illustrations helps to avoid confusion. For example, color provides a visual reminder to students not to confuse acceleration vectors with force vectors.

SUPPLEMENTS

Quality supplements make a difference. Each of the supplements that accompany this text has been carefully developed to create a unique set of high-quality, tested tools that provide the means by which instructors and students can enliven and reinforce the teaching and learning of physics.

For the Instructor

Instructor's Solution Manual. Written by Professor Irv Miller of Drexel University, this manual contains complete solutions to all problems in the text. Considerable effort has been put into checking the accuracy of the solutions.

Transparencies. Approximately 200 four-color acetates are available in the transparency set.

Test-Item File and PH Test. The test questions that accompany this text have been developed by Professor Charles Scherr of the University of Texas and have been used at Austin by thousands of engineering and science students over the course of a decade. Approximately 500 questions have been constructed in algorithmic form with a broad range of input data, which allows each question to generate a significant and reliable number of versions. PH Test, the testing program, provides the means to create a variety of student testing devices, homework and drill sheets, and answer sheets in minutes. PH Test is available in Macintosh and DOS versions.

"Physics You Can See" Demonstration Experiments Video. This video is a collection of brief demonstrations for classroom use.

For the Student

The Princeton Learning Guide. This study guide for our text has been adapted from guides developed at Princeton University during the past 20 years. The guide features complete problem assignments with aids in the form of leading questions and keyed answers. Help is given in the form of increasingly detailed hints and references to the text. The emphasis in this guide is on conceptual understanding, especially understanding of the links in the physical world.

Student Solution Manual. This manual provides complete solutions to many of the end-of-chapter problems. It also provides a discussion of answers to some of the end-of-chapter questions.

The Portable TA: A Problem-Solving Guide (Volumes I and II). A collection of problems with carefully detailed solutions to provide students with extra practice in problem-solving techniques. This collection is the result of two years of class testing.

The New York Times/Themes of the Times. A semiannual newspaper containing articles from *The New York Times*, this supplement demonstrates the connection between what goes on in the classroom and in the world around us. It is available in quantity, *free* for each student.

TECHNOLOGY-BASED SUPPLEMENTS

LOGAL Physics Explorer. Available only from Prentice Hall, this interactive "run time" version of LOGAL's award-winning simulation software contains over 100 simulations based on examples and problems from the text. Physics Explorer allows students to compare predictions with results by altering parameters of a problem. The simulation models cover mechanics, waves, diffraction and interference, and optics. Extension Exercises expand the boundaries of the initial problem, and Concept Checks offer immediate on-screen feedback. Available in Windows and Macintosh versions.

The Interactive Physics Player™. Available only from Prentice Hall, this "player" version of Knowledge Revolution's highly acclaimed Interactive Physics II computer program brings the text to life with real-time simulations and animations of over 100 of the examples and problems. The program allows the user to change values and view the outcomes, creating and observing "what if" scenarios. Tools showing numerical results, graphs, and vectors make it easy to analyze the simulation. The Interactive Physics Player provides students with a way to visualize the concepts they are trying to learn. Instructors can use it as a demonstration tool in class or in lab. The program is available in both Macintosh and Windows versions.

Interactive Physics II Workbook. Written by Cindy Schwarz of Vassar College, this highly interactive workbook/software package contains 40 simulation projects of varying degrees of difficulty. Each contains a physics review, simulation details, hints, explanation of results, math help, and a self-test. Available in Windows and Macintosh versions.

Mathematica Projects for Scientists and Engineers. This supplement, written by Rodney Varley of Hunter College, includes approximately 40 detailed modules that contain an explanation of the Mathematica model, reviews of the physical and mathematical concepts, "Learn by Doing" exercises that ask you to build or modify Mathematica simulations, and additional exercises with suggested hints.

Matlab Projects for Scientists and Engineers. This supplement, written by Alejandro Garcia of San Jose State University, contains approximately 20 detailed modules based on the numerical processing software Matlab. Each module contains reviews of the primary physical and mathematical concepts, an explanation of the Matlab model, "Learn by Doing" exercises that allow you to build or modify Matlab simulations, and additional exercises with suggested hints.

TEXT VERSIONS AND TEACHING ALTERNATIVES

This text provides a complete introduction to physics. It is available in a 41-chapter version, which ends with a chapter on special relativity and a chapter on quantum physics. For those who want more detailed coverage, a 46-chapter extended version containing additional coverage of quantum physics, nuclear physics, particles, and cosmology is also available. The text is published in split volumes as well. Volume I includes Chapters 1 to 21, covering mechanics, waves, and thermodynamics. Volume II includes Chapters 22 to 46 of the extended version, covering electricity and magnetism, light and optics, and modern physics.

The question of length is a knotty one; there is widespread agreement that a text such as this one is too long, but very little agreement on just what should be eliminated. For that reason, we have written this book in such a way that any instructor can make his or her own decisions on cuts. We nevertheless give two sorts of guidance:

- ***Optional Material.*** An asterisk (*) indicates material that only the most thorough courses will typically have time to cover. For the most part, such material is either self-contained modern physics or an in-depth discussion of material already introduced.
- ***A Lean Alternative.*** At many institutions, the constraints of time coupled with the degree of preparation of the students necessitate that the introductory physics course be compressed or that more time be spent on fewer topics. We propose here one possible way—out of many—in which this text could be used in a highly streamlined course. The criteria we use for suggesting what to retain and what to delete are associated with the degree to which the material is self-contained or will be studied later in a more applications-oriented course, not with how "hard" that material is. We would eliminate or de-emphasize the following topics or sets of topics: relative motion; the more complicated aspects of rotational motion; statics; the more advanced aspects of harmonic motion; the Doppler effect; the more advanced aspects of wave superposition, including diffraction; fluids and solids; much of statistical physics, including probability distributions and entropy; the effects of materials in electromagnetism; applied aspects of time-dependent circuits, including alternating currents; aspects of electromagnetic radiation; mirrors and lenses; the Lorentz transformations; and the more detailed applications of quantum mechanics.

  Thus our compressed course would consist of the following:

Chapter	Sections	Chapter	Sections
1	1,2,3,5,6	20	1,2,3,4
2	1,2,3,4,5	22	1,2,3,4
3	1,2,3,4,5	23	1,2,3,4
4	1,2,3,4,5	24	1,2,3,4
5	1,2,3,4	25	1,2,3,4,6,7
6	1,2,3,4,5	26	1,2,3
7	1,2	27	1,2,3,7
8	1,3,4,6	28	1,2,3
9	1,2,3,5,6	29	1,2,3,4
10	1,2,3,4	30	1,2,3,5
12	2,3,4	31	2,3,4,5,6
13	1,3,4,5	33	1,2,3
14	1,2,3,5,6	35	1,2,3,4
15	1,3,4	36	1,2,3
17	1,2,3	38	1,2,3
18	1,2,3,4,5,6	40	1,2,3,4,7
19	1,2,3	41	1,2,3,4

ACKNOWLEDGMENTS

It is clear that a project of this magnitude cannot be accomplished by three authors alone. We are grateful to the many people who have contributed to making this a better text.

A special thanks goes to Professor Irv Miller of Drexel University. In addition to writing the Instructor's Solution Manual, Irv was an invaluable resource in creating, checking, and refining the problem sets and answers. Working with the input of Professor T. S. Venkataraman, also of Drexel, Irv coordinated the independent critique and feedback of the problems and answers from the following: Edward Adelson (Ohio State University); Howard Bale (University of North Dakota); John Barach (Vanderbilt University); William R. Cochran (Youngstown State University); George Dixon (Oklahoma State University); Brent Foy (Wright State University); Rex Gandy (Auburn University); Mark Heald (Swarthmore College); Paul Heckert (Western Carolina University): Larry Hmurcik (University of Bridgeport); Rex Joyner (Indiana Institute of Technology); Brij M. Khorana (Rose-Hulman Institute of Technology); Douglas A. Kurtze (North Dakota State University); David Markowitz (University of Connecticut); George Miner (University of Dayton); R. J. Peterson (University of Colorado, Boulder); and James H. Smith (University of Illinois at Urbana-Champaign). Equal thanks go to Laszlo Takacs of the University of Maryland at Baltimore County. Laszlo supplied many new problems and modified many first-edition problems for this edition. We are also exceedingly grateful to Tony Buffa, who was tremendously helpful in providing suggestions for updating and redrafting the art program. Special thanks to John Broadhurst for invaluable help with the application boxes.

We want to thank John Malone of the teaching laboratory at the University of Virginia for his help with demonstrations and photographs. John is the kind of person that no lecturer in physics should be without.

We would also like to acknowledge and thank the 70 physics instructors who provided valuable feedback. We took all of their comments very seriously and only regret that we could not satisfy them all.

V. K. Agarwal
Moorhead State University

Thomas Armstrong
University of Kansas

Philip S. Baringer
University of Kansas

John E. Bartelt
Vanderbilt University

William Bassichis
Texas A & M University

Benjamin F. Bayman
University of Minnesota

Michael Browne
University of Idaho

Timothy Burns
Leeward Community College

Alice Chance
Western Connecticut State University

Robert Clark
Texas A & M University

Lucien Cremaldi
University of Mississippi

W. Lawrence Croft
Mississippi State University

Chris L. Davis
University of Louisville

Robin Davis
University of Kansas

Jack Denson
Mississippi State University

John DiNardo
Drexel University

Robert J. Endorf
University of Cincinnati

Arnold Feldman
University of Hawaii at Manoa

Rex Gandy
Auburn University

Alexander B. Gardner
Howard University

Robert E. Gibbs
Eastern Washington University

Wallace L. Glab
Texas Tech University

James R. Goff
Pima Community College

Alan I. Goldman
Iowa State University

Phillip Gutierrez
University of Oklahoma

Frank Hagelberg
SUNY-Albany

Robert F. Harder
George Fox College

Warren W. Hein
South Dakota State University

Jerome Hosken
City College of San Francisco

Joey Houston
Michigan State University

Francis L. Howell
University of North Dakota

Evan W. Jones
Sierra College

Alain E. Kaloyeros
SUNY-Albany

Robert J. Kearney
University of Idaho

Arthur Z. Kovacs
Rochester Institute of Technology

Claude Laird
University of Kansas

Vance Gordon Lind
Utah State University

A. Eugene Livingston
University of Notre Dame

B. A. Logan
University of Ottawa

David Markowitz
University of Connecticut

Erwin Marquit
University of Minnesota

Marvin L. Marshak
University of Minnesota

Charles R. McKenzie
Salisbury State University

Norman McNeal
Sauk Valley Community College

Forrest Meiere
Indiana University-Purdue University

Irvin A. Miller
Drexel University

Thomas Muller
University of California-Los Angeles

Richard Murphy
University of Missouri-Kansas City

Peter Nemethy
New York University

David Ober
Ball State University

Gottlieb S. Oehrlein
SUNY-Albany

Micheal J. O'Shea
Kansas State University

Dan Overcash
Auburn University

Patrick Papin
San Diego State University

Robert A. Pelcovits
Brown University

Don D. Reeder
University of Wisconsin-Madison

John Lewis Robinson
University of British Columbia

Ernest Rost
University of Colorado

Mendel Sachs
SUNY-Buffalo

Francesca Sammarruca
University of Idaho

William G. Sturrus
Youngstown State University

Richard E. Swanson
Sandhills Community College

Laszlo Takacs
University of Maryland, Baltimore County

Leo H. Takahashi
Pennsylvania State University

Robert Tribble
Texas A & M University

Rod Varley
Hunter College

Gianfranco Vidali
Syracuse University

John Wahr
University of Colorado

Fa-chung Wang
Prairie View A & M University

Gail S. Welsh
Salisbury State University

We are also grateful to the many reviewers of our first edition:

Maris Abolins *(Michigan State University)*, Ricardo Alarcon *(Arizona State University)*, Bradley Antanaitis *(Lafayette College)*, Carl Bender *(Washington University)*, Hans-

Uno Bengtsson *(University of California, Los Angeles),* Robert Bowden *(Virginia Polytechnic Institute and State University),* Bennet Brabson *(Indiana University),* Edward Chang *(University of Massachusetts),* Albert Claus *(Loyola University),* Robert Coakley *(University of Southern Maine),* James Dicello *(Clarkson University),* N. John DiNardo *(Drexel University),* P. E. Eastman *(University of Waterloo),* Gabor Forgacs *(Clarkson University),* A. L. Ford *(Texas A & M University),* William Fickinger *(Case Western Reserve University),* Rex Gandy *(Auburn University),* Simon George *(California State University, Long Beach),* James Gerhart *(University of Washington),* Bruce Harmon *(Iowa State University),* Joseph Hemsky *(Wright State University),* Alvin Jenkins *(North Carolina State University),* Karen Johnston *(North Carolina State University),* Garth Jones *(University of British Columbia),* Leonard Kahn *(University of Rhode Island),* Charles Kaufman *(University of Rhode Island),* Thomas Keil *(Worcester Polytechnic University),* Carl Kocher *(Oregon State University),* Karl Ludwig *(Boston University),* Robert Marande *(Pennsylvania State University),* David Markowitz *(University of Connecticut),* Roy Middleton *(University of Pennsylvania),* George Miner *(University of Dayton),* Lorenzo Narducci *(Drexel University),* Jay Orear *(Cornell University),* Patrick Papin *(San Diego State University),* Kwangjai Park *(University of Oregon),* R. J. Peterson *(University of Colorado),* Frank Pinski *(University of Cincinnati),* Lawrence Pinsky *(University of Houston),* Stephen Pinsky *(Ohio State University),* Richard Plano *(Rutgers University),* Hans Plendl *(Florida State University),* Shafigur Rahman *(Allegheny College),* Peter Riley *(University of Texas, Austin),* L. David Roper *(Virginia Polytechnic Institute and State University),* Richard Roth *(Eastern Michigan University),* Carl Rotter *(West Virginia University),* Charles Scherr *(University of Texas),* Eric Sheldon *(University of Lowell),* Charles Shirkey *(Bowling Green State University),* Marlin Simon *(Auburn University),* Robert Simpson *(University of New Hampshire),* James Smith *(University of Illinois),* J. C. Sprott *(University of Wisconsin),* Malcolm Steuer *(University of Georgia),* Thor Stromberg *(New Mexico State University),* Smio Tani *(Marquette University),* William Walker *(University of California, Santa Barbara),* George Williams *(University of Utah).*

Finally, we would like to thank the publishing team at Prentice Hall who have carried this project through. Our editors, Ray Henderson and Tim Bozik, who directed the project, have been a constant source of ideas, encouragement, and material help. Carol Trueheart, our development editor, insisted on reminding us that this book is meant for students first, and that a failure to communicate at the appropriate level is the worst kind of failure. She has been encouraging from beginning to end, and we cannot imagine completing this work without her help. Director of Marketing Gary June has provided constant support and ideas, and we thank him. Last but not least, the production of a book such as this one is an enormous task demanding the most careful attention to detail. We are grateful to Tony Buffa, who was tremendously helpful in updating and redrafting the art program to increase clarity and improve perspective. David George and Karen Malley did an impossibly careful job of proofreading. We want to thank Rolando Corujo, who created many of the complex pieces of art for this edition. We especially want to thank Barbara Mack, Joanne Jimenez, and Jennifer Wenzel for the elaborate juggling job they did to bring this book to press. Finally, we would like to thank, in advance, Alison Reeves for the work she is doing to insure the continued success of this book. To all the individuals listed above, and to the many others at Prentice Hall who have worked to make this book a success, we extend our heartfelt thanks.

The cumulative and accelerating nature of science and technology make it more imperative than ever that our emerging scientists and engineers understand how few and how solid are the pillars of the enterprise. From this view, the distinctions between "science" and "engineering," and between "classical physics" and "modern physics," melt. We want in this book to make evident the pillars of physics as well as the highly interconnected structure that has been erected on those pillars.

About the Authors

Paul M. Fishbane

Paul Fishbane has been teaching undergraduate courses at the University of Virginia, where he is Professor of Physics, for some 25 years. He received his doctoral degree from Princeton University in 1967 and has published some 100 papers in his field, theoretical high energy physics. He has held visiting appointments at the State University of New York at Stony Brook, Los Alamos Scientific Lab, CERN laboratory in Switzerland, Amsterdam's NIKHEF laboratory, France's Institut de Physique Nucleaire, the University of Paris-Sud, and the Ecole Polytechnique. He has been active for many years at the Aspen Center for Physics, where current issues in physics are discussed with an international group of participants. His other interests include biking, music, and the physics of the kitchen. All of the rest of his time is spent trying to keep up with his family, especially his youngest son Nicholas.

Stephen Gasiorowicz

Stephen Gasiorowicz was born in Poland and received his Ph.D. in physics at the University of California, Los Angeles in 1952. After spending 8 years at the Lawrence Radiation Laboratory in Berkeley, California, he joined the faculty of the University of Minnesota, where his field of research is theoretical high energy physics. As a visiting professor, he has traveled to the Niels Bohr Institute, NORDITA in Copenhagen, the Max Planck Institute for Physics and Astrophysics in Munich, DESY in Hamburg, Fermilab in Batavia, and the Universities of Marseille and Tokyo. He has been a frequent visitor and an officer of the Aspen Center for Physics. He has written books on elementary particle physics and quantum physics. A relatively new occupation is that of grandfather, which still leaves some time for reading (history), biking, canoeing, and skiing.

Stephen T. Thornton

Stephen Thornton performed his doctoral research at Oak Ridge National Laboratory while completing his Ph.D. at the University of Tennessee. He joined the faculty at the University of Virginia in 1968, and became the first Director of the Institute of Nuclear and Particle Physics. He has held two Fulbright fellowships and has performed nuclear experiments at accelerators throughout the United States and Europe. His recent interests include teaching physical science to school teachers and performing science outreach to children. He helped establish the Center for Science Education at Virginia. He has published over 100 research papers as well as three textbooks. He has two grown sons and is married to Dr. Kathryn Thornton, NASA astronaut. They have three daughters. He owns and operates a Christmas tree farm, and his interests include snow skiing and scuba diving.

A Guided Tour

We hope more than anything else that we can convey to you the excitement of a vast body of knowledge. Like anything worthwhile, scientific knowledge takes some work to acquire, but the effort will be repaid manyfold. The most important characteristic of physics is that it describes and explains a vast set of phenomena in terms of just a few basic principles. You will have succeeded if you understand these principles well enough to use them to solve problems. The conceptual basis for solving those problems must also be present, however, and if it is, you can go very far with your knowledge.

This book is a tool, which, with the aid of your instructor and your fellow students, will allow you to acquire an understanding of physics. Tools half-used give only half value, and we encourage you to make use of every aspect of this book and of the ancillary materials that are available with it. For now, let's take a brief walk through the book itself.

Chapter Opener
Each chapter opens with an illustration that indicates how thinking about physics can explain an immense range of activities in the world around us.

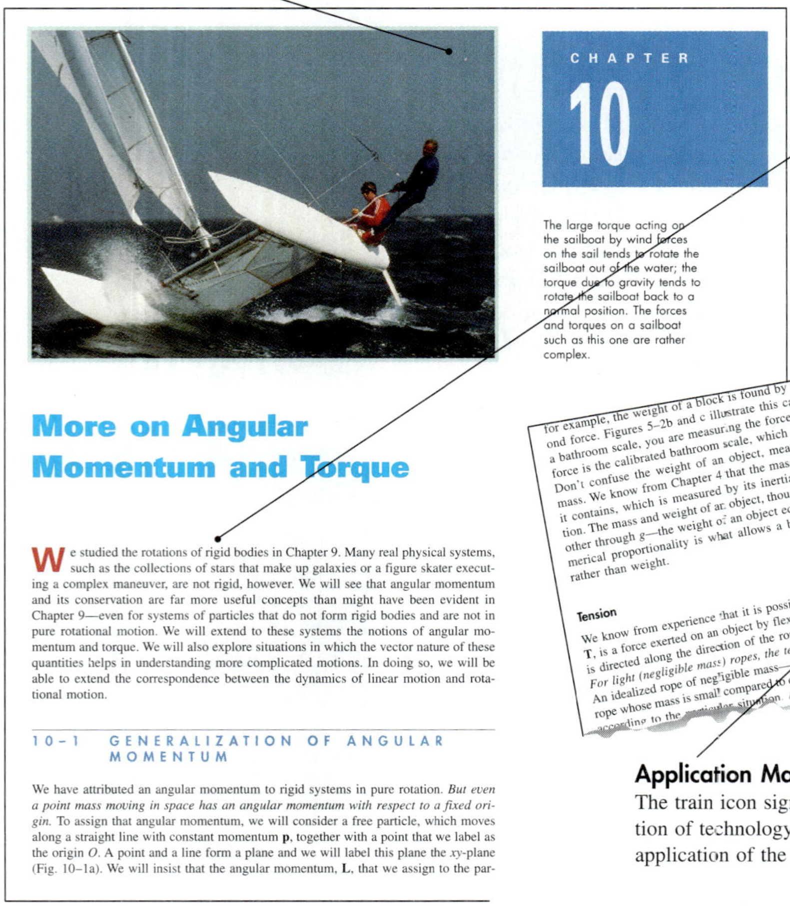

CHAPTER 10

The large torque acting on the sailboat by wind forces on the sail tends to rotate the sailboat out of the water; the torque due to gravity tends to rotate the sailboat back to a normal position. The forces and torques on a sailboat such as this one are rather complex.

More on Angular Momentum and Torque

We studied the rotations of rigid bodies in Chapter 9. Many real physical systems, such as the collections of stars that make up galaxies or a figure skater executing a complex maneuver, are not rigid, however. We will see that angular momentum and its conservation are far more useful concepts than might have been evident in Chapter 9—even for systems of particles that do not form rigid bodies and are not in pure rotational motion. We will extend to these systems the notions of angular momentum and torque. We will also explore situations in which the vector nature of these quantities helps in understanding more complicated motions. In doing so, we will be able to extend the correspondence between the dynamics of linear motion and rotational motion.

10–1 GENERALIZATION OF ANGULAR MOMENTUM

We have attributed an angular momentum to rigid systems in pure rotation. *But even a point mass moving in space has an angular momentum with respect to a fixed origin.* To assign that angular momentum, we will consider a free particle, which moves along a straight line with constant momentum **p**, together with a point that we label as the origin *O*. A point and a line form a plane and we will label this plane the *xy*-plane (Fig. 10–1a). We will insist that the angular momentum, **L**, that we assign to the par-

Introductory Text
Each chapter introduces the topic being presented and its connection to the physical concepts covered in previous chapters and in chapters to come.

Problem Solving Margin Note
This type of note contains references to useful techniques for problems.

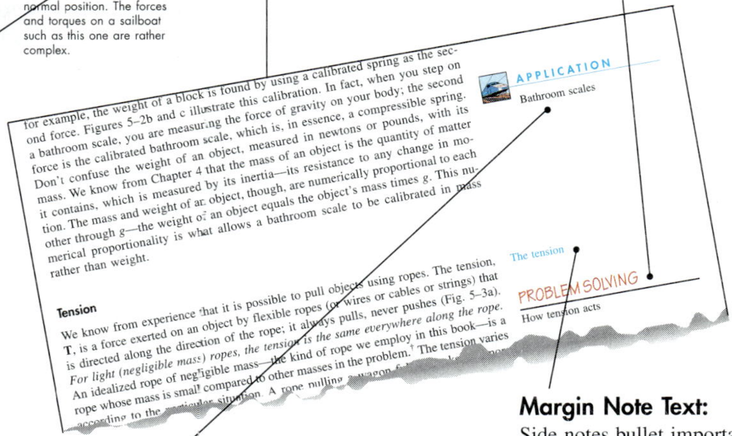

for example, the weight of a block is found by using a calibrated spring as the second force. Figures 5–2b and c illustrate this calibration. In fact, when you step on a bathroom scale, you are measuring the force of gravity on your body; the second force is the calibrated bathroom scale, which is, in essence, a compressible spring. Don't confuse the weight of an object, measured in newtons or pounds, with its mass. We know from Chapter 4 that the mass of an object is the quantity of matter it contains, which is measured by its inertia—its resistance to any change in motion. The mass and weight of an object, though, are numerically proportional to each other through *g*—the weight of an object equals the object's mass times *g*. This numerical proportionality is what allows a bathroom scale to be calibrated in mass rather than weight.

APPLICATION
Bathroom scales

Tension

We know from experience that it is possible to pull objects using ropes. The tension, **T**, is a force exerted on an object by flexible ropes (or wires or cables or strings) that is directed along the direction of the rope; it always pulls, never pushes (Fig. 5–3a). *For light (negligible mass) ropes, the tension is the same everywhere along the rope.* An idealized rope of negligible mass—the kind of rope we employ in this book—is a rope whose mass is small compared to other masses in the problem.[†] The tension varies

The tension
PROBLEM SOLVING
How tension acts

Margin Note Text:
Side notes bullet important results presented in the text.

Application Margin Note
The train icon signals the presentation of technology or a relevant application of the text.

Examples

Worked examples illustrate the use of concepts presented in the text. Most are accompanied, as this one is, by a figure or a student sketch. Solutions clearly show *how* and *why* an answer is reached.

EXAMPLE 10–7 A spool of mass M, rotational inertia I about its axis, and radius R unwinds under the force of gravity (Fig. 10–23). By using energy considerations, find the speed of the spool's center of mass after it has unwound a length h of thread.

Solution: The total kinetic energy of the spool is the sum of the kinetic energy of motion of the center of mass and the kinetic energy of rotation about the center of mass. Thus

$$K = \frac{1}{2}Mv^2 + \frac{1}{2}I\omega^2.$$

The relation between the angular velocity and the velocity of the center of mass is identical to that of a cylinder rolling without sliding, so that

$$v = R\omega.$$

This implies that

$$K = \frac{1}{2}MR^2\omega^2 + \frac{1}{2}I\omega^2 = \frac{1}{2}(MR^2 + I)\omega^2.$$

Note that a simple application of the parallel-axis theorem shows that the quantity in parentheses is just the rotational inertia about the point at which the thread separates from the surface.

Another contribution to the energy is the potential energy of gravity. We can measure the potential energy from the initial height of the spool so that if the spool falls a distance h, the potential energy is $-Mgh$ and the spool has gained a kinetic energy Mgh. With a zero initial angular velocity, the kinetic energy after the spool has fallen a distance h is

$$K = Mgh.$$

Thus

$$\frac{1}{2}(MR^2 + I)\omega^2 = Mgh,$$

so

$$v = R\omega = R\sqrt{\frac{2Mgh}{MR^2 + I}}.$$

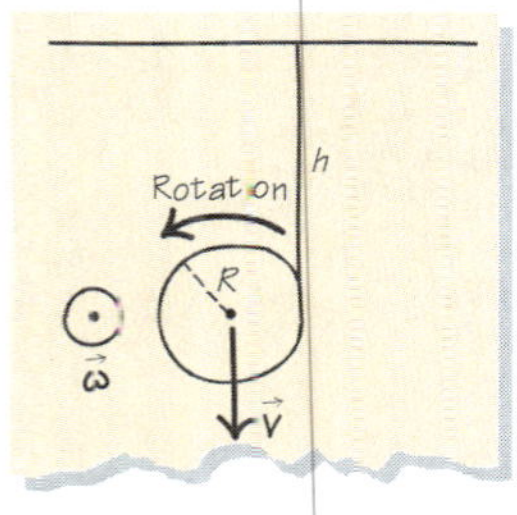

FIGURE 10–23 Example 10–7.

More Parallels Between Rotational and Linear Motion

Despite all the subtleties that appear when the full vector nature of angular momentum is involved, it is important to stress that there are no new laws of physics that describe rotational motion on the classical (the non-quantum-mechanical) level. *Every result on classical physics obtained in these last two chapters is ...*

musement park ride $= c_1t^2 + c_2t + c_3$ 0.5 m, $d_1 = -1.0$ he locations of the time in Figs. 3–3b

∞ The unit vectors **i**, **j**, and **k** are discussed in Chapter 1.

Cross-Reference Marginal Note

Margin cross references are signaled by this icon and indicate where the referenced material has been presented earlier.

Chapter Review Text

At the end of each chapter, students have a chance to review the most important results.

APPLICATION

Electronic devices based on tunneling

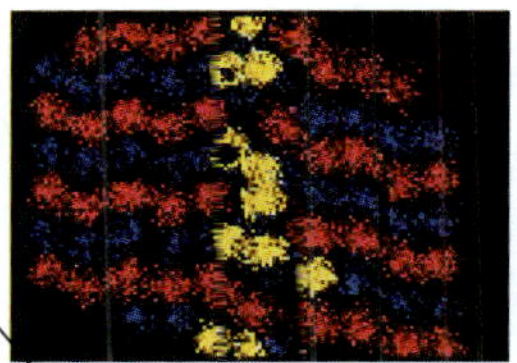

FIGURE 7–15 Individual atoms in a solar cell, shown here in false color, are made visible by using a scanning tunneling microscope. This image is part of a project to improve solar cells.

smaller. When the ^{2}H nuclei are raised to high temperatures (billions of degrees), the nuclei become agitated, move rapidly, have an increased energy E, and thus they have a smaller value of $U_{av} - E$. Effectively, by raising the energy of the nuclei, they are brought closer to the top of the barrier and both $U_{av} - E$ and the thickness of the barrier are reduced. Much research and development has been done since the early 1950s to bring the ^{2}H nuclei to these high temperatures. The technical difficulties are enormous but the reward for success would be great. The fusion reaction could yield a virtually unlimited amount of energy because seawater contains ^{2}H nuclei in abundance

Other Applications

Several other manifestations of tunneling have technological application. The scanning tunneling microscope is a device that detects electrons that tunnel through a potential barrier at the surface of a piece of metal. It is used to study variations of atomic size in the surface of the metal in great detail (Fig. 7–15). Tunneling has also been used in studies of superconductivity. One of the most important applications is in the tunnel diode, a device with wide application in circuits. Quantum tunneling is not just some esoteric effect of interest to a small class of physicists. Many modern electronic devices rely in one way or another on electron tunneling. In later chapters, we will return to these applications.

SUMMARY

Conservative forces can be specified in terms of a single scalar function called the potential energy. In one-dimensional motion, the potential energy $U(x)$ depends on one coordinate only and is defined by

$$\text{for conservative forces: } U(x) \equiv U(x_0) - \int_{x_0}^{x} F\,dx', \tag{7–6}$$

where $U(x_0)$ ($= U_0$) is an arbitrary constant and x_0 is an arbitrary point. In turn, the force is given in terms of U by

$$F(x) = -\frac{dU(x)}{dx}. \tag{7–4}$$

Two important examples of potential energy are gravity and the spring. For gravity,

$$U(y) = mgy + U_0, \tag{7–11}$$

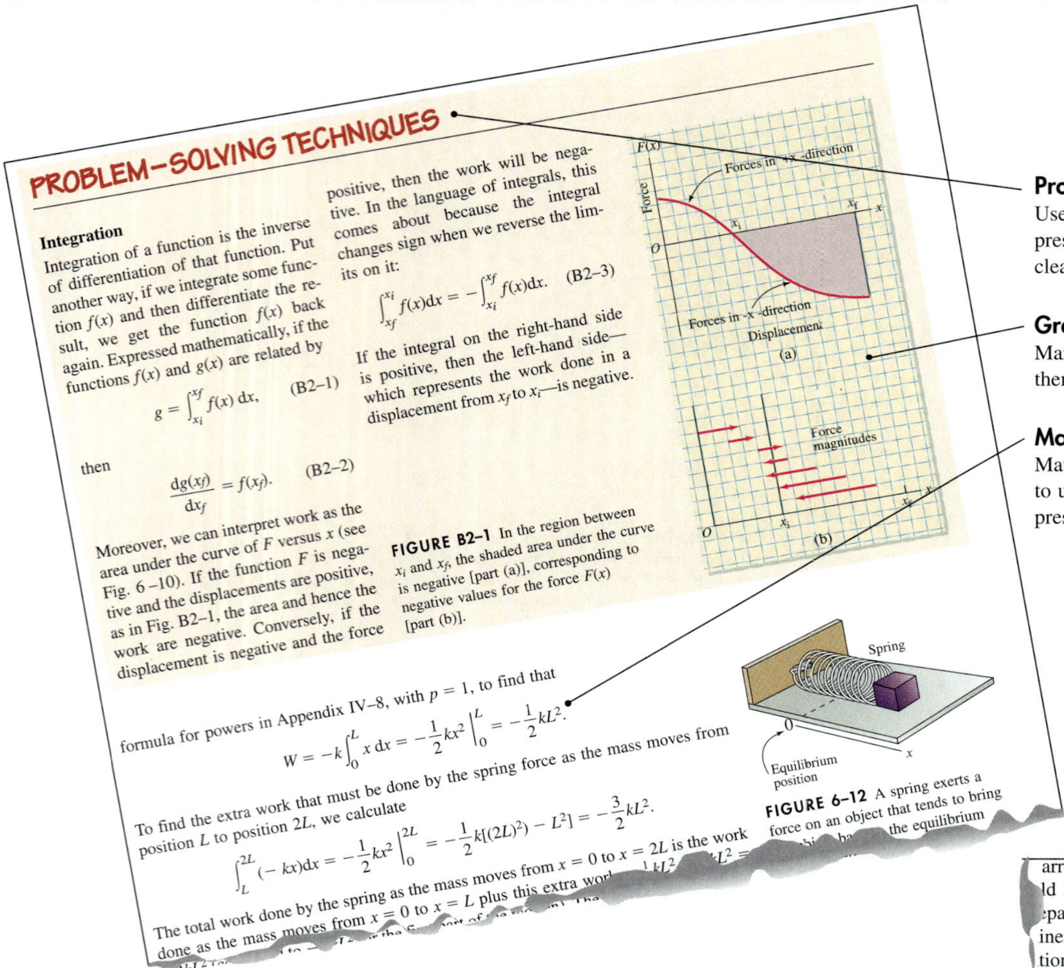

PROBLEM–SOLVING TECHNIQUES

Integration

Integration of a function is the inverse of differentiation of that function. Put another way, if we integrate some function $f(x)$ and then differentiate the result, we get the function $f(x)$ back again. Expressed mathematically, if the functions $f(x)$ and $g(x)$ are related by

$$g = \int_{x_i}^{x_f} f(x)\,dx, \quad \text{(B2–1)}$$

then

$$\frac{dg(x_f)}{dx_f} = f(x_f). \quad \text{(B2–2)}$$

Moreover, we can interpret work as the area under the curve of F versus x (see Fig. 6–10). If the function F is positive and the displacements are positive, as in Fig. B2–1, the area and hence the work are negative. Conversely, if the displacement is negative and the force is positive, then the work will be negative. In the language of integrals, this comes about because the integral changes sign when we reverse the limits on it:

$$\int_{x_f}^{x_i} f(x)dx = -\int_{x_i}^{x_f} f(x)dx. \quad \text{(B2–3)}$$

If the integral on the right-hand side is positive, then the left-hand side—which represents the work done in a displacement from x_f to x_i—is negative.

FIGURE B2–1 In the region between x_i and x_f, the shaded area under the curve is negative [part (a)], corresponding to negative values for the force $F(x)$ [part (b)].

formula for powers in Appendix IV–8, with $p = 1$, to find that

$$W = -k\int_0^L x\,dx = -\frac{1}{2}kx^2\Big|_0^L = -\frac{1}{2}kL^2.$$

To find the extra work that must be done by the spring force as the mass moves from position L to position $2L$, we calculate

$$\int_L^{2L}(-kx)dx = -\frac{1}{2}kx^2\Big|_0^{2L} = -\frac{1}{2}k[(2L)^2) - L^2] = -\frac{3}{2}kL^2.$$

The total work done by the spring as the mass moves from $x = 0$ to $x = 2L$ is the work done as the mass moves from $x = 0$ to $x = L$ plus this extra work ...

FIGURE 6–12 A spring exerts a force on an object that tends to bring ... the equilibrium ...

Problem-Solving Techniques
Useful strategies for solving problems are presented, working through a problem in a clear step-by-step manner.

Graphs
Many graphs contain grid lines to make them easy to use.

Mathematics
Mathematics is introduced when you need to use it, in the context of the physics being presented.

APPLICATION: CONSTRUCTION CRANES

The tower crane used at construction sites (Fig. 11AB–1) consists of a very flimsy tower constructed from three planes arranged in a triangle. Each plane consists of a lattice of triangular bars (this is called a geodetic construction). At the top of this tower is a normal crane boom, which can lift large masses, such as furnaces, air conditioners, and steel girders. The tower can rotate to place the mass at the correct angle, while the crane can move the mass radially inward or outward to position the object at the correct location; the object is finally set down. The strength of the tower is completely insufficient to withstand the bending moment of a large mass at the end of the boom, so there is a balance weight at the far side of the crane that moves in an opposite direction from the lifting equipment and the object being moved. The motion of the counterweight is controlled by strain sensors at the top of the tower and ensures that the crane boom produces no net torque at the top of the tower.

A similar idea is applied to the portable cranes that are carried by trucks. These cranes are provided with sen- ...

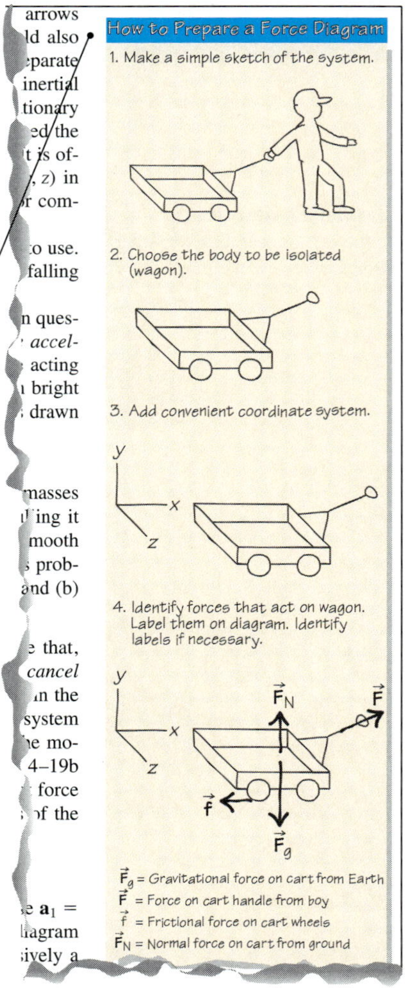

How to Prepare a Force Diagram

1. Make a simple sketch of the system.
2. Choose the body to be isolated (wagon).
3. Add convenient coordinate system.
4. Identify forces that act on wagon. Label them on diagram. Identify labels if necessary.

$\vec{F}_g$ = Gravitational force on cart from Earth
$\vec{F}$ = Force on cart handle from boy
$\vec{f}$ = Frictional force on cart wheels
$\vec{F}_N$ = Normal force on cart from ground

Application Box
This edition provides increased coverage of engineering-oriented material, most noticeably in the new application boxes.

How to Draw
Illustration strategies for problem solving are presented in the clear and easily reproduced style of student sketches.

Problems

End-of-chapter problems answer the question: "How much have I learned?"

Student Sketch

All figures in the end-of-chapter material are drawn as student sketches and represent the first step in problem solving.

Level of Difficulty

A problem's level of difficulty is indicated: Level I problems are the easiest, Level II problems are usually multistep, and Level III problems are especially challenging.

PROBLEMS

3–1 Position and Displacement

1. (I) A car travels 21 km to the northeast, then 15 km to the east, before it travels 28 km to the north. Express the position vector from where the car starts to the point at which each turn occurs. What is the car's total displacement?

2. (I) A particle is located by the position described by the vector $\mathbf{r} = (c_1 - c_2t)\mathbf{i} + (d_1 + d_2t + d_3t^2)\mathbf{j}$, where $c_1 = 12$ m, $c_2 = 3$ m/s, $d_1 = -20$ m, $d_2 = -3$ m/s, and $d_3 = 0.5$ m/s^2. At what time(s) does the particle pass through the position $x = 0$ m? At what time(s), and where, does the particle cross the line $x = y$? Sketch the particle's trajectory from $t = -8$ s to $t = 10$ s.

3. (I) A gym teacher organizes a series of indoor races, which follow along the walls of the gym; the race starts from corner A, continues to corner B, and so forth. The gym is a rectangle with the distance $AB = 25$ m and the distance $BC = 35$ m. Suppose that the origin of a coordinate system is at point A, leg AB is in the $+x$-direction, and leg BC is in the $+y$-direction. Express the position vector of a running student at each of the four corners.

4. (I) A runner races with a uniform speed of 27.0 km/h around a circular track of radius 172 m. Draw the track and the runner's displacement vector after 20 s, 40 s, 60 s, and 2 min, assuming that at $t = 0$ s the runner is at the three o'clock position. Assume counterclockwise motion.

5. (I) A treasure map locates the site of a treasure by reference to two starting points, A and B. A is chosen to be the origin, and B is at the point (2.5 km)**i**. The instructions state that the treasure lies at the intersection of two lines. One line starts at A and passes through the point (2.0 km)**i** + (4.0 km)**j**; the other line, starting at B, passes through the point (6.0 km)**i** − (8.0 km)**j**. Sketch the instructions and find the location of the intersection graphically as well as algebraically. Express the vector that gives the displacement of the intersection point from point C, whose location relative to the coordinate system specified by A and B is given by (1.2 km)**i** − (2.2 km)**j**.

6. (II) The position of a particle in a given coordinate system is $\mathbf{r}(t) = (5 - 2t^2)\mathbf{i} + (-4 + 3t)\mathbf{j}$ where the distances are in meters when t is in seconds. At what time will the particle cross the y-axis? At what time will it cross the x-axis? Can you find an equation that relates the y-coordinate to the x-coordinate and therefore gives the trajectory in the xy-plane? Where would the x- and y-axes have to be moved so that at $t = 0$ s the tra-

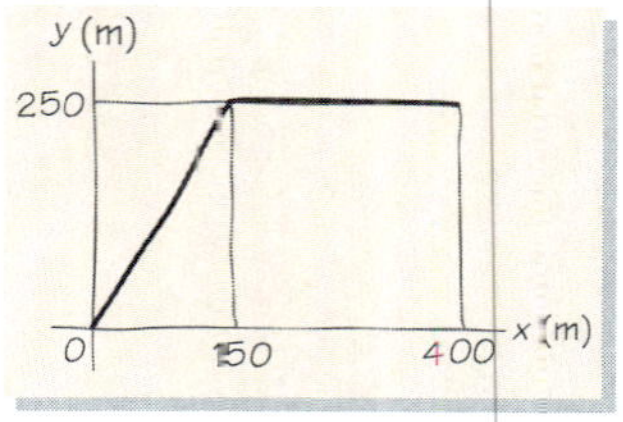

FIGURE 3–29 Problem 7.

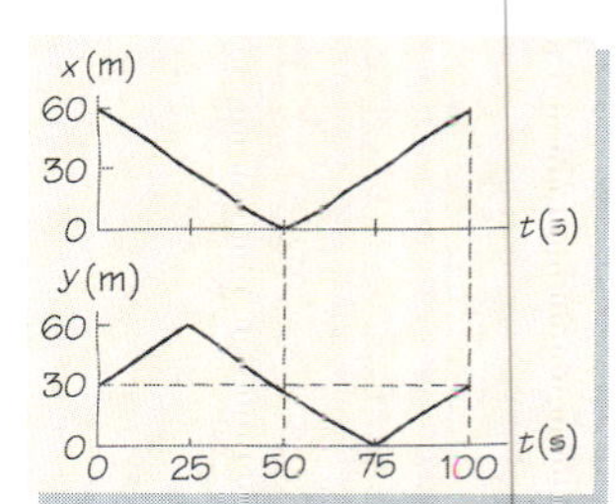

FIGURE 3–30 Problem 8.

10. (III) In a "shoot the coconut" lecture demonstration, the position of the coconut is given by the vector $(h_0 - \frac{1}{2}gt^2)\mathbf{j}$, whereas that of the projectile, aimed at the coconut at time $t = 0$, is given by $(-L + ut)\mathbf{i} - [(h_0ut/L) - (\frac{1}{2}gt^2)]\mathbf{j}$. Show that the two will always collide, and find the time at which this takes place (Fig. 3–32). Express the displacement vector of the coconut relative to the projectile.

Conceptual Questions

These questions check a student's understanding of the ideas that underlie a situation.

A law of physics has a wide range of applicability. Laws such as Newton's laws provide a framework for detailed predictions about what will happen under certain conditions. In their breadth and predictive power, Newton's laws are an ideal illustration of what constitutes a law of physics.

UNDERSTANDING KEY CONCEPTS

1. A small but dense mass is swinging freely at the end of a light string. A very sharp knife cuts the string when the mass is at the bottom of its swing; the knife does not disturb anything else. What is the subsequent flight of the mass?
2. Someone pushes on a wall. What experiment can you propose to determine the force with which the person pushes?
3. If you were in a freely falling elevator, the contact force on you due to the floor would drop to zero. If the elevator were to accelerate rapidly upward, the contact force between you and the floor would increase. Why?
4. An astronaut is working while in orbit. When the astronaut assembles a piece of equipment, will he or she notice a difference between working with components of large mass as op- ... of small mass?
5. When a satellite travels around Earth in a circular orbit it moves at a constant speed. Does Newton's first law apply in this situation? Is the velocity constant? Is there a force present?
6. There is a well-known parlor trick in which a tablecloth is pulled sharply from beneath a dinner setting, leaving the setting in place. Why does this work?
7. Is a spaceship heading from Earth to the Moon traveling in a force-free environment? Explain. What about a spaceship traveling from Earth to Mars that is currently in a region far away from either planet?
8. An adult sits on a child's table. The table is about to break. Is it correct to say that it is the weight of the adult that is causing the table to break?

General Problems

62. (I) Neutron stars are incredibly dense objects, compared to the Sun. One such (hypothetical) star has a radius of 40 km and rotates with a period of 26 s. What are the velocity and acceleration of a particle on the surface due to the rotation? Compare this acceleration with the acceleration of gravity on Earth.

63. (II) A golfer wants to land a golf ball on the green located 105 m away horizontally but 8.0 m down. The golfer chooses an eight iron that he knows will result in the ball leaving the tee at an elevation angle of 65°. (a) With what velocity should the ball leave the tee? (b) What is the maximum height of the ball above the green?

64. (II) A softball player hits the ball when it is 0.9 m above home plate. The ball leaves the bat at an elevation angle of 30°. What initial speed must the ball have to clear a fence 1.8 m high lo-

barrel makes with the ground? Unlike the projectile treated in Section 3–4, the cannonball is fired from the edge of a cliff of height h_0 above the level plain at which it is aimed. Show that the angle θ that gives the largest horizontal range is given by $\sin^2\theta = v_0^2/2(v_0^2 + 2gh_0)$. [*Hint*: In calculus, you learn that a function of a variable such as θ has a maximum (or minimum) at an angle θ_0, the angle for which the derivative of the function with respect to θ is zero.]

69. (III) A juggler is able to handle four balls simultaneously. He takes 0.5 s to cycle each ball through his hands, throw the ball, and be ready to catch the next ball. (a) With what velocity must he throw each ball vertically? (b) What is the position of the other three balls when he has just caught one of the balls? (c) How high must he throw the balls if he is to juggle five balls?

70. (III) A wheel 72 cm in diameter rolls along a road, with the ... speed of 18 km/h.

General Problems

A comprehensive check of understanding involves problems that are not tied to a particular section of text.

CHAPTER 1

Tooling Up

A wide range of physical phenomena is presented to us by this rock-climbing scene. We can say that the force of gravity acts on the climber, who is saved by using the friction between himself and the rock. If we were to look at where friction comes from on an atomic scale, we would also discover a range of electromagnetic phenomena. Why does the rock have the properties it does? What makes the sky blue? Physics can help us to understand a huge variety of things in the world around us.

Why physics? What is there about today's world that makes an understanding of the fundamentals of science so essential? To see forward, let's look backward. Curiosity and a need to make sense of what we see are distinctly human traits. The first steps toward the sciences arose out of the recognition of patterns of regularity. Many of these patterns—the sequence of day and night, the seasons, and the regular motion of the heavenly bodies—raised questions about their causes. It was not the proposed answers to these questions that moved us toward science, *it was the attention given to observation.* The accuracy of the observations of both position and time by ancient astronomers was surprisingly high. The Babylonians could predict the motion of planets. The description of the appearance of a brilliant "new star" in A.D. 1054—now known to be the Crab supernova—was recorded by Chinese astronomers in day-by-day detail. This description is accurate enough to confirm today's nuclear physics calculations of the star's brightness (Fig. 1–1).

Blind reverence for authority impedes scientific progress, and for a long period the work of the ancient Greeks was regarded with crippling reverence. For example, the idea that gravity makes heavier objects fall more quickly than lighter ones persisted from Greek times all the way to the Renaissance. An important component of the Renaissance was a departure from too much reverence for authority, and a reintroduction of the role of experiment in natural science. Among those who must be given credit for solid reestablishment of experiment are William Gilbert, for magnetism, and Galileo Galilei, for motion. Galileo had a more immediate impact on the rise of physics because of his influence on Isaac Newton, whose laws of motion form the underpinning of almost all of physics. The give and take between experimental observation, the mathematical formulation of descriptive and predictive theories, and further experi-

The scientific method

mental tests of predictions of these theories form the *scientific method.* In the scientific method, human imagination, through intuition and curiosity, is subject to the checks and balances of experiment, thought, and further testing. The scientific method has led to accelerated progress in our understanding of the physical world.

This progress has moved along two fronts. On one side, new concepts are built—sometimes in small steps, sometimes in large ones—on earlier concepts. These concepts summarize an ever-increasing body of experimental information, as well as permit an extrapolation of physical ideas into areas where patterns of regularity had not previously been seen or even suspected. We speak of opening up new fields. On the other side, progress is driven by an improvement of experimental techniques, which have allowed scientists to probe nature in domains hitherto inaccessible to experimentation. For example, much of modern technology can be traced to the development of pumps efficient enough to allow the creation of near-vacuum conditions in sealed vessels.

FIGURE 1–1 We know from historical record that the crab nebula is the result of a great stellar explosion. At its heart is a very compact remnant—a neutron star—of the original star. Objects such as these allow us to test our understanding of many aspects of the natural world.

Scales of the Observable World

Physics underwent a series of revolutionary developments in the period 1900–1925, and the field is sometimes divided into classical (pre-1900) and modern (post-1900) physics. This division seems artificial to us, and we prefer to think of physics in terms of *scales.* The laws of physics and the structure of matter appear to have a hierarchical structure. Newton's laws of motion were thought for a long time to be universally applicable. Subsequent observation has shown, however, that there are extreme circumstances in which these laws must be replaced by a different picture of motion. We need quantum mechanics to describe matter at very small distances, and we need special relativity to describe motion at speeds close to that of light. The behavior of atoms and their constituents cannot be understood without these new laws of physics. These laws have the characteristic that for objects that are large on an atomic scale, or move slowly in comparison with the speed of light, they reduce to the simpler laws of motion discovered by Newton. Thus for the description of most of the world, and for many—but not all—engineering applications, Newton's laws apply.

This middle ground in the description of nature includes the great subjects of classical physics—mechanics, waves, thermodynamics, and electricity and magnetism—and forms the content of most of this book. In the middle ground, the effects of the detailed underlying structure of matter appear in the form of some numerical constants that can be taken from experiment. Numbers such as the conductivity of electricity or the index of refraction of light summarize a great deal of complexity in the structure of matter; as far as classical physics is concerned, their numerical value is the only connection with that structure, and these numbers can be taken from experiment.

To understand the realm of the very small, or the very fast, we need quantum mechanics, and we need relativity. This realm is what we might call the frontier, where we are continuing to build our understanding. A deeper understanding of the frontiers is one of the permanent goals of science. Even when we treat middle-ground physics, we shall indicate the underlying frontier physics. The reason that this is useful is obvious: technology is pushing engineering practice into these frontiers, and quantum physics—the physics of small scales—is becoming more and more essential to modern engineering. In the coming decades, few engineers will be able to function without some understanding of quantum mechanics and relativity. Devices such as the laser have moved from the physics research laboratory into everyday use in engineering practice.

Students sometimes feel that physics is somehow separate from engineering. In reality, it is only the goals of the two disciplines that differ. The aim of much of engineering is to use knowledge of the basic structure and functioning of materials to advance technology. That knowledge is the domain of physics. Both the advancement of technology and the attainment of the needed knowledge are extremely important as we move toward the twenty-first century. That is why a physics education is so essential to future engineers or to anyone who wishes to function successfully in the emerging world.

1–1 SCIENTIFIC NOTATION

One language of engineering and physics is that of numbers. The range of numbers that appear in the physical world is truly enormous. For example, the mass of Earth is about 5,980,000,000,000,000,000,000,000 kilograms (kg), and the diameter of a proton is about 0.000000000000001 meter (m). So many zeros are inconvenient, and we employ a shorthand method of writing very large and very small numbers. By using powers of 10, Earth's mass is more easily written as 5.98×10^{24} kg, and the diameter of a proton is written as 10^{-15} m. In this notation, 10^3 represents 1000 and 10^{-4} means 0.0001. We shall use this standard *scientific notation* throughout this book.

A considerable advantage of scientific notation is that multiplication and division are easily performed by adding and subtracting exponents of 10. Thus the product $100 \times 100 = 10{,}000$ can be written as $10^2 \times 10^2 = 10^{2+2} = 10^4$. The awkward multiplication $0.00000055 \times 24{,}000$ can be done more easily as $(5.5 \times 10^{-7}) \times (2.4 \times 10^4) = (5.5 \times 2.4) \times 10^{-7+4} = 13.2 \times 10^{-3} = 1.32 \times 10^1 \times 10^{-3} = 1.32 \times 10^{-2}$. Where division is involved, we simply change the sign of an exponent and use the multiplication rules. For example,

$$\frac{7.5 \times 10^{-3}}{2.5 \times 10^{-4}} = \frac{7.5}{2.5} \times 10^{-3} \times 10^{+4} = 3 \times 10 = 30.$$

1–2 LENGTH, TIME, AND MASS

In our exploration of the physical world, we have found that length, time, and mass play a fundamental role in measurement (Fig. 1–2). As Tables 1–1, 1–2, and 1–3 show, these quantities cover an enormous range of values in our universe. These three quantities are already intuitively familiar, and the idea that there is a system of units for specifying them is not a strange one. The tables contain a particular, standardized system of units. The importance of standardization is evident. You may wear a size 7 shoe in the United States, but this unit would not be of much use if you were to travel to Europe, where a different system is used; there, your shoe size would be 38.

Hundreds of years ago, people used what was readily available as standards for measurement. Length measurements such as the foot came into use in this manner. Over time, measurement systems have become both more precise and more universal. For an early example, French scientists established the forerunner of the International System of measurements in 1791. They defined the meter, the second, and the kilo-

(a) (b) (c)

FIGURE 1–2 (a) Measuring distance. The measurements of length needed to study a traffic accident have very different kinds of precision requirements than those involved in, say, the fabrication of integrated circuits. (b) It has become commonplace to see clocks and watches that display time in a digital fashion. (c) We must have an objective, repeatable way to measure mass.

gram: The meter—roughly one yard—was defined as one ten-millionth (10^{-7}) of the distance along Earth's surface between the equator and the North Pole; the second as 1/86,400 of a mean solar day; and the kilogram as the mass of a certain quantity of water. In 1889, an international organization called the General Conference on Weights and Measures was formed to meet periodically to refine these units of measure. In 1960, this organization decided to name a system of units based on the meter, kilogram, and second the International System, with the abbreviation **SI** (for the French words **Système International**). We use the term SI in this book. This system is also known as the *metric system* or *mks system* (after *m*eter, *k*ilogram, and *s*econd).

The International System of units, or SI

Length

The definition of the meter has been changed several times. In 1889, one meter was defined as the distance between two finely engraved marks on a bar of platinum–iridium that was kept in a vault outside Paris. Even though several copies of this bar were distributed throughout the world, such a standard of length had many shortcomings. For instance, with progress in optical techniques, the scratches on the bar were seen to be fuzzy and imprecise. In 1960, the standard of length was changed to depend upon an atomic constant—the wavelength of a particular orange-red light emitted by an isotope of krypton (^{86}Kr) gas. Because our ability (and need) to measure length has led us to require even greater accuracy, this standard also became insufficiently precise. Therefore, in 1983, the 17th General Conference on Weights and Measures established a standard of length based on the speed of light in vacuum (denoted by the letter c). A **meter** (m) is now defined as the distance light travels in vacuum during 1/299,792,458 second. Some orders of magnitude for lengths are given in Table 1–1.

The definition of the meter

Time

The second was originally defined as 1/86,400 of the mean solar day, which is the time interval, averaged over a year, from noon of one day to noon of the next. This definition is insufficient because Earth's rotation is both slightly irregular and gradually slowing down from year to year. Therefore, in 1967, a definition of the second was adopted that depends on an atomic standard. The **second** (s) is now defined as the duration of 9,192,631,770 periods of a particular vibration of a cesium atom isotope (^{133}Cs). Clocks based on this standard are, in effect, identical because all atoms of ^{133}Cs are indistinguishable and because frequency can be measured in the laboratory to an accuracy of about four parts in 10^{13}. Some orders of magnitude for time are given in Table 1–2.

The definition of the second

Mass

The kilogram was originally defined as the mass of 1 liter of water under certain conditions of temperature and pressure. In 1901, the standard **kilogram** (kg) was defined as the mass of a particular cylinder of platinum–iridium alloy kept at the International

The definition of the kilogram

TABLE 1–1 Orders of Magnitude for Length

Parameter	*Length (m)*	*Parameter*	*Length (m)*
Proton	10^{-15}	Diameter of solar system	10^{13}
Hydrogen atom	10^{-10}	Distance to nearest star (Proxima Centauri)	10^{17}
Flu virus	10^{-7}	Diameter of our galaxy (Milky Way)	10^{21}
Raindrop	10^{-3}	Distance to nearest galaxy	10^{22}
Height of person	10^{0}	Distance to edge of observable universe	10^{26}
One mile	10^{3}		
Diameter of Earth	10^{7}		
Earth–Moon distance	10^{9}		
Earth–Sun distance	10^{11}		

TABLE 1–2 Orders of Magnitude for Time

Parameter	*Time (s)*	*Parameter*	*Time (s)*
Time for light to cross proton	10^{-23}	Period of human heartbeat	10^{0}
Time for light to cross atom	10^{-19}	Class lecture	10^{3}
Period of visible light wave	10^{-15}	One Earth day	10^{5}
Period of vibration for standard cesium clock	10^{-10}	One Earth year	10^{7}
		Age of Greek antiquities	10^{11}
Half-life of muon	10^{-6}	Age of Earth	10^{17}
Period of highest audible sound	10^{-4}	Age of universe	10^{18}

Bureau of Weights and Measures in France. Duplicate copies of the cylinder made of this particularly stable alloy are kept in laboratories such as the National Institute of Standards and Technology in Maryland. Although the standards of time and length can be reproduced to precisions of one part in 10^{12}, the standard of mass can be reproduced only to perhaps one part in 10^8 or 10^9. This standard of mass leaves much to be desired. We would like to find an atomic or natural standard for mass but, even though we know that all atoms of the same type have the same mass, nobody knows how to count atoms with the required accuracy. Some orders of magnitude for mass are given in Table 1–3.

Other Systems of Units

The International System of units is by far the most important and widely accepted system used in the world today. Two other systems, however, are still in common use.

cgs. The *cgs system* is based on the centimeter, gram, and second and is a metric system derived directly from SI. This system finds its major use in measurements of electricity and magnetism. Densities (mass per unit volume) normally are still quoted in grams per cubic centimeter (g/cm^3) because most densities are nearer unity in this system than in SI. For example, the density of water is 1 g/cm^3 in cgs but is 10^3 kg/m^3 in SI. The definition of the units of the cgs system is based on those of SI:

$$1 \text{ cm} \equiv 0.01 \text{ m and } 1 \text{ g} \equiv 0.001 \text{ kg},$$

where we have used the symbol $\equiv$ to indicate a definition.

British Engineering System. The *British engineering system*, or British system, is based on units of the inch, pound, and second. This system is used only in the United States and in parts of the British Commonwealth, where it is in the process of being replaced by SI. Even in the United States, scientists seldom use the British system, but existing technology based on this system may require its use in engineering applications. Economic considerations—the desirability of international trade—suggest that even this limited use will eventually disappear.

The British system of units is now *defined* in terms of SI units. The unit of length, the **inch** (in), is defined as 1 in $\equiv$ 2.54 cm (Fig. 1–3). In the British system we must distinguish between "mass" and "weight," a topic we will discuss later. It is sufficient

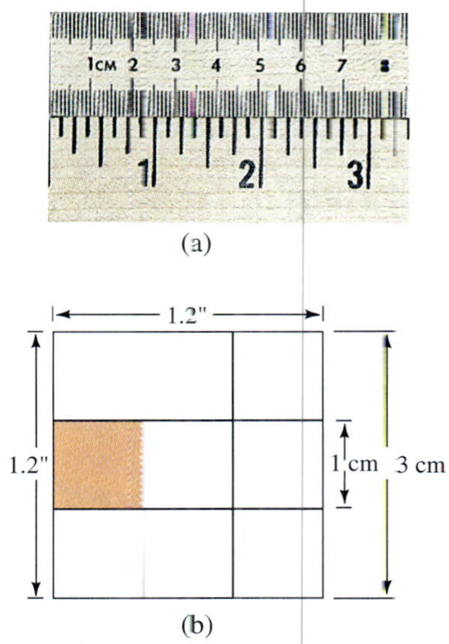

FIGURE 1–3 (a) One inch is exactly 2.54 cm. (b) A square 3 cm × 3 cm in size has an area 9 cm^2, or $(3/2.54)^2$ in^2 = 1.4 in^2.

TABLE 1–3 Orders of Magnitude for Mass

Parameter	*Mass (kg)*	*Parameter*	*Mass (kg)*
Electron	10^{-30}	Battleship	10^{8}
Hydrogen atom	10^{-27}	Moon	10^{23}
Uranium atom	10^{-24}	Earth	10^{25}
Dust particle	10^{-13}	Sun	10^{30}
Raindrop	10^{-6}	Our galaxy (Milky Way)	10^{41}
Piece of paper	10^{-2}	Observable universe	10^{52}
Man	10^{2}		

APPLICATION: THE GLOBAL POSITIONING SYSTEM

Why do we need high precision in our standards of length and time? As a practical application of high-precision time measurements, consider the *Global Positioning System* (Fig. 1AB–1). The system consists of a network of 24 satellites orbiting Earth at an altitude of 20,000 km. The satellites carry atomic clocks—accurate to about four parts in 10^{13}—that emit regular signals, which travel with the speed of light. Because the speed of light is finite, signals from different satellites arrive at a receiver at slightly different times, depending on the position of the receiver. The receiver's *position* can then be determined because the position of the satellites is very accurately tracked. The accuracy with which the receiver's position is determined has to do with the uncertainty in the time measurement and the uncertainty in the speed of light, known to three parts in 10^{10}. There are limitations on this accuracy, since the speed of light in air differs a little from the absolute speed of light; it is the meteorological variations in the speed in air that set limits on the accuracy of the position measurement. At present, the civilian users of hand-held receivers may determine their position to an accuracy of ± 15 m. The Federal Aviation Authority has decided to use satellite positioning to track airplanes to airports, and the system has become indispensible for navigation. It is easy to imagine a host of applications of this technology in general transportation and geography.

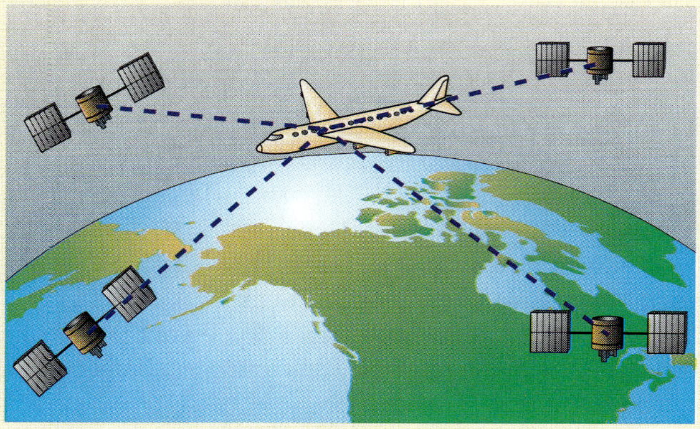

FIGURE 1AB–1 With the GPS system, travelers—from airplane pilots to hikers—can confirm their positions with remarkable accuracy.

for now to state that a **weight** of 1 pound (lb) is a force (a push or pull) equivalent to the force exerted by a mass of exactly 0.45359237 kg on a bathroom scale—such scales read the weight of an object—at a particular point on Earth's surface. This means that a mass of 1 kg has the same weight as about 2.2 lb. (Notice that we do *not* write 1 kg = 2.2 lb, because mass and weight are *not* the same.) A ton is 2000 lb, but a metric ton is 1000 kg, which is equivalent to a weight of 2204.6 lb. The unit of mass in the British system is called a *slug*, but it is seldom used. A slug is equal to 14.5939 kg. Finally, the unit of time—the *second*—is the same in the British system as it is in SI.

TABLE 1–4 Unit Prefixes for Powers of 10

Prefix	*Symbol*	*Multiple*
exa†	E	10^{18}
peta†	P	10^{15}
tera	T	10^{12}
giga	G	10^{9}
mega	M	10^{6}
kilo	k	10^{3}
hecto†	h	10^{2}
deka†	da	10^{1}
deci†	d	10^{-1}
centi	c	10^{-2}
milli	m	10^{-3}
micro	μ	10^{-6}
nano	n	10^{-9}
pico†	f	10^{-15}
atto†	a	10^{-18}

†Except for centi, the prefixes near 1 (10^0) are rarely used in the United States. Similarly, the very small and very large multiples (powers of ± 15, ± 18) are not normally used or recognized. You should become familiar with all others.

Some Shorthand and History Concerning Units

A useful set of prefixes in SI replace certain powers of ten. These prefixes are listed in Table 1–4. For example, the distance between New York City and Los Angeles is 4,483,000 m, or 4.483×10^6 m. Neither of these quantities is convenient for everyday use. However, we can use the prefix *kilo-*, which stands for 10^3, to state this distance as 4483 kilometers (km). A finger's width, for example, would not normally be written as 0.015 m or as 1.5×10^{-2} m, but rather as 1.5 centimeters, or 1.5 cm. The standard of mass, the kilogram (kg), is often more convenient to use than the gram (g). In most countries of the world, meat and produce prices are quoted per kilogram because that is a convenient mass unit. You will soon appreciate the simplicity of the International System and obtain a feel for its units of length and mass.

As in every discipline, some special units are present because of historical precedent or simply because they are easy to use. For example, the quantity 10^{-6} m is equivalent to a micrometer, or μm, in SI, but is also known as a micron. This double name for the same quantity causes little confusion. Another example involves a unit often used in atomic physics for the wavelength of light, which is the angstrom, Å. One angstrom has the value 10^{-10} m or 0.1 nm (1 nanometer (nm) = 10^{-9} m). The range of visible light is about 3000–7000 Å, or 300–700 nm. Both the nanometer and the angstrom are commonly used.

At the other extreme of the length scale, astronomers and astrophysicists find the meter to be too small for practical use, and they employ three other distance measures:

(1) the *astronomical unit* (AU), which is the mean distance between Earth and the Sun; (2) the *light-year* (ly), the distance that light travels in one year; and (3) the *parsec* (pc), which is 3.0857×10^{16} m. The astronomical unit ($1 \text{ AU} = 1.496 \times 10^{11}$ m) is useful for distances within our solar system. The light-year, 0.95×10^{16} m, is useful when dealing with interstellar distances, as is the parsec, equal to about 3.26 ly or 206,265 AU.

Unit Relations and Conversions

The angstrom and the astronomical unit are examples of *derived units,* units defined in terms of the basic set (the meter, the kilogram, and the second). These units are simple changes in scale of the basic units. A more complicated type of derived unit involves combinations of the fundamental units. Sometimes these combinations are given names, sometimes not. Speed is measured in meters per second (m/s), but it is not given a separate name. When we study force, however, we will find it to be measured in kilogram-meters per second squared ($\text{kg} \cdot \text{m/s}^2$); this more complicated unit is given a name, the **newton** (N). The newton is a derived unit in SI.

In deriving relations between physical quantities within a *single* system of units, the units on both sides of such relations must match. This provides us with a very useful tool for checking results that we may derive. If units on both sides of an equation cannot be made to match, there is an error somewhere. Further, if a problem is stated using more than one unit for a particular quantity—such as meters and inches for length—then we must find a simple way to *convert* between different systems of units to solve the problem. As we will now see, this can be done conveniently and systematically with the primary conversion equations that relate one set of units to another. Examples of such primary equations are the expressions

When a single unit system is employed, the units on one side of an equation must be the same as those on the other side.

PROBLEM SOLVING

Conversion of units

$$1 \text{ in} = 2.54 \text{ cm and } 100 \text{ cm} = 1 \text{ m}.$$

The first of these can be rewritten as

$$1 = \frac{2.54 \text{ cm}}{1 \text{ in}} \quad \text{or} \quad 1 = \frac{1 \text{ in}}{2.54 \text{ cm}}.$$

Any equation can be multiplied by the pure number 1 without change; by judicious choice of the factor 1, units can be canceled and replaced by others. For example,

$$15 \text{ in} = (15 \text{ in})(1) = (15 \cancel{\text{in}})\left(\frac{2.54 \text{ cm}}{1 \cancel{\text{in}}}\right).$$

The unit "in" cancels, and

$$15 \text{ in} = (15)(2.54 \text{ cm}) = 38.1 \text{ cm}.$$

Some other examples of conversions of this type are

$$1 \text{ yd} = 1 \cancel{\text{yd}}\left(\frac{36 \cancel{\text{in}}}{1 \cancel{\text{yd}}}\right)\left(\frac{2.54 \cancel{\text{cm}}}{1 \cancel{\text{in}}}\right)\left(\frac{1 \text{ m}}{100 \cancel{\text{cm}}}\right) = 0.9144 \text{ m}$$

and

$$1 \text{ mi} = 1 \cancel{\text{mi}}\left(\frac{5280 \cancel{\text{ft}}}{\cancel{\text{mi}}}\right)\left(\frac{12 \cancel{\text{in}}}{\cancel{\text{ft}}}\right)\left(\frac{2.54 \cancel{\text{cm}}}{\cancel{\text{in}}}\right)\left(\frac{1 \cancel{\text{m}}}{100 \cancel{\text{cm}}}\right)\left(\frac{1 \text{ km}}{1000 \cancel{\text{m}}}\right) = 1.609 \text{ km}.$$

The technique described here is applicable to more complicated examples.

EXAMPLE 1–1 Given that the speed of light is 2.998×10^8 m/s, what is the distance that light travels in 1 yr? (This distance is the light-year.)

Solution: The primary relations involve time in this case; for example,

$$60 \text{ s} = 1 \text{ min}.$$

We can write this as

$$1 = \frac{60\ \text{s}}{1\ \text{min}}.$$

Similarly,

$$1 = \frac{365.25\ \text{d}}{1\ \text{yr}}.$$

For more precision, we use 365.25 d as an average year instead of just 365 d. The distance 1 ly is the speed of light times the time duration, 1 yr:

$$1\ \text{ly} = (2.998 \times 10^8\ \text{m/s})(1\ \text{yr}).$$

This equation has mixed units, a situation we can rectify by converting the units of time from years to seconds:

$$\begin{aligned} 1\ \text{ly} &= \left(2.988 \times 10^8\ \frac{\text{m}}{\cancel{\text{s}}}\right)(1\ \cancel{\text{yr}})\left(\frac{365.25\ \cancel{\text{d}}}{1\ \cancel{\text{yr}}}\right)\left(\frac{24\ \cancel{\text{h}}}{1\ \cancel{\text{d}}}\right)\left(\frac{60\ \cancel{\text{min}}}{1\ \cancel{\text{h}}}\right)\left(\frac{60\ \cancel{\text{s}}}{1\ \cancel{\text{min}}}\right) \\ &= 9.461 \times 10^{15}\ \text{m}. \end{aligned}$$

We have calculated that 1 ly $= 9.461 \times 10^{15}$ m. The units for second, minute, hour, day, and year all appear in both the numerator and denominator and, therefore, cancel. The only unit left is the meter.

1–3 ACCURACY AND SIGNIFICANT FIGURES

Uncertainty in Measurement

Physics rests on experiment, and experiment requires measurement. But measurements are, at best, only approximate. An **uncertainty** is an indication of the accuracy of a measurement. The uncertainty depends on the accuracy and calibration of the instrument that is making the measurement and on how well the instrument can be read. We can best illustrate the meaning of uncertainty with an example. If the width of a page of paper is measured with a ruler to be 21.6 cm with an uncertainty of 1 mm (or 0.1 cm), it would be correct to say that the width is 21.6 cm ± 0.1 cm, or 21.6 ± 0.1 cm. (The ± is read as "plus or minus.") Here, 21.6 cm is called the *central value* and 0.1 cm the *uncertainty* around that central value. In this case, the basis of the uncertainty lies in how well our eyes can read the ruler and on the precision with which the ruler was made. We often use the term *percentage uncertainty* as a measure of the ratio of the uncertainty of a quantity to its central value. The percentage uncertainty is found by multiplying this ratio by 100. The percentage uncertainty of our paper measurement is thus

$$100\left(\frac{0.1\ \cancel{\text{cm}}}{21.6\ \cancel{\text{cm}}}\right) = 0.5\%.$$

We can find the area of the paper by measuring the length and multiplying by the width. Suppose that we measure the paper's length to be 27.9 ± 0.1 cm. The percentage uncertainty for the length is 0.4%. We find the area by multiplying (21.6 cm)(27.9 cm) = 603 cm^2. Because the measurements of width and length both contain uncertainties, the area is also uncertain. But what is this uncertainty? To get a rough idea, just add the percentage uncertainties when multiplying or dividing numbers.[†] In our

[†]A more precise measure is that if P_1 and P_2 are the percentage uncertainties of two quantities being multiplied, the net uncertainty is $\sqrt{P_1^2 + P_2^2}$.

case, the total area uncertainty is about 0.5% + 0.4%, or 0.9%. This means an uncertainty of $(0.009)(603\ cm^2) = 5\ cm^2$, and the area of the paper is $603 \pm 5\ cm^2$.

Significant Figures

Physical quantities are never known with exact precision (unless they are merely definitions). We imply a certain degree of uncertainty in a quantity when we assign a certain number of digits to its numerical value. Thus, when we say that an object is 2.00 m long, we mean that it is between 1.995 m and 2.005 m long. If we wanted to say that the length is somewhere between 1.9995 m and 2.0005 m, we would say that the length is 2.000 m. In the first case, 3 significant figures are used to describe the object's length; in the second case, the number of significant figures is 4. When we say that a sheet of paper has an area of 603 cm^2 we are using 3 significant figures, which means that the area lies between 602.5 cm^2 and 603.5 cm^2.

Zeros that are used only to set a decimal point are not part of our count of significant figures. Thus 0.00035 has 2 significant figures, not 6. To take a more extreme example, we mentioned in Section 1–1 that the mass of Earth is 5,980,000,000,000,000,000,000,000 kg. Surely we do not know Earth's mass to 25 significant figures! Scientific notation provides a way to avoid this ambiguity. When we write the mass of Earth as 5.98×10^{24} kg, we indicate unambiguously that we know the mass to 3 significant figures; if we knew the mass to only 2 significant figures, we would write 6.0×10^{24} kg.

There is one point you should keep in mind: *Carry a calculation only to as many significant figures as are contained in the input parameter with the fewest significant figures.* In other words, the quantity in a calculation that contains the most uncertainty largely dictates the accuracy of the final result. Watch out! Hand-held calculators make it all too easy to violate this rule. For example, if you are asked to calculate the ratio of 3.0 to 11.0 on your calculator, you might be tempted to write 0.27272727272. . . . But in this case, the number 3.0 has the fewest significant figures—2—and thus you should carry your calculation to only 2 significant figures, 0.27.

Resist the temptation to keep all the digits your calculator can supply.

PROBLEM SOLVING

Use good sense with significant figures

1–4 DIMENSIONAL ANALYSIS

Dimensions

Experience has shown that there are three basic ways to describe any physical quantity: the space it takes up, the matter it contains, and how long it persists. All descriptions of matter, relationships, and events are combinations of these three basic characteristics. All measurements can be reduced ultimately to the measurement of length, time, and mass. Any physical quantity, no matter how complex, can be expressed as an algebraic combination of these three basic quantities. Speed, for example, is a length per time.

Length, time, and mass therefore have a significance far beyond that of providing the basis of a system of units. They specify the three **primary dimensions**. We use the abbreviations $[L]$, $[T]$, and $[M]$ for these primary dimensions. The **dimension** of a physical quantity is the algebraic combination of $[L]$, $[T]$, and $[M]$ from which the quantity is formed. The speed v provides an example. The dimension of v is

Any physical quantity has a dimension that can be formed from the three primary dimensions of length, mass, and time.

$$[v] = [L/T] \quad \text{or} \quad [LT^{-1}].$$

Do not confuse the dimension of a quantity with the units in which it is measured. A speed may have units of meters per second, miles per hour, or, for that matter, light-years per century. All of these different choices of units are consistent with the dimension $[LT^{-1}]$. In what follows, the square brackets, as used here, indicate that we are dealing with dimensions.

The dimension of a quantity should not be confused with the units in which it is measured.

Any physical quantity has dimensions that are algebraic combinations $[L^q T^r M^s]$ of the primary dimensions, where the superscripts q, r, and s refer to the order (or power) of the dimension. Thus, for example, an area has dimension $[L^2]$. If all of the exponents q, r, and s are zero, the combination will be dimensionless. The exponents q, r, and s can be positive integers, negative integers, or even fractional powers.

Dimensional Analysis

The dimensions on one side of an equation must be the same as those on the other side.

Study of the dimensions of an equation—*dimensional analysis*—is an important exercise with several different uses in physics. Any equation that relates physical quantities must have consistent dimensions; that is, *the dimensions on one side of an equation must be the same as those on the other side.* This provides a valuable check for any calculation. Dimensional analysis can also reveal *scaling laws* (see Section 1–7), which describe how changing one quantity in a physical situation requires changes in others. Finally, when there is reason to believe that only certain physical quantities can enter into a physical situation, dimensional analysis can provide us with powerful insights.

Let's look at some examples of dimensional analysis. In Chapter 7, we derive a relation between the height h of a dropped object and the speed of that object. This relation involves the *acceleration of gravity*, g, a quantity whose dimension is $[g] = [LT^2]$. The relation reads

$$gh = \frac{1}{2}v^2.$$

Let's compare the dimensions on each side of this equation. The dimension of h is $[L]$, so the left-hand side has dimensions $[LT^{-2}][L] = [L^2T^{-2}]$. The right-hand side has the dimensions of speed squared, $[LT^{-1}]^2 = [L^2T^{-2}]$. Thus the dimensions match. If, through error, we had written a relation $gh^2 = \frac{1}{2}v^2$, then this check would have revealed the error. Note that dimensional analysis does not help us understand the numerical factor $\frac{1}{2}$.

EXAMPLE 1–2 Newton's law of universal gravitation gives the force between two objects of mass, m_1 and m_2, separated by a distance r, as

$$F = G\left(\frac{m_1 m_2}{r^2}\right).$$

Use dimensional analysis to find the units of the gravitational constant, G.

Solution: First, the dimensions of the two sides of the equation must match. In the previous section, we learned that the unit of force is the newton, equivalent to $\text{kg}\cdot\text{m/s}^2$. Using these units, the dimensions of force must be $[MLT^{-2}]$. We now know the dimensions of every quantity in the equation for gravitational force except G. Writing the dimensions for both sides gives

$$[MLT^{-2}] = [G][M][M][L^2] = [G][M^2L^{-2}].$$

Note that the individual dimensions can be consolidated inside the square brackets or left within their own brackets—whichever is easiest. We solve for the dimension of G as

$$[G] = \frac{[MLT^{-2}]}{[M^2L^{-2}]} = [MLT^{-2}][M^{-2}L^2] = [M^{-1}L^3T^{-2}].$$

FIGURE 1–4 The square base of the great pyramid of Giza is 232 m on a side, and the peak is 138 m above the base. Napoleon's engineers estimated that the material that forms the pyramid could be used to build a wall 1 m high and 0.1 m thick all the way around France. Can you estimate the length of the perimeter of France?

1–5 HOW A LITTLE REASONING GOES A LONG WAY

Sometimes we must make a quick calculation. We may want to check a complicated numerical calculation to see if the answer is reasonable. Or we may not have access to all the data needed, and an estimate, or a very rough approximation, is in order (Fig. 1–4). In these cases, we perform an *order-of-magnitude* calculation, in which variables are rounded off to the nearest power of 10 or to some other easily handled number. The final result of a calculation with variables so dramatically rounded off is accurate only within an order of magnitude, but such an estimate can be extremely useful.

Some Examples

The sight of a Christmas tree lot may make us wonder how many natural trees are sold at Christmas time. There are roughly 100 million families in the United States, but perhaps only half of them—50 million—have a Christmas tree. It is reasonable to assume that about half of those families with a tree buy a natural tree. We thus arrive at an estimate of about 25 million natural trees sold.

Now, let's say that we have a friend who owns a 200-acre Christmas tree farm (Fig. 1–5), and we wonder how many trees he can plant. We reasonably suppose that trees are planted about 6 ft apart, so that each tree takes up about 36 ft^2. Considering space occupied by roads, buildings, or other uses, we might change this estimate to 1 tree/50 ft^2. But how big is 200 acres? Most of us do not remember the precise size of an acre, but we might hazard a guess that a typical suburban house lot is about a quarter-acre, and that it is perhaps 100 ft across by 100 ft deep, or 10^4 ft^2. This indicates that an acre is about 40,000 ft^2. (In fact, 1 acre = 43,560 ft^2, so our guess is not far off.) If each tree requires about 50 ft^2, then our friend can plant 40,000/50 = 800 trees—let's take 1000 for simplicity—per acre, for a total of 200,000 trees. This is a reasonable order-of-magnitude calculation.

FIGURE 1–5 Images such as this one allow us to make useful estimates—in this case, a set of questions revolving around the number of Christmas trees that are planted.

EXAMPLE 1–3 Estimate the average area available to each person in the United States and then to each person on Earth.

Solution: There are some 250 million people in the United States. The United States is roughly a rectangle about 3000 mi (about 5000 km) from east to west and 2000 mi (about 3000 km) from north to south, giving an approximate total area of 15×10^6 km^2. With $(1\text{ km})^2 = (10^3\text{ m})^2 = 10^6\text{ m}^2$, this translates to 15×10^{12} m^2. Dividing 15×10^{12} m^2 by 3×10^8 people, we obtain 5×10^4 m^2 per person. This is a square roughly 200 m on a side.

For Earth, we might estimate a total of about 5 billion people. We remember from geography that the circumference of Earth is 25,000 mi, or about 40,000 km. We divide 40,000 km by 6 (circumference = $2\pi r$, and we will approximate 2π as 6) to obtain a radius of 7000 km. The total surface area of Earth is $4\pi r^2$, $12 \times 7 \times 7 \times 10^6$ km^2, or roughly 5×10^8 km^2. However, only about one-third of Earth's surface area is land—roughly 2×10^8 km^2, or 2×10^{14} m^2. Dividing 2×10^{14} m^2 by 5×10^9 people, we estimate an area on Earth of about 5×10^4 m^2 per person. The United States has about the same population density as does Earth.

EXAMPLE 1–4 A human body consists mainly of water. One mole (mol) of water, which consists of about 6×10^{23} molecules, has a mass of 18 g. Assuming that the molecules of water in your body are closely packed together, make a rough estimate of the size of a molecule.

Solution: We want to find the number of molecules N in your body, then divide your total volume by N to determine the volume of a single molecule.

The number N is determined by finding out how many moles you contain.† For a mass, let's use 60 kg, which is equivalent to a weight of about 132 lb. Because 60 kg = 60×10^3 g = 6×10^4 g, the number of moles n is

$$n = \frac{\text{total mass}}{\text{mass of 1 mol}} = \frac{6 \times 10^4\ \cancel{\text{g}}}{18\ \cancel{\text{g}}} = 3.3 \times 10^3\text{ mol}.$$

†A mole is about 6×10^{23} molecules. It is the quantity of a substance whose weight in grams equals the substance's molecular weight.

The total number of molecules is found by multiplying the number of moles by 6×10^{23}:

$$N = (6 \times 10^{23} \text{ molecules/}\cancel{\text{mol}})(3.3 \times 10^3 \cancel{\text{mol}}) = 2 \times 10^{27} \text{ molecules.}$$

To find the volume of one molecule, we must divide your total volume by N. We might estimate that you form a solid with height 2 m, width 0.5 m, and depth 0.3 m, for a total volume

$$V = (2 \text{ m})(0.5 \text{ m})(0.3 \text{ m}) = 0.3 \text{ m}^3.$$

Thus the volume V_1 of a molecule is

$$V_1 = \frac{V}{N} = \frac{0.3 \text{ m}^3}{2 \times 10^{27} \text{ molecules}} = 0.15 \times 10^{-27} \text{ m}^3\text{/molecule.}$$

We can estimate the linear size of a molecule by assuming that it forms a little cube with sides of length ℓ; its volume is then $V_1 = \ell^3$. Thus

$$\ell = \sqrt[3]{V_1} = \sqrt[3]{0.15 \times 10^{-27} \text{ m}^3} \simeq 0.5 \times 10^{-9} \text{ m},$$

or about 5×10^{-10} m. Because 1 Å is 10^{-10} m, the size of a water molecule is estimated to be several angstroms. In fact, the separation between the hydrogen atoms in a water molecule is about 2 Å.

1–6 SCALARS AND VECTORS

Scalars have no direction associated with them.

The mathematical descriptions of physical systems in this book deal with two types of quantities. One type is an ordinary algebraic quantity called a **scalar**. A scalar does not have a direction associated with it. The statement that the mass of a ball is $\frac{1}{4}$ kg specifies all we need to know about its mass. In particular, no direction is specified when a mass is given. The same is true of the time it takes the ball to travel a certain distance. Some other examples of scalar quantities are temperature, the energy of a moving body, and electric charge. Some scalar quantities, such as mass, are always positive, whereas others, such as electric charge, can be positive or negative.

Vectors have both magnitude and direction.

There are physical quantities, however, that cannot be described by scalars. Experiment shows that direction does matter at times: Objects fall down and not sideways. Specifying the velocity of a ball requires specifying not only the speed (how fast it is going), but also the direction in which it is traveling. Quantities that must be described by both a magnitude—always positive—and a direction are called **vectors**. Vectors describe velocity, force, electric fields, and numerous other quantities. They play an important role in physics, and this section summarizes their properties.

An Example: The Displacement Vector

The definition of displacement

One important vector quantity is **displacement**, the difference between two positions of an object. We shall use displacement to describe many of the properties of vectors. To understand displacement, imagine a treasure hunt in which we must proceed from some point K to a second point L, which is 30 paces northeast of K. The displacement from K to L may be drawn as an arrow on a map (Fig. 1–6). That arrow is the pictorial representation of the displacement vector from K to L.

We can give the vector a name—**B** in Fig. 1–6. The vector **B** has two attributes: a length, or **magnitude** (30 paces), and a direction (to the northeast). We refer to the tail of the vector as the point where it starts; the tip of the vector represents the point where it ends. Although the length of the displacement vector and the direction in which it points are fixed, the position of the tail (or the tip) of the vector is not. We can shift a vector by moving it to another location in such a way that the vector retains its original direction. Thus vector **B** represents the displacement from *any* point to any other point that is 30 paces away from, and to the northeast of, the starting point (Fig. 1–7).

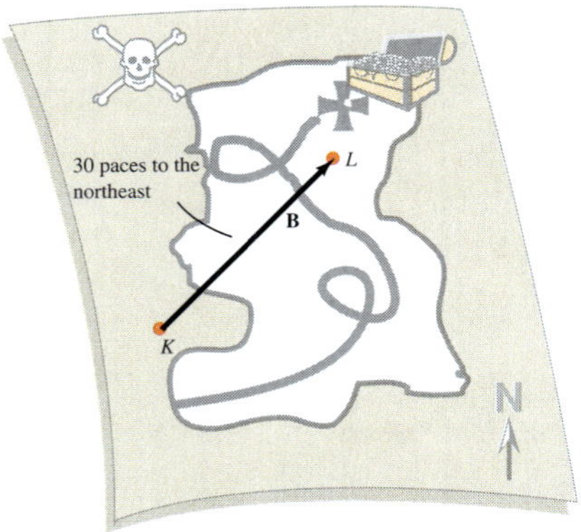

FIGURE 1–6 The displacement vector **B** from some point K to a second point L on the map for a treasure hunt.

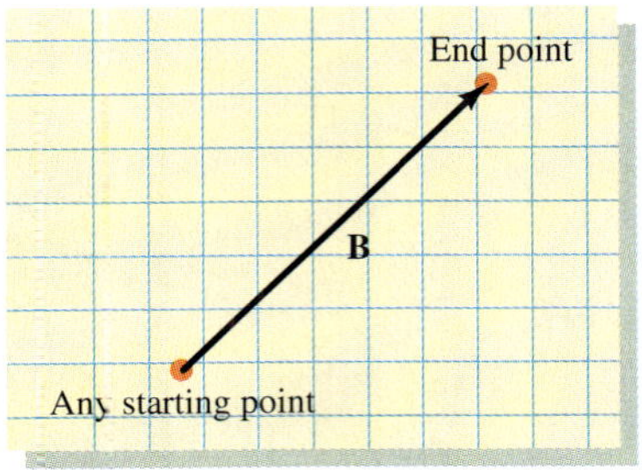

FIGURE 1–7 The same vector **B** represents the displacement 30 paces to the northeast from *any* point. Vectors can be shifted about without changing their characteristics, as long as their magnitude (length) and direction (orientation) are unchanged.

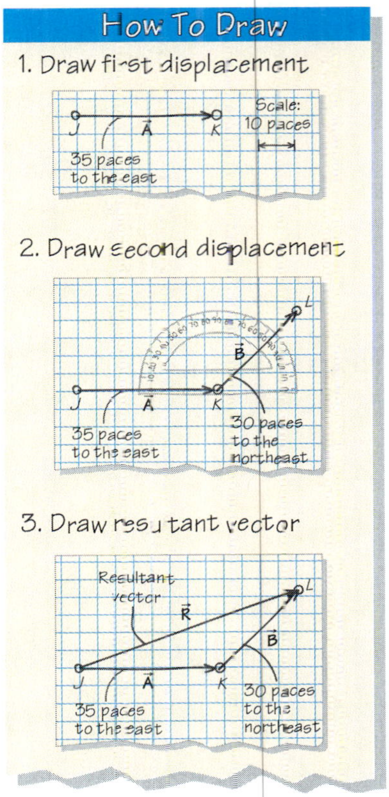

FIGURE 1–8 The treasure hunt specifies a series of displacement vectors, which leads to a net displacement. Here, we use a series of steps to outline how to add vectors to arrive at a resultant vector.

In the text, we denote all vectors with a boldface letter, such as **B**. An alternative notation, useful for writing and lecturing, is to put an arrow over the quantity, $\vec{B}$. The magnitude of **B** is sometimes denoted by $|\mathbf{B}|$ but more usually by the unadorned symbol B.

Addition and Subtraction of Vectors

Addition. The result of two successive displacements is also a displacement, which we call a *net displacement.* For example, our treasure hunt may specify an initial displacement from the point J to the point K, which is 35 paces due east of J. We call this displacement **A** in Fig. 1–8. Our second displacement would be vector **B**, which starts at point K and proceeds 30 paces northeast to point L. The net displacement, then, takes us from point J to point L, and we denote this by the vector **R**. The vector **R** is the *sum of the two vectors* **A** and **B**:

$$\mathbf{R} = \mathbf{A} + \mathbf{B}. \tag{1–1}$$

The sum of the two vectors, which is known as the **resultant vector**, is formed as follows. Draw vector **A**, then place the tail of vector **B** on the tip of vector **A**. The line from the tail of **A** to the tip of **B** is vector **R**. The addition of vectors is *commutative*—that is, the order of the vectors does not matter, so

$$\mathbf{A} + \mathbf{B} = \mathbf{B} + \mathbf{A}. \tag{1–2}$$

This is easy to see in Fig. 1–9 and using our prescription of placing the tail of the second vector at the tip of the first. The figure contains the sum in both orders, and the result is the same.

FIGURE 1–9 (a) We can find the sum **R** of **A** and **B** by placing **A** and **B** tail to head *in either order.* (b) By combining the two graphs in a parallelogram, the equality of the sum in either order is apparent. We can pass either through point K or point I.

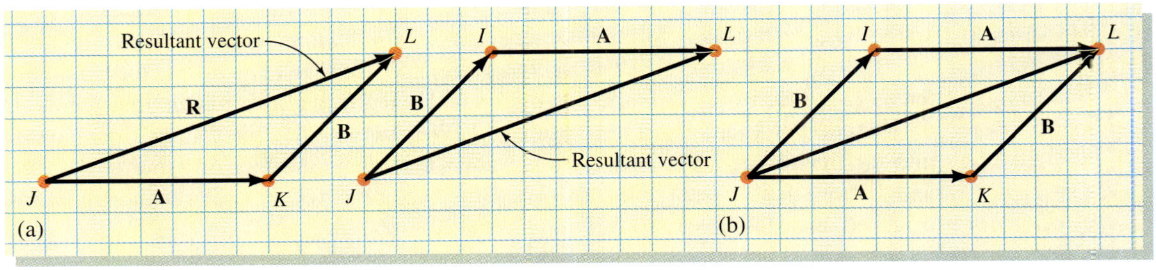

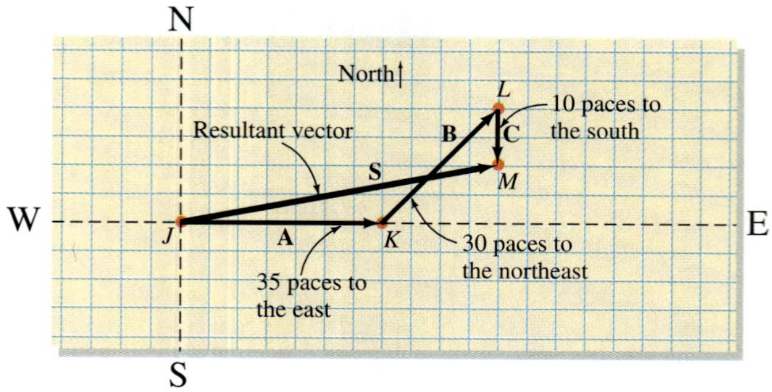

FIGURE 1–10 One more displacement vector, **C**, is added to the displacements in Fig. 1–9. The net displacement is **S**.

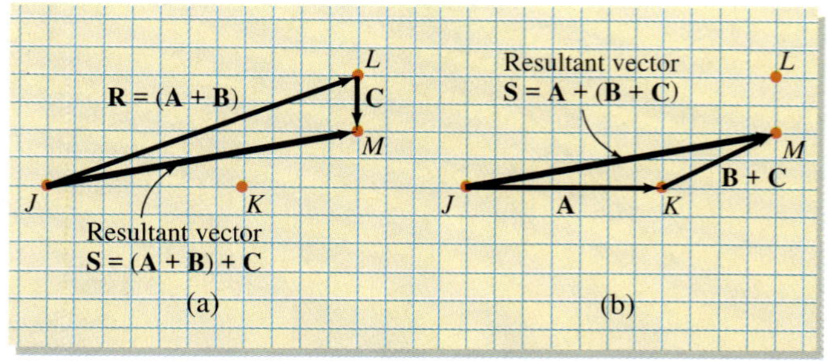

FIGURE 1–11 (a) The net displacement **S** is found by first adding **A** to **B**, then adding the result to **C**. (b) The net displacement **S** is found by first adding **B** to **C**, then adding the result to **A**. The net displacement is the same in both cases, which shows that vector addition is associative.

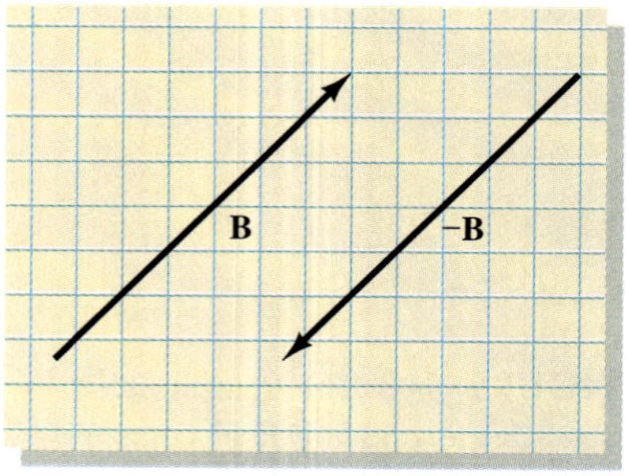

FIGURE 1–12 The vector $-\mathbf{B}$ is obtained from **B** by reversing the direction of **B** while leaving its magnitude unchanged.

The addition of vectors does not depend on how the vectors are grouped, or on their order.

Figure 1–10 shows the result of adding one more displacement vector to the series of displacements in Fig. 1–8: The vector **C** takes us from point L to point M on the figure. Vector **C** has magnitude 10 paces and points in a southerly direction. When **C** is added to $\mathbf{R} = \mathbf{A} + \mathbf{B}$, we obtain a new vector, **S**, which represents the net displacement from point J to point M:

$$\mathbf{S} = \mathbf{A} + \mathbf{B} + \mathbf{C}. \tag{1–3}$$

With this displacement vector, we can see that vector addition is *associative*. This term means that we can group the addition in any way we find convenient (Fig. 1–11):

$$\mathbf{S} = (\mathbf{A} + \mathbf{B}) + \mathbf{C} = \mathbf{A} + (\mathbf{B} + \mathbf{C}). \tag{1–4}$$

Subtraction. The *null vector* **0** is a special vector with zero magnitude. With the help of this vector, we can define the negative, $-\mathbf{B}$, of a vector **B**: When $-\mathbf{B}$ is added to the vector **B**, the sum is the null vector:

$$\mathbf{0} = \mathbf{B} + (-\mathbf{B}). \tag{1–5}$$

This means, as shown in Fig. 1–12, that $-\mathbf{B}$ is a vector that has the same magnitude as **B**, but $-\mathbf{B}$ points in the opposite direction. The subtraction $\mathbf{A} - \mathbf{B}$ of two vectors is simply the addition of **A** and $-\mathbf{B}$:

$$\mathbf{T} = \mathbf{A} - \mathbf{B} = \mathbf{A} + (-\mathbf{B}). \tag{1–6}$$

This vector is shown in Fig. 1–13.

Scalar Multiplication and Unit Vectors

The product of a vector and a scalar is a vector.

When vectors are multiplied by scalars, the result is a vector. Twice the displacement **B** is just a displacement with twice the magnitude of **B**, but **B** continues to point in the same direction. More generally, $b\mathbf{B}$ has a length b times that of the vector **B** (Fig. 1–14). The vector $4\mathbf{B}$ is formed by the *scalar multiplication* of 4 and **B**.

Unit vectors

This allows us to define *unit vectors* as follows (Fig. 1–15): Any vector **U** is written as $\mathbf{U} = U\hat{\mathbf{u}}$, where U is the magnitude of **U**, and $\hat{\mathbf{u}}$ is a **unit vector** that points in the direction of **U** and has a magnitude of 1. We place all the units (m, m/s, and so on) into the magnitude U, so that the unit vector $\hat{\mathbf{u}}$ is *dimensionless*. In other words, the unit vector simply specifies a direction. Other notations for the unit vector will be introduced as needed.

The definition of the position vector

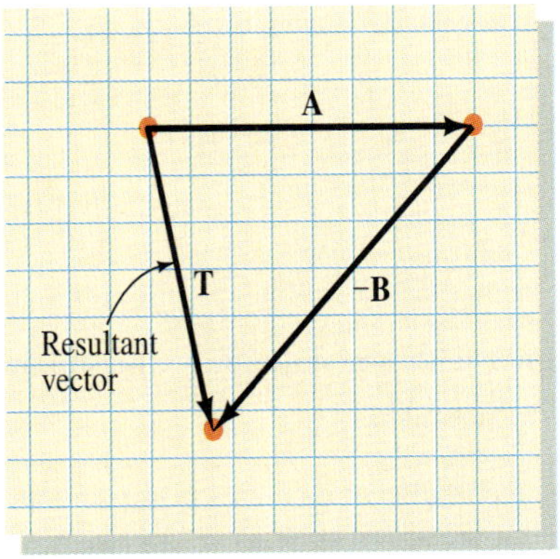

FIGURE 1–13 The vector difference $\mathbf{T} = \mathbf{A} - \mathbf{B}$.

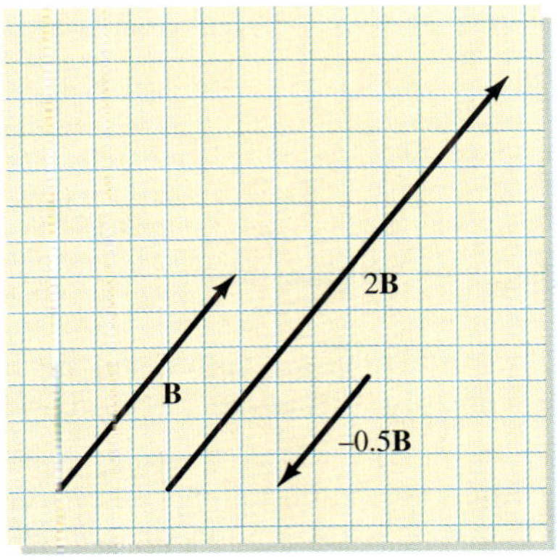

FIGURE 1–14 Scalar multiplication of a vector **B** by a scalar quantity b is a vector with the same (or opposite) direction as **B** but a magnitude or length scaled by the factor b.

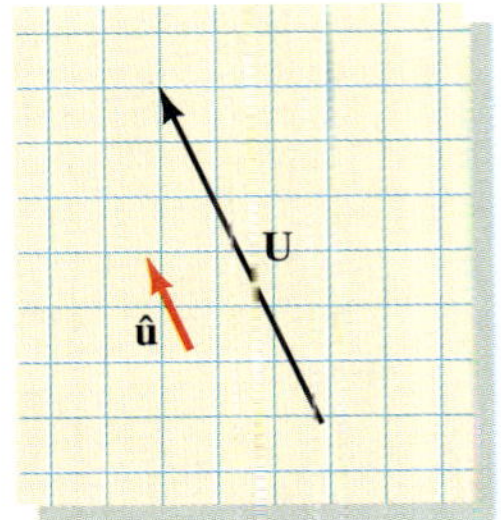

FIGURE 1–15 Any vector **U** can be broken into a scalar magnitude, U, times a unit vector (here labeled $\hat{\mathbf{u}}$), that points in the same direction as **U** $(\mathbf{U} = U\hat{\mathbf{u}})$. All units are placed into U so that $\hat{\mathbf{u}}$ is dimensionless.

Components

It is convenient to use coordinate axes to describe the location of points. In two dimensions, as on the surface of a table, we shall often specify two perpendicular axes, the x-axis and the y-axis (Fig. 1–16a). A point P can be located by giving its coordinates (x_1, y_1); that is, how far it is along the x-direction or the y-direction from the origin O. The point P may also be described with a *position vector* **D** that extends from the origin to P. We use the term **position vector** to denote the displacement *as measured from the origin of a particular coordinate frame.*

With our knowledge of vector addition and scalar multiplication, we arrive at this displacement in two steps: We first make a displacement of magnitude x_1 along the x-axis, then we follow it with a displacement of magnitude y_1 along the y-axis (Fig. 1–16b). To write this in vectorial form, we employ a unit vector **i** that points in the + x-direction and another unit vector **j** that points in the + y-direction. Then $x_1\mathbf{i}$ is a vector pointing in the x-direction whose magnitude is the absolute value of x_1, whereas $y_1\mathbf{j}$ is a vector pointing in the y-direction whose magnitude is the absolute value of y_1. Thus

$$\mathbf{D} = x_1\mathbf{i} + y_1\mathbf{j}. \tag{1–7}$$

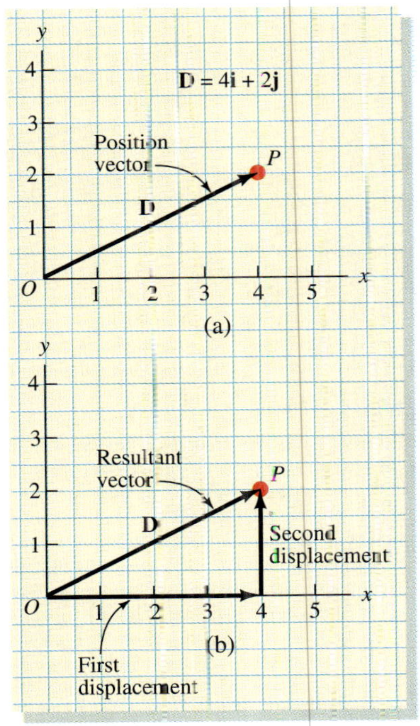

FIGURE 1–16 (a) A point P can be located in a two-dimensional coordinate system by specifying the two coordinates $x_1 = 4$ and $y_1 = 2$ along their x- and y-axes, respectively. The position of point P with respect to the origin, O, is the vector **D**. (b) The displacement **D** from O to P can be produced with two successive displacements, taken in either order: a displacement in the x-direction and a displacement in the y-direction.

Figure 1–17 applies this kind of reasoning to a more general vector **V**. We draw the vector in the xy-plane. The vector has **components** V_x and V_y, the elements into which the vector can be resolved, defined such that

$$\mathbf{V} = V_x\mathbf{i} + V_y\mathbf{j}. \tag{1–8}$$

The **component vectors** of **V** are the vectors $V_x\mathbf{i}$ and $V_y\mathbf{j}$. The coordinate frame is specified in the figure, *as it always must be if we refer to particular components.* Figure 1–17 also shows another way to describe the vector **V**. It follows from Pythagoras's theorem that the length of **V** is

$$V = \sqrt{V_x^2 + V_y^2}. \tag{1–9}$$

We see from the figure that **V** makes an angle θ with the + x-direction. The angle θ is given by

$$\tan\theta = \frac{V_y}{V_x} \quad \text{or} \quad \theta = \tan^{-1}\left(\frac{V_y}{V_x}\right). \tag{1–10}$$

Unit vectors in the x- and y-directions

Equivalently, the vector components are described in terms of V and θ by

$$V_x = V\cos\theta \tag{1–11a}$$

The definitions of components and component vectors

and

$$V_y = V\sin\theta. \tag{1–11b}$$

We have therefore shown that

In vector addition, the components add algebraically to form the components of the resultant vector.

In two dimensions, a vector may be described either with a magnitude V and an angle θ measured from the x-axis, or with components V_x and V_y.

PROBLEM SOLVING

Use components with vector addition and subtraction.

Addition (and subtraction) are especially simple to handle in terms of components. The component vectors add independently, so the components do too. Thus, for example, if **V** and **W** are two vectors with components (V_x, V_y) and (W_x, W_y), then

$$\mathbf{V} + \mathbf{W} = (V_x + W_x)\mathbf{i} + (V_y + W_y)\mathbf{j}. \tag{1–12}$$

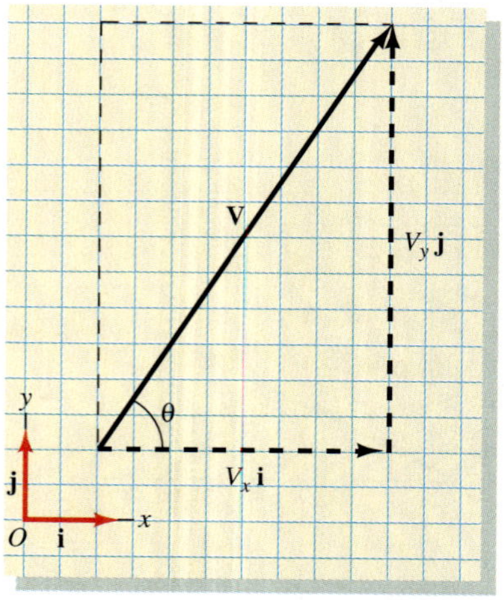

FIGURE 1–17 A vector **V** in two dimensions is described either by a magnitude V and an angle θ or by the components V_x and V_y.

EXAMPLE 1–5 The displacement from a point A to point B describes a displacement vector **D**. The (x, y) coordinates of point A (measured in meters) are (2.0, 6.0); the (x, y) coordinates of B are (5.0, 10.0). (a) Describe **D** in terms of the position vectors **A** and **B** of the points A and B, respectively. (b) Calculate the length of **D** and the angle it makes with the x-axis.

Solution: First, we draw a figure of the situation, as shown in Fig. 1–18. The figure includes the specified coordinate system as well as the two points A and B, the position vectors **A** and **B**, and the displacement vector **D**.

(a) The figure tells us immediately that **A** + **D** = **B**. We can now solve this equation for **D**:

$$\mathbf{D} = \mathbf{B} - \mathbf{A}.$$

We have specified the coordinates of both **A** and **B**:

$$\mathbf{A} = (2.0\ \text{m})\mathbf{i} - (6.0\ \text{m})\mathbf{j} \quad \text{and} \quad \mathbf{B} = (5.0\ \text{m})\mathbf{i} + (10.0\ \text{m})\mathbf{j}.$$

Then, from Eq. (1–12),

$$\mathbf{D} = (5.0\ \text{m})\mathbf{i} - (2.0\ \text{m})\mathbf{i} + (10.0\ \text{m} - 6.0\ \text{m})\mathbf{j} = (3.0\ \text{m})\mathbf{i} + (4.0\ \text{m})\mathbf{j}.$$

The minus signs appear because, in **B** − **A**, the components of **A** are *subtracted* from those of **B**.

(b) Once we know the components of **D**, we can use Eqs. (1–9) and (1–10) to find the length D and the angle θ that **D** makes with the x-axis. We have, from Eq. (1–9),

$$D = \sqrt{D_x^2 + D_y^2} = \sqrt{(3.0\ \text{m})^2 + (4.0\ \text{m})^2} = 5.0\ \text{m},$$

and, from Eq. (1–10),

$$\tan\theta = \frac{D_y}{D_x} = \frac{4.0\ \text{m}}{3.0\ \text{m}} = 1.33;$$

that is, $\theta = 53°$.

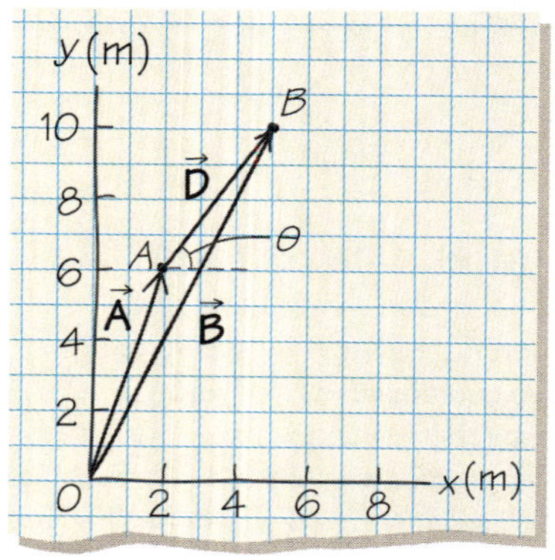

FIGURE 1–18 Example 1–5. The two points A and B have position vectors **A** and **B**, respectively. The vector **D** is the displacement from point A to point B.

EXAMPLE 1–6 Figure 1–10 specifies a set of points J, K, L, and M, which refer to a treasure hunt. In the hunt, distances are measured in paces. Express the vectors **A**, **B**, and **C** by describing the displacements from J to K, from K to L, and from L to M, respectively. Describe these displacements in component form relative to a set of axes in which the x-axis points to the east and the y-axis points north. Use your description to calculate the sum of the three vectors. Give the length and inclination with respect to the x-axis of the resultant vector.

Solution: Figure 1–19 adds a set of x- and y-axes to Fig. 1–10 drawn in as specified in the problem. The vector **A** lies along the x-direction; thus, it has no

y-component and is given by

$$\mathbf{A} = (35 \text{ paces})\mathbf{i}.$$

The vector **B** points in the northeast direction; that is, it makes an angle of $\phi = 45°$ with the x-axis. Thus $\sin\phi = \cos\phi = 1/\sqrt{2} \simeq 0.71$. From Eqs. (1–11a and b) we have $B_x = B\cos\phi = (30 \text{ paces})(0.71) = 21$ paces. B_y has the same value. Thus

$$\mathbf{B} = B_x\mathbf{i} + B_y\mathbf{j} = (21 \text{ paces})\mathbf{i} + (21 \text{ paces})\mathbf{j}.$$

Finally, the vector **C** has a y-component only. Because it points to the $-y$-direction, it has a component that is -10 paces:

$$\mathbf{C} = (-10 \text{ paces})\mathbf{j}.$$

Now that we have the three vectors, we can find their sum **S** by adding the components:

$$\mathbf{S} = \mathbf{A} + \mathbf{B} + \mathbf{C} = (35 \text{ paces})\mathbf{i} + (21 \text{ paces})\mathbf{i} + (21 \text{ paces})\mathbf{j} - (10 \text{ paces})\mathbf{j}$$

$$= (35 \text{ paces} + 21 \text{ paces})\mathbf{i} + (21 \text{ paces} - 10 \text{ paces})\mathbf{j} = (56 \text{ paces})\mathbf{i} + (11 \text{ paces})\mathbf{j}.$$

To find the length and angle of **S**, given its components, we use Eqs. (1–9) and (1–10):

$$S = \sqrt{(56 \text{ paces})^2 + (11 \text{ paces})^2} = 57 \text{ paces};$$

$$\tan\theta = \frac{11 \text{ paces}}{56 \text{ paces}} = 0.20.$$

or $\theta = 11°$.

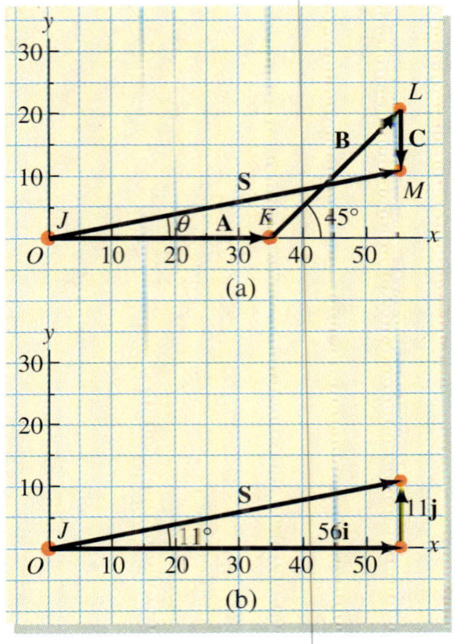

FIGURE 1–19 Example 1–6. (a) A set of coordinate axes has been added to the treasure hunt specified in Fig. 1–10. (b) **S** is constructed by adding the components of **A**, **B**, and **C**.

Vectors in Three-Dimensional Space

The vectors we have considered thus far are two-dimensional. Vectors can also represent quantities in three dimensions. A vector in three dimensions can be specified with a *Cartesian* or *Euclidean* set of axes x, y, and z. (The two-dimensional x- and y-axes are also called Cartesian.) Figure 1–20 illustrates such a three-dimensional system, showing the conventional orientation of the three axes. The orientation of the axes is best described using a *right-hand rule*. Start with the usual x- and y-axes (x to the east, say, and y to the north). To find the direction of the z-axis:

> **Point the fingers of your right hand along the x-axis. Now curl these fingers in the direction of the y-axis. Your thumb will point along the $+z$-axis.**

This right-hand rule has become a well-established convention and will appear in many places in this book. Figure 1–21 shows how a vector **V** is decomposed into three components along the three axes.

The three unit vectors for the three axes are denoted by **i**, **j**, and **k**; the unit vector **k** points in the $+z$-direction (Fig. 1–22). A point P in three-dimensional space is

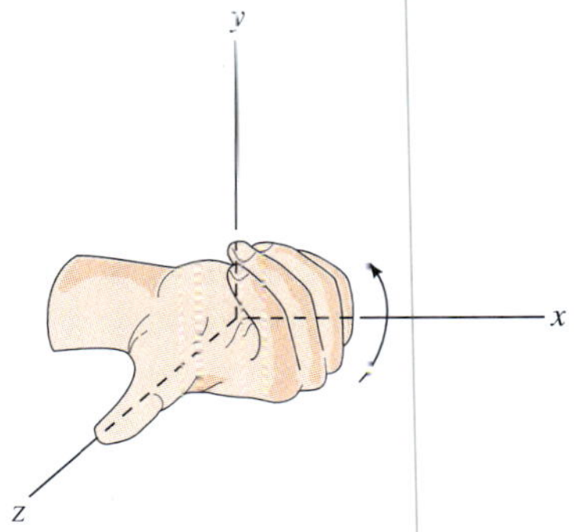

FIGURE 1–20 A Cartesian coordinate system for three dimensions. A right-hand rule specifies the orientation of the three axes. The z-axis points out of the plane of the paper. The thumb points along the z-axis; the fingers curl from the x-axis toward the y-axis.

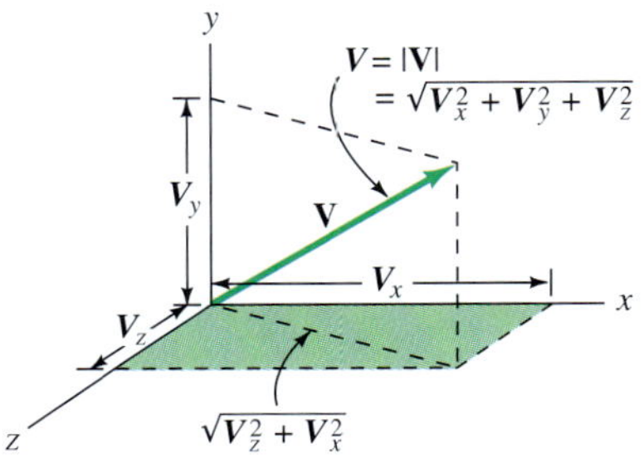

FIGURE 1–21 A vector **V** decomposed into three components in three-dimensional space.

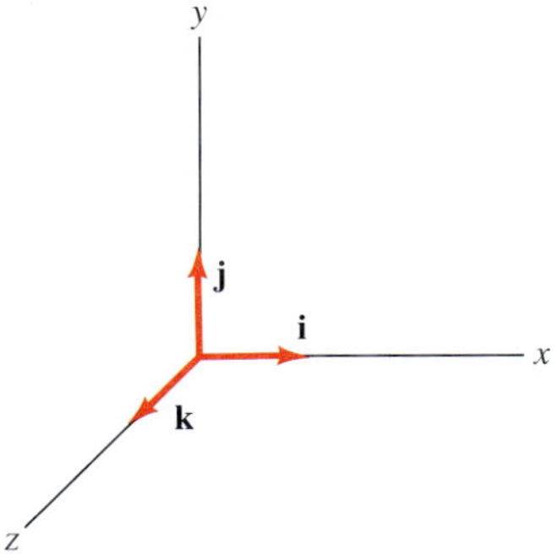

FIGURE 1–22 The unit vectors **i**, **j**, and **k** are aligned along the three Cartesian axes x, y, and z, respectively. Each has unit magnitude.

now assigned the coordinates (x, y, z) in a given frame, and its displacement vector from the origin—its position vector—is

$$\mathbf{D} = x\mathbf{i} + y\mathbf{j} + z\mathbf{k}. \tag{1–13}$$

Similarly, any vector $\mathbf{V}$ has the components (V_x, V_y, V_z), such that

$$\mathbf{V} = V_x\mathbf{i} + V_y\mathbf{j} + V_z\mathbf{k}. \tag{1–14}$$

Pythagoras's theorem again tells us the magnitude, or length, of $\mathbf{V}$:

$$V = \sqrt{V_x^2 + V_y^2 + V_z^2}. \tag{1–15}$$

Vector Equations

Vector equations are shorthand for a set of equations, one for each component.

Vector equations are equations in which vector quantities appear on both sides of the equal sign. For example, we may want to express the velocity of a projectile in terms of its position. Velocity is a vector, as is position. The equality in such an equation means either that the magnitudes and direction of the net vector on each side are the same, or, equivalently, that the respective components of the vector equation are equal. In two dimensions, a vector equation is really shorthand for two separate equations for the components; in three dimensions, a vector equation represents three separate equations.

* 1–7 THE USES OF DIMENSIONAL ANALYSIS[†]

Scaling Laws

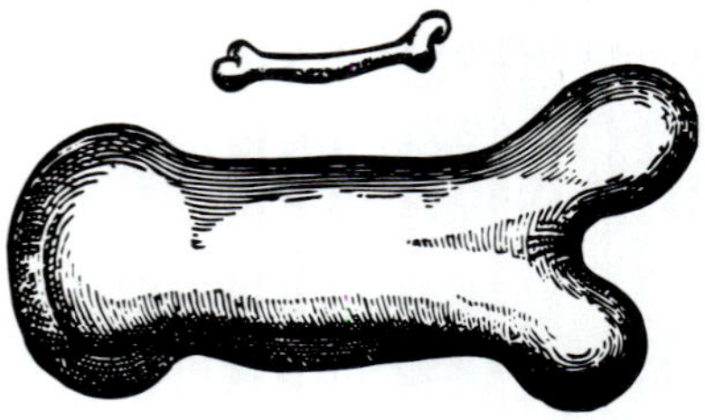

FIGURE 1–23 The scaling of bones. (From Galileo, *Dialogue Concerning Two New Sciences*, 1638.)

In Section 1–4, dimensional analysis was used as a check in problem solving. There are two other important applications of dimensional analysis. In the first of these, we can discover scaling laws. *Scaling laws* reveal how a change in one dimensional parameter of a problem leads to changes in another parameter with the same dimensions. Let us consider a problem that Galileo studied in his book *Dialogue Concerning Two New Sciences*; namely, the scaling properties of animals: What must giants look like? Consider a bone (Fig. 1–23) whose length is ℓ, characterizing the height, or linear size, of an animal. The width of the bone is b. It is a fact that the strength of such a bone is proportional to its cross-sectional area; that is,

$$\text{strength} = c_1 b^2.$$

The constant factor c_1 is characteristic of the material, and it does not change if the size of the animal changes. This strength must be such as to support the weight or, equivalently, the *mass* m of the animal. But the mass is proportional to the animal's volume, and the volume is, in turn, proportional to ℓ^3, just as the volume of a cube is equal to the length of the side cubed. Thus

$$m = c_2 \ell^3.$$

The precise value of c_2 depends on the precise shape of the animal and need not concern us. Now, if the strength of the bone must be proportional to the mass, we have

$$\text{strength} = c_3 m.$$

Combining these three equations, we have

$$c_1 b^2 = c_3 m = c_3 c_2 \ell^3;$$

$$c = \frac{\ell^3}{b^2}, \quad \text{where } c = \frac{c_1}{c_2 c_3}. \tag{1–16}$$

The constant c is characteristic of our animal's internal structure, but not of its size.

[†]An asterisk (*) indicates optional material that may be skipped without loss of continuity.

Now what happens if we try to change the size parameter ℓ by a factor f? The constant c cannot change, so b must change to compensate for the change in ℓ. Equation (1–16) can be satisfied when we change ℓ to $f \times \ell$ only if b changes by a factor f'. We find f' by insisting that the ratio ℓ^3/b^2 be constant:

$$\frac{f^3}{f'^2} = 1 \quad \text{or} \quad f' = f^{3/2}. \tag{1–17}$$

If the length of the bone changes by a factor f, then its width must change by a factor $f^{3/2}$. For example, if the length is changed by a factor 3, as in Fig. 1–23, then according to Eq. (1–17), the width of the bone must change by a factor $(3)^{3/2} \simeq 5$. The scaling argument demonstrates that the giant animal cannot have the same proportions as its smaller model and helps us to understand why hippopotamuses will never be mistaken for overgrown chihuahuas (Fig. 1–24).

FIGURE 1–24 The giant ants of the 1950s movie *Them* have the same proportions as their normal counterparts do. Galileo's scaling analysis shows that such creatures could not exist because their structure could not support their weight.

Understanding the scaling behavior of physical systems is of primary importance for engineering. Such understanding allows filmmakers to film ripples in a bathtub in slow motion and convince us that they are mighty waves on the ocean, or tells us how the stiffness of a beam varies with its length and width. Scaling arguments tell us how to strengthen structures or how to test small models of large aircraft in wind tunnels.

Relations Based on Dimensional Analysis

The second major application of dimensional analysis lies in the derivation of new relations between physical quantities. Consider the simple pendulum as an example. The simple pendulum consists of a small bob of mass m on the end of a light string of length ℓ. A pendulum swings because, when it is displaced away from the vertical direction, gravity pulls it back down. It overshoots the minimum, swings to the other side, then repeats its motion. One full cycle of this motion takes a time τ that we can call the *period*. The problem is this: How does the period depend on the mass of the pendulum bob?

To answer this question, we must gather a list of physical parameters that might be relevant to the period. This requires some knowledge. Could the period of the pendulum depend on the *internal* structure of either the bob or the string? We must answer no because if we change the string or change the bob from lead to wood (of the same mass) the period does not change. Nor do we expect air resistance to play a large role. The list we gather includes the bob mass m, with dimension $[M]$; the length ℓ of the string, with dimension $[L]$; and the acceleration of gravity, g, discussed in Section 1–4. The latter quantity has dimension $[g] = [LT^{-2}]$. There are no other dimensional quantities on which the period of the pendulum should depend.

The dimension of the period is time $[T]$. We now look for an algebraic combination of m, ℓ, and g that has the dimension of τ. We want to find q, r, and s so that

$$[\tau] = [m^q][\ell^r][g^s],$$

or, in terms of the dimensions,

$$[T] = [M^q][L^r][L^s T^{-2s}].$$

Now there are no powers of $[M]$ on the left-hand side, so $q = 0$. We see that *the mass does not enter at all into the period.* In Problem (1–53), we continue this treatment and show that, in fact, the only combination of the parameters with the same dimension as that of the period is

$$\tau \propto \sqrt{\frac{\ell}{g}}.$$

(The symbol $\propto$ indicates proportionality.) This is an illuminating result. It does not tell us the dimensionless numerical coefficient by which the square root is multiplied to give an equality for τ rather than a proportionality. But it does reveal the dependence of τ on ℓ and g, and it does say that the mass does not enter into the result. One measurement of the period of a pendulum with known length would determine the unknown numerical coefficient. We might even hope that the coefficient is not too far from unity. In fact, this is very often the case!

SUMMARY

The range of quantities relevant to our understanding of the physical world is so large that it is useful to employ scientific notation. In this notation, any number can be represented by a decimal number from 1 to 10 multiplied by a power of 10.

The quantities that appear in physics and engineering have units as well as sizes. The International System of units, or SI, provides reproducible and precise definitions of mass, length, and time. The SI units are, respectively, kilograms (kg), meters (m), and seconds (s). Some quantities are used so often that their units are given a special name (for example, force is measured in newtons, N, in SI), but these units are derived units: They can always be expressed in terms of the three primary units (for example, $1\ \text{N} = 1\ \text{kg} \cdot \text{m/s}^2$). Units that appear in equations can be manipulated algebraically; after conversion to a single unit system, the units on both sides of any correct equation will match.

Mass, length, and time are the quantities with the three primary dimensions, abbreviated M, L, T. Dimensions should not be confused with units, which refer to a particular choice of unit system. Any physical quantity has dimensions that are rational combinations of the primary dimensions. Dimensions can be manipulated algebraically, and both sides of any correct equation will have the same dimensions. When we analyze the dimensions of an equation, we are performing a dimensional analysis. Dimensional analysis is useful for checking the answers to problems, for learning scaling laws, and for discovering relations between physical quantities.

Numbers that represent physical quantities can be measured only to a certain accuracy. An explicit way to indicate this accuracy is to write a physical quantity x as a central value $\pm$ an uncertainty. Calculations involving physical quantities are meaningful only to within the known accuracy of those quantities. When several numbers of different accuracies are involved in a calculation, the least accurate quantity primarily determines the accuracy of the result. A second way to indicate the known accuracy of a physical quantity is through the use of significant figures; that is, the number of digits between 1 and 10, which are then multiplied by the power of 10 in scientific notation.

The ability to estimate is one that should be cultivated. An educated first guess is a valuable start to the solution of any problem. An order-of-magnitude calculation is such a guess. Similarly, when you arrive at an answer to a physical problem, it is always wise to ask yourself if it makes sense.

Some, but not all, physical quantities include directional information. Temperature and time do not. Such quantities are scalars: They have magnitude (including the appropriate dimensions and units) only. Displacement and velocity do include directional information; such quantities are represented by vectors. The vectors **A** and **B** are mathematical objects with both magnitude and direction. They obey the rule

$$\mathbf{A} + \mathbf{B} = \mathbf{B} + \mathbf{A}. \tag{1–2}$$

Vectors can be expressed in graphical form; we draw them as arrows of length equal to their magnitude within a particular coordinate system. The simplest such system is the Cartesian system, with mutually perpendicular x-, y-, and z-axes for three-dimensional space. We can express any vector **V** in terms of the unit vectors—vectors of unit length—for a given coordinate system. Thus the unit vectors **i**, **j**, and **k** point along the x-, y-, and z-axes, respectively. Then

$$\mathbf{V} = V_x\mathbf{i} + V_y\mathbf{j} + V_z\mathbf{k}. \tag{1–14}$$

The quantities V_x, V_y, and V_z are the components of **V**. The magnitude of **V** is

$$V = \sqrt{V_x^2 + V_y^2 + V_z^2}. \tag{1–15}$$

Vector equations are relations that equate different vectors. In such an equation, the components of the vectors are equal on both sides, so that, in three dimensions, a vector equation stands for three separate equations.

UNDERSTANDING KEY CONCEPTS

1. If space were somehow four-dimensional instead of three-dimensional, would the concept of displacement vectors still make sense? In what way could the idea of vector displacement be generalized?
2. How might the measurement of distance in centimeters, meters, and kilometers rather than in inches, feet, yards, and miles be more convenient? How might it be less convenient?
3. What is your height in centimeters and meters?
4. What is your mass (notice that we do not say *weight*) in kilograms?
5. The human pulse and the swing of a pendulum are possible time units. Are they ideal ones?
6. How can length be defined in terms of the speeds of light and time?
7. Three vectors all have the same magnitude. Is it possible to add them together to obtain a null vector?
8. Three vectors all have different magnitudes, given by 4 m, 5 m, and 7 m. Is it possible for the three vectors to add to zero? Prove your result by drawing a figure.
9. Is the vector sum of the two unit vectors **i** + **j** also a unit vector? Under what conditions is it possible for two unit vectors to add to a resultant vector that also has unit length?
10. Suppose we were in radio contact with some distant civilization. How would we communicate how large we are? What assumptions are you making in providing your suggestions?
11. A useful estimate is that there are $\pi \times 10^7$ s in a year. How accurate is this?
12. The next time you weigh yourself, consider the uncertainty of the numbers you read. What would you estimate the percentage uncertainty in this measurement of your weight to be?
13. A small worm absorbs the oxygen it needs through its surface. Assume that the oxygen needed by an animal is proportional to its mass, and estimate how much the absorption of oxygen per unit area would have to increase if the worm were to increase each of its dimensions by a factor of 10. You may find it interesting to know that the lungs of a human being have about 100 m^2 of surface available for the absorption of oxygen.
14. A mouse eats the equivalent of about one-quarter of its mass in food every day. You do not. Why?

PROBLEMS

1–1 Scientific Notation

1. (I) Ten thousand jelly beans are in a jar, and 25 percent of them are green. Express in scientific notation the number of green jelly beans.
2. (I) What is the product of 10^5 and 10^{-4}? The ratio $10^{-4}/10^5$?
3. (II) Calculate the cube root of the number 10^{21}, as well as the square of the resulting number.

1–2 Length, Time, and Mass

4. (I) The Empire State Building is 1472 ft high. Express this height in both meters and centimeters.
5. (I) What is your height in atomic diameters (see Table 1–1)?
6. (I) A 5-ft, 5-in-tall skier should use skis 5 cm longer than her height. How long should her skis be? Skis are made in 5-cm intervals (150 cm, 155 cm, and so on). What length skis should she buy if she rounds off to the nearest 5 cm?
7. (I) Grapes sell in Italy for 1000 Lire per kilogram. If the conversion rate between dollars and Lire is 1400 Lire = \$1, what is the price of grapes in dollars per kilogram?
8. (II) Gasoline is heavily taxed in Europe, and at one time in West Germany cost 1.90 Deutschmark (DM) per liter. What price in dollars per gallon is this if the currency conversion is 1.7 DM = \$1 US? One gallon (gal) = 3.8 liters (L).
9. (II) The acceleration due to gravity, g, is 9.80 m/s^2 in SI. Convert this to the British system, where length is measured in feet rather than meters.
10. (II) The gravitational constant G is $6.67 \times 10^{-11}\ m^3 \cdot s^{-2} \cdot kg^{-1}$. What is G in units of $cm^3 \cdot s^{-2} \cdot g^{-1}$?
11. (II) The radius of the Moon (assumed spherical) is 1.74×10^3 km, and its mass is 7.35×10^{22} kg. What is the density of the Moon in grams per cubic centimeter?
12. (II) A neutron star has a radius of 15 km and a mass of 1.4×10^{31} kg. What is the density of the neutron star in metric tons per cubic centimeter?
13. (II) Gasoline consumption in Europe is measured in liters per 100 km. For example, a small Opel uses 7.0 L per 100 km while the gasoline consumption of a large Mercedes is 23 L per 100 km. Convert these to mi/gal.

1–3 Accuracy and Significant Figures

14. (I) A student wishes to make a measurement of the road distance from his dormitory to the physics building of his university. He uses his car's trip odometer, which measures distance only in units of a tenth of a mile. (a) He makes one trip, and the odometer reads 0.3 mi. What can he say is the distance, and, in particular, to how many significant figures?

(b) One day he has nothing better to do with his time and, adding trips both to and from, he makes 100 trips; his odometer measures 27.2 mi. What can he now say is the distance? How is this result consistent with the result of part (a)? (c) A friend challenges the student on his measurement and, upon reflection, the student is not sure whether he made 99, 100, or 101 trips. How should he modify his statement of the distance?

15. (II) A well-known approximation to π is $\pi \simeq \frac{22}{7}$. What percentage error does this result have? How much better is the approximation 355/113?

16. (II) The force F on a mass m moving at speed v in a circular path of radius r has magnitude $F = mv^2/r$. The mass is measured to be 0.00535 kg, the radius is 0.3 m, and the speed is 1.1 m/s. Give the magnitude of the force. Pay attention to the number of significant figures.

17. (II) A rectangular box is stated to have width $w = (1.25 \pm 0.03)$ m, length $\ell = (0.5 \pm 0.1)$ m, and height $h = (0.137 \pm 0.028)$ m. What is the volume, stated to the appropriate number of significant figures with an uncertainty?

18. (II) You wish to determine your density (mass/unit volume) by two measurements: by weighing yourself on a digital scale, and by submerging yourself in a tank of water with vertical sides and noting the rise in water level on a scale marked in centimeters (Fig. 1–25). The surface area of water in the tank is 1.5 m^2. The digital scale gives the weight in pounds at 0.5-lb intervals, and you cannot make a reading of the water level to better than 0.5-cm accuracy. Suppose that the weight reading is 213 lb, and the water level changes from 152 cm to 158.5 cm. What is your density? Express your result as a central value with a percentage error.

FIGURE 1–25 Problem 18.

19. (III) If you want to know the area of a circle to 5-percent accuracy, how accurately should you measure the diameter of the circle?

20. (III) Consider the infinite series $\sum_{n=0}^{\infty}(x^n/n!)$. The symbol $n!$ means the product $1 \times 2 \times 3 \times \cdots \times n$. By definition, $0! = 1$. If $x = 0.1$, how many terms in the series suffice to give a result correct to 6 significant figures?

1–4 Dimensional Analysis

21. (I) The kinetic energy of a baseball is denoted by $mv^2/2 = p^2/2m$, where m is the baseball's mass and v is its speed. This relation can be used to define p, the baseball's momentum. Use dimensional analysis to find the dimensions of momentum.

22. (I) One of Einstein's most famous results is contained in the formula $E = mc^2$, where E is the energy content of the mass m, and c is the speed of light. What are the dimensions of E?

23. (I) A length L that appears in atomic physics is given by the formula $L = h/m_ec$, where m_e is the mass of an electron, c is the speed of light, and h is a constant known as Planck's constant. What are the dimensions of h?

24. (II) What are the dimensions of h^2/m^3G, where h is a constant called Planck's constant, m is a mass, and G is the gravitational constant? The dimensions of the constants in this formula can be found in the list of physical constants given in Appendix II.

25. (III) A force F acting on a body of mass m a distance r from some origin has magnitude $F = Ame^{-\alpha r}/r^3$, where A and α are both constants. The constant $e = 2.718\ldots$ is Euler's constant. Given that the force has dimensions kg·m/s^2, what are the dimensions of (a) the constant α? (b) the constant A?

1–5 How a Little Reasoning Goes a Long Way

26. (I) How many times does an average person's heart beat in a lifetime? Estimate the number of times an automobile tire rotates in a trip across the United States.

27. (II) A criminal, posing as a tourist, wants to smuggle $100 million in gold across the U.S. border in his station wagon. Is he likely to make it? Use estimates, and explain your reasoning.

28. (II) Estimate the volume of concrete used for the construction of the tunnel under the English Channel. The system consists of two railway tunnels and a service tunnel, and is 30 km long.

29. (II) Estimate the area used for the storage of one bit on a 1.44 MB $3\frac{1}{2}''$ floppy disk (1 byte = 8 bits).

30. (II) What is the weight of water in a full 5-gal can? About what volume of water can a typical person carry?

31. (II) Make separate estimates of the number of automobile mechanics in the United States based on (a) your total bill for automobile repairs compared with a reasonable average salary for a mechanic; (b) how many hours it takes to repair a car; (c) the number of people you know who are automobile mechanics compared with the total number of people you know.

32. (II) The Sun is 93 million mi from Earth. What is the diameter of the Sun? (A dime held at arm's length will just about cover the surface of the Sun [Fig. 1–26].)

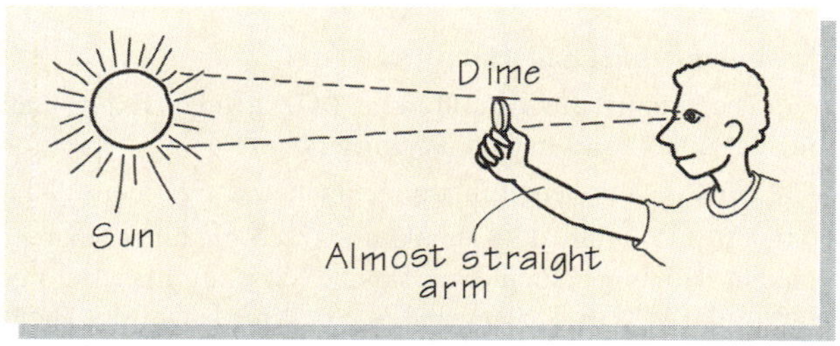

FIGURE 1–26 Problem 32.

33. (II) A typical cloud contains droplets of water with an average radius of 0.5×10^{-4} m. How many droplets are needed for a cloud that provides a rainfall of 1 cm in your city?

34. (II) A strong radioactive point source emits 10^8 gamma particles per second uniformly in all directions. What fraction of these particles will hit a circular detector of diameter 4 cm that is 2 m away from the source?

35. (II) There are over 100 million automobiles in the United States, and each one is driven about 20,000 mi/yr. Estimate the number of automobiles on the road at any one moment.

36. (II) Suppose that all the eighteen-wheeler trucks in the United States line up bumper-to-bumper on I-80. Can they form a continuous line from New York to San Francisco?

37. (II) Suppose that the circumference of Earth is a perfect circle of exactly 25,000 mi. Somebody prepares a wire that is supposed to go around the equator completely, but makes it 1 m too long by mistake. If this 1-m-too-long wire were placed around the equator in a perfect circle with the ends of the wire just touching each other, by how much would the wire be off the ground?

38. (II) Make a rough estimate of the number of apples produced each year in a 10-acre orchard.

39. (II) The mass of one atom of hydrogen is 1.6×10^{-27} kg. Given that the mass of the Sun is 2×10^{30} kg, how many atoms of hydrogen would it contain if it consisted purely of hydrogen? Actually, it consists of 70 percent hydrogen and 30 percent helium by mass, and the mass of a helium atom is 6.4×10^{-27} kg. How many hydrogen atoms does the Sun have?

40. (II) Suppose that oil consists of molecules of CH_2. This means that 1 mol of oil, consisting of 6×10^{23} molecules, has a mass of 14 g. One milliliter (mL) of oil has a mass of 0.95 g. When this much oil is poured on water, it does not spread forever but spreads until it makes a circular film of area 1.5×10^{7} cm^2. How does the fact that the oil does not spread forever support the idea of atoms and molecules? Assuming that the oil slick is one molecule thick (a *monomolecular layer*), with the molecules touching, and that the molecules are spheres, estimate the size of an oil molecule.

1–6 Scalars and Vectors

41. (I) A girl jogs around a circular lake. Devise a simple coordinate system to describe her position and direction of travel at any time, assuming that she begins at the south end of the lake and jogs clockwise at a speed of 2 m/s.

42. (I) Draw the vector $3\mathbf{i} + 4\mathbf{j}$ by first drawing the x-component vector, then the y-component vector, then adding them graphically. Multiply the vector by a factor of two and repeat the exercise.

43. (I) A drunken sailor stumbles 4 paces north, 6 paces northeast, 2 paces east, and 5 paces west. Describe the final location from the initial position by a single displacement vector.

44. (I) What is the resultant vector when the vectors $\mathbf{A} = 6\mathbf{i} - 5\mathbf{j}$ and $\mathbf{B} = 8\mathbf{i} + 3\mathbf{j}$ are added together? When $\mathbf{B}$ is subtracted from $\mathbf{A}$?

45. (II) A football player catches the kickoff on the 5-yd line and runs straight up the field for 20 yd, turns left for 15 yd, goes straight up the field for 10 yd, turns right for 25 yd, reverses his field (makes a 180° turn) for 10 yd, and then streaks straight up the field for a touchdown. Define a coordinate system and list the entire path in vector form.

46. (II) Draw a vector $\mathbf{V}$ that points in the northwesterly direction, making an angle α with the northerly direction, as in Fig. 1–27. If north is chosen as the $+y$-direction and east as the $+x$-direction, what is the x-component of $\mathbf{V}$?

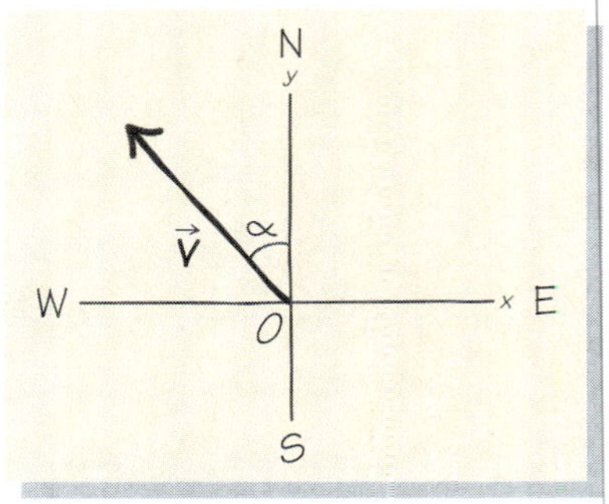

FIGURE 1–27 Problem 46.

47. (II) Suppose that in Problem 46 you choose north as the $+x$-direction and west as the $+y$-direction. What is the x-component of $\mathbf{V}$ in this case?

48. (II) Refer to the situation outlined in Problems 46 and 47. Choose the $+x$-axis as the line that makes an angle of 45° with the northerly direction and is inclined to the east, and the $+y$-axis as the line that makes a 45° angle with the westerly direction and is inclined to the north. What is the x-component of $\mathbf{V}$ in this case?

49. (II) In computer-aided drafting programs, lines can be specified in either rectangular or polar coordinates (Fig. 1–28). In such programs, if the coordinates (x, y) are given, the cursor draws a line from its current position a distance x to the right and a distance y up. If polar coordinates (r, θ) are given, the line is drawn from its current position through an angle θ in the counterclockwise direction from the positive x-axis, through a distance r. Give the instructions both in terms of (x, y) and (r, θ) for the drafting of the triangle shown in the figure, starting at the point A.

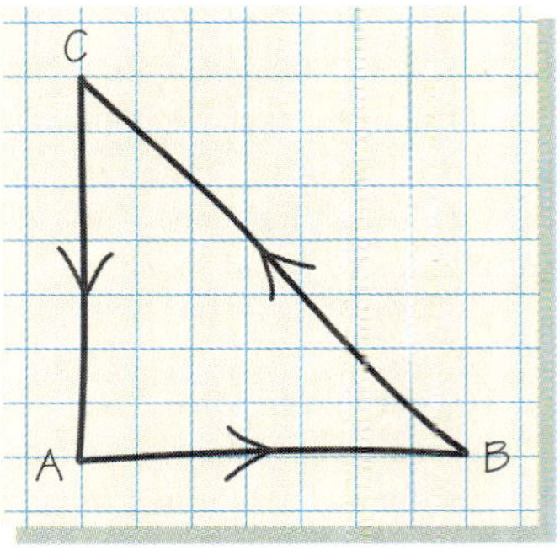

FIGURE 1–28 Problem 49.

50. (II) Consider the following vectors: $\mathbf{A} = -2\mathbf{i} - 3\mathbf{j}$; $\mathbf{B} = \mathbf{i} + 2\mathbf{j} + 3\mathbf{k}$; $\mathbf{C} = 3\mathbf{j} + 3\mathbf{k}$; and $\mathbf{D} = -2\mathbf{i} - \mathbf{k}$. Find (a) $\mathbf{A} + \mathbf{B} + \mathbf{C} + \mathbf{D}$; (b) $\mathbf{A} - \mathbf{D}$; (c) $\mathbf{A} + \mathbf{D} - \mathbf{B}$ and (d) $|\mathbf{A} + \mathbf{C}|$.

51. (II) Vectors $\mathbf{A}$, $\mathbf{B}$, $\mathbf{C}$, and $\mathbf{D}$ are shown in Fig. 1–29. (a) Give the vectors in component form. (b) Determine the following quantities both algebraically and graphically: $2\mathbf{A} - \mathbf{C} - \mathbf{D}$, $\mathbf{B} + \mathbf{C}/2$, $|\mathbf{D} - \mathbf{B}|$.

52. (II) Suppose that you have three vectors, $\mathbf{A} = 3\mathbf{i} + 4\mathbf{j}$, $\mathbf{B} = 2\mathbf{i} - 2\mathbf{j} + 4\mathbf{k}$, and $\mathbf{C} = -\mathbf{i} + 5\mathbf{j} - 3\mathbf{k}$. Show that the sum of these three vectors can alternatively be computed by first summing $\mathbf{A}$ and $\mathbf{B}$ and then summing the resultant with $\mathbf{C}$, or by first summing $\mathbf{B}$ and $\mathbf{C}$ and then summing the resultant with $\mathbf{A}$.

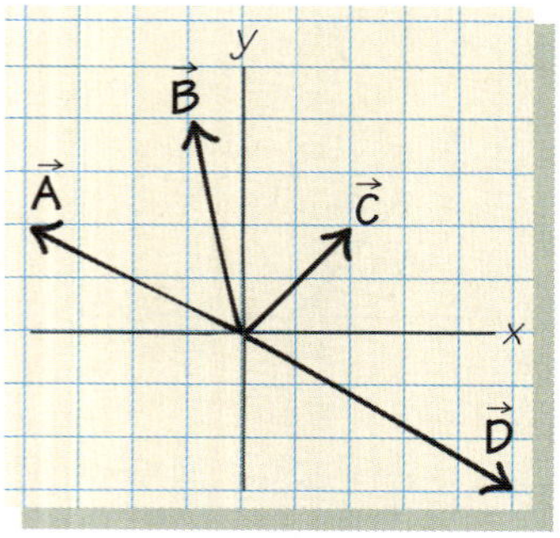

FIGURE 1–29 Problem 51.

*1–7 *The Uses of Dimensional Analysis*

53. (II) We have seen in the text that the period of a simple pendulum is independent of the mass of the pendulum bob. Further, we have seen that the dimensional relation between the period, τ, the length, ℓ, of the pendulum, and the acceleration of gravity, g, takes the form

$$[\tau] = [\ell^r][g^s].$$

Use the fact that the dimension of τ is [T], that of ℓ is [L], and that of g is [L/T^2] to show that

$$\tau = \sqrt{\frac{\ell}{g}}.$$

54. (II) In quantum mechanics, the fundamental constant called Planck's constant, h, has dimensions of $[ML^2T^{-1}]$. Construct a quantity with the dimensions of length from h, a mass m, and c, the speed of light.

55. (II) It is known that the quantity Ke^2/hc is dimensionless (K is a numerical constant; h and c are as discussed in Problem 54). (a) What are the dimensions of e? (b) What are the dimensions of e^2/R, where R is a length?

56. (II) You are told that the speed of sound in a metal depends only on the density ρ ($[ML^{-3}]$) and on the bulk modulus of the metal, B, which has dimensions $[ML^{-1}\,T^{-2}]$. Express the sound speed in terms of ρ and B.

General Problems

57. (I) There are 100 cm in a meter, 1000 kg in a metric ton, and 1000 g in a kilogram. (a) How many metric tons are in 1 g? (b) How many cubic meters are in 1 cm^3?

58. (I) A human hair has a diameter of about 10^{-4} m. Given that all of the hair on a head can be gathered into a lock of diameter 4 cm, estimate the fraction of the human scalp area from which hairs spring, and the number of hairs on the head.

59. (I) Given that the speed of light is 2.998×10^8 m/s and that the distance from the Moon to Earth is 2.4×10^5 mi, how long does it take light to travel from the Moon to Earth?

60. (I) Measure the height, width, and thickness of this book in centimeters. Estimate its mass. Calculate its density. An object will float in water if its density is less than that of water. Without doing the experiment, would you expect this book to float in water?

61. (I) Eighteen grams of water is known to contain 6.02×10^{23} molecules of H_2O. What is the mass in kilograms of one molecule of H_2O?

62. (II) The density of a human body is approximately 1 g/cm^3, which is also the density of water. Use the result of the calculation in Problem 61 to estimate the number of molecules of water that a typical human body would contain if it were made up entirely of water.

63. (II) Imagine that molecules of H_2O are stacked up in a cubic array, like a large number of cubical boxes, with a water molecule at the center of each cube. Let the side of each cube be L. Given that the density of water is 1 g/cm^3, estimate the distance L from the data given in Problem 61.

64. (II) A silver nucleus consists of 108 closely packed nucleons (protons and neutrons) and has a radius of 5×10^{-15} m. A neutron star is basically an overgrown nucleus, with neutrons only, closely packed in the same way as nucleons are in the silver nucleus. If the radius of a neutron star is 12 km, how many neutrons does it contain?

65. (II) A typical star has a mass of about 2×10^{30} kg, and there are about 10^{11} stars in a galaxy. What is the mass of this typical galaxy? Assume that stars are made primarily of hydrogen; the mass of a hydrogen atom is 1.67×10^{-27} kg. How many hydrogen atoms are there in a galaxy?

66. (II) The gasoline usage rate required to propel an automobile is very roughly proportional to the mass of the automobile. Assuming that the proportions and types of materials of an automobile do not change, calculate the percentage gasoline savings that would be realized if cars were reduced by 12 percent in all their dimensions.

67. In aquatic animals, the energy E available for motion is proportional to the mass of the animal, and the friction F with their skin is proportional to the surface area. All such animals have the same density, very close to that of water. If the maximum speed v such an animal can reach varies as $\sqrt{E/F}$, show that v is also proportional to $\sqrt{L}$, where L is some length characterizing the animal's size.

68. (II) Determine the thickness of a page of this book to an accuracy better than 10 percent. Explain your method and give your uncertainty.

69. (II) The water supply of Pittsburgh is contaminated by an oil spill. Make some reasonable assumptions in order to estimate how many trucks per day are needed to bring in a minimum supply of water. What if each person were allowed to take a bath every 3 days?

70. (II) Assume that houses are set on quarter-acre lots and each house receives four pieces of mail a day. Estimate how far a postal carrier walks in 1 day, and how much mail he or she carries. What is the mass of all this mail?

71. (II) According to Kepler's third law of planetary motion, the square of the period of a planet is proportional to the cube of its mean distance from the Sun. Given that Earth, whose period is 1 yr, is 1.5×10^8 km away from the Sun, calculate the distance from the Sun to Venus, whose period is 0.61 yr, and the period of Saturn, which is 14×10^8 km away from the Sun.

72. (II) A mouse is 10 cm in length, whereas an elephant is 4 m in length. The amount of food an animal must eat is proportional to its heat loss, and the heat loss is proportional to its surface area. Compare the percentage of body weight that a mouse and an elephant must eat each day. Ignore the detailed differences in shape between an elephant and a mouse.

73. (II) The mass of a cylinder cut from the entire atmosphere of area 1 cm^2 is approximately the same as the mass of a cylinder of the same area of water that is 30 ft high. Use this to estimate the number of "molecules of air" in the atmosphere, given the fact that, on average, one molecule of air is 1.6 times as massive as one molecule of water. (See Example 1–4.)

74. (II) Show that $(\mathbf{A} + \mathbf{B}) + \mathbf{C} = \mathbf{A} + (\mathbf{B} + \mathbf{C})$, where **A**, **B**, and **C** are vectors.

75. (II) A vector **u** in the xy-plane has x- and y-components $u \cos \theta$ and $u \sin \theta$, respectively, where θ is the angle that **u** makes with the $+x$-axis. A second vector, **v**, also lies in the xy-plane, and it is perpendicular to **u**. (a) Draw a figure to show that there are two possibilities for the vector **v**: either its x- and y-components are $-v \sin \theta$ and $v \cos \theta$, or else they are $v \sin \theta$ and $-v \cos \theta$. (b) Show that $v_x u_x + v_y u_y = 0$, independent of θ.

76. (II) For objects that move in a circle about an origin O, it can be convenient to use the mutually perpendicular unit vectors $\mathbf{i}_r$ and $\mathbf{i}_t$, defined as in Fig. 1–30. If a Cartesian coordinate system has its origin at O, with an x-axis chosen so that the angle between it and the line OP is θ, then (a) show that $\mathbf{i}_r = \cos \theta\ \mathbf{i} + \sin \theta\ \mathbf{j}$; (b) calculate the y-component of $\mathbf{i}_r$; and (c) express $\mathbf{i}_t$ as a combination of **i** and **j**.

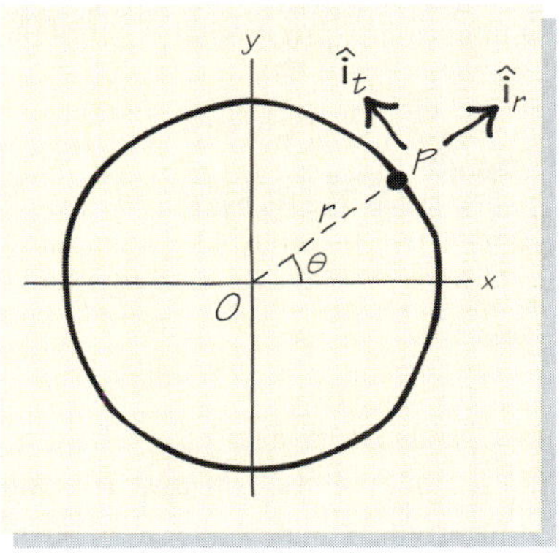

FIGURE 1–30 Problem 76.

77. (II) A vector **r** has length r and points in the direction shown in Fig. 1–31. The angles θ and ϕ are drawn in the figure. (a) Show that the x-component of **r** is $r \sin \theta \cos \phi$. (b) Show that the z-component of **r** is $r \cos \theta$. (c) Find the y-component of **r** in terms of r, θ, and ϕ.

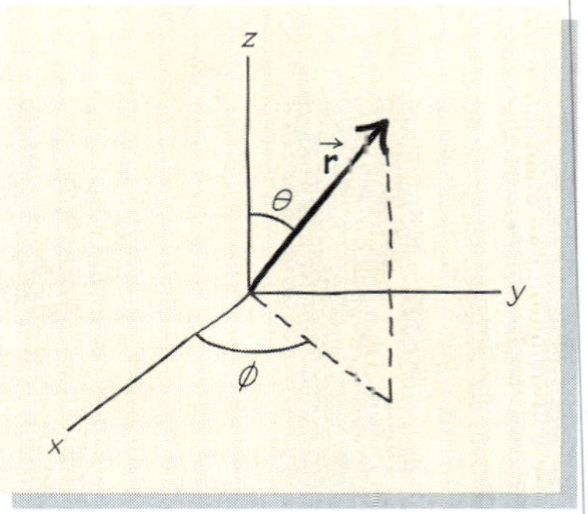

FIGURE 1–31 Problem 77.

78. (II) The number of molecules in 22.4 L of air at the surface of Earth is around 6×10^{23}. A liter is $10^{-3}\ m^3$. Let us say that the volume of air you take in and then expel with each breath is 2.5 L. Let us also say that the air around Earth is a uniform blanket 8 km thick. (Actually, it drops as height increases, but our assumption is not far off.) Finally, Michelangelo Buonarotti lived for 91 yr and died sufficiently long ago that all the "molecules of air" he ever breathed are thoroughly dispersed throughout the atmosphere. Under all these conditions, how many molecules of air breathed by Michelangelo are in your lungs right now?

79. (II) A long-legged bird, the stilt, has legs 20 cm long and a mass of 150 g. A flamingo has a mass of 2.5 kg. How long would you estimate the flamingo's legs to be, assuming that the density of the two birds is the same?

80. (II) An elephant has a mass of 4000 kg. What would you estimate the diameter of an elephant's leg to be, given that yours is around 12 cm? Does your result make sense?

81. (III) A stretched wire has three physical attributes: the density λ, or mass per unit length; the total length ℓ; and the tension τ. The latter is related to how hard the wire is being pulled to keep it stretched, and has dimensions of $[MLT^{-2}]$. Show by dimensional analysis that if the time t_0 of one back-and-forth vibration of the wire in a direction perpendicular to its length depends only on these three quantities, then t_0 has the form $t_0 = (\text{a constant})\ \ell \sqrt{\lambda/\tau}$.

CHAPTER

2

The Super High Altitude Research Project of the Lawrence Livermore National Laboratory is based on a two-stage gas gun meant to accelerate a projectile to very high speeds in very short periods of time. The white pipe in the center of the photo is the launch tube that directs the projectile; burning hydrogen gas that makes the projectile accelerate is seen here escaping from each end of the tube.

Straight-Line Motion

Understanding motion is one of the key goals of physical law. That is why we begin with a study of **mechanics**, the science of motion and its causes. The *description* of motion is the domain of **kinematics**, which is the subject of this chapter and the next. Once we know how to describe motion, we can explore the *causes* of motion, which come under the heading of **dynamics**. In this chapter, we study motion in a straight line; that is, one-dimensional motion. In the next chapter, we shall extend this study to two and three dimensions—motion on a plane and motion in space.

2–1 DISPLACEMENT

Consider the motion of athletes running a 100-m dash. Some runners are able to attain a tremendous advantage right at the beginning of the race—they leap ahead of the others. Other runners have a late kick that allows them to take the lead at the end of the race (Fig. 2–1). Suppose that we set up an electronic timing system along the path of the 100-m dash that records the times of the runners every 5 m. Data provided by this experiment would yield the distance each runner travels as a function of time.

We first want to quantify the notion of *distance traveled.* To do this, we construct a coordinate system with an x-axis, using the variable x to indicate the distance traveled. The choice of origin on the x-axis is up to us, but it is convenient to label the distance markers from the starting line, which we therefore set at $x = 0$ m. We also choose the direction that the runners take to be the $+x$-direction. We have invented a set of reasonable data points, shown in Fig. 2–2a, that represent the distance traveled by one runner—actually, the distance of a point on the runner's chest—plotted against the time it takes the runner to reach that x-value. We choose the starting time to be $t = 0$ s, again for convenience.

Figure 2–2b shows a curve that interpolates the data points of Fig. 2–2a to reasonable accuracy. It can be used to read the time taken to reach any distance x, even those values not given directly by the data. The curve, which we call $x(t)$, represents the distance from the starting point, x, as a function of time t. To understand this curve better, we use the notion of *displacement*, a concept we met in Chapter 1. The displacement is the *change* in the position of an object. If we denote the position at time t_1 by x_1, and the position at time t_2 by x_2, then the displacement is the difference between these two positions; this is defined mathematically by

$$\Delta x \equiv x_2 - x_1. \qquad (2\text{–}1)$$

FIGURE 2–1 The 100-m dash provides us with a test case in which we can refine our ideas of motion.

We use the capital Greek letter Δ (delta) to indicate a change (or difference) in a variable from one value to another. The time interval is, similarly,

$$\Delta t = t_2 - t_1. \qquad (2\text{–}2)$$

There is an important distinction between Δx and Δt: the displacement Δx is really a *vector*, while the time interval Δt is a *scalar*. In our example of the runner, x increases steadily with time, and the vectorial property of the displacement does not play an important role. If there were motion in both the $+x$- and $-x$-directions, then we could have a negative displacement during some time interval and a positive one during some other time interval. For one-dimensional motion, just keeping track of the sign of x as well as its magnitude takes care of its specification. For two- or three-dimensional motion, however, the displacement has both a magnitude and a direction, and the full vectorial description is important. In anticipation, we define the vector displacement by

Vectors and scalars were introduced in Chapter 1.

The displacement vector is independent of the origin.

$$\mathbf{\Delta x} \equiv \mathbf{x}_2 - \mathbf{x}_1, \qquad (2\text{–}3)$$

where $\mathbf{x}_2$ is the *position vector* at time t_2 (the vector from the starting line to the position of the runner at time t_2) and $\mathbf{x}_1$ is the position vector at t_1.

The vector aspect of our one-dimensional case is handled neatly with the unit vector $\mathbf{i}$ pointing along the $+x$-axis (Fig. 2–3a). The vector $\mathbf{x}_1$, for example, is written as $x_1\mathbf{i}$, where x_1 represents the distance from the origin, including the sign.

While the definition of the position vector depends on the choice of origin, the vector displacement $\mathbf{\Delta x}$ does not. Suppose that the origin were placed 10 m back of

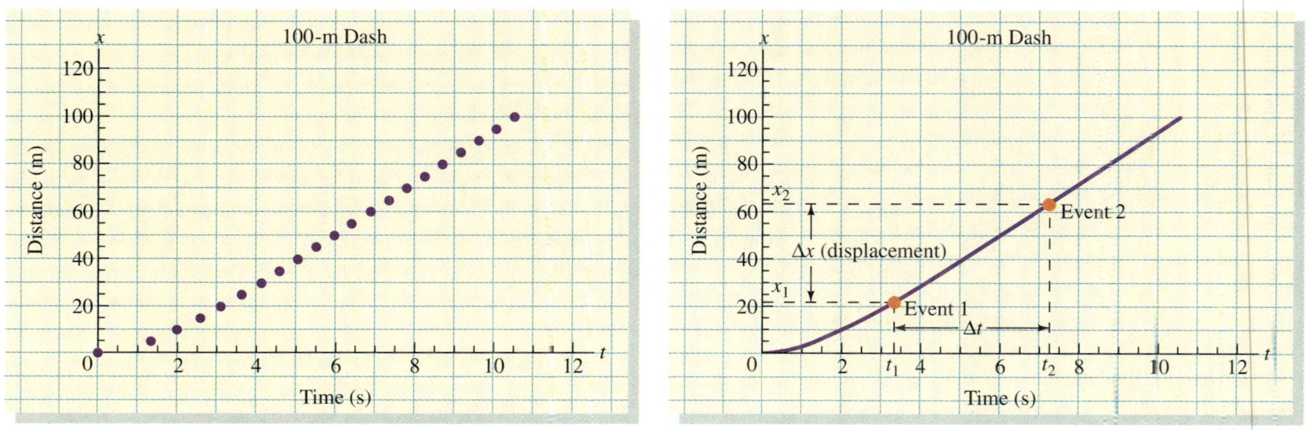

(a)

(b)

FIGURE 2–2 (a) A set of data (invented for this illustration) of distance versus time at 5-m intervals for the 100-m dash. (b) The curve is an interpolation of the data shown in part (a). The runner is at position x_1 at time t_1 (event 1) and at position x_2 at time t_2 (event 2). In the time interval $\Delta t = t_2 - t_1$, the runner has had a displacement $\Delta x = x_2 - x_1$.

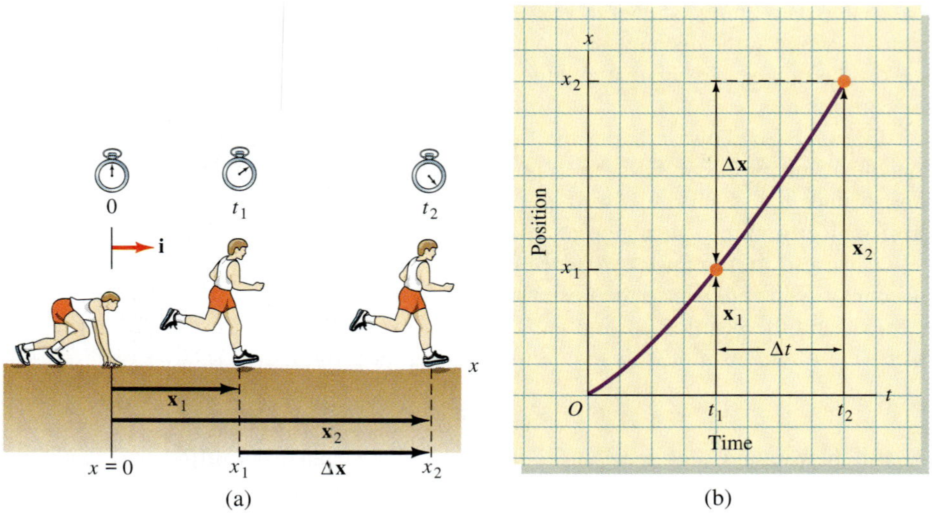

FIGURE 2–3 (a) The position vectors $\mathbf{x}_1$ and $\mathbf{x}_2$ run from the origin to the positions x_1 and x_2, respectively, of the runner. The displacement vector $\Delta\mathbf{x} = \mathbf{x}_2 - \mathbf{x}_1$ is also drawn. (b) The information contained in part (a) can be plotted as a curve of position versus time.

the starting blocks. Then the position at time t_1 would be given by $(x_1 + 10\text{ m})\mathbf{i}$, and that at time t_2 would be given by $(x_2 + 10\text{ m})\mathbf{i}$. The displacement would still be given by

$$\Delta\mathbf{x} = (x_2 - x_1)\mathbf{i}. \tag{2–4}$$

This relationship is shown in Fig. 2–3.

EXAMPLE 2–1 Use Fig. 2–2 to find the position vectors for the dash at times 2 s and 5 s. What is the displacement vector between these two times?

Solution: We can use the diagram in Fig. 2–2 to read the distance traveled for the two times 2 s and 5 s. To determine the displacement vector, defined in Eq. (2–3), we first find the position vectors for 2 s and 5 s. Because the origin is at the beginning of the race, the position vector at 2 s is $\mathbf{x}_1 = 10\mathbf{i}$ m, and at 5 s, $\mathbf{x}_2 = 40\mathbf{i}$ m. Substitution of the position vectors into Eq. (2–3) gives us the displacement vector $\Delta\mathbf{x}$:

$$\Delta\mathbf{x} = \mathbf{x}_2 - \mathbf{x}_1 = (40\text{ m})\mathbf{i} - (10\text{ m})\mathbf{i} = (30\text{ m})\mathbf{i}.$$

We put the appropriate numerical values in at the last step of the problem to determine the displacement vector. The magnitude of the displacement vector between 2 s and 5 s is the magnitude of $\Delta\mathbf{x}$, or 30 m.

2–2 SPEED AND VELOCITY

Speed

The quantities *speed* and *velocity* describe how fast the position of an object changes. The speed of an automobile or airplane is a familiar concept; we want to refine it here. The **average speed** for a particular motion is defined as the total distance traveled divided by the time taken to travel that distance:

$$\text{average speed} \equiv \frac{\text{total distance traveled}}{\text{time interval}}. \tag{2–5}$$

The speed, a scalar quantity, is always positive. As the time interval in the definition of average speed is changed, the average speed may also change. For example, during the first 5.01 s of the 100-m dash, the runner has traveled 40 m. The average speed

PROBLEM-SOLVING TECHNIQUES

1. Problem solving is important. The techniques we present here represent the basis for a plan of action on *any* problem. As we proceed, other techniques will fit into the general scheme. Read the problem, then read it again. Failure to read the problem carefully is perhaps the source of more false starts and wrong answers than is any other cause.
2. Draw a sketch or diagram of the problem that will help you visualize the situation presented by the problem. We illustrate this aspect of the technique in Fig. B1–1.
3. Write down the given and known quantities.
4. Make sure you understand which quantities are to be found.
5. There are generally only a few principles applicable to the solution of a problem. Think about which principles link the quantities to be determined to those that are known.
6. Use the principles that apply to the situation to guide you to the equation or equations that contain the quantities in the problem. Pay attention—at times, certain equations apply to a given situation and others do not. The rest is mathematics! Several equations may need to be manipulated together at times. Count the number of equations available to see if there are enough equations to determine the unknowns.
7. When you solve for an unknown in terms of the known quantities, use symbols, not numbers. Wait until the end to replace symbols with numbers and units. It is important to include units; the answer may require them and the proper cancellation of units provides a valuable accuracy check.
8. When you arrive at a number, think about it. Does it make sense? If you find that it takes 3 min to drive from New York City to Los Angeles, you have probably made a mistake!
9. Use *any* checks you can find for your result.

How To Draw Sketches

1. **Problem Statement:** A girl starts from rest at $x = -10$ m (time = 0), walks to $x = 20$ m at $t = 25$ s, turns around, and walks back to $x = -10$ m where she stops at 45 s. Sketch a plot of position versus time.

2. **Thinking Process:** Let's first make a sketch of the motion along the x-axis and mark the start, turn around, and stop positions as 1, 2, and 3.

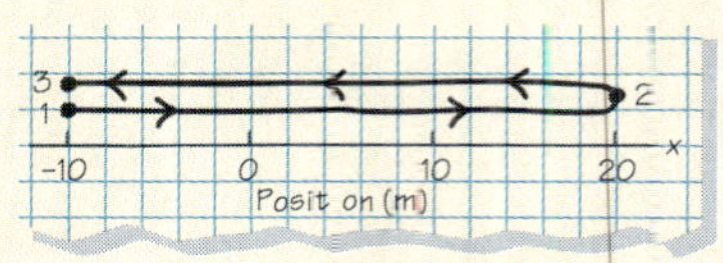

(a)

Note that the positions 1, 2, and 3 are "at rest." For these three positions, the speed is zero and their slope on the x versus t plot must be zero.

3. **Make a Sketch:**

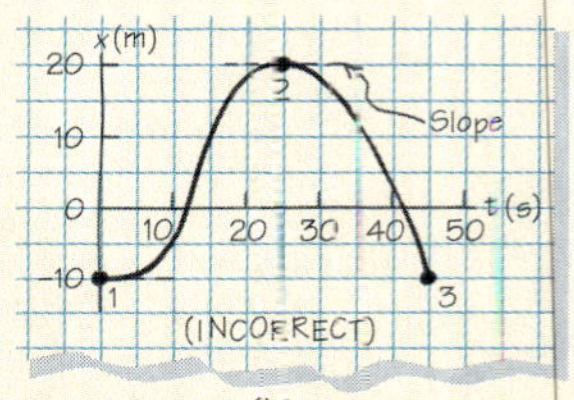

(b)

4. **Check Sketch:** The curve goes through all three known positions. However, the slope (remember the speed) at position 2 is not zero so we must fix that.

5. **Redraw Sketch:**

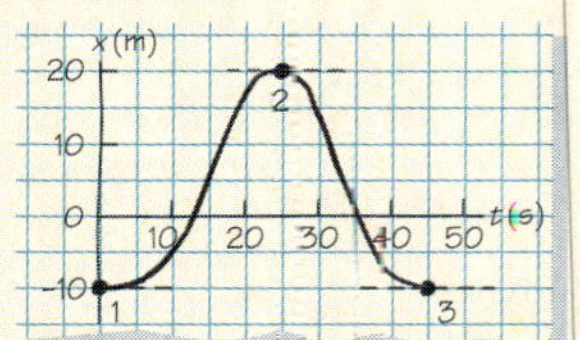

(c)

FIGURE 2B1–1 Some suggestions for illustrations.

over this interval is

$$\text{for 0 s to 5.01 s:}\quad \text{average speed} = \frac{40\ \text{m}}{5.01\ \text{s}} = 8.0\ \text{m/s}.$$

For the last 5.49 s, however, the runner progresses 60 m, and the average speed is

$$\text{for 5.01 s to 10.5 s:}\quad \text{average speed} = \frac{60\ \text{m}}{5.49\ \text{s}} = 10.9\ \text{m/s}.$$

The runner completes the 100-m dash in 10.5 s. Over the entire 100-m dash, then, the runner's average speed is (100 m)/(10.5 s), or 9.5 m/s. Although the notion of speed may be useful for sports, it is of limited utility in mechanics, where velocity plays a much more important role.

Velocity

Whereas speed refers to the total distance traveled, **velocity** refers to how fast the *displacement* changes. Velocity, like speed, is measured over a certain time interval. If a car has a displacement $\Delta\mathbf{x}$ in a particular time interval Δt, then the car's **average velocity**, $\mathbf{v}_{av}$, over that time interval is defined by

$$\mathbf{v}_{av} \equiv \frac{\text{displacement}}{\text{time interval}} \tag{2–6}$$

$$= \frac{\mathbf{x}_2 - \mathbf{x}_1}{t_2 - t_1} \tag{2–7}$$

$$= \frac{\Delta\mathbf{x}}{\Delta t}. \tag{2–8}$$

Note that $\mathbf{v}_{av}$ is the ratio of a vector to an ordinary number (scalar), and *it is therefore a vector*. We write it in boldface for that reason. The dimensions of velocity are $[LT^{-1}]$, with SI units of meters per second.

Average Velocity. The average velocity, $\Delta\mathbf{x}/\Delta t$, provides us with only limited information. For example, consider a car being driven along a straight road for 1 h (so $\Delta t =$ 1 h). If $\Delta\mathbf{x}$ is a displacement from the starting point to a point 30 mi down the road, then the above expression will read

$$\mathbf{v}_{av} = (30 \text{ mi/h})\mathbf{i} \tag{2–9}$$

no matter how the car traveled during that hour. It could have traveled at 60 mi/h to a point 45 mi down the road, turned around and traveled 15 mi back to the finishing point at 60 mi/h. It could have traveled a steady 30 mi/h, or it could have traveled the 30 mi at 120 mi/h, arriving in 15 min, and simply parked at the finish point until the hour was up. All that the definition in Eq. (2–8) has given us is the *average velocity* over the time interval in question. It is perfectly consistent to have a zero average velocity over a finite time interval even though a considerable distance may have been covered. For example, if an automobile starts from a given point and returns to that point at the end of a given time interval, the net displacement is zero, and the average velocity over the time interval is zero.

Instantaneous Velocity. The definition of the average velocity includes a time interval. We learn more about the motion when smaller time intervals are used. For example, if we had asked about the average velocity during 15-min intervals in the one-hour trip just discussed, we would already have a more detailed picture of the car's motion, even if the car is not moving steadily during those 15-min intervals. Dividing the hour into 60 1-min intervals and finding the average velocity during each minute would give us still more information: We would find the total displacement vector for a given minute and could then divide by 1 min to find the average velocity during that minute.

It is possible to continue to make the time interval Δt progressively smaller, finding the average velocity for each interval. Suppose that we look at the particular time interval from some time t to a time $t + \Delta t$. In this time interval, the displacement $\Delta\mathbf{x} = \mathbf{x}(t + \Delta t) - \mathbf{x}(t)$ occurs. The average velocity during this interval is the ratio of $\Delta\mathbf{x}$ to Δt. Now, if Δt shrinks to zero, so does $\Delta\mathbf{x}$, *but their ratio remains finite*! (In the example of the car trip, we could imagine that there is some finite average velocity in 1-min intervals, just as there is in 1-s intervals, or in intervals of 1/100 s.) We say that we are taking the *limit* as Δt approaches zero, symbolized by $\Delta t \to 0$. This limit refers to a particular time t and gives us the average velocity over a shorter and shorter time interval around that time. The **instantaneous velocity** at time t is the velocity of an object at that given instant of time, and it is defined as the limit of the average velocity as $\Delta t \to 0$:

The definition of instantaneous velocity, a vector

PROBLEM SOLVING

See Appendix IV–7 for rules for differentiation.

$$\mathbf{v}(t) \equiv \lim_{\Delta t \to 0} \frac{\mathbf{x}(t + \Delta t) - \mathbf{x}(t)}{\Delta t}. \tag{2–10}$$

APPLICATION: TIMED LIGHTS

You may have traveled roads with signs indicating that the traffic lights are timed for a certain average speed, v_{av}. To arrange such timing in any one direction, a traffic engineer must know only the distance between lights, and then arrange the timing of the lights accordingly. If the distance between the first light and a second light is Δx, then the lights should be timed so that the second light turns green at a time $\Delta x/v_{av}$ after the first has done so (Fig. 2AB–1). Note that it is the *average* vehicle speed that enters into such an arrangement. For the design of this system, it is irrelevant whether a driver achieves the average speed by driving rapidly for a portion of the distance and slowly for another portion, or whether he or she maintains v_{av} throughout. (Although it is more fuel efficient to do the latter!)

FIGURE 2AB–1 A close look at this photograph reveals several red lights in sequence, followed by green lights, followed by red lights. A motorist driving at the proper speed will meet only green lights.

Notice that we refer to the instantaneous velocity without subscript in Eq. (2–10). When we use the term "*velocity*," we mean instantaneous velocity, unless we state otherwise. The right-hand side of Eq. (2–10) is the *definition* of a derivative in calculus. Velocity is the derivative of **x** with respect to the variable t:

The instantaneous velocity is the derivative of the displacement vector with respect to time.

$$\mathbf{v} = \lim_{\Delta t \to 0} \frac{\Delta \mathbf{x}}{\Delta t} = \frac{d\mathbf{x}}{dt}. \tag{2–11}$$

We shall soon see, in our discussion of an automobile's velocity, that the derivative of a function has a simple graphical interpretation in terms of the graph of the function.

We define the instantaneous speed v to be the magnitude of the instantaneous velocity:

$$v \equiv |\mathbf{v}|. \tag{2–12}$$

This definition shows that there are some subtle points in the notion of speed. For example, the average speed is generally not equal to the magnitude of the average velocity. In the case of a car traveling at a uniform 60 mi/h for 1 h—30 mi straight out and 30 mi straight back—the magnitude of the average velocity for the period from 15 min to 45 min is zero, while the average speed is 60 mi/h. But in the first 15-min time interval, the quantities are identical. You may feel by now that the actual use of the term *speed* is confusing even if the definitions are precise. You are correct, and we will attempt to limit the use of the word *speed* in our discussion of mechanics to a synonym for the magnitude of the instantaneous velocity, as given in Eq. (2–12).

The derivative is the slope of a curve.

An Automobile's Velocity. Figure 2–4 shows an automobile's position as a function of time as it moves in a straight line from one traffic light to the next. The automobile starts when the first light turns green and comes to a halt at the second light, which is red, although this is perhaps not obvious from the curve. To learn such information, we must find the automobile's velocity as a function of time. Marked on Fig. 2–4 are two displacements, Δx, centered around the particular time $t = 3$ s. The larger interval corresponds to a time interval $\Delta t = 6$ s, while the smaller corresponds to the time interval $\Delta t = 4$ s. (The motion is one-dimensional, so we can dispense with the boldfaced notation here.) For each time interval, there is a corresponding displacement, Δx. In each case, the ratio $\Delta x/\Delta t$ gives us the value of v_{av} over the corresponding time interval Δt. This value is the slope of the line that connects the two points marking the

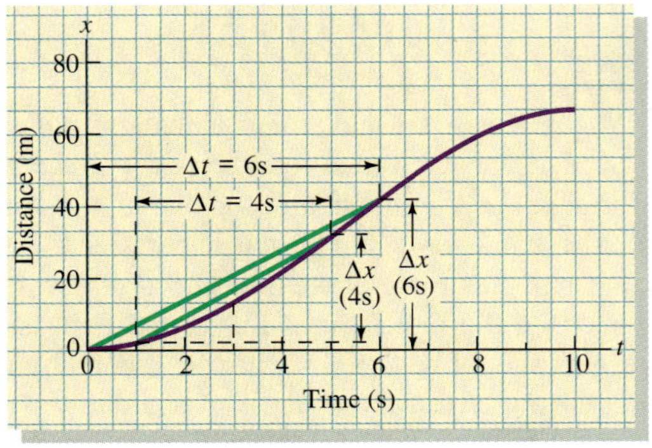

FIGURE 2–4 The displacements Δx corresponding to the two time intervals $\Delta t = 6$ s and $\Delta t = 4$ s, centered around the central time $t = 3$ s. The magnitude of the average velocity for each interval is given by the ratio $\Delta x/\Delta t$ and differs for the two intervals. The slopes of the straight, solid lines (in green) are the average velocities.

interval. You can see that, as the time interval Δt around the central value of $t = 3$ s becomes smaller, the slope of the line connecting the two points of the interval more nearly matches the slope of the tangent to the curve at that point (Fig. 2–5a). In the limit that Δt becomes infinitesimally small, we arrive at the instantaneous slope dx/dt at the particular time of interest. This is the instantaneous velocity at that point.

PROBLEM SOLVING

The derivative is a slope.

All this is completely consistent with what we know from calculus: *The derivative* dx/dt *at any given time* t *is the slope of the function* $x(t)$ *at that time.* Thus the slope of the tangent at any time is the instantaneous velocity of the automobile at that time. We can apply this idea to find the velocity of the car at any time t along its trajectory between the two lights. In Fig. 2–5b, we show the tangents to the curve for four separate time points. The slope of each of these four lines gives the instantaneous velocity of the automobile at the precise times when the lines are tangent to the curve. We can see, for example, that the car has come to rest at the second light, at roughly 65 m, because the slope is zero at that point. We had already seen how to find the derivative as an algebraic limit. Graphical methods like this give us an alternative way to find the derivative—in this case, the instantaneous velocity.

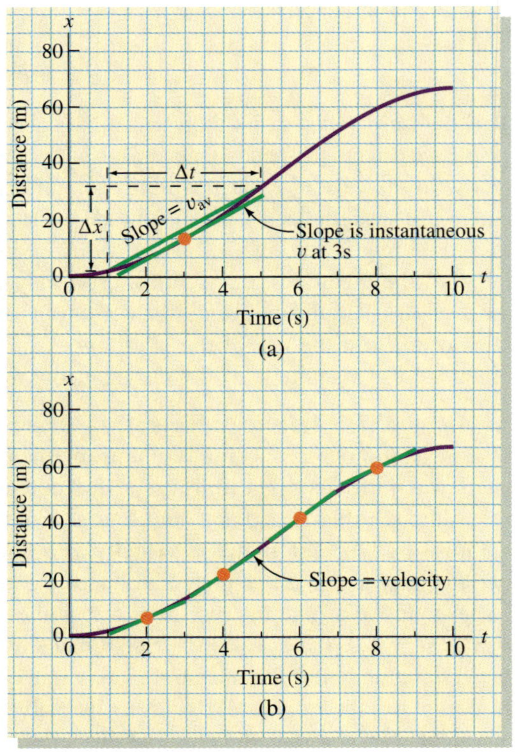

FIGURE 2–5 (a) In the limit that the time interval around a particular time becomes zero, the average velocity for that interval becomes the slope of the curve of displacement versus time at that time. (b) The slope of a line (in green) tangent to the curve of displacement versus time gives the instantaneous velocity at a time corresponding to the tangent point. For example, at the end of the curve, the line tangent to the curve is flat (has zero slope), so that the instantaneous velocity there is zero. The automobile has come to a stop at the light.

EXAMPLE 2–2 Using the data in Table 2–1, calculate the average velocity of a runner over the time intervals $\Delta t = 3.91$ s and $\Delta t = 8.20$ s, centered around $x = 30.0$ m. Also calculate the average velocity over the smallest time interval available near $x = 30.0$ m.

Solution: We must determine the runner's average velocity centered around the distance $x = 30.0$ m using time intervals of 3.91 s and 8.20 s. The average velocity can be calculated by using Eqs. (2–6) through (2–8). The direction of the average velocity is the direction of the runner's motion. According to Table 2–1, we find that the time points on either side of 30.0 m that give a time interval, Δt, of 3.91 s are $t_1 = 2.01$ s and $t_2 = 5.92$ s; for $\Delta t = 8.20$ s, the times are $t_1 = 0$ s and $t_2 = 8.20$ s. We obtain the distances that correspond to each of these time points from Table 2–1. The average velocities are then

$$\text{for } \Delta t = 3.91 \text{ s:} \quad \mathbf{v}_{av} = \frac{\mathbf{x}_2 - \mathbf{x}_1}{t_2 - t_1} = \left(\frac{50 \text{ m} - 10 \text{ m}}{5.92 \text{ s} - 2.01 \text{ s}}\right)\mathbf{i} = (10.2 \text{ m/s})\mathbf{i};$$

$$\text{for } \Delta t = 8.20 \text{ s:} \quad \mathbf{v}_{av} = \frac{\mathbf{x}_2 - \mathbf{x}_1}{t_2 - t_1} = \left(\frac{75 \text{ m} - 0 \text{ m}}{8.20 \text{ s} - 0 \text{ s}}\right)\mathbf{i} = (9.2 \text{ m/s})\mathbf{i}.$$

The smallest available time interval around 30.0 m in Table 2–1 is the interval from 25 m (3.60 s) to 35 m (4.55 s), and

$$\text{for } \Delta t = 0.95 \text{ s:} \quad \mathbf{v}_{av} = \frac{\mathbf{x}_2 - \mathbf{x}_1}{t_2 - t_1} = \left(\frac{35 \text{ m} - 25 \text{ m}}{4.55 \text{ s} - 3.60 \text{ s}}\right)\mathbf{i} = (10.5 \text{ m/s})\mathbf{i}.$$

TABLE 2–1 Times for a 100-m Dash

Distance (m)	*Time(s)*
0	0
5	1.36
10	2.01
15	2.57
20	3.09
25	3.60
30	4.09
35	4.55
40	5.01
45	5.47
50	5.92
55	6.37
60	6.83
65	7.28
70	7.74
75	8.20
80	8.65
85	9.11
90	9.57
95	10.04
100	10.50

EXAMPLE 2–3 Figure 2–6 displays the data shown in Table 2–1. Using graphical techniques, determine the velocity at times $t = 2$ s.

Solution: The instantaneous velocity is given by the slope of the curve shown in Fig. 2–6 at the given time $t = 2$ s. We carefully draw a line tangent to the curve at $t = 2$ s in Fig. 2–6. Using the scales of the axes, we are able to measure the slope by examining any two points on the straight line. We see that $x = 90$ m when $t = 12$ s and $x = 0$ m when $t = 0.8$ s. Therefore we have $\Delta x = 90$ m for $\Delta t = 11.2$ s. The slope is

$$\frac{\Delta x}{\Delta t} = \frac{90 \text{ m}}{11.2 \text{ s}} = 8.0 \text{ m/s}.$$

The instantaneous velocity is now given by

$$\mathbf{v} = (\text{slope})\mathbf{i} = (8.0 \text{ m/s})\mathbf{i}.$$

Note that the slope at $t = 2$ s is positive, which is why the coefficient of **i** in the expression for the velocity is positive.

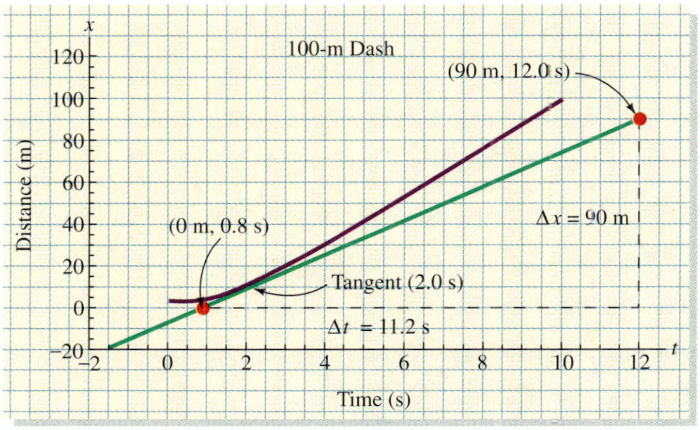

FIGURE 2–6 The straight green line is the tangent to the curve of position versus time at $t = 2$ s. The points at $x = 90$ m ($t = 12$ s) and $x = 0$ m ($t = 0.8$ s) determine the slope of this line and therefore the instantaneous velocity at $t = 2$ s.

2–3 ACCELERATION

Average Acceleration

The dimension and SI unit of acceleration

The term **acceleration** refers to the rate of change in velocity of an object with respect to time. The runner in the 100-m dash of Fig. 2–2 starts off with $v = 0$ m/s at $t = 0$ s. Two seconds later, he is moving with $v = 8$ m/s. His velocity has changed by $\Delta v = 8$ m/s over a time interval $\Delta t = 2$ s. We define the **average acceleration**, $\mathbf{a}_{av}$, in terms of velocity $\mathbf{v}_1$ at time t_1 and velocity $\mathbf{v}_2$ at time t_2:

$$\mathbf{a}_{av} \equiv \frac{\mathbf{v}_2 - \mathbf{v}_1}{t_2 - t_1} = \frac{\Delta \mathbf{v}}{\Delta t}. \qquad (2\text{–}13)$$

Note that because velocity is a vector, so is acceleration. But because $\mathbf{v}$ and $\Delta\mathbf{v}$ need not point in the same direction, *the velocity vector and the acceleration vector are not necessarily parallel*. The dimensions of acceleration are $[LT^{-2}]$, with SI units of meters per second squared (m/s^2).

EXAMPLE 2–4 A runner in the 100-m dash accelerates uniformly to 10 m/s at 4 s and maintains this velocity for the next 4 s. She then realizes that she is going to win and slows uniformly over the next 4.7 s to reach a velocity of 8 m/s at the end of the race. She has run the 100-m dash in 12.7 s. What is the runner's average acceleration over the time periods 0 s to 4 s, 4 s to 8 s, and 8 s to 12.7 s?

Solution: The positive direction of the acceleration is the direction of the runner's motion, which is described by the unit vector **i.** We must be careful to keep track of signs. The magnitudes of all required velocities are given in the problem except for the additional magnitude $v = 0$ m/s at $t = 0$ s. We represent the speeds during the various time intervals in Fig. 2–7, where the changes in velocity between the times given are linear; that is, the acceleration is uniform. Equation (2–13) is used to find the average acceleration. We give the coefficient of **i** in each case:

$$\text{for 0 s to 4 s:} \quad a_{av} = \frac{v_2 - v_1}{t_2 - t_1} = \frac{10\ \text{m/s} - 0\ \text{m/s}}{4\ \text{s} - 0\ \text{s}} = 2.5\ \text{m/s}^2;$$

$$\text{for 4 s to 8 s:} \quad a_{av} = \frac{v_2 - v_1}{t_2 - t_1} = \frac{10\ \text{m/s} - 10\ \text{m/s}}{8\ \text{s} - 4\ \text{s}} = 0\ \text{m/s}^2;$$

$$\text{for 8 s to 12.7 s:} \quad a_{av} = \frac{v_2 - v_1}{t_2 - t_1} = \frac{8\ \text{m/s} - 10\ \text{m/s}}{12.7\ \text{s} - 8\ \text{s}} = -0.42\ \text{m/s}^2.$$

The greatest average acceleration occurs at the beginning of the race, when the runner is attempting to reach her greatest speed. Although she is running at her highest velocity during the middle part of the race, her average acceleration during this period is zero. During the time interval when she is slowing down at the end of the race, her average acceleration is negative.

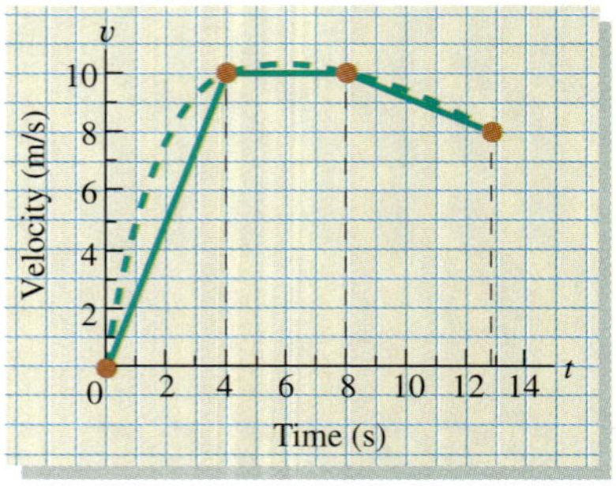

FIGURE 2–7 Example 2–4. The dashed line is the actual velocity; the segmented straight lines are the means of finding the average acceleration.

In Example 2–4, the fact that the runner's average acceleration at the end of the race was negative means that the magnitude of the velocity is decreasing rather than increasing. You will sometimes see the term **deceleration** to describe situations in which the *magnitude* of the velocity decreases. As for any one-dimensional vector, a negative value for the average acceleration means that the direction of the average acceleration is along $-\mathbf{i}$ rather than $+\mathbf{i}$.

Instantaneous Acceleration

The instantaneous acceleration is the derivative with respect to time of the velocity vector.

In physical phenomena, it is the instantaneous values of quantities that are most important. The velocity provides one example. We shall also want to work with the **instantaneous acceleration**. In fact, when we use the term *acceleration*, we shall be referring to the instantaneous acceleration, unless otherwise stated. We define the instantaneous acceleration as the limit of $\Delta\mathbf{v}/\Delta t$ as the time interval Δt goes to zero. This is again a derivative:

$$\mathbf{a} \equiv \lim_{\Delta t \to 0} \frac{\Delta \mathbf{v}}{\Delta t} = \frac{d\mathbf{v}}{dt}. \tag{2–14}$$

The instantaneous acceleration is the second time derivative of the displacement vector.

What is the relation between acceleration and displacement? If we examine the derivatives presented in Eqs. (2–11) and (2–14), we find that

$$\mathbf{a} = \frac{d\mathbf{v}}{dt} = \frac{d}{dt}\left[\frac{d\mathbf{x}}{dt}\right] = \frac{d^2\mathbf{x}}{dt^2}. \tag{2–15}$$

The acceleration is the time derivative of the velocity; equivalently, the acceleration is *the second time derivative of the displacement.*

We have discovered that we can find the velocity of a runner from a plot of position versus time; analogously, we can determine the acceleration from a plot of velocity versus time (Fig. 2–8a). We obtain the acceleration at any time t by finding the slope of the tangent to the curve of v as a function of t at the particular time t. In Fig. 2–8b, we show the tangent to the curve of v versus t at times 2 s and 4 s. We can determine the acceleration of our runner either by finding the slopes of tangents to the velocity curve or by taking the algebraic time derivative of the function $v(t)$. We show the acceleration curve in Fig. 2–8c. This curve was derived from the velocity curve in Fig. 2–8b. Notice that the acceleration is initially very high as the runner gains speed, but by 4 s, when the runner moves at a steady speed, the acceleration drops to zero.

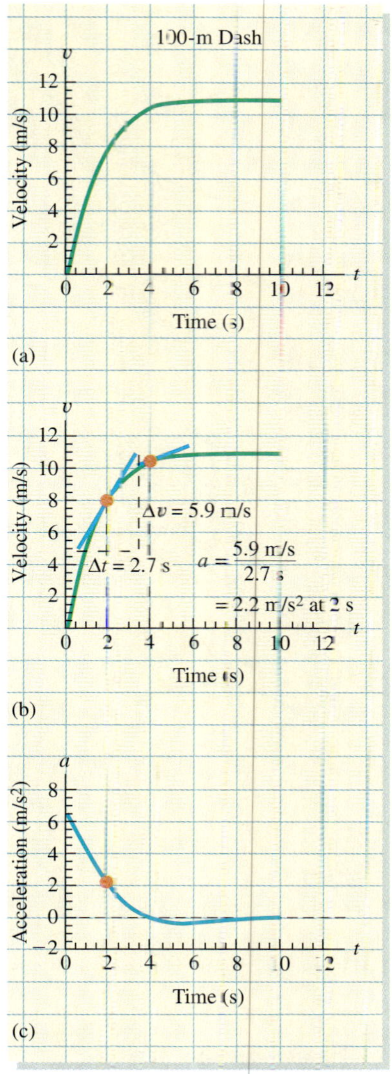

FIGURE 2–8 (a) The velocity of the runner whose position is shown in Fig. 2–2. This curve can be determined, for example, by finding the slope at each point along the position versus time curve and plotting it. Figure 2–6 illustrates the procedure for $t = 2$ s. (b) The instantaneous acceleration of the runner is found by measuring the slope of the tangents to the curve of velocity versus time. Two such tangents are drawn in blue; their slope gives the acceleration at $t = 2$ s and $t = 4$ s. The slope of the tangent at $t = 2$ s is calculated here. (c) The acceleration of the runner as a function of time. This curve can be found by plotting the slope of the tangents to the velocity versus time curve as a function of time.

EXAMPLE 2–5 The position x of an experimental rocket moving along a long rail is measured to be $x(t) = (5 \text{ m/s})t + (8 \text{ m/s}^2)t^2 + (4 \text{ m/s}^3)t^3 - (0.25 \text{ m/s}^4)t^4$ over the first 10 s of its motion, where t is in seconds and x is in meters. Find the velocity and acceleration of the rocket and display the results graphically.

Solution: The directions of vectors **x**, **v**, and **a** are taken as positive along the rail (which is one-dimensional, so we do not need boldfaced notation). The plot of the position x as a function of time is shown in Fig. 2–9a. The velocity and acceleration of the rocket can be determined by taking the time derivatives in Eqs. (2–11) and (2–15), respectively:

$$v = \frac{dx}{dt} = (5 \text{ m/s}) + (16 \text{ m/s}^2)t + (12 \text{ m/s}^3)t^2 - (1 \text{ m/s}^4)t^3;$$

$$a = \frac{dv}{dt} = (16 \text{ m/s}^2) + (24 \text{ m/s}^3)t - (3 \text{ m/s}^4)t^2.$$

Velocity and acceleration are measured in m/s and m/s^2, respectively, if time is measured in seconds. We plot these results over the time period 0 s to 10 s in Fig. 2–9b and c, respectively. Note that although the position x is zero at time $t = 0$, neither the velocity nor the acceleration is zero at this time.

If we wished, we could next discuss the rate of change of acceleration, $d\mathbf{a}/dt$. However, *we do not need to discuss further derivatives* for reasons that will become clear in Chapter 4. We shall see that it is acceleration, not its changes, that plays a primary role in Newton's laws of motion.

2–4 MOTION WITH CONSTANT ACCELERATION

The simplest example of acceleration is constant acceleration (Fig. 2–10). It is a physically important case for a variety of reasons, not the least of which is that we experience a constant acceleration near Earth's surface—that associated with gravity.

The velocity as a function of time for constant acceleration

Although we might correctly argue that we do not accelerate when we are standing on the ground, we certainly accelerate under the influence of gravity when, for example, we dive into a pool.

In our discussion of displacement, velocity, and acceleration, we concentrated on the *differences* between average and instantaneous quantities. But a *constant* acceleration means that $\mathbf{a} = \mathbf{a}_{av}$. Constant acceleration means also that the speed changes steadily; that is, it increases linearly with time. Mathematically,

$$\text{for constant acceleration:} \quad \mathbf{a} = \mathbf{a}_{av} = \frac{\mathbf{v}(t) - \mathbf{v}(t_0)}{t - t_0}, \qquad (2\text{–}16)$$

where t_0 is *any* fixed time and $\mathbf{v}(t_0)$ is the velocity at that time. It is often convenient to label the initial velocity $\mathbf{v}(t_0)$ as $\mathbf{v}_0$. Because we are dealing with only one dimension, we shall drop the boldfaced notation for vectors. We use a minus sign when the vector is along the $-x$-direction. We also set $t_0 = 0$. Equation (2–16) can then be solved for v:

$$\text{for constant acceleration: } v = at + v_0. \qquad (2\text{–}17)$$

This result for the velocity has the linear dependence on time that we had foreseen. (See also Section 2–6.)

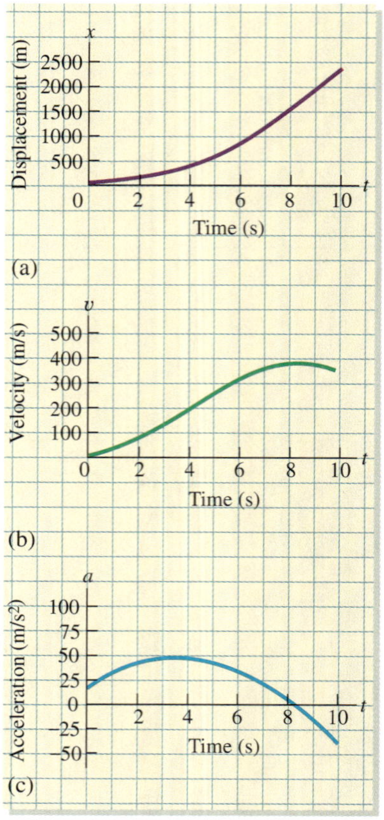

FIGURE 2–9 Example 2–5. We have plotted the displacement, the instantaneous velocity, and the instantaneous acceleration as functions of time.

When the velocity is constant, the displacement changes linearly with time. Here, the velocity changes linearly with time, so we would expect the displacement to change even more rapidly. To see how this is realized mathematically, we can write Eq. (2–7) as

$$v_{av} = \frac{x(t) - x(t_0)}{t - t_0}.$$

If we let $x_0 = x(t_0)$, $x = x(t)$, and then let $t_0 = 0$ for convenience, we have

$$v_{av} = \frac{x - x_0}{t}.$$

This equation is solved for x to yield

$$x = v_{av}t + x_0. \qquad (2\text{–}18)$$

Figure 2–11 shows a plot of velocity as a function of time for the case of constant acceleration. In this special case, the average velocity is simply the average of the initial and final velocities over the total time period t. The initial velocity is v_0 and the final velocity at time t is $v = v(t)$. The average velocity for the case of constant acceleration is then

$$\text{for constant acceleration:} \quad v_{av} = \frac{v_0 + v}{2}. \qquad (2\text{–}19)$$

FIGURE 2–10 The diver—or more exactly, one special point in the diver—accelerates uniformly under the influence of gravity.

For this case, v_{av} is just the velocity at time $t/2$. When we substitute this result for v_{av} into Eq. (2–18), we find that

$$\text{for constant acceleration:} \quad x = \frac{(v_0 + v)t}{2} + x_0 = \frac{vt}{2} + \frac{v_0 t}{2} + x_0. \qquad (2\text{–}20)$$

We now substitute v from Eq. (2–17) into this equation and determine the position as a function of time:

The position as a function of time for constant acceleration

$$\text{for constant acceleration:} \quad x = \frac{at^2}{2} + v_0 t + x_0. \qquad (2\text{–}21)$$

As we had expected, the position varies more rapidly than linearly with time: it varies quadratically.

We should discuss one more useful relation. The three boxed equations, which all contain the time, allow us to find a relationship between displacement, velocity, and acceleration that does not involve time. To do so, we first solve for the time t from Eq. (2–17):

$$\text{for constant acceleration:} \quad t = \frac{v - v_0}{a}. \tag{2–22}$$

If we substitute this expression for time into the first part of Eq. (2–20), we have

$$x = \left(\frac{v_0 + v}{2}\right)\left(\frac{v - v_0}{a}\right) + x_0$$

for constant acceleration, or

$$x - x_0 = \frac{v^2 - v_0^2}{2a}. \tag{2–23}$$

This equation can be written as

$$\text{for constant acceleration:} \quad v^2 = v_0^2 + 2a(x - x_0). \tag{2–24}$$

This result gives us the velocity at any position x from the constant acceleration a and the initial velocity and position. The time does not enter into this result. Of course, if we want to find the time t, we can find it by using Eq. (2–22).

The four boxed equations of this section are particularly useful, but you need not memorize them. We can get all the information in these equations from the simple statement that, in this case, the acceleration is constant; alternatively, simply remember that the position changes quadratically with time, Eq. (2–21).

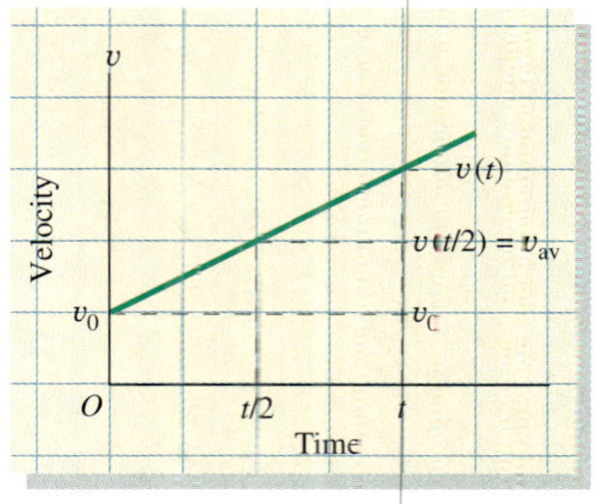

FIGURE 2–11 The velocity changes linearly with time for constant acceleration.

EXAMPLE 2–6 An amateur bowler releases a ball with an initial velocity of 2 m/s; the ball slows down with a constant negative acceleration of $-0.2\ \text{m/s}^2$. How far does the ball roll before stopping, and how long does it take to stop?

Solution: Choose a coordinate system with $x = 0$ m at the point where the ball leaves the bowler's hand, and start the clock at $t = 0$ s when the ball leaves his hand (Fig. 2–12). The x-axis is along the direction of the ball's motion. The initial conditions are then $x_0 = 0$ m, $t_0 = 0$ s, and $v_0 = 2$ m/s. The acceleration is negative along the $+x$-direction and has the constant value $-0.2\ \text{m/s}^2$. In Fig. 2–12b, we have sketched the constant acceleration, as well as the linearly decreasing velocity and the quadratically varying position that correspond to the acceleration.

We know the initial and final velocities (zero) as well as the acceleration. We do not know the final x-position of the ball or the time elapsed. If we look at the available equations, we see that we know all the variables in Eq.(2–24) except x, which is the displacement of the ball. We can use Eq. (2–24) to find the x-value when the final velocity v has become zero. The solution for x is, from that equation,

$$x = x_0 + \frac{v^2 - v_0^2}{2a}.$$

All the quantities on the right-hand side are known, giving the value of x as:

$$x = (0\ \text{m}) + \frac{(0\ \text{m/s})^2 - (2\ \text{m/s})^2}{2(-0.2\ \text{m/s}^2)} = 10\ \text{m}.$$

Use Eq. (2–22), or Eq. (2–17), rearranged, to determine the time of motion:

$$t = \frac{(0\ \text{m/s}) - (2\ \text{m/s})}{-0.2\ \text{m/s}^2} = 10\ \text{s}.$$

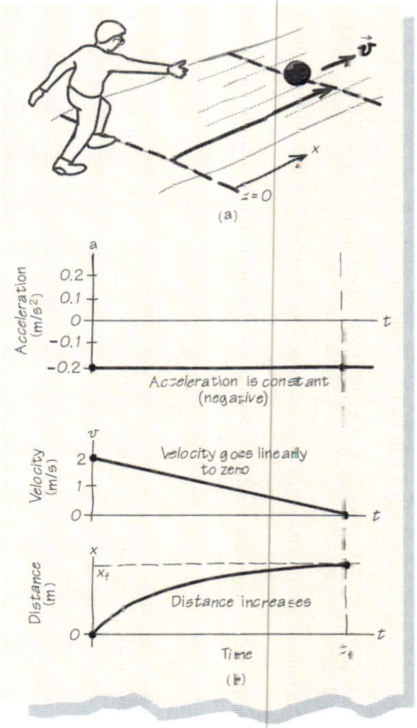

FIGURE 2–12 (a) Example 2–6. The origin O marks the spot ($x = 0$ m) where the ball leaves the bowler's hand. (b) Sketches of acceleration, velocity, and displacement as functions of time.

Any checks you can find for your answer are helpful. In particular, use Eq. (2–21) with $t = 10$ s to determine the displacement again:

$$x = \frac{(-0.2 \text{ m/s}^2)(10 \text{ s})^2}{2} + (2 \text{ m/s})(10 \text{ s}) + (0 \text{ m}) = 10 \text{ m}.$$

So the displacement is again 10 m.

EXAMPLE 2–7 A runner bursts out of the starting blocks 0.1 s after the gun signals the start of a race. She runs at constant acceleration for the next 1.9 s of the race. If she has gone 8.0 m after 2.0 s, what are her acceleration and velocity at this time?

Solution: The runner is not moving during the first 0.1 s of the race, and thus her acceleration is zero during this period of time. During the next 1.9 s, she has an acceleration that is not zero. We can use the four boxed equations of this section *only if the acceleration is constant* during the entire time. Consider a time $t' = t - 0.1$ s, which is the interval of constant acceleration. For the times $t = 0.1$ s and 2.0 s, $t' =$ 0 s and 1.9 s, respectively. Over the time period $t' = 0$ s to 1.9 s, we can use the four boxed equations because the acceleration is constant and the initial time t'_0 is 0. The initial conditions of the problem are therefore $t'_0 = 0$ s, $x_0 = 0$ m, $v_0 = 0$ m/s, and we want the acceleration and velocity at $t' = 1.9$ s and $x = 8.0$ m. An examination of the four boxed equations shows that there are two unknowns (v and a) in Eqs. (2–17) and (2–24). Equation (2–19) does not allow us to determine either v or a. However, Eq. (2–21) allows us to determine the acceleration. We insert the known values (with the primed values of time) into Eq. (2–21) and solve for the acceleration:

$$8.0 \text{ m} = \frac{a(1.9 \text{ s})^2}{2} + (0 \text{ m/s})(0 \text{ s}) + (0 \text{ m});$$

$$a = \frac{16 \text{ m}}{3.6 \text{ s}^2} = 4.4 \text{ m/s}^2.$$

Now we can use Eq. (2–17) to determine the velocity at $t' = 1.9$ s ($t = 2.0$ s) after the runner starts:

$$v = at' + v_0 = (4.4 \text{ m/s}^{\not{2}})(1.9 \not{\text{s}}) + 0 \text{ m/s} = 8.4 \text{ m/s}.$$

General Relations between Position, Velocity, and Acceleration for Constant Acceleration

There is nothing very special about the time $t = 0$ in motion under constant acceleration; in fact, Example 2–7 illustrated the difficulty with Eqs. (2–17) through (2–24) if the initial time must always be zero. That is why it is useful to allow t_0 to be an arbitrary time in the equations relating acceleration, velocity, and displacement. In this case, the four boxed equations of this section become

for constant acceleration:

$$v = a(t - t_0) + v_0, \qquad \text{(2–25a)}$$

$$v_{av} = \frac{v_0 + v}{2}, \quad \text{(unchanged)} \qquad \text{(2–25b)}$$

$$x = \frac{a(t - t_0)^2}{2} + v_0(t - t_0) + x_0, \qquad \text{(2–25c)}$$

$$v^2 = v_0^2 + 2a(x - x_0). \quad \text{(unchanged)} \qquad \text{(2–25d)}$$

The only difference between the four boxed equations (2–17, 2–19, 2–21, and 2–24) and Eqs. (2–25a–d) is that t is replaced by $t - t_0$ in the latter equations. Remember,

Eqs. (2–25a–d) are valid only when the acceleration is constant between the times t_0 and t. When $t_0 = 0$ in Eqs. (2–25a–d), we obtain the four boxed equations. Example 2–7 can now be worked more easily by letting $t_0 = 0.1$ s and $t = 2$ s because the acceleration is constant during this time interval.

EXAMPLE 2–8 On a given flight, a T-38 training jet has an acceleration of 3.6 m/s^2 that lasts 5.0 s during the initial phase of takeoff (Fig. 2–13). The afterburner engines are then turned up to full power for an acceleration of 5.1 m/s^2. The speed needed for takeoff is 164 knots (1 m/s = 1.94 knots). Calculate the length of runway needed and the total time of takeoff.

FIGURE 2–13 Example 2–8.

Solution: There are two different constant accelerations in this example; therefore, we need to divide the problem into two parts. Let us first look at the initial 5.0 s of takeoff. For this period, we have the values $x_0 = 0$ m and $v_0 = 0$ m/s at $t_0 = 0$ s. We use Eqs. (2–25a–d) to find the velocity and distance at 5.0 s. Equation (2–25a) gives the velocity,

$$\text{for } t = 5.0\text{ s:}\quad v(t) = (3.6\text{ m/s}^2)(5.0\text{ s} - 0\text{ s}) + (0\text{ m/s}) = 18\text{ m/s}.$$

Next we use Eq. (2–25c) to determine the distance the jet has traveled:

$$\text{for } t = 5.0\text{ s:}\quad x(t) = \frac{(3.6\text{ m/s}^2)(5.0\text{ s} - 0\text{ s})^2}{2} + (0\text{ m/s})(5.0\text{ s} - 0\text{ s}) + 0\text{ m} = 45\text{ m}.$$

We now move to the second phase of takeoff, where full power is applied. We have a new set of initial conditions beginning with the time $t_0 = 5.0$ s; namely, $x_0 = 45$ m and $v_0 = 18$ m/s. We want to find the time and distance corresponding to a final velocity of 164 knots. Let us first convert this value into SI units. The final velocity of 164 knots is

$$(164\text{ knots})\frac{1\text{ m/s}}{1.94\text{ knots}} = 84.4\text{ m/s}.$$

We use Eq. (2–25a) to find the time t at takeoff; the acceleration is now 5.1 m/s^2. From Eq. (2–25a) we have

$$a(t - t_0) = v - v_0,$$

$$t = \frac{v - v_0}{a} + t_0 = \frac{84.4\text{ m/s} - 18\text{ m/s}}{5.1\text{ m/s}^2} + 5.0\text{ s} = 18\text{ s}.$$

Note that we waited until we had solved for the variable t before inserting the numerical values (with units) for v, v_0, a, and t_0. This technique also serves as a check when the cancellation of units gives the expected result, seconds in this case.

Equation (2–25c) can be used directly to determine the takeoff distance because all the variables for the second phase, except x, are now known:

$$x = \frac{(5.1\text{ m/s}^2)(18\text{ s} - 5.0\text{ s})^2}{2} + (18\text{ m/s})(18\text{ s} - 5.0\text{ s}) + 45\text{ m} = 710\text{ m}.$$

Because we have included $x_0 = 45$ m from the first phase, 710 m (or 2330 ft) is the total amount of runway used.

2–5 FREELY FALLING OBJECTS

Near Earth's surface, objects fall with constant acceleration.

In Section 2–4, we mentioned an important example of constant acceleration: gravity. The **acceleration due to gravity** is given the symbol **g**, and its magnitude is approximately 9.8 m/s^2.[†] Ignoring the effects of air resistance, any object dropped in the vicin-

[†]There are variations of the order of one percent in the magnitude and direction of **g** over Earth's surface.

ity of Earth's surface will move with constant acceleration **g**. The direction of **g** is down, toward Earth's center, a direction easily found using a plumb bob (a string with a mass at its end).

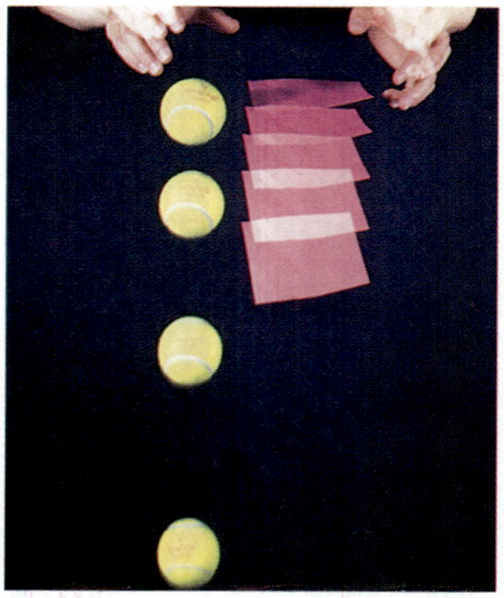

(a)

(b)

FIGURE 2–14 (a) The rubber ball falls directly to the floor while the sheet of paper floats down more slowly. (b) After the sheet of paper is wadded up and the experiment is repeated, the ball and the paper fall almost together.

Experimental Tests of the Acceleration of Gravity

Galileo Galilei, who could be considered the first modern physicist, systematically investigated the motions of falling objects. Centuries earlier, Aristotle had suggested (incorrectly) that the speed of a falling object depends on the weight of the object, and that this speed is proportional to the distance fallen. For Galileo, such assertions were no longer sufficient. His approach was to test by experiment. Galileo performed precise measurements in planned experiments and mathematically described the results. He determined that the distance that objects fall after starting from rest is proportional to the square of the time; equivalently, the speed of a falling object is proportional to the square root of the distance fallen. As Eqs. (2–21) and (2–24) show, this type of motion is characteristic of constant acceleration.

If we drop a hard rubber ball and a sheet of paper simultaneously from the same height, we observe that the paper floats down and the ball reaches the floor first (Fig. 2–14a). The ball experiences a greater acceleration and a larger final velocity than does the sheet of paper. This is because air resistance affects the sheet of paper to a much greater degree than it affects the ball. If we wad the paper up and repeat the experiment (Fig. 2–14b), the effect of air resistance on the paper is decreased and the falling times for the paper and the ball in this experiment are more nearly equal. If the same experiment is done in a vacuum, the falling times are, as best as we can measure, equal. Although air resistance can be an important effect, we shall assume for now that we are dealing with small, heavy objects ("particles") with negligible air resistance. A falling particle near Earth's surface undergoes a constant acceleration that is the same for *all* particles.

Quantitative Description of the Motion of a Freely Falling Object

To study the effects of gravity, it is easiest to set up a coordinate system with a direction perpendicular to Earth's surface. Let us align the y-axis with the vertical direction. We have two choices for this direction. If we choose the positive direction toward the center of the Earth, then the acceleration of gravity is $\mathbf{g} = g\mathbf{j}$. (Recall that $\mathbf{j}$ is the unit vector in the y-direction.) If we choose the positive direction of the y-axis away from

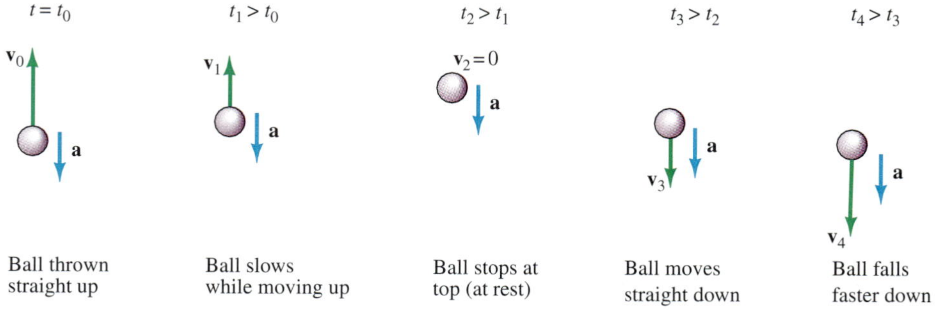

FIGURE 2–15 A ball tossed in the air slows, stops, and falls back. Although position and velocity change with time, the acceleration is constant.

Earth's center, then $\mathbf{g} = -g\mathbf{j}$. We could work problems by using either choice of axes, but let's choose the latter because it will facilitate our future discussion of motion in two or three dimensions. The vertical motion of any freely moving object for which air resistance can be ignored is then summarized by the four boxed equations of Section 2–4 for constant acceleration with $a = -g$,

PROBLEM SOLVING

A good choice of axes simplifies problem solving.

for a freely falling body:

$$v = -gt + v_0, \tag{2–26a}$$

$$v_{av} = \frac{v_0 + v}{2}, \quad \text{(unchanged)} \tag{2–26b}$$

$$y = \frac{-gt^2}{2} + v_0 t + y_0, \tag{2–26c}$$

$$v^2 = v_0^2 - 2g(y - y_0). \tag{2–26d}$$

Similar changes can be made to Eq. (2–25a–d). Remember that it is only the near constancy of the acceleration of gravity near the Earth's surface that makes these equations applicable. Figure 2–15 represents the up-and-down motion of an object, with the corresponding velocity and acceleration vectors.

EXAMPLE 2–9 How long does it take a ball to hit the ground if it is dropped from rest from a height of 100 m (Fig. 2–16)? What is the ball's velocity just before it hits the ground?

Solution: In Fig. 2–16, we indicate that the ground is at a level $y = 0$ m, while $y = 100$ m is the position from which the ball is dropped. Our initial conditions at $t_0 = 0$ s are $y_0 = 100$ m and $v_0 = 0$ m/s. We use Eq. (2–26c) to determine t, the time of flight:

$$0\text{ m} = \frac{-(9.80\text{ m/s}^2)t^2}{2} - (0\text{ m/s})t + 100\text{ m};$$

$$t^2 = \frac{200\text{ m}}{9.80\text{ m/s}^2} = 20.4\text{ s}^2,$$

$$t = 4.52\text{ s}.$$

We use Eq. (2–26a) to calculate the velocity:

$$\mathbf{v} = -(9.80\text{ m/s}^2)(4.52\text{ s})\mathbf{j} + 0\text{ m/s} = -(44.3\text{ m/s})\mathbf{j}.$$

Notice that the final velocity has a large magnitude and is *negative;* that is, it points downward. The negative sign appears because we chose the direction of the y-axis to be up, not down.

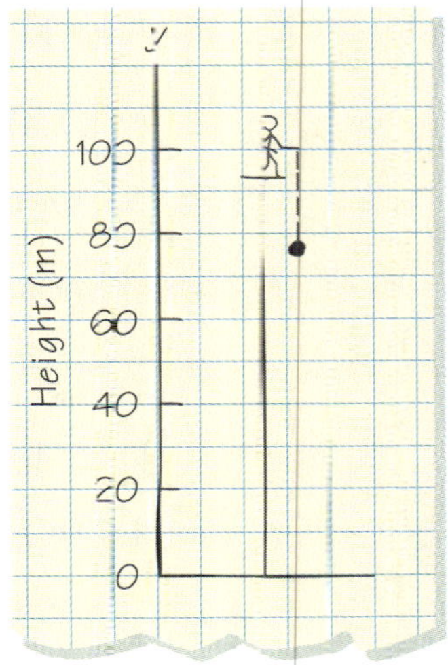

FIGURE 2–16 Example 2–9.

In Example 2–9, we calculated the velocity of an object falling freely from rest. Assume that an object starts at $y_0 = h$ with $v_0 = 0$ at $t_0 = 0$. What is the magnitude of the velocity when the object hits the ground ($y = 0$)? We use Eq. (2–26d) to determine

$$v^2 = 0 - 2g(0 - h) = 2gh;$$

$$\text{for constant acceleration:} \quad v = \sqrt{2gh}. \tag{2–27}$$

The positive sign of the square root is appropriate because it refers to the *magnitude* of the velocity. If the y-axis is up, the velocity is $\mathbf{v} = -(\sqrt{2gh})\mathbf{j}$.

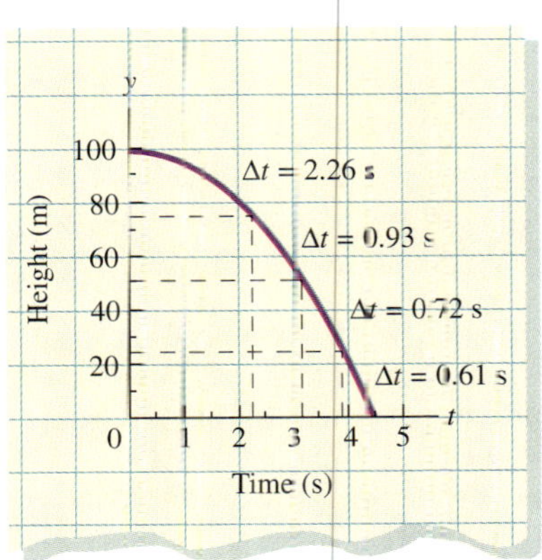

FIGURE 2–17 Example 2–10. The time for a ball to drop a given distance decreases as the ball gains velocity.

EXAMPLE 2–10 Calculate the time elapsed for the ball of Example 2–9 to drop from 100 m to 75 m, from 75 m to 50 m, from 50 m to 25 m, and from 25 m to the ground (Fig. 2–17).

Solution: Let's calculate the times at 75 m, 50 m, and 25 m with value $y_0 = 100$ m at $t_0 = 0$ s. By the same procedure used in Example 2–9, we find, using Eq.

(2–26c), that

$$y = \frac{-gt^2}{2} + y_0,$$

$$t^2 = \frac{2(y_0 - y)}{g};$$

$$\text{fall from 100 m to 75 m:} \quad t^2 = \frac{2(100\ \text{m} - 75\ \text{m})}{9.80\ \text{m/s}^2} = 5.10\ \text{s}^2;$$

$$t = 2.26\ \text{s};$$

$$\text{fall from 100 m to 50 m:} \quad t^2 = \frac{2(100\ \text{m} - 50\ \text{m})}{9.80\ \text{m/s}^2} = 10.2\ \text{s}^2;$$

$$t = 3.19\ \text{s};$$

$$\text{fall from 100 m to 25 m:} \quad t^2 = \frac{2(100\ \text{m} - 25\ \text{m})}{9.80\ \text{m/s}^2} = 15.3\ \text{s}^2;$$

$$t = 3.91\ \text{s};$$

$$\text{fall from 100 m to 0 m:} \quad t = 4.52\ \text{s (from Example 2–9)}.$$

Thus the times of flight are

$$100\ \text{m to 75 m}, \quad t = 2.26\ \text{s} - 0\ \text{s} = 2.26\ \text{s};$$

$$75\ \text{m to 50 m}, \quad t = 3.19\ \text{s} - 2.26\ \text{s} = 0.93\ \text{s};$$

$$50\ \text{m to 25 m}, \quad t = 3.91\ \text{s} - 3.19\ \text{s} = 0.72\ \text{s};$$

$$25\ \text{m to 0 m}, \quad t = 4.52\ \text{s} - 3.91\ \text{s} = 0.61\ \text{s}.$$

Due to the increasing velocity, the time needed for the ball to cover a distance of 25 m becomes shorter (Fig. 2–17).

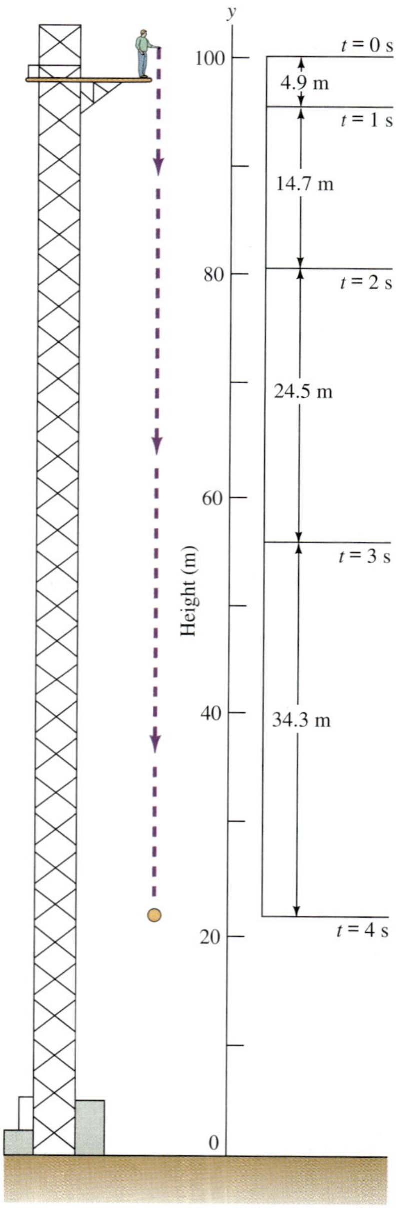

FIGURE 2–18 Example 2–11. When an object falls freely under the influence of gravity alone, the distance traveled in a given time interval increases as the square of the total time elapsed.

EXAMPLE 2–11 Calculate the position of the ball at $t = 1$ s, 2 s, 3 s, and 4 s under the conditions stated in Example 2–9.

Solution: The initial conditions are the same as before; Fig. 2–18 illustrates the situation. We use Eq. (2–26c) to find y for each t:

$$y = \frac{-gt^2}{2} + 100\ \text{m};$$

$$t = 1\ \text{s}, \quad y = \frac{-(9.80\ \text{m/s}^2)(1\ \text{s})^2}{2} + 100\ \text{m} = 95.1\ \text{m};$$

$$t = 2\ \text{s}, \quad y = \frac{-(9.80\ \text{m/s}^2)(2\ \text{s})^2}{2} + 100\ \text{m} = 80.4\ \text{m};$$

$$t = 3\ \text{s}, \quad y = \frac{-(9.80\ \text{m/s}^2)(3\ \text{s})^2}{2} + 100\ \text{m} = 55.9\ \text{m};$$

$$t = 4\ \text{s}, \quad y = \frac{-(9.80\ \text{m/s}^2)(4\ \text{s})^2}{2} + 100\ \text{m} = 21.6\ \text{m}.$$

We show these distances in Fig. 2–18 for the 1-s equal time intervals. During the first second, the ball travels 100 m − 95.1 m = 4.9 m, but during the fourth second, the ball travels 55.9 m − 21.6 m = 34.3 m. Figure 2–19 clearly illustrates the increasing distance traveled in each time interval.

EXAMPLE 2–12 A ball is thrown straight up with a speed of 10.0 m/s from a third-floor window that is located 15.0 m above the ground. Calculate the maximum height of the ball, the ball's velocity when it hits the ground, and the total time it takes to reach the ground (Fig. 2–20).

Solution: A ball thrown upward reaches its maximum height when its velocity is zero. At this point, the ball's direction of motion changes. For the initial conditions at $t_0 = 0$ s, we know that $y_0 = 15.0$ m and $v_0 = +10.0$ m/s. We use Eq. (2–26a) to determine the time when the ball reaches its maximum height:

$$v = 0 \text{ m/s} = -(9.80 \text{ m/s}^2)t + 10.0 \text{ m/s};$$

$$t = \frac{10.0 \cancel{\text{m/s}}}{9.80 \cancel{\text{m}}/\text{s}^2} = 1.02 \text{ s}.$$

The maximum height above the ground, y_{max}, is found by using this value of t in Eq. (2–26c):

$$y_{max} = \frac{-(9.80 \text{ m/s}^2)(1.02 \text{ s})^2}{2} - (10.0 \text{ m/s})(1.02 \text{ s}) + 15.0 \text{ m} = 20.1 \text{ m}.$$

We have two choices to work the remainder of this example. We can either take a new initial condition to be at the maximum height of the ball, or we can keep the original initial conditions. Let's work the example using the second choice so the initial conditions are as given in the problem

We use Eq. (2–26d) to calculate the final velocity v. When the ball hits the ground, the value of y is zero:

$$\begin{aligned} v^2 &= (10.0 \text{ m/s})^2 - 2(9.80 \text{ m/s}^2)(0 \text{ m} - 15.0 \text{ m}) \\ &= 100 \text{ m}^2/\text{s}^2 + 294 \text{ m}^2/\text{s}^2 = 394 \text{ m}^2/\text{s}^2; \\ v &= -19.9 \text{ m/s}. \end{aligned}$$

Notice that the final velocity does not depend on whether the initial direction of the ball is up or down. The final value of the velocity, however, is negative, because the direction of the ball is down. We can choose either the + or − solution of the square root of v^2, whichever is physically sensible.

Equation (2–26a) is used to find the flight time:

$$-19.9 \text{ m/s} = -(9.80 \text{ m/s}^2)t + 10.0 \text{ m/s};$$

$$t = \frac{19.9 \cancel{\text{m/s}} + 10.0 \cancel{\text{m/s}}}{9.80 \cancel{\text{m}}/\text{s}^2} = 3.05 \text{ s}.$$

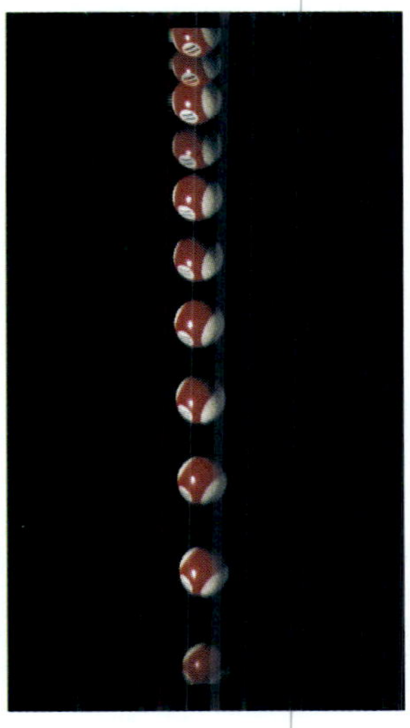

FIGURE 2–19 This multiple image of a ball falling under the influence of gravity is made by flashing a stroboscope 22 times at equal time intervals over 0.5 s.

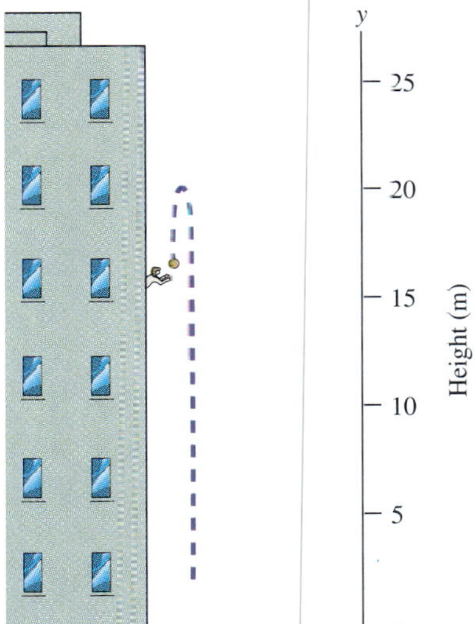

FIGURE 2–20 Example 2–12.

* 2–6 MOTION WITH NONCONSTANT ACCELERATION

We have learned how to find the velocity if the displacement function $x(t)$ is known ($v = dx/dt$); we can also determine the acceleration if the velocity $v(t)$ is known ($a = dv/dt$). (Again we suppress the boldfaced notation for one-dimensional motion.) What about the inverse? Can we determine the displacement if the velocity is known or the velocity if the acceleration is known? What happens if the acceleration is not constant? You will learn in your calculus course (if you have not already) that the techniques of integration allow us to find the displacement if we are given the velocity and to find the velocity if we are given the acceleration.

Displacement as an Integral of Velocity over Time

Assume that we know how the velocity depends on time, as shown in Fig. 2–21a. In this example, the velocity is not changing linearly with time, so we know that the acceleration is not constant. We divide the total time interval from an initial time t_i to the final time t_f into many small equal time intervals Δt. These new time intervals are so short that the velocity cannot change very much over their duration. Let's concentrate

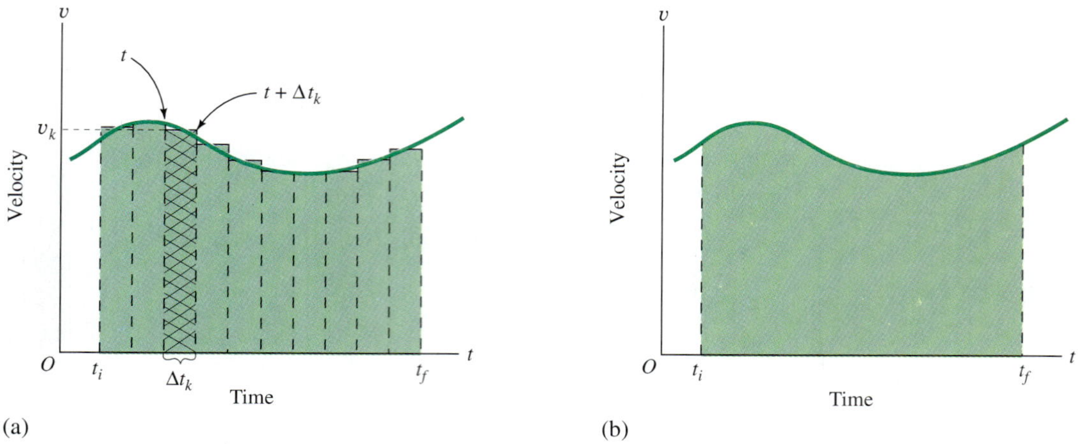

FIGURE 2–21 (a) The displacement over the crosshatched area is $\Delta x_k = v_k\,\Delta t_k$, where Δt_k is a small time interval and v_k is the average velocity in this interval. If the interval is small enough, v_k is the value of the instantaneous velocity at time t. If all the Δx_k are summed, we find the net displacement between initial time t_i and final time t_f, which is the shaded area under the curve of velocity versus time. (b) This area is also found by integration, $x_f - x_i = \int_{t_i}^{t_f} v\,\mathrm{d}t$.

on the tiny time interval from t to $t + \Delta t$, where t is some intermediate time between t_i and t_f. The velocity $v(t)$ at the beginning of the time interval Δt must be similar to the velocity $v(t + \Delta t)$ at the end; further, it should be nearly the same as the average value in the time interval. We therefore write this average value as just $v(t)$ itself. By definition, it is the net displacement Δx in the time interval divided by Δt, $v(t) = \Delta x/\Delta t$. Thus, for small intervals,

$$\Delta x = v(t)\Delta t.$$

Let's now label the tiny time intervals with an index k, and make the time intervals equal. Then the displacement of an object during the time period Δt_k is $\Delta x_k = v_k\,\Delta t_k$ and is shown graphically by the crosshatched area of the shaded curve in Fig. 2–21a.

The total displacement $x_f - x_i$ over the total time interval t_i to t_f is then a sum over the small displacements $\Delta x_k = v_k\,\Delta t_k$. There must be N equal tiny time intervals Δt_k to span the time interval from t_i to t_f.

$$N\,\Delta t_k = t_f - t_i;$$

$$x_f - x_i = \sum_{k=1}^{N} \Delta x_k = \sum_{k=1}^{N} v_k\,\Delta t_k. \tag{2–28}$$

The shaded area under the curve in Fig. 2–21a represents the displacement $x_f - x_i$. If we are concerned with accuracy because v changes significantly over the time interval Δt, we can simply make Δt smaller and smaller. In the limit that each time interval Δt goes to zero, we have an exact result:

$$x_f - x_i = \lim_{\Delta t \to 0} \sum_{k=1}^{N} v_k\,\Delta t_k. \tag{2–29}$$

The right-hand side of this expression is the definition of the integral of $v(t)$ over time:

$$x_f - x_i = \int_{t_i}^{t_f} v(t)\,\mathrm{d}t. \tag{2–30}$$

This expression is called a **definite integral** because the limits of integration—here t_f and t_i—are specified. The symbol $\int$ replaces the summation sign Σ when we sum over an infinite number of infinitesimal intervals of Δt. The area under the curve of velocity versus time between t_i and t_f in Fig. 2–21b is exactly the displacement $x_f - x_i$.

If the integration limits are not specified, we have

$$x = \int v(t)\,\mathrm{d}t + C, \tag{2–31}$$

where C is a constant of integration determined from the initial conditions discussed earlier (that is, x_0 and v_0 at t_0). Equation (2–31) is called an **indefinite integral**. You will find a table of some indefinite integrals in Appendix IV–8.

In Eqs. (2–30) and (2–31), we speak of taking the integral of $v(t)$ over the time. The function $v(t)$ in this case is called the *integrand*. The integral of a function between two limits is the area under the curve that represents a graph of the function versus its variable. For example, Fig. 2–21 shows that the integral of $v(t)$ over t is the total area under the curve $v(t)$ versus t.

Velocity as an Integral of Acceleration over Time

We can perform an analysis for the velocity and acceleration just as we have done for displacement and velocity. We find that velocity is an integral of the acceleration over time:

$$v_f - v_i = \int_{t_i}^{t_f} a(t)\, \mathrm{d}t; \tag{2–32}$$

$$v = \int a(t)\, \mathrm{d}t + C, \tag{2–33}$$

We have again written both the definite and indefinite integral forms.

More on Integration

Differentiation and integration are inverse processes. Take the derivative of a function. The integral of that derivative is the original function (except for an additive constant). To illustrate this, consider the function $f(t) = C_n t^n$, where C_n is a constant and n is an integer. The derivative of $f(t)$ is

$$\frac{\mathrm{d}f(t)}{\mathrm{d}t} = \frac{\mathrm{d}}{\mathrm{d}t}(C_n t^n) = nC_n t^{n-1}. \tag{2–34}$$

A polynomial such as $C_0 + C_1 t + C_2 t^2 + \cdots + C_n t^n$ can be differentiated by means of this equation. The integral of $C_n t^n$ is

$$\int f(t)\, \mathrm{d}t = \int C_n t^n\, \mathrm{d}t = \frac{C_n t^{n+1}}{n+1} + C, \tag{2–35}$$

where C is an indefinite constant to be determined by the initial conditions. As a check, let us verify that the derivative of the right-hand side of Eq. (2–35) gives us $C_n t^n$; that is, perform the inverse operation. Using Eq. (2–34), we have

$$\frac{\mathrm{d}}{\mathrm{d}t}\left(\frac{C_n t^{n+1}}{n+1} + C\right) = \frac{(n+1)C_n t^{(n+1)-1}}{n+1} + 0 = C_n t^n. \tag{2–36}$$

Therefore, our results are consistent. Notice that the derivative of the constant C in Eq. (2–36) is zero, an application of Eq. (2–34) for $n = 0$.

The Special Case of Constant Acceleration

The value of Eqs. (2–30) and (2–32) is that they can be used to find the position and the velocity of an object whether its acceleration is constant or not. We can demonstrate the consistency of these integration techniques by using them to rederive the earlier constant-acceleration results. Suppose that the magnitude of the constant acceleration is a. Beginning with Eq. (2–32), with the initial conditions x_0 and v_0 for $t = t_0$, we have

$$v - v_0 = \int_{t_0}^{t} a\, \mathrm{d}t.$$

The integral of a constant C_1 is, according to Eq. (2–35), $C_1 t$, so

$$\text{for constant acceleration:}\quad v - v_0 = at\Big|_{t_0}^{t} = a(t - t_0),$$

where the vertical line indicates that the total function is to be evaluated at the two limits of integration (t_0 and t) and the results subtracted. Rearranging this result gives Eq. (2–25a),

$$\text{for constant acceleration:}\quad v = v_0 + a(t - t_0).$$

We integrate this result once more to determine x:

for constant acceleration:

$$x - x_0 = \int_{t_0}^{t} v\,\mathrm{d}t = \int_{t_0}^{t} (v_0 + at - at_0)\,\mathrm{d}t = \left(v_0 t + \frac{at^2}{2} - at_0 t\right)\Bigg|_{t_0}^{t}$$

$$= v_0(t - t_0) + \frac{a}{2}(t^2 - t_0^2) - at_0(t - t_0)$$

$$= v_0(t - t_0) + \frac{at^2}{2} - at_0 t + \frac{at_0^2}{2} = v_0(t - t_0) + \frac{a}{2}(t - t_0)^2;$$

$$x = \frac{a(t - t_0)^2}{2} + v_0(t - t_0) + x_0,$$

which is Eq. (2–25c). Integration techniques allow us to obtain these results much more quickly than previously, and they allow us to dispense with the use of the average velocity.

SUMMARY

Kinematics is the mathematical description of motion. Quantities of particular physical significance are displacement, instantaneous velocity, and instantaneous acceleration. These are all vector quantities. The displacement $\mathbf{\Delta x}$ measures the change in an object's position; where an object moves from position $\mathbf{x}_1$ to position $\mathbf{x}_2$, its displacement is

$$\mathbf{\Delta x} \equiv \mathbf{x}_2 - \mathbf{x}_1. \tag{2–3}$$

The instantaneous velocity at any given time is defined as the limit of the average velocity as $\Delta t \rightarrow 0$:

$$\mathbf{v}(t) \equiv \lim_{\Delta t \rightarrow 0} \frac{\mathbf{x}(t + \Delta t) - \mathbf{x}(t)}{\Delta t}. \tag{2–10}$$

It is thus the rate of change, or derivative, of the displacement at that time:

$$\mathbf{v} = \lim_{\Delta t \rightarrow 0} \frac{\mathbf{\Delta x}}{\Delta t} = \frac{\mathrm{d}\mathbf{x}}{\mathrm{d}t}. \tag{2–11}$$

The instantaneous acceleration is defined as the rate of change of velocity:

$$\mathbf{a} \equiv \lim_{\Delta t \rightarrow 0} \frac{\mathbf{\Delta v}}{\Delta t} = \frac{\mathrm{d}\mathbf{v}}{\mathrm{d}t}. \tag{2–14}$$

We can also define the average speed and average velocity, but these concepts are used less frequently.

Useful relations between displacement, velocity, and the acceleration that are valid for constant acceleration in one dimension are for constant acceleration:

$$v = a(t - t_0) + v_0, \qquad (2\text{–}25a)$$

$$v_{av} = \frac{v_0 + v}{2}, \qquad (2\text{–}19, 2\text{–}25b)$$

$$x = \frac{a(t - t_0)^2}{2} + v_0(t - t_0) + x_0, \qquad (2\text{–}25c)$$

$$v^2 = v_0^2 + 2a(x - x_0). \qquad (2\text{–}24, 2\text{–}25d)$$

Graphical techniques are useful in determining velocity. Graphically, velocity is the slope of a curve of displacement versus time. Similar analysis can determine acceleration from a curve of velocity versus time.

The acceleration due to gravity, $\mathbf{g} = -g\mathbf{j}$, is an important example of a constant acceleration. In this case, the acceleration points to Earth's center. A particle falling a distance h from rest has the final velocity

$$\text{for constant acceleration:} \quad v = \sqrt{2gh}. \qquad (2\text{–}27)$$

By using integration techniques, the displacement and velocity can be determined from the velocity and acceleration, respectively:

$$x_f - x_i = \int_{t_i}^{t_f} v(t)\, dt. \qquad (2\text{–}30)$$

$$v_f - v_i = \int_{t_i}^{t_f} a(t)\, dt. \qquad (2\text{–}32)$$

UNDERSTANDING KEY CONCEPTS

1. Why is it a good idea to increase the space between your car and the car in front of you when the speed of the cars increases?

2. A piece of chalk is thrown straight up; at some point it stops and begins to drop. Can there be a nonzero acceleration at this point, even though the velocity falls to zero?

3. In a series of thought experiments, an object is dropped from rest from a given height on a variety of planets. Each of these planets has a different acceleration due to gravity, g_x. Describe how the time of fall varies with g_x. How does the speed of the object at the end of the fall vary with g_x?

4. You are in the unfortunate position of being in an elevator with 20 bowling balls when the elevator cable breaks, causing both you and the elevator to fall under the acceleration of gravity. The emergency brake has not yet cut in. What is happening inside the elevator?

5. If an object that is restricted to moving along a straight line has a positive initial velocity, and if the acceleration is always negative, can the velocity remain positive?

6. What is the role of an air bag placed where a falling object is expected to land? How can an air bag prevent injury to someone who jumps from a height?

7. For the data shown in Table 2–1, will there be any difference between the average speed over some interval and the magnitude of the average velocity over the same interval? How would you answer the same question if the motion were not on a straight track?

8. What should the velocity of the runner in Fig. 2–2 be at $t = 0$ s? Do the data justify your conclusion? Explain.

9. "Zeno's paradox" comes to us from ancient Greece. It concerns the difficulty that a runner might have in catching a tortoise near the finish line of a race if the tortoise is ahead of the runner at one point, as follows. At some time, the tortoise is a distance of L in front of the runner. After a time interval Δt, the runner is $L/2$ behind the tortoise. After a later time $\Delta t/2$, the runner is $L/4$ behind the tortoise. After a time $\Delta t/4$, the runner is $L/8$ behind the tortoise. The runner always appears to be behind! Where did the Greeks go wrong? By the way, the correct answer to this question was given only in Newton's time and lies behind the crucial concepts of calculus.

10. If the velocity of an object is positive, is its acceleration necessarily positive? Is there any connection between the sign of the velocity and the sign of the acceleration?

11. A juggler tosses a beanbag straight up with initial speed v_0 under the influence of gravity, lets a second beanbag drop from rest, and tosses a third straight down with initial speed v_0. Compare the subsequent accelerations of the three beanbags.

12. A beanbag is tossed straight up. It rises, reaches a maximum height, then falls back down. What is the acceleration of the beanbag at its maximum height?

13. Given a stopwatch and a measuring rod, how would you determine the average acceleration experienced by someone jumping on a trampoline?

14. You are given a measuring rod and a movie camera, with a rather precisely known speed of the motion of the film. How would you use this to determine (to some degree of accuracy) the instantaneous velocity of a person jumping up and down on a trampoline? How would you use your apparatus to measure the instantaneous acceleration?

PROBLEMS

2–1 Displacement

1. (I) A grasshopper jumps along a groove aligned with the x-axis. Starting at the origin, the grasshopper's first jump has a displacement $+32$ cm, the second jump has a displacement -27 cm, the third a displacement -23 cm, and the fourth a displacement $+39$ cm. What is the net displacement? At what position is the grasshopper after all four jumps?
2. (I) Using the data in Table 2–1, draw position vectors to the runner for 40 m and 80 m. Write and draw the displacement from 40 m to 80 m.
3. (I) A gym teacher organizes a series of indoor races in the gym, which is 42 m in length. The students run from one end to the other and back again. After three round trips, what is the distance traveled by each student, and what is the displacement vector? Draw a graph of the magnitude of the displacement vector as a function of time if it takes 7 s to run each 42-m leg. Assume that the speed is constant.

2–2 Speed and Velocity

4. (I) In 1991, Carl Lewis edged out Leroy Burrell in the World Championships in Tokyo to set a new world record in the 100-m dash. Their times at 10-m intervals are given here. Calculate the average velocity for Lewis's world record for 0 m to 50 m, 50 m to 100 m, and 0 m to 100 m.

	Time(s)	
Distance(m)	***Lewis***	***Burrell***
10	1.88	1.83
20	2.96	2.89
30	3.88	3.79
40	4.77	4.68
50	5.61	5.55
60	6.46	6.41
70	7.30	7.28
80	8.13	8.12
90	9.00	9.01
100	9.86	9.88

5. (I) (a) Plot the path of an automobile that travels from a starting point to a point 15 km along a straight road at 75 km/h. It stops for 25 min, then continues on the same straight road for 40 km at 100 km/h. After a 5-min stop, it returns to its starting point at 60 km/h. Draw your position axis as horizontal and your time axis as vertical. (b) On the same plot, draw the path of an automobile that starts from the same spot 25 min after the first one and travels at 74 km/h in the original direction of the first automobile. Where, and how often, will the two cars meet?
6. (I) A car moving at 60 mi/h passes a pickup truck moving at 50 mi/h. The car goes on for 20 mi, then stops at a rest stop for 40 min. The car resumes its journey, again at 60 mi/h. Assuming the truck did not stop and maintained its speed, how long after the initial passing does it take for the car to catch the truck again? Solve this problem by graphical means.
7. (I) Redraw the same paths for the two vehicles of Problem 5 on a plot in which the horizontal axis is the time axis and the position, x, is along the vertical axis. Suppose that you took this new plot and simply relabeled the axes, so that the vertical direction represents time and the horizontal direction represents position. Interpret the paths of the two vehicles.
8. (I) An automobile travels north, covering a distance of 30 mi in 40 min, stops for 20 min, and then continues north for 20 mi, taking 1/2 h. Assume that the car moves uniformly during each segment of the trip. Calculate the average velocity of the total trip. Calculate the average velocity for the first half (by time) and the last half of the trip.
9. (II) Using Fig. 2–5, find the average velocity at 60 m for $\Delta t =$ 1 s and 2 s. Find the instantaneous velocity at 60 m.
10. (II) The position of a falling particle is given by $x = x_0 + v_0 t - 1/2\, gt^2$. What is the velocity of the particle as a function of time? Calculate the average velocity during the time intervals $t = 0$ s to 1 s, $t = 1$ s to 2 s, and $t = \tau$ s to $(\tau + 1)$ s.
11. (II) Use the velocity of a particle as a function of time, tabulated below, to calculate the position of the particle at each of the times; assume that at $t = 0$ s, the particle was at the origin and at rest. [*Hint*: A graph is simplest.]

Time(s)	*Velocity(m/s)*
0.5	0.75
1.5	1.75
2.5	8.75
3.5	21.75
4.5	39.25
5.5	62.75
6.5	90.75
7.5	122.75

12. (II) The height of a bungee jumper above ground level is given as a function of time t by $y = (30 \text{ m})\cos[\pi t/(6 \text{ s})] + (42 \text{ m})$. (a) Sketch the function $y(t)$ from $t = 0$ s to $t = 9$ s. (b) Calculate the average velocity of the jumper between $t = 2$ s and $t =$ 4 s and between $t = 4$ s and $t = 8$ s. (c) What is the instantaneous velocity of the jumper when he is closest to the ground?
13. (II) An old brain teaser reads as follows: Two trains leave different stations 80 km apart and travel toward each other on a straight track. One train has a speed of 80 km/h, and the other has a speed of 160 km/h. A very fast insect leaves the slower train and heads toward the faster train at a speed of 240 km/h. Upon encountering the second train, it turns around and just as rapidly returns to the first train. It continues these maneuvers until it is squashed between the two trains when they collide. Graph what has happened, and use your graph to determine how far the insect will have traveled. Can you think of a way to estimate your result or to calculate it rapidly?
14. (II) Traffic signals are placed along a straight road at positions $x = 0$ m, 600 m, and 1200 m (Fig. 2–22). The time intervals during which the signals are green are shown by the thick lines in the figure. (a) Draw the displacement versus time curves (fastest and slowest) for a car that passes through all the lights, when the car moves with constant speed. (b) Draw a similar set of lines for a car traveling in the opposite direction. (c) Assuming that the lights are timed such that a car passes

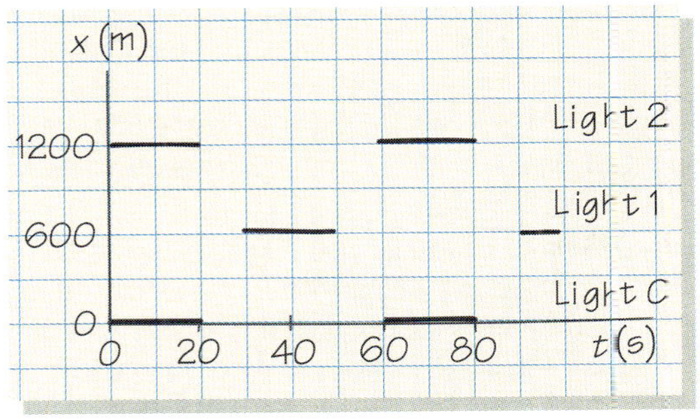

FIGURE 2–22 Problem 14.

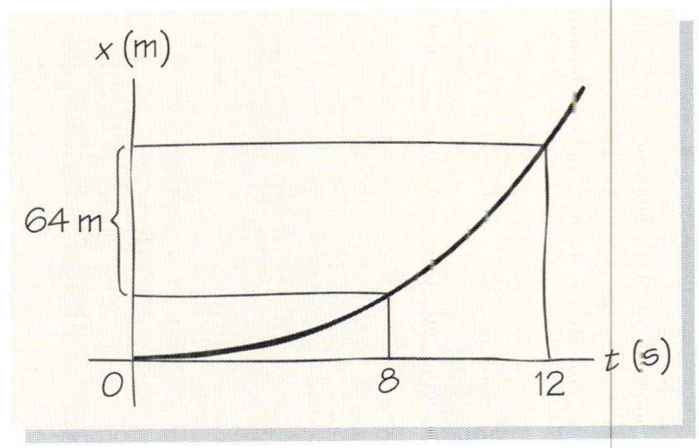

FIGURE 2–23 Problem 21.

through all lights in the middle of the time interval, what is the speed that the lights are timed for? (d) What is the fastest constant speed of a car that makes it through all the signals, assuming it arrives at the first light at the optimal moment?

15. (III) The distance an ant moves is given by $x = 0.02t^3 - 0.1t^2 + 2t$ cm, where t is in seconds. Calculate the velocities for $t = 1$ s, 5 s, and 10 s. What is the average velocity for the first 10 s? Why is the formula unrealistic for long times?

16. (III) The displacement of a particle as a function of t is described by the equation $x = \sqrt{(2\text{ m}^2/\text{s})(t + 1\text{ s})}$, where t is measured in seconds. (a) Plot $x(t)$ between $t = 0$ s and $t = 5$ s. (b) Calculate the average velocity between $t = 1$ s and $t = 5$ s; between $t = 2$ s and $t = 4$ s, and between $t = 2.8$ s and 3.2 s. (c) Compare these results with the instantaneous velocity at $t = 3$ s.

2–3 Acceleration

17. (I) An automobile badly in need of repairs is able to accelerate only in the forward direction at a constant value of 0.6 m/s^2. How long does it take the automobile to get to 40 mi/h?

18. (I) A bicyclist is pedaling at a constant speed of 10 m/s when she decides to slow down. She stops pedaling and sits up, and the combined effects of wind resistance and road friction cause a negative acceleration of -0.3 m/s^2. If this acceleration does not change, how long would it take her to slow to 5 m/s?

19. (I) A car is said to go from rest to 60 mi/h in 9.0 s. Assuming that the acceleration is uniform, what is its value in units of g?

20. (II) Car A leaves a city and travels along a straight road for 1.5 min at 60 km/h. It then accelerates uniformly for 0.25 min until it reaches a speed of 80 km/h. It proceeds at that speed for 2.0 min, then decelerates uniformly for 0.5 min until it comes to rest. Car B leaves the same city along the same road and accelerates uniformly for 1.6 min until it reaches a speed of 120 km/h. It then decelerates uniformly until it comes to rest again after 1.6 min. (a) Plot the curve of the cars' motions on a graph in which the vertical axis is the speed v and the horizontal axis is the time t. (b) Plot the trajectories of the cars on a graph in which the vertical axis is the distance x from the city and the horizontal axis is time t. (c) How far will the two cars have traveled during the different stages?

21. (II) An automobile starting from rest at $t = 0$ s undergoes constant acceleration on a straight line (Fig. 2–23). It is observed to pass two marks separated by 64 m, the first at $t = 8$ s and the second at $t = 12$ s. What is the value of the acceleration?

22. (II) Inclined planes are convenient tools to study motion under a constant acceleration. The time of passage of a ball rolling on an inclined plane is measured by three light gates positioned 60 cm apart. The ball passes the light gates at 0.30 s, 1.15 s, and 1.70 s. Find the acceleration of the ball.

23. (II) Suppose the position of a particle is described by $x = A\sin(\pi t/12)$. Calculate the velocity and acceleration of the particle as a function of time.

24. (II) Consider the motion of the particle whose velocity is tabulated in Problem 11. Use these data to make a table of approximate values of the acceleration for $t = 1$ s, 2 s, . . . , 7 s.

25. (II) The position x of a block attached to a spring as a function of time is given by the formula $x = A\sin(\pi t/12)$, as shown in Fig. 2–24. Describe in words the motion of the block.

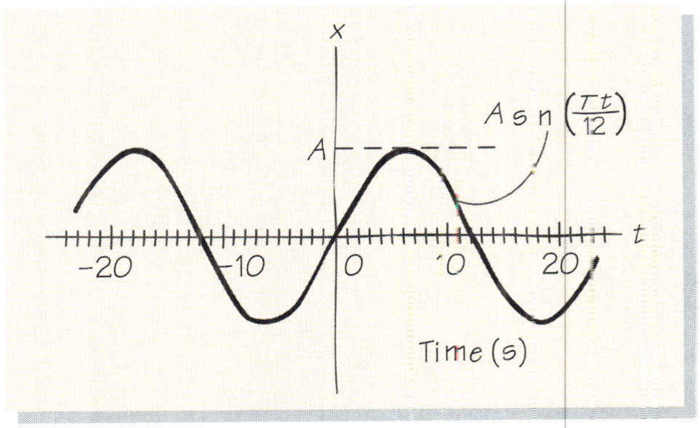

FIGURE 2–24 Problem 25.

26. (II) The position of a particle is given by $x = At^2 - Be^{\alpha t}$. The particle is initially at $x = -2.0$ m with $v = 4.0$ m/s at $t = 0$ s. After 0.2 s, the velocity is observed to be 5 m/s. What is the acceleration after 1.0 s?

27. (II) Consider an object whose acceleration is determined by its velocity, as in the equation $a = A - (v/t_0)$. Here t_0 and A are constants with the dimensions of time and acceleration, respectively. Assume that the object starts out at $t = 0$ s with an initial velocity v_0. Sketch the behavior of the acceleration and of the velocity. Describe the motion after a long time.

2–4 Motion with Constant Acceleration

28. (I) A drag racer reaches 128 mi/h in a $\frac{1}{4}$-mi race. Assuming a constant acceleration, what was the elapsed time?

29. (I) A car traveling 25 mi/h must reach a minimum of 50 mi/h within a 1000-ft access lane. What must the car's constant acceleration be?

30. (I) An airplane starting from rest reaches its takeoff velocity of 212 mi/h over a runway of 6000 ft. How long did this take if the plane rolled with a constant acceleration?

31. (I) A ball undergoes a constant acceleration of 1.5 m/s^2. What is the average velocity over the period 1 s to 2 s, assuming the ball started from rest at $t = 0$ s?

32. (I) A rocket accelerates uniformly from rest to a speed of 4.2×10^3 mi/h in 125 s. Over what distance does the rocket accelerate?

33. (I) A soccer ball rolls with an initial velocity of 8.0 m/s in an easterly direction across a flat field. Friction slows the ball down at the rate of 0.5 m/s^2. (a) Express the ball's velocity as a function of time. (b) Where is the ball 5.0 s after it starts to roll?

34. (II) The speed of a landing airplane is 60 m/s. After touching ground it rolls a distance of 300 m on the runway at a constant velocity. It then decelerates at 2.5 m/s^2 until it stops. (a) Sketch the displacement–time and velocity–time curves. (b) Calculate the distance traveled on the ground, and the time interval between touch-down and full stop.

35. (II) A car travels at a constant velocity of 20 m/s toward an intersection. When the car is 80 m from the intersection, the traffic light turns yellow. The driver continues with constant velocity for 1.2 s and then applies the brakes, with a constant acceleration such that the car stops just at the intersection. (a) Sketch the displacement–time and velocity–time curves for the motion. (b) Determine the acceleration of the car during the braking period.

36. (II) A car accelerates from rest at 3 m/s^2 for 4 s, travels at a constant speed for 7 s, accelerates at 1 m/s^2 for 15 s, and then decelerates to rest at 2.5 m/s^2 (Fig. 2–25). How far has the car traveled?

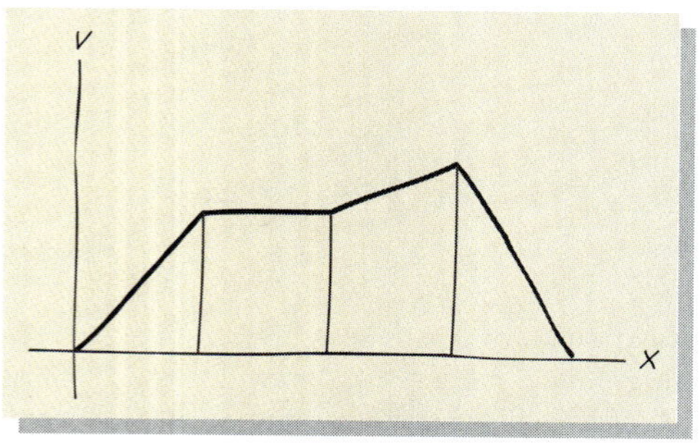

FIGURE 2–25 Problem 36.

37. (II) A child is in an open-cage elevator facing out on a hotel lobby. The elevator is descending at a constant speed of 1 m/s. The child lets a penny drop from his hand when the elevator is 20 m above the floor of the lobby. How much time does the penny spend in the air? Ignore air resistance.

38. (II) Suppose that a runner were capable of a constant acceleration of 3 m/s^2 for the entire length of a 100-m dash. (a) How long would it take the runner to run the first 10 m? (b) How long for the first 50 m? (c) For the second 50 m? (d) For the entire 100 m? (e) Compare to the times of Problem 4.

39. (II) In 1979, the Japanese tested a magnetically levitated train. The train is both suspended and propelled by magnetic forces. The train traveled on a straight 7000-m-long track starting from rest; it reached a peak speed of 144 m/s before it came to rest again. Both the acceleration and deceleration were constant and of the same magnitude. The entire length of the track was used. (a) What was the magnitude of the acceleration (and deceleration)? (b) How much time was spent on the trip from one end of the track to the other?

40. (II) Your bus is leaving the stop, accelerating at a constant rate of 0.6 m/s^2. You turn the corner to see the bus pulling out of the stop 30 m ahead of you. What is the minimum steady speed with which you must run to catch the bus? Olympic sprinters can run at 10 m/s.

41. (II) A speeder is traveling at 75 mi/h. He passes a standing police car, which starts to chase him. The police car accelerates from 0 mi/h to 85 mi/h in 13 s and travels at 85 mi/h thereafter. (a) Sketch the positions of both cars on the same x-versus-t graph. (b) How far from its starting point does the police car overtake the speeder? (c) What is the elapsed time?

42. (II) A car is moving at 40 mi/h when the driver sees a light turn red. She hits the brake pedal when she is 100 ft from the light, and the deceleration of the automobile is $0.5g$, where g is the acceleration due to gravity. Does she stop before she arrives at the light? How far does she travel before stopping?

43. (II) In Problem 42, calculate the time it takes to stop from the moment the brakes are applied.

44. (II) Two automobiles are geared quite differently and are to be used for a drag race over a distance of 400 m. Car A accelerates at a constant value of 5.0 m/s^2 for the first 200 m, then at a constant value of 2.5 m/s^2 for the remaining 200 m. Car B accelerates at a constant rate of 4.5 m/s^2 for the first 200 m but at 3.0 m/s^2 for the remaining distance. (a) Give the value of the speed of each automobile at the 200-m mark and the time it took each to get there. (b) What are the finishing times for the race and the values of the respective speeds at the end of the race?

45. (II) An electron in the picture tube of a TV set, traveling in a straight line, accelerates uniformly from speed 3×10^4 m/s to 5×10^6 m/s along a length of 2 cm. (a) How much time does the electron spend in this 2-cm region? (b) What is the magnitude of the electron's acceleration?

46. (II) You have an old, heavy automobile that does not accelerate very rapidly, but can maintain acceleration for a long period. Suppose that your car has a maximum acceleration that takes it from 0 mi/h to 50 mi/h in 20 s. What would the speed be if this average acceleration were maintained for 1 min? How far does your car go during the first 20 s, and how far would it travel in 1 min under the above conditions?

47. (II) An elevator accelerates from the ground with a uniform acceleration a. After 3 s, an object is dropped out of an opening in the floor of the elevator and that object hits the ground 3.5 s later. How large is the acceleration? How high was the elevator when the object was dropped?

48. (II) Two small objects A and B are suspended from the ends of a rope thrown over a pulley (Fig. 2–26). A is 1.2 m above B when the system is released from rest. A descends with a downward acceleration of 0.3 m/s^2 and, because of the rope, B accelerates upward at the same rate. How much time elapses before the objects bump into each other?

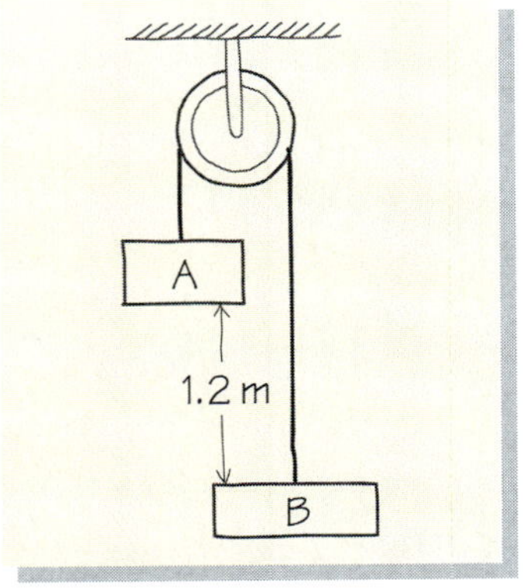

FIGURE 2–26 Problem 48.

49. (II) A car moving at 60 mi/h can be brought to rest in 4 s. Assuming that the deceleration is uniform, how far will the car travel between the time the brakes are applied and the time the car stops?

50. (II) A mountain climber is attached to a rope. She slips, and after she has fallen 5 m, the rope starts to decelerate her. If the constant deceleration is $5g$ (five times the acceleration due to gravity), how much will the rope have to stretch? (In reality, the deceleration depends on the stretching of the rope.)

51. (II) A bullet traveling at 600 m/s penetrates a block of wood and comes to rest with a constant deceleration after traveling 20 cm (Fig. 2–27). What is the magnitude of the deceleration? How long does it take the bullet to stop?

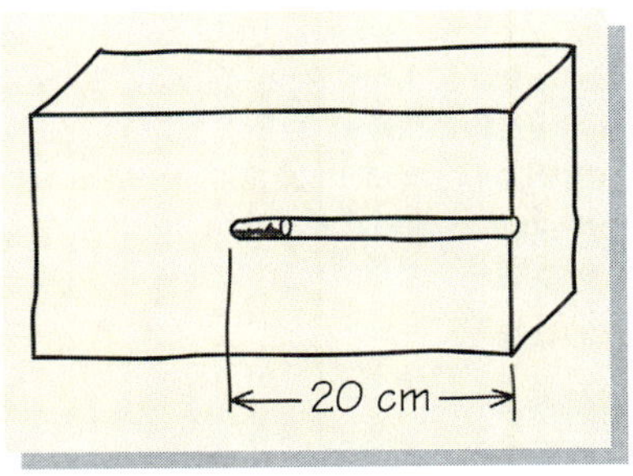

FIGURE 2–27 Problem 51.

2–5 Freely Falling Objects

52. (I) The tower of Pisa is 54.5 m tall. Assuming that Galileo dropped his object from rest from the top of the tower and that the effects of air resistance were negligible, how long would it have taken the object to fall?

53. (I) A story claims that someone who fell off New York City's Empire State Building (which has approximately 100 floors) was overheard to say "so far, so good" as he passed a third-floor window. Make some estimates to see if this is possible.

54. (I) A string is to have a series of lead sinkers tried to it. The first is tied at the bottom, and the second is tied 10 cm up from the bottom. The string can be held at its top and dropped from a height onto the top of a drum on which the first sinker already rests; each time a sinker hits the drum, a tap is heard. How far above the bottom sinker must the third, fourth and fifth sinkers be tied so that the series of four taps is spaced by equal time intervals when the string is dropped?

55. (I) The acceleration due to gravity on the surface of the Moon is only about one-sixth the acceleration due to gravity on Earth's surface. In the celebrated experiment of the dropped feather performed by an astronaut on the Moon, how long did it take for the feather to drop 1 m to the surface if it started from rest?

56. (I) The acceleration due to gravity on the surface of Jupiter is 25.9 m/s^2. If it were possible to perform the experiment, how long would it take for an object that is initially at rest to fall a distance of 10 m on Jupiter's surface? Ignore any effects due to "air" resistance.

57. (II) A rock is thrown upward from the top of a building at an initial speed of 10 m/s. How much later must a second rock be dropped from rest at the same initial height of 20 m so that the two rocks hit the ground at the same time?

58. (II) An astronaut shipwrecked on a distant planet with unknown characteristics is on top of a cliff, which he wishes to descend. He does not know the acceleration due to gravity on the planet, and he has only a good watch with which to make measurements. He wants to learn the height of the cliff, and to do this he makes two measurements (Fig. 2–28). First, he lets a rock fall from rest off the cliff edge; he finds that the rock takes 4.15 s to reach the distant ground. Second, he releases the rock from the same spot but tosses it upward so that it rises a height of what he estimates to be 2 m before it falls to the ground below. This time the rock takes 6.30 s to reach the ground. What is the height of the cliff?

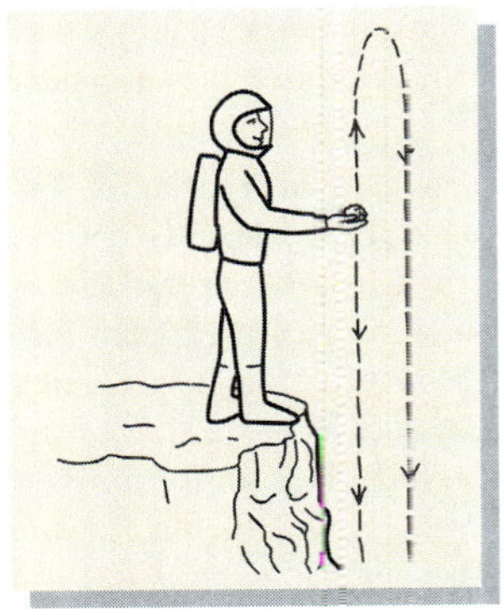

FIGURE 2–28 Problem 58.

59. (II) A ball is dropped from the roof of a 25-m-tall building. What is the velocity of the object when it touches the ground? Suppose that the ball is a perfect golf ball, and it bounces such that the velocity as it leaves the ground has the same magnitude but the opposite direction as the velocity with which it reached the ground. How high will the ball bounce? Now suppose instead that the ball bounces back to a height of 20 m. What was the velocity with which it left the ground?

60. (II) A ball is thrown upward from the ground. It passes a window 10 m above the ground and is seen to descend past the window 2.2 s after it went by on its way up (Fig. 2–29). It reaches the ground 3.6 s after it was thrown. Use this information to calculate the acceleration due to gravity, g.

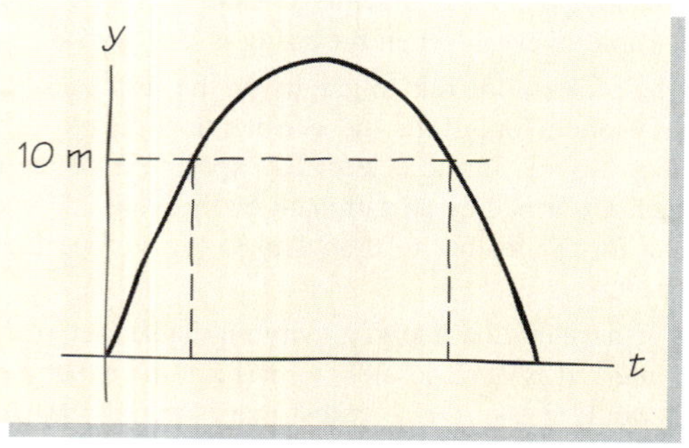

FIGURE 2–29 Problem 60.

*2–6 Motion with Nonconstant Acceleration

61. (II) An object moves with an acceleration that depends on time and has the form $a = -4\sqrt{t}$ m/s^2. Its velocity at $t = 0$ s is 15 m/s. How far will the object travel before it comes to a stop?

62. (I) A machine causes an object's speed to increase exponentially, $v(t) = v_0\, e^{at}$, where $a = 0.5$ s^{-1} and $v_0 = 1$ m/s. If the object starts from the origin, how far has it traveled after 2 s?

63. (II) The height of a mass suspended from a ceiling by a spring at different times t is given by the formula $y = (0.2\text{ m})\sin(4\pi t)$, where t is measured in seconds. (a) Plot the height as a function of time for times $t = 0$ s to $t = 1.0$ s (b) Use your graph to determine the instantaneous velocity at time $t = 0.15$ s. (c) A measurement of the slopes at different times shows that the instantaneous velocity can be represented by the formula $\mathbf{v} = A\cos(4\pi t)\mathbf{j}$, where again t is measured in seconds. Use your measurement from part (b) to determine A. (Do not forget the units of A.) (d) Plot v as a function of t, and use that graph to determine the instantaneous acceleration at time $t = 0.15$ s.

64. (III) When the effect of air resistance is taken into account, the acceleration of a falling object is described by the equation $a = ge^{-bt}$, where g is the acceleration due to gravity, $b = 0.5$ s^{-1}, and t is the time measured from the moment of release. (a) Calculate the velocity and displacement (from the place of release) of the object as a function of time. (b) Sketch these curves. (c) How long does it take for the object to fall 50 m? (d) Compare this time with the time it would take to fall 50 m without air resistance, i.e., with $b = 0$.

65. (III) A powerful rocket moves for a short time with an acceleration that grows with time according to the formula $a = \alpha t^2$. If the rocket is to accelerate from rest in this way until it reaches a speed v_f, how long must the acceleration be maintained?

General Problems

66. (I) A test of your reaction time is to catch a 12-in ruler held vertically by another person. Put your thumb and one finger near the bottom of the ruler, and as soon as the other person releases the ruler, squeeze your thumb and finger together to prevent the ruler from falling. Suppose that in such a test the ruler is grabbed after 5 in of it have passed your hand. What is the time interval between the visual detection of movement and the squeezing together of the fingers?

FIGURE 2–30 Problem 67.

67. (I) The simplest juggling act involves two objects, one of which is transferred from one hand to the other when the second object is tossed upward (Fig. 2–30). Perform an experiment that will tell you how fast you can transfer an object from one hand to the other (for example, by transferring something back and forth 20 times while a friend times you). Estimate from this how high you would have to toss the second object to perform the juggling act. How high would you have to toss the objects if you wanted to juggle three of them?

68. (I) You could probably jump off an 8-ft wall without hurting yourself (but do not try it!). Estimate what your deceleration would be when you hit the ground.

69. (I) A high jumper can jump 2 m on Earth. All other things being equal, how high could the same jumper jump on the surface of the Moon, where the acceleration of gravity is 1/6 that of Earth?

70. (II) Two long-separated friends, June and Bill, spot each other in an airport terminal from a distance of 30 m. They start to run toward each other (Fig. 2–31). Bill accelerates at a constant rate of 0.9 m/s^2, and June at a constant rate of 1.2 m/s^2. How far from June's initial position do they meet?

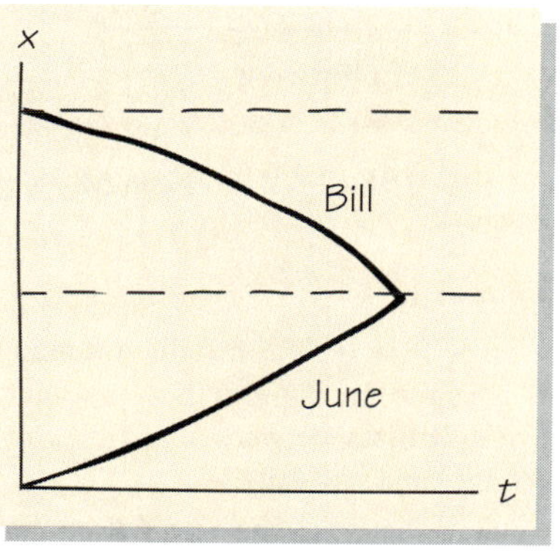

FIGURE 2–31 Problem 70.

71. (II) A Moon rock is thrown upward with velocity 7 m/s. After 7 s, it has a downward velocity of 4 m/s. What is the acceleration due to gravity on the Moon? How high above the starting point did the rock go before it began to fall?

72. (II) A water balloon is dropped from the top of a tower, 200 m off the ground. An alert archer at the base of the tower sees the balloon and shoots an arrow straight up toward the balloon 5 s after the balloon is dropped. The arrow's initial velocity is 40 m/s. Where does the arrow intercept the balloon?

73. (II) There are several known cases of paratroopers whose chutes did not open as they fell, but who survived by falling into brush or snow or onto a steep hillside. It is possible to survive a fall when the deceleration on impact is some 500 m/s^2, equivalent to about 50g. What is the distance traveled within a snowbank while a paratrooper comes to a stop if the deceleration is constant with a magnitude of 50g? The speed with which a paratrooper would enter the snowbank is about 40 m/s. Over how much time does the paratrooper decelerate?

74. (III) One test for the effects of the acceleration of gravity is to tie a set of weights to a string, with the second lowest separated from the lowest by L_0, the third lowest from the second lowest by L_1, and so forth, and to then drop the string (Fig. 2–32). Supposing that free fall corresponds to motion with constant acceleration, how should the separations $L_1, L_2, \ldots, L_n$ (where n is the number of weights) be related to L_0 if the sounds made by the weights as they land form a steady beat? The lowest weight starts at the surface onto which the other weights fall.

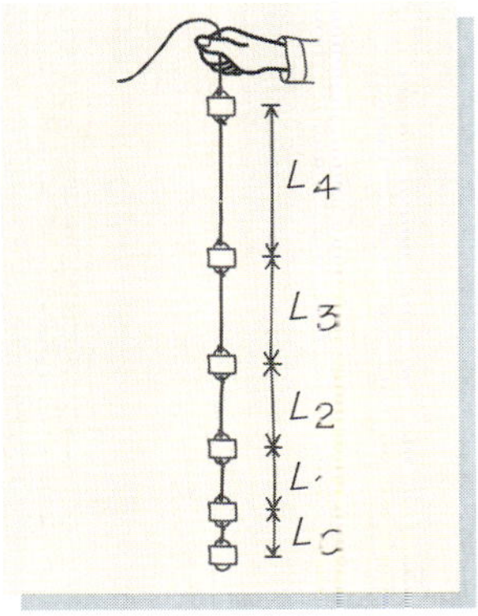

FIGURE 2–32 Problem 74.

CHAPTER

3

The water droplets that fly from this fountain follow parabolic paths. Such paths are a consequence of the constant vertical acceleration associated with local gravity.

Motion in a Plane

We explored one-dimensional motion in Chapter 2. There, the vector nature of displacement, velocity, and acceleration plays a limited role. There are only two directions in straight-line motion—right or left—and these are adequately represented by assigning a plus or minus sign to the quantities used to describe such motion. In this chapter, we extend our study of motion into two and three dimensions—motion in a plane and motion in space, respectively—where vectors play a more elaborate role. We will concentrate on motion in two dimensions. This type of motion is simpler and, in most cases, is the motion of interest to us. For example, a rock thrown from a cliff moves in a vertical plane, as does a simple pendulum, and planetary orbits form planes.

3–1 POSITION AND DISPLACEMENT

The motion of a planet circling the Sun traces out a path, or an orbit, through space. Similarly, a rock thrown off a cliff follows a path, which is known as its **trajectory**. In fact, any pointlike object traces its own trajectory as it moves through space. Figure 3–1 depicts a particle moving in a two-dimensional plane. We label the plane as the xy-plane and introduce a Cartesian coordinate system that contains an origin and x- and y-axes. The particle is at the position P at time t; this position is described by the position vector $\mathbf{r}_P$, which points from the origin to the point P. At a later time $t + \Delta t$ (where Δt is not necessarily small), the particle is located at position Q and is described by the position vector $\mathbf{r}_Q$. This change in the particle's position between times t and

$t + \Delta t$ can be described by the displacement vector $\Delta \mathbf{r}$; this vector is defined by

$$\Delta \mathbf{r} \equiv \mathbf{r}_Q - \mathbf{r}_P. \tag{3–1}$$

The vector $\Delta \mathbf{r}$ points from the tip of vector $\mathbf{r}_P$ to the tip of vector $\mathbf{r}_Q$. Whereas the position vectors $\mathbf{r}_P$ and $\mathbf{r}_Q$ depend on the choice of origin, the displacement vector $\Delta \mathbf{r}$ is independent of the choice of origin. To see this clearly, let's imagine that there is a new origin O' such that the vector from O' to O is the fixed vector $\mathbf{b}$ (Fig. 3–2). The position vector of point P in the new coordinate system is $\mathbf{r}_P + \mathbf{b}$, and that of point Q in the new system is $\mathbf{r}_Q + \mathbf{b}$. If we calculate the displacement $\Delta \mathbf{r}$, which is the difference $[\mathbf{r}_Q + \mathbf{b}] - [\mathbf{r}_P + \mathbf{b}]$, the vector $\mathbf{b}$ cancels. In other words, we have again arrived at the displacement vector defined in Eq. (3–1). This can also be seen in the graphical representation in Fig. 3–2. The choice of origin for the coordinate system did not affect our calculation of the change in the particle's position.

As the particle moves, its components with respect to the coordinate axes change. In other words, the components of the position vector $\mathbf{r}$ along Cartesian coordinate axes change with time:

$$\mathbf{r} = x(t)\mathbf{i} + y(t)\mathbf{j}. \tag{3–2}$$

For three-dimensional motion, we would set up three axes, define three mutually perpendicular unit vectors $\mathbf{i}$, $\mathbf{j}$, and $\mathbf{k}$, and write a position vector in the form

$$\mathbf{r} = x(t)\mathbf{i} + y(t)\mathbf{j} + z(t)\mathbf{k}. \tag{3–3}$$

The fact that there is more than one component is the only real difference between one-dimensional motion and two- or three-dimensional motion.

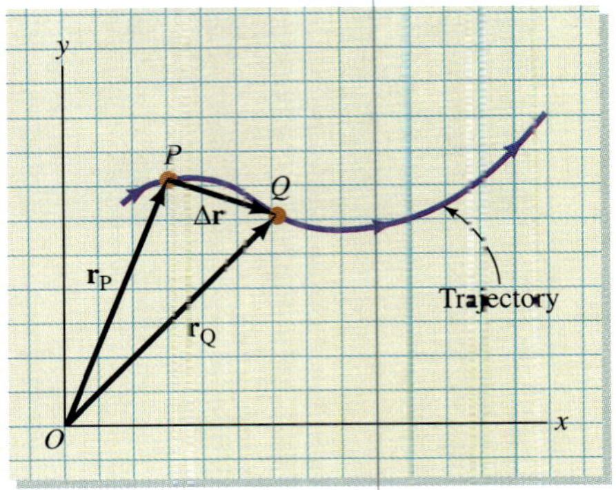

FIGURE 3–1 Position vectors $\mathbf{r}_P$ and $\mathbf{r}_Q$ point from the origin to the positions P and Q at the two times t and $t + \Delta t$, respectively, along the particle's trajectory. The displacement vector between these times is $\Delta \mathbf{r} = \mathbf{r}_Q - \mathbf{r}_P$.

PROBLEM SOLVING

The displacement is independent of the choice of origin.

The unit vectors **i**, **j**, and **k** are discussed in Chapter 1.

EXAMPLE 3–1 The position of a small bumper car in an amusement park ride (Fig. 3–3a) is described as a function of time by the coordinates $x = c_1 t^2 + c_2 t + c_3$ and $y = d_1 t^2 + d_2 t + d_3$, where $c_1 = 0.2\ \text{m/s}^2$, $c_2 = 5.0\ \text{m/s}$, $c_3 = 0.5\ \text{m}$, $d_1 = -1.0\ \text{m/s}^2$, $d_2 = 10.0\ \text{m/s}$, and $d_3 = 2.0\ \text{m}$. These functions—that is, the locations of the car along the x-axis and the y-axis—are plotted as a function of time in Figs. 3–3b and 3–3c, respectively. Find the position vectors of the car at $t = 3.0$ s and $t = 6.0$ s, and the displacement vector between these times.

Solution: First, we find the x- and y-coordinates of the car at the two times—3.0 s and 6.0 s—by simply inserting the two values of time into the equations given for x and y:

$$\text{for } t = 3.0\text{ s:}\quad x(t) = (0.2\ \text{m/s}^2)(3.0\ \text{s})^2 + (5.0\ \text{m/s})(3.0\ \text{s}) + 0.5\ \text{m} = 17\ \text{m},$$

$$y(t) = (-1.0\ \text{m/s}^2)(3.0\ \text{s})^2 + (10.0\ \text{m/s})(3.0\ \text{s}) + 2.0\ \text{m} = 23\ \text{m};$$

$$\text{for } t = 6.0\text{ s:}\quad x(t) = (0.2\ \text{m/s}^2)(6.0\ \text{s})^2 + (5.0\ \text{m/s})(6.0\ \text{s}) + 0.5\ \text{m} = 38\ \text{m},$$

$$y(t) = (-1.0\ \text{m/s}^2)(6.0\ \text{s})^2 + (10.0\ \text{m/s})(6.0\ \text{s}) + 2.0\ \text{m} = 26\ \text{m}.$$

Now that we have the components, Eq. (3–2) gives us the position vectors of the car at the two times:

$$\text{for } t = 3.0\text{ s:}\quad \mathbf{r}(t) = (17\mathbf{i} + 23\mathbf{j})\ \text{m};$$

$$\text{for } t = 6.0\text{ s:}\quad \mathbf{r}(t) = (38\mathbf{i} + 26\mathbf{j})\ \text{m}.$$

Finally, the displacement vector of the car between 3 s and 6 s is [Eq. (3–1)]

$$\begin{aligned}\Delta \mathbf{r} &= \mathbf{r}(t_{6s}) - \mathbf{r}(t_{3s}) \\ &= (38\mathbf{i} + 26\mathbf{j})\ \text{m} - (17\mathbf{i} + 23\mathbf{j})\ \text{m} \\ &= (21\mathbf{i} + 3\mathbf{j})\ \text{m}.\end{aligned}$$

One other graphical representation of this motion is useful here; we can plot x against y, moment by moment (Fig. 3–3d). This curve is the trajectory of the car.

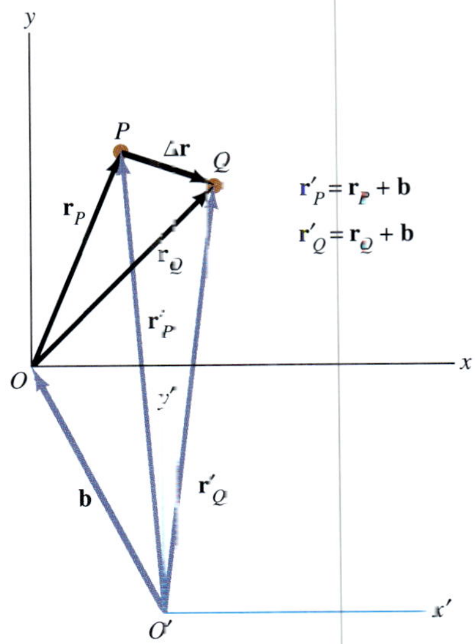

FIGURE 3–2 A displacement vector $\Delta \mathbf{r}$ is independent of the origin. O and O' are two origins and, although the initial and final position vectors to points P and Q do depend on the origins, the difference between these position vectors does not.

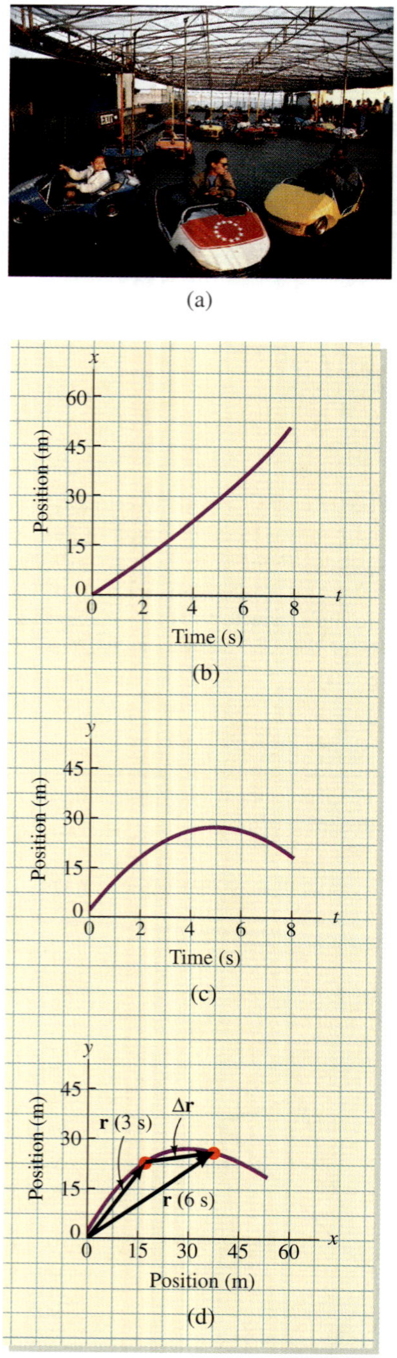

FIGURE 3–3 (a) Example 3–1. An xy-coordinate system can be laid out on the floor. (b) The location along the x-axis of the car's motion from 0 s to 5 s. (c) The car's location along the y-axis. (d) By combining the results of part (b) and part (c), we can plot the trajectory, a graph of the y-position versus the x-position.

3–2 VELOCITY AND ACCELERATION

Velocity

As in Chapter 2, we want the velocity of a particle to describe the rate of change in the position of the particle as it moves on its trajectory. This description can be obtained using Eq. (3–1) for the particle's displacement as our starting point. The average velocity, $\mathbf{v}_{av}$, over the finite time interval from t to $t + \Delta t$ is then defined by

$$\mathbf{v}_{av} \equiv \frac{\mathbf{r}(t + \Delta t) - \mathbf{r}(t)}{\Delta t} = \frac{\Delta \mathbf{r}}{\Delta t}. \tag{3–4}$$

Equation (3–4) reveals that the direction of $\mathbf{v}_{av}$ points in the same direction as the displacement vector $\Delta\mathbf{r}$.

As the time interval Δt around a given time t decreases, the displacement vector $\Delta\mathbf{r}$ becomes tangent to the particle's trajectory at the location of the moving particle. Then, as in Eq. (2–11), the *instantaneous velocity* $\mathbf{v}(t)$ is obtained by letting Δt become infinitesimally small:

The velocity vector

$$\mathbf{v}(t) \equiv \lim_{\Delta t \to 0} \frac{\mathbf{r}(t + \Delta t) - \mathbf{r}(t)}{\Delta t} = \frac{d\mathbf{r}}{dt}. \tag{3–5}$$

The instantaneous velocity $\mathbf{v}$ at time t is defined as the limit of the average velocity as $\Delta t \to 0$; in other words, the instantaneous velocity $\mathbf{v}$ is the time derivative of the position vector. The direction of $\mathbf{v}$ at time t is tangent to the trajectory curve at that time (Fig. 3–4).

We can write the velocity vector in terms of components by using Eqs. (3–2) and (3–5):

$$\mathbf{v} = \frac{d}{dt}\mathbf{r}(t) = \frac{d}{dt}(x\mathbf{i} + y\mathbf{j}) \tag{3–6}$$

$$= \frac{dx}{dt}\mathbf{i} + \frac{dy}{dt}\mathbf{j}. \tag{3–7}$$

This result follows because the unit vectors $\mathbf{i}$ and $\mathbf{j}$ are constant. We write Eq. (3–7) in the form

$$\mathbf{v} = v_x\mathbf{i} + v_y\mathbf{j} \tag{3–8}$$

$$= \mathbf{v}_x + \mathbf{v}_y, \tag{3–9}$$

where

$$v_x = \frac{dx}{dt}, \tag{3–10a}$$

$$v_y = \frac{dy}{dt}, \tag{3–10b}$$

and the component vectors $\mathbf{v}_x$ and $\mathbf{v}_y$ are

The component vectors of velocity

$$\mathbf{v}_x = \frac{dx}{dt}\mathbf{i}, \tag{3–11a}$$

$$\mathbf{v}_y = \frac{dy}{dt}\mathbf{j}. \tag{3–11b}$$

The component vectors $\mathbf{v}_x$ and $\mathbf{v}_y$ of the velocity vector $\mathbf{v}$ are drawn on Fig. 3–4. The magnitude of the velocity $\mathbf{v}$ can be written in terms of the components of $\mathbf{v}$:

$$v = |\mathbf{v}| = \sqrt{v_x^2 + v_y^2}. \tag{3–12}$$

The direction θ of the velocity vector **v** with respect to the x-axis is determined in terms of the components by

$$\tan\theta = \frac{v_y}{v_x}. \tag{3–13}$$

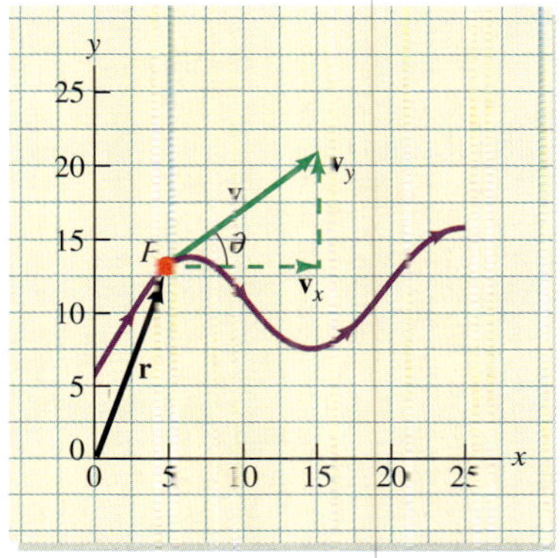

FIGURE 3–4 The velocity vector **v** at point P is tangent to the particle's trajectory at that point. The component vectors $\mathbf{v}_x$ and $\mathbf{v}_y$ of the velocity vector **v** at that point are also included.

EXAMPLE 3–2 Use the data presented in Example 3–1 to find the bumper car's average velocity over the period from 3.0 s to 6.0 s, and the car's instantaneous velocity at $t = 3.0$ s.

Solution: We found in Example 3–1 that the displacement vector of the bumper car between $t = 3.0$ s and $t = 6.0$ s was

$$\Delta\mathbf{r} = (21\mathbf{i} + 3\mathbf{j})\text{ m}.$$

The average velocity is given by Eq. (3–4),

$$\mathbf{v}_{\text{av}} = \frac{\Delta\mathbf{r}}{\Delta t} = \frac{(21\mathbf{i} + 3\mathbf{j})\text{ m}}{3.0\text{ s}}$$
$$= (7\mathbf{i} + \mathbf{j})\text{ m/s}.$$

Now, we can use Eq. (3–7) to evaluate the car's instantaneous velocity at $t =$ 3.0 s by differentiating x and y given in Example 3–1:

$$\frac{dx}{dt} = \frac{d}{dt}(c_1t^2 + c_2t + c_3) = 2c_1t + c_2;$$

$$\frac{dy}{dt} = \frac{d}{dt}(d_1t^2 + d_2t + d_3) = 2d_1t + d_2.$$

Substituting the numerical values of c_1, c_2, d_1, and d_2 at $t = 3.0$ s from Example 3–1, we have

$$\frac{dx}{dt} = 2(0.2\text{ m/s}^2)(3.0\text{ s}) + 5.0\text{ m/s} = 6.2\text{ m/s};$$

$$\frac{dy}{dt} = 2(-1.0\text{ m/s}^2)(3.0\text{ s}) + 10.0\text{ m/s} = 4.0\text{ m/s}.$$

From these velocity components, we calculate the velocity at $t = 3.0$ s to be

$$\mathbf{v} = \frac{dx}{dt}\mathbf{i} + \frac{dy}{dt}\mathbf{j}$$
$$= (6.2\mathbf{i} + 4.0\mathbf{j})\text{ m/s}.$$

This velocity vector is shown in Fig. 3–5.

Acceleration

Acceleration describes how rapidly velocity changes. Our discussion of how acceleration is found from velocity is analogous to the discussion of how velocity is found from displacement. For a finite time interval Δt, the average acceleration is defined as

$$\mathbf{a}_{\text{av}} \equiv \frac{\mathbf{v}(t + \Delta t) - \mathbf{v}(t)}{\Delta t} = \frac{\Delta\mathbf{v}}{\Delta t}. \tag{3–14}$$

The instantaneous acceleration at time t is the limit of the average acceleration as Δt approaches zero, which is a derivative:

$$\mathbf{a} \equiv \lim_{\Delta t\to 0}\frac{\mathbf{v}(t + \Delta t) - \mathbf{v}(t)}{\Delta t} = \frac{d\mathbf{v}}{dt}. \tag{3–15}$$

The acceleration vector

The components of acceleration

As we did for velocity, we can express acceleration in terms of its components:

$$\mathbf{a} = \frac{dv_x}{dt}\mathbf{i} + \frac{dv_y}{dt}\mathbf{j} \tag{3–16}$$

$$= a_x\mathbf{i} + a_y\mathbf{j}. \tag{3–17}$$

Here, the components of the acceleration vector are

$$a_x = \frac{dv_x}{dt} = \frac{d^2x}{dt^2}, \tag{3–18a}$$

$$a_y = \frac{dv_y}{dt} = \frac{d^2y}{dt^2}. \tag{3–18b}$$

EXAMPLE 3–3 Calculate the instantaneous acceleration of the bumper car in Example 3–1 at $t = 1.0$ s and $t = 3.0$ s. What are the magnitude and direction of the acceleration at these times?

Solution: Equations (3–18a and b) express the car's acceleration for any time t. The values for x and y as functions of time were given in Example 3–1, and we found the velocity components in Example 3–2 (see Fig. 3–5). Equations (3–18) then give

$$a_x = \frac{dv_x}{dt} = \frac{d}{dt}(2c_1t + c_2) = 2c_1$$

and

$$a_y = \frac{dv_y}{dt} = \frac{d}{dt}(2d_1t + d_2) = 2d_1.$$

Once we find the acceleration components, we can use Eq. (3–17) to determine the acceleration:

$$\mathbf{a} = 2c_1\mathbf{i} + 2d_1\mathbf{j}.$$

In this case, the car's acceleration is a constant—it is independent of time—and so it is exactly the same for $t = 1.0$ s and for $t = 3.0$ s. Given the values of c_1 and d_1 (0.2 m/s^2 and -1.0 m/s^2, respectively), the numerical value of the acceleration is

$$\mathbf{a} = (0.4\mathbf{i} - 2.0\mathbf{j})\ \text{m/s}^2.$$

The magnitude of the acceleration is

$$a = |\mathbf{a}| = \sqrt{a_x^2 + a_y^2}$$

$$= \sqrt{(0.4)^2 + (-2.0)^2}\ \text{m/s}^2 = \sqrt{4.2}\ \text{m/s}^2 = 2.0\ \text{m/s}^2.$$

The acceleration vector makes an angle θ with the x-axis, which is shown in Fig. 3–5; the angle θ is

$$\tan\theta = \frac{a_y}{a_x} = \frac{-2.0\ \text{m/s}^2}{0.4\ \text{m/s}^2} = -5.0,$$

$$\theta = -79°.$$

Representing Trajectories

Figure 3–5 shows the trajectory of the bumper car discussed in Examples 3–1, 3–2, and 3–3; this trajectory is a curve representing the car's x-position versus its y-position. In Fig. 3–5, we show a position vector **r**, a velocity vector **v**, and an acceleration vector **a**, each at $t = 3$ s. Although the figure shows **a** at the particular point corresponding to $t = 3.0$ s, it is, in fact, the same vector everywhere along the curve. This is because we calculated that the acceleration vector of the car is independent of time.

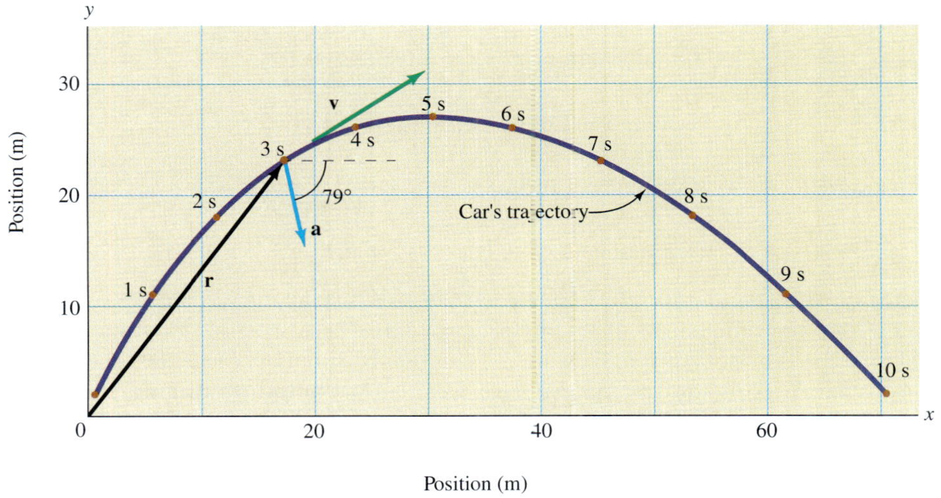

FIGURE 3–5 The trajectory of the car's path in Example 3–1 is plotted for times up to 10 s: the position vector **r**, velocity **v**, and acceleration **a** at $t = 3.0$ s.

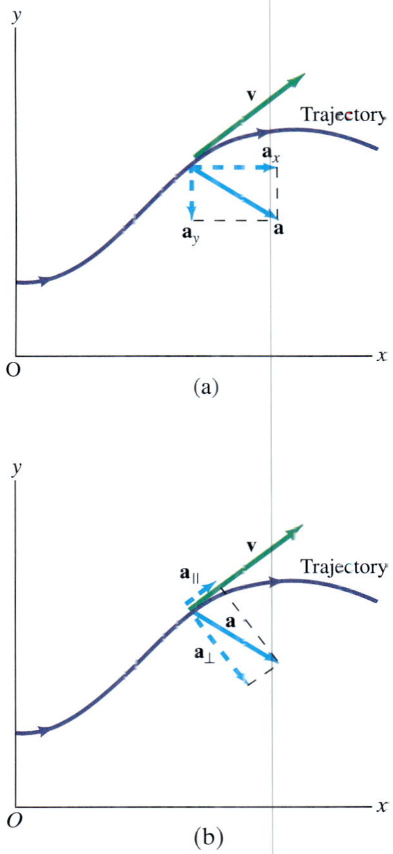

FIGURE 3–6 The velocity **v** and acceleration **a** of a particle following some trajectory: (a) The acceleration of the particle is separated into x- and y-components; (b) the acceleration of the particle is separated into components parallel and perpendicular to the path.

We can apply a graphical representation like that of Fig. 3–5 to any motion. For example, consider a trajectory, formed by the locus of the position vector **r** as a function of time. The velocity vector **v** at any time t is a vector of magnitude $|\mathbf{v}|$ that is *tangential* to the trajectory at time t. A plot of the "locus" of **v** can similarly then be drawn; it is the curve of the points whose *horizontal coordinate* at time t is v_x and whose *vertical coordinate* at that time is v_y. The acceleration at time t is given by a vector whose magnitude is $|\mathbf{a}|$ and whose direction is tangential to the velocity curve at time t. Note: *The acceleration vector **a** is tangent to the velocity curve but not to the trajectory itself.*

Figure 3–6 shows the path of an object with **v** and **a** indicated at a particular time. In Fig. 3–6a, the acceleration is separated into its a_x and a_y components. Alternatively, we can separate the acceleration **a** into components that are parallel (tangential) and perpendicular (normal) to the velocity vector (Fig. 3–6b). We label these components $a_{\parallel}$ and $a_{\perp}$, respectively. The component $a_{\parallel}$ of **a** that is parallel to **v** affects the magnitude but not the direction of **v**. Similarly, the $a_{\perp}$ component changes the direction but not the magnitude of **v**. It is useful to refer separately to the parallel and perpendicular components of an object's acceleration because they affect the velocity differently.

3–3 MOTION WITH CONSTANT ACCELERATION

When an object moves under the effects of *constant acceleration*, it can move only in a plane. The plane is formed by the initial velocity vector and the acceleration vector; the object cannot acquire a component of velocity perpendicular to this plane. As we have noted, the classic case of constant acceleration is gravity. Because there are any number of examples of objects moving under the influence of gravity, a discussion of constant acceleration is one of the simplest and most important for us to undertake. When we say that the acceleration **a** remains constant, we mean it is constant in both magnitude and direction. That means that both components a_x and a_y must be constant. When both $\mathbf{a}_x$ and $\mathbf{a}_y$ are constant, our one-dimensional results from Chapter 2 *independently* determine the x- and y-components of position **r** and velocity **v** in terms of the constant acceleration. In other words, we can think of the x- and y-motions as separate ones, governed by separate constant accelerations. We use Eqs. (2–17) and (2–21)

PROBLEM SOLVING

We can use the results of Chapter 2 when the acceleration vector is constant.

The components of position and velocity for a constant acceleration

to find:

$$x\text{-component of } \mathbf{r}: \quad x = x_0 + v_{0x}t + \frac{1}{2}a_xt^2, \tag{3–19}$$

$$x\text{-component of } \mathbf{v}: \quad v_x = v_{0x} + a_xt; \tag{3–20}$$

$$y\text{-component of } \mathbf{r}: \quad y = y_0 + v_{0y}t + \frac{1}{2}a_yt^2, \tag{3–21}$$

$$y\text{-component of } \mathbf{v}: \quad v_y = v_{0y} + a_yt. \tag{3–22}$$

Here, x_0 and y_0 are the specified components of $\mathbf{r} = \mathbf{r}_0$ at an initial time $t = 0$, and v_{0_x} and v_{0_y} are the components of $\mathbf{v} = \mathbf{v}_0$ at time $t = 0$. These specified values at the initial time $t = 0$ must be given; they are not determined simply because the acceleration is constant. We can now say that we have specified the **initial conditions**. In component form, the initial conditions are

$$\mathbf{r}_0 = x_0\mathbf{i} + y_0\mathbf{j} \tag{3–23}$$

and

$$\mathbf{v} = v_{0x}\mathbf{i} + v_{0y}\mathbf{j} \tag{3–24}$$

at $t = 0$.

Equations (3–19) through (3–22), which address position and velocity for motion with constant acceleration $\mathbf{a}$, can be written more compactly in vector form:

$$\mathbf{r} = \mathbf{r}_0 + \mathbf{v}_0t + \frac{1}{2}\mathbf{a}t^2 \tag{3–25}$$

$$\mathbf{v} = \mathbf{v}_0 + \mathbf{a}t. \tag{3–26}$$

These results are valid *only* when $\mathbf{a}$ is constant.

EXAMPLE 3–4 A wayward golf ball rolls off the edge of a vertical cliff overlooking the Pacific Ocean. The golf ball has a horizontal velocity component of 10 m/s and no vertical component when it leaves the cliff. Describe the subsequent motion.

Solution: Here, the displacement, the velocity, and the acceleration all lie in the same plane, which we assign to be the xy-plane. The ball's constant acceleration $\mathbf{a}$ is equal to that of gravity and thus $\mathbf{a} = \mathbf{g}$. The vector $\mathbf{g}$ points toward the center of Earth and has a magnitude of 9.8 m/s^2. Now we can draw a coordinate system as in Fig. 3–7a, placing the origin at the point where the ball leaves the cliff. For constant acceleration, we apply Eqs. (3–19) through (3–22).

We must specify the golf ball's initial position and velocity by using the given information. We set $t = 0$ s at the time when the motion starts; at that moment, then, we have $x_0 = 0$ m, $v_{0x} = 10$ m/s, $y_0 = 0$ m, and $v_{0y} = 0$ m/s. The acceleration due to gravity has components $a_x = 0$ m/s^2 and $a_y = -9.8$ m/s^2. Because there is no component of acceleration in the x-direction, the horizontal velocity is constant and will remain at 10 m/s. We determine the velocity components from Eqs. (3–20) and (3–22):

$$v_x = 10 \text{ m/s};$$

$$v_y = 0 \text{ m/s} + (-9.8 \text{ m/s}^2)t = (-9.8 \text{ m/s}^2)t. \tag{3–27}$$

The position of the golf ball is determined from Eqs. (3–19) and (3–21):

$$x = 0 \text{ m} + (10 \text{ m/s})t + \frac{1}{2}(0 \text{ m/s}^2)t^2 = (10 \text{ m/s})t;$$

$$y = 0 \text{ m} + (0 \text{ m/s})t + \left[-\frac{1}{2}(9.8 \text{ m/s}^2)t^2\right] = (-4.9 \text{ m/s}^2)t^2. \tag{3–28}$$

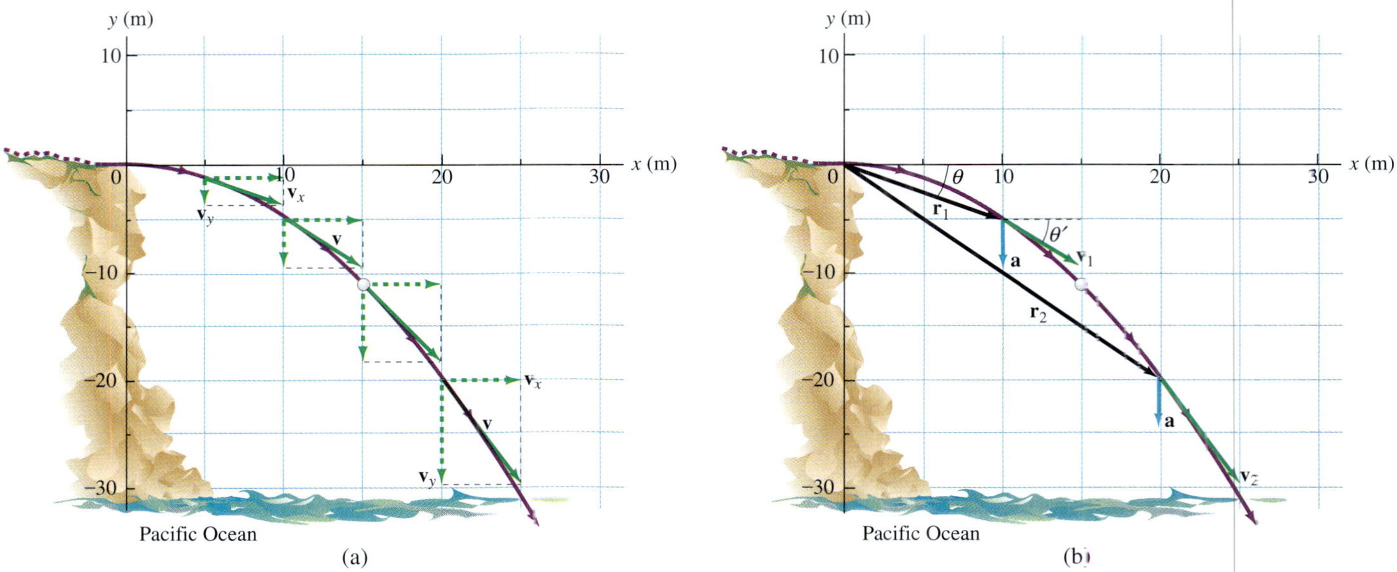

FIGURE 3–7 (a) Example 3–4. The velocity vector **v** and components $\mathbf{v}_x$ and $\mathbf{v}_y$ of the golf ball are shown at 0.5-s intervals up to 2.0 s. (b) The position **r**, velocity **v**, and acceleration **a** of the golf ball for $t_1 = 1$ s and $t_2 = 2$ s.

Figure 3–7a shows the trajectory of the golf ball. It also shows the velocity vector and its components every 0.5 s up to 2 s. While the horizontal component of the velocity stays constant, the vertical component changes linearly with time. Further, the total velocity vector is tangent to the ball's path of motion at each point along its trajectory.

Figure 3–7b shows the position vector **r**, velocity **v**, and acceleration **a** at $t = 1$ s and $t = 2$ s. While **a** remains constant, **r** and **v** change with time. The three vectors **r**, **v**, and **a** do not generally point in the same direction as one another at a given time during the golf ball's motion. The directions of **r** and **v** are determined by

$$\tan\theta = \frac{y}{x} = \frac{(-4.9\ \text{m/s}^2)t^2}{(10\ \text{m/s})t} = (-0.49\ \text{s}^{-1})t$$

and

$$\tan\theta' = \frac{v_y}{v_x} = \frac{(-9.8\ \text{m/s}^2)t}{(10\ \text{m/s})} = (-0.98\ \text{s}^{-1})t.$$

Both angles vary with time.

3–4 PROJECTILE MOTION

Let's consider the motion of a golf ball. The golf ball is an example of a *projectile* that moves under the effect of gravity. In the absence of air resistance, what is the trajectory of such a projectile? The motion is that of constant acceleration, which is due to gravity in this case. This constant acceleration, **g**, has only a vertical component. We calculate the ball's trajectory by separating its motion into horizontal and vertical components and then applying the kinematic equations for constant acceleration (Sec. 3–3).

Our coordinate system is best designed by placing the origin at the starting point, assigning the y-direction vertically and the x-direction along the horizontal (Fig. 3–8). The initial position of the ball is $x_0 = y_0 = 0$; the initial velocity at $t = 0$ is $\mathbf{v}_0$. The trajectory of the golf ball makes an initial angle with the horizontal that is referred to as the elevation angle θ_0. The components of $\mathbf{v}_0$ are thus

$$v_{0x} = v_0\cos\theta_0 \quad\text{and}\quad v_{0y} = v_0\sin\theta_0. \tag{3–29}$$

The components of the acceleration are the constants

$$a_x = 0 \quad\text{and}\quad a_y = -g. \tag{3–30}$$

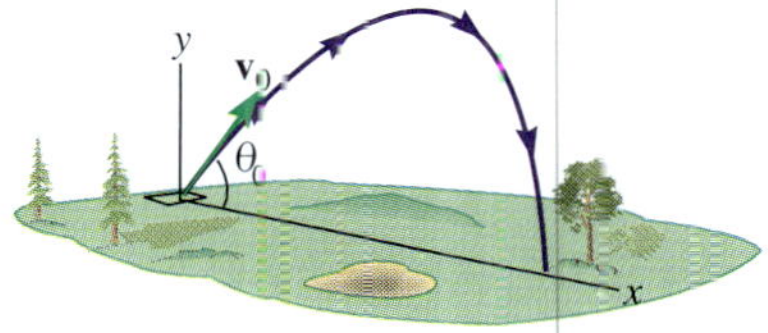

FIGURE 3–8 A golf ball leaves a tee with an initial velocity of magnitude v_0 at an elevation angle θ_0.

Using Eqs. (3–19) through (3–22), we find the components of **r** and **v** (the position and velocity of the ball, respectively) to be

$$x = 0 + (v_0 \cos \theta_0)t + \frac{1}{2}(0)t^2 = (v_0 \cos \theta_0)t, \tag{3–31}$$

$$y = 0 + (v_0 \sin \theta_0)t - \frac{1}{2}gt^2 = (v_0 \sin \theta_0)t - \frac{1}{2}gt^2, \tag{3–32}$$

and

$$v_x = v_0 \cos \theta_0 + (0)t = v_0 \cos \theta_0, \tag{3–33}$$

$$v_y = v_0 \sin \theta_0 - gt. \tag{3–34}$$

The Trajectory

We can find the trajectory of the golf ball by plotting its height y versus its x-position. We know both x and y as functions of time, and we can eliminate the time dependence by using Eq. (3–31) to find the time t as a function of x. We then insert the result for t into Eq. (3–32):

$$t = \frac{x}{v_0 \cos \theta_0}; \tag{3–35}$$

$$y = (v_0 \sin \theta_0)\frac{x}{v_0 \cos \theta_0} - \frac{1}{2}g\left(\frac{x}{v_0 \cos \theta_0}\right)^2$$

$$= (\tan \theta_0)x - \left(\frac{g}{2v_0^2 \cos^2 \theta_0}\right)x^2. \tag{3–36}$$

The coefficients of x and x^2 in Eq. (3–36) are both constants, so this equation has the form

$$y = C_1 x - C_2 x^2, \tag{3–37}$$

Objects moving with constant acceleration follow a parabolic trajectory.

This is the general equation of a parabola with its axis parallel to the y-axis. *The trajectory of all objects moving with constant acceleration is parabolic.* Take a look at the parabolic motion illustrated in the chapter-opening photograph or in Fig. 3–9; this motion is examined at various points of the trajectory in Fig. 3–10.

FIGURE 3–9 The path followed by a projectile is parabolic.

Range. We define the **range** R of a projectile launched from the ground ($y = 0$) to be the horizontal distance that the projectile travels over level ground. The quantity R is the value of x when the projectile has returned to the ground; that is, when y again equals zero. From Eq. (3–37), then, we have

$$0 = R(C_1 - C_2 R). \tag{3–38}$$

The value $R = 0$ clearly satisfies the condition $y = 0$ in this equation, but this is the starting point of the projectile's motion—it is launched from the ground and its x-position is zero at launch time. y can alternatively be zero if the factor $C_1 - C_2 R = 0$ in Eq. (3–38), or $R = C_1/C_2$; this case corresponds to the projectile having landed back on the ground after its flight. Inserting the values of C_1 and C_2 from Eq. (3–36) yields

$$R = \frac{C_1}{C_2} = \frac{\tan \theta_0 \,(2v_0^2 \cos^2 \theta_0)}{g} = \frac{2v_0^2}{g}\left(\frac{\sin \theta_0}{\cos \theta_0}\right)\cos^2 \theta_0 = \frac{v_0^2}{g} 2 \sin \theta_0 \cos \theta_0.$$

From trigonometry, $\sin(2\theta_0) = 2 \sin\theta_0 \cos\theta_0$, and we find

$$\text{for motion over level ground:} \quad R = \frac{{v_0}^2 \sin(2\theta_0)}{g}. \tag{3–39}$$

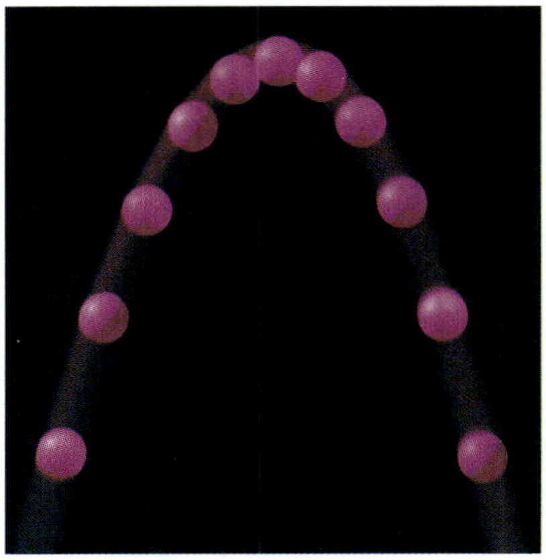

FIGURE 3–10 The motion of a ball thrown in the air and moving under the influence of gravity. Air resistance is ignored. The ball moves with constant acceleration, which in this case is directed downward due to gravity. The velocity vector changes throughout the motion, but the vector component in the horizontal direction does not. The vector component in the vertical direction changes linearly with time. The resulting trajectory forms a parabola.

The range varies with the initial angle of the projectile. If, for example, $\theta_0 = 0°$, then $R = 0$; when a projectile moves off horizontally from ground level, the slightest drop makes it plow into the ground. If $\theta_0 = 90°$, Eq (3–39) again gives $R = 0$; when a projectile is launched straight up, it comes back straight down. In fact, as θ_0 increases from 0° to 45° and then to 90°, $\sin(2\theta_0)$ first increases from 0 to 1, then decreases back down to 0, respectively. In other words, there are *two* initial angles at which to launch a projectile that give the same range for a given initial speed (Fig. 3–11). For example, a pop fly and a line drive in softball can both be caught by the shortstop. The range reaches a maximum value when the factor $\sin(2\theta_0)$ reaches its maximum value of 1. This occurs for $2\theta_0 = 90°$, or $\theta_0 = 45°$, in which case

$$R_{\text{max}} = \frac{v_0^2}{g}. \tag{3–40}$$

If the projectile is shot at an angle higher or lower than 45°, the range is shorter (Fig. 3–11).

Flight Time. Let T be the total flight time of a golf ball. Figure 3–12 shows that the golf ball reaches its maximum height exactly halfway through its motion. At this point, its motion is horizontal; that is, the vertical component of velocity is zero. This occurs at time $t = T/2$. We can find $T/2$ by setting $v_y = 0$ in Eq. (3–34):

$$0 = v_0 \sin\theta_0 - \frac{gT}{2}.$$

We solve for T to find that

$$T = \frac{2v_0}{g} \sin\theta_0. \tag{3–41}$$

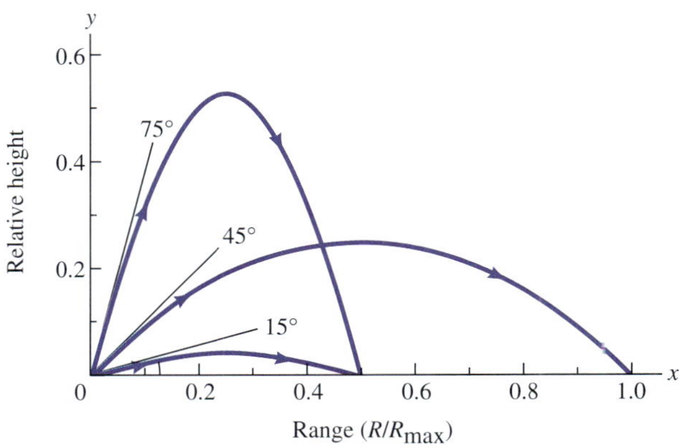

FIGURE 3–11 If the initial speed is fixed and if air resistance is ignored, a projectile's trajectory will have a maximum range for an elevation angle of 45°. The range is the horizontal distance the projectile travels over level ground.

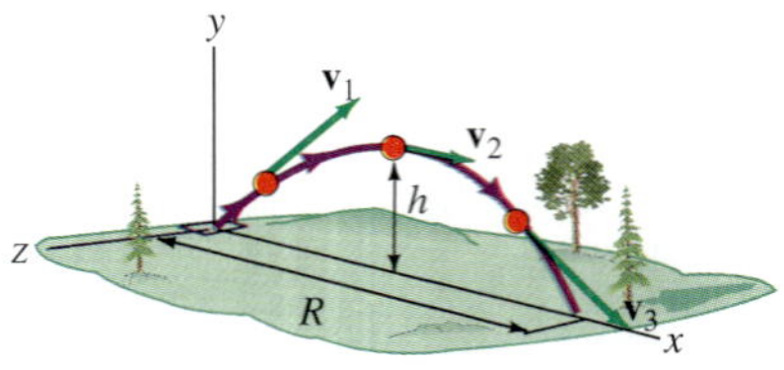

FIGURE 3–12 A projectile (a golf ball) moving under the force of gravity is at its maximum height when $v_y = 0$. At that moment, the ball is traveling horizontally.

Maximum Height. The maximum height $y_{max} = h$ is reached at time $T/2$. From Eq. (3–32), we find the height at this time is

$$h = (v_0 \sin\theta_0)\frac{2v_0}{2g}\sin\theta_0 - \frac{1}{2}g\left(\frac{2v_0}{2g}\sin\theta_0\right)^2 = v_0^2\,\frac{\sin^2\theta_0}{g} - gv_0^2\,\frac{\sin^2\theta_0}{2g^2}$$

$$= \frac{v_0{}^2\sin^2\theta_0}{2g}. \tag{3–42}$$

We use Eqs. (3–36), (3–39), (3–41), and (3–42) to determine a projectile's trajectory, range, flight time, and maximum height, respectively. The range and flight time refer to the special case where the golf ball returns to its original height. These equations need not be memorized; however, it is important to understand how these equations were obtained. We apply these methods again in Examples 3–5 through 3–8.

EXAMPLE 3–5 In order to win a bet that he can drive a golf ball a horizontal distance of 250 m, an amateur golfer goes to a cliff overlooking the ocean (Fig. 3–13). The cliff is 52 m above the ocean. The golfer strikes the golf ball so that the ball's initial speed is 48 m/s and the elevation angle (from the horizontal) is 36°. Does he win his bet? What is the horizontal distance actually covered by the ball?

Solution: We use the trajectory equation to find the value of x at which the golf ball reaches the ocean surface. We again place the origin of our coordinate system at the tee, letting $+y$ extend upward. The initial conditions are $x_0 = 0$ m, $y_0 = 0$ m, $v_0 = 48$ m/s, and $\theta_0 = 36°$. We cannot use Eq. (3–39) to calculate the range because that result applies to level ground. We want to determine the distance from the tee to the point at which the golf ball reaches the ocean ($y = -52$ m), not the distance when the ball returns to $y = 0$ m. However, we can still use Eq. (3–36) to find the value of x when $y = -52$ m. This value of x is the desired range R'. Equation (3–36) becomes

$$y = -52\text{ m} = R'\tan\theta_0 - R'^2\,\frac{g}{2v_0{}^2\cos^2\theta_0}.$$

Rearrange this equation to find that

$$R'^2 - \frac{2v_0^2\cos^2\theta_0\tan\theta_0}{g}R' + \frac{2yv_0^2\cos^2\theta_0}{g} = 0;$$

$$R'^2 + bR' + c = 0,$$

where $$b = -\frac{2v_0^2\cos^2\theta_0\tan\theta_0}{g} \quad\text{and}\quad c = \frac{2yv_0^2\cos^2\theta_0}{g}.$$

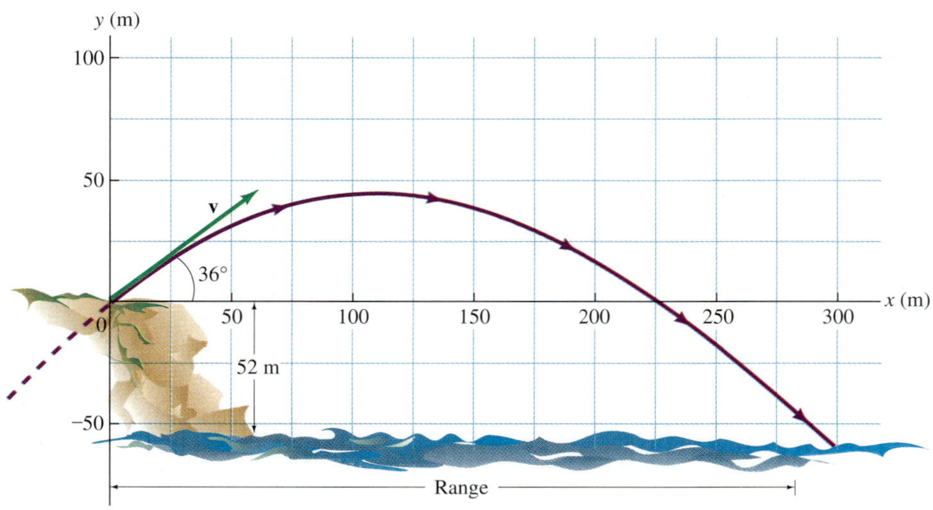

FIGURE 3–13 Examples 3–5 and 3–6.

Solve this quadratic equation to find R':

$$R' = \frac{-b \pm \sqrt{b^2 - 4c}}{2}.$$

Insert the values of b and c to obtain

$$R' = \frac{v_0^2 \cos^2\theta_0 \tan\theta_0}{g} \pm \frac{1}{2}\sqrt{\frac{4v_0^4 \cos^4\theta_0 \tan^2\theta_0}{g^2} - \frac{8yv_0^2\cos^2\theta_0}{g}}.$$

Now insert the value of $y = -52$ m and the values of the constants to determine the range:

$$R' = \frac{(48\ \text{m/s})^2(\cos 36°)^2(\tan 36°)}{9.8\ \text{m/s}^2}$$

$$\pm \frac{1}{2}\sqrt{\frac{4(48\ \text{m/s})^4(\cos 36°)^4(\tan 36°)^2}{(9.8\ \text{m/s}^2)^2} - \frac{8(-52\ \text{m})(48\ \text{m/s})^2(\cos 36°)^2}{9.8\ \text{m/s}^2}}$$

$$= 281\ \text{m or } -57\ \text{m}.$$

Now, did the golfer drive the ball a distance of 281 m or −57 m? The positive value must be correct. The negative result arises because it is a suitable solution to the quadratic equation that produces a parabola before $t = 0$ s (see the dashed line on the left in Fig. 3–13). The golfer wins his bet.

EXAMPLE 3–6 What was the maximum height above the ocean of the golf ball in Example 3–5, and how long was the golf ball in flight?

Solution: Let's refer again to Fig. 3–13. The maximum height of the golf ball occurs when the vertical component of the velocity is zero, and Eq. (3–42) gives the correct result, h, for the height above the original location. Notice, however, that the answer needed is $h + 52$ m, because we want to determine the maximum height *above the ocean*:

$$h = \frac{v_0^2 \sin^2\theta_0}{2g} = \frac{(48\ \text{m/s})^2(\sin 36°)^2}{2(9.8\ \text{m/s}^2)} = 41\ \text{m}.$$

The answer is 41 m + 52 m = 93 m.

We can find the total time T of the trip by using Eq. (3–31). We have already determined that the total horizontal distance traveled is 281 m. From Eq. (3–31):

$$281\ \text{m} = (48\ \text{m/s})(\cos 36°)T;$$

$$T = \frac{281\ \text{m}}{(48\ \text{m/s})(\cos 36°)} = 7.2\ \text{s}.$$

EXAMPLE 3–7 A group of engineering students constructs a slingshot device that lobs water balloons at a target. The device is constructed so that the launching speed is 12 m/s; the target is 14 m away at the same elevation on the other side of a fence. How can they accomplish their mission?

Solution: In this case, sketched in Fig. 3–14, the range equation for horizontal ground is relevant. The range varies with the initial angle, so the students need to find a value of θ_0 that will give a range of 14 m. Equation (3–39) applies here, and the known parameters are $R = 14$ m and $v_0 = 12$ m/s:

$$R = 14\ \text{m} = \frac{(12\ \text{m/s})^2 \sin(2\theta_0)}{9.8\ \text{m/s}^2};$$

$$\sin(2\theta_0) = 0.95.$$

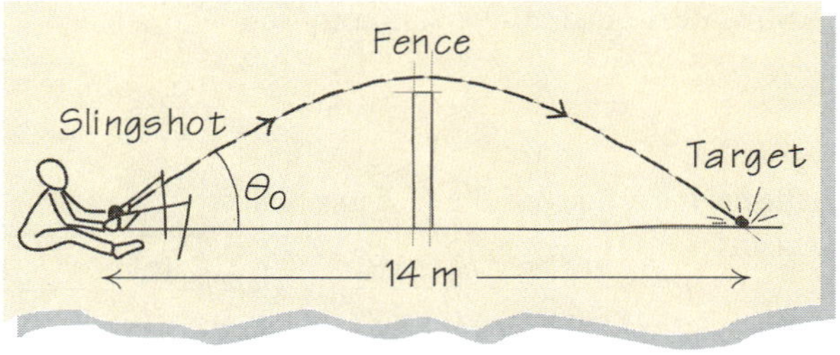

FIGURE 3–14 Example 3–7. The students can orient their slingshot in two ways to get the same range for the same initial speed. One is shown here.

This equation has *two* solutions, $2\theta_0 = 72°$ and $2\theta_0 = 108°$, or $\theta_0 = 36°$ and $54°$. These are the two possible initial angles that result in a given range, as in Fig. 3–11. We have drawn one of these trajectories (Fig. 3–14). A reminder: There will always be two initial angles that generate the same range, except the 45° initial angle, which produces the longest range. Figure 3–15 shows a typical projectile trajectory.

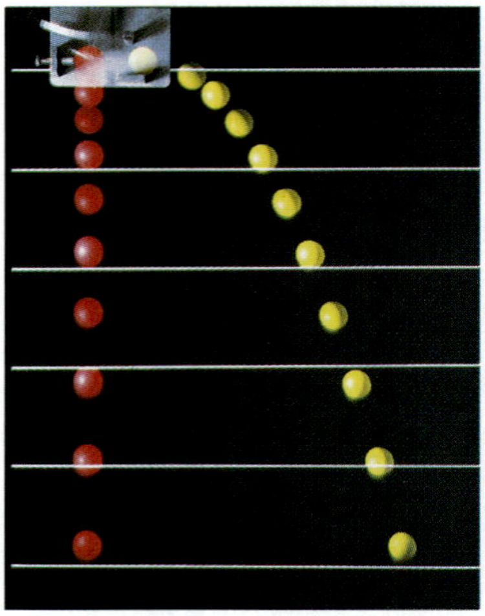

FIGURE 3–15 Two balls released simultaneously have two different trajectories. The difference lies in the (constant) x-component of velocity. But in a given time, each moves the same vertical distance.

EXAMPLE 3–8 A boy would rather shoot coconuts down from a tree than climb the tree or wait for the coconuts to drop. The boy aims his slingshot at a coconut, but just when his rock leaves the slingshot, the coconut falls from the tree. Show that the rock will hit the coconut.

Solution: Let's suppose that the rock starts out at the point $(x, y) = (0, 0)$ and that the coconut is at the point (x_0, y_0). The rock will have initial velocity (v_{x0}, v_{y0}). First let's see what would happen if there were no gravity acting. Without gravity, the properly aimed rock would follow a straight-line path that would place it at the point $(x_0 = v_{x0}t, y_0 = v_{y0}t)$ after a time t. This time is just the time necessary for the rock to reach the coconut, which is still at the tree since gravity has been ignored so far. Now let's factor in the effect of gravity. During the time t that the rock travels toward the coconut, the coconut falls the distance $gt^2/2$ (Fig. 3–16). In other words, the height of the coconut after time t is [Eq. (3–21)]

$$y = y_0 - \frac{1}{2}gt^2.$$

The rock's horizontal velocity component remains constant at v_{x0}. However, the vertical velocity component of the rock is changing under the effect of gravity and, after time t, Eq. (3–21) shows us that the rock's height is not $v_{y0}t$ but rather:

$$y = v_{y0}t - \frac{1}{2}gt^2.$$

The rock is a height $gt^2/2$ below the height it would have if it followed a straight-line motion, which is precisely the distance the coconut falls, as shown in Fig. 3–16. Thus the rock will hit the coconut at this point. This effect can be demonstrated in the lecture room or the laboratory. In effect, the parabolic path of the rock "tracks" the falling coconut.

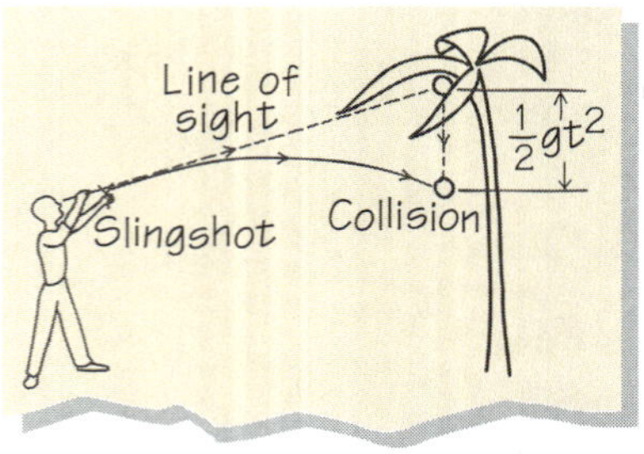

FIGURE 3–16 Example 3–8.

3–5 CIRCULAR MOTION

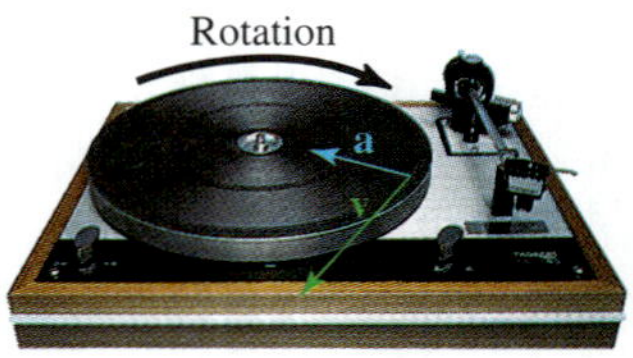

FIGURE 3–17 A phonograph turntable provides an example of uniform circular motion.

An object following a circular path is accelerating, even when its speed is constant.

The steady motion of any object on a spinning phonograph turntable is called **uniform circular motion** (Fig. 3–17). An object is experiencing uniform circular motion if it travels at a constant speed along a circular path. Perhaps surprisingly, this object is actually accelerating. Velocity is a vector quantity; although the object is not changing speed—the magnitude of the velocity remains the same—the *direction of its velocity is changing*. This means that an object undergoing uniform circular motion accelerates. The x- and y-motions are connected in uniform circular motion, so we will not want to separate these motions as we did for projectile motion.

APPLICATION: ACHIEVING CONTROLLED FREE FALL

An exciting application of simple parabolic motion in Earth's gravity is a NASA service that produces "zero gravity" for up to 20 seconds (Fig. 3AB–1). The NASA service is needed to test satellite and shuttle components for proper operation in a zero-gravity environment. Note that "zero gravity" does not mean that there is no force of gravity; rather, it means that there are none of the associated "contact forces" that otherwise act on an object (you normally feel such forces on your seat when you are in a chair, for example). Thus an object that falls in a vacuum (to eliminate air resistance) is said to be in an environment of zero gravity, as is an orbiting satellite. NASA achieves the zero-gravity effect with an airplane (suitably furnished for whatever experiment is desired) that is flown to a reasonable height and then allowed to dive to reach its maximum safe speed. At that point, control of the airplane is taken over by a computer system that provides an accurate parabolic flight path; the top of the flight path's parabola is limited by the lowest safe air speed. The motion of the airplane tracks the motion of a projectile under the influence of gravity, and so is that of an object in free fall. Anything inside the airplane also moves on a parabolic projectile trajectory, so it is as if the airplane were not present for the object: The object is in free fall.

FIGURE 3AB-1 The interior of the airplane used by NASA for a zero-gravity environment, during that part of the flight where the plane follows the same parabolic path taken by a projectile in free fall. In this photograph, astronauts in training are experiencing some of the same effects they will feel during a shuttle flight.

Plane Polar Coordinates

PROBLEM SOLVING

Use of plane polar coordinates can simplify problem solving.

The description of circular motion is very much simplified if we use **plane polar coordinates**. Figure 3–18 shows the two polar coordinates that locate a point in a plane, the radial coordinate ρ, and the angular coordinate ϕ. Conventionally, the angle ϕ is chosen to be measured from the $+x$-axis and to increase in the counterclockwise direction. We can measure it in *radians* (rad), defined by the relation 2π radians equals the circumference of any circle divided by the radius of that circle. If the angle ϕ is measured in radians, then the arc length intercepted by any circle is simply the product of the radius and the angle:

$$\text{arc length} = R\phi. \tag{3–43}$$

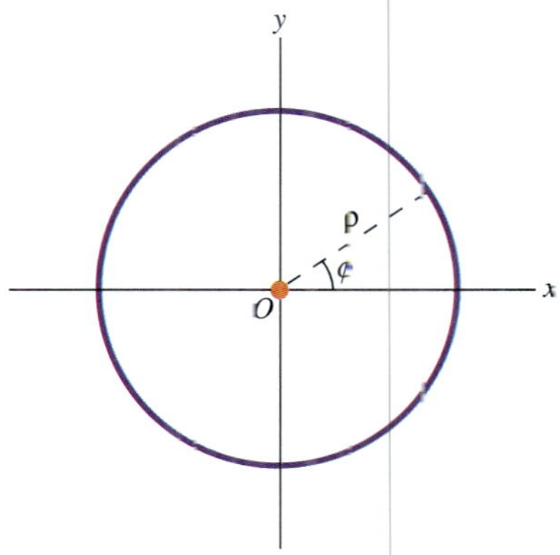

FIGURE 3–18 The radial variable ρ and angular variable ϕ of plane polar coordinates.

Notice that radians are dimensionless because they represent the ratio (arc length)/(radius); that is, the ratio of two lengths. Nevertheless, we will use radians as a unit. You may want to note that the relation $360° = 2\pi$ rad gives us the relation between degrees and radians: $1° \cong 0.0175$ rad and 1 rad $\cong 57.3°$.

For an object moving along a circle, the radial coordinate is fixed, $\rho = R$; thus, the motion is described by a single variable, the angle ϕ, which may have a time dependence $\phi(t)$. Suppose that during a time interval dt, the change in the angle is $d\phi$ (Fig. 3–19). The arc length ds traversed during that time interval is given by

$$ds = R\,d\phi. \tag{3–44}$$

Division by the time interval dt gives us an expression for the speed of motion:

$$\frac{ds}{dt} = v = R\frac{d\phi}{dt}. \tag{3–45}$$

PROBLEM SOLVING

Only one variable is necessary in circular motion.

We are thus led to define the **angular speed**, ω, of the object moving along the circle as *the rate of change of the angle ϕ*, or

The definition of angular speed

$$\omega \equiv \frac{d\phi}{dt}. \tag{3–46}$$

In terms of ω, we have, for the object's speed, v:

$$v = R\omega. \tag{3–47}$$

The quantity ω has dimensions $[T^{-1}]$ and is measured in radians per second (rad/s) in SI.

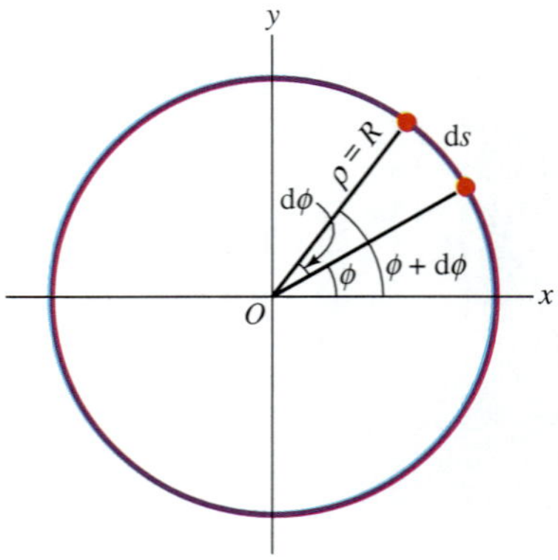

FIGURE 3–19 During the time interval dt, a particle in circular motion traverses an arc length ds and an angle $d\phi$.

Uniform Circular Motion

We say that circular motion is *uniform* when the speed v and the angular speed ω of an object moving along the circle are constant. However, if these two quantities are not constant—as in the case of an old phonograph turntable that varies in angular speed—then we refer to nonuniform circular motion. An important quantity that characterizes uniform circular motion is the time the moving object takes to make one complete revolution. This time is called the **period**, T. Because the distance traveled in one revolution is $2\pi R$, the period is given by

$$2\pi R = vT; \tag{3–48}$$

$$T = \frac{2\pi R}{v} = \frac{2\pi R}{\omega R} = \frac{2\pi}{\omega}. \tag{3–49}$$

The relations between period, frequency, and angular frequency

The **frequency**, f, counts the number of revolutions that the particle makes per unit time. The frequency is the inverse of the period T:

$$f \equiv \frac{1}{T}; \tag{3–50}$$

and the relation between angular speed and frequency is

$$\omega = 2\pi f. \tag{3–51}$$

SI unit of frequency

The SI unit of frequency is the hertz (Hz), defined as one cycle per second (cps). Another and perhaps more familiar unit, revolutions per minute (rev/min or rpm), is also useful; 60 rev/min = 1 Hz.

Acceleration in Uniform Circular Motion. Although the speed of an object in uniform circular motion is constant, the *direction* of the velocity changes, and thus the acceleration is not zero. Let's first consider the directional aspects of uniform circular motion. Specifically, we want to study the direction of the acceleration. We already know that acceleration is the change in velocity per unit time, so we first examine how the velocity changes. In Fig. 3–20a, we imagine a particle located at the point determined by angle ϕ at time t. At a later time $t + \Delta t$, the particle's angular position is given by $\phi + \Delta\phi$. The *direction* of the velocity vector is always tangential to the circle and therefore changes continuously with time. Figure 3–20a shows the velocity $\mathbf{v}(t)$ at time t. At time $t + \Delta t$, the velocity vector is $\mathbf{v}(t + \Delta t)$. These two vectors have the

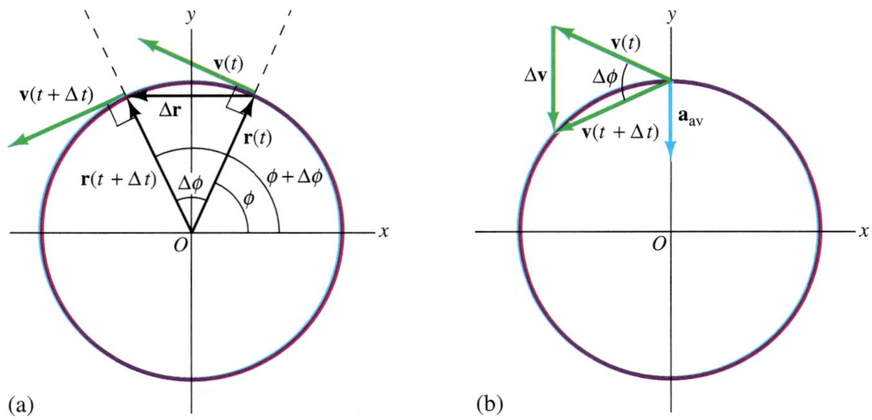

FIGURE 3–20 (a) During time Δt, a particle in motion has changed its position vector from $\mathbf{r}(t)$ to $\mathbf{r}(t + \Delta t)$ and its velocity vector from $\mathbf{v}(t)$ to $\mathbf{v}(t + \Delta t)$. For uniform circular motion, the magnitude of $\mathbf{v}$ is constant; however, the direction is always perpendicular to $\mathbf{r}$ and is therefore changing continuously. (b) In order to form the vector difference $\Delta\mathbf{v} = \mathbf{v}(t + \Delta t) - \mathbf{v}(t)$, we translate the vector $\mathbf{v}(t + \Delta t)$ so that its tail meets the tail of the vector $\mathbf{v}(t)$. We do this at the midway point in the particle's path from time t to time $t + \Delta t$. In part (a), the angle between $\mathbf{r}(t)$ and $\mathbf{r}(t + \Delta t)$ is $\Delta\phi$, the same angle as that between $\mathbf{v}(t)$ and $\mathbf{v}(t + \Delta t)$. We see from the figure that the average change in velocity, which is proportional to $\mathbf{a}_{av}$, points toward the center of the circle.

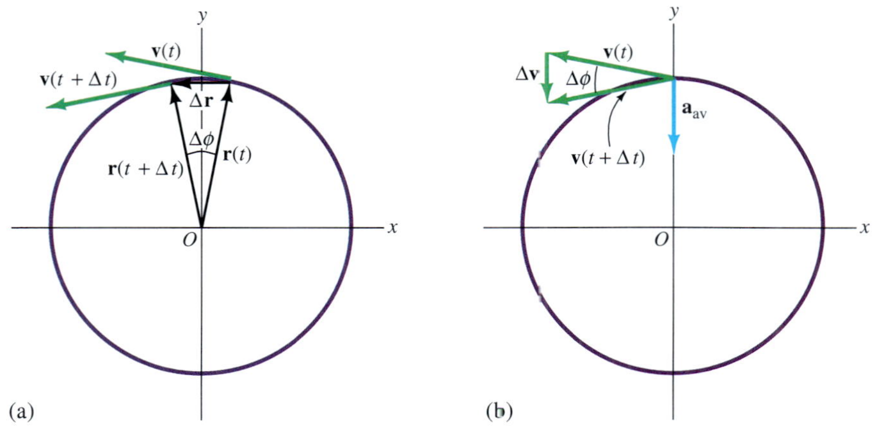

FIGURE 3–21 (a) The time difference Δt is smaller than in Figs. 3–20a and 3–20b; so too are $\Delta\phi$, $\Delta\mathbf{r}$, and $\Delta\mathbf{v}$. (b) The resulting $\mathbf{a}_{av}$ continues to point to the center of the circle. In the limit that $\Delta t \to 0$, $\mathbf{a}_{av}$ becomes the instantaneous acceleration $\mathbf{a}(t)$. It points to the center of the circle.

same magnitude; the difference between them lies in the fact that they point in different directions. The *change* in the velocity vector over the time period Δt is given by $\Delta\mathbf{v} = \mathbf{v}(t + \Delta t) - \mathbf{v}(t)$. This vector difference, which we have drawn to occur at the midway point (time $t + \Delta t/2$) is shown in Fig. 3–20b. We see that $\Delta\mathbf{v}$, and hence $\mathbf{a}_{av} = \Delta\mathbf{v}/\Delta t$, points toward the center of the circle. Let's now make the time interval Δt smaller, as in Fig. 3–21a. The points corresponding to ϕ and $\phi + \Delta\phi$ move closer together, and $\Delta\mathbf{v}$—or equivalently, $\mathbf{a}_{av}$—continues to point to the center of the circle (Fig. 3–21b). In the limit in which Δt goes to zero, the ratio $\Delta\mathbf{v}/\Delta t$ gives us the instantaneous acceleration. *The instantaneous acceleration points precisely to the center of the circle.*

To obtain the magnitude of the acceleration, notice that the angle between $\mathbf{v}$ and $\mathbf{v} + \Delta\mathbf{v}$ is the same as the angle between $\mathbf{r}$ and $\mathbf{r} + \Delta\mathbf{r}$, so they form similar triangles (see Fig. 3–20b). We then have

$$\frac{\Delta v}{v} = \frac{\Delta r}{r}. \qquad (3\text{–}52)$$

We write this in the form $\Delta v = (v/r)\,\Delta r$, and, after dividing by Δt, we obtain

$$\frac{\Delta v}{\Delta t} = \frac{v}{r}\frac{\Delta r}{\Delta t}.$$

If we continue this process for smaller and smaller time intervals Δt, then $\Delta\phi$, $\Delta\mathbf{r}$, and $\Delta\mathbf{v}$ all become smaller and smaller. When we use $\Delta r = v\,\Delta t$, we are led automatically to the instantaneous acceleration:

$$a = \lim_{\Delta t \to 0}\frac{\Delta v}{\Delta t} = \frac{v}{r}\lim_{\Delta t \to 0}\frac{\Delta r}{\Delta t}$$

$$= \frac{v}{r}v = \frac{v^2}{r}. \qquad (3\text{–}53)$$

An alternative form for the acceleration is expressed in terms of ω rather than v:

$$a = \frac{v^2}{r} = \frac{(\omega r)^2}{r}$$

$$= r\omega^2. \qquad (3\text{–}54)$$

The acceleration of an object in uniform circular motion

As previously noted, the instantaneous acceleration of an object undergoing uniform circular motion is directed exactly inward toward the center of the circle. We can summarize this information with a unit vector in the radial direction, defined by

The acceleration of an object in uniform circular motion is always directed toward the center of the circle.

$$\hat{\mathbf{r}} \equiv \frac{\mathbf{r}}{r}, \qquad (3\text{–}55)$$

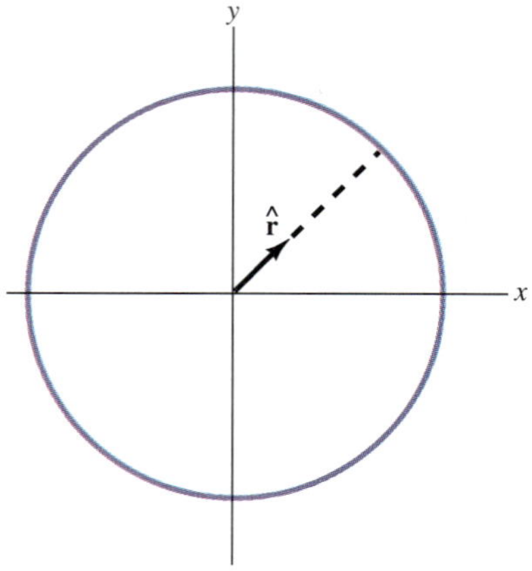

FIGURE 3–22 The unit vector $\hat{\mathbf{r}}$ has length one and points away from the origin.

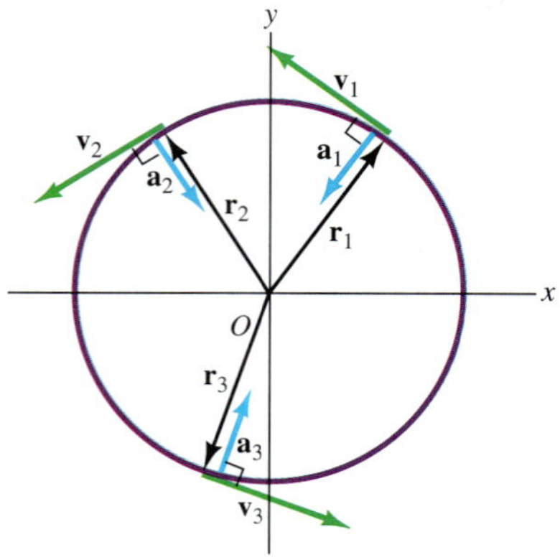

FIGURE 3–23 The instantaneous position **r**, instantaneous velocity **v**, and instantaneous acceleration **a** are shown for several positions around the path of a particle in uniform circular motion.

where the caret over the unit vector $\hat{\mathbf{r}}$ distinguishes this vector from the position vector **r**. In contrast to the Cartesian unit vectors **i** and **j**, which are fixed in space, the unit vector in the radial direction varies with the angle ϕ. Notice that the unit vector $\hat{\mathbf{r}}$ points outward from the origin (Fig. 3–22). In terms of this vector, the centripetal acceleration is

$$\mathbf{a} = -\frac{v^2}{r}\hat{\mathbf{r}}. \tag{3–56}$$

We should reemphasize the sign here: The unit vector $\hat{\mathbf{r}}$ points *away* from the origin, so the acceleration points *toward* the origin.

The relationship of **r**, **v**, and **a** is shown in Fig. 3–23 for various times during the particle's circular orbit. (Note that **a** is always perpendicular to **v**.) All the vectors are constant in magnitude but *vary continuously in direction.* Quantities that point to the center of a circle, such as the acceleration in circular motion, are described as **centripetal**, meaning center-seeking.

EXAMPLE 3–9 A child's top spins uniformly at 16 Hz. What is the centripetal acceleration on the outside surface if the radius of the top is 3 cm?

Solution: The angular speed is determined from $\omega = 2\pi f$:

$$\omega = (2\pi \text{ rad})(16 \text{ Hz}) = 101 \text{ rad/s}.$$

The acceleration a has magnitude

$$a = \omega^2 r = (101 \text{ rad/s})^2(3 \text{ cm})\left(\frac{1 \text{ m}}{100 \text{ cm}}\right) = 306 \text{ m/s}^2.$$

Remember that the unit *radian* is a dimensionless quantity. It is a measure of angle and we use it in the measurement of angular speed. However, the unit is dropped in this example during the calculation of the acceleration; it would not be appropriate to write the answer for acceleration as 306 $\text{rad}^2 \cdot \text{m/s}^2$.

3–6 RELATIVE MOTION

When we drive with velocity **v** past a tree, we normally think of the tree as stationary and ourselves as moving. Yet this way of thinking can be reversed: If we measure the position of the tree with respect to ourselves, it is also perfectly correct to think of the tree moving past us with a velocity $-\mathbf{v}$. At the same time, someone picnicking in the shade of the tree would say that the tree is stationary; that is, the picnicker measures the tree to have zero velocity. It is all a question of the **frame of reference**, and it is necessary for us to be able to see how descriptions can vary according to the frame of reference of the observers. Although it is not quite so obvious in the example of the tree, we shall discuss situations in which the laws of physics are more easily understood in one frame of reference than in another. The laws of planetary orbits and those of electricity and magnetism are good examples of such situations that we'll be exploring in later chapters. Our aim here is to see how the observations of some simple situations made from different reference frames are related to one another.

Kinematic Quantities as Seen from Two Frames of Reference

An object may appear to have one motion to one observer and a different motion to a second observer, depending on how the two observers are moving with respect to one another. The two observers are said to have a **relative motion**. Figure 3–24 illustrates how the view of a scene depends on the observer. The position of an object—for example, the person marked as A—is measured differently by the two observers B and

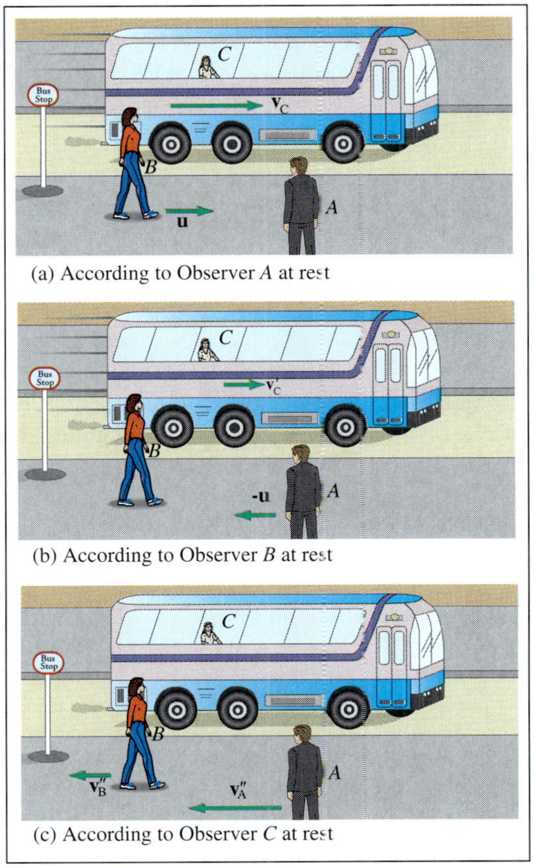

FIGURE 3–24 The scene in part (a) is observed from the point of view of (b) observer B and (c) observer C, who are in relative motion.

C, who are each in a different frame of reference. If we specify the relative motion of the observers, then we also specify the velocity at which one observer sees the other moving. For example, if observer A sees observer B moving with velocity $\mathbf{u}$, then B sees A moving with velocity $-\mathbf{u}$ (in the opposite direction at the same speed). Now suppose that observer A sees object P moving with velocity $\mathbf{v}_A$ and observer B sees object P moving with velocity $\mathbf{v}_B$. Velocities $\mathbf{v}_A$ and $\mathbf{v}_B$ must be related through the relative velocity $\mathbf{u}$. From Fig. 3–24, which refers to one-dimensional motion,

How two observers each measure the velocity of an object

$$\mathbf{v}_B = \mathbf{v}_A - \mathbf{u}. \tag{3–57}$$

As the vector notation indicates, Eq. (3–57) holds even when we refer to more than one dimension. We can see this more formally as follows. Suppose the position of an object were measured from reference frame A as $\mathbf{r}_A$ and from frame B as $\mathbf{r}_B$. As Figure 3–25 shows, the relation is

$$\mathbf{r}_B = \mathbf{r}_A - \mathbf{R}, \tag{3–58}$$

where $\mathbf{R}$ is the position of the origin of frame B as measured in frame A. The relative movement of the two frames is specified according to

$$d\mathbf{R}/dt = \mathbf{u}. \tag{3–59}$$

Then, by measuring the time rate of change of Eq. (3–60), we see how the velocities in the two frames compare:

$$d\mathbf{r}_B/dt = d\mathbf{r}_A/dt - d\mathbf{R}/dt = d\mathbf{r}_A/dt - \mathbf{u}. \tag{3–60}$$

This is exactly Eq. (3–57).

Will our two observers agree on the *acceleration* of an object P? That depends on

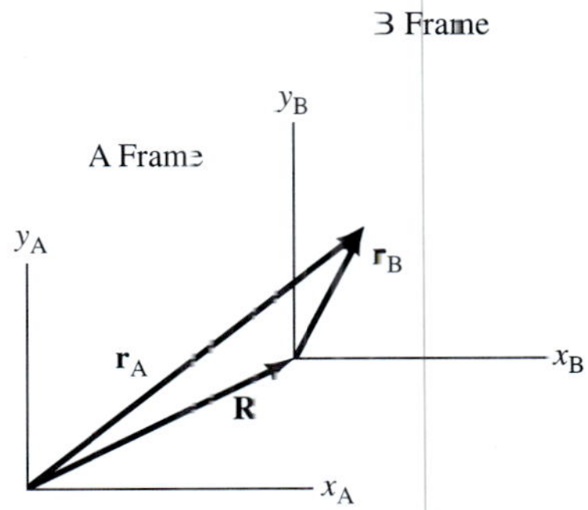

FIGURE 3–25 The position of an object can be measured by an observer in the A frame or by an observer in the B frame. The positions in the two frames can be related once the relative position of the two frames is known.

Will our two observers agree on the *acceleration* of an object *P*? That depends on whether the relative motion of the two observers has a constant velocity or not. If we differentiate Eq. (3–57) with respect to time, we obtain the relation between the two accelerations of the object as seen by the two observers. We have

$$\frac{d\mathbf{v}_B}{dt} = \frac{d\mathbf{v}_A}{dt} - \frac{d\mathbf{u}}{dt},$$

or

How two observers measure the acceleration of an object

$$\mathbf{a}_B = \mathbf{a}_A - \frac{d\mathbf{u}}{dt}, \tag{3–61}$$

where $\mathbf{a}_A$ and $\mathbf{a}_B$ are the accelerations of the object as seen by observers A and B, respectively. If $\mathbf{u}$ is a constant—and thus its time derivative is zero—then the two observers agree on the object's acceleration if not on its velocity. However, it is not always the case that $\mathbf{u}$ is a constant. Consider a tricycle standing motionless relative to a woman we will call observer A, while her child (observer B) spins on a nearby merry-go-round (Fig. 3–26). The child observes the tricycle to be moving—but not in a straight line. Motion in which the direction and/or the speed of an object changes is accelerated motion. Because $\mathbf{v}_A$ (the velocity of the tricycle as measured by the mother) is zero at all times, so too is $\mathbf{a}_A$. Equation (3–61) shows that the child would measure the acceleration of the tricycle to be

$$\mathbf{a}_B = -\frac{d\mathbf{u}}{dt}, \tag{3–62}$$

where $\mathbf{u}$ is the velocity that the mother measures her child to have on the merry-go-round; this velocity is not constant, so its time derivative is not zero.

EXAMPLE 3–10 A boat must cross a river that is 150 m wide. The river has a current of 3 km/h, and the boat can be rowed through the water with a uniform speed of 4 km/h (Fig. 3–27a). Set up two coordinate systems in which to describe the displacement of the boat: one fixed to the bank and the other fixed to a spot moving with the current of the river. Using these coordinate systems, express the position vector of the boat at time *t*; assume that the boat leaves the dock at the angle θ with respect to a point moving with the water, as shown in Fig. 3–27b. Calculate θ such that the boat lands at a point exactly opposite the starting point. How long will the trip take?

Solution: The two frames of reference in this case refer to a frame whose origin, *O*, is fixed on the dock on the bank of the river (Fig. 3–27a), and a frame whose

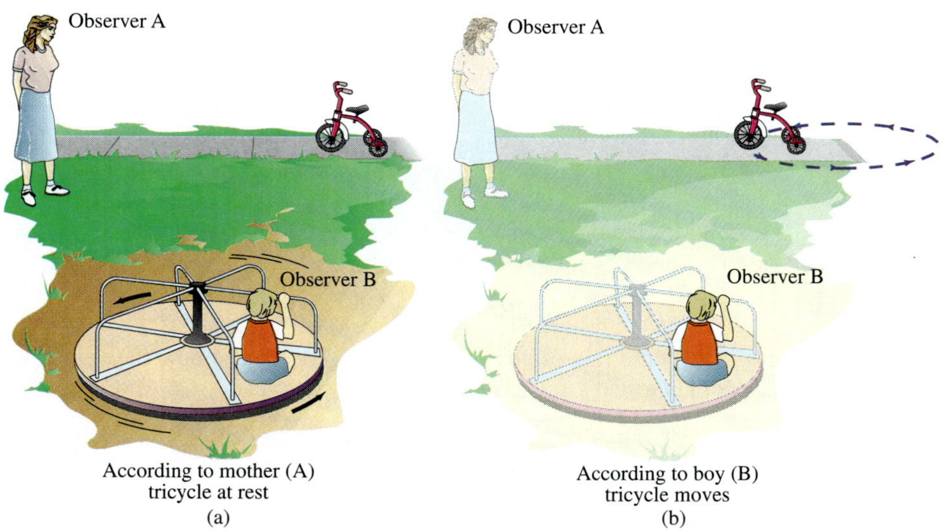

FIGURE 3–26 A tricycle and a mother are at rest relative to each other. A child is sitting on a rotating merry-go-round. The mother and child, observers A and B, respectively, are not moving in steady straight-line motion with respect to one another; they measure the tricycle as having different accelerations. (a) The mother sees the tricycle at rest. (b) From the point of view of a frame at rest relative to the child, the tricycle moves in circles—hence it accelerates.

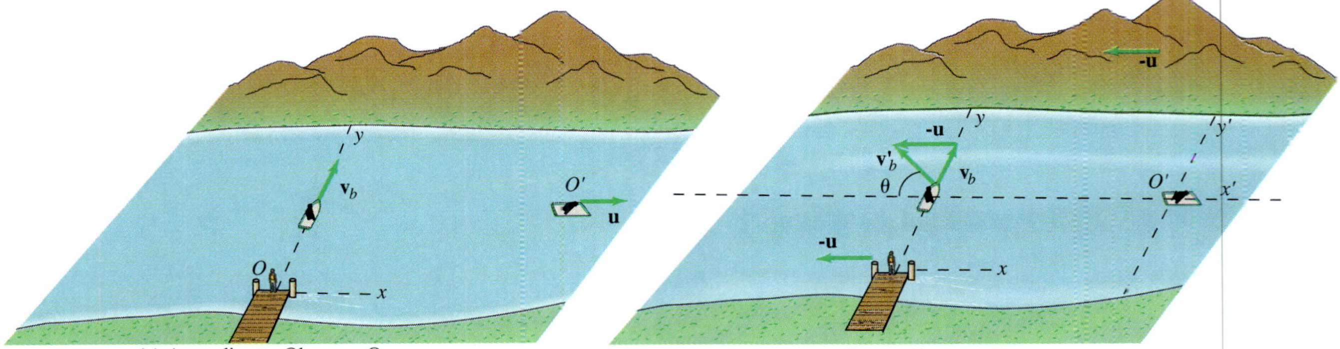

(a) According to Observer O

(b) According to Observer O'

FIGURE 3–27 Example 3–10. (a) As seen by an observer O standing on the dock, the boat moves straight across the river. (b) As seen by an observer O′ moving with the water, the boat moves at an angle θ.

origin, O', is fixed to a hypothetical spot in the water—a raft that moves with the current (Fig. 3–27b). In each frame of reference, we set the x-direction in the direction of the current, with the y-direction pointing directly across the river. The origin O' is seen by the observer on the fixed dock to move with the velocity

$$\mathbf{u} = u\mathbf{i},$$

where u is the speed of the current (3 km/h).

If the boat can move at speed $v = 4$ km/h with respect to the water, then, as observed in the frame moving with the water, the boat moves with velocity

$$\mathbf{v}'_{\mathrm{B}} = -v\cos\theta\ \mathbf{i} + v\sin\theta\ \mathbf{j}.$$

As long as the rower does not look at the bank and fixes his vision on a spot in the water, he sees himself moving with speed v at the angle θ through the water. An observer on the dock, however, sees the boat move with velocity

$$\mathbf{v}_{\mathrm{B}} = \mathbf{v}'_{\mathrm{B}} + \mathbf{u} = (-v\cos\theta + u)\mathbf{i} + v\sin\theta\ \mathbf{j},$$

which is consistent with Eq. (3–57). All velocities are constant, so the position of the boat as seen by the observer on the dock is

$$\mathbf{r}_{\mathrm{A}} = (-v\cos\theta + u)t\mathbf{i} + vt\sin\theta\ \mathbf{j}.$$

If angle θ is chosen such that the dockside observer sees no horizontal velocity (x-component of velocity), he will see the boat moving straight across the water to the point on the opposite shore. This occurs when the angle is given by

$$\cos\theta = \frac{u}{v} = \frac{3\ \text{km/h}}{4\ \text{km/h}} = 0.75;$$

$$\theta = 41°.$$

The boat is then seen by the dockside observer to move uniformly across the river with speed $v\sin\theta$, and the trip takes a time

$$t = \frac{y}{v\sin\theta} = \frac{0.15\ \text{km}}{(4\ \text{km/h})(\sin 41°)} = 0.06\ \text{h} \simeq 3\ \text{min}.$$

With no river current, the boat could be rowed directly across the water; in this case, the time to cross would be (0.15 km)/(4 km/h) = 0.04 h ≃ 2 min.

SUMMARY

A particle moving in space follows a trajectory, or path. In three dimensions, the position vector of such a particle is

$$\mathbf{r} = x(t)\mathbf{i} + y(t)\mathbf{j} + z(t)\mathbf{k}. \qquad (3\text{–}3)$$

The displacement vector between times t and $t + \Delta t$ is the difference between the position vectors at these times. It is given by

$$\Delta\mathbf{r} = \mathbf{r}(t + \Delta t) - \mathbf{r}(t). \tag{3–1}$$

The instantaneous velocity $\mathbf{v}$ is found from the displacement vector $\Delta\mathbf{r}$ over small time intervals:

$$\mathbf{v} \equiv \lim_{\Delta t \to 0} \frac{\mathbf{r}(t + \Delta t) - \mathbf{r}(t)}{\Delta t} = \frac{d\mathbf{r}}{dt}. \tag{3–5}$$

In two dimensions, the velocity is expressed in terms of its component vectors as

$$\mathbf{v} = \mathbf{v}_x + \mathbf{v}_y, \tag{3–9}$$

where, in terms of unit vectors,

$$\mathbf{v}_x = \frac{dx}{dt}\,\mathbf{i}, \tag{3–11a}$$

$$\mathbf{v}_y = \frac{dy}{dt}\,\mathbf{j}. \tag{3–11b}$$

In terms of its components, the magnitude of the velocity $\mathbf{v}$ is

$$v = |\mathbf{v}| = \sqrt{v_x^2 + v_y^2}, \tag{3–12}$$

Further, the angle θ that $\mathbf{v}$ makes with the x-axis is given by

$$\tan\theta = \frac{v_y}{v_x}. \tag{3–13}$$

The instantaneous acceleration is a derivative of the velocity:

$$\mathbf{a} \equiv \lim_{\Delta t \to 0} \frac{\mathbf{v}(t + \Delta t) - \mathbf{v}(t)}{\Delta t} = \frac{d\mathbf{v}}{dt}. \tag{3–15}$$

Like velocity, acceleration can also be expressed in vector components. The acceleration has components a_x and a_y, which are derivatives of the x- and y-components of the velocity.

For constant acceleration in the xy-plane, we have

$$x = x_0 + v_{0x}t + \frac{1}{2}a_x t^2, \tag{3–19}$$

$$v_x = v_{0x} + a_x t; \tag{3–20}$$

$$y = y_0 + v_{0y}t + \frac{1}{2}a_y t^2, \tag{3–21}$$

$$v_y = v_{0y} + a_y t, \tag{3–22}$$

which can be written more concisely as

$$\mathbf{r} = \mathbf{r}_0 + \mathbf{v}_0 t + \frac{1}{2}\mathbf{a}t^2, \tag{3–25}$$

$$\mathbf{v} = \mathbf{v}_0 + \mathbf{a}t. \tag{3–26}$$

In the absence of air resistance, a projectile moves under the influence of gravity with a constant acceleration vector $\mathbf{a} = \mathbf{g}$. The trajectory of such a projectile is a parabola. The range (the horizontal distance a projectile launched from the ground travels over level ground) and maximum height of the trajectory can be calculated, as can the projectile's flight time.

Circular motion is most simply described by means of plane polar coordinates, with ρ as the radial coordinate and ϕ as the angular coordinate. Angles are measured in radians (which are dimensionless), so that the arc length formed by the angle ϕ in a circle of radius R is

$$\text{arc length} = R\phi. \tag{3–43}$$

The angular speed of an object in circular motion is

$$\omega \equiv \frac{d\phi}{dt}. \qquad (3\text{–}46)$$

The speed of a particle moving in circular motion with angular speed ω is

$$v = R\omega. \qquad (3\text{–}47)$$

The period T is the time to make one complete revolution, and the frequency f is the inverse of T. The relation between angular speed and frequency is

$$\omega = 2\pi f. \qquad (3\text{–}51)$$

The velocity of a particle in uniform (or constant) circular motion is tangent to the circle. The acceleration of this particle has magnitude

$$a = \frac{dv}{dt} = \frac{v^2}{r} = r\omega^2. \qquad (3\text{–}53,\ 3\text{–}54)$$

The direction of the particle's acceleration is centripetal; that is, it points inward along a radius toward the center of the circle.

Suppose that two observers, A and B, are in relative motion. Observer A measures the velocity of observer B to be **u**, the relative velocity. Then, if the velocities of an object as measured by observers A and B are $\mathbf{v}_A$ and $\mathbf{v}_B$, respectively, these velocities are related by

$$\mathbf{v}_B = \mathbf{v}_A - \mathbf{u}. \qquad (3\text{–}57)$$

The acceleration of the object according to the two observers obeys the relation

$$\mathbf{a}_B = \mathbf{a}_A - \frac{d\mathbf{u}}{dt}. \qquad (3\text{–}61)$$

UNDERSTANDING KEY CONCEPTS

1. What factors determine how far an athlete travels in the long jump? Which factor do you think must be the most important?
2. Why is it important to have limits on wind speed for world-record track and field events?
3. Under what conditions is it possible to have a constant speed yet a nonzero acceleration?
4. You wish to row a boat to a point directly opposite your starting point on the bank of a swiftly moving river. The best speed you can make through the water—meaning with respect to the water—is 2 mi/h. The downstream current of the river is also 2 mi/h. Can you reach your desired goal in a finite time without doing some walking? Explain your answer.
5. There are two initial angles for which the range of a projectile is the same (except in the instance where the initial angle is 45°). What are the differences in the two trajectories that might lead a football quarterback to choose one of them over the other in throwing a pass? If you are familiar with football, you will know that the quarterback usually chooses one of these two trajectories.
6. A ball is thrown straight up and comes straight down. What is the value of its velocity at the top of its flight? the value of its acceleration?
7. Your lecture demonstrations may include a car that, even while moving, shoots a ball vertically off a platform on the car. If the car is moving with constant speed on a horizontal surface, the ball ends up at the same position on the car that it left. What would happen if the car were decelerating as it released the ball?
8. Suppose that you measure the period for successive rotations of a phonograph turntable as being unchanging. Is this enough to assure you that the turntable has uniform circular motion?
9. With a given initial speed, there are two initial firing angles for which a projectile has the same range. How many initial velocities give the same maximum height? The same time of flight?
10. You are running in the rain. At what angle should you hold your umbrella for the most protection? Assume that no wind is blowing, so that the raindrops fall straight down. You may need to assume some value for the raindrops' speed as well as your own.
11. You are on the rotating platform of a merry-go-round and throw a ball straight up. Describe the motion of the ball as you see it.
12. Earth is a rotating platform because it turns on an axis. If a projectile initially aimed at a particular point on the equator is shot from the North Pole, will it arrive at that point, assuming that it could travel such a great distance? Describe the motion of the projectile as seen by someone on Earth.
13. Suppose a projectile is fired from Washington, D.C. toward the equator. How does this situation differ from the one discussed in Question 12?
14. A motorcyclist rides on a large, rotating platform in an amusement park. The cyclist starts at the center of the rotating platform and travels in a radial direction with a constant speed. Is it correct to say that, when he or she has traveled some distance from the platform center, the acceleration due to the circular motion (which is directed in the inward radial direction) will stop the motorcycle? Analyze the motion.

PROBLEMS

3–1 *Position and Displacement*

1. (I) A car travels 21 km to the northeast, then 15 km to the east, before it travels 28 km to the north. Express the position vector from where the car starts to the point at which each turn occurs. What is the car's total displacement?

2. (I) A particle is located by the position described by the vector $\mathbf{r} = (c_1 - c_2 t)\mathbf{i} + (d_1 + d_2 t + d_3 t^2)\mathbf{j}$, where $c_1 = 12$ m, $c_2 = 3$ m/s, $d_1 = -20$ m, $d_2 = -3$ m/s, and $d_3 = 0.5$ m/s^2. At what time(s) does the particle pass through the position $x = 0$ m? At what time(s), and where, does the particle cross the line $x = y$? Sketch the particle's trajectory from $t = -8$ s to $t = 10$ s.

3. (I) A gym teacher organizes a series of indoor races, which follow along the walls of the gym; the race starts from corner A, continues to corner B, and so forth. The gym is a rectangle with the distance $AB = 25$ m and the distance $BC = 35$ m. Suppose that the origin of a coordinate system is at point A, leg AB is in the $+x$-direction, and leg BC is in the $+y$-direction. Express the position vector of a running student at each of the four corners.

4. (I) A runner races with a uniform speed of 27.0 km/h around a circular track of radius 172 m. Draw the track and the runner's displacement vector after 20 s, 40 s, 60 s, and 2 min, assuming that at $t = 0$ s the runner is at the three o'clock position. Assume counterclockwise motion.

5. (I) A treasure map locates the site of a treasure by reference to two starting points, A and B. A is chosen to be the origin, and B is at the point (2.5 km)$\mathbf{i}$. The instructions state that the treasure lies at the intersection of two lines. One line starts at A and passes through the point (2.0 km)$\mathbf{i}$ + (4.0 km)$\mathbf{j}$; the other line, starting at B, passes through the point (6.0 km)$\mathbf{i}$ − (8.0 km)$\mathbf{j}$. Sketch the instructions and find the location of the intersection graphically as well as algebraically. Express the vector that gives the displacement of the intersection point from point C, whose location relative to the coordinate system specified by A and B is given by (1.2 km)$\mathbf{i}$ − (2.2 km)$\mathbf{j}$.

6. (II) The position of a particle in a given coordinate system is $\mathbf{r}(t) = (5 - 2t^2)\mathbf{i} + (-4 + 3t)\mathbf{j}$ where the distances are in meters when t is in seconds. At what time will the particle cross the y-axis? At what time will it cross the x-axis? Can you find an equation that relates the y-coordinate to the x-coordinate and therefore gives the trajectory in the xy-plane? Where would the x- and y-axes have to be moved so that at $t = 0$ s the trajectory passes through the origin?

7. (II) Figure 3–28 shows the trajectory of a bicyclist traveling with a constant speed of 25.2 km/h. Express the position vector of the bicyclist as a function of time.

8. (II) The x- and y-coordinates of the position of a land surveyor are shown on Fig. 3–29 as a function of time. (a) Plot the trajectory of the surveyor. (b) Calculate the area of the plot that she surveyed.

9. (II) The position of a particle is given in a certain coordinate system by the vector $\mathbf{r}(t) = (4\text{ m})\cos(\pi t/T)\mathbf{i} - (4\text{ m})\sin(\pi t/T)\mathbf{j}$. Find the displacement vector at times $t = (T/3)$, $t = (T/2)$, and $t = 2T$. In each case, compute the distance to the origin. What is the angle that the position vector makes with the $+x$-axis for arbitrary t?

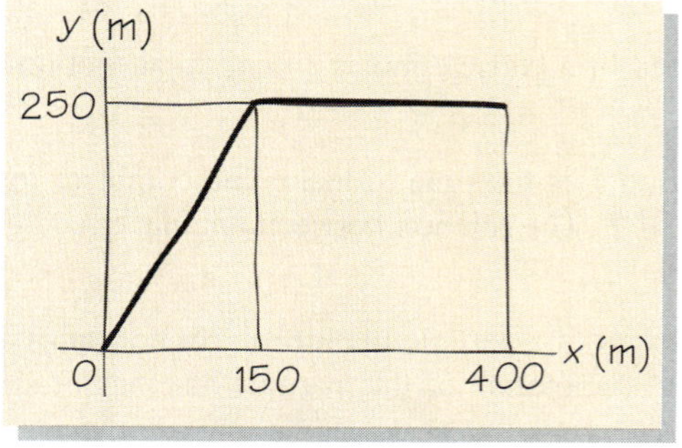

FIGURE 3–28 Problem 7.

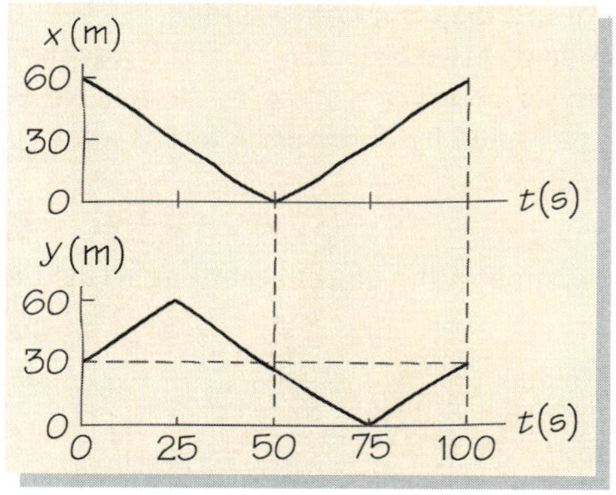

FIGURE 3–29 Problem 8.

10. (III) In a "shoot the coconut" lecture demonstration, the position of the coconut is given by the vector $(h_0 - \frac{1}{2}gt^2)\mathbf{j}$, whereas that of the projectile, aimed at the coconut at time $t = 0$, is given by $(-L + ut)\mathbf{i} + [(h_0 ut/L) - (\frac{1}{2}gt^2)]\mathbf{j}$. Show that the two will always collide, and find the time at which this takes place (Fig. 3–30). Express the displacement vector of the coconut relative to the projectile.

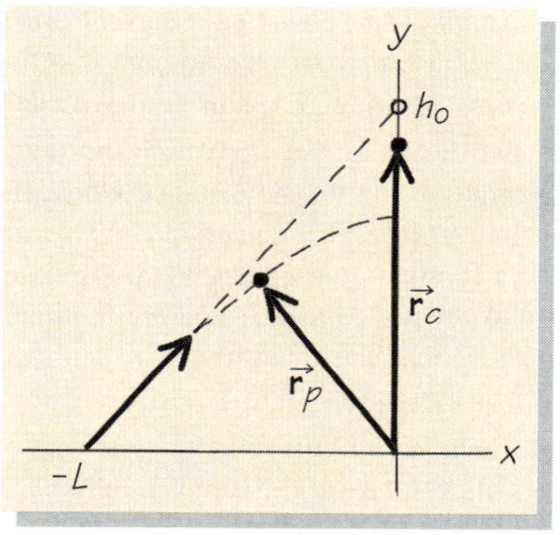

FIGURE 3–30 Problem 10.

3–2 *Velocity and Acceleration*

11. (I) A tennis ball leaves a racquet with an initial speed of 15 m/s. After 2, it has slowed down to 13 m/s. What is the mag-

nitude of the average acceleration? What would you need to know to determine the direction of the acceleration?

12. (I) A particle is observed to move with the coordinates $x(t) = 4 \text{ m} + (3 \text{ m/s})t + (1 \text{ m/s}^2)t^2$ and $y(t) = 6 \text{ m} - (4 \text{ m/s})t + (0.5 \text{ m/s}^2)t^2$. What are the particle's position, velocity, and acceleration? At what time are the velocity's horizontal and vertical components equal?

13. (I) At a given moment, a fly moving through the air has a velocity vector that changes with time according to $v_x = 2.2$ m/s, $v_y = (3.7 \text{ m/s}^2)t$ and $v_z = (-1.2 \text{ m/s}^3)t^2 + 3.3$ m/s, where t is measured in seconds. What is the fly's acceleration?

14. (I) A particle moves in such a way that its coordinates are

$$x(t) = A \cos \omega t; \; y(t) = A \sin \omega t$$

Calculate the x- and y-components of the velocity and the acceleration of the particle.

15. (II) A whale traveling southwest at 7.0 km/h is spotted 5.0 km to the northwest off the coast of Malibu. Photographers jump into a boat that can move at 30 km/h. With what velocity will the photographers intercept the whale, assuming that their boat travels a straight-line path? What is the position vector of the whale from the original point on the coast when the photographers reach the whale?

16. (II) A lifeguard standing on a tower throws a buoy to a swimmer 20 m from the tower (Fig. 3–31). The lifeguard, positioned 3 m above the water, pulls in the rope at a speed of 1 m/s. How fast is the swimmer coming to the shore when he is (a) 15 m and (b) 5 m from the water's edge?

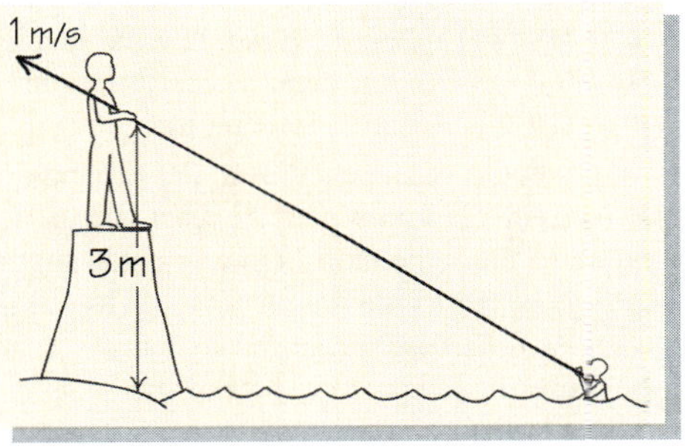

FIGURE 3–31 Problem 16.

17. (II) An engineering student holds her open compass perpendicular to the drafting board, touching the board with both tips of the compass (Fig. 3–32). She slowly closes the compass so that the tips move toward each other with a speed of $2v_0 = 0.06$ m/s. Initially, the angle between the arms of the compass is $2\theta = 60°$. The arms of the compass are $L = 15$ cm long. (a) Calculate the velocity of the joint of the compass as a function of time. (b) Give the velocity of the joint at the moment that she starts to close the compass, and at the moment that the two tips reach each other.

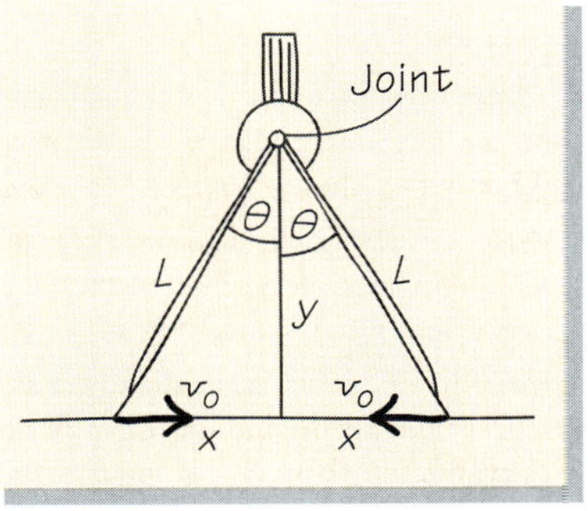

FIGURE 3–32 Problem 17.

18. (II) Calculate the velocity vectors for the coconut and projectile in Problem 10. Express the difference between the two velocity vectors (the relative velocity). What is the magnitude of the relative velocity? What is the angle that its direction makes with the $+x$-axis?

19. (II) Calculate the velocity vector for the particle described by the position vector given in Problem 9. Calculate the angle that the velocity vector makes with the $+x$-axis, and show that the velocity vector is always perpendicular to the position vector.

20. (II) A particle moving with an initial velocity $\mathbf{v} = (50 \text{ m/s})\mathbf{j}$ undergoes an acceleration $\mathbf{a} = [35 \text{ m/s}^2 - (2 \text{ m/s}^5)t^3]\mathbf{i} + [4 \text{ m/s}^2 - (1 \text{ m/s}^4)t^2]\mathbf{j}$. What are the particle's position and velocity after 3 s, assuming that it starts at the origin?

21. (II) The motion of a planet about a star may be described by the vector $\mathbf{r}_1 = R\cos(2\pi ft)\mathbf{i} + R\sin(2\pi ft)\mathbf{j}$. The position vector of another planet about the same star is given by $\mathbf{r}_2 = 4R \cos(\pi ft/4)\mathbf{i} + 4R\sin(\pi ft/4)\mathbf{j}$. Show that both of these describe circular motion with constant speed. Calculate the position vector of the second planet relative to the first planet. Sketch the path by noting that one circular motion is eight times faster than the other. This is like the motion of a planet whose year is 8 Earth years.

22. (II) Calculate the acceleration vectors for the two planets in Problem 21. Express the vector that describes the acceleration of the second planet relative to the first one.

23. (II) A bag is dropped from a hot-air balloon. Its height is given by the formula $h = H - ut - (u/B)e^{-Bt}$. What are the dimensions of B? What is the initial velocity? What is the velocity as $t \to \infty$? Calculate the accelerations at $t = 0$ and at $t = \infty$.

3–3 Motion with Constant Acceleration

24. (I) An airplane is flying due south on a level course at a speed of 400 km/h. At an altitude of 2.50×10^3 m directly above a hospital, the airplane meets severe turbulence and descends with a vertical acceleration of 5.00 m/s² for 8 s. (a) What is the total displacement of the plane in the 8 s? (b) What is the plane's velocity at the end of the 8 s? (c) What is the plane's position vector with respect to the hospital at the end of the 8 s?

25. (I) A launching mechanism accelerates a baseball horizontally at 24 m/s² for 0.5 s. The baseball's initial velocity is zero. With what velocity will it leave the launcher? The baseball leaves the launcher at the same speed when the launcher is turned in the vertical direction. How high will the baseball go?

26. (I) A gymnast works out on a trampoline. At the instant that she leaves the trampoline, a point on her waist is 1.1 m above the floor and at the center of the trampoline. At that instant, the point has an upward velocity of 7.3 m/s and a horizontal velocity of 3.0 m/s. Write equations that describe the subsequent motion of that point, and find its maximum height.

27. (II) A boy shoots a rock with an initial velocity of 21 m/s straight up from his slingshot. He quickly reloads and shoots another rock in the same way 3.0 s later. (a) At what time and (b) at what height do the rocks meet? (c) What is the velocity of each rock when they meet?

28. (II) A man in the crow's nest of a sailing ship moving through smooth seas at a steady 12 km/h accidentally lets a cannonball drop from his station, which is 8.5 m above the deck at the top of the mainmast. (a) Assuming that he dropped the ball from a position immediately adjacent to the vertical mast, where does the ball land with respect to the mast? (b) How long does it take for the ball to fall to the deck? (c) In the time it takes the ball to fall, how far has the ball moved with respect to an observer fixed on the shore?

3–4 Projectile Motion

29. (I) A runner attempting a broad jump leaves the ground with a horizontal velocity of magnitude 9.0 m/s. Assuming the horizontal component of velocity is unaffected, what vertical component of velocity must the runner acquire to jump 9.5 m?

30. (I) An engineering student wants to throw a ball out a third-story dorm window (10 m off the ground) onto a target on the ground placed 8.0 m away from the building. (a) If the student throws the ball horizontally, with what velocity must it be thrown? (b) What must the velocity of the ball be if it is thrown up at an elevation angle of 29°? (c) What is the ball's time of flight in case (b)?

31. (I) A projectile is shot at an angle of 34° to the horizontal with an initial speed of 225 m/s. What is the speed at the maximum height of the trajectory?

32. (I) At what points in a projectile's trajectory above level ground is the magnitude of the velocity a maximum and a minimum? What are these velocities in terms of the initial speed v_0 and elevation angle θ_0?

33. (I) A projectile is shot at an angle of 40° to the horizontal over level ground. Assuming air resistance plays no role, what angle does the projectile make with the horizontal when it lands?

34. (I) Find the initial angle if the range of a projectile is twice its maximum height.

35. (II) (a) Show that the range R can be expressed in terms of the maximum height h, and in particular that $R = 4h \cot \theta_0$. (b) Show that, when the range is a maximum, $h = R/4$.

36. (II) Galileo throws a rock from the top of the Leaning Tower of Pisa at an upward angle of 45° with speed v_0. The rock is in flight for 4.0 s and hits the ground 20 m from the base of the building. Ignore air resistance and ignore the fact that the tower tilts a bit. (a) What is the speed v_0? (b) How high off the ground is the top of the tower? (c) What is the speed of the rock just before it hits the ground?

37. (II) In the Battle of Hastings in A.D. 1066, during which the Normans of France defeated the Saxons in England, an important role was played by Norman archers who shot arrows over a wall of shields erected by the Saxons. If the Norman bows had a maximum range of 350 m and the arrows were shot at an elevation of 55°, how close were the Normans to the Saxons? Assume that the arrows reached their target.

38. (II) A punter kicks a football during a critical football game. The ball leaves his foot from ground level with a speed of 28 m/s at an angle of 50° to the horizontal. At the very top of its flight, the ball hits a wandering seagull. The ball and the seagull each stop dead and fall vertically from the point of collision (Fig. 3–33). In the following, ignore air resistance.

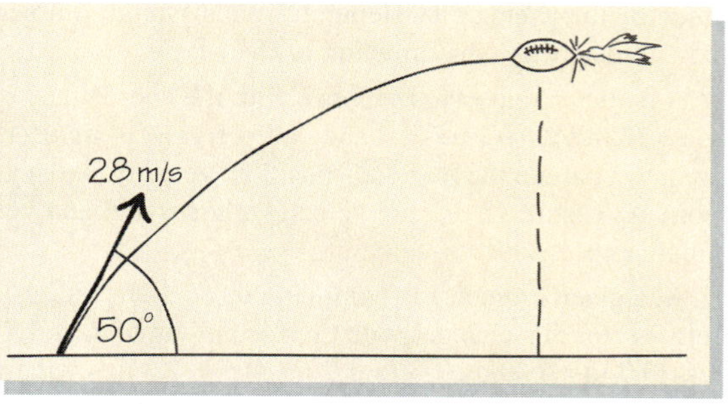

FIGURE 3–33 Problem 38.

(a) With what speed is the ball moving when it strikes the seagull? (b) How high was the unfortunate seagull when it met the ball? (c) What is the speed of the seagull when it hits the ground?

39. (II) A place kicker attempts an extra point, giving the ball an initial velocity of 15 m/s at an angle of 37° with the field. The uprights are 15 m from the point at which the ball is kicked, and the horizontal bar is 4.0 m from the ground. (a) At what time after the kick will the ball pass the goal posts? (b) Is the kick successful, and by how many meters does the ball clear or pass beneath the bar?

40. (II) Astronaut John Q. Hero had the good fortune to play golf on the Moon. The acceleration of gravity on the surface of the Moon is only about one-sixth of that on the surface of Earth. Assuming that Hero was not noticeably hampered by his space suit, and assuming that his best drive on Earth (unhampered by air resistance) is some 210 yd, how far can he drive a ball on the Moon? Derive a general answer to this question for planet X if g_X is a known fraction of g.

41. (II) A projectile is launched over flat ground and the effects of air resistance are minimal. At what angles with respect to the ground should the launcher be oriented so that the projectile's range is half its maximum range? Why are there two possible angles? What are the angles so that the range is zero; that is, the projectile lands at the foot of the launcher?

42. (II) You must throw a baseball to hit a target on the ground 50 m from the base of a building that is 20 m in height. You are standing at a point on the edge of the roof nearest the target. (a) With what velocity must you throw the baseball if it is to leave the hand horizontally? (b) With what velocity must you throw the baseball if it is to leave the hand at an angle of 45° up from the horizontal? (c) What is the horizontal component of the initial value of the velocity in case (b)?

3–5 Circular Motion

43. (I) The Space Shuttle is in a circular orbit 220 km above Earth's surface and completes an Earth revolution every 89 min. (a) What is the Shuttle's speed? (b) Acceleration?

44. (I) The Moon circles Earth at a distance of 3.84×10^5 km. The period is approximately 28 d. What is the magnitude of the moon's acceleration, in units of g, as the Moon orbits Earth?

45. (I) A runner in the 200-m dash must make part of the dash around a curve that forms the arc of a circle. This arc has a radius of curvature of 30 m. Assuming that she runs at a steady speed and completes the 200 m in 24.7 s, what is her centripetal acceleration while she is running the curve?

46. (I) A rock placed in a plastic bag is tied to a rope 1.2 m long. The rock is whirled in a horizontal circle. (a) What is the rock's centripetal acceleration if the period of motion is 1.8 s? (b) The plastic bag will break if the radial acceleration exceeds 56 m/s^2. With what speed must the rock be whirled if the plastic bag is to be broken?

47. (I) A passenger on the outer edge of a merry-go-round, 7.5 m from the central pivot, learns that when the merry-go-round is in steady motion, his centripetal acceleration is 3.3 m/s^2. How long does it take to make one revolution?

48. (II) The shaft of the engine of a car rotates at 4000 rev/min. A flywheel, 20 cm in diameter, rotates with the shaft. Calculate the centripetal acceleration of a point on the rim of the flywheel, and express it in units of g, the acceleration of gravity.

49. (II) Safety requires that the centripetal acceleration of cars traveling along highway curves may not exceed one-tenth of the acceleration of gravity, even when traveling at the posted speed limit of 55 mi/h. How small can the radius of curvature of a curve be?

50. (II) Suppose that a point object is in uniform circular motion, moving steadily at a distance R from some central point. The time for one revolution is T. Use dimensional analysis to find the dependence of the centripetal acceleration on T and R. Compare this result to the acceleration derived from a detailed analysis of uniform circular motion performed in Section 3–6.

51. (II) A mass is tethered to a post and moves in a circular path of radius $r = 0.5$ m on an air table—friction free—at a constant speed $v = 6.3$ m/s. We employ the coordinate system shown in Fig. 3–34. (a) If at $t = 0$ s, the mass is at $\theta = 0°$, what are the coordinates (x, y) of the mass at $t = \frac{1}{2\pi}$ s? (b) What is the acceleration vector of the mass at $t = 0$ s? (c) What is the acceleration vector of the mass when $\theta = 90°$?

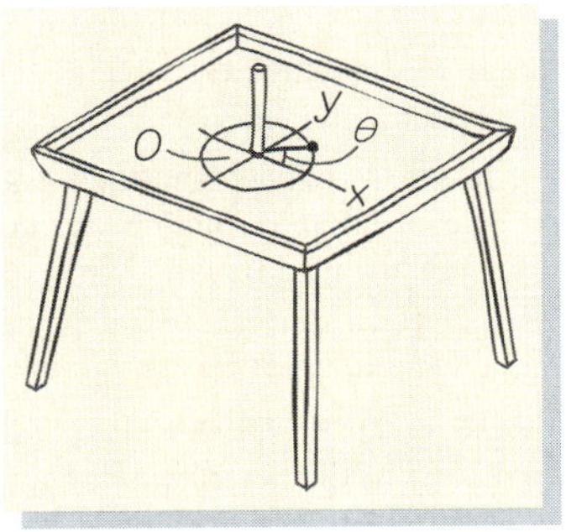

FIGURE 3–34 Problem 51.

52. (II) The Space Shuttle is moving in a circular orbit with a speed of 7.8 km/s and a period of 87 min. In order to return to Earth, the Shuttle fires its retro engines opposite to its direction of motion. The engines provide a deceleration of 6 m/s^2 that is constant in magnitude and direction. What is the total acceleration of the Shuttle?

53. (II) An electron in a research apparatus follows a circular path. On the electron's first circuit of the apparatus, its speed is v_0, and the radius of its circular path is R. Each time it makes one circuit, it passes a short region where it receives a "kick" and gains an additional speed of $v_0/100$. The electron follows a circular path such that the magnitude of its acceleration is always the same. What is the radius of the circular path after the electron has received 10 kicks?

54. (II) An automobile moves on a circular track of radius 1.00 km. It starts from rest from the point $(x, y) = (1.00 \text{ km}, 0 \text{ km})$ and moves counterclockwise with a steady *tangential* acceleration such that it returns to the starting point with a speed of 30.0 m/s after one lap. (The origin of the Cartesian coordinate system is at the center of the circular track.) What is the car's velocity (magnitude and direction) when it is one-eighth of the way around the track? Express the position and velocity at this point in terms of the unit vectors along the x- and y-axes.

3–6 Relative Motion

55. (I) A sailor wants to travel due east from Miami at a velocity of 15 km/h with respect to a coordinate system fixed on land. The sailor must contend with the Gulf Stream, which moves north at 5 km/h. With what velocity with respect to the water should the sailboat proceed under sail?

56. (I) During an uphill portion of a bicycle race, a cyclist reads a message on a board informing him that the leader is 15 s ahead and that the leader is traveling at 4 m/s. The cyclist's speedometer informs him that he is traveling at 3 m/s. (a) What is the speed of the leader with respect to the cyclist? (b) How far in front of the cyclist is the leader, assuming that the speeds have not changed in the last half minute or so?

57. (I) Rain is falling steadily, but there is no wind. You are in an automobile that moves at 80 km/h, and you see from the drops on a side window that the rain makes streaks at a 58° angle with respect to the vertical. What is the vertical velocity of the raindrops?

58. (II) A cyclist's top speed on a flat road is v. This is an "air speed" because the limiting speed for a cyclist is determined by the wind resistance. In other words, this is her top speed with respect to the air. She cycles a flat course straight north for a distance L, turns around, and cycles straight south for the same distance. In the following, ignore the time it takes her to turn around, and assume that the cyclist can maintain her top speed with respect to the air. (a) Write a formula for the total course time t_0 in terms of L and v. (b) There is a north wind blowing at speed v_w. Write a new formula for the course time t_1 of the cyclist, including the effect of v_w. (c) Show that for $v_w << v$, the course time can be approximated by $t_1 = t_0[1 + (v_w^2/v^2)]$. To show this result you may want to use the approximation $(1 - x)^{-1} \simeq 1 + x$ for $x << 1$. (d) Plot the time t_1 as a function of v_w and show that it is always greater than t_0. What happens at $v_w = v$, and why?

59. (II) An airplane flies due south with respect to the ground at an air speed of 900 km/h for 2 h before turning and moving southwest with respect to the ground for 3 h. During the entire trip, a wind blows in the easterly direction at 120 km/h. (a) What is the plane's average speed with respect to the ground? (b) What is the plane's average velocity with respect to the ground? (c) What is the final position vector?

60. (II) An airplane is to fly due north from New Orleans to St. Louis, a distance of 673 mi. On that day and at the altitude of the flight, a wind blows from the west at a steady speed of 65 mi/h. The airplane can maintain an air speed of 180 mi/h. Ignore the periods of takeoff and landing. (a) In what direction must the airplane fly in order to arrive at St. Louis without changing direction? Draw a diagram and label this direction with an angle. Would this calculation change if the distance between the cities were twice as great? (b) What is the flying time for this flight? (c) Recalculate the flying time if the airplane heads due north until it reaches the latitude of St. Louis, and then flies due west, into the wind to reach the city.

61. (III) Earth has a radius of 6.4×10^6 m, and its orbit around the Sun has a radius of some 1.5×10^{11} m. Earth simultaneously rotates about its own axis and moves around the Sun (Fig. 3–35). Assume a circular orbit and that Earth's axis of rotation is perpendicular to its orbital plane. (a) What is the speed with respect to the Sun of the point on Earth's equator nearest the Sun? (b) of the point on Earth's equator farthest from the Sun? (c) of the two points on Earth's equator, midway between the points in parts (a) and (b)?

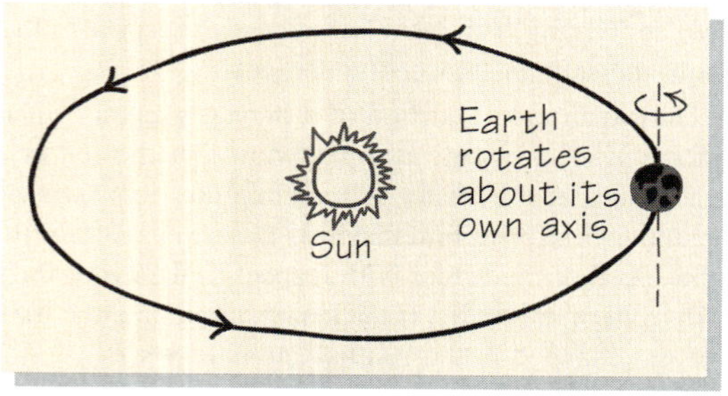

FIGURE 3–35 Problem 61.

General Problems

62. (I) Neutron stars are incredibly dense objects, compared to the Sun. One such (hypothetical) star has a radius of 40 km and rotates with a period of 26 s. What are the velocity and acceleration of a particle on the surface due to the rotation? Compare this acceleration with the acceleration of gravity on Earth.

63. (II) A golfer wants to land a golf ball on the green located 105 m away horizontally but 8.0 m down. The golfer chooses an eight iron that he knows will result in the ball leaving the tee at an elevation angle of 65°. (a) With what velocity should the ball leave the tee? (b) What is the maximum height of the ball above the green?

64. (II) A softball player hits the ball when it is 0.9 m above home plate. The ball leaves the bat at an elevation angle of 30°. What initial speed must the ball have to clear a fence 1.8 m high located 65 m away in left field?

65. (II) A boat is required to traverse a river that is 150 m wide. The current in the river moves with a speed of 6 km/h. The boat can be rowed on still water with a speed of 10 km/h. Set up a convenient coordinate system in which to describe the various displacements. Using this coordinate system, write down the position vector of the boat at time t, assuming that the boat moves with uniform speed and that it leaves one side with the velocity vector making an angle θ with the direction of the river. Calculate θ such that the boat lands at a point exactly opposite the starting point. How long will the trip take?

66. (II) A possible way to measure g is to send a mass traveling horizontally with a known speed off the edge of a vertical drop, such as a table. The table height can be measured, as can the distance from the base of the table where the mass strikes the floor. (a) If the speed of the mass just as it leaves the table is 2.50 m/s, and the distance from the table base to the point at which the mass strikes the floor is 108 cm, how long was the mass in flight? (b) The table height is 86 cm. What is the value of g obtained from this data? (c) Use $g = 9.80\ \text{m/s}^2$ and neglect the air resistance to calculate the speed of the mass when it arrives at the floor.

67. (II) A sailor on top of a mast 26 m high drops a hammer. The ship is rolling with a maximum angle away from the vertical of no more than 15°. At the moment that the hammer is dropped, the mast is exactly vertical while the top of the mast is moving laterally at a speed of 3.6 m/s. Will the hammer fall into the sea or onto the deck, given that the ship is 19 m wide?

68. (III) A cannon can project a cannonball from its barrel with a certain muzzle speed v_0. Ignoring the effects of air resistance, what formula expresses the distance the cannonball travels before it reaches the ground as a function of the angle θ that the barrel makes with the ground? Unlike the projectile treated in Section 3–4, the cannonball is fired from the edge of a cliff of height h_0 above the level plain at which it is aimed. Show that the angle θ that gives the largest horizontal range is given by $\sin^2\theta = v_0^2/2(v_0^2 + 2gh_0)$. [*Hint*: In calculus, you learn that a function of a variable such as θ has a maximum (or minimum) at an angle θ_0, the angle for which the derivative of the function with respect to θ is zero.]

69. (III) A juggler is able to handle four balls simultaneously. He takes 0.5 s to cycle each ball through his hands, throw the ball, and be ready to catch the next ball. (a) With what velocity must he throw each ball vertically? (b) What is the position of the other three balls when he has just caught one of the balls? (c) How high must he throw the balls if he is to juggle five balls?

70. (III) A wheel 72 cm in diameter rolls along a road, with the center moving in a straight line at a uniform speed of 18 km/h. What are the position vector, the velocity vector, and the acceleration vector of a fixed point on the rim of the wheel, relative to a fixed point on the straight line followed by the wheel on the road?

CHAPTER 4

The idea that the apples on these trees might have something in common with the Moon originated with Isaac Newton, who showed that the same force governs the falling of an object and the Moon's movement around Earth. This observation, popularly linked with Newton watching an apple fall, is one of the truly significant discoveries in science.

Newton's Laws

In the previous chapters, we explored quantities that *describe* motion, such as velocity and acceleration. Such a description is called kinematics. We cannot, however, *predict* the motion of an object using kinematics alone. As we shall see in this chapter, forces cause changes in motion and enable us to predict the subsequent motion of an object. The study of the *causes* of motion is called **dynamics**.

The laws that govern the motion of an object were described by Isaac Newton in 1687. Now known as Newton's laws, they are based on careful and extensive observation of motion and its changes. These laws provide an extraordinarily accurate description of the motions of all objects—small or large, simple or complicated—on the everyday scale. There are, however, some motions that cannot be characterized using these laws. We must go beyond Newton's laws to describe motion within systems as small as the atom and for motion at speeds near the speed of light (300,000 km/s). Nevertheless, Newton's laws represent a tremendous achievement in their simplicity and breadth. Much of this book treats Newton's laws as applied to the motion of the heavenly bodies, fluids, springs, projectiles, electric charges, and many other systems. We use Newton's laws to calculate the motion of an object given the forces acting on it.

4–1 FORCES AND NEWTON'S FIRST LAW

Newton's first law of motion describes what happens to atoms, apples, asteroids, and any other objects—*moving or at rest*—when they are left alone. We might at first think that when a moving object is left alone, it will eventually come to rest. Indeed, the an-

Newton's first law

cient Greeks believed this; however, they were wrong. Galileo challenged this prejudice in the first half of the seventeenth century with experiments on the motion of objects on smooth planes. Isaac Newton built upon Galileo's work and, with great insight and power of abstraction, correctly and simply stated what happens:

FIGURE 4–1 The force **F** acts on the wagon. Other forces act on the wagon as well.

When an object is left alone, it maintains a constant velocity.

This law is *Newton's first law*, or the *law of inertia*. Notice that an object at rest is a special case of an object with constant velocity!

Constant velocity means that an object has both a constant speed and an unchanging direction of motion,

$$\mathbf{v} = \mathbf{v}_0 = \text{constant vector.} \tag{4–1}$$

The subscript on $\mathbf{v}_0$ indicates the value of the velocity $\mathbf{v}$ at time $t = 0$ or any other starting time. It is important to realize that Eq. (4–1) is a vector equation, which means that each vectorial *component* of the velocity is a constant. We know from Chapter 2 that this means that the acceleration $\mathbf{a}$ of the object is zero. Mathematically, we can write this as

$$\mathbf{a} = \frac{d\mathbf{v}}{dt} = 0. \tag{4–2}$$

Further, we know from Chapter 3 that constant velocity means that an object's position vector $\mathbf{r}$ changes linearly with time; that is, $d\mathbf{r}/dt$ = a constant. In other words, all three position coordinates change linearly with time:

$$\mathbf{r} = \mathbf{r}_0 + \mathbf{v}_0 t. \tag{4–3}$$

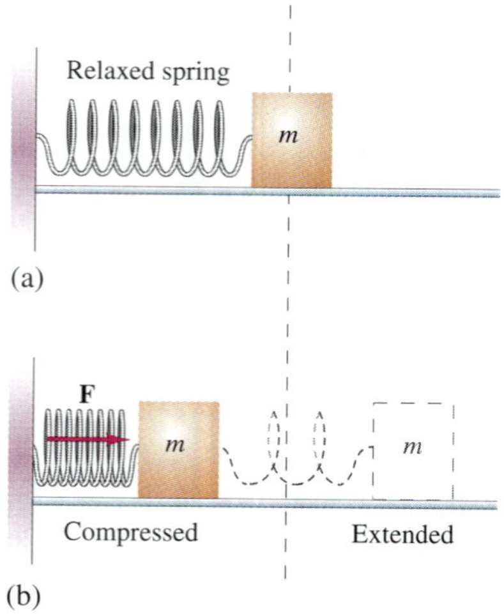

FIGURE 4–2 (a) A mass attached to the end of a relaxed spring is at rest. (b) The spring has been compressed to the left and now pushes the mass to the right.

Again, an object at rest is just a special case, with $\mathbf{v}_0 = 0$, of this general result.

According to Newton's approach, an object that is left alone—whether it is moving or at rest—has a very specific relation to the concept of **force**. Although we will investigate a precise definition of force in Section 4–2, for now we will use the intuitive notion that a force is something that acts to push or pull an object. Let's take a look at Fig. 4–1, in which an adult is pulling a child's wagon. Forces acting on the object will have an identifiable source—in this case, the adult. A good example of a force source is a compressed spring. In Fig. 4–2, we show a mass attached to the end of a relaxed spring. The end of the spring is then pushed to the left and released. It exerts a force on the mass and pushes it to the right.

A force has a magnitude and a direction and is thus described by a *vector*, denoted by **F**. The **net force** exerted on an object is the overall push or pull that comes from separate pushes or pulls; the net force is the vector sum of all the different forces that act on that object. We will use the notation $\mathbf{F}_{net}$ when it is important to distinguish the net force from a particular force. For example, a cart is being pulled by two ropes in Fig. 4–3. If we ignore gravity, there are two forces acting on the cart: $\mathbf{F}_1$ and $\mathbf{F}_2$. The net force exerted on the cart is the vector sum of $\mathbf{F}_1$ and $\mathbf{F}_2$; that is, $\mathbf{F}_{net} = \mathbf{F}_1 + \mathbf{F}_2$. Of course, more than two forces can act on an object; each push and pull on an object can be labeled as an individual force $\mathbf{F}_i$. These forces add vectorially to yield a net force, $\mathbf{F}_{net}$, acting on the object:

$$\mathbf{F}_{net} = \sum_i \mathbf{F}_i. \tag{4–4}$$

Newton's concept of an object being left alone is that there is no net force acting on it.

It is an experimental fact that forces add as vectors, and experiment is the ultimate judge of statements in science. Forces that are equal in magnitude but oppositely directed add vectorially to zero. In Fig. 4–3, if the magnitude of the forces exerted by the ropes were equal and if the forces were oppositely directed, there would be no net force on the cart. In fact, any number of forces can add vectorially to zero net force. In Newton's approach, an object is left alone when there is zero *net* force acting on it; thus, an object that is left alone means more than it may seem at first. Zero net force could mean either that there are *no* forces acting on an object at all or that there are

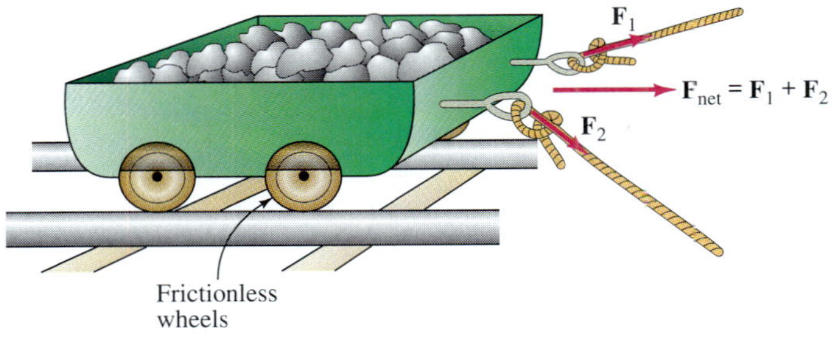

FIGURE 4–3 Two forces, $\mathbf{F}_1$ and $\mathbf{F}_2$, act on a mass, resulting in the net force, $\mathbf{F}_{net}$. We are ignoring gravity here.

several forces acting on the object that add vectorially to zero. If two ropes exerting equal and opposite forces on a cart were simultaneously cut, there would be no discernable effect on the cart.

A question may spring to mind: Is an object ever really left alone? Is there a "force-free" environment, in which there is *no net force* and where we can really perform experiments to test Newton's first law? In the special case where the constant velocity of an object is zero and the object is at rest, the answer is yes. The object remains at rest because there is zero net force acting on it. The answer, however, becomes less clear when the velocity of an object is not zero. In fact, our common experience appears to contradict Newton's first law; we continually observe moving objects gradually slow down and come to a halt. This is because a moving object is seldom left alone. Try as we might to eliminate the overall forces acting on an object, there are always small residual forces—called *frictional forces*—that act on an object to slow it down and ultimately bring it to rest (Fig. 4–4). For example, a brick sliding on a table comes to a stop quite rapidly. We can try to approximate a force-free environment and reduce the frictional forces for the brick by pouring a thin layer of oil on the table. The brick might travel farther but would nevertheless still come to a stop fairly rapidly; obviously, this is not a very good approximation to the ideal "force-free" environment. Sliding a car with smooth tires on an icy road brings us much closer to friction-free motion. As we approach closer and closer to the idealization of a force-free environment, experiment confirms Newton's first law, which we can rephrase as

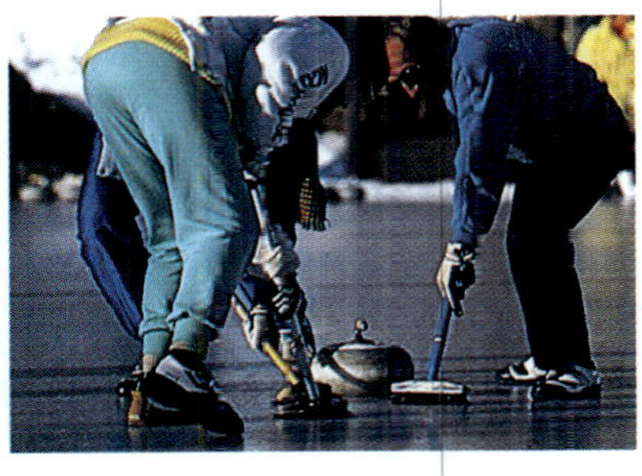

FIGURE 4–4 In the game of curling, the broom melts a layer of ice and reduces the friction for the sliding rock, but friction eventually brings the rock to rest.

A more mathematical form of Newton's first law

When the net force, $\mathbf{F}_{net}$, acting on an object is zero, that object moves with constant velocity.

There's Nothing Special about Being at Rest

The formulation of Newton's first law includes its converse:

A consequence of Newton's first law

An object moving with constant velocity has no net force acting on it.

This version of the first law has some interesting consequences. We have already indicated that "being at rest" is just a special case of moving with a constant velocity. No net force is acting in either situation; thus, the object retains its initial velocity. To approach this situation from another perspective, let's consider how observers in different reference frames see the motion of an object.

Suppose that a bus moves with uniform velocity. An observer A watches the bus moving with what he measures to be velocity $\mathbf{v}_{bus} = \mathbf{v}_1$. Let's also imagine an observer B who is walking parallel to the street at the same velocity as the bus. Observer A measures observer B to have velocity $\mathbf{v}_1$ as well. Figure 4–5a shows the situation from the point of view of observer A, whereas Fig. 4–5b depicts observer B's point of view. We can see from Fig. 4–5b that observer B measures the velocity of the bus relative to her to be $\mathbf{v}_{bus} = 0$. Both observers see uniform velocities, albeit different ones. At a given time, both observers report uniform velocities and therefore agree that there is no net force acting on the bus according to Newton's first law.

(a) According to Person *A* at rest

(b) According to Person *B* (at rest)

FIGURE 4–5 Observer A stands watching a bus moving down a street, while observer B moves at the same speed as the moving bus, but on the sidewalk. In (a) we see things from the point of view of observer A's frame; in (b) we see things from the point of view of observer B's (and C's) frame.

(a)

(b)

FIGURE 4–6 A parachutist falling to Earth is acted upon by gravity and by a drag force. In (a) these forces are not equal and opposite, whereas in (b) they are.

The reference frame of an observer for whom Newton's first law holds is an **inertial reference frame**. What we have realized is that there is no fundamental way to distinguish between different inertial reference frames. Observer A and observer B are both in inertial reference frames and they both agree that there is no net force on the bus—even if they disagree on the bus's velocity. So, what is the bus's true velocity: $\mathbf{v}_1$ (A's measurement) or 0 (B's measurement)? Which observer is truly at rest: observer A, who is at rest relative to the street, or observer B, who is at rest relative to the bus? The first law provides no clue as to who is at rest. *When observers A and B try to determine which one of them is at rest or exactly what his absolute velocity is by looking at the bus, they cannot do so.* Observer B might go so far as to say that the street is also moving; in fact, there is no real way to make the distinction. If you have ever had the experience of sitting in a very slowly moving train leaving a station on a track adjacent to another train, you will recall a disorientation as to whether your own train or the other train was moving.

In summary, when we say that there is no way to distinguish between inertial frames in the absence of forces, we are also saying that there is nothing sacred about the particular inertial frame in which an object is at rest. In other words, an object can be said to be at rest only in a particular inertial frame, not at rest in an absolute sense. Further, there is no way to say if any particular inertial frame is at rest in an absolute sense. We shall see in Section 4–2 that, *even in the presence of forces*, there is no fundamental way to distinguish between different inertial frames.

Net Forces

As we remarked earlier, the net force acting on an object may be zero either because there are no forces acting at all or because several forces are acting that add vectorially to zero. In fact, these two situations amount to the same thing in determining the object's motion. We observe all around us that constant motion is normally the result of the cancellation of many forces. When a parachutist first leaps from a plane, as in Fig. 4–6a, there is a net force accelerating him downward because the upward force of air drag, $\mathbf{F}_d$, does not cancel the downward force of gravity, $\mathbf{F}_g$. When the parachute is fully deployed, as in Fig. 4–6b, the parachutist falls at a constant velocity because $\mathbf{F}_d$ and $\mathbf{F}_g$ cancel. Similarly, an apple resting motionless in the palm of your hand remains there because the downward-directed gravitational force acting on it is exactly canceled by an upward contact force from your hand.

EXAMPLE 4–1 Three children each tug at the same sled (Fig. 4–7a). (All the forces are in the horizontal plane.) The three forces on the sled have the vectorial decomposition $\mathbf{F}_1 = -5\mathbf{k}$ units, $\mathbf{F}_2 = 5\mathbf{i}$ units, and $\mathbf{F}_3 = (-5\mathbf{i} + 5\mathbf{k})$ units. What is the net force on the sled? What can you say about its consequent motion? Ignore the force of gravity on the sled.

Solution: The net force on the sled is the vectorial sum of the individual forces (Fig. 4–7b). By adding the components, we see that the three forces add vectorially to zero:

$$\mathbf{F}_{\text{net}} = \mathbf{F}_1 + \mathbf{F}_2 + \mathbf{F}_3 = -5\mathbf{k} + 5\mathbf{i} + (-5\mathbf{i} + 5\mathbf{k})$$
$$= (-5 + 5)\mathbf{k} + (5 - 5)\mathbf{i} = 0.$$

The net force is zero; according to the first law, then, the sled's velocity is unchanging. If it is at rest when the three children apply their forces, it remains at rest.

Note: we have not specified the units of force in this example. When the net force is zero, the units are irrelevant. We shall return to the question of dimensions and units in the next section.

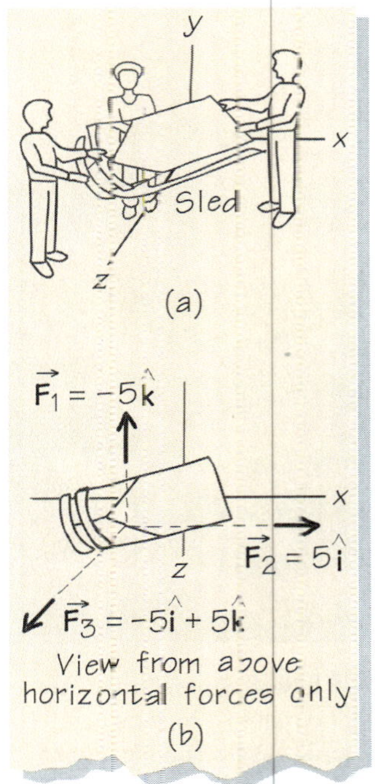

FIGURE 4–7 (a) Example 4–1. Three children pull on the sled in the *xz*-plane (horizontal).(b) Force diagram for forces on the sled. We have ignored the force of gravity.

Contact Forces and Forces at a Distance

In Example 4–1, the children must grab the sled in order to tug on it. Our qualitative notions about pushes and pulls arise to a large extent from *contact forces* that are associated with *physical* contact. Let's consider the wagon shown in Fig. 4–1. The child sitting in the wagon exerts a force on the wagon due to contact with it. The child's mother pulls the wagon by a handle that is in contact with both the mother's hand and the wagon. We may think of the wagon as being an isolated object with external forces acting on it due to the mother, the sitting child, its contact with the ground, and the pull of Earth. The first three of these forces are contact forces.

Physical contact, however, is not necessary for forces to act. Forces can also act from a distance. Gravity, the fourth force acting on the wagon in our example, is not a contact force; a softball thrown into the air is acted upon by gravity throughout its path of motion. One bar magnet can repel or attract another without actually touching it. Interestingly, a closer examination of forces reveals that *all* forces act at a distance. The distance involved in the contact forces we have discussed just happens to be of atomic size. The atoms of the mother's hand and those of the wagon handle interact—forces act between them—over imperceptibly small distances. Contact forces refer to forces that act over distances too short to be visible to the eye.

Now, we know that there are forces acting at a distance upon objects, so how can we be *sure* that there is no net force acting on an object moving with constant velocity? If we answer that there is no net external force acting on the object because the object's motion is constant, we present a circular argument. In fact, the first law makes sense only because we have performed many experiments with forces in different situations. These results depend heavily on Newton's second law, which describes *changes* in motion as a response to forces. Physicists know a good deal about the sources of force: We know that magnets and electric charges, for example, lead to forces that act at a distance, and we know how to detect contact or friction forces of various kinds. We know that there is a force acting on a mass when that mass is attached to a compressed spring. For the first law to be useful to us, we must have *independent* means of determining the forces.

PROBLEM SOLVING

We must identify the forces to apply Newton's laws.

4–2 NEWTON'S SECOND LAW OF MOTION

Newton's first law expresses the fact that when the net force acting on an object is zero, the object will maintain a constant velocity. Simply put, an object maintains a constant velocity when it is left alone. What happens to this object when it is not left alone; that is, when a net force acts on it? Newton's second law answers this question:

> **An object acted upon by a net force accelerates.**

Further, the object will accelerate, or change its velocity, in the same direction as the net force.

We mentioned in Section 4–1 that a car on a horizontal surface of slick ice approaches a friction-free environment. If a car were at rest under these conditions and you could push it with a given force—say by bracing your back against a tree and pushing with your feet—the car would move with a given acceleration. In fact, the acceleration of the car is proportional to the magnitude of the push you give it; if somehow you were to push the car with twice as much force, its acceleration would double.

FIGURE 4–8 An enormous force would be required to provide substantial acceleration and change this oil tanker's direction or bring it to a stop quickly. Because its engines (or those of tugs) can exert relatively limited forces, the tanker's acceleration will be small. It is necessary to think far ahead when operating such tankers.

As our own experience tells us, the specific response of an object to a given net force also depends on the mass of the object. Mass is related to the property known as *inertia*, which describes an object's resistance to a change in its motion; an object's resistance to change can be either its resistance to slowing down if the object is already moving or its resistance to starting to move if it is at rest. For this reason, the mass that appears in Newton's second law is often called the **inertial mass**. If the mass of the car in our example were doubled and you were to push it with the same force you used the first time, the car would move with half of the initial acceleration. The greater the object's mass, the less the acceleration (Fig. 4–8); the less the mass, the greater the acceleration (Fig. 4–9). In fact, for a given force, experiment describes an *inverse* relation between mass and acceleration.

The basic relation we have just described between the net force on an object, $\mathbf{F}_{\text{net}}$, the mass of the object, m, and the object's acceleration, $\mathbf{a}$, was discovered by Newton and has come to be known as **Newton's second law**. It is stated mathematically as[†]

$$\mathbf{F}_{\text{net}} = m\mathbf{a}. \tag{4–5}$$

FIGURE 4–9 The fisherman causes the fish to accelerate out of the water with the help of the tension in the fishing line.

Earlier, we discussed how acceleration is the rate of change of velocity, $\mathbf{a} = \text{d}\mathbf{v}/\text{d}t$. In terms of the velocity, then, Newton's second law becomes

$$\mathbf{F}_{\text{net}} = m\frac{\text{d}\mathbf{v}}{\text{d}t}. \tag{4–6}$$

We must emphasize that the force appearing here is the net force; that is, the vector sum of all the forces acting on the object. As with any vectorial expression, Newton's second law is equivalent to three scalar equations for the components:

$$F_{x,\text{net}} = ma_x = m\frac{\text{d}v_x}{\text{d}t}; \tag{4–7a}$$

$$F_{y,\text{net}} = ma_y = m\frac{\text{d}v_y}{\text{d}t}; \tag{4–7b}$$

$$F_{z,\text{net}} = ma_z = m\frac{\text{d}v_z}{\text{d}t}. \tag{4–7c}$$

[†]When the object has a mass that changes with time (as in rocket motion), we shall modify the second law.

We shall refer interchangeably to Newton's second law in its vector form, Eq. (4–5) or (4–6), or in its equivalent component form, Eqs. (4–7).

More About Mass

Under the influence of a given force, it seems intuitively reasonable that an object with a small mass accelerates more briskly than does an object with a large mass. Mass is an *intrinsic* property of an object; a given object has a given mass—regardless of the forces acting on it. Let's look at an example in which mass enters into Newton's second law: Imagine that you are again braced against a tree while you push a sleigh that is sitting on a horizontal icy surface (Fig. 4–10). This surface is so smooth that we can consider it to be friction-free. You push the sleigh three times, each time with exactly the same effort; once when the sleigh is empty (Fig. 4–10a), again when it contains your little sister (Fig. 4–10b), and a final time when it contains your entire family (Fig. 4–10c). With each push, the sleigh will undergo successively smaller accelerations. The empty sleigh will show the largest acceleration while the fully loaded sleigh will have the smallest. In each case, the sleigh responds quite differently to your efforts even though the force you exert on it is the same. Now let's consider the mathematical expression of this property. Let the mass of the empty sleigh be m_1 and the mass of the fully loaded sleigh be m_2. When they are subject to the same net force, the respective accelerations a_1 and a_2 of the empty and fully loaded sleigh are related to their masses by

$$\text{for a given force:} \quad \frac{m_1}{m_2} = \frac{a_2}{a_1}. \tag{4–8}$$

FIGURE 4–10 By bracing yourself against a tree, you can exert a fixed force against a sleigh. You do so three times, as follows: (a) the sleigh is empty; (b) the sleigh contains your little sister; (c) the sleigh contains your whole family. Even though you push the same way each time, the results are quite different. The empty sleigh leaves with a much greater speed than does the full sleigh.

Mass is a scalar quantity; it has no direction associated with it. It is always positive. As a consequence, the acceleration of an object is in the same direction as the force exerted on it. Furthermore, masses are additive; that is, if an object contains two parts with masses m_1 and m_2, the total mass of that object is[†]

$$m = m_1 + m_2. \tag{4–9}$$

Is the Second Law Merely a Definition of Force?

In Chapter 5 and beyond, we will explore many applications of Newton's second law. At this point, however, we must consider what the second law really means. Does Eq. (4–5) do more than simply define the force? We can measure an object's acceleration with a meter stick and a watch. Do we use that measurement merely to define the quantity **F**/*m*? The power of the second law lies in the fact that it goes beyond this mere definition. In this subsection, we shall explain how forces and masses are defined and how Newton's second law makes predictions.

Suppose that we have a large supply of lumps of gold. We will label one of them A and call it our standard mass, $m_A = 1$ kg. We also have available a spring and a very smooth (frictionless) table. We compress the spring, place lump A against it, release the spring, and measure the acceleration of the lump as it is pushed away by the spring (Fig. 4–11). Suppose that when the spring is compressed by 1 cm, lump A has an acceleration of $a_A = 2\ \text{m/s}^2$, as in Fig. 4–11b. At this point, *Newton's second law provides a precise definition of the force exerted by this spring* when it is compressed by 1 cm. Because $F = m_A a_A$, the force is $F = (1\ \text{kg})(2\ \text{m/s}^2) = 2\ \text{kg}\cdot\text{m/s}^2$. Every time the spring is compressed by 1 cm, it will exert this force.

We can now calculate the masses of our other lumps of gold. Place a second lump, B, in front of the spring and compress the spring by the same 1 cm (Fig. 4–11c). When the spring is released, the acceleration of lump B is measured to be $a_B = 3\ \text{m/s}^2$. By

[†]When the forces holding the system together are strong, special relativity implies a modification. In most of the situations we meet, the effect is very small.

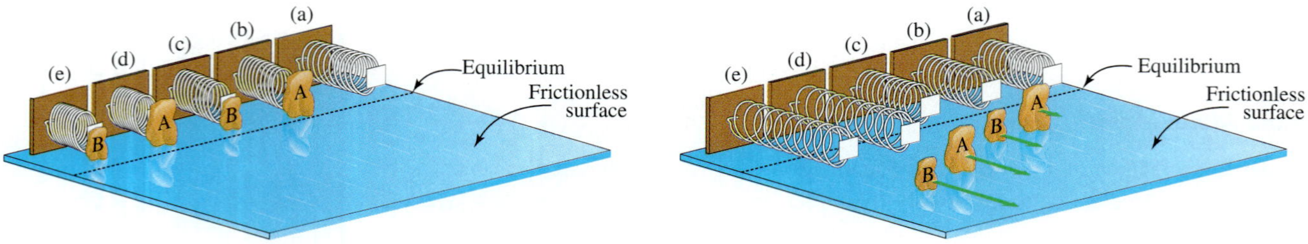

FIGURE 4–11 (a)–(c) A series of experiments with a spring and a set of masses allows us to construct a mass scale and to learn about the forces exerted by the spring. (d)–(e) Once we have established masses and learned the force law, we can predict the motion of a mass.

A force law describes the force due to one source and how that force can vary.

Newton's second law, the mass m_B is

$$m_B = \frac{F}{a_B} = m_A \frac{a_A}{a_B} = m_A \frac{2 \text{ m/s}^2}{3 \text{ m/s}^2} = \frac{2}{3} m_A = \frac{2}{3} \text{ kg}.$$

In this way, we can determine the masses of each of our lumps of gold—or of any object for that matter.

The force of the spring is always the same whenever it is compressed by 1 cm. Let's now compress the spring by a different amount, place lump A in front of it, and measure the acceleration of lump A when the spring is released (Fig. 4–11d). This time, we measure $a'_A = 5 \text{ m/s}^2$. We can again use Newton's second law to define the force that the spring exerts when it is compressed by the new amount; it is $F' = m_A a'_A = (1 \text{ kg})(5 \text{ m/s}^2) = 5 \text{ kg} \cdot \text{m/s}^2$. We can continue this process to define the force exerted by the spring for each different amount of compression. We have found a *force law*, which in this case is just a catalog of how much force the particular spring exerts at any given compression. A force law describes the force due to a particular source and how that force depends on variables such as position or time. Throughout this book, we shall study the force laws describing friction, gravity, springs, and electric charges, among others.

At this point, Newton's second law enables us to make predictions for the first time. If we compress the spring to the position shown in Fig. 4–11d and place lump B against it, we can predict lump B's acceleration. We have already determined that the force exerted by the spring in this second position has magnitude $F' = 5 \text{ kg} \cdot \text{m/s}^2$. When the spring is released, as in Fig. 4–11e, the second law *predicts* the acceleration to be

$$a'_B = \frac{F'}{m_B} = \frac{5 \text{ kg} \cdot \text{m/s}^2}{2/3 \text{ kg}} = 7.5 \text{ m/s}^2.$$

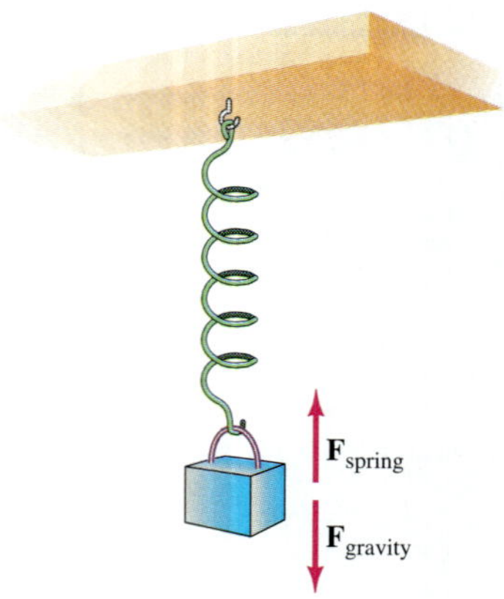

FIGURE 4–12 By using Newton's first law and a known force, such as that exerted by the stretched spring, we can measure other forces—in this case, the force of gravity acting on the (stationary) block.

Let's suppose that we know just how much force is exerted by the spring for a given compression or stretch. We can then use the spring to learn about *other* forces. Instead of studying the motions these other forces cause, we can just balance them against the spring and use the first law. For example, if we turn our spring to the vertical and suspend a block of known mass from it so that the block is motionless (Fig. 4–12), the force of gravity acting on the block must exactly balance the upward force of the stretched spring. By observing how much the spring is stretched, we have measured the force of gravity on the block. We could now predict how the block would accelerate if there were no spring and only gravity acted on it. In this book, we are most interested in using the second law for its ability to predict motion.

EXAMPLE 4–2 You must deliver a box of bowling balls to a bowling alley. The balls will be placed in a box that is initially at rest (its mass is very small compared to even one bowling ball). You start with one ball in the box, exert a given force upon the box, and, at the end of a time period Δt, the box moves at a speed of 3.2 m/s. You then add more bowling balls to the box; you exert the same amount of force on the box for the same period of time (Δt) and find the box to have a final speed of 0.40 m/s. How many balls are in the box now?

Solution: In the two cases described (which we will call 1 and 2), the box containing the bowling ball(s) is subject to the same force. Moreover, we can ignore the mass of the box. The mass of the box with one ball in it is then approximately $m_1 = m$, and its mass with the unknown number of balls is $m_2 = nm$, where m is the mass of one bowling ball and n is the number of balls, which we are trying to find. Using Newton's second law, we can find the accelerations a_1 and a_2 during the period Δt when the force (of constant magnitude F) operates. These accelerations are

$$a_1 = \frac{F}{m_1} = \frac{F}{m} \quad \text{and} \quad a_2 = \frac{F}{m_2} = \frac{F}{nm}.$$

Although we do not know the numerical values of the two accelerations, we do know the speeds v_1 and v_2 after a fixed period of acceleration. Further, we learned in Chapter 2 that an object that starts at rest and undergoes a fixed acceleration $\mathbf{a}$ for a given period of time Δt has the velocity $\Delta\mathbf{v} = \mathbf{v} = \mathbf{a}\,\Delta t$. In our one-dimensional case, then, we have

$$v_1 = a_1\,\Delta t = \frac{F\,\Delta t}{m} \quad \text{and} \quad v_2 = a_2\,\Delta t = \frac{F\,\Delta t}{nm}.$$

We now have enough information to solve for the unknown, n. We can solve for the ratio F/m in terms of v_1 and Δt and substitute it into the equation for v_2, which we could solve for n. Alternatively, we can simply take the ratio of the two speeds:

$$\frac{v_1}{v_2} = \frac{\cancel{F}\,\cancel{\Delta t}/\cancel{m}}{\cancel{F}\,\cancel{\Delta t}/n\cancel{m}} = n.$$

Numerical substitution gives $n = (3.2\text{ m/s})/(0.40\text{ m/s}) = 8$ bowling balls.

Units of force

Units. We first discussed the dimensions and units of mass, length, and time in Chapter 1. Because acceleration has dimensions of $[LT^{-2}]$ and units of m/s^2 in SI, force has dimensions of $[M \cdot LT^{-2}]$ and, in SI, units of kg · m/s^2, or **newtons** (N):

$$1\text{ N} \equiv 1\text{ kg}\cdot\text{m/s}^2. \qquad (4\text{–}10)$$

In other words, a 1-N force exerted upon an object with a mass of 1 kg will produce an acceleration of 1 m/s^2.

In the cgs system, the force 1 g · cm/s^2 is called the *dyne*:

$$1\text{ dyne} \equiv 1\text{ g}\cdot\text{cm/s}^2 = 10^{-5}\text{ N}. \qquad (4\text{–}11)$$

Here again, a force of 1 dyne acting on a mass of 1 g causes an acceleration of 1 cm/s^2.

Another force unit in everyday use is the pound (lb); 1 lb = 4.448 N. The pound is used in the British engineering system, in which mass is measured in slugs.

How Do Different Observers See Newton's Second Law?

Let's revisit the two streetside observers A and B (Fig. 4–5). They are in different inertial frames, moving with fixed velocity with respect to one another. To be more general, let's label this fixed velocity $\mathbf{u}$. These observers measure a moving object to have different velocities, $\mathbf{v}_A$ and $\mathbf{v}_B$. But what if there is a net force on the object? Do each of the observers see Newton's second law in the same way? We have already seen in Eq. (3–57) how the two observers see the velocity of a bus:

$$\mathbf{v}_B = \mathbf{v}_A - \mathbf{u}$$

Let's now see how observers A and B measure the *rate of change* in the bus's velocity. The rate of change is a derivative with respect to time, so we take the derivative with respect to time of both sides of the equation:

$$\frac{d\mathbf{v}_B}{dt} = \frac{d\mathbf{v}_A}{dt} - \frac{d\mathbf{u}}{dt}. \qquad (4\text{–}12)$$

APPLICATION: SAFE EJECTION SEATS

The ability to construct devices that provide constant acceleration has an important application in the design of the aircraft emergency ejection seat. Early ejection seats consisted of an explosive charge placed under the seat assembly. In an emergency, the crew member would send an electric current to the detonator, which fired the charge in milliseconds. Because of the short duration of the acceleration, it had to be very large, in order to provide the crew member with enough height to clear the tail of the airplane. Collapsed spinal disks, crushed vertebrae, and heart tears were the not uncommon by-products of this method of ejection.

In modern ejection assemblies, the explosive charge is replaced by a small rocket engine, which provides a smaller acceleration over longer periods (Fig. 4AB–1). In this way, the crew member can gain the needed height without the danger of the high accelerations of the explosive force. The crew member not only clears the tail assembly, but can gain enough height to deploy a parachute before the rocket has burned out. This means that the seat can even be used at ground level in the event something goes wrong on take-off or landing.

FIGURE 4AB–1 Test ejection of a pilot from F-14 Tomcat cockpit. Note the jet plumes below the seat and the release of the parachute.

The observers are in inertial frames, and thus their relative velocity $\mathbf{u}$ is constant and its derivative is zero. Therefore

$$\frac{d\mathbf{v}_B}{dt} = \frac{d\mathbf{v}_A}{dt}. \qquad (4\text{–}13)$$

Because the mass m of the object under observation—the bus—does not depend on the velocity of the observer,[†] we multiply both sides of Eq. (4–13) by m to find that

$$m\frac{d\mathbf{v}_B}{dt} = m\frac{d\mathbf{v}_A}{dt},$$

or, according to the second law,

$$\mathbf{F}_B = \mathbf{F}_A. \qquad (4\text{–}14)$$

$\mathbf{F}_B$ is the force on an accelerating object, such as bus, as measured by observer B, and $\mathbf{F}_A$ is the force measured by observer A.

Observers in different inertial frames agree on the net force acting on an object.

A different way of stating this result is that *two observers in different inertial frames cannot by experiment tell which of them is moving and which of them is at rest.* Remarkably, this remains true in situations where speeds are comparable to the speed of light and Newton's second law in the form we have stated is no longer useful (see Chapter 40).

INTERIM SUMMARY

We have learned how to measure masses and forces in terms of a set of standard masses and forces. Force has units of newtons in SI. Newton's second law states that $\mathbf{F}_{net} = m\mathbf{a}$. This law predicts the acceleration of an object of known mass if the net force acting on it is known. We can view Newton's first law as a special case of the second law:

[†] Actually, experiment shows that at speeds close to the speed of light these remarks must be reexamined.

When the net force acting on an object is zero, the object's acceleration is zero, and it maintains a constant velocity.

The second law is used initially to calibrate mass and to define a single force law, such as the force exerted by a particular spring at different compressions or stretches. After that, the first law can be used to define all other forces. Once these forces are known, the second law allows us to predict the motion of any object of known mass. Neither the first nor the second law provides the means to tell observers in different inertial frames which one is moving and which is standing still. (See Fig. 4–13 for an example of these two laws.)

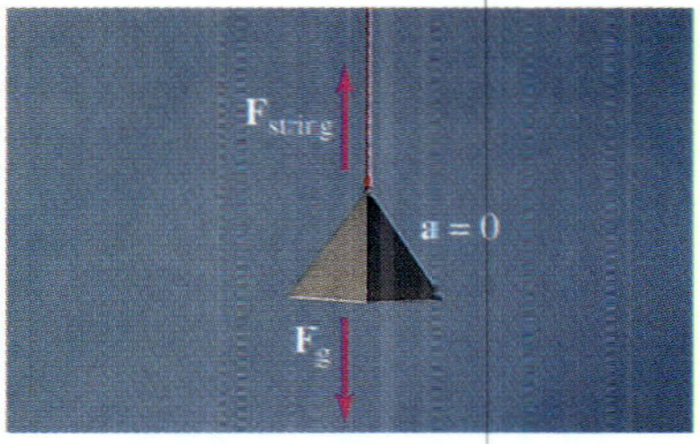

(a)

(b)

FIGURE 4–13 (a) Two equal and opposite forces act on this object, the tension due to the string and gravity. The net force is zero, and the object maintains its velocity, which has the value zero in this case. (b) When the string is cut, only gravity acts, and the object accelerates toward Earth according to Newton's second law.

4–3 NEWTON'S THIRD LAW OF MOTION

The first and second laws describe forces acting *on* objects, and such forces have a source. For example, the force acting on the lumps of gold in Section 4–2 was due to the compression of a spring, and the Sun is the source of the gravitational force that causes Earth to circle the Sun. *When a force due to object B acts on object A, then a force due to object A also acts on object B* (Fig. 4–14). When you push on a wall, the wall pushes back on you. It is less obvious that when Earth tugs on an apple, causing it to fall (the force is gravity), the apple also tugs on Earth, causing Earth to accelerate toward the apple. Earth exerts a force on the Moon, making it orbit us overhead, while the Moon exerts a gravitational pull on Earth. In fact, ocean tides are associated closely with the force the Moon exerts on Earth. Earth and the Moon exert gravitational forces *on each other.*

Newton's third law

Newton quantified this phenomenon in his third law. According to Newton's third law, the force on Earth due to the apple is equal in magnitude but opposite in direction to the force on the apple due to Earth. Forces do not simply act *on* objects; rather, forces act *between* two objects or between an object and its surroundings. Objects are said to *interact* when forces act between them. This effect does not depend on the observer's inertial frame of reference.

A mathematical statement of the third law is the following: If the force on object A due to object B is $\mathbf{F}_{AB}$, then there exists an equal and opposite force $\mathbf{F}_{BA}$ that acts on object B due to object A. The mathematical relation is[†]

$$\mathbf{F}_{BA} = -\mathbf{F}_{AB}. \tag{4–15}$$

PROBLEM SOLVING

Equal action and reaction does not mean equal motion.

The third law is sometimes called *the law of equal action and reaction*. Do not let this particular phrasing mislead you into believing that the *accelerations* of the two objects are the same. Each object accelerates according to the second law, which means that the response depends on the mass of the object.

The third law is illustrated in Examples 4–3 and 4–4. In these examples, we imagine an outer space environment where we can isolate astronauts and satellites from all forces except the forces they exert on one another. Such a situation is more difficult to arrange on Earth.

FIGURE 4–14 A demonstration of Newton's third law. Action causes reaction; in this case, the release of CO_2 from a fire extinguisher causes the initially stationary cart to be propelled in a direction opposite that of the released gas.

EXAMPLE 4–3 An astronaut and a satellite are in an environment where they can be considered to form an isolated system with no external forces acting on that system. The astronaut tugs on the satellite with a force of 10.0 N to the right (and down) (Fig. 4–15). What is the force on the astronaut?

Solution: We label the two objects involved as the astronaut A and the satellite S. The given force on the satellite due to the astronaut is $\mathbf{F}_{SA}$ = 10.0 N to the right (and down). According to the third law, the force on the astronaut due to the satellite, $\mathbf{F}_{AS}$, is

$$\mathbf{F}_{AS} = -\mathbf{F}_{SA} = 10.0 \text{ N to the left.}$$

[†]We shall see that Newton's third law must be reformulated when we study electromagnetic forces.

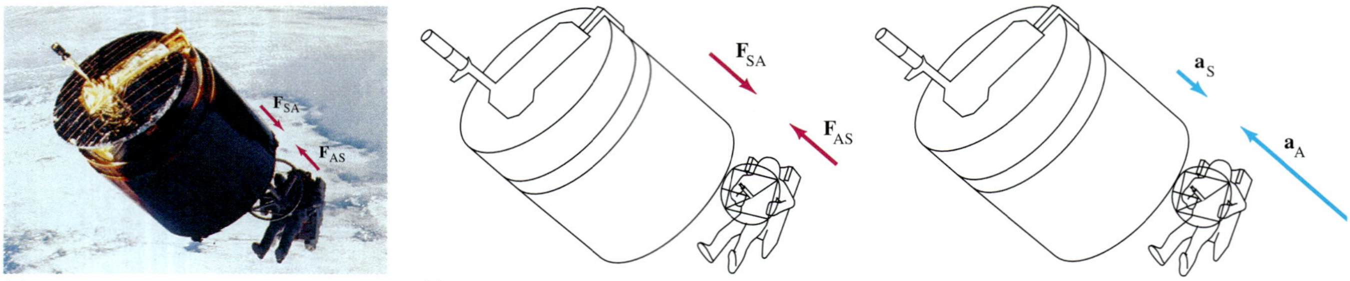

(a) (b) (c)

FIGURE 4–15 (a) Example 4–3. (b) According to Newton's third law, the force the astronaut exerts on the satellite is equal and opposite to the force the satellite exerts on the astronaut. (c) Even though the force on the astronaut has the same magnitude as the force on the satellite, the accelerations of astronaut and satellite are quite different; they have unequal masses.

The force on the astronaut due to the satellite has a magnitude of 10.0 N, which is the same magnitude as the force exerted on the satellite due to the astronaut; however, the force on the astronaut is directed to the left (and up), whereas the force on the satellite is directed to the right (and down) (Fig. 4–15b).

EXAMPLE 4–4 Let's now assume that the mass of the astronaut in Example 4–3 is 75.5 kg and that of the satellite is 755 kg. What is the acceleration of each?

Solution: The force, and thus the acceleration, is along the line between the astronaut and satellite and therefore we can drop the vector notation. We use the second law to determine the acceleration of each object. According to this law, the satellite has an acceleration of magnitude

$$a_S = \frac{F_{SA}}{m_S} = \frac{10.0\ \text{N}}{755\ \text{kg}} = 0.0132\ \text{m/s}^2.$$

This acceleration of the satellite is directed to the right (and down). For the astronaut, the acceleration has magnitude

$$a_A = \frac{F_{AS}}{m_A} = \frac{F_{SA}}{m_A} = \frac{10.0\ \text{N}}{75.5\ \text{kg}} = 0.132\ \text{m/s}^2.$$

This acceleration is directed to the left (and up). Thus, as Fig. 4–15c shows, the astronaut experiences an acceleration whose magnitude is 10 times larger than that of the satellite. This difference is the result of the difference in the masses of the satellite and the astronaut.

These examples demonstrate why the force that the apple exerts on Earth is not directly observable. With a small apple of mass $m_a = 0.1$ kg, the force on the apple due to Earth ($\mathbf{F}_{aE}$) is approximately 1 N, directed downward. This leads to an acceleration of magnitude $F_{aE}/m_a = 10\ \text{m/s}^2$ when the apple falls from the tree. According to Newton's third law, the upward force $\mathbf{F}_{Ea}$ that the apple exerts on Earth also has magnitude 1 N. Because Earth's mass, m_E, is approximately 6×10^{24} kg, its upward acceleration has magnitude $F_{Ea}/m_E \simeq 1.7 \times 10^{-25}\ \text{m/s}^2$, much too small to be observable. The evidence that led Newton to the third law involved the Earth–Moon system, where the law's effects are much more significant.

*4–4 THE EFFECTS OF NONINERTIAL FRAMES

Let's return to our streetside observers one more time and reconsider observers A and B moving with respect to one another. We found that if they are in inertial frames—so that their relative velocity **u** is constant—they agree on the forces acting on *any object* they observe.

FIGURE 4–16 The woman is at rest with respect to the stop sign. Because she does not see the sign accelerating, she would say that there is no net force acting on it. The boy on the bicycle, however, is in an accelerating frame. As seen from his frame of reference, the stop sign is accelerating, and he may think that forces of unknown origin act on the stop sign.

Suppose, however, that **u** is not a constant velocity. One or both of the observers are in a **noninertial frame of reference**. According to Eq. (4–12), observer B will measure an object to have a different acceleration from the acceleration observer A measures the object to have, and the two observers will disagree on the forces acting. To take a specific example, consider a stop sign at rest relative to a woman standing nearby; however, the sign is not at rest relative to a boy on a bicycle, as shown in Fig. 4–16. Assume the boy could somehow block out the surroundings and concentrate solely on the stop sign. As he goes around the corner, the boy would observe the sign to be moving—and not in a straight line. A change in either the observed direction of motion of the stop sign or in its speed (relative to the boy) implies acceleration. The boy could say, "I am accelerating, so the stop sign I see only *seems* to be accelerating, and there is actually no net force on it." Or he could say, "I am at rest, but some unusual force acts on the stop sign to make it move and accelerate in the odd way that I observe. I cannot identify the source of this odd force on the stop sign."

The observant child's second interpretation is an interesting one. His failure to find and identify the source of the "force"† is an indication that the acceleration is associated with an observation that was made from a *noninertial frame of reference*. In effect, Newton's second law does not apply to motions observed from noninertial frames. A certain amount of experience is necessary to determine whether all possible force sources acting on an object have been identified, and whether an observer is in an inertial or noninertial frame. We shall normally deal with real forces in inertial frames and give ample warning when accelerating (noninertial) frames are involved.

EXAMPLE 4–5 Professor A is at rest on a train platform; her friend, Professor B, is leaving the station in a train with acceleration α in the $+x$-direction (Fig. 4–17). Professor A considers herself to be at rest and states that there is no net force acting on her. What does Professor B observe Professor A's motion to be, and how might he interpret that motion?

Solution: The motion in this example is in one dimension and so we can drop our boldfaced notation for vectors. We then say that Professor B is accelerating in the positive direction, which is to the right. Professor B wants to measure the mo-

†"Forces" associated with noninertial frames—forces with no identifiable sources—are referred to as *fictitious forces, pseudo-forces, or noninertial forces.*

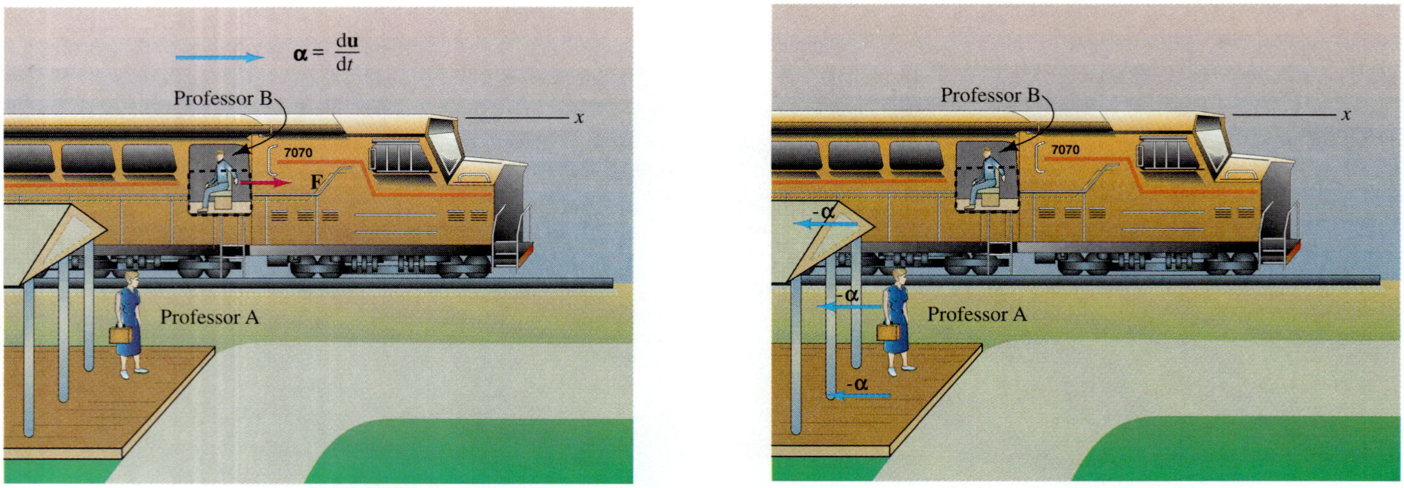

(a) According to Professor A

(b) According to Professor B

FIGURE 4–17 (a) Example 4–5. (b) Because Professor B is in a noninertial frame, Newton's second law does not apply to his measurements of Professor A's motion, and he sees Professor A accelerating.

tion of Professor A. The variables are

$$v_A = \text{velocity of Professor A as measured by A;}$$

$$v_B = \text{velocity of Professor A as measured by B;}$$

$$u = \text{velocity of Professor B as measured by A.}$$

We first applied Eq. (4–12) in the case of the bus's velocity, and it continues to relate the rates of change in the various velocities in this example:

$$\frac{dv_B}{dt} = \frac{dv_A}{dt} - \frac{du}{dt}.$$

Because Professor A measures herself to be standing still, $dv_A/dt = 0$. In addition, the acceleration of the train as seen by Professor A is given by $du/dt = \alpha$. Thus

$$\frac{dv_B}{dt} = 0 - \alpha = -\alpha.$$

Professor B sees Professor A accelerating *backward* (minus sign) with an acceleration of magnitude α. He would then say that, according to the second law, there must be a force on Professor A of $-m\alpha$ that is responsible for giving her this motion. He would not, however, be able to find an identifiable agent for this force and might in this way decide that he is in a noninertial frame. Indeed, Professor B sees the entire train platform accelerating backward!

4–5 USING NEWTON'S LAWS I: FORCE DIAGRAMS

Ordinarily, we take two different steps to apply Newton's second law. In this section, we will discuss the first step, in which we identify all the forces that act on an object and we express the second law explicitly. In the second step, the topic of Section 4–6, we find the subsequent motion of the object.

PROBLEM SOLVING

How to prepare a force diagram

Newton's second law, $\mathbf{F}_{net} = m\mathbf{a}$, relates the mass of an object, the forces acting on it, and its acceleration. We must be able to distinguish these quantities in order to use the second law. Therefore, *we must know exactly what object we are talking about*. This step may be less obvious than it seems. For example, if we want to analyze the motion of a wagon being pulled by a child, we must consider *only* the forces acting *on the wagon*—not those forces that act on the child. Further, if we want to find the motion of the wagon, it is only the forces on the wagon and the mass of the wagon itself that enter Newton's second law. We want to *isolate* the wagon.

The wagon is most easily isolated in a sketch that we shall refer to as a **force diagram** for the wagon. The force diagram indicates each individual force that acts on the wagon and indicates the direction of these forces. These forces will appear as arrows along the directions of the force vectors (Fig. 4–18). The force diagram should also indicate a set of coordinate axes so that we can use this diagram to help us separate the vectors into their vector components. In order to avoid the effects of noninertial frames, place the axes in an inertial frame—usually attached to some fixed, stationary point—rather than attached to an accelerating point. In Fig. 4–18, we have placed the origin of the coordinate system at a spot on the ground adjacent to the wagon. It is often convenient to use a set of Cartesian axes: (x, y) for planar figures or (x, y, z) in space. Newton's second law breaks down into separate equations for the vector components along these mutually perpendicular axes.

Any orientation of the axes is acceptable, but certain choices will be easier to use. For example, it is often convenient to orient the y-axis vertically in the study of falling objects so that the force of gravity has only a y-component.

Sometimes it is convenient to draw the resulting acceleration of the object in question on the force diagram, but *it is most important that we do not confuse the acceleration with a force*. The acceleration is the object's *response* to the net force acting on it; *it is not a force itself*. In figures, we single out the acceleration vector in bright blue when we include it in the force diagram, whereas force vectors are always drawn in bright pink.

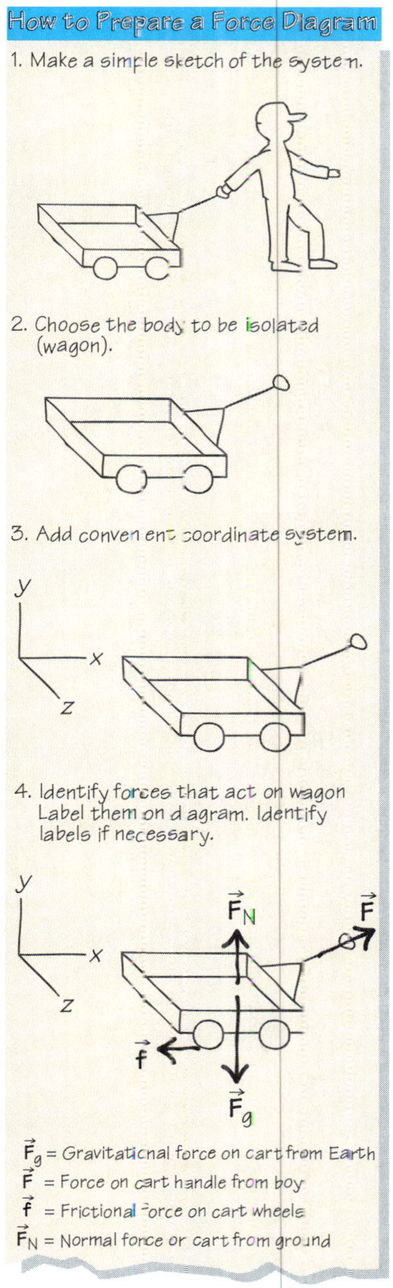

FIGURE 4–18 How to draw a force diagram for a wagon being pulled by a child.

EXAMPLE 4–6 Block 1 is glued to the top of block 2 (Fig. 4–19a). The masses of the blocks are m_1 and m_2, respectively. A rope is attached to block 2, pulling it horizontally to the right with a force of constant magnitude T along a perfectly smooth horizontal surface. What equation describes the motion of block 1? Solve this problem in two ways: (a) Consider the system of the two blocks glued together; and (b) consider block 1 isolated.

Solution: Before we find the equation describing the motion, let's note that, whatever forces act in the vertical direction (gravity, for example), *they must cancel out entirely* because there is no motion of either block 1 or the whole system in the vertical direction. Thus we ignore all vertical forces. (a) We can consider the system of two blocks glued together to find the motion of block 1 alone, because the motion of block 1 will also be the motion of the two blocks together. Figure 4–19b shows the system isolated, together with the net force that acts on it. That net force is the force $\mathbf{T}$ due to the rope because any vertical forces cancel. The mass of the system is $m_1 + m_2$. Thus Newton's second law is, for the system,

$$(m_1 + m_2)\mathbf{a}_{sys} = \mathbf{T}.$$

This equation allows us to find the acceleration of the system; hence, because $\mathbf{a}_1 = \mathbf{a}_{sys}$, it also gives the acceleration of block 1. (b) Figure 4–19c is a force diagram for block 1 alone. This time, the net (horizontal) force on block 1 is exclusively a contact force due to block 2. This contact force is written as $\mathbf{F}_{12}$, where the subscript specifies that we have a force *on* block 1 *due to* block 2. Thus we have

$$m_1\mathbf{a}_1 = \mathbf{F}_{12}.$$

This certainly does not appear to be equivalent to our result in part (a) yet. To go further, let's look at the force diagram for block 2 (Fig. 4–19d). From that diagram, we find

$$m_2\mathbf{a}_2 = \mathbf{T} + \mathbf{F}_{21}.$$

Here, $\mathbf{F}_{21}$ is the force on block 2 due to block 1. According to Newton's third law, however, $\mathbf{F}_{21} = -\mathbf{F}_{12}$; moreover, the expression for Newton's second law applied to block 2 is

$$m_2\mathbf{a}_2 = \mathbf{T} - \mathbf{F}_{12}.$$

We then solve for the unknown force $\mathbf{F}_{12}$, with result $\mathbf{F}_{12} = \mathbf{T} - m_2\mathbf{a}_2$. We substitute this back into our equation for block 1:

$$m_1\mathbf{a}_1 = \mathbf{F}_{12} = \mathbf{T} - m_2\mathbf{a}_2.$$

Finally, the blocks move together and thus $\mathbf{a}_2 = \mathbf{a}_1 = \mathbf{a}_{sys}$, so that

$$m_1\mathbf{a}_1 = \mathbf{T} - m_2\mathbf{a}_1,$$

or $(m_1 + m_2)\mathbf{a}_{sys} = \mathbf{T}$. We can now recognize the equation we found in part (a) and discover that the two methods give the same answer.

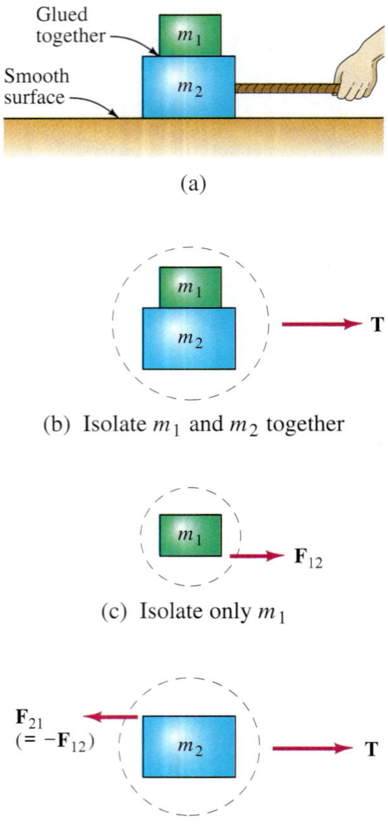

(a)

(b) Isolate m_1 and m_2 together

(c) Isolate only m_1

(d) Isolate only m_2

FIGURE 4–19 (a) Example 4–6. Two blocks that are glued together are pulled by a rope. Force diagrams for (b) the two-block system, (c) block 1 alone, and (d) block 2 alone.

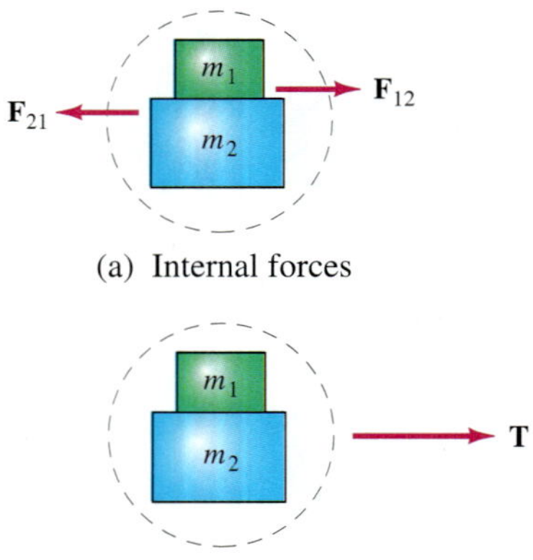

(a) Internal forces

(b) Net external force

FIGURE 4–20 (a) The internal forces within the system of two blocks studied in Example 4–6. (b) The net external force on that same system. Some vertical forces act (gravity and contact forces), but these do not contribute to the net external forces because they cancel.

External and Internal Forces

Example 4–6 illustrates the important difference between *external* and *internal* forces. Internal forces are those that act within a system we are isolating. These forces may act to hold the system rigidly together or they may simply act between different parts of the system. If we consider the two blocks in Example 4–6 as a single, isolated system, the internal forces would be associated with the glue holding the two masses together. We can examine these internal forces, $\mathbf{F}_{12}$ and $\mathbf{F}_{21}$ in Fig. 4–20a.

External forces, in contrast, are forces that act on the isolated system from outside. In Example 4–6, the net force $\mathbf{T}$ is an external force acting on the two-block system (Fig. 4–20b). Once a system has been isolated, only the external forces acting on it influence its overall motion. As we shall see in more detail in Chapter 8, internal forces do not enter into the second law because by Newton's third law they cancel in pairs. For example, we saw that the internal forces $\mathbf{F}_{12}$ and $\mathbf{F}_{21}$ of the two-block system canceled each other and did not enter into the solution of the problem. *In the statement of the second law, the net force is actually the net external force.* All of the objects we deal with are, in fact, complicated systems composed of many atoms held together by internal forces. But even though these internal forces act within the object, they cannot influence its overall acceleration.

If a given system is broken up into separate pieces, the internal forces for the original system may become external forces for the pieces. The motion of the smaller pieces is then governed by these external forces. In Example 4–6, the isolation of block 1 compelled us to treat $\mathbf{F}_{12}$ as an external force on block 1. In that same example, we made two different choices of the objects to be isolated; hence, we made different choices as to which forces to treat as external and which ones as internal. Each time, we found that only the external forces affected the motion of the blocks. We saw that a judicious choice of which object to isolate can simplify the calculation considerably. Take a careful look at the force diagrams in Fig. 4–19 to learn how to isolate a given object and recognize the forces on it.

Some Common Forces

We will use a familiar example to illustrate the preparation of force diagrams and introduce some forces that enter into many physical situations. Let's consider a sled of mass m moving down a snow-covered hill. What is the force diagram for the sled?

Three forces act on the sled, two of which involve contact between the sled and the hill (Fig. 4–21). One of the forces, however, is not a contact force. It is the force due to Earth—the **force of gravity**, $\mathbf{F}_g$. It acts vertically downward in Fig. 4–21.

Figure 4–21 shows a second force, one exerted by the hill on the sled. We know that this force must be present because, in its absence, the force of gravity would cause the sled to accelerate down into the surface of the hill. We call this force the **normal force** and label it $\mathbf{F}_N$. It is a contact force and acts in a direction *normal*, or perpen-

PROBLEM-SOLVING TECHNIQUES

Let's refine some aspects of the general problem-solving strategy laid out in Chapter 2. To prepare a force diagram for a given object (see Fig. 4–18):

1. Identify and isolate the object in question. Make a sketch with the object clearly labeled.
2. Identify all the forces acting on the isolated object. Draw each force on the force diagram as a labeled arrow; include rough approximations of the direction and magnitude of each force.
3. If more than one direction is involved, draw a set of coordinate axes with the origin at a fixed point of the diagram; that is, the origin should not be attached to the object itself. Choose these axes so that you can easily pick out the components of the various forces along them.
4. Separate all forces into their components with respect to the coordinate axes you have chosen.
5. If you include an arrow that represents your guess as to the acceleration of the object, distinguish it clearly from the arrows representing forces. Remember, the acceleration is not a force, it is the response to a force.

Figure 4B1–1 summarizes these steps and gives some further hints for finding the net force.

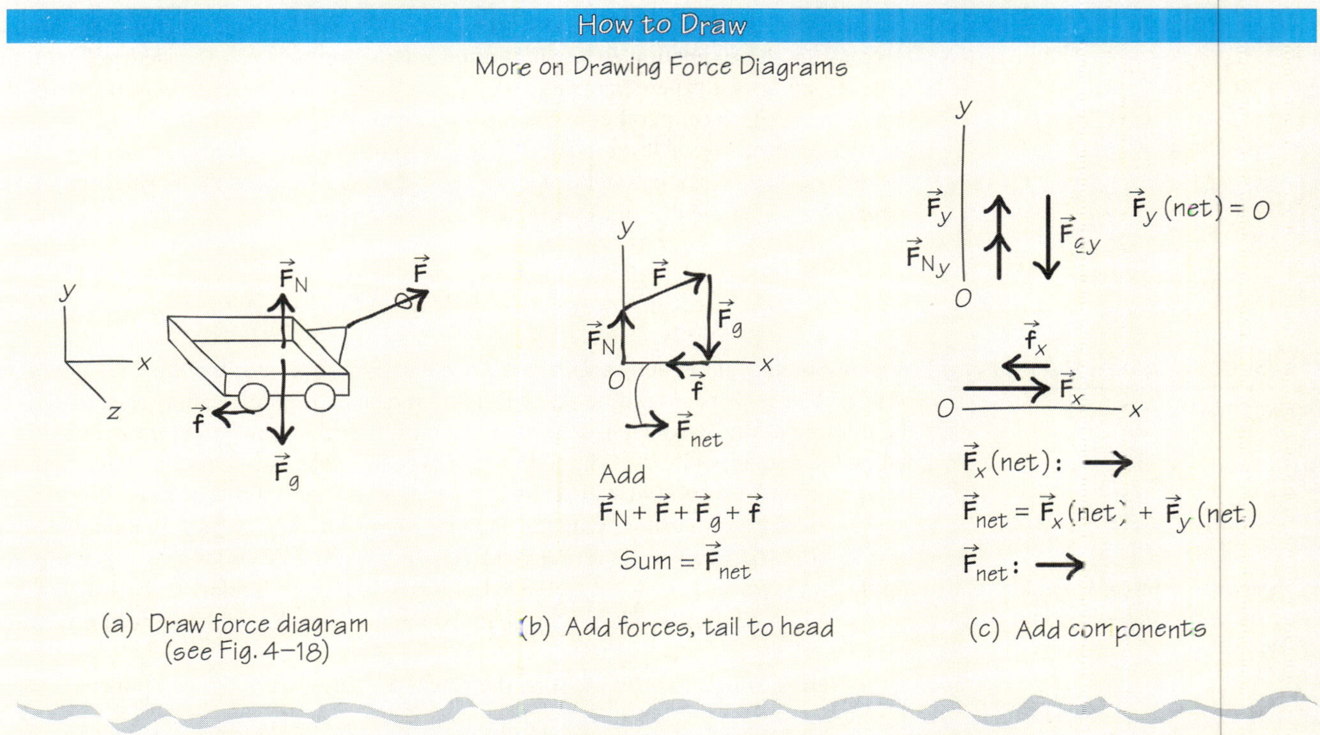

FIGURE 4B1–1 More hints on drawing force diagrams. Parts (b) and (c) show two ways to find the net force, the first a graphical method and the second a method in which components are used.

dicular, to the surface of the hill. This force prevents an object from penetrating into another surface; in this case, the hill. The magnitude of the normal force is not a prefixed constant, but adjusts itself to cancel any forces that might make the sled penetrate the hill. The normal force is simple to use in practice—just as simple as gravity. In contrast to gravity, though, which is a direct and fundamental force between two massive objects (here, Earth and the sled), a normal force actually results from the addition of many complicated intermolecular forces within and between the materials making up the object and the surface on which it rests. Fortunately, we do not have to worry about these complications.

The net force in Newton's second law comes from external forces only.

PROBLEM SOLVING

Your choice of which object to isolate can simplify problem solving.

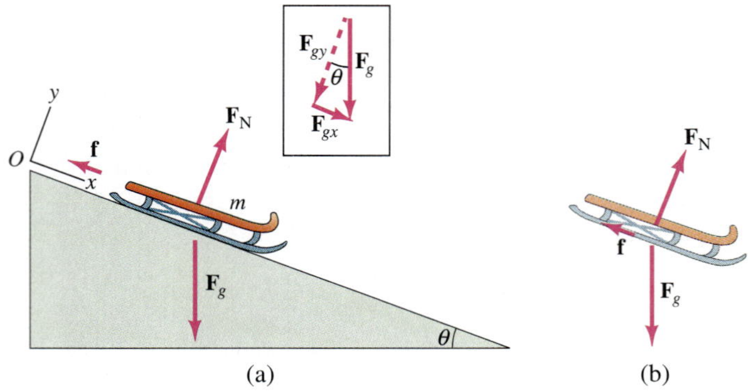

FIGURE 4–21 Force diagram for a mass (sled) sliding on an inclined plane. In the inset, the force of gravity, $\mathbf{F}_g$, is decomposed into components perpendicular and parallel to the plane. Part (b) is a simplified sketch that isolates the forces.

The force of gravity, $\mathbf{F}_g$

The normal force, $\mathbf{F}_N$

The friction force, $\mathbf{f}$

The third force is the **friction force**, which we label **f**. Friction, like the normal force, is an approximation to a complicated interaction between the sled rails and the snow. Friction is a type of contact force as well. Its direction here is opposite to the motion at the surface, parallel to the hill and uphill (Fig. 4–21). If the sled were moving uphill, the force of friction would act downhill.

Friction is always parallel to the surface—perpendicular to the normal force. The second law as applied to the sled, $\mathbf{F}_{\text{net}} = m\mathbf{a}$, becomes

$$\mathbf{F}_g + \mathbf{F}_N + \mathbf{f} = m\mathbf{a}. \tag{4–16}$$

PROBLEM SOLVING

Choose your axes carefully.

The unit vectors **i**, **j**, and **k** in the x-, y-, and z-directions were introduced in Chapter 1.

We now break this equation into component form. To do so, we must choose a set of axes. These axes should be placed in an inertial (nonaccelerating) frame to avoid the complications associated with noninertial frames. A convenient choice is one in which the decomposition of the force vectors is simplest. Another convenient choice, often equivalent, is one in which the acceleration is parallel to one axis. When there are several forces and they cannot *all* point along an axis, as in this example, it is best to align the axes with as many forces as possible. A good choice for this example is shown in Fig. 4–21, where the y-axis is perpendicular to the hill and the x-axis points downhill. Then two of the three forces—$\mathbf{F}_N$ and **f**—are along these axes.

In terms of our coordinate axes, the forces $\mathbf{F}_N$ and **f** have the component form

$$\mathbf{F}_N = F_N\mathbf{j},$$

$$\mathbf{f} = -f\mathbf{i}.$$

Finally, we must find the components of $\mathbf{F}_g$ in the x- and y-directions. To do this, we have drawn a useful diagram for $\mathbf{F}_g$ inset into Fig. 4–21. From Fig. 4–21, then, we find that

$$\mathbf{F}_g = F_g \sin\theta\,\mathbf{i} - F_g \cos\theta\,\mathbf{j},$$

where F_g stands for the (positive) magnitude of the force of gravity. Because all three forces have only x- or y-components, the interesting part of the motion occurs entirely in the xy-plane. Any motion in the z-direction (perpendicular to the page) is unchanging.

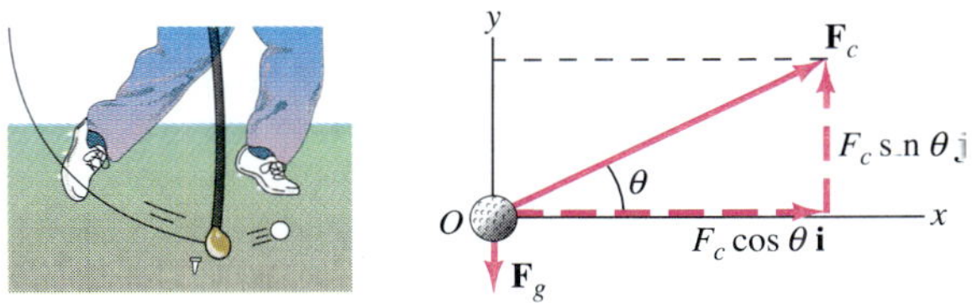

(a) (b)

FIGURE 4–22 (a) Example 4–7. (b) Force diagram for the golf ball being hit by a golf club. The ball has already left the tee, so there is no contact force from the tee.

Equation (4–16) can now be written as

$$F_g \sin\theta\,\mathbf{i} - F_g\cos\theta\,\mathbf{j} + F_N\mathbf{j} - f\mathbf{i} = ma_x\mathbf{i} + ma_y\mathbf{j}.$$

The equations for the separate components of this equation are

$$\text{in the } x\text{-direction:} \qquad F_g \sin\theta - f = ma_x; \tag{4–17}$$

$$\text{in the } y\text{-direction:} \qquad -F_g\cos\theta + F_N = ma_y. \tag{4–18a}$$

We can immediately simplify Eq. (4–18a) by setting $a_y = 0$: We know that the sled does not leave the hill surface. It is this constraint that determines the magnitude of $\mathbf{F}_N$. With $a_y = 0$, Eq. (4–18a) becomes

$$\text{in the } y\text{-direction:} \qquad -F_g\cos\theta + F_N = 0. \tag{4–18b}$$

Equation (4–17) is the dynamical equation for the unknown sled acceleration a_x. That equation depends on the magnitude of the friction force, f, and as we shall learn in Chapter 5, f depends on the value of F_N, which, in turn, we can find from Eq. (4–18a). We shall return to this in Chapter 5.

EXAMPLE 4–7 A swinging golf club strikes a golf ball. During contact, the force of the club (magnitude F_c) on the ball makes an angle θ with respect to the horizontal. You may assume that the ball is no longer in contact with the tee. Draw a force diagram for the golf ball and specify the forces in the coordinate system that you choose.

Solution: We must sketch a force diagram for the golf ball. The ball in Fig. 4–22a has left the tee, so there is no contact force from it. The only forces acting on the ball are therefore the force due to the club, $\mathbf{F}_c$—this is a contact force—and the force of gravity, $\mathbf{F}_g$. Figure 4–22b isolates the ball and shows the forces that act on it. It also contains a coordinate system for which the forces are decomposed as

$$\mathbf{F}_c = F_c\cos\theta\,\mathbf{i} + F_c\sin\theta\,\mathbf{j}$$

and

$$\mathbf{F}_g = -mg\mathbf{j},$$

where m is the mass of the golf ball.

4–6 USING NEWTON'S LAWS II: FINDING THE MOTION

We have learned how to use force diagrams to write Newton's second law for an object that is acted on by forces. This is a very important step along the way to finding the motion of that object. That motion is described by *solving* the equations expressing Newton's second law. It is for this reason that we call those equations the *equations of motion*. We can solve them only if we know the particular *force law*; that is, the force expressed as a function of variables such as position, velocity, time, or any other parameters of the problem. Force laws are determined in experiments that measure these forces in controlled conditions. Many of the most common forces depend only on the position of the object—and the simplest case of all occurs when the force is constant. Let's take a look at how we solve the equations of motion for a constant force in Example 4–8.

EXAMPLE 4–8 The constant net force **F** shown in Fig. 4–23 acts on a nugget of gold whose mass is m. What is the motion of the nugget?

Solution: Because **F** is a constant vector, it points in a fixed direction. For convenience, we choose the x-axis to point in the same direction as the force, so that

$$\mathbf{F} = F\mathbf{i}. \tag{4–19}$$

The force has no components in the y- and z-directions so the nugget of gold will not have a component of acceleration in these directions. The nugget can, of course, have motion with constant velocity in these directions. Motion of constant velocity in the y- and z-directions means that these components of the nugget's position vector change linearly with time:

$$y(t) = y_0 + v_{0y}t; \tag{4–20a}$$

$$z(t) = z_0 + v_{0z}t. \tag{4–20b}$$

The quantities v_{0y} and v_{0z} are the constant components of the velocity vector in the y- and z-directions, and y_0 and z_0 are the values of the y- and z-components of the position at time $t = 0$.

We must still describe the motion in the x-direction, though. The x-component of Newton's second law is

$$F_x = F = ma_x, \tag{4–21}$$

or, equivalently,

$$F = m\frac{d^2x}{dt^2}. \tag{4–22}$$

We want to find the function $x(t)$ that satisfies this equation. We have already done so in Chapter 2, and we can turn to that chapter to find the answer. Once the *initial conditions* are stated, the answer is given by Eq. (2–16). We restate the result below in Eq. (4–23). For the initial conditions, we place the location of the object at $t = 0$ to be $x = x_0$, and the x-component of the velocity at that time is $dx/dt = v_{0x}$. The result is

$$x(t) = x_0 + v_{0x}t + \frac{F}{2m}t^2. \tag{4–23}$$

This result obeys both the equation of motion—Eq. (4–22)—and the initial conditions at $t = 0$. We can verify this by direct differentiation and direct substitution.

The motion described in Example 4–8 is merely a review of our study of kinematics under constant acceleration from Chapters 2 and 3. What is new here is the idea

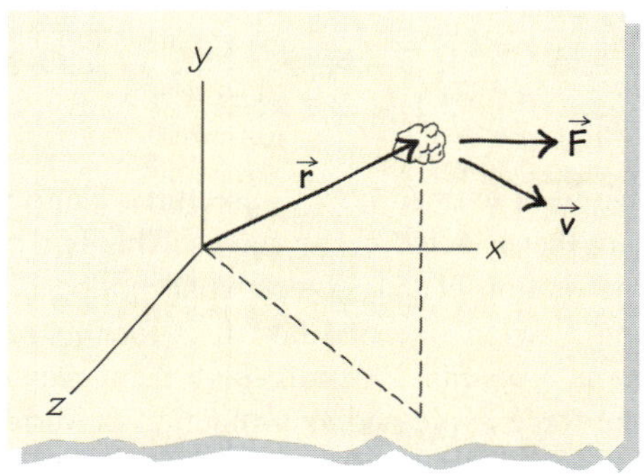

FIGURE 4–23 Example 4–8. A gold nugget is located at a displacement **r** from the origin of a coordinate system and moves with velocity **v**.

that constant acceleration is associated with a constant force. We will be returning to this example in Chapter 5. It is especially important because it is applicable to the force law of *gravity* near Earth's surface.

When the force on an object is not constant but depends explicitly on its position, the equations of motion become more complex. In fact, we can find simple expressions for such motion only for a few important cases of the force law. For example, we shall see later that the force law for a spring that has been compressed or stretched by an amount x from a relaxed position is proportional to x. The larger the compression or stretch, the stronger the spring force (Fig. 4–24). Mathematically, the proportionality is expressed as[†]

$$\text{force law:} \qquad F \propto x.$$

We also have Newton's second law,

$$F = m\frac{d^2x}{dt^2}.$$

The equation of motion therefore has the form

$$x \propto \frac{d^2x}{dt^2},$$

where we have left off all the constants. When the other components of the force are known, and when all the constants and signs are properly included, we are left with a differential equation for x. The meaning of the equation in this case is that x must be a function of time whose second derivative is proportional to the function itself. Only certain functions will satisfy such an equation[‡].

FIGURE 4–24 The stretched bungee cord acts like a spring; the more it is stretched, the stronger the force it exerts on the jumper. Here, the force on the jumper is upward

When the equations of motion have simple mathematical solutions, as for the spring, we say we have an *analytic solution*. In fact, there are not many problems that we can solve in this way. Problems with analytic solutions populate textbooks such as this one. You might get the mistaken impression that all problems have such solutions. Although there are indeed important, real-life problems that can be solved analytically, the equations of motion frequently have to be solved numerically with the aid of computers. While the idealized motion of a rock thrown up in the air is simple, the problem of the exact motion of a rocket launched from Cape Canaveral is quite a different matter. The force laws cannot be written so simply because air friction, the acceleration of gravity, and wind forces all vary with altitude, and the propulsion forces need to be adjusted to compensate for them. Sophisticated computer calculations are needed to learn just how the Space Shuttle will behave as it rises into orbit.

APPLICATION
Real projectile motion

Chaos

Some problems are quite different from those we can solve in terms of simple functions. One consequence of the computer revolution has been a much deeper understanding of systems with motions that are quite different from the very regular ones that appear in analytic solutions. These systems are called **chaotic**, and examples occur all around us: from planetary atmospheres—the great red spot of Jupiter as well as Earthly hurricanes exhibit chaotic phenomena—to aspects of the flying capability of advanced aircraft to the turbulent flow of fluids. Systems with chaotic behavior still obey Newton's second law, but their motions are highly dependent on initial conditions; that is, their behavior may change radically if the initial conditions change just a little. Such systems have a rich and complex behavior (Fig. 4–25).

Prediction of the behavior of our chaotic atmosphere from Newton's laws must be done numerically. Here we face the problem that we may not have sufficient knowledge of the initial conditions. Even if we do, the computer may be unable to solve the

FIGURE 4–25 (a) The great red spot of Jupiter, a gigantic storm within Jupiter's atmosphere, exhibits chaotic behavior. The spot is about two Earth diameters across. This picture, with computer-enhanced colors, was taken from one of the Voyager satellites.

[†]Here we are assuming one-dimensional motion.

[‡]We shall see later that these solutions are sinusoidal: the object moves back and forth.

problem rapidly. Although we are sure that Newton's laws apply to the problem of predicting the weather, no one has succeeded in finding either an analytical or a numerical solution that will accurately predict tomorrow's weather today!

4–7 WHAT IS A LAW OF PHYSICS?

All natural phenomena are governed by the laws of physics. But what do we mean by a *law of physics*? The word "law" is loosely applied to many statements; for example, we used it when we talked of a "force law" describing how a particular force varies as a function of position or velocity. In this context, we could equally well have used the word "prescription." When we explore electricity in Chapter 27, we will study "Ohm's law," which relates voltage, V, to current, I, in a resistor. For many materials, these two quantities are a constant multiple of one another ($V = IR$), and this isolated fact is given the name of a "law." It is not valid, however, for many materials; it is generally not true when temperatures vary; further, it is certainly not true for currents that vary rapidly in time. Thus Ohm's law or the spring force law are not "laws" in the same sense as Newton's laws. What characterizes a true physical law such as Newton's second law is the breadth of its applicability. But even Newton's laws cease to be accurate for systems that move with extremely high speeds (near the speed of light, when special relativity applies) or for systems that are very tiny (where the laws of quantum mechanics apply). In those cases, however, there is still a clear connection to the laws postulated by Newton.

Newton's descriptions have proven to be so universally applicable that they have repeatedly earned their label as *laws of physics*. They played a crucial role in the explanation of most of physical science up to the latter half of the nineteenth century.

The events associated with the discovery of the planet Neptune provide us with a spectacular example of the power of Newton's laws. The motion of astronomical objects is sufficiently frictionless that the fundamental forces can be studied under ideal conditions. Newton himself deduced the force law for the interaction between the Sun and the planets. This law is called the *law of universal gravitation*. The motion of a planet is accurately predicted by taking into consideration the presence of the Sun alone. Even more accurate predictions use the same fundamental law but also take into account the presence of other planets. In the nineteenth century, elaborate mathematical methods were developed to solve the full equations of motion for astronomical objects—remember, this was before the era of computers. Calculations of the motion of the planet Uranus did not agree with observations. By this time, the laws of motion and the law of gravitation had been tested in so many cases that the most conservative view of the discrepancy in Uranus's position was that there must be an additional, unobserved astronomical object somewhere in the solar system that interacts with Uranus by gravitation and thereby perturbs Uranus's orbit. This view was preferable to the alternative, in which the well-established laws of physics were violated.

In 1845, the astronomers Urbain Le Verrier and John Adams independently and simultaneously showed that the tiny discrepancy in the orbit of Uranus could indeed be explained by the presence of an undiscovered planet that was about twice as far from the Sun as was Uranus. Their calculations were based only on Newton's laws and on the law of universal gravitation. Both men predicted the details of the new planet's orbit. When, in 1846, Le Verrier convinced the astronomer Johann Galle to look for the new planet at a location he had predicted, Neptune was discovered.

SUMMARY

Newton's three laws express the dynamics of motion: how forces acting between objects determine the subsequent motion of those objects. The first law states what happens to an object—moving or at rest—when it is left alone:

$\mathbf{F}_{net}$ is the vector sum of any individual forces that act on an object. When $\mathbf{F}_{net}$ is zero, the object moves with constant velocity.

Forces, which behave as vectors, act on objects and cause them to accelerate. For a given force, this acceleration is inversely proportional to the mass m of the object in question. This is expressed in Newton's second law,

$$\mathbf{F}_{net} = m\mathbf{a}. \tag{4–5}$$

In SI, the force is measured in newtons, abbreviated N, where $1\ \mathrm{N} \equiv 1\ \mathrm{kg \cdot m/s^2}$.

Forces act between objects. If objects A and B interact—if there are forces acting between them—then Newton's third law states that the force on object A due to object B, $\mathbf{F}_{AB}$, is equal and opposite to the force on object B due to object A, $\mathbf{F}_{BA}$:

$$\mathbf{F}_{BA} = -\mathbf{F}_{AB}. \tag{4–15}$$

Observers in reference frames moving with respect to one another observe the motion of a given object differently. An observer who verifies that Newton's second law holds, with known or identifiable sources of forces, is said to be in an inertial frame. If a second observer moves with constant velocity relative to the first, the second observer is also in an inertial frame; if there is nonconstant relative motion, the second observer is in a noninertial frame. Observers in inertial frames agree on the forces they see acting on an object. There is no experiment they can perform to decide who is moving in an absolute sense. Observers in noninertial frames disagree on the forces that act on an object and, in effect, Newton's second law fails for an accelerating observer.

Newton's laws help us to determine the motion of an object if we know the nature of the forces that act on it. Conversely, the laws allow us to measure forces acting on an object by measuring the object's motion. If we want to determine the motion of an object, we must independently know the forces that act on it. A force law describes the force due to a particular source and how that force varies with position, time, or other variables. We must also distinguish between external forces that act on an object and internal forces that act within the object. Only the external forces determine an object's motion. To use Newton's laws, we draw force diagrams that conceptually isolate the object. We include all the forces acting on it, keeping in mind that a force is a vector. Once a force diagram has been prepared, we choose a convenient set of axes to write the three components of the second law. These equations can be solved, either analytically or numerically, to find the object's motion.

A law of physics has a wide range of applicability. Laws such as Newton's laws provide a framework for detailed predictions about what will happen under certain conditions. In their breadth and predictive power, Newton's laws are an ideal illustration of what constitutes a law of physics.

UNDERSTANDING KEY CONCEPTS

1. A small but dense mass is swinging freely at the end of a light string. A very sharp knife cuts the string when the mass is at the bottom of its swing; the knife does not disturb anything else. What is the subsequent flight of the mass?
2. Someone pushes on a wall. What experiment can you propose to determine the force with which the person pushes?
3. If you were in a freely falling elevator, the contact force on you due to the floor would drop to zero. If the elevator were to accelerate rapidly upward, the contact force between you and the floor would increase. Why?
4. An astronaut is working while in orbit. When the astronaut assembles a piece of equipment, will he or she notice a difference between working with components of large mass as opposed to components of small mass?
5. When a satellite travels around Earth in a circular orbit, it moves at a constant speed. Does Newton's first law apply in this situation? Is the velocity constant? Is there a force present?
6. There is a well-known parlor trick in which a tablecloth is pulled sharply from beneath a dinner setting, leaving the setting in place. Why does this work?
7. Is a spaceship heading from Earth to the Moon traveling in a force-free environment? Explain. What about a spaceship traveling from Earth to Mars that is currently in a region far away from either planet?
8. An adult sits on a child's table. The table is about to break. Is it correct to say that it is the weight of the adult that is causing the table to break?

9. From your experience with forces, which of these are contact forces: friction, the force of gravity, the normal force on an object on the floor, the force due to a magnet, the pull on the rope used in a pulley?

10. A box is placed on a table. The box's weight and the normal force on it are equal and opposite. Is this a good example of Newton's third law?

11. Shortly after jumping from an airplane, a parachutist will descend with constant velocity. Why is this?

12. Using Newton's second law, devise a system (other than the compressed spring described in Section 4–3) that could be used to measure masses.

13. Newton stated that a reference frame at rest with respect to the distant stars would be a good inertial system. Comment on this. Suggest systems that would be good inertial systems for experiments conducted in, or on (a) a physics lab; (b) the Space Shuttle; (c) a ship at sea; and (d) Mars.

14. A rubber ball and a golf ball have the same mass but the rubber ball has a larger radius than the golf ball. If they are identically accelerated with the same initial force, why might the golf ball go farther in the atmosphere?

15. Describe the forces that are responsible for the acceleration of an automobile. In particular, in what way is friction between the tires and the road responsible for the acceleration? What, then, is the role of the engine?

16. If you tie a rock to one end of a piece of string and swing it in a circle, you experience a force on your hand. Why?

17. When a baseball hits a bat, a force must be exerted on the ball to change its direction of motion. Describe the forces between the ball and the bat. What effect does the ball have on the bat? Does the batter feel the effects of any of these forces?

18. Give three examples of noninertial forces.

19. A marble is placed on the side of a bowl. It rolls down into the bowl, then continues to roll up and down the bowl's side until it finally comes to rest at the bottom. Explain this in terms of the first and second laws.

20. Describe qualitatively how you might go about constructing an accelerometer, which measures the acceleration of a ship. [*Hint*: Think about how a spring placed between your back and the back of a seat would react when an automobile in which you are sitting accelerates forward. Such devices are of great importance for "blind" navigation, which might be required for a submarine. They allow us to follow the path when we know the acceleration as a function of time and then reconstruct the velocity and the position as functions of time. (This procedure is described in Chapter 2.) It is, in fact, difficult to construct an accurate version of such a device without knowing about rotational motion.]

21. You are standing on a scale in an elevator. The elevator suddenly starts to move upward. What happens to the reading of the scale, and why?

22. A diver jumps from a high platform and experiences a feeling of weightlessness. Is the force of gravity no longer acting on the diver?

23. A 5-kg mass is placed on a table. A professor states that the normal force on the mass due to the table is (5 kg) (9.8 m/s^2) and that this is a consequence of Newton's third law. Is this a sound analysis?

24. In Example 4–5, we discussed Professor A, who is at rest on a train platform, and Professor B, in the accelerating train, to illustrate the consequences of the noninertial nature of Professor B's reference frame. How do we know that Professor A's frame is inertial?

25. Consider a horse that pulls on a cart. By Newton's third law, the cart pulls on the horse with a force of equal magnitude but in the opposite direction. How can there be any motion?

26. What makes a car go forward when the engine is turned on and the transmission is engaged?

27. You are standing in a stationary bus and suddenly find yourself thrown backward. What does Newton's second law say about that?

28. A car is stationary on a flat parking lot. The force of gravity acts downward and an equal and opposite normal force acts upward. Is it correct to say that these forces are equal and opposite because of Newton's third law?

29. In a tug of war, one side is stronger than the other. Assuming that both sides exert themselves to the maximum, discuss the motion of a handkerchief tied to the rope being pulled.

30. A fellow student states that forces cause an object to move. Criticize this statement.

31. Baron Munchausen claimed that it is possible for a very strong man to pull himself off the ground and rise into the air by pulling on his bootstraps. Discuss this mode of liftoff in the cold light of Newton's laws.

PROBLEMS

4–1 Forces and Newton's First Law

1. (I) In applying Newton's laws, we must identify the forces acting on an object. Are there any forces acting on the following objects? If so, list them: (a) the Space Shuttle in Earth orbit; (b) an ice skater coasting on ice; (c) the *Voyager I* spacecraft far past the orbit of the planet Pluto.

2. (I) The forces acting on an airplane are the following:

gravity: $F_g = 4 \times 10^5$ N down;

engine thrust: $F_E = 10^5$ N forward;

lift: $F_L = 5 \times 10^5$ N up;

air drag: $F_D = 5 \times 10^4$ N backward.

What is the net force on the airplane?

3. (I) In a tug of war, a red ribbon tied around a point on the rope between the two teams moves with a uniform velocity of 0.1 m/s in the y-direction. One team exerts a force on the rope of 600 N in the $+y$-direction. What force does the other team exert on the rope?

4. (I) A boat sailing in the northeasterly direction with constant speed experiences a wind force of magnitude 10^4 N from the east. What is the force on the sailboat due to the resistance of the water to motion through it?

5. (I) In a classic demonstration, Otto von Guericke used 16 horses—8 on each side—to try to pull apart two hemispheres forming a sphere from which air had been evacuated. Could he as well have used only 8 horses on one side, with the other side tied to a sturdy tree?

6. (II) Three nonzero forces act on a particle at the origin of a coordinate system. $\mathbf{F}_1$ is in the z-direction, whereas $\mathbf{F}_2$ and $\mathbf{F}_3$ lie in the xy-plane. Can you arrange the magnitudes and directions of $\mathbf{F}_2$ and $\mathbf{F}_3$ (keeping them always in the xy-plane) so that the particle does not accelerate?

7. (II) (a) A spider is suspended from a single vertical thread; the spider has a mass of $m = 30$ mg. The spider is acted upon by the force of gravity, which is directed downward and has magnitude 3×10^{-4} N, and by the *tension*, T, in the thread, a common type of contact force that always acts in the direction of the thread, and is directed away from the point at which it is attached. In this case, the tension acts in the upward direction. What is the magnitude of the tension? (b) The spider is now attached to two threads of equal length, which make a 120° angle with each other, as in Fig. 4–26. The spider is motionless, waiting for a victim. What is the tension in each thread? [*Hint*: The tension is a vector directed away from the attachment point. Newton's first law must be satisfied in its vector form.]

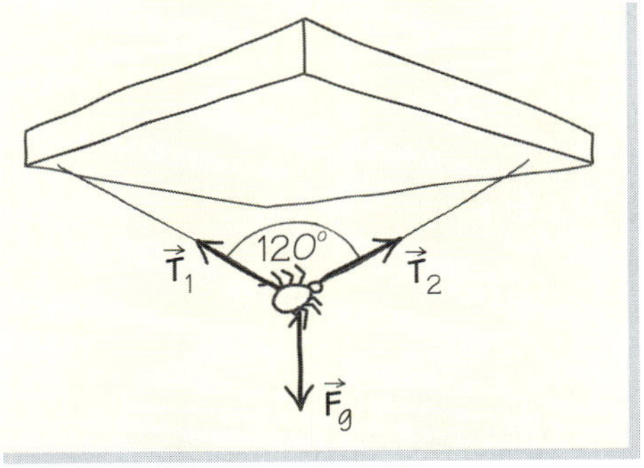

FIGURE 4–26 Problem 7.

8. (II) A very large—consider it to be infinitely large—mesh of stiff wires makes a horizontal plane and large electric charges are placed at the intersections of the wires. A mass with another charge is placed below one of the charges on the mesh. Electric charges exert forces on one another that are proportional to the strength of the charges. Each charge on the mesh repels the lone charge with a force that varies with the distance between the charges. The forces are directed along the line between the charges. Show that the net force on the lone charge is a repulsion directed straight downward.

4–2 *Newton's Second Law of Motion*

9. (I) A car coasts along a road with initial velocity $\mathbf{v}_0$. It inevitably slows down and finally comes to rest. (a) Describe why this is so. (b) An observer traveling with uniform velocity $\mathbf{v}_0$ starts out at rest relative to the car. What does she see with the passage of time? How does she explain what happens?

10. (I) The force of gravity on an apple (mass 0.1 kg) has a magnitude of about 1 N. What is the acceleration of the apple as it falls toward Earth? How large is the force of gravity on a falling automobile, mass 1000 kg, if it has the same acceleration as the apple? In each case, assume that only the force of gravity acts.

11. (I) The force of gravitation attracts two masses m_1 and m_2 to each other. If the masses are separated by a distance d, the magnitude of the force on each mass is

$$F = Gm_1m_2/d^2,$$

where G is a constant. Suppose $m_2 = 3m_1$. Make a sketch of the two masses with vectors that indicate the direction of the forces on the two bodies. Draw the lengths of the vectors to correspond to the magnitude of the two forces. Repeat the sketch but replace the force vectors with acceleration vectors.

12. (II) A spring exerts a force when it is compressed or extended. The force is proportional to the distance x by which it is compressed or stretched away from its equilibrium position. The direction of the spring's force is toward its equilibrium position. Draw diagrams with the spring compressed and extended. Choose a direction for x. Write mathematical expressions for the force in both situations.

13. (II) Five forces, all of the same magnitude F, act on an object of mass m at the origin of the coordinate system shown in Fig. 4–27. Two of the forces are aligned along the x-axis, one is oriented in the $+x$-direction and one in the $-x$-direction. Two other forces are similarly aligned with the z-axis. The fifth force points in the $+y$-direction. What are the direction and magnitude of the acceleration of the object? If we had specified the forces in a different order, would the answer have been different?

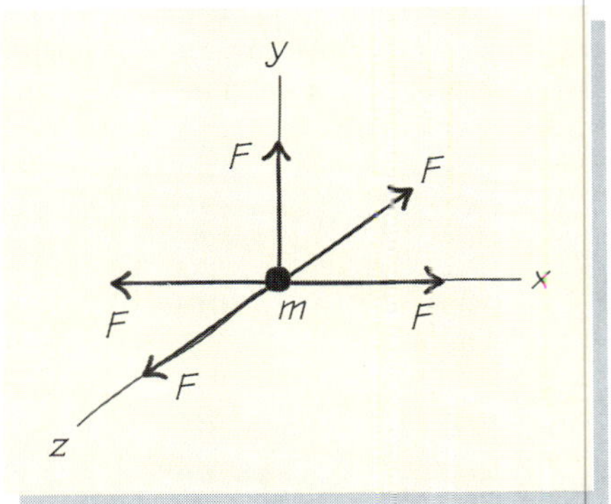

FIGURE 4–27 Problem 13.

14. (II) A sports car of mass 820 kg accelerates from 0 mi/h to 60 mi/h in 6.00 s. What is the average force, in newtons, that the road exerts on the wheels of the car? A huge station wagon of mass 1900 kg can accelerate from 0 mi/h to 60 mi/h in 9.5 s. What force does the road exert on the station wagon? What would the station wagon's acceleration be if it were acted on by the same external force as the sports car experiences?

15. (II) A car of mass 1150 kg accelerates from rest to 100 km/h in 11 s. With additional streamlining, the same car undergoes acceleration to the same speed in 9 s. What is the difference in the force exerted by the air (the *drag force*) on the car in the two cases? For this problem, assume the drag force is constant. (This assumption is a poor one in practice.)

16. (II) A common type of contact force is that provided by a taut rope. Suppose that a taut rope is attached to an object. In what direction can the rope apply a force to the object? This force is known as the *tension*. (a) A cart being pulled horizontally by a light rope (the word "light" in this context means that you can ignore the mass of the rope in all your considerations) has a mass of 25 kg and accelerates in the horizontal direction at 2.4 m/s^2. What is the tension in the rope? If the same cart is pulled so that it accelerates at 0.65 m/s^2, what is the tension? Assume that the only force acting on the cart is the tension. (b) Suppose that the rope passes over a fixed pulley (Fig. 4–28) and the cart accelerates horizontally at 1.4 m/s^2. What is the upward force on the pulley?

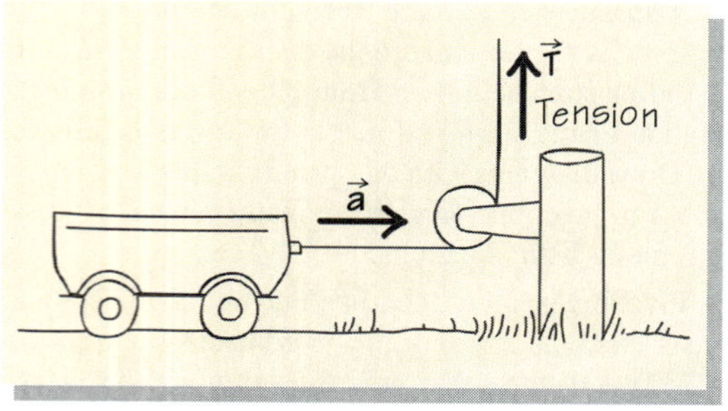

FIGURE 4–28 Problem 16.

17. (II) A father pulls his identical twins on identical sleds tied one after the other (Fig. 4–29). He exerts a force **F** that makes an angle of 30° with the horizontal and that leads to an acceleration of the two sleds. What is the tension in the rope that he is pulling? What is the tension in the rope that connects the front sled to the rear sled? The mass of each sled plus its twin is m. Assume that there is no friction between the sleds' runners and the snow surface.

FIGURE 4–29 Problem 17.

18. (II) Electrons, mass about 10^{-30} kg, are constituents of atoms and respond to electrical forces. Such forces, of varying strengths, can be generated in the laboratory. Suppose that a constant electrical force of 10^{-15} N acts on an electron. What is the speed of the electron after 10^{-10} s? after 10^{-9} s? after 10^{-7} s?

4–3 Newton's Third Law of Motion

19. (I) A hook is screwed into a ceiling; one end of a string is tied to the hook and the other end has an 8-kg mass attached to it. Assuming that the force of gravity acting on a mass m has magnitude mg, where $g \simeq 9.8$ m/s^2, what force does the hook exert on the string?

20. (I) A 3000-kg pickup truck pulls a 1200-kg boat on a trailer, and they are accelerating together at 1.2 m/s^2. What is the horizontal force that the truck and boat trailer exert on the road?

21. (I) A falling automobile, mass 950 kg, has an acceleration of magnitude 9.8 m/s^2 when only the force of gravity acts on it. What is the magnitude of the upward acceleration of Earth? Take the mass of Earth to be 6.0×10^{24} kg.

22. (I) The force exerted on a satellite by an astronaut is $(0.5\mathbf{i} + 0.7\mathbf{j} + 5.1\mathbf{k})$ N. What is the force exerted by the satellite on the astronaut? What is the magnitude of this force?

23. (II) In Problem 1, you were asked to find the forces acting on certain objects. By Newton's third law, these original objects are the sources for forces that act on other objects. What are these other objects for each of the following original objects: (a) the Space Shuttle in an Earth orbit, (b) an ice skater coasting on ice, (c) the *Voyager I* spacecraft far past the orbit of the planet Pluto?

24. (II) A force of magnitude 8.0 N pushes on a horizontally stacked set of blocks on a frictionless surface (Fig. 4–30) with masses $m_1 = 2.0$ kg, $m_2 = 3.0$ kg, and $m_3 = 4.0$ kg. (a) What is the acceleration of the stack? (b) What are the forces on block 1, as well as the net force on this block? (c) Repeat part (b) for block 2. (d) Repeat part (b) for block 3.

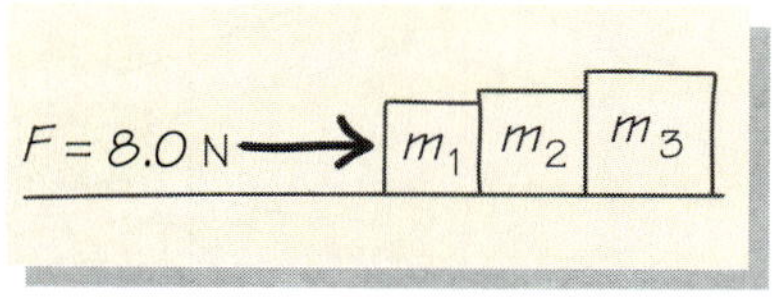

FIGURE 4–30 Problem 24.

25. (II) The engine of a train pulls five cars, each of mass $m =$ 30,000 kg. During a period of acceleration, the force between the engine and the first car is 45,000 N. (a) What is the acceleration of the train? (b) Sketch the forces acting on the first car. (c) How large is the tension in the hook between the first and second cars?

26. (II) Repeat Problem 24, but this time with the blocks stacked in the reverse order; that is, block 3 to the left and block 1 to the right.

27. (II) Three charges move through space with no forces acting on them except the electric forces that they exert on each other. In an appropriate coordinate system, some of the forces can be broken down as follows: The force that charge 1 exerts on charge 2 is $\mathbf{F}_{21} = (2\mathbf{i} - 3\mathbf{j} + \mathbf{k})$ N; and the force that charge 1 exerts on charge 3 is $\mathbf{F}_{31} = (-3\mathbf{i} + 2\mathbf{j} - 3\mathbf{k})$ N. What is the total, or net, force on charge 1?

**4–4 The Effects of Noninertial Frames*

28. (I) An observer inside an elevator that is in free fall will see any object that was initially at rest in midair inside the elevator remain in that position. How does he explain this fact, assuming that he knows about the existence of gravity?

29. (I) Consider the situation described in Example 4–5, with $v_A = 0$ and $v_B = 10$ m/s. The train is accelerating in the x-direction at 2 m/s^2, according to Professor A. Can Professor B tell that he is accelerating? Does he think that Professor A is accelerating? Do our physical senses help us solve these questions? What is the acceleration of Professor A according to Professor B? Is this a real acceleration?

30. (II) An observer sits in a cylindrical bathtub that can rotate about a vertical axis. At first, the tub is at rest—relative to a neighboring tree for example. A little later, the tub is rotating at high speed around its vertical axis. The observer is still at rest relative to the tub, but the water level is no longer flat; it runs up on the sides of the tub. How does the observer explain this?

31. (II) A simple demonstration of the effect of a noninertial frame can be made by dangling a watch from a chain while you are taking a trip in a jet. What is the angle of the chain with respect to a vertical window edge when (a) the jet is parked at the gate; (b) the jet is accelerating with some acceleration a' along the runway just before takeoff; (c) the jet is at cruising speed at 30,000 ft? You can try this type of experiment in an accelerating automobile.

32. (II) A mass of 2 kg lies on a horizontal table that is placed in the back of a truck (Fig. 4–31a). The mass is held in position by a string and a force of 6 N acts on it in the $-y$-direction. The truck then accelerates in the $+x$-direction. As a result, an observer in the truck sees that the string holding the mass on the table makes an angle of 20° with respect to its initial orientation, as shown in Fig. 4–31b. What is the acceleration of the truck? Describe what the observer in the truck sees and how he might interpret his observations.

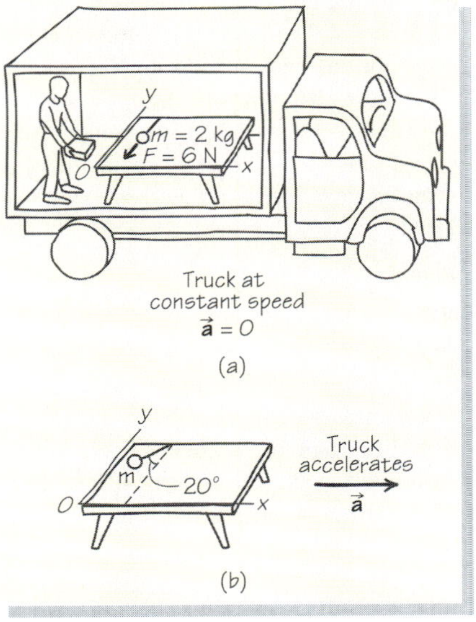

FIGURE 4–31 Problem 32.

4–5 Using Newton's Laws I: Force Diagrams

33. (I) A car is accelerating straight ahead on a flat road. Draw the force diagram for the car.

34. (I) A child slides down a water slide at an amusement park. What forces are present? Draw a force diagram for the child.

35. (I) Consider an astronaut sitting in a rocket just after blastoff from Cape Canaveral. Draw a force diagram showing the forces present on the astronaut sitting in her seat.

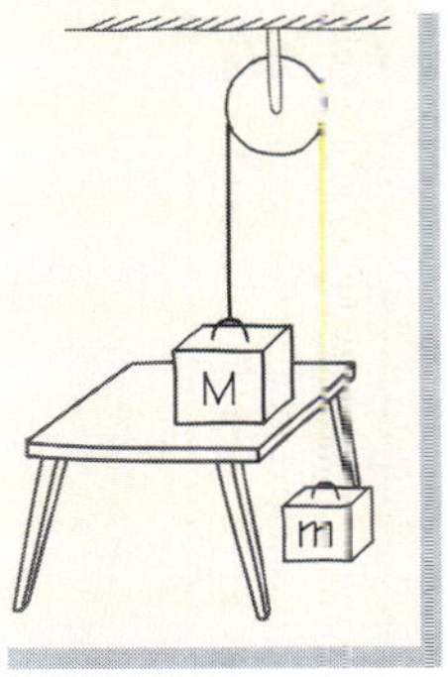

FIGURE 4–32 Problem 36.

36. (I) Consider the arrangement shown in Fig. 4–32. The whole system is at rest. Draw all the forces, draw the force diagram for each mass, and give the equations of motion for each mass.

37. (I) A person throws a medicine ball into the air. Draw separate force diagrams for the ball and for the person (a) for the time just before the ball leaves the hands, and (b) after the ball has been thrown.

38. (I) A piano sits within a large freight elevator. (a) Draw a force diagram for the piano alone. (b) Draw a force diagram for the elevator alone.

39. (II) Using a rope, a horse pulls a wagon that is initially at rest. (a) Draw a force diagram for the horse. Include friction, Earth's gravity, the influence of the wagon, and any other forces that may be operating. Draw the direction of the forces as accurately as possible. (b) Repeat part (a) for the wagon. (c) Repeat part (a) for Earth, including only the horse and wagon as influences. (d) Why doesn't the horse accelerate backward toward the wagon? In answering, think about the different forces acting on horse and wagon.

40. (II) A block of mass M sits on a rough horizontal surface. A rope inclined upward at an angle θ with the horizontal exerts a force of magnitude T on the block. The block remains stationary. Draw a force diagram and express all the forces on the block in terms of M, g, θ, and T.

41. (II) Consider a brick sliding down an inclined plane of 30°. Draw a force diagram for the brick that includes the gravity, friction, and normal forces. Choose a coordinate system and give the force components in equation form.

42. (II) Draw two force diagrams for a tug of war in which the strengths of the two sides are unbalanced. (The two objects in the diagram will be the two groups of contestants.) Be sure to include all forces!

43. (II) A heavy man stands on a ladder that is leaning against a rough-surfaced wall and rests on the ground (Fig. 4–33). Draw the force diagram for the system of man and ladder. Ignore the mass of the ladder. If the ground is very smooth, the ladder may start to slip. Why?

44. (II) Two astronauts, Joe and Moe, are taking a space walk. They are connected to the ship only by a slack rope. They have a light rope between them and decide to have an outer space tug of war. In their space suits, Joe has a mass of 100 kg, whereas Moe has a mass of 110 kg. The loser of the tug of war is the first astronaut to cross an imaginary line midway between them. (a) Draw a force diagram for each astronaut.

FIGURE 4–33 Problem 43.

What is the constraint of Newton's third law? (b) Show that Moe wins the tug of war. (c) Miffed at his loss, Joe decides to challenge Moe to a tug of war with their feet firmly planted on Earth. Draw the force diagram for each astronaut. (d) Joe wins the tug of war this time. How is this possible?

45. (II) An electrically charged mass, when placed between two charged parallel plates, experiences a constant force F_e perpendicular to the plates. Suppose that such a mass moving vertically with some speed v_y enters the region between two such plates, which are oriented vertically. Draw a force diagram for the mass while it is between the plates. Include the force of gravity. Note that there are two possibilities for the direction of F_e; choose one or the other.

46. (II) Objects made of iron are attracted by an electromagnet when a switch is thrown activating it. Suppose that a small piece of iron is suspended by a thread in the vicinity of an electromagnet, as shown in Fig. 4–34. (a) Draw a force diagram for the iron object immediately after the switch is closed and the electromagnet is activated, and (b) for the situation when the switch has been closed for a long time.

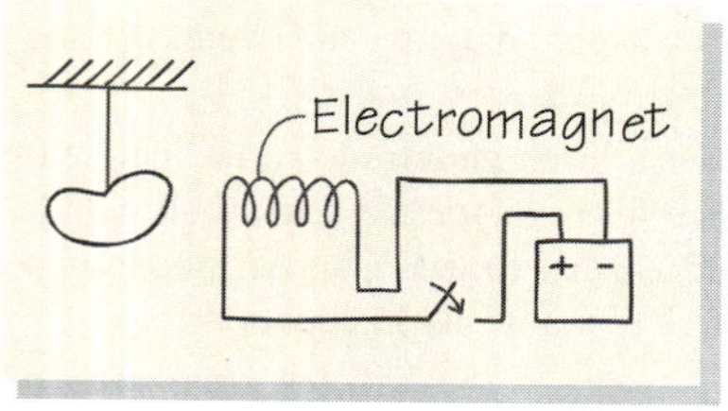

FIGURE 4–34 Problem 46.

47. (II) A person is standing in an elevator that is moving upward with a constant speed. (a) What are the forces acting on the person? (b) Suppose that the elevator is accelerating upward with an acceleration g that is equal in magnitude to the acceleration that the person would have if the only force acting on him were gravity. What are the forces acting on the person, and how large are they? (c) Suppose that the cable breaks and the elevator, together with its unfortunate passenger, is falling freely. What forces are acting on the person?

4–6 Using Newton's Laws II: Finding the Motion

48. (I) An object of mass 25.43 kg that can move only in the xy-plane is at rest at the origin of the xy-coordinate system at time $t = 0$ s. Constant forces $\mathbf{F}_1 = (1.715, 0, 0)$ N and $\mathbf{F}_2 = (0, 1.128, 0)$ N act on the object. (a) Draw the force diagram for the object. (b) State the initial conditions; that is, the position and velocity vectors at $t = 0$ s. (c) What are the position and velocity of the object at time $t = 1.200$ s? (d) at $t = 3.600$ s?

49. (II) An object of mass M is subject to two constant forces: $\mathbf{F}_1$, pointing in the x-direction, and $\mathbf{F}_2$, pointing in the y-direction. Show that the motion is one of constant acceleration. Find the magnitude and direction of this uniform acceleration.

50. (II) A third constant force $\mathbf{F}_3 = (F_{3x}, F_{3y}, 0)$ acts on the object in Problem 48 at $t = 0$ s in addition to the forces $\mathbf{F}_1$ and $\mathbf{F}_2$. (a) What must F_{3x} and F_{3y} be so that the object does not accelerate? (b) What must F_{3x} and F_{3y} be so that at $t = 5.000$ s, the object is at position $\mathbf{r} = (1.000, 1.000, 0)$ m?

51. (III) After some time, the displacement of an object of $m = 2.0$ kg is measured to be $x = At^{3/2}$, where $A = 6.0\ \text{m/s}^{3/2}$. Determine the net force acting on this object. Note that the force will depend on the time.

52. (III) When a small sphere, starting from rest, falls through a liquid, it experiences a so-called linear drag force $\mathbf{F}_D$, in addition to the force of gravity mg directed downward. $\mathbf{F}_D$ is directed upward and has a magnitude proportional to the velocity, $F_D = bv$, where b is a constant. The equation of motion—Newton's second law—for motion in the vertical direction z (measured with the +direction downward) reads $m\,dv/dt = mg - bv$. (a) Show that this equation is solved when the velocity takes the form $v(t) = mg(1 - e^{-bt/m})/b$. (b) Sketch the speed as a function of time. (c) Sketch the position as a function of time.

53. (III) The force on a mass m that can move along the x-axis is given by $F = -kx$. (This force is characteristic of the spring; k is a property of this particular spring and is called the spring constant.) (a) Show that, with an appropriate choice of the constant c, $x = A\cos(ct) + B\sin(ct)$ is a solution of the equation of motion. What is the value of c? (b) If the object is at $x = 0$ at $t = 0$ and if its velocity at that time is v_0, what are A and B?

General Problems

54. (I) The gravitational force that the Sun exerts on Earth has a magnitude of 3.5×10^{22} N and is directed toward the Sun. What force does Earth exert on the sun?

55. (II) A load of mass 53.2 kg is to be lifted by two ropes, each going over pulleys. The ropes are separated in such a way that both ropes initially make an angle of 65° with the horizontal (Fig. 4–35). The magnitude of the forces exerted by each of

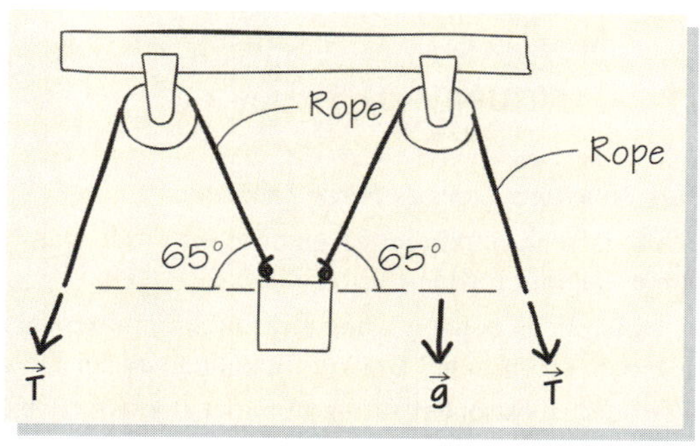

FIGURE 4–35 Problem 55.

the ropes is 333 N. (a) Draw a force diagram for the load. The only forces acting on it are the forces of the ropes, directed along the ropes, and the force of gravity, which points down and has a magnitude given by the mass of the load times the constant acceleration $g = 9.80\ \mathrm{m/s^2}$. (b) What is the net force on the load? (c) What is the initial acceleration (magnitude and direction) of the load?

56. (II) An object of weight $F_g = 20$ N is suspended as shown in Fig. 4–36. The rope and pulleys are very light and friction is negligible. Determine the weight of the object that maintains the equilibrium, given that the angle between the rope and the horizontal is 35° on both sides of the movable pulley.

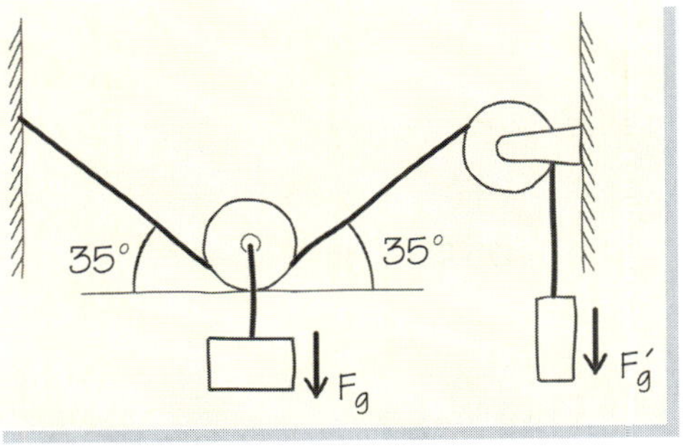

FIGURE 4–36 Problem 56.

57. (II) A window washer sits on a board supported by two ropes that go over pulleys attached to a scaffold. He holds the other ends of the ropes in his hands (Fig. 4–37). He pulls on them in such a way that he moves upward at constant speed. Draw a force diagram for the window washer only and give all the equations; assume that the tension in a given rope is the same everywhere in the rope, even after it wraps around a pulley. Suppose that the effect of the window washer's efforts leads to an upward acceleration of both the window washer and the board. How are the equations modified?

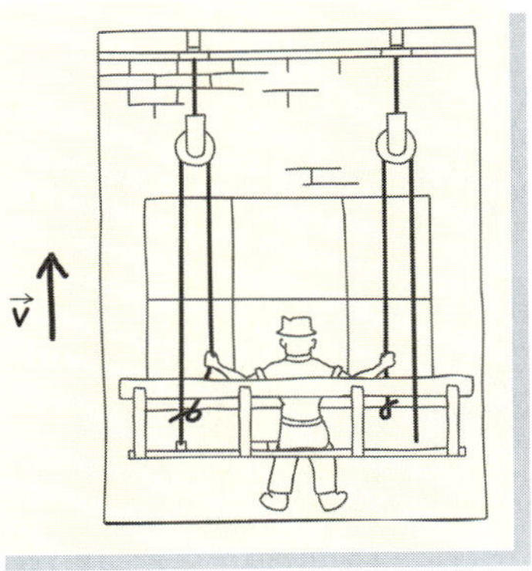

FIGURE 4–37 Problem 57.

58. (II) A tennis ball is dropped onto a granite slab and it bounces back to about 90 percent of the height from which it was dropped. What forces acted on the tennis ball for it to exhibit this behavior? In what direction do these forces point? Estimate their magnitude.

59. (II) A train consists of an engine and three cars tied closely together. The mass of the whole train is 10^5 kg; the engine alone has a mass of 4×10^4 kg and each car has a mass of 2×10^4 kg. Through its wheels, the engine can exert a horizontal force of 2×10^4 N on Earth. (a) When the engineer wishes to accelerate forward, in which direction should the engine exert its force on Earth? What is the magnitude of the force exerted by Earth on the engine then? (b) When the engineer acts to accelerate the train, what is the force on the set of three cars and what is the subsequent acceleration of these cars? (Treat the three cars as a unified whole.) (c) When the engineer acts to accelerate the train, what are all the forces on, and subsequent acceleration of, the second car of the train?

60. (II) A boat is being pulled up the middle of a canal at a steady speed by two horses, one on each side of the canal (Fig. 4–38). The ropes tying the boat to the horses each make an angle of 30° with the lengthwise direction of the canal. (a) Draw a force diagram for the boat. Include the friction force due to the boat sliding through the water. (b) If the force exerted on each horse by its rope is 2000 N, what is the force of friction that the boat experiences due to the water?

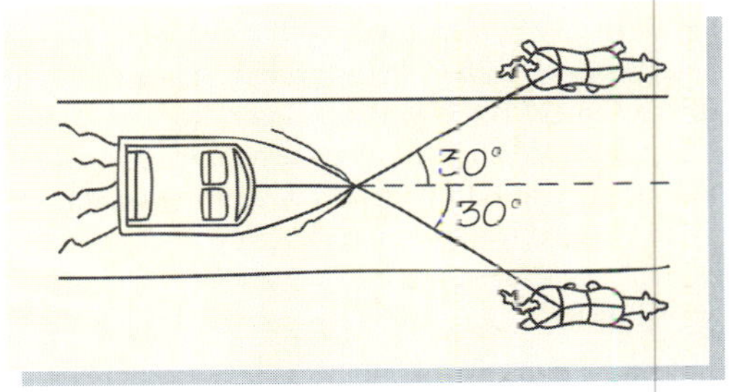

FIGURE 4–38 Problem 60.

61. (II) A passenger in a stationary elevator is acted on both by the force of gravity, which is directed downward, and by a contact force, directed upward by the floor. The contact force from the floor cancels the force of gravity, and in this way there is no net force on the passenger, which explains why she measures herself to be at rest with respect to the walls. (a) What is the force diagram on the passenger when the elevator descends at constant velocity? (b) The elevator cable breaks. (Do not be alarmed, because the emergency brake will act if you finish this problem in time.) As seen from the outside, the passenger and elevator fall together and accelerate under the force of gravity. A force diagram for the passenger would show the single force of gravity acting. What does the passenger observe, and how would she draw a force diagram for herself?

62. (II) A parachutist has jumped from a plane. (a) What are the forces acting on the parachutist when the parachute opens? Suppose that the force due to air resistance opposes the parachutist's motion and is proportional to the speed of the system; that is, $\mathbf{F}_{\text{air resistance}} = -A\mathbf{v}$. (b) What are the dimensions of the constant A? (c) Sketch the downward acceleration of the parachutist as a function of velocity. Show that the velocity of the system becomes constant at some point. This constant velocity is called the *terminal velocity*. (d) If the force of gravity on the system is 1000 N and the terminal velocity is 6.0 m/s, what is the value of the constant A?

63. (II) A parachutist experiences two forces: One is the force of gravity, which is of the form $\mathbf{F}_g = -mg\mathbf{j}$. Here, m is the mass of the parachutist and g is the acceleration due to gravity. The other force is a drag force, and, in contrast to Problem 62, it has the form $\mathbf{F}_d = Av^2\mathbf{j}$ in this problem, where the velocity of the parachutist is given by $-v\mathbf{j}$. (a) What are the dimensions (and units) of A? (b) Complete the equation $dv/dt = \dots$. (c) At some point, the parachutist reaches a terminal velocity that is constant. What is it?

64. (II) An egg of mass 60 g is in a stiff box of mass 500 g, surrounded on all sides by padding that keeps the egg in place. A force of $\mathbf{F} = (3.2\text{ N})\mathbf{i} + (2.8\text{ N})\mathbf{j}$ acts on the outside of the box, in addition to the force of gravity (which is oriented in the $-y$-direction). (a) Draw force diagrams for the egg–box system and for the egg. (b) Express Newton's second law as it applies to the egg. (c) What are the acceleration of the egg–box system and net force on the egg?

65. (II) A bicyclist rides a flat course at constant speed v with respect to the ground into a stiff but steady headwind of speed v_w, also with respect to the ground. (a) Draw a force diagram for the cyclist–cycle system. List the forces acting on this system, including the wind resistance, which pushes in his face, and the frictional forces with the ground, which propel him forward. (b) If you are observing the cyclist from a reference frame at rest with respect to the ground, what is the vector sum of all the forces? (c) If you are observing the cyclist from a frame at rest with respect to the bicycle, what are the speed of the cyclist, the speed of the wind, and the vector sum of all the forces? (d) If you are observing the cyclist from a frame at rest with respect to the air, what are the speed of the cyclist, the speed of the wind, and the vector sum of all the forces?

66. (II) An electron is placed between two charged parallel plates, oriented vertically. The electron experiences a constant force perpendicular to the plates. Suppose that the electron, whose mass is 9×10^{-31} kg, is moving vertically with $v_y = 2 \times 10^6$ m/s, and enters the region between the two plates (Fig. 4–39). This region has a vertical height of 1 cm, and while in it, the electron experiences a force of 3×10^{-18} N. (a) How long does the electron stay in the region? (b) After the electron has passed through the region, what is its horizontal velocity component?

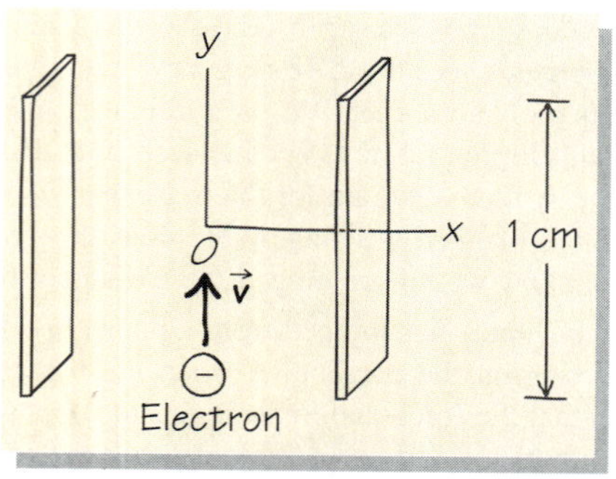

FIGURE 4–39 Problem 66.

67. (III) As shown in Fig. 4–40, a plumb bob suspended from a frame settles to a steady position as the frame slides down a frictionless inclined plane. What is the angle the plumb bob makes with the vertical during the slide?

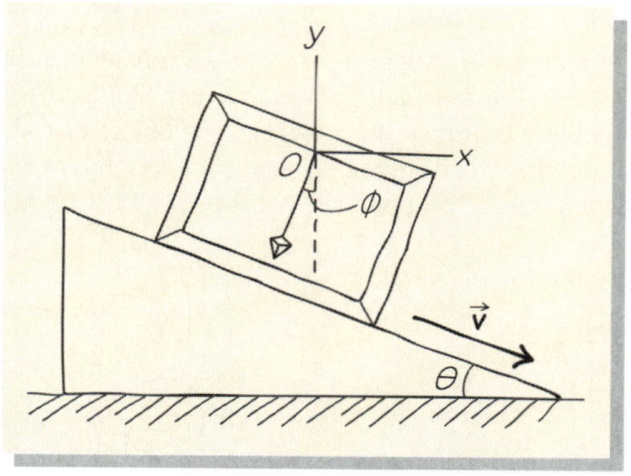

FIGURE 4–40 Problem 67.

68. (III) The acceleration of an object dropped from a height is smaller at the equator than at the poles, even when the force of gravity is assumed to be the same everywhere on the surface of Earth. (a) Assuming that the difference has something to do with the rotation of Earth, obtain an expression for the difference between the accelerations in terms of the radius of Earth and its rotation frequency. (b) Given that the radius of Earth is $R = 6.38 \times 10^6$ m and that the acceleration at the poles is $g_0 = 9.80\text{ m/s}^2$, calculate the acceleration of a falling object at the equator. (c) Calculate the acceleration of a falling object as a function of the angle of latitude, in terms of g_0. (d) If experiment shows a deviation from your prediction, what effects could be responsible for that deviation?

CHAPTER

5

The crawler that transports the Space Shuttle to its launch pad uses friction between its treads and the ground to move the mammoth assembly. The combined weight of the crawler and shuttle assembly is some 30 million Newtons.

Applications of Newton's Laws

In this chapter, we apply Newton's laws to a variety of situations in which several forces act, including gravity, tension, normal and contact forces, friction, and drag. These forces are ones that frequently act on us and the objects all around us. We shall also look at the role of forces in circular motion, as well as the features of motion in rotating frames. Finally, we shall take the time to see how the forces we look at in this chapter are ultimately described in terms of more fundamental forces—in particular, ones that act at a microscopic level.

5–1 SOME SIMPLE CONSTANT FORCES

Constant forces—forces whose magnitude and direction are independent of time and position—describe a wide variety of phenomena. We discussed some examples in Chapter 4: the force of *gravity*, $\mathbf{F}_g$, which acts on an apple falling from a tree, and the *normal force*, $\mathbf{F}_N$, which acts on the apple when it sits on a kitchen table. Although a normal force is not necessarily constant, it is indeed constant in the situations to be illustrated here.

Gravity

The cause of projectile motion near Earth is the force of gravity. As we saw in Chapter 3, projectile motion is motion with a constant acceleration **a**, which points downward and has magnitude g. The acceleration vector provides a complete description of projectile motion, which is generally parabolic (Fig. 5–1). Having now studied the *causes* of acceleration, we can attribute this constant acceleration to the force of gravity—a force that is constant and directed down toward the center of Earth. Newton's second law tells us that an object accelerates if a force acts on it. According to Newton's second law, then, the acceleration **a** of an object on which gravity alone acts can be written in terms of the force of gravity, $\mathbf{F}_g$, and the inertial mass m of an object:

$$\mathbf{a} = \frac{\mathbf{F}_g}{m}. \tag{5–1}$$

All objects at the same location accelerate under gravity in the same way, regardless of their mass.

A very special characteristic of the force of gravity is that, at any given location, *it causes all objects to accelerate in the same way—no matter what their mass.* In other words, the right-hand side of Eq. (5–1) is independent of the mass m; *the only way this can happen is for the force itself to be proportional to the mass*, so that the factor m in the denominator cancels with another such factor in the force itself.

FIGURE 5–1 A number of forces act on the child playing on a rope swing, including gravity and the tension of the rope.

Let's look more carefully at the meaning of this remarkable cancellation. That an object has constant acceleration under the influence of gravity means only that this force takes the form

$$\mathbf{F}_g = m_g\mathbf{g}, \tag{5–2}$$

where **g** is a constant vector with dimensions of acceleration, pointing to the center of Earth.[†] The "gravitational mass" m_g is the property of the object that determines the strength of the gravitational force acting on it. However, the acceleration of *all* objects under the influence of gravity is *precisely the same*; mathematically, this statement is

$$m_g = m. \tag{5–3}$$

This relation is remarkable because we have no reason to think that the force of gravity has anything special to do with the inertial mass. We recall from Chapter 4 that the inertial mass determines the *response* to a force, and tests to determine the inertial mass are possible without using gravity at all. Conversely, tests to determine the force of gravity on an object can be performed without using motion at all—say, by observing the equilibrium stretch of a spring while the object hangs from it under the influence of gravity. Nevertheless, the equality of Eq. (5–3) has been experimentally verified to a very high degree of accuracy.[‡] Thus the force of gravity has the simple form

The force of gravity

$$\mathbf{F}_g = m_g\mathbf{g} = m\mathbf{g}. \tag{5–4}$$

The acceleration **a** of any object at Earth's surface under the influence of gravity alone is given by the constant vector **g**:

$$\mathbf{a} = \frac{\mathbf{F}_g}{m} = \mathbf{g}. \tag{5–5}$$

Do not confuse weight with mass. Weight is a force.

As described in Chapter 3, the magnitude of **g** on the surface of Earth is roughly

$$g = 9.8\ \text{m/s}^2. \tag{5–6}$$

[†]This force is present near any large spherical object, but the magnitude of **g** is different.

[‡]The so-called Eötvös experiments verify that the inertial and gravitational masses are the same to one part in 10^{12}.

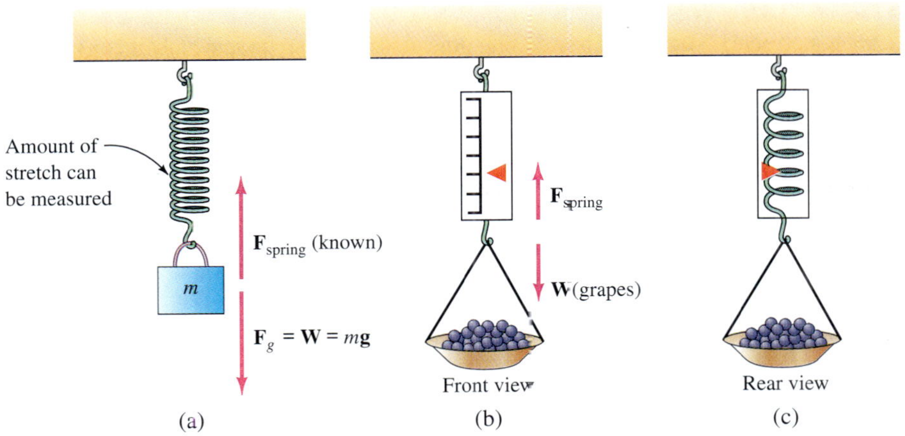

FIGURE 5–2 The weight of an object is found by balancing it against a spring, for which the force law is known. (b) and (c) illustrate the calibration.

Experiments show that this value varies by about 1 percent over Earth's surface, with the higher values occurring at the poles. This observed variance is due to irregularities in the shape and density of Earth and to Earth's rotation.

The force of gravity on an object is commonly called the object's **weight**, **W**, where

$$\mathbf{W} = \mathbf{F}_g = m\mathbf{g}. \tag{5–7}$$

APPLICATION

Bathroom scales

We can experimentally determine the weight of an object by balancing the force of gravity against a second calibrated force that is acting on the object. In Fig. 5–2a, for example, the weight of a block is found by using a calibrated spring as the second force. Figures 5–2b and c illustrate this calibration. In fact, when you step on a bathroom scale, you are measuring the force of gravity on your body; the second force is the calibrated bathroom scale, which is, in essence, a compressible spring. Don't confuse the weight of an object, measured in newtons or pounds, with its mass. We know from Chapter 4 that the mass of an object is the quantity of matter it contains, which is measured by its inertia—its resistance to any change in motion. The mass and weight of an object, though, are numerically proportional to each other through g—the weight of an object equals the object's mass times g. This numerical proportionality is what allows a bathroom scale to be calibrated in mass rather than weight.

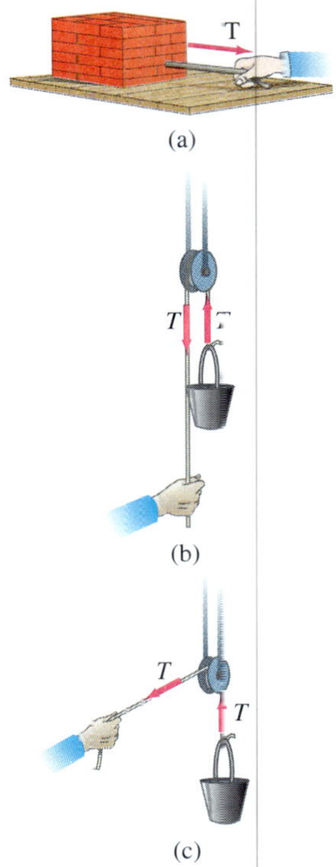

FIGURE 5–3 (a) The tension force pulls but cannot push. (b, c) A rope of negligible mass with a given tension can maintain that tension even if the direction of the rope is changed: Because the rope is passed over an ideal pulley, the tension can be transmitted around corners.

Tension

The tension

We know from experience that it is possible to pull objects using ropes. The tension, **T**, is a force exerted on an object by flexible ropes (or wires or cables or strings) that is directed along the direction of the rope; it always pulls, never pushes (Fig. 5–3a). *For light (negligible mass) ropes, the tension is the same everywhere along the rope.* An idealized rope of negligible mass—the kind of rope we employ in this book—is a rope whose mass is small compared to other masses in the problem.[†] The tension varies according to the particular situation. A rope pulling a wagon full of rocks has more tension than does the same rope pulling an empty wagon. Tension originates in the molecular forces that hold a rope together and give it flexibility. Note that the tension is the same everywhere in the rope, even when the rope turns corners, provided that it turns the corner by means of (idealized—negligible mass) pulleys turning on frictionless bearings (Figs. 5–3b, c).

PROBLEM SOLVING

How tension acts

[†]When the mass of the rope is not negligible, then the tension will vary along the rope.

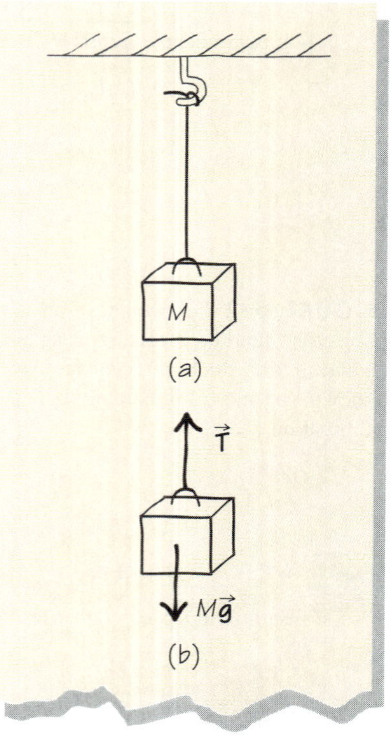

FIGURE 5–4 (a) Example 5–1. (b) Force diagram for a 2.0-kg package suspended from a line.

EXAMPLE 5–1 Let's consider Fig. 5–4a, which shows a fishing line hanging from a stationary support. This fishing line is 10-lb-test line, which means that it may break when the tension within it exceeds 10 lb. A package of mass $M = 2.0$ kg is suspended from the line. Find the tension in the line. Will the line hold? What will happen with a package of mass 5.0 kg?

Solution: We want to find the tension, T, in the line for the two different masses. We draw a force diagram for mass M (Fig. 5–4b) to determine the tension: As long as the line holds, $\mathbf{T}$ and $M\mathbf{g}$ balance, and there is no acceleration of the package. Without acceleration, Newton's second law reads

$$\mathbf{F}_{\text{net}} = \mathbf{T} + M\mathbf{g} = 0.$$

The tension is directed upward while the vector $\mathbf{g}$ is directed downward, so the second law is represented by a one-dimensional equation:

$$T - Mg = 0.$$

Thus, when the mass is suspended, the tension in the line has magnitude

$$T = Mg.$$

When $M = 2.0$ kg, we have

$$T = Mg = (2.0\text{ kg})(9.8\text{ m/s}^2) = 19.6\text{ N} = (19.6\,\cancel{\text{N}})\left(\frac{1\text{ lb}}{4.45\,\cancel{\text{N}}}\right) = 4.4\text{ lb}.$$

The line will hold.

When $M = 5.0$ kg,

$$T = (5.0\text{ kg})(9.8\text{ m/s}^2) = 49\text{ N} = (49\,\cancel{\text{N}})\left(\frac{1\text{ lb}}{4.45\,\cancel{\text{N}}}\right) = 11\text{ lb}.$$

The line may not hold with the heavier package suspended from it because the tension within it exceeds its breaking tension.

How could we measure the tension within a rope at a given point? One way would be to cut the rope at the given point and insert a calibrated spring scale. The scale will stretch by an amount corresponding to the tension in the rope.

Normal Force

The normal force

Let's take a look at the forces acting on an apple that has been placed on a table (Fig. 5–5). In addition to the force of gravity, a *normal force*, $\mathbf{F}_N$, acts on the apple. In Chapter 4, we were introduced to this force and acknowledged that such a normal force must exist—or gravity would cause the apple to simply accelerate into the surface of the table. We also learned that the normal force acts perpendicular to, and away from, the surface at the object's point of contact. We know that the normal force acts in a direction perpendicular to the table top because the apple does not accelerate to one side of the table or the other. In fact, whatever the tilt of the table, the apple does not move into its surface because *the normal force always acts perpendicular to, and away from, the surface.* The normal force pushes but never pulls.

PROBLEM SOLVING

Normal forces are perpendicular to the surface.

As for tension, the magnitude of $\mathbf{F}_N$ is determined by the situation, often by the use of Newton's first law. An apple of 0.25 kg sitting on a table must experience an upwardly directed normal force that just cancels the downward force of gravity; the magnitude of this force would be: $mg = (0.25\text{ kg})(9.8\text{ m/s}^2) = 2.45$ N.

How can we measure the normal force? An ordinary bathroom scale suffices. The spring in the scale opposes the force of gravity in exactly the same way as does the normal force and therefore gives the value of the normal force. In Fig. 5–6, a scale measures the normal force of 2.5 N that acts on the apple (although the force is often

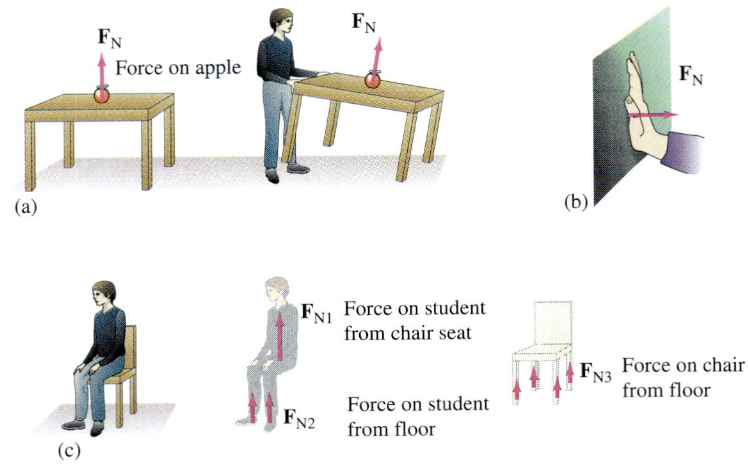

FIGURE 5–5 Some normal forces. These forces are always perpendicular to the surface involved.

expressed in units other than newtons!). The scale could equally well be placed on other surfaces that exert normal forces. For example, the normal force exerted on a hand by a wall can be measured directly if a scale is inserted between the wall and the hand.

There are other types of forces that are similar to the tension and the normal force; namely, the support forces supplied by hooks, connection points, bearings, and so forth. The direction and magnitude of such forces are determined by the requirement that the attachment point in question does not move. For example, a hook from which a 20-lb weight is suspended must supply an upward force of 20 lb if the mass is not to fall.

Some Examples

In many everyday situations, the tension and the normal force are constant. Examples involving constant tensions, constant normal forces, and gravity provide us with solvable equations of motion. We shall investigate several such examples here.

It may be helpful for us to recall the motion that results when *any* constant force **F** acts on an object of mass m. The object's velocity and position as a function of time represent the solutions to Newton's second law, $\mathbf{F} = m\mathbf{a}$. In one dimension, where the acceleration, the velocity, and the position have only x-components, these solutions are

$$x = x_0 + v_0 t + \frac{1}{2}at^2 = x_0 + v_0 t + \frac{F}{2m}t^2, \qquad (5\text{–}8)$$

for constant force:

$$v = v_0 + at = v_0 + \frac{F}{m}t. \qquad (5\text{–}9)$$

The constants x_0 and v_0 refer here to the values of position and velocity at time $t = 0$ and are usually specified in the statement of the problem. Sometimes measurements of position and velocity at two different times allow us to determine x_0 and v_0. For a force with three constant components, we must apply the solution to each component separately. Remember that these formulas apply only when **F** is constant.

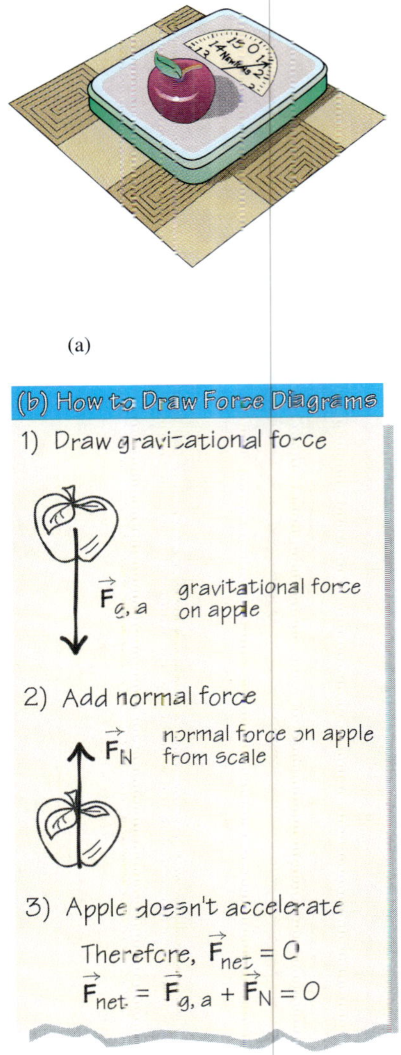

FIGURE 5–6 (a) A scale can be used to measure the normal force on an apple. (b) We can draw the forces acting on the apple.

EXAMPLE 5–2 A coffee cup of mass 75 g is placed on a slippery (frictionless) ramp tilted at an angle $\theta = 20°$ (Fig. 5–7a). The coffee cup starts from rest at $t = 0$ s and slides down the ramp. How far down the ramp has the cup moved after 2.0 s?

Solution: As emphasized in Chapter 4, we first prepare a force diagram, choose a convenient set of axes, and decompose the forces into their components along these

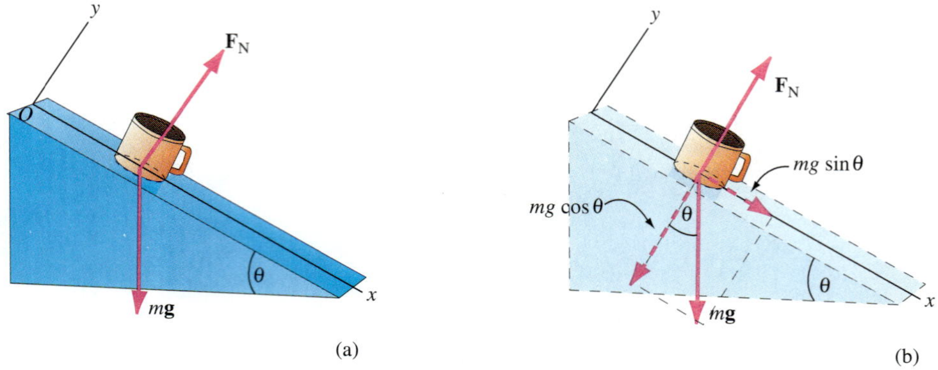

FIGURE 5–7 (a) Example 5–2. A coffee cup on a frictionless, tilted ramp. (b) Force diagram for the coffee cup, including the vector decomposition of the forces.

axes (Fig. 5–7b). We then write Newton's second law and solve the equations of motion. Because there is no friction, the only forces acting on the cup are gravity, $\mathbf{F}_g$, and the normal force, $\mathbf{F}_N$. *Note*: $\mathbf{F}_N$ is perpendicular to the surface of the inclined plane and thus it is *not* oriented in the vertical direction.

We choose the axes shown in Fig. 5–7b because we know from experience that the cup will not "jump" off the ramp; in other words, the acceleration **a** of the cup will be along the ramp and will not have a component perpendicular to the ramp. Accordingly, the x-direction points down along the ramp and the origin is placed at the starting point of the coffee cup. The acceleration **a** will have only an x-component a_x:

$$\mathbf{a} = a_x\,\mathbf{i}.$$

This choice of axes is also convenient for the normal force, which has only a y-component:

$$\mathbf{F}_N = F_N\,\mathbf{j}.$$

The magnitude F_N is as yet unknown. We will find it below from the requirement that the acceleration has no y-component. The force of gravity has both x- and y-components (Fig. 5–7b):

$$\mathbf{F}_g = mg\sin\theta\,\mathbf{i} - mg\cos\theta\,\mathbf{j}.$$

where m is the mass of the cup.

Newton's second law has the vector form

$$\mathbf{F}_g + \mathbf{F}_N = m\mathbf{a},$$

or

$$mg\sin\theta\,\mathbf{i} - mg\cos\theta\,\mathbf{j} + F_N\,\mathbf{j} = ma_x\,\mathbf{i}.$$

Each component of this equation is a separate equation of motion:

$$\text{for the } x\text{-component:} \quad mg\sin\theta = ma_x; \tag{5–10}$$

$$\text{for the } y\text{-component:} \quad -mg\cos\theta + F_N = 0. \tag{5–11}$$

Now we solve the equations of motion. Equation (5–10) gives

$$a_x = g\sin\theta. \tag{5–12}$$

Equation (5–11) shows that $\mathbf{F}_N$ has magnitude $mg\cos\theta$.

To find how the position of the coffee cup varies with time, we use the kinematical equation for constant acceleration in one dimension: Eq. (5–8). In this equation, the cup starts at the origin so $x_0 = 0$; moreover, $v_0 = 0$ because the cup starts at rest. We then insert Eq. (5–12) into Eq. (5–8), with the result

$$x = \frac{1}{2}a_x t^2 = \frac{1}{2}g\sin\theta\,t^2.$$

At $t = 2.0$ s,

$$x = (0.5)(9.8 \text{ m/s}^2)(\sin 20°)(2.0 \text{ s})^2 = 6.7 \text{ m}.$$

We find the cup has moved 6.7 m after 2.0 s. As is typical for problems involving gravity, the answer is independent of the mass of the cup. The angle $\theta = 0°$ presents a special limit; in this case, the plane is flat, and there should be no acceleration whatsoever. This is ensured by the $\sin\theta$ factor in the acceleration, which is zero when $\theta = 0°$. The normal force has magnitude mg, as expected. A second limit is the case $\theta = 90°$, when $\sin\theta = 1$. Here, $a_x = g$ and $F_N = 0$, as expected.

EXAMPLE 5–3 Because of a wager, a woman wishes to lift a professional football player off his feet. The player is a huge interior lineman with a mass of 149 kg. (He weighs 328 lb.) The woman has devised a system for the task, which is shown in Fig. 5–8a. We will assume that all pulleys, ropes, and miscellaneous gear in the apparatus have negligible mass and are frictionless. What is the magnitude of the downward force the woman must exert on the end of the rope in order to lift the lineman?

Solution: Even a smooth lift at constant velocity is sufficient; we do not ask that the lineman accelerate continuously. This condition is met when the net force on the lineman balances his weight. The tension in the rope is transmitted all the way through the rope, and the tension is the force that must be supplied by the woman pulling on the rope. Thus, we want to find the tension, T.

If we examine Fig. 5–8a, we see that all four rope segments, 1 through 4, pull upward on the lineman. Figure 5–8b is the force diagram for the lineman and the two lower pulleys. As the rope becomes taut when the woman begins to pull, the sum of the tensions $T_1 + T_2 + T_3 + T_4$ increases from zero. As long as this sum is less than the weight of the lineman, Mg, he will remain on the ground. But when this sum becomes equal to Mg, there is no net force on the lineman, and any additional tension—no matter how small—will start him accelerating upward. Thus the condition to lift the lineman is

$$T_1 + T_2 + T_3 + T_4 = Mg.$$

Because T is the same everywhere in the rope,

$$4T = Mg.$$

$$T = \frac{Mg}{4} = \frac{(149 \text{ kg})(9.80 \text{ m/s}^2)}{4} = 365 \text{ N}.$$

This is the magnitude of the smallest downward force that the woman must apply to her end of the rope. Provided that she weighs more than 365 N (82.1 lb), which corresponds to a mass of $m = W/g = 37.3$ kg, she can apply at least this force just by hanging on the rope.

The arrangement described here is called a *block and tackle.*

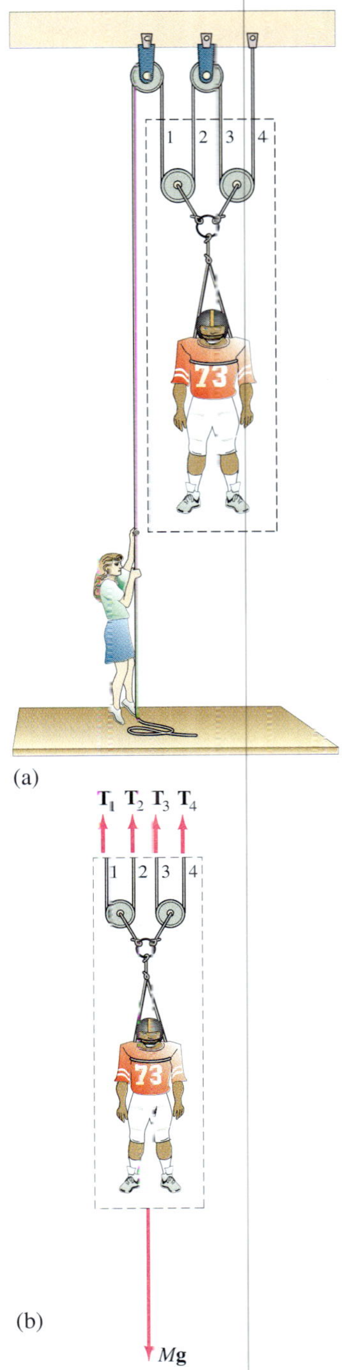

FIGURE 5–8 (a) Example 5–3. (b) Force diagram for the lineman.

APPLICATION

Block and tackle

In Examples 5–4 and 5–5, we will investigate all three forces discussed so far in this section. These examples also introduce something new: Two masses are involved. Each mass requires an identification of the forces acting on it, a force diagram, and an equation of motion.

EXAMPLE 5–4 Masses $m_1 = 1.0$ kg and $m_2 = 1.1$ kg are attached to opposite ends of a massless rope draped over a pulley (Fig. 5–9a). (This device is called an *Atwood machine.*) Mass m_2 rests on a scale that measures the normal force exerted on m_2. There is no motion. What is the reading on the scale, and what is the tension in the rope?

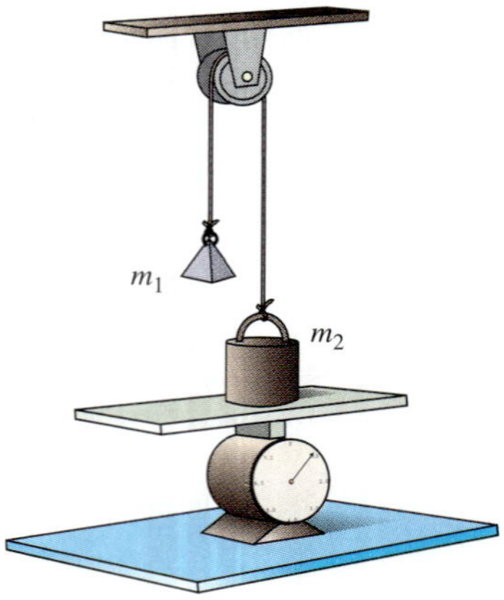

(a)

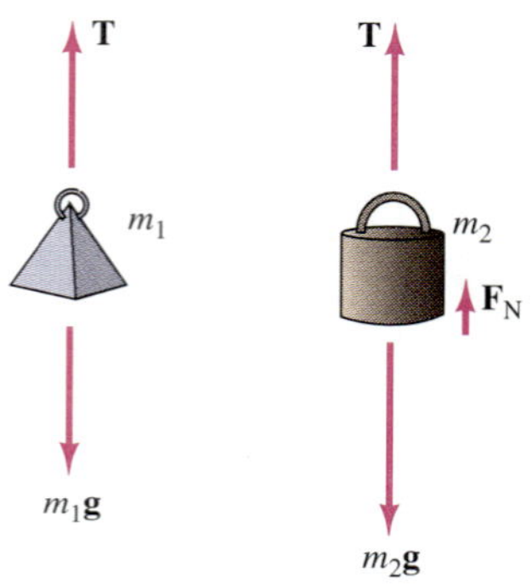

(b)

FIGURE 5–9 (a) Example 5–4. An Atwood machine. (b) Force diagrams for the masses.

PROBLEM SOLVING

It is often useful to choose different coordinates for each part of a system.

Solution: Let's first prepare a force diagram for each mass (Fig. 5–9b) and find the equations of motion for each mass. Because there is no acceleration, the net force on each mass is zero. All three forces—gravity, the tension, and the normal force—are vertical, so we have only one component (y-component) to consider; thus, we drop the vector (boldface) notation.

Take the upward direction to be positive. The forces acting on m_1 are gravity ($-m_1g$) and the rope tension (T). Newton's first law gives

$$F_{\text{net}} = -m_1g + T = 0. \tag{5–13}$$

Three forces act on mass m_2: $-m_2g$, T, and F_N. Notice that the same value of T acts on each mass, pulling (acting upward) in each case. Newton's first law for mass m_2 is then

$$F_{\text{net}} = -m_2g + T + F_N = 0. \tag{5–14}$$

We can immediately solve Eq. (5–13) for the tension:

$$T = m_1g = (1.0\ \text{kg})(9.8\ \text{m/s}^2) = 9.8\ \text{N}.$$

The tension is one of the quantities we were asked to find. If we insert this value of T into Eq. (5–14), we can solve for the normal force and hence the scale reading:

$$F_N = m_2g - T = m_2g - m_1g = (m_2 - m_1)g$$

$$= (1.1\ \text{kg} - 1.0\ \text{kg})(9.8\ \text{m/s}^2) = (0.1\ \text{kg})(9.8\ \text{m/s}^2) \simeq 1\ \text{N}.$$

For comparison, the weight of m_2 alone is $(1.1\ \text{kg})(9.8\ \text{m/s}^2) = 10.8\ \text{N}$.

EXAMPLE 5–5 Consider two masses $m_1 = 1.00$ kg and $m_2 = 2.00$ kg connected by a rope that passes over an ideal pulley. Mass m_1 hangs straight down, while m_2 slides without friction on a ramp inclined at an angle θ. At $t = 0$, the system is started from rest in the position shown in Fig. 5–10a. Describe the motion of the two masses for $\theta = 25.0°$. Find the angle θ' for which the system remains motionless.

Solution: In this problem, we again have two masses for which we must write the equations of motion: $\mathbf{F}_{\text{net on 1}} = m_1\mathbf{a}_1$ and $\mathbf{F}_{\text{net on 2}} = m_2\mathbf{a}_2$. Connected as they are by a rope, the motion of one mass is correlated with the motion of the other. *The magnitudes of the accelerations are the same.* We begin by drawing a force diagram for each mass (Fig. 5–10b). A force with common magnitude acts on each mass; this force is the tension, $\mathbf{T}$. $\mathbf{T}$ and gravity $m_1\mathbf{g}$ act on m_1. Three forces act on m_2: the normal force, $\mathbf{F}_N$, tension, $\mathbf{T}$, and gravity, $m_2\mathbf{g}$.

Once the force diagrams are drawn, we choose coordinate systems and write Newton's second law. *We are not compelled to choose the same coordinate system for the two masses.* In fact, choosing different systems makes this problem easier to solve (Fig. 5–10b). We choose the origin of each coordinate system to be at the location of the respective mass at $t = 0$. Because m_1 moves only in the vertical direction, labeled y_1, its acceleration $\mathbf{a}_1$ is aligned with y_1. For m_1, then, we need look only at the component of Newton's second law along y_1,

$$T - m_1g = m_1a_1. \tag{5–15}$$

Although the mass m_2 has forces acting in two directions, it moves only in the direction labeled x_2 in Fig. 5–10b. Thus its acceleration vector $\mathbf{a}_2$ is aligned with x_2. We must now decompose the forces acting on m_2 into components. $\mathbf{F}_N$ is in the $+y_2$-direction and $\mathbf{T}$ is in the $-x_2$-direction. The third force, gravity, has two components; we can determine these components by recalling from geometry that the angle θ indicated in Fig. 5–10b is the same as the ramp angle θ in Fig. 5–10a. Then the force of gravity $m_2\mathbf{g}$ has x_2-component $m_2g \sin\theta$ and y_2-component $-m_2g \cos\theta$. Thus Newton's second law for mass m_2, $\mathbf{F}_{\text{net}} = \mathbf{F}_N + \mathbf{T} + m_2\mathbf{g} = m_2\mathbf{a}_2$, breaks down into the two component equations for the x_2-component: $$-T + m_2 g \sin\theta = m_2 a_2; \tag{5–16}$$

for the y_2-component: $$F_N - m_2 g \cos\theta = 0. \tag{5–17}$$

The three equations (5–15), (5–16), and (5–17) are not enough to solve for the four unknowns T, F_N, a_1, and a_2. The missing fourth equation expresses the connection between the motion of m_1 and that of m_2; that is, the motion of m_1 in the $+y$-direction is matched by the motion of m_2 in the $+x$-direction, so

$$a_1 = a_2 \equiv a. \tag{5–18}$$

The single acceleration magnitude a is then substituted in Eqs. (5–15), (5–16), and (5–17), which become three equations for the three unknowns a, T, and F_N. Equation (5–17) gives the normal force,

$$F_N = m_2 g \cos\theta.$$

The tension cancels in the sum of Eqs. (5–15) and (5–16):

$$T - m_1 g - T + m_2 g \sin\theta = m_1 a + m_2 a.$$

We are left with an equation for a with solution

$$a = \frac{m_2 \sin\theta - m_1}{m_1 + m_2} g. \tag{5–19}$$

Now that we have found the accelerations, we can find positions as a function of time by using the constant acceleration equations: Eq. (5–8). Each mass starts from rest at the origin of its respective coordinate system, so $v_0 = 0$ for both masses, and both y_{10} ($\equiv y_1$ at $t = 0$) and x_{20} ($\equiv x_2$ at $t = 0$) are zero. Thus

for m_1: $$y_1 = \frac{1}{2}at^2,$$

for m_2: $$x_2 = \frac{1}{2}at^2.$$

The masses move only in these directions. Together with Eq. (5–19), these equations describe the motion fully.

For the special case $\theta = 25.0°$, we use $\sin 25.0 = 0.423$, so

$$a = \frac{(2.00\ \text{kg})(0.423) - 1.00\ \text{kg}}{1.00\ \text{kg} + 2.00\ \text{kg}}(9.80\ \text{m/s}^2) = -0.503\ \text{m/s}^2.$$

Thus $y_1 = (-0.503\ \text{m/s}^2)t^2/2 = x_2$. Note the minus sign in the acceleration: The sign indicates that m_1 drops and m_2 moves up the ramp. Equation (5–19) shows that the acceleration—including its sign—depends on the masses and the ramp angle.

We must also find the angle for which the acceleration is zero given m_1 and m_2. According to Eq. (5–19), the acceleration is zero at an angle θ' for which

$$m_2 \sin\theta' - m_1 = 0, \tag{5–20}$$

or $\sin\theta' = m_1/m_2$. For this particular problem, $\sin\theta' = \dfrac{1.00}{2.00}$, or $\theta' = 30.0°$.

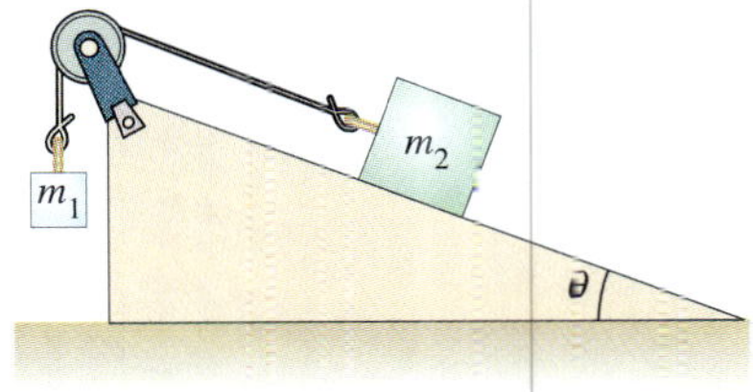

(a)

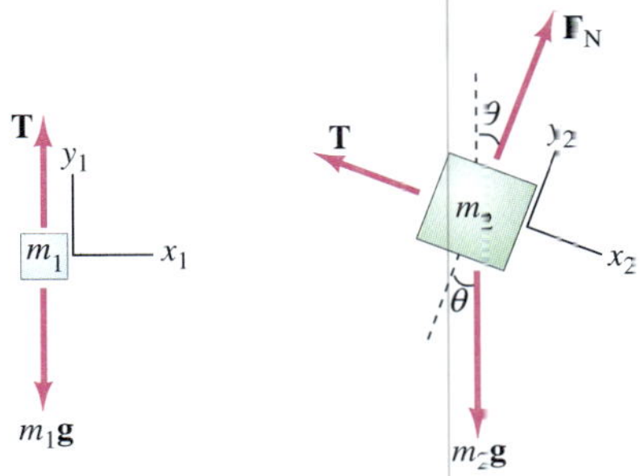

(b)

FIGURE 5–10 (a) Example 5–5. (b) Force diagrams for masses m_1 and m_2.

5–2 FRICTION

Friction is a familiar concept to all of us. It is a contact force that impedes sliding. It allows us to walk, it allows cars to move along the road, and it even holds nails and screws in place (Fig. 5–11). The need to reduce friction in machinery requires us to learn how to lubricate rubbing surfaces. Some 20 percent of the power of an automobile is spent overcoming internal friction—and where there is friction, there is surface wear.

FIGURE 5–11 Nails are held in place by the force of friction, which can be quite substantial.

PROBLEM-SOLVING TECHNIQUES

When more than one mass appears in a problem, there are two useful techniques to keep in mind:

1. There may be a constraint equation that connects the motion of the masses, as for the two masses connected by a taut rope in Example 5–5.
2. You can choose different coordinate systems for the different masses. As in Example 5–5, a problem can be made simpler if the coordinate systems are well chosen.

Let's suppose that you want to slide a crate from one place to another. You push on it with a small horizontal force, but nothing happens. Even when you push as hard as you can, the crate does not move. Why not? **Static friction** acts between the floor and the crate in the absence of motion in such a way as to *prevent* motion. This force must be variable because it balances each of your own different pushes. Suppose that you finally get the crate moving with the help of another person. The combined force overcame the static friction because *static friction has a maximum magnitude*.

Once the crate is moving, it is easier to keep it moving at a constant speed. There is still friction opposing your push, but it is now **kinetic** (or **sliding**) **friction**; that is, friction associated with motion. The magnitude of kinetic friction is smaller than the maximum value of static friction. The entire sequence of getting the crate started and keeping it moving is illustrated in Fig. 5–12. Both static and kinetic friction act along the surface (Fig. 5–13). Static friction acts in a direction opposite to the component of an applied force along the surface; sliding friction acts opposite to the direction of the velocity of a sliding object at its point or points of contact.

The Force of Friction: A Quantitative Treatment

Studies of the nature of friction were made by Leonardo da Vinci some 200 years before Newton's work on dynamics. Leonardo experimented with a set of blocks of varying sizes sliding on table tops (Fig. 5–14) and learned some surprising facts. He found that *both static and kinetic friction are independent of the apparent contact area of the block on the table top. Moreover, both static and kinetic friction are proportional to the magnitude F_N of the normal force exerted by the table top on the blocks.* To determine that the force is independent of the apparent area that is in contact with the surface, Leonardo simply took a given block and turned it so that faces of different areas were in contact with the table top. He found that the friction force on the block was the same no matter what face of the block was in contact with the table top.

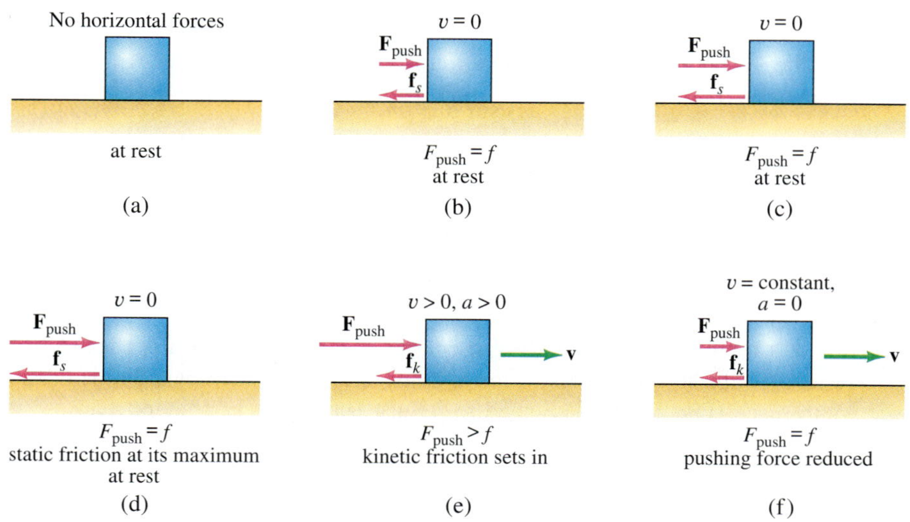

FIGURE 5–12 A sequence in which an increasing pushing force is opposed by an equal and opposite static friction force. Static friction can increase only to a certain point, after which the crate accelerates. Kinetic friction, which applies when the crate actually moves, is smaller than the maximum size of static friction, so it takes less pushing force to move the crate at that point.

The proportionality constant that relates the friction force and the normal force is the **coefficient of friction, μ**. The (unitless) constant μ is determined experimentally. The maximum value of static friction is generally not equal to the force of kinetic friction, so we distinguish two coefficients: μ_s for static friction and μ_k for kinetic friction.

If we write the force of static friction as $\mathbf{f}_s$ and that of kinetic friction as $\mathbf{f}_k$, their magnitudes are given by

$$\text{static friction:} \quad 0 \le f_s \le \mu_s F_N \tag{5–21}$$

and

$$\text{kinetic friction:} \quad f_s = \mu_k F_N. \tag{5–22}$$

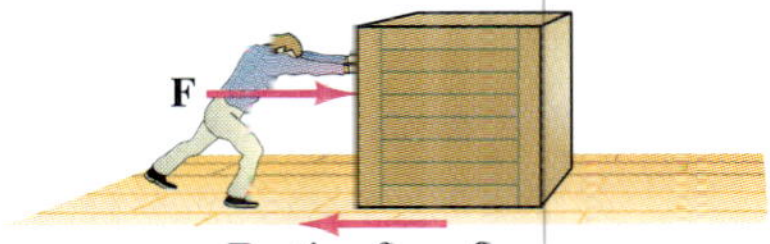

FIGURE 5–13 Friction opposes the motion of the crate.

The forces of static and kinetic friction

Equation (5–21) expresses a range because static friction is variable. The experimental fact that the maximum value of static friction exceeds kinetic friction implies the inequality

$$\mu_s > \mu_k. \tag{5–23}$$

We also make the assumption, reasonably well satisfied, that μ_k *is independent of the relative speed of the two surfaces.*

The coefficients of friction depend on the two surfaces involved. We know from everyday experience that a basketball shoe on a basketball court involves a larger coefficient of friction than does the blade of an ice skate on a frozen lake. A lubricating material—such as sweat—between the basketball shoe and the court will drastically reduce the coefficient of friction. Table 5–1 shows some typical values of coefficients of static friction. The materials listed are generally unlubricated ("dry"). Coefficients of kinetic friction can be anywhere from roughly 25 percent to 100 percent of the corresponding coefficients of static friction.

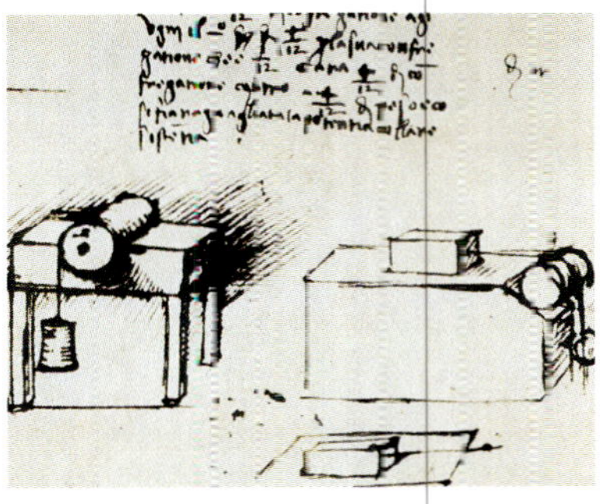

FIGURE 5–14 A page from Leonardo's notebooks. By simply turning the faces of blocks in contact with a table top, Leonardo showed that the friction between the blocks and the table top is independent of the apparent surface area in contact.

The values in Table 5–1 are meant only to be indicative—the coefficients of friction are sharply dependent on such things as the cleanliness of the surfaces, their roughness, and so forth. Two very rough objects may have a large coefficient of friction that can be reduced once the objects are smoothed. But if they are smoothed too much and are free of dirt and oxidation as well, the coefficient of friction may rise virtually to infinity because the surfaces weld together!

In Example 5–6, we will learn one way to measure the coefficient of static friction.

TABLE 5–1
Some Coefficients of Static Friction†

Materials	μ_s
Brake material on a brake drum	1.2
A dry tire on dry asphalt	1.0
Hard steel on hard steel	0.3
Oak on oak, parallel to the grain	0.6
A book on a table	0.3
A wet tire on wet asphalt	0.2
Ice on wood	0.05
Teflon on steel	0.04

†The coefficients of kinetic friction are from 25 percent to 100 percent of the corresponding coefficients of static friction.

EXAMPLE 5–6 A mass m is set at rest on a horizontal ramp; there is friction between the mass and the ramp. The ramp is slowly raised from its horizontal orientation. By analyzing the forces on the mass, find the critical ramp angle θ_c at which the mass will start to slide.

Solution: In Fig. 5–15, we can see the ramp and the mass. The forces on the mass are drawn and a suitable coordinate system is shown, with the origin at the initial position of the mass. The $+x$-direction is oriented down along the ramp, and the y-direction is perpendicular to the ramp. This figure thus serves as a force diagram. The three forces acting on the mass before it starts to move are the force of gravity (vertically down), the normal force, $\mathbf{F}_N$ (perpendicular to the surface), and the force of static friction, $\mathbf{f}_s$ (up the ramp). The friction force opposes the motion that would take place if there were no friction. The force magnitudes F_N and f_s are each determined by the fact that the mass remains motionless for a sufficiently shallow ramp angle. Note that we must find F_N because the maximum value of static friction depends on F_N [see Eq. (5–21)].

If the mass is stationary, Newton's second law with zero acceleration applies: $\mathbf{F}_g + \mathbf{F}_N + \mathbf{f}_s = 0$, or, in Cartesian form,

$$mg \sin\theta\, \mathbf{i} - mg \cos\theta\, \mathbf{j} + F_N \mathbf{j} - f_s\, \mathbf{i} = 0.$$

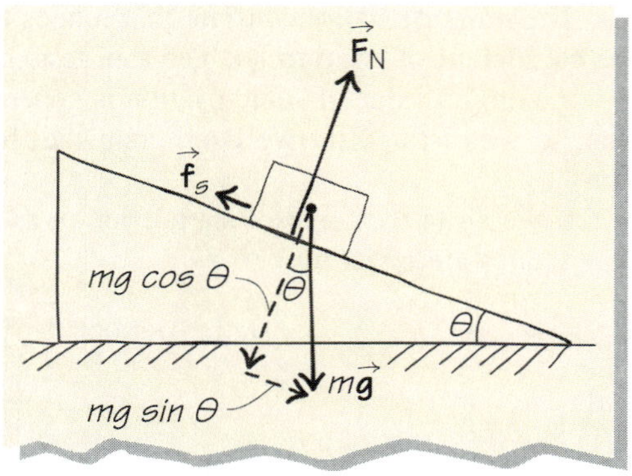

FIGURE 5–15 Example 5–6. A mass m experiencing a friction force on a ramp. The angle at which the mass starts to slide tells us the coefficient of static friction.

In component form, the equations of motion are

$$\text{for the } x\text{-component:} \qquad mg\sin\theta - f_s = 0, \tag{5–24a}$$

$$\text{for the } y\text{-component:} \qquad F_N - mg\cos\theta = 0. \tag{5–24b}$$

The x-component equation relates the force of static friction to the force of gravity,

$$f_s = mg\sin\theta. \tag{5–25}$$

The y-component equation determines F_N as a function of θ:

$$F_N = mg\cos\theta. \tag{5–26}$$

As θ (and $\sin\theta$) increases, the force of friction from Eq. (5–25) required to hold the mass in static equilibrium also increases. Eventually, static friction reaches its maximum value, $\mu_s N$. Beyond that point, the mass will start to slide. When we set static friction to its maximum value, $f_s = \mu_s F_N$, Eq. (5–25) determines the critical angle θ_c at which the mass starts to slide. Equation (5–25) then becomes

$$\mu_s F_N = mg\sin\theta_c,$$

or, from Eq. (5–26),

$$\mu_s mg\cos\theta_c = mg\sin\theta_c.$$

Cancel the factor mg from this equation to find θ_c:

$$\frac{\sin\theta_c}{\cos\theta_c} = \tan\theta_c = \mu_s. \tag{5–27}$$

Thus if we measure the angle at which the mass begins to slip, we measure μ_s.

With an extension of the reasoning used in Example 5–6, we can figure out how to measure the coefficient of kinetic friction. Once the mass begins to slip, kinetic friction acts, given by $f_k = \mu_k F_N = \mu_k mg\cos\theta$. If we use Newton's second law and apply it along the x-direction, then, instead of Eq. (5–24a), we find:

$$mg\sin\theta - \mu_k mg\cos\theta = ma_x. \tag{5–28}$$

Because μ_k is smaller than μ_s, the ramp can be lowered back down, decreasing θ while the mass is still sliding. There is a second critical value of θ—call it θ'_c—for which the forces in Eq. (5–28) cancel and the object no longer accelerates. Instead, the object slides at constant velocity. This critical angle is given by

$$\frac{\sin\theta'_c}{\cos\theta'_c} = \tan\theta'_c = \mu_k, \tag{5–29}$$

Thus θ'_c measures the coefficient of kinetic friction.

Let's now take a look at Example 5–7, which is a straightforward application of the concept of static friction.

EXAMPLE 5–7 A professor with a light eraser in her hand leans against a blackboard (Fig. 5–16a). Her arm makes an angle of 30° with the horizontal, and the force $\mathbf{F}_{\text{prof}}$ exerted by her arm on the eraser has magnitude $F_{\text{prof}} = 50$ N. The coefficient of static friction between the eraser and the blackboard is $\mu_s = 0.15$. Does the eraser slip?

Solution: Figure 5–16b is a force diagram for the eraser, which also indicates a useful coordinate system. If nothing moves, Newton's first law applies: $\mathbf{F}_{\text{N}} + \mathbf{f}_s + \mathbf{F}_{\text{prof}} = 0$. In the coordinate system on the figure,

$$-F_N \mathbf{i} - f_s \mathbf{j} + F_{\text{prof}} \cos\theta\, \mathbf{i} + F_{\text{prof}} \sin\theta\, \mathbf{j} = 0.$$

The two component equations are

$$\text{for the } x\text{-component:} \qquad -F_N + F_{\text{prof}} \cos\theta = 0,$$

$$\text{for the } y\text{-component:} \qquad -f_s + F_{\text{prof}} \sin\theta = 0.$$

The x-component equation determines F_N from the requirement that it balances the perpendicular component of the force the professor exerts. Once we have found that $F_N = F_{\text{prof}} \cos\theta$, we can determine the *maximum* value of static friction, $f_{s,\text{max}} = \mu_s F_N = \mu_s F_{\text{prof}} \cos\theta$. When this maximum value is exceeded, the eraser begins to slip. Thus, when we substitute the maximum value of static friction into the y-component equation, we find a condition for the critical angle θ_c for which the eraser begins to slip:

$$-\mu_s F_{\text{prof}} \cos\theta_c + F_{\text{prof}} \sin\theta_c = 0;$$

$$\frac{\sin\theta_c}{\cos\theta_c} = \tan\theta_c = \mu_s.$$

Note the striking feature that the critical angle is independent of the force the professor exerts! Numerical substitution yields $\tan\theta_c = 0.15$, or $\theta_c = 8.5°$. This angle is less than the 30° angle made by the arm, so the eraser indeed slips.

(a)

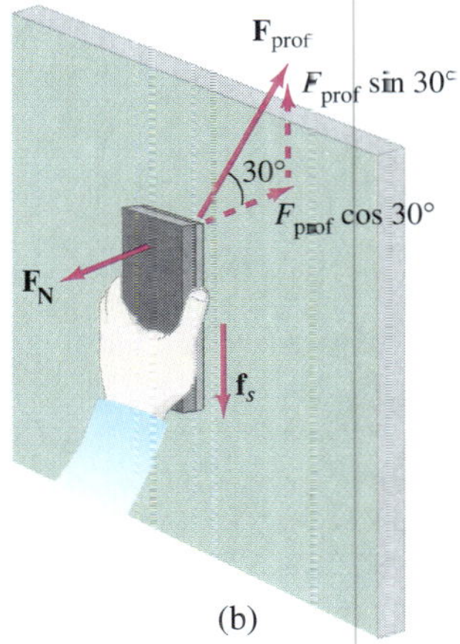

(b)

FIGURE 5–16 (a) Example 5–7. A professor leaning against an eraser at a blackboard. (b) Force diagram for the eraser at the blackboard.

Static friction does not always act to impede motion. If we are not dealing with a pointlike object or with a rigid system, then static friction can act to accelerate a system. Walking is perhaps the most familiar example of this phenomenon (Fig. 5–17): Static friction between our feet and Earth is the force that propels us forward when we walk. By Newton's third law, there is a corresponding backward-directed force on the planet. Without friction, our feet would fruitlessly slip *backward* when we tried to walk, an experiment that is easy to perform on a frozen puddle. This point is illustrated by Example 5–8.

FIGURE 5–17 A walker uses the force of static friction between shoe and ground to move forward.

EXAMPLE 5–8 An automobile with four-wheel drive and a powerful engine has a mass of 1000 kg. Its weight is evenly distributed on its four wheels, whose coefficient of static friction with the dry road is $\mu_s = 0.8$. If the car starts from rest on a horizontal surface, what is the greatest forward acceleration that it can attain without spinning its wheels?

Solution: The automobile has its greatest forward acceleration when the static friction between the wheels and the road is at its maximum. The engine of the car serves to impart the rotational motion shown in Fig. 5–18a to the wheels. If there were no friction between the wheels and the road, the wheels would simply spin. Because the wheels do not slip, it is *static* friction that acts here, and μ_s, not μ_k, enters the problem. The motion of the tires at the point of contact with the road is to the rear in the absence of friction. The force of friction opposes this rearward motion, so the direction of $\mathbf{f}_s$ is toward the front of the car. The forward frictional force is the *only* external horizontal force acting on the car and hence determines its forward acceleration. We draw the forces acting on the car in Fig. 5–18b.

Figure 5–18b is a force diagram for the car, and Newton's second law is

$$\mathbf{F}_{\text{net}} = \mathbf{F}_{\text{N}} + m\mathbf{g} + \mathbf{f}_s = m\mathbf{a},$$

where m is the mass of the car. (Actually, there is a separate normal force and friction force at each wheel, but because the weight is distributed evenly over each wheel, all the normal and friction forces at each wheel are equal.) With the coordinate system in the figure, $\mathbf{F}_{\text{N}} = F_N\mathbf{j}$, $m\mathbf{g} = -mg\,\mathbf{j}$, $\mathbf{f}_s = f_s\,\mathbf{i}$, and the acceleration is forward, so $\mathbf{a} = a\mathbf{i}$. Newton's second law is then

$$F_N\,\mathbf{j} - mg\,\mathbf{j} + f_s\,\mathbf{i} = ma\,\mathbf{i},$$

and the component equations are

for the x-component: $\quad f_s = ma,$

for the y-component: $\quad N - mg = 0.$

The acceleration is a maximum a_{max} when static friction reaches its maximum value $\mu_s F_N = \mu_s mg$, where we have used $F_N = mg$ from the y-component equation. Thus

$$\mu_s mg = ma_{\text{max}}.$$

The mass of the car cancels out of this expression, leaving

$$a_{\text{max}} = \mu_s g = (0.8)(9.8\ \text{m/s}^2) \simeq 8\ \text{m/s}^2.$$

This is quite a significant forward acceleration. Note that as μ_s decreases, the maximum acceleration decreases; in other words, when μ_s is zero, the automobile can only spin its wheels.

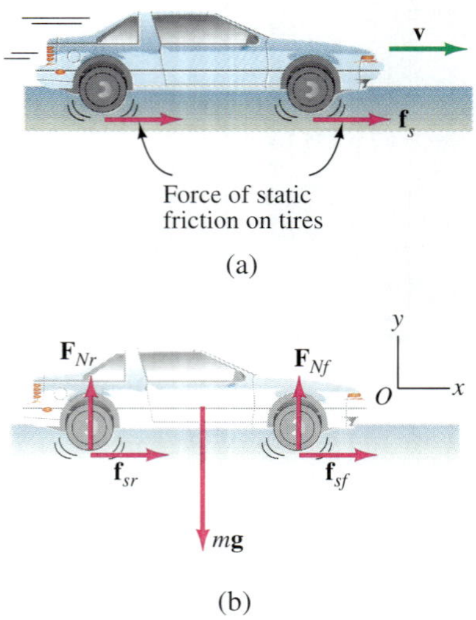

FIGURE 5–18 (a) Example 5–8. A car accelerates forward under the influence of static friction. (b) Force diagram for the car.

Where Does Friction Come From?

To understand the origins of friction, let's look closely at two surfaces that rub against one another. Figure 5–19a shows a microscopic view of two such surfaces in contact. Because of the hills and valleys present on any rough surface—and all surfaces are rough when viewed closely enough (Fig. 5–19b)—the area on two surfaces that actually touches together is a small fraction of the area that appears to be in contact. Friction forces are due to three major effects: the interlocking of surface irregularities; the attraction between the contact points due to forces between the molecules of the two objects (the objects "adhere"); and the "plowing out" of softer materials by harder ones. The coefficient of static friction can be greater than that of kinetic friction because the materials have a longer time to "settle in" together.

This description helps us understand why the friction force is independent of the apparent surface area that is in contact, while it is dependent on the normal force. The normal force is a measure of how strongly the two surfaces are pressed together; when the normal force is large, the two surfaces are pressed strongly together. The rough surfaces shown in Fig. 5–19a mesh more closely when pressed strongly together, and the

FIGURE 5–19 (a) A microscopic view of the contact region between two rough surfaces. (b) At lower magnification, the same surfaces are apparently smooth.

actual surface area in contact increases. In fact, the normal force is a good measure of the actual surface contact area! Whether we place the broad side or the narrow end of a brick on a table, approximately the same surface areas are in actual contact, even if the apparent contact areas are vastly different. So the friction force is, finally, proportional to the real contact area.

The study of friction, wear, and lubrication is called *tribology*. Despite much effort, a truly fundamental understanding of friction remains elusive. The discovery of Teflon™ was a happy accident, not the result of a planned development program. The discoverer has stated that he was lucky he was not blown up in the process.

5–3 DRAG FORCES

A spoon dropping through molasses, an automobile moving at highway speeds, and a plane using a parachute to slow down are all subject to a substantial *drag force*, which is a resistive force somewhat like friction. However, in motion *through* media such as gases or liquids, there is no normal force, yet there is still a drag force acting to impede the motion. Drag forces act like sliding friction in that *they act in a direction opposite to that of the motion*, but they differ from sliding friction in that they depend on the *speed* v of the object that is moving through the medium.

In many everyday situations—an automobile moving on a highway is one—the drag force, $\mathbf{F}_D$, is found by experiment to have magnitude

$$F_D = \frac{1}{2}\rho A C_D v^2, \tag{5–30}$$

where ρ is the mass density (the mass per unit volume) of the medium through which the object moves, A is the maximum cross-sectional area presented by the moving object, and C_D is the *drag coefficient*. The area A is the area found by cutting across the object in a plane perpendicular to the direction of motion. The drag coefficient, C_D, is dimensionless and depends on the shape of the object. A very highly streamlined object might have a drag coefficient as small as 0.1, whereas a particularly awkward shape will have a drag coefficient greater than 1. The most streamlined automobiles have drag coefficients around 0.25.

Terminal Velocity

The fact that the drag force on an object increases with the speed has an important consequence. Consider a falling parachutist who is acted on by both gravity and a drag force like that of Eq. (5–30) due to the air (Fig. 5–20). When the parachutist first starts to fall, his or her speed is slow, so the drag force is small and there is an acceleration of $\mathbf{g}$. As the parachutist's speed increases, so does the drag; at some point, the drag will actually cancel the force of gravity. Then the parachutist no longer accelerates and his or her speed remains constant at a maximum value. The maximum speed v_t is called the *terminal velocity*.[†] Figure 5–20 illustrates a parachutist who has reached terminal velocity.

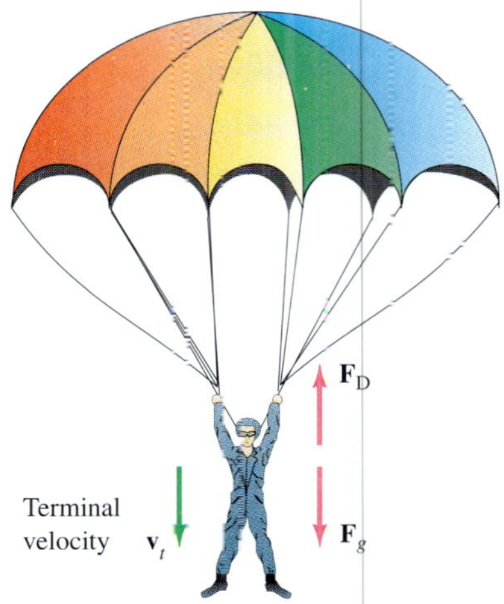

FIGURE 5–20 When the drag force and the force of gravity acting on a parachutist cancel, the parachutist has reached his terminal velocity.

Let's calculate the terminal velocity in the case of falling objects, where the applied force $\mathbf{F}_1$ is the force of gravity. We assume that the drag force is given by Eq. (5–30), which is a good approximation for all but very small falling objects. The terminal velocity is found from the condition that the forces of gravity and drag cancel: $F_g = F_D$, or

$$mg = \frac{1}{2}\rho A C_D v_t^2.$$

[†]Some typical values are: for a parachutist, 5 m/s; for a falling raindrop, 8 m/s; for a baseball, 40 m/s; for a skydiver, 60 m/s.

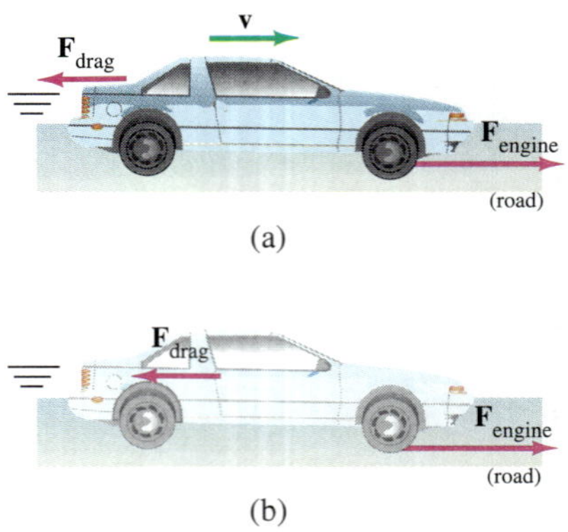

FIGURE 5–21 (a) Example 5–9. (b) Force diagram for an automobile that moves under the influence of the force provided by the engine and the drag of the air through which it moves.

We solve for the terminal velocity:

$$v_t = \sqrt{\frac{2mg}{\rho A C_D}}. \tag{5–31}$$

We can apply similar reasoning to the calculation of the terminal velocity for other external forces.

EXAMPLE 5–9 The maximum force with which a certain automobile engine can accelerate a car is 3200 N. At highway speeds, Eq. (5–30) is a good approximation to the true drag force. The density of air is about 1.2 kg/m^3, the cross-sectional area of the automobile is 3.4 m^2, and the drag coefficient is 0.50. Assuming that the drag force is the only force resisting the motion, what is the maximum speed of the automobile?

Solution: Figure 5–21 shows the forces and a force diagram for the automobile. It includes the forces $\mathbf{F}_{\text{engine}}$ and $\mathbf{F}_{\text{drag}}$. (Friction from the road actually propels the car, but this occurs because of forces provided by the engine so we will refer to this force as $\boldsymbol{F}_{\text{engine}}$.) The maximum speed, or terminal velocity v_t, of the car occurs when the net force on the car is zero. This occurs when the vector sum of the engine and drag forces is zero:

$$\mathbf{F}_{\text{engine}} + \mathbf{F}_{\text{drag}} = 0.$$

Because the two forces are equal in magnitude but opposite in direction, we can write $F_{\text{engine}} = F_{\text{drag}}$. We then use Eq. (5–30) for the drag force to find that

$$F_{\text{engine}} = \frac{1}{2}\rho A C_p v_t^2.$$

We solve this equation for the terminal velocity squared:

$$v_t^2 = \frac{2F_{\text{engine}}}{\rho A C_D} = \frac{(2)(3200\ \text{N})}{(1.2\ \text{kg/m}^3)(3.4\ \text{m}^2)(0.50)} = 3100\ \text{m}^2/\text{s}^2.$$

The terminal velocity of the automobile is then

$$v_t = 56\ \text{m/s, or } 125\ \text{mi/h}.$$

In a more realistic situation, road friction, sometimes referred to as rolling friction, also plays an important role in opposing the motion of an automobile. Road friction is fairly constant over a large range of speeds and, as a rule of thumb, road friction is as important as drag at about 40 mi/h. As the speed increases, the relative size of drag compared to road friction increases.

A Better Approximation to the Drag Force

The drag force

The experimental force law for the drag force, $\mathbf{F}_\text{D}$, depends in a more complicated way on the medium through which an object moves and on the shape and size of the moving object than Eq. (5–30) suggests. At low speeds, the drag force is linearly dependent upon the speed. As the speed increases, however, new effects involving the onset of turbulence in the medium set in, and the drag force becomes proportional to the square of the object's speed. Thus, a better approximation to the drag force is

$$F_D = bv + cv^2. \tag{5–32}$$

The coefficients b and c contain information on the shape of the moving object as well as on the medium in which it moves. The first term always dominates for sufficiently low speeds, while the second term dominates for higher speeds.

EXAMPLE 5–10 A marble of mass 5 g falls into a jar of oil (Fig. 5–22). The drag force on the marble is given by Eq. (5–32) with $b = 0.2$ kg/s and $c = 0.1$ kg/m. (These values are typical for real fluids such as oil and for an object the size and shape of a marble.) Find the value of the speed for which the two terms in the drag

APPLICATION: DRAG FORCES IN AN EVERYDAY DEVICE

A simple application of the use of drag forces in low-level technology is the door closer used on storm doors. The device consists of a piston fitting into a closed cylinder with a very small hole at the closed end. A strong spring is placed behind the piston to move it to the closed end of the cylinder (Fig. 5AB–1). The device is attached to the storm door so that, when the door is opened, the piston is pulled to the bottom of the cylinder. There is a valve-like seal between the piston and the cylinder that allows air to fill the cylinder; this makes it easy to open the door. However, the seal does not allow air to leave the cylinder. When the open door is released, the spring force tries to push the piston into the cylinder, thus closing the door. This force is opposed by the drag force of the air leaving by the very small hole in the closed end of the cylinder. This causes the door to close quietly and smoothly.

FIGURE 5AB–1 Drag forces are effectively used in this common device.

force are equal. Which of the two terms is dominant when the drag force is comparable to the force of gravity on the marble?

Solution: Let the value of the speed for which the two terms in the drag force are equal be v'. Then v' is determined by

$$bv' = cv'^2.$$

This equation is solved by

$$v' = \frac{b}{c} = \frac{0.2\ \text{kg/s}}{0.1\ \text{kg/m}} = 2\ \text{m/s}.$$

At terminal speed, magnitudes of the force of gravity and the drag force are equal; that is, $F_D = mg$. For now, let's compare only the first term in the drag, bv:

$$bv = mg;$$

$$v = \frac{mg}{b} = \frac{(5 \times 10^{-3}\ \text{kg})(9.8\ \text{m/s}^2)}{0.2\ \text{kg/s}} = 0.25\ \text{m/s}.$$

When the values of the two terms in the drag force are equal—when $bv = cv^2$—the speed of the marble is 2 m/s. However, when the gravity and drag are the same size, the b term in the drag force dominates; this yields a speed of 0.25 m/s.

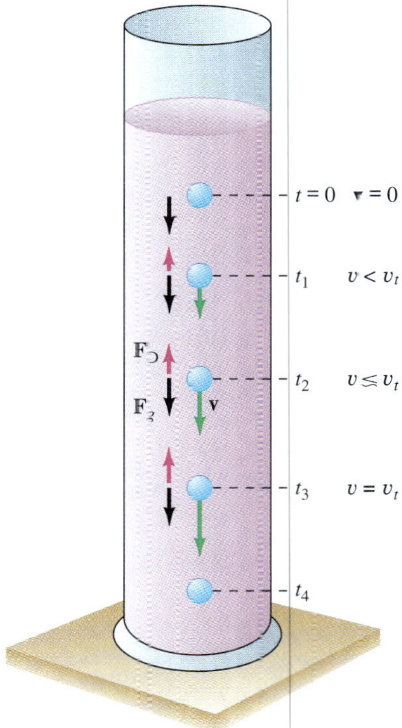

FIGURE 5–22 Example 5–10. When the marble has reached the terminal velocity, the magnitudes of the drag force and gravity are equal.

5-4 FORCES AND CIRCULAR MOTION

If you were to tie this book to the end of a rope and swing it smoothly over your head in a nearly horizontal circle, keeping your hand as close to the center of the circle as possible (Fig. 5–23a), you would feel the rope become taut. This tautness is just the tension in the rope. As we discussed in Section 5–1, a rope under tension pulls on whatever it is connected to—in this case, your hand at one end and the book at the other. In fact, the pull of the rope would be the only horizontal force on the book. Thus, there would be a net horizontal force on the book, and the book would accelerate in the horizontal plane. This isn't surprising, though; in Chapter 3, we realized that an object that is in circular motion is accelerating—even an object in circular motion at constant speed (*uniform circular motion*). There is acceleration because the *direction* of the velocity vector is changing. If the constant speed of the uniform circular motion of the book is v and the radius of its circular motion is r, then Eq. (3–56) tells us that the acceleration vector **a** of the book is

$$\mathbf{a} = -\frac{v^2}{r}\hat{\mathbf{r}}. \qquad (5\text{–}33)$$

Newton's second law as applied to circular motion

The unit vector $\hat{\mathbf{r}}$ points outward from the center of the circular motion. The minus sign in Eq. (5–33) indicates that the acceleration is *toward* the center of the circle.

Newton's second law, $\mathbf{F} = m\mathbf{a}$, tells us that the force **F** required to keep a mass m in uniform circular motion is

$$\mathbf{F} = -\frac{mv^2}{r}\hat{\mathbf{r}}. \qquad (5\text{–}34)$$

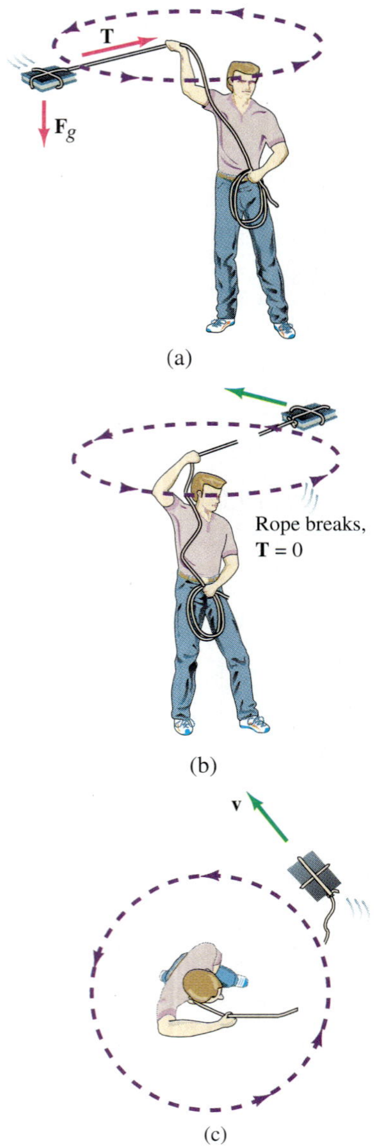

FIGURE 5–23 (a) The tension, **T**, in the rope holding a whirling book is centripetal force, which maintains the circular motion of the book. (b) If the rope breaks, there is no force on the book, which then moves in a straight line tangent to the circle. (c) Top view of (b).

Again, the minus sign reminds us that the force vector points to the center of the circle. Forces pointing in toward the center of a circle are called *centripetal forces*; the acceleration toward the center of a circle is a *centripetal acceleration.* In the case of the swinging book, tension is the force that keeps the book in its circular path and Eq. (5–34) tells us the *magnitude* of the tension in the rope. Because the rope leads to the center of the circle, the *direction* of the tension force on the book is indeed toward the center.

If the rope were cut, the tension that provided the centripetal force would no longer be present. According to the first law, then, the book would move without acceleration in the horizontal plane; that is, in a straight line (Fig. 5–23b).

Centripetal forces that lead to uniform circular motion *are not constant forces.* They are constantly changing in direction, just like the centripetal acceleration.

The tension of a rope is not the only centripetal force that can maintain an object in uniform circular motion. A planet in circular orbit about the Sun or a satellite in circular orbit about Earth are maintained in their circular motion by gravitational forces. Friction between the wheels and the road is the force that takes an automobile around a curve or, as Example 5–11 illustrates, that acts on a die resting on a rotating turntable.

EXAMPLE 5–11 A small die sits 0.15 m from the center of a rotating horizontal turntable (Fig. 5–24a). If the coefficient of static friction between the die and the turntable is 0.55, what is the largest possible angular speed such that the die will not slide off?

Solution: Figure 5–24b is a force diagram for the die. The centripetal force maintaining the die in uniform circular motion is static friction, $\mathbf{f}_s$, between the die and the turntable. The largest rotation frequency that is possible without the die slipping occurs when the friction force reaches its maximum value, $f_s = \mu_s F_N$.

Newton's second law, $\mathbf{F}_{\text{net}} = \mathbf{F}_{\text{N}} + \mathbf{F}_g + \mathbf{f}_s = m\mathbf{a}$, must be decomposed into equations for the vertical direction and for the horizontal plane. For uniform circular motion, the horizontal motion equations are represented by a single radial component equation, Eq. (5–34).

There is no acceleration component in the vertical direction; that is, the direction in which $\mathbf{F}_{\text{N}}$ and $\mathbf{F}_g = m\mathbf{g}$ act. Thus the vertical component of Newton's second law is

$$F_N - mg = 0.$$

In turn, the maximum value of static friction is

$$f_s = \mu_s F_N = \mu_s mg.$$

The maximum radial acceleration $\mathbf{a}_{\text{max}}$ occurs for the maximum value of static friction. The radial component of Newton's second law, the coefficient of $\hat{\mathbf{r}}$ in Eq. (5–34), is then

$$\mu_s mg = \frac{m(v_{\text{max}})^2}{r}.$$

Note that the mass of the die cancels. We want to express this result in terms of a frequency of rotation. From our work in Chapter 3, we know that if the turntable is rotating with angular speed ω, then a point on the turntable that is a distance r from the center moves with speed $v = \omega r$. Thus our radial equation of motion becomes

$$\mu_s g = \frac{(v_{\text{max}})^2}{r} = \frac{(\omega_{\text{max}} r)^2}{r} = (\omega_{\text{max}})^2 r.$$

We solve for ω_{max} and find that

$$\omega_{\text{max}} = \sqrt{\frac{\mu_s g}{r}} = \sqrt{\frac{(0.55)(9.8\ \text{m/s}^2)}{0.15\ \text{m}}} = 6.0\ \text{rad/s}.$$

Normal forces can also act centripetally, as in Example 5–12.

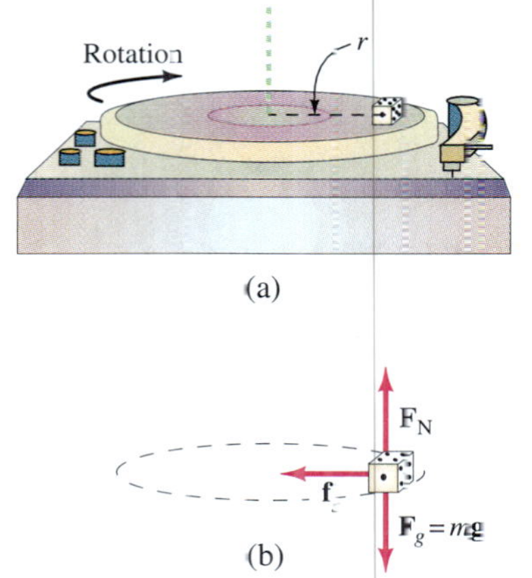

FIGURE 5–24 (a) Example 5–11. (b) Force diagram for the die.

EXAMPLE 5–12 A space station designed to keep its inhabitants comfortable consists of a hollow circular tube that is rotating around its central axis (Fig. 5–25a). The astronauts move on the inner surface of the outermost wall. If the distance from the outer wall to the axis is 50 m, what must be the speed v of a point on the outer wall such that a bathroom scale will read what it would read on Earth?

SOLUTION: Rather than setting up Newton's second law in what might be a rather complicated coordinate system, let's try to think through this problem to gain more insight. An astronaut of mass m who stands on a scale in the space station moves in a circular path of radius R at constant speed v. He therefore accelerates toward the center of the circle (the axis) with magnitude v^2/R from Eq. (5–33). The force that causes this acceleration is the normal force, $\mathbf{F}_N$, of magnitude mv^2/R, which is supplied by the inner surface of the tube (Fig. 5–25b). The bathroom scale indicates this force, just as it would indicate the magnitude of the (upwardly directed) normal force on the astronaut if he were standing on the scale on Earth. Because the normal force on Earth balances gravity, it would have magnitude mg. Thus the scale on the space station reads as it would read on Earth provided that

$$\frac{mv^2}{R} = mg.$$

The mass m cancels out from this equation and we can solve for the speed v:

$$v = \sqrt{gR} = \sqrt{(9.8\ \text{m/s}^2)(50\ \text{m})} = 22\ \text{m/s}.$$

This is a fairly considerable speed—about 50 mi/h. Motion such as this involves internal forces that the station must be constructed to withstand. These internal forces are proportional to v^2. It would thus be a great luxury to have a large enough rotation to give the astronauts the illusion of being on Earth. Actual space stations will have to make do with much less of a normal force—if any at all.

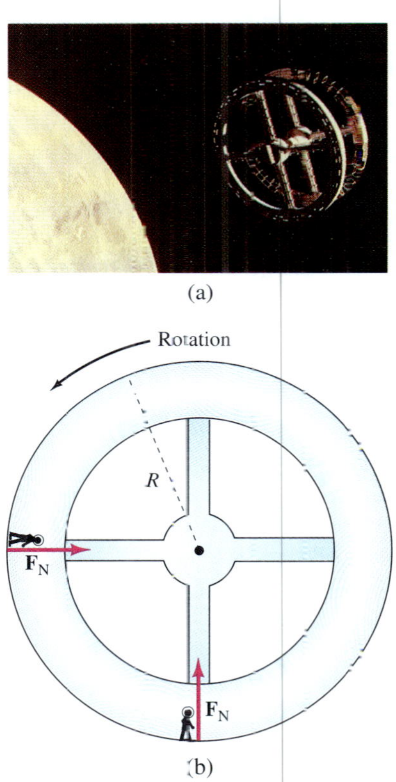

FIGURE 5–25 (a) Example 5–12. The space station from the movie *2001, A Space Odyssey*. (b) A normal force, $\mathbf{F}_N$, pointing to the circle's center acts on each astronaut. The existence of this force gives the astronauts the illusion of weight.

We have discussed how tension, a normal force, and a friction force can each act as the centripetal force responsible for circular motion. In particular, static friction (the tires holding to the road rather than skidding) is the centripetal force that causes an automobile to make a turn. But static friction is limited to a maximum value. Above this value, sliding friction takes over and the automobile starts to skid. Thus the speed and/or radius with which an automobile can make a turn is limited. This fact explains why it is useful to "bank" a curve. In a banked curve, the road is tilted so that the normal force, perpendicular to the road surface, has a component pointing to the center of the circle of the curve. In this way, the normal force acts together with static friction to make up the centripetal force that accelerates the automobile through the curve. Example 5–13 illustrates this point in the extreme case where there is no friction.

APPLICATION

Banking turns

EXAMPLE 5–13 It is a dark and stormy night, and a driver advancing along an icy road must negotiate a turn. The road traces out an arc of a circle with radius

320 m and is banked at an angle of 5.1°. We will assume that the friction between the tires and the road is zero. At what speed should the driver take the curve to avoid sliding off the road?

Solution: Figure 5–26 is a force diagram for the car as seen from the front, with the center of the circle toward the $+x$-direction. In the absence of friction, there are only two forces acting on the car. The force of gravity is purely in the y-direction, and the normal force, perpendicular to the banked road, has components in the y- and x-directions. Newton's second law is $\mathbf{F}_{\text{net}} = \mathbf{F}_\text{N} + \mathbf{F}_g = m\mathbf{a}$, or

$$F_N \cos\theta\,\mathbf{j} + F_N \sin\theta\,\mathbf{i} - mg\,\mathbf{j} = \frac{mv^2}{r}\mathbf{i}.$$

In component form, we have

$$\text{for the } x\text{-component:} \qquad F_N \sin\theta = \frac{mv^2}{r}, \tag{5–35}$$

$$\text{for the } y\text{-component:} \qquad F_N \cos\theta - mg = 0. \tag{5–36}$$

Only the x-component of F_N acts as the centripetal force to carry the car around the curve. When the automobile moves without sliding, the vertical component of F_N just balances the force of gravity, and the horizontal component provides the centripetal force, as in Eq. (5–34).

Equation (5–36) determines F_N:

$$F_N = \frac{mg}{\cos\theta},$$

and this is inserted into Eq. (5–35), which is solved for v:

$$\frac{mg}{\cos\theta}\sin\theta = \frac{mv^2}{r};$$

$$v = \sqrt{gr\tan\theta}.$$

With $r = 320$ m and $\theta = 5.1°$ ($\tan 5.1° = 0.089$),

$$v = \sqrt{(9.8\text{ m/s}^2)(320\text{ m})(0.089)} = 17\text{ m/s} \simeq 60\text{ km/h}.$$

This system is balanced quite delicately. For a given banking, curvature, and speed, Eq. (5–35) uniquely determines F_N. If F_N is too small—if the speed is too slow—the product $F_N \cos\theta$ cannot balance gravity, and the car slips *down* the slope. If F_N is too large—if the speed is too fast—$F_N \cos\theta$ would exceed mg, and the car would skid *up* the slope and off the road. The driver must negotiate the curve just right! When there is some static friction, there is some leeway for error because static friction can take on an appropriate value within a range and hold the car on the road (see Problem 49).

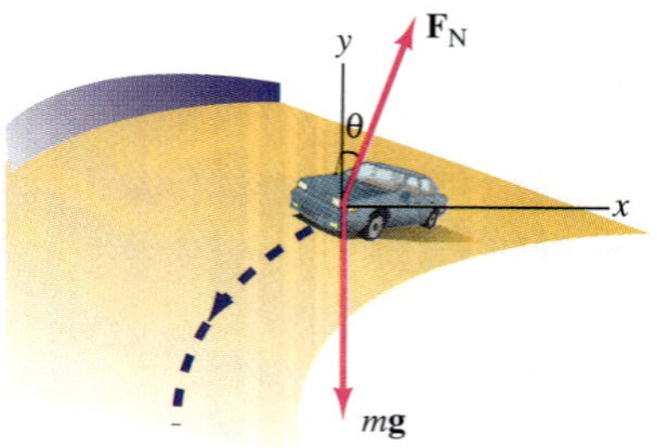

(a)

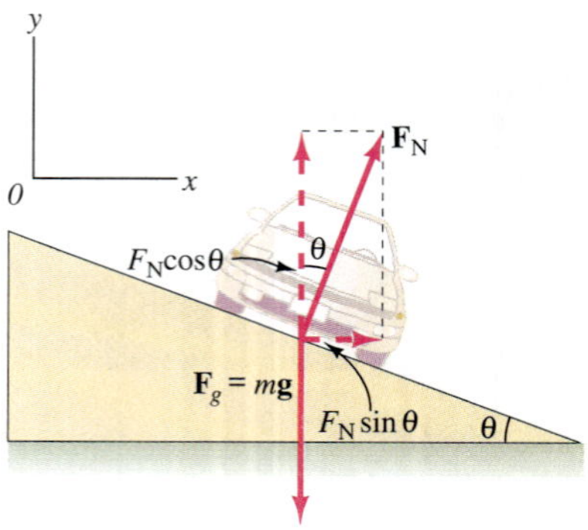

(b)

FIGURE 5–26 (a) Example 5–13. (b) The automobile is seen face on. The center of the circle described by the road curve is to the right.

Circular Motion with Changing Speed

What happens when the speed changes in circular motion? Figure 5–27a illustrates a *simple pendulum*: A point mass m suspended from a string of negligible mass that has a length ℓ. The pendulum swings in a plane under the influence of gravity. During this swing, the mass on the end follows a circular path determined by the length of the string, but it is *not* uniform circular motion because the speed is not constant. To understand this, we draw a force diagram (Fig. 5–27b). Two forces act on the mass: gravity and tension. The mass accelerates according to Newton's second law, $\mathbf{F}_{\text{net}} = \mathbf{F}_g + \mathbf{T} = m\mathbf{a}$, because these two forces are unbalanced. When the mass is at the particular angle θ shown, the forces can be broken into components perpendicular to the path (radial forces), and components along the path (tangential forces). The tension is purely

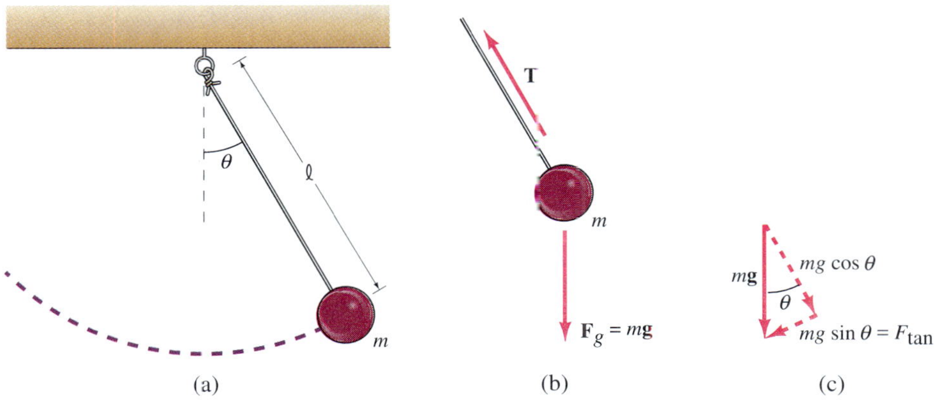

FIGURE 5–27 (a) A simple pendulum of mass m and length ℓ. The path followed traces a circle, but the speed along this path varies with angle θ. (b) Force diagram for the simple pendulum. (c) The components of the force of gravity on the simple pendulum.

radial and inwardly directed, whereas the force of gravity has an outward radial component $mg\cos\theta$ and a tangential component $mg\sin\theta$ (Fig. 5–27c). Thus the component breakdown of the second law is

$$\text{for the radial (inward) component:}\quad T - mg\cos\theta = \frac{mv^2}{r},$$

$$\text{for the tangential component:}\quad mg\sin\theta = ma_{\tan} = m\frac{dv}{dt},$$

where the tangential component of the acceleration, $a_{\tan}$, is the rate of change of speed. *The speed of the pendulum changes because a tangential force is present.* The radial forces simply guide the mass on a circular path and *do not change its speed.* Figure 5–28 enables us to see that the pendulum changes its speed.

Tangential forces are required to change the speed of an object.

For *any* curving path, there will generally be forces acting with components perpendicular to the path and tangential to the path. Force components perpendicular to the path change the direction of the object's motion but not its speed. The tangential components of the net force change the magnitude of the velocity but not its direction.

EXAMPLE 5–14 A skater of mass $m = 52$ kg is swinging in a circle of radius $r = 2.0$ m at a speed $v = 3.0$ m/s. He maintains this motion by holding onto a rope attached through a frictionless bearing to a fixed pole at the center of the circle (Fig. 5–29). The tension of the rope provides the centripetal force for his circular motion and we can regard the ice as being frictionless. As he swings around the pole, however, he runs into a sandy patch of ice 48 cm long, where the coefficient of friction is $\mu_k = 0.10$. (a) What is the tension, T, in the rope before he hits the sandy patch? (b) What is his speed just after he passes the sandy patch? (c) How must the tension change to keep him on the same circular path after he has passed the sandy patch?

Solution: (a) Before he reaches the sandy patch, the skater is in uniform circular motion. The tension is the unique centripetal force,

$$T = \frac{mv^2}{r} = \frac{(52\text{ kg})(3.0\text{ m/s})^2}{2.0\text{ m}} = 230\text{ N}.$$

Compare this to his weight, which is $mg = (52\text{ kg})(9.8\text{ m/s}^2) = 510$ N.

(b) The friction of the sandy patch is a constant tangential force, which slows the skater down. The patch is small enough so that we can assume that the motion of the skater through the patch is a straight line and use our formulas for one-dimensional motion. The magnitude of the friction force is

$$f_k = \mu_k F_N = \mu_k mg = (0.10)(510\text{ N}) = 51\text{ N}.$$

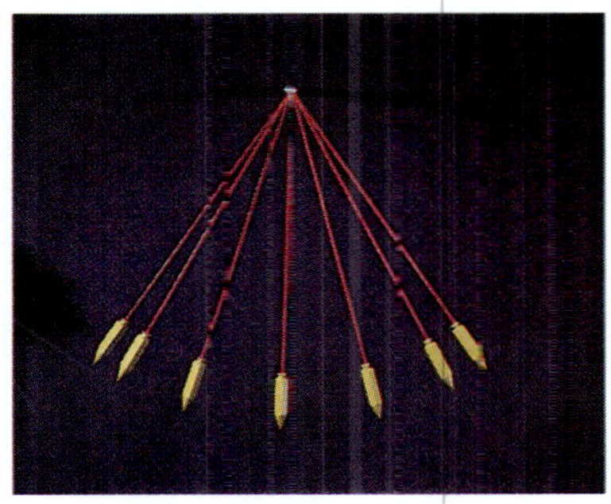

FIGURE 5–28 As the pendulum bob moves under the influence of gravity and the string tension, it speeds up to a maximum at the bottom of the swing and moves most slowly at the top. This is revealed here by photographing the swing with light flashes at equal time intervals, here 1/10 s.

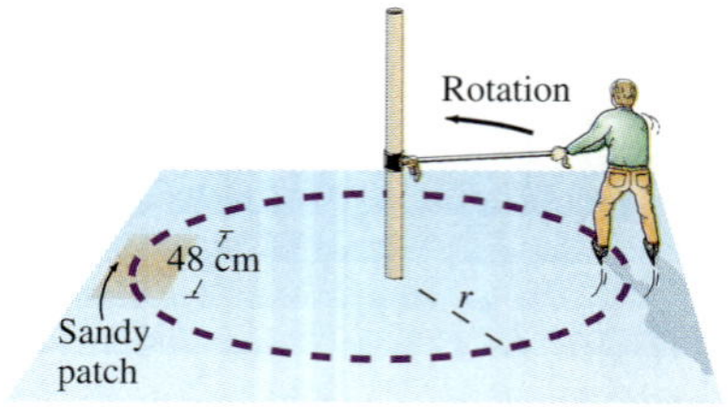

FIGURE 5–29 Example 5–14.

Then, we use the one-dimensional formula relating speed change to distance traveled for a constant force, Eq. (2–24):

$$v^2 = v_0^2 + 2a\,\Delta x = v_0^2 - \frac{2f_k}{m}\,\Delta x$$

$$= (3.0\ \text{m/s})^2 - \frac{2(51\ \text{N})}{52\ \text{kg}}(48\ \text{cm})\left(\frac{1\ \text{m}}{10^2\ \text{cm}}\right) = 8.1\ \text{m}^2/\text{s}^2.$$

Note that the acceleration is negative because friction slows the skater down. Thus, after passing the sandy patch, the skater has slowed to

$$v = \sqrt{8.1\ \text{m}^2/\text{s}^2} = 2.8\ \text{m/s}.$$

(c) We simply follow the first procedure and compute the tension with the new speed:

$$T = \frac{mv^2}{r} = \frac{(52\ \text{kg})(8.1\ \text{m}^2/\text{s}^2)}{2.0\ \text{m}} = 210\ \text{N}.$$

This is 20 N less than the tension in the rope before the sandy patch is encountered.

Circular Motion and Noninertial Frames

We have examined the centripetal forces illustrated in the doughnut-shaped space station of Example 5–12. This system also illustrates the effects of noninertial frames, first mentioned in Section 4–4. An astronaut in the space station is an observer in a noninertial frame. Seen from outside, the astronaut is in uniform circular motion and is therefore accelerating. The centripetal force responsible for this acceleration is the normal force, which can be read by a scale at the astronaut's feet. As the astronaut perceives it, however, everything around him within the station is at rest. The scale, of course, continues to read the same normal force. This force is directed toward the center of the circle. Thus Newton's first law no longer holds for the astronaut: He is at rest and there is no identifiable source for the force that keeps him on the scale. That "force" is directed *outward* ("centrifugal," meaning "directed away from the center") because it balances the inward-directed normal force that the scale reads.

While the notion of a centrifugal force is a commonplace one, we should keep in mind that it is not a force in the same sense as the forces that appear in Newton's laws. It is merely the consequence of an observer being in a rotating reference frame. We use the language of the centrifugal force to interpret the invisible hand that forces us against the door when an automobile makes a high-speed turn. Seen from outside the automobile, a passenger is accelerating centripetally. A centripetal normal force from the door of the automobile pushing on the passenger is the force responsible for moving the passenger in his curving path. The automobile passenger views things differently. The passenger perceives herself to be at rest with respect to the car, yet she feels the normal force from the door. The passenger then invents a "centrifugal force" to balance the normal force from the door.

5–5 FUNDAMENTAL FORCES

Certain forces are *fundamental forces.* They differ from what we might call secondary forces: weight, the normal force, friction, drag, tension, spring forces, and so forth. The fundamental forces play a privileged role among the various forces that we investigate. In fact, the secondary forces are ultimately derived from these forces. For example, the force of gravity (the term we normally use when an object moves near Earth's surface) is just a special case of the fundamental force of *universal gravitation* (which describes the force on objects located near or far from any mass) first understood by Newton. The force exerted by a stiff coil of wire forming a spring depends on the fundamental electric forces between the molecules that make up the material of the spring. It is easy to derive the connection between gravity and universal gravitation. Even with the largest computers at our disposal, however, we can only dream of calculating the spring force from fundamental forces. In any case, such a calculation would not clarify any aspect of physics.

What we refer to today as the fundamental forces are not necessarily the forces that were called fundamental 200 or 100 or even 20 years ago. Our progress on the understanding of what is truly fundamental has been steady. There have been a series of *unifications* in the historical development of fundamental forces: What were previously thought to be distinct fundamental forces have become just different aspects of the same fundamental force.

Three fundamental forces

We now believe that there are only three fundamental forces in nature. They are the force of **universal gravitation**, the **electroweak** force, and the **strong** force (often called the **nuclear** force). The expression of the force of universal gravitation (see Chapter 12) was one of Newton's greatest discoveries. Albert Einstein replaced Newton's expression in his law of general relativity, making some new predictions of tiny experimental effects that are not predicted by Newtonian gravitation. Einstein also made persistent but unsuccessful efforts to unify gravitation with the other fundamental forces. The electroweak force is a recent discovery, dating from a theory proposed by Sheldon L. Glashow, Abdus Salam, and Steven Weinberg in the mid-1970s that was verified experimentally in the early 1980s. This work unified two forces formerly thought to be independent and fundamental: the *weak* force, responsible mainly for some types of radioactive processes in nuclei and important in the evolution of the universe, and the forces of *electromagnetism* (Chapters 22–35). On the scale appropriate to the secondary forces that we deal with every day, the electromagnetism aspect of the electroweak force is dominant, and it is often convenient to refer simply to the electromagnetic force. Electromagnetism is itself the result of a nineteenth-century unification of the forces of *electricity* and *magnetism*; these two forces had previously been thought to be independent. The strong force is responsible for holding together the nuclei of atoms. Both the strong force and the electroweak force have been the object of more recent attempts at unification, which have yet to bear fruit. Since the 1980s, there has also been a new effort to bring gravitation into the unification program. Unification continues to be a fascinating beacon that will allow us to describe more of the universe in a simpler way.

Except for gravity, all the forces discussed in this chapter are electromagnetic in origin.

On the everyday scale of the secondary forces that we have discussed in this chapter, only the force of gravity is a direct aspect of a fundamental force: universal gravitation. All the other forces that we have investigated are ultimately due to the electromagnetic force, which binds atoms and molecules together into ordinary matter. We do not directly see the strong force on our scale; it holds nuclei together so tightly that, for most practical purposes, we can think of them as indivisible lumps of matter. Only when we discuss nuclear energy or the composition of stars do strong forces come into practical play.

It is customary to speak of the relative size of the fundamental forces. This cannot be done without talking about a particular situation; that is, placing the fundamental forces in context. We can say the following: Within the nucleus, the typical strong force is about 100 times larger than the electromagnetic forces. The gravitational force between two protons in the nucleus is many, many orders of magnitude smaller than either the strong or the electroweak force. Thus the forces are ordered in decreasing strength as strong, electroweak, and gravitational. It is only because the strong force acts over such a limited range that the electroweak and gravitational forces ever come into view. The gravitational force dominates on the astronomical scale because matter is arranged into an electrically neutral combination, in which the electromagnetic forces cancel out to zero.

SUMMARY

This chapter is devoted to describing various forces that occur in nature and to applying the problem-solving techniques developed in Chapter 4. These techniques are very important and we will use them throughout the book.

The force of gravity (or weight), tension, and the normal force are all common forces for which it is often possible to solve the equations of motion given by Newton's second law. The force of gravity is expressed as

$$\mathbf{F}_g = m\mathbf{g}, \tag{5–4}$$

where $\mathbf{g}$ is a constant vector that points down toward Earth's center.

The tension, $\mathbf{T}$, is a variable force that is determined according to the circumstance and exerted by ropes (or wires or cables or strings). Tension always pulls on a mass in the direction taken by the rope; it is transmitted everywhere along the rope, taking a single constant value when the mass of the rope is negligible.

The normal force, $\mathbf{F}_N$, is also variable. It is directed perpendicularly to a surface and acts to cancel any other forces that might make a mass accelerate into the surface.

The friction force acts when two surfaces slide or attempt to slide across one another. Static friction, $\mathbf{f}_s$, is variable and acts in a direction that would oppose any sliding that would occur if there were no friction. It can increase up to a maximum value proportional to the magnitude of the normal force:

$$0 \le f_s \le \mu_s F_N. \tag{5–21}$$

Kinetic friction, $\mathbf{f}_k$, acts when sliding actually is occurring and is also proportional to F_N:

$$f_k = \mu_k F_N. \tag{5–22}$$

The constants μ_s and μ_k are the coefficients of friction, and generally

$$\mu_s > \mu_k. \tag{5–23}$$

Another type of friction, the drag force, occurs when objects move through media. For all but small objects, the drag most frequently encountered varies with the speed squared:

$$F_D = \frac{1}{2}\rho A C_D v^2. \tag{5–30}$$

It implies that objects accelerating within media can be accelerated only up to a terminal velocity.

Forces are responsible for accelerating objects moving in a curved path. When the motion is uniform circular motion, then Newton's second law takes the form

$$\mathbf{F} = -\frac{mv^2}{r}\hat{\mathbf{r}}. \tag{5–34}$$

The vector $-\hat{\mathbf{r}}$ is directed to the center of the circle and the force is said to be centripetal. When both the direction and the magnitude of the velocity change, then $\mathbf{F}$ has a component tangential to the motion. This component causes the magnitude of the velocity to change, while a force component perpendicular to the motion causes the direction of the motion to change.

All the forces of nature are ultimately described in terms of three fundamental forces: the force of universal gravitation, the electroweak force, and the strong force. The electromagnetic force, which is part of the electroweak force, is responsible for most of the secondary forces, including tension, friction, drag, normal forces, and spring forces.

UNDERSTANDING KEY CONCEPTS

1. A tightrope walker moves to the center of a thin wire that was initially stretched taut to a horizontal position. Why is it that the wire cannot remain horizontal?
2. We have said that it is actually the force of friction that is responsible for both the acceleration of automobiles and our ability to walk. What is the role of the engine or of the muscles in these processes?
3. An observer sees a mass hanging motionless from a vertical string. Under what circumstances is the tension in the string greater than or less than the weight of the mass?
4. What are some factors that could limit how fast a hot rod can go in a $\frac{1}{4}$-mi race?
5. How does the fact that a rope has mass complicate solving a problem about lifting a load with a pulley? What is the effect of friction in the pulley?
6. Why is it helpful for an automobile with an engine in the front to have front-wheel drive? Why is it useful to put sand in the trunk of your car in winter if your car has rear-wheel drive?
7. Why do bicyclists or motorcyclists "lean into" a curve? In explaining why, make use of the fact that we are able to balance ourselves best when we feel that the net force on us is coming from directly beneath our feet. Also think about the forces that friction must cancel.
8. How might the result of Example 5–8 vary if the weight was not evenly distributed over the wheels and/or the car was equipped with rear-wheel drive instead of four-wheel drive?
9. Suppose that a rope has tension because a mass is suspended from its end. Let's say that the rope is now cut, a spring is inserted at the cut, and we observe the stretch of the spring as a measure of the rope's tension. If the mass of the rope is negligible, does the observed tension of the rope depend on where the cut is made along the string? Does your answer change if the rope cannot be considered massless?
10. A bowl of water with floating ice cubes is placed on a scale. The ice cubes melt. Does the reading of the scale change? Why or why not?
11. What is the role of the keel, which runs along the center of the bottom of a sailboat? Some sailboats have centerboards rather than keels. These are simply large boards that can be lowered or raised in the position of a keel. What is the role of a centerboard? Why might you prefer a keel to a centerboard or vice versa?
12. We have referred to massless ropes and the tension in them. What physical considerations allow you to think of a rope as massless?
13. The speed of the boats (shells) used in scull racing is, to a good approximation, independent of the number of people rowing (provided that the number is larger than three or four). At first, this might appear strange: The more people there are, the larger the forward propelling force available to overcome the drag of the water. Can you explain this seeming contradiction?
14. Describe a series of experiments that would have allowed Leonardo da Vinci to decide that the force of friction depends only on the normal force on an object.
15. At the moment a car in a loop-the-loop roller coaster is at the top of the loop (directly below the track), can there be a normal force on it? Such a force would point straight down, in the same direction as that of gravity.
16. The riders on the loop-the-loop of Question 15 experience a feeling of near-weightlessness close to the top of the loop. Why? Would the coins in their pockets fall out?
17. Suppose that it were possible for a ship to sail all the way around the world along a great circle. Is a centripetal force necessary to keep the ship moving in this circle? What force or forces would act centripetally?
18. We have said that the tension in a string is the result of the intermolecular forces within it. Suppose that you had some idea of the forces between individual, separate molecules within the string. How could you estimate the tension at which the string breaks?
19. When you sit in the passenger seat of a car that makes a tight turn to the left, you could be thrown out of the car if the door should open (and you are not belted in). Why?
20. At the beginning of Section 5–4, we spoke of swinging a book on the end of a rope in a nearly horizontal circle. Could the plane of the circle be *perfectly* horizontal?
21. Imagine that you and your partner are on skates on a perfectly frictionless ice surface. You hold on to opposite ends of a rope and pull toward each other. What happens if you each have exactly the same mass? If you have half your partner's mass?
22. Why is it hard to run when the ground is icy?
23. Tarzan swings from tree to tree on a vine. At which point in his swing is the vine most likely to break, and why?
24. A person on a rapidly moving Ferris wheel feels that she is about to fly off the seat when her seat reaches the top of the circular path. Why is that?
25. A hemispherical bowl is placed open end up on a table and rotates around its own vertical axis. A die is allowed to slide down from the edge into the bowl. Describe the motion of the die as seen by a hypothetical observer at the bottom of the bowl, assuming that there is no friction between the bowl and the die.

PROBLEMS

5–1 Some Simple Constant Forces

1. (I) An 8-g bullet that travels at 500 m/s is fired into a rigidly fixed block of wood. The bullet is found 7 cm into the wood. What was the average force exerted by the wood opposing the bullet's motion? Assume that the deceleration was uniform.
2. (I) In Example 5–3, a woman lifts a football lineman (mass 149 kg) by pulling down on one end of a rope with 365 N of force. Suppose that the woman has a mass of 50 kg. If a bathroom scale were beneath her feet as she lifted the lineman, what would the scale read in pounds? Analyze the force dia-

gram for the woman and remember that a scale of this type reads the upward normal force that the floor exerts on her.

3. (I) A helium balloon just manages to lift 100 kg (including the mass of the balloon, the helium it contains, and its payload) off the ground and then hovers 1 m off the ground. The upward force that maintains the balloon is *buoyancy* (we shall treat this force in Chapter 16). What is the magnitude of the buoyancy in this case?

4. (I) A woman of mass 56 kg sits in a racing car. When she depresses the accelerator, the car accelerates in a straight line to 170 km/h in 9.3 s. What are the direction and magnitude of the force she experiences? Where is the force applied?

5. (I) A hockey puck of mass 0.10 kg slides without friction on ice. In an appropriate coordinate system, its velocity is $\mathbf{v}_1 = (1.4\mathbf{i} + 3.0\mathbf{j})$ m/s. A constant force $\mathbf{F} = 4.0\mathbf{i}$ N is then applied to the puck. After how many seconds will the puck have a speed of 6 m/s?

6. (II) A man of mass 85 kg is escaping a burning building using a rope that will break if the tension exceeds 700 N. (a) With what acceleration must he slide down the rope if it is not to break? (b) How far down the rope is he, and what is his velocity, after 4.0 s, assuming he drops with the minimum acceleration of part (a)?

7. (II) A metal rod of mass 5.6 kg and length 3.5 m is suspended from the ceiling. (a) What is the tension in the rod at a distance of 2.0 m from the top? (b) What is it at a distance of 3.0 m from the top?

8. (II) A brick hangs from a string attached to the ceiling. When a horizontal force of 12 N is applied to the brick, the string makes an angle of 25° with the vertical (Fig. 5–30). What is the mass of the brick?

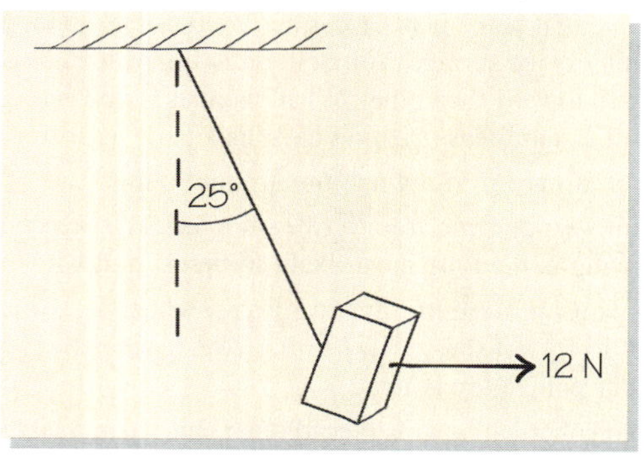

FIGURE 5–30 Problem 8.

9. (II) Figure 5–31 shows a person applying a horizontal force in trying to push a 25-kg block up a frictionless plane inclined at an angle of 15°. (a) Calculate the force needed just to keep the block in equilibrium. (b) Suppose that she applies three times that force. What will be the acceleration of the block?

10. (II) Two blocks, of mass m_1 and m_2, are placed in contact on a smooth surface, with the more massive one (m_1) on the left. A force of magnitude F, pointing to the right, is applied to the block on the left. (a) What is the acceleration of the system? (b) What force acts on the block on the right?

FIGURE 5–31 Problem 9.

11. (II) An automobile of mass 1200 kg pulls another automobile of mass 1400 kg with a tow rope. (a) In order to pull out onto a highway, the automobile must accelerate to 55 mi/h in an access lane that is only 120 m long. What must its acceleration be? (b) What is the tension in the tow rope?

12. (II) An Atwood machine consists of a massless string connecting two masses over a massless, frictionless pulley (Fig. 5–32). In this case, the masses are 0.800 kg and 0.650 kg. The system is released from rest with the 0.8-kg mass 2.15 m above the floor and the 0.65-kg mass on the floor. (a) What is the acceleration of the 0.8-kg mass? of the 0.65-kg mass? (b) What is the speed of the 0.8-kg mass just before it hits the floor? (c) How long does it take the 0.8-kg mass to reach the floor?

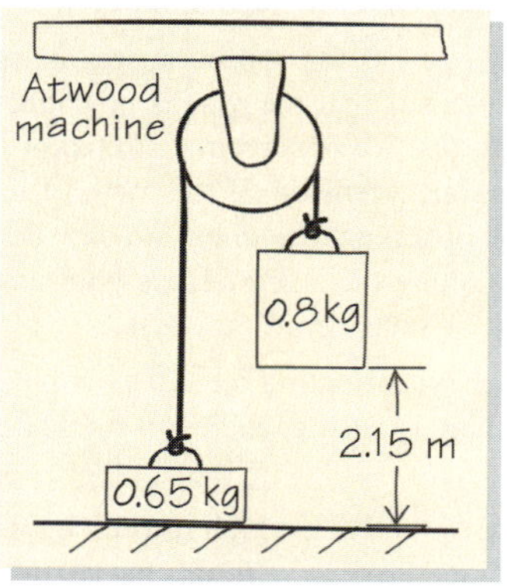

FIGURE 5–32 Problem 12.

13. (II) Two blocks of masses M and m are connected by a light rope which passes over a frictionless pulley. Mass M sits on

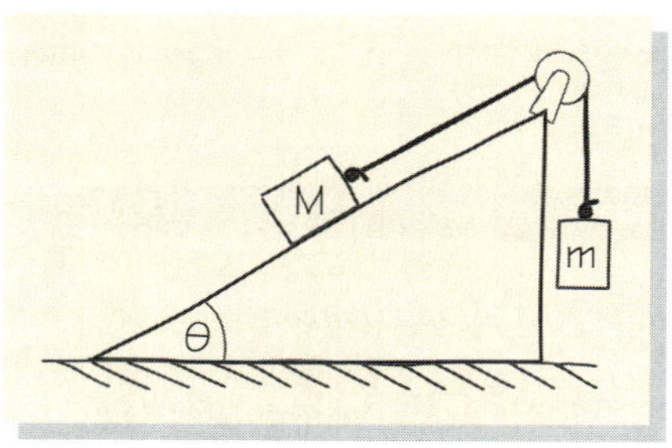

FIGURE 5–33 Problem 13.

an inclined plane with an angle of inclination of 30° (Fig. 5–33). The coefficient of static friction between mass M and the inclined plane is 0.20, while $m = 3.0$ kg. Determine the largest and smallest possible values of M for which the system remains in equilibrium. Calculate the force of static friction on the block of mass M if $M = 6$ kg.

14. (II) Consider a variation on an Atwood machine in which the masses are each on an incline (Fig. 5–34). The mass sliding on incline 1, m_1, is 1.50 kg, and the angle of this incline is $\theta_1 = 62°$. If the mass on the second incline, m_2, is 2.50 kg, what is the angle θ_2 so that the system does not accelerate?

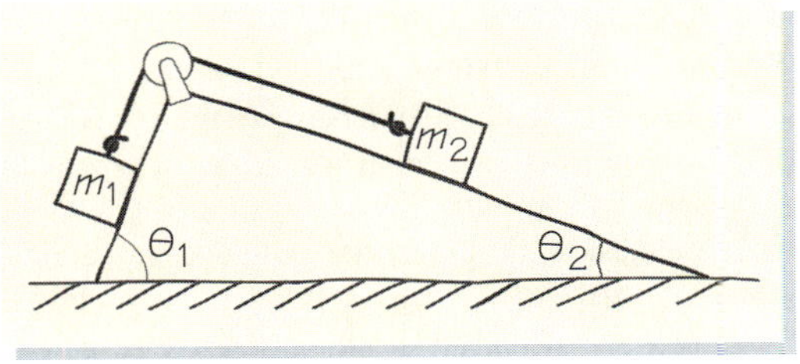

FIGURE 5–34 Problem 14.

15. (III) Consider the three-pulley arrangement shown in Fig. 5–35. The three masses m_1, m_2, and m_3 have the values 2.00 kg, 5.00 kg, and 4.00 kg, respectively. All the pulleys are frictionless, and the strings are massless. What are the tensions in all the strings, and what are the accelerations of the masses?

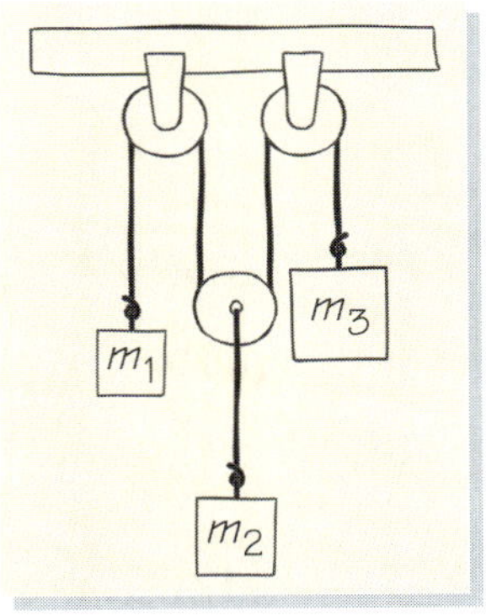

FIGURE 5–35 Problem 15.

16. (III) In the text, we stated that the tension is the same everywhere in a massless rope. Give the equation that describes the tension as a function of position for a stationary hanging rope, as in Fig. 5–36a, with constant mass density $\lambda =$ mass/unit length. Figure 5–36b shows the forces on an element of rope. For this problem assume that no separate mass is attached at the bottom. [*Hint*: Measure the height h up from the bottom point of the rope to any point along the rope. The bottom of an isolated slice is at height h, and its top is at height $h + \Delta h$. This means that its mass Δm is given by $\lambda\,\Delta h$. The tension T_{up} acting at the top of the slice is $T(h + \Delta h)$, and the tension T_{down} acting at the bottom of the slice is $T(h)$. The resulting equation can be written as a *differential equation* for T as a function of h by dividing by Δh and recognizing that in the limit of $\Delta h \to 0$, $d[T(h)]/dh = [T(h + \Delta h) - T(h)]/\Delta h$. The equation's solution is the *function* $T(h)$ that satisfies the equation.]

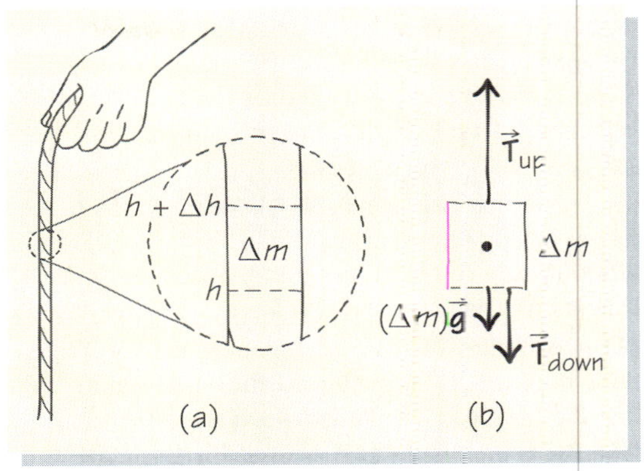

FIGURE 5–36 Problem 16.

17. (III) A compound Atwood machine is constructed by replacing one of the masses of a conventional Atwood machine (see Problem 12) with the pulley of another Atwood machine (Fig. 5–37). Altogether there are three masses, two ropes, and two pulleys; the ropes and pulleys are to be considered massless and friction free. Describe the method by which the motions of the masses of this machine can be analyzed. (*Hint*: There are two independent tensions and two independent accelerations, which must be found by analyzing the equations of motion for the three masses simultaneously.) Solve these equations for arbitrary masses. What happens if $m_2 = m_3 \neq m_1$? [*Hint*: If m_1 accelerates, so does the pulley, so m_2 and m_3 share an additional acceleration.]

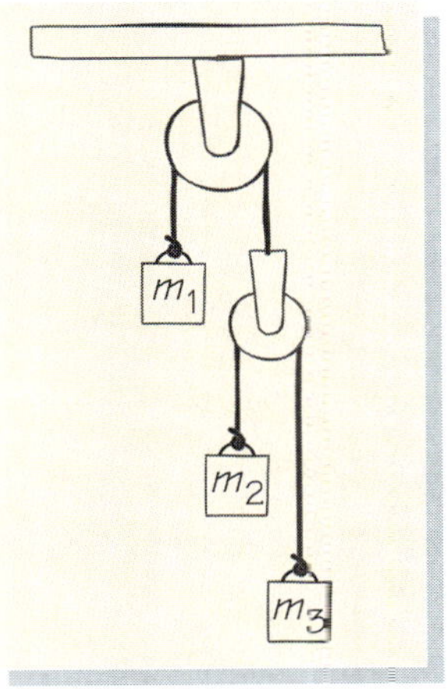

FIGURE 5–37 Problem 17.

18. (III) A double-pulley system—like that described in Problem 17—has masses $m_1 = 2.00$ kg, $m_2 = 1.20$ kg, and $m_3 = 0.800$ kg. (a) What are the accelerations of all the masses? (b) What are the tensions in all the ropes?

5–2 *Friction*

19. (I) A truck of mass 5000 kg accelerates on a straight road at 1.2 m/s^2. Assuming that air resistance is negligible, what is the coefficient of static friction between the road and the tires? (Don't worry about differences in normal forces, on each wheel, and assume 4-wheel drive.)

20. (I) A boy of mass 35 kg runs into an ice-skating rink in street shoes and begins a smooth slide. He starts with a speed of 4.5 m/s and comes to a stop after sliding for 3.4 s. What is the coefficient of kinetic friction of his shoes on ice?

21. (I) A car of mass 1200 kg is moving at 25 m/s. The driver suddenly sees a dog crossing the road, slams on the brakes, and manages to stop the car in 4.2 s. What is the coefficient of static friction between the tires and the road? (Assume that the acceleration is constant and that there is no skidding).

22. (I) The coefficient of static friction between a worker's shoes and the floor is 0.81, while the coefficient of static friction between the floor and a crate is 0.43. The worker, mass 80 kg, pushes the crate, which has a mass of 140 kg. What is the friction force on the worker due to friction between the shoes and the floor when the crate starts to slide?

23. (I) A worker must push an 85-kg crate across a floor. The coefficient of kinetic friction between the crate and the floor is $\mu_k = 0.4$. What is the minimum force that the worker must exert to keep the crate moving once the crate starts moving?

24. (I) A rope is connected to a 25-kg cement block placed on a board leaning against a wall at an angle of 25° with respect to the horizontal (Fig. 5–38). The coefficient of kinetic friction between the cement block and board is $\mu_k = 0.4$. (a) What is the tension in the rope if it is pulled at constant speed straight up the board? (b) What is the tension if the rope is pulled up at constant speed at an angle 40° from the horizontal?

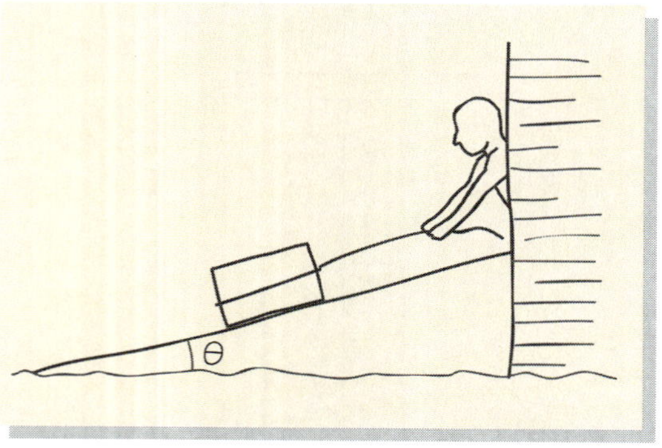

FIGURE 5–38 Problem 24.

25. (II) A person learning to snow ski will use the snowplow position; it is a rudimentary way of keeping one's skiing speed under control. Let's imagine that a beginning skier finds herself on an icy slope of 22°. Only by setting her skis in a good snowplow position, with the tips of both skis pointed inward and the inner edges dug in, is she able to keep from accelerating. Effectively, what coefficient of sliding friction is created by the snowplow?

26. (II) A man wants to push a package of shingles of total mass 25 kg up a roof being built at an angle of 35°. The coefficient of kinetic friction between the package and the roofing paper already in place is $\mu_k = 0.32$. (a) How much force does the man have to exert on the package directly along the slope of the roof to cause the package to accelerate at 0.20 m/s^2? (b) If the coefficient of static friction is 0.42, will the package remain on the roof?

27. (II) The coefficient of static friction between a car of mass 1500 kg and an asphalt road is $\mu_k = 0.7$. (a) What is the shortest distance over which the car can accelerate from rest to a speed of 96 km/h? (b) How long will this take? (Because we neglect drag and rolling friction, the distance and time will be unusually short.)

28. (II) A crate of mass 250 kg is loaded on the back of a truck. The coefficient of static friction between the crate and the truck bed is μ_s. The truck decelerates such that it comes to a stop from a speed of 60 mi/h (26.7 m/s) in a distance of 140 m. How large must μ_s be so that the crate does not slide forward on the truck bed?

29. (II) A 500-g box is placed on a board at a 30° incline and accelerates from rest down the board at 0.3 m/s^2. (a) How long does the box take to travel down the board of length 1.2 m? (b) What frictional force opposes the motion of the box? (c) What is the coefficient of friction between box and board?

30. (II) A block of mass $m = 2.0$ kg is placed on a horizontal surface. The coefficient of static friction between the block and the surface is $\mu_s = 0.4$. A light rope is tied to the block and thrown over a frictionless pulley (Fig. 5–39). The free end of the rope is pulled with a slowly increasing force T. At what value of T will the block start to move if the angle that the rope makes with the horizontal is 50°?

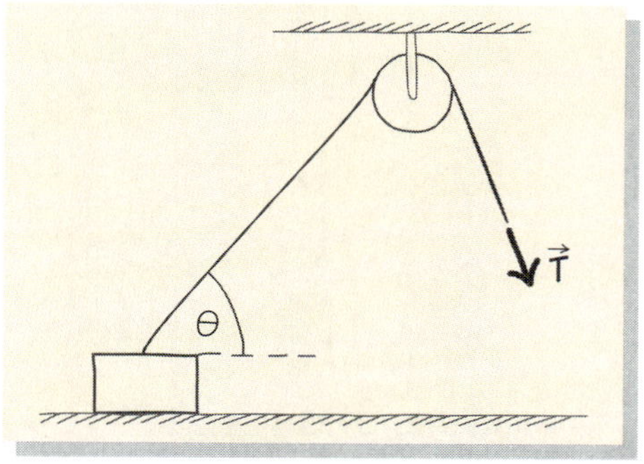

FIGURE 5–39 Problem 30.

31. (II) A 50-kg box rests on a rough horizontal surface with which the box has a large coefficient of static friction, $\mu_s = 0.75$. The box is pulled by means of a light rope with a force of magnitude F, making an angle θ with the horizontal. (a) Find the magnitude of the force F that will just start the box moving horizontally as a function of θ. (b) Show that there is some angle θ for which F takes a minimum value. What is this value for our case, and what is the force F corresponding to this value? Explain why, physically, there is such a minimum value.

32. (II) A block of mass 1.2 kg rests on top of another block of mass 1.8 kg, which rests on a frictionless surface. The coefficient of static friction between the blocks is $\mu_s = 0.3$. What is the maximum horizontal force that can be applied to the upper block so that the blocks accelerate together, without the upper block sliding on the lower one? If the horizontal force is applied to the lower block instead, what is the maximum force that will give rise to the same motion?

33. (II) Consider again the professor of Example 5–7. Her mass is 55 kg, and the coefficient of static friction between her and the floor is μ_1. What is the minimum value of μ_1 for which she will not slip on the floor?

34. (II) A pile of snow at the crest of a roof with a slope of 40° from the horizontal starts to slide off. The distance from the crest to the edge of the roof is 8 m, and the coefficient of kinetic friction for the snow on the roof is 0.1. (a) What is the speed of the pile of snow when it reaches the edge of the roof? (b) Assuming that it is 6 m from the edge of the roof to the ground, how far out from the base of the building does the snow land?

35. (III) A mass m_1 rests on top of another mass, m_2, which in turn rests on a frictionless horizontal surface (Fig. 5–40). A light cord is attached to m_2, which is used to pull on it with a force F. (a) Find the acceleration of each object when the surface between the two objects is frictionless. (b) Find the acceleration of each object when the surface between the two objects is rough enough to ensure that m_1 does not slide on m_2. (c) What are the magnitude and direction of the contact forces exerted by the lower object on the upper one, assuming that the upper object is sliding on the lower object with a nonzero coefficient of kinetic friction μ_k? (d) Find the acceleration of each object if the surface between the two objects is such that the upper object is sliding on the lower one under the influence of kinetic friction, with a coefficient of kinetic friction μ_k.

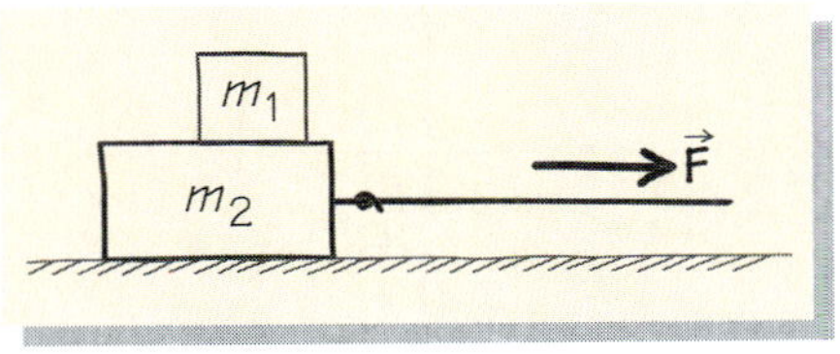

FIGURE 5–40 Problem 35.

5–3 *Drag Forces*

36. (I) A parachute is rigged so that a parachutist of total mass 116 kg with full gear reaches the ground at a terminal velocity of 4.9 m/s. Assuming that the drag force on the parachutist, moving with speed v, has a magnitude equal to kv^2, what is the value of k?

37. (I) Estimate the drag force on an automobile cruising at 55 mi/h. Assume that the drag coefficient, C_D, is 0.3 and that the car's cross-sectional area is 5 m^2. Take air to have a density of 1.25 kg/m^3.

38. (I) A race car of mass 800 kg has a maximum acceleration from rest of 4.8 m/s^2. Assume that the car's engine is such that the force on the tires is constant and that the car's effective cross-sectional area into the air is 1.8 m^2. If the car's top speed is observed to be 90 m/s, what is its drag coefficient, C_D? Take the density of air as 1.25 kg/m^3.

39. (I) A ball of mass 500 g is observed to reach its terminal velocity of 18 m/s after being dropped from the top of a tall building. Assume that the density of air is 1.25 kg/m^3 and the drag coefficient, C_D, is 0.4. What is the effective cross-sectional area of the ball? You can use Eq. (5–30) here.

40. (II) A sphere of radius r_1 and mass m_1 falls through the air and is found to have a terminal velocity v_1. Equation (5–30) applies. (a) What is the terminal velocity v_2 for a sphere with twice the radius and the same mass density as the first sphere? (b) Generalize the result of part (a) to find the terminal velocity for a sphere with a radius z times the first radius but the same mass density.

41. (II) A barge, moving at uniform speed, is pulled by two horses moving on opposite sides of a canal in which the barge floats. The ropes connecting the horses to the barge make an angle of 28° with the line of motion of the barge. If the resistance to the motion of the barge is characterized by a frictional force given by $F = -(160\ \text{N}\cdot\text{s/m})v$, where v is the speed of the barge in meters per second and the tension in each of the ropes is 42 N, what is the speed of the barge?

42. (II) Assume that the drag force on a parachute is given by Eq. (5–30) in the text. The effective area of the parachute is 30 m^2 and the density of air is 1.25 kg/m^3. If a 90-kg parachutist finds that his terminal velocity is 6 m/s, what is the drag coefficient?

43. (II) The terminal velocity of a skydiver of mass 75 kg can be controlled by the orientation of her body and can range from 40 m/s to 60 m/s. Assume that she can change the area presented to the ground by a factor of 1.5 in going from the minimum to the maximum terminal speed, and that the larger area presented to the ground slows the skydiver down. Now, how does the drag coefficient change? (Express your answer with a proportion.)

44. (II) A marble of mass m falls through a fluid and is subject to the drag force $\mathbf{F}_D = -A\mathbf{v}$, where $\mathbf{v}$ is the velocity of the marble. The marble will reach a terminal velocity given by $\mathbf{v}_t = m\mathbf{g}/A$. Use dimensional analysis to estimate how long it will take to reach the terminal velocity. [*Hint*: A characteristic "time" can be constructed from A, g, and m.]

5–4 *Forces and Circular Motion*

45. (I) An airplane of mass 2×10^4 kg executes a banked turn of radius 30 km while flying at 200 m/s. What acceleration will the passengers feel as a result of the turn?

46. (I) A rock swings in a nearly horizontal circle at the end of a string whose breaking tension is 12 N. The circular path is 0.25 m in radius, and the rock's mass is 150 g. What is the maximum speed the rock can have before the string breaks?

47. (I) A man of mass 65 kg stands at the edge of a merry-go-round of radius 5.3 m. The merry-go-round turns at 6.0 rev/min. What are the magnitude and direction of the net force on the man?

48. (I) An accelerometer shows that an airplane flying at 850 km/h undergoes a vertical acceleration of 0.17 g's ($1\ g = 9.8$ m/s^2) at a certain moment. What is the radius of curvature of the airplane's (horizontal) path at that point?

49. (II) An automobile makes a turn whose radius is 150 m (Fig. 5–41). The road is banked at an angle of 18°, and the coefficient of friction between the wheels and road is 0.3. Find the maximum and minimum speeds for the car to stay on the road without skidding up or down the banked road.

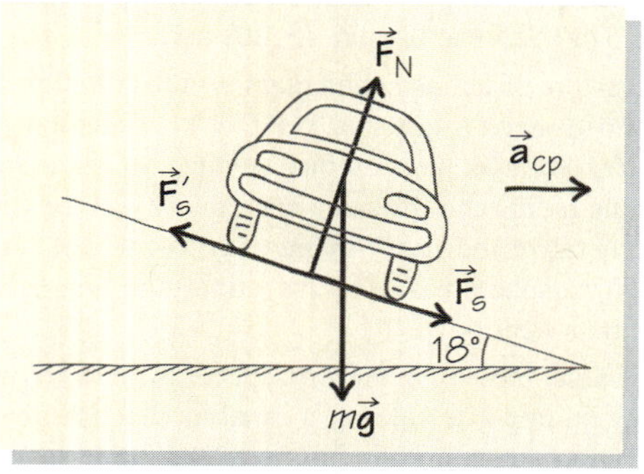

FIGURE 5–41 Problem 49.

50. (II) A merry-go-round has a circular platform that is 1 m from the central axis at its inner edge and is 5 m from the central axis at its outer edge. The ride turns at a rate of one full rotation every 10 s. A passenger holds himself to the surface with a pair of very sticky shoes and is most comfortable when he orients his body length along the line of the net force on him. What angle does his body make to the vertical (a) 1 m from the axis? (b) 3 m from the axis? (c) 5 m from the axis?

51. (II) The coefficient of static friction between a small stone and a horizontal turntable is measured in the following way. The stone is placed on the turntable at a distance R from the axis, and the speed of rotation is slowly increased to 45 rev/min. When the experiment is repeated for several different values of R, it is found that the stone remains on the turntable if $R <$ 26 cm, and that it slides off with increasing speed of rotation if $R >$ 26 cm. Determine μ_s from these data.

52. (II) A Ferris wheel in an amusement park has a radius of 30 m and makes one complete turn every 75 s. Calculate the normal force that a passenger of mass 60 kg experiences through the seat of the pants (the seat bottom is parallel to the ground) when the passenger is (a) at the bottom of the path, nearest the ground; (b) at the maximum height of the path.

53. (II) Assume that the acceleration of the Moon due to Earth's gravity is 0.0027 m/s^2. What is the velocity of the Moon with respect to Earth if the period of the Moon's motion around Earth is 28 d? Do not look up the distance between Earth and the Moon; instead, calculate it. Compare this result with the distance given in Appendix III–1.

54. (II) A fighter pilot makes a dive almost vertically down and pulls up while traveling at 1500 km/h in a turn of radius 2.50 km. How many g's will the fighter pilot feel at the bottom of the dive? Because the pilot will black out if the number of g's is greater than 11, is this a safe maneuver? (The number of g's is the acceleration in units of g = 9.80 m/s^2.)

55. (II) A mass of 1.00 kg hangs from a rope placed through a hole in a smooth, frictionless table. At the other end of the rope is attached a puck of mass 400 g, 80 cm from the hole in the table. The puck swings in a circular orbit around the hole (Fig.

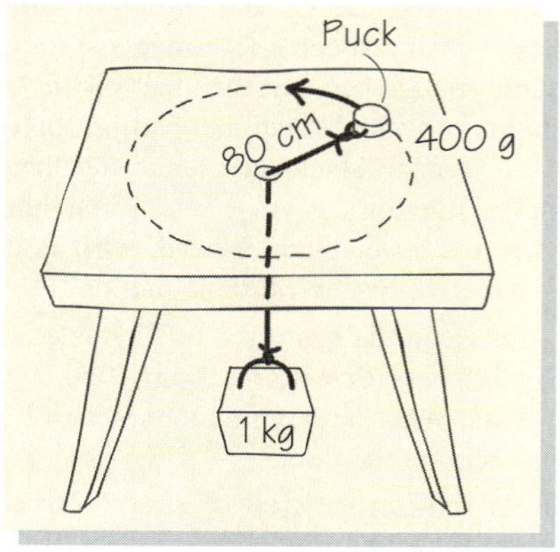

FIGURE 5–42 Problem 55.

5–42). With what speed must the puck rotate to keep it 80 cm from the hole?

56. (II) A mass m at the end of a string of length r moves at constant speed v in a circle on a frictionless table. (a) Find the tension in the string in terms of m, r, and v. (b) A second mass identical to the first is attached at the midpoint of the string and the two are whirled; the speed of the outer mass is again v. Draw force diagrams for the two masses and calculate the tensions in terms of m, r, and v.

57. (II) Consider the *conical pendulum*, a mass on the end of a massless string, with the other end of the string fixed on a ceiling. Given the proper push, this pendulum can swing in a circle at a given angle θ, maintaining the same height h throughout its swing, as shown in Fig. 5–43. (a) What is the force diagram for such a pendulum? (b) If the mass of the pendulum is m = 0.2 kg, the length of the pendulum is ℓ = 0.5 m, and the angle at which it swings is θ = 10°, what is the speed of the mass as it swings?

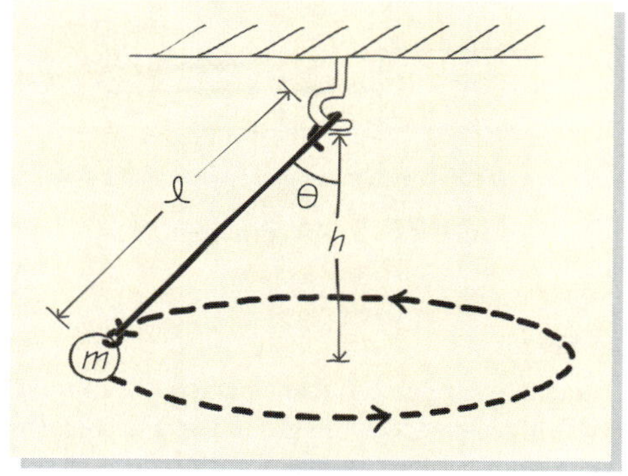

FIGURE 5–43 Problem 57.

58. (II) A small mass slides without friction in a horizontal circular path around the sides of a circular bowl. The bottom of the bowl may be described as a parabola, with the height h above the bottom varying quadratically with the distance r from the axis: $h = br^2$. The mass is observed to move in its circular path with a speed v. What is the height of the path?

59. (II) A small block slides in a horizontal circle on the inside of a conical surface, with the cone making an angle of 50° with the vertical (Fig. 5–44). Assuming that there is no friction between the block and the surface, and the block slides with an angular speed of 6 rad/s, at what vertical height above the apex of the cone does the block slide?

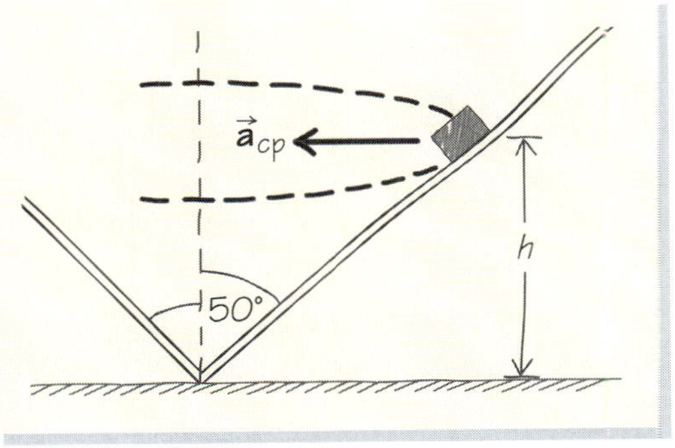

FIGURE 5–44 Problem 59.

60. (II) Two light strings 1.0 m in length are attached to a vertical support 1.0 m apart, and a mass of 5.0 kg at the end of the two strings is whirled about the vertical z-axis (Fig. 5–45). Both strings are taut, so that they and the vertical support form an equilateral triangle. The tension in the upper string is measured to be 150 N. (a) What is the tension in the lower string? (b) How much time does it take for the apparatus to make one complete circuit around the vertical support?

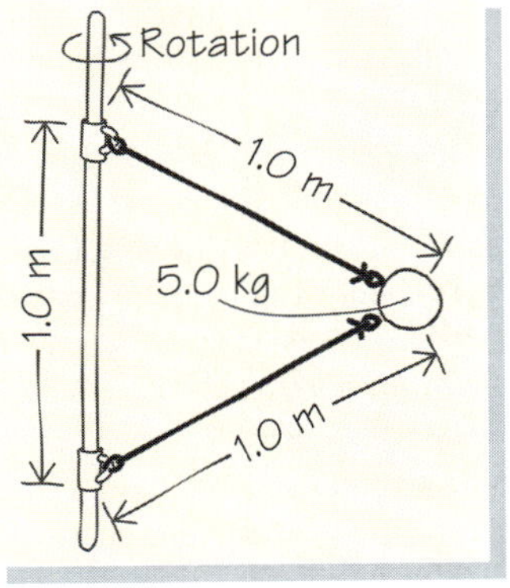

FIGURE 5–45 Problem 60.

61. (II) The road in Example 5–13 is now in better condition, and the coefficient of static friction between the tires and the road is 0.8. With what maximum constant speed can the motorist now negotiate the curve? Friction acts here both to provide a centripetal force and to help cancel the vertical forces. Is there one and only one speed with which the curve can be taken, as in Example 5–13? Why or why not?

62. (II) A train of mass 1.5×10^5 kg is traveling horizontally at 80 km/h and rounding a bend whose radius of curvature is 2 km. At the same time it is decelerating at a rate of 0.2 m/s^2. The length of the train is negligible compared with the size of the bend, and the train can be treated as a point. What net force does the track exert on the train? Give an approximate answer in which the net change in speed is small compared to the speed itself. Such questions help engineers decide how "robustly" a track must be constructed or how much to bank it.

63. (II) A small puck of mass $m = 0.2$ kg moves in a circle of radius 0.5 m on a table top; the puck is tied with a massless string to a tether at the origin. The coefficient of kinetic friction between the puck and the table top is $\mu_k = 0.2$. At $t = 0$ s, the puck is at the point shown with a velocity in the $+y$-direction of magnitude 10 m/s. (a) What is the tension in the string at $t = 0$ s? (b) What is the tension in the string at the end of one revolution?

64. (II) A pendulum hangs at rest from a hook in a ceiling of a building. The building is located at a latitude such that the radius vector from the center of Earth to the building makes an angle θ with Earth's axis of rotation. Assume that the force of gravity is the same everywhere and points directly to the center of Earth. Where will the pendulum point? (By definition, Earth rotates about its axis once a day.)

65. (III) A satellite of mass 3000 kg travels in a circular orbit 180 km above Earth, where the acceleration due to gravity is 5 percent smaller than on Earth's surface. Assume that, in a year, the satellite loses 5 km in altitude because of the drag of the extremely thin atmosphere at that altitude. What would you estimate the density of air to be at that altitude, given that the effective area of the satellite is 6 m^2 and the drag coefficient, C_D, in Eq. (5–30) is 1.0?

66. (III) Consider a ball thrown outward from the center of a platform that rotates counterclockwise with uniform angular velocity ω (Fig. 5–46). An observer standing off the platform (in an inertial reference frame) will describe the ball as moving with uniform velocity in a straight line. (Ignore the effect of gravity; imagine looking down on the platform from above, so that you do not see the up-and-down motion of the ball.) The inertial observer will see that the ball reaches a horizontal distance r from the center in time $t = r/v$, where v is the speed of the ball. (a) Show that, in time t, a point at a radius r on the platform will have moved a distance $d = v\omega t^2$. (b) An observer moving with the platform will see the ball curve away to the right (as seen from the center of the platform). Show that the perceived acceleration is perpendicular to the velocity vector, and that its magnitude is $2\omega v$. (c) What is the direction of the

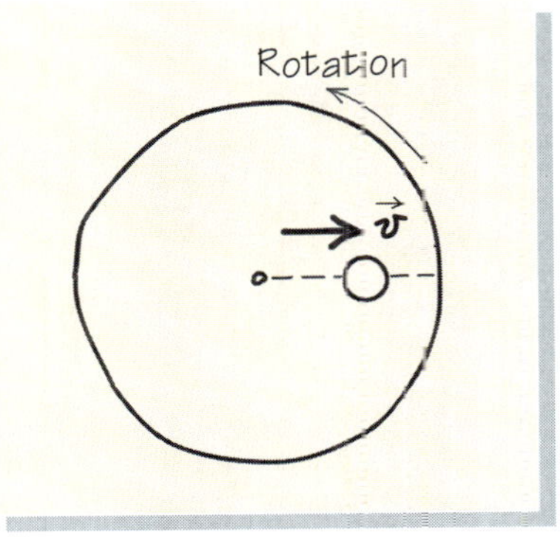

FIGURE 5–46 Problem 66.

perceived acceleration of a ball thrown by a person on the platform from the rim toward the center? It is not surprising that an observer in an accelerated frame sees force-free motion that nevertheless undergoes acceleration. Under certain circumstances—for example, in the study of global wind motion on a rotating Earth—it is convenient to study motion as seen in an accelerating frame. The frame-imposed acceleration is then attributed to a purely fictitious force, which, for rotating systems like the one treated in this problem, is called the *Coriolis force*.

General Problems

67. (I) A 5.0-kg sphere at the end of a 1.2-m cable swings in a horizontal circle on a frictionless surface at the rate of one revolution every 1.4 s. What is the tension in the cable?

68. (II) A small die that is placed 5 in from the center of a turntable begins to slide. The turntable is rotating at $33\frac{1}{3}$ rev/min. What is the coefficient of static friction between the die and the turntable?

69. (II) Three masses (from left to right: 0.3 kg, 0.4 kg, and 0.2 kg) are connected by light cords to make a "train" sliding on a frictionless surface. They are accelerated by a constant horizontal force $F = 1.5$ N that pulls the rightmost mass to the right. What is the tension, T, in the cord (a) between the 0.3-kg and the 0.4-kg masses? (b) between the 0.4-kg and 0.2-kg masses?

70. (II) A string 6.95 m long is strung between two pegs (4.96 m apart) on a ceiling. A mass of 3.88 kg is attached to a point 2.96 m along the string. What are the tensions in the two segments of the string?

71. (II) Consider a system of masses connected by light ropes that pass over massless and frictionless pulleys (Fig. 5–47). (a) When m_1 is displaced vertically by Δx_1, what is the displacement Δx_2 of m_2? (b) For $m_1 = 1.2$ kg and $m_2 = 1.8$ kg, calculate the respective accelerations of the two masses. (c) What is the tension in the string for the masses given in part (b)?

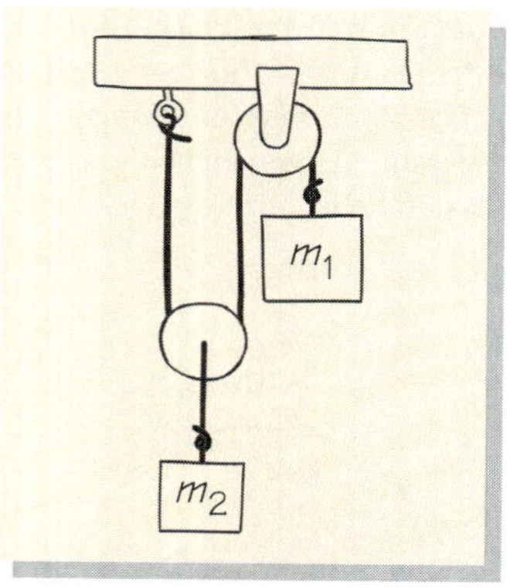

FIGURE 5–47 Problem 71.

72. (II) Masses $m_1 = 0.6$ kg and $m_2 = 1.2$ kg are connected by a taut rope. Mass m_2 is just over the edge of a ramp inclined at an angle of $\theta = 35°$, as in Fig. 5–48, and the masses have a coefficient of kinetic friction of $\mu_k = 0.25$ with the surface. At $t = 0$ s, the system is given an initial speed of $v_0 = 1.5$ m/s, which starts mass m_2 down the ramp. (a) Draw the force diagram for each mass. (b) Solve the equations of motion to predict the motion of the system with time. Assume that the rope is long enough so that mass m_1 does not go over the edge.

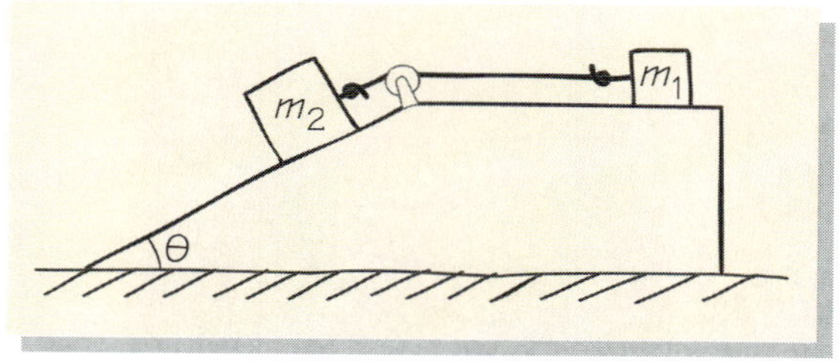

FIGURE 5–48 Problem 72.

73. (II) One of the entertainments at the carnival is a rotating cylinder. The participants step in and place themselves against the interior wall. The cylinder starts to rotate more and more rapidly, and at some point the floor falls away, leaving the customers stuck like so many flies to the wall. If the cylinder were to slow down without the floor coming back up, the participants would begin to slip down. In terms of the relevant parameters, express the rotational speed, ω, at which this happens.

74. (II) A motorcycle moves in a horizontal circular path on the inside surface of a vertical cylinder of radius 8 m. Assuming that the coefficient of static friction between the wheels of the motorcycle and the wall is 0.9, how fast must the motorcycle move so that it stays in the horizontal path?

75. (II) Consider the conical pendulum described in Problem 57. Express the angular velocity in terms of the string angle θ and the string length ℓ.

76. (II) A tractor of mass 1200 kg is pulling a sled loaded with 450 kg of hay bales. The coefficient of kinetic friction between the sled and the ground is 0.50. (a) What horizontal force must the tractor exert to move at constant speed? (b) What is the tension in the rope between the tractor and sled? (c) If the tractor stops, how much horizontal force must it exert to get the sled moving again if $\mu_s = 0.60$?

77. (II) A bicyclist traveling at 10 m/s rides around an unbanked curve. If the coefficient of friction between the tires and the road is $\mu = 0.4$, what is the shortest turn the bicyclist can safely make? Is the coefficient of friction here static or kinetic?

78. (II) Two cars are traveling at 60 mi/h, one behind the other. The driver of the second car reacts by braking 0.8 s after she observes the sudden braking of the car ahead of her. The front car has a mass of 1200 kg, and the coefficient of friction between the tires and the road with the brakes applied is 0.8. The second car has mass 1600 kg, and the coefficient of friction with the brakes applied is 0.7. How far behind must the second car have been in order to avoid hitting the first car?

79. (II) A stunt motorcyclist rides with uniform speed on the inside rim of a vertical circular ramp of radius 12 m (Fig. 5–49). How fast must the motorcyclist travel to avoid leaving the surface at the top of the loop?

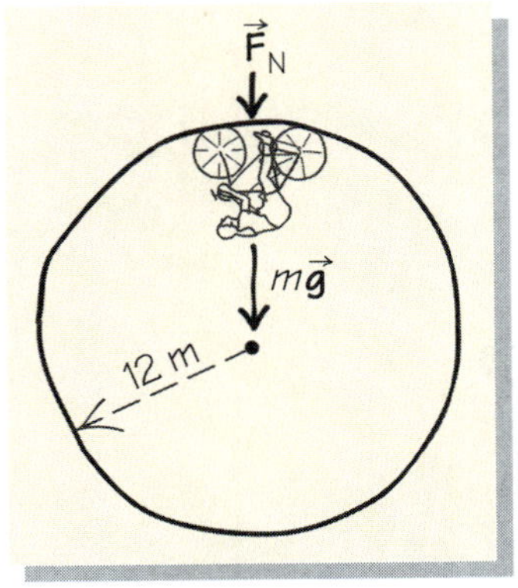

FIGURE 5–49 Problem 79.

80. (III) The new white belt of a long horizontal conveyor is moving with a constant speed of $v = 3.0$ m/s. A small block of carbon is placed on the belt with zero initial velocity relative to the ground. The block will slip a bit before moving with the belt, leaving a black mark on the belt (Fig. 5–50). How long is that mark if the coefficient of kinetic friction between the carbon block and the belt is 0.20 and the coefficient of static friction is 0.30?

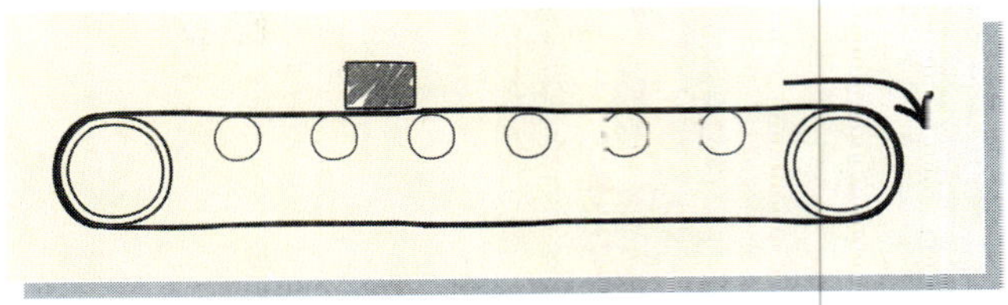

FIGURE 5–50 Problem 80.

CHAPTER

6

Work is done to lift these crushed stones against gravity as they move through this conveyor belt assembly. As the contact force from the conveyor belt does positive work on the stones, gravity is doing an equal magnitude of negative work on them.

Work and Kinetic Energy

A baseball flying through the air will change speed due to the forces of gravity and air drag. We have solved the equations of motion—Newton's second law—when gravity alone acts on the ball. If we included air drag in that same approach, however, it would present formidable technical obstacles. We can use another approach to this problem that is much simpler. It involves two new concepts that are the subjects of this chapter: work and kinetic energy. The notion of energy is one of the most basic concepts in physics and indeed in all the sciences. Energy takes many forms and, in this chapter, we will focus on the energy contained in a moving object, which we call its energy of motion, or *kinetic energy*. The *work* done on an object involves the force acting on it as it moves. We can relate the change in the kinetic energy of an object to the work done on it as it moved; this relation is called the work–energy theorem. The change in speed of a baseball, which is related to its change in kinetic energy, may thereby be related to the work done on it by the forces of gravity and air drag. Once we learn how to calculate the work done on an object, we have a powerful tool for the understanding of motion.

Nowadays, no matter how complex a force might be, a computer can generate solutions to Newton's second law with any given set of initial conditions. Powerful computers were not available in the nineteenth century, however, and it is fortunate that the scientists of that time made an effort to extract information from the general structure of the equations of motion. Their legacy, which includes the notion of energy, is a much deeper understanding of the properties of systems in which forces act than reams of computer output could possibly provide.

6–1 KINETIC ENERGY AND WORK

In this section, we want to introduce the idea that moving objects have kinetic energy and we want to relate that kinetic energy to the forces acting on the object. There are two reasons for exploring this idea. First, the energy approach can provide a significant shortcut for many aspects of problem solving. Second, the idea of energy turns out to be a truly fundamental one in physical science. Let's consider a one-dimensional problem in which a *constant* net force of magnitude F_{net} acts on an object of mass m in the $+x$-direction. In this case, the acceleration (a) of the object is also a constant, with magnitude F_{net}/m. Equation (2–24) gives us a relation between the object's acceleration and its displacement:

We introduced this relation in Eq. (2–24).

$$v^2 = v_0^2 = 2a(\Delta x) = 2(F_{net}/m)(\Delta x). \tag{6–1}$$

Here, $\Delta x = x - x_0$ represents the change in the object's position—the displacement from an initial position x_0 to the present position x. Similarly, $v^2 - v_0^2$ is the change in the speed squared, where v_0 is the initial speed.

If we multiply by m and divide by 2, we find

$$F_{net}\Delta x = \frac{1}{2}mv^2 - \frac{1}{2}mv_0^2. \tag{6–2}$$

If an object is subjected to a constant net force, this type of relation is all we need to determine how the speed of the object has changed after it has traveled a certain distance. Notice that the time dependence of the motion does not appear here.

Work

The quantity on the left-hand side of Eq. (6–2) is the **net work**, W_{net}, done by the net force on the object:

The definition of the work done by a constant force in one dimension

$$\text{for constant net force in one dimension:} \quad W_{net} \equiv F_{net}\Delta x. \tag{6–3}$$

The SI unit of work

The unit of work is so important that a special name has been reserved for it: the **joule** (J). One joule is the work done by a force of 1 N in pushing an object a distance of 1 m. Thus 1 J = 1 N·m. In the cgs system, the unit of work is the **erg**, which is the work done by a force of 1 dyne in moving an object a distance of 1 cm. We have

$$1 \text{ erg} = (1 \text{ dyne})(1 \text{ cm}) = (1 \text{ g}\cdot\text{cm/s}^2)(1 \text{ cm}) = (1 \text{ g}\cdot\text{cm}^2/\text{s}^2)\left(\frac{1 \text{ kg}}{10^3 \text{ g}}\right)\left[\frac{(1 \text{ m})^2}{(10^2 \text{ cm})^2}\right]$$
$$= 10^{-7} \text{ kg}\cdot\text{m}^2/\text{s}^2 = 10^{-7} \text{ J}. \tag{6–4}$$

The word *work* has several connotations in our language; the more colloquial uses of the word should not be confused with the precise definition presented in Eq. (6–3). We all agree that pushing a crate across a rough floor requires us to do work on the crate. This is because a force must be applied for work to be done on an object. Less in accord with our everyday use of the word is the idea that *there must be displacement if there is to be work* (Fig. 6–1). In holding a bag of groceries stationary, no work is ever done, even though colloquial usage of the word suggests that it is a lot of "work" to hold the bag stationary.

(a)

(b)

FIGURE 6–1 In raising the barbell from a position (a) near the floor to a position (b) above his head, this weightlifter has done positive work. When holding the barbell above his head, the weightlifter is doing no work.

Kinetic Energy

The definition of kinetic energy

The quantity $mv^2/2$, whose change is expressed on the right-hand side of Eq. (6–2), is the **kinetic energy**, K:

$$K \equiv \tfrac{1}{2}mv^2. \tag{6–5}$$

Notice that the kinetic energy cannot be a negative number. There is no directionality in the kinetic energy; it is a scalar quantity with dimensions of [mass][velocity]^2:

$$[K] = [M][L^2T^{-2}] = [ML^2T^{-2}].$$

The dimensions and units of energy

The combination of SI units with this dimension is $\text{kg} \cdot \text{m}^2/\text{s}^2$. According to Eq. (6–2), kinetic energy has the same dimensions as work. Thus we also measure kinetic energy (and, as we shall see, all forms of energy) in joules. Note that work has the units $\text{N} \cdot \text{m}$, whereas kinetic energy has the units $\text{kg} \cdot \text{m}^2/\text{s}^2$. We can verify the consistency of these units by recalling that $1\ \text{N} = 1\ \text{kg} \cdot \text{m/s}^2$, so $1\ \text{N} \cdot \text{m} = 1\ \text{kg} \cdot \text{m}^2/\text{s}^2$. See Table 6–1 for typical values of kinetic energy.

The Work–Energy Theorem

Let's look again at Eq. (6–2) in terms of our definitions of kinetic energy and work. Suppose that an object moves under the influence of a constant net force from an initial point x_i to a final point x_f (and thus undergoes a displacement $\Delta x = x_f - x_i$). Its speed changes, and hence so does its kinetic energy. This change ΔK in kinetic energy is given by

$$\Delta K = K_f - K_i. \tag{6–6}$$

In terms of ΔK and W_{net}, Eq. (6–2) becomes

$$W_{\text{net}} = \Delta K. \tag{6–7}$$

The work–energy theorem

This equation is known as the **work–energy theorem**.

The *sign* of work and therefore of the change in kinetic energy in Eq. (6–7) can be either positive or negative. We see from Eq. (6–3) that the sign of W_{net} is determined by the relative signs of F_{net} and Δx. If the force is directed in the *same* direction as the displacement (they have the same sign), the net work done during that displacement is positive. If the force is directed in a direction *opposite* to the displacement (they have opposite signs), the work done is negative. Suppose that a baseball moves upward. The displacement is upward, with a sign opposite to that of the force of gravity acting on it. The product of force and displacement is negative. Thus the net work done by gravity on the ball as it rises is negative. According to the work–energy theorem, the *change* in kinetic energy is then negative; that is, the baseball loses speed. Similarly, when the ball falls, the work done on it is positive, and it gains kinetic energy.

TABLE 6–1
Some Orders of Magnitude for Kinetic Energies

System	*Kinetic Energy (J)*
Electron in orbit around a nucleus	10^{-18}
Molecule of air at room temperature	10^{-17}
Electron in a TV tube	10^{-15}
Falling raindrop	10^{-3}
Baseball pitch	10^2
A running human	10^3
Automobile on a highway	10^5
Large earthquake	10^{17}
Earth in orbital motion around the Sun	10^{33}

EXAMPLE 6–1 A car of mass 1200 kg falls a vertical distance of 24 m starting from rest (Fig. 6–2). What is the work done by the force of gravity on the car? Use the work–energy theorem to find the final velocity of the car just before it hits the water. (Treat the car as a pointlike object.)

Solution: We measure the position of the car along a vertical y-axis, conventionally pointing upward. Then the downward displacement is $\Delta y = -24$ m. The force of gravity is the only force acting, so it makes the only contribution to the net work. Gravity has magnitude mg and is also oriented downward, so the y-component of the net force is $F_{\text{net}} = -mg$. Therefore the net work is

$$W_{\text{net}} = F_{\text{net}}\Delta y = (-mg)\,\Delta y = (-(1200\ \text{kg})(9.8\ \text{m/s}^2))(-24\ \text{m}) = 2.8 \times 10^5\ \text{J}.$$

We can now use the work–energy theorem to find the final speed v. The initial kinetic energy, K_i, is zero because the car starts from rest. Thus Eqs. (6–5), (6–7), and (6–8) give

$$\Delta K = K_f = W_{\text{net}} = mg\,\Delta y;$$

$$\tfrac{1}{2}mv^2 = mg\,\Delta y.$$

The mass cancels and we find that

$$v = \sqrt{2g\,\Delta y} = \sqrt{2(9.8\ \text{m/s}^2)(24\ \text{m})} = 22\ \text{m/s, nearly 50 mi/h.}$$

(This problem is one we could have solved in Chapter 3; we treat it here to bring in the concepts of work and kinetic energy.)

FIGURE 6–2 Example 6–1.

Calculating the Work

It is possible to calculate the work done by each force acting on an object. For example, if each force were constant and acted in one dimension, then the work done on an object by a particular force F as the object moves through a displacement Δx is given by

$$W = F\Delta x.$$

Remember, however, that the work–energy theorem specifies the *net* work. Often, several forces act on an object. When a piano is slowly lowered at constant speed by a rope, the force of gravity is canceled by the rope tension; that is why there is no acceleration. Here, the net force is zero, so the *net work* is also zero. This is consistent with the fact that there is no change in kinetic energy and no change in speed.

Because the net force is the sum of the individual forces acting, the net work can be decomposed into a sum of all the work done by individual forces. *This is a result that will hold for the most general definition of work.* It is often simpler to find the work W_i done by individual forces F_i and take the algebraic sum of the W_i to find the net work. In the case of the piano lowered by a rope at constant speed, positive work is done by gravity as the piano is lowered. The tension of the rope, however, which points upward, does the same magnitude of *negative* work on the piano. The forces cancel and the net work is zero in this case.

PROBLEM SOLVING

The work done by the sum of forces is equal to the sum of the work done by individual forces.

EXAMPLE 6–2 A box of books of mass 100 kg is pushed with constant speed in a straight line across a rough floor with a coefficient of kinetic friction $\mu_k = 0.2$. Find the work done by the force that pushes the box if the box is moved a distance $d = 3$ m.

Solution: We approach this problem by drawing a force diagram (Fig. 6–3). With no vertical displacement, no work is done by gravity or by the normal force. The forces in the vertical direction must cancel and so $F_N = mg$. Because the box moves with a constant velocity, the *net* horizontal force must vanish. Thus the pushing force **F** must be equal in magnitude but opposite in direction to the force of friction **f**, whose magnitude is given by $f = \mu_k F_N = \mu_k mg$. Hence the magnitude of **F** is also $\mu_k mg$. The direction of **F** is the same direction as the displacement d. Thus the work done by the pushing force is positive. We have for the work done by the pushing force

$$W = Fd = \mu_k mgd = (0.2)(100\ \text{kg})(9.8\ \text{m/s}^2)(3\ \text{m}) = 6 \times 10^2\ \text{J}.$$

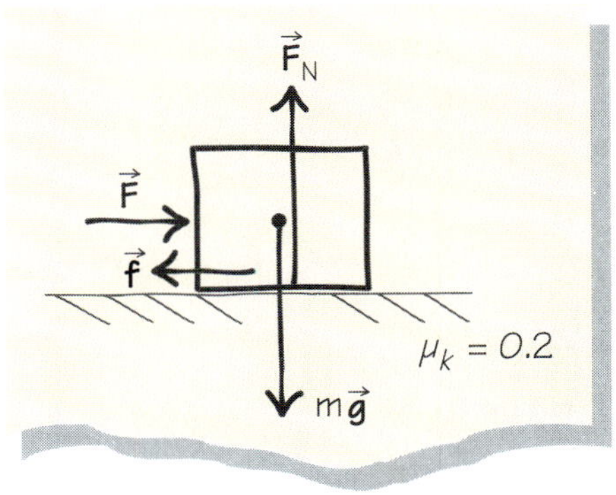

FIGURE 6–3 Example 6–2. Force diagram for a crate being pushed across a floor.

APPLICATION: JET ENGINES-KINETIC ENERGY SYSTEMS

A jet engine is a good example of a system that works by employing the work–energy theorem (Fig. 6AB–1). In the reference frame of the engine, air of mass m is brought into the engine at velocity v by the intake fans, which compress the air—indeed, the fans are also called compressors. The air, consisting mainly of oxygen and nitrogen molecules, is then mixed with fuel. Several things then happen to the air as it burns. One chemical reaction that occurs during combustion produces two water molecules ($2 \times H_2O$) for each oxygen molecule (O_2). This doubles the volume of that part of the oxygen from the air that combines with the hydrogen. In another chemical reaction, part of the oxygen (O_2) combines with carbon to form CO_2; this portion of the oxygen maintains its volume. The nitrogen does not burn. Because of the water-producing reaction, a bigger volume of gas leaves the engine than entered it. To be able to keep up the continuous action of the engine, the outgoing gas must therefore leave with a velocity greater than the incoming velocity. Furthermore, the mass of the departing gas is larger since the mass of the fuel that has been burned has been added to it. Thus the outgoing gas has substantially higher kinetic energy than does the incoming gas. By the work–energy theorem, the engine has done work to provide this kinetic energy increase. In effect, this work is associated with a force that is applied to the outgoing gas to give it increased kinetic energy; by the third law, then, a corresponding force acts on the airplane to move it forward against the frictional drag forces of the air. In this way, the work done by the engine propels the airplane.

FIGURE 6AB–1 The jet engines you see here are devices that use energy to do work and thereby propel the airplane.

EXAMPLE 6–3 A crate of mass 96 kg is pushed across a horizontal floor by a force **F**. The coefficient of kinetic friction between the crate and the floor is $\mu_k = 0.27$ (Fig. 6–4a). The crate moves with uniform velocity. What is the magnitude of force **F**? Suppose that at some point the crate passes onto a new section of floor, where $\mu_k' = 0.085$. The pushing force on the crate is unchanged. After 1.25 m on the new section of the floor, the crate moves with a speed of $v_f = 2.3$ m/s. What was the original speed of the crate, v_i?

Solution: We start with the force diagram drawn in Fig. 6–4b. This diagram permits us to find the force, which must balance the force of friction because the crate is moving with constant speed on the first section of floor. The force of fric-

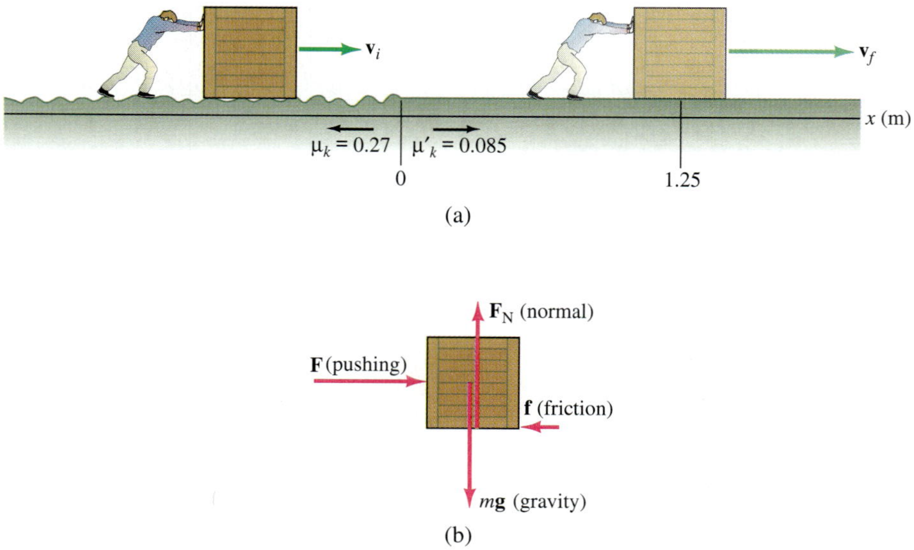

FIGURE 6–4 (a) Example 6–3. The floor changes composition from left to right to a more slippery surface at $x = 0$ m. (b) Force diagram for the crate.

tion that opposes **F** has magnitude

$$f = \mu_k mg = (0.27)(96\ \text{kg})(9.8\ \text{m/s}^2) = 2.5 \times 10^2\ \text{N}.$$

Force **F** must have just this magnitude: $F = 2.5 \times 10^2$ N.

The force of friction is decreased on the new section of floor, whereas the pushing force remains the same. Thus there is a (constant) net force in the direction of motion and the crate accelerates uniformly. The work done on the crate can be calculated. Then the work–energy theorem will allow us to find how much the crate's kinetic energy has changed and hence to find the original speed.

On the new section of floor, the force of friction is given by

$$f' = \mu_k' mg.$$

The net force on the crate then has magnitude

$$F_{\text{net}} = F - f' = (\mu_k - \mu_k')mg$$

and acts in the direction of motion of the crate. The net work done on the crate when it moves a distance Δx on the new section of floor is then

$$W_{\text{net}} = F_{\text{net}}\Delta x = (\mu_k - \mu_k')mg\ \Delta x.$$

According to the work–energy theorem, this is the increase in kinetic energy of the crate as it moves over the new section of floor:

$$W_{\text{net}} = K_f - K_i = \tfrac{1}{2}mv_f^2 - \tfrac{1}{2}mv_i^2.$$

We solve this equation for the initial speed:

$$v_i^2 = v_f^2 - \frac{2W_{\text{net}}}{m} = v_f^2 - \frac{2(\mu_k - \mu_k')mg\ \Delta x}{m}$$

$$= (2.3\ \text{m/s})^2 - 2(0.27 - 0.085)(9.8\ \text{m/s}^2)(1.25\ \text{m}) = 0.76\ \text{m}^2/\text{s}^2;$$

$$v_i = 0.87\ \text{m/s}.$$

Note that we did not need to find the kinetic energy of the crate over the entire distance where it was changing.

FIGURE 6–5 A simple machine is a kind of force amplifier. It does the same work that a large force would do over a small distance by applying a small force over a large distance. Thus a lever lifts a large mass a height h by applying a force much less than the mass's weight over a distance much greater than h. Here, we can see a lever, a ramp, a wedge, and various screw-driven devices.

We can do a fixed amount of work by applying a large force over a small distance or a small force over a large distance. We are all familiar with the classic simple machines: the ramp, the lever, the block and tackle (see Example 5–3), the screw thread, and so forth (Fig. 6–5). The simple machines are devices that do a given amount of work by employing a small force over a compensatingly longer distance. For example, the lever lifts a very heavy object a small distance by applying a force (a small force compared to the object's weight) to the other end of the lever, which must move a good deal. The woman in Example 5–3 pulls quite a bit of rope with a relatively small force to lift a football player a short distance. In each case, we say we are using a "mechanical advantage," or that we are using "leverage." The simple machines are therefore force "amplifiers." They represent engineering at its very best: simple, elegant, and practical.

APPLICATION
Simple machines

6–2 CONSTANT FORCES IN SPACE

Thus far, we have discussed the concepts of kinetic energy, work, and the work–energy theorem as they apply to one direction only. We now want to extend these concepts to two and three dimensions. We continue to assume that the net force **F** is a constant, but now it is a vector, as are the displacement, velocity, and acceleration.[†] (For simplicity, we have dropped the subscript "net" on the net force).

[†]A constant force is a vector that does not change with time and has the same value at every point in space.

The application of energy and work to two and three dimensions is a straightforward one; we know that Newton's laws include the information that motion and the effects of forces on motion are *independently applicable to the different Cartesian directions.* Let's apply these concepts to two dimensions for simplicity. Newton's second law, $\mathbf{F} = m\mathbf{a}$, now represents two separate equations for the x- and y-directions. If we label the components of all vector quantities with the appropriate axes, then we can derive an equation like the work–energy theorem—Eq. (6–2)—for each direction:

$$F_x\,\Delta x = \tfrac{1}{2}mv_x^2 - \tfrac{1}{2}mv_{x0}^2; \qquad (6\text{–}8a)$$

$$F_y\,\Delta y = \tfrac{1}{2}mv_y^2 - \tfrac{1}{2}mv_{y0}^2. \qquad (6\text{–}8b)$$

Here, the displacement vector $\Delta\mathbf{r} = \Delta x\mathbf{i} + \Delta y\mathbf{j}$. The velocity $\mathbf{v}$ and the initial velocity $\mathbf{v}_0$ have each been separated into their components, as has the net force $\mathbf{F}$.

FIGURE 6–6 The tugboat is doing positive work on the freighter even though the direction of the tugboat force on the freighter is different than the direction of the freighter's velocity and displacement.

Let's now take the sum of these two equations. We recognize that the square of the *magnitude* of the velocity—speed squared (v^2)—is the sum of the x- and y-components of the velocity squared:

$$v^2 = v_x^2 + v_y^2.$$

The summed equations then form the generalization of the work–energy theorem:

$$F_x\,\Delta x + F_y\,\Delta y = \tfrac{1}{2}mv^2 - \tfrac{1}{2}mv_0^2. \qquad (6\text{–}9)$$

If we continue to define kinetic energy as in Eq. (6–5)—$K = mv^2/2$—then the change in kinetic energy appears on the right-hand side of this equation.

The left-hand side of Eq. (6–9) generalizes the definition of the work done on an object in two dimensions. The work done easily generalizes to three dimensions:

The scalar product of two vectors

$$\text{for constant } \mathbf{F}\text{:} \quad W \equiv F_x\,\Delta x + F_y\,\Delta y + F_z\,\Delta z. \qquad (6\text{–}10)$$

This definition for the work contains the components of the two vectors $\mathbf{F}$ and $\Delta\mathbf{r}$ multiplied, dimension by dimension, with the resulting products added together. This combination of two vectors occurs frequently and is called the **scalar product** (or **dot product**) of these two vectors. The work, then, is the scalar product of $\mathbf{F}$ and $\Delta\mathbf{r}$. A scalar product, although it is the product of two vectors, is itself a scalar quantity. If you are not already familiar with this concept, refer to the separate box explaining the scalar product.

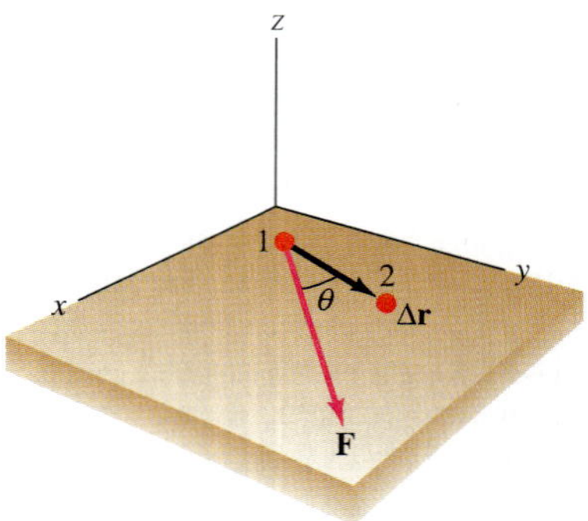

FIGURE 6–7 The two vectors $\Delta\mathbf{r}$ and $\mathbf{F}$ always form a plane. We have labeled that plane as the xy-plane. The angle between these vectors determines the work done by the constant force on an object undergoing displacement $\Delta\mathbf{r}$, namely $W = F\,\Delta r \cos\theta$.

From Eq. (6–10) and the expression Eq. (B1–6) for the scalar product, the work W done by a constant force $\mathbf{F}$ acting on an object that moves through a displacement $\Delta\mathbf{r}$ is given by

$$W = \mathbf{F}\cdot\Delta\mathbf{r}. \qquad (6\text{–}11)$$

With this definition of work, the work–energy theorem takes exactly the same form as before—$W_{net} = \Delta K$—as long as the net force is a constant vector. It does not matter that the force and the displacement point in different directions (Fig. 6–6).

Suppose the vectors $\mathbf{F}$ and $\Delta\mathbf{r}$ at any given position have an angle θ between them (Fig. 6–7). From Eq. (B1–1) in the Problem-Solving Techniques box on page 151 and Eq. (6–11), the work then takes the form

The work done by a constant force

$$W = F\,\Delta r\cos\theta. \qquad (6\text{–}12)$$

Equivalently, the work done is the product of the magnitude of force $\mathbf{F}$ and the magnitude of the displacement $\Delta\mathbf{r}$ *in the direction of the force.* Put another way, *only the component of the displacement along the force counts in the work.* In particular, a force perpendicular to the motion does no work. A frequently occurring example is the normal force, $\mathbf{F}_N$, which is perpendicular to the surface on which an object moves; $\mathbf{F}_N$ does no work on that object. Another important example of a force that does no work in this way is the centripetal force responsible for uniform circular motion (see Section 6–4).

PROBLEM SOLVING

Only the component of the displacement along the force counts in the work.

PROBLEM-SOLVING TECHNIQUES

The Scalar Product

In Chapter 1, we discussed the definition of vectors and the multiplication of vectors by scalars. The product bA of a scalar b and a vector **A** is a vector. It points in the same direction as **A** and has magnitude bA, where A is the magnitude of **A**. One way to multiply two vectors—**A** and **B**, for example—is the **scalar product A · B**. The scalar product is a scalar quantity whose value is

$$\mathbf{A}\cdot\mathbf{B} \equiv AB\cos\theta. \quad \text{(B1–1)}$$

Here, θ is the angle between the directions of the two vectors (Fig. 6B1–1a). The scalar product has the properties that

$$\mathbf{A}\cdot\mathbf{B} = \mathbf{B}\cdot\mathbf{A} \quad \text{(B1–2)}$$

and

$$\mathbf{A}\cdot(\mathbf{B}+\mathbf{C}) = (\mathbf{A}\cdot\mathbf{B}) + (\mathbf{A}\cdot\mathbf{C}). \quad \text{(B1–3)}$$

If two vectors are perpendicular (*orthogonal*) to each other, then $\theta = 90°$ and $\cos\theta = 0$, and their scalar product is zero. If the vectors are parallel to each other, then the scalar product takes on its maximum value; that is, the product of the magnitudes of the two vectors. The scalar product of a vector with itself is the square of its magnitude, $\mathbf{A}\cdot\mathbf{A} = A^2$. The unit vectors **i, j**, and **k** along some set of orthogonal axes x, y, and z have the property that

$$\mathbf{i}\cdot\mathbf{i} = \mathbf{j}\cdot\mathbf{j} = \mathbf{k}\cdot\mathbf{k} = 1. \quad \text{(B1–4)}$$

Because they are orthogonal to each other,

$$\mathbf{i}\cdot\mathbf{j} = \mathbf{j}\cdot\mathbf{k} = \mathbf{i}\cdot\mathbf{k} = 0. \quad \text{(B1–5)}$$

Two vectors **A** and **B** can be decomposed into their vector components: $\mathbf{A} = A_x\mathbf{i} + A_y\mathbf{j} + A_z\mathbf{k}$ and $\mathbf{B} = B_x\mathbf{i} + B_y\mathbf{j} + B_z\mathbf{k}$. The rules in Eqs. (B1–4) and (B1–5) allow us to write the scalar product of **A** and **B** as

$$\mathbf{A}\cdot\mathbf{B} = (A_x\mathbf{i} + A_y\mathbf{j} + A_z\mathbf{k})\cdot(B_x\mathbf{i} + B_y\mathbf{j} + B_z\mathbf{k})$$

$$= A_xB_x + A_yB_y + A_zB_z. \quad \text{(B1–6)}$$

Thus the scalar product of two vectors is *the sum of the product of the components of the two vectors.*

The scalar product is a scalar quantity, so it remains the same even if the axes of our coordinate system are rotated. If we consider two vectors **A** and **B**, we may choose our coordinate frame in such a way that **A** lies along the x-axis, $\mathbf{A} = A\mathbf{i}$. The other axes can be arranged so that the vector **B** has only x- and y-components, $\mathbf{B} = B_x\mathbf{i} + B_y\mathbf{j}$. These vectors are shown in Fig. 6B1–1b, which is a view looking down on the plane formed by **A** and **B**. According to Eq. (B1–6), the scalar product is then given by

$$\mathbf{A}\cdot\mathbf{B} = AB_x.$$

Thus the scalar product of two vectors may be described as the *product of the length of one vector and the projection of the other vector along the direction of the first one.* Because $B_x = B\cos\theta$, we recover here our original definition: Eq. (B1–1).

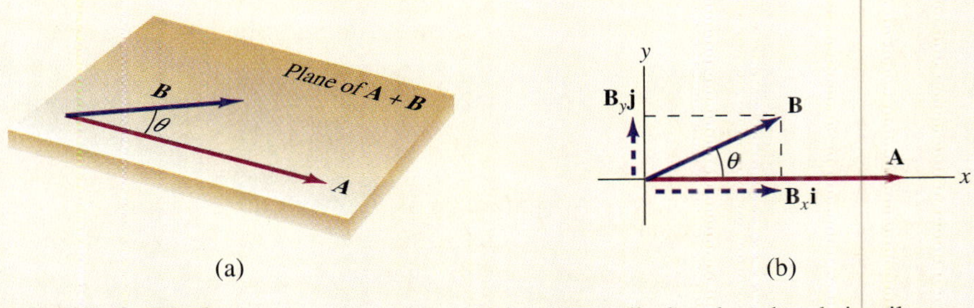

FIGURE 6B1–1 (a) Two vectors **A** and **B** have been displaced so that their tails meet at the same point. They are oriented in space with the angle θ between them. (b) The x-axis of our coordinate system has been redefined so that **A** lies along the $+x$-direction. The scalar product is independent of the orientation of the axes.

EXAMPLE 6–4 A block of mass m is placed on an inclined plane that makes an angle θ with the horizontal (Fig. 6–8a). A horizontal force **F** is applied to the block. Find the magnitude of **F** such that the block moves up the plane with acceleration **a**. What is the work done by **F**? (Assume that the plane is rough, with coefficient of kinetic friction μ_k, and that the distance the block is moved along the plane is d).

Solution: Start with a force diagram for the block (Fig. 6–8b). We will first determine the magnitude F of the horizontal force. Once we have found this value, the work done by this force on the block will be

$$W = Fd\cos\theta.$$

In Fig. 6–8c, we choose axes and decompose all the forces into components. The direction up along the plane is the x-direction, and the direction pointing directly away

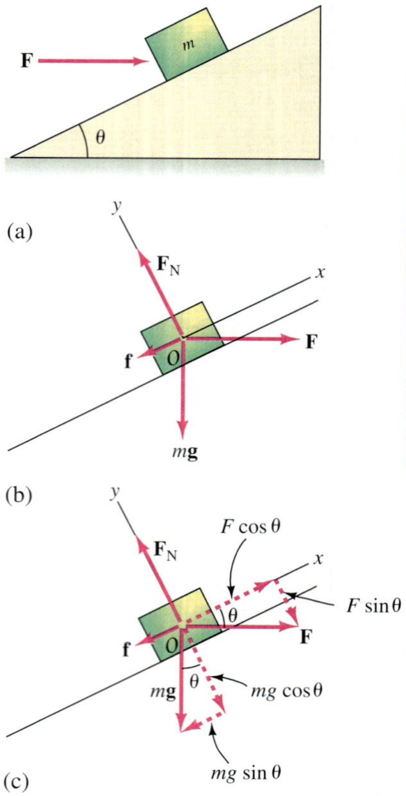

FIGURE 6–8 (a) Example 6–4. (b) Force diagram for the block on the inclined plane. (c) The forces are decomposed into components along the plane surface and perpendicular to it.

from the surface of the plane is the y-direction. The y-components of the forces must add up to zero because there is no acceleration in the direction perpendicular to the plane. We thus have

$$F_N - F\sin\theta - mg\cos\theta = 0.$$

From this equation, $F_N = F\sin\theta + mg\cos\theta$. There is also an acceleration up the plane, which is determined by Newton's second law applied to the x-direction:

$$ma = F\cos\theta - mg\sin\theta - \mu_k F_N$$

$$= F\cos\theta - mg\sin\theta - \mu_k(F\sin\theta + mg\cos\theta).$$

Hence

$$F = \frac{ma + mg\sin\theta + \mu_k mg\cos\theta}{\cos\theta - \mu_k\sin\theta}.$$

The work done by the force in accelerating the block is then

$$W = Fd\cos\theta = md\frac{a + g\sin\theta + \mu_k g\cos\theta}{1 - \mu_k\tan\theta}.$$

Application of the Work–Energy Theorem to Projectile Motion

A particularly simple application of the work–energy theorem for constant forces concerns projectile motion. Let's suppose that an object of mass m is thrown with initial velocity $\mathbf{v}_1 = v_{1x}\mathbf{i} + v_{1y}\mathbf{j}$ off a tall building. We choose the upward vertical direction to be the y-direction. After the object has fallen a vertical distance h (if we choose the top of the building to be $y = 0$, then the mass falls from $y = 0$ to $y = -h$), it has acquired a velocity $\mathbf{v}_2 = v_{2x}\mathbf{i} + v_{2y}\mathbf{j}$ (Fig. 6–9). The speed v_2 at this point is determined from the kinetic energy K_2, and the work–energy theorem gives K_2:

$$K_2 = K_1 + W. \qquad (6\text{–}13)$$

Here, W is the work done by gravity on the falling object as it moves from its initial point to its final point. The force of gravity is given by $\mathbf{F} = F_y\mathbf{j} = -mg\mathbf{j}$. The work done is therefore

$$W = \mathbf{F}\cdot\Delta\mathbf{r} = (F_y\mathbf{j})\cdot[(\Delta x)\mathbf{i} + (\Delta y)\mathbf{j} + (\Delta z)\mathbf{k}]$$

$$= F_y\,\Delta y = (-mg)(-h) = mgh. \qquad (6\text{–}14)$$

The force has only a y-component, so only the y-component of the displacement enters into the work. Substituting into Eq. (6–13), we have

$$\tfrac{1}{2}mv_2^2 = \tfrac{1}{2}mv_1^2 + mgh. \qquad (6\text{–}15)$$

We have arrived at this simple expression without solving Newton's second law. Note that the increase in speed depends only on the vertical distance fallen; it does *not* depend on the horizontal distance covered, even though the speed has contributions from both the vertical and horizontal components of the velocity.

6–3 FORCES THAT VARY WITH POSITION

Thus far in our study of the work–energy theorem, the force acting on an object has been constant. In contrast, we may have a force whose magnitude varies from point to point, one whose direction varies from point to point, or one for which both the magnitude and direction vary. We want to generalize the definition of work to include these cases so that the work–energy theorem continues to hold.

Variable Forces in One Dimension

Let's again consider motion in one dimension—we will drop the vector notation temporarily—but this time, let's suppose that the magnitude of the net force varies with position. We can use our earlier development of the work–energy theorem by subdividing the total interval over which the object moves (Fig. 6–10a). We make the subintervals of the object's displacement small enough so that the force may be considered to be constant in each of them. For example, if a force changes smoothly by some factor over a distance of 1 m, then it may change by only 1 percent of this factor over distances of 1 cm. But if, as in Fig. 6–10b, the force changes more abruptly during one portion of the object's movement, we can always make the intervals smaller in this region so that we can regard the force as constant in each interval.

Let's take a look at how we might partition a varying force into intervals (Fig. 6–11). In each interval shown in Fig. 6–11, the force (acting in one dimension) may be viewed as constant. We suppose that the force takes on the approximately constant value F_1 in the interval from initial position x_i to x_1, the value F_2 in the interval x_1 to x_2, and, generally, the value F_k in the interval x_{k-1} to x_k. There are a total of N intervals. Let's calculate the total work done by the force in moving an object of mass m from initial position x_i to final position x_f; we will use the expression for work appropriate for constant force in each interval. In each of the intervals, we apply Eqs. (6–2) and (6–3):

$$W_1 = F_1(x_1 - x_1) = \tfrac{1}{2}m(v_1^2 - v_i^2) = K_1 - K_i, \qquad (6\text{–}16a)$$

$$W_2 = F_2(x_2 - x_1) = \tfrac{1}{2}m(v_2^2 - v_1^2) = K_2 - K_1, \qquad (6\text{–}16b)$$

$$W_3 = F_3(x_3 - x_2) = \tfrac{1}{2}m(v_3^2 - v_2^2) = K_3 - K_2, \qquad (6\text{–}16c)$$

and so on, finishing with the last interval, which is expressed as

$$W_N = F_N(x_f - x_{N-1}) = \tfrac{1}{2}m(v_f^2 - v_{N-1}^2) = K_f - K_{N-1}. \qquad (6\text{–}16d)$$

Now let's sum all these equations. The work done in moving the object through each interval of displacement is placed on the left-hand side of the equation; this is the total work done. On the right-hand side, all of the terms that involve $v_1^2, v_2^2, \ldots, v_{N-1}^2$ cancel, leaving

$$W = \sum_{k=1}^{N} W_k = \tfrac{1}{2}m(v_f^2 - v_i^2)$$

$$= K_f - K_i = \Delta K. \qquad (6\text{–}17)$$

We find that Equation (6–17) is the work–energy theorem again.

The left-hand side of Eq. (6–17) is just the sum of the areas of individual rectangles—each one of the form $F_k \Delta x_k$. In the limit that the intervals Δx_k become infinitesimally small, the sum of their areas is the total area under the curve of $F(x)$ versus x. This area is the *integral* of $F(x)$ over x. We denote it mathematically as

$$W = \lim_{N \to 0} \sum_{j=1}^{N} F(x_j)\Delta x_j = \int_{x_i}^{x_f} F(x)\, dx. \qquad (6\text{–}18)$$

Equation (6–17) is useful to find ΔK only if we can evaluate the work, and this is done by using the techniques of **integration**. If you are not familiar with the properties of integrals and the techniques of integration, you may want to refer back to Section 2–6 or consult the Problem-Solving Techniques box included here on page 155.

Just as work is given by an integral of the force, force is given by a derivative of the work, because differentiation and integration are inverse processes. In particular, Eq. (B2–2) from the Problem-Solving Techniques box allows us to find the force if the work done by that force is known:

$$\frac{dW(x_f)}{dx_f} = F(x_f). \qquad (6\text{–}19)$$

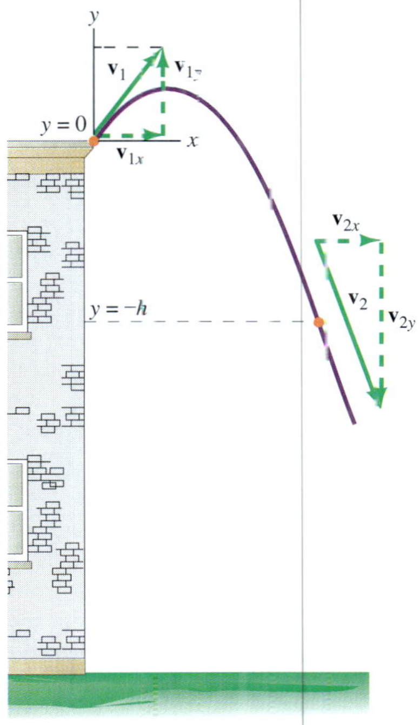

FIGURE 6–9 A projectile is launched off the top of a building. The work that gravity does on the object as it moves along its trajectory determines its kinetic energy and thus its speed at various points on the trajectory.

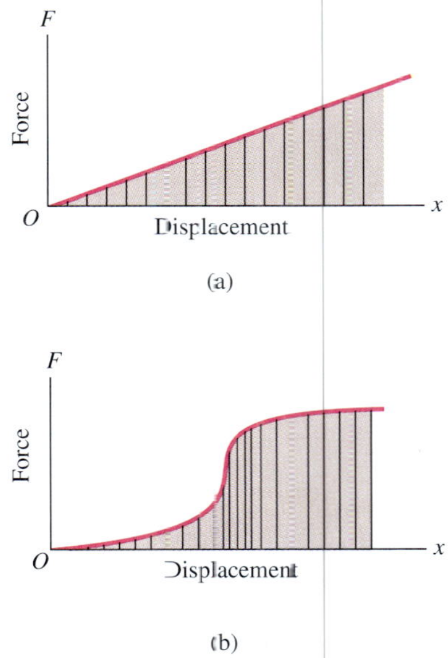

FIGURE 6–10 (a) A smoothly varying force is approximately constant in small, equally spaced intervals of x. (b) If there is a region where the force changes more rapidly, the intervals in that region must be made smaller, so that the force can be thought of as constant in each interval.

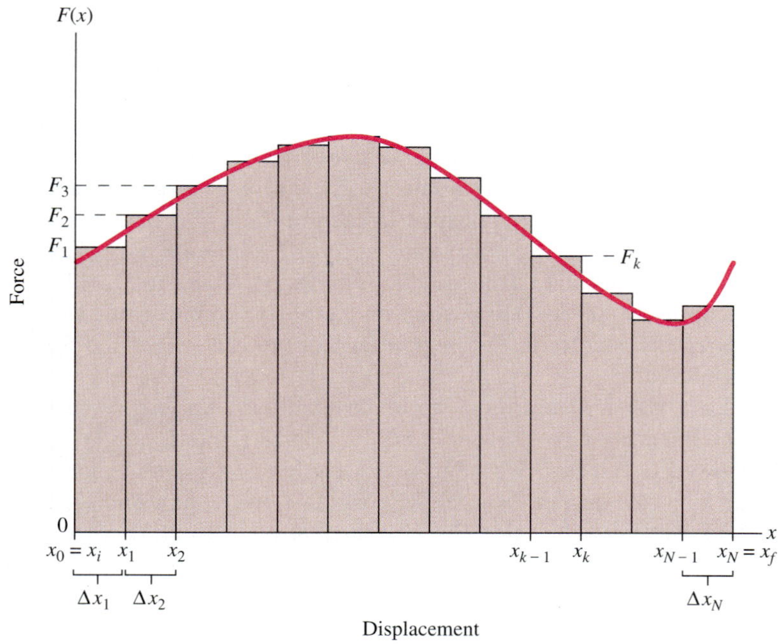

FIGURE 6–11 A variable force $F(x)$ has an approximately constant value ($F(x_j) = F_j$) within each of the small intervals Δx_j in the region between x_i and x_f. The work done by the force on an object as the object moves between these limits is the sum of the work W_k done in each interval, for which we can use the constant force formula for work, $W_k = F_k \Delta x_k$. This sum, W, is the integral of $F(x)$ over x, and it is equal to the area under the curve of F versus x.

Finally, we can recover our earlier result for constant force, Eq. (6–4). Multiplicative constants (in this case, F) can be removed from beneath an integral sign, so that

$$\text{for constant } F: \quad W = F\int_{x_i}^{x_f} \mathrm{d}x = F(x_f - x_i) = F\,\Delta x.$$

The Work Done by a Spring

One of the most important examples of a one-dimensional variable force is the force exerted on a mass by a spring attached to it (Fig. 6–12). This force takes the form

$$F = -kx \tag{6–20}$$

Hooke's law

Here, x measures the displacement of the mass from an equilibrium position; k is a constant characteristic of the particular spring known as the **spring constant.** This force law is known as **Hooke's law,** after its seventeenth-century discoverer, Robert Hooke. Note the sign here: When x is positive, the mass is on the right side and the force points to the left; when x is negative, the mass is on the left side and the force points to the right. The spring force always acts to bring the mass back to $x = 0$.

Let's now calculate the work done on the mass by the spring force when the mass moves from the equilibrium position to the position $x = L$ (Fig. 6–13a), then calculate the extra work that is done as the mass moves still further to a position $2L$ beyond the equilibrium point. (Note that if the mass is at rest, it cannot move away from $x = 0$ under the influence of the spring force alone; the spring force is zero there. But the mass may have been set in motion by other forces. We assume that is the case.) To calculate the work, we must integrate the (nonconstant) force over the displacement, as in Eq. (6–18):

$$W = \int_0^L (-kx)\,\mathrm{d}x.$$

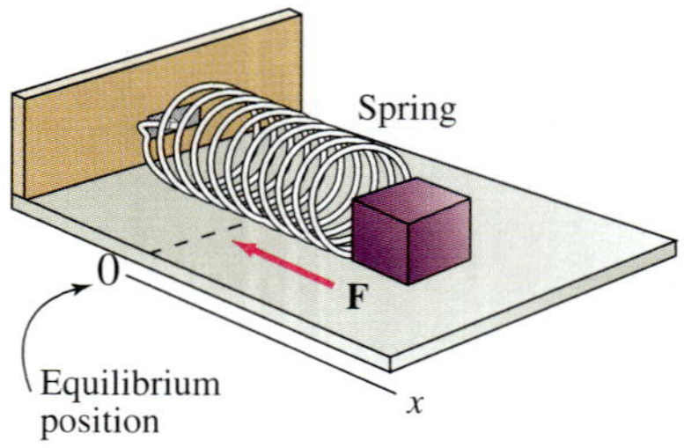

FIGURE 6–12 A spring exerts a force on an object that tends to bring the object back to the equilibrium position. Here the spring is stretched out past the equilibrium position.

Note the sign: The spring force is in the direction opposite to the displacement of the mass, so the work done by the spring will be negative. We use the general integration

PROBLEM-SOLVING TECHNIQUES

Integration

Integration of a function is the inverse of differentiation of that function. Put another way, if we integrate some function $f(x)$ and then differentiate the result, we get the function $f(x)$ back again. Expressed mathematically, if the functions $f(x)$ and $g(x)$ are related by

$$g = \int_{x_i}^{x_f} f(x)\mathrm{d}x, \qquad \text{(B2–1)}$$

then

$$\frac{\mathrm{d}g(x_f)}{\mathrm{d}x_f} = f(x_f). \qquad \text{(B2–2)}$$

Moreover, we can interpret work as the area under the curve of F versus x (see Fig. 6–11). If the function F is negative and the displacements are positive, as in Fig. 6B2–1, the area and hence the work are negative. Conversely, if the displacement is negative and the force positive, then the work will be negative. In the language of integrals, this comes about because the integral changes sign when we reverse the limits on it:

$$\int_{x_f}^{x_i} f(x)\mathrm{d}x = -\int_{x_i}^{x_f} f(x)\mathrm{d}x. \qquad \text{(B2–3)}$$

If the integral on the right-hand side is positive, then the left-hand side—which represents the work done in a displacement from x_f to x_i—is negative.

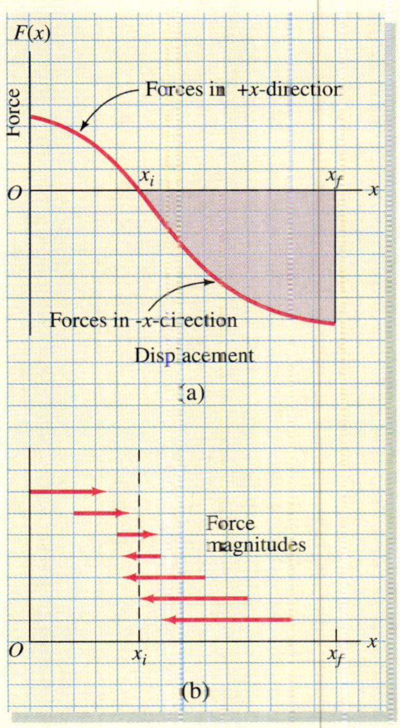

FIGURE 6B2–1 In the region between x_i and x_f, the shaded area under the curve is negative [part (a)], corresponding to negative values for the force $F(x)$ [part (b)].

formula for powers in Appendix IV–8, with $p = 1$, to find that

$$W = -k\int_0^L x\,\mathrm{d}x = -\frac{1}{2}kx^2\Big|_0^L = -\frac{1}{2}kL^2.$$

To find the extra work that must be done by the spring force as the mass moves from position L to position $2L$, we calculate

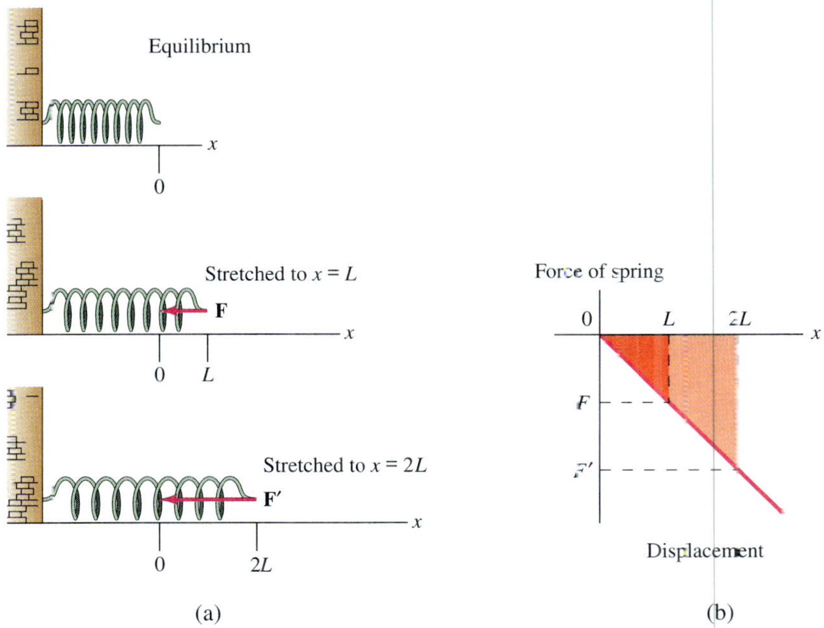

FIGURE 6–13 (a) The force exerted by a spring changes as the spring stretches (or compresses). Twice the stretch corresponds to twice the force. (b) The area under the curve of force versus x is the work the spring force performs on a mass attached to its end. The work done as the spring stretches from $x = 0$ to $x = 2L$ is four times the work done as the spring stretches from $x = 0$ to $x = L$.

$$\int_{L}^{2L}(-\,kx)\mathrm{d}x = -\frac{1}{2}kx^2\Big|_{0}^{2L} = -\frac{1}{2}k[(2L)^2) - L^2] = -\frac{3}{2}kL^2.$$

The total work done by the spring as the mass moves from $x = 0$ to $x = 2L$ is the work done as the mass moves from $x = 0$ to $x = L$ plus this extra work: $-\frac{1}{2}kL^2 - \frac{3}{2}kL^2 = -2kL^2$ (compared to $-\frac{1}{2}kL^2$ for the first part of the motion). The spring does four times as much work when the mass moves twice the distance.

It is easy to understand these results graphically. Figure 6–13b is a graph of the spring force F as a function of x. The work done by this force as the mass moves from one value of x to another is the area under the curve of F versus x, between the two values of x. In particular, we will consider the shaded triangle under the force curve (here, a straight line with slope k) from $x = 0$ to $x = L$ and a second shaded triangle under the force curve from $x = 0$ to $x = 2L$. These triangles are similar but the second has twice the base length and twice the height. The second triangle has four times the area of the first, so the spring force does four times as much work when the mass moves twice as far.

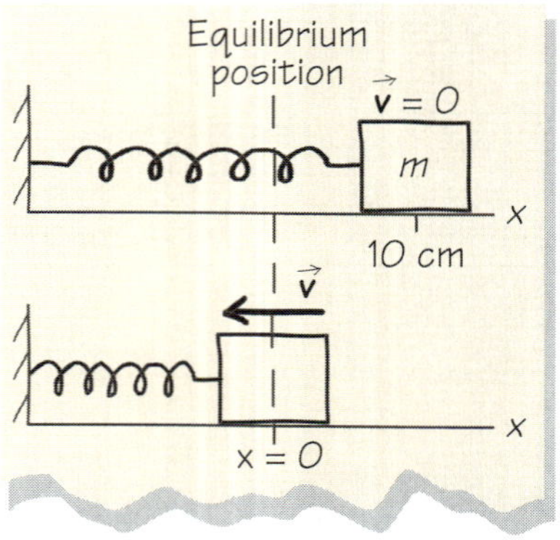

FIGURE 6–14 Example 6–5.

EXAMPLE 6–5 A mass $m = 50$ g is placed on the end of a spring with a spring constant k = 6.0 N/m. The mass is released from rest at $x = 10$ cm (Fig. 6–14). Opposing the force of the spring is an unknown drag force. Find the work done on the mass by the drag force if the mass's velocity at the equilibrium position $x = 0$ is 0.85 m/s.

Solution: We must put several elements into place to be able to use the work–energy theorem and find the work done by the drag force. The net work is the difference between the final kinetic energy K_f and the initial kinetic energy K_i; we know both K_i (0 because the mass starts from rest) and K_f. The net work, however, is also the algebraic sum of the work done by the spring, W_s, and the work done by the drag force, W_D. Thus if we can find W_s we can solve for W_D.

We know that the force exerted by the spring on the mass is $F = -kx$. The work done by the spring is then

$$W_s = \int_{x_0}^{0}(-kx)\mathrm{d}x = -\int_{0}^{x_0}(-kx)\mathrm{d}x = \frac{1}{2}kx^2\Big|_{0}^{x_0} = \frac{1}{2}kx_0^2.$$

We now apply the work–energy theorem. We denote the speed of the mass at the equilibrium point as v_1, and thus

$$W_{\text{net}} = W_s + W_D = K_f - K_i$$

or

$$W_D = K_f - K_i - W_s = \tfrac{1}{2}mv_1^2 - 0 - \tfrac{1}{2}kx_0^2$$

$$= (0.5)(50 \times 10^{-3}\ \text{kg})(0.85\ \text{m/s})^2 - (0.5)(6.0\ \text{N/m})(0.10\ \text{m})^2 = -0.012\ \text{J}.$$

Why is the sign negative? It is because the drag acts in a direction opposite to the displacement of the mass.

Forces that Vary in Both Magnitude and Direction

We must make one further generalization in order to define work that will satisfy the work–energy theorem. It is the case where the force acts not only with varying magnitude but also with varying direction. The net displacement $\mathbf{\Delta r}$ of an object under the influence of a net force may result from a rather complicated *path* in space (Fig. 6–15). (Here we have drawn the path in a plane, but the path could move through three-dimensional space.) The path is described with a changing position vector $\mathbf{r}$. At the distance r from the origin, the force on the object is approximately a constant vector

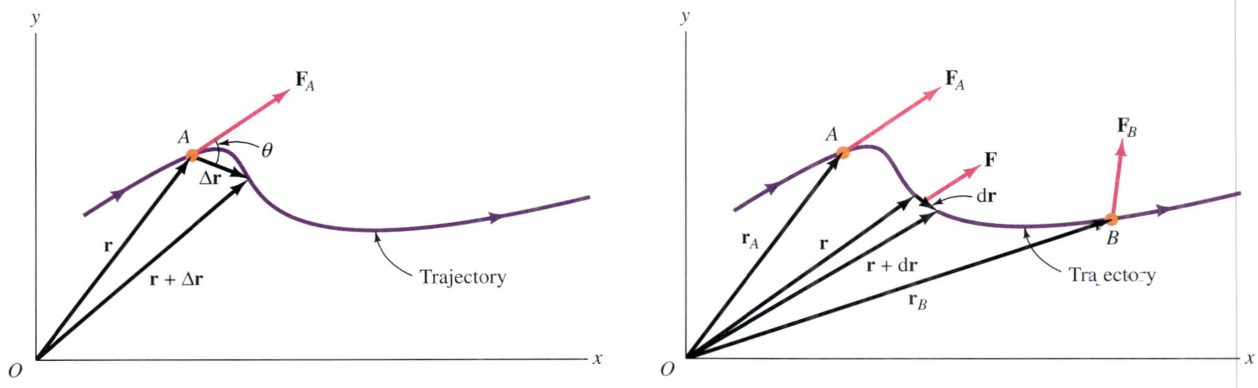

(a) (b)

FIGURE 6–15 (a) An object moves from position $\mathbf{r}$ at time t to position $\mathbf{r} + \Delta\mathbf{r}$ at time $t + \Delta t$, while a force $\mathbf{F}_A$ acts on it. The work done on the object by the force over the displacement $\Delta\mathbf{r}$ is $\mathbf{F}_A \cdot \Delta\mathbf{r}$. (b) In a finite time, there is a net displacement from point A to point B.

$\mathbf{F}_A$: It does not change very much as the object moves through a small displacement $\Delta\mathbf{r}$ from the point labeled by position vector $\mathbf{r}$ to an adjacent point labeled by position vector $\mathbf{r} + \Delta\mathbf{r}$ (Fig. 6–15a). As $\Delta\mathbf{r} \to 0$, this displacement is *tangential* to the curve at the tip of $\mathbf{r}$, while the force vector points in some other direction. The work done on the object in moving through the small interval is then $\mathbf{F}_A \cdot \Delta\mathbf{r}$. We find the total work done for a displacement from point A with position vector $\mathbf{r}_\mathrm{A}$ to point B with position vector $\mathbf{r}_\mathrm{B}$ (Fig. 6–15b) by summing these infinitesimal contributions:

$$W = \int_{\mathbf{r}_\mathrm{A}}^{\mathbf{r}_\mathrm{B}} \mathbf{F} \cdot d\mathbf{r}. \tag{6–21}$$

The work done by a force that varies over space

This formula is the general definition of work and is consistent with all our earlier definitions. The integral that appears here is called a **line integral** because it depends not only on the beginning and ending points A and B, but, in general, on the path, or line, taken to move between these points. We shall explore the properties of this integral in Section 6–4.

No Work Is Done in Uniform Circular Motion

We conclude this section with the important observation that *no net work is done on a particle that undergoes uniform circular motion* (Fig. 6–16). This remark holds for *any* part of a circular trajectory at constant speed. To see this result, it is enough to recall from Chapter 5 that an object undergoing uniform circular motion experiences an acceleration that is directed along the radius to the center of the circle. Thus the force is directed in the (negative) radial direction and is always *perpendicular* to the direction of motion, which is tangential. If the infinitesimal displacement along an arc is $d\mathbf{s}$, as in Fig. 6–17, then the scalar product $\mathbf{F} \cdot d\mathbf{s}$ is zero, and thus no work is done.

If the motion is circular with varying speed, a tangential force $\mathbf{F}_t$ must be present (one that is parallel to the direction of the displacement). Work is done for such motion. Because the force is directed along (or opposite to) the displacement $d\mathbf{s}$, the scalar product $\mathbf{F}_t \cdot d\mathbf{s}$ for the infinitesimal work done has nonzero magnitude $F_t ds$. We integrate this quantity to find the total work.

FIGURE 6–16 The net work done on a Ferris wheel passenger in moving from the bottom to the top and back again is zero. In particular, no work is done by the forces, a combination of friction and normal forces, that uniformly move the passenger in a circle.

EXAMPLE 6–6 A metallic sphere with a mass of 8.0 kg is attached to a vertical shaft by a rod 1.5 m long of negligible mass that is perpendicular to the shaft (Fig. 6–18). A tangential force $\mathbf{F}$ is applied to the sphere for one-quarter turn. As a result, the sphere rotates about the frictionless shaft with an angular speed of 1.2 rev/s. If the sphere was initially at rest, what was the magnitude of the tangential force?

Angular variables were introduced in Chapter 3.

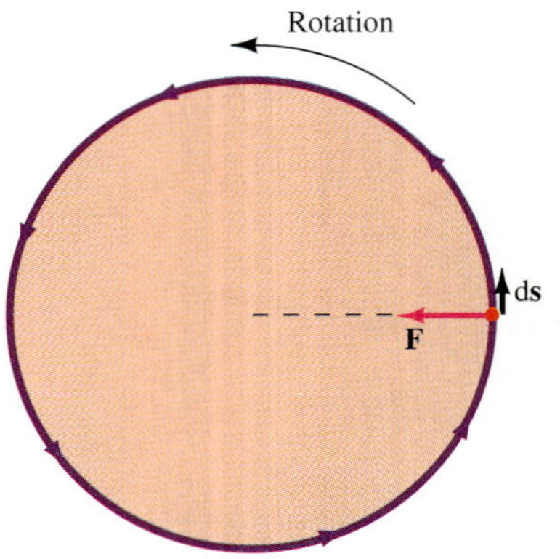

FIGURE 6–17 In the uniform circular motion of an object, the infinitesimal displacement is tangent to the circle. The force responsible for this motion is directed toward the center of the circle and is perpendicular to the displacement. The work done by this force on the object is therefore zero.

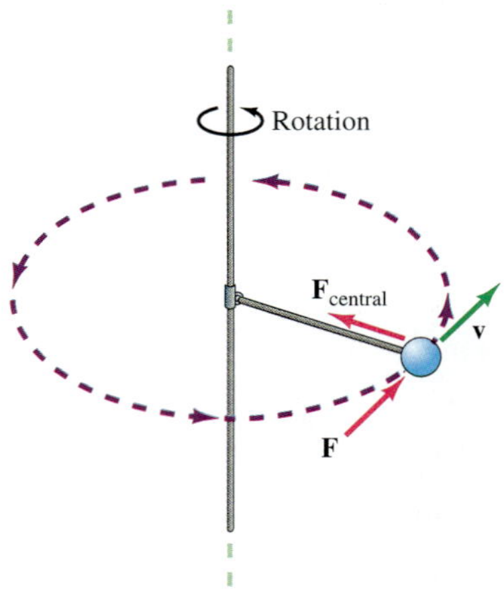

FIGURE 6–18 Example 6–6.

Solution: The angular speed after the force has acted is 1.2 rev/s = (1.2 rev/s)(2π rad/rev) = 7.5 rad/s, so the speed of the sphere at the end of the push is $v = \omega r = (7.5\ \text{s}^{-1})(1.5\ \text{m}) = 11$ m/s. From the work–energy theorem, then,

$$K_f - K_i = \tfrac{1}{2}mv_f^2 - \tfrac{1}{2}mv_i^2 = \tfrac{1}{2}(8.0\ \text{kg})(11\ \text{m/s})^2) = 4.8 \times 10^2\ \text{J}.$$

The force **F**, applied for a quarter-turn, acts over a distance $\frac{1}{4}(2\pi r) = \frac{1}{4}\pi(3.0\ \text{m}) = 2.4$ m. Thus the force has magnitude

$$F = \frac{K_f - K_i}{\text{distance}} = \frac{4.8 \times 10^2\ \text{J}}{2.4\ \text{m}} = 2.0 \times 10^2\ \text{N}.$$

6–4 CONSERVATIVE AND NONCONSERVATIVE FORCES

One of the most important questions that arises when an object undergoes a certain displacement is: How does the work done depend on the path taken by the object during its displacement? We call this a question of *path dependence.* We illustrate what is involved by considering, in order, gravity, friction, and the spring force.

Gravity. Let us consider the work that gravity does in moving an object of mass m from the top of a ramp of length L to the bottom. Figure 6–19a shows the geometry of the ramp, which makes an angle θ with the horizontal. The force of gravity acts vertically downward with magnitude mg on the object. This force can be decomposed into a component $mg \cos \theta$ perpendicular to the ramp and a component $mg \sin \theta$ pointing along the ramp toward its bottom. When an object slides down the ramp from an initial point to a final point, the motion is perpendicular to the force $mg \cos \theta$, and the work done by that component of the force is therefore zero. The work done by the force of gravity to move the object the total length L along the direction of motion is thus

$$W = (mg \sin \theta)L. \tag{6–22}$$

Consider now a second path from the top of the ramp to the bottom. This second displacement is achieved by dropping the object vertically off the back of the ramp and then moving it horizontally to the previous end point (Fig. 6–19b). The force downward is mg, but the distance through which the force acts is $L \sin \theta$. Thus the work done for this part of the second path of the displacement is $(mg)L \sin \theta$. The object is then moved horizontally; no work is done by gravity during this portion of the second displacement because the force of gravity is perpendicular to this motion. Thus the total work done by gravity is the same as that calculated in Eq. (6–22). We would find the same result for other paths: The work done by gravity depends only on the difference between the final and initial heights of the object being moved; it does not depend on the path by which this height difference was reached.

Friction. Let us next consider the work done by friction on a coffee cup of mass m that is pushed across a rough horizontal surface. In this case, friction is not the only force acting on the cup, but it is not the work–energy theorem that is at issue here—that theorem involves the *net* work—rather, it is only the work done by friction that interests us here. The force of friction on the cup has magnitude $\mu_k mg$ and it is directed *opposite* to the motion. Suppose that the cup in Fig. 6–20 is moved (with the aid of some external force) along the $+x$-axis from $x = 0$ to $x = x_{\text{max}}$ and then back again. For the first half of the motion, the displacement of the cup is $+x_{\text{max}}\mathbf{i}$, while the force of friction is directed back toward $x = 0$: $\mathbf{f} = -\mu_k mg\mathbf{i}$. In the motion back to the origin, the displacement is $-x_{\text{max}}\mathbf{i}$, while the force of friction is directed to $+x$: $\mathbf{f} = \mu_k mg\mathbf{i}$. Then

$$W = W_{\text{out}} + W_{\text{back}} = (-\mu_k mg)(x_{\text{max}}) + (\mu_k mg)(-x_{\text{max}}) = -2\mu_k mgx_{\text{max}}.$$

Rather than canceling, the same amount of negative work is done on the cup by friction both out and back, and there is work done even though the net displacement is zero. Moreover, the amount of work depends on x_{max}; that is, it depends on the path taken by the cup.

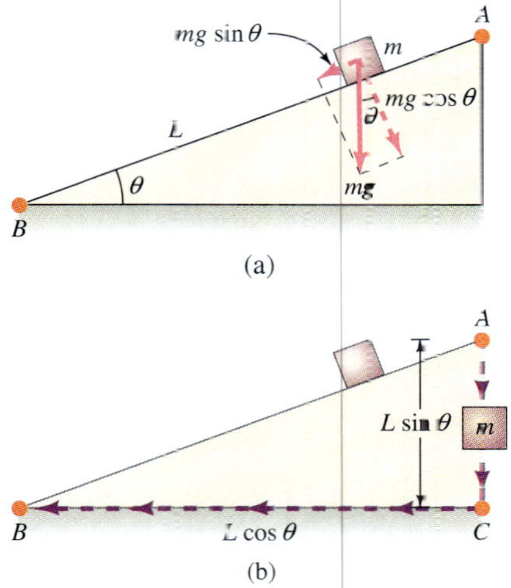

FIGURE 6–19 (a) An object of mass m moves from point A to point B on an inclined plane under the influence of gravity. Gravity does positive (or negative) work on the object as it moves down (or up) the plane. (b) The object now moves from point A to point B by a different path: a vertical motion from point A to point C, followed by a horizontal movement from C to B. The work done by gravity is exactly the same as in part (a).

Spring Force. Equation (6–18) is an expression for the work done that applies to any one-dimensional force that depends only on the position of the object. It is a general property of integration that integrals such as these depend only on x_f and x_i, not on any intermediate points. This is illustrated with the spring force in the following example.

EXAMPLE 6–7 A mass m is attached to the end of a spring with spring constant k (Fig. 6–21a). The mass moves from the point $x = L$ to the point $x = 0$ in two ways: (1) Path 1 is the direct motion (Fig. 6–21b), and (2) path 2 is the displacement from $x = L$ to $x = a$, with $a < 0$, followed by a displacement to $x = b$ with $b > L$, completed with a final displacement to $x = 0$ (Fig. 6–21c). Find the work done on the mass by the spring for each path. (You should find the same answer in each case.)

Solution: Unlike gravity, the spring exerts a force that varies with position. For the work, we must therefore use the form $W = \int F\,dx$ between suitable limits. In the case of path 1, we have

$$W = \int_L^0 (-kx)\,dx = -\frac{kx^2}{2}\Big|_0^L = -\left(0 - \frac{kL^2}{2}\right) = \frac{kL^2}{2}.$$

The work done by the spring on the mass is positive.

In the case of path 2, there are three contributions to W—corresponding to the three displacement steps described—and their sum is

$$W = \int_L^a (-kx)\,dx + \int_a^b (-kx)\,dx + \int_b^0 (-kx)\,dx = -\frac{kx^2}{2}\Big|_L^a - \frac{kx^2}{2}\Big|_a^b - \frac{kx^2}{2}\Big|_b^0$$

$$= -\frac{ka^2}{2} + \frac{kL^2}{2} - \frac{kb^2}{2} + \frac{ka^2}{2} - \frac{k0^2}{2} + \frac{kb^2}{2} = \frac{kL^2}{2},$$

the same work as for path 1.

Conservative and Nonconservative Forces

We have seen that the work done is independent of the path taken by the object for both gravity and the spring force; however, that is not the case for friction. In two and three dimensions, it is not enough to say that the work done on an object during displacement will be path independent if the force depends solely on the object's position. The integral in Eq. (6–21) may or may not depend on the path. In any case, we can generally divide forces into two classes: Forces for which the work is independent of the path are called **conservative forces**; those for which the work depends on the path are called **nonconservative forces**. We have seen that gravity and the spring force are conservative, whereas friction is nonconservative.

The meaning and definition of conservative forces

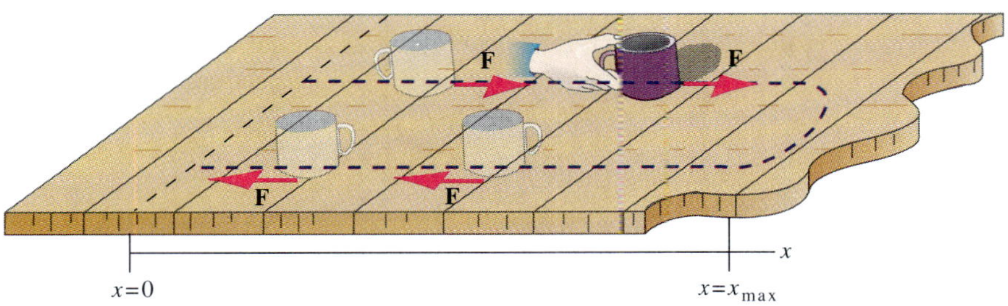

FIGURE 6–20 A cup moves along a horizontal surface from $x = 0$ to $x = x_{max}$ and back again. One of the forces acting is friction. The work done by friction in the back-and-forth motion is not zero.

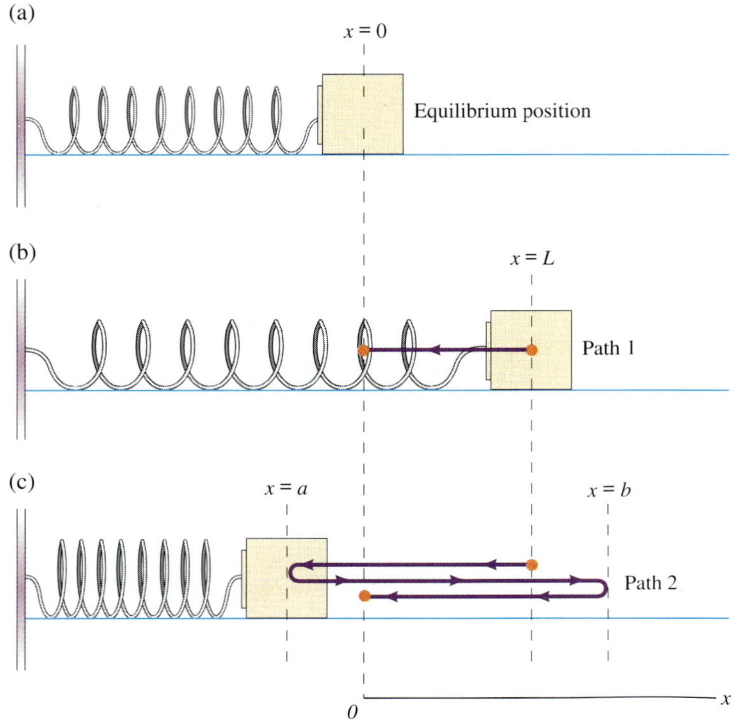

FIGURE 6–21 (a) Example 6–7. The equilibrium position of the mass attached to the spring. (b) The spring is pulled to $x = L$ and then moves from this point back to $x = 0$. (c) A more complicated path for motion from $x = L$ to $x = 0$.

An important property of conservative forces

The work done by a conservative force in moving an object from one position to another is independent of the path taken by the object.

The statement that conservative forces do work that is independent of the path of the object can be reformulated in terms of closed paths, paths that end at the same point at which they start:

The work done by a conservative force in moving an object along any closed path is zero.

The relation between these two statements is easy to understand. One possible closed path is the path in which the object does not move at all, in which case $W = 0$. But the statement that the work done is path independent means that the same amount of work is done along *any* path for which the object starts and finishes at the same point. Thus no work is done in moving along any closed path. This statement may be used as a definition of conservative forces.

Any one-dimensional force that depends only on position is conservative.

We can now restate our conclusions about forces in one dimension, which were earlier stated in terms of path independence. *Any* force that can be written in the form $F = F(x)$ is a conservative force. This includes constant forces as a special case.

Friction is Nonconservative

You might think from our discussion that forces with constant magnitude are always conservative, a conclusion that friction shows is patently false. This apparent difficulty is resolved by noting that a nonconservative force such as friction cannot, in fact, be represented by a function of the form $F(x)$. In fact, friction depends on the direction of the velocity as well as the position (Fig. 6–22):

$$\mathbf{F} = \mu_k \mathbf{F}_N \text{ in the direction opposite to } \mathbf{v},$$

which means that, strictly speaking, $\mathbf{F} = \mathbf{F}(x, \mathbf{v})$. The work done by friction when an object moves out cannot possibly cancel the work done by friction as the object comes back along a reversed path, as was the case for the spring.

FIGURE 6–22 The snowplow position increases the effect of kinetic friction and thereby controls the speed of the skier. This force is doing negative work on the skier.

The importance of conservative forces is twofold. First, we shall see in Chapter 7 that we can dispense with the work itself for these forces and instead use the notion of a *potential energy*. Second, *most of the forces we deal with are conservative.* This is true of all the fundamental forces in nature and, in particular, for those forces known as *central forces,* in which the force is directed along a line from a fixed center. The gravitational force between the Sun and the planets is one example.

6–5 POWER

Up to this point, we have said nothing about how *fast* work is done. **Power**—symbol P— is *the rate at which work is done*:

The definition of power

$$P \equiv \frac{dW}{dt}. \tag{6–23}$$

It is easy to calculate the power for a constant force. In one dimension, we have

$$W = F\,\Delta x, \tag{6–24}$$

where F is the force and Δx is the displacement from some fixed starting point. In two or three dimensions, we have

$$W = \mathbf{F}\cdot\Delta\mathbf{r}. \tag{6–25}$$

Let's now calculate the power. The amount of work done is a function of time because the displacement changes with time. The general calculus rule states that $d(fg)/dt = g(df/dt) + f(dg/dt)$; therefore, we have in one dimension

$$P = \frac{dW}{dt} = F\frac{d(\Delta x)}{dt} - \Delta x\frac{dF}{dt}.$$

As long as the force is constant in time, the second term on the right is zero. The rate of change of displacement $d(\Delta x)/dt$ is the velocity, so we find

$$P = Fv. \tag{6–26}$$

In two or three dimensions, we can generalize this result to

$$P = \mathbf{F}\cdot\frac{d\Delta\mathbf{r}}{dt} = \mathbf{F}\cdot\mathbf{v}. \tag{6–27}$$

We have derived this result for a constant force but for a velocity that can change with time. The power P in Eq. (6–27) is therefore more appropriately called the **instantaneous power**. It can be shown with a little more calculus that Eq. (6–27) is the correct expression for the instantaneous power even when the forces depend on time.

The instantaneous power

The units of power are joules per second (J/s) and the unit has been given its own name. A **watt** (W for short; do not confuse this unit with the algebraic symbol W that we use for work!) is the power generated when a force of 1 N displaces an object with a speed of 1 m/s. Another commonly used unit of power is the **horsepower** (hp):

The SI unit of power

$$1\text{ hp} = 746\text{ W}. \tag{6–28}$$

A useful measure of *electrical energy* is based on the watt or, more conveniently, the kilowatt (kW). The **kilowatt-hour** (kWh) is the amount of work done when 1 kW of power is generated for 1 h.[†] Because there are 3600 s in 1 h,

$$1\text{ kWh} = 3.6 \times 10^{6}\text{ J}. \tag{6–29}$$

[†]A typical large hydroelectric project generates a power of 10^{10} W; household energy consumption in the northern United States typically runs from 250 to 1000 kWh per month.

FIGURE 6–23 As this nineteenth-century representation shows, James Watt's steam engine could do the same work as many horses.

EXAMPLE 6–8 Early in the nineteenth century, James Watt wanted to market his newly discovered steam engine to a society that until then had relied heavily on horses. So Watt invented a unit that made it clear how useful a steam engine could be. He conducted a demonstration in which a horse lifted water from a well over a certain period of time and called the corresponding power expended "one horse-power" (Fig. 6–23). Assume that water has a mass density of 1.0×10^3 kg/m^3, the well was 20.0 m deep, and the horse worked for 8.0 h. How many liters of water did the horse raise from the well?

Solution: Suppose that the mass density of water is denoted by ρ. This means that a volume V of water has mass $m = \rho V$. Then the work done in lifting a mass m of water from the bottom of the well is

$$W = F\,\Delta y = mg\,\Delta y,$$

where Δy is the depth of the well. Thus the work done in lifting a volume V from the well in a time t is $\rho V g\,\Delta y$, and the power is

$$P = \frac{\text{work}}{\text{time}} = \frac{\rho V g\,\Delta y}{t}.$$

The unknown here is the volume V, and we can solve for it:

$$V = \frac{Pt}{\rho g\,\Delta y}.$$

All the quantities on the right-hand side are known. The power is, by definition, 1 hp = 746 W, and

$$V = \frac{(746\text{ W})(8.0\text{ h} \times 3600\text{ s/h})}{(1.0 \times 10^3\text{ kg/m}^3)(9.8\text{ m/s}^2)(20.0\text{ m})} = 1.1 \times 10^2\text{ m}^3.$$

Because there are 10^3 L in 1 m^3, the number of liters lifted by the horse is 1.1×10^5 L.

* 6–6 KINETIC ENERGY AT VERY HIGH SPEEDS

We mentioned in Chapter 5 that Newton's laws cease to be applicable in two domains. One is the domain in which speeds approach the speed of light and the other is the domain of quantum physics, which applies largely to atoms and smaller entities. The first domain is discussed in Chapter 40, whose subject is special relativity. In Chapter 40, we will show that the maximum speed attainable by any particle is the speed of light itself. We will also show that the precise expression for the kinetic energy of a particle of mass m is

$$K = mc^2\left(\frac{1}{\sqrt{1 - (v/c)^2}} - 1\right). \tag{6–30}$$

Here, c is the speed of light with a value $c = 3 \times 10^8$ m/s. In this section, we explore the expression for K.

Equation (6–30) reduces to the usual expression $K = mv^2/2$ when $(v/c)^2$ is very small (see Problem 71). We call a quantity x "very small" when x^2 can be neglected in comparison with x, and we denote this by $x \ll 1$. For example, if $v/c = 10^{-2}$, then $(v/c)^2 = 10-4$; this is indeed much less than 10^{-2} if we maintain 1-percent accuracy. Under normal conditions, v/c is much smaller than 10^{-2}. The speed v_E of Earth around the Sun is a large number, but it is still true that $v_E/c \simeq 10^{-4}$. Molecular speeds in air are also of this order of magnitude.

For particles in cosmic-ray showers and in particle accelerators (atom smashers), v can be very close to c, and the factor $1/\sqrt{1 - (v/c)^2}$ can be very large. In the highest energy accelerators, this factor is as large as 10^4, and the relativistic kinetic energy is very much larger than the Newtonian kinetic energy formula would indicate. We note that the formula for K ceases to make sense when $(v/c) \geq 1$. It is indeed an essential tenet of the theory of relativity, which improves on Newton's laws for high-speed particles, that no information—that is, no particles—can be sent at a speed greater than the speed of light. The case of $v = c$ is delicate: With $m = 0$, the expression for K in Eq. (6–30) is ambiguous but not manifestly wrong. The theory of relativity states that massless particles not only can move with the speed of light, but always *must* move with the speed of light. An example of such a particle is the elementary particle called the *neutrino,* which plays an important role in the fundamental structure of matter, in astrophysics, and in cosmology. Experiment shows the neutrino has a mass at most 10^{-5} times the electron mass. A second example is the photon, which is the particle that represents light itself in the quantum physics description of nature.

SUMMARY

The central result of this chapter is the work–energy theorem as it applies to an object of mass m under the influence of one or more forces. It is given by

$$W_{\text{net}} = \Delta K, \tag{6–7}$$

where W_{net} is the net work done by the net force on the object. In this expression, the quantity K is the kinetic energy of the object:

$$K \equiv \tfrac{1}{2}mv^2. \tag{6–5}$$

This is a scalar quantity formed from the velocity vector, with v^2 (the speed squared) $= v_x^2 + v_y^2 + v_z^2$. The change ΔK is the change in the kinetic energy of the object as it moves from an initial position $\mathbf{r}_i$ to a final position $\mathbf{r}_f$ through a displacement $\Delta\mathbf{r} = \mathbf{r}_f - \mathbf{r}_i$:

$$\Delta K = K_f - K_i. \tag{6–6}$$

Here, K_i is the kinetic energy at the initial position and K_f is the kinetic energy at the final position. The net work, W_{net}, depends on the net force acting on an object as well as its displacement and, in general, on how the displacement is made. The net work is the work done by the net force; equivalently, it is the sum of the works done by the individual forces that make up the net force. The work done by a force can be expressed in different ways according to the form the force takes. We can enumerate these forms:

for a constant net force in one dimension:

$$W = F\,\Delta x.$$

for a constant force in three dimensions:

$$W = F_x\,\Delta x + F_y\,\Delta y + F_z\,\Delta z \tag{6–10}$$

$$= \mathbf{F}\cdot\Delta\mathbf{r} \tag{6–11}$$

$$= F\,\Delta r\cos\theta, \tag{6–12}$$

where θ is the angle between the vectors $\mathbf{F}$ and $\Delta\mathbf{r}$. For a nonconstant force in one dimension:

$$W = \int_{x_i}^{x_f} F(x)\,dx. \tag{6–18}$$

For nonconstant, space-dependent force in three dimensions,

$$W = \int_{\mathbf{r}_A}^{\mathbf{r}_B} \mathbf{F} \cdot d\mathbf{r}. \quad (6\text{–}21)$$

In Eq. (6–21), the displacement is between points A and B. This form for work is the most general one and reduces to the other forms in the appropriate limit. By definition, we conclude that no work is done on an object that is in uniform circular motion. Both work and kinetic energy are measured in units of joules (J) in SI.

In some cases, the work done depends on the path the object takes; in others, the work done is independent of the path. When the work done is path dependent, we say that the force is nonconservative; friction provides an example. When the work done is path independent, we say that the force is conservative; gravity or the spring force provide examples. Conservative forces are important because all the fundamental forces of nature are conservative.

The work–energy theorem allows us to calculate the speeds of objects when the work done by the forces is known, and it allows us to calculate the net work that must be done if a certain speed is to be achieved. This theorem is often much simpler to use for these purposes than is Newton's second law.

Power is the rate at which work is done:

$$P \equiv \frac{dW}{dt} \quad (6\text{–}23)$$

From the definition of the instantaneous power, we find that

in one dimension: $P = Fv$; (6–26)

in two or three dimensions: $P = \mathbf{F} \cdot \mathbf{v}$. (6–27)

The unit of power in SI is the watt (W).

UNDERSTANDING KEY CONCEPTS

1. We mentioned in the introduction that the work–energy theorem might help us to analyze the motion of a baseball under the influence of both gravity and drag force from the air. How would you do so?
2. Does it make sense to refer to a force doing negative work when an object moves under its influence? What does negative work mean? In answering, consider what happens when an object is stopped by a force.
3. It certainly seems like work to us when we hold a bag of groceries for a long period of time. Are we expending energy when we hold a bag of groceries for a long period of time? How is the answer to this question consistent with the work–energy theorem?
4. The centripetal forces that cause uniform circular motion do no work because they are perpendicular to the motion. How do such forces fit into the work–energy theorem?
5. A piano can be lifted to the third story of a building by having a crew carry it up the stairs or by using some type of pulley system. Is the same work done in both cases? Assume that friction can be neglected.
6. If the moving crew of Question 5 uses a rope and pulley, it pulls on the rope in the same direction as the force of gravity. Because the crew pulls the rope in the direction opposite to the displacement of the load, is the crew doing negative work on the load?
7. A man pushes against the smokestack on a cruise boat. When the ship is stationary, he does no work. When the ship starts to move in the direction in which he is pushing, he appears to be doing work, yet he experiences no change in the level of his exertion. Why is this? Keep in mind that the man does not fall down because the force of friction keeps his shoes from sliding backward. What is the work done on the man by the rough deck?
8. You do no net work when you walk at a constant speed. Why do you get tired?
9. A parachutist jumps from a plane and lands safely in a field. Does the net work done on the parachutist depend on the height from which he or she jumps? (The work done by gravity does depend on that height.)
10. No work is done in uniform circular motion. Suppose that you observe circular motion in which the moving object first speeds up and then slows down to its original speed. Is any net work done?
11. No work is done on a bag of groceries while you are holding it stationary. Is work done on the same bag if you are holding it while you move steadily upward in an elevator? What is the difference in the two situations? Are your hands still the origin of the force that does work on the bag?
12. A car with cruise control transports you at a constant speed. The engine does work. How much of that work is done on you?

13. One of the entertainments at the carnival is a rotating (vertical) cylinder. The participants step in and place themselves against the interior wall. The cylinder starts to rotate more and more rapidly, and at some point the floor falls away, leaving the customers stuck like so many flies to a wall. Is any work done on the participants? If so, what force does the work?

14. Tarzan swings from tree to tree on a jungle vine. Is any work done on him during the motion? If so, what forces do the work?

15. A stunt consists of one acrobat standing on the short end of a seesaw whose pivot point is not at its midpoint. A second acrobat leaps down on the long end of the board and flips the first acrobat several meters into the air. How would you determine the work done by the second acrobat in flipping the first one into the air?

16. Two identical twins work side-by-side as butchers. They use identical motions and identical hatchets. One brings a hatchet down on some very tender meat, and the other on a large bone. Which one does more work per swing?

17. You tow a small child on a sled at a constant speed by pulling the sled with a rope. The rope makes an angle θ with respect to the horizontal. What forces act on the sled and which ones do work?

18. A one-dimensional force acts on an object, changing its velocity from 0 to **v**. By the work–energy theorem, the work done is $W = \frac{1}{2}mv_f^2 - \frac{1}{2}mv_i^2 = \frac{1}{2}mv^2$. An observer moving with velocity **v** with respect to the original system sees the initial velocity as $-\mathbf{v}$ and the final velocity as 0. This observer would conclude that $W = -\frac{1}{2}mv^2$. What accounts for the difference?

19. A diver plunges from a 10-m high diving board into water. How would you determine the average force of resistance of the water that slows down and stops the diver? (Neglect the force of gravity on the diver while she is in the water. We shall see in Chapter 16 how the water's buoyancy takes care of that.)

20. A parachutist jumps off a tower. What measurements would you have to make to determine the work done by the drag force of the air during the entire fall? The drag force is a rather complicated function of the velocity of the jumper.

21. Are the following forces conservative or nonconservative? (a) Air drag on a parachute. (b) The force opposing the fall of a steel ball bearing in a beaker of water. (c) The explosive force causing a bullet to leave a rifle barrel. (d) The force of an ideal trampoline that propels you into the air.

22. How do you know that the drag forces you experience when you swim are not conservative?

PROBLEMS

6–1 Kinetic Energy and Work

1. (I) An automobile of mass 10^3 kg moves at 1.0 km/h $= 0.28$ m/s. (a) What is its kinetic energy? (b) At what speeds must a person of mass 80 kg and a bullet of mass 10 g move to have the same kinetic energy as the automobile? (c) What would the speed of the automobile be if the kinetic energy doubled?

2. (I) A construction worker of mass 93 kg rides in an elevator up to the tenth floor, which is 33 m above the ground. The elevator travels with uniform speed. (a) What is the net work done on the worker? (b) What is the work done on the worker by the contact force of the elevator? (c) What is the work done on the worker by gravity?

3. (I) A person lifts a suitcase of mass 10 kg from the floor. Ignore the initial acceleration of the suitcase, and suppose that it moves upward at a constant speed between a height $h = 0$ m and $h = 1$ m. (a) What are the forces acting on the suitcase, as well as the net force? (b) What is the net work done on the suitcase? (c) Find the work done on the suitcase by the person.

4. (I) A bedroom bureau of mass 42 kg is moved from the first floor of an apartment building to the penthouse on the ninth floor, 32 m higher. (a) What is the work done on the bureau by three men in carrying it up the steps? (b) If the three men take it up on an elevator, how much work is done on the bureau by the normal force of the floor of the elevator?

5. (I) A truck carrying a 66-kg crate accelerates uniformly from rest to 63 km/h in 15 s. Calculate the work done on the crate by the truck.

6. (I) An old piano of mass 120 kg is removed from an apartment building being converted into condominiums. The previous owners found it too much trouble to remove and left it. The workmen decide the easiest thing to do is to drop it out of a double-width window to the ground 30 m below. (a) How much work do the workmen do if they just push it out the window? (b) If the men slowly lower the piano by rope, what is the work done on the piano by the rope's tension? (c) How much work does gravity do in each case?

7. (I) A man pushes a refrigerator of mass 40 kg at uniform speed for a distance of 1.5 m to the kitchen wall. The coefficient of friction between the refrigerator and the floor is $\mu_k = 0.4$. (a) How much work does the man do in moving the refrigerator? (b) What other sources of work done are there? (c) What is the net work done in this process?

8. (I) A person pulls a heavy load of mass 37 kg up the side of a building by using a frictionless pulley. The load travels up a distance of 7.5 m. Take the load to move with constant velocity and ignore any acceleration at the beginning or end of the move. (a) How much work is done on the load by gravity? (b) by the tension of the rope? (c) by the person?

9. (II) Consider the woman who lifts the huge interior lineman in Example 5–3. How much work does she do while pulling down a 2 m length of rope? What is the work done by gravity on the lineman while this is going on?

10. (II) The mass $M = 60$ kg is lifted to a height of $h = 2$ m using the system of pulleys shown in Fig. 6–24. The motion is slow and the initial acceleration is negligible. (a) Find the force that must be applied at the free end of the rope. (b) Find the work done on the mass by this force. (c) Calculate the work done on the mass by gravity during the process.

11. (I) A baseball of mass 145 g leaves a pitcher's hand at 96.6 mi/h, but, due to air resistance, it arrives at home plate 60.0 ft away traveling at 95.3 mi/h. Assume that the magnitude of the

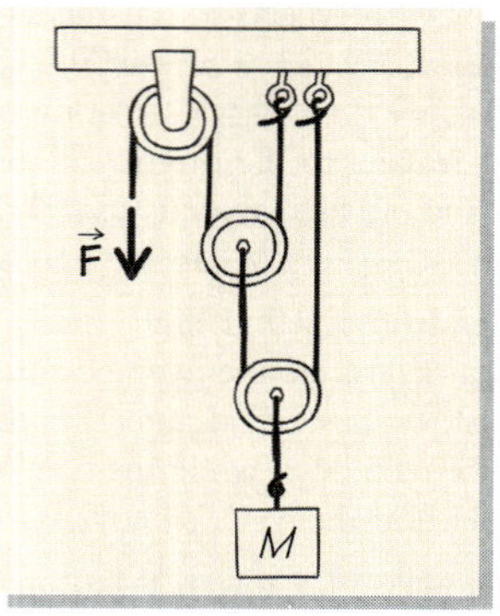

FIGURE 6–24 Problem 10.

ball's acceleration is constant and that the ball travels in a straight line (ignore gravity). How much work is done by friction during the flight of the ball?

12. (II) A ball of mass 240 g is dropped from a height of 2 m. (a) What is the work done on the ball by gravity? (b) Suppose that the ball bounces to a height of only 1.5 m. How much work is done by gravity on the ball as it moves from ground level to 1.5 m?

13. (II) A construction worker of mass 75 kg hoists a load of bricks of mass 42 kg by throwing a rope attached to the load over a pulley and letting his weight lift the load. Assuming that there is no friction, what is the work done by gravity during a 2.0-s period?

14. (II) Two masses are connected by a light string over a light, frictionless pulley, as in Fig. 6–25. The table surface is also frictionless. (a) Apply the work–energy theorem for this system to calculate the speed of the masses after the masses have moved a distance Δx, starting from rest. Note that the work of the tensions drops out. (b) Use this result to obtain the acceleration of the system.

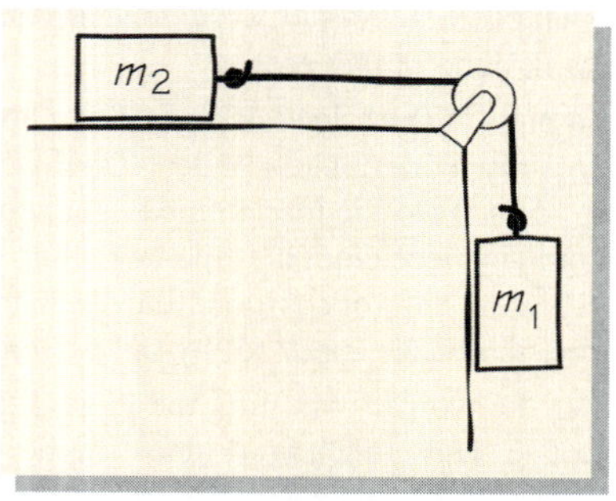

FIGURE 6–25 Problem 14.

15. (II) A waterfall, whose height is 40 m, has 200 m^3 of water falling every second. How many joules of work are done by gravity every hour? (The mass of 1 m^3 of water is 10^3 kg.)

16. (II) A construction crew is required to pull up a load of mass 58 kg by means of a rope thrown over a pulley. They are to lift the load from rest on the ground to a height of 12 m, and the load should arrive at the end point with a speed of 2.0 m/s. (a) Calculate the work done by the crew if it accelerates the load uniformly over the whole distance. (b) Repeat the calculation for the case in which the acceleration takes place in the first 5 m and the load is pulled with uniform speed the rest of the way.

17. (II) A child has three different sets of cubical blocks (Fig. 6–26). The first set consists of 3 blocks, each 12 cm on a side and of mass 36 g; the second set consists of 6 blocks, each 6 cm on a side and of mass 18 g; the third is a set of 12 blocks, each 3 cm on a side and of mass 9 g. For each set, what is the work the child must do to stack the blocks into a tower 36 cm high? The blocks can be treated as point objects at their *centers* in calculating the work.

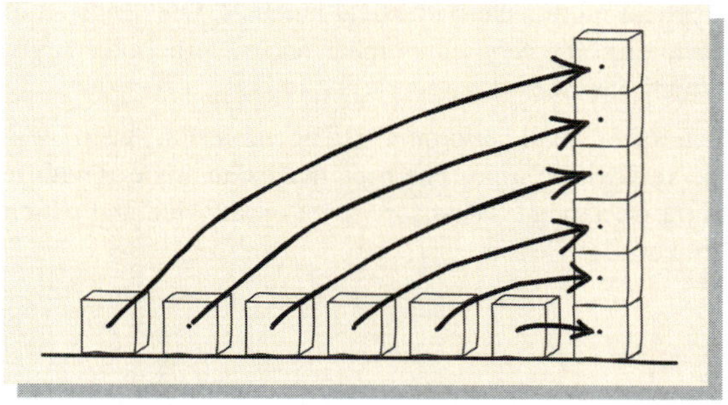

FIGURE 6–26 Problem 17.

18. (III) Find the work that must be done by a force lifting against gravity to raise a coiled rope of length L and mass M entirely off a level surface. [*Hint*: Use the method of Problem 17 and divide the rope into more and more segments.]

6–2 Constant Forces in Space

19. (I) What is the scalar product of $\mathbf{A} = -2\mathbf{i} + 3\mathbf{j} - 5\mathbf{k}$ and $\mathbf{B} = 5\mathbf{i} + \mathbf{j} - 2\mathbf{k}$?

20. (I) A force $\mathbf{F} = (2.7\text{ N})\,\mathbf{i} - (6.8\text{ N})\,\mathbf{j}$ is used to displace an object of mass 2.6 kg by an amount $\mathbf{r} = (3.4\text{ m})\,\mathbf{i} + (1.2\text{ m})\,\mathbf{j}$. What is the work done by the force on the object?

21. (I) Assuming that the object of Problem 20 is initially at rest, what is its final speed?

22. (I) Show that the vector $\mathbf{v} = -y\mathbf{i} + x\mathbf{j}$ is always perpendicular to the vector $\mathbf{u} = x\mathbf{i} + y\mathbf{j}$.

23. (I) Consider two vectors, $\mathbf{u} = 3\mathbf{i} - 4\mathbf{j} + 7\mathbf{k}$ and $\mathbf{v} = -2\mathbf{i} + 3\mathbf{j} + z\mathbf{k}$. What must z be so that $\mathbf{u}$ and $\mathbf{v}$ are orthogonal?

24. (I) A person puts a suitcase of mass 15 kg into a van, moving the suitcase a total distance of 1.6 m: 0.6 m up and 1 m horizontally. How much work is done by the person?

25. (I) A man pulls a sled by a rope, moving his two daughters to the top of a 15° slope. He holds the rope parallel to the slope. If the daughters and the sled have a total mass of 43 kg and the length of the slope is 36 m, how much work does the man do on the sled, assuming that he pulls the sled with uniform velocity? Ignore all friction on the sled.

26. (I) A block of material with mass 560 kg is used in the construction of a building. During one part of the process of setting the block in place, a complex network of cables acts on it and its motion is transformed from a horizontal motion with speed 10 cm/s to a vertical motion with speed 38 cm/s. What is the net work done on the block during this motion?

27. (II) A skier of mass 72 kg (including skis), starting from rest, slides down a slope at an angle of 18° with the horizontal. The coefficient of kinetic friction is $\mu_k = 0.12$. What is the net work done on the skier in the first 7.0 s of descent?

28. (II) Consider a vector **A** in the xy-plane. Its x- and y-components are A_1 and A_2, respectively. Show that any vector **B** in the same plane that points in a direction perpendicular to **A** must have components $-cA_2$ and cA_1, respectively, where the magnitude of c is the ratio of the lengths ($|c| = B/A$).

29. (II) Sketch the direction of the vector $\mathbf{e} = \cos\theta\mathbf{i} + \sin\theta\mathbf{j}$. Show that it has unit length, and use your sketch to give an expression for the unit vectors **f** that are perpendicular to **e** and that lie in the xy-plane. How many such vectors **f** are there?

30. (II) Consider the vector $\mathbf{A} = 3\mathbf{i} - 4\mathbf{j} + 5\mathbf{k}$. Find the most general vector in the yz-plane that is perpendicular to **A**.

31. (II) Consider the unit vector $\mathbf{e} = -0.6\mathbf{i} + 0.8\mathbf{j}$. What is the magnitude of the projection of vector $\mathbf{A} = 3\mathbf{i} - 2\mathbf{j}$ onto the line along which **e** points?

32. (II) A stone is thrown from a height h_0 above a level field, leaving the hand at a 40° angle. Ignore all effects of air resistance. (a) Compute the work done by gravity as the stone follows its trajectory back to the height h_0. Recall that the motion can be divided into motion in the vertical direction and motion in the horizontal direction. (b) Show, by applying the work–energy theorem, that the speed of the stone when it reaches h_0 again is identical to the speed it had when it left the hand.

33. (II) A force given by $\mathbf{F} = (2\mathbf{i} - 5\mathbf{j})$ N acts on an object that moves from $\mathbf{r}_1 = (7\mathbf{i} - 8\mathbf{j} + 2\mathbf{k})$ m to a new position $\mathbf{r}_2 = (5\mathbf{i} - 4\mathbf{j} + 5\mathbf{k})$ m. How much work does this force do on the object?

34. (II) A 25-kg crate slides down a plane that makes an angle of 30° with the horizontal, starting from rest at the top. The speed of the crate when it reaches the bottom of the 10-m-long slide is 8 m/s. What is the coefficient of friction? How much work is done by the force of friction?

6–3 Forces that Vary with Position

35. (I) A small gizmo is confined to a groove that is aligned with the x-direction. A rod pulls the gizmo in the $+x$-direction. The rod is attached to an apparatus such that the pulling force is 0.3 N when the gizmo is to the left of a point in the groove we label as the origin and 0.7 N when the gizmo is to the right of the origin. What is the work the pulling force does on the gizmo as the gizmo moves from $x = -6$ cm to $x = +7$ cm?

36. (I) A spring with spring constant $k = 12$ N/m is attached to a wall at ground level. The end of the relaxed spring is on the floor at a location that we take to be the origin. A mass of 3.0 kg is attached to the end of the spring, and the spring is stretched by 50 cm and released. How much work has the spring done on the mass by the time the mass passes through the origin?

37. (I) A one-dimensional force F depends on the position x of a particle on which it acts as $F = g_1x - g_2x^3$, where g_1 and g_2 are constants. What is the work done in moving the particle from the origin to $x = 2.0$ m?

38. (I) A one-dimensional force on a particle is given by $F = \alpha x$, where $\alpha = -5.00$ N/m for $x < 0$ and $\alpha = +3.00$ N/m for $x > 0$ (Fig. 6–27). Calculate the work done by the force on a block when the block is moved from $x = -3.00$ m to $x = +4.00$ m.

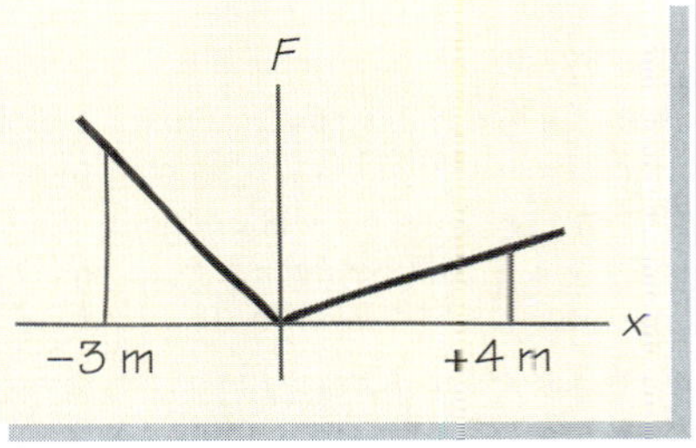

FIGURE 6–27 Problem 38.

39. (I) A spring gun is made by compressing a spring (assumed to be perfect) and latching it. A spring of constant $k = 60$ N/m is used and the latch is located at a distance of 7 cm from equilibrium. The pellets have mass 4 g. What is the muzzle velocity of the gun?

40. (II) A man pushing a 50-kg crate up a slope that makes an angle of 30° with the horizontal exerts a force parallel to the plane. The coefficient of kinetic friction varies along the slope and is given by $\mu_k = \mu_1 + [(\mu_2 - \mu_1)s/L]$, where s is the distance along the slope starting at the bottom. The largest value of s is L, where $L = 10$ m is the length of the slope; $\mu_1 = 0.2$, and $\mu_2 = 0.3$. What is the work done on the crate by the force as a function of s if the force varies such that the crate is pushed at a constant speed?

41. (II) A nonstandard spring exerts a force $F = -k_1x - k_2x^3$ to restore itself to equilibrium, where x is the distance from equilibrium. The values of k_1 and k_2 are 5.0 N/m and 15 N/m^3, respectively. Calculate the work done to stretch the spring from 0.10 m to 0.20 m.

42. (II) A rocket is scheduled to blast off from Cape Canaveral to study a neighboring solar system. Earth's gravitational force is $F = K/r^2$, where r is the distance from Earth's center and K is a negative constant. What is the minimum work the rocket engine must do so that the rocket leaves the gravitational force of Earth? Assume that the mass of the rocket does not vary in the process. (This assumption is actually very poor, but a more exact treatment must wait until Chapter 8.)

6–4 Conservative and Nonconservative Forces

43. (I) A child swings a streamer toy over her head in a nearly horizontal plane. The toy, of mass 85 g, is at the end of a massless string of length 1.5 m. She starts twirling the toy from rest while she slowly lets out the string to full length, and gets the angular speed up to 2 rev/s. How much work has she done?

44. (I) A 100-g ball is tossed straight up in the air, rises to a maximum point, then falls back until it is 1 m below the position of the hand that tossed it up. A second ball is simply dropped from the same hand position, and also lands 1 m below that position. What is the net work done by gravity in the two cases?

45. (I) You are in the process of moving. A 54-kg bed can be brought from ground level to the second floor (4.0 m above ground level) either by pulling it straight up by means of a rope or by dragging it up a frictionless plane inclined at 30° to the horizontal. Calculate the work done in each case by those who move the bed.

46. (I) An object of mass 1.5 kg is suspended from a vertical spring that is attached to the ceiling. Without the mass, the spring is 30 cm long. When the mass is attached to it, the spring is extended to a length of 110 cm. What is the work done by the force of gravity during the extension of the spring?

47. (I) A constant force of 10 N pushes a particle along the x-axis. The position of the particle is represented by $x = 11\text{ m} - (2\text{ m/s})t + (0.5\text{ m/s}^2)t^2$. Find the work done by the force between $t = 0$ s and $t = 1$ s, and between $t = 1$ s and $t = 2$ s. Is the force conservative?

48. (II) A force **F** has components $F_x = axy - by^2$, $F_y = -axy + bx^2$, where $a = 2\text{ N/m}^2$ and $b = 2\text{ N/m}^2$ (Fig. 6–28). Calculate the work done on an object of mass 4 kg if it is moved in a closed path from $(x, y) = (0, 1)$ to $(4, 1)$, to $(4, 3)$, to $(0, 3)$, and back to $(0, 1)$. The path between the points is always the shortest straight one, and all the distances are given in meters.

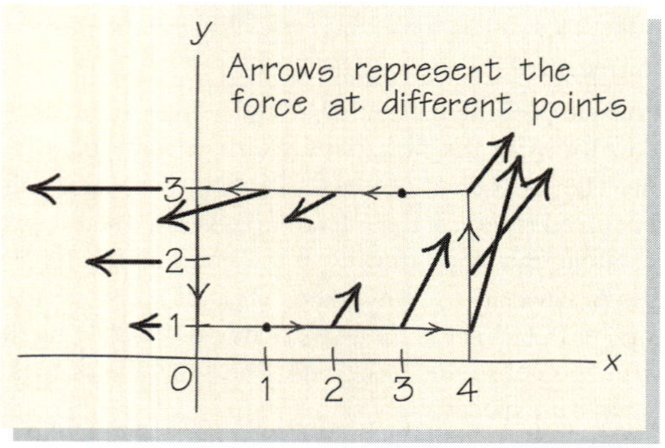

FIGURE 6–28 Problem 48.

49. (II) A force with an x-component acts on a mass with a strength that varies only with the position x of the mass on the x-axis, according to $|F(x)| = Ax^2$. The sign of the force is negative when x is positive and positive when x is negative, indicating that the force always attracts the mass toward the origin. Compute the work done by this force when $A = 1500\text{ N/m}^2$ and the mass moves along the x-axis (a) from $x = -5.0$ cm to the origin; (b) from $x = -5.0$ cm to $+5.0$ cm; (c) from $x = +5.0$ cm to $+2.0$ cm; (d) from $x = -2.0$ cm to -5.0 cm.

50. (II) A child's playground ride consists of four seats, of mass 5 kg each, connected to a vertical axle with spokes of small mass. The seats are placed equidistant in a circle of radius 1.4 m and rotate about the vertical axle. A child of mass 15 kg sits in one of the seats, and his friend pushes the ride to accelerate him from rest to 0.4 rev/s. How much work does the friend do?

51. (II) The net force acting on a particle depends on the position of the particle on the x-axis according to the relation $F = F_0 + Cx$, where $F_0 = 5$ N and $C = -2$ N/m. The particle is initially at rest at the point $x = 0$ m when the force begins to act. (a) Calculate the work done by the force when the particle reaches $x = 1$ m, 2 m, 3 m, and 4 m. (b) Determine any positions (other than at $x = 0$ m) where the work done is zero. (c) Is the force conservative?

52. (II) An object of mass m is to be moved from the top of a building of height h to a point on the ground a horizontal distance h from its original location, so that the position vector at the beginning may be chosen to be $h\mathbf{j}$, and at the end, $h\mathbf{i}$. Two possible paths are: (a) the object is lowered at constant speed by rope and, after it reaches the ground, it is moved horizontally to the final location; and (b) the object is allowed to slide along a straight support that runs from the initial point to the final point. Show that the work done by the force of gravity is the same in both cases.

53. (III) Prove that a force acting in one dimension is conservative if it is a function of position only, and not a function of any other information about the motion of an object under its influence. Does this include forces with constant magnitude? In view of your answer, how does the friction force manage to be nonconservative?

54. (III) A small object of mass m is moved up along a track that forms one quarter of a circle of radius R in the vertical plane (Fig. 6–29). The object moves with a small uniform speed maintained by a tangential force (that varies with the angle) along the track. Calculate the work done by this force in moving the object through a 90° arc from the lowest point to the highest point (a) by direct use of the definition of work, and (b) by using the work–energy theorem.

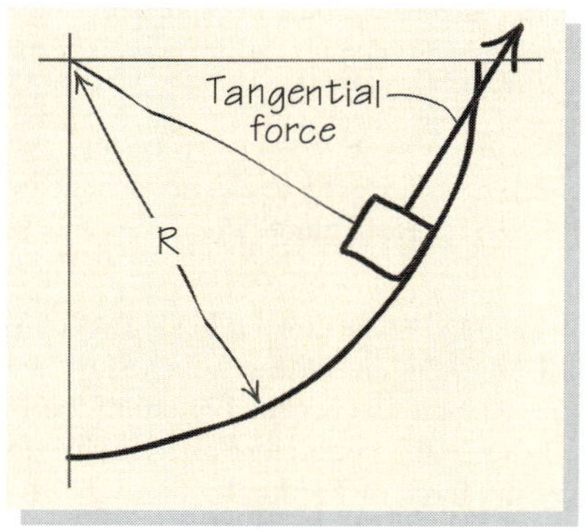

FIGURE 6–29 Problem 54.

55. (III) The force acting on an object of mass m is given by the expression $F(x) = C|x|$, where $|x|$ is measured in centimeters. (a) How much work is done by the force when the object is moved from $x = -4.0$ cm to $x = +4.0$ cm? Compare your result with the amount of work done if the form of the force law were $F(x) = Cx$. (b) Repeat the calculation if the object moves from $x = 0.0$ cm to $x = 8.0$ cm.

56. (III) Consider a force that acts on an object of mass m, which moves in the xy-plane. The force is given by $\mathbf{F}(x, y) = k_1x\mathbf{i} + k_2y\mathbf{j}$. Calculate the work done by the object if it moves in a circle of unit radius (given by $x^2 + y^2 = 1\text{ m}^2$) starting at $x = 1$ m, $y = 0$ m, and ending at a point that makes an angle of (a) 90°, (b) 180°, and (c) 360° with the original direction of the position radius vector. [*Hint*: The problem is simplified with polar coordinates r and θ, where $x = r\cos\theta$ and $y = r\sin\theta$.]

6–5 Power

57. (I) Electricity costs about \$0.08 per kWh. Your monthly electric bill is \$26.00. Assuming that your only use of electricity is for light and that you keep your house lit 5 h/d, how many 100-W bulbs do you keep going?

58. (I) How much energy is used by running a fleet of 100 80-hp cars around the clock for 1 month?

59. (I) How long does it take you to climb four flights of steps? Assume that your body is 20 percent efficient, and estimate the power you have to generate.

60. (I) Two engines are used to move a mass of 18 kg, starting from rest, a distance of 20 m in a straight line along a frictionless flat surface. Engine 1 exerts a constant force of 2.0 N, and engine 2 exerts a constant force of 3.0 N. (a) What is the work done by each engine? (b) What is the average power expended by each engine during the process?

61. (I) A test car of mass 700 kg is moving at a speed of 15 mi/h when it crashes into a wall to test its bumper. If the car comes to rest in 0.3 s, how much average power is expended in the process? (To find the average power, simply imagine Eq. (6–23) for a finite time interval: $P_{av} = \Delta W/\Delta t$, where ΔW is the work done during the time interval Δt.)

62. (I) Assume that a car of mass 1200 kg has an engine with power output of 80 hp. How long would it take to accelerate such a car to a speed of 100 km/h? (Neglect air resistance, which would make this time much larger.)

63. (II) The maximum power of a particular horse is 1 hp. With what speed can this horse pull a sled on level ground if the weight of the sled with its load is 5000 N and if the coefficient of kinetic friction is $\mu_k = 0.03$? What is the maximum speed on a 5° upward incline?

64. (II) An accelerator accelerates a proton to $0.999c$, where c is the speed of light. If 1.35×10^{10} protons are accelerated every minute, how much power is expended by the accelerator, assuming 9.76 percent efficiency? (For this problem, with $v \simeq c$, use Eq. (6–30) for the kinetic energy. The *efficiency* is the fraction of the total power that goes into changing the kinetic energy of the protons.)

65. (II) Consider the waterfall in Problem 15. If the waterfall is used to produce electricity in a power station, and if the efficiency of conversion of kinetic energy of falling water to electrical energy is 60 percent, what is the power production of the station?

66. (II) An escalator moves people from one floor up to another. The height difference between the floors is 5.2 m, and the angle that the escalator makes with the horizontal is 33°. The speed of the escalator is 1.5 m/s, and it is supposed to carry a maximum of 60 passengers, with an average mass of 75 kg. How much power must be generated by the motor that runs the escalator?

67. (II) Trained athletes can exert power for their movements ranging from around 5 hp for a second to 0.4 hp or less for periods extending over several hours. (a) A bicyclist is limited by wind resistance, which is roughly of the form $F = Av^2$, where $A \simeq 0.08$ kg/m. Estimate the speed a cyclist can maintain for 1 h. (b) Estimate the time it takes a weight lifter to lift 100 kg a distance of 2 m. (c) Assuming that not too much time is taken up turning the corners, estimate how fast it is possible to climb three flights of steps, a vertical distance of 12 m. You can easily try this one!

*6–6 Kinetic Energy at Very High Speeds

68. (I) A proton is accelerated from rest to a final speed of $0.85c$, where c is the speed of light. How much work is done by the accelerator on the proton, given that a proton's mass is 1.6×10^{-27} kg?

69. (I) What is the kinetic energy of an electron (mass $= 9.1 \times 10^{-31}$ kg) moving at a speed of $0.9999c$? of $0.9999999999c$?

70. (I) (a) How much work does it take to accelerate an electron from a speed of $0.1c$ to $0.8c$? (b) from $0.8c$ to $0.998c$? (c) from $0.998c$ to $0.99998c$?

71. (II) In order to show that Eq. (6–30) reduces to the usual form for small v/c, we need an approximation for $1/\sqrt{1-x}$ for small values of x, namely, $1/\sqrt{1-x} \approx 1 + \frac{1}{2}x + \frac{3}{8}x^2$. Check this approximation by calculating both sides of the expression on your calculator. For what value of x is the approximation correct to within 10 percent? 1 percent?

General Problems

72. (I) A grocery store pays a monthly power bill of \$475. Electricity costs \$0.09/kWh. How many joules of energy were used in the month?

73. (I) A ball with mass 100 g is set in motion inside a bowl, with initial speed $v = 2$ m/s. Due to friction, the ball ultimately comes to rest. How much net work was done by the external forces acting on the ball?

74. (I) A pile driver works by lifting a large mass and dropping it to the ground. A driver used to put pilings in the ground for a tall building has a mass of 1400 kg and is raised to a height of 6 m in 4 s. (a) How much work is done by the engine each time the weight is lifted? (b) What horsepower engine must be used to run the pile driver?

75. (I) The so-called Domesday Book recorded a general census carried out in England in the year 1086. It catalogued all 5000 waterwheels in the country, each of which had a power output of roughly 2 hp. Waterwheels were the main source of nonanimal energy at the time. By contrast, the power of a Boeing 747 jet airplane at maximum thrust is approximately 1 MW, and at cruise level the power output is roughly 0.3 MW. What fraction of the total nonanimal power of eleventh-century England does a cruising 747 represent?

76. (II) A mass $M = 3$ kg moving without friction in the xy-plane starts at the point labeled by the position vector $\mathbf{r}_i = 0\mathbf{i} + 0\mathbf{j}$ with velocity $\mathbf{v}_i = (2\mathbf{i} + 1\mathbf{j})$ m/s. Two forces, $\mathbf{F}_1 = (2\mathbf{i} + 7\mathbf{j})$ N and $\mathbf{F}_2 = (2\mathbf{i} - 5\mathbf{j})$ N, act on the mass as it moves in a straight line to the point labeled by the position vector $\mathbf{r}_f = (10\mathbf{i} + 5\mathbf{j})$ m (Fig. 6–30). (a) How much work is done by $\mathbf{F}_1$

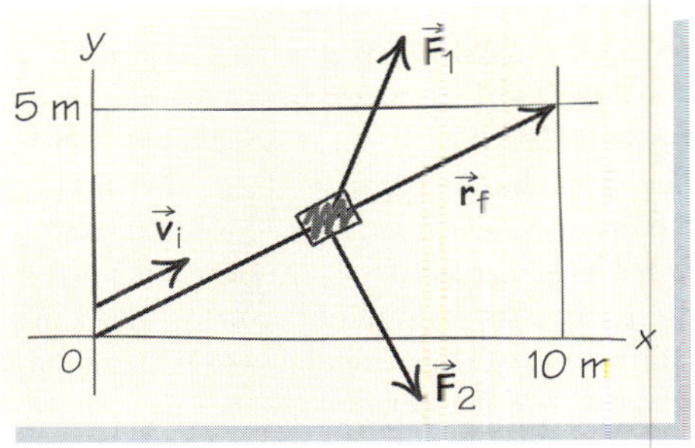

FIGURE 6–30 Problem 76.

as the mass moves from $\mathbf{r}_i$ to $\mathbf{r}_f$? (b) What power is provided by $\mathbf{F}_1$ at the instant the mass is at $\mathbf{r}_i$? (c) What is the kinetic energy of the mass when it reaches $\mathbf{r}_f$?

77. (II) A rocket in outer space, initially at rest, is accelerated uniformly at the rate of 2.0 m/s^2 for 33 s. (Neglect gravity throughout.) (a) If the mass of the rocket, assumed here to be constant, is 1.0×10^4 kg, how much work is done by the rocket engine? (b) Suppose that the rocket decelerates at such a rate that it comes to rest in 55 s. How much work is done on the rocket during the deceleration?

78. (II) A worker pushes a box of mass $m = 40.5$ kg in a straight line along a rough floor. The applied force **F** has magnitude 285 N and acts downward at an angle $\theta = 25.5°$ with respect to the horizontal. The box is initially at rest at the position $x_1 = 0$ m, and it has speed $v_2 = 1.72$ m/s at position $x_2 = 3.50$ m. (a) Use the data to calculate the coefficient of friction. (b) What is the net work done? (c) How much work is done to overcome friction? (d) What is the instantaneous power generated by the worker at $x = x_2$?

79. (II) A block of mass $m = 2.6$ kg is placed on an inclined plane that makes an angle $\theta = 32°$ with the horizontal. It is given an initial speed v_0 up the ramp and slides a distance 1.3 m up the ramp before it comes to a stop. The coefficient of kinetic friction between the block and the ramp is $\mu_k = 0.25$. (a) What are the three forces acting on the block? Give the magnitude and direction of each. (b) What is the work done by each of the three forces during the motion of the block? (c) What was the initial speed v_0?

80. (II) The engine of an automobile requires 45 hp to maintain a constant speed of 80 km/h. (a) What is the resistive force against the automobile? (b) If the resistive force is proportional to the velocity, what must the engine power be to drive at constant speeds of 60 km/h? (c) at 140 km/h?

81. (II) A mass of 4.0 kg is attached to a string tied to a hook in the ceiling. The length of the string is 1.0 m, and the mass is released from rest in an initial position in which the string makes an angle of 30° with the vertical. Calculate the work done by gravity by the time the string is in a vertical position for the first time. [*Hint*: Recall that you can use any path you like to get from the initial to the final point, because the work done by gravity is path independent.]

82. (II) A mass of 1 kg is accelerated by applying a force. In each case, calculate the work, in joules, that this force must do on the mass. (a) The mass is brought from rest to a speed of 1 m/s. (b) The mass is accelerated from 100 m/s to 101 m/s. (c) The speed of the mass is increased from v m/s to $(v + 1)$ m/s. (d) If $v \gg 1$ m/s, find an approximation that simplifies the calculation in part (c).

83. (II) A puck of mass $m = 0.2$ kg moves in a circle of radius 0.8 m on a table top and is tied with a massless rope to a tether at the origin. The coefficient of kinetic friction between the puck and the table top is $\mu_k = 0.02$. At $t = 0$ s, the puck is at the point shown in Fig. 6–31 with a velocity in the $+y$-direction of magnitude 10 m/s. (a) How much work is done by the rope on the first revolution? (b) How much work is done by friction on the first revolution? (c) What is the kinetic energy at the end of one revolution?

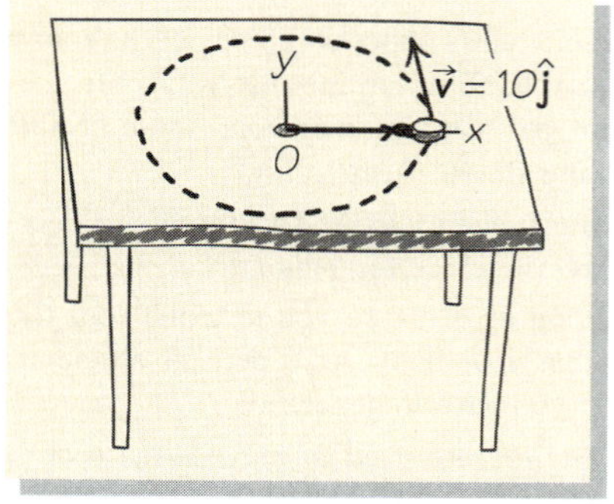

FIGURE 6–31 Problem 83.

84. (II) An automobile of mass 1100 kg is brought to a halt by applying the brakes, which lock the wheels. The coefficient of kinetic friction between the wheels and the road is 0.55. The car leaves skid marks 48 m long. (a) What is the force of friction between the car and the road? (b) What is the work done on the car by friction in bringing the car to a halt? Include the sign of the work. (c) What was the speed of the automobile when the brakes were first applied?

85. (II) A mass m is hauled from ground level up an inclined plane that makes an angle θ with the horizontal by means of a rope passing over a frictionless pulley. The mass is pulled along until it reaches a height H. The building of the Egyptian pyramids and many other ancient construction jobs used such ramps. Logs were sometimes used as rollers to reduce friction. (a) Show that if the contact between the mass and the ramp is frictionless, the work done by the tension in the rope (or, equivalently, by the person hauling the rope) is independent of the angle θ. (b) Calculate the work done by the tension in the rope as a function of the ramp angle θ if the coefficient of kinetic friction between the mass and the surface is μ_k.

86. (II) A large laser designed for nuclear fusion can produce 2×10^4 J of energy over a time period of 10^{-9} s. (a) How much power can such a laser produce during its discharge? (b) How much power is required to reenergize it over a period of 45 min?

87. (II) A pendulum of length L and mass m starts from an initial position in which it makes an angle θ_i with the vertical. Calculate the work done by the force of gravity as the mass moves from θ_i to θ_f and use your result to calculate the speed of the pendulum at the bottom of the swing. Why is it possible to ignore the tension in the string of the pendulum? [*Hint*: The work done by gravity is path independent.]

88. (II) A small dam produces electrical power. The water falls a distance of 18 m to turn a turbine. If the efficiency to produce electrical energy is only 68 percent, at what rate must water flow over the dam to produce 850 kW of electrical energy? (See Problem 64 for the meaning of efficiency.) The power system costs \$3.5 million. How many years will it take for the power plant to pay for itself if electricity can be sold for \$0.10/kWh? Ignore effects such as inflation and the cost of borrowing money.

89. (II) The force the Sun exerts on a planet in a circular orbit of radius x around the Sun is given by an expression of the form $F(x) = -mK/x^2$, where m is the mass of the planet and K is a constant. How much work must be done by a passing celestial body to move the planet to a radius that is 1 percent larger?

90. (II) Early twentieth-century investigations of the structure of atoms carried out by Ernest Rutherford involved the collisions of alpha-particles (α-particles) and gold atoms. The most interesting results concern the cases in which an α-particle is scattered at a large angle from a gold nucleus. Assume that an α-particle approaches a gold nucleus (at rest) head on and that the force between these two objects is repulsive, with magnitude $F = k/r^2$, where $k = 3.65 \times 10^{-26}\ \mathrm{N \cdot m^2}$. Suppose that the α-particle moves in toward the nucleus from far away under the sole influence of the repulsive force of the nucleus. What is the minimum kinetic energy the α-particle must have initially in order for it to get as close as 1.00×10^{-14} m?

CHAPTER

7

The motion of a roller coaster car illustrates the conversion of potential energy into kinetic energy. As the car rises, it slows; as it drops, it speeds up. The total energy of the car—the sum of the kinetic and potential energies—is conserved if there is no friction.

Potential Energy and Conservation of Energy

A roller coaster gains kinetic energy as it descends. Where did the energy come from? When the roller coaster rises from a low point, it loses kinetic energy. What happened to that energy? Conservative forces such as gravity have a special property: the work they do to make an object gain or lose energy can be recovered. The work is "stored" and we can think of it as a form of energy. That energy, *potential energy*, is a function of the position of the object, and the sum of the potential energy and the kinetic energy—the total energy—remains constant throughout the object's motion. As we will see, energy is conserved when conservative forces act. Thus when the roller coaster descends, potential energy is converted into kinetic energy; that kinetic energy is converted back into potential energy when the roller coaster climbs. Conservation of energy is a fundamental property that can also significantly simplify the solution of problems involving conservative forces. What about nonconservative forces such as friction? Nonconservative forces cannot be characterized by a potential energy. When such forces are present, energy is lost—converted into heat, abrasion, and noise.

A close look at the mechanisms of energy loss for nonconservative forces reveals that we can account for *all* the energy at the microscopic level. Thermal energy, for example, is actually related to the kinetic energy of molecular motion. Indeed, *all* the fundamental forces (see Chapter 5) are conservative. The concept of the conservation of energy is one of the central concepts of science.

7–1 ENERGY CONSERVATION

Here, we develop the concepts of potential energy and total energy. The results of this section *apply only to conservative forces*. The work–energy theorem shows us how the kinetic energy of an object changes when there is net work done on it. Now we shall see that we can reexpress the work done by conservative forces in terms of a simple function of position called the potential energy.

By definition, the work done on a given object by a conservative force when the object moves from point A to point B depends only on those points; it does not depend on any points in between: $W = W(\mathrm{A}, \mathrm{B})$. We can actually say more than this. Suppose we go from A to B, then from B to C. In this case, the total work does not depend on the intermediate point B and thus

$$W(\mathrm{A}, \mathrm{C}) = W(\mathrm{A}, \mathrm{B}) - W(\mathrm{B}, \mathrm{C}).$$

But this, in turn, can be true only if

$$W(\mathrm{A}, \mathrm{B}) = U(\mathrm{A}) - U(\mathrm{B}), \tag{7–1}$$

where U is some scalar function that depends only on position. Not only does W depend on the end points only, but it does so as a difference of two values of a function that depends only on position. If we know this function for a given conservative force, then we know once and for all how much work that force does as an object moves from one spot to another. We have already worked out some examples of this result. One occurs when an object is moved from height y_1 to height y_2. As we saw in the last chapter, the work done by gravity is

$$W(y_1, y_2) = mg(y_1 - y_2).$$

Comparing the work done by gravity with Eq. (7–1) we find

$$\text{for gravity:} \quad U(y) = mgy + U_0, \tag{7–2}$$

where U_0 is any constant. We call the function U the **potential energy**.

We can easily see how this idea works for a general one-dimensional conservative force $F(x)$. The work done in moving from point x_0 to point x_1 is

$$W(x_0, x_1) = \int_{x_0}^{x_1} F(x)\,\mathrm{d}x. \tag{7–3}$$

As we show next, we can guarantee that $W(x_0, x_1) = U(x_0) - U(x_1)$ as stated in Eq. (7–1) if and only if

$$F(x) = -\frac{\mathrm{d}U(x)}{\mathrm{d}x}. \tag{7–4}$$

The force in one dimension in terms of the potential energy

The function U is the potential energy. To check that Eq. (7–4) leads to Eq. (7–1), use Eq. (7–4) in Eq. (7–3) along with the fact that integration and differentiation are inverse processes to find

$$W(x_0, x_1) = \int_{x_0}^{x_1} \left[-\frac{\mathrm{d}U(x)}{\mathrm{d}x}\right] \mathrm{d}x = -U(x_1) + U(x_0). \tag{7–5}$$

This is indeed Eq. (7–1).

Note that we now have a way to find the one-dimensional conservative force from the corresponding potential energy function, $U(x)$; we simply use Eq. (7–4). We can check that this works for the case of gravity, where we have already found $U(y) = mgy + U_0$, or Eq. (7–2).

$$F(y) = -\frac{\mathrm{d}U}{\mathrm{d}y} = -\frac{\mathrm{d}}{\mathrm{d}y}(mgy) - \frac{\mathrm{d}U_0}{\mathrm{d}y} = -mg.$$

The Definition of Potential Energy

If we compare Eq. (7–3) with Eq. (7–5), we find an expression for the potential energy $U(x)$:

The definition of potential energy

$$\text{for conservative forces:} \quad U(x) = U(x_0) - \int_{x_0}^{x} F(x')\,dx'. \tag{7–6}$$

Note that the dependence on the initial point x_0 does not appear on the left-hand side of the equation; that dependence cancels on the right-hand side. Usually, we simply write $U_0 = U(x_0)$. The value of U_0, as well as the point x_0 itself, can be chosen for convenience.

PROBLEM SOLVING

U_0 is arbitrary.

Equation (7–6) gives us a way to interpret the potential energy. The change in the potential energy associated with a given force as an object moves from one point to another is the negative of the work done by that force over that displacement. The potential energy represents the possibility of getting that work back (Fig. 7–1).

FIGURE 7–1 The stretched bow and arrow system contains potential energy, much of which can be converted into the kinetic energy of the arrow.

The Conservation of Energy

Let's now apply our expression for work in terms of the potential energy to the work–energy theorem. We have

$$\frac{1}{2}mv_1^2 - \frac{1}{2}mv_0^2 = W(x_0, x_1) = -U(x_1) + U(x_0).$$

Simple rearrangement gives

$$\frac{1}{2}mv_1^2 + U(x_1) = \frac{1}{2}mv_0^2 + U(x_0). \tag{7–7}$$

This is a remarkable relation: It expresses the conservation of energy. Because this is true for any two points along the path of an object that is moving under the influence of the conservative force $F(x)$, Eq. (7–7) shows that *the quantity* $\frac{1}{2}mv^2 + U(x)$ *must have the same value at every point in the trajectory of the object.* We call this quantity E:

$$E \equiv \frac{1}{2}mv^2 + U(x). \tag{7–8}$$

The principle of the conservation of energy

The quantity E is called the **total energy** (or more simply, the **energy**) of the object, and we have shown that it is *conserved* during the motion, no matter how that energy is transformed from potential energy to kinetic energy and back again. A thrown ball starts with a certain kinetic energy. This kinetic energy is converted to potential energy as the ball rises to its maximum height; the potential energy is converted back to kinetic energy as the ball falls. Throughout the ball's flight, the total energy is always the same. The idea that the total energy is conserved is one of the most powerful ideas in physics and is one of the most useful (Fig. 7–2).

At this point, you should feel more comfortable with the idea that the potential energy is the negative of the work. As a force does positive work, which increases the kinetic energy, the potential energy decreases; as the force does negative work, which decreases the kinetic energy, the potential energy increases.

Applications of the Conservation of Energy

The conservation of E means that any change ΔK in the kinetic energy is compensated by an equal and opposite change ΔU in the potential energy:

$$\Delta K = -\Delta U. \tag{7–9}$$

This equation explains the origin of the term potential energy. Potential energy is energy that has the potential to be converted into energy of motion; that is, kinetic energy.

Energy conservation is a very useful tool for determining the speed of an object. If the function $U(x)$ is known and the initial conditions are given, we can determine the speed at any point x on the trajectory of an object. We simply use the fact that the total energy is constant.

PROBLEM SOLVING

We use the conservation of energy to find speeds.

It is clear from Eq. (7–7) that we can add an arbitrary constant U_0 to the potential energy $U(x)$ because this constant would cancel on both sides of Eq. (7–7). This arbitrary constant is chosen for convenience in practical applications. This choice does not affect *changes* in the potential energy or the total energy. The process of choosing a particular point x_0 where $U(x_0)$ is defined to be zero is sometimes referred to as "choosing the zero of the potential energy."

The point where the potential energy is zero does not affect predictions about observable quantities.

Application to Gravity. We have already seen from Eq. (7–2) that for a mass m under the influence of gravity

$$U(y) - U(y_0) = mgy - mgy_0. \tag{7–10}$$

(Here, we measure y with the positive direction up.) If we choose the reference height to be ground level in this case ($y_0 = 0$), and write $U(y_0) = U(0)$ as the arbitrary constant U_0, we recover Eq. (7–2):

$$U(y) = mgy - U_0. \tag{7–11}$$

We may further define the potential energy at $y = 0$ to be $U_0 = 0$; we have chosen the zero of the potential energy to be at height zero. Then we find the simple result

The potential energy of a mass acted on by gravity

$$U(y) = mgy, \tag{7–12}$$

and the conserved energy is

$$E = \frac{1}{2}mv^2 + mgy. \tag{7–13}$$

Let's see how energy conservations help us to find the ground-level speed of a ball dropped from rest. At the beginning of the ball's motion, the height is y and $v = 0$, so $E_{\text{beginning}} = mgy$. At the end, $y = 0$ and the speed is v, so $E_{\text{end}} = \frac{1}{2}mv^2$. Equating the two values of E—that is, $E_{\text{end}} = E_{\text{beginning}}$—we have

$$\frac{1}{2}mv^2 = mgy;$$

$$v = \sqrt{2gy}.$$

We derived this result using Newton's equation of motion in Chapter 2; here, we found the same result in a much simpler way.

We defined U to be zero at height $y = 0$. Alternatively, we can define U at, say, the top of a building ($y = H$) so that E is zero there. To do so, choose the arbitrary constant U_0 so that

$$E = 0 = mgH + U_0;$$

(Notice there is no kinetic energy at $y = H$.) This equation determines U_0:

$$U_0 = -mgH.$$

With this choice, the total energy is given at any height y by

$$E = \frac{1}{2}mv^2 + mgy - mgH = \frac{1}{2}mv^2 + mg(y - H). \tag{7–14}$$

In this equation, $y - H$ is the displacement of the ball from a starting point on the roof of our building to its new position at height y. Whatever the choice of U_0, the total energy is unchanged during the fall (Fig. 7–3).

FIGURE 7–2 The conservation of energy gives us an immediate way to calculate the speed of each ball as a function of height, given its initial speed.

FIGURE 7–3 The total energy of this Mexican cliff diver is conserved when he jumps from rest off the top of the cliff, where his kinetic energy is zero, to the water below, where his potential energy is taken to be zero. As he drops, his kinetic energy grows as his potential energy drops.

If we use the form of Eq. (7–14), an object starting at $y = H$ with $v = 0$ has $E = 0$. Thus $E = 0$ later; that is,

$$\frac{1}{2}mv^2 + mg(y - H) = 0;$$

$$v^2 = -2g(y - H).$$

If we take into account that $y - H$ is always negative for an object falling from rest from an initial height H, we see that this result is also in agreement with the formula obtained in Chapter 2.

EXAMPLE 7–1 A brick is thrown nearly straight up from 18 m above the ground by a person standing at the edge of a roof (Fig. 7–4). Use the fact that energy is conserved to determine the velocity with which the brick is thrown, given that it reaches the ground with a velocity of 24 m/s.

Solution: Let's choose the potential energy to be zero at ground level. Using Eq. (7–11), we have

$$U(y) = mgy + U_0 = mgy.$$

where y is the height measured from ground level. The total energy, E, is unchanging and it is

$$E = \frac{1}{2}mv^2 + mgy.$$

In particular, if the final speed is v_f, the final height is $y_f = 0$ m, and the initial velocity v_i is set when the brick is at an initial height $y_i = 18$ m, then we find

$$\frac{1}{2}mv_f^2 + mgy_f = \frac{1}{2}mv_f^2 + 0 = \frac{1}{2}mv_i^2 + mgy_i.$$

We divide out the unknown mass m and multiply by 2, solving for v_i:

$$v_i^2 = v_f^2 - 2gy_i = (24\text{ m/s})^2 - 2(9.8\text{ m/s}^2)(18\text{ m}) = 220\text{ m}^2/\text{s}^2;$$

$$v_i = \pm\ 15\text{ m/s}.$$

Note that energy conservation does not determine the sign of v_i. It does not matter whether the brick is thrown upward or downward with that velocity. In fact, when the brick is thrown upward, it reaches the top of its trajectory and then comes down, passing the roof with the same speed v_i with which it was thrown, except that now the velocity is negative instead of positive.

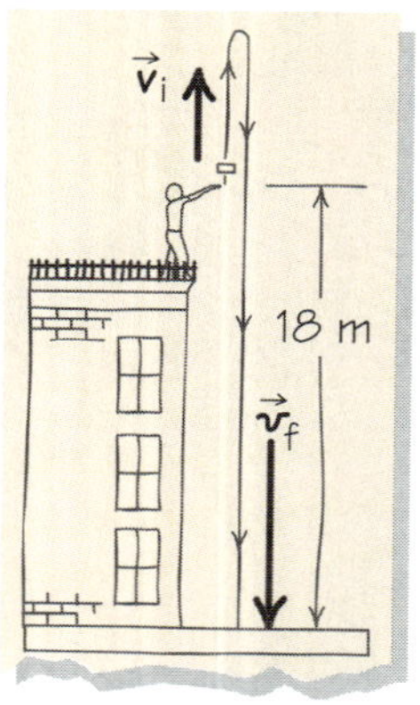

FIGURE 7–4 Example 7–1.

Application to a Spring Force. An ideal spring provides another example of a conservative force. The displacement from the equilibrium position of a mass m attached to the spring is labeled as x. The force on the mass is a restoring force, pulling the mass back to equilibrium:

$$F(x) = -kx, \qquad (7\text{–}15)$$

where k is the spring constant. Equation (7–10) takes the form

$$U(x) = U(x_0) - \int_{x_0}^{x} (-kx)\,dx = U(x_0) + \frac{1}{2}kx^2 - \frac{1}{2}kx_0^2.$$

The potential energy of a mass at the end of a spring

We can now proceed as we did for gravity. We set the arbitrary point x_0 to be zero, and denote $U(0)$ by U_0. We have

$$U(x) = \frac{1}{2}kx^2 + U_0. \qquad (7\text{–}16)$$

It is customary to take U_0 to be zero, which is equivalent to choosing the zero of potential energy at $x = 0$. With this choice, Eq. (7–16) becomes

$$U(x) = \frac{1}{2}kx^2. \tag{7–17}$$

The total conserved energy is given by the sum of the kinetic and potential energies: $E = (mv^2/2) + (kx^2/2)$.

EXAMPLE 7–2 One end of a massless spring is placed on a flat surface; the other end points upward (Fig. 7–5a). A mass of 1.0 kg is gently set down on top of the spring, until the spring is compressed by 17 cm to a new equilibrium position (Fig. 7–5b). What is the spring constant?

Now, the 1.0-kg mass is removed and a 2.0-kg mass is set on top of the spring. The spring is then compressed by hand so that the end of the spring is 42 cm lower than the position of the spring with no mass on top (Fig. 7–5c). The spring is then suddenly released. What is the maximum kinetic energy of the 2.0-kg mass?

Solution: Use the coordinate y to measure the position of the masses. The top of the unloaded spring defines $y = 0$ (Fig. 7–5a), with $+y$ directed upward. We set both the potential energy of gravity and the potential energy of the spring to be zero at that point. Thus

$$U_g = mgy, \quad U_{\text{spring}} = \frac{1}{2}ky^2,$$

where k is the spring constant.

We find the spring constant by using Newton's first law. As usual, we drop vector notation for this one-dimensional problem. When the 1.0-kg mass, m_1, is resting motionless on the spring, the net force on the mass is zero. The force due to the spring acts upward (in the positive direction), $F_{\text{spring}} = -ky_1$, where the compression is $y_1 = -17$ cm (Fig. 7–5b). Thus

$$F_{\text{net}} = -ky_1 - m_1g = 0.$$

We then solve for the unknown spring constant k:

$$k = -\frac{m_1g}{y_1} = -\frac{(1.0\text{ kg})(9.8\text{ m/s}^2)}{-0.17\text{ m}} = 58\text{ kg/s}^2 = 58\text{ N/m}.$$

We can determine the potential energy of the mass because we know the compressed spring position $y_2 = -0.42$ m (Fig. 7–5c) with the 2.0-kg mass, m_2. To find the mass's speed at any position, we then use the conservation of total energy, E,

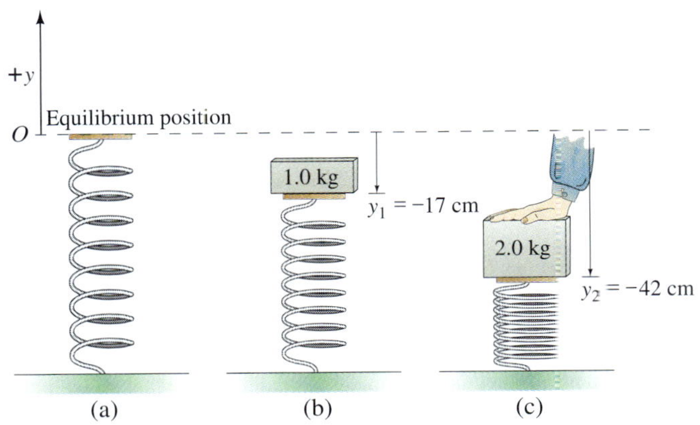

FIGURE 7–5 (a) Example 7–2. In our coordinate system, the top of the relaxed spring is at $y = 0$. (b) When a 1.0-kg mass is placed gently on top of the spring, the new equilibrium position is $y = -17$ cm. (c) A 2.0-kg mass is attached to the end of the spring, and the mass is then pushed down to $y = -42$ cm. The mass is then released.

which is the sum of the kinetic energy, the spring potential energy, and the gravitational potential energy:

$$E = \frac{1}{2}m_2v^2 + \frac{1}{2}ky^2 + m_2gy.$$

We evaluate E by noting that $v = 0$ m/s at $y = y_2$. Thus

$$E = \frac{1}{2}(2.0\text{ kg})(0\text{ m/s})^2 + \frac{1}{2}(58\text{ N/m})y_2^2 + (2.0\text{ kg})(9.8\text{ m/s}^2)y_2$$

$$= (29\text{ N/m})(-0.42\text{ m})^2 + (19.6\text{ kg}\cdot\text{m/s}^2)(-0.42\text{ m}) = -3.1\text{ J}.$$

The potential energy has a minimum—and therefore the kinetic energy has a maximum—when the force is zero. This occurs at y_3, where the spring force and the force of gravity cancel; namely, at $-ky_3 - m_2g = 0$, or

$$y_3 = -\frac{m_2g}{k} = -\frac{(2.0\text{ kg})(9.8\text{ m/s}^2)}{58\text{ N/m}} = -0.34\text{ m}.$$

At this point, where we have seen that the kinetic energy has a maximum, $E = K_{max} + \frac{1}{2}ky_3^2 + m_2gy_3$. Because E is constant, we just set this to the value of E we have already calculated, $E = -3.1$ J, or

$$K_{max} = -\frac{1}{2}ky_3^2 - m_2gy_3 + E$$

$$= -\frac{1}{2}(58\text{ N/m})(-0.34\text{ m})^2 - (2.0\text{ kg})(9.8\text{ m/s}^2)(-0.34\text{ m}) - 3.1\text{ J}$$

$$= 0.21\text{ J}.$$

The Energy of Systems

Up to this point, we have concentrated on how energy applies to problems involving single objects. The scalar nature of energy makes it easy to apply it to entire systems—as long as those systems involve conservative forces. Let's look at a problem involving a system.

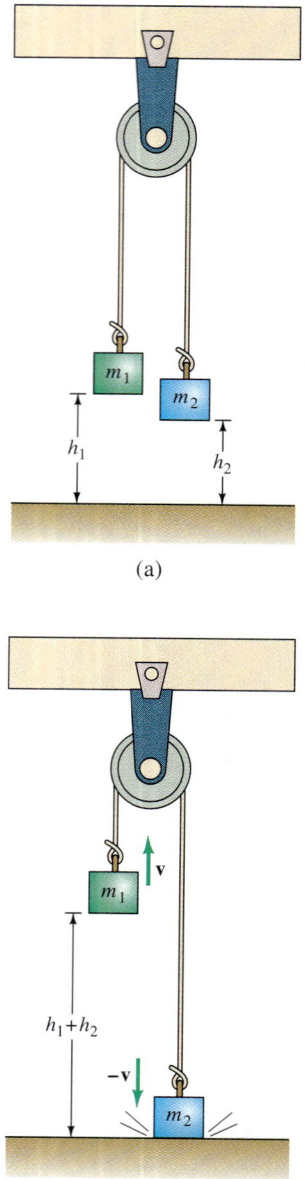

FIGURE 7–6 (a) Initial conditions for Example 7–3. (b) The situation just before mass m_2 hits the floor.

EXAMPLE 7–3 Consider the Atwood machine introduced in Example 5–4 (Fig. 7–6a). The masses have the values $m_1 = 1.37$ kg and $m_2 = 1.51$ kg. The system is released from rest with m_2 84 cm from the floor. Use energy conservation to find the speed of m_2 just before it hits the floor.

Solution: We consider the two masses tied together as a whole. Only gravity acts on this system, and gravity is conservative, so we can use the conservation of energy with the potential energy due to gravity. Figure 7–6a shows the system in its initial state, with no kinetic energy:

$$E_{init} = m_1gh_1 + m_2gh_2$$

The final state of the system is illustrated in Fig. 7–6b. There is movement at this time, so the system contains kinetic energy as well as potential energy. Each mass has the same speed, v, because they are tied together. Thus

$$E_{final} = m_1g(h_1 + h_2) + \frac{1}{2}m_1v_1^2 + m_2g(0) + \frac{1}{2}m_2v_2^2 = m_1g(h_1 + h_2) + \frac{1}{2}(m_1 + m_2)v^2.$$

When we set the final and initial energies equal we find

$$(m_2 - m_1)gh_2 = \frac{1}{2}(m_2 + m_1)v^2.$$

Solving for v, we find

$$v = \sqrt{\frac{2(m_2 - m_1)gh_2}{(m_2 + m_1)}} = \sqrt{\frac{2(1.51\text{ kg} - 1.37\text{ kg})(9.81\text{ m/s}^2)(0.84\text{ m})}{(1.51\text{ kg} + 1.37\text{ kg})}} = 0.895\text{ m/s}.$$

APPLICATION

Balancing energy

Conservation of energy is a powerful tool in the study of complicated systems, when accompanied by the realization that energy comes in many forms. It may be possible to account for energy entering or leaving the system by certain routes and at certain rates. For example, if a certain amount of electric lighting is necessary in a building, how does that affect the heating or air conditioning that must be installed? In constructing a refinery, how do the heights at which various pipes must be placed affect the necessary pumping apparatus? The conservation of energy is a bookkeeping tool that allows engineers to design and analyze the behavior of a system without knowing every last detail.

INTERIM SUMMARY

The work done by a conservative force acting on an object as it moves from a point x_0 to a point x_1 is independent of the path taken by the object. For such forces, the work done can be expressed as a function of one end point of the path minus a function of the other. We can then express this work in terms of a scalar function of position: the potential energy—again, *only* for conservative forces. For one-dimensional conservative forces, the potential energy is defined by

$$\text{for conservative forces:} \qquad U(x) = U(x_0) - \int_{x_0}^{x} F(x')\,dx'. \qquad (7\text{–}6)$$

The potential energy difference between two points is thus the negative of the work done between those points. Because we only ever deal with differences of potential energy, it is only defined up to an arbitrary constant, which is expressed here by the fact that both x_0 and the value U_0 of U at the point x_0 are arbitrary. If the potential energy is known, then the (one-dimensional) force is found by taking a derivative:

$$F(x) = -\frac{dU(x)}{dx}. \qquad (7\text{–}4)$$

The work–energy theorem takes a simple useful form for conservative forces. The energy E, defined by

$$E = \frac{1}{2}mv^2 + U(x), \qquad (7\text{–}8)$$

is conserved, meaning it has the same value everywhere along the trajectory of an object that is acted on by that force only. The conservation of energy is a powerful tool for solving problems.

7–2 ENERGY CONSERVATION AND ALLOWED MOTION

Let's take a look at the energy conservation equation for a mass attached to a spring with spring constant k. To do so, we will use an **energy diagram** of the system. For such a diagram, we place the position of the mass (the object whose motion is of interest to us) on the horizontal axis and we place any of the energies—potential energy, $U = kx^2/2$; kinetic energy, $K = mv^2/2$; or total energy, $E = U + K$—on the vertical axis. Suppose that the spring is initially compressed by an amount x_0, then released

The energy diagram

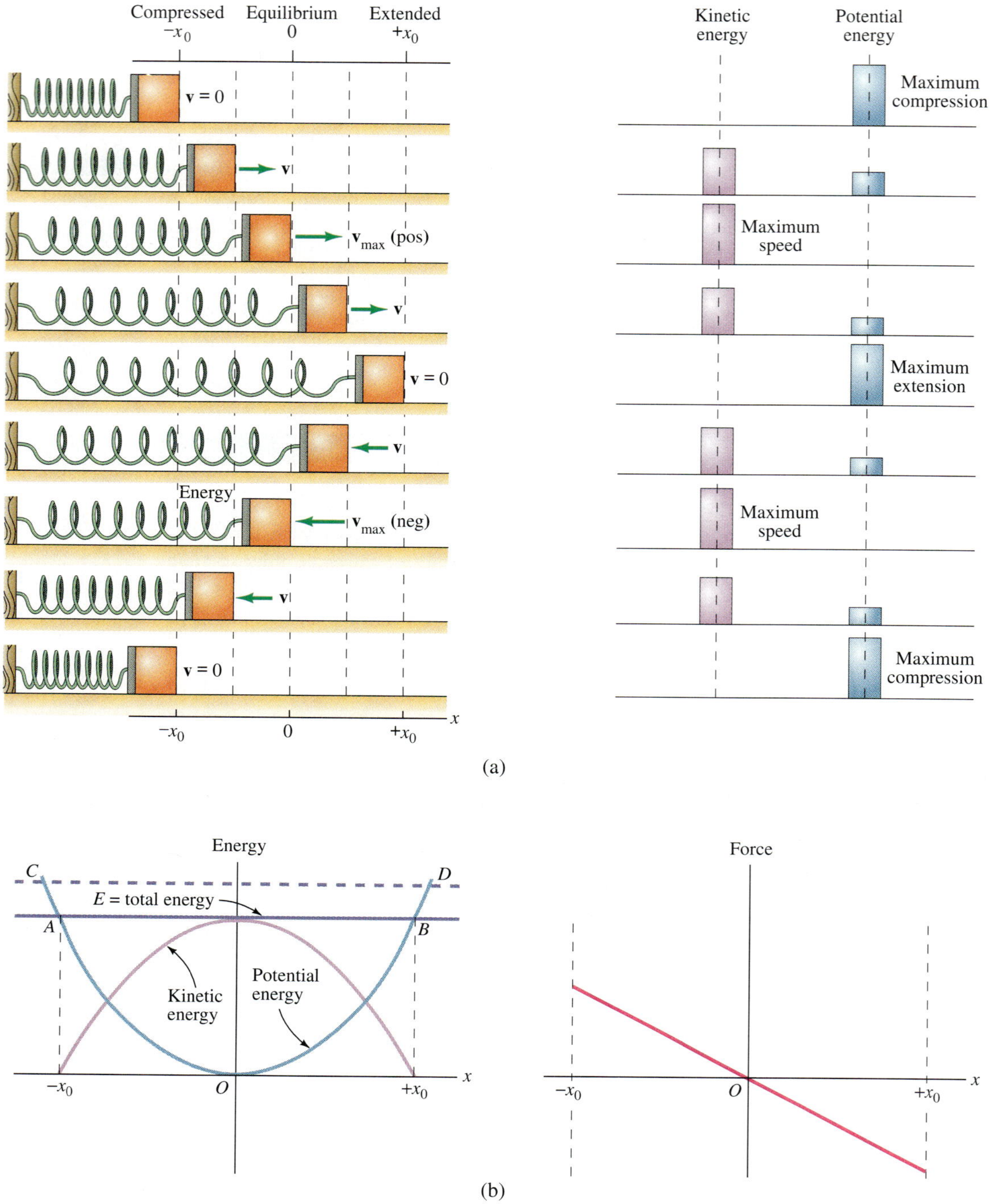

FIGURE 7–7 A mass on the end of a spring; equilibrium is $x = 0$. (a) At the top, the spring is compressed by an amount x_0. During one period (or cycle), the mass moves through $x = 0$ to an extension $x = x_0$ and back to $x = -x_0$. As the mass moves, the kinetic and potential energies associated with the mass-spring system change: One increases as the other decreases. (b) A plot of the force of a spring on a mass attached to its end, together with an energy diagram for the mass. The potential energy $U = \frac{1}{2}kx^2$ is drawn. The straight line represents the constant total energy, E; the kinetic energy, K, is the difference between E and U. The curve of K is determined purely by the conservation of energy. Points $+x_0$ and $-x_0$ are turning points for the motion, where the kinetic energy and therefore the speed are zero. For x outside the range $-x_0 < x < +x_0$, the kinetic energy would be negative, so the mass cannot go outside this range.

(Fig. 7–7a). Initially, $v = 0$ and $x = -x_0$; at this point, then, $E = kx_0^2/2$. Because E is unchanging, this is the value of E once and for all. Figure 7–7b is an energy diagram for the system, in which the spring at its initial, compressed position corresponds to $-x_0$. Equation (7–4) states that the force is the negative of the slope of the potential energy. The slope of $U(x)$ at $-x_0$ is negative, which means that the force is positive according to Eq. (7–4); the mass thus accelerates to the right, back to the equilibrium position. As it moves to the right past $x = 0$, the potential energy curve falls below the horizontal line that represents the constant total energy, E.

Energy conservation, as expressed in Eq. (7–8), implies that the difference between E and the potential energy curve is equal to the kinetic energy—$K = \frac{1}{2}mv^2$—as seen in Fig. 7–7b. The kinetic energy and the potential energy always add up to the total energy. In other words, if the kinetic energy increases, the potential energy decreases by an equal amount. The kinetic energy increases to a maximum at $x = 0$ as the mass moves to the right. The mass slows down as it continues to the right because the potential energy curve rises and the kinetic energy decreases. Finally, at $+x_0$, $U(x)$ is equal to E again. The mass cannot go any farther to the right; there, $U(x) > E$ and this would require a negative kinetic energy, an impossibility for the quantity $\frac{1}{2}mv^2$. Thus $+x_0$ and (for exactly the same reason) $-x_0$ are *turning points* for the motion. **Turning points** are those points where the speed drops to zero and the mass changes direction. The mass must therefore remain between $-x_0$ and $+x_0$. If the energy were larger, then the horizontal line at E in Fig. 7–7b would just lie higher and the turning points would correspond to points C and D. A turning point is always characterized by the fact that $E = U(x)$ at that point so that $K = \frac{1}{2}mv^2 = 0$.

Let's suppose that an object moves from $-x_0$ to $+x_0$ and then back to $-x_0$. It then restarts its motion from $-x_0$ and there is nothing to distinguish this second traversal from the first. The motion of the mass attached to the end of a spring is repetitive or, to use more technical language, it is **periodic**. Periodic motion appears everywhere in the physical world and is characteristic of the motion near *any* minimum of a potential energy function.

Energy Conservation and Equilibrium

We can use the energy diagram technique to explore the stability of systems. The mass on a spring is a stable system because the spring force always accelerates the mass back toward the equilibrium point. Let's now look at a rather different system—a marble balanced on top of a hill. Physically, we know that once the marble starts rolling, it rolls all the way down the hill. Suppose that, at any point on the hill, the height $y = y_0 - cx^2$, where x is the horizontal distance from the top of the hill, y_0 is the height of the top of the hill, and c is a constant. Because the potential energy of the marble is that of gravity and is therefore proportional to y, the marble has a potential energy superficially like that of the spring—$U = U(y) \propto -x^2$—*where the coefficient of x^2 is negative.* This reflects a force whose magnitude grows with the marble's displacement but whose sign, unlike the spring force, tends to send the marble *away* from $x = 0$. This type of force is relevant for *unstable* situations such as the marble on the top of the hill; its potential has the form

$$U(x) = -\frac{1}{2}\kappa x^2, \tag{7–18}$$

where the new constant κ is positive. The total energy is then $E = (mv^2/2) - (\kappa x^2/2)$.

How do we see the behavior of the marble on the hill in this language? We draw an energy diagram that includes the potential energy of the marble (Fig. 7–8). We observe that if the marble starts from rest at some positive point x_0, then the energy is negative: $v = 0$ there, so $E = -\frac{1}{2}\kappa x_0^2$. (A negative total energy is perfectly acceptable, because we have seen that we can add or subtract any constant amount to the energy

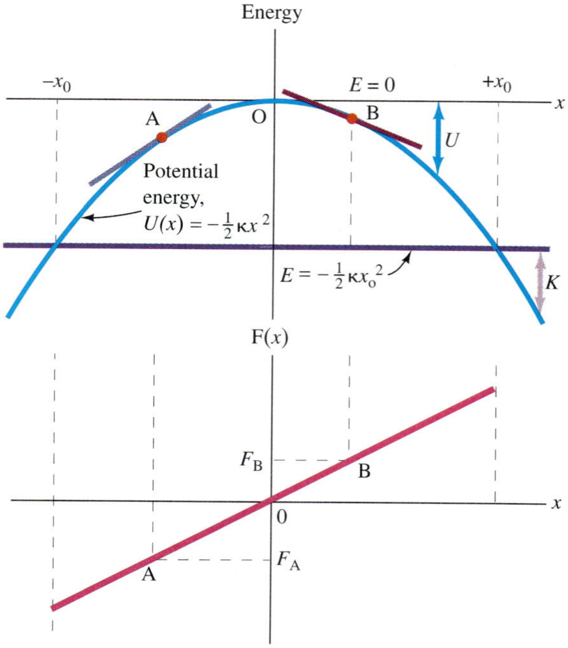

FIGURE 7–8 Energy diagram for the potential energy function $U(x) = -\frac{1}{2}\kappa x^2$, where κ is positive. We have also drawn the force that leads to this potential energy. For the total energy $E = -\frac{1}{2}\kappa x_0^2$, the particle can exist only where $x \geq x_0$ or $x \leq -x_0$; otherwise, the kinetic energy would be negative. For $E = 0$, the position $x = 0$ is an unstable equilibrium point. A slight displacement either way from $x = 0$ will send the particle hurtling away from the equilibrium point.

Stable and unstable equilibrium points

without affecting the behavior of the marble.) We notice that the slope of the potential energy is negative at $x = x_0$, so the force is positive and the marble accelerates in the direction of increasing x. As it moves to the right, the potential energy becomes more and more negative and thus the kinetic energy (given by the difference between the total and potential energies) becomes larger and larger. We call this *runaway motion*. Once started, the marble accelerates down the hill.

If the marble were to start at the very top of the hill ($x = 0$), there would be no acceleration because the slope of the curve $U(x)$ vanishes there—and the force is the slope of the curve $U(x)$. With an initial velocity of zero, nothing happens. In principle, the marble could stay there forever. Points such as this occur frequently in physical systems and we call them **equilibrium points**. Because the slightest perturbation would send the marble rolling one way or another, $x = 0$ is an **unstable** equilibrium point. Unstable equilibrium points are characterized by peaks, or maxima, in the potential energy. We can contrast this with the potential energy curve $U(x) = kx^2/2$ for the spring (see Fig. 7–7b). There, too, $x = 0$ is an equilibrium point; in this case, however, a slight displacement from $x = 0$ brings a restoring force into play that accelerates the particle *back* toward $x = 0$. A perturbation from this equilibrium point leads to the periodic behavior discussed previously, and not to runaway motion. When the potential energy has a trough, or minimum—as for a mass on a spring, a skateboarder on a ramp, or a marble in a bowl—the equilibrium point is **stable** (Fig. 7–9).

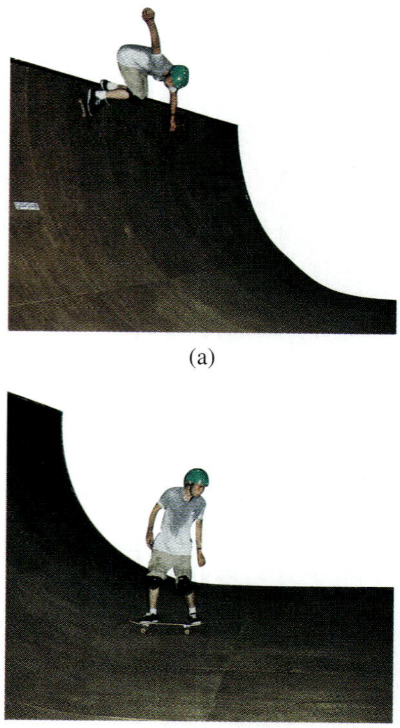

(a)

(b)

FIGURE 7–9 This skateboarding trough has a stable equilibrium point at the lowest height. (a) The forces on the skateboarder tend to move him to the equilibrium point. (b) When the skateboarder moves farther to the right, he will find a rise that will move him back to the equilibrium point.

More on Energy Diagrams

To illustrate some new features of motion under the influence of conservative forces, let us return to the potential energy $U(x)$ of Eq. (7–18) but *add* the term qx^4. This potential energy is shown in the energy diagram in Fig. 7–10. The qx^4 term may be negligible for small $|x|$, but as $|x|$ becomes larger, it begins to dominate the negative x^2 term of Eq. (7–18). We analyze the motion of a particle with such a potential energy by first choosing a total energy, this time E_0.

The motion of a particle with energy E_0 that starts from rest at x_0 (point B) is periodic. The motion is described by the kinetic energy $K = E_0 - U(x)$. Starting from zero at x_0, K grows until it reaches a maximum above the right-hand-side minimum ($x = x_1$) in $U(x)$. The kinetic energy then decreases, reaching a minimum at $x = 0$; it

PROBLEM-SOLVING TECHNIQUES

Assuming that we know the potential energy, the energy diagram is a useful tool for analyzing motion in one dimension under the influence of a conservative force. The choice of the zero of the potential energy should already have been made.

1. Motion in regions of the energy diagram where the kinetic energy would be negative are excluded.
2. A single point where the object is known to be at rest (zero kinetic energy) determines the total energy E, which is a constant (by conservation of energy) and can be drawn as such on the diagram. The object is then at rest at other places—turning points—where the flat line of E intersects the curve of potential energy. Two such points, A and B, are shown in Fig. 7B1–1.
3. With both U and E known, the speed at any point—such as x_0 on Fig. 7B1–1—is determined by finding the kinetic energy there, which is the difference $E - U$.
4. The shape of the curve of U versus position carries useful information. If the curve has a minimum, and if the total energy is such that the turning points are on either side of that minimum, then that minimum is a stable equilibrium point. If the curve has a maximum, and if the total energy is such as to allow the object to start from rest at that point, then the maximum is an unstable equilibrium point.
5. For motion in more than one dimension, the energy diagram is more complicated but can still prove useful. Figure 7B1–2 shows a possible potential energy for an object that can move in two dimensions; this plot looks like a Mexican hat. If the total energy is less than the peak of the hat, then the motion of the object will be restricted to lie within the circle formed by the well of the potential energy curve.

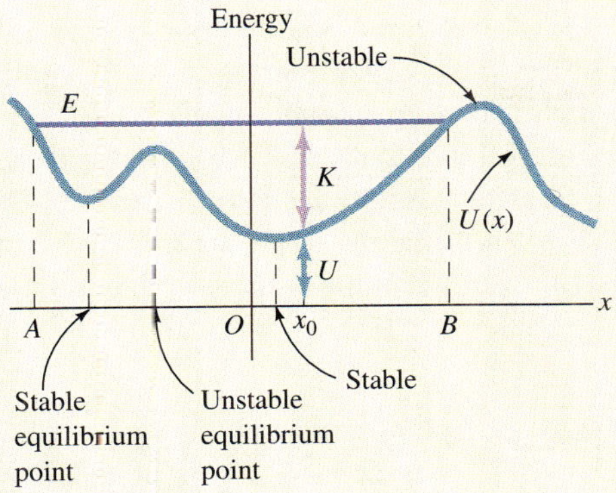

FIGURE 7B1–1 An energy diagram for one-dimensional motion: The solid curve is a plot of potential energy, U, versus position. The total energy, E, is a constant as the position changes, expressing conservation of energy. The kinetic energy, K, is then the difference between E and U. The points A and B represent turning points for the given total energy. Stable and unstable equilibrium points are also shown.

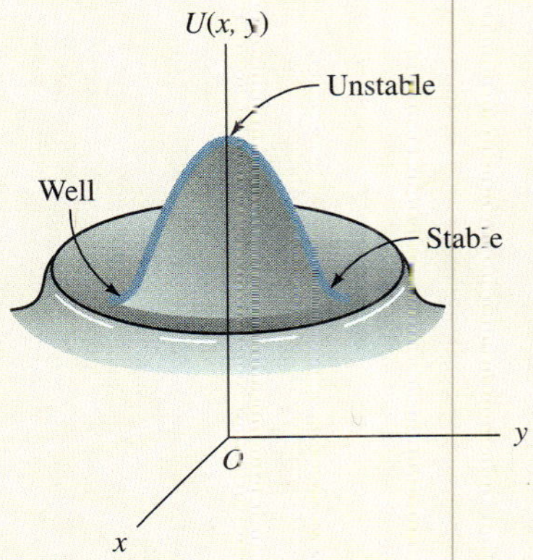

FIGURE 7B1–2 An energy diagram for two-dimensional motion, with a potential energy curve known as the Mexican hat potential. The circular well is a locus of stable equilibrium points for motion toward or away from the origin. The tip of the hat is an unstable equilibrium point.

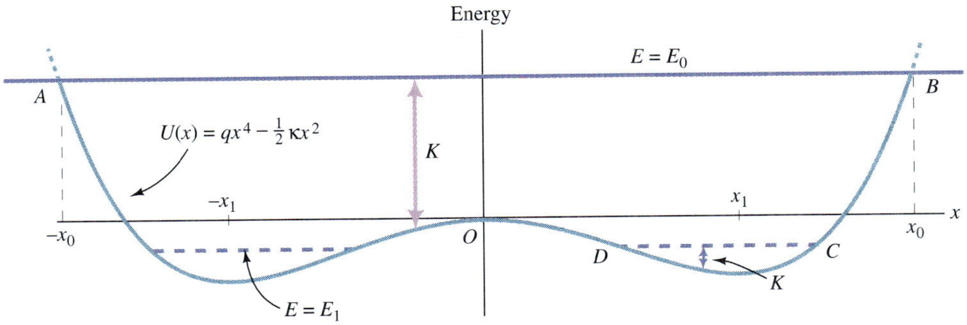

FIGURE 7–10 The term qx^4 is added to the potential energy $U(x)$ of Fig. 7–8. The graph of U shows that two minima of potential energy occur at $x = \pm x_1$. A particle with energy E_1 can move only within either of these pockets.

increases again until the left-hand-side minimum in $U(x)$ is reached ($x = -x_1$), and then decreases to zero at the point $x = -x_0$, where the line $E = E_0$ intersects the potential energy curve. The particle then moves to the right, retracing its motion, speeding up near $x = -x_1$, slowing down near $x = 0$, speeding up as x approaches x_1, and coming to a stop at $x = x_0$; starting from this time, the whole motion is repeated.

Potential energy barriers

If the energy of the particle is E_1 and the particle starts between C and D, the particle moves between the two turning points C and D. The particle never reaches the left side of the vertical axis. We describe the peak between the two troughs as a **potential energy barrier**—so called because it may act to keep a particle in one trough from getting to the other. For an energy value such as E_1, $x = 0$ is inaccessible to the particle. We see that there is a stable equilibrium point located at the bottom of each trough. There, a particle displaced from the bottom of the trough will experience a restoring force that pulls it back toward the bottom.

PROBLEM SOLVING

Uses of the potential energy curve

Our discussion has shown that the potential energy curve $U(x)$ contains a great deal of information about the motion of the particle. Energy diagram techniques are useful even when the equation of motion is a very complicated differential equation.

7–3 MOTION IN TWO OR THREE DIMENSIONS

Just as for one dimension, there exists a potential energy that is a function of position for conservative forces acting in two or three dimensions. This potential energy is again defined so that the change in its value in going from one point to another is the negative of the work done by the force between those points. The potential energy is now a function of all three variables, x, y, and z; equivalently, it is a function of the position vector $\mathbf{r}$: $U(x, y, z) = U(\mathbf{r})$. The total energy,

$$E \equiv \frac{1}{2}m(v_x^2 + v_y^2 + v_z^2) + U(x, y, z)$$

$$= \frac{1}{2}mv^2 + U(\mathbf{r}), \qquad (7\text{–}19)$$

is then independent of time; that is, it is a constant of the motion. The value of the constant E is fixed by the initial value of U (at the starting point) and by the initial velocity. The potential energy function $U(\mathbf{r})$ determines the force in a manner analogous to Eq. (7–4), which gives $F(x) = -\mathrm{d}U/\mathrm{d}x$ for a force acting in one dimension. In particular, the force may have two (or three) components in two- (or three-) dimensional space but, *for conservative forces, all components of the force—the complete force vector—can be found from a single potential energy function $U(x, y, z)$*. The potential energy is certainly a more economical way for us to describe the force than is a listing of three components.

The conservation of energy holds for conservative forces in any number of dimensions.

Potential Energy for Projectile Motion

The potential energy due to gravity for projectile motion

In Section 7–1, we derived and applied the expression $U = mgy + U_0$ [Eq. (7–11)] for up-and-down motion. *This expression is valid also for arbitrary projectile motion under the influence of gravity*. We studied the motion of a mass on a ramp in Section 6–4 to show that the force of gravity is conservative in that the work does not depend on the path taken—even when that path contains horizontal components. Thus there is a potential energy. To find that potential energy, we must evaluate the work done when gravity acts on an object. We have seen that this work, and hence the potential energy change, depends only on the change in height of the object; it does not depend on the object's horizontal motion. For example, $W = mgL \sin\theta$ [Eq. (6–22)] is the work done by gravity in moving a mass a total length L along a ramp that makes an angle θ with the horizontal. But the quantity $L \sin\theta$ is the height through which the mass drops (see Fig. 6–22), h. The work done is thus mgh, which is just the expression that holds when we consider vertical motion only. We conclude that *the potential energy function $U(y) = mgy + U_0$ applies to general projectile motion under the influence of gravity*.

EXAMPLE 7–4 An inexperienced golfer hits a bad shot and the ball leaves the tee with an initial speed of 28 m/s at an angle of 84° with respect to the horizontal. A bee is cruising innocently at a height of 37 m when it has the bad luck to meet the golf ball. What is the speed of the ball when it hits the bee? Ignore all effects of air resistance.

Solution: We use the conservation of energy to find the speed of the golf ball at any height given an initial speed v_i of 0 m/s. We let the potential energy function be zero at ground level height; that is $h = 0$ m, so $U_0 = 0$ and $U(h) = mgh$. The golf ball has an initial energy of $E_i = K_i + U(0) = mv_i^2/2 + mg(0) = mv_i^2/2$, where m is the mass of the ball and v_i is the initial speed. The energy of the ball when it meets the bee at height h is $E_f = K_f + U(h) = mv_f^2/2 + mgh$, where v_f is the desired information. Conservation of energy allows us to equate these two energies:

$$E_i = \frac{mv_i^2}{2} = E_f = \frac{mv_f^2}{2} - mgh.$$

Cancel the factor m and solve for v_f^2:

$$v_f^2 = v_i^2 - 2gh = (28 \text{ m/s})^2 - 2(9.8 \text{ m/s}^2)(37 \text{ m}) = 58.8 \text{ m}^2/\text{s}^2;$$

$$v_f = 7.67 \text{ m/s}.$$

The golf ball is fairly close to the top of its trajectory and has slowed down considerably. Note that we do not need to use the detailed information about the initial angle of the ball nor calculate the angle at which the ball is moving when it hits the bee. That is part of the utility of working with the conservation of energy.

Central Forces

There is a most interesting case that occurs frequently in nature: a force with a magnitude that depends only on the radial distance from a fixed point and has a direction aligned along the corresponding radius vector. For example, the force of gravitational attraction on Earth due to the Sun depends on the distance between Earth and the Sun and is aligned along the line between the two objects. Such forces are called **central forces** and *all central forces are conservative*.† Such a force takes the form

PROBLEM SOLVING

Central forces are conservative.

$$\mathbf{F(r)} = F(r)\left(\frac{\mathbf{r}}{r}\right) = F(r)\hat{\mathbf{r}}, \tag{7–20}$$

where $\hat{\mathbf{r}}$ is a unit vector pointing away from the origin. This force is reminiscent of a one-dimensional force. The potential energy function of a central force depends only on the radial distance r; the expression for the force in terms of a potential energy function $U(r)$ is

A central force is expressed in terms of a potential energy that depends only on the radial distance.

$$\mathbf{F} = -\frac{dU(r)}{dr}\hat{\mathbf{r}}. \tag{7–21}$$

Example 7–5 illustrates how we can apply Eq. (7–21).

EXAMPLE 7–5 The potential energy describing the gravitational interaction (the *gravitational potential energy*) between two point masses m_1 and m_2 is given by the expression

$$U(r) = -\frac{Gm_1m_2}{r},$$

†More precisely, a conservative central force must depend *only* on r.

where G is a constant and r is the distance between the two point masses. Calculate the force experienced by mass m_2 due to the presence of m_1. Is the force attractive or repulsive?

Solution: We define the vector **r** to be in the direction from m_1 (taken to be located at the origin) to m_2 (Fig. 7–11). With the help of Eq. (7–21) and $(d/dr)(1/r) = -1/r^2$, we find

$$\mathbf{F} = -\frac{Gm_1m_2}{r^2}\hat{\mathbf{r}}.$$

The force is proportional to $1/r^2$ (an inverse-square form), and it is directed opposite to the position vector **r** from m_1 to m_2. It therefore acts to pull m_2 back toward m_1 and is an *attractive* force.

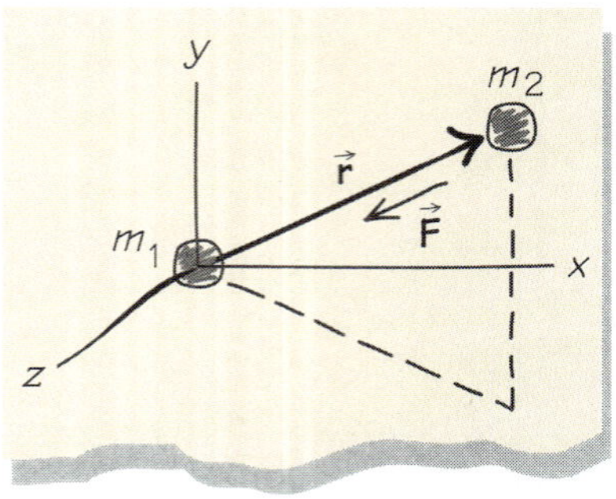

FIGURE 7–11 Example 7–5. The vector **r** pointing outward radially from the origin (the location of mass m_1) is the position vector of mass m_2. The force vector on m_2 acts inward radially.

*A General Form

If the force is not a central force, it will still be a conservative force if the work it does is path independent. There is, then, a potential energy function that depends on x, y, and z separately. The rule that takes us from the potential energy function to the force itself is a simple generalization of Eq. (7–4) or (7–21). To express this rule, we use **partial derivatives**. If we have a function $U(x, y)$ of two variables, we define partial derivatives—denoted by $\partial U/\partial x$ and $\partial U/\partial y$—as follows: The partial derivative $\partial U/\partial x$ is obtained by differentiating $U(x, y)$ with respect to *x while treating y as if it were a constant*; the partial derivative $\partial U/\partial y$ is obtained by differentiating $U(x, y)$ with respect to *y while treating x as if it were a constant.* Given this simple definition, the force **F** has components given by

PROBLEM SOLVING

Partial derivatives

$$F_x = -\frac{\partial U}{\partial x},\ F_y = -\frac{\partial U}{\partial y},\ \text{and}\ F_z = -\frac{\partial U}{\partial z}. \tag{7–22}$$

In particular, if the potential energy depends on, say, the variable y alone, then Eq. (7–22) shows that the force has a component only in the y-direction. The force of gravity provides a good example: In this case, U depends on y (the height) alone, and **F** has only a y-component.

7–4 IS ENERGY CONSERVATION A GENERAL PRINCIPLE?

At first glance, it appears that most systems involve nonconservative forces because forces that dissipate energy, such as friction or drag, are almost everywhere. In the presence of friction, energy does not appear to be conserved. Actually, the energy that is lost in friction can be identified through experiment as **thermal energy**; that is, the type of energy that raises the temperature of two objects when they are rubbed together. When thermal energy is measured and included as still another component of the total energy, *the total energy is then found to be conserved.* Thermal energy is not a form of potential energy because it is not completely retrievable. A system that contains thermal energy therefore differs from a system for which kinetic and potential energies can be freely converted back and forth. The ways in which mechanical and thermal energy can be converted back and forth are important, and we shall devote a great deal of effort to this subject when we study thermodynamics.

The conservation of energy is a general principle.

PROBLEM SOLVING

Conservation of energy and nonconservative forces

In fact, when we look closely enough, *all the fundamental forces in nature are conservative*, and conservation of energy always holds. On the atomic level, there is no such thing as friction. *The conservation of energy is one of the most fundamental principles of physics.* It has received experimental support from all fields of physics. We shall encounter energy conservation in a variety of applications, including thermodynamics, electromagnetism, relativity, and quantum physics (Fig. 7–12).

The work–energy theorem when both conservative and nonconservative forces are present

Energy Conservation and Nonconservative Forces

While it is important to realize that all fundamental forces are conservative, we do not often operate at the level of the fundamental forces. How then do we deal with the nonfundamental and often nonconservative forces, such as friction, that appear in everyday situations?

Suppose both conservative and nonconservative forces act on an object. A skier descending a mountain moves under the influence of both gravity and friction. In an example such as this, we can first construct the potential energy as if there were no friction. We then ask how much energy is lost due to friction; the loss of energy is attributable to any work done by friction. All this is possible because we can divide the work as it appears in the work–energy theorem into a part that can be associated with a potential energy (the work of the conservative forces) and another part that cannot. Thus, if we write W_{nc} as the work done by the net nonconservative force and U as the potential energy of the net conservative force, the work–energy theorem becomes

$$W_{nc} = \Delta E = \Delta(K + U). \qquad (7\text{–}23)$$

Here, we have continued to think of E as $K + U$, but E is no longer conserved.

Let's consider our skier. If friction is ignored, we have $E = K + mgy =$ constant, where $K = mv^2/2$ is the kinetic energy at a height y above some zero level. When friction is introduced, the initial and final energies, E_i and E_f respectively, differ, with $E_f < E_i$. The deficit is due to the work done by friction. We have

$$W_{\text{friction}} = E_f - E_i = (K_f + mgy_f) - (K_i + mgy_i). \qquad (7\text{–}24)$$

The work done by friction is negative, so that the final energy is less than the initial energy.

FIGURE 7–12 A fireworks display represents the conversion of chemical energy into the kinetic energy of the shooting components and the energy of the excited atoms, providing electromagnetic energy—light

APPLICATION: CONVERSION OF ENERGY IN ELECTRICAL POWER GENERATION

The conversion of one form of energy into another finds an application in the electrical power industry; it is necessary for economic reasons to smooth out the load on electrical generating plants. One way in which this is done is to build an upper and a lower water storage reservoir, separated by a reasonable vertical distance and connected with a hydroelectric generating plant (Fig. 7AB–1). The plant operates by converting the mechanical energy of falling water to electrical energy; this is done through fundamental laws of electricity and magnetism. During periods when there is a low requirement for electrical energy, the excess electrical energy generated by the plant can be used in reverse, driving motors that pump water from the lower to the upper reservoir. This work done on the water by the motors gives it potential energy. During periods of high demand for electrical energy, the water is allowed to flow down from the upper reservoir through the electrical generation apparatus to the lower reservoir. This transfers the water's potential energy back again to electrical energy. What is important about this process is that it is easily and entirely reversible on a very large scale, in contrast to the conversion of chemical energy to electrical energy in the burning of fossil fuel. Such a scheme has been implemented in several places, most notably in Luxembourg.

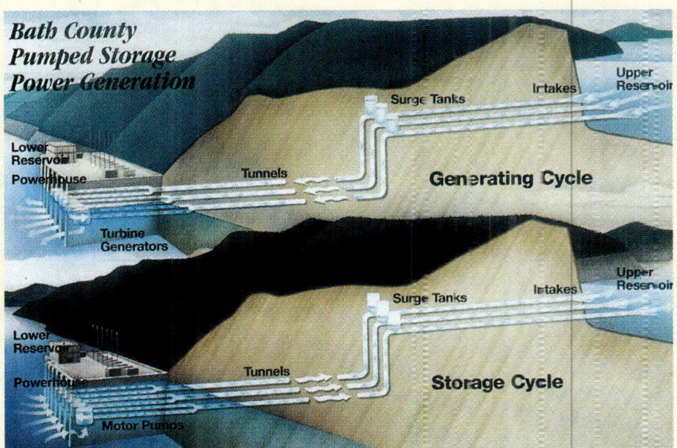

FIGURE 7AB–1 The Bath County pumped storage power generation system in Virginia illustrates how the energy associated with lifting water can later be re-used in the form of electric energy. Water is pumped into the upper reservoir during times of low demand for electricity and is released during times of high demand to generate hydroelectric power.

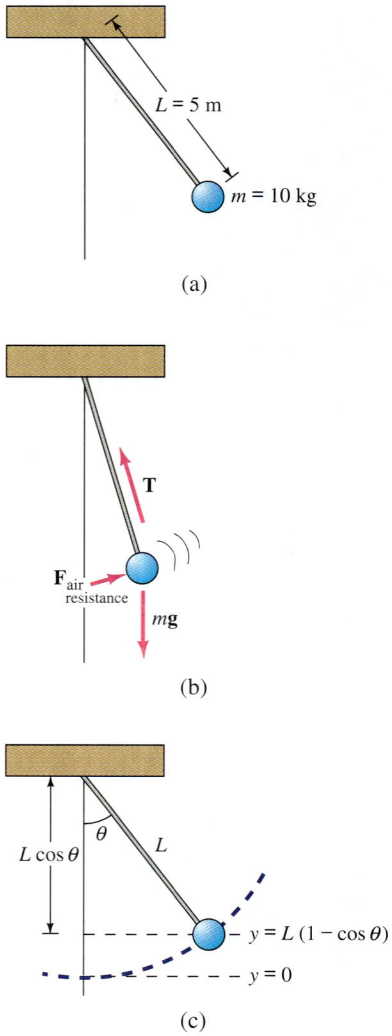

FIGURE 7–13 (a) Example 7–6. (b) Force diagram for the mass. (c) Geometry for the determination of the potential energy.

EXAMPLE 7–6 A ball of mass 10 kg is attached to a 5-m-long wire that swings freely from a support. In other words, we have a *simple pendulum* (Fig 7–13a). The ball is pulled aside so that the wire makes an angle of 31° to the vertical and it is released from rest. After 10 swings, the maximum angle that the ball reaches is 25° from the vertical. What is the work done by air resistance and any other nonconservative forces acting on the ball during these 10 swings?

Solution: As Fig. 7–13b shows, the forces acting are gravity, the tension of the wire, and the air resistance, which is nonconservative. Tension acts in a direction perpendicular to the motion, so it does no work and plays no role in any energy considerations. Gravity is conservative and so it has a potential energy. We will first express conservation of energy as if air resistance were not present. If we take the zero of potential energy at the minimum point of the swing, and measure the height y from that level, then $U(y) = mgy$. In terms of the angle θ of the wire, we see from Fig. 7–13c that

$$y = L(1 - \cos\theta),$$

where L is the wire's length. There is no kinetic energy at the top of the swing, so the total energy is

$$E = U(y_{\max}) = mgy_{\max} = mgL(1 - \cos\theta_{\max}).$$

The *loss* of energy over the ten swings is then

$$E_i - E_f = mgL(1 - \cos 31°) - mgL(1 - \cos 25°) = mgL(\cos 25° - \cos 31°)$$
$$= (10\text{ kg})(9.8\text{ m/s}^2)(5\text{ m})(\cos 25° - \cos 31°) = 24\text{ J}.$$

This is the work done by air resistance.

In this discussion and example, we used the energy loss to find the work done by nonconservative forces. But, of course, we can use this approach in a different way. If we can calculate the work done by friction—we can do this if we know its magnitude and direction and the distance over which it acts—then we can *predict* the energy loss.

Energy Conservation as One of the Underpinnings of Physics

In our discussion of energy conservation, we saw that principle as a powerful shortcut to information that could be obtained only laboriously from the equations of motion. It is important to understand that the conservation of energy is entirely equivalent to Newton's laws where ordinary mechanical systems are concerned. Indeed, energy conservation is now generally viewed as a more fundamental formulation of dynamics than is Newton's second law, even though they are equivalent for mechanical systems.

To demonstrate the equivalence, let's concentrate for simplicity on motion in one dimension. We write

$$E = \frac{1}{2}mv^2 + U(x). \tag{7–25}$$

E is a constant and both x and v are functions of time; they are the position and speed of the object in motion. Because energy is conserved, we have

$$\frac{dE}{dt} = 0. \tag{7–26}$$

We apply this to Eq. (7–25), using the chain rule when needed:

$$\frac{dE}{dt} = \frac{1}{2}m\frac{dv^2}{dt} + \frac{dU(x)}{dt} = \frac{1}{2}m\frac{dv^2}{dv}\frac{dv}{dt} + \frac{dU(x)}{dx}\frac{dx}{dt}$$
$$= \frac{1}{2}m(2v)\frac{dv}{dt} + \frac{dU(x)}{dx}v = v\left[m\frac{dv}{dt} + \frac{dU(x)}{dx}\right] = 0.$$

Unless $v = 0$ at all times, the contents of the square bracket must be zero; that is,

$$m\frac{dv}{dt} = -\frac{dU(x)}{dx} = F(x). \tag{7–27}$$

This is just our original formulation of Newton's second law.

It is interesting to note that although physicists and mathematicians studied Newton's equations almost from the time of their discovery, the principle of energy conservation—formulated by Julius Robert Mayer (1842), Hermann von Helmholtz (1847), and James Prescott Joule (1850)—grew out of experimental research on the properties of heat and the conversion of mechanical energy into thermal energy; this principle did not develop from Newtonian mechanics.

*7–5 BARRIER TUNNELING IN QUANTUM PHYSICS

Our energy-diagram approach has shown that if the total energy of a particle is less than the potential energy in some region of space, then the particle cannot cross through that region. We must modify this statement for atoms and molecules.

For extremely small systems (in the range of 10^{-9} m or less), the description of motion by Newton's second law is no longer adequate. A proper treatment of such systems is provided by *quantum mechanics*, which predicts certain phenomena that are impossible according to Newton's laws. One such phenomenon is **barrier tunneling**, or **barrier penetration**. Let's consider a particle whose potential energy is of the form given in Fig. 7–14. Newtonian mechanics states that a particle with energy E_1 in the right-hand trough can never appear in the left-hand trough because there is a potential energy barrier between the two regions (see Section 7–1). Inside the region between points D and D' in the figure—inside the barrier—a particle with energy E_1 must have negative kinetic energy because $E_1 < U(x)$ there. The mass of a particle is positive and therefore $mv^2/2$ must be positive; thus, the motion of a particle initially in the right-side well is forever restricted to the region CD. Not so, according to quantum theory. It is possible for a particle of atomic dimensions to **tunnel** through, or cross, the barrier and thus traverse the forbidden region.

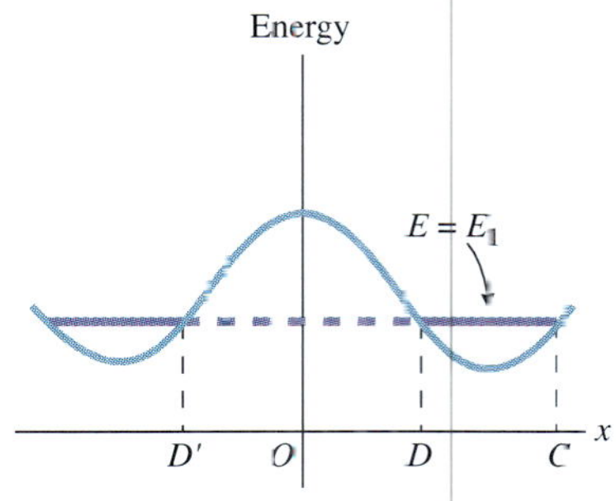

FIGURE 7–14 A potential energy barrier prevents a particle on the right with energy E_1 from entering the region on the left.

Quantum mechanics predicts quantitatively the probability of a particle crossing a barrier. This rate is proportional to a factor e^{-S}, where S depends on both the width of the barrier and the difference between the barrier height (on an energy diagram) and the energy of the tunneling particle. The greater the height and the greater the width, the larger the value of S and the less likely the particle is to cross the barrier.

The factor S also contains a controlling parameter for this process. This parameter is a fundamental physical constant determined by Max Planck that is, appropriately, known as **Planck's constant**, h, whose value is $h = 6.6 \times 10^{-34}$ J·s. This number is a small one; when it is used in the evaluation of S, it dictates that tunneling is a phenomenon that becomes important only in systems of atomic size. For example, the typical height of a barrier encountered in atomic systems is on the order of 10^{-19} J, and the typical width is 10^{-10} m. For an electron encountering a barrier of this type, S is typically on the order of unity (1). The factor e^{-S} is also on the order of unity, and the electron is likely to tunnel through the barrier. Compare this case with the case of a marble rolling with a speed of 10 cm/s and meeting a bump in the road that is 1 cm high and 1 mm wide. For this case, S takes a value on the order of 10^{28}. The factor e^{-S} is so small that we would never see the marble tunnel through the barrier. Thus quantum physics does not violate our everyday experience: Macroscopic particles do not tunnel through regions that are expected to be impassable.

Tunneling in Nuclear Fusion

Barrier penetration manifests itself in a number of phenomena at the atomic and nuclear levels. Nuclear fusion is one important example. Hydrogen comes in several forms:

There is the familiar hydrogen nucleus, ^{1}H, which is really a proton; there is also a heavy form of hydrogen, ^{2}H (deuterium), whose nucleus is called the deuteron, which is a proton and a neutron bound together by nuclear forces. There is another, still heavier, form of hydrogen, ^{3}H (tritium), whose nucleus is called the triton, which is one proton and two neutrons bound together by nuclear forces. All of these nuclei are electrically charged because the proton is electrically charged. If it were possible to bring these particles together, then a nuclear reaction could occur.

One example of such a reaction is a so-called fusion reaction, $^2\text{H} + {}^2\text{H} \rightarrow {}^3\text{H} + {}^1\text{H} + (6.4 \times 10^{-13}\ \text{J})$. The energy on the right-hand side is energy released during the reaction—energy that can be converted into electric energy, for example.

APPLICATION

Nuclear fusion

The reaction is held back because the two nuclei ^{2}H each carry a positive charge and, as we shall learn in Chapter 22, two positive charges repel each other. This means that when one ^{2}H nucleus approaches the other, there is a potential barrier that should totally prohibit contact between these nuclei. Quantum physics allows barrier penetration but, under ordinary circumstances, $U_{av} - E$ is so large (remember, the classical barrier is present when $U - E$ is positive) that the reaction cannot happen. The only way to improve the rate or, equivalently, to reduce the number S, is to make $U_{av} - E$ smaller. When the ^{2}H nuclei are raised to high temperatures (billions of degrees), the nuclei become agitated, move rapidly, have an increased energy E, and thus they have a smaller value of $U_{av} - E$. Effectively, by raising the energy of the nuclei, they are brought closer to the top of the barrier and both $U_{av} - E$ *and* the thickness of the barrier are reduced. Much research and development has been done since the early 1950s to bring the ^{2}H nuclei to these high temperatures. The technical difficulties are enormous but the reward for success would be great. The fusion reaction could yield a virtually unlimited amount of energy because seawater contains ^{2}H nuclei in abundance.

APPLICATION

Electronic devices based on tunneling

Other Applications

Several other manifestations of tunneling have technological application. The scanning tunneling microscope is a device that detects electrons that tunnel through a potential barrier at the surface of a piece of metal. It is used to study variations of atomic size in the surface of the metal in great detail (Fig. 7–15). Tunneling has also been used in studies of superconductivity. One of the most important applications is in the tunnel diode, a device with wide application in circuits. Quantum tunneling is not just some esoteric effect of interest to a small class of physicists. Many modern electronic devices rely in one way or another on electron tunneling. In later chapters, we will return to these applications.

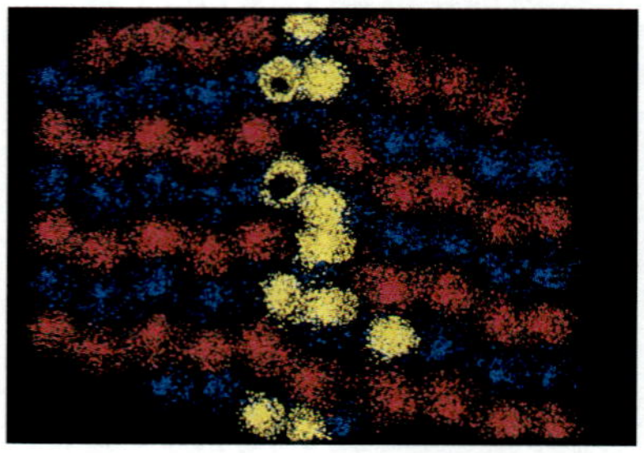

FIGURE 7–15 Individual atoms in a solar cell, shown here in false color, are made visible by using a scanning tunneling microscope. This image is part of a project to improve solar cells.

SUMMARY

Conservative forces can be specified in terms of a single scalar function called the potential energy. In one-dimensional motion, the potential energy $U(x)$ depends on one coordinate only and is defined by

$$\text{for conservative forces:} \quad U(x) \equiv U(x_0) - \int_{x_0}^{x} F\, dx', \tag{7–6}$$

where $U(x_0)$ ($= U_0$) is an arbitrary constant and x_0 is an arbitrary point. In turn, the force is given in terms of U by

$$F(x) = -\frac{dU(x)}{dx}. \tag{7–4}$$

Two important examples of potential energy are gravity and the spring. For gravity,

$$U(y) = mgy + U_0, \tag{7–11}$$

where y is the height above ground level. For the spring,

$$U(x) = \frac{1}{2}kx^2. \tag{7–17}$$

where x is the distance from the equilibrium point of the spring.

For central forces (forces aligned in the radial direction), the potential energy depends only on r, and

$$\mathbf{F} = -\frac{dU(r)}{dr}\hat{\mathbf{r}}. \tag{7–21}$$

For conservative systems, the total energy—the sum of the kinetic and potential energies—does not change during the motion; thus, the total energy is a constant whose value may be determined at any time during the motion:

$$E = \frac{1}{2}mv^2 + U(x, y, z). \tag{7–19}$$

The fact that E is constant means that any change in the kinetic energy is compensated by an equal but opposite change in the potential energy during the motion. *The total energy is conserved.* The utility of this conservation law is enormous: We can determine the speed when the position of an object is known. Energy diagrams provide us with a systematic way to approach problems that involve energy conservation. With them, we can understand such useful concepts as turning points, potential energy barriers, and stable and unstable equilibria.

Although dissipative forces are always present in the world as we see it, the law of conservation of energy holds for all the basic laws of physics. Its general validity allows us to find the energy dissipated through friction or drag forces, often without being able to calculate this energy loss directly.

Energy conservation is also a characteristic of atomic systems that must be described by quantum mechanics. There, under certain very specific circumstances, particles can act as if they have negative kinetic energy and can thereby tunnel through potential barriers. While this phenomenon may violate our intuition, it nevertheless occurs, and it conserves energy.

UNDERSTANDING KEY CONCEPTS

1. Are drag forces such as air resistance conservative?

2. Discuss why it is possible for the total mechanical energy to be negative, even if the kinetic energy cannot be negative.

3. If we add a constant term to the potential energy of an object, why doesn't this change the object's motion?

4. Explain why a rubber ball seems never to bounce back to its original height when dropped from rest. What happens to the energy? Explain why the ball can bounce back to a much greater height when it is thrown down.

5. Is energy conserved when you ingest sugar and then go out and exercise? In what ways, and in which of the various processes for converting food into muscular activity, might the energy conservation principle come into play?

6. Does a real spring exert a truly conservative force? What experiment might you perform to check the answer to this question?

7. How, if at all, does experiment demonstrate that the gravitational force is conservative?

8. A golf ball is dropped with zero initial velocity from a height of 20 m. It bounces on a concrete pad and rises to a maximum height of 19 m. Are any of the forces that the ball experiences during the motion nonconservative?

9. Consider motion with a potential energy of the shape given in Fig. 7–7 (see page 180). Let $U(x)$ describe the height of a smooth slide above the ground. In the absence of friction the motion is easily predicted to repeat with a uniform repetition time. Discuss the motion, starting from the right or the left side, when a small amount of friction is present.

10. When a force acts in two dimensions, are there two separate laws for the conservation of energy that correspond to motion in each direction?

11. The potential energy is zero at a given point. Is the force necessarily zero or nonzero at that point?

12. An object moves on a rough inclined plane. Can you still use the concept of energy conservation to relate the height and the speed? How?

13. The motion of the pendulum of a grandfather clock is slowed by air resistance and bearing friction, yet such clocks can run without stopping for years. What types of mechanisms are typically used to supply the necessary energy?

14. You are watching fireworks. You see a glowing point streak into the air. This is followed by a loud noise and, about a second later, by the appearance of many scattered dots of light, which sink in graceful curves before disappearing. Account for all of the energy supplied and spent at each stage of the spectacle.

15. A steel marble is bounced off the top step of a staircase. It rises to 1 m above the step, then falls onto the second step, bounces off, falls onto the third step, and so on for 39 steps. Assuming that there is no air resistance or other source of energy loss, how high will the marble rise after the last bounce? (This is a question, so do not do 39 calculations!)

PROBLEMS

7–1 Energy Conservation

1. (I) An unknown constant force F pushes a 10-kg body from rest on the ground vertically upward. At a height of 2 m, the velocity of the object is $\mathbf{v} = 2.4\mathbf{j}$ m/s. (a) Find the change in the potential energy associated with gravity. (b) What is the net work done, and what is the work done by the unknown force?

2. (I) Consider the expression $U = mgy$ for an object of mass m under the influence of gravity, where y is measured from the ground up. Express U in terms of z, where z is measured in a downward direction from a rooftop 30 m above the ground.

3. (I) A baseball pitcher throws a ball at 95 mi/h off a roof that is 80 m above the ground. How high will the ball be when it is traveling at 120 mi/h?

4. (I) A rock is thrown straight down into a deserted quarry from the edge, which is 18 m above the bottom. The rock has a speed of 21 m/s when it reaches the bottom. What was the rock's initial speed?

5. (I) The spring of a toy gun launches rubber-tipped projectiles with a spring constant of 5 N/m. The spring is compressed by 7 cm with the projectile in place. How much kinetic energy is imparted to the projectile?

6. (II) The potential energy of an archery bow is measured to be $U(x) = bx^2 + cx^3$, where x is the distance the bow string is pulled back from its equilibrium position. When an archer pulls the string a distance x, what force does the archer exert on the string and what force does the string exert on the archer?

7. (II) A package of mass 5 kg is subject to a constant force of 8 N pointing in the $+x$-direction. (a) Calculate the potential energy of the package as a function of its position x, defining it such that $U(x)$ at $x = 0$ m is zero. (b) Assuming that the package has a velocity of 2 m/s at $x = -1$ m, calculate the total mechanical energy of the package. (c) What is the speed of the package at $x = 3$ m?

8. (II) A spring has a spring constant of 42.38 N/m and obeys Hooke's law. How far must the spring be pulled back if its potential energy is to be 1.565 J? What is the mass of a ball at the end of the spring if the maximum speed of the ball is observed to be 2.402 m/s when the spring is released?

9. (II) The energy of a harmonic oscillator (a mass moving on the end of a spring) is given by $E = \frac{1}{2}mv^2 + \frac{1}{2}kx^2$. Plot contours of constant E on a graph in which x is measured along one axis and v is measured along the perpendicular axis. Choose the parameters $E = 16.0$ J, $m = 2.0$ kg, and $k = 8.0$ J/m^2. Such a plot is called a *phase plot*; the motion of a system is restricted to the curve corresponding to the energy E.

10. (II) Sketch the potential energy for a 5-kg mass that can move between the ground and a height of 10 m. (a) Assume that zero potential energy is at the ground, (b) at a height of 10 m, (c) at 4 m.

11. (II) A block of mass 0.528 kg slides with uniform velocity of 3.85 m/s on a horizontal frictionless surface. At some point, it strikes a horizontal spring in equilibrium. If the spring constant is $k = 26.7$ N/m, by how much will the spring be compressed by the time the block comes to rest? What is the amount of compression if the surface is rough under the spring, with coefficient of kinetic friction $\mu_k = 0.411$?

12. (II) An archery bow acts much like a spring displaced from equilibrium when the bow is drawn. Suppose that an archer displaces the string from equilibrium by 60 cm and exerts a force of 150 N. (a) What is the "spring constant"? (b) What is the speed of an arrow of mass 25 g that leaves the bow as the string reaches the equilibrium position? [*Hint*: Use the expression for energy for a mass moving under the influence of a spring.]

13. (II) A cannonball of mass 15 kg is dropped from rest from a height of 6.0 m. It falls onto a large vertically oriented spring, which is compressed from its relaxed position when the cannonball lands on it. The spring has spring constant $k = 10^4$ N/m. What is the maximum compression of the spring? How much would the spring be compressed if a man of mass 60 kg jumped onto it from a height of 1.5 m?

14. (II) A mass $m = 0.3$ kg slides along the x-axis of a horizontal frictionless surface with speed $v_x = 1.7$ m/s. It runs into a relaxed spring oriented along the x-axis. This spring had previously been observed to stretch by 0.8 cm when it was oriented vertically with the mass suspended from it. (a) What is the maximum compression of the spring when the mass runs into it? (b) The mass rebounds as a result of having compressed the spring. What is its velocity when it leaves contact with the spring?

15. (II) A spring with spring constant $k = 200$ N/m is used as a launcher for a small block whose mass is 10 g. The block is placed against the compressed spring in a horizontal arrangement on a smooth horizontal surface. The spring, with the block, is compressed 5 cm and then released. (a) Find the speed of the block just as it leaves the spring. (b) The block encounters a rough surface as it leaves the spring. How much work does friction do in bringing the block to an eventual stop? (c) The block slides a distance of 3.5 m before stopping. What is the coefficient of kinetic friction between block and surface?

7–2 Energy Conservation and Allowed Motion

16. (II) For a conservative one-dimensional force, show that the sign of the slope of the potential energy function at a position x determines the direction, positive or negative, in which the force acts.

17. (II) When two atoms on a line are far apart, there is no force between them. As they start to move closer, there is an attraction between them, which, at very close distances, turns into a strongly repulsive force. Sketch the potential energy as a function of the distance between the atoms.

18. (II) Figure 7–16 shows the force $F(x)$ that acts on a particle moving along the x-axis. (a) Plot the potential energy of the particle as a function of x. (b) The particle starts its motion at $x = -2$ m, with zero initial velocity. How far to the right will the particle travel?

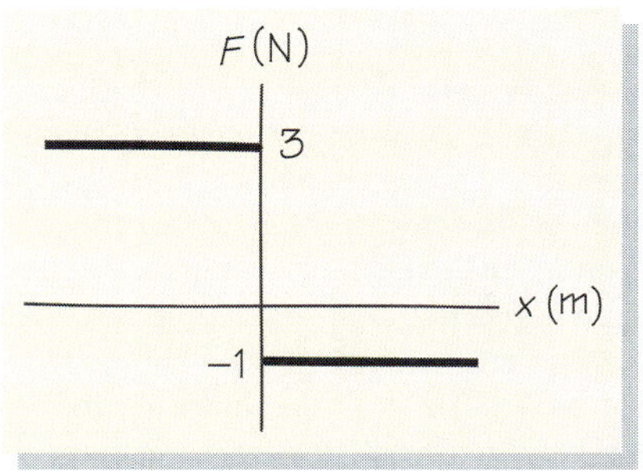

FIGURE 7–16 Problem 18.

19. (II) Consider a force $\mathbf{F}(x)$, acting along the x-axis, that is opposite to the spring force (Hooke's law). In other words, the force has the single vector component $F(x) = +kx$. A mass m is under the influence of this force and of no other forces. Suppose that the mass is placed at rest just to the right of the origin. (a) Which way will the mass move, if at all? (b) Find the speed of the mass as a function of its distance from the origin. (c) Repeat the problem but assume that the mass had been placed just to the left of the origin.

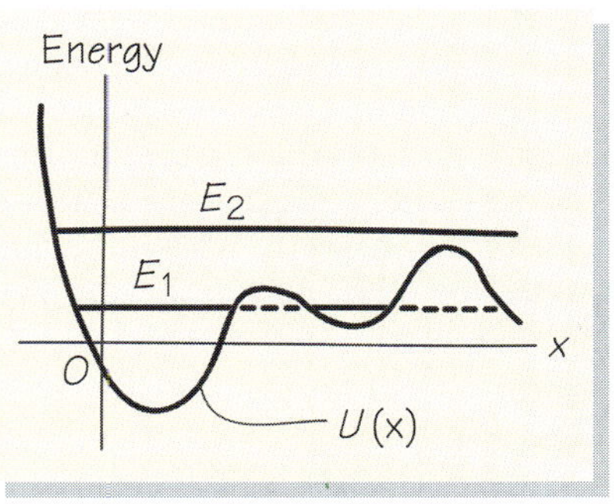

FIGURE 7–17 Problem 20.

20. (II) Consider the energy diagram in Fig. 7–17. (a) What are the limits of motion for energies E_1 and E_2? Redraw the figure and label it as necessary. (b) Describe the circumstances under which the particle is always at rest. (c) Find the energies and positions for which motion within turning points is possible. (d) Find the equilibrium positions on your drawing. Are they stable or unstable?

21. (II) Draw a one-dimensional potential energy $[U(x)]$ diagram with the following characteristics: (a) The particle can never reach negative x. (b) There are three regions in x where the particle can move within turning points. (c) The particle can never reach infinity. (d) The particle has unstable equilibrium positions at 1 nm and 2 nm.

22. (II) Consider the potential energy $U(x)$ shown in Fig. 7–18. (a) What is the sign of the force at the positions 1 through 6? (b) Which positions have the most positive, most negative, and zero force? (c) Find the equilibrium positions and indicate whether they are stable or unstable.

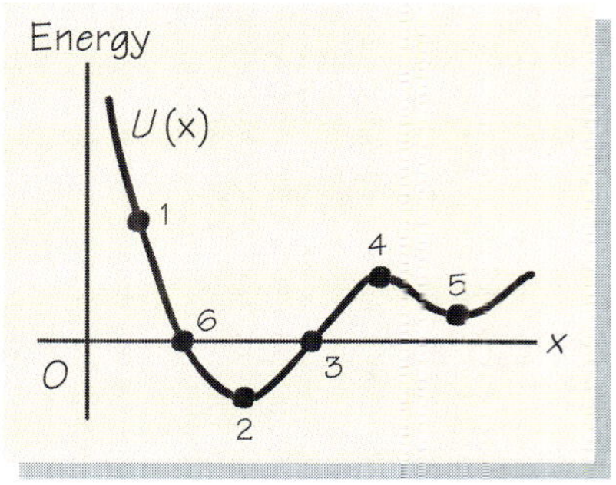

FIGURE 7–18 Problem 22.

23. (II) The potential energy of an object constrained to move in the x-direction is given by $U(x) = \alpha x^4 - \beta x^2$, where $\alpha = 26$ J/m^4 and $\beta = -3.0$ J/m^2. Find the equilibrium points and state whether they are stable or unstable.

24. (II) A particle is moving in a potential well described by the potential energy

$$U(x) = -\frac{3\ \mathrm{J\cdot m^2}}{1\ \mathrm{m}^2 + (x + 2\ \mathrm{m})^2} - \frac{3\ \mathrm{J\cdot m^2}}{1\ \mathrm{m}^2 + (x - 2\ \mathrm{m})^2},$$

where x is measured in meters. (a) Sketch the shape of the potential energy for $-4\ \mathrm{m} < x < 4\ \mathrm{m}$. (b) The speed of a particle of mass 0.20 kg at $x = -2$ m is 4.2 m/s. Can the particle reach $x = 3.0$ m?

7–3 Motion in Two or Three Dimensions

25. (I) A projectile fired from a gun leaves the barrel at a speed of 500 m/s. The gun is placed 180 m above a level plain. Use energy conservation to calculate the speed of the projectile when it is 16 m above the plain. Neglect all drag effects.

26. (I) What is the gravitational potential energy of the Earth–Moon system? See the appendices for the data you need.

27. (I) A cannonball is fired horizontally with an initial speed of 125 m/s from the top of a cliff that is 68 m above the sea. What is the speed of the cannonball when it hits the water? How is that changed if the cannon is inclined at a 32° angle with the horizontal without any change in initial speed? Ignore all effects of air resistance.

28. (I) A particle has potential energy that depends only on the distance r from some central point. This potential energy has the form $U(r) = U_0 - k/r^2$. What is the corresponding force law?

29. (II) A ball is thrown with initial speed v_0 in a trajectory that makes an initial angle θ with the ground. Air resistance is small. (a) By using the principle of conservation of energy, show that the speed of the ball when it reaches the height h above the ground is $v_h = \sqrt{v_0^2 - 2\,gh}$, independent of the angle of the throw. (b) Use the result of part (a) to find the initial speed v_0 that is required if a vertically thrown ball is to reach height h just before turning back. (c) Use the results of part (a) to find the initial speed v_0 required so that a ball thrown at 45° reaches a maximum height h.

30. (II) An object of mass m is subject to two forces: One force acts only in the x-direction and is due to a spring of spring constant k; the other force acts only in the y-direction and is due to a spring with the same spring constant k. (a) What is the potential energy? (b) Use your result to show that the net force is proportional to the distance from the equilibrium point and points toward that point. (c) Using your result from part (b), show that one possible motion is uniform circular motion in the xy-plane.

31. (II) A ball of mass m is thrown with a speed v, at an angle θ from the horizontal, off the top of a tall building at height h. (a) Show that, in general, it is not possible to tell whether the angle θ is above or below the horizontal when a ball hits the ground just by a measurement of the speed of the ball when it hits the ground. (b) By knowing the horizontal distance that the ball travels, show that the angle θ can be determined.

32. (II) A skier slides down a hill, starting with zero velocity at a height of 95 m above the bottom of the hill. The shape of the terrain is shown in Fig. 7–19. What is the velocity of the skier on top of the second, intermediate hill, whose height is 32 m? What is the skier's velocity at the bottom of the hill? Neglect all frictional effects. Is this neglect reasonable?

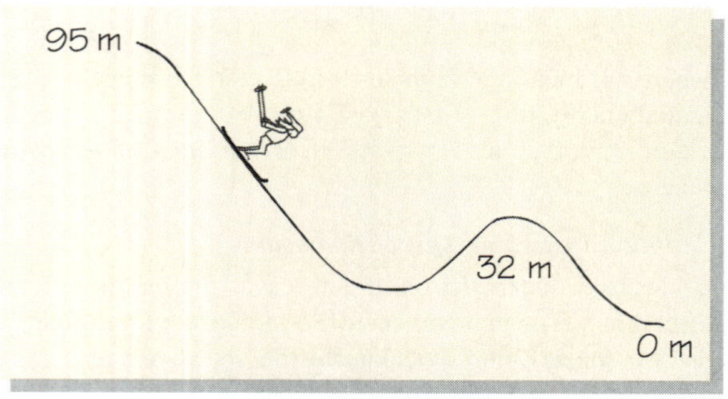

FIGURE 7–19 Problem 32.

33. (II) Suppose that the skier of Problem 32 reaches the bottom of the hill at a speed of 23 m/s. Assuming that the skier, including equipment, has a mass of 75 kg, how much work is done by the resistive forces of friction and drag?

34. (II) A ski jumper starts from rest and follows, with its several ups and downs, the rather bumpy ski jump shown in Fig. 7–20. The jump track starts at a height $h = 20$ m above the eventual landing point, and the jumper leaves the track while he moves horizontally at a height above the landing point that is exactly one-half h. Assuming that the effects of air resistance are negligible (in real ski jumping they are not), what is the horizontal distance of the edge of the track from the landing point?

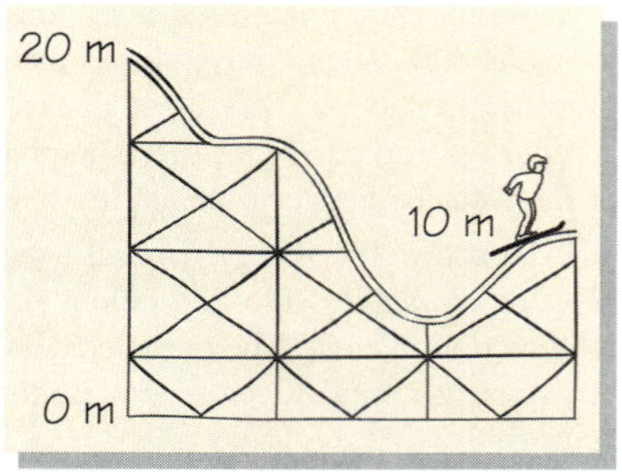

FIGURE 7–20 Problem 34.

35. (II) A particle of mass $m = 30$ g slides inside a bowl whose cross section has circular arcs at each side and a flat horizontal central portion between points a and b of length 30 cm (Fig. 7–21). The curved sides of the bowl are frictionless, and for the flat bottom the coefficient of kinetic friction $\mu_k = 0.21$. The particle is released from rest at the rim, which is 15 cm above the flat part of the bowl. (a) What is the speed of the particle at a? (b) What is the speed of the particle at b? (c) Where does the particle finally come to rest?

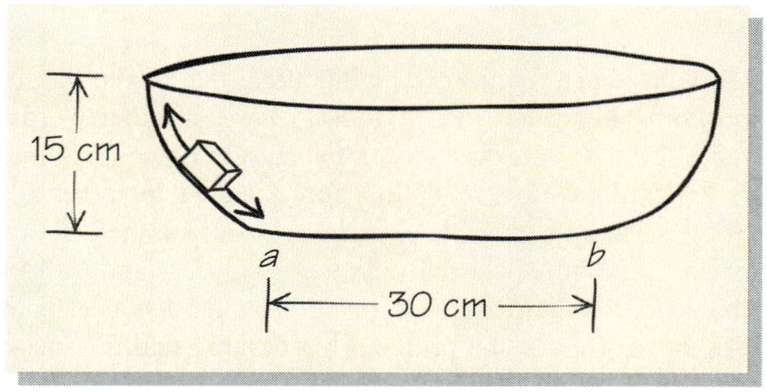

FIGURE 7–21 Problem 35.

36. (II) A toy car of mass M slides down a frictionless track that makes a circular loop of radius R at the bottom (Fig. 7–22). Suppose that the car starts from rest at a height H, with $H > 2R$. (a) What is the car's speed at the bottom of the circle? (b) at the top of the circle? (c) What is the force exerted by the track at the top of the circle? (d) What is the minimum

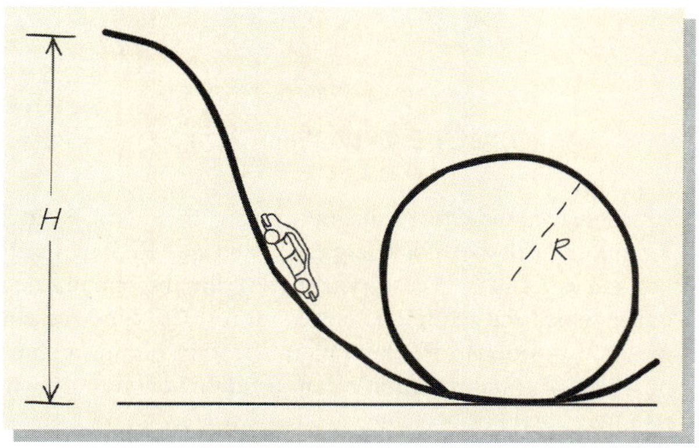

FIGURE 7–22 Problem 36.

value of H such that the car goes around the loop without falling off under the force of gravity?

37. (II) A block is constrained to move without friction along the x-axis. The block is attached to a spring of equilibrium length L. The other end of the spring is fixed at a point $x = 0, y = h$ (Fig. 7–23), where $h < L$. (a) What is the potential energy of the system? (b) What is the net force acting on the block? Sketch both $U(x)$ and $\mathbf{F}(x)$.

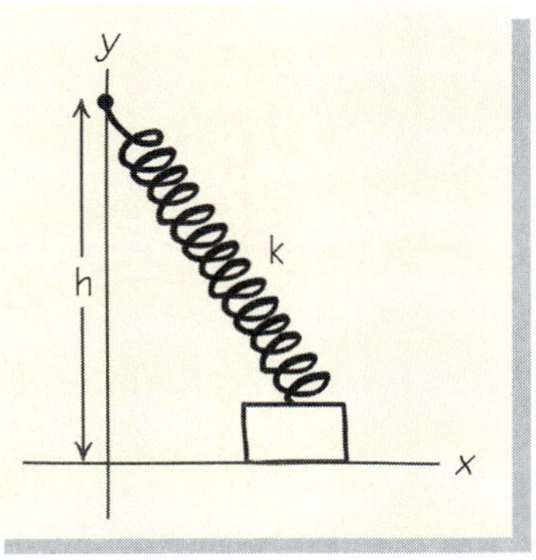

FIGURE 7–23 Problem 37.

38. (II) The potential energy of two atoms separated by a distance r may be written in the form $U(r) = U_0[(r_0/r)^{12} - 2(r_0/r)^6]$ (Fig. 7–24). Find the separation r at which there is no force between the atoms. What is the magnitude of the potential energy there?

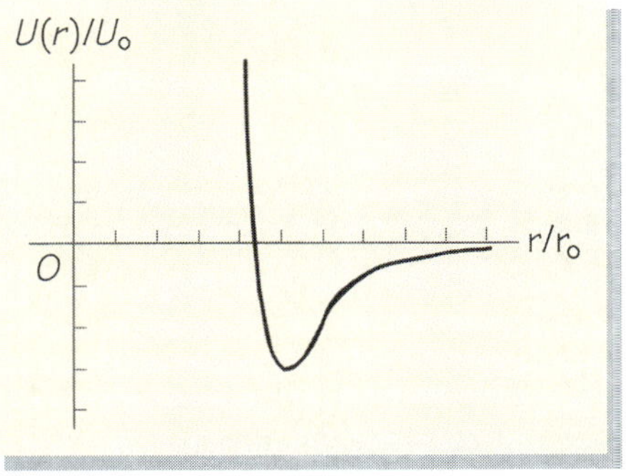

FIGURE 7–24 Problem 38.

39. (II) The potential energy of a satellite of mass m, moving in a circular orbit of radius r about a planet of mass M, is given by $U(r) = -GMm/r$, where G is a universal constant. (a) Calculate the force on the satellite. (b) From the equation $\mathbf{F} = m\mathbf{a}$, calculate the kinetic energy for a particle in this orbit. (c) What is the total energy?

40. (II) The potential energy of a particle moving in the xy-plane is given by $U(x, y) = a_1x^2 + a_2xy + a_3y^2$, where $a_1 = 5\ \text{J/m}^2$, $a_2 = -8\ \text{J/m}^2$, and $a_3 = 3\ \text{J/m}^2$. Calculate the force vector.

41. (III) The ski jumper of Problem 34 has survived the jump and prepares for another jump on a track of a different design: The track takes the jumper through a vertical distance H from its start to its take-off point, but the jumper can choose the angle at which he leaves the take-off point. The lip at the edge can be varied so that the jumper leaves the jump at any angle—from taking off horizontally to leaving vertically upward. Show that the angle θ with the horizontal that leads to the maximum-distance jump onto a horizontal plane a distance D below the lip is given by $\sin^2 \theta = 1/[2 + (D/H)]$. Ignore all friction and drag effects. (A real landing area is, in fact, sloped downward to allow the jumper to make a smooth landing.)

7–4 Is Energy Conservation a General Principle?

42. (I) A 10 g ping-pong ball is dropped with zero initial velocity from a height of 1.0 m, and it bounces back to a height of 0.90 m. What is the work done by the nonconservative forces in this process?

43. (I) A parachutist jumps off a training tower that is 85 m high. She starts at rest and reaches the ground with a vertical speed of 5.0 m/s. How much work was done by the drag forces acting on her, given that her mass is 75 kg?

44. (II) A track consists of a descending ramp, a straight track, and an ascending ramp. The smooth ramps both make an angle of 30° with the horizontal. The coefficient of kinetic friction on the horizontal surface is $\mu_k = 0.20$. An object starts from rest at a vertical height of 1.85 m on the descending ramp. It slides down the ramp, across the horizontal stretch, and up the ascending ramp. It reaches a vertical height of 0.75 m before coming to rest. (a) How long is the horizontal part of the track? (b) The object starts sliding back from the 1.22 m height. How far along the horizontal stretch does it slide?

**7–5 Barrier Tunneling in Quantum Physics*

45. (I) Suppose that the fraction of particles that hit a potential energy barrier and tunnel through it is given by e^{-S}. What is the value of S if one particle in a million tunnels through?

46. (I) Suppose that, in Problem 45, S has the value 20. How many particles would have to strike the barrier for 10,000 of them to tunnel through?

47. (II) An interpretation of the factor S, described in Section 7–5, is that if a single particle can hit a potential energy barrier over and over again, then the fraction of times it will tunnel through is e^{-S}. Suppose that a particle is trapped between two barriers, 10^{-14} m apart, for which $S = 72$. The particle moves at a speed of 10^6 m/s back and forth between the barriers. How long, on average, will it take the particle to get out of the confined space between the barriers? This is similar to what happens to an alpha particle in a radioactive nucleus.

General Problems

48. (I) Assume that 1 kWh of electric energy costs 12 cents. Estimate the cost of lighting a three-room apartment per day.

49. (I) A furniture mover pushes a crate of mass 60 kg up a rough slope through a vertical distance of 1.0 m at a uniform speed. What is the change in the potential energy of the crate?

50. (I) A diver jumps off a rigid diving platform 6.0 m above the water with an initial upward velocity of 3.2 m/s. Assuming that his takeoff is very nearly in a vertical direction and that there are no drag forces on the diver, with what velocity will the diver hit the water? (Treat the diver as a point particle.)

51. (I) The force on an object of mass m moving along the x-axis is given by $F(x) = -ax + bx^2$, where $a = 3$ N/m and $b = 0.2$ N/m^2. (a) Calculate the potential energy function $U(x)$, letting $U(x) = 3$ J at $x = 0$ cm. (b) Sketch $U(x)$ as a function of x from $x = 0$ m to $x = 4.0$ m in steps of 0.5 m.

52. (II) A massless spring hangs vertically in equilibrium with no mass at its end. When a 2.0-kg mass is connected to the bottom, the new equilibrium position is 5.0 cm lower. The mass is then pulled down and released. It is observed that the speed of the mass is 2.1 m/s when the mass passes the original equilibrium position (before the mass was attached). How far down were the mass and spring pulled together when released?

53. (II) Are the following forces conservative or nonconservative? (a) The force $F(x) = ax + bx^3 + cx^4$. (b) The force $\mathbf{F}(x, y) = Ax^2\,\mathbf{i} + Bxy\,\mathbf{j}$.

54. (II) A particular spring exerts a force, as a function of displacement from equilibrium, given by $F = -(0.2\text{ N/m})x - (0.4\text{ N/m}^3)x^3$ (Fig. 7–25). This force acts on an object of mass 2 kg and displaces it from $x = 0$ m to $x = 3$ m. Consider the object being displaced along two different paths: (a) directly from $x = 0$ m to $x = 3$ m, and (b) from $x = 0$ m to $x = -2$ m, then to $x = 7$ m, and back to $x = 3$ m. Show that the work done by the force is the same for both paths.

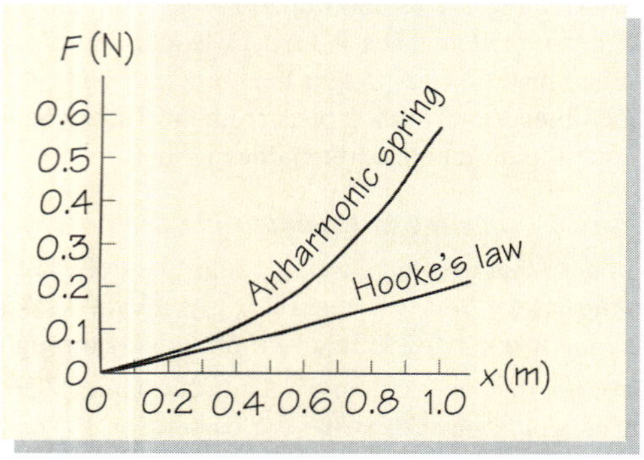

FIGURE 7–25 Problem 54.

55. (II) A conservative force does 2 J of work in moving a particle from point A to point C via path ABC (Fig. 7–26). The force does -1 J of work to move the particle from D to F, 3 J for E to B, 1 J for E to F, and 1 J for B to C. How much work does the force do as the object moves from C to A, from A to E, from D to C?

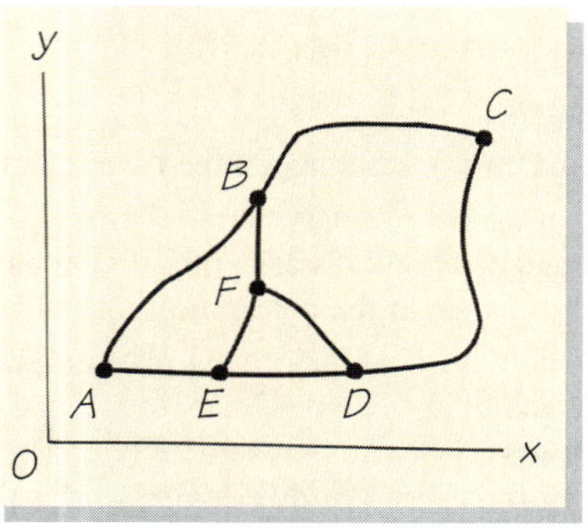

FIGURE 7–26 Problem 55.

56. (II) A particle of mass 50 g leaves, from rest, point a on a loop-the-loop. The heights of the points a, b, c, and d as measured from the table level are 10 cm, 0 cm, 8 cm, and 12 cm, respectively (Fig. 7–27). Ignore friction. (a) What are the speeds of the particle at points b, c, and d? (b) How high up on the other side does the particle rise?

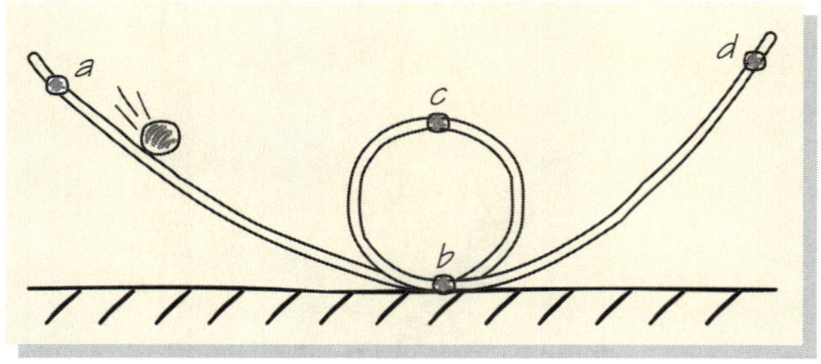

FIGURE 7–27 Problem 56.

57. (II) A skier skis from rest from a vertical height $h_1 = 18$ m over two successively lower hills of vertical heights $h_2 = 15$ m and $h_3 = 7$ m (Fig. 7–28). The summit of the third hill fits a circle of radius h_3 centered at height 0 m. Friction with the snow and air resistance are negligible. (a) Find her speeds at x_1, x_2, and x_3. (b) Does the skier leave the surface at x_3? If not, what should h_1 be so that she just leaves the surface at x_3?

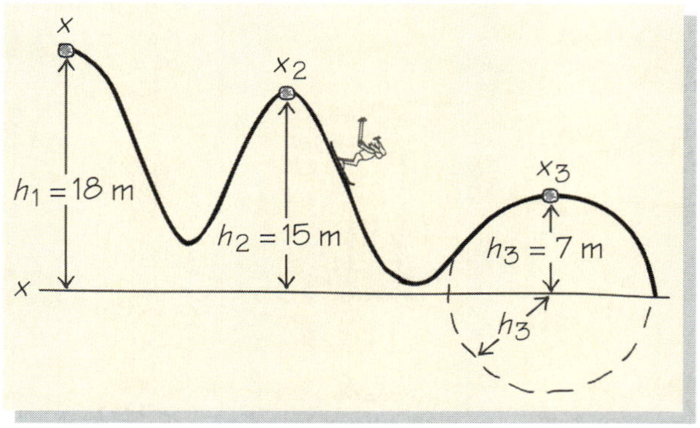

FIGURE 7–28 Problem 57.

58. (II) A pile driver works by lifting a large mass and dropping it to the ground. The mass is 2500 kg, and it is raised to a height of 20 m above the pile for each stroke. The pile driver encounters resistance of a constant force of 8×10^6 N on each stroke. Use a combination of energy conservation and the work–energy theorem to determine how many strokes it takes to drive a pile 2 m into the ground. For which forces involved is it possible to use energy conservation, and for which forces is this not possible?

59. (II) The mass of a simple pendulum of length $L = 1$ m is released with the string originally in a horizontal position. (a) Calculate the speed of the mass at its lowest position. (b) What is the speed when the string makes an angle of 45° with the vertical? (c) Determine the tension in the string in both positions if the mass is 0.2 kg.

60. (II) A simple pendulum of length $L = 1.0$ m, mass $= 0.20$ kg is released from the horizontal position. When the mass is at its lowest point, the string hits a nail a distance h above the mass, so that the mass loops around the nail (Fig. 7–29). How large can h be so that the string of the pendulum remains taut even when the mass loops to a point right above the nail? [*Hint*: The string is taut as long as there is tension in the string].

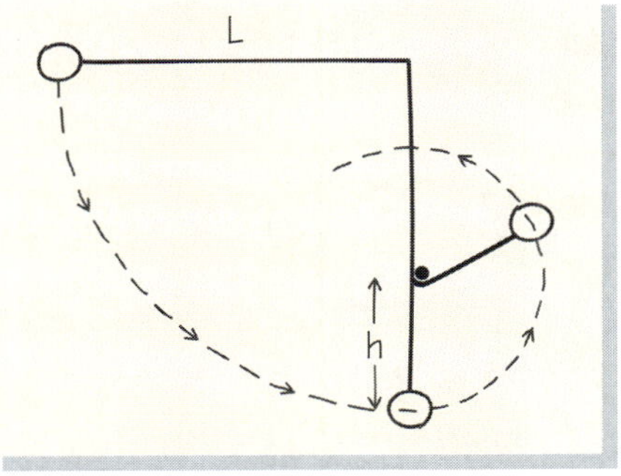

FIGURE 7–29 Problem 60.

61. (II) An electron is attracted to a proton (the latter is much heavier than the former) with a central force whose magnitude is given by Coulomb's law, $F = C/r^2$, where $C = 2.3 \times 10^{-28}$ $\text{kg} \cdot \text{m}^3/\text{s}^2$. (a) Is this a conservative force? (b) Write an expression for the potential energy associated with this system. (c) An electron is very far away from a proton and starts from rest. It falls straight toward the proton under the influence of the force. What is the speed of the electron when it is 1.2×10^{-12} m from the proton?

62. (II) A batter hits a baseball and the baseball leaves the bat making a 38° angle with the ground. Air resistance has negligible effect on the trajectory of the baseball, which travels a total horizontal distance of 115 m. (a) What is the speed of the baseball just after it leaves the bat? (b) Use the conservation of the total energy of the baseball to calculate the maximum height to which the baseball rises. (c) What is the speed of the baseball when it has first risen to half its maximum height? (d) when it falls back to half its maximum height?

63. (II) A 3.0-kg block is held against a spring with spring constant $k = 25$ N/cm, compressing the spring 3 cm from its relaxed position. When the block is released, the spring expands and pushes the block upward along a rough surface inclined at a 20° angle (Fig. 7–30). The coefficient of kinetic friction between the block and the surface is $\mu_k = 0.1$. (a) What is the work done on the block by the spring as it extends from its compressed position to its equilibrium position? (b) by friction while the block moves 3 cm as in part (a)? (c) by gravity during the same motion? (d) What is the speed of the block when the spring reaches its equilibrium position? (e) If the block is not attached to the spring, how far up the incline will it slide before it comes to rest? (f) Suppose that the block is attached to the spring so that the spring is extended when the block slides past the equilibrium point. By how much will the spring be extended before the block comes to rest?

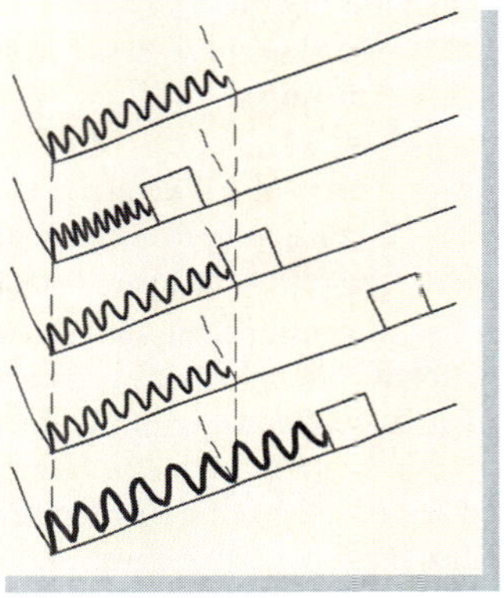

FIGURE 7–30 Problem 53.

64. (II) A block of wood of mass 3.0 kg is placed on a horizontal table and attached by a massless rope of length 6.0 cm to a vertical axis that passes through the table. The rope is initially swung around the axis, causing the wood to have a tangential velocity of 5.0 m/s. After one revolution the wood is observed to have a speed of only 3.0 m/s. (a) How much work has friction done on the block of wood during the first revolution? (b) What is the coefficient of kinetic friction between the block of wood and the table? (c) What is the potential energy of the block of wood at the beginning, after one revolution, and when the block of wood comes to rest? (d) How many revolutions does the block of wood make before stopping?

65. (II) Two blocks of mass $m_1 = 5.0$ kg and $m_2 = 2.0$ kg are supported by the system of light frictionless pulleys and massless strings shown in Fig. 7–31. Mass m_1 is at rest at a height $h = 0.8$ m above the ground when the system is released. Use the conservation of energy to determine the speed of m when it hits the ground.

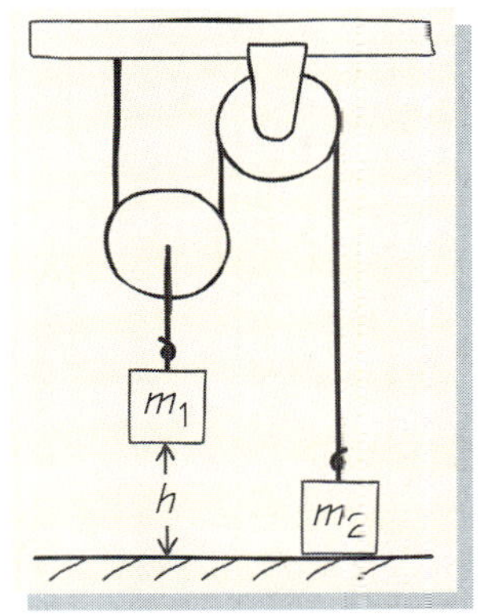

FIGURE 7–31 Problem 55.

66. (II) A representation of the nuclear force between two nucleons (neutron or proton) is given by the *Yukawa potential energy function*, $U(r) = -Ae^{-kr}/r$, where $1/k$ has the approximate value of 10^{-15} m and A is a constant. (a) Plot $U(r)$ versus r in steps of 0.2×10^{-15} m up to 2.4×10^{-15} m. Plot U in units of $A \times 10^{15}$. (b) At what distance is the potential energy a minimum? (c) Determine the force $F(r)$. (d) Determine the force at $r = 0.1 \times 10^{-15}$ m and 10×10^{-15} m.

67. (II) Given the relationship between the force and the potential energy, show that Newton's third law is satisfied if the potential energy has the form $U(x_1 - x_2)$ for two particles located at x_1 and x_2, respectively.

68. (III) A chain with a mass of 3.0 kg and a length of 3.0 m lies on a table, with 1.2 m hanging over the edge. How much energy is required to get all of the chain back on the table?

69. (III) Consider two-dimensional motion. You are given the components of the force **F** in the x- and y-directions, respectively, as $F_x(x, y)$ and $F_y(x, y)$. Show that if the force is conservative, then $\partial F_x/\partial y = \partial F_y/\partial x$. [*Hint*: Use the fact that the forces can be represented as derivatives of a single potential energy function $U(x, y)$ when the force is conservative.] What are the analogous conditions on the forces for three-dimensional motion?

70. (III) The surface of water in a bucket is determined by the condition that the potential energy per unit mass, as determined in the frame in which the water is at rest, is constant at all places on the surface. Assume that the bucket is set into rotation at an angular speed ω (introduced in Chapter 3) about its vertical central axis and that the water rotates with the bucket. Write a formula that gives the shape of the water surface. [*Hint*: Note that the normal force of the liquid must account for the centripetal acceleration.]

71. (III) A ball at the end of a pendulum of length L is released at rest from an initial position in which the pendulum string is horizontal (Fig. 7–32). The floor is just beneath the low point of the swing. The string is cut after the ball has passed the low point, with the string at an angle $\alpha = \theta - 90°$ to the vertical. Find the horizontal distance that the ball travels from the low point before it bounces on the floor.

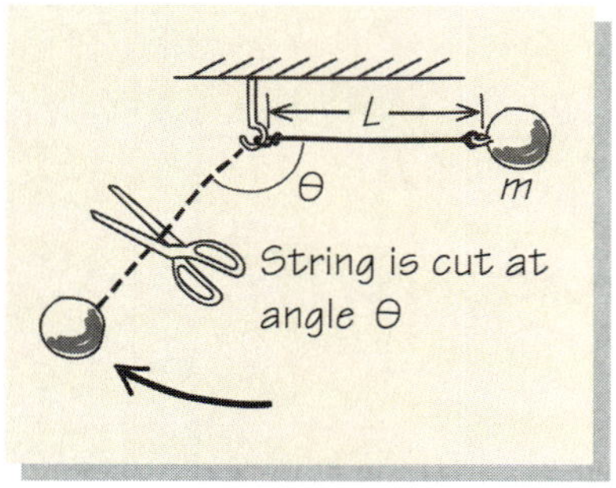

FIGURE 7–32 Problem 71.

CHAPTER

8

The collision between two baseball players has severely disrupted each of their trajectories. This collision is complicated by the fact that the players do not form an isolated system: Friction between their cleats and the ground also plays an important role.

Linear Momentum, Collisions, and the Center of Mass

The idea that energy is conserved is an enormously useful concept. In this chapter, we will identify another quantity that is conserved in isolated systems: momentum. The momentum of an object plays an important role in Newton's second law: A force produces a change in momentum. When a system of particles is isolated, the total momentum of the system is constant. This principle, the conservation of momentum, is particularly useful for understanding the behavior of colliding objects. We can use the conservation of momentum principle even when we know no details about the force or when the forces are very complicated.

As we study the momentum of a system, we will learn that there is a particular point of the system—the *center of mass*—which is distinguished because its motion is so simple. For an isolated system, the center of mass moves as if it were free of all forces. When external forces act on the system, the center of mass accelerates according to Newton's second law just as a point object would.

8–1 MOMENTUM AND ITS CONSERVATION

Newton's second law describes how forces change the motion of objects. We have expressed this law in terms of the mass and the acceleration of an object. A more general form of the second law ($\mathbf{F} = m\mathbf{a}$) is applicable when the mass changes, as in a rocket. This form of the second law is

$$\mathbf{F} = \frac{d(m\mathbf{v})}{dt}. \tag{8–1}$$

The definition of momentum

The combination mass times velocity, $m\mathbf{v}$, is called the **linear momentum**, or just **momentum**. We denote it by $\mathbf{p}$:

$$\mathbf{p} \equiv m\mathbf{v}. \tag{8–2}$$

Newton's second law states that the force on an object equals the rate of change of the object's momentum.

The momentum of an object is a vector whose direction is that of the velocity. Its dimensions are those of a mass times a velocity, namely, $[M \cdot LT^{-1}]$; in SI, the units of momentum are kg · m/s.

In terms of momentum, the second law [Eq. (8–1)] has the general form

$$\mathbf{F} = \frac{d\mathbf{p}}{dt}. \tag{8–3}$$

The kinetic energy of an object can also be expressed in terms of the momentum:

$$K = \frac{1}{2}mv^2 = \frac{p^2}{2m}, \tag{8–4}$$

where p is the magnitude of the momentum.

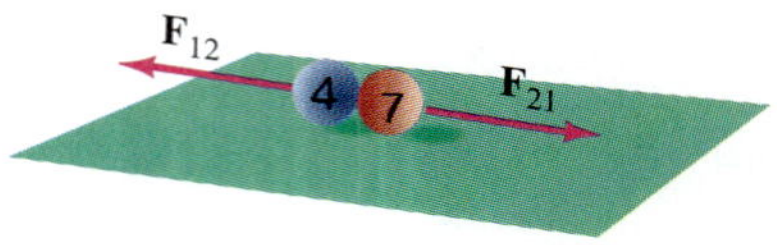

(a)

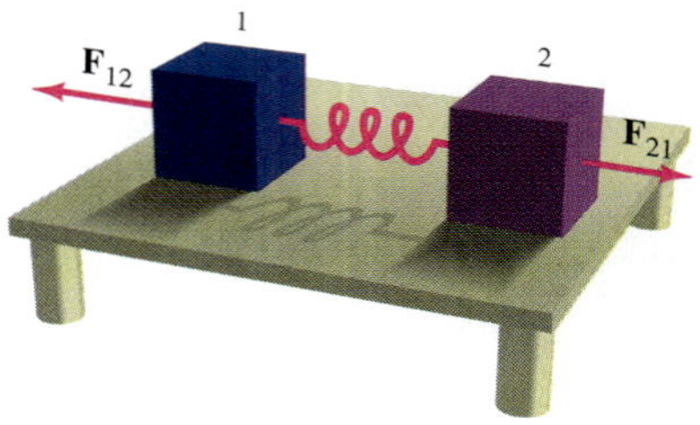

(b)

FIGURE 8–1 (a) Two billiard balls collide and (b) two masses interact because they are connected by a spring. In each case, the objects exert forces on one another. $\mathbf{F}_{12}$ is the force exerted on object 1 by object 2, and $\mathbf{F}_{21}$ is the force exerted on object 2 by object 1. In these collisions, the forces are contact forces although, in general, physical contact is not necessary for two objects to exert forces on one another.

Conservation of Momentum

When objects exert a force on one another, we say they interact. Let's consider the interaction between objects 1 and 2 in Fig. 8–1. The two objects may be in contact, as in a collision of two billiard balls; they may exert a force on each other at a distance, as in the gravitational attraction between Earth and the Moon; or they may simply be connected by a spring. Let $\mathbf{F}_{12}$ denote the force exerted on object 1 by object 2 and let $\mathbf{F}_{21}$ denote the force exerted on object 2 by object 1 (the first subscript always labels the object that is acted *upon*). Then, Newton's third law states that

$$\mathbf{F}_{12} = -\mathbf{F}_{21}. \tag{8–5}$$

When it is expressed in terms of momentum, Newton's second law [Eq. (8–3)] tells us that the rate of change of each object's momentum is the force acting on it:

$$\frac{d\mathbf{p}_1}{dt} = \mathbf{F}_{12}, \tag{8–6}$$

$$\frac{d\mathbf{p}_2}{dt} = \mathbf{F}_{21} = -\mathbf{F}_{12}. \tag{8–7}$$

Addition of these two equations leads to

$$\frac{d\mathbf{p}_1}{dt} + \frac{d\mathbf{p}_2}{dt} = \mathbf{F}_{12} + \mathbf{F}_{21};$$

$$\frac{d(\mathbf{p}_1 + \mathbf{p}_2)}{dt} = \mathbf{F}_{12} - \mathbf{F}_{12} = 0. \tag{8–8}$$

Newton's third law is treated in Chapter 4.

As a consequence,

$$\text{for zero external forces:} \quad \mathbf{p}_1 + \mathbf{p}_2 = \text{a constant}. \tag{8–9}$$

The conservation of momentum

In other words, Newton's third law implies that *the sum of the momenta of an isolated system of two objects that exert forces on one another is a constant, no matter what form the forces take.* This is called the **principle of conservation of momentum**. We will see that this result is not confined to two objects. Like the principle of the conservation of energy, it holds even when we deal with forces whose details we don't understand. The conservation laws are important both as general principles and as powerful tools for solving problems.

EXAMPLE 8–1 A billiard ball moves with a velocity of 1.20 m/s in the $+y$-direction on a billiard table and strikes an identical ball that was initially at rest (Fig. 8–2a). The rolling ball is deflected so that its velocity has a component of 0.80 m/s in the $+y$-direction and a component of 0.56 m/s in the $+x$-direction (Fig. 8–2b). What are the final velocity and final speed of the struck ball?

Solution: We will denote the initial velocity of the rolling ball as $\mathbf{v}_{1i}$ and the final velocity of that ball by $\mathbf{v}_{1f}$. Let the final velocity of the struck ball be $\mathbf{v}_{2f}$. If m is the mass of each ball, $\mathbf{p}_i = m\mathbf{v}_{1i}$ because the struck ball has an initial velocity of zero. Then

$$m\mathbf{v}_{1i} = m\mathbf{v}_{1f} + m\mathbf{v}_{2f},$$

this simplifies to $\mathbf{v}_{1i} = \mathbf{v}_{1f} + \mathbf{v}_{2f}$, or $\mathbf{v}_{2f} = \mathbf{v}_{1i} - \mathbf{v}_{1f}$. We have

$$\mathbf{v}_{1i} = (1.20 \text{ m/s})\mathbf{j} \quad \text{and} \quad \mathbf{v}_{1f} = (0.56 \text{ m/s})\mathbf{i} + (0.80 \text{ m/s})\mathbf{j},$$

so the final velocity of the struck ball is

$$\begin{aligned}\mathbf{v}_{2f} = \mathbf{v}_{1i} - \mathbf{v}_{1f} &= (1.20 \text{ m/s})\mathbf{j} - [(0.56 \text{ m/s})\mathbf{i} + (0.80 \text{ m/s})\mathbf{j}] \\ &= (-0.56 \text{ m/s})\mathbf{i} - (0.40 \text{ m/s})\mathbf{j}.\end{aligned}$$

The final speed of the struck ball is then

$$v_{2f} = \sqrt{(v^2{}_{2f,x} + v^2{}_{2f,y})} = \sqrt{(-0.56 \text{ m/s})^2 + (0.40 \text{ m/s})^2} = 0.69 \text{ m/s}.$$

Conservation of Momentum for a System of Many Objects. The conservation of momentum is not confined to a system of two interacting objects. Suppose that there are three objects in a system on which no external forces act. Then the total force $\mathbf{F}_1$ on object 1 is given by the sum of the forces on object 1 due to objects 2 and 3:

$$\mathbf{F}_1 = \mathbf{F}_{12} + \mathbf{F}_{13}. \tag{8–10a}$$

Similarly, the total force $\mathbf{F}_2$ on object 2 is given by the sum of the forces on object 2 due to objects 1 and 3, so

$$\mathbf{F}_2 = \mathbf{F}_{21} + \mathbf{F}_{23}; \tag{8–10b}$$

and similarly

$$\mathbf{F}_3 = \mathbf{F}_{31} + \mathbf{F}_{32}. \tag{8–10c}$$

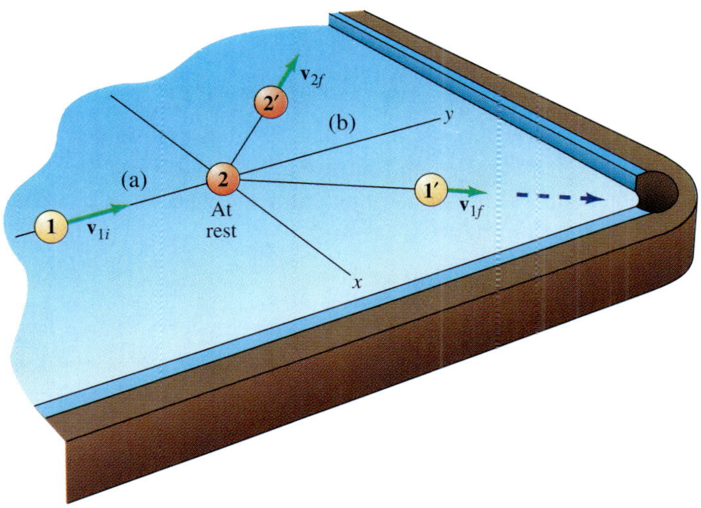

FIGURE 8–2 Example 8–1. (a) Before the collision. (b) After the collision.

Addition of these three equations, with the help of $\mathbf{F}_{12} = -\mathbf{F}_{21}$, $\mathbf{F}_{13} = -\mathbf{F}_{31}$, and $\mathbf{F}_{23} = -\mathbf{F}_{32}$, yields

$$\frac{d\mathbf{p}_1}{dt} + \frac{d\mathbf{p}_2}{dt} + \frac{d\mathbf{p}_3}{dt} = \mathbf{F}_1 + \mathbf{F}_2 + \mathbf{F}_3$$

$$= \mathbf{F}_{12} + \mathbf{F}_{13} + \mathbf{F}_{21} + \mathbf{F}_{23} + \mathbf{F}_{31} + \mathbf{F}_{32} = 0. \quad (8\text{–}11)$$

Consequently, the sum of the momenta of the three objects $\mathbf{P} = \mathbf{p}_1 + \mathbf{p}_2 + \mathbf{p}_3$ is constant throughout the motion:

$$\mathbf{P} = \mathbf{p}_1 + \mathbf{p}_2 + \mathbf{p}_3 = \text{a constant.} \quad (8\text{–}12)$$

We can extend this demonstration to N interacting objects and prove that the sum of the objects' momenta is constant throughout the motion.

PROBLEM SOLVING

A good choice of reference frames makes collision problems simpler.

Conservation of Momentum in Different Inertial Frames

We showed in Chapter 4 that Newton's laws of motion look the same to all observers in inertial frames of reference; that is, in all frames moving with uniform velocity with respect to the system being observed. It is equally true that if momentum is conserved in one inertial reference frame, it is conserved in all inertial frames. This fact may be used to great advantage because calculations involving colliding objects may be simpler in one reference frame than in another.

For example, let's consider the collision of two objects of equal mass: One of them is at rest and the other is moving with momentum $\mathbf{p} = p\,\mathbf{i}$ (Fig. 8–3a). It simplifies the problem to go to a second reference frame[†] by adding the momentum $-\mathbf{p}/2$ to the momentum of each mass. In this second frame, then, the objects have equal and opposite momentum $\mathbf{p}/2$ and $-\mathbf{p}/2$ (Fig. 8–3b). The total momentum is zero and we can immediately say that the final momenta must be $\mathbf{p}'$ and $-\mathbf{p}'$ in this frame (Fig. 8–3c). We now return to the original reference frame by adding the momentum $+\mathbf{p}/2$ to every momentum. Thus, in the original frame, the final momenta must be of the form $\mathbf{p}' + \mathbf{p}/2$ and $\mathbf{p}' - \mathbf{p}/2$, respectively (Fig. 8–3d). This way of finding the possible momenta is simpler than working entirely in the original reference frame.

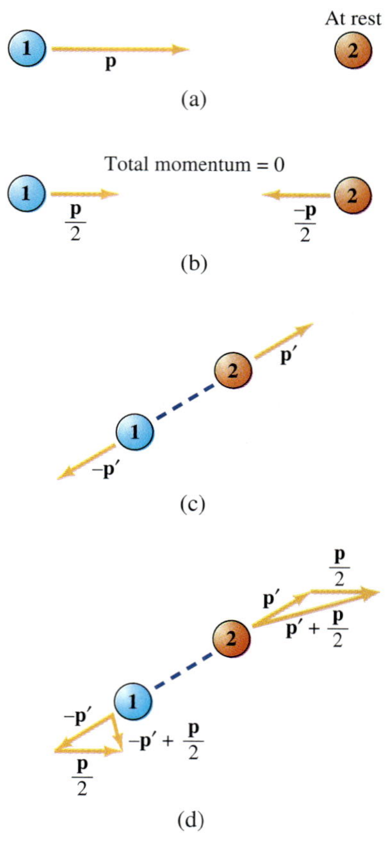

FIGURE 8–3 (a) View before collision from the frame in which ball 2 is at rest. (b) View before collision from the frame in which the total momentum is zero. (c) View after collision from the frame in which the total momentum is zero. (d) View after collision from the frame in which ball 2 was originally at rest.

8–2 IMPULSE AND ONE-DIMENSIONAL COLLISIONS

What happens when objects *collide?* The word "collision" evokes the image of a sharp contact, such as the collision between two billiard balls. In order to apply the conservation of momentum to collisions, we will first discuss these brief, sharp forces in more detail.

Impulsive Forces

During a collision, the force that alters the motion of the two objects is generally active for only a short time. This force is called an **impulsive force**. Let's see how such a force is related to the momentum change of an object on which the force acts (Fig. 8–4). We define the **impulse, J**, according to

$$\mathbf{J} \equiv \int \mathbf{F}\,dt, \quad (8\text{–}13)$$

where $\mathbf{F}$ is the (impulsive) force that acts on the object. From Newton's second law,

$$\mathbf{J} = \int_{t_1}^{t_2} \left(\frac{d\mathbf{p}}{dt}\right) dt = \mathbf{p}_{t_2} - \mathbf{p}_{t_1} = \mathbf{p}_f - \mathbf{p}_i = \Delta\mathbf{p}. \quad (8\text{–}14)$$

[†]We will see later that this is the center-of-mass reference frame.

FIGURE 8–4 The contact force that sends the football on its way is impulsive. The force acts over a limited time.

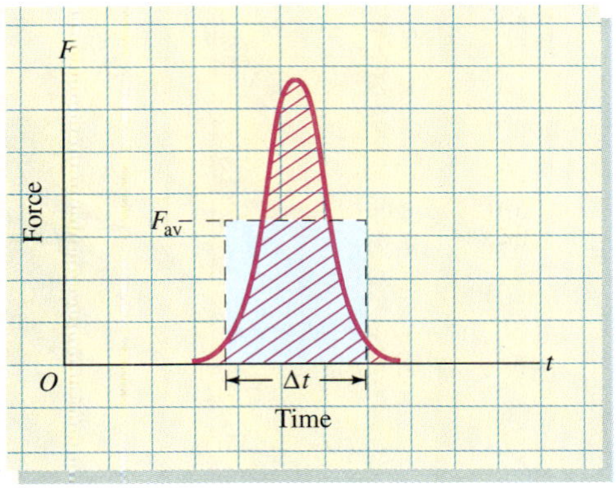

FIGURE 8–5 The momentum of an object is changed by a time-dependent force **F** that acts on it. If that force acts over a time period Δt, then the same change in momentum is produced by its average value $\mathbf{F}_{av}$ over that time period.

Here, the force acts for the time interval $\Delta t = t_2 - t_1$. Because the integral of force over time is the area under a curve of force versus time, we can also represent the impulse by the product of the time interval Δt and a quantity that we may call the *average force* $\mathbf{F}_{av}$. The average force refers to the average value of the force over the time interval Δt. In this case, the impulse or change in momentum can be written in the form

$$\Delta \mathbf{p} = \mathbf{J} = \int \mathbf{F}\, dt = \mathbf{F}_{av}\, \Delta t, \tag{8–15}$$

as depicted in Fig. 8–5. Remember that this equation is a *vector* equation. In Example 8–2, however, the vector aspect does not play a crucial role.

EXAMPLE 8–2 The average force exerted by a bat on a baseball during the time of contact, which lasts 2.00×10^{-3} s, is 6660 N. The mass of the baseball is 0.145 kg and its speed is 33.5 m/s just before the bat collides with it. What is the velocity of the ball when it leaves the bat? Assume that the ball leaves the bat along the same line of direction from which it is pitched.

Solution: Figure 8–6 illustrates this problem. To find the momentum change of the ball, we use the impulse, Eq. (8–15):

$$\Delta \mathbf{p} = \mathbf{J} = \int \mathbf{F}\, \mathrm{d}t = \mathbf{F}_{av}\, \Delta t.$$

Both the average force, $\mathbf{F}_{av}$, and the time interval are given. The motion can be described in one dimension because all motion takes place along a line; thus, we can drop the boldfaced notation. We must, however, keep track of the signs. We do so by aligning the motion along the x-axis, with the original motion of the ball—the direction of the pitch—in the $+x$-direction. The momentum change Δp is $p_f - p_i$ and the initial momentum $p_i = mv_0$, where m is the baseball's mass and v_0 is the (positive) velocity of the ball when it reaches the bat. Thus

$$p_f = p_i + F_{av}\, \Delta t = mv_0 + F_{av}\, \Delta t.$$

The final momentum is the product of m and the final velocity, v_f, and hence

$$v_f = \frac{p_f}{m} = \frac{mv_0 + F_{av}\, \Delta t}{m} = v_0 + \frac{F_{av}\, \Delta t}{m}.$$

We note that the *sign* of the impulse received by the ball is negative because the force is directed in the $-x$-direction. Then

$$v_f = +33.5 \text{ m/s} + \frac{(-6660 \text{ N})(2.00 \times 10^{-3} \text{ s})}{0.145 \text{ kg}}$$

$$= 33.5 \text{ m/s} - 91.9 \text{ m/s} = -58.4 \text{ m/s}.$$

The minus sign indicates that the ball moves in the negative direction back toward the pitcher.

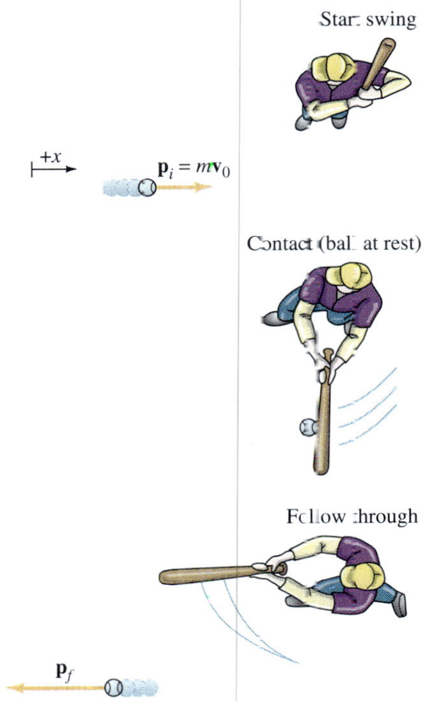

FIGURE 8–6 Example 8–2. The direction of the impulse imparted to a ball hit by a bat is to the left, in the $-x$-direction.

APPLICATION: THE PILE DRIVER

Momentum and impulse find a practical application in the pile driver (Fig. 8AB–1). The pile driver uses a vertical cylinder closed at the lower end and containing a piston. The upper side of this piston is attached to the center of a massive M-shaped piece of steel—a yoke. The outer legs of the M reach down beyond the bottom of the cylinder and engage with the top of the pile being driven. The whole assembly, complete with guides to allow the yoke to remain aligned with the cylinder, is hoisted by a crane and positioned on top of the pile. Compressed air and fuel is supplied to the cylinder below the piston and ignited. The resultant explosion drives the piston and the steel yoke upward. The pressure is then released from the cylinder and the yoke falls. During the fall, it gains kinetic energy and, with it, momentum. When the yoke reaches the top of the pile, it is brought to rest in an exceedingly short time (determined by the compressibility of the steel yoke and the steel pile). Thus dp/dt is very large and a large impulsive force is transmitted to the pile. This process generates forces large enough to break up small rocks in front of the pile as it is being driven into the ground. More importantly, the process we have described here does not require a fixed fulcrum such as would be needed for a static force.

FIGURE 8AB–1 The pile driver is one of the devices that enables us to build large structures.

Momentum Conservation in Collisions

We can clearly see the manifestations of the conservation of momentum in the study of collisions within an isolated system of objects. Attempts to understand collisions were carried out by Galileo and his contemporaries. The laws that describe collisions in one dimension were formulated by John Wallis, Christopher Wren,† and Christian Huygens in 1668.

Let's start with collisions between two objects. The two objects are free before the collision and each may be characterized by its constant momentum. During the brief interaction, their momenta change because each object experiences an impulsive force due to the other object. The impulses felt by the two objects are equal and opposite because the forces are equal and opposite. The increase in the momentum of one object is equal to the loss of momentum in the other. After the collision, the two objects are again free but have different momenta. However, *the sum of their momenta is unchanged.*

In one dimension, the vector aspect is reduced to the fact that the momenta can be negative as well as positive. Let's suppose that two objects are moving along the x-axis. Object 1 has mass m_1 and velocity v_1, whereas object 2 has mass m_2 and velocity v_2. There will be a collision if these objects approach each other from opposite directions or if one overtakes the other as they move in the same direction. The total initial momentum is given by

$$p_i = m_1 v_1 + m_2 v_2. \tag{8–16}$$

Several things can happen to objects during a collision:

1. The two objects can shatter into several pieces.

†Best remembered today as an architect.

2. Mass can be transferred from one object to the other; after the collision, then, the two objects would have masses m_6 and m_7.

3. The two masses can coalesce into one, as in the collision of two blobs of putty.

4. The masses can remain unchanged. Even in this case, though, there are different possibilities. The objects can be completely unaltered, as in the collision of two billiard balls, or they can be deformed, as in the collision of two automobiles.

Figure 8–7 shows the various final configurations of a two-object collision in one dimension. *In all cases, momentum will be conserved.* In the first case, if the objects happen to break up into three objects of masses m_3, m_4, and m_5 moving with velocities v_3, v_4, and v_5, respectively, then

$$m_3v_3 + m_4v_4 + m_5v_5 = m_1v_1 + m_2v_2. \quad (8–17)$$

In the second case, if the final velocities are v_6 and v_7, respectively, then

$$m_6v_6 = m_7v_7 = m_1v_1 + m_2v_2. \quad (8–18)$$

In the third case, if the mass of the single final object is M and its velocity is v, then

$$Mv = m_1v_1 + m_2v_2. \quad (8–19)$$

Mass conservation gives us the additional information that $M = m_1 + m_2$.[†] Finally, in the last case, if the final velocities of the two objects are v_8 and v_9, respectively, then

$$m_1v_8 + m_2v_9 = m_1v_1 + m_2v_2. \quad (8–20)$$

Only in the case of Eq. (8–19) is the final motion (meaning all final velocities) determined by this equation alone. In all the other cases, more information is required.

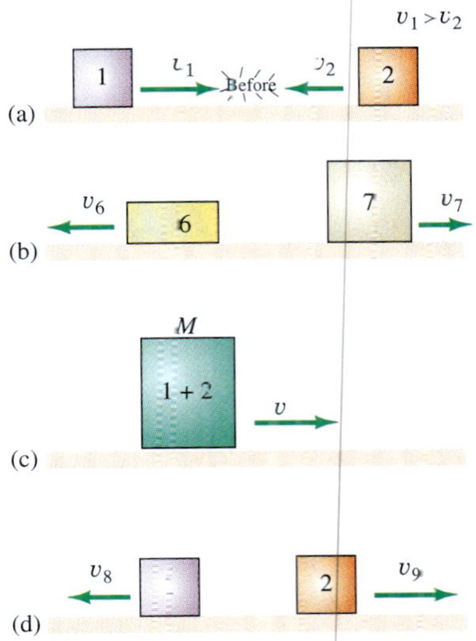

FIGURE 8–7 A collision between two objects, as shown before the collision in part (a), can have several outcomes. The two objects shatter into several pieces (not shown). (b) Mass is transferred between the objects (c) The two masses combine as the objects coalesce into one object. (d) The original masses do not change.

Energy Conservation in Collisions

We may not be able to determine *all* the unknowns in a collision—the final velocities if given the initial ones, for example—simply by using the conservation of momentum. Another important tool is the *conservation of energy.* Strictly speaking, energy is always conserved, but we cannot always take advantage of that fact in our calculations of collisions. For example, when friction is important, all of the energy does not all go into energy of motion; instead, we say that the energy is *dissipated.* For example, in a collision between two cars, some energy goes into the crumpling of metal and as a result there is less energy available for motion—the kinetic energy after the collision. If the energy decrease is known (as in Example 8–3), then we can use this information.

EXAMPLE 8–3 A 14,000-kg truck and a 2000-kg car have a head-on collision (Fig. 8–8). Despite attempts to stop, the truck has a speed of 6.6 m/s in the $+x$-direction when they collide and the car has a speed of 8.8 m/s in the $-x$-direction. If 10 percent of the initial total kinetic energy is dissipated through damage to the vehicles, what are the final velocities of the truck and the car after the collision? Assume that all motion takes place in one dimension.

[†]This need not be the case in nuclear or subnuclear collisions.

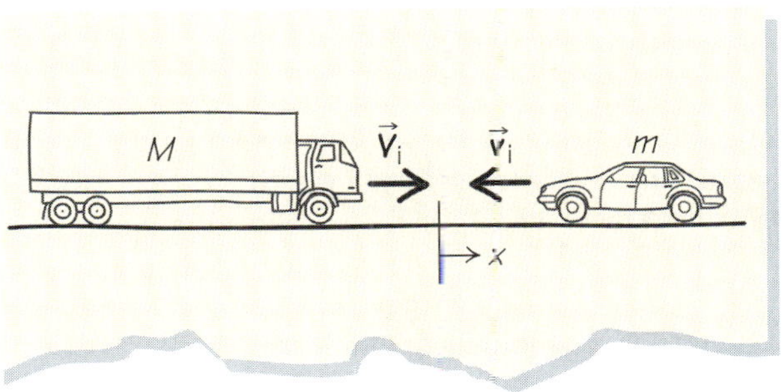

FIGURE 8–8 Example 8–3.

Solution: Let the masses of the truck and car be M and m, respectively. The initial velocities of the truck and the car are denoted V_i and v_i, respectively, and their respective final velocities are V_f and v_f. With our coordinate system, V_i, is positive and v_i is negative. By momentum conservation, we have

$$MV_f + mv_f = MV_i + mv_i \equiv p_1.$$

The initial total energy is the sum of the kinetic energies of the two vehicles, $(MV_i^2/2) + (mv_i^2/2)$. The sum of the final kinetic energies is 90 percent of this quantity, so

$$\frac{1}{2}MV_f^2 + \frac{1}{2}mv_f^2 = 0.9(\frac{1}{2}MV_i^2 + \frac{1}{2}mv_i^2) \equiv 0.9\ K_i.$$

These are two equations for the two unknowns, V_f and v_f. We have defined the initial momentum p_i and the initial kinetic energy K_i because both are known.

The first equation directly yields

$$V_f = \frac{p_i - mv_f}{M}.$$

Substituting this value into the second equation, we have a quadratic equation for vf:

$$\frac{1}{2}M\left(\frac{p_i - mv_f}{M}\right)^2 + \frac{1}{2}mv_f^2 - 0.9\ K_i = 0;$$

$$\frac{1}{2}M\frac{p_i^2}{M^2} - \frac{1}{2}M\left(\frac{2p_imv_f}{M^2}\right) + \frac{1}{2}M\frac{m^2v_f^2}{M^2} + \frac{1}{2}mv_f^2 - 0.9\ K_i = 0;$$

$$(m^2 + Mm)v_f^2 - 2mv_fp_i + p_i^2 - 2(0.9)M\ K_i = 0.$$

This equation has solutions

$$v_f = \frac{2mp_i \pm \sqrt{(2mp_i)^2 - 4(m^2 + Mm)[p_i^2 - 2(0.9)M\ K_i]}}{2(m^2 + Mm)}.$$

When numbers are inserted for the two possible solutions represented by the $\pm$ sign, the minus sign gives a negative velocity for the car and a positive velocity for the truck. This means that the car continues its motion to the left, going "through" the truck, while the truck similarly goes "through" the car. Although this scenario can happen for the collision of elementary particles, it cannot happen for motor vehicles; thus, the only physical solution corresponds to the plus sign. Inserting numbers, we find

$$v_f = 17.2\text{ m/s} \quad \text{and} \quad V_f = 2.9\text{ m/s}.$$

The truck continues in the $+x$-direction with a speed less than its initial speed; the car has completely reversed its direction and is moving even faster than its initial speed.

Classification of Collisions

Elastic, inelastic, and perfectly inelastic collisions

The cases described so far provide a way to classify collisions; namely, by whether kinetic energy is lost or not. If no mass is transferred between the objects and if all the initial kinetic energy of the two colliding objects goes into the kinetic energy of the objects after the collision, we call the collision **elastic**; in other words, the kinetic energy is conserved. If the kinetic energy is not conserved in a collision, it is called an **inelastic** collision. Another alternative is the situation in which two objects collide and coalesce. This kind of collision is called **perfectly inelastic** because, as we shall see in Section 8–3, it corresponds to the maximum loss of kinetic energy.

8–3 PERFECTLY INELASTIC COLLISIONS; EXPLOSIONS

Perfectly Inelastic Collisions

We have described a range of possible collisions from elastic to inelastic to perfectly inelastic. The simplest to treat is the case of *perfectly inelastic collisions* in one dimension, which are those in which the objects coalesce as a result of the collision. These collisions are both described by Eq. (8–19). Mass conservation implies that $M = m_1 + m_2$. If we divide Eq. (8–19) by the factor $M = m_1 + m_2$, we find the velocity of the coalesced object:

$$v_1 = \frac{m_1 v_1 + m_2 v_2}{m_1 + m_2}. \tag{8–21}$$

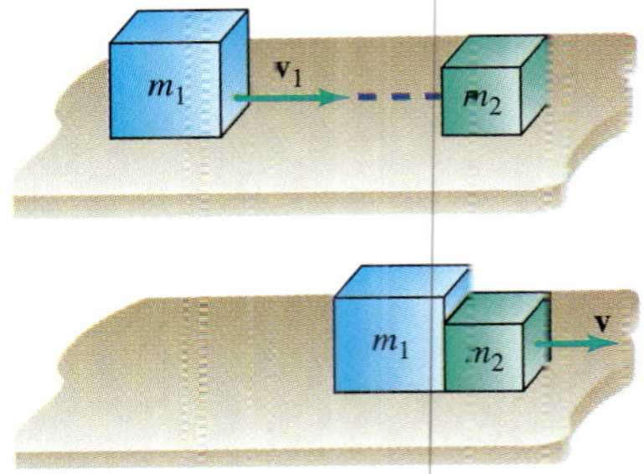

FIGURE 8–9 An object with mass m_1 and velocity $\mathbf{v}_1$ collides with an object of mass m_2 at rest. The two objects stick together and move on together with velocity $\mathbf{v}$.

Let's analyze this result for some special cases.

Suppose that one of the objects is at rest and the other object smashes into it (Fig. 8–9). In this case, $v_2 = 0$, and we find

$$v = \frac{m_1 v_1}{m_1 + m_2} = \left(\frac{m_1}{M}\right) v_1. \tag{8–22}$$

If $m_1 >> m_2$, the "composite" object will move with a velocity nearly equal to that of the initially moving object; a train "catching a cow" does not slow down very much. In contrast, when $m_1 << m_2$, as when a stationary athlete catches a ball, we get the opposite effect; that is, the athlete will recoil with only a low velocity, just the fraction m_1/m_2 of the velocity of the ball.

Next, consider the case of a head-on collision in which the two objects have equal and opposite velocities; that is, $v_2 = -v_1$. In this case, Eq. (8–21) becomes

$$v = \frac{m_1 - m_2}{m_1 + m_2} v_1. \tag{8–23}$$

In the special case that $m_1 = m_2$, the two objects have equal and opposite *momenta* because

$$m_1 v_1 + m_2 v_2 = m_1 v_1 + m_1 v_2 = m_1(v_1 + v_2) = 0. \tag{8–24}$$

In that case, the final momentum must be zero and thus $v = 0$, as Eq. (8–23) verifies. The objects collide and stay there, as when a fullback is tackled head-on by a linebacker (Fig. 8–10).

(a)

(b)

FIGURE 8–10 (a) A collision in the making. (b) When a ball carrier is tackled head-on by a defensive player, a violent collision takes place. If their masses are equal and their velocities are equal and opposite, their final velocities will be zero.

Energy Loss in Perfectly Inelastic Collisions. Let's find the change in energy for a collision in which the objects coalesce. Before the collision, the total energy E_i is the sum of the kinetic energies of the two objects. The final energy E_f is the kinetic energy of the composite object of mass $M = m_1 + m_2$. The change in energy $\Delta E = E_f - E_i$ is, using Eq. (8–21),

$$\begin{aligned}
\Delta E &= \frac{1}{2}Mv^2 - \left(\frac{1}{2}m_1 v_1^2 + \frac{1}{2}m_2 v_2^2\right) \\
&= \frac{1}{2}\frac{(m_1 + m_2)(m_1 v_1 + m_2 v_2)^2}{(m_1 + m_2)^2} - \left(\frac{1}{2}m_1 v_1^2 + \frac{1}{2}m_2 v_2^2\right) \\
&= \frac{\frac{1}{2}\left[m_1^2 v_1^2 + 2m_1 m_2 v_1 v_2 + m_2^2 v_2^2 - (m_1 + m_2)(m_1 v_1^2 + m_2 v_2^2)\right]}{m_1 + m_2} \\
&= \frac{\frac{1}{2}\left[(m_1 m_2)(-v_1^2 - v_2^2 + 2v_1 v_2)\right]}{m_1 + m_2} = -\frac{m_1 m_2}{2(m_1 + m_2)}(v_1 - v_2)^2.
\end{aligned}$$

The right-hand side of Eq. (8–25) is always negative, corresponding to an energy loss. We say the collision is inelastic. When there is only one final composite object, there is a frame of reference in which the composite object is at rest. In this frame, there is no final kinetic energy; thus the collision is called *perfectly inelastic.* In this frame of reference, the total momentum is zero, and *all the energy of motion* in the initial state of the system goes into the "sticking together" of the objects.

EXAMPLE 8–4 A dog who jumps into the interior of a stationary ice boat is moving at $v_0 = 32$ km/h when he enters the boat (Fig. 8–11). The dog's mass is 14 kg and that of the boat plus boater is 160 kg. (a) Assuming that the ice surface is frictionless, what is the speed of the boat after the collision? (b) What is the ratio of the energy loss to the initial energy, and where did the energy go?

Solution: (a) The initial momentum is that of the dog because the boat is at rest. If m is the dog's mass and v_0 is the dog's speed as it enters the boat, then

$$p_i = mv_0.$$

The final momentum is given by $p_f = Mv$, where v is the unknown speed and M is the *total* mass of the boat, boater, and dog; namely, 174 kg. Equating p_f and p_i, we find

$$v = \frac{p_f}{M} = \frac{p_i}{M} = \frac{mv_0}{M} = \frac{(14\text{ kg})(32\text{ km/h})}{174\text{ kg}} = 2.6\text{ km/h} = 0.72\text{ m/s}.$$

(b) The initial energy is the kinetic energy of the dog:

$$K_i = \frac{1}{2}mv_0^2.$$

The final energy is again all in the form of kinetic energy:

$$K_f = \frac{1}{2}Mv^2 = \frac{1}{2}M\left(\frac{mv_0}{M}\right)^2 = \frac{1}{2}\left(\frac{m}{M}\right)mv_0^2 = \frac{m}{M}K_i.$$

Thus the energy loss is given by

$$\Delta E = K_i - K_f = K_i - \frac{m}{M}K_i = K_i\left(1 - \frac{m}{M}\right),$$

and the ratio of the energy loss to the initial energy is

$$\frac{\Delta E}{K_i} = 1 - \frac{m}{M},$$

FIGURE 8–11
Example 8–4.

a number less than 1. The energy has decreased. Numerically,

$$\frac{\Delta E}{K_i} = 1 - \frac{14\text{ kg}}{174\text{ kg}} = 0.92.$$

Energy is lost as the boater "gives" in order to bring the dog to rest.

Explosions

Imagine that we were to film a perfectly inelastic collision in a frame of reference in which the total momentum is zero. In this reference system, the two objects approach each other and merge, leaving a composite object at rest. If we ran the film in reverse, it would look like a film of an explosion. The "initial" object of mass $M = m_1 + m_2$, at rest, breaks up into two objects, m_1 and m_2, moving with velocities such that the momentum is still zero; that is, with

$$m_1v_1 + m_2v_2 = 0. \tag{8–26}$$

Energy conservation suggests that an explosion is possible as long as there is an initial potential energy U such that

$$U = \frac{1}{2}m_1v_1^2 + \frac{1}{2}m_2v_2^2. \tag{8–27}$$

An explosive does have potential energy stored in its molecules and this potential energy is released by the detonation. Let's next take a look at a different type of "explosion," one in which an unstable atomic nucleus disintegrates.

EXAMPLE 8–5 One type of polonium nucleus (symbol ^{210}Po), with mass 3.49×10^{-25} kg, can decay into an α-particle (actually a helium nucleus), mass 6.64×10^{-27} kg, and a certain type of lead nucleus (symbol ^{206}Pb), mass 3.42×10^{-25} kg:

$$^{210}\text{Po} \rightarrow \alpha + {}^{206}\text{Pb}.$$

In this process, the decay products have a kinetic energy of 8.65×10^{-13} J above and beyond any kinetic energy possessed by the polonium nucleus itself. Consider such a decay with the polonium nucleus at rest. Find the speeds of the α-particle and the lead nucleus.

Solution: Let's denote by Q the kinetic energy of the decay products. Then conservation of momentum and the expression for the kinetic energy read, respectively,

$$M_\alpha v_\alpha = M_{\text{Pb}}v_{\text{Pb}};$$

$$Q = \frac{1}{2}M_\alpha v_\alpha^2 + \frac{1}{2}M_{\text{Pb}}v_{\text{Pb}}^2.$$

Here, v is the speed of the respective particles. These two equations can be solved for the two variables v_α and v_{Pb}. We find

$$v_\alpha = \sqrt{\frac{2Q}{M_\alpha\,(1 + M_\alpha/M_{\text{Pb}})}}$$

$$v_{\text{Pb}} = \sqrt{\frac{2Q}{M_{\text{Pb}}\,(1 + M_{\text{Pb}}/M_\alpha)}}.$$

With $Q = 8.65 \times 10^{-13}$ J, numerical evaluation gives $v_\alpha = 1.60 \times 10^7$ m/s and $v_{\text{Pb}} = 3.10 \times 10^5$ m/s. Note that the speed of the heavier of the two decay products is much less than the speed of the lighter, a result already visible in the conservation of momentum equation.

8-4 ELASTIC AND INELASTIC COLLISIONS

Let's continue to work in one dimension. In an *elastic collision*, there is no mass transfer from one object to another. Further, *all the kinetic energy in the initial state goes into kinetic energy in the final state.* If the final velocities of objects 1 and 2 are denoted by v_3 and v_4, then, in addition to the momentum conservation equation for one dimension,

$$m_1v_1 + m_2v_2 = m_1v_3 + m_2v_4, \tag{8–28}$$

we have the energy conservation equation

$$\frac{1}{2}m_1v_1^2 + \frac{1}{2}m_2v_2^2 = \frac{1}{2}m_1v_3^2 + \frac{1}{2}m_2v_4^2. \tag{8–29}$$

With this information we can find the final velocities of the colliding objects if their initial velocities are known. We rewrite the momentum conservation equation, Eq. (8–28), as

$$m_1(v_1 - v_3) = -m_2(v_2 - v_4). \tag{8–30}$$

We use the fact that $v_1^2 - v_3^2 = (v_1 - v_3)(v_1 + v_3)$ and $v_2^2 - v_4^2 = (v_2 - v_4)(v_2 + v_4)$ to rewrite the energy conservation equation, Eq. (8–29), in the form

$$\frac{1}{2}m_1(v_1 - v_3)(v_1 + v_3) = -\frac{1}{2}m_2(v_2 - v_4)(v_2 + v_4). \tag{8–31}$$

Dividing both sides of Eq. (8–31) by the two sides of Eq. (8–30) leads to the equation

$$v_1 + v_3 = v_2 + v_4. \tag{8–32}$$

If we use the letter u to denote the *relative velocity* of the two colliding objects, then

$$u_i = v_1 - v_2;$$

$$u_f = v_3 - v_4.$$

Using these quantities, Eq. (8–32) can be written in the form

$$u_i = -u_f. \tag{8–33}$$

Relative speeds remain unchanged in an elastic collision.

Equation (8–33) shows that *when the collision is elastic, the relative velocity of the colliding objects changes sign but does not change in magnitude.* A simple way to remember this result is that the relative velocity behaves like the velocity of a perfectly elastic rubber ball hitting a brick wall.

We may solve Eq. (8–32) for one of the unknown variables, v_4, for example,

$$v_4 = v_1 - v_2 + v_3,$$

and substitute this value into the momentum conservation equation, Eq. (8–28). We then have

$$m_1v_1 + m_2v_2 = m_1v_3 + m_2(v_1 - v_2 + v_3),$$

which may be rewritten in the form

$$(m_1 + m_2)v_3 = (m_1 - m_2)v_1 + 2m_2v_2;$$

$$v_3 = \frac{m_1 - m_2}{m_1 + m_2}v_1 + \frac{2m_2}{m_1 + m_2}v_2. \tag{8–34}$$

A similar calculation leads to the formula

$$v_4 = \frac{2m_1}{m_1 + m_2}v_1 + \frac{m_2 - m_1}{m_1 + m_2}v_2. \tag{8–35}$$

These equations are rather complicated and it is useful to consider some special cases that simplify them.

1. **Object 2 is initially at rest.** We set $v_2 = 0$, so Eqs. (8–34) and (8–35) become

$$v_3 = \frac{m_1 - m_2}{m_1 + m_2} v_1 \tag{8–36a}$$

and

$$v_4 = \frac{2m_1}{m_1 + m_2} v_1. \tag{8–36b}$$

Let's consider the following situations (in which object 2 is initially at rest):

1a. The objects have equal masses (Fig. 8–12a). In this case, $v_3 = 0$ and $v_4 = v_1$. The two objects in effect change roles: The moving object comes to rest and the object that was initially at rest moves with the initial velocity of the first object. This effect can be seen vividly in hard billiard shots along a line.

1b. Mass $m_1 >> $ mass m_2 (Fig. 8–12b). In this case, Eqs. (8–36) yield $v_3 \simeq v_1$ and $v_4 \simeq 2v_1$. The velocity of the moving object decreases a little, while the object that was at rest picks up almost twice the velocity of the incoming object.

1c. Mass $m_2 >> $ mass m_1 (Fig. 8–12c). In this case, Eqs. (8–36) yield $v_3 \simeq -v_1$ and $v_4 \simeq (2m_1/m_2)v_1$. The moving object very nearly reverses its velocity, while the object initially at rest recoils with a very small velocity. In the limit that m_2 approaches infinity, the recoil velocity can be neglected and the final velocity of the first object is equal and opposite to its incident velocity. This is what happens when a tennis ball is bounced off a wall.

2. **The initial total momentum is zero.** The two objects approach each other with velocities such that the initial total momentum is zero:

$$m_1 v_1 + m_2 v_2 = 0. \tag{8–37}$$

Thus

$$v_2 = -\frac{m_1}{m_2} v_1.$$

When this value is substituted into Eq. (8–34), we find

$$v_3 = \frac{m_1 - m_2}{m_1 + m_2} v_1 + \left(\frac{2m_2}{m_1 + m_2}\right)\left(-\frac{m_1}{m_2}\right) v_1 = \left(\frac{m_1 - m_2 - 2m_1}{m_1 + m_2}\right) v_1 = -v_1. \tag{8–38}$$

To find the value of v_4, we can recall that the initial total momentum was zero. Thus, by momentum conservation, the final total momentum $(m_1 v_3 + m_2 v_4)$ is also zero and

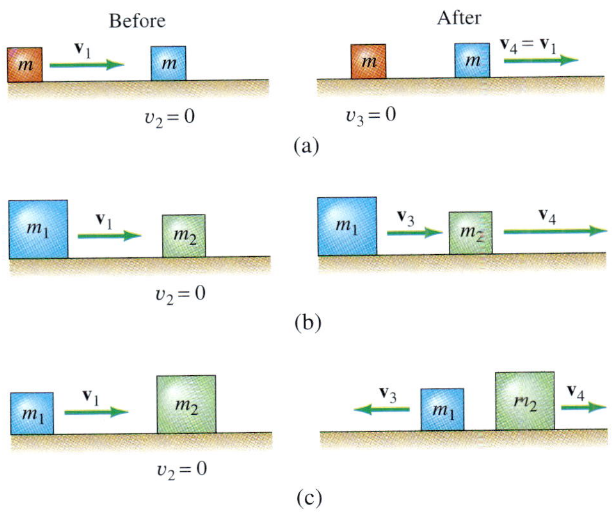

FIGURE 8–12 Two objects collide elastically in one dimension, with the second object initially at rest. (a) If the two masses are equal, the objects simply exchange velocities. (b) If $m_1 >> m_2$, both objects move off to the right. (c) If $m_2 >> m_1$, mass m_1 reverses its direction and m_2 moves slowly off to the right.

$$v_4 = -\frac{m_1}{m_2}v_3 = \frac{m_1}{m_2}v_1 = -v_2. \tag{8–39}$$

Therefore, in the case where the total momentum is zero, the velocities of each of the objects are unchanged in magnitude but they change sign. In effect, under these circumstances, each of the objects acts as if it hit an infinitely massive brick wall.

Examples 8–6 and 8–7 illustrate some features of collisions in one dimension.

EXAMPLE 8–6 Two spheres with masses of 1.0 kg and 1.5 kg hang at rest at the ends of strings that are both 1.5 m long. These two strings are attached to the same point on the ceiling. The lighter sphere is pulled aside so that its string makes an angle $\theta = 60°$ with the vertical (Fig. 8–13a). The lighter sphere is then released and the two spheres collide elastically. When they rebound, what is the largest angle with respect to the vertical that the string holding the lighter sphere makes?

Solution: We label the light mass m_1 and the heavy mass m_2 (Fig. 8–13). To find the velocities $\mathbf{v}_1'$ and $\mathbf{v}_2'$ resulting from the collision, we must first find the velocity $\mathbf{v}_1$ of m_1 just before the collision. Conservation of momentum and energy give us its velocity just after the collision. We can then use energy conservation to discover the height to which the lighter sphere rebounds.

We begin by calculating the initial potential energy, which is converted to kinetic energy when sphere 1 swings down to the minimum point (neglecting air resistance). From its kinetic energy, we can find its speed, v_1, just before the collision. The mass m_1 is raised a distance $L(1 - \cos\theta) = L/2$ above the minimum point. We take the potential energy to be zero when the spheres are hanging free. The initial potential energy of m_1 is then $U_i = m_1gL/2$. The kinetic energy of m_1 just before the collision is this potential energy:

$$\frac{1}{2}m_1v_1^2 = \frac{1}{2}m_1gL$$

$$v_1 = \sqrt{gL}.$$

The collision takes place at the bottom of the swing, where $\mathbf{v}_1$ is horizontal. Let the positive direction be to the right. Thus sphere 1 swinging from left to right has a velocity $+v_1$. (The collision takes place in one dimension so we drop the boldfaced notation.) The initial momentum is

$$p_i = m_1v_1 + m_2v_2 = m_1v_1.$$

To find the velocities v_1' and v_2' we use conservation of momentum, which states that

$$p_f = m_1v_1' + m_2v_2' = p_i = m_1v_1.$$

As in Eq. (8–33), energy conservation implies that the initial relative velocity, v_{rel}, and the final relative velocity are equal in magnitude but opposite in sign. With

$$v_{rel} = v_1 - v_2 = v_1,$$

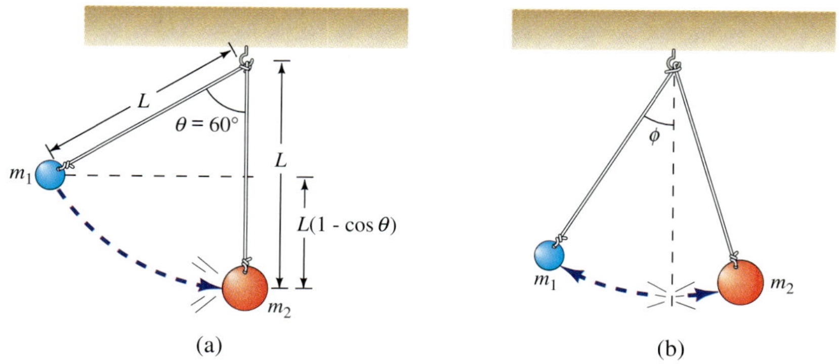

FIGURE 8–13 Example 8–6. Two spheres hang from strings of equal length. (a) Sphere m_1 is pulled back to the left at an angle θ and released. It collides with sphere m_2, which is at rest. (b) The balls recoil, with sphere m_1 reaching a maximum height characterized by the angle ϕ.

we have a final relative velocity of

$$v'_1 - v'_2 = -v_{rel} = -v_1.$$

We now have two simultaneous equations that can be solved for v'_1 and v'_2, namely, $m_1v'_1 + m_2v'_2 = m_1v_1$ (momentum conservation) and $v'_1 - v'_2 = -v_1$ (energy conservation). We are interested only in the velocity v'_1 of the lighter sphere, and the solution for this quantity is

$$v'_1 = \frac{m_1 - m_2}{m_1 + m_2} v_1 = \frac{m_1 - m_2}{m_1 + m_2} \sqrt{gL}.$$

Because m_2 is larger than m_1, v'_1 is negative, which indicates that the sphere recoils back to the left.

We next find the height to which m_1 recoils. The kinetic energy right after the collision is $m_1(v'_1)^2/2$. This energy will be converted into a potential energy, U_f, at the end of the swing. The expression for U_f is $m_1gL(1 - \cos\phi)$, where ϕ is the desired angle of recoil (Fig. 8–13b). Thus the conservation of energy gives

$$m_1gL(1 - \cos\phi) = \frac{1}{2}m_1(v'_1)^2.$$

We solve for ϕ:

$$1 - \cos\phi = \frac{(v'_1)^2}{2gL} = \frac{(m_1 - m_2)^2}{(m_1 + m_2)^2} gL \frac{1}{2gL} = \frac{(m_1 - m_2)^2}{2(m_1 + m_2)^2}$$

$$= \frac{(1.0\text{ kg} - 1.5\text{ kg})^2}{2(1.0\text{ kg} + 1.5\text{ kg})^2} = 0.020,$$

or $\phi = 11°$.

Friction is discussed in Chapter 5.

EXAMPLE 8–7 A 10-g bullet is fired in the $+x$-direction into a stationary block of wood that has a mass of 5 kg (Fig. 8–14). The speed of the bullet before entry into the wood block is $v_0 = 500$ m/s. What is the speed of the block just after the bullet has become embedded? What distance d will the block slide on a surface with a coefficient of friction equal to 0.50?

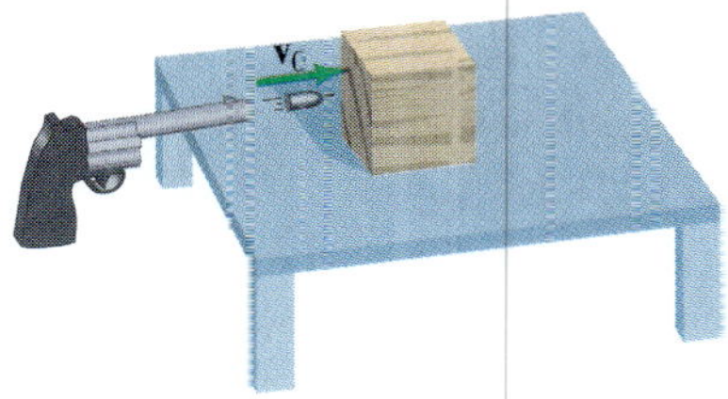

(a)

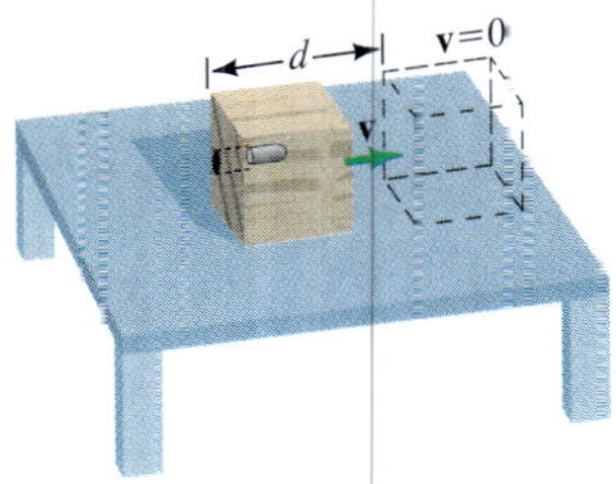

(b)

FIGURE 8–14 Example 8–7.

Solution: Let m be the mass of the bullet and v_0 be its velocity before it enters the wood block (Fig. 8–14a). The initial momentum is then $p_i = mv_0$. If M is the combined mass of the bullet plus the block and v is the velocity of the block and bullet just after the bullet enters the block, then momentum conservation implies that

$$mv_0 = Mv;$$

$$v = \frac{m}{M}v_0 = \frac{10\text{ g}}{5010\text{ g}}(500\text{ m/s}) = 1\text{ m/s}.$$

Note that we can ignore the mass of the bullet compared to that of the block and still attain good accuracy.

The force of friction is

$$\mu N = -\mu Mg.$$

(The minus sign indicates that friction points to the left.) This has constant magnitude and leads to a constant acceleration a of the block, according to Newton's second law:

$$-\mu Mg = Ma;$$

$$a = -\mu g.$$

The negative sign means the block slows down, traveling a distance d before it stops (Fig. 8–14b). Because the acceleration is uniform, we can use the relation $v_f^2 - v_i^2 = 2ad$ [from Eq. (2–22)]. With $v_f = 0$, we have

$$d = -\frac{v_i^2}{2a} = \frac{1}{2}\frac{v_i^2}{\mu g} = \frac{1}{2}\frac{(1\text{ m/s})^2}{(0.5)(9.8\text{ m/s}^2)} = 0.1\text{ m}.$$

8-5 ELASTIC COLLISIONS IN SPACE

In one dimension, the possible motion of the colliding objects is severely limited. When collisions occur in space, however, as when billiard balls collide on a billiard table, there is a greater range of allowed motion and the vector nature of the mathematical equations becomes important.

The law of conservation of momentum [Eq. (8–9)] for the collision of two objects of masses m_1 and m_2, with initial velocities $\mathbf{v}_1$ and $\mathbf{v}_2$ and final velocities $\mathbf{v}_3$ and $\mathbf{v}_4$, reads:

$$m_1\mathbf{v}_1 + m_2\mathbf{v}_2 = m_1\mathbf{v}_3 + m_2\mathbf{v}_4. \tag{8–40}$$

The energy conservation law is

$$\frac{1}{2}m_1v_1^2 + \frac{1}{2}m_2v_2^2 = \frac{1}{2}m_1v_3^2 + \frac{1}{2}m_2v_4^2. \tag{8–41}$$

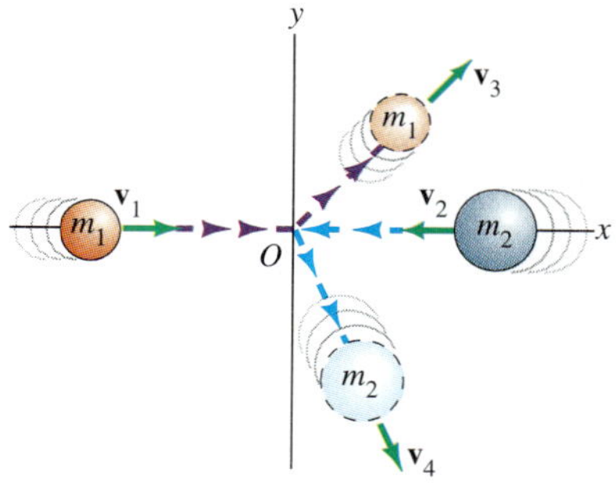

FIGURE 8–15 An object of mass m_1 and velocity $\mathbf{v}_1$ collides with another object of mass m_2 and velocity $\mathbf{v}_2$ in two dimensions.

Let's first consider collisions in two dimensions, where everything happens in a plane, such as the xy-plane shown in Fig. 8–15. A familiar example is the collision of billiard balls on a pool table (but the masses of the colliding objects need not be equal). We want to find the magnitudes of the final velocities and the angles they make with the x-axis. Equivalently, we want the x- and y-components of the final velocities of objects 1 and 2. There are four unknowns but only three equations. [Equation (8–40) is a vector equation and actually comprises two equations: one for the x-components and one for the y-components.] Therefore, for a given set of initial velocities, there is no unique solution for the final velocities and the final objects can move in a variety of directions.

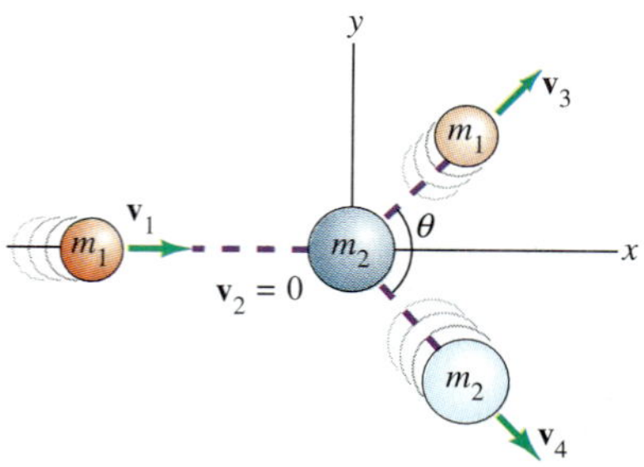

FIGURE 8–16 An object of mass m_1 and velocity $\mathbf{v}_1$ collides with an object of mass m_2 at rest. The objects move off after the collision.

It is useful to see some of the constraints that these equations impose in the special case where one of the objects is initially at rest. Figure 8–15 shows the geometry. It is possible to solve this problem completely for the final velocities once the angles with which the objects leave the collision are specified. Let's look, though, at an interesting, if limited, result. If we choose object 2 to be at rest, then $v_2 = 0$. We take the square of Eq. (8–40) in the sense that the square of a vector equation $\mathbf{A} = \mathbf{B}$ implies $\mathbf{A}\cdot\mathbf{A} = \mathbf{B}\cdot\mathbf{B}$. With $v_2 = 0$, we find

$$m_1^2v_1^2 = m_1^2v_3^2 + 2m_1m_2\mathbf{v}_3\cdot\mathbf{v}_4 + m_2^2v_4^2. \tag{8–42}$$

The energy conservation equation multiplied by $2m_1$ reads

$$m_1^2v_1^2 = m_1^2v_3^2 + m_1m_2v_4^2. \tag{8–43}$$

If θ is the angle between the final velocity vectors $\mathbf{v}_3$ and $\mathbf{v}_4$, so that $\mathbf{v}_3\cdot\mathbf{v}_4 = v_3v_4\cos\theta$, then the difference of Eqs. (8–42) and (8–43) is

$$m_1m_2v_4^2 = 2m_1m_2v_3v_4\cos\theta + m_2^2v_4^2.$$

Dividing by m_2v_4 leads to

$$m_1v_4 - m_2v_4 = 2m_1v_3\cos\theta. \tag{8–44}$$

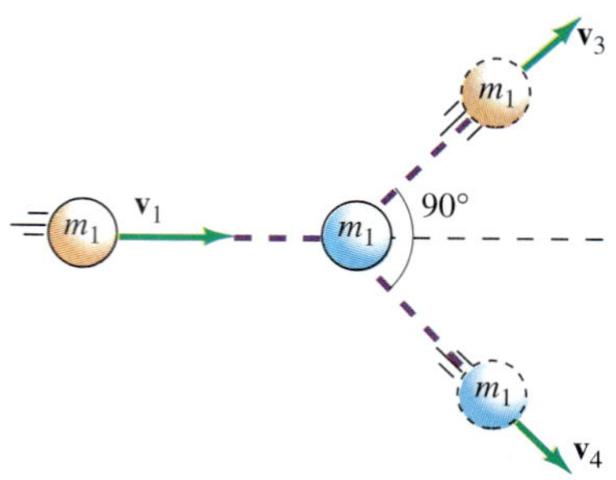

FIGURE 8–17 An object of mass m_1 and velocity $\mathbf{v}_1$ collides elastically with an object of the same mass at rest. The angle between the final velocities is 90°.

In the special case where $m_1 = m_2$, Eq. (8–44) implies that $\cos\theta = 0$, or $\theta = 90°$, independent of the velocities, which is a fact known to billiards players (Fig. 8–17). The angle between a ball and the recoiling target ball is always a right angle. Energy conservation, as in Eq. (8–42), requires in this equal-mass case that

$$v_1^2 = v_3^2 + v_4^2. \tag{8–45}$$

The final velocities are constrained to form the sides of a right triangle whose hypotenuse is given by v_1.

EXAMPLE 8–8 Two billiard balls of equal mass approach each other along the x-axis; one is moving to the right with a speed of $v_1 = 10$ m/s and the other is moving to the left with a speed of $v_2 = 5$ m/s. After the collision, which is elastic, one of the balls moves in the direction of the y-axis (Fig. 8–18). What are the velocities $\mathbf{v}_3$ and $\mathbf{v}_4$ of the balls after the collision?

Solution: The fact that the masses of the two balls are equal simplifies the algebra. If we write the momentum conservation equation and divide by the factor of m that appears everywhere, we deal with a conservation of velocities.

Dividing by the mass factor in the energy conservation equation gives a conservation law for the squares of the velocities.

The initial velocities of the balls are

$$\mathbf{v}_1 = (10\ \text{m/s})\mathbf{i}$$

and

$$\mathbf{v}_2 = (-5\ \text{m/s})\mathbf{i}.$$

After the collision one ball moves in the y-direction; its velocity is given by $\mathbf{v}_3 = v_3\,\mathbf{j}$. Let the velocity of the second ball after the collision be $\mathbf{v}_4 = v_{4x}\,\mathbf{i} + v_{4y}\,\mathbf{j}$. Momentum conservation now reads

$$(10\ \text{m/s})\mathbf{i} + (-5\ \text{m/s})\mathbf{i} = v_3\,\mathbf{j} + v_{4x}\,\mathbf{i} + v_{4y}\,\mathbf{j}.$$

We now separately equate the coefficients of $\mathbf{i}$ and those of $\mathbf{j}$; that is, we use the fact that conservation of momentum holds in the x-direction and in the y-direction separately. We find

$$v_{4x} = (10\ \text{m/s}) + (-5\ \text{m/s}) = 5\ \text{m/s} \quad \text{and} \quad v_{4y} = -v_3.$$

Kinetic energy is conserved because the collision is elastic and this result leads to

$$\frac{1}{2}(10\ \text{m/s})^2 + \frac{1}{2}(-5\ \text{m/s})^2 = \frac{1}{2}v_3^2 + \frac{1}{2}(v_{4x}^2 + v_{4y}^2);$$
$$100\ \text{m}^2/\text{s}^2 + 25\ \text{m}^2/\text{s}^2 = v_3^2 + 25\ \text{m}^2/\text{s}^2 + v_3^2.$$

Thus we have $100\ \text{m}^2/\text{s}^2 = 2v_3^2$, or $v_3 = \sqrt{50}$ m/s. We also have $v_{4y} = -\sqrt{50}$ m/s.

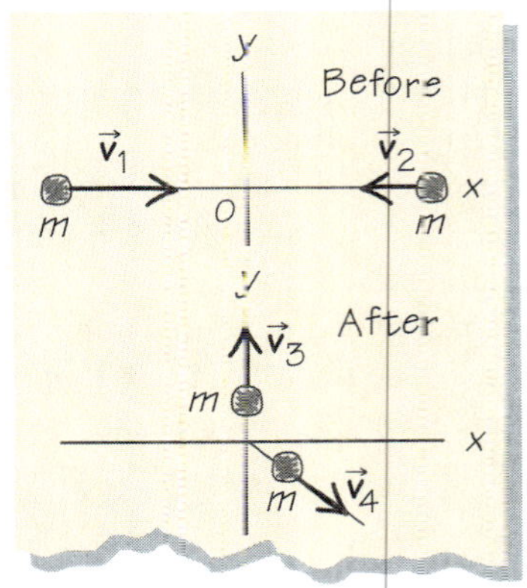

FIGURE 8–18 Example 8–8.

What about collisions in three dimensions? Conservation of momentum allows us to show that there is always an inertial reference frame in which elastic collisions between two objects take place in a plane (Fig. 8–19). The results established for two dimensions are thus directly applicable to three dimensions.

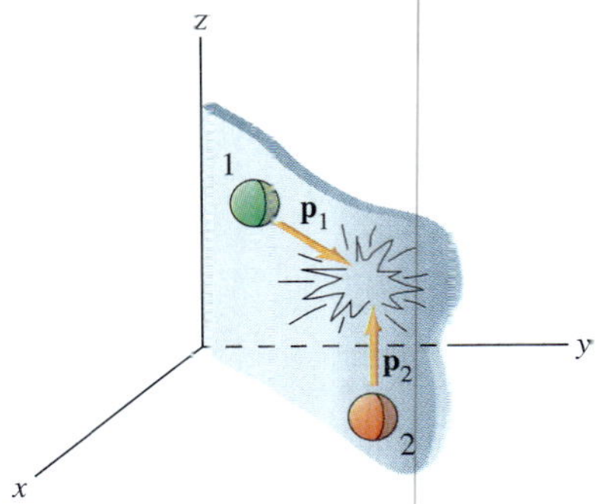

FIGURE 8–19 The elastic collisions of two bodies take place in a plane which is formed by the two momentum vectors of the incoming masses (or of the outgoing masses).

INTERIM SUMMARY

The momentum $\mathbf{p} \equiv m\mathbf{v}$ of an object of mass m moving with velocity $\mathbf{v}$ plays an important role in the interactions of objects. In particular, Newton's second law states that the net force acting on an object is the rate of change of its momentum; that is, $\mathbf{F} = d\mathbf{p}/dt$. Using Newton's second law, we find the change in momentum in terms of the impulse, $\mathbf{J}$:

$$\Delta\mathbf{p} = \mathbf{J} = \int \mathbf{F}\,dt = \mathbf{F}_{av}\,\Delta t. \tag{8–15}$$

Impulses are most useful when the force acts briefly, as in a collision.

The concept of momentum is useful for understanding isolated systems of interacting objects, which are systems on which no external force acts. The total momentum of an isolated system of objects is conserved; importantly, this can be applied to collisions of two objects. Collisions involve only interparticle forces between the colliding objects, and the effects of external forces can normally be ignored during a collision. Thus the total momentum of the system of objects is conserved whether the collisions are elastic, inelastic, or perfectly inelastic. In elastic collisions, the sum of the kinetic energies is conserved. During inelastic collisions, the sum of the kinetic energies changes. The change is generally a decrease due to the presence of dissipative forces, such as friction. In perfectly inelastic collisions, the objects that collide merge into a single object whose mass is the sum of the masses of the colliding objects.

PROBLEM-SOLVING TECHNIQUES

Collision problems involve an interaction between objects that is limited in space (and time). At some early time, the objects do not exert any forces on one another; at some later time, they again exert no forces on one another. Typically, we must find some parameters of the final (or initial) motion—the final velocity of one of the objects, for example. The following set of steps can be useful.

1. We must clearly identify the relevant interaction between the objects involved in a collision. This means that the objects are isolated or that, during the collision, external forces are small compared to the impulsive forces.
2. We must identify the objects involved before the collision and those involved after the collision. The objects that result from a collision may not be the same as the objects before the collision. Explosions or collisions in which the colliding objects coalesce are such examples.
3. Identify the quantities that are known and those unknown quantities that are to be found. In particular, a count of the number of known and unknown quantities, including the number of vector components, is helpful.
4. Remember that the conservation of momentum, which is a vector relation, is always applicable. If the motion is in one, two, or three dimensions, the conservation of momentum gives one, two, or three relations among the momenta (or velocities), respectively.
5. If the collision is known to be elastic, the conservation of energy provides an additional equation that involves the speeds of the objects. If the collision is not elastic, then an equation for the energy is available only if information about the energy loss is given.
6. The number of equations that include the unknowns must match the number of unknowns; if so, we can solve the problem by using algebra.

8–6 THE CENTER OF MASS

In a system of many objects, there is a particular point with special properties, called the *center of mass.* As we shall see, the importance of this point is that it moves under Newton's second law as if the total mass of the system were concentrated there.

To find this point, recall from Section 8–1 that, in the absence of external forces, the sum of the momenta of three objects is a constant of the motion; the procedure we used is easily generalized to an isolated N-object system. We find that the total momentum, $\mathbf{P}$, of the system does not change in the absence of external forces:

$$\mathbf{P} = m_1\mathbf{v}_1 + m_2\mathbf{v}_2 + \cdots + m_N\mathbf{v}_N. \tag{8–46}$$

The momentum of the system is a constant vector—one that *does not change in magnitude or direction as a function of time.* If we also assume that the masses do not change as a function of time, we have

$$\mathbf{P} = m_1\frac{d\mathbf{r}_1}{dt} + m_2\frac{d\mathbf{r}_2}{dt} + \cdots + m_N\frac{d\mathbf{r}_N}{dt} = \text{a constant};$$

$$\frac{d}{dt}(m_1\mathbf{r}_1 + m_2\mathbf{r}_2 + \cdots + m_N\mathbf{r}_N) = \text{a constant}. \tag{8–47}$$

The position vector of the center of mass

The quantity in parentheses has the dimensions $[ML]$. We divide it by the total mass, $M = m_1 + m_2 + \cdots + m_N$, to obtain a vector that determines the position vector $\mathbf{R}$ of the **center of mass**, the average position of all of the particles in a system, weighted by their mass:

$$\mathbf{R} \equiv \frac{m_1\mathbf{r}_1 + m_2\mathbf{r}_2 + \cdots + m_N\mathbf{r}_N}{M}. \tag{8–48}$$

From its definition, we see that the center-of-mass vector, $\mathbf{R}$, has components X, Y, and Z, which are

$$X = \frac{m_1x_1 + m_2x_2 + \cdots + m_Nx_N}{M}, \tag{8–49a}$$

$$Y = \frac{m_1 y_1 + m_2 y_2 + \cdots + m_N y_N}{M}, \tag{8–49b}$$

$$Z = \frac{m_1 z_1 + m_2 z_2 + \cdots + m_N z_N}{M}. \tag{8–49c}$$

Figure 8–20 illustrates several examples of the center of mass position for two and three objects.

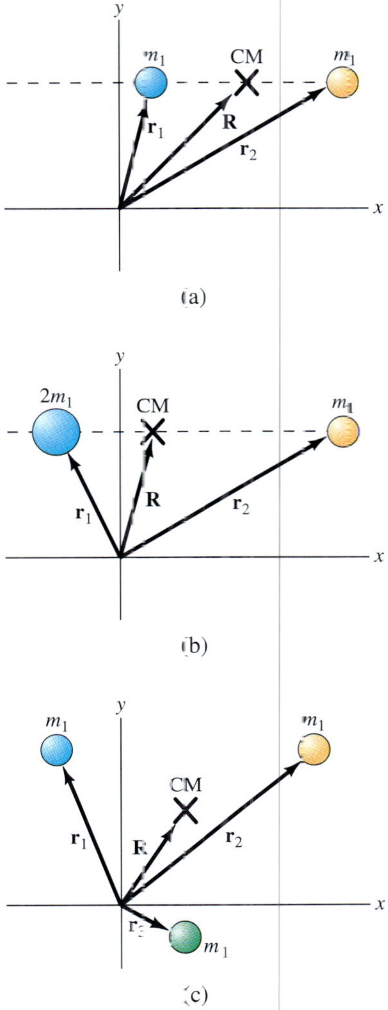

FIGURE 8–20 Several examples of the center of mass. (a) Both balls have equal masses. (b) One ball has twice the mass of the other. (c) Three balls of equal mass. We shall look below at several explicit calculations of the center of mass.

Velocity and Acceleration of the Center of Mass

We can find the velocity, **V**, of the center of mass by taking a derivative of its position **R** [Eq. (8–48)]:

$$\mathbf{V} \equiv \frac{d\mathbf{R}}{dt} = \frac{m_1 \mathbf{v}_1 + m_2 \mathbf{v}_2 + \cdots + m_N \mathbf{v}_N}{M}. \tag{8–50}$$

The numerator is the sum of the momenta of the individual pieces, so we may write this result as

$$\mathbf{V} = \mathbf{P}_{\text{tot}}/M. \tag{8–51}$$

Differentiating once more with respect to time gives us the acceleration, **A**, of the center of mass:

$$\mathbf{A} \equiv \frac{d\mathbf{V}}{dt} = \frac{d^2\mathbf{R}}{dt^2}. \tag{8–52}$$

Multiplying by M gives the relation

$$M\frac{d^2\mathbf{R}}{dt^2} = \frac{d\mathbf{p}_1}{dt} + \frac{d\mathbf{p}_2}{dt} + \cdots + \frac{d\mathbf{p}_N}{dt} = \frac{d\mathbf{P}_{\text{tot}}}{dt}. \tag{8–53}$$

The importance of the center of mass

The center of mass is useful for several reasons. Equation (8–51) shows that, *in the absence of external forces, the center of mass moves with constant velocity*. No matter how complicated the motion of the constituent objects, the motion of its center of mass is simple. We shall soon see how Eq. (8–53) shows that when the system is not isolated—so that external forces act on its constituents—its *center of mass moves according to Newton's second law, with an acceleration equal to the net external force divided by the total mass*. For example, the center of mass of a diver moves with the same, simple parabolic motion that a rock with the same initial velocity would have if it were thrown from the diving board. It is important to be able to calculate the center of mass, as in Examples 8–9 and 8–10.

EXAMPLE 8–9 Two pointlike masses are placed on a massless rod that is 1.5 m long. The masses are placed as follows: 1.6 kg at the left end and 1.8 kg 1.2 m from the left end. (a) What is the location of the center of mass? (b) By moving the 1.8-kg mass, can you arrange to have the center of mass in the middle of the rod?

Solution: (a) Figure 8–21 illustrates this one-dimensional problem. We align the rod with length L along the x-axis, and we place its left end at the origin. We

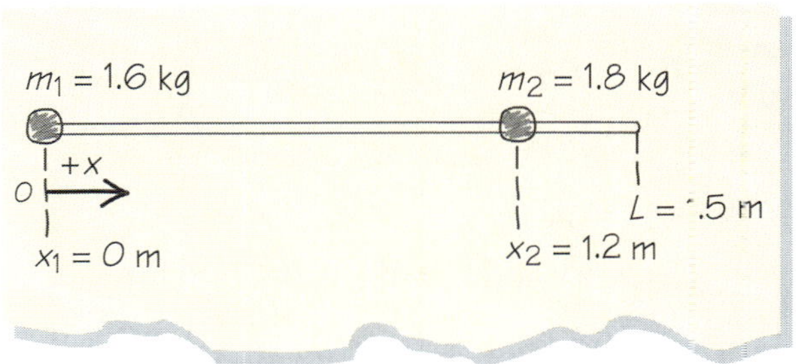

FIGURE 8–21 Example 8–9. The rod is aligned with the x-axis, with its left end at the origin.

call the 1.6-kg mass m_1, placed at $x_1 = 0$ m. The 1.8-kg mass is m_2, and it is placed at $x = 1.2$ m. The location of the center of mass is given by Eq. (8–49a):

$$X = \frac{m_1x_1 + m_2x_2}{m_1 + m_2} = \frac{(1.6\text{ kg})(0\text{ m}) + (1.8\text{ kg})(1.2\text{ m})}{1.6\text{ kg} + 1.8\text{ kg}} = 0.64\text{ m}.$$

(b) The midpoint of the rod is $x_{\text{mid}} = 0.75$ m. Let's place m_2 at a new position x_2 and ask if there is a solution for x_2 along the rod ($x_2 < L$) such that $X = x_{\text{mid}}$. We have

$$X = x_{\text{mid}} = \frac{m_1x_1 + m_2x_2}{m_1 + m_2} = \frac{m_1(0\text{ m}) + m_2x_2}{M} = \frac{m_2x_2}{M}.$$

We can solve this equation for x_2:

$$x_2 = \frac{Mx_{\text{mid}}}{m_2} = \frac{(1.6\text{ kg} + 1.8\text{ kg})(0.75\text{ m})}{1.8\text{ kg}} = 1.4\text{ m}.$$

This value is indeed less than L, so if we place the 1.8-kg mass at this point, the center of mass is at the midpoint of the rod.

EXAMPLE 8–10 Add a third mass of 2.3 kg to the masses of Example 8–9 at the point shown in Fig. 8–22. Find the center of mass.

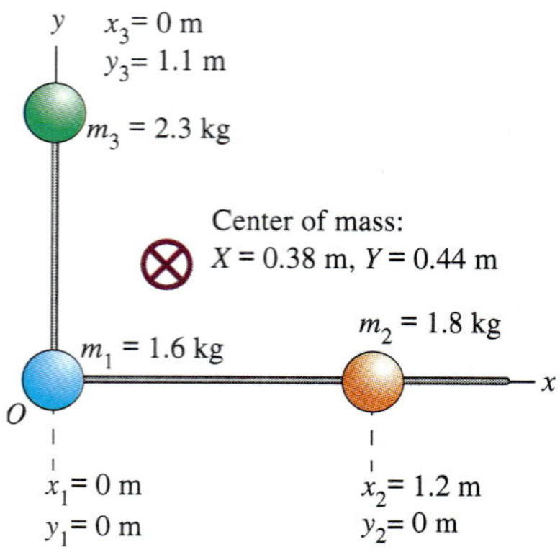

FIGURE 8–22 Example 8–10. The location of the center of mass is marked with a circled X.

Solution: By adding the third mass, m_3, the problem has become a two-dimensional one. We again align the first two masses along the x-axis and place the y-axis as shown in Fig. 8–22, so that the third mass is at $x_3 = 0$ m, $y_3 = 1.1$ m. We must now use Eqs. (8–49) to find both X and Y. With $M = 1.6\text{ kg} + 1.8\text{ kg} + 2.3\text{ kg} = 5.7\text{ kg}$, we have

$$X = \frac{m_1x_1 + m_2x_2 + m_3x_3}{M}$$

$$= \frac{(1.6\text{ kg})(0\text{ m}) + (1.8\text{ kg})(1.2\text{ m}) + (2.3\text{ kg})(0\text{ m})}{5.7\text{ kg}} = 0.38\text{ m},$$

$$Y = \frac{m_1y_1 + m_2y_2 + m_3y_3}{M}$$

$$= \frac{(1.6\text{ kg})(0\text{ m}) + (1.8\text{ kg})(0\text{ m}) + (2.3\text{ kg})(1.1\text{ m})}{5.7\text{ kg}} = 0.44\text{ m}.$$

This point is indicated in Fig. 8–22.

The Center of Mass and External Forces

What happens when there are external forces present in addition to interparticle forces? Let's again consider the three-object system shown in Fig. 8–23 and write

$$\frac{d\mathbf{p}_1}{dt} = \mathbf{F}_{12} + \mathbf{F}_{13} + \mathbf{F}_{1,\text{ext}},$$

$$\frac{d\mathbf{p}_2}{dt} = \mathbf{F}_{21} + \mathbf{F}_{23} + \mathbf{F}_{2,\text{ext}},$$

$$\frac{d\mathbf{p}_3}{dt} = \mathbf{F}_{31} + \mathbf{F}_{32} + \mathbf{F}_{3,\text{ext}}.$$

Here $\mathbf{F}_{i,\text{ext}}$ is the external force on object i. If we add these equations, use

$$\mathbf{P}_{\text{tot}} = \mathbf{p}_1 + \mathbf{p}_2 + \mathbf{p}_3,$$

and use Newton's third law—according to which $\mathbf{F}_{12} + \mathbf{F}_{21} = 0$, $\mathbf{F}_{13} + \mathbf{F}_{31} = 0$, and $\mathbf{F}_{23} + \mathbf{F}_{32} = 0$—we find that

$$\frac{d\mathbf{P}_{tot}}{dt} = \mathbf{F}_{1,ext} + \mathbf{F}_{2,ext} + \mathbf{F}_{3,ext} = \mathbf{F}_{tot,ext}. \quad (8\text{–}54)$$

This procedure can be extended to N objects, and Eqs. (8–53) and (8–54) give the general result:

$$\mathbf{F}_{tot,ext} = \frac{d\mathbf{P}_{tot}}{dt} = M\frac{d^2\mathbf{R}}{dt^2}. \quad (8\text{–}55)$$

For any system with total mass M in which there are both internal and external forces, we have shown that *the center of mass moves like a single point mass of mass M subject to the total external force on the system.* In other words, the system obeys the equation

$$\mathbf{F}_{tot,ext} = M\mathbf{A} \quad (8\text{–}56)$$

where $\mathbf{F}_{tot,ext}$ is the total external force on the system and $\mathbf{A}$ is the acceleration of the center of mass.

This result is very useful. Consider a bowling ball rolling down an inclined plane. The bowling ball consists of some 10^{27} molecules and these interact with each other by means of electric forces. Despite this huge complexity, it is possible to treat a bowling ball as a point mass (or as a simple rigid object, as we shall see in Chapters 9 and 10) and study its motion without being aware of the underlying atomic structure of the ball or even of the laws that govern the motion of atoms. Without this possibility, the development of physics would not have occurred.

Figure 8–24 shows that the motion of a system's center of mass is simple even though there is a rather complex motion of parts of the system.

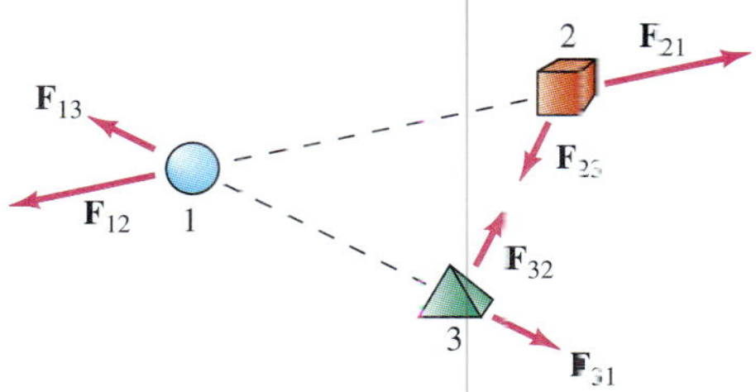

FIGURE 8–23 There are altogether six forces in a three-object system. These forces are labeled both by the object that causes them and the object on which they act. According to Newton's third law, these forces, whether they are attractive or repulsive, are equal and opposite in pairs.

The center of mass moves under the influence of external forces according to Newton's second law.

The Center of Mass of a Continuous Mass Distribution

We know that matter is composed of discrete masses; if we look closely enough, we will find evidence for the presence of atoms. But to an instrument that is not sufficiently acute to see atoms (such as our eyes), a solid object looks and behaves like a continuum of matter; that is, a *continuous mass distribution*. We may define the center of mass of a continuous distribution of mass by a simple generalization of Eq. (8–49). Consider a thin rod of mass M lying along the x-axis. We may take one end to lie at $x = a$ and the other at $x = b$. Now let's divide the interval between a and b into a large number N of tiny segments (Fig. 8–25). The segments are of equal length, Δx, but they are not necessarily of equal mass, because the rod may not have a uniform

FIGURE 8–24 The center of mass of the dancer moves as a point mass under the influence of an external force. In this case, the external force is gravity so the center of mass has a parabolic trajectory.

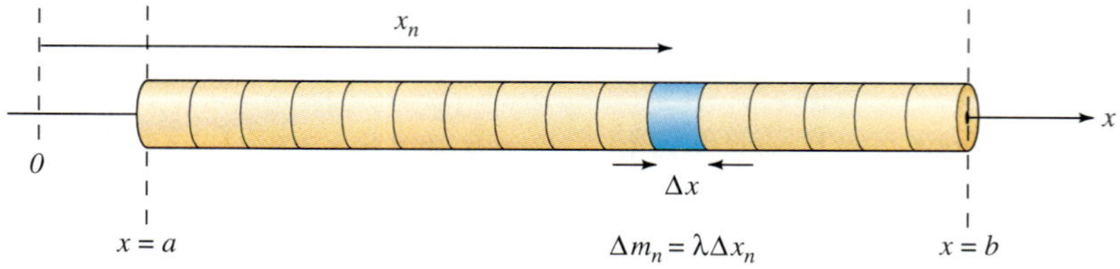

FIGURE 8–25 A continuous mass is divided into N tiny mass segments of equal length Δx (where N is large). This technique allows us to find the center of mass. The nth segment, x_n, has mass Δm_n.

mass distribution. The position of the nth segment is labeled x_n, and we call the mass of this tiny segment Δm_n. We may write this mass in terms of the local mass density:

$$\Delta m_n = \left(\frac{\Delta m_n}{\Delta x}\right)\Delta x. \tag{8–57}$$

The masses of all the segments add up to the total mass of the rod, so that we have

$$\Delta m_1 + \Delta m_2 + \cdots + \Delta m_n = \left(\frac{\Delta m_1}{\Delta x} + \frac{\Delta m_2}{\Delta x} + \cdots + \frac{\Delta m_n}{\Delta x}\right)\Delta x = M. \tag{8–58}$$

If we were to use infinitesimal intervals, this equation would read

$$M = \int_a^b \left(\frac{\mathrm{d}m}{\mathrm{d}x}\right)\mathrm{d}x. \tag{8–59}$$

According to Eq. (8–49a), we have

$$\begin{aligned} X &= \frac{\Delta m_1 x_1 + \Delta m_2 x_2 + \cdots + \Delta m_n x_n}{M} \\ &= \frac{1}{M}\left[\left(\frac{\Delta m_1}{\Delta x}\right)x_1 + \left(\frac{\Delta m_2}{\Delta x}\right)x_2 + \cdots + \left(\frac{\Delta m_n}{\Delta x}\right)x_n\right]\Delta x. \end{aligned}$$

As $N \to \infty$ this equation becomes an integral:

$$X = \frac{1}{M}\int_a^b \left(\frac{\mathrm{d}m}{\mathrm{d}x}\right)x\,\mathrm{d}x.$$

It is usual to write the mass density $\mathrm{d}m/\mathrm{d}x$ of a one-dimensional object as λ. If λ is not a constant, it will depend on x. In this notation, the position of the center of mass is

$$X = \frac{1}{M}\int_a^b x\lambda\,\mathrm{d}x. \tag{8–60}$$

Equation (8–60) allows us to calculate the center of mass of a one-dimensional object, as in Example 8–11.

EXAMPLE 8–11 Consider a rod of length L, whose mass density ($\mathrm{d}m/\mathrm{d}x$) is given by $\lambda = C(1 + ax^2)$, where x is the distance from the light end and C is a constant with dimensions of mass per length (Fig. 8–26). Calculate the center of mass of the rod.

Solution: We first note that the total mass of the rod is given by

$$M = \int_0^L \lambda\,\mathrm{d}x = \int_0^L C(1 + ax^2)\mathrm{d}x = C\left(x + \frac{ax^3}{3}\right)\Bigg|_0^L = C\left(L + \frac{aL^3}{3}\right).$$

Now, from Eq. (8–60),

$$X = \frac{1}{M}\int_0^L xC(1 + ax^2)\mathrm{d}x = \frac{1}{M}C\left(\frac{x^2}{2} + \frac{ax^4}{4}\right)\Bigg|_0^L$$

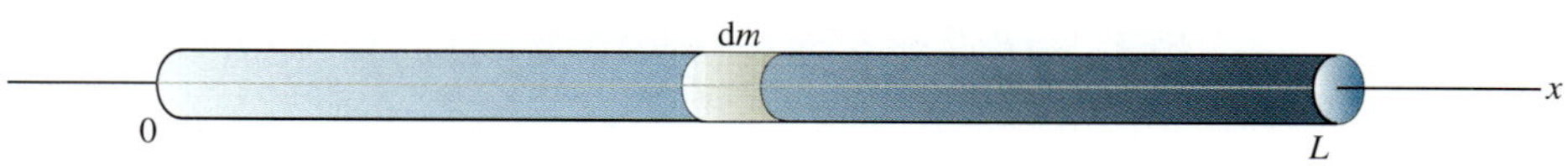

FIGURE 8–26 Example 8–11.

$$= \frac{1}{M} C\left(\frac{L^2}{2} + \frac{aL^4}{4}\right)$$

$$= \frac{C[(L^2/2) + (aL^4/4)]}{C[L + (aL^3/3)]} = \frac{(L/2) + (aL^3/4)}{1 + (aL^2/3)} = \left(\frac{L}{2}\right)\frac{1 + (aL^2/2)}{1 + (aL^2/3)}.$$

If $a = 0$, then the rod is uniform, and the center of mass is at $L/2$, as expected. In the case of nonuniformity, the center of mass is closer to the more massive end. Note that the parameter C of the mass density cancels.

Continuous Objects in Two and Three Dimensions. Suppose that instead of a one-dimensional object, we deal with a two-dimensional one such as a sheet of metal with an area A. We divide the object into tiny elements of area $\Delta A = \Delta x\,\Delta y$ and mass Δm (Fig. 8–27). Suppose that for the element located at the point (x, y) the mass density is $\sigma(x, y)$. This quantity is analogous to the one-dimensional mass density λ; σ has dimensions of mass/unit area. If we take the area element small enough so that σ is a constant within that area, then the mass of that area element is

$$\Delta m = \sigma\,\Delta x\,\Delta y.$$

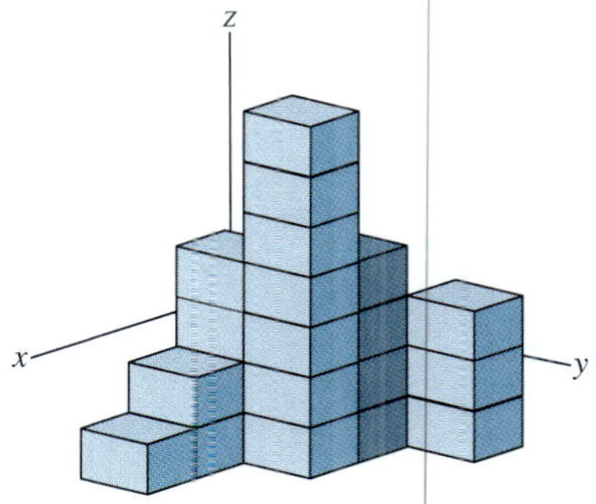

FIGURE 8–27 We have placed a two-dimensional object in the xy-plane and divided it up into squares, each of area $\Delta x\,\Delta y$. The mass of a square at a given point is the two-dimensional density σ times the area. The height of the towers in this figure represents the value of σ; we see it can vary from point to point.

Remember, the value of σ can vary from place to place, as the varying height within Fig. 8–27 indicates. We can find the total mass by summing the masses over all the elements, and, in the calculus limit, this summation takes the form[†]

$$M = \lim \sum \Delta m = \lim \sum \sigma\,\Delta x\,\Delta y = \int_{\text{Surface}} \sigma(x, y)\,dx\,dy. \quad (8\text{–}61)$$

Similar reasoning tells us that the coordinates of the center of mass are

$$MX = \lim \sum x\,\Delta m = \int_{\text{Surface}} x\sigma(x, y)\,dx\,dy; \quad (8\text{–}62\text{a})$$

$$MY = \lim \sum y\,\Delta m = \int_{\text{Surface}} y\sigma(x, y)\,dx\,dy. \quad (8\text{–}62\text{b})$$

When we want to learn about the center of mass of a three-dimensional object of volume V, we must know its three-dimensional mass density $\rho(x, y, z)$, measured in mass/unit volume. The dependence on x, y, z is merely an expression of the fact that the mass density can vary from point to point. The total mass is

$$M = \int_{\text{Volume}} \rho(x, y, z)\,dx\,dy\,dz, \quad (8\text{–}63)$$

while the center of mass is located at the position

$$M\mathbf{R} = \int_{\text{Volume}} \mathbf{r}\rho(x, y, z)\,dx\,dy\,dz. \quad (8\text{–}64)$$

8–7 ROCKET MOTION

A rocket is a system that undergoes a kind of continuous explosion. To understand rocket motion, imagine an astronaut who steps out of a space vehicle to do some repairs and, for some reason, finds himself or herself at rest with respect to the vehicle but separated from it. No amount of bodily contortions will bring the astronaut closer to the vehicle because the center of mass of the astronaut, if initially at rest, will remain at rest in the absence of external forces. If the astronaut is holding something that

[†] The subscript "Surface" on the single integral sign denotes a two-dimensional integration over the area A.

can be thrown away, then he or she is in business. For example, if the astronaut throws a hammer in a direction away from the space vehicle, he or she will start to move toward the space vehicle and safety, albeit with a small speed (Fig. 8–28). Momentum conservation lies behind this motion. If the astronaut and the hammer are initially at rest, the total momentum of the astronaut–hammer system is zero, and this must still be the case after the hammer is thrown. A rocket moves because of the same principle used by the astronaut. The ejected fuel of the rocket acts as a whole series of thrown hammers. The presence or absence of an atmosphere has nothing to do with the basic principle of operation.

APPLICATION

Rocket propulsion

FIGURE 8–28 An astronaut can get a small push back toward the Space Shuttle by throwing a hammer in the opposite direction.

A rocket is propelled by the ejection of hot gases from burning chemical fuel. Two important parameters for describing the characteristics of a rocket are the exhaust speed[†] of the hot gases, u_{ex}, and the rate at which the gases are expelled by the rocket; that is, the rate of change of mass dm/dt of the rocket (Fig. 8–29). Keep in mind that the exhaust speed involves a burning reaction within the rocket and that it is measured *with respect to the rocket*. Suppose the initial mass of the rocket with all its fuel is m_0; at some later time t, its velocity with respect to some ground-based (inertial) observer is v while its mass has been reduced to m. In the next small time interval Δt, the ground-based observer sees a small bit of exhaust gas, mass Δm, is ejected backward relative to the rocket; the ground-based observer sees this bit of gas moving with velocity $v - u_{ex}$, while the observer sees the rocket itself now traveling at velocity $v + \Delta v$ and with the decreased mass $m - \Delta m$. In the absence of external forces, the total momentum at time t equals the total momentum at time $t + \Delta t$. We have $P(t) = mv$, while $P(t + \Delta t) = (\Delta m)(v - u_{ex}) + (m - \Delta m)(v + \Delta v)$. (The last two terms are the momentum of the exhaust gas and the rocket, respectively.) Thus

$$\begin{aligned} mv &= (\Delta m)(v - u_{ex}) + (m - \Delta m)(v + \Delta v) \\ &= (\Delta m)v - (\Delta m)u_{ex} + mv + m\Delta v - (\Delta m)v - (\Delta m)(\Delta v), \end{aligned}$$

or, neglecting the very small term $(\Delta m)(\Delta v)$,

$$m\Delta v + (\Delta m)u_{ex} = 0. \tag{8–65}$$

This equation describes the fact that the momentum given up by the rocket is taken up by the gas (Fig. 8–30). Let us divide Eq. (8–65) by the time interval Δt and take the limit $\Delta t \to 0$. We have

$$-u_{ex}\frac{dm}{dt} = m\frac{dv}{dt}. \tag{8–66}$$

This equation looks just like the usual $F = ma$ equation. The left-hand side of Eq. (8–66) is called the **thrust** of the rocket. Notice that because dm/dt is negative (the rocket loses mass), the thrust is a positive quantity. Equation (8–66) may be written in a more convenient form by dividing both sides of the equation by mu_{ex}. It then takes the form

$$\frac{1}{u_{ex}}\frac{dv}{dt} = -\frac{1}{m}\frac{dm}{dt}. \tag{8–67}$$

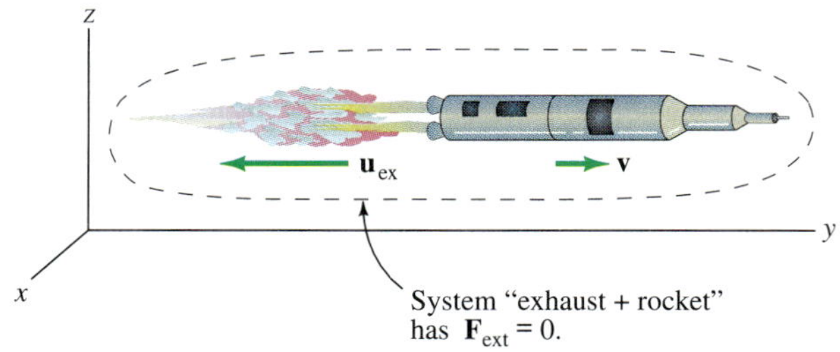

FIGURE 8–29 A rocket expels gases that have a speed u_{ex} with respect to the rocket. If we know the rate at which the gases are expelled, conservation of momentum allows us to find the velocity **v** of the rocket.

[†]Sometimes called the exhaust velocity. We prefer the term speed because it is always positive.

We now use

$$\frac{d(\ln m)}{dt} = \left(\frac{1}{m}\right)\left(\frac{dm}{dt}\right). \tag{8–68}$$

[This formula follows from the calculus results in Appendix IV–7. We use

$$\frac{d(\ln x)}{dx} = \frac{1}{x}.$$

We must replace x by m, and use the chain rule of calculus.] It follows from Eqs. (8–67) and (8–68) that

$$\frac{d}{dt}\left(\ln m + \frac{v}{u_{ex}}\right) = 0.$$

This equation is equivalent to

$$\ln m = -\frac{v}{u_{ex}} + \text{a constant}. \tag{8–68}$$

At $t = 0$, we have $m = m_0$, the initial mass of the rocket, and $v = 0$, for example. In this case, we find that the constant in Eq. (8–68) is given by $\ln m_0$. Equation (8–68) is rearranged to give

$$\frac{v}{u_{ex}} = (\ln m_0) - (\ln m). \tag{8–69}$$

Finally, we use the fact that $(\ln a) - (\ln b) = \ln(a/b)$ to obtain

$$v = u_{ex} \ln\left(\frac{m_0}{m}\right). \tag{8–70}$$

FIGURE 8–30 The momentum of the rocket plus the fuel it has just ejected is constant, but the rocket has a forward momentum because the fuel has a momentum toward the rear.

The speed of a rocket

This equation for the rocket speed is a fundamental one in rocket propulsion.

Let's look more closely at Eq. (8–70). First of all, we note that the speed v is proportional to u_{ex}. This dependence is to be expected because u_{ex} is the only quantity with the dimensions of speed that appears in the problem. Next, we observe that the coefficient of u_{ex} must be a dimensionless quantity, and therefore it must be a function of m/m_0, which is the fraction of the rocket mass left over after time t. Because the logarithm is a very slowly varying function of its argument, the most productive way to get a sizable speed is by making u_{ex} large. Chemical rockets can generate gas exhaust speeds up to a maximum of about 4000 m/s, and thus the typical speeds of rockets are also in the range of 10^3 m/s. Large values of m_0/m help only moderately; for example, $\ln 10 = 2.3$, so changing the mass ratio by a factor of 10 improves the speed by a factor of 2.3. Much higher speeds can be achieved with nuclear reactions, but rockets that use nuclear energy for propulsion in this way would present enormous environmental problems!

Rocket Motion in the Presence of Gravity

The previous discussion involves a rocket free of all external forces. When gravity acts on the rocket, Eq. (8–70) is modified so that, in addition to the term proportional to $u_{ex} \ln(m_0/m)$ due to the fuel exhaust, the velocity has a falling motion due to gravity superimposed on it. As we already know from Chapter 2, the velocity in free fall is $-gt$. Thus Eq. (8–70) is modified to

$$v = u_{ex} \ln\left(\frac{m_0}{m}\right) - gt. \tag{8–71}$$

EXAMPLE 8–12 A Saturn V rocket (the vehicle that sent humans to the Moon) of mass 2.5×10^6 kg takes off from Earth in a vertical direction. It burns fuel at a uniform rate of 1.6×10^4 kg/s for a duration of 2 min. If the exhaust speed of the gas is given by 3.0 km/s, what is the speed of the rocket immediately after the combustion ceases?

Solution: We use the formula in Eq. (8–71) to find the rocket's speed. The mass m in this formula is the final mass of the rocket after the gas has been burned. We can calculate this mass from the initial mass, m_0, and the rate of combustion, dm/dt:

$$m = m_0 - \frac{dm}{dt}t = 2.5 \times 10^6\ \text{kg} - (1.6 \times 10^4\ \text{kg/s})(120\ \text{s}) = 0.58 \times 10^6\ \text{kg}.$$

We also have

$$\ln\left(\frac{m_0}{m}\right) = \ln\left(\frac{2.5 \times 10^6\ \text{kg}}{0.58 \times 10^6\ \text{kg}}\right) = \ln\left(\frac{2.5}{0.58}\right) = 1.5.$$

Thus, from Eq. (8–71),

$$v_f = (3.0 \times 10^3\ \text{m/s})(1.5) - (9.8\ \text{m/s}^2)(120\ \text{s}) = 3.2 \times 10^3\ \text{m/s}.$$

Our calculations are correct only near the surface of Earth. Earth's gravitational force on the rocket decreases as the rocket moves away from the surface, so in effect g decreases as the rocket moves away.

* 8–8 MOMENTUM TRANSFER AT HIGH ENERGIES

High-energy physics is the study of the properties of the elementary particles that make up matter at the most fundamental level. These properties are studied with the aid of high-energy accelerators. There, projectile particles such as protons or electrons are accelerated to high speeds and allowed to collide with target particles of the same type. The results of such collisions are then analyzed by means of sophisticated detectors in which the energies and momenta of the particles that emerge after the collisions are precisely measured. The patterns revealed in these measurements are used to probe the underlying structure of the colliding particles.

As an example of how such studies are made, imagine a BB gun fired at two targets: One is cotton candy, the other is cotton candy with a tiny steel pellet implanted at its center. A BB shot (the projectile) will hardly be deflected by the first target. Most of the time, the same will be true for the second target. Occasionally, however, the projectile will penetrate the second target close to its center and, upon colliding with the steel pellet, will be deflected through a large angle. An analysis of the collisions in the two cases would allow us to deduce that the second target had a hard, compact center. The pattern of many projectile deflections would even allow us to say whether the steel pellet is spherical or cubical. In this example, there is no benefit in increasing the power of the BB gun. All we would need to improve the experiment is to be able to aim more accurately at the center of the cotton candy.

In the case of high-energy collisions, though, there is indeed a benefit in increasing the speed—or momentum—of the projectile. For collisions of atomic or sub-atomic particles, the rules of *quantum mechanics* apply. This is a field of study developed in the first part of the twentieth century, with some surprising consequences. One consequence of quantum mechanics is that high momenta (or high energy) are required to study the regions close to the center of a target. More precisely, the **Heisenberg uncertainty relation**, discovered by physicist Werner Heisenberg, describes a relation between the magnitude of the momentum change, $\Delta\mathbf{p}$, of a target particle in a collision and the size of the region around the center of the target that can be studied. Let us denote by r_{coll} the smallest radius of a spherical region around the center of a target that can be studied in a collision. Then the uncertainty relation is

$$|\Delta \mathbf{p}| r_{coll} \simeq \hbar, \qquad (8\text{–}72)$$

where $\hbar \simeq 10^{-34}\ \mathrm{J \cdot s}$ is Planck's constant divided by 2π. According to Eq. (8–72), the larger the value of $\Delta \mathbf{p}$, the smaller r_{coll} can be.

Experiments with elementary particles—like the BB gun fired at cotton candy—have been carried out and have revealed much about the structure of these particles. Ernest Rutherford's experiments in the first decade of the twentieth century were the first to reveal the presence of the nucleus in atoms, which plays a role like the steel pellet in the cotton candy. More recently, experiments involving protons colliding at momenta that could probe down to distances of 10^{-15} m revealed that the proton itself contains "pellets" known as *quarks*. When a next generation of accelerators is built, we will discern even more detail. Judging by the enormous difference between the physics of atoms and the physics of nuclei, we can expect some surprises as we probe more and more deeply.

SUMMARY

The momentum (or linear momentum) of an object of mass m moving with velocity $\mathbf{v}$ is defined by

$$\mathbf{p} = m\mathbf{v}. \qquad (8\text{–}2)$$

The rate of change of momentum with time is given by Newton's second law:

$$\mathbf{F} = \frac{d\mathbf{p}}{dt}, \qquad (8\text{–}3)$$

where $\mathbf{F}$ is the net force acting on the object. The momentum of a collection of objects with momenta $\mathbf{p}_i$ (where $i = 1, 2, \cdots, N$) is their vector sum, and the net external force on the system determines the rate of change of the total momentum. In particular, if the only forces present are the internal forces exerted by the objects on each other, then the total momentum does not change with time; that is, momentum is conserved. In the case of two objects, this relation is

$$\mathbf{p}_1 + \mathbf{p}_2 = \text{a constant}. \qquad (8\text{–}9)$$

One consequence of Eq. (8–3) is that the change in momentum of an object influenced by a force that acts over a limited time (an impulsive force) is an integral over the force and is called the impulse, $\mathbf{J}$:

$$\mathbf{J} \equiv \int \mathbf{F}\, dt = \mathbf{F}_{av}\, \Delta t. \qquad (8\text{–}15)$$

The impulse does not involve the detailed time dependence of the force.

Collisions involve interparticle forces, so that the total momentum of the system of objects is conserved whether the collisions are elastic, inelastic, or perfectly inelastic. During elastic collisions, the sum of the kinetic energies is conserved. Inelastic collisions involve a change in the sum of the kinetic energies. The change is generally a decrease due to the presence of dissipative forces, such as friction. In perfectly inelastic collisions, the objects that collide merge into a single object whose mass is the sum of the masses of the colliding objects.

The center of mass of a collection of objects with masses m_i and position vectors $\mathbf{r}_i$ ($i = 1, 2, \cdots, N$) is the mean position of a system's mass, which is defined by

$$\mathbf{R} \equiv \frac{m_1\mathbf{r}_1 + m_2\mathbf{r}_2 + \cdots + m_N\mathbf{r}_N}{M}. \qquad (8\text{–}48)$$

In the absence of external forces, the center of mass moves with uniform velocity **V**, given by

$$\mathbf{V} = \frac{\mathbf{P}_{\text{tot}}}{M} = \frac{m_1\mathbf{v}_1 + m_2\mathbf{v}_2 + \cdots + m_N\mathbf{v}_N}{M}. \tag{8–50}$$

The total mass of the system is $M = m_1 + m_2 + \cdots + m_N$. The center of mass for a continuous object may be calculated in terms of the mass density. For a one-dimensional object with mass density λ, it is given by

$$X = \frac{1}{M}\int \lambda x \,\mathrm{d}x, \tag{8–60}$$

where the mass is related to $\lambda \equiv \mathrm{d}m/\mathrm{d}x$ by

$$M = \int\left(\frac{\mathrm{d}m}{\mathrm{d}x}\right)\mathrm{d}x. \tag{8–59}$$

For a two-dimensional object with mass density σ, the corresponding expressions are

$$X = \frac{1}{M}\int_A \sigma\, x \,\mathrm{d}x\,\mathrm{d}y \tag{8–62a}$$

and

$$Y = \frac{1}{M}\int_A \sigma\, y \,\mathrm{d}x\,\mathrm{d}y, \tag{8–62b}$$

where

$$M = \int_A \sigma \,\mathrm{d}x\,\mathrm{d}y. \tag{8–61}$$

Similar expressions hold for a three-dimensional object with mass density ρ.

When a net external force $\mathbf{F}_{\text{net}}$ acts on a system, the center of mass continues to move in a simple way, as described by Newton's second law: The position of the center of mass **R** obeys the equation

$$\mathbf{F}_{\text{tot,ext}} = M\mathbf{A} = M\frac{\mathrm{d}^2\mathbf{R}}{\mathrm{d}t^2}. \qquad (8–55), (8–56)$$

Here **A** is the acceleration of the center of mass.

The concept of momentum conservation is useful for the description of rocket motion. For a rocket that is not influenced by external forces, whose mass is initially m_0, and whose mass at time t is m, the speed is

$$v = u_{\text{ex}} \ln\left(\frac{m_0}{m}\right). \tag{8–70}$$

Here, u_{ex} is the exhaust speed of the gas relative to the rocket.

According to the Heisenberg uncertainty relation, when collisions are used to study the structure of a target, the momentum of an incoming projectile must increase as the distance to be probed decreases.

UNDERSTANDING KEY CONCEPTS

1. Discuss momentum conservation for a comet that enters the solar system, is deflected, and then leaves the solar system in a direction different from the direction by which it entered.
2. A common toy consists of a series of five balls that touch and form a line; each ball is suspended as a pendulum. When ball 1 is pulled away and then released, it strikes the line and ball 5 rises in a motion like the reverse of ball 1. How can you explain this motion?
3. Describe a physical object for which the center of mass is not actually inside the object.
4. A diver leaps off the diving board, performing a difficult series of maneuvers. Can he cause his center of mass to perform a midair loop?
5. The center of mass of a championship-level high jumper passes below the bar even though the jumper passes above the bar. How is this possible?

6. A particle collides with another particle at rest. If there are two particles observed to come out of the collision, do the momenta of the two final state particles and the initial incoming particle have to lie in a plane? Why or why not? What if the final system consists of three particles?

7. A vase falls to the floor and shatters. Is momentum conserved in the collision? What objects need to be taken into account in describing the conservation of momentum?

8. As a tennis racket hits a tennis ball, the racket continues to move forward. Is this consistent with the conservation of momentum?

9. If a tennis player wants more power, should he or she choose a racket with more or less tension on the strings? If the player wants more control? [*Hint*: Think about how long the ball is on the strings in each case.]

10. The mass on the end of a pendulum swings due to the effects of gravity on the mass. Does momentum have to be conserved in the interaction of this mass with Earth? If so, does this mean that Earth moves back and forth along with the mass?

11. We have said that an impulsive force acts for only a short time. What decides whether a time is "short" or not?

12. Cricket players catch balls hit as hard as baseballs but they do not use padded mitts. How do they avoid injuring their hands?

13. If you have the misfortune to be in an automobile collision, you are better off in a more massive car (all other things being equal). Why?

14. If you have the misfortune to be in an automobile collision, you are better off in a car that tends to crumple on impact rather than a car that holds together stiffly. Why?

15. A closed railroad car is at rest on a flat stretch of track. A cannon located inside the car at the front end points to the rear. What is the motion of the railroad car when the cannon is fired, and the shell is absorbed by the rear wall?

16. A large, closed crate contains many pigeons that sit on the floor of the crate. A sudden noise makes them all fly up and hit the top of the crate at the same time. Will they be able to lift the crate off the ground?

17. Drag forces act when a parachute opens. The parachute and its load will slow down to a terminal velocity. Why is the drag force velocity dependent? Think of the parachute as colliding with a lot of tiny air molecules.

18. A billiard ball strikes the cushion of a billiard table at an angle of 45°. Assuming that there is no energy loss (and no effects due to the spin of the ball), what can you say about the angle at which the billiard ball bounces off? What would happen if there were some energy loss?

PROBLEMS

8–1 Momentum and Its Conservation

1. (I) Calculate the magnitudes of the momenta of (a) a 40-g arrow traveling at a speed of 110 km/h; (b) a 145-g baseball traveling at a speed of 35 m/s; (c) a 72-kg sprinter running at 22 mi/h; (d) a 95-kg tackler running 100 m in 12.5 s.

2. (I) A 0.800-kg mass falls vertically downward from a roof 36.5 m high. What is the momentum of the object after 2.0 s, given that the initial velocity is zero?

3. (I) Calculate the magnitudes of the momenta of (a) a man of mass 70 kg, running 6 m/s; (b) a freight car of 100,000 kg, moving 60 m/s; (c) a car of 1100 kg, moving 25 mi/h; (d) a proton moving at 2×10^5 m/s; (e) a feather of 10 g in an airless container that has fallen 10 cm due to gravity.

4. (II) An object of mass m is constrained to move in a circle of radius R by a central force F. (a) What is the magnitude of the momentum of the object? (b) Suppose that the force has magnitude $F = Kv$, where K is a constant and v is the speed of the object. Calculate R in terms of K and the momentum of the object. (This kind of force acts on a charged object in a uniform magnetic field.)

5. (II) A 7-kg rifle is used to fire a 10-g bullet, which travels with a speed of 700 m/s. (a) What is the speed of recoil of the rifle? (b) How much energy does it transmit to the shoulder of the person using the rifle as it stops?

6. (II) Two objects, of masses m and M, respectively, move in circular orbits that have the same center (Fig. 8–31). If the force that gives rise to this motion is a force of attraction between the two objects acting along a line joining them, using momentum conservation and Newton's second law, (a) show that they move with the same angular speed (see Chapter 3), and (b) calculate the ratio of the radii of the two circular orbits.

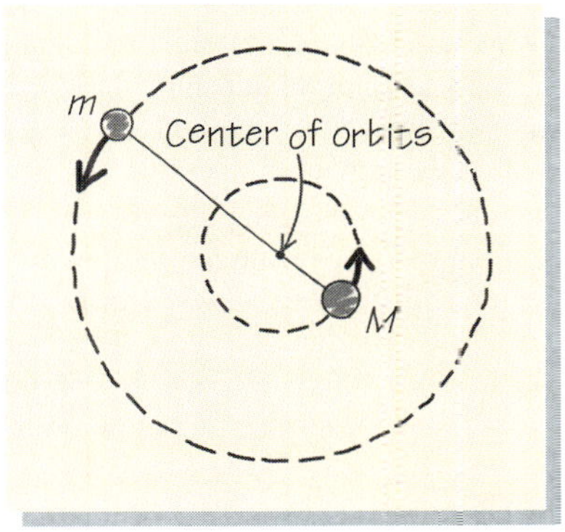

FIGURE 8–31 Problem 5.

7. (II) Two objects are moving in the xy-plane. The first, with mass 2.4 kg, has a velocity $\mathbf{v}_1 = (-2.0\text{ m/s})\mathbf{i} + (-3.5\text{ m/s})\mathbf{j}$; the second object, with mass 1.6 kg, has velocity $\mathbf{v}_2 = (1.8\text{ m/s})\mathbf{i} + (-1.5\text{ m/s})\mathbf{j}$. (a) What is the total momentum of the system? (b) If the system observed at a later time shows that the 2.4-kg object has $v_1' = (2.5\text{ m/s})\mathbf{i}$, what is the velocity of the 1.6-kg object? (c) Consider again the initial situation. Now suppose that there has been a mass transfer, so that the first object now has a mass of 2.1 kg. The total mass is conserved.

What is v_1' if the velocity of the second object is $(-2.5\text{ m/s})\mathbf{j} + (1.3\text{ m/s})\mathbf{k}$? (d) Calculate the sum of the kinetic energies in the initial configuration and in the configurations of parts (b) and (c). Compare and discuss them.

8. (III) (a) A billiard ball collides head-on with a second billiard ball that is touching a third billiard ball placed directly behind it. Describe the subsequent motion of all three billiard balls. Assume that all collisions are elastic. (b) A moving billiard ball hits the two balls at rest precisely between the two touching balls at rest. The velocity of the moving ball is perpendicular to the line between the centers of the two balls at rest. Again, all collisions are elastic. Describe the subsequent motion now.

8–2 Impulse and One-Dimensional Collisions

9. (I) A baseball of mass 0.15 kg, moving horizontally with a momentum of 4.0 kg · m/s, is struck head-on by a baseball bat with an impulse of 10 N · s. What is the speed of the baseball after it is struck?

10. (I) If a 25-g arrow moving at 90 km/h penetrates a block of wood suspended by a rope, what impulse is delivered to the block? What velocity will the wood block acquire if its mass is 5 kg? Ignore the mass of the arrow compared to that of the wood block.

11. (I) A 145-g baseball traveling at a speed of 36 m/s hits a bat and moves back along its incoming trajectory with a speed of 45 m/s. What is the impulse delivered to the ball by the bat? If the duration of the bat–ball collision is 7.0×10^{-4} s, what is the average force exerted by the bat on the ball during this period?

12. (I) A 120-g ball is dropped from a height of 4.8 m onto a hard floor and bounces back to exactly that height. (a) What is the impulse received by the ball? (b) What is the average force on the ball during the 0.013 s that, an independent measurement shows, it was in contact with the floor?

13. (I) A fire rescue unit uses a tightly woven net to catch an 80-kg person who jumps out of a burning building from a height of 11 m. What is the impulse transmitted to the net? If the net sinks 70 cm as it slows down the jumper, what is the average force exerted on the jumper by the net?

14. (II) (a) Estimate the possible impulse received by a nail when it is hit by a hammer. (b) Estimate the possible impulse received by a football when it is kicked in a kickoff. The ball has mass of about 250 g and travels a horizontal distance of some 50 m. Note the difference between parts (a) and (b); in the latter, you have additional information on the momentum the ball receives.

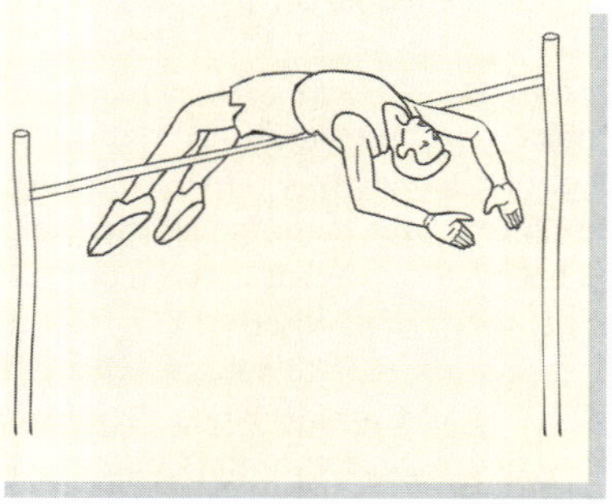

FIGURE 8–32 Problem 15.

15. (II) A high jumper of mass 55 kg clears 6′7″ (Fig. 8–32). What is the impulse transmitted to the jumper by the ground? (It has been found that the center of mass of the high jumper remains *below* the bar. In this problem, assume that the center of mass starts at 3′10″ and reaches the height 6′7″.)

16. (II) A communications satellite of mass 4000 kg can be ejected from the cargo bay of the Space Shuttle by means of springs. A particular satellite is ejected at 0.3 m/s. (a) What impulse does the spring provide? (b) If the spring operates over a time period of $\Delta t = 0.2$ s, what average force does the spring provide?

17. (II) A ball of mass 260 g is dropped from a height of 2.0 m. It hits the ground and rebounds to a height of 1.4 m. Assuming that the ball is in contact with the ground for 0.004 s, what is the average force exerted on the ball during the contact?

18. (II) A golf ball of mass 0.05 kg, placed on a tee, is struck by a golf club. The speed of the golf ball as it leaves the tee is 100 m/s. (a) What is the impulse? (b) If the time of contact between the club and the ball is 0.02 s, what is the average force? (c) If the force decreases to zero linearly with time during the 0.02 s, what is the value of the force at the beginning of the contact?

19. (II) A catcher catches a 45-m/s pitch and his glove recoils 0.25 m. The mass of the ball is 0.14 kg. Assume that the deceleration of the ball during the catching time, Δt, is constant and equal to some a_{av}. (a) Find the average force, F_{av}, that the catcher exerts. (b) How much work does he do? (c) Find Δt.

20. (II) Superman rushes to save Lois Lane, who has fallen (has been pushed?) from a window 100 m above a crowded street. Superman swoops down in the nick of time, arriving when Lois is 1.0 m above the street and stopping her just at ground level. Lois has been on a diet and has a mass of 50 kg. Ignore air resistance throughout. (a) What is the impulse that Lois receives as Superman catches her? (b) If the force that Superman supplies in stopping Lois is constant, how long does it take for Lois to come to rest? (c) What is the average force Superman applies to Lois? Compare this to the force of gravity on her. Draw your own conclusions about these last-minute rescues.

21. (II) A ball of mass 150 g is dropped from a height of 60 cm onto the first step at the top of a staircase. Each step is 20 cm above the next. The ball bounces perfectly elastically, but it has a small horizontal velocity, so it hits the second step on the next bounce, and then the third, and so on. Assume that the size of each step is such that the ball always bounces onto the next step down. What is the impulse transmitted to the nth step?

22. (II) A brother and sister live near a lake. In the middle of winter, when the lake is solidly and smoothly frozen, they play catch by sliding a rock back and forth between them across the ice. Ignore all friction. The mass of the two children is 40 kg each, and the rock has a mass of 2.0 kg. Each time the rock is given a push by the child who has just caught it, it leaves the hand with a speed of 1.0 m/s (relative to the hand). (a) The children are initially at rest, and the sister first slides the rock toward her brother. What is the initial motion of the sister after she releases the rock? (b) The brother receives the rock. What is his initial velocity after he has received it? (c) The brother then slides the rock to his sister. What is his velocity now? (d) The sister catches the rock. What is her velocity now?

(e) Will either child ever have a speed such that the rock will never reach the other player? In other words, can the game go on forever?

8–3 *Perfectly Inelastic Collisions; Explosions*

23. (I) A 250-g cart moves on an air track (a one-dimensional system in which motion is friction free) at 1.2 m/s. It collides with and sticks to another cart of mass 500 g, which was moving in the opposite direction at 0.80 m/s before the collision. What is the velocity of the composite cart after the collision?

24. (I) An object of mass m_0 with speed v_0 hits another object, of mass m, at rest. The two masses stick together in the process. Find the fractional change in kinetic energy.

25. (I) A snowball of mass 400 g is thrown with a speed of 10 m/s and hits the loosely placed head of a snowman. If the snowball sticks to the head, which is of mass 5 kg, with what initial speed will the now enlarged head recoil?

26. (I) A fireworks rocket is shot straight up in the air. Its "payload," mass 0.80 kg, explodes just as it reaches its peak height and separates into two fragments before eventually exploding again into spiraling colors. The first fragment, of mass 0.30 kg, heads straight down with an initial speed of 18 m/s. What is the velocity immediately after the explosion of the second fragment?

27. (I) Two objects each move with speed v in opposite directions along a line. They meet and have a completely inelastic collision. After the collision, the composite object moves along the same line with a speed of $v/2$. What is the ratio of the masses of the two objects?

28. (II) An ^{241}Am nucleus at rest emits an alpha particle (a ^{4}He nucleus). The energy released during the process is 6 MeV $= 9.6 \times 10^{-13}$ J. (a) What are the speeds of the alpha particle and the remaining ^{237}Np nucleus? (b) What is the kinetic energy of the ^{237}Np nucleus? (The masses of the nuclei may be taken to be $241m_0$, $4m_0$, and $237m_0$, respectively, where $m_0 = 1.66 \times 10^{-27}$ kg.)

29. (II) In a *ballistic pendulum,* a bullet of mass m and speed v embeds in a block of mass M suspended by a string (Fig. 8–33). The block and bullet, of total mass $m + M$, then moves as a pendulum and the maximum height h that it reaches can easily be measured. This apparatus can be used to measure the speed of the bullet. (a) What fraction of the bullet's kinetic energy is lost in the collision? (b) Give a formula for measuring v in terms of m, M, g, and h.

30. (II) A bullet strikes a ballistic pendulum (see Problem 29) whose target block has a mass of 5 kg. The block is observed to rise a distance of 3 cm, and the bullet is weighed and found to have a mass of 20 gm. What is the speed of the bullet when it strikes the block?

31. (III) Two persons, one of mass m_1 and the other of mass m_2, are sitting initially at rest on a bobsled of mass M on a frozen lake. Answer the following questions for a frame fixed to the lake. (a) Determine the final velocity of the bobsled if the person of mass m_1 jumps off the rear of the bobsled with horizontal velocity v, followed a few seconds later by the second person, who also jumps off in the same way with the same horizontal velocity with respect to the sled. (b) Find the velocity of the bobsled if the person of mass m_2 jumps off first. (c) What is the velocity if both jump off at the same time?

8–4 *Elastic and Inelastic Collisions*

32. (I) An object with velocity (1.4 m/s)**i** and mass 0.30 kg collides with an object whose velocity is (−2.5 m/s)**i** and whose mass is 0.15 kg. The motion takes place in one dimension. (a) What are the final velocities of the objects if the collision is elastic? (b) What is the total initial kinetic energy in the collision?

33. (I) On an air track, a 0.4-kg mass m_1 moves at 3 m/s in the positive direction. It approaches a stationary mass m_2 of 0.8 kg. They collide and, after the collision, the velocity of mass m_2 is 1.6 m/s in the positive direction. (a) What is the velocity of mass m_1 after the collision? (b) Is this collision elastic or inelastic? If the latter, what percentage of the maximum possible kinetic energy loss occurs?

34. (I) A bullet of mass 20 g is moving horizontally at a speed of 600 m/s when it strikes a 3-kg block at rest. (a) The block is made of wood; the bullet sinks in and sticks. What is the initial speed of the bullet–block combination? (b) The block is made of a very hard steel, as is the bullet, and the collision is perfectly elastic. What is the resulting velocity of the block? [*Hint*: Use the approximation that the mass of the bullet is much less than the mass of the block.]

35. (II) A block of mass 126 g is moving along the $+x$-axis with a speed of 0.875 m/s. Just ahead of it is a 9.66-kg mass, moving in the same direction with the same speed. At some point, the large mass hits a wall and bounces off the wall perfectly elastically (Fig. 8–34). What is the return speed of the small mass after its elastic collision with the large mass?

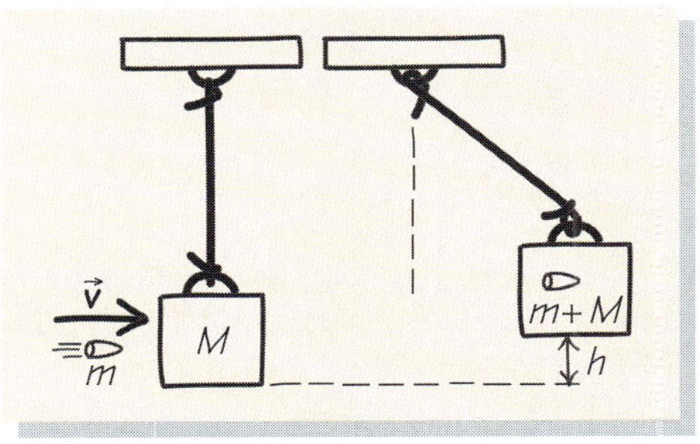

FIGURE 8–33 Problem 29.

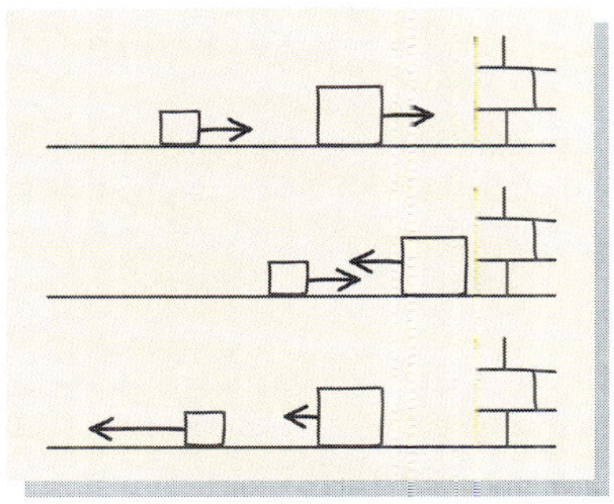

FIGURE 8–34 Problem 35.

36. (II) A machine gun in automatic mode fires 20-g bullets with $v_{\text{bullet}} = 300$ m/s at 60 bullets/s. (a) If the bullets enter a thick wooden wall, what is the average force exerted against the wall? (b) If the bullets hit a steel wall and rebound elastically, what is the average force on the wall?

37. (II) Two spheres of mass 400 g and 600 g, respectively, are suspended from the ceiling by massless strings 1 m long. The lighter sphere is pulled aside through an angle of 80° and let go. It swings and collides elastically with the second sphere at the bottom of the swing. How high will the lighter of the spheres swing (in terms of an angle)?

38. (II) A pendulum of mass $m = 0.1$ kg and length $l = 1.5$ m is released from a horizontal position. At the bottom of its swing, it collides elastically with a mass $M = 0.5$ kg. (a) What is the velocity of mass M right after the collision? (b) To what height above its low point does mass m rebound?

39. (II) A superball has collisions that are nearly perfectly elastic. A superball of mass M is dropped from rest from a height h (where $h >>$ the size of the superball) together with a smaller marble of mass m; the marble is initially just a little above the top of the superball and remains right over it throughout the fall. The superball hits the floor first and immediately rebounds elastically, colliding with the marble. (a) What is the speed of the superball and the marble just before the superball hits the floor? (b) Just after the superball rebounds from the floor but before it hits the marble? (c) What is the velocity of the marble after the superball hits it in the head-on collision? (d) How high does the marble go after its collision with the superball, assuming the marble has stayed in line with the superball, so that all the motion is vertical? (e) What is the answer to part (d) in the limit $M >> m$?

40. (II) Consider a perfectly elastic collision in one dimension between a ball of mass M and another of mass m, where $M >> m$. The light ball is initially at rest, and the heavy ball has a given initial velocity v_i and final velocity v_f. What is the fractional velocity change $(v_i - v_f)/v_i$ of the large mass? Use $M >> m$ to find an approximate expression.

8–5 Elastic Collisions in Space

41. (I) A ball of mass 1.2 kg moves along the positive x-axis with a speed of 2.4 m/s. Another ball (mass 0.8 kg) moves along the negative x-axis with a speed of 3.6 m/s. After colliding with each other, the lighter ball moves at a speed of 1.8 m/s along a line that makes an angle of 60° with the positive x-axis. What the speed and direction of the heavier ball? Is the collision elastic?

42. (I) Two objects of equal mass approach each other with equal but opposite velocities along the x-axis. After a collision, one particle has velocity $v_1\mathbf{i} + v_2\mathbf{j}$. What is the velocity of the other particle?

43. (I) A billiard ball with velocity $\mathbf{v} = (2.50 \text{ m/s})\mathbf{i}$ strikes a stationary billiard ball of the same mass. After the collision, the first billiard ball has velocity $\mathbf{v}_1 = (0.50 \text{ m/s})\mathbf{i} + (-1.00 \text{ m/s})\mathbf{j}$. What is the velocity of the second ball? Is the collision elastic?

44. (I) Suppose there are two billiard balls of equal mass; one is at rest and one is moving with a speed of 1.5 m/s. They collide elastically. After the collision, one of the two balls is measured to be moving with speed 1.2 m/s. What is the speed of the other ball?

45. (I) A billiard ball moving at 3.0 m/s collides with another billiard ball at rest. The balls move off at right angles to one another. If the first ball continues with a speed of 1.5 m/s, what is the speed of the ball that was initially at rest?

46. (II) Two objects with masses 2.0 kg and 3.0 kg move toward each other, both with speeds $v_0 = 5$ m/s. They collide head on and stick together. (a) Calculate their final velocity. (b) Calculate the amount of kinetic energy lost during the process. (c) Suppose the two masses approach each other at 90° before the collision (e.g., along the x- and y-axes). What will be the kinetic energy loss in this case?

47. (II) In a target shooting game, wooden blocks are thrown into the air and shot in flight. A block of 0.80 kg has a speed of 10 m/s at the top of its trajectory when it is hit by a bullet from below at an angle of 60° from the horizontal (Fig. 8–35). The mass of the bullet is 5.0 g and its speed is 550 m/s when it hits the block. The bullet is embedded in the block. What is the velocity of the block immediately after impact?

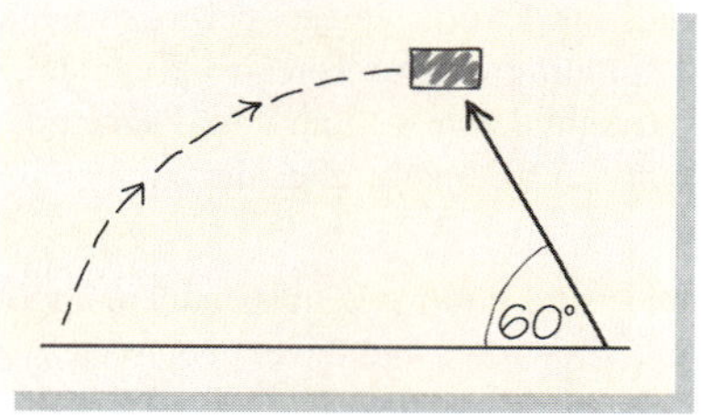

FIGURE 8–35 Problem 47.

48. (II) An air puck with mass 0.3 kg and velocity $(-1.5 \text{ m/s})\mathbf{i} - (2.0 \text{ m/s})\mathbf{j}$ on a frictionless table collides with a second air puck, of mass 0.2 kg and velocity $(5.0 \text{ m/s})\mathbf{i}$ (Fig. 8–36). As a result of the collision, the first air puck comes to rest. What is the kinetic energy of the second air puck after the collision?

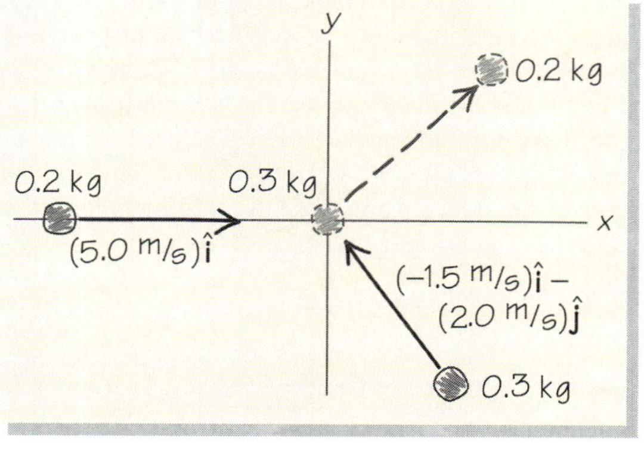

FIGURE 8–36 Problem 48.

49. (II) In a court hearing, a police expert reconstructs an accident in the following way (Fig. 8–37): A sports car of mass $m_1 = 1000$ kg collided with a parked pickup truck of mass $m_2 = 1500$ kg. From the skid marks, it is estimated that the speed of the pickup immediately after the collision was 21.6 m/s, at an angle of 33.7° with the direction of the road. The

sports car left a mark at an angle of 60° with the road, but did not stop, leaving the scene badly damaged. Can the expert determine from these data if the sports car was speeding and its driver fully responsible?

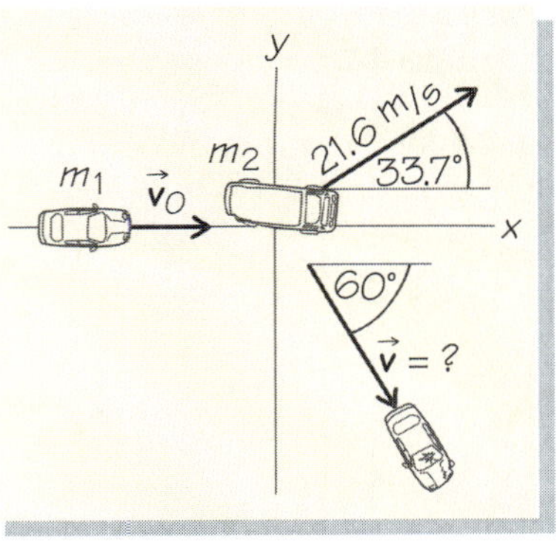

FIGURE 8–37 Problem 49.

50. (III) A proton of mass m and kinetic energy K scatters elastically from an alpha particle (with mass four times that of the proton). The proton is deflected at 30° from its original direction. (a) At what angle does the alpha particle recoil? (b) Draw a momentum diagram of the collision. (c) What are the final energies of the two particles?

8–6 The Center of Mass

51. (I) A massless, rigid bar that is 2.24 m long connects two tiny spheres (of negligible radii compared to the length of the bar) of masses 895 g and 478 g, respectively. Locate the system's center of mass.

52. (I) Point masses of 0.2 kg, 0.1 kg, and 0.3 kg are located in the xy-plane at the origin, at $\mathbf{r}_2 = (0.3\text{ m})\mathbf{j}$, and at $\mathbf{r}_3 = (0.2\text{ m})\mathbf{i} + (0.2\text{ m})\mathbf{j}$, respectively. Where is their center of mass?

53. (I) Redo Example 8–10 by placing the origin at the location of mass m_3, which was added to the original system.

54. (I) What is the velocity of the center of mass of the system in Problem 8, both before and after the collision (both parts), if the speed of the moving billiard ball is v?

55. (I) Two children of masses 25 kg and 30 kg, respectively, stand 2 m apart on skates on a smooth ice rink. The lighter of the children holds a 3-kg ball, and throws it to the heavier child. After the throw the lighter child recoils at 2.0 m/s. With what speed will the center of mass of the two children and the ball move?

56. (II) A uniform iron bar that is 0.40 m long, with mass 1.5 kg, is placed along the y-axis. From the midpoint of that bar a 0.65-m-long massless bar extends perpendicular to it in the x-direction; a 0.5-kg point mass is placed at its end. Where is the center of mass of this system? [*Hint*: First calculate the center of mass of the iron bar, $\mathbf{R}_1$, then treat that bar as a point mass of mass 1.5 kg at $\mathbf{R}_1$.]

57. (II) A mallet forms a symmetric T-shape. The top of the T is a uniform iron block of mass 4.0 kg. The wooden handle is uniform, 1.2 m long, and has a mass of 1.8 kg. Where is the mallet's center of mass? (See the hint in Problem 56.)

58. (II) Where is the center of mass of a uniform, L-shaped iron rod of sides 0.4 m and 0.7 m, respectively? (Assume that the rod is so narrow that the dimensions of the outer bend are the same as those of the inner bend of the L.)

59. (II) By using arguments based on symmetry, show that the center of mass of a uniform, thin, spherical shell made of any material of uniform density is at the center of the shell. Using this result, show that the center of mass of any sphere whose density varies arbitrarily only with distance from the center of the sphere but not with angle (in other words, a spherically symmetric density) is at the center of the sphere.

60. (II) Calculate the center of mass of the T-shaped object shown in Fig. 8–38. Assume that the mass density is uniform.

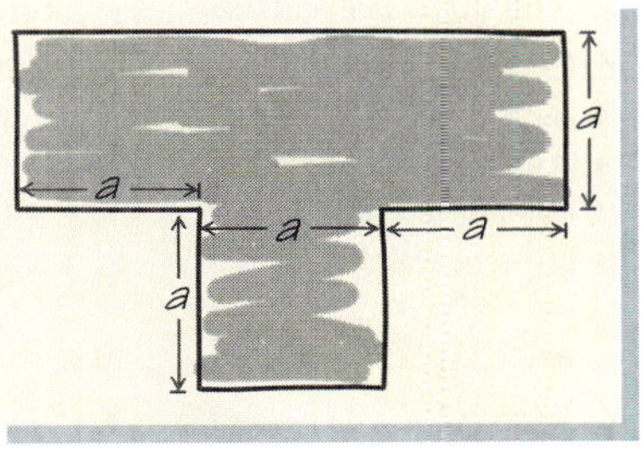

FIGURE 8–38 Problem 60.

61. (II) Find the center of mass of a flat, semicircular object of radius R. Assume that the density is uniform.

62. (III) Show that the center of mass of a planar object can be found by dividing the object into a convenient number of sections of known mass M_i, finding the center of mass of each section, and then treating each section as a collection of point masses M_i located at the center of mass of that section.

63. (III) Calculate the center of mass of a stick extending from $x = 0$ to $x = L$; assume that the density of the mass distribution is given by $\lambda(x) = K/(x^2 + a^2)$, where K is a constant with dimensions $[ML]$.

64. (III) Find the center of mass of a flat triangular object of uniform density with sides of length a, b, c.

8–7 Rocket Motion

65. (I) The thrust of a rocket is 10 million N. If the speed of its engine's exhaust gas is 3.0 km/s, how fast is the rocket's mass changing?

66. (I) An orbiter must fire its thruster rockets to increase its speed. If the mass of the orbiter is about 80,000 kg and the exhaust gas velocity is 2800 m/s, how much mass must be discarded to increase the speed from 7900 m/s to 8100 m/s?

67. (I) After ejecting a communication satellite, the Space Shuttle must make a correction to account for the change in momentum. One of the thrusters with $u_{ex} = 10^3$ m/s is used to increase the orbital velocity by 10 m/s. What percent of the mass of the Space Shuttle must be discarded?

68. (I) A rocket that uses a fuel with an exhaust speed of 2700 m/s is shot straight up and reaches a final speed $v = 3400$ m/s when the fuel is exhausted, which happens 120 s after the launch. What fraction of the initial mass of the rocket with all its fuel is the mass of the portion of the rocket that remains? Do not ignore gravity, but assume the force of gravity to be the same throughout the burn period.

69. (II) The *burnout velocity* is the final velocity of a rocket when all the fuel is burned away. If m_0 is the total initial mass of the rocket and m_{fuel} is the mass of the fuel, show that the burnout velocity is given by

$$v = u_{\text{ex}} \ln\left(\frac{m_0}{m_0 - m_{\text{fuel}}}\right) - gt,$$

where g is assumed to be constant. If the ratio m_{fuel}/m_0 is 70 percent, $t = 90$ s, and u_{ex} is 2800 m/s, what is the burnout velocity?

70. (II) The rocket considered in Problem 69 carries human passengers. Consequently, its initial upward acceleration may not exceed $2g$. What is the largest initial burnout rate (the mass expelled over unit time) that satisfies this condition? (The mass of the rocket is 80,000 kg.)

71. (II) To test certain aspects of rocket propulsion, a student stands initially motionless on frictionless ice. He has 100 watermelon seeds that he can spit horizontally in the $-x$-direction, thereby propelling himself in the $+x$-direction. The speed he can give one seed with respect to himself is u, but if he tries to spit two seeds simultaneously, he can give them only speed $u/2$. More generally, if he tries to spit n seeds simultaneously, he can give them speed u/n. The student has mass M and each seed has mass m. (a) To maximize his speed after spitting all 100 seeds, is the student better off spitting out all 100 seeds at once, spitting out the seeds one at a time, or something in between? Assume that the ratio m/M and even $100m/M$ can be taken to be small. (b) Let $M = 50$ kg, $m = 1$ g, and $u = 3$ m/s. What is the maximum final speed of the student?

72. (III) Because the final velocity of rockets is limited by the exhaust velocity and by the mass ratio of total mass to the rest of the rocket minus fuel, engineers have developed multistage rockets in which the fuel tank of each stage is discarded before the next stage is ignited. Consider a two-stage rocket for which m_0 is the total mass of the rocket at liftoff, m_1 is the mass of the fuel in the first stage, m_1' is the mass of the first stage that is discarded, and m_2 is the mass of the second-stage fuel. Find an equation that gives the final velocity of the rocket; assume that gravity is constant.

General Problems

73. (I) A 48-kg figure skater (skater 1) moving at 6 m/s is picked up by her 82-kg partner (skater 2), who is skating in the same direction at 9 m/s. They then coast off together. Ignore all friction. (a) What is the velocity of their center of mass before he picks her up? (b) After he picks her up? (c) What were the velocities of skater 1 and 2 in the center of mass frame before he picked her up? (d) During a brief period of 0.05 s, they crash into the barrier and collapse together. What average force is exerted on the barrier?

74. (II) A railroad car, of mass 20 Mg, is released from rest in a railway switchyard and rolls to the bottom of a slope 5 m below its original height. At the low point, it collides with and sticks to another car of 10 Mg. The two cars roll off together up another slope to a position a height h above the low point, where they come to a stop. Ignoring the effects of friction, find h.

75. (II) A cannon of mass 800 kg shoots a 5-kg projectile at an angle of 35° from the horizontal. The projectile has an initial speed of 800 m/s. The cannon is on wheels. (a) Draw a diagram showing the momentum vectors. (b) With what velocity does the cannon recoil along the horizontal ground? (c) What happens to the recoiling cannon's component of momentum perpendicular to the ground?

76. (II) A small bag of sand is suspended by a thread of length $L = 1.2$ m. A bullet of mass $m = 8$ g, moving with speed $v_0 = 600$ m/s in the horizontal direction, hits the bag and leaves with a speed $v = 250$ m/s (Fig. 8–39). The bag swings to a maximum angle of $\theta = 40°$. What is the mass of the sandbag?

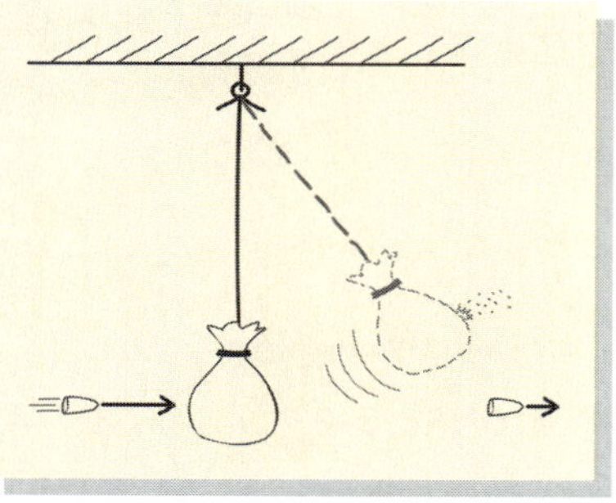

FIGURE 8–39 Problem 76.

77. (II) Two masses slide without friction down the sides of a hemispherical bowl. They each start with zero velocity at the lip, which is a height h above the bottom. They move in opposite directions and collide elastically at the bottom of the bowl. Suppose that the masses are m and M, respectively, with $m < M$. (a) By how much will the lighter mass overshoot the lip after the collision? (b) How high will the masses move if they coalesce on contact?

78. (II) A boxcar of length 8 m and height 2.4 m is at rest on frictionless rails. Inside the boxcar, whose mass is 3200 kg, a tank containing 2400 kg of water is located at one end. The tank is 1 m long and 2.4 m high. At some point, the walls of the tank start to leak and the water fills the floor of the boxcar uniformly. Assuming that all the water stays in the boxcar, describe how the breakage of the tank manifests itself to an outside observer.

79. (II) A sphere of styrofoam has radius R. A cavity of radius $R/2$, centered a distance $R/2$ directly above the center of the sphere, is hollowed out and filled with a solid material of density five times the density of styrofoam (Fig. 8–40). Where is the center of mass of the new sphere?

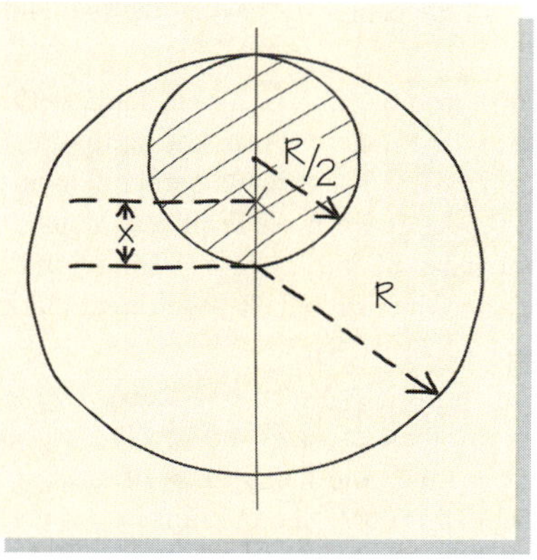

FIGURE 8–40 Problem 79.

80. (II) In the Olympic Games of the year 2020, a new sport is introduced—cyclo-shooting. A bicyclist moving in the $+x$-direction fires a gun in the y-direction in an attempt to hit a target lined up with the position $x = 0$ m. The mass of the cyclist plus the bicycle is 65 kg, and the cyclist is originally moving with speed 35 km/h. The gun shoots a projectile, with mass 15 g, at a speed of 1600 km/h. The target is a distance 30 m from the bicycle track. (a) At what x-value should the cyclist fire in order to hit the target? Ignore air resistance. (b) Immediately after she fires, the cyclist moves in a new direction due to the recoil. What is the new value of her velocity, both magnitude and direction?

81. (II) Two students sit on opposite ends of a 4-m-long sled that is initially at rest on frictionless ice. Each student has a mass of 65 kg; the sled's mass is 30 kg. The student at one end slides a 3-kg object on the sled across to the other at a uniform speed of 6 m/s relative to the sled (the object moves friction free on the sled). (a) What is the sled's speed relative to the ice before the second student catches the object? (b) After the second student catches the object? (c) Over what distance does the sled move while the object slides across? (d) Over what distance does the center of mass move while the object slides across?

82. (III) Two identical spherical billiard balls, each with a radius of 3 cm, move toward each other with velocities $(0.6\text{ m/s})\mathbf{i}$ and $(-0.2\text{ m/s})\mathbf{i}$. The center of one of them moves along the $+x$-axis ($y = 0$ m), and the center of the other one moves in the $-x$-direction, 1.5 cm below the x-axis ($y = -1.5$ cm). Assuming that the collision is perfectly elastic, what are the final velocities of the two billiard balls? Pretend that the objects slide without friction. Note that the impulse is directed along the line that joins the two centers at the moment of collision.

CHAPTER

9

This water wheel provides a time-honored example of rotational motion. The weight of the water falling on the wheel supplies the dynamical means for centripetal acceleration of the wheel, and the result is motion at a constant angular speed.

Rotations of Rigid Bodies

To this point our discussion of dynamics has treated very different objects such as crates and planets as if they were all point objects (particles). Further, our study of an extended object, or of a system of particles, has only reached the point where we can show that, under the influence of a net external force, the center of mass moves as if the entire system were concentrated at that single point. But there is more to the motion of an extended object than the motion of its center of mass. Consider the vast cloud of dust that circled the Sun and ultimately condensed into a number of planets, including Earth. Computer simulations indicate that the gravitational attractions between the dust particles led the great dust swirls to break up into isolated clusters and subsequently condense into planets orbiting the Sun and spinning on their own axes. The description of this motion is certainly more complicated than the description of the motion of the center of mass of the solar system. Even with a relatively simple object of fixed form, such as a thrown blackboard eraser, a complete description of its motion requires us to include its rotations.

In this chapter, we study the rotational motion of **rigid bodies**, which are objects that do not change volume or shape. Blackboard erasers, rocks, baseball bats, and heavy sheets of metal are all rigid bodies. We explore the rotation of rigid bodies as we did the motion of point objects: We start with a simple description (kinematics) and then continue with the causes of rotations (dynamics). Even though rotational motion in classical physics appears to encompass many new concepts, this motion is in fact described by the application of Newton's laws to collections of particles. No new physical laws are involved.

9–1 MOTION OF A RIGID BODY

Let's consider the possible motions of a rigid body in which one point of the object is permanently fixed in space. Any other point P will always remain at a constant distance from the fixed point; moreover, the point P will also remain at a fixed distance from any other point Q. For example, if a clock hand is moving with one end fixed at the center of the clock's face, then all the points of the hand will move in circles about the fixed point (Fig. 9–1). The motion of the hand consists of a **rotation** about an axis that passes through the fixed point and is perpendicular to the surface swept out by the hand as it moves.

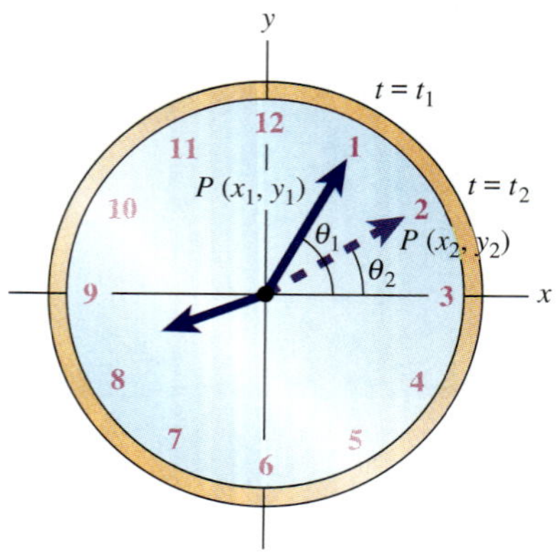

FIGURE 9–1 A flat object in the xy-plane rotates about a fixed axis perpendicular to the plane. Here, the object is the hand of a clock, which rotates through the angle $\theta_2 - \theta_1$. The fixed point is at the center of the clock face.

Let's look more closely at Fig. 9–1. We have placed the origin at the fixed point and have drawn a set of x- and y-axes that are fixed in space. At some time t_1, point P has coordinates (x_1, y_1); at a later time, point P has moved to where its coordinates are now (x_2, y_2). *Both* x and y change with time. In terms of polar coordinates, we can describe the angle θ measured from the x-axis to the hand and the distance R of the hand from the origin. *The only variable that changes with time is the angle* θ. Our point P has undergone an *angular displacement.*

Polar coordinates and the angular description of circular motion were introduced in Section 3–5.

The special feature of a rigid body is that *when P moves through a certain angle in the rotational motion about the origin, all points in the rigid body turn through exactly the same angle.* The motion with one fixed point is thus described by the circular motion of an arbitrary line that is drawn from the origin to a point P in the object. That motion need not be uniform, but it *is* just circular motion. For the description of a two-dimensional rigid body on a flat surface, this means that we need only one variable—the angle of circular rotation θ.

In rigid-body rotations about an axis, a single angle is sufficient to describe the motion.

In the case of a three-dimensional rigid body, once one point in the object is fixed, every other point must move in such a way that it is at a constant distance from that fixed point. However, it is sufficient for us to consider the situation in which we fix a *second* point of the object in space. The line joining the two fixed points is fixed in space; we call this the *axis.* The points lying along the axis will *always* be fixed in space. A point P that is off the axis will always be at a fixed distance from the two fixed points and at a fixed perpendicular distance from the axis (Fig. 9–2). With two points fixed, the motion of the three-dimensional object—here, a pot on a potter's wheel—consists of rotations about the axis. And as the discussion of the clock hand shows, to describe rotations about a fixed axis we need only specify a single angle. Other examples of this limited motion are provided by Earth's rotation about its axis or by a turning merry-go-round.

In Section 3–5, we showed how angle θ acts as a convenient variable for the description of circular motion.

In considering the motion of a rigid object about a fixed axis, we will not insist that the axis pass through the center of mass; rather, we are considering circular motion about *any* axis.

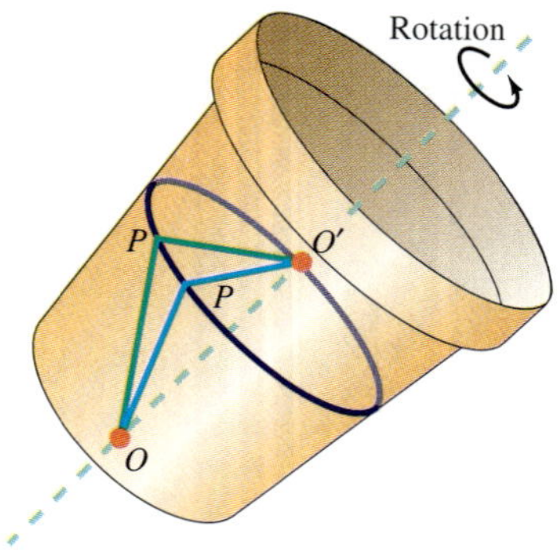

FIGURE 9–2 A three-dimensional object—a clay pot on a potter's wheel—rotates about the axis defined by the line between the two fixed points O and O'. The point P remains the same distance from these points or from any point on the axis as the pot rotates about the axis.

9–2 ROTATIONS ABOUT AN AXIS

Angular Velocity and Angular Acceleration

Suppose that a rigid body rotates about a fixed axis; you may want to think of the pot on the potter's wheel as an example. In such motion, the angle of any point P in the object relative to the axis changes with time. We want to give a mathematical description of this motion. Let's orient the z-axis along the rotation axis (Fig. 9–3a). Consider the position vector $\mathbf{r}$ from the origin to point P. Take the projection of $\mathbf{r}$ on the xy-plane to go to point P' and let θ_0 be the angle this projection makes with the x-axis (Fig. 9–3b). Angle θ, which varies with time, is measured in *radians* (2π rad $= 360°$).[†] Figure 9–3b depicts a view down the z-axis, where the projection point P' has moved from angle θ_0 to angle θ in the time interval Δt.

[†]In Section 3–5, we used the notation ϕ for this angle.

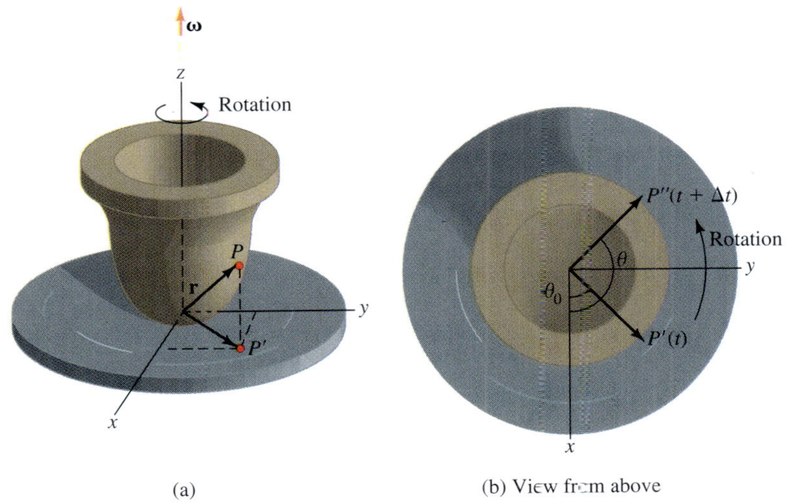

FIGURE 9–3 (a) A pot on a potter's wheel rotates about the z-axis. (b) The pot as viewed looking down the z-axis. The angle of rotation θ, which varies with time, is defined to be the angular displacement from the x-axis. It is positive when the pot rotates counterclockwise.

The average angular velocity, ω_{av}, over a finite time interval Δt is obtained from the change in the angle over that time interval:

$$\omega_{av} = \frac{\theta(t + \Delta t) - \theta(t)}{\Delta t}. \tag{9–1}$$

The angular velocity and its units

We find the *instantaneous angular velocity* (or just the *angular velocity*) $\omega(t)$, given by Eq. (3–46), by taking the limit $\Delta t \to 0$:

$$\omega = \frac{d\theta}{dt}. \tag{9–2}$$

PROBLEM SOLVING

Using a right-hand rule

The units that describe angular velocity are *radians per second* (rad/s). Because a radian is a dimensionless measure (it is the ratio of an arc length to the circumference), the dimensions of ω are $[T^{-1}]$. It is conventional to measure the angle in a counterclockwise direction from the x-axis, using a *right-hand rule: If the extended thumb of the right hand points in the +z-direction (along the axis of rotation), then the fingers curve in the positive direction of rotation,* as shown in Fig. 9–4. Thus an angle that increases in the counterclockwise direction corresponds to a *positive* angular velocity.

Suppose that the angular velocity is a *constant;* that is, $\omega(t) = \omega_0$. Then the angle θ grows linearly with time:

$$\theta = \theta_0 + \omega_0 t, \tag{9–3}$$

where θ_0 is the value in radians of the angle θ at time $t = 0$ (Fig. 9–3b). As noted in our discussion of the clock hand, any point P in the object moves in a circle. According to Section 3–5, this point makes a complete circle during a time T, which is the *period* of the motion. Because a full circle is 2π rad,

$$T = \frac{2\pi}{\omega_0}. \tag{9–4}$$

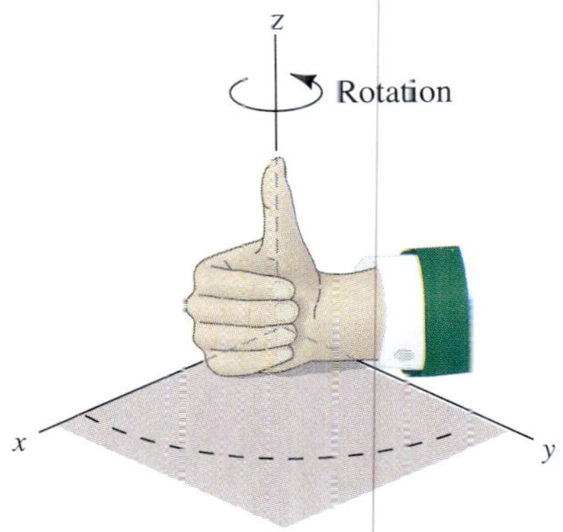

FIGURE 9–4 A right-hand rule. The thumb of the right hand points along the axis of rotation (the z-axis) when the fingers curl in the direction of rotation (from the x-axis to the y-axis).

The *frequency*, f, measures the number of times per second that point P returns to its original position in uniform rotation:

$$f = \frac{1}{T} = \frac{\omega_0}{2\pi}. \tag{9–5}$$

Relations among frequency, angular velocity, and period

For nonuniform motion, we also speak of an instantaneous frequency, $f(t) = \omega(t)/2\pi$.

If the angular velocity is not constant, we may define an average angular acceleration, α_{av}:

$$\alpha_{av} = \frac{\omega(t + \Delta t) - \omega(t)}{\Delta t}. \tag{9–6}$$

The angular acceleration

The instantaneous angular acceleration $\alpha(t)$ or, more simply, the **angular acceleration**, is defined as the instantaneous rate of change of the angular velocity:

$$\alpha = \frac{d\omega}{dt} = \frac{d^2\theta}{dt^2}. \tag{9–7}$$

Note that this equation has the same structure as the equation for linear motion,

$$a = \frac{dv}{dt} = \frac{d^2x}{dt^2}.$$

The angular acceleration has units of radians per second squared, and the dimension of α is $[T^{-2}]$ because radians are dimensionless.

Constant Angular Acceleration

A constant angular acceleration means that the angular velocity has a constant time derivative. Hence the angular velocity changes linearly with time; that is,

$$\omega = \omega_0 + \alpha t, \tag{9–8}$$

where ω_0 is the angular velocity at $t = 0$. The angle is then given by

The change in angle under constant angular acceleration

$$\theta(t) = \theta_0 + \omega_0 t + \frac{1}{2}\alpha t^2. \tag{9–9}$$

This solution corresponds to the following equation for linear motion:

$$x(t) = x_0 + v_0 t + \frac{1}{2}at^2.$$

We may eliminate the time from Eqs. (9–8) and (9–9) to find that

$$\omega^2 = \omega_0^2 + 2\alpha(\theta - \theta_0), \tag{9–10}$$

which is the analog of the linear motion expression

$$v^2 = v_0^2 + 2a(x - x_0).$$

EXAMPLE 9–1 A pulley is rotating at the rate of 32 rev/min. A motor speeds up the wheel so that, 30.0 s later, it is turning at 82 rev/min. (a) What is the average angular acceleration in radians per second? (b) How far will a point 0.30 m from the center of the pulley have traveled during the acceleration period, assuming that the acceleration is uniform?

Solution: (a) Each revolution consists of an angular displacement of 2π rad. Thus 32 rev/min corresponds to (32 rev/min)(2π rad/rev) = 64π rad/min, or 64π rad/60 s = 3.35π rad/s = ω_0. After 30 s, the angular velocity ω_f is

$$\omega_f = \left(\frac{82\text{ rev}}{\text{min}}\right)\left(\frac{2\pi\text{ rad}}{\text{rev}}\right)\left(\frac{1\text{ min}}{60\text{ s}}\right) = 8.6\pi\text{ rad/s}.$$

Thus the average angular acceleration is

$$\alpha_{av} = \frac{\omega_f - \omega_0}{\Delta t} = \frac{(8.6\pi - 3.35\pi)\text{ rad/s}}{30.0\text{ s}} = 0.17\text{ rad/s}^2.$$

(b) We calculate the angular displacement $\Delta\theta$ during the period by using the for-

$$\theta = \theta_0 + \omega_0 t + \frac{1}{2}\alpha t^2,$$

mula applicable for uniform acceleration. With $\omega_0 = 3.35\pi$ rad/s and $\alpha_{av} =$

$0.17\ \text{rad/s}^2$,

$$\Delta\theta = \theta - \theta_0 = (3.35\ \text{rad/s})(30.0\ \text{s}) + \frac{1}{2}(0.17\ \text{rad/s}^2)(30.0\ \text{s})^2 = 180\ \text{rad}.$$

The distance traveled is

$$D = r\,\Delta\theta = (0.30\ \text{m})(180\ \text{rad}) = 54\ \text{m}.$$

The unit "radian" is deleted from the final value because radians are dimensionless.

Acceleration of a Rotating Rigid Body

Consider the rotating clock hand, and, more specifically, a point located at a distance r from the axis of rotation. The point moves in a circle; its velocity has a magnitude of $r\omega$ and a direction that is tangential to the circle of radius r. We learned in Chapter 3 that a point moving in a circle of radius r with speed v undergoes an acceleration in the radial direction toward the center of the circle. The magnitude of that acceleration is v^2/r. In terms of the angular speed—$\omega = v/r$—this acceleration may be expressed as

$$a = \frac{v^2}{r} = \omega^2 r. \tag{9–11}$$

If there is an angular acceleration α, then there is also a linear acceleration in the tangential direction, whose magnitude is

$$a_{\text{tan}} = \alpha r. \tag{9–12}$$

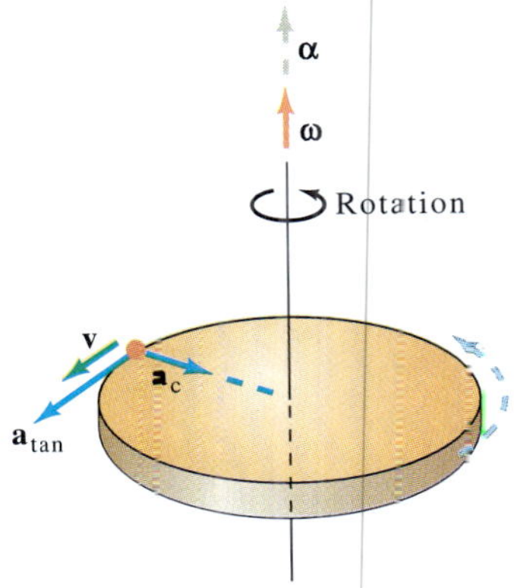

FIGURE 9–5 A rotating potter's wheel undergoes an angular acceleration. A point on its margin must have a centripetal acceleration component—this component keeps the point on its circular path—and a tangential component.

Figure 9–5 illustrates these acceleration components for a point on a potter's wheel that is undergoing an angular acceleration. Whenever a rigid body accelerates, there must be a force involved. The radial acceleration is ultimately due to the interatomic forces that fix the relative positions of the atoms of the rigid body. The tangential acceleration must involve an external force. We'll discuss this in more detail later.

Vectorial Properties of Angular Motion

We have seen that there is a parallel between position and angle, between velocity and angular velocity, and between acceleration and angular acceleration. The difference appears to be that the quantities that describe the kinematics of linear motion are vectors, whereas the notion of a vectorial description of rotations has not played any real role in our discussion. Does rotational motion have vectorial properties? In the description of rotation about an axis, there is a special direction—the direction of the axis. Can we say that the angle, angular velocity, and angular acceleration are vectors just by stating that they point in the direction of the axis? It is not quite so simple. Vectors are not just numbers with an associated direction; vectors must satisfy certain addition rules, one of them being the condition of commutativity:

$$\mathbf{A} + \mathbf{B} = \mathbf{B} + \mathbf{A}.$$

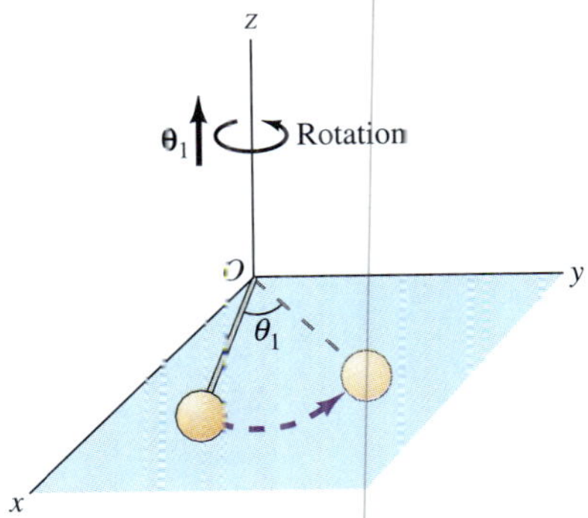

FIGURE 9–6 A proposal for defining a vector angle. When the rotation is counterclockwise in the xy-plane as seen from above, the vector angle θ_1 points in the z-direction.

The angular velocity is a vector with a direction defined by the right-hand rule.

Analogous to our definition of a position vector, suppose we were to try to define a *vector angle* by associating the magnitude of the angle with the direction of the axis about which the rotation takes place. This specification includes a right-hand rule for the direction of the axis to resolve the question of which way the "vector" really points. We call this attempt at a vector angle θ_1. Let the axis for this angle, and hence its direction, be the z-axis (Fig. 9–6). We similarly define an angle θ_2 for a rotation about the y-axis. In order to check commutativity for these proposed vector angles, using the book illustrated in Fig. 9–7, we perform the operation $\theta_1 + \theta_2$ by performing the rotation θ_1 first and subsequently performing the rotation θ_2. We compare that with the operation $\theta_2 + \theta_1$, in which the rotation θ_2 is performed first (Fig. 9–8). Both angles will have magnitude 90°. If the final positions of the book are not the same for the two

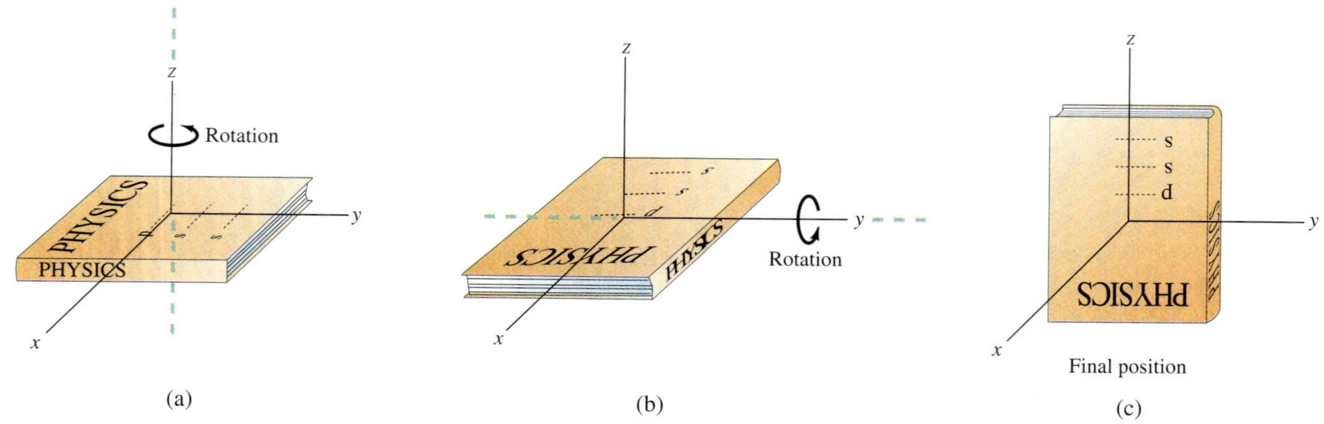

FIGURE 9–7 (a) A book is initially oriented with its cover facing the z-axis and the book spine facing the x-axis. (b) The book first rotates from its position in part (a) by an angle 90° about the z-axis (θ_1) and then (c) 90° about the y-axis (θ_2).

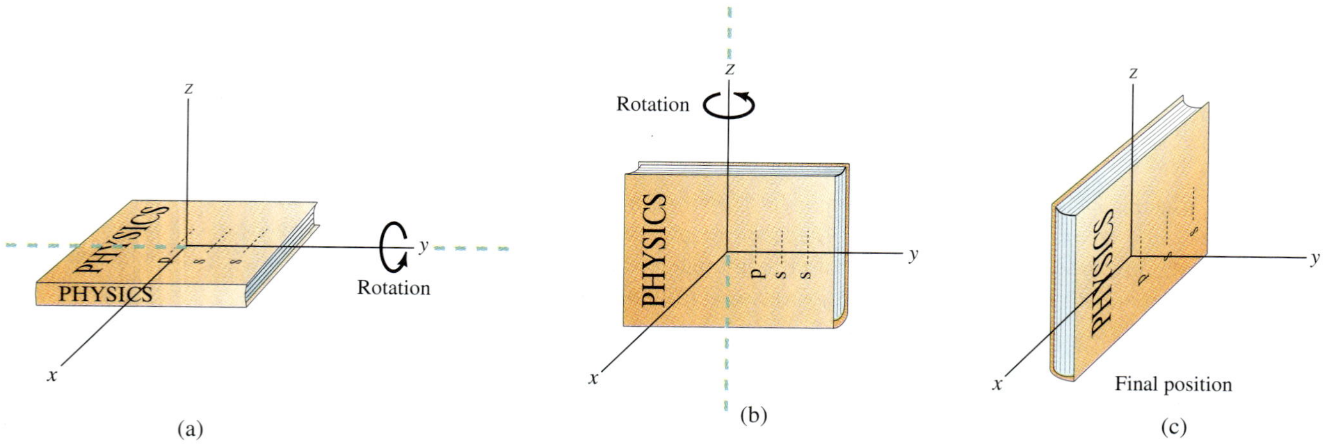

FIGURE 9–8 The order of rotations of the book is reversed from the order taken in Fig. 9–7. (a) In this case, the book starts from the position of Fig. 9–7a and (b) rotates with the rotation θ_2, followed by (c) the rotation θ_1. The final position of the book is different than the final position in Fig. 9–7c.

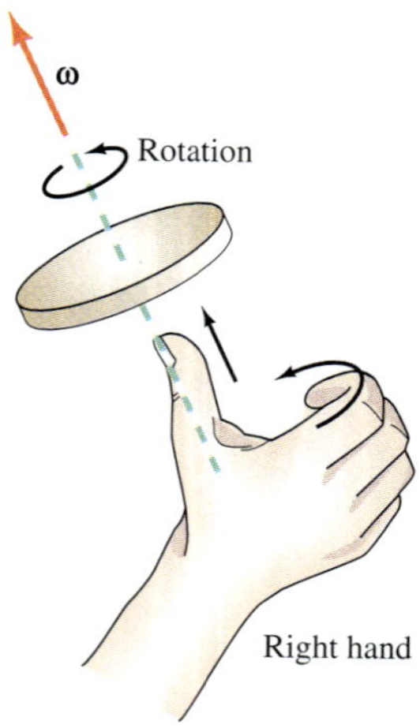

FIGURE 9–9 A right-hand rule is used to determine the direction of the angular velocity, $\boldsymbol{\omega}$. With the fingers of the right hand curling in the sense of the rotation, the thumb points in the direction of $\boldsymbol{\omega}$.

operations, then commutativity is not satisfied. The final positions of the book in Figs. 9–7c and 9–8c are clearly not the same. We conclude that commutativity fails and that the angles as defined above *are not vectors*!

All is not lost, however: The commutativity test *does* work with infinitesimal angles. You can check for yourself with this book by letting θ_1 and θ_2 specify rotations of only a few degrees. Although this fact does not help with large-angle rotations, it does help with angular velocities. If a small angle $\Delta\theta$ is a vector, then $\Delta\theta/\Delta t$ is a vector. This defines the instantaneous angular velocity; thus, *the angular velocity is a vector*. This vector has the magnitude ω. The direction of $\boldsymbol{\omega}$ is defined by a right-hand rule: If the fingers curl with the direction of the rotation, then the extended thumb of the right hand points in the direction of the angular velocity (Fig. 9–9). The angular acceleration vector $\boldsymbol{\alpha}$, with magnitude α, is defined as the rate of change of the angular velocity vector, exactly as the linear acceleration vector is the rate of change of the linear velocity vector.

9–3 ROTATIONAL KINETIC ENERGY

A spinning ball consists of moving mass. This ball, or any rotating object, must have **rotational kinetic energy** (Fig. 9–10). Consider the object shown in Fig. 9–11, which rotates about the z-axis with angular velocity ω. We have chosen the origin of the coordinate system arbitrarily and divide the object into elements labeled by the indices i, with each element having mass Δm_i. In other words, we are treating the object as a system of discrete particles. The total kinetic energy of the object is given by

$$K = \sum_i K_i = \frac{1}{2}\sum_i \Delta m_i\, v_i^2.$$

Because each mass element is rotating about the z-axis with angular velocity ω, the speed of each element is given by $v_i = R_i\omega$. The length R_i is the *perpendicular distance* between the mass element and the rotation axis, as shown in Fig. 9–11. Substitution of this value of v_i into the expression for the kinetic energy gives

$$K = \frac{1}{2}\sum_i \Delta m_i\, R_i^2\omega^2. \tag{9–13}$$

The kinetic energy of our rotating object is thus proportional to ω^2. We may write

$$K = \frac{1}{2}I\omega^2, \tag{9–14}$$

where

$$I \equiv \sum_i \Delta m_i\, R_i^2. \tag{9–15}$$

FIGURE 9–10 The flying sparks make it evident that the rotating wheel has kinetic energy.

The expression of the kinetic energy of rotation and the definition of the rotational inertia

The quantity I is a property of the rigid object called the **rotational inertia**, or the **moment of inertia**. It has dimensions of $[ML^2]$ and is measured in units of kg · m^2 in SI. From its definition, we see that I sums the contributions of each element of the object; for a single element or single particle, then,

The rotational inertia is defined with respect to an axis.

$$I = mR^2, \tag{9–16}$$

where R is the perpendicular distance from the single element to the axis of rotation. According to Eq. (9–15), the rotational inertia can be defined with respect to any axis we choose, not just the axis of rotation. Moreover, *I depends on the axis from which R_i is measured.*

The rotational inertia has a role in rotational motion that is analogous to the role of the mass in linear motion, except for the fact that it is defined in relation to a particular axis. The kinetic energy of a pointlike object is one-half its mass times the velocity squared; the kinetic energy of a rotating object is one-half its rotational inertia times its angular velocity squared; that is, $K_{\text{point}} = mv^2/2$ and $K_{\text{rotation}} = I\omega^2/2$. Just as the mass measures the resistance of an object to changes in velocity, we shall see that the rotational inertia measures the resistance of an object to changes in its angular velocity.

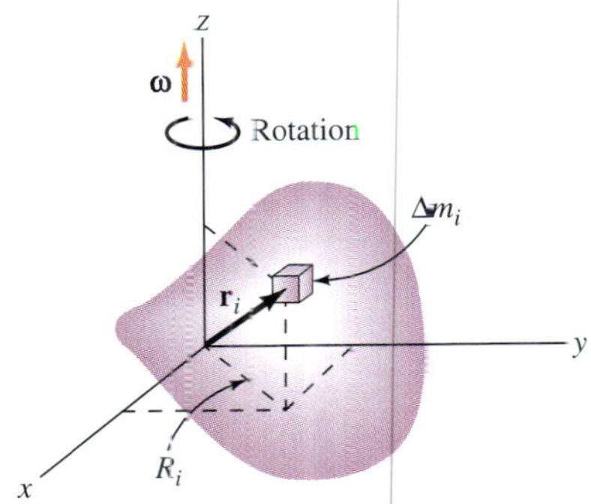

FIGURE 9–11 A rigid body rotates about the z-axis. It is divided into tiny segments labeled with the index i, of mass Δm_i, located at position $\mathbf{r}_i$. The perpendicular distance of segment i from the z-axis is R_i.

A *flywheel* is a rotating disk with a large rotational inertia and the ability to reach a large angular velocity; a flywheel thereby stores energy. In Zurich, the Swiss government operates a bus powered by a flywheel electrical system. A flywheel mounted in the bus powers an electrical generator; the electricity runs electric motors that drive the wheels. At each terminus of the bus run, the bus is connected to Zurich's electric power distribution net, and the bus's electrical generator is run in reverse, acting as a motor to speed up the flywheel and thereby store enough energy for the return trip.

APPLICATION

Flywheels

We have defined the rotational inertia for a system of discrete elements. For an object that is a continuum of mass, we can take the limit in which the elements have

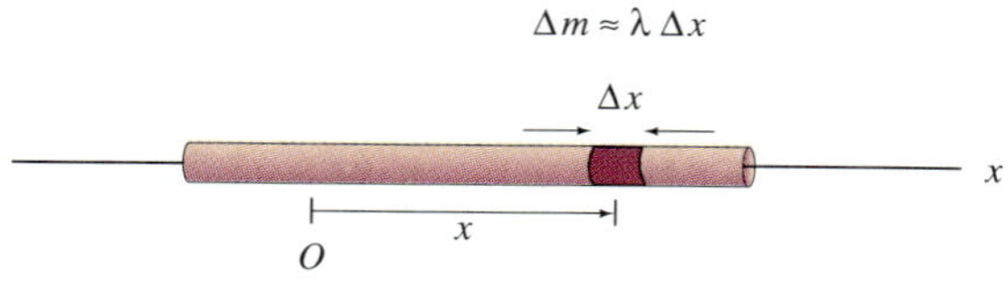

FIGURE 9–12 A one-dimensional object is aligned with the x-axis. A segment of width Δx is a displacement x from the origin.

infinitesimal volume and infinitesimal mass. In this case, the sum becomes an integral over the volume of the object. A symbolic way to write this continuum expression is to label the mass element by the differential notation dm, and thus

$$I = \int R^2\, dm. \tag{9–17}$$

This formal expression is not useful as it stands because m is not a useful variable of integration. We can remedy this and the possibility that the mass element dm can vary from point to point by the use of the *mass density* function ρ. The mass density $\rho(\mathbf{r})$ at point $\mathbf{r}$, which we suppose is the point labeled by i, is defined as

$$\rho(\mathbf{r}) \equiv \lim_{\Delta V_i \to 0} \frac{\Delta m_i}{\Delta V_i} = \frac{dm}{dV}.$$

Here, both the volume of the element and the mass of that element approach zero. Now we can write the mass element as $dm = \rho\, dV$. The total mass is found from the mass density by an integration over the volume V (the single integral sign with the subscript "Volume" denotes a three-dimensional integration over the volume):

$$M = \sum_i \Delta m_i \to \int dm = \int_{\text{Volume}} \rho\, dV. \tag{9–18}$$

The rotational inertia of a continuous object is therefore

$$I = \int R^2\, dm = \int_{\text{Volume}} \rho R^2\, dV. \tag{9–19}$$

Don't panic when we refer to volume or surface integrals. In cases of interest in this book, we make use of symmetry to reduce all such integrals to single integrals. An equation such as Eq. (9–19) is also applicable for one-dimensional (linear) objects or for two-dimensional (flat) objects. In these cases, we may replace the mass per unit volume ρ by a mass per unit length $\lambda \equiv dm/dx$, or a mass per unit area $\sigma \equiv dm/dA$. Thus

for a one-dimensional object: $dm = \lambda\, dx$;

for a two-dimensional object: $dm = \sigma\, dA$;

for a three-dimensional object: $dm = \rho\, dV$.

The densities λ, σ, and ρ may vary from point to point of the object.

To evaluate the rotational inertia in the one-dimensional case, we can align our object with the x-axis, and

$$I = \int \lambda x^2\, dx. \tag{9–20}$$

The distance x is measured from the origin, as in Fig. 9–12, and the integral is carried out over the length of the object. When the object is two-dimensional, we have

$$I = \int_{\text{Surface}} \sigma r^2\, dA. \tag{9–21}$$

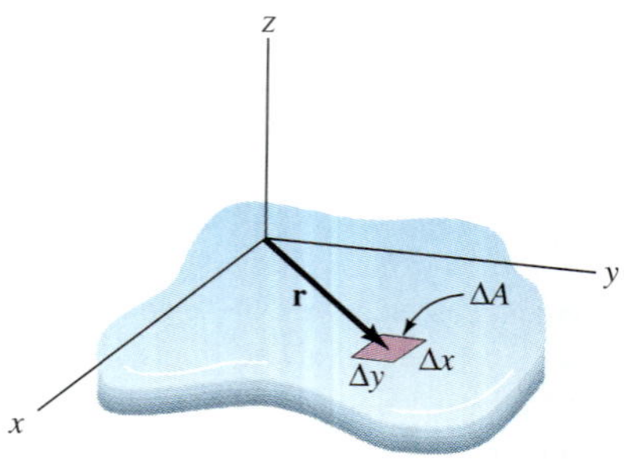

FIGURE 9–13 A two-dimensional object is aligned in the xy-plane. A segment of area ΔA is a position $\mathbf{r}$ from the origin.

We have taken the axis perpendicular to the object and integrated over its entire area (Fig. 9–13). The quantity r^2 is the square of the distance of the integration element to the axis about which I is measured.

9–4 EVALUATION OF ROTATIONAL INERTIA

A description of the rotational motion of an object requires us to evaluate its rotational inertia. For continuous objects, this evaluation involves integrations. For objects that have circular symmetry about the axis of rotation, it is easiest to label the points by the distance from that axis and by an angle variable rather than by Cartesian coordinates. We will see how this works in a series of examples. Table 9–1 contains a summary of the results, as well as rotational inertias of other objects for which the calculations are not carried out in this section.

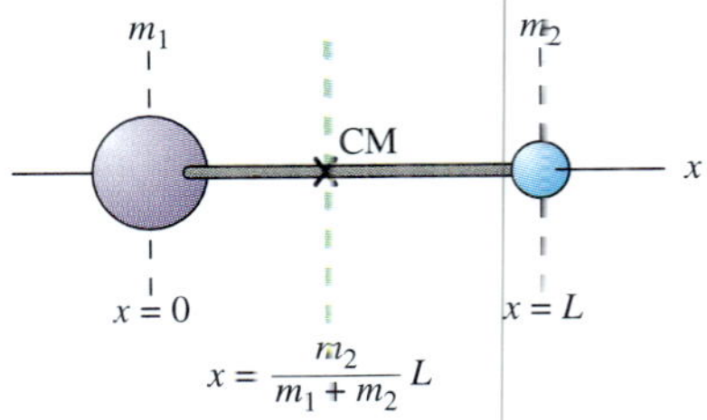

FIGURE 9–14 A dumbbell consists of two point masses, m_1 and m_2, connected by a rigid massless rod of length L. The object is aligned along the x-axis, with m_1 at the origin. The axis of rotation is perpendicular to the rod, passing through the center of mass.

Dumbbell. The simplest rotating object that we can contemplate is a dumbbell consisting of two point masses, m_1 and m_2, connected by a massless rigid rod of length L (Fig. 9–14). The total mass is $M = m_1 + m_2$. If we put mass m_1 at the origin and mass m_2 at $x = L$, we find from Eq. (8–49a) that the center of mass is at

$$X = \frac{(m_1)(0) + (m_2)(L)}{m_1 + m_2} = \frac{m_2 L}{M}.$$

Now let's consider the case in which the axis of rotation goes through the center of mass; that is, through the point $x = m_2L/M$. The axis is taken perpendicular to the rod. As measured from the center of mass, the coordinates of m_1 and m_2 are $-m_2L/M$ and $L - (m_2/M) = m_1L/M$. As in Fig. 9–14, the rotational inertia about an axis passing through the center of mass and perpendicular to the axis of the dumbbell is given by

$$I = m_1\left(-\frac{m_2 L}{M}\right)^2 + m_2\left(\frac{m_1 L}{M}\right)^2 = \frac{L^2(m_1 m_2^2 + m_2 m_1^2)}{M^2}$$

$$= \frac{m_1 m_2 L^2 (m_2 + m_1)}{(m_1 + m_2)^2}$$

$$= \frac{m_1 m_2}{m_1 + m_2} L^2. \tag{9–22}$$

TABLE 9–1 Rotational Inertias of Simple Solids

$\frac{1}{12}ML^2$	MR^2	$\frac{1}{2}MR^2$	$\frac{2}{5}MR^2$
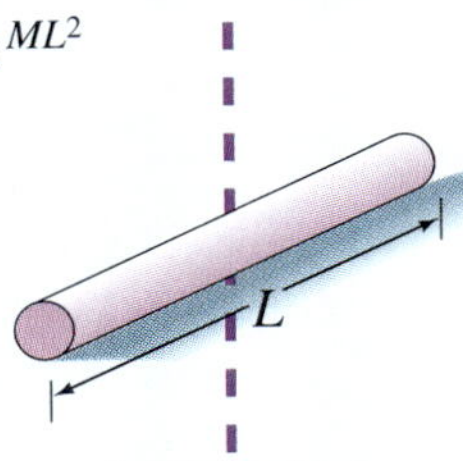	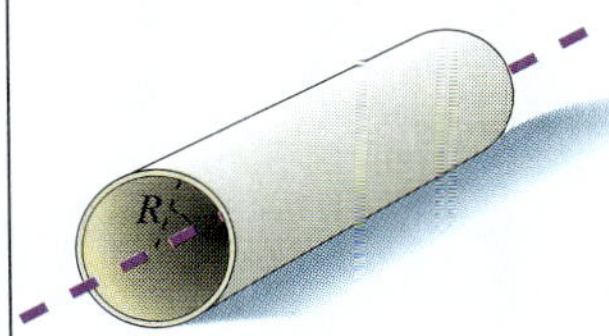		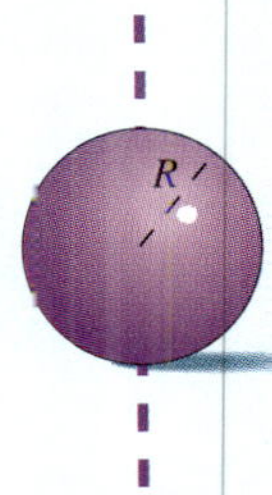
Solid rod about perpendicular axis through center	Cylindrical shell about central axis	Solid cylinder about central axis	Solid sphere about any diameter
$\frac{1}{3}ML^2$	$\frac{1}{2}M(R_1^2 + R_2^2)$	$\frac{1}{4}MR^2 + \frac{1}{12}ML^2$	$\frac{2}{3}MR^2$
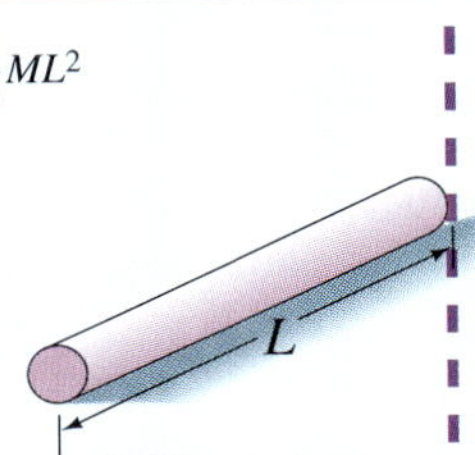	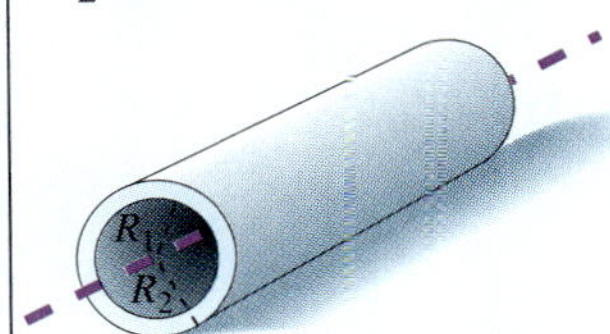	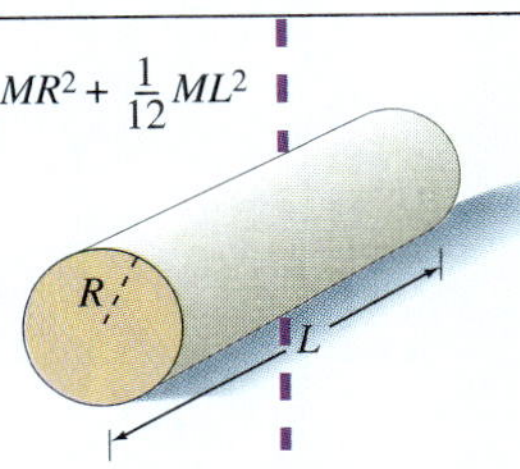	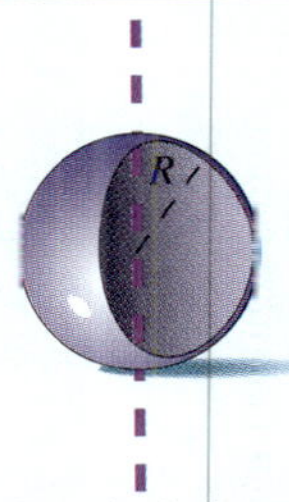
Solid rod about perpendicular axis through end	Hollow cylinder about central axis	Solid cylinder about diameter through center	Thin spherical shell about any diameter

If the masses are equal, then $m_1m_2/(m_1 + m_2) = m/2$; if $m_2 >> m_1$, then it is a good approximation to write $m_1m_2/(m_1 + m_2) = m_1$, the mass of the light point mass. These results are easy to understand physically. In the case of equal masses m, these two point masses rotate about an axis at a distance $L/2$ from each of them, and the rotational inertia is then $I = 2m(L/2)^2 = mL^2/2$. In the case of one very light mass, $m_1 << m_2$, then, to a good approximation, the axis passes through the position of the heavier mass m_2, and the mass m_1 rotates at a distance L from the axis, so $I = m_1L^2$. Note that if the axis of rotation is the line connecting two point masses, then $r_1^2 = 0$, and the rotational inertia about this axis is zero.

EXAMPLE 9–2 A dumbbell consists of point masses $m_1 = 2.0$ kg and $m_2 = 1.0$ kg attached by a rigid massless rod of length $L = 0.6$ m. Calculate the rotational inertia of the dumbbell (a) about an axis perpendicular to the rod of the dumbbell and going through the center of mass; and (b) about an axis perpendicular to the rod of the dumbbell and going through the 2.0-kg mass.

Solution: (a) In Fig. 9–14, let the larger mass be at the origin, $x_1 = 0$ m, and the smaller mass at $x_2 = L$. The center of mass is located at

$$X = \frac{m_1x_1 + m_2x_2}{m_1 + m_2} = \frac{[(2.0\text{ kg})(0\text{ m}) + (1.0\text{ kg})(0.6\text{ m})]}{2.0\text{ kg} + 1.0\text{ kg}} = 0.2\text{ m}.$$

Now

$$I_{\text{cm}} = m_1(x_1 - X)^2 + m_2(x_2 + X)^2$$
$$= (2.0\text{ kg})(0\text{ m} - 0.2\text{ m})^2 + (1.0\text{ kg})(0.6\text{ m} - 0.2\text{ m})^2 = 0.24\text{ kg}\cdot\text{m}^2.$$

(b) Mass m_1 does not contribute to the rotational inertia because it is on the axis. Mass m_2 is a distance L from the axis. Thus

$$I = m_2L^2 = (1.0\text{ kg})(0.6\text{ m})^2 = 0.4\text{ kg}\cdot\text{m}^2.$$

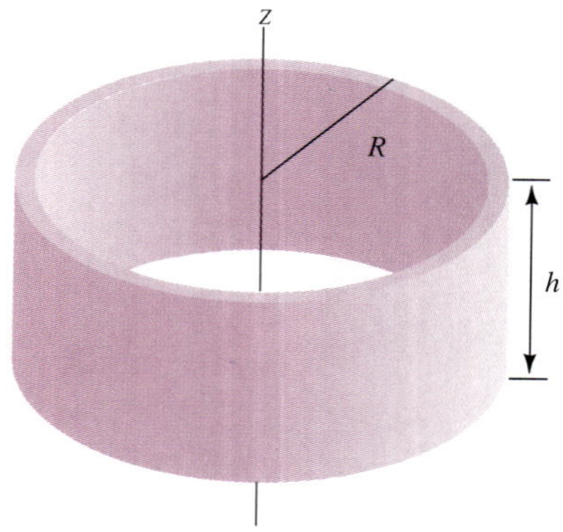

FIGURE 9–15 A thin cylinder of mass M has height h and radius R. We want to find its rotational inertia about the z-axis, which is the symmetry axis of the cylinder.

Thin Cylinder. The thin cylinder shown in Fig. 9–15 has radius R, height h, and total mass M, distributed uniformly around the ring. Let's find the rotational inertia about the central axis, which runs along the z-axis. The word "thin" means that we can consider all of the mass to be the same distance R from the axis. Thus, the rotational inertia is simply

$$I = MR^2. \tag{9–23}$$

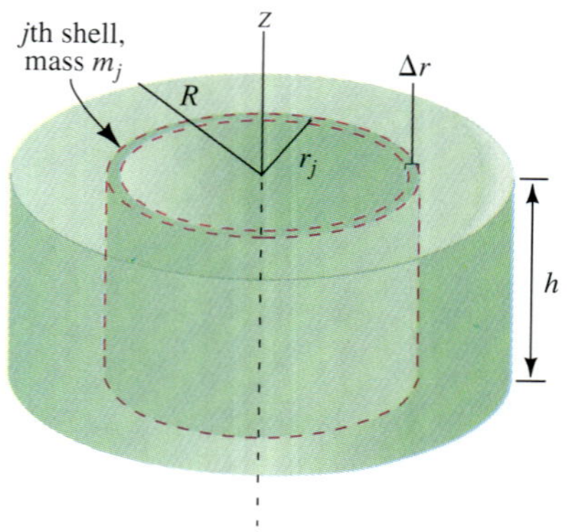

FIGURE 9–16 To find the rotational inertia of a thick cylinder of mass M, height h, and radius R about its symmetry axis, we break it into a series of thin concentric cylinders. We know the rotational inertia of the thin cylinders, and can sum these inertias to find the rotational inertia of the solid cylinder.

Solid Cylinder. The rotational inertia of a uniform solid cylinder about its central axis can be calculated by using our result for the thin cylinder. The solid cylinder has exterior radius R, height h, and mass M. The mass density ρ is given by

$$\rho = \frac{M}{V} = \frac{M}{\pi R^2h}. \tag{9–24}$$

We can very often simplify the calculation of the rotational inertia of an object by breaking the object up into appropriate pieces. In this case, we can divide the solid cylinder into thin cylindrical shells. The jth shell is a distance r_j from the axis and has thickness Δr and mass Δm_j (Fig. 9–16). The cylindrical shell labeled j in Fig. 9–16 contributes an amount $m_jr_j^2$ to the rotational inertia, I, of the solid cylinder. We must sum these contributions to find I:

$$I = \sum_j (\Delta m_j)\, r_j^2. \tag{9–25}$$

The sum in Eq. (9–25) is a formal one. We can carry it out once we find the mass of each shell. The mass of a given volume is ρ times that volume; for the thin shell labeled j, the volume is approximately the shell thickness times the shell circumference times the height, $V_j = (\Delta r)(2\pi r_j)h$. Thus $\Delta m_j = \rho(\Delta r)(2\pi r_j)h$, and

$$I = \sum_j (\Delta m_j)r_j^2 = \sum_j \rho(\Delta r)(2\pi r_j)hr_j^2 = 2\pi h\rho\sum_j (\Delta r)r_j^3.$$

In the limit that the cylindrical shells are very thin, the sum, which includes shells from $r = 0$ to $r = R$, becomes an integral:

$$I = 2\pi h\rho\int_0^R r^3\,\mathrm{d}r = 2\pi h\rho\frac{R^4}{4}.$$

When we substitute Eq. (9–24) for ρ, we find

$$I = 2\pi h\frac{M}{\pi R^2 h}\frac{R^4}{4} = \frac{1}{2}MR^2.$$

We have reduced the formal three-dimensional integral for I to a one-dimensional integral. This sort of simplification is very often possible for multiple integrals. Note that the value of I is less than that of the thin cylinder of the same mass and radius. The weighted sum for the solid cylinder includes mass pieces that give a smaller contribution because they are closer to the axis. Note too that the result is independent of the height and applies for a thin disk as well as a tall cylinder.

PROBLEM SOLVING

Reducing multi-dimensional integrals to one-dimensional integrals

Rod. Next we consider the rotational inertia of a uniform rod of length L about an axis that is perpendicular to it and passes through its center of mass. The center of mass is the midpoint of the rod and is the point at which we will place the origin of our coordinate system (Fig. 9–17). The mass density is the total mass divided by the rod's length, $\lambda = M/L$. Consider the mass $\mathrm{d}m$ contained in a slice of the rod of length $\mathrm{d}x$ at a distance x from the axis. We have $\mathrm{d}m = \lambda\,\mathrm{d}x$, so the rotational inertia about the axis is given by

$$I = \int_{-L/2}^{L/2} x^2\lambda\,\mathrm{d}x = \lambda\int_{-L/2}^{L/2} x^2\,\mathrm{d}x = \lambda\frac{x^3}{3}\Big|_{-L/2}^{L/2}$$
$$= \frac{ML^2}{12}.$$

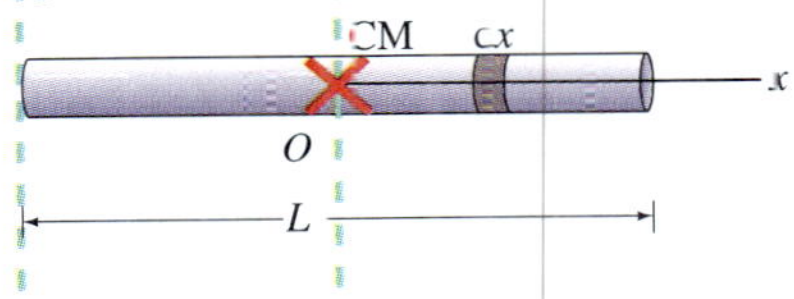

FIGURE 9–17 To find the rotational inertia of a massive rod of length L about an axis perpendicular to it and passing through an end point, we first find the rotational inertia about the parallel axis through the center of mass, which is at the midpoint if the rod's density is constant.

If λ is not a constant but is symmetric about the midpoint, so that the midpoint is still the center of mass, then λ is replaced by $\lambda(x)$, and the rotational inertia is the integral

$$I = \int_{-L/2}^{L/2} x^2\lambda(x)\,\mathrm{d}x.$$

The rotational inertia for rotations about an axis perpendicular to the rod and passing through one end of the uniform rod is given by

$$I = \int_0^L x^2\lambda\,\mathrm{d}x = \frac{\lambda}{3}x^3\Big|_0^L = \frac{\lambda L^3}{3} = \frac{1}{3}ML^2. \qquad (9\text{–}26)$$

This second axis is also drawn in Fig. 9–17.

The Parallel-Axis Theorem

PROBLEM SOLVING

Thanks to the parallel-axis theorem it is often sufficient to calculate rotational inertia about a symmetry axis.

We have worked out the rotational inertia for several simple systems and for several specific axes. Yet how useful are such exercises? After all, there is a different answer for every axis. However, there is a natural set of axes for every object, and these are axes that go through the center of mass. It is sufficient to calculate the rotational inertia of objects about these axes because the rotational inertia about any other axis can be obtained from the **parallel-axis theorem**. This theorem, whose proof we will not detail, states that *the rotational inertia of an object about any axis is given by the sum of the rotational inertia about an axis that goes through the center of mass and is parallel to the given axis, and of the product of the total mass M of the object and the*

The parallel-axis theorem

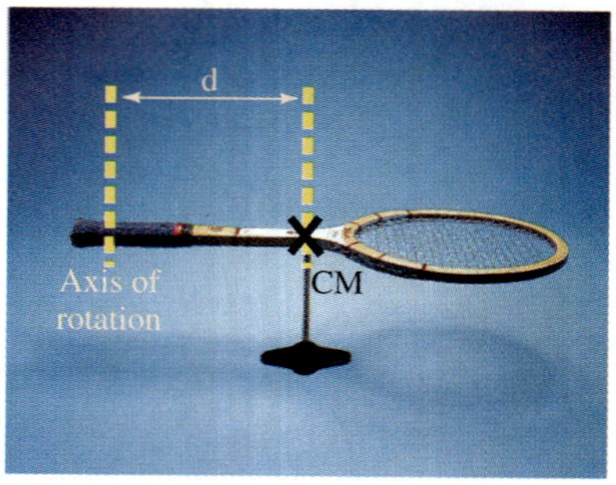

FIGURE 9–18 To find the rotational inertia of any three-dimensional object about a given axis of rotation, first find the rotational inertia about the parallel axis through the center of mass. The parallel-axis theorem gives a general relation between these two rotational inertias. Here, the object in question is a tennis racket.

square of the perpendicular distance d between the two axes (Fig. 9–18). Mathematically, the parallel-axis theorem states that

$$I = I_{cm} + Md^2. \tag{9–27}$$

As a simple example, let's take a dumbbell with equal masses m and consider two parallel axes: one passing through the center of mass and one through one of the masses (Fig. 9–19). For the axis through the center of mass, we have $I_{cm} = m(L/2)^2 + m(L/2)^2 = mL^2/2$, while direct calculation for the axis through the end gives $I_{end} = mL^2$. What does the parallel axis theorem give for I_{end}? The distance d in Eq. (9–27) is $L/2$, while $M = 2m$, so the parallel axis theorem gives $I_{end} = I_{cm} + M(L/2)^2 = mL^2/2 + (2m)(L/2)^2 = mL^2$, the same result given by the direct calculation.

The theorem shows that the rotational inertia of any object is smallest if the axis of rotation goes through the object's center of mass ($d = 0$). This means that the kinetic energy of rotational motion of an object for fixed angular velocity ω is smallest if the rotation is about an axis containing the center of mass.

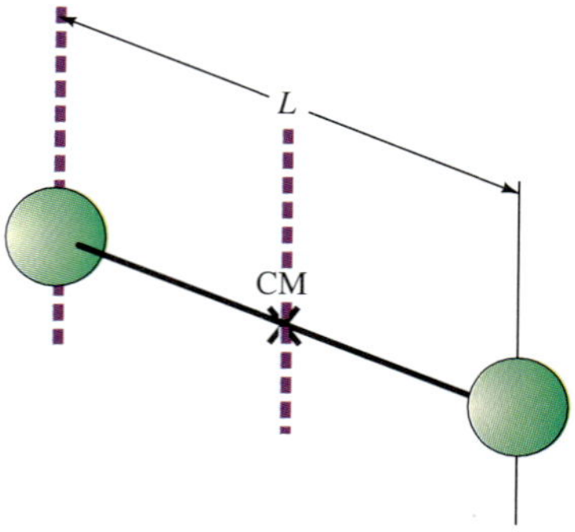

FIGURE 9–19 Two parallel axes, one through the center of mass and the other through one end, can be drawn through this dumbbell of equal masses. By using the rotational inertia for the center of mass axis and the parallel-axis theorem, we can easily find the rotational inertia about the axis through one end.

EXAMPLE 9–3 Consider a uniform, thin, rectangular sheet of metal of mass M, with width a and length b. Calculate the rotational inertia about an axis that is perpendicular to the sheet and passes through one of the corners.

Solution: Choose the corner through which the axis passes as the origin and line up the width along the x-axis and the length along the y-axis (Fig. 9–20). We have $0 \le x \le a$ and $0 \le y \le b$. The two-dimensional mass density σ is uniform and given by

$$\sigma = \frac{M}{ab}.$$

Thus

$$I_{corner} = \int_0^a dx \int_0^b dy\, \sigma(x^2 + y^2) = \frac{M}{ab}\int_0^a dx \int_0^b dy\,(x^2 + y^2)$$

$$= \frac{M}{ab}\left(\int_0^a x^2\,dx \int_0^b dy + \int_0^a dx \int_0^b y^2\,dy\right) = \frac{M}{ab}\left[\left(\frac{a^3}{3}\right)b + \left(\frac{b^3}{3}\right)a\right]$$

$$= \frac{1}{3}M(a^2 + b^2).$$

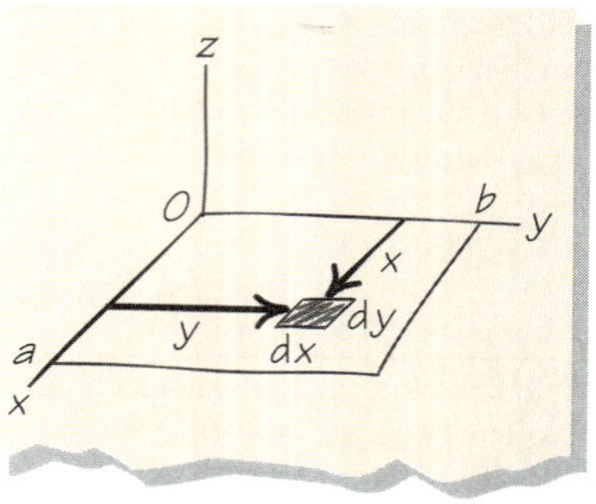

FIGURE 9–20 Example 9–3. We align the width of a rectangular plate with the x-axis and the length with the y-axis. We calculate the rotational inertia about the z-axis.

EXAMPLE 9–4 Use the result of Example 9–3 to calculate the rotational inertia of the same sheet of metal about an axis that is perpendicular to the plane of the sheet and passes through the sheet's center of mass.

Solution: The center of mass is located at the midpoint of the rectangle; that is, at $x = a/2$, $y = b/2$. The distance to the corner is given by

$$d^2 = \left(\frac{a}{2}\right)^2 + \left(\frac{b}{2}\right)^2.$$

Thus, by the parallel-axis theorem, $I_{corner} = I_{cm} + Md^2$:

$$I_{cm} = I_{corner} - Md^2 = \frac{1}{3}M(a^2 + b^2) - M\left[\left(\frac{a}{2}\right)^2 + \left(\frac{b}{2}\right)^2\right]$$

$$= M(a^2 + b^2)\left(\frac{1}{3} - \frac{1}{4}\right) = \frac{1}{12}M(a^2 + b^2).$$

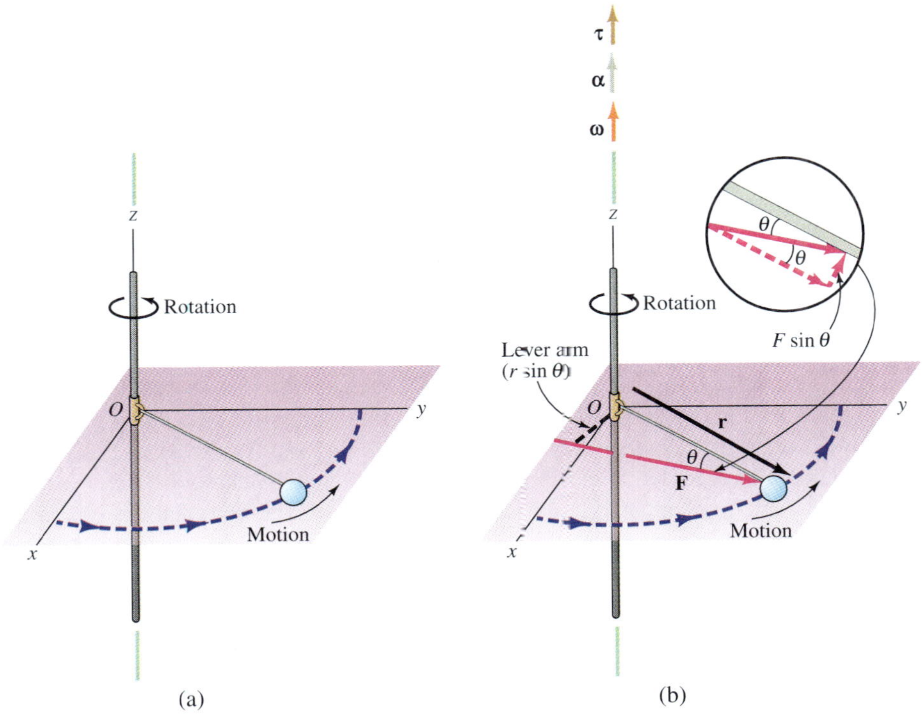

FIGURE 9–21 (a) A rod with a bob at its end is attached to a shaft and can rotate in the plane perpendicular to the shaft. The shaft acts as a rotation axis. (b) A force lying in the plane in which the rod can move is applied to the rod, causing a rotation. The effectiveness of the force in causing a rotation depends on the strength of the force, on where along the rod it is applied, and on its direction, specified by the angle θ. The lever arm is $r \sin \theta$.

9–5 TORQUE

We have described rotations; now, let's turn to the question of what causes them. The analogies we have established between rotational motion and the motion of point particles (we refer to this as *linear motion*) will be useful here. Newton's second law describes the dynamics of linear motion by the equation $\mathbf{F} = m\mathbf{a}$. We'll see here that there is in rotational motion an equation with a quality analogous to force called **torque**.

It will be helpful to keep a definite example of a rotation and its cause in mind. For example, we might think of the rotation of a rod with a bob at its end; the rod is attached at a point O to a shaft that can rotate (Fig. 9–21a). The rod and bob together have a mass M. The rod, of length R, makes a right angle with the shaft. The motion will be entirely in the plane perpendicular to the shaft. Suppose now that a force $\mathbf{F}$ acts in that plane somewhere along the rod (Fig. 9–21b). When this force causes the system to undergo rotational motion, we say that there is a torque.

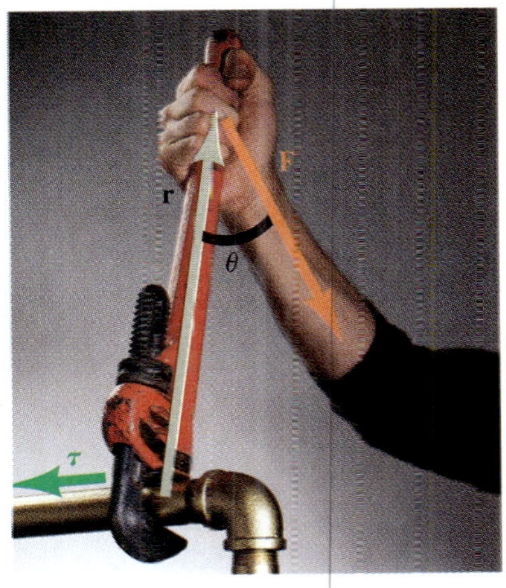

FIGURE 9–22 A plumber taking advantage of a long lever arm. A long wrench is more effective than a short one for freeing a stubborn pipe joint.

We can intuitively think of a torque as a *twist*, just as a force is intuitively a push or a pull. The force will be effective in setting an extended system (such as that of Fig. 9–21) into rotational motion about a point O, depending on two features. First, a larger force will be more effective in exerting torque than would an otherwise similar force of smaller magnitude. Second, a force is most effective at producing a torque if it is applied at a point P that is *far* from point O and is at *right angles* to the line OP that connects O and the point of application of the force. In other words, the torque must be proportional to $r_\perp \equiv r \sin \theta$ in Fig. 9–21b, where θ is the angle between $\mathbf{r}$ and $\mathbf{F}$. This factor is known as the *lever arm*, or *moment arm*. Plumbers who use long wrenches to free stubborn pipes are quite familiar with the value of a large lever arm (Fig. 9–22).

We can summarize these features further by writing torque, denoted by the symbol τ, as

$$\tau = rF \sin \theta. \tag{9–28}$$

Torque and its dimensions and units

The dimensions of torque are $[ML^2T^{-2}]$, and in SI it is measured in N · m. For any angle, we can interpret the factor $r \sin \theta$ as the length of the projection of $\mathbf{r}$ perpendicular to the line of action of $\mathbf{F}$. This is just the lever arm through which the force acts.

Equivalently, we can interpret the factor $F \sin\theta$ as the length of the projection of **F** perpendicular to the vector **r**. Only the component of **F** perpendicular to **r** is effective at producing a twist.

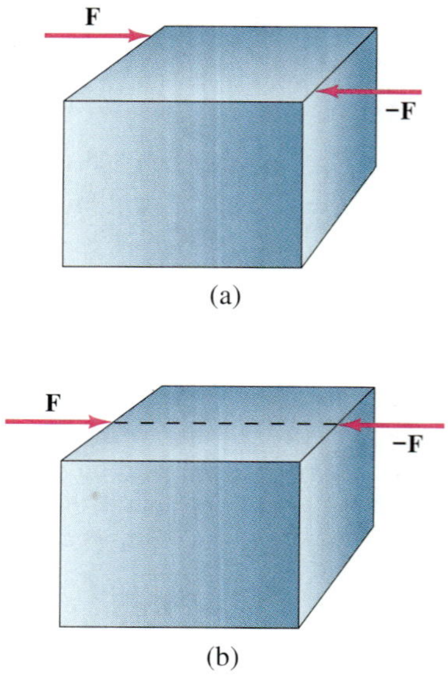

FIGURE 9–23 (a) A pair of equal and opposite forces that do not act along the same line exert a torque on an object. (b) There is no torque only if the same pair of forces act along the same line.

Dynamical Equation for Rotational Motion

To this point, our expression for torque is no more than an attempt to include the factors we know intuitively are effective at producing a twist. To make this concept a more useful one, we want to connect it *quantitatively* to a dynamical equation for rotations. We do this by applying Newton's second law.

Both internal and external forces act on a given element of the rod. In simple linear motion, internal forces do not cause any acceleration because of Newton's third law. In the same way, we can see that internal forces cannot cause a twist because they are aligned. As Fig. 9–23a shows, a pair of equal and opposite forces can produce a twist; however, they cannot if they are aligned (Fig. 9–23b). Any twist due to a force on the element labeled due to element j is canceled by an opposite twist due to a force on element j due to element i. From this point on, we will assume that the force producing torque is an external force.

Suppose, then, that a net force $\mathbf{F}_i$ acts on an element of the rod labeled by i (Fig. 9–24). We take this force perpendicular to the rod: Any component along the rod has zero lever arm and is ineffective at producing a twist. The direction of the force is *tangential.* The ith element is a distance r_i from the fixed point of attachment to the shaft and has mass m_i. The torque on element i about the origin therefore has magnitude $\tau_i = r_iF_i$. At the same time, we know from the second law that F_i produces an instantaneous tangential acceleration a_i of element i whose magnitude is $a_i = F_i/m_i$. Because the acceleration a_i is tangential, we can use Eq. (9–12), $a_i = \alpha r_i$. In other words, because the force is perpendicular to the rod, the instantaneous acceleration of the mass element is associated with an *angular* acceleration of the mass element. Because of internal forces within the (rigid) rod, *the entire rod undergoes the same angular acceleration,* so there is no subscript i on the quantity α. Combining, we have Newton's second law:

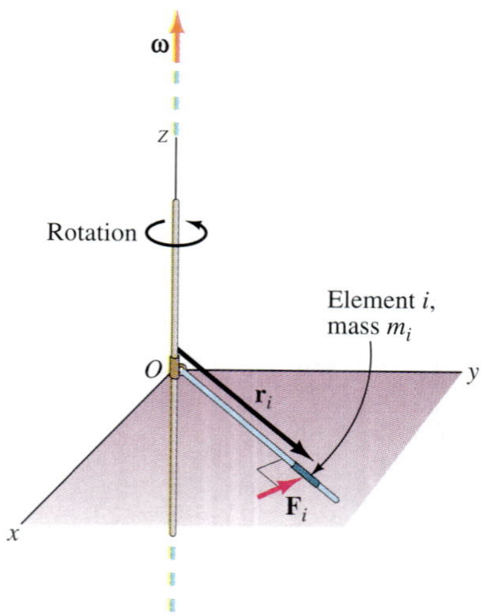

FIGURE 9–24 A force of magnitude F_i acts on an element i of a rod attached to a shaft. The force is perpendicular to the rod.

$$F_i = m_ir_i\alpha;$$

hence, the torque on the element i has magnitude

$$\tau_i = r_iF_i = m_ir_i^2\alpha.$$

The *net* torque, τ, therefore has magnitude

$$\tau = \sum_i m_ir_i^2\alpha = \left(\sum_i m_ir_i^2\right)\alpha,$$

where we have used the fact that α is the same for each element. The quantity in parentheses is the rotational inertia, I, about the shaft [Eq. (9–15)], so we have found the dynamical relation we were looking for:

$$\tau = I\alpha. \tag{9–29}$$

When there is an angular acceleration, α, about an axis perpendicular to the plane of rotation, it is dynamically determined by the torque about the same point according to Eq. (9–29). *This equation is analogous to Newton's second law, which determines the linear acceleration of an object in terms of the object's inertial mass and the force acting on the object.* The major difference is that all the quantities in Eq. (9–29)—the torque, the angular acceleration, and the rotational inertia—refer to a given axis.

The dynamical equation of rotational motion

Newton's second law is a vector relation and we have already stated that the angular acceleration is a vector. Where is the vector aspect of the torque? Figure 9–25 shows that we can use a right-hand rule to define correctly the direction of the torque:

If the fingers of the right hand are aligned along the direction of the line perpendicular to the axis extending to the point where the force is applied (in other words, along the vector **r**) and then curled in the direction of the force, the thumb points in the direction of the torque vector $\boldsymbol{\tau}$. The procedure is illustrated in Fig. 9–26, and we can see that the torque is in the same direction as the angular acceleration. The vector equation of motion takes the form

$$\boldsymbol{\tau} = I\boldsymbol{\alpha}. \tag{9–30}$$

In this chapter, the vector nature of torques will not play a very important role because we are restricting ourselves to rotations about a single axis. It is important, however, to become used to thinking of the dynamics of rotational motion in terms of vectors. In the more general treatment given to rotations in Chapter 10, the vector nature of torque is critical.

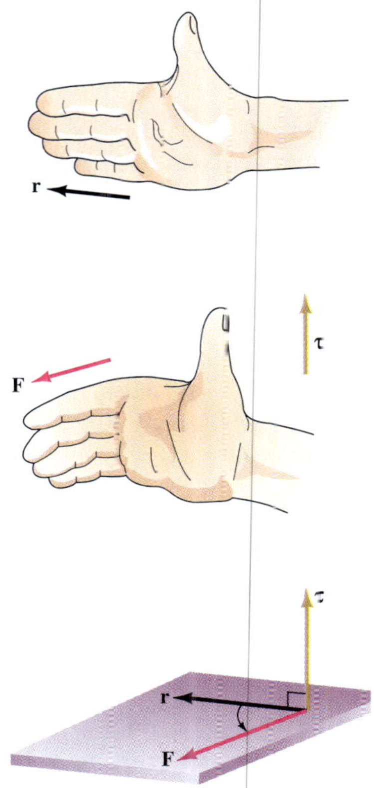

FIGURE 9–25 The direction of the torque can be determined by a right-hand rule applied to the vectors **r** and **F**. The fingers of the right hand are aligned along **r** and then curled toward **F**, with the angle between **r** and **F** always taken to be less than 180°. The torque is then along the thumb.

EXAMPLE 9–5 A massless rod of length $L = 0.83$ m connects two small spheres of mass $m = 0.25$ kg each. The rod is constrained to rotate about an axis perpendicular to the rod and passing through its midpoint. The initial angular velocity has magnitude $\omega_0 = 2.1$ rad/s, and the rod rotates counterclockwise. A tangential force of magnitude $F = 9.6$ N is applied to one of the spheres (Fig. 9–27). If the force is applied for $t = 2.0$ s, what is the final angular velocity of the rod?

Solution: Choose the axis of rotation to be the z-axis such that the initial angular velocity ω_0 about the axis of rotation points in the $+z$-direction (Fig. 9–27). The rotational inertia of the two-sphere system about the axis is given by

$$I = m\left(\frac{L}{2}\right)^2 + m\left(\frac{L}{2}\right)^2 = \frac{mL^2}{2}.$$

The angular acceleration about the axis points in the same direction as the torque about that axis, whose magnitude is given by

$$\tau = \frac{L}{2}F.$$

Equation (9–29) gives a constant angular acceleration for a constant torque:

$$\alpha = \frac{\tau}{I} = \frac{LF/2}{mL^2/2} = \frac{F}{mL}.$$

The angular velocity then has a linear dependence on the time:

$$\omega = \omega_0 + \alpha t = \omega_0 + \frac{F}{mL}t.$$

At time $t = 2.0$ s, we have

$$\omega = 2.1 \text{ rad/s} + \frac{9.6 \text{ N}}{(0.25 \text{ kg})(0.83 \text{ m})}(2.0 \text{ s}) = 95 \text{ rad/s}.$$

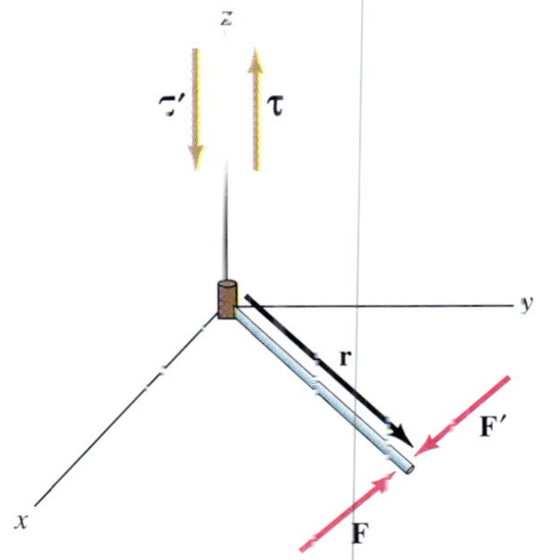

FIGURE 9–26 The direction of the angular acceleration of the rod, α, and of the torque on the rod, τ, depends on the direction of the force. When the force is in one direction (**F**), the torque $\boldsymbol{\tau}$ and angular acceleration point along $+z$; when the force is in the other direction (**F**′), the torque $\boldsymbol{\tau}'$ and angular acceleration point along $-z$.

Force Diagrams Revisited

The discussion of torque suggests that in the force-diagram technique, we should take into account the location at which forces act. A description of such an extended force diagram and some advice on how to approach problems that involve rigid-body motion are given in the Problem-Solving Techniques box.

Example 9–6 shows the role played by the force diagram in clarifying the forces and the torques that act on objects when rotations can occur. It also illustrates a useful energy technique.

PROBLEM-SOLVING TECHNIQUES

1. Identify and isolate the rigid body on which external forces act.
2. Identify the external forces that act on the object as well as where they act.
3. Prepare an extended force diagram: All forces should be included *plus information as to where they act* on the rigid body. Where the force acts is of paramount importance in understanding torques. Note that gravity acts as if it were applied to the center of mass of the object (see Eq. (9–32) and what follows it).
4. Identify a single convenient axis, which will normally be the axis about which the rigid body rotates.
5. Find the torques about this axis that result from the acting forces. The direction of positive torque should be clearly understood.
6. Express the dynamical equations of motion that correspond to the *net* torque. This step generally requires knowledge (or calculation) of the rotational inertia about the axis of rotation. The solutions of the dynamical equations describe the rotational motion of the rigid body.
7. The net force governs the linear motion of the center of mass.

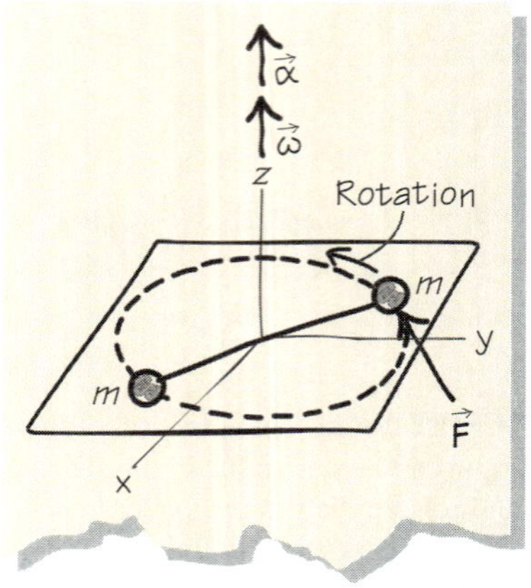

FIGURE 9–27 Example 9–5.

EXAMPLE 9–6 A bucket of water of mass $m = 12$ kg is connected to a rope of negligible mass. The rope is wrapped around a uniform solid cylindrical flywheel of mass $M = 88$ kg and radius $R = 0.50$ m (Fig. 9–28a). The flywheel is free to rotate about its axis, which is horizontal. The bucket is brought to a height $h = 28$ m and dropped, making the flywheel turn. Use the dynamical equation of motion to find the angular velocity of the flywheel after the bucket has fallen for 5 s.

Solution: We start with a force diagram drawn according to our new rules. Figure 9–28b shows a view of the flywheel from along the horizontal axis. This central axis is the one from which we measure torque. The only torque on the flywheel about this axis comes from the tension T of the rope. The angular velocity in Fig. 9–28b will point along the axis of rotation; that is, perpendicular to the page. Similarly, all forces are vertical. From now on, we will drop the boldfaced notation for these quantities.

We take the upward direction to be the $+y$-direction. The forces acting on the flywheel are the normal force, $\mathbf{F}_N$, at the axis (due to the supports that hold the rod horizontal), the force of gravity, and the tension. Because the center of mass of the flywheel does not accelerate, the net force acting on it must be zero, such that

$$\text{second law for flywheel:} \quad F_N - T - Mg = 0.$$

The forces acting on the bucket are the tension and the force of gravity. The mass will accelerate down with (as yet unknown) magnitude a, so Newton's second law gives

$$\text{second law for bucket:} \quad T - mg = -ma.$$

The torque acting on the flywheel has magnitude TR, so the dynamical torque equation is

$$TR = I\alpha.$$

Application of the right-hand rule shows that the torque is directed out of the page. Then α will also be out of the page, corresponding to an increasing angular speed in the sense shown in Fig. 9–28b.

We now have three dynamical equations and four unknowns: F_N, T, a, and α. There is, however, a fourth equation, which is a *kinematic* relation between a and α. This relation comes from the fact that if the bucket moves at a speed v, then any point on the rope moves at the same speed, and thus the outer rim of the flywheel moves at that speed. This means that the flywheel rotates about its axis at an angular speed ω such that

$$v = \omega R.$$

If we take a time derivative of this equation, we find the fourth relation:

$$a = \alpha R.$$

Now we solve our equations for the unknowns. When we substitute $\alpha = a/R$ into the dynamical equation for rotation, we find that

$$T = \frac{I\alpha}{R} = \frac{Ia}{R^2}.$$

Because the rotational inertia of a flywheel about its axis is $I = \frac{1}{2}MR^2$ [Eq. (9–26)], this becomes

$$T = \frac{1}{2}Ma.$$

We substitute this result into Newton's second law for the bucket:

$$\frac{1}{2}Ma - mg = -ma.$$

This equation can be solved for the magnitude of the bucket's acceleration:

$$a = \frac{mg}{m + M/2}.$$

In turn, the angular acceleration of the flywheel is

$$\alpha = \frac{a}{R} = \frac{mg}{[m + (M/2)]R}.$$

This angular acceleration is constant, implying that the angular velocity changes linearly with time:

$$\omega = \omega_0 + \alpha t = \omega_0 + \frac{mgt}{[m + (M/2)]R}.$$

By the initial conditions of the problem, $\omega_0 = 0$ rad/s, and after 5 s,

$$\omega = \frac{(12\text{ kg})(9.8\text{ m/s}^2)(5\text{ s})}{\{(12\text{ kg}) + [(88\text{ kg})/2]\}(0.50\text{ m})} = 21\text{ rad/s}.$$

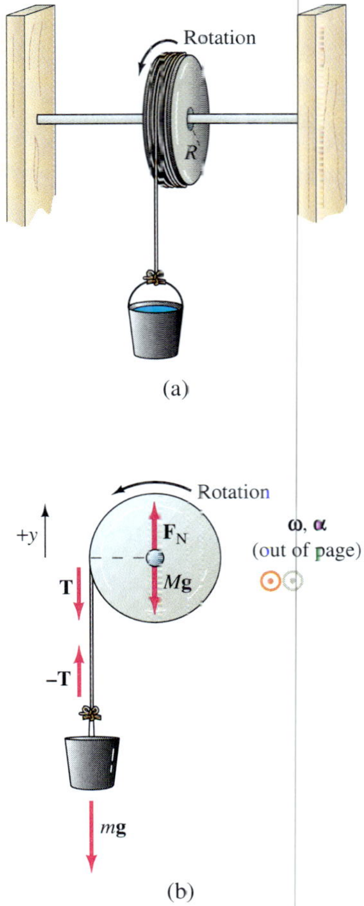

FIGURE 9–28 (a) Example 9–6. (b) Combined force diagram for the suspended mass m and for the cylindrical flywheel, including the locations where the various forces act. The circled dots indicate that the vectors (here, $\boldsymbol{\omega}$ and $\boldsymbol{\alpha}$) are directed out of the plane of the page.

Gravity and Extended Objects

Gravity acts on every point of an object at once. Here we demonstrate that the torque on an extended object due to gravity can be computed as if the entire mass of the object were concentrated at its center of mass. Let's consider a rod of length L and mass M fixed to a pivot at its upper end (Fig. 9–29a). We divide the rod into a series of pieces labeled by the subscript i. The pieces, which have mass Δm_i, are small enough

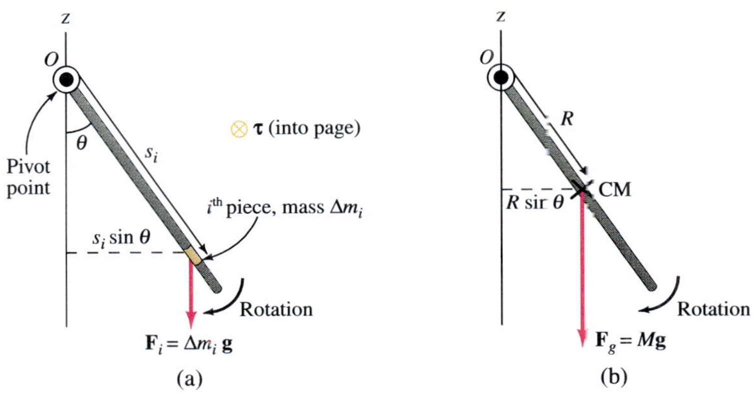

FIGURE 9–29 The force of gravity rotates a rod about the point O. (a) The differential element of force $\mathbf{F}_i$ acts on a differential mass Δm_i located a distance s_i down the rod. The circled cross indicates that the torque τ is directed into the plane of the page. (b) The net effect of the sum of the torques acting on each piece is a net torque that is the same as would be present if the entire mass of the rod were concentrated at the rod's center of mass, here marked with an ×.

so that we can consider each of them to be a given distance s_i down the rod. We will compute the net torque due to gravity about the pivot point, assuming the rod makes an angle θ to the vertical. The net torque, $\boldsymbol{\tau}$, is the vector sum of the torques on each piece, $\boldsymbol{\tau}_i$. All the torques are directed into the page, so this sum is the simple algebraic sum of the magnitudes τ_i. In turn, the moment arm for the force of gravity on the piece labeled by i is $s_i \sin\theta$, so the torque τ_i is

$$\tau_i = (s_i \sin\theta)\Delta m_i g.$$

The net torque then has magnitude

$$\tau = \sum_i \tau_i = \sum_i (s_i \sin\theta)(\Delta m_i g) = \left[\sum_i (s_i)(\Delta m_i)\right](\sin\theta)g. \tag{9–31}$$

By definition (see Chapter 8), the quantity in brackets is MR, where R is the distance of the center of mass from the reference point. Thus the net torque due to gravity is (Fig. 9–29b)

$$\tau = R(Mg)\sin\theta. \tag{9–32}$$

The torque due to gravity

Compare this result with Eq. (9–28), which is our general form for the torque. *The torque due to gravity on an extended object of total mass M may be represented by the torque due to gravity acting on a particle of mass M located at the object's center of mass.* This important and very useful result is a general one that holds for any object, not just a rod (see Chapter 11 for a more general proof).

The same reasoning can be applied to energy considerations. When an extended object moves in any way, we can find the change in its gravitational potential energy by finding the change in height, Δh, of the center of mass. The change in gravitational potential energy is then $Mg(\Delta h)$.

9–6 ANGULAR MOMENTUM AND ITS CONSERVATION

We have found a dynamical equation for rotational motion that is analogous to Newton's second law. The analogue between linear and rotational motion can be of further use to us. In our original formulation of Newton's second law, the quantity $m\,d\mathbf{v}/dt$ appears. We saw in Chapter 8 that it is useful to define the linear momentum $\mathbf{p} \equiv m\mathbf{v}$ and to formulate a more general form of Newton's second law:

$$\mathbf{F} = \frac{d\mathbf{p}}{dt}.$$

The definition of angular momentum

Similarly, it is useful to define the **angular momentum**, **L**, of a symmetrical object that rotates about its symmetry axis (or an axis parallel to the symmetry axis) with angular velocity $\boldsymbol{\omega}$ by

$$\mathbf{L} \equiv I\boldsymbol{\omega}. \tag{9–33}$$

Here, I is the rotational inertia of the object with respect to the rotation axis.

EXAMPLE 9–7 Calculate the angular momentum of Earth's motion about its axis of rotation, given that Earth's mass is 6×10^{24} kg and its radius is 6.4×10^6 m. Assume that the mass density is uniform.

Solution: Earth makes a single revolution about its axis in 24 h. Thus, the period of rotation is

$$T = (24\text{ h})\frac{60\text{ min}}{1\text{ h}}\,\frac{60\text{ s}}{1\text{ min}} = 86{,}400\text{ s}.$$

Hence, from the connection between ω and T [Eq. (9–4)],

$$\omega = \frac{2\pi}{T} = \frac{6.28\ \text{rad}}{86{,}400\ \text{s}} = 7.3 \times 10^{-5}\ \text{rad/s}.$$

The rotational inertia of a uniform sphere is given in Table 9–1:

$$I = \frac{2}{5}MR^2 = (0.4)(6 \times 10^{24}\ \text{kg})(6.4 \times 10^6\ \text{m})^2 = 10 \times 10^{37}\ \text{kg}\cdot\text{m}^2.$$

Thus

$$L = I\omega = (10 \times 10^{37}\ \text{kg}\cdot\text{m}^2)(7.3 \times 10^{-5}\ \text{rad/s}) = 7 \times 10^{33}\ \text{kg}\cdot\text{m}^2\text{/s}.$$

The assumption that Earth is a sphere of uniform density is not correct, however, which is why our calculated value of I is some 20 percent larger than the correct value of $7.9 \times 10^{37}\ \text{kg}\cdot\text{m}^2$.

In terms of angular momentum, the dynamical equation for rotational motion takes the more general form

$$\boldsymbol{\tau} = \frac{d\mathbf{L}}{dt}. \qquad (9\text{–}34)$$

The torque determines the rate of change of the angular momentum.

Angular momentum for rotations of the type just described is a vector that points in the same direction as $\boldsymbol{\omega}$. For uniform rotational motion about an axis, the angular momentum does not change in either magnitude or direction. Angular momentum shares with momentum† the important property that it is independent of time for a system that is "left alone"; that is, a system on which there is no torque due to external forces. *Note that it is possible that the external torque is zero even when the external force is not zero, depending on where the external force is applied and on its direction.* (Similarly, it is possible to have a net torque even though the net force is zero.) When the net torque is zero, the angular momentum is independent of time and is *conserved.* For rigid bodies, the rotational inertia is constant, and the conservation of angular momentum means that the angular velocity is constant in time. When the rotational inertia can vary because the system considered can vary its shape, then the conservation of angular momentum is a very important and useful principle, as we will see in Chapter 10.

Because angular momentum is so fundamental to rotational motion, it is useful to write the energy in terms of it. We have

$$K = \frac{1}{2}I\omega^2 = \frac{1}{2}\frac{(I\omega)^2}{I} = \frac{L^2}{2I}. \qquad (9\text{–}35)$$

Parallels between Rotational and Linear Motion

Throughout this chapter, we have emphasized the similarities and the differences between rotational motion and linear motion. The major differences are that the vectors that describe the parameters of rotational motion are all measured with respect to an axis, and a right-hand rule applies to specify their direction. The similarities, or parallels, between these two types of motion are many. Although the description of rotational motion may appear to be complicated, keep in mind that no fundamentally new physical laws are involved. Everything we have discussed to this point is derived from application of Newton's laws to extended systems of point masses.

†The simple term "momentum" refers to linear momentum, not to angular momentum.

APPLICATION: ANGULAR MOMENTUM CONSERVATION IN DRAG RACING

Angular momentum conservation plays a crucial role in drag racing. Unlike the normal operation of a car, the aim in drag racing is to have the clutch fully engaged—and therefore the rear tires spinning—before the vehicle crosses the starting line (Fig. 9AB–1). The reason for this is that when the engine turns at its maximum possible rate, it has a maximum amount of angular momentum. The transmission shafts and the rear wheels, which start at rest before the clutch begins its engagement, have a large rotational inertia. As the clutch is engaged, some of the engine's angular momentum is transferred to the shafts and wheels; thus the angular momentum of the engine is reduced. The total energy of the engine/transmission-wheel system is also reduced. (This is analogous to the collision of a moving object with a stationary one, in which momentum is conserved, but kinetic energy is reduced). If this lost energy can be replaced by burning fuel before the vehicle crosses the starting line, then the maximum energy is available to propel the automobile along the drag racing track. But if the clutch is engaged after the start of the race, then the lost energy has to be replaced *during* the race and this results in a longer time to the finishing post. Note that, because the wheels spin at the start, it is kinetic rather than static friction that propels the car forward, and kinetic friction is weaker than the maximum value of static friction. Interestingly, this is less significant than the need to replace the energy lost in what we might call the "energy-transmission angular collision."

FIGURE 9AB–1 The spinning rear wheels of this drag racer raise clouds of smoke from hot rubber. This is a very inefficient way to drive your car on the road.

9–7 ROLLING

Kinematics of Rolling

Our discussion so far has dealt with the kinematics of pure rotation. *Rolling* is a type of motion that is more than a pure rotation. When a bicycle wheel rolls, rotation occurs *along with* linear motion. The wheel undergoes pure rotation about its axis but, to an observer on the ground, that axis is moving; as a consequence, the observer sees a point on the bicycle rim undergo motion that is more than a simple rotation (Fig. 9–30). Rolling involves a connected linear and rotational motion. Our first task is to describe rolling and discover the connection between the linear and rotational motion that it implies.

The connection between the linear and rotational motion in rolling without slipping

Consider a wheel of radius R that rolls without slipping or skidding in a straight line on a horizontal surface (Fig. 9–31a). The center of the wheel moves at uniform speed v. In rolling without slipping, the connection between the speed v and the magnitude of the angular velocity, ω, of the wheel's rotation is

$$v = R\omega. \tag{9–36}$$

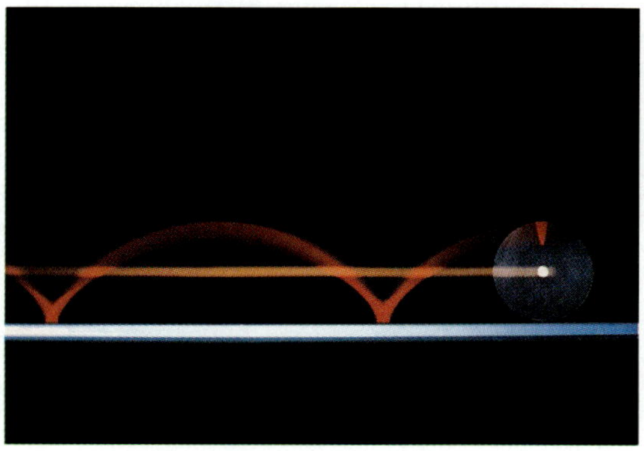

FIGURE 9–30 A rolling cylinder. A small light at the center and one at the edge show that the center of mass moves linearly, whereas a point along the edge has rotational motion and traces out a *cycloid*.

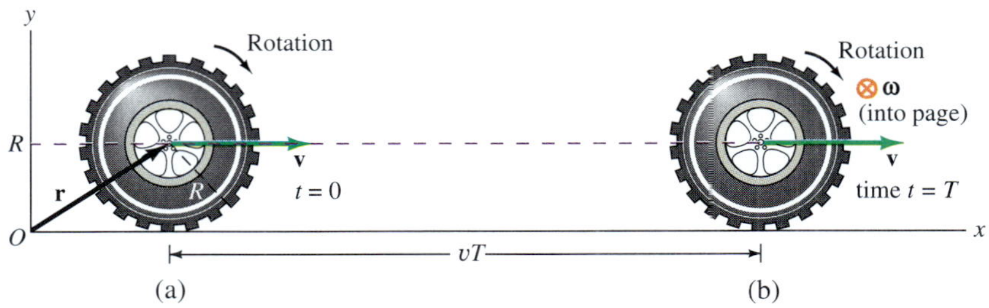

FIGURE 9–31 (a) A wheel of radius r rolls without slipping in the x-direction. Its plane lies in the xy-plane, and its center is at the position $\mathbf{r}$. (b) If the center of the wheel is at $x = 0$ at $t = 0$ and moves at speed v along the $+x$-direction, then the angular speed about the center has magnitude $\omega = v/r$.

A simple way to see this result is to imagine a chalk mark made on the rim of the wheel that rolls along a plane. The wheel leaves chalk marks on the plane as it rolls, and the distance between chalk marks made in one revolution is $2\pi R$ (the wheel's circumference). If the time taken to move this distance is T, then the center of mass will have traveled a distance vT (Fig. 9–31b). The distances are equal, so $vT = 2\pi R$, or $v = 2\pi R/T$. But $2\pi/T$ is the angular speed, ω, from which Eq. (9–36) follows.

Note that the motion of the wheel may be viewed as a rotation about the contact point with the road (Fig. 9–32a). The angular velocity about that point is determined by the fact that the wheel hub, at a distance R from the point of contact, has speed v (Fig. 9–32b). Thus this angular speed is also $\omega = v/R$. Because the top of the wheel is at a distance $2R$ from the point of contact, $v_{top} = 2R\omega = 2v$.

(a)

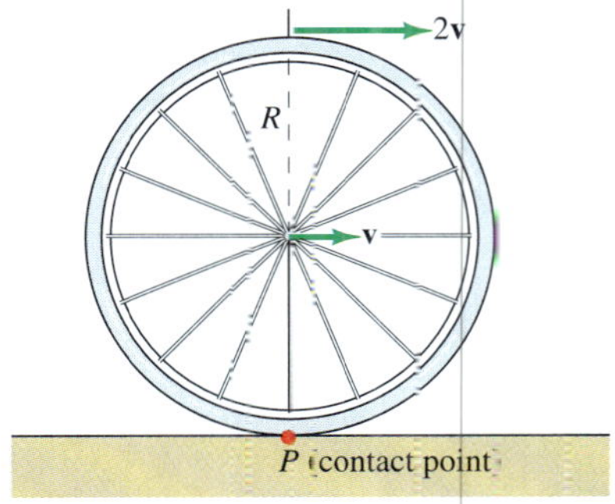

(b)

FIGURE 9–32 (a) A rotating bicycle wheel. The rest frame with respect to the road is the frame of reference in which the point of contact between the rolling bicycle wheel and the road is at rest. (b) In this frame, the center of the wheel moves with velocity $\mathbf{v}$ and the top of the wheel with velocity $2\mathbf{v}$. Both points rotate about the contact point.

Energy in Rolling

With respect to the point of contact, the motion of the wheel is a *pure* rotation about the point of contact, and the total kinetic energy is

$$K = K_{rot} = \frac{1}{2} I_{contact}\, \omega^2, \tag{9–37}$$

where $I_{contact}$ is the rotational inertia of the wheel about an axis through the point of contact. By the parallel-axis theorem, the rotational inertia $I_{contact}$ may be written in terms of the rotational inertia about the central axis through the center of mass:

$$I_{contact} = I_{cm} + MR^2, \tag{9–38}$$

where M is the mass of the rolling object. Thus the total kinetic energy of the rolling object is

$$K = \frac{1}{2}(I_{cm} + MR^2)\omega^2 = \frac{1}{2} I_{cm}\omega^2 + \frac{1}{2} Mv^2, \tag{9–39}$$

where we have used $v = R\omega$. Equation (9–39) expresses the important result that *the kinetic energy of an object rolling without slipping is the sum of the kinetic energy of rotation about its center of mass* ($I_{cm}\omega^2/2$) *and the kinetic energy of the linear motion of the object as if all the mass were at the center of mass.*

The total kinetic energy of a rolling object in terms of the center of mass

EXAMPLE 9–8 Calculate the total kinetic energy of a solid ball of mass 2 kg and radius 10 cm that rolls without slipping on a flat surface at a speed of 0.8 m/s.

Solution: The rotational inertia of a solid ball about an axis through its center is $\frac{2}{5}MR^2$, as can be seen in Table 9–1. Thus, from Eq. (9–39), the total kinetic energy is

$$K = \frac{1}{2}\left(\frac{2}{5}MR^2\right)\omega^2 + \frac{1}{2}Mv^2 = \frac{1}{5}Mv^2 + \frac{1}{2}Mv^2 = \frac{7}{10}Mv^2$$

$$= 0.7(2\text{ kg})(0.8\text{ m/s})^2 = 0.9\text{ J}.$$

The kinetic energy associated with the rolling is 40 percent of the energy associated with the linear motion.

Dynamics of Rolling

Consider a round shape of mass M and radius R that rolls without slipping down a plane making an angle θ with the horizontal (Fig. 9–33). We will insist that the object be symmetric about the central axis but it is otherwise unrestricted; for example, it could be a solid cylinder, a hollow cylinder, or a sphere. Figure 9–33 assumes that the object is a solid cylinder, but our analysis will not depend on that. We will calculate the angular velocity of the object after its center of mass has traveled a distance ℓ down the incline, starting from rest.

The forces are included in Fig. 9–33, which thereby serves as a force diagram. The force of gravity, $M\mathbf{g}$, may be separated into a component normal to the plane, $Mg\cos\theta$, and a component parallel to the plane, $Mg\sin\theta$. Both components act on the center of mass of the rolling object, as we learned in Section 9–5. There is also the contact force $\mathbf{F}_N$, normal to the plane and acting at the point of contact, which just cancels the component of gravity normal to the plane. Finally, there is the force of friction, $\mathbf{f}$, which acts at the point of contact and points back up along the plane. The existence of the frictional force is crucial if there is to be rolling rather than sliding.

To apply the equations of motion to the rolling object, we want to find the net torque acting on the object. We will use the axis through the center of the object as a reference axis, as in Fig. 9–33. Friction is the *only* force of all the forces acting on the object that exerts a torque about the center of mass; both the normal force and gravity have no lever arm through this point.

The net force along the plane determines the linear acceleration, magnitude a in this direction. According to Newton's second law,

$$Ma = Mg\sin\theta - f. \tag{9–40}$$

Because it is static friction that is involved, its magnitude, f, is not uniquely determined until we include the dynamical equation of rotation. The torque due to friction has magnitude

$$\tau = fR. \tag{9–41}$$

By the right-hand rule, the torque's direction is along $\boldsymbol{\omega}$ in Fig. 9-33. The dynamical equation that determines the rate of change of angular momentum about the axis of the cylinder comes from Eq. (9–29) with the torque given by Eq. (9–41):

$$I\alpha = \tau = fR. \tag{9–42}$$

This equation implies that the friction force is given by $f = I\alpha/R$. Substituting into Newton's second law, Eq. (9–40), we get

$$Ma = Mg\sin\theta - \frac{I\alpha}{R}. \tag{9–43}$$

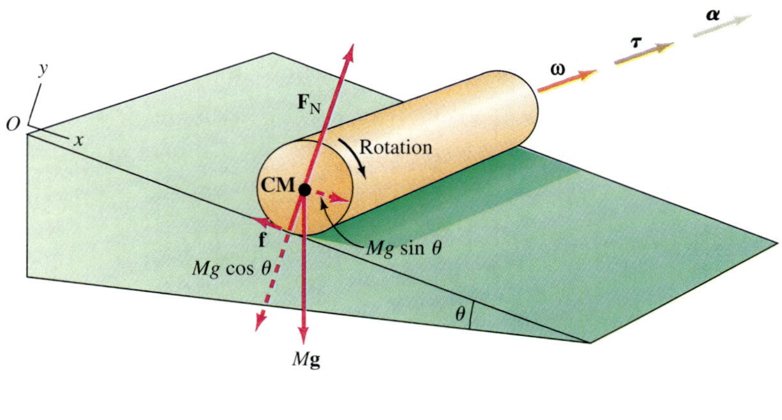

FIGURE 9–33 A cylinder of radius R rolls down an inclined plane. Static friction causes the cylinder to roll without slipping. The rotational motion can be analyzed about any axis. One good one passes through the geometrical center of the cylinder and the other is formed by the line of contact between the cylinder and the plane. The forces acting are friction $\mathbf{f}$, the normal force $\mathbf{F}_N$, and gravity $M\mathbf{g}$. We have decomposed the force of gravity into components along, and perpendicular to, the inclined plane.

Finally, we use Eq. (9–36), $v = R\omega$, to derive a relation between a and α. The time derivative of that equation immediately gives us $a = R\alpha$. Substitution into Eq. (9–43) then gives us a single equation for the angular acceleration:

$$M\alpha R = Mg \sin\theta - \frac{I\alpha}{R},$$

with the algebraic solution

$$\alpha = \frac{MgR \sin\theta}{MR^2 + I}. \tag{9–44}$$

Only geometric factors appear on the right, and the angular acceleration is a constant. With a constant angular acceleration, we find the usual linear dependence of angular velocity on time:

$$\omega = \omega_0 + \frac{MgR \sin\theta}{MR^2 + I} t. \tag{9–45}$$

Here, the angular velocity at $t = 0$ is ω_0. If the object starts from rest, then $\omega_0 = 0$. Similarly, the linear acceleration is constant, implying a center-of-mass velocity that increases linearly with time according to

$$v = \omega R = v_0 + \frac{(MgR^2 \sin\theta)}{(MR^2 + I)} t. \tag{9–46}$$

We can also determine the speed gained after the object has rolled a total distance ℓ down the plane. According to Eq. (2–25d), we have

$$V_{\text{cm}}^2 = 2a\ell = 2\frac{(MgR^2 \sin\theta)}{(MR^2 + I)}\ell. \tag{9–47}$$

For an object of radius R and mass M that is symmetric about a central axis, dimensional analysis tells us that the rotational inertia will equal CMR^2, where C is a pure number determined by the geometry. Thus Eq. (9–45) becomes

$$\omega = \omega_0 + \frac{(g/R)\sin\theta}{1 + C} t,$$

or, equivalently,

$$V_{\text{cm}} = v_0 + \frac{g \sin\theta}{1 + C} t. \tag{9–48}$$

Similarly,

$$V_{\text{cm}}^2 = 2al = 2\frac{(g \sin\theta)}{1 + C}\ell. \tag{9–49}$$

It is instructive to compare this last result with the speed gained by a point object that moves down the same ramp without friction: $v^2 = 2g\ell \sin\theta$. The rolling object moves less rapidly because *some of the potential energy goes into rotational motion.* Note also the striking result that *the speed of the center of mass is independent of both M and R.*

Let's look at a few special cases. The smallest possible value of C is zero, corresponding to an object with all its mass concentrated at its central axis; this is just the case of a small sliding object since there is no rotational energy in this case. The case $C = 1/2$ is the solid cylinder (see Table 9–1). The case $C = 1$ corresponds to a thin hollow cylinder with all its mass concentrated at the outside. Still larger values of C are possible if we consider yo-yo-like objects, where the rolling takes place on the inner cylinder. Thus the factor $1 + C$ increases from a minimum value of 1, and the rate at which ω, or v, increases with time for symmetric rolling objects is always less than that of a sliding object.

EXAMPLE 9–9 A ramp of length $d = 1.5$ m is set at an angle of 5.0° to the horizontal. Two objects are initially at rest at the top of the ramp and simultaneously start rolling without slipping. Object 1 is a solid cylinder of mass 0.65 kg and radius $R = 4.7$ cm. Object 2 is a hollow, thin-walled tube of the same radius R and of mass 0.13 kg. How much time does it take for each object to arrive at the bottom of the ramp?

Solution: Because there is no slipping, we can use the direct kinematic connection between linear motion and angular motion of the rolling objects; moreover, we may use the results of our discussion with different values of I for the two different objects. We therefore know the values of the linear acceleration in each case. Because we want to know the time required to travel a given straight-line distance d under constant acceleration a, we use the constant-acceleration equation $d = at^2/2$, or $t = \sqrt{2d/a}$, which applies when the initial velocity is zero. Equation (9–44) with $a = \alpha R$ tells us that

$$a = \frac{MgR^2 \sin\theta}{MR^2 + I},$$

where M is the object's mass. For the case of the solid cylinder, $I = MR^2/2$. The center of the solid cylinder then has linear acceleration

$$a_1 = \frac{g \sin\theta}{1 + \frac{1}{2}} = \frac{2}{3} g \sin\theta.$$

The hollow cylinder has a rotational inertia about its axis of $I = MR^2$; thus, its linear acceleration has magnitude

$$a_2 = \frac{g \sin\theta}{1 + 1} = \frac{1}{2} g \sin\theta.$$

In each case, the linear acceleration is independent of the mass and the radius, so this information is not needed. The only relevant difference between the solid cylinder and the hollow tube is that they have different rotational inertias because their masses are distributed differently. Numerically,

$$a_1 = \frac{2}{3} g \sin\theta = (0.67)(9.8\ \text{m/s}^2)(\sin 5.0°) = 0.57\ \text{m/s}^2;$$

$$a_2 = \frac{1}{2} g \sin\theta = (0.5)(9.8\ \text{m/s}^2)(\sin 5.0°) = 0.43\ \text{m/s}^2.$$

The time for descent of the cylinder is then

$$t_1 = \sqrt{\frac{2d}{a_1}} = \sqrt{\frac{2(1.5\ \text{m})}{0.57\ \text{m/s}^2}} = 2.3\ \text{s};$$

for the tube,

$$t_2 = \sqrt{\frac{2(1.5\ \text{m})}{0.43\ \text{m/s}^2}} = 2.6\ \text{s}.$$

It is easy to try the experiment and see that a solid cylinder reaches the bottom of a ramp faster than a hollow one. Figure 9–34 illustrates this race.

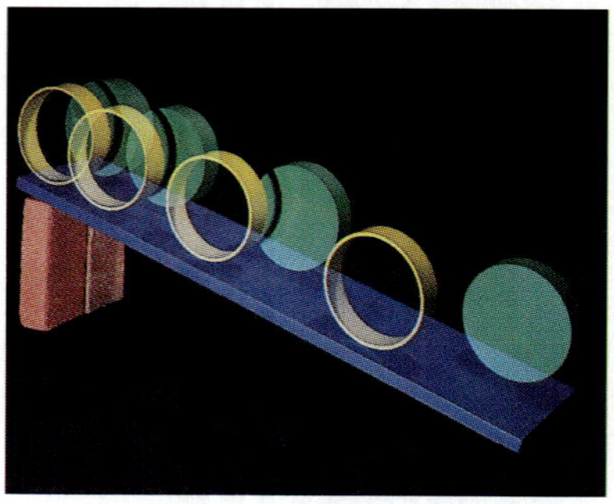

FIGURE 9–34 Example 9–9. The rate at which a rolling object accelerates depends on its rotational inertia. In a race between a solid cylinder and a hollow cylinder, both starting from rest and rolling down an inclined plane, the solid cylinder comes in first.

SUMMARY

A rigid body can undergo rotations as well as linear motion. The linear motion can be described by specifying the motion of the center of mass. Rotations about fixed axes can be described with a single angle. Changes in this angle define the angular velocity vector, $\boldsymbol{\omega}$, and the angular acceleration vector, $\boldsymbol{\alpha}$. The directions of these vectors are determined by a right-hand rule. The period, T, and frequency, f, of rotational motion are related by

$$f = \frac{1}{T} = \frac{\omega}{2\pi}. \tag{9–5}$$

The rotational kinetic energy of a rigid body rotating about an axis is

$$K = \frac{1}{2} I\omega^2, \tag{9–14}$$

where I is the rotational inertia of the rotating object with respect to that axis. For a discrete system of point masses Δm_i that is a perpendicular distance R_i from an axis,

$$I \equiv \sum_i \Delta m_i R_i^2. \tag{9–15}$$

In particular, for a single mass m that is a perpendicular distance from an axis that can serve as a rotation axis, the rotational inertia is

$$I = mR^2. \tag{9–16}$$

The definition in Eq. (9–15) can be extended to a continuous object; thus, for an object with mass density ρ,

$$I = \int_{\text{Volume}} \rho R^2 \, dV, \tag{9–19}$$

where R^2 is the perpendicular distance of an internal point of the object to the axis.

Rotational inertia plays a role in rotational motion analogous to the role played by mass in linear motion. A useful tool for its evaluation is the parallel-axis theorem. This theorem states that the rotational inertia of an object about a given axis is

$$I = I_{\text{cm}} + Md^2 \tag{9–27}$$

In this equation, I_{cm} is the rotational inertia about an axis that goes through the center of mass and is parallel to the given axis, M is the mass of the object, and d is the perpendicular distance between the axes.

Torque, $\boldsymbol{\tau}$, is the cause of rotational motion and is analogous to force. For a rigid body rotating about a fixed axis with rotational inertia I about that axis, the equation of motion is analogous to Newton's second law:

$$\boldsymbol{\tau}_{\text{net}} = I\boldsymbol{\alpha}, \tag{9–30}$$

where $\boldsymbol{\alpha}$ is the angular acceleration. If the torque is constant, the angular velocity grows linearly with time.

The net torque is expressed in terms of the net force by the expression

$$\tau = rF \sin\theta, \tag{9–28}$$

where r is the distance from the point of application of the force to the axis of rotation and θ is the angle between the net force and r. The combination $r \sin\theta$ is the lever arm, or perpendicular distance from the axis to the line along which the force acts. The direction of the torque is specified by a right-hand rule. The torque due to gravity acts as if it were applied to the center of mass.

The technique of force diagrams for solving linear motion can be extended to rotational motion. Note that the point of application of the force must be specified.

The equation of motion also takes the form

$$\boldsymbol{\tau} = \frac{d\mathbf{L}}{dt}, \tag{9–34}$$

where $\mathbf{L}$ is the angular momentum—a quantity analogous to the linear momentum in linear motion. For rotations of a symmetric, rigid body about a symmetry axis, the angular momentum is

$$\mathbf{L} = I\boldsymbol{\omega}. \tag{9–33}$$

When the net torque is zero, the angular momentum is constant; thus, the angular momentum is conserved for isolated systems.

In rolling, linear motion is combined with rotational motion. Objects with radius r that roll without slipping have an angular velocity about their axes, ω, that is related to the speed v of the center of mass of the object:

$$v = r\omega. \tag{9–36}$$

The kinetic energy of a rolling object is the sum of its rotational kinetic energy about its axis of rotation, $I_{\text{cm}}\omega^2/2$, and the kinetic energy of its linear motion:

$$K = \frac{1}{2}(I_{\text{cm}} + Mr^2)\omega^2 = \frac{1}{2}I_{\text{cm}}\omega^2 + \frac{1}{2}Mv^2. \tag{9–39}$$

UNDERSTANDING KEY CONCEPTS

1. A record turntable rotates in the clockwise sense when seen from above. In what direction is the angular velocity of the turntable? If you had used a left-hand rule rather than a right-hand rule, which direction would you have chosen for the direction of the angular velocity vector? Is the choice between a right-hand or a left-hand rule purely conventional?
2. In the discussion of rolling without slipping, we saw that the kinetic energy of the rolling object is greater than the kinetic energy associated with the linear motion alone. Are there cases of rolling, with or without slipping, for which this is not true?
3. Suppose that you have a set of spherical objects with the same total mass but different radial distributions of mass. Which objects have the larger rotational inertias about an axis through the center: the spheres with more mass at the center or the spheres with more mass toward the outer surface?
4. A paddle wheel that propels a Mississippi River boat dips down into the water. Discuss the direction of motion of the portion of the paddle wheel under the water, according to an observer on the shore, then according to an observer on the boat.
5. A ball starts from rest down the inside of a parabolic bowl and rolls without slipping. At the bottom, the surface is a frictionless surface. The ball then moves up the other side of the bowl to a certain height. Is this height higher than, lower than, or the same as the height from which the ball started?
6. Solid and hollow cylinders both roll from rest down an inclined plane. Explain the difference in speeds at the bottom.
7. Devise a method to determine the rotational inertia of a sphere whose density can vary with the distance from its center.
8. Two skaters holding on to opposite ends of a rope circle a common point between them. In order to double their angular velocity about that point, should they lengthen or shorten the distance between them, and by how much?
9. You have a flat outline of the continental United States cut out in $\frac{1}{4}$-in plywood. Devise a method to find the center of mass of this outline.
10. When canoeing in rapidly flowing water, it is possible for the opposite ends of the canoe to become pressed against rocks, such that the canoe is aligned across the flow of the stream. Why is this bad?
11. Why can't flywheels be made to store arbitrarily large amounts of energy in their rotational motion?
12. A solid and a hollow cylinder are rolled from rest down an inclined plane. At the bottom, the center of the solid cylinder is moving more rapidly, even if the two cylinders have the same radius and mass. Is this a violation of the conservation of energy?
13. Why did the wheels used to steer large sailing ships have large radii?
14. When you switch to a longer wrench, you have not suddenly become stronger. So why is it easier to loosen a pipe with a long wrench than with a short wrench?
15. Attach one end of a given spring to a fixed pivot point on a frictionless table. Attach a mass to the other end, and start the mass in circular motion about the pivot point (Fig. 9–35). What happens when the revolution time is decreased by some external means, and why?
16. When a wheel of radius R rolls without slipping, the relationship between the speed of the center of the wheel, v, and the angular speed, ω, is $v = R\omega$. How would this relationship change if there were some slipping?

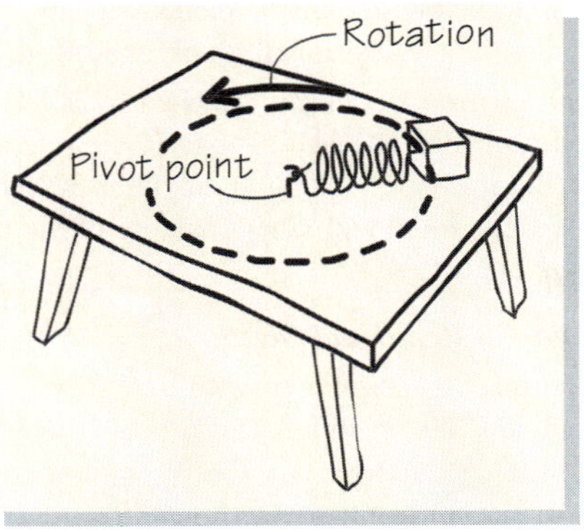

FIGURE 9–35 Question 15.

17. Suppose that you swing a rock at the end of a string in a circle in a vertical plane. What happens to the rock if the string suddenly breaks?

18. You have two wheels with rotational inertias in the ratio 2:1. Initially, the wheels are at rest. If the same torque is applied to both of them for the same length of time, what will be the ratio of their angular momenta at the end of that time interval?

19. You have two wheels with rotational inertias in the ratio 2:1. Initially, the wheels are at rest. If the same torque is applied to both of them for the same length of time, what will be the ratio of their rotational kinetic energies at the end of that time interval?

20. A wheel rotates about a central axis. When the angular speed increases beyond a certain value, the wheel breaks. Why?

PROBLEMS

9–2 Rotations about an Axis

1. (I) When a phonograph turntable is switched off, it comes to rest from its original $33\frac{1}{3}$ rev/min in 5.15 s. What is the magnitude of the angular deceleration, assuming that it is constant while the turntable comes to rest?

2. (I) How many revolutions does the turntable of Problem 1 make between the moment it is switched off and the moment it stops?

3. (I) A skater does a pirouette at the rate of 2.0 rev/s and then stops within 3/4 rev. Assume that the angular deceleration is constant, and calculate its magnitude.

4. (I) A turntable slows down from 2.0 rev/s to rest in 350 s. What is the average angular acceleration of the turntable?

5. (I) What is the angular velocity of Earth's rotation about its axis? What is the angular velocity of Earth in its orbital motion around the Sun?

6. (I) A carousel has a 7-m radius and requires 8 s for a single revolution at full speed. A carousel pig sits at a distance of 3 m from the axis, and a carousel horse at a distance of 6 m. (a) What is the period T for a single revolution of the pig? (b) of the horse? (c) What is the angular frequency of the motion of the pig? (d) of the horse? (e) What is the velocity of the pig? (f) of the horse? (g) What is the centripetal acceleration of the pig in its motion around the axis? (h) of the horse?

7. (II) A centrifuge whose maximum rotation rate is 10,000 rev/min can be brought to rest in 4.00 s. (a) What is the average angular acceleration of the centrifuge? (b) What is the distance that a point on the rim travels during the deceleration time, assuming that the radius of the centrifuge is 8 cm and that the acceleration is uniform?

8. (II) A phonograph turntable is rotating at 33.33 rev/min in a clockwise direction, viewed from above. (a) What is its angular velocity, $\boldsymbol{\omega}$, both direction and magnitude, in rad/s? (b) The turntable is switched off and comes to rest in 4.791 s. What is the average value of the angular acceleration, direction, and magnitude, during that period?

9. (II) A carousel initially at rest has an angular acceleration of 0.4 rad/s^2 and accelerates for 5 s. It then rotates at a constant angular velocity for 30 s before slowing down at the same rate with which it accelerated. (a) What is the average acceleration during the first 20 s? (b) How many total revolutions does it make? (c) How far does a child sitting on a horse 3 m from the center travel?

10. (II) A thread is wrapped around a cylindrical spool, of radius 1.5 cm, whose central axis is fixed on a support (Fig. 9–36). The thread is pulled off at a constant rate, causing the spool to spin at a constant rate; it takes 2 s to pull off 3 m of thread. What is the angular velocity of the spool while the thread is pulled off?

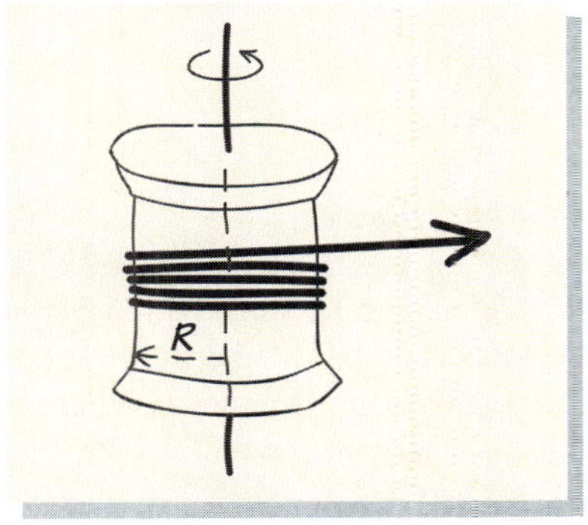

FIGURE 9–36 Problem 10.

11. (II) A more careful measurement of the unwrapping of the thread in Problem 10 shows that the spool accelerates from rest at a steady rate in the 2 s it takes to pull off 3 m of thread. (a) Give a formula for the position of the hand that pulls the thread as a function of time. What is the value of the constant (linear) acceleration? (b) What is the value of the angular acceleration of the spool? (c) Give a formula for the magnitude of the angular velocity as a function of time.

12. (II) A vacuum pump is connected to its electric motor by a belt drive (Fig. 9–37). The motor rotates at a rate of 4200 rev/min and the diameter of the motor shaft is 4 cm. How large should the pulley be if it is designed for a speed of 15 rev/s?

FIGURE 9–37 Problem 12.

13. (II) The path of the tip of a needle on a phonograph record may be described by the formula $r = r_0 - (\theta\rho/2\pi)$, where r_0 is the outer starting radius, θ is the angle (in radians) that a fixed radial line in the record makes with the needle arm, and ρ is the spacing between grooves. Assuming that the record turns at a rate of $33\frac{1}{3}$ rev/min and that the radial distance traveled by the needle is 9 cm in 20 min, what is ρ?

14. (II) The angular acceleration of a wheel starting from rest has magnitude $C_1 t + C_2 t^3$, where $C_1 = 48$ rad/s^3 and $C_2 = -9.5$ rad/s^5. (a) What is its angular velocity at 3.0 s? (b) How many revolutions has the wheel made after 2.0 s? (c) When will the wheel be at rest again?

15. (III) A rigid solid undergoes rotational motion about an axis. Its angular velocity has magnitude $\omega = \alpha t$, where α is constant. As the angular speed, ω, increases with time, the period, T, decreases. (a) Show that the rate of change of the period is described by the equation $dT/dt = -2\pi/\alpha t^2$. (The period is infinite at $t = 0$ because the rotation has not yet started, and a measurement of the time for 1 rev at that rate would be infinite!) (b) Show that the change in the period between $t = t_1$ and $t = t_2$ is given by

$$T(t_2) - T(t_1) = \frac{2\pi}{\alpha}\left(\frac{1}{t_2} - \frac{1}{t_1}\right).$$

9–3 *Rotational Kinetic Energy*

16. (I) A metal ball of mass 5.0 kg at the end of a 0.92-m-long wire rotates with an angular speed of 125 rev/min ($125 \times 2\pi$ rad/min). What is the rotational kinetic energy of the ball?

17. (I) Two identical balls are spinning on a flat surface. Ball #1 has three times the angular speed of ball #2. What is the ratio of their kinetic energies?

18. (I) Measurements of the amount of energy used by an electric motor to speed up a wheel from rest show that to bring the wheel from rest to an angular speed of 3.7 rad/s, the motor expends 7600 J. In a second use of the motor and wheel, the motor expends 9200 J to bring the wheel up from rest to an unmeasured angular speed. The motor and wheel have practically negligible amounts of friction or other type of damping. What is the unmeasured angular speed?

19. (II) A string is wrapped around a cylindrical spool of radius 1 cm. The axis of the spool is fixed. A length of string of 0.8 m is pulled off in 1.5 s at a constant tension of 20 N. What is the rotational inertia of the spool?

20. (II) What is the rotational kinetic energy of a dumbbell consisting of two equal (compact) masses of 2.5 kg each, connected by a massless rod of length 0.80 m, when the dumbbell rotates about an axis through the center of, and perpendicular to, the rod at 400 rev/min? What is the rotational kinetic energy if the dumbbell rotates with the same angular velocity about a parallel axis through one of the masses?

21. (II) A ball of mass 0.75 kg is attached by a 1.5-m-long rope to the top of a rod. The ball swings in a circle at the rate of 25 rad/s with the rope making an angle of 30° with the vertical. What is the rotational kinetic energy of the ball? What is it when the angle is 60°?

9–4 *Evaluation of Rotational Inertia*

22. (I) A pipe made of aluminum with a density of 2.7 g/cm^3 is a right cylinder 20 cm long, whose outer diameter is 10 cm and whose inner diameter is 7 cm. What is the rotational inertia about the central axis of the pipe? Note that the rotational inertia of the thick cylinder can be expressed as the rotational inertia for a solid cylinder of radius R_2 minus the rotational inertia of the solid cylinder of radius R_1.

23. (I) What is the rotational inertia of a uniform 4.0-kg iron rod, 0.25 m long, about (a) an axis through its center point and perpendicular to the rod? (b) an axis through an end point and perpendicular to the rod?

24. (I) *Estimate* the rotational inertia of a tennis ball about an axis through its center.

25. (II) A neutron star has a constant density of 6×10^{17} kg/m^3 and a mass five times that of our Sun. Compare its rotational inertia with that of Earth (assume constant density). In both cases the reference axis is an axis through the center of the sphere; Table 9–1 gives the rotational inertia for such an axis.

26. (II) Mass m_1 sits at the point $(x, y, z) = (0$ m, 0 m, 0 m$)$, and mass m_2 at point (0 m, 1 m, 0 m). (a) Where is the center of mass? (b) What is the rotational inertia about an axis through the center of mass and parallel to the z-axis? (c) parallel to the y-axis? (d) parallel to the x-axis? (e) What is the rotational inertia about an axis through the origin and along the z-axis? (f) Verify the parallel-axis theorem for this system, using the results of parts (b) and (e).

27. (II) Find the rotational inertia about the symmetry axis of a thick cylinder of mass M. Take the inner radius to be R_1 and outer radius to be R_2.

28. (II) Calculate the rotational inertia of a section of a right circular cylinder of radius R that subtends an angle θ_0 at the origin (Fig. 9–38), when the reference axis is at the origin and perpendicular to the section.

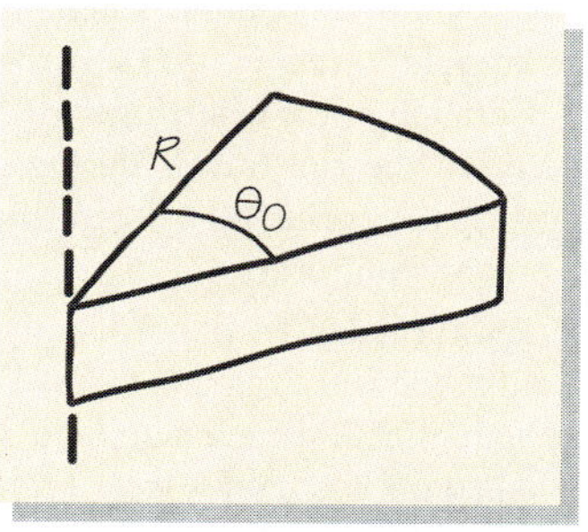

FIGURE 9–38 Problem 28.

29. (II) Calculate the rotational inertia about the central axis of the solid cone of mass M illustrated in Fig. 9–39, of opening half-angle α and height H.

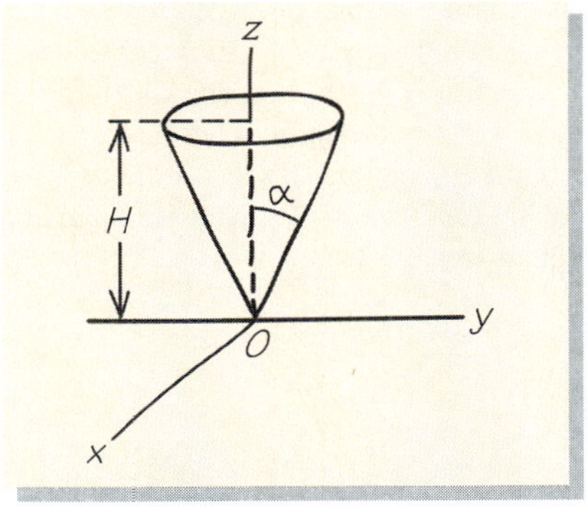

FIGURE 9–39 Problem 29.

30. (II) A thin stick of length $L = 2$ m is denser at one end than at the other: Its mass density is $\lambda = (0.42\ \text{kg/m}) - (0.15\ \text{kg/m}^2)x$, where x measures the distance from the heavier end of the stick. The stick rotates with period $T = 0.3$ s about an axis perpendicular to the stick through the heavy end. Determine the rotational kinetic energy of the stick.

31. (II) Calculate the rotational inertia of a sphere of radius R and mass M about an axis through the center of the sphere; assume that the density is not uniform but is given by ρ_1 for $0 \leq r \leq R_1$ and by ρ_2 for $R_1 \leq r \leq R$ (Fig. 9–40).

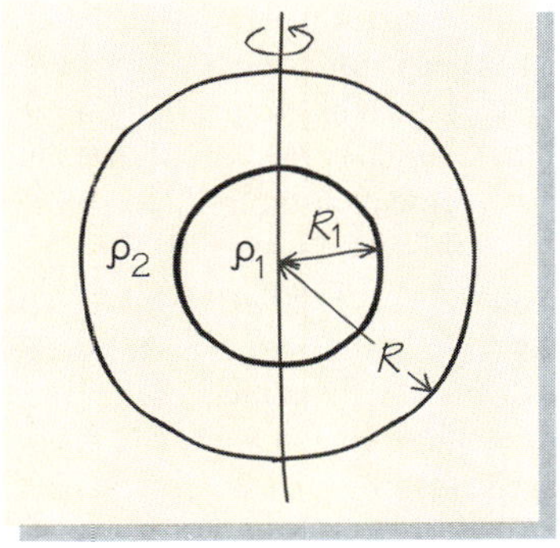

FIGURE 9–40 Problem 31.

32. (III) Use Eq. (9–19) to show that the rotational inertia about an axis through the center of a uniform, constant-density sphere of mass M and radius R is $\frac{2}{5}MR^2$.

33. (III) In Problem 79 of Chapter 8, we described a styrofoam sphere of radius R. A cavity of radius $R/2$, centered a distance $R/2$ directly above the center of the sphere, was hollowed out and filled with a solid material of density five times the density of styrofoam. In that problem, the location of the center of mass of the composite sphere was determined. What is the rotational inertia of the composite sphere about a horizontal axis through the center of mass? Express your result in terms of the total mass M of the composite sphere and its radius R. [*Hint*: You may view the mass as consisting of a large sphere of radius R and density ρ and a small sphere, off-center, of radius $R/2$ and density 4ρ. Use the parallel-axis theorem.]

34. (III) A square of mass m is made of thin wire with sides of length a. Calculate the rotational inertia about the axes shown in Fig. 9–41.

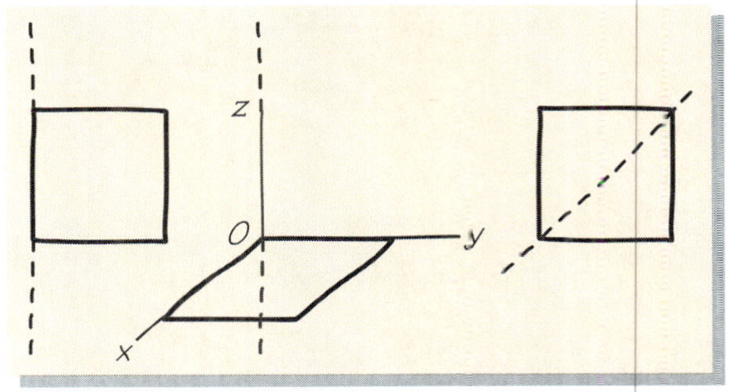

FIGURE 9–41 Problem 34.

9–5 *Torque*

35. (I) An instruction manual calls for a bolt to be tightened to 20 ft · lb. All you have is a 10-in wrench. How much force do you need to apply to the end of the wrench to tighten the bolt as required?

36. (I) A plumber of mass 74 kg just loosens a rusted-in bolt with the help of a 45-cm-long wrench. He places the wrench in a horizontal position and applies torque by hanging from the end of the wrench. What is the torque applied?

37. (I) The assistant of the plumber in Problem 36 meets a similar situation but aligns the wrench at an angle of 50° from the vertical (Fig. 9–42). What must his mass be in order for him to use the same technique to loosen the same bolt?

FIGURE 9–42 Problem 37.

38. (I) A flywheel of rotational inertia $I = 25\ \text{kg} \cdot \text{m}^2$ rotates with angular speed 6.0 rad/s. A tangential force of 10.0 N is applied at a distance of 0.30 m from the center in such a way that the angular speed decreases. How long will it take for the wheel to stop?

39. (II) A uniform rod of length L lies along the x-axis. A force F_{1y} is applied to one end of it, and a force $-F_{2y}$ is applied to the other end of it. How large is the torque on the rod about its center of mass?

40. (II) A uniform rod 1 m long with mass 0.6 kg is pivoted at one end, as shown in Fig. 9–43, and released from a horizon-

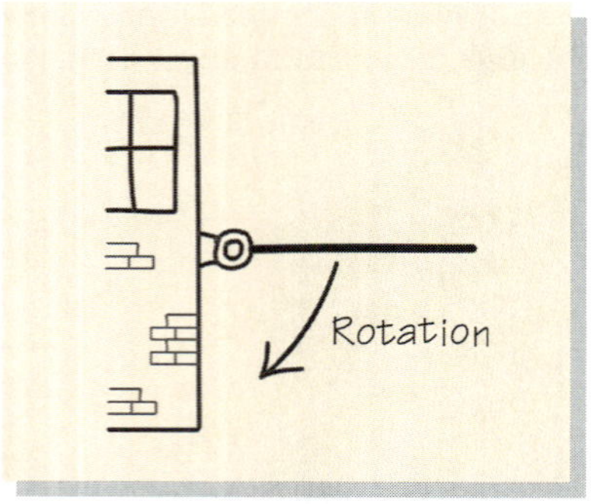

FIGURE 9–43 Problem 40.

tal position. Find the torque about the pivot exerted by the force of gravity as a function of the angle that the rod makes with the horizontal direction.

41. (II) A seesaw pivots as shown in Fig. 9–44. (a) What is the net torque about the pivot point? (b) Give an example for which the application of three different forces and their points of application will balance the seesaw. Two of the forces must point down and the other one up.

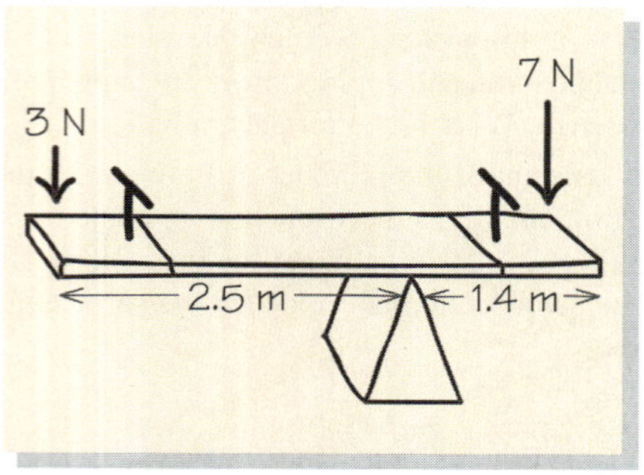

FIGURE 9–44 Problem 41.

42. (II) A two-dimensional object placed in the xy-plane has several forces acting on it. Find the torques about points A and B in Fig. 9–45.

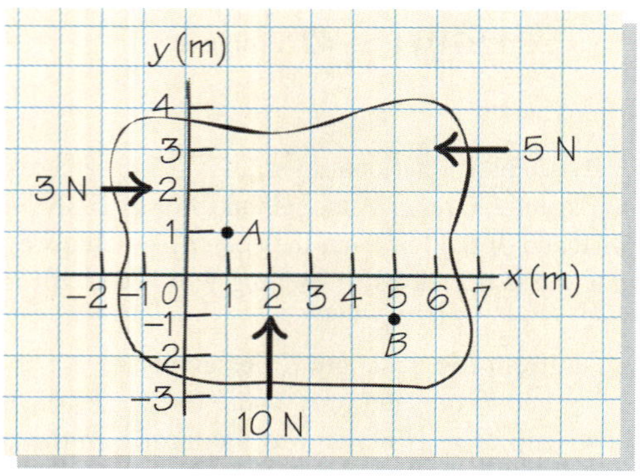

FIGURE 9–45 Problem 42.

43. (II) An aluminum casting is finished with a band sander. The sandpaper is stretched by two cylindrical rollers, one of them driven by a motor (Fig. 9–46). The rollers have diameter of 16 cm. How large is the torque that has to be applied to the driven cylinder, if the coefficient of kinetic friction between aluminum and sandpaper is 1.2, and the force applied to push the casting against the sandpaper is 4 N?

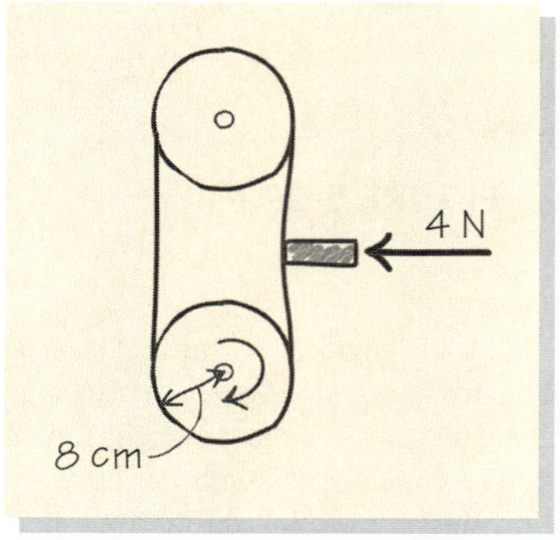

FIGURE 9–46 Problem 43.

44. (II) A massless rope is wrapped around a hollow cylinder of radius 25 cm whose central axis is fixed in a horizontal position. A mass of 10 kg hangs from the rope and, starting from rest, moves 80 cm in 2.0 s. What is the mass of the cylinder?

45. (II) A wheel of radius 24.6 cm whose axis is fixed starts from rest and reaches an angular velocity of 4.15 rad/s in 2.68 s due to a force of 13.4 N acting tangentially on the rim. (a) What is the rotational inertia of the wheel? (b) What is the change in the angular momentum during the 2.68 s? (c) How many revolutions does the wheel make? (d) How much rotational kinetic energy does the wheel have after 2.68 s?

46. (III) A motorcycle has a mass of 500 kg, the wheels have a diameter of 60 cm, and the centers of the wheels are separated by 1.5 m. Assuming that the weight is distributed uniformly over the wheels, that the wheels roll without sliding, and that the coefficient of friction between the wheels and the road is $\mu = 0.5$, calculate the torque about the center of the front wheel exerted by the forces between the road and the wheels when there is a maximum braking.

9–6 Angular Momentum and Its Conservation

47. (I) A student sits on a piano stool and holds the axle of a bicycle wheel that rotates with angular velocity $\boldsymbol{\omega}$, of magnitude 4π rad/s, pointing upward. The wheel's axis of rotation goes through the axis of the stool, which is at rest. The rotational inertia of the wheel about its axis is $1.2\ \text{kg}\cdot\text{m}^2$, and the rotational inertia of the student and stool about the stool's axis is $8\ \text{kg}\cdot\text{m}^2$. The student suddenly flips the shaft of the wheel, so that its angular velocity points down. How fast and in what direction will the student and stool rotate? Ignore friction.

48. (I) Two identical tops spin with angular velocities 10π rad/s up and 12π rad/s down, respectively, about vertical axes on a table. The tops bump into one another and separate. After the collision, one of the tops has an angular velocity of 5π rad/s in its original direction. What is the angular velocity of the other top?

49. (I) A spherically symmetric celestial object rotates at 3.24592 rev/s. Through some mechanism, which does not involve an application of external torque or a change in the spherical symmetry, its radius decreases rapidly and uniformly. As a consequence, the rate of revolution changes to 3.24608 rev/s. What is the fractional change in the radius?

50. (II) A child of mass 25 kg stands at the edge of a rotating platform of mass 150 kg and radius 4.0 m. The platform with the child on it rotates with an angular speed of 6.2 rad/s. The child jumps off in a radial direction. (a) What happens to the angular speed of the platform? (b) What happens to the platform if, a little later, the child, starting at rest, jumps back onto the platform? (Treat the platform as a uniform disk.)

51. (II) Compact disks and long-playing records are made from the same material. The former have a diameter of about 12 cm; the latter, about 32 cm. When in use, records spin at $33\frac{1}{3}$ rev/min, and compact disks spin at, say, 400 rev/min. What is the ratio of the angular momentum of a compact disk in use to that of a record? Assume that a compact disk has half the thickness of a record.

9–7 Rolling

52. (I) Bicycle racers sometimes use solid wheels in order to cut down the drag force between the air and the spokes of an ordinary wheel. This can be an important effect because drag forces rise quite rapidly as the speed of an object through the air increases. If the radius of a wheel is 35 cm and the speed of the bicycle relative to the ground is 30 mi/h (13.4 m/s), what is the speed relative to the ground of the end of the spoke closest to the rim for (a) a spoke leading to the contact point with the ground; that is, a spoke pointing vertically down? (b) a spoke pointing vertically up? (c) a spoke that is horizontal and points forward? (d) a spoke that is horizontal and points backward?

53. (II) A cylindrical shell starting from rest rolls down an inclined plane that makes an angle of 20° with the horizontal. How far will the shell travel in 4 s? How far would a solid cylinder travel in the same time?

54. (II) The following objects all roll without slipping, have uniform density, mass M, and radius R, and the speed of the center of mass in each case is v. Find the ratio of the rotational kinetic energy to the total kinetic energy for (a) a solid cylinder, (b) a hollow cylinder, and (c) a sphere.

55. (II) Figure 9–47 shows a disk attached to an axle placed on an incline made of two parallel bars. The radius of the disk is 0.12 cm and its mass is 0.8 kg; the radius of the axle is 0.020 m and its mass is 0.10 kg, not including the part inside the disk. Calculate the acceleration of the system if the incline mades an angle of 5° with the horizontal and the axle rolls without slipping.

56. (II) A hollow cylinder moves down an inclined plane of length ℓ and angle θ. The cylinder has uniform density, mass M, and radius R. It is initially at rest at the top of the plane. Calculate and compare the times taken to reach the bottom if the cylinder rolls without slipping as opposed to the case in which the cylinder slips all the way down the plane without rolling. In the latter case, the coefficient of friction must be zero.

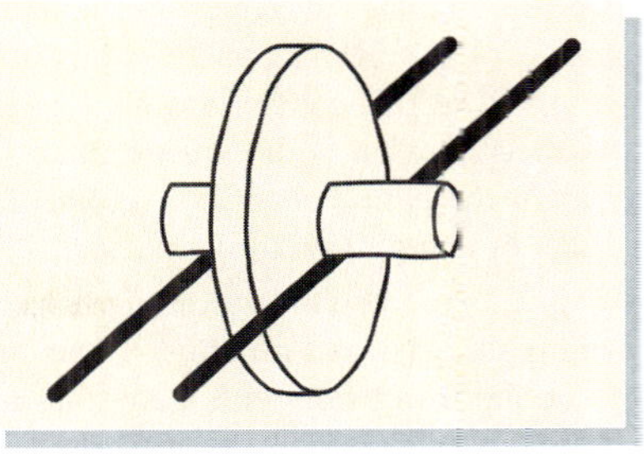

FIGURE 9–47 Problem 55.

General Problems

57. (I) A basketball player shoots a desperate last shot, spinning the ball at an angular speed of 15 rad/s to give it "action." The ball is shot with an initial velocity of 4.1 m/s at an elevation angle of 45° and leaves the player's hands 1.7 m off the floor. Unfortunately, the shot misses the backboard, the rim and the net. How many revolutions has the ball made when it hits the floor? Ignore air resistance.

58. (I) *Estimate* the angular momentum of a spinning ice skater.

59. (II) Earth's radius is 6.4×10^6 m. Assume that its density is not uniform; that is, the inner core has a density of 8.0×10^3 kg/m^3 and the outer mantle has a density of 3.0×10^3 kg/m^3. Given that the rotational inertia of Earth is 8.3×10^{37} kg $\cdot$ m^2, calculate the radius R at which the density changes. [*Hint*: Go back to Problem 31.]

60. (II) A solid cylinder of mass 20 kg and radius 50 cm rotates at 300 rev/min about its central axis. What is the rotational kinetic energy of this motion? Suppose that a 0.2-kg mass is attached at one point on the rim of the cylinder. If the additional mass can be treated as a point mass, and the rotational speed is unchanged, what is the percentage change in the rotational kinetic energy?

61. (II) A solid uniform cylinder of mass M and radius R is projected up an incline of angle θ. It rolls without slipping from an initial speed v_0 of the center of mass. What distance s does the center of the cylinder travel before it starts to fall back?

62. (II) When a bicycle rider accelerates, he must accelerate his own and his bicycle's linear motion as well as the angular motion of the wheels. Suppose that the cyclist has mass 55 kg; the bicycle (not counting wheels), 8 kg; and both wheels together, 1.8 kg. Assume that the wheels, each of radius 30 cm, have all their mass concentrated in the (thin) rim. (a) At 25 km/h, what fraction of the kinetic energy of the rider plus his bicycle is in linear motion, and what fraction is in rotational motion? (b) Suppose that the cyclist loses 3 kg on a diet. What percentage of the original force is required to accelerate the system uniformly from 0 km/h to 25 km/h in 10 s? (c) Suppose that instead of going on a diet, the cyclist replaces his wheels with ones of total mass 1.2 kg. Now what is the percentage of the original force required to accelerate the system uniformly from 0 km/h to 25 km/h in 10 s?

63. (II) A hollow cylinder of radius 15 cm and mass 3.0 kg rolls without slipping at a constant speed of 1.6 m/s. (a) What is its angular momentum about its symmetry axis? (b) What is its rotational kinetic energy? (c) What is its total kinetic energy?

64. (II) A pulsar (the remnant of a star after a supernova explosion) has a mass of 4×10^{30} kg and a radius of 10 km. It rotates with a period of 0.03 s. (a) Assuming a spherical shape and constant density, what is the kinetic rotational energy of the pulsar? (b) If the period changes by 1 part in 10^8 in 1 yr, what is the rate of energy loss of the pulsar?

65. (II) A yo-yo has mass M and external radius R. The central stem has negligible mass and a radius r. The string is pulled horizontally on the lower side with a constant force F, while the yo-yo rests on a rough horizontal surface (Fig. 9–48). What is the maximum value of F for which the yo-yo will roll without slipping, assuming that the coefficient of friction between the yo-yo and the surface is μ?

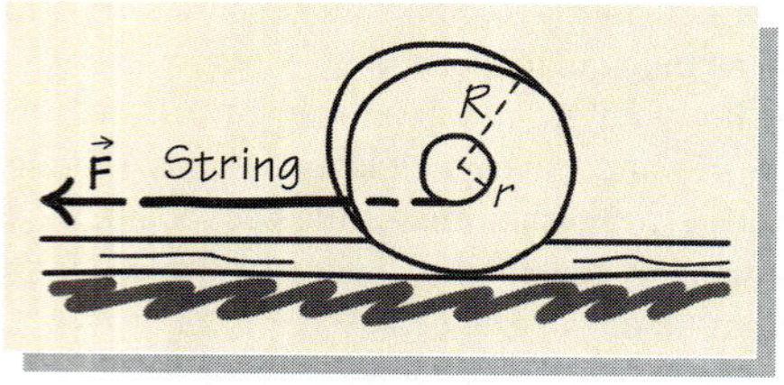

FIGURE 9–48 Problem 65.

66. (II) A cylinder of known mass M, radius R, and rotational inertia I is placed on an inclined plane with angle θ (Fig. 9–49). A string is wound around the cylinder and pulled up with a tension T parallel to the inclined plane. The coefficient of static friction is large enough to prevent slipping. (a) Find the tension T_0 needed to keep the cylinder in equilibrium. (b) Find the acceleration of the cylinder if the tension is known and is different from T_0.

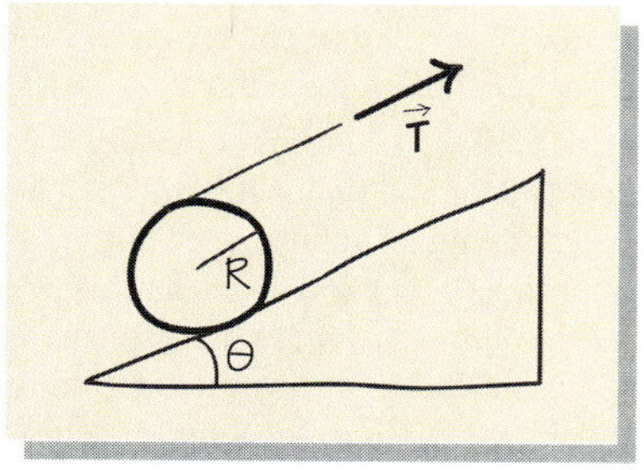

FIGURE 9–49 Problem 66.

67. (II) Suppose that the yo-yo described in Problem 65 is released from rest while the upper end of the cord is held steady. What is the tension in the cord during the yo-yo's downward motion?

68. (II) A thin rod of mass M and length ℓ is lying on a frictionless table. It is given an impulse of 1 N·s by a force **F** on one end, at an angle of 45° to the rod. How far will the center of mass of the rod travel when the rod has completed 2 rev?

CHAPTER 10

The large torque acting on the sailboat by wind forces on the sail tends to rotate the sailboat out of the water; the torque due to gravity tends to rotate the sailboat back to a normal position. The forces and torques on a sailboat such as this one are rather complex.

More on Angular Momentum and Torque

We studied the rotations of rigid bodies in Chapter 9. Many real physical systems, such as the collections of stars that make up galaxies or a figure skater executing a complex maneuver, are not rigid, however. We will see that angular momentum and its conservation are far more useful concepts than might have been evident in Chapter 9—even for systems of particles that do not form rigid bodies and are not in pure rotational motion. We will extend to these systems the notions of angular momentum and torque. We will also explore situations in which the vector nature of these quantities helps in understanding more complicated motions. In doing so, we will be able to extend the correspondence between the dynamics of linear motion and rotational motion.

10–1 GENERALIZATION OF ANGULAR MOMENTUM

We have attributed an angular momentum to rigid systems in pure rotation. *But even a point mass moving in space has an angular momentum with respect to a fixed origin.* To assign that angular momentum, we will consider a free particle, which moves along a straight line with constant momentum **p**, together with a point that we label as the origin O. A point and a line form a plane and we will label this plane the xy-plane (Fig. 10–1a). We will insist that the angular momentum, **L**, that we assign to the par-

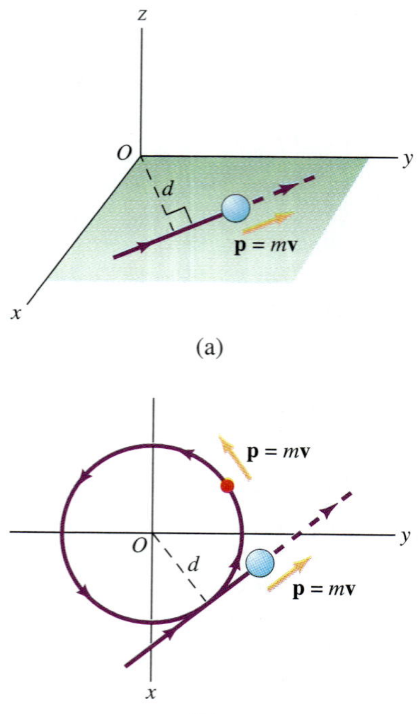

FIGURE 10–1 (a) A mass m moves freely in a plane with velocity **v**. (b) To find its angular momentum with respect to a point O we match its motion when it is closest to O to a particle of the same mass moving in a circle about O.

ticle is constant because there are no forces—thus no torques—acting on the particle. As the particle moves, its point of closest approach to O is a distance d from the origin. At this point d, the line of motion is perpendicular to the line from O to the origin; further, the particle's motion at point d is *instantaneously* identical to that of a second particle of the same mass moving in a circular path in the xy-plane about point O (Fig. 10–1b). We already know how to find the angular momentum of the second particle: Its angular momentum is in the $+z$-direction and has magnitude

$$L = I\omega = (md^2)(v/d) = mvd = pd.$$

Therefore this is the angular momentum at all times that we assign to the freely moving particle:

$$\mathbf{L} = pd\,\mathbf{k}. \tag{10–1}$$

As Fig 10–2 illustrates, the freely moving particle may at some time be at point P, a distance r from the origin. Here, the momentum makes an angle θ with respect to the line from O. Geometry shows that $d = r \sin\theta$ so that, in term of these variables,

$$\mathbf{L} = pd\,\mathbf{k} = pr\sin\theta\,\mathbf{k}. \tag{10–2}$$

We can now simply take Eq. (10–2) as the basis for a more general definition of the angular momentum with respect to a point O of a particle with momentum p that is a distance r from O, with the angle θ between r and p. It no longer matters that the particle be free. The difference between a free particle and one on which a torque acts about point O is that for the free particle, Eq. (10–2) is the angular momentum *for all time*; whereas for the particle on which a torque acts, Eq. (10–2) is only the value of **L** at the moment picked out in Fig. 10–2. It does not matter if the particle is moving in a circle, in a straight line, or on any other trajectory (Fig. 10–3). Note that a force may act, but if that force exerts no torque on the particle about point O, Eq. (10–2) remains the constant value of **L**. Notice that we now determine angular momentum with respect to a point rather than an axis.

The angular momentum of a point mass moving in the xy-plane

There is angular momentum even if there is no circular motion.

Reduction of Angular Momentum to Earlier Form

How does our result reduce to a symmetric rigid body rotating about point O with angular velocity ω in the $+z$-direction—a case we studied in Chapter 9? To find the net angular momentum of any system, we sum the angular momenta of the individual pieces of the system. In particular, let's look at the angular momentum about the origin of a dumbbell rotating around the z-axis, where the dumbbell is off the xy-plane (Fig. 10–4). Once we understand this case, the generalization to any three-dimensional symmetric object rotating about the z-axis follows by decomposing the object into pairs of tiny masses forming dumbbells like the one we consider here. At the moment shown in Fig. 10–4, the masses are above the y-axis. $\mathbf{L}_1$ and $\mathbf{L}_2$ have magnitudes r_1mv_1 and r_2mv_2, and are perpendicular to $\boldsymbol{r}_1$ and $\boldsymbol{r}_2$, respectively; $\mathbf{L}_1$ and $\mathbf{L}_2$ are also shown in that figure. The horizontal components of the $\mathbf{L}_i$ (the subscript i stands for 1 and 2 here) cancel each other, while the z-components add. To find the z-component, we observe from the figure that the angle ϕ equals the angle θ. Thus the mass labeled with the subscript i has z-component

$$(L_i)_z = L_i \sin\phi = r_imv_i \sin\phi = r_imv_i \sin\theta.$$

The combination $r_i \sin\theta$ is equal to d, which is the radius of the dumbbell, so that $(L_i)_z = mv_id$. Finally, the velocity of the mass labeled i has magnitude $v_i = \omega d$. Thus the z-component of the angular momentum of each of the two masses is

$$(L_i)_z = mvd = m\omega d^2.$$

The net angular momentum of the dumbbell is then in the $+z$-direction, with magnitude

$$L = (L_1)_z + (L_2)_z = 2(m\omega d^2).$$

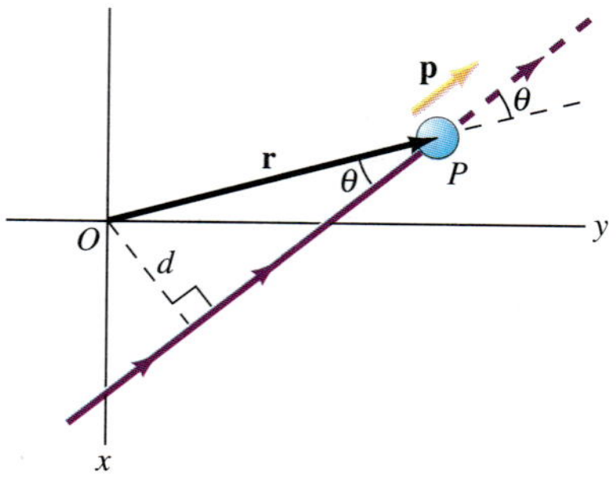

FIGURE 10–2 At a later time mass m has moved to point P. Its angular momentum, however, is the same as it was in Fig. 10–1.

We can break a more general symmetric rigid body into pairs of masses in this way; the subscript i now runs over many values, and we must retain the subscript on each mass m and each distance d. However, the angular speed ω is the same for each piece of the object, so it would have no index. Thus

$$L = \left[\sum_i (m_i d_i^2)\right]\omega = I\omega. \qquad (10\text{–}3)$$

We have recognized the sum in Eq. (10–3) as the rotational inertia about the z-axis. We can conclude that our new definition of the angular momentum about a point—Eq. (10–2)—reduces to the earlier form, Eq. (9–33), for the angular momentum about an axis in the case of a symmetric rigid body.

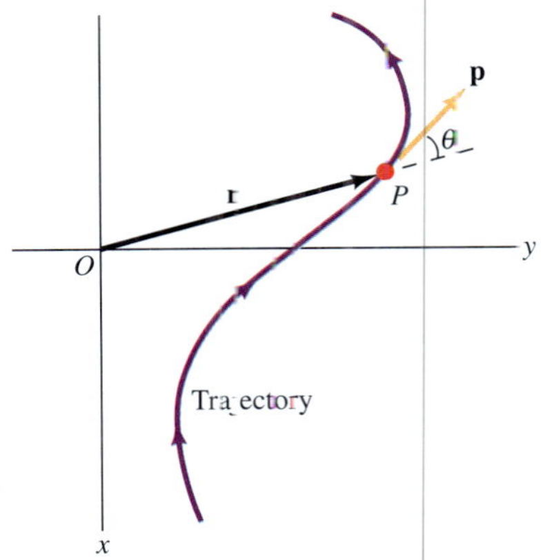

FIGURE 10–3 A particle of the same mass and velocity as that of the free particle in Fig. 10–2 at the same point P. Its angular momentum at that point is the same as that of the freely moving particle.

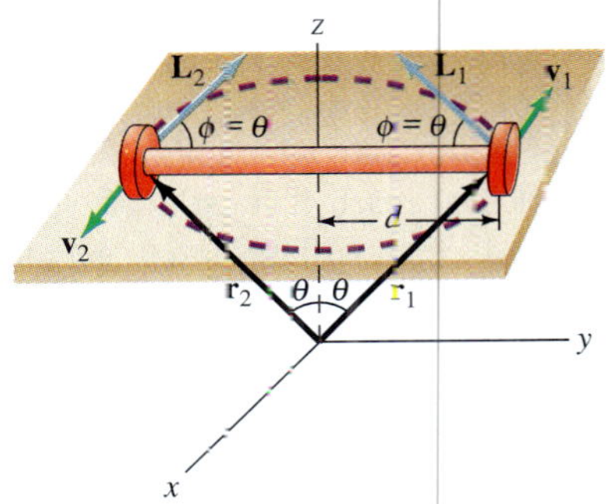

FIGURE 10–4 To find the angular momentum of a symmetric rigid body about the symmetry axis in terms of our results for the angular momentum of point masses, we break up the rigid body into pairs of symmetric mass elements.

Angular Momentum and Central Forces

Central forces—forces directed along a line from a given source—provide examples of forces with no torque about a special point. In this case, the point is the source of the force itself. The Sun is responsible for a central force on the planets, and this force is directed toward the Sun itself. Thus the gravitational force due to the Sun exerts no torque on the planets, and the angular momentum of the planets as they circle the Sun is constant. Because both gravitation and electrostatic forces are central, the case of central forces is of particular importance. Example 10–1 illustrates angular momentum for an object under the influence of a central force.

EXAMPLE 10–1 A comet, of mass $M = 10^{15}$ kg, moves in a highly eccentric orbit about the Sun, as shown in Fig. 10–5. At its closest approach to the Sun, $d = 10^6$ km, the comet is measured to be moving at a speed of 6×10^6 m/s. What is the angular momentum of the comet with respect to the Sun if the comet is treated as a point mass?

Solution: The momentum of the comet has magnitude $p = Mv$, where v is the speed of the comet. The angular momentum is given by the momentum times the distance of closest approach. Therefore it has magnitude

$$L = Mvd = (10^{15}\ \text{kg})(6 \times 10^6\ \text{m/s})(10^6\ \text{km})\,\frac{10^3\ \text{m}}{1\ \text{km}} = 6 \times 10^{30}\ \text{kg}\cdot\text{m}^2/\text{s}.$$

The direction of the angular momentum is given by a right-hand rule. In Fig. 10–5, this direction is up out of the plane of the orbit, in what is labeled the $+z$-direction.

FIGURE 10–5 Example 10–1. The direction of the angular momentum is found from a right-hand rule to be up out of the plane of the orbit, in the $+z$-direction.

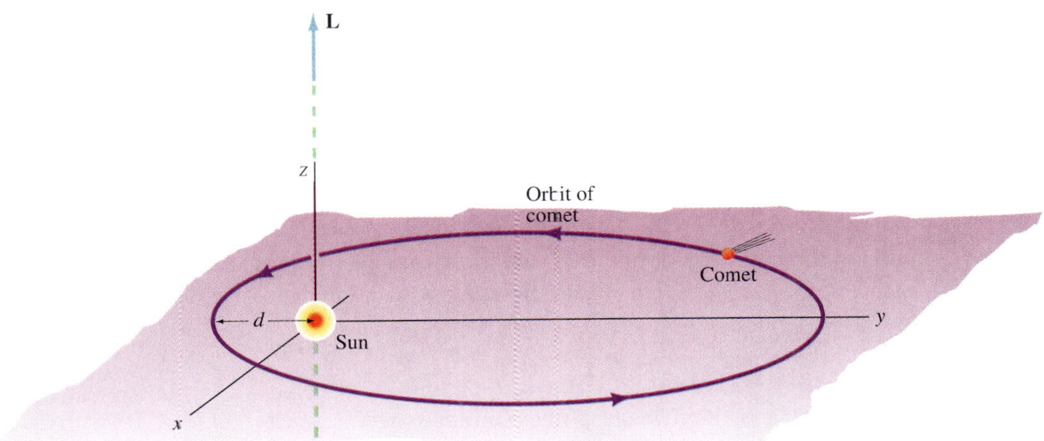

PROBLEM–SOLVING TECHNIQUES

The Vector Product

In addition to the scalar product discussed in Chapter 6, we can define a product of two vectors—**A** and **B**—that is itself a vector. The **vector product** (or **cross product**) of **A** and **B** is defined to be perpendicular to both **A** and **B**; it is denoted **A** × **B**. Any two vectors form a plane; for example, we can choose the x-axis along **A** and then define the xy-axis so that **B** has only x and y components. The direction of the vector product of **A** and **B** is then perpendicular to this plane. A right-hand rule is used to define the direction of the vector product because there is an ambiguity as to whether the direction is into or out of the plane. The direction of **A** × **B** is specified as follows (Fig. 10B1–1): Point the fingers of your right hand in the direction of **A** and curl toward **B** according to the rule that the angle through which the fingers are curled to reach **B** must always be less than 180° (it will be hard to get the fingers to follow the opposite rule!). The direction of your thumb then indicates the direction of the vector product **A** × **B**. The magnitude of the vector product is given by

$$|\mathbf{A} \times \mathbf{B}| = AB \sin\theta, \qquad \text{(B1–1)}$$

where θ is the angle (less than 180°, so that the sine is positive) between **A** and **B**.

An immediate consequence of the right-hand rule is that

$$\mathbf{A} \times \mathbf{B} = -\mathbf{B} \times \mathbf{A}. \qquad \text{(B1–2)}$$

Thus the vector product is not commutative. It also follows from Eq. (B1–1) that *the vector product of two vectors that are parallel (or antiparallel) to each other is zero*: When $\theta = 0°$ or 180°, $\sin\theta = 0$. The vector product attains its maximum magnitude when **A** and **B** are perpendicular to each other because then $\sin\theta = 1$. Contrast this to the scalar product, which is zero when the two vectors are perpendicular and has a maximum magnitude when they are parallel.

The unit vectors **i**, **j**, and **k** along the x-, y-, and z-axes, respectively, obey the relations

$$\mathbf{i} \times \mathbf{j} = -\mathbf{j} \times \mathbf{i} = \mathbf{k}, \qquad \text{(B1–3a)}$$

$$\mathbf{j} \times \mathbf{k} = -\mathbf{k} \times \mathbf{j} = \mathbf{i}, \qquad \text{(B1–3b)}$$

$$\mathbf{k} \times \mathbf{i} = -\mathbf{i} \times \mathbf{k} = \mathbf{j}. \qquad \text{(B1–3c)}$$

Also $\mathbf{i} \times \mathbf{i} = \mathbf{j} \times \mathbf{j} = \mathbf{k} \times \mathbf{k} = 0$. We can expand the vectors **A** and **B** into their components and find

$$\begin{aligned}\mathbf{A} \times \mathbf{B} &= (A_x\mathbf{i} + A_y\mathbf{j} + A_z\mathbf{k}) \\ &\quad \times (B_x\mathbf{i} + B_y\mathbf{j} + B_z\mathbf{k}) \\ &= A_xB_y(\mathbf{i} \times \mathbf{j}) + A_xB_z(\mathbf{i} \times \mathbf{k}) \\ &\quad + A_yB_x(\mathbf{j} \times \mathbf{i}) + A_yB_z(\mathbf{j} \times \mathbf{k}) \\ &\quad + A_zB_x(\mathbf{k} \times \mathbf{i}) + A_zB_y(\mathbf{k} \times \mathbf{j}).\end{aligned}$$

With the results of Eqs. (B1–3), we find that

$$\begin{aligned}\mathbf{A} \times \mathbf{B} &= (A_yB_z - A_zB_y)\mathbf{i} \\ &\quad + (A_zB_x - A_xB_z)\mathbf{j} \\ &\quad + (A_xB_y - A_yB_x)\mathbf{k}. \qquad \text{(B1–4)}\end{aligned}$$

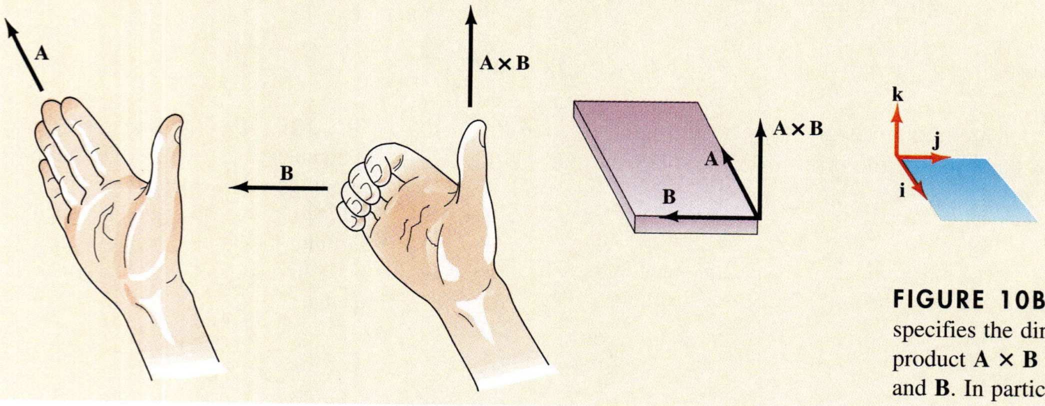

FIGURE 10B1–1 A right-hand rule specifies the direction of the vector product **A** × **B** between two vectors **A** and **B**. In particular, **k** = **i** × **j**.

Angular Momentum as a Vector Product

The angular momentum of a point mass about an origin

Vector products (see "The Vector Product" box) are useful for many relationships involving rotation. One of the most important examples is provided by the angular momentum *about some origin O* of a point mass with momentum **p**, which is given by

$$\mathbf{L} \equiv \mathbf{r} \times \mathbf{p}, \qquad \text{(10–4)}$$

where **r** is the position vector of the point mass with respect to O (Fig. 10–6). We can reduce this result to that previously given in Eq. (10–2) and described there. The magnitude of **L** is $rp \sin\theta$, where θ is the angle between **r** and **p**, and $r \sin\theta$ is the perpendicular distance d of Eq. (10–2). By the property of the vector product, the vector **L** is perpendicular to *both* **r** and **p**, with the direction given by the right-hand rule. Note that **L** is zero when **r** and **p** are parallel; that is, when the straight-line extension of the vector **p** passes through the point of reference, O. Examples 10–2 and 10–3 illustrate the use of Eq. (10–4).

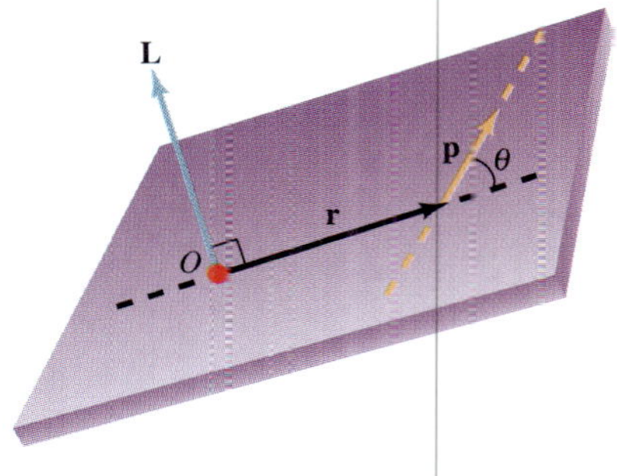

FIGURE 10–6 Vector relation for the angular momentum **L** of a point particle about a point O.

EXAMPLE 10–2 In an engineering design, a light but stiff rod of length R is attached at an angle θ to a shaft along the z-axis; it is used to rotate a mass M about the shaft (Fig. 10–7). The mass moves with speed v. Describe the angular momentum, **L**, of the mass with respect to the attachment point of the rod.

Solution: The angular momentum vector is perpendicular both to the vector from point O to the mass, which runs along the rod, and to the mass's momentum. Thus **L** has a component both in the z-direction and in the horizontal plane, pointing to the shaft—that is, radial. **L** has magnitude Rmv. To find its components, we can refer directly to Fig. 10–4, which treated a similar case with a pair of masses; here we are interested in the angular momentum for one of the pair. Replacing r_1 in Fig. 10–4 with R, we have

$$L_z = L \sin\theta = L\frac{d}{R} = (RMv)\frac{d}{R} = dMv;$$

$$L_{\text{radial}} = -L\cos\theta = -L\frac{\sqrt{R^2 - d^2}}{R} = -\sqrt{R^2 - d^2}\, Mv.$$

Note that because the angular momentum is not constant (the horizontal component rotates so as to point to the shaft), there must be a torque about point O. This classic engineering problem tells us that the attachment to the shaft must be appropriately constructed to supply the torque.

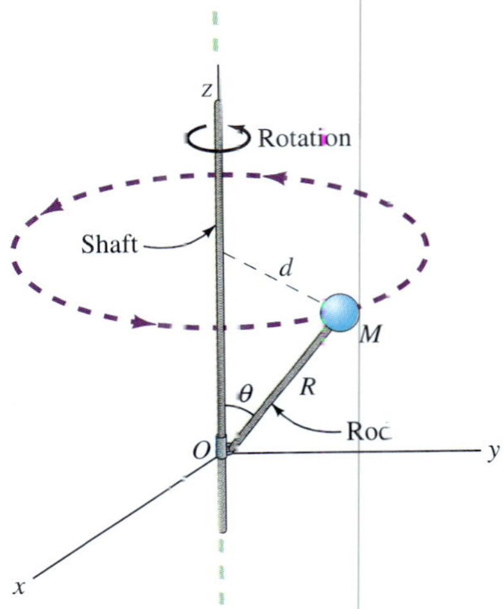

FIGURE 10–7 Example 10–2.

EXAMPLE 10–3 A tether ball is tied by a rope to a central pole. As the ball whirls around the pole in a (nearly) horizontal plane, the rope gradually winds around (or unwinds from) the pole, shortening (or lengthening) the amount of rope between the ball and the pole (Fig. 10–8). The motion is spiral rather than circular: Near the time t, when the ball is at a distance nearly equal to R and rotating around the pole with angular speed nearly equal to ω, the distance is more precisely $r = R - A\omega t$, where A is a constant with dimensions of length and $A\omega t$ is small compared to R. Find the ball's angular momentum with respect to the pole at time t.

Solution: Because the tether ball does not follow a circular path, we must use the more general form $\mathbf{L} = \mathbf{r} \times \mathbf{p}$ for the angular momentum. We take the horizontal plane to be the xy-plane. We first write out the x- and y-coordinates of the ball as a function of time, so as to be able to calculate the velocities and thus the momentum. We have

$$x = r\cos\theta = (R - A\omega t)\cos(\omega t);$$

$$y = r\sin\theta = (R - A\omega t)\sin(\omega t).$$

From this expression, we can also calculate the momentum components of the ball:

$$p_x = Mv_x = M\frac{dx}{dt} = M[(R - A\omega t)(-\omega\sin(\omega t)) + (-A\omega)\cos(\omega t)];$$

$$p_y = Mv_y = M\frac{dy}{dt} = M[(R - A\omega t)(\omega\cos(\omega t)) + (-A\omega)\sin(\omega t)].$$

We can now use Eq. (B1–4) in "The Vector Product" box on page 270 to find **L**:

$$\begin{aligned}\mathbf{L} &= (xp_y - yp_x)\mathbf{k}\\ &= M\begin{bmatrix}(R - A\omega t)^2\omega\cos^2(\omega t) + (R - A\omega t)(-A\omega)\cos(\omega t)\sin(\omega t)\\ -(R - A\omega t)^2(-\omega)\sin^2(\omega t) - (R - A\omega t)(-A\omega)\sin(\omega t)\cos(\omega t)\end{bmatrix}\mathbf{k}\\ &= M(R - A\omega t)^2\omega[\cos^2(\omega t) + \sin^2(\omega t)]\mathbf{k} = M(R - A\omega t)^2\omega\mathbf{k}.\end{aligned}$$

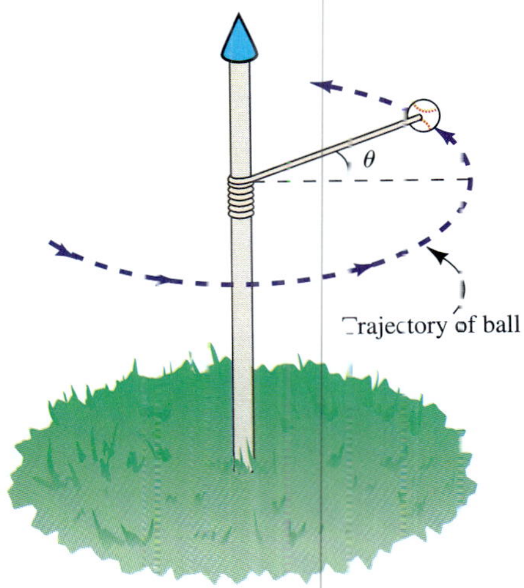

FIGURE 10–8 Example 10–3. As the rope winds around the pole, it shortens.

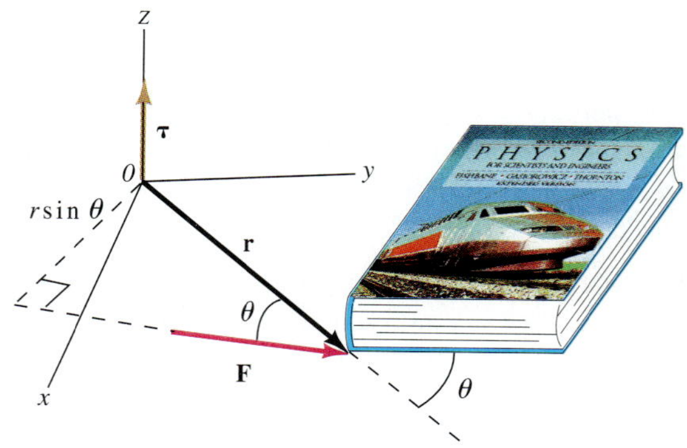

FIGURE 10–9 Vector relation for the torque about a point O. The force is applied at a specific point on an extended object.

$\mathbf{L}$ always lies in the z-direction because the motion is, by assumption, always in the xy-plane. But the angular momentum changes with time—albeit slowly ($R >> A\omega t$). As a check, note that if $A = 0$ (no change in length), the angular momentum has the familiar magnitude $MR^2\omega = MvR$.

FIGURE 10–10 The tugboat can rotate the large ship by exerting a force on it. Notice that the force is perpendicular to the large ship and acts at the bow; this maximizes the torque.

10–2 GENERALIZATION OF TORQUE

Torque, introduced in Chapter 9, has a magnitude that is given by a lever arm times the magnitude of force; it has a direction specified by a right-hand rule. The torque, $\boldsymbol{\tau}$, with respect to any reference point O due to a force $\mathbf{F}$ applied at some point P of an object is therefore compactly written by using a vector product:

The torque on an object about a reference point

$$\boldsymbol{\tau} = \mathbf{r} \times \mathbf{F}, \tag{10–5}$$

where $\mathbf{r}$ is the position vector from O to the point of application (P) of the force (Fig. 10–9). The object in question can be an extended system (Fig. 10–10) or it can even be a point mass located at P. The reference point O about which the torque is defined can be located inside or outside the object. Aligning the fingers of the right hand along $\mathbf{r}$ and curling them toward $\mathbf{F}$, the torque is in the direction of the thumb.

The test as to whether this is the proper definition of torque is whether the dynamical relation between torque and angular momentum, Eq. (9–34), continues to be satisfied:

The rate of change of angular momentum is given by the torque.

$$\boldsymbol{\tau} = \frac{d\mathbf{L}}{dt}. \tag{10–6}$$

We can check that Eq. (10–6) is indeed satisfied if we take Eq. (10–4), which is our generalized expression for the angular momentum of a point mass, and compute its rate of change. Keep in mind that we can always use Newton's second law. The rate of change of angular momentum can be found by using the chain rule of calculus (Appendix IV–7):

$$\frac{d\mathbf{L}}{dt} = \left(\frac{d\mathbf{r}}{dt} \times \mathbf{p}\right) + \left(\mathbf{r} \times \frac{d\mathbf{p}}{dt}\right). \tag{10–7}$$

The first part of Eq. (10–7) drops out because it is of the form $\mathbf{v} \times (m\mathbf{v})$, and the vector product of two parallel vectors is zero. For the second part of Eq. (10–7) we use Newton's second law in the form $d\mathbf{p}/dt = \mathbf{F}$ and obtain the equation

$$\frac{d\mathbf{L}}{dt} = \mathbf{r} \times \mathbf{F} = \boldsymbol{\tau},$$

which is just the desired Eq. (10–6).

Having shown that the general form $\mathbf{L} = \mathbf{r} \times \mathbf{p}$ reduces to the form $I\omega$ for symmetric rigid bodies, we can use the fact that $\boldsymbol{\tau} = d\mathbf{L}/dt$ to see that the general form $\boldsymbol{\tau} = \mathbf{r} \times \mathbf{F}$ similarly reduces to the form $I\alpha$.

We may use the properties of the unit vectors **i** and **j** in Eqs. (B1–3) to express torque in terms of the components of **r** and **F**.

If **r** is written as $\mathbf{r} = x\mathbf{i} + y\mathbf{j} + z\mathbf{k}$, Eq. (B1–4) shows that torque is given in terms of components by

$$\boldsymbol{\tau} = (yF_z - zF_y)\mathbf{i} + (zF_x - xF_z)\mathbf{j} + (xF_y - yF_x)\mathbf{k}. \tag{10–8}$$

EXAMPLE 10–4 A stone of mass $m = 1$ kg is dropped from an outstretched arm of length $\ell = 0.8$ m (Fig. 10–11). For the time when the stone has dropped exactly $d = 1$ m, find the net torque on the stone about the shoulder, labeled point O. Ignore air resistance.

Solution: The net torque is determined by the net force, and in this case the only force acting is gravity, $\mathbf{F} = -mg\mathbf{k}$. We are using the coordinate system with the shoulder at the origin. Although the distance of the rock from point O increases with time, the lever arm for the force of gravity about point O is constant, given by the arm length ℓ. The magnitude of the torque will therefore be ℓmg, and its direction will be given by the right-hand rule to be in the $+y$-direction. Equivalently, we can use the vector product definition in terms of components. We have, from Eq. (10–8),

$$\boldsymbol{\tau} = (\ell\,\mathbf{i}) \times (-F_g\mathbf{k}) = \ell F_g\,\mathbf{j} = \ell mg\,\mathbf{j}.$$

Note that the torque is constant over time and independent of the distance the stone has fallen.

The magnitude of the torque is

$$\tau = (0.8\ \text{m})(1\ \text{kg})(9.8\ \text{m/s}^2) = 8\ \text{N}\cdot\text{m}.$$

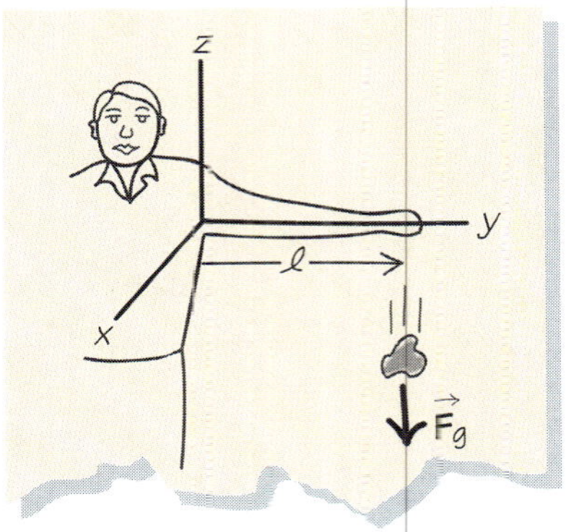

FIGURE 10–11 Example 10–4. The origin of the coordinate system is placed at the shoulder. $\mathbf{F}_g$ is the force of gravity on the falling stone.

10–3 THE DYNAMICS OF ROTATION

In Section 10–2, we saw that the dynamical equation of motion [Eq. (10–6)] holds for a point mass whose angular momentum is defined with respect to a particular point. To see that the same kind of equation of motion applies to extended objects, we will now consider a system that is a collection of point particles. Our more general expressions for the angular momentum and torque will provide some new and powerful ways to think about rotations and the motions of extended objects, both rigid and nonrigid.

To find our new expressions, let's consider a collection of particles with masses m_i located at positions $\mathbf{r}_i$ with respect to a point O, as in Fig. 10–12. If the momenta of the particles are given by $\mathbf{p}_i$, then the particle with the label i has an angular momentum $\mathbf{L}_i = \mathbf{r}_i \times \mathbf{p}_i$, and if the force acting on the ith particle is $\mathbf{F}_i$, then

$$\boldsymbol{\tau}_i = \mathbf{r}_i \times \mathbf{F}_i = \frac{d\mathbf{L}_i}{dt} \tag{10–9}$$

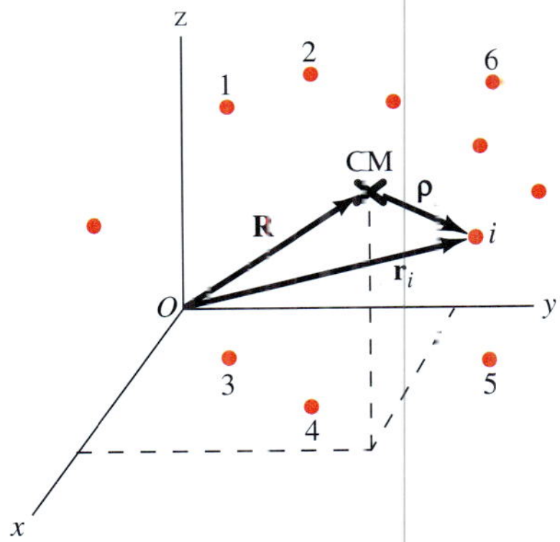

FIGURE 10–12 A system of many particles labeled by the index i. The position of particle i with respect to an origin O is $\mathbf{r}_i$, and the displacement of particle i from the center of mass is $\boldsymbol{\rho}_i$.

The total angular momentum of the collection of particles is

$$\mathbf{L} = \sum_i \mathbf{L}_i. \tag{10–10}$$

Similarly, the total torque is the sum of the individual torques on the particles:

$$\boldsymbol{\tau} = \sum_i \boldsymbol{\tau}_i. \tag{10–11}$$

The rate of change of the total angular momentum is obtained by summing Eq. (10–9) over the component particles, and we obtain Eq. (10–6):

The rate of change of the total angular momentum of a system is given by the total torque.

$$\boldsymbol{\tau} = \frac{d\mathbf{L}}{dt}. \tag{10–12}$$

Equation (10–12) expands the rotational analogue of Newton's second law to any extended system.

The Role of Interparticle Forces in Torque

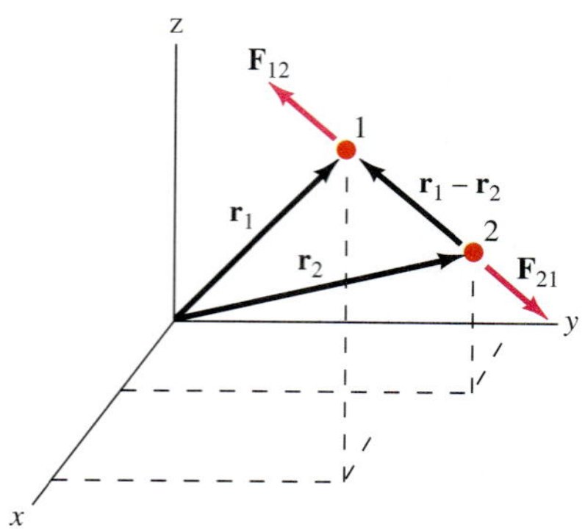

FIGURE 10–13 Two mass points, labeled 1 and 2, within an extended object, and the internal forces they exert on one another.

What role do the forces between the individual particles (or between different regions of a continuous object) play in the dynamics of rotations? Let's consider just two particles (or two small mass regions), with labels 1 and 2; this result can be generalized. For the torque on particle 1, we have (Fig. 10–13)

$$\boldsymbol{\tau}_1 = \mathbf{r}_1 \times (\mathbf{F}_1^{\text{ext}} + \mathbf{F}_{12}), \tag{10–13a}$$

where $\mathbf{F}_{12}$ is the (internal) force exerted on particle 1 due to particle 2 and $\mathbf{F}_1^{\text{ext}}$ is the external force on particle 1. We also have

$$\boldsymbol{\tau}_2 = \mathbf{r}_2 \times (\mathbf{F}_2^{\text{ext}} + \mathbf{F}_{21}). \tag{10–13b}$$

We must assume that the forces between 1 and 2 are directed along a line between the two particles.

The interparticle forces contribute to the total torque through the terms

$$(\mathbf{r}_1 \times \mathbf{F}_{12}) + (\mathbf{r}_2 \times \mathbf{F}_{21}) = (\mathbf{r}_1 \times \mathbf{F}_{12}) - (\mathbf{r}_2 \times \mathbf{F}_{12})$$
$$= (\mathbf{r}_1 - \mathbf{r}_2) \times \mathbf{F}_{12}. \tag{10–14}$$

Interparticle forces play no role in the torque on a system.

We have used Newton's third law. Observe that $(\mathbf{r}_1 - \mathbf{r}_2)$ is the vector leading from particle 2 to particle 1; further, observe that the interparticle force $\mathbf{F}_{12} = -\mathbf{F}_{21}$ is a vector *along the direction between the two particles.* Thus the vector product that appears in Eq. (10–14) is zero, and *only external forces appear in the total torque* $\boldsymbol{\tau}$ that is to be used in Eq. (10–12). This result is similar to the result derived in Chapter 8, where we showed that interparticle forces do not contribute to the change in linear momentum of a system composed of many interacting particles. As in the case of the momentum equation, this allows us to discuss the motion of extended bodies without knowing the details of how their individual parts interact with each other. Only the torques due to external forces enter.

Dynamical Equations of Motion with Reference to Any Axis

Features of rotations about any axis

We showed in Eq. (9–33) that, for a symmetric rigid body with some rotational inertia I about its center of mass axis, the dynamical equation for rotation takes the form

$$\boldsymbol{\tau} = \frac{d\mathbf{L}}{dt} = I\frac{d\boldsymbol{\omega}}{dt} = I\boldsymbol{\alpha}. \tag{10–15}$$

In solving equations that describe the angular motion of an extended object, however, it is often useful to consider rotations about a second axis A, parallel to the axis through the center of mass. We can derive three important results concerning the rotations of a rigid body about any axis A:

1. The angular momentum about A is the sum of two terms: (a) the angular momentum $\mathbf{R} \times \mathbf{P}$ about A of a point mass that is carrying the whole mass (M) of the object, as though it were located at the center of mass and moving with the velocity ($\mathbf{V}$) of the center of mass; and (b) the angular momentum $\mathbf{L}_{\text{cm}}$ of the object about the center of mass:

$$\mathbf{L} = (\mathbf{R} \times \mathbf{P}) + \mathbf{L}_{\text{cm}}. \tag{10–16}$$

Here, $\mathbf{P} = M\mathbf{V}$.

2. The total torque about A is the sum of two terms: (a) the torque about A due to the total external force applied to the center of mass; and (b) the torque about the center of mass:

$$\boldsymbol{\tau} = (\mathbf{R} \times \mathbf{F}_{\text{tot}}) + \boldsymbol{\tau}_{\text{cm}}. \tag{10–17}$$

3. The rate of change in each term of the total angular momentum in Eq. (10–16) is equal to the corresponding term of the total torque; that is,

$$\mathbf{R} \times \mathbf{F}_{\text{tot}} = \frac{\mathrm{d}}{\mathrm{d}t}(\mathbf{R} \times \mathbf{P}) \quad \text{and} \quad \boldsymbol{\tau}_{\text{cm}} = \frac{\mathrm{d}}{\mathrm{d}t}\mathbf{L}_{\text{cm}}. \tag{10–18}$$

How to Get Equations (10–16) through (10–18) for Angular Momentum and Torque

We break our extended object into a set of discrete pieces labeled with the subscript i. The center of mass position $\mathbf{R}$ with respect to an origin O was defined in Eq. (8–48) by

$$\mathbf{R} = \frac{1}{M}\sum_i^N m_i\mathbf{r}_i,$$

where $M = \sum_i^N m_i$ is the total mass, N is the number of pieces, and $\mathbf{r}_i$ is the position of the ith piece. We introduce the vector $\boldsymbol{\rho}_i$ (Fig. 10–12), the position vector of particle i as measured from the center of mass:

$$\mathbf{r}_i = \mathbf{R} + \boldsymbol{\rho}_i.$$

The angular momentum with respect to point O may now be written in the form

$$\mathbf{L} = \sum_i \mathbf{r}_i \times \mathbf{p}_i = \sum_i (\mathbf{R} + \boldsymbol{\rho}_i) \times \mathbf{p}_i = \left(\mathbf{R} \times \sum_i \mathbf{p}_i\right) + \left(\sum_i \boldsymbol{\rho}_i \times \mathbf{p}_i\right)$$
$$= (\mathbf{R} \times \mathbf{P}) + \left(\sum_i \boldsymbol{\rho}_i \times \mathbf{p}_i\right),$$

where $\mathbf{P} = \sum \mathbf{p}_i$ is the total momentum of the system. The term $\sum \boldsymbol{\rho}_i \times \mathbf{p}_i$ is the angular momentum $\mathbf{L}_{\text{cm}}$ about the center of mass. We have thus proven Eq. (10–16).

There is a similar decomposition of the total torque:

$$\boldsymbol{\tau} = \sum_i \mathbf{r}_i \times \mathbf{F}_i = \sum_i (\mathbf{R} + \boldsymbol{\rho}_i) \times \mathbf{F}_i = \left(\mathbf{R} \times \sum_i \mathbf{F}_i\right) + \left(\sum_i \boldsymbol{\rho}_i \times \mathbf{F}_i\right)$$
$$= (\mathbf{R} \times \mathbf{F}_{\text{tot}}) + \left(\sum_i \boldsymbol{\rho}_i \times \mathbf{F}_i\right).$$

The second term in the final equality is the torque about the center of mass, and so we have demonstrated Eq. (10–17).

To derive the third result, we take the rate of change of the $\mathbf{R} \times \mathbf{P}$ term in Eq. (10–18). We find that it is the same as the first term of Eq. (10–17):

$$\frac{\mathrm{d}(\mathbf{R} \times \mathbf{P})}{\mathrm{d}t} = \left(\frac{\mathrm{d}\mathbf{R}}{\mathrm{d}t} \times \mathbf{P}\right) + \left(\mathbf{R} \times \frac{\mathrm{d}\mathbf{P}}{\mathrm{d}t}\right) = \left[\left(\frac{1}{M}\right)\mathbf{P} \times \mathbf{P}\right] + (\mathbf{R} \times \mathbf{F}_{\text{tot}})$$
$$= \mathbf{R} \times \mathbf{F}_{\text{tot}}.$$

This, in turn, implies that the rate of change of the total angular momentum about the center of mass is equal to the torque about the center of mass due to all the forces:

$$\frac{\mathrm{d}\mathbf{L}_{\text{cm}}}{\mathrm{d}t} = \boldsymbol{\tau}_{\text{cm}}.$$

Why It Is Useful to Be Able to Choose the Reference Point

The results just summarized are of practical importance because they provide us with alternative ways to approach problems with both linear and rotational motion. We can choose points of reference for torque, angular momentum, and rotational inertia for convenience. This is illustrated by taking a second look at a cylinder rolling down an incline, which is a problem first examined in Chapter 9 (Fig. 10–14a). There we used a point on the symmetry axis of the cylinder as a reference. Let's now choose our ref-

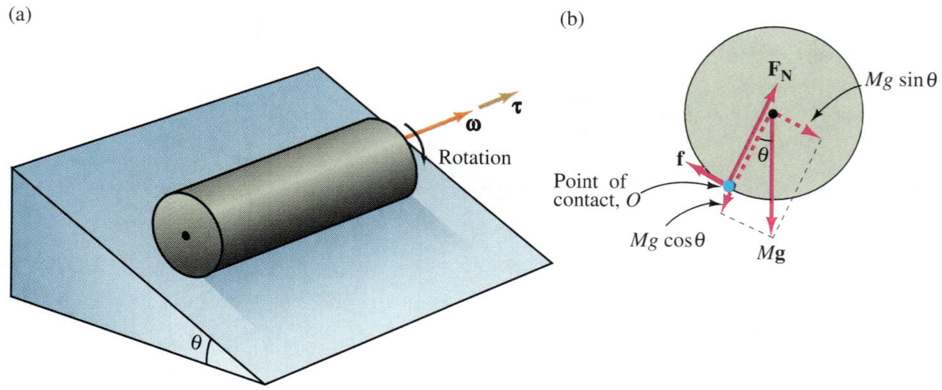

FIGURE 10–14 (a) A cylinder rolls without slipping down an inclined plane. (b) Extended force diagram for the rolling cylinder.

erence point to be point O, the point of contact between the cylinder and the plane. It is most efficient to choose as a reference point the point through which the largest number of individual forces act. For such a point the lever arm is zero; thus, these forces do not contribute to the torque (Fig. 10–14b). In this case, only gravity produces a torque about the contact point; moreover, only the component of gravity that is parallel to the plane—of magnitude $mg\sin\theta$—contributes. Using the right-hand rule, we see that the corresponding torque about O points into the plane of the page and has magnitude

$$\tau = (Mg\sin\theta)R. \tag{10–19}$$

Having calculated the torque about the contact point, let's look at the angular momentum about this same point. As indicated in Eq. (10–16), there are two terms. First, there is the angular momentum of the total mass as though that mass were placed at the object's center of mass, $\mathbf{R}\times\mathbf{P} = \mathbf{R}\times(M\mathbf{V})$. $\mathbf{R}$ and $\mathbf{V}$ are perpendicular to each other, so this term has magnitude

$$MVR = M(\omega R)R = MR^2\omega. \tag{10–20}$$

Second, there is the angular momentum about the symmetry axis passing through the center of mass, $L = I_{cm}\omega$. Thus the total angular momentum about the point of contact is the sum of Eq. (10–20) and $I_{cm}\omega$:

$$L = (MR^2 + I_{cm})\omega. \tag{10–21}$$

We have chosen the same reference point for both the torque and the angular momentum. The dynamical equation $dL/dt = \tau$ is therefore applicable and, from Eqs. (10–19) and (10–21), it takes the form

$$(MR^2 + I_{cm})\frac{d\omega}{dt} = MgR\sin\theta;$$

$$\frac{d\omega}{dt} = \frac{MgR\sin\theta}{MR^2 + I_{cm}}.$$

This equation states that the angular acceleration $d\omega/dt$ is constant, so the angular velocity increases linearly with time:

$$\omega(t) = \omega_0 + \frac{MgR\sin\theta}{MR^2 + I_{cm}}t. \tag{10–22}$$

Here, the angular velocity at $t = 0$ is ω_0. For a symmetric rolling object, I_{cm} will take the form $I_{cm} = CMR^2$, where C is a numerical constant. Then the velocity $v = \omega R$ of the center axis becomes

$$v(t) = v_0 + \frac{g\sin\theta}{1 + C}t. \tag{10–23}$$

This result is the same one found in Eq. (9–48), yet we have used a different reference point. The choice of reference point does not affect the result and can be chosen for convenience.

PROBLEM SOLVING

Choose a convenient axis. Physical results won't depend on that choice.

Angular Impulse

The definition of angular impulse

The angular impulse is as useful for the motion of extended systems as the "ordinary," or linear, impulse is for the motion of point masses. The linear impulse, $\mathbf{J} = \mathbf{F}\,\Delta t$, describes the change $\Delta\mathbf{p}$ in momentum when a force $\mathbf{F}$ acts for a short duration Δt. Similarly, the dynamical equation for rotational motion tells us that the change in the angular momentum of a system, $\Delta\mathbf{L}$, when a torque, $\boldsymbol{\tau}$ acts for a duration Δt is given by the angular impulse $\mathbf{J}_\tau$:

$$\Delta\mathbf{L} = \boldsymbol{\tau}\,\Delta t \equiv \mathbf{J}_\tau. \tag{10–24}$$

Here, both the torque and angular momentum are measured with respect to the same reference point.

Let the reference point for torque and angular momentum be O. The force, $\mathbf{F}$, is applied briefly at $\mathbf{r}$. The torque is then $\boldsymbol{\tau} = \mathbf{r} \times \mathbf{F}$; multiplying by Δt, we see that the angular impulse is given by

$$\mathbf{J}_\tau = \boldsymbol{\tau}\,\Delta t = \mathbf{r} \times (\mathbf{F}\,\Delta t) = \mathbf{r} \times \mathbf{J}. \tag{10–25}$$

EXAMPLE 10–5 A court expert models a collision in which a stationary automobile is struck from the side by treating the automobile as a rod of mass M and length ℓ at rest on a frictionless surface. An impulse $F\,\Delta t$ is applied at right angles at a distance $\ell/3$ from one end of the rod (Fig. 10–15). Describe the subsequent motion of the rod.

Solution: The motion may be decomposed into motion of the center of mass and rotation about the center of mass. The initial momentum of the center of mass is zero, and therefore its final momentum is given by the impulse:

$$\Delta P = P_f - P_i = P_f = MV_f = F\,\Delta t.$$

Similarly, the angular impulse is equal to the change in angular momentum in time Δt. The initial angular momentum is zero, and the distance from the point of application of the force to the center of mass is $(\ell/2) - (\ell/3)$. Thus, if ω_f is the final angular velocity of the rod about the center of mass,

$$\Delta L = L_f - L_i = I\omega_f = (F\,\Delta t)\left(\frac{\ell}{2} - \frac{\ell}{3}\right) = \frac{(F\,\Delta t)\ell}{6}.$$

Because $I = M\ell^2/12$ (see Table 9–1), this relation reads

$$\frac{M\ell^2\,\omega_f}{12} = \frac{(F\,\Delta t)\ell}{6};$$

$$\omega_f = \frac{2}{M\ell}(F\,\Delta t) = \frac{2}{\ell}V_f.$$

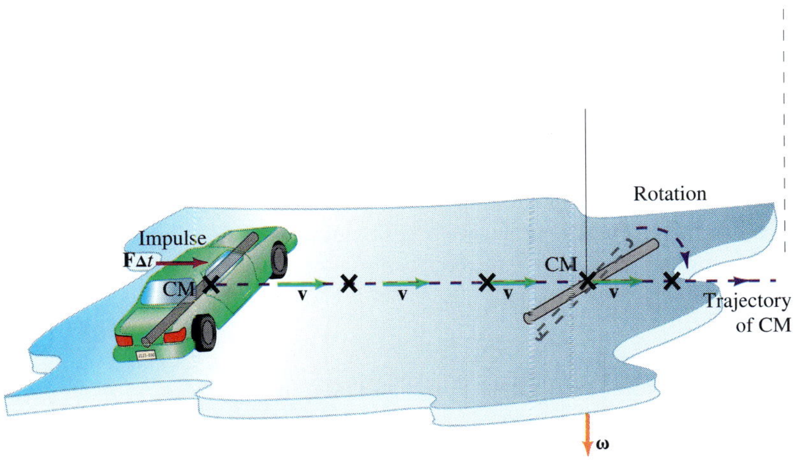

FIGURE 10–15 Example 10–5. The automobile is modeled by a uniform rod. (a) Before: An impulse is given to the rod a distance $\ell/3$ from the top. (b) After: The rod moves with a combination of a linear center-of-mass motion and a rotation about the center of mass.

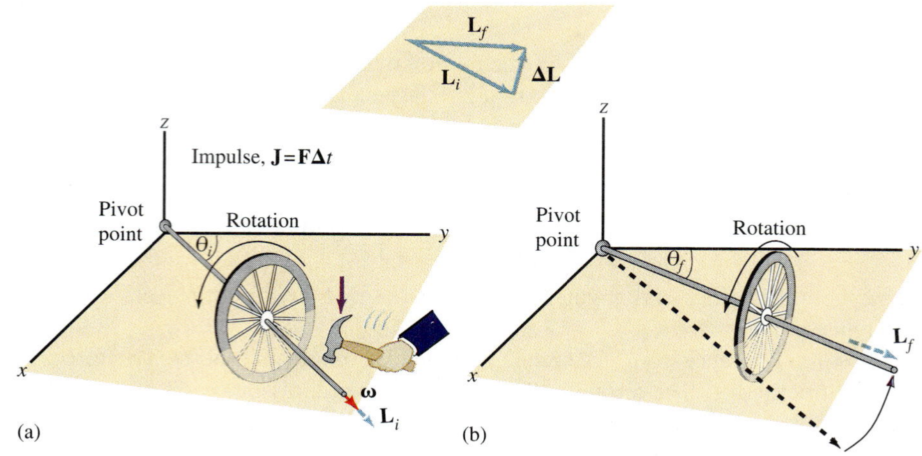

FIGURE 10–16 (a) A rapidly rotating bicycle wheel is spinning with its angular velocity, $\boldsymbol{\omega}$, and its initial angular momentum, $\mathbf{L}_i$, aligned as shown. One end of the axis is fixed on a pivot point, and the other end is given an impulse $\mathbf{J} = -J\mathbf{k}$. (b) When a corresponding angular momentum change, $\Delta\mathbf{L}$, is added to the initial angular momentum, the result is that the free end of the wheel tends to rotate toward the y-axis.

As in Fig. 10–15, the motion of the rod consists of a linear center-of-mass motion, with velocity $V = F\,\Delta t/\mathrm{m}$ along the direction of the impulse, plus a rotation about the center of mass, with angular velocity $\omega = 2V/\ell$.

The Vectorial Nature of Angular Impulse. The angular impulse illustrates one of the most surprising aspects of the dynamics of rotation, one in which the *vector* nature of angular momentum and torque plays a crucial role. Consider an object such as a bicycle wheel rotating about its axle, with its angular velocity $\boldsymbol{\omega}$ and initial angular momentum $\mathbf{L}_i$, aligned horizontally in the xy-plane (Fig. 10–16a). The wheel is pivoted about one end of the axle and the other end is struck by a hammer blow downward, receiving an impulse $\mathbf{J} = -J\mathbf{k}$. What is the immediate response of the wheel as a result of the impulse?

Let's compute the angular impulse, $\mathbf{J}_\tau$, using the fixed end (pivot point) of the axle as a reference point. If the vector from the fixed end to the point where the impulse is applied is $\boldsymbol{\ell}$, then the angular impulse is

$$\mathbf{J}_\tau = \boldsymbol{\ell} \times \mathbf{J}.$$

If we use a right-hand rule on this vector product, we see that its direction, and hence the direction of the instantaneous angular momentum change $\Delta\mathbf{L}$, points in the direction indicated in Fig. 10–16b. Even though the blow comes from above, *the free end of the wheel tends to rotate back toward the y-axis*! The wheel moves in a direction quite different from the direction of the blow. This is a surprising consequence of the vector nature of angular momentum and torque that will be echoed by a similar phenomenon in the motion of rotating objects (gyroscopes) under the effect of gravity (Section 10–7).

10–4 CONSERVATION OF ANGULAR MOMENTUM

The general dynamical equation for rotational motion is $\boldsymbol{\tau}_{\text{net}} = \mathrm{d}\mathbf{L}/\mathrm{d}t$, where the torque always refers to *external* forces. If there is no net external torque on a system—rigid or otherwise—the angular momentum of the system is constant or *conserved,*

$$\frac{\mathrm{d}\mathbf{L}}{\mathrm{d}t} = 0, \tag{10–26}$$

so that the angular momentum is constant during the motion of the system. For a nonrigid system, the rotational inertia may change, and the conservation of angular momentum is a powerful tool for study of the motion.

APPLICATION: NAVIGATION THROUGH THE CONSERVATION OF ANGULAR MOMENTUM

The rotation axis of a rotating object that has no torques acting on it will point in a fixed direction. A device such as this with a built-in fixed direction is a self-contained tool for navigation. An interesting version consists of a small metal sphere, engraved with black and white marks and levitated by a magnetic field (Fig. 10AB–1). The whole assembly is placed inside a vacuum enclosure mounted on gimbals. The sphere can then be sped up to a very high angular velocity with the aid of special magnetic fields. The rotating sphere is held in a horizontally stable position by an electrostatic field; any drifting up or down in the enclosure is corrected by servomechanisms. Once the rotation is established, the sphere's rotation axis remains aligned in a constant spatial direction. As there are no mechanical bearings, there are no vestigial torques that can cause the sphere's axis to rotate. The whole assembly can then be placed in an airplane. With the aid of photocells, the markings on the sphere can be read; the position of these marks relative to the casing, which is attached to the airplane, then measures the exact rotation of the sphere relative to the plane. This allows for precise control of the flight path. The particular device described here provided an accuracy of 200 m on a transatlantic crossing. Note that this device does not require any communication from the ground in order to work.

FIGURE 10AB–1 Two versions of the heart of a navigating device based on conservation of momentum for a rotating sphere.

Central Forces

The torque on a particle is zero if the force and the displacement vector **r** from the point of reference (the origin) to the point of application of the force are *parallel* (or antiparallel) as for central forces; this is because the vector product between **r** and **F** is zero. Here, the angular momentum about the origin is conserved, as illustrated in Example 10–1.

Consider an object moving under the influence of a central force, and take the origin of the coordinate system to be at the source of this force. Let the initial velocity of the object be $\mathbf{v}_0$. The initial direction of the angular momentum is given by $\mathbf{r} \times \mathbf{v}_0$, that is, perpendicular to the plane formed by **r** and $\mathbf{v}_0$. Because the angular momentum is constant and its direction does not change, *the motion is always confined to the initial plane formed by* **r** *and* $\mathbf{v}_0$. To obtain the magnitude of the angular momentum, consider the path shown in Fig. 10–17a. At any given moment t, the particle is at position **r** and is moving with a velocity **v** tangent to the trajectory. The magnitude of the angular momentum is given by

$$L = |\mathbf{r} \times \mathbf{p}| = |\mathbf{r} \times m\,\mathbf{v}| = rmv \sin\alpha, \tag{10–27}$$

where α is the the angle between **r** and **v**. The velocity $\mathbf{v} = d\mathbf{r}/dt$ so that, if the position changes from **r** to $\mathbf{r} + \Delta\mathbf{r}$ in a small time interval Δt as in Fig. 10–17b, and if we replace **v** by $\Delta\mathbf{r}/\Delta t$, we find

$$L = mr\frac{|\Delta\mathbf{r}|\sin\alpha}{\Delta t} = mr\frac{r\,\Delta\theta}{\Delta t}. \tag{10–28}$$

We have used the geometric relation; $|\Delta\mathbf{r}| \sin\alpha = r\,\Delta\theta$. In the limit $\Delta t \to 0$, Eq. (10–28) becomes

$$L = mr^2\frac{d\theta}{dt} = mr^2\omega, \tag{10–29}$$

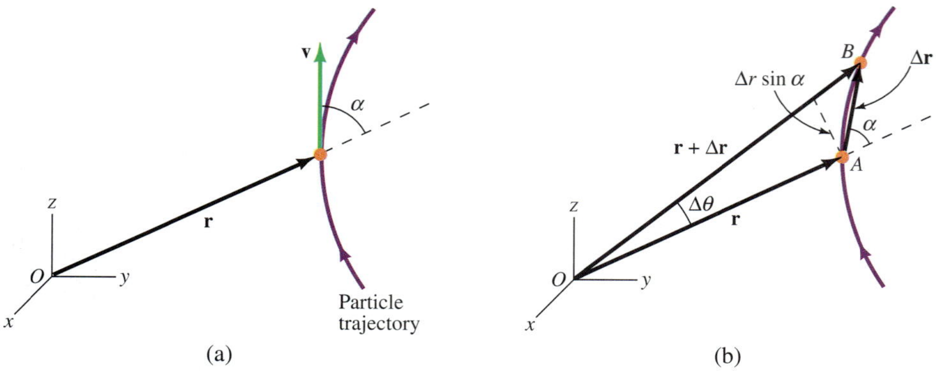

FIGURE 10–17 The trajectory of a particle with constant angular momentum. (a) The velocity at any given time is tangent to the trajectory and makes an angle α with the position vector. (b) In a small time interval, the particle moves from point A to point B and is displaced by $\Delta\mathbf{r}$. The angle $\Delta\theta$ is approximately $\Delta\theta = (|\Delta\mathbf{r}| \sin\alpha)/r$.

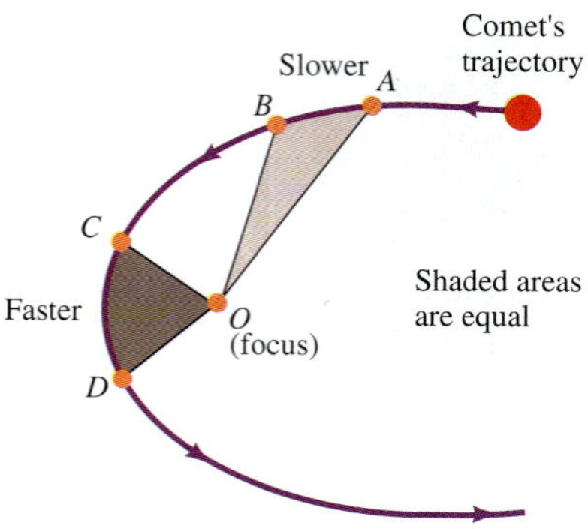

FIGURE 10–18 If a particle moves with constant angular momentum, then its trajectory sweeps out equal areas in equal times. If the motion of a comet from A to B takes time Δt and the motion from C to D takes the same amount of time, then the particle must move more quickly in traveling from C to D.

where ω is the instantaneous angular velocity about the origin. Thus $r^2\omega$ is constant throughout the motion when L is constant. For uniform circular motion, both r and ω are constant and there is nothing new in Eq. (10–29). If r is not constant, then Eq. (10–29) can be interpreted as follows: Consider the time interval Δt and the geometry of Fig. 10–17b. For infinitesimal displacements, the area of the triangle OAB is given by

$$\Delta A = \frac{1}{2}(r\,\Delta\theta)r = \frac{1}{2}r^2\,\Delta\theta, \tag{10–30}$$

from which it follows that

$$dA = \frac{1}{2}r^2\frac{d\theta}{dt}\,dt = \frac{1}{2}r^2\omega\,dt.$$

From Eq. (10–29),

$$\frac{dA}{dt} = \frac{1}{2}r^2\omega = \frac{L_0}{2m}, \tag{10–31}$$

Kepler's second law

where L_0 is the constant value of the angular momentum. Whatever the trajectory of the particle (which depends on the detailed form of the central force), a consequence of angular momentum conservation is that *the rate at which the radius vector sweeps out an area is constant.* In the context of the gravitational force (Chapter 12), this is known as **Kepler's second law**. For example, a comet, whose motion about the Sun is an ellipse with the Sun located at one of the focal points, sweeps out equal areas in equal lengths of time. This means that it must move very rapidly when it is near the Sun compared with its speed at large distances from the Sun (Fig. 10–18).

Note that a freely moving particle experiences no torque. Thus it too satisfies the condition of Eq. (10–31) that its position vector, as measured from any point, sweeps out each area at a uniform rate, as shown in Fig. 10–19.

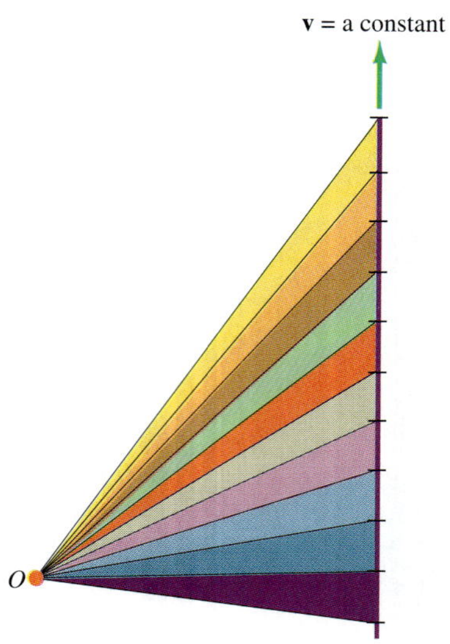

FIGURE 10–19 The triangles that a uniformly moving particle sweeps out with respect to any point O have equal areas.

Nonrigid Objects

When an object is not rigid and its angular momentum is constant, the rotational inertia and the angular velocity can each change in such a way that their product remains constant. This possibility is commonly realized by divers or figure skaters (Fig. 10–20). In another example, water going down the drain holds its angular momentum constant by swirling more quickly as it approaches the drain.

EXAMPLE 10–6 In a common classroom experiment, a student sits on a spinning stool with weights in his hands. As an idealized version of this experiment, consider a solid cylinder of diameter $d = 0.5$ m and mass $M = 50$ kg (Fig. 10–21). The cylinder is oriented vertically and spins freely about its axis with a period of 3 s. Two massless rods are attached horizontally to the cylinder, with their ends $L = 1$ m from

(a)

(b)

FIGURE 10–20 (a) Angular momentum is conserved even for nonrigid objects. The cat is not a rigid body. By rotating the central part of the body one way and the outer part of the body (legs and head) the opposite way, the cat can land on its feet even though its angular momentum must remain constant at zero. (b) Although a figure skater spins more slowly when her arms are extended than when her arms are close to her body, her angular momentum remains constant.

the surface, and there is a mass $m = 2$ kg at the end of each rod (Fig. 10–21a). The rods are drawn into the cylinder by an internal mechanism until the 2-kg masses are at the surface of the cylinder (Fig. 10–21b). What are the initial and final angular velocities?

Solution: The angular momentum, L, about the symmetry axis of the cylinder is constant because there are no external torques. (The force pulling in the "arms" is *internal* and *radial*.) Thus the product $I\omega$ is constant. As the masses are pulled in, I decreases; therefore, ω must increase. If I_0 and ω_0 refer to the initial conditions and I and ω refer to the final conditions, we find

$$I_0\omega_0 = I\omega;$$

$$\omega = \omega_0 \frac{I_0}{I}.$$

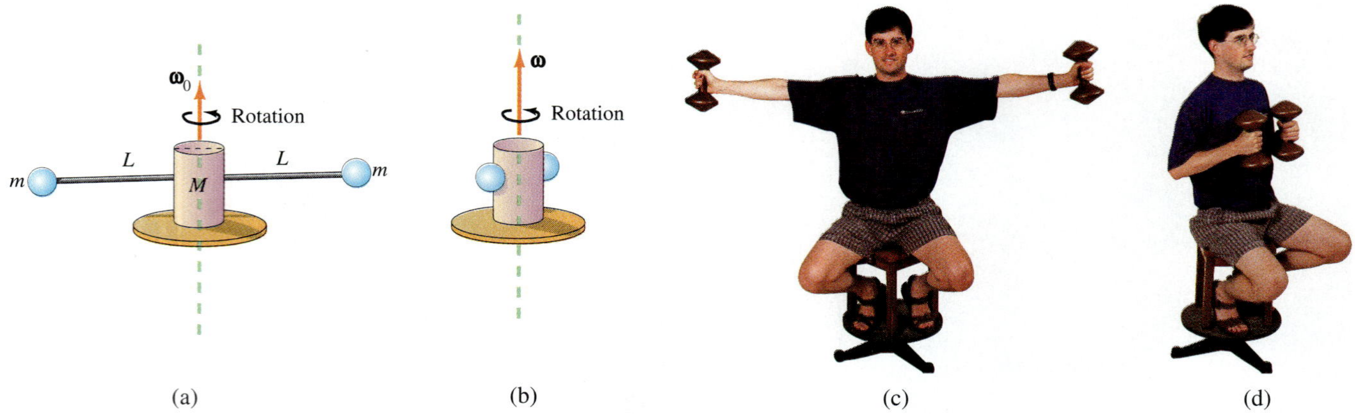

FIGURE 10–21 (a) Example 10–6. (b) When the masses are brought in closer to the rotation axis, the conservation of angular momentum requires the rotation to speed up. (c), (d) A demonstration of this.

I_0 is the rotational inertia of a solid cylinder ($MR^2/2$) of mass $M = 50$ kg and radius $R = d/2 = 0.25$ m, plus the rotational inertia of the two masses of 2 kg a distance $L + R = 1.25$ m from the axis:

$$I_0 = \frac{MR^2}{2} + 2m(L + R)^2.$$

I is found the same way, except the two masses are a distance R from the axis:

$$I = \frac{MR^2}{2} + 2mR^2 = \left(\frac{M}{2} + 2m\right)R^2.$$

The initial angular velocity is found from the period:

$$\omega_0 = \frac{2\pi\,\text{rad}}{T} = \frac{6.28\,\text{rad}}{3\,\text{s}} = 2\,\text{rad/s}.$$

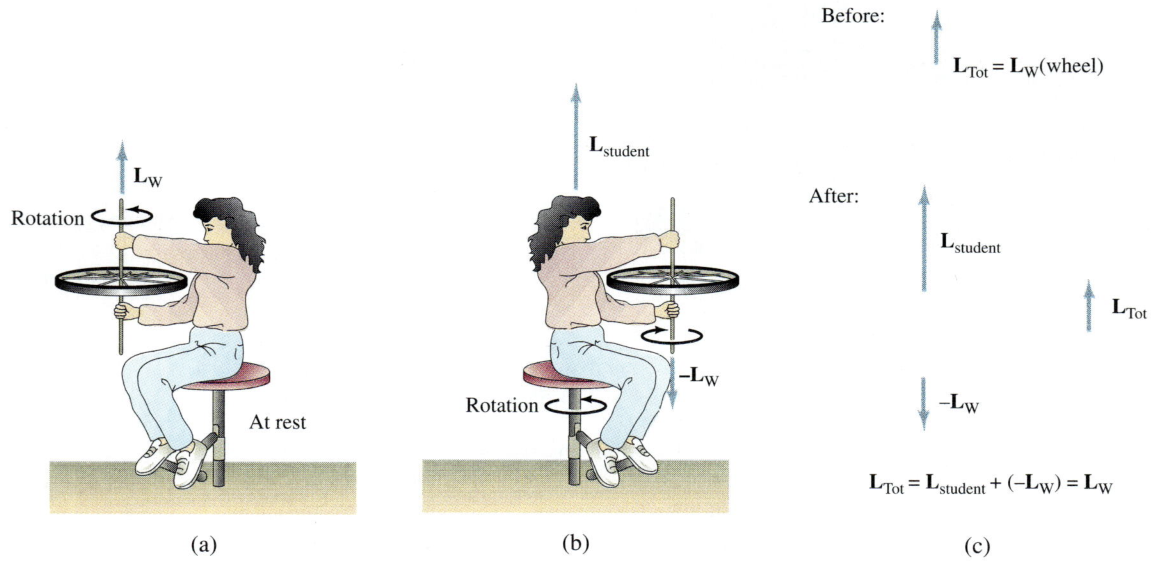

FIGURE 10–22 (a) A student on a stool that can rotate holds a freely rotating wheel. (b) When the wheel is inverted, the stool begins to turn in order to hold the total angular momentum constant. (c) To the now negative angular momentum of the wheel must be added a positive angular momentum from the student. Only in this way can the total angular momentum be conserved.

Thus

$$\omega = \omega_0 \frac{(MR^2/2) + 2m(L+R)^2}{[(M/2) + 2m]R^2}$$

$$= (2\ \text{rad/s}) \frac{(0.5)(50\ \text{kg})(0.25\ \text{m})^2 + 2(2\ \text{kg})(1.25\ \text{m})^2}{(0.5)(50\ \text{kg})(0.25\ \text{m})^2 + 2(2\ \text{kg})(0.25\ \text{m})^2}$$

$$= (2\ \text{rad/s}) \frac{7.8\ \text{kg}\cdot\text{m}^2}{1.8\ \text{kg}\cdot\text{m}^2} = 9\ \text{rad/s}.$$

The angular velocity has increased by a factor of 4.5. The movement of the end weights into the axis is remarkably effective at decreasing the rotational inertia! You can perform this demonstration in class (Figs. 10–21c and 10–21d).

The conservation of angular momentum is the conservation of a vector quantity. Suppose that a spinning bicycle wheel with its angular momentum pointing up (the wheel spins counterclockwise as seen from above) is handed to a student perched on a stool that can itself rotate (Fig. 10–22a). If the student now turns the wheel upside down, so that its angular velocity points downward (negative), he or she will begin to spin counterclockwise—the original direction of rotation of the wheel (Fig. 10–22b). This new spin direction is generated to conserve the total angular momentum (Fig. 10–22c).

An Astrophysical Example. A supernova provides a dramatic astrophysical illustration of the conservation of angular momentum. When a massive star stops burning because it has used up its fuel, it undergoes a collapse due to gravitational forces. There is a stupendous explosion and some of the stellar material is blown off in a radial direction. What is left behind is a *neutron star*: an extremely dense sphere of matter with a mass of several solar masses and a diameter of perhaps 10 km (compared with 10^6 km for the original star). This large ratio leads to very fast rates of rotation for neutron stars, some of which have periods on the order of milliseconds, compared to days for the original star. Many such rapidly rotating neutron stars have been observed.

10–5 WORK AND ENERGY IN ANGULAR MOTION

Rotating systems of particles—even those that are nonrigid—must have energy, and work must be done on them to change their energy. Now we want to explore the energy of rotating systems of many objects. In Chapter 9, we studied the rotational kinetic energy of a symmetrical rigid body whose rotational inertia about the axis of rotation is I and whose angular velocity is ω. We found that this energy is

$$K = \frac{1}{2} I\omega^2. \tag{10–32}$$

The angular velocity is a vector: if it is given by

$$\omega = \omega_x\mathbf{i} + \omega_y\mathbf{j} + \omega_z\mathbf{k}, \tag{10–33}$$

then

$$\omega^2 = \boldsymbol{\omega}\cdot\boldsymbol{\omega} = \omega_x^2 + \omega_y^2 + \omega_z^2. \tag{10–34}$$

If a torque is present, the rotational kinetic energy will change because the angular speed changes. We calculate the rate of change of energy; that is, the *instantaneous power*

$$\frac{dK}{dt} = \frac{1}{2} I \frac{d}{dt}(\omega^2) = \frac{I\,d\boldsymbol{\omega}}{dt}\cdot\boldsymbol{\omega} = I\boldsymbol{\alpha}\cdot\boldsymbol{\omega} = \boldsymbol{\tau}\cdot\boldsymbol{\omega}, \tag{10–35}$$

Power in rotating systems

where we have used Eq. (10–15), $I\boldsymbol{\alpha} = \boldsymbol{\tau}$. Equation (10–35) is the analogue of the linear motion equation for power that we saw in Chapter 6: $dK/dt = \mathbf{F}\cdot\mathbf{v}$.

The Work–Energy Theorem for Rotations

It is fairly straightforward to derive a work–energy theorem for rotational motion about a fixed axis. To define the work, we follow the analogue with linear motion that has proven so useful. Just as the infinitesimal work in linear motion is defined as $dW = \mathbf{F} \cdot d\mathbf{x}$, the infinitesimal work done in rotating a rigid body through an infinitesimal angle $d\boldsymbol{\theta}$ about the axis is defined to be

The work–energy theorem for rotations

$$dW \equiv \boldsymbol{\tau} \cdot d\boldsymbol{\theta}. \tag{10–36}$$

Note that an infinitesimal angle *is* a vector. When the torque, $\boldsymbol{\tau}$, points along the axis of rotation, this reduces to the form

$$dW = \tau\, d\theta. \tag{10–37}$$

This will be a reasonable definition for the work if it leads to an appropriate work–energy theorem. We have, using $\tau = I\, d\omega/dt$ and $d\theta = \omega\, dt$.

$$W = \int_{\theta_0}^{\theta} \tau\, d\theta = \int_0^t I \frac{d\omega}{dt}\, \omega\, dt = \int_0^t I\left(\frac{1}{2}\right)\frac{d\omega^2}{dt}\, dt$$

$$= \frac{1}{2} I \int_{\omega_0}^{\omega} d\omega^2 = \frac{1}{2} I(\omega^2 - \omega_0^2) = K - K_0. \tag{10–38}$$

Thus the work–energy theorem for rotational motion about a fixed axis is that *the work done on an object with rotational inertia I by a torque about an axis of rotation is equal to the increase in the rotational kinetic energy of the object.*

The Energy of an Extended Object in Motion

We next consider the energy of an extended object—rigid or not. We do this as we did in the subhead in which we derived Eqs. (10–16) and (10–17). First, we break up the object into a number of individual pieces with masses m_i located at $\mathbf{r}_i$ relative to some origin. We introduce the center of mass position $\mathbf{R}$ and the coordinates $\boldsymbol{\rho}_i$ relative to the center of mass according to $\mathbf{r}_i = \boldsymbol{\rho}_i + \mathbf{R}$, as in Fig. 10–12. When we take a derivative with respect to time, we have

$$\mathbf{v}_i = \frac{d\mathbf{r}_i}{dt} = \frac{d\mathbf{R}}{dt} + \frac{d\boldsymbol{\rho}_i}{dt} = \mathbf{V} + \mathbf{u}_i, \tag{10–39}$$

where $\mathbf{u}_i \equiv d\boldsymbol{\rho}_i/dt$ are the velocities of the masses m_i with respect to the center of mass and $\mathbf{V}$ is the velocity of the center of mass. We can then show that the kinetic energy of the system takes the form

$$K = \sum_i \frac{1}{2} m_i \mathbf{v}_i^2 = \frac{1}{2} MV^2 + \frac{1}{2}\sum_i m_i \mathbf{u}_i^2. \tag{10–40}$$

(We do not prove this result here.) The total kinetic energy consists of two parts: the kinetic energy of the total mass of the object moving with the velocity of the center of mass, and the kinetic energy of the motion relative to the center of mass.

The kinetic energy of Eq. (10–40) takes on a familiar form for a *rigid* body. In that case, the vectors $\boldsymbol{\rho}_i$ have a fixed magnitude, and only the angular variable changes. For rotations with angular speed ω about an axis through the center of mass, the speed of any point that is a radial distance r from the axis is given by $v = \omega r$. Thus $u_i = \omega \rho_i$, and $u_i^2 = \omega^2 \rho_i^2$. We can substitute this expression into Eq. (10–40) to find that

$$K = \frac{1}{2} MV^2 + \frac{1}{2} I_{cm}\omega^2. \tag{10–41}$$

The total kinetic energy of a rotating rigid body consists of a center-of-mass term and a term for rotations about the center of mass.

Equation (10–41) shows that *the total kinetic energy of a rigid body consists of the kinetic energy of the total mass moving with the velocity of the center of mass together with the rotational kinetic energy of the object rotating about an axis passing through the center of mass.* We had already derived this result [see Eq. (9–39)] for the special case of rolling objects. Now we see that it is more general. This result is useful in calculations involving a combination of rotational and linear motion, as Example 10–7 shows.

EXAMPLE 10–7 A spool of mass M, rotational inertia I about its axis, and radius R unwinds under the force of gravity (Fig. 10–23). By using energy considerations, find the speed of the spool's center of mass after it has unwound a length h of thread.

Solution: The total kinetic energy of the spool is the sum of the kinetic energy of motion of the center of mass and the kinetic energy of rotation about the center of mass. Thus

$$K = \frac{1}{2}Mv^2 + \frac{1}{2}I\omega^2.$$

The relation between the angular velocity and the velocity of the center of mass is identical to that of a cylinder rolling without sliding, so that

$$v = R\omega.$$

This implies that

$$K = \frac{1}{2}MR^2\omega^2 + \frac{1}{2}I\omega^2 = \frac{1}{2}(MR^2 + I)\omega^2.$$

Note that a simple application of the parallel-axis theorem shows that the quantity in parentheses is just the rotational inertia about the point at which the thread separates from the surface.

Another contribution to the energy is the potential energy of gravity. We can measure the potential energy from the initial height of the spool so that if the spool falls a distance h, the potential energy is $-Mgh$ and the spool has gained a kinetic energy Mgh. With a zero initial angular velocity, the kinetic energy after the spool has fallen a distance h is

$$K = Mgh.$$

Thus

$$\frac{1}{2}(MR^2 + I)\omega^2 = Mgh,$$

so

$$v = R\omega = R\sqrt{\frac{2Mgh}{MR^2 + I}}.$$

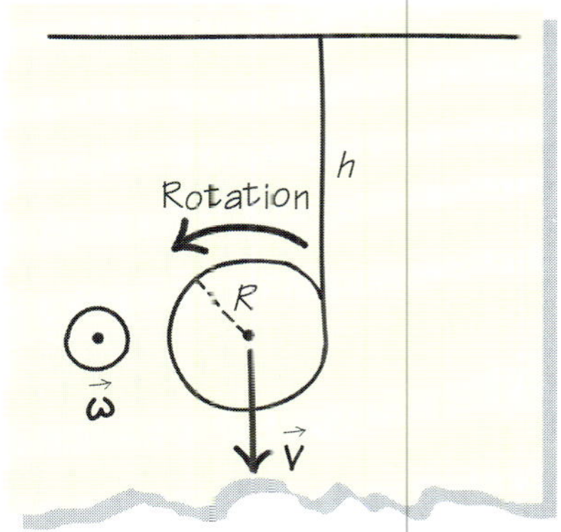

FIGURE 10–23 Example 10–7.

More Parallels Between Rotational and Linear Motion

Despite all the subtleties that appear when the full vector nature of angular momentum is involved, it is important to stress that there are no new laws of physics that describe rotational motion on the classical (the non-quantum-mechanical) level. *Every result on classical physics obtained in these last two chapters is a direct consequence of the application of Newton's laws of motion to aggregates of particles.* The rotational motion of both rigid and nonrigid objects has so many simplifying features that it can be described in terms of derived quantities such as angular velocity, angular momentum, and torque, but the basic laws are not new. Thus the analogies between quantities that appear in classical linear motion and those that appear in classical rotational motion are not accidental; instead, they are consequences of the fact that aggregate systems are made up of individual pieces that separately obey the laws of linear motion discovered by Newton (Table 10–1).

*10–6 QUANTIZATION OF ANGULAR MOMENTUM

When systems with atomic dimensions or smaller are involved, the effects of quantum physics become important. One such effect is that *angular momentum can have only certain discrete values*—we say that it is *quantized*. Consider an electron that is or-

TABLE 10–1 Analogies Between Linear and Rotational Motion

Linear Motion	*Rotational Motion*
Infinitesimal linear displacement: $\mathbf{dr}$	Infinitesimal angular displacement: $d\theta$
Velocity: $\mathbf{v}$	Angular velocity: $\boldsymbol{\omega}$
Acceleration: $\mathbf{a} = \dfrac{d\mathbf{v}}{dt}$	Angular acceleration: $\boldsymbol{\alpha} = \dfrac{d\boldsymbol{\omega}}{dt}$
Momentum: $\mathbf{p} = m\mathbf{v}$	Angular momentum: $\mathbf{L} = \mathbf{r} \times \mathbf{p}$
Force: $\mathbf{F} = \dfrac{d\mathbf{p}}{dt}$	Torque: $\boldsymbol{\tau} = \mathbf{r} \times \mathbf{F} = \dfrac{d\mathbf{L}}{dt}$
Impulse: $\Delta\mathbf{p} = \mathbf{F}\,\Delta t$	Angular impulse: $\Delta\mathbf{L} = \boldsymbol{\tau}\,\Delta t$
Kinetic energy: $\frac{1}{2}mv^2$	Kinetic energy: $\frac{1}{2}I\omega^2$
Work: $\int \mathbf{F} \cdot \mathbf{dr}$	Work: $\int \boldsymbol{\tau} \cdot d\boldsymbol{\theta}$
Power: $\mathbf{v} \cdot \mathbf{F}$	Power: $\boldsymbol{\omega} \cdot \boldsymbol{\tau}$

The quantization of angular momentum

biting the nucleus of an atom, under the influence of the central force due to the nucleus. The electron has an angular momentum, **L**, about an axis perpendicular to the plane in which it orbits, much like a planet orbiting the Sun. In Newtonian physics, angular momentum can have any magnitude at all. Niels Bohr proposed in 1913 that the angular momentum can only have a component in the direction perpendicular to the plane of motion given by

$$L_z = n\hbar \tag{10–42}$$

(in the z-direction, here). In this quantum mechanical expression, n has *only* the integer values 0, ± 1, $\pm 2, \ldots$, and $\hbar$ is Planck's constant, h, divided by 2π, with value $\hbar \simeq 10^{-34}\ \mathrm{J \cdot s}$.

Equation (10–42) must apply to macroscopic as well as microscopic systems. However, it is not in conflict with our experience about macroscopic systems because $\hbar$ is so small. A particle of mass 0.001 mg rotating about an axis at a distance of 1 μm with angular velocity 1 rad/s has a value of L_z given by

$$L_z = mvr = m\omega r^2 = (10^{-9}\ \mathrm{kg})(1\ \mathrm{rad/s})(10^{-6}\ \mathrm{m})^2 = 10^{-21}\ \mathrm{J \cdot s}.$$

This value corresponds to 10^{13} units of $\hbar$. The quantization is not observable because its observation would require a determination of the radius or the angular velocity to an accuracy of one part in 10^{13}. In molecular or atomic systems, however, the masses and distances are very tiny, and the values of the classical angular momentum are of the same order of magnitude as $\hbar$. For example, consider a molecule of nitrogen, N_2. Its classical angular momentum is given by $I\omega$. For the purpose of determining I, we can consider the molecule to form a tiny dumbbell made of two nuclei of mass M a distance d apart.† The rotational inertia for rotations about an axis perpendicular to, and halfway down, the line connecting the nuclei is $I = Md^2/2$. The mass of a nitrogen nucleus is the mass of 14 protons, $M = 14\,(1.6 \times 10^{-27}\ \mathrm{kg}) = 2.2 \times 10^{-26}\ \mathrm{kg}$. The typical distance between the nuclei of any molecule is about 10^{-10} m, so $I \simeq 10^{-46}\ \mathrm{kg \cdot m^2}$. To determine L, we use the expression for the kinetic energy of the system, from Eq. (9–35): $K = L^2/2I$. The typical kinetic energy of a molecule is in the range of $10^{-2}\ \mathrm{eV} \simeq (10^{-2}\ \mathrm{eV})\,(1.6 \times 10^{-19}\ \mathrm{J/eV}) \simeq 1.6 \times 10^{-21}\ \mathrm{J}$. Thus we estimate a classical angular momentum

$$L = \sqrt{2IK} \simeq \sqrt{2(10^{-46}\ \mathrm{kg \cdot m^2})(1.6 \times 10^{-21}\ \mathrm{J})} \simeq 5.6 \times 10^{-34}\ \mathrm{J \cdot s}.$$

This value of the angular momentum is between 5 and 6 units of $\hbar$. The difference between 5 and 6 units of $\hbar$ corresponds to a 10- to 20-percent difference in energy and the fact that the angular momentum is quantized—it has only the discrete values dic-

†The electrons are some 10,000 times lighter than the nuclei and can be ignored.

tated by quantum mechanics—is easily detectable in experiments that measure the energies of the molecule, such as those described next.

Quantization of Energy

Equation (9–35) shows that if the angular momentum is quantized, then *the allowed values of energy of a system such as the N_2 molecule or of an electron orbiting an atomic nucleus are quantized as well.* That is because the energy of an orbiting electron can be expressed in terms of L.

When the system has one of its possible values of energy, corresponding to different values of n, it is said to be in an **allowed state**. The possible values of energy can then be labeled by the particular value of n—as E_n—and we then refer to a particular **energy level**. When a molecule or an orbiting electron absorbs or emits energy as a result of collisions or other interactions, then this change can occur only in discrete jumps corresponding to changes from one state to another. For example, the electron of an atom may lose energy by jumping from a state labeled n to a state labeled $n - 1$. When it does so, the conservation of energy requires that the energy loss be compensated by an energy gain somewhere else. This energy appears as radiation (light), with a frequency (color) given by

$$f_{\text{rad}} \sim E_n - E_{n-1}. \tag{10–43}$$

Thus if the allowed energies are discrete, the frequencies of Eq. (10–43) take on only discrete values. Equation (10–43) then provides us with the means to detect the effects of the quantization of angular momentum. Atoms or molecules initially excited (moved into an allowed state with an increased energy) by collisions or by impinging radiation will jump spontaneously to lower energy levels, one step at a time, emitting radiation (light) in the process. Thus light is emitted with characteristic frequencies determined by equations such as Eq. (10–43). The frequency of this radiation can be measured. It was the observation that atoms and molecules emit light with discrete frequencies that provided one of the keys to the development of quantum mechanics.

The consequences of the discreteness of atomic and molecular energies are momentous. For example, there is a minimum energy—an *energy gap*—required to move an atomic system such as a hydrogen atom up the "quantum ladder" from the lowest state, characterized by $n = 1$, to states with higher n. It is the relatively large size of the energy gap that explains the stability of atoms. It is difficult to excite a hydrogen atom. The fact that even large molecular systems have energy gaps is crucial to the stability of biological systems. If there were no energy gaps, even the slightest perturbation would change molecular systems in important ways.

*10–7 PRECESSION

Recall the motion of a top: There is a rapid rotation about the symmetry axis: **precession**, which is a rotation of the symmetry axis itself around a vertical axis through the point of the top; and *nutation*, the name given to the up-and-down bobbing motion of the symmetry axis while it rotates about the vertical. The general motion of tops and of gyroscopes is beyond the scope of this book, but we can indicate with an example how precession arises.

In linear motion, there are two rather different ways that Newton's second law can be satisfied. First, the momentum can change its magnitude; this leads to changes in speed. Second, if the applied force is not parallel to the momentum, the momentum can change *direction* rather than magnitude, as in circular motion. The same is true for rotational motion. If the applied torque is parallel to the angular momentum, then we have changes in the magnitude of the angular momentum, and this situation represents the majority of the cases we have studied. But if the torque is in some other direction, then we can have changes in the direction of the angular momentum.

EXAMPLE 10–8 In making a science-fiction movie about an asteroid that collides with Earth, a studio constructs a model Earth attached firmly to a support fixed at the South Pole and spinning rapidly with ω in the $+z$-direction (Fig. 10–24a). A model asteroid approaches horizontally from the $+y$-direction and strikes a grazing blow with the model Earth just at the North Pole. Describe the new motion of the model Earth.

Solution: Because the model Earth is fixed at its pivot at the South Pole, we calculate the torque on the model Earth about this point. The collision delivers an impulse $\mathbf{J} = -J\mathbf{j}$ to the model Earth in the given coordinate system. According to

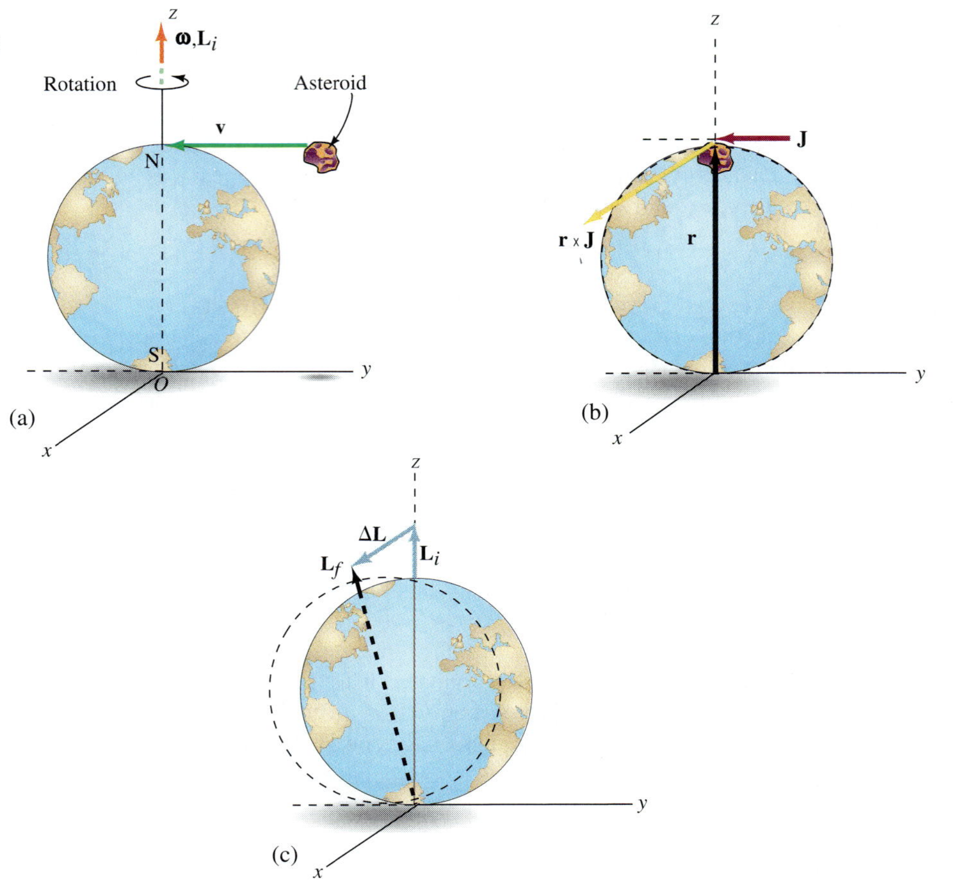

FIGURE 10–24 (a) Example 10–8. (b) The collision of the model asteroid along the $-y$-direction causes the spinning model Earth to tilt toward the $+x$-direction.

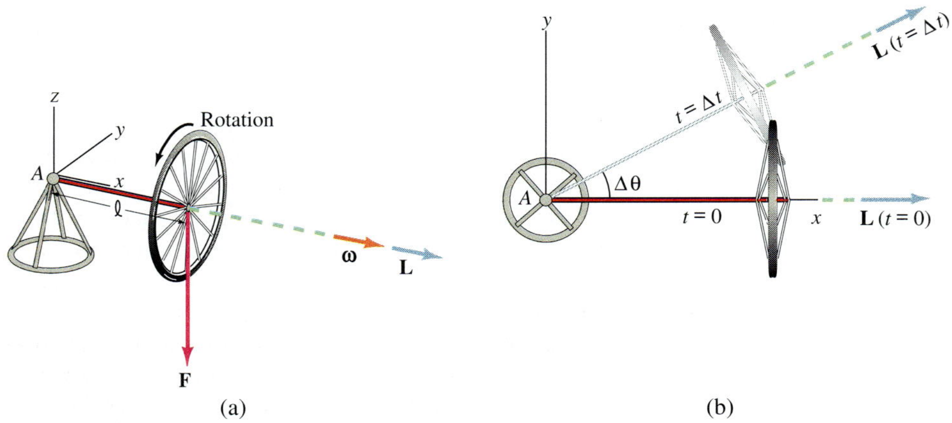

FIGURE 10–25 (a) A wheel rotates with angular velocity $\boldsymbol{\omega}$ about a massless horizontal shaft that can pivot about point A. The force of gravity, $\mathbf{F}_g = m\mathbf{g}$, acts at the center of the wheel. (b) As seen from above, the shaft has rotated an angle $\Delta\theta$ after a time Δt. The angular momentum vector has therefore changed direction.

Eq. (10–25), the angular impulse about the support point is $\mathbf{J}_\tau = \mathbf{r} \times \mathbf{J}$, where $\mathbf{r}$ is the vector from the fixed point to the point where the impulse is delivered. Because $\mathbf{r}$ is in the $+z$-direction, the vector product $\mathbf{r} \times \mathbf{J}$ is in the $+x$-direction. But this describes the change in angular momentum, $\Delta\mathbf{L}$, produced by the angular impulse. The initial angular momentum $\mathbf{L}_i$ is in the $+z$-direction, and when $\Delta\mathbf{L} = (+\Delta L)\mathbf{i}$ is added to this, the new angular momentum $\mathbf{L}_f$ corresponds to Earth having tilted toward the $+x$-axis (Fig. 10–24b). Note that this is *not* the direction from which the model asteroid comes.

Torque on a Spinning Wheel

As another example of the consequence of the vector nature of the dynamical equations of motion, consider a wheel of rotational inertia I rotating with angular velocity $\boldsymbol{\omega}$ about a horizontal shaft of length ℓ, which has one end pivoted at point A, and the other end free. The system is shown at $t = 0$ in Fig. 10–25a. If the sense of rotation of the wheel is as shown, then the angular momentum about the pivot initially has magnitude $L = I\omega$ and points in the x-direction. Now recall that a gravitational force of magnitude Mg, pointing in the $-z$-direction, acts on the center of mass of the wheel. The torque about point A due to this force has magnitude

$$\tau = Mg\ell, \qquad (10\text{–}44)$$

and it points in *the* $+y$*-direction* by the right-hand rule. Because

$$\Delta L = \tau\,\Delta t,$$

the angular momentum after a short time interval Δt is given by

$$\mathbf{L} = I\omega\mathbf{i} + Mg\ell(\Delta t)\mathbf{j}. \qquad (10\text{–}45)$$

Thus the angular momentum vector has rotated slightly in a counterclockwise direction, as seen from above. As Fig. 10–25b shows, the new angular momentum makes an angle

$$\Delta\theta = \frac{Mg\ell\,\Delta t}{I\omega} \qquad (10\text{–}46)$$

with the original direction. This is the angle of the shaft, and so we have seen that the direction of the shaft changes with time in a steady way. When the shaft of the spinning wheel steadily rotates with the pivot point fixed, we say that the shaft *precesses* (Fig. 10–26a). The angular velocity of the precession of the angular velocity vector determined from Eq. (10–46) is

$$\omega_p = \frac{\Delta\theta}{\Delta t} = \frac{Mg\ell}{I\omega} \qquad (10\text{–}47)$$

and is known as the *angular frequency of precession.* The precession frequency is, in fact, a vector that points in the $+z$-direction due to the right-hand rule. The precession frequency turns out to be independent of the angle the initial value of $\mathbf{L}$ makes with the vertical (see Problem 43). This formula applies for any top—not just a rotating wheel—if ℓ is the distance of the top's center of mass from the support point (Fig. 10–26b).

Our description of the precession of the wheel's shaft is only approximately correct. Energy considerations give rise to a perturbation to the simple precession just described, which consists of an up-and-down bobbing motion of the shaft below the xy-plane: a nutation (Fig. 10–27). The effect is small if the angular velocity of the wheel is large, or more precisely, if $\omega_p << \omega$. In this section. we have been able to point out only the simplest features of a top's motion. Nevertheless, the laws of rotational motion described in this chapter are sufficient to account for all these features.

(a)

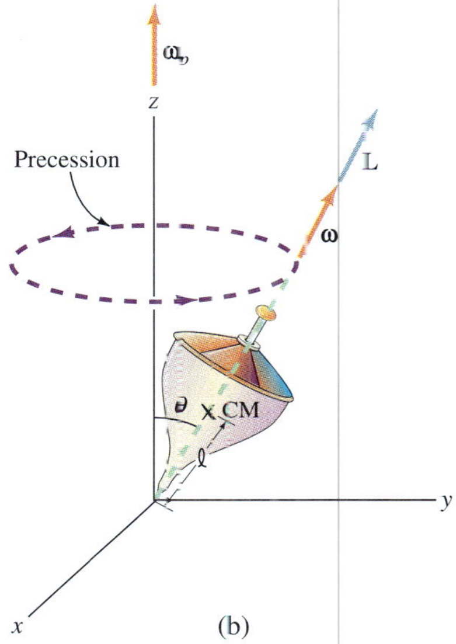

(b)

FIGURE 10–26 (a) Simple precession. (b) The precession of a top is governed by Eq. (10–45), where ℓ is the distance between the pivot point and the top's center of mass.

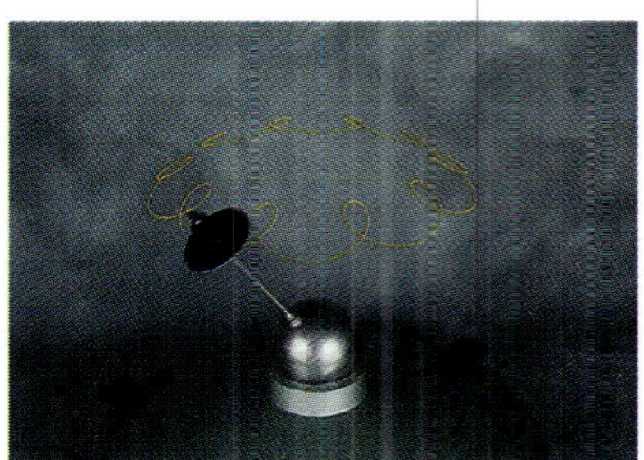

FIGURE 10–27 The interplay of gravity, angular momentum, and torque lead to perturbations on the precession of a top, resulting in the up-and-down bobbing motion called nutation.

SUMMARY

For a particle moving with momentum $\mathbf{p}$, the general form of the angular momentum about a point is

$$\mathbf{L} = \mathbf{r} \times \mathbf{p}, \tag{10–4}$$

where $\mathbf{r}$ is the position vector from the point to the particle. This result involves the vector product of the vectors $\mathbf{r}$ and $\mathbf{p}$ with angle θ between them. The vector product has magnitude $rp \sin \theta$, and its direction—perpendicular to both $\mathbf{r}$ and $\mathbf{p}$—is determined by a right-hand rule. The angular momentum of a system of particles, rigid or otherwise, about some point is a sum over terms such as Eq. (10–4).

A dynamical equation obeyed by extended systems follows from consideration of Newton's second law. It states that the rate of change of angular momentum about some point O is given by

$$\boldsymbol{\tau} = \frac{d\mathbf{L}}{dt}, \tag{10–12}$$

where $\boldsymbol{\tau}$ is the torque applied to the object. The torque is generally defined by

$$\boldsymbol{\tau} = \mathbf{r} \times \mathbf{F}, \tag{10–5}$$

where $\mathbf{r}$ is the vector from the point of reference to the point where the force, $\mathbf{F}$, is applied.

If a rigid body rotates about an axis A that does *not* go through its center of mass, then (1) the angular momentum about A is the sum of the angular momentum about A of a point mass carrying the whole mass of the object, as if it is located at the center of mass and moves with the velocity of the center of mass, plus the angular momentum of the object about the center of mass,

$$\mathbf{L} = (\mathbf{R} \times \mathbf{P}) + \mathbf{L}_{cm}; \tag{10–16}$$

(2) the total torque about A is the sum of the torque about A due to the total external force applied to the center of mass and the torque about the center of mass,

$$\boldsymbol{\tau} = (\mathbf{R} \times \mathbf{F}_{tot}) + \boldsymbol{\tau}_{cm}; \tag{10–17}$$

and (3) the rate of change of each part of the total angular momentum is equal to the corresponding part of the total torque.

In the absence of torque, angular momentum is constant throughout the motion, and this is an important tool in solving problems of motion of extended systems, especially nonrigid ones. It also helps us understand the motion of particles. The torque will naturally be zero in the absence of external forces, but it also is zero for central forces. As a consequence, in the orbit of an object moving under the influence of a central force, the areas swept out by position vectors from the source of the force to the moving object in equal times are equal.

The kinetic energy of rotational motion is given by $K = I\omega^2/2$, and the change in K equals the work done—just as in linear motion. For rotational motion about an axis, this relation (the work–energy theorem) reads

$$W = \int_{\theta_0}^{\theta} \tau \, d\theta = K - K_0, \tag{10–38}$$

where K_0 is the initial rotational energy at the angle θ_0. In many problems both rotational and linear motion are present. The total energy of a rigid body may be written as the sum of the linear kinetic energy, calculated as if all the mass of the object were concentrated at the center of mass, and the rotational energy of the object about the center of mass:

$$K = \frac{1}{2}MV^2 + \frac{1}{2}I_{cm}\omega^2. \tag{10–41}$$

Angular momentum in atomic systems is quantized in the form

$$L_z = n\hbar = n\frac{h}{2\pi}, \qquad (10\text{–}42)$$

where h is Planck's constant and $n = 0, \pm 1, \pm 2, \ldots$. The consequences of this quantization include the result that not all energy values are permitted for atoms and molecules.

Many subtle effects are associated with the effect of torque on angular momentum, and we illustrated one of them: the precession of angular momentum for a top. If the top is spinning while inclined from the vertical, then gravity produces a torque on the top that causes its axis to precess about the vertical with a precession frequency

$$\omega_p = \frac{Mg\ell}{I\omega}. \qquad (10\text{–}47)$$

UNDERSTANDING KEY CONCEPTS

1. Angular momentum is sometimes given in units of joule-seconds (J · s). Is this a correct SI unit for angular momentum?
2. A comet is heading at high speed straight into the center of the Sun. Why is the angular momentum of the comet with respect to the Sun zero?
3. When a quarterback throws a football, he tries to put quite a bit of spin its axis. Since some of the energy put into the ball then goes into rotational motion, this means that there is less energy for translational motion. Why is it nevertheless done?
4. In taking a fast corner on a bicycle, it is safer if you crouch as low as possible. Why is that?
5. Films from Skylab show the astronauts reorienting themselves as they float weightlessly by spinning their arms. Is this consistent with the conservation of angular momentum?
6. A long, flexible, heavy bar can be very useful to a tightrope walker. Why?
7. Why is it easier on your back for you to lift heavy objects by bending your knees, keeping your back straight, and straightening your knees, rather than by picking up the object as you bend over from your waist with your legs straight?
8. An astronaut floating in a space station holds the axle of a rotating wheel. When the astronaut is vertical and the axle points away from her, she suddenly rotates the axle to point up. What happens to the astronaut?
9. A diver prepares for a complex set of midair maneuvers. In springing off an elastic diving board, the diver wishes to acquire angular momentum relative to her center of mass. What should she strive for in the take-off?
10. In Example 10–3, we studied a tether ball and found that as the rope tying it to the pole changes, so does the angular momentum. Where does the corresponding torque come from?
11. It is possible to tell whether an egg is hardboiled without cracking it open by setting it in rotation on a table. Discuss what you would find, and why. Compare the behavior of the hardboiled egg to that of a raw one.
12. If you were to tie a rock to a rope and swing it in a horizontal circle above your head, you could rather easily do so without spinning around yourself. Is this a violation of the conservation of angular momentum?
13. The center of mass of an object accelerates as the result of an impulse (a brief force). If the object is extended, the impulse may also be an angular impulse. Is it possible to have an angular impulse without a linear impulse?
14. A diver executes a series of midair maneuvers. To do so, is it necessary for the diver to give herself some angular momentum about her center of mass?
15. We saw that the torque–angular momentum relation is a consequence of Newton's laws. Is it also true in Newtonian mechanics that the conservation of angular momentum follows from Newton's laws?
16. You are given a stool of known rotational inertia that can rotate with minimal friction. You are also given a stopwatch, a very light meter stick, and two known masses that can slide along the meter stick. How would you use this apparatus to measure the rotational inertia about the rotational axis of a person sitting on the stool?
17. The propeller of a single-engine propeller airplane rotates clockwise, as seen from the cockpit. The plane makes a slow turn to the right. What else happens?
18. If only one set of hand brakes on your bicycle works when you descend a steep downhill slope, which set would you prefer?
19. A woman stands on the edge of a freely rotating platform. She walks toward the center along a radius. Will the speed of rotation of the platform change? If so, in what way, and what is the source of the torque?
20. A horizontal platform is rotating at a certain speed. A boy jumps onto the platform from an overhanging tree branch. He lands with both feet straddling the center and remains standing. Will the platform speed up, slow down, or neither?
21. A boat can be stabilized against rolling by attaching a large flywheel to the sides (Fig. 10–28). The attachment point is above the waterline. Both ω and the rotational inertia, I, about the axis are large. (a) A wave hits the boat on the side. The wave would tend to rock the boat or rotate it about its longitudinal axis without the flywheel. With the flywheel installed, how does the boat react to this wave? (b) A wave comes straight at the bow (the front). It would lift the bow of the boat without the flywheel. With the flywheel, what happens?

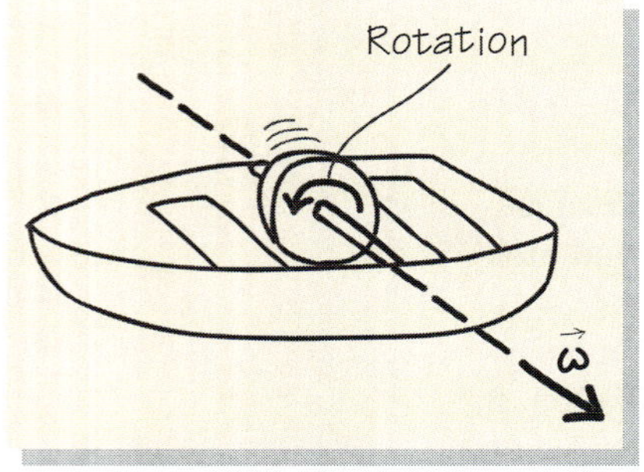

FIGURE 10–28 Question 21.

22. When jugglers perform juggling acts, the props they use are frequently put into a spinning motion when they are tossed into the air. Why is this a good idea?
23. A baseball is thrown horizontally, hits a rough floor, and bounces. The ball is spinning in the direction of its motion (like a rolling bicycle wheel). Which of the following is true: When the ball leaves the floor, it does so at (a) the same angle at which it struck the floor; (b) a larger angle with respect to the floor; (c) a smaller angle with respect to the floor?
24. Think about the demonstration in which someone spinning on a stool pulls in his or her arms and speeds up because of angular momentum conservation. In the process, does the energy of rotation decrease, remain constant, or increase? Explain.

PROBLEMS

10–1 Generalization of Angular Momentum

1. (I) An airplane of mass 2000 kg located 100 km north of New York City is flying 200 km/h in an easterly direction. (a) What is its angular momentum with respect to New York City? (b) What if it is flying in a northeasterly direction?
2. (I) What is the angular momentum about the origin of a particle of mass 200 g at position $\mathbf{r} = (2\mathbf{i} - 3\mathbf{j} + \mathbf{k})$ m, moving with a velocity of $\mathbf{v} = (5\mathbf{i} + 15\mathbf{j} - 10\mathbf{k})$ m/s?
3. (I) You are standing on the corner of Main Street and Elm, watching the cars on Main pass at a steady 10 m/s. You stand 5 m from the line of traffic. You watch a red convertible, whose mass is about 1000 kg, from the moment it is one block away, a distance of 200 m. Treat the car as pointlike, and assume that Main Street is straight. (a) What are the car's angular momentum, magnitude, and direction, with respect to you when it is one block away? (b) What is its angular momentum with respect to you when it passes your position on the corner?
4. (I) Assuming that each object is pointlike, how large is the angular momentum (a) of Earth about the Sun; (b) of the Moon about Earth? (c) Compare these results to the angular momentum of Earth (no longer pointlike) about its own axis; assume constant density.
5. (I) A bicycle travels east. The mass of the wheel, 1.8 kg, is uniformly distributed along the rim, with the mass of the hub and spokes negligible. If the radius of the wheel is 18 cm and the wheel rotates at the rate of 4.2 rev/s, what is the direction and the magnitude of the angular momentum of the wheel about its axis? (Neglect the width of the rim.)
6. (II) Calculate the angular momentum about the origin of a particle of mass m moving along the trajectory $y = ax + b$ with uniform speed v.
7. (II) A rock of mass 60 g is thrown with initial horizontal speed $v_x = 25$ m/s off a building from a height of 30 m. Calculate the angular momentum of the rock about the line along the edge of the roof as a function of time.
8. (II) A unicycle has a wheel of mass 3.5 kg, rotational inertia $0.45\ \text{kg}\cdot\text{m}^2$ about the axle, and radius 0.4 m. What is its angular momentum with respect to a point on the road if the wheel rolls without slipping with an angular velocity of 8 rad/s?
9. (II) The position vector of an object of mass m subject to two constant forces that act at right angles is given by $\mathbf{r} = (\frac{1}{2}at^2)\mathbf{i} + (vt)\mathbf{j} + (\frac{1}{2}bt^2 - wt)\mathbf{k}$. Calculate the angular momentum of this object about the origin.
10. (II) An object of mass M moves in a path given by $\mathbf{r} = (x_0 + \rho\cos[\omega t])\mathbf{i} + (y_0 + \rho\sin[\omega t])\mathbf{j}$. What is the angular momentum of the object about the origin?
11. (II) Consider two objects whose position vectors are given by $\mathbf{r}_1$ and $\mathbf{r}_2$ and whose momenta are given by $\mathbf{p}_1 = m_1\mathbf{v}_1$ and $\mathbf{p}_2 = m_2\mathbf{v}_2$, respectively (Fig. 10–29). Show that in the special case that the center of mass of the two bodies is at rest at the origin (that is, $\mathbf{P} = \mathbf{p}_1 + \mathbf{p}_2 = 0$, and the position of the center of mass is $\mathbf{R} = 0$), the sum of the angular momenta of the two objects about the center of mass equals the angular momentum of a single object of mass $\mu = m_1m_2/(m_1 + m_2)$, rotating in circular motion about the origin at a distance $r = r_2 - r_1$. The quantity μ is called the *reduced mass*. [*Hint*: Introduce $\mathbf{r} \equiv \mathbf{r}_2 - \mathbf{r}_1$, and express $\mathbf{p}_1$ in terms of $d\mathbf{r}/dt$.]

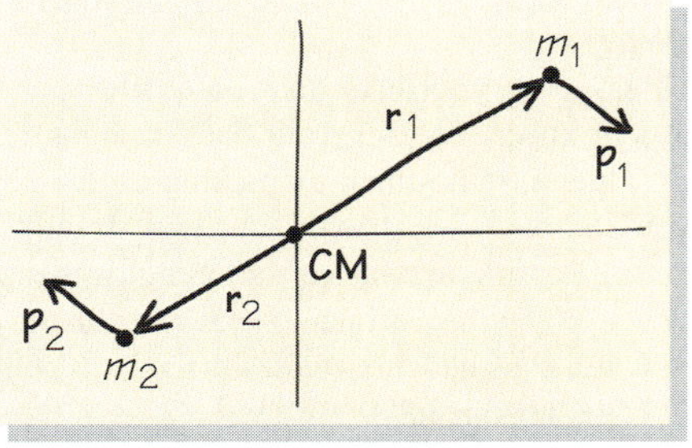

FIGURE 10–29 Problem 11.

12. (II) Earth is not a point object but is a sphere with a rotational inertia of $9.8 \times 10^{37}\ \text{kg}\cdot\text{m}^2$ about its axis. Assume that Earth's axis of rotation is parallel to the axis of the orbital motion of Earth around the Sun and that the Sun is so massive that it can be considered to be fixed. Calculate (a) the rotational inertia of Earth about the axis of its orbital motion around the Sun; (b) the total angular momentum of Earth about that same axis. (c) Calculate the fractional difference between your result for part (b) and the angular momentum you would find for part (b) if Earth were pointlike; explain why the difference is small.
13. (II) A square, 20 cm on the side, is made of very light sticks. Four identical masses of $m = 0.1$ kg form the corners of the

square. The square rotates with an angular velocity of 8 rad/s about an axis perpendicular to its plane through the center of the square. (a) Calculate the rotational inertia of the system about the rotation axis and use it to find the angular momentum about this axis. (b) Use the general definition of angular momentum to calculate the angular momentum of each mass with respect to the center of the square, and add these up. Compare the results of (a) and (b).

14. (II) Consider the square studied in Problem 13. Calculate the angular momentum of each particle about a point on the axis of rotation 25 cm below the plane of the square. Compare the total angular momentum calculated in this way with the results of Problem 13.

15. (II) Three identical masses m are attached to the corners of an equilateral triangle of sides d. Calculate the angular momentum (a) if the triangle rotates at an angular velocity ω about the center of mass around an axis perpendicular to the plane of the triangle; (b) if the triangle rotates with angular velocity ω about one of its sides; (c) if the triangle rotates about an axis going through one of its vertices to the midpoint of the opposite side.

10–2 Generalization of Torque

16. (I) A construction worker of mass 72 kg stands at the end of a 3.4-m-long (massless) horizontal mast attached to a building. What is the magnitude of the torque exerted on the hinge that holds the mast fixed in position?

17. (I) What is the torque about the origin on a particle positioned at $\mathbf{r} = (3\mathbf{i} - \mathbf{j} - 5\mathbf{k})$ m, exerted by a force of $\mathbf{F} = (2\mathbf{i} + 4\mathbf{j} + 3\mathbf{k})$ N?

18. (I) A flagpole 2.4 m long is attached to a building. It makes an angle of 30° with the horizontal. A mass of 65 kg is suspended from the end. What is the torque acting on the point of attachment to the building due to the suspended mass?

19. (I) A 1.0-m-long massless stick lies on a table. A force of 200 N is applied to one end of the stick for 0.1 s at an angle of 45° to the stick, and in a direction pointing away from the center of the stick. What is the torque about the other end of the stick during the brief period that the force is acting?

20. (II) Consider two forces **F** and −**F** that act at different points on an extended object. Show that the net force of this combination is zero and that the torque about any point P is independent of the location of P, and depends only on the separation of the two points at which the forces act.

10–3 The Dynamics of Rotation

21. (I) A point mass M is attached to a turntable at a distance R from the center (Fig. 10–30). The turntable rotates with constant angular speed ω about its axis. If the axis is horizontal, so that the turntable rotates in a vertical plane, what is the torque that the force of gravity on the mass exerts about the axis as a function of time? Assume that the mass is at the topmost position at $t = 0$.

22. (II) The position of a ball of mass m thrown from a building is given by $\mathbf{r} = (v \cos \theta t)\mathbf{i} + (v \sin \theta t - gt^2/2)\mathbf{j}$, measured from the point from which the ball was thrown. What is the torque about the origin that the force of gravity exerts on the ball?

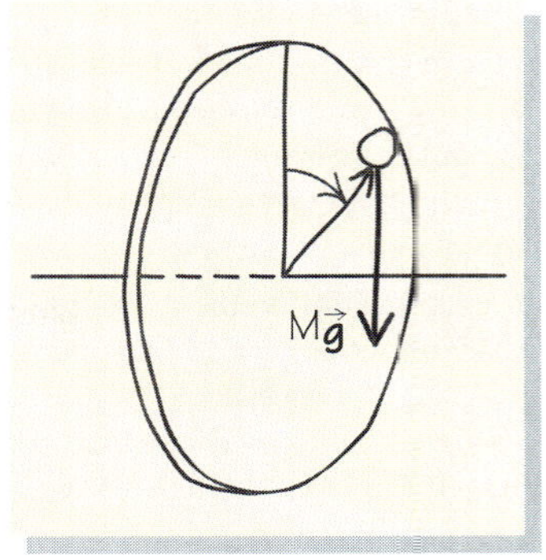

FIGURE 10–30 Problem 21.

23. (II) A ball of mass m slides at speed v on a frictionless horizontal surface and bounces elastically from a wall. The initial path of the ball makes an angle θ with the wall (Fig. 10–31). Find the initial and final angular momenta of the ball about the point A. What causes the change in angular momentum?

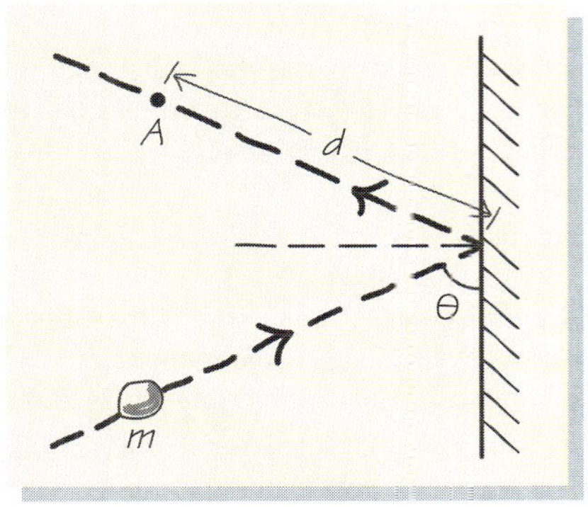

FIGURE 10–31 Problem 23.

24. (II) A pulley system is used to lift a heavy mass. How much force must be applied to lift the object in Fig. 10–32 at a steady speed? Neglect friction at the axle.

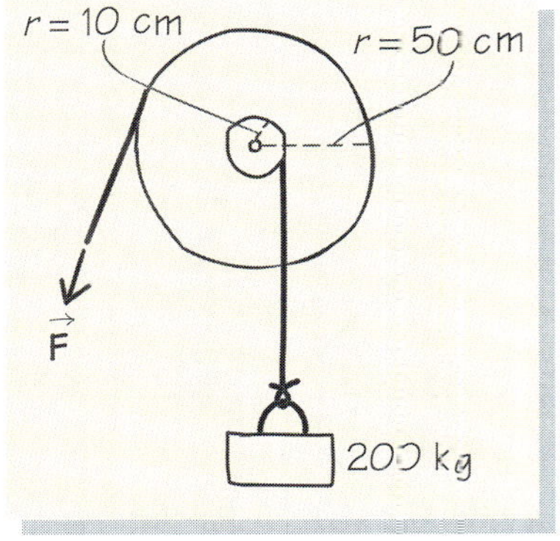

FIGURE 10–32 Problem 24.

25. (II) What is the vector product of $\mathbf{A} = 2\mathbf{i} - 4\mathbf{j} + 5\mathbf{k}$ and $\mathbf{B} = \mathbf{i} + 3\mathbf{j} - 2\mathbf{k}$?

26. (II) Show that the magnitude of the vector product of two vectors is the area of the parallelogram for which the two vectors form adjacent sides.

10–4 Conservation of Angular Momentum

27. (I) A playground merry-go-round of diameter 3.4 m and rotational inertia 120 kg · m^2 is pushed with no one on it by three children to an angular speed of 2.5 rad/s. Two of the children, of mass 25 kg each, jump on the edge of the merry-go-round, coming radially in. What is the new angular speed?

28. (I) A uniform disk rotating without friction about its (vertical) central axis, with total mass 138 kg and radius 1.5 m, acts as a turntable. Its angular speed is $\omega = 1.2$ rad/s. A person of mass 74 kg jumps straight down onto the rotating turntable. The person lands 1.2 m from the axis. What is the new angular speed of the turntable?

29. (I) A firetruck, mass 6000 kg, passes a parked car on a straight street at $t = 0$ with speed 15 m/s. A physics graduate student finds that, 10 s later, the angular momentum of the firetruck with respect to the parked car is twice its value at $t = 0$. What is the speed of the truck at $t = 10$ s?

30. (I) A skater twirls at 1.0 rev/s with her arms extended and holds a 5-kg mass in each hand; each mass is 0.9 m from the axis of rotation. She pulls the masses in along the radial direction until they are 0.3 m from the axis of rotation. Assuming that the rotational inertia of the arms is negligible and that the rotational inertia of the skater without the masses is 2.9 kg · m^2, what is the speed of rotation after the masses have been pulled in?

31. (II) A bug of mass $m = 2.0$ g walks around a horizontal turntable, which may be viewed as a uniform cylinder of mass $M = 0.24$ kg. If both the turntable and the bug are initially at rest, how much does the turntable rotate relative to the ground while the bug makes one full circle relative to the turntable?

32. (II) A small mass of 0.5 kg slides down a frictionless slope starting from rest at 1.6 m above the ground level. When the slope reaches the bottom, it levels off, and the mass strikes the bottom of a vertical uniform bar of mass 3.2 kg and length 1.0 m, pivoted at its midpoint, and sticks to it (Fig. 10–33). With what angular speed will the bar start its rotation?

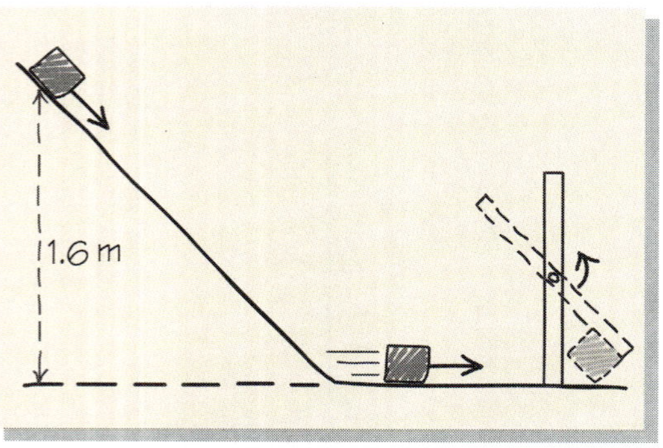

FIGURE 10–33 Problem 32.

10–5 Work and Energy in Angular Motion

33. (I) An airplane engine develops 240 hp while turning the propeller at 3200 rev/min. What is the torque exerted on the propeller axis by the engine?

34. (I) A flywheel has radius 1.2 m and mass 680 kg, almost all of which is concentrated on the rim. It is spinning at $\omega = 4.5$ rad/s about its axis when a torque is applied along the axis, producing an angular acceleration of 0.3 rad/s^2. How much time does it take for ω to increase to 6 rad/s, and how much work is done by the torque during this time?

35. (I) The work done to slow down a spool used in a manufacturing process during a certain period is 1200 J. The spool is fixed on an axis, and has rotational inertia of 0.033 kg · m^2 about that axis. The spool is rotating at an angular speed of 490 rad/s before the specified period. What is the spool's angular speed at the end of the period?

36. (II) Repeat the problem of the cylinder rolling down the inclined plane treated in Section 9–7, but this time use energy techniques to learn the speed of the cylinder as a function of distance traveled.

37. (II) A cylinder of mass 0.2 kg rolls without slipping down an inclined plane of 15°. What is its rotational kinetic energy after it rolls 80 cm?

38. (II) A particular flywheel used for the storage of energy is a solid steel cylinder of density 8 g/cm^3. The cylinder has a radius of 1.4 m, is 30 cm thick, and spins about its axis at a frequency of 320 rev/min. (a) What is the rotational inertia of the flywheel? (b) What is the work the flywheel can do in being brought to a halt, if there is no energy loss to frictional forces?

*10–6 Quantization of Angular Momentum

39. (II) Consider an object of mass m that moves in a circular orbit caused by a central force given by $F = -kr$. Suppose that the Bohr quantization condition is applied to this motion. What are the allowed quantized radii, velocities, and kinetic energy values? [*Hint*: The acceleration for circular motion is v^2/r, and $E = (mv^2/2) + U(r)$.]

40. (II) Repeat the calculation of Problem 39 for an object that moves in a circular orbit caused by a central constant attractive force, obtained from the potential energy $U(r) = Cr$.

41. (II) The energy of the hydrogen atom, when quantized, is given by $E_n = -(13.6\text{ eV})/n^2$, where $n = 1, 2, 3, \ldots$. Particles of energy 2.0 eV repeatedly pass through a gas of hydrogen atoms in the $n = 1$ state but never excite them out of this lowest-energy state. Explain. What energies would excite the hydrogen atoms?

42. (II) A proton has mass 1.67×10^{-27} kg and "radius" 1.3×10^{-15} m. (We put the radius in quotation marks because the proton is not a classical object with a radius, like a baseball.) It also can be thought of as having angular momentum with respect to an internal axis of 0.5×10^{-34} kg · m^2/s. Take as a model that the proton is a uniform sphere. (a) What, according to the model, is the angular frequency of the proton's rotational motion? (b) What is the speed with which the outermost portion of the proton, on its equator, moves? Compare this to the speed of light, 3×10^8 m/s. (c) What is the energy associated with the rotational motion of the proton?

*10–7 Precession

43. (II) By repeating the derivation of the precession frequency of a top, but with the top making an angle ϕ with respect to the vertical, show that the precession frequency is *independent* of the angle ϕ.

44. (II) A student sits on a piano stool that is not rotating. She holds a vertical shaft on which a bicycle wheel of rotational inertia I is mounted and rotates with an angular speed ω oriented upward. She wants to tilt the wheel away from herself in a radial direction. What is the direction of the torque that she must exert? Suppose she succeeds in reversing the direction of the shaft by 180°. What will the speed and direction of rotation of the student and the piano stool be, assuming that the rotational inertia of the student and stool together is I^*?

45. (II) A wheel with massless spokes has mass 1 kg and radius 10 cm and is mounted on one end of a massless axle (Fig. 10–34). The axle rests on a pivot at a point 16 cm from the mounting point and 10 cm from the wheel. At the other end a mass of 0.8 kg is attached. The wheel spins at an angular frequency of 10 rad/s. What is the rate of precession?

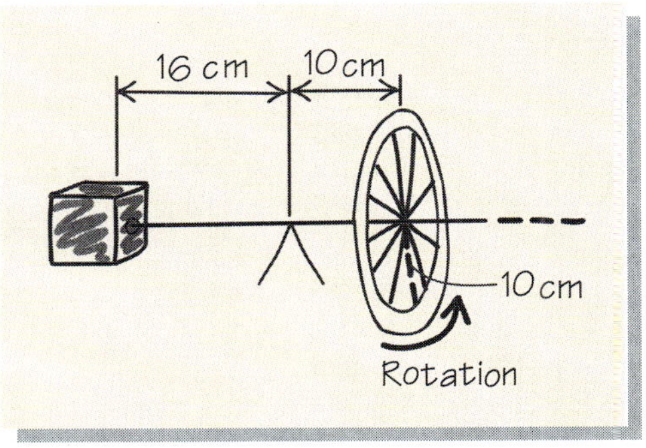

FIGURE 10–34 Problem 45.

General Problems

46. (I) An 80-g ball is thrown out of a second-story window 8.0 m above the ground. The initial velocity of the ball is horizontal, with magnitude 2.4 m/s. What is the angular momentum of the ball, as a function of time, about the point on the ground directly below the window?

47. (II) A solid cylinder of mass 0.85 kg and radius 4.2 cm initially at rest rolls down a plane inclined at 28° with the horizontal and 1.5 m long. Use energy conservation to calculate the angular velocity of the cylinder at the bottom of the ramp; assume that all the kinetic energy of the cylinder is in rolling motion (that is, there is no sliding).

48. (II) A constant-density cylinder of mass 0.5 kg and radius 4 cm can rotate freely about an axis through its center. It has thread wound around an attached axle of radius 0.5 cm that also runs through its center (Fig. 10–35). The thread is attached to a mass of 1 kg, which slides down an inclined plane with an acceleration of 0.1 m/s^2. What is the coefficient of kinetic friction between the block and plane?

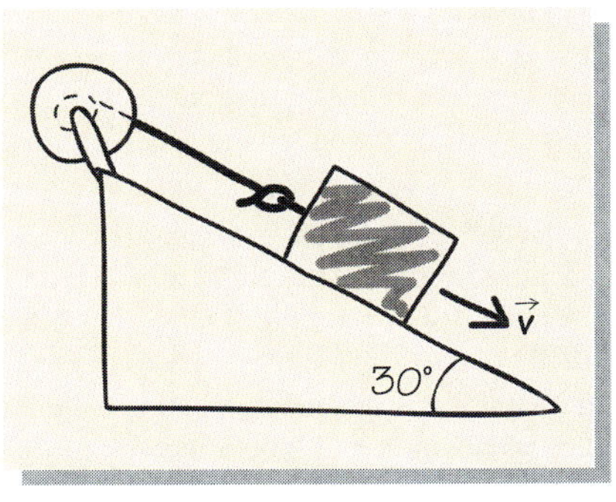

FIGURE 10–35 Problem 48.

49. (II) A child of mass 32 kg stands at the center of a platform of radius 2 m and rotational inertia 450 kg · m^2. The circular platform rotates about a frictionless shaft with angular speed of 0.8 rad/s. The child walks in a radial direction until he reaches the rim. What will the angular velocity of the platform be when that happens? What is the change in energy of the platform plus child? Identify the source of the work responsible for the change in rotational kinetic energy.

50. (II) A bullet of mass 5.0 g and velocity 330 m/s passes through a wheel at rest (Fig. 10–36). The wheel is a solid disk of mass 2.0 kg and radius 18 cm. The bullet passes through the wheel at a perpendicular distance of 15 cm from the center, and the bullet's final velocity is 220 m/s. What are the wheel's angular velocity, angular momentum, and kinetic energy? Is energy of motion conserved?

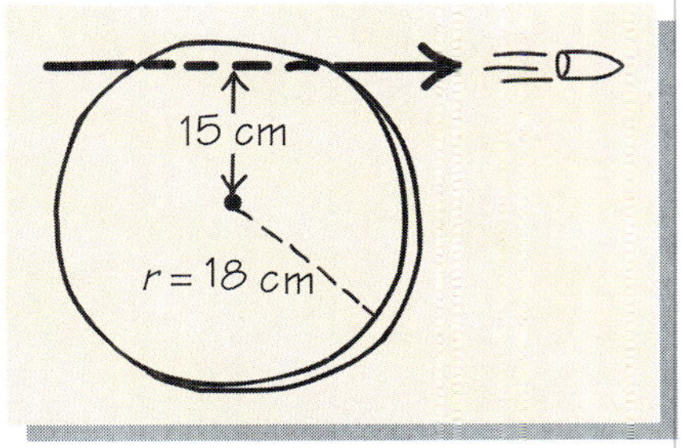

FIGURE 10–36 Problem 50.

51. (II) Figure 10–37 shows the Atwood's machine treated in the problems of Chapter 5, with a rope of negligible mass 2 m long. Earlier the pulley was treated as massless. Now suppose that the pulley can be approximated by a solid disk of radius 0.1 m and mass 2 kg. The system is released from rest with the 4-kg mass 1.5 m from the floor and the 1-kg mass on the floor. (a) What is the speed of either block just before the 4-kg mass hits the floor? (b) How long does it take the 4-kg mass to reach the floor?

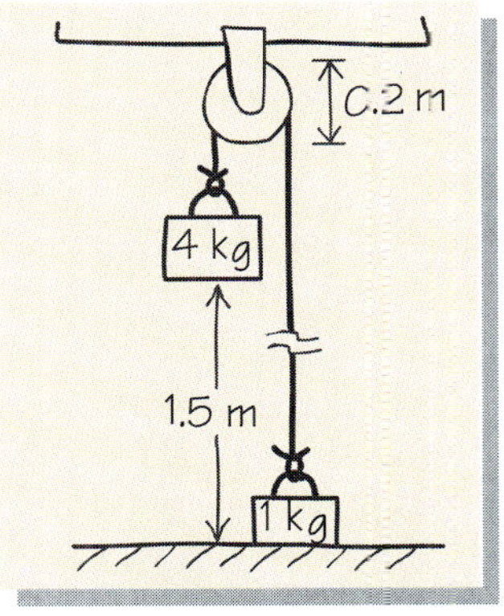

FIGURE 10–37 Problem 51.

52. (II) Two weights of mass 1.2 kg and 0.85 kg, respectively are connected by a massless string that passes over a pulley. The pulley is a hollow cylinder of radius 18 cm and mass 0.45 kg. What are (a) the acceleration of the system and (b) the time that it takes for the larger mass to descend a distance of 1.6 m, if the weights start from rest?

53. (II) A ball of mass M, radius R, and uniform density falls from rest from the top of a hemispherical bowl (Fig. 10–38). The left side of the bowl is frictionless, but the right side has a large coefficient of friction with the ball, and for all practical purposes the ball immediately rolls without slipping. How far up the right side of the bowl does the ball reach? Explain your answer in light of our assumptions about the very short distance in which sliding changes to rolling.

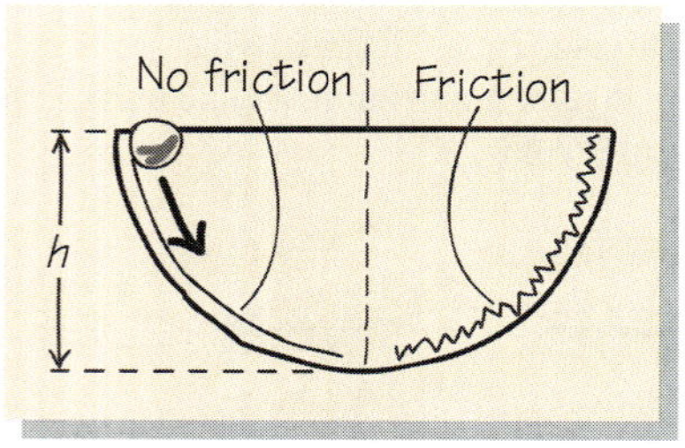

FIGURE 10–38 Problem 53.

54. (II) A solid door of mass $M = 20$ kg and width $\ell = 1$ m is hit at a right angle by a mud ball of mass $m = 0.5$ kg, which, as Fig. 10–39 shows, hits the door at the edge with speed $v = 10$ m/s and sticks. (a) What is the rotational inertia of the door about the hinges? (b) What is the angular velocity of the door after having been struck? (c) What fraction of the initial energy does the moving door–mud ball system retain?

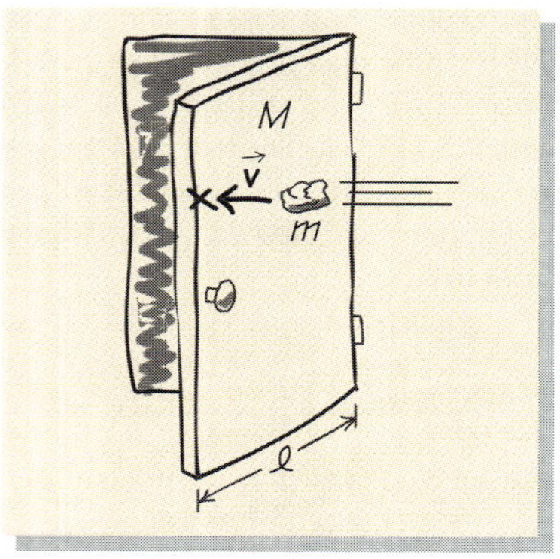

FIGURE 10–39 Problem 54.

55. (II) A 75-kg bank robber is escaping on a 450-kg motorcycle, runs out of gas, and thereafter coasts (friction free) at 80 km/h. As he passes under an overpass, a 75-kg policeman drops vertically onto the back of the seat and hangs on. Ignore the impulse due to the sudden change of the force of friction on the road on the translational and rotational motion. (a) Find the final velocity of the motorcycle. (b) What fraction of the initial kinetic energy (motorcycle plus robber) is lost? Ignore the rotational energy in the wheels. (c) Redo part (b) but include the effect of the rotational energy of the wheels, which each have a rotational inertia about their axes of 3 kg · m^2 and a radius $r = 0.5$ m.

56. (II) Electric power is used to speed up a centrifuge whose rotational inertia is 0.2 kg · m^2; 2.7 kW of power were used to make the centrifuge accelerate at a steady rate from rest to 24,000 rev/min. If the electricity use were 100 percent efficient, how much time is required to speed up the centrifuge?

57. (II) A point mass $m = 0.2$ kg is attached to a string, which passes through a hole in a table and rotates in a circle of radius $r = 0.8$ m with an angular velocity of 40 rad/s. What mass M must be attached to the end of the string under the table to maintain this motion? Suppose that mass M is slowly increased by an amount that makes it descend a distance 0.1 m. What is the amount of the increase of M? What will the new angular velocity of the point mass be? (*Hint*: Use angular momentum conservation.)

58. (II) Show that $(\mathbf{r} \times \mathbf{p}) \cdot (\mathbf{r} \times \mathbf{p}) = r^2p^2 - (\mathbf{r} \cdot \mathbf{p})^2$. (*Hint*: It is convenient, without any loss of generality, to assume that both $\mathbf{r}$ and $\mathbf{p}$ lie in the xy-plane.) Use this result to express the kinetic energy of a particle in terms of the momentum in the radial direction and of the square of the angular momentum.

59. (II) A cylindrical shaft of radius 5 cm is connected by a band to a solid cylindrical flywheel of mass 300 kg and of radius 0.35 m (Fig. 10–40). A motor brings the shaft up to a rotational rate of 1400 rev/min. Calculate the amount of work done by the motor, neglecting the rotational inertia of the shaft.

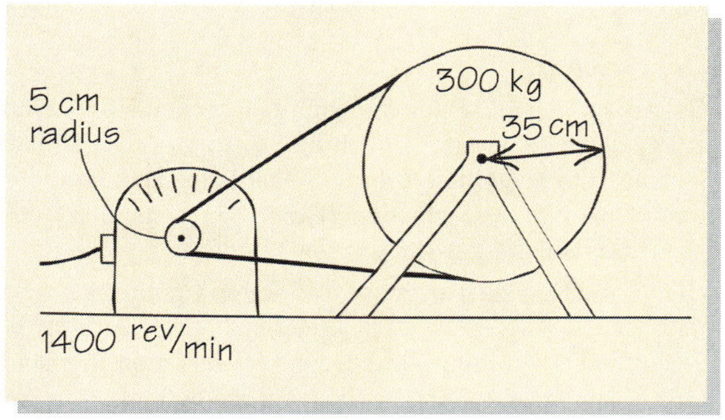

FIGURE 10–40 Problem 59.

60. (II) A thin rod of mass M, length ℓ, and constant density is standing on end on a rough table that forms the xy-plane. The rod begins to fall, with its top moving in the $+x$-direction, but as it falls, its point of contact does not move. As the rod hits the table, what are its (a) angular velocity, (b) angular momentum, and (c) kinetic energy?

61. (II) An object of mass M moves in a circular planar orbit about a center of gravitational attraction. The force of attraction has magnitude $F = K/r^2$, where r is the radius of the circle, and it is directed toward the center. Calculate (a) the velocity, (b) the radius, (c) the period, T, and (d) the acceleration of the object, all in terms of the angular momentum L, M, and K.

62. (III) A putty ball of mass $m = M/3$ is thrown with velocity $\mathbf{v} = v\mathbf{i}$ and hits the top of the thin rod in Problem 60 as the rod stands vertically. If the putty ball makes a completely inelastic collision, and if again the point of contact between the rod and the table does not move, what are the angular velocity, angular momentum, and kinetic energy of the system as it hits the table?

63. (III) A particular top can be approximated as a solid cylinder of mass 100 g and radius 2 cm. A string of negligible mass and length 1 m is wound around the top, which is started by pulling horizontally on the string with a constant force of magnitude 0.6 N. The top starts from rest at point O, and the string is pulled off. Ignore all friction between the top and the table on which it moves. (a) What is the final velocity of the center of mass of the top? (b) the final angular velocity of the top about its center of mass?

64. (III) A uniform solid cylinder of radius R and mass M rests against a vertical curb of height h, where $h < R$ (Fig. 10–41). The cylinder is mounted through its axis on a frictionless horizontal axle. You exert a horizontal force of magnitude F on the axle, pushing the cylinder against the curb. What is the minimum value of F that will cause the cylinder to roll up over the curb?

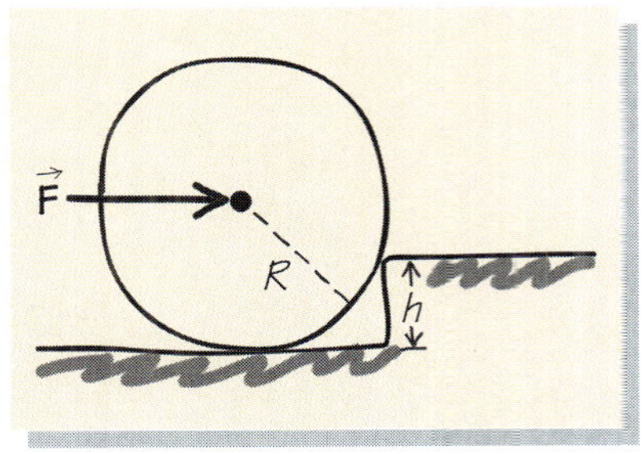

FIGURE 10–41 Problem 64.

CHAPTER

11

The tightrope artist Philippe Petit, here seen in Frankfort, uses the large bar to control the position of his center of mass and minimize the torque acting on the system formed by him and the bar.

Statics

As Chapters 9 and 10 show, the motions of extended objects present a rich array of possibilities because both linear and rotational motion are possible. In the area of physics called **statics**, we are interested in the conditions that leave rigid bodies motionless. The problems presented by statics for the motion of extended objects must include the possibility that there are torques acting as well as forces. Statics has its most notable use in the field of civil engineering, as in the design of bridges or buildings. Applications extend to the analysis of the role of muscle, tendon, and bone in living systems.

11–1 STATIC CONDITIONS FOR RIGID BODIES

In principle, it is enough to say that a building or a bridge will be static if the net force on every component is zero. A full calculation of the net force on any one component must include forces due to other components, which are sometimes very difficult to include. For example, it is not an easy problem to calculate the bending of a beam supported at two ends. However, a discussion of the conditions under which the building is static is greatly simplified if the building is *rigid*. The external forces acting on such an object have two effects: First, no matter where they are actually applied on the object, their vectorial sum produces a linear acceleration of the center of mass. Second, depending on where they are applied, they may produce torques that act to rotate the object. In the previous two chapters, we were interested in the resulting motion. In this chapter, we will look at the conditions for *no center-of-mass acceleration and no angular acceleration about any axis*; in most practical applications, there is also *no center-of-mass velocity and no angular velocity about any axis.* This area of study is known as statics (Fig. 11–1).

FIGURE 11–1 Among the forces and torques acting on this rock are those due to contact forces and to gravity. This static marvel has remained balanced for a very long time, suggesting that the equilibrium is stable.

Suppose that we have a set of external forces $\mathbf{F}_i$, labeled by the index i, that act at various points described by the position vectors $\mathbf{R}_i$ on a rigid body of mass M. Then the full dynamical equations for the object are

$$\mathbf{F}_{\text{net}} = \sum_i \mathbf{F}_i = M\mathbf{A} \tag{11–1}$$

and

$$\boldsymbol{\tau}_{\text{net}} = \sum_i (\mathbf{R}_i \times \mathbf{F}_i) = I\boldsymbol{\alpha}. \tag{11–2}$$

The points at which the forces act have position vecters $\mathbf{R}_i$ measured with respect to any convenient origin. Here, $\mathbf{A}$ is the acceleration of the center of mass due to the net force, $\mathbf{F}_{\text{net}}$; $\boldsymbol{\alpha}$ is the angular acceleration of the object body due to the net torque, $\boldsymbol{\tau}_{\text{net}}$, *about the point chosen as the origin.* (Ordinarily, the rotation about a point also corresponds to rotation about an axis, which is why we have included the form $I\boldsymbol{\alpha}$ in Eq. (11–2). In that case, the rotational inertia, I, must be calculated about the axis of rotation.)

The calculation of torque is treated in Chapter 10.

The conditions under which a rigid body is static

We can now write down the conditions for statics. The condition under which the linear acceleration drops out is contained in Eq. (11–1) with $\mathbf{A} = 0$, which states that the net force is zero:

$$\sum_i \mathbf{F}_i = 0. \tag{11–3}$$

The condition that the net torque drops out is obtained from Eq. (11–2) with the angular acceleration $\boldsymbol{\alpha} = 0$:

$$\sum_i (\mathbf{R}_i \times \mathbf{F}_i) = 0. \tag{11–4}$$

Equations (11–3) and (11–4) are all we need to study the statics of rigid bodies.

Recall from Chapter 9 that a torque has a magnitude given by the product of the force and a length known as the *moment arm.* To find the moment arm about a certain point O, extend the line of the force. The closest, or perpendicular, distance between the extension line and point O is the moment arm.

The Condition of No Torque Is Independent of the Choice of Axis

"No torque about any one axis" means "No torque about any axis."

When we say that there is no angular acceleration [Eq. (11–4)], to which axis are we referring? We can show that if an object has no angular acceleration about any one axis, then, assuming that the object is not in linear acceleration, it is not in angular acceleration about another axis. Equation (11–4) applies to *any* choice of origin when Eq. (11–3) holds.

In order to prove this important result, we consider a new origin displaced by $\mathbf{D}$ from the old origin and compute the angular acceleration about this new origin. Any angular acceleration is due to a series of forces, labeled by an index i, applied to the extended object. If the force labeled i is applied at point $\mathbf{R}_i$ with respect to the original origin, then it is applied at the point $\mathbf{R}_i'$ in the new system:

$$\mathbf{R}_i' = \mathbf{R}_i + \mathbf{D}.$$

The condition for no rotational acceleration is

$$\begin{aligned} 0 = \sum_i (\mathbf{R}_i \times \mathbf{F}_i) &= \sum_i [(\mathbf{R}_i' - \mathbf{D}) \times \mathbf{F}_i] = \sum_i (\mathbf{R}_i' \times \mathbf{F}_i) - \sum_i (\mathbf{D} \times \mathbf{F}_i) \\ &= \sum_i (\mathbf{R}_i' \times \mathbf{F}_i) - \left(\mathbf{D} \times \sum_i \mathbf{F}_i\right). \end{aligned}$$

From Eq. (11–3), the sum over the forces in the second term of the right-hand side is zero. Thus

$$\sum_i (\mathbf{R}_i \times \mathbf{F}_i) = \sum_i (\mathbf{R}_i' \times \mathbf{F}_i) = 0,$$

which shows that if there is no torque about one origin, then there is no torque about another origin.

Because a static object has no torque about any axis, we can choose the axis about which we find the torques that compose the net torque. A good choice leads to considerable simplification; the calculation of the torques about some points may be especially simple. In particular, the torque about the point where a force is applied, or anywhere along the line of that force, is zero because the moment arm is zero for that force. For example, let's consider a hard ball being pushed against a curb by a horizontal force **F** that is applied at precisely curb height (Fig. 11–2a). The ball itself has a radius equal to the curb height. A horizontal force can make the ball climb the curb if it causes the ball to rotate about the point P. This occurs when there is a net torque about P. But the contact force, $\mathbf{F}_C$, and the external force, **F**, are both directed so that extensions of them pass through point P. These forces therefore produce no torque about P, as the extended force diagram Fig. 11–2b shows. Thus the external force cannot make the ball climb the curb and the ball is static.

PROBLEM SOLVING

The calculation of torque can be simplified by a good choice of axis.

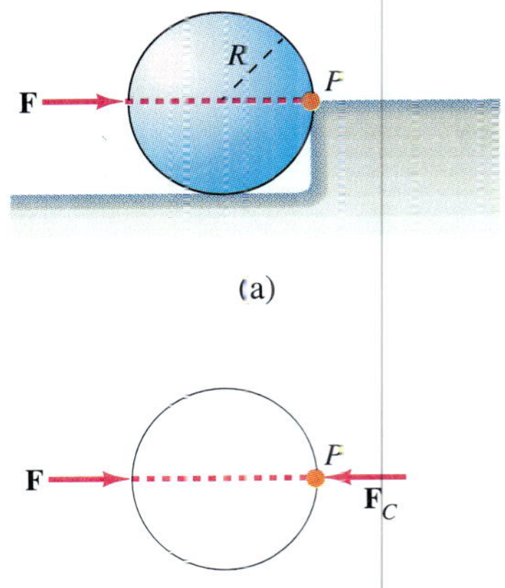

FIGURE 11–2 (a) The ball has a radius equal to the height of the curb, and a horizontal force **F** is applied at exactly curb height. (b) Extended force diagram (we have ignored the vertical forces). The forces shown can have no torque about point P.

11–2 GRAVITY AND RIGID BODIES

We often apply the static conditions to mechanical structures such as bridges, where gravity is an external force. Unless the structure is so huge that the acceleration of gravity differs over its length or height (an unusual situation, to say the least), gravity acts equally on all its different portions. In Section 9–5, however, we showed that there is a considerable simplification: *Gravity acts as though it were applied to a concentrated (point) mass of the total mass M at the center of mass of the extended object.* Thus, gravity produces a torque as though the gravitational force acts on the center of mass (which is sometimes called the *center of gravity* in this context).

Torque due to gravity was first discussed in Chapter 9.

Gravity produces a torque as though it acts on the center of mass.

EXAMPLE 11–1 Consider a rectangular book with a uniform mass density and a width L. The book is lying on a table with one side parallel to the table edge, but the book is hanging off the edge by an amount ℓ (Fig. 11–3a). What is the largest possible value of ℓ for which the book does not rotate off the edge?

Solution: Figure 11–3b shows a force diagram with the force of gravity, $m\mathbf{g}$, acting on the book's center of mass at a point midway along its width. If the book falls off the table, it will do so by rotating about the table edge. Thus we calculate the torque about this axis and place the origin at the edge of the table. We can satisfy the first equilibrium condition, Eq. (11–3), if

$$\mathbf{F}_N = -m\mathbf{g}.$$

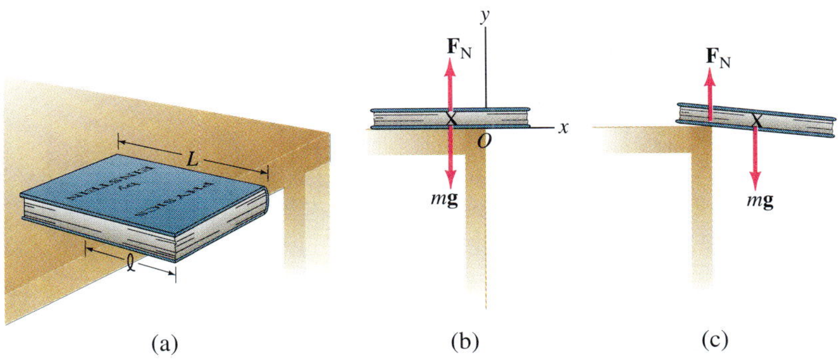

FIGURE 11–3 (a) Example 11–1. (b) The center of mass of the book is over the table surface. (c) The center of mass of the book is beyond the table edge.

FIGURE 11–4 In an extension of Example 11–1, it is possible to show that, with as few as four books, the topmost book in a stable pile can lie entirely over the edge of the table.

When the center of mass of the book lies above the table surface, we can satisfy the second equilibrium condition—Eq. (11–4)—for rotations about the edge of the table if $\mathbf{F}_N$ acts at the same point as the force of gravity, as shown in Fig. 11–3b. Note that at the point where the book is ready to rotate off the table, $\mathbf{F}_N$ does indeed act only at the edge of the table.

If the book's center of mass is beyond the edge of the table, then the second equilibrium condition can no longer be satisfied. The normal force will be acting at the edge of the table; at this point, the lever arm for the torque of the normal force is zero. The force of gravity, however, does have a torque for rotations about this point. Thus the equilibrium condition can no longer be satisfied, and the book falls (Fig. 11–3c). The equilibrium conditions *can* be satisfied if the center of mass of the book lies above the table surface and the largest possible value of ℓ is $L/2$.

Structures Without Mortar

APPLICATION

Stable construction

An interesting generalization of Example 11–1 involves the static conditions for a stack of several books of mass m on a table. In Problem 13, we consider the conditions under which one book overhangs a second, identical book on the edge of a table. If you stack enough books, it is possible to make a stable pile with the topmost book lying as far off the table as desired. Figure 11–4 shows such a stack. Even with a stack of only four books, the topmost book can be made to lie completely off the table edge. These results can be applied to other constructions; for example, they suggest that if building blocks are used judiciously, it is possible to build stable structures without worrying about mortar. A crucial added ingredient is the keystone of an arch, which, in effect, allows the joining of the topmost edges of two stacks of books, such as the one in Fig. 11–5. With the keystone, neither pile need be stable by itself. Mortar often represents the weak point of a structure and many ancient building structures that have lasted through the ages have made good use of the possibility of equilibrium without mortar. Among these we may list Egyptian, Greek, and Roman monuments, as well as the mighty cathedrals erected during the middle ages in Europe. Modern buildings, too, use principles like these; Frank Lloyd Wright's Falling Water is a splendid example.

FIGURE 11–5 The arch is a stable architectural element that has been used since ancient times.

An Experimental Method to Find the Center of Mass

The fact that gravity effectively acts on the center of mass shows that a suspended object will be in stable equilibrium when the center of mass lies directly below the suspension point. Figure 11–6a shows a suspended object with its center of mass indicated. The forces acting on the object are the tension, T, of the suspending rope, acting at the suspension point, and the object's weight, acting at the center of mass. The equi-

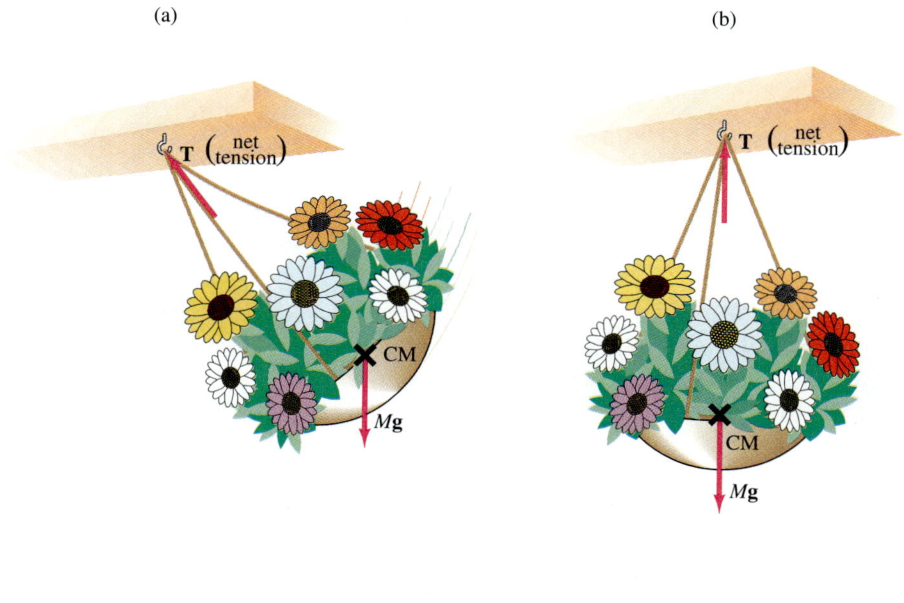

FIGURE 11–6 (a) If a suspended object does not line up with its center of mass directly below the pivot point, there will be a net torque about some point, and the object cannot be in static equilibrium. (b) With the center of mass directly below the pivot point, there is no net torque about any point, and the object is in static equilibrium.

(a) (b) (c)

FIGURE 11–7 Here's a technique to find the center of mass of this cutout map of the United States. (a) Suspend the map from any point, and drop a plumb line from that point. The center of mass lies somewhere along the line. (b) The same is true for a second plumb line dropped from a second point, and the intersection of the two lines is the location of the center of mass. (c) The plumb line dropped from a third point does indeed pass through the center of mass.

librium condition for the force is that T is equal and opposite to the weight. However, if the center of mass does not lie directly below the suspension point as in Fig. 11–6a, there will be a net torque and the object cannot be in equilibrium. For example, take the origin at the center of mass; in this case, the tension will have a torque about this point. If the center of mass lies directly below the suspension point, then neither force exerts a torque about the center of mass (Fig. 11–6b).

The center of mass of a flat object can be determined by using a plumb bob. The object is suspended three times in succession, each time from a different point on the object. A plumb bob is dropped from each suspension point and a vertical line is drawn on the object. In each case, the center of mass lies along these lines, and thus the intersection of these lines is the center of mass. Figure 11–7 illustrates the method.

11–3 APPLICATIONS OF STATICS

APPLICATION

The limits of safe construction

How many beams of a given size are required to support a roof of a given weight? Can the beams be reduced in size? The equations for statics are often used to determine the forces ("loads") on such components, which in turn are used to find the size of components needed for construction on various scales, or to determine the limits of safe construction (Fig. 11–8). In Examples 11–2 and 11–3, we treat some of these aspects.

FIGURE 11–8 The Statue of Liberty forms a complex system whose successful construction required knowledge of many static forces. Interestingly, the raised arm was originally designed to lie along the black dashed line. The additional torque caused by the misplaced arm required an extensive repair, completed in 1986.

EXAMPLE 11–2 In order to handle a very hot object of mass 5 kg, a worker hooks it with a light but stiff pole of length $L = 3$ m, which he then carries horizontally with two hands (Fig. 11–9). His right hand is at one end of the pole (point A), while his left hand is at point B, a distance $\ell = 0.75$ m farther down the pole, and the mass is at the other end. What forces must the worker's hands exert on the pole?

Solution: We want to use the static conditions to determine the forces exerted by the hands on the pole. To do so, we first note that the extended object on which the forces act is the pole along with the hot object at its end; the external forces acting on it are gravity and contact forces from each hand. In the force diagram of Fig. 11–9b, we have used our intuition that the hands will exert upward forces but, as is usual with force diagrams, it is not necessary that the forces drawn correspond precisely in direction and magnitude to the forces to be determined; the static conditions will themselves determine these forces.

PROBLEM-SOLVING TECHNIQUES

Suggestions for Statics Problems

1. Prepare an extended force diagram that includes all the forces and where they act. Follow the first three steps described in the Problem-Solving Techniques box in Chapter 9.
2. Choose a single convenient origin and think about the choice of axes. In particular, the torques due to the forces may be simpler to compute about some origins than about others; further, the more forces are aligned with coordinate axes, the better.
3. Be clear about the direction you have chosen for positive torque.
4. Write down the conditions for statics that follow from steps 1 through 3—Eqs. (11–3) and (11–4).
5. Count the resulting equations as well as the unknown forces to be determined to make sure that the number of equations corresponds to the number of unknowns and the problem does indeed have a unique solution.
6. In solving the equations for statics, carry through an algebraic rather than a numerical solution as far as possible. In this way, you can make checks in various limits or in special cases and you can see if the results are reasonable.

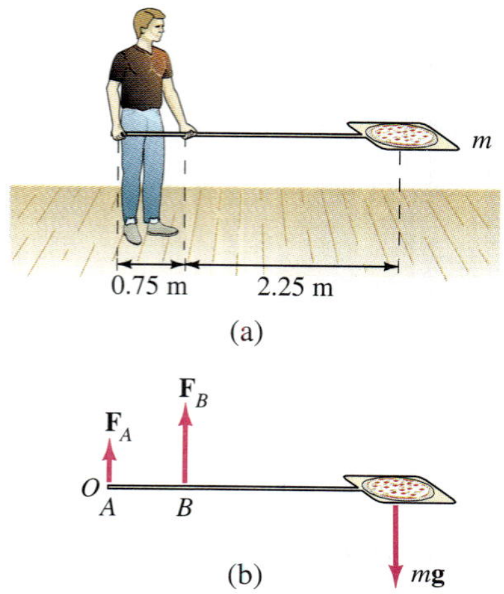

FIGURE 11–9 (a) Example 11–2. (b) Extended force diagram. The object on which the forces act is the massless rod with the hot, massive object at the end. The problem solution shows that the actual direction of $\mathbf{F}_A$ points down.

Let's choose point A to be the origin. Equation (11–3) reduces to one equation for the vertical components of the force; Eq. (11–4) is also one equation because the torques about point A are all into or out of the page. Two is the right number of equations to determine the two unknown forces exerted by the hands. Then our static conditions are

$$F_A + F_B - mg = 0$$

and

$$\ell F_B - Lmg = 0.$$

Note that F_B and mg must exert torques in opposite directions and that there is no torque due to F_A with our choice of A as origin.

These equations can be solved for the unknown forces:

$$F_B = mg\left(\frac{L}{\ell}\right);$$

$$F_A = mg - mg\frac{L}{\ell} = mg\left(\frac{\ell - L}{\ell}\right).$$

Because $\ell < L$, F_A is negative: Hand A must exert a downward force. In effect, point B, which lies between the two ends of the pole, acts as a pivot point about which the pole may rotate. The force at A must point in the *same* direction as the force of gravity on the mass to allow the torque to be zero. $\mathbf{F}_B$ must then counteract $m\mathbf{g}$ and $\mathbf{F}_A$. Note the magnification factor in the determination of the unknown forces. For example, F_B is much larger than mg if L is much larger than ℓ. It is difficult to hold a mass at the end of a horizontal pole!

Inserting numbers, we find

$$F_A = (5\ \text{kg})(9.8\ \text{m/s}^2)\frac{0.75\ \text{m} - 3\ \text{m}}{0.75\ \text{m}} = -150\ \text{N};$$

$$F_B = (5\ \text{kg})(9.8\ \text{m/s}^2)\frac{3\ \text{m}}{0.75\ \text{m}} = 200\ \text{N}.$$

EXAMPLE 11–3 A crane whose cabin and engine are effectively fixed to Earth is used to lift a mass m of 5300 kg (Fig. 11–10a). The arm of the crane is supported at its base, at point B, by a strong but friction-free pivot, and at its top, at point A,

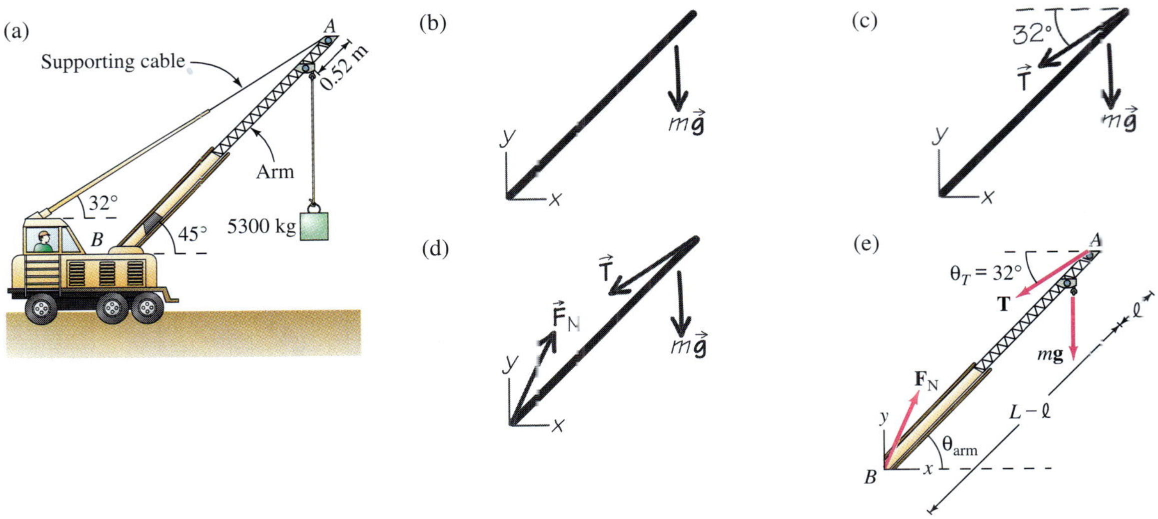

FIGURE 11–10 (a) Example 11–3. (b–d) Sketches illustrating the preparation of the extended force diagram. (e) Extended force diagram. The object on which the forces act is the arm of the crane. The relevant angle between the gravitational force and the position vector from point B along the crane's arm is given by $\theta_{arm} + 90°$. The relevant angle between the tension and the position vector from point B along the crane's arm is given by $-\theta_{arm} + 180° + \theta_T$. The sines of these angles enter into the expression for the torque about point B due to the forces of gravity and tension, respectively.

by a cable. The arm and the cable make angles of 45° and 32°, respectively, with the horizontal, and the arm is 10.0 m long. The mass is lifted from a point on the arm 0.52 m from point A. Ignoring the mass of the arm, compute the tension in the cable.

Solution: We want to find the magnitude of one of the forces (the cable tension) that acts on the crane arm and therefore we take the arm to be the object for which a force diagram is drawn (Fig. 11–10b). The forces acting on the arm are: (1) the tension, $\mathbf{T}$, acting at A along the cable; (2) the weight of the 5300-kg mass; and (3) an unknown contact force, $\mathbf{F}_N$, at the bottom pivot point (Fig. 11–10c, d). The contact force points in an arbitrary direction to be determined. We must choose an origin to write the equations of motion. If the origin is at the application point of any of the forces, then there will be no torque due to that force. Aside from this consideration, there is no special advantage to choosing any one of the three candidate points, and we choose point B as shown (Fig. 11–10c). There are three equations: two for the x- and y-components of the net force and one for the torque. There are three unknowns: the two components of $\mathbf{F}_N$ and the magnitude of $\mathbf{T}$. The number of equations will match the unknowns, hence the problem has a solution.

The net force equation has the two vector components:

$$F_{Nx} - T\cos\theta_T = 0, \tag{11–5}$$

$$F_{Ny} - T\sin\theta_T - mg = 0. \tag{11–6}$$

The torques are perpendicular to the page, so the net torque equation has one vector component:

$$-(L-\ell)mg\sin(\theta_{arm} + 90°) + LT\sin(-\theta_{arm} + 180° + \theta_T) = 0, \tag{11–7}$$

where L is the arm length (AB) and ℓ is the distance away from point A that the mass hangs. Equation (11–7) is sufficient to solve for T because the choice of B as the origin removes F_N from the net torque equation:

$$T = \left(\frac{L-\ell}{L}\right)\frac{\sin(\theta_{arm} + 90°)}{\sin(-\theta_{arm} + 180° + \theta_T)}mg$$

$$= \left(\frac{10.0 \text{ m} - 0.52 \text{ m}}{10.0 \text{ m}}\right)\frac{\sin(45° + 90°)}{\sin(-45° + 180° + 32°)}(5300 \text{ kg})(9.8 \text{ m/s}^2)$$

$$= 1.5 \times 10^5 \text{ N}.$$

A useful check on this result is that if $\ell = L$, the mass is hung directly from point B and, from physical considerations, we expect no tension to be required; this is indeed the case. It is also true that if $\theta_T = \theta_{arm}$—so that the cable is also attached at the pivot point at B—then there is no way that T can provide a torque of the opposite sign from the torque coming from the weight, and the arm cannot be stabilized. Indeed, we cannot solve for T when the angles are equal.

FIGURE 11–11 A crane is held rigid by internal forces. External forces, such as those provided by cables and bolts, hold the crane static.

Example 11–3 is one you might use if you were going to design a crane and had to choose the size of the supporting cable (Fig. 11–11). Notice that in this design, the tension in the cable must be much greater than the weight itself. It may also be necessary to solve for the contact force to ensure that the pivot itself is sufficiently strong.

It is a good exercise to choose some other point—say, point A—to be the origin and verify that the answer is independent of the choice of origin.

EXAMPLE 11–4 A ladder of length 3.0 m is leaning against a wall at an angle of 58° as in Fig. 11–12a. Its eight rungs are spaced 0.33 m apart. The ladder's mass is insignificant compared to the mass $m = 85$ kg of a window washer who is climbing the ladder. The coefficient of static friction between the rubber feet of the ladder and the floor is $\mu_s = 0.51$, but assume that the wall is smooth and there is no friction between the top of the ladder and the wall. Is the ladder safe from slipping if the washer climbs to the seventh rung?

Solution: The extended object on which the forces act is the ladder; the question of interest concerns the magnitude of the friction force, and hence of the normal force at the floor. The normal forces are labeled $\mathbf{F}_{floor}$ and $\mathbf{F}_{wall}$ in Fig. 11-12b, which is an extended force diagram. The friction force at the floor acts to the left. To find out if the ladder slips, it is necessary to note that the static friction force takes on its *maximum value* just before slipping. The magnitude of this force is

$$f_{floor} = \mu_s F_{floor}. \tag{11–8}$$

The angle of the ladder is θ and the ladder length is L. The weight, $m\mathbf{g}$, acts at a distance d up the ladder, and points directly downward.

There will be three equations: two for the cancellation of forces and one for the torque. The three unknowns will be the two contact forces and the friction force. If the friction force that we determine exceeds the maximum static value, Eq. (11–8), then the ladder will slip. The bottom of the ladder is a useful choice for the origin

FIGURE 11–12 (a) Example 11–4. The window washer is a thin person standing vertically, so his center of mass is over the rung on which he stands. (b) Extended force diagram for the ladder. (c) The lever arm for the force of gravity about the origin is $d \cos \theta$, and the lever arm for the normal force of the wall is $L \sin \theta$.

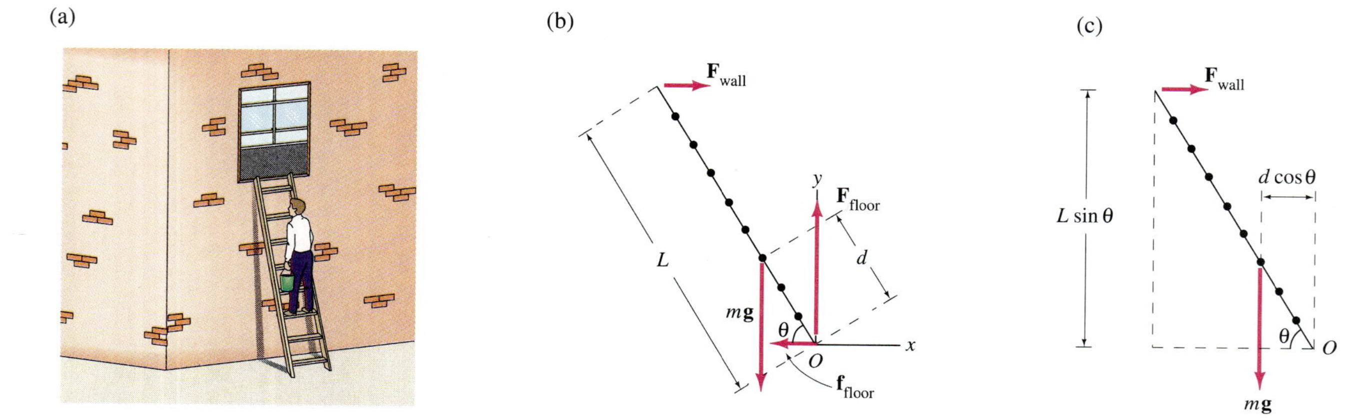

because two forces act at this point. We let $+x$ be to the right and $+y$ be upward. The statics equations for the forces are

$$F_{\text{floor}} - mg = 0, \tag{11–9}$$

$$F_{\text{wall}} - f_{\text{floor}} = 0. \tag{11–10}$$

For the torque, we see in Fig. 11–12c that the lever arm for the force of gravity about the origin is $d\cos\theta$, whereas the lever arm for the normal force of the wall is $L\sin\theta$. Thus

$$mgd\cos\theta - F_{\text{wall}}L\sin\theta = 0. \tag{11–11}$$

The contact force at the wall tends to rotate the ladder clockwise, while the weight tends to rotate the ladder counterclockwise.

Equations (11–10) and (11–11) can be solved for the contact force at the wall and the friction, and we find that

$$f_{\text{floor}} = \frac{mgd\cot\theta}{L} \tag{11–12}$$

Equation (11–9) gives the contact force at the floor directly. The condition that the friction force at the floor not exceed the maximum value of static friction is

$$\frac{mgd\cot\theta}{L} \le \mu_s F_{\text{floor}} = \mu_s mg,$$

$$d \le \mu_s L\tan\theta.$$

The higher the washer climbs, the greater the danger! For the seventh rung, $d = 2.33$ m, and with $\theta = 58°$, the condition becomes

$$2.33\text{ m} \le 0.51(3.0\text{ m})(\tan 58°) = 2.4\text{ m}.$$

The condition is satisfied, but barely. One more rung and the ladder would slip.

The reason the ladder becomes more likely to slip as the washer climbs is that, as the washer moves up the ladder, the lever arm of the washer's weight with respect to the contact point with the ground increases. The torque due to the washer's weight about the contact point with the ground increases, until the torque about this point due to the normal force with the wall can no longer counteract the effect of the torque due to the weight.

EXAMPLE 11–5 The biceps muscle is responsible for bending your arm. It acts through a kind of lever system, as in Fig. 11–13a. Some typical values for a, the elbow–hand distance, and x, the distance from the biceps attachment point to the

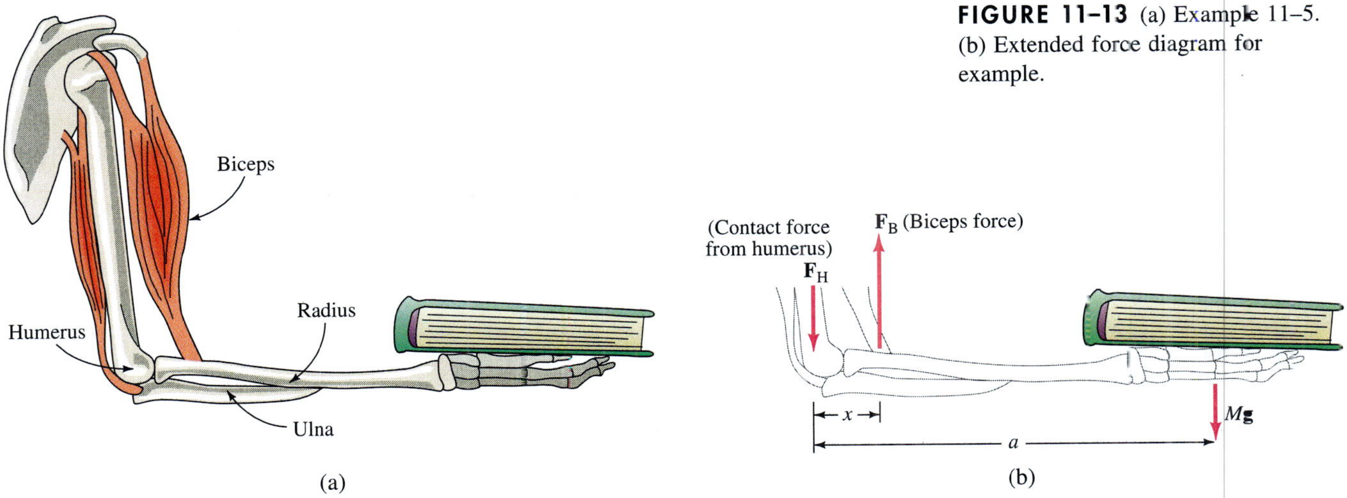

FIGURE 11–13 (a) Example 11–5. (b) Extended force diagram for example.

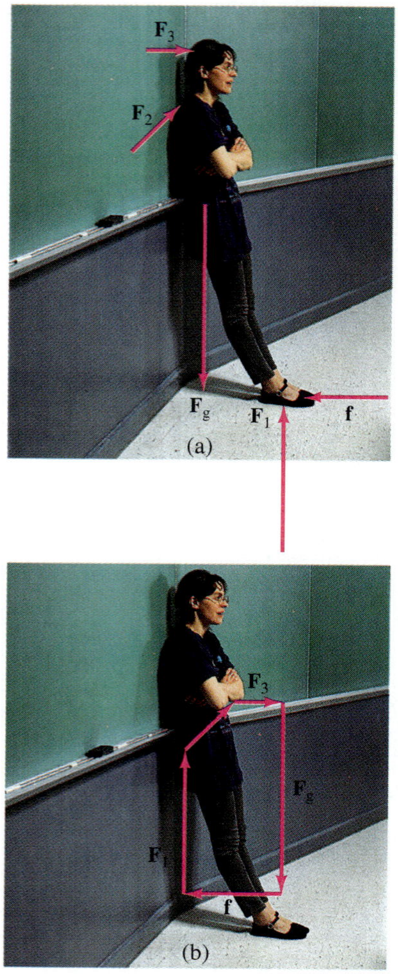

FIGURE 11–14 Biological systems can change the relative orientations of their different parts to allow them to be stable. (a) At least five different extenal forces act on the professor. (b) These forces add to a net force of zero.

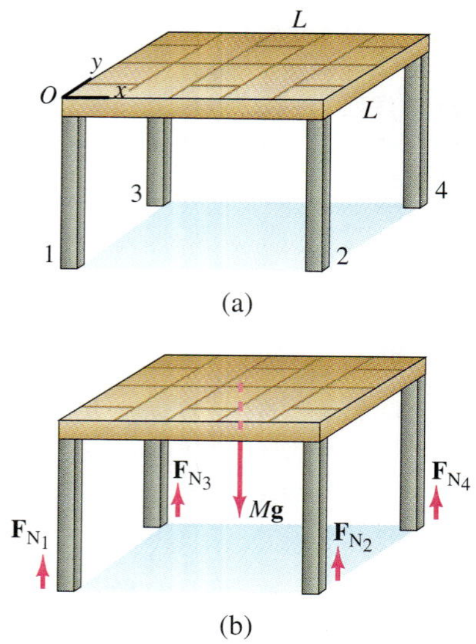

FIGURE 11–15 (a) A uniform four-legged table. (b) Extended force diagram for the table.

elbow, are $a = 30$ cm and $x = 4$ cm. If a mass M is held in your hand with your forearm horizontal and your upper arm vertical, what upward force does the biceps have to exert on the forearm bones? Ignore the weight of the forearm bone itself, and assume the hand-forearm forms a single system.

Solution: If the forearm weight is ignored, there are three vertical forces on the forearm: the contact force from the upper arm acting at the elbow; the force $\mathbf{F}_\mathrm{B}$ of the biceps on the forearm; and the weight of the mass. These are shown in the extended force diagram in Fig. 11–13b. The condition that there is no net force on the forearm determines the contact force from the upper arm. The condition that there is no net torque about the elbow determines F_B in terms of the distances and the mass's weight. Note that the contact force from the upper arm does not enter because that force exerts no torque about the elbow. We have

$$0 = (Mg)a - (F_B)x,$$

$$F_B = Mg\,\frac{a}{x}.$$

The force exerted by the contracting muscle must be a large factor greater than the weight itself. For typical arm dimensions, this factor is (30 cm)/(4 cm) = 7.5.

The physiological example described here is representative of the adaptable structures that occur throughout the kingdom of living organisms. For the considerations of this chapter, we must think of such structures as a collection of rigid parts—the bones—with variable orientation (Fig. 11–14).

Underdetermined Systems

We sometimes face problems in static systems where the forces cannot be uniquely determined by the conditions of force and torque equilibrium. There are fewer equations for the forces to be determined than there are forces. Such systems are said to be *underdetermined.* When a system is underdetermined, small incidental effects can become important.

As an explicit example, consider a uniform square table of mass M with four light legs on a horizontal surface (Fig. 11–15a). The top has sides of length L and, if we take the axes shown with the origin at leg 1, legs 1, 2, 3, and 4 make contact with the ground at the respective points $(x, y) = (0, 0)$, $(L, 0)$, $(0, L)$, and (L, L). The center of mass of the table is at its geometric center, the point $(x, y) = (L/2, L/2)$.

The forces acting on the table are the four contact forces $\mathbf{F}_{\mathrm{N}_1}, \ldots, \mathbf{F}_{\mathrm{N}_4}$ acting upward on the legs, and the force of gravity, $M\mathbf{g}$, acting on the center of mass. Figure 11–15b is the force diagram. The forces all act in the vertical, or z-, direction. We calculate the torque about the origin. The torque vectors due to the forces are perpendicular to these forces and therefore lie in the xy-plane. Our static conditions thus consist of one force equation (for the z-component) and two torque equations (for the x- and y-components). These equations are, respectively,

$$F_{N_1} + F_{N_2} + F_{N_3} + F_{N_4} - Mg = 0, \tag{11–13}$$

$$LF_{N_3} - \frac{L}{\sqrt{2}}\,Mg\sin 45^\circ + \sqrt{2}LF_{N_4}\sin 45^\circ = 0,$$

$$-LF_{N_2} + \frac{L}{\sqrt{2}}\,Mg\sin 45^\circ - \sqrt{2}LF_{N_4}\sin 45^\circ = 0.$$

Sin 45° = $1/\sqrt{2}$, so the last two equations simplify to

$$F_{N_3} - \frac{1}{2}Mg + F_{N_4} = 0, \tag{11–14}$$

$$F_{N_2} - \frac{1}{2}Mg + F_{N_4} = 0. \tag{11–15}$$

APPLICATION: CONSTRUCTION CRANES

The tower crane used at construction sites (Fig. 11AB–1) consists of a very flimsy tower constructed from three planes arranged in a triangle. Each plane consists of a lattice of triangular bars (this is called a geodetic construction). At the top of this tower is a normal crane boom, which can lift large masses, such as furnaces, air conditioners, and steel girders. The tower can rotate to place the mass at the correct angle, while the crane can move the mass radially inward or outward to position the object at the correct location; the object is finally set down. The strength of the tower is completely insufficient to withstand the bending moment of a large mass at the end of the boom, so there is a balance weight at the far side of the crane that moves in an opposite direction from the lifting equipment and the object being moved. The motion of the counterweight is controlled by strain sensors at the top of the tower and ensures that the crane boom produces no net torque at the top of the tower.

A similar idea is applied to the portable cranes that are carried by trucks. These cranes are provided with sensors that measure the downward force of the crane and truck body on each wheel. If this force reaches a lower limit, indicating that a wheel is lifting off the ground, movement of the crane boom is halted, with operator controls being overridden. Thus the crane remains in static equilibrium and does not tip over.

FIGURE 11AB–1 This type of construction crane has become a familiar feature of the urban landscape

We have only three equations to determine the four unknown forces $\mathbf{F}_{Ni}$. Indeed, any one leg could be removed completely. In this case, Eqs. (11–13) through (11–15) are three equations for the three remaining unknown contact forces. Here we see how a small change could make a big difference in the situation. If a teacup were placed near the location of the missing leg, just off the center of the table, the table would tip over toward the missing leg.

Finite-Element Methods

In this chapter, we have treated a rigid body as a collection of perfectly rigid parts. In some of the examples given, we can alter the relation between the orientations of the various rigid components of an extended object—like the crane of Example 11–3.

In reality, the components of extended objects are more complicated than such idealizations. Beams stretch, compress, bend, or twist. It is important to try to understand this behavior because the behavior of construction components is of paramount importance in the beauty, cost, and safety of a structure. The behavior of a beam depends not only on the external forces that act on it, but also on the forces within it. Many of these internal forces are the result of the action of external forces. These internal forces are in general referred to as **stresses**, and the response of extended objects—their stretching or bending or deforming—to stresses are generally called **strains**.[†] Because of the complexity of a real system, we do not have a mathematically exact way of computing, for example, the way in which a real beam will bend under the influence of a suspended weight.

[†]A more complete treatment of stresses and strains is given in Chapter 21.

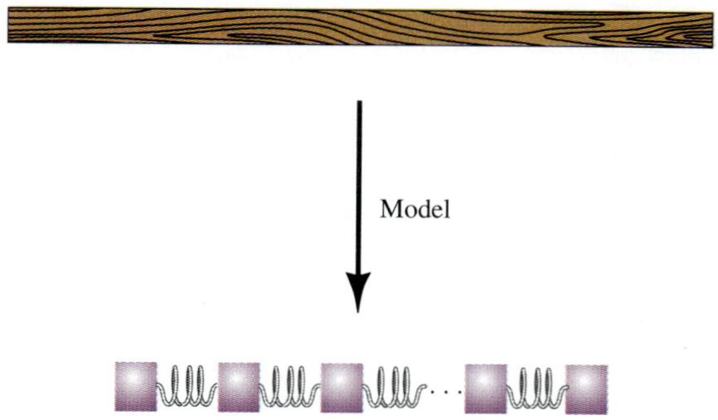

FIGURE 11–16 A thin beam is not truly rigid and may stretch or compress when it is subjected to tension or compression forces. It may be modeled by a series of masses connected by springs.

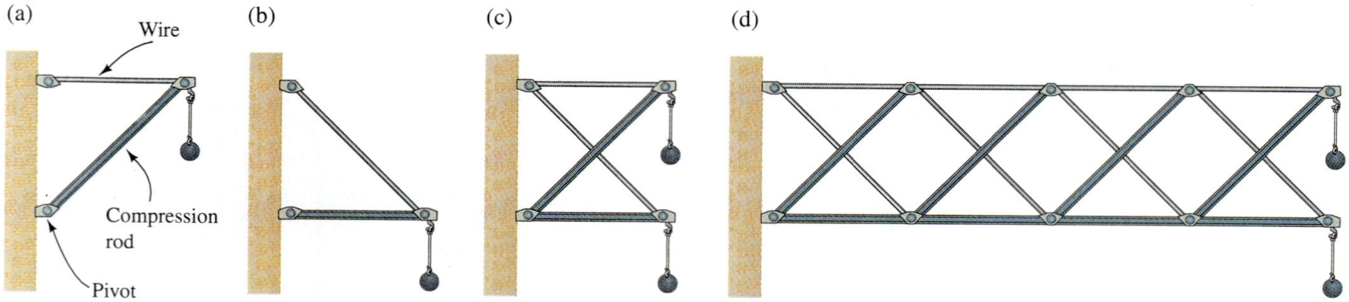

FIGURE 11–17 A finite-element model of a cantilever beam. (a), (b) Two basic parts are combined to provide (c) the primitive elements for the beam. (d) Together these elements make a model for a beam. (After J. E. Gordon, *The New Science of Strong Materials*, Princeton University Press, Princeton, New Jersey, 1976.)

APPLICATION

Applying finite-element methods

There is, however, a numerical approximation technique that springs directly from the methods we developed earlier. We can imagine breaking a continuous object into tiny elements, each of which can be thought of as an idealized rigid component. The accuracy of this method, known as the **finite-element method**, improves systematically as the number of elements grows larger. Consider as an example a one-dimensional system: a thin beam subject to forces in this one direction. We idealize the beam as a series of N rigid and massive components connected by springs (Fig. 11–16). The springs model the internal forces in the beam by resisting compression and stretching. In the model, the components are spaced and assigned masses so that the average mass density of the model matches that of the real beam. The strength of the individual springs is determined by matching some measured properties of the real beam, such as how the beam stretches when a known tension is applied. The model can now be used to see how the beam reacts to some external longitudinal force. There will be a static condition for each of the components and an array of N simultaneous equations will have to be solved. In other words, in solving problems with finite-element methods, we repeat the same relatively simple steps for many elements, a method that is well suited for numerical solution on a computer.

A more realistic type of example treats a *cantilever*, which is a projecting beam of finite thickness supported at one end only, as a succession of repeated components. The components pictured in Figs. 11–17a and 11–17b can each act as cantilevers and, when they are combined (Fig. 11–17c) and strung together in succession (Fig. 11–17d), they form a finite-element model of the cantilever beam. An element to the right of a second element acts as a weight for that second element; the succession of elements is treated as a succession of coupled statics problems. In this way, we can determine the size of beam necessary to support a certain weight at the end with a certain tolerated amount of bending (Fig. 11–18) and the amount of bending that results.

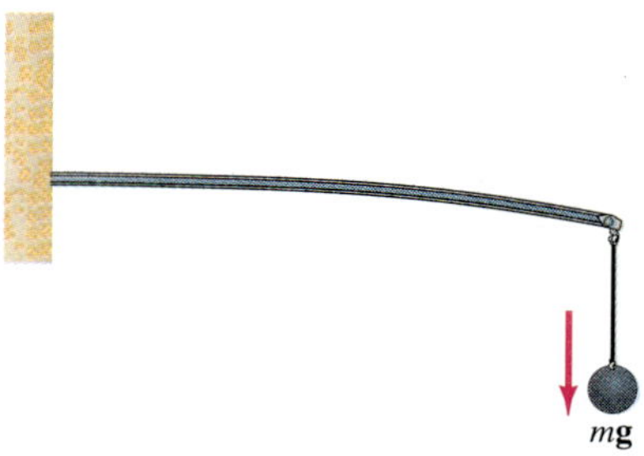

FIGURE 11–18 By using the model described in Fig. 11–17, it is possible to calculate various characteristics of a cantilever beam, such as how it will strain (bend) under stress.

Finite-element methods, in which a continuous system is approximated by numerous small elements, have uses in many areas of the physical sciences. Among numerous examples are supercomputer studies of the turbulent behavior of fluids, thermonuclear fusion research, and the quantum mechanics of atomic and molecular systems.

SUMMARY

An extended object is static when the net external force and the net external torque about any convenient origin are both zero:

$$\sum_i \mathbf{F}_i = 0 \tag{11–3}$$

and

$$\sum_i (\mathbf{R}_i \times \mathbf{F}_i) = 0. \tag{11–4}$$

In order to apply these conditions, it is necessary to know both what the external forces acting on an object are and *where* they act. Gravity acts as though it were applied to the center of mass of a rigid body. We can say that the weight of an object acts on its center of mass.

The static conditions are most often used to learn under what conditions a structure can be held static or to learn what forces various components of a structure must endure. Some structures are underdetermined, which means that the static conditions do not by themselves determine the forces acting

UNDERSTANDING KEY CONCEPTS

1. It is a common (and quite useful) piece of advice for those who are learning mountaineering techniques to "stand away" from the mountain on steep slopes rather than to follow one's instincts and "hug" the mountainside. Explain why this is so, using your knowledge of torques and forces on a climber standing on the steep slope of a mountain.
2. A baby pulls straight down with all his might on the flush handles of the closed drawer of a bureau (Fig. 11–19). Can he cause the bureau to tip over?

FIGURE 11–19 Question 2.

3. Why does a rope from which an object is suspended line up with the vertical?
4. Give at least three examples for which an object is not in equilibrium even though the net force on the object is zero
5. You are sitting quietly in a porch rocker that is suspended by chains from the ceiling. Are you in a stable or unstable equilibrium?
6. Bridges and buildings are not really rigid. Does this mean that nothing we have said in this chapter is relevant to them? How can an object be "approximately" rigid?
7. You have a flat map of the continental United States cut out of plywood. Devise two separate methods for determining the map's center of mass, one of them using only one finger.
8. Is the height of the four-legged table discussed in Section 11–3 irrelevant to the calculation of the equilibrium conditions for the table?
9. A pendulum suspended from the roof of an accelerating rail car makes a nonzero angle with the vertical. The pendulum is not swinging. Is this a case of stable equilibrium?
10. For some objects (U-shaped objects, for example), the center of mass is outside the object itself. For such objects, can we still think of gravity as though it acts on the center of mass?
11. The doughnut-shaped space station in Stanley Kubrick's film *2001, A Space Odyssey* rotates about its axis of symmetry (see Fig. 5–25a). Is the rotating station in stable equilibrium, unstable equilibrium, or neither?
12. Does the method of finding the center of mass of a flat object, as discussed in this chapter, work even if the density of the object is not constant?
13. You want to hold a beam with its end against a frictionless wall. Is it possible to do so by running a rope from the far end of the beam to any attachment point on the wall?
14. Will a ladder placed against a rough vertical wall remain standing when the floor is so smooth that there is no friction between it and the feet of the ladder?
15. Identical twins are placed at opposite ends of a seesaw pivoted about its midpoint. No forces other than those due to the twins, the pivot, and gravity act on the seesaw. What determines the inclination to the horizontal made by the seesaw when it is balanced? Is the equilibrium stable, unstable, or neutral?
16. Suppose that the seesaw of Question 15, still pivoted about its midpoint, has a sharp downward bend at the midpoint. If there is an equilibrium, will it be stable, unstable, or neutral?

PROBLEMS

11–2 *Gravity and Rigid Bodies*

1. (I) A uniform board of mass 80 kg and length 3.6 m is placed on top of a pivot 1.2 m from one end. What mass must be put at that end to allow the board to balance?

2. (I) A 30-kg board 4 m long is supported in a horizontal position at the two ends. A 70-kg worker stands 1.5 m from one end. What forces are exerted by the board on the two support points?

3. (I) Two workmen each carry one end of a 2.2-m-long ladder of mass 24 kg. The ladder is tapered so that its center of mass is 0.9 m from the wider end. What are the forces exerted by the ladder on the two workmen?

4. (I) A rail of length 2.0 m and mass 5.0 kg runs horizontally between two scales; a bowling ball of mass 5.5 kg is allowed to roll at a steady speed of 0.2 m/s from the left scale to the right scale (Fig. 11–20). During the 10 s the ball is moving, how do the readings on the two scales change?

5. (I) A projectile has broken into two parts. At time t_0, the two parts are located at the following points: m_1 at $(x, y, z) = (3, 0, 0)$ and m_2 at $(0, 0, 3)$, where all distances are measured in meters. The mass m_1 is twice the mass m_2. What is the location at time t_0 of the point that follows a parabolic trajectory (assuming that there is such a point)?

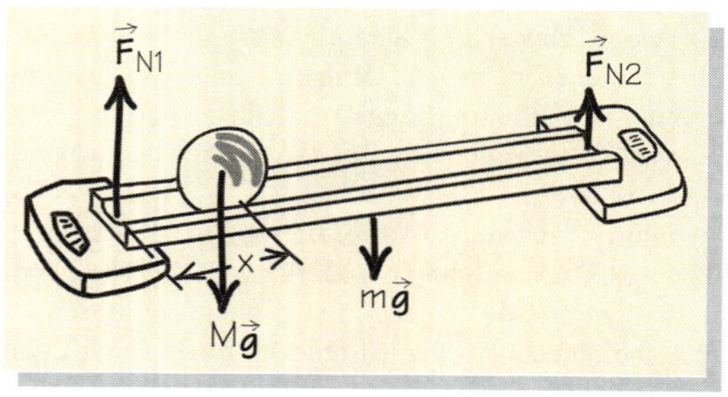

FIGURE 11–20 Problem 4.

6. (II) Two people of unequal strength must carry a uniform beam of length L while holding it horizontal. The weaker of the two holds the beam at one end. (a) How far from the other end must the stronger person hold the beam in order to support three-quarters of the weight? (b) Is there a way in which the stronger person can carry the beam at one end and still support more than half the weight of the beam?

7. (II) A playground seesaw is balanced at its midpoint. Two children, weighing 25 kg and 40 kg, respectively, want to balance on the seesaw. If the children are separated by a distance of 2.8 m, how far from the pivot point will the lighter child sit?

8. (II) Consider a seesaw whose total mass is 20 kg and total length is 4.0 m. Suppose the seesaw is placed off center on the pivot point so that the pivot point is 30 cm from the center of the seesaw (Fig. 11–21). How far from the center will the children of Problem 7 have to sit if the lighter child sits on the longer part of the seesaw? (Their separation is still 2.8 m.)

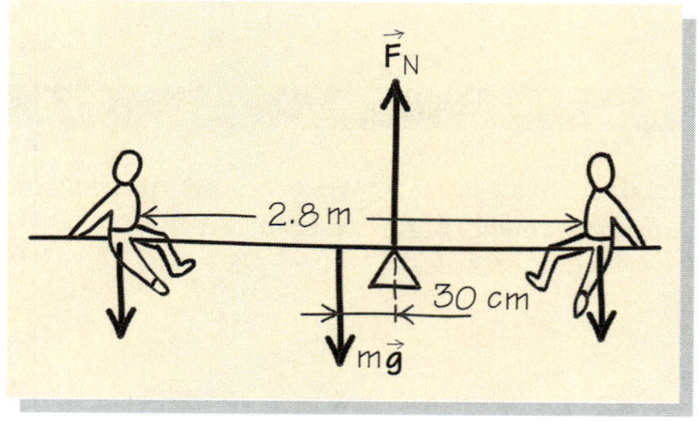

FIGURE 11–21 Problem 8.

9. (II) In order to hang a load of mass $M_1 = 30$ kg from the horizontal, flat roof of a building, a plank of length $L = 2.4$ m is placed on the roof (Fig. 11–22). One end is held in place with a chunk of concrete of mass $M_2 = 15$ kg, and the other supports the load M_1 with a light rope. How far can the end of the plank reach without tipping over? Neglect the mass of the plank.

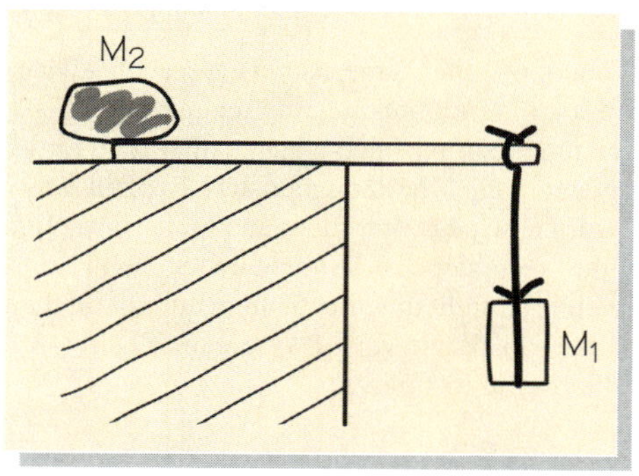

FIGURE 11–22 Problem 9.

10. (II) A uniform beam of mass 80 kg and length 3 m rests on two pivots, one at the left edge and one 2.4 m from the left edge. How far to the right of the right pivot can a mass of 150 kg be placed without the beam tipping?

11. (II) A rectangular piece of plywood (60 cm × 120 cm) lies in the horizontal xy-plane. The surface mass density of the plywood is 3 kg/m^2. Calculate the torque about one of the corners due to the force of gravity.

12. (II) A book of mass 1 kg is placed such that 2/3 of the book is hanging over the edge of a table. A paperweight of mass m is placed centrally on top of the 1/3 of the book that is on the table. What is the minimum mass of the paperweight such that the book doesn't fall off the table?

13. (II) Two books are stacked at the edge of the table, with their lengths perpendicular to the table edge. If the width of each book is L, how far out from the table edge can the top book's extreme edge be placed without the books falling down? How does this generalize to three books?

11–3 Applications of Statics

14. (I) A door 90 cm by 195 cm of mass 14 kg hangs on two hinges: One is attached to the bottom of one side of the door, and the other to the top of the same side. What are the horizontal forces exerted on the door by the hinges?

15. (I) A football player is at the top of a pushup. The angle that the (rigid) torso makes with the floor is 25°. His arms are perpendicular to his torso and his center of mass is located at a point 3/8 of the distance from the shoulders to the feet. Assuming that the mass of his head can be neglected, what is the force, in terms of the player's weight, along his arms?

16. (I) A student wants to place a flower pot on a board that juts out from a window so it will get more sunlight (Fig. 11–23). The flower pot has a mass of 2.0 kg and needs to be 1.0 m from the window sill. The student can only place a nail into the sill 4 cm from the edge. Neglect the mass of the board and find out how much force the nail must exert to hold the board in place.

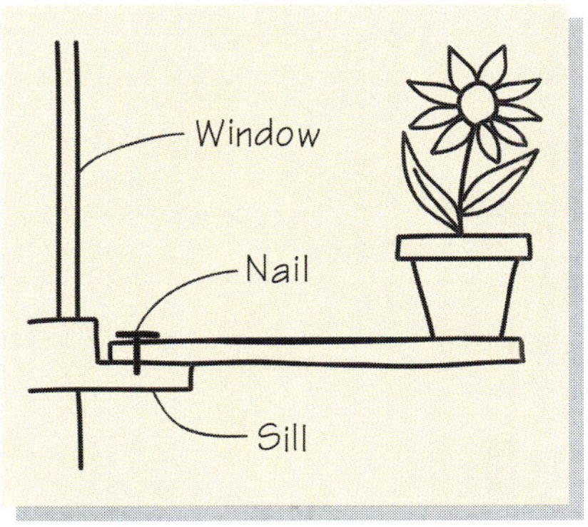

FIGURE 11–23 Problem 16.

17. (I) A uniform rod, mass 12 kg and length 1.5 m, rests on two points, one at its left end and one at the center point. What are the contact forces on the rod at these points? Comment on the stability of the situation.

18. (II) An 8.5-m extension ladder of mass 26 kg is propped up against a wall, touching at a point 8.0 m above the level ground. A man of mass 75 kg climbs 7 m up the ladder to repair a window. The ladder rests against a frictionless wall, but the ground has friction. Determine all the forces on the ladder.

19. (II) Consider a ladder of mass 10 kg and length 4 m, leaning against a vertical wall at an angle of 30° with the vertical. The coefficient of friction between the ladder and the floor is $\mu_s = 0.4$, and there is no friction between the wall and the ladder. A man of mass 80 kg climbs up the ladder. (a) How high can he climb before the ladder begins to slip? (b) Work out your calculation by taking the torques about the three points A, B, and O in Fig. 11–24, and show that the resulting equations are independent of the choice of point.

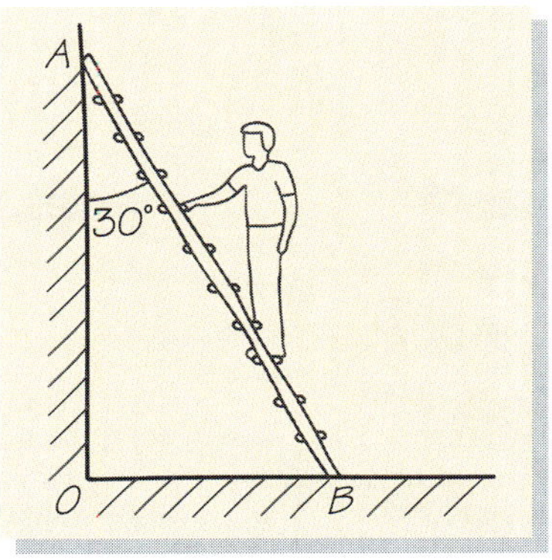

FIGURE 11–24 Problem 19.

20. (II) Consider the ladder in Example 11–4 (see Fig. 11–12a). The 3.0-m-long ladder is placed against a frictionless wall, making the same 58° angle to a different horizontal surface; this time, much to his dismay, the same 85-kg window washer finds himself starting to slip when he steps to the second rung. What is the coefficient of static friction between the ladder and floor?

21. (II) Using a uniform strut, a rigid brace hinged at the floor, a person holds a 30-kg engine in equilibrium while it is being repaired. The strut has a mass of 12.5 kg. A smooth rope passes over a pulley at the end of the strut (Fig. 11–25). (a) What is the force exerted on the rope by the person? (Specify the direction of this force by calculating the angle θ that the rope makes with the horizontal.) (b)What forces are exerted by the strut?

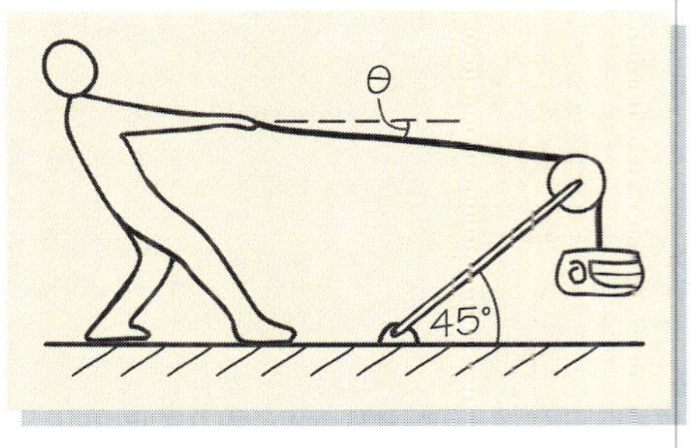

FIGURE 11–25 Problem 21.

22. (II) A piece of plywood leans against a wooden wall with which it has a coefficient of static friction of $\mu_s = 0.3$. (a) If the coefficient of static friction between the board and the floor is also 0.3, what is the minimum angle that the board makes with the floor? (b) What happens if the coefficient of friction between the board and the floor is zero?

23. (II) A stepladder consisting of two ladders of mass M and length L is held together by a crossbar attached to the midpoints of the two ladders (Fig. 11–26). What force is exerted on the crossbar by each ladder if the length of the crossbar is $L/2$? Assume that any friction between the ladders and the floor is negligible.

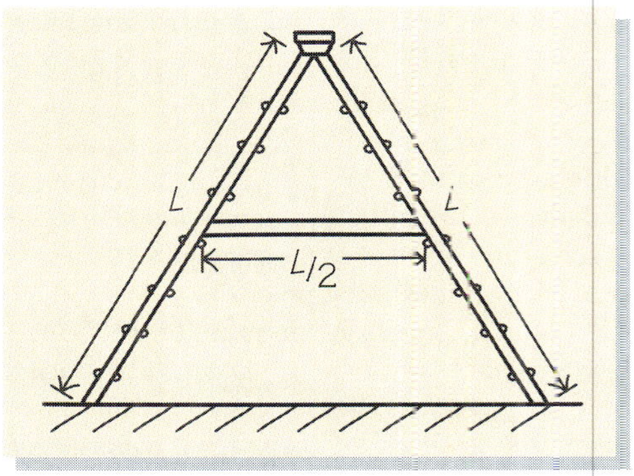

FIGURE 11–26 Problem 23.

24. (II) A flagpole of mass 6 kg and length 2.4 m is hinged at a wall and supported in a horizontal position by a cable attached to the free end (Fig. 11–27). The cable makes an angle of 25° with the horizontal. What is the tension in the cable? What is the vector force exerted on the hinge at the wall?

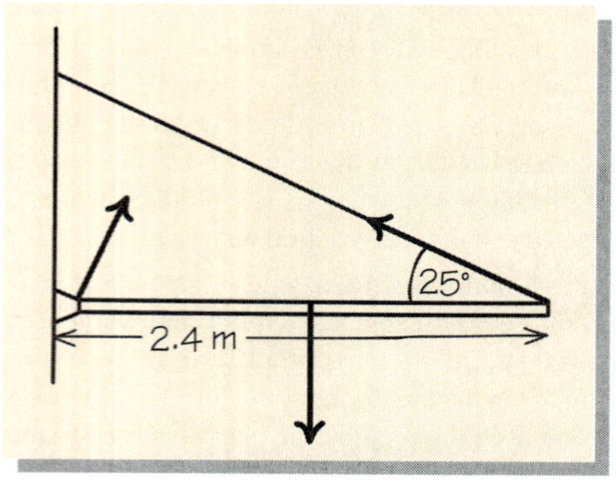

FIGURE 11–27 Problem 24.

25. (II) A desk of height 0.82 m, length 1.54 m, and mass 43 kg is pushed across a horizontal floor at a steady speed with a horizontal force **F** applied at the top (Fig. 11–28). The coefficient of kinetic friction between the legs and the floor is $\mu_k = 0.45$. What is the friction force at each leg, and what is **F**? Assume that the two right legs each support the same forces, as do the left legs.

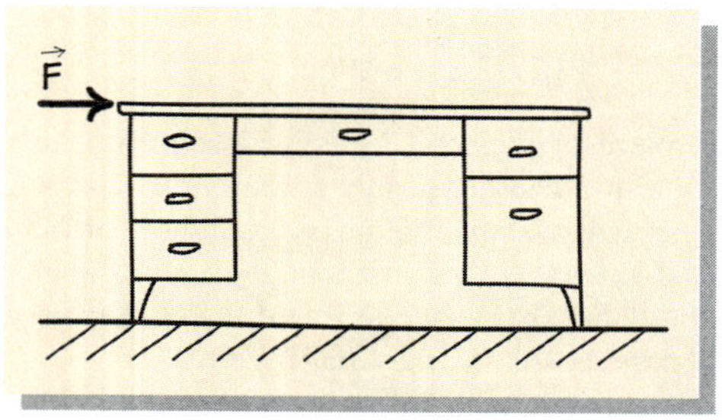

FIGURE 11–28 Problem 25.

26. (II) Repeat Problem 25, but this time assume that the desk is being pushed down a slope of 5.5°. The force applied is parallel to the sloping floor. The center of mass of the desk is in the middle of its long dimension and 0.28 m down from its top surface.

27. (II) Figure 11–29 is a side view of a seat used for babies. It enables a baby to sit at the edge of a table by means of four points of contact; in the side view shown, two of these points are visible—point A at the top of the table, and point B beneath the table. The other two points are aligned directly behind. The center of mass of the baby plus the seat can be approximated by a mass m at point C. (a) Taking into account only forces in the xy-plane, the plane of the figure, calculate the forces *on the table* at points A and B. (Remember that there are four contact points, not two. Assume that the symmetric legs share the force equally.) (b) What happens for $\ell_2 \to 0$? for $\ell_1 \to 0$? (c) Work out numerical values for the forces for $m = 10$ kg, $\ell_1 = 20$ cm, and $\ell_2 = 30$ cm.

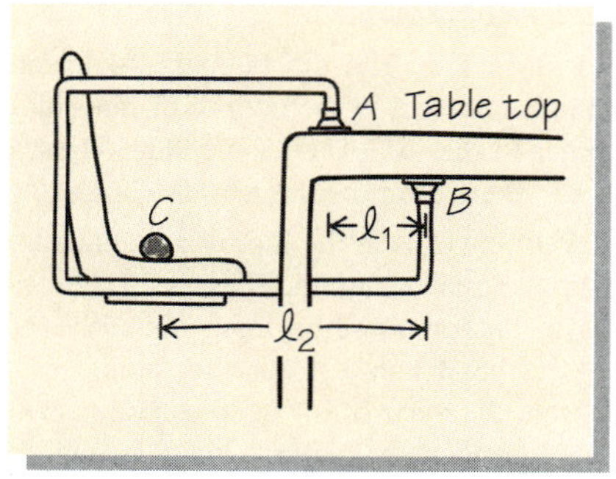

FIGURE 11–29 Problem 27.

28. (II) Consider the baby seat of Problem 27. The table has a total mass $M = 18$ kg, all concentrated in the (uniform) top, and the center of the table is 70 cm from the vertical extension of point P. For the geometry of the baby seat and the table top shown in Fig. 11–30, what mass m of the baby plus the seat will cause the table to tip over by a rotation at point P? Can the possible movements of a baby seriously destabilize the situation?

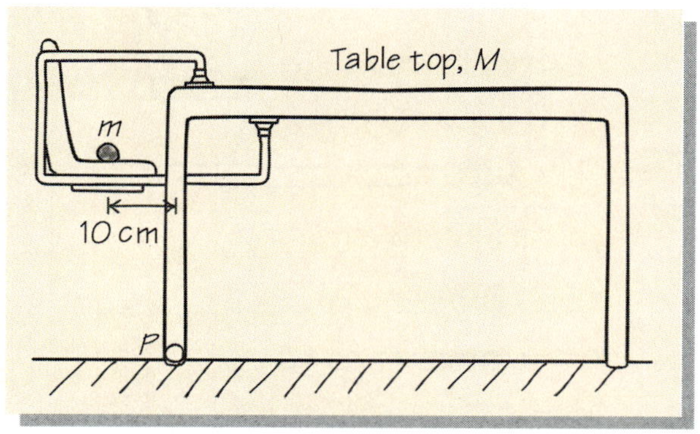

FIGURE 11–30 Problem 28.

29. (II) A trap door 1.8 m square with mass 20 kg is hinged at one edge and is attached to a rope at the opposite edge (Fig. 11–31). The trap door makes an angle of 55° with the horizontal, and the rope is perpendicular to the door. (a) What is the tension in the rope? (b) What is the force vector acting on the door at the hinge?

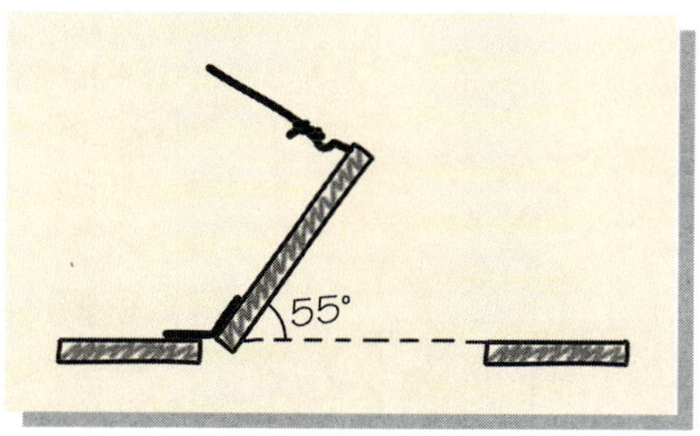

FIGURE 11–31 Problem 29.

30. (II) A beam of mass 10 kg is hinged at the top, and its bottom end is pulled to the side by a horizontal rope. The beam makes an angle of 30° with the vertical. What is the tension in the rope, and what is the force exerted on the beam at the suspension point?

31. (II) A variant of the crane in Example 11–3 is shown in Fig. 11–32. The pivot at point B is frictionless, and the beam, of length 3 m, has mass 100 kg. The rope makes an angle of 30° with the horizontal and can withstand a tension of 10,000 N before breaking. If this arrangement is used to lift masses from the point shown on the beam, what is the maximum mass that can be so raised?

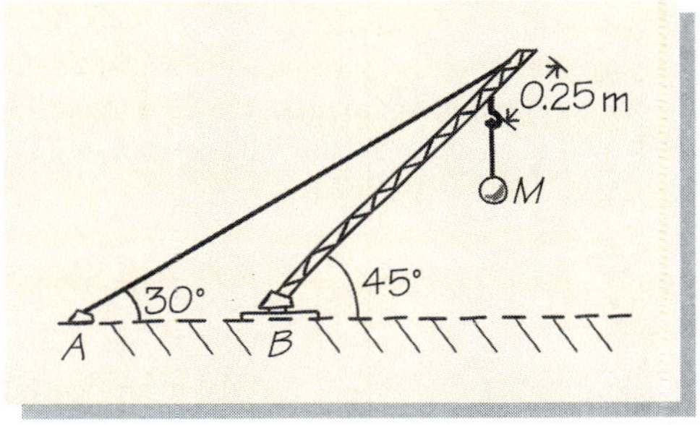

FIGURE 11–32 Problem 31.

32. (II) A uniform door is attached to the wall by two hinges. Draw a force diagram for the door and describe the various forces that can act. Describe the conditions for static equilibrium, and find the forces that are solvable. Is this an underdetermined system? If so, which forces are underdetermined?

33. (II) Consider the four-legged table discussed in Section 11–4, with leg 1 slightly shorter than the others. Suppose that a teacup were placed on the diagonal between legs 1 and 4, at location

$$(x, y) = \left(\frac{L}{2} + x, \frac{L}{2} + x\right).$$

(a) Find a satisfactory solution to the magnitude of the contact forces on the legs if $x > 0$. Neglect the masses of the legs. (b) Repeat part (a) for $x < 0$.

34. (II) A large, spherical satellite is held in place in the bay of a space station with six ropes: four are attached and equally spaced around the equator and the other two are attached at the poles. (a) What equations describe the situation in which the satellite is held motionless with respect to the station? (b) Is this an underdetermined system? (c) Is the system underdetermined when all the ropes are under identical tension?

35. (III) A seaside tower is supported by a cable (Fig. 11–33). A horizontal wind that increases with height often blows from the left, exerting a horizontal force to the right on the tower. The horizontal force on a unit length of the tower increases with the height h according to force/unit length $= \alpha h$, where h is the height from ground level, and, if h is measured in meters, $\alpha = 50\ \text{N/m}^2$. The total height of the tower is 20 m. What is the tension in the cable?

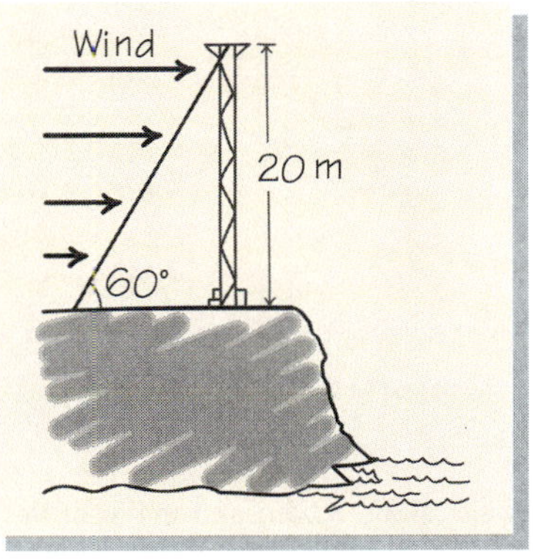

FIGURE 11–33 Problem 35.

36. (III) The mythical Greek king Sisyphus, pushing a large, round rock up a mountain, wishes to take a rest and supports the rock with a horizontal rope attached to the top of the rock (Fig. 11–34). The coefficient of static friction between the rock and the slope is $\mu_s = 0.6$. (a) What is the largest value of θ for which this method of support is possible? (b) If the mass of the rock is 288 kg, what tension must the rope be able to support on this maximum slope? (c) How does the tension vary as a function of θ for angles less than the maximum angle? [*Hint*: It is easiest to take moments about the point of contact.]

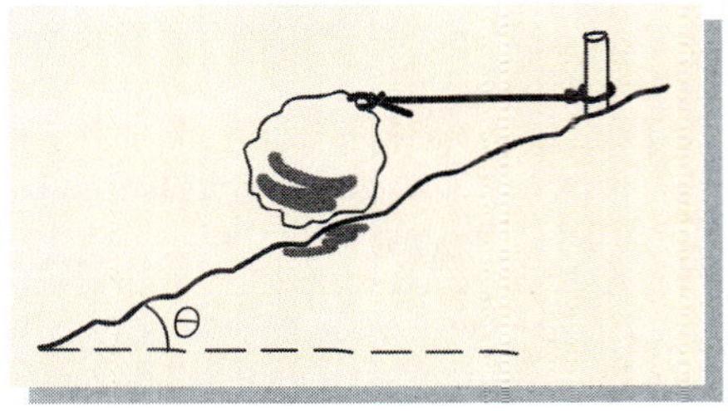

FIGURE 11–34 Problem 36.

General Problems

37. (II) A 30-cm-wide shelf is supported at the wall by an L-shaped bracket and a cable at each end of the shelf placed at 45° to the wall (Fig. 11–35). A 20-kg sack of potatoes is placed and centered on the shelf. What is the tension of the cable if we ignore the mass of the shelf?

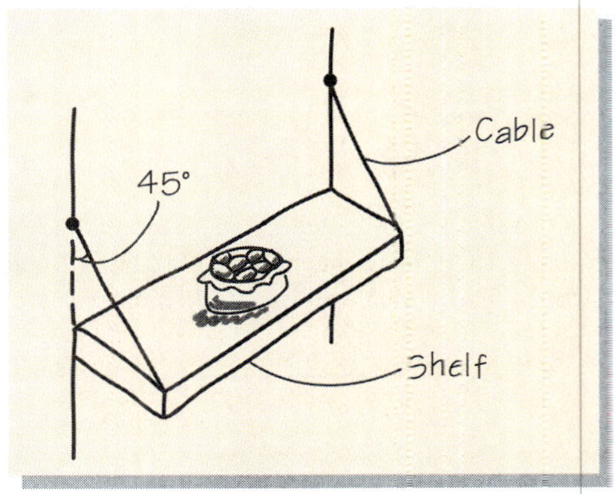

FIGURE 11–35 Problem 37.

38. (II) The cross section of an A-frame house is shown in Fig. 11–36. The total height of the apex is 5.0 m, and the 1.5-m-long crossbeam is two-thirds of the way up the roof line. The crossbeam must support two roof beams, each of mass 3000 kg. (a) Does the crossbeam push the roof beams out or does it pull them in? (b) What is the force exerted by the roof beams on the crossbeam?

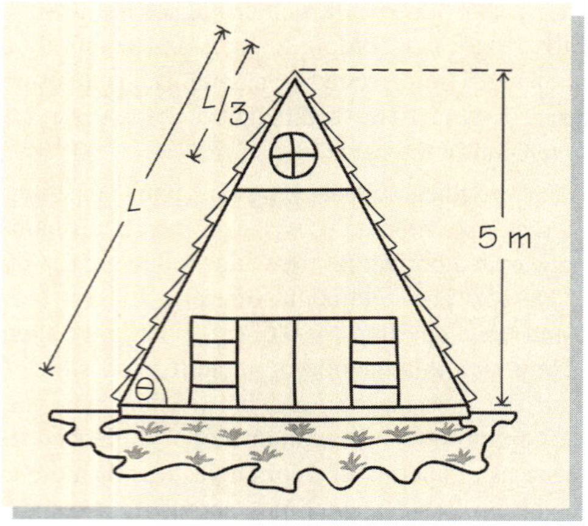

FIGURE 11–36 Problem 38.

39. (II) A uniform beam of length L and mass M is freely pivoted at one end about an attachment point in a wall. The other end is supported by a horizontal cable also attached to the wall, so that the beam makes an angle θ_0 with the horizontal (Fig. 11–37). (a) What is the tension in the cable? (b) The cable snaps. What is the angular acceleration of the beam about its pivot point immediately afterward? (c) What is the angular velocity of the beam as it falls through the horizontal position?

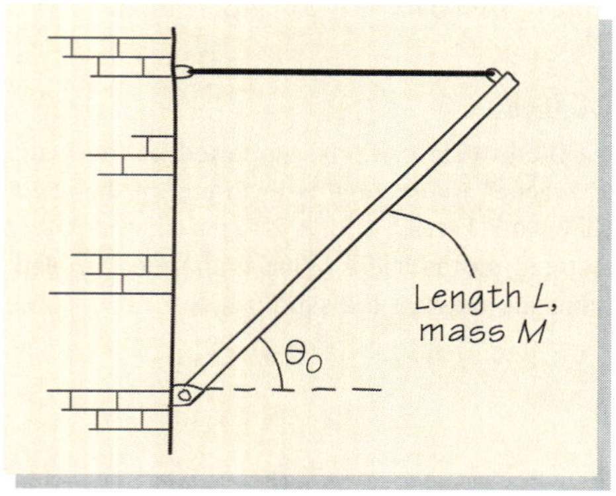

FIGURE 11–37 Problem 39.

40. (II) A uniform board of length 2.4 m and weight 47 N has one end on the ground. With the aid of a horizontal force applied at the upper end by means of an attached horizontal rope, the board is held at an angle θ with respect to the vertical (Fig. 11–38). The coefficient of static friction between the end of the board and the ground is $\mu_s = 0.32$. What is the range of angles the board can make with the vertical and still be in static equilibrium? How does the tension in the rope vary with the angle within the angle's possible range?

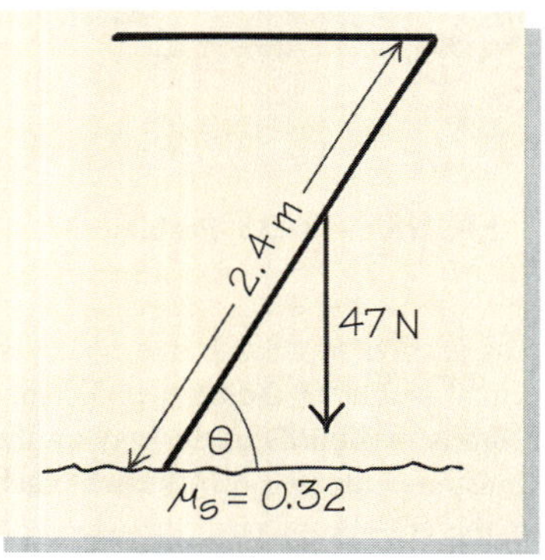

FIGURE 11–38 Problem 40.

41. (II) A sign is to be constructed from a piece of plywood in the shape of a 30° right triangle of mass 15 kg. It is to be attached to a wall, as shown in Fig. 11–39. The lower attachment point is a frictionless pivot, and the upper point is a rope that can be reeled in or out to make the bottom of the sign and the rope itself horizontal. What is the tension in the rope?

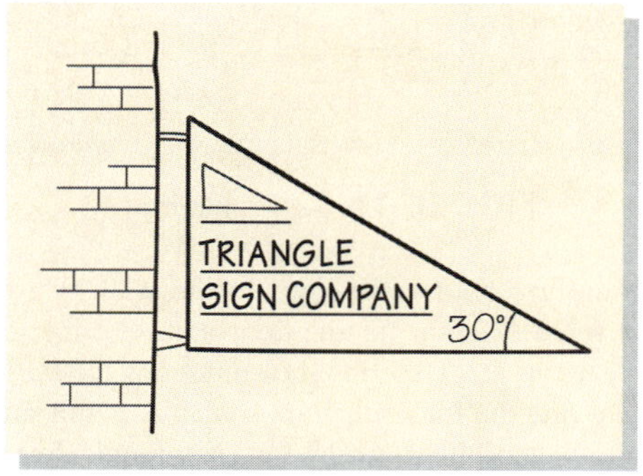

FIGURE 11–39 Problem 41.

42. (II) A car is lifted vertically by a jack placed at the car's rear end 40 cm off the central axis, so that the weight of the car is supported by the jack and the two front wheels (Fig. 11–40). The distance between the front wheels is 1.60 m, the distance from the axis connecting the two wheels to the center of mass of the car is 80 cm, and the distance from the rear of the car to the center of mass is 2.10 m. What fraction of the car's weight is carried by each of the wheels, and what fraction is carried by the jack? (Note that the weight on the wheels will *not* be symmetrically distributed.)

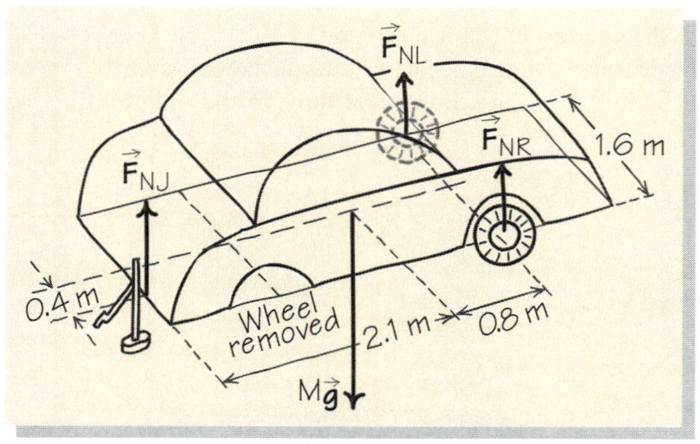

FIGURE 11–40 Problem 42.

43. (II) A roller of radius 30 cm and mass 80 kg is pulled by a force F applied horizontally to the axle. How large must F be in order for the roller to climb a step 15 cm high?

44. (II) Consider the roller of Problem 43. Assume that the force F can be applied in any direction. What direction minimizes the magnitude of F necessary to make the roller climb the step?

45. (II) A chest of drawers 58 cm wide, 1.6 m long, and 1 m high is pulled by a horizontal rope attached at a height of 58 cm to the midpoint of the long side. The force is such that the chest moves with uniform velocity. (a) Express the force on the legs on the side with the rope, and the force on the legs on the opposite side, in terms of the coefficient of kinetic friction μ_k. (b) For what value of μ_k will the chest topple over?

46. (II) A box 0.8 m long and 1.2 m tall is placed on a flat-bed truck. The truck accelerates at a rate of 2 m/s^2. Will the box topple over?

47. (II) A cylinder of mass M and radius R rests on an inclined plane (Fig. 11–41). It is held in place by a horizontal string that is attached to the edge of the cylinder. If the angle that the plane makes with the horizontal is θ, and the coefficient of static friction between the cylinder and the plane is μ_s, what is the smallest value of μ_s that will maintain this position as an equilibrium position?

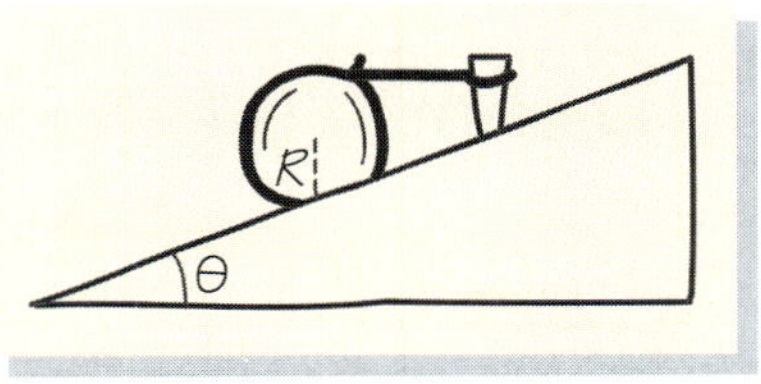

FIGURE 11–41 Problem 47.

48. (II) A centrifugal governor consists of light rods pivoted at points A, B and C, as shown in Fig. 11–42. They are loaded with masses M, m, and m respectively, and the whole apparatus rotates about the vertical axis. Depending on the angular velocity, the mass M slides up or down, and its position can be used to control the flow of steam in a steam engine. Derive a relationship between the angular velocity and the angle the arms make with the vertical.

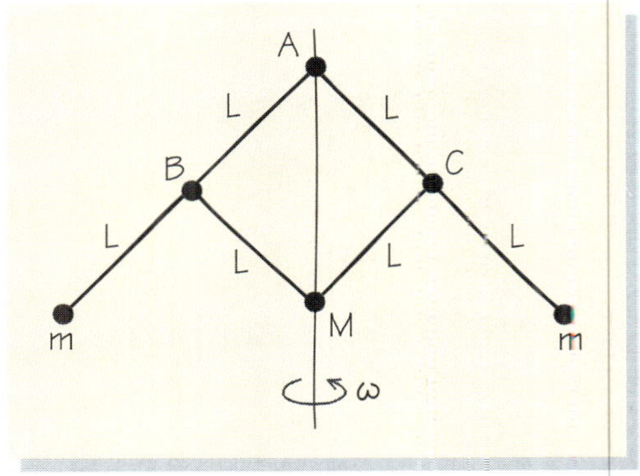

FIGURE 11–42 Problem 48.

49. (III) Two steel ball bearings of mass m and radius 2 cm are placed in a vertical tube of diameter 6 cm that is closed on the bottom. Find all the forces on the two balls.

50. (III) Three steel ball bearings of mass m and radius 2 cm are placed in a closed vertical tube of diameter 6 cm that is closed on the bottom. Find the forces on the middle ball bearing.

CHAPTER

12

This montage of photos from the Voyager I satellite shows four of Jupiter's largest moons. The force that is largely responsible for their motions—gravitation—is the same force that governs the motion of a falling apple.

Gravitation

Most of the forces that we see every day are disguised forms of the more basic forces that govern the interactions of matter. Examples include friction, tension, and contact forces, which are all ultimately explained by the forces holding the components of matter together. There is, however, one important exception, and that is gravity. Gravity is simply a special limit of the force of universal gravitation; as far as we know today, gravitation is one of the truly fundamental forces. Gravitation governs not only the motion of a falling apple but also the majestic orbital motions of the Moon around Earth, the planets around the Sun, and the stars in their voyage around the galaxy. In 1687, Isaac Newton published the realization in his *Principia* that the same force governs both the falling apple and the orbiting Moon; this was one of the great intellectual leaps in human history.

Happily, the law of gravitation is a force law for which the equations of motion can be solved. We can thus find a full description of the motion of astronomical bodies. Like the other fundamental forces, gravitation is conservative and lends itself to a description in terms of potential energy. As we will see in this chapter, the variety of motions possible when gravitation acts presents a rich and beautiful design.

12-1 FUNDAMENTAL FORCES

The forces that act on an object determine its motion. We can distinguish between fundamental (or basic) forces and derived forces. Fundamental forces are those for which we cannot find (or have not yet found) an underlying force from which they are derived. Derived forces result from the operation of some underlying fundamental force.

Gravitation is a fundamental force.

For example, friction and the spring force are both derived forces. In the final analysis, these forces result from the forces between molecules, and *these forces turn out to be electric forces*, which *are* fundamental. The fundamental forces are all conservative, and that is why energy conservation is such an important principle.

In the realm of classical physics and in most parts of atomic physics, there are only two domains in which the fundamental forces appear undisguised. For both the gravitational force between masses and the electromagnetic forces between charges, *the force that is observed is the basic force.*

The development of the theory of gravitation is one of the important achievements of physics, both in its Newtonian form and in its later elaboration by Einstein. It is the Newtonian form of the law, combined with advances in computing technology, that allows us to plan interplanetary journeys for probes that can pass the planets Jupiter, Uranus, and Neptune on one long trajectory—even with our present chemical sources of fuel (Fig. 12–1).

12–2 EARLY OBSERVATIONS OF PLANETARY MOTION

The earliest astronomical observations of the night sky led ancient peoples to divide the points of light they saw into two classes: the so-called fixed stars, which move in seemingly perfect circles around Earth, and the planets (in Greek, the "wanderers"), which appear to move in complicated, erratic patterns in the night sky. The first interpretations of these observations placed Earth at the center of the universe, in a **geocentric** frame. Earth was pictured as being surrounded by a rotating spherical shell in which the fixed stars were embedded. The planets, the Sun, and the Moon seemed to have more complicated motions—motions explained by placing the planets upon moving, transparent shells within the outer, fixed-star shell. These shells had to move in complicated ways if the observations were to be consistent with this frame. In the second century A.D., Claudius Ptolemy made the most detailed formulation of the notions just outlined. To explain the various paths taken by planets, he had to depart from the simple spherical shell description and construct the paths of motion to form circles that themselves lay on circles around Earth (Fig. 12–2). His theory was limited by a culturally imposed belief that perfect, circular motion describes celestial motion. His *epicy-*

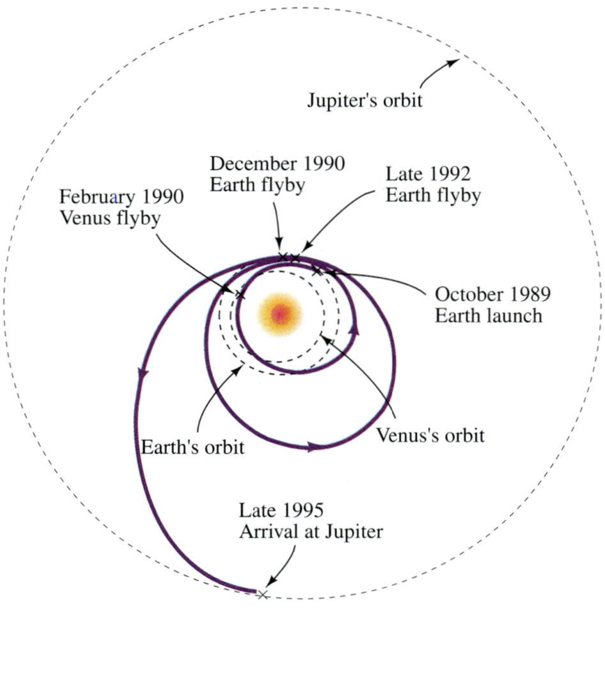

FIGURE 12–1 An understanding of the inverse-square force law permits us to calculate numerically rather complicated satellite trajectories. The satellite *Galileo* was launched in 1989 by NASA to study Jupiter and its atmosphere.

cles (the circles on circles used to describe the movements of the planets) gave a reasonable description of the motion of the planets, and the idea proved serviceable for over 1400 years.

The Copernican Picture

In 1543, Nicolaus Copernicus introduced the **heliocentric** frame, with the Sun at the center of the solar system. Unfortunately, Copernicus continued to insist on describing all motions with circles and, because the true motions of the planets about the Sun are not circles, epicycles continued to be needed in the Copernican description to accommodate the observations of planetary motion (Fig. 12–3). The apparent immobility of the stars was ascribed to their great distance. Partly because the epicycles meant that the Copernican description was not really more economical than the Ptolemaic description, the iconoclastic Copernican hypothesis was not accepted for almost a century. During this transitional period, the construction of more refined instruments—there were still no telescopes—allowed Tycho Brahe, toward the end of the sixteenth century, to improve the knowledge of planetary **orbits**, or paths, to an accuracy of less than half a minute of arc.

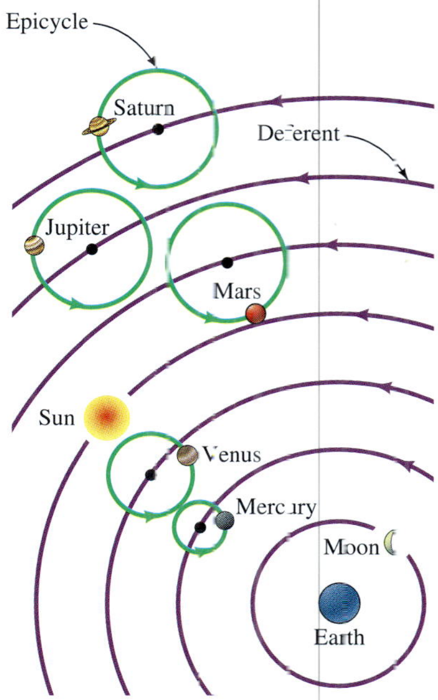

FIGURE 12–2 Diagram of planetary motion based on the Ptolemaic system. According to Ptolemy, Earth is the center of the world system. Planets move in small circular paths called epicycles, and the centers of the epicycles move around Earth on large circles called deferents. (After D. Scott Birney, *Modern Astronomy,* Allyn and Bacon, 1969.)

Kepler's Laws

At Brahe's death in 1601, his assistant Johannes Kepler inherited the data that Brahe had accumulated. Kepler spent some 20 years analyzing these data, looking for mathematical regularities. He came to the conclusion that the idea of circular orbits should be discarded and replaced with elliptical orbits. This was one of the crucial breakthroughs in the data analysis. **Kepler's laws** summarize Kepler's most important conclusions:

Kepler's laws of planetary motion

1. Planets move in planar elliptical paths with the Sun at one focus of the ellipse (Fig. 12–4a).
2. During equal time intervals, the radius vector from the Sun to a planet sweeps out equal areas (Fig. 12–4b).
3. If T is the time that it takes for a planet to make one full revolution around the Sun, and if R is half the major axis of the ellipse (R reduces to the radius of the planet's orbit if that orbit is circular), then

$$\frac{T^2}{R^3} = C, \qquad (12\text{–}1)$$

where C is a constant *whose value is the same for all planets.*

Kepler's laws were so simple that it was no longer possible for scientists to cling to the pre-Copernican ideas (Fig. 12–5).

Kepler's second law (the equal-area law) follows from the conservation of angular momentum.

We have already found the origin of Kepler's second law in Chapter 10: Eq. (10–31). The law follows from the conservation of angular momentum and it is a consequence of the fact that *the gravitational force between the Sun and each planet is central*; that is, the force acts along the line between the Sun and the planet. In fact, Kepler's second law can be taken as evidence that the gravitational force is central. Conservation of angular momentum also implies that the trajectories of planets must lie in a plane—the plane that is perpendicular to the direction of the fixed angular momentum vector.

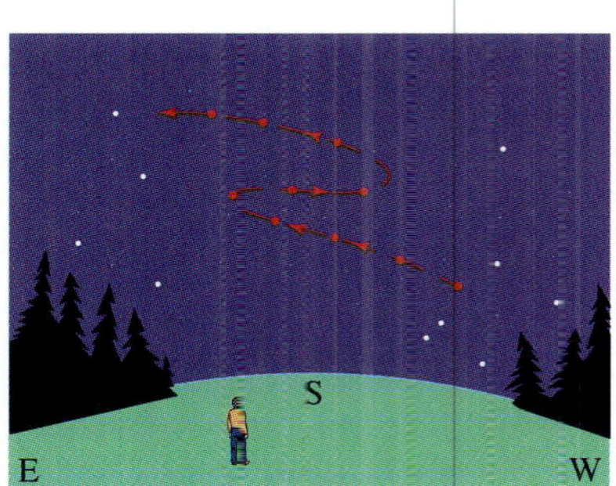

FIGURE 12–3 The Ptolemaic universe has a difficult time explaining the retrograde motion of Mars shown here. The Sun and Moon move west to east relative to fixed stars from night to night, but planets do a retrograde "loop" from east to west every once in a while. Only the Keplerian description explains these observations.

12–3 NEWTON'S INVERSE-SQUARE LAW

Newton understood the importance of Kepler's laws. Because the planets do not move in straight lines, Newton realized that they must be subject to a net force. He concluded from Kepler's second law that the net force on a planet must point from the planet to

Kepler's first law (elliptical orbits) follows from a central inverse-square force law.

the Sun. He also found that the elliptical paths described by Kepler are the consequence of an *inverse-square force law:* If the source of the force is at the origin O and the mass on which the force acts is at position $\mathbf{r}$, then the force has magnitude k/r^2, where k is a constant.

More generally, Newton showed that the paths of the mass under the influence of this force had to be conic sections. **Conic sections** are the shapes obtained if a cone is sliced by a plane. As can be seen in Fig. 12–6a, if a cone is sliced by a plane parallel to the cone's base, the conic section is a **circle**. If the plane is tilted a little, the conic section is an **ellipse**. If the slice is cut parallel to the slope of the cone, the conic section turns out to be a **parabola**. If the slice is steeper than the slope of the cone, the conic section is a **hyperbola**. The planets from which Kepler discovered his first law move in closed elliptical orbits, but there are astronomical bodies that move along trajectories described by all these mathematical curves (Fig. 12–6b).

By postulating a central inverse-square force law, Newton was able to explain not only Kepler's first two laws, but Kepler's third law as well. To deal with Kepler's third law, let's consider the very special case of a circular orbit of radius R. The force has magnitude[†] k/R^2. We know from Chapter 3 that the acceleration for an object moving with speed v in a circle of radius R is v^2/R. We can now apply Newton's law of motion $F = ma$ for the case of a planet of mass m, so that F is replaced by k/R^2 and ma is replaced by mv^2/R:

$$\frac{k}{R^2} = \frac{mv^2}{R}. \tag{12–2}$$

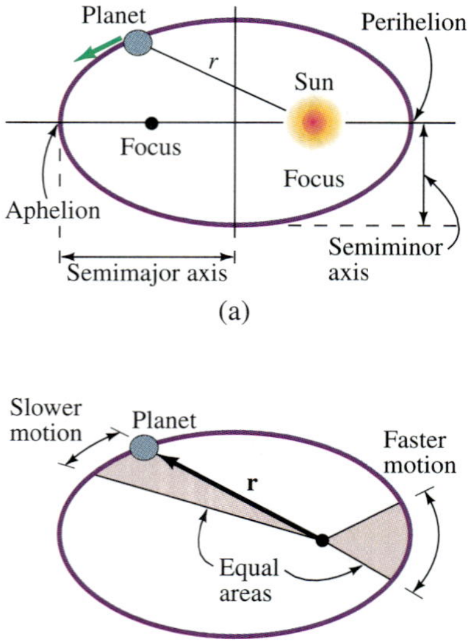

FIGURE 12–4 (a) Kepler's first law specifies that the planets move in elliptical orbits with the Sun at one focus. The perihelion is the distance of closest approach and the aphelion is the farthest distance between a planet and the Sun. (b) Kepler's second law specifies that the radius vector from the Sun to a planet sweeps out equal areas in equal times. Thus the two shaded areas shown are equal if the same time is taken to sweep out each of them. This means that the planet must move more quickly over a part of the orbit closer to the Sun than over a more distant part.

FIGURE 12–5 Galileo's early improvements to the telescope allowed him to make critical observations—including observing the moons of Jupiter and the phases of Venus—that helped destroy the pre-Copernican point of view. Unfortunately, the Church treated Galileo harshly for propagating his findings.

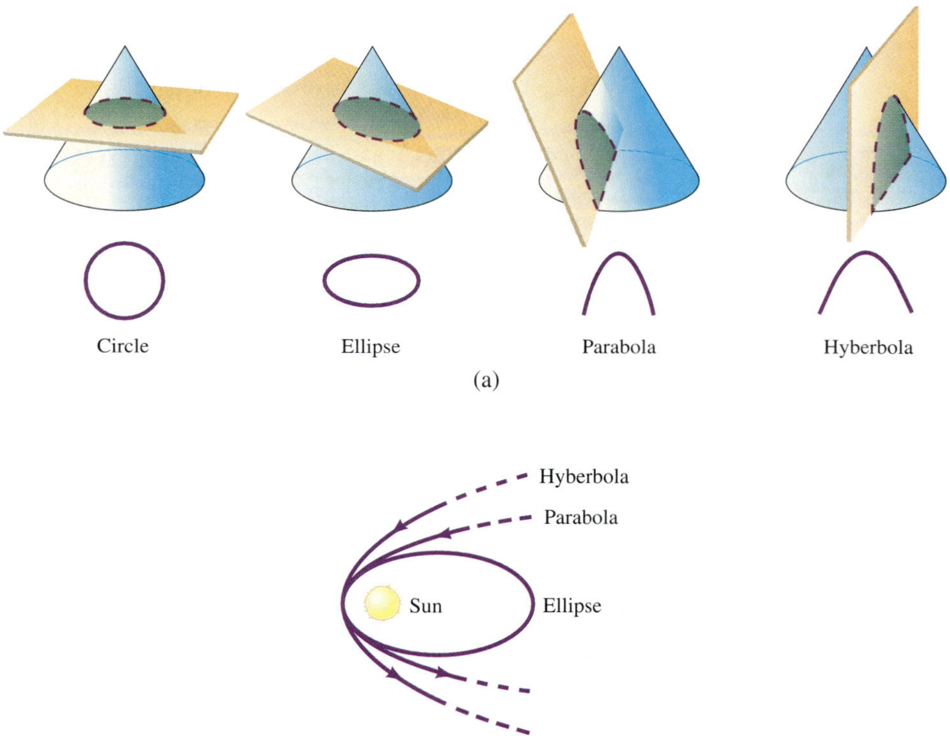

FIGURE 12–6 (a) Slicing a cone with a plane produces conic sections. Several familiar geometric curves are shown here. (b) The orbits of astronomical objects such as planets can follow all the curves drawn in (a).

[†]We are assuming that the Sun is (nearly) fixed, an assumption justified by the Sun's large mass.

In terms of the *period*, T, of the planet's orbital rotation, we have v = circumference/T, so that $T = 2\pi R/v$. Squaring, we have

$$T^2 = \frac{4\pi^2 R^2}{v^2}.$$

If we now substitute $v^2 = kR/mR^2 = k/mR$ from Eq. (12–2), we find

$$T^2 = \frac{4\pi^2 R^2}{k/mR} = \left(\frac{4\pi^2 m}{k}\right) R^3. \tag{12–3}$$

This is just the statement of Kepler's third law as it applies to a circle, with the additional condition that the constant $C = (4\pi^2 m/k)$ in Eq. (12–1) should be the same for all planets—as it will be if k is proportional to the mass of the planet.

Kepler's third law follows from a central inverse-square force law.

The Law of Universal Gravitation

The constant k just described involves the force between two objects; by Newton's third law, there is a symmetry between these objects as they appear in the force law. If k is proportional to one of the masses, it must also be proportional to the other one. Then k must have the form GmM, where m is the mass of one object, M is the mass of the other, and G is a proportionality constant called the **gravitational constant**. Newton put all this together into what is now known as **Newton's law of universal gravitation**: The force of gravitation acting on a point mass of mass m due to another point mass of mass M is an attractive force with an inverse-square form,

Newton's law of universal gravitation

$$\mathbf{F} = -\left(\frac{GmM}{r^2}\right)\hat{\mathbf{r}}. \tag{12–4}$$

Here $\hat{\mathbf{r}}$ is the unit vector in the direction from the mass M to the mass m, and r is the distance separating them. The minus sign indicates that the force is attractive. Of course, Newton's third law states that if Eq. (12–4) is the force exerted on m by M, then its negative is the force exerted on M by m. Mass M is attracted to m with the same magnitude of force that m is attracted to M.

We can now return briefly to Kepler's third law. According to Newton, the constant C in Kepler's third law is given by

$$C = \frac{4\pi^2 m}{k} = \frac{4\pi^2 m}{GmM} = \frac{4\pi^2}{GM}. \tag{12–5}$$

Kepler's third law, Eq. (12–3), thus takes the form

$$T^2 = \frac{4\pi^2 R^3}{GM}. \tag{12–6}$$

The Gravitational Constant. The constant G must be determined from experimental data. It characterizes the strength of the gravitational force. Its dimensions are $[L^3M^{-1}T^{-2}]$, and its units in SI are $\mathrm{N \cdot m^2/kg^2}$.

The constant G in Eq. (12–4) cannot be measured independently unless the two masses involved are known; for this reason, G cannot be measured for astronomical objects such as Earth, the Moon, or the Sun. The value of G was first determined by Henry Cavendish in 1798. The basic idea of his experiment is shown in Fig. 12–7. Two masses m at the end of a rod of negligible mass and length $2L$ are suspended by a quartz fiber. Large masses M are placed near the masses m, with their centers of mass separated by a distance d. The magnitude of the torque about the point of suspension due to the force of gravitation on the masses is counterbalanced by a second torque due to the resistance of the fiber to twisting and proportional to the angle of deflection

The gravitational constant

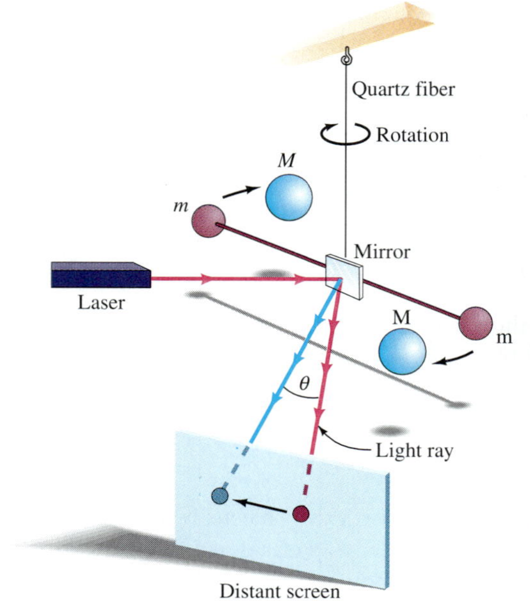

(a)

(b)

FIGURE 12–7 (a) Schematic diagram of Cavendish's experiment. The attraction of the smaller masses to the larger ones twists the fiber holding the rod. A light shining on a mirror on the rod indicates the small rotation by reflection on a distant screen. Switching the positions of the large masses reverses the rotation. (b) A Cavendish experiment apparatus.

θ. When this resistance is known, the torque due to the gravitational force, and hence G, can be determined. The most recent measurements lead to the value

$$G = 6.673 \times 10^{-11}\ \mathrm{N \cdot m^2/kg^2}. \tag{12–7}$$

with an uncertainty of about 0.06 percent. The Cavendish experiment is not a simple one. The magnitude of the gravitational force between two 10-kg masses separated by 10 cm is only about 7×10^{-7} N.

The Potential Energy Associated with the Gravitational Force

Potential energy is discussed in Chapter 7.

The potential energy of a system of two masses that interact with each other through the gravitational force

The gravitational force is central and depends only on the distance of the influenced object from the force center. It is therefore *conservative* and can be derived from a potential energy function. We show here that the potential energy of a system of two point masses interacting with each other through the gravitational force (Fig. 12–8) is

$$U(r) = -\frac{GmM}{r}. \tag{12–8}$$

We have chosen the potential energy to be zero at infinity. To demonstrate this result, we recall the definition of potential energy from Chapter 7, Eq. (7–9):

$$U(r) - U(\infty) = -\int_{\infty}^{r} \mathbf{F}(\mathbf{r}') \cdot \mathrm{d}\mathbf{r}',$$

where we have chosen the point x_0 in Eq. (7–9) at ∞. The force points from the location of mass m to the origin (the location of M), and we choose the path to go directly along a radial direction so that $\hat{\mathbf{r}} \cdot \mathrm{d}\mathbf{r}' = \mathrm{d}r'$. (Any path gives the same result: The force is conservative.) We thus obtain

$$U(r) - U(\infty) = -\int_{\infty}^{r} \frac{-GmM}{r'^2}\,\mathrm{d}r' = -\frac{GmM}{r'}\bigg|_{\infty}^{r} = -\frac{GmM}{r}.$$

By choosing $U(\infty) = 0$, Eq. (12–8) follows.

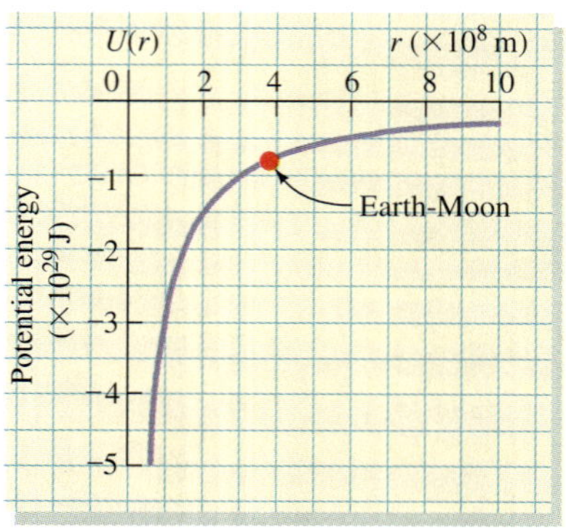

FIGURE 12–8 The potential energy of the gravitational force is negative and has a $1/r$ dependence. We define it to be zero at $r = \infty$.

The Superposition Principle

The gravitational force obeys the **principle of superposition**: When several particles of varying masses exert gravitational forces on a point mass m, the net force on m is the (vectorial) sum of the individual forces. The same idea holds for potential energies.

The potential energy of a point mass m at position $\mathbf{r}$ due to a point mass M_1 at position $\mathbf{r}_1$ is $U_1 = -GmM_1/|\mathbf{r} - \mathbf{r}_1|$, and the potential energy of m due to a point mass M_2 at position $\mathbf{r}_2$ is $U_2 = -GmM_2/|\mathbf{r} - \mathbf{r}_2|$. Then, if both M_1 and M_2 are present in their original positions, the potential energy U of m in the presence of the two masses is the *sum* of the potential energies U_1 and U_2. If the source of the gravitational force or potential energy for a point mass is an extended continuous object, we can apply the principle of superposition through integration.

The superposition principle applied to gravitation. See also Section 4–1.

PROBLEM SOLVING

Use superposition for a continuous object.

EXAMPLE 12–1 A satellite is to be sent to the position between the Moon and Earth where there is no net gravitational force on an object due to those two bodies. Locate that point.

Solution: Figure 12–9 shows us that the force on the satellite due to the Moon, $\mathbf{F}_M$, is to the right, while the corresponding force due to Earth, $\mathbf{F}_E$, is to the left. Let's write the Earth–Moon distance as d, while the Earth–satellite distance is x, and the Moon–satellite distance is $(d - x)$. The net force on the satellite then has magnitude

$$F_{\text{net}} = F_M - F_E = Gm\left[\frac{M_E}{(x)^2} - \frac{M_M}{(d-x)^2}\right].$$

The condition that this equal zero determines x

$$\left[\frac{M_E}{(x)^2} - \frac{M_M}{(d-x)^2}\right] = 0,$$

or

$$(d-x)^2 = x^2R,$$

where we have defined the ratio $R = M_M/M_E = 0.0123$ (see Appendix III–1.1). This quadratic equation for x has the following solution with x between 0 and d:

$$x = \frac{1-\sqrt{R}}{1-R}d = \frac{1-\sqrt{0.0123}}{1-0.0123}d = 0.900\,d.$$

The desired position is thus 9/10ths of the way to the Moon. The result is independent of the mass of the satellite.

FIGURE 12–9 Example 12–1. This view of the Moon from some 17,000 km was taken by the Apollo 11 astronauts on their way home. The whole Earth view was taken by the Apollo 17 astronauts on their way to the Moon. The satellite feels gravitational forces from both the Moon and Earth.

Measuring Earth's Mass

Local gravity is just a manifestation of the general law of gravitation. Later, we will discuss the fact that the gravitational force due to a spherically symmetric object—Earth is a good approximation of one—acts as if the mass of the object were con-

centrated at its center. Given this statement, we can determine Earth's mass, M_E, from the size of the acceleration due to gravity, g, and Earth's radius, R. The force exerted on an object of mass m on Earth's surface by Earth's gravitational attraction is given by

Finding Earth's mass

$$F = \frac{GmM_E}{R^2} = mg. \qquad (12\text{–}9)$$

We thus obtain

$$M_E = \frac{gR^2}{G} = \frac{(9.80\ \text{m/s}^2)(6.37 \times 10^6\ \text{m})^2}{6.673 \times 10^{-11}\ \text{N}\cdot\text{m}^2/\text{kg}^2} = 5.96 \times 10^{24}\ \text{kg},$$

where we have used the measured values of g, R, and G. This mass determination shows that Earth has an average density of 5.5 g/cm^3.

In making this calculation, we are using what might be called the *central dogma* of physics: The laws of nature are the same everywhere, be it Cavendish's laboratory, the environs of Earth, the solar system, or extragalactic space.

12–4 PLANETS AND SATELLITES

The force of gravitation [Eq. (12–4)] determines the trajectories (or orbits) followed by astronomical bodies and other properties of their motion. Newton's second law for the orbit of a single object subject to the gravitational force law due to a single source can be solved to give an *analytic form for the orbit.* (This is certainly not true of *every* force law!)

Escape Speed

It is simplest to approach the orbit of a single object—we refer to such objects as satellites—from the point of view of energy. The energy is the sum of the kinetic and potential energies and is thus given by

$$E = K + U = \frac{1}{2}mv^2 - \frac{GmM}{r}. \qquad (12\text{–}10)$$

The escape speed

The **escape speed**, v_{esc}, of a projectile launched from Earth's surface is the minimum speed with which the projectile must leave the surface in order to leave the vicinity of Earth forever; that is, travel an infinite distance from Earth. In other words, the escape speed is the speed at $r = R_E$ that gives zero speed at infinite r. The object will have zero kinetic energy at $r = \infty$, and the potential energy is defined to be zero at that point, so the total energy will be $E = 0$ at infinity. Because the total energy is conserved, we can set $E = 0$ everywhere; then, the condition that $E = 0$ at the surface is the condition that determines v_{esc}. Let the mass of the projectile be m and Earth's mass be M. At Earth's surface, $r = R_E$, and

$$E = \frac{1}{2}mv^2_{\text{esc}} - \frac{GMm}{R_E} = 0.$$

If we solve this equation for v_{esc}, the factor m cancels, and we find that

$$v_{\text{esc}} = \sqrt{\frac{2GM}{R_E}}. \qquad (12\text{–}11)$$

With the known values of G, M, and R_E, we find that the escape speed from Earth's surface is $v_{\text{esc}} = 1.12 \times 10^4$ m/s $= 11.2$ km/s.

It is interesting that the escape speed is independent of the direction in which the object leaves the surface. (Of course, the object cannot pass through Earth itself!) Even if the object leaves in an initially horizontal direction, it will still escape.

Types of Orbits

We can classify orbits according to the sign of the energy in Eq. (12–10). If the energy is positive, then as r becomes infinite, $|v| > 0$. This corresponds to an orbit that never closes and it is, in fact, a hyperbolic orbit. When the energy of the object is exactly zero, we have the special case of a parabolic orbit. The object starts at infinity with a speed of zero, swings past the Sun in a parabolic orbit, and slows down as it moves away from the Sun, ending up with zero velocity when r becomes infinitely large. For negative energies, r cannot become too large because then v^2 would become negative, which is impossible. In that case, the orbit is limited and it is elliptical (Fig. 12–10a) or circular. Halley's comet, like other comets, follows a very elongated elliptical orbit (Fig. 12–10b). Circular orbits are the simplest of all, as Examples 2 through 4 illustrate.

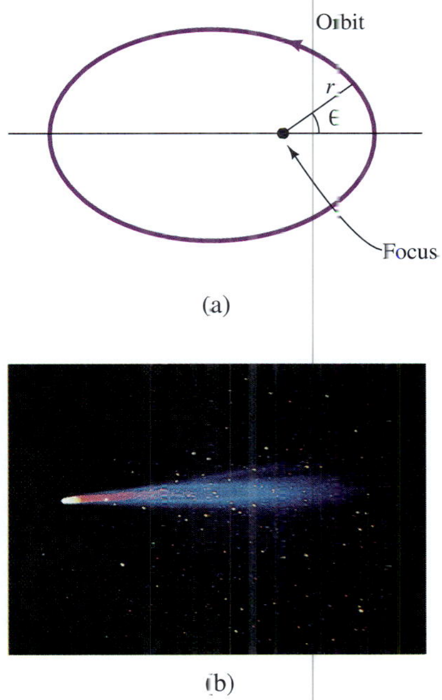

FIGURE 12–10 (a) For negative total energies, the value of r is limited and the orbit is elliptical (or circular). (b) Halley's comet has an elongated elliptical orbit with a period of 76 yr.

EXAMPLE 12–2 A particle of mass m moves in a circular orbit of radius r under the influence of the gravitational force due to an object of mass $M >> m$ (Fig. 12–11). Calculate the total energy of the particle as a function of r.

Solution: Equation (12–10) gives the total energy, but it is a function of both v and r. We want to eliminate the speed. This can be done by using Newton's second law, $\mathbf{F} = m\mathbf{a}$. For a circular orbit, the acceleration is centripetal and has the form $a = v^2/r$, directed toward the center. The force has the magnitude GmM/r^2 and is also directed toward the center. The second law therefore has the form

$$\frac{GmM}{r^2} = \frac{mv^2}{r},$$

or, equivalently,

$$v^2 = \frac{GM}{r}.$$

This equation can be used to eliminate the speed. The total energy is

$$E = \frac{1}{2}mv^2 - \frac{GmM}{r} = \frac{1}{2}m\frac{GM}{r} - \frac{GmM}{r} = -\frac{1}{2}\frac{GmM}{r}.$$

The total energy is just one-half the potential energy for a circular orbit. It is negative, as is appropriate for a closed orbit.

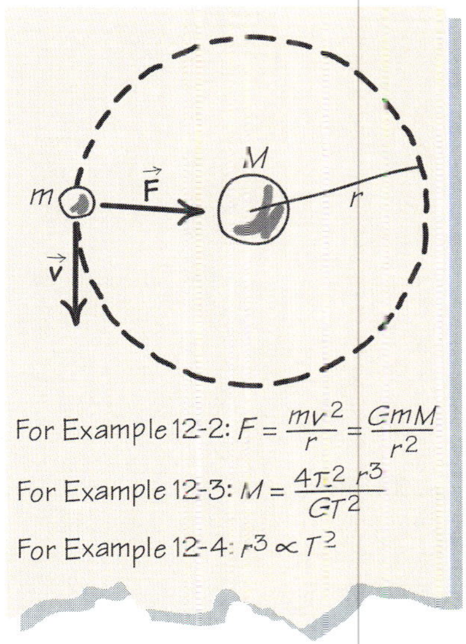

FIGURE 12–11 This simple sketch can be used to help with Examples 12–2, 3, and 4.

EXAMPLE 12–3 Calculate the mass of the Sun, assuming that Earth's orbit around the Sun is circular, with radius $r = 1.5 \times 10^8$ km.

Solution: Kepler's third law relates the distance of Earth from the Sun, the period of Earth's orbit, and the mass of the Sun. This law is expressed by Eq. (12–6), which may be rewritten as

$$M = \frac{4\pi^2 r^3}{GT^2}.$$

Given the value of G, the radius of the orbit, r, and the period, T, which is 365 days = (365 d)(24 h/d)(60 min/h)(60 s/min) = 3.15×10^7 s, we find that

$$M = \frac{4(3.14)^2(1.5 \times 10^{11}\ \text{m})^3}{(6.67 \times 10^{-11}\ \text{N}\cdot\text{m}^2/\text{kg}^2)(3.15 \times 10^7\ \text{s})^2} = 2.0 \times 10^{30}\ \text{kg}.$$

This is a factor of 3×10^5 larger than the mass of Earth.

EXAMPLE 12–4 A satellite moves in a circular orbit around Earth, taking 90 min to complete 1 rev (Fig. 12–12). We are given the following information: The distance from the Moon to Earth is $d_{ME} = 3.84 \times 10^8$ m; the Moon's orbit is circular; the period of the Moon's rotation about Earth is $T_M = 27.32$ d; Earth's radius is

$R_E = 6.37 \times 10^6$ m; and Earth's gravitational force acts as if all of Earth's mass were concentrated at its center. With this information, calculate the height of the satellite above Earth.

Solution: A satellite in circular orbit around Earth behaves like a planet in circular orbit around the Sun. We can thus use Kepler's third law in the simplifying case that the ellipse reduces to a circle: The cube of the radius of the orbit is proportional to the square of the orbital period. To make use of this result, the period of the satellite must be in the same units as that of the Moon. We thus have $T_S = 90$ min $= 1.5$ h $= (1.5 \text{ h})/(1 \text{ d}/24 \text{ h}) = 0.0625$ d. If we call the distance from the satellite to the center of Earth D, then Kepler's third law [Eq. (12–1)] as applied to the satellite is

$$\left(\frac{D}{d_{ME}}\right)^3 = \left(\frac{T_S}{T_M}\right)^2;$$

$$\left(\frac{D}{3.84 \times 10^8 \text{ m}}\right)^3 = \left(\frac{0.0625 \text{ d}}{27.32 \text{ d}}\right)^2;$$

that is,

$$D = (3.84 \times 10^8 \text{ m})(0.00229)^{2/3} = 6.67 \times 10^6 \text{ m}.$$

To find the height above the surface, we must subtract Earth's radius,

$$(6.67 - 6.37) \times 10^6 \text{ m} = 3 \times 10^5 \text{ m} = 300 \text{ km}.$$

Many satellites use this orbit, which is one that lies just above Earth's atmosphere.

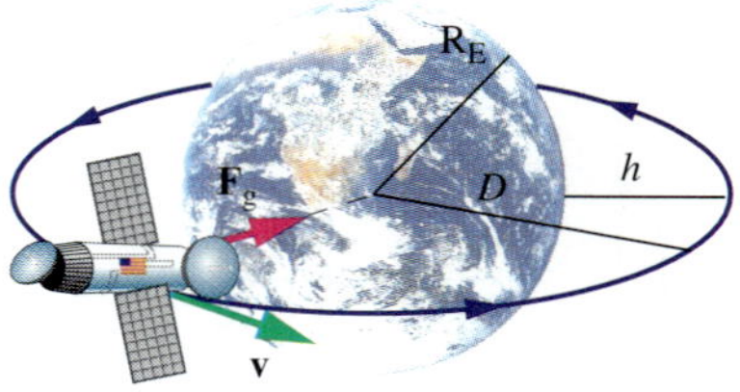

FIGURE 12–12 Example 12–4. A satellite in orbit about Earth.

Properties of Noncircular Orbits. The hyperbolic, parabolic, and elliptical orbits corresponding to different signs of total energy are closely related; as stated in Section 12–3, they are all *conic sections* (see Fig. 12–6). Let's focus our attention on the solar system, although we should keep in mind that our results will be applicable to, say, satellites orbiting Earth as well. The Sun, which is very heavy in comparison to the planets, plays a special role in the orbits. For closed orbits in particular, the Sun sits at one *focus* of the ellipse, a fact noted by Kepler in his first law and illustrated in Fig. 12–4a. The point where the planet makes its closest approach to the Sun ($r = r_{min}$) is called the **perihelion**; the point where the farthest distance between a planet and the Sun is attained ($r = r_{max}$) is the **aphelion**.† The **semimajor axis**, a, is a length that sets the overall size of the orbit (Fig. 12–13). From Fig. 12–4a, we have

$$\text{semimajor axis} = a = \frac{1}{2}(r_{min} + r_{max}). \tag{12–12}$$

The **eccentricity**, e, of the orbit is a dimensionless measurement of the elongation of the orbit and is proportional to the difference between r_{max} and r_{min}:

$$e = \frac{r_{max} - r_{min}}{2a}. \tag{12–13}$$

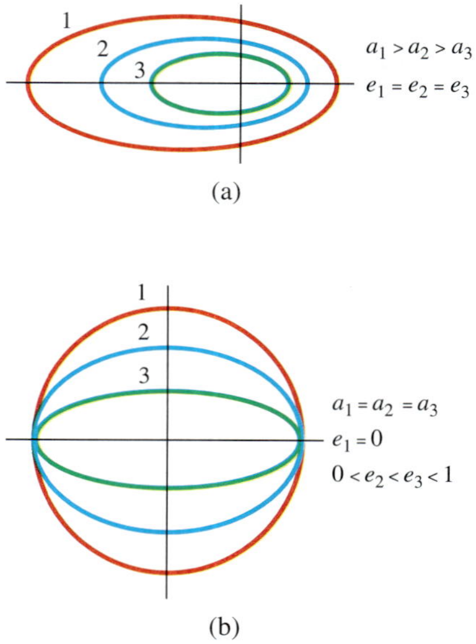

FIGURE 12–13 Two ellipses with (a) the same eccentricity and different semimajor axes; and (b) the same semimajor axes but different eccentricities.

Note that when $r_{max} = r_{min}$, the orbit reduces to a circular one of radius a and the eccentricity is zero. The most extreme possible orbit corresponds to $r_{min} = 0$, when $r_{max} = 2a$, and

$$e_{extreme} = \frac{2a}{2a} = 1.$$

The eccentricity therefore varies between zero and one.

†For Earth satellites, the corresponding words are perigee and apogee.

EXAMPLE 12–5 Consider the motion of a comet in an elliptical orbit around a star. The eccentricity of the orbit is given by $e = 0.20$ and the distance between the perihelion and the aphelion is 1.0×10^8 km. (a) Find the distances of nearest and farthest approaches of the comet. (b) If the speed of the comet is 81 km/s at perihelion, what is its speed at aphelion?

Solution: (a) We are given e, and $2a$ is the sum of the aphelion and perihelion distances. Equations (12–12) and (12–13) are two equations that can be solved for r_{max} and r_{min} in terms of a and e. If we add Eq. (12–12) to a times Eq. (12–13), r_{min} cancels, and

$$r_{max} = a + ae = a(1 + e). \tag{12–14}$$

In turn,

$$r_{min} = 2\left(a - \frac{r_{max}}{2}\right) = 2\left[a - \frac{a(1 - e)}{2}\right] = a(1 - e). \tag{12–15}$$

Numerically, with $e = 0.20$ and $2a = 1.0 \times 10^8$ km,

$$r_{max} = \frac{(1.0 \times 10^8 \text{ km})(1.20)}{2} = 6.0 \times 10^7 \text{ km};$$

$$r_{min} = \frac{(1.0 \times 10^8 \text{ km})(0.80)}{2} = 4.0 \times 10^7 \text{ km}.$$

This would be an unusually circular orbit for a solar comet.

(b) At both perihelion and aphelion, the orbit is at right angles to the vector $\mathbf{r}$ from the star to the comet. The angular momentum is therefore just the product of r with mv, where v is the speed. The angular momentum is conserved in motion under the influence of gravitation, so we find

$$mv_{perihelion}r_{min} = mv_{aphelion}r_{max}.$$

We can solve to find the speed at aphelion:

$$v_{aphelion} = v_{perihelion}\frac{r_{min}}{r_{max}} = (81 \text{ km/s})\left(\frac{4.0 \times 10^7 \text{ km}}{6.0 \times 10^7 \text{ km}}\right) = 54 \text{ km/s}.$$

We used the conservation of angular momentum to relate the speed at perihelion to the speed at aphelion in this example. The use of energy conservation would also allow us to find the star's mass (see Problem 36).

12–5 GRAVITATION AND EXTENDED OBJECTS

We have viewed the planets and the Sun on the scale of the entire solar system and have treated them as point masses. But they are not really pointlike. This problem is especially evident if we think about how the gravitational force from Earth acts on an apple. To an apple, Earth most certainly is an extended object! We can study how extended objects behave by using the principle of superposition (Sec. 12–3).

The Gravitational Force Due to a Spherically Symmetric Object

We suggested in Section 12–3 that when an extended object is spherically symmetric, the force it exerts on a point mass outside the extended object is the same as the force that would be exerted if the entire mass of the extended object were concentrated at its center. It is this assertion that allowed us to relate the acceleration due to gravity at Earth's surface, g, to Earth's mass and radius, as in Eq. (12–9). In fact, for any spher ically symmetric object† of mass M and radius R, the acceleration due to gravity at its surface is

$$g = \frac{GM}{R^2}. \tag{12–16}$$

Thus we would expect the acceleration of gravity to be different at the surfaces of the different planets (Fig. 12–14).

PROBLEM SOLVING

The gravitational effects of spherically symmetric objects

Many mass distributions—such as Earth—are at least approximately spherically symmetric. For such systems, the mass density depends only on the distance from the center of the distribution. The mass density at the center could be larger or smaller than that at the surface but, at a given radial distance from the center, it must be the same. Earth conforms well to these requirements. Earth's core is denser than the outer layers—it most likely consists primarily of iron. The small deviations from spherical symmetry come from the fact that, because of its rotation, Earth bulges a little in the equatorial region; there is an additional distortion that makes the planet look a little pear-shaped. In addition, there are small local regions of greater or lesser density. Here, we will summarize the principal conclusions about the gravitational force due to spherical systems. The same results hold for the gravitational potential energy.

FIGURE 12–14 The force of gravity—the weight—is less on the Moon's surface than on Earth's surface. Astronaut John Young, on the Apollo 16 mission, was able to jump to quite a height despite the massive spacesuit.

1. Suppose that a point mass is *outside* a spherically symmetric object. The gravitational force experienced by the point mass is identical to the force that would arise if the whole mass of the spherical object were concentrated at its center. In other words, the force exerted on the point mass is the same as the force that would be exerted on it if the uniform spherical object (total mass M) were a point mass M located at the center of the sphere.
2. Suppose that a point mass is *inside* a thin spherical shell of constant density. Then there is no gravitational force on the mass. This conclusion holds for a point mass inside an arbitrarily thick shell as well, as long as the mass density of the shell depends *only* on the distance from the geometric center of the shell.

These two results show that the gravitational force on a point mass m within Earth at a distance r from the center, for example, would be due to a mass M' concentrated at Earth's center; M' is only the mass contained within Earth up to the radius r (Fig. 12–15a).

EXAMPLE 12–6 Suppose that a tunnel is drilled through our planet along a diameter. Assume that Earth's mass density is uniform and is given by ρ. Describe the force on a point mass m dropped into the hole as a function of the distance of the mass from the center.

Solution: The gravitational force on the point mass m is due only to the mass of the material contained within a radius r, where r is the distance from the point mass m to the center of Earth. The force is attractive, toward Earth's center, and it is given by

$$F = -\frac{GmM'}{r^2},$$

where the mass M' that attracts the point mass is the total mass *inside* radius r (Fig. 12–15a). Mass M' is given by (volume inside) × (density):

$$M' = \left(\frac{4\pi r^3}{3}\right)\rho.$$

†This means its density varies only with distance from the center.

Thus, as in Fig. 12–15b,

$$F = -\left(\frac{4\pi Gm\rho}{3}\right)r.$$

The result that F is proportional to r shows that, inside Earth, the point mass acts as if it were moving under the influence of a spring with spring constant $k = 4\pi Gm\rho/3$. This motion is oscillatory, and the point mass moves from one end of the tunnel to the other and back (Chapter 13).

The results for spherically symmetric objects will not be derived mathematically until Chapter 24, where we give a simple but indirect proof. Direct proof, which would follow by integrating the contributions of the force from different pieces of our extended object, would take us too far afield now. However, we can give a physical picture of just how it is that the mass *outside* a point mass exerts no force on our mass.† Figure 12–16 shows our point mass placed within a shell. The point mass is taken to be off-center. Consider a double cone making an opening angle θ at the point mass and, in particular, the gravitational effects of the two circular sections that the cone cuts on the shell. The point mass is a distance r_1 from each point in the area on the right, A_1, and a different distance r_2 from each point in the area on the left, A_2. The shell is so thin that these areas do not change as we go from the inner to the outer surface of the shell. If the density of the shell is ρ and its thickness is τ, then the force F_1 that attracts the point mass to the right-hand area is

$$F_1 = \frac{Gm}{r_1^2}\rho\tau A_1,$$

and the force F_2 that attracts the point mass to the left-hand side is

$$F_2 = \frac{Gm}{r_2^2}\rho\tau A_2.$$

Now we note that areas A_1 and A_2 are proportional to r_1^2 and r_2^2, respectively, so that

$$\frac{A_1}{r_1^2} = \frac{A_2}{r_2^2}.$$

This means that the two forces are independent of the distance of the point mass to the circular sections and the forces cancel because they pull in opposite directions. In effect, as the point mass moves closer to one side, the inverse-distance-squared factor in the force increases, but the amount of mass seen decreases by the same factor; thus, the pull of every sector of the shell is the same and cancels.

This curious effect is very special to the inverse-square law—a law that also happens to hold for electric charges. Indeed, it was Benjamin Franklin who noticed that an electric charge surrounded by a shell of the opposite charge feels no net force, and it was on the basis of this observation that Joseph Priestley in the eighteenth century first suggested that electric forces obey the inverse-square law.

How *g* Varies with Altitude

Equation (12–16), together with the result that a symmetric, spherical Earth behaves like a centered point mass, implies that g varies with altitude. Suppose that we measure altitude, h, from sea level, and that Earth's radius at sea level is R_E. From Eq. (12–16), we find

$$g(h) = \frac{GM}{(R_E + h)^2} = \frac{GM}{R_E^2[1 + (h/R_E)^2]}. \quad (12\text{–}17)$$

†Strictly speaking, our arguments hold only for small values of θ. But generalization is possible.

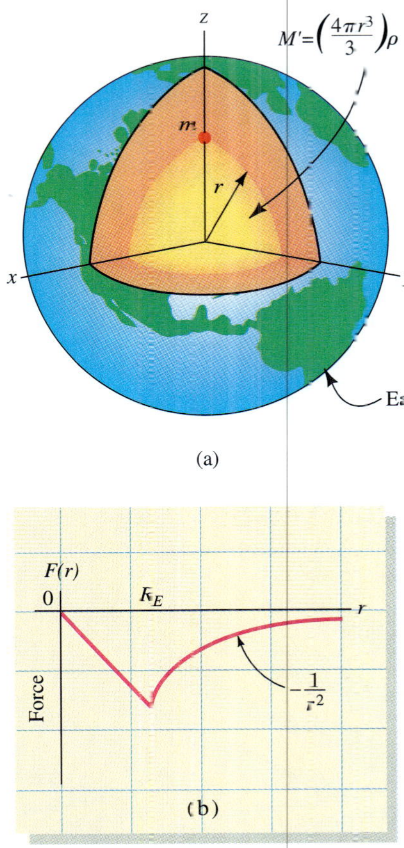

FIGURE 12–15 (a) The gravitational force on an object with mass m located inside Earth depends on the total amount of mass M' inside the sphere whose density is ρ and whose radius r is the distance of mass m from the center. (b) A graph of the variation of the force with r.

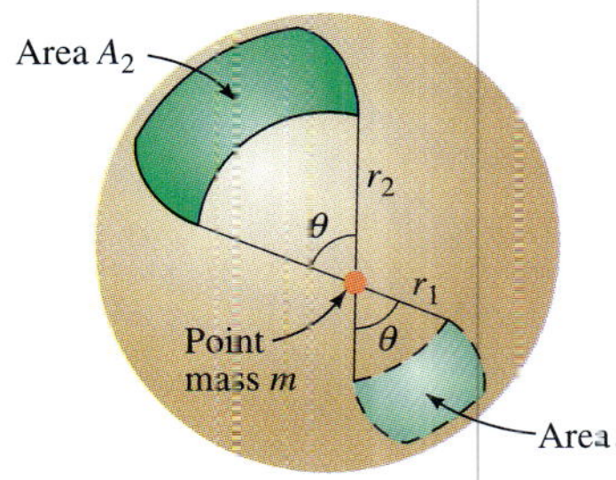

FIGURE 12–16 The total attraction that a point mass m experiences inside a spherical shell of uniform density is zero, because the force due to the mass in area A_1 is equal to that from the area A_2.

Now, if the ratio $h/R_E << 1$ (even for the top of Mount Everest, $h/R_E \simeq 10^{-3}$), then we can use the approximation $1/(1+x)^2 \simeq 1 - 2x$ (for $x << 1$; see Appendix IV–9), so $g(h) \simeq (GM/R_E^2)[1 - (2h/R_E)]$, or

$$\frac{g(h)}{g(0)} \simeq \frac{(GM/R_E^2)[1 - (2h/R_E)]}{(GM/R_E^2)} = 1 - \frac{2h}{R_E}. \qquad (12\text{–}18)$$

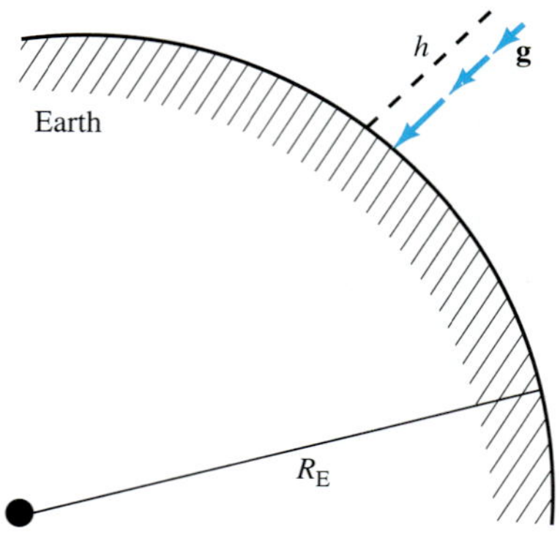

FIGURE 12–17 As we move away from the surface, the value of g decreases.

To estimate the size of the effect, we can see from Eq. (12–18) that, at the top of the Sears Tower in Chicago, with $h = 443$ m, g is 99.99 percent of its value at sea level; even at the top of Mount Everest, with $h = 8848$ m, g is 99.74 percent of its value at sea level (Fig. 12–17).

Other effects modify Eq. (12–16). In Chapter 5, we mentioned the effect on g of Earth's rotation. This effect is largest at the equator, where g is 99.57 percent of what it would be if Earth did not rotate. This effect decreases with latitude and it is not present at the poles. Finally, Earth is not a perfectly uniform sphere. Not only is it not spherical, but it contains lumps of higher or lower density. Measurements of g are routinely made to many significant figures, more than enough to reveal all the effects we have discussed.

*12–6 A CLOSER LOOK AT GRAVITATION

When we have applied the law of gravitation, it has always been in the context of the attraction exerted on a "small" object by a "large" object. We have considered planets or comets moving around the Sun, tennis balls falling on Earth's surface, and so on. We assumed that the massive object was at rest and that the orbits being studied (for example, the ellipses) were those of the light object. We know that, just as Earth exerts a force on the Moon, the Moon exerts a force on Earth. The Moon moves in an orbit around Earth, but does Earth move around the Moon? The answer to this question is yes. When two objects move under the influence of the gravitational forces they mutually exert on each other, they move about their common center of mass. This is a property we first observed in the context of momentum. If we regard the two objects as an isolated system, then there is no external force on that system, and its center of mass remains unaccelerated. When one of the two objects is much heavier than the other, the center of mass is close to the heavier object and may even be within that object. In the Earth–Sun system, the Sun's motion is a very small orbit about a point very close to the center of (and well within) the Sun. Cases in which the center of mass is well separated from the two objects in question occur in double star systems with stars of comparable masses. These stars each move in a circular orbit around a point between them (Fig. 12–18).

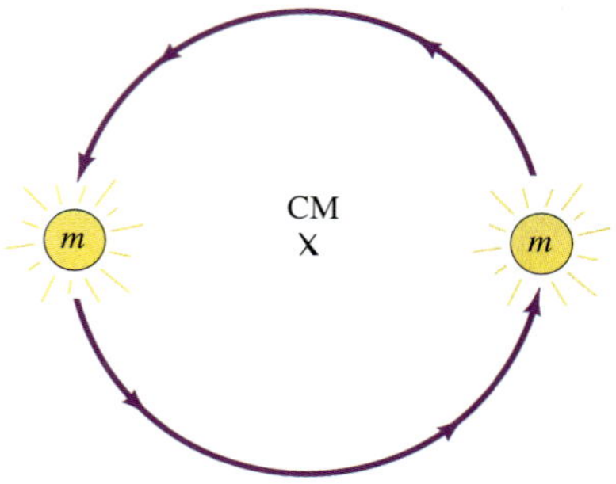

FIGURE 12–18 In a double star system with stars of equal mass, the center of mass is midway between the stars. Each star orbits around this point.

Ocean Tides

In the absence of a moon, the oceans would form a concentric envelope of water around Earth (Fig. 12–19a). With the Moon added to this picture, a superficial view of the system has the Moon attracting water on the near side of Earth more strongly than water on the far side. This would lead to a bulge in the water distribution which would peak

FIGURE 12–19 (a) If there were no Moon, the ocean could surround Earth in a uniform manner. (b) The Moon's gravitational attraction is stronger on the portion of the ocean nearer the Moon. If this were the only effect acting, there would be only one tide a day. Newton correctly pointed out that the orbiting of Earth about the center of mass of the Earth–Moon system leads to two bulges and two high tides per day.

APPLICATION: MAPPING g

There is a great deal of interest in mapping the exact details of Earth's gravity for prospecting purposes. Although Earth's average density is 5.5×10^3 kg/m^3, the density near the surface varies significantly. Water has a density of 1.0×10^3 kg/m^3, oil less than this; iron-containing rocks have a higher density than the average. Thus an accurate map of Earth's gravity would provide a global mineral prospecting tool. This is relatively easy to do because of the inverse-square law, which implies that irregularities close to the region being measured have a large effect on the size of g. NASA proposed a global project that involved a satellite in the form of an evacuated hollow sphere, containing a smaller dense sphere. Once in flight, the position of the smaller sphere within the larger would be sensed optically. This sensing would control gas jet thrusters on the large sphere and keep the spheres concentric. Normally a satellite is not in exact free fall because there are weak nongravitational forces acting on it. The most significant of these are the frictional drag due to the very dilute atmosphere at the satellite's altitude and momentum transfer produced by gases escaping from the satellite. (That is why the Space Shuttle is described as a microgravity rather than zero gravity environment.) The interior gravitation-sensing sphere is not acted on by any forces except the radiation pressure of the light from the optical sensors, and this is negligible. By following the precise path of the interior satellite, variations in the strength of the gravitational force acting on it can be measured.

in the direction of the Moon (Fig. 12–19b). Earth would rotate within this stationary, distorted water envelope, and thus the water level would be seen to rise once every 24 h. This argument, due to Galileo, fails to account for the fact that there are *two* high tides per day, not one.

Newton showed that, just as the Moon attracts the water nearest it more strongly than Earth as a whole, making a bulge toward the Moon, so the Moon also attracts Earth as a whole more strongly than the water on the *far* side of Earth, leaving behind a second bulge on the far side. The result is two high tides per day (Fig. 12–20).

Let's look at this argument in more detail. In addition to experiencing a gravitational force due to Earth itself, the surface points are attracted to the Moon. Because of Earth's finite size, the Moon exerts slightly different forces on different parts of the surface; points nearer the Moon experience stronger forces than points farther from the Moon. In particular, we want to see how these forces *differ* from the "average" force due to the Moon, the force that the Moon exerts on Earth treated as a point at its own center. A mass of water on the surface moves differently from the overall motion according to how much the force on it *differs* from the average force. Water nearest the Moon experiences *a little bit more* than the average force due to the Moon. Thus the force there differs from the average force by a vector pointing *toward* the Moon. Similarly, water farthest from the Moon experiences *a little bit less* than the average force. There, the force differs from the average force by a vector pointing *away from* the Moon. The force differences form what are called *tidal forces* (Fig. 12–21a). Earth's shell of water is accordingly distorted by the tidal forces to form the two bulges shown in Fig. 12–21b.

The Sun also contributes to tide formation, although less than half as much as does the Moon. This is not because the gravitational force of the Sun is smaller than that of the Moon. On the contrary, it is about 175 times larger. However, the *difference* in the Sun's gravitational force from one side of Earth to the other is smaller due to the fact that Earth's radius is so much less than the Earth–Sun distance. When the Moon and

(a)

(b)

FIGURE 12–20 Alma beach in the Fundy National Park, New Brunswick, Canada, at (a) high and (b) low tide. The Bay of Fundy has some of the largest tidal differences in the world.

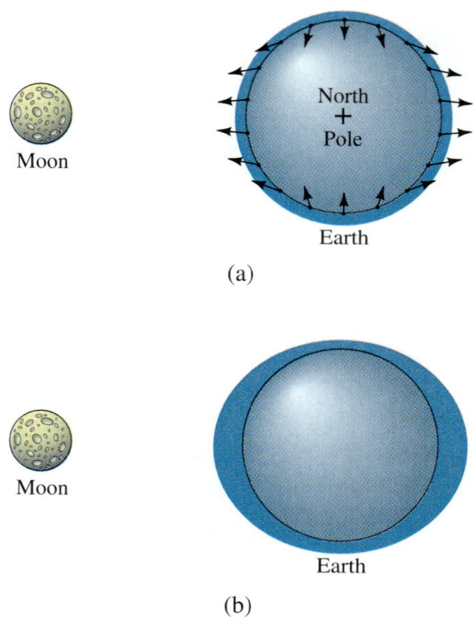

FIGURE 12–21 (a) The arrows show how the forces from the Moon over Earth differ from the average value of the force on Earth. (b) The resulting water distribution has two bulges.

the Sun are lined up, at a new or full Moon, an especially large tide results; when they are at right angles, the tidal effects partially cancel.

The friction between the water and Earth moving beneath the bulges leads to a dissipation of energy, and Earth's rotation about its axis slows down. Calculations show that the day is thereby lengthened by about 10^{-3} s every century, a prediction verified by experiment.

Effects of Other Objects

The previous discussion involved the gravitational effects of more than one object. This has implications for planetary orbits. The superposition principle tells us that the orbit of any planet is affected not only by the Sun, but also by the presence of all other planets—although to a much smaller extent. The largest effects are due to the most massive planets, Jupiter and Saturn. One consequence of this is that the potential energy of a planet of mass m will have the form

$$U = -Gm\left(\frac{M_S}{r_S} + \frac{M_1}{r_1} + \frac{M_2}{r_2} + \cdots\right), \qquad (12\text{–}19)$$

where the terms in parentheses represent the contribution of the Sun and those of the other planets. Thus the net force on a planet is no longer a pure $1/r^2$ force directed exactly at the Sun; there are small corrections to it. As a consequence, the orbits are no longer exact ellipses that close on themselves, but instead, the orbits *precess* (Fig. 12–22). Almost all major figures in nineteenth-century mathematics worked on the problem of computing orbits subject to perturbations from additional masses; in fact, accurate orbits were computed even before the advent of large-scale computing machines. Two interesting historical events are worth mentioning in this connection. First, the calculation of the orbit of Uranus, with the inclusion of all the perturbations, did not fit the observed orbit. In 1845, both John Adams (an undergraduate at Cambridge University) and Urbain Le Verrier in France calculated the potential effects of a hypothetical new planet and published their results. Adams's work was ignored, whereas Le Verrier was more successful in mounting a search for the new planet, which culminated in the discovery of the planet Neptune in 1846. Second, the calculation of the precession of the perihelion of the planet Mercury was also carried out to great accuracy by Le Verrier. The result of the comparison of observation and theory left a discrepancy in the rate of precession of the perihelion that amounted to only 43″ of arc

FIGURE 12–22 An orbit that does not close on itself. The orbit is said to precess.

per century (out of an observed total of some 5600″ of arc per century). Both theory and experiment were so good that there was no doubt of the existence of this discrepancy, whose explanation had to wait for Einstein's theory of gravitation.

Equality of Inertial and Gravitational Masses

The parameter m in $\mathbf{F} = m\mathbf{a}$ describes a property of an object that is properly called the *inertial mass* [see Eq. (5–1)]. It is a constant that characterizes the object, and it appears as a coefficient of the acceleration in response to *any* force. The parameter m that appears in the expression for the gravitational force that is exerted on that object is the *gravitational mass*, and there is no a priori reason why the inertial and gravitational masses should be equal. Their equality to one part in 10^{11} has been demonstrated by Robert Dicke and Vladimir Borisovitch Braginsky. Newton had already measured the equality of these quantities to an accuracy of one part in 10^3, and Loránd von Eötvös carried out measurements in the period from 1890 to 1922 to an accuracy of one part in 10^9. The equality of inertial and gravitational masses made a great impression on Einstein and led toward his formulation of the equivalence principle, a cornerstone of the general theory of relativity (see Section 12–7).

How Accurate Is Newton's Law of Gravitation?

With what accuracy do we know that the inverse-square law suggested by Newton holds? It is well verified that celestial phenomena obey an inverse-square law, but these phenomena occur on scales ranging from hundreds of kilometers to much greater distances. The Cavendish experiment and its modern versions verify the inverse-square law at distances ranging from centimeters to meters. But because there is no precise, direct confirmation of the law on distances ranging from meters to hundreds of kilometers, tests of the force of gravitation in this range are still proceeding. It is useful to remember that Newton's predictions are subject to the same experimental verification as are those of any other scientist.

*12–7 EINSTEIN'S THEORY OF GRAVITATION

As remarkably accurate as it is, Newton's theory of gravitation was superseded in 1915 by a still more accurate description: Albert Einstein's theory of gravitation. The Einstein theory, also known as the **general theory of relativity**, reduces to Newton's theory for objects that move with a speed v that is small compared to the speed of light, c ($v \ll c$), and when the gravitational potential energies are small compared to mc^2. These conditions are satisfied except in regions close to very large masses; this is why Newton's theory was so very successful in virtually all its applications.

Einstein's theory arose out of his attempts in 1915 to combine Newton's theory with the special theory of relativity, which will be discussed in Chapter 40. While working on this problem, Einstein had what he described as "the happiest moment of my life": the idea that a freely falling person does not feel (and has no way to measure) his or her own weight. This idea was generalized in 1907 to form the **equivalence principle**. According to this principle, no experiment can distinguish between the following two situations: (1) a physical system at rest that is subject to a uniform gravitational force; and (2) a physical system that is uniformly accelerating in the absence of gravity. A simple example will impart the flavor of what is implied. Suppose an observer stands in an elevator and experiences a force on his feet. He can interpret this effect as being due to an upward acceleration of the elevator. Because the observer, of mass M, experiences an acceleration, a, he must be subject to an upward force of magnitude $F = Ma$. In fact, this force is the contact force that the elevator floor exerts on his feet. A second interpretation is that the elevator is at rest,

The equivalence principle

but there is a uniform gravitational force acting to pull him downward. He again feels the upward normal force, F_N, of the floor and, because the elevator and thus the observer are at rest, that force must just cancel the gravitational force. Thus $F_N = Mg$, where g is the acceleration due to gravity. Notice that we have used the same mass M in both descriptions. Strictly speaking, the mass M in $F = Ma$ is the *inertial* mass, whereas it is the *gravitational* mass in the force-identity equation $F_N = Mg$. The gravitational and inertial masses were first observed by Newton to be identical, but he was unable to draw deep conclusions from this observation. Einstein's equivalence principle states that if g has the same numerical value as a, there is no way for the observer to distinguish the two cases; thus, the inertial mass and the gravitational mass are *required* to be equal.

Predictions of the Equivalence Principle

Light Falls under the Influence of Gravity. Light falls just like matter does. To see this, let's consider the elevator again. A beam of light shines across the elevator in a horizontal direction from one side just as the elevator accelerates upward (Fig. 12–23). Because it takes light some time to travel across the elevator, it will hit the opposite wall closer to the floor than a horizontal line parallel to the floor would indicate. More importantly, for constant acceleration, the amount by which the elevator moves upward is proportional to the square of the time. That means that a *series* of measurements made by an occupant of the elevator that located the light beam relative to the floor would show that the light beam follows a parabolic path, just as a falling object would under the effect of gravity. If we are to interpret this observation from the point of view that the elevator and its contents feel a gravitational force, then the observed deflection of the light must be due to the gravitational force. Light, in this sense, does not behave any differently than matter, except that it moves faster. The first observation of the deflection of light due to gravitation was made during a solar eclipse in 1919. At such a time, pairs of stars whose light passes very close to the Sun become visible. The angular spread between the stars when their light comes around the two sides of the Sun can be compared with their observed angular spread when they are seen away from the Sun (Fig. 12–24). The confirmation of the predicted effect did more than anything else to bring Einstein's theory to the public's attention.

Gravitational Lenses. Recently, astronomers have observed two identical *quasars* (extraordinarily bright sources of light) very close to one another as seen from Earth (Fig. 12–25). Because there is good evidence that quasars are billions of light-years

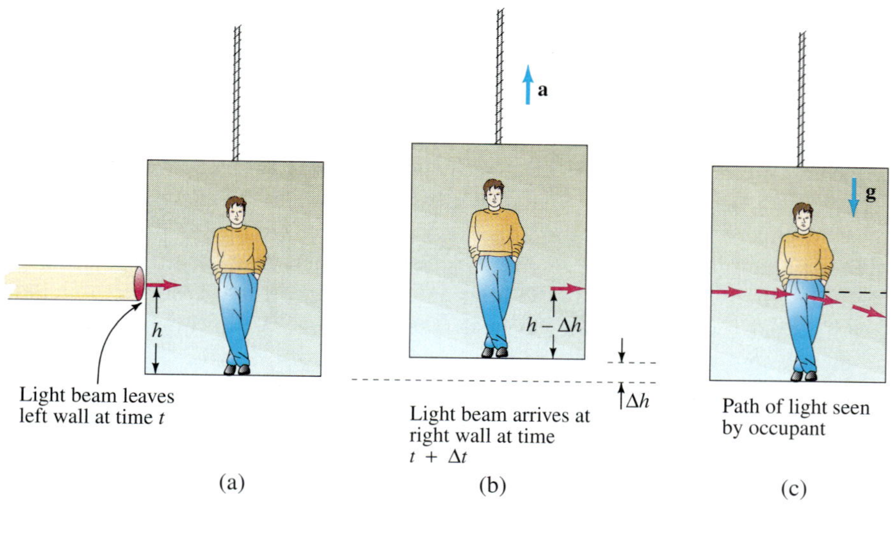

FIGURE 12–23 (a) A beam of light shines across an elevator at a time t when the elevator starts to rise. (b) At time $t + \Delta t$ the light hits the opposite side of the elevator at a point lower than it would have hit had the elevator not been moving. (c) The observer in the elevator cannot know whether the elevator is accelerating upward at **a** or whether gravity **g** is present. We conclude from thought experiments like this that light must be bent when gravitational forces are present.

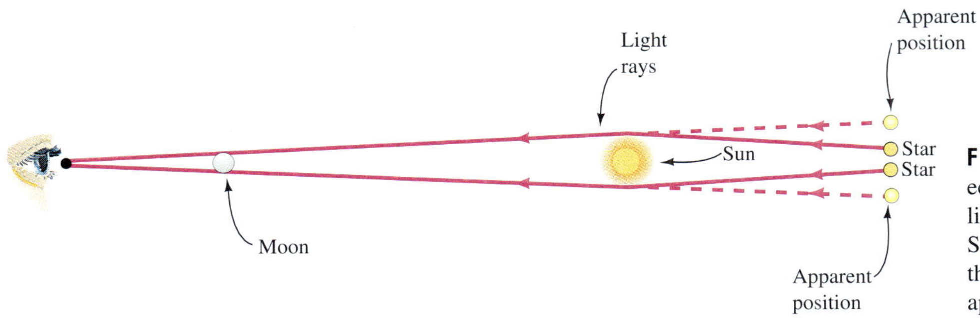

FIGURE 12–24 During a solar eclipse, it is possible to observe the light of two stars on either side of the Sun. Because the Sun's gravity bends the light, the stars appear to be farther apart than they actually are.

away, any two quasars are literally quite disconnected from one another; thus, it is highly improbable that any two could be as close to identical as the observed pair appear to be. The generally accepted interpretation is that there is only one quasar, but its light passes close to an extremely massive object, which bends light going around it just as a lens would; this gives rise to a double image; the effect is said to be due to a *gravitational lens*. An even more spectacular demonstration of this effect is seen in a second discovery, where we can detect a ring image of the source (Fig. 12–26). (Can you see how the ring is produced?) To produce such a large effect, the light must be bent by a galaxy whose mass equals that of about 300 billion suns.

Black Holes. The fact that light falls when it passes near masses implies that it is possible to imagine a large enough mass, localized in a small enough region, for which the speed of light is smaller than the escape speed from the surface of the mass. Such a mass would not be directly visible because light or matter could not escape from it. It would manifest itself only through the gravitational force it exerts. Such a mass forms a *black hole*. Recent observations have provided indirect evidence for black holes. There exist pairs of stars—only one of which can be seen directly. The other is invisible, but its presence and mass can be deduced from the motion of the visible star. In addition, some such pairs emit X-rays copiously. The characteristics of the X-rays demand the interpretation that they come from matter falling into a black hole. The strong gravitational force necessary to produce such dramatic effects can come only from a black hole.

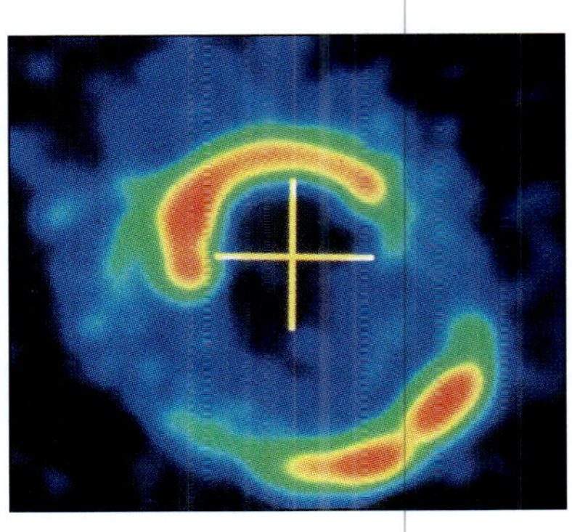

FIGURE 12–26 This photograph, taken by the Hubble Space Telescope, of a massive, compact galactic cluster illustrates gravitational lensing. The lensing effects are evident in the arc-like pattern. They are caused when light from an object far beyond the cluster passes near the cluster, which magnifies, distorts, and brightens that light on its way to our eyes. The image tells us a great deal both about the matter far beyond the cluster and about the cluster itself.

Precession of Planetary Orbits. The equivalence principle forms the foundation of the general theory of relativity, which is the full theory of gravitation. The mathematical application of the theory leads to subtle corrections to Newton's gravitational force law. The corrections to the Newtonian form predict, among other things, that the perihelion of Mercury should precess by 43″ of arc per century, an amount that is within observational error when the effects of other planets on the motion of Mercury are taken into account. Other predictions of the general theory of relativity are being confirmed by recent experiments.†

†See Clifford M. Will, *Was Einstein Right?*, Basic Books, New York, 1986.

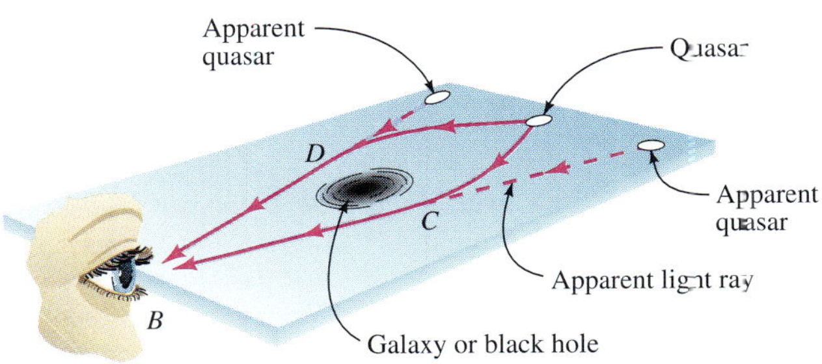

FIGURE 12–25 A double image results when light from a single quasar is bent as it passes on either side of an object with strong gravity, such as a galaxy or a large black hole. The effect is greatly exaggerated here.

SUMMARY

Astronomical observations led Kepler to three laws for planetary motion:

1. Planets move in planar elliptical paths with the Sun at one focus of the ellipse.

2. During equal time intervals, the radius vector from the Sun to a planet sweeps out equal areas.

3. If T is the time that it takes for a planet to make one full revolution around the Sun, and if R is half the major axis of the ellipse (R reduces to the radius of the orbit of the planet if that orbit is circular), then

$$\frac{T^2}{R^3} = C, \tag{12–1}$$

where C is a constant whose value is the same for all planets.

Newton showed that these laws are a consequence of a law of universal gravitation, which states that any two point masses m and M, separated by a distance r, will attract each other with a force that is along the radius vector connecting the masses. The force on object 2 due to object 1 has the inverse-square form

$$\mathbf{F}_{21} = -\left(\frac{GmM}{r^2}\right)\hat{\mathbf{r}}, \tag{12–4}$$

where $\hat{\mathbf{r}}$ is the unit vector pointing from object 1 to object 2. The constant G has been measured, and its value is $G = 6.673 \times 10^{-11}\ \text{N}\cdot\text{m}^2/\text{kg}^2$. The gravitational force is conservative and may be derived from a potential energy:

$$U(r) = -\frac{GmM}{r}. \tag{12–8}$$

Masses that exert gravitational forces are not always pointlike. The most important case occurs when there is a spherically symmetric mass distribution, such as Earth or the Sun. In this case, the gravitational force is the same as if all the mass of the extended object were concentrated at the center of the spherical distribution. Another consequence of the formula given in Eq. (12–8) is that the gravitational force exerted on an object anywhere inside an arbitrarily thick spherical shell, with a mass distribution that depends only on the distance from the center, is zero.

The gravitational force acts on both objects; for example, not only does Earth revolve around the Sun, but the Sun also revolves around Earth. More precisely, both objects revolve around their center of mass. When this phenomenon is taken into account in the Earth–Moon system, the tides can be understood.

The Newtonian theory of gravity is a limiting case of a more accurate and fundamental theory of gravity: Einstein's general theory of relativity. That theory is based on the equivalence principle, which states that no experiment can distinguish between a uniform acceleration and the effects of a uniform gravitational force. One consequence is the equality between the inertial mass (m_i in $F = m_i a$) and the gravitational mass (the masses that appear in the law of universal gravitation). Another consequence is the fact that light falls just like ordinary matter when it is subject to gravity.

UNDERSTANDING KEY CONCEPTS

1. What are some forces, other than those mentioned in this chapter, that are not fundamental? From what fundamental forces are they derived?

2. Any projectile fired with enough initial speed will eventually escape Earth, regardless of the direction in which it is fired. How do you reconcile this statement with the fact that the height above Earth to which a cannonball will rise depends on the angle at which it is fired?

3. To a good approximation, the Sun and the Moon both move with respect to Earth in the plane of Earth's equator. Does this mean there would be little or no tide at the North Pole if there were surface water at the North Pole?

4. The European satellite launch area is in French Guiana, because less energy is required to launch rockets into orbit from there than from a point in Europe. Explain why this is so.
5. How can Earth's rotational motion be used to minimize the fuel needed to boost a satellite into a given orbit around the planet? How can Earth's orbital motion be used to minimize the fuel needed to boost a satellite into a given orbit around the Sun?
6. The satellites of Jupiter follow Kepler's third law: The square of their periods, divided by the radius of their orbits cubed, is a constant. Is this the same constant as for the planets moving around the Sun?
7. Newton performed a thought experiment when he described an artillery shell fired horizontally from the top of a high mountain. Ignoring air resistance, the shell will land farther and farther out as the speed with which the shell is fired increases. Eventually, the curvature of Earth becomes significant, and the surface falls away from beneath the shell. At last, for a sufficiently high initial speed, Earth's surface falls away from the path of the shell as the shell circles Earth, and this is a circular orbit. How is this result consistent with the statement that an artillery shell follows a parabolic path rather than the arc of a circle?
8. Describe the path of a celestial object whose angular momentum with respect to the Sun is zero.
9. If the gravitational force were a central force proportional to $1/r^3$ rather than $1/r^2$, the planetary orbits would no longer be closed (unless they are circular). Would the planets still sweep out equal areas in equal times?
10. We often hear that Earth satellites burn up when they leave their orbit and return to Earth. Why don't satellites burn up as they go up into orbit?
11. Communication satellites are placed in *geosynchronous* orbit, which means that they are always above a given point on Earth. Is it possible to place such a satellite above the North Pole? Above what points on Earth is it possible to have a geosynchronous orbit?
12. When astronauts float in the bay of their spacecraft, they are in outer space, above Earth's atmosphere. Is gravity acting on them?
13. Very careful measurements of the orbits of satellites can help teach us about Earth's internal structure. How can we use such measurements to study regions with a mass density that is higher or lower than that of Earth's average mass density?
14. If Earth were a perfect sphere, would you weigh more or less at the equator than at the poles?
15. Earth is not a perfect sphere. What types of observations might we use to learn this fact?
16. When a satellite is in circular orbit around Earth, there is a direct relationship between its angular momentum and the radius of the orbit. Suppose that such a satellite collides elastically with a meteor that was heading directly toward the center of Earth. Because the impulse is in a radial direction, the angular impulse is zero. What happens? Draw some diagrams, using the conservation laws that you know. (If you get stuck, review Kepler's first law.)
17. In the text, we spoke of friction between Earth and its shell of water as slowing down the rotation frequency of Earth. Would this still occur if the Moon were not present?
18. Is it possible for friction from the tides to slow down Earth's rotation and still conserve Earth's angular momentum (with its oceans)? The origin of the tides involves the gravitational force due to the Moon on Earth's oceans, so Earth and its oceans do not form an isolated system.
19. A satellite is in circular orbit around Earth. How much work is done on the satellite by the gravitational force of Earth during one orbit?
20. Assume that Earth is perfectly spherical and that its density depends only on the distance from its center. A large asteroid makes a close pass to Earth. Could such an asteroid (which is nevertheless small compared to Earth itself) change the rate of Earth's rotation without actually colliding?
21. You are in a spaceship very far away (say, 100 Earth radii) from Earth. Could you move to a region where the gravitational force due to Earth is less by ejecting some material from the spaceship in the proper direction?
22. Edgar Rice Burroughs's character Tarzan discovers that Earth is hollow. Tarzan finds an entry and discovers a whole new civilization, with modernistic buildings, people walking around or driving modernistic vehicles, and so on, in Earth's interior. What is wrong with this picture?

PROBLEMS

12–2 *Early Observations of Planetary Motion*

1. (I) Use the data listed in the appendices to calculate C in Eq. (12–1).
2. (I) In planetary tables, we find that Jupiter's satellites Io and Europa each follow nearly circular orbits: Io's orbit has a mean radius of 422,000 km and Europa's orbit has a mean radius of 671,400 km. The period of Io is 152,854 s. What is Europa's period?
3. (II) Consider an object of mass m, moving in a circular orbit, subject to a central attractive force whose magnitude is given by $F(r) = h/r^3$. (a) What are the dimensions of h? (b) Show that the angular momentum for the motion is uniquely determined by h and m. (c) What is the resulting relation between period and radius analogous to Kepler's third law for this force?
4. (III) Angular momentum is conserved for a radial, or central, force. Show that the orbits due to a radial force lie in a plane.

12–3 *Newton's Inverse-Square Law*

5. (I) A man of mass 95 kg is dancing with his wife, who has a mass of 68 kg. Assume that each person's mass is concentrated at their respective centers of mass, which are separated by 48 cm. (a) What is the gravitational attraction between them? (b) Which person has the greater gravitational attraction toward the other?
6. (I) Calculate the gravitational attraction between a proton and an electron in a hydrogen atom if the radius of the atom is 0.6×10^{-10} m. The masses can be found in Appendix V.
7. (I) A Cavendish experiment involves the force between two spheres of 1 kg each whose centers are separated by 40 cm. Using the known value of G, find the gravitational force between these spheres. Compare this force to the weight of a fly.
8. (I) What is the acceleration due to gravity on the surface of (a) the Moon ($R = 1.74 \times 10^3$ km, $m = 7.35 \times 10^{22}$ kg);

(b) Mars ($R = 3.40 \times 10^3$ km, $m = 6.42 \times 10^{23}$ kg); (c) Jupiter ($R = 7.14 \times 10^4$ km, $m = 1.90 \times 10^{27}$ kg); (d) the Sun ($R = 6.96 \times 10^5$ km, $m = 1.99 \times 10^{30}$ kg)?

9. (I) A satellite orbits Earth in 90 minutes. What is the radius of its motion around the center of Earth?

10. (I) What is the period of a satellite circling the Moon at a height of 300 km above the Moon's surface?

11. (I) What is the period of a satellite circling Earth at a height of 300 km above Earth's surface?

12. (I) What is the surface gravity (the value of g) on an asteroid of diameter 150 km and density 4800 kg/m^3?

13. (I) Two identical satellites move in circular orbits around Earth. One has twice the kinetic energy of the other. The radius of the faster one's orbit is three Earth radii. What is the radius of the slower one's orbit?

14. (II) A weight lifter can lift 115 kg on Earth. What mass could the same weight lifter lift on (a) the Moon, (b) the Sun (use the data in Problem 8)?

15. (II) The height achieved in a jump is determined by the initial vertical velocity that the jumper is able to achieve. Assuming that this is a fixed number, how high can an athlete jump on Mars if she can clear 1.85 m on Earth?

16. (II) There is a point on the line joining two astronomical bodies where there is no gravitational force on a rocket. Find this point for (a) the Earth–Sun system, (b) the Earth–Moon system.

12–4 Planets and Satellites

17. (I) Pluto has the most eccentric orbit of all the planets, with $e = 0.25$. Its semimajor axis is 39.5 AU. What is $(r_{max} - r_{min})/(r_{max} + r_{min})$ for Pluto's orbit?

18. (I) The semimajor axis of Earth's orbit is 149.6×10^6 km, while its eccentricity is 0.017. What is the maximum distance between Earth and the Sun as Earth traces out its orbit?

19. (I) Determine the escape speed of an object from the Sun's surface.

20. (I) What are the escape speeds on the surface of (a) the Moon, (b) Mars, and (c) Jupiter? (Use the data in Problem 8).

21. (I) Consider the asteroid of Problem 12. What is the escape speed from the surface of that satellite?

22. (I) The radius of a neutron star is 600 times smaller than Earth's radius, and its mass is 4×10^5 times larger than Earth's mass. What is the escape velocity from the surface of a neutron star? (Ignore the fact that, at high speeds, one should not really use $mv^2/2$ for the kinetic energy.)

23. (II) If the asteroid of Problem 12 rotates with an angular speed ω about an axis, material on the equator will have a tendency to be thrown off. How slowly must it rotate so that material that is not attached permanently just barely stays on the surface?

24. (II) A rocket is sent vertically upward from Earth's surface with an initial speed of 7 km/s. How far above Earth's surface will it go before falling back? Ignore atmospheric friction.

25. (II) Astronomers discover a meteorite at a distance of 80,000 km from the center of Earth. The meteorite is moving directly toward Earth with a velocity of 2000 m/s. What will be the velocity of the meteorite when it hits Earth's surface? Ignore all drag effects.

26. (II) Determine the minimum energy needed to allow an unmanned rocket of mass 1200 kg to leave the Moon's surface and arrive at a point very far away. Ignore the effect of Earth.

27. (II) Calculate the distance from Earth's center to a satellite in circular orbit with a period that is (a) one-third the Moon's period; (b) three times the Moon's period; (c) one-thousandth of the Moon's period.

28. (II) Using the information in Example 12–3, determine a satellite's time of revolution about Earth if it is in a circular orbit 250 km above Earth. How would your answer change if the mass of the satellite were to double?

29. (II) Determine the velocity with respect to Earth of a satellite of mass 500 kg in a circular orbit 200 km above Earth's surface. What is its kinetic energy? What is its angular momentum?

30. (II) A geosynchronous communications satellite orbits Earth, always positioned above the same point on the equator. (a) What is the period and angular velocity of the satellite? (b) What is the radius of the orbit? (c) Show that Kepler's third law applies to the orbits of the satellite and the Moon.

31. (II) The mass of Mars is 6.42×10^{23} kg, and its radius is 3393 km. What is the period of a satellite in a circular orbit 95 km above the surface of Mars?

32. (II) The Moon goes around Earth once in 27.3 d. What is the distance between the Moon and Earth?

33. (II) Use the mass of the Sun, given in Problem 8, to estimate the distance from Earth to the Sun. Assume that the orbit is circular.

34. (II) A satellite is fired off horizontally with an initial speed $v_0 = 9.3$ km/s from the North Pole. Ignore air resistance. (a) What is the maximum distance from Earth's center attained by the satellite? Use both the conservation of energy and the conservation of angular momentum about the center of Earth. (b) What is the maximum distance attained if the satellite is fired vertically with the same speed? (c) if the satellite is fired at an angle of 30° with the same speed? (d) Sketch the motion of the satellite in the three cases.

35. (II) A small package is fired off Earth's surface with a speed v at a 45° angle. It reaches a maximum height h above the surface at $h = 6370$ km, a value equal to Earth's radius itself (Fig. 12–27). What is its speed when it reaches this height? Ignore any effects that might come from Earth's rotation.

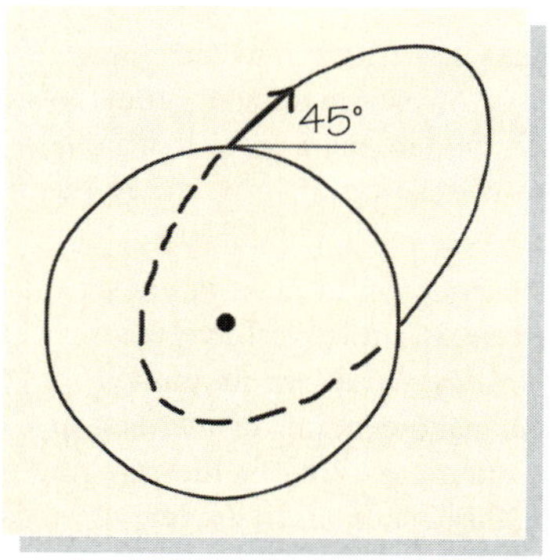

FIGURE 12–27 Problem 35.

36. (II) In Example 12–4, we used the conservation of angular momentum to find the speed of a comet at perihelion and aphelion around a star. Use the fact that the energy at these two points must be equal to find the mass of the star.

37. (II) An object of mass 3×10^{15} kg approaches the solar system (Fig. 12–28). When it is very far away—where the gravitational potential energy can be neglected in comparison with its kinetic energy—the object moves with a velocity of 12 km/s in a straight line. By straight-line extrapolation, the closest this line would come to the Sun is 3×10^8 km. The point of the object's nearest approach to the Sun is characterized by the fact that the radius vector from the object to the Sun is perpendicular to the tangent to the path at that point. (a) Sketch the orbit of the object. (b) Use conservation of energy and of angular momentum to calculate the velocity of the object at the point of nearest approach. (c) Calculate the distance of nearest approach.

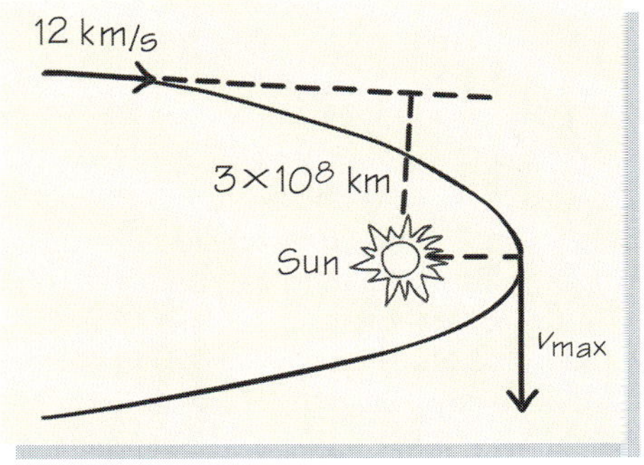

FIGURE 12–28 Problem 37.

38. (II) A satellite is in a circular orbit of radius R_1 around Earth. Small rockets aboard the satellite change its direction so that it has an elliptical orbit. The change causes the satellite to lose half its orbital angular momentum, but the total energy remains constant. In terms of R_1, what are the perigee and apogee distances of the new orbit with respect to the center of Earth?

39. (II) A satellite of mass 300 kg is in circular orbit 2000 km above Earth's surface ($M_E = 6 \times 10^{24}$ kg, $r_E = 6370$ km). (a) What is the orbital speed of the satellite? (b) What is the angular momentum of the satellite? (c) A rocket engine is fired, reducing the speed of the satellite to half its initial value, but leaving the direction of motion unchanged. What is the new angular momentum? (d) Does the satellite crash as a result of the maneuvers in part (c)? Explain your answer.

40. (III) The distance of closest approach of Halley's comet to the Sun is 8.9×10^{10} m. Its period is 76 yr. What is the nature of its orbit? Calculate the following: (a) semimajor axes; (b) eccentricity; (c) aphelion distance (farthest distance from Sun).

41. (III) A satellite of mass 2000 kg is in circular orbit about Earth at a distance of 300 km above the surface. (a) What is the speed of the satellite in its orbit? (b) What is the angular momentum of the satellite about the center of Earth? (c) What is the total energy of the satellite in its orbit? (d) Controllers back on Earth wish to move the satellite to a new orbit 500 km above the surface. They propose to do this by briefly firing a rocket engine on the satellite for several seconds in the direction of the center of Earth; that is, the force on the satellite is directly away from Earth's center. What is the torque on the satellite about the center of its orbit? (e) Can the new orbit be circular?

12–5 Gravitation and Extended Objects

42. (I) What is the approximate difference between the value of g at sea level and the value on top of a 10,000-ft-high mountain? Assume that Earth has a constant density, and take Earth's radius as the radius at sea level.

43. (I) Equation (12–19) is an approximate form. The corresponding exact form can be derived from Eq. (12–18): $g(h)/g(0) = (R_E^2)/(R_E + h)^2$. Verify the accuracy of the approximation by calculating the ratio $g(h)/g(0)$ for $h = 10{,}000$ m (the altitude of a cruising passenger jet) according to the approximate and the exact forms.

44. (II) A deep hole in Earth reaches a depth of one-half of Earth's radius (Fig. 12–29). How much work is done when a 1-kg mass is slowly lifted from the bottom of the hole to Earth's surface?

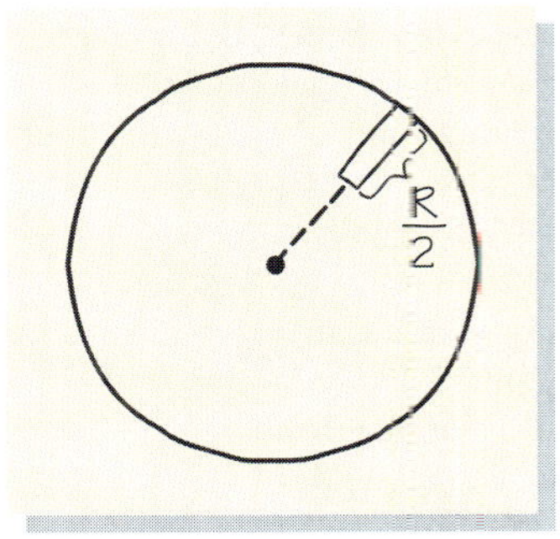

FIGURE 12–29 Problem 44.

45. (II) What is the speed of the mass in the tunnel through Earth in Example 12–5 as it passes through the center of Earth?

46. (III) Rather than a tunnel through Earth's center, as in Example 12–5, consider a tunnel drilled along a chord of the earth, meaning that it passes a perpendicular distance d away from the center of Earth (Fig. 12–30). Find the potential energy of a mass placed in such a tunnel as a function of (a) its distance r from the center of Earth and (b) its distance x from the midpoint of the tunnel.

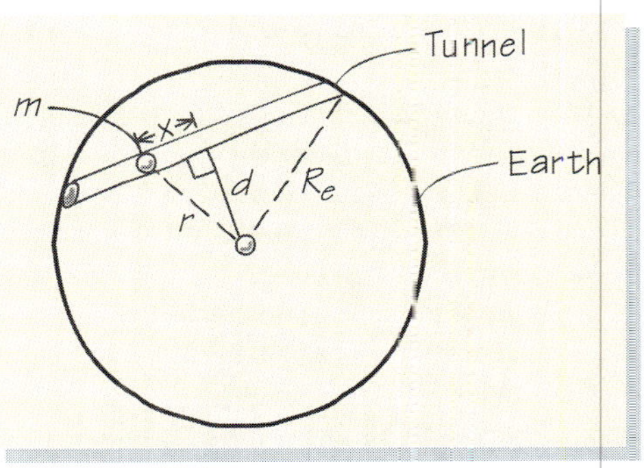

FIGURE 12–30 Problem 46.

12–6 A Closer Look at Gravitation

47. (I) The mean Earth–Sun separation is 1.50×10^8 km, Earth's radius is 6.37×10^3 km, the mass of the Sun is 1.99×10^{30}

kg, and Earth's mass is 5.98×10^{24} kg. Given these data, locate the center of mass of the Earth–Sun system.

48. (I) At a certain moment, the Sun, Jupiter, and Saturn are lined up, with Jupiter between the Sun and Saturn. Suppose the orbits of Jupiter and Saturn were circular, with mean distances from the Sun of 7.78×10^8 km and 1.42×10^9 km, respectively. The mass of Saturn is 0.029 percent of that of the Sun. What is the ratio of the gravitational force on Jupiter due to Saturn to the gravitational force on Jupiter due to the Sun, at the specified moment?

49. (II) A binary star system consists of two stars, each of mass M, orbiting around their common center of mass with radii R from the center of mass. Determine the period of revolution. [*Hint*: See Problem 11 of Chapter 10.]

*12–7 Einstein's Theory of Gravitation

50. (I) Fighter pilots are able to withstand accelerations up to $7g$ for a short period. A jet dives toward Earth and pulls up at the last second in a parabolic orbit. Draw a force diagram showing the various forces on the pilot at the bottom of the orbit. If the dive was at night, could the pilot tell the difference between an increased value of gravitational force and the effect of the contact forces on him at the bottom of the dive?

51. (II) A very sharply defined laser beam, directed horizontally, enters a hotel room at height h. At what height does the light beam hit the opposite wall, which is 8 m from the first wall? Compare the difference in heights to the size of an atom. Is this a feasible experiment with which to test Einstein's theory of gravitation?

52. (II) An elevator of width w in free space is accelerated upward with acceleration g. A ray of light, traveling with speed c, enters through a pinhole on one side of the elevator, at right angles to the side at the moment the elevator starts to accelerate. It will strike the opposite wall at a somewhat lower height. What is the angle of deflection of the light? According to the equivalence principle, a passenger in the elevator could not distinguish this bending of light from a bending due to the effects of gravity.

General Problems

53. (I) A spaceship of the future is cylindrical in shape, with a radius of 60 m. In order to simulate terrestrial gravity on the inside surface of the cylinder, the spaceship is made to rotate about its axis. What is the angular velocity of the spaceship about its axis?

54. (I) (a) What is the acceleration g due to gravity on a planet with the same density as Earth but with twice the radius? (b) The orbital period and radius of Jupiter's Moon Ganymede are 7.16 d and 660,000 mi, respectively. What is the period of the Moon Io, whose orbital radius is 262,000 mi? (c) Planets A and B are both in circular orbits around a star. Planet A has half the orbital speed of planet B. What is the radius of A in terms of the radius of B?

55. (II) The Little Prince (a character in a book by Antoine de Saint-Exupery) lives on the spherically symmetric asteriod B-612 (Fig. 12–31). The density of asteroids, including B-612, is 5.2×10^3 kg/m^3. Assume that the asteroid does not rotate. The Little Prince noticed that he felt lighter whenever he walked quickly around his asteroid. In fact, he found that he became weightless and started to orbit the asteroid like a satellite whenever he speeded up to 2 m/s. (a) Estimate the radius of the asteroid from these data. (b) What is the escape speed for the asteroid? (c) Suppose that B-612 does rotate about an axis such that the length of the day there is 12 h. Can the Little Prince take advantage of this rotation when he wants to orbit his asteroid?

FIGURE 12–31 Problem 55.

56. (II) A neutron star has a mass of 4×10^{30} kg and a radius of 10 km. (a) Calculate the acceleration due to gravity at the surface of the neutron star. (b) What is the difference between the gravitational forces acting on the top and the bottom of a dumbbell held vertically on the surface (that is, with one end on the surface of the neutron star and the other 1 cm above the surface)? The dumbbell consists of two 1-g point masses connected by a massless connector of length 1 cm.

57. (II) Suppose that, instead of a $1/r^2$ dependence, an attractive central force varied with distance as $1/r^n$. (a) Would such forces support a circular orbit? (b) Find the resulting relation between period and radius analogous to Kepler's third law for this force.

58. (II) Consider a cluster of galaxies that fills a sphere of radius R and average mass density ρ. (There are so many galaxies that you can assume uniform density.) There is a galaxy of mass M at the edge of this sphere (Fig. 12–32). (a) Write an expression for the energy of the galaxy. (b) In the big-bang model of the origin of the universe, the velocity of the galaxy is directed radially outward from the center of the sphere; the galaxy's speed is $v = HR$, where $H = (15 \text{ km/s})/(10^6 \text{ ly})$ is the Hubble parameter. For what critical density ρ_c of the large cluster will the galaxy be able to escape to infinity with a final velocity of zero?

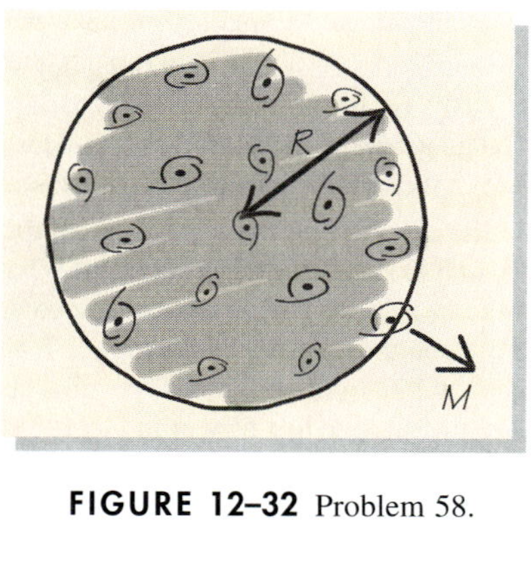

FIGURE 12–32 Problem 58.

59. (II) A satellite in low circular Earth orbit is subject to a very small constant friction force, f, due to the thin atmosphere. As it spirals in, it slowly decreases its radius. Find the decrease in radius per revolution under the assumption that the orbit is approximately circular with radius r. Find the changes in potential energy, total energy, and kinetic energy per orbit. [*Hint*: If you find the kinetic energy increasing, you are on the right track!]

60. (II) The electric force between an electron and a (much more massive) proton is attractive and of magnitude e^2/r^2, where $e^2 = 2.3 \times 10^{-28}\ \mathrm{N \cdot m^2}$. The electron circles the proton in a circular orbit of total energy $E = -2.2 \times 10^{-18}$ J (We have taken zero potential energy to be at infinite separation.) (a) What is the radius of the orbit? (b) What is the period of the orbit? (c) If another electron were in another circular orbit around the proton, with an orbit radius four times as large as the first, what would its total energy and period be?

61. (II) An astronaut of mass 115 kg (including equipment) finds himself drifting away from his orbiting space ship at 0.05 m/s. He throws a 3-kg wrench in the direction of his drift and comes to rest relative to the ship 1 m from its surface. The ship is a sphere of radius 12 m and mass 10^5 kg. (a) At what speed does he throw the wrench? (b) How many hours must he wait for the gravitational attraction of the ship to pull him to its surface, assuming that the force of gravity is approximately constant in the region of interest?

62. (II) Suppose that an object of mass m is placed at the point at which the gravitational attraction of the Moon is just canceled by that of Earth; further, suppose that the object is displaced by a small distance x along a line perpendicular to the line connecting the centers of Earth and the Moon (Fig. 12–33). What are the magnitude and direction of the net force on the object as a function of x? Calculate your answer by using the approximation $(r^2 + x^2)^n = r^{2n}[1 + (nx^2/r^2) + \cdots]$, valid for $x^2/r^2 << 1$.

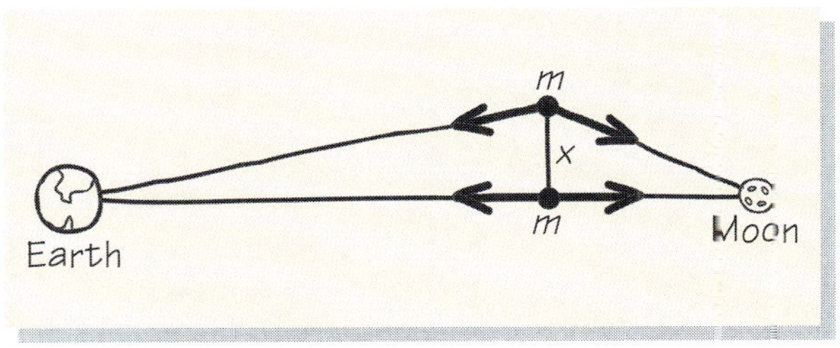

FIGURE 12–33 Problem 62.

63. (III) Three stars, each the mass of the Sun, form an equilateral triangle. Each moves in a circular orbit about the center of mass of the system because of the gravitational force exerted by the other two stars. (a) Is such an arrangement possible? (b) If so, what is the period of the motion, assuming that the side of the triangle is an Earth–Sun distance? (c) Is the system stable?

64. (III) Astrologers claim that a person's life is influenced by the position of the planets at the moment of that person's birth. To check whether this influence could be due to gravity, compare the following two quantities: the change in the gravitational force on a baby in a hospital due to the change in the position of Jupiter from one day to the next, and the change in the gravitational force due to the presence or absence of a 2-ton truck parked near the hospital at a distance of 100 m. Jupiter has a mass of 1.90×10^{27} kg; its mean distance from the Sun is 0.78×10^9 km, and its period is 11.9 yr. Assume a circular orbit for Jupiter, and a circular orbit of radius 1.5×10^8 km for Earth. Choose the region of closest approach of the two planets for convenience.

65. (III) When the first nuclear weapons were detonated, concern was expressed in some quarters that a huge nuclear chain reaction would be set up, blowing Earth to pieces. Show that the energy that would be required to disassemble Earth completely into pieces totally separate from each other is $\frac{3}{5}GM^2{}_E/R_E$. [*Hint*: Imagine that layers of Earth are peeled off one by one like layers of an onion.]

CHAPTER

13

An earthquake in Oakland, California caused this highway bridge to collapse under the influence of harmonic resonant forces. By studying the behavior of systems—such as the soil underlying the highway—under harmonic forces, we can learn to build structures that can resist damage.

Simple Harmonic Motion

Rhythmic motion—also known as periodic motion—is an important feature of the physical world. The very concept of time may have arisen from the observation that certain motions, such as the human heartbeat and the cycling of the seasons, repeat themselves in a reliable and regular way. The most basic type of rhythmic motion appears over and over again: simple harmonic motion. Examples are provided by the motion of a mass on the end of a spring and by the motion of a pendulum. *Simple harmonic motion is motion in which the position of a point varies with time as a sine or a cosine.* Such motion occurs when we have restoring forces—forces that tend to bring an object back to a point—that vary linearly with a position variable. While the spring force is an example that we will use repeatedly, simple harmonic motion is of universal importance because *virtually any stable equilibrium situation involves a linear restoring force.* We will discover that simple harmonic motion is a fundamental part of the physical world.

13–1 THE KINEMATICS OF SIMPLE HARMONIC MOTION

Simple harmonic motion—the motion followed by a mass on the end of a spring or a pendulum—is a basic form of **periodic**, or **oscillatory**, motion: motion that repeats. The word "harmonic," signifying agreement and accord, reveals that we have always seen beauty in this motion. In the back-and-forth of simple harmonic motion, the position $x(t)$ of an object is of the form $\sin(\omega t)$ or $\cos(\omega t)$. Both sines and cosines repeat themselves periodically as time passes. The trigonometric functions are functions of a dimensionless argument, an angle measured in degrees or radians. Thus the coefficient ω of the time must have the dimensions $[T^{-1}]$. We'll see later that ω is a fundamental property of the motion, determined by the inertia of the moving objects and the restoring forces acting on them.

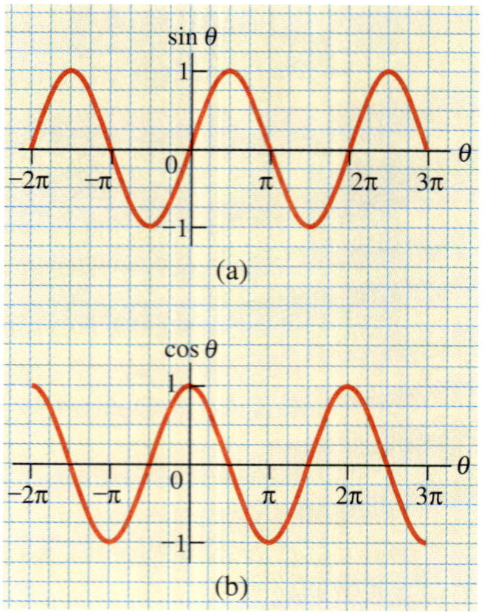

FIGURE 13–1 Plots of (a) sin θ and (b) cos θ, both as a function of θ.

How do we decide if the motion of a mass on the end of a spring has a sine or a cosine dependence? Let's look at a graph of sin θ versus θ next to a graph of cos θ versus θ (Fig. 13–1). Both functions repeat every time the angle θ changes by 2π rad. When $\theta = 0$, the sine function is zero, whereas the cosine function is $+1$, but this is only a matter of placing the axis. Indeed, the functions are *identical* if the origin of the θ axis is shifted; if θ is shifted by an angle we call the **phase**, δ, the curves drawn in Fig. 13–1 shift along the horizontal axis and move between the sine and the cosine functions. Thus, if δ is chosen properly, the function $\sin(\omega t + \delta)$ can represent $\sin(\omega t)$, $\cos(\omega t)$, or anything in between.

The sine function is dimensionless and, moreover, varies between -1 and $+1$. But $x(t)$ has dimensions of length. We therefore multiply the sine function by an **amplitude**, A, which has the proper dimensions of length. The amplitude is defined to be positive and describes the magnitude of the maximum excursion away from the point of zero displacement (Fig. 13–2). The resulting expression for the position of an object in simple harmonic motion is

$$x(t) = A\sin(\omega t + \delta). \tag{13–1}$$

The mathematical definition of simple harmonic motion, amplitude, and phase

Properties of Simple Harmonic Motion

Three independent parameters appear in simple harmonic motion: the amplitude, A; the phase, δ; and the constant ω. The amplitude and the phase are determined by specifying the value of position $x(t)$ and velocity $v(t) = dx/dt$ at an initial time $t = 0$. We say that A and δ are determined by the **initial conditions** for the motion. It follows from Eq. (13–1) that, at $t = 0$,

$$x = A\sin\delta$$

and

$$\frac{dx}{dt} = \omega A\cos\delta.$$

These equations are enough to specify both A and δ once ω is known.

The period of simple harmonic motion

The constant ω has to do with the repetition time for the motion. The sine function repeats itself either when the angle increases by 2π rad (see Fig. 13–1) or, because

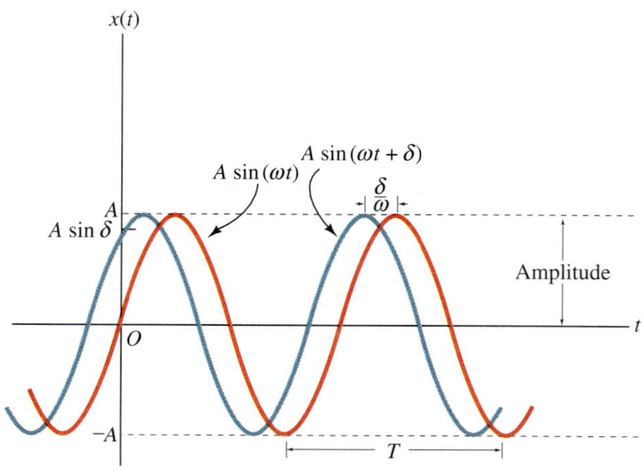

FIGURE 13–2 The phase, δ, corresponds to a sliding of the curve of displacement versus time to earlier or later times for simple harmonic motion. The amplitude and period are also indicated.

δ is a constant, when ωt increases by 2π. The time that is required for this is by definition T, so that $\omega T = 2\pi$. We can solve for T, the **period** of simple harmonic motion:

$$T = \frac{2\pi}{\omega}. \tag{13–2}$$

This equation provides us with an interpretation of ω, whose value specifies the value of T. In Chapter 3, where we described uniform circular motion, we defined the **frequency**, f, as the inverse of the period. A period of 5 s means a frequency of one every five seconds, while a period of 0.5 s means a frequency of two per second, and so forth:

The frequency of simple harmonic motion

$$f = \frac{1}{T}. \tag{13–3}$$

If the period is measured in seconds, the frequency is measured in s^{-1}. In SI, the unit s^{-1} is the **hertz** (Hz), after the physicist Heinrich Hertz:

The SI unit of frequency

$$1\ \text{Hz} = 1\ \text{s}^{-1}. \tag{13–4}$$

By comparing Eqs. (13–2) and (13–3), we find that

$$f = \frac{\omega}{2\pi}. \tag{13–5}$$

The constant ω is referred to as the **angular frequency**. We will see in Section 13–2 that ω can be identified with the angular speed, a quantity defined and used in Chapters 3 and 9. Inversion of Eq. (13–2) or (13–5) gives

The angular frequency

$$\omega = \frac{2\pi}{T} = 2\pi f. \tag{13–6}$$

When the position is specified as a function of time, the velocity and the acceleration are determined by taking successive derivatives. As a consequence of Eq. (13–1), we have (see Appendix IV–7)

$$v = \frac{dx}{dt} = \omega A \cos(\omega t + \delta). \tag{13–7}$$

One further derivative gives the acceleration as a function of time:

$$a = \frac{dv}{dt} = -\omega^2 A \sin(\omega t + \delta) = -\omega^2 x. \tag{13–8}$$

The acceleration is proportional to the displacement, with a crucial minus sign. We will argue that virtually all stable equilibrium situations are associated with simple harmonic motion; thus, the proportionality of the acceleration and the displacement is a universal property of motion near equilibrium.

In simple harmonic motion, the acceleration is proportional to the position.

Relations among Position, Velocity, and Acceleration in Simple Harmonic Motion

In Fig. 13–3, we plot the position, velocity, and acceleration of an object in simple harmonic motion over two full periods, starting with $x(t) = A \sin(\omega t)$. (For convenience, the phase has been taken to be zero. The relations discussed here are not affected by the phase.) Here, we take the $+x$-direction to the right and the $-x$-direction to the left.

Let's follow Fig. 13–3. The object starts at the origin at $t = 0$. Here, the velocity is maximum in magnitude and is positive. The acceleration is zero, so that the velocity is not changing. After one-quarter of the period ($\omega t = \pi/2$), the object has moved

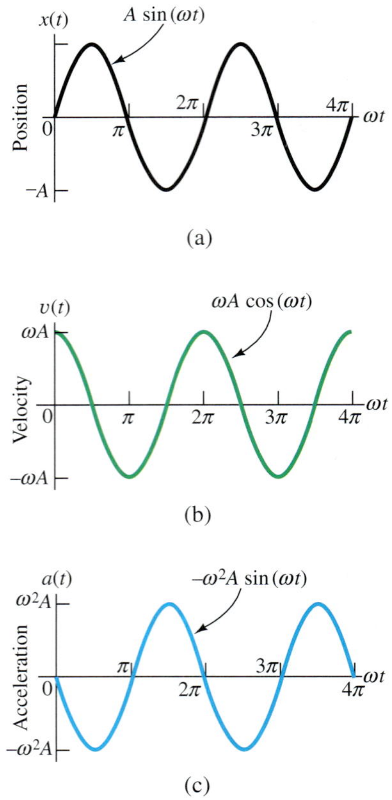

FIGURE 13–3 Starting with (a) position $x(t) = A\sin(\omega t)$, a single derivative gives (b) the velocity $v(t) = \omega A\sin(\omega t)$. (c) One further time derivative gives the acceleration $a(t) = -\omega^2 A\sin(\omega t)$. We have in each case plotted two cycles, or periods, for zero phase; the curves repeat after every period.

to the right-hand extreme and is ready to turn around. The velocity is zero at this turnaround point, but the acceleration has actually reached a maximum in magnitude and is negative, indicating that the velocity will be turning to the left and will become negative. (Think of a ball thrown in the air; at the maximum height—the turnaround point—the velocity is zero even if the acceleration is nonzero and directed toward Earth.) After one-half the period ($\omega t = \pi$), the object once again passes through the origin, this time moving to the left. The acceleration is again zero. The three-quarter mark ($\omega t = 3\pi/2$) is at another turnaround, characterized by a maximum negative value of x—the object is at its left-hand extreme—and zero velocity. The acceleration is maximum and positive, meaning that the velocity is becoming positive, and the object will subsequently move back to the right. Finally, after one full period ($\omega t = 2\pi$), the object has come back to its starting point, moving to the right through the origin with its largest positive velocity and zero acceleration. The situation at $t = 2\pi/\omega$ is identical to what it was at $t = 0$.

EXAMPLE 13–1 A cork floating on a pond moves in simple harmonic motion, bobbing up and down over a range of 4 cm (Fig.13–4). The period of the motion is $T = 1.0$ s, and a clock is started at $t = 0$ s when the cork is at its minimum height. What are the height and velocity of the cork at $t = 10.5$ s?

Solution: We suppose that the cork moves along the z-axis, and we take the origin $z = 0$ to be the midpoint of the motion. Thus the maximum value of z is $z_{max} = 2$ cm, and the minimum value is $z_{min} = -2$ cm. The motion takes the general form $z(t) = A\sin(\omega t + \delta)$. We know the period, T, and from Eq. (13–6), $\omega = 2\pi/T$. The constants A and δ must be determined from other information, namely, the initial conditions. The amplitude, A, is the maximum excursion from equilibrium, and is given by $A = z_{max} = |z_{min}|$. The phase, δ, is then determined by the initial condition that the height is a minimum when $t = 0$ s. Thus the equation determining δ is

$$z_{min} = A\sin(\omega t + \delta)\big|_{t=0} = A\sin\delta.$$

When we substitute $A = |z_{min}|$, this equation becomes $z_{min} = |z_{min}|\sin\delta$. Because z_{min} is negative, this result implies

$$\sin\delta = -1.$$

When the sine function is -1, its argument is $-\pi/2$ or $3\pi/2$. In fact, any integer multiple of 2π can be added to or subtracted from $-\pi/2$, and it is just a matter of convenience to choose the phase to be $-\pi/2$. When a simple phase such as this occurs, it is often worthwhile to expand the sine function with trigonometric identities. In particular, from Appendix IV–4,

$$\sin(\omega t + \delta) = \sin\left(\omega t - \frac{\pi}{2}\right) = \sin(\omega t)\cos\left(\frac{\pi}{2}\right) - \cos(\omega t)\sin\left(\frac{\pi}{2}\right) = -\cos\omega t.$$

We have used the fact that $\cos(\pi/2) = 0$ and $\sin(\pi/2) = 1$. Then, instead of $\sin(\omega t + \delta)$, we have $-\cos(\omega t)$ appearing in the expression for $z(t)$. We gather our results:

$$z = -A\cos\left(\frac{2\pi t}{T}\right),$$

where $A = 2$ cm and $T = 1$ s.

The velocity is the time derivative of this expression:

$$v = \frac{dz}{dt} = -A\left(\frac{-2\pi}{T}\right)\sin\left(\frac{2\pi t}{T}\right) = \frac{2\pi A}{T}\sin\left(\frac{2\pi t}{T}\right).$$

We now want to evaluate z and v at $t = 10.5$ s, or 10.5 periods. Both z and v repeat themselves every period, so the values of z and v at 10.5 s are the same as at

0.5 s (0.5 period):

$$\cos\left[\frac{(2\pi)(10.5\text{ s})}{1.0\text{ s}}\right] = \cos[(2\pi)(10.5)] = \cos[2\pi(10) + 2\pi(0.5)]$$

$$= \cos[2\pi(0.5)] = -1;$$

$$\sin\left[\frac{(2\pi)(10.5\text{ s})}{1.0\text{ s}}\right] = \sin[(2\pi)(10.5)] = \sin[2\pi(10) + 2\pi(0.5)]$$

$$= \sin[2\pi(0.5)] = 0.$$

Thus, for $t = 10.5$ s,

$$z = -A(-1) = A = +2\text{ cm}$$

and

$$v = \frac{2\pi A}{T}(0) = 0\text{ cm/s}.$$

It is simple to deduce these results from physical reasoning. We are interested in where the cork is after exactly one-half a period. Because the cork starts at its minimum height, half a period later it is at its maximum height, +2 cm in this case. That is a point where the cork stops and starts back down, so the velocity is zero there.

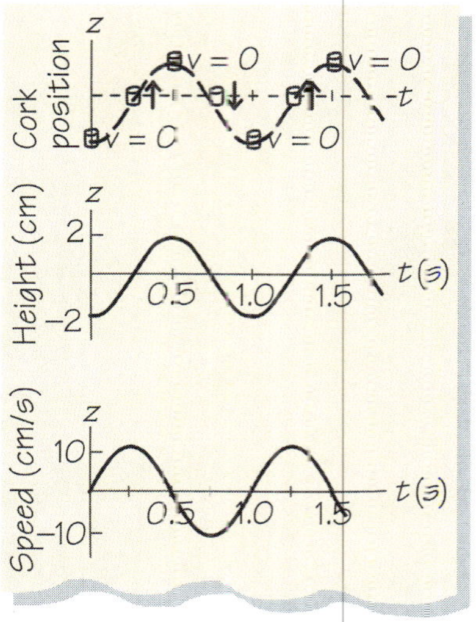

FIGURE 13–4 Example 13–1.

13–2 A CONNECTION WITH CIRCULAR MOTION

In Chapter 3, we discussed another kind of periodic motion: uniform circular motion. We now demonstrate that this motion is intimately related to simple harmonic motion (Fig. 13–5). Figure 13–6 recalls uniform circular motion for a point moving in the xy-plane a constant distance R from the origin. The motion is described by an angle θ, measured from the x-axis, that varies linearly with time:

$$\theta = \omega t + \delta. \qquad (13\text{–}9)$$

PROBLEM SOLVING

Each period, the motion repeats.

The phase, δ, is just the value of θ at time $t = 0$.

If we were to look at a side view of the uniform circular motion of a thumbtack stuck on a rotating turntable, we would see the thumbtack oscillate in simple harmonic motion. Figure 13–6 indicates the *projection* of the circular motion on the x- and y-axes. Simple trigonometry gives us these projections:

$$x = R\cos\theta = R\cos(\omega t + \delta); \qquad (13\text{–}10)$$

$$y = R\sin\theta = R\sin(\omega t + \delta). \qquad (13\text{–}11)$$

FIGURE 13–5 The relation between uniform circular motion and simple harmonic motion is evident in the piston-linkage connection on the wheel.

Thus uniform circular motion corresponds to simple harmonic motion in both of the x- and y-directions. A cosine rather than sine appears in x but, as we have already discussed, this is just the standard form with a different phase. We can use the trigonometric identity $\sin[\theta + (\pi/2)] = \sin\theta\cos(\pi/2) + \cos\theta\sin(\pi/2) = \cos\theta$ to replace the cosine in Eq. (13–10) with a sine function, and we thereby obtain

$$x = R\sin\left(\omega t + \delta + \frac{\pi}{2}\right). \qquad (13\text{–}12)$$

Projections of uniform circular motion are simple harmonic motion.

Both the x- and y-motions are now in the standard form of Eq. (13–1). The two motions have a phase that differs by exactly $\pi/2$ (90°), and the sign of this extra phase specifies the direction—clockwise or counterclockwise—of the corresponding uniform circular motion (see Problem 22).

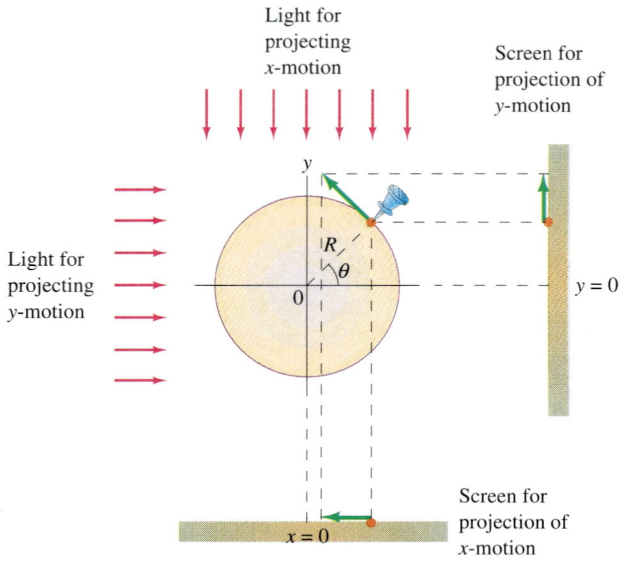

FIGURE 13–6 Uniform circular motion in the xy-plane, and its projection onto the x- and y-axes. These projections are simple harmonic motion.

13–3 SPRINGS AND SIMPLE HARMONIC MOTION

Springs give rise to simple harmonic motion. Let's restrict ourselves to one-dimensional motion and dispense with vector notation. If the spring is not overly stretched or compressed, the spring force on a mass displaced by x from the equilibrium position of the spring has the form known as *Hooke's law*:

$$F = -kx. \tag{13–13}$$

The spring force. Hooke's law is discussed in Chapter 6.

Here k is the *spring constant.* Thus the force is linear in the displacement. The minus sign in Eq. (13–13) makes the force a *restoring force.* A displacement in the $+x$-direction gives rise to a force that acts in the $-x$-direction and vice versa (Fig. 13–7). Newton's second law provides us with the connection between the force and the acceleration; namely, $F = -kx = ma$. Thus the acceleration of a mass on the end of a spring is proportional to its displacement:

$$a = -\frac{k}{m}x. \tag{13–14}$$

The angular frequency of simple harmonic motion

This is indeed simple harmonic motion because the acceleration is proportional to the position. Comparison of Eqs. (13–8) and (13–14) yields the important result that the angular frequency is determined by the mass and the spring constant:

$$\omega^2 = \frac{k}{m}; \tag{13–15}$$

$$\omega = \sqrt{\frac{k}{m}}. \tag{13–16}$$

The period of simple harmonic motion does not depend on the amplitude.

In turn, Eqs. (13–2) and (13–3) give the period and the frequency of the oscillations:

$$T = 2\pi\sqrt{\frac{m}{k}} \quad \text{and} \quad f = \frac{1}{2\pi}\sqrt{\frac{k}{m}}. \tag{13–17}$$

Remarkably, *the period of the motion is independent of the amplitude.*

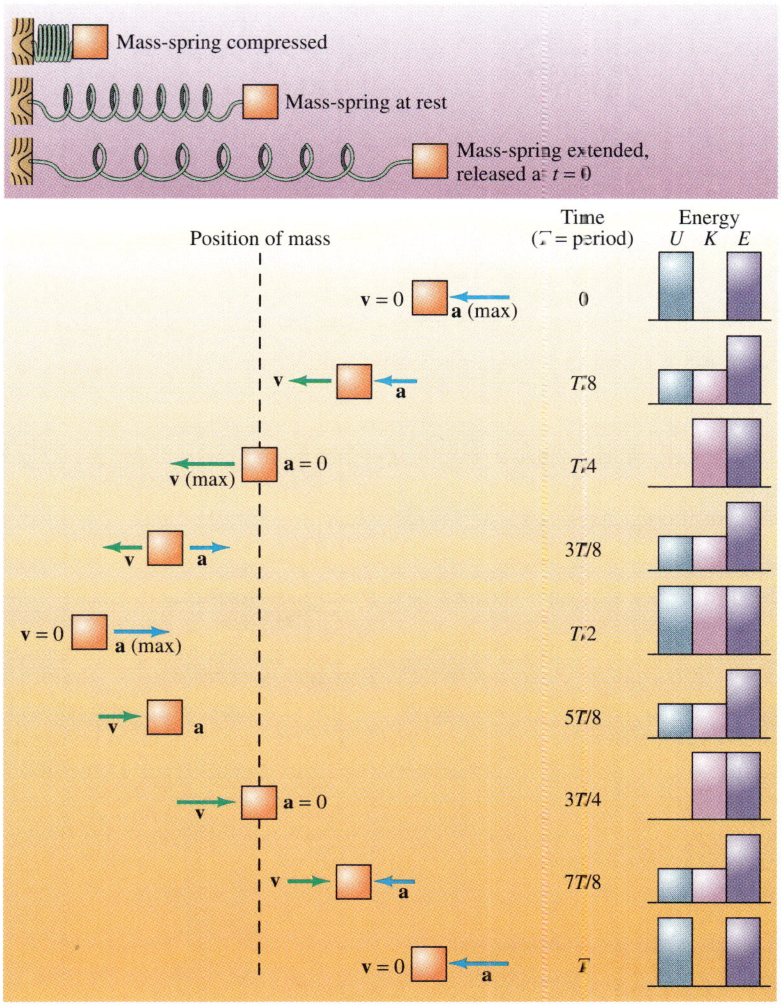

FIGURE 13–7 We can follow the velocity and acceleration of a mass that moves in simple harmonic motion on the end of a spring. The speed is lowest (and the acceleration is highest) when the displacement from equilibrium is a maximum, and the speed is highest (and the acceleration is lowest) when the displacement is a minimum. We can also see the play between kinetic and potential energy; one is large where the other is small.

EXAMPLE 13–2 A mass of 0.50 kg moves along the x-direction under the influence of a spring with spring constant $k = 2.0$ N/m. The origin of the x-axis is at the equilibrium point of the mass. At $t = 0$ s, the mass is at the origin and moving with a speed of 0.5 m/s in the $+x$-direction. (a) At what time t_1 does the mass first arrive at its maximum extension? (b) What is this maximum extension?

Solution: (a) The information given includes k, m, the position at $t = 0$ s, and the velocity at $t = 0$ s. This is enough information to calculate ω, the period, the amplitude, and the phase. From Eq. (13–16), the angular frequency, ω, is

$$\omega = \sqrt{\frac{k}{m}} = \sqrt{\frac{2.0\text{ N/m}}{0.50\text{ kg}}} = \sqrt{4.0\text{ s}^{-2}} = 2.0\text{ rad/s}.$$

The period T is

$$T = \frac{2\pi}{\omega} = \frac{2\pi\text{ rad}}{2.0\text{ rad/s}} = 3.1\text{ s}.$$

The mass travels from the equilibrium position to its maximum extension in precisely one-quarter period, so

$$t_1 = \frac{T}{4} = 0.79\text{ s}.$$

(b) We use the information about x and v at $t = 0$ s to find the amplitude. Writing $x(t) = A\sin(\omega t + \delta)$, we get

$$t = 0\ \text{s}, \qquad x = A\sin\delta = 0.$$

This implies that $\delta = 0$. We use this, in turn, for the value of v:

$$t = 0\ \text{s}, \qquad v = A\omega\cos(0) = A\omega.$$

The argument of the cosine is zero because both t and δ are zero. Thus

$$A = \frac{v|_{t=0}}{\omega} = \frac{0.50\ \text{m/s}}{2.0\ \text{rad/s}} = 0.25\ \text{m},$$

which is the maximum excursion of the mass from the origin.

The spring is the prototype of dynamic systems that have simple harmonic motion. Many other dynamic systems also exhibit simple harmonic motion. All such systems are reducible in one way or another to a spring, meaning that the *form* of the force is the same as that of the spring: a restoring force linear in some variable.

The Hanging Spring: How Constant Forces Affect Simple Harmonic Motion

When a constant force such as gravity influences a mass that is already attached to a spring, remarkably little changes. The motion of this mass is exactly as it would be with no gravity, except that the equilibrium point changes. To show this, suppose that a mass is suspended vertically from a spring (Fig. 13–8). The mass would be at height $y = y_0$ if there were no gravity, where y is measured up from ground level. In the presence of gravity, the mass hangs at a new, lower equilibrium position, $y_1 = y_0 - \Delta y$. The positive value Δy is determined by a cancellation between the downward (negative) force of gravity, whose magnitude is $F_g = mg$, and the upward (positive) force exerted by the spring,

$$F_{\text{spring}} - F_g = 0.$$

Because $F_{\text{spring}} = -k(-\Delta y) = k\,\Delta y = k(y_0 - y_1)$, we have

$$k(y_0 - y_1) - mg = 0. \tag{13–18}$$

Suppose that we now move the mass to a new position, y. The stretch of the spring from its original equilibrium position (no gravity) is $y - y_0$, and the spring force is

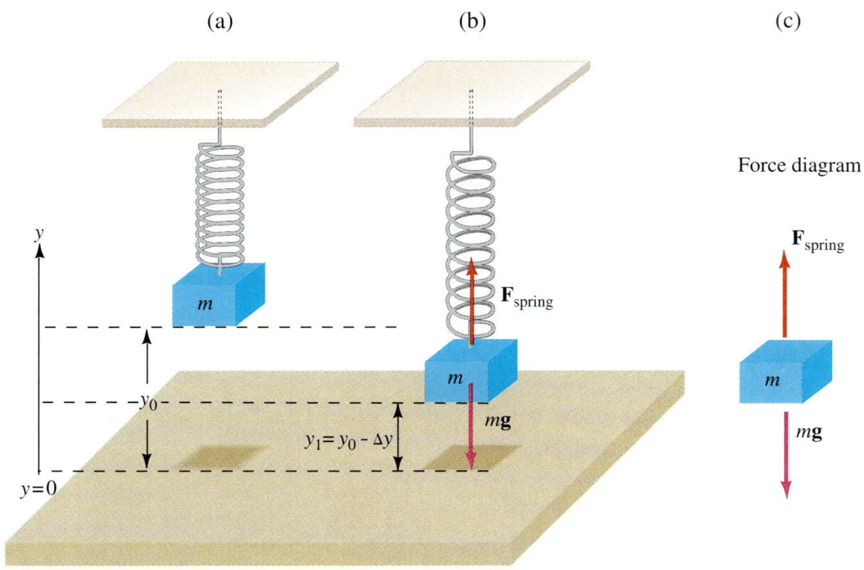

FIGURE 13–8 A mass on the end of a spring is suspended vertically. (a) If its equilibrium length would place it at height y_0 in the absence of gravity, then (b) it will be stretched an additional amount, Δy, to a new equilibrium position, y_1, under the influence of gravity. (c) Force diagram for the mass.

$F_{spring} = -k(y - y_0)$. The signs are satisfactory because if y is higher than y_0, then $y - y_0$ is positive and the spring force is negative (or downward). If y is below y_0, then $y - y_0$ is negative and the spring force is positive (or upward). The force on the mass is then

$$F_{net} = F_{spring} - F_g = -k(y - y_0) - mg. \tag{13–19}$$

According to Eq. (13–18), we can replace the quantity mg by $k(y_0 - y_1)$, and the net force is

$$F_{net} = -k(y - y_0) - k(y_0 - y_1) = -k(y - y_1). \tag{13–20}$$

Equation (13–20) is again the equation for a spring force with exactly the same spring constant, and hence *the motion of a mass on the end of a vertical spring has exactly the same frequency as it does when it is on the end of a horizontal spring.* The only change is that the displacement is measured from the new equilibrium position, y_1, rather than the old one, y_0.

13–4 ENERGY AND SIMPLE HARMONIC MOTION

We extensively examined questions involving energy as they apply to the spring force in Chapter 7, where the topic was the conservation of energy. In particular, we found that the work done by a spring force in moving a mass from one position to another is independent of the path taken by the mass. This means that *the spring force is conservative.* A conservative force depending on a position variable x has a potential energy function $U(x)$ associated with it; the total energy, E (the sum of kinetic energy, K, and potential energy), is *conserved* throughout any motion.

The computation of the potential energy $U(x)$ of an object attached to a spring was treated in Section 7–2. We found

$$U(x) = \frac{1}{2}kx^2. \tag{13–21}$$

The potential energy associated with a spring

In Eq. (13–21), zero potential energy has been chosen at the equilibrium position of the spring, $x = 0$. The kinetic energy is simply

$$K = \frac{1}{2}mv^2. \tag{13–22}$$

Because both x and v are known for simple harmonic motion from Eqs. (13–1) and (13–7), the variation in time of U and K can be plotted. If we write the argument of $\omega t + \delta$ as θ, we have

$$U = \frac{1}{2}kA^2 \sin^2\theta, \tag{13–23}$$

$$K = \frac{1}{2}mA^2\omega^2 \cos^2\theta = \frac{1}{2}kA^2 \cos^2\theta. \tag{13–24}$$

In writing this form for K, we have used $\omega^2 = k/m$, Eq. (13–15). Figure 13–9 is a plot of the potential and kinetic energy functions as θ varies between 0 and 2π, which corresponds to a complete cycle. Both $\sin^2\theta$ and $\cos^2\theta$ vary between 0 and 1; when $\sin^2\theta$ is a minimum, $\cos^2\theta$ is a maximum and vice versa. Thus U and K each vary between 0 and $kA^2/2$. Suppose that an object attached to a spring starts at the origin and moves to the right. At the origin, the potential energy is zero and K is a maximum. As the mass moves to the right, it slows until it has reached its turnaround point at one-quarter cycle, where the velocity and hence K are zero. Because x is at its maximum here, U is also a maximum. The mass now moves to the left, gaining speed until the speed is a maximum as it passes through the origin once more. Here, after one-half cycle, K is a maximum and U is a minimum. Finally, at the left-hand turnaround point, K is a minimum and U is a maximum. *The energy flows back and forth between U and K.*

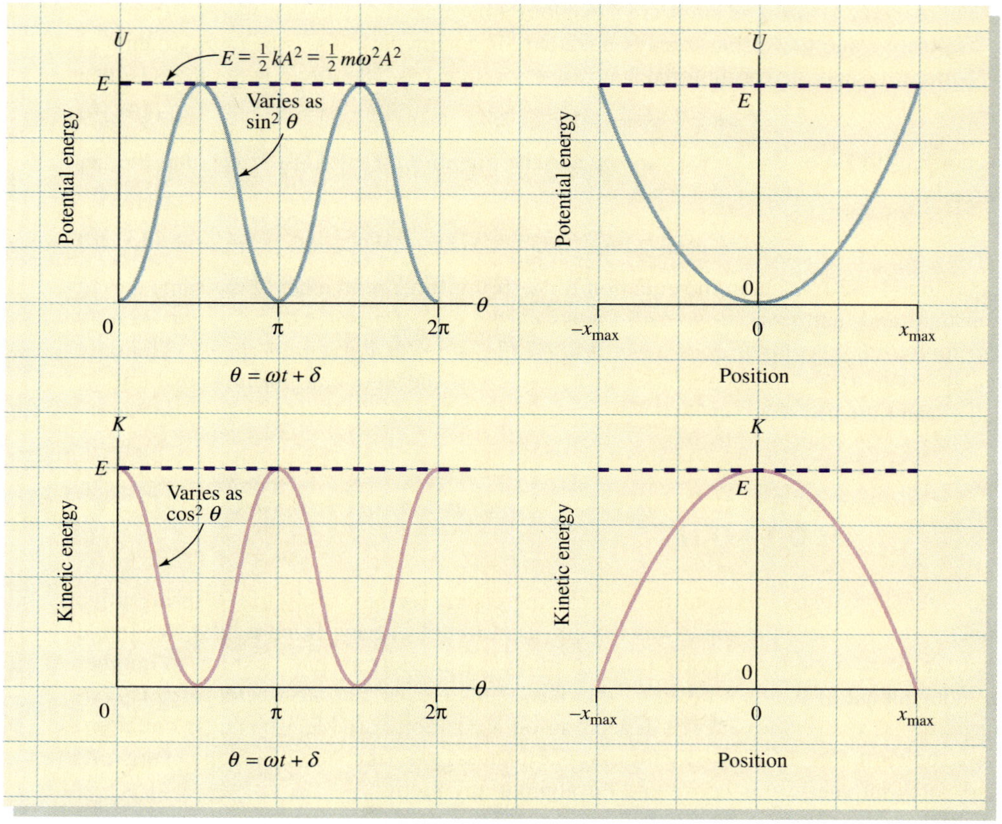

FIGURE 13–9 (a) The potential energy and (b) the kinetic energy of a mass in simple harmonic motion plotted over one cycle. When one is a maximum, the other is a minimum, and their sum, the total energy, is conserved.

The Total Energy

The total energy, $E = U + K$, must be constant. We have

$$E = \frac{1}{2}kA^2 \sin^2(\omega t + \delta) + \frac{1}{2}kA^2 \cos^2(\omega t + \delta)$$

$$= \frac{1}{2}kA^2[\sin^2(\omega t + \delta) + \cos^2(\omega t + \delta)]. \quad (13\text{–}25)$$

The total energy in simple harmonic motion

Because the sum of $\sin^2\theta$ and $\cos^2\theta$ is unity for any θ, E is indeed constant in time:

$$E = \frac{1}{2}kA^2. \quad (13\text{–}26)$$

The quadratic dependence of energy on amplitude is typical of simple harmonic motion.

EXAMPLE 13–3 A mass m attached to a spring of spring constant k is stretched a length X from its equilibrium position and released with no initial motion. (a) What is the maximum speed attained by the mass in the subsequent motion? (b) At what time is this speed first attained?

Solution: (a) The conservation of energy is a useful tool in this type of problem. Just before the mass is released from rest, all of its energy is potential energy; that is, the total energy is

$$E = \frac{1}{2}kX^2.$$

This agrees with Eq. (13–26) because the maximum excursion in space of the motion is, by definition, the amplitude of the motion. E is the value of the energy for all time. When the maximum speed is attained, all the energy is in the form of kinetic energy:

$$\frac{1}{2}mv_{\max}^2 = E = \frac{1}{2}kX^2.$$

We solve for $v_{\max}$:

$$v_{\max} = \sqrt{\frac{k}{m}}X = \omega X.$$

(b) The conservation of energy is not very helpful when we want to learn about time dependence. To answer the second question, we must make use of our knowledge about the time dependence of simple harmonic motion. The maximum speed is attained here when $x = 0$. Because the mass is released at the maximum value of x, the first time the mass passes through the origin is one-quarter period later:

$$t = \frac{T}{4} = \frac{1}{4}2\pi\sqrt{\frac{m}{k}} = \frac{\pi}{2}\sqrt{\frac{m}{k}}.$$

PROBLEM SOLVING

For time dependence, energy is less useful.

13–5 THE SIMPLE PENDULUM

The simple pendulum

Simple harmonic motion occurs throughout nature. A particularly important example is the ordinary pendulum, such as the swinging pendulum in some clocks. For centuries, such clocks were the most accurate way to measure time. When we idealize the form of the pendulum to a point mass suspended from a massless string of length ℓ, as in Fig. 13–10, we have the **simple pendulum**. The mass moves along the arc of a circle traced out by the end of the taut string (Fig. 13–11a).

APPLICATION

Pendulum clocks

Let's first look at how Newton's second law applies. Suppose that the string makes an angle θ with the vertical. The force diagram for the mass, Fig. 13–11b, includes the force of gravity, $m\mathbf{g}$, and the tension of the string, $\mathbf{T}$. The tension is perpendicular to the path of the motion. Its only role in the motion is to constrain that motion to lie along the arc of a circle of radius ℓ. The position s of the mass along the arc of the circle is given by

$$s = \ell\theta, \qquad (13\text{–}27)$$

where s is measured from $\theta = 0$. The angle θ varies with time and is the quantity we wish to determine dynamically. Accordingly, we are interested only in components of forces along the arc of the path.

To obtain the velocity along the arc of the circle, we differentiate s with respect to time. Because ℓ is a constant, we find

$$v = \frac{ds}{dt} = \ell\frac{d\theta}{dt}. \qquad (13\text{–}28)$$

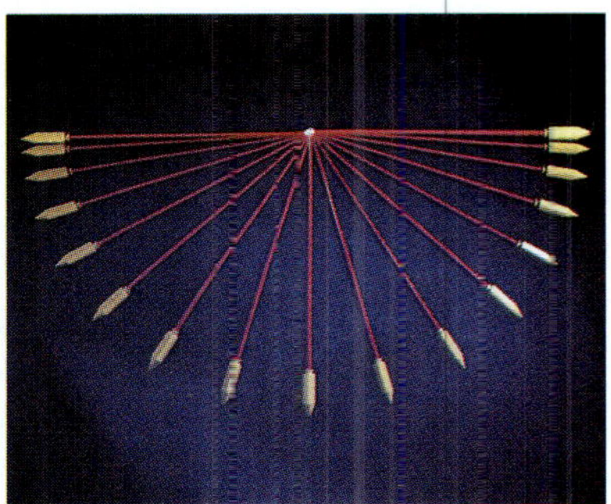

FIGURE 13–10 The simple pendulum, strobed at equal time intervals. The pendulum bob moves faster near the bottom of its swing and more slowly near the ends.

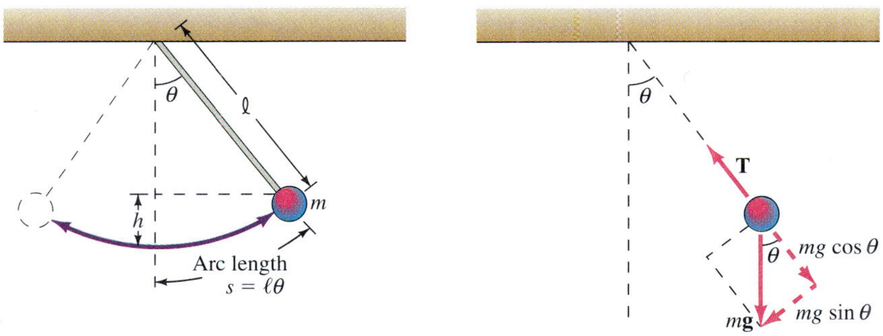

(a) (b)

FIGURE 13–11 (a) Geometry of the simple pendulum. (b) Force diagram for the simple pendulum. The tension, $\mathbf{T}$, is along the radial direction, but $m\mathbf{g}$ has components along both the radial direction and tangent to the arc of the circle traced out by the motion of the mass.

The tangential acceleration, which is the component of the total acceleration along the arc of the circle, is, in turn, associated with changes in the magnitude of this velocity:

$$a = \frac{dv}{dt} = \ell \frac{d^2\theta}{dt^2}. \tag{13–29}$$

This component of the total acceleration (which should *not* be confused with the centripetal component of the acceleration) is due to the tangential force component

$$F_t = -mg \sin \theta. \tag{13–30}$$

The sign of this force component is important. It is in the negative direction when θ is itself positive—when the mass is on the right side of the vertical—and in the positive direction when the mass is on the left side of the vertical. This means that *the force of gravity always acts to bring the mass back to the vertical.* This is enough to ensure that the motion will be oscillatory but not enough to guarantee simple harmonic motion: In simple harmonic motion, the force must be *linear* in the dynamical variable itself—in this case, the angle.

Using Eq. (13–29) for the acceleration and Eq. (13–30) for the force, we see that Newton's second law takes the form

$$m\ell \frac{d^2\theta}{dt^2} = -mg \sin \theta.$$

Canceling the mass from this equation, we get

$$\ell \frac{d^2\theta}{dt^2} = -g \sin \theta. \tag{13–31}$$

Equation (13–31) would satisfy our requirement for simple harmonic motion if, instead of $\sin\theta$, θ itself appeared on the right-hand side. But this will indeed happen when the angle θ is small. To see this, refer to Fig. 13–12a for $\sin \theta$ and θ, which gives

$$\sin \theta = \frac{x}{\ell} \quad \text{and} \quad \theta = \frac{s}{\ell}, \tag{13–32}$$

where s is the arc length corresponding to angle θ. In Fig. 13–12b, note that x approaches s more and more closely as θ becomes smaller. Thus, for small θ, $\sin \simeq \theta$. This can also be seen from the expansion for the sine function:

$$\sin \theta = \theta - \frac{\theta^3}{3!} + \frac{\theta^5}{5!} - \cdots. \tag{13–33}$$

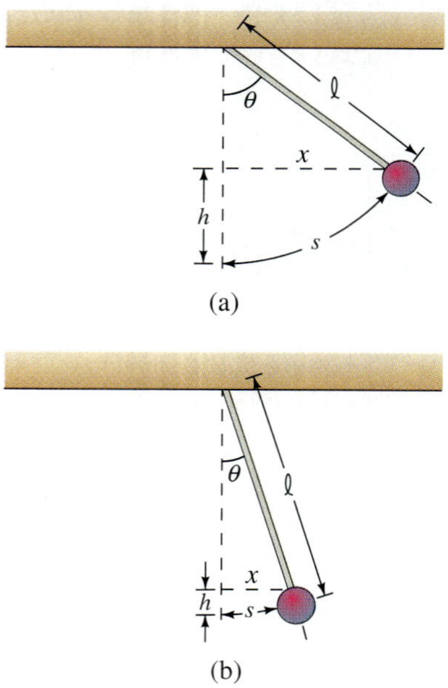

FIGURE 13–12 Comparing the horizontal displacement x of the mass of a simple pendulum with the arc length s traced out by the motion. These two lengths differ by a larger percentage (a) when θ is large than (b) when θ is small.

When θ is small, the terms of order θ^3, θ^5, and so forth can be ignored, justifying our approximation. For example, when $\theta = 0.2$ rad (about 11°), the difference between $\sin \theta$ and θ is about 1 percent, but when $\theta = 0.1$ rad (about 6°), the difference is only about 0.1 percent.

The small-θ approximation means that a pendulum will have true simple harmonic motion only for small excursions. This is because, with this approximation, Eq. (13–31) becomes

$$\text{for small } \theta\text{:} \qquad \ell \frac{d^2\theta}{dt^2} = -g\theta. \tag{13–34}$$

Equation (13–34) is *precisely* the equation for simple harmonic motion, as shown by comparison with Newton's second law for the spring, $m\, d^2x/dt^2 = -kx$. We must change only some variable names and constants to go from spring to pendulum:

$$x \to \theta, \qquad k \to g, \qquad \text{and} \qquad m \to \ell. \tag{13–35}$$

(Here, the horizontal arrow simply indicates a name substitution.) The solution to the motion of the simple pendulum for small angles is then taken directly from the solution for the motion of the spring:

$$\theta = \theta_0 \sin(\omega t + \delta) \quad \text{with} \quad \omega = \sqrt{\frac{g}{\ell}}. \tag{13–36}$$

Again, dynamics has determined the frequency. Here, θ_0 is the amplitude of the angular motion—the maximum angle attained. The period and frequency of the pendulum's motion come from Eqs. (13–5) and (13–6):

$$T = 2\pi\sqrt{\frac{\ell}{g}} \quad \text{and} \quad f = \frac{1}{2\pi}\sqrt{\frac{g}{\ell}}. \tag{13–37}$$

The period and frequency of the simple pendulum

As for the mass on the spring, *the period of the pendulum is independent of the amplitude*. This explains why a pendulum clock can be accurate: The amplitude does not enter into the period of the swing!

EXAMPLE 13–4 A simple pendulum 2.00 m long is suspended in a region where $g = 9.81\ \text{m/s}^2$. The point mass at the end is displaced from the vertical and given a small push, so its maximum speed is 0.11 m/s. What is the maximum horizontal displacement of the mass from the vertical line it makes when at rest? Assume that all the motion takes place at small angles.

Solution: The angle that the string makes with the vertical varies harmonically, $\theta = \theta_0 \cos(\omega t + \delta)$, where ω is the angular frequency. The horizontal displacement from the vertical is $x = \ell\theta$ (where ℓ is the string's length), as long as θ remains small (see Fig. 13–12b). Thus x also varies harmonically: $x = A \cos(\omega t + \delta)$, where $A = \ell\theta_0$. This is the quantity we want to find. Another good small-angle approximation is that the vertical component of the velocity is small, so $v \simeq dx/dt$. Thus we have

$$v \simeq \frac{d}{dt}[A\cos(\omega t + \delta)] = A\frac{d}{dt}[\cos(\omega t + \delta)] = -A\omega\sin(\omega t + \delta).$$

From this expression, we see that v varies harmonically, with amplitude $A\omega$. The maximum value of v occurs at $\theta = 0$ and is given by $v_{max} = A\omega = 0.11$ m/s. Finally, we can find ω from Eq. (13–36):

$$\omega = \sqrt{\frac{g}{\ell}} = \sqrt{\frac{9.81\ \text{m/s}^2}{2.00\ \text{m}}} = 2.21\ \text{rad/s}.$$

Thus, from $v_{max} = A\omega$,

$$A = \frac{v_{max}}{\omega} = \frac{0.11\ \text{m/s}}{2.21\ \text{rad/s}} = 0.05\ \text{m}.$$

This horizontal displacement of 5.0 cm is indeed small compared to the length, so our small-angle approximations are good. Figure 13–13 illustrates the motion.

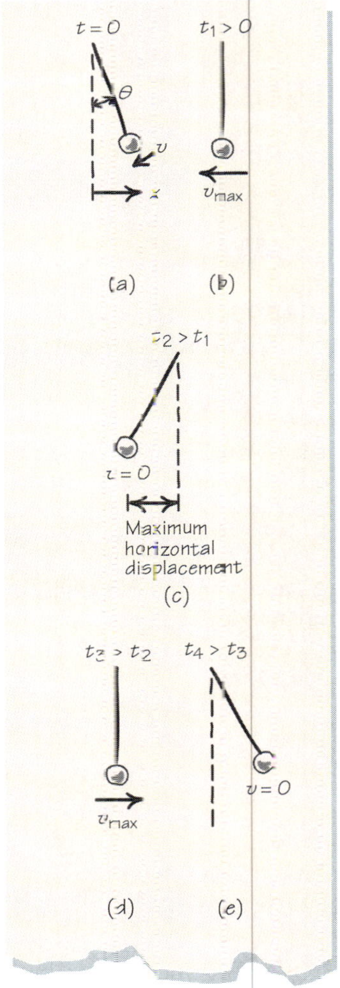

FIGURE 13–13 Example 13–4.

The Energy of the Simple Pendulum

Let's now consider the simple pendulum from the point of view of energy. The kinetic energy, K, which is a function of θ, is found by expressing the speed as a function of θ. Equation (13–28) gives us what we require, and

$$K(\theta) = \frac{1}{2}mv^2 = \frac{1}{2}m\ell^2\left(\frac{d\theta}{dt}\right)^2. \tag{13–38}$$

The only force with a component along the motion of the mass is gravity. This force is conservative and to find the associated potential energy function, U, express the height h gained by the mass in terms of θ. From Fig. 13–11a, we see that $h = \ell - \ell\cos\theta$, so that

$$U = mgh = mg\ell\,(1 - \cos\theta). \tag{13–39}$$

Here, we have taken zero potential energy to be at the bottom of the swing, at $\theta = 0$. In the small-angle approximation, $\cos\theta \simeq 1 - (\theta^2/2)$, and

$$U(\theta) = mg\ell(1 - 1 + \frac{1}{2}\theta^2)$$
$$= \frac{1}{2}mg\ell\theta^2. \qquad (13\text{–}40)$$

The potential energy of the simple pendulum

Comparing Eqs. (13–38) and (13–40) for the kinetic and potential energies with their counterparts for the spring, Eqs. (13–22) and (13–21), we see that they are formally the same—with the name changes listed in Eq. (13–35).

The period of oscillation of a simple pendulum provides a highly accurate method for the measurement of g because the time of many oscillations gives a very precise measure of the period of one oscillation. The period of a pendulum has played an important historical role in the definition of time. The grandfather clock, now more a curiosity or a work of art than a scientific instrument, was crucial in setting standards for time (Fig. 13–14).

FIGURE 13–14 The pendulum clock was the best timekeeper available for many years.

13–6 HARMONIC MOTION AND EQUILIBRIUM

The universal importance of the motion described in this chapter stems from the fact that *almost all small departures from stable equilibrium exhibit simple harmonic motion.* Everyday experience bears this out, at least in a qualitative way. A marble nudged a little from its stable equilibrium at the bottom of a bowl rolls back and forth, as does a mass hanging from a string or an automobile rocking on its worn shock absorbers. It is for this reason that simple harmonic motion occurs so often in physics. Table 13–1 gives a sampling of the range of periods of mechanical systems that move in simple harmonic motion.

In the spring, the pendulum, and indeed every case of stable equilibrium, some mass is trapped in a *potential energy well.* This term refers to a *minimum* in the curve of potential energy versus some position variable (Fig. 13–15). In the spring case, the position variable is the stretch of the spring and the minimum is at zero stretch. For the spring, we have seen that the potential energy forms a parabola. For a pendulum, the appropriate variable is the pendulum's angle θ; and the potential energy curve is also a parabola in θ near $\theta = 0$.

Simple harmonic motion nearly always occurs near points of stable equilibrium.

The Taylor expansion (App. IV–8) is a very general mathematical result that allows us to see why the minimum of a well forms a parabola and hence why simple harmonic motion is universal near equilibrium. Suppose we apply the Taylor expansion to a potential energy function near a minimum, which characterizes a stable equilibrium point. Let's label the position of the minimum as the origin, $x = 0$. Then, with each differentiation with respect to x labeled with a prime, we have

The concept of the potential energy well is treated in Chapter 7.

$$U(x) = U(0) + xU'(x)|_{x=0} + (x^2/2)U''(x)|_{x=0} + \cdots .$$

TABLE 13–1 Periods of Mechanical Systems in Simple Harmonic Motion

Mechanical System	*Period (s)*
Sloshing of water in a tidal basin or large lake	10^2 to 10^4
Large structures (bridges, buildings)	> 1
Strings or air columns of musical instruments	5×10^{-2} to 10^{-4}
Fluid machinery	10^{-4} to 10^{-5}
Piezoelectric crystals, ultrasound generators	10^{-5} to 5×10^{-1}
Vibrations in molecules	10^{-14}

The constant $U(0)$ plays no physical role and, as we know, we can always replace it by 0. The first derivative of U at $x = 0$ is zero because that is a minimum point. Thus, if we keep the first nonzero term in the Taylor expansion, we find

$$U(x) \simeq (x^2/2)U''(0). \tag{13–41}$$

(Here, we have used the notation $U''(x)|_{x=0} = U''(0)$.) This is indeed in the form of a spring force, with $U''(0)$ playing the role of the spring constant. Thus the force takes the general form near the equilibrium point

$$F(x) = -\frac{dU}{dx} \simeq -U''(0)x, \tag{13–42}$$

the familiar linear restoring force.

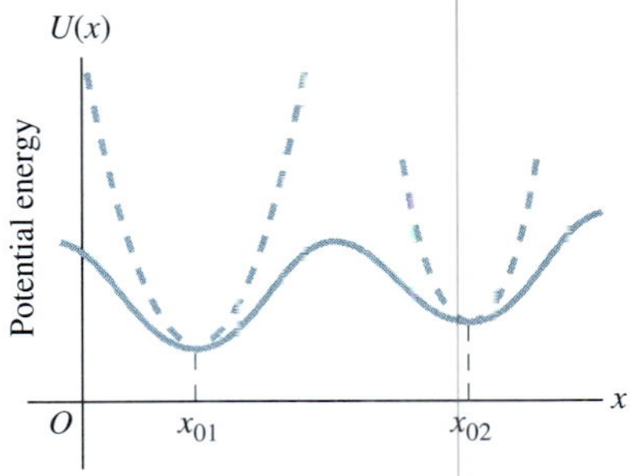

FIGURE 13–15 A potential well, in which potential energy has a minimum at $x = x_0$. This point is a point of stable equilibrium. The dashed line is a parabola that matches the minimum of the well.

The Physical Pendulum

A pendulum need not consist of a massless string with a pointlike mass at the end of it. When the suspended, swinging object has some other form, we call it a *physical pendulum.* Any object can be suspended from any point on the object and act as a physical pendulum. This illustrates the point that simple harmonic motion is a general characteristic of motion about a stable equilibrium. The appropriate approach to such problems is to recast them as a spring problem, through analysis of either the potential energy or the dynamical equation, and then use the results we have already found for the spring.

PROBLEM SOLVING

Try to restate stable equilibrium situations to resemble the spring problem.

We have the necessary tools to handle the physical pendulum: We can study the torque, τ, on it and the corresponding angular acceleration. Figure 13–16 illustrates a physical pendulum allowed to pivot through some horizontal axis, called y, which defines the vertical plane of the swinging. We take the y-axis to be into the page and through the oscillation point. We need to know the rotational inertia, I, about the pivot axis, the total mass M of the object, and the distance r from the center of mass to the pivot axis. The stable equilibrium point for this object is $\theta = 0$, when the center of mass hangs directly below the axis. When $\theta \neq 0$, only the force of gravity, which acts on the center of mass, exerts a (restoring) torque. From Eq. (9–28), the equation of motion governing the behavior of the object is

$$\tau = I\alpha, \tag{13–43}$$

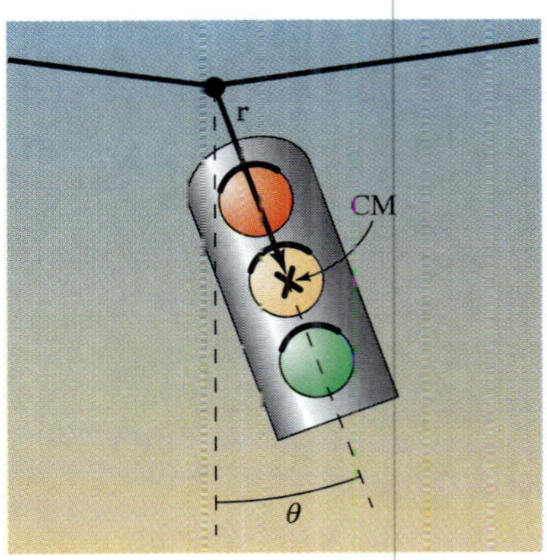

FIGURE 13–16 A swinging traffic light is an example of a physical pendulum. The center of mass of the oscillating light executes simple harmonic motion about the stable equilibrium position, $\theta = 0$.

where α is the angular acceleration about the pivot axis. The magnitude of α is $d^2\theta/dt^2$ and, because it acts to *decrease* θ, its sign is negative.

The torque is given by $\boldsymbol{\tau} = \mathbf{r} \times \mathbf{F}$, where $\mathbf{r}$ is the vector from the pivot axis to the center of mass and $\mathbf{F}$ is the force of gravity acting on the center of mass. According to the right-hand rule, this torque is along the $+y$-axis (when θ is positive, as shown); α must therefore also be along the $+y$-axis, which corresponds in Fig. 13–16 to an angular acceleration that brings the object back to equilibrium. (Both $\boldsymbol{\tau}$ and $\boldsymbol{\alpha}$ would change signs if the object were drawn with θ on the other side of the vertical axis.) The magnitude of $\boldsymbol{\tau}$ is

$$\tau = Fr\sin\theta = Mgr\sin\theta, \tag{13–44}$$

so the equation of motion, Eq. (13–44), becomes

$$Mgr\sin\theta = -I\frac{d^2\theta}{dt^2}. \tag{13–45}$$

Equation (13–45) has exactly the same form as the equation governing the motion of the simple pendulum, which is Eq. (13–31). For small angles, the sine of θ can be replaced with θ itself and, as for the simple pendulum, simple harmonic motion follows: $\theta = \theta_0 \sin(\omega t + \delta)$. To find ω, it is necessary only to replace ℓ and g for the simple pendulum with I and Mgr, respectively, for the physical pendulum. Thus, from Eqs. (13–36) and (13–37),

$$\omega = \sqrt{\frac{Mgr}{I}}, \tag{13–46a}$$

$$T = 2\pi\sqrt{\frac{I}{Mgr}}. \tag{13–46b}$$

For a physical pendulum of a given shape, the period of the small amplitude oscillations are independent of the mass.

Because I is always M times some length squared, M will cancel from Eqs. (13–46). *The mass of a physical pendulum does not enter into its period.* One check on the period of the physical pendulum as expressed in Eq. (13–46b) is to verify that it does reduce to the simple pendulum when the swinging object is a point mass m on a massless string of length ℓ. In that limit, $I \to m\ell^2$, $r \to \ell$, $M \to m$, and T reduces to the appropriate value,

$$T \to 2\pi\sqrt{\frac{m\ell^2}{mg\ell}} = 2\pi\sqrt{\frac{\ell}{g}}.$$

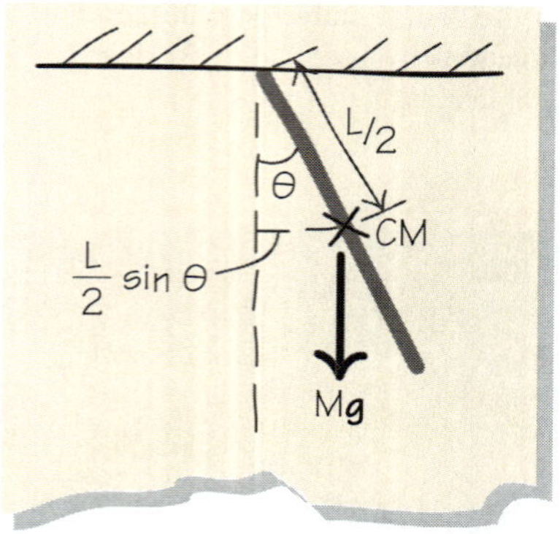

FIGURE 13–17 Example 13–5.

EXAMPLE 13–5 A thin, uniform rod of mass M and length L swings from its end as a physical pendulum (Fig. 13–17). What is the period of the oscillatory motion for small angles? Find the length ℓ of the simple pendulum that has the same period as the swinging rod.

Solution: Equation (13–46b) gives the period of a physical pendulum swinging in small angles and thus is directly applicable here. We need to find the rotational inertia of a rod about an axis through its end. This quantity was calculated in Eq. (9–24) and is listed in Table 9–1. It is

$$I = \frac{1}{3}ML^2.$$

The distance of the center of mass from the end is $r = L/2$, and Eq. (13–46b) then gives the period of the motion as

$$T_{\text{physical pendulum}} = 2\pi\sqrt{\frac{I}{Mgr}} = 2\pi\sqrt{\frac{ML^2/3}{MgL/2}} = 2\pi\sqrt{\frac{2}{3}\left(\frac{L}{g}\right)}.$$

The length ℓ of a simple pendulum of the same period is determined by

$$T_{\text{simple pendulum}} = 2\pi\sqrt{\frac{\ell}{g}} = 2\pi\sqrt{\frac{2}{3}\left(\frac{L}{g}\right)};$$

$$\ell = \frac{2}{3}L.$$

13–7 DAMPED HARMONIC MOTION

Almost all physical systems, including springs and other oscillating systems, are affected by friction or drag (resistive) forces (Fig. 13–18). These forces tend to remove energy from a moving system and thereby slow it down, or damp its motion. Their universality and importance are evident in the world around us. The pendulum of a clock does not swing forever. There is an energy loss due to damping; as can be seen from Eq. (13–26), a decrease in energy implies a decrease in amplitude. For the pendulum to maintain a given amplitude, energy must be supplied from, say, a wound spring or falling weights.

What are the *quantitative* effects of these nonconservative, or dissipative, forces? We can answer in the case of a drag force, $\mathbf{F}_d$, proportional to velocity:

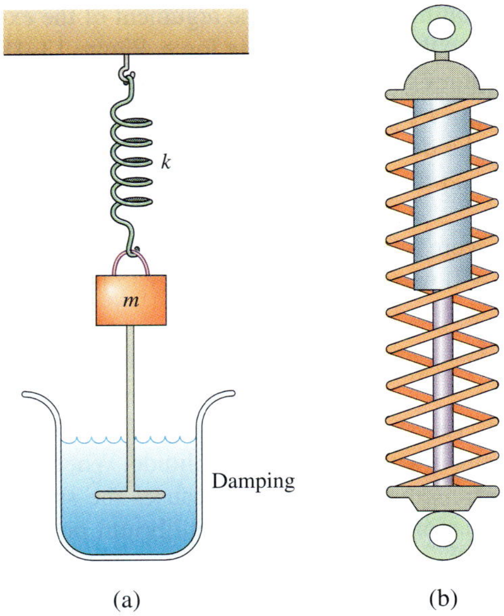

FIGURE 13–18 (a) A damped oscillator. (b) An automobile shock absorber is a damped harmonic oscillator.

$$\mathbf{F}_d = -b\mathbf{v} = -b\frac{d\mathbf{x}}{dt}, \tag{13–47}$$

where b is the *damping coefficient* (or *damping parameter*). The minus sign indicates that this force is always opposite to the direction of motion. For this case the equations of motion are *solvable* in analytic form, and we can at least take these solutions as a guide to the behavior of damped systems. The equation of motion is

$$-kx - b\frac{dx}{dt} = m\frac{d^2x}{dt^2}. \tag{13–48}$$

Just how does the damping modify simple harmonic motion? To solve differential equations such as Eq. (13–48), we attempt trial solutions that are educated guesses and see if they work. As a guess to the solution of Eq. (13–48), we keep a sinusoidal component but also incorporate a decreasing term due to damping. Our trial solution is

PROBLEM SOLVING

Solve differential equations with trial solutions.

$$x = Ae^{-\alpha t}\sin(\omega' t + \delta), \tag{13–49}$$

where A, δ, and α are constants to be determined. We use a frequency ω' rather than ω_0, the frequency in the absence of drag, because we want to allow for the possibility that the frequency is changed by the damping. We check whether the trial solution is a good one by seeing if the equation of motion, in this case, Eq. (13–48), is satisfied when the trial solution is plugged into it. This procedure is left to Problem 71. The result is that the trial solution is satisfactory provided that

$$\alpha = \frac{b}{2m} \tag{13–50}$$

and that

$$\omega' = \sqrt{\frac{k}{m} - \frac{b^2}{4m^2}} = \sqrt{\omega_0^2 - \frac{b^2}{4m^2}}. \tag{13–51}$$

Thus *the damping factor α and the modified angular frequency ω' are determined by the equations of motion.* (*A* and δ remain undetermined.) It is easy to check that Eq. (13–49) reduces to standard simple harmonic motion when $b = 0$.

The falling exponential function in Eq. (13–49) is a kind of *envelope* (shown by the dashed lines in Fig. 13–19a) that modulates what would otherwise be simple harmonic motion; this causes the amplitude of the motion to decrease as time goes on.

(a)

(b)

FIGURE 13–21 The Tacoma Narrows Bridge in Tacoma, Washington. (a) In November 1940, strong winds drove the bridge into oscillatory motion. (b) It may have been a resonant phenomenon that increased the amplitude of the oscillations to the point of collapse.

(a)

(b)

FIGURE 13–22 An example of resonant driven periodic motion. (a) A push in rhythm with the swing's natural frequency drives the swing to large amplitudes. (b) The child can rhythmically "pump," a more complex way of bringing in the resonant effect.

driving force has gone on and will go on forever—we mean, of course, for a long time—and the mathematical solution must show that *the motion of the mass is simple harmonic motion with the frequency of the driving force.* In other words, the mass will follow the time dependence of the driving force. Any exponential effects due to friction will have long since died out—exponentially, of course. Imagine, for example, a spring with a mass on its end suspended from a harmonically moving hand. In time, the mass will move with the frequency of your hand motion, even if the motion of the mass and your hand are not in phase. After long times, the solution to the equation of motion is then

$$x = A \sin(\omega t + \delta). \tag{13–56}$$

This solution can be verified by substitution into Eq. (13–55), and the amplitude A and phase δ are determined in this substitution. This is a complicated exercise in algebra, which we forego. The physically interesting amplitude of oscillation is in this case determined by Eq. (13–55):

$$A = \frac{F_0}{\sqrt{m^2(\omega^2 - \omega_0^2)^2 + b^2\omega^2}}. \tag{13–57}$$

This amplitude has the remarkable property of *resonance: It is peaked when the driving frequency, ω, nears the natural frequency, ω_0.* In fact, if there were no damping ($b = 0$), the amplitude would become infinite when $\omega = \omega_0$. This is not a realistic physical situation because it corresponds to the spring being stretched to infinite length. A real spring will snap rather than accept an infinite stretch; in other words, some form of damping will ultimately occur. But it does illustrate that, at resonance, the response of a harmonic system to a driving force can be catastrophically large.

Properties of Resonance

We plot amplitude as a function of ω in Fig. 13–23 for several values of b. We choose $F_0 = 0.01$ N, the natural frequency $\omega_0 = 1$ rad/s, and $m = 1$ kg. The damping coefficient b is given the values 0.01, 0.05, and 0.2 kg/s. From the plot, we can see that the position of the amplitude shifts slightly from $\omega_0 = 1$ rad/s when b increases from a small value. From Eq. (13–57), the peak amplitude generally occurs at

$$\omega_{\max}^2 = \omega_0^2 - \frac{1}{2}\left(\frac{b^2}{m^2}\right). \tag{13–58}$$

That the peak is less sharp for larger b is sometimes expressed as the fact that the *width* of the peak is a measure of b. To be more precise, it is possible to compute the *total width at half-maximum* $\Delta\omega$ of the peak, found by evaluating the amplitude where it is one-half its peak value and measuring the spread of frequencies to which this corresponds, $\Delta\omega$ (Fig. 13–24). For small b, this width can be shown to be

$$\Delta\omega \simeq \frac{2b}{m}. \tag{13–59}$$

Thus the sharpness $\Delta\omega$ of the resonance peak is a direct measure of the damping coefficient, b, divided by m: The smaller the damping, the sharper the resonance peak.

*Resonance and Uncertainty

We have seen that the amplitude for a driven, damped harmonic oscillator has a resonance peak. We have also seen that the undriven damped harmonic oscillator has a characteristic decay time. These phenomena are closely related, and this relation has important consequences for our ability to construct systems with resonances, such as radio tuners or filters for screening electronic noise.

Equation (13–49) can be used to find the rate at which the energy is dissipated in an undriven damped oscillator. The energy is proportional to the amplitude *squared*

APPLICATION: QUARTZ WATCHES

The most obvious application of simple harmonic motion can be seen in clocks. The pendulum clock is the simplest example. A more modern example is the quartz crystal that provides the time standard in a wristwatch (Fig. 13AB–1). The crystal takes the form of a bar, end mounted and cut in relation to the "piezo axis" of the crystal. The "piezo" property of a crystal is an electrical property; in this case, it refers to the fact that an applied electric field causes the crystal to flex, and a harmonically varying electric field leads to harmonic flexing. The amplitude of flexure at the resonant frequency of the crystal is very much larger than at other frequencies, and the energy loss due to damping passes through a maximum at the resonant frequency. The electric circuitry that provides the driving frequency senses where the energy loss is greatest, so there is a feedback to the fixed resonant frequency. In a cheap watch, the crystal is driven only at its resonant frequency, and the accuracy of the period is on the order of one part in 10^5 to 10^6. Better watches can adjust the driving frequency, so that the quartz bar can be operated slightly away from resonance. Such watches have accuracies of one part in 10^7. Time standards that can keep the oscillating bar at a constant temperature achieve stabilities of better than one part in 10^{10}, not as good as atomic clocks, but much more convenient.

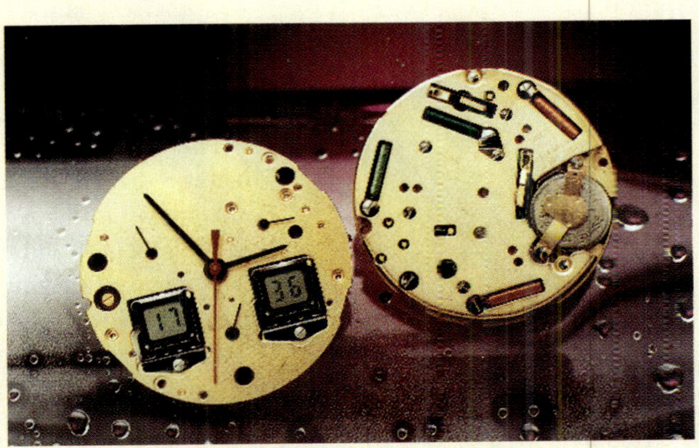

FIGURE 13AB–1 The quartz watch has put accurate timekeeping within the reach of everyone.

and therefore falls as $e^{-t/\tau}$, where $\tau = m/b$ is the lifetime. The width in frequency of the driven harmonic oscillator is given in Eq. (13–59), and we see that it is inversely proportional to τ. From Eqs. (13–53) and (13–59), we have

$$\tau\Delta\omega \text{ is on the order of } 1. \tag{13–60}$$

This equation is known as the **uncertainty principle**: It expresses the impossibility of measurements that are arbitrarily precise in both time and frequency. Strictly speak-

The uncertainty principle

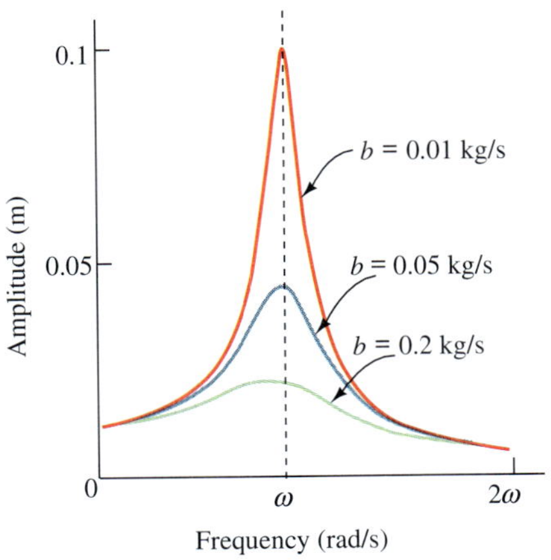

FIGURE 13–23 The amplitude of a harmonic system under the influence of a harmonic driving force with frequency ω. The natural frequency of the harmonic system is $\omega_0 = 1$ rad/s. There is a drag force with damping coefficient b. Resonance occurs at the peak (near $\omega = \omega_0$) and is strongest when b is smallest.

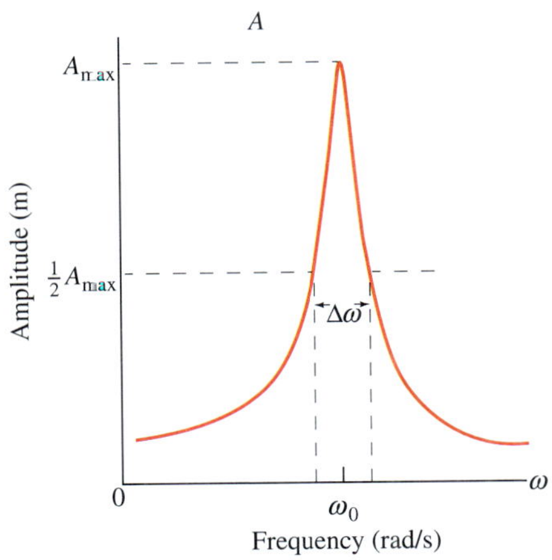

FIGURE 13–24 The width at half-maximum of a resonant peak near ω_0 is $\Delta\omega$.

ing, we have derived it only for a special damping force. But it actually represents a very general property. It states that if the damping time of a harmonic oscillator is large, then the resonance width is small, and vice versa. The weaker the damping of a harmonic oscillator, the more sharply it responds to, or selects, a harmonic driving force of the right frequency. What is the significance of this result? One of the important uses of the resonance phenomenon, both in mechanical systems and in electric circuits, is that it allows us to select, or filter, certain frequencies by letting these frequencies act as a driving force on our selector. For example, the tuner on a radio selects a given station by letting in only frequencies near the central frequency of the station. Equation (13–60) sets strong limits on our ability to design filters that can respond to only a narrow range of driving frequencies. The transient effects of such filters die away more slowly with time as the frequencies they select are more and more limited. This is a very general result that cannot be avoided with clever design.

APPLICATION
Filters

SUMMARY

Simple harmonic motion is characterized by a particular type of periodic behavior; namely, sinusoidal time dependence of some position variable of a moving object. In one dimension, this dependence can be written as

$$x(t) = A\sin(\omega t + \delta). \tag{13–1}$$

The amplitude, A, measures the maximum displacement from equilibrium. The phase, δ, is the angle by which the motion is shifted from $x = 0$ at $t = 0$; the angular frequency, ω, measures how rapidly the motion repeats. The period, T, and frequency, f, of the simple harmonic motion are related to ω by

$$T = \frac{2\pi}{\omega} \tag{13–2}$$

and by

$$f = \frac{1}{T}. \tag{13–3}$$

In simple harmonic motion, the velocity and the acceleration are also sinusoidal. In particular, the acceleration is proportional to the displacement but opposite in sign:

$$a = -\omega^2 x. \tag{13–8}$$

Uniform circular motion is closely related to simple harmonic motion. It is just the result of simultaneous simple harmonic motion of equal amplitudes and frequencies in the x- and y-directions, 90° out of phase.

Because acceleration is proportional to displacement in simple harmonic motion, the force that leads to this motion is a restoring force proportional to the displacement. The prototype of this is the force exerted on a mass by a spring, Hooke's law:

$$F = -kx, \tag{13–13}$$

where k is the spring constant and x is the displacement of the mass from the equilibrium position of the spring. The resulting angular frequency of the motion of a mass m subject to this force is

$$\omega = \sqrt{\frac{k}{m}}. \tag{13–16}$$

The period of the motion is independent of the amplitude.

The potential energy of a mass subject to a spring force is

$$U(x) = \frac{1}{2}kx^2. \tag{13–21}$$

It oscillates with time as $\sin^2(\omega t + \delta)$, whereas the kinetic energy oscillates with time as $\cos^2(\omega t + \delta)$, and the total energy, E, which is their sum, remains constant with time:

$$E = \frac{1}{2}kA^2. \quad (13\text{–}26)$$

The simple pendulum, consisting of a mass on the end of a light string of length ℓ, also undergoes simple harmonic motion when it is not allowed to swing too far. In this case, it is the angle with respect to the vertical that varies sinusoidally, and the period of the motion is independent of the mass as well as the amplitude:

$$T = 2\pi\sqrt{\frac{\ell}{g}}. \quad (13\text{–}37)$$

Almost all small departures from stable equilibrium situations exhibit simple harmonic motion. Two examples are a bead rolling back and forth across the bottom of a bowl and a physical pendulum. The period of a physical pendulum depends on the geometric factor contained in its rotational inertia about the axis through which it swings, but its total mass cancels from the period.

The presence of additional velocity-dependent drag, or resistive, forces causes the amplitude of a particle that moves under the influence of springlike forces to decrease. When a system that, by itself, would move in simple harmonic motion is driven by a force with sinusoidal time dependence, the system moves with the frequency of the driving force. The amplitude of the resulting motion of the system shows resonant behavior when the frequency of the driving force equals the natural frequency of the system. The width of the resonance peak is inversely related to the exponential rate of falloff of the undriven system due to damping—a result known as the uncertainty principle.

UNDERSTANDING KEY CONCEPTS

1. Describe some examples of simple harmonic motion that are not discussed in the chapter.
2. Describe an experiment to determine the spring constant of a spring.
3. If the amplitude of simple harmonic motion is doubled, what happens to the maximum kinetic energy?
4. Will a pendulum clock lose time or gain time when it is taken from sea level to the top of a mountain? Ignore all damping effects.
5. In the previous question, we suggested that all damping effects be ignored. Suppose that you could not ignore damping effects due to the air, which becomes less dense as you gain in altitude. Why might you find it difficult to predict whether a pendulum clock gains or loses time when it is taken from sea level to the top of a mountain?
6. A damped harmonic oscillator driven by a harmonic external force maintains a steady oscillatory motion. Is energy lost to friction in the motion? If so, what keeps the oscillator moving?
7. The length of a simple pendulum is doubled, and the mass at the end is halved. What happens to the period?
8. What is the phenomenon that allows you to increase the amplitude of your motion when you swing on a swing?
9. Suppose you stand on a swing instead of sitting on it. Will your frequency of oscillation increase or decrease?
10. It can be shown that if the frequency, the energy, or the amplitude of a harmonic oscillator changes very slowly then, although the energy E is no longer constant, the ratio E/ω, where ω is the angular frequency, does not change. Suppose you have a pendulum whose maximum angle of deflection is θ_0. If the bob at the end of the pendulum consists of fine sand in a spherical container, and the sand leaks very slowly out the bottom, what will happen to the motion?
11. Make use of the assertion in Question 10 to discuss what happens to the period and amplitude of a pendulum in which the length is very slowly decreased by a factor of 1.5.
12. Discuss if and why each of the following is an example of simple harmonic motion, damped harmonic motion, and/or driven harmonic motion: (a) leaves blowing on a tree limb; (b) children seesawing; (c) a child swinging on a playground swing; (d) a car bouncing up and down after hitting a large pothole in the road; and (e) water sloshing back and forth in a tub.
13. A mass is suspended from two ropes of equal length, attached to different points on a ceiling. Would small oscillations about the point of stable equilibrium represent simple harmonic motion?
14. A simple pendulum acts as a *conical pendulum* when the mass moves in a horizontal circle (Fig. 13–25). What force keeps it moving in a circle? Does the total energy consist of potential and kinetic energies that each vary in this case?

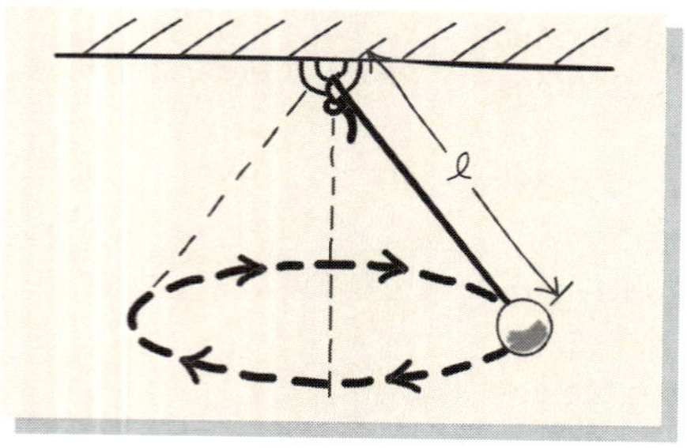

FIGURE 13–25 Question 14.

15. Suppose that there were a harmonic driving force on a mass attached to a horizontal spring and that very little damping occurs. What could supply the energy that would allow the amplitude to become enormously large near resonance?

16. What are you doing when you adjust your bounce up and down on a diving board to get a big boost for your dive?

17. A large sled is sliding down a snowy hill. Resting on the sled are a mass and a spring; one end of the spring is attached firmly to the sled and the other end is attached to the mass. There is little friction between the mass and the sled and the mass is set into oscillation. Will the sled accelerate smoothly down the hill?

PROBLEMS

13–1 The Kinematics of Simple Harmonic Motion

1. (I) What phase, δ, is necessary if $\sin(\theta + \delta) = \cos(\theta)$? This can be shown graphically, with Fig. 13–1, or by using a trigonometric identity for the sine of the sum of two angles.

2. (I) A simple harmonic motion along the x-direction has the following properties: The maximum amplitude = 0.15 m, the time between the maximum and minimum x-values = 0.4 s, and $x = 0.1$ m for $t = 0.1$ s. Find the period, the angular frequency, and the general equation of motion.

3. (I) A spring has the speed $v = 0.4\sin(\omega t + \pi)$ m/s, where $\omega = 2.00$ rad/s. Plot x, v, and a as functions of time for three periods of motion.

4. (I) The angular frequency of a mass on the end of a spring in simple harmonic motion is 3.827 rad/s. What is the period of the motion?

5. (I) The amplitude of the motion of a mass attached to a spring is $A = 2.84$ m, while the maximum speed of the mass is 4.36 m/s. What is the period of the motion?

6. (I) The maximum speed of a mass attached to a spring is $v_{max} = 0.486$ m/s, while the maximum acceleration is 1.66 m/s^2. What is the maximum displacement of the mass?

7. (I) What is the position of the mass in Problem 6 at time $t =$ 1 s, given that the mass is precisely at the origin ($x = 0$ m) at $t = 0$ s?

8. (II) The expression for the position of a simple harmonic oscillator is given by $x(t) = B\cos(\omega t) + C\sin(\omega t)$. (a) Show that this can be written in the form given in Eq. (13–1) and express the constants B and C in terms of the constants A and δ. (b) Express B and C in terms of the position x_0 and the speed v_0 at time $t = t_0$. (c) What is the maximum speed in terms of B and C?

9. (II) A harmonic oscillator operates at a frequency of 813.52 Hz. What is the amplitude for which the maximum acceleration is 183.25 m/s^2?

10. (II) A small object is placed on a horizontal platform, which vibrates vertically with an amplitude of 1 cm. The frequency of the vibration is slowly increased. At what frequency will the object start bouncing on the platform?

11. (II) The motion of a mass can be described by the function $x(t) = A\sin(\omega t + \delta)$, where $\omega = 2.0$ rad/s and $\delta = 0.40$ rad. Express the motion as a cosine function.

12. (II) A professor pacing back and forth is observed by the class to move back and forth along the x-axis in a rough approximation to simple harmonic motion. Relative to the center of the classroom, the motion is between the two extremes +3 m and −3 m. The professor was at $x = -0.3$ m at $t = 0$ s, and the motion is observed to repeat itself 6 times in 90 s. What are the amplitude, phase, period, frequency, and angular frequency of the motion?

13. (II) A spring has spring constant 0.5 N/m and a 0.2-kg mass on its end, which has a maximum speed of 2.0 m/s. (a) What are the angular frequency and period of the system? (b) What is the amplitude of the motion?

14. (II) A sailing ship rolls sideways in simple harmonic motion, with the period given by $T = 8$ s (Fig. 13–26). The tip of a 25-m mast travels a maximum of 3 m from the vertical position. What is the speed of the tip of the mast at the instant it is in a vertical position?

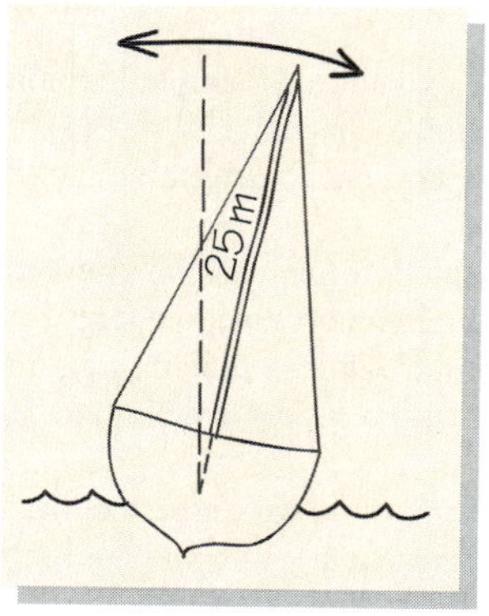

FIGURE 13–26 Problem 14.

15. **(II)** A particle undergoing simple harmonic motion travels a total distance of 6.98 cm during one cycle of 1.71 s. (a) What is the average speed of the particle? (b) What are its maximum speed and acceleration?

16. (II) A mass $m = 0.3$ kg is attached to an ideal spring of spring constant $k = 15$ N/m. All motion takes place in the (horizontal) x-direction, and the equilibrium position of the mass is defined to be $x = 0$ m. The mass is then displaced to $x = 0.050$ m and released from rest. (a) Write an expression for the function $x(t)$ that describes the subsequent motion. (b) At what position x does the maximum positive acceleration of the mass occur? (c) What is the magnitude of the maximum acceleration? (d) What is the maximum speed of the mass?

17. (II) An object oscillates with an angular frequency of 3.0 rad/s. Its initial displacement from equilibrium is +3.0 cm, and the initial velocity is 5 cm/s in the direction of the equilibrium point. (a) Find the displacement as a function of time. (b) How soon will the displacement be +3 cm again? (c) At what times will be object move with a speed of 5 cm/s? (d) Sketch x and v as functions of t.

13–2 A Connection with Circular Motion

18. (I) Find the phase angle δ for uniform circular motion when $x = -R$ and $y = 0$ at $t = 0$.

19. (I) A particle moves in the xy-plane so that its x and y coordinates are described by Eq. (13–10) and Eq. (13–11); namely,

$$x = R\cos(\omega t + \delta), \qquad y = R\sin(\omega t + \delta).$$

Show that the distance of the particle from the origin is a constant, and find that constant.

20. (I) When a certain uniform circular motion in the xy-plane is projected onto the x- and y-axes, this projection gives $x(t) = R\sin[\omega t + \delta - (\pi/2)]$ and $y(t) = R\sin(\omega t + \delta)$. (a) Show that the uniform circular motion is clockwise. (b) What would happen if instead of an extra phase $-\pi/2$ in $x(t)$, there was an extra phase $+\pi/2$ in $y(t)$?

21. (I) Write down a formula that describes the motion of Earth around the Sun as seen by a distant observer in the plane of Earth's orbit. Assume a circular orbit.

22. (II) A small object is placed at the outer edge of an LP record of diameter 12 in. The record rotates clockwise at $33\frac{1}{3}$ rev/min. (a) What is the projection of the object's motion on the x-axis, assuming that at $t = 0$ the projection of the motion places the object at $x = 0$? (The point $x = 0$ corresponds to the center of the record.) (b) Give the amplitude, angular frequency, and largest speed of the projection of the motion.

23. (II) Show that if the motion along the x-and y-axes of a mass moving in the xy-plane is $x(t) = R_1\cos(\omega t + \delta)$ and $y(t) = R_2\sin(\omega t + \delta)$, then this motion traces out an ellipse of axis lengths R_1 and R_2. (An ellipse is the curve described by $[x^2/a^2] + [y^2/b^2] = 1$, where a and b are the axis lengths.)

13–3 Springs and Simple Harmonic Motion

24. (I) A 10-kg ball is suspended motionless under the influence of gravity from a spring with a force constant of $k = 2 \times 10^3$ N/m. (a) How much is the spring stretched? (b) The same spring is sent to the Moon and suspended in the same way. By how much is the spring stretched in this case?

25. (I) A student has a spring with $k = 200$ N/m and wants to build a horizontal mass–spring system with period 1.0 s. What mass should the student use at the end of the spring?

26. (I) The spring on a scale is compressed by 2.45 cm when a 60-kg man stands on it. What is the spring constant of the spring?

27. (I) The period of a 45-g mass attached to the end of a spring is 3.1 s. Find the spring constant of the spring.

28. (I) A mass moves along the x-axis under the influence of a spring whose equilibrium position is at the origin. The mass moves between limits of -2 cm and $+2$ cm, and the period of its motion is 0.5 s. (a) What is the period if the mass is doubled? halved? (b) What is the period if the amplitude is doubled to 4 cm? halved to 1 cm?

29. (II) The plate at the base of a floor sander moves back and forth in simple harmonic motion at a frequency of 20 oscillations/s, and the amplitude of the motion is 0.8 cm. If the mass of the oscillating plate is 1200 g, what is the maximum value of the driving force?

30. (II) A spring is placed in a vertical position by suspending it from a hook at its top. A similar hook on the bottom of the spring is measured to be 45 cm above a table top. A mass of 180 g, of negligible size, is then suspended from the bottom hook, which is now found to be 20 cm above the table top. The mass is then pulled down a distance of 15 cm and released (Fig. 13–27). Find (a) the spring constant, k; (b) the angular frequency, ω; (c) the position of the bottom hook after 5 s.

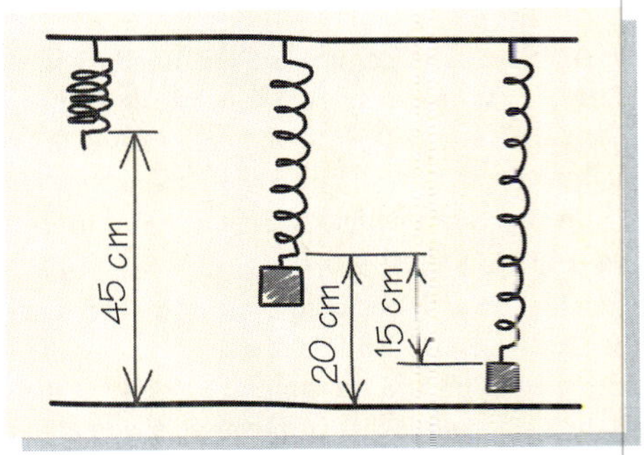

FIGURE 13–27 Problem 30.

31. (II) A spring is suspended from the ceiling. A mass of 25 g is attached to it, and the spring stretches by 12 cm. Ignore damping. What is the period of the oscillation of a 75-g mass attached to the spring?

32. (II) Two identical springs of spring constant k are attached end to end to make one longer spring. Show that this new spring has spring constant $k/2$. The springs are said to be attached *in series*. The case of n springs attached in series gives a spring n times as long, with a spring constant $k_n = k/n$.

33. (II) A small object of mass $m = 0.06$ kg is held in place by two springs (Fig. 13–28). The one acting on the left has spring constant $k_1 = 100$ N/m; the one acting on the right has spring constant $k_2 = 200$ N/m. The object is moved away from its equilibrium position by 1 cm to the right, and released at time $t = 0$. What is the displacement of the object as a function of time?

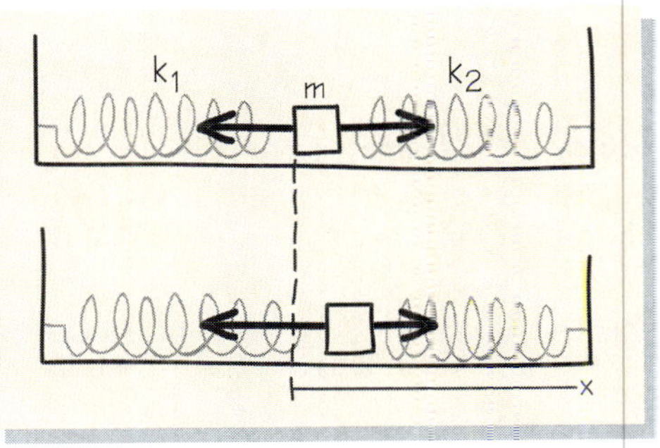

FIGURE 13–28 Problem 33.

13–4 *Energy and Simple Harmonic Motion*

34. (I) What is the energy of a 0.5-kg mass that moves with amplitude 5 cm on a flat, frictionless table and is attached to a spring of spring constant 860 N/m?

35. (I) A trampoline acts like a spring of spring constant 1200 N/m. A mass is placed on the trampoline. The trampoline surface is depressed by a total of 30 cm. What is the period of oscillation of the system if this mass is pushed beyond the equilibrium displacement?

36. (I) A mass of 1.2 kg, attached to a spring, is in simple harmonic motion along the x-axis, and its period is $T = 2.5$ s. If the total energy of the spring and mass is 2.7 J, what is the amplitude of the oscillation?

37. (I) A spring with a 1-kg fish at its end and with spring constant $k = 2$ N/m is compressed 3 cm from equilibrium and then released. Use the conservation of energy to find the maximum speed of the fish.

38. (I) A puck of mass 582 g moves horizontally with speed $v = 2.5$ m/s on a frictionless surface toward the end of a relaxed spring for which $k = 840$ N/m, and compresses the spring. By how much is the spring compressed? How would your answer change if the mass of the puck were increased by 100 percent?

39. (I) A mass at the end of a string moves in a circle in a vertical plane. The only forces acting on it are the tension of the string, which is central, and the force of gravity. If the length of the string is 1.2 m and the angular velocity at the top of the circle is $\omega = 2.2$ rad/s, what is the angular velocity at the bottom of the circle?

40. (II) Consider a mass m, moving along the x-axis, with potential energy given by $U(x) = \frac{1}{2}m\omega^2x^2$. Show that the motion of this mass is simple harmonic motion with angular frequency ω by using $dE/dt = 0$, where E is the (constant) total energy of the object.

41. (II) A point mass m on a turntable that rotates with angular frequency ω is located a distance d from the center of the turntable. Show that the energy of the point mass is the sum of the energies of the harmonic motions in the x- and y-directions.

42. (II) A mass m of 0.2 kg is attached to the end of a spring and released from rest at $t = 0$ s from an extended position $x_{max} = 6$ cm. All other forces acting on the spring cancel, so that the motion of the mass is due entirely to the effect of the spring. After its release, the mass's speed drops back to zero for the first time after 0.3 s. Find the maximum speed of the mass.

43. (II) A mass m attached to the end of a spring is released from rest at $t = 0$ s from an extended position x_{max}. The mass $m = 0.2$ kg, and $k = 1$ N/m. After 0.5 s, the speed of the mass is measured to be 1.5 m/s. Calculate x_{max}, the maximum speed of the motion, and the total energy.

13–5 *The Simple Pendulum*

44. (I) A simple pendulum 1.20 m long is suspended in a location where g is 9.82 m/s^2. What is the period of the pendulum?

45. (I) A simple pendulum has a frequency of 0.342 Hz. The length of the pendulum string is 2.12 m. What is the local value of g?

46. (I) A thin wire 7.58 m long is attached to the ceiling of a large lecture room. A lead ball is attached at the bottom. The wire is displaced by an angle of 0.075 rad and released. Express the angular displacement of the lead ball as a function of time, given that $g = 9.80$ m/s^2.

47. (I) You need to measure the height of a room. You have a watch but no meter stick. You also have your human brain, not to mention an opposable thumb. A long pendulum with a point mass at the end extends from the ceiling to the floor and has a period $T = 3$ s. (a) What is the height of the room? (b) You take the same pendulum to the top floor of a skyscraper and measure a period $T = 3.0002$ s. What is the height of the skyscraper?

48. (II) The difference in temperature between summer and winter causes the length of the pendulum in a clock to change by one part in 30,000. What time-difference error will this make in one week?

49. (II) A small lead ball of mass 2 kg is suspended at the end of a light string 1 m in length. A small peg, 0.5 m below the suspension point, catches the string in its swing (Fig. 13–29). The ball is set swinging through small angles. (a) What is the period of this pendulum? (b) The ball is started swinging on the side that does not catch the peg, at an initial height 0.05 m above the low point. How high does it rise on the side where the peg restricts the pendulum length to 0.5 m?

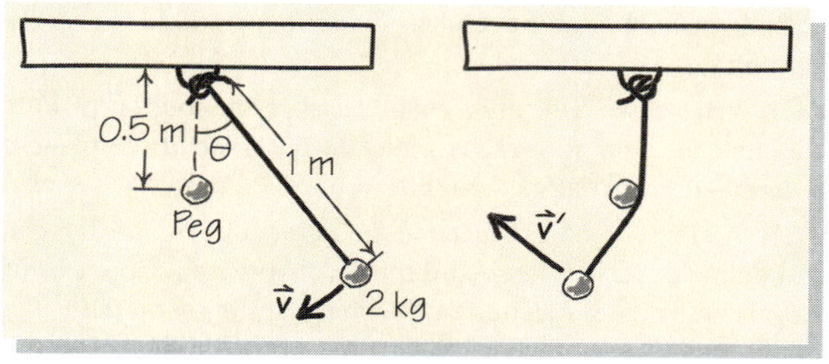

FIGURE 13–29 Problem 49.

13–6 *Harmonic Motion and Equilibrium*

50. (I) A small door to allow a dog to pass in and out of the house has a mass of 0.6 kg and is 25 cm wide and 35 cm tall. The door is hinged along the top of the 25-cm width. What is the frequency of oscillation of the door?

51. (I) A student wants to build a pendulum out of a circle of plywood as shown in Fig. 13–30. The circle has a radius of $R = 10$ cm and the plywood has a mass of 200 g. What is the period of the motion?

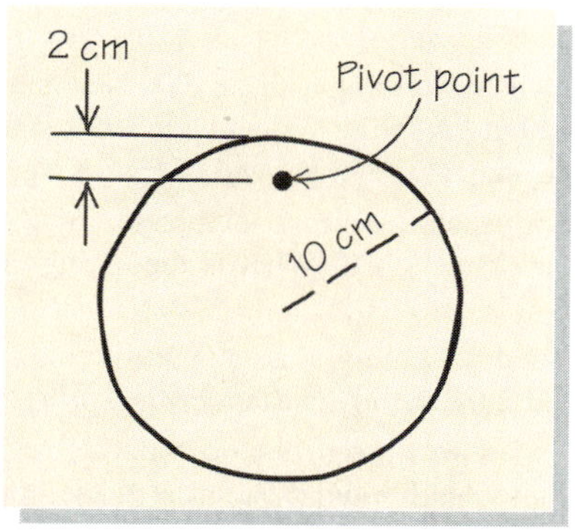

FIGURE 13–30 Problem 51.

52. (II) A uniformly dense book of dimensions 5 cm × 12 cm × 20 cm has a mass of 3 kg. You pick it up close to one corner with a pair of calipers and let it swing. What is the angular frequency? You can do this experiment with your own textbook.

53. (II) A thin, uniform rod of mass M and length L oscillates with small amplitude in a vertical plane about a pin that pierces the rod a distance y from the upper end. Calculate the period of the motion as a function of y. [*Hint*: You need to use the parallel-axis theorem to calculate the rotational inertia to be used in Eq. (13–46b).]

54. (II) An object of mass m has potential energy given by $U(x) = U_0(x^2 - a^2)^2$. (a) What are the stable equilibrium positions of the object? (b) What is the angular frequency of its motion about a stable equilibrium point, when the object is displaced by an amount that is small compared with a? [*Hint*: Let $x = x_{eq} + z$, and keep only up to z^2 terms in $U(z)$.]

55. (II) A straight wire 80 cm long is bent in the middle into an L-shape and balanced with the two ends down on a knife edge. With what frequency will it oscillate about its equilibrium position?

56. (II) Imagine a tunnel that has been drilled through Earth: a smooth, straight tunnel with a frictionless interior (Fig. 13–31). The deepest point of the tunnel is at depth d, and the coordinate x measures the distance along the tunnel from its deepest point to an arbitrary point P a distance ℓ from Earth's center. The known parameters are the mass and radius of Earth, M_E and R_E, respectively, as well as G. Earth is assumed to have uniform density. (a) What is the total mass of that portion of Earth that lies within the distance ℓ in terms of ℓ, M_E, R_E, and G? (b) What is the gravitational force, in direction and magnitude, acting on a ball of mass m at point P? (c) What is the *total* force, in direction and magnitude, acting on the ball *as a function of* x and of the constants of the problem? Why is there no net force acting perpendicular to the tunnel? (d) What is the period of the motion if the ball is released at rest at an entrance to the tunnel? Ignore air resistance. (e) What is the period of a satellite in circular orbit around Earth at a radius equal to Earth's radius? Ignore air resistance.

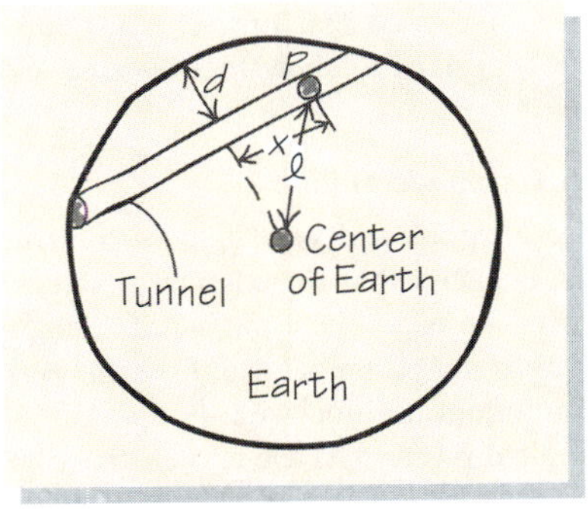

FIGURE 13–31 Problem 56.

57. (II) A thin uniform disk of mass M and radius R hangs from a nail driven straight into the disk at a distance ℓ from the center. (a) What is the rotational inertia of the disk about the nail? (b) What is the equation of motion for small oscillations of this pendulum about the point where the nail is driven in? (*Hint*: Use Newton's law for the torque about the point in question.) (c) What is the period, T, of oscillations about the point of suspension? (d) What is T in the limit where ℓ goes to zero?

58. (II) A uniform circular disk of radius R and mass M is attached at its center to one end of a massless, rigid rod of length L. The other end of the rod is attached to the ceiling and pivots freely about that point. The system thus makes a physical pendulum. What is the period of small oscillations of the pendulum?

59. (II) A *torsion pendulum* consists of a dumbbell suspended from its center by a wire that resists being twisted (Fig. 13–32). The dumbbell has length ℓ and has masses m on each end. It remains in the horizontal plane as it twists back and forth about the equilibrium position, $\theta = 0$ rad. For small oscillations about this equilibrium position, the torque exerted by the wire has the form $\tau = -\gamma\theta$, where γ is the *torsion constant*. (a) For $m = 80$ g, $\ell = 30$ cm, and $\gamma = 2 \times 10^5\ \text{g}\cdot\text{cm}^2/\text{s}^2$, what is the period of the oscillation? (b) If the system is started from rest at $\theta = 0.1$ rad, what is the total energy of the system? (c) What is the maximum speed of either mass, given the initial conditions of part (b)?

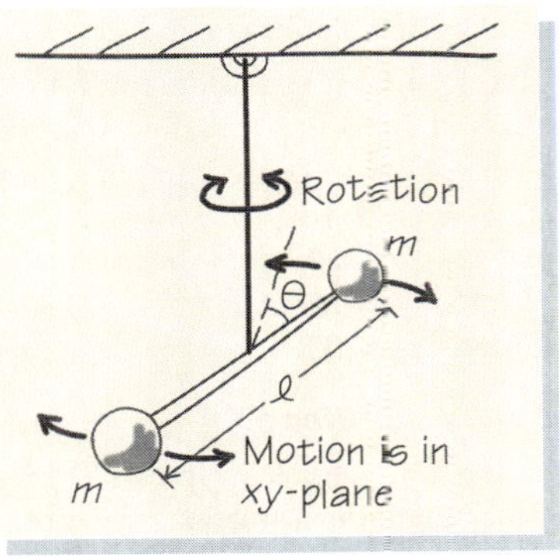

FIGURE 13–32 Problem 59.

60. (II) A slender, uniform rod of mass m and length ℓ can be pivoted on a frictionless horizontal support at any point along the rod's length. The rod then moves as a physical pendulum for small oscillations about the vertical equilibrium position. Suppose that the pivot point is at a point $z\ell$ from the end of the rod, where z is a fraction between 0 and 1. Using an equation for the angular acceleration of the rod as a function of the angle θ, which measures the departure from the vertical, find the period of small-angle oscillations. Check the answer by specifying $z = 0.5$, where the period should become large.

61. (III) A small metal block of mass m is placed on a smooth horizontal table and constrained to move along a frictionless, rectilinear groove. The block is attached to one end of a spring (of spring constant k) whose other end is fastened to a pin P (Fig. 13–33). Let length ℓ be the equilibrium length of the spring and the perpendicular distance to the groove. The spring is now pulled a distance x_0 from the equilibrium position and released. Show that, if the displacement along the groove $x << \ell$, the restoring force on the block is proportional to x^3, so the motion is not simple harmonic (although it will still be periodic). [*Hint*: For $x << \ell$, $\sqrt{x^2 + \ell^2} \simeq \ell + (x^2/2\ell)$.]

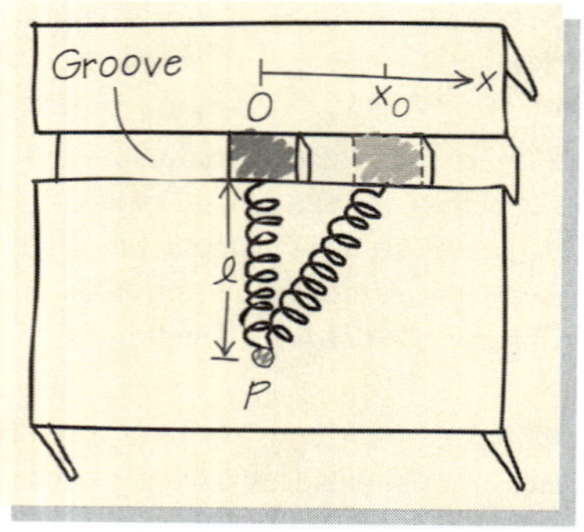

FIGURE 13–33 Problem 61.

62. (III) A bead of mass $m = 50$ g slides without friction at the bottom of a bowl whose bottom traces out the arc of a circle of radius $R = 20$ cm. If the bead moves in a plane that passes through the bottom of the bowl, its position can be described by an angle θ measuring its angular displacement from the bottom. (a) Express the potential energy as a function of θ for very small angles θ. [*Hint*: The potential energy is a linear function of height. Reexpress the energy in terms of θ, and use an approximation suitable for small θ.]

At $t = 0$ s the mass is released from rest from a very small angle $\theta_0 = 0.1$ rad. Parts (b)–(e) refer to this initial condition. (b) Give an expression $\theta(t)$ describing the subsequent motion. In particular, what is the frequency of small oscillations at the bottom of the bowl? (c) What is the total energy of the system? (d) the velocity of the bead at $t = 0.1$ s? (e) the acceleration of the bead at $t = 0.2$ s?

13–7 Damped Harmonic Motion

63. (I) Consider the gong struck in Example 13–6. What is the loudness of the sound (as a fraction of the original loudness) after 2 min? How much time does it take for the sound to have a loudness 1/10,000 of the original loudness? Your ear is a very good detector of loudness, and you would have no trouble hearing the gong for this entire time.

64. (I) A mass on a spring with a natural angular frequency $\omega_0 =$ 2.5 rad/s is placed in an environment in which there is a damping force proportional to the speed of the mass. If the amplitude is reduced to 0.65 times its initial value in 5.4 s, what is the angular frequency of the damped motion?

65. (I) The damping coefficient of a damped harmonic oscillator can be adjusted. Two measurements are made: First, when the damping coefficient is zero, the angular frequency of motion is 3880 rad/s. Second, a static measurement shows that the effective spring constant of the system is 184 N/m. To what value should the damping coefficient be set in order to have critical damping?

66. (II) A mass m is attached to a horizontal spring, with spring constant k, and the mass can move under the spring's influence along the x-axis on a frictionless tabletop. It is initially at rest at the equilibrium position $x = 0$. At $t = 0$, the mass is struck a brief blow that gives it a speed v_0 in the $-x$-direction. (a) What is the position x of the mass as a function of time? (b) Suppose that there is a small drag force $f_D = -bv$, where b is a constant and v is the speed of the mass. What is the position x of the mass as a function of time? (c) Suppose that, rather than a drag force proportional to speed, there is a small force f of kinetic friction, constant in magnitude but always opposing the motion, acting on the object. How, qualitatively, does the motion differ from that described in part (b)?

67. (II) A harmonic oscillator with natural period $T = 2.0$ s is placed in an environment where its motion is damped, with a damping force proportional to its speed. The amplitude of the oscillation drops to 70 percent of its original value in 5.0 s. What is the period of the oscillator in the new environment?

68. (II) Consider a damped harmonic oscillator. The damping, proportional to the speed, is sufficiently weak so that it is a good approximation to view the amplitude as constant over the duration of a cycle. What is the energy of the oscillator at time t if its original energy at time $t = 0$ is E_0? [*Hint*: For one cycle, the usual energy formula involving the square of the amplitude can be used.]

69. (II) A spring with $k = 12.0$ N/m and an attached bob oscillates in a viscous medium (Fig. 13–34). A given maximum, of +6.0 cm from the equilibrium position, is observed at $t = 1.5$ s, and the next maximum, of +5.6 cm, occurs at $t = 2.5$ s. What will the position of the bob be at 3.0 s and at 4.8 s? What was its position at $t = 0$ s?

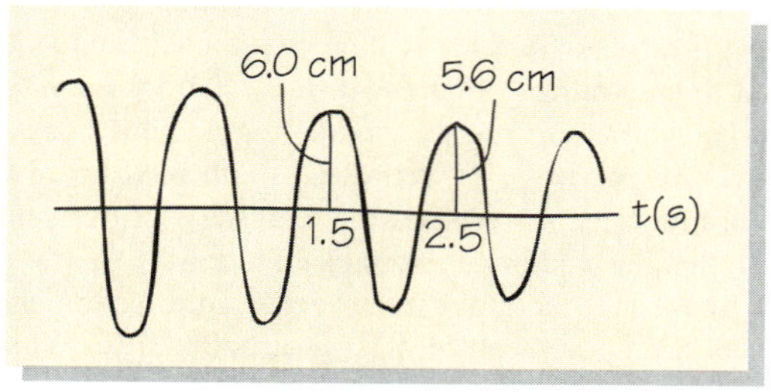

FIGURE 13–34 Problem 69.

70. (II) For what mass m of the bob in Problem 69 will the spring system have its critical value? What are the lifetime and Q factor in this case?

71. (II) Show by direct substitution that Eq. (13–49) is a solution to the equations of motion of the damped harmonic oscillator, Eq. (13–48), provided that Eqs. (13–50) and (13–51) hold. [*Hint*: You will need to show that the sine and cosine terms *separately* vanish.]

13–8 Driven Harmonic Motion

72. (I) A long, flexible, and very light strip of metal, such as a saw blade, is clamped to the edge of a table, extending horizontally over the edge like a miniature diving board. When a gob of putty is placed on the end, the strip sags and comes to a new equilibrium position with its tip a distance 1.0 cm below its original position. If the end of the strip is then lightly tapped, its tip will oscillate in simple harmonic motion. With what frequency should the tip be tapped to make the strip oscillate with maximum amplitude?

73. (I) A mass of 0.5 kg is suspended from a spring, which stretches by 8 cm. The support from which the spring is suspended is set into sinusoidal motion. At what frequency would you expect resonant behavior?

74. (II) Show that resonance occurs for driven harmonic motion when $\omega = \sqrt{\omega_0^2 - (b^2/2m^2)}$. The resonant frequency occurs when the amplitude has a maximum as a function of frequency.

75. (II) A particular spring has a spring constant of 85 N/m and a mass of 0.548 kg at its end. When the spring is driven in a viscous medium, the resonant motion occurs at an angular frequency of 12.2 rad/s. What is (a) the damping parameter due to the viscous medium? (b) the lifetime of the system? (c) the sharpness of the resonance peak?

76. (II) Consider the driven, damped harmonic oscillator discussed in Section 13–8. (a) Calculate the power dissipated by the damping force. (b) Calculate the average power loss, using the fact that the average of $\sin^2(\omega t + \delta)$ over a cycle is one-half. [*Hint*: Recall that $P = Fv$.]

General Problems

77. (II) When you bounce up and down on a diving board of negligible mass, you find yourself and the board bouncing with the maximum amplitude when your bounce frequency is once every 1.2 s. How much will the board deflect vertically when you stand on the end without bouncing? Assume that the damping that must be present is small.

78. (II) Consider simple harmonic motion of a mass on the end of a spring, $x = (0.5 \text{ m}) \sin(\omega t + \delta)$. At $t = 0$ s the position is –0.1 m, and the velocity is 1 m/s in the $-x$-direction. The total energy of the motion is 5 J. What is the value of the (a) phase, δ; (b) frequency, f; (c) acceleration at $t = 0$ s; (d) spring constant, k; (e) mass, m?

79. (II) A mass m on the end of a spring oscillates with angular frequency ω. The mass is removed, the spring is cut in two, and the mass is reattached. What is the new angular frequency?

80. (II) A spring of equilibrium length 30 cm has one end anchored, and a mass is attached to the other end. The mass is set in uniform circular motion in a plane. The angular frequency of the rotational motion is two-thirds the natural angular frequency of the spring, ω_0. (a) By how much is the spring extended by the motion? (b) Derive the general result for $\omega = \alpha\omega_0$.

81. (II) In the spectacular sport of bungee jumping, a light elastic cord (a bungee cord) is tied tightly around the ankles of someone who jumps from a bridge of height H (the other end of the cord is attached to the bridge). The length of the cord is calculated so that the jumper, of mass m, will not quite reach the surface of the water below the bridge before he or she springs back up. Suppose that the cord behaves like a spring of spring constant $10mg/H$, where g is the acceleration due to gravity. (a) How long must the cord be so that a jumper just touches the water before being pulled back up? Neglect the height of the jumper and any effects due to friction. (b) Friction damps the up-and-down motion of the jumper that results after the initial jump. How far above the water would the jumper be when the oscillations have ceased? Express your answer as a fraction of H.

82. (II) A block of mass $m = 0.3$ kg is dropped onto a horizontal platform supported by a spring with spring constant $k = 5$ N/m. The mass of the platform and spring are negligible compared with m. The speed of the block when it hits the platform is 2 m/s. (a) Find the maximum compression of the spring. (b) How long does it take to reach the lowest point after the block hits the platform?

83. (II) The following systems, illustrated in Fig. 13–35, exhibit simple harmonic motion. What is the period of each motion? (a) A toy tightrope walker, with geometry as shown, whose body has mass much less than that of the barbell weights, each of mass 50 g. The toy sways from side to side in harmonic motion. (b) A mass m attached to two parallel springs, each of spring constant k. Ignore gravity. (c) A spring with spring constant k and a mass m attached to each end. (d) A mass m hanging vertically from a spring, of spring constant k, under the influence of gravity.

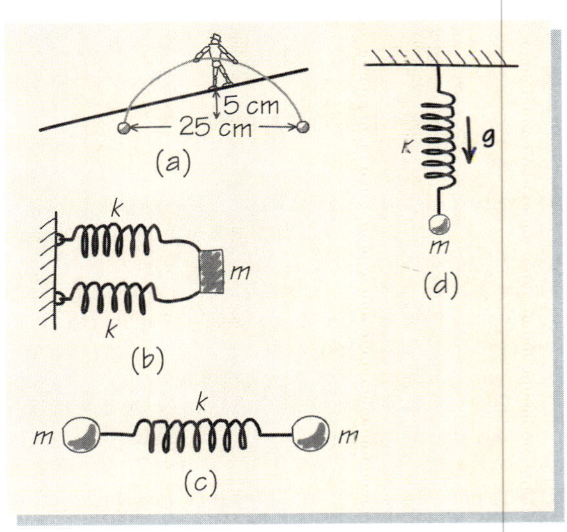

FIGURE 13–35 Problem 83.

84. (II) A simple pendulum of mass $m = 0.5$ kg and length $L = 0.8$ m is attached to a cart of mass $M = 1$ kg (Fig. 13–36). The mass of the pendulum support is negligible. The cart can roll freely on a horizontal surface. At $t = 0$, the pendulum bob is released from rest when the string makes an angle of 10° with the vertical. Assume that the resulting motion of the cart relative to the ground is simple harmonic motion. Determine the amplitude of the motion of the cart. Why would you expect the cart to move in simple harmonic motion?

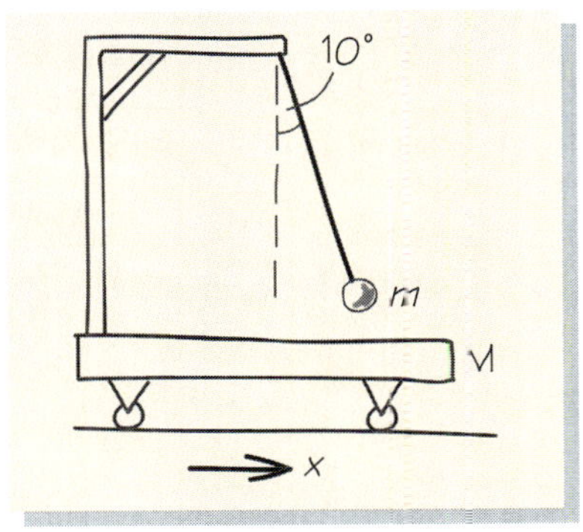

FIGURE 13–36 Problem 84.

85. (II) A bar 5.0 m long with a mass of 12 kg has a sharp bend (totaling 14°) at its midpoint (Fig. 13–37). The bar rests on a pivot placed under the bend. Twins of mass 32 kg each are seated at opposite ends of the bar. What is the period of small oscillations of this modified seesaw?

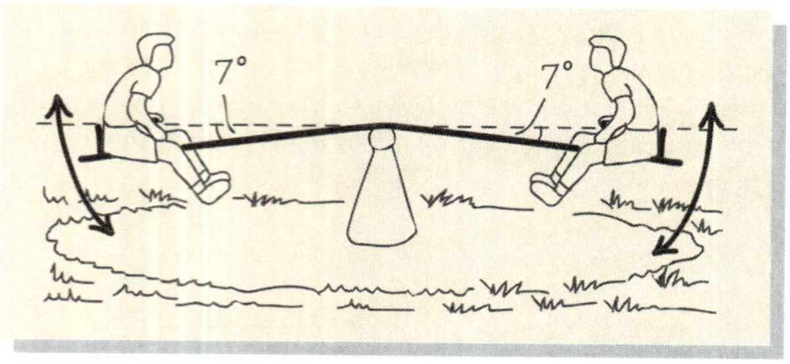

FIGURE 13–37 Problem 85.

86. (II) When a rope of length L is stretched by an amount x, the rope acts like an imperfect spring that exerts the force $F = -f_1(x/L) - f_2(x/L)^2$, where $f_1 > 0$, $f_2 > 0$, and $f_1 >> f_2$. The motion is restricted to the x-direction. (a) Describe qualitatively what happens when a large bucket hanging vertically at the end of the rope is slowly filled with water. (b) Find the potential energy function for the rope and discuss what happens when the bucket, at rest in equilibrium, is set in motion with an initial vertical velocity $\mathbf{v}_0$. What happens when the initial velocity is increased?

87. (II) The potential energy of a diatomic molecule whose two atoms have the same mass, m, and are separated by a distance r is given by the formula $U(r) = (A/r^2) - (e^2/r)$, where A and e^2 are positive constants. (a) Find the equilibrium separation r_0 of the two atoms. (b) Show that if the atoms are slightly displaced, so that their separation is $r_0(1 + x)$, then they will undergo simple harmonic motion about the equilibrium position. Calculate the angular frequency of the harmonic motion. Use $(1 + x)^{-2} = 1 - 2x + 3x^2 - \cdots$ and $(1 + x)^{-1} = 1 - x + x^2 - \cdots$ for $x << 1$. (Remember the reduced mass.)

88. (II) A small mass m is placed midway on the line between two large masses M. The large masses are separated by a distance L. (a) What gravitational force does the small mass experience? (Assume that the system is isolated from other masses.) (b) The small mass is displaced a distance y in a direction perpendicular to the line between the masses. What is the direction of the net force on that mass? (c) Evaluate the magnitude of the net force for $y << L$. Give an expression that incorporates the inequality. (d) What is the motion of m when it is displaced from the midway point?

89. (II) A thin, circular hoop of mass m and radius R hangs on a small peg (Fig. 13–38). What is the frequency of small oscillations with the center of mass moving back and forth along the x-direction?

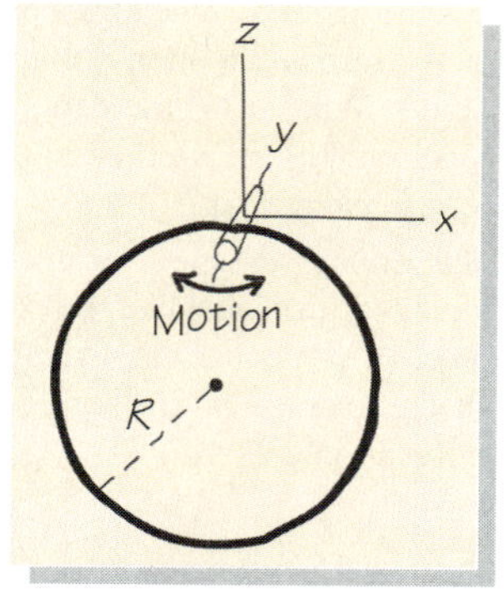

FIGURE 13–38 Problem 89.

90. (II) A machine gun fires bullets in a direction perpendicular to the plane of a target 250 m away from the gun. The bullets leave the muzzle at a speed of 410 m/s. The machine gun is set up on a platform that is vibrating with a frequency of 5.0 Hz and an amplitude of 10.0 cm in a horizontal direction parallel to the plane of the target. Describe the distribution of bullet impact points on the target. The bullets are fired with a frequency much greater than 5 Hz.

91. (III) A solid, uniform cylinder of mass m and radius R is fitted with a frictionless axle along the cylinder's long axis. A horizontal spring (spring constant k) is attached to this axle (Fig. 13–39). Under the influence of the spring, the cylinder rolls back and forth without slipping on a horizontal surface. What is the frequency of this motion?

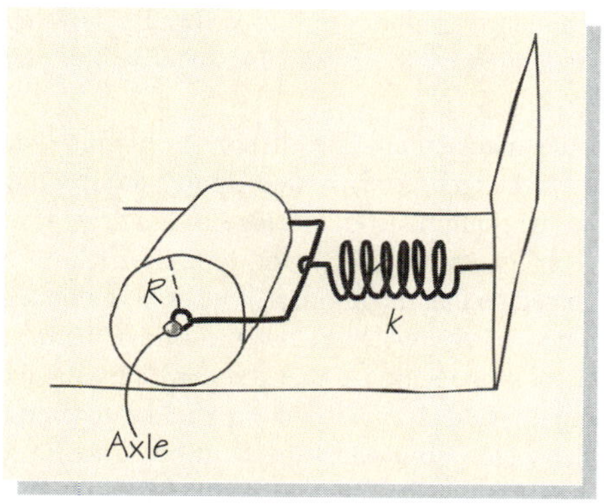

FIGURE 13–39 Problem 91.

CHAPTER

14

These steel band musicians in Trinidad are making complex disturbances in the air. We perceive these disturbances as sound. Sound is only one example of the widespread set of phenomena that we classify under the heading of mechanical waves.

Waves

Waves are familiar to all of us. We can see ocean waves by means of light waves. Music may originate as waves on piano strings, be transported as waves of electrical current in a stereo system, form waves on the cone of a loudspeaker, and reach our ears as sound waves in air, which, in turn, form waves within the mechanical structure of our ears and are carried as waves of electrical impulses to our brains. Waves may appear as traveling waves, which move in some direction, as ocean surf moves toward a beach. They may appear as standing waves, like the vibrations of a guitar string. They may appear in one-dimensional systems, such as a Slinky spring; two-dimensional surfaces, such as the surface of a pond; or three-dimensional media, such as the air around us.

The point masses and perfectly rigid bodies that we have considered up to now are an idealization: We assumed that the internal forces for a rigid body are so strong that its different parts cannot move relative to each other. Waves form because continuous media *can* distort in response to internal forces. In Chapters 14 and 15, we study waves on a guitar string, on the sea, or in the air, all of which involve continuous media. Eventually, we shall encounter (in Chapter 35) electromagnetic waves such as light waves, in which no medium at all is involved!

14–1 ORGANIZED OSCILLATIONS

A guitar string under tension oscillates, or vibrates, in an organized way when it is plucked. The atoms comprising the guitar string are bound together by interatomic forces. Undisturbed, the atoms are in stable equilibrium positions and the whole string is in equilibrium. When the string is plucked, a group of atoms is displaced from its equilibrium position, exerting forces on the neighboring atoms and displacing them. This displacement will, in turn, lead to a displacement of the next neighboring set of atoms, and so on. Thus the disturbance propagates through the length of the entire string, distorting the string. It is the whole string that moves *in an organized way* to form what we call **waves**. We talked earlier about the motion of either point masses (one or several) or rigid extended bodies, in which the relative motion of different parts of the body is unimportant. In wave motion, however, it is the relative motion of different parts of an extended body that interests us.

To understand better the origin of wave motion, think of a model in which an extended medium such as a solid is represented by a collection of point masses connected by springs. Such a medium can be distorted, but if the distortion is not too great, the medium can return to its original shape, and it will support waves. We call such a medium an **elastic medium**. A stretched guitar string is formed from an elastic medium. Elastic media include materials that we might not normally think of as elastic. The ocean surface responds elastically to distortions, and the surface supports waves. Molecules connected by springs make a poor model of air, yet air responds elastically when a small region of it is abruptly compressed; the result is sound. In all of these cases, the elastic response of the medium leads to an organized and collective motion: waves that move and spread. If waves require matter to propagate, as is true of waves on strings, water waves, or sound waves, we call them **mechanical waves**. Such waves are the subject of this chapter and the next.

Although the microscopic model of a guitar string as a succession of point masses connected by springs can be used for a quantitative treatment of waves, we will take a more descriptive approach. For example, we employ the tension in the string without worrying about the microscopic explanation of tension—namely, that tension comes from the forces holding the molecules of the string together.

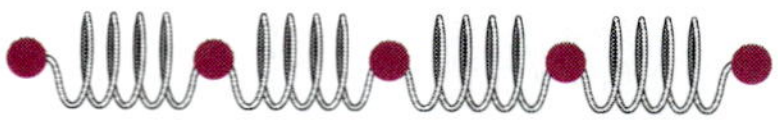

FIGURE 14–1 A series of point masses connected by springs in a line. In equilibrium, the entire system is at rest or is moving uniformly as a whole.

Transverse Motion and Longitudinal Motion

Let's look a little more closely at the kinds of waves that occur on a string under tension. A mechanical model of this system, shown in Fig. 14–1, is a series of point masses connected by springs and attached at the ends. Such a system exhibits the organized oscillatory motion of waves. This model illustrates a way to place wave motion into two important classes, whose significance goes beyond the stretched string.

Transverse waves versus longitudinal waves

The first class of wave motion is illustrated in Fig. 14–2a: In **transverse motion**, all the point masses move *perpendicular* to the original, equilibrium line of masses and springs, with a maximum displacement called the **amplitude**. This describes one possible **mode of vibration** of the string. This type of wave is easily seen on a Slinky toy

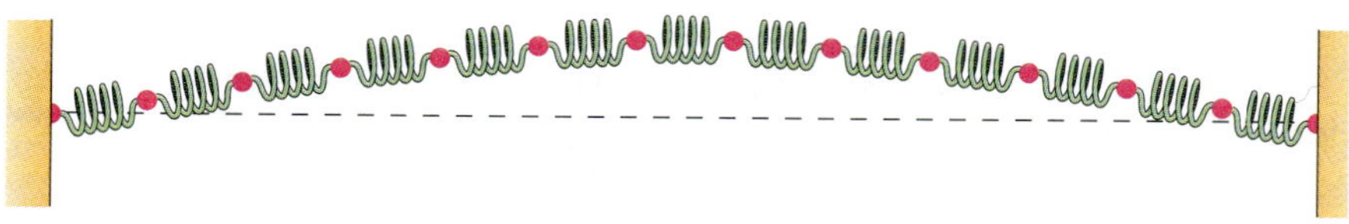

(a)

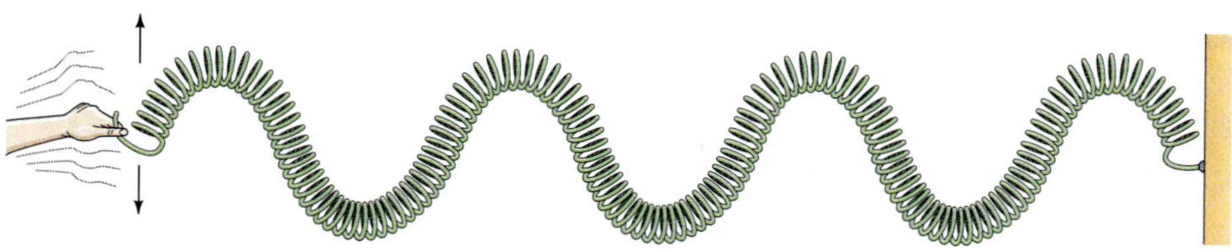

(b)

FIGURE 14–2 (a) The point masses of the system illustrated in Fig. 14–1 are displaced transverse to the equilibrium line (dashed). All the springs are stretched and act to bring the system back to equilibrium. (b) Transverse waves on a Slinky toy.

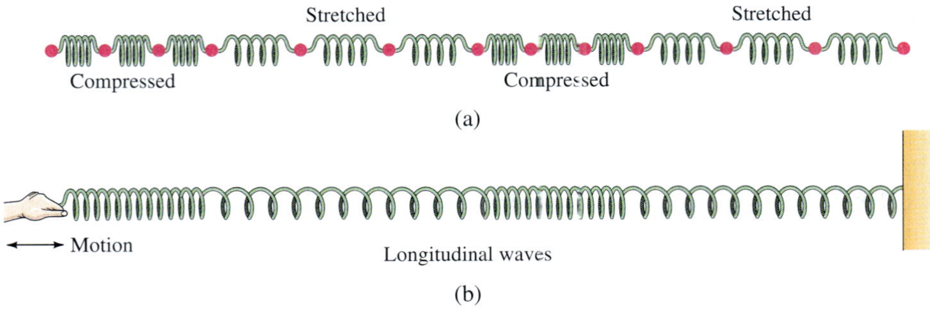

FIGURE 14–3 (a) The point masses of the system illustrated in Fig. 14–1 are displaced longitudinally along the equilibrium line. Some springs are stretched, and some are compressed. (b) Longitudinal waves on a Slinky toy.

or a rope. If one end of the Slinky is whipped from side to side, transverse waves will be generated (Fig. 14–2b).

The second class of wave motion—**longitudinal motion**—is illustrated in Fig. 14–3a and consists of displacements *along* the line of the springs. Some of the springs are stretched and some are compressed, but the spring forces ultimately act to bring each point back to its equilibrium position. The Slinky also shows this type of wave. If the end is pumped back and forth along the direction of the Slinky, longitudinal waves are generated (Fig. 14–3b).

Some waves include both transverse and longitudinal motion. An example is provided by shore waves. If a cork is put into the ocean to mark the passage of a wave, it will move partly up and down and partly toward and away from the shore. In doing so, it traces out a nearly circular path (Fig. 14–4). This is because water is incompressible; it cannot be squeezed. Therefore the trough of a water wave can be formed only by pushing water forward or backward, toward a neighboring crest. This is the origin of the horizontal (longitudinal) motion that accompanies the up-and-down (transverse) motion.

Standing Waves and Traveling Waves

Standing waves versus traveling waves

In addition to the transverse or longitudinal classification of waves, a string under tension also illustrates a different classification: standing (or stationary) waves and traveling (or progressive) waves. In **standing waves**, all the different parts of the medium move together with the same time dependence (Fig. 14–5). A vibrating guitar string is an example of a standing wave. Standing waves are typically present when the ends of a string are fixed or otherwise constrained.

In **traveling waves**, energy is propagated in a definite direction and with a definite speed. If we whip one end of a rope back and forth, a traveling wave moves down the rope (Fig. 14–6). The intriguing feature of this wave is that *all the actual motion is transverse to the overall alignment of the rope, even though the disturbance—the wave—moves along the rope.* Think of a wave formed by a crowd in a stadium, in which each spectator successively moves his or her arms up and down in response to a neighbor's motion, propagating a "wave" around the stadium perpendicular to the direction of the arm motion. The wave travels from one side of the stadium to the other much faster than any spectator could.

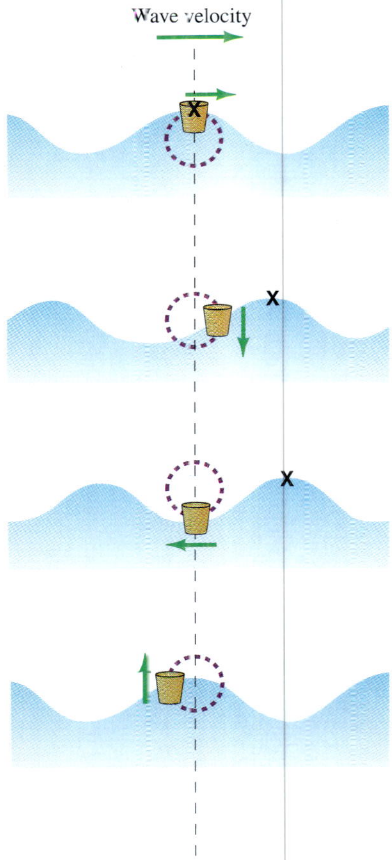

FIGURE 14–4 A cork floats on the ocean surface, where the water is not too deep, as a wave passes. The x marks the crest of the passing wave; the dashed circle marks the path followed by the cork; and the arrow, the direction of motion of the cork at the particular point in the wave's passage.

We have illustrated the concepts of standing waves and traveling waves with transverse waves, but longitudinal waves can also form both types. Sound waves, which are longitudinal, can form both standing waves and traveling waves. For example, standing sound waves are generated within an organ pipe, whereas a Swiss yodel carrying from one mountain top to another is a traveling sound wave.

Multidimensional Waves

Waves on a string, a one-dimensional system, will continue to be our primary example, whether they are standing or traveling waves. But a two-dimensional system, such as the surface of the ocean, can also support waves. Just as standing waves can be gen-

FIGURE 14–5 Vibrating strings of a guitar are examples of standing waves, which can be established on strings fixed at both ends.

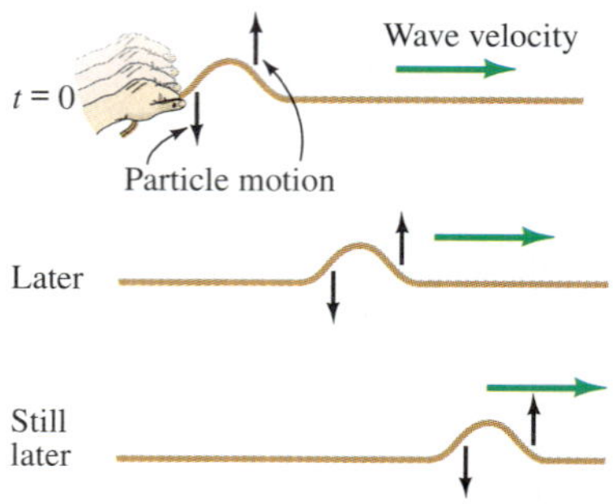

FIGURE 14–6 A traveling wave on a string. At time $t = 0$, a hand generates a wave that travels down the taut string.

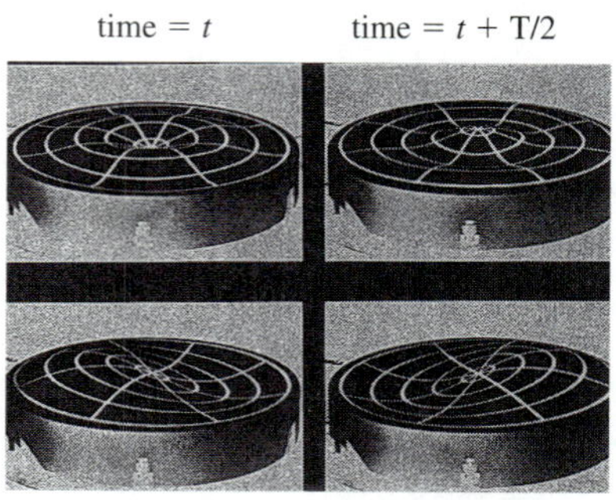

FIGURE 14–7 Standing waves on a drumhead. The *nodes,* points where the drumhead is not in motion, form lines rather than individual points as they did for the string. The lines shown are not necessarily nodes. Can you determine the nodes for each pair of vibrations?

erated on a string, they can also be generated on a drumhead. These waves consist of an organized harmonic motion transverse to the drumhead itself (Fig. 14–7). Traveling waves in two-dimensional systems form **wave fronts**, which are lines perpendicular to the direction of motion of the wave. This is a familiar phenomenon, illustrated by the spreading ripples generated by throwing a rock into a pond, as in Fig. 14–8. The ripples move radially away from a central point and the wave front is an expanding circle. For a three-dimensional system, such as open air, the wave fronts formed by traveling waves move out as spheres from a central point.

14–2 STANDING WAVES

As we have seen, a guitar string with two fixed ends supports standing waves. The fact that the two ends are fixed is referred to as a set of *boundary conditions* for the wave motion. The variables needed to understand standing waves on a string with fixed ends are the tension, T, the length L, and the mass density, μ.† The mass density, or mass per unit length, is such that the mass of any segment of length ℓ is $m = \mu\ell$. These three variables can be changed. For example, we tighten or loosen a guitar string by turning its peg, we change the length of the string when we press it against a fret, and we change the mass density when we replace a light string with a heavy one.

If we pluck the guitar string, it vibrates in a transverse standing wave. Some possible vibrations are shown in Fig. 14–9. The transverse displacement of a standing wave may differ at different points along the string; however, the *time dependence* of the transverse motion is the same everywhere along the string. For example, the center point on the string might move up and down with an amplitude of 1 cm, while a point halfway between one end and the center point might move up and down with an amplitude of only 0.5 cm; but the two points move together.

Mathematical Description of Standing Waves

To describe a standing wave of a guitar string mathematically, suppose that the string is aligned along the x-direction, with one end at $x = 0$ and the other end at $x = L$. The position along the string is labeled x, and the transverse displacement of a point on the string is labeled $z(x, t)$. Then every point along the string moves with the same time dependence if

$$z(x, t) = f(x)g(t). \tag{14–1}$$

In view of the close relation between waves and springlike forces discussed in Section 14–1, we expect harmonic time dependence,

$$g(t) = \cos(\omega t + \phi), \tag{14–2}$$

where ω is just like the angular frequency of simple harmonic motion and ϕ is a phase. Thus any particular point on the string is in simple harmonic motion. For the space dependence, a simple function that fits the kind of situation we are discussing is

$$f(x) = z_0 \sin(kx + \delta). \tag{14–3}$$

The constant z_0 is the amplitude, which measures the maximum displacement; δ is another phase. The constant k, which is positive, should not be confused with a spring constant. It is just a constant analogous to ω, whose dimensions make the argument of the sine dimensionless. Substituting for the functions $f(x)$ and $g(t)$ in Eq. (14–1), we get a standing wave of the form

$$z(x, t) = z_0 \sin(kx + \delta) \cos(\omega t + \phi). \tag{14–4}$$

We can call this a *harmonic standing wave.*

†The mythical massless strings we have so often used could not support waves.

Properties of Harmonic Standing Waves

Let's study the various features of the standing wave expressed by Eq. (14–4). The *angular frequency,* ω, is familiar from Chapter 13. The *frequency* of the repetition is $f = \omega/2\pi$, and the *period* of the repetition is $\tau = 1/f = 2\pi/\omega$.† The function $\sin(kx + \delta)$ repeats itself in space, and the distance over which it repeats is $2\pi/k$. We refer to this distance as the **wavelength**, λ, and k is called the **wave number**. These quantities are related by

$$\lambda = \frac{2\pi}{k}. \qquad (14\text{–}5)$$

FIGURE 14–8 The spreading waves resulting from a disturbance of a water surface form circles centered on the disturbance.

Boundary Conditions Determine the Wavelength. The possible values of k are determined by boundary conditions. For this particular case, the boundary conditions are that the ends of the string are fixed: $z = 0$ at $x = 0$ and at $x = L$ for all times t. In order to satisfy $z(0, t) = 0$, we take $\delta = 0$. In order to satisfy $z(L, t) = 0$, we require that

A harmonic standing wave on a string

The wavelength and wave number

$$\sin(kL) = 0. \qquad (14\text{–}6)$$

Equation (14–6) can be satisfied if the argument of the sine function is any positive integer multiple of π (negative integer multiples only reverse overall sign of the wave), or if k takes on the discrete values

Boundary conditions fix the possible values of the wavelengths of standing waves.

$$k_n = \frac{n\pi}{L} \qquad \text{for } n = 1, 2, 3, \ldots . \qquad (14\text{–}7)$$

(The case $n = 0$ gives just a flat or undistorted string—we say the string is *unexcited.*) Equation (14–5) then implies that the allowed wavelengths also carry an index n:

$$\lambda_n = \frac{2\pi}{k_n} = \frac{2}{n}L \qquad \text{for } n = 1, 2, 3, \ldots . \qquad (14\text{–}8)$$

PROBLEM SOLVING

A way to remember the possible values of the wavelength.

When the string is set into motion with values of k that correspond to a certain standing wave, we say the string is *excited* into those waves. Equation (14–8) can be rewritten as $L = \frac{1}{2}n\lambda_n$, providing a convenient way to recall the possible values of the wavelength: *The total length of a string can be divided into an integral number of half-wavelengths.* The value of n labels the **modes**, or possible vibration patterns, of the possible standing waves; $n = 1$ labels the *fundamental mode* (see Fig. 14–9), also called the *first harmonic.* $n = 2, 3, \ldots$ are called the second, third, … harmonics. Note that for odd values of n, the number of wavelengths that can fit on the string is a whole number plus one-half, while for even values of n, a whole number of wavelengths fits into the string. Note also that, for $n > 1$, there are *nodes* where the string rests motionless with no transverse displacement. Although these nodes are not caused by a physical clamping of the string as at the end points, they might as well be because these points do not move. Not counting the end points, there are $n - 1$ of these nodes—for example, one node in the $n = 2$ case of Fig. 14–9. We can also see the presence of *antinodes,* locations where the displacement is a maximum.

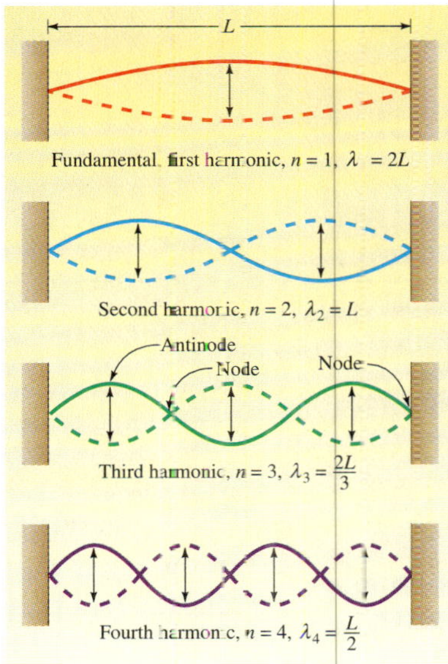

FIGURE 14–9 Some transverse standing waves of a vibrating string with fixed ends. We see some of the possible modes, together with nodes and antinodes. The string moves back and forth between the extremes shown here, as the arrows indicate.

EXAMPLE 14–1 Standing waves can be excited on a string of length $L = 1.00$ m whose ends are fixed at $x = 0$ and $x = L$. Find the two smallest values of the wave number and the two longest wavelengths allowed for standing waves. Sketch the space dependence of these waves.

Solution: The possible wave numbers, k_n, are proportional to n, whereas the possible wavelengths, λ_n, are proportional to n^{-1}. Thus the smallest values of k and the largest values of λ are the two cases $n = 1$ and $n = 2$ of Eqs. (14–7) and (14–8),

†We use τ for the period because we do not want to confuse it with the tension, T, of the string.

respectively. These are

$$k_1 = \frac{\pi}{L} = 3.14\ \text{m}^{-1}, \qquad k_2 = \frac{2\pi}{L} = 6.28\ \text{m}^{-1}$$

and

$$\lambda_1 = 2L = 2.00\ \text{m}, \qquad \lambda_2 = \frac{2L}{2} = 1.00\ \text{m}.$$

The space dependence of the waves is given by the function $\sin(k_n x)$ times an arbitrary amplitude z_0. Figure 14–9 shows the possibilities $n = 1$ through $n = 4$, with z_0 left unspecified.

Dynamics Determines the Frequency. We have seen how boundary conditions determine k. Initial conditions, which are a type of boundary condition in time, determine the phase ϕ. *But boundary conditions do not determine the angular frequency* ω. We will next show that, as in Chapter 13, ω *is determined by Newton's second law, not by the boundary conditions.*

The Wave Equation

Just how do we apply $F = ma$ to the string? In other words, what is the dynamical equation of motion? We find the equation of motion by studying the forces on a small segment of the string as it is stretched and distorted (see the next subsection "How To Get the Wave Equation"). One crucial approximation leads to a simplified equation of motion. This is the *assumption of small distortion*, in which the stretch of the string is kept small. As a result of the small-distortion approximation, we find the dynamical equation of motion for transverse displacements of the string to be

$$T\frac{\partial^2 z}{\partial x^2} = \mu\frac{\partial^2 z}{\partial t^2}. \tag{14–9}$$

A wave equation

Relations such as this are important in many areas of science and engineering and are known as **wave equations**. The symbol ∂ is used for partial differentiation. The transverse displacement z is a function of *two* variables (x and t). The partial-derivative notation simply means that for $\partial z/\partial x$, the variable t is held fixed while the derivative is taken with respect to x. The wave equation (or its generalization to two or three dimensions) describes not only transverse and longitudinal mechanical waves—both standing and traveling—but also electromagnetic waves, such as light (see Chapter 35).

PROBLEM SOLVING

Partial derivatives

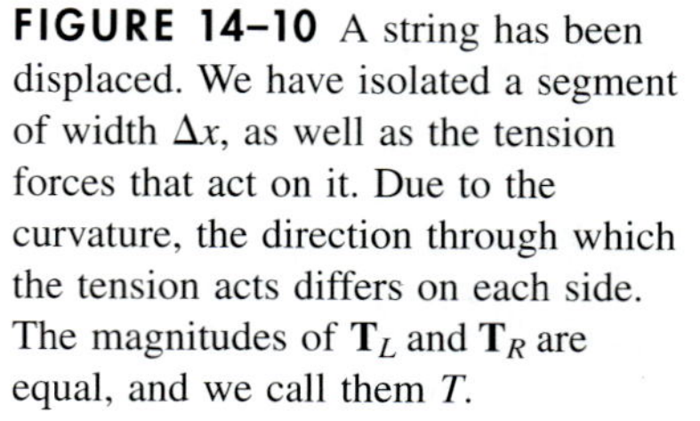

FIGURE 14–10 A string has been displaced. We have isolated a segment of width Δx, as well as the tension forces that act on it. Due to the curvature, the direction through which the tension acts differs on each side. The magnitudes of $\mathbf{T}_L$ and $\mathbf{T}_R$ are equal, and we call them T.

*How To Get the Wave Equation

Consider a small segment of length Δx and mass $\mu\Delta x$ of a uniform string, where μ is the mass density of the string. The ends of the segment are at x and $x + \Delta x$. We suppose the string has been displaced in the transverse direction (the z-direction) to take the shape shown in Fig. 14–10. We look for the tranverse forces acting on this segment.[†] We assume the distortion of the string is small, meaning that (a) the tension is the same throughout the string, and (b) all angles are small.

The forces acting on the segment are the tensions on the left and right sides. The *directions* of the tensions are slightly different because the string is curved, and that is what gives a net vertical force on the segment. On the left side the vertical component of the tension is downward and on the right it is upward. For small angles, the magnitudes of the vertical components are $T\sin\theta \simeq T\tan\theta$; that is, the tension times the

[†]For waves other than transverse waves on a string, only the details would change.

slope of the string at the respective points. In turn, this slope is the derivative of z with respect to x. Thus the vertical component of the tension on the left side is

$$F_L = -T\frac{\partial}{\partial x}z(x, t).$$

The slope is usually denoted by dz/dx but, because z depends on two variables, x and t, we use the partial-derivative notation described in the text. The force on the right end of the segment points upward. This time, the slope involved is the slope at $x + \Delta x$:

$$F_R = T\frac{\partial}{\partial x}z(x + \Delta x, t).$$

Combining, the net vertical force on our element is

$$\begin{aligned} F = F_R + F_L &= T\frac{\partial}{\partial x}z(x + \Delta x, t) - T\frac{\partial}{\partial x}z(x, t) \\ &= T\left[\frac{\partial}{\partial x}z(x + \Delta x, t) - \frac{\partial}{\partial x}z(x, t)\right]. \end{aligned} \tag{14–10}$$

Now recall that by definition of the derivative, in the limit that Δx approaches zero,

$$\frac{f(x + \Delta x) - f(x)}{\Delta x} = \frac{df}{dx}$$

for any function $f(x)$. We apply this to Eq. (14–10) for $f(x) = \partial z/\partial x$:

$$F = T(\Delta x)\frac{\partial}{\partial x}\left(\frac{\partial z}{\partial x}\right) = T(\Delta x)\frac{\partial^2 z}{\partial x^2}.$$

This equation is the net force on our small segment.

The mass times the vertical acceleration of the segment is simply given by

$$ma = \mu(\Delta x)\frac{\partial^2 z}{\partial t^2}. \tag{14–11}$$

The dynamical equation of motion results from the equality $F = ma$. In equating F and ma, the factor Δx cancels, and we are left with Eq. (14–9).

How the Wave Equation Determines the Frequency

We want to verify that our trial standing-wave solution, Eq. (14–4), is indeed a solution of the wave equation. In so doing, we determine the angular frequency, ω. We test the solution by taking the necessary derivatives of Eq. (14–4) and substituting them into the wave equation, Eq. (14–9). The first time derivative of Eq. (14–4) gives

$$\frac{\partial z}{\partial t} = z_0 \sin(kx + \delta)\omega \sin(\omega t + \phi)$$

and hence

$$\frac{\partial^2 z}{\partial t^2} = z_0 \sin(kx + \delta)(-\omega^2)\cos(\omega t + \phi).$$

For the left-hand side of Eq. (14–9) with Eq. (14–4) substituted, the derivatives with respect to x leave the time dependence unchanged. Therefore the $\cos(\omega t + \phi)$ term cancels from both sides of Eq. (14–9) with Eq. (14–4) substituted, and we find that

$$T\frac{\partial^2[z_0 \sin(kx + \delta)]}{\partial x^2} = -\mu\omega^2 z_0 \sin(kx + \delta); \tag{14–12}$$

$$-Tk^2 z_0 \sin(kx + \delta) = -\mu\omega^2 z_0 \sin(kx + \delta).$$

Finally, we can cancel the factor $z_0 \sin(kx + \delta)$ from both sides, leaving

$$Tk^2 = \mu\omega^2. \tag{14–13}$$

The angular frequency is determined by the dynamics.

Recall that the wave number, k, is related to the wavelength, λ, by $k = 2\pi/\lambda$. Thus, from Eq. (14–13), the angular frequency is given by

$$\omega = \frac{2\pi}{\lambda}\sqrt{\frac{T}{\mu}}. \tag{14–14}$$

When we substitute $\omega = 2\pi/\tau$, Eq. (14–14) becomes a relation between the period of the oscillations of the string, τ, and the wavelength of the standing waves:

$$\tau = \lambda\sqrt{\frac{\mu}{T}}. \tag{14–15}$$

What can we learn about the motion of the string from Eqs. (14–14) and (14–15)?

1. As the wavelength of possible standing waves on a particular string decreases, so does the period of the motion.
2. A less dense string vibrates more rapidly than a dense one.
3. A string under great tension vibrates more rapidly than a string under less tension.

Harmonics

For the boundary conditions previously discussed, $\lambda_n = (2/n)L$, and the frequency, angular frequency, and period can have the discrete values

$$f_n = \frac{n}{2L}\sqrt{\frac{T}{\mu}}, \tag{14–16a}$$

$$\omega_n = \frac{n\pi}{L}\sqrt{\frac{T}{\mu}}, \tag{14–16b}$$

$$\tau_n = \frac{2L}{n}\sqrt{\frac{\mu}{T}}. \tag{14–16c}$$

The frequency f_1, the lowest frequency, is known as the **fundamental frequency**, and $\lambda_1(= 2L)$ is the **fundamental wavelength**. The series of frequencies that are integral multiples of the lowest possible frequency are called the **harmonic series**. The frequency $f_n = nf_1$ is called the nth *harmonic* of the string; Fig. 14–9 depicts some harmonics. (We will return to these topics in Section 14–6.)

FIGURE 14–11 This piano tuner adjusts the natural frequencies of the strings by changing their tensions.

EXAMPLE 14–2 The time for 100 full vibrations (100 periods) in the fundamental mode of a piano wire (Fig. 14–11) is 0.5 s. The wire length is $L = 2$ m and the total mass of the wire is 25 g. What is the tension on the wire? By how much must the tension be increased in order to halve this time?

Solution: The fundamental mode is labeled $n = 1$. For the period of one vibration, Eq. (14–16c) gives

$$\tau_1 = 2L\sqrt{\frac{\mu}{T}}.$$

This may be solved for the tension, T:

$$T = \frac{4\mu L^2}{\tau_1^2}.$$

We know that $L = 2$ m, $\tau_1 = (0.5\ \text{s})/100 = 5 \times 10^{-3}$ s, and the mass density, μ, is $(0.025\ \text{kg})/(2\ \text{m}) = 0.0125$ kg/m. Thus the tension is

$$T = \frac{4(0.0125\ \text{kg/m})(4\ \text{m}^2)}{2.5 \times 10^{-5}\ \text{s}^2} = 800\ \text{N},$$

which is equivalent to the weight of an average man.

From Eq. (14–16c), it can be seen directly that the period will be halved if the square root of the tension is doubled or if the tension is increased by a factor of 4. The new tension will be 3200 N.

Standing Waves in More than One Dimension

The boundary conditions determining the allowed values of k and the dynamics determining the allowed values of ω are far more complicated when the system is a surface or a volume rather than a one-dimensional system like the string. For example, the boundary conditions for a drumhead require that the vibrations vanish on the circular rim of the drum. Figure 14–7 showed a few of the modes possible on a circular drumhead. The modes on an elliptical drumhead differ from those on a circular drumhead. The common feature of all these systems is that the wavelengths and frequencies are discrete, so they can be labeled by integers.

14–3 TRAVELING WAVES

Suppose that a hand shakes one end of a string and disturbs it. If the string is so long that we do not need to worry about what is happening at the other end, then we can have traveling waves on the string. Traveling waves can move over large distances even though the material that comprises the propagating medium moves in a very limited way.

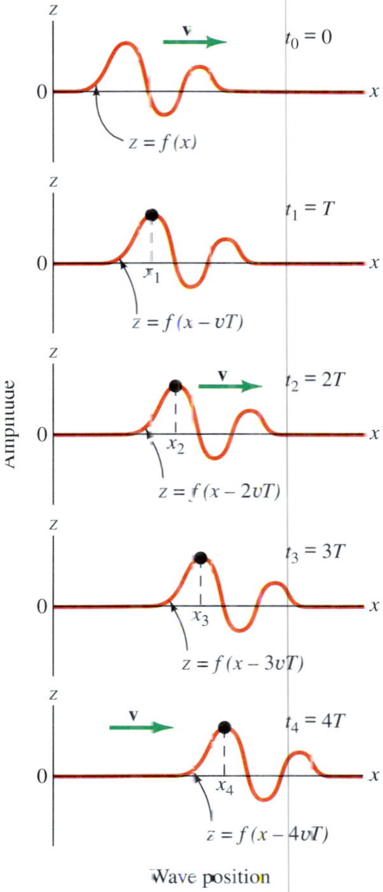

FIGURE 14–12 The curve is some function f of the variable $x - vt$. As time goes on ($t_2 > t_1$), the curve moves to the right with speed v.

Mathematical Description of Traveling Waves

Picture a string oriented along the x-axis, with mass density μ and tension T. We want to characterize the transverse displacement $z(x, t)$ of a traveling wave. We require a mathematical form that describes a shape moving *as a whole* with speed v to the right, for example. We will now show that this can be done with *any function* of the quantity $x - vt$:

$$z(x, t) = f(x - vt), \tag{14–17}$$

where the choice of function f corresponds to a particular moving shape.

To see why this works, consider a function such as that in Fig. 14–12. We single out a particular point on the function, such as a peak, and mark this point with a black dot. At time t_1, this point is at an x-value of x_1. If t_1 increases to t_2, the argument of f will have the same value—the point will still be at the black dot—only if x increases to compensate for the decreased value of the term $-vt$. In order for the point to stay at the same argument, we must have

$$x_2 - vt_2 = x_1 - vt_1;$$

$$x_2 = vt_2 + x_1 - vt_1 = x_1 + v(t_2 - t_1). \tag{14–18}$$

Because Eq. (14–18) describes a point moving to the right with speed v, $f(x - vt)$ describes some form moving in the same way. Similarly, $f(x + vt)$ represents a shape moving to the left. In summary, $f(x - vt)$—or $f(x + vt)$—*represents a shape that moves to the right—or left—with speed* v.

Any function of $x - vt$ (or $x + vt$) describes a traveling wave moving in the positive (negative) x-direction with speed v.

Again, let's be guided by our experience in simple harmonic motion and use a sinusoidal form for f. We cannot, however, use the simple form $\sin(x - vt)$ because the argument of the sine must be dimensionless. To make a dimensionless argument, we multiply $x - vt$ by the wave number, k, as in Eq. (14–3). Thus, the following equation represents a harmonic traveling wave on a string:

A harmonic traveling wave on a string

$$z(x, t) = z_0 \sin[k(x - vt)] = z_0 \sin(kx - kvt). \qquad (14\text{–}19)$$

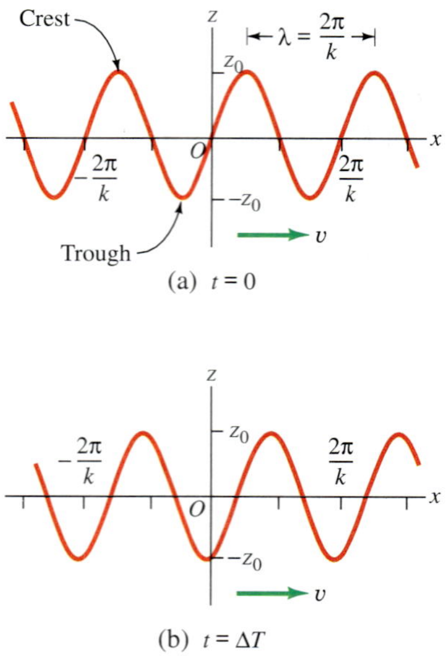

FIGURE 14–13 (a) The harmonic traveling wave of Eq. (14–19) is plotted at the fixed time $t = 0$. (b) As time passes, this pattern will move to the right.

Properties of Harmonic Traveling Waves

If we were to take a snapshot of $z(x, t)$ in Eq. (14–19) at $t = 0$ and plot $z(x, t) = z_0 \sin(kx)$, we would get Fig. 14–13a. The sine curve is zero at $x = 0$ and at every value of x equal to an integer multiple of π/k. It repeats every time x increases by $2\pi/k$. The repetition length is the wavelength, λ; *as before, the relation between* λ *and* k *is* $\lambda = 2\pi/k$. The maxima of the sine curve, $z = z_0$, are known as **crests** of the wave, and the minima, $z = -z_0$, are **troughs**.

We have already shown that the wave described by Eq. (14–19) moves to the right with speed v (Fig. 14–13b). How much time does this wave take to move by one wavelength? The wave has the same configuration when kvt changes by 2π,

$$kv\tau = 2\pi;$$

$$\tau = \frac{2\pi}{kv}. \qquad (14\text{–}20)$$

The time τ is the period of the wave. We have a frequency f and an angular frequency ω with the usual relation to the period:

$$f = \frac{1}{\tau}, \qquad \omega = 2\pi f = \frac{2\pi}{\tau}. \qquad (14\text{–}21)$$

Equation (14–21) states *exactly the same relations as those that hold in simple harmonic motion.* Comparison of Eqs. (14–20) and (14–21) shows that the wave speed is related to the wave number, wavelength, frequency, and angular frequency by

The relations between frequency, wavelength, wave number, angular frequency, and wave speed

$$v = \lambda f = \frac{\lambda}{\tau} = \frac{\omega}{k}. \qquad (14\text{–}22)$$

The important relation $v = \lambda f$ among speed, wavelength, and frequency is easily understood if we think about a passing train possessing cars of equal length. The frequency is measured by counting the number of cars that pass in a given time and the wavelength is the length of one car. The speed of the train is then the length of a car times the number of cars that pass per unit time. If we measure three cars passing every 5 s, then $f = 3/(5\text{ s}) = 0.6$ Hz, and if a car has a length of $\lambda = 30$ m, the speed is $v = \lambda f = (30\text{ m})(0.6\text{ Hz}), = 18$ m/s.

Equation (14–22) suggests alternative useful ways to write Eq. (14–19) for the transverse displacement:

$$z(x, t) = z_0 \sin(kx - \omega t) \qquad (14\text{–}23)$$

$$= z_0 \left[\sin 2\pi \left(\frac{x}{\lambda} - \frac{t}{\tau} \right) \right]. \qquad (14\text{–}24)$$

If we examine the traveling wave at some fixed position x_0, then the kx_0 term in Eq. (14–23) acts as a phase, and the z-motion at x_0 is simple harmonic motion.

EXAMPLE 14–3 A person standing at the narrow entrance to a harbor sees sinusoidal water waves moving into the harbor. He counts 50 wave crests in 1.0 min and he estimates the distance between the crests to be 3.0 m (with the help of an anchored boat of known size). Write an expression for the form of the wave height where the person is standing. What are the wavelength, wave number, frequency, angular frequency, and speed of these waves?

Solution: If $+x$ is taken as the direction into the harbor, and if h is the height of the waves above the average water level, then an expression for h is

$$h(x, t) = h_0 \sin(kx - \omega t).$$

The amplitude, or maximum height h_0, is not specified in the problem, but the wave number and the angular frequency are specified. The wavelength, λ, is just the distance between crests, $\lambda = 3.0$ m. The wave number, k, is given in terms of λ by Eq. (14–5):

$$k = \frac{2\pi}{\lambda} = \frac{2\pi}{3.0}\,\text{m}^{-1} = 2.1\ \text{m}^{-1}.$$

The frequency, f, of the waves is 50 min^{-1}, or

$$f = (50\ \text{min}^{-1})\frac{1\ \text{min}}{60\ \text{s}} = 0.83\ \text{s}^{-1}.$$

The angular frequency, ω, and the frequency, f, are related by Eq. (14–21):

$$\omega = 2\pi f = (2\pi\ \text{rad})(0.83\ \text{s}^{-1}) = 5.2\ \text{rad/s}.$$

Finally, the speed v is given by Eq. (14–22),

$$v = \lambda f = (3.0\ \text{m})(0.83\ \text{s}^{-1}) = 2.5\ \text{m/s}.$$

Dynamics Determines the Speed. Unlike the standing waves of Section 14–2, traveling waves have no boundary conditions for determining wavelength—even though the wavelength is related to the frequency and the wave speed by $v = \lambda f$. In fact, *the dynamical equations determine the speed of a traveling wave.* We showed in Section 14–2 that the wave equation, or dynamical equation of motion, takes the form of Eq. (14–9): $T\,\partial^2 z/\partial x^2 = \mu\,\partial^2 z/\partial t^2$. It is not difficult to check that our traveling wave satisfies this equation by calculating the derivatives. We leave this exercise for the problems; here, we simply state that if we use one of the forms for $z(x, t)$, such as Eq. (14–23), then we find an algebraic relation that must be satisfied after taking the appropriate derivatives:

$$Tk^2 = \mu\omega^2. \qquad (14\text{–}25)$$

The quantities k and ω are already kinematically related by the wave speed, $v = \omega/k$. Thus Eq. (14–25) gives the speed of the wave in terms of the mass density and tension of the string:

$$v = \sqrt{\frac{T}{\mu}}. \qquad (14\text{–}26)$$

The speed of a traveling wave on a string

Equation (14–25) is exactly the same as Eq. (14–13) of the standing wave problem. The only difference is that we did not interpret the ratio ω/k as a speed for the case of the standing wave.

EXAMPLE 14–4 A very long wire is held under tension by suspending weights from an end that passes over a pulley. The speed of transverse traveling waves on the wire is 51 m/s. The wire is replaced by another with three times the mass density and twice as much weight is suspended from the end of the new wire. What is the speed of the traveling waves now?

Solution: Equation (14–26) gives the original speed v_1 in terms of the original tension T_1 and mass density μ_1:

$$v_1 = \sqrt{\frac{T_1}{\mu_1}}.$$

When Newton's second law is applied to the wire in equilibrium—with zero acceleration—it implies that the tension is given precisely by the suspended weight. Thus T has been doubled and the new tension is $T_2 = 2T_1$. The new mass density is $\mu_2 = 3\mu_1$. Then the new speed is

$$v_2 = \sqrt{\frac{T_2}{\mu_2}} = \sqrt{\frac{2T_1}{3\mu_1}} = \sqrt{\frac{2}{3}}v_1.$$

Thus the speed has been decreased to

$$v_2 = (\sqrt{0.67})(51 \text{ m/s}) = 42 \text{ m/s}.$$

Increasing the tension tends to increase the speed, whereas increasing the mass density tends to decrease the speed. In this case, the mass density is increased by a larger factor than is the tension, so the speed decreases.

14–4 ESTIMATING THE SPEEDS OF TRAVELING WAVES

The speed of a traveling wave is one of its most important features. If the time interval between a lightning flash and the thunderclap is 5 s, how far away was the lightning strike? How much time does it take for a tsunami (a tidal wave) to travel from the point at which an earthquake generates it to a distant shore? Here we shall examine some techniques for estimating wave speeds.

A General Form of the Wave Equation

A string of mass density μ under tension T obeys the wave equation [Eq. (14–9)]. Using the fact that $v = \sqrt{T/\mu}$, we can rewrite the wave equation as

$$\frac{\partial^2 z}{\partial x^2} = \frac{1}{v^2}\frac{\partial^2 z}{\partial t^2}. \qquad (14\text{–}27)$$

In this form, the wave equation takes on a more universal aspect than the special situation of transverse waves on a string. The variable z now represents any physical quantity whose variation describes a wave; for example, the height of a water wave, the back-and-forth motion of the coils of a Slinky, the compression of a metal, or the variation of air pressure in a sound wave. The direction of motion of the wave is represented by x and v is the speed with which the wave moves.

All of these waves can take the very general form $z(x, t) = f(x \pm vt)$ and this form will automatically satisfy the wave equation. To demonstrate this, we need only take the derivatives of $f(x \pm vt)$. For example, denote $y = x - vt$ so that $f(x - vt)$ is written $f(y)$. Then use the chain rule for derivatives (Appendix IV–7):

$$\frac{\partial f(y)}{\partial x} = \frac{df(y)}{dy}\frac{\partial y}{\partial x} = \frac{df(y)}{dy}.$$

In the last step, we have used $\partial y/\partial x = \partial(x - vt)/\partial x = (\partial x/\partial x) - [\partial(vt)/\partial x] = 1$. By following this procedure and inserting the appropriate derivatives in the wave equation—Eq. (14–27)—we can verify that the wave equation is satisfied (see Problem 17).

APPLICATION: SOLDERING CIRCUIT BOARDS

Surface waves, like small amplitude water waves, are used in an interesting fashion in the preparation of the printed circuit boards that form the central parts of computers, radios, televisions, and VCRs. A printed circuit is a fiberglass plastic board with many holes punched into it. The electronic components of the circuit are placed at the correct location on one side of the board, and wire legs that enable electrical contact to these components to be made are placed through the appropriate holes. The attachment of the component wires is by soldering; i.e., by melting a metal alloy at the appropriate intersections. Previously, this was done by hand—one joint at a time. Today, the board is suspended over a rectangular container of molten solder. A traveling surface wave of solder is then started at one end of the container. The height of this wave is very carefully controlled. When the wave reaches the board, it passes just below the surface of the board, touching each of the joints, and depositing solder at each. The bath temperature is chosen so that a properly fused connection occurs at every appropriate location. This method of assembly, which is called "wave soldering," is fast, and it does not "forget" any connections. In addition, and very importantly, wave soldering subjects the component wires of the board to a controlled temperature rise because the solder contacts the joint for a brief controlled period; excessive heating of the components is thereby avoided.

How the Wave Speed Depends on the Medium

Once we have established that a medium is elastic and obeys a wave equation, we can try to determine the wave speed, which depends on the nature of the medium. For the string ($v = \sqrt{T/\mu}$), the tension, T, acts as a restoring force that tends to bring the string back to its equilibrium position. The mass density, μ, describes the reaction of the string to this restoring force. Generally, this suggests that the speed of mechanical waves is a function of an internal force factor (an elastic restoring force) divided by a mass factor characteristic of the system.

A dimensional analysis helps us push further. The tension is a force and has dimensions $[MLT^{-2}]$. The mass density (mass per unit length) has dimensions $[ML^{-1}]$. Therefore

Dimensional analysis is described in Chapter 1.

$$\left[\frac{\text{tension}}{\text{mass density}}\right] = \left[\frac{MLT^{-2}}{ML^{-1}}\right] = [L^2T^{-2}].$$

Thus, if v is a function of the ratio of the restoring force factor to the mass factor, *the particular function must be a square root*:

$$\text{wave speed} = \sqrt{\frac{\text{restoring force factor}}{\text{mass factor}}}. \qquad (14\text{–}28)$$

A general form for the wave speed

If we restrict ourselves to bulk (macroscopic) properties of the string, Eq. (14–26) is the *only* dimensionally correct relation for wave speed. Indeed, the *only* bulk properties of the string that we know are the tension, the total mass, the length, and the thickness; the first three of these are figured into the wave speed. (For the true mass density, or mass per unit *volume*, the thickness of the string would also enter.) As we discussed in Chapter 1, this analysis does not permit a calculation of constant factors such as 2 or π, but it will give the system's dependence on the bulk properties, and it will give a correct estimate of orders of magnitude.

The Speed of Sound in Solids

To illustrate the dimensional approach, let's estimate a longitudinal wave speed in solids. To do so, we need to know that the *Young's modulus*, Y, of a solid determines the (small) stretch of the solid under the influence of an external force (see Chapter 21). In particular, Young's modulus describes the *fractional* longitudinal stretch, $\Delta L/L$, of a given material of length L when a force per unit area, F/A, is applied to its end: $(\Delta L/L)Y = F/A$. Young's modulus therefore has dimensions of force per area.

Let's now suppose that one end of a metal bar is vibrated rhythmically. Waves will form and travel along the bar. As with the string, we expect that the speed of these waves is independent of the exact dimensions of the bar. Then only two mechanical properties of the bar can determine the wave speed: Young's modulus and the mass density (mass per unit volume), ρ. The first factor is a measure of restoring force and the second is a mass factor. In addition, the ratio Y/ρ has dimensions of force/area divided by mass/volume, or $[(MLT^{-2}L^{-2})/(ML^{-3})] = [L^2T^{-2}]$. The dimensions are indeed those of speed squared. We have thus estimated the speed of mechanical waves in a metal bar to be

$$v = (\text{a constant})\sqrt{\frac{Y}{\rho}}. \tag{14–29}$$

(We expect the constant to be on the order of 1.) The waves whose speed we have estimated here are longitudinal waves because Young's modulus measures the resistance of the bar to longitudinal compression or longitudinal stretching. These waves are known as **sound waves**.

A more careful analysis, based on the wave equation, shows that the simple form $v = \sqrt{Y/\rho}$ is *exact* (the constant = 1); this result is confirmed by experiment. Young's moduli are given in Table 21–1 for several solids. For silver, for example, whose density is roughly 10^4 kg/m^3 and whose Young's modulus is 7.5×10^{10} N/m^2, the quantity $\sqrt{Y/\rho} = 2700$ m/s, which is a very good approximation of the measured value of the speed of sound in silver.

INTERIM SUMMARY

Mechanical waves are organized motions within elastic media. Standing waves are waves for which all the motion within the medium has the same time dependence. For sinusoidal standing waves, the displacement of any point in the medium has the form

$$z(x, t) = z_0 \sin(kx + \delta)\cos(\omega t + \phi). \tag{14–4}$$

The important parameters of this wave are the amplitude z_0, the angular frequency, ω, and the wave number, k. The wavelength, λ (the length over which the wave repeats itself) is related to the wave number by $\lambda = 2\pi/k$. We also use the frequency, f, and the period, τ. These parameters are related by $\lambda f = \lambda/\tau = \omega/k$. Possible values of the wavelength are restricted by the boundary conditions of the medium.

Traveling waves are characterized by a series of internal motions that propagate progressively with some definite speed through the medium. These waves can travel over large distances even though the material that makes up the medium moves in only a limited way. The wavelength, wave number, frequency, and period are also useful parameters for traveling waves. In addition, traveling waves move at a definite speed v, which is related to the parameters already introduced according to $v = \lambda f = \lambda/\tau = \omega/k$. When the particles of the medium move parallel to the direction of wave propagation, longitudinal waves form; when the particles move perpendicular to the wave direction, transverse waves form. Standing waves can also be classified into longitudinal and transverse types.

The angular frequency of waves is determined by the dynamical wave equation, obtained by the application of Newton's second law to an elastic medium:

$$\frac{\partial^2 z}{\partial x^2} = \frac{1}{v^2}\frac{\partial^2 z}{\partial t^2}. \tag{14–27}$$

This equation applies to all types of mechanical waves. The wave speed v is a property of the medium:

$$v = \sqrt{\frac{\text{restoring force factor}}{\text{mass factor}}}. \tag{14–28}$$

In the case of a string of mass density μ under tension T, $v = \sqrt{T/\mu}$.

14–5 ENERGY AND POWER IN WAVES

In waves, masses are in motion, so energy must be present (Fig. 14–14). In traveling waves, energy can be delivered from one end of the system to the other even though no mass is transferred in the motion. It is an important observation that *energy can be transmitted without the transport of mass.*

Traveling waves transmit energy.

FIGURE 14–14 That waves carry energy is a fact well known to anyone who has spent time at the seashore.

The Power Delivered by Waves

Let's think about right-moving transverse traveling waves on a uniform string that has a mass per unit length μ and tension T. The power P delivered at a point x along the string is given by $P = \mathbf{F}\cdot\mathbf{v}_{\text{transverse}}$, where $\mathbf{v}_{\text{transverse}}$ is the (transverse) velocity of the string at that point and $\mathbf{F}$ is the force acting on the portion of the string to the right of x. Only the transverse component of the force enters into the scalar product. To find the transverse component of the force, refer to Fig. 14–15, which shows that the transverse component of force at x acts downward (negative sign) with magnitude

$$T\sin\theta \simeq T\tan\theta = T\frac{\partial z}{\partial x}.$$

We have used the small distortion approximation here—the angle θ is small—and we have remembered to use the partial-derivative notation again, as a reminder that z depends on two variables, x and t. The velocity of the string at x is given by

$$v_{\text{transverse}} = \frac{\partial z}{\partial t}. \tag{14–30}$$

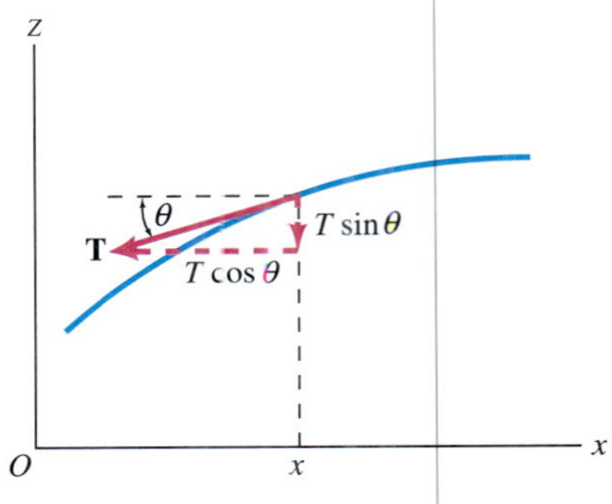

FIGURE 14–15 A string that is distorted is subject to a tension force from the left whose direction depends on the string distortion. We are interested in this case in the vertical component.

Thus the power delivered to the string at point x is

$$P = \mathbf{F}\cdot\mathbf{v}_{\text{transverse}} = -T\frac{\partial z}{\partial x}\frac{\partial z}{\partial t}. \tag{14–31}$$

It is simple to calculate the derivatives involved here for a harmonic traveling wave, $z = z_0\sin(kx - \omega t)$. We find

$$P = Tk\omega z_0^2\cos^2(kx - \omega t) = \mu v\omega^2 z_0^2\cos^2(kx - \omega t). \tag{14–32}$$

To find the last form, we have used Eq. (14–26), which gives $T = \mu v^2 = \mu(\omega/k)v$, where v is the wave velocity. (Don't confuse v with the transverse velocity of a portion of the string.)

It is often useful to think about the time average of the power delivered. Let us denote this average with angular brackets. To find the time average, we need the average value taken by the cosine squared over a long time period, namely 1/2. Thus

$$\langle P\rangle = \frac{1}{2}\mu\omega^2 z_0^2 v. \tag{14–33}$$

The average power delivered by a harmonic wave

Energy Transport

There is no mechanism for energy loss in a string such as we have described. Thus the power supplied by whatever generates the waves in the first place is transmitted from one piece of the string to another. This power is thus the rate at which energy is carried by the system, or the rate of energy transport. We can also think of the vibrating string as having an energy density u—an energy per unit length. This energy density must also take the form of a traveling wave. How is the energy density u related to the rate P at which energy is delivered? Suppose we knew the energy density. Then note that the quantity $v\,\Delta t$ is the length of wave that moves past a given point in a brief time interval Δt. All the energy in that length, namely the energy density times the length, or $uv\,\Delta t$, will be delivered; the rate of energy delivery P is then found by dividing this quantity by Δt. In other words,

A traveling harmonic wave transports energy at the wave speed.

The power transmitted by a wave is proportional to the squares of both the amplitude and the frequency.

$$u(x, t) = P/v = \mu\omega^2 z_0^2\cos^2(kx - \omega t). \tag{14–34}$$

We have shown that the power delivered and the energy density on the string are each traveling waves moving with speed $v = \omega/k = \sqrt{(T/\mu)}$. These quantities are always positive. Moreover, *they are proportional to the squares of both the amplitude and the frequency*. Although we have derived our results for traveling waves on a string, the quadratic dependence of the rate of energy transport, of power, or of energy density on amplitude and frequency is a general feature of *all* waves.

EXAMPLE 14–5 A long rope of mass density 150 g/m is tied at one end. A person holds the other end with a horizontal force of 95 N so that the equilibrium position of the rope is horizontal. The person then shakes the held end of the rope up and down (Fig. 14–16) such that sinusoidal traveling waves are generated with a frequency of 1.0 s^{-1} and an amplitude of 5.0 cm. What power must be delivered by the hand that is shaking the rope?

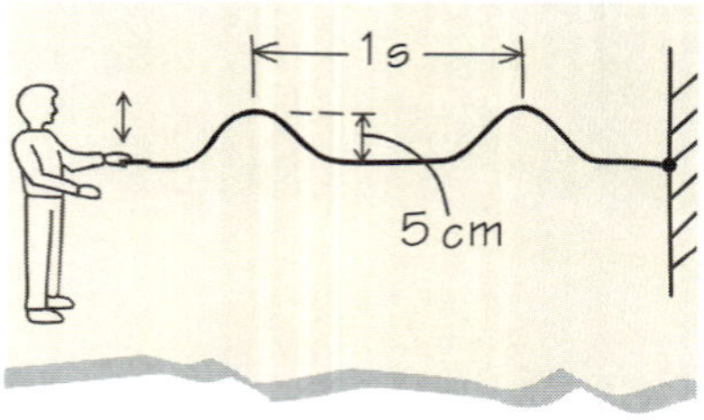

FIGURE 14–16 Example 14–5.

Solution: The power delivered to the rope is the power carried by the rope in its wave motion. Thus Eq. (14–33) can be used to calculate the power. We are given the tension $T = 95$ N, the mass density $\mu = 0.15$ kg/m, the frequency $f = 1.0$ Hz, and the amplitude $z_0 = 0.050$ m. To apply Eq. (14–33), we need $\omega = 2\pi f = 6.28$ rad/s, and $v = \sqrt{T/\mu} = \sqrt{(95\text{ N})/(0.15\text{ kg/m})} = 25.2$ m/s. Then Eq. (14–33) gives

$$P = \frac{1}{2}(0.15\text{ kg/m})(6.28\text{ s}^{-1})^2(0.050\text{ m})^2(25.2\text{ m/s}) = 0.19\text{ W}.$$

This is a small fraction of 1 hp and is easily delivered by the hand holding the rope.

The energy density of standing waves has the same dependence on amplitude and frequency as the total energy density of traveling waves. However, no power is delivered by standing waves because such waves do not propagate through space.

Consequences of Energy Conservation

We can use our results for the energy density, together with the conservation of energy, to determine how the amplitude (wave height) of a periodic wave originating at a point varies with distance from that point. For example, a toe periodically dipped into a pond is the source of the energy contained in the ripples that spread outward in circles from the toe. If a given ripple has radius r, the energy in the ripple is spread evenly over the circle, which has length $2\pi r$. Thus the total energy density falls off as $1/r$; further, the amplitude falls off as $1/\sqrt{r}$ because the energy density is proportional to the square of the wave's amplitude. Similarly, a horn sounding a note in open air is the source of a wave that spreads outward from the source as a sphere. Sound energy is spread over the spherical surface of area $4\pi r^2$ when a particular part of the wave has spread to a radius r. Thus the energy density at a point decreases as $1/r^2$, and the amplitude of the sound waves from the horn decreases as $1/r$.

14–6 SOUND

In Section 14–4, we referred to the compressional waves traveling down a metal bar as *sound*. Another type of sound wave is generated in air by the mechanical motion of some object in contact with the air—the vibration of vocal cords, an oscillating violin string, or the collapsing hot air column formed by a lightning stroke. A traveling sound wave transports energy, which can then be picked up by another mechanical system such as an eardrum.

The Nature of Sound Waves

Let's look at sound in air on a microscopic level. The molecules in air are widely spaced so the intermolecular forces that act as springs in solid or liquid bodies play a very limited role here. Except for brief, random collisions with each other, air molecules move freely. How, then, can a wave propagate? Consider a pipe oriented along the x-axis as in Fig. 14–17. At one end of the pipe is a speaker. When the speaker diaphragm moves to the right, the air molecules in a thin slice to the right of the diaphragm within the pipe are subject to a force that acts along the $+x$-direction. These air molecules acquire a net momentum in the $+x$-direction in addition to their random motion. Some of the molecules in the original slice will then move to an adjacent region, increasing the number of molecules there and creating a region of *compression*, where the molecular density is higher. When the diaphragm moves back to the left, fewer molecules remain in this region, which corresponds to a region of *rarefaction* where the molecular density is lower. We thus have adjacent regions of compression and rarefaction, as in Fig. 14–17. In the region of compression, there are more molecules and therefore more molecular collisions. The air molecules that recoil from this region of more collisions will fill in the rarefied region. The molecules to which the net momentum had been transferred will move in the $+x$-direction, so the region of compression moves to the right in the form of a pulse. If the diaphragm undergoes periodic motion, regions of compression and rarefaction are propagated in the $+x$-direction as a periodic wave in air. Figure 14–17 shows the molecular pattern of a sound wave in air. The displacement of the molecules is aligned with the propagation direction of the wave. Sound waves in air are therefore *longitudinal* waves.

Sound waves in air are longitudinal waves.

Note that there is no net movement of molecules in the direction of the wave. It is the alteration of the density from the average (or equilibrium) density that propagates. If ρ is the density of air (or any other gas), then sound waves may be described in terms of $\Delta\rho = \rho - \rho_0$, the deviation from the equilibrium density ρ_0. A harmonic sound wave traveling in the $+x$-direction would be described by

$$\Delta\rho = A_\rho \sin(kx - \omega t). \tag{14–35}$$

where A_ρ is the maximum deviation of the density from ρ_0. Generally, $A_\rho << \rho_0$, so we may speak of a nearly constant mass density, even in the presence of sound waves.

The Speed of Sound

We can use the techniques of Section 14–4—in particular Eq. (14–28)—to estimate the speed of sound waves in a gas (such as air). We must find a restoring force factor and a mass factor for the gas; these factors should not depend on the total amount of gas because the speed is not dependent on the total amount of gas. The mass factor is

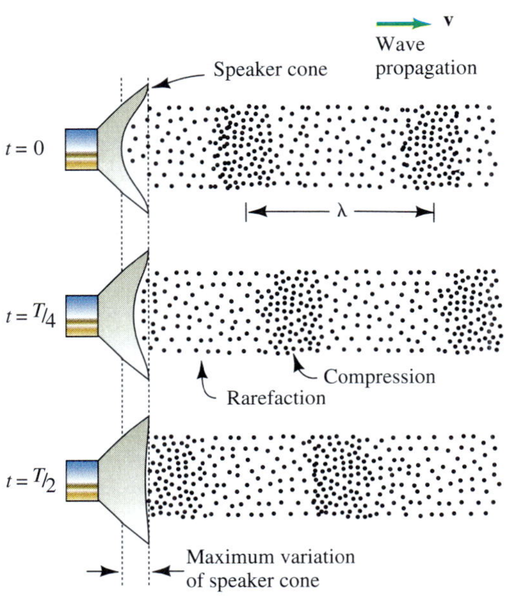

FIGURE 14–17 When a piston—here, a speaker diaphragm—at one end of a pipe filled with air moves along the $+x$-direction, there is a force on the air molecules to the right of the piston. If the piston moves back and forth, the result is alternate regions of higher and lower density propagating along the pipe.

easy: The only quantity that fits is the mass density ρ_0. The restoring force factor is the pressure P_0 (see Chapter 16). If we were to insert a tiny membrane into the gas and observe one side of the membrane, we would see it being bombarded by a stream of molecules. When these molecules collide elastically with the membrane, they transfer momentum to it. A transfer of momentum is equivalent to a force. Therefore, we may characterize the gas by a force per unit area, which is the **pressure**. Pressure has dimensions of force per area, or $[MLT^{-2}][L^{-2}] = [ML^{-1}T^{-2}]$, whereas density has dimensions of mass per volume, or $[ML^{-3}]$. The ratio P_0/ρ_0 has dimensions $[L^2T^{-2}]$, a speed squared. We thus estimate for the speed of sound in a gas

$$v_{\text{sound}} \simeq \sqrt{\frac{P_0}{\rho_0}}. \tag{14–36}$$

This result was first found by Newton, and it is appropriately called Newton's formula.

A more detailed calculation gives a better value for the speed of sound,

The speed of sound

$$v_{\text{sound}} = \sqrt{\frac{\gamma P_0}{\rho_0}}. \tag{14–37}$$

Here γ is a constant with the approximate value 1.4 for air. The numerical correction to our estimate due to the factor γ is small. For normal atmospheric conditions at sea level, $P_0 \simeq 10^5\ \text{N/m}^2$, $\rho_0 \simeq 1.3\ \text{kg/m}^3$,

$$v_{\text{sound}} \simeq 330\ \text{m/s} = 740\ \text{mi/h}. \tag{14–38}$$

In air, sound travels 1 mi in about 5 s. Sound waves move at different speeds in different materials. Table 14–1 gives the speed of longitudinal sound waves in various media.

TABLE 14–1
The Speed of Sound Waves in Various Media†

Medium	*Speed of Sound (m/s)*
Hydrogen	1284
Air	330
Liquid mercury (20°C)	1450
Methyl alcohol	1189
Water	1402
Polyethylene	920
Lead	1210
Silver	2700
Aluminum	5000
Beryllium	12,870

†Longitudinal waves. Temperature = 0°C, pressure = 1 atm unless otherwise stated.

Standing Sound Waves

Standing sound waves are generated if the necessary boundary conditions exist. A pipe much narrower than the wavelength of standing waves set up in it provides a suitable one-dimensional environment. If the pipe is closed at an end, then there can be no displacement of gas molecules across the end. The closed end of the pipe is then like the fixed end of a string. If both ends are closed, we have the equivalent of a string fixed at both ends. As Fig. 14–18 shows, the wavelength for these standing waves takes on only the discrete values allowed by Eqs. (14–7) and (14–8). The frequencies are determined in terms of the wavelengths by Eq. (14–22),

$$v_{\text{sound}} = \lambda f. \tag{14–39}$$

As for the vibrating string, the discrete integer n labels the modes [Eqs. (14–16)]. When $n = 1$, we have the fundamental frequency (discussed in Section 14–2); when $n > 1$, we speak of the nth harmonic.

EXAMPLE 14–6 A pipe closed at both ends has length $L = 4.0$ m. Compute the fundamental frequency for standing sound waves as well as the frequencies of the next two modes.

Solution: As Eq. (14–8) states, the fundamental wavelength, λ_1, for a system with two closed ends is twice the length of the system: $\lambda_1 = 2L$. Figure 14–18 shows how the standing wave fits within the pipe with a node at each end. Higher modes have shorter wavelengths, enumerated in this case by $n = 2, 3$: $\lambda_2 = \lambda_1/2$, $\lambda_3 = \lambda_1/3$. The relation between frequency and wavelength is given by Eq. (14–39). Thus the frequencies in question are determined by

$$f_n = \frac{v_{\text{sound}}}{\lambda_n} = \frac{v_{\text{sound}} n}{2L}.$$

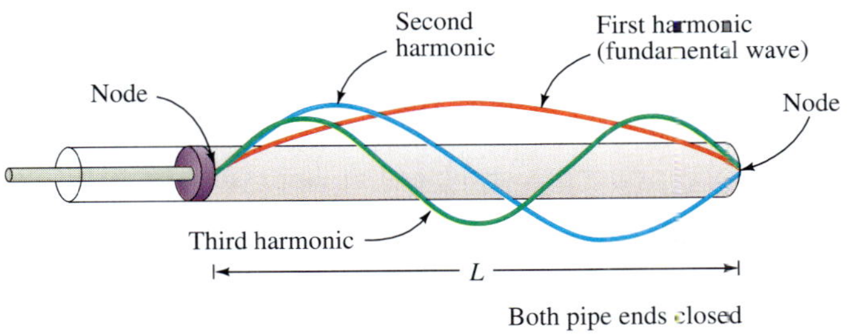

FIGURE 14–18 Example 14–6. There must be a node of molecular displacement at each end of a closed pipe. Thus one-half of the standing sound wave with the longest possible wavelength fits in the pipe. The longest wavelength represents the fundamental frequency, which corresponds to the first harmonic. The next two harmonics are also shown. The curves represent the amplitudes (not to scale) of the density wave.

Taking v_{sound} from Eq. (14–38), we find

$$f_1 = \frac{330 \text{ m/s}}{2(4.0 \text{ m})} = 41 \text{ Hz},$$

$$f_2 = 2f_1 = 83 \text{ Hz},$$

$$f_3 = 3f_1 = 124 \text{ Hz}.$$

Figure 14–18 also shows these second and third harmonics.

Other boundary conditions are possible for standing sound waves. When one end of a pipe is open, the pressure at that end takes on the value of the pressure of the exterior. There is no variation of the pressure there, which means that there is a pressure node. This can be shown to mean that the displacement has an amplitude maximum (or *antinode*) rather than a node at the open end (Fig. 14–19).

EXAMPLE 14–7 An organ pipe with the dimensions of the pipe in Example 14–6 is closed at one end and open at the other. Find the fundamental frequency and next harmonic in this case.

Solution: When the organ pipe is closed at one end, then the longest possible standing wave has a wavelength four times the length of the pipe, as shown in Fig. 14–19. This is because the wave at the open end is a maximum, or an antinode. Thus $\lambda_1 = 4L$, and the fundamental frequency is

$$f_1 = \frac{v_{\text{sound}}}{\lambda_1} = \frac{330 \text{ m/s}}{4(4.0 \text{ m})} = 21 \text{ Hz}.$$

This is just half of the fundamental frequency when both ends are closed.

The next harmonic is not the harmonic with a frequency twice the fundamental frequency; such a wave would have a node at the open end. Rather, it is the harmonic with three times the fundamental frequency, also illustrated in Fig. 14–19. For this wave, three-quarters of one wavelength fits into the pipe.

Figure 14–20 shows the first few harmonics for an organ pipe with both ends open. The condition that there is an antinode at each end produces a different harmonic series.

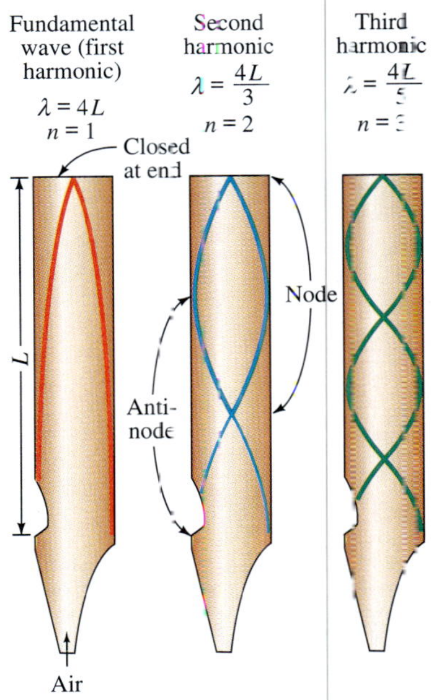

FIGURE 14–19 The amplitude of a standing sound wave in an organ pipe open at one end has a maximum at the open end. In this situation, one-fourth of the standing wave with the longest possible wavelength fits within the pipe. The next harmonic is also shown. The curves represent the amplitudes (not to scale) of the density wave.

Excitation of Standing Waves. We can excite standing waves in an organ pipe if we vibrate the air within at the standing wave frequency; for example, with externally generated sound waves of that frequency. In addition, these standing waves must be able to generate traveling sound waves of the same frequency if the distant members of an audience are to perceive the standing waves within the organ pipe as sound. The couplings between these different types of waves normally involve rather complicated mechanisms. The many different ways that standing waves in musical instruments are excited (bow, single and double reed, mouthpiece, and so forth) indicates the rich variety of physical phenomena involved.

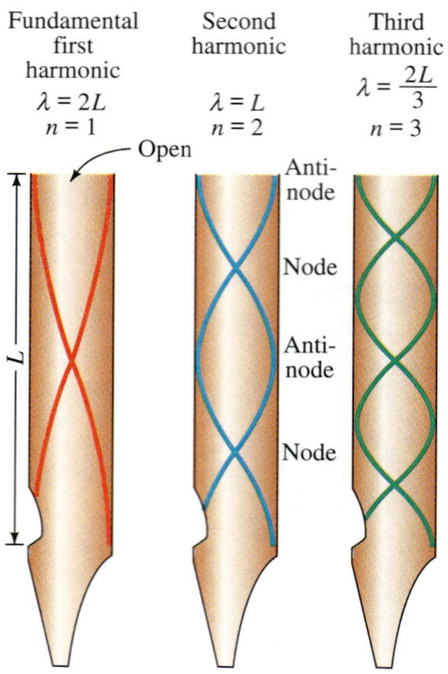

FIGURE 14–20 An organ pipe open at each end has standing waves with antinodes at each end, hence a different harmonic series than the organ pipe closed at one end.

PROBLEM SOLVING

Exciting sound waves

The intensity is the average power per unit area.

FIGURE 14–21 The goblet breaks because a resonance phenomenon acts that allows energy to be transferred easily from sound waves in air to the glass. The sound frequency corresponds to a natural frequency of the goblet.

In each case, the excitation of waves involves a transfer of energy. This transfer of energy, from the energy of an external source—such as the air forced across a clarinet mouthpiece—to the energy of the standing wave within the tube of a clarinet, takes place in ways closely related to the *resonance* phenomenon discussed in Chapter 13. Thus, for example, if a tuning fork is placed near the end of an organ pipe and if the frequency of the tuning fork is the same as the fundamental frequency of the pipe, strong standing waves will rapidly be established within the pipe. Perhaps you have seen a resonance phenomenon act in the opposite direction: When a trumpet plays a note whose frequency matches the natural frequency of a delicate goblet, standing waves can be induced in the goblet strong enough to make it break (Fig. 14–21).

Hearing Sounds

The combination of ear and brain acts as a very sensitive instrument with many interesting properties. For present purposes, it suffices to note that the combination can detect pitch and loudness. *Pitch* is a measure of frequency. The human ear can hear frequencies ranging from about 20 Hz to 20,000 Hz. *Loudness* is a measure of the power carried in a wave.

Because sound waves in air form two-dimensional fronts, it is appropriate to define the average power per unit area rather than the total power in the wave. This average power per unit area is called the **intensity**, I. The intensity measures energy per unit time per unit area, and has SI units of watts per meters squared (W/m^2). Intensity, like power, is proportional to the square of both the amplitude and the frequency. Intensity is not a particularly convenient measure physiologically; a measure better suited to the human ear is the *decibel scale,* in which the measure of intensity is a dimensionless quantity, β, with units of decibels (dB). It is defined by

$$\beta \equiv 10 \log_{10}\left(\frac{I}{I_0}\right), \tag{14–40}$$

where

$$I_0 = 10^{-12}\ W/m^2.$$

A logarithmic scale such as this is designed to cover a large range. The smallest detectable intensity is I_0 itself, for which $\beta = 0$, and the largest intensity still perceived as sound is about $I = 1\ W/m^2$, or $\beta = 120$ dB. Higher intensities are just painful. The range of intensities detectable as sound by the ear–brain combination is thus an impressively large factor of 10^{12} (Fig. 14–22). Curiously enough, the range of sound intensities detectable by the human ear is about the same as the range of light intensities detectable by the human eye.

14–7 THE DOPPLER EFFECT

Traveling waves move with a finite velocity. It is therefore possible for an observer who measures the waves (the receiver) to move relative to them or for the generator of traveling waves (the source) to move relative to these waves. The movement of the source or of the receiver affects the measured frequencies. An extreme case that illustrates this is a receiver who moves along with the wave crests. The receiver measures a wavelength, just as if he or she were stationary with respect to the medium, by extending a yardstick from one crest to another. But the receiver measures no frequency at all, for the frequency measures the rate at which crests pass the observer.

The perception of a sudden change in the pitch of an automobile horn (a source) as a vehicle passes is a common experience. Related effects occur when the listener (a receiver) moves. In 1842, Christian Doppler gave an explanation for this phenomenon,

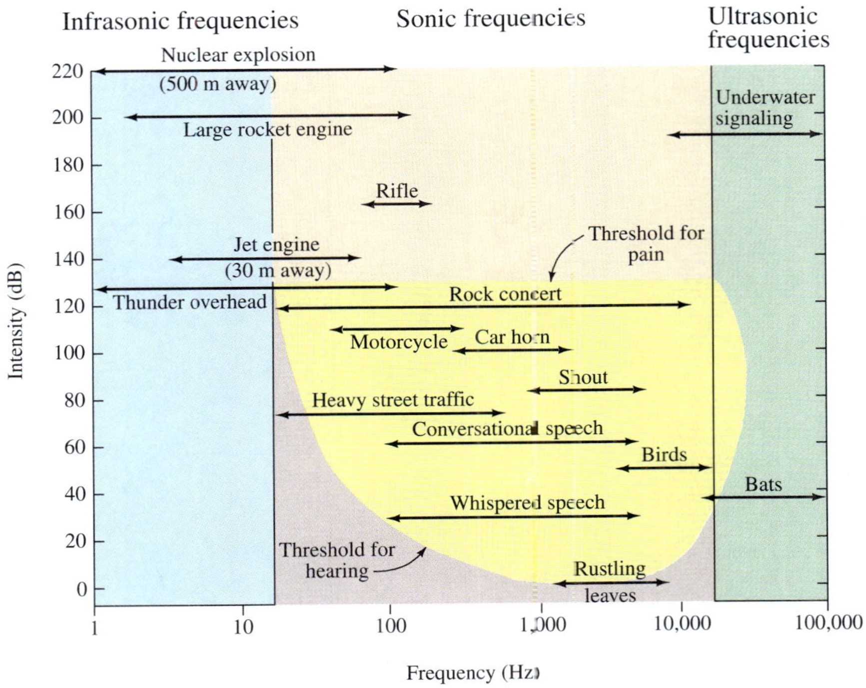

FIGURE 14–22 Typical noise levels at the source: frequency versus β, the intensity measure on the decibel scale. In acoustics—the study of sound—frequency ranges of sound are categorized as infrasonic (low frequency), sonic (or audible, medium frequency), or ultrasonic (high frequency).

as well as a corresponding phenomenon for light. The **Doppler effect** is the shift in frequency and wavelength of waves that results from a source moving with respect to the medium, receiver moving with respect to the medium, or even a moving medium. We'll treat the various possibilities in order.

Case A: A Moving Source. Consider a point source of traveling waves, such as a siren wailing at some frequency f_0 (period $\tau_0 = 1/f_0$). These waves spread away from the source at a wave speed v that is characteristic of the medium alone and is unaffected by the source's motion. If the source is stationary with respect to the medium, these waves are symmetric on all sides of the source, and the wavelength is $\lambda = v/f_0$ [from Eq. (14–22)]. But if the source moves with velocity $\mathbf{v}_s$ with respect to the medium, then the wave fronts will be squeezed together in the direction of $\mathbf{v}_s$ and spread apart in the opposite direction (Figs. 14–23a and b). This is because larger circles correspond to emission at some earlier time, when the source was farther back in the direction $-\mathbf{v}_s$. Thus the wavelengths are shorter along the direction $+\mathbf{v}_s$ and longer in the opposite direction.

For simplicity, consider the source to be moving either directly toward or directly away from the observer (Fig. 14–23c). If the source is moving toward the observer, then during one period τ_0 it moves a distance $v_s\tau_0 = v_s/f_0$. The wavelength is then decreased by this amount:

$$\lambda' = \lambda - \frac{v_s}{f_0} = \frac{v - v_s}{f_0}. \tag{14–41}$$

Because the wave speed remains unaffected, the frequency with which the observer receives the waves becomes

$$f' = \frac{v}{\lambda'} = f_0\left(\frac{v}{v - v_s}\right)$$
$$= \frac{f_0}{1 - (v_s/v)}. \tag{14–42}$$

When the source moves toward the observer, the wavelength *decreases*, and the frequency *increases*—the pitch will be higher in the case of sound. To find the result when the source moves away from the observer, simply reverse the sign of v_s:

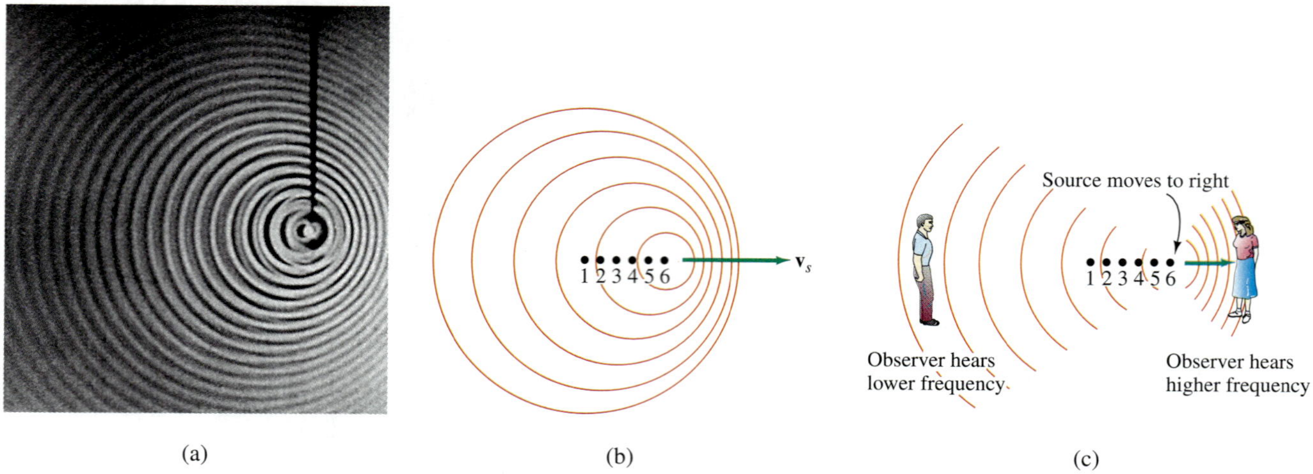

FIGURE 14–23 (a) The Doppler effect as seen in the waves of a ripple tank. (b) A moving source of waves. The dots along the direction of the velocity represent the position of the source when the wave centered on them was emitted. The dot at the farthest left corresponds with the outermost wave, and so forth. (c) Here we show two observers at rest, on either side of the source.

$$\lambda' = \lambda + \frac{v_s}{f_0} = \frac{v + v_s}{f_0}; \tag{14–43}$$

$$f' = \frac{f_0}{1 + (v_s/v)}. \tag{14–44}$$

When the source moves away from the observer, the wavelength *increases*, and the frequency *decreases*—for sound, the pitch drops. A manifestation of this phenomenon is familiar to anyone who has heard a siren drop to a lower pitch as the vehicle with the siren approaches, then passes.

Case B: A Moving Observer. Figure 14–24 depicts an observer moving with velocity $\mathbf{v}_r$ with respect to a wave-carrying medium, within which a stationary source is exciting waves with frequency f_0. We have illustrated the case of an observer moving directly *toward* the source. Such an observer will measure exactly the same wavelength λ as if he or she were at rest but the observer sees the wave crests pass by more quickly. The observer measures a *modified wave speed*

$$v' = v + |v_r|. \tag{14–45}$$

The wavelength and wave speed provide the modified wave frequency that the moving observer measures:

$$f' = \frac{v'}{\lambda} = \frac{v + |v_r|}{\lambda} = f_0 + \frac{|v_r|}{\lambda} = f_0\left(1 + \frac{|v_r|}{v}\right). \tag{14–46}$$

When an observer moves *toward* the source of traveling waves, the wavelength is unchanged, the wave speed *increases*, and the frequency *increases*. We reverse the sign of v_r and repeat the derivation to learn that when an observer moves *away from* a wave source, the wavelength is again unchanged and the wave speed and the frequency both *decrease*:

$$v' = v - |v_r|; \tag{14–47}$$

$$f' = \frac{v - |v_r|}{\lambda} = f_0 - \frac{|v_r|}{\lambda} = f_0\left(1 - \frac{|v_r|}{v}\right). \tag{14–48}$$

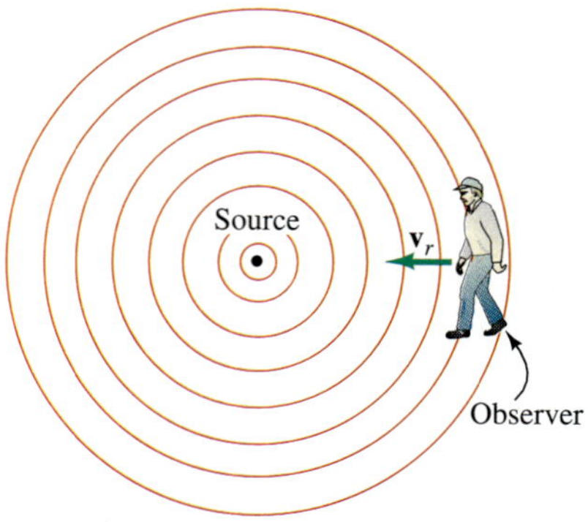

FIGURE 14–24 An observer moving with velocity $\mathbf{v}_r$ receives waves from a fixed source. In this situation, the observer hears an increased frequency.

EXAMPLE 14–8 While moving along a straight road, you first approach, and then pass, a stationary car just at the side of the road whose one-note horn is stuck on. Having perfect pitch, you recognize that the frequency you hear approaching the car is 853 Hz, while you hear a note of 741 Hz receding from the car. Use the value 330 m/s for the speed of sound in air to find the frequency the unfortunate occupant of the car hears and your own speed.

Solution: As you move toward the car, the frequency you hear, f_+, is increased over the at-rest frequency f_0 according to Eq. (14–46), $f_+ = f_0\left(1 + \dfrac{v_r}{v}\right)$, where v_r is your speed and v is the speed of sound. Similarly, as you move away, you hear a decreased frequency $f_- = f_0\left(1 - \dfrac{v_r}{v}\right)$. These are two equations for the two unknowns f_0 and v_r. We solve them by taking their sum and difference:

$$f_+ + f_- = 2f_0$$

$$f_+ - f_- = 2f_0(v_r/v) = (f_+ + f_-)(v_r/v).$$

The first equation immediately gives us f_0 as the average of the two frequencies,

$$f_0 = \frac{1}{2}(f_+ + f_-) = \frac{1}{2}(853\text{ Hz} + 741\text{ Hz}) = 797\text{ Hz}.$$

This is the frequency the car's occupant hears. The second equation gives us v_r:

$$v_r = v\frac{f_+ - f_-}{f_+ + f_-} = (330\text{ m/s})\frac{853\text{ Hz} - 741\text{ Hz}}{853\text{ Hz} + 741\text{ Hz}} = 23.2\text{ m/s}.$$

Case C: A Moving Source and a Moving Observer. If all motion is kept to the same line, which we label the x-axis, then we can simply combine the results of cases A and B. *Any velocity, including that of the wave itself, is positive if it is to the right and negative if it is to the left.* With this reminder about the sign of the velocity, we shall drop the absolute-value signs. These velocities refer to motion with respect to the medium. The effect of the moving source is to change the wavelength but not the wave speed, and the effect of the moving observer is to change the wave speed but not the wavelength. This can be expressed as

$$\lambda' = \frac{v - v_s}{f_0}; \qquad (14\text{–}49)$$

$$v' = v - v_r.$$

The modified frequency f' is then

$$f' = \frac{v'}{\lambda'} = \left(\frac{v - v_r}{v - v_s}\right)f_0. \qquad (14\text{–}50)$$

The Doppler effect for a moving source and a moving receiver

Source and observer approach one another when v_s and v have the same sign and v_r has the opposite sign. In this case, f' is *increased* over f_0. When the source and observer are moving away from each other, v_r and v have the same sign and v_s has the opposite sign. The perceived frequency is *decreased.* Equations (14–49) and (14–50) include cases A and B as special cases. Note that Eq. (14–50) is *not* symmetric between the source and the observer. If we know the relative speed, we can tell what is moving—the source or the observer—by measuring the at-rest and in-motion frequencies.

When the speeds of the observer and the source are *small* compared to v, as is often the case when the Doppler effect applies to sound waves, then Eq. (14–50) can be shown to be approximately

$$f' \simeq \left(1 + \frac{v_s - v_r}{v}\right)f_0. \qquad (14\text{–}51)$$

In this approximation, the Doppler shift (the change in frequency) depends only on the relative velocity between source and observer, and it is no longer possible to say who is moving—the source or the observer.

Case D: A Moving Medium. We can consider a fourth possibility: that the medium is itself in motion with respect to some fixed reference frame. Waves generated in a stream provide an example. The source is fixed with respect to the bank, and the stream is moving with speed v_m (Fig. 14–25). Let's take a look at how an observer on the bank downstream from the source sees the waves. There are two effects: First, the observer sees the crests moving by at an increased speed $v' = v + v_m$. Second, the wavelength is also increased because a downstream crest moves an additional distance before a second crest is emitted. From Fig. 14–25, it is easy to deduce that the wavelength increases from λ to $\lambda' = \lambda[1 + (v_m/v)]$. The frequency measured by the observer is $f' = v'/\lambda' = (v + v_m)/\lambda[1 + (v_m/v)] = (v/\lambda)[(v + v_m)/(v + v_m)] = v/\lambda = f$.

Similarly, an observer on the bank upstream from the source will see both the wave speed and the wavelength decreased, and in the same ratio, $[1 + (v_m/v)]$, so the frequency $f' = v'/\lambda'$ is unchanged. In other words, because both the speed and the wavelength are changed in the same way by the movement of the medium, *the frequency is unchanged by the movement of the medium.* The fact that the frequency is unchanged is the reason why it is possible to give outdoor concerts even under windy conditions. The intensity of the sound may be decreased if the wind is blowing in the wrong direction, but at least the pitch remains true.

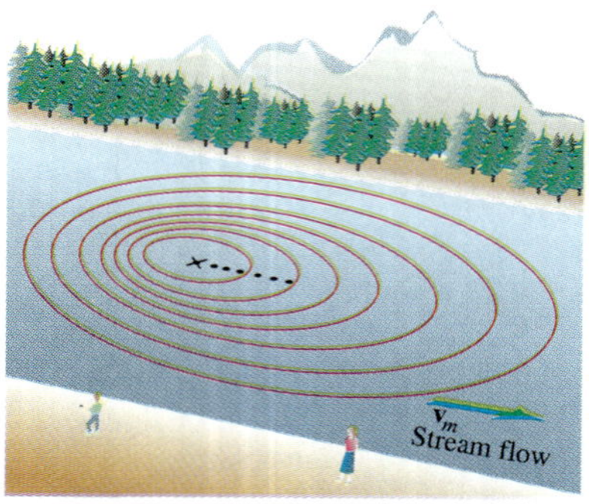

FIGURE 14–25 Traveling waves generated at a fixed point, marked by a cross, in a moving medium. The centers of the circles move downstream.

*14–8 SHOCK WAVES

When a source of traveling waves moves with a speed exceeding the wave speed, we encounter a new situation in which *shock waves* can form. This can be seen from Eqs. (14–42) or (14–50), which show that when $v_s = v$, the observed frequency becomes infinite. When $v_s = v$, the waves emitted directly in front of the source pile up on top of one another, so the wavelength is zero. Thus the time separation between the waves is zero, which corresponds to an infinite frequency. When v_s exceeds v, the source outruns the motion of the waves themselves; thus, instead of the squeezed wave fronts that occur for $v_s < v$ (see Fig. 14–23), the wave fronts are actually left behind.

The geometry of this situation can be analyzed as follows. Suppose that the wave source is at point x_0 at $t = 0$ and emits a crest (Fig. 14–26). In a time t_2, the source has traveled to x_2, a distance $v_s t_2$ from x_0. In this same time, the crest has moved out to a circular wave front a distance vt_2 from x_0. The perpendicular to the line drawn from x_2 tangent to this circle makes an angle θ_0 with respect to the line perpendicular to the motion of the source, and

$$\cos\theta_0 = \frac{vt_2}{v_s t_2} = \frac{v}{v_s}. \tag{14–52}$$

What is special about this line is that *all* the wave fronts emitted as the source moves

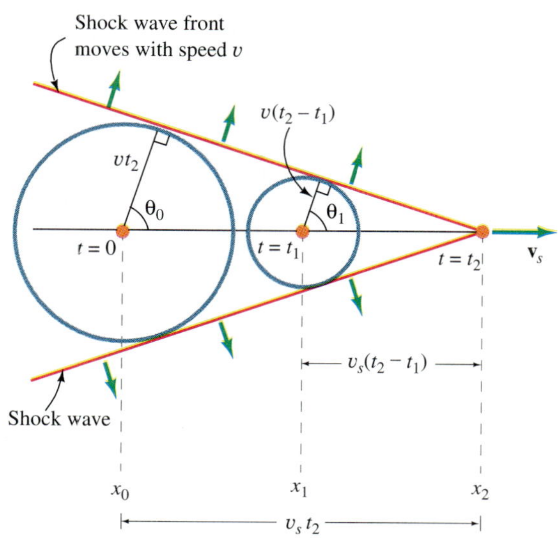

FIGURE 14–26 When v, the wave speed, is less than v_s, the speed of the source, a linear wave front forms along the red lines. This front is called a shock wave. The source continuously emits disturbances that generate circular waves. At $t = 0$, the source is at the center of the larger circle. At $t = t_1$, the source is at the center of the smaller circle, and at $t = t_2$, the source is about to emit another wave front. Successive circles form the continuous line of the shock wave.

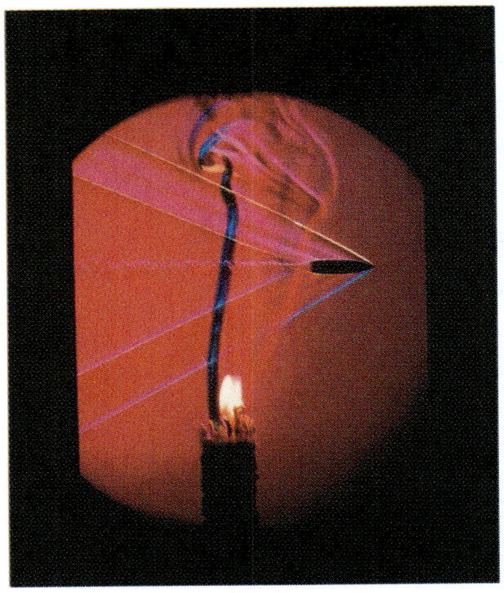

FIGURE 14–27 A bullet passing through the hot gases above a candle flame produces a shock wave because its speed is faster than the speed of sound in air. The ratio of these speeds determines the angle of the shock wave.

FIGURE 14–28 The blue glow is Cerenkov radiation that is produced when a charged particle moves through the water at a speed greater than the speed of light in water. In this case, the particles are the result of radioactivity in the reactor elements suspended in the water.

lie along this tangent. For instance, consider the crest emitted at t_1, when the source is at x_1, a distance $v_s t_1$ from x_0. The geometry of the figure gives

$$\cos\theta_1 = \frac{v(t_2 - t_1)}{v_s(t_2 - t_1)} = \frac{v}{v_s}.$$

Thus $\theta_1 = \theta_0$ and the tangent is formed by all the waves emitted during the voyage of the source. This line—a linear wave front called a **shock wave**—moves out with a speed v, maintaining the angle given by Eq. (14–52). When the wave is propagating in a three-dimensional medium, such as air, a cone is formed instead. It must be stressed that *there is only one such wave front.* A cone whose apex is at any location on the trajectory of the source represents just an earlier or later location of the cone that appears in Fig. 14–26. The wave front will pass an observer located on the ground just once (Fig. 14–27).

When $v < v_s$, Eq. (14–52) determines without ambiguity the angle of the shock wave. The ratio v_s/v is called the *Mach number;* the larger the Mach number, the closer θ_0 comes to 90°; that is, the more acute is the angle of the line or cone. The sonic boom that occurs whenever an airplane moves faster than the speed of sound is a manifestation of the arrival of a shock wave of sound at an observer's ear. A second type of shock wave is associated with the movement of an electrically charged particle through some medium such as a gas or Plexiglas. The speed of light in this medium can be less than the speed of light in a vacuum, and a charged particle, whose speed cannot exceed that of light in a vacuum, can nevertheless move through the medium faster than the speed of light in that medium. Light waves then propagate along a shock wave front whose angle is given by Eq. (14–52). These waves are called *Cerenkov radiation,* after Pavel Cerenkov (Fig. 14–28). When the speed of light in a given medium is known, measurement of the angle of the shock wave gives the speed of the charged particle. This is a useful measurement tool.

FIGURE 14–29 The wakes of these motorboats on the Connecticut River are a consequence of the fact that the speed of water waves depends on their wavelengths.

Wakes

The *wake,* or track, that extends out behind an ocean liner, a row boat, or a paddling duck (Fig. 14–29) is unlike the shock wave that is responsible for a sonic boom (see Fig. 14–26). Upon observation, we see that the wake forms the same angle for *all* these examples, even though the ocean liner moves much more swiftly than the duck. This wake forms a vee with an angle of 39° between the linear wave fronts. A wave in deep water (the depth is much greater than the wavelength) behaves differently from all the other waves we have studied in this chapter, because *its speed depends on its wavelength.* The wake of a ship, and other differences between that wake and the shock waves discussed here, depend on the variation of speed with wavelength, a phenomenon called *dispersion.*

SUMMARY

Extended bodies that respond elastically to external and internal forces support organized motions called waves. Waves can be classified into standing waves, in which the motion of the entire medium has the same sinusoidal time dependence, and traveling waves, in which limited internal motion propagates progressively with some definite speed and in some definite direction through the medium. Both traveling and standing waves are further classified into either longitudinal waves or transverse waves. In transverse waves, the motion within the medium is perpendicular to the direction in which the waves are organized. In longitudinal waves, the motion is parallel to the wave direction.

For harmonic standing waves, the motion of any point in the medium is described by the equation

$$z(x, t) = z_0 \sin(kx + \delta) \cos(\omega t + \phi). \tag{14–4}$$

The parameters of this wave are the amplitude z_0, the angular frequency ω, and the wave number k. The wavelength, λ, the length over which a standing wave repeats itself, is related to the wave number by Eq. (14–5), $\lambda = 2\pi/k$. We also use the frequency, f, and the period, τ. These parameters are all related by $\lambda f = \lambda/\tau = \omega/k$, from Eq. (14–22).

For standing waves, possible values of k and of λ are determined by the boundary conditions of the medium and are generally discrete. The angular frequency, however, is determined by the dynamics of the internal motion of the medium. For example, for transverse waves on a string, the dynamical equation of motion is a form of the wave equation:

$$T\frac{\partial^2 z}{\partial x^2} = \mu\frac{\partial^2 z}{\partial t^2}, \tag{14–9}$$

where T is the string's tension and μ is its mass density. The frequency is determined from this equation to be

$$\omega = \frac{2\pi}{\lambda}\sqrt{\frac{T}{\mu}}. \tag{14–14}$$

More generally, the dynamical wave equation takes the form

$$\frac{\partial^2 z}{\partial x^2} = \frac{1}{v^2}\frac{\partial^2 z}{\partial t^2}. \tag{14–27}$$

Here, the quantity v is given by

$$\text{wave speed} = \sqrt{\frac{\text{restoring force factor}}{\text{mass factor}}}. \tag{14–28}$$

For a sinusoidal traveling wave along the x-direction, the displacement z of a point in the medium has the form

$$z(x, t) = z_0 \sin(kx - \omega t). \tag{14–23}$$

The speed of the traveling wave is

$$v = \lambda f = \frac{\lambda}{\tau} = \frac{\omega}{k}. \tag{14–22}$$

This speed is determined by the wave equation. In the case of a string of mass density μ under tension T, the speed is

$$v = \sqrt{\frac{T}{\mu}}. \tag{14–26}$$

Waves carry energy, and traveling waves carry energy and momentum along the direction of their motion, even if the medium has no net motion in this direction. For both standing waves and traveling waves, the total energy density is proportional to the amplitude squared. The total energy density of a traveling wave in particular is

$$\frac{\mathrm{d}E}{\mathrm{d}x} = \mu\omega^2 z_0^2 \cos^2(kx - \omega t). \tag{14–32}$$

This energy density moves along with the wave, delivering an average power

$$\langle P \rangle = \frac{1}{2}\mu\omega^2 z_0^2 v. \tag{14–33}$$

Sound waves are longitudinal waves in solids, liquids, or gases. The speed of sound in air is given by

$$v_{\text{sound}} = \sqrt{\frac{\gamma P_0}{\rho_0}}, \tag{14–37}$$

where γ is a constant close to 1, and P_0 and ρ_0 are the pressure and the density of the medium, respectively. For normal atmospheric conditions at sea level, the speed of sound in air is roughly 330 m/s.

The Doppler effect describes what happens to the frequency and wavelength of traveling waves emitted by a source moving with velocity $\mathbf{v}_s$ and detected by a receiver moving with velocity $\mathbf{v}_r$. If v is the speed of a wave within the medium, then the frequency perceived by the receiver is

$$f' = \left(\frac{v - v_r}{v - v_s}\right) f_0. \tag{14–50}$$

Here, v_r and v_s are positive (negative) when the corresponding motion is along (against) the motion of the wave.

UNDERSTANDING KEY CONCEPTS

1. Give some examples of transverse waves and longitudinal waves and discuss the nature of the mechanical motion in each wave.
2. When a gong is sounded, it produces sound waves in air. If the same gong were struck in a vacuum, would it continue to vibrate for less time, for the same amount of time, or for more time than it does when struck in the same way in air? Explain your reasoning.
3. In old cowboy movies, you have probably seen someone place his ear on railroad rails to tell if a train is coming. Why is this done, and how does it work?
4. Would you expect the speed of sound (a) in hydrogen and (b) in argon to be greater than or less than the speed of sound in oxygen if all gases are under the same pressure?
5. When a lecturer in a classroom demonstration takes a lungful of helium and speaks, his or her voice is comically higher in pitch. Why is that?
6. Captain Kirk is somewhere between galaxies. When a gong sounds in a neighboring spaceship, Kirk reacts to the sound. What is wrong with this scenario?
7. Suppose that a string supports a standing wave with a node at a particular location. How would things change if the point of the node were clamped physically?
8. What prevents sound waves from being transverse waves in a dilute gas such as air? In solids, transverse "sound" waves are generally possible. What can we learn about the nature of the intermolecular forces in such a solid?
9. Are the nodes actually motionless when a longitudinal standing wave is established on a Slinky?

10. The speed of deep water waves of wavelength λ is given by the formula $v = \sqrt{g\lambda/2\pi}$. When there is a storm somewhere on the ocean, it manifests itself by large breakers (breaking waves) at the shore. Initially, you observe that the breakers arrive every 20 s at the shore. Somewhat later, you observe that they are coming every 10 s. How do you interpret this?

11. A child concocts a telephone with two cans that are open at one end and connected by a string attached through a small hole to the closed ends. When the string is made taut by pulling on the cans at opposite ends, two children talking into the open ends can carry out a soft conversation over rather large distances. How does this happen?

12. A common toy consists of a series of five balls that touch and form a line; each ball is suspended as a pendulum. When ball 1 is pulled away and then released, it strikes the line and ball 5 rises in a motion like the reverse of ball 1. If there is a small gap between the balls at equilibrium, then a series of collisions occurs in motion, each with an exchange of velocities, until the last ball swings up. The phenomenon does not change even if the balls touch when at rest. If we describe the phenomenon as a collision of the first ball with an object four times its mass (this is how we describe the four touching balls), we do not get the observed motion of the balls. Why is the last description inadequate?

PROBLEMS

14–2 *Standing Waves*

1. (I) A guitar string has end points 54 cm apart. What are the three largest wavelengths of standing waves this string can support? Sketch the standing waves for these wavelengths.

2. (I) The third harmonic of a certain piano wire ($\mu = 0.01$ kg/m, $T = 500$ N) has a frequency of 600 Hz. What is the wavelength of the fundamental?

3. (I) A standing wave is formed on a string for which three and one-half wavelengths fit into the total length of the string, which is 2.7 m. The wave has a period of 0.10 s. If the string has a mass density of 220 g/m, what is the tension in the string?

4. (I) A string that is fastened at the two end points, $x = \pm 60$ cm, oscillates in its fundamental mode (no nodes except at the walls). The period of the oscillation is 1 s, the displacement at $x = 0$ cm at $t = 0$ s is 2 cm, and at $t = 0.1$ s, 3 cm. Express $z(x, t)$ for the displacement.

5. (I) Two wires of the same length and under the same tension have fundamental frequencies that are in the ratio 3:1. What is the ratio of their masses?

6. (I) The tension on a wire is 2840 N. The wire is 2.6 m long and the fundamental frequency is 310 Hz. What is the mass of the wire?

7. (II) A steel string under tension vibrates in its lowest mode at a frequency of 1200 Hz. A string of the same material and length but three times the thickness vibrates in its lowest mode at a frequency of 400 Hz. What is the ratio of the tensions of the two strings?

8. (II) A string with fixed ends is in a standing wave that vibrates in five segments (Fig. 14–30). The frequency of this mode is 120 Hz. (a) Find the fundamental frequency of the string. (b) The tension in the string is reduced by a factor of 9. What is the new fundamental frequency?

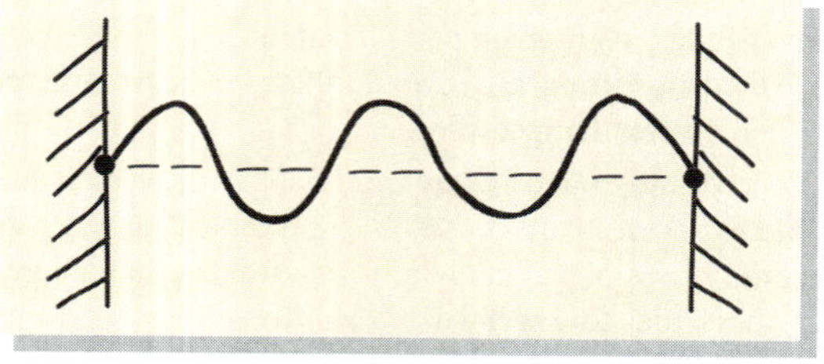

FIGURE 14–30 Problem 8.

9. (II) A rope has a mass density of 40 g/cm and is under tension of 200 N. The rope is fixed at one end and is connected to a rod at the other end that sets up standing waves by moving slightly. The rope is 1 m long. Determine the three lowest possible frequencies and the corresponding angular frequencies and periods.

10. (II) Figure 14–31 shows a pulse at $t = 0$ traveling to the right with a speed of 0.5 m/s. (a) Sketch, properly scaled, the shape of the rope at $t = 4$ s. (b) Plot the displacement of the point at $x = 5$ m as a function of time.

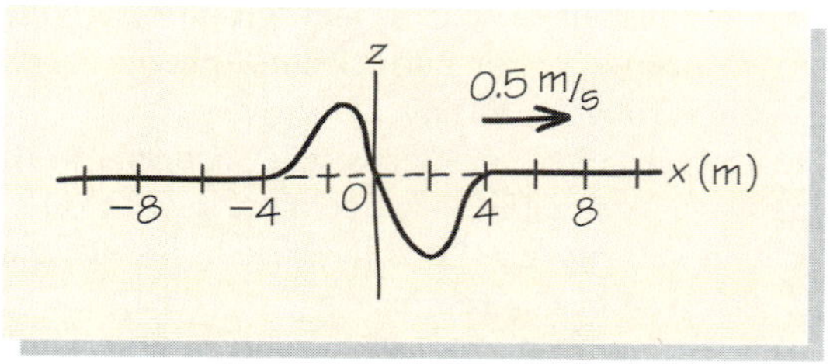

FIGURE 14–31 Problem 10.

11. (II) The mass density of copper is 8.92×10^3 kg/m^3. What is the tension on a copper wire 1.00 mm in diameter and 60.0 cm long, fastened at each end, when the frequency of vibration in the $n = 3$ mode is 870 Hz?

12. (II) A string with length L and linear density μ is under tension T. One end is fixed; the other end remains horizontal while it moves periodically up and down. Standing waves with an antinode (a point of maximum displacement) can be induced in this way, as in Fig. 14–32. Find the possible wavelengths and frequencies.

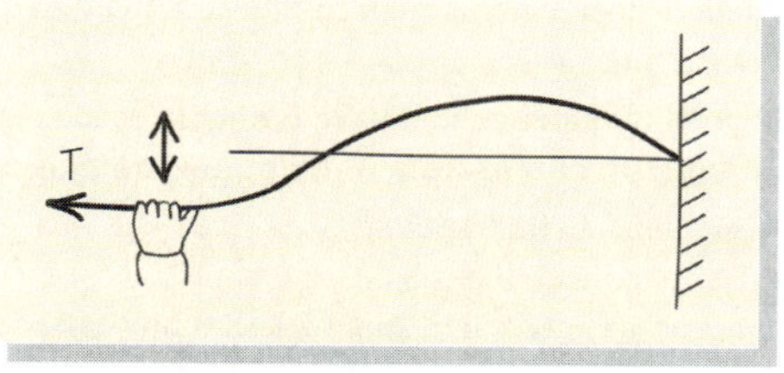

FIGURE 14–32 Problem 12.

14–3 *Traveling Waves*

13. (I) A wave is described by a displacement function $y = A \sin(ax - bt)$, where $a = 0.3\ \text{m}^{-1}$ and $b = 0.02\ \text{s}^{-1}$. Plot the displacement versus t at both $x = 1$ m and $x = 12$ m.

14. (I) A wave is described by $y = A \sin(kx + \omega t)$, where $k = 3.0\ \text{m}^{-1}$, $\omega = 2.0\ \text{s}^{-1}$, and $A = 0.40$ m. Determine (a) the wavelength, (b) the frequency, (c) the maximum amplitude, (d) the period, and (e) the wave number.

15. (II) Use Eq. (14–23) for the traveling wave $z(x, t)$ and the wave equation to show that $Tk^2 = \mu\omega^2$.

16. (II) A perfect triangular pulse travels along a string with a velocity of 4 m/s. The shape of the pulse at $t = 0$ is shown in Fig. 14–33. Plot the displacement and velocity of the segment at $x = 3$ m as a function of time. Is such a pulse physically possible?

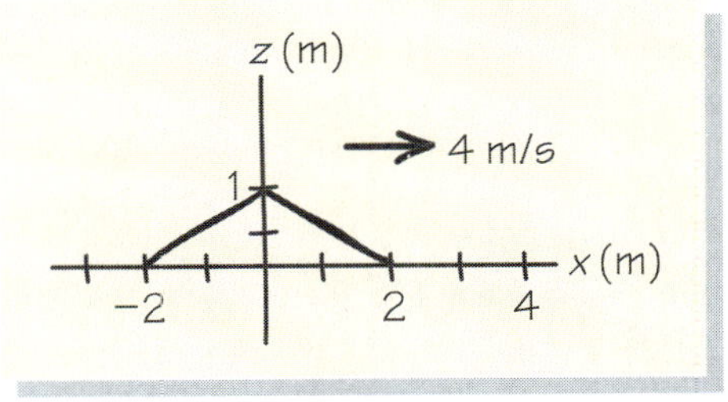

FIGURE 14–33 Problem 16.

17. (II) Show by substitution that the general form of a traveling wave, $f(x - vt)$, satisfies the wave equation.

18. (II) Between $x = 0$ cm and $x = 10$ cm, a very long wire (so long that the boundary effects do not matter) is distorted from its straight-line equilibrium form into a triangular shape like that of Fig. 14–33 with the peak 4 cm above the original position. The tension and density of the wire are such that waves travel along the wire at a speed of 400 m/s. (a) What is the shape of the wire at the initial time $t = 0$? (b) What is the shape of the wire at an arbitrary time t, assuming that the wave travels in the $-x$-direction?

19. (III) Suppose you make a little hump in a hall rug with linear mass density 1.5 kg/m and that the hump propagates with speed 2 m/s to the right. In this case, the segments of the rug do not return to their original position; rather, the pulse results in the displacement of the whole rug. Assuming that the hump involves 0.6 m of the rug and extends 0.5 m along the floor, calculate the momentum carried by the pulse. Similar disturbances, called dislocations, play an important role in the plastic deformation of solids. [*Hint*: Consider the global effect of the pulse on a long rug.]

14–4 *Estimating the Speeds of Traveling Waves*

20. (I) The French train known as the TGV travels at 300 km/hr. What is the frequency with which the cars of this train pass a given point at that speed, given that the cars are 30 m in length?

21. (I) Consider a long length of the wire used in Example 14–2. The wire is taut under a tension of 500 N. If the wire is plucked, what will be the speed of the traveling wave?

22. (I) A rope of density μ_1 is attached to another rope, of density μ_2 (Fig. 14–34). A traveling wave of speed v_1 is sent down the first rope. What will the speed in the second rope be?

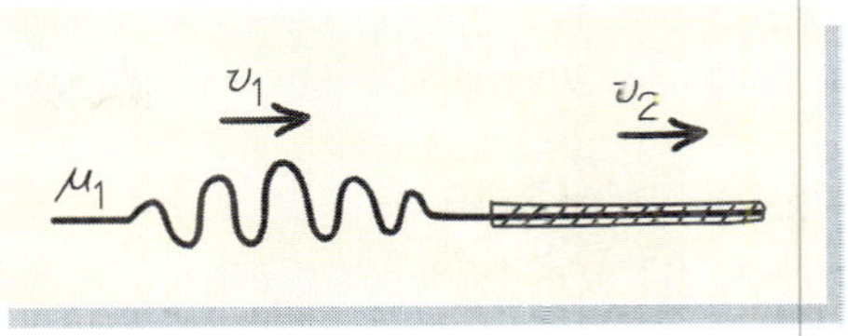

FIGURE 14–34 Problem 22.

23. (I) A harmonic wave with wavelength 0.27 m moves with speed 13 m/s. Find the wave number k and the frequency f of this wave.

24. (I) A mountain climber of mass 63 kg hangs from a nylon rope 28 m below the top of a cliff. The total mass of the rope is 1.3 kg; assume that this mass is small enough so that the variation in equilibrium tension along the rope can be ignored. How long does it take a wave pulse to travel up the rope from the man to the top of the cliff?

25. (I) The values of Young's modulus, Y, for aluminum and copper are $Y_{Al} = 7.0 \times 10^{10}\ \text{N/m}^2$ and $Y_{Cu} = 11 \times 10^{10}\ \text{N/m}^2$. Their mass densities are $2.70 \times 10^3\ \text{kg/m}^3$ and $8.96 \times 10^3\ \text{kg/m}^3$, respectively. Calculate the speed of elastic wave propagation in these two media.

26. (II) Assume that experiment has shown that the velocity of waves in shallow water is independent of the wavelength of these waves. Use dimensional analysis to show that the speed of these waves is proportional to the square root of the depth D. Express the speed in terms of the relevant constants.

27. (II) Assume that the expression for the velocity of shallow water waves in terms of g and depth D that you derived in Problem 26 is exact. A speed boat that travels close to the speed of such waves tends to build up a large wave just at its bow and thus will have trouble going any faster. At what speed in kilometers per hour will this happen in a harbor in which the water is 5 m deep?

28. (II) The considerations of Problem 26 apply to all water waves for which the wavelength is much larger than the depth. For tidal waves (or *tsunami*: waves caused by underwater earthquakes), the wavelengths are enormous, many hundreds of kilometers. What is the speed of a tidal wave in water that is 3 km deep?

29. (III) When n springs, each of spring constant k and equilibrium length ℓ, are attached in a line, the resulting system is a spring with spring constant $k_n = k/n$ and length $L = n\ell$. If there is a mass m at the connection point between each spring (the springs themselves are assumed to be massless), then the resulting system can support longitudinal or compressional, waves (see Fig. 14–3b). By imagining an arbitrarily long series, you can see that the speed of these waves is independent of the total length of the spring, and hence of n. Use the available constants k, ℓ, and m to estimate the speed of longitudinal waves. This system is an excellent model for compressional waves on a real spring. [*Hint*: Dimensional analysis can be useful here.]

30. (III) A set of tiny beads is connected in a line by springs, as in Problem 29. The equilibrium length of the springs is 3 cm. The beads have mass 5 g each, and the spring constant of the springs is 20 N/m. Longitudinal waves can be generated on the string of beads by moving the end of the string along the direction of the string. Estimate the speed of such longitudinal pulses. [*Hint*: Dimensional analysis can be useful here.]

14–5 Energy and Power in Waves

31. (I) A wave of wavelength 0.7 m is sent moving down a rope of density 12 g/cm, under a tension of 800 N. The amplitude of the wave is 5 cm. What is the average power transported by the rope? Suppose that the wavelength is doubled; what is the average power transported by the rope now?

32. (I) Calculate the kinetic energy and potential energy contained in one wavelength of a harmonic traveling wave on a string whose mass density is μ and whose tension is T.

33. (I) The power delivered per unit area by a sound wave is given by an expression similar to that of Eq. (14–33). In this case, μ is replaced by the density, ρ, of the gas in which the sound propagates: $P/\text{area} = \text{intensity} = \frac{1}{2}\rho\omega^2 A^2 v$, where A is the amplitude of the displacement of air molecules from their average positions. Given an 8600-Hz sound wave with an intensity of 5.0×10^{-7} W/m^2, what is A? The density of air is 1.30 kg/m^3, and the speed of sound in air is 330 m/s.

34. (II) A sound wave has an intensity of 1.7×10^{-6} W/m^2. Given that the speed of sound in air is 330 m/s, what is the energy density of the sound wave?

35. (II) Obtain an expression analogous to Eq. (14–32) for the rate of energy transport in a left-moving wave. In particular, show that the power remains positive.

36. (II) Harmonic water waves of a given amplitude are generated on a water surface by rhythmic dipping of a pointer at one spot. The waves thus generated spread in circular wave fronts. If R is the distance from the generation point, show that conservation of energy implies that the amplitude z_0 must vary with R, and that this variation is of the form $z_0 \approx 1/\sqrt{R}$.

37. (III) We derived the power or rate of energy transport for a sinusoidal wave [Eq. (14–32)]. Obtain an expression for the power delivered by an arbitrary right-moving wave of the form $f(x - vt)$.

14–6 Sound

38. (I) Sound waves in air travel at 330 m/s, whereas light travels at 3×10^8 m/s. You can determine how far away a lightning strike is by counting the number of seconds between the time you see the lightning flash and the time you hear the thunderclap. How far are you from the lightning bolt for each 1-s difference between the flash and the thunderclap?

39. (I) Using the information in Table 14–1, compute the densities of air and hydrogen.

40. (I) A telephone receiver is at your left ear, with the sound of the conversation at a level of 58 dB. At the same time, your little sister's screams, at a level of 93 dB, enter your right ear. What is the ratio of the sound intensities that enter your two ears?

41. (I) A series of noises have intensities 8.3×10^{-4} W/m^2, 3.5×10^{-6} W/m^2, and 7.2×10^{-9} W/m^2, respectively. What is the sound level of each in decibels?

42. (I) A sound suddenly appears to the ear to be twice as intense as usual. How has the amplitude of the pressure wave (the magnitude of the maximum pressure in the wave) changed?

43. (II) A sound source emits sound with equal intensity in all directions. What is the power output of the source if the intensity of the sound measured at 5 m from the source is 83 dB?

44. (II) Two sounds have levels of 35 dB and 53 dB, respectively. What is the ratio of the power of the two noises? What is the ratio of the amplitudes of the pressure waves they make?

45. (II) In an ideal fluid, the speed of sound is given by $v = \sqrt{B/\rho}$, where B is the *bulk modulus*, a measure of how easy it is to compress the fluid. Sound waves in a fluid are only longitudinal because an ideal fluid cannot bounce back from the application of transverse forces, sometimes called shearing forces. Using the information given in Table 14–1, calculate the bulk modulus of water.

46. (II) A source of sound waves is operating in midair, far from any reflecting surfaces. If, 1 m from the source, the sound intensity level is 120 dB, what is the intensity 1000 m from the source? What is the total power output of the source?

14–7 The Doppler Effect

47. (I) A horn emits sound at a frequency of 160 Hz. What frequency is heard by an observer moving away from the source at 26.0 m/s?

48. (I) A musician with perfect pitch—the capacity to identify pitch by ear—has run a red light and is being pursued by a police car whose siren emits at the single frequency $f = 400$ Hz. The musician's car is moving east at 60 mi/h (27 m/s), and the police car follows at 80 mi/h. The speed of sound is 330 m/s, and the air is still. What does the musician say the frequency of the siren is?

49. (I) A whistle emits sound with a frequency of 333 Hz. The whistle is mounted on a train that travels at 140 km/h in the $-x$-direction toward an observer at rest. (a) What frequency is heard by the observer? (b) Suppose that the whistle is at rest, and the observer moves in the $+x$-direction at 140 km/h. What frequency is heard by the observer?

50. (I) A child is blowing a whistle while riding her bicycle toward you on the sidewalk. You have set up an accurate audio receiver that measures the frequency of the whistle sound to be 793 Hz. You estimate the bicycle to be moving at 7 mph (3.1 m/s). What would you measure the frequency of the whistle to be at rest?

51. (II) The frequency of light emitted by a star moving away from us (the observer) with speed v will be *red shifted* (its frequency will be lowered) by an amount properly described by the treatment in Section 14–7, provided $v/c << 1$, where c is the speed of light ($c = 3 \times 10^8$ m/s). How fast is a galaxy receding from us if the observed frequency is 0.95 times the natural frequency of the light?

52. (II) A truck is traveling at 60 mi/h (27 m/s) down the interstate highway where you are changing a flat tire. As the truck approaches, the truck driver sounds an air horn, which has a frequency of 185 Hz. (a) What frequency do you hear? (b) Suppose that you were driving your car at 50 mi/h while the truck driver blows his air horn after he passes you. What frequency would you hear?

53. (II) An ambulance siren is the source of sound emitted with a frequency of 1600 Hz. The ambulance is driving toward the base of a large cliff at 95 km/h (Fig. 14–35). What is the frequency of the reflected sound wave heard at the ambulance?

FIGURE 14–35 Problem 53.

54. (II) You are moving southward on a divided highway and, heading northward on the other side, a fire engine passes with its (one-note) siren screaming. You are close enough to each other as you pass to assume that you and the fire engine are moving on the same line. Your speedometer reads 45 mi/h, and, being a trained musician, you recognize that the frequency you hear before you pass the fire engine is 910 Hz and 580 Hz after you pass. Use the value 330 m/s for the speed of sound in air to find the frequency emitted by the siren and the speed of the fire engine.

55. (III) A tuning fork with frequency 440 Hz is dropped from a tower of height 100 m by an enterprising student. She can hear the fork sounding as it falls. Given that the speed of sound in air is 330 m/s and that $g = 10 \text{ m/s}^2$, what frequency does the student hear as a function of time? Do not forget that it takes time for sound to travel back to the student.

56. (III) Show that, for $v << c$, where c is the speed of a wave in a certain medium, the frequency shift for an observer at rest and a source moving away at speed v is the same as that for a source at rest and the observer moving at speed v in the opposite direction.

*14–8 Shock Waves

57. (I) The angle made by the shock wave associated with the sonic boom of a low-flying airplane is $\theta_0 = 70°$. What is the speed of the airplane?

58. (I) An airplane moving at three times the speed of sound (Mach 3) flies horizontally and at a very low altitude. It emits sound continuously with a frequency that the pilot hears as 300 Hz. (a) What angle does the plane's shock wave cone make with the vertical? (b) What is the frequency of the sound heard by a stationary observer after the plane has passed? (Assume that the plane is moving directly away from the observer.)

General Problems

59. (II) A string is fastened to a wall at $x = 0$ cm and is free to move on another wall at $x = 60$ cm, with the constraint that it is always perpendicular to the wall there (Fig. 14–36). What are the permissible values of the wavelength?

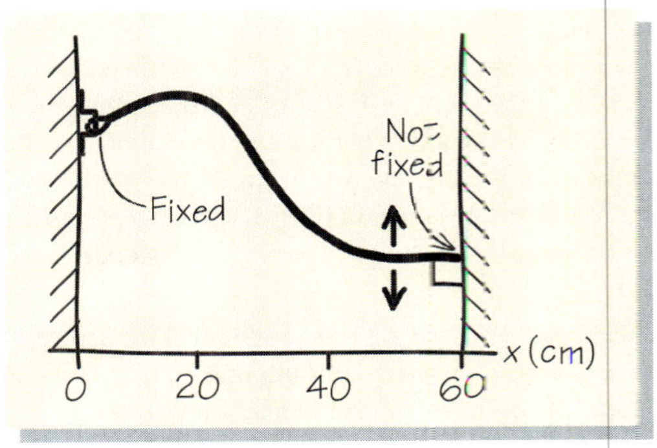

FIGURE 14–36 Problem 59.

60. (II) A string is stretched between a pulley and a wave generator consisting of a plate vibrating up and down with small amplitude and frequency 120 Hz (Fig. 14–37). A standing wave with two intermediate nodes is generated when the string has a mass of 215 g attached to it. How large a load is needed to generate standing waves with one and with four nodes?

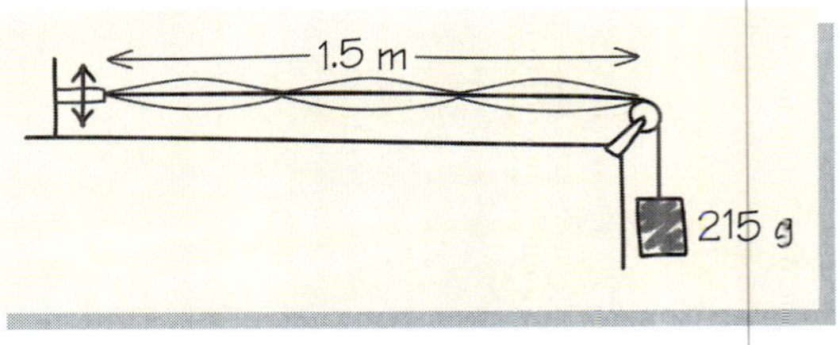

FIGURE 14–37 Problem 60.

61. (II) The sound intensity at a rock concert reaches the pain threshold of 120 dB. Normal conversation has an intensity of 55 dB. What is the ratio of the power delivered to the ear at a rock concert compared to that delivered by a typical conversation?

62. (II) Transverse traveling waves are generated on a uniform wire of constant tension and a given mass. By what factor will the required power be increased or decreased if (a) the given mass of wire is doubled in length and the angular frequency remains constant? (b) the amplitude is doubled and the angular frequency is halved? (c) both the wavelength and the amplitude are doubled? (d) the given mass of wire is doubled in length and the wavelength is halved?

63. (II) The density of neon is approximately 0.9 kg/m³ at standard atmospheric pressure and temperature. The value of γ for neon is 5/3. A loudspeaker emits a note with a frequency of 400 Hz in air. What will the frequency and wavelength of that note be if the loudspeaker is put into a chamber filled with neon?

64. (II) The left-hand end of a long, taut string is moved harmonically up and down with amplitude 0.14 m. This motion produces a wave of wavelength $\lambda = 2.5$ m that travels in the $+x$-direction at speed $v = 245$ m/s. (a) What is the frequency, f, of the traveling wave? (b) What is the maximum transverse velocity of a point along the string? (c) What is the maximum transverse acceleration of a point along the string?

65. (II) A tuning fork with frequency $f = 512$ Hz is held by someone who swings it vigorously in a circle in the horizontal plane. The radius of the circle is 1.0 m, and the frequency of revolution is 3.0 rev/s. (a) What are the maximum and minimum frequencies that a second person would hear? (b) Which part of the rotation corresponds to the highest frequency the second person hears, and which part corresponds to the lowest?

66. (II) A wire is attached to two walls 1 m apart. The wire oscillates in such a way that a point on the wire 25 cm from one end is held fixed at its equilibrium position. What are the possible wavelengths for the first three allowed modes?

67. (II) A tuning fork with a frequency of 440 Hz is held just above the top of a uniform tube containing water (Fig. 14–38). The tube can excite the column of air above the water, whose level can be changed by a spigot at the bottom of the tube. As the water is drained out, the sound intensity of the fork is enhanced when the air column has a length of 0.6 m and again when the air column has a length of 1 m. Use these data to find the speed of sound in air.

FIGURE 14–38 Problem 67.

68. (III) A square sheet of metal with sides of length L is clamped on all four sides. It is convenient to put these sides at $x = 0$, $x = L$, and $y = 0$, $y = L$ (Fig. 14–39). The vertical displacement at any point on the sheet is given by $z(x, y, t)$, and it satisfies the wave equation

$$\frac{\partial^2 z}{\partial x^2} + \frac{\partial^2 z}{\partial y^2} = \frac{1}{v^2}\frac{\partial^2 z}{\partial t^2}.$$

Show that a standing wave of the form $z(x, y, t) = A\sin(kx)\sin(qy)\cos(\omega t)$ satisfies the wave equation, and find the values of k and q that satisfy the boundary conditions and the constraints of the wave equation.

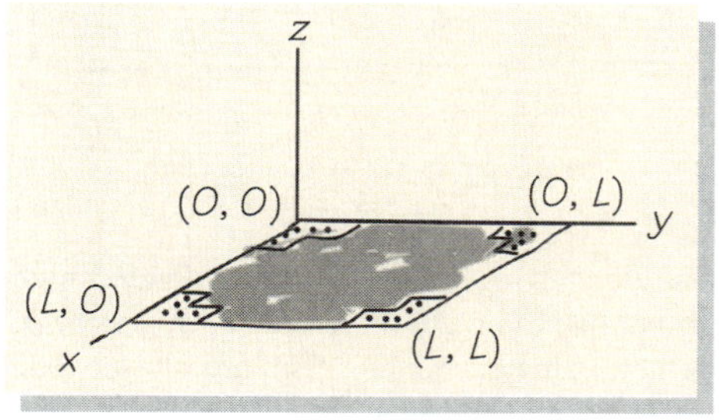

FIGURE 14–39 Problem 68.

69. (III) The relation between angular frequency and wave number for a stiff wire takes the form $\omega^2 = (T/\mu)k^2 + \alpha k^4$, where α is a measure of the stiffness. Suppose that such a stiff wire is clamped at $x = 0$ and $x = L$. Express the frequency for the two lowest modes of the wire. For which frequency is the stiffness a more important effect?

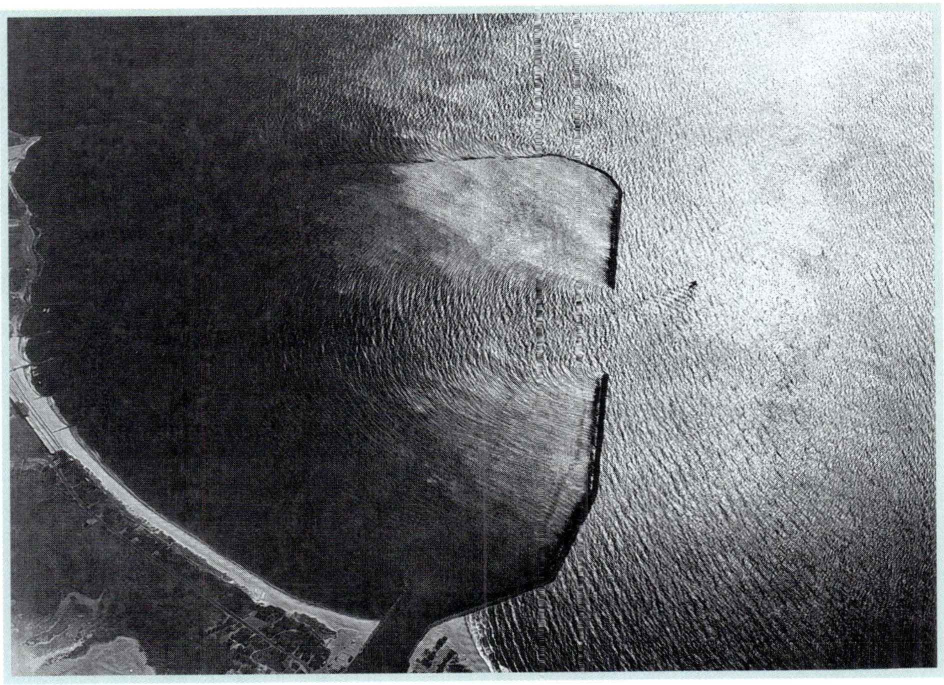

CHAPTER

15

As waves enter the protected harbor through a narrow opening, they spread in opening circles. This is a diffraction phenomenon that can be most easily understood in terms of Huygens' principle.

Superposition and Interference of Waves

The harmonic waves discussed in Chapter 14 are idealizations. Waves observed in nature, such as water waves, sound waves, waves on strings, or light waves, are rarely as simple as harmonic waves. The surface of the ocean is a chaotic welter of waves moving in various directions. A guitar string is set into a very complicated motion when it is plucked. A single shake of the end of a rope sends a single pulse rather than a sine function traveling down the rope.

Nevertheless, we can understand more complicated waves entirely in terms of simple harmonic waves: Harmonic waves can be combined, or superposed, to produce the full variety of waves observed in nature. Superposed waves, which sometimes reinforce and sometimes cancel each other, are said to interfere. This interference produces regular patterns of reinforcement and cancellation; such patterns are seen most easily in light waves as the colors in oil spots or in the pattern of ocean waves as they enter a harbor.

Our goal in this chapter is to understand the different ways in which waves can combine. We will see how standing waves can be formed by superpositions of traveling waves, how beats arise, how there can be well-organized interference patterns in space, and how waves of any shape can be formed by the superposition of harmonic waves.

15–1 THE SUPERPOSITION PRINCIPLE

The superposition principle

The waves that appear in nature are often complicated ones. Nevertheless, our work on harmonic waves is relevant because elastic media that undergo small displacements exhibit **linearity**. Linearity means that if two waves can travel through a medium, then a third wave whose form is the algebraic sum of these waves is possible. Stated mathematically, *the sum of two solutions of the equations of motion of these systems is also a solution.* When linearity applies, the sum of any number of waves can add to form an allowed wave, and a wave multiplied by any constant is also an allowed wave. This fact is known as the **superposition principle**.

To understand how a sum of two solutions of an equation can also be a solution, let's reconsider the dynamical equations that determine the solutions. Newton's second law, applied to large systems with elastic restoring forces, led us to the wave equation in Chapter 14. For transverse waves on a string, Eq. (14–27) took the form

$$\frac{\partial^2 z}{\partial x^2} = \frac{1}{v^2}\frac{\partial^2 z}{\partial t^2}, \tag{15–1}$$

where x is the distance along the string and $z(x, t)$ is the transverse displacement of the string at point x and time t. Equation (15–1) is *linear* because the function z appears linearly; that is, as z rather than as z^2 or as z to some other power. If $z_1(x, t)$ and $z_2(x, t)$ are two different solutions satisfying some linear equation such as the wave equation, then the sum $z_1(x, t) + z_2(x, t)$ also satisfies that equation. Superposition works only for waves of small amplitude because when a medium is stretched or distorted too far, the restoring force is no longer purely springlike, and the dyamical equation becomes nonlinear.

Interference

The superposition principle allows us to add sinusoidal waves of different amplitudes, wavelengths, and frequencies, each of them moving in different directions. The sums considered are *algebraic* sums, complete with the possibilities of algebraic cancellations (Fig. 15–1). The results of adding different waves present a rich palette of patterns called **interference phenomena**.

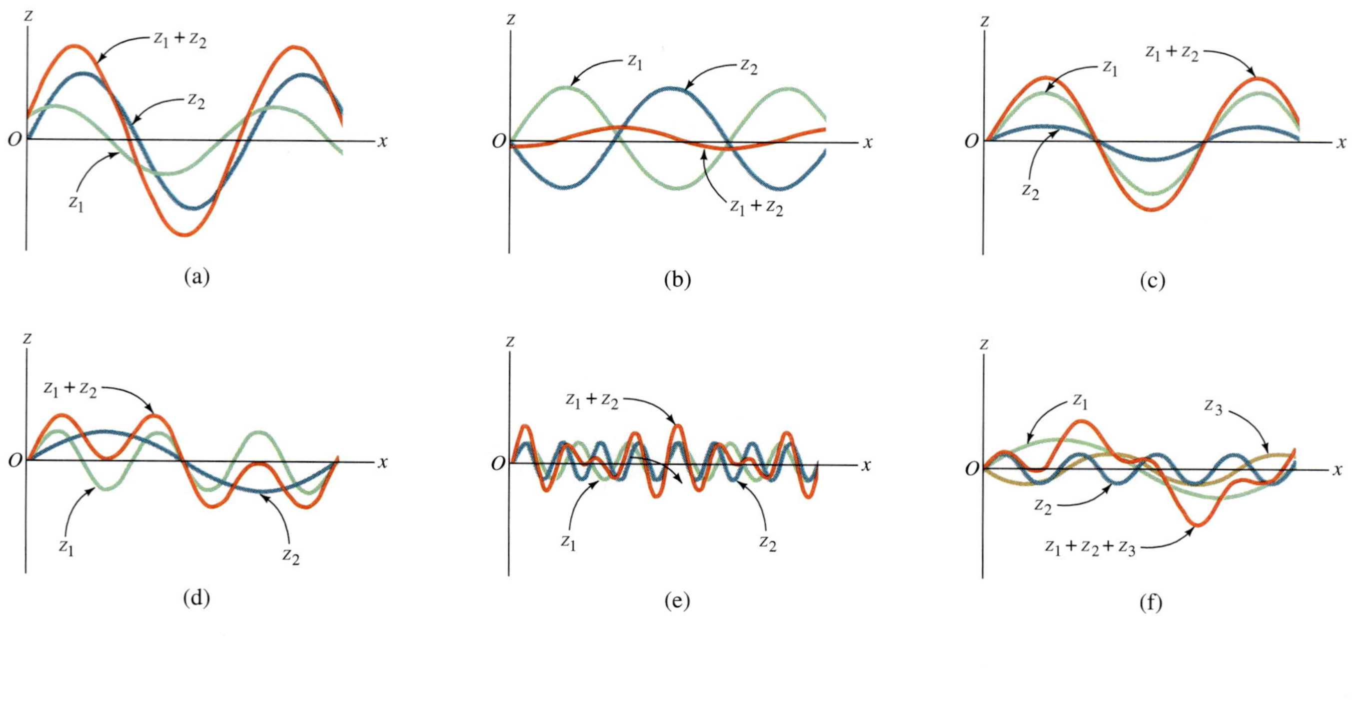

FIGURE 15–1 The superposition of (a) two waves of the same frequency, with a slight phase difference between them; (b) two waves of the same amplitude and frequency, with a phase difference of nearly 180° between them; (c) two waves of the same frequency and phase but different amplitudes; (d) two waves of the same amplitude but frequencies that differ by a factor of 3; (e) two waves of the same amplitude but slightly different frequencies, resulting in beats; (f) three waves of different amplitudes and frequencies.

A simple and striking example of interference is the case of two transverse waves that travel to the right on a string. Let the waves have the same amplitude and be 180° (π rad) out of phase. The two waves have the form

$$z_1(x, t) = z_0 \sin(kx - \omega t) \tag{15–2}$$

and

$$z_2(x, t) = z_0 \sin(kx - \omega t + \pi) = -z_0 \sin(kx - \omega t). \tag{15–3}$$

Destructive and constructive interference

The superposition of these two transverse waves adds to zero! This is an extreme case of **destructive interference**, which is interference that occurs at any point where two waves have the opposite sign. Although interference may be *fully* destructive (waves adding to zero at a given point), keep in mind that intermediate levels of cancellation are possible. **Constructive interference** occurs when the waves add in such a way that they reinforce. If, for example, our two waves z_1 and z_2 had exactly the same phase, we would find a summed solution with the same x- and t-dependence of either wave, but with amplitude $2z_0$.

Coherence

Coherence. Only coherent waves can exhibit stable interference patterns.

What conditions are necessary for waves from two different sources (for example, the sound waves coming from two loudspeakers) to exhibit an interference pattern? If the two sources put out a series of waves with different frequencies and with phases that change randomly, then there is no consistent interference pattern. Any pattern would change so rapidly that we could not recognize it; *an interference pattern will only appear if the waves have a definite, stable relation between their frequencies and phases.* The waves are then said to be **coherent**. When we have waves without a definite, stable relation between their frequencies and phases, the waves are **incoherent**. Sources that are not point sources, such as piano strings or light bulbs or the human voice, are generally incoherent with other sources because different parts of the waves produced by those sources generally have different phases

15–2 STANDING WAVES THROUGH INTERFERENCE

Standing waves can be viewed as the result of the superposition of traveling waves moving in opposite directions. Imagine a very long string with two waves that move in opposite directions but are otherwise identical, each a solution of the wave equation:

$$\text{for the wave moving to the left:} \quad z_l(x, t) = z_0 \sin(kx + \omega t); \tag{15–4a}$$

$$\text{for the wave moving to the right:} \quad z_r(x, t) = z_0 \sin(kx - \omega t). \tag{15–4b}$$

We have taken the wavelengths, frequencies, phases, and amplitudes to be equal. According to the principle of superposition, the resultant displacement is

$$z(x, t) = z_r(x, t) + z_l(x, t) = z_0[\sin(kx - \omega t) + \sin(kx + \omega t)]. \tag{15–5}$$

If we use the trigonometric identity 5.3 of Appendix IV–4, with $a = kx - \omega t$ and $b = kx + \omega t$, then we obtain

$$\begin{aligned}\sin(kx - \omega t) &+ \sin(kx + \omega t)\\ &= 2 \sin\left[\frac{(kx - \omega t) + (kx + \omega t)}{2}\right] \cos\left[\frac{(kx - \omega t) - (kx + \omega t)}{2}\right]\\ &= 2 \sin(kx) \cos(\omega t).\end{aligned}$$

In the last line, we have used the fact that $\cos(-a) = \cos(a)$. Thus,

$$z(x, t) = 2z_0 \sin(kx) \cos(\omega t). \tag{15–6}$$

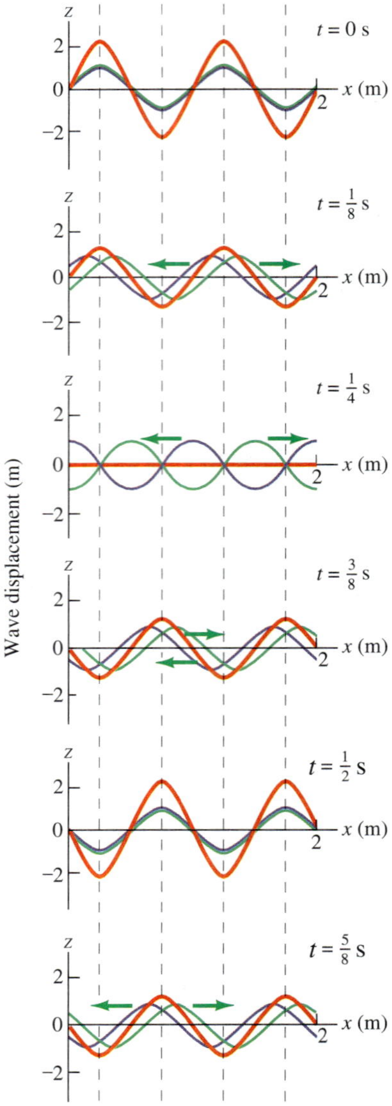

FIGURE 15–2 Superposition of two traveling waves produces a standing wave, plotted in red. Shown are the waves $\sin(kx - \omega t)$, moving rightward, and $\sin(kx + \omega t)$, moving leftward. The amplitude of each wave is 1 m, the wavelength is 1 m, and the period is 1 s. Thus, $k = 2\pi\ \text{m}^{-1}$ and $\omega = 2\pi$ rad/s. Arrows indicate the direction of motion of the two traveling waves. Time progresses in $\frac{1}{8}$-s intervals starting at $t = 0$ s and ending at $t = \frac{5}{8}$ s. Note that the peaks of the standing wave remain at exactly the same x-values.

This expression is identical to our expression for standing waves, Eq. (14–4), except for an (unimportant) phase difference and a factor of 2 that can be taken as part of the amplitude. It is now easier to understand why the wave speed v appeared in the equation for a standing wave: It is the speed of the two traveling waves that are superposed to form the standing wave. In Fig. 15–2, we plot two traveling waves and their superposition. The superposition produces a standing wave whose period and wavelength are the same as those of the traveling waves that compose it.

EXAMPLE 15–1 Two traveling waves move in opposite directions in a one-dimensional medium. They have the form given by Eqs. (15–4a, b). For each wave, the wave speed is 1.8 m/s and the angular frequencies of the traveling waves are 29 rad/s. Find the wavelength of the resulting standing wave.

Solution: We have seen that the wave number k, and hence the wavelength λ, of the standing wave formed by two waves traveling in opposite directions such as those of this example are the *same* as the wave number and wavelength of the traveling waves. This result is expressed in Eq. (15–6). In turn, we can find the wavelength of the traveling waves from the wave speed and the frequency:

$$\lambda = v/f = v/(\omega/2\pi) = 2\pi v/\omega = (6.28)(1.8\ \text{m/s})/(29\ \text{rad/s}) = 0.39\ \text{m}.$$

Interference of Incident and Reflected Waves

Let's look at an *incident,* or incoming, traveling wave that encounters a wall; this will help us to understand the production of standing waves (Fig. 15–3). We can attach the two ends of a string to two walls so that the displacement at the walls ($x = 0$ and $x = L$) is always zero. An incident wave that moves to the right and arrives at $x = L$ will be *reflected* from the wall. The energy carried to the right by the traveling wave bounces back and travels to the left as a reflected wave. The reflected wave has the same frequency, f, as the incoming wave because collisions with the wall of any part of the incoming wave, such as the crests, also have frequency f, and it is these collisions that generate the reflected wave. The reflected wave moves with the same speed as the original traveling wave because the wave speed v is just a property of the string. Figure 15–4a shows the incoming wave; Fig. 15–4b, the reflected wave. The reflected wave travels to the left and can now interfere with the rightward-moving incident wave to produce a standing wave (Fig. 15–4c) with the same frequency and wavelength—exactly as in the derivation of Eq. (15–6). It is thus correct to think of standing waves on a string fixed at both ends as a superposition of traveling waves and of their reflections at the two ends.

One feature of the reflection deserves special attention. The standing wave formed on the string must have nodes at the two fixed ends. This means that the incident wave and its reflection must *cancel* at each end (and hence form a node). This requires that *the sinusoidal wave that reflects from a fixed end must have a displacement with a sign opposite to that of the incoming wave and must therefore be inverted* (see also Section 15–6).

15–3 BEATS

Beats occur when two waves of nearly the same frequency are superposed; a long wavelength (or small frequency) interference pattern is produced. Musicians use beats to tune their instruments, and the phenomenon is of scientific importance because it provides a sensitive technique for measuring frequency (or period) differences. A simple visual analogue to beats can be realized with the help of two combs with slightly different tooth spacings; the slightly different spacings are analogous to slightly different frequencies. In Fig. 15–5, two combs are superposed. Where the teeth lie directly on top of one another, the spaces are visible. Some distance away, the teeth will lie out of phase and no spaces are visible. Thus, the superposed combs exhibit a pattern that repeats over a much greater distance than the separation of the teeth. The very large spacing of the interference pattern observed when the two combs are superposed is like the very large period of the interference pattern observed when waves with two slightly different periods are superposed.

Consider two traveling waves of slightly different frequencies, as in Figs. 15–6a and 15–6b. The two waves have equal amplitudes of 1 m. The first wave has $\omega_1 = 1$ Hz and $k_1 = 1.0\ \text{m}^{-1}$, and the second has $\omega_2 = 0.9$ Hz and $k_2 = (1/0.9)\ \text{m}^{-1} = 1.1\ \text{m}^{-1}$. To the eye, the two waves are barely distinguishable from each other. Yet their superposition has a striking visual signature: *A very different wave—one with a very long wavelength—is formed* (Fig. 15–6c).

What has happened is the following: Both waves start out together at $x = 0$ m, where their similarity causes them to interfere constructively. The intensity of the superposed wave is large there. Because their wavelengths differ slightly, the two waves begin to differ from one another as we move along x; at some point along x (at about 30 m), they are almost perfectly out of phase. At this point, they interfere destructively. The intensity of the superposed wave is small there. The cycle repeats itself as the waves slowly come together again, at about 55 m, as the combs of Fig. 15–5 demonstrated.

APPLICATION

Beats are used to measure frequency differences.

FIGURE 15–3 The parallel waves generated at the bottom of this photograph are reflected from the 45° barrier and move to the right.

Calculation of the Beat Frequency

To analyze this situation mathematically, start with the two waves

$$z_1(x, t) = z_0 \sin(k_1 x - \omega_1 t) \qquad \text{and} \qquad z_2(x, t) = z_0 \sin(k_2 x - \omega_2 t). \tag{15–7}$$

A convenient way to account for the fact that the frequencies of the two waves are not very different is to deal with the sum and the difference of the frequencies. The frequency difference will be a small quantity. If $\omega_1 > \omega_2$, and $k_1 > k_2$, we define

$$\delta\omega \equiv (\omega_1 - \omega_2) \qquad \text{and} \qquad \Omega \equiv \frac{1}{2}(\omega_1 + \omega_2), \tag{15–8}$$

and similarly,

$$\delta k \equiv (k_1 - k_2) \qquad \text{and} \qquad K \equiv \frac{1}{2}(k_1 + k_2). \tag{15–9}$$

It is easy to show that the ratios $\delta\omega/\delta k$ and Ω/K are given by the wave speed $v = \omega_1/k_1 = \omega_2/k_2$ in the medium. Both $\delta\omega$ and δk are positive and small.

We can easily reexpress the original frequencies and wave numbers in terms of their sums and differences:

$$\omega_1 = \Omega + \frac{1}{2}\delta\omega \qquad \text{and} \qquad \omega_2 = \Omega - \frac{1}{2}\delta\omega, \tag{15–10}$$

$$k_1 = K + \frac{1}{2}\delta k \qquad \text{and} \qquad k_2 = K - \frac{1}{2}\delta k. \tag{15–11}$$

When Eqs. (15–10) and (15–11) are substituted into the waves of Eq. (15–7), we find that

$$\begin{aligned} z_1(x, t) &= z_0 \sin\left[\left(K + \frac{\delta k}{2}\right)x - \left(\Omega + \frac{\delta\omega}{2}\right)t\right] \\ &= z_0 \sin\left\{(Kx - \Omega t) + \frac{1}{2}[(\delta k)x - (\delta\omega)t]\right\}, \end{aligned} \tag{15–12}$$

$$\begin{aligned} z_2(x, t) &= z_0 \sin\left[\left(K - \frac{\delta k}{2}\right)x - \left(\Omega - \frac{\delta\omega}{2}\right)t\right] \\ &= z_0 \sin\left\{(Kx - \Omega t) - \frac{1}{2}[(\delta k)x - (\delta\omega)t]\right\}. \end{aligned} \tag{15–13}$$

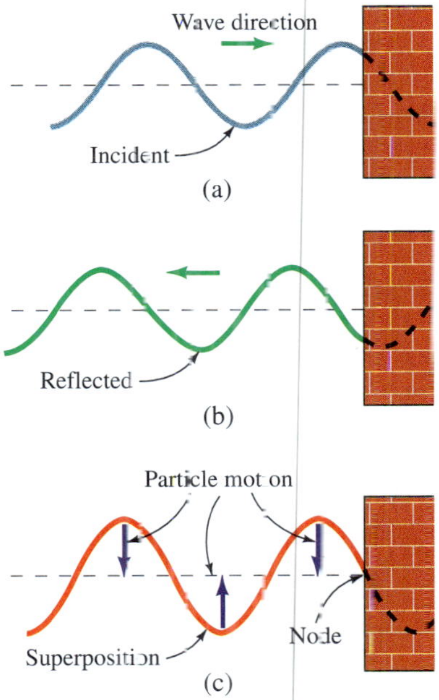

FIGURE 15–4 The reflection of a transverse traveling wave on a string from a wall (the second wall is far to the left). (a) The incident wave $\sin(kx - \omega t)$ moves to the right. (b) The reflection of that wave has the form $-\sin(kx + \omega t)$ and moves to the left. (c) The superposition of the traveling wave and its reflection is a standing wave with nodes at the walls. The arrows in parts (a) and (b) represent the direction of the waveform; in (c) the arrows indicate the actual motion of the string itself.

To sum these two terms and find the resultant superposition, we expand the trigonometric functions with identity 5.3 of Appendix IV–4. The result is

$$z(x, t) = z_1(x, t) + z_2(x, t)$$
$$= 2z_0 \sin(Kx - \Omega t) \cos\left[\left(\frac{\delta k}{2}\right)x - \left(\frac{\delta\omega}{2}\right)t\right]. \quad (15\text{–}14)$$

We have used the fact that $\cos(-a) = \cos(a)$ to arrive at this result.

Equation (15–14) states that the sum of the original two waves can be reformulated as a *product* of two traveling waves. The wavelength and frequency of the first traveling wave, which is the sine term in Eq. (15–14), are given by the *average* of corresponding quantities for the original two waves. The envelope formed by the second wave is the dashed curve in Fig. 15–6c. This envelope is a wave traveling with the same speed as the sine wave but with a much longer wavelength. The envelope wave is called a *beat* of the original two waves. The coefficient of x in the cosine factor in Eq. (15–14), $\delta k/2$, gives the wave number of the beats, whereas the coefficient of the time, $\delta\omega/2$, gives the angular frequency. In turn, we can use Eqs. (15–8) and (15–9) to find the wavelength and frequency of the beats:

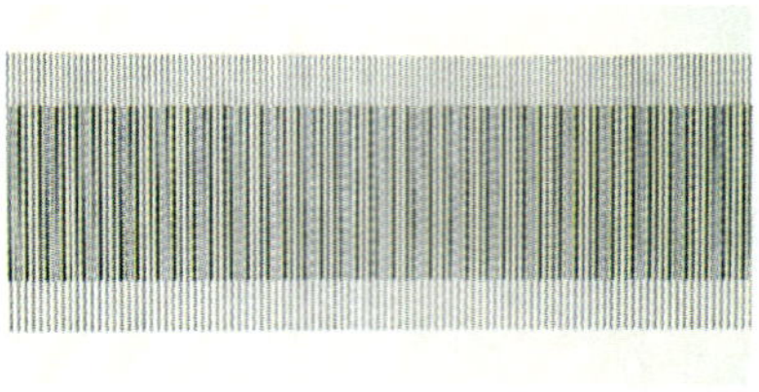

FIGURE 15–5 Two superposed combs whose tooth spacing is slightly different illustrate beats by analogy. They produce what is known as a moiré pattern.

The beat frequency

$$\frac{1}{\lambda_{\text{beat}}} = \frac{k_{\text{beat}}}{2\pi} = \frac{k_1 - k_2}{4\pi} = \frac{1}{2}\left(\frac{1}{\lambda_1} - \frac{1}{\lambda_2}\right); \quad (15\text{–}15)$$

$$f_{\text{beat}} = 2\pi\omega_{\text{beat}} = \pi(\omega_1 - \omega_2) = \frac{1}{2}(f_1 - f_2). \quad (15\text{–}16)$$

The beat has a wave number and frequency that are 1/2 the *differences* of the original wave numbers and frequencies. Because this difference is small by assumption, the wave number of the envelope is small and the wavelength is *large*. Similarly, the frequency of the beat is *small*. In deriving this result, we tacitly assumed $f_1 > f_2$, so that f_{beat} was positive. Note, however, that the beat factor in Eq. (15–14) is a cosine factor, and the cosine is unchanged if its argument changes sign. This means that if $f_2 > f_1$, f_{beat} would be given by $(f_2 - f_1)/2$ rather than $(f_1 - f_2)/2$: the beat frequency is always positive.

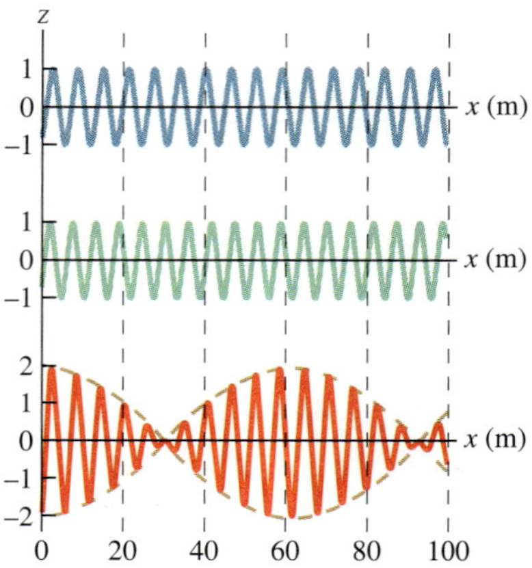

FIGURE 15–6 Two traveling waves of identical amplitude, $z_0 = 1$ m, and nearly the same frequency and wavelength, at $t = 1$ s. (a) The first traveling wave has $\omega = 1$ rad/s and $k_1 = 1\ \text{m}^{-1}$. (b) The second traveling wave has $\omega_2 = 0.9$ rad/s and $k_1 = 1.1\ \text{m}^{-1}$. (c) Beats are visible in their superposition, formed by the envelope shown as a dashed line. The vertical lines trace the regions of constructive and destructive interference.

Figure 15–7 is an enlarged version of the superposed waves plotted in Fig. 15–6c. Recall that the original two waves have $k_1 = 1.0\ \text{m}^{-1}$ and $k_2 = 1.1\ \text{m}^{-1}$. From Eq. (15–15), the beat wavelength is then

$$\lambda_{\text{beat}} = \frac{4\pi}{|(k_1 - k_2)|} = \frac{4\pi}{|(1\ \text{m}^{-1} - 1.1\ \text{m}^{-1})|} = 110\ \text{m}.$$

Only a little over half a wavelength is plotted in Fig. 15–7, and we can indeed see that $\lambda_{\text{beat}}/2 \simeq 55$ m. The frequency of the beat is, in this case,

$$f_{\text{beat}} = \pi(\omega_1 - \omega_2) = \pi(1.0\ \text{Hz} - 0.9\ \text{Hz}) = 0.31\ \text{Hz}.$$

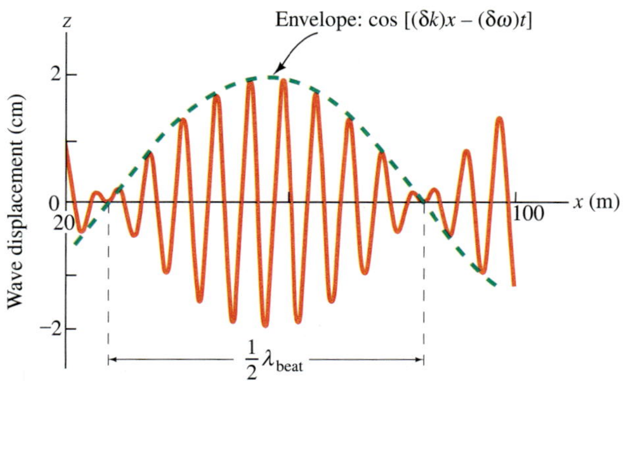

FIGURE 15–7 An enlarged look at the superposition of Fig. 15–6c. The dashed envelope is a traveling wave of wavelength $\lambda_{\text{beat}} \simeq 110$ m.

Beats in Sound. When we hear beats of sound, the combination of ear and brain has a definite physiological response. In addition to the note of average frequency, we hear a second note with the pitch of the beat frequency if the beat frequency exceeds 25 Hz or so. If the beat frequency is less than this, however, we hear regular pulses in amplitude. Because the ear interprets loudness as proportional to the wave intensity, which is in turn proportional to the amplitude squared, the ear hears a pulse every time the beat passes through either a maximum or minimum. Thus *the frequency of the perceived pulse is twice the beat frequency.*

PROBLEM SOLVING

The pulse frequency differs from the beat frequency.

EXAMPLE 15–2 Two tuning forks are each labeled 440 Hz. Yet when they are struck simultaneously, there is a 1-Hz pulsing in the strength of the sound. What is the difference in the frequencies of the two tuning forks? Is it possible to tell which one has the higher pitch from the information given?

Solution: The frequency of the *perceived* pulses occurs with twice the beat frequency:

$$f_{\text{pulse}} = 2f_{\text{beat}} = 2\pi(\omega_1 - \omega_2) = f_1 - f_2.$$

Thus $f_1 - f_2 = 1$ Hz, and this is the difference in frequencies of the forks. The frequency of the beat depends only on the absolute value of the frequency differences, so there is no way to tell which fork has the higher pitch from the information given.

FIGURE 15–8 Two sources that produce waves in a ripple tank form an interference pattern, visible as regions of constructive and destructive interference radiating out from the sources.

15–4 SPATIAL INTERFERENCE PHENOMENA

Sound waves, light waves, and water waves all exhibit *interference patterns* in space.† For example, if a laser beam is separated into two beams by passing it through two holes, it is possible to produce a regular series of intense spots where the two beams interfere constructively. A ripple tank is another useful tool for demonstrating interference patterns (Fig. 15–8). *Any* wave that propagates in two or three dimensions can show such spatial interference phenomena. Spatial interference phenomena are the basis of many precision measurements of distance, as we shall see when we study the interference of light waves in Chapters 38 and 39.

In order to see how spatial interference phenomena arise, imagine a single source of harmonic traveling waves with wavelength λ and frequency f, such as a harmonically driven loudspeaker and two pipes that carry the sound waves away from the loudspeaker. The pipes fix the path that the waves can follow, and the two paths do not necessarily have the same length even if they end up at the same place (Fig. 15–9). (This is what is special about two or three dimensions.) At the beginning of the respective paths, the waves start in phase. For example, if there is a wave crest at the start of path 1, then the same crest is at the start of path 2. These two crests each travel down their paths at the same speed. If the paths are the same length, $L_1 = L_2$, then they will both arrive at the observation point P at the same time. They will still be in phase at P, and they will interfere constructively. Under these circumstances, the adjacent troughs will arrive together, as will the next crests, and so forth. The interference will always be constructive.

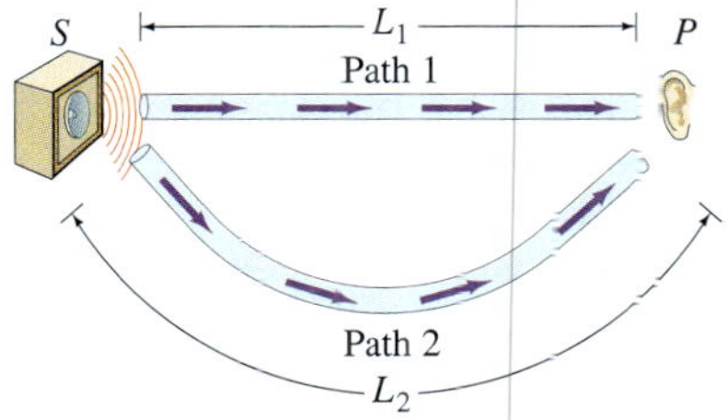

FIGURE 15–9 Unless they propagate on a string or on other one-dimensional media, waves from a source S may follow different paths to get to a given observation point P. Here two pipes of different lengths determine two different paths between S and P.

PROBLEM SOLVING

Determining the interference pattern

Conversely, if the path lengths are different, then these same crests will not arrive at the same time. Whether the interference is constructive or destructive depends on the *difference* in path length, $\Delta L = L_2 - L_1$. Suppose path 2 is longer, so that ΔL is positive. If ΔL is one full wavelength, $\Delta L = \lambda$, then the second crest will arrive at P along path 1 just when the first crest arrives at P along path 2. The two waves will be

†Not to be confused with beats, in which time dependence is primary.

in phase and the interference will be constructive. If $\Delta L = \lambda/2$, then when the crest arrives at P from path 2, it meets a trough from path 1. In this case, the interference is destructive, and no disturbance is recorded at P. The situation can be summed up as follows:

The conditions for constructive and destructive interference

for constructive interference: $\Delta L = n\lambda,\ n = 0, \pm1, \pm2, \ldots;$ (15–17)

for destructive interference: $\Delta L = (n + \frac{1}{2})\lambda,\ n = 0, \pm1, \pm2, \ldots.$ (15–18)

The effect just described also occurs with two sources at different distances from an observation point. Imagine two separate sources emitting harmonic waves of the same wavelength in a given medium, as though we were making a rhythmic disturbance at two different points on a water surface. At each point, a series of spreading ripples that interfere with one another is started. In Fig. 15–10, we locate two separated sources, S_1 and S_2, each producing waves with wavelength λ and frequency f. We assume that these waves are *in phase*; that is, a crest emerges from S_1 at the same moment that a crest emerges from S_2. These waves are *coherent*. An observation point P is located a distance L_i from source S_i; the path difference is $\Delta L = L_2 - L_1$. Then the conditions for constructive interference are simply that the crests, or the troughs, of the wave arrive at P together. For this two-source situation, the same conditions respectively apply for constructive and destructive interference as before; namely, Eqs. (15–17) and (15–18).

Locations of Maxima and Minima

The observation point P can be anywhere that the waves propagate. Some locations will have constructive interference, some destructive, and some points will have an intermediate level of interference, depending only on the distances from the sources. The locations where the interference is constructive are said to exhibit interference *maxima*, and the locations where the interference is destructive exhibit interference *minima*. Finding the spatial pattern of the maxima and minima is an exercise in geometry—of the type sketched out in Fig. 15–11.

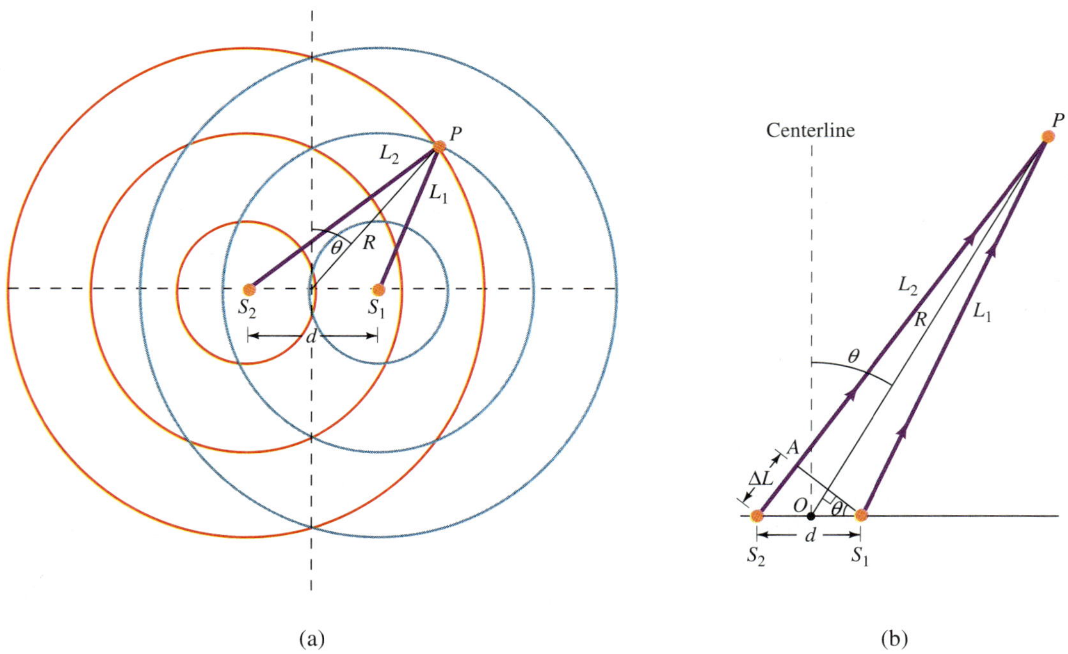

FIGURE 15–10 (a) The water waves generated by two coherent sources in phase spread away in circles from these sources. Whether the interference is constructive or destructive at observation point P depends on the difference in path lengths. The circles are (arbitrarily) drawn on the wave crests. (b) The geometry necessary to determine the path difference $\Delta L = L_2 - L_1$ and hence for the pattern of interference maxima and minima. The observation point P is far away compared to the separation of the two sources, d.

We can get a feeling for patterns of this type with a visual analogue to wave interference. Figure 15–12a shows a set of gray concentric circles around a point representing a wave source. Think of the gray circles as simplified crests and the clear areas between the circles as simplified troughs. When two such wave sources are present, as in Fig. 15–12b, we can think of constructive interference as an overlap of the gray areas (crests) with themselves or of the clear areas (troughs) with themselves; we can think of destructive interference as a uniform gray where the gray areas and the clear areas overlap. In fact, bands of maxima and minima are clearly visible in the figure.

To find the patterns of maxima or of minima defined by Eqs. (15–17) and (15–18), the necessary geometric information is the wavelength, λ, and the separation of the two sources, d. It is convenient to put an origin, O, midway between the sources and to measure the distance R from point P to the origin, as well as the angle θ between the centerline and the line from the origin to P. The centerline is the line perpendicular to and bisecting the line of the sources. The geometry is shown in Fig. 15–10b. Note that at the angle $\theta = 0$, where $L_1 = L_2$, we have $\Delta L = 0$; *there is a maximum everywhere along the centerline.* The regions of the other maxima and minima are simple to see when $R >> d$; that is, when P is very far from the pair of sources. This case is illustrated in Fig. 15–10b, where, to a good approximation, all three lines—the line from S_1 to P (S_1P), the line from the origin to P, and the line from S_2 to P (S_2P)—make the same angle θ with the centerline. Then ΔL can be measured by dropping a line from S_1 perpendicular to S_2P. This line meets S_2P at point A in the figure. S_2S_1 is perpendicular to the centerline, and AS_1 is perpendicular to S_2P, so the angle formed by lines AS_1 and S_2S_1 is θ. Because $AP = S_1P$, we have $\Delta L = L_2 - L_1 = S_2A$, and $\sin\theta = \Delta L/d$, or

$$\Delta L = d\sin\theta. \tag{15–19}$$

For $R >> d$, we can restate Eqs. (15–17) and (15–18) for the maxima and minima in terms of the angle θ by using Eq. (15–19):

$$\text{for maxima:}\quad \sin\theta = n\frac{\lambda}{d},\ n = 0, \pm1, \pm2, \ldots; \tag{15–20}$$

$$\text{for minima:}\quad \sin\theta = \left(n + \frac{1}{2}\right)\frac{\lambda}{d},\ n = 0, \pm1, \pm2, \ldots. \tag{15–21}$$

These equations determine a set of angles for which there are interference maxima and minima. Positive θ values are to the right of the centerline and negative values are to the left. The array of angles is symmetric about the centerline, so we can concentrate on the right side. The case $n = 0$ is a maximum at $\theta = 0$, which is along the centerline itself. As we increase θ, the first minimum occurs at $\sin\theta = \lambda/2d$. The second maximum is at $\sin\theta = \lambda/d$, the second minimum at $3\lambda/2d$, and so forth. In Fig. 15–12b, these angles are marked on the pattern made by overlaying two sets of concentric circles, one set slightly displaced from the other. In the original pattern from which these figures were made, $\lambda = 5$ mm and $d = 16$ mm. The first three minima should then be

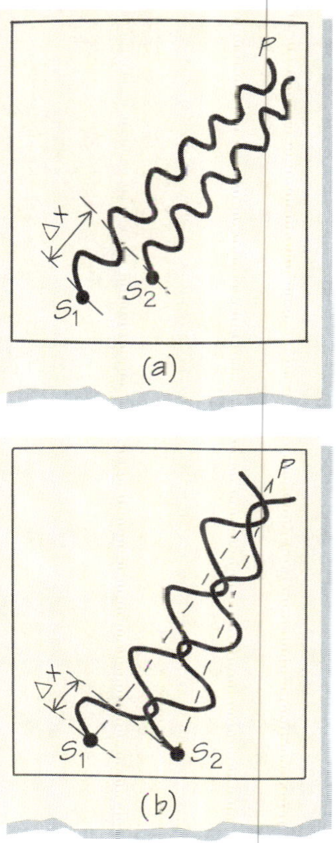

FIGURE 15–11 The conditions for maxima and minima in the interference of coherent waves of the same wavelength emitted by two sources. (a) Maxima correspond to the difference in path length equal to an integer multiple of λ, while (b) minima occur when the path difference is an integral multiple of λ plus $\lambda/2$.

The angle locations of the maxima and minima in the interference pattern of two sources in phase

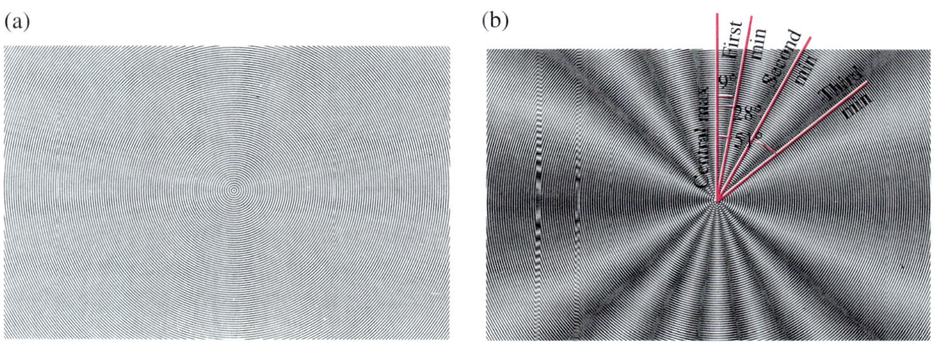

FIGURE 15–12 (a) Concentric circles representing waves propagating out from a source. (b) Overlay of the circles that spread from two adjacent sources. Regions of constructive and destructive interference, marked as max (maxima) and min (minima), are clearly visible. We have measured λ = 5 mm and d = 16 mm on the original of this reproduction, so the ratio λ/d = 0.32. The first minimum should occur at $\sin\theta_1 = \lambda/2d = 0.16$, or $\theta_1 = 9°$, and this is the value of the angle found in the figure.

at $\sin\theta_1 = 0.16$, $\sin\theta_2 = 0.47$, and $\sin\theta_3 = 0.79$, or $\theta_1 = 9°$, $\theta_2 = 28°$, and $\theta_3 = 51°$, in agreement with the figures.

Any constant phase difference between two harmonic waves—coherent waves—will lead to a definite interference pattern. Example 15–3 describes the pattern that occurs when two waves of the same wavelength have phases differing by π.

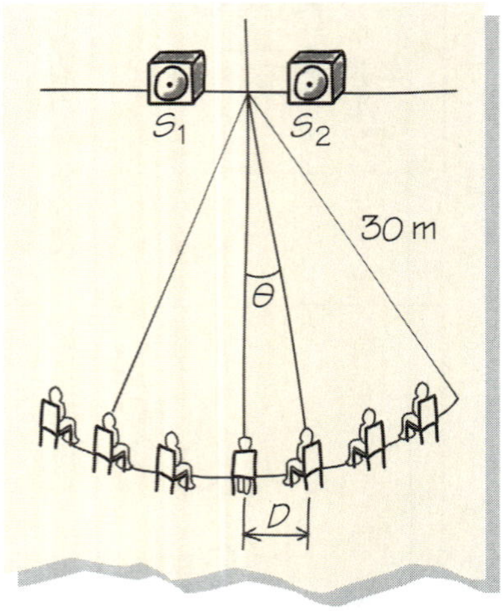

FIGURE 15–13 Example 15–3. The geometry is the same as that illustrated in Fig. 15–10b.

EXAMPLE 15–3 Two loudspeakers, S_1 and S_2, are separated by $d = 0.50$ m. These speakers form sound waves with the same amplitude at a frequency of precisely $f = 4400$ Hz. The amplifier emits the two waves 180° out of phase. A set of chairs is arranged in a semicircle 30.0 m from the speakers. Find the amplitude of the wave at the chair on the perpendicular bisector (centerline) of the line between the speakers. At what distance D to the right of this central chair is there a first maximum in sound intensity?

Solution: As long as we are not too far from the centerline, the distance D along the semicircle is close to the horizontal distance, and the geometry used for the discussion of the in-phase sources can be used here (Fig. 15–13). The fact that the sources have a phase difference of 180° means that when one speaker emits a wave crest, the other emits a wave trough. Think again about the centerline, for which the difference ΔL of the path lengths from the speaker to the observation point is zero. Along this line, the 180° phase difference implies that the wave generated by S_1 is the negative of the wave generated by S_2, and the waves cancel along the centerline. The same reasoning implies that all the results given by Eqs. (15–20) and (15–21)—for sources far from the observation point—are switched: *The maxima become minima, and vice versa*:

$$\text{for minima:} \quad \sin\theta = n\frac{\lambda}{d},\ n = 0, \pm 1, \pm 2, \ldots;$$

$$\text{for maxima:} \quad \sin\theta = \left(n + \frac{1}{2}\right)\frac{\lambda}{d},\ n = 0, \pm 1, \pm 2, \ldots.$$

The first maximum ($n = 0$) is at an angle given by

$$\sin\theta = \frac{\lambda}{2d}.$$

If the semicircle of chairs is a distance L from the speakers, then $\theta = D/L$. For small θ, $\sin\theta \simeq \theta$, and the value of D for which the first maximum occurs is

$$\sin\theta = \frac{D}{L} = \frac{\lambda}{2d};$$

$$D = \frac{\lambda}{2d}L = \frac{v}{2fd}L.$$

We have used the relation $\lambda f = v$, where v is the wave speed. The speed of sound is roughly 330 m/s, so

$$D = \frac{(330\ \text{m/s})}{2(4400\ \text{Hz})(0.50\ \text{m})}(30.0\ \text{m}) = 2.2\ \text{m}.$$

We see that $D << L$, so our small-angle approximation is justified.

Broadening and Narrowing

Interference patterns broaden as wavelengths increase and/or source separations decrease.

Note that as λ increases and/or d decreases, λ/d increases, and the angles given by Eqs. (15–20) and (15–21) increase. The pattern *spreads*, or is *broadened*. The pattern *shrinks*, or *narrows*, as λ decreases and/or d increases. For example, as the separation of our two loudspeakers decreases, the pattern of interference maxima and minima spreads in angle.

APPLICATION: SILENCE BY SOUND WAVE CANCELLATION

One modern use of the interference between waves is in the design of noise-canceling headphones in aviation (Fig. 15AB–1). Aircraft cockpits are noisy environments; even with the type of headphones that fully cover the ears, sufficient noise penetrates to cause hearing discomfort. This is particularly true for aircraft such as helicopters, which have a high power-to-weight ratio. The most recent headsets have a speaker at the outside of the headphone and a microphone located close to the entrance of the ear. Sound received at the microphone is amplified and supplied to the speaker with a phase difference, such that the sound from the speaker cancels the original signal as completely as possible just beyond the entrance to the ear, nominally at the surface of the eardrum. True communication from the radio or intercom of the aircraft is not subject to cancellation and is therefore much more audible to the crew members.

FIGURE 15AB–1 An electronic noise cancellation head set such as this one can reduce noise in the low frequency range by some 10–15 dB over conventional head sets.

15–5 PULSES

Real waves normally appear as **pulses**, which are disturbances with a definite beginning and a definite end—like a train passing in the night. Pure harmonic waves (cosines or sines) are an idealization in that they are infinitely long. Any truly periodic wave is also an idealization. However, pulses do not present any special difficulty as far as the wave equation is concerned. We saw in Chapter 14 that *any* function of the form $f(x \pm vt)$ satisfies the wave equation. The function f can describe a traveling pulse as easily as a sine or cosine.

Pulses are of interest to us because they occur throughout the physical world. Our messages, of whatever type, are communicated through pulses. The sound of a drum, a flash of light, and a ripple caused by a pebble dropped into a pond are all pulses. We will see in Section 15–7 that pulses are themselves formed from the superposition of harmonic waves. As we will now see, the superposition of pulses also provides us with insight into wave interference.

Collisions of Pulses

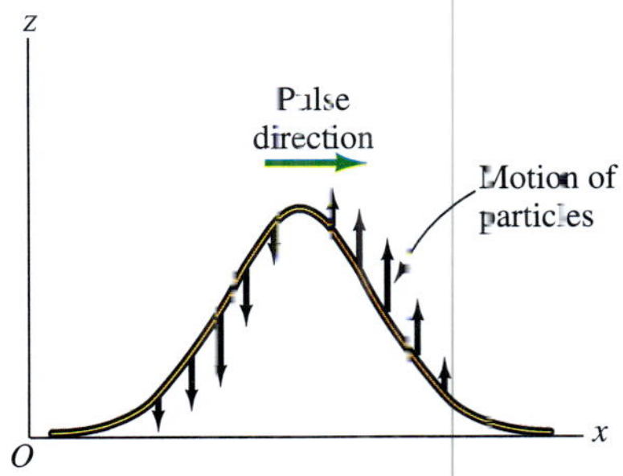

FIGURE 15–14 A pulse moving to the right (horizontal arrow) on a string. The vertical arrows indicate the velocities of elements of the string

Suppose that there is a pulse that moves to the right (Fig. 15–14). There is upward transverse motion along the leading edge of the illustrated pulse. The transverse motion vanishes at the peak, and there is downward transverse motion along the trailing edge until the string again comes to rest after the pulse has passed. The velocity of the transverse motion is indicated in the figure by vertical arrows.

Consider the superposition of two pulses of identical shape on the string: Pulse 1 is positive and moves to the right; pulse 2 is negative and moves to the left. Figure 15–15 illustrates a sequence in which the pulses approach each other, superpose, then continue on. At the moment when their respective peaks would be at exactly the same position, $t = 0$ s, the superposition of the two pulses gives a flat string. This situation is in fact no more peculiar than that of a ball that has been thrown vertically and is

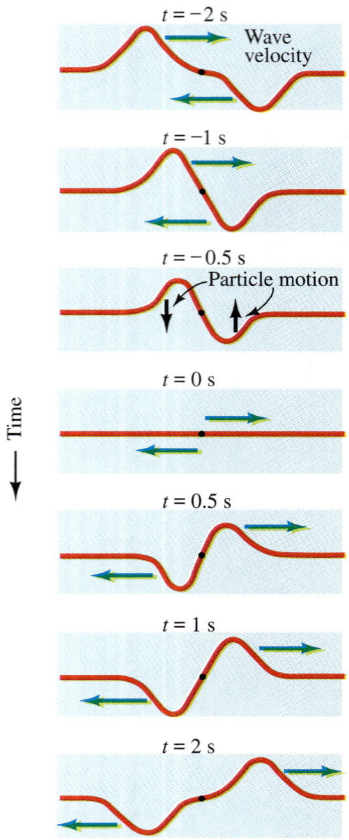

FIGURE 15–15 The superposition of two pulses, one moving to the right and one to the left. The center point, at the dot, remains motionless throughout.

A wave is inverted when it reflects from a boundary where the medium is fixed.

momentarily at rest at the peak of its flight. Even if the ball is momentarily at rest, its acceleration remains constant. Similarly, even though the string is flat momentarily, it may still be moving. Not only do the displacements of the pulses superpose, but so do their time derivatives, which measure their transverse velocities. The velocities of the string are indicated in sequence in Fig. 15–15. When the peaks superpose, the leading edge of pulse 1 is superposed on the trailing edge of pulse 2. Both edges are moving upward, so there is a region just right of the center of the superposed peaks moving upward rapidly. Similarly, the region just left of the center of the superposed peaks, consisting of the trailing edge of pulse 1 and the leading edge of pulse 2, is moving downward rapidly. Even though the string is flat for a brief moment, it is not at rest!

15–6 REFLECTION AND TRANSMISSION

When an incident ocean wave meets a wall, the wave rebounds and forms a reflected wave that moves back toward the sea. If the wall is a low one, part of the incident wave can pass over it and continue as a transmitted wave. We are familiar with the fact that waves that meet obstacles can reflect from, and be transmitted across, boundaries.

Reflection

In our discussion of two pulses of opposite sign that meet and interfere, the central point between the two pulses—marked on Fig. 15–15 by a dot—never moves at all. At this point, the displacements of the two waves have equal magnitude but are opposite in sign. This depiction helps us to think about what happens when an incident wave or pulse reflects from the fixed end of a string (or rope). The point that is stationary in the collision of two pulses can be considered to correspond to the fixed end of the string. The two colliding pulses must be of opposite sign to leave a central point stationary. Similarly, as we remarked in Section 15–2, the sinusoidal wave that reflects from a fixed wall must be inverted or else there would be a displacement at the wall. Thus, we can think of reflection as the meeting of two waves in which one wave is coming from the other side of the wall. The string and the inverted pulse or wave on the other side of the wall are purely imaginary, just an aid to picturing the reflected wave, which is not at all imaginary. Figure 15–16 shows a sequence of a pulse reflecting from the fixed end of a string. The pulse moving away from the boundary is the *reflected* pulse.

Suppose that, instead of being fixed at the wall, the end of a rope under tension were *free to move in the vertical direction.* This could be arranged by fixing the rope to a support that allows it to slide freely, as in Fig. 15–17a. The analogous sequence of reflection is shown in Fig. 15–17b. As before, the reflected pulse matches the incident pulse in shape. This time, however, the reflected pulse must be right-side up rather than inverted because the end of the string has been thrown upward by the incident pulse. In this case, we can also imagine a traveling wave on the other side of the slide support moving to the left. The two pulses meet and combine at the slide support, and the pulse height at the support is twice as high as the original amplitude. What has happened here is that the free end gets a vertical impulse when the pulse arrives. Because there is no mass to be accelerated farther down the string (there is no string to the right of the spot!), the end can move an additional amount. That the reflected wave is right-side up is confirmed by experiment.

EXAMPLE 15–4 A harmonic sound wave travels in the $+x$-direction. We can describe this wave by using the quantity z that measures the longitudinal displacement of a layer of air from its equilibrium position. z will be a function of the longitudinal position x and will take the form $z_i = A_i \cos(kx - \omega t + \phi_i)$. The wave encounters a brick wall at $x = 0$ and, because no air can cross the brick wall, the boundary condition $z = 0$ at $x = 0$ holds. What is the form of the reflected wave, and what is the form of the superposition of the incident and reflected waves?

Solution: The reflected wave will travel to the left and thus it takes the general form $z_r = A_r \cos(kx + \omega t + \phi_r)$. The boundary condition stipulates that the superposition of the two waves (the net wave) must be zero at $x = 0$: $z_i + z_r = 0$ at $x = 0$ or, for all t,

$$A_i \cos[k(0) - \omega t + \phi_i] + A_r \cos[k(0) + \omega t + \phi_r]$$
$$= A_i \cos(-\omega t + \phi_i) + A_r \cos(\omega t + \phi_r) = 0.$$

We can expand the cosine functions by using the identities in Appendix IV–4—such as $\cos(\omega t + \phi) = \cos(\omega t)\cos\phi - \sin(\omega t)\sin\phi$—and separate the coefficients of $\sin(\omega t)$ and $\cos(\omega t)$ that result:

$$(A_i \cos\phi_i + A_r \cos\phi_r)\cos(\omega t) + (A_i \sin\phi_i - A_r \sin\phi_r)\sin(\omega t) = 0.$$

If this equation is to hold for all t, the coefficients of both $\cos(\omega t)$ and $\sin(\omega t)$ must be zero. For example, if the expression is to be zero at $t = 0$, where $\sin(\omega t)$ is zero automatically, the coefficient of $\cos(\omega t)$ must be zero; if the expression is to be zero at $t = \pi/2\omega$, where $\cos(\omega t)$ is zero automatically, the coefficient of $\sin(\omega t)$ must be zero. Thus

$$A_i \cos\phi_i = -A_r \cos\phi_r;$$
$$A_i \sin\phi_i = A_r \sin\phi_r.$$

We suppose that all signs are taken care of by phases; that is, the amplitudes are positive. If we square these two equations, add them (use $\sin^2\phi + \cos^2\phi = 1$), and take the square root, we find that

$$A_i = A_r.$$

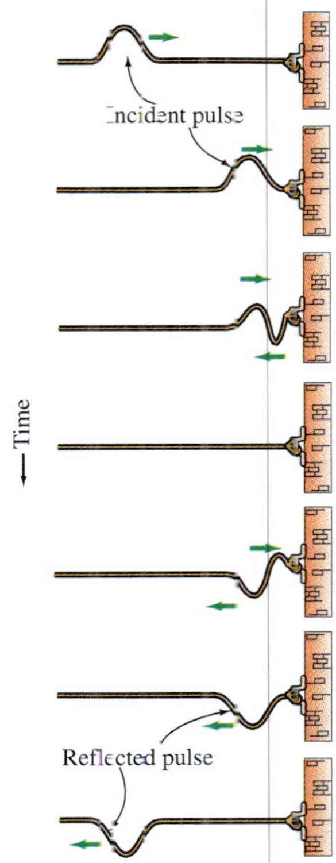

FIGURE 15–16 An incident pulse reflecting from the fixed end of a string is inverted. If the pulse is symmetric, then at some point the string is flat, because the reflected pulse cancels the incoming pulse by destructive interference.

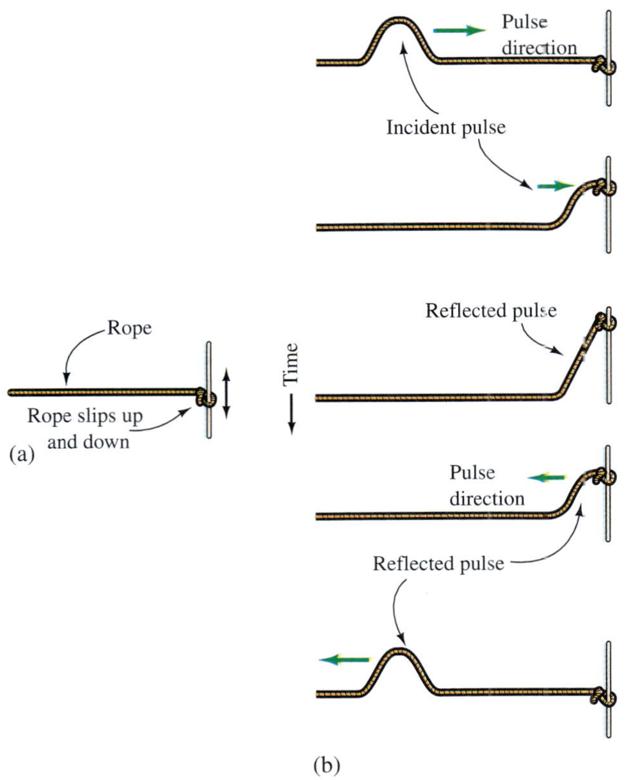

FIGURE 15–17 (a) The right-hand end of a string that is free to slide vertically. (b) The reflection of an incident pulse is upright when the end of the string is free.

This gives us the amplitude of the reflected wave. To learn about the phase, note that we can choose $\phi_i = 0$, because this is only a matter of setting the clock. In that case, $\cos\phi_i = 1$, and we have $A_i = -A_r \cos\phi_r = -A_i \cos\phi_r$, or $-1 = \cos\phi_r$. Thus $\phi_r = \pi$, and the reflected wave takes the form

$$z_r = A_i \cos(kx + \omega t + \pi) = -A_i \cos(kx + \omega t),$$

where we have again used the same identity from Appendix IV–4.

Finally, with $\phi_i = 0$, the total (superposed) wave is the algebraic sum of the incident and reflected waves:

$$z_{\text{total}} = z_i + z_r = A_i[\cos(kx - \omega t) - \cos(kx + \omega t)]$$
$$= 2A_i \sin(kx)\sin(\omega t).$$

This is a standing wave, a result we have already derived for waves on a string.

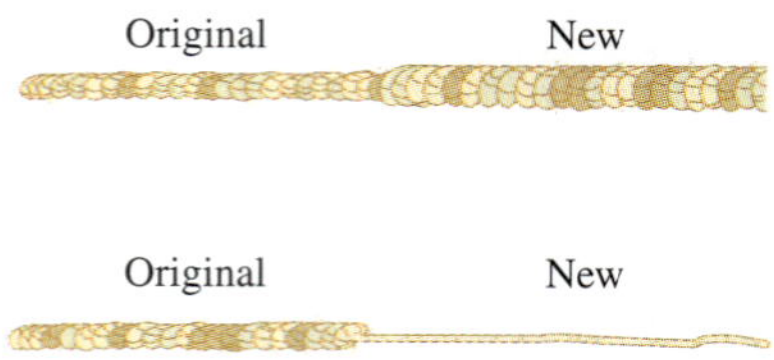

FIGURE 15–18 A string is attached to a heavier string (top) and to a lighter string (bottom).

Transmission

In addition to being reflected at boundaries, waves and pulses can also pass through boundaries separating two media. This is the process of *transmission*. Up to this point, we have supposed that one end of a string was either tightly fixed or completely free. Imagine instead that our string is attached to a new string that is heavier or lighter than the original one (Fig. 15–18). If the new string is sufficiently heavy, then it acts just like the wall; if the new string is sufficiently light, then it serves as a device that allows the end of the original string to move freely. This arrangement gives us a new feature: a *transmitted wave*, a wave carried by the new string.

Properties of Reflected Waves and Transmitted Waves

Figure 15–19a shows what happens when a pulse is incident on a heavier string and Fig. 15–19b shows what happens when a pulse is incident on a lighter string. Consider first the reflected pulse. In view of our discussion of the string with fixed or free ends,

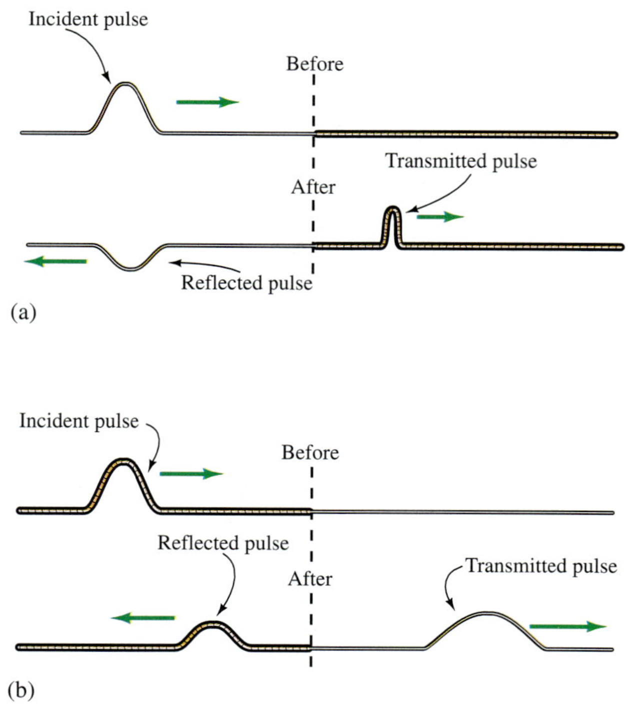

FIGURE 15–19 (a) A pulse is incident on a heavy string from a lighter one. The reflected pulse is reduced in amplitude and inverted. The transmitted pulse has an amplitude less than that of the incident pulse and moves more slowly than either the incident or reflected pulse. (b) A pulse is incident on a light string from a heavier one. The reflected pulse is reduced in amplitude and right-side up. The transmitted pulse has an amplitude less than that of the incident pulse and moves more rapidly than either the incident or reflected pulses.

it is not surprising that when the new string is heavier than the original string, the reflected pulse is inverted; when the new string is lighter, the reflected pulse is right-side up. The amplitude of the reflected pulse decreases to zero in the limit where the new string has a mass density equivalent to that of the original string because, in that case, there is no boundary. This behavior is both smooth and reasonable.

Consequences of the Conservation of Energy. If we have an incident pulse that is partly reflected and partly transmitted, then the sum of the energies in these two pulses equals the energy of the incident pulse. If there is no transmitted pulse, as is the case when the new string is either infinitely heavy or infinitely light, then the energy in the reflected pulse matches the energy in the incoming pulse. Otherwise, both the transmitted and reflected waves must share the total amplitude of the incident wave and each therefore must have an amplitude reduced from that of the incident wave.

We can state all this in a slightly different way. If the wave speed in the medium containing the incident wave is greater than the wave speed in the medium containing the transmitted wave, then the reflected wave is inverted but is otherwise similar in shape to the incident wave. The transmitted wave is always right-side up, but its shape will be stretched if the wave speed is higher in the second medium and compressed if the wave speed is lower in the second medium. This information is further summarized in Table 15–1.

EXAMPLE 15–5 A string with mass density μ_1 is attached at $x = 0$ to a second string, with mass density μ_2. The strings are under tension T. An incident harmonic traveling wave of amplitude A_i arrives on the first string. A reflected harmonic wave of amplitude A_r results on the first string, and a transmitted harmonic wave of amplitude A_t results on the second string (Fig. 15–20). The strings behave elastically so there is no mechanism for energy loss. Use the fact that energy is conserved to relate the three amplitudes.

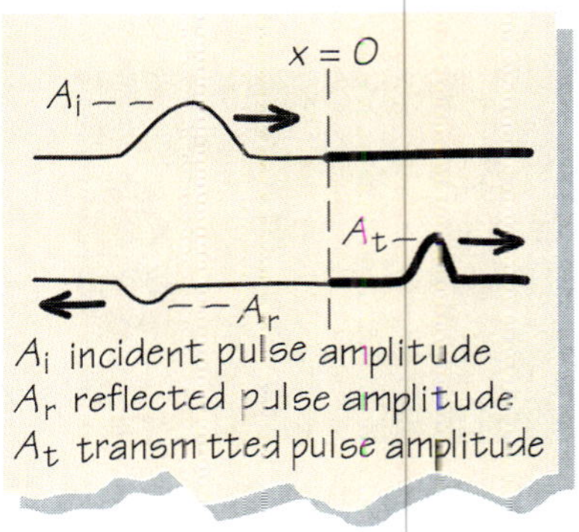

FIGURE 15–20 Example 15–5.

Solution: For traveling waves, energy conservation dictates that the rate of energy delivery that leaves the boundary in the form of reflected and transmitted waves must on average equal the rate at which energy reaches the boundary; if not, energy would be lost somewhere. This is equivalent to saying that the sum of the power delivered by the reflected and transmitted waves must equal the power delivered by the incident wave. The power delivered by traveling waves was discussed in Chapter 14; Eq. (14–33) gives the average power in a harmonic wave of amplitude A (equivalent to z_0): $P = \frac{1}{2}\mu\omega^2A^2v$. The speed of both the incident and reflected waves is v_1, whereas the speed of the transmitted wave is v_2. Thus, energy conservation gives

$$\frac{\mu_1\omega_i^2A_i^2v_1}{s} = \frac{\mu_1\omega_r^2A_r^2v_1}{2} + \frac{\mu_2\omega_t^2A_t^2v_2}{2},$$

where ω_i, ω_r, and ω_t are the frequencies of the three waves. These frequencies must all be the same because they simply represent a counting of the rate at which crests reach and leave the boundary. (If the frequency changed, wave crests would be lost or would pile up somewhere.) Finally, we can substitute the wave speeds $v_1 = \sqrt{T/\mu_1}$

TABLE 15–1 Properties of Reflected Waves and Transmitted Waves†

A Wave in Medium 1 Is Incident on a Boundary that Separates Medium 1 from Medium 2		
Wave	***Reflected Wave***	***Transmitted***
Medium 1 is less dense than medium 2 (wave speed in 1 is higher than in 2)	Inverted	Right-side up, Compressed
Medium 1 is denser than medium 2 (wave speed in 1 is lower than in 2)	Right-side up	Right-side up, Stretched

†See Fig. 15–19.

and $v_2 = \sqrt{T/\mu_2}$. The frequencies and the factors of $\sqrt{T}$ and of $\frac{1}{2}$ cancel, and we are left with the relation between the three amplitudes:

$$\sqrt{\mu_1}A_i^2 = \sqrt{\mu_1}A_r^2 + \sqrt{\mu_2}A_t^2.$$

Note that when there is no transmitted amplitude—either because there is a wall at $x = 0$ or because the end of the first string is free—then the relation is $A_i^2 = A_r^2$. Because this relation depends on the *squares* of the amplitudes, it tells us nothing about whether the reflected wave is right-side up (positive A_r) or inverted (negative A_r).

APPLICATION
Imaging with waves

Uses of Reflected and Transmitted Waves

We use reflected light waves constantly: This is how we see the world around us. Modern technology uses the phenomena of reflection and transmission in a variety of different ways. Geophysicists use sound waves to explore Earth's interior. On a small scale, explosions are set near Earth's surface; the sounds from these explosions travel outward from the source. Their arrival times are monitored at a variety of locations after they have been reflected and attenuated by their passage through different materials and across boundaries. Analysis of these measurements can reveal the presence of ore bodies or of geologic formations that may contain oil. On a large scale, earthquakes initiate waves of such intensity that they can travel, by a variety of different paths, all the way through Earth. Seismographs measure and record the arrival time and amplitude of such waves. These waves, called seismic waves, are reflected, transmitted, or both, during their travel; thus, they carry information both about the boundaries, or layers, they encounter within Earth's interior and about the nature of the material between these boundaries. Fourier analysis of the detected waves (a wave-decomposition technique that we will discuss in Section 15–7) helps geophysicists form an image from these signals.

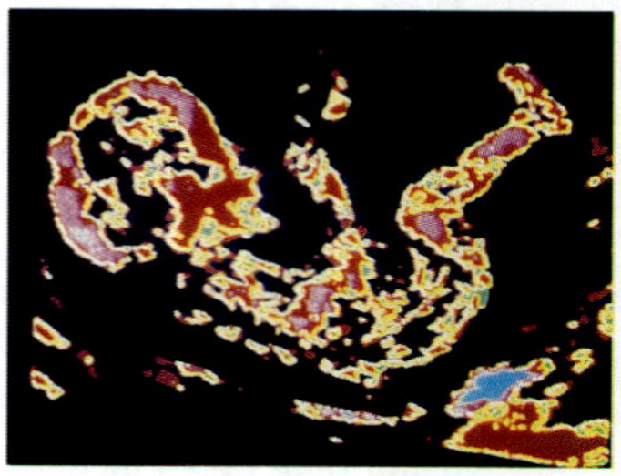

FIGURE 15–21 The colors in this ultrasound image of a human fetus are the result of computer processing of the image.

Doctors use ultrasound (sound of very high frequencies) to explore the human body (Fig. 15–21). These same techniques are also used for investigating the imperfections in materials. Radar and sonar devices emit electromagnetic waves or sound waves, then detect the arrival of their reflections, allowing us to learn about the reflecting object. Finally, Doppler shifts of the reflected wave tell us about the speed of the reflecting object.

*15–7 FOURIER DECOMPOSITION OF WAVES

Fourier's theorem

We have now studied two types of waves: single pulses, which are apparently not associated with periodic phenomena, and harmonic (or sinusoidal) waves, which are periodic. Another important category of waves are periodic waves that are not harmonic (Fig. 15–22). In fact, the superposition principle, together with a remarkable mathematical result called *Fourier's theorem*, shows that waves that fall into these three categories are very closely related. The substance of the theorem is that *any wave can be approximated by a superposition of purely harmonic (or sinusoidal) waves with different frequencies.*

The theorem states rules for finding just what harmonic components are necessary to make up a wave—what we call the *decomposition* of the wave. Moreover, it states how only a few harmonic components can form an approximation to the wave. Suppose that the wave we wish to fit with harmonic forms has period T or, equivalently, frequency $f = 1/T$, and angular frequency $\omega = 2\pi f$ (where f and ω are the *fundamental* frequency and *fundamental* angular frequency of the wave). For such waves, the period in time is related by properties of the medium (in particular, the wave speed) to a period in space—the wavelength. Thus, for a periodic wave, we could refer either to the periodic time dependence at a particular point or to a snapshot in time of a wave in space. Let's look at the periodic time dependence at a particular point, $f(t)$, of the wave we want to fit. This fit is known as a *Fourier expansion* and takes the general form

$$f(t) = \sum_{n=1}^{\infty} A_n \sin(n\omega t + \phi_n). \tag{15–22}$$

The only frequencies that enter into the sum are integer multiples of the fundamental frequency, $f_n = nf$. These frequencies are called harmonics of the fundamental frequency, with second harmonic referring to $n = 2$, third harmonic referring to $n = 3$, and so on.

If, instead of concentrating on the time dependence, we take a snapshot of a periodic wave $g(x)$ in space, then that wave, however complex in form, has a fundamental wavelength λ or a fundamental wave number $k = 2\pi/\lambda$. The Fourier expansion of our wave takes the form

$$g(x) = \sum_{n=1}^{\infty} A'_n \sin(nkx + \theta_n). \tag{15–23}$$

Here, the higher harmonics have wave numbers k_n that are integer multiples of the fundamental wave number. The Fourier expansion specifies rules for the calculation of the constants in Eqs. (15–22) and (15–23).

If the infinite sum in Eq. (15–22) or (15–23) is truncated (cut off), the resulting finite sum forms an approximation to the original waveform. Such approximations are quite useful. We may want to learn about the important frequencies in a particular waveform so that we can suppress or enhance certain modes. For example, in the construction of bridges or buildings, it is important to suppress certain types of movement by mechanical reinforcements, so that frequencies that are resonant with external forces are eliminated and the system cannot be driven to large amplitude oscillations.

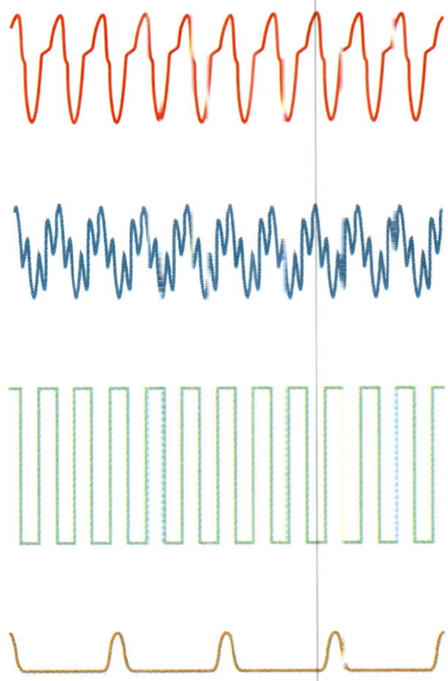

FIGURE 15–22 Some nonharmonic periodic waves.

APPLICATION

Using the Fourier expansion

Fourier Decomposition of a Triangular Wave

To demonstrate Fourier decomposition, let's try to fit the triangular wave illustrated in Fig. 15–23 with harmonic functions. (Such waves might be used in electronics applications where they are generated precisely by the superposition of harmonic waves.) The triangular wave has a period T and an angular frequency $\omega = 2\pi/T$. The phases ϕ_n in the Fourier expansion for the triangular wave in the figure are equal to $\pi/2$, which means that only cosines appear in the expansion. The triangular waveform is symmetric about $t = 0$, and the cosine also has this symmetry. The cosine is thus a "natural" expansion function for the triangular wave. Cosines that reach negative maxima at $t = \pm T/2$ would make it easy to build the sharp negative trough of the triangular wave at these points. This feature is shared by $\cos(\omega t)$, $\cos(3\omega t)$, $\cos(5\omega t)$, and so forth, but *not* by $\cos(2\omega t)$, $\cos(4\omega t)$, and so on. The calculation of A_n bears this out: $A_2, A_4, \ldots$ all turn out to be zero. When the expansion rules are applied, the exact Fourier expansion of our wave is

$$z(t) = \frac{8}{\pi^2} z_0 \sum_{n=\text{odd}} \frac{1}{n^2} \cos(n\omega t)$$

$$= (0.81) z_0 \left[\cos(\omega t) + \frac{1}{9}\cos(3\omega t) + \frac{1}{25}\cos(5\omega t) + \cdots \right]. \tag{15–24}$$

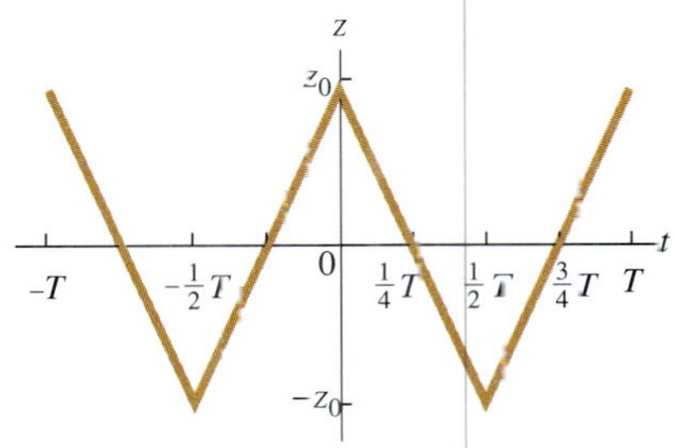

FIGURE 15–23 A triangular wave of period T. The time dependence of the transverse displacement z of a string is plotted at some fixed point in space.

Figure 15–24 illustrates this Fourier expansion. Figure 15–24a shows the individual terms $z_1 = 0.81 z_0 \cos(\omega t)$ and $z_2 = (0.81/9) z_0 \cos(3\omega t)$. Figure 15–24b shows how

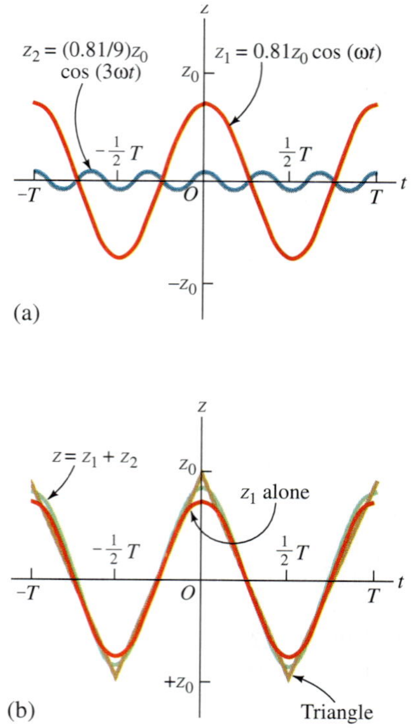

FIGURE 15–24 Fourier analysis of the triangular waveform of Fig. 15–22. (a) The first two contributing harmonics, $z_1 = 0.81z_0 \cos(\omega t)$ and $z_2 = (0.81/9)z_0 \cos(3\omega t)$. (b) The triangular waveform, the expression z_1 alone, and an approximation containing the first two terms of Eq. (15–24). The approximation with z_1 alone is already a good one, and the two-term expression is an even better one.

first one term then two terms of Eq. (15–24) match the triangular wave. The $\cos(\omega t)$ term alone does quite well in approximating the triangular wave. By the time the second term is included, the approximation is excellent.

The Fourier theorem is of great importance in engineering and physics because of its relevance to the solution of many different problems. In addition to the applications in vibrations and wave motion that are simply sketched out here, heat flow problems and the solutions of a great class of problems in electricity and magnetism require the Fourier expansion.

Fourier Analysis and the Ear

Our ears work very well as Fourier analysis devices. When the note A is struck on a tuning fork designated A, a pure harmonic wave of period $T = 1/(440 \text{ Hz}) = 2.27 \times 10^{-3}$ s emerges. The same note played on a piano, a violin, a trombone, an oboe, or a saxophone contains a harmonic wave of the same fundamental frequency. But each instrument emits its own complicated waveform (Fig. 15–25). In the language of Fourier analysis, this means that a note played by any of those instruments contains vast numbers of differing harmonic components. The ear is capable of picking out the fundamental frequency; we can all agree when the same note is played by a piano and an electric guitar. It is the higher components by which the ear differentiates between different instruments or between different voices. Even different violins have very different higher harmonic contents, a very subtle matter as far as the construction of violins is concerned. Although the physical differences between Stradivarius violins and others of lesser value are so slight that attempts to duplicate "Strads" fail, the sounds of Strads are generally more appealing than those of ordinary violins. A Fourier analysis shows that better violins have a far greater component of higher harmonics. Our ears can even distinguish between individual violins made by the same master!

*15–8 PULSES AND THE UNCERTAINTY PRINCIPLE

We saw a different version of the uncertainty principle in Chapter 13.

When we treat pulses with Fourier's theorem, we discover a fundamental relation between the width of the pulse and the frequencies of the periodic waves of which it is composed. This relation is known as the **uncertainty principle**. In Section 15–7, we used Fourier's theorem to treat general periodic waves. At first sight, pulses would seem to have little to do with such periodic waves. Pulses are not periodic. Actually, however, there is a way to think of pulses as periodic, and this is illustrated in the last drawing in Fig. 15–22. Here we see a periodic wave that is a series of short, repeated pulses, with a very long period, T, between the pulses, or, equivalently, separated by a very long wavelength, λ. This means that the fundamental angular frequency, ω (or the fundamental wave number, k) is very small, because $\omega = 2\pi/T$ (or $k = 2\pi/\lambda$).

A true pulse in time is like the periodic sequence shown in the limit that $T \to \infty$, or $\omega \to 0$. Similarly, a true pulse in space is like the periodic sequence shown in the limit that $\lambda \to \infty$, or $k \to 0$. In the Fourier expansion, the sum is taken over integer multiples of angular frequencies (wave numbers). Therefore, *all* frequencies enter into the sum and *the sum in the Fourier expansion is converted into an integral.*

Fourier Decomposition of Pulses

The range of frequencies that enter into the Fourier expansion of a pulse is limited.

What frequencies are involved in the Fourier expansion of a pulse? Although the exact answer to this question depends on the shape of the pulse, we can state an important principle. *If a pulse is strictly limited in time (or space), then it is strictly limited in the range of angular frequencies (or wave numbers) that enter into it.* In other words, the angular frequencies that contribute are in a narrow range, and we can estimate that range by the following argument. The different frequencies that contribute to a pulse interfere constructively in the region where the pulse is strong and interfere destructively in the region where the pulse drops to zero. Suppose that a pulse lasts a time Δt; we say it has a **width** in time of Δt. Suppose also that the contributing frequencies go from a minimum angular frequency ω_{min} to a maximum ω_{max}, with $\Delta\omega = \omega_{\text{max}} - \omega_{\text{min}}$. We want these waves to be out of phase at the beginning of the pulse, to be in phase at the center of the pulse, and then to be out of phase once more at the end of the pulse. If, in time span Δt, the number of periods of the waves of frequency ω_{max} differs by just one from the number of periods of the waves of frequency ω_{min}, then waves that are out of phase at the beginning of the pulse will be in phase at the center and out of phase at the end. Now, the number of periods of a wave of angular frequency ω contained in a time span Δt is

$$\frac{\Delta t}{T} = \Delta t\frac{\omega}{2\pi}.$$

Our condition is then

$$\frac{\Delta t}{2\pi}(\omega_{\text{max}} - \omega_{\text{min}}) \simeq 1;$$

$$(\Delta t)(\Delta\omega) \simeq 2\pi. \tag{15–25}$$

This equation is a **reciprocal relation**; if Δt is large, $\Delta\omega$ is small—and vice versa—because 2π is a constant. A more detailed mathematical study shows that this equation holds for all pulses, even if the value of the constant on the right-hand side of Eq. (15–25) might be slightly different for different mathematical pulse forms. A very sharp pulse in time contains a very broad range of harmonic components. Similarly, a pulse that is very broad in time will have only a limited range of angular frequencies.

A more commonly used reciprocal relation follows if we use the frequency, $f = \omega/2\pi$, in place of the angular frequency, ω:

The reciprocal relation for frequency and time

$$(\Delta t)(\Delta f) \simeq 1. \tag{15–26}$$

It is also worthwhile to see how the reciprocal relation works when we study a pulse in space rather than time. Here, the breakdown of the pulse is in terms of a continuous range of wavelengths rather than angular frequencies. Suppose that the pulse has width Δx in space. Then the same method of estimating the range of wavelengths, $\Delta\lambda$, of the contributing harmonic waves can be used. The condition would then be

$$\Delta x\left(\frac{1}{\lambda_{\text{min}}} - \frac{1}{\lambda_{\text{max}}}\right) \simeq 1,$$

or, in terms of the wave number, $k = 2\pi/\lambda$,

The reciprocal relation for width in space and for wave number

$$(\Delta x)(\Delta k) \simeq 2\pi. \tag{15–27}$$

Here, $\Delta k = k_{\text{max}} - k_{\text{min}}$ is the range of wave numbers that must enter into the construction of a pulse whose width in space is Δx. Again, the narrower Δx is, the larger the range of wave numbers must be.

There is a very real place where the reciprocal relation is relevant: communication by signal construction. Pure harmonic waves cannot convey information because they have neither a beginning nor an end. Signals and information transfer are necessarily in the form of pulses. For example, all movement of information within and between

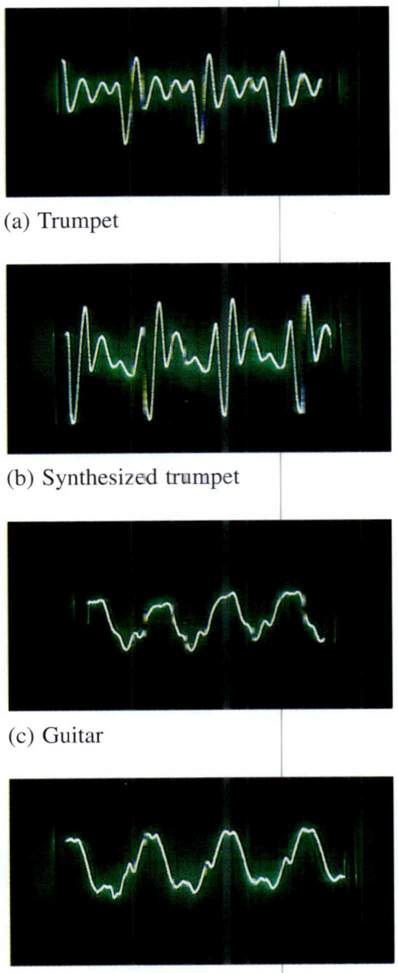

(a) Trumpet

(b) Synthesized trumpet

(c) Guitar

(d) Synthesized guitar

FIGURE 15–25 The characteristic sounds of different instruments are associated with frequencies above that of the sound of the "note" being played. These higher frequencies are known as overtones, and their presence makes the waveform of the sound different than the purely harmonic sine or cosine form. A synthesizer electronically generates the characteristic pattern of overtones to trick the ear. Shown in Figs. (a) to (d) are the note of F, played, respectively, by a trumpet, a synthesized trumpet, a guitar, and a synthesized guitar.

computers is done with sequences of pulses in binary form. Rapid communication of information requires that the pulses carrying the information be very narrow. If not, and, in particular, if pulses overlap to make wider pulses, information is lost in the confusion between the pulses. According to the reciprocal relation, this means that a very broad spectrum of frequencies must go into the construction of these narrow pulses: The narrower the pulse, the broader the range of frequencies must be. But this costs money—in a very literal sense. The energy density and power in harmonic waves of frequency ω are proportional to ω^2, as discussed in Section 14–5. The reciprocal relation forces a trade-off: A sharper pulse, which is more effective in transferring information, is more expensive.

APPLICATION
The costs of communication

Uncertainty Relations

In the context of the microscopic world of atoms, the reciprocal relations we have discussed are called **uncertainty relations**; the fact that such relations come into play is the *uncertainty principle.* In atomic physics, quantum mechanics is important and the key idea in quantum mechanics is that all matter has wavelike characteristics. Although the words "uncertainty principle" are often used only in connection with quantum mechanics, the principle really applies to any wavelike phenomenon. The trade-off in the context of quantum mechanics involves the position and speed of microscopic particles. If you measure the position of a particle at any given time with high precision, then you can know the speed—or the momentum—only with poor precision, and vice versa. The momentum is closely related to a wave number in quantum mechanics. We will discuss this further in Chapter 41.

SUMMARY

The algebraic sum, or superposition, of two solutions of the wave equation is also a solution. The collection of physical phenomena that follow from this fact fall under the general heading of wave interference. When two waves that are superposed have opposite algebraic signs at a particular point or a particular moment of time, they interfere destructively. When they have the same sign, they interfere constructively.

Standing waves can be considered to be the result of the superposition of traveling waves moving in opposite directions. Standing waves on a string can be decomposed into the superposition of traveling waves with their own reflections from the ends.

When two waves with nearly equal frequencies, f_1 and f_2, are superposed, the result is that the original wave is modulated by a wave of much smaller frequency and larger wavelength called a beat. The beat frequency is

$$f_{\text{beat}} = \frac{1}{2}(f_1 - f_2). \qquad (15\text{–}16)$$

Two waves of the same wavelength are coherent if there is a constant phase difference between them. If the distances that the waves travel differ by ΔL, an interference pattern that depends only on ΔL and on the phase difference is traced out. If the phase difference at the source is zero, the interference pattern is

$$\text{for constructive interference:}\quad \Delta L = n\lambda,\ n = 0, \pm 1, \pm 2, \ldots; \qquad (15\text{–}17)$$

$$\text{for destructive interference:}\quad \Delta L = (n + \frac{1}{2})\lambda,\ n = 0, \pm 1, \pm 2, \ldots. \qquad (15\text{–}18)$$

Suppose that two coherent wave sources of wavelength λ and zero phase difference are separated by a distance d that is much smaller than the distance to the observation point. Then there is an interference pattern that is a function of the angle θ made with the perpendicular bisector of the line separating the wave sources:

$$\text{for maxima:} \quad \sin\theta = n\frac{\lambda}{d}, \quad n = 0, \pm 1, \pm 2, \ldots; \tag{15–20}$$

$$\text{for minima:} \quad \sin\theta = \left(n + \frac{1}{2}\right)\frac{\lambda}{d}, \quad n = 0, \pm 1, \pm 2, \ldots. \tag{15–21}$$

The pattern described by these equations broadens (the angle θ corresponding to a particular minimum or maximum increases) as the wavelengths involved increase and/or the separation between the sources decreases. This is a very general feature of interference patterns.

Waves reflect from boundaries and are transmitted past boundaries that separate one wave-carrying medium from another. A reflected wave is inverted or right-side up, compared to the original wave, according to whether the medium from which the wave reflects is denser or less dense than the original medium.

Fourier's theorem shows that any wave, both periodic and nonperiodic (pulses), can be arbitrarily approximated as well as we like by a superposition of purely harmonic waves with different frequencies. This result allows for the construction of arbitrary waveforms from harmonic waves and for the analysis of physical systems in terms of harmonic waves.

A pulse can also be treated by Fourier analysis. If a pulse has a time span of Δt, then the range of frequencies, Δf, that comprise that pulse is limited:

$$(\Delta t)(\Delta f) \simeq 1. \tag{15–26}$$

There is a similar relation between the range in space of the pulse, Δx, and the range in wave numbers, Δk, that enter into the pulse's decomposition:

$$(\Delta x)(\Delta k) \simeq 2\pi. \tag{15–27}$$

These relations hold whenever wave phenomena are present and explain why sharp information-carrying signals are expensive to generate.

UNDERSTANDING KEY CONCEPTS

1. A standing wave on a string can be constructed from two traveling waves that move in opposite directions. Can a traveling wave be constructed from the sum of two standing waves?
2. When transverse positive and negative pulses that have the same symmetric shape and size but travel in opposite directions meet, is it necessary that there be a moment when the string or wire on which they move is flat? If so, how do the pulses "know" to continue moving on the string?
3. Two people on each end of a long rope send off a wave pulse. If both wave pulses are on the same side of the rope, describe what happens when the pulses meet. What about when the pulses are on opposite sides? What happens when one pulse is oriented at a 90° angle to the second pulse?
4. Listening to the beat frequency between an unknown and a standard (known) tuning fork tells you only about the difference between the two frequencies. The unknown frequency could be larger or smaller than the known frequency. With a piece of chewing gum at your disposal, can you devise a method to determine the unknown frequency?
5. A wave with truly sharp edges, such as the triangular wave discussed in Section 15–7, can never exist on a string in nature. What might limit how sharp the changes can be?
6. Consider a sinusoidal traveling wave that moves down a sequence of successively lighter strings tied together end to end. The wave starts at the end containing the heaviest string. Qualitatively describe what happens to the frequency and wavelength of the wave by the time it reaches the lightest end of the system.
7. A pebble dropped into still water produces a pulse. The speed of water waves of different wavelengths differ from one another. Why does a dropped pebble produce a series of spreading concentric waves rather than a single wave front?
8. How do you reconcile the principle of conservation of energy with the observation that, in one region, the wave medium is not in motion when waves interfere destructively?
9. In Section 15–1, we referred to two waves that move to the right and cancel; they give no wave at all. Yet each wave alone has an energy density. What happens to the energy? [*Hint:* What are the implications of the superposition principle for the energy density?]
10. Musicians often use beats to tune to some standard, such as an oboe. Can a trumpet player tell from the beat frequency whether one particular note is sharp (at a higher frequency) or flat (at a lower frequency) compared to the same note played on the oboe?

PROBLEMS

15–1 *The Superposition Principle*

1. (I) Show that the superposition of the two traveling waves $\sin(kx - \omega t)$ and $\cos(kx - \omega t)$ can be written in the form $A \sin(kx - \omega t + \phi)$, and find A and ϕ.

2. (I) Two traveling waves have the forms $z_1 = z_0 \sin(kx - \omega t)$ and $z_2 = z_0 \sin(kx - \omega t + \pi/2)$, respectively. Sketch the superposition of these two waves.

3. (I) Work out the result of the superposition of the two waves of Problem 2 in the form $z = z_1 + z_2 = A \sin(kx - \omega t + \delta)$.

4. (II) Consider the superposition of two harmonic waves, $\psi(x, t) = A \sin(kx - \omega t) + 2A \cos(kx - \omega t + \theta)$. (a) For what values of θ will this superposition result in the maximum constructive interference? (b) the maximum destructive interference? (c) If this superposition is written in the form $B \sin(kx - \omega t + \alpha)$, express B and α in terms of A and θ.

5. (II) Two sinusoidal waves with the same frequency travel down a long rope (Fig. 15–26). The waves have amplitudes 3 cm and 2 cm, respectively, but the second wave trails the first wave by a phase of $\pi/2$. Determine the amplitude of the resultant motion.

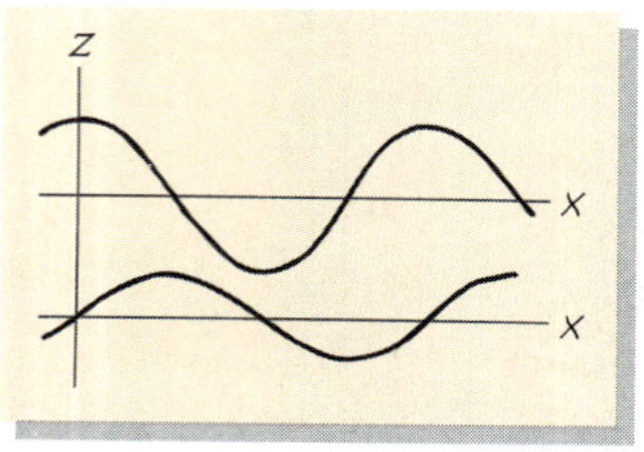

FIGURE 15–26 Problem 5.

6. (II) Four sinusoidal waves have the same frequency, but the first two waves have amplitudes twice that of the other two waves. The phases of the four waves are 0, $\pi/2$, π, and $3\pi/2$, respectively. Plot the resulting superposed wave, and describe its motion.

7. (II) Two sine waves, $A \sin(\omega t + kx)$ and $A \sin(\omega t + kx + \delta)$, combine to form a sine wave with the same amplitude A. For what values of δ of the second wave is this possible?

15–2 *Standing Waves through Interference*

8. (I) Two waves are described by $\psi_1 = (5.0\text{ cm}) \sin(kx - \omega t)$ and $\psi_2 = (5.0\text{ cm}) \sin(kx + \omega t)$, where $k = \pi/2\text{ cm}^{-1}$ and $\omega = 3\pi\text{ s}^{-1}$. (a) Show that the addition of the two waves produces a standing wave. (b) Determine the amplitude and frequency of the resulting standing wave.

9. (I) Consider traveling waves moving in opposite directions: one has the form $z_r(x, t) = z_0 \sin(kx - \omega t)$ and the other has the form $z_l(x, t) = z_0 \cos(kx + \omega t)$. Do these waves add to a standing wave? If so, what is the expression for the standing wave, and what are the locations of the nodes? Note that $\cos\theta$ can be written as $\sin[\theta + (\pi/2)]$.

10. (II) Show by a specific construction that two standing waves can add up to give a traveling wave. What standing wave would you have to add to $A(\cos kx)(\cos \omega t)$ to get a wave traveling to the left?

11. (II) Two traveling waves that move in opposite directions along the x-axis interfere to produce a standing wave of the form $6 \sin(kx) \cos(\omega t)$, where $k = 3\pi\text{ m}^{-1}$ and $\omega = 22\pi\text{ s}^{-1}$. (a) What are the frequencies and speeds of the traveling waves? (b) What are the forms of the traveling waves?

12. (II) One harmonic of an organ pipe has a single node at 2/3 of the length of the pipe and an antinode at the open end (see Fig. 14–20). Write an expression to describe this mode as a standing wave. Find two traveling waves that produce the same result. The length of the pipe is 1.2 m and the speed of sound is 330 m/s.

13. (II) Show that the sum of a wave that travels to the right and a wave that travels to the left with the same amplitude, wave number, and frequency is a standing wave, independent of phase. [*Hint*: Consider $\psi(x, t) = A \sin(kx - \omega t) + A \sin(kx + \omega t + \theta)$; define $u \equiv kx + (\theta/2)$ and $v \equiv \omega t + (\theta/2)$, and use the appropriate identities.]

14. (II) The most general forms of two traveling waves that move in opposite directions along the x-axis with the same amplitude, wavelength, and frequency are $z_r(x, t) = z_0 \sin(kx - \omega t + \phi_1)$ and $z_l(x, t) = z_0 \sin(kx + \omega t + \phi_2)$. What is the form of the standing wave that results from their superposition?

15. (II) A harmonic wave traveling to the right along the x-axis has its maximum amplitude at the time $t = 0$ s at the point $x = 0$ m. Superimposed on this is a wave traveling to the left with equal amplitude, wavelength, and frequency, such that the net standing wave has nodes at $x = 3$ m and at $x = 6$ m. What is the largest possible value of the wavelength?

15–3 *Beats*

16. (I) You have a tuning fork of unknown frequency and a tuning fork of frequency 264 Hz. Combined, the two produce a pulse frequency ($f_{\text{pulse}} = 2f_{\text{beat}}$) of 2 Hz. What are the possible values of the frequency emitted by the first tuning fork?

17. (I) You have a tuning fork of frequency 512 Hz and look for another tuning fork of smaller frequency so that there is a perceived beat frequency of 6 Hz between them. What frequency should the other tuning fork have?

18. (I) Two violinists play their (identical) A-strings, nominally 440 Hz. A beat is heard at a pulsing of 2 Hz. What is the magnitude of the fractional difference in wave speed, $(v_1 - v_2)/v_1$, for the two strings? [*Hint*: The length L of a violin string is the same from violin to violin.]

19. (I) Two strings of the same material on a pair of guitars are in tune with a frequency of 414 Hz. The string of one of them is tightened slightly, and you hear a beat frequency of 8 Hz. What is the frequency of the tightened string?

20. (I) Two speakers are wired so that they emit sound coherently. Because of a manufacturing error, the same electrical impulse that leads one to emit sound of wavelength 80.10 cm leads the other one to emit sound of wavelength 79.71 cm. What is the pulse frequency ($f_{\text{pulse}} = 2f_{\text{beat}}$)?

21. (II) Two identical piano wires have a frequency of 512 Hz when under the same tension. As a consequence of slippage,

the tension of one of the wires changes slightly, so that its frequency decreases slightly. If 4 beats/s are heard when the wires both vibrate, what is the fractional change in the tension of the lower-frequency string?

15–4 *Spatial Interference Phenomena*

22. (I) Two speakers that emit waves with the same wavelength λ are placed at $x = L$ and $x = -L$, respectively. Assuming that the speakers emit waves with equal amplitudes A and that the phase of the wave on emission is zero at $t = 0$ in both cases, what is the form of the traveling waves that travel toward a microphone located between them at $x = 0$?

23. (I) Find the waveform picked up by the microphone in Problem 22; that is, the superposition of the two waves arriving at the microphone.

24. (I) Two speakers separated by 30 cm are fed with the same pure sinusoidal wave of 1500 Hz. A microphone is placed 6.0 m away along the centerline between the speakers (Fig. 15–27). It picks up an intensity maximum. It is then moved slowly out to the side. (a) How far out to the side, in meters, does the microphone have to be moved to pick up a first minimum? (b) How much farther does the microphone now have to be moved to pick up a second maximum? (c) How do the answers to parts (a) and (b) change if the speakers are now separated by just 10 cm? The approximation $\sin\theta \simeq \theta$ is not valid here, although the angles of the lines from the speakers to the microphone are nearly equal.

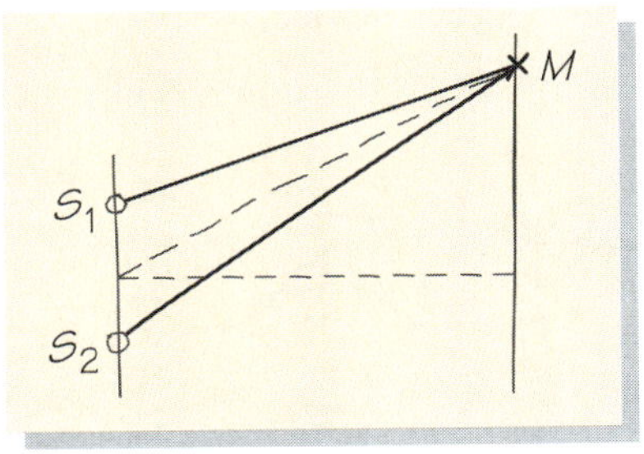

FIGURE 15–27 Problem 24.

25. (I) Two coherent sources are 3.0 cm apart and make harmonic ripples of the same frequency (see Fig. 15–8). Consider the ripples along a straight line L_1 parallel to the line L_2 that connects the sources and 40 cm away from line L_2. If the distance between the central maximum at L_1 and the next maximum on L_1 is 8.0 cm, what is the wavelength of the ripples?

26. (I) Sound of frequency 600 Hz is emitted in phase from two speakers 1.4 m apart (Fig. 15–28). A microphone can move along a line parallel to the line that joins the two speakers and is 10 m from the base of the speakers; it measures sound intensity, which can then be plotted versus distance. The interference produced is constructive in the center. Where is the first minimum in the sound intensity curve along the track of the microphone?

27. (II) Two coherent wave sources 5 cm apart make ripples in a water tank (see Fig. 15–8). The wavelength of the waves is 3 cm. (a) Determine the angle between the perpendicular bisector of the line segment connecting the two wave sources and

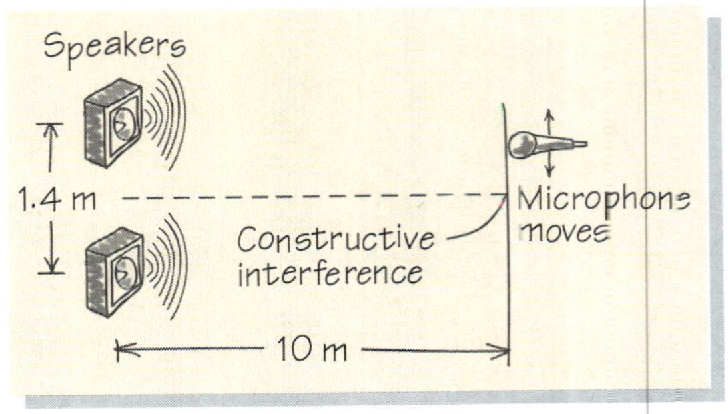

FIGURE 15–28 Problem 26.

the direction of the first noncentral maximum. (b) What is the direction of the first maximum if the two sources operate with opposite phases? (c) How will the answers to (a) and (b) change if the wavelength is changed to 6 cm?

28. (II) Harmonic light waves are emitted by a source equidistant between two narrow vertical slits 0.2 mm apart in a screen (Fig. 15–29). The light waves detected at a screen parallel to the screen containing the slits but 25 cm away have the usual maximum at the point opposite the midpoint of the two slits, and the first minimum occurs at a point 0.4 mm from the maximum. What is the wavelength of the light?

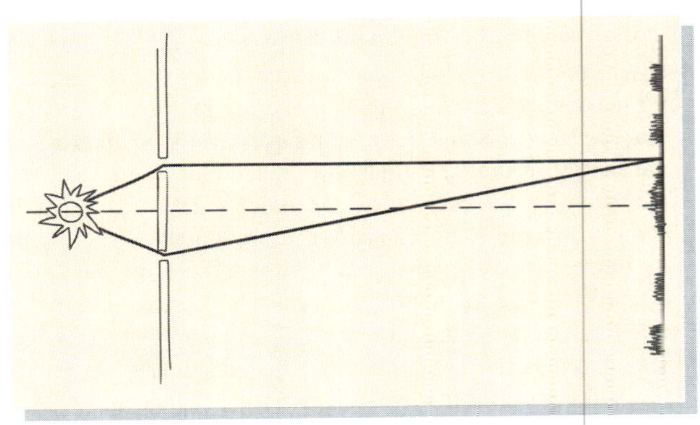

FIGURE 15–29 Problem 28.

29. (II) Coherent waves emerge from two slits separated by a distance $2a$. A screen is placed at a distance L away from the slits and the first interference minimum is found at a distance d from the center point. Use the small-angle approximation to show that the relationship between the wavelength (λ) and the other lengths is given by $\lambda = 4ad/L$.

30. (II) Two tuning forks vibrating in phase with frequency 440 Hz are placed 2 m apart. A detector is moved in a circle of radius 40 m around the center of the tuning forks (Fig. 15–30). Plot the approximate variation of the square of the net amplitude as a function of angle. Neglect the small variation in the magnitude of the amplitude that results from the sources not being at the same distance from the detector. Take the velocity of sound to be 330 m/s.

31. (II) An oscillator, located at $(x, y) = (-2\text{ cm}, 0\text{ cm})$ that dips in and out of water in a tank produces a water wave of the form $\psi_1 = (3\text{ cm})\sin(\pi r_1 - \pi v t)$, where r_1 is the distance from the oscillator; a second oscillator, located at $(x,$

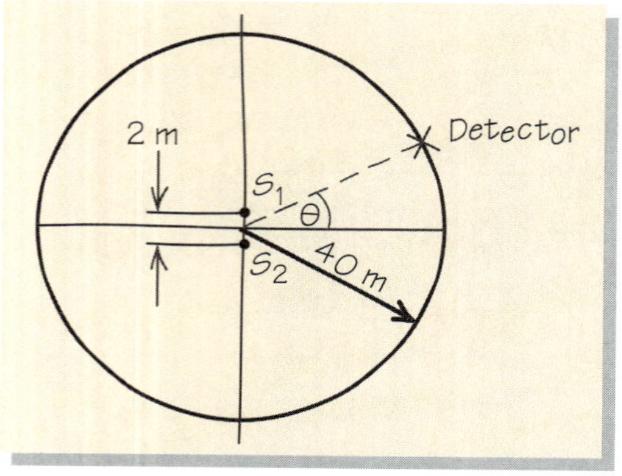

FIGURE 15–30 Problem 30.

$y) = (+2$ cm, 0 cm), produces a water wave of the form $\psi_2 = (3 \text{ cm}) \sin(\pi r_2 - \pi v t)$, where r_2 is the distance from the second oscillator. (a) Where on the line between the sources (the x-axis) are the interference maxima located? (b) What is the form of the wave along the line $x = 0$ cm? (Take all times in seconds.)

32. (II) Two tuning forks that vibrate coherently (in phase) at 512 Hz are placed 0.5 m apart on a horizontal surface. A circular wall of radius 10 m, centered on the midpoint between the sources, is placed around them. Describe the interference pattern of maxima and minima at the wall.

33. (II) Suppose that two in-phase sources of surface waves in water generate water waves that spread out in circles with equal amplitudes. (a) Show that the intensity of the disturbance, or the energy generated in the water, is, at a maximum, four times the intensity for each wave taken separately. (b) Similarly, show that there is no intensity at a minimum. (c) Show that the intensity at any given point if there were no wave interference is twice the intensity from each wave. (d) Do the different answers to parts (a), (b), and (c) mean that the conservation of energy principle must be abandoned for waves? Explain your answer.

15–5 Pulses

34. (I) A student pulls a rope taut and shakes it once, sending a wave pulse down the rope. The pulse is described by a *Gaussian function*,

$$z(x, t) = z_0 \exp[-(x - (0.5 \text{ m/s})t)^2/(0.1 \text{ m})^2].$$

Where is the center of the pulse at time t?

35. (I) At $t = 0$, a sawtoothed pulse has the shape $z = 0$ for $x < -a$ and for $x > 0$; $z = k(x + a)$ for $-a < x < 0$. Sketch this pulse. Suppose that the pulse moves at speed v in the $+x$-direction. What is the algebraic expression for the moving pulse?

36. (II) Consider a train of pulses of the type described in Problem 35. Here, at $t = 0$ the displacement is $z = 0$ for $x < -a + nb$ and for $x > nb$; $z = k(x + a - nb)$ for $-a + nb < x < nb$; n is any integer—positive, negative, or zero—and $b > a$. (a) Sketch this wave train of pulses. (b) Suppose that the entire train of pulses moves at speed v in the $+x$-direction. What is the algebraic expression for the train? (c) What are the period and wavelength of this nonharmonic wave?

37. (II) Consider a transverse wave that moves to the right on a string, in the form of a *Gaussian function*, namely

$$z(x, t) = z_0 e^{[-(x - vt)^2/\alpha^2]}.$$

By taking a time derivative, compute the transverse velocity of this pulse, and sketch it as a function of $(x - vt)/\alpha$. Do the same for the acceleration. Note that any time derivative is automatically a function of $x - vt$ and is itself therefore a quantity that forms a traveling wave.

38. (II) By using the results of Problem 37, *estimate* the energy in a Gaussian pulse that moves to the right. What would the total energy in a harmonic wave be? One of the reasons that real waves have beginnings and ends and are therefore, strictly speaking, pulses, is that pulses involve only a finite amount of energy.

39. (II) Two transverse pulses travel along a string at speed v in opposite directions. The pulse that moves to the right approximates a square with sides of length ℓ projecting above the equilibrium level of the string. The pulse that moves to the left approximates a square with sides of length $\ell/2$ projecting below the equilibrium level. Assume that the superposition principle holds, and describe the approximate shape of the string as the pulses collide.

15–6 Reflection and Transmission

40. (I) A pulse shaped like an upright two-dimensional pyramid, or an isosceles triangle, travels to the left along a string under tension (Fig. 15–31). The left-hand end of the string is free to move up and down. Sketch a sequence of drawings that show how the pulse reflects from the free end.

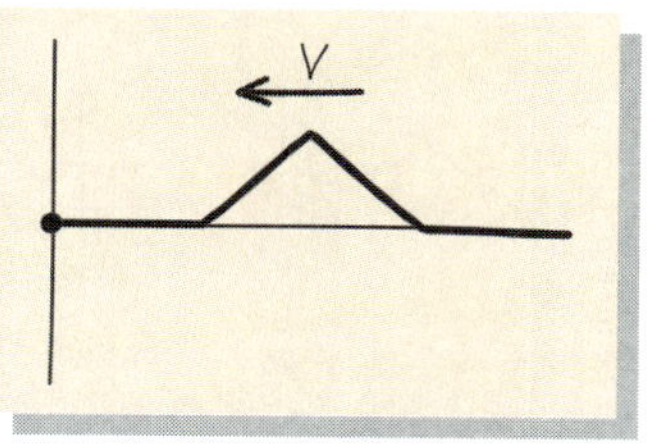

FIGURE 15–31 Problem 40.

41. (I) A string with linear mass density μ_1 is attached to another string, with linear mass density μ_2. A wave with amplitude A that travels along the μ_1 string will at the attachment point give rise to a reflected wave with amplitude B and a transmitted wave with amplitude C, where

$$B = \frac{\sqrt{\mu_2} - \sqrt{\mu_1}}{\sqrt{\mu_2} + \sqrt{\mu_1}} A, \qquad C = \frac{2\sqrt{\mu_1}}{\sqrt{\mu_2} + \sqrt{\mu_1}} A.$$

For $\mu_1 = 5.0 \times 10^{-2}$ kg/m and $\mu_2 = 3.8 \times 10^{-2}$ kg/m, what are B and C in terms of the amplitude of the incoming wave?

42. (I) Use the formulas in Problem 41 to show that the average power of the reflected wave plus the average power of the transmitted wave add to the average power of the incident wave.

43. (I) A string with mass density μ_1 is attached to a string with mass density μ_2. The amplitude of a reflected pulse is 85 percent of the amplitude of the incident pulse and is inverted. What is the ratio of μ_1 to μ_2? (See Problem 41.)

44. (II) When a symmetric pulse is incident on the fixed end of a rope, the reflected and incident pulse interfere, and there is a moment when the rope is completely flat. Is this a violation of the conservation of energy? (The energy of waves is discussed in Chapter 14.) Where is the energy at the moment the rope is flat?

45. (III) A string with mass density μ_1 is attached at the point $x = 0$ to a string with mass density μ_2. A tension T is maintained in both strings. A transverse wave of the form $A\cos(kx - \omega t)$, moving to the right, is generated far to the left in the first string. Use the fact that the string is unbroken at $x = 0$ and that no sharp kinks can develop there to obtain the forms of the reflected and transmitted waves. [*Hint*: The stated boundary conditions are sufficient to determine all the parameters of the reflected and transmitted waves. These conditions are (a) that the sum of the incident and reflected waves at $x = 0$ on string 1 equals the transmitted wave at $x = 0$ (unbroken string), and (b) that the derivative with respect to x of the sum of the incident and reflected waves at $x = 0$ on string 1 equals the derivative with respect to x of the transmitted wave at $x = 0$ (no kinks).]

*15–7 Fourier Decomposition of Waves

46. (III) Figure 15–32 shows a square wave $f(t)$ with period $T = 2\pi$ (or angular frequency $\omega = 1$) and amplitude $= 1$ unit. The wave's Fourier expansion is

$$f(t) = \frac{4}{\pi}\left[\sin(\omega t) + \frac{1}{3}\sin(3\omega t) + \frac{1}{5}\sin(5\omega t) + \cdots\right].$$

Plot an approximation to the square wave that corresponds to the first, the first two, and the first three terms in this series. (It helps in this to have access to computer software or to a calculator that can plot graphs.)

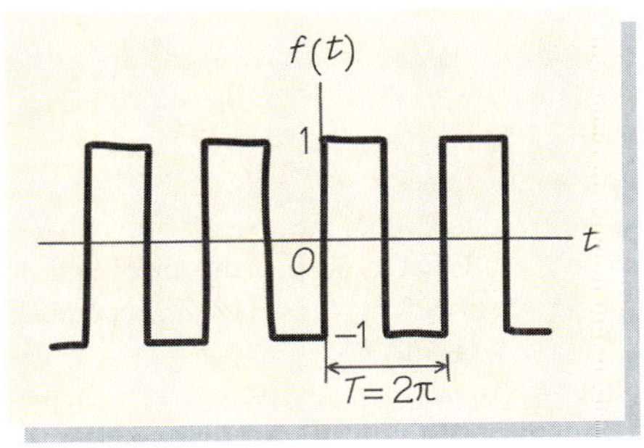

FIGURE 15–32 Problem 46.

*15–8 Pulses and the Uncertainty Principle

47. (I) A synthesizer, which is an electronic device for generating and combining harmonic waves, generates an electromagnetic pulse of width 10^{-6} s. Use Eq. (15–26) to find the range of frequencies that the synthesizer must utilize to generate the pulse.

48. (II) Waves are usually not infinite, but come in wave pulses caused by the turning on and turning off of the source. Suppose a laser with wavelength 650 nm is switched on for one second. What is the average frequency, and the width of the range of frequencies in this pulse?

General Problems

49. (I) Two harmonic waves are given by $y_1 = A\cos(kx - \omega t)$ and $y_2 = A\sin[kx - \omega t + (\pi/3)]$, where $k = 5\pi\ \text{m}^{-1}$, $\omega = 800\pi\ \text{s}^{-1}$, and $A = 4$ cm. (a) What is the frequency of each wave? What is the wavelength of each wave? What is the speed of each wave? (b) What is the amplitude of the wave that is the superposition, $y_1 + y_2$, of these two waves?

50. (I) Two rectangular pulses 6 cm long and of amplitude 4 cm but opposite sign are traveling down a string from opposite directions. At $t = 0$ s, the pulses approach each other with their centers 20 cm apart. The wave speed on the string is 1 cm/s (a) Describe and sketch the shape of the string at $t = 0$, 2, 5, 8, 10, and 15 s. (b) What has happened at $t = 10$ s? What is the energy at this time?

51. (I) Two violinists play their (identical) A-strings, nominally 440 Hz. A beat is heard at a pulsing of 2 Hz. What is the fractional difference in tensions $(T_1 - T_2)/T_1$ of the two strings? [*Hint*: The length of a violin string is the same from violin to violin; when a violinist tunes his or her instrument, the string tension is changed.]

52. (II) The police are busy on the day that you hear two police cars sounding their one-note sirens as you wait at a light (Fig. 15–33). The police cars are moving at 67.1 mi/h (30.0 m/s) and their drivers hear their own sirens emitting at a frequency of 650 Hz. One police car passes you while the other is still approaching. (a) What frequency do you hear for the siren that has passed? (b) for the siren that is approaching?

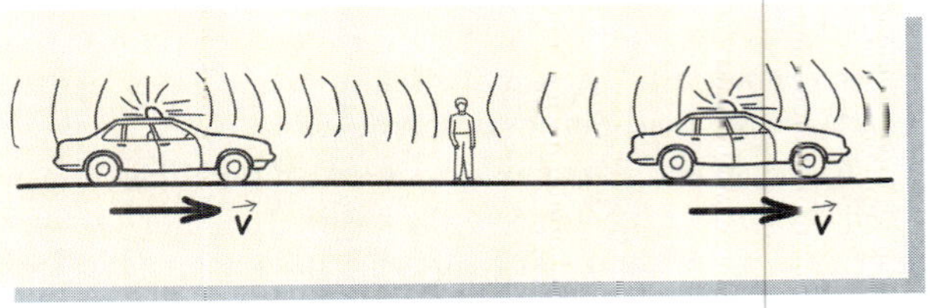

FIGURE 15–33 Problem 52.

53. (II) Consider two sources of sound that both emit waves with a frequency of 1800 Hz. These sources are placed on a turntable of radius 50 cm that rotates with an angular speed of 120 rev/min. Calculate the pulse frequency ($f_{pulse} = 2f_{beat}$) that results from the opposite Doppler shifts of the two sources, heard at a large distance from the turntable as a function of time. Ignore the fact that the sources are not quite equidistant from the detector of the sound. The speed of sound in air is 330 m/s.

54. (II) A guitar string has a fundamental frequency of 400 Hz. Another apparently identical guitar string is stretched beside the first one. (a) During successive stretching, each time tightening the second guitar string, the beat frequencies are 10, 5, and 1 Hz, consecutively. During the tightening, the beat frequency never passes through zero. What are the fundamental frequencies of the second guitar string for each case? (b) By what percentage should the tension be increased for the final adjustment to make the guitar strings have identical frequencies?

55. (II) A loudspeaker (source) that emits sound waves uniformly in all directions is 10 m from a microphone (receiver). Both the source and the receiver are 20 m from a smooth rigid wall that reflects the sound waves from the source (Fig. 15–34). As the frequency of the waves from the source is smoothly increased from zero, there is a first interference maximum detected by the receiver and subsequently a series of further maxima. List the frequencies of the three maxima that occur above 80 Hz. Assume that sound reflects from the walls as would a ball in an elastic collision.

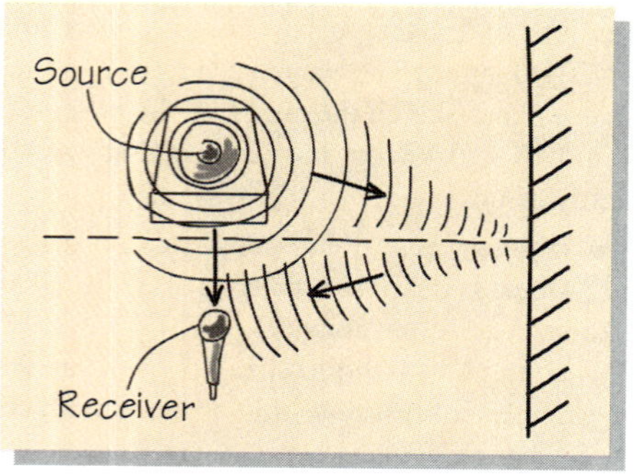

FIGURE 15–34 Problem 55.

56. (II) Consider two wave sources on a water surface. They generate identical waves of $\lambda = 1$ cm. What possible values of the separation of the sources, d, can you take to ensure that there is no disturbance to the side of the sources (that is, on the extension of the line joining the sources)? When this happens, is there a disturbance on that same line but *between* the sources? If so, describe it; if not, explain why not. Neglect the change in amplitude as a function of distance from the source; that is, treat this as a one-dimensional problem.

57. (II) The speed of the propagation of light in different media depends on the properties of the media. The formula is given by $v = c/n$, where $c = 3 \times 10^8$ m/s and n is the *index of refraction* ($n > 1$ for all media). Consider light of frequency 0.5×10^{15} Hz entering from empty space, with $n = 1$, into glass, with $n = 1.52$. What is the wavelength of the light in empty space? What is the wavelength of the light in glass?

58. (II) Waves are generated in phase in a water tank. The waves move along the x- and y-axes, respectively, with equal amplitude, so that the total wave is given by $\psi(x, y, t) = \sin(kx - \omega t) + \sin(ky - \omega t)$. Show that the pattern of interference maxima is given by the line $x - y = 2n\pi/k$ (where $n = 0, \pm 1, \pm 2, \ldots$).

59. (II) Which of the following equations are linear? Explain your reasoning:

$$\text{(a)}\ \frac{\partial^2\psi}{\partial t^2} = v^2\frac{\partial^2\psi}{\partial x^2}.$$

$$\text{(b)}\ \frac{\partial\psi}{\partial t} = -K\frac{\partial^2\psi}{\partial x^2}.$$

$$\text{(c)}\ \frac{\partial^2\psi}{\partial t^2} = a\psi^2 + b\frac{\partial\psi}{\partial x}.$$

$$\text{(d)}\ \frac{\partial\psi}{\partial t} = 2\psi\frac{\partial\psi}{\partial x} + 3\psi.$$

60. (II) Two sources emit sound coherently and in phase at a frequency of 845 Hz with the same intensity. They are 35.0 cm apart on a horizontal surface. A circular wall of radius 5.3 m (centered on the midpoint between the sources) is placed around them. Describe systematically the interference pattern of maxima and minima at the wall.

61. (III) Two waves of the form $z_1 = z_0 \sin(k_1x - \omega_1 t)$ and $z_2 = z_0 \cos(k_2x - \omega_2 t)$, respectively, propagate at speed v in a nondispersive medium. The frequency difference $\Delta f = f_1 - f_2$ is small. What function describes the envelope of the resulting superposition of the two waves? Express this envelope as a wave, and give its frequency and wavelength.

62. (III) Sinusoidal waves are generated on an infinitely long rope. One wave, $y_1(x, t)$, moves to the left and has amplitude y_0, wave number k_1, and angular frequency ω_1; the other, $y_2(x, t)$, moves to the right with amplitude y_0, wave number k_2, and angular frequency ω_2. Each wave has the same phase. (a) Express $y_i(x, t)$ for each of the waves. (b) Assuming that the rope is a nondispersive medium, what is ω_2 in terms of k_1, k_2, and ω_1? In parts (c) and (d), assume that $k_2 = k_1 + \delta k$ and $\omega_2 = \omega_1 + \delta\omega$. Take the speed of traveling waves on the rope to be v. (c) What is $\delta\omega$ in terms of δk and v? (d) Superpose these waves, assuming that δk and $\delta\omega$ are small. Use the notation $k_{av} = (k_1 + k_2)/2$ and $\omega_{av} = (\omega_1 + \omega_2)/2$. Show that the result of the superposition is two waves, one moving at the low speed $v_l = v\,\delta k/2k_{av}$ and wavelength $2\pi/k_{av}$, and one moving at the high speed $v_r = 2vk_{av}/\delta k$ and the long wavelength $4\pi/\delta k$.

63. (III) Imagine that sinusoidal waves are generated on an infinitely long rope. One wave, for which the displacement is $y_1(x, t)$, moves to the left and has amplitude y_0, wave number k_1, and angular frequency ω_1. The other wave, with displacement $y_2(x, t)$, moves to the right with amplitude y_0, wave number k_2, and angular frequency ω_2. Each has the same phase, which you may take to be zero. (a) Write the appropriate displacement function $y_i(x, t)$ for each of these two waves. (b) What is ω_2 in terms of k_1, k_2, and ω_1?

Suppose now that $k_2 = k_1 + \delta k$ and $\omega_2 = \omega_1 + \delta\omega$, where δk and $\delta\omega$ are small. Take the speed of traveling waves on the rope to be v. (c) What is $\delta\omega$ in terms of δk and of the speed v of waves on the rope? (d) Consider the superposition of these waves. In the limit of $\delta k \to 0$ and $\delta\omega \to 0$, show that the result of the superposition is a standing wave. In particular, what are the wavelength and angular frequency of the standing wave? (e) Superpose these waves again, but this time assume only that δk and $\delta\omega$ are small. Use the notation $k_{av} = (k_1 + k_2)/2$ and $\omega_{av} = (\omega_1 + \omega_2)/2$. Show that the result of the superposition is the product of two waves, one moving at the low speed $v_l = v\,\delta k/2k_{av}$ and short wavelength $2\pi/k_{av}$, and one moving at the high speed $v_r = 2vk_{av}/\delta k$ and long wavelength $4\pi/\delta k$.

CHAPTER

16

Rafting down a highly turbulent section of a river. The motion of the water is very complex and depends on many properties that vary from fluid to fluid.

Properties of Fluids

Waves on the sea, smoke rings, and a whirlpool of water flowing down a drain are all part of the fascinating variety of responses that fluids have to forces. The motion of individual particles and of rigid, extended bodies is completely predictable and relatively simple if the forces are known. As soon as we leave these idealized forms of matter, however, the description of motion becomes more complicated. We have learned that forces that induce only small departures from stable-equilibrium configurations lead universally to simple harmonic motion and, in extended systems, to waves. But the response of fluids to forces goes well beyond wave motion.

In this chapter, we discuss density and pressure, general properties of fluids, which leads us to the notion of buoyancy. We also study the consequences of the conservation of mass and energy for fluids in motion. A good deal of idealization is necessary here; however, even with idealization, we can still arrive at a useful description of the behavior of real fluids.

16–1 STATES OF MATTER

In between rigid solids and dilute gases, matter comes in liquid form. The particular form that matter takes is determined ultimately by the forces between its molecules. Figure 16–1 is a curve that depicts the typical force between two molecules as a function of their separation, r. At very small separations, the force is large and positive; it is strongly repulsive. It becomes attractive at a separation r_0, whose value is approximately 10^{-10} m, and then tapers off to zero at a distance of a few times 10^{-10} m. Close to the distance $r = r_0$, the curve is almost a straight line with a negative slope, and $F = -k(r - r_0)$ is just the typical restoring spring force that leads to oscillatory behavior, as discussed in Chapters 13 and 14.

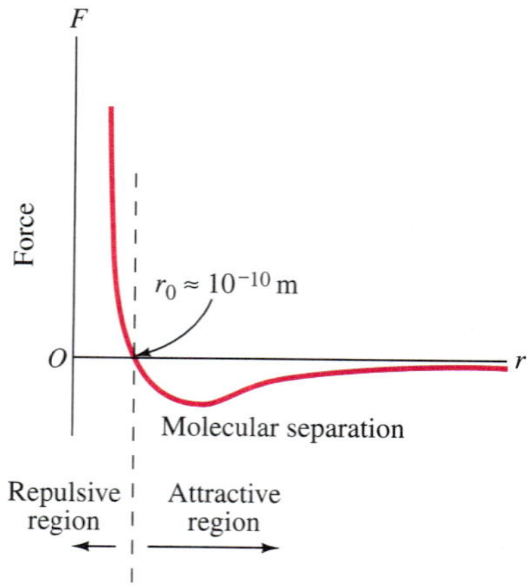

FIGURE 16–1 Qualitative behavior of the force between a pair of molecules as a function of their separation, r. Near $r = r_0$, the point at which the force is zero, the force is springlike (restoring). For $r < r_0$, the force is strongly repulsive; for $r > r_0$, it is attractive.

Gases

If molecules are, on average, sufficiently far apart, then they experience the force shown in Fig. 16–1 only at large values of r, where the force is small. In effect, their only interaction occurs when the molecules occasionally collide with each other. The molecules behave as a **gas**. We will study gases further in Chapters 17 through 19 and only note here that gases are easily compressed and that the shape of a body of gas is easily deformed.

Solids

If molecules are sufficiently close together that the forces due to neighboring molecules are significant, matter will take on a form determined by the nature of the molecules and by the detailed properties of the forces. In some cases, the forces act to organize the molecules into a **lattice**, which is a three-dimensional structure in which the molecules or atoms appear to be connected by rigid bonds and have little freedom to move, except for small displacements (Fig. 16–2). This is a typical **solid**: a substance whose structure resists forces that would deform its shape.

Liquids

In liquids, the molecules are closely spaced and the intermolecular forces between the molecules are strong, but they do not lead to an organized lattice structure. For such aggregates of molecules, there is no resistance to deformation. Nevertheless, unlike the molecules of gases, the molecules of a liquid are close to each other, and there are repulsive forces that tend to resist compression.

Density

The mass density

A quantity of matter has a mass, M, and a volume, V. The average **mass density**, ρ, of the matter is given by

$$\rho = \frac{M}{V}. \qquad (16\text{–}1)$$

Table 16–1 gives the densities of some typical substances. Densities vary with temperature; the density of water at sea level is 0.9998×10^3 kg/m^3 at 0°C and 0.9584×10^3 kg/m^3 at 100°C. The densities of liquids also can change under compression, although the effect is often small. The fractional density change of water in a cylinder is only about 1.5×10^{-8} when a force per unit area of 3000 N/m^2 acts on a piston. This is equivalent to the force per unit area produced by a 75-kg person standing on a piston of area 0.25 m^2. It is therefore a very good approximation to treat water as incompressible. Finally, we can define the **specific gravity** of a substance as the ratio of the mass density of that substance to the mass density of water; thus, for example, the specific gravity of glycerin is 1.26.

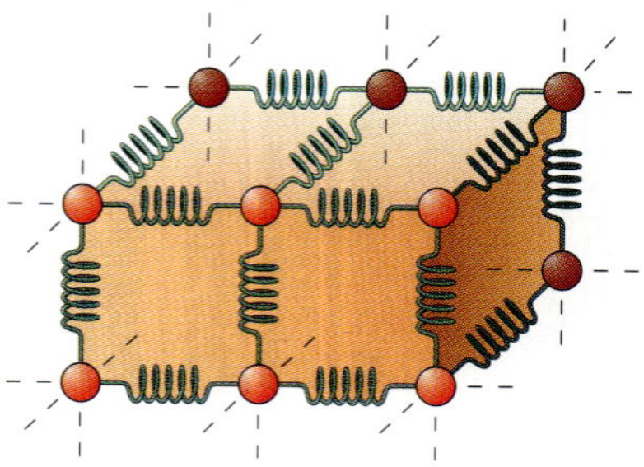

FIGURE 16–2 A lattice structure characteristic of solids. For cubic lattices, the molecules or atoms reside at the corners. The forces between the molecules are springlike for small displacements from the equilibrium positions, as shown here schematically.

EXAMPLE 16–1 A neutron star is a star that is much smaller than our Sun and has the density of an atomic nucleus. A typical neutron star has a radius of 10 km and a mass of 2×10^{30} kg (the mass of the Sun). How much would a 1-cm^3 volume of such a star weigh under the influence of Earth's gravity?

Solution: To find the weight of 1 cm^3 of a neutron star on Earth's surface, we must find the mass of this volume. To determine this, we need the density, which we

TABLE 16–1 Mass Densities

Phase	Material	Density (kg/m^3)
Gas	Hydrogen (0°C, 1 atm)	0.090
Gas	Air at sea level	1.29
Gas	Chlorine (0°C, 1 atm)	3.21
Liquid	Ethyl alcohol	0.79×10^3
Liquid	A good olive oil	0.92×10^3
Liquid	Water	1.00×10^3
Liquid	Glycerin	1.26×10^3
Liquid	Earth's core	9.5×10^3
Liquid	Liquid mercury	13.6×10^3
Liquid	Sun's core	1.6×10^5
Solid	Cork	0.25×10^3
Solid	Applewood	0.745×10^3
Solid	Ice	0.917×10^3
Solid	Quartz	2.65×10^3
Solid	Steel alloys	$7.6 \times 10^3 - 8.9 \times 10^3$
Solid	Iridium	22.4×10^3

can calculate from the given information. The star's radius is 10 km = 10^4 m. Thus the density is

$$\rho = \frac{\text{star mass}}{\text{star volume}} = \frac{M}{4\pi R^3/3} = \frac{2 \times 10^{30}\ \text{kg}}{4\pi(10^4\ \text{m})^3/3} = 0.5 \times 10^{18}\ \text{kg/m}^3$$

$$= 0.5 \times 10^{12}\ \text{kg/cm}^3.$$

Thus 1 cm^3 of the material of a neutron star has a mass 0.5×10^{12} kg, and it weighs $(0.5 \times 10^{12}\ \text{kg})g = (0.5 \times 10^{12}\ \text{kg})(9.8\ \text{m/s}^2) = 0.5 \times 10^{13}$ N. For comparison, 1 cm^3 of water on Earth's surface has a mass of 1 g and weighs 0.01 N. A neutron star is very dense indeed!

We stress that the distinction among the three forms of matter is not absolute. If the forces are large enough, even rock will flow. Gases share many properties with liquids; one major difference between gases and liquids is in the compressibility. A collection of molecules can form a solid, a liquid, or a gas, depending on external circumstances. Water is solid at low temperatures, but it can become a liquid at 0°C and a gas at 100°C. We say that matter appears in different **phases**; the particular phase depends on external conditions. At very high temperatures, the molecules themselves may dissociate into negatively charged electrons and positively charged ions. That kind of matter has very different properties because electrical forces between the constituents play an important role. Matter that is electrically dissociated into positive ions and negatively charged electrons is called a **plasma**.

We shall define a **fluid** to be a substance that is not resistant to deformation. This description involves how a material responds to external forces and applies to both liquids and gases. Therefore, except for phenomena in which compressibility is important, our description of the behavior of fluids applies to gases as well as to liquids.

16–2 PRESSURE

The first part of this chapter concerns **hydrostatics**—the study of fluids at rest. Density and pressure are the major variables of interest for such fluids. We have already discussed density. **Pressure** is a measure of the force transmitted by fluids, even fluids at rest. Swimmers know from experience that liquids can transmit forces. If you dive too

The definition of pressure

deeply into water, your eardrum may hurt or even rupture. This is due to the water pressure on the eardrum. But a sheet of glass does not shatter under water, because there is as much pressure on one side of it as on the other. One way to determine the force exerted by a fluid is to balance that force against a calibrated spring (Fig. 16–3). The plate on the spring balance is in contact with the fluid, while the spring is inside a fluid-free enclosure, so that there is no force on the plate from the inside due to the fluid. The average pressure, p, on any surface of area A is defined as the *force per unit area* acting perpendicular to that surface:

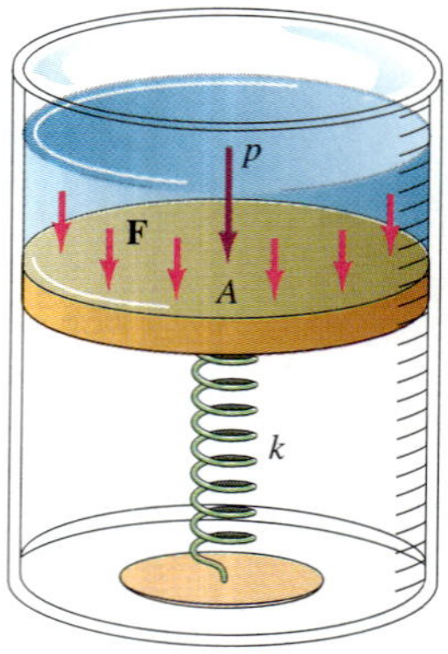

FIGURE 16–3 A device that would allow the measurement of pressure in a liquid. The fit between the piston and cylinder walls allows the piston to move but is tight enough so that a near vacuum can be maintained on the side with the spring. The compression of the spring provides a measurement of the force, from which the pressure is determined.

$$p = \frac{F}{A}. \tag{16–2}$$

Under the influence of gravity, pressure will vary as a function of depth. Thus, for a more precise definition, we consider a small area ΔA located at point $\mathbf{r}$, and take the limit $\Delta A \to 0$. Denoting the force perpendicular to this area as ΔF, we find

$$p = \lim_{\Delta A \to 0} \frac{\Delta F}{\Delta A} = \frac{\mathrm{d}F}{\mathrm{d}A}. \tag{16–3}$$

The pressure is the same in all directions.

It is important to note that *when a fluid is at rest, the pressure at any given point must be the same in all directions.* The microscopic origin of pressure is the bombardment of the surface on which the pressure acts by the individual molecules of the substance (Chapter 19). The pressure is the same in all directions because the molecular movements are oriented randomly. For fluids at rest, the pressure is simply a positive scalar quantity. The pressure at a static interface between two fluids is the same on both sides of the interface; if it were not, there would be a net force at the interface and a resulting motion of the interface.

Units of pressure

In SI, pressure is measured in *pascals* (Pa), named after Blaise Pascal, who studied pressure in the mid-seventeenth century:†

$$1 \text{ Pa} = 1 \text{ N/m}^2. \tag{16–4}$$

16–3 VARIATION OF PRESSURE IN A FLUID AT REST

The gravitational force causes the pressure of fluids at rest to be higher at ocean depths than at Earth's surface. In order to establish exactly how pressure varies with depth for an incompressible fluid, consider a liquid of fixed density ρ at rest in a container. Imagine isolating a very thin cylinder of that liquid (Fig. 16–4). (The top, base, and walls of the cylinder are purely imaginary; their sole purpose is to help us visualize the forces acting on an element of the fluid.) The cylinder's top, of area A, is horizontal and lies a distance y below the surface of the liquid. The bottom of the cylinder is at a depth $y + \Delta y$. Note that we measure the depth y downward from the surface, so it increases as we go down. We denote the pressure at the top of our cylinder by p and the pressure at the bottom by $p + \Delta p$. We want to find an expression for Δp.

The downward force on the cylinder, which is applied at the top surface by pressure from the surrounding fluid, is pA; the upward force, which is applied at the bottom surface by the pressure there, is $(p + \Delta p)A$. Additionally, there is a downward force on the cylinder due to gravity; namely, the weight of the liquid inside the cylinder. That force has a magnitude of $(\Delta m)g$, and the mass of the cylinder, Δm, is the density times the volume of the cylinder. Thus, the downward force of gravity is $\rho(A\ \Delta y)g$.

†Other common units for pressure are pounds per square inch (lb/in²): 1 lb/in² $= 6.9 \times 10^3$ Pa; the atmosphere (atm), which is the atmospheric pressure at sea level: 1 atm $= 1.01 \times 10^5$ Pa $= 14.7$ lb/in²; the bar: 1 bar $= 10^5$ Pa; and the torr: 1 torr $= 133.32$ Pa.

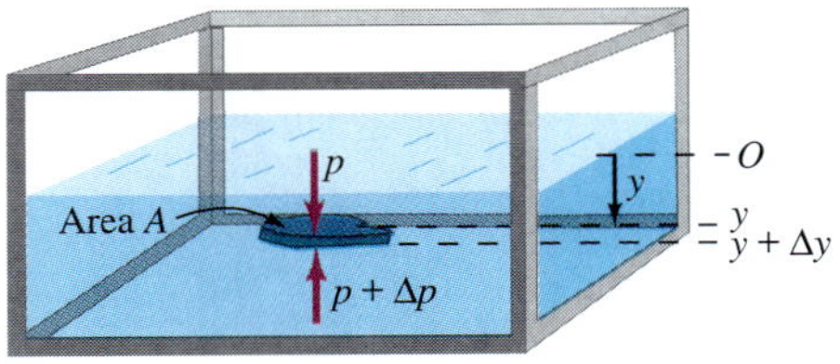

FIGURE 16–4 An imaginary thin cylinder of fluid is isolated to show the forces that act to keep it in equilibrium.

All together, we find

$$F_{up} = (p + \Delta p)A,$$

$$F_{down} = pA + (\Delta m)g = pA + \rho(A\ \Delta y)g.$$

Because the cylinder does not accelerate, the net force must cancel:

$$F_{net} = F_{up} - F_{down} = (p + \Delta p)A - pA - \rho(A\ \Delta y)g = 0.$$

We cancel the pA terms, divide by A, and rearrange terms to find that

$$\frac{\Delta p}{\Delta y} = \rho g.$$

In the limit $\Delta y \to 0$, this result becomes

$$\frac{dp}{dy} = \rho g. \qquad (16\text{–}5)$$

FIGURE 16–5 To equalize the pressures of the surrounding water and the sinus cavities, a diver blows air into his or her sinuses.

According to Eq. (16–5), the pressure change is positive for positive Δy. Pressure therefore increases with increasing depth (Fig. 16–5). In physical terms, this phenomenon is a measure of the increased amount, and hence increased weight, of the liquid above any given point as the depth of the point increases. There is a greater weight of water above a submarine at an ocean depth of 500 m than above a submarine at a depth of 20 m.

The pressure of an incompressible fluid under the influence of gravity

Equation (16–5) looks just like the formula for the constant acceleration g of an object acted on by the constant force of gravity, $dv/dt = g$. When we studied projectile motion in Chapters 2 and 3, we saw that this equation has a solution $v = v_0 + gt$. By comparison, we can write the solution to Eq. (16–5), which gives the pressure as a function of depth, as

Pressure varies only with depth.

$$p = p_0 + \rho g y. \qquad (16\text{–}6)$$

Alternatively, we can integrate Eq. (16–5) to find this result. The constant p_0 is the value of the pressure at the surface, where $y = 0$; for a body of fluid such as the ocean, this is just atmospheric pressure. The pressure increase with depth y is equal to the weight per unit area of the overlying fluid. The total mass of the fluid above the cylinder is $\rho V = \rho A y$, so its weight is $\rho A y g$, and thus the weight per unit area is $W/A = \rho y g$.

Note that our derivation is completely independent of the horizontal location of the cylinder within the liquid. The pressure is *independent* of the horizontal position and is dependent only on depth. Thus the pressure of a liquid in a series of open, connected vessels at the same depth is the same. The pressure at the fluid surfaces is equal to atmospheric pressure (the surface is in contact with the atmosphere and is therefore at the same pressure), and so the heights of the fluid surfaces in these vessels must be the same (Fig. 16–6).

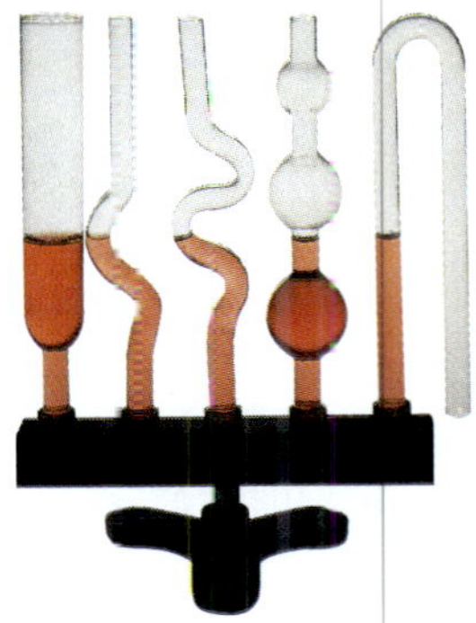

FIGURE 16–6 The fluid level in a series of connected vessels is the same because the pressure at the surface is the same.

An immediate consequence of Eq. (16–6) is that if p_0 is changed as a result of some external effect, p is changed by exactly the same amount. In other words, *the same change in pressure applied to any point in a fluid at rest is transmitted to every part of the fluid.* This statement is commonly known as **Pascal's principle**.

Pascal's principle

EXAMPLE 16–2 A hydraulic press (Fig. 16–7) is used to lift an object of mass 400 kg. The supporting cylindrical column has a diameter of 20 cm. The object is lifted by an operator using foot power to push on a piston. The height of the opera-

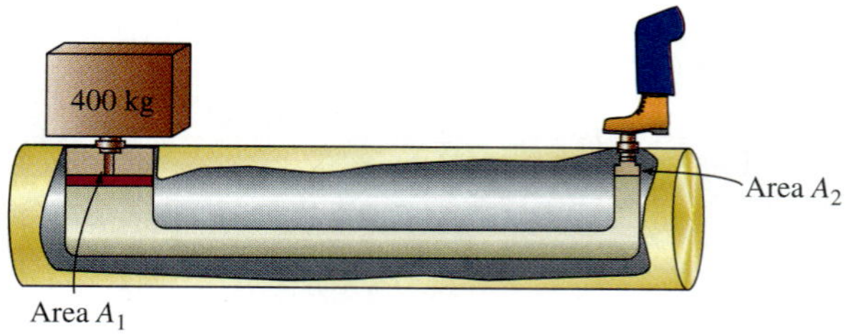

FIGURE 16–7 Example 16–2.

tor's foot and the height at which the supporting column contacts the fluid are the same. Assuming that the force applied by the operator is 150 N, what is the diameter of the piston to which the foot is applied?

Solution: Pascal's principle tells us that the pressure p at the foot piston equals the pressure at the supporting column. We can find the force at the supporting column by multiplying p by the area A_1 of the column; this force must equal the weight of the mass to be lifted, mg. In turn, we know the pressure at the piston is the applied force, F_{foot}, divided by the piston area, A_2. Thus

$$mg = pA_1 = (F_{\text{foot}}/A_2)A_1.$$

By using the geometrical relation $A = \pi r^2 = \pi d^2/4$, where d is the diameter, we find instead

$$d_1^2/d_2^2 = mg/F_{\text{foot}},$$

or

$$d_2 = d_1\sqrt{(F_{\text{foot}}/mg)} = (0.20\text{ m})\sqrt{[(150\text{ N})/((400\text{ kg})(9.8\text{ m/s}^2))]} = 3.9\text{ cm}.$$

EXAMPLE 16–3 A column of mercury ($\rho_m = 13.6 \times 10^3\text{ kg/m}^3$) in a vertical tube that is open at the bottom rests in an open pool of mercury (Fig. 16–8). The column is sealed off at the top after all the air is evacuated from the empty region, creating a vacuum. What is the height H of the column of mercury?

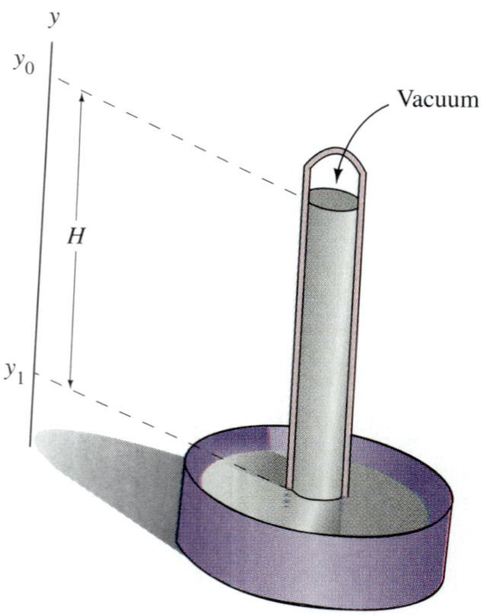

FIGURE 16–8 Example 16–3. A column of mercury balanced by atmospheric pressure.

Solution: The pressure at height y_1 (as measured up from ground level) on the open part of the pool of mercury is the atmospheric pressure, $p_1 = 1.01 \times 10^5$ Pa. The pressure at the same height y_1 within the column must be the same. But this pressure can be calculated from Eq. (16–6). Let the vertical coordinate of the surface inside the tube be denoted by y_0. Because we have evacuated the column, the value of p_0 for Eq. (16–6), which is the pressure at y_0 and represents the weight of the gas above the column, is zero (in reality, a little mercury will have evaporated, but the pressure that it exerts is negligible). The equality of the pressures at the level of the open pool then gives, with $H = y_0 - y_1$,

$$p_1 = p_0 + \rho g(y_0 - y_1) = p_0 + \rho gH = \rho gH.$$

Thus

$$H = \frac{p_1}{\rho g}.$$

We find that, numerically,

$$H = \frac{1.01 \times 10^5\text{ Pa}}{(13.6 \times 10^3\text{ kg/m}^3)(9.8\text{ m/s}^2)} = 0.76\text{ m}.$$

This example describes the working principle of a **barometer**, a device used to measure atmospheric pressure. The pressure at the bottom of a column of mercury exactly 0.760 m = 760 mm high is 1 atm at sea level.

APPLICATION

The barometer

EXAMPLE 16–4 A U-shaped tube of uniform cross section is open at both ends, as in Fig. 16–9. First, amyl bromide is poured into the tube; amyl bromide is a colorless fluid that does not mix with water. Then, water is carefully added to one side of the U, so there is water at the top of one column of fluid and amyl bromide is at the top of the other. The two columns of fluid meet at a point that we label $h = 0$. What is the ratio of the heights of the fluids above $h = 0$ in each column? The density of water is $\rho_w = 1.00 \times 10^3\ \text{kg/m}^3$ and that of amyl bromide is $\rho_{ab} = 1.26 \times 10^3\ \text{kg/m}^3$. Repeat the calculation for heptyl ether, a fluid that does not mix with water and has a mass density of $0.81 \times 10^3\ \text{kg/m}^3$.

Solution: The atmospheric pressure is the same at the open surface of both columns (we neglect the weight of any amount of air that adds to the pressure if one column has a lower fluid height than the other). The pressure in the two fluids at $h = 0$, where the two liquids meet, is also equal when the system has come to rest (otherwise there would be movement!). Let h_{ab} be the height of the column with the amyl bromide above level $h = 0$ and h_w be the height of the water column above level $h = 0$. For the pressure at $h = 0$, we find

for the amyl bromide column: $p_{ab} = p_0 + \rho_{ab} h_{ab} g$;

for the water column: $p_w = p_0 + \rho_w h_w g$.

Setting these pressures equal to one another, we have

$$\rho_{ab} h_{ab} g = \rho_w h_w g.$$

Rearranging,

$$\frac{h_{ab}}{h_w} = \frac{\rho_w}{\rho_{ab}} = \frac{1.00 \times 10^3\ \text{kg/m}^3}{1.26 \times 10^3\ \text{kg/m}^3} = 0.79.$$

The water column is higher than the amyl bromide column. This makes sense physically because amyl bromide is denser than water; the shorter column of amyl bromide weighs the same as the taller column of water. Note that this result is independent of the amount of either fluid.

We repeat the calculation with heptyl ether ($\rho_{he} = 0.81 \times 10^3\ \text{kg/m}^3$) to find that

$$\frac{h_{he}}{h_w} = \frac{\rho_w}{\rho_{he}} = \frac{1.00 \times 10^3\ \text{kg/m}^3}{0.81 \times 10^3\ \text{kg/m}^3} = 1.23.$$

The height of the column with heptyl ether is higher than the water column.

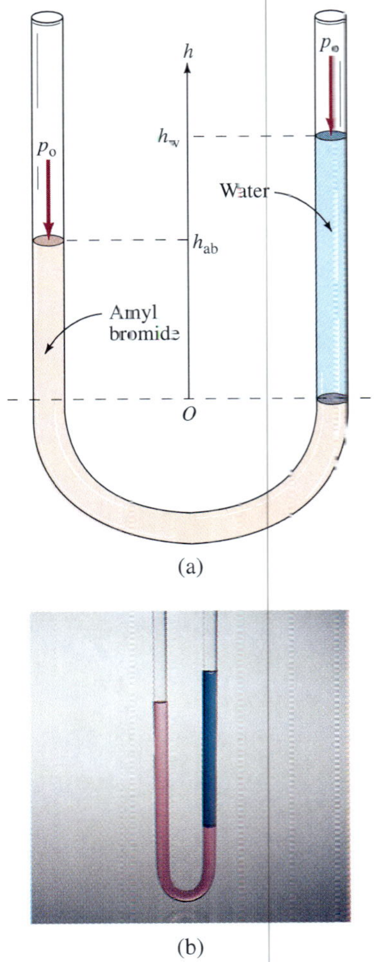

FIGURE 16–9 Example 16–4. The water has been dyed blue, and the amyl bromide has been dyed pink.

EXAMPLE 16–5 A closed container of liquid is connected to the exterior by two pistons: a small one of area $A_1 = 1\ \text{cm}^2$ and a large one of area $A_2 = 100\ \text{cm}^2$ (Fig. 16–10). Both pistons are at the same height. When a force of magnitude $F_1 = 100\ \text{N}$ is applied downward on the small piston, what mass m can be lifted by the large piston? Show that the work done in pushing the small piston down a distance d_1 is equal to the work done in lifting the mass m (assume that the liquid is incompressible).

Solution: When force F_1 is imposed on the small piston, the pressure p in the liquid at that point is given by

$$p = \frac{F_1}{A_1} = \frac{100\ \text{N}}{1\ \text{cm}^2} = \frac{10^2\ \text{N}}{10^{-4}\ \text{m}^2} = 10^6\ \text{Pa}.$$

Because the large piston is at the same height as the small one, according to Pascal's principle the pressure at the large piston will have the same value, $p = 10^6$ Pa. A force F_2 acts upward at this piston, given by the pressure times the area:

$$F_2 = pA_2.$$

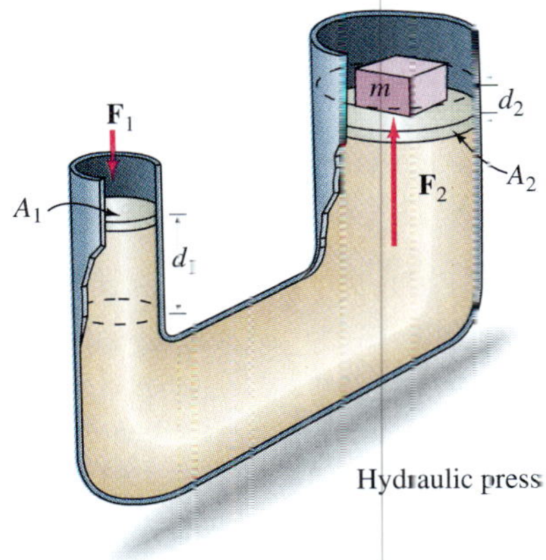

FIGURE 16–10 Example 16–5. A closed container of liquid with two pistons. An external force $\mathbf{F}_1$ is imposed on the small piston, giving an amplified upward force $\mathbf{F}_2$ due to pressure on the larger piston.

APPLICATION

The hydraulic press

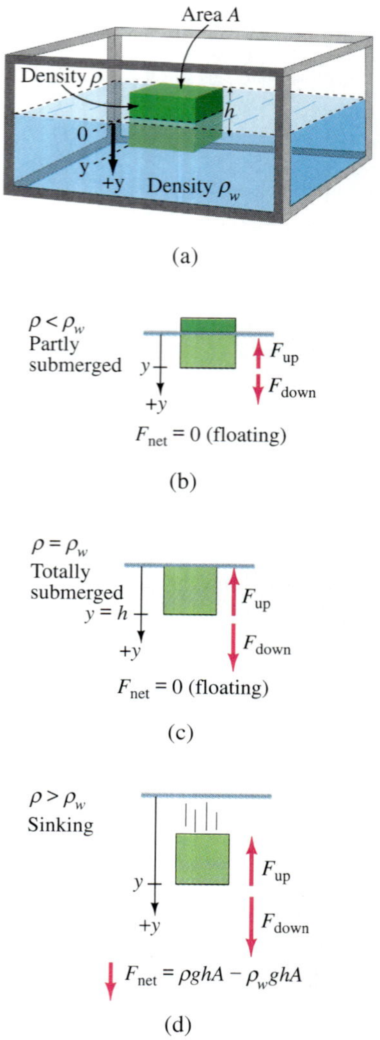

FIGURE 16–11 (a) A block of material of density ρ is partly immersed in water of density ρ_w. The volume of the block is $V = Ah$, and the volume of water displaced by the block is $V_d = Ay$. (b) Force diagram of the forces acting on the lower surface of the block, which is partly immersed, as in part (a). The net force is $F_{net} = 0$, and the block floats. (c) When $\rho = \rho_w$, the block is totally submerged, and the block floats because $F_{net} = 0$. (d) When $\rho > \rho_w$, the block sinks because $F_{net} > 0$.

Correspondingly, the mass that can be lifted is given by $mg = F_2$, or

$$m = \frac{pA_2}{g} = \frac{(10^6\ \text{Pa})(100\ \text{cm}^2)}{9.8\ \text{m/s}^2} = 1020\ \text{kg}.$$

For the second part of the problem, we compare the work done in pushing down the small piston, W_p, to the work done in lifting the mass, W_m. We find

$$W_p = F_1 d_1.$$

At the same time, the work done in lifting the mass m is

$$W_m = F_2 d_2,$$

where d_2 is the height through which m is lifted. To find d_2, note that as long as the fluid is incompressible, the volume change in the movement of the small cylinder is matched by the volume change in the large cylinder. The volume of the liquid displaced by the small piston is $A_1 d_1$. The volume displaced in the movement of the large piston is $A_2 d_2$. Equating these volumes, we have $A_1 d_1 = A_2 d_2$; hence,

$$d_2 = \frac{A_1 d_1}{A_2}.$$

The work done in raising the mass is then

$$W_m = F_2 d_2 = pA_2\left(\frac{A_1 d_1}{A_2}\right) = pA_1 d_1.$$

But the quantity pA_1 is the original force F_1, and the right-hand side is the original work $W_p = F_1 d_1$ done in pushing down the small piston. The work done on both sides is equal, but the forces are considerably different: $F_2/F_1 = d_1/d_2 = A_2/A_1 = 100\ \text{cm}^2/1\ \text{cm}^2 = 100$ because $F_1 d_1 = F_2 d_2$.

The device treated here describes the *hydraulic press* schematically. Pascal's principle is similarly used in the design of hydraulic brakes.

16–4 BUOYANCY AND ARCHIMEDES' PRINCIPLE

An ocean-going liner floats when it is launched. When a brick is put into water, however, it sinks. Why? Experience tells us that this has something to do with the density of the material (and not the mass; a small brick sinks just like a large one). Let's consider, as in Fig. 16–11a, a block of material of volume $V = Ah$ and density ρ (the mass is thus $m = \rho V$) partly immersed in water, which has a density of ρ_w. Figure 16–11b shows the forces acting on the lower surface of the partly submerged block. The upward force is due to the pressure of the liquid, as given by Eq. (16–6): $p = p_0 + \rho_w g y$, where y is the vertical submerged distance of the block. The upward force on the block is therefore

$$F_{up} = pA = p_0 A + \rho_w g y A.$$

The downward force has two components: the atmospheric pressure above the block, p_0, and the weight of the block. The first downward component is $F_{atm} = p_0 A$, and the second is the force of gravity on the block, $F_g = mg = (\rho A h)g$. Thus the total downward force is

$$F_{down} = p_0 A + \rho g h A.$$

Therefore,

$$F_{net} = F_{down} - F_{up} = p_0 A + \rho g h A - p_0 A - \rho_w g y A = \rho g h A - \rho_w g y A. \quad (16\text{–}7)$$

A state of equilibrium—floating—demands that the forces balance, $F_{net} = 0$. If we set

Eq. (16–7) equal to zero and divide by gA, we get

$$\text{for floating:}\qquad \rho h = \rho_w y; \tag{16–8}$$

$$\frac{\rho}{\rho_w} = \frac{y}{h}.$$

If $\rho < \rho_w$, then $y/h < 1$ and only a fraction of the block is submerged. In the limiting case that $\rho = \rho_w$, the block is totally submerged and $y = h$ (Fig. 16–11c). In this case, the block would float just under the surface because the upward and downward forces cancel. If $\rho > \rho_w$, then the block must sink (Fig. 16–11d) because the two forces can no longer cancel completely. In this case, once the block is fully submerged ($y \geq h$), the net (downward) force is

$$\text{for sinking:}\qquad F_{\text{net}} = \rho g h A - \rho_w g h A > 0. \tag{16–9}$$

We can interpret the difference between the weight of the block $\rho g h A$ and F_{net} as an *upward* **buoyant force**:

$$F_{\text{buoy}} = F_g - F_{\text{net}}. \tag{16–10}$$

For the case when the block is partly submerged, we have $F_{\text{net}} = 0$, and Eq. (16–10) gives

$$\text{for partial submergence:}\qquad F_{\text{buoy}} = \rho g h A - 0 = \rho g h A = \rho_w g y A,$$

where we have used Eq. (16–8) in the last step. When the block is totally submerged (at any depth where $y \geq h$), F_{net} is given by Eq. (16–9), and we find

$$\text{for total submergence:}\qquad F_{\text{buoy}} = \rho g h A - (\rho g h A - \rho_w g h A) = \rho_w g h A.$$

If we think of V as the volume of the object beneath the water ($V = yA$ or hA, according to whether the object is partly or totally submerged), then we can combine our results into the single expression (Fig. 16–12).

$$F_{\text{buoy}} = \rho_w g V. \tag{16–11}$$

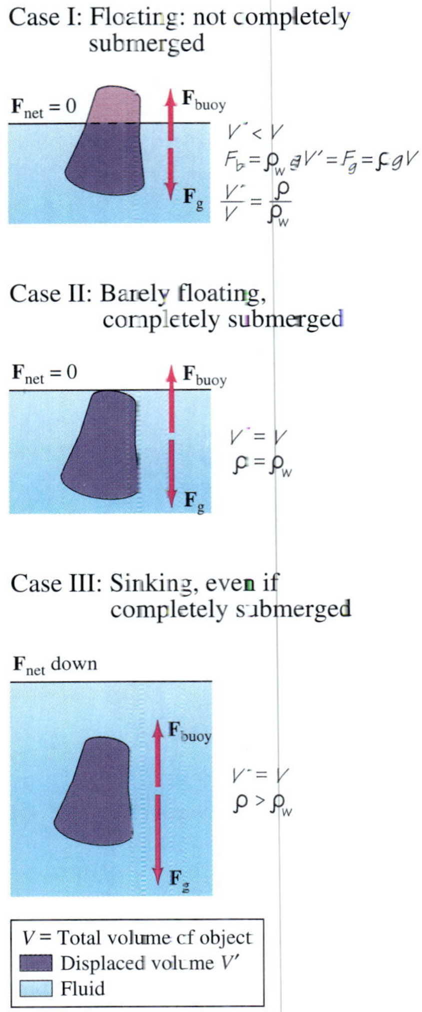

FIGURE 16–12 By comparing the buoyant force F_{buoy} to the force of gravity F_g, we determine whether a body sinks, just floats, or pops up.

The buoyant force opposes the downwardly directed force of gravity, $\rho g h A$, on the block. For a floating body, the buoyant force and the force of gravity cancel exactly. When $\rho > \rho_w$, the buoyant force opposes the force of gravity even though it does not cancel the force of gravity completely; a sinking object descends with less acceleration than it would under the influence of gravity alone. When a metal sphere is immersed in a bucket of water and suspended from a spring balance, the balance will indicate a lower weight than it would for the same sphere in air. The difference in these weights is the buoyant force. The buoyant force is not limited to water; it is what allows a balloon to rise in air (Fig. 16–13a), and even a steel ball to float in mercury (Fig. 16–13b).

Archimedes' Principle

The buoyant force

Regardless of whether the block is partly or totally submerged, Eq. (16–11) shows that the buoyant force is equal to the weight of the liquid (here, water) displaced by the block:

$$W_{\text{liq}} = m_{\text{liq}} g = \rho_{\text{liq}} V_{\text{liq}} g = \rho_{\text{liq}} y A g,$$

where $y = h$ when the block is totally submerged. This fact was discovered in the third century B.C. by Archimedes, a great scientist and mathematician of ancient Greece, and is known as **Archimedes' principle**:

Archimedes' principle

The buoyant force on an immersed object equals the weight of displaced liquid.

We may check the validity of this principle by considering the behavior of a block of iron, with a mass M and a density $\rho_{\text{Fe}} = 7.86 \times 10^3\ \text{kg/m}^3$, in water. The upwardly

(a)

(b)

FIGURE 16–13 (a) This hot-air balloon takes advantage of the fact that hot air is less dense than cold air at the same pressure. Buoyancy accordingly makes the balloon rise. (b) The buoyant force allows a steel ball to float on the surface of liquid mercury, whose density is some 70 percent greater than that of steel (iron).

directed buoyant force is the weight of water displaced, $F_{\text{buoy}} = mg = \rho_w Vg$, where m is the mass of displaced water and V is its volume. The downward force of gravity on the iron block is $F_g = Mg = \rho_{\text{Fe}} Vg$. Because $\rho_{\text{Fe}} > \rho_w$, the net force is downward and the block sinks. The net downward force—effectively the weight of iron in water—is

$$F_{\text{net}} = F_g - F_{\text{buoy}} = \rho_{\text{Fe}} Vg - \rho_w Vg = \left(1 - \frac{\rho_w}{\rho_{\text{Fe}}}\right)\rho_{\text{Fe}} Vg$$

$$= \left(1 - \frac{\rho_w}{\rho_{\text{Fe}}}\right) F_g. \tag{16–12}$$

We can check the magnitude of the net force by weighing an object in water (Fig. 16–14). Such experiments verify Archimedes' principle. Archimedes is said to have discovered this when he was asked to ascertain that a crown was made entirely of gold—without marring the crown; he did this by weighing the crown in air and in water, thereby measuring its density [Eq. (16–12)]. It is said that he found the density of the crown to be less than that of gold, thus the gold in the crown must have been diluted with another metal.

Although our calculations were carried out for a rectangular block, the results do not depend on the shape of the block. An object of any shape can be thought of as being composed of small rectangular blocks. Similarly, we performed our calculations for an object of constant density, but all that matters is the weight of the liquid that is displaced by the object. Our considerations apply equally well to objects in which the density varies, as Examples 16–6 through 16–8 show.

EXAMPLE 16–6 An air-filled lead balloon of radius $R = 0.1$ m is totally submerged in a tank of water (Fig. 16–15). What is the thickness t of the lead skin of the balloon if the balloon neither rises nor sinks? The density of lead, ρ_{Pb}, is $11.3 \times 10^3\ \text{kg/m}^3$.

Solution: We must calculate the weight of the water displaced by the balloon. To do this, we need the volume of lead and the volume of air. In order for the balloon not to sink, we expect that the thickness of the balloon must be quite small compared to the radius R: $t << R = 0.1$ m. If this is the case, we can use a simple ap-

FIGURE 16–14 By Archimedes' principle, the buoyant force on the sphere, $\mathbf{F}_{\text{buoy}}$, equals the weight of the water the sphere displaces. The scale measures the difference between the weight of the sphere and the buoyant force, $\mathbf{F}_{\text{scale}} = \mathbf{F}_g - \mathbf{F}_{\text{buoy}}$. (a) No water is displaced; the scale measures the weight of the sphere. (b) Half the volume of the sphere is displaced. (c) The entire volume of the sphere is displaced. The buoyancy force in this case is 9 N.

(a) (b) (c)

proximation. The volume of lead, V_{Pb}, is the volume of the outer sphere, of radius R, minus the volume of the inner sphere, whose radius is $R - t$. For $t << R$, this difference is approximately the area of a sphere of radius R, or $4\pi R^2$, times the thickness t: $V_{\text{Pb}} \simeq 4\pi R^2 t$. The error made in using this approximation is small (of order Rt^2). Thus, the weight of the lead sphere is approximately

$$W_{\text{Pb}} = mg = \rho_{\text{Pb}} V_{\text{Pb}} g = 4\pi R^2 t \rho_{\text{Pb}} g.$$

The weight of the enclosed air is negligible; air has a density some 1000 times less than that of water.

The weight of the displaced water is

$$W_w = \rho_w V_s g = \frac{4}{3}\pi R^3 \rho_w g,$$

where V_s is the volume of the (outer) sphere. The weights are equal if the sphere neither floats above the surface nor sinks. We set the two weights equal to each other and solve for the unknown thickness t:

$$W_{\text{Pb}} = 4\pi R^2 t \rho_{\text{Pb}} g = W_w = \frac{4}{3}\pi R^3 \rho_w g;$$

$$t = \frac{\rho_w}{\rho_{\text{Pb}}}\frac{R}{3} = \frac{10^3\ \text{kg/m}^3}{11.3 \times 10^3\ \text{kg/m}^3}\ \frac{0.1\ \text{m}}{3} \simeq 0.003\ \text{m} = 3\ \text{mm}.$$

The fact that 0.003 m << 0.1 m ($t << R$) justifies our approximation.

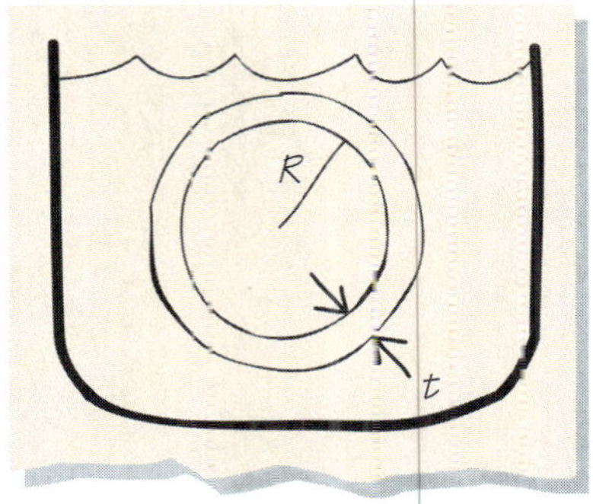

FIGURE 16–15 Example 16–6. A lead balloon, with a skin of thickness t, submerged in water.

EXAMPLE 16–7 A metal sphere weighs 29.4 N in air and 18.5 N in water. What is its average density?

Solution: The weight of the sphere in water, W_w, is the weight in air, W_{air}, minus the buoyant force. Let the volume of the sphere be V and its (unknown) average density be ρ. We have $W_{\text{air}} = \rho g V$, while the buoyant force is the weight of the displaced water, $\rho_w g V$, or

$$F_{\text{buoy}} = \rho_w g V = (\rho_w/\rho)\rho g V = (\rho_w/\rho) W_{\text{air}}.$$

But the buoyant force has magnitude $F_{\text{buoy}} = W_{\text{air}} - W_w = 29.4\ \text{N} - 18.5\ \text{N} = 10.9\ \text{N}$. Thus,

$$\frac{\rho_w}{\rho} = \frac{F_{\text{buoy}}}{W_{\text{air}}} = \frac{10.9\ \text{N}}{29.4\ \text{N}} = 0.37;$$

$$\rho = \frac{\rho_w}{0.37} = 2.7 \times 10^3\ \text{kg/m}^3.$$

EXAMPLE 16–8 A rectangular tub made of a thin shell of poured cement has length $\ell = 1$ m, width $w = 80$ cm, depth $d = 60$ cm, and mass $M = 200$ kg. The tub floats in a lake. How many people of mass $m = 80$ kg each can stand in the tub before it sinks?

Solution: Let the distance that the tub sinks be given by y (Fig. 16–16). The volume of displaced water is $\ell w y$. The weight of the displaced water, and therefore the buoyant force, is

$$F_{\text{buoy}} = \rho(\ell w y)g,$$

where the density ρ of water is 1 g/cm^3 = 10^3 kg/m^3. The downward force due to the weight of the tub and X people in it is

$$F_{\text{down}} = (M + Xm)g.$$

FIGURE 16–16 Example 16–8. A tub sinks to a water depth y. When $y > d$, the depth of the tub, the tub sinks.

The tub floats at depth y as long as the balance equation $F_{\text{buoy}} = F_{\text{down}}$ holds. The tub will finally fill with water and sink when $y = d$. Just before this point, the balance equation is

$$\rho(\ell wd)g = (M + Xm)g.$$

The solution of this equation for the number of people the tub can hold just before sinking, X, is

$$X = \frac{\rho \ell wd - M}{m}.$$

We insert numerical values to find that

$$X = \frac{(10^3\ \text{kg/m}^3)(1\ \text{m})(0.8\ \text{m})(0.6\ \text{m}) - 200\ \text{kg}}{80\ \text{kg}} = 3.5.$$

With three people, the tub does not sink; with four people, she goes down. Even the three tubbers should be careful not to rock the boat!

FIGURE 16–17 Under the right conditions, soil can act like a fluid (mud) and flow.

16–5 FLUIDS IN MOTION

The marvelous fluid flow patterns observed in nature go far beyond the particle and rigid body motions we have studied to this point. **Hydrodynamics** describes fluids in motion, including the flow of placid streams, eddies in a pond, mud slides, smoke rings, ocean waves and tides, arterial blood flow, Jupiter's (and Earth's) atmosphere, the rotational flow of the galaxy, and nuclear fission (Fig. 16–17). Numerical calculations performed by supercomputers are necessary to solve the most complex types of behavior. What is remarkable is the variety of phenomena that can be described with a very simplified set of equations.

Assume incompressibility.

We assume, as we have discussed, that the fluid under consideration is incompressible. This does not mean that we exclude phenomena that involve gases, such as air flow; it just means that we will discuss these phenomena only under circumstances in which compressibility does not play a role.

Assume constant temperature.

We also assume that we are dealing with a fluid at constant temperature. (We will learn more about temperature and heat flow in Chapters 17 through 20.) Phenomena such as the **convection** of fluids—in which a liquid in the bottom of a vessel is heated, rises, cools, and falls in a circulating pattern—are thus excluded from our consideration.

Assume steady flow.

In general, the velocity of a fluid at a given point, $\mathbf{v}(x, y, z, t)$, as well as the pressure, $p(x, y, z, t)$, may vary with time, but we will consider only **steady flow**: fluid movement for which both $\mathbf{v}$ and p do not depend on time. The pressure and the velocity may vary from point to point, but we assume that any changes are smooth ones.

FIGURE 16–18 The water moves faster near the center of the vortex than it does far from the center. The water is in rotational flow.

Consider a fluid in motion. We could represent the direction of the velocity at some point in Fig. 16–18 by an arrow at that point, with the length of the arrow proportional to the magnitude of the velocity. For steady (time-independent) flow, there is a *local velocity* $\mathbf{v}(\mathbf{r})$. If we start at some point, then a tiny element of the fluid at that point moves in the direction of the velocity vector. An instant later, the element is at an adjacent point, which has a slightly different velocity vector. This vector dictates the element's direction and speed of motion in the next instant, and determines the next step of the motion. In this way, the motion of a tiny element of fluid can be traced. The line of motion of a small fluid element that maintains its integrity is called a **streamline** (Fig. 16–19).

Assume laminar flow.

Streamlines cannot cross because a velocity vector cannot have two directions. Flow described by streamlines is called **laminar flow**; in laminar flow, each element of fluid travels along a smooth, well-defined path. Under certain circumstances, which we will describe in Section 16–9, fluid flow ceases to be laminar and becomes turbulent. **Turbulent flow**, as in the flow of a white-water river, involves rapidly fluctuating velocities and pressures. Turbulence is a characteristic of many flow patterns that we observe in nature, but there is no simple theory that describes it. The study of turbulence is an exciting area of research in engineering and physics.

Assume irrotational flow.

We do *not* consider flow in which streamlines close in on themselves, such as those shown in Fig. 16–20; that is, we consider only **irrotational flow**. As a consequence, we do not discuss vortices, or whirlpools, such as those that arise in water running out of a bathtub, smoke rings, or even the much simpler motion of coffee stirred by a spoon in a cup. A mathematical expression of the statement that the flow is irrotational is that the **circulation**, defined as the integral of the velocity along any *closed* path in the fluid, is zero.

To simplify matters further, we make one more assumption: We neglect the analogue of friction in the fluid. That there is such an analogue is obvious from simple observations. The flow of honey in Fig. 16–21 is different than the flow of water. The friction that enters here is an internal friction, which is called **viscosity** (see Section 16–9). In neglecting viscosity, we are neglecting internal energy losses. The conservation of energy can then be used profitably in the description of fluids.

To summarize, in the next two sections we make the following assumptions:

1. The fluid is incompressible.
2. The temperature does not vary.
3. The flow is steady, so that the velocity and pressure do not depend on time.
4. The flow is laminar rather than turbulent.
5. The flow is irrotational, so there is no circulation.
6. There is no viscosity in the fluid.

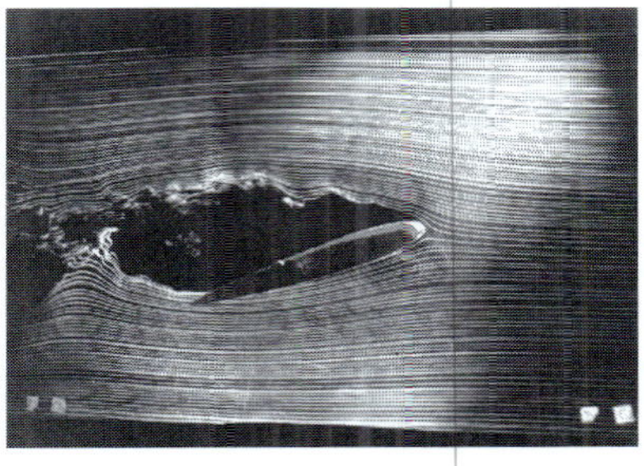

FIGURE 16–19 The use of smoke in a wind tunnel allows a direct visualization of the motion of air. Here we see that turbulent flow has been induced behind the wing section.

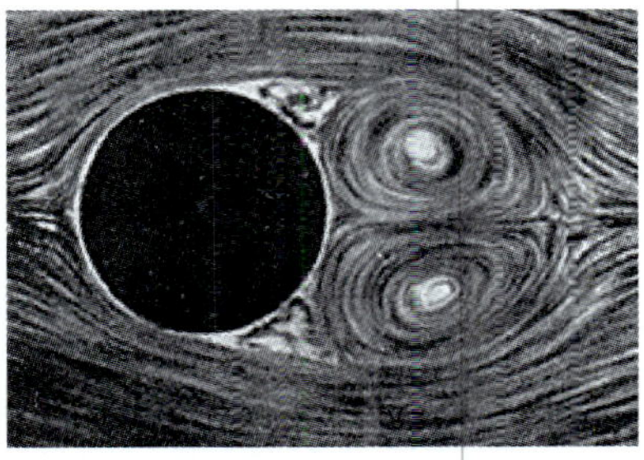

FIGURE 16–20 Streamlines for a cylinder pulled steadily through a fluid, showing circulation in the fluid.

FIGURE 16–21 Honey flows slowly because its viscosity is high.

16–6 THE EQUATION OF CONTINUITY

For a fluid in motion, we can pick a group of adjacent streamlines, which we say form a thin bundle, and follow them (Fig. 16–22). Because streamlines never cross, the bundle forms a thin tube with nonuniform cross section, known as a **streamtube**, in which fluid flows; no fluid passes from the inside to the outside of a streamtube. We can choose a tube thin enough so that the velocity across a section drawn at right angles to the tube is the same across the whole cross section. The volume of fluid that crosses the section of area A_1 in a given time t is the volume of a cylinder whose base is A_1 and whose length is $\ell_1 = v_1 t$, where v_1 is the speed across that section; the volume of fluid is thus $v_1 t A_1$. If the density of the fluid at this section is ρ_1, then the mass of fluid crossing the section in the given time t is

$$m_1 = \rho_1 v_1 t A_1. \qquad (16\text{–}13)$$

Similarly, the mass of fluid crossing a section of area A_2 at right angles to the bundle of streamlines with speed v_2 and density ρ_2 in the same time t is

$$m_2 = \rho_2 v_2 t A_2.$$

But because fluid neither enters nor leaves the tube by the side walls, the *conservation of mass* implies that $m_1 = m_2$, or, if we cancel the time t,

$$\rho_1 v_1 A_1 = \rho_2 v_2 A_2. \qquad (16\text{–}14)$$

The conservation of mass applied to fluids, also known as the equation of continuity

This equation, the **equation of continuity**, is true even for fluids whose densities can vary with position. When we restrict ourselves to incompressible flow, we have $\rho_1 = \rho_2$ and derive the very useful equation

$$v_1 A_1 = v_2 A_2. \qquad (16\text{–}15)$$

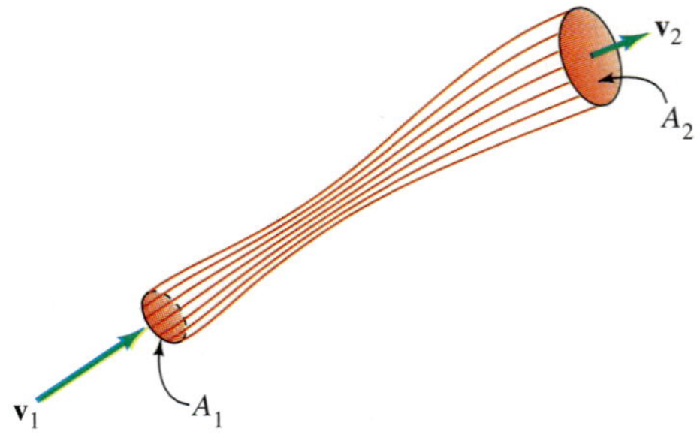

FIGURE 16–22 Flow contained in a bundle of streamlines (streamtube). The velocity perpendicular to the surface of area A_1 is $\mathbf{v}_1$, and that perpendicular to the surface of area A_2 is $\mathbf{v}_2$.

The conservation of mass applied to incompressible fluids

This result states that the product of speed and area of any bundle of streamlines is the same anywhere along the bundle. The speed of fluid flow thus varies inversely with the area of the streamtube. The product (velocity perpendicular to an area) × (area) is the **flux**, Φ. Equation (16–15) describes the **conservation of flux**. Perhaps you have observed the conservation of flux while you have entered or left a building on a windy day: The wind speed is greatest in the doorway.

PROBLEM SOLVING

Conserved quantities are very helpful in problem-solving.

Note that flux can be interpreted as the volume of fluid that passes through an area per second. If the velocity of a fluid is perpendicular to a given area A, the volume of fluid that passes through that area in time t is a length $L = vt$ (the distance the fluid moves in time t) multiplied by A, or $V = vtA$. Thus *the flux*, $\Phi = V/t = vA$, *is the volume of flow per second.*

Flux is the volume of fluid that passes through a given area per unit time.

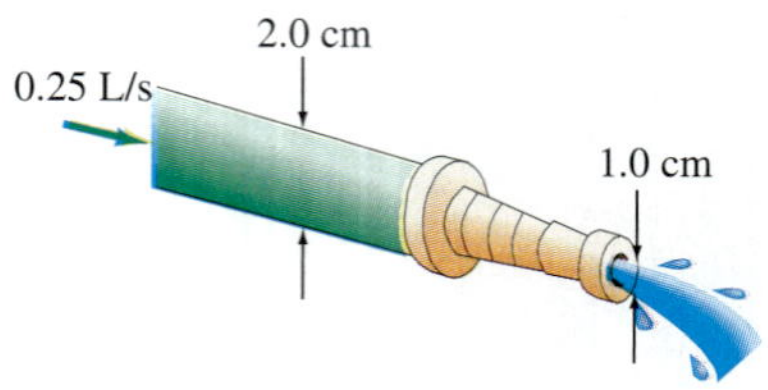

FIGURE 16–23 Example 16–9.

EXAMPLE 16–9 Figure 16–23 shows water that flows at a rate of 0.25 L/s through a garden hose of inside diameter 2 cm. What is the speed of the water in the hose? The nozzle of the hose is a circular opening of diameter 1.0 cm. What is the speed of the water when it emerges?

Solution: We are given the volume of water that passes through the hose per second. To find the speed v of the water in the hose, whose cross section is A, we use the fact that the volume that passes per second is the flux, $\Phi = vA$. Thus the speed in the hose is $v_{\text{hose}} = \Phi/A_{\text{hose}}$. From Eq. (16–15), the conservation of flux, we know that the volume per second that passes the nozzle must be the same as the volume per second that passes through the hose, or

$$A_{\text{noz}}v_{\text{noz}} = A_{\text{hose}}v_{\text{hose}}.$$

We solve this equation for the desired quantity, v_{noz}:

$$v_{\text{noz}} = \frac{A_{\text{hose}}}{A_{\text{noz}}}\, v_{\text{hose}} = \frac{A_{\text{hose}}}{A_{\text{noz}}}\, \frac{\Phi}{A_{\text{hose}}} = \frac{\Phi}{A_{\text{noz}}}.$$

The radius of the nozzle is half the diameter, or 0.50 cm, and the area $A = \pi r^2$. Thus we have

$$v_{\text{noz}} = \frac{0.25\ \text{L/s}}{\pi(0.50\ \text{cm})^2} = 320\ \text{cm/s} = 3.20\ \text{m/s}.$$

Note that because the radius of the nozzle is half that of the hose, and because the area is proportional to the radius squared, the nozzle has one-fourth the area of the hose; the velocity of the emerging stream is four times that of the water in the hose: $v_{\text{hose}} = \frac{1}{4}v_{\text{noz}} = 0.8$ m/s.

This example illustrates how a small nozzle causes the water to exit a hose with a large velocity and thus spray over a long distance.

APPLICATION

The use of nozzles

16–7 BERNOULLI'S EQUATION

In principle, the motion of a fluid can be obtained by breaking the fluid down into small elements and applying Newton's second law to each of them. We saw in our previous study of motion that the conservation of energy is a powerful alternative to the second law when we have only conservative forces. If there is no viscosity in a fluid, so that there is no energy dissipation, and if the fluid is incompressible, it becomes straightforward to formulate the conservation of energy for fluids. This will give us a simple and important relation to describe the motion.

We again consider a streamtube—a bundle of streamlines of fluid—this time under the influence of gravity. Concentrate on the fluid between two cross sections of the tube: one at height h_1 with cross-sectional area A_1, the other at height h_2 with area A_2 (Fig. 16–24). The velocity perpendicular to the cross sections will have magnitudes v_1 and v_2, respectively, and we take our tube to be thin enough so that the velocity does not vary across the slices. There will be flow to the right provided that the pressure p_1 at region A_1 is larger than the pressure p_2 at region A_2.

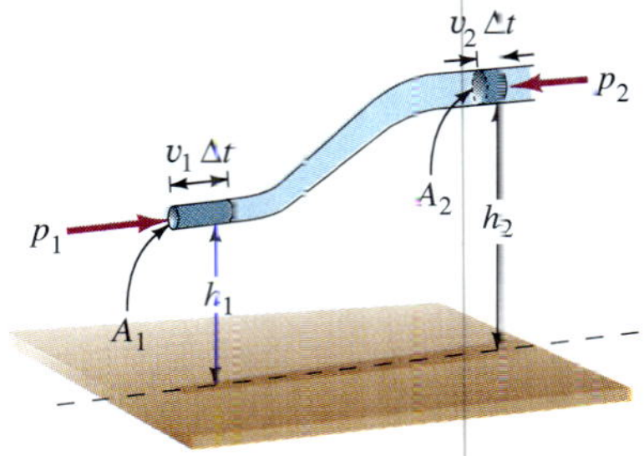

FIGURE 16–24 A streamtube of fluid at different heights. In time Δt, the fluid moves to the right under the influence of a difference in pressures between p_1 and p_2. In effect, the shaded slice in region 1 is transferred to the shaded slice in region 2.

Calculating the Work to Move a Fluid Segment

Let's examine what happens during a time interval dt. The fluid that starts out in the region between A_1 and A_2 at $t = 0$ will have advanced. The distance by which it advances at height h_1 is $v_1\,dt$, and the distance by which it advances at height h_2 is $v_2\,dt$. We calculate the net work done in advancing the fluid in the tube by looking at the forces acting at either end. The work done at the left-hand region by pressure p_1 is given by (force) × (displacement). This can be written as

$$W_1 = (p_1A_1)(v_1\,dt). \qquad (16\text{–}16)$$

The work done at the right-hand region by pressure p_2 can be calculated similarly, but there is an additional minus sign because the force (and thus the pressure) acts to the left while the displacement is to the right, so that $\mathbf{F}\cdot\mathbf{s} = -Fs$. Thus

$$W_2 = -(p_2A_2)(v_2\,dt).$$

The net work done on the streamtube between regions 1 and 2 is

$$W_{net} = W_1 + W_2 = (p_1 - p_2)\,vA\,dt. \qquad (16\text{–}17)$$

where in the last line we have made use of Eq. (16–15), which expresses the constancy of the product vA at any point along the tube:

$$vA = v_1A_1 = v_2A_2.$$

Applying the Work–Energy Theorem

We have determined the amount of work done in moving the fluid in the tube section during the time dt. To apply the work–energy relation, we now want to find the change in energy. Examining Fig. 16–24, we can see that what has happened in the time interval dt is the transfer of the small slice of fluid from region 1 to region 2. This transfer has two effects on the energy: The potential energy changes as the slice is lifted from h_1 to h_2, and the kinetic energy changes as the speed changes from v_1 to v_2. The change in potential energy is

$$\Delta U = mg\,\Delta h = (\rho vA\,dt)g(h_2 - h_1). \qquad (16\text{–}18)$$

Again we have used the constancy of vA. The change in kinetic energy is

$$\Delta K = \frac{1}{2}(\rho vA\,dt)(v_2^2 - v_1^2). \qquad (16\text{–}19)$$

Because $W_{net} = \Delta K + \Delta U$, we find, from Eqs. (16–17) through (16–19),

$$(p_1 - p_2)vA\,dt = \frac{1}{2}(\rho vA\,dt)(v_2^2 - v_1^2) + (\rho vA\,dt)g(h_2 - h_1).$$

After the factor $vA\,dt$ divides out,

$$p_1 - p_2 = \frac{1}{2}\rho(v_2^2 - v_1^2) + \rho g(h_2 - h_1).$$

PROBLEM-SOLVING TECHNIQUES

Bernoulli's equation expresses conservation of energy for a fluid; as such, its use is similar to the use of all conservation laws. Keep in mind the following points:

1. When we say the right-hand side of Eq. (16–21) is a constant, we mean that it takes on the *same* value for all points within the fluid. If we know the value of $p + \frac{1}{2}\rho v^2 + \rho gh$ at one point, then that is the value it takes at all points.
2. In its typical application, we can find the value of the constant in Bernoulli's equation because we know the value of all the variables at some point. Then we can use the value of the constant to find the unknown value of a variable at another point.
3. When one of the three terms in the expression $p + \frac{1}{2}\rho v^2 + \rho gh$ is the same at all points, then we can ignore it and concentrate on using Bernoulli's equation to learn about the other variables. For example, in Section 16–8, we first apply Bernoulli's equation to the case of a fluid at rest. In that case, v is the same everywhere ($v = 0$), and we can use Bernoulli's equation as a conservation law for the quantity $p + \rho gh$.

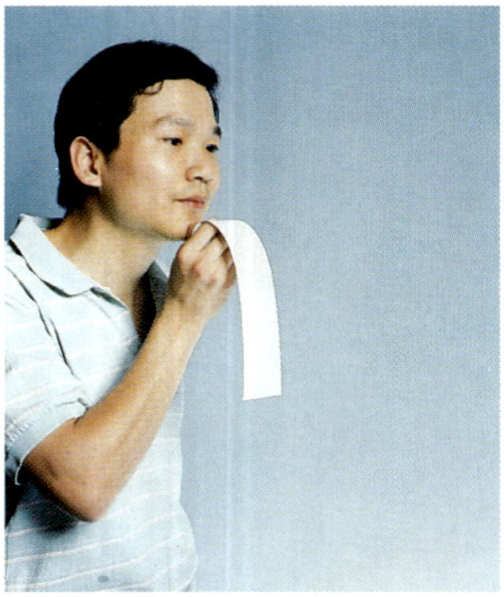

(a)

(b)

FIGURE 16–25 By blowing over the top of the paper, the pressure above the paper is reduced according to Bernoulli's principle. Below the paper, the pressure is unchanged, and there is a net force that makes the paper rise.

This equation can be rearranged into a conservation law, with one side dependent on the quantities defined in region 1 and the other side dependent on the quantities defined in region 2:

$$p_1 + \frac{1}{2}\rho v_1^2 + \rho gh_1 = p_2 + \frac{1}{2}\rho v_2^2 + \rho gh_2. \qquad (16–20)$$

In other words,

$$\text{along a streamline:} \quad p + \frac{1}{2}\rho v^2 + \rho gh = \text{a constant.} \qquad (16–21)$$

Our derivation deals with the motion of an element of fluid, and thus our conclusions hold only for motion along a streamline. It can be shown in general that *for irrotational flow, the constant in Eq. (16–21) is the same for all streamlines.* Equation (16–21) is **Bernoulli's equation**, derived in 1738 by Daniel Bernoulli, one of a remarkable family of scientists and mathematicians. In Section 16–8, we discuss applications of this formula.

16–8 APPLICATIONS OF BERNOULLI'S EQUATION

Bernoulli's equation is useful for describing a variety of phenomena. First, we observe that if a fluid is at rest, so that $v = 0$ in Eq. (16–21), then

$$p + \rho gh = \text{a constant.} \qquad (16–22)$$

This is just the hydrostatic relation between pressure and height that we obtained in Eq. (16–6).

The Bernoulli Effect

For flow in a horizontal (constant height) tube, the conservation of energy [Eq. (16–21)] implies that

$$p + \frac{1}{2}\rho v^2 = \text{a constant.} \qquad (16–23)$$

Thus the pressure must be lower in a region in which a fluid is moving faster. You can observe this phenomenon—known as the **Bernoulli effect**—in a simple experiment. If you blow air across the top of a single sheet of paper (Fig.16–25), the pressure above the paper is reduced. However, the pressure is unchanged below the paper; thus, the paper rises. The Bernoulli effect also enters into the operation of a chimney. When the wind blows over the top of a chimney, the pressure is decreased at the top by the Bernoulli effect, and the air in the tube rises. This is an important element in a chimney's "draw."

APPLICATION

The Venturi flowmeter

A Measuring Device. Equation (16–23) is applied in the *Venturi flowmeter*, a device designed to measure the speed of fluids in a pipe (Fig. 16–26a). Figure 16–26b is a simplified version of the flowmeter. The large pipe, with cross-sectional area A_1, is constricted smoothly to a smaller area A_2. A U-shaped tube containing a different fluid connects the two segments of pipe. The speeds in the larger and smaller sections of the pipe are v_1 and v_2, respectively; similarly, the pressures in the two sections are p_1 and p_2, respectively. Then Eq. (16–23) tells us that

$$p_1 + \frac{1}{2}\rho v_1^2 = p_2 + \frac{1}{2}\rho v_1^2;$$

$$p_1 - p_2 = \frac{1}{2}\rho(v_2^2 - v_1^2).$$

The continuity equation, Eq. (16–15), tells us that $v_2 = v_1 A_1/A_2$, so

$$p_1 - p_2 = \frac{1}{2}\rho v_1^2\left[\left(\frac{A_1}{A_2}\right)^2 - 1\right]. \qquad (16\text{–}24)$$

The difference in pressures is determined from the difference in heights, h, of the two columns of fluid in the U-shaped tube (see Section 16–3). Equation (16–24) is then used to find speed v_1 in terms of this pressure difference.

Fluid Motion with Constant Speed

In the Bernoulli effect, we did not change the height but the speed of the flowing fluid did change. If the area of a pipe is fixed and the fluid is incompressible, then the equation of continuity, Eq. (16–15), shows that the speed cannot change. Such a pipe will force the fluid moving inside to change pressure if the height changes: In this case, Bernoulli's equation becomes $p + \rho g h$ = a constant, or Eq. (16–22).

EXAMPLE 16–10 A giraffe needs a strong heart because of its long neck (Fig. 16–27). Suppose that the difference between the aortic valve (the place where arterial blood comes out of the heart) and the head of a giraffe is 2.50 m, and that the

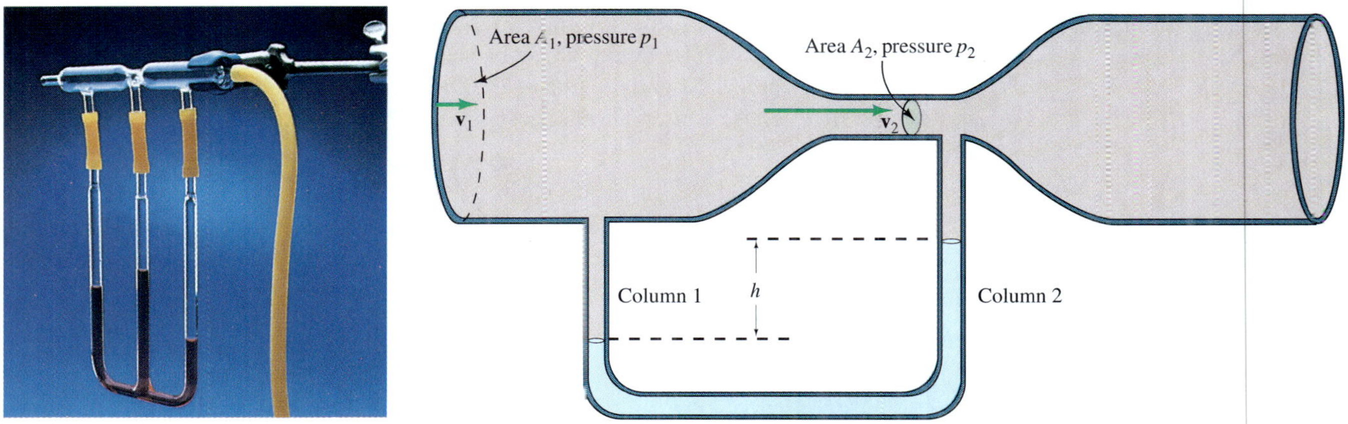

(a) (b)

FIGURE 16–26 (a) The Venturi flowmeter. (b) The speed v_1 of a fluid in a pipe is measured by the difference in heights, h, of column 1 and column 2 of a different fluid. The number of measuring tubes that tap the pipe through which the fluid flows can be varied according to the measurement desired.

artery leading from near the aortic valve to the head has constant cross section. What is the minimum pressure at the aortic valve? Blood is an incompressible fluid with density $\rho = 1.0\ \text{g/cm}^3$.

Solution: The minimum pressure at the heart is the pressure that will allow the blood to arrive at the head with zero pressure. Because the blood is incompressible, the equation of continuity implies that the speed of the blood will be the same at the head as at the heart. We can then employ Bernoulli's equation to find the pressure at the heart. Let's use subscripts 1 and 2 for the heart and head, respectively. Then, with $v_1 = v_2$, the terms proportional to v^2 in Bernoulli's equation cancel, and Eq. (16–22) applies. With $p_2 = 0$, we have

$$p_1 + \rho g h_1 = \rho g h_2$$
$$p_1 = \rho g(h_2 - h_1).$$

Numerically, $p_1 = (1.0\ \text{g/cm}^3)(10^2\ \text{cm/m})^3(10^{-3}\ \text{kg/g})(9.80\ \text{m/s}^2)(2.50\ \text{m}) = 2.45 \times 10^4\ \text{Pa}$. Various factors, in particular the loss of energy due to friction between the blood and the arterial walls, make this a low estimate. This number can be compared to the peak output pressure of the human heart, 1.6×10^4 Pa.

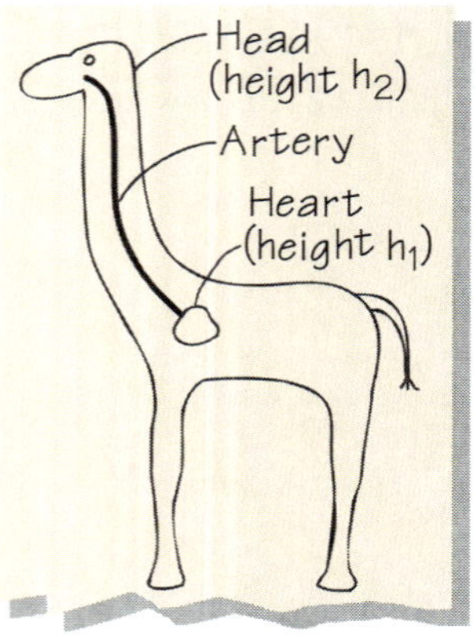

FIGURE 16–27 Example 16–10.

Flow from a Tank

Another interesting phenomenon of practical importance in fluid flow is the speed of flow of a liquid from an opening in a tank (Fig. 16–28a). Let's make a small hole in the side of an open tank of water at a depth h below the surface. We apply Bernoulli's equation, Eq. (16–21), to the fluid at different locations. Denote the height above the hole by the variable y, so the height at the water surface is $y = h$ and at the opening $y = 0$. Equation (16–21) states that the value of $p + (\rho v^2/2) + \rho g y$ is the same at point 1 (on the upper surface, where the pressure $p = p_0$) and point 2 (just outside the hole, where $p = p_0$ because the sides of the stream of fluid flowing from the tank are in equilibrium with the atmosphere). The speed v_2 at point 2 is to be determined. We equate the values of $p + (\rho v^2/2) + \rho g y$ at the two points and obtain

$$p_0 + \frac{1}{2}\rho v_1^2 + \rho g h = p_0 + \frac{1}{2}\rho v_2^2 + \rho g(0).$$

We can neglect the small velocity v_1 of the liquid at the top of the tank. We solve for the speed at point 2 (the hole) to find that

$$v_2 = \sqrt{2gh} \tag{16–25}$$

(Fig. 16–28b). Notice that this is just the speed of a single particle falling through a height h under the influence of gravity, as in Eq. (2–27). We find the same result for the falling liquid as for a falling object because they both are consequences of the conservation of energy. To find the rate of fluid flow from the hole, we use the expression for the flux, whose conservation is described by Eq. (16–15):†

$$\Phi = vA. \tag{16–26}$$

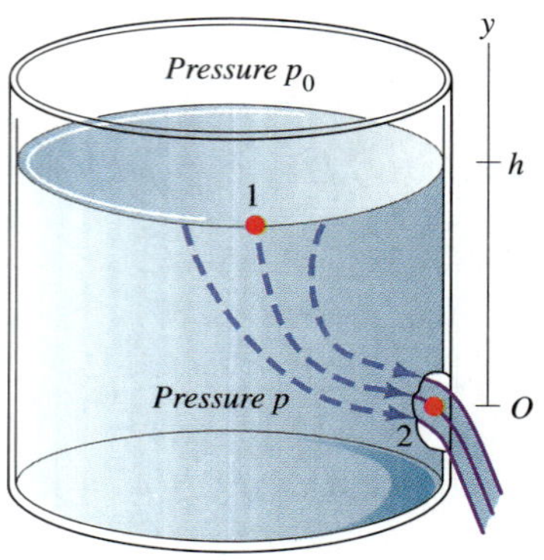

(a)

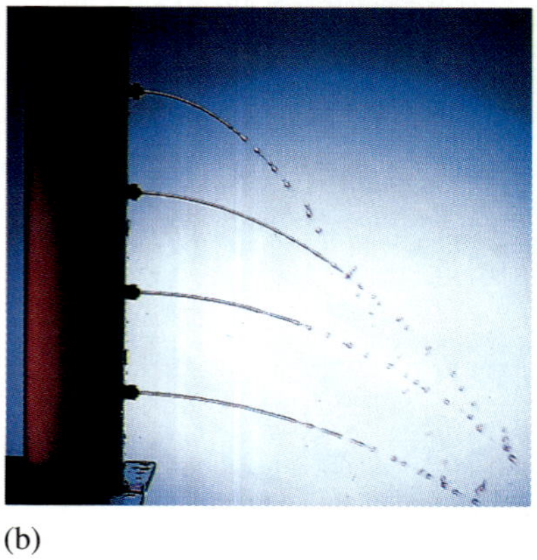

(b)

FIGURE 16–28 The escape of fluid through a hole in a tank. (a) The curved lines are streamlines for real fluid flow. (b) The pressure at the bottom of the tank is greater than the pressure at the top, so a stream of water leaving the tank has greater speed if it leaves from a hole that is lower on the tank.

EXAMPLE 16–11 A tank of cross-sectional area $A = 0.07\ \text{m}^2$ is filled with water. A tightly fitting piston, with a total mass $m = 10$ kg, rests on top of the water (Fig. 16–29). A circular hole of diameter 1.5 cm is opened at a depth of 60 cm below the water level of the tank. What is the initial rate of flow of water out of the hole?

Solution: At surface level, the total pressure $P = p_0 + p_1$, where p_0 is the atmospheric pressure and p_1 is the pressure due to the piston resting on the water, $p_1 = mg/A$. As in our discussion, the pressure when the fluid has left the hole is p_0. An

†A more careful treatment replaces the right-hand side of Eq. (16–26) with the expression kvA, where the coefficient k is less than 1. Its correct value for a simple small hole is 0.62. In the examples and problems, we use $k = 1$ for simplicity.

APPLICATION: REEDED INSTRUMENTS

Both single-reed (clarinet and saxophone) and double-reed (oboe and bassoon) instruments rely on the Bernoulli effect to produce their sounds. Let's consider an oboe. The two strips of cane that form the double reed are tightly joined at one end to a brass tube, which is connected with a cork joint to the top of the oboe. The other end of the strips is free; here, the sides of the two strips touch but they are open in the center (Fig. 16AB–1). This end of the double reed is placed between an oboe player's lips; the player then pressurizes his or her mouth cavity and air is blown between the two reeds, producing sound. By the Bernoulli effect, the movement of the air between the reeds lowers the pressure there relative to the static pressure in the mouth cavity. The resulting net force presses the reeds together and the entrance hole closes. The air flow then ceases and the pressure differential becomes zero, allowing the reeds to return to their natural open position. This cycle repeats at the natural mechanical resonance frequency of the reeds. A reed assembly blown without being connected to the oboe produces a musical note at this resonance, called the "crow tone." The pulses of air generated by the reed act as a driving force for an air column oscillator made from the oboe pipe. Standing waves are set up in the air column, at frequencies that can be changed by opening and closing holes in the pipe. The starting transient state of the instrument contains frequencies associated with the crow tone, giving the characteristic "pauk" sound of the instrument.

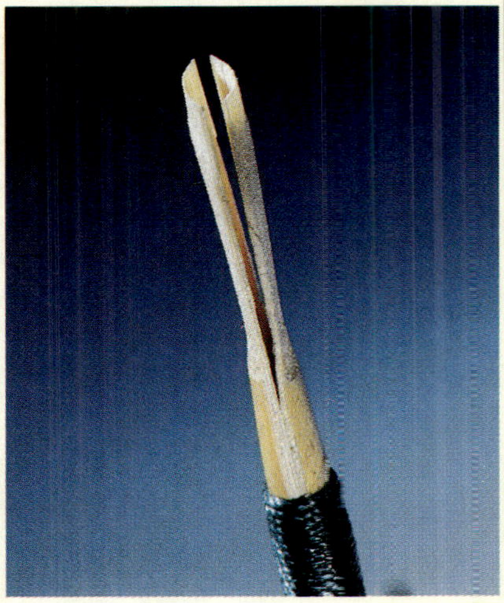

FIGURE 16AB–1 An oboe reed.

application of Bernoulli's equation at the top of the water and at the opening leads to the equality

$$p_0 + p_1 + \rho g h = p_0 + \frac{1}{2}\rho v^2.$$

We cancel p_0 from both sides of the equation and solve for v^2:

$$v^2 = \frac{2(p_1 + \rho g h)}{\rho} = \frac{2(mg/A + \rho g h)}{\rho} = 2g\left(\frac{m}{\rho A} + h\right).$$

Numerically, with $\rho = 10^3\ \text{kg/m}^3$, we find

$$v^2 = 2(9.8\ \text{m/s}^2)\left(\frac{10\ \text{kg}}{(10^3\ \text{kg/m}^3)(0.07\ \text{m}^2)} + 0.6\ \text{m}\right) = 15\ \text{m}^2/\text{s}^2,$$

or $v \approx 4$ m/s. From our discussion of flux, we know that the rate of flow is the flux $\Phi = vA_{\text{hole}} = v\pi r_{\text{hole}}^2$. The hole has a radius of 0.75 cm = 7.5×10^{-3} m. Thus $\Phi = (4\ \text{m/s})\pi(7.5 \times 10^{-3}\ \text{m})^2 = 7 \times 10^{-4}\ \text{m}^3/\text{s} = 0.7$ L/s.

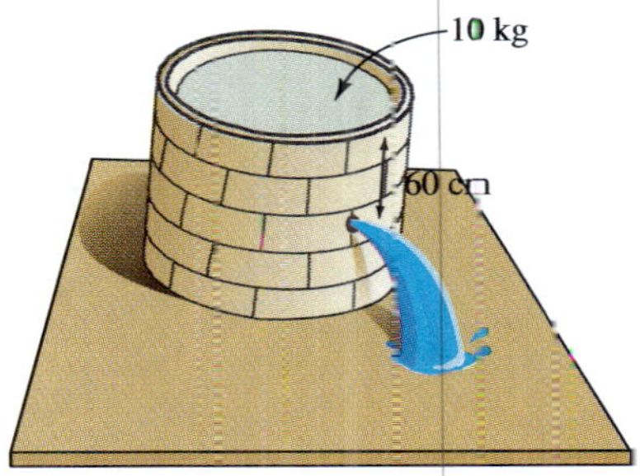

FIGURE 16–29 Example 16–11.

*Lift

The lift

The Bernoulli effect is ultimately responsible for **lift**, an upward net force on the wings of an airplane. A correct description of this phenomenon requires that we relax some of the simplifying assumptions listed at the end of Section 16–5. Circulation, viscosity, and turbulence play an important, though indirect, role in lift. Consider the cross section of a wing shown in Fig. 16–30a. We visualize the situation from the reference frame of the wing, so that we think of the air moving toward the wing, which is at rest, rather than the wing moving through stationary air. (This is, in fact, just what happens in a wind tunnel in which problems such as this one are studied.) Air approaches the wing with some speed v along a set of streamlines. Some of the streamlines go below the wing, some go above. The wing is designed so that the air speed above the wing is greater than that below the wing. This can happen only if there is circulation in the

APPLICATION

Wing design

clockwise direction (Fig. 16–30b).[†] Circulation occurs only around the wing. The flow elsewhere is irrotational, which means that the constant appearing in the Bernoulli equation, Eq. (16–21), is the same for all the streamlines that flow near the wing, below or above. From Eq. (16–23), we see that the pressure on top of the wing, p_t, must therefore be less than the pressure below, p_b, and there is a net upward force, or lift, on the wing. The net upward pressure, $p_{\text{net}} = p_b - p_t$, is determined from Eq. (16–23):

$$p_b + \frac{1}{2}\rho v_b^2 = p_t + \frac{1}{2}\rho v_t^2,$$

where ρ is the density of air. We thus find that

$$p_{\text{net}} = p_b - p_t = \frac{1}{2}\rho(v_t^2 - v_b^2) = \rho\left(\frac{v_t + v_b}{2}\right)(v_t - v_b) \simeq \rho v(v_t - v_b), \quad (16\text{–}27)$$

where v is the average of the top and bottom speeds. Because the difference between the top speed and the bottom speed is small, both are nearly equal to the average speed, which we call the *airstream speed* v.

The expression in Equation (16–27) can be used to calculate the lift. If the wings have a width w and a tip-to-tip span S, then their total area is wS (we are ignoring the fuselage), and the lift is given by

$$L = wSp_{\text{net}} = wS(p_b - p_t) \simeq wS\rho v(v_t - v_b). \quad (16\text{–}28)$$

Experiments show that for speeds much slower than the speed of sound, $v_t - v_b$ is proportional to the airstream speed v, with a proportionality constant K (in general, $K \ll 1$), which depends on the wing design and the wing angle relative to the airplane direction. A typical value of the constant K is 0.12 so, for a small plane with a total wing area of 30 m^2 and an air speed of 200 km/h = 56 m/s and with the air density of about 1.3 kg/m^3, the lift is given by

$$L = (0.12)(30\ \text{m}^2)(1.3\ \text{kg/m}^3)(56\ \text{m/s})^2 = 1.6 \times 10^4\ \text{N}.$$

Thus the mass that can be lifted is $(1.6 \times 10^4\ \text{N})/(9.8\ \text{m/s}^2) \simeq 1600$ kg. The fact that air is compressible plays little role if the relative speed of the wing and the air is much smaller than the speed of sound in air, $v_s \simeq 330$ m/s. (The design and operation of the wings of supersonic aircraft presents a different set of problems.)

We remarked previously that there must be circulation in the problem. To give the flow circulation, it is necessary to have some turbulence, and this is induced by the viscosity of the air and further enhanced by the placement of small protuberances (ailerons) on the wings.

Racing cars are designed so that the air beneath the car moves faster than the air above it, and a downward pressure is exerted on the car. This increases the friction between the wheels and the track, and this friction is ultimately responsible for the propulsion of the car.

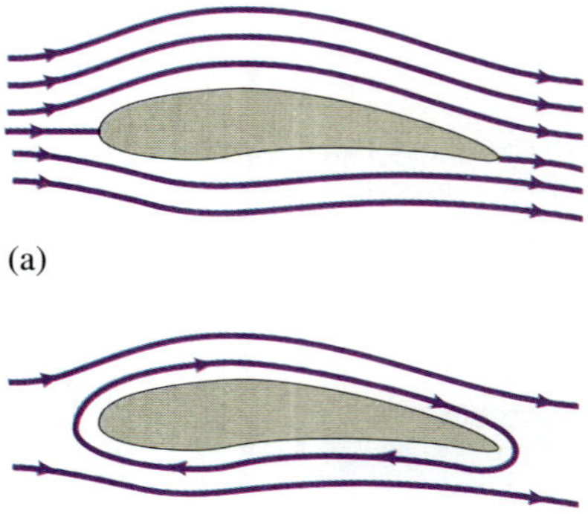

(a)

(b)

FIGURE 16–30 (a) Schematic version of streamlines over an airplane wing section like that in Fig. 16–19. There is a streamline that ends on the front of the wing at the stagnation point and a streamline that starts at the rear edge of the wing. (b) Because the compressibility of the air can be neglected in this application, the equation of continuity shows that there must be a nonzero circulation around the wing, which in effect forms a closed loop.

*16–9 COMMENTS ON REAL FLUIDS

Our discussion has been limited by assuming idealized kinds of flow with idealized fluids. Some of our assumptions apply only approximately, at best, to real fluids. Nonviscous flow, for example, has sometimes been called the flow of "dry water." Here, we describe some qualitative notions of how real fluids behave under ordinary circumstances.

Viscosity

Figure 16–31a is a picture of the velocity profile of pipe flow for fluids idealized as we have described. For real fluids, the liquid near the surface of the pipe is at rest, and

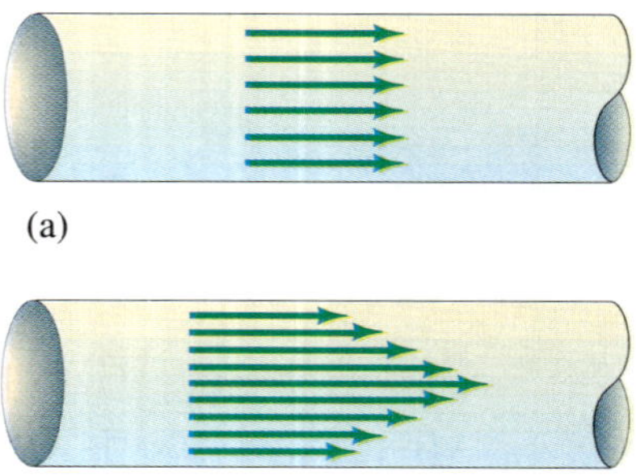

(a)

(b)

FIGURE 16–31 Velocity profiles for laminar flow in a pipe: (a) idealized fluid flow that does not take into account viscosity; (b) realistic fluid flow (Poiseuille flow).

[†]No air actually moves forward under the wing; rather, the speed beneath the wing is less than the speed above the wing.

the velocity profile is actually parabolic (Figure 16–31b). Such flow is referred to as **Poiseuille flow**, after Jean Poiseuille, who studied blood circulation. The reason for this effect is the internal friction of the fluid, or *viscosity*. Viscosity comes from intermolecular collisions. The molecules of a fluid have a random motion, to be described in Chapter 19. If two adjacent layers of fluid flow with slightly different speeds, the random sidewise intrusion of some slower molecules into the faster stream will tend to slow down the faster stream, whereas intrusion of faster molecules into the slower stream will tend to speed up the slower stream.

Viscous flow can be analyzed if we take two glass plates with a film of fluid between them and slide one plate over the other. The easier it is to slide the plates, the less viscous is the fluid between them. To quantify this effect, take the smaller plate to have an area A; it is separated from the other plate by a distance y (Fig. 16–32). If the upper plate is to move with a uniform velocity $\mathbf{v}$, a force must be applied—just as if a brick is made to slide along a rough surface at constant speed. This force, F, is proportional to A and inversely proportional to y. In addition, the force increases linearly with speed (like the damping force on a mass at the end of a spring or like air drag). The proportionality constant η in the formula

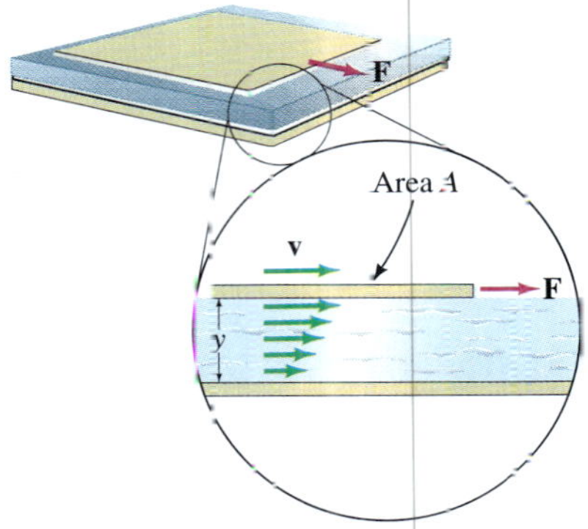

FIGURE 16–32 Two glass plates have a film of fluid between them; one slides against the other. The top plate moves with speed v. Fluid near the top plate moves with the same speed as that plate, while fluid near the bottom plate is at rest

$$F = \frac{\eta v A}{y} \tag{16–29}$$

is called the **coefficient of viscosity**. The coefficient of viscosity is measured in newton · seconds per square meter = kilograms per meter · second. Typical values are 10^{-3} kg/(m · s) for water and 830×10^{-3} kg/(m · s) for glycerin, both at 20°C. We mention the temperature because, as anyone who has poured syrup over pancakes knows, viscosity is strongly temperature dependent.

Turbulence

A flow regime marked by violent and random movement is referred to as turbulent. At the onset of turbulence, the whole notion of neatly separated streamlines ceases to make sense (Fig. 16–33). A dimensionless parameter known as the **Reynolds number**, *Re*, determines this regime. It is defined as

$$Re = vL\left(\frac{\rho}{\eta}\right). \tag{16–30}$$

(a)

(b)

FIGURE 16–33 Turbulence, an important feature of fluid flow, is seen here in (a) the air above a candle and (b) the atmosphere of Venus. The blue seen in the image results from computer processing. The large swirl at the bottom is at the planet's South Pole.

where ρ is the density of the fluid, v is its speed, L is a length associated with the flow, such as the pipe's diameter for flow in a pipe, and η is the viscosity coefficent. When the Reynolds number increases past about 2000 to 3000, the flow becomes turbulent. For water flowing through a pipe 1 cm in diameter, $Re \simeq (10^4\ \text{s/m})v$, so the flow becomes turbulent when v reaches merely 0.3 m/s. This shows that the kind of flow we have treated in this chapter is not very common. Fortunately, a small amount of turbulence does not change the applicability of Bernoulli's equation for short distances of flow (just as a small amount of friction does not disturb energy conservation over short times), so Eq. (16–21), although not exact, is still a very good and useful approximation.

SUMMARY

Liquids are distinguished from gases in that the molecules that form liquids are closer together and impose strong forces on each other. Liquids are distinguished from solids in that the molecules in liquids do not form a rigid lattice. Both liquids and gases deform in response to external forces and therefore flow as fluids. The density of a material of mass M and volume V is unchanging in incompressible fluids:

$$\rho = \frac{M}{V}; \tag{16–1}$$

this is the case for the fluids we study.

The pressure, measured in SI in pascals, expresses the way that fluids transmit forces. The pressure is the force per unit area on any area A:

$$p = \frac{F}{A}; \tag{16–2}$$

it is perpendicular to A. When a fluid is at rest, the pressure at any given point is the same in all directions. For an incompressible fluid in Earth's gravity, the pressure increases with the depth y according to

$$p = p_0 + \rho g y, \tag{16–6}$$

where p_0 is the pressure at $y = 0$. A change in pressure applied to a fluid at rest is transmitted without change to every part of the fluid, a statement known as Pascal's principle. This principle is the basis of hydraulic lifts and brakes.

For an object immersed in a fluid, there is an upward buoyant force equal to the weight of the displaced fluid,

$$F_{\text{buoy}} = \rho g V. \tag{16–11}$$

This is Archimedes' principle.

Nonviscous (friction-free) fluids in smooth (nonturbulent) motion follow paths traced out by streamlines. The conservation of mass for a fluid moving in this way is stated as the continuity equation,

$$\rho_1 v_1 A_1 = \rho_2 v_2 A_2. \tag{16–14}$$

Here, ρ and v are the density and speed, respectively, at some point in a tube of streamlines; A is the corresponding cross section of the tube. If the fluid is incompressible, the (constant) density cancels, and the resulting expression is the conservation of flux:

$$v_1 A_1 = v_2 A_2. \tag{16–15}$$

The conservation of energy for a moving fluid under the influence of gravity is called Bernoulli's equation:

$$\text{along a streamline:} \quad p + \frac{1}{2}\rho v^2 + \rho g h = \text{a constant}. \tag{16–21}$$

When the fluid is at rest, this reduces to Eq. (16–6); when the fluid flow is horizontal, this reduces to

$$p + \frac{1}{2}\rho v^2 = \text{a constant.} \tag{16–23}$$

This equation is the basis of the Venturi flowmeter, which measures the speed of fluids in pipes, as well as the basis for understanding lift forces on wings. The full Bernoulli equation helps us to understand the flow of fluids from holes in fluid-filled tanks.

UNDERSTANDING KEY CONCEPTS

1. How does a hot-air balloon get off the ground?
2. We weigh a brick twice with a bathroom scale. The first time, the scale and the brick are inside a tub of water, and the second time, they are outside the tub. Water can enter the scale. Does the scale indicate a difference in weight in the two cases?
3. One accurate method of measuring the fat content of your body is to measure your weight twice: once while you are immersed in a tank of water, and once when you are out of the water. Explain how such a method might work.
4. If an empty balloon is weighed on a scale, a certain value is obtained. If a child now blows up the balloon and waits for several minutes so that the temperature of the air in the balloon is the same as the room temperature, what value will the scale give for the balloon's weight?
5. Suppose that a certain fluid has no viscosity. Will the acceleration of objects falling under gravity within this fluid be independent of the mass of the objects?
6. Water-storage tanks for communities normally are placed high off the ground rather than underground. This is obviously a safety hazard, so why is it done?
7. Mercury is poisonous. Why, then, is it typically used in barometers to measure atmospheric pressure?
8. The "antilift" effect for a racing car, the opposite of the lift effect on airplanes, is said to be so strong that a car could race on an upside-down track. Is this a plausible statement? Make some rough estimates of the surface area of a racing car, its mass, and its velocity, and assume that the difference between the air speeds under the car and over the car is about 20 m/s.
9. One way to get the water out of a stopped-up sink is to use a long piece of rubber tubing. You put one end in the water and suck on the other end until you have filled the tubing. You then close off your end, bring that end to a point below the bottom of the sink, and reopen that end. The water will flow out through the tubing until the sink is empty. You have made use of the *siphon mechanism* (see Problem 65). Explain how it works.
10. The siphon shown in Fig. 16–34 and described in Question 9 is limited in the height h_0 that it can overcome. What determines this limit?
11. Do sailboats make use of any of the aspects of Bernoulli's laws, such as those we described for airplanes? In particular, do sails need to be curved in order to work?
12. Which weighs more, a ton of Styrofoam or a ton of lead? Which of the two has the larger volume? How might you estimate the density of each?
13. People find it very easy to float in the Dead Sea. Why?

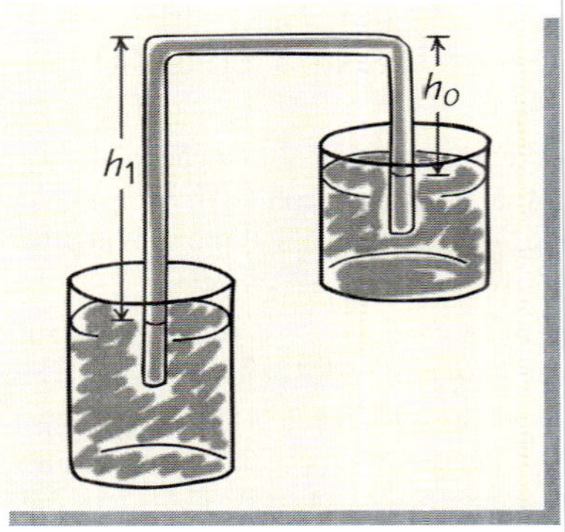

FIGURE 16–34 Question 10.

14. The point of a cone-shaped bottle (pointed end up) is removed to provide an opening. The pressure of the liquid at the very bottom of the bottle is the sum of the atmospheric pressure and the weight of the liquid above. If we consider a part of the bottom that is not directly below the opening, we might argue that there is less pressure because, above that point, there is less liquid and no atmosphere. This argument violates Pascal's principle. What has the argument left out?
15. Suppose that you put a mixture of oil and water into a centrifuge that rotates at high speeds (the oil is less dense than the water and does not mix with it). The two liquids separate. Why, and which component will be farthest from the center of the centrifuge?
16. An old proposal for transcontinental travel along a fixed latitude is the following: Take a balloon high into the stratosphere. Wait until Earth has rotated the desired distance beneath you, then descend. Comment on the feasibility of this mode of transport.
17. A snorkel is a breathing tube meant to be used while you swim under water; it runs from your mouth to the water surface. Why is the length of a snorkel limited?
18. In the course of a lazy summer morning, ice cubes floating in a pitcher of water melt. What happens to the water level in the pitcher? (Ignore evaporation.)
19. Suppose that you put a kilogram weight and a jar half-filled with water on a kitchen scale. The scale reads 2.5 kg. Now you place the kilogram weight *inside* the jar, and no water spills out. You might argue that because the kilogram weight weighs less in water than in air, the scale would read less than 2.5 kg, but the reading remains 2.5 kg. What is wrong with your argument?

16–1 *States of Matter*

1. (I) Jupiter has a radius $R = 7.14 \times 10^4$ km, and the acceleration due to gravity at the surface is $g_J = 22.9$ m/s^2. Use these data to calculate Jupiter's average density.
2. (I) The numbers given in Problem 1 have the following values for some of the other planets: Venus: $R = 6.05 \times 10^6$ m, $g = 8.86$ m/s^2; Mars: $R = 3.39 \times 10^6$ m, $g = 3.73$ m/s^2; Uranus: $R = 2.54 \times 10^7$ m, $g = 9.12$ m/s^2; Neptune: $R = 2.48 \times 10^7$ m, $g = 12.0$ m/s^2. Calculate the densities of these bodies.
3. (I) The density of a nucleus is about 2×10^{17} kg/m^3. The amount of water in a large lake is 10^{13} m^3. If this amount of water were compressed to nuclear density, how many liters of water would there be?
4. (I) A platinum sphere has a diameter of 4.05 cm. What is the diameter of an aluminum sphere of the same mass, given that the densities of platinum and aluminum are 21.4×10^3 kg/m^3 and 2.70×10^3 kg/m^3, respectively?
5. (I) Atoms of an imaginary two-dimensional solid are arranged in a square lattice. The distance of neighboring atoms is a. Assume that the solid is acted upon by a shear force and deformed by a *small* angle f without any change in the nearest neighbor distance (Fig. 16–35). Calculate the change in the distance between atoms situated at opposite corners of a square.

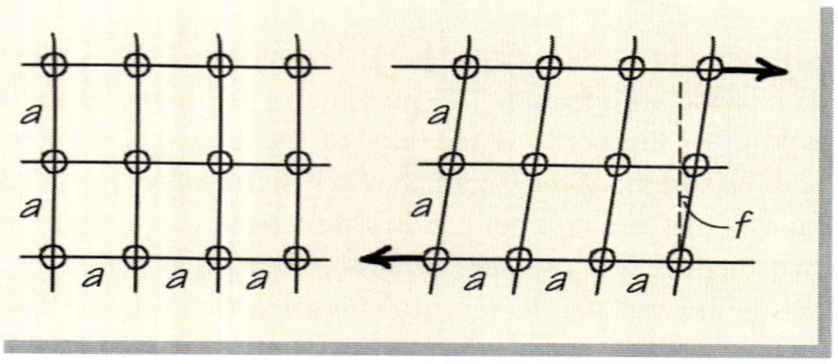

FIGURE 16–35 Problem 5.

6. (II) A wedge of weight 20 N and angle 60° is floating on the surface of water in the symmetrical position shown in Fig. 16–36. Calculate the force of the water that acts on each of the two surfaces.

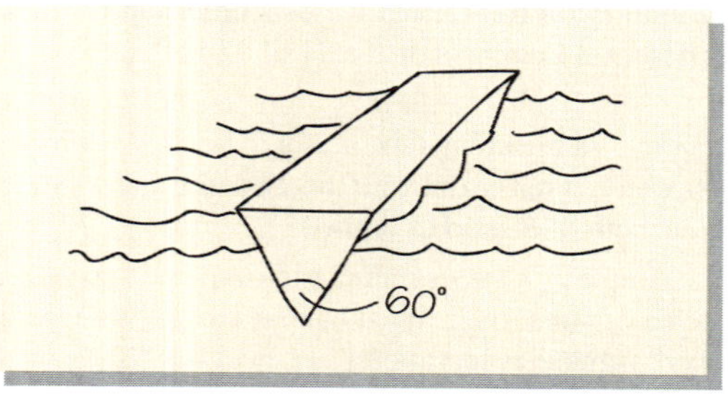

FIGURE 16–36 Problem 6.

16–2 *Pressure*

7. (I) A circus clown stands on a pair of stilts that each have a square cross section of 4.0 cm per side. If the mass of the clown plus the stilts is 68 kg, what pressure is exerted on the floor?
8. (I) The density of mercury is 13.6 times that of water. Compare the height of a column of water to that of a column of mercury, assuming that the pressure exerted by the weights of the respective liquids at the bottom of each column are the same.
9. (II) A hollow stainless steel sphere of radius 20 cm is evacuated so that there is a vacuum inside. (a) What is the sum of the magnitudes of the forces that act to compress the sphere? (b) There is a circular hole of diameter 4 cm on the side of the sphere for access to the inside. Calculate the force needed to pull a flat plate off the hole when the sphere is evacuated. Do you think that you could remove such a plate by pulling on it?
10. (II) During a hurricane, the atmospheric pressure changes dramatically. Explain why it is recommended that house windows be kept slightly open during a hurricane. What is the net force on a wall that is 300 ft^2 in area when the pressure on one side is 14.7 lb/in^2 and the pressure on the other is 14 lb/in^2?
11. (II) A hollow metal pyramid with a square base that is 15 cm on each side has a mass of 1.8 kg. What is the pressure exerted by the pyramid on the table on which it stands? Suppose that the temperature in the room rises so that the metal expands. Will the pressure increase or decrease as a result of the expansion?

16–3 *Variation of Pressure in a Fluid at Rest*

12. (I) Using the densities in Table 16–1, what is the height of a fluid column in a barometer that has water as its fluid, in an air pressure of 1 atm? One that has alcohol? (Neglect the vapor pressure of the fluid above the column.)
13. (I) What is the pressure at depths of 1 m, 10 m, 100 m, and 10 km under the surface of the ocean? Take $\rho = 1.03 \times 10^3$ kg/m^3 for the density of sea water and $p_0 = 1.01 \times 10^5$ Pa for the atmospheric pressure at the ocean surface. (Ignore the fact that, to this accuracy, density varies with depth.)
14. (I) Recreational scuba divers rarely go deeper than 100 feet (30 m). What is the water pressure at this depth? By what factor does your result differ from sea-level water pressure?
15. (I) What is the pressure 1 cm below the surface of a column of mercury exposed to air? 3 cm below the surface?
16. (I) An oceanic research vehicle operates at a depth of 1.20 km under water. Assuming that the pressure inside the vehicle is 1.0 atm, what is the force on a small window of dimensions 10 cm $\times$ 10 cm?
17. (II) Assume that a spaceship lands on Venus with the cabin temperature maintained such that the pressure is the same as the surface pressure on Earth—1 atm. What would be the height of a column of mercury in a barometer in the spaceship on Venus? Repeat for a spaceship landing on Neptune. (Use the data given in Problem 2.)
18. (II) In a pressure test chamber, a person starts acting strange when the gauge pressure exceeds 40 lb/in^2. *Gauge pressure* is the pressure in excess of atmospheric pressure. This is a well-known effect that limits the depth at which scuba divers should breathe pure air. In sea water, whose density is 1.03 g/cm^3, to what depth should a diver be limited?
19. (II) A hydraulic jack is used to lift a 1200-kg car on a piston of diameter 30 cm (Fig. 16–37). How large a force is needed

to push down the smaller piston if its diameter is 2 cm? By how much is the car lifted with a single push, assuming that the small piston moves 0.5 m?

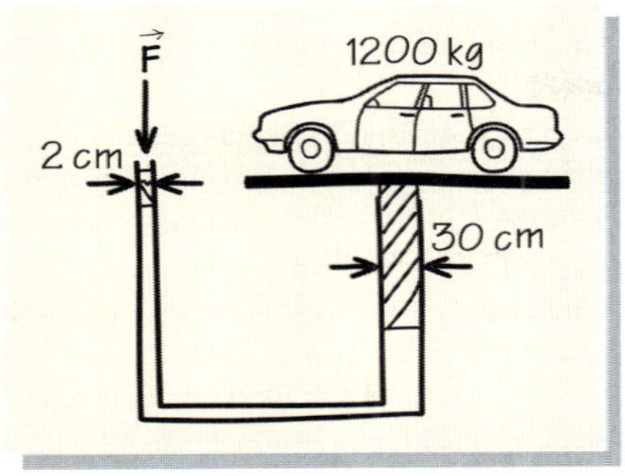

FIGURE 16–37 Problem 19.

20. (III) The vertical wall of a small dam is 20 m long and it holds back a body of water that is 6 m deep (Fig. 16–38). Calculate the force acting on the wall due to the pressure of the water. Determine the effective point of action of this force. [*Hint*: Recall the method used to calculate the center of mass of a nonuniform object in Chapter 8].

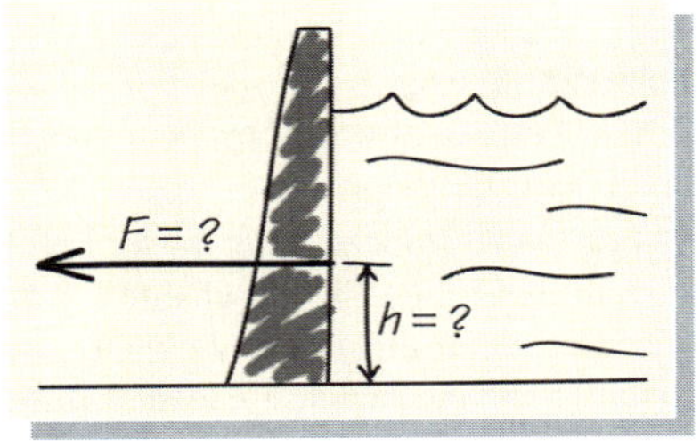

FIGURE 16–38 Problem 20.

16–4 Buoyancy and Archimedes' Principle

21. (I) A bathysphere (a spherical, watertight research facility) 2.6 m in diameter has a mass of 9400 kg. It is released from a submarine at a depth of 20 m below the surface. Will the bathysphere sink or rise?

22. (I) Use the densities in Table 16–1 to show that ice floats in water. What fraction of the volume of a floating ice cube will be above the surface?

23. (I) A cubical box, contents unknown, floats in water; 16 percent of the volume of the box is above water. What is the average density of the box and its contents?

24. (II) A hemispherically shaped bowl is floating on the surface of water; the mass of the bowl is 1.8 kg (Fig. 16–39). Water is poured into the bowl; when $6.9 \times 10^3\ \text{cm}^3$ of water has been added, the bowl just sinks. What is the outer radius of the bowl?

25. (II) A group of Scouts intend to make a raft to float down the Ohio River. The mass of four of them plus their equipment is 400 kg. They find trees with an average diameter of 20 cm and a specific gravity of 0.8. Determine the minimum area of the log raft that would keep them dry.

26. (II) A temporary bridge is supported on floating empty drums that are 1 m in diameter and 1.8 m long. The design criterion is that the drums may only be submerged to 2/3 of their diameter under a maximum load of 5000-kg trucks, 7.5 m long, that follow each other bumper to bumper in a single line. How far apart should the drums be placed? Neglect the weight of the bridge and of the drums.

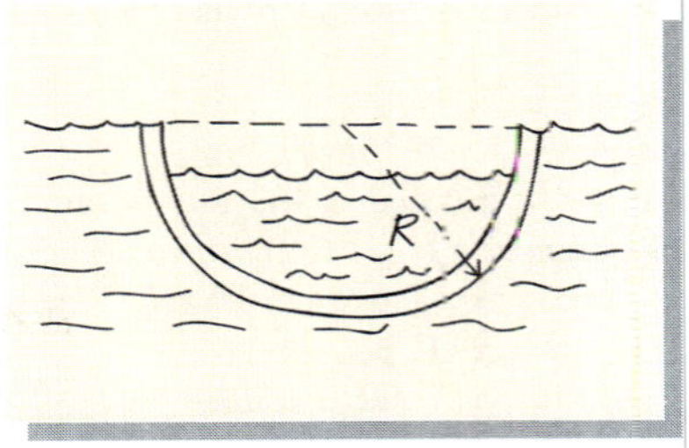

FIGURE 16–39 Problem 24.

27. (II) A valuable preserved biological specimen is weighed by suspending it from a spring scale. It weighs 0.45 N when it is suspended in air and 0.081 N when it is suspended in a bottle of alcohol. What is its density?

28. (II) Consider a spherical balloon filled with helium, with a density of $0.18\ \text{kg/m}^3$. The density of air is $1.3\ \text{kg/m}^3$. What must the radius of the balloon be in order to lift a load of 100 kg (this includes the mass of the balloon)?

29. (II) Olive oil floats on water. Take ρ_1 to be the density of the oil and ρ_2 to be the density of the water. Consider an oil–water interface across which a bouillon cube of density ρ_3 floats; use the geometry shown in Fig. 16–40. We know that for the oil to float on the water, ρ_1 must be less than ρ_2. What is the condition on ρ_3 so that the cube floats at the oil–water interface? How much of the cube will be in the water?

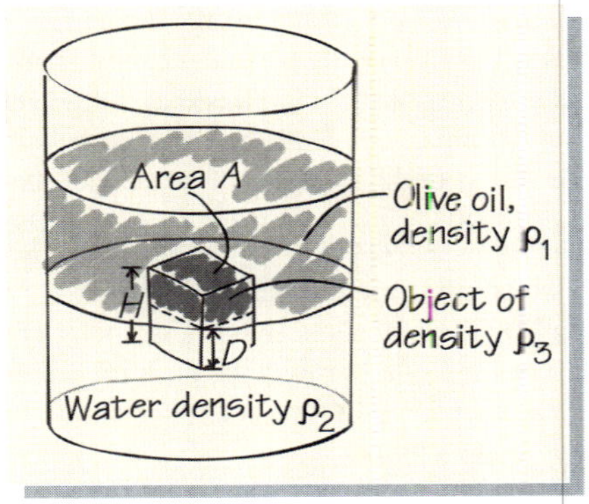

FIGURE 16–40 Problem 29.

30. (II) A beaker of mass 350 g contains 2.8 kg of water and rests on a bathroom scale. A 5.2-kg block of aluminum, of specific gravity 2.7, is suspended from a spring scale and completely submerged in the water. Find the readings on both scales.

31. (II) Balloons of mass 3.5 g can be filled with helium to make spheres of radius 21 cm. How many of these balloons must a child of mass 32 kg hold in order to float off Earth's surface? Assume that helium has a density of $0.18\ \text{kg/m}^3$.

32. (III) A sphere of radius R of material with an average density of $0.75\ \text{g/cm}^3$ is immersed in a body of water. What is the height of that portion of the sphere that projects above the surface?

16–6 The Equation of Continuity

33. (I) Water flows at a speed of 1.3 m/s through a hose of diameter 1.5 cm and emerges from a nozzle of radius 0.5 cm. With what speed does the water emerge? How much water emerges from the nozzle per second?

34. (I) Water flows down a channel 1.0 m deep and 0.5 m wide at a rate of 2 metric tonnes/s. At some point the channel widens to 0.8 m. How fast does the water flow in the wider channel?

35. (I) A steady stream of automobiles moves along the eastbound side of a highway in a region where there is neither exit nor entrance. Where two eastbound lanes are available, the automobiles are spaced so that there are 66 cars per mile in each lane, and the average speed is 45 mi/h. In a region to the west, where three eastbound lanes are available, the automobiles are spaced to give 35 cars per mile in each lane. What is the average speed in the region where three eastbound lanes are available?

36. (II) A lawn sprinkler is connected to a garden hose of inside diameter 2 cm that has a water flow rate of 0.3 L/s. The sprinkler has 10 nozzle openings, each with a diameter of 3 mm. (a) Calculate the speed of the water that emerges from the nozzles. (b) Explain how the nozzles can be arranged to cause the sprinkler to rotate. What principle could you use to calculate the torque on the sprinkler? What are the limitations, if any, on the speed of rotation of the sprinkler?

37. (II) Students would like to spray water 18 m across their neighbor's yard onto a patio. They have a hose of inside diameter 1.5 cm that can spray a distance of only 1.5 m. What size nozzle do they need?

38. (II) When water drains out of a hole in the bottom of a tank, the stream radius contracts as the speed of the water in the stream increases. Assume that the stream starts with zero water velocity at the top of the tank, at a height H above the hole, and ignore the various instabilities that will eventually cause the stream to break up. Find the distance below the hole at which the radius of the stream is 80 percent of the hole radius.

16–8 Applications of Bernoulli's Equation

39. (I) A bilge pump is used to pump water out of a leaking ship (Fig. 16–41). The pump's hose has a diameter of 3.0 cm, and the pump moves water through the hose up and out of a porthole 5.0 m above the waterline at a speed of 4.0 m/s. Calculate the power of the pump.

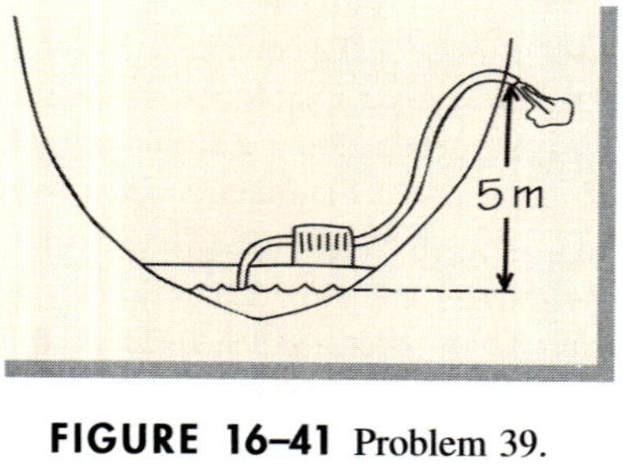

FIGURE 16–41 Problem 39.

40. (I) What pressure must a pump generate to get a jet of water to leave a nozzle at a speed of 3 m/s and a height of 25 m above the pump?

41. (I) Wind gusts at 60 mi/h past a ship's cabin in which the air is at rest and the pressure is 1 atm. What are the pressure outside the cabin and the net pressure on the walls past which the wind blows?

42. (I) A stream of water sprays like a fountain vertically from a small hole in a pipe, reaching a height of 2.4 m. Assuming that the water in the pipe is static, what is the gauge water pressure in the pipe? (See Problem 8.)

43. (I) A folded piece of notebook paper can exhibit lift in the air flow from a fan of a window air conditioner. A plastic straw is taped perpendicular to the plane of the paper through the paper's middle, and a thread runs through the straw allowing the paper to move up and down the thread. The folded paper has an area of 300 cm^2. The air speed is about 4 m/s and the proportionality constant K in the approximate lift formula $L = KA\rho v^2$ is taken to be $K = 0.14$. Here, A is the surface area of the wings, ρ is the air density, and v is the air speed. How much lift is possible? Given the mass of the paper and straw to be 8 g, will the paper rise?

44. (II) A small garden fountain shoots a vertical jet of water at a rate of 0.10 L/s to a height of 0.50 m. (a) What is the initial speed of the jet, and what is the radius of the hole out of which the jet passes? (b) What pressure must the pump of the fountain supply? (Assume that it sits just below the emerging jet.) (c) At height 0.25 m, what is the speed of the jet, and what is the radius of the column of water? Ignore effects of turbulence, such as the breakup of the jet.

45. (II) Heavy rains have flooded your favorite professor's basement to a depth of 15 cm. This basement is 7.5 m × 12 m in area. You have a hose of diameter 1.2 cm and can run the hose from the water in the basement to a level 3 m below the level of the basement floor, into a low portion of the yard. When the hose is filled with water, it acts as a siphon (see Question 9) and empties the basement. (a) What is the speed of the stream of water leaving the hose? (b) Approximately how long does it take to empty the basement?

46. (II) A liquid of density 1.4×10^3 kg/m^3 flows in a horizontal pipe. The cross-sectional area in one part of the pipe is 75 cm^2. When the liquid enters another part of the pipe with cross-sectional area 150 cm^2, the pressure as measured by a gauge is 2.0×10^4 Pa higher than it was in the first part. Calculate the velocities of the liquid in the two parts of the pipe.

47. (II) A large horizontal pipe of diameter 10 cm contains water flowing with a speed of 0.3 m/s. The pipe branches into four pipelets, each of diameter 2 cm. The four pipelets each run horizontally after an initial height change of 3.5 m (Fig. 16–42). (a) What are the speeds of the fluid in the horizontal portion of each pipelet? (b) If the pressure in the large pipe is $p = 2.5$ atm, what is the pressure in the pipelets?

48. (II) A uniform glass tube with internal cross-sectional area A is formed into a U-shape and partly filled with a volume V of an incompressible fluid of density ρ. The two ends of the tube are open to the atmosphere. At equilibrium, the level in each vertical segment of the tube is $h = 0$. By applying additional pressure to one end of the tube, the levels of the liquid are displaced, so one side is at height h_0 and the other is at height

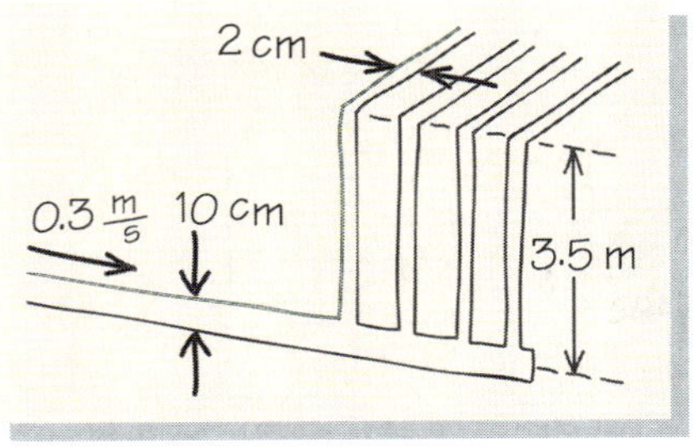

FIGURE 16–42 Problem 47.

$-h_0$. The additional pressure is suddenly released, and the liquid level starts to oscillate. (a) Use Bernoulli's equation to show that the force on the fluid in the tube is proportional to the displacement of the fluids from their equilibrium levels, so the force is harmonic. (b) Find the "spring constant" of the harmonic motion described in part (a) and use it, together with the mass in motion (ρV), to calculate the oscillation frequency.

49. (II) Water flows through a large horizontal pipe of diameter D at speed v_0. The pressure in the pipe is p_0. The pipe branches into two horizontal pipelets of diameter d_1 and d_2, respectively, both at the same height as the large pipe. Write the equations that would allow you to find the pressures p_1 and p_2 and the speeds v_1 and v_2 in the two pipes. Show that one of these four quantities must be known to be able to solve for the others.

50. (II) An open water tank stands on a plane surface. The water surface in the tank is a height h above the plane. A small hole is opened up at a depth y below the surface of the water. (a) Show that the jet of water will hit the plane surface a distance D from the tank, where $D = \sqrt{4y(h - y)}$. (b) Show that the hole should be placed at a depth $y = h/2$ for the jet to cover a maximum horizontal distance.

*16–9 *Comments on Real Fluids*

51. (II) The retarding force on a sphere of radius R that moves with speed v through a fluid of viscosity η is proportional to η and depends on R and v, in the form $R^a v^b$. Use dimensional analysis to determine the powers a and b.

52. (III) The result of the solution of Problem 51, the retarding force on a sphere of radius R that moves with speed v through a fluid of viscosity η, is $F = 6\pi\eta Rv$, known as *Stokes's law*. Such a sphere, with mass density ρm, falls through glycerin, whose density is ρ. The forces on the sphere are the retarding Stokes's law force, the buoyant force, and the force of gravity. (a) Derive a formula for the value of the velocity v_t for which the net force is zero. (b) How does v_t depend on the radius of the sphere? (c) If the viscosity of glycerin is 8.3 kg/(m · s) and its density is $\rho = 1.26 \times 10^3$ kg/m^3, what is the terminal velocity of a sphere of iron, which has $\rho = 8.5 \times 10^3$ kg/m^3 and a diameter of 5 cm? Of an iron sphere with a radius of 1 mm?

General Problems

53. (II) A rectangular container is divided into two parts by a movable vertical partition. This partition is sealed so that no water can seep from one side to the other. Initially, the wall is held in the middle of the 30-cm length of the container. One side is filled with 2 kg of water to a depth of 10 cm; the other side is attached to a spring in equilibrium (unstretched, uncompressed). The spring constant is 180 N/m. The partition is now released and the spring compresses. What is the position of the partition at maximum compression? Assume that the process takes place without turbulence so that no energy is dissipated.

54. (II) A hollow sphere of radius 0.2 m and total mass 1.4 kg is placed in water. One very small lead pellet of mass 0.2 kg is attached to the inside surface of the sphere. When the sphere is floating in equilibrium, the lead pellet is oriented on the bottom. If the sphere is rotated slightly, so that the pellet is moved off to the side, and then released, the sphere will rock, and the pellet will move like a pendulum. Calculate the period of small oscillations for this motion. Assume that the center of the sphere does not move.

55. (II) A depth of about 100 ft ($\approx$ 30 m) is as deep as a scuba diver should go without special precautions because of an effect that causes disorientation. What is the water pressure at a depth of 30 m? Does a scuba diver take in the same mass of air in each breath at 30 m that he or she would at a depth of 5 m? Explain.

56. (II) A cylindrical bucket of fluid of density ρ sits at the center of a turntable that rotates with some fixed angular velocity, ω. Show that the pressure variation in the fluid at a horizontal distance r from the axis is given by $p = p_{\text{axis}} + \frac{1}{2}\rho\omega^2 r^2$. [*Hint*: If you place yourself within the rotating fluid and at rest relative to it, the problem is one of hydrostatics. The only force in the horizontal direction is a fictitious force away from the axis of the rotation, due to the fact that your frame of reference is not inertial. This force is the centrifugal force.]

57. (II) Late in the seventeenth century, King Louis XIV of France had a series of fountains constructed at Marly, some 20 km from his chateau at Versailles. These fountains were fed from the Seine River, whose altitude is some 150 m below that of Marly, through a remarkable series of pumps. The tallest of these fountains was the Grand Jet, which is calculated to have risen a height of 37 m from its base. (a) At what speed does this jet leave the orifice at the fountain's base? (b) The rate of flow in the Grand Jet is measured to be 0.051 m^3/s. Compare this to an estimate of the rate of flow from a garden hose. (c) Given the rate of flow in part (b), calculate the area of the orifice from which the Grand Jet comes. (d) What is the pressure of the water just behind the orifice from which the Grand Jet comes? Beyond the crude calculations outlined here, the height to which a jet of water will rise can be affected by the addition of a short length of conical pipe above the level of the orifice at the base.

58. (II) A cylindrical tank is filled with water to a height h_1 above its base. Unfortunately, someone has cut a hole in the bottom through which fluid escapes in a vertical stream into your room. (a) What is the speed of the escaping stream at the hole? (b) You insert a flexible hose into the hole. If the other end of the tube is held at the same height as the base of the tank and is directed horizontally, what is the speed of the water as it leaves the hose? (c) You hold the tube, whose end is still at the height of the base of the tank, so that the stream is directed straight up. How high does the escaping stream shoot?

59. (II) An inventor proposes to you the following perpetual motion machine (a machine that can do work "forever" without your having to add energy to it) (Fig. 16–43). A cylindrical tower of height h filled with water rests on the ground. Blocks of wood are inserted at the bottom of the tower. They float to the top, where they are removed. The potential energy that they acquired in floating to the top can be converted into work. They can, for example, be dropped down for re-use, with the kinetic energy they acquire in falling used to turn the wheels of a turbine. When they reach the bottom, the heavier blocks are again inserted, and the process is repeated. How can you prove to the mad inventor that at best this machine can do no net work? [*Hint*: Calculate the work required to insert a block, which requires in effect that a mass of water whose volume is that of the block be raised to the top. Compare this to the energy the block acquires in being accelerated to the top under the influences of gravity and the buoyant force. Neglect any drag caused by the water.]

Do not buy this machine! Perpetual motion machines are impossible to construct (see Chapters 18 and 20). A close analysis of the details of such machines always reveals this fact.

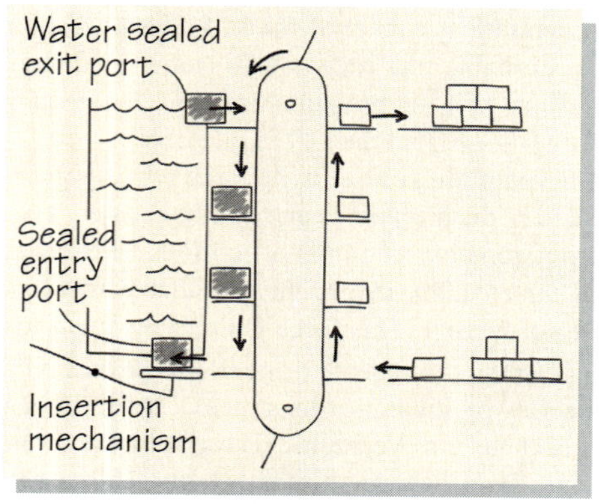

FIGURE 16–43 Problem 59.

60. (II) Consider the machine discussed in Problem 59. Why does the machine still fail if you replace the entry port by an exchange chamber (Fig. 16–44) with two sliding doors, so that the insertion of the blocks does not require that work be done?

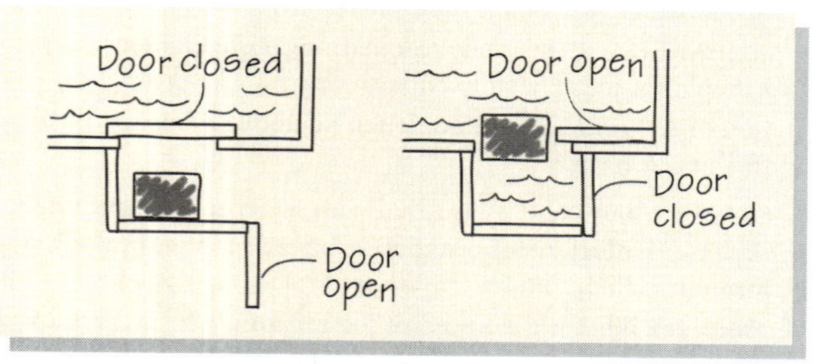

FIGURE 16–44 Problem 60.

61. (II) A cubical block of wood 0.20 m on each side floats in water, with 8.0 cm of the block above the surface of the water (Fig. 16–45). (a) What is the density of the wood? (b) Suppose that the block of wood is pushed down below the equilibrium level by a distance Δ. What is the magnitude of the restoring force that pushes the wood back toward the equilibrium level? (c) In general, if dissipation of energy is ignored, the block will overshoot the equilibrium level and rise to a height Δ above that level, and is then subjected to a downward force. How large is that force? (d) Recall from Chapter 13 that a force that is proportional to the displacement Δ from equilibrium (of the form $F = -k\Delta$) and acts on an object of mass m leads to oscillations with the frequency $f = \sqrt{k/4\pi^2 m}$. Calculate that frequency for the block that bobs in the water.

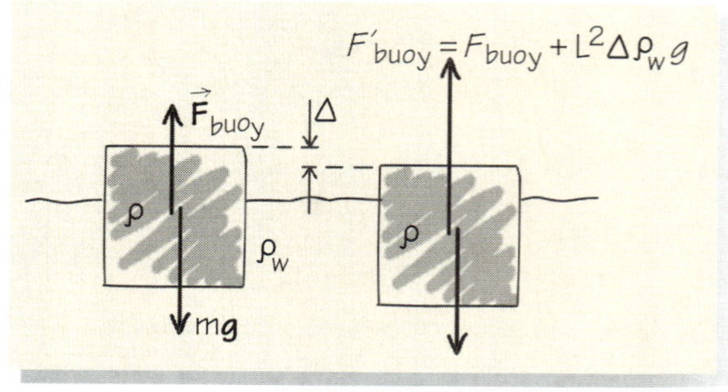

FIGURE 16–45 Problem 61.

62. (II) The ticket-window opening of a movie theater, where money goes into the cashier's hands and tickets emerge, is a rectangle 8 cm × 16 cm. A breeze of speed 5 m/s comes out of this opening. Sometimes, as a result, paper money goes flying. (a) What is the difference in air pressure between the inside and the outside of the theater? (b) Suppose that the opening were the size of a door, 0.9 m × 2.1 m. What would the speed of the breeze be in this case, assuming the air pressure difference were the same as in part (a)?

63. (II) A water-storage tank is an upright cylinder of height 8 m and radius 6 m. The storage tank is full of water but is vented to the atmosphere. The bottom of the tank is placed 30 m off the ground. A pipe of diameter 12 cm runs vertically down from the tank and goes 1 m underground before turning horizontal. (a) What is the water pressure in the bottom of the tank? (b) in the horizontal pipe underground? (c) The water flow in the pipe is 80 L/s. How fast does the water level in the tank drop? (d) A saboteur drills a hole of diameter 6 mm near the bottom of the tank. How fast does the water shoot out? (e) What volume flow of water is lost out the hole?

64. (II) A *Pitot tube* is a device for measuring flow velocities (Fig. 16–46). It is as useful for measuring the flow velocity of blood in an artery as for measuring the air speed of a jet airplane. The fluid passes the opening at point B, creating a pressure p_B inside. The fluid enters the opening of the tube and comes to a halt, creating a pressure p_A. Apply Bernoulli's equation as needed to show that the flow velocity is given by $v = \sqrt{2(p_A - p_B)/\rho} = \sqrt{2gh\rho_\ell/\rho}$, where ρ is the density of the fluid being measured, ρ_ℓ is the density of the fluid shown in the device, and h is the difference in height of the liquids.

65. (II) With a siphon, it is possible to transfer fluids from one container to another. Figure 16–34 shows a siphon that must be filled with fluid to start the transfer. If the speed with which the fluid leaves the higher container is zero, and if the fluid has a fixed density ρ, find an expression for the speed of the fluid as it enters the lower container, in terms of the heights h_0 and h_1.

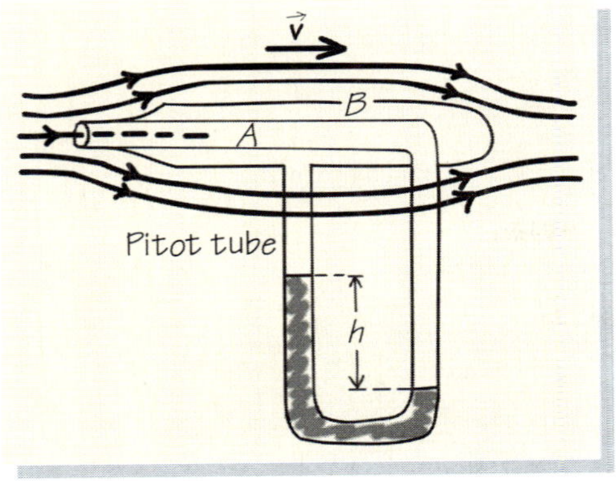

FIGURE 16–46 Problem 64.

66. (III) A tank of surface area A is filled with water to a height x above the position of a hole of surface area σ, and the water flows out. Assume throughout that the fluid speed is zero at the top surface. (a) How much water flows out of the hole in the first time interval Δt after the hole is opened? (b) How large is the drop in the water level, Δx, in that time interval? (c) Use the results of parts (a) and (b) in the limit of small Δt to express a relation of the form $dx/dt = f(x)$. This is a differential equation for the height as a function of time. Take care to get the right signs, noting that x decreases with time, so that Δx is a negative quantity. (d) Show that this equation is solved by the formula

$$x(t) = \left[\sqrt{x_0} - \frac{1}{2}\left(\frac{\sigma}{A}\right)t\sqrt{2g}\right]^2.$$

(e) How long will it take for the water to drop to the level of the hole?

67. (III) The ice cream cone shown in Fig. 16–47 is filled with melted ice cream of density 1.2 g/cm^3. The cone has a diameter of 6 cm at the larger end and is 10 cm long. Find the pressure at the bottom of the cone. If a small hole of diameter 1 mm is opened at the bottom, the ice cream starts to run out. Ignoring the viscosity of the melted ice cream, find the amount of time it takes the ice cream to run out. Assume that the fluid speed is zero at the top.

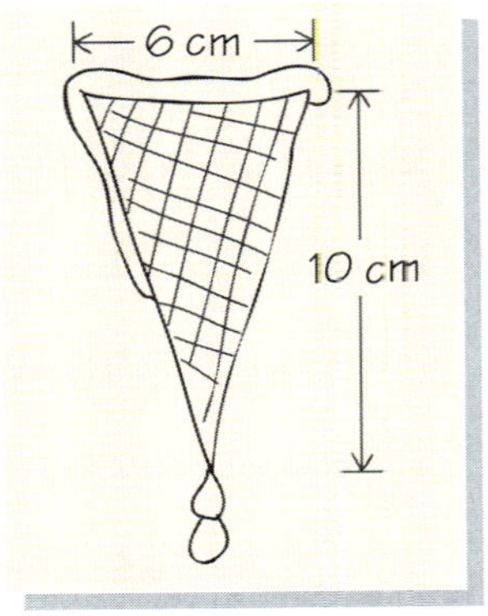

FIGURE 16–47 Problem 67.

CHAPTER

17

The surface of the Sun is an active region. Here, we see patterns of magnetism that hold hot streamers of ionized gases above the surface. The patterns of temperature are an important clue to deciphering the dynamics of this activity.

Temperature and Ideal Gases

Although temperature has not been a very important part of our discussions to this point, many properties of matter depend on it. Water behaves quite differently at temperatures below its freezing point than it does at temperatures above its boiling point. Temperature has something to do with energy, as we can see from the fact that hot steam can lift a piston. How much energy does hot steam contain, and how does that energy get into the steam? How can we quantify "hotness," and just how hot does steam need to be to lift a piston? By studying the behavior of gases, we can learn how to measure temperature and how temperature is a measure of energy.

Although thermal phenomena are ultimately explained by a statistical treatment of the behavior of matter on an atomic scale, the historical development of thermal physics was based on a description of bulk matter. The study of thermal phenomena received its strongest impetus from the engineers who studied the steam engines that powered the industrial revolution of the nineteenth century. The treatment of thermal phenomena that they developed—the subject of Chapters 17 through 20—is known as **thermodynamics**. Thermal phenomena are pervasive in the physical world and our understanding of them has had a profound effect on the way we live.

FIGURE 17–1 The vendor's stand is *not* in thermal equilibrium with its environment. The steam is evidence of an interaction with the surroundings that would eventually cool the stand down to where it would be serving frozen hot dogs.

17–1 TEMPERATURE AND THERMAL EQUILIBRIUM

We call a physical system a **thermal system** whenever we are interested in its temperature-dependent properties. A piston and a cylinder full of hot steam form a good example to keep in mind as we proceed. We all have an intuitive feeling for temperature—for example, we can usually tell when one system is hotter than another. Our aim here is to develop a more precise notion of temperature.

Central to the idea of temperature is the notion of **thermal equilibrium** (Fig. 17–1). Suppose that we pour cold water into a bucket of hot water. Experience shows that, after a short time, the water becomes lukewarm. Until that point, we cannot say that the water has a particular temperature. It is only when thermal equilibrium has been reached, meaning that temperature-dependent quantities are no longer changing, that the temperature becomes a variable that characterizes the whole bucket of water; that is, the whole thermal system. We can say that if two thermal systems are in thermal equilibrium with one another, then they have the same temperature. Finally, we can add that if two systems are each in thermal equilibrium with a third, then they are in thermal equilibrium with one another.

If we put an ice cube into a bathtub full of hot water, the temperature of the tubful of water changes minimally; if we add a teaspoon of hot water to a swimming pool, the temperature change is virtually undetectable. These observations allow us to define a **thermal reservoir** (or **heat bath**) as a thermal system so large that it maintains a constant temperature when it interacts with other thermal systems. A thermal reservoir is of great importance because it allows us to bring a given system to a predetermined temperature, the temperature of the reservoir. To do so, all we need to do is bring the given (smaller) system into **thermal contact** with the reservoir: Any two thermal systems in contact with each other *will reach the same equilibrium temperature.*

Radiation, conduction, and convection are three ways to achieve thermal contact.

Thermal contact is established between thermal systems in three important ways. We can illustrate them with two metal plates in an enclosure, one red hot and the other cold. In thermal contact by *radiation*, the hot plate emits electromagnetic waves (the "radiation" in question), and the cold plate absorbs them. This form of thermal contact operates even with the plates separated by some distance. Thermal contact by *conduction* occurs when the two plates actually touch. Finally, thermal contact occurs by *convection* when there is a fluid between the plates. The fluid acts as a third thermal system whose temperature increases in the region near the hot plate through its own thermal contact with the plate. The fluid then circulates in large currents to the colder plate, increasing its temperature. Conversely, if two fluids at different temperatures are separated by a double-walled, silvered glass flask with a vacuum between the two glass surfaces, the fluids will take a very long time to reach the same temperature. The vacuum prevents thermal contact by conduction or convection, and the silvered surfaces of the glass walls reflect radiation back to the heat source and prevent thermal contact by radiation. The liquids are **thermally isolated** from one another.

Thermometers

A **thermometer** is a device that measures the temperature of a thermal system quantitatively—in general, by coming to thermal equilibrium with that system. A thermometer must satisfy several criteria. It should be small enough to have a minimal ef-

(a)

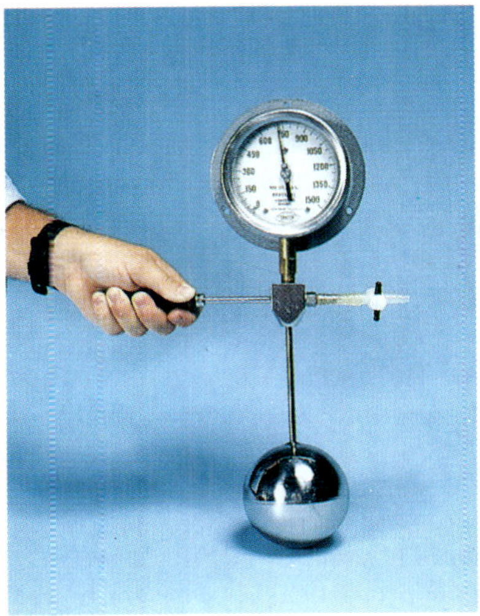

(b)

(c)

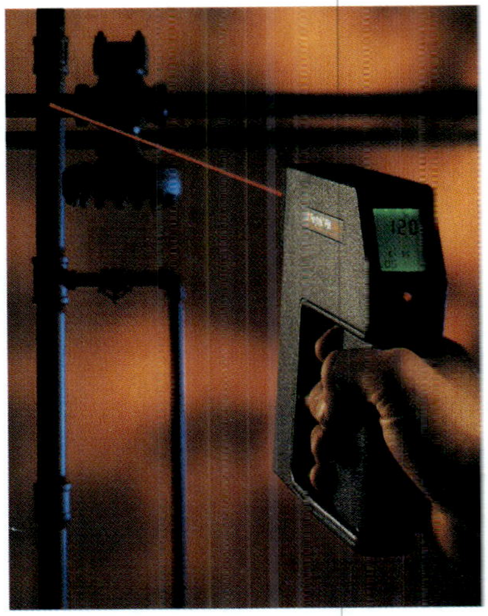

(d)

FIGURE 17–2 (a) This thermometer contains a (red) fluid that expands at higher temperatures and contracts at lower temperatures. (b) This thermometer measures temperature in terms of the pressure of the gas in the bulb. (c) This digital thermometer uses the temperature dependence of electronic properties of materials. It is calibrated in the Celsius scale, in which the temperature of ice melting in pure water is 0 degrees. (d) This device measures temperature using emitted radiation.

fect on the system being measured. (A small room thermometer brought inside from the cold will have a minimal effect on the room temperature.) The thermometer must make good thermal contact with the system being measured. Like a meter stick, the scale of the thermometer should be reproducible, easily read, and universally available.

It is easy to find properties of matter that are suitable for thermometry. Ordinary fever thermometers rely on the fact that fluids (and solids) expand and contract when their temperatures change. We can therefore make a thermometer by marking numbers on a column of mercury that measure the mercury's volume. The pressures and volumes of gases, the volumes of liquids or solids, reflective properties of so-called liquid crystals, electrical and magnetic properties of solids, and the color of light emitted by hot objects are all properties that vary with temperature and can be used as thermometers (Fig. 17–2). Not all of the properties of matter that depend on temperature are suitable for making thermometers: The brittleness of many materials depends on temperature, but we would not make a thermometer that works by seeing how easy it is to smash a material with a hammer. Table 17–1 gives a representative list of some useful thermometers.

See Chapter 21 for a discussion of *thermal expansion.*

TABLE 17–1 Some Thermometers

Thermometer	*Physical Property Measured*	*Comments*
Ideal gas	Pressure and volume of dilute gas	See Section 17–2
Mercury bulb	Expansion or contraction of fluid	Good where fluid does not change phase
Bimetallic strip	Difference in expansion of two metals	Often used in solids
Resistance	Electrical resistance	See Chapter 27
Thermocouple	Electrical voltage across different metals	Most widely used thermometer in industry
Paramagnetic	Magnetic properties of matter	Useful at ultracold temperatures; see Section 32–5
Optical pyrometer	Color of emitted light	Useful at high temperatures; see Section 17–4

17–2 IDEAL GASES AND ABSOLUTE TEMPERATURE

Pressure was defined in Chapter 16.

Consider a gas in a bottle with a sliding piston (Fig. 17–3). We can adjust the volume of gas by changing the position of the piston and we can adjust the pressure by changing the weight on the piston. A valve allows us to control the *amount* of gas—in other words, its mass.

If we dip the bottle into thermal reservoirs at various temperatures, we find that, as the gas becomes hotter, its pressure increases if the volume is kept constant, and its volume increases if the pressure is held fixed. The volume and the pressure correspondingly decrease when the gas becomes colder (Fig. 17–4). In fact:

A technique to define temperature

> **The product of pressure and volume, divided by the mass of the gas in the bottle, increases or decreases according to whether the thermal reservoir in contact with the bottle is hotter or colder, respectively. We use this property to define temperature as some proportionality constant times the product of pressure and volume, divided by mass.**

Ideal gases

The gas thermometer we construct in this way comes extremely close to being a universal thermometer because experiment shows that *the proportionality constant is the same for all gases if their densities are sufficiently low.* Dilute (or low-density) gases, which we call **ideal gases**, will thus enable us to give a universal definition of temperature. Just how dilute the gas needs to be is simply a question of the accuracy demanded of the thermometer. Gas thermometers accurate enough for laboratory use may contain gas hundreds of times less dense than air at sea level.

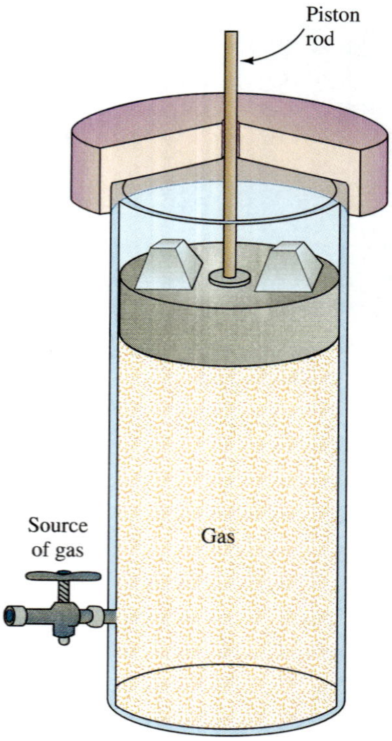

FIGURE 17–3 A bottle of gas whose pressure and volume can be measured. A tightly fitting piston can be moved to change the volume of gas, and weights on the piston can be changed to vary the pressure.

Figure 17–5 plots the pressure of a fixed amount of a dilute gas at fixed volume against our temperature scale. (In this case, we say that we have a constant-volume ideal-gas thermometer.) The part of the curve where T and p drop to zero is drawn as a dashed line. Temperatures here cannot be measured with our thermometer because gases liquefy or freeze in this region of extreme cold. The dashed line is thus an extrapolation.

Let's define the temperature $T = 0$ as the point *where the pressure, p, would become zero* (if that were possible), so

$$\text{for constant volume: } p = (\text{a constant})\, T. \tag{17–1}$$

We still need to specify the constant in this relation; we do so by defining T at some other point. This second point, called a *defined point*, was chosen in 1968 by international agreement to be the **triple point** of water. The triple point of water is the state at which water vapor, liquid water, and frozen water coexist. The temperature at the triple point, where the pressure is p_{tp} (= 4.58 mm of mercury), is defined as 273.16 K. The letter K stands for kelvin, and the temperature scale so defined is the **Kelvin scale**. (During the mid- to late-nineteenth century, Sir William Thomson, Lord Kelvin, did important work on thermal phenomena and on electricity and magnetism.) To summarize, temperature in the Kelvin scale, T, is defined by a constant-volume ideal-gas thermometer as

The Kelvin temperature scale [see Eq. (17–11)]

$$T = \left(\frac{p}{p_{tp}}\right) 273.16\ \text{K}, \tag{17–2}$$

where p is the pressure of the ideal gas in the thermometer. Figure 17–6 illustrates Eq. (17–2). The assignment of 273.16 K as the triple point makes the Kelvin scale compatible with the Celsius scale, to be described below.

Table 17–2 gives us a feeling for the Kelvin scale by listing some temperatures that occur in nature. Remarkably, the numbers that refer to the origin of our universe in the so-called Big Bang, which occurred some 1.8×10^{10} years ago, can be estimated quite reliably from laboratory data. Although experiments have come quite close to 0 K, this temperature cannot in fact be reached.

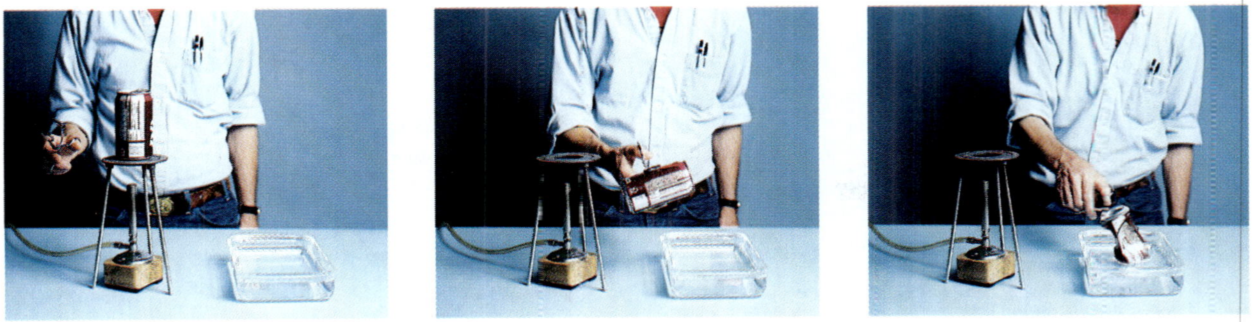

(a) (b) (c)

FIGURE 17–4 (a) A small amount of water is poured into a can, then heated to boiling, filling the can with water vapor. (b) The can is turned upside down and plunged into a container of water at room temperature. (c) The steam inside the can condenses, causing the inside pressure to suddenly drop. The outside pressure then crushes the can.

EXAMPLE 17–1 Estimate the pressure in a hypothetical bottle of air that, starting from sea level (surface temperature and pressure), has been heated to the temperature of the surface of the Sun. Express your answer in atmospheres.

Solution: We assume that the gas in the bottle is ideal over the desired range. The temperature must be in kelvins. From Table 17–2, we estimate that a typical temperature on Earth, where the pressure is 1 atm, is 300 K; the temperature at the surface of the Sun is approximately 6000 K. According to Eq. (17–1), the ratio of the two temperatures in question is the same as the ratio of the two pressures. If we refer to the temperature and pressure on Earth and the Sun with the subscripts E and S, respectively, we have

$$\frac{p_E}{T_E} = \frac{p_S}{T_S};$$

$$p_S = p_E\left(\frac{T_S}{T_E}\right) = 1\text{ atm}\left(\frac{6000\text{ K}}{300\text{ K}}\right) = 20\text{ atm}.$$

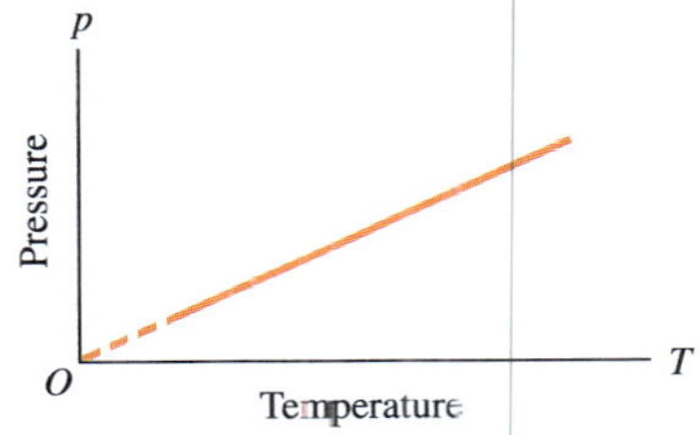

FIGURE 17–5 As p varies in an ideal gas of constant volume, T is *defined* by drawing a straight line.

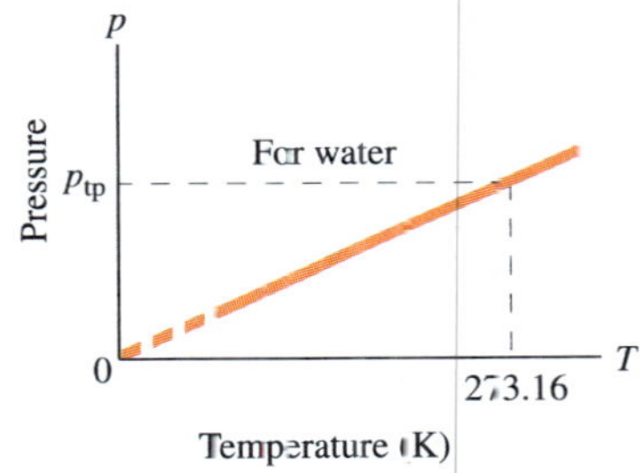

FIGURE 17–6 The Kelvin temperature scale is defined by a constant-volume ideal-gas thermometer. The two fixed points are $T = 0$ K located where p extrapolates to zero, and $T = 273.16$ K, located at the pressure of the triple point of water, p_{tp}.

TABLE 17–2 Temperatures Occurring in Nature

Physical System	*Temperature (K)*
Lowest temperature reached in the lab	$\simeq 10^{-6}$
Temperature of background radiation that fills the universe	3
Liquid helium at 1 atm	4.2
Coldest recorded outdoor temperature on Earth	185
Average surface temperature on Mars	218
Freezing point of water	273.15
Body temperature	310
Hottest recorded outdoor temperature	331
Boiling point of water	373
Melting point of gold	1335
Surface of the Sun	6000
Interior of the Sun	10^7
Helium fusion reaction	10^8
Interior of hottest stars	10^9
Universe at 3 min after the Big Bang	5×10^8
Universe at 1 s after the Big Bang	10^{10}
Universe at 10^{-16} s after the Big Bang	10^{14}
Universe at 10^{-12} s after the Big Bang	10^{16}

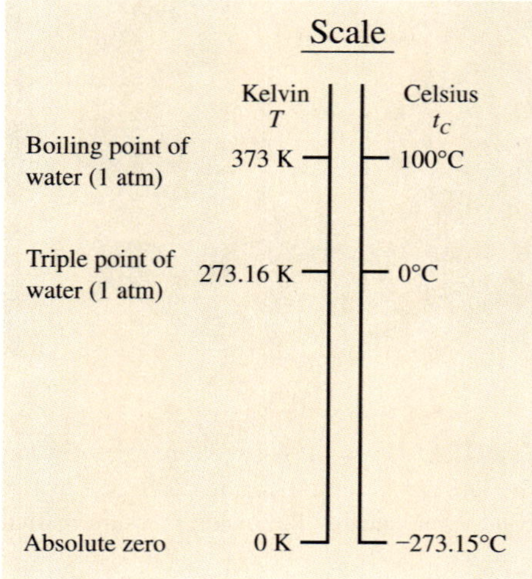

FIGURE 17–7 Comparison of the Kelvin and Celsius temperature scales.

The difficulty of finding containers that can withstand high temperatures is one reason why ideal-gas thermometers are not very practical at high temperatures. It is not easy to construct a suitable container for temperatures much above 2000 K, even at room pressures.

Other Temperature Scales

By changing both the slope and the intercept in Fig. 17–4, we can construct other temperature scales. Equivalently, if two points on the linear curve are assigned to special temperatures, a new scale emerges. Different scales are useful in different regimes. The temperature in the **Celsius scale**—once known as the centigrade scale—is denoted by t_C and measured in units of °C. It assigns the value $t_C = 0°\text{C}$ to the freezing point of water at 1 atm of pressure (the *ice point*) and $t_C = 100°\text{C}$ to boiling water at 1 atm. From Fig. 17–7, we see that

The Celsius scale

$$t_C = T - 273.15, \tag{17–3}$$

where T is the temperature in the Kelvin scale.† Note that the slopes of the Kelvin and the Celsius scales are exactly the same.

Other scales differ from the Kelvin scale in slope as well as in zero level. The **Fahrenheit scale**, in which the temperature t_F is measured in °F, is an example. It was invented with a human scale in mind. The ice point of water is 32°F, the temperature of the human body is about 100°F, and the steam point of water is 212°F. The relation between the Fahrenheit scale and the Kelvin scale is

$$t_F = \frac{9}{5}T - 459.67. \tag{17–4}$$

Again, T is measured in kelvins. We can easily find the relation between any two scales; see, for example, Fig. 17–8.

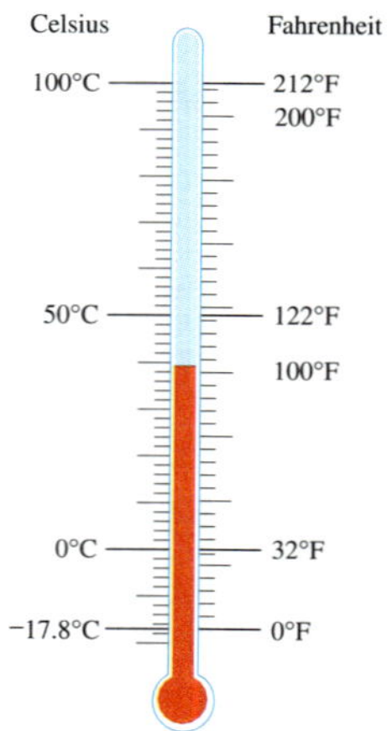

FIGURE 17–8 Comparison of the Celsius and Fahrenheit temperature scales.

17–3 THE EQUATION OF STATE

The pressure p, temperature T, volume V, and the amount (or mass) of material are the variables that describe a thermal system such as a bottle of gas with a piston. We have already discussed the first three variables. One possible unit for the amount of material in a thermal system is the **mole** (abbreviated mol), which represents either an amount of gas with a certain mass (according to the chemical species of the material) or a count of the number of molecules. One mole of a given gas is defined to be the amount of

Avogadro's number

†The difference of 0.01 between 273.16 in Eq. (17–2) and 273.15 here is the difference between the temperature of water at its triple point and water at its ice point.

gas with a mass in grams equal to the *atomic* (or *molecular*) *weight* of the gas. For example, 1 mol of helium gas has a mass of 4 g; the atomic weight of helium is 4 units of atomic mass. One mole of gas always contains *Avogadro's number* of molecules—$N_A = 6.022 \times 10^{23}$ molecules. The total number of molecules in a container, N, can be written in terms of the number of moles, n, as

$$N = nN_A. \tag{17–5}$$

Variables such as p, T, V, and n that are used to describe thermal systems are called **thermodynamic variables**.

Ideal gases obey *Boyle's law*, discovered in 1662 by Robert Boyle. Boyle's law states that, for a dilute gas held at constant temperature, the product of pressure, p, and volume, V, divided by the number of moles, n, is a constant:

$$\text{for constant } T\text{:} \quad \frac{pV}{n} = \text{a constant.} \tag{17–6}$$

Boyle's law

We previously defined temperature as a linear function of pressure if volume is constant. We can combine this result [Eq. (17–1)] with Boyle's law into the **ideal gas law**:

$$pV = nRT, \tag{17–7}$$

The ideal gas law

where R is a proportionality constant. The ideal gas law is an example of a relation between the thermodynamic variables of a thermal system, called the **equation of state**. Such a relation holds for any thermodynamic system and ensures that the variables are not independent of one another. The ideal gas law is the equation of state for an ideal gas. Measurements of ideal gases give

$$R \simeq 8.314 \frac{\text{N}\cdot\text{m}}{\text{mol}\cdot\text{K}} = 8.314 \text{ J/mol}\cdot\text{K}. \tag{17–8}$$

The universal gas constant

R is known as the **universal gas constant**. If we count the number of molecules of a gas, N, rather than the number of moles, n, we can use Eq. (17–5) in order to rewrite Eq. (17–7) as

$$pV = NkT, \tag{17–9}$$

where **Boltzmann's constant**, k, is given by

$$k = \frac{R}{N_A} \simeq 1.381 \times 10^{-23} \text{ J/K}. \tag{17–10}$$

Boltzmann's constant

We shall see that Boltzmann's constant, named for Ludwig Boltzmann—one of the important figures in nineteenth-century thermal physics—provides the bridge between temperature and energy. This gives it a particularly important role in thermal phenomena.

EXAMPLE 17–2 Calculate the volume occupied by 1 mol of an ideal gas at 20.0°C and at atmospheric pressure. Use this volume to calculate the mass density of air, whose molecular weight is 29.2 g/mol.

Solution: It follows from the ideal gas law that for 1 mol of gas, $V = RT/p$. Under the given conditions (T in kelvins is $20.0 + 273 = 293$ K),

$$V = \frac{(1 \text{ mol})(8.31 \text{ J/mol}\cdot\text{K})(293 \text{ K})}{1.01 \times 10^5 \text{ Pa}} = 2.41 \times 10^{-2} \text{ m}^3.$$

The mass density of air under these conditions is then the mass of 1 mol of air divided by this volume:

$$\rho = \frac{29.2 \text{ g}}{2.41 \times 10^{-2} \text{ m}^3} = \frac{(29.2 \text{ g})(10^{-3} \text{ kg/g})}{2.41 \times 10^{-2} \text{ m}^3} = 1.21 \text{ kg/m}^3.$$

Changing the Thermodynamic Variables of a Gas

Thermal systems change as the thermodynamic variables change. (We say that a *thermal transformation*, or, alternatively, a *thermal process*, occurs.) For example, when the cylinder of gas in Section 17–2 was put in contact with a hotter thermal reservoir, the gas's temperature changed; so did the gas's volume as the piston moved out. In this transformation, the temperature, the volume, and the pressure all changed subject to the constraint of the equation of state. There are many different ways to change a thermal system. For example, in an **isothermal** transformation, the temperature remains fixed while the pressure and volume both change. In an **isobaric** transformation, the pressure is held fixed while both temperature and volume change. Such transformations are most easily shown on diagrams in which one variable is plotted against another. For example, in a p–V diagram, pressure is plotted against volume. When a diagram shows various phases of a material, it is referred to as a *phase diagram.* Figure 17–9 is a p–T phase diagram for water.

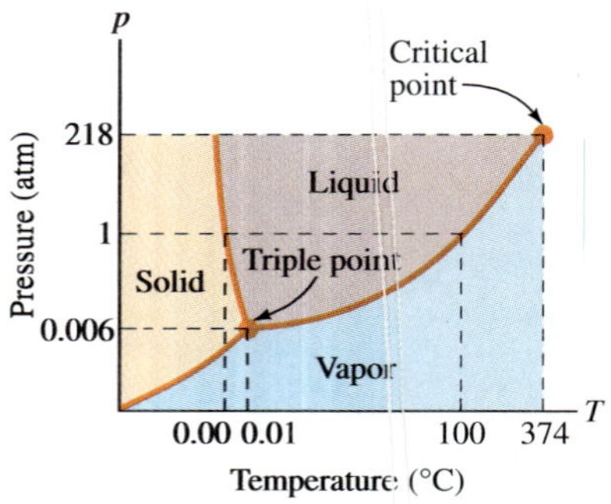

FIGURE 17–9 Phase diagrams are useful for the different phases of a material, such as those for water, shown here: liquid, solid (ice), and vapor (steam). The intersection of these phases is the triple point.

EXAMPLE 17–3 Exactly 1 mol of an ideal gas is taken through the sequence of changes shown on the p–V diagram in Fig. 17–10. The change $A \to B$ is an isothermal transformation. In $B \to C$, the transformation is isobaric: The gas is compressed in volume in such a way that the pressure remains fixed. (a) If $p_A = 5.00$ atm and $V_A = 8.00$ L, what is the temperature T_A? (b) If $V_B = 40.0$ L, what is p_B? (c) If $T_C = T_A/8$, what is V_C?

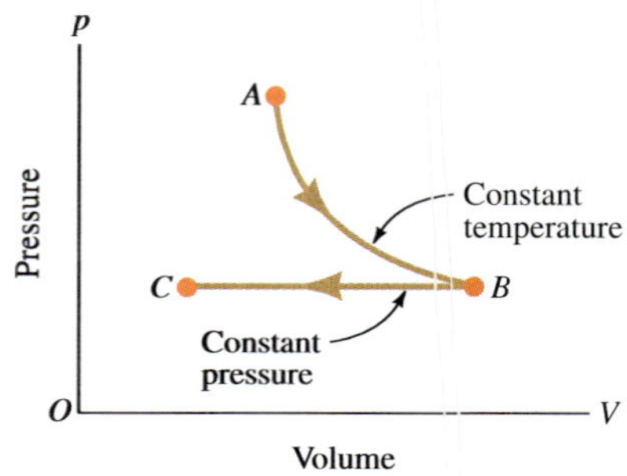

FIGURE 17–10 Example 17–3. A p–V diagram of the path followed by an ideal gas in undergoing first an isothermal transformation and then an isobaric transformation.

Solution: The starting point for all the parts of this problem is the ideal gas law, $pV = nRT$. For this sample of gas, $n = 1$ mol.

(a) We know p and V at point A; the temperature T_A is then given by

$$T_A = \frac{p_A V_A}{nR} = \frac{(5.00\ \text{atm})(8.00\ \text{L})}{(8.134\ \text{N} \cdot \text{m/mol} \cdot \text{K})(1\ \text{mol})}.$$

We make some unit conversions in order to cancel units:

$$T_A = \frac{(5.00\ \text{atm})(1.01 \times 10^5\ \text{Pa/atm})(8.00\ \text{L})(10^{-3}\ \text{m}^3/\text{L})}{(8.314\ \text{N} \cdot \text{m/mol} \cdot \text{K})(1\ \text{mol})} = 486\ \text{K}.$$

(b) $T_B = T_A$, from the conditions stated in the problem. We know V and T at point B, and hence we can compute p_B:

$$p_B = \frac{nRT_B}{V_B} = \frac{nRT_A}{V_B} = \left(\frac{nR}{V_B}\right)\left(\frac{p_A V_A}{nR}\right) = p_A\left(\frac{V_A}{V_B}\right)$$

$$= (5.00\ \text{atm})\left(\frac{8.00\ \text{L}}{40.0\ \text{L}}\right) = 1.00\ \text{atm}.$$

(c) The change $B \to C$ is one in which pressure is held constant. The quantities $p_C = p_B$ and T_C are known. The volume at C is given by

$$V_C = \frac{nRT_C}{p_C} = \frac{nR}{p_C}\left(\frac{T_A}{8}\right) = \frac{nR}{8p_C}\left(\frac{p_A V_A}{nR}\right) = \frac{p_A V_A}{8p_C} = \frac{p_A V_A}{8p_B}$$

$$= \frac{(5.00\ \text{atm})(8.00\ \text{L})}{8(1.00\ \text{atm})} = 5.00\ \text{L}.$$

How Close Do Real Gases Come to Being Ideal?

When is a real gas "sufficiently dilute" to obey the ideal gas equation of state? To answer this question, let's look at how well Boyle's law is satisfied. Consider a series of identical closed tubes. A pressure gauge can measure the pressure of any gas in the tubes. Put different amounts (a different number of moles, n) of the same gas in each tube and immerse them in the same thermal reservoir. If Boyle's law is satisfied ex-

APPLICATION: CARBONATED BEVERAGES

The properties of phase diagrams suggest ways in which the presence of different phases of a material can be used to advantage. An example of such an application of a phase diagram is the delivery of carbon dioxide to bars and restaurants for the production of carbonated beverages (Fig. 17AB–1). At a pressure of one atmosphere—and depending on the temperature—carbon dioxide exists only in the solid phase (low temperatures) and in the gas phase (high temperatures). Both are inconvenient to use; the gas is bulky and the solid is difficult to incorporate in mixing systems. Carbon dioxide is therefore transported and delivered at pressures greater than 40 atmospheres (4×10^6 Pa), pressures at which the room-temperature phase is liquid. This liquid can then be conveniently stored, transported efficiently through pipes, controlled by automatic valves, and so forth. When it is needed for incorporation into soft drinks near room temperature, it is released to one atmosphere of pressure, at which point it changes to the gas phase.

FIGURE 17AB–1 A carbonated beverage dispenser.

actly, the (measured) ratio pV/n will be the same in each tube, independent of the pressure in the tube.

Experiment shows that for monatomic gases, such as helium, the ratio pV/n is independent of the pressure over a wider range of pressures, and therefore fits the ideal gas limit better than do polyatomic gases.[†] But *all gases give precisely the same value of pV/n when extrapolated to zero pressure or to zero density.* All gases are indeed ideal in this limit; a more appropriate definition of temperature than that presented in Eq. (17–2) would be to *extrapolate measurements linearly* with such a thermometer to zero pressure:

A gas of any substance is ideal if the gas is dilute enough.

$$T = (273.16\ \text{K}) \lim_{p \to 0} \frac{p}{p_{tp}}. \qquad (17\text{–}11)$$

The Kelvin temperature scale redefined

Equation (17–11) provides us with the desired universal way to define temperature because it holds in the domain where *every* gas is ideal.

The van der Waals Equation of State

In 1873, Johannes D. van der Waals proposed an equation of state that works well for many real gases. His formula, known as the *van der Waals equation*, takes the form

The van der Waals equation of state

$$\left[p + a\left(\frac{n}{V}\right)^2\right]\left(\frac{V}{n} - b\right) = RT. \qquad (17\text{–}12)$$

The constants a and b are both positive. Van der Waals was motivated to write this equation based on a molecular model of gases (see Chapter 19). We can verify that the van der Waals equation reduces, as it must, to the ideal gas law when the gas is very dilute: In the limit as $n \to 0$, the term $a(n/V)^2$ is small compared to p, and the term V/n is large compared to b. Thus, in this limit, Eq. (17–12) reduces to $pV/n = RT$.

The van der Waals equation of state is a considerable empirical improvement over the ideal gas law for polyatomic gases that are about as dilute as the air we breathe. In fact, it describes all gases well, even when a gas is nearly a liquid; that is, at very low temperatures and at densities approaching those of real liquids.

[†]The molecule of a monatomic gas consists of one atom; the molecule of a polyatomic gas consists of several.

PROBLEM SOLVING

Quantities with the same dimensions may well be related.

A Connection Between Temperature and Energy?

An important fact follows from the ideal gas law presented in Eq. (17–7) or Eq. (17–9). The quantities pV/n and pV/N have the dimensions of (force/length^2)(length^3) = (force)(length), or $[MLT^{-2}][L] = [ML^2T^{-2}]$; n or N simply counts the number of moles or molecules, so that these quantities are dimensionless. The factor (force)(length) has the dimensions of work or, equivalently, energy. *The dimensions of RT and kT are those of energy.* This is already reflected in the units of R and k [Eqs. (17–8) and (17–10)]. The significance of this observation is that temperature, as defined by the ideal gas law, must be related closely to the energy of a thermal system. This *thermal*, or *internal*, energy is the subject of Chapter 18.

*17–4 THE BACKGROUND TEMPERATURE OF THE UNIVERSE

APPLICATION

The optical pyrometer

Most of us have enjoyed an evening before a roaring blaze that slowly dies down during the course of the evening. As the fire dies, its color changes from brilliant white to bright yellow, to orange, and finally to a dull red. The change is more marked if you look into a cavelike hole within the embers, a spot where the temperature is steadier because it is protected from the vagaries of drafts. This phenomenon is the basis of a thermometer called the *optical pyrometer* (Table 17–1). This thermometer works because the color or, rather, the range of colors emitted by a hot object is characteristic of the temperature of that object. This thermometer works best for relatively hot objects with temperatures in the range of several hundred kelvins.

The energy states of atoms are discussed in Chapter 10.

The light emitted by hot objects is **electromagnetic radiation**. Radiation exists in the form of waves, and all frequencies are possible in principle. All matter, at any temperature, emits such radiation. Atoms go into excited states when the temperature of the matter they comprise is raised. They radiate energy when they fall back into their normal energy states. For example, the atoms heated by an electric current in a lightbulb filament radiate visible light, which we detect as a range of colors: from red to orange to yellow to green to blue to violet, as the frequency of the radiation increases from 0.5×10^{15} Hz to approximately 1.0×10^{15} Hz.

A body at temperature T emits electromagnetic radiation that is characteristic of that temperature.

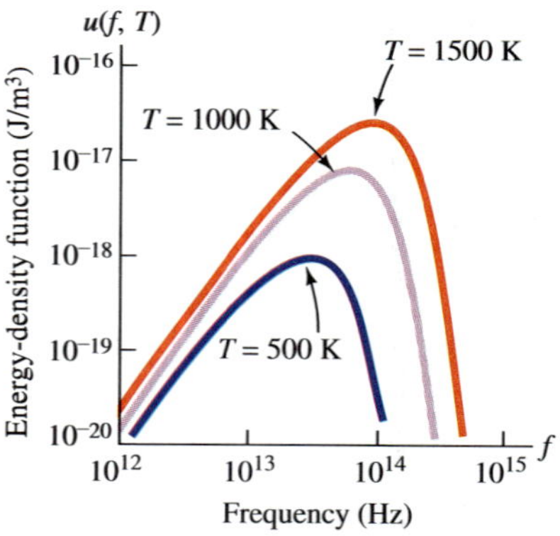

FIGURE 17–11 The energy density of electromagnetic radiation in thermal equilibrium at the indicated temperatures. The maximum energy density occurs for increasing frequency as the temperature increases. The maximum energy density increases *very* rapidly with T for a given f; note that this is a log–log plot.

In some ways, radiation in an enclosure acts like a dilute gas in a bottle; in particular, the radiation has a temperature just as a gas does. However, there is an important difference: Radiation achieves thermal equilibrium through absorption and reemission of radiation by the atoms of the walls of the enclosure rather than through collisions among its components (as the molecules that make up a dilute gas do). The proportion of high frequencies to low frequencies of the radiation at thermal equilibrium in an enclosure rises with temperature. The average frequency emitted by a white-hot object is greater than the average frequency the object emits when it is cherry red. We can give a function that describes the radiation emitted at *different* frequencies by an object at a given temperature. Such a function is denoted by $u(f, T)$, where *$u(f, T)$ df is the radiation energy per unit volume, or energy density, with a frequency between f and $f + df$.* Moreover, the radiation is said to be at a given temperature—the temperature of the object—when it and the object are in thermal equilibrium, meaning that the radiation interacts with the object over a long period and that the system of object and radiation is isolated from its surroundings. Radiation of this type is known as **blackbody radiation**, and the body that emits and absorbs such radiation is known as a **blackbody**. Figure 17–11 shows this function for different values of temperature. *The measurement of $u(f, T)$ provides a measurement of T.*

The form of the energy–density function $u(f, T)$ was a subject of intense study in the latter part of the nineteenth century. The work of Gustav Robert Kirchhoff, Wilhelm Wien, Lord Rayleigh (Fig. 17–12), and James Jeans set the stage for the breakthrough of Max Planck, who determined in 1900 that

$$u(f, T) = \frac{8\pi h}{c^3} \frac{f^3}{e^{hf/kT} - 1}. \qquad (17\text{–}13)$$

This formula, known as the **Planck formula**, was found to fit perfectly the data on the energy density of radiation for a wide range of frequencies and temperatures. In addition to Boltzmann's constant, k [introduced in Eq. (17–10)], and the Kelvin temperature, T, the formula contains the speed of light, $c = 3 \times 10^8$ m/s, which is the speed of electromagnetic waves of *every* frequency. A new fundamental constant also appears in this formula. The new constant h, called *Planck's constant*, was found by fitting the formula to the observed energy density, and its value is $h = 6.625 \times 10^{-34}$ J·s.

We introduced Planck's constant in Chapter 7.

The discovery of the Planck formula was the first step on the path to the basic theory of matter and radiation, called *quantum theory*. This theory is based on a set of ideas that revolutionized the physical sciences in the twentieth century. (The other great revolution in twentieth-century physics—Einstein's theory of relativity—also has its origin in the investigation of electromagnetic radiation.)

An interesting consequence of the Planck formula is that the total power radiated through a small hole in an enclosure containing radiation in thermal equilibrium at temperature T must be proportional to the fourth power of the temperature, T^4. The **Stefan–Boltzmann formula**, which was discovered experimentally by Josef Stefan in 1879 and inferred from thermodynamics by Boltzmann in 1884 before the discovery of the Planck formula, states that the radiated power (energy per unit time) per unit area emitted through the hole is given by

$$P(T) = \sigma T^4, \qquad (17\text{–}14)$$

where $\sigma = 5.68 \times 10^{-8}$ W/(m^2·K^4). The T dependence in Eq. (17–14) can be obtained by integrating Eq. (17–13) over all frequencies f (see Problem 62).

FIGURE 17–12 Lord Kelvin and Lord Rayleigh, July 1900.

Let's examine the Planck formula for the energy–density function, Eq. (17–13). The quantity hf has dimensions of energy as does the quantity kT. (Indeed, these quantities must have the same dimensions because their ratio is the argument of the exponential function.) If the frequencies are small so that $hf < kT$, then the exponential function in $u(f, T)$ has a small argument, and we can use the expansion

$$\text{for small } x: \qquad e^x \simeq 1 + x.$$

In this case, the denominator of $u(f, T)$ is

$$e^{hf/kT} - 1 \simeq 1 + \frac{hf}{kT} - 1 = \frac{hf}{kT},$$

and the function $u(f, T)$ takes the approximate form

$$\text{for } hf < kT: \qquad u(f, T) \simeq \frac{8\pi h}{c^3}(f^3)\frac{kT}{hf} = \frac{8\pi f^2}{c^3}kT. \qquad (17\text{–}15)$$

This expression is independent of Planck's constant and is thus independent of quantum physics. It was, in fact, first found by Lord Rayleigh in 1900 by using a derivation based on classical physics. The energy density in Eq. (17–15) rises with increasing frequency as f^2 and linearly with temperature.

If, however, the frequencies are large, so that $hf > kT$, then the exponential argument in $u(f, T)$ is large, and the exponential function plays a major role. The denominator then becomes

$$e^{hf/kT} - 1 \simeq e^{hf/kT}.$$

(Here we have used the fact that for $hf/kT > 1$, $e^{hf/kT} > 1$.) Then the energy density is approximately

$$\text{for } hf > kT: \qquad u(f, T) \simeq \frac{8\pi h}{c^3} f^3 e^{-hf/kT}. \qquad (17\text{–}16)$$

Because the exponential function falls with f much faster than the polynomial factor f^3 grows, $u(f, T)$ decreases as f grows in this range of f. Note that h plays a crucial role in this part of the energy–density function, indicating that quantum physics is important.

As we have seen, the energy–density function increases with f for $hf < kT$ and decreases with f for $hf > kT$. It is not surprising, then, that the function has an intermediate maximum, which we can estimate as occurring at about $hf \simeq kT$. The underlying physics of the high-frequency side of the curve of the energy–density function in Fig. 17–11 is quantum physics because it involves Planck's constant, but the low-frequency side can be explained by classical physics. The maximum moves to higher values of f as T increases. We can therefore say that the importance of quantum physics decreases as the temperature grows; in fact, this is generally true.

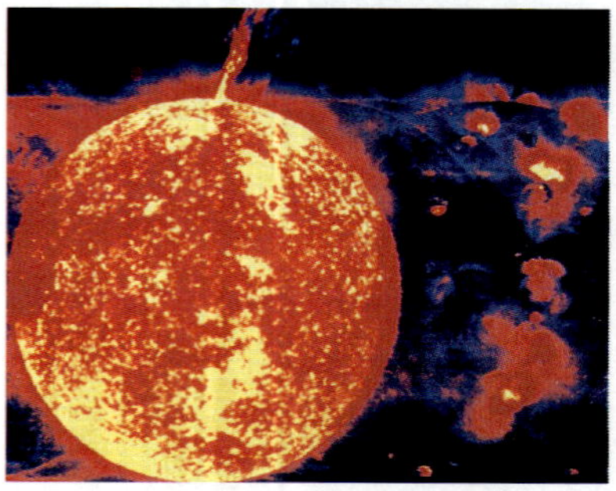

FIGURE 17–13 Example 17–4. The temperature of the Sun's surface is not constant but varies considerably in active areas, such as regions with sun flares.

EXAMPLE 17–4 The surface of the Sun is observed to emit light most strongly at a frequency of 2×10^{14} Hz. Estimate the temperature of the Sun's surface (Fig. 17–13).

Solution: The strongest emission occurs at the maximum of the energy–density function. An appropriate estimate follows from the rough value of this maximum, at a frequency given by $hf/kT \simeq 1$. The temperature is then approximated by

$$T \simeq \frac{hf}{k} = \frac{(6.63 \times 10^{-34}\ \mathrm{J \cdot s})(2 \times 10^{14}\ \mathrm{Hz})}{1.38 \times 10^{-23}\ \mathrm{J/K}} = 10{,}000\ \mathrm{K}.$$

A more precise value can be found by looking for the maximum of the energy–density function. Because it is wavelength rather than frequency that is measured, it is the energy density as a function of wavelength that is maximized. The more accurate expression (see Problem 66) is

$$\lambda_{max} T = 2.9 \times 10^{-3}\ \mathrm{m \cdot K}.$$

An estimate from this formula yields 6000 K for the temperature at the Sun's surface.

A measurement of the value of the energy–density function, Eq. (17–13), at any particular frequency is equivalent to a measurement of the temperature. A crude way to measure this quantity is to hold your hand near a fire and feel the radiant heat. A pyrometer does this job more precisely: The color of the visible radiation emitted from a hot cavity is a measure of the frequency distribution. An antenna that receives and measures electromagnetic radiation is another accurate method of measuring the energy–density function. The size and nature of such antennas determine the particular frequency or range of frequencies they receive.

The Discovery and Measurement of the Background Radiation of the Universe

In 1964, Arno Penzias and Robert Wilson used a large radio antenna, shown in Fig. 17–14, to study the radio waves emitted within our galaxy. They found a puzzling anomaly, which they at first thought might be a problem with their antenna: A background electromagnetic radiation that comes from no particular source is present. Their antenna was sensitive only to the particular frequency $f = 4.08 \times 10^9$ Hz, but later measurements over other frequencies showed that this background radiation was consistent with the Planck formula for $T \simeq 3$ K.

FIGURE 17–14 Arno Penzias (right) and Robert Wilson in front of the antenna they used in detecting the background radiation of the universe.

Penzias and Wilson were not aware of work by Ralph Alpher, George Gamow, and Robert Herman in the late 1940s, or of work by James Peebles that was contemporary with their own, in which a cosmological, or large-scale, model of the universe predicted the existence of this radiation. As described by this model, the early universe expanded from an initial catastrophic event—the Big Bang. As the universe expanded, it cooled, maintaining thermal equilibrium between electromagnetic radiation and matter up to a point, some 100,000 years after the Big Bang, when the temperature was

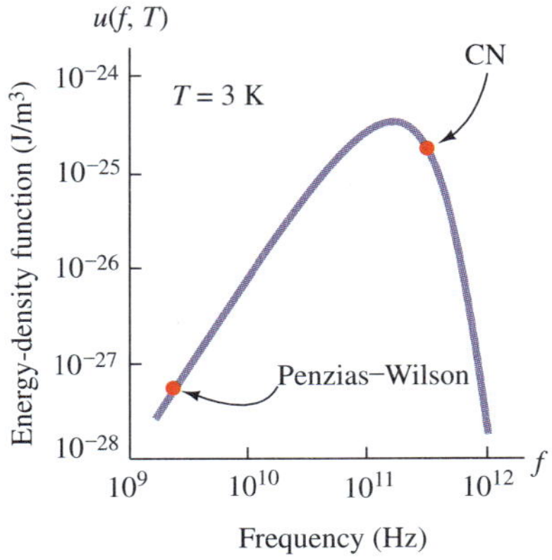

FIGURE 17–15 The energy density of electromagnetic radiation in thermal equilibrium at $T = 3$ K. The Penzias–Wilson frequency measurement and the frequency associated with interstellar cyanogen, labeled CN, are marked by red dots.

about 3000 K. At that point, the model shows that thermal equilibrium between electromagnetic radiation and matter was broken off Remarkably, as the universe has continued to expand, the Planck formula still describes the radiation, but with a temperature that drops as the universe expands. *A measurement of the temperature of the radiation is equivalent to a measurement of the rate of expansion of the universe; hence, it is equivalent to the time that has elapsed since the radiation was no longer in thermal equilibrium with matter.* Presently, the temperature of the radiation, as characterized by the Planck formula, is about 3 K, and this radiation is the oldest directly observable relic of the explosion in which the universe is widely believed to have formed.

Penzias and Wilson's frequency measurement occurs on the classical, or low-frequency, side of the Planck formula, as Fig. 17–15, the energy–density function for $T =$ 3 K, indicates. Since 1964, many other Earth-based measurements have been made by radio antennas sensitive to a great variety of frequencies. Because Earth's atmosphere absorbs radiation of high frequency, these measurements are all made on the classical side of the Planck formula. Fortunately, one other technique can make a measurement on the quantum, or high-frequency, side of the peak of energy density. The measurement in question is the amount of light emitted by interstellar cyanogen, the CN molecule. This light is emitted with a strength that can be explained only if the cyanogen is in the presence of background radiation of a certain temperature. The frequency measurement of CN, first made in 1941, is marked on Fig. 17–15. In early 1990, a satel-

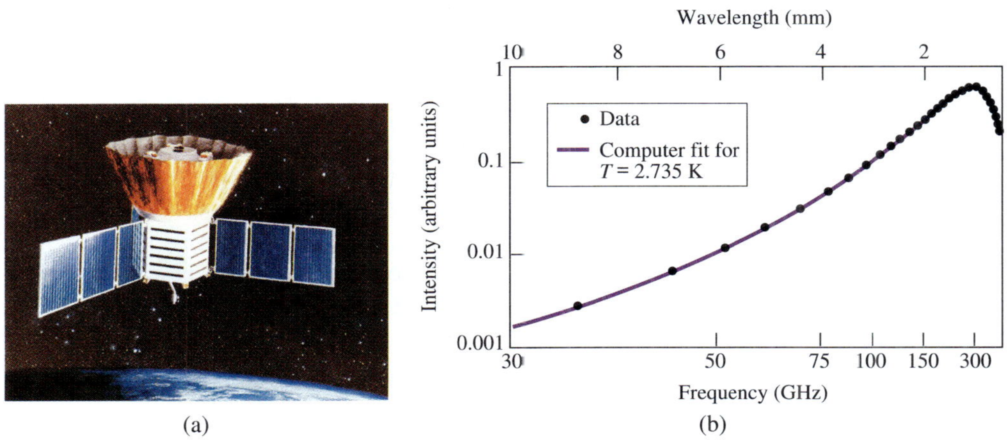

FIGURE 17–16 (a) The Cosmic Orbiting Background Experiment (COBE) satellite measured cosmic background radiation. (b) The measurements show remarkable agreement with the Planck formula. The data shown for $T = 2.735$ K fit the smooth curve to remarkable accuracy.

lite was launched to measure the background radiation (Fig. 17–16a). The satellite measures a range of frequency from roughly 6×10^{10} Hz to 6×10^{11} Hz with extraordinary precision. This range includes the maximum of the energy–density curve. We show the results in Fig. 17–16b; a temperature of about 2.7 K fits the curve perfectly. (The actual temperature can be determined to six significant figures!) Still other measurements reveal that the temperatures in different directions differ ever so slightly from one another. Some regions are slightly hotter, while others are slightly cooler. These temperature inhomogeneities may well be relics of the instabilities that led to the formation of the great clusters of galaxies present today. All these results provide a vivid verification of the Big Bang model of the universe, which predicts the presence of this background blackbody radiation.

SUMMARY

The temperature of a system in thermal equilibrium is defined operationally on the Kelvin scale by the behavior of the pressure, p, of a dilute gas in a constant-volume gas thermometer:

$$T = (273.16) \lim_{p \to 0} \frac{p}{p_{tp}}, \tag{17–11}$$

where p_{tp} is the pressure of the gas in the thermometer at the triple point of water (where the gaseous, liquid, and solid phases coexist); thus 273.16 K is chosen to be the temperature at the triple point. The other fixed point of the Kelvin scale is built into the thermometer: $T = 0$ K is the point where the pressure extrapolates to zero.

Two other temperature scales in common use are the Celsius scale and the Fahrenheit scale. The Celsius temperature is given in terms of the Kelvin scale by

$$t_C = T - 273.15. \tag{17–3}$$

Any sufficiently dilute gas behaves as an ideal gas, meaning that its thermodynamic variables are related by the ideal gas law, an equation of state:

$$pV = nRT = NkT, \tag{17–7, 17–9}$$

where the universal gas constant is

$$R \simeq 8.314 \text{ J/mol} \cdot \text{K}. \tag{17–8}$$

and Boltzmann's constant is

$$k = \frac{R}{N_A} \simeq 1.381 \times 10^{-23} \text{ J/K}, \tag{17–10}$$

where NA is Avogadro's number. The quantities kT and RT have dimensions of energy.

Real gases are ideal only in the limit that they are very dilute. The small departures from ideal behavior are accurately reflected for many real gases by the empirical van der Waals equation of state,

$$\left[p + a\left(\frac{n}{V}\right)^2\right]\left(\frac{V}{n} - b\right) = RT. \tag{17–12}$$

Electromagnetic radiation displays the characteristics of a thermal system. When such radiation is in equilibrium with another thermal system at temperature T, it is known as blackbody radiation and contains a range of frequencies f that have an energy per unit volume (or energy density) per unit frequency, given by

$$u(f, T)\frac{8\pi h}{c^3} = \frac{f^3}{e^{hf/kT} - 1}, \tag{17–13}$$

a result known as the Planck formula. The constant h is Planck's constant.

Electromagnetic radiation with this frequency range is observed to occur throughout the universe at a characteristic temperature of 3 K. This is strong evidence for the Big Bang theory of the origin of the universe.

UNDERSTANDING KEY CONCEPTS

1. Estimate to within an order of magnitude the number of molecules in a breath of air.
2. Why is mercury used in most thermometers rather than a more common liquid such as water?
3. If a scuba diver takes the same volume of air for each breath and breathes at a constant rate, why is the tank of air used up much faster at greater water depths?
4. How can a triple point of water exist? If ice, water, and vapor all existed together, would the ice cool the vapor, making it water? Or would the vapor and water melt the ice?
5. Why is it advisable to measure the pressure in automobile tires when the car has not been driven for some time?
6. Devise an experiment to prove the experimental fact that hot objects radiate more energy per unit area than cold objects do.
7. Why is it important to check the pressure in automobile tires before the winter season?
8. All gases obey the ideal gas law when they are made sufficiently dilute. Is this true when the temperature is lowered sufficiently?
9. In recent years, an attempt has been made to convert Americans to the use of the metric system, so that, for example, two sets of tools would not be necessary for a mechanic who works on both American and European cars. Would it be of equal practical importance to convert Americans from the use of Fahrenheit to Celsius temperature scales?
10. When we observe a fire, we are observing radiation that has escaped the region of the hot fire. Is this a mechanism for cooling the fire? In what way could a fire be enclosed so that it remains hot?
11. Suppose that we have no idea that gases consist of molecules. Could we still use the low-pressure limit of a gas to define temperature?
12. Substances in thermal contact reach thermal equilibrium after some time. Can you think of any experimental test that will indicate just when that time will have arrived?
13. If the molecules of gases move around a lot, why do we never see a container of gas shake?
14. The pressure of a gas produces an outward force on the walls of a container. Why is there not a sideways force as well?
15. The Chamber of Commerce of a town plagued by temperature extremes devises a way to avoid bad publicity: It defines a new temperature scale, the G scale, such that (a) $-40°F$ is defined to be $+20°G$; (b) $+120°F$ is defined to be $-40°G$; and (c) the scale is linear. Is there anything wrong with this scale?

PROBLEMS

17–1 Temperature and Thermal Equilibrium

1. (II) Which of these pairs of systems are in thermal equilibrium with one another? (a) A roast reaching the rare stage and the oven in which it sits. (b) The point of a meat thermometer in the roast of part (a) and the roast. (c) A sunbather getting a tan and the air around him. (d) A bather in a hot tub and the water around her. (e) An ice cube in a glass of water at 0°C. (f) Molten iron ore in a blast furnace and the walls of the furnace.
2. (II) You must measure the different temperatures of a series of bottles that contain 1 L of an unknown fluid. Which one of the following techniques would make a suitable thermometer, and why or why not? (a) The time it takes 1 L of ice cubes dumped into the fluid containers to melt. (b) The melting time for ice cubes sealed in full 1-L plastic bags that are dumped into the bottles. (c) The melting time for ice sealed in full 1-mL plastic bags that are dumped into the bottles. (d) All of the above techniques, but the fluids in the bottles are gently stirred during the measurement.
3. (II) Thermodynamic variables are either extensive or intensive. A variable is *extensive* if, when two identical thermodynamic systems are combined into one, the variable of the combined system is double its original value in each system; otherwise, it is *intensive*. Which of the following variables are extensive, and which are intensive: (a) volume, (b) temperature, (c) pressure, (d) number of moles?
4. (II) A physicist needs to measure room temperature. Not having anything better at hand, she decides to make use of the thermal expansion of water to construct a thermometer (Fig. 17–17). She calibrates her apparatus to a dependable thermometer at 10°C and 40°C, and divides the range between these temperatures into 30 equal intervals. What will be the true temperature when her equipment reads 25°C? She knows that the volume of water can be described in this temperature range by the formula

$$V = V_0(1 - 2.525 \times 10^{-4} + 4.98 \times 10^{-6}\,T + 4.94 \times 10^{-6}\,T^2),$$

where T is the temperature in °C.

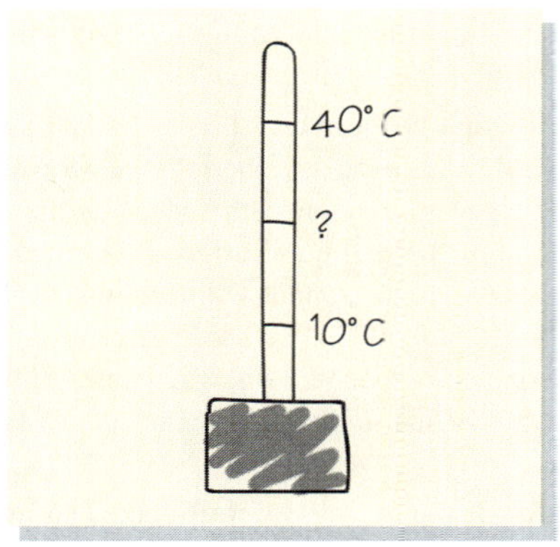

FIGURE 17–17 Problem 4.

17–2 Ideal Gases and Absolute Temperature

5. (I) At what temperature values are the following scales the same: (a) the Fahrenheit and the Celsius, (b) the Celsius and the Kelvin, (c) the Fahrenheit and the Kelvin?

6. (I) The title of Ray Bradbury's science-fiction novel *Fahrenheit 451* refers to the temperature at which paper ignites. What is that temperature on the Celsius scale?

7. (I) An oral fever thermometer ranges from 35.5°C to 42.5°C. What is the corresponding range on the Fahrenheit scale?

8. (I) An ideal gas in a container is heated from 20°C to 215°C. By what factor does the pressure increase?

9. (I) A very dilute gas confined to a closed one-liter container is put on the burner of a stove and the temperature is raised from 271 K to 349 K. By how much does the pressure, which is originally at 1 Pa, change?

10. (I) If an ideal gas thermometer could be constructed to measure the temperature of a hot star (see Table 17–2), by what factor would the pressure in the thermometer differ from *ptp*?

11. (II) (a) Determine the temperature of the surface of Venus (730 K) in the Celsius and the Fahrenheit scales. (b) Determine the temperature of the boiling point of liquid nitrogen (77 K) in the Celsius and the Fahrenheit scales. (c) A comfortable room temperature is 75°F. Determine this temperature in the Kelvin and Celsius scales. (d) Heat pumps become less efficient than furnaces for heating homes when the temperature drops below 36°F. New laws may require that all temperatures be posted in the Celsius and the Kelvin scales as well as the Fahrenheit scale. What is this temperature in these other systems?

12. (II) In 1701, Isaac Newton proposed a temperature scale in which ice water has a temperature of 0°, whereas the human body in good health is assigned a temperature of 12°. In this scale, what is the temperature of someone with a slight fever? What is the temperature of boiling water?

13. (II) A temperature scale—the Reaumur scale—developed in the early eighteenth century sets the ice point at 0°R and the boiling point of water at 80°R. Give the conversion formula from the Reaumur scale to the Kelvin scale and to the Fahrenheit scale.

14. (II) For the new temperature scale described in Question 15, what are the general conversion rules from °G to °F and vice versa? What is the conversion rule from °C to °G?

15. (II) The Rankine scale for temperature, named for William Rankine, is used in engineering applications. It has units of °R and is defined by $t_R = 9T/5$, where T is the temperature in the Kelvin scale. Find the value of t_R for the temperatures 100°C, 4.2 K, 6000 K, 32°F, −30°F, and −25°C.

16. (II) A bottle that has a freely sliding vertical piston and contains an ideal gas is used as a thermometer for water: The bottle is submerged in a water bath (whose temperature is to be measured) until thermal equilibrium is reached (Fig. 17–18). When the bottle is in contact with ice water, the volume inside the bottle is 0.50 L. The bottle is removed from the ice water and put in a second bath. This time, the enclosed volume is 0.55 L. What is the temperature of the second bath, in °C?

17. (II) One Sunday morning, a family takes an automobile trip to Grandma's. At the start of the trip, the temperature is 288 K (15°C), and the gauge pressure in the tires is 32 lb/in² (psi).

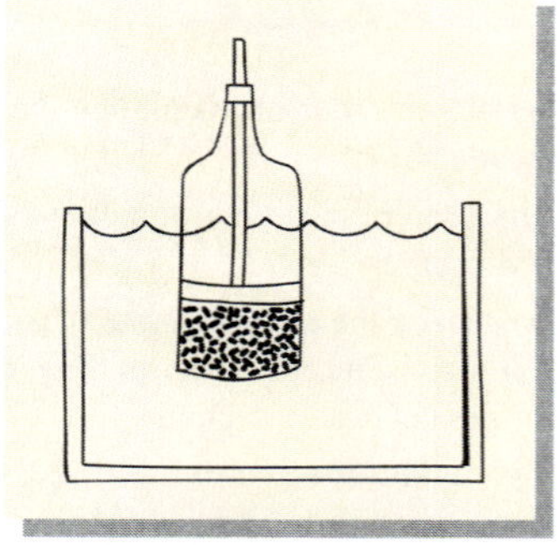

FIGURE 17–18 Problem 16.

(The gauge pressure is the excess over 14.5 psi, the exterior air pressure.) After an hour's ride over an interstate highway, the gauge pressure in the tires is 38 psi. What is the temperature of the air in the tires, assuming that air behaves as an ideal gas? Neglect any changes in volume of the tires.

18. (II) When the temperature of mercury changes by ΔT, the fractional volume change, $\Delta V/V$, is given by $\Delta V/V = \beta\Delta T$, where $\beta = 1.8 \times 10^{-4}\ \mathrm{K}^{-1}$. The quantity β is called the *coefficient of thermal expansion.* Consider a thermometer containing a bulb attached to a thin, cylindrical capillary tube. If the bulb and capillary tube contain 0.2 cm³ of mercury, how large must the diameter of the capillary tube be so that a 1-K change in temperature corresponds to a 2-mm change in the mercury level in the capillary tube? (Neglect the change in volume of the glass bulb when the temperature is raised by 1 K.)

19. (II) A copper bowl of volume 1500 cm³ is filled to the brim with water. Assume that the coefficient of thermal expansion, β, defined in Problem 18, is $5.1 \times 10^{-5}\ \mathrm{K}^{-1}$ for copper and $2.07 \times 10^{-4}\ \mathrm{K}^{-1}$ for water. How much water will spill out when the temperature of the system is raised from 20°C to 50°C?

17–3 The Equation of State

20. (I) A container holds gas at a pressure of 1.0 atm and a temperature of 300 K. Half the gas leaks out while the temperature is raised to 340 K. What is the pressure in the container?

21. (I) A dilute gas in a container fitted with a piston occupies a volume of 2500 cm³. If the pressure is increased by 50 percent and the temperature in degrees kelvin is decreased by 15 percent, what volume will the gas occupy?

22. (I) A 1000-L container is filled with an ideal gas at 1.0 atm and 300 K. How many moles are in the container? How many gas molecules?

23. (I) What is the volume occupied by three moles of an ideal gas at 30 K and 10^{-7} atm in the upper atmosphere?

24. (I) What is the mass density of steam at a temperature of 100°C and a pressure of 1 atm, given that the molecular weight (the number of grams of 1 mol of the material) of water is 18 g/mol?

25. (I) What is the mass density of air (average molecular weight 29 g/mol), at a pressure of 0.31 atm and a temperature of −43°C? (These conditions correspond roughly to those at the top of Mt. Everest.)

26. (I) A helium balloon is filled with 10,000 cm³ of helium gas at *STP* (standard temperature and pressure: 0°C and 1 atm).

How many moles and molecules of helium (molecular weight 4 g/mol) are contained in the balloon?

27. (I) A container of ideal gas whose volume can be changed by the movement of a plunger is placed in a large bath that maintains the temperature of the gas at a fixed value of 295 K. The plunger is pulled out so that the volume is increased by a factor of 1.78. By what factor is the pressure changed in the vessel, if at all?

28. (I) The atomic weight of an elemental substance, A, is closely related to the mass of the atoms of that substance. Atoms consist of electrons, of negligible mass, and a total of A protons and neutrons, which each have nearly the same mass M. The total mass of an atom is therefore approximately AM, and N_A atoms have a total mass AMN_A. By the definition of Avogadro's number, N_A, this mass is A g. Given the value of N_A, what is M in grams?

29. (I) A container at 0°C holds a mixture of gases: 80 percent nitrogen (N_2 molecular weight 28 g/mol) and 20 percent oxygen (O_2 molecular weight 32 g/mol) by weight. Assuming that for a dilute gas each constituent behaves as if it alone occupied the volume, what is the partial pressure due to each constituent of the mixture, given that the total mass of gas in the container is 8.0 g, and that the volume is 4.0 L. What is the total pressure in the container?

30. (I) Consider a container of an ideal gas at a fixed pressure. The temperature is 280 K. What is the fractional change in volume if the temperature changes to 281 K? Can you derive a general formula that holds for all ideal gases?

31. (II) Given that the molecular weight of water (H_2O) is 18 g/mol and that the volume occupied by 1 g of water is 10^{-6} m^3, use Avogadro's number to find the distance between neighboring water molecules. Assume for simplicity that the molecules are stacked like cubes.

32. (II) The lowest pressure achievable with a rotary vane pump is on the order of 10^{-2} N/m^2. What is the number density of molecules of the gas in a chamber that holds gas at this pressure and at a temperature of 20°C?

33. (II) The pressure of an ideal gas in a closed container is 0.6 atm at 35°C. The number of molecules is 5×10^{22}. (a) What are the pressure in pascals and the temperature in kelvins? (b) What is the volume of the container? (c) If the container is heated to 120°C, what is the pressure in atmospheres?

34. (II) The molecular weight of H_2 is 2 g/mol. Consider H_2 gas in a container at STP (see Problem 26). (a) What volume is occupied by 1 kg? (b) What mass and volume are occupied by 1 mol and 1 kmol?

35. (II) Use the ideal gas law to calculate the volume occupied by 1 mol of ideal gas at 1 atm pressure and 0°C. Given that the average molecular weight of air is 28.9 g/mol, calculate the mass density of air, in kg/m^3, at the above conditions.

36. (II) Use the result of Problem 35 and Avogadro's number for the number of molecules in 1 mol, 6.02×10^{23}, to estimate the number of molecules in your lecture hall.

37. (II) What is the mass density of helium gas (atomic weight 4 g/mol) at 1 atm pressure and a temperature of 8 K?

38. (II) A child takes 12 puffs of air to blow up a spherical balloon. The diameter of the balloon is 26 cm. The temperature is 21°C, and the exterior atmospheric pressure is 1 atm. (a) How many molecules are in each of the child's puffs? (b) If the surrounding air and the balloon are heated to 35°C, what size will the balloon be?

39. (II) In an experiment, a vacuum of 10^{-10} atm is achieved in a bottle. If the bottle is at room temperature (30°C), what is the number of molecules in the bottle per cubic centimeter?

40. (II) At altitudes above 14,000 ft, pilots must breathe air with enriched oxygen because the density of the atmosphere is decreased. The pressure of air is only about 0.25 atm at a typical height for flying of 32,000 ft. If the airplane is not pressurized, what must the fraction of oxygen in the air be in order for a pilot to breathe the same amount of oxygen as at sea level?

41. (II) An ideal gas is contained in a tank at 120 atm of pressure and 263 K. (a) If half the mass of gas is drawn off and the temperature then rises by 50 K, what is the new pressure? (b) Suppose instead that the temperature first rises by 50 K and then half the mass of the gas is drawn off. What is the new pressure?

42. (II) For safety reasons, compressed gases for laboratory use are often stored outdoors and transferred to where they are needed via a system of tubes and valves (Fig. 17–19). Suppose that a 200-liter tank contains hydrogen at 87 atm at −5°C at the beginning of an experiment. If 800 L are used at 25°C and 1 atm pressure, what is the pressure of the remaining hydrogen? The temperature outside does not change during the measurement.

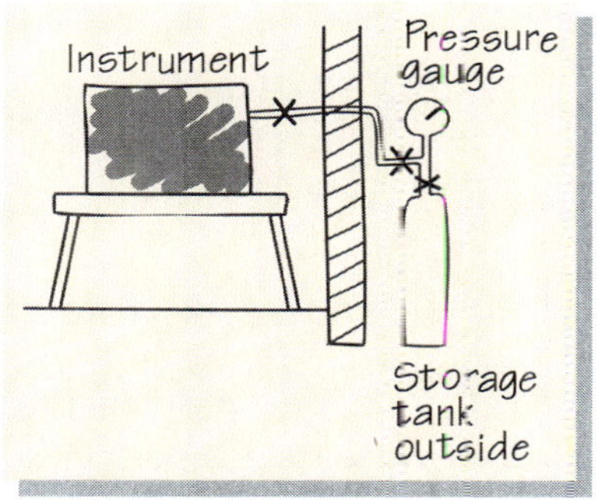

FIGURE 17–19 Problem 42.

43. (II) The cylinder of a manual bicycle tire pump is 35 cm long; its inside diameter is 2.5 cm. When the piston is pulled up, the valve to the tire is closed and air at atmospheric pressure enters the cylinder through the gap between the piston and the wall of the cylinder, which acts like another valve (Fig. 17–20). When the piston is pushed down, the air in the cylinder is compressed. The valve to the tire opens when the pressure in the tire is reached, and air is transferred to the tire during the rest of the stroke. The last 5 percent of the volume of air remains in the cylinder and the hose. Suppose that you are pumping up a tire that already contains air at 2.5 atm gauge pressure. How much air is transferred to the tire at this pressure? What is the volume of this air at atmospheric pressure? Sketch a p–V diagram of the process. Assume that the temperature is constant.

44. (II) Figure 17–21 represents the variation of the volume and pressure of gas contained in a cylinder with a movable piston. The curved sections of the curve are isotherms. Calculate the

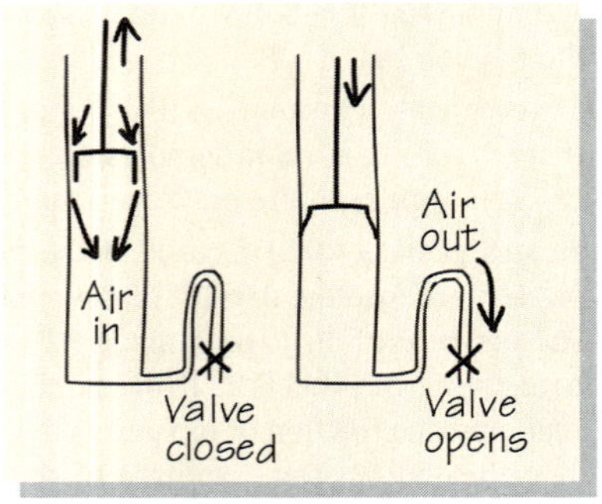

FIGURE 17–20 Problem 43.

temperature, pressure, and volume at the points 1, 2, 3, and 4 given that $p_1 = 1$ atm, $V_1 = 1$ L, $T_1 = 0$°C, $V_2 = 3.5\ V_1$, and $p_4 = 3p_1$.

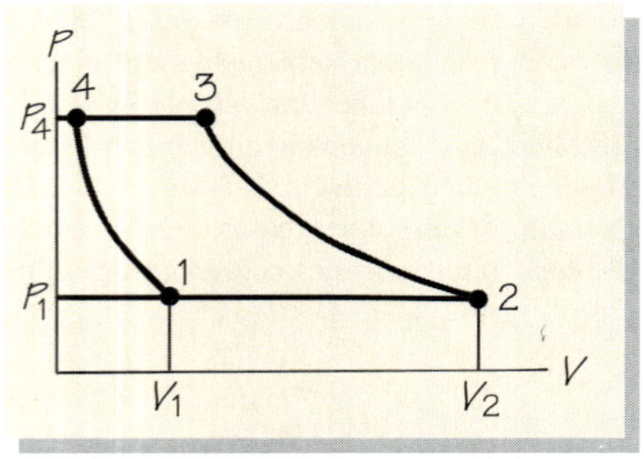

FIGURE 17–21 Problem 44.

45. (II) The specific volume of a gas is its volume per unit mass; that is, the inverse of the density. Compute the specific volume and density of oxygen gas, O_2, at $T = 15$°C and 1 atm of pressure, assuming that oxygen gas is ideal.

46. (II) An ideal gas is confined in a steel cylinder at 20°C and a pressure of 4.0 atm. (a) If the cylinder is surrounded by boiling water and allowed to come to equilibrium, what will the pressure of the gas be? (b) If the gas is allowed to escape until the pressure again reaches 4.0 atm, what fraction of the original gas, by weight, will escape? (c) If the temperature of the remaining gas now returns to 20°C, what is its pressure?

47. (II) A cylinder is closed at one end by a movable piston. The cylinder contains 300 cm^3 of air at 20°C and 10^5 Pa pressure. The cylinder undergoes the following changes: (a) the gas is heated in such a way that its volume doubles but its pressure remains constant; (b) the volume is kept constant and the temperature is changed until the pressure increases by 30 percent; (c) the gas cools, and the piston position is adjusted to keep the pressure constant until the initial volume is reached. Calculate the volume, pressure, and temperature after each step.

48. (II) The law of Joseph Gay-Lussac and Jacques Charles (1802) states that any ideal gas undergoes the same fractional increase in the product of pressure, p, and volume, V, when the gas is brought from one temperature to another. What is the fractional increase in pV for an ideal gas brought from the temperature of ice melting at 1 atm of pressure to the temperature of water boiling at the same pressure?

49. (II) Calculate the volume of 24 g of the gaseous form of ethyl ether, $C_4H_{10}O$, at a temperature of 120 K and pressure of 0.080 atm. Assume that the gas is ideal.

50. (II) A tank of volume 0.20 m^3 contains 250 mol of helium gas at 25°C. Assuming that the helium behaves like an ideal gas, what is the pressure in the tank?

51. (II) It is a scorching summer day with an air temperature of 308 K. What is the value, in joules, of RT for 1 mol of air? Repeat this for a cold winter's night when the air temperature is 258 K. Compare these results to the kinetic energy that a ball with the mass of 1 mol of air (28.9 g) acquires when it falls 1.0 m under the influence of gravity.

52. (II) At the beginning of a compression stroke, the cylinder in a combustion engine contains 1 L of air at atmospheric pressure and a temperature of 20°C (Fig. 17–22). At the end of the stroke, the air has been compressed to a volume of 60 cm^3, and the total pressure is 35 atm. What is the temperature of the air when it is so compressed?

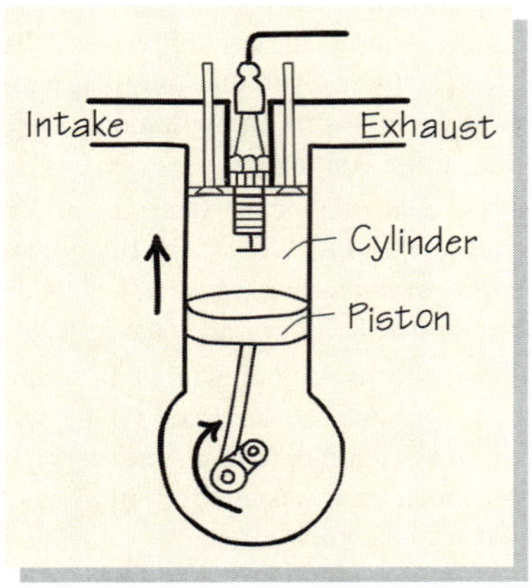

FIGURE 17–22 Problem 52.

53. (II) The steam point of water is the point where water vapor (steam) and boiling water coexist at a pressure of 1 atm. When a constant-volume gas thermometer calibrated to read a temperature of 273.16 K at the triple point of water is used to measure the temperature of the steam point, the result depends slightly on the pressure of the gas in the thermometer because the gas is not precisely ideal (Fig. 17–23). In particular, if oxygen is used in the thermometer, the temperature of the steam point is determined to be 373.35 K when the gas pressure is 0.4 atm, and 373.25 K when the pressure is reduced to 0.2 atm. What is the temperature of the steam point?

54. (II) Carbon dioxide obeys the van der Waals equation of state with $a = 5.96 \times 10^6$ atm·cm^6/mol^2 and $b = 98.6$ cm^3/mol. We know that 0.2 moles of CO_2 occupies 2×10^3 cm^3 at 2.5 atm of pressure. (a) Find the temperature of the gas. (b) Redo your calculation, but this time assume that CO_2 is an ideal gas. What is the percentage difference between the two answers?

55. (II) There is a certain temperature T_B (the *Boyle temperature*) for which a van der Waals gas behaves as if it were ideal. (a) Show that an expression for this temperature is $T_B = (a/bR)[1 - (bn/V)]$. (b) The constants a and b of the van der Waals equation of state are $a = 0.140$ m^6·Pa/mol^2 and $b = 4.00 \times 10^{-5}$ m^3/mol for argon gas. What is the value of the Boyle

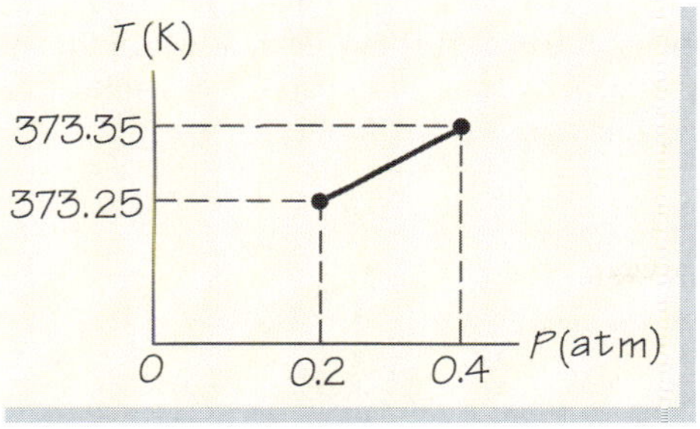

FIGURE 17–23 Problem 53.

temperature of argon gas when the gas is so dilute that the term bn/V in the expression for the Boyle temperature can be ignored?

56. (III) An isotherm of 1 mol of a gas that obeys the van der Waals equation of state and is plotted on a p–V curve falls monotonically for large T. As T is decreased and new isotherms are plotted, the curve begins to show a kink, and at a temperature known as the *critical temperature*, T_c, there is a flat spot in the curve, where the slope is zero (Fig. 17–24). At this point, $p = p_c$ and $V = V_c$. Find T_c, p_c, and V_c in terms of the constants a and b.

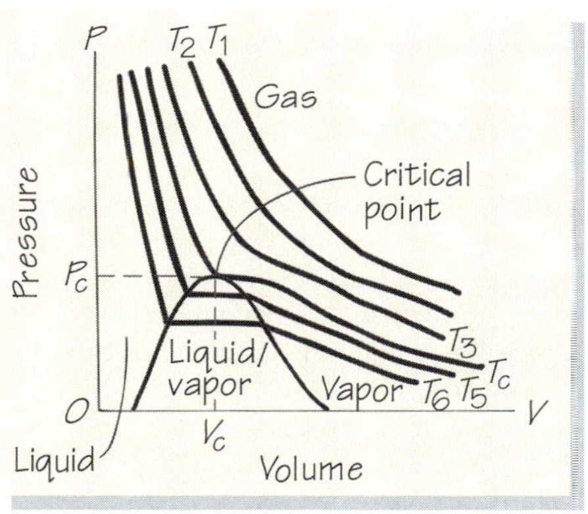

FIGURE 17–24 Problem 56.

*17–4 The Background Temperature of the Universe

57. (I) According to the Stefan–Boltzmann formula, the energy of radiation emitted by a blackbody per unit area per unit time is proportional to T^4, where T is the temperature in the Kelvin scale. Compare the energy emitted by a tungsten filament at 3200°C and the same filament at room temperature (25°C).

58. (I) The intensity of radiation from a small source decreases with the square of the distance from the source. Consider the energy of radiation that reaches your face after being emitted by solid aluminum at room temperature (25°C). At what distance will this energy of radiation be the same as the energy emitted by the same amount of molten aluminum (800°C) that reaches your face when you are 6 m from the aluminum? (Your answer will not agree with your experience. Can you explain why? A look at Fig. 17–11 should help.)

59. (I) The dominant frequency of radiation emitted by an object is related to its temperature by $hf \simeq kT$. Find the dominant radiation frequency emitted by (a) an object in interstellar space at 3 K; (b) a body of water at 280 K; (c) an electric stove heating unit (800 K); (d) melting tantalum (3000 K).

60. (I) The human eye is most sensitive to yellow light with a wavelength of approximately 550 nm. What is the temperature of an incandescent bulb filament that radiates most of its energy as a blackbody at this wavelength?

61. (II) The temperature of your skin is about 35°C. Calculate the wavelength at which the Planck radiation curve has a maximum for this temperature, and therefore the wavelength at which your body radiates the most energy. The correct relation between λ_{max} and T is given in Example 17–4.

62. (II) Show that the integral over all frequencies of the Planck formula given by

$$U(T) = \int_0^\infty \left(\frac{8\pi h f^3}{c^3}\right)\frac{1}{e^{hf/kT} - 1}\,df$$

gives a result that is of the form (a constant) $\times\, T^4$. [*Hint*: Change variables from f to hf/kT.] The energy emitted per unit area per unit time, $P(T)$, is proportional to $U(T)$, and thus $P(T)$ is also proportional to T^4, as in the Stefan–Boltzmann formula, Eq. (17–14).

63. (II) The surface of the Sun is at a temperature of 6000 K. At what rate is energy radiated from the whole surface of the Sun, given that the radius of the Sun is $R = 6.95 \times 10^8$ m?

64. (II) Assume that the radiation emitted from the Sun moves radially outward from the sun, and that no radiation is absorbed between the Sun and Earth. How much energy in the form of radiation will fall per second on an area of 1 m^2 on Earth, if that area is perpendicular to the straight-line path of the radiation? The distance from the sun to Earth is 1.5×10^{11} m.

65. (II) The average surface temperature of Earth is 290 K. How much energy per second is radiated by Earth's surface, assuming that Earth simulates a blackbody. Compare this result with the amount of radiative energy that reaches Earth from the Sun. Use the result of Problem 64.

66. (III) The energy density per unit frequency in the frequency range from f to $f + df$ in blackbody radiation is $u(f, T)\,df$, where $u(f, T)$ is given by Eq. (17–13). An alternative way to express the blackbody radiation is to give $u'(\lambda, T)\,d\lambda$, the energy density per unit wavelength in the wavelength range from λ to $d\lambda$. (a) Use the fundamental wave relation $c = \lambda f$, where c is the speed of light, to show that $u(f, T)\,df = u[(c/\lambda), T]c\,d\lambda/\lambda^2$, so $u' = u[(c/\lambda), T]c/\lambda^2$. (b) Assuming that the temperature is fixed, use the results of part (a) to find an equation for the λ_{max} for which $u'(\lambda, T)$ has a maximum. (c) The result of part (b) is a transcendental equation. Solve it for $\lambda_{max}T$. For what range of temperatures does λ_{max} fall within the visible spectrum, $\lambda \simeq 450$ nm to 650 nm?

67. (III) Find the constant coefficient of T^4 in the Stefan–Boltzmann formula, Eq. (17–14), given that the relation between $P(T)$ and the total energy density $U(T)$ calculated in Problem 62 is $P(T) = cU(T)/4$, where c is the speed of light. [*Hint*: An appropriate entry in a table of integrals is

$$\int_0^\infty [x^3/(e^x - 1)]\,dx = \pi^4/15.]$$

General Problems

68. (II) Some astronomers estimate that toward the center of the galaxy, the average interstellar gas is molecular hydrogen, at a density of roughly 1 molecule per cubic centimeter and a temperature of between 10 and 20 K. Estimate the pressure of this gas. Compare this with atmospheric pressure (1 atm = 760 torr) and very good laboratory vacuums (10^{-10} torr) that are achieved only with some difficulty.

69. (II) Use Archimedes' principle (Section 16–4) to estimate how much helium is needed for a balloon to lift a payload of 230 kg (including the mass of the balloon) in air at 1 atm pressure and 10°C. You will need the molecular weight of helium (4 g/mol) and that of air (28.9 g/mol).

70. (II) A constant-volume gas thermometer is placed in contact with ice water and the pressure in the thermometer is observed to be 0.285×10^5 N/m^2. If the same thermometer is then placed in a pot of boiling water, the pressure rises to a new value of 0.394×10^5 N/m^2. The thermometer is calibrated in °C. What temperature will this thermometer read if it has a pressure reading of 0.315×10^5 N/m^2?

71. (II) A cylindrical vessel with a tight but movable piston is placed in a vertical position, so that the piston, whose area is 70 cm^2, is subject to atmospheric pressure. When the gas in the vessel is heated from a temperature of 20°C to 80°C, a 0.5-kg mass must be placed on top of the piston to hold the piston at the position it occupied at the lower temperature (Fig. 17–25). What is the volume of 0.2 mol of the gas?

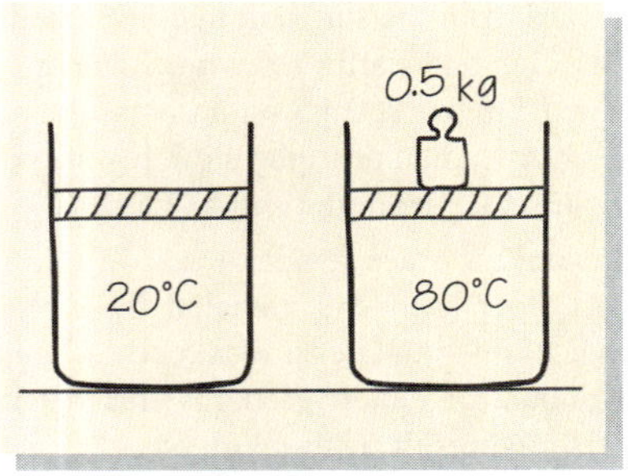

FIGURE 17–25 Problem 71.

72. (II) An air bubble of volume 20 cm^3 rises from the bottom of a lake 30 m deep. The temperature at the bottom is 4°C. The bubble rises to the surface, which is at 20°C and at atmospheric pressure. Assuming that the bubble is in thermal equilibrium with the surroundings at all times, calculate the volume of the bubble at the surface.

73. (II) In Chapter 14, we stated that the speed of sound in air is given by $v_{\text{sound}} = \sqrt{\gamma p/\rho_0}$, where the constant $\gamma \simeq 1.4$ for air near STP (see Problem 26), p is the pressure, and ρ_0 is the density of the air. Reexpress the speed of sound in terms of the temperature. What is the difference in the speed of sound at the coldest and hottest times of day in a desert?

74. (II) A bottle containing air is closed with a watertight yet smoothly moving piston. The bottle with its air has a total mass of 0.50 kg. At the surface of a body of water whose temperature is a uniform 288 K throughout, the volume of air contained in the bottle is 2.0 L (Fig. 17–26). Recall that the pressure of water increases with depth below the surface, D, as $p_{H_2O} = p_0 + \rho g D$, where p_0 is the surface pressure and $\rho = 1.0$ kg/L. The bottle is submerged. (a) What is the volume of the air in the bottle as a function of depth? (b) Calculate the buoyant force on the bottle as a function of depth. (c) At what depth do the buoyant force and the force of gravity cancel? (d) Is the depth at which the force on the bottle is zero, calculated in part (c), a stable or an unstable equilibrium?

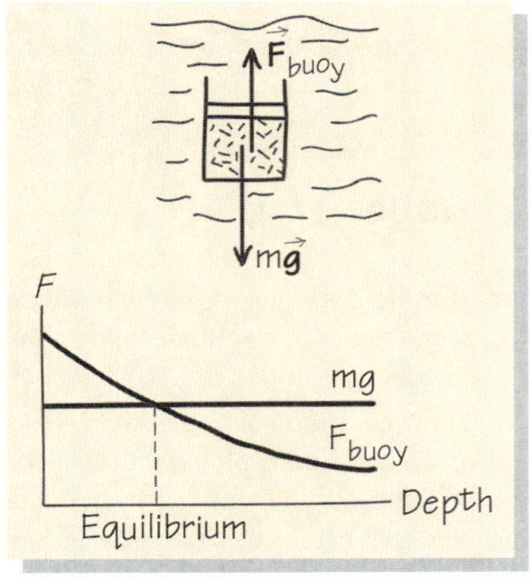

FIGURE 17–26 Problem 74.

75. (II) Suppose that, instead of the Kelvin scale, an alternative temperature scale—call it the Kelvin′ scale, with the units K′—were defined in which the zero level remains unaffected but the temperature at the triple point of water is taken to be 500 K′. Calculate the value of Boltzmann's constant in units of J/K′.

76. (II) Consider the radiation emitted from a furnace through a hole of area 1 cm^2. Assume that all the radiation is absorbed by a beaker that contains 50 g of water. Given that it takes 4.2 J of energy to raise the temperature of 1 g of water by 1°C, how long will it take for the water to be heated by 20°C if the furnace temperature is 2000 K? Assume that there are no energy losses due to the beaker.

77. (III) The van der Waals equation of state, Eq. (17–12), can be put into a form that is an expansion in powers of the density. Dividing by RT and then by $[(V/n) - b]$ in Eq. (17–12), we can write

$$p + \frac{a(n/V)^2}{RT} = \left(\frac{V}{n} - b\right)^{-1} = \frac{n}{V}\left(1 - b\,\frac{n}{V}\right)^{-1}.$$

(a) Show that, for $bn/V < 1$, this equation has the approximate form

$$\frac{p}{RT} = \frac{n}{V} + b\left(\frac{n}{V}\right)^2 - \frac{a}{RT}\left(\frac{n}{V}\right)^2.$$

(b) The result in part (a) is of the general form

$$\frac{p}{RT} = \frac{n}{V} + B_2(T)\left(\frac{n}{V}\right)^2,$$

where, in this case, the function $B_2(T)$ is $B_2(T) = b - (a/RT)$. This is an example of a *virial expansion* for the equation of state, and the function $B_2(T)$ is known as a *virial coefficient.* Show that $B_2(T)$ starts out negative as T increases from some small value, passes through zero, and then becomes positive. The value of T at which $B_2(T)$ is zero is the Boyle temperature, where a real gas satisfying the van der Waals equation behaves as though it were ideal (see Problem 55).

CHAPTER 18

This large plant near Geyserville in northern California uses thermal energy from within Earth to do work and generate electric power. While such a plant does not require a separate fuel source that must be replenished, any electric energy that it has produced has depleted at least that much thermal energy from the underground thermal source.

Heat Flow and the First Law of Thermodynamics

Hot steam contains thermal energy. Steam can lift a piston or turn a turbine and can thereby transfer energy to a mechanical system. Here, we will see how this form of energy is related to the forms of energy with which we are familiar, and how it can be used to do mechanical work. We will see that, as the temperature of a thermal system increases, so does the amount of thermal energy in the system. The energy of thermal systems (thermal energy) and its relation to the energy of the systems' surroundings is summarized in two important laws of physics. The first law of thermodynamics, which we examine here, describes thermal energy as just another form of energy—like kinetic energy or potential energy. The principle of the conservation of energy must now include the thermal energy of a system. The second law of thermodynamics, to be discussed in Chapter 20, states limitations on how much of a system's thermal energy can be converted to other forms of energy.

18–1 CHANGES IN THERMAL SYSTEMS

Thermal systems in equilibrium, such as a bottle of hot steam, are described by only a few thermodynamic variables. For a gas, these variables are the temperature, T, the pressure, p, the volume, V, and the number of moles, n (or the number of molecules, N). Thermodynamic variables describe the *state* of the thermal system. As discussed in Chapter 17, an equation of state relates these variables. For example, the equation of state for an ideal gas [Eqs. (17–7) and (17–9)] is $pV = nRT = NkT$, where R is the universal gas constant and k is Boltzmann's constant. Let's assume that n is fixed. Because the remaining three variables are connected by an equation of state, it is enough to describe the state on a two-dimensional plot, such as a p–V diagram, a p–T diagram, or a V–T diagram; in fact, this is what we usually do. Changes, or *transformations*, in the state of a thermal system may be represented as curves on such diagrams—at least if thermal equilibrium is established at every stage of the transformation. We have already described and used such curves in Chapter 17 (see Example 17–4).

PROBLEM SOLVING

Use diagrams to describe transformations.

To maintain thermal equilibrium at each step of a transformation, changes must be made slowly. Just how slowly depends on the internal dynamics that lead to equilibrium. As a piston compresses a gas, the molecules near the piston receive a push, are speeded up (thereby gaining energy), and distribute their additional energy to other molecules through collisions. The time required for the effect of the push to be evenly spread through the gas is characterized by the size of the container divided by the *speed of sound* in the gas. This time is typically short, so it is not difficult to compress a gas while maintaining thermal equilibrium at each stage.

Reversible processes

Irreversible processes

Reversible and Irreversible Processes

A change in which thermal equilibrium is maintained throughout, such as the slow compression just described, is called a **reversible process**. Such changes can always be reversed—the thermodynamic variables can return to their original values—by reversing the external conditions. The volume of a cylinder of gas held at a fixed temperature is increased from V_1 to V_2 in a controlled way when a piston is slowly pulled out of the cylinder, as in Fig. 18–1a. If the piston is slowly pushed back in, as in Fig. 18–1b, the external conditions are reversed, and the thermodynamic variables of the gas recover their initial values. These changes are reversible, and Fig. 18–1c shows the transformation as a curve on a p–V diagram. Because the thermal system remains in equilibrium throughout a reversible process, reversible processes always connect equilibrium states.

Conversely, a gas expanding to fill a vacuum through an open stopcock (Fig. 18–2a) provides an example of an irreversible process, a process in which the thermal system's changes cannot be retraced. The expansion from V_1 to V_2 occurs rapidly and without control. Nothing you do with the stopcock will cause the released gas to reverse itself and reconcentrate within the original volume. We can mark only the original and final points on a p–V diagram (Fig. 18–2b); we cannot mark the points in between, however, because the system is not in thermodynamic equilibrium during the expansion. It is useful to mark a curve with dots, as in Fig. 18–2b, but the dotted curve is meant only to imply that there has been a change, *not* to specify the state of the system during the change. The particular irreversible process described here, in which the gas expands freely, is known as **free expansion**.

Unless we state otherwise, we will consider only reversible changes of state. Thermodynamics is based on our ability to treat reversible processes. There are situations (such as free expansion) in which equilibrium thermodynamics does not apply. The physics of *nonequilibrium thermodynamics* describes processes such as rapid chemical reactions, but that subject is still developing.

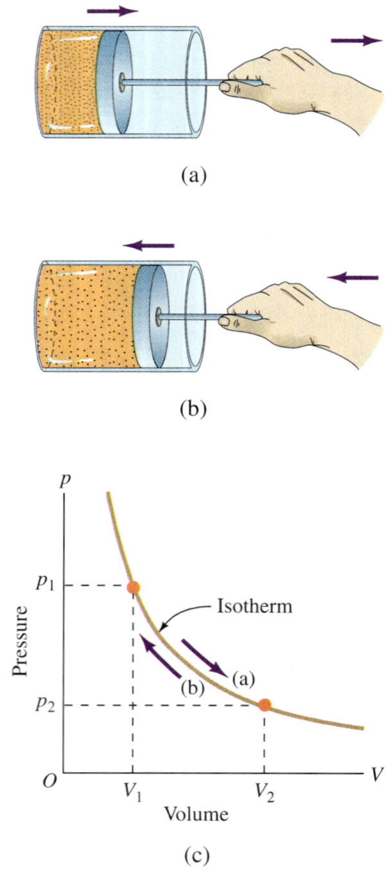

FIGURE 18–1 The volume of a gas in a cylinder is reversibly changed when a piston is (a) slowly pulled or (b) slowly pushed. The temperature is held fixed during the transformations. (c) Constant-temperature reversible transformation of a gas between V_1 and V_2 on a p–V diagram. The curve is an *isotherm*.

EXAMPLE 18–1 A gas in contact with a thermal reservoir at temperature T is expanded reversibly from V_1 to V_2. The trace of points made by this change on a p–V diagram is an isotherm. Describe the isothermal curve when an ideal gas is used.

Solution: The ideal gas law, $pV = nRT$, shows that the product of p times V is constant for a fixed temperature, T. The curve that corresponds to $p \times V =$ a constant or, equivalently, p proportional to V^{-1} is a hyperbola. Figure 18–3 shows a family of hyperbolas. According to the ideal gas law, the different curves correspond to larger or smaller values of T.

We suggested in Section 17–3 that the temperature of a system is closely related to its **internal energy**, or **thermal energy**. We will learn here that the internal energy of a thermal system is a function of the thermodynamic variables; in particular, the higher the temperature of a system, the higher the internal energy. Changes in the internal energy of a thermal system can manifest themselves in only a few ways: *The system can transfer its internal (thermal) energy by changing the temperature (or phase) of another system, or it can use its internal energy to do mechanical work on its surroundings, or both.*

Up to the middle of the nineteenth century, it was thought that when the temperature of a system changes, a type of conserved fluid—known as *caloric*, or more simply, heat—flows into or out of the system. In other words, as much caloric was thought to exit (or enter) the surroundings of the system as entered (or exited) the system itself. Today we know that this is wrong.

When the temperature of a thermal system changes and, in the process, the temperature of a neighboring system changes (Fig. 18–4), we say that there has been a **heat flow** into or out of the system.† We will use the term *heat flow* rather than *heat* because heat flow is more like work than like energy: Work is done *on* or *by* a mechanical system, but we do not speak of such a system as containing a certain amount of work. Heat flow into, and/or mechanical work done on, a system may each contribute to raising the temperature of a thermal system. We employ the symbol Q for heat flow.

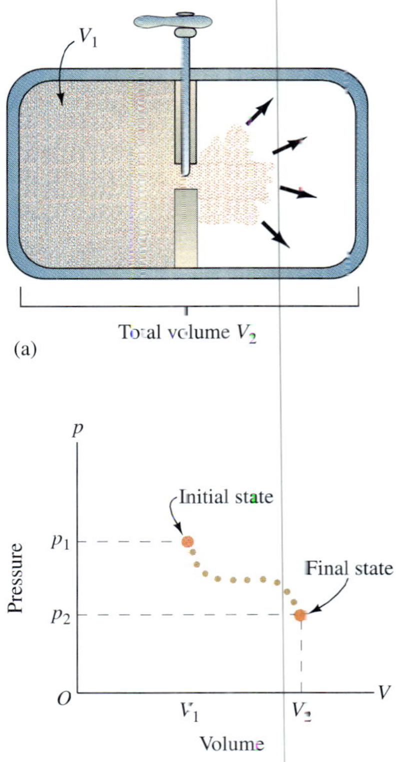

FIGURE 18–2 (a) A gas initially confined to the left-hand side of a container expands to fill the container's total volume when a stopcock is opened. (b) Irreversible expansion of a gas from V_1 to V_2. The path is not well defined, and the dots serve only to indicate that there has been some irreversible change

The unit of heat flow is the calorie (cal). Provisionally, 1 cal is defined to be the amount of heat flow required to raise the temperature of 1 g of water at atmospheric pressure from 14.5°C to 15.5°C. We must wait until later to relate this quantity to the SI units with which we are familiar. Another unit for heat flow used in industrial applications is the *British thermal unit* (Btu), which is defined as the heat flow required to raise the temperature of a 1-lb weight of water by 1°F, averaged over the temperature range 32°F to 212°F. The British thermal unit is related to the calorie by 1 Btu = 252.02 cal.

A provisional definition of the calorie

Heat Capacity

The definition of heat capacity

The quantitative connection between heat flow and temperature change is the **heat capacity**, C. An infinitesimal heat flow, dQ, *into* a thermal system will change the temperature by an infinitesimal amount, dT, with a proportionality constant C:

$$dQ = C\,dT. \qquad (18\text{–}1)$$

In this equation, dQ can be positive or negative. The quantity C is always positive; thus, the sign of the temperature change indicates the sign of the heat flow. Equation (18–1) states that there is a positive heat flow into a system when its temperature increases and there is a negative heat flow into the system when its temperature decreases. The heat capacity is characteristic of the type of material and is proportional to the amount of material; it may also depend on the temperature. A substance with a large heat capacity undergoes a small temperature change for a given heat flow, or requires a large heat flow for a given temperature change. Water has a relatively large heat capacity, which is why large bodies of water have moderating effects on the climate in their vicinity.

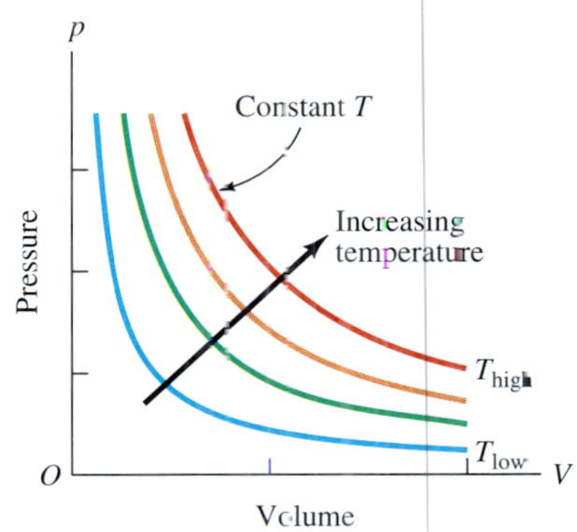

FIGURE 18–3 Example 18–1. An isotherm for an ideal gas is a hyperbola, $p \propto TV^{-1}$. The isotherms corresponding to different temperatures are indicated.

We can eliminate the dependence of heat capacity on the amount of material by defining the **specific heat**, c, as the heat capacity of 1 g of the material—the heat capacity per unit mass of the material—and the **molar heat capacity**, c', as the heat ca-

†We will see later that we can also have a heat flow without a temperature change when the system does work or undergoes a phase change.

FIGURE 18–4 In this steel manufacturing scene, thermal energy is being supplied to steel rods.

pacity of 1 mol of the material. If the mass in grams of a thermal system consisting of n mol is m, then

$$C = mc = nc', \tag{18–2}$$

and if A is the molecular weight (the mass of 1 mol, in grams), then

$$c' = Ac. \tag{18–3}$$

The units of heat capacity are cal/K; those of specific heat are cal/g · K; and of molar specific heat, cal/mol · K.

Path Dependence of Heat Flow

PROBLEM SOLVING

Heat flow depends on the path of temperature change.

The heat capacity is defined in Eq. (18–1) in terms of an infinitesimal temperature change. But temperature can change in different ways: The volume or the pressure—or neither—can be held constant during the temperature change. The heat capacity generally depends on how the temperature change is made and on the values of the thermodynamic variables p, V, and T near their respective values where that (small) change is made. This dependence is indicated by writing $C = C(p, T)$ or $C = C(V, T)$. We cannot simply integrate Eq. (18–1) to find a net heat flow unless we know more about how either pressure or volume changes as temperature changes. Generally, the changes will describe a path on a p–V diagram, and heat flow *depends on the path* along which temperature changes. This result recalls the calculation of the work done by a force in moving a mass from a point A in space to a point B. As we learned in Chapter 6, the work done may or may not be path dependent, depending on the force.

In the case of heat flow, there are two cases of special interest: Either the volume or the pressure is kept constant. For these cases, we label the heat capacity with the subscript V or p, and Eq. (18–1) becomes

Heat capacities at constant volume or constant pressure

$$\text{for constant volume:} \quad dQ = C_V\, dT; \tag{18–4}$$

$$\text{for constant pressure:} \quad dQ = C_p\, dT. \tag{18–5}$$

(Note that the subscript must be used on c and c' as well. The volume change with temperature for liquids is insignificant compared with that of gases; for solids, pressure is not a useful variable. Therefore, the subscript is not used for these cases.) Generally, C_V differs from C_p. We will see an example when we derive C_V and C_p for an ideal gas.

For an ideal gas, the heat capacities turn out to be independent of temperature. The integration of Eqs. (18–4) and (18–5) to find the total heat flow for finite temperature changes then becomes simple. We illustrate this calculation in Example 18–2.

EXAMPLE 18–2 An ideal gas undergoes transformations that take it from point A to point B on a p–V diagram by the two different paths, α and β, shown in Fig. 18–5. The figure depicts the various thermodynamic variables. Points A and B have the same temperature, $T = T_0$ (they lie on an isotherm). Assuming that C_V and C_p are constants, what is the heat flow into the gas (a) for path α (through point 1) and (b) for path β (through point 2)?

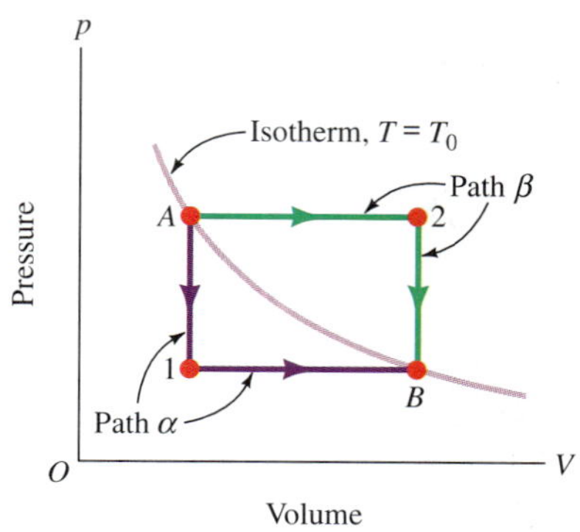

FIGURE 18–5 Example 18–2.

Solution: Because $T_A = T_B = T_0$ (there is no net temperature change of the gas), a tempting answer to this question is that there is no heat flow at all. However, the change $A \rightarrow B$ is not infinitesimal, and so we must integrate along the respective paths to find the correct answer.

(a) Along path α, there are two contributions: from point A to point 1, and from point 1 to point B. Point 1 is at temperature T_1:

$$\Delta Q_\alpha = \int_{T_0}^{T_1} C_V\, dT + \int_{T_1}^{T_0} C_p\, dT.$$

The quantity C_V appears in the first term because the first leg of path α is at constant volume, whereas C_p appears in the second term because the second leg is at constant pressure. C_V and C_p are constants, so they come out of the integrals and $\Delta Q_\alpha = (T_0 - T_1)(C_p - C_V)$.

(b) For path β, we pass in two steps through point 2 at temperature T_2:

$$\Delta Q_\beta = \int_{T_0}^{T_2} C_p \, dT + \int_{T_2}^{T_0} C_V \, dT = C_p \int_{T_0}^{T_2} dT + C_V \int_{T_2}^{T_0} dT$$

$$= C_p(T_2 - T_0) + C_V(T_0 - T_2) = (T_2 - T_0)(C_p - C_V).$$

You can see in Fig. 18–5 that, for an ideal gas, the factor $(T_0 - T_1)$ is not equal to the factor $(T_2 - T_0)$. Thus the two answers are different, and the heat flow depends on the path that is followed.

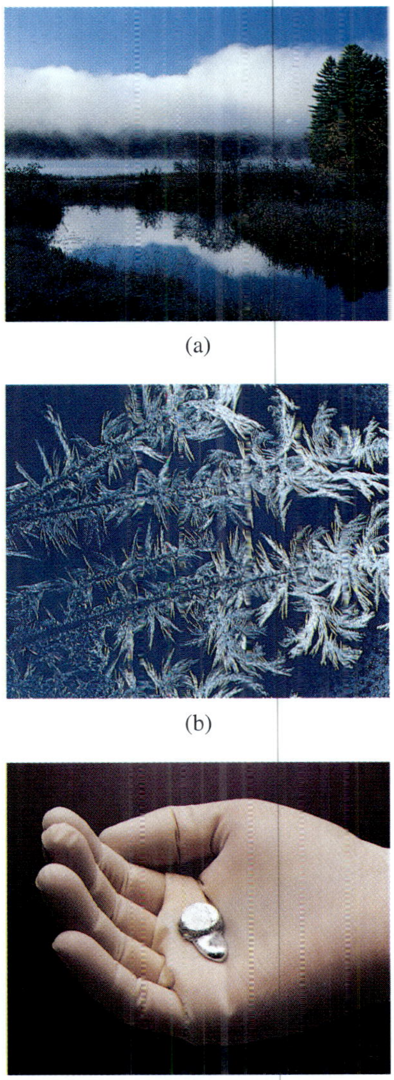

(a)

(b)

(c)

FIGURE 18–6 (a) Mist forms when water droplets condense from water vapor. (b) Ice crystals form on the window when liquid water gives up heat flow to the exterior environment. (c) Gallium melts at the temperature of the human body. When it does so, it absorbs heat flow from its surrounding environment.

Phase Changes and Heat Flow

When there is a **change of phase** in a system, there is no temperature change even though there is a heat flow. For example, a solid state may change to a liquid state (fusion or melting), or a gaseous state may change to a liquid state (condensation) (Fig. 18–6). It takes a certain heat flow to convert 1 g of ice at its melting temperature of 0°C to water at 0°C (Fig. 18–7). The necessary heat flow is called the **latent heat of fusion**, L_f. Measurements indicate that $L_f = 79.6$ cal/g for ice. During the melting process, there is no change of temperature and, to a good approximation, there is no change of pressure or volume. The heat flow goes into changing the molecular structure of the substance. It also takes a certain heat flow to convert 1 g of water at its vaporization temperature, 100°C (at 1 atm pressure), to 1 g of steam. The heat flow necessary for this is called the **latent heat of vaporization**, L_v, whose value for water in particular is 540 cal/g. Most of this heat flow goes into overcoming the intermolecular forces that make a liquid so much denser than a gas. Some latent heats are given in Table 18–1.

EXAMPLE 18–3 One hundred grams of ice at 0°C is dropped into 200 g of water at 49°C. The system is thermally isolated (see Chapter 17). After a period of time, the ice is entirely melted, leaving 300 g of water at 6°C. Assume that the specific heat of water has the constant value $c = 1$ cal/g · °C. Calculate the latent heat of fusion of water.

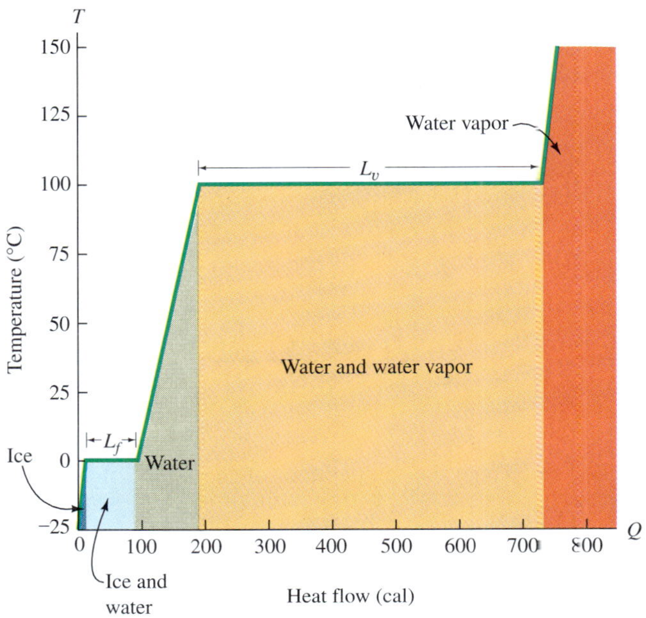

FIGURE 18–7 Phase changes for 1 g of water. Note that the temperature remains constant as heat flow is added during melting and vaporization.

TABLE 18–1 Some Latent Heats

Material	*Melting Temperature (K)*	*Latent Heat of Fusion (kJ/mol)*
H_2	14	0.12
O_2	54	0.44
Water (H_2O)	273	6.0
Ethyl alcohol (CH_3CH_2OH)	159	4.8
Mercury	234	2.3
Lead	600	4.8
Copper	1357	13
Material	*Boiling Temperature*† *(K)*	*Latent Heat of Vaporization (kJ/mol)*
H_2	20	0.92
O_2	90	6.8
Water (H_2O)	373	41
Ethyl alcohol (CH_3CH_2OH)	351	39
Mercury	630	59
Lead	2023	178
Copper	2839	300

†Boiling takes place at atmospheric pressure.

Solution: Because this system is thermally isolated, the temperature cannot change as a result of thermal contact with the surroundings.† Any heat flow must occur within the system between the water and the melting ice, and the net heat flow to the outside must be zero. We can summarize this condition as

$$\Delta Q_1 + \Delta Q_2 + \Delta Q_3 = 0,$$

where ΔQ_1 is the heat flow needed to melt $m_1 = 100$ g of ice, ΔQ_2 is the heat flow necessary to raise the temperature of 100 g of water (from the melted ice) from $T_i = 0°\text{C}$ to $T_f = 6°\text{C}$, and ΔQ_3 is the heat flow necessary to cool $m_2 = 200$ g of water initially at $T'_i = 49°\text{C}$ to T_f. By definition, L_f is the heat flow necessary to melt 1 g, so if m_1 is measured in grams, we find

$$\Delta Q_1 = m_1 L_f,$$

$$\Delta Q_2 = cm_1(T_f - T_i),$$

$$\Delta Q_3 = cm_2(T_f - T'_i).$$

Thus

$$m_1 L_f + cm_1(T_f - T_i) + cm_2(T_f - T'_i) = 0.$$

Solving for L_f, we find that

$$L_f = \frac{-cm_1(T_f - T_i) - cm_2(T_f - T'_i)}{m_1}$$

$$= \frac{-(1\ \text{cal/g}\cdot°\text{C})(100\ \text{g})(6°\text{C} - 0°\text{C}) - (1\ \text{cal/g}\cdot°\text{C})(200\ \text{g})(6°\text{C} - 49°\text{C})}{100\ \text{g}}$$

$$= 80\ \text{cal/g}.$$

Note that the heat flow needed to change the temperature of the 200 g of water is negative; the water is cooled.

†We will see in Section 18–4 that the system must also do no work on its surroundings.

Calorimetry

The measurement of heat capacities is called **calorimetry**. Calorimetry is based on the fact that *when a system is mechanically and thermally isolated, heat flow is conserved within the system* (Fig. 18–8). (A system is said to be *mechanically isolated* if it cannot do mechanical work on its surroundings.) A heat flow into one part of such a system must be matched with a corresponding heat flow out of another part. In such circumstances, heat flow is present only as long as there is a temperature difference between parts of the system; it ceases when equilibrium is established throughout the system. The fact that the pattern of heat flow represents an irreversible process is not important: We are comparing initial and final equilibrium states, not finding infinitesimal changes along paths.

There is no heat flow to or from a mechanically and thermally isolated system.

Our definition of the calorie shows that the specific heat for water at 15°C and at 1 atm of pressure is

$$C_{H_2O} = 1 \text{ cal/(g} \cdot {}^\circ\text{C)}.$$

The measurement of the specific heats of other substances is based on that of water, as shown in Example 18–4.

FIGURE 18–8 Thermally isolated systems were described in Chapter 17. This type of container is used to keep water cold.

EXAMPLE 18–4 A block of iron (mass 1.0×10^2 g) is heated in an oven to 5.0×10^2 K and then plunged into a closed, thermally insulated container of 0.50 kg of water at 292 K. The block and the water come to an equilibrium temperature of 297 K. What is the specific heat of iron? Assume that the specific heats of water and iron do not vary significantly over the temperature ranges in question.

Solution: There is a positive heat flow into the water, ΔQ_{H_2O}, because it becomes hotter. This must be matched by a negative heat flow into the iron block, ΔQ_{Fe}, as follows:

$$\Delta Q_{\text{system}} = 0 = \Delta Q_{H_2O} + \Delta Q_{Fe}.$$

The finite heat flow, ΔQ, is found from the infinitesimal heat flow, dQ, by summing, or integrating, over the infinitesimal temperature changes. The heat capacities are assumed to be constant, so this integration is simple:

$$\Delta Q = \int_{T_i}^{T_f} mc \, dT = mc \int_{T_i}^{T_f} dT = mc(T_f - T_i),$$

where m is the mass of material involved.

We have

$$\begin{aligned} 0 &= \Delta Q_{H_2O} + \Delta Q_{Fe} \\ &= (0.50 \times 10^3 \text{ g})(1 \text{ cal/g} \cdot \text{K})(297 \text{ K} - 292 \text{ K}) \\ &\quad + (1.0 \times 10^2 \text{ g})c_{Fe}[297 \text{ K} - (5.0 \times 10^2 \text{ K})] \\ &= 2.5 \times 10^3 \text{ cal} + (-2.03 \times 10^4 \text{ g} \cdot \text{K})c_{Fe}. \end{aligned}$$

We solve for c_{Fe} to find that

$$c_{Fe} = \frac{2.5 \times 10^3 \text{ cal}}{2.03 \times 10^4 \text{ g} \cdot \text{K}} = 0.12 \text{ cal/g} \cdot \text{K}.$$

Example 18–4 shows how the specific heat of iron (or of many materials) can be measured. Calorimetry is used most often in the opposite sense: Given the specific heats, temperature changes are calculated in situations where there is a heat flow from one system to another. Table 18–2 gives the molar heat capacities and specific heats *at constant volume* for a variety of materials at room temperature. In this table, the gases are at standard pressure. In Chapter 19, we will see that the values in Table 18–2 can be explained in terms of the microscopic constituents and structure of the material.

PROBLEM SOLVING

Uses of calorimetry

TABLE 18–2 Molar Heat Capacities and Specific Heats of Some Materials at Constant Volume[†]

Material	*Molar Heat Capacity (cal/mol · K)*	*Specific Heat (cal/g · K)*
Helium gas	3.00	0.75
Argon gas	30.00	0.75
N_2 gas	4.94	0.176
O_2 gas	4.97	0.155
SO_2 gas	7.52	0.118
Diamond	1.46	0.12
Beryllium	3.85	0.43
Silicon	4.78	0.17
Aluminum	5.82	0.22
Copper	5.85	0.092
Lead	6.32	0.031
Water	18.0	1.00
Mercury	6.7	0.033

[†]Gases at standard temperature and pressure.

C_V and C_p for ideal gases can be calculated when the first law of thermodynamics is known. A general prediction (to be shown in Section 18–6) is that the heat capacity for constant pressure is larger than the heat capacity for constant volume:

$$C_p \geq C_V. \tag{18–6}$$

This inequality holds even for nonideal gases.

18–3 THE MECHANICAL EQUIVALENT OF HEAT

Placing a thermal system in thermal contact with a second thermal system at a different temperature is one way to produce a temperature change in the first system. There is a heat flow in this case. For example, we can heat water by placing it in a pot on a hot stove. A major discovery of the first half of the nineteenth century was the observation that the temperature of a thermal system can also be raised by performing mechanical work on that system. For example, when an object is rubbed so that work is done on it by friction, its temperature is raised (Fig. 18–9).

FIGURE 18–9 The presence of friction allows the mechanical energy of rubbed sticks to be converted into heat energy. Fire is the result.

These observations and the understanding of their significance took place through the slow buildup of rather complicated data rather than through a single brilliant experiment. The history is thus rather tangled, with many participants. We cite in particular the work of Count Rumford (Benjamin Thompson) and of James Joule. While supervising the boring of cannon barrels in Bavaria in 1798, Rumford observed and quantified the continual heating of the metal (Fig. 18–10). He reasoned that if the heating were due to the transfer of a fluid—as proposed by the caloric theory—then the fluid eventually would be used up and no more heating should occur. In the mid-nineteenth century, Joule's careful experiments established that a given amount of mechanical work done on a thermal system produces a temperature rise that corresponds to a given heat flow, as defined by Eq. (18–1).

Joule's Experiments

An idealized version of Joule's experiments is described here. In these experiments, the temperature of a thermally isolated quantity of water is measurably raised by a set of paddles that churn the water (Fig. 18–11). The water resists the motion of the turning paddles, and work must be performed to keep them rotating. The amount of work done can be measured precisely by attaching the paddles to a falling mass (which keeps the paddles moving), and then measuring the distance the mass has fallen.

When the paddles have done a certain amount of work on the water, the temperature of the system (water plus paddles) will have risen by a definite amount. Once the temperature has been raised, *there is no way to tell* whether the temperature increase occurred because work had been done on the system or because of heat flow into it. A series of experiments of this type establishes that there is a specific temperature rise for a specific amount of work; therefore, there is a precise equivalent between work done on a system, W, and the heat flow into it, ΔQ. Mathematically,

$$\text{for a thermally isolated system:} \quad \Delta Q = W. \tag{18–7}$$

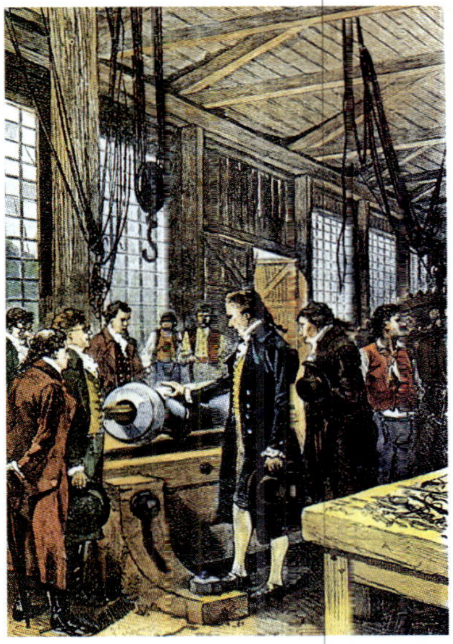

FIGURE 18–10 Count Rumford observed that the temperature of cannon barrels rose steadily as they were being hollowed out. He concluded that the caloric theory could not be correct.

Because *no* experiment can be performed that can tell whether a given temperature change was produced by mechanical work or by heat flow, we can assert that *heat flow, like work, is an energy transfer.* This result is known as the **mechanical equivalent of heat**. Equation (18–7) can be translated into a relation between the SI unit of work, the joule, and the unit of heat flow, the calorie:

$$1 \text{ cal} = 4.185 \text{ J}. \tag{18–8}$$

This relation may seem to be a simple equivalence of units; however, it is far more. Until now, the calorie was unrelated to any other units familiar to us. Now we can state that the calorie has the dimensions of energy, and that the SI unit of heat flow is the joule. Equation (18–8) thus states the crucial result of a basic experiment. It should not be confused with, say, the statement that 1 Btu is 252.02 cal, which is merely the relation between different units.

The mechanical equivalent of heat

Today, the *numerical definition* of the calorie is taken to be Eq. (18–8); it is no longer based on a measurement of the temperature change of water, as it was in Section 18–2. Such a definition is possible only because of the physical equivalence between heat flow and work. The calorie continues to be a widely used unit.

The definition of the calorie

EXAMPLE 18–5 The paddles of the apparatus shown in Fig. 18–11 are driven by a mass m of 0.50 kg falling at a constant speed. The mass of the thermally isolated water in the apparatus is 250 g. How much is the temperature of the water raised if the driving mass falls a distance of 2.0 m? Ignore the mass of the paddles.

Solution: According to the work–energy theorem, the work, W, done on the water is the potential energy change of mass m when it falls a distance Δy. Because this potential energy change is given by $mg\,\Delta y$, the work done is

$$W = mg\,\Delta y.$$

According to Eq. (18–7), this work is equivalent to a heat flow, $\Delta Q = W$. The specific heat of water has the constant value $c_{H_2O} = 1\ \text{cal/g}\cdot{}^\circ\text{C}$, so we get $m_{H_2O}c_{H_2O}\,\Delta T = mg\,\Delta y$, or

$$\Delta T = \frac{mg\,\Delta y}{m_{H_2O}c_{H_2O}} = \frac{(0.50\text{ kg})(9.8\text{ m/s}^2)(2.0\text{ m})}{(250\text{ g})(1\text{ cal/g}\cdot{}^\circ\text{C})}\left(4.185\,\text{cal/J}\right)$$
$$= 0.0094^\circ\text{C}.$$

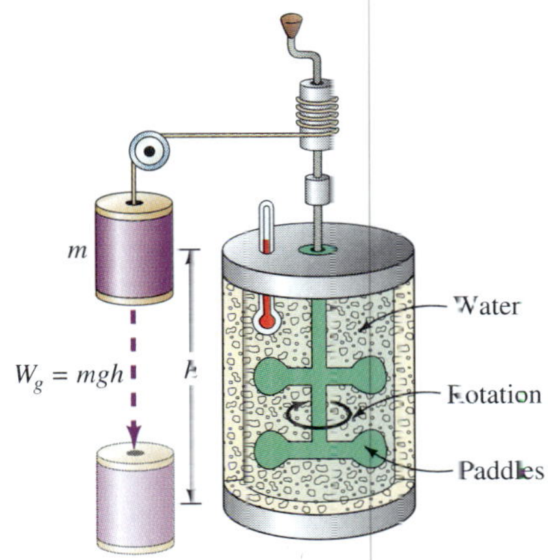

FIGURE 18–11 Schematic diagram of the apparatus used in Joule's classic experiments. Churning paddles raise the temperature of a thermally isolated container of water.

A Microscopic View

Thermal (or internal) energy is associated with the chaotic motion of the molecular constituents of matter. This motion is more and more violent as thermal energy increases, and temperature increases with thermal energy. Rapidly moving molecules transfer their energy to more slowly moving molecules by collision, so it is not too surprising that heat flow is an energy transfer. By the same reasoning, if thermal energy is the energy of the constituent molecules of a thermal system, then this energy is a function of the state of the gas; that is, of its thermodynamic variables, p, V, and T. This point of view will be described quantitatively in Chapter 19. From the thermodynamic (macroscopic) point of view, however, the observation that heat flow is a kind of energy transfer—like work—is not at all obvious. The principle of conservation of energy is one of the key principles of physics, and the discovery of the mechanical equivalent of heat was one of the truly important discoveries of the nineteenth century.

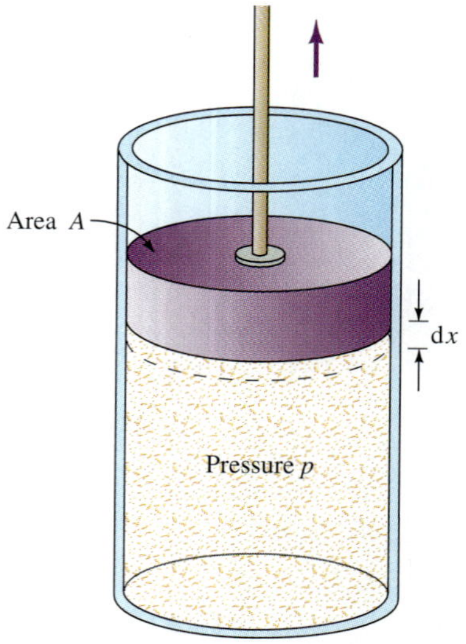

FIGURE 18–12 The volume of a cylinder increases when a piston of area A moves out a distance dx under the influence of the pressure, p, of the gas contained in the cylinder.

18–4 WORK DONE BY THERMAL SYSTEMS

In Section 18–3, we saw that work can be done on a thermal system. By Newton's third law, this means that the thermal system can perform work on its surroundings—negative work, in the case of the churning paddles of Joule's experiments. Thermal systems, particularly gases, which can expand greatly, can also do *positive* work on their surroundings.

Consider a piston of area A pushed out by the pressure p of the gas contained in a cylinder (Fig. 18–12). By the definition of pressure, the magnitude of the force F on the piston is pA. Moreover, if the piston moves out, this force acts in the direction of the displacement, which we take to have infinitesimal magnitude dx. The infinitesimal work done by the gas in moving the piston is

$$dW = \mathbf{F} \cdot \mathbf{dx} = pA\, dx. \tag{18–9}$$

Because $A\,dx = dV$ (the infinitesimal change in the volume of the gas), this expression can be rewritten as

The work done by a gas when its volume changes

$$dW = p\, dV. \tag{18–10}$$

Equation (18–9) is specific to the linear motion of a piston with the geometry of Fig. 18–12, but Eq. (18–10) is a more general form that holds when a gas changes its volume by dV in pushing against its surroundings in any direction. Note the sign carefully: dW, the work the gas does *on* its surroundings, will be positive or negative according to whether dV is positive or negative.

The work done by a changing thermal system depends on how the change is made.

The pressure, p, in Eq. (18–10) is a function of V and T. As for the heat flow, we must specify the way in which the finite volume change is made in the integration of Eq. (18–10). Thus, *the net work done by the gas depends on the path in a p–V, p–T, or V–T diagram.* The integral expressing the work done in going from state 1 to state 2 is

PROBLEM SOLVING

Work is an area beneath a curve of p versus V.

$$W = \int_{V_1}^{V_2} p(V, T)\, dV. \tag{18–11}$$

This integral is defined only when the path $1 \rightarrow 2$ is specified.

If the path of the transformation $1 \rightarrow 2$ is drawn on a p–V diagram, as in Fig. 18–13a, then the work done by the gas has a simple interpretation: *It is the area under the curve of p versus V*, which is shaded in the figure. The sign of the work done by the gas is determined by whether V_2 is larger than V_1, as in Fig. 18–13a, or smaller than V_1, as in Fig. 18–13b. In Fig. 18–13b, the net work done *by* the gas is negative; in other words, there is positive work done *on* the gas to compress it. All this follows the normal rules of calculus, where $\int_{x_1}^{x_2} F(x)\, dx$ is the area under a curve of $F(x)$ versus x.

If we think of the work done by a thermal system as the area under a curve of p versus V, the fact that the work done depends on the path of the change becomes evident. The path describes the curve between two states, 1 and 2, and the area under different curves may certainly differ.

Cyclic Transformations

The operation of the cylinders in an automobile engine is repetitive. This is an example of a **cyclic transformation**: a change in which a thermal system follows a path that takes it back to its original state. The importance of such processes rests on the fact that all engines work in cycles. The net work done by the thermal system in a cyclic transformation can be calculated by starting anywhere along the path (Fig. 18–14a). Let's divide the cycle of Fig. 18–14a into two parts, as in Fig. 18–14b and c. The first part is the change $B \to A$, along path 1 (Fig. 18–14b), and the second part is $A \to B$, along path 2 (Fig. 18–14c). The net work W_{cycle} adds algebraically (Fig. 18–14d) and is thus given by

$$W_{\text{cycle}} = \underset{\text{path 1}}{W_{B\to A}} + \underset{\text{path 2}}{W_{A\to B}}.$$

As we saw in Chapter 6 during our discussion of work, $\int_{x1}^{x2} f(x)\,dx = -\int_{x2}^{x1} f(x)\,dx$. Thus the integral over path 2, which goes from larger to smaller values of x, is the negative of the area under that path. The net work is then the area under path 1 minus the area under path 2:

$$W_{\text{cycle}} = \text{area enclosed by cyclic path.} \tag{18–12}$$

Because the cycle is traced in a clockwise direction, the work done by the gas along path 1 is larger than the work done on the gas along path 2. The net work done on the surroundings is positive; if this cycle were traced in a counterclockwise direction, the work done would be negative.

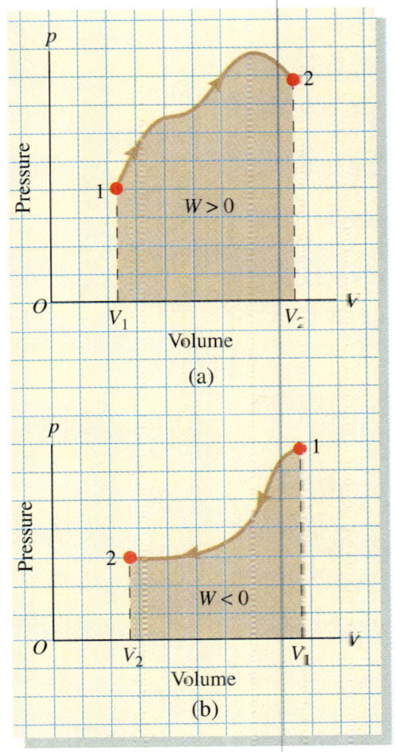

FIGURE 18–13 The work done by a gas on its surroundings (like that of the piston–cylinder system of Fig. 18–12) in moving reversibly from point 1 to point 2 is the area under the curve of p versus V. (a) If the volume V_2 is greater than V_1 (the gas expands), the work the gas does on its surroundings is positive. (b) If the volume V_1 is greater than V_2, the gas is compressed, and the work it does on its surroundings is negative.

The work done by a gas during a cyclic transformation is the area enclosed by the curve that describes the changes on a p–V diagram.

EXAMPLE 18–6 A thermal system is taken around the cycle shown in Fig. 18–15. Along the paths $A \to B$ and $C \to D$, the volume is held fixed at V_1 and V_2, respectively. From $B \to C$ and from $D \to A$, the pressure is held fixed at p_2 and p_1, respectively. Calculate the work done by the thermal system in tracing out one cycle.

Solution: This problem can be solved by two methods: by performing the integration over each leg explicitly; or by evaluating the area enclosed by the cycle.

Using the first method,

$$W_{\text{cycle}} = \int_{V_A}^{V_B} p\,dV + \int_{V_B}^{V_C} p\,dV + \int_{V_C}^{V_D} p\,dV + \int_{V_D}^{V_A} p\,dV. \tag{18–13}$$

The first term on the right-hand side of Eq. (18–13) runs from $V_A = V_1$ to $V_B = V_1$. The volume is fixed at each step; that is, $dV = 0$ all along path $A \to B$, and the integral drops out. The same is true for the third term, over path $C \to D$. To evaluate the second term of Eq. (18–13), over path $B \to C$, we note that the pressure takes the constant value p_2 and thus comes out of the integral:

$$W_{B\to C} = p_2 \int_{V_B}^{V_C} dV = p_2 \int_{V_1}^{V_2} dV = p_2(V_2 - V_1).$$

For the fourth term of Eq. (18–13), over path $D \to A$, p has the constant value p_1, and

$$W_{D\to A} = p_2 \int_{V_D}^{V_A} dV = p_1 \int_{V_2}^{V_1} dV = p_1(V_1 - V_2).$$

Collecting these results,

$$\begin{aligned} W_{\text{cycle}} &= p_2(V_2 - V_1) + p_1(V_1 - V_2) = p_2(V_2 - V_1) - p_1(V_2 - V_1) \\ &= (p_2 - p_1)(V_2 - V_1). \end{aligned}$$

By the second method, the area enclosed in a rectangle is the width $(V_2 - V_1)$ times the height $(p_2 - p_1)$:

$$W_{\text{cycle}} = (p_2 - p_1)(V_2 - V_1).$$

Because the cycle is traced out in a clockwise manner, W_{cycle} is positive.

As expected, the two methods give the same result.

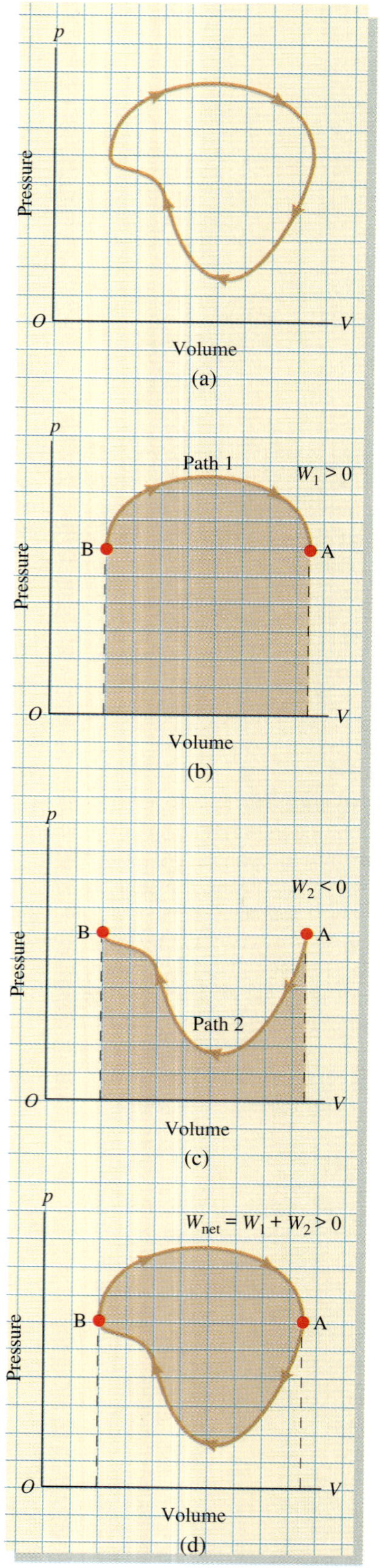

FIGURE 18–14 In a closed cycle, the work done by a gas on its surroundings is the area enclosed by the curve of p versus V. To show this, the full cycle is broken in two paths.

Work Done in Various Transformations

Constant Pressure. An isobaric (constant-pressure) transformation is drawn on a p–V diagram in Fig. 18–16a. For such transformations, the work done by the gas is just the volume change multiplied by the constant value of the pressure. How can an isobaric transformation be arranged? Generally, if the volume of a gas is to increase without a change in pressure, the temperature must increase. Let's calculate this temperature change for an ideal gas. Suppose that the initial temperature, pressure, and volume are T, p, and V, and that V is increased by dV while p is held constant. It follows from the ideal gas law, $T = pV/nR$, that

$$\text{for constant } p: \qquad dT = \frac{p\,dV}{nR}. \tag{18–14}$$

Substituting T/V for p/nR, Eq. (18–14) becomes

$$\text{for constant } p: \qquad dT = \frac{T}{V}\,dV. \tag{18–15}$$

If the volume change dV is accompanied by a heat flow into the gas, giving precisely the temperature change in Eq. (18–15), then the pressure will be maintained at a constant value. The heat flow in this case must be $dQ = C_p\,dT$.

Constant Volume. When the volume is kept fixed, the curve of the transformation is said to be an **isochore** (Fig. 18–16b). Thermal systems do no work—positive or negative—in an isochoric transformation because there is zero area under the curve on a p–V diagram. The pressure and temperature both change, however, and so such a transformation must be accompanied by a heat flow, just as for the isobaric and the isothermal transformations. In this case, the heat flow is $dQ = C_V\,dT$.

Adiabatic. Finally, there is a particular curve that will be followed if the gas does work while it is thermally isolated from its environment. Imagine a cylinder of gas enclosed in a thermal insulator, such as Styrofoam, with only the mechanical movements of a piston connecting it to its surroundings. Because of the thermal isolation, there is no heat flow dQ into or out of the system. When $dQ = 0$, the temperature can change only if work is done on the system. Reversible transformations of a thermal system in which there is no heat flow to the system are called **adiabatic** transformations. In Section 18–7, we will explore the precise curve followed by adiabatic transformations for ideal gases. These transformations will be important when we study engines in more detail.

Internal Energy

In an adiabatic transformation, a thermal system does work on its surroundings but has no thermal contact with those surroundings. Simple conservation of energy, as formulated in Chapters 6 and 7, can be applied to it. We can think of the system as having a thermal (or internal) energy that can change if work is done on or by the system. We call this thermal energy U.† In particular, when a system does work on its surroundings in an adiabatic transformation, *it changes its internal (thermal) energy*, U, by the amount of work it does:

$$\text{for an infinitesimal adiabatic transformation:} \qquad dU = -dW; \tag{18–16}$$

$$\text{for a finite adiabatic transformation:} \qquad U_B - U_A = -W_{A\to B}. \tag{18–17}$$

†Don't confuse U with a potential energy!

In an adiabatic transformation, a thermal system such as the insulated cylinder behaves as an ordinary mechanical system; any changes in its internal energy are matched by mechanical work done. This means that if we want to study how the internal energy changes when we go from one thermodynamic state to another, we should use adiabatic transformations because then we can avoid the complications of heat flow. *The adiabatic process allows us to learn just how the internal energy depends on the thermodynamic variables.*

Adiabatic transformations

The thermal (internal) energy is a function of p and V, p and T, or T and V. The thermal energy is associated with the energy of the molecules of a thermal system; it is simply a function of the thermodynamic variables of the system, not of how those variables might have changed. Changes in internal energy, like changes in the potential energy of a conservative mechanical system, are *not* path dependent. This contrasts sharply with work or heat flow, which *are* path dependent.

The thermal energy is a function of state, and any changes in it are independent of how the changes are made.

In Eq. (18–16), both dU and dW are written as differentials. The symbol dU has its usual meaning as a differential, in that a finite change in internal energy is

$$\int_A^B dU = U_B - U_A, \tag{18–18}$$

which is independent of the path between A and B. There is no equivalent expression for work because the integral of dW *does* depend on the path taken. In a strict mathematical sense, dW is not a true differential and is sometimes denoted by $đW$. There is no need for the more complicated notation, however, provided we keep in mind that the integral of dW is path dependent.

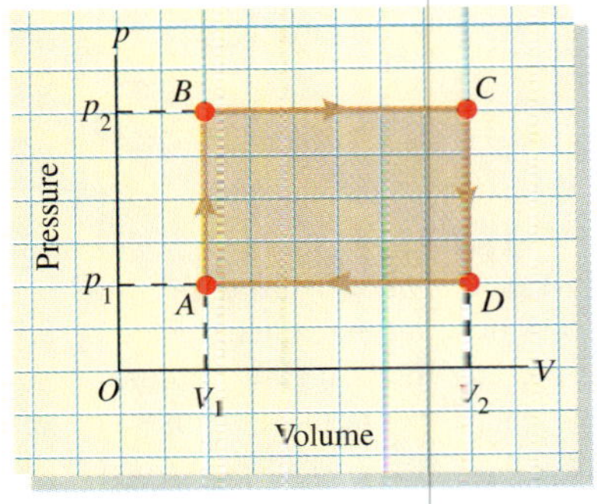

FIGURE 18–15 Example 18–6.

The minus signs on the right-hand sides of Eqs. (18–16) and (18–17) occur because W always denotes the work done *by* the system *on* its surroundings. When the system does positive work in an adiabatic transformation, its internal energy U decreases; when the work done is negative, the internal energy increases.

Let's now reconsider the Joule experiments that determined the mechanical equivalent of heat (Section 18–3). The water whose temperature is raised by the churning paddles is thermally isolated. The transformation is irreversible; it is not possible to "unchurn" the water and use it to raise the fallen weight. Nevertheless, the (negative) net work done by the water does indeed determine the (positive) change in the internal energy of the water. *How* a change is made is irrelevant for the internal energy. It is purely a function of state. Thus Eq. (18–17) holds for both reversible and irreversible transformations of thermally isolated systems.

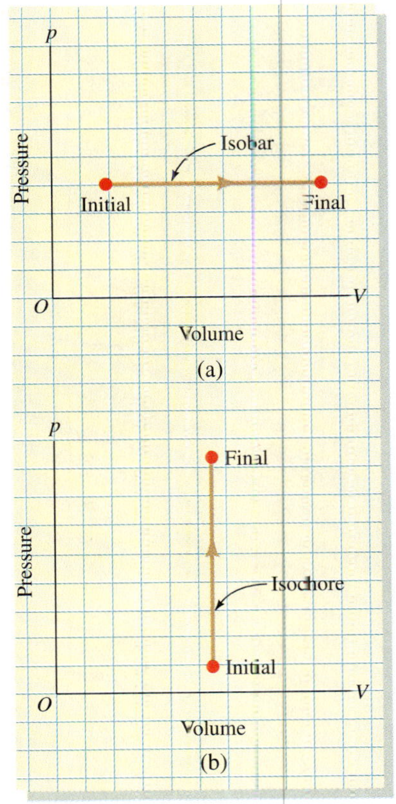

FIGURE 18–16 (a) An isobar, which represents a constant-pressure transformation. (b) An isochore, which represents a constant-volume transformation.

18–5 THE FIRST LAW OF THERMODYNAMICS

The **first law of thermodynamics** is a statement of the conservation of energy for thermal systems. It is essential in understanding any thermal system—particularly those in which fuel is burned and work is done, such as engines or power plants (Fig. 18–17). Because heat flow and work are equivalent in terms of internal energy, the change in the internal energy of a thermal system in going from state A to state B is the negative of the work that the system does on its surroundings, plus the heat flow into the system:

$$\Delta U = U_B - U_A = -W_{A\to B} + Q_{A\to B}. \tag{18–19}$$

Equation (18–19) is the primary result of this chapter. The minus sign in front of W indicates that W is the work done *by* the thermal system, and the plus sign in front of Q indicates that Q is the heat flow *into* the system. In Fig. 18–17, we give another view of the meaning of these terms.

Because U is a function of state only, ΔU is the same whether the transformation $A \to B$ is reversible or irreversible. Similarly, the work done by the system can be mea-

The first law of thermodynamics

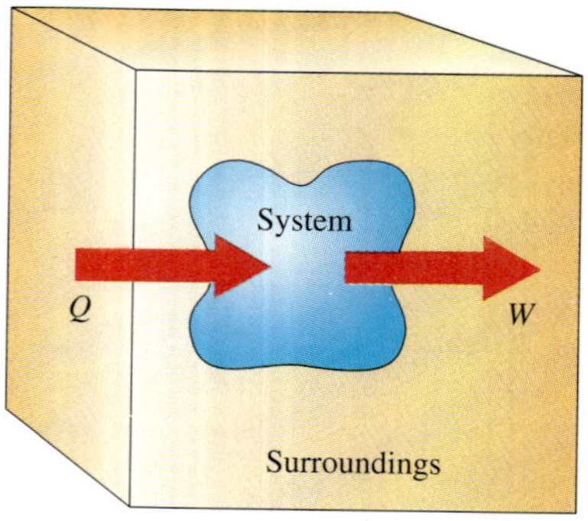

FIGURE 18–17 In the first law of thermodynamics, W represents the work done by the system, while Q represents the heat flow into the system. Either term can be negative or positive, depending on the transformations involved.

In closed cycles, work done equals heat flow.

FIGURE 18–18 This engine uses thermal energy—burning fuel—to do work in cycles and produce electrical energy.

sured by observing its effect on its surroundings—for example, how high the system lifts a mass—independently of whether the transformation $A \to B$ is reversible or irreversible. The heat flow $Q_{A \to B}$ can be determined from the first law of thermodynamics if the transformation is irreversible or determined independently by making a calorimetric measurement.

The differential form of the first law of thermodynamics is

$$dU = -dW + dQ. \tag{18–20}$$

While both dW and dQ depend on the path, dU is independent of path. Remarkably, the path dependence of the *difference* $-dW + dQ$ cancels.

The First Law of Thermodynamics in Closed Cycles

Engines operate and perform work in closed cycles. Thus it is of practical importance to be able to apply the first law of thermodynamics to a closed cycle (Fig. 18–18). There is no change in internal energy over a cycle because the system comes back to its original state. The work done by the thermal system over the cycle must then match the heat flow into the system, by the first law:

$$Q_{\text{cycle}} = W_{\text{cycle}}. \tag{18–21}$$

EXAMPLE 18–7 Suppose that a thermal system follows the cycle $A \to B \to C \to D \to A$, as drawn in Fig. 18–15 and treated in Example 18–6. What is the total heat flow into the system after one cycle is completed?

Solution: We can use the fact that the internal energy depends only on the state to recognize that the internal energy change in the cycle is zero because the initial point is the same as the final point. The fact that the system passes through a complicated succession of intermediate states is irrelevant. Then Eq. (18–21) tells us immediately that

$$Q_{\text{cycle}} = W_{\text{cycle}}.$$

The heat flow into the system is precisely the work done by the system, so, according to Example 18–6,

$$Q_{\text{cycle}} = (p_2 - p_1)(V_2 - V_1).$$

Note again that the work is just the area enclosed by the cycle.

How the Energy Changes in Constant-Volume and Constant-Pressure Transformations

No work is done in reversible changes under constant volume, and Eq. (18–20) becomes

$$\text{for constant volume:} \quad dU = dQ_V = C_V\,dT. \tag{18–22}$$

We have emphasized the fact that the volume is constant by the subscript on dQ. No subscript is necessary for dU because the change in U is independent of the path taken. Similarly, we can express the first law of thermodynamics for reversible changes under constant pressure as

$$\text{for constant pressure:} \quad dU = -dW_p + dQ_p = -p\,dV + C_p\,dT. \tag{18–23}$$

Equations (18–22) and (18–23) can both be integrated over paths with constant V and constant p, respectively, to find the net internal energy change under a finite volume or pressure change. Keep in mind that in such an integration the heat capacities may not necessarily be constants that can be removed from under the integral sign.

18–6 INTERNAL ENERGY OF IDEAL GASES

Joule performed an experiment in 1845 to study the properties of U for ideal gases. He demonstrated that, *for ideal gases, the internal energy is a function of temperature alone*: $U = U(T)$. Equivalently, the equation of state implies that we can say that U is a function only of the product pV.

Joule's experiment is as follows: Fill one part of a two-part container with a dilute sample of gas (if it is sufficiently dilute, we can be sure it is ideal). The other part of the container is evacuated, and the two parts are connected by a tube with a stopcock. The container as a whole is sealed and does no work on the outside world. The container is submerged in a thermally isolated container of water or some other fluid whose temperature can be monitored. This fluid, initially at temperature T_1, acts as a calorimeter: If its temperature changes because of some change in the state of the gas in the container, then there has been a heat flow into or out of the gas. This setup is shown schematically in Fig. 18–19a.

The stopcock is now opened and the gas undergoes free expansion into the previously empty part of the container. The crucial feature of free expansion is that the gas does no work because it does not cause mechanical movement in its surroundings. The new temperature of the fluid, T_2, is then measured. When the gas is sufficiently dilute, the experiment shows that $T_2 = T_1$ (Fig. 18–19b). *The temperature of an ideal gas undergoing free expansion remains constant.* This result means that there has been no heat flow into the gas. There is no work done and no heat flow in free expansion. Thus, according to the first law of thermodynamics, the internal energy, U, remains unchanged. But because the volume of the gas does change, U must be independent of volume. Therefore, for an ideal gas, U is a function only of temperature:

The internal energy of an ideal gas depends only on the temperature.

$$U_{\text{ideal}} = U(T). \tag{18–24}$$

This result follows directly from experiment, but it can also be derived from the second law of thermodynamics, which we will explore in Chapter 20.

For real gases—which are not perfectly ideal—there is a small deviation from the result of Eq. (18–24), and temperature changes slightly during free expansion. In other words, U is a function of density as well as temperature. We typically ignore this effect.

Let's now couple the fact that the internal energy of an ideal gas depends only on the temperature with Eqs. (18–22) and (18–23). Because U depends *only* on T, the integral of Eq. (18–22) gives

$$U(T_B) - U(T_A) = \int_{T_A}^{T_B} C_V \, dT. \tag{18–25}$$

This equation must hold for *all* values of the volume because the left-hand side is independent of volume. Hence Joule's experiment shows that C_V must itself be independent of volume. Equation (18–25) expresses only a *change* in energy, so it is equivalent to the form

$$U(T) = \int_0^T C_V \, dT + \text{a constant}. \tag{18–26}$$

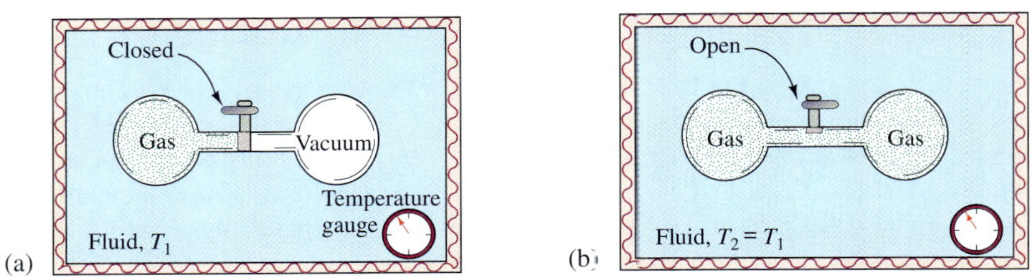

FIGURE 18–19 Schematic diagram of Joule's experiment. When Joule opened the stopcock to allow free expansion, he observed no temperature change for a dilute (ideal) gas.

The constant is arbitrary because only *changes* in U matter, not the actual values of $U(T)$ and $U(0)$. We therefore take the constant to be zero. Finally, we can use the experimental fact that, over a wide temperature range, C_V is independent of temperature. We can remove C_V from behind the integral sign and find

The internal energy of an ideal gas as a function of temperature

$$\text{for } C_V \text{ independent of temperature: } U(T) = C_V T. \qquad (18\text{–}27)$$

The internal energy of an ideal gas is simple indeed, as long as C_V is constant with temperature.

EXAMPLE 18–8 An ideal gas in contact with a thermal reservoir at temperature $T = T_0$ does work by expanding against a piston. That work is 10 J. How much heat flow goes into the gas during its expansion?

Solution: Application of the first law of thermodynamics gives us the heat flow, Q, in terms of the work done, W, and the change in internal energy, ΔU:

$$Q = \Delta U + W.$$

Because the gas is ideal, U is a function only of temperature; because the transformation is isothermal, the change in U is zero. Thus

$$Q = W = 10 \text{ J}.$$

This heat flow comes from the thermal reservoir.

A Relation between C_p and C_V for Ideal Gases

We can express C_p in terms of C_V. Equations (18–22) and (18–23) are two differential expressions for the energy change, dU. Because Eq. (18–23) specifies constant pressure, the term $p\,dV$ that occurs on the right-hand side can also be written for an ideal gas as $p\,dV = nR\,dT$. Then both Eqs. (18–22) and (18–23) involve only terms proportional to dT on the right-hand side. Because the energy shift dU is independent of whether the infinitesimal transformation is made at constant V or at constant p, the two expressions for dU must be equal:

The relation between the heat capacities at constant pressure and constant volume for an ideal gas

$$C_V\,dT = -nR\,dT + C_p\,dT. \qquad (18\text{–}28)$$

We cancel the common factor dT from this equation to find that

$$C_p = C_V + nR, \qquad (18\text{–}29)$$

which is a result that applies to ideal gases. Because C_V is a function of T only, C_p is also; to the extent that C_V is in fact constant with T, C_p is as well. The prediction of Eq. (18–29) is indeed satisfied by experiment for ideal gases and is well approximated for real gases. Although relations such as this are useful predictions typical of thermodynamics, it is necessary to use the microscopic dynamics of gases to compute quantities such as C_V. This calculation will be made in Chapter 19.

18–7 MORE APPLICATIONS FOR IDEAL GASES

In order to understand the behavior of engines or, more generally, of gaseous thermal systems, we want to express the changes that occur in terms of the thermodynamic variables. In principle, this is possible as long as the transformations are reversible. Better, if we can approximate real gases by ideal gases, then all the transformations we have discussed can be expressed mathematically, as can the corresponding work done, heat flow, energy changes, and related quantities. Some of these mathematical expressions have already been given. For example, the shape of the p–V curve that represents an isothermal transformation at $T = T_0$ is a hyperbola, $p = nRT_0/V$. The relation $C_p = C_V + nR$ [Eq. (18–29)] is a consequence of the fact that the gas is ideal.

Of the four particular reversible transformations discussed earlier—constant-volume (isochoric), constant-pressure (isobaric), constant-temperature (isothermal), and adiabatic transformations—the first two are simple to treat. Let's now look at the other two in more detail.

Isothermal Transformations of an Ideal Gas

When the temperature is constant, the work done by an ideal gas is simple to compute. We see from Fig. 18–20 that

$$W = \int_{V_1}^{V_2} p\,\mathrm{d}V = \int_{V_1}^{V_2} \frac{nRT_0}{V}\,\mathrm{d}V = nRT_0 \int_{V_1}^{V_2} \frac{\mathrm{d}V}{V}. \qquad (18\text{–}30)$$

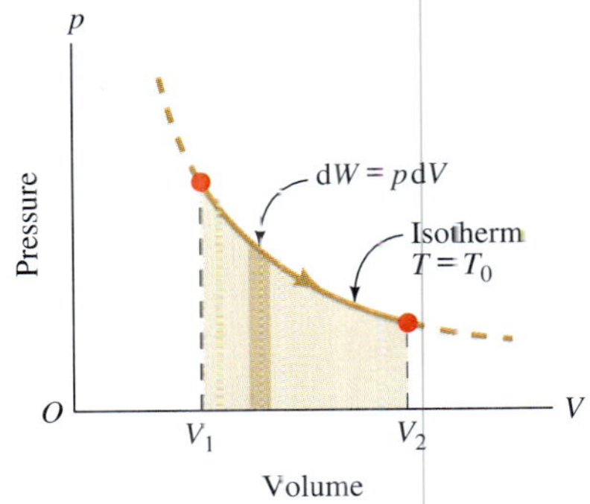

FIGURE 18–20 An ideal gas expands from V_1 to V_2 while its temperature is held fixed, and in doing so does work on its surroundings.

The integral of $1/x$ is the natural logarithm, $\ln x$ (see Appendix IV–6). Because $\ln x_1 - \ln x_2 = \ln(x_1/x_2)$, we have, for the work done by an ideal gas in an isothermal transformation:

$$W = nRT_0 \ln\left(\frac{V_2}{V_1}\right). \qquad (18\text{–}31)$$

As expected, W is positive when the gas expands ($V_2 > V_1$) and negative when the gas contracts ($V_2 < V_1$).

EXAMPLE 18–9 Suppose that 0.12 mol of an ideal gas in contact with a thermal reservoir that holds the temperature fixed at $T_0 = 9.8°\mathrm{C}$ has an initial volume of 1.3 L and does 14 J of work. What is the final volume, and what is the final pressure?

Solution: Equation (18–31)—the work done in an isothermal transformation—can be solved for the final volume V_f because every other quantity in the equation is known. T_0 must be converted to the Kelvin scale: $T_0 = 9.8 + 273.15 \simeq 283$ K. Note that nRT_0 has the dimensions of pV and hence of work. We have

$$\ln\left(\frac{V_f}{V_i}\right) = \frac{W}{nRT_0} = \frac{14\ \mathrm{J}}{(0.12\ \mathrm{mol})(8.314\ \mathrm{J/mol\cdot K})(283\ \mathrm{K})} = 0.050.$$

Thus

$$\frac{V_f}{V_i} = e^{0.050} = 1.051,$$

and the final volume is

$$V_f = V_i(1.051) = (1.3\ \mathrm{L})(1.051) = 1.4\ \mathrm{L}.$$

The final pressure is given by the equation of state:

$$p_f = \frac{nRT_0}{V_f} = \frac{(0.12\ \mathrm{mol})(8.314\ \mathrm{J/mol\cdot K})(283\ \mathrm{K})}{(1.4\ \mathrm{L})(10^{-3}\ \mathrm{m^3/L})} = 2.1 \times 10^5\ \mathrm{Pa} = 2.0\ \mathrm{atm}.$$

The pressure has decreased from an initial value of $p_i = p_f(V_f/V_i) = 2.2$ atm.

The internal energy of an ideal gas is unchanged in an isothermal transformation. Thus, by the first law of thermodynamics, Eq. (18–31) also expresses the heat flow into the gas during an isothermal transformation.

Adiabatic Transformations of an Ideal Gas

In an adiabatic expansion, the first law of thermodynamics becomes $dU = -dW$. Because dW is positive, dU is negative. U is proportional to T, so the temperature will fall in an adiabatic expansion and rise in an adiabatic compression. When a gas expands in a thermally isolated container, it cools; when it is compressed, it heats up (as in a compressed bicycle pump).

The curve describing an adiabatic transformation

In a separate subsection that follows, we find that the curves that describe this transformation are given by

$$pV^\gamma = p_0V_0^\gamma = \text{a constant}, \tag{18–32}$$

where (p_0, V_0) are the initial values of pressure and volume, respectively, of the ideal gas that undergoes an adiabatic transformation to new values (p, V). The constant γ is defined to be

$$\gamma \equiv \frac{C_p}{C_V} = \frac{C_V + nR}{C_V} = \frac{c'_V + R}{c'_V}, \tag{18–33}$$

where c' is the molar heat capacity, from Eq. (18–2). We have used Eq. (18–29), which relates C_p to C_V. Because $C_p > C_V$, γ is greater than unity (a result that is generally true). In terms of V and T, the transformation follows the curves

$$TV^{\gamma-1} = \text{a constant}. \tag{18–34}$$

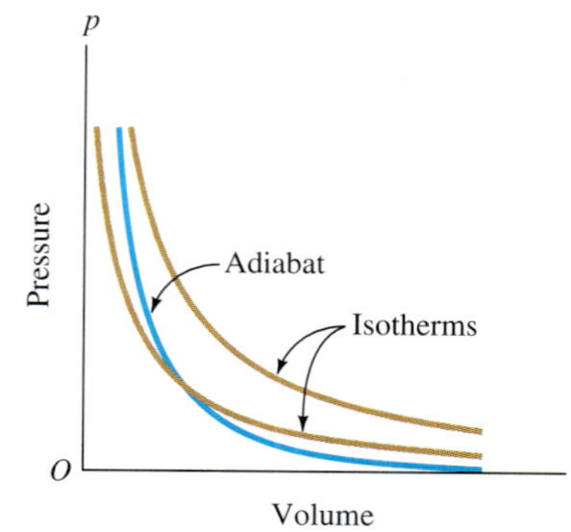

FIGURE 18–21 For ideal gases, adiabatic curves ($p \propto V^{-\gamma}$), called adiabats, are steeper than isotherms ($p \propto V^{-1}$) on a p–V diagram.

Compare Equation (18–32) with the curve that describes an isothermal transformation, pV = a constant, or $p \propto V^{-1}$. The latter curve is a hyperbola, symmetric in p and V. The adiabatic curve $p \propto V^{-1}$ falls more steeply as V increases, because $\gamma > 1$. Figure 18–21 shows the two families of curves for a monatomic ideal gas, for which $\gamma = 1.67$. The constant in Eq. (18–32), which distinguishes the two adiabatic curves in Fig. 18–21, is determined when one point on the curve is known. If we know that the pressure equals p_0 and the volume equals V_0 at some point on an adiabatic curve, then the constant equals $p_0V_0^\gamma$.

*How to Describe the Adiabatic Transformation of an Ideal Gas on a *p–V* Diagram

To derive the curves that are described by Eq. (18–32), we start with the first law of thermodynamics in its differential form [Eq. (18–20)], with the heat flow, dQ, set to zero: $dU = -dW$. The infinitesimal work done is still given by $dW = p\,dV$. We also know that dU is proportional to the temperature change, dT. In particular, dU can be calculated by differentiating Eq. (18–26) with respect to the temperature, T. Because C_V is constant for an ideal gas, we have $dU/dT = C_V$, or

$$dU = C_V\,dT.$$

Thus the adiabatic transformation satisfies

$$C_V\,dT = -p\,dV. \tag{18–35}$$

All three thermodynamic variables appear here, but, because of the ideal gas law, $nRT = pV$, only two are independent. To find the curve of p versus V, we eliminate T in favor of p and V. The ideal gas law implies that

$$nR\,dT = p\,dV + V\,dp$$

(see Appendix IV–7). Equation (18–35) can thus be rewritten as

$$C_V\frac{p\,dV + V\,dp}{nR} = -p\,dV.$$

Solving for $V\,dp$, we find

$$V\,dp = -\frac{C_V + nR}{C_V}p\,dV. \tag{18–36}$$

The numerator on the right-hand side of Eq. (18–36) is the heat capacity at constant pressure, C_p, as Eq. (18–29) shows. Thus the right-hand side may be written as $-(C_p/C_V)p\,dV = -\gamma p\,dV$, with γ defined in Eq. (18–33).

With the definition of γ, we rewrite Eq. (18–36) in the form

$$\frac{dp}{p} = -\gamma\frac{dV}{V}.$$

Each side of this equation may be integrated separately, with the lower limits on the integrals determined by the initial state (p_0, V_0, T_0). Thus

$$\int_{p_0}^{p}\frac{dp}{p} = -\gamma\int_{V_0}^{V}\frac{dV}{V}.$$

Each integral gives a logarithm, and

$$\ln\left(\frac{p}{p_0}\right) = -\gamma\ln\left(\frac{V}{V_0}\right).$$

Because $A \ln x = \ln(x^A)$, it follows that

$$\ln\left(\frac{p}{p_0}\right) + \ln\left[\left(\frac{V}{V_0}\right)^{\gamma}\right] = 0.$$

We use $\ln x_1 + \ln x_2 = \ln(x_1x_2)$ to find

$$\ln\left[\left(\frac{p}{p_0}\right)\left(\frac{V}{V_0}\right)^{\gamma}\right] = 0.$$

We may now take the exponential of both sides of the equation to find

$$pV^{\gamma} = p_0V_0^{\gamma} = \text{a constant}.$$

This is Eq. (18–32). When we substitute the ideal gas law, we can show that the corresponding equation in terms of T and V is Eq. (18–34).

Variation of Atmospheric Temperature with Height

Adiabatic transformations explain the variation of atmospheric temperature with height rather well. Air is transported from high altitude to low altitude and vice versa by large atmospheric wind currents. Air is a sufficiently good thermal insulator so that the transport of air between different altitudes is, to a good approximation, adiabatic. When air rises, the pressure decreases, so that the air undergoes adiabatic expansion; its temperature therefore falls. When air is carried to lower altitudes, it undergoes adiabatic compression, and its temperature rises.

If we assume that air is an ideal gas, we can quantify these concepts. Suppose that pressure and temperature are related by an adiabatic transformation. The adiabatic curve for p versus T is derived from, for example, Eq. (18–32) with $V \propto T/p$. For the p-versus-T adiabatic curve, this gives the equation $p^{1-\gamma}T^{\gamma}$ = a constant. If we raise both sides to the power $1/\gamma$ and solve for T, we find that

$$T = (\text{a constant})p^{(\gamma-1)/\gamma}. \qquad (18\text{–}37)$$

If we know how pressure varies with height, then Eq. (18–37) can tell us how temperature varies with height. But how does pressure vary with height? In Chapter 16, we found that pressure decreases linearly with increasing height for an *incompressible* fluid, such as water. Air, which is assumed to obey the ideal gas law in this case, is certainly not incompressible, but we can use the methods of Section 16–3 to find an equation for the way pressure varies with height. There, we considered an imaginary horizontal slab cut out of the fluid, with its base at height h, where the pressure is p. The top of the slab, where the pressure is $p + dp$, is at height $h + dh$. Note that we expect dp to be negative because pressure should decrease with height. According to Eq. (16–5),

APPLICATION: ADIABATIC TRANSFORMATIONS IN DIESEL ENGINES

The diesel engine is a well-known application of an adiabatic transformation (Fig. 18AB–1). In its operation, a cylinder is filled with air at atmospheric pressure and temperature, and then compressed to less than 1/20 of its original volume. This compression is nearly adiabatic; if it were truly adiabatic and if the air were an ideal gas, then the ratio of the final to initial temperatures would be $T_f/T_i = (V_i/V_f)^{(\gamma-1)}$. Although some heat is lost to the cylinder walls, piston, and cylinder head, and the ideal gas equation of state is not exact for air, the final temperature of the air is greater than the ignition temperatures of most hydrocarbons. When, at this point, a combustible hydrocarbon is introduced into the pressurized air, it will burn, transferring its chemical energy into heat energy and raising the pressure of the gases in the cylinder. The hot gas is allowed to expand, driving down the piston and doing work. In practice, diesel engines are very tolerant to the type of fuel burned; everything from flour to powdered coal has been used. Note that because the end temperature of the compressed gas is proportional to the incoming temperature, diesel engines do not work as well in cold climates.

FIGURE 18AB–1 The modern diesel engine is a highly efficient and durable one. Adiabatic transformations play an important part in its operation.

$$\frac{dp}{dh} = -\rho g,$$

where ρ is the density of the fluid. From this equation, we see that as h decreases, p increases due to the increasing weight of the overlying fluid (in this case, air). For water, ρ is constant. For air, ρ is not constant but is given by

$$\rho = \frac{m}{V} = \frac{nM}{V} = M\left(\frac{p}{RT}\right),$$

where M is the mass of 1 mol of air (its molecular weight). The equation for pressure as a function of height becomes

$$\frac{dp}{dh} = -\left(\frac{Mg}{RT}\right)p. \tag{18–38}$$

Rather than solving this equation directly (see Problem 45), we can convert it into an equation for temperature as a function of height (see Problem 57). This is done by using the adiabatic curve, Eq. (18–37). Temperature is thus found to vary with height as

$$T = T_0 - \left(\frac{Mgh}{R}\right)\left(\frac{\gamma - 1}{\gamma}\right). \tag{18–39}$$

Because $\gamma > 1$, temperature does indeed drop with increasing height—as anyone who has been to the mountains can appreciate.

How does Eq. (18–39) compare with observation? The molecular weight of air is approximately 0.029 kg/mol, $\gamma_{air} \simeq 1.4$, and $g \simeq 9.8\ m/s^2$. Then if we take the derivative of T [Eq. (18–39)] with respect to h, we find that

$$\frac{dT}{dh} = -\frac{(0.029\ \text{kg/mol})(9.8\ \text{m/s}^2)}{8.3\ \text{J/°C}\cdot\text{mol}}\left(\frac{1.4 - 1}{1.4}\right) = 0.010\text{°C/m}.$$

We have estimated that temperature drops about 10°C/km. The experimental result is approximately 6.5°C/km. The main sources of error in our calculation are the neglect of the effects of water condensation and, closer to the surface, the effects of local topography, such as mountain ranges. In the vicinity of flat deserts and high in the atmosphere, the value calculated is a fairly good approximation.

SUMMARY

Reversible changes of state are changes that are made sufficiently slowly that the temperature, T, the pressure, p, and the volume, V—the thermodynamic variables that describe a thermal system such as a gas—are well defined at all times. The three variables are related by the equation of state of the gas. Reversible transformations, in contrast with irreversible transformations, can be traced out on a plot of any one of the three variables versus either of the other two.

When the temperature of a system changes, the internal energy of the system has changed. This can happen because the system has done work on its surroundings and/or because there has been a heat flow into the system. Infinitesimal heat flow into a system, dQ, is related to the infinitesimal temperature change of the system, dT, by the heat capacity, C, which depends not only on the system's composition, but on how the temperature change is made:

$$\text{for constant volume:} \quad dQ = C_V\,dT \tag{18–4}$$
$$\text{for constant pressure:} \quad dQ = C_p\,dT. \tag{18–5}$$

It is generally true that

$$C_p \geq C_V. \tag{18–6}$$

Because dQ depends on how a transformation is made, the net heat flow in a finite transformation is path dependent. Heat flow is measured in calories. Calorimetry, the measurement of heat capacities, is based on the fact that heat flow is conserved in a mechanically and thermally isolated system.

The specific heat of a material, c, is the heat capacity of 1 g of the material; its molar heat capacity, c', is the heat capacity of 1 mol of the material.

Heat flow, like work, is an energy transfer. Any transformation of a thermal system caused by a heat flow can be made equally well by doing work on the system. This means that heat has an exact mechanical equivalent,

$$\text{for a thermally isolated system:} \quad \Delta Q = W. \tag{18–7}$$

The equivalence of work and heat means that the calorie can be expressed as an energy unit:

$$1\ \text{cal} = 4.185\ \text{J}. \tag{18–8}$$

A gas undergoing an infinitesimal transformation does an infinitesimal amount of work on its surroundings, given by

$$dW = p\,dV. \tag{18–10}$$

As for heat flow, the net work done in a finite transformation is path dependent. If the path is adiabatic ($dQ = 0$), the work done by the system is the negative of the internal energy change,

$$\text{for an infinitesimal adiabatic transformation:} \quad dU = -dW; \tag{18–16}$$
$$\text{for a finite adiabatic transformation:} \quad U_B - U_A = -W_{A\to B}. \tag{18–17}$$

An adiabatic transformation is a reversible transformation in which there is no heat flow between a system and its surroundings.

The internal energy (thermal energy), U, is a function of state. Any changes U undergoes in thermal transformations are path independent. The first law of thermodynamics specifies that the change in the internal energy of a thermal system in going from state A to state B is the negative of the work it does on its surroundings, plus the heat flow into the system:

$$\Delta U = U_B - U_A = -W_{A\to B} + Q_{A\to B}. \qquad (18\text{–}19)$$

The work in this equation, W, is the work done *by* the thermal system, and Q is the heat flow *into* the system. The differential form of the first law of thermodynamics is written as

$$dU = -dW + dQ. \qquad (18\text{–}20)$$

The internal energy of ideal gases is a function of temperature alone:

$$U_{\text{ideal}} = U(T). \qquad (18\text{–}24)$$

Because C_V is practically independent of temperature, the dependence is linear,

$$U(T) = C_V T. \qquad (18\text{–}27)$$

It is also true that, for an ideal gas,

$$C_p = C_V + nR. \qquad (18\text{–}29)$$

The curve of an adiabatic transformation can be calculated for an ideal gas by using the ideal gas law. It is given on a p–V diagram by

$$pV^\gamma = \text{a constant}, \qquad (18\text{–}32)$$

where

$$\gamma \equiv \frac{C_p}{C_V} = \frac{C_V + nR}{C_V} = \frac{c'_V + R}{c'_V}. \qquad (18\text{–}33)$$

This equation can be used to predict the variation of temperature with height in the atmosphere.

UNDERSTANDING KEY CONCEPTS

1. When a bicycle tire is pumped up, the end of the pump near the valve feels hotter. What is happening?
2. A gas expands adiabatically, doing 500 J of work. By how much does the internal energy of the gas change?
3. If a bucket of water is carried uphill, raising the water's potential energy, will its temperature change?
4. Which of the following are reversible processes: (a) the slow inflation of a balloon with a bicycle pump; (b) the heating up of a drill used to bore a hole in a log; (c) the slow stretching of a wire by an external force, carried out at a constant temperature?
5. Why is free expansion not an example of an adiabatic process?
6. Would it make sense for you to shake a container of soup in order to heat it? Would the fact that the soup is in an insulating container change your answer?
7. Should the cooling fluid for an engine have a large or a small heat capacity?
8. What use can you think of for fluids with very large specific heats? With very small specific heats?
9. When a mass m_1 of water at temperature T_1 is mixed with a mass m_2 of water at temperature T_2, with $T_1 > T_2$, the temperature change of the hotter mass, ΔT_1, is related to the temperature change of the colder mass, ΔT_2, by $\Delta T_1/\Delta T_2 = -m_2/m_1$. In an experiment in which water and mercury are mixed, it was found that 100 g of mercury acts like 3.3 g of water. Explain both the formula and the observation about mercury.
10. A well-insulated container is divided into two separate compartments by an adiabatic wall. One compartment contains gas at a pressure p_1 and temperature T_1; the other contains gas at a pressure p_2 and temperature T_2. What happens when the partition is removed?
11. Why is the latent heat of vaporization for water so much larger than the latent heat of fusion?
12. When a freeze threatens, how can spraying fruit trees with water protect the fruit from frost damage?
13. Why does perspiring during exercise help you to cool off?
14. For an ideal gas, the heat capacities are independent of temperature, and it is easy to relate temperature change to heat flow. Would it be impossible to relate temperature change to heat flow if the heat capacities varied with temperature?
15. Can all thermodynamic transformations be plotted on a graph of one thermodynamic variable versus another?
16. A change in a thermal system is reversible if (a) there is no friction and (b) the change is carried out slowly. Explain these constraints.
17. In the evening, after a hot summer day, why is it much cooler in the suburbs than downtown?
18. The center of a continent has more temperature extremes than do coastal areas at the same latitude. What effect contributes significantly to this?
19. People who drive cars in cold climates observe that the car's engine sometimes overheats if they keep the engine idling. This does not happen while they are driving, even though the engine puts out much more power then. Why?

PROBLEMS

18–1 Changes in Thermal Systems

1. (II) Are the following thermal transformations reversible or irreversible? Show the changes on a diagram when one is requested. (a) The sudden release of a gas from one portion of a container, of volume V_1, into a second portion, previously under vacuum. The total final volume, the volume of the whole container, is V_2 (p–V diagram). (b) As in part (a), but rather than a sudden release, the release is made a little at a time until there is no further flow of gas (p–V diagram). (c) The compression of a gas in thermal contact with a heat reservoir by loading weights, a little at a time, onto a piston connected to the gas (V–T diagram). (d) The slow heating of a gas by the release, a little at a time, of a warmer gas into the original sample of gas (p–T diagram). (e) The slow heating of a gas by placing it in thermal contact with a series of ever hotter thermal reservoirs (p–T diagram). (f) The slow rusting of a nail. (g) The release of electrical energy as lightning. (h) The release of chemical energy in the operation of a rechargeable battery to lift a weight.

18–2 Thermal Energy and Heat Flow

2. (I) (a) Using Table 18–2, calculate the heat capacity of a 5.0-kg aluminum rod. (b) Assuming that heat capacity is constant with temperature, how much heat flow is required to raise the temperature of the rod from 25°C to 87°C?

3. (I) A water heater heats 0.8 L of water/min from 20°C to 50°C. If the efficiency of the heater is 80 percent, how much power is consumed by the heater?

4. (I) The latent heat of fusion of oxygen (O_2) is 0.44 kJ/mol. Given that the atomic weight of molecular oxygen is 32 g/mol, find the latent heat of fusion in units of (a) kJ/kg and (b) cal/g.

5. (I) Consider an ideal gas initially in a state (p, V) and finally in a state (p, $2V$). If the transformation occurs at a constant pressure, what is the heat flow into the gas in terms of p and V?

6. (II) A 10-kg lead brick is at room temperature, 25°C. The melting point of lead is 327°C. Using the specific heat for lead, given in Table 18–2, calculate the heat flow to the brick needed to bring the brick to the melting temperature. Will the entire lead brick melt at this point? Explain.

7. (II) A 300-cm^3 glass is filled with 100 g of ice at 0°C and 200 g of water at 25°C. (a) Characterize the content of the glass after equilibrium has been reached. Neglect heat transfer to and from the environment. (b) Repeat your calculations for 50 g of ice and 250 g of water.

8. (II) In Example 18–4, an experiment in calorimetry was described in which the assumption is made that the heat capacity of water and iron are constant over the temperature range of the experiment. Describe a series of experiments that would establish whether or not these heat capacities are indeed constant.

9. (II) The quantity 0.6 kg of thermally isolated water at an initial temperature of 291 K serves as a calorimeter. Into it can be placed 1.2 kg of metal that has been heated to 373 K. Find the specific heats and molar heat capacities of the metals if the equilibrium temperature is (a) 316 K; the metal is aluminum. (b) 305 K; the metal is iron. (c) 297 K; the metal is lead.

10. (II) A calorimeter consists of 400 g of water at 25°C A 500-g piece of copper at 100°C is thrown into the water, and the equilibrium temperature is found to be 28.3°C. What are the specific heat of copper in cal/g · K and the molar heat capacity, given that 1 mol of copper has a mass of 63.5 g?

11. (II) The quantity 200 g of thermally isolated water at an initial temperature of 293 K serves as a calorimeter. Into it are placed, *successively*, 200 g of iron, then 200 g of silver, each of which has been heated to 400 K. Using the known specific heats of iron and silver, find the final equilibrium temperature. (Use $c_{Ag} = 0.0557$ cal/g · K and $c_{Fe} = 0.112$ cal/g · K.)

12. (II) An axe head consisting of 1 kg of iron is left outdoors one cold winter's night and is brought indoors when the outside temperature is a brisk 250 K. The room into which it is brought is initially at a nice, comfortable 293 K and 1 atm of pressure. The volume of the room, which is well insulated, is 40 m^3. Assuming that the axe head comes to thermal equilibrium with the air in the room, by how much is the temperature of the room lowered? (Ignore the thermal interaction with furniture, walls, and so forth. Use $c_{air} = 0.172$ cal/g · K $c_{Fe} = 0.112$ cal/g · K, and 28.8 g/mol for the molecular weight of air.)

13. (II) A thermally isolated system consists of 1 mol of helium gas at 100 K and 2 mol of a solid at 200 K; the gas and the solid are separated by an insulating wall of negligible mass. If the wall is removed, what is the equilibrium temperature of the system? Assume that at the given temperatures, the heat capacity of the solid is given by $3R/2$ per mole.

18–3 The Mechanical Equivalent of Heat

14. (I) How many kilojoules of energy are required to raise the temperature of a 5-kg bar of aluminum by 10°C?

15. (I) One hundred grams of ice at 0°C is added to 500 grams of thermally insulated water at 24°C. What is the temperature of the water after the ice melts?

16. (I) A 50-g lead bullet traveling at 200 m/s stops in a target. How much would the bullet's temperature rise if 80 percent of the energy were to go into heating it?

17. (I) A primitive demonstration of the mechanical equivalent of heat consists of repeatedly dropping a bag of lead shot from a given height and measuring the temperature change in the bag after a number of drops. Given that the heat capacity of lead is 128 J/kg · K, how many times would the bag have to be dropped from a height of 2.0 m to change its temperature by 5°C? Neglect all heat flows from the lead.

18. (II) Suppose that a kilogram of water drops, starting from rest, from a height of 5.2 m. When the water hits the ground suppose that half of the kinetic energy it attains in falling is converted into thermal energy within the water. By how much will the temperature of the water rise? Neglect the effect of air resistance.

19. (II) Julius Robert Mayer, who first enunciated the principle of the conservation of energy, estimated the mechanical equivalent of heat by considering a quantity of air confined in a cylinder by a movable piston. If the volume is fixed, 0.172 cal must be supplied to raise the temperature of 1 g of air by 1°C. If the piston is not fixed but exerts a constant pressure of 1 atm ($= 1.01 \times 10^5$ N/m^2), 0.241 cal are needed to raise the tem-

perature of 1 g of air by 1°C. If the volume change experienced by 1 g of air when the temperature changes by 1°C at constant pressure is 2.83×10^{-6} m^3/g · K, what is the relation between the calorie and the joule?

20. (II) The amount of solar energy delivered to a horizontal surface in Washington, D.C., averaged over a full year, is 160 W/m^2. Assuming a 10 percent efficiency for absorption and conversion to usable heating, how many liters of water can be heated from 0°C to 20°C in a 24-h period by a 3-m^2 solar panel?

21. (II) Angel Falls, in Venezuela, has the highest free waterfall in the world, at 2,648 ft. The water at the top of the falls is at a temperature of 10°C. (a) What is the temperature of the water in the pool at the base of the falls? As a first approximation, ignore the effects of air resistance, so that there is no terminal velocity. Also, ignore the effects of evaporation. [*Hint*: The kinetic energy that the water gains in falling and losing potential energy goes into turbulent motion at the base of the falls.] The exact terminal velocity is a complicated function of the precise configuration of the waterfall, such as whether it forms a tight tube or whether it is a broad, dispersed spray. (b) Repeat part (a) for Niagara Falls, which is approximately 50 m high. (c) Return to Angel Falls. Discuss the effect of air resistance and the fact that there is a terminal velocity. Would you expect the water to be heated as much as in part (a)? (d) It takes a positive heat flow into a liquid, the latent heat of vaporization, to cause it to evaporate. What is the sign of the temperature change in the waterfall, when you take into account the effect of evaporation of the falling droplets?

22. (II) Work is done on a thermally isolated container of 2 kg of water under 1 atm of pressure, initially at a temperature of 291 K. Assume that the specific heat of water does not vary with temperature, and that churning paddles do work on the water at a rate of 50 W. How long must the paddles continue to turn to bring the water to its boiling point, 373 K?

18–4 Work Done by Thermal Systems

23. (I) The volume of a gas in a cylinder slowly increases by a factor of 3 while its pressure decreases from 9×10^5 Pa to 1×10^5 Pa, as shown in Fig. 18–22. How much work is done on the piston by the gas during the expansion?

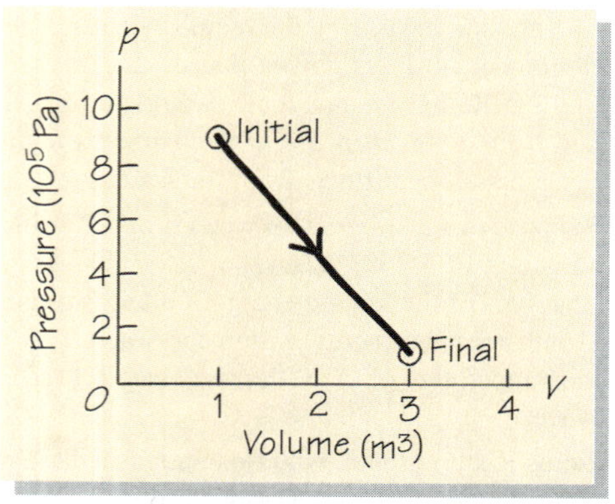

FIGURE 18–22 Problem 23.

24. (I) Express the work done by an ideal gas at constant temperature T_0 as a function of the initial pressure, p_i, and final pressure, p_f.

25. (I) If a small amount of water is evaporated at atmospheric pressure until a volume of 3.0 m^3 is occupied, how much work is done? Ignore the volume taken up by any remaining liquid water.

26. (I) Gas in a cylinder is allowed to expand from a volume of 0.2×10^{-3} m^3 to a volume of 0.7×10^{-3} m^3, while the pressure changes linearly with the volume from an initial pressure of 5.2 atm to a final pressure of 2.2 atm. What is the work done by the gas?

27. (II) Half a mole of an ideal gas is compressed isothermally at 293 K so that the pressure is increased from 1.0 atm to 3.6 atm. What are the initial and final volumes of the gas, and how much work is done by the gas during the compression?

28. (II) Two liters of nitrogen gas at an initial temperature of 320 K expand at a constant pressure of 5 atm against a piston, thereby doing work to raise a total mass of 100 kg. The gas is allowed to expand to a volume of 5 L. (a) How high can the mass be lifted? (b) What is the final temperature of the gas?

29. (II) A gas is compressed from V_1 to V_2 by reversibly following the path shown in Fig. 18–23, $p = p_0 - \beta V$, where β is a constant. (a) What is the dimension of β? (b) Determine the work done by the gas during the transformation by integrating the infinitesimal work along the path, $dW = p\,dV$. (c) Check your answer by evaluating the area under the p–V curve in the figure.

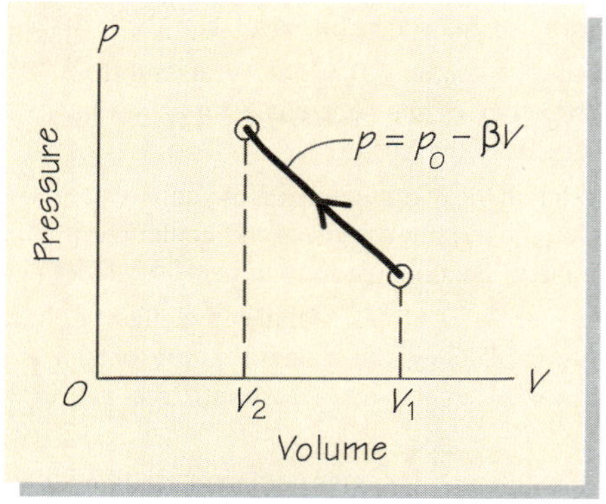

FIGURE 18–23 Problem 29.

30. (II) Figure 18–24 shows a reversible path for a gas being taken from state A to state B. Both states have the same pressure. (a) What is the work done in taking this path? (b) Draw a second

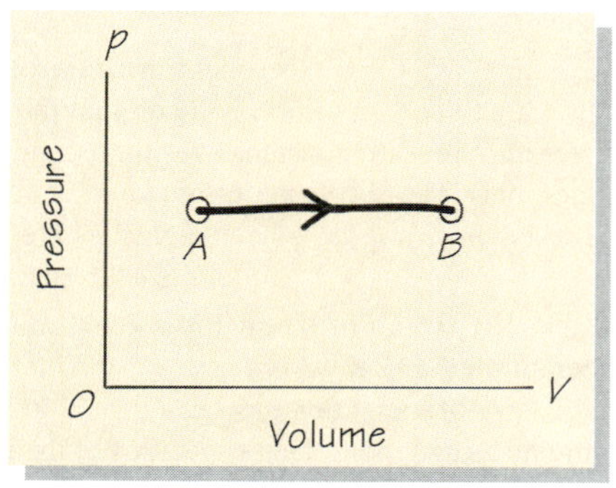

FIGURE 18–24 Problem 30.

reversible path that connects the same points by which no work is done. If the gas is ideal, to what minimum temperature is the gas taken on this path? (Real gases cannot be taken to this temperature.) (c) Draw a third path in which the work done is greater than the work done in following the original path.

31. (II) Two moles of an ideal gas are carried around the thermodynamic cycle shown in Fig. 18–25. The cycle consists of (1) an isothermal expansion $A \rightarrow B$ at a temperature of 700 K, with the pressure at A given by $p_A = 8$ atm; (2) an isobaric compression $B \rightarrow C$ at 3 atm; and (3) an isochoric pressure increase $C \rightarrow A$. What work is done by the gas per cycle?

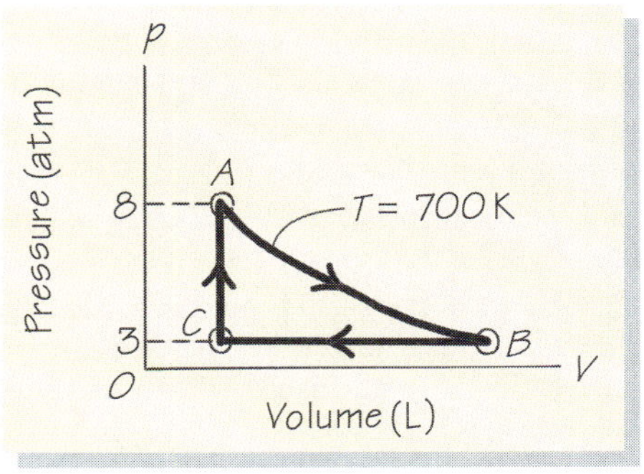

FIGURE 18–25 Problem 31.

32. (III) An external force compresses 0.10 mol of an ideal gas in thermal isolation. The curve of this transformation on a p–V diagram is $pV^{1.4}$ = a constant. The gas initially has a volume of 2.5 L and a temperature of 350 K. When the compression is finished, the temperature has increased to 400 K (Fig. 18–26). (a) What are the final volume and final pressure? (b) How much work was done on the gas to compress it?

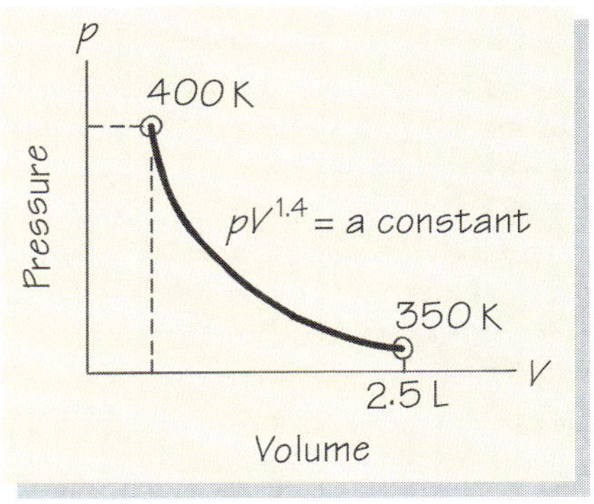

FIGURE 18–26 Problem 32.

18–5 The First Law of Thermodynamics

33. (I) A certain engine follows a closed thermodynamic cycle in which the heat flow into the system is measured to be 633 cal. Assuming that all transformations of the cycle can be approximated as reversible ones, what is the net work done, in joules, by the engine in one cycle?

34. (I) An ideal gas undergoes a reversible transformation from an initial state (p = 1.0 atm, V = 350 cm^3) to a final state (p = 2.0 atm, V = 175 cm^3). The transformation is carried out in two steps: The gas is compressed at constant pressure, and when the final volume is achieved, the pressure is increased while the volume is held fixed. Use the first law of thermodynamics to calculate the heat flow into the gas during this transformation. Assume (as is shown in Section 18–6) that, for an ideal gas, the internal energy U is a function of T only.

35. (II) Consider the p–V diagram shown in Fig. 18–27. If the system is taken from point A to point B via path $A \rightarrow C \rightarrow B$, 40,000 cal of heat flows into the system, and the system does 20,000 cal of work. (a) How much heat flows into the system along path $A \rightarrow D \rightarrow B$ if the work done is 7000 cal? (b) When the system returns from point B to point A along the curved path, the work done is 15,000 cal. Does the system absorb or liberate heat, and how much?

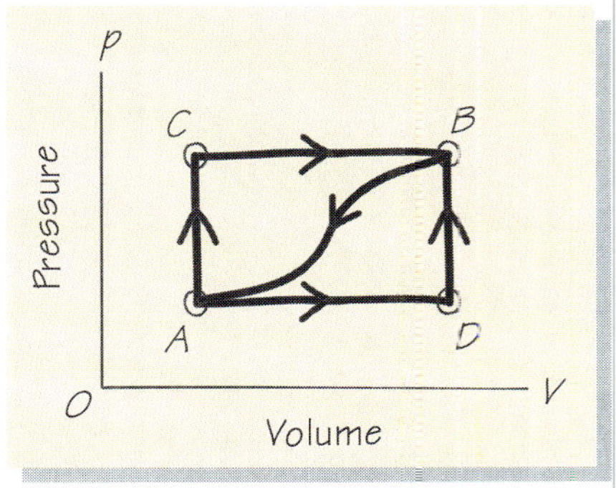

FIGURE 18–27 Problem 35.

36. (II) The latent heat of vaporization for water at 1 atm and 100°C is 540 cal/g. How much of that heat is converted into the mechanical work needed to change the volume of 1 g of water into 1 g of steam? [*Hint*: Calculate the volume of 1 g of steam.]

18–6 Internal Energy of Ideal Gases

37. (I) A thermal system consisting of 1 mol of an ideal gas forms a cycle that contains a leg AB in which pressure decreases linearly as the volume increases, a leg BC in which the gas is compressed at constant pressure until the initial volume is attained, and a leg CA in which the volume is held fixed as the pressure increases to its initial value (Fig. 18–28). (a) Express the internal energy at A, B, and C in terms of p_A, p_B, V_A, and V_B. (b) How much work does the gas do over the cycle?

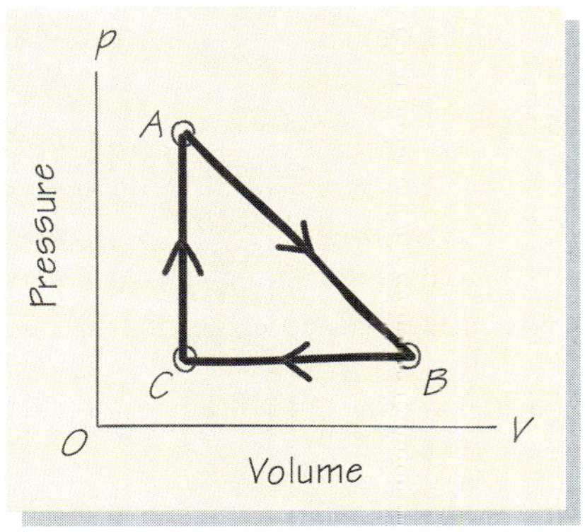

FIGURE 18–28 Problem 37.

38. (II) An ideal monatomic gas, with $c'_V = 3R/2$, undergoes a change of state. On a plot of pressure versus volume, it follows straight lines from an initial point (p_0, V_0) to $(p_0, 3V_0)$ to $(4p_0, 3V_0)$. (a) Draw the p–V plot. (b) By how much does the internal energy change? (c) How much work is done by the gas on its surroundings? (d) How much heat is added to the gas?

39. (II) Five grams of N_2 gas (molecular weight 28 g/mol) are allowed to expand from an initial state $(p_i, V_i) = (1.2 \times 10^5\ \text{N/m}^2, 3.5 \times 10^{-3}\ \text{m}^3)$ to a final state $(p_f, V_f) = (1.8 \times 10^5\ \text{N/m}^2, 5.0 \times 10^{-3}\ \text{m}^3)$ along a path that forms a straight line on a p–V plot (Fig. 18–29). (a) What are the initial and final temperatures? (b) What is the change in the internal energy of the gas? (c) What is the work done by the gas? (d) What is the heat flow into or out of the gas during the process?

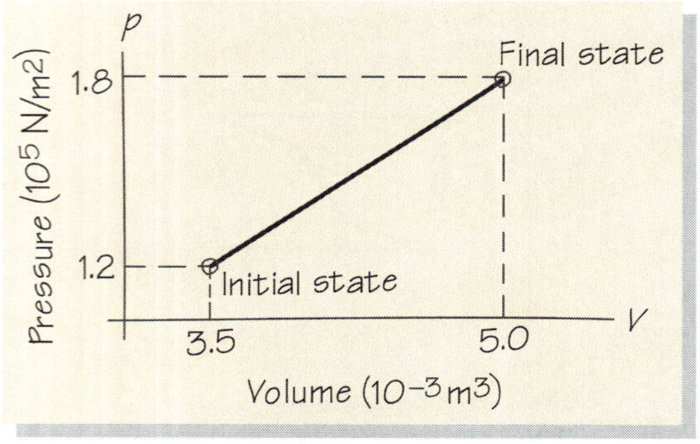

FIGURE 18–29 Problem 39.

40. (II) One mole of an ideal gas in contact with a thermal reservoir at 350 K is under 10 atm of pressure. The pressure is slowly increased to 25 atm. (a) What are the initial and final volumes of the gas? (b) How much work is done *on* the gas? (c) What is the change in internal energy of the gas? (d) What is the heat flow into the gas?

41. (II) According to a calorimeter in contact with 0.2 mol of an ideal gas, the gas does 8 J of work while 5 J of heat is added to it. If the molar specific heat of the gas, c'_V, is 20.8 J/mol · K, independent of temperature, by how much is the temperature of the gas changed in the process described? Is it warmed or cooled?

42. (II) One mole of an ideal gas, $\gamma = 5/3$, is at STP, and 500 J of heat is added at constant pressure. (a) What are the initial and final internal energies of the gas? (b) What is the work done by the gas? (c) Repeat parts (a) and (b) for the same amount of heat added at constant volume.

18–7 More Applications for Ideal Gases

43. (I) How much work does it take to compress 30 mol of an ideal gas at a fixed temperature of 15°C to half its volume?

44. (I) One mole of ideal gas undergoes an adiabatic transformation in which its temperature changes by 25 K. How much work is required to bring about this transformation, given that the constant-volume molar heat capacity of the gas is $c'_V = 30.00$ cal/mol · K?

45. (I) Equation (18–38) is an equation for pressure as a function of height for the atmosphere, which is treated as an ideal gas. Show that, if T is independent of height, this equation has the solution $p(h) = p_0 e^{-Mg(h-h_0)/RT}$, where p_0 is the pressure at some reference height $h = h_0$.

46. (I) A monatomic ideal gas with $\gamma = 1.67$ undergoes an adiabatic compression starting from an initial temperature of 251 K and an initial volume of 1.73 L. The final volume is 1.12 L. What is the final temperature?

47. (II) The temperature of the (dry) air at the bottom of a mountain valley is 26°C. The air moves as a brisk wind to the ridge 600 m above the valley floor. What is the temperature of the air on the ridge?

48. (II) The wild fires of the California coast are often fed by desert winds, which heat up to about 100°F in the desert at an altitude of 500 m, then rush down to sea level. What is the temperature of the air in these winds at sea level?

49. (II) Consider two thermal states of an ideal gas, A and B, together with the possible reversible paths that connect them. Is there an adiabatic path that connects these states when (a) $V_A = V_B$? (b) $p_A = p_B$? (c) $T_A = T_B$?

50. (II) A diatomic ideal gas such as air, for which $\gamma = 1.4$, expands adiabatically to 10 times its original volume. (a) By what factor does the temperature change? (b) By what factor does the pressure change?

51. (II) One liter of gas characterized by $\gamma = 1.4$ is allowed to expand adiabatically to twice its volume. How does the pressure change compare with the pressure change that would occur for an isothermal expansion?

52. (II) How much work is done by 1.00 L of helium when it expands adiabatically from sea-level pressure (1.00 atm) to the point where the pressure is 0.10 atm? Helium is monatomic, with $c'_V = 3R/2$.

53. (II) A particular adiabatic transformation of an ideal gas occurs as the pressure increases by a factor of 5 while the volume decreases by a factor of 3. What is the ratio of C_V to C_p? How much work is done on the gas during the compression, in terms of the initial pressure and volume? By how much does the internal energy of the gas change?

54. (II) The compression ratio (the ratio of the maximum volume to the minimum volume) of the cylinder of a diesel engine is 17:1. The working gas is air (for which $\gamma = 1.4$), which enters the cylinder at room temperature, approximately 310 K. The compression is so rapid that there is no heat flow through the cylinder walls, and the compression can be said to be adiabatic. To what temperature is the air heated?

55. (II) In Problem 17–43, the operation of a bicycle pump was studied on the assumption that the air in the pump is in thermal equilibrium with the environment. Repeat the analysis for the case when the pump is pushed down so quickly that no heat exchange occurs and the process is adiabatic. Assume that the temperature of the environment is 22°C. (The real process actually lies between the very slow isothermal and the very rapid adiabatic case).

56. (III) Consider the *Carnot cycle* shown in Fig. 18–30. One mole of an ideal gas with volume V_A at temperature T_1 undergoes an isothermal expansion to a volume V_B. This is followed by an adiabatic expansion to a volume V_C and then an isothermal compression at temperature T_2 to volume V_D; the cycle is closed by an adiabatic compression ending with volume V_A at temperature T_1. You will need to know the form of the transformation curve followed on each leg, and the heat flow to the gas for each leg. For the adiabatic legs, you know the heat flows, but, for the isothermal legs, you may want to find the work done directly and use your knowledge of the energy change of an ideal gas when the temperature is constant. You also need to know the starting and ending points of the curves. What is the net heat, expressed in terms of the given variables, supplied to the gas during the cycle?

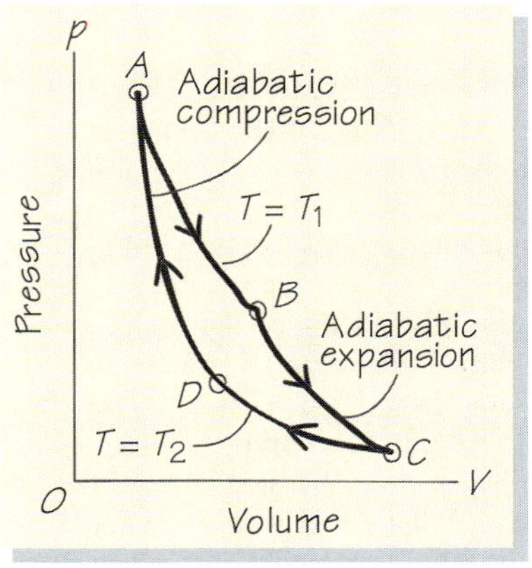

FIGURE 18–30 Problem 56.

57. (III) In Eq. (18–39) we state that by following an adiabatic curve, the temperature of Earth's atmosphere drops linearly with height h from the surface, and give the coefficient of h. Here, this result is derived. The starting points are Eq. (18–37), which expresses the relation between T and p in an adiabatic transformation of an ideal gas, as well as Eq. (18–38), which expresses the pressure of the atmosphere as a function of height. (a) Take the derivative with respect to height of Eq. (18–37), using the chain rule of differentiation (see Appendix IV–7) for the right-hand side, to show that

$$\frac{dT}{dh} = \text{a constant}\left(\frac{\gamma - 1}{\gamma}\right)p^{[(\gamma-1)/\gamma]-1}\frac{dp}{dh}.$$

(b) Show that this result can be rewritten as

$$\frac{dT}{dh} = \left(\frac{\gamma - 1}{\gamma}\right)Tp^{-1}\frac{dp}{dh}.$$

(c) Substitute Eq. (18–38) for dp/dh into the result derived in part (b) to find that

$$\frac{dT}{dh} = -\left(\frac{Mg}{R}\right)\left(\frac{\gamma - 1}{\gamma}\right).$$

(d) Show that Eq. (18–39) is the solution of the differential equation derived in part (c).

58. (III) An experiment shows that the amount of work done is 88.9 J when a certain ideal gas is compressed adiabatically from an initial pressure of 1.00 atm and an initial volume of 1.00 L to a final volume of 0.500 L. In a second experiment, the same amount of gas is again compressed from 1.00 L to 0.500 L, but this time it is in thermal contact with a thermal reservoir at $T = 20°C$, and 189 J of work is required for the compression. (In the second experiment, the initial pressure is not necessarily 1 atm.) (a) How many moles of gas are there? (b) What is γ for the gas? [*Hint*: Calculate the work done in the adiabatic compression for several values of γ between 1.2 and 2.] (c) What is C_V for the gas?

General Problems

59. (I) Five liters of oxygen at one-half atmospheric pressure and 273 K is heated at constant pressure until its volume has tripled. It is then compressed isothermally back to 5 L. (a) Draw the process on a p–V diagram. (b) What is the final pressure of the gas? (c) How much work is done by the gas during the entire process?

60. (II) A system that consists of 0.6 mol of helium gas undergoes an isobaric compression from a volume of 1.5 L to 0.1 L at a pressure of 6.0 atm. Is there heat flow in this transformation? If so, what are its magnitude and sign?

61. (II) How much heat flow must there be into a system of 10 L of oxygen initially at STP if (a) the gas triples in volume while the pressure stays constant and (b) the gas doubles in pressure at constant volume?

62. (II) A camper carries a 5-gal plastic container of water up a 40-m-high hill to her campsite. (a) How much heat flow does it take to raise the temperature of 2 qt of the water by 65°C? (b) Compare this energy with the energy expended to carry the entire 5 gal up the hill. (c) How much heat flow will it take to boil the 2 qt of water if it is initially at 23°C?

63. (II) Consider the adiabatic and isothermal processes that each pass through the point (p_0, V_0) on a p–V diagram (Fig. 18–31). Which of the processes has the greater slope at this point? Compute the ratio of the slopes at this point.

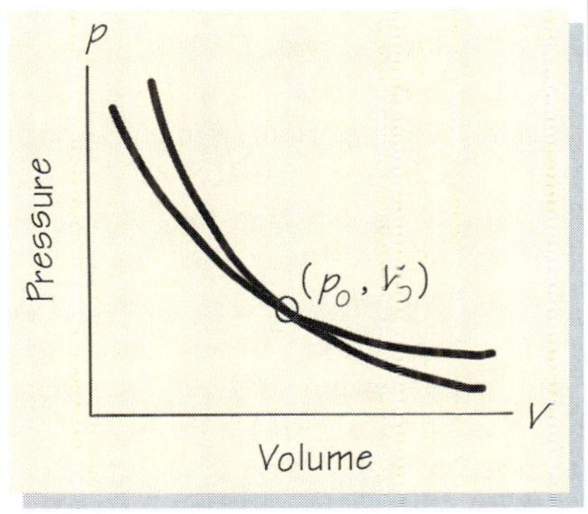

FIGURE 18–31 Problem 63.

64. (II) A 1-mol sample of an ideal gas for which $c'_V = \frac{3}{2}R$ follows the closed thermodynamic cycle shown in Fig. 18–32. There are three legs in the cycle: an isothermal expansion, $A \to B$; an isobaric compression, $B \to C$; and a constant-volume increase in pressure, $C \to A$. The temperature during the isothermal leg is $T = 300$ K, $p_A = 5$ atm, and $p_B = p_C = 2$ atm. What are (a) V_A? (b) V_B? (c) the work done by the system during leg $A \to B$? (d) Q for leg $A \to B$? (e) T_C? (f) the net work done by the system over the complete cycle?

65. (II) A 100-g piece of copper is heated from 0°C to 100°C at atmospheric pressure. What is the change in its internal energy?

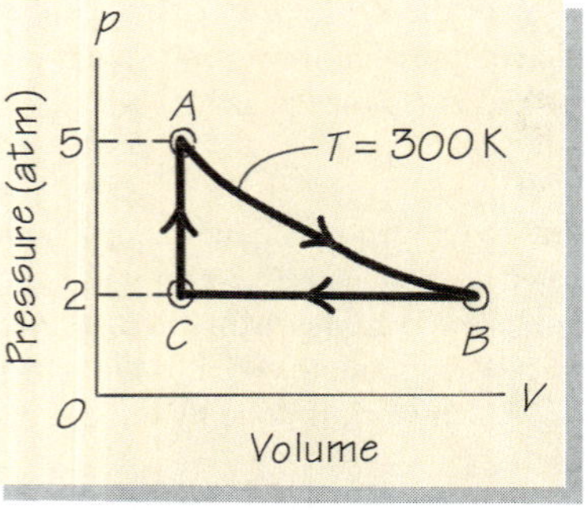

FIGURE 18–32 Problem 64.

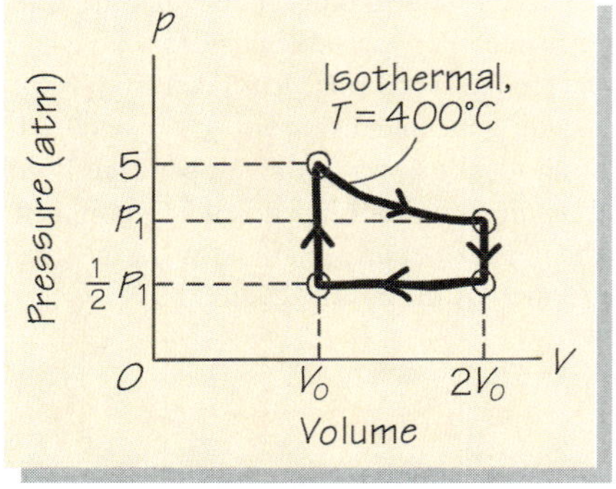

FIGURE 18–33 Problem 73.

66. (II) A pan contains 0.2 kg of water at 20°C. (a) How much heat does it require to raise the temperature of the water to 100°C and vaporize it at 1 atm? (b) How much work is done by the water as it changes to the gaseous phase? (c) What is the total change in internal energy of the water?

67. (II) A nuclear power plant requires 1000 MW (megawatts) of cooling. If environmental concerns limit the temperature rise of the water to 12°C, how much water flow is required to cool the plant?

68. (II) Heat flow is transferred to 0.3 mol of an ideal gas at a temperature of 500 K and pressure of 2.5 atm, resulting in an isothermal expansion of the volume by a factor of 5. How many calories are involved in the heat flow?

69. (II) One mole of N_2 gas initially at a temperature of 300 K and pressure of 1 atm is compressed isothermally until its volume is reduced by a factor of 10. (a) What is the final pressure of the gas? (b) What is the work done by the gas? (c) How much heat flow is transferred?

70. (II) An electric immersion heater is placed in a coffee cup to heat the water to boiling. The cup initially contains 0.33 L of water at 70°F. The heater is rated for 350 W. Neglect heat loss from the water and the cup. How long does it take to heat the water up to boiling? How long after this does it take to boil the water away completely?

71. (II) A glass contains 150 cm^3 of water at 70°F. Four ice cubes of 25 g each of temperature 10°F are dropped into the water. Neglecting heat-flow loss, what is the final temperature of the water? The specific heat of ice in this temperature range is 2.04 J/g · K.

72. (II) The temperature of espresso coffee (mostly water) can be increased by blowing 100°C steam through it. How much steam (in grams) is needed to heat up a 50-cm^3 cup of espresso from 45°C to 70°C? What is the volume of this quantity of steam, assuming that the steam is an ideal gas?

73. (II) One mole of an ideal gas is carried around the thermodynamic cycle shown in Fig. 18–33. The cycle consists of an isothermal expansion at a temperature of 400°C with an initial pressure of 5 atm, leading to a doubling of the initial volume. This is followed by a pressure drop of a factor of 2 at constant volume, and then by an isobaric compression until the initial volume is restored. The cycle is completed by a constant-volume pressure increase to the initial value of 5 atm. Calculate the values of p, T, and V not given, and the work done by the gas during one cycle.

74. (II) In experiments carried out between 1759 and 1762, Joseph Black started with a glass cup of mass 32 g, containing 467 g of water at 88°C. Black took a piece of ice of mass 404 g at 0°C and put it in the cup. The final equilibrium temperature was measured to be 12°C. How many calories are needed to melt 1 g of ice; that is, convert it to water at 0°C? That quantity is the latent heat of fusion. What would the equilibrium temperature be if no energy were required to convert ice to water at 0°C? Ignore the heat capacity of the glass.

75. (II) An ideal gas for which $c'_V = 5R/2$ is carried around a cycle $a \to b \to c$. There are 2.4 mol of the gas in the cycle. The expansion $a \to b$ is a straight line on a p–V diagram with $T_b = T_a = 540$ K, $b \to c$ is a constant-pressure segment with $p = 1.8$ atm, and $c \to a$ is a constant-volume segment at 15 L. (a) What is p_a? (b) What is V_b? (c) What is the work done in segment $a \to b$? (d) What is T_c? (e) What is the change in internal energy in segment $c \to a$? (f) What is the net work done during the entire cycle?

76. (III) A cyclic process with 0.4 mol of ideal gas is represented by a circle on the p–V diagram (if the appropriate scale is chosen) (Fig. 18–34). Calculate (a) the amount of heat transferred to the gas between the minimum and maximum volumes, as given in the figure; and (b) the largest and smallest internal energies during the cycle, given that $c'_V = 5R/2$.

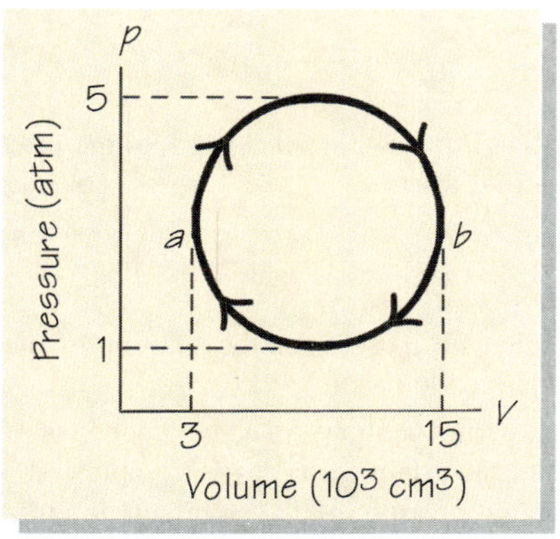

FIGURE 18–34 Problem 76.

CHAPTER

19

The properties of the molecules that make up the gases within these lights determine the colors produced when electrical effects act. Indeed, a more complete understanding of the thermodynamic properties of these and other systems critically depends on an understanding of these systems at the molecular level.

Molecules and Gases

A hypothetical film of the molecular motion of a dilute gas would show the molecules rushing around at different velocities, bouncing elastically off each other and off the walls in a three-dimensional game of billiards. Huge numbers of molecules make up even the smallest measurable quantity of a gas. The behavior of such a vast number of molecules can be expressed only in terms of averages. The kinetic theory of ideal gases, the focus of this chapter, is based on this kind of statistical treatment. In this chapter, we will see how ideal—that is, dilute—gases can be explained in kinetic theory. Along the way, we will learn about some distributions and their properties, and about transport phenomena. We will see that the kinetic theory of gases is remarkable because, with so little input, it can explain so much of the thermodynamics we have already studied.

Statistical techniques for the study of large collections of particles have been the subject of intensive work in physics and other fields. Early work was done by James Clerk Maxwell in the 1850s and by Ludwig Boltzmann and Josiah Willard Gibbs in the latter half of the nineteenth century. Research in this area falls under the general heading of **statistical physics**, a vast field that also includes the physics of liquids and solids. Statistical techniques allow us to explain precisely a great many of the physical properties observed in bulk matter. Statistical physics has been used to explain many new and often unexpected phenomena: phase changes, superconductivity, complex crystalline structure, and chaotic behavior, to name just a few.

19–1 A MICROSCOPIC VIEW OF GASES

There are some 6×10^{23} molecules (Avogadro's number) in 1 mol of any gas, and such incomprehensibly large numbers evoke large crowds. But just how crowded are the molecules? It follows from the ideal gas law that, at standard temperature and pressure (STP), 0°C and 1 atm, the volume of 1 mol of gas is 22.4 L = 22.4×10^{-3} m^3. This means that, on average, each molecule occupies a volume of $(22.4 \times 10^{-3}\ \text{m}^3)/(6 \times 10^{23}) \simeq 4 \times 10^{-26}\ \text{m}^3$. Using a radius r of 10^{-10} m for a typical molecule, we calculate the molecular volume to be $\frac{4}{3}\pi r^3 \simeq 4 \times 10^{-30}\ \text{m}^3$. A molecule generally takes up only $(4 \times 10^{-30}\ \text{m}^3)/(4 \times 10^{-26}\ \text{m}^3) = 10^{-4}$ of the volume available to it! The cube root of this number, namely, about 0.05 or $\frac{1}{20}$, gives the ratio of the typical molecular radius to the mean spacing between molecules. If we were to scale molecules to the size of students, with a radius of 0.5 m, the molecules would be spaced some 10 m, or 30 ft, apart.

As a gas becomes more dilute, molecular collisions occur less frequently. A dilute (or ideal) gas is a collection of independent molecules that move about with differing velocities, only rarely interacting with one another. The rare collisions do, however, play an important role. They are the means by which gases come to thermal equilibrium. When a hot gas (consisting of rapidly moving molecules) is mixed with a cold gas (composed of more slowly moving molecules), elastic collisions between fast molecules and slow molecules slow down the fast molecules and speed up the slow ones (see Chapter 8). The entire system comes to equilibrium at an intermediate temperature.

Why do averages tell us anything at all about a gas? After all, if we toss 10 coins, the average number of heads is 5, but we have very little confidence that we will actually get heads 5 times in just 10 tosses. (Try it!) It is only because the number of molecules we deal with is so large that we can be confident that average quantities really do represent the behavior of gases. Suppose that the number of coin tosses (N) is very large. The average number of heads is $0.5N$. In a given trial, the probability that we will get heads more than 51 percent of the time is about 38 in 100 for $N = 100$. This probability drops to about 2 in 100 for $N = 10^4$, and is on the order of 1 in 10^{88} for a million tosses. It is generally true that, for systems with a large number of independent components, large deviations from the average value of a variable are unlikely. When a gas that contains 10^{24} or so molecules in thermal equilibrium is sampled, the chances that the measured value of any randomly varying quantity will differ from its average value are tiny indeed.

(a)

(b)

FIGURE 19–1 After Otto von Guericke invented the air pump, he performed an experiment in 1657 in Magdeburg, Germany that allowed the effect of air pressure to be observed directly. He pumped the air out of a sphere formed from two hemispheres. Even horses could not pull the two halves apart because of the pressure of air molecules outside the sphere. In this modern reenactment, (a) the two halves of the sphere are easily separated before it has been pumped out, but (b) it cannot be pulled apart once it has been emptied of air.

19–2 PRESSURE AND MOLECULAR MOTION

In kinetic theory, a container of dilute gas consists of N independently moving molecules. These molecules move with a variety of speeds in a variety of directions. In this microscopic model, pressure is due to molecular collisions with the walls of the container (Fig. 19–1). Each time a molecule hits and rebounds from the wall, there is a momentum transfer to the wall—in other words, a force on the wall. We shall calculate the *average* momentum transfer to calculate the pressure.

Some Average Values in a Gas

A dilute gas consists of N molecules of mass m, and it is in thermal equilibrium at temperature T. The gas is in a box of volume V oriented in a Cartesian coordinate system, as shown in Fig. 19–2a. The molecules in the box are moving in random directions and *the average velocity of the molecules is zero* because as many move in one direction as move in another. The same can be said for the average x-, y-, or z-components of the velocity. We will use angular brackets, $\langle\ \rangle$, to indicate the average value, so

$$\langle \mathbf{v} \rangle = \langle v_x \rangle = \langle v_y \rangle = \langle v_z \rangle = 0. \tag{19–1}$$

The average velocity may be zero, but the same is not true for the average *speed.* The speed is always positive, and it must have an average value that is likewise positive. We will concentrate on the average value of the speed squared (which is the same as the velocity squared) $\langle v^2 \rangle$. Note that this is *not* the same as the square of the average value of the velocity: $\langle v^2 \rangle \neq \langle \mathbf{v} \rangle^2$; in fact, the latter quantity is zero because $\langle \mathbf{v} \rangle$ is zero. We refer to $\sqrt{\langle v^2 \rangle}$ as the **root-mean-square (rms) velocity**.

We can relate $\langle v^2 \rangle$ to the internal energy of a dilute gas by kinetic theory. Forces between molecules are important only when the molecules are close together, and this happens relatively rarely in dilute gases. Thus the internal energy, U, consists mainly of the kinetic energies of the molecules. In other words, U is given by

The internal energy of a dilute gas is the sum of the kinetic energies of its molecules.

$$U = N\langle K \rangle = N\left(\frac{1}{2} m \langle v^2 \rangle\right), \tag{19–2}$$

where $\langle K \rangle$ is the average kinetic energy per molecule and N is the total number of molecules. This equation provides us with a link to the temperature of the gas, because, as we learned in Chapter 18, the internal energy of an ideal gas is a function only of temperature.

The average of a sum of terms is the sum of the average of those terms, so

$$\langle v^2 \rangle = \langle v_x^2 + v_y^2 + v_z^2 \rangle = \langle v_x^2 \rangle + \langle v_y^2 \rangle + \langle v_z^2 \rangle. \tag{19–3}$$

The gas as a whole is not moving, so there must be an equivalence among the three directions. The average value of the x-component of the velocity squared must be the same as that of the y-component and that of the z-component. Thus,

$$\langle v_x^2 \rangle = \langle v_y^2 \rangle = \langle v_z^2 \rangle. \tag{19–4}$$

Using this relation in Eqs. (19–3) and (19–2), it is possible to relate the average of the components of the velocity squared to the thermal energy of the gas:

$$\langle v^2 \rangle = 3\langle v_x^2 \rangle;$$

$$\langle v_x^2 \rangle = \frac{1}{3}\langle v^2 \rangle = \frac{2}{3}\frac{U}{mN}. \tag{19–5}$$

Next we will relate $\langle v_x^2 \rangle$ to the pressure of the gas.

The Origin of Pressure

Pressure arises from the multiple collisions the molecules of a gas have with the walls that contain the gas. For a quantitative calculation, we compute the momentum transfer to a wall due to a single collision, and then find the number of molecules that strike the wall per unit time. We use this to find the average momentum transfer to the wall per unit time, which is the force on the wall. Pressure is then the force per unit area. Figure 19–2b shows a molecule colliding elastically with the right-hand wall of the box of Fig. 19–2a. The wall is oriented in the yz-plane, so only the x-component of the velocity changes. If the velocity before the collision is

$$\mathbf{v}_i = v_x\mathbf{i} + v_y\mathbf{j} + v_z\mathbf{k},$$

then the velocity after the collision is

$$\mathbf{v}_f = -v_x\mathbf{i} + v_y\mathbf{j} + v_z\mathbf{k}.$$

The momentum change of the molecule, $\Delta\mathbf{P}_{\text{mol}}$, is

$$\Delta\mathbf{P}_{\text{mol}} = m\mathbf{v}_f - m\mathbf{v}_i = -2mv_x\mathbf{i};$$

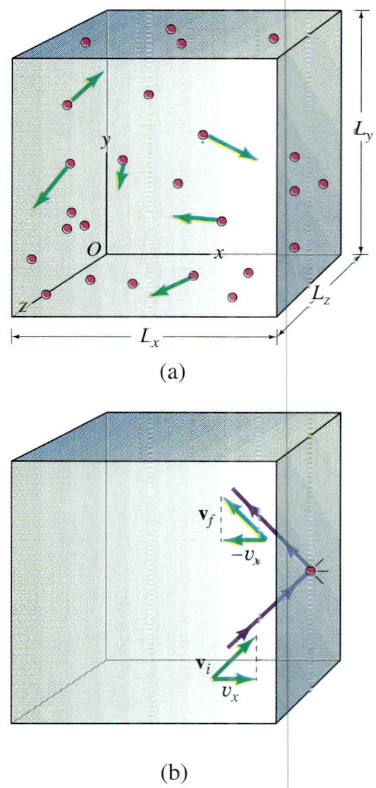

FIGURE 19–2 (a) The dimensions and orientation of a box that contains a gas. (b) A typical molecule, whose x-component of velocity is v_x, has an elastic collision with the right-hand wall. The x-component of the velocity changes its sign in the collision, but the y- and z-components are unchanged.

the momentum transfer to the wall, $\Delta\mathbf{P}$, is the negative of this:

$$\Delta\mathbf{P} = 2mv_x\mathbf{i}. \tag{19–6}$$

This positive result is reasonable: the wall is pushed to the right by the collision.

We have found the momentum transfer from a single molecule. As we continue, we think of this molecule as an "average" molecule; it has the average x-component of velocity, which we continue to write as v_x for now. The next question is: How many collisions with the wall occur per unit time? The number of molecules with an x-component of velocity of magnitude v_x that strike an area A in a time interval dt is the number of molecules contained in a cylindrical volume whose base is the area A and whose length is $v_x\,dt$ (Fig. 19–3). Molecules that travel toward this area and are farther away than the cylinder height will not reach the wall in the infinitesimal time interval dt. Note also that by making the time interval short enough, we don't have to worry about the possibility that the molecules collide with each other before they reach the wall. If the number of molecules per unit volume (the number density) is $n = N/V$ (not the number of moles here!), then the total number of molecules in our cylindrical volume is $(n)(v_x\,dt)(A)$. Thus the number of collisions suffered by the wall is

$$\frac{1}{2}(n)(v_x\,dt)(A) = \frac{N}{2V}(v_x\,dt)A.$$

The factor $\frac{1}{2}$ is present because only half of the molecules are moving to the right; the other half are moving to the left. Only the ones that move to the right contribute to the pressure in this case. The number of collisions must be multiplied by the individual collision momentum transfer ΔP_x to find the total momentum dP_x transferred in the time interval dt:

$$dP_x = (2mv_x)\left(\frac{N}{2V}\right)(v_x\,dt)A = mv_x^2\frac{N}{V}\,dt\,A. \tag{19–7}$$

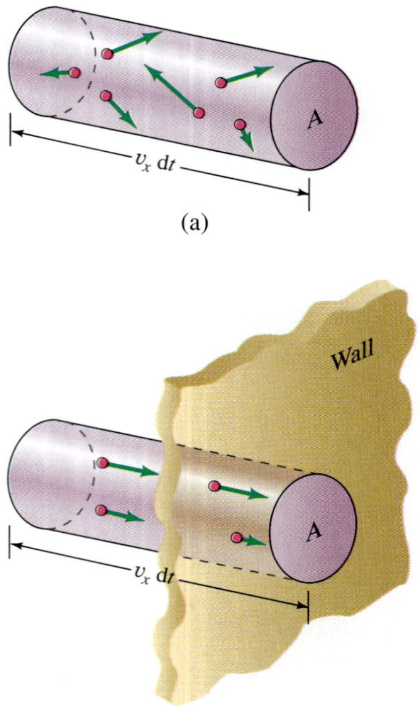

FIGURE 19–3 (a) All the molecules that will collide with an area A of a wall in a time interval dt are contained in a cylinder of area A and length $v_x\,dt$. (b) Only the x-component of velocity is indicated, and only those with x-component of velocity directed toward the wall.

Thus the total momentum transfer per unit time—the force exerted on area A—is

$$\frac{dP_x}{dt} = F_x = mv_x^2\frac{N}{V}A. \tag{19–8}$$

The pressure on the wall, p, is the force per unit area:

$$p = \frac{F_x}{A} = mv_x^2\frac{N}{V}. \tag{19–9}$$

Finally, recall that we have employed the average x-component of velocity squared for each molecule. We make this explicit by employing the notation $\langle v_x^2\rangle$ rather than v_x^2. Then, with the help of Eq. (19–5), we find

The pressure of a dilute gas in terms of its internal energy

$$p = m\langle v_x^2\rangle\frac{N}{V} = m\left(\frac{2}{3}\frac{U}{mN}\right)\frac{N}{V} = \frac{2}{3}\frac{U}{V};$$

$$pV = \frac{2}{3}U. \tag{19–10}$$

This derivation is an important one. We have used the dynamical variables of mechanics to find a relation between thermodynamic variables.

EXAMPLE 19–1 Exactly 1 mol of helium gas has an internal energy of 3600 J. It is contained within a cube of sides 0.50 m. (a) Compare helium under these conditions with air at standard temperature and pressure (STP). If air is ideal at STP, can the same be said for helium? (b) Approximately how many times per second do the walls suffer collisions from the molecules of the helium gas?

Solution: (a) We compute the pressure and temperature of the helium and compare them with those of air at STP. Equation (19–10) can be used to compute the pressure:

$$p = \frac{2}{3}\frac{U}{V} = \frac{2}{3}\frac{3600\ \text{J}}{(0.50\ \text{m})^3} = 1.9 \times 10^4\ \text{N/m}^2,$$

which is about 0.2 atm. The temperature can be computed from the ideal gas law:

$$T = \frac{pV}{nR} = \frac{(1.9 \times 10^4\ \text{N/m}^2)(0.50\ \text{m})^3}{(1\ \text{mol})(8.31\ \text{J/mol})} = 290\ \text{K}.$$

This is in the range of room temperature. The helium has pressure and temperature similar to those of air at sea level and can be treated as ideal.

(b) We have seen that the number of molecules that collide with a wall in an infinitesimal time dt is

$$\text{number of collisions} = \frac{N}{2V}(v_x\,dt)A.$$

This equation is approximately true for small but finite time differences, so we replace dt by $\Delta t = 1$ s. The area A is the area of one wall, $A = (0.50\ \text{m})^2$. For the typical value of the magnitude of one component of the velocity, v_x, we use $\langle v_x^2\rangle$ and Eq. (19–5):

$$\sqrt{\langle v_x^2\rangle} = \sqrt{\frac{1}{3}\langle v^2\rangle} = \sqrt{\frac{2}{3}\left(\frac{U}{mN}\right)}.$$

The quantity mN, the mass of one molecule times the number of molecules, is the total mass of a gas, M. There is 1 mol of helium so the mass is 4 g $= 4 \times 10^{-3}$ kg. A measure of speed v_x is then

$$v_x = \sqrt{\frac{2}{3}\left(\frac{3600\ \text{J}}{4 \times 10^{-3}\ \text{kg}}\right)} \simeq 780\ \text{m/s}.$$

Finally, because there are six walls and all six directions are equivalent, we must multiply our result by 6 to find the total number of collisions with the walls:

$$\begin{aligned}\text{total number of collisions} &= 6\,\frac{N}{2V}Av_x\,\Delta t\\ &= 6\,\frac{6.0 \times 10^{23}}{(2)(0.50\ \text{m})^3}[(0.50)^2\ \text{m}^2](780\ \text{m/s})(1\ \text{s})\\ &= 2.8 \times 10^{27}\ \text{collisions in 1 s.}\end{aligned}$$

19–3 THE INTERPRETATION OF TEMPERATURE

How can we interpret the temperature of a gas in terms of a microscopic model? The relation

$$U = N\langle K\rangle = \frac{1}{2}Nm\langle v^2\rangle \tag{19–11}$$

is a link provided by the kinetic theory between the macroscopic and microscopic properties of a monatomic ideal gas. Equation (19-10) gives $U = \frac{3}{2}pV$. But $pV = nRT = NkT$, where k is Boltzmann's constant. Thus

$$U = \frac{3}{2}nRT = \frac{3}{2}NkT. \tag{19–12}$$

Equations (19–11) and (19–12) together give a microscopic interpretation of temperature:

$$kT = \frac{2}{3}\frac{U}{N} = \frac{2}{3}\langle K \rangle. \tag{19–13}$$

The temperature of an ideal gas is proportional to its energy.

The number of molecules has canceled from this expression, so T is (correctly) independent of the amount of gas. *The temperature of an ideal gas is a measure of the average kinetic energy of the constituents.* See Fig. 19–4 for another microscopic interpretation.

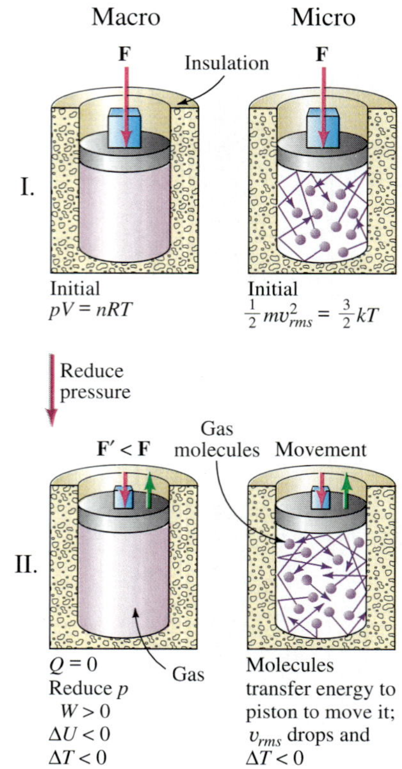

FIGURE 19–4 Macroscopic and microscopic views of the work done by an insulated gas. Molecules colliding with the recoiling piston lose energy. Thus the temperature drops.

EXAMPLE 19–2 Exactly 1 mol of helium gas in a large volume is cooled to 31.5 K. What is the internal energy of the gas as well as the average value of the x-component of velocity squared, $\langle v_x^2 \rangle$?

Solution: Because the gas is in a large volume, it is dilute and behaves as an ideal gas. The internal energy of the gas is therefore given by Eq. (19–12). For 1 mol,

$$U = \frac{3}{2}nRT = (1.5)(8.31\ \text{J/K})(31.5\ \text{K}) = 393\ \text{J}.$$

To compute $\langle v_x^2 \rangle$, we note that $\langle v_x^2 \rangle = \frac{1}{3}\langle v^2 \rangle$, and that $\langle v^2 \rangle$ is given in terms of U by Eq. (19–11):

$$kT = \frac{2}{3}\left(\frac{1}{2}\right)(m\langle v^2 \rangle) = \frac{m}{3}3\langle v_x^2 \rangle = m\langle v_x^2 \rangle;$$

$$\langle v_x^2 \rangle = \frac{kT}{m}.$$

The mass of a helium molecule is $m = (4\ \text{g/mol})(10^{-3}\ \text{kg/g})/(6.0 \times 10^{23}\ \text{molecules/mol}) = 6.7 \times 10^{-27}\ \text{kg}$, and

$$\langle v_x^2 \rangle = \frac{(1.38 \times 10^{-23}\ \text{J/K})(31.5\ \text{K})}{6.7 \times 10^{-27}\ \text{kg}} = 6.2 \times 10^4\ \text{m}^2/\text{s}^2.$$

Thus $\langle v_x \rangle$ is about 250 m/s. Even at temperatures as low as 31.5 K, the helium atoms are moving quite rapidly.

The zero of the Kelvin temperature scale has a simple meaning: It is the point at which the pressure drops to zero. In the microscopic view of an ideal gas, the meaning of this result is easily understood. From Eq. (19–13), the temperature is zero when the average kinetic energy of the ideal gas is zero. Pressure vanishes because the molecules no longer move around and bounce against the walls.

Interpretation of the van der Waals Gas

Knowing how the microscopic properties of an ideal gas relate to its macroscopic behavior enables us to understand the origin of the van der Waals equation of state, Eq. (17–12):

$$\left[p + a\left(\frac{n}{V}\right)^2\right]\left(\frac{V}{n} - b\right) = RT.$$

This equation of state is more accurate than the ideal gas law away from the dilute gas limit. It reduces to the ideal gas law when the constants a and b are zero. These constants have a simple meaning in the kinetic theory. Gas molecules are not truly like pointlike billiard balls. Billiard balls exert a force on one another only during a very brief time period when they touch. Real molecules have forces that act between them over distances larger than their radii. These forces are repulsive at short range, so that the molecules bounce off one another, but are slightly attractive at longer distances.

The attractive component of the intermolecular force acts to make a gas more compact. This translates into a reduced pressure: The term $a(n/V)^2$ represents the effect of the long-range attraction between molecules. If we solve Eq. (17–12) for p, we see it is reduced by $a(n/V)^2$. As for the term proportional to b, molecules, like billiard balls, take up some space. This term is present because of the strong repulsion between molecules at a characteristic radius. By appearing as a term subtracted from V, the constant b measures the volume unavailable for the motion of molecules because it is space already occupied by other molecules. Example 19–3 illustrates how we can use our interpretation of the constant b to estimate the size of molecules.

EXAMPLE 19–3 Measurements show that nitrogen gas obeys the van der Waals equation of state with the constant $b = 3\ 94 \times 10^{-5}\ \text{m}^3/\text{mol}$. What is the size of a nitrogen molecule?

Solution: Because b is subtracted from the term that represents the total volume per mole (V/n) in the van der Waals equation of state, b represents the volume occupied by 1 mol of molecules. Therefore each molecule has volume

$$V_m = \frac{b}{N_A} = \frac{3.94 \times 10^{-5}\ \text{m}^3/\text{mol}}{6.02 \times 10^{23}\ \text{molecules/mol}} = 6.54 \times 10^{-28}\ \text{m}^3/\text{molecule}.$$

If the molecule is spherical with radius r, then $V_m = \frac{4}{3}\pi r^3$, and

$$r = \left(\frac{3}{4\pi} V_m\right)^{1/3} = \left[\frac{3(0.654 \times 10^{-27}\ \text{m}^3)}{4(3.14)}\right]^{1/3} = 5.39 \times 10^{-10}\ \text{m}.$$

This radius is in good agreement with more direct atomic measurements.

19–4 PROBABILITY DISTRIBUTIONS

Just as every student in a class does not receive the average grade on a given test, the molecules in a gas do not each have the average velocity or the average position. Instead, each molecule has its own velocity and position. Two functions describe the probability that a molecule has a particular velocity or a particular position. The **velocity distribution function** describes how many molecules have one velocity, how many have another, and so forth. The **position distribution function** similarly describes how the molecules are distributed in space. We need to know something about these distributions in order to understand more about the thermal properties of gases.

We all have a certain familiarity with the notions of probability and averages (Fig. 19–5). In this section, we will sharpen these notions so that we can use them in the analysis of gases and other thermal systems.

FIGURE 19–5 Roulette is a well-known application of the laws of probability.

PROBLEM SOLVING

Using probability notions

Probabilities in a Set of Grades

Consider a quiz taken by a class of 100 students. The grades can run from 0 to a maximum score of 50. The number of students who receive each grade is shown in Fig. 19–6, where the number of students with grade g, N_g, is plotted as a function of the grade g. The total number of students here is $N = 100$, and if we add up all the N_g, we get all of the students taking the quiz:

$$\sum_g N_g = N. \tag{19–14}$$

Given such a distribution of grades, we can find the average grade. We take the number of people with a given grade, multiply it by the grade, add these products for each grade, and divide by the total number of students. We denote the average grade by $\langle g \rangle$. According to the above technique, we find

$$\langle g \rangle = \frac{1}{N}\left(\sum_g g N_g\right). \tag{19–15}$$

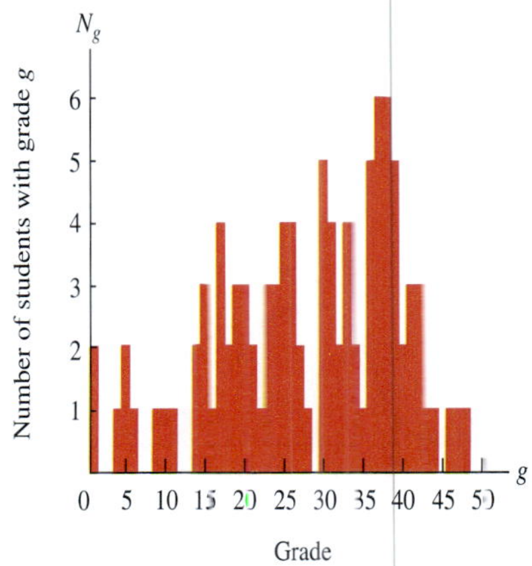

FIGURE 19–6 Bar graph of quiz grades in a class.

In the case illustrated, $\langle g \rangle$ is 28. We could also ask for the average of the *square* of the grade $\langle g^2 \rangle$, given by

$$\langle g^2 \rangle = \frac{1}{N}\left(\sum_g g^2 N_g\right). \tag{19–16}$$

In the case illustrated, $\langle g^2 \rangle = 908.6 = (30.1)^2$. Note that $\langle g^2 \rangle \neq \langle g \rangle^2$! The quantity $\langle g^2 \rangle$ contains different information than does $\langle g \rangle$. You would get the same value for $\langle g \rangle$ if half the class had 6 points and half had 50 points but, in that case, $\langle g^2 \rangle = 1268 = (35.6)^2$.

What is the probability that any one student, chosen at random, has a grade of 38 points? This probability is the number of students with a grade of 38 divided by the total number of students—in this case, $\frac{6}{50}$. More generally, the probability of finding a student who received a grade g is

$$P_g = \frac{1}{N} N_g, \tag{19–17}$$

and P_g is the **probability distribution** for the grades. As a consequence of Eq. (19–14), the probability distribution is such that

$$\sum_g P_g = 1. \tag{19–18}$$

That the sum of probabilities is 1 is known as a **normalization condition**. As we see from Eqs. (19–15) and (19–16), averages are obtained from weighted sums over P_g:

$$\langle g \rangle = \sum_g g P_g, \tag{19–19}$$

$$\langle g^2 \rangle = \sum_g g^2 P_g. \tag{19–20}$$

We can ask one more type of question: If we pick one student at random, what are the odds that his or her grade lies between 15 and 19? The probability of finding that the student has a grade between 15 and 19 is the number of students with grades in that range divided by the total number of students. We write this as follows:

$$P(15 \le g \le 19) = \frac{1}{N}\sum_{g=15}^{19} N_g = \sum_{g=15}^{19} P_g.$$

For our example, this is equal to $\frac{13}{100} = 0.13$. We will see that this type of question is especially relevant to quantities that vary continuously.

Continuous Distributions

PROBLEM SOLVING

Using probability for quantities that can vary continuously

Some distributions do not involve discrete quantities, such as the grades on an exam, but instead involve continuous quantities, such as the heights of students. We do not do things much differently, but we must recognize that the probability that any student will have a *particular* height x—say, $x = 1.73000055$ m—is zero. Instead, we must measure the probability $P(x)\,\Delta x$ that the outcome lies within some small interval Δx around the value x (Fig. 19–7). In our example, this could be the probability that a student has a height within an interval $\Delta x = 0.01$ m around $x = 1.73$ m. The normalization condition now expresses the fact that all students must have some height, so that the sum of all probabilities adds to 1:

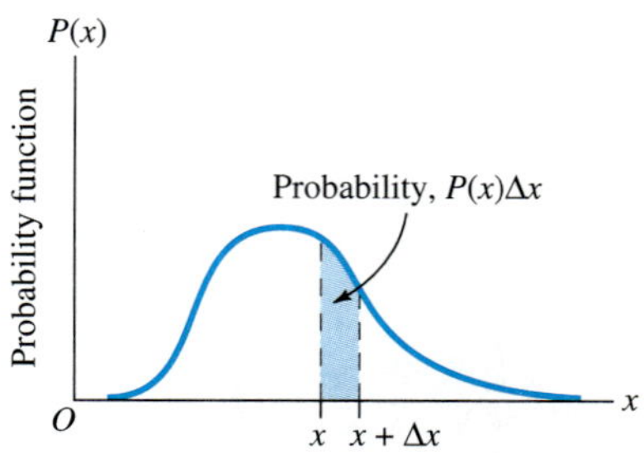

FIGURE 19–7 The probability $P(x)\,\Delta x$ associated with a continuous distribution is the area underneath the curve between x and $x + \Delta x$.

$$\sum P(x)\,\Delta x = 1. \tag{19–21}$$

A sum appears here because—even if x takes on a continuous range of values—the intervals Δx have divided the region into a *finite* number of segments.

The concept we have described can be refined by making the interval smaller and smaller, thereby finding the probability that an outcome near a particular x-value occurs with more and more accuracy. Calculus notation is appropriate in this limit. The quantity Δx is written as dx, and the sum becomes an integral whose upper and lower limits are the upper and lower values of the variable in question, x_2 and x_1:

$$\int_{x_1}^{x_2} P(x)\, dx = 1. \tag{19–22}$$

This is the normalization condition in the case of continuous variables.

The averages of functions of x are found from formulas similar to those of the discrete case. Thus

$$\langle x^2 \rangle = \int_{x_1}^{x_2} x^2\, P(x)\, dx. \tag{19–23}$$

EXAMPLE 19–4 Suppose that all heights between 1.5 m and 1.8 m are equally likely among a group of students and that no students are outside this height range. Find the probability distribution and use it to calculate the average height.

Solution: Let x represent height. The probability $P(x)$ is zero for $x < 1.5$ m and for $x > 1.8$ m. It is a constant P in between. We calculate P by using the normalization condition, Eq. (19–22):

$$\int_0^\infty P(x)\, dx = \int_{1.5\text{ m}}^{1.8\text{ m}} P\, dx = P\int_{1.5\text{ m}}^{1.8\text{ m}} dx = P(1.8\text{ m} - 1.5\text{ m}) = P(0.3\text{ m}) = 1,$$

so the constant $P = 1/(0.3\text{ m})$. Then $\langle x \rangle$ is given by

$$\langle x \rangle = \int_{1.5\text{ m}}^{1.8\text{ m}} xP\, dx = P\int_{1.5\text{ m}}^{1.8\text{ m}} x\, dx = P\frac{x^2}{2}\Bigg|_{1.5\text{ m}}^{1.8\text{ m}}$$

$$= \frac{1}{2(0.3\text{ m})}[(1.8\text{ m})^2 - (1.5\text{ m})^2] = 1.65\text{ m}.$$

Not surprisingly, this is the midpoint of the range of heights.

19–5 THE VELOCITY DISTRIBUTION OF GASES

In this section, we extend our ideas about probability distributions to molecules in a gas. The velocity distribution function $F(\mathbf{v})$ is a probability distribution for the velocities of the gas molecules. Velocities form a continuum, so we use a probability function appropriate for continuous variables. The velocity distribution function for ideal gases was first described by James Clerk Maxwell in 1859—a date when the molecular model of matter was by no means universally accepted. Let's start with a number distribution $N(\mathbf{v})$, such that

$N(\mathbf{v})d^3v$ = number of gas molecules with a velocity between $\mathbf{v}$ *and* $\mathbf{v} + d\mathbf{v}$.

The notation requires a little explanation. As was true of the height distribution of Section 19–4, the probability of finding any particular velocity $\mathbf{v}$ is zero. However, we can ask about the probability that a velocity will be somewhere in the neighborhood of $\mathbf{v}$. This "neighborhood" is a box of volume

$$d\mathbf{v} = d^3v = dv_x\, dv_y\, dv_z, \tag{19–24}$$

centered about the tip of the vector $\mathbf{v}$ (Fig. 19–8). The total number of molecules is N, and thus

$$\int N(\mathbf{v})\, d^3v = N. \tag{19–25}$$

The distribution $N(\mathbf{v})$ leads directly to a *probability* distribution $F(\mathbf{v})$:

$$F(\mathbf{v}) = \frac{1}{N}N(\mathbf{v}). \tag{19–26}$$

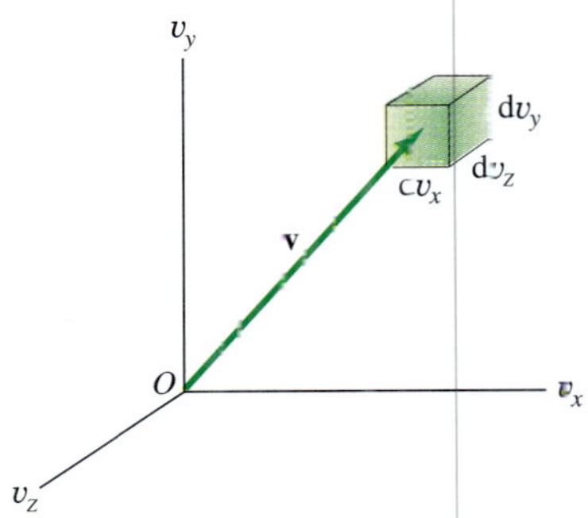

FIGURE 19–8 A molecule has a certain probability of having a velocity $\mathbf{v}$ within the "volume" $d\mathbf{v} = d^3v = dv_x\, dv_y\, dv_z$. Here we show the small box d^3v within which the velocity vector is located.

The meaning of this function is that

$F(\mathbf{v})\mathrm{d}^3v =$ the probability that the velocity of a gas molecule is between $\mathbf{v}$ and $\mathbf{v} + \mathrm{d}\mathbf{v}$.

From $F(\mathbf{v})$, we can calculate the averages for continuously varying quantities. In particular, the average of the velocity squared is

$$\langle v^2 \rangle = \int v^2 F(\mathbf{v})\, \mathrm{d}^3v. \tag{19–27}$$

To determine the probability distribution $F(\mathbf{v})$, we must make some *physical* assumptions. The model of molecules darting about and occasionally colliding with other molecules suggests that a given molecule may have any velocity. This leads to a fundamental physical postulate: Any way in which a gas's total energy and total momentum (which is zero) can be shared among the molecules is equally likely. Once this postulate is accepted, we are in the same position as someone asking for the probability of getting any particular number between 100 and 600 when throwing 100 dice. What counts is the *number of ways* of getting that number. There is only one way of getting 100 or 600, but there are many ways of getting 400; thus, the probability of getting 400 is much greater than the probability of getting 100 or 600. When this sort of counting is done properly for molecules whose individual kinetic energy is $\frac{1}{2}mv^2$ and whose average kinetic energy is $\frac{1}{2}m\langle v^2 \rangle = \frac{3}{2}kT$, the distribution function is

The velocity distribution function for an ideal gas

$$F(\mathbf{v}) = \left(\frac{m}{2\pi kT}\right)^{3/2} e^{-mv^2/2kT}. \tag{19–28}$$

The velocity distribution function is properly normalized, $\int F(\mathbf{v})\, \mathrm{d}^3v = 1$, meaning that the sum over all probabilities is 1. $F(\mathbf{v})$ is a function only of the magnitude of the velocity (that is, the speed), not the direction. This expresses the physically reasonable fact that there is no preferred direction in a container of gas in which the gas as a whole is not moving. All directions are equally likely. It is possible to make a direct experimental verification of Eq. (19–28) by cutting a tiny hole in a container of gas and measuring the different speeds of the molecules that leak out.

The Average of the Velocity Squared

To determine the internal energy, we need the average of the velocity squared, $\langle v^2 \rangle$, which we can now calculate. From the probability distribution $F(\mathbf{v})$, we have

$$\langle v^2 \rangle = \int v^2 F(\mathbf{v})\, \mathrm{d}^3v = \left(\frac{m}{2\pi kT}\right)^{3/2} \int v^2 e^{-mv^2/2kT}\, \mathrm{d}^3v. \tag{19–29}$$

PROBLEM SOLVING

Scaling technique for integrals

This integral can be simplified by using a technique known as **scaling**. This scaling technique is of great importance to scientists and engineers because it allows the *dependence* of physical quantities on other physical parameters to be extracted. Both kT and mv^2 have the dimensions of energy, so v^2 has the dimensions of kT/m. We therefore change the integration variable in Eq. (19–29) from $\mathbf{v}$ to the dimensionless vector $\mathbf{u}$ by the definition

$$\mathbf{v} \equiv \mathbf{u}\sqrt{\frac{2kT}{m}}. \tag{19–30}$$

This definition means that each component of the vector $\mathbf{v}$ is transformed—that is, it stands for three equations. With this definition, the volume element d^3v has the form

$$\mathrm{d}^3v = \left(\frac{2kT}{m}\right)^{3/2} \mathrm{d}^3u.$$

The factor $\sqrt{2kT/m}$ is cubed because d^3v stands for $\mathrm{d}v_x\, \mathrm{d}v_y\, \mathrm{d}v_z$. Thus

$$\langle v^2 \rangle = \left(\frac{m}{2\pi \mathrm{kT}}\right)^{3/2} \left(\frac{2kT}{m}\right)^{5/2} \int u^2 e^{-u^2}\, \mathrm{d}^3u = \frac{2kT}{m}\frac{1}{\pi\sqrt{\pi}} \int u^2 e^{-u^2}\, \mathrm{d}^3u. \tag{19–31}$$

APPLICATION: ISOTOPE SEPARATION

The fact that the faster molecules of a gas escape in greater quantity through a tiny hole than do the slower ones is the practical basis for the separation of uranium isotopes.† The uranium isotope ^{238}U is more common, but the relatively rare isotope ^{235}U is more useful for nuclear reactors such as those used for generating electric power. These isotopes cannot be separated by chemical means because they have the same chemical properties. However, the fact that ^{238}U is a bit heavier than ^{235}U means that when they are mixed in a gas—uranium hexafluoride, UF_6, is used—the ^{238}U component moves, on average, at a lower speed. This is because the factor $1/\sqrt{m}$ appears in any of the measures of speed, such as v_{rms}. The UF_6 gas is allowed to escape through many tiny holes from one container at a given temperature into another container that is initially empty. The percentage of ^{235}U in the gas that has escaped is slightly higher than that in the original container. Thousands of repetitions of this process of isotope separation lead to significantly increased percentages of ^{235}U.

Equation (19–31) shows that $\langle v^2 \rangle$ is a dimensionless constant times $2kT/m$. We have already estimated [Eqs. (19–11) and (19–13)] that

$$\langle v^2 \rangle = \frac{2U}{Nm} = \frac{3kT}{m}. \tag{19–32}$$

Thus, Eq. (19–31) is in accord with Eq. (19–32)—at least in terms of dimensions. We need to integrate Eq. (19–31) to show that it is in exact agreement with Eq. (19–32). In fact, we can show with the help of a table of integrals that

$$\frac{1}{\pi\sqrt{\pi}} \int u^2 e^{-u^2}\, d^3u = \frac{3}{2}.$$

Thus Eq. (19–32) follows.

We can get a feel for the speeds of molecules by considering the root-mean-square (rms) speed v_{rms}, the square root of the average value of the speed squared:

$$v_{rms} = \sqrt{\langle v^2 \rangle} = \sqrt{\frac{3kT}{m}}. \tag{19–33}$$

This result follows from Eq. (19–32). For helium gas, which has an atomic mass $m = 6.7 \times 10^{-27}$ kg, the rms speed is 556 m/s, 1362 m/s, and 2486 m/s for $T = 50$ K, $T = 300$ K, and $T = 1000$ K, respectively. Contrast this with argon gas, which has a molecular mass 10 times greater. The corresponding rms speeds are $\sqrt{10}$ times smaller: 176 m/s, 431 m/s, and 786 m/s.

*19–6 THE MAXWELL–BOLTZMANN DISTRIBUTION

An ideal gas may consist of molecules that have just one atom (monatomic) such as helium (He) or argon (Ar); it may consist of molecules with two (diatomic) or more atoms, such as nitrogen (N_2), oxygen (O_2), or water (H_2O) (Fig. 19–9). Is there a difference between these types of ideal gases? The answer came in the last part of the nineteenth century from the work of Ludwig Boltzmann.

The velocity distribution of a monatomic ideal gas [Eq. (19–28)] has the general form

$$F(\mathbf{v}) = \frac{1}{Z} e^{-mv^2/2kT} = \frac{1}{Z} e^{-K/kT}, \tag{19–34}$$

where K is the kinetic energy ($K = mv^2/2$); the factor $1/Z$ takes into account the normalization.

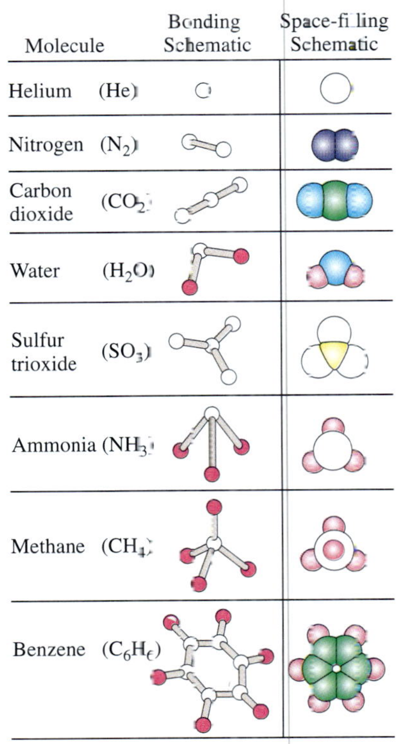

FIGURE 19–9 Molecules come in different shapes, as shown here in two schematic representations

†Elements are characterized by their chemical properties, which depend on the number of electrons they carry, and hence on the number of (positively charged) protons in their nuclei. *Isotopes* of an element differ in the number of (electrically neutral) neutrons in the nucleus of the element, and thus differ in atomic weight. For uranium, the atomic weights 235 and 238 refer to the total number of protons plus neutrons.

Boltzmann generalized the velocity distribution to a distribution that describes the probability that any one molecule has a given energy E. He proposed that in the original velocity distribution, Eq. (19–34), *the total energy of the molecule, E, should replace the kinetic energy, K, for a point mass*:

The Maxwell–Boltzmann distribution, which describes the probability that a molecule has a given energy

$$F = \frac{1}{Z} e^{-E/kT}. \tag{19–35}$$

This distribution function is known as the **Maxwell–Boltzmann distribution**. Equation (19–35) is one of the most frequently used formulas in the physics of the behavior of matter. This distribution function depends on all the dynamical variables that enter into the energy of a molecule. It is only for free, pointlike molecules that the energy depends on **v** alone. The energy of a rotating molecule with rotational inertia also depends on the angular velocities of rotation about the various axes. Later in this section, we will see how this affects the specific heat of gases.

The Energy Distribution for Diatomic Molecules

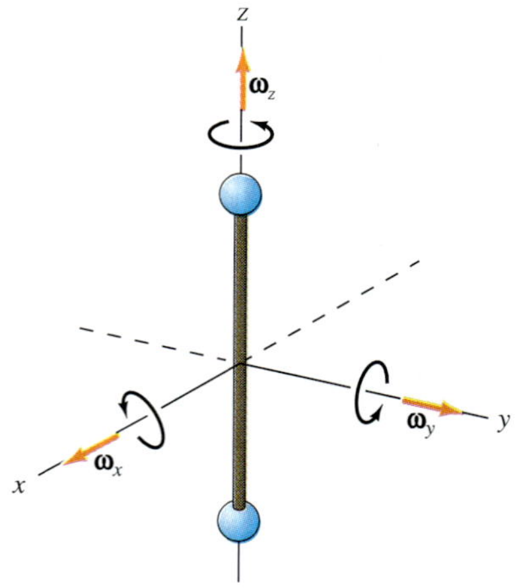

FIGURE 19–10 A dumbbell-like molecule can rotate about three axes. If the masses at the end of the rod are pointlike, the rotational inertia about the z-axis is zero. The rotational inertia about the x- and y-axes is I.

To see the effect of structure on the thermal properties of a gas, let's consider a gas composed of diatomic molecules. To describe a symmetric diatomic molecule, such as O_2, we use a classical model in which the two molecules act as though they are connected by a rigid rod. The rotational inertia has the same value, I, about each of the two axes that pass through the center of the rod and are perpendicular to it (Fig. 19–10). The energy of the molecule is a sum of the overall translational kinetic energy from linear motion, or translation (Fig. 19–11a), and of the rotational motion of the atoms about the center of mass (Fig. 19–11b; see Chapters 9 and 10). The rotational energy is

$$R_{\text{rot}} = \frac{1}{2} I\omega_x^2 + \frac{1}{2} I\omega_y^2, \tag{19–36}$$

where ω_x and ω_y are the components of the angular velocity along the x- and y-axes, respectively (Fig. 19–10). The ω_z term is missing from Eq. (19–36) because the rotational inertia about the z-axis is zero. The probability distribution F now refers to the probability that the velocity of a molecule lies in a region d^3v about the velocity vector **v**, and that the molecule's angular velocity lies in a region; $d^2\omega = d\omega_x\, d\omega_y$ about the angular velocity vector ω whose components are ω_x and ω_y. The total energy of a molecule is now $E = K + E_{\text{rot}}$, and, from Eq. (19–35), the Maxwell–Boltzmann distribution is

$$F = \frac{1}{Z} e^{(-mv^2/2kT)-(I\omega_x^2/2kT)-(I\omega_y^2/2kT)}. \tag{19–37}$$

By integrating this function times the energy per molecule over d^3v, $d\omega_x$, and $d\omega_y$, we find the average energy, $\langle E\rangle$:

$$\langle E\rangle = \frac{1}{2} m\langle v_x^2\rangle + \frac{1}{2} m\langle v_y^2\rangle + \frac{1}{2} m\langle v_z^2\rangle + \frac{1}{2} I\langle \omega_x^2\rangle + \frac{1}{2} I\langle \omega_y^2\rangle. \tag{19–38}$$

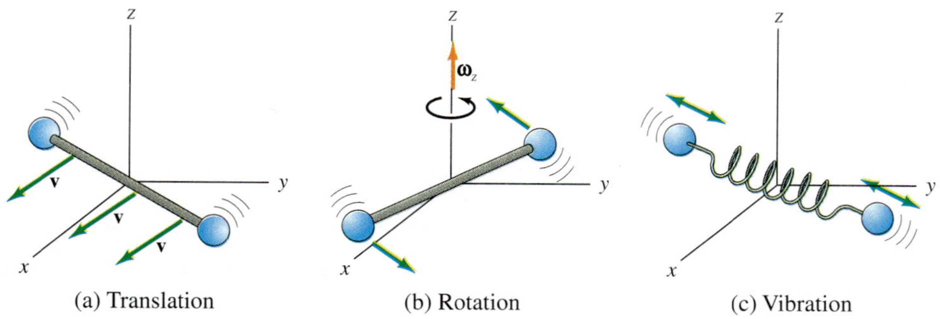

FIGURE 19–11 Schematic diagram of (a) translation, (b) rotation, and (c) vibration of a diatomic molecule.

When the integrations are performed to find the averages, each of these five terms is precisely $kT/2$. Finally, the internal energy of the gas is given by $U = N\langle E\rangle$, where N is the number of molecules. Thus, for our rigid-rod diatomic molecule,

$$\langle E\rangle = \frac{U}{N} = \frac{5}{2}kT. \quad (19\text{–}39)$$

This result should be compared to the monatomic case, in which there are only the three terms of the kinetic energy of linear motion and $\langle E\rangle = \frac{3}{2}kT$, which is a result already given in Eq. (19–12). If the diatomic molecules are not rigidly attached, the bond between them acts like a spring with spring constant k (Fig. 19–11c). There are then two additional vibrational contributions to the energy, of the form $E_{vib} = (mV^2/2) + (kx^2/2)$. Here, V is the speed of the atoms relative to the center of mass and x is the displacement of the vibrating atoms from an equilibrium position. Each of these terms also gives a contribution of $kT/2$ to the average energy and, for these molecules, $\langle E\rangle = \frac{7}{2}kT$.

Equipartition

The counting we just performed shows that the total average energy per molecule depends on how many independent motions a molecule can have. If a molecule acts like a point mass, all it can do is move in the x-, y-, and z-directions. In this case, the energy has the three terms, proportional to v_x^2, v_y^2, and v_z^2, corresponding to linear motion in the x-, y-, and z-directions, respectively. If, in addition, the molecule is diatomic with a rotational inertia about axes x and y, we found two new terms in the energy—this time proportional to ω_x^2 and ω_y^2—making five terms in all. If vibration is possible, then we found two more terms, this time proportional to the squares of the relative speed and of the atomic separation; thus there are in all seven such terms. Every term in the energy expression that is quadratic in an independent dynamical variable designates a **degree of freedom**. Generally, *the contribution of each degree of freedom to the average energy of a molecule is kT/2*. This result is called the **equipartition theorem**. If s is the number of degrees of freedom, then

$$\langle E\rangle = \frac{s}{2}kT. \quad (19\text{–}40)$$

The equipartition theorem

A dilute gas—even one whose molecules have a complicated structure—continues to obey the ideal gas law. The internal energy is just

$$U = N\langle E\rangle = \frac{s}{2}NkT \quad (19\text{–}41)$$

The internal energy continues to be linearly dependent on temperature, as for any ideal gas. The term by which the temperature is multiplied in Eq. (19–41) is the constant-volume heat capacity of the gas, C_V. The heat capacity therefore depends on the number of degrees of freedom of the molecules through the relation

The heat capacity of an ideal gas is proportional to the number of degrees of freedom.

$$C_V = \frac{s}{2}nR. \quad (19\text{–}42)$$

Molecular nitrogen, for example, should have a *molar heat capacity*, c', or heat capacity per mole ($n = 1$), of $\frac{7}{2}R$, at constant volume: $\frac{3}{2}R$ is from the overall motion of the center of mass, $\frac{2}{2}R$ is from rotations about two axes, and $\frac{2}{2}R$ is from the kinetic and potential energies of the "spring" that connects the two nitrogen atoms. How does this compare with experimental results? Figure 19–12 plots c_V'/R, where c_V' is the constant-volume molar heat capacity, as a function of temperature. At lower temperatures, hydrogen gas (H_2) behaves like a monatomic system, with a molar heat capacity of $\frac{3}{2}R$, which explains why we can treat air at room temperature as a monatomic gas. The

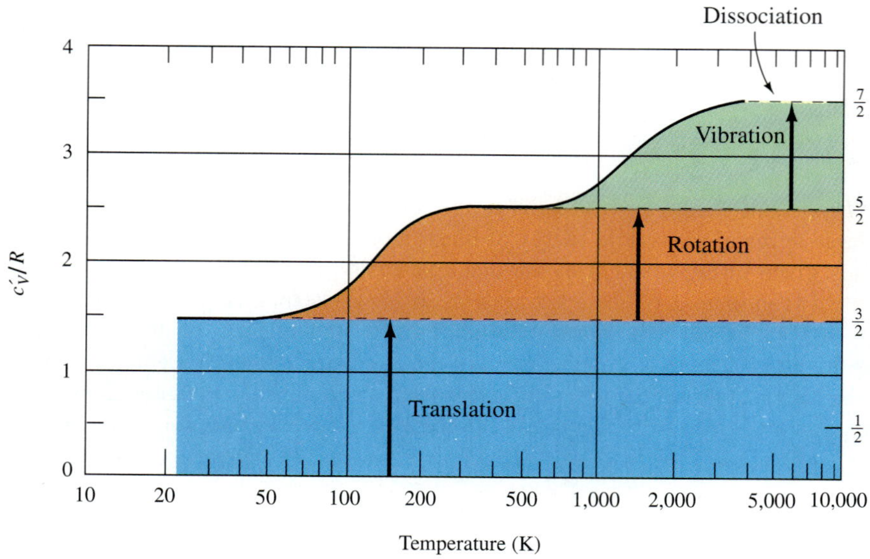

FIGURE 19–12 The constant-volume molar heat capacity divided by the gas constant, c_V'/R, for H_2 as a function of temperature.

full factor of $\frac{7}{2}R$ comes in only at high temperatures. The steplike nature of the curve suggests that, somehow, the effects enter one at a time. The reason for this cannot be found in the realm of classical physics, but instead requires quantum mechanics. Quantum physics sets certain minimum temperatures for the excitation of the different degrees of freedom.

*19–7 COLLISIONS AND TRANSPORT PHENOMENA

Molecules, even in a dilute ideal gas, follow a tortuous path within their container. They frequently collide with one another and the walls, changing in direction and speed with each collision (Fig. 19–13). An air molecule in a room undergoes billions of collisions per second. If the air is diluted to one-millionth of 1 atm, it still suffers thousands of collisions per second. Even so, the collisions are brief, and most of the time the molecule is free of the influence of other molecules. The *average* distance a molecule travels between collisions is a statistical quantity that can be calculated because of the large number of molecules and collisions involved.

Molecules that are characterized by some special property, such as having higher speeds due to local heating, will, through collisions, carry that special property away from their initial location. In effect, they will carry thermal energy from one place to another. If some of the molecules affect the sense of smell (molecules that make up perfume, for example), they will carry the odor as they move and collide. The movement of such properties is called **transport** (of thermal energy, odor, and so forth).

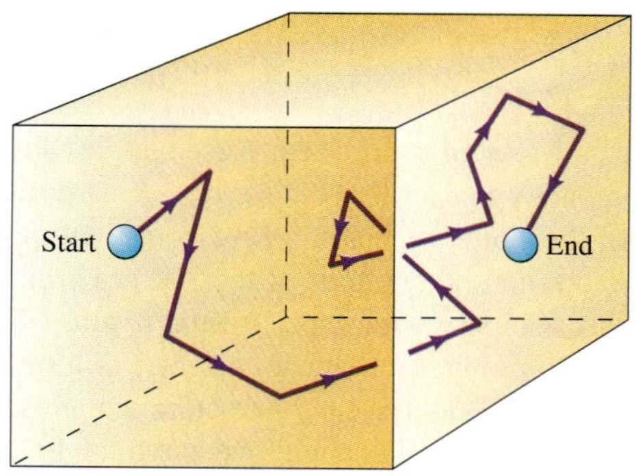

FIGURE 19–13 A molecule follows a tortuous zigzag path because of its multiple collisions with other molecules.

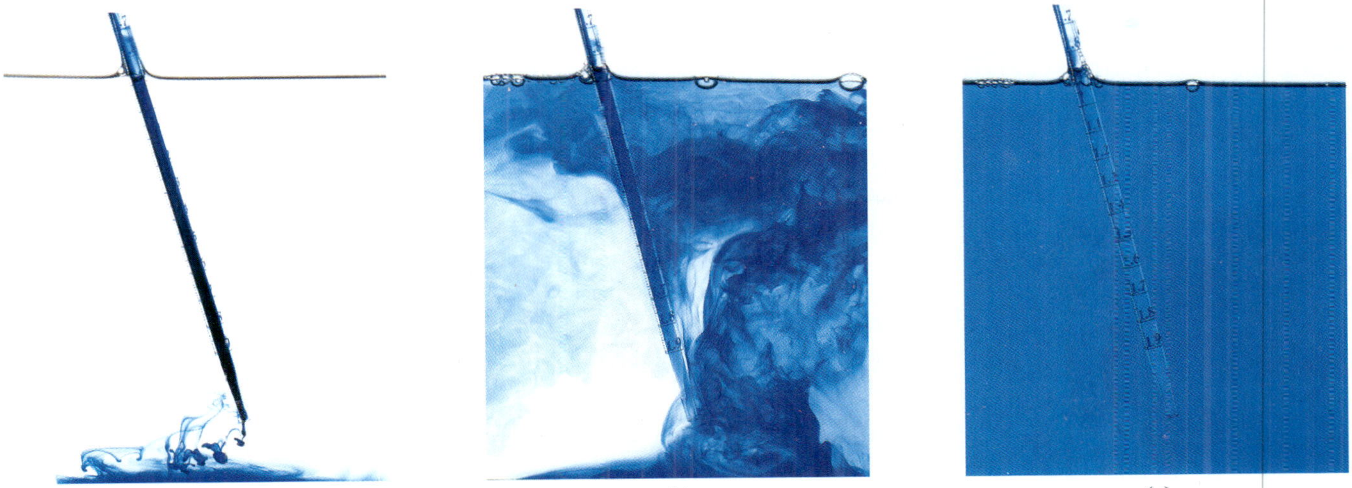

(a) (b) (c)

FIGURE 19–14 (a) Dye is placed with the help of a pipette at the bottom of a flask of water. (b) A combination of convection and diffusion has moved the dye into other regions of the flask. (c) Dye has diffused throughout the flask of water to make a uniform distribution. While stirring can speed the process, diffusion will make this happen if given enough time.

Figure 19–14 illustrates transport in a liquid. Ordinary thermodynamics has nothing to say about transport phenomena, but kinetic theory can explain them.

Collisions and Movement in a Gas

For a given gas density, the distance that a molecule travels before it has a collision depends on the size of the molecules. If molecules were infinitely small, they would never run into each other; if they were very large, they would always be in each other's way. Molecules of diameter D will collide when the path of the center of one molecule lies within an area πD^2 presented by the second molecule (Fig. 19–15). This area is the **collision cross section**, σ:

$$\sigma = \pi D^2. \tag{19–43}$$

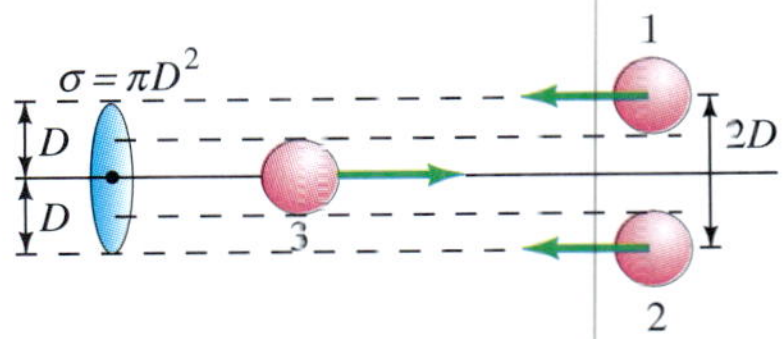

FIGURE 19–15 If the projected centers of two molecules of diameter D that move together are less than or equal to a distance D apart, the molecules will collide. An area $\sigma = \pi D^2$ of one molecule is available for collisions with another molecule.

Even if the molecules do not behave like billiard balls, there is some effective distance D that characterizes the collision, and the collision cross section is still given by Eq. (19–43).

Consider now a molecule that moves with speed v and sweeps out an area σ. If no other molecules were present, then in time t the molecule would travel a distance $d = vt$, sweeping out a volume $V = \sigma d = \sigma vt$. However, our molecule is not alone: Within the volume V, there are $N_V = nV$ target molecules, where n is the number density of molecules (not the number of moles). The traveling molecule therefore suffers N_V collisions. Even though the path is bent with each collision, the volume, V, remains unchanged if the speed is not changed by the effect of multiple collisions; and, on average, it is not. Because the number of collisions in time t is N_V, there is on average a collision every t/N_V seconds. This is the **mean collision time**, τ. Since we are averaging over many molecules, v must be replaced by some average value. It is reasonable to replace it with v_{rms}, so that

$$\tau = \frac{t}{N_V} = \frac{t}{nV} = \frac{t}{n\sigma v_{rms}t} = \frac{1}{n\sigma v_{rms}}. \tag{19–44}$$

A better calculation takes into account the fact that the target molecules are also moving, and Eq. (19–44) is modified to

$$\tau = \frac{1}{\sqrt{2}n\sigma v_{rms}}. \tag{19–45}$$

The average time between collisions

The average distance a given molecule travels before it is involved in a collision is the **mean free path**, λ. In the same spirit in which we found τ, λ is the rms speed times the mean collision time:

The mean free path

$$\lambda = \tau v_{\text{rms}} = \frac{1}{\sqrt{2}n\sigma}. \tag{19–46}$$

As is reasonable, λ is inversely proportional to both the density and the collision cross section.

EXAMPLE 19–5 What is the mean free path in air at sea level when the temperature is 300 K? By how much does the mean free path change when the temperature drops to 275 K? Take $r \simeq 10^{-10}$ m as a typical molecular radius, and treat the atmosphere as ideal.

Solution: For pressure at sea level, we use 1 atm $\simeq 10^5$ Pa. The atmospheric number density n can be calculated from the ideal gas law:

$$n = \frac{N}{V} = \frac{p}{kT} = \frac{10^5\ \text{Pa}}{(1.38 \times 10^{-23}\ \text{J/K})(300\ \text{K})} = 2.4 \times 10^{25}\ \text{m}^{-3}.$$

The collision cross section, σ, is given by Eq. (19–43). From Eq. (19–46), the mean free path is

$$\lambda = \frac{1}{\sqrt{2}n\sigma} = \frac{1}{\sqrt{2}n\pi(2r)^2}$$

$$= \frac{1}{\sqrt{2}(2.4 \times 10^{25}\ \text{m}^{-3})(3.14)(2 \times 10^{-10}\ \text{m})^2} = 2.3 \times 10^{-7}\ \text{m},$$

which is much greater than the average molecular spacing.

The mean free path changes with temperature because the number density of molecules is inversely proportional to T. The mean free path is inversely proportional to n, hence it is directly proportional to T:

$$\frac{\lambda_{275\ \text{K}}}{\lambda_{300\ \text{K}}} = \frac{275\ \text{K}}{300\ \text{K}} \simeq 0.9.$$

The Random Walk and Diffusion

How far does a molecule move, *on average*, from its initial position in a given amount of time? This problem is similar to a classic problem in mathematics called the **random walk**, known historically as the drunkard's walk. In this problem, a drunkard starts at a lamppost and takes steps that are equal in length but random in direction (Fig. 19–16). Compare Fig. 19–16 to Fig. 19–13, a typical molecular path in a gas. If the length of the drunkard's step is replaced by the mean free path, and if the time between steps is replaced by the mean collision time, then the problems are very similar. Of course, the molecule moves in three dimensions rather than two.

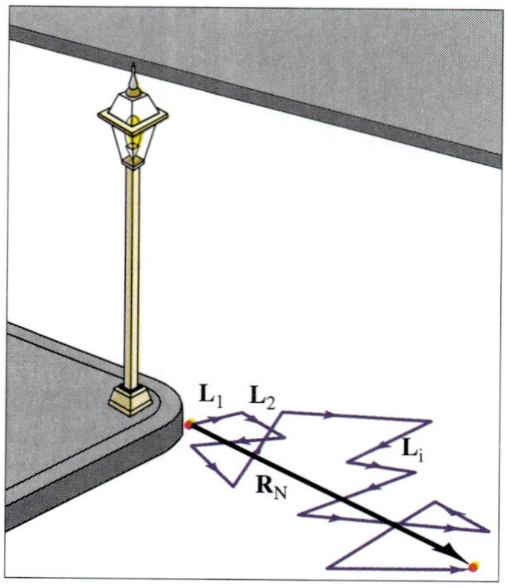

FIGURE 19–16 The random walk (a drunkard's walk around a lamppost).

We can find the average displacement of a molecule after N steps as follows. Let the successive displacements of the molecules be $\mathbf{L}_1, \mathbf{L}_2, \ldots, \mathbf{L}_N$. These have random directions, but their magnitudes are all the same value, L. After N steps, the net displacement of a molecule is $\mathbf{R}_N = \mathbf{L}_1 + \mathbf{L}_2 + \cdots + \mathbf{L}_N$. Squaring this quantity, we find that

$$R_N^2 = L_1^2 + L_2^2 + \cdots L_N^2 + 2\mathbf{L}_1 \cdot \mathbf{L}_2 + 2\mathbf{L}_1 \cdot \mathbf{L}_3 + \cdots + 2\mathbf{L}_{N-1} \cdot \mathbf{L}_N.$$

Because the directions of the vectors $\mathbf{L}_i$ are random, all the dot products, such as $\mathbf{L}_1 \cdot \mathbf{L}_2$, average to zero, whereas the terms L_i^2 all equal L^2. We are left with

$$\langle R_N^2 \rangle = NL^2. \tag{19–47}$$

To connect this result to the properties of the gas, we replace L^2 by the mean free path squared. The time interval between steps is the mean collision time, τ, so after N steps, a time $t = N\tau$ has elapsed. N can thus be replaced by t/τ. Therefore, Eq. (19–47) states that, after time t, a molecule will have moved on average a distance squared given by

$$\langle r^2 \rangle = \frac{t}{\tau}\lambda^2. \tag{19–48}$$

It is typical of random-walk problems that the displacement *squared* is linear in time.

The calculations leading to Eq. (19–48) can be refined, leading to changes in some of the details. More importantly, the result can be verified experimentally. Molecules that have been marked in one way or another (for example, by using some radioactive molecules) can be traced as they move. The movement of molecules by random collisions is called **diffusion**. The actual motion of the random walk is observable when particles larger than molecules, such as smoke particles, are added to a gas. This motion is called **Brownian motion**.

SUMMARY

Molecules are spread rather sparsely in gases under normal conditions. The kinetic theory of ideal gases is a statistical description in which molecules bounce randomly and relatively infrequently off each other and off the walls of their container. This theory accurately predicts many thermodynamic properties.

The internal energy, U, is given in this model by the number of molecules, N, times the average energy of a molecule:

$$U = N\left(\frac{1}{2}m\langle v^2 \rangle\right), \tag{19–2}$$

where $\langle v^2 \rangle$ is the average of the quantity velocity squared. The pressure of a gas arises from multiple elastic collisions between the walls and the molecules. Pressure is related to internal energy by

$$pV = \frac{2}{3}U, \tag{19–10}$$

where V is the volume of gas. In the ideal gas law, pV is proportional to temperature, so temperature can be interpreted as a measure of the internal energy, or of the average energy of molecules:

$$kT = \frac{2}{3}\frac{U}{N} = \frac{2}{3}\langle K \rangle. \tag{19–13}$$

Probability distributions describe the probability that various outcomes will be represented in a large statistical sample. For gases, $F(\mathbf{v})\,d^3v$ describes the probability that a given molecule will have a velocity $\mathbf{v}$ in the range from $\mathbf{v}$ to $\mathbf{v} + d\mathbf{v}$. Averages are computed from the probability distribution by integration. For example, the average of v^2 is

$$\langle v^2 \rangle = \int v^2 F(\mathbf{v})d^3v. \tag{19–27}$$

The probability distribution $F(\mathbf{v})$ for the velocity of the molecules of a gas is

$$F(\mathbf{v}) = \left(\frac{m}{2\pi kT}\right)^{3/2} e^{-mv^2/2kT}. \tag{19–28}$$

Boltzmann generalized the velocity distribution to the cases in which molecules have internal energy as well as a center-of-mass kinetic energy, or in which the molecules are subject to external forces. The Maxwell–Boltzmann distribution is given by

$$F = \frac{1}{Z}e^{-E/kT}, \tag{19–35}$$

where E is the total energy of a molecule. Every variable that appears quadratically in the expression for the energy of a single molecule—such variables are called degrees of freedom—contributes $\frac{1}{2}kT$ to the average energy. If the number of degrees of freedom is labeled s, then the average energy of a molecule is

$$\langle E \rangle = \frac{s}{2}kT. \qquad (19\text{–}40)$$

This result is the equipartition theorem. The internal energy is $U = N\langle E \rangle$.

Transport phenomena describe the motion of molecules, momentum, energy, and so forth through a gas as a result of multiple molecular collisions. Molecules have a collision cross section, σ, which describes the area available for collisions with another molecule. If the number of molecules per unit volume is n, then the mean time τ between collisions, called the mean collision time, is

$$\tau = \frac{1}{\sqrt{2}n\sigma\langle v \rangle}. \qquad (19\text{–}45)$$

The average distance a given molecule travels before it suffers a collision is the mean free path, λ:

$$\lambda = \tau\langle v \rangle = \frac{1}{\sqrt{2}n\sigma}. \qquad (19\text{–}46)$$

As a result of multiple collisions with other molecules, an individual molecule within a gas will on average move. In a time t, the molecule moves a distance squared that is proportional to t. Such movement is called diffusion.

UNDERSTANDING KEY CONCEPTS

1. Consider an ideal gas in an insulated cylinder outfitted with a piston. The piston is moved slowly, so that the volume of the gas is decreased. What happens to the temperature of the gas? How do you explain your answer in terms of the motion of molecules?

2. Which travel faster, on average, the oxygen molecules or the nitrogen molecules in your room?

3. In our discussion of pressure, we used the fact that the collisions of molecules with the walls are elastic. If there are diatomic molecules, they could be set into rotation by the collision, and thus their translational kinetic energy after the collision could be smaller than before the collision. How can these statements be reconciled? What is the effect on the pressure?

4. We made the connection between temperature and the average kinetic energy of a molecule without taking into account the large number of intermolecular collisions estimated in Section 19–1. Why do you think this approximation holds?

5. In an extremely dilute gas, the rate of interatomic collisions is very small. As the gas becomes less dense, the rate of interatomic collisions becomes much less than the collision rate with the container's walls. Can we still make a connection between temperature and the average kinetic energy of a molecule? What are the possible problems in making the connection in this limit?

6. Very few hydrogen molecules are present in Earth's atmosphere today. Is this fact in conflict with the possibility of their abundance a long time ago? In explaining your answer, ignore the possibility that hydrogen is removed by chemical reactions.

7. Does temperature depend on the rotational kinetic energy of molecules?

8. Can we assign a temperature to a single molecule? Explain.

9. Why, from the point of view of kinetic theory, does the air near a hot stove become heated?

10. In Section 19–1, we stated that the typical distance a molecule travels before it suffers a collision in 1 mol of a gas at standard temperature and pressure is roughly 3×10^{-7} m. This will be much less than the size of a container of gas. How is it that the approximation that each molecule in the cylinder collides freely with the wall, without worrying about intermediate collisions between molecules, leads to a satisfactory result for the pressure, Eq. (19–10)?

11. At standard temperature and pressure, a molecule of one mole of an ideal gas has a mean free path of about 3×10^{-7} m, some 100 times greater than the average intermolecular spacing. How is it possible for the mean free path to be greater than the average spacing?

12. You have a container of a dilute gas at a certain temperature and inject a small quantity of a different type of molecule into that container. What processes determine how long it takes the new molecules to come to thermal equilibrium with the original molecules?

13. Gravity keeps the atmosphere of Earth close to the surface. Why, then, is there no atmosphere on the Moon, which also exerts a gravitational force?

14. In Newtonian mechanics, energy is added to a system when work is done on it. In the language of classical mechanics, how is energy added to a gas in a container when a piston is pushed in and the container's volume is changed?

15. Given our explanation of the origin of pressure, why is there no net force on a pane of glass placed in a container of gas, even though there may be a net force on a wall of the container?

16. Will the air just above the surface of hot water in a bowl be at the same temperature as the water? If so, what molecular mechanism is responsible for the equality of the two temperatures?

17. Why will a cup of hot water cool off much more rapidly when air is blown across its surface than when the air above the cup is stationary?

18. The smoke from the burning end of a cigarette rises; the flame of a candle points upward. Why?

19. When a candle burns, carbon combines with the oxygen in air to form carbon dioxide, a compound used in fire extinguishers to put out fires. Why does the candle not snuff itself out? Could you burn a candle in a satellite that is in free fall?

PROBLEMS

19–1 *A Microscopic View of Gases*

1. (I) We can define a volume ratio R_V of a gas to be the ratio of the volume taken up by the molecules to the total volume of the container; similarly, we can define a spacing ratio R_L to be the ratio of the linear size of a molecule to the mean spacing between molecules. Compute R_V and R_L for a gas with molecules of diameter 2×10^{-10} m. Assume that the gas obeys the ideal gas law for the following values of temperature and pressure: (a) $T = 300$ K, $p = 1$ atm; (b) $T = 5$ K, $p = 1$ atm; (c) $T = 300$ K, $p = 10^{-8}$ atm; (d) $T = 5$ K, $p = 10^{-8}$ atm.

2. (I) For an ideal gas at 0°C and 1 atm, the ratio of the typical molecular radius (10^{-10} m) to the intermolecular spacing is about $\frac{1}{20}$. What is this ratio in liquid water (H_2O, molecular weight 18 g/mol)?

3. (II) Assume that the carbon and hydrogen atoms that make up oil behave like spheres of diameter 10^{-8} cm, and that the fundamental molecular component of oil, CH_2, would take up a corresponding area of 10^{-15} cm^2 on a water surface. How many such components are there in 0.1 L of oil if the density of oil is 90 percent of the density of water? What would the area of a one-molecule-thick layer of this amount of oil on water be? (This type of experiment, said to have been performed by Benjamin Franklin, among others, is sometimes used to compute the size of molecules.)

19–2 *Pressure and Molecular Motion*

4. (I) A rubber ball, confined to move in the x-direction, bounces elastically back and forth between two walls. The ball, which has a mass of 0.125 kg, moves at a speed of 12 m/s, and the spacing of the walls is such that the ball strikes the right-hand wall 48 times in a ten-second period. What is the average force on the right-hand wall due to the ball during that ten-second span?

5. (I) Grains of sand of mass 3×10^{-3} g each, fall from a height of 0.8 m on a sticky surface at a rate of 50 grains per second per cm^2. What pressure does this shower of sand exert on the surface, assuming that air resistance can be neglected?

6. (II) Use dimensional analysis to estimate the number of air molecules that strike the 12-in × 12-in screen of a television set during a 1-h program.

7. (II) One mol of a monatomic ideal gas is placed in a chamber under 5 atm pressure. The volume of the chamber is 5000 cm^3. (a) What is the internal energy of the gas? (b) What is the temperature of the gas? (c) Assuming that the mass of a molecule of the gas is 3.36×10^{-26} kg, what is the value of $\langle v^2 \rangle$ of a gas molecule? (d) What is the root-mean-square (rms) velocity?

8. (II) A cubic box of volume 0.05 m^3 contains helium gas at 0.20 atm pressure and 27°C. (a) How many total collisions do the molecules make with the walls per second? (b) What is the total internal energy of the gas? (c) If the temperature is doubled, what happens to the rate of collisions?

9. (II) The apparatus shown in Fig. 19–17 is designed to demonstrate aspects of the kinetic theory of gases. It consists of a transparent cubical box, 20 cm on each side, containing 100 steel balls of diameter 5 mm. The density of steel is 7.8 g/cm^3. The bottom of the box vibrates so that the steel balls bounce around. The top of the box consists of a movable piston of mass 1 kg. What is the rms velocity of the steel balls if the top of the box is in dynamic equilibrium with the "gas" of steel balls? Ignore gravity in your treatment of the motion of the steel balls. Is this approximation justified?

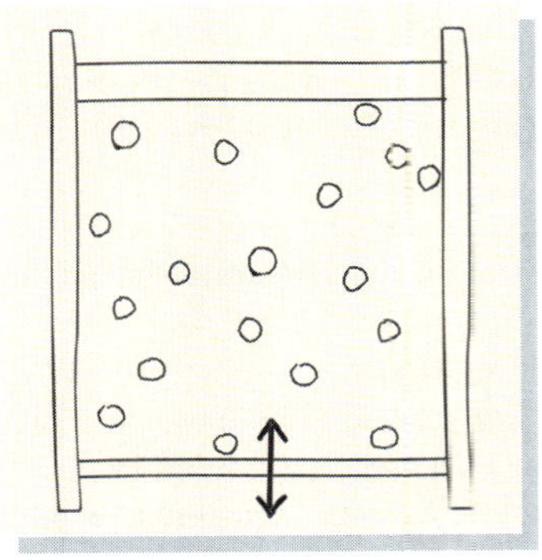

FIGURE 19–17 Problem 9.

10. (II) Particles of mass 1.3×10^{-17} kg, when suspended in a liquid at room temperature, have an rms speed of 0.030 m/s. Use this to determine Avogadro's number. (You will need to look up the gas constant R.)

11. (II) The rms velocity of galaxies in a large part of the visible universe is roughly 100 km/s. The number density of these galaxies is 3×10^{-20}/ly^3, and the average mass of a galaxy is 3×10^{41} kg. What is the pressure of a gas of such galaxies?

12. (II) Molecules are confined to move in a plane, for example, by putting them between two glass plates separated by a distance that is small compared to the mean free path. What is the relation between the pressure and the internal energy? What is the relation between internal energy and temperature for an ideal gas?

19–3 *The Interpretation of Temperature*

13. (I) Consider the mole of helium gas with an internal energy of 3600 J that we discussed in Example 19–1. (a) What is the temperature of the gas? (b) If 1/1000 of this sample of gas had the same internal energy as the total gas, what would its temperature be? (c) What would the internal energy of 1/1000 of the original sample of gas be, at its original temperature?
14. (I) A gas is heated from 15 K to 300 K. What is the change in the rms velocity?
15. (I) What is the average kinetic energy per molecule of a gas at room temperature, 293 K? How fast would a baseball of mass 0.1 kg be moving to have this same kinetic energy?
16. (I) An ideal gas is contained in a vessel of volume 4×10^{-4} m^3 that is under a pressure of 8.6 atm. What is the internal energy of the gas?
17. (I) The temperature of hydrogen in a dilute plasma—a state of matter in which atoms are broken apart into electrons and nuclei—inside a nuclear fusion reactor needs to be on the order of 20 million K in order to initiate nuclear fusion. What is the rms velocity of the hydrogen nuclei in the plasma?
18. (II) A mixture of nitrogen gas and sulfur dioxide gas is in equilibrium at 27°C. What is the rms velocity of the sulfur dioxide molecules (SO_2, molecular weight 64 g/mol), and that of the nitrogen molecules (N_2, molecular weight 28 g/mol)?
19. (II) Calculate the rms velocity at 300 K of the principal components of air: O_2 molecules (molecular weight 32 g/mol), N_2 (molecular weight 28 g/mol), CO_2 (molecular weight 44 g/mol) and H_2 (molecular weight 2 g/mol).
20. (II) The rms speed of 1 mol of argon atoms (atomic weight 40 g/mol) in a box is 400 m/s. (a) What is the temperature inside the box? (b) What is the internal energy? (c) If the box has a volume of 0.1 m^3, what is the pressure? Treat the gas as ideal.
21. (II) Estimate v_{rms} for hydrogen atoms (a) on the surface of the Sun, where the temperature is 6000 K; (b) on the surface of the Moon, where the temperature at one point is 150 K in the shade.
22. (II) A proton of kinetic energy 5×10^9 eV comes to a stop because of molecular collisions in a thin tube that contains 0.01 mol of oxygen at STP, as in Fig. 19–18. By how much is the temperature in the tube increased at equilibrium?

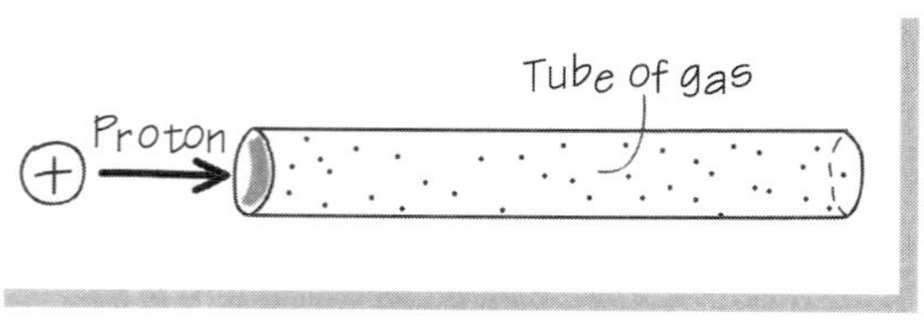

FIGURE 19–18 Problem 22.

23. (II) A "gas" of water droplets 10^{-6} m in diameter is in equilibrium with air molecules at 300 K. Given that the density of water is 1.0×10^3 kg/m^3, what is the value of v_{rms} for the droplets?
24. (II) A relativistic gas is one in which a significant fraction of the constituents have speeds that are some finite fraction of the speed of light, $c \simeq 3 \times 10^8$ m/s. Suppose that the rms speed of the constituents of a gas of atomic hydrogen is just 1 percent of c. What is the temperature? The energy that binds the electron and the proton together into a hydrogen atom is 2.18×10^{-18} J. Would this gas break down into a gas of electrons and protons? In the interior of stars, where temperatures comfortably exceed the temperature you have calculated, atoms cannot exist; they are broken into their components.
25. (II) If their kinetic energy greatly exceeds the average gravitational potential energy, galaxies in the sky may be viewed as an ideal gas. Given that the rms velocity of galaxies is 100 km/s and the average mass is 3×10^{41} kg, what is the temperature of a gas of galaxies?

19–4 *Probability Distributions*

26. (I) A lottery that costs \$1 to play has a prize of \$1,000,000. The state that runs the lottery is obligated to tell the players that the probability of winning is one in 10,000,000. Is it a good idea—from a financial point of view—to play this lottery? Answer this question by calculating the ratio of your outlay to your revenue under the assumption that you play the game an unlimited number of times.
27. (I) Nine cars are measured at a given spot on a highway to have speeds v of 52.3, 54.5, 57.0, 57.2, 57.9, 63.6, 63.6, 68.1, and 82.2 mi/h. Calculate (a) the average speed, and (b) the rms speed.
28. (I) Take a deck of cards. Draw cards from that deck, one by one, without putting any cards back in the deck. What is the probability that the first card is a spade? What is the probability that the second one is also a spade? What is the probability that the third one is also a spade? (You may extend this procedure to calculate the probability that when you draw thirteen cards, they are all spades, by multiplying all the thirteen probabilities together).
29. (I) Four dice are tossed. What are the probabilities of getting (a) four sixes; (b) three sixes and a five?
30. (II) A cage of 100 Ping-Pong balls has air blowing through it to keep the Ping-Pong balls moving. Ten Ping-Pong balls have the number 0 painted on them, ten have the number 1, and so forth, finishing with ten Ping-Pong balls painted with the number 9. The apparatus is used to choose the numbers for a lottery, and the balls are not returned to the cage after they are chosen. (a) The first drawing is for three balls to win \$100. Your number is 337. Before any balls are drawn, what is the probability that you will win? (b) The first ball drawn is 3. What are your chances now? (c) The second ball drawn is 3. Now what is the probability that you will win?
31. (II) Consider the distribution of grades g discussed in Section 19–4. (a) Plot the distribution of the numbers D for the students, where D is defined as the deviation of the grade from the average. (D can be positive or negative.) (b) Calculate the average value of D^2. (c) Show that $\langle D \rangle = 0$ and $\langle D^2 \rangle = \langle g^2 \rangle - \langle g \rangle^2$.

32. (II) Find an expression for the probability of getting 13 spades in a bridge hand. (In bridge, the entire deck of 52 cards is dealt to four players.)

33. (II) In a true-or-false test, a correct answer is awarded +1 point and an incorrect answer is awarded −1 point. In a test involving 87 students, the following grade distribution is found (Fig. 19–19):

−50 to −30: 3 students;
−30 to −10: 18 students;
−10 to +10: 29 students;
+10 to +30: 22 students;
+30 to +50: 15 students.

(a) Calculate the average grade. (b) Calculate $\langle D^2 \rangle$, where D is defined in Problem 31. (c) Suppose that the distribution was such that every student has a grade between +10 and +30. With this information, what would the average be?

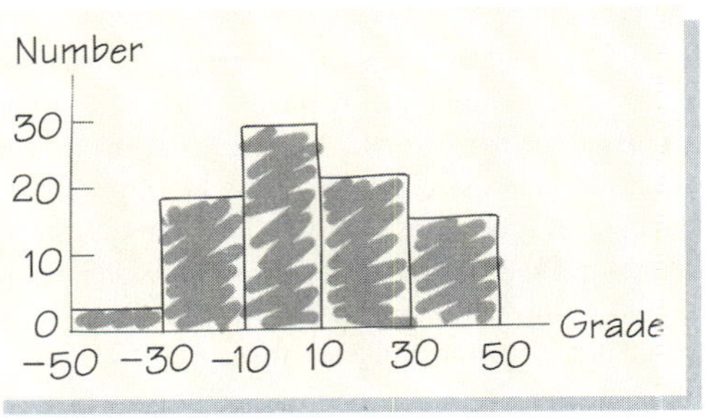

FIGURE 19–19 Problem 33.

34. (II) Suppose that the distribution of grades (the number of test takers, N, with a grade x) in a national test is approximated by a continuous curve of the form (Fig. 19–20)

$$\begin{aligned} N(x) &= 25x && \text{for } 0 < x < 60 \\ &= 6000 - 75x && \text{for } 60 < x < 80 \\ &= 0 && \text{for } 80 < x < 100. \end{aligned}$$

(a) What is the total number of students who took the test? (b) What is the average grade?

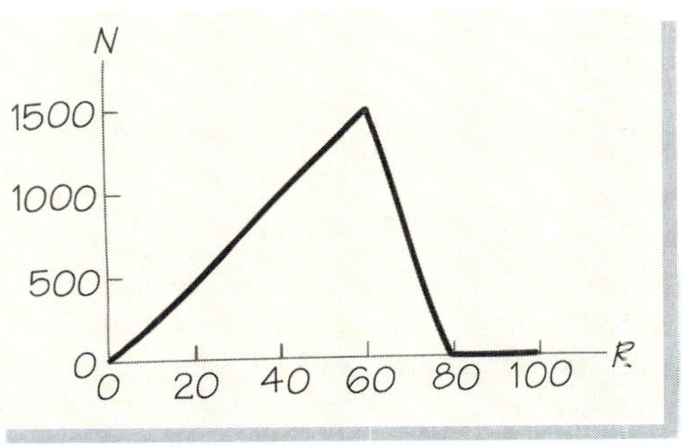

FIGURE 19–20 Problem 34.

35. (III) A way to find the value of π (*Buffon's method*) is the following: Drop a needle of length L repeatedly onto a grid of parallel lines separated by a distance L (Fig. 19–21). Show that the probability of the needle falling on a line is $2/\pi$. [*Hint*: Take the lines as parallel to the x-axis. Calculate the average value of the component of the needle's length along the y-axis.]

FIGURE 19–21 Problem 35.

19–5 The Velocity Distribution of Gases

36. (I) What is the rms speed of a helium atom in the vapor above liquid helium at 4 K? (The system is enclosed in a Thermos bottle.)

37. (I) Oxygen atoms come in a variety of isotopes: ^{16}O, ^{17}O, and ^{18}O (with the superscript representing A). The latter two isotopes are present only in small quantities. A gas consists of diatomic molecules of oxygen (mostly $^{16}O_2$, with small admixtures of ^{16}O–^{17}O and ^{16}O–^{18}O). Calculate the ratios of the rms speeds for the latter two to that of the dominant (^{16}O–^{16}O) form. Does your answer depend on temperature?

38. (I) If the rms speed of molecules of gaseous H_2O is 200 m/s, what will be the rms speed of CO_2 molecules at the same temperature? Assume that both of these are an ideal gas.

39. (II) Calculate $\langle v \rangle$, the average speed for an ideal gas. In doing this calculation with the help of the velocity distribution $F(\mathbf{v})$, use the fact that for a spherically symmetric integrand you may replace d^3v by $4\pi v^2\ dv$. You will need the integral

$$\int_0^\infty z^3 e^{-z^2}\, dz = \tfrac{1}{2}.$$

[*Hint*: This problem illustrates the use of dimensional analysis as an intermediate step.]

40. (II) Use the result of Problem 39 to find the numerical value of the average speed for an ideal gas of CO_2 at 300 K, and compare it with v_{rms} at that temperature.

41. (III) Consider a gas being blown along at a velocity $\mathbf{u}$, so that its velocity distribution is given by

$$F(\mathbf{v}) = \frac{1}{Z}\, e^{-m(\mathbf{v}-\mathbf{u})^2/2kT}.$$

(a) Show that Z is the same as for the usual velocity distribution, as in Eq. (19–28). (b) Calculate $\langle \mathbf{v} \rangle$ and $\langle \mathbf{v}^2 \rangle$ and interpret your results.

**19–6 The Maxwell–Boltzmann Distribution*

42. (I) The value of c'_p for carbon monoxide (CO) is 29.2 J/mol · K. What value do you expect for c'_V for the diatomic CO molecule?

43. (I) One mole of hydrogen gas is determined to have an internal energy of 1.87×10^4 J at 900 K. How many degrees of freedom are there for the hydrogen molecule? Which degrees of freedom are likely to be available?

44. (I) A gas of N_2 molecules is in equilibrium at a temperature of 700 K. Calculate the rms angular momentum $I\sqrt{\omega^2}$ for an

N_2 molecule, given that the mass of a nitrogen atom is 2.33×10^{-26} kg and that the average separation of the two atoms in an N_2 molecule is approximately 0.7 nm (Fig. 19–22). (You may assume that the rotational—but not the vibrational—degrees of freedom participate in the equipartition theorem.)

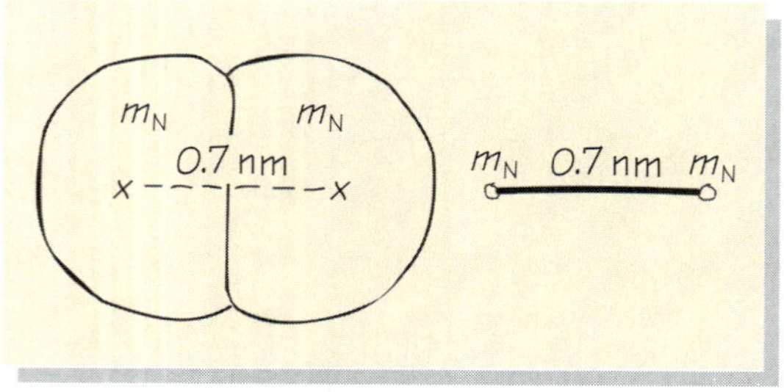

FIGURE 19–22 Problem 44.

45. (II) The following data can be found in a table of thermodynamic properties of gases:

$$\text{Argon Ar: } c_p = 0.214 \text{ cal/g}\cdot\text{K}; \ c_V = 0.074 \text{ cal/g}\cdot\text{K}$$
$$\text{Oxygen } O_2\text{: } c_p = 0.219 \text{ cal/g}\cdot\text{K}; \ c_V = 0.157 \text{ cal/g}\cdot\text{K}$$
$$\text{Water } H_2O\text{: } c_p = 0.445 \text{ cal/g}\cdot\text{K}; \ c_V = 0.335 \text{ cal/g}\cdot\text{K}$$
$$\text{Carbon dioxide } CO_2\text{: } c_p = 0.201 \text{ cal/g}\cdot\text{K};$$
$$c_V = 0.156 \text{ cal/g}\cdot\text{K}$$

What can you infer about the shape of the molecules from these data?

46. (II) A gas of diatomic molecules is in equilibrium at a temperature of 300 K. Assuming that the rotational inertia of the molecule [as used in Eq. (19–36)] is 3×10^{-40} kg·m^2, what is the rms angular speed of the molecules? What is the mean angular momentum of the molecules?

47. (III) Write down an integral for the probability that a molecule is found to have a velocity larger than 90 percent of v_{rms}. Show that the probability is independent of the temperature. Show that this is not the case if we ask for a velocity larger than some fixed number; for example, the escape velocity from a planet.

48. (III) A crystal can be thought of as a collection of N atoms arranged in a lattice, with all the atoms connected by springs (Fig. 19–23). Thus we can think of each atom as a mass at the end of three springs, aligned respectively along the x-, y-, and z-directions. Using this model of the lattice, count the number of degrees of freedom for each atom, then use the equipartition theorem to compute the molar heat capacity of the crystal at temperature T.

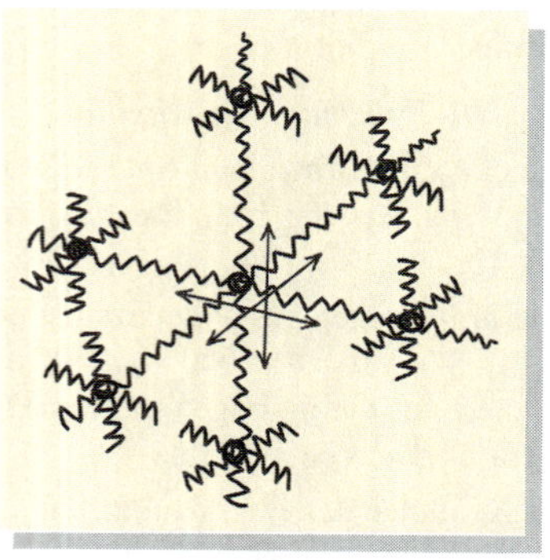

FIGURE 19–23 Problem 48.

*19–7 *Collisions and Transport Phenomena*

49. (I) For a collision between two molecules of radii R_1 and R_2, respectively, show that the collision cross section is $\sigma = \pi(R_1 + R_2)^2$ (Fig. 19–24).

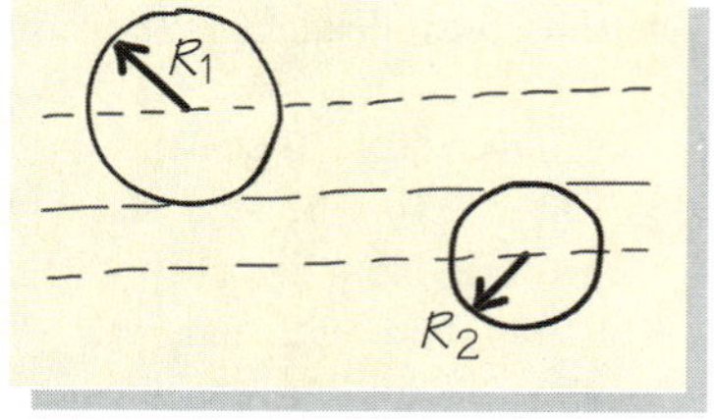

FIGURE 19–24 Problem 49.

50. (I) Compute the pressure of a container of air molecules of diameter 2.5×10^{-10} m if the temperature is 273 K and the mean free path is (a) 1 in, (b) 0.1 μm. The gas should be treated as ideal.

51. (II) One can treat the flow of gas through a vacuum hose as if the gas were a fluid if the diameter of the hose is much larger than the mean free path. If, however, the mean free path is much larger than the diameter of the hose, the motion of the individual free molecules has to be considered. Consider a hose of diameter 1 cm. At what pressure is the mean free path of air at 300 K shorter than 0.01 cm? Longer than 100 cm? Take the diameter of an air molecule to be 3×10^{-10} m.

52. (II) Using a reasonable value for the collision cross section or, more simply, a model of a gas in which the molecules are like tiny billiard balls, make an estimate of the density at which the mean free path becomes equal to the mean spacing between molecules.

53. (II) Ten tennis balls are rolling around randomly on a tennis court (Fig. 19–25). Estimate the mean free path for collisions.

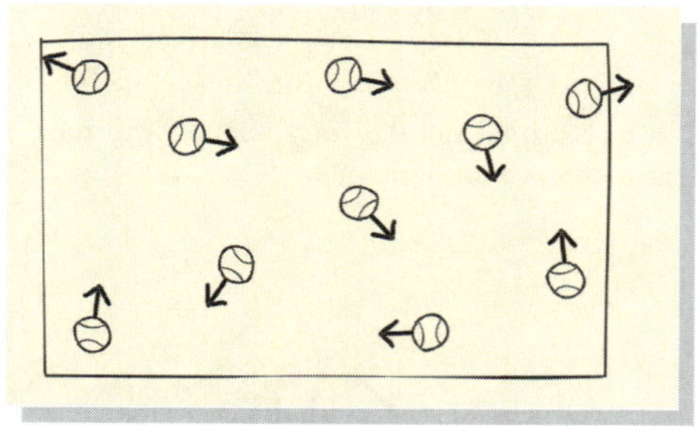

FIGURE 19–25 Problem 53.

54. (II) A student walks across a field that is 100 m $\times$ 50 m. Estimate the student's mean free path if there are (a) 10, (b) 100, (c) 1000 other students on the field.

55. (II) A cubic box of sides 10 cm contains helium atoms (diameter $\simeq 10^{-10}$ m) at a pressure of 3×10^4 Pa. If the mean free path in the box is 4×10^{-4} m, how many helium atoms are in the box? How many are there if the mean free path is 40 cm? What will the pressure be if the temperature is held fixed?

56. (III) According to one theory, planets are formed as a consequence of near collisions between stars. The number density of stars in our part of the galaxy is 10^{-51} m^{-3}. The velocity of the Sun relative to these stars can be taken to be 25 km/s, and the size of the stars is roughly 10^9 m. If we assume that a passage of two stars at a distance of 100 radii is close enough to set off the tidal forces that give rise to planet formation, what is the probability that the Sun would have acquired a planetary system in 10 billion yr?

General Problems

57. (I) Argon and methane (CH_4) gas are mixed together at 400 K. What is the ratio of the rms speeds of the argon and methane atoms? What happens to this ratio if the temperature is halved?

58. (II) The density as well as the temperature of material in intergalactic space is very uncertain. Suppose that the density and temperature were one hydrogen atom per cubic meter and 3 K, respectively. (a) What is the pressure? (b) What is the mean free path if the diameter of a hydrogen atom is 10^{-10} m?

59. (II) The acceleration due to gravity on the surface of Mercury is 3.5 m/s^2, and Mercury's radius is 2.4×10^6 m. (a) What is the escape speed of a particle on Mercury? (b) Suppose that the atmosphere of Mercury were pure H_2 gas. What would the temperature be so that, if the molecules had the velocity distribution given by Eq. (19–28) for that temperature, the rms speed of the H_2 molecules matched the escape speed? Qualitatively, what is the effect on the temperature of the remaining gas? (c) If the temperature were less than the result of part (b), would there be a similar effect, and why or why not? (d) Suppose that Mercury's atmosphere has two or more molecular components. What happens to the composition of the atmosphere over time as a result of the effects discussed here?

60. (II) The surface of the Sun contains both hydrogen and helium atoms at 6000 K. (a) What is the average kinetic energy of each type of atom? (b) What is the rms speed of each? (c) Which is most likely to escape from the sun?

61. (II) The regions in which the beams of accelerated particles circulate in accelerators have high vacuums (10^{-9} torr) to allow the particles to circulate freely. (a) What are the mean free paths of air molecules inside the accelerator at room temperature? (b) By how much does the mean free path decrease if the pressure inside increases to 10^{-6} torr?

62. (II) Suppose that molecules were confined to move in a plane. Calculate the velocity distribution $F(\mathbf{v})$, assuming that it is of the general form $(1/Z)e^{-(mv^2/2kT)}$. In other words, calculate Z so that $\int_{\text{surface}} F(\mathbf{v})\, d^2\mathbf{v} = 1$.

63. (II) Brownian motion is the random walk followed by individual molecules as well as by larger objects such as smoke particles. The Brownian motion of smoke particles in a room can be observed with the aid of a microscope. If the mass of the smoke particles is typically 3×10^{-15} kg, what is the rms speed of the particles at room temperature?

64. (II) Peter and Paul are standing in opposite corners of a room, 7 m apart. The air in the room does not move. Peter holds a small vial full of H_2S while Paul holds a vial containing Cl_2. They open their vials at the same time (Fig. 19–26). Paul smells H_2S after 50 s. When will Peter smell the Cl_2?

65. (II) Calculate an expression for the pressure of a gas of massless particles, using the following facts: Massless particles always move with the speed of light, so that $v = c$; the energy of a massless particle, E, is related to its momentum by $E = |\mathbf{p}|c$. Use this expression to show that for a gas of massless particles, $pV = \frac{1}{3}U$. This result is applicable to the blackbody radiation discussed in Chapter 17.

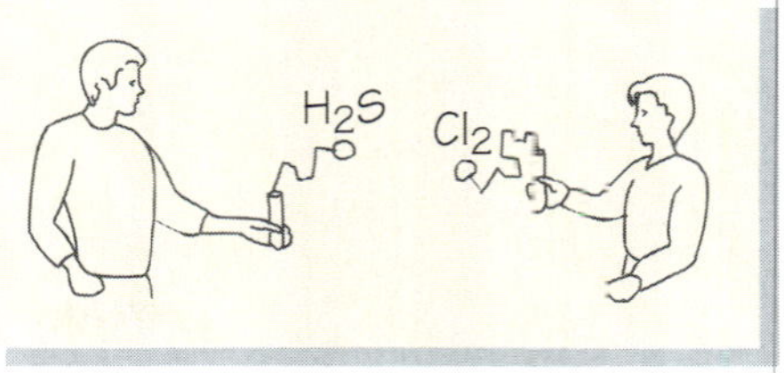

FIGURE 19–26 Problem 64.

66. (II) The relation $pV = \frac{2}{3}U$ is correct only for a monatomic gas. Show that the more general relation is $pV = (\gamma - 1)U$, where $\gamma = c'_p/c'_V$.

67. (II) Consider a tank of volume 0.30 m^3 containing 2.5 mol of helium gas at 20°C. If we add 1 mol of O_2 at 0°C to the helium and let the whole system come to equilibrium, what will the equilibrium temperature be? What will the pressure be?

68. (II) Assume that an H_2O molecule has three rotational degrees of freedom in addition to its three linear degrees of freedom (Fig. 19–27). Calculate the molar heat capacity of its vapor. Compare this with the experimental value at low pressures of 0.48 cal/g · K. What role might vibrational motion play?

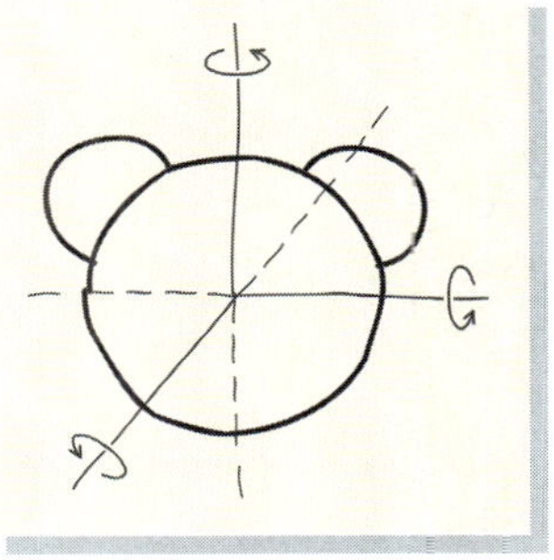

FIGURE 19–27 Problem 68.

69. (II) A vessel is pumped out to a high vacuum and then heated to get rid of water vapor. It is then filled with dry helium and pumped out again to a very high vacuum of 10^{-11} torr, so that it now contains only helium atoms at 300 K. Helium atoms have a diameter of about 10^{-10} m. (a) What is the density of helium atoms? (b) the mean collision time? (c) the mean free path?

70. (II) The *law of partial pressures* states that the total pressure of a mixture of gases is the sum of the pressures that each gas would exert if it were present alone in the same amount as in the mixture. Use the kinetic theory to demonstrate this fact. Could this law be true if the gases were not ideal?

71. (III) The escape speed of a particle from Earth is 11 km/s. Assuming that the oxygen molecules at Earth's surface are at a temperature of 300 K, how would you use the velocity distribution to calculate the probability that a particular molecule at the surface would escape Earth's gravitational attraction if the molecule were to suffer no further molecular collisions? Express this probability in the form of an integral, and manipulate the integral into a form that involves the ratio v_{rms}/v_{escape}, but do not attempt to perform the integration. [*Hint*: This problem is analogous to finding the probability that a given student in a class has a grade of 35 or more on a test.]

CHAPTER

20

The surroundings of this power plant, which converts thermal energy to electrical energy, betray some of the unavoidable consequences of such operations: The water acts as a sink for unconverted energy associated with an exhaust stage of the conversion cycle, energy that the second law of thermodynamics states must be present.

The Second Law of Thermodynamics

The first law of thermodynamics does not specify the direction of thermal energy transfer. Nevertheless, we know that the spontaneous transfer of thermal energy always proceeds from a hotter system to a cooler one—ice cubes melt when they are placed in hot water. This directionality in nature is the essence of the second law of thermodynamics and this law has some surprising consequences. It implies that there are limits as to how much of the thermal energy in a system can be used to do mechanical work when this work is done cyclically, as in an engine. The directionality of the spontaneous transfer of thermal energy can be understood from a molecular point of view by using the notions of probability developed in Chapter 19. These notions will allow us to develop the idea of order in a thermal system and to see that a spontaneous process proceeds from a more ordered to a less ordered state.

20–1 BEYOND ENERGY CONSERVATION

The first law of thermodynamics brings the idea of energy and its conservation into thermal systems. However, there are some facts about thermal systems that this law cannot explain. First, it is impossible to devise an engine that extracts all the thermal energy from a thermal system and turns that energy into work. Such an engine, if it could be constructed, could power New York City while it cooled Long Island Sound. Yet such an engine is completely consistent with the first law of thermodynamics.

A second fact about thermal systems looks superficially different. When ice cubes are put into hot tea, the ice cubes melt and the tea cools. The first law of thermodynamics requires only that the heat flow into the ice cubes must match the heat flow out of the hot tea. According to the first law, heat flow can proceed in either direction. Nevertheless, lukewarm tea never *spontaneously* turns into hot tea with a couple of ice cubes floating in it. Another example of the second fact is the observation that gas released into an evacuated chamber through a small opening spontaneously spreads throughout the available space. We never see the gas spontaneously concentrating in a corner of the chamber, leaving most of the chamber evacuated.

These phenomena can be summarized as follows:

1. *Not all the thermal energy in a thermal system is available to do work.*
2. *Thermal systems spontaneously change only in certain ways.* In particular, spontaneous heat flow always occurs from a hotter body to a colder one.

These experimental facts are summarized in the **second law of thermodynamics**. We will formulate this law as it applies to point 1 in Section 20–2. Point 2, which describes the direction of spontaneous processes, introduces the concept of *entropy*, a new thermodynamic variable. We will address this point in Section 20–5.

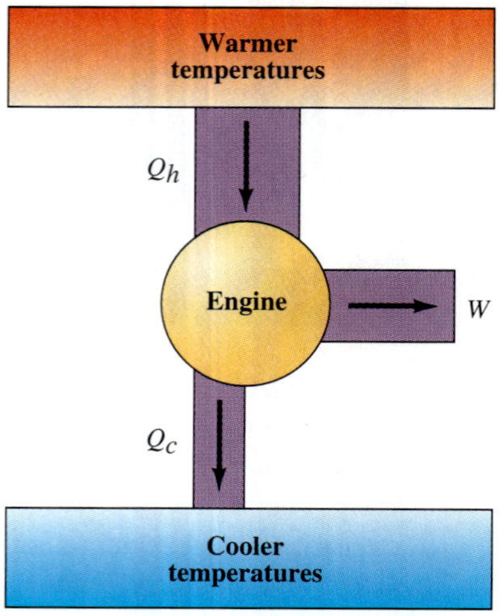

FIGURE 20–1 Abstract representation of an engine, which takes energy—here represented by a system at higher temperatures—to perform work. It is not possible to convert thermal energy entirely into work, so there is some thermal energy left over.

Engines

A young engineer, Sadi Carnot, gave an explicit statement of the second law of thermodynamics in 1824. He was interested in improving the efficiency of steam engines and found that there are limits to the possible efficiency of any engine. Just what is an engine? An engine is a device that transforms thermal energy into mechanical energy. In an automobile engine, the thermal energy of burning fuel is converted into the energy of the automobile's motion. In all known engines, however, the conversion of thermal energy into mechanical energy is accompanied by the emission of exhaust gases, which carry off some thermal energy. This means that only part of the thermal energy of the burning fuel is converted into mechanical energy (Fig. 20–1).

A useful engine exhibits two crucial features:

Cycles were introduced in Chapter 18.

1. An engine must work in **cycles** if it is to be useful. To see why a *noncyclic* operation is less useful, consider a gas that undergoes isothermal expansion. The gas is confined to a cylinder with a movable piston. The cylinder is in thermal contact with a thermal reservoir, such as a kitchen hot plate kept at a fixed temperature. During the expansion, the energy required by the piston to do work is supplied by the thermal reservoir. In this process, we convert thermal energy into mechanical energy. But we do not have an engine because the process operates in only one direction. The volume of the cylinder cannot increase indefinitely. The pressure, which is inversely proportional to the volume, decreases so, at some point, the expanding gas will no longer be able to push the piston. To be a part of a functioning engine, *the piston must eventually be recompressed and restored to its original position, ready to do work again.*

2. A cyclic engine must include more than one thermal reservoir. To understand this, let's reconsider the cylinder and piston. If the piston is returned to its original position while the cylinder remains in thermal contact with the original thermal reservoir, then all the work that the gas did while it expanded must be used to recompress the gas. The original expansion is reversed and nothing has been gained. In Chapter 18, we learned that the area enclosed by a closed path on a p–V diagram is the work done during the cycle. In this case, the area enclosed is zero, so the net work done by the system is also zero (Fig. 20–2). For a cycle to do finite work, thermal contact with the original thermal reservoir must be broken, and *temperatures other than the temperature of the original reservoir must come into play.* In this example, if we compress the gas in the cylinder at a lower temperature (so that its internal pressure will be lower)

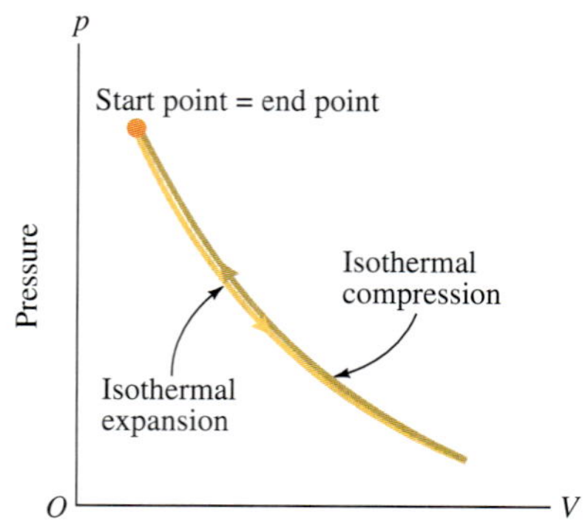

FIGURE 20–2 A cycle in which a gas expands isothermally and is then compressed isothermally at the same temperature. The work done is zero.

than was present during the expansion, less work will be needed for the compression than was produced in the expansion.

The definition of efficiency

Efficiency. Engines have an **efficiency**, which is a measure of how well the heat flow from the hotter thermal reservoir (such as a burning fuel) is converted into work (Fig. 20–1). If the work done in one complete cycle is W, then we define the efficiency, η, as the ratio of the work done to the total positive heat flow supplied by the burning fuel, Q_h:

$$\eta \equiv \frac{W}{Q_h}. \quad (20\text{–}1)$$

According to the first law of thermodynamics, this efficiency can run from 0 to 1. When the efficiency is 1, or 100 percent, all the thermal energy taken from the thermal reservoir is converted to mechanical work; this is consistent with the conservation of energy (the first law). We will see, however, that an efficiency of 100 percent is impossible according to the second law. Figure 20–3 illustrates how the efficiency is calculated.

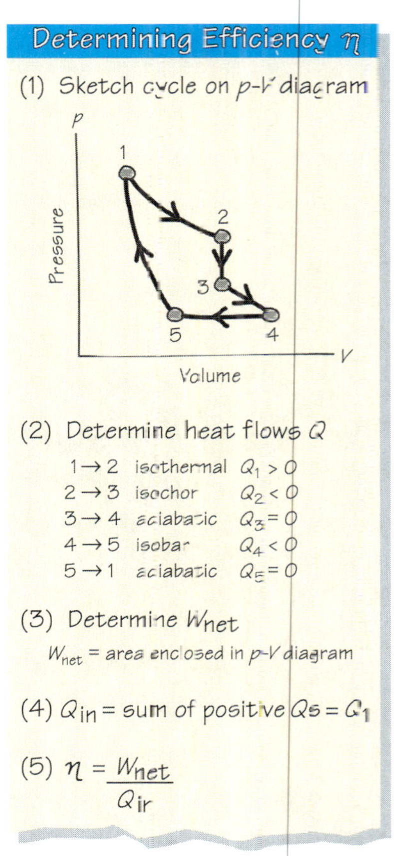

FIGURE 20–3 An illustration of the technique of calculating efficiency.

PROBLEM SOLVING

How to calculate the efficiency

Spontaneous Processes

We have mentioned two examples (hot tea plus ice cubes and an expanding gas) that illustrate a directionality in certain physical processes. Let's concentrate on the expanding gas example (free expansion). We can discuss this phenomenon most directly in terms of molecules. When a valve is opened and gas escapes from a small tank into a large empty chamber, the gas spreads and fills the chamber until the density and temperature are uniform throughout that chamber. Now imagine a film of this process made on a scale in which individual molecules can be seen. The molecules move rapidly and collide frequently in the small tank. When the valve is opened, molecules that reach the opening escape. They enter a vast space, empty of molecules, and the first batch of molecules encounter nothing in their way. Their first collisions will be with the walls of the chamber. As more and more molecules escape into the large chamber, they will begin to collide with other molecules as well as with the walls. After a short time, their velocities will be distributed according to Eq. (19–28).

Let's now suppose that we were to run the film backward. Because each collision is an elastic collision, it is as acceptable viewed backward as it is viewed forward. Thus each action in the reversed film is perfectly possible; in fact, the whole film, run backward, is possible in theory. It should be possible for the air to evacuate the large chamber spontaneously and concentrate into the small tank. So why do we never see this happen?

The explanation lies in the number of ways in which events involving many possibilities can happen. Consider the two poker hands in Fig. 20–4. In terms of probabilities, both hands are equally unlikely. The probability of each is approximately 2×10^{-7}, yet one looks much more special than the other. The reason is that hand I is a uniquely interesting one, whereas hand II is of a very undistinguished type because there are many, many ways to get 1 spade, 2 hearts, 1 diamond, and 1 club. Thus what we are really comparing is one distinct hand from a huge number of "ordinary" hands. An ordinary hand is much more probable because an immense number of individual (all equally rare) hands fit into this class.

Molecular distributions are like distributions of possible poker hands. Each molecule has an equal probability of being anywhere in the chamber, including the opening to the small tank, and each molecule has an equal probability of moving in any direction, including the one that will send it back into the small tank. Yet if we count the huge number of possible distributions for the huge number of molecules, the chance of them all ending up in the small tank is comparatively like being dealt a royal straight flush, although far less likely numerically. Thus the directionality of a spontaneous process such as free expansion is no miracle; it is just an almost sure bet.

20–2 THE SECOND LAW OF THERMODYNAMICS

There are two equivalent formulations of the second law of thermodynamics that are useful in understanding the conversion of thermal energy to mechanical energy.

The Kelvin form. ***It is impossible to construct a cyclic engine whose only effect is to convert thermal energy from a body at a given temperature into an equivalent amount of mechanical work.***

The Kelvin form of the second law of thermodynamics

The Clausius form of the second law of thermodynamics

There is no way to extract thermal energy from the ocean and use that energy to run an electrical generator *without further effect*, such as heating the atmosphere. In the example discussed in Section 20–1 (a gas expanding against a piston), thermal energy from a thermal reservoir is indeed converted into work, but the engine is not cyclic—the gas increases in volume. In an engine that takes thermal energy from a thermal reservoir cyclically and converts that energy to work, the additional effect is typically a heat flow from the engine into its lower-temperature surroundings. This brings us to the second formulation of the second law of thermodynamics:

The Clausius form.† ***It is impossible to construct a cyclic engine whose only effect is to transfer thermal energy from a colder body to a hotter body.***

In other words, there is no cyclic engine that can freeze water and use the released energy to boil more water *with no further effect.* In this case, the further effect is associated with the fact that the cyclic combination freezer–boiler *must* use some energy (such as electrical) from the surroundings.

In this book, we will use mainly the Kelvin form, which was first stated by Lord Kelvin. Detailed analysis, however, shows that the two formulations of the second law of thermodynamics are entirely equivalent.

I

II

FIGURE 20–4 Has poker hand I been dealt honestly, or would you suspect a stacked deck? Hand I is just as likely as hand II.

20–3 THE CARNOT CYCLE

We stated earlier that thermal reservoirs at two (or more) different temperatures are necessary for the operation of a reversible thermodynamic cycle that performs work; that is, a reversible engine. A minimal version of such an engine, one that requires exactly two such reservoirs, was invented by Sadi Carnot in 1824. The reversible cycle followed by this engine—known as a **Carnot cycle**—is of great importance in thermodynamics.

The four steps of the Carnot cycle

The Carnot cycle, demonstrated in Figure 20–5 with a gas-filled cylinder and a piston, consists of four reversible steps. Figure 20–5a is a p–V diagram of these four steps and Fig. 20–5b depicts the cycle schematically. Step I (AB in Fig. 20–5a) is an isothermal expansion: The expansion takes place with gas in the cylinder in thermal contact with a thermal reservoir—say, a kitchen hot plate—at temperature T_h. Step II (BC) is an adiabatic step: The volume slowly expands further, while the pressure as well as the temperature of the gas decrease. During step II, the cylinder of gas has been insulated with styrofoam from thermal contact with its surroundings so that there is no heat flow ($\Delta Q = 0$). The temperature of the gas is monitored throughout this adiabatic expansion. When the temperature has dropped to T_c (the temperature of a second, colder, thermal reservoir—for example, a bathtub full of cold water), it is placed in thermal contact with this second reservoir. In step III (CD), an isothermal compression of the gas at T_c takes place. The final pressure of step III is determined so that when the gas is removed from contact with the bathtub and again placed in thermal isolation, the adiabatic compression of step IV (DA) increases the temperature and pressure of the gas until it returns to the starting point, A.

†Rudolf Clausius is best known for introducing the concept of entropy.

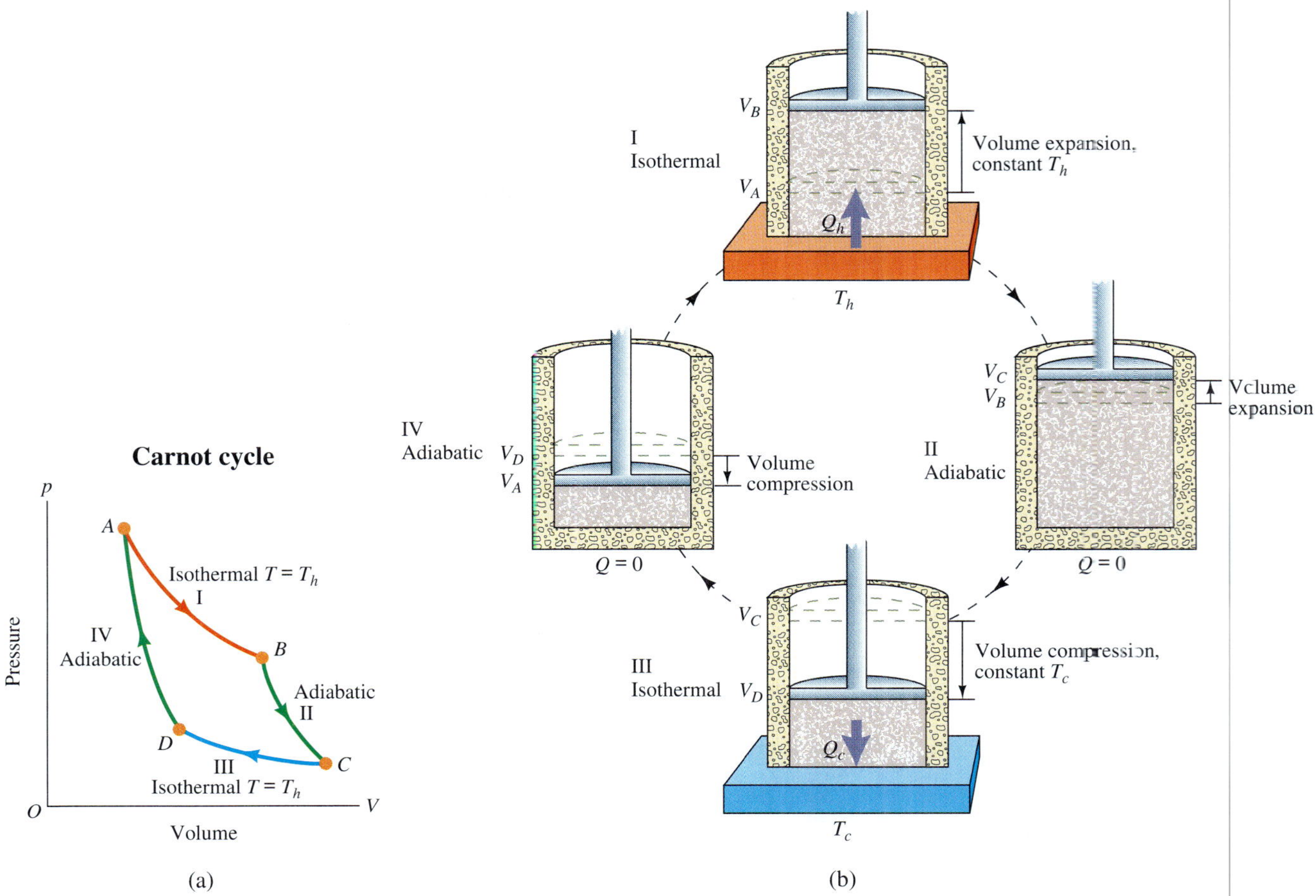

FIGURE 20–5 (a) The Carnot cycle, which consists of two adiabatic and two isothermal transformations. (b) Schematic diagram of the Carnot cycle.

In Fig. 20–5a, the isothermal legs are drawn as hyperbolas, and the adiabatic legs follow the curve p = (a constant) $V^{-\gamma}$ for $\gamma < 1$—as would be appropriate for an ideal gas (see Chapter 18). However, there is no reason to insist that the working fluid must be an ideal gas.† *Any* gas or other compressible fluid would be suitable because isothermal or adiabatic changes are possible for any substance. The crucial features of the Carnot cycle are independent of whether the working fluid is an ideal gas or not.

The work done in tracing out a complete Carnot cycle is the area enclosed by the curve in the p–V diagram. The cycle shown is clockwise, so the work done is positive. There is also heat flow associated with the Carnot cycle. It cannot occur during the adiabatic steps II and IV because, by definition, the system is thermally isolated for these steps. During step I, there is a positive heat flow Q_h from the hotter thermal reservoir into the thermal system. (For an ideal gas, $\Delta U = 0$ during an isothermal process, and then $Q_h = W_{AB}$, the positive work done in the expansion.) Similarly, there is a negative heat flow $-Q_c$ from the colder thermal reservoir to the system during step III. In other words, there is a positive heat flow Q_c from the thermal system to the colder reservoir. Thus the Carnot engine absorbs a heat flow Q_h during step I, gives up a heat flow Q_c during step III, and does a total amount of mechanical work W. Figure 20–6 is a schematic representation of the result of the operation of this cycle. The first law of thermodynamics can be applied to the entire cycle. Because the engine returns to

The work done in a cycle was discussed in Chapter 18.

†The "working fluid" is the material that expands and contracts, allowing the engine to do work.

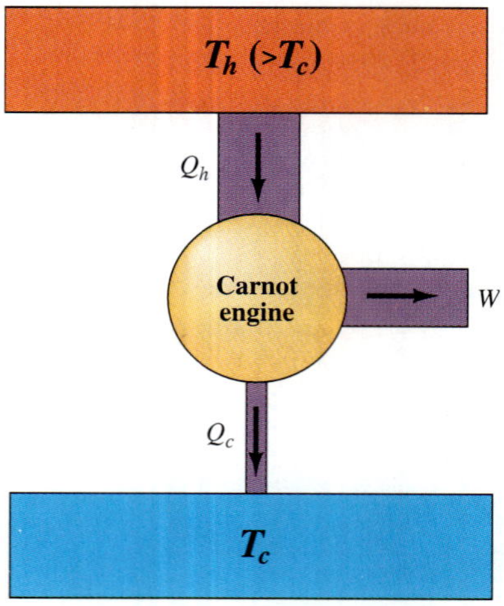

FIGURE 20–6 Schematic diagram of the result of the operation of one cycle of a Carnot engine. Heat flow Q_h is partly converted into mechanical work, W, and partly released as heat flow Q_c.

its original state, the internal energy is unchanged. This procedure holds even if the gas is not ideal. Then, according to the first law,

$$0 = W - Q_h + Q_c. \tag{20–2}$$

Equation (20–2) allows us to find the efficiency of the Carnot cycle, $\eta_C = W/Q_h$:

$$\eta_C = \frac{Q_h - Q_c}{Q_h} = 1 - \frac{Q_c}{Q_h}. \tag{20–3}$$

When the working gas of the Carnot cycle is ideal, the efficiency can be calculated in terms of the temperatures of the thermal reservoirs:

$$\eta_C = 1 - \frac{T_c}{T_h}. \tag{20–4}$$

This result follows from an analysis of the Carnot cycle with an ideal working gas. The analysis that gives this result is possible because the curves that describe both the adiabatic and isothermal curves for an ideal gas were found in Chapter 18.

How to Find the Efficiency of the Carnot Engine

Equation (20–4) is an important one. Let's see how to get it. Recall that for an ideal gas, the isothermal and adiabatic curves are

for an isothermal transformation: $\quad pV = nRT = \text{a constant},$

for an adiabatic transformation: $\quad pV^\gamma = \text{a constant},$

where the quantity $\gamma = C_p/C_V$ is greater than 1.

The Carnot cycle is labeled in Fig. 20–5a. For an ideal gas, the thermal energy is a function of temperature only, so the thermal energy, U, is constant during the isothermal steps. For these steps, $\Delta U = 0$, and the first law of thermodynamics, $\Delta U = \Delta Q - \Delta W$, implies that

for step I: $\quad Q_h = Q_{A\to B},$

and

for step III: $\quad Q_c = -W_{C\to D}.$

There is a minus sign in the second equation because we have defined Q_c as the heat flow *from* the thermal system to the cold reservoir, whereas in the first law the heat flow is always *to* the thermal system.

The work done during the isothermal steps is

$$W_{A\to B} = \int_{V_A}^{V_B} p\,dV = \int_{V_A}^{V_B} \frac{nRT_h}{V}\,dV = nRT_h \ln\left(\frac{V_B}{V_A}\right);$$

$$W_{C\to D} = \int_{V_C}^{V_D} p\,dV = \int_{V_C}^{V_D} \frac{nRT_c}{V}\,dV = nRT_c \ln\left(\frac{V_D}{V_C}\right).$$

Using these results we find the efficiency of the cycle:

$$\eta = 1 - \frac{Q_c}{Q_h} = 1 - \frac{-W_{C\to D}}{W_{A\to B}} = 1 + \frac{nRT_c}{nRT_h}\frac{\ln(V_D/V_C)}{\ln(V_B/V_A)}$$

$$= 1 - \frac{T_c}{T_h}\frac{\ln(V_C/V_D)}{\ln(V_B/V_A)}. \tag{20–5}$$

We have not yet used any information about the shape of the adiabatic curves in this expression. These curves will relate the various volumes. Points B and C lie on the same adiabatic curve, so $p_B V_B^\gamma = p_C V_C^\gamma$. Similarly, for points A and D, $p_A V_A^\gamma = p_D V_D^\gamma$. The ratio of these two equations gives

$$\frac{p_B V_B^\gamma}{p_A V_A^\gamma} = \frac{p_C V_C^\gamma}{p_D V_D^\gamma}. \qquad (20\text{–}6)$$

Next, we substitute for the ratio of the pressures on each side of this equation. Because A and B lie on an isotherm, $\frac{p_B}{p_A} = \frac{nRT_h/V_B}{nRT_h/V_A} = \frac{V_A}{V_B}$ and, similarly, because C and D also lie on an isotherm, $\frac{p_C}{p_D} = \frac{nRT_c/V_C}{nRT_c/V_D} = \frac{V_D}{V_C}$. We insert these ratios into Eq. (20–6) to find that

$$\frac{V_B^{\gamma-1}}{V_A^{\gamma-1}} = \frac{V_C^{\gamma-1}}{V_D^{\gamma-1}}.$$

The same power, $\gamma - 1$, appears on both sides of this equation, so the arguments must be equal:

$$\frac{V_B}{V_A} = \frac{V_C}{V_D}.$$

This result considerably simplifies the expression for efficiency. The arguments of the logarithms in Eq. (20–5) are the same, so the logarithms cancel and we finish with Eq. (20–4).

EXAMPLE 20–1 In a Carnot engine with an ideal gas, calorimetry reveals that the heat flow from the hot thermal reservoir in one cycle equals 38 J, and the heat flow to the cold thermal reservoir is 28 J. The temperature of the cold reservoir is 290 K. (a) What is the mechanical work done during one cycle? (b) What is the efficiency of the cycle? (c) What is the temperature of the hot reservoir?

Solution: (a) To find the mechanical work done during one cycle, we can find the area enclosed by the cycle on a p–V diagram. This evaluation can be done by integration, but only with great difficulty because all the parameters of the cycle are not yet known. A much simpler way to compute the work done is to use the first law of thermodynamics, which, for a closed cycle, states that the work done plus the heat flow out of a thermal system equals the heat flow into the system: $W + Q_c = Q_h$, from Eq. (20–2). Thus

$$W = Q_h - Q_c = 38\text{ J} - 28\text{ J} = 10\text{ J}.$$

(b) Equation (20–3) expresses the efficiency directly in terms of the given heat flows:

$$\eta = 1 - \frac{Q_c}{Q_h} = 1 - \frac{28\text{ J}}{38\text{ J}} = 0.26.$$

(c) Equation (20–4) gives an alternative expression for the efficiency in terms of the temperatures of the two thermal reservoirs. We know the efficiency and the temperature of the cold reservoir, so we can solve for the temperature of the hot reservoir. From Eq. (20–4),

$$T_h = \frac{T_c}{1 - \eta} = \frac{290\text{ K}}{1 - 0.26} = 390\text{ K}.$$

The Importance of the Carnot Engine

The second law of thermodynamics implies two important results:

1. *All Carnot cycles that operate between the same two temperatures have the same efficiency.* In particular, the efficiency of a Carnot cycle *does not depend on the use of an ideal gas*, and the expression for the efficiency is the one we have found for an ideal gas.

2. *The Carnot engine is the most efficient engine possible that operates between the two given temperatures.* (We will demonstrate this in Section 20–4.)

We can demonstrate the first result by showing that we would violate the second law of thermodynamics if it were not true. Suppose that we have two Carnot engines, A and B, that operate between the two same temperatures. Further, suppose that somehow the efficiency $\eta_A = W_A/Q_{hA}$ of cycle A were 75 percent, whereas the efficiency $\eta_B = W_B/Q_{hB}$ of cycle B were 74 percent. Let's adjust the amount of working gas in the cycles so that $Q_{hA} = Q_{hB}$, say, 100 J. This does not change the efficiencies. The work done by the two engines, which is the product of efficiency and heat flows, would then be different; namely, $W_A = (75\%)(100\text{ J}) = 75$ J, whereas $W_B = (74\%)(100\text{ J}) = 74$ J. Cycle A does more work than cycle B by an amount $\Delta W = W_A - W_B = 1$ J. We can now connect these two cycles in a way that violates the second law. Like any Carnot engine, engine B is reversible. We connect it to engine A as in Fig. 20–7 and run it in reverse.

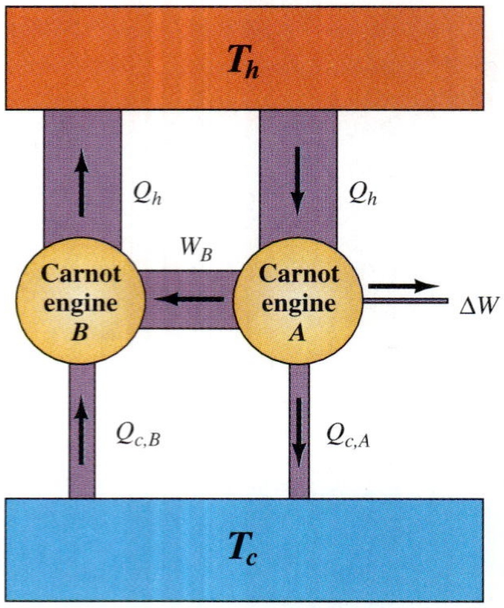

FIGURE 20–7 Carnot engines A and B are connected together, with B run in reverse. If engine A is more efficient than engine B, the net effect of the combination is to circumvent the Kelvin form of the second law of thermodynamics.

Running a Carnot engine—or any reversible engine—in reverse means that the direction of the transformation in each step is reversed, so that the path runs counterclockwise on the p–V diagram. For example, step I in Figure 20–5a becomes an isothermal compression rather than an isothermal expansion. The work done during one cycle is then negative—positive work must be supplied to the cycle from the exterior to make it run—while heat flow Q_c is taken from the cold reservoir and heat flow Q_h goes to the hot reservoir. This result is indicated on Fig. 20–7 for Carnot engine B.

We can divide the 75 J of work done by engine A in Fig. 20–7 into two parts: We use 74 J to run engine B and 1 J is left over. An equivalent amount of heat flow goes into the hot reservoir as is taken out, so this reservoir plays no role. Thus the net effect of this combination engine is to employ only one reservoir to extract 1 J of work with each cycle. But such an arrangement violates the Kelvin form of the second law of thermodynamics. This result is quite general, and it is a remarkable one because Carnot engines can be constructed with many different working fluids.

The Carnot cycle, with any working fluid, is the most efficient cycle that operates between two temperature extremes.

We have shown that the efficiency of *any* Carnot cycle that operates between temperatures T_c and T_h is given by Eq. (20–4), whether or not the gas is ideal:

The efficiency of any Carnot cycle

$$\text{for any Carnot cycle:} \qquad \eta_C = 1 - \frac{T_c}{T_h}. \tag{20–7}$$

We have found a universal expression for the efficiency of a Carnot cycle in terms of the temperatures of the thermal reservoirs. Note that the temperature scale here is the ideal-gas temperature scale—the Kelvin scale. For a Carnot engine to be efficient, the temperatures of the two reservoirs should be quite different from one another ($T_c < T_h$), and the temperature of the cold reservoir should be as low as possible (Fig. 20–8). If T_c were to be zero, the efficiency would be 1 (100 percent), independent of T_h. This is the only situation that could give an efficiency of 1. Such a reservoir does not exist, and thus *a perfectly efficient Carnot engine is not possible.*

FIGURE 20–8 The cyclic conversion of thermal energy into other forms of energy cannot be perfectly efficient; some thermal energy must be rejected. The cooling towers of a power plant, whether non-nuclear or nuclear, are the sites at which thermal energy is rejected. The clouds you see are condensing steam.

What happens if instead of the Carnot engine, with its idealized reversible cycle, we have an engine with some element of irreversibility? For example, there might be friction between the piston and the cylinder walls. We could reasonably expect that such an engine is less efficient than the Carnot engine that operates between the same two temperatures. We can, in fact, prove that this must be the case (see Problem 15).

20–4 OTHER TYPES OF ENGINES

Practical work-producing engines differ from the Carnot cycle in two ways. First, the path they trace out in a p–V diagram may not have the form of the Carnot cycle (two isothermal transformations and two adiabatic transformations), even though the path is reversible. These engines thus involve heat flows at temperatures other than the two temperatures of the Carnot cycle. Second, real engines generally involve some friction; that is, some irreversibility. We ignore this second possibility in the following discussion.

The Stirling Engine

As an example of a reversible non-Carnot cycle, consider the ideal Stirling cycle (Fig. 20–9), and compare this cycle to the Carnot cycle in Fig. 20–5a. Steps I and III are isothermal steps, just as in the Carnot cycle. The adiabatic steps of the Carnot cycle, however, are replaced by changes in pressure and temperature that maintain a constant volume. No work is done during the constant-volume steps. However, because the temperature changes, the internal energy also changes so that, by the first law of thermodynamics, there is a heat flow to or from the working gas of the engine during the constant-volume steps. This heat flow occurs at decreasing temperatures during step II and at increasing temperatures during step IV. It is as though a series of ever-colder thermal reservoirs enter into step II and ever-hotter thermal reservoirs enter into step IV. Increasingly hotter fuel must be used throughout step IV. Under these conditions, we note that the efficiency of an engine is really a measure of the work done for the total fuel burned; thus, the denominator of the efficiency should be the entire positive heat flow to the thermal system, which we write as Q_{pos}.

Here, we work out the efficiency of the ideal Stirling engine, η_S, when the working fluid is 1 mol of an ideal gas. We find that

$$\eta_S = \frac{W}{Q_{pos}} = \frac{R(T_h - T_c)\ln(V_2/V_1)}{RT_h\ln(V_2/V_1) + C_V(T_h - T_c)}. \tag{20–8}$$

This result would be the corresponding Carnot efficiency if it were not for the term proportional to C_V in the denominator (due to step IV). Because this extra term is positive, the denominator of η_S is greater than the denominator of η_C, and thus *the efficiency of the Stirling cycle is less than that of the Carnot cycle.*

Stirling cycle

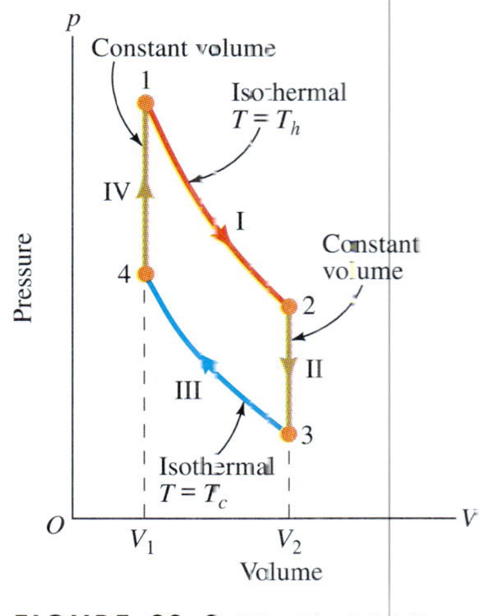

FIGURE 20–9 The ideal Stirling cycle. There is a positive heat flow to the engine in steps I and IV, and a heat flow from the engine in steps II and III. This cycle was described by Robert Stirling almost 20 years before Carnot discussed his own ideas. The engine Stirling patented in 1816 may be useful in outer space and under other conditions and has become the subject of active research.

APPLICATION

The Stirling engine

How to Get the Efficiency of the Stirling Engine

We assume that the working gas in a Stirling engine is ideal. We need the total positive heat flow $Q_{pos} = Q_I + Q_{IV}$ as well as W, the work done during one cycle (Fig. 20–9). During step I, the temperature is constant and $\Delta U = 0$. Thus, by the first law of thermodynamics, the heat flow to the engine is the work done during the transformation:

$$Q_I = W_I = \int_{V_1}^{V_2} p\,dV = RT_h\int_{V_1}^{V_2}\frac{dV}{V} = RT_h\ln\left(\frac{V_2}{V_1}\right).$$

According to the first law, the heat flow to the engine during step IV is the change in internal energy $U(T_h) - U(T_c)$ because no work is done by the engine during this portion of the cycle. Thus,

$$Q_{IV} = U(T_h) - U(T_c) = C_V(T_h - T_c).$$

Combining these two equations, we see that the total heat flow to the engine is

$$Q_{pos} = Q_I + Q_{IV} = RT_h\ln\left(\frac{V_2}{V_1}\right) + C_V(T_h - T_c).$$

The total work done in the cycle is the area enclosed by the cycle on the p–V diagram:

$$\begin{aligned} W = W_I + W_{III} &= RT_h\ln\left(\frac{V_2}{V_1}\right) + RT_c\ln\left(\frac{V_1}{V_2}\right) = RT_h\ln\left(\frac{V_2}{V_1}\right) - RT_c\ln\left(\frac{V_2}{V_1}\right) \\ &= R(T_h - T_c)\ln\left(\frac{V_2}{V_1}\right). \end{aligned}$$

The efficiency of this cycle, W/Q_{pos}, is then given by Eq. (20–8).

The Efficiencies of Other Types of Engines

It is possible to decompose an arbitrary reversible cycle into a series of Carnot cycles. By doing so, we can prove that *any* non-Carnot cycle (even if it is reversible) that operates between two temperature extremes T_h and T_c is less efficient than the Carnot cycle that operates between those same temperatures.

Let's apply this technique to the Stirling cycle. Figure 20–10a shows the path of the Stirling cycle approximated by a series of adiabatic and isothermal transformations. We can add a transformation as shown in Fig. 20–10b that will modify the approximation into one that consists of two Carnot cycles. This added transformation is an isothermal transformation from E_3 to E_2 and back again. Because this new transformation is a reversible round trip along the same line, it has no net effect on the overall efficiency of the approximation shown in Fig. 20–10a. At this point, we can think of the original approximation as consisting of two separate Carnot cycles. In Fig. 20–10c, these two cycles are labeled I and II.

Now, the combination of two Carnot cycles is less efficient than a single Carnot cycle that operates between T_h and T_c. To see why, suppose that the area under the approximating cycles is the same as the area under the Stirling cycle. Then the work done, which appears in the numerator of the expression for the efficiency, is unchanged. However, the denominator, Q_h, of the approximating cycle is bigger than the corresponding denominator for a single Carnot cycle that operates between T_h and T_c. This is because Carnot cycle I takes its incoming heat from two sources. For the step E_2E_3, it takes up the released heat from Carnot cycle II. However, along E_1E_2, it must take

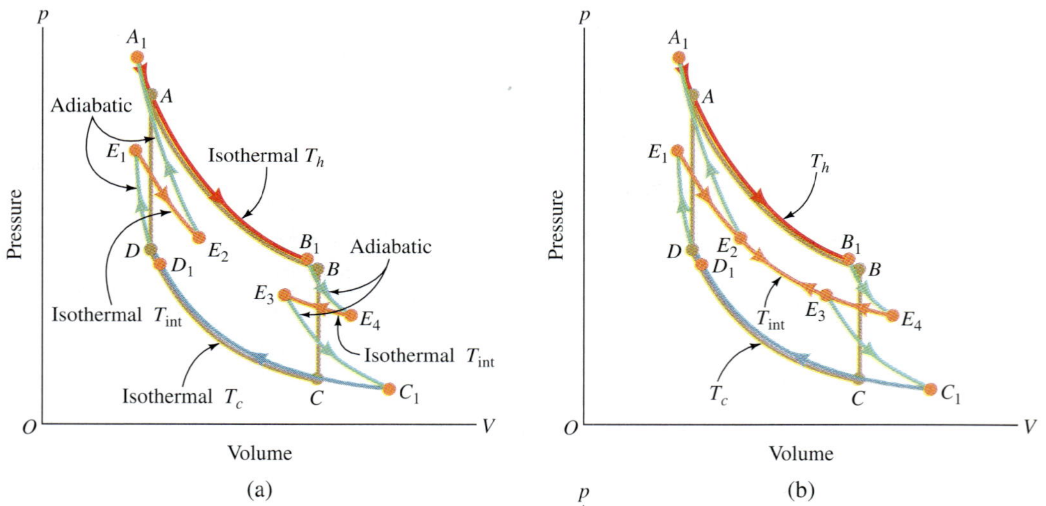

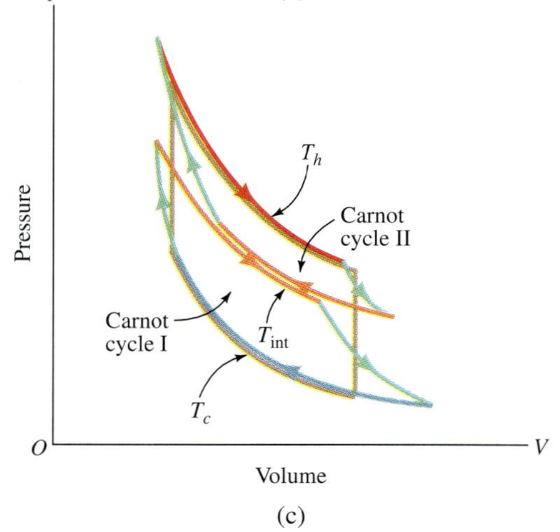

FIGURE 20–10 (a) The Stirling cycle is approximated by a series of adiabatic and isothermal transformations. (b) By adding an isothermal back-and-forth segment at T_{int}, the Stirling cycle is approximated by two Carnot cycles. (c) The two Carnot cycles that approximate the Stirling cycle are labeled I and II.

APPLICATION: MANDATING EFFICIENCIES

In addition to the Stirling cycle that we have discussed, other reversible cycles represent idealized versions of real engines. A short list might include the diesel engine (Fig. 20AB–1); the Otto cycle (Fig. 20AB–2), which is similar to an automobile engine; and the Brayton cycle (Fig. 20AB–3). Detailed treatment of these cycles will be left to end-of-chapter problems. Keep in mind that, in addition to the fact that these cycles are inherently less efficient than the Carnot cycle, there is always additional inefficiency in real engines associated with irreversible portions of the cycle.

From the end of the 1960s, the U.S. government has requested that the efficiency of internal combustion engines be improved and that air pollution be reduced. Engines of that period used an excess of fuel, and the residue of unburnt hydrocarbons was a significant pollution source. The pollution problem was alleviated by adding air to the products of combustion as they left the cylinder, thereby burning them. What about the improvement in engine efficiency? The easiest way to accomplish this is to increase the maximum temperature at which the fuel is burned. But a higher temperature leads to a secondary reaction: the oxidation of the nitrogen in the added air. The resulting oxides of nitrogen added very significantly to "acid rain" problems. Thus, the government-mandated increase in efficiency resulted in a new pollution problem. The solution to this built-in conflict is a complicated one, involving recycled gases and catalysts. It is not yet completely resolved. Perhaps the moral of this tale is that having scientifically literate people in government is not a luxury but a necessity.

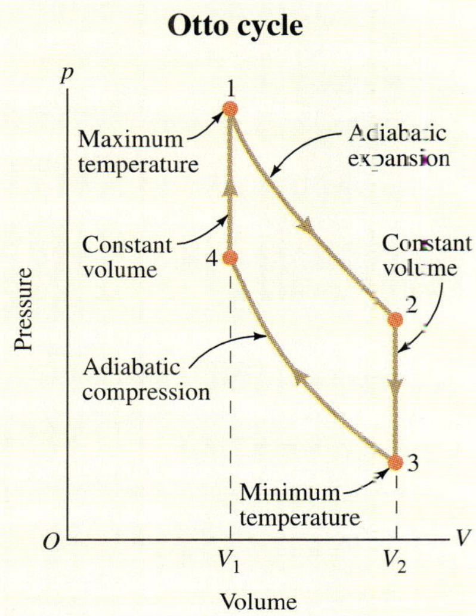

FIGURE 20AB–2 The Otto cycle, for the engine designed by Nikolaus Otto in 1876.

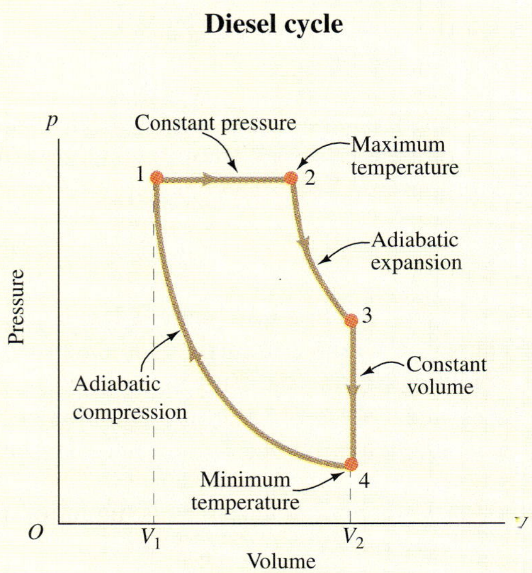

FIGURE 20AB–1 The Diesel cycle, for the engine patented by Rudolf Diesel in 1892.

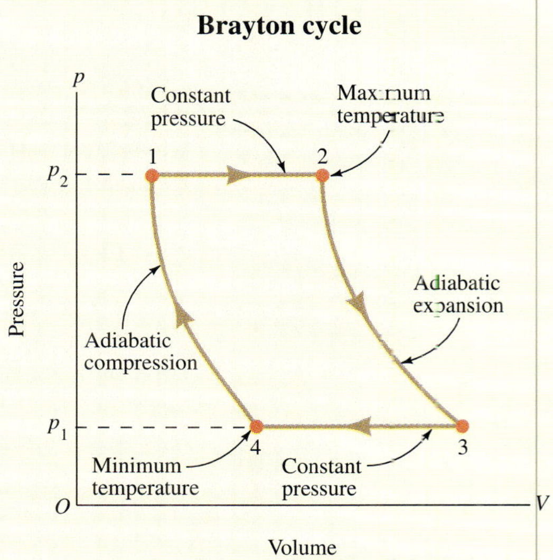

FIGURE 20AB–3 The Brayton cycle, named for George B. Brayton.

an additional amount of heat from the intermediate reservoir at T_{int}. This intermediate use of heat is *in addition to* the heat that cycle II takes at T_h. The same amount of work requires more heat, and the double Carnot approximation is less efficient. Thus, to the extent that the double Carnot cycle approximates the Stirling cycle, the Stirling cycle is less efficient than the single Carnot cycle.

The Carnot engine is more efficient than any other type of engine.

The two-cycle approximation shown is not, in fact, a very good approximation to the Stirling cycle. However, we can always improve the approximation by the inclusion of more Carnot cycles that operate between more intermediate temperatures. If we repeat the reasoning we applied for two cycles, we will always show that the approximation is less efficient than a single Carnot engine operating between the same temperature extremes; the intermediate Carnot cycles always involve some additional heat flow at intermediate temperatures.The technique described here can be applied to any reversible cycle. We learn the same lesson: *Any reversible cycle operating between given temperature extremes can be approximated by a series of Carnot cycles, and its efficiency will always be less than that of the single Carnot cycle that operates between the given temperatures.*

FIGURE 20–11 A heat pump, in which mechanical energy is used to transfer thermal energy from the ground to the interior of a building.

Heat Pumps and Refrigerators

APPLICATION

Heat pumps and refrigerators

Heat pumps and **refrigerators** are nothing more than work-producing engines run in reverse. A heat engine run in reverse—counterclockwise on a p–V diagram—transfers thermal energy from colder to hotter thermal reservoirs. For this to occur, work must be performed *on* the system. Heat pumps and refrigerators are similar to one another in operation, but they are used differently. A refrigerator removes thermal energy from a colder reservoir (which keeps a freezer cold). In doing so, the refrigerator is responsible for a heat flow at a higher temperature into its surroundings, something you can feel when you put your hand at the back of the refrigerator. A heat pump heats a house by taking thermal energy from a colder reservoir (for example, cool ground below a house) and dumping this thermal energy into the house at a higher temperature (Fig. 20–11). In each case, it is necessary to supply additional energy through the performance of work on the thermal system.

The definition of the coefficient of performance of a heat pump

We can analyze heat pumps or refrigerators by the same method that we used to analyze heat engines. Because of the different ways these devices are used, their efficiencies are defined differently. For a heat pump, where the total positive heat flow rejected from the pump into a building, Q_{rej}, is of interest, we use the *coefficient of performance* K_{hp}, which is defined as

$$K_{\text{hp}} \equiv \frac{Q_{\text{rej}}}{W}, \tag{20–9}$$

where W is the total work that must be supplied in one cycle. For a refrigerator, where we are interested in the total positive heat flow absorbed from the interior of the refrigerator, Q_{abs}, the coefficient of performance K_{ref} is defined by

$$K_{\text{ref}} \equiv \frac{Q_{\text{abs}}}{W}, \tag{20–10}$$

where W is the work supplied to the refrigerator. For a heat pump, a larger coefficient of performance means that a given amount of work performed by the motor leads to a larger transfer of heat flow into the building. By the same token, for a refrigerator, a larger coefficient of performance corresponds to a larger heat flow from the interior of the refrigerator to its surroundings when the motor performs a given amount of work.

EXAMPLE 20–2 The Carnot cycle shown in Fig. 20–5a is run in reverse and used as a heat pump. Show that K_{hp} is given by $T_h/(T_h - T_c)$.

Solution: In solving this problem, all formulas given previously for the operation of the Carnot cycle can be used, except that the signs must be changed because the cycle runs in reverse. These signs may be easily remembered with a schematic drawing of the operation, as in Fig. 20–12, where W, Q_c, and Q_h are all taken to be positive. It is also necessary to interpret the total heat flow rejected, Q_{rej}, as the heat flow Q_h transferred to the thermal reservoir at T_h. By the first law of thermodynamics,

$$W = Q_h - Q_c,$$

so K_{hp} is given by

$$K_{hp} = \frac{Q_h}{Q_h - Q_c} = \frac{1}{1 - (Q_c/Q_h)},$$

according to Eq. (20–9). The ratio Q_c/Q_h is, in fact, the ratio of temperatures—see Eq. (20–7) and the accompanying discussion. It is not important that the signs of the heat flows are reversed because we use only the ratio of these quantities. With $Q_c/Q_h = T_c/T_h$, we find

$$K_{hp} = \frac{1}{1 - (T_c/T_h)} = \frac{T_h}{T_h - T_c}. \quad (20\text{–}11)$$

The coefficient of performance of the Carnot heat pump is a number larger than 1 and it increases, or improves, if the difference between the temperatures of the two thermal reservoirs is reduced: It is easier to transfer thermal energy from the cold ground to a warm house when the temperature difference is smaller. An ideal heat pump used for heating a house in winter could operate between the temperatures $T_c = -5°\text{C} = 268$ K and $T_h = 45°\text{C} = 318$ K. (Although 45°C is much hotter than the interior of a house, if the size of the heat pump is not to dwarf the house itself or if the cost of insulating the house is not to be prohibitive, the temperature of the exhaust must be somewhat higher than the desired interior temperature.) The ideal K_{hp} of a Carnot cycle that operates between these temperatures is then $T_h/(T_h - T_c) =$ 318 K/50 K = 6.36. In contrast, an actual heat pump that operates between the temperatures 45°C and −5°C has a value of K_{hp} that is only about half this value.

The coefficient of performance, as defined in Eq. (20–10), for a Carnot cycle run as a refrigerator (see Problem 22) is given by

$$K_{ref} = \frac{T_c}{T_h - T_c}. \quad (20\text{–}12)$$

Just as for the heat pump, the coefficient of performance is improved if the temperature range across which the refrigerator runs is decreased.

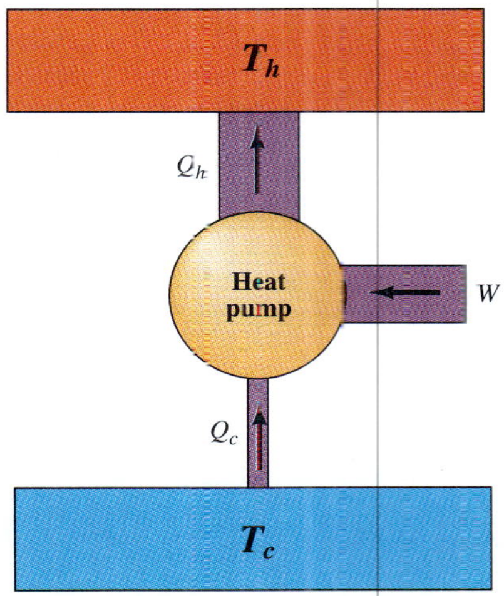

FIGURE 20–12 Example 20–2. The Carnot cycle operated as a heat pump.

20–5 ENTROPY AND THE SECOND LAW

Entropy as a Thermodynamic Variable

In this section, we will use the Carnot cycle to identify a new thermodynamic variable: entropy. This variable is like internal energy in that it is a function of state. On a macroscopic level, we will see that the entropy is a measure of the energy unavailable to do useful work; on a microscopic level, we will see that the entropy is a measure of disorder.

In a Carnot cycle, the efficiency—defined by $1 - (Q_c/Q_h)$—is given by Eq. (20–7): $\eta_c = 1 - (T_c/T_h)$. Thus, Q_c/Q_h must equal T_c/T_h. This relation can be expressed as

$$\frac{-Q_c}{T_c} + \frac{Q_h}{T_h} = 0. \quad (20\text{–}13)$$

Recall that Q_c is the heat flow from the Carnot engine to the cold thermal reservoir. The expression $-Q_c$ in the first term of Eq. (20–13), minus sign and all, is then the heat flow to the engine at the cold temperature, just as Q_h is the heat flow *to* the engine at the high temperature.

Is there a relation comparable to Eq. (20–13) for another reversible cycle? Let's consider the Stirling cycle as approximated by two Carnot cycles. In Fig. 20–10c, the Carnot cycle that operates between T_c and T_{int} is Carnot cycle I; the Carnot cycle that operates between T_{int} and T_h is Carnot cycle II. Equation (20–13) holds for each of the two Carnot cycles, and the sum of the equations for each cycle is

$$\frac{-Q_{c,\,\text{I}}}{T_c} + \frac{W_{\text{int, I}}}{T_{\text{int}}} + \frac{-Q_{\text{int, II}}}{T_{\text{int}}} + \frac{Q_{h,\,\text{II}}}{T_h} = 0.$$

The interpretation of the heat flow in the first term (including the minus sign) is as before: the heat flow to the Stirling cycle at its minimum temperature. The numerator of the second term is the *net* heat flow to the cycle, Q_{int}, at temperature T_{int}. The third term is the heat flow to the engine at the maximum temperature.

This result is generalized if the cycle is more closely approximated by more Carnot cycles that operate at finer temperature divisions. We can add Eq. (20–13) for each of these approximating Carnot cycles. The heat flows at the intermediate temperatures T_i will cancel in part, leaving, in each case, the *net* heat flow to the cycle, Q_i, at temperature T_i. The first and last terms, $-Q_{\text{min}}/T_{\text{min}}$ and $Q_{\text{max}}/T_{\text{max}}$, remain. The net effect is

$$\frac{-Q_{\text{min}}}{T_{\text{min}}} + \frac{Q_1}{T_1} + \frac{Q_2}{T_2} + \cdots + \frac{Q_{\text{max}}}{T_{\text{max}}} = 0. \qquad (20\text{–}14)$$

The heat flows in Eq. (20–14) represent heat flows *to* the thermal system. We mean here that a heat flow to the system is positive, and a heat flow from the system is negative. If we let ΔQ_i be the net heat flow at temperature T_i, Eq. (20–14) becomes

$$\sum_i \frac{\Delta Q_i}{T_i} = 0, \qquad (20\text{–}15)$$

where the sum is taken over the full cycle. This expression depends on the original cycle being approximated by a finite number of Carnot cycles that operate at the successive temperatures T_i. In the limit in which the temperature divisions become finer and finer, the sum in Eq. (20–15) becomes an integral. In integral notation, Eq. (20–15) is known as **Clausius's theorem**:

Clausius's theorem for reversible cycles

$$\text{for a reversible cycle:} \quad \oint \frac{dQ}{T} = 0. \qquad (20\text{–}16)$$

The circle placed on the integral is a reminder that the integral is taken over a closed cycle. Because any reversible cycle can be approximated by a series of Carnot cycles, Clausius's theorem is a general result that holds for *any reversible cycle, which means any closed path on a p–V diagram.*

We have already encountered a result like Eq. (20–16). In the discussion of conservative forces in Chapter 7, we saw that the integral $\oint \mathbf{F} \cdot d\mathbf{s}$ equals zero, which allowed us to define a potential energy **U**. This function depends on position alone; moreover, the integral of dU along a path that connects two points is independent of the path. The same situation applies here. There is a function S, which we call **entropy**, that is analogous to U. The analogue of $dU = -\mathbf{F} \cdot d\mathbf{s}$ is the expression that defines the change in entropy as

The definition of the change in entropy

$$\text{for a reversible transformation:} \quad dS \equiv \frac{dQ}{T}. \qquad (20\text{–}17)$$

According to Eq. (20–16), the integral of dS around a closed, reversible path on a p–V diagram is zero. The analogue to the expression for the change in U, which is $U_B - U_A = \int_A^B dU = -\int_A^B \mathbf{F} \cdot d\mathbf{s}$, is an expression for the change in entropy when a thermal system undergoes a transformation between any two thermodynamic states A and B:

$$S(B) - S(A) = \int_A^B \frac{dQ}{T}, \qquad (20\text{–}18)$$

where the integral is taken over *any* reversible path between these states because the integral is independent of path. As for U, the entropy depends only on the position on a p–V diagram; that is, *the entropy, S, is a function of state.*

Entropy is a function of state.

The fact that an integral of dS between two points along some path is independent of the path means that S behaves like the thermal energy U. A differential quantity whose integral is independent of the path is said to be a perfect differential. The infinitesimal heat flow dQ is *not* a perfect differential because its integrals are path dependent. But when dQ is divided by the temperature at which it occurs, the differential becomes perfect.

As is true for potential energy, only *changes* in entropy have any physical meaning—as Eq. (20–18) shows. The choice of zero for the entropy function is a matter of convenience.

EXAMPLE 20–3 Calculate the entropy change of 1 mol of an ideal gas that undergoes an isothermal transformation from an initial state of pressure 1.50 atm and volume 500 cm^3 to a final state of pressure 0.90 atm.

Solution: To calculate the entropy change along the specified isothermal path, we must know the heat flow along that path. For the isothermal transformation of an ideal gas, the internal energy does not change, so the first law of thermodynamics tells us that along our path, $dQ = dW + dU = dW = p\,dV$. Thus, from Eq. (20–18) and the ideal gas law, $pV = nRT$, we find

$$\Delta S = \int \frac{p\,dV}{T} = \int \frac{nRT}{VT}\,dV = nR\int \frac{dV}{V} = nR\ln\left(\frac{V_f}{V_i}\right) = -nR\ln\left(\frac{p_f}{p_i}\right)$$

$$= -(1\text{ mol})(8.3\text{ J/mol}\cdot\text{K})\ln\left(\frac{0.90\text{ atm}}{1.50\text{ atm}}\right) = 4.2\text{ J/K}.$$

How Entropy Changes for Irreversible or Spontaneous Processes

Irreversible processes (processes with friction, for example) lead to increased inefficiency in any otherwise reversible engine. Let's see what this implies for heat flows to such a cycle. The efficiency of an arbitrary cycle is

$$\eta = \frac{W}{Q_{abs}} = 1 - \frac{Q_{rej}}{Q_{abs}},$$

where Q_{rej} is the total heat flow rejected by the cycle and Q_{abs} is the total heat flow absorbed. We have used the relation $W = Q_{abs} - Q_{rej}$ from the first law of thermodynamics. Each of these heat flows is positive. The efficiency can be reduced in two ways. First, the heat flow *to* the system, Q_{abs}, can be reduced. Second, the heat flow rejected, Q_{rej}, can be increased (which is equivalent to a reduced heat flow to the system). Thus, the efficiency is reduced if the heat flow *to* the system is algebraically reduced anywhere along the cycle. In other words, the effect of irreversibilities or of spontaneous processes is to reduce the algebraic heat flow to the system, dQ. In turn, the integral in Clausius's theorem, Eq. (20–16), is reduced algebraically below zero: Clausius's inequality

$$\text{for an irreversible cycle:} \quad \oint \frac{dQ}{T} < 0. \qquad (20\text{–}19)$$

This result, known as **Clausius's inequality**, is a general result that holds for heat pumps as well as for engines.

Equations (20–16) and (20–19) can be combined into

$$\oint \frac{dQ}{T} \leq 0. \qquad (20\text{–}20)$$

The equality applies when the path is reversible; the inequality applies when the transformations include some irreversible portion.

Equation (20–20) can be used to learn about entropy changes in irreversible transformations. Consider the cycle shown in Fig. 20–13, in which leg $A \to B$ is irreversible. From Eq. (20–20),

$$0 > \oint \frac{dQ}{T} = \int_{A\,\text{irrev}}^{B} \frac{dQ}{T} + \int_{B\,\text{rev}}^{A} \frac{dQ}{T}.$$

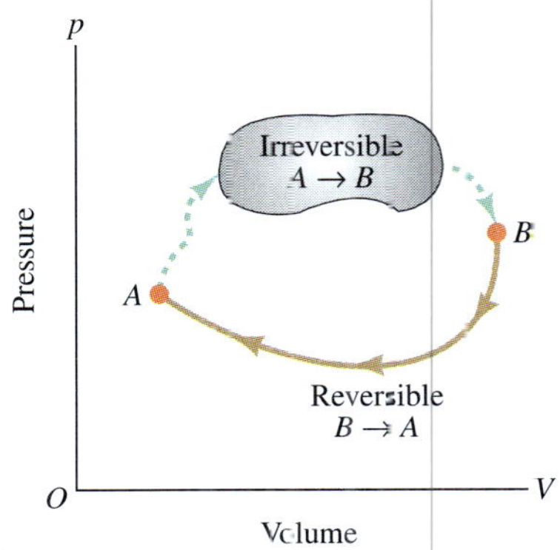

FIGURE 20–13 A cycle that contains an irreversible path from A to B and a reversible path from B back to A.

The right-hand integral is given by $S(A) - S(B)$ because it is taken over a reversible path. When we move this quantity to the left-hand side, we find that

$$S(B) - S(A) > \int_{A\ \text{irrev}}^{B} \frac{dQ}{T}. \tag{20–21}$$

Thus, *the difference in entropy between two points is greater than the integral of* dQ/T *over an irreversible change.*

How do we compute the increase in entropy when there is an irreversible transformation from thermodynamic state A to thermodynamic state B? We have seen that entropy is a function of state, so we can compute the change $S(B) - S(A)$ by any means we like; that is, *we can use a reversible path to find the change in entropy as long as that path connects the two given states.* We will illustrate this procedure in Section 20–6.

The Entropy of an Isolated System Never Decreases

When an isolated system undergoes an irreversible process, its entropy increases.

One type of irreversible change is special: the irreversible change with no heat flow. Free expansion into a thermally isolated container is an example of such a spontaneous transformation. If there is no heat flow, then dQ is zero every step of the way, and

$$\text{for a thermally isolated system:} \quad S(B) - S(A) \geq 0. \tag{20–22}$$

The equality holds when a reversible, or adiabatic, transformation takes place; the inequality holds when the transformation from A to B is irreversible, as is true for free expansion. Note that the insulated container into which a gas expands in free expansion is a small, isolated system. But even if a given system is not thermally isolated, combinations of systems may well be. We may have a small system such as a cylinder of gas with a movable piston in thermal contact with a reservoir. Neither the cylinder nor the reservoir is isolated, but the combination of the two may be. *A spontaneous process in an isolated system increases the system's entropy.* Because the universe—our entire surroundings—is in contact with no other system, we say that irreversible processes increase the entropy of the universe.

The fact that the final entropy is greater than or equal to the initial entropy means that every spontaneous process increases the entropy of an isolated thermal system. Only the state of maximum entropy is a stable one: For a state of maximum entropy, no spontaneous changes occur. For an isolated thermal system, *the state of maximum entropy is the state of stable equilibrium.*

20–6 ENTROPY AND IDEAL GASES

When a gas is ideal, we can calculate its entropy change in reversible transformations. In fact, we can use a combination of constant-volume and isothermal transformations to calculate the entropy of an ideal gas. In this way, we can acquire a feeling for the nature of entropy as a thermodynamic variable. Also, we can use the results in Section 20–7 when we investigate the physical nature of entropy.

Consider the isothermal transformation from a point A at temperature T_0 and volume V_0 to a point B at temperature T_0 but at volume V. In this transformation, the internal energy—which for an ideal gas is a function only of temperature—is unchanged. Therefore, according to the first law of thermodynamics, $dW = dQ$ along the isotherm. Then, according to Eq. (20–18),

$$S(B) - S(A) = \int_A^B \frac{dQ}{T_0} = \int_A^B \frac{dW}{T_0}. \tag{20–23}$$

The infinitesimal work done at any point along the path is $dW = p\,dV = (nRT_0/V)\,dV$. The factor T_0 cancels, and

$$\text{for a fixed temperature:} \quad S(B) - S(A) = \int_A^B \frac{nRT_0}{V}\,dV\left(\frac{1}{T_0}\right)$$

$$= NR\int_A^B \frac{dV}{V} = nR\ln\left(\frac{V}{V_0}\right). \tag{20–24}$$

This expression is the change in entropy for a fixed temperature. There will be other transformations in which the temperature is changed. To keep such changes distinct from the one we have just used, we next perform a transformation at fixed volume. This transformation takes an ideal gas from point $A = (T_0, V_0)$ to point $B = (T, V_0)$. For a constant-volume transformation, the work done is zero and, according to the first law of thermodynamics, the heat flow $dQ = dU = C_V\,dT$. Thus

$$S(B) - S(A) = \int_A^B \frac{dQ}{T} = \int_A^B C_V\frac{dT}{T}. \tag{20–25}$$

C_V is a constant, so

$$\text{for a fixed volume:} \quad S(B) - S(A) = C_V\int_A^B \frac{dT}{T} = C_V\ln\left(\frac{T}{T_0}\right). \tag{20–26}$$

Equation (20–24) gives the entropy change for fixed T and Eq. (20–26) gives it for fixed V. These are independent state variables so, to find the entropy change in going from any standard state (T_0, V_0) to state (T, V), simply add the two expressions:

$$S(T, V) - S(T_0, V_0) = C_V\ln\left(\frac{T}{T_0}\right) + nR\ln\left(\frac{V}{V_0}\right). \tag{20–27}$$

The entropy of an ideal gas

This expression specifies the function $S(T, V)$ up to a constant value $S_0 = S(T_0, V_0)$. If we want S in terms of other variables, such as pressure and volume, the ideal gas law can be used for the substitution.

Note that we used reversible paths to calculate the entropy change. Because entropy is a function of the thermodynamic state, once we know the entropy change between given thermodynamic states, it makes no difference how the transformation between those states is made. The entropy change is the same whether that transformation is reversible or irreversible.

EXAMPLE 20–4 You have 3.0 mol of an ideal gas in thermal contact with a thermal reservoir at $T = 300$ K. The gas expands spontaneously—the gas does not push against a piston to do work—from 12 L to 24 L. What is the entropy change of the gas? of the universe? Repeat the exercise for an expansion from 12 L to 120 L.

Solution: The entropy of the gas, being a state function, can be computed (up to a constant) when the state is known. In this case, we can use Eq. (20–27) with fixed $T = T_0$. Because $\ln(1) = 0$,

$$\Delta S = C_V\ln\left(\frac{T_0}{T_0}\right) + nR\ln\left(\frac{V}{V_0}\right) = C_V(0) + nR\ln\left(\frac{V}{V_0}\right)$$

$$= nR\ln\left(\frac{V}{V_0}\right) = (3.0\text{ mol})(8.3\text{ J/mol}\cdot\text{K})\ln\frac{24L}{12L}$$

$$= 25\ln(2)\text{ J/K} = 17\text{ J/K}.$$

We have found that the entropy change of the gas is positive. To determine the entropy change in the rest of the surroundings, we must determine the heat flow to the thermal reservoir. The gas does no work in its expansion, so the heat flow out is equal to the change in internal energy. But this change is zero because the gas is ideal and because its internal energy depends only on the (unchanging) temperature. There is thus no heat flow to the thermal reservoir and its entropy change, $\Delta Q/T_0$, is zero. The entropy change of the universe is therefore the entropy change of the gas. The fact that it is positive indicates that the transformation involved—in this case, free expansion—is irreversible.

When the expansion is from 12 L to 120 L, the entropy change is as just described, except that ln(2) is replaced by ln(10):

$$\Delta S = 25 \ln(10)\ \text{J/K} = 57\ \text{J/K}.$$

The natural logarithm changes fairly slowly with its argument, so, compared to the ratio of volume changes in the two cases, the difference in entropy changes is not very great.

20–7 THE MEANING OF ENTROPY

The entropy of a system and all its surroundings—the universe—increases when an irreversible process takes place. In effect, increased entropy corresponds to a lost heat flow, and hence a lost energy. The loss of heat flow represents energy that can never be turned into work because, by the nature of irreversible processes, the process cannot be reversed to recapture this heat flow without making other changes. Of course, the energy is not really lost; energy is always conserved. But it has been turned into the random motion of the microscopic constituents of the thermal system and cannot be sufficiently well "organized" to allow all of it to go into, say, lifting a weight. By organization, we mean order; if all the molecules were moving in precisely the same direction, for example, they would be well ordered. They might have the same total energy as a disorganized gas, but *all* of their energy could be used to lift a weight. Any degree of irreversibility reduces our ability to do work because it increases the disorder.

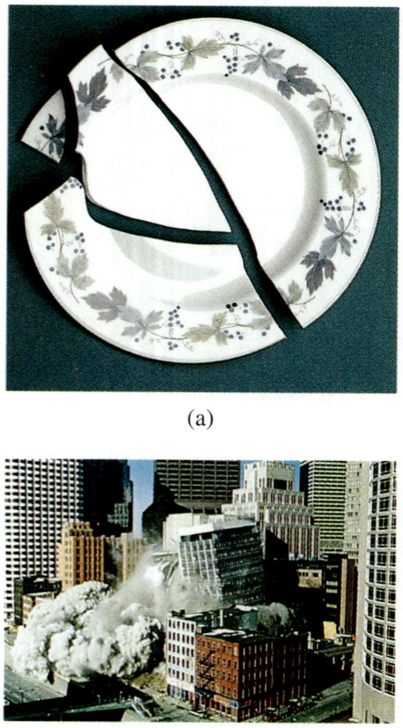

(a)

(b)

FIGURE 20–14 (a) When a plate breaks, it is more disordered—the entropy of the "plate system" increases. (b) The entropy of the demolished building has rather suddenly increased.

The preceding paragraph contains the seeds of a deeper understanding of entropy based on a microscopic picture. The idea of microscopic order and disorder is the key to entropy (Fig. 20–14). Let's consider this idea more closely.

We indicated in Section 20–1 that, when a gas expands into an empty space, its molecules move from a configuration that has a low probability (all the molecules are in one corner) to one that has the largest possible probability (the molecules are spread uniformly throughout the allowed volume). Equivalently, the molecules go from a configuration that has few ways of occurring (like the poker hand in Fig. 20–4 with a royal straight flush) to one that has many ways of occurring (like the "ordinary" hand). Initially, all the molecules are confined by walls to a small tank. This initial state is more *ordered* than the final state because the individual molecules are more localized. A loss of order is associated with a change of the system from a state of low probability to one of higher probability. Equation (20–24) illustrates how entropy increases for an ideal gas when the volume increases.

We can further illustrate this idea by mixing two gases that are initially separated by a partition (Fig. 20–15). Consider a large container of volume V divided by a partition into volumes V_1 and V_2. There are n_1 mol of gas 1 in V_1 and n_2 mol of gas 2 in V_2. Each gas has the same temperature and pressure. The partition is opened, and the gases in the two parts mix. Before the partition is opened, the degree of order is greater: All the molecules of each gas are on their respective sides. After the partition is opened, the disorder is greater: Any one molecule is somewhere in the greater volume, rather than being on one side or the other. It is more probable that each gas will spread separately throughout the greater volume than that it will stay on its respective side. Thus, we would expect the entropy to increase when the partition is opened, which is in line with the idea that the mixing of the two sides is an irreversible process. This increased entropy is the **entropy of mixing**.

EXAMPLE 20–5 A quantity of 0.20 mol of argon gas contained in a volume of 5.0 L mixes with 0.50 mol of neon gas contained in a volume of 12.5 L, making a total volume of 17.5 L. Both gases are at the same temperature and can be regarded as ideal. All volumes are thermally isolated. What is the change in entropy?

Solution: The pressure of each gas must be the same because the pressure of an ideal gas is given by $p = nRT/V$, and the temperatures are the same. The ratios n/V are also the same. The only change in entropy comes from the changes in volume of the two gases:

$$\Delta S = \Delta S_{Ar} + \Delta S_{Ne}.$$

For the change in the entropies of each gas, use Eq. (20–24), which is appropriate for a fixed temperature:

$$\Delta S_{Ar} = n_{Ar} R \ln\left(\frac{V_{Ar}}{V_{0,\,Ar}}\right) = (0.20 \text{ mol})(8.3 \text{ J/mol}\cdot\text{K})\ln\left(\frac{17.5 \text{ L}}{12.5 \text{ L}}\right)$$

$$= 17 \ln(3.5) \text{ J/K} = 21 \text{ J/K};$$

$$\Delta S_{Ne} = n_{Ne} R \ln\left(\frac{V_{Ne}}{V_{0,\,Ne}}\right) = (0.50 \text{ mol})(8.3 \text{ J/mol}\cdot\text{K})\ln\left(\frac{17.5 \text{ L}}{12.5 \text{ L}}\right)$$

$$= 42 \ln(1.4) \text{ J/K} = 14 \text{ J/K}.$$

The entropy of mixing is thus

$$\Delta S = \Delta S_{Ar} + \Delta S_{Ne} = 35 \text{ J/K}.$$

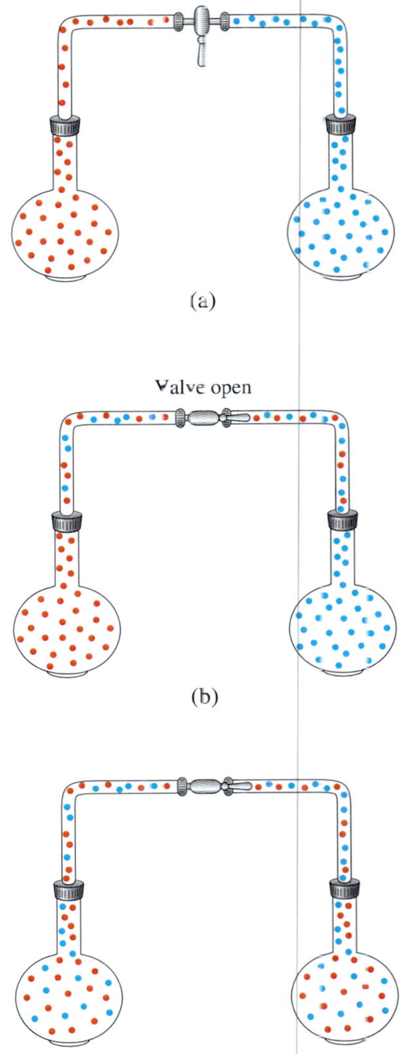

FIGURE 20–15 (a) Two gases are initially unmixed. (b) A valve is opened, and each gas diffuses into the other's flask. (c) The final state is more disordered than the initial state. The system's entropy has increased

Entropy is a measure of disorder.

We can similarly interpret the direction of spontaneous heat flow, in which systems with different temperatures tend to the same intermediate temperature when they are placed in thermal contact—ice cubes melt when they are placed in hot tea. In this case, the more ordered system is the cooler one. On average, the velocity vectors of the molecules of steam at a lower temperature are confined to a smaller range than are those of steam at a higher temperature. The heated system is more disordered because, on average, the velocity vectors spread over a greater range. Equation (20–26) illustrates how entropy increases for an ideal gas when temperature increases.

More advanced treatments of statistical physics quantify the idea that entropy is a measure of disorder. In fact, *all the consequences of the second law of thermodynamics follow from the treatment of entropy as a measure of disorder.*

Engines and Entropy

The principle that entropy never decreases is a way of formulating the second law of thermodynamics that emphasizes the directionality of thermal processes. The application of this formulation to engines is less obvious because the entropy in a reversible cycle does not change. Here, we show how the entropy formulation of the second law applies to engines and leads to our earlier formulations of the second law.

Consider a thermal reservoir at temperature T_h. Can we construct an engine that, in a single cycle, extracts thermal energy through a (positive) heat flow Q_h from that reservoir to lift a weight and does nothing else? Let's suppose that we could do so. There is no change in the entropy of the weight that has been lifted and there is no change in the engine after a full cycle. Thus, in the isolated system that consists of the weight, the engine, and the reservoir, the only entropy change is that of the reservoir, as given by $\Delta S = -Q_h/T_h$. This entropy change is negative and therefore violates the entropy formulation of the second law. We deduce that, in order to satisfy the entropy formulation of the second law, there must be a second thermal reservoir (at temperature T_c) such that the net entropy change is positive. If this second reservoir is to have a positive entropy change, then there must be a heat flow Q_c into it. This gives rise to an entropy change $\Delta S' = Q_c/T_c$.

The isolated system, which now consists of the weight, the engine, and the two reservoirs, must satisfy the entropy formulation of the second law, namely, $\Delta S_{\text{tot}} = \Delta S + \Delta S' \geq 0$, or

$$-\frac{Q_h}{T_h} + \frac{Q_c}{T_c} \geq 0;$$

$$\frac{Q_c}{Q_h} \geq \frac{T_c}{T_h}.$$

Because the work done to lift the weight is $W = Q_h - Q_c$ and the efficiency of the engine is defined as $\eta \equiv W/Q_h = (Q_h - Q_c)/Q_h = 1 - (Q_c/Q_h)$, the entropy formulation of the second law leads to

$$\eta \leq 1 - \frac{T_c}{T_h}.$$

We are now on familiar ground. The equality applies for the reversible Carnot cycle, where there is no heat flow except at the two temperature extremes. In this case, ΔS_{tot} is zero, which requires that $(Q_c/Q_h) = (T_c/T_h)$, and the efficiency is $\eta = 1 - (T_c/T_h)$. For cycles with irreversibility, or for cycles such as the Stirling cycle in which there is heat flow at intermediate temperatures, the inequality sign holds. Thus, the entropy formulation of the second law implies that the Carnot engine is the most efficient type of engine.

We have now connected the two apparently dissimilar aspects of the second law of thermodynamics: the impossibility of constructing a perfectly efficient engine and the inevitable direction of spontaneous processes.

*The Arrow of Time

Drop an egg and film the event. (Choose your laboratory carefully.) This process is a typical, if particularly spectacular, macroscopic event. Do the same for the collision of two billiard balls. The collision of two billiard balls resembles the collision of two molecules, and the film of this event represents fairly the fundamental interactions between the molecular constituents of matter.

If you were to see these two films, could you tell whether they were run forward or backward? The film of the ruins of the egg reassembling into a perfect ovoid would appear to be a cheap comic trick that could not possibly be confused with a real event (Fig. 20–16). But the two billiard balls colliding in reverse would be absolutely indistinguishable from the real thing—at least in the short time period around the collision.

FIGURE 20–16 In this stroboscopic photograph, the direction of increasing time follows the egg descending and breaking, not reassembling and rising.

The elementary forces between the microscopic constituents of matter are said to be indifferent to the direction of time, or *invariant under time reversal.* This result is fundamental, and it seems to hold (with one very tiny exception discussed in Chapter 46) for all the fundamental forces in nature, including the electromagnetic forces that govern the collisions of gas molecules. If the fundamental forces are invariant under time reversal, and if they are the forces that act in ordinary materials, why does there seem to be a preferred direction of time—an **arrow of time**—in which eggs break rather than reassemble? Why do spontaneous processes, in which entropy increases, proceed in one direction and not the other?

We have seen that the reason is statistical. Individual collisions between molecules do not depend on the direction of time. The fundamental laws that govern the reformation of the egg are thus independent of the direction of time—it is not against any physical laws for a broken egg to reassemble itself. It is just highly unlikely. The overwhelming majority of collisions between gas molecules lead to more disordered situations. The arrow of time expresses not a definite preferred direction of time in a fundamental sense but an overwhelming likelihood that out of a random sequence of microscopic interactions, less ordered situations will result. Irreversibility means only that the spontaneous reversal of the "irreversible" process is so unlikely as not to be worth considering.

SUMMARY

The second law of thermodynamics limits the efficiency of cyclic engines and governs the direction of spontaneous processes in thermal systems. Cyclic engines do work in repeated cycles by extracting a heat flow from hotter external systems, converting part of that heat flow into work, and rejecting a heat flow to cooler thermal systems. The efficiency of such an engine is defined by

$$\eta \equiv \frac{W}{Q_h}, \tag{20–1}$$

where W is the work done in one complete cycle and Q_h is the total positive heat flow the cycle extracts from its sources of fuel.

Two equivalent ways to state the second law of thermodynamics are the Kelvin form and the Clausius form. The Kelvin form states that it is impossible to convert thermal energy from a body at a given temperature into an equivalent amount of mechanical work with no other effect; the Clausius form states that it is impossible to transfer thermal energy from a colder body to a hotter body with no other effect.

In the Carnot cycle, which involves four reversible steps that occur at just two temperatures, a working fluid undergoes alternate adiabatic and isothermal transformations. It is the most efficient cycle possible that operates between two temperatures T_h and T_c. No matter what the nature of the working fluid, the efficiency of the Carnot engine is

$$\text{for any Carnot cycle:} \quad \eta_C = 1 - \frac{T_c}{T_h}. \tag{20–7}$$

The Carnot cycle that operates between two temperatures is more efficient than any other cycle that operates between the same temperatures extremes. Any element of irreversibility also decreases the efficiency.

When the reversible cycles of work-producing engines are run in reverse, they act as refrigerators or heat pumps. Work is done on both of these machines while they extract a heat flow from a cold reservoir and release a heat flow to a hot reservoir.

A new thermodynamic variable, entropy, S, measures the amount of disorder in a thermal system. Like thermal energy, it is a function of state, and only *changes* in entropy have physical significance. Entropy changes are path independent. For infinitesimal reversible transformations,

$$dS \equiv \frac{dQ}{T}. \tag{20–17}$$

For an ideal gas, entropy can be expressed as a function of T and V as

$$S(T, V) - S(T_0, V_0) = C_V \ln\left(\frac{T}{T_0}\right) - nR \ln\left(\frac{V}{V_0}\right). \tag{20–27}$$

Note that entropy decreases as temperature decreases.

The path independence of entropy changes is equivalent to the fact that, for a reversible cycle,

$$\text{for a reversible cycle:} \quad \oint \frac{dQ}{T} = 0, \tag{20–16}$$

a result known as Clausius's theorem. If a cycle has any element of irreversibility, Clausius's theorem is modified to Clausius's inequality,

$$\text{for an irreversible cycle:} \quad \oint \frac{dQ}{T} < 0. \tag{20–19}$$

If the change $A \to B$ is irreversible, the following relation holds:

$$S(B) - S(A) > \int_{A \atop \text{irrev}}^{B} \frac{dQ}{T}. \qquad (20\text{–}21)$$

In particular, for transformations of thermally isolated systems, we find

$$\text{for a thermally isolated system:} \qquad S(B) - S(A) \geq 0. \qquad (20\text{–}22)$$

A spontaneous process in an isolated system increases the system's entropy, and the state of maximum entropy is the state of stable equilibrium. The increase of entropy expresses the inevitable increase in the disorder of thermally isolated systems.

UNDERSTANDING KEY CONCEPTS

1. If the ocean is viewed as a reservoir at one temperature and the atmosphere as a reservoir at another temperature, when can thermal energy flow spontaneously from the ocean to the air?
2. When the brakes are applied, a car decelerates and then stops. A film of the process run in reverse obviously looks fake. Does the process in reverse violate the second law of thermodynamics? If so, in what way?
3. The historian Herman Daly has called the second law of thermodynamics "the law of random, ravage, and rust." Is he right?
4. Why is it not possible for an engine—even a Carnot engine—to have an efficiency of one?
5. Suppose that a gas obeys the van der Waals equation of state. What is the efficiency of a Carnot engine that uses this gas as a working fluid in terms of the temperatures of the hot and cold reservoirs?
6. In what climates are heat pumps most likely to be used?
7. A professor spends a couple of hours straightening up his office—putting books away, filing papers, throwing old exam papers away, and so on. He has created a less probable configuration out of a more probable configuration. Has he violated the second law of thermodynamics?
8. Imagine a microscopic device at the boundary between two containers of a gas that could sense the arrival of a particularly fast molecule heading from container A toward container B and could open a tiny door to let that molecule through. The same device, which is known generically as a *Maxwell demon,* could sense the arrival of a particularly slow molecule heading from container B toward container A and could let the molecule pass through. How would such a device violate the second law of thermodynamics? Is such a device feasible?[†]
9. Give examples of things you have observed recently that exemplify the second law of thermodynamics. Discuss the entropy change associated with them.
10. Would heat pumps that use an underground water reservoir be more efficient than those that use the outside air as a thermal reservoir if the pumps are employed to (a) heat in the winter? (b) cool in the summer?
11. List and discuss some processes not already mentioned that will not violate the first law of thermodynamics but will violate the second law.
12. When gasoline is burned in the chamber of a piston, chemical energy is turned into mechanical energy that is used to propel an automobile. How is this process consistent with the second law of thermodynamics?
13. If you were to run a film of an egg that drops to the floor, you would have little trouble determining if the film were running backward or not. Yet if the film were in extreme closeup, so that you were observing the interaction of molecules in the dropped egg, you would be hard pressed to say if the film were running forward or backward. How are these two statements consistent?
14. Is it possible to cool a house by leaving a refrigerator door open? What would be the net effect if you were to leave the door open?
15. Is the entropy of a gas an extensive quantity (such as volume) or an intensive quantity (such as pressure or temperature)? (See Problem 3 in Chapter 17.)
16. Biological systems use energy (ultimately from the Sun) in organizing highly ordered organisms, such as humans. Does this mean that biological systems violate the second law of thermodynamics?

PROBLEMS

20–1 Beyond Energy Conservation

1. (I) An engine produces 3.8×10^5 J of mechanical energy accompanied by a total heat flow of 7.6×10^5 J into the engine. What is its efficiency?
2. (I) Suppose that you toss a pair of dice. Make a list of all the possible outcomes, and calculate the probability of getting totals of 2, 3, 4, … , 12, respectively.
3. (II) Six tokens, numbered one to six, are put into a hat. The tokens with even numbers are white, the ones with odd numbers are black. Suppose you draw four tokens at random, without putting them back into the hat. (a) List all the possible outcomes, without regard to order (for example, 1234 and 2134 do not count as separate outcomes). (b) What is the probability of drawing the tokens with 1, 2, 3, 4 on them? (c) What is the probability of drawing two white and two black tokens?

[†]For an interesting discussion of Maxwell demons, see Charles H. Bennett, "Demons, Engines, and the Second Law," *Scientific American,* Nov. 1987, pp. 108–116.

4. (II) The N molecules of a gas are placed in the left half of a chamber and kept from entering the right half by a membrane. The membrane is then removed and the molecules distribute themselves uniformly throughout the whole volume (Fig. 20–17). What is the probability that, at some time during their subsequent random motion, they will all collect in the left half of the chamber, for the cases $N = 2$, 100, 10^6, and 10^{23}? [*Hint*: This is equivalent to the probability of getting nothing but heads in 10^{23} coin tosses. To get an idea of this number in scientific notation, use the fact that $\frac{1}{2} \simeq 10^{-0.3}$.]

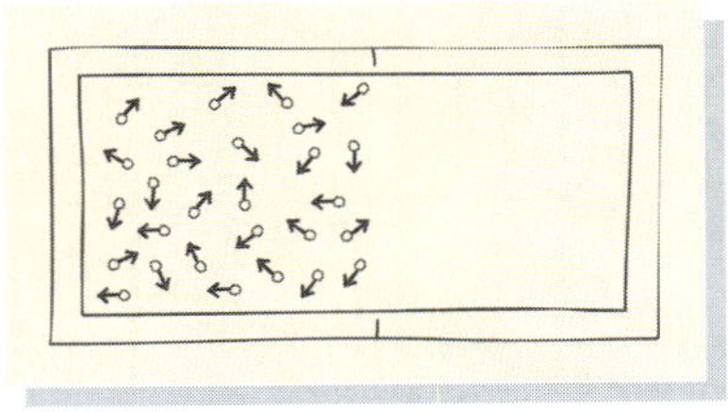

FIGURE 20–17 Problem 4.

20–3 *The Carnot Cycle*

5. (I) What is the maximum efficiency of a steam engine whose boiler (the hottest-temperature reservoir in contact with the engine) is at 160°C and whose condenser (the coldest-temperature reservoir in contact with the engine) is at 35°C?

6. (I) A heat engine absorbs heat from a thermal reservoir at 400 K, but the engine's maximum efficiency is only 31 percent. What must the temperature of the reservoir be to raise the efficiency to 38 percent?

7. (I) A heat engine receives heat from a thermal reservoir at 750 K and transfers it to a thermal reservoir at 340 K. The efficiency of the engine is only 70 percent of the maximum possible. (a) What is the engine's actual efficiency? (b) How much work does the engine produce for each calorie of heat input?

8. (I) Temperature differences in the ocean have been proposed as a possible energy source in the tropics. Surface water at 30°C could act as a hot reservoir, and deep water at 4°C could serve as a cold reservoir. Ammonia gas could be a working fluid in a heat engine that runs between the two thermal reservoirs. What is the maximum efficiency of such an engine?

9. (II) A Carnot engine absorbs thermal energy from a reservoir at 450°C. Its efficiency is 0.55, and the work delivered by the engine is 5 kWh. (a) What is the temperature of the cold reservoir? (b) How much thermal energy flows into the system, and how much is rejected?

10. (II) What is the maximum efficiency of a heat engine that operates between 450°C and 145°C? If this engine generates 2000 J of mechanical energy, how many calories does it absorb from the hot reservoir, and how many calories does it transfer into the cold reservoir?

11. (II) An inventor claims to have built an engine that takes in 3.0×10^8 J of thermal energy at 450 K, rejects 1.4×10^8 J of thermal energy at 250 K, and delivers 1.0×10^8 J of work in 1 h of cyclic operation. Is there anything wrong with this claim?

12. (II) A power plant generates 440 MW of electric power. At what rate does the plant generate waste heat if its efficiency is 35 percent? Assuming that the plant operates between 480°C and 40°C, what is the maximum efficiency possible?

13. (II) An ideal diatomic gas is used in a Carnot engine with thermal reservoirs at 77 K and 300 K. What is the ratio of the maximum volume to the minimum volume if the pressure drops by a factor of 2 during the isothermal expansion? [*Hint*: For a diatomic gas, $\gamma = 1.4$.]

14. (II) Prove that two reversible adiabatic paths cannot intersect. Do this by assuming that they do, constructing a cycle that consists of the two paths and another reversible transformation, and showing that such a cycle would violate the second law of thermodynamics (Fig. 20–18). (This result is simply shown for an ideal gas. Your proof should apply to a nonideal gas as well.)

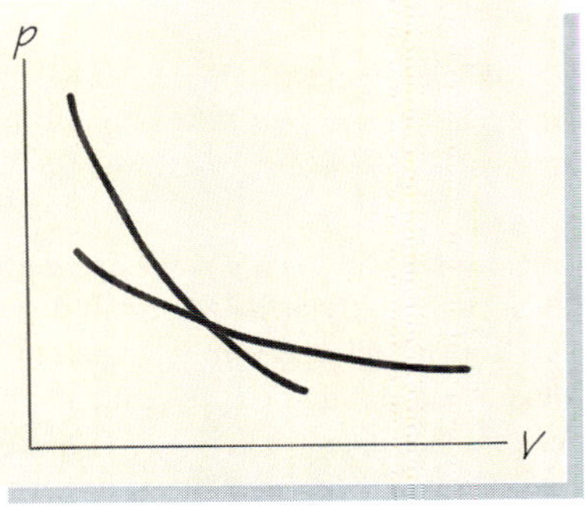

FIGURE 20–18 Problem 14.

15. (III) Any element of irreversibility makes a Carnot engine, and, in fact, any engine, less efficient. Let engine A be a pseudo-Carnot engine, meaning that it has some irreversibility somewhere along the cycle; let engine B be an ideal Carnot engine—the two arranged in size so that they exhaust the same heat flow Q_c. If engine A is *more* efficient than engine B, then by combining the two engines such that A runs normally and B runs backward, as in Figure 20–7, show that you arrive at a contradiction of the Kelvin form of the second law of thermodynamics. Now that you have shown that engine A must be less efficient than engine B, why can the preceding argument not be reversed, running the Carnot cycle forward and the pseudo-Carnot cycle backward to get a contradiction of the second law?

16. (III) Show that if it is possible to build an engine that violates the Kelvin form of the second law of thermodynamics, it is also possible to build an engine that violates the Clausius form of the second law. [*Hint*: Consider a Carnot engine driven by the Kelvin-violating engine.]

20–4 *Other Types of Engines*

17. (I) An air conditioner is rated at 8000 Btu/h. Assuming maximum possible efficiency, an exhaust temperature of 38°C, and an interior temperature of 22°C, what is the electrical power consumption? (The work done on the air conditioner is supplied by the wall socket.)

18. (I) What is the coefficient of performance for an ideal Carnot heat pump operated between outside air at 30° C and a deep water well at 15° C? Would the coefficient of performance be larger or smaller if your heat pump followed any other type of cycle?

19. (I) Two refrigerators can each be approximated as Carnot cycles run in reverse. Each refrigerator works with the same cold temperature reservoir $T_c = 273$ K. Refrigerator 1 evacuates to a hot reservoir at temperature $T_h = 293$ K, while refrigerator 2 evacuates to a hot reservoir at a temperature of $T_h = 303$ K. Compare the coefficients of performance of these refrigerators. (Real refrigerators will have coefficients of performance substantially less than these.)

20. (I) A heat pump acts like a refrigerator that cools the outside of a house and gives rise to a heat flow to the inside of the house. If the outside temperature is $+5°C$ and the inside temperature is 22°C, what is the maximum amount of British thermal units a pump can deliver per kilowatt-hour of electric energy supplied? (The work done on the pump has electric energy as its source.)

21. (I) Consider the diesel cycle in Figure 20AB–1. In which portions of the cycle is there heat flow into or out of the engine? What is the work done in the transformation $2 \rightarrow 3$ in terms of T_2 and T_3?

22. (II) An "ideal refrigerator," a Carnot engine that operates in reverse, has a *compressor* as its engine. It does work W to extract heat Q_c from a cold reservoir at temperature T_c and dumps heat Q_h into a hot reservoir at T_h. Use the known efficiency of a Carnot engine that operates between T_h and T_c and the conservation of energy for one cycle to show that the coefficient of performance $K_{ref} \equiv Q_c/W$ for the compressor is given by $K = T_c/(T_h - T_c)$. Note that K_{ref} is large when the cold and hot reservoirs have nearly the same temperature.

23. (II) A refrigerator with coefficient of performance $K_{ref} = 5.1$ (see Problem 22) gives rise to a heat flow out of the cooling compartment at a rate of 400 cal/min. What is the required power of the motor that operates this refrigerator?

24. (II) To compensate for a given energy loss when a refrigerator is opened too often, the refrigerator's compressor must do work. Take $T_c = 0°C$ and $T_h = 25°C$, and suppose that the refrigerator's coefficient of performance is 20 percent of its ideal value. What is the rate that the compressor does work; that is, what is the power consumption for a given rate of heat loss dQ/dt?

25. (II) The specifications of a freezer claim that it can remove 80 cal/s heat flow from the compartment at $-20°C$ and release 94 cal/s into the room at 25°C while using 60 W of electrical energy to drive the compressor. Can you trust this statement?

26. (II) In order to make an ice cube from water, a refrigerator must extract 1400 cal of heat at the temperature of the freezer, $-12°C$. The room temperature is 30°C. What minimum work must an ideal refrigerator do?

27. (II) You can buy a 5-lb bag of ice for about $2. Estimate the fraction of this price that pays for the electric energy needed to run the ice-making machine. Assume that the inside of the machine is at $-10°C$, that the outside temperature is 30°C, that the initial temperature of the water is 20°C, and that the coefficient of performance is 30 percent of the ideal. Electrical energy costs $0.10/kWh.

28. (II) The temperature inside a Carnot refrigerator placed in a room at 72°F is 40°F. The heat flow into the refrigerator from the room is 0.40×10^8 J/h. How many watts are needed to operate the refrigerator: that is, to remove the heat flow into the refrigerator?

29. (II) A heat engine operates on the cycle shown in Fig. 20–19, with heat absorbed or ejected on each leg as indicated. Its operating fluid is an ideal gas with heat capacity at constant volume C_V. (a) What are temperatures T_3 and T_4 at points 3 and 4, respectively, in terms of the indicated pressures and volumes? (b) What is the internal energy change along leg IV, ΔU_{IV}? Include the sign. (c) What is Q_{IV} in terms of ΔU_{IV} and the indicated pressures and volumes? (d) What is the efficiency of the engine in terms of the indicated pressures and volumes and C_V?

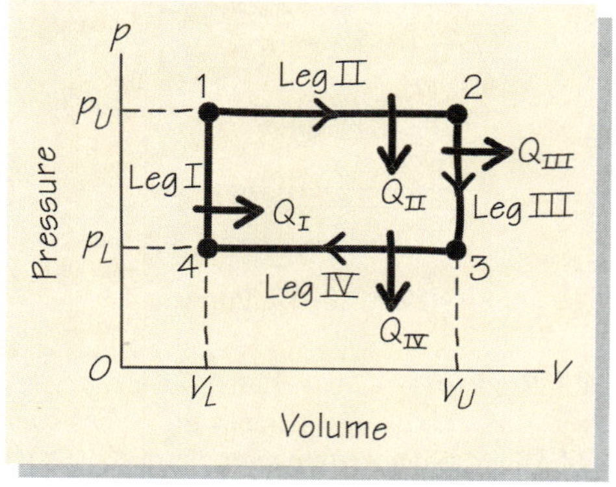

FIGURE 20–19 Problem 29.

30. (II) Consider the ideal Stirling cycle shown in Fig. 20–9, working between a maximum temperature T_h and a minimum temperature T_c, and a minimum volume V_1 and a maximum volume V_2. Suppose that the working gas of the cycle is 0.1 mol of an ideal gas with $c'_V = 5R/2$. (a) What are the heat flows to the cycle during *each* leg? Be sure to give the sign. For which legs is the heat flow positive? (b) What work is done by the cycle during *each* leg? Again, be sure to include the sign. (c) If, in the definition of the efficiency of this cycle, $\eta = W/Q_{pos}$, where Q_{pos} is the *total* positive heat flow to the engine, what is the efficiency of the cycle when $T_h = 700$ K and $T_c = 400$ K? $V_1 = 0.5$ L and $V_2 = 1.5$ L. Compare this efficiency to the efficiency of a Carnot cycle that operates between the same temperature extremes.

31. (II) The Otto cycle, represented in Figure 20AB–2, runs between minimum and maximum volumes V_1 and V_2 and minimum and maximum temperatures T_1 and T_2, respectively. (a) Show that the efficiency is given by $\eta = 1 - (T_2 - T_3)/(T_1 - T_4)$. (b) Show that if the working fluid of the cycle is an ideal gas, the efficiency of this cycle can alternatively be written as $\eta = 1 - (V_1/V_2)^{\gamma - 1}$, where $\gamma = C_p/C_V$. The efficiency of this cycle is thus *independent* of the temperatures between which it operates, depending instead only on γ and on geometry. The ratio V_2/V_1 is the compression ratio. A typical compression ratio is 8, and $\gamma = 1.4$, which gives a predicted efficiency of 56 percent.

32. (III) The Brayton cycle is represented by Figure 20AB–3. Show that when its working fluid is an ideal gas, the cycle's efficiency is given by the same formula as that for the Otto cycle in Problem 31. In this case, V_1 and V_2 are not the extreme volumes of the engine but are the volumes at the start and finish of the adiabatic expansion step, respectively.

33. (III) Find the efficiency of the diesel cycle illustrated in Fig. 20AB–1 in terms of $\gamma = C_p/C_V$ and the temperatures at points 1, 2, 3, and 4. Assume that it contains 1 mol of an ideal gas. (Rudolf Diesel tried to produce an engine based on the Carnot cycle. He was not able to produce such an engine, but he came close: A diesel engine nearly matches a Carnot engine in efficiency. Diesel's engine is one of the most efficient in use today.)

20–5 *Entropy and the Second Law*

34. (I) Consider a large, isolated thermal system in two parts. One part is a thermal reservoir at temperature T_1, labeled I, and the other part is a thermal reservoir at temperature T_2, labeled II. Reservoir I is colder than reservoir II. These two reservoirs are briefly connected, and there is a heat flow Q from II to I. (a) What are the entropy changes for I and for II? (b) Compute the entropy change for the whole system, and show that it is positive.

35. (I) One end of a metal rod is in contact with a thermal reservoir at 1273 K, and the other end is in contact with a reservoir at 293 K. If the rate at which heat flow passes from the hot end to the cold end is 30 J/min, what is the rate of total entropy change? Is the process reversible?

36. (II) Find the entropy change for (a) 1 g of ice melting at 0°C; (b) 1 g of water evaporating at 100°C. You will need to look up the heats of melting and of vaporization.

37. (II) One kilogram of iron at 80°C is dropped into 0.5 L of water at 20°C. Given that the specific heat of water is 1 cal/g · K and that of iron is 0.107 cal/g · K, calculate the final equilibrium temperature of the system and the increase in entropy.

38. (II) Forty grams of water at 100°C is changed reversibly into steam at 100°C. (a) Describe such a process. (b) What is the change in entropy of the water and of the universe? (c) How would part (b) change if the process were irreversible?

39. (II) Calculate the change in entropy of the universe if 0.3 kg of water at 70°C is mixed with 0.2 kg of water at 15°C in a thermally insulated container. [*Hint*: Even though this process is irreversible, the change in entropy can be calculated by devising a reversible process that leads from the given initial state to the expected final state.]

40. (III) Thermodynamic tables often give the specific heat of materials in the form of a list of coefficients a, b, c, d in the expression $c_p = a + bT + cT^2 + d/T$. How can you use these data to find the entropy change of the material as it undergoes a transformation from T_i to T_f?

20–6 *Entropy and Ideal Gases*

41. (I) Five moles of an ideal gas is in thermal isolation and undergoes free expansion from 35 L to 100 L. What is the entropy change of the gas? of the universe?

42. (I) The heat capacity at constant volume of a sample of gas is determined to be 20 J/K. If this amount of gas is heated in a constant volume container from 273 K to 2200 K, what is the change in entropy?

43. (I) One mole of an ideal gas expands at constant pressure from an initial volume of 250 cm^3 to a final volume of 650 cm^3. What is the change in entropy, assuming that the molar specific heat is $c'_V = 3R/2$?

44. (I) One mole of ideal gas with molar specific heat $c'_V = 3R/2$ is heated from 273 K to 350 K at constant volume. What is the change in entropy?

45. (I) An ideal gas undergoes an isothermal compression. Does the entropy increase?

46. (II) Ten moles of a monatomic gas held in a tank at 20 atm escape from the tank isothermally at 23°C. What is the change in the entropy of the universe?

47. (II) One mole of a monatomic ideal gas at an initial pressure of 30 atm and a temperature of 600 K is allowed to undergo a rapid (adiabatic) free expansion from a small vessel into a vessel of 50 times greater volume. Find the change in temperature and the increase in entropy.

48. (II) One mole of a monatomic ideal gas is carried in two steps from (p_1, V_1) to $(p_1, 5V_1)$ to $(0.3p_1, 5V_1)$ by reversible thermodynamic processes that are straight lines on a p–V diagram (Fig. 20–20). (a) What is the overall change in the gas's internal energy? (b) What is the overall work done by the gas? (c) How much net heat flow is added overall? (d) What is the net change in the entropy?

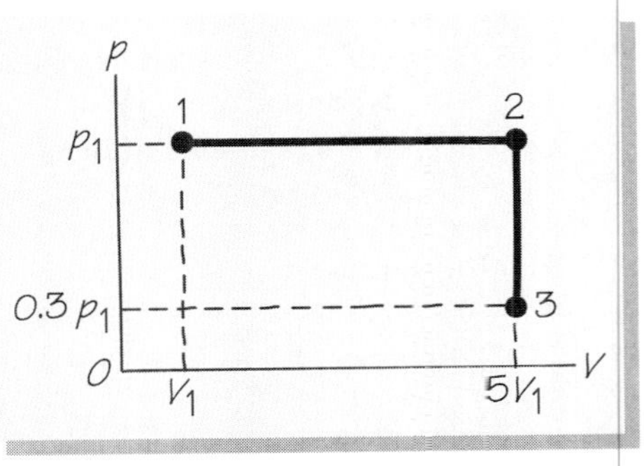

FIGURE 20–20 Problem 48.

49. (II) A gas obeys the well-known equation of state $pV =$ (a constant) T. The gas expands, doubling in volume. (a) Plot p versus V when the expansion is isobaric and when it is isothermal. (b) What work is done by the gas on its surroundings for both cases above? (c) What is the entropy change of the gas for both cases above?

50. (II) Show that, in terms of temperature and pressure, the entropy of an ideal gas with constant heat capacities is $S(T, p) = C_p \ln(T/T_0) - nR \ln(p/p_0) + S(T_0, p_0)$.

51. (III) Dilute helium gas is taken through the following reversible cycle on a p–V diagram: AB is an isobaric compression; BC is an adiabatic expansion; CD is an isothermal expansion; DA is constant-volume pressure increase (Fig. 20–21). Express the heat flow, work done on the system, change in internal energy, change in temperature, and change in entropy for each leg of the cycle in terms of the quantities T_A, T_B, T_C, and p_A. Pay attention to the signs.

20–7 *The Meaning of Entropy*

52. (I) A total of 12 molecules are in a large box, which we mentally divide into two halves, a left half and a right half. Any one molecule is equally likely to be in either of the two halves. What is the probability that all 12 molecules will be found in the left half? How does your answer change if there are 120 molecules?

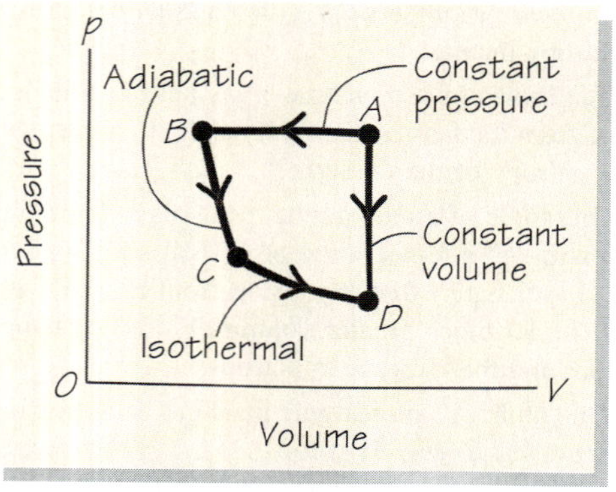

FIGURE 20–21 Problem 51.

53. (II) An ideal gas in a cylinder with a movable piston is placed in thermal contact with a thermal reservoir, and the gas is slowly compressed. How does the entropy of the gas in the cylinder change? How does the entropy of the reservoir change? Is the answer to these questions in conflict with the second law of thermodynamics? Explain your answer.

54. (II) A box of volume 5 L contains 3 L of He in one side and 2 L of O_2 in the other, separated by a partition. The temperature is 273 K, and the pressure in both sides of the box is 1 atm. The partition is removed. (a) Describe the new equilibrium configuration. (b) What is the change in entropy?

55. (II) A thermally isolated container of volume $3V_0$ is divided by a partition. In one portion, of volume V_0, there are n mol of a monatomic ideal gas at temperature T_0. The other portion is empty. The partition is then broken, and the gas expands to fill the entire container (Fig. 20–22). (a) What is the final temperature of the gas? (b) the change in entropy of the gas? (c) the change in entropy of the universe? (d) Suppose that the gas is originally confined to volume V_0 by means of a piston instead of a partition. The piston is then slowly withdrawn until a final volume of $3V_0$ is obtained. What is the change in entropy of the gas?

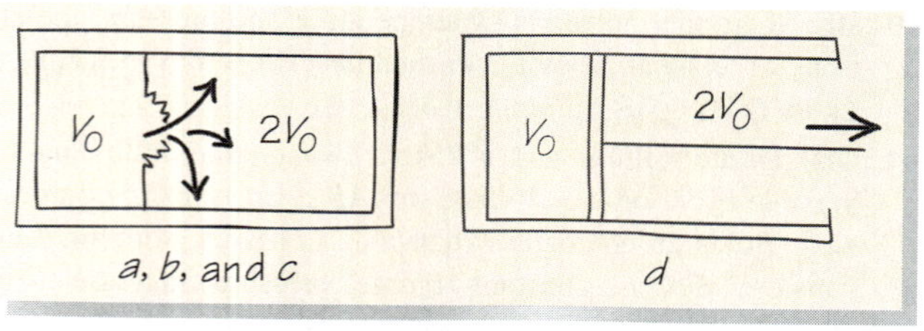

FIGURE 20–22 Problem 55.

General Problems

56. (II) A nuclear power plant requires 600 MW of cooling. If environmental concerns limit the temperature rise of the cooling water to 15°C, how much water flow is required to cool the heat generated by the plant? If the water is at an initial temperature of 285 K, what is the entropy change per kilogram of water?

57. (II) Fifty grams of oxygen gas at 320 K does 80 J of work while 40 cal of heat flow is absorbed by the gas. (a) What is the change in internal energy? (b) in the temperature of the gas? (c) in entropy, assuming that the change is isobaric?

58. (II) Calculate the entropy change when 1 kg of water is cooled from 100°C to 0°C. Calculate the entropy change when the water turns to ice at 0°C (Fig. 20–23). The latent heat of fusion is 80 cal/g.

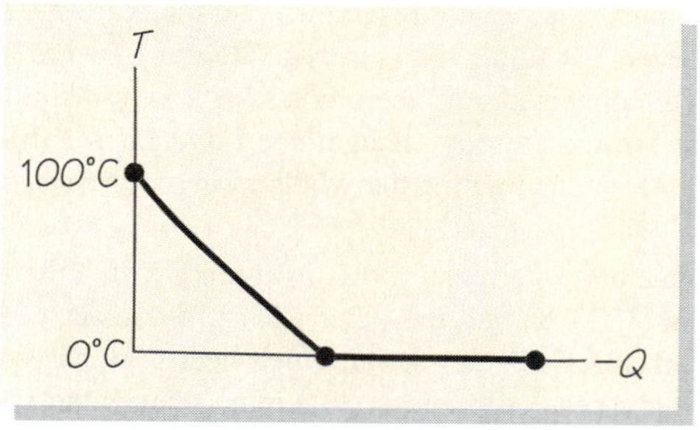

FIGURE 20–23 Problem 58.

59. (II) Two moles of helium at STP is compressed in an isothermal process to a pressure of 2 atm. (a) What is the final volume? (b) the final temperature? (c) the change in entropy?

60. (II) A Carnot engine for which the working fluid is 1 mol of an ideal gas with $c'_V = (5/2)R$ operates between two thermal reservoirs of temperatures $T_h = 800$ K and $T_c = 300$ K. When the gas is in contact with the hot reservoir, its minimum volume is $V_1 = 1.2$ L, and when it is in contact with the cold reservoir, its maximum volume is $V_3 = 15$ L. (a) What is the efficiency of this engine? (b) For one cycle, what is the net energy change of the gas? (c) the net entropy change of the gas? (d) the net change in the entropy of the universe? (e) What is V_2, the maximum volume when the gas is in contact with the hot reservoir?

61. (II) Three moles of oxygen at STP undergoes an adiabatic compression to a final pressure of 3 atm. (a) What is the final volume? (b) the final temperature? (c) the change in entropy?

62. (II) One liter of nitrogen at STP is adiabatically compressed to a volume of 0.5 L when a membrane breaks and the nitrogen gas undergoes free expansion back to 1 L. (a) What is the final pressure? (b) the final temperature? (c) the change in entropy?

63. (II) A heat engine operates between the temperatures of 400 K and 300 K, doing 100 J of work. This work is used to run a refrigerator between the same temperatures. (a) Calculate the heat transfer between each reservoir and the engine or the refrigerator, assuming that both of them have ideal efficiency/coefficient of performance. (b) Repeat the calculation if the heat engine's efficiency is only 80 percent of the ideal. (c) Repeat the calculation for the case of an ideal heat engine but for a refrigerator whose coefficient of performance is 80 percent of the ideal.

64. (II) One Carnot engine drives another in series, as shown in Fig. 20–24. The heat released from the first engine is absorbed by the second. Find the overall efficiency, which is determined by the total work produced divided by the heat input to the first engine.

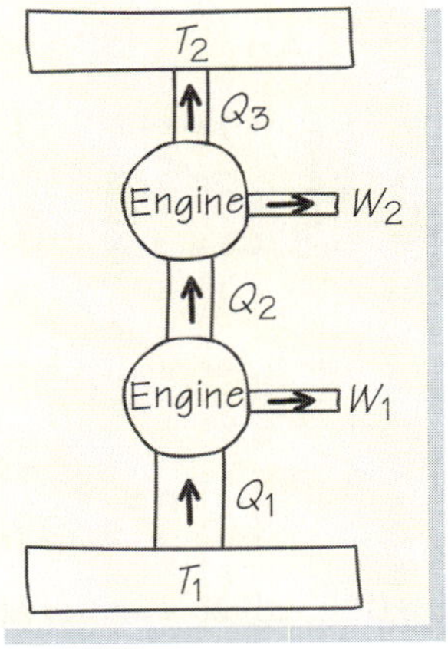

FIGURE 20–24 Problem 64.

65. (II) A coal-fired, 300-MW generating plant operates between 750 K and 400 K, with an efficiency 60 percent of the maximum efficiency possible. At what rate is waste heat flow produced? Suppose that water is used to carry off the waste heat flow, and it does so by being heated by 12°C. How much water must flow through the machinery per second?

66. (II) Consider the reversible thermodynamic cycle of the following form (Fig. 20–25): (i) One mole of a monatomic ideal gas at temperature T_a and pressure p_a is allowed to expand isobarically until the volume has increased by a factor r. (ii) The gas is then cooled so that its absolute temperature is reduced by a factor of 2, with the cooling performed at constant volume. (iii) The gas is then compressed isobarically to its original volume. (iv) Finally, the gas is heated to its starting temperature, T_a, at constant volume. (a) What is the mechanical work done by the engine? (b) the heat transfer in the four legs of the cycle? (c) the efficiency of the cycle?

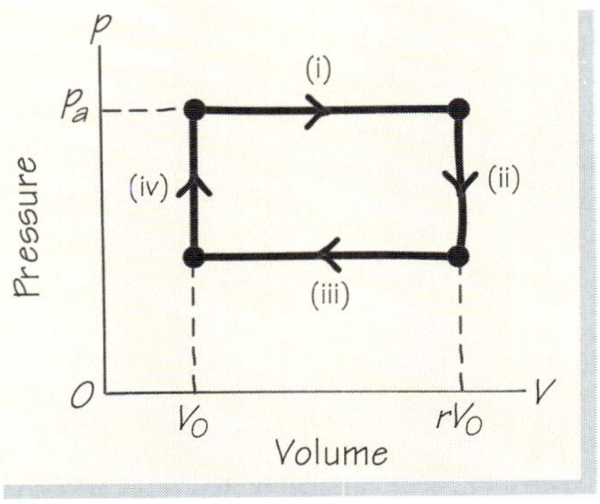

FIGURE 20–25 Problem 66.

67. (II) A heat engine operates in a reversible cycle with the following steps (Fig. 20–26): (i) An ideal gas characterized by $\gamma = 1.4$ expands adiabatically, so that it cools from an initial temperature of 330°C to 30°C; (ii) it is compressed isothermally until it reaches its initial volume; and (iii) it is then heated at constant volume until it reaches its initial temperature. Calculate the efficiency of the cycle, and compare it with the Carnot efficiency that corresponds to these temperatures.

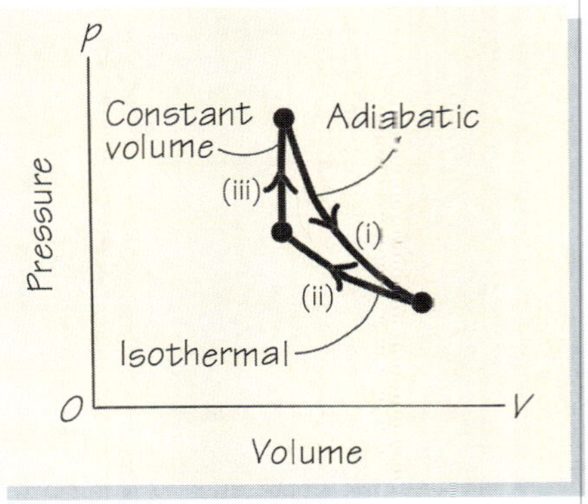

FIGURE 20–26 Problem 67.

68. (II) The variables T and S may be used instead of the thermodynamic variables p and V. Plot a T–S diagram for the Carnot cycle. What does the area within the cycle represent? [*Hint*: Use the fact that the first law of thermodynamics, which states that $dQ = dU + p\,dV$ may be written in the form $T\,dS = dU + p\,dV$, and integrate around the cycle.]

69. (II) In engineering, a useful quantity called *enthalpy*, H, is defined by $H = U + pV$. Calculate the enthalpy of an ideal gas as a function of temperature.

70. (II) Suppose that there is steady flow of a mass M of gas through a turbine. Gas enters the turbine at an elevation z_1, pressure p_1, internal energy U_1, and speed v_1, and leaves at an elevation z_2, pressure p_2, internal energy U_1, and speed v_2. Thermal energy Q is supplied to the turbine. Work W is done by the turbine, and V_1 and V_2 are the respective volumes occupied by mass M on entering and leaving the turbine (a) Use energy conservation to show that the change in enthalpy (see Problem 69) in terms of the specified quantities can be written in the form $H_2 - H_1 = Q - W + \frac{1}{2}mv_1^2 + Mgz_1 - (\frac{1}{2}mv_2^2 + Mgz_2)$. (b) Use this result to show that there is no change in enthalpy in the Joule experiment described in Section 18–6 (in which there is no change in elevation, and the change in the speed of the gas is so small that $v_2^2 - v_1^2$ can be neglected).

71. (III) The Brayton cycle is used to extract work from a high-temperature, gas-cooled nuclear reactor with helium (a monatomic ideal gas) as the working fluid. The helium enters a compressor, and its pressure is raised adiabatically. It is then heated at constant pressure to a high temperature in the reactor core. The energy is supplied by the fission of uranium or plutonium. The helium then enters a turbine, in which it is allowed to expand adiabatically. It then passes through a precooler, in which heat is rejected at constant pressure until it reaches the initial conditions. Suppose that helium enters the compressor at 30°C and a pressure of 20 atm. It is compressed to 50 atm and then heated to 1200°C in the reactor core (Fig. 20–27). (a) What are the temperatures at the end of compression and expansion? (b) What are the heat flows per kilogram of helium in each cycle? (c) How much net work is done per kilogram of helium in each cycle?

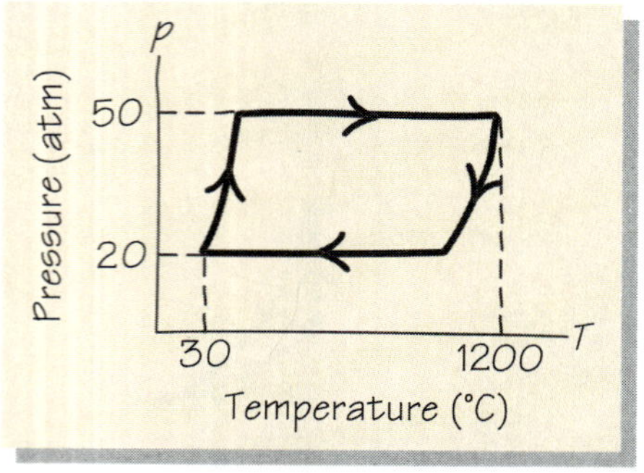

FIGURE 20–27 Problem 71.

72. (III) Three reversible paths, legs i, ii, and iii, connect points (p_2, V_2) and (p_1, V_1) in Fig. 20–28. (a) Calculate $\int_1^2 dW$ and $\int_1^2 dQ$ for the three paths. (b) Calculate $\int_1^2 dQ/T$ for the three paths.

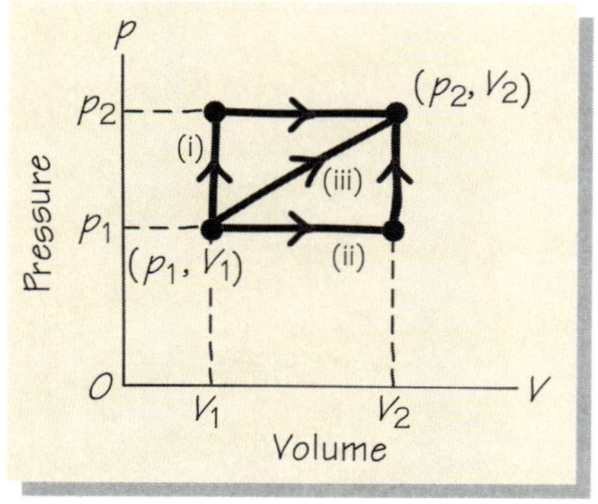

FIGURE 20–28 Problem 72.

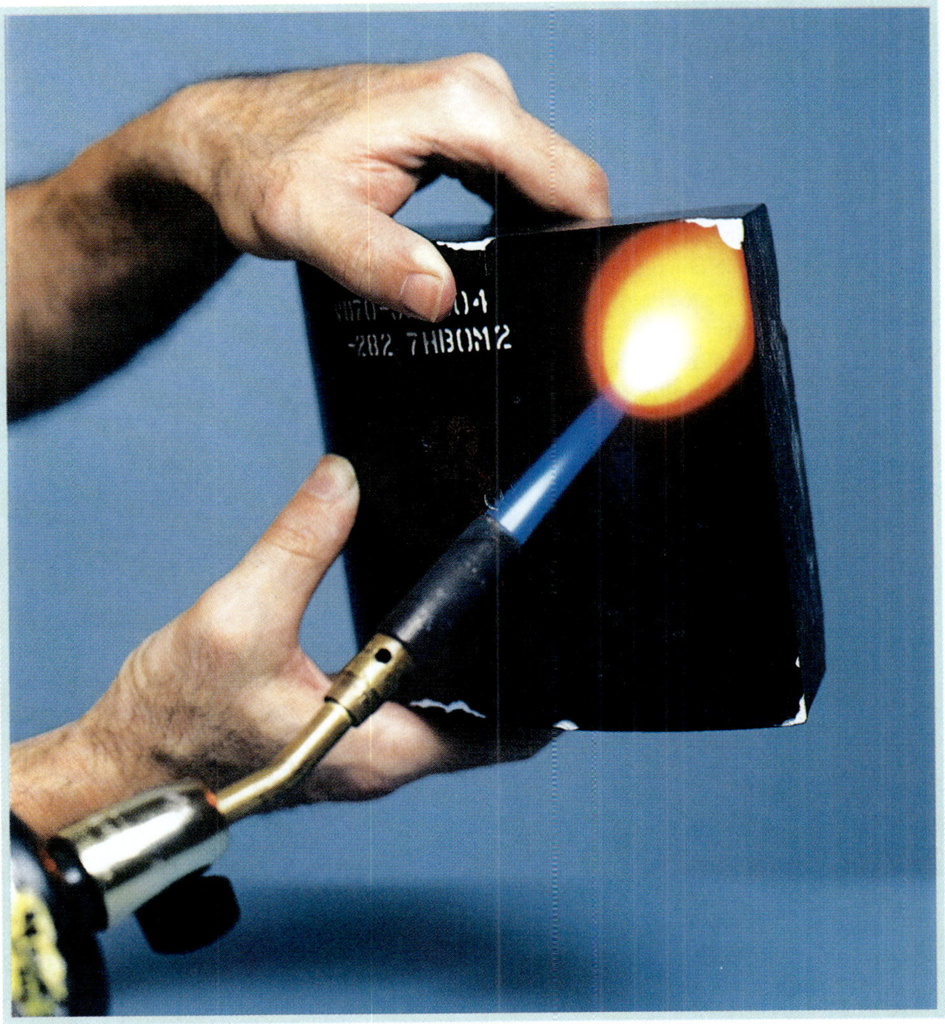

CHAPTER

21

We now know a great deal about how to manufacture materials with desirable thermal properties. This tile for the exterior skin of the Space Shuttle has a coefficient of thermal conductivity that is so low that it can be touched only centimeters from a spot that is red hot. In addition, the tile changes its dimensions very little when its temperature changes.

Properties of Solids

The properties of solids reflect a wide range of the physical laws we have discussed to this point. Solids deform in response to forces that act on them, propagate sound waves, respond to temperature changes by expanding or contracting, and conduct thermal energy. All these properties are explained ultimately by the atomic structure of materials.

21-1 THE NATURE OF SOLIDS

Solids are a form of matter that may appear at first glance to resist any kind of deformation. Our everyday experience suggests that solids move rigidly when forces act upon them and rotate rigidly under the influence of torques. Upon closer examination, though, solids exhibit more complicated responses to external influences. When com-

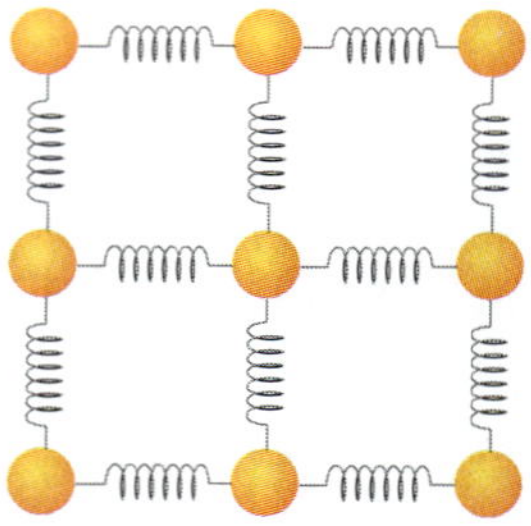

FIGURE 21–1 Schematic diagram of atoms connected by bonds in a lattice structure.

Defects, or dislocations, are important in solids.

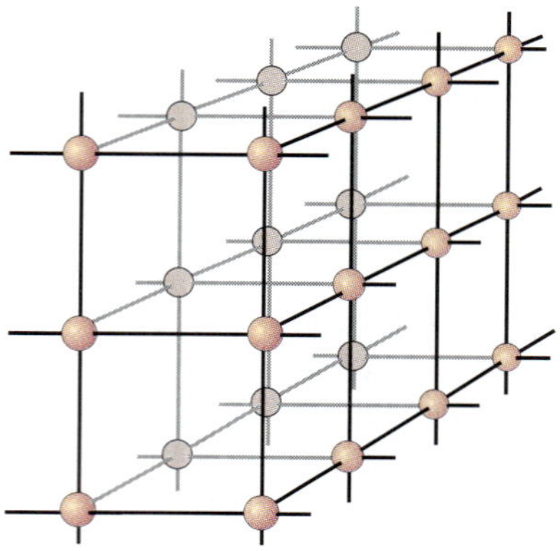

FIGURE 21–2 Sketch of a cubic lattice.

FIGURE 21–3 These beautiful crystals are characteristic of the mineral known as amethyst, a type of quartz.

pressed, their volumes change slightly; when put under tension or heated, they stretch; and when heated at one end, they conduct thermal energy to the other end. All these properties can be understood in terms of the underlying atomic structure of matter. As indicated in Section 16–1, **solids** are aggregates of atoms or molecules for which the interatomic forces lead to an organized, three-dimensional grouping of atoms, called a *lattice structure*. Figure 21–1, a schematic two-dimensional representation of such a structure, depicts the interatomic forces as springs. The lattice structure is one of stable equilibrium; therefore, the use of spring forces is at least approximately correct. The possible lattice structures are called **crystals**. Different crystals are determined by the nature of the interatomic forces between atoms or molecules and by the shapes and orientations of these constituents; that is, how they fit together. Figure 21–2 shows the three-dimensional structure of one particularly simple type of crystal, the *cubic lattice*.

If geometrically complicated molecules rather than simple atoms make up a crystal, or if several different kinds of atoms are present, the number of possible crystal structures increases rapidly (Fig. 21–3). The way different atoms assemble into crystals depends also on the conditions under which they are formed. Both diamond and graphite—two solids with very different properties—are formed from pure carbon. In a certain sense, the formation of a crystal by atoms is like a three-dimensional jigsaw puzzle with identical pieces, *which has more than one solution*!

Almost as important as crystalline structure to the macroscopic behavior of a solid are its lattice imperfections, called **defects**, or **dislocations**. Defects are closely associated with the way a crystal grows. Figure 21–4 shows a simple two-dimensional crystal that is easy to grow and observe: a raft of bubbles that float on the surface of a liquid. Figure 21–4a is a perfect crystalline form, whereas a defect has been introduced in Fig. 21–4b. Can you recognize it? A more general type of defect is associated with grains and grain boundaries. A **grain** is a region in a solid where the crystal structure is perfect. Because chance plays a role in the organization of atoms when crystal growth starts, the crystalline structure of one grain is oriented differently from that of another, and the boundaries where these grains meet involve special kinds of defects. Figure 21–5a shows such a boundary. A real solid is made up of many grains (Fig. 21–5b). Another class of defects occurs when single lattice sites are empty or are occupied by an impurity (an atom chemically different from those that make up the bulk of the crystal), or when impurities occupy spaces between lattice sites. Defects are related to a crystal's color, luminescence, transport properties, and mechanical properties. The study of the nature of defects is a major field of research and has taught us much about the behavior of solids.

The division between solids, liquids, and gases is not a very precise one. Substances can be distinguished by what we call **long-range order** or **short-range order**, determined by whether the arrangements of the atoms and molecules persist over distances of many atomic spacings or not. Crystals have long-range order, whereas most fluids

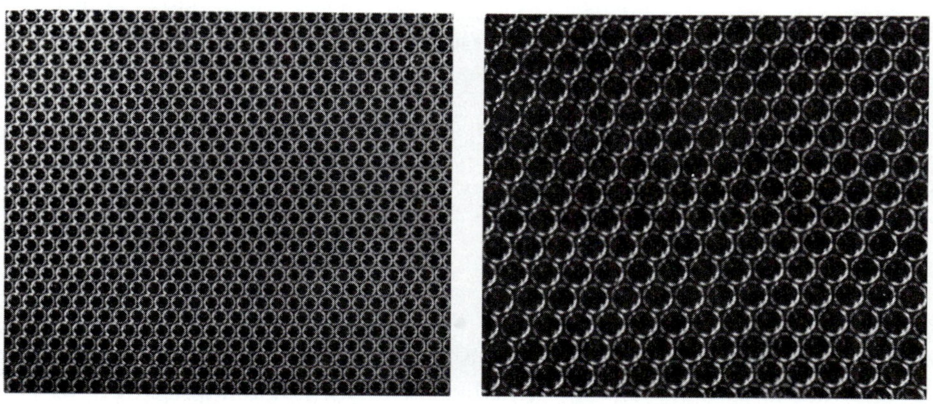

(a) (b)

FIGURE 21–4 (a) A perfect crystalline raft of bubbles. Bubbles will arrange themselves in crystalline patterns (in this case, two-dimensional), just like the atoms in real solids. (b) A dislocation in a raft of bubbles.

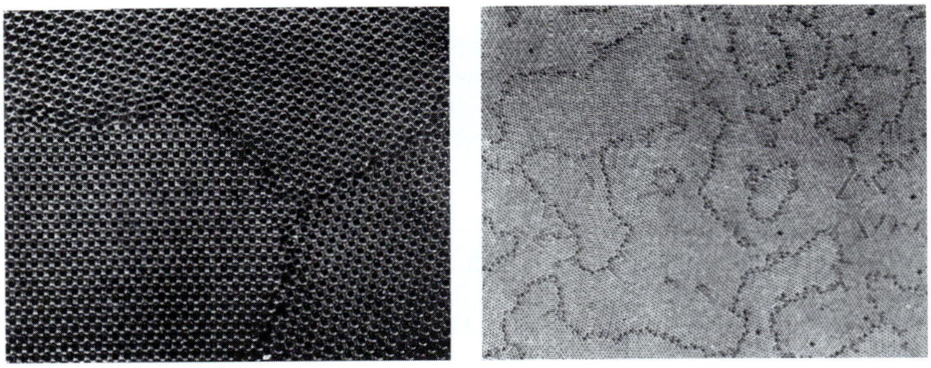
(a) (b)

FIGURE 21–5 (a) A close look at the grain boundaries in a crystal, as illustrated by a raft of bubbles. (b) An overview of numerous grains in a crystal.

have only short-range order. But, again, the division is artificial, for there is a small class of substances, called *amorphous solids*, that are like solids in some respects and like liquids in other respects. Substances exist at each of the gradations in between the boundaries we have constructed thus far. Consider *liquid crystals*, which are materials that contain rodlike molecules arranged parallel to one another. These molecules are in crystalline order if viewed from one end, along their axial direction, but they are free to move, in the fashion of molecules in liquids, along that direction. It is almost as though a liquid crystal is a solid in two directions but a liquid in the third direction. Liquid crystals are used in electronic displays. Many researchers feel that the materials that cannot be cleanly labeled as either solids or liquids are the most interesting ones, because they open avenues to new understanding and new applications.

Let's return to our simple model of lattice structure. Figure 21–6 suggests how a cubic structure might react to compression (Fig. 21–6a), to stretching (Fig. 21–6b), and to a *shear* (Fig. 21–6c). Just how hard it is to compress, stretch, or shear a solid depends on its interatomic forces. In the final analysis, it is these forces, along with the grain and defect structure, that differentiate steel from a pie crust. The lattice structure and the springlike bonds that connect the atoms or molecules also explain how waves propagate in solids. If successive layers of interatomic bonds are compressed, they will spring back to the equilibrium length and overshoot it, thus propagating the compression. Similarly, a small shear force on one cell gives rise through interatomic forces to a shear force in the following cell, and so forth, leading to transverse waves associated with the resistance to shear forces.

The concept of shear was introduced in Chapter 16.

Conduction of thermal energy is qualitatively similar. When one end of a rod is heated, the atoms become agitated and acquire kinetic energy. The bonds permit this kinetic energy to be transferred to neighboring layers of atoms; in this way, thermal energy travels along the rod. When the external conditions undergo a more extreme change, the very character of the solid changes. Under extreme compression, structural changes are possible—graphite can be changed into diamond; under strong tension, solids break; and when solids are heated to a sufficiently high temperature, they turn into liquids.

FIGURE 21–6 Schematic diagram of lattice distortions: (a) compressed in the horizontal direction; (b) stretched in the horizontal direction (c) subjected to shear forces, which act in opposite directions at the top and bottom.

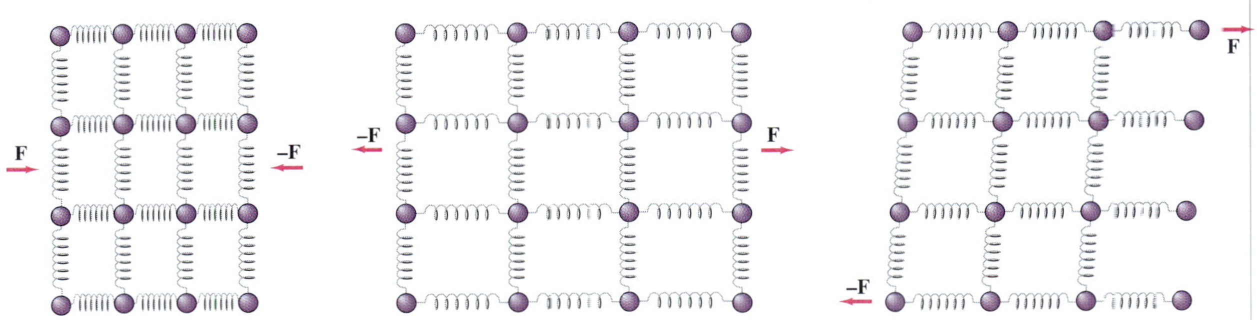

(a) (b) (c)

In principle, all properties of solids can be described by describing the atomic structure of solids, but in practice it is simpler to describe the properties of solids on a macroscopic scale. In developing the behavior of solids (or other materials) in terms of empirical observation, keep in mind that we are taking an approach that, at best, can give us "laws" that are more properly described as rules of thumb, valid only under restricted conditions.

21–2 STRESSES AND STRAINS

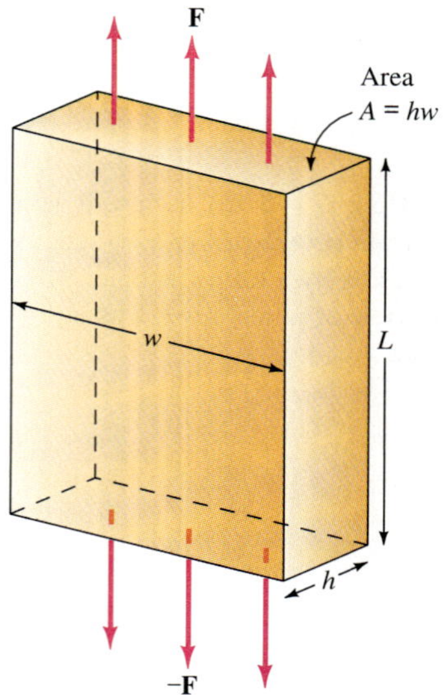

FIGURE 21–7 Equal but opposite forces applied evenly to opposite ends of a piece of material cause elongation.

Suppose that a force acts on a solid bar and tends to stretch it (Fig. 21–7). We have referred to such a force as a *tension*. **Stress** is a measure of the force that acts on the bar, whereas **strain** is a measure of the solid's response to a stress such as tension. We shall assume that the applied force is small, which implies that the response is small. (If the force is too large, the concept of a small response breaks down; when a heavy hammer strikes a piece of ice, the ice shatters.) We define the stress on a solid, S, as the external force per unit area that is exerted on it:

The definition of stress

$$S \equiv \frac{F}{A}, \tag{21–1}$$

where F is the magnitude of the force that acts per area A of the solid (Fig. 21–8). Tension gives rise to positive stress, and compression gives rise to negative stress. (The stress should contain a reference to the direction of the applied force, but we shall work with magnitudes and simple signs unless we specify otherwise.) Stress has the same dimensions as pressure and has SI units of newtons per square meter. Actually, this unit is rather small for ordinary applications, and a more practical unit for stress is a meganewton per square meter (MN/m^2).

When tension is applied to a rod, the rod stretches; when the rod is subjected to a small compressive force, its length decreases. The *response* to the stress associated with either a tension or a compression is called the *compressional strain*, e. It is defined as *the fractional change in the length of the solid along the direction of the compressional force*:

The definition of compressional strain

$$e \equiv \frac{\Delta L}{L}. \tag{21–2}$$

Note that e is dimensionless. It is negative if the stress is due to compression, and positive if it is due to tension.

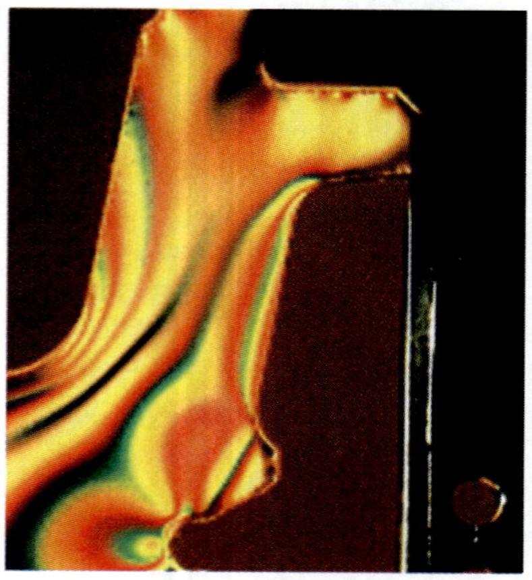

FIGURE 21–8 The passage of polarized light through materials such as this plastic cross section of a truck's axle shaft is stress dependent. The stress is greatest where the lines are most closely spaced.

The bar in Fig. 21–7 does more than just stretch as a result of a tension; when it stretches, it also *shrinks* in its lateral dimensions—the width h and depth w. Similarly, it will bulge in these directions if the force is a compression. The fractional shrinking or bulging is proportional to the compressional strain by an amount characteristic of the material, and it is described by the positive constant called *Poisson's ratio*, σ, after Siméon Poisson:

$$\frac{\Delta h}{h} = \frac{\Delta w}{w} = -\sigma e. \tag{21–3}$$

A typical value of σ for many solids is 0.3. This constant is always less than 0.5: If it were to exceed 0.5, the solid would expand when a uniform pressure is applied to it (see Problem 18).

If an external force F of constant magnitude is applied everywhere perpendicular to the surface of a solid, the volume, V, of that solid changes. The volume decreases if the force is inward and increases if the force is outward. The *volume stress*, p, is defined as stress was defined in Eq. (21–1): $p \equiv F/A$, where A is the *total* surface area of the solid. We have used the same symbol as that for pressure because the volume stress is identical to the pressure if the force is directed inward. For solids, though, the volume stress can equally well refer to an outward force on the body. We may also define the *volume strain*, e_V, by the fractional change in volume due to a volume stress:

$e_V \equiv \Delta V/V$. *When Poisson's ratio is small*, the volume strain is just a special case of the compressional strain; a bar that stretches while its cross-sectional area A remains constant has a volume $V = AL$, and when the bar is stressed longitudinally, the change in volume is $\Delta V = A\,\Delta L$. When these quantities are substituted in the definition for e_V, we obtain Eq. (21–2) for e.

We want to mention one more kind of stress and its corresponding strain: shear stress and shear strain. Suppose that a uniform force acts *along* a face of a solid (Fig. 21–9a). This is a shear force by definition. If the bottom surface of the solid is fixed (by some kind of contact force), the result is distortion of the solid (Fig. 21–9b). The quantity F/A is the *shear stress* on this solid, and the *shear strain* is the fractional amount by which the upper surface moves, $\Delta L/L$. Note that the volume is essentially unchanged under shear stress, in contrast to the case of longitudinal stress.

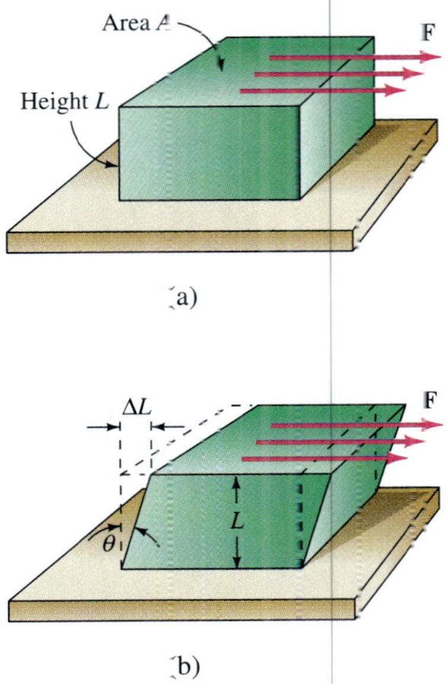

FIGURE 21–9 (a) An object fixed along its bottom surface has a force applied. (b) The effect of shear stress from uniform force distorts the object.

The Relation between Stress and Strain

Let's turn now to the relation between stress and strain. As early as the seventeenth century, Robert Hooke realized that stress and compressional strain are proportional to each other. A clear understanding of this proportionality was first provided by Thomas Young in 1800, who wrote down the relation

$$\frac{S}{e} = Y. \qquad (21\text{–}4)$$

This equation applies to the relation between (small) compression or tension and the corresponding strain. We introduced Y in Chapter 14 as **Young's modulus**, or the *elastic modulus*. It is a constant whose value varies with the material; it measures the "stiffness" of the material. Y is analogous to the *spring constant*, k, in the spring force $F = -kx$. In fact, Young's modulus is a constant only for *small* strains because the atomic bonds behave just like simple springs. Because e is dimensionless, Young's modulus has the same dimensions as S and therefore has units of force per area—meganewtons per square meter, for instance. Note that Young's modulus is a constant only for small stresses; for large stresses, the strain ceases to be proportional to the stresses, as experimentation with licorice or bread dough shows. Table 21–1 lists values of Young's modulus for various solids.

Young's modulus gives the relation between stress and compressional strain.

TABLE 21–1
Young's Moduli for Various Solids

Material	*Y* (MN/m^2)
Rubber	7
Wood	14,000
Concrete	17,000–30,000
Bone	9,000–21,000
Glass	70,000
Aluminum	73,000
Steel	210,000
Diamond	1,200,000

For small stresses and strains, the value of Young's modulus is generally independent of whether the material is under tension (positive strain) or under compression (negative strain). In some materials, however, the response to compression is not always the same as the response to tension. This is due to imperfections, such as tiny cracks in the material. Compression tends to reduce the cracks, whereas tension magnifies them.† Thus, for materials such as cast iron or cement, Young's modulus is larger for compression than for tension. The relative weakness of concrete under tension is remedied by the addition of iron rods to make what is called *reinforced concrete*. The rods are stiff under tension, though less so under compression, which causes them to buckle. Conversely, concrete is stiff under compression. Reinforced concrete is therefore stiff under both tension and compression.

APPLICATION
Reinforced concrete

EXAMPLE 21–1 A vertical steel rod of length 2.0 m and diameter 2.0 cm is fixed at the top and has a mass m of 9500 kg hanging from its lower end (Fig. 21–10). Given that Young's modulus for this particular type of steel is 250,000 MN/m^2, calculate the elongation of the rod. What is the strain?

Solution: The stress is given by $S = mg/A$, where A is the cross-sectional area of the rod. Equations (21–2) and (21–4) express the amount of stretch ΔL of the rod of length L and Young's modulus Y in terms of the stress:

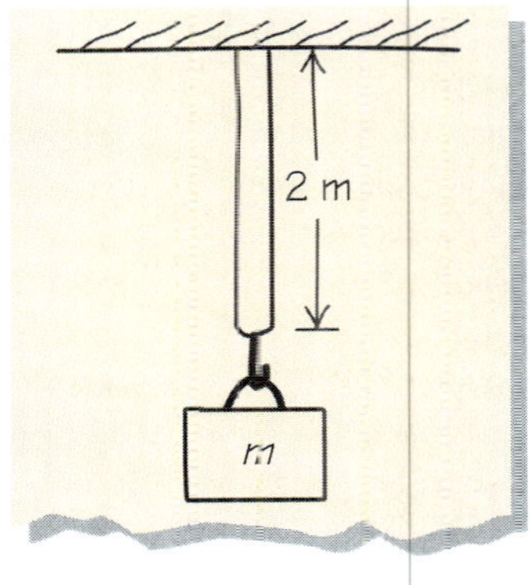

FIGURE 21–10 Example 21–1.

†A stack of bricks may be viewed as one brick with a number of cracks going right through it; the stack has no resistance to tension.

$$\Delta L = eL = \frac{S}{Y}L = \frac{mg}{AY}L.$$

The area of the rod is $A = \pi r^2$, where $r = \frac{1}{2}(2.0 \text{ cm}) = 1.0$ cm is the radius of the cross section. Thus we find

$$\Delta L = \frac{mg}{\pi r^2 Y}L = \frac{(9.5 \times 10^3 \text{ kg})(9.8 \text{ m/s}^2)}{3.14(1.0 \times 10^{-2} \text{ m})^2(2.5 \times 10^{11} \text{ N/m}^2)}(2.0 \text{ m})$$

The elongation is just 2.4 mm, and the strain is

$$\frac{\Delta L}{L} = \frac{2.4 \times 10^{-3} \text{ m}}{2.0 \text{ m}} = 1.2 \times 10^{-3}.$$

The bulk modulus gives the relation between volume stress and volume strain.

The relation between volume stress and volume strain takes exactly the same form as the relation between ordinary stress and strain for small strains. The proportionality constant $B =$ (volume stress)/(volume strain), which is analogous to Young's modulus, is called the **bulk modulus**. We have

$$B \equiv -\frac{F/A}{\Delta V/V} = -\frac{p}{e_V}. \qquad (21\text{–}5)$$

The minus sign means that an *inward* pressure on a solid, which is positive, implies a *decrease* in volume, $\Delta V < 0$. There is a relation between Young's modulus and the bulk modulus that must involve Poisson's ratio. Only two of these three constants are independent of one another (see Problem 17). Finally, there is a *shear modulus*, G, which is given by the ratio of shear stress to shear strain. Liquids have a shear modulus of zero, meaning that they have an arbitrarily large shear strain for even small shear stresses. Liquids do not resist shear at all.

The Strength of Materials

An important property of solids is their strength, or, more specifically, their *tensile strength*, which is also known as *fracture stress*. This quantity is the stress required to break a material into two pieces—by pulling it along an axis, for example. The tensile strength of different materials varies a great deal. Steel may break when the tension is 3000 MN/m^2, whereas a piece of concrete may be pulled apart by a stress of only 5 to 10 MN/m^2. The tensile strengths for a variety of materials are listed in Table 21–2. The values of tensile strength are a factor of 100 to 1000 times smaller than the values of Y (Table 21–1) for the same materials. There is useful information in this huge discrepancy, and it involves the importance of cracks and imperfections in a solid.

TABLE 21–2
Tensile Strengths for Various Solids

Material	*Tensile Strength (MN/m^2)*
Steel piano wire	3000
Steel	400–1500
Cast iron	70–250
Aluminum (pure)	70
Aluminum alloys	140–550
Copper	140
Titanium alloys	700–1400
Spruce, along grain	100
Spruce, across grain	3
Glass	70–170
Brick	5
Cotton	350
Spider silk	240
Human tendon	100
Rope	80

The tensile strength is a kind of *critical stress*. We may define a corresponding *critical strain* as $(\Delta L/L)_c$, such that

$$\text{tensile strength} = Y\left(\frac{\Delta L}{L}\right)_c.$$

The factor of 100 to 1000 difference between Young's modulus and the tensile strength for a given solid means that a fractional extension of any rod by 1 percent, or even by 0.1 percent, will break the rod. This is because the pervasive existence of cracks weakens the material (Figure 21–11). Without cracks, the stress on a section of rod between two layers of atoms is carried by many bonds. When cracks are present, though, there are fewer bonds to carry the stress. The force F in the expression F/A for the stress is the same, but the real area A over which the force acts is very much smaller than the apparent area. This means that the stress experienced by the material is much larger than the stress obtained by using the apparent area. For extremely thin—micrometer-size—fibers, which are nearly perfect crystals, measurements give a tensile strength of

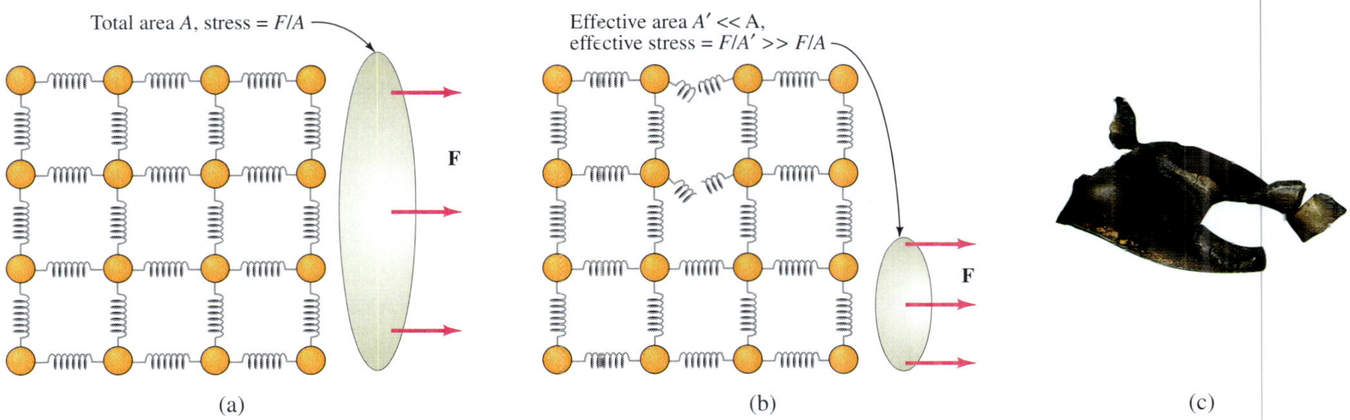

FIGURE 21–11 (a) An uncracked sample of material. (b) Cracks in the material lead to an increased susceptibility to fracture under stress. (c) This piece of metal has fractured under stress.

around $0.2Y$. This value is what we would expect upon evaluating $(\Delta L/L)_c$ from our understanding of interatomic forces. The discrepancy between the values of the tensile strength for bulk materials and tiny perfect crystals shows again the importance of imperfections in the structure of solids.

21–3 WAVES IN SOLIDS

The speed of sound in solids

We used dimensional analysis in Chapter 14 to estimate the speed of the propagation of waves in a solid with density ρ and Young's modulus Y:

$$v = \sqrt{\frac{Y}{\rho}}. \tag{21–6}$$

Young's modulus refers to compressive and tensile forces, so the wave being discussed is a longitudinal wave composed of alternate compressions and extensions of the solid. Sound waves in air are also longitudinal and, accordingly, longitudinal waves in solids are referred to as sound waves.

In Chapter 14, we pointed out that the constant in front of the square root in Eq. (14–27) [Eq. (21–6)] cannot be estimated by dimensional analysis, so we set it equal to 1. In order to find the constant (which, in fact, is 1 to a good approximation), we need to find a dynamical wave equation for the longitudinal waves in a given solid. Let x be the position in the longitudinal direction and let $z(x, t)$ be the longitudinal displacement from equilibrium of the material at x. We must show that $z(x, t)$ varies with position x and time t in such a way as to satisfy the wave equation first discussed in Chapter 14. This equation takes the general form

$$\frac{\partial^2 z}{\partial x^2} - \frac{1}{v^2}\frac{\partial^2 z}{\partial t^2} = 0. \tag{21–7}$$

(Recall that the partial derivative is, for all practical purposes, just an ordinary derivative that acts on a quantity that depends on several independent variables; in this case, x and t.) Equation (21–7) has the general wave solution

$$z(x, t) = z(x \pm vt), \tag{21–8}$$

in which v is the wave speed. Just as Hooke's law implies the harmonic motion of a mass attached to a spring, it follows from the stress–strain relationship that z obeys the wave equation and that the speed of sound in Eq. (21–6) is indeed correct, at least for a thin rod. When a time-varying stress is imposed on one end of a material from the outside, there is a time-varying strain at that point. The displacement at that point causes a stress on a neighboring point, which is in turn displaced, and so on. The details are worked out in the following subsection.

How to Get the Wave Equation for Longitudinal Waves in Solids

We find the wave equation for longitudinal waves in solids in a manner that is analogous to how we arrived at Eq. (14–9). Consider a rod of cross-sectional area A that lies along the x-axis. Concentrate on a section of the rod between x and $x + \mathrm{d}x$ (Fig. 21–12). The stress (force per unit area) will vary with position and also with time. The *net* stress acting on the section of the rod is

$$\Delta S = S(x + \mathrm{d}x) - S(x) = \left[\frac{\partial S(x)}{\partial x}\right]\mathrm{d}x.$$

Thus the net force on the section, ΔF, is given by

$$\Delta F = A\ \Delta S = A\left[\frac{\partial S(x)}{\partial x}\right]\mathrm{d}x.$$

Let's denote the longitudinal displacement of the material within the rod from its equilibrium position by $z(x, t)$. The force given above leads to an acceleration of the material. The mass of material in the section of the rod of length $\mathrm{d}x$ is given by the density ρ multiplied by the volume, so that $m = \rho A\ \mathrm{d}x$; the acceleration of the material is given by the second time derivative of the displacement $z(x, t)$. Thus $a = \partial^2 z(x, t)/\partial t^2$, and Newton's second law, $ma = \Delta F$, yields

$$\rho A\ \mathrm{d}x\left(\frac{\partial^2 z}{\partial t^2}\right) = A\left(\frac{\partial S}{\partial x}\right)\mathrm{d}x;$$

$$\rho\left(\frac{\partial^2 z}{\partial t^2}\right) = \frac{\partial S}{\partial x}. \qquad (21\text{–}9a)$$

Let's now relate the stress to the strain. The strain, given by the fractional displacement, is proportional to the stress. The total length of the section of the rod is $\mathrm{d}x$, and the net increase in length is the difference between the displacements measurable at the two ends of the section, $z(x + \mathrm{d}x) - z(x)$ (Fig. 21–12b). Thus the strain is given by the ratio of the difference of displacements to the original length, $\mathrm{d}x$:

$$e = \frac{z(x + \mathrm{d}x, t) - z(x, t)}{\mathrm{d}x} = \frac{\partial z}{\partial x}.$$

Equation (21–4) states that $S = eY$, or

$$S = Y\frac{\partial z}{\partial x}.$$

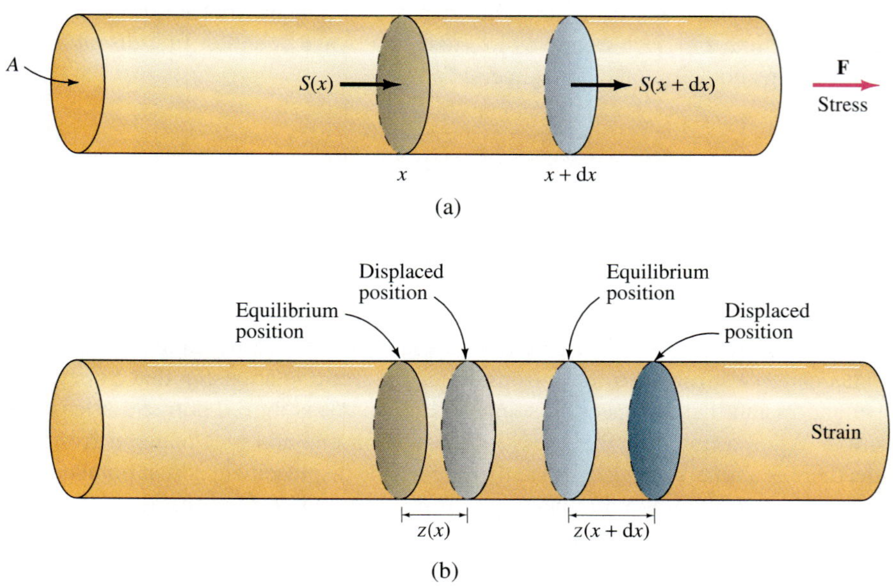

FIGURE 21–12 The (a) stresses on and (b) resulting elongation of a section of a rod that lies between the positions x and $x + \mathrm{d}x$.

When this result is substituted into Eq. (21–9a), we find that

$$\rho\left(\frac{\partial^2 z}{\partial t^2}\right) = Y\left(\frac{\partial^2 z}{\partial x^2}\right). \tag{21–9b}$$

This result has the form of the wave equation, Eq. (21–7), and the speed can be identified to be that of Eq. (21–6).

We can get some feeling for the speed of sound in solids by calculating this quantity in a steel rod. The density of steel is $\rho = 7.9 \times 10^3\ \text{kg/m}^3$ while Young's modulus is $Y = 200{,}000\ \text{MN/m}^2$. Then, using Eq. (21–6), we find

$$v = \sqrt{\frac{Y}{\rho}} = \sqrt{\frac{2 \times 10^{11}\ \text{N/m}^2}{7.9 \times 10^3\ \text{kg/m}^3}} = \sqrt{0.25 \times 10^8\ \text{m}^2/\text{s}^2} = 5 \times 10^3\ \text{m/s}.$$

This number, more than ten times larger than the speed of sound in air, is in agreement with experimental data.

Other Waves in Solids

Just as small longitudinal displacements within a solid lead to longitudinal waves, small transverse displacements within a solid lead to transverse waves. This is because solids resist shear forces, and the wave speed of transverse waves—also called *shear waves*—is described by using the shear modulus, G, of Section 21–2:

$$v_{\text{shear}} = \sqrt{\frac{G}{\rho}}. \tag{21–10}$$

In most cases, the shear modulus is lower than Young's modulus. The speed of shear waves is thus less than that of longitudinal waves.

Probing Earth

APPLICATION
Geological exploration

Earth supports both sound waves and shear waves and there are many boundaries within it—the mantle/core transition, for example—where these waves are reflected or transmitted (see Problem 27). This phenomenon is the basis for an important tool used in geological exploration and research. Earthquakes are responsible for initiating both kinds of waves within the planet; the varying arrival times of these waves at a seismograph—a device that detects such vibrations—gives information about the location of the earthquake. Earthquake, or seismic, waves can follow many paths, reflecting from various internal layers and boundaries between different materials; this fact is helpful in pinpointing the location and strength of the earthquake (Fig. 21–13). More important, information on Earth's *internal* structure comes from the arrival time and

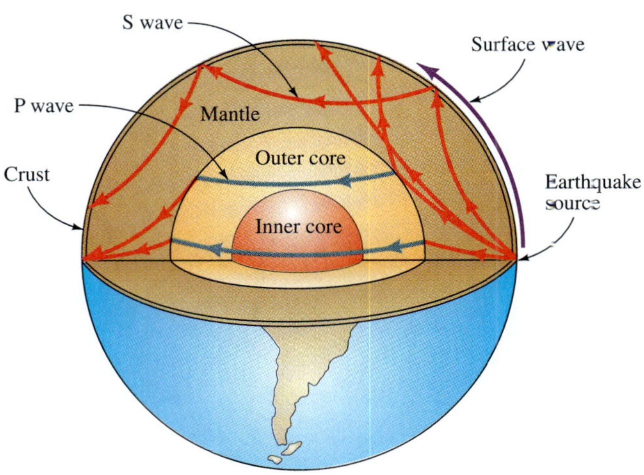

FIGURE 21–13 Several types of waves are possible in solids. Seismic waves, produced by events such as earthquakes, include all types. (After S. Judson and M. Kauffman, *Physical Geology,* Prentice Hall, 1990.)

strength of waves initiated either by earthquakes or by explosions that are set intentionally. Such data teach us not only about the existence of reflecting boundaries, but also about varying densities within Earth. For example, we know that Earth's central core is liquid because shear waves do not pass through it. The search for petroleum relies heavily on evidence from wave propagation within Earth.

21–4 THERMAL EXPANSION

The definition of the coefficient of thermal expansion

We know that a solid expands when it is heated. A simple demonstration of this fact is shown in Fig. 21–14. Empirically, this expansion is described by the **coefficient of thermal expansion**, α, defined by

$$\alpha \equiv \frac{1}{L}\frac{dL}{dT}. \tag{21–11}$$

The quantity $\alpha\, dT$ is the fractional change in the length L of a solid due to a change in its temperature, dT. In principle, the coefficient α is a function of temperature T but tends to vary quite slowly with T. The analogous **coefficient of volume expansion**, β, is simply related to α. $\beta\, dT$ is the fractional change in the volume V due to a temperature change dT:

$$\beta \equiv \frac{1}{V}\frac{dV}{dT}. \tag{21–12}$$

We can show that $\beta = 3\alpha$ by considering a cube of volume $V = L^3$. We have

$$\beta = \frac{1}{V}\frac{dV}{dT} = \frac{1}{L^3}\left(3L^2\frac{dL}{dT}\right) = \frac{3}{L}\frac{dL}{dT} = 3\alpha. \tag{21–13}$$

This result applies as long as the material is *isotropic*; that is, its properties are the same in any direction within the material. Thermal expansion also occurs in liquids. Liquids are more correctly described as isotropic than are solids, whose crystal structure implies preferred directions. For this reason, the volume expansion coefficient β is usually given when liquids are involved. Both β and α are measured in units of K^{-1}. Table 21–3 lists some thermal expansion coefficients.

TABLE 21–3
Coefficients of Thermal and Volume Expansion (at 20°C)

Material	*α (K^{-1})*
Aluminum	2.30×10^{-5}
Copper	1.67×10^{-5}
Pyrex glass	3.2×10^{-6}
Fused quartz	4.2×10^{-7}
Steel	1.05×10^{-5}
	β (K^{-1})
Ethyl alcohol	1.12×10^{-3}
Gasoline	9.5×10^{-4}
Mercury	1.82×10^{-4}
Water	2.07×10^{-4}

EXAMPLE 21–2 A steel bridge is 600 m long. How much allowance must be made for linear expansion between the temperatures of +40°C and −40°C?

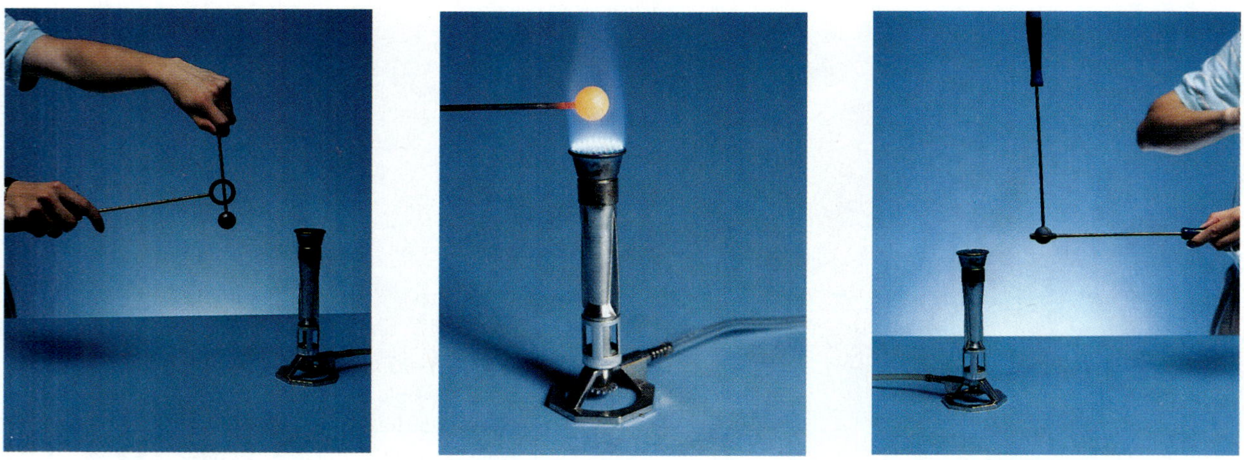

(a) (b) (c)

FIGURE 21–14 Thermal expansion: The linear dimensions of a material expand proportionally with the temperature increase. In this instance, the linear dimension of the steel ball is its radius. (a) It just fits through a ring at room temperature, (b) is heated, and (c) becomes too big to fit through the ring.

Solution: Equation (21–11) may be written as

$$\Delta L = \alpha L \, \Delta T. \tag{21–14}$$

We take α from Table 21–3 to be $1.05 \times 10^{-5}\ \mathrm{K}^{-1}$ for steel (α varies little with temperature in the range +40°C to −40°C). Numerical evaluation gives

$$\Delta L = (1.05 \times 10^{-5}\ \mathrm{K}^{-1})(600\ \mathrm{m})(80\ \mathrm{K}) = 5 \times 10^{-1}\ \mathrm{m} = 50\ \mathrm{cm}.$$

Expansion joints that allow for this much thermal expansion are distributed at regular intervals along bridges (Fig. 21–15).

APPLICATION

Expansion joints

EXAMPLE 21–3 You must design a gasoline container to store gasoline under conditions in which the annual temperature extremes are −20°C to +40°C. You want to be able to fill the container at any time of the year and not lose more than 0.1% (a fraction 0.001) of the gasoline due to overflow. In choosing the material for your container, what range of possible values of the coefficient of thermal expansion α should delimit your search?

Solution: Both container and liquid will change their volumes as the temperature changes. If the container material has a coefficient of thermal expansion α that is one-third β ($\beta = 9.5 \times 10^{-4}\ \mathrm{K}^{-1}$ is the coefficient of volume expansion for gasoline), then both container volume and gasoline volume will change in the same way, and a full container will remain full. If the expansion coefficient of the container exceeds that of the fluid, then fluid will be squeezed out upon cooling; if the expansion coefficient of the container is lower than that of the fluid, then fluid will be squeezed out upon heating. Thus the maximum possible value of α is found by filling the container at the maximum temperature; the container then shrinks more than the gasoline when the temperature descends to its lowest value. We require

$$\left.\frac{\Delta V}{V}\right|_{\text{container}} - \left.\frac{\Delta V}{V}\right|_{\text{gasoline}} = -0.001.$$

(The minus sign is present because the container shrinks more than the gasoline in this case.) Thus

$$3\alpha_{\max}\,\Delta T - \beta\,\Delta T = -0.001,$$

$$\alpha_{\max} = \frac{-0.001 + \beta\,\Delta T}{3\,\Delta T} = \frac{-0.001 + (9.5 \times 10^{-4}\ \mathrm{K}^{-1})(-20^\circ\mathrm{C} - 40^\circ\mathrm{C})}{3(-20^\circ\mathrm{C} - 40^\circ\mathrm{C})} = 3.2 \times 10^{-4}\ \mathrm{K}^{-1}.$$

Similarly, the minimum value of α can be found by filling the container at the minimum temperature, then insisting that the gasoline expands only a little more than the container:

$$\left.\frac{\Delta V}{V}\right|_{\text{gasoline}} - \left.\frac{\Delta V}{V}\right|_{\text{container}} = 0.001.$$

Thus

$$\beta\,\Delta T - 3\alpha_{\min}\,\Delta T = 0.001,$$

$$\alpha_{\min} = \frac{0.001 - \beta\,\Delta T}{-3\,\Delta T} = \frac{0.001 - (9.5 \times 10^{-4}\ \mathrm{K}^{-1})(40^\circ\mathrm{C} - (-20^\circ\mathrm{C})}{-3(40^\circ\mathrm{C} - (-20^\circ\mathrm{C}))} = 3.1 \times 10^{-4}\ \mathrm{K}^{-1}.$$

In fact, it would be hard to find solids with values of α this large.

FIGURE 21–15 The expansion joint in this bridge allows the material of the bridge to undergo expansion in hot weather and contraction in cold weather.

Water has the very special property that it decreases in density—it expands—as it solidifies on cooling (Fig. 21–16). This means that ice floats on water.

FIGURE 21–16 The plastic bottle of water on the left has been placed in a freezer, whereas the one on the right has not. Water expands when it freezes, meaning that ice floats on liquid water. This fact has important consequences for Earth's weather and for life on Earth.

APPLICATION: REINFORCEMENT OF MASONRY BUILDINGS

An interesting use of thermal expansion is the restoration of old masonry buildings. Such buildings often have bulging walls that result from the outward pressure of a pitched roof. To straighten such a wall, a hole is drilled in each side of the building and a metal bar passed through the building. This bar is usually concealed in the floorbeam region. The ends of the bar are threaded and metal plates are placed over the protruding bar ends (Fig. 21AB–1). Nuts are then tightened on the threads to force the plates against the walls. The bar is heated along its length, traditionally with flames, but now by electric energy. When the bar lengthens, the nuts are tightened as much as possible. When the bar cools and contracts, there is a large compression force against the walls, straightening them, and resisting future outward forces due to roof loads. Such plates, together with the nuts in their centers, can be seen on many old European buildings.

FIGURE 21AB–1 The star is a metal plate holding the old brick wall in place. A metal rod at the center of the star extends through the wall to the other side, where a second star is placed. There are several such devices placed on this old warehouse wall.

21–5 THE CONDUCTION OF THERMAL ENERGY

Another important property of solids has to do with the transport of thermal energy. When one end of a solid beam of length L is maintained at a temperature T_1 and the other end is kept at a lower temperature, T_2, then there will be a steady transport of thermal energy from the higher-temperature end to the lower-temperature end. Experiment shows that the rate of transfer of thermal energy is proportional to the area of the beam, A, and to the temperature difference between the two ends, $\Delta T = (T_1 - T_2)$; it is inversely proportional to the length of the beam, L.

The rate of thermal energy transfer, written as $\mathrm{d}Q/\mathrm{d}t$, is thus described by an equation of the type

$$\frac{\mathrm{d}Q}{\mathrm{d}t} \propto A\frac{\Delta T}{L}. \tag{21–15}$$

Actually, the linear dependence on the temperature difference ΔT is true only for small ΔT. This is not a difficulty if we think about the rate of energy transfer for a thin slice of length Δx rather than for the entire beam at once. The temperature difference across a thin slice is very small, and we can safely assume that the relation between ΔT and the rate of thermal energy flow is linear. We can then replace L by Δx and rewrite Eq. (21–15) in derivative form as

$$\frac{\mathrm{d}Q}{\mathrm{d}t} = -\kappa A\frac{\mathrm{d}T}{\mathrm{d}x}. \tag{21–16}$$

The definition of thermal conductivity

The coefficient κ is the **thermal conductivity**, a constant that varies from material to material and may have a temperature dependence. The SI units of κ are $\mathrm{W/(m \cdot K)}$, or $\mathrm{J/(m \cdot s \cdot {}^\circ C)}$. (Note that a difference of 1°C is identical to a difference of 1 K.) We give some typical values of thermal conductivity in Table 21–4. When κ is large, the material is said to be a good thermal conductor, and when κ is small, the material is a poor thermal conductor. The minus sign in Eq. (21–16) indicates that, if the *temperature gradient* $\mathrm{d}T/\mathrm{d}x$ is negative—so the temperature is lower for larger values of x—then thermal energy flows toward larger x. The differential equation (21–16) applies to liquids and gases as well as to solids, and Table 21–4 lists values of κ for materials of all three states.

When Eq. (21–16) applies, we say that thermal energy is transported by *conduction.*

EXAMPLE 21–4 An igloo, a hemispherical enclosure built of ice ($\kappa = 1.67$ J/m · s · °C), has an inner radius of 2.5 m. The thickness of the ice is 0.30 m. At what rate must thermal energy be generated to maintain the air inside the igloo at 5°C when the outside temperature is −40°C? Ignore all thermal energy losses through the ground or by air currents.

Solution: The rate at which energy must be supplied is just the rate of loss of thermal energy by heat conduction, so we want to calculate the right-hand side of Eq. (21–16). The temperature gradient is given by $(T_{in} - T_{out})$/thickness of ice. The area A is that of half a sphere. Because the ice thickness is small compared to either the inner or outer radius, it does not matter very much which radius we use, so we use the median radius, $R_{med} = 2.5\text{ m} + 0.15\text{ m} = 2.65\text{ m}$. From Eq. (21–16),

$$\frac{dQ}{dt} = -\kappa A\frac{dT}{dx} = -\kappa(2\pi R_{med}{}^2)\frac{T_{in} - T_{out}}{\text{thickness of ice}}$$

$$= -(1.67\text{ J/m}\cdot\text{s}\cdot{}^\circ\text{C})(2)(3.14)(2.65\text{ m})^2\frac{(5^\circ\text{C}) - (-40^\circ\text{C})}{0.30\text{ m}}$$

$$= -11 \times 10^3\text{ J/s} = -11\text{ kW}.$$

For comparison, a room-size kerosene heater puts out some 5 kW.

TABLE 21–4
Thermal Conductivities for Various Materials

Material	κ *(kW/m · K)*
Aluminum	0.21
Copper	0.39
Lead	0.035
Silver	0.42
Water (at 0°C)	5.65×10^{-4}
Water (at 20°C)	5.99×10^{-4}
Air	2.5×10^{-5}
Asbestos	0.80×10^{-4}
Brick	6.3×10^{-4}
Glass	10.5×10^{-4}

How Is Thermal Energy Conducted in Materials?

We know that thermal energy is the energy of motion of the constituents of matter. Let's now explore the physical origins of thermal conductivity. Metals conduct thermal energy better than nonmetals do. Within the atoms of metals, the outermost electrons that orbit the nucleus are attached very loosely (they are *weakly bound*), and a slight disturbance of the system can dislodge them. The practically free electrons of metals are efficient in transferring energy from one end of a piece of metal to the other. The same electrons also transfer electric charge (as we shall see in Chapter 24), and thermal and electrical conductivity in metals are closely related.

Nonmetals also conduct thermal energy, but less well than metals do. The outermost electrons of the atoms of nonmetals are attached tightly and are therefore not readily available to conduct thermal energy, and the molecules cannot move freely in a solid. Compressions and rarefactions of a nonmetal enter into the transport of thermal energy, but purely elastic waves traverse a material without depositing energy along the way. Only the scattering of waves by impurities, defects, and nonelasticities in the medium lead to the transfer of thermal energy.

In gases, thermal energy conduction is a transport phenomenon much like others discussed in Section 19–8. Molecules that move more rapidly because they are in a region of higher temperature collide with molecules in a neighboring region, giving the adjacent molecules more kinetic energy and consequently more thermal energy. This model explains some unique features of thermal transport in gases; for example, the thermal conductivity, κ, of gases varies with temperature as $\sqrt{T}$.

Thermal energy transport by convection and radiation are described in Chapters 16 and 17.

Table 21–4 shows that air—and this is generally true for gases—has a small thermal conductivity. Despite that fact, a house built of double walls with air between them will be poorly insulated. The reason is *convection* in fluids, a phenomenon that involves the large-scale movement of gases or liquids. Thermal energy is transported efficiently by convection. Fiberglass insulation is designed to trap air so that it cannot flow, or convect, easily. Thermal energy can also be transported by *radiation*. In such transport, thermal energy crosses transparent spaces—spaces in which electromagnetic radiation can travel.

APPLICATION

Insulating walls

Thermal Resistance

The definition of thermal resistance

The **thermal resistance**, or ***R* value**, is used as an alternative to the thermal conductivity rating; it is inversely proportional to thermal conductivity. For a slab of a given material of thickness L, R is defined by

$$R \equiv \frac{L}{\kappa}. \tag{21–17}$$

Whereas κ depends only on the type of material, R *depends also on the thickness of the material*. Therefore, R characterizes the effectiveness of a thermal barrier of given thickness. A large R value corresponds to a small thermal conductivity and/or a large thickness; a piece of material with such characteristics is a good *thermal insulator*, whereas a piece of material with a low R value is a poor insulator. The R value has SI units of $m^2 \cdot K/W$. When it is used in connection with insulation in construction in America, however, it is often quoted in units of $ft^2 \cdot h \cdot {}^\circ F/Btu$, where $1\ ft^2 \cdot h \cdot {}^\circ F/Btu = 0.18 m^2 \cdot K/W$. The R value of a 6-in piece of fiberglass insulation is $22\ ft^2 \cdot h \cdot {}^\circ F/Btu$, compared to $0.32\ ft^2 \cdot h \cdot {}^\circ F/Btu$ for a $\frac{3}{8}$-in piece of plasterboard.

In terms of R, the relation between thermal energy flow and temperature difference from Eqs. (21–16) and (21–17) is

$$\frac{\Delta Q}{\Delta t} = \frac{1}{R} A\, \Delta T, \tag{21–18}$$

where A is the area across which thermal energy is transported. Because R refers to a given material of a given thickness, it is interesting to see how the R values of pieces of material combine.

Thermal Resistance in Series. Let's first consider combining two conducting solids of the same area A *in series*, as shown in Fig. 21–17. Suppose that the left-hand piece has an R value of R_1 and the right-hand piece R_2, and that the temperature varies from T_c (cold) on the left to T_h (hot) on the right. We want to find the effective R value of the combined piece, R_{eff}. When thermal energy flows at a steady rate through the two materials, as shown in Fig. 21–17, then energy conservation implies that the rate of transport of thermal energy across one piece must equal the rate of transport across the second piece, and hence also the rate across the combined piece. Application of this idea (see Problem 49) implies that

PROBLEM SOLVING

When thermal resistances are combined in series, the R values are additive.

$$\text{for two solids in series:} \quad R_{eff} = R_1 + R_2. \tag{21–19}$$

This result generalizes for many slabs of material that are combined in series.

Thermal Resistance in Parallel. Now suppose that instead of being placed in series, two slabs are placed *in parallel*, as shown in Fig. 21–18. In this case, the temperature difference on the right-hand and left-hand sides of both slabs is the same, and the total rate of transport of thermal energy is the sum of the transport rates through the two slabs. In Problem 51, this fact is applied, and the result is

PROBLEM SOLVING

When thermal resistances are combined in parallel, the values of area/R are additive.

$$\text{for two solids in parallel:} \quad \frac{1}{R_{eff}} = \frac{1}{A_1 + A_2}\left(\frac{A_1}{R_1} + \frac{A_2}{R_2}\right). \tag{21–20}$$

This result generalizes for more than two slabs placed in parallel.

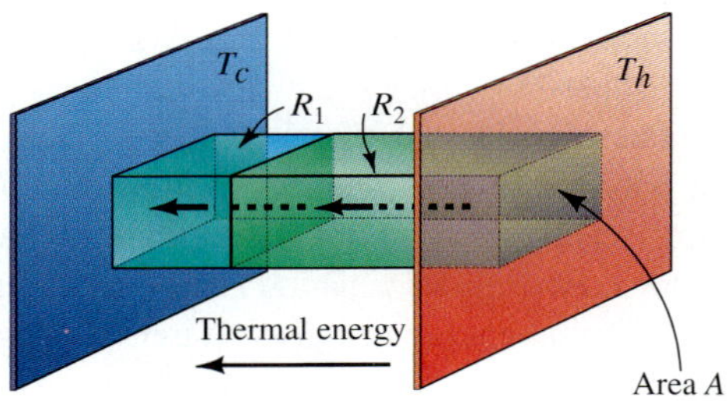

FIGURE 21–17 Thermal resistance, or R values, for two materials placed "in series"—one behind the other. Thermal energy passes through the materials as shown.

EXAMPLE 21–5 A hole in a wall has been temporarily patched on the inside with a 1-square-ft piece of plywood for which $R = 1\ \text{ft}^2\cdot\text{h}\cdot{}^\circ\text{F/Btu}$. The rest of the wall is insulated with fiberglass insulation for which $R = 22\ \text{ft}^2\cdot\text{h}\cdot{}^\circ\text{F/Btu}$. What is the R value of the repaired wall, whose total area is $100\ \text{ft}^2$?

Solution: A sketch of this situation, such as Fig. 21–18, shows that we have two materials placed in parallel, and Eq. (21–20) applies. Note that because the R value already takes the thickness of the material into account, the plywood is not necessarily the same thickness as the fiberglass insulation. In fact, plywood is only a fraction of an inch thick, whereas the fiberglass is 6 in thick. The rest of the thickness may be taken up by air, but because this air will be continuous with the outside environment, we ignore its insulating properties. If A_1 and R_1 are taken to be the area and R value of the plywood and A_2 and R_2 are the area and R value of the fiberglass, respectively, we find

$$\frac{1}{R_{\text{eff}}} = \frac{1}{A_1 + A_2}\left(\frac{A_1}{R_1} + \frac{A_2}{R_2}\right) = \frac{1}{100\ \text{ft}^2}\left(\frac{1\ \text{ft}^2}{1\ \text{ft}^2\cdot\text{h}\cdot{}^\circ\text{F/Btu}} + \frac{100\ \text{ft}^2 - 1\ \text{ft}^2}{22\ \text{ft}^2\cdot\text{h}\cdot{}^\circ\text{F/Btu}}\right)$$

$$= \frac{1}{18\ \text{ft}^2\cdot\text{h}\cdot{}^\circ\text{F/Btu}}.$$

Thus

$$R_{\text{eff}} = 18\ \text{ft}^2\cdot\text{h}\cdot{}^\circ\text{F/Btu}.$$

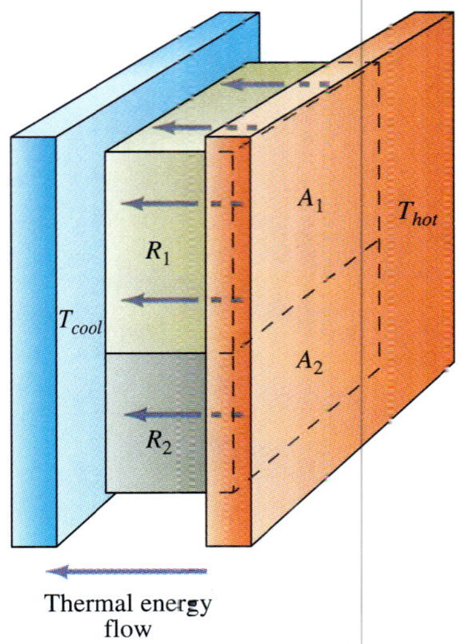

FIGURE 21–18 Thermal resistance, or R values, for two materials placed "in parallel." Thermal energy passes through the materials as shown.

In Example 21–5, a hole in a wall that is patched with little insulation and takes up only 1 percent of the area of the wall reduces the R value by some 17 percent. This effect is typical of thermal insulators combined in parallel. In particular, if one of the elements of a set of insulators combined in parallel has a very small R value, Eq. (21–20) shows that the effective R value is very small. Small holes can be very costly sources of thermal energy loss. Indeed, the best window has a thermal resistance of about one-tenth of normal wall and ceiling insulation. Thus once wall and ceiling energy losses become small, the only way to conserve energy further is to reduce window area. This is already practiced in commercial buildings in the Midwest.

SUMMARY

Solids consist of atoms arranged in regular crystalline structures, with springlike forces between the atoms. As a consequence, solid bodies may stretch or be compressed when subjected to external forces. The force per unit area, or stress, S, is related to the fractional change in length, or strain, e, by Y, Young's modulus:

$$\frac{S}{e} = Y. \tag{21–4}$$

A similar relation applies to volume stress, p, and volume strain, e_V, which are related by the bulk modulus, B:

$$B = -\frac{p}{e_V}. \tag{21–5}$$

The real tensile strength of solids, which is a measure of the amount of strain at which they fracture, is smaller than one would expect because of the many imperfections that occur in the crystalline structure of solids.

The springlike nature of the forces that hold solids together permits the propagation of longitudinal waves (sound), with the velocity of propagation given (approximately) by

$$v = \sqrt{\frac{Y}{\rho}}, \tag{21–6}$$

where ρ is the density of the material. Solids also support transverse shear waves, which move more slowly than sound waves.

Materials expand when heated, and they are characterized by a coefficient of thermal expansion, α, defined by

$$\alpha \equiv \frac{1}{L}\frac{dL}{dT}. \tag{21–11}$$

Materials also conduct thermal energy. The rate of transport of thermal energy depends on the temperature drop per unit length and the area of the material through which the thermal energy flows. The thermal conductivity, κ, is defined by

$$\frac{dQ}{dt} = -\kappa A \frac{dT}{dx}. \tag{21–16}$$

where the left-hand side represents the flow of thermal energy per unit time, A is the area, and dT/dx is the temperature drop per unit length. The reciprocal of κ is related to the thermal resistance, or R value, of the insulating material, with

$$R \equiv \frac{L}{\kappa}, \tag{21–17}$$

where L is the thickness of the material. The R values of two insulating slabs of the same area are additive when the slabs are connected in series, whereas the quantities area/R are additive when the slabs are connected in parallel.

UNDERSTANDING KEY CONCEPTS

1. Why is cement unsuitable as a construction material for a boiler?

2. In Chapter 5 of the Book of Exodus, the Israelites in captivity complain to Pharaoh that they are being asked to make bricks without straw. Why was it a good idea to put straw in the clay that was allowed to dry in the hot sun?

3. Why is it inadvisable to wear wet clothing on a cold day? In view of this question, how does a diver's wet suit work?

4. How can seismographs in the United States record earthquakes in other continents?

5. Thermopane windows are windows made of two glass panes with air between them. Explain why they are better thermal insulators than a simple window pane of twice the thickness.

6. Why does the amount of stretch of a bar on which a tension acts depend on the overall length of the bar? [*Hint*: Think of breaking the bar into two pieces of equal length and of how much each piece would stretch under the same force.]

7. If it is possible to make diamonds from graphite by using high pressure, how would it be possible to make graphite from diamonds?

8. If a doughnut-shaped piece of a solid that expands under heat is heated, does the hole expand or shrink?

9. On a macroscopic level, solids are distinguishable from liquids by their resistance to shear. Is this distinction likely to be a sharp one?

10. What kind of crystal structure might have little resistance to shear forces in some directions and much resistance in other directions? Would such a crystal have direction-dependent resistance to stretching?

11. To survive a cold night in the woods, you are advised to make a thick bed of dry leaves to sleep on and a thick cover of dry leaves to sleep under. Why does this work?

12. When you place your hand on a smooth block of wood, it feels warm; when you place your hand on a smooth block of marble, it feels cool. Why is this statement consistent with the statement that the wood and marble are both at the same temperature?

13. Water is one of the relatively rare substances that becomes less dense rather than more dense when it freezes. Describe what might happen to the ocean if ice were denser than liquid water.

14. Styrofoam, which has a very low density, is a good thermal insulator. Are all (or most) good thermal insulators low-density materials? Why or why not?

15. Why are cross-country skiers told to wear several layers of clothing rather than a single, thick coat?

16. You glue together two strips of metals that have different coefficients of thermal expansion, back-to-back. What will happen to such a composite strip when it is heated or cooled? Can you propose some technological use for such strips?

PROBLEMS

21–1 The Nature of Solids

1. (I) The crystal structure of copper is face-centered cubic; i.e., the structure is built of cubes of edge length a, with copper atoms at the corners as well as the face centers of the cube (Fig. 21–19). The lattice length $a = 0.361$ nm. (a) Calculate the diameter of a copper atom. (Assume that atoms are described by spheres that are centered on the corners and centers, and that they just touch). (b) Show that the structure can also be described as a stacking of triangular layers of atoms, perpendicular to the maximum diagonal of the cube.

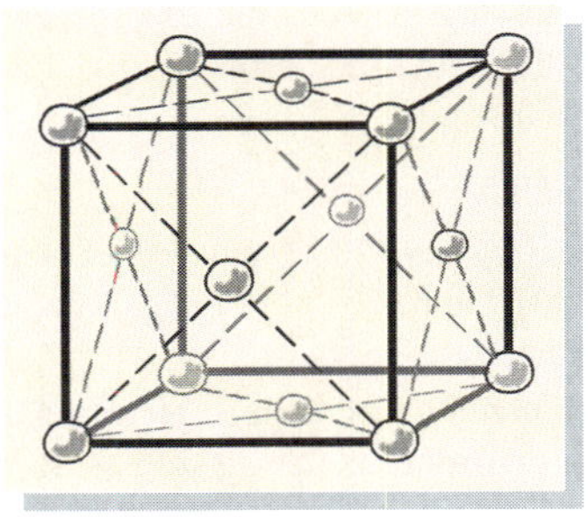

FIGURE 21–19 Problem 1.

2. (I) The building block of a simple cubic lattice is a cube with atoms at the corners. If the atoms are represented by spheres of diameter d that are centered on the corners and just touch, what is the diameter of the largest impurity that will not displace any of the existing atoms that can be put at the center of the cube? [Figure 21–20 shows a section through two opposite face diagonals. The figure has a hint: The longer side of the rectangle is $\sqrt{2}$ times the shorter side.]

FIGURE 21–20 Problem 2.

3. (II) The volume filling of a crystal structure can be defined as the ratio of the volume of the atoms as represented by touching spheres to that of the crystal unit. Calculate this ratio for simple cubic and face-centered cubic lattices. [Refer to Problems 1 and 2.]

21–2 Stresses and Strains

4. (I) A concrete pier of area 0.25 m^2 and height 3 m is built to hold up a bridge (Fig. 21–21). The load on the pier is 300 tons (2.73×10^5 kg). $Y = 17{,}000$ MN/m^2 for the concrete. How far will the concrete compress?
5. (I) A mine elevator is supported by a steel cable 2.0 cm in diameter. The mass of the elevator and its contents is 800 kg. By how much is the cable stretched when the elevator is 250 m below Earth's surface?
6. (I) A steel bolt must withstand forces up to 1000 N. If a safety factor of 3 is to be taken into account, what must the minimum diameter of the bolt be? (With a safety factor of 3, the bolt should be able to support 3×1000 N before it snaps.)

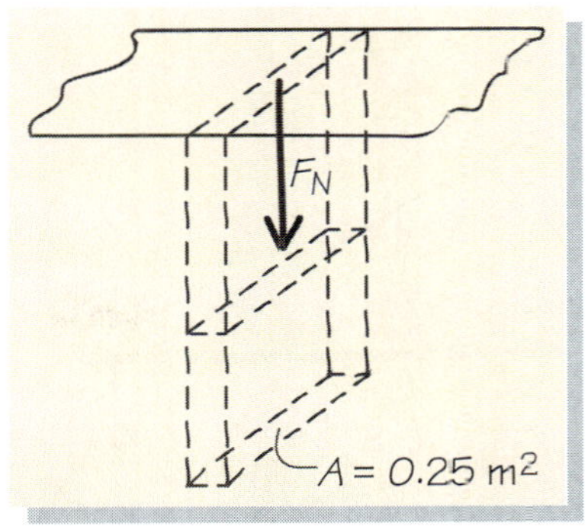

FIGURE 21–21 Problem 4.

7. (I) Calculate the critical strain for steel piano wire. You will need to consult the tables. How far would a wire 1 m long stretch before breaking?
8. (I) Assume that the piano wire of Problem 7 is 1 mm in diameter. Calculate the weight it could hold before fracturing.
9. (II) A crate of mass 30 kg slides across the ground. The coefficient of kinetic friction between the crate and the ground is 0.3. The physical contact area between the crate and the ground (which is less than the area of the whole crate) is 0.35 m^2. Calculate the shear stress on the crate.
10. (II) A steel wire of diameter 5.0 mm and length 2.0 m stretches 0.30 mm when a load of 60.0 kg is hung from it. What is its Young's modulus? How much mass can the wire hold before it may fracture? Use the tensile strength of steel from the tables.
11. (II) The 3-km-long cables on a large suspension bridge are stretched from their equilibrium length by 3 m. Estimate the change in the equilibrium separation between any two adjacent atoms along the cable.
12. (II) A steel bar of length 5 m is placed in a structure where it is subject to extreme stress under tension. The area of the bar is 40 cm^2, and it has room to stretch by 4 mm. It cannot stretch any more than this without butting up against a much stronger part of the structure. Does the bar break before it stretches the 4 mm? Use the Young's modulus and tensile strength from the tables.
13. (II) A uniform beam of length 2.0 m and mass 10 kg is freely pivoted at one end about a point fixed to the wall (Fig. 21–22). It is held in a horizontal position by a steel cable of diameter 2.0 mm. The cable makes an angle of 30° with the horizontal. A load of 30 kg is suspended from the end of the beam. What will be the angle of the beam relative to the horizontal? [*Hint*: The distortion is small, so you may calculate the tension in the cable without taking into consideration the slight change in angle.]
14. (II) A 0.5-m-long piece of metal is compressed longitudinally by 0.1 mm. Its Poisson's ratio is 0.32. Calculate its volume strain.
15. (II) Poisson's ratio for steel is 0.3. What is the new diameter of the steel rod used in Example 21–1 after the rod has been stretched?

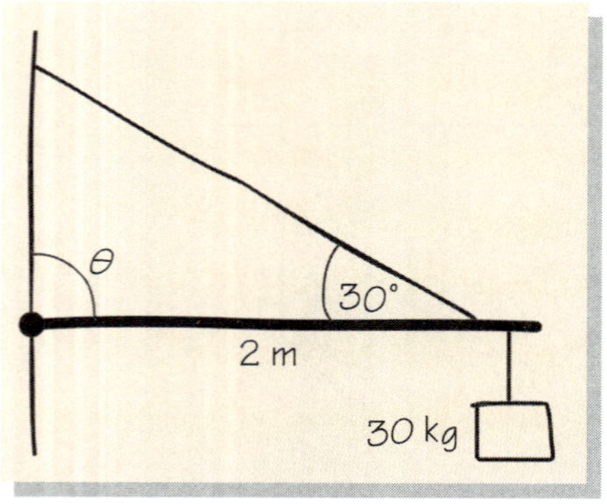

FIGURE 21–22 Problem 13.

16. (III) According to our discussion of Poisson's ratio (Section 21–2), a bar of length L and cross-sectional area A under compression at its ends expands at the sides. Find the amount of pressure that must be exerted on the sides to keep them fixed when there is a compression F on the two ends. Find the total amount of compression in the longitudinal direction under these conditions, and show that a bar with constrained sides is stiffer under compression than the same bar with unconstrained sides. (Keep the terms up to order σ^2 only.)

17. (III) Let the strain on a rod along the direction of applied stress be e_1. Because the stretching produces a slight reduction in the diameter of the rod, we may speak of an induced transverse strain, given by $e_{tr} = -\sigma e_1$, where σ is Poisson's ratio (Fig. 21–23). Recall that Y is defined in terms of longitudinal stress and strain. A cube of the rod's material immersed in a liquid, so that the pressure on all sides is p, suffers a volume change given by $\Delta V/V = 3p(1 - 2\sigma)Y$. (a) Derive this result. (b) Show that when Poisson's ratio is not zero, the relation between Young's modulus and the bulk modulus under uniform pressure is given by

$$B = \frac{1}{3}\frac{Y}{1 - 2\sigma}.$$

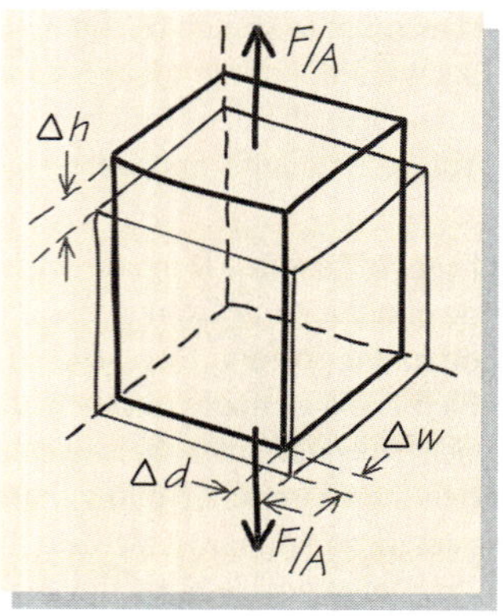

FIGURE 21–23 Problem 17.

18. (II) By using the results of Problem 17, show that Poisson's ratio, σ, must be less than 0.5 because, if it exceeds 0.5, the total volume of the solid would increase when a uniform pressure is applied to it.

19. (III) *Compressibility* is defined by $-\Delta V/pV$, the reciprocal of the bulk modulus. Using the equation for $\Delta V/V$ in Problem 17, calculate the compressibility of silver, given that $\sigma = 0.38$ and $Y = 7.9 \times 10^4$ MN/m^2.

20. (III) Show by analogy with a simple spring, for which the force is $F = -kx$, that a volumetric potential energy associated with strain—that is, potential energy per unit volume of a wire or rod of length L and cross-sectional area A—is given by $u = \frac{1}{2}$ (stress)(strain).

21–3 Waves in Solids

21. (I) The speed of sound in a material of density 2.7×10^3 kg/m^3 is measured to be 5.1×10^3 m/s. What is Young's modulus for that material?

22. (I) The shear modulus of copper is $G = 4.2 \times 10^{10}$ N/m^2. Given that the density of copper is 8.93×10^3 kg/m^3, what is the speed of a shear wave in copper?

23. (I) The density of pure gold is 19.3 g/cm^3, and the speed of sound in a small sample is measured to be 2.04×10^5 cm/s. What is Young's modulus for gold in dynes/cm^2?

24. (I) Given that the ratio of densities for steel and lead is 7.86: 11.3, and that the ratio of Y values is 200,000 MN/m^2: 16,000 MN/m^2, what is the ratio of the speeds of sound in these two materials?

25. (I) A scuba diver has to send a message to save an oil platform in the Gulf of Mexico. He realizes that sound travels faster in steel than in air or water. With only one second left for the message to be received on the platform, he finds a steel mooring cable that is 5 km long, which runs directly to the platform. Will the message reach the platform in time? (The density of steel is 7.9×10^3 kg/m^3 and $Y = 210{,}000$ MN/m^2 for steel.)

26. (II) Young's modulus for Earth can be approximated by 5.5×10^5 MN/m^2, and the corresponding shear modulus by 2.0×10^5 MN/m^2. A first signal of an earthquake arrives at a detection station at $t = 0$ min, and a second signal arrives 2 min later. Assume that the successive signals are those of longitudinal waves and shear waves, respectively, that follow the same straight-line path through the planet. Assume further that the average density of Earth is 5.5 g/cm^3. How far from the station did the earthquake take place?

27. (II) Earthquake waves within Earth are of two types: S (transverse) waves, which can be supported within Earth's mantle but not in the liquid part of the core, and P (longitudinal) waves, which propagate in both the mantle and the core (see Fig. 21–13). The speeds of S waves and P waves in the mantle are approximately 4 km/s and 8 km/s, respectively. The speeds of these waves can be calculated from the Young's and shear moduli. Take the density of Earth's mantle to be 5 g/cm^3. Calculate the Young's and shear moduli. If the difference between the time of arrival of the two waves is 27 min, how far away was the earthquake? Ignore Earth's thin crust, and assume that the waves travel in a straight line.

28. (II) A steel rod of length 0.8 m is rigidly supported at one end. Calculate the frequencies of the lowest three longitudinal modes,and the lowest three transverse (shear) modes, given that $Y = 2.1 \times 10^{11}$ N/m^2 and $G = 8.1 \times 10^{10}$ N/m^2.

21–4 Thermal Expansion

29. (I) How much farther will a 1-m-long aluminum rod expand than a steel one when each undergoes a temperature increase of 40°C?

30. (I) We learned that the steel rod in Example 21–1 extended by only 0.25 mm under a tensile load of about 10^5 N. By how much will the same steel rod elongate if it is heated 100°C?

31. (I) The area of the circle formed by a circular loop of metal is found to increase by 1.6 percent when the temperature changes by 10°C. What is the coefficient of thermal expansion of the metal?

32. (I) The coefficient of thermal expansion for Invar is $0.7 \times 10^{-6}\ \mathrm{K}^{-1}$. (Invar is a steel alloy.) Given that an Invar pendulum clock keeps perfect time at a room temperature of 20°C, how much time will the clock gain or lose per day when it is in a room at 30°C? (For the time-keeping properties of a pendulum, see Chapter 13).

33. (I) In the design of a mercury thermometer, it is desirable for each °C change to correspond to a 1 cm expansion of the metal. What is the cross section of the mercury column if the mercury at the lowest reading on the thermometer is contained in a bulb whose volume is 0.7 cm^3? Ignore any possible expansion of the glass tube.

34. (II) A metal bar is 2.00 m long with a coefficient of thermal expansion of $1.54 \times 10^{-5}\ \mathrm{K}^{-1}$. It is rigidly held between two fixed beams. When the temperature rises, the metal bar takes on the shape of the arc of a circle (Fig. 21–24). What is the radius of curvature of the circle when the temperature rises by 50°C? [*Hint*: You can use the small-angle approximation $\sin\theta \simeq \theta - (\theta^3/6)$.]

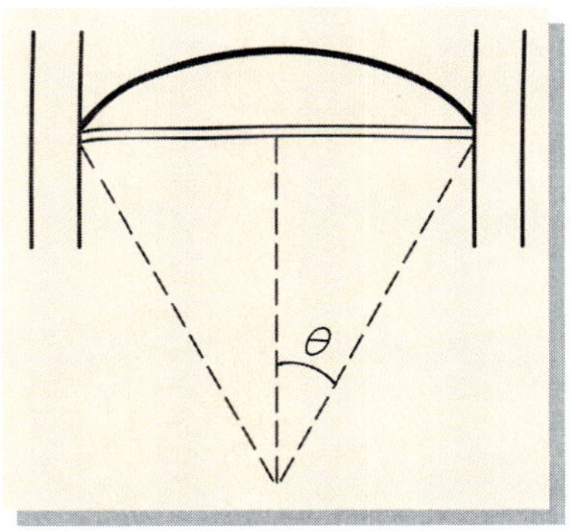

FIGURE 21–24 Problem 34.

35. (II) As we know from Chapter 13, the time expressed by a grandfather clock depends on the length of the pendulum. (a) Which is the best material for the pendulum of a clock to be used outdoors: aluminum, copper, or steel? (See Table 21–3.) (b) If the temperature of the clock ranges from −20°C to +30°C over the course of a year, and if the clock ticks with frequency $f = 1.00000$ Hz at 5°C, what is the yearly frequency range for each of the three materials?

36. (II) In space, where heat transfer by air is missing, large temperature differences can develop. A space telescope (Fig. 21–25) consists of a main mirror and an instrument unit, held together by titanium rods. The optical axis of the mirror (perpendicular to the mirror, through its center) hits the middle of the receiving window of the instrument unit. Suppose that the temperature of two neighboring rods increases by 20°C due to direct sunlight, with the temperature of the other two remaining unchanged. How far will the optical axis of the mirror move on the instrument unit, given that the rods are 5 m long and that $\alpha = 8.5 \times 10^{-6}\ \mathrm{K}^{-1}$?

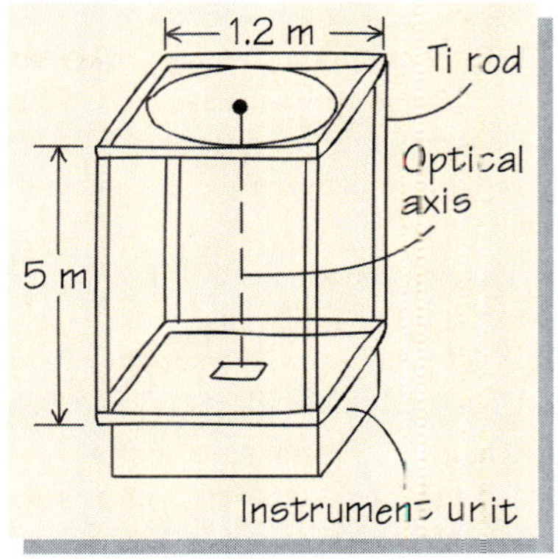

FIGURE 21–25 Problem 36.

37. (II) The coefficient of volume expansion, β, for ethyl alcohol is $1.12 \times 10^{-3}\ \mathrm{K}^{-1}$. Suppose that a cylindrical copper container, with a coefficient of thermal expansion of $1.67 \times 10^{-5}\ \mathrm{K}^{-1}$, is filled to the very brim with ethyl alcohol at 5°C. What percentage of the alcohol will spill if the temperature is raised to 25°C?

38. (III) A strip of brass 2.0 mm thick and 10 cm long is glued back-to-back to a strip of Invar with the same dimensions. When the temperature is increased by ΔT, the composite strip tends to curl into an arc of a circle (Fig. 21–26). Find the radius of curvature of this circle for $\Delta T = 5$°C, given that the coefficient of thermal expansion is $19 \times 10^{-6}\ \mathrm{K}^{-1}$ for brass and $0.7 \times 10^{-6}\ \mathrm{K}^{-1}$ for Invar.

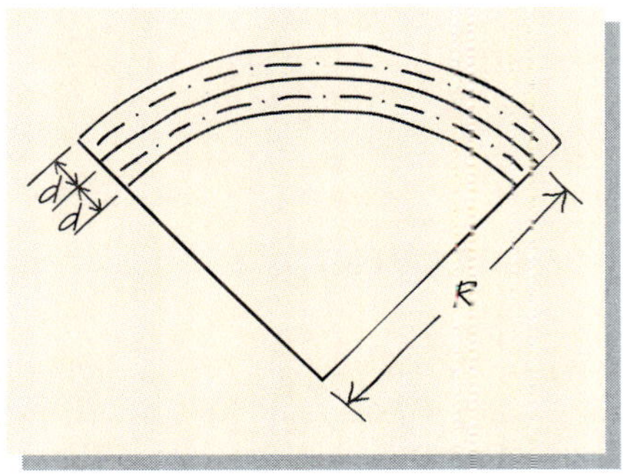

FIGURE 21–26 Problem 38.

21–5 The Conduction of Thermal Energy

39. (I) Compare the rate of thermal energy transfer across a piece of asbestos that is 1 m^2 in area and 1 cm thick with that of a piece of glass of the same dimensions for a temperature difference of 60°C.

40. (I) A single window pane of thickness 6.0 mm has an area of 1.0 m^2. The inside and outside temperatures are 73°F and 13°F, respectively. Calculate the rate of thermal energy loss through the window. (See Table 21–4.)

41. (I) A brick wall 5 m by 5 m in size and 0.1 m thick separates the inside of a house from the outside, where the temperature is 285 K (12°C). A heater supplies energy inside at a rate of 0.35 kW. What is the inside temperature? Ignore all surfaces except the brick wall in question.

42. (I) The glass from a basement window is replaced by a sheet of asbestos that is 2 cm thick. The window is 80 cm wide and 40 cm high, and there is no air leakage around the edges. Given that the heat loss through the surface is 0.06 J/s, what is the temperature difference between the outside and the inside of the basement?

43. (I) A house of dimensions 36 ft × 21 ft × 9 ft is built of bricks that are 9 in thick. The inside and outside temperatures are 76°F and 25°F, respectively. Calculate the rate of thermal energy loss through the brick wall. (See Table 21–4.)

44. (I) Your refrigerator can be thought of as a box with six sides of total area 2.5 m^2. The effective R value of the walls is 1.5 $m^2 \cdot K/W$. The temperature inside is 5°C, while the temperature outside is 25°C. Calculate the rate of heat loss.

45. (II) Pine wood 1 in thick has an R value of 1.3, compared with 3.0 for fiberglass insulation of the same thickness. (The units of R are $ft^2 \cdot h \cdot °F/Btu$.) Calculate the rate of thermal energy loss of a 6-in thickness of these same materials across a 4 ft × 8 ft section of wall for a temperature difference across the wall of 35°F. If 10 percent of the wall section is made of wood studs and the space between studs is filled with fiberglass, calculate the effective R value of the wall section (Fig. 21–27).

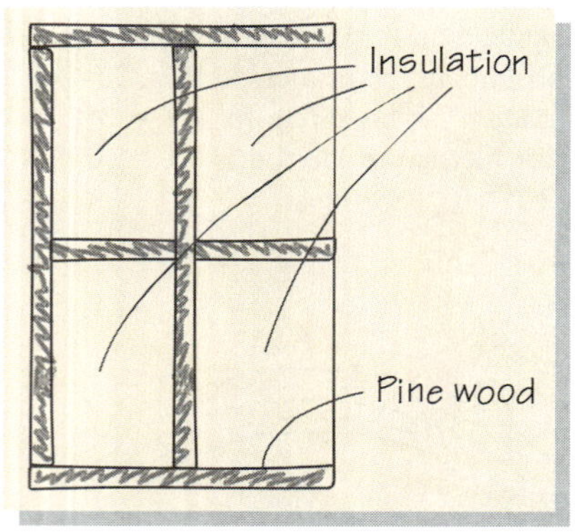

FIGURE 21–27 Problem 45.

46. (II) The R value of brick is about 1.1 $ft^2 \cdot h \cdot °F/Btu$ across the brick's width. A dog who expends 50 W of energy lives in a brick doghouse of total effective area 2.0 m^2. How much temperature variation between the inside and outside of the doghouse can the dog withstand if only its body heat warms the house? Assume that the door locks out drafts and that the heat conduction through the door is identical to that through the walls.

47. (II) A wall of area A consists of two slabs: one of thickness L_1 and thermal conductivity κ_1, the other of thickness L_2 and thermal conductivity κ_2, as shown in Fig. 21–28. The temperatures are T_1 and T_2, respectively, on the left-facing side and the right-facing side of the wall. (a) Calculate the effective thermal conductivity of the wall in terms of the parameters given. (b) Calculate the temperature at the interface of the two slabs.

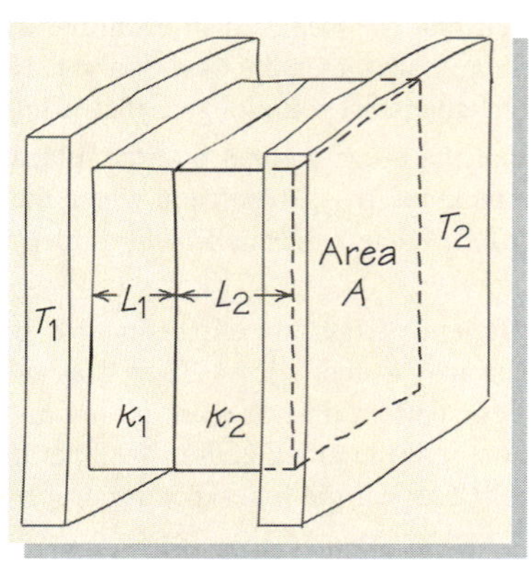

FIGURE 21–28 Problem 47.

48. (II) Consider a wall of thickness L and area A. Part of the area, A_1, consists of a material with thermal conductivity κ_1 and the remaining area, $A - A_1$, consists of a material with thermal conductivity κ_2 (Fig. 21–29). Calculate how much thermal energy flows through each part of the wall, and calculate an effective thermal conductivity for the wall as a whole.

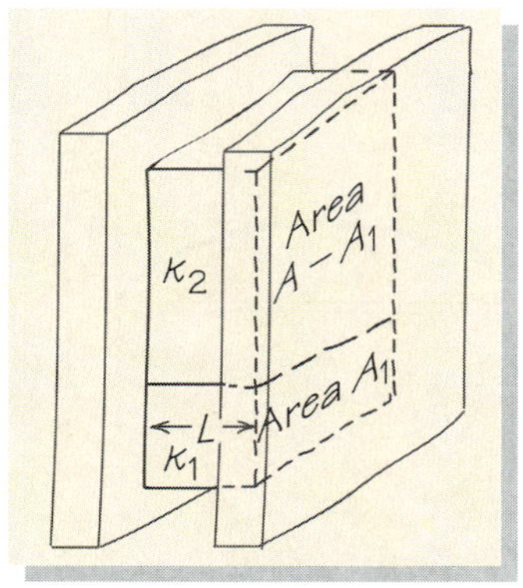

FIGURE 21–29 Problem 48.

49. (II) Equation (21–19) states that when two slabs of conducting material are combined in series, as pictured in Fig. 21–17, their R values are additive. Show that this is so by using the principle that the rate of flow of thermal energy is the same for each slab and for the two slabs combined:

$$\frac{dQ}{dt} = \frac{1}{R_1} A(T_1 - T_{\text{int}}) = \frac{1}{R_2} A(T_{\text{int}} - T_2)$$
$$= \frac{1}{R_{\text{eff}}} A(T_1 - T_2).$$

50. (II) Two pieces of material are arranged in series as in Fig. 21–17. If the temperature to the left is T_1 and the temperature to the right is T_2, find the temperature between the pieces, T_{int}, as a function of T_1, T_2, and the R values of the two pieces.

51. (II) Equation (21–20) gives the effective R value for two slabs of conducting material combined in parallel, as pictured in Fig. 21–18. Show that this equation is valid by using the principle that the temperature difference on opposite sides of both slabs is the same, and that the rate of flow of thermal energy across the combined slabs is the sum of the rate of flow of thermal energy across each slab.

52. (III) Repeat Example 21–4 but do not assume that the igloo's thickness is small compared to its radii. You may assume that the surfaces of constant temperature form concentric hemispheres. [*Hint*: Break up the hemisphere of ice into thin hemispherical shells and find the R value of all these hemispheres combined in series. Note that the area of the hemispherical shells is not constant, so Eq. (21–16) must be modified appropriately.]

53. (III) A long, cylindrical hot-water pipe of radius R_1 is at temperature T_1. It is wrapped with insulation of thickness a and of thermal conductivity κ; the outer surface of the insulation is at temperature T_2. Derive an expression for the rate of heat loss per unit length of insulated pipe. [*Hint*: The temperature within the insulation varies with radial position r. Break the insulation into thin cylindrical shells and find the areas of these shells as a function of their radii.]

General Problems

54. (II) When a 3000-ton submarine dives into deep waters, the compressive volume strain on the submarine is 2.5 percent. What is the loss of buoyancy for the submarine? In order to maintain the same buoyancy that it has just below the surface, the submarine must lose some mass, presumably by dumping water from its ballast tanks. How much water must be dumped?

55. (II) A trapper lives in a cave. At the cave's entrance, the trapper has constructed an exterior wall from lumber 6 cm thick. The wood has a thermal conductivity, κ, of 0.35 W/m · K. Assume that all thermal energy loss occurs through the wall, which forms a rectangle of dimensions 2.5 m × 2.0 m. This wall, together with a stove, is capable of maintaining the cave's interior at a temperature of 8°C when the outside temperature is −25°C. What is the power output of the stove?

56. (II) The trapper of Problem 55 tires of looking at the blank entrance wall and decides to cut a window in it. The window will have a sheet of mica 2.0 mm thick with a thermal conductivity of 0.7 W/m · K. How large a window can he construct if he keeps the same stove and plans to keep the temperature inside the cave at 6°C?

57. (II) Cold air blows through a crack in the wall of Problem 55, so the trapper covers the crack with a piece of wood 5 mm thick. The thermal conductivity of the wood is 25×10^{-2} W/m · K. If the area of the crack is 6 cm^2 and the temperature difference between outside and inside is 30°C, what is the heat loss through the covered crack?

58. (II) A brass ring has to be fit onto a solid steel axle. To do this, the ring is made somewhat smaller than the axle and it is heated to 100°C until it fits onto the axle (Fig. 21–30). Calculate the stress in the ring when it cools back to room temperature (25°C). Neglect the deformation of the axle. For brass, $\alpha = 19 \times 10^{-6}\ \text{K}^{-1}$, $Y = 106{,}000\ \text{MN/m}^2$.

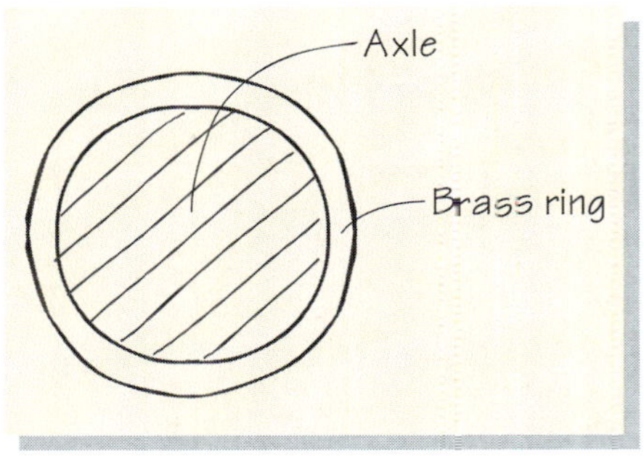

FIGURE 21–30 Problem 58.

59. (II) The coefficient of thermal expansion of the iron rods inside a bar of reinforced concrete is $12.5 \times 10^{-6}\ \text{K}^{-1}$, whereas the coefficient of thermal expansion of concrete is 12.0×10^{-5} K^{-1}. Calculate the additional stress on a 10-m-long bar of reinforced concrete for a temperature increase from 10°C to 30°C; assume that the new length of the bar is that of the iron rods within. The fact that the coefficients of thermal expansion are practically the same for the right variety of iron bar and concrete allows the fabrication of reinforced concrete that remains useful under large temperature extremes.

CHAPTER

22

Investigations of the nature of lightning were important in the understanding of electrical phenomena. Here, Benjamin Franklin is performing one of his famous kite experiments, the results of which he published between 1751 and 1753.

Electric Charge

Electricity and magnetism is a subject with ramifications throughout the physical world. Electromagnetic forces control the structure of atoms and of all materials, and light and other electromagnetic waves are pervasive. The understanding of these forces is one of the great success stories of science. In this chapter, we shall introduce electric charge, a new property carried by the constituents of atoms, and the fundamental law of the interaction of two charges at rest, Coulomb's law. This force law is as fundamental as the universal law of gravitation. In fact, it has the same space dependence as gravitation. However, the force described by Coulomb's law can be either attractive or repulsive.

22–1 PROPERTIES OF CHARGED MATTER

A Brief History of the Study of Electricity and Magnetism

Most people have at least some acquaintance with electric charges, the forces between them, and the fact that magnetism has something to do with electricity. As simple as these things may seem, the experimental evidence and the understanding of that evidence developed only over a long time period.

The word *electricity* has its roots in the Greek word for "amber" (*electrum*), and the first written mention of the curious effects of rubbed amber dates from the fifth century B.C. Far earlier, ancient peoples surely observed the crackling and sparking of rubbed fur. It was not until the eighteenth century that the critical discovery that electric forces can be either repulsive or attractive was made. Over time, the idea developed that a quantity (which we now call electric charge) is associated with electric forces. Among the many important names associated with these discoveries are Stephen Gray, Charles Dufay, and Benjamin Franklin.

Benjamin Franklin is best known for his exploitation of the existing idea that electrical phenomena were associated with a kind of fluid contained in matter. Repulsion and attraction were the result of an excess or deficiency of the fluid. Implicit in this model is what we now recognize as the phenomenon of the conservation of charge: If the fluid were to flow out of an object, it would leave behind a deficiency. Franklin introduced the terms "positive" and "negative" for the two types of charge, and also set the standard sign convention in which the electron, the actual particle that moves in conductors, has negative charge. Franklin was known for his spectacular (and dangerous) experiments with lightning, which he recognized as an electrical effect. Franklin and his friend Joseph Priestley, as well as Henry Cavendish, are linked with the discovery that the fundamental force between electric charges is proportional to the inverse square of the distance between them. This law was confirmed more directly by John Robison and then by Charles Coulomb in the mid- and late-eighteenth centuries, respectively. This inverse-square law is now known as Coulomb's law.

By the first part of the nineteenth century, magnetism, which was initially thought to be unrelated to electricity, became the object of intensive experimentation. The nature of magnetism and its relation to electricity began to be clarified around 1820, primarily through the work of Hans Christian Oersted, André-Marie Ampère, and Michael Faraday. James Clerk Maxwell carried the unification to completion in the 1860s. Electricity and magnetism were aspects of the single fundamental subject electromagnetism.

The real nature of charged matter was revealed only with the experimental exploration of atoms. Quantum mechanics is an additional element needed to explain the properties of atoms. All the electrical properties of matter can now be understood within the framework of quantum theory.

Matter and Electric Charge

In most of our discussions to this point, we have characterized bulk matter—and the atoms that make up matter—by a single attribute: mass. When we probe the structure of atoms more deeply, we find that atoms are made up of electrons and nuclei. Electrons and nuclei can be characterized by another attribute, **electric charge** (usually labeled q). Electric charges exert (electric) forces on one another proportional to the product of their charges, just as masses exert gravitational forces on one another proportional to the product of their masses. Electric forces hold the atom together. However, a new element enters into electric forces that does not occur in gravitation: Charges come with two signs, and, depending on the signs of the interacting charges, the forces between charges can be attractive *or* repulsive. We speak of *net* charge as the algebraic sum of charges. The set of phenomena associated with the forces between stationary charges form the subject of **electrostatics**, or **static electricity**.

Atoms are **electrically neutral** (or just "neutral"); that is, an atom has no net electric charge. We know this because the electric forces between atoms are small. However, an atom's electrons (symbolized by e) each carry the same unit of negative charge, $q_{\text{electron}} = -e$. The electrons orbit in shell-like regions around the much heavier nucleus, which consists of electrically neutral neutrons (abbreviated n) and positively charged protons (abbreviated p). The proton charge is equal in magnitude but opposite in sign to that of the electron, $q_{\text{proton}} = +e$. Because the number of protons in an atom equals the number of electrons, the atom is electrically neutral. Although the nucleus possesses 99.95 percent of the mass of an atom, the nuclear radius is only about 10^{-5} of the atomic radius.

Chemical elements differ in the number of protons—equal to the number of electrons—in the nuclei of their atoms. Chemical properties have to do with the behavior of the electrons. The electrons that are on average closer to the nucleus are more difficult to dislodge from the atom because of the strength of their attraction to the positively charged nucleus. The outermost electrons—farther away from the nucleus—are attracted less strongly and are more easily dislodged. The ease with which this happens greatly influences the properties of the element that contains those electrons. An atom that has lost one or more electrons, and is thus left with a positive charge, is called a *positive ion*. A *negative ion* is an atom that has gained electrons.

The evidence that led to the discovery of electric charge and electric forces depended on the electrical properties of bulk matter and only indirectly on the fact that matter is made of atoms. For that reason, we want to look briefly at the electrical properties of bulk matter. If the outer electrons of atoms in bulk matter are especially easy to dislodge (are *weakly bound* to their nucleus), they behave as though they are almost free, and can move through the material nearly unimpeded. Such materials are good **conductors**. Metals are good conductors; some metals, such as copper, silver, and aluminum, are better conductors than others. A certain class of materials—**superconductors**—have electrons that, in effect, move with *no* inhibition when the material is made sufficiently cold. The electrons of most nonmetallic solids do not travel easily; such solids, including rubber, glass, and plastics, are **insulators**. Silicon, germanium, and an increasing number of synthetic combinations of materials are substances that we can make into insulators or conductors by controlling the electric forces on them or the temperature. Such substances are called **semiconductors**, and they play an important role in technology.

Conductors, superconductors, insulators, and semiconductors

The ease with which charges move through matter is closely related to our ability to transfer charges back and forth between different materials. When we transfer charges between different materials, we say that we have either *charged* or *discharged* them. Rubbing a material in which the outer electrons are loosely bound, such as certain plastics, may carry those electrons elsewhere, to be ultimately deposited on some object. The original material then has an excess positive charge: It has lost some electrons. The object to which the electrons have been carried will have an excess of electrons and is now negatively charged. When charge is carried from one object to another in this way, the objects are said to be **charged by conduction**. Note that both the original material and the object have acquired a charge.

Charging by conduction

We can gain another measure of control over the charge on an object by connecting that object to Earth with a good conductor. When a negatively charged object is connected to the ground in this way, electrons flow from the object to the ground and leave the object neutral. If, instead, an object has an excess positive charge, then electrons flow from the ground and neutralize the object. Why does the charge flow? Earth itself is a good conductor; in effect, the conducting line from the object to Earth allows the charge on the object to be shared with Earth, but the planet is so large that the remaining additional charge on the object is undetectable. Such an object is said to be **grounded** (Fig. 22–1). By walking across a carpet on a dry winter day, we may slowly accumulate a charge. When we become grounded by touching a conductor connected to Earth, such as a radiator, we suddenly discharge our electric charge to Earth. The resulting spark can be startling.

FIGURE 22–1 The large copper rod that is being pounded into the ground will serve as the electrical ground for the household electrical service box.

Evidence that Charges Are of Two Types

Some important properties of electric charge can be demonstrated by performing experiments with materials that are readily available in an elementary physics laboratory (Fig. 22–2a). We can transfer electric charge by rubbing a Teflon rod on a piece of fur, or by rubbing a glass rod on silk. The Teflon acquires a charge, and the fur acquires an equal but opposite charge (Fig. 22–2b); similarly, the glass acquires a charge and

Like charges repel, unlike charges attract.

the silk, an equal but opposite charge (Fig. 22–2c). (Actually, the Teflon now has an excess of electrons and the fur has a deficiency of electrons, whereas the glass has a deficiency of electrons and the silk has an excess. Thus the Teflon rod becomes negative, and the glass rod becomes positive.)

To study the effects of forces between charges, we want to transfer our charges to small (initially neutral) masses because small masses react to forces most visibly. We can do this by using small cork balls coated with a conducting paint, which allows charge to move around easily on the surface. A cork ball is hung by a thin insulating thread (Fig. 22–3a). If we touch a negatively charged Teflon rod to a cork ball, the ball is immediately repelled by the rod (Fig. 22–3b). If we touch the negatively charged Teflon rod to *two* suspended (neutral) cork balls, the balls strongly repel each other (Fig. 22–3c). Similar behavior occurs between two cork balls that have been touched by a positively charged glass rod. However, if we touch the Teflon rod to one cork ball and the glass rod to one cork ball, the balls attract each other (Fig. 22–3d).

We conclude from these experiments that the electric charges on the Teflon and glass rods are different, and that

Like charges repel, and unlike charges attract.

These conclusions are the simplest explanation for what has been observed. For example, while the Teflon rod touches the cork ball, some of the rod's negative charge is transferred to the ball. Now both the ball and the rod have a negative charge. The ball, which has been charged by conduction, immediately jumps away from the rod. Our other observations are similarly explained by the rule that like charges repel and unlike charges attract (Fig. 22–4).

Another, more subtle, effect is also present. Before the negatively charged Teflon rod actually touches the neutral cork ball, the ball is *attracted* to the rod, not repelled by it. How can we explain this initial attraction? Because we have coated the cork ball with conducting paint, there are mobile electrons on the surface of the ball. When the negatively charged Teflon rod comes near, the mobile electrons are repelled and move to the far side of the cork ball (Fig. 22–5). That leaves an equal amount of excess positive charge on the area of the ball near the rod. Those positive charges are attracted

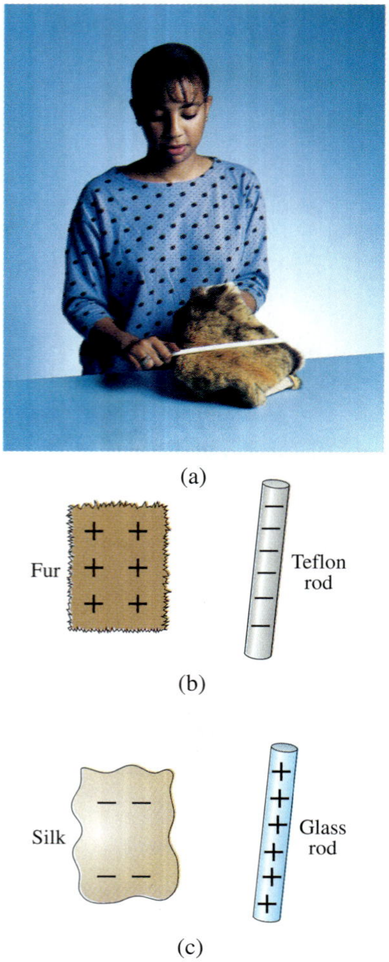

FIGURE 22–2 When (a,b) Teflon is rubbed against fur and (c) glass is rubbed against silk, electric charge is transferred.

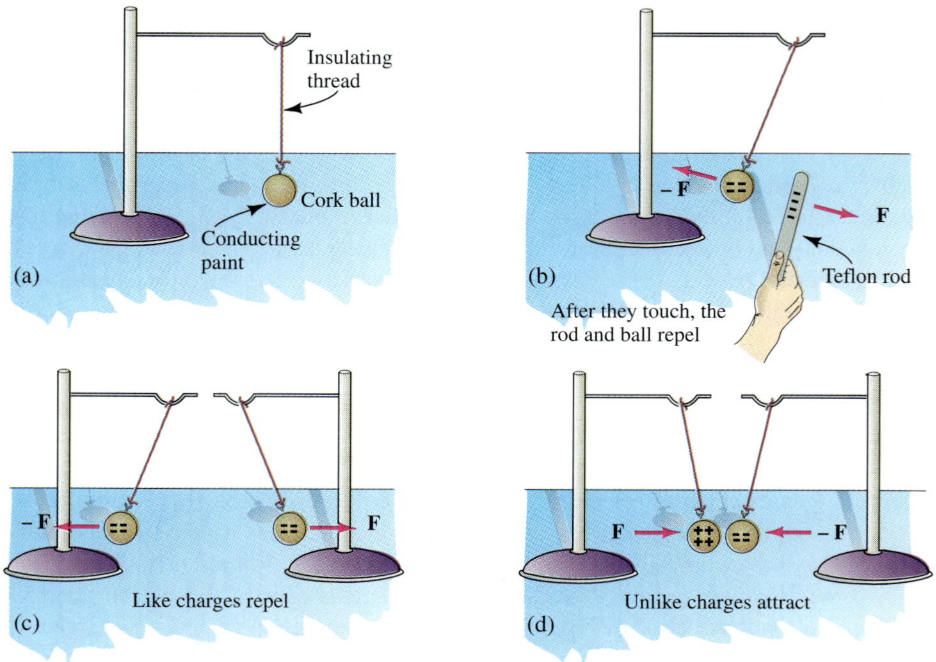

FIGURE 22–3 (a) An insulated cork ball covered with a thin layer of conducting paint can indicate the presence of small electric charges. (b) A negatively charged Teflon rod first approaches the neutral coated cork ball, which is initially attracted to the rod. After the rod touches the ball, the ball becomes charged and strongly repels the charged rod. (c) If we touch two initially neutral cork balls with a negatively charged Teflon rod, the two balls repel each other: Like charges repel. (d) If we touch one initially neutral cork ball with a negatively charged Teflon rod and a similar ball with a positively charged glass rod, the two balls attract each other: Unlike charges attract.

FIGURE 22–4 An eighteenth-century experiment on static electricity by Stephen Gray. The boy, suspended in air, was charged electrostatically; bits of paper were then attracted to him.

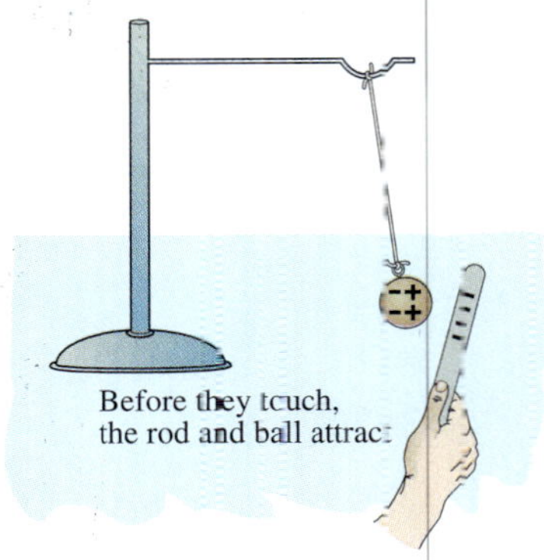

FIGURE 22–5 The neutral cork ball is initially attracted to the charged Teflon rod because some electrons on the ball move to the far side due to the repulsive force from the rod. The positive charges on the ball are on average closer to the rod, so the attractive force on them due to the rod is greater than the repulsive force on the shifted electrons.

to the rod more strongly than the negative charges on the other side of the ball are repelled. In other words, when the positive charges on the cork ball are closer than the ball's negative charges to the Teflon rod, the *net* force is attractive. The initial attraction can therefore be understood if the electric force weakens when the distance between the charges increases. We call this phenomenon, in which charges within an object are redistributed due to the presence of external charge, **charge polarization**. The fact that electrical forces weaken with distance between the interacting charges is of great importance, and we shall return to it.

Charge by Induction

Charging by induction

Another experiment explains how initially neutral conductors can obtain a *charge by induction*, or an *induced charge*. Consider two neutral metal spheres that each stand on an insulated post and are in side-by-side contact (Fig. 22–6a). If we bring a negatively charged Teflon rod very close to one sphere, mobile electrons in the sphere move to the opposite side of the far sphere, leaving opposite charges on the two spheres (Fig. 22–6b). The spheres have a total charge of zero, but one is positive and the other negative. While the Teflon rod is still near, we separate the two spheres, leaving them oppositely charged (Fig. 22–6c). Even when we remove the Teflon rod, the charges induced by the rod remain on the two metal spheres (Fig. 22–6d). We say that the spheres have been **charged by induction**. These charges can then be transferred to cork balls coated with conducting paint. The cork balls attract demonstrating that the charges are opposite in sign. Note that only conductors can be charged by induction.

Units of Charge

The coulomb is the SI unit of charge.

Just how much charge an electron carries depends on how the scale for charge is defined. The SI unit of charge is called the **coulomb** (C). We can determine the value of the coulomb by specifying the magnitude of the force between two objects separated by a distance of 1 m, with each object carrying 1 C of charge.

The electron charge

The magnitude of the charge on the electron—the smallest charge found in nature—has been measured to high precision. An approximation sufficient for our purposes is

$$e \simeq 1.60 \times 10^{-19}\ \text{C}. \qquad (22\text{–}1)$$

The mass and charge of the neutron, proton, and electron are given in Table 22–1.

EXAMPLE 22–1 A glass rod rubbed with silk has a charge of $+110$ nC (110×10^{-9} C). By how many electrons is the rod deficient?

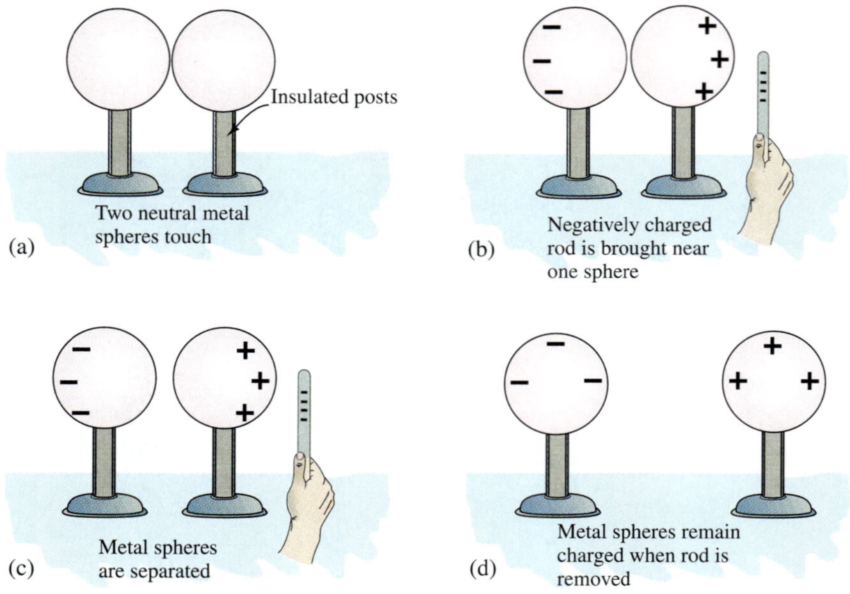

FIGURE 22–6 (a) Two neutral metal spheres on insulated posts touch. (b) A negatively charged Teflon rod polarizes the metal spheres. (c) If the metal spheres are separated while the Teflon rod is nearby, the spheres are charged oppositely. (d) When the Teflon rod is removed, the two metal spheres are still charged oppositely. Note that the total charge of the two spheres remains zero throughout.

Solution: Electrons were transferred from the glass rod when it was rubbed with silk, leaving an excess positive charge on the rod. Each electron has charge of magnitude e, so the number of transferred electrons must be

$$\text{transferred electrons} = \frac{\text{net charge}}{\text{charge on each electron}}$$

$$= \frac{110 \times 10^{-9}\ \text{C}}{1.6 \times 10^{-19}\ \text{C/electron}} = 6.9 \times 10^{11}\ \text{electrons.}$$

EXAMPLE 22–2 The largest American Eagle gold coin has a mass of 28.4 g. The atomic number of gold—the number of protons in the nucleus of an atom of gold—is 79, and thus the number of electrons in a neutral gold atom is also 79. The atomic mass of gold is 197, which means that 1 mol of gold has a mass $m_{\text{Au}} = 197$ g. How many electrons are contained in one pure-gold coin? What is the total negative charge contained in the coin?

Solution: The number of gold atoms in a mass of 28.4 g is

$$\frac{mN}{m_{\text{Au}}} = \frac{(28.4\ \text{g})(6.02 \times 10^{23}\ \text{atoms/mol})}{197\ \text{g/mol}} = 8.68 \times 10^{22}\ \text{atoms,}$$

where $N = 6.02 \times 10^{23}$ atoms/mol is Avogadro's number, the number of atoms in 1 mol of any substance. Each neutral Au atom (Au is the chemical symbol for gold) contains 79 electrons, so the total number of electrons is

$$\text{number of electrons} = (79\ \text{electrons/atom})(8.68 \times 10^{22}\ \text{atoms})$$

$$= 6.85 \times 10^{24}\ \text{electrons.}$$

TABLE 22–1 Mass and Charge of Atomic Constituents

	Mass (kg)	*Charge (C)*
Neutron, n	1.675×10^{-27}	0
Proton, p	1.673×10^{-27}	1.602×10^{-19}
Electron, e^-	9.11×10^{-31}	-1.602×10^{-19}

The total charge of these electrons is

$$\text{total electron charge} = (\text{number of electrons})(\text{charge per electron})$$
$$= (6.85 \times 10^{24}\ \text{electrons})(-1.6 \times 10^{-19}\ \text{C/atoms})$$
$$= -1.1 \times 10^{6}\ \text{C}.$$

The gold is neutral, so a positive net charge of the same magnitude is present due to the protons. Notice that the number of electrons transferred by rubbing the glass rod in Example 22–1 with silk is 10^{13} times smaller than the number of electrons contained in the gold coin; the rod of Example 22–1 would have a comparable number of electrons.

The Electroscope The *electroscope* is a device used to detect excess free charge (Fig. 22–7a). It can be used in two ways. First, when free charge is transferred to the electroscope, the charge is distributed through the external ball at the top of the electroscope as well as the metal rod and the gold leaf inside the case. The leaf and the rod then each contain a net charge of the same sign, and they repel (Fig. 22–7b). The leaf moves away from the rod until the vertical component of the electrostatic repulsion is balanced by the gravity on the leaf. Addition of more charge moves the leaf still more, and the angle made by the leaf is a measure of the amount of free charge involved. Second, if a charged object is brought to a given distance from the ball at the top of the electroscope, a charge is induced on the ball, leaving the rod and leaf inside the case with a charge of the same magnitude; again they repel (Fig. 22–7c). (For an example of how this approach can be made more quantitative, see Problem 8.)

22–2 CHARGE IS CONSERVED AND QUANTIZED

Charge conservation

The simple experiments described in Section 22–1 strongly suggest that *charge is conserved.* Further experiments show that the **conservation of charge** is a fundamental physical law: *Net* charge is the same before and after any interaction.

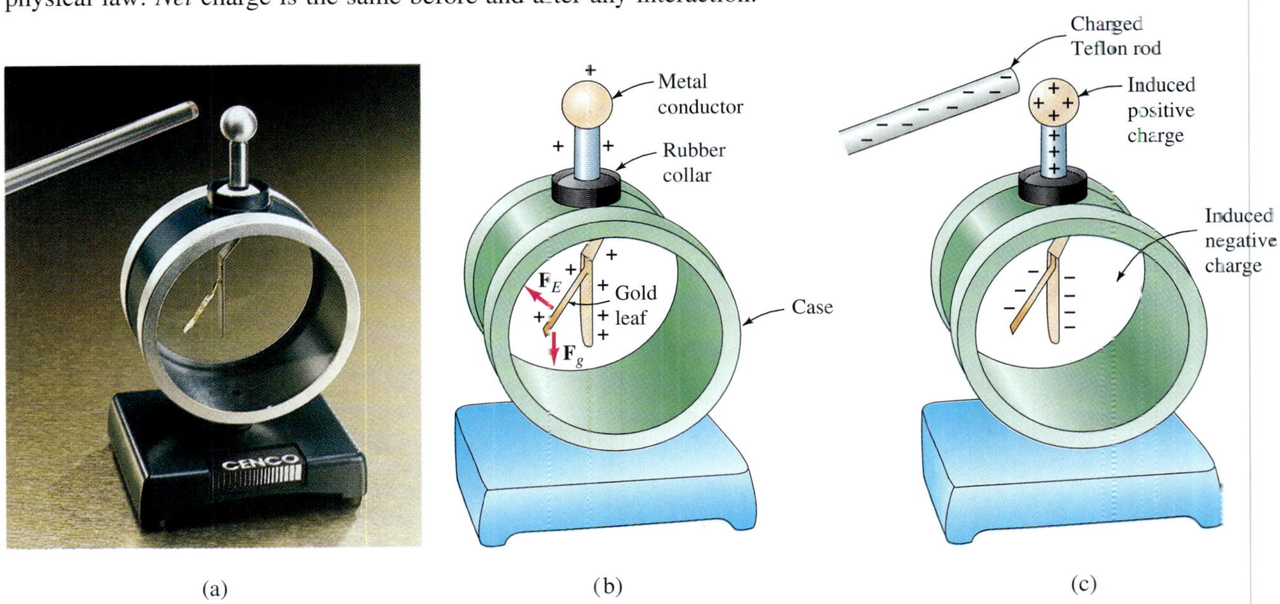

FIGURE 22–7 (a) An electroscope, a device that detects the presence of charge. (b) Schematic diagram of an electroscope. When free charge is added to the metal conductor, the gold leaf and the metal rod repel, causing the gold leaf to move away from the rod. (c) It is not necessary to add free charge to the electroscope. When a charged object is brought close to the metal ball at the top, a charge is induced on the ball, leaving a charge of the opposite sign on the rod and leaf. They again repel.

Evidence of Charge Conservation

The idea that charge conservation was subject to experimental test and verification did not really take root until the twentieth century. The classical view implicit in the picture of charge as a kind of fluid is that, just as the quantity of a fluid such as water is conserved, charge conservation is automatic. Thus, little was done with the specific aim of checking charge conservation.

Once microscopic reactions could be observed more directly, it became easier to check charge conservation directly. Let's look at some of these reactions. One of the reactions between atomic nuclei that takes place in a nuclear reactor is

$$n\,{}^{235}_{92}\mathrm{U} \rightarrow {}^{143}_{56}\mathrm{Ba} + {}^{90}_{16}\mathrm{Kr} + 3n + \text{energy}.$$

Here, the total number of protons (92) is the same on both "sides" of the reaction.†

Even when the number of electrons or protons changes during a reaction, the total charge remains unchanged. Thus another reaction that can take place in a nucleus is *electron capture*,

$$e^- + p \rightarrow n + \nu,$$

where ν stands for a neutral particle called the *neutrino*. (The neutrino, unlike the neutron, has no mass, as far as we can tell.) In this reaction, the numbers of both protons and electrons change, but charge is still conserved.

Another type of charged particle is called a *pion*, denoted by π, with a superscript to indicate the sign of the charge it carries. In the reaction

$$\gamma + p \rightarrow n + \pi^+,$$

a *photon*, γ, which is a small (neutral) packet of electromagnetic radiation, impinges on a proton and produces a neutron and a pion. To the high accuracy of the experiments that investigate this reaction, the charge of the pion is *exactly* the same as that of the proton. Other particles, called *positrons*, are practically identical to electrons, except for the *sign* of the charge, and are denoted by e^+. In the reaction

$$\gamma + p \rightarrow p + e^+ + e^-,$$

an electron is produced, but then only in partnership with a positron, whose charge has exactly the same magnitude (Fig. 22–8). In fact, in observed reactions involving the so-called elementary particles, *no one has ever seen a single case of net charge appearing or disappearing.*

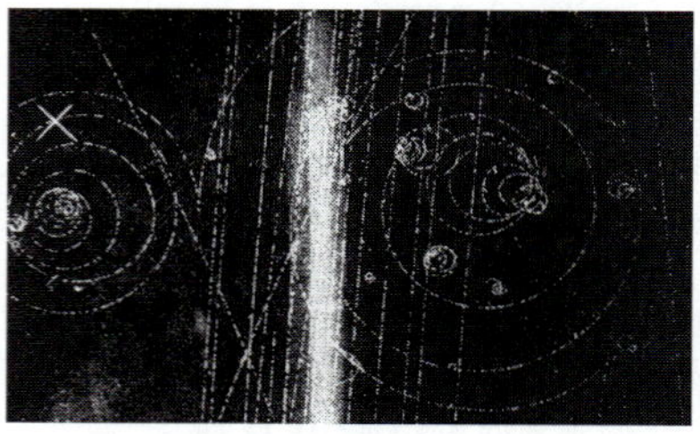

FIGURE 22–8 Pair production of an electron–positron pair. The event took place in a magnetic field, and the electron and positron spiral in opposite directions in this field as they lose energy.

Is it possible for a little of the charge on an electron or a proton to wear off, like paint? Again, all the evidence points to the fact that the electron and the proton charges are always the same, no matter where or when they are measured. In looking at quasars (distant and powerful sources of light), we are looking at matter that existed billions of years ago (it took that long for the light to reach Earth). Observations of the color of the light quasars emit suggest that, to a very high accuracy, the properties of their atoms are identical to the properties of atoms here on Earth. This implies that the charge of electrons has remained constant over billions of years.

Charge Quantization

Charge quantization

We have already indicated that charges appear to be organized in discrete bundles. The magnitude of the charge of such a bundle is that of one electron. Greater charges are always multiples of these values. The facts that, within experimental accuracy, charge occurs in integral multiples of the electron charge, known as **charge quantization**, and that charges are never observed with values smaller than the electron charge were first established in 1909 by the pioneering experiments of Robert Millikan.

†The superscript on the element symbol is the atomic mass, the sum of the numbers of protons and neutrons in one atom; the subscript is the number of protons.

APPLICATION: SMOKE DETECTORS

The smoke detectors common in many homes depend on the separation of charges. They contain a small radioactive source centered within a cylinder that is open at one end (Fig. 22AB–1). The source steadily emits so-called alpha (α) particles, which have enough energy to knock electrons from the molecules of the air when they collide with those molecules. The outer cylinder contains a net charge, and the positive ions that are produced when the air molecules are broken up are attracted to it. The rate at which the ions arrive at the outer wall is sensed electronically. When there is a fire in the house, large organic molecules—bacon fat or pinewood resin, for example—enter the cylinder. These large molecules present a better target for the α particles and, because they are more loosely bound they are also more likely to be ionized in the collision. The increased number of ions arriving at the cylinder sets off the alarm. The only disadvantage to this system is that any large molecule—for example, from paint solvents or ordinary cooking smoke—will set off the alarm. The latest detectors have disable buttons that temporarily turn off the detector and allow the turkey to roast in peace.

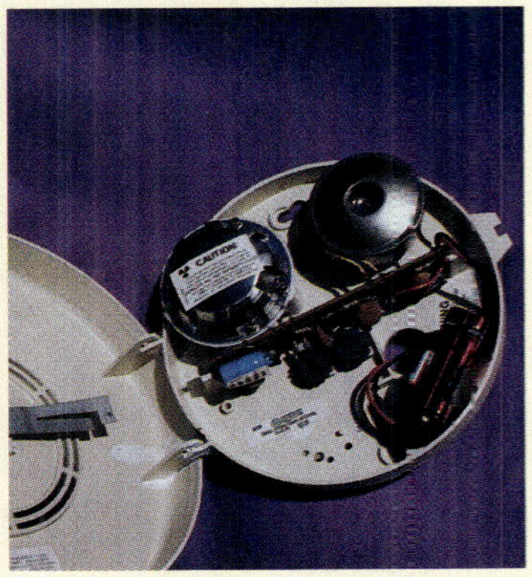

FIGURE 22AB–1 A typical household smoke detector.

In 1964, Murray Gell-Mann and George Zweig proposed that protons and neutrons are composed of even more fundamental particles, called **quarks**, whose charges are either $2e/3$ or $-e/3$. There is strong experimental evidence that quarks really do make up particles such as protons, but that quarks cannot be isolated. Despite many searches, such fractional electron charges have never been observed on freely moving particles. Most physicists now believe that only combinations of quarks possessing a net charge that is an integer multiple of e can ever be isolated and independently observed. We refer to any charge that can be isolated as **free charge**.

In summary, we can say that

Charge is conserved absolutely

and that

Free charge is quantized in positive or negative integral multiples of *e*.

22–3 COULOMB'S LAW

Encouraged by Benjamin Franklin, Joseph Priestley concluded in the mid-eighteenth century from Franklin's and his own experiments that the electric force between two charged objects varies as the inverse square of the distance between the objects. Priestley made this deduction after he observed that there is no charge on the inside surface of a closed or nearly closed metal vessel—all the charge is on the outside surface—and that the force on a charged object placed inside such a vessel is zero. This is like the phenomenon we discussed in Chapter 12: There is no gravitational force on an object inside a uniform spherical shell of matter. In gravitation, this result is a direct consequence of the $1/r^2$ nature of the force law. By analogy, Priestley argued that the electric force responsible for his observations must have a $1/r^2$ dependence.

In 1785, Charles Coulomb directly determined the force law for electrostatics. He performed the relevant experiments with a torsion balance similar to the one Henry Cavendish used in 1798 to measure the gravitational constant, G (Fig. 22–9). In Coulomb's work, small charged balls replaced the massive ones of the Cavendish ap-

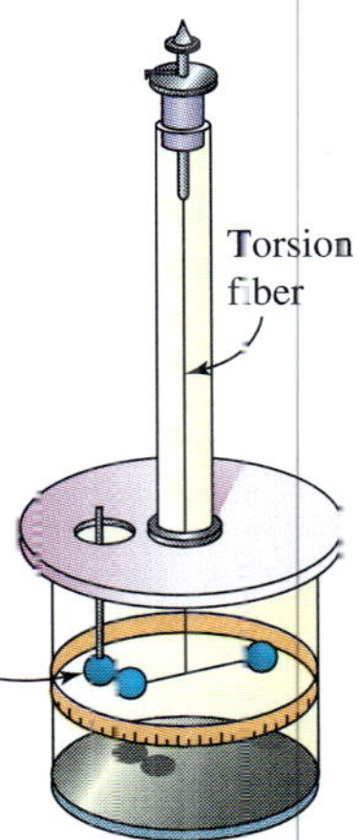

FIGURE 22–9 Coulomb's torsion balance, used to verify the inverse-square form of the force between electric charges.

The Cavendish apparatus is described in Chapter 12.

paratus. Coulomb showed that the electrostatic force is central—directed on the line between the charges—and varies as

$$F \propto \frac{1}{r^2}, \tag{22–2}$$

where r is the distance between the centers of the charge sources. By changing the charge on the balls, possibly as we discussed in Section 22–1, Coulomb inferred that the force is proportional to the product of the charges q_1 and q_2 on the balls:

$$F \propto q_1 q_2. \tag{22–3}$$

To demonstrate the results of Eq. (22–3), we can ground one cork ball, neutralizing it, and charge another identical ball, giving it net (unknown) charge q. After we touch the two balls together, they each have a charge of $q/2$. Then we measure the force between these two balls. Next, we ground one ball again to neutralize it, and touch the balls together once more. Each then has a charge of $q/4$, and we measure the force between them to have decreased by a factor of 4 for the same amount of separation. This set of results is consistent with Eq. (22–3): In the first case, $F \propto (q/2)(q/2) = q^2/4$; in the second, $F \propto (q/4)(q/4) = q^2/16$.

Combining Eqs. (22–2) and (22–3) gives us a first view of **Coulomb's law**, the electrostatic force law. The magnitude of the force is

$$F = \frac{k|q_1 q_2|}{r^2}, \tag{22–4}$$

where k is a proportionality constant. The force is attractive when the charges have opposite sign and repulsive when they have the same sign.

The constant k plays the same role that the constant G plays in Newton's law of universal gravitation. The magnitude of k depends on the units used for charge. It is then possible to define the coulomb by assigning a value to k:

$$k = \frac{1}{4\pi\epsilon_0}, \tag{22–5}$$

where ϵ_0 is known as the *permittivity of free space*. (We shall see later that the value of ϵ_0 follows directly from the defined value of the speed of light, so in this sense ϵ_0 is itself defined.) To four significant figures, the permittivity is

$$\epsilon_0 \simeq 8.854 \times 10^{-12}\ \mathrm{C^2/N\cdot m^2}. \tag{22–6}$$

The value of k (to four significant figures) follows from Eqs. (22–5) and (22–6):

$$k = 8.988 \times 10^{9}\ \mathrm{N\cdot m^2/C^2}. \tag{22–7}$$

In this book, we will usually round k off to $9 \times 10^9\ \mathrm{N\cdot m^2/C^2}$. Now that we have assigned a value to k, we can tentatively define the coulomb. From Eqs. (22–4) and (22–7), we say that *when the force between two particular charges separated by 1 m is equal to the numerical value of k in newtons* (8.988×10^9 N), *these charges are each 1 C.*

The definition of the coulomb

Note that Coulomb's law expresses the force between charged *pointlike* objects. We mentioned in Chapter 12 (and shall prove in Chapter 24) that because the gravitational force has a $1/r^2$ dependence for pointlike objects, the force between spherically symmetric distributions of mass is the same as the force between pointlike objects of the same mass placed at the centers of the spheres. The same behavior holds for electric charges because the $1/r^2$ spatial dependences of the gravitational and electrical forces are the same. This is why Coulomb was able to measure a $1/r^2$ force even though the objects he used were not point charges. All that is required is that the charge on the balls used in the experiment be distributed in a spherically symmetric way—for example, uniformly over their surfaces. Further, the balls should not be conducting to avoid the possibility that charge polarization would redistribute the charges on the balls.

The electric force, often called the **Coulomb force**, is associated with a direction and is therefore a vector. We write Coulomb's law as

$$\mathbf{F}_{12} = \frac{1}{4\pi\epsilon_0}\left(\frac{q_1 q_2}{r_{12}^2}\right)\hat{\mathbf{r}}_{12}. \tag{22–8}$$

Coulomb's law

where $\mathbf{F}_{12}$ is the force exerted on point charge q_1 by point charge q_2 when they are separated by a distance r_{12}. The unit vector $\hat{\mathbf{r}}_{12}$ is directed from q_2 to q_1 along the line between the two charges (Fig. 22–10). Note that if q_1 and q_2 have opposite signs, Eq. (22–8) indicates that the force is attractive, along $-\hat{\mathbf{r}}_{12}$. But rather than remembering the subscripts on $\mathbf{F}$ and the unit vector $\hat{\mathbf{r}}$, it is easier to remember that like charges repel and unlike charges attract.

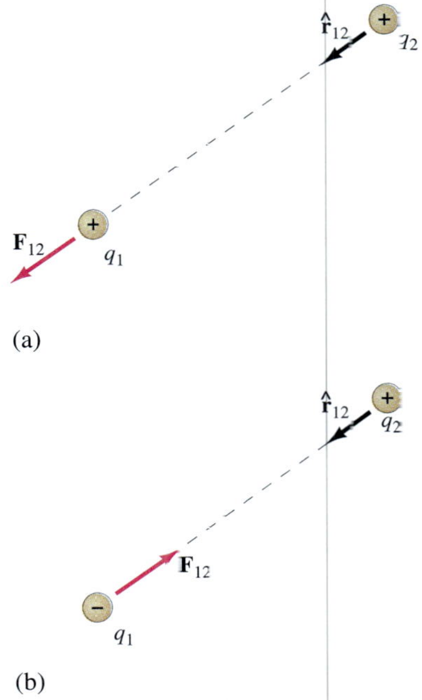

FIGURE 22–10 $\mathbf{F}_{12}$ is the force on q_1 due to q_2. The force is in the direction of the unit vector $\hat{\mathbf{r}}_{12}$ for (a) like charges and $-\hat{\mathbf{r}}_{12}$ for (b) opposite charges.

EXAMPLE 22–3 Compare the electric force and the gravitational force for the single proton and single electron in a hydrogen atom. Assume a purely classical model of the hydrogen atom, in which the electron moves in a circular orbit around the proton, which is at the atom's center. The radius of a hydrogen atom is about 5×10^{-11} m.

Solution: First we calculate the gravitational force, obtaining the masses (m_e for the electron, m_p for the proton) from Table 22–1. Both the gravitational and electric forces are attractive in this case, so we must calculate only the magnitudes. Using Eq. (12–4), we have

$$F_g = \frac{G m_e m_p}{r^2}.$$

When we insert the values into this equation, we find that

$$F_g = \frac{(6.67 \times 10^{-11}\ \mathrm{N\cdot m^2/kg^2})(9.11 \times 10^{-31}\ \mathrm{kg})(1.67 \times 10^{-27}\ \mathrm{kg})}{(5 \times 10^{-11}\ \mathrm{m})^2}$$

$$= 4 \times 10^{-47}\ \mathrm{N}.$$

The electric force, from Eq. (22–8), is

$$F_E = \frac{(9 \times 10^{9}\ \mathrm{N\cdot m^2/C^2})(1.6 \times 10^{-19}\ \mathrm{C})(1.6 \times 10^{-19}\ \mathrm{C})}{(5 \times 10^{-11}\ \mathrm{m})^2} = 9 \times 10^{-8}\ \mathrm{N}.$$

$$\frac{F_E}{F_g} = \frac{9 \times 10^{-8}\ \mathrm{N}}{4 \times 10^{-47}\ \mathrm{N}} \simeq 2 \times 10^{39}.$$

This result is independent of r. It shows that, on the atomic scale, the electric force is much greater than the gravitational force and justifies ignoring gravitation at that level.

EXAMPLE 22–4 Two small cork balls are both charged to 40 nC and placed 4 cm apart. What is the magnitude of the electric force between them? Each cork ball has a mass of 0.4 g. Compare the electric force and the cork ball's weight.

Solution: The electric force is

$$F_E = \frac{k q_1 q_2}{r^2} = \frac{(9 \times 10^{9}\ \mathrm{N\cdot m^2/C^2})(40 \times 10^{-9}\ \mathrm{C})(40 \times 10^{-9}\ \mathrm{C})}{(4 \times 10^{-2}\ \mathrm{m})^2} = 0.01\ \mathrm{N}.$$

The weight of each cork ball is

$$W = mg = (0.4 \times 10^{-3}\ \mathrm{kg})(9.8\ \mathrm{m/s^2}) = 0.004\ \mathrm{N}.$$

Therefore, the (repulsive) electric force is strong enough to lift the upper cork ball if one is placed 4 cm above the other (Fig. 22–11). (A charge of 40 nC is somewhat larger than that which could realistically be placed on a ball with a diameter of 1 cm.)

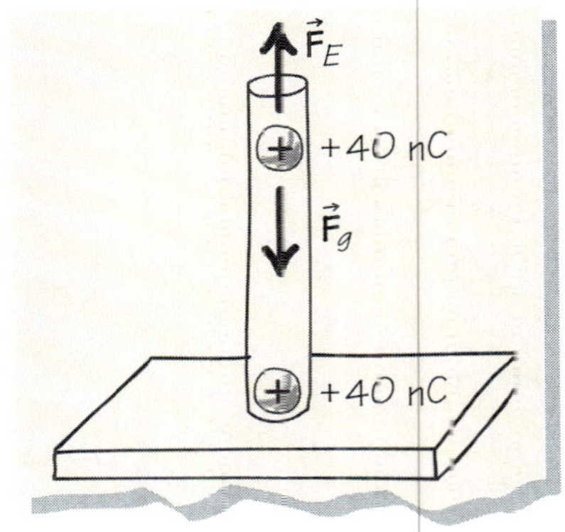

FIGURE 22–11 Example 22–4. A small object suspended in space because equal but opposite gravitational and electric forces act on it.

22–4 FORCES INVOLVING MULTIPLE CHARGES

The superposition principle applies to Coulomb's law.

What happens if multiple charges are present? Experiment shows that the **principle of superposition** applies: The force on any one charge due to a collection of other charges is the vector sum of the forces due to each individual charge. In this respect, the Coulomb force is again like the gravitational force.

The superposition principle allows us to find the force due to a set of charges on another charge (or, for that matter, on another set of charges). It does not matter if the set of charges is discrete or continuous; we can always use integration techniques to break the continuous charge distribution into infinitesimally small pieces and treat it as discrete.

As an example of how superposition is applied, consider four charges, numbered 1, 2, 3, 4 (Fig. 22–12). The total force on charge q_2 is the *vector sum* of the forces due to the other individual charges, q_1, q_3, and q_4:

$$\mathbf{F}_{2,\,\text{total}} = \mathbf{F}_{21} + \mathbf{F}_{23} + \mathbf{F}_{24}. \tag{22–9}$$

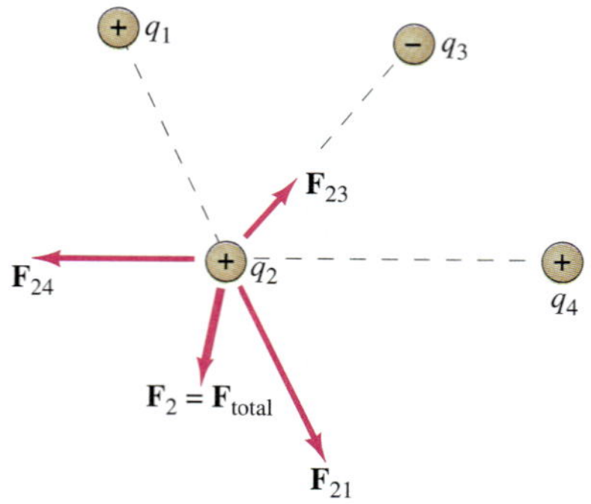

FIGURE 22–12 The superposition principle applies for multiple charges. The total force on charge q_2 is the vectorial sum of the individual forces on q_2 due to charges q_1, q_3, and q_4.

If there are N charges—$q_1, q_2, \ldots, q_N$—all acting on a charge q, the total force $\mathbf{F}$ on charge q is the vector sum of the individual forces F_i on charge q due to charge q_i:

$$\mathbf{F} = \sum_{i=1}^{N} \mathbf{F}_i = \frac{q}{4\pi\epsilon_0} \sum_{i=1}^{N} \frac{q_i}{r_i^2}\hat{\mathbf{r}}_i. \tag{22–10}$$

The vector $\hat{\mathbf{r}}_i$ is the unit vector from charge q_i to charge q. We have moved the common factor $q/4\pi\epsilon_0$ out of the sum.

EXAMPLE 22–5 Consider three point charges $q_1 = q_2 = 2.0$ nC and $q_3 = -3.0$ nC, which are placed as in Fig. 22–13. Find the forces on q_1 and q_3.

Solution: The force on q_1 is due to the presence of charges q_2 and q_3. We want to find the vector forces on q_1 due to each of the charges q_2 and q_3 separately, then add them vectorially to find the net force on q_1. Similar remarks hold for the calculation of the force on q_3.

The force on q_1 is

$$\mathbf{F}_1 = \mathbf{F}_{12} + \mathbf{F}_{13} = \frac{q_1}{4\pi\epsilon_0}\left[\left(\frac{q_2}{r_{12}^2}\right)\hat{\mathbf{r}}_{12} + \left(\frac{q_3}{r_{13}^2}\right)\hat{\mathbf{r}}_{13}\right].$$

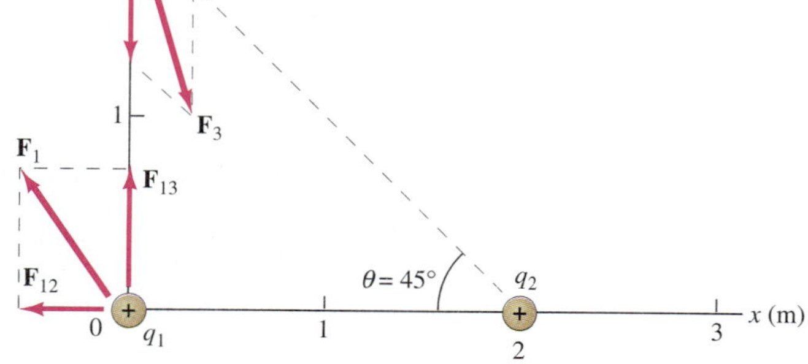

FIGURE 22–13 Example 22–5. The positions of three point charges are indicated. Charges q_1 and q_2 are positive, while q_3 is negative. Forces $\mathbf{F}_{12}$ and $\mathbf{F}_{13}$ on charge q_1, and their resultant, $\mathbf{F}_1$, as well as forces $\mathbf{F}_{31}$ and $\mathbf{F}_{32}$ on charge q_3, and their resultant, $\mathbf{F}_3$, are drawn.

PROBLEM-SOLVING TECHNIQUES

We often need to calculate electric forces on a given charge when several other fixed charges or continuous distributions of charges are present. In these cases, keep in mind the following techniques:

1. Draw a clear diagram of the situation. Be sure to distinguish between the fixed external charges and the charges on which the forces must be found. The diagram should contain coordinate axes for reference.
2. Do not forget that the electric force that acts on a charge is a vector quantity; when many charges are present, the net force is a vector sum. It is usually simplest to use unit vectors in a Cartesian coordinate system.
3. Search for symmetries in the distribution of charges that give rise to the electric force. When symmetries are present, the net force along certain directions will be zero. For example, if a point charge is midway between two identical charges, we know without performing any calculations that the net force on it will be zero.

From Fig. 22–13, we can deduce that $\hat{\mathbf{r}}_{12} = -\mathbf{i}$ and $\hat{\mathbf{r}}_{13} = -\mathbf{j}$. Thus

$$\mathbf{F}_1 = (9.0 \times 10^9\ \mathrm{N \cdot m^2/C^2})(2.0 \times 10^{-9}\ \mathrm{C}) \times \left[\frac{(2.0 \times 10^{-9}\ \mathrm{C})}{(2.0\ \mathrm{m})^2}(-\mathbf{i}) + \frac{(-3 \times 10^{-9}\ \mathrm{C})}{(2.0\ \mathrm{m})^2}(-\mathbf{j})\right]$$
$$= (-9.0 \times 10^{-9}\ \mathrm{N})\mathbf{i} + (13.5 \times 10^{-9}\ \mathrm{N})\mathbf{j}.$$

The direction of force $\mathbf{F}_1$ is shown in Fig. 22–13.

The force on q_3 is calculated in much the same way, with the caution that the unit vector $\hat{\mathbf{r}}_{32}$, which points from q_2 to q_3, is given by $-\cos\theta\,\mathbf{i} + \sin\theta\,\mathbf{j}$:

$$\mathbf{F}_3 = \mathbf{F}_{31} + \mathbf{F}_{32} = \frac{q_3}{4\pi\epsilon_0}\left[\left(\frac{q_1}{r_{31}^2}\right)\hat{\mathbf{r}}_{31} + \left(\frac{q_2}{r_{32}^2}\right)\hat{\mathbf{r}}_{32}\right]$$
$$= (9.0 \times 10^9\ \mathrm{N \cdot m^2/C^2})(-3.0 \times 10^{-9}\ \mathrm{C})$$
$$\left[\frac{(2.0 \times 10^{-9}\ \mathrm{C})}{(2.0\ \mathrm{m})^2}\mathbf{j} + \frac{(2.0 \times 10^{-9}\ \mathrm{C})}{\sqrt{(2.0^2\ \mathrm{m}^2 + 2.0^2\ \mathrm{m}^2)^2}}(-\cos\theta\,\mathbf{i} + \sin\theta\,\mathbf{j})\right].$$

The angle θ is 45°, or $\pi/4$ rad, so $\mathbf{F}_3$ becomes

$$\mathbf{F}_3 = (-13.5 \times 10^{-9}\ \mathrm{N})\mathbf{j} + (4.8 \times 10^{-9}\ \mathrm{N})\mathbf{i} - (4.8 \times 10^{-9}\ \mathrm{N})\mathbf{j}$$
$$= (4.8 \times 10^{-9}\ \mathrm{N})\mathbf{i} - (18.3 \times 10^{-9}\ \mathrm{N})\mathbf{j}.$$

Continuous Distributions of Charges

The fact that charge is quantized will have no physical consequence when we deal with charges that are much larger than e. Such charges are actually composed of large numbers of electrons or of protons. We can normally treat a large collection of point charges as a *continuous distribution* of charge. To analyze this situation, we can follow the approach suggested in Chapter 12, where we discussed continuous distributions of mass. Let's consider first the interaction of a point charge q with a large continuous charge distribution (Fig. 22–14). The force on q due to the tiny element of volume shown, which contains charge Δq and is a distance r from q, is

$$\Delta\mathbf{F} = \frac{q}{4\pi\epsilon_0}\frac{\Delta q}{r^2}\hat{\mathbf{r}}.$$

In turn, the net force on q is the sum over the forces due to the elements Δq:

$$\mathbf{F} = \sum \Delta\mathbf{F} = \sum \frac{q}{4\pi\epsilon_0}\frac{\Delta q}{r^2}\hat{\mathbf{r}} = \frac{q}{4\pi\epsilon_0}\sum \frac{\Delta q}{r^2}\hat{\mathbf{r}}. \qquad (22\text{–}11)$$

A formal expression for the force on a point charge due to a continuous distribution of charge

This sum is a vector sum, so care must be exercised when it is used.

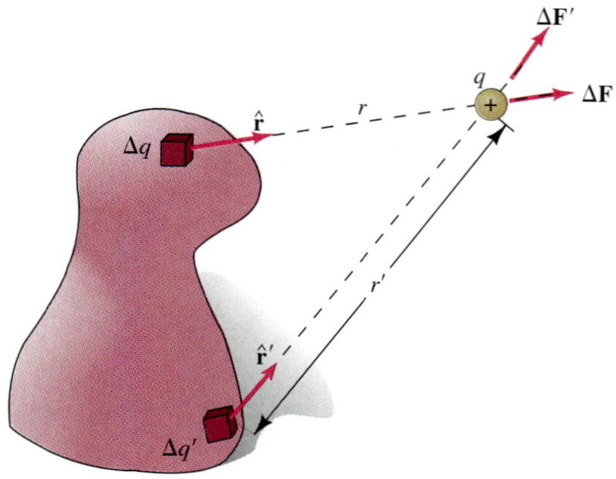

FIGURE 22–14 To find the total force on a point charge q due to a continuous charge distribution, integrate over the tiny charge elements Δq. We show the forces $\Delta\mathbf{F}$ and $\Delta\mathbf{F}'$ due to two of the tiny charge elements Δq and $\Delta q'$. Notice that the vector $\hat{\mathbf{r}}$ will change as we move through the distribution.

PROBLEM SOLVING

Techniques for finding the force due to continuous charge distributions

At this point, the sum is really only symbolic. We can find a more concrete form in terms of a *charge density*. It is also convenient to write this form according to whether the continuous distribution of charge is distributed *along a line*, spread *over a plane*, or spread *throughout a volume*. These three cases correspond to one-, two-, or three-dimensional space. In each case, we replace the finite charge Δq in a small length, area, or volume with the infinitesimal charge dq, and replace the sum in Eq. (22–11) with an integral.

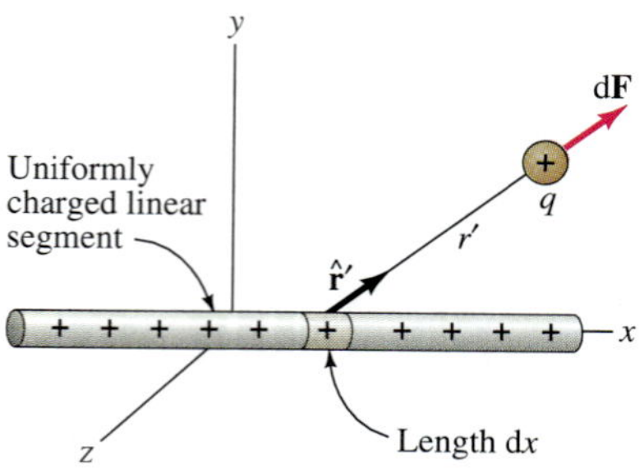

FIGURE 22–15 A one-dimensional charge distribution and the infinitesimal force on a point charge due to an infinitesimal piece of it.

One Dimension: Line Segment. If the charge distribution is distributed along a line that we label as the x-axis, we denote the *linear charge density* (charge/unit length) by $\lambda(x)$. The charge on an infinitesimal length dx of the line is

$$\text{for a charged line:} \qquad dq = \lambda(x)\, dx. \qquad (22\text{–}12)$$

Note that λ is a function of x; that is, the charge density can vary along the line. The force on the point charge q is then (Fig. 22–15)

$$\mathbf{F} = \frac{q}{4\pi\epsilon_0}\int \hat{\mathbf{r}}'\,\frac{\lambda(\mathbf{x})\, dx}{r'^2}. \qquad (22\text{–}13)$$

Notice the meaning of the integration: We move along the line of charge, and each point along the line is a different distance r' and at a different direction $\hat{\mathbf{r}}'$ from the charge q. We shall look at some examples to see how this works in practice.

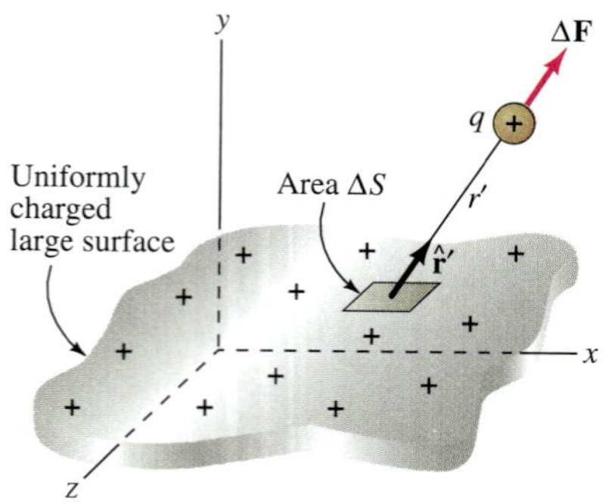

FIGURE 22–16 A two-dimensional charge distribution and the infinitesimal force on a point charge due to an infinitesimal piece of it.

Two Dimensions: Surface. Here the charge is distributed across a surface. We denote the *surface charge density* (charge/unit area) by σ. The density σ can vary from point to point on the surface. The charge on an infinitesimal area dS of the surface is (Fig. 22–16)

$$\text{for a charged surface:} \qquad dq = \sigma\, dS. \qquad (22\text{–}14)$$

The force on the point charge q is (Fig. 22–16)

$$\mathbf{F} = \frac{q}{4\pi\epsilon_0}\int_{\text{Surface}} \hat{\mathbf{r}}'\,\frac{\sigma\, dS}{r'^2}. \qquad (22\text{–}15)$$

Here, we are integrating over all the elements of the surface, as indicated by the subscript of the integral sign. In practice, such integrals often reduce to simple one-dimensional integrals.

Three Dimensions: Volume. When the charge is distributed through a volume, we write the *volume charge density* of the distribution as $\rho(\mathbf{r}')$, which means that the infinitesimal charge dq contained in the infinitesimal volume dV is

$$dq = \rho\, dV. \qquad (22\text{–}16)$$

In terms of the charge density of the continuous charge distribution, the *net* force due to the volume element shown in Fig. 22–14 is

$$\mathbf{F} = \frac{q}{4\pi\epsilon_0}\int_{\text{Volume}} \hat{\mathbf{r}}\,\frac{\rho\,dV}{r^2}. \qquad (22\text{–}17)$$

The integration is over the entire volume of the charge distribution, and that is why we have used the subscript. Again, in practice such integrals often reduce to simpler one-dimensional integrals.

In each of these cases, the argument of the charge distribution is the vector displacement $\mathbf{r}'$ because what counts is the vector displacement from an element of the charge distribution to the point charge on which the force acts. We may have, however, a *uniform* charge distribution, in which charge is distributed evenly throughout a region. In that case, λ is the total charge on the line divided by the length of the line, σ is the total charge on the surface divided by the area of the surface, and ρ is the total charge in the three-dimensional region divided by the volume of the region. All three quantities are constants that can be removed from the integral for the net force. Keep in mind that a fixed uniform charge distribution is not possible with a conductor, within or on which charges are free to move.

The integrals that express the force may well be simple to perform. Conversely, the integral may be difficult to perform, particularly if there is no symmetry in the distribution. Numerical integration on a computer can always be used if the integration is too difficult to perform analytically. Example 22–6 illustrates a typical integration and how symmetry can simplify a problem.

EXAMPLE 22–6 Find the force on a point charge q_1 located on the axis of a uniformly charged ring of total charge Q (Fig. 22–17). The radius of the ring is R, and q_1 is located a distance L from the center of the ring.

Solution: The ring has a continuous charge distribution, but it is spread along a (curved) line, so the integration must be one-dimensional. A small segment of the ring contains charge dq (Fig. 22–17a). All such segments are located a distance $r' = \sqrt{L^2 + R^2}$ from charge q_1, and the line to any segment on the ring makes the angle θ with the y-axis.

Next, look at the components of the force on q_1. Because every segment of the ring is the same distance r' from q_1, the *magnitude* of the infinitesimal force from each infinitesimal slice is the same. This is not true for the direction. The force from segment dq at the top of the ring ($z = 0$, $y = R$) is $d\mathbf{F}_{dq}$, and this force has components in the $+x$-direction and the $-y$-direction (Fig. 22–17b). The force from seg-

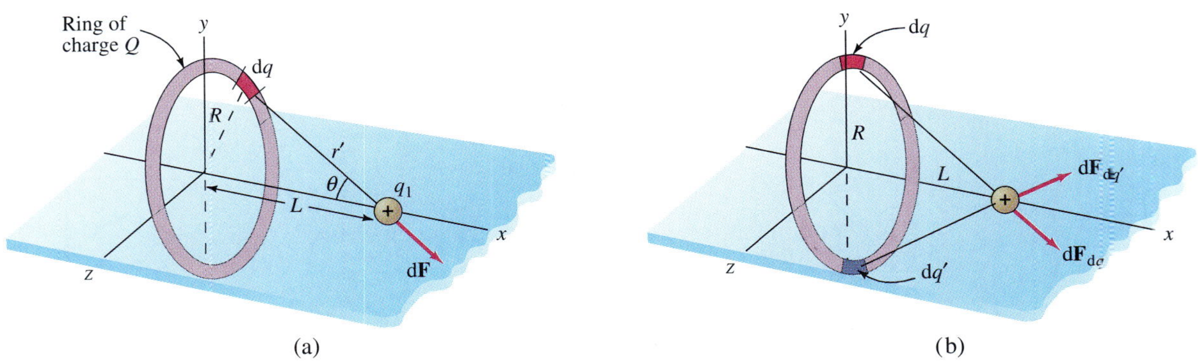

FIGURE 22–17 (a) Example 22–6. The force on a point charge q_1 due to a ring with total charge Q. First we find the force between the point charge and a tiny ring segment with charge dq. (b) Only the x-component of the force needs to be determined, because the y- and z-components will cancel due to symmetry.

ment dq' at the bottom of the ring ($z = 0$, $y = -R$) is $d\mathbf{F}_{dq'}$, and this force has components in the $+x$-direction and the $+y$-direction. If the magnitude dq equals the magnitude dq', the y-components of the force will cancel each other, and the x-components of the force will be additive. The y-components are the components perpendicular to the axis of the ring. This cancellation will hold for every perpendicular component of the force because we can always consider the charge elements in pairs. Thus we need compute only the component F_x. The x-component from the element shown in Fig. 22–17a is

$$dF_x = \frac{q_1}{4\pi\epsilon_0}\frac{dq}{r'^2}\cos\theta = \frac{q_1}{4\pi\epsilon_0}\frac{\cos\theta}{R^2 + L^2}\,dq.$$

The net force has only an x-component and is the sum over the infinitesimal x-components:

$$F_x = \int dF_x = \int \frac{q_1}{4\pi\epsilon_0}\frac{\cos\theta}{R^2 + L^2}\,dq.$$

The entire coefficient of dq is a constant and can be placed outside the integral sign. Thus

$$F_x = \frac{q_1}{4\pi\epsilon_0}\frac{\cos\theta}{R^2 + L^2}\int dq = \frac{q_1 Q}{4\pi\epsilon_0}\frac{\cos\theta}{R^2 + L^2}.$$

We have used the fact that $\int dq = Q$. Finally, from trigonometry we find

$$\cos\theta = \frac{L}{\sqrt{R^2 + L^2}},$$

so

$$F_x = \frac{q_1 Q}{4\pi\epsilon_0}\frac{L}{(R^2 + L^2)^{3/2}}. \tag{22–18}$$

A check is always desirable, and we can make two checks on this result. When point charge q_1 is very far from the ring, the ring should appear as a distant point of total charge Q, and the force should take on the Coulomb form $q_1Q/(4\pi\epsilon_0 L^2)$; this is indeed the limit of Eq. (22–18) when $L >> R$. When $L = 0$, the charge is at the middle of the ring, and, by symmetry, the net force should be zero. This limit also fits our result.

EXAMPLE 22–7 A straight rod of length L is aligned along the x-axis, with the ends at $x = \pm L/2$ (Fig. 22–18). The total charge on the rod is zero but the charge density is not; it is given by $\lambda(x) = 2\lambda_0 x/L$ (positive to the right of the origin, negative to the left). Find the force on a charge q located at a point $x = R$ on the x-axis, to the right of the right-hand end of the rod.

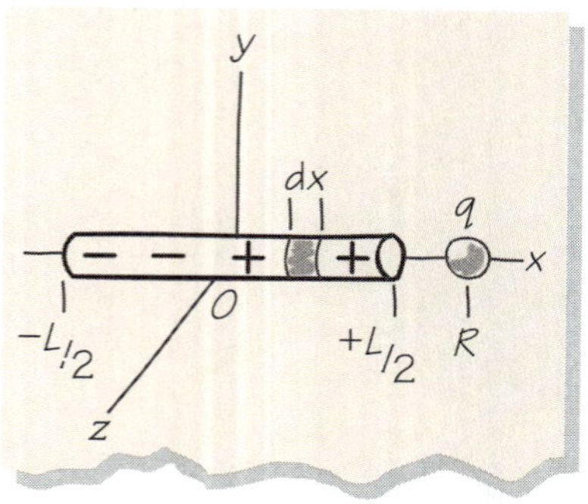

FIGURE 22–18 Example 22–7. A nonuniform charge density.

Solution: Consider a thin slice of the rod located at point x, with thickness dx. The charge on that slice is given by

$$dQ = \lambda\,dx = \frac{2\lambda_0}{L}x\,dx.$$

The infinitesimal force exerted by this charge on charge q is given by

$$d\mathbf{F} = \frac{q}{4\pi\epsilon_0}\frac{2\lambda_0}{L}x\,dx\frac{1}{(R - x)^2}\mathbf{i},$$

and the total force on the charge is the integral over $d\mathbf{F}$,

$$\mathbf{F} = \int d\mathbf{F} = \int_{-L/2}^{L/2}\frac{q}{4\pi\epsilon_0}\frac{2\lambda_0}{L}\frac{x\,dx}{(R - x)^2}\mathbf{i} = \frac{2q\lambda_0}{4\pi\epsilon_0 L}\mathbf{i}\int_{-L/2}^{L/2}\frac{x\,dx}{(R - x)^2}$$

$$= \frac{2q\lambda_0}{4\pi\epsilon_0 L}\,\mathbf{i}\int_{-L/2}^{L/2}\left[-\frac{1}{R-x}+\frac{R}{(R-x)^2}\right]dx$$

$$= \frac{2q\lambda_0}{4\pi\epsilon_0 L}\left\{\ln\left[\frac{R-(L/2)}{R+(L/2)}\right]+R\left[\frac{1}{R-(L/2)}-\frac{1}{R+(L/2)}\right]\right\}\mathbf{i}.$$

Notice that we set the limits on the integral according to where the charge distribution tells us. In this case, charge was present from $x = -L/2$ to $x = L/2$.

The Force Due to a Spherically Symmetric Charge Distribution

A charge distribution that is *spherically symmetric* is both important physically and easy to treat. Such a distribution is in the form of a sphere centered at, say, point P, and the charge density has a constant value at a given distance from P. Notice that the charge density could vary with the distance from P, but if there is spherical symmetry, the charge density must look the same in any direction as seen from P. This case was discussed extensively in Chapter 12 for gravitational force. Those results depend only on the fact that the force varies inversely with the distance squared, so we can use the results here. The force of the spherically symmetric charge distribution on a point charge q outside the distribution (Fig. 22–19a) is the same as though the entire charge of the distribution were concentrated at P (Fig. 22–19b). If, as in Fig. 22–19c, the point charge q is inside any part of the distribution, then the force on q due to the part of the distribution that lies outside q is zero (Fig. 22–19d).

22–5 THE SIGNIFICANCE OF ELECTRIC FORCES

We have now introduced a second basic force of nature: We have added the electrical interaction represented by the Coulomb force to the law of universal gravitation. Both the electric and gravitational forces have an inverse-square dependence on the distance between the interacting pointlike objects. Both forces are also proportional to the product of a characteristic attribute of the two objects—either mass or charge.

On the cosmic scale, the gravitational law looms large. It is the force that keeps Earth rotating around the Sun and the Moon rotating around Earth. The reasons why gravitation dominates electric forces on the astronomical scale are twofold. First, astronomical bodies have a great deal of mass. Second, astronomical bodies are almost exactly charge neutral, so the electric forces between them are relatively small. On anything less than an astronomical scale, however, the electric forces are normally much larger than the gravitational ones; apart from the direct effects of Earth's gravity, our

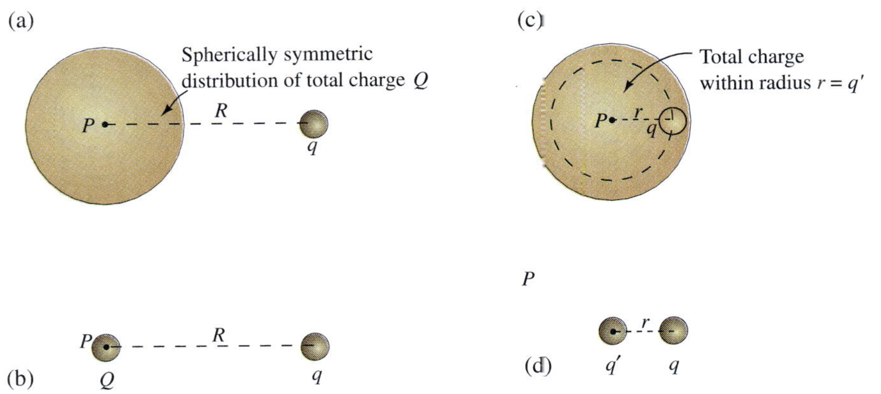

FIGURE 22–19 (a) A spherically symmetric charge distribution of total charge Q is centered on the point P. The force on a point charge q outside the distribution a distance R from P is the same as (b) the force it would experience if a point charge Q were located at P. (c) If q lies inside the distribution a distance r from P, and q' is the total charge that lies within a sphere of radius r centered on P, then it experiences the same force it would have (d) if there were a point charge q' at P.

everyday experience depends far more on the electric force than on the gravitational one.

As we have seen for the hydrogen atom, the electric force dominates the gravitational force on a microscopic scale. Even though a full explanation requires quantum physics, we can now state that the electric force is responsible for

1. electrons binding to a positive nucleus, forming a stable atom;
2. atoms binding together into molecules;
3. atoms or molecules binding together into liquids and solids;
4. all chemical reactions; and
5. all biological processes.

The electric force is behind such nonfundamental forces as friction and other contact forces. Electric energy fuels our homes, starts our cars, and runs our factories.

SUMMARY

Electric charge occurs in two forms, which we label as positive and negative charge. Charges of the same sign repel each other, and charges of unlike sign attract each other. In SI units, charge is measured in coulombs.

Much of the behavior of materials under the influence of electric forces is characterized by the ease with which electrons are dislodged from their constituent atoms and molecules and move through the material. Metals are normally good conductors of electric charge, whereas most nonmetals are not and are called insulators.

The basic electric charge is that of the electron. The electron has a charge of $-e$, and the proton has a charge of $+e$, with $e = 1.602 \times 10^{-19}$ C. Electric charge in matter is quantized in multiples of e. Charge is conserved in all interactions, meaning that the net charge before an interaction is the same as the net charge after the interaction.

The electric force between point charges q_1 and q_2 separated by a distance r_{12} is

$$\mathbf{F}_{12} = \frac{1}{4\pi\epsilon_0}\left(\frac{q_1 q_2}{r_{12}^2}\right)\hat{\mathbf{r}}_{12}, \tag{22–8}$$

where the factor $1/4\pi\epsilon_0$ characterizes the strength of the force. This equation is Coulomb's law.

The principle of superposition applies when multiple charges are present. The forces on a point charge q due to all other charges add together vectorially. For continuous charge distributions, we must integrate over the charges, and the force of such a distribution on q depends on the charge distribution. For charges distributed on a line, over a surface, or through a volume, the force is, respectively,

$$\mathbf{F} = \frac{q}{4\pi\epsilon_0}\int \hat{\mathbf{r}}'\,\frac{\lambda(\mathbf{x})\,dx}{r'^2} \tag{22–13}$$

$$\mathbf{F} = \frac{q}{4\pi\epsilon_0}\int_{\text{Surface}} \hat{\mathbf{r}}'\,\frac{\sigma\,dS}{r'^2} \tag{22–15}$$

and

$$\mathbf{F} = \frac{q}{4\pi\epsilon_0}\int_{\text{Volume}} \hat{\mathbf{r}}'\,\frac{\rho\,dV}{r'^2}. \tag{22–17}$$

Here λ, σ, and ρ are the one-, two-, and three-dimensional charge densities, respectively.

On all but the astronomical scale, electric forces tend to be much stronger than gravitational forces. The electric force is responsible for making atoms, molecules, solids, and liquids stable, and for producing all chemical reactions and biological processes.

UNDERSTANDING KEY CONCEPTS

1. Two identical positive charges are placed on a table and fixed there. Find all the places on the table where the net force on a test charge due to the two charges is zero.
2. Particles of opposite charge attract with an inverse-square law. Are there analogues of Kepler's laws for a system composed of such a pair, and what are they?
3. When you walk across a carpet, you often pick up enough electric charge to cause a spark when you touch a doorknob. In climates that are dry in winter, this phenomenon is much more common in the winter than in the summer. Why?
4. Two metallic spheres on insulating stands are placed on an air-track. The mass of one sphere is five times larger than the other, and the charges are both positive in the ratio 3:1. The two objects are held at rest, and then let go. What determines how far the two objects each move in a short time interval? How would you find the ratio of the distances that they travel in that interval?
5. By using the apparatus discussed in Section 22–1, how could you determine what charge you accumulate by walking across a wool rug?
6. Atoms consist of negatively charged electrons bound to the positively charged nucleus by the Coulomb force. The electrons are rearranged when two different chemicals are brought together. Would you expect the electrons that are closer to the nucleus or the ones that are farther from the nucleus to be more involved in chemical reactions?
7. Neutrons and protons are believed to be made of two types of charged particles called quarks, having charge $-\frac{1}{3}e$ and $\frac{2}{3}e$, as mentioned in Section 22–2. List the possible combinations of only three quarks that make up neutrons and protons.
8. Some materials lose electrons easily by rubbing, so why are many of the objects around us not charged at all times?
9. *Earnshaw's theorem* states that a point charge cannot be in stable equilibrium while purely electrostatic forces act on the point charge. Consider a ring that is uniformly positively charged, with a positive charge at the center. It appears that the center charge suffers an identical repulsive force from every direction. How can the theorem be true?
10. How is the existence of a battery, which sends negative charges out of one of its contact points, consistent with the conservation of charge?
11. You have a cork ball with a charge of -4.8×10^{-19} C and three uncharged cork balls. Can you devise a method of touching cork balls together in sequence that will give a charge of -0.8×10^{-19} C to one of the balls?
12. We spoke of generating a spark on a winter's day when we touch a conducting line to Earth and become grounded. Automobile tires are such good insulators that a car body is not connected to Earth by a conductor. How do you explain the spark that occurs when you touch a car door after you have rubbed the car upholstery?
13. Suppose that the electric charge of a fundamental particle such as an electron depends on the speed v of the particle, so that $e = e_0[1 + (\kappa v^2/c^2)]$, where e_0 is the particle's "rest charge," c is the speed of light, and κ is some tiny number. Discuss ways in which you might measure κ. Is there any experimental reason why κ must be small, if not zero?
14. Does the modification of the electric charge proposed in Question 13 necessarily violate the principle that it should not be possible to detect the absolute velocity of an object by means of any experiment?
15. The color of the light emitted by quasars is evidence that the charge on electrons has not changed over billions of years. Is saying that the charge on electrons and protons is unchanged equivalent to the statement that charge is conserved?
16. Suppose that electrons had charge $-e$ and protons had charge $+e(1 + \delta)$, with δ very small. Would there necessarily be an additional repulsive $1/r^2$ force between the Moon and Earth, for example, that could overpower the gravitational attraction between these bodies?
17. Consider a uniform, spherical positive charge distribution. A negative charge is placed at the center. Discuss the net force on a negative point charge placed at the center of the sphere, and discuss what happens to the point charge if it is placed a bit off center.

PROBLEMS

22–1 Properties of Charged Matter

1. (I) A cork ball is charged to +1 nC. How many less electrons than protons does the ball have?
2. (I) A uranium atom has undergone a violent collision that has stripped off 18 of its electrons. What is the charge of the resulting atom? If a uranium nucleus contains 92 protons and 146 neutrons, what is the charge of the nucleus?
3. (I) What is the total charge of all the electrons in 1 g of CO_2?
4. (I) Three identical metallic spheres are connected by wires, and a charge Q is placed on one of them. The wires are then removed. One of the spheres is then connected by wire to the ground. That wire is then removed. This particular sphere is then connected by a wire to one of the other spheres. What is the charge on each of the spheres when the process is completed?
5. (II) A cork ball that is covered with conducting paint and charged to -4×10^{-10} C is touched by an identical but uncharged cork ball; the balls then separate. This second cork ball is then touched by a third uncharged cork ball, and they separate. What is the charge of each ball at the end, and how many excess electrons does each ball have?
6. (II) A cork ball covered with conducting paint is charged to -1.6×10^{-2} C. You have three similar but uncharged cork balls. Describe a method by which to produce a cork ball with a charge of -0.2×10^{-12} C. Do you need all three extra balls? Explain.
7. (II) An aluminum ball of mass 0.1 g is given a negative charge of 1 μC. What is the fractional increase in the number of electrons the ball contains?

8. (II) Two cork balls of mass 0.2 g hang from the same support point by massless insulating threads of length 20 cm (Fig. 22–20). A total positive charge of 3.0×10^{-8} C is added to the system. Half this charge is taken up by each ball, and the balls spread apart to a new equilibrium position. (a) Draw a force diagram for each cork ball. (b) What is the tension in the threads before the charge is added, and what is it after? (c) What is the value of angle θ in the figure? This device is a type of electroscope, or *electrometer*, a meter that measures electric charge. Angle θ measures the amount of charge on the balls if we can be sure that the charge is divided between them equally. This constraint is circumvented when the electrometer is made of a single strip of conducting material draped at its midpoint over a hook; the charge is then distributed over the strip equally, and half the strip repels the other half.

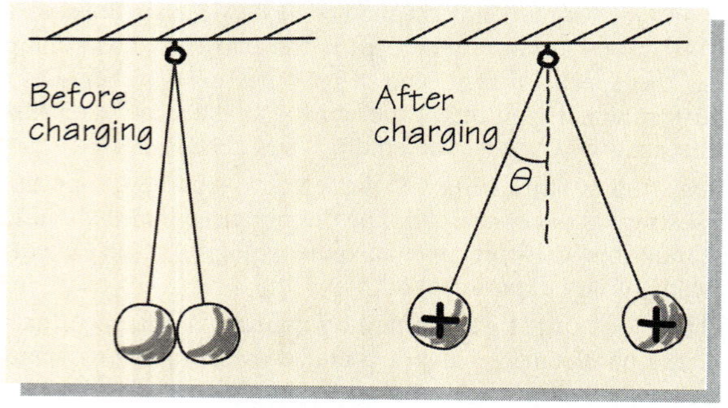

FIGURE 22–20 Problem 8.

9. (II) Silicon is the most abundant material on Earth's surface. (a) Assume that Earth is made of silicon (28 g/mol), and calculate the total number of negative charges contained within Earth. (b) When we neutralize a cork ball that has a charge of 1 μC by grounding it to Earth, what fractional change are we making in the total negative charge contained within Earth?

22–2 Charge is Conserved and Quantized

10. (I) One possible result of the high-energy collision of two protons is the reaction $p + p \to X + p$. What is the electric charge of particle X?

11. (I) *Antiparticles* have the same mass as their counterpart particles but have an opposite charge. For example, the antiparticle of an electron, e^- is the positron, e^+. Most antiparticles are denoted by a bar over the particle, so $\bar{p}$ is the antiparticle of the proton, and it has a charge of $-e$. Which of the following reactions satisfy the conservation of charge: (a) $p + \bar{p} \to e^+ + e^- + e^+ + e^- + 2n$; (b) $e^+ + e^- \to 2p + n + 2\gamma$; (c) $e^+ + e^- \to e^+ + e^- + p + \bar{p} + 2\gamma$; (d) $n + p \to e^- + p + \bar{p}$?

12. (I) How much charge is contained in 1 mg of electrons?

13. (II) The electric charge of an object is independent of the object's motion. Suppose that this were not true, but that the charge of a particle such as an electron or a proton that moves at speed v has the form $e = e_0[1 + (v^2/c^2)]$, where e_0 is the particle's charge when at rest and $c \simeq 3 \times 10^8$ m/s is the speed of light. What would the net charge on a hydrogen atom be, assuming that the atom consists of a proton at rest and an electron orbiting the proton at average speed $(v/c) \simeq (1/137)$?

22–3 Coulomb's Law

14. (I) How far apart must two protons be for the force on each other to be the same as the weight of one proton on Earth's surface?

15. (I) A proton is believed to consist of two "up" quarks of charge $+\frac{2}{3}e$ and one "down" quark of charge $-\frac{1}{3}e$. Assume that all three quarks are equidistant from each other at the distance of 1.5×10^{-15} m. What are the electrostatic forces between each pair of the three quarks?

16. (I) Two small balls, each of mass 2 g, are each charged with +12 nC. What distance apart must they be if the force on one of them has the same magnitude as the weight of that ball?

17. (I) Two identical charged sodium ions separated by 4.5×10^{-9} m have a force between them of 1.1×10^{-11} N. What is the charge of each ion, and how many electron charges does this represent?

18. (I) Two small cork balls have the same charge. When their centers are placed 2 cm apart, the force between them is observed to be 0.18 N. What is the cork balls' charge? Why do we have to assume that the size of the cork balls is small compared to 2 cm?

19. (I) Two tiny cork balls, both of mass 0.10 g, each have just one electron charge, $q = -1.6 \times 10^{-19}$ C. They are separated by 15 cm, which is much greater than their sizes. What is the ratio of the magnitudes of the Coulomb force between them to the gravitational force they exert on each other? Why is this result so different from that of Example 22–3?

20. (II) The experiment of Cavendish to determine the gravitational constant (see Chapter 12) relies on the measurement of a force of about 7×10^{-7} N between two masses separated by a distance of 0.1 m. One possible source of error is a small electric charge on the balls. Assuming the charges are equal, what is the magnitude of the largest allowed charge, if the force is to be measured to at least a 0.1 percent accuracy?

21. (II) Suppose that we were to measure a charge in some new unit, which we will call the esu, so defined that Coulomb's law reads, in magnitude, $F = q_1q_2/r^2$, and so that $F = 1$ dyne (10^{-5} N) when $q_1 = q_2 = 1$ esu and $r = 1$ cm. (a) How many esu are there in 1 C? (b) What is the charge of the electron in esu? (The esu is an actual unit, the *electrostatic unit.*)

22. (II) An electron and a proton attract each other with a $1/r^2$ electric force, just like the gravitational force. Suppose that an electron moves in a circular orbit about a proton. (a) If the period of the circular motion is 24 h, what is the radius of the orbit? (b) If the period is 4×10^{-16} s, as it is in a hydrogen atom, what is the radius of the orbit?

23. (II) A charge q is split into two parts, $q = q_1 + q_2$. In order to maximize the repulsive Coulomb force between q_1 and q_2, what fraction of the original charge q should q_1 and q_2 have?

24. (II) An alpha particle (a helium nucleus, composed of 2 protons and 2 neutrons) is directed onto a particular tungsten nucleus (^{184}W, with 74 protons and 110 neutrons). The alpha particle stops and turns around at a distance of 2×10^{-13} m from the tungsten nucleus (Fig. 22–21). Ignore the effects of electrons, and treat the alpha particle and tungsten nucleus as pointlike. What is the Coulomb force on the alpha particle at its closest approach to the nucleus?

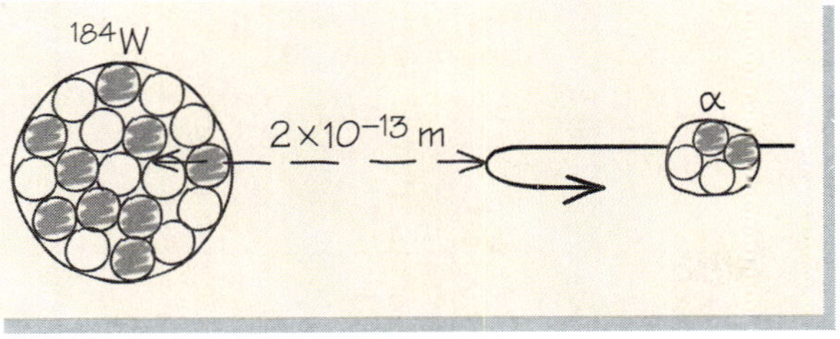

FIGURE 22–21 Problem 24.

25. (II) An electron orbits in uniform circular motion about a much heavier—and therefore nearly stationary—proton at a distance of 3×10^{-10} m. (a) What are the magnitude and direction of the Coulomb force exerted on the electron by the proton? (b) What is the speed of the electron in its circular orbit? (c) the frequency of the circular orbit? (d) Calculate the spring constant of a spring with an electron at its end and the frequency of part (c).

26. (II) Two pointlike objects are placed 8.75 cm apart and are given equal charge. The first object, of mass 31.3 g, has an initial acceleration of 1.93 m/s^2 toward the second object. (a) What is the mass of the second object if its initial acceleration toward the first is 5.36 m/s^2? (b) What is the charge of each object?

27. (II) Two cork balls, each of mass 0.20 g, are hung by insulating threads 20.0 cm long from a common point. The cork balls are given an equal charge by a Teflon rod. The balls repel and deflect as shown in Fig. 22–22. What charge q was given to each cork ball? Assume uniform charge.

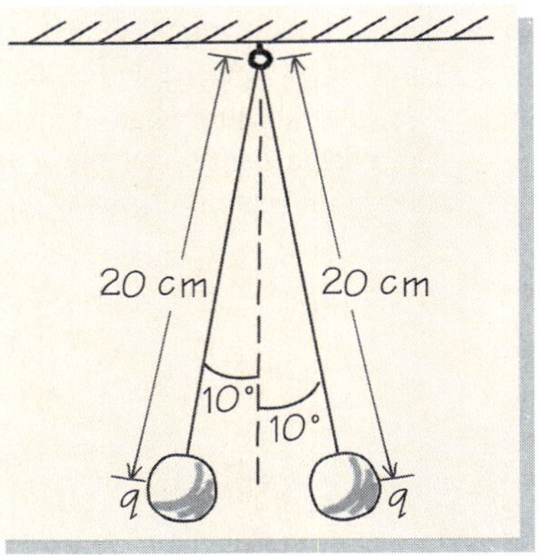

FIGURE 22–22 Problem 27.

28. (II) Astronomical data tell us that Earth's radius is 6.4×10^6 m, that its mass is 5.98×10^{24} kg, that the Moon's mass is 7.36×10^{22} kg, and that the mean Earth–Moon separation is 3.8×10^8 m. Suppose that, instead of being electrically neutral, as we believe, Earth and the Moon each have an excess positive charge of 5.7×10^{13} C. (a) What is the magnitude of the electrical repulsion between Earth and the Moon? (b) What is the ratio of this repulsive force to the attractive gravitational force? (c) If the charge on Earth were distributed uniformly throughout its volume, what would the excess charge density be, in coulombs per cubic meter (C/m^3)? (d) Assume that the excess positive charge is due to excess protons, which have an electric charge of 1.6×10^{-19} C. Calculate the density of protons, in units of protons per cubic meter, that corresponds to the conditions in part (c). (e) Earth's mean density is 5.52×10^3 kg/m^3, and a proton has mass 1.67×10^{-27} kg. Protons account for about half of Earth's mass. Compute the density of all protons in Earth, and compare this to your answer in part (d).

29. (II) Three unknown charges q_1, q_2, and q_3 exert forces on each other. When q_1 and q_2 are 15.0 cm apart (q_3 is absent), they attract each other with a force of 1.4×10^{-2} N. When q_2 and q_3 are 20.0 cm apart (q_1 is absent), they attract with a force of 3.8×10^{-2} N. When q_1 and q_3 are 10.0 cm apart (q_2 is absent), they repel each other with a force of 5.2×10^{-2} N. Find the magnitude and sign of each charge.

30. (II) An electron has a mass of 0.9×10^{-30} kg and a charge of -1.6×10^{-19} C. Earth's mass is 6×10^{24} kg, and its radius is 6.4×10^6 m. Suppose that Earth has a net negative charge Q at its center. (a) How large would Q have to be for the charge repulsion on an electron to cancel the gravitational attraction at Earth's surface? (b) Suppose that this net charge is due to a discrepancy between the positive proton charge and the negative electron charge. Assume that half of Earth's mass is due to protons, each of which has a mass of 1.5×10^{-27} kg (the rest is neutrons, assumed to be neutral; electrons do not contribute much to the mass). What is the size of the charge discrepancy, compared to the electron charge?

31. (III) Use the similarity between Coulomb's law and the law of universal gravitation to calculate the distance of closest approach between a point charge of $+10^{-6}$ C, which starts at infinity with kinetic energy of 1 J, and a fixed point charge of $+10^{-4}$ C. Assume that the moving point charge is aimed straight at the fixed point charge. [*Hint*: The similarity to gravity consists of using the notions of potential energy and energy conservation.]

22–4 Forces that Involve Multiple Charges

32. (I) A charge $-5q$ is placed at $x = 0$ and a charge $-3q$ is placed at $x = 10$. Is there a point on the x-axis at which the net force on a charge Q is zero, and if so, where is it?

33. (I) A charge $-5q$ is placed at $x = 0$ and a charge $+3q$ is placed at $x = 10$. Where, on the x-axis, is the net force on a charge Q zero?

34. (I) Six identical charges of magnitude 2.4×10^{-7} C are placed on a straight line at $5 =$ cm intervals, starting at $x = 0$. What is the force on the charge at $x = 25$ cm?

35. (I) A positive point charge q sits at the center of a circle of radius R on which a total negative charge Q is uniformly distributed. What is the net force on q?

36. (II) A charge of q is fixed on a plane at the origin (0, 0) of an xy-coordinate system, and a charge $3q$ is fixed at (3 cm, −3 cm). Where must a charge of $-2q$ be placed at rest for it to be in equilibrium (that is, so that it remains at rest)? Is the equilibrium stable?

37. (II) What is the total force on each of the three quarks in Problem 15 due to the other two quarks?

38. (II) Three negative charges of magnitude 0.8 μC are placed at the corners of an equilateral triangle of sides 14 cm. What is the net force on a charge of 1.2 μC placed at the midpoint of one of the sides?

39. (II) Four positive charges $+q$ sit in a plane at the corners of a square whose sides have length d, as in Fig. 22–23. A negative charge, $-q$, is placed in the middle of the square. (a) What is the net force on the negative charge? (b) Is the equilibrium point at the center a stable equilibrium for motion of the negative charge in the plane of the square? (c) for motion of the negative charge perpendicular to the plane of the square?

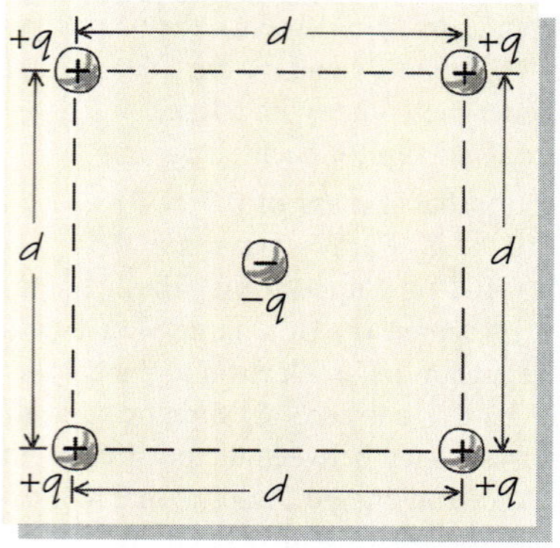

FIGURE 22–23 Problem 39.

40. (II) Calculate the force between two identical dipoles consisting of dumbbells with equal and opposite charges q and $-q$ at the end of a rigid rod of length $2d$. The dipoles are parallel, as shown in Figure 22–24, and they are a distance x apart. Derive a first-order approximation for $d << x$. [*Hint*: Use $(1 + y)^n \simeq 1 + ny + \cdots$ for $y << 1$.]

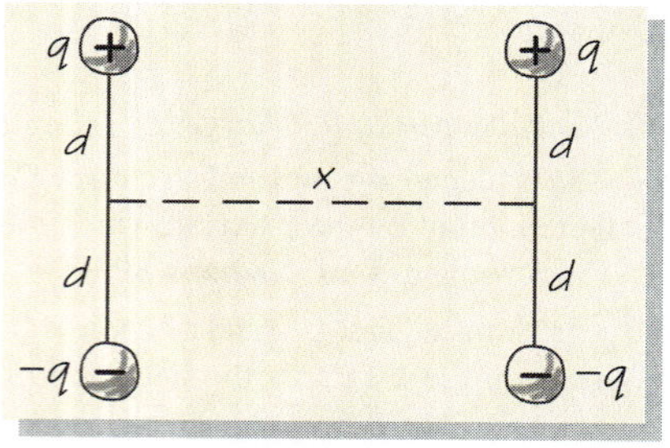

FIGURE 22–24 Problem 40.

41. (II) Charges q, $2q$, $-4q$, and $-2q$ (q is positive) occupy the four corners of a square of sides $2L$, centered at the origin of a coordinate system (Fig. 22–25). (a) What is the net force on charge q due to the other charges? (b) What is the force on a new charge Q placed at the origin?

42. (II) A charge Q is distributed uniformly along a rod of length $2L$, extending from $y = -L$ to $y = L$ (Fig. 22–26). A charge q is placed on the x-axis at $x = D$. (a) In what direction is the force on q, given that Q and q have the same sign? (b) What is the charge on a segment of the rod of infinitesimal length dy? (c) What is the force vector on charge q due to the small segment dy? (d) Express an integral that describes the total force in the x-direction. (e) Compute the integral in order to find the total force in the x-direction.

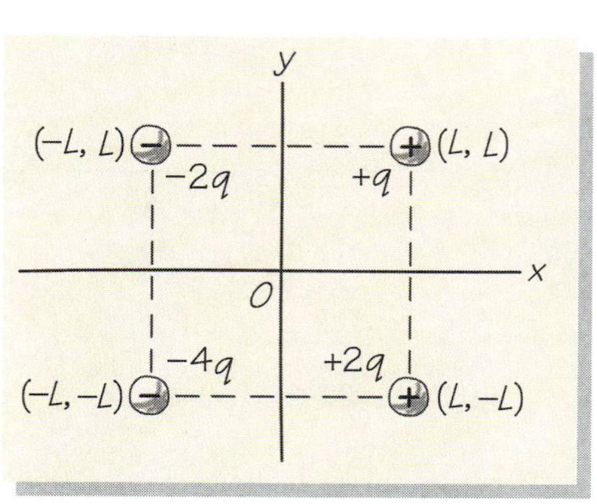

FIGURE 22–25 Problem 41.

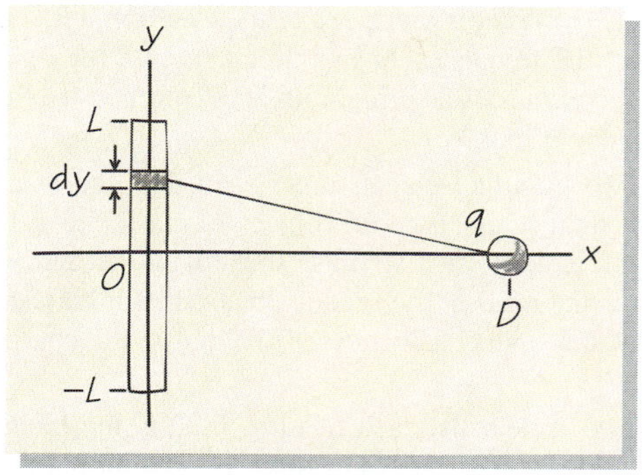

FIGURE 22–26 Problem 42.

43. (II) A charge is spread uniformly along the y-axis, stretching infinitely far in both the positive and negative directions. The charge density (charge per unit length) on the y-axis is λ. Find the force on a point charge q placed on the x-axis at $x = x_0$.

44. (II) A charge is spread uniformly along the y-axis from $y = 0$ to $y = +\infty$. The charge density on the y-axis is λ (Fig. 22–27). Find the force on a point charge q placed on the x-axis at $x = x_0$.

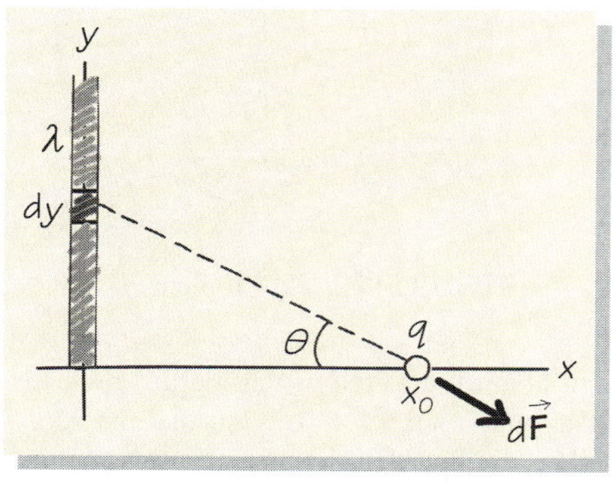

FIGURE 22–27 Problem 44.

45. (II) A long, thin rod of length L that contains a uniform distribution of charge Q points away from a point charge q. The nearest part of the rod is a distance d from the point charge. What is the electric force exerted on the charge q by the rod?

46. (II) A charge Q is distributed uniformly over a thin ring of radius R. The ring is oriented in the xy-plane, with its center at the origin. Find the force on a charge q located at the origin, and discuss the stability of its motion in the xy-plane. How does this compare with the case of a point charge placed at the center of a sphere whose surface is uniformly charged?

47. (II) Use the results of Example 22–6 to calculate the force on a positive point charge of magnitude 0.65 μC located 5 cm above the center of a uniformly charged solid plate of radius 8 cm that carries a total positive charge of 1.6 μC. [*Hint*: Break the disk into concentric rings, use the results of Example 22–6 for each ring, and sum over the forces due to the rings.]

48. (II) Calculate the force exerted on a charge q by an infinite plane sheet with surface charge density (charge per unit area) σ. [*Hint*: Break up the plane into concentric rings centered below the charge, use the results of Example 22–6 for the force from each ring, then sum over the forces due to the rings.]

49. (II) Two rigid plates of equal size, made of different plastics, are rubbed against each other. This results in equal and opposite charges on the two plates. How large are these charges if it takes 0.1 N to separate the two plates? The area of each plate is 0.05 m^2 and the charge distribution may be assumed to be uniform. [*Hint*: Use the result of Problem 48.]

50. (II) Consider an infinite vertical sheet that carries a charge of 2×10^{-6} C/m^2. A cork ball of mass 2 g is suspended by a string 50 cm long at a distance of 10 cm from the charged sheet. What is the string orientation (a) if a charge $q = 8 \times 10^{-8}$ C is placed on the cork ball? (b) if instead a charge $q = -5 \times 10^{-8}$ C is placed on the ball?

51. (II) A total charge of 0.75 μC is distributed uniformly over a thin, semicircular wire of radius 5.0 cm. What is the force on a charge of 0.30 μC located at the center of the circle?

52. (II) A succession of $n + 1$ alternating positive and negative charges q are located along the x-axis at the points $x = 0, x = d, x = 2d, \ldots, x = nd$. An isolated charge Q is placed as shown in Fig. 22–28 at the point $x = D$ a very long distance away from the origin ($D >> nd$). (a) Write a general expression for the electric force on charge Q. (b) Approximate your result, using the condition $D >> nd$. Keep only leading and next-to-leading terms. (*Hint*: Use $(1 + x)^{-2} \simeq 1 - 2x$ for $x << 1$.]

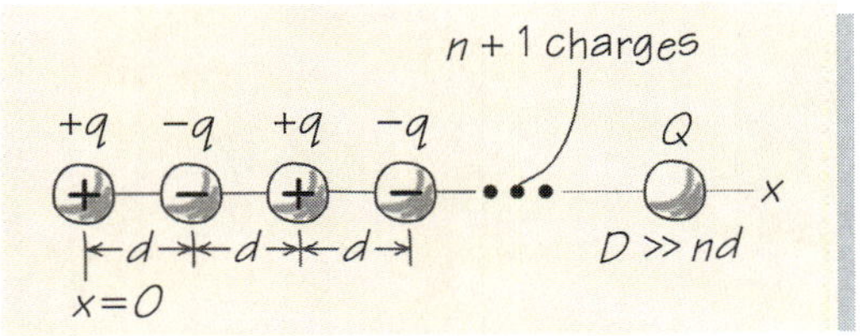

FIGURE 22–28 Problem 52.

53. (III) What is the force per unit area between two infinite uniformly charged plates with a surface charge density of $+10^{-5}$ C/m^2 and -10^{-5} C/m^2, respectively, when the distance between the plates is 10 cm? What if the distance between the plates is doubled? [*Hint*: You may use the result of Problem 48.]

General Problems

54. (II) How much charge $+Q$ should be distributed uniformly over a square, horizontal plate of dimensions 30 cm × 30 cm if a ball of mass 0.1 g and charge 0.1 μC is to remain suspended 1 mm over the surface of the plate? Take gravity into account in this problem. How would your answer change if the ball were to be suspended 2 mm over the plate? *Qualitatively*, how would your answer change if the ball were to be suspended 1 m over the plate?

55. (II) A single charge $q_1 = +2 \times 10^{-8}$ C is fixed at the base of a plane that makes an angle θ with the horizontal direction. A small ball of mass $m = 0.5$ g and charge $+2 \times 10^{-8}$ C is placed in a smooth, frictionless groove in the plane that extends directly to the fixed charge (Fig. 22–29). It is allowed to move up and down until it finds a stable position $\ell = 8$ cm from the fixed charge. What is θ?

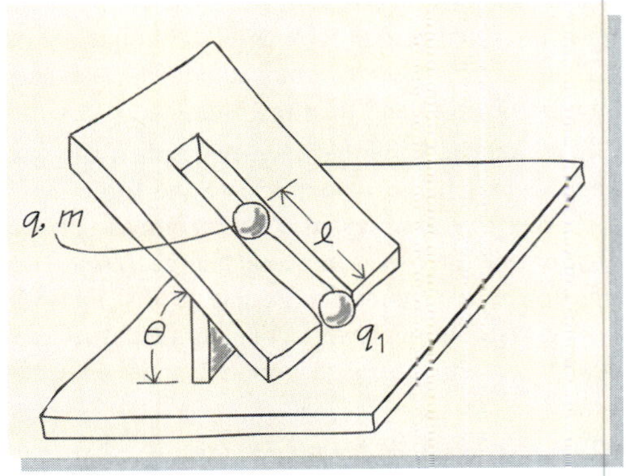

FIGURE 22–29 Problem 55.

56. (II) The nucleus of an iron atom contains 26 protons within a sphere of radius 4×10^{-15} m. What is the Coulomb force between two protons at opposite sides of this nucleus? The answer to this problem illustrates that the force that holds the nucleus together against the Coulomb repulsion of its constituents must be strong indeed.

57. (II) An electron moves in a circular planetary orbit around a proton. (a) If the centripetal force is the attractive Coulomb force, what is the speed of the electron in terms of the charge e and the radius of the circular orbit? (b) What is the angular momentum, L, of the electron in the orbit? (c) Express the speed in terms of e and L. (d) Express the radius of the orbit in terms of e and L. (e) Express in terms of e and L the time it takes for the electron to go around the circle once. (f) Evaluate all these quantities, given that $L = 1.05 \times 10^{-34}$ kg · m^2/s. This corresponds to a simplified version of the hydrogen atom.

58. (II) Suppose that the proton charge were slightly larger than the electron charge, so that $q_{proton} = (1 + \delta)e$ and $q_{electron} = -e$, where $0 < \delta << 1$. (a) Given that there are approximately 1.25×10^{57} protons (and electrons) in the Sun, and approximately 1.15×10^{44} protons and electrons in Earth, what is the upper limit on δ set by the fact that the resultant Earth–Sun electric repulsion cannot be large enough to cancel the attraction due to gravity? The mass of the Sun is approximately 2×10^{30} kg, that of Earth is approximately 6×10^{24} kg, and $G = 6.7 \times 10^{-11}$ N · m^2/kg^2. (b) How would your value of δ change the weight of a football player who contains 3×10^{28} protons and electrons?

59. (II) Two fixed positive charges q are separated by a length ℓ. A third positive charge q of mass m is constrained to run on a line between the two fixed charges. (a) When the third charge is placed a distance x from the left-hand fixed charge, what is the net force on the third charge? Where is this force zero; in other words, where is the equilibrium point? (b) What is the net force as a function of the displacement of the third charge from the equilibrium point of part (a)? (c) For *small* values of the displacement from the equilibrium point, the third charge behaves as if a spring were acting on it. What is the value of the oscillation frequency?

60. (II) A positive charge q and a negative charge $-\alpha q$ ($\alpha > 1$) are fixed at a distance ℓ apart. Another positive charge q of mass m is constrained to move on the line connecting the two fixed charges. (a) Calculate the net force on the moving charge when it is at a distance x from the fixed positive charge. (b) Where is the force zero? (c) What is the frequency of oscillations if the moving charge is moved a small distance from its equilibrium position and then released, and if $\alpha = 4$?

61. (II) Show that the force between two spherically symmetric distributions of charge is identical to the force between two point charges that are located at the geometric center of each distribution and have the same total charge. [*Hint*: Use the fact that the force on a point charge due to distribution 1 is the same as if distribution 1 were concentrated at its center, then use similar reasoning for distribution 2, and then use Newton's third law.]

62. (III) Two rods, each of length $2L$, are placed parallel to one another a distance R apart. Each carries a total charge Q, distributed uniformly over the length of the rod. Give an integral for the magnitude of the force between the rods, but do not evaluate it. Without working out any integrals, can you determine the force between the rods for $R >> L$?

63. (III) Consider an infinite number of identical point charges q located at equally spaced points on the x-axis at the locations $x = na$ (n takes on integer values that range from $-\infty$ to $+\infty$) (Fig. 22–30). (a) Write an expression for the force on a charge Q, located at $x = 0$ and $y = R$, due to all the point charges q, and show the direction of the net force. (b) Take the limit of your result when the intercharge spacing $a \rightarrow 0$ and the charge $q \rightarrow 0$ such that $q/a = \lambda$ (a fixed charge density). Show that your expression can be written as an integral, and use dimensional analysis to determine the R-dependence of the force on charge Q.

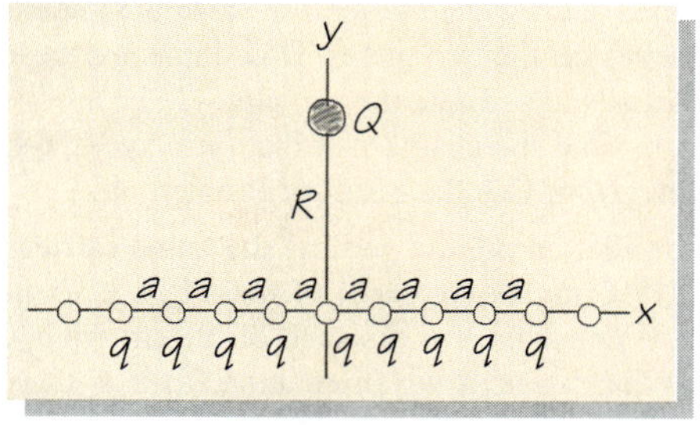

FIGURE 22–30 Problem 63.

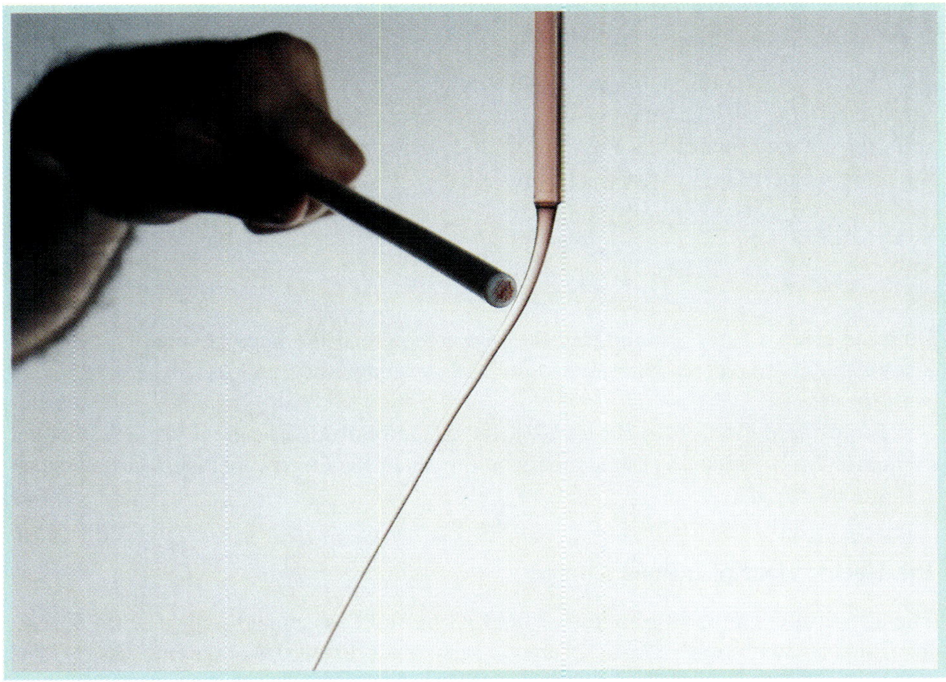

CHAPTER

23

An electric field forms around a teflon rod when it is rubbed with fur. The field induces a net dipole moment in a stream of water when it is brought close to the water, and the dipole moment is automatically oriented so that it is attracted to the rod.

Electric Field

Charges exert forces on one another over large distances and across empty space. This idea of "action at a distance" seems difficult to accept. Action at a distance suggests that the object responsible for the force on a second object somehow reaches out, measures the distance to the second object, and then acts. Michael Faraday suggested a way around this conceptual difficulty: The first object influences the surrounding space by setting up a *field* around itself that is present whether there is a second object or not. When the second object is located at a given point, the field at this point acts on that object. This important idea can be developed quantitatively and, like any really good idea, leads to further ideas that go far beyond the original concept in utility and insight. In this chapter, we introduce and develop the concept of an electric field produced by static charges and learn some of the ways in which it can be useful. We shall continue to use the field concept in future chapters because it forms a basis for understanding many electrical and magnetic effects.

23–1 ELECTRIC FIELD

It is useful to think of a distribution of charges as giving rise to an **electric field**, which acts on any charge placed in that field. We can detect the electric field at any particular point by placing a small positive **test charge** q_0 at that location and seeing if it experiences a force. A test charge is only a probe: It does not *produce* the electric field that we are trying to measure; the field is due to other charges. The electric field, **E**, can be defined—this is what is called an operational definition—by measuring the mag-

The definition of the electric field

nitude and direction of the electric force **F** acting on the test charge. The definition of the field is

$$\mathbf{E} \equiv \frac{\mathbf{F}}{q_0}. \tag{23–1}$$

We use a *small* test charge q_0 because a large charge might cause the charges responsible for the electric field to move (Fig. 23–1). This would affect the field itself. Thus, we more properly define the electric field by

The SI units of electric field

$$\mathbf{E} \equiv \lim_{q_0 \to 0} \frac{\mathbf{F}}{q_0}. \tag{23–2}$$

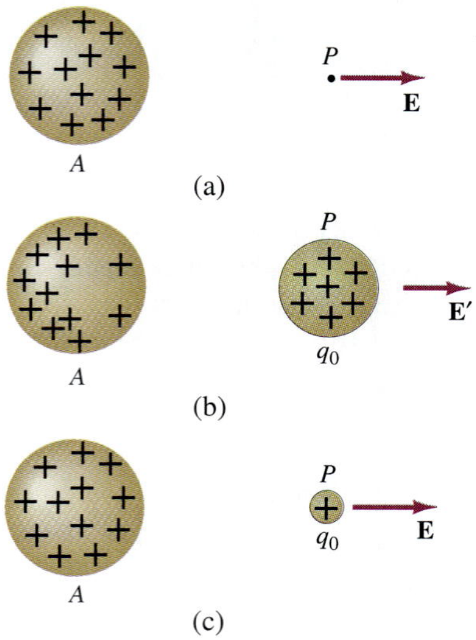

FIGURE 23–1 (a) An electric field exists at a point P due to charges on the sphere A. (b) The test charge q_0 repels the charges on sphere A. A new electric field, $\mathbf{E}'$, is produced at P by sphere A, because the charges on A have been redistributed. (c) The test charge q_0 is now so small that it hardly affects the charges on sphere A. The electric field produced by A at point P is now the same as in part (a). In each case the electric field is due to the charge on sphere A.

Both the electric force and the electric field are vectors. We know a vector such as the electric field completely when we know both its magnitude and its direction *at every point in space.*

From the definition in Eq. (23–2), the SI units of electric field are newtons per coulomb (N/C). Table 23–1 gives the magnitudes of the electric fields in various physical situations.

The Electric Field of a Point Charge

The simplest example of an electric field is the field associated with a point charge, q_1. Consider two point charges, q_1 and q_0, located a distance r apart (Fig. 23–2). The Coulomb force on q_0 due to q_1 is

$$\text{for a point charge:} \qquad \mathbf{F}_{01} = \frac{q_1 q_0}{4\pi\epsilon_0 r^2}\,\hat{\mathbf{r}}_{01}, \tag{23–3}$$

after Eq. (22–8). If we take q_0 as our test charge, we can use Eqs. (23–1) and (23–3) to find the electric field due to q_1:

$$\text{for a point charge:} \qquad \mathbf{E}_1 = \frac{\mathbf{F}_{01}}{q_0} = \frac{q_1}{4\pi\epsilon_0 r^2}\,\hat{\mathbf{r}}_{01}. \tag{23–4}$$

The value of the test charge has canceled, so the limiting process in Eq. (23–2) introduces no complications. Equation (23–4) specifies that $\mathbf{E}_1$ is in the same direction as $\mathbf{F}_{01}$, that of the unit vector $\hat{\mathbf{r}}_{01}$, which points from q_1 to q_0. Figure 23–3 shows the direction of $\mathbf{E}_1$ as determined by moving our test charge to various points a distance r away from q_1. This field is radial (Fig. 23–3a), and we have used the radial unit vector $\hat{\mathbf{r}}$ (measured from q_1) to specify completely the electric field **E** due to a point charge q (note that we drop the subscripts on $\hat{\mathbf{r}}_{01}$, $\mathbf{E}_1$, and q_1):

The electric field due to a point charge q is directed away from a positive charge and toward a negative charge.

$$\text{for a point charge:} \qquad \mathbf{E} = \frac{q}{4\pi\epsilon_0 r^2}\,\hat{\mathbf{r}}. \tag{23–5}$$

TABLE 23–1 Values of Electric Fields (N/C)

Interplanetary space	10^{-3}–10^{-2}
Atmosphere at Earth's surface in clear weather	100–200
Electrical breakdown of dry air	3×10^6
Just outside large sphere of a Van de Graaff accelerator	10^6
In the Fermilab particle accelerator	1.2×10^7
In atoms within the radius of an electron orbit	10^9
In the electromagnetic radiation of the most intense laser	10^{12}
Outside a uranium nucleus, at a distance from the center of twice the nucleus's radius	5×10^{20}

The electric field points *away* from a positive charge, as shown in Fig. 23–3b. When the charge is negative, the electric field has the same magnitude but is opposite in direction. The electric field due to a negative charge points *toward* that charge, as in Fig. 23–3c.

The Usefulness of the Field Concept

Once we know the electric field, **E**, produced by a point charge q, we can find the force on any point charge q' placed in that field by using Eq. (23–1); that is,

$$\mathbf{F} = q'\,\mathbf{E}. \tag{23–6}$$

More importantly, *any* distribution of charges—not simply a point charge—produces an electric field throughout space. We use the subscript "ext" (for external) on **E** to emphasize that this field is present *independent* of the charge q' on which the force acts. Once we know $\mathbf{E}_{ext}$, *the force on any point charge q' in the field* is the generalization of Eq. (23–6):

The force on a point charge in an electric field

$$\text{for a point charge in an external electric field:} \quad \mathbf{F} = q'\,\mathbf{E}_{ext}. \tag{23–7}$$

From a purely practical point of view, the field concept is a very useful one. We calculate the field from a charge distribution once and for all, and we then use Eq. (23–7) to find the effect of that distribution on other charges. Alternatively, we can use a test charge to *measure* electric fields so that we can know the subsequent effects of those fields.

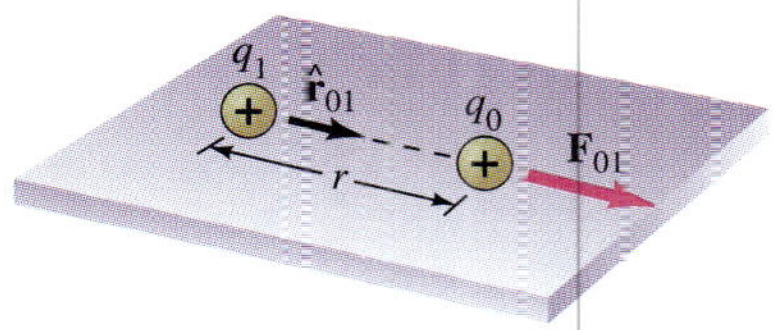

FIGURE 23–2 The force $\mathbf{F}_{01}$ exerted on point charge q_0 by point charge q_1; both charges are positive.

But why do we bother to introduce fields? Why not deal instead with forces between charges? In the chapter introduction, we mentioned the role the field plays in resolving the conceptual difficulties of action at a distance. We shall see in Chapter 25 that the field carries energy. To preserve the important idea of energy conservation, the field is a *necessary* concept. But the real power of the field concept appears when the field arises from *accelerating charges*. Even if these charges are limited to a small region (for example, within the arms of an antenna), the fields they produce—both electric and magnetic fields—spread through all of space at the speed of light. The supernova known as 1987A took place approximately 163,000 years ago; electric fields caused by the violent motion of many charges within and around the exploding star reached Earth on February 23, 1987. These traveling fields caused electrons in terrestrial antennas to move; this was the signal that supernova 1987A had occurred. This description of the process is easy to grasp. Contrast it with the idea that an electrical force exists between the charges in the supernova and those in a terrestrial detector that depends not only on the separation between the charges but also on a time lag between their respective motions. Such a force law is both difficult to express and difficult to work with.

FIGURE 23–3 (a) Threads floating in oil become aligned with the electric field of this point charge. (b) The direction of the electric field **E** due to q_1 is radial. The charge is positive, and the field points away from it. (c) The charge is negative, and the field points toward it.

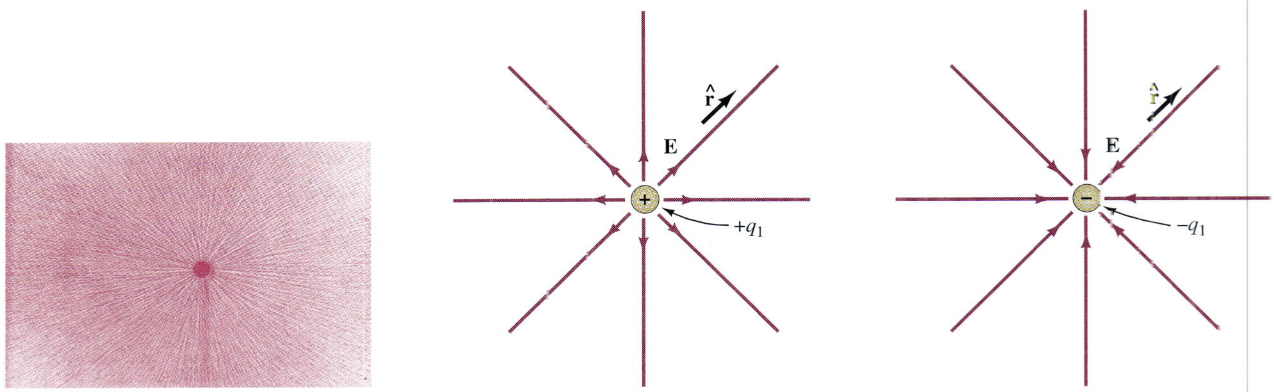

(a) (b) (c)

The notion of a field is useful in many areas of physical science. We can think of a mass distribution having a gravitational field analogous to the electric field of a charge distribution. We employ a velocity field in hydrodynamics; this field describes the velocity $\mathbf{v}$ at all points where fluid flow occurs, such as in the pipes of a city water system. In thermal physics, a temperature field describes the temperature at all points in a room. In the last case, there is no directionality to the field; we have a scalar field instead of a vector field.

EXAMPLE 23–1 Find the electric field due to a point charge $q = +1.4\ \mu\text{C}$ at a distance of 0.10 m from the charge (Fig. 23–4a). What is the force on a charge $q' = -1.2\ \mu\text{C}$ that is placed this far from q (Fig. 23–4b)?

Solution: The electric field of a point charge is given directly by Eq. (23–5):

$$\mathbf{E} = \left(\frac{q}{4\pi\epsilon_0 r^2}\right)\hat{\mathbf{r}} = \left[\frac{(9.0\times10^9\ \text{N}\cdot\text{m}^2/\text{C}^2)(1.4\times10^{-6}\ \text{C})}{(0.10\ \text{m})^2}\right]\hat{\mathbf{r}} = (1.3\times10^6\ \text{N/C})\hat{\mathbf{r}}.$$

The electric field is directed radially outward from the position of the 1.4-μC charge (Fig. 23–4b). If the charge were negative rather than positive, the field would point radially inward rather than radially outward.

To find the force on q', we treat the electric field that we have determined as an external field and use Eq. (23–7). The magnitude of the force is the magnitude of the field times the magnitude of q':

$$F = |q'|E = (1.2\times10^{-6}\ \text{C})(1.3\times10^6\ \text{N/C}) = 1.5\ \text{N}.$$

The sign of q' is negative; when it is multiplied by the magnitude of the outward radial field, the resulting force on q' acts radially inward. It is no surprise that opposite charges attract.

Alternatively, we can use Eq. (23–7) directly, including the vector notation. We get the same result as before:

$$\mathbf{F} = (-1.2\times10^{-6}\ \text{C})[(1.3\times10^6\ \text{N/C})\ \hat{\mathbf{r}}] = (-1.5\ \text{N})\ \hat{\mathbf{r}}.$$

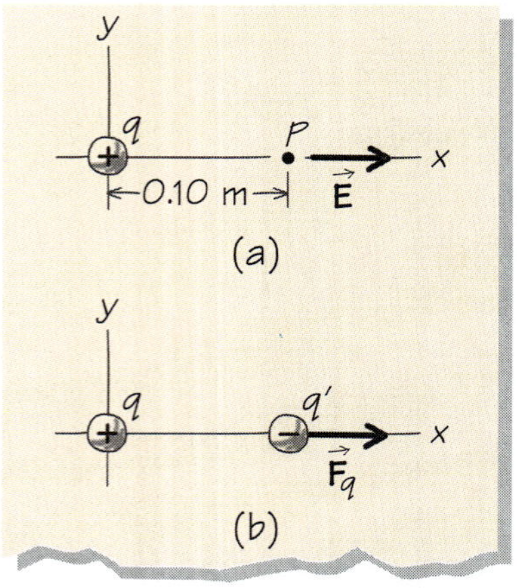

FIGURE 23–4 Example 23–1.

If more than one point charge is responsible for producing a net electric field, that field is determined by the principle of superposition. The superposition principle states that the net electric force on an object is the vector sum of the forces due to individual point charges. Therefore *the net electric field is the vector sum of the fields of individual charges present.* The net force exerted on our test charge q_0 due to all the other charges in the region is

$$\mathbf{F}_{\text{net}} = \mathbf{F}_{01} + \mathbf{F}_{02} + \mathbf{F}_{03} + \cdots = \sum_i \mathbf{F}_{0i}. \tag{23–8}$$

Thus

$$\mathbf{E}_{\text{net}} = \frac{\mathbf{F}_{\text{net}}}{q_0} = \frac{\mathbf{F}_{01}}{q_0} + \frac{\mathbf{F}_{02}}{q_0} + \frac{\mathbf{F}_{03}}{q_0} + \cdots \tag{23–9}$$

$$= \mathbf{E}_1 + \mathbf{E}_2 + \mathbf{E}_3 + \cdots = \sum_i \mathbf{E}_i. \tag{23–10}$$

In Eq. (23–10), $\mathbf{E}_2$, for example, is the electric field due solely to the charge q_2 at the point in space where we have placed q_0. By using Eq. (23–5) for each point charge q_i, we find that

The net electric field produced by a group of point charges

$$\text{for a group of point charges:}\quad \mathbf{E}_{\text{net}} = \frac{1}{4\pi\epsilon_0}\sum_i \frac{q_i}{r_i^2}\hat{\mathbf{r}}_i. \tag{23–11}$$

In this equation, the unit vector $\hat{\mathbf{r}}_i$ is directed from the position of charge q_i to the position where the field is being measured.

EXAMPLE 23–2 Consider three charges placed on a line: $q_1 = +2\ \mu C$ at $x_1 = -2$ cm, $q_2 = +3\ \mu C$ at $x_2 = +4$ cm, and $q_3 = -2\ \mu C$ at $x_3 = +10$ cm (Fig. 23–5). Find the electric field at point A, which is the origin of the coordinate system.

Solution: We solve this from a straightforward application of Eq. (23–11). Although it is generally important to remember that the required sum is vectorial, all the positions lie along a straight line in this case. The electric field at point A is

$$\mathbf{E}_A = \mathbf{E}_1 + \mathbf{E}_2 + \mathbf{E}_3,$$

where $\mathbf{E}_j$ (j = 1, 2, or 3) is the field due to charge q_j at point A. Application of Eq. (23–11) for the individual electric fields gives

$$\mathbf{E}_A = \frac{1}{4\pi\epsilon_0}\left(\frac{q_1(+\mathbf{i})}{x_1^2} + \frac{q_2(-\mathbf{i})}{x_2^2} + \frac{q_3(-\mathbf{i})}{x_3^2}\right). \tag{23–12}$$

We must pay careful attention to signs. The unit vectors $\pm\,\mathbf{i}$ in parentheses indicate the direction of the unit vector $\hat{\mathbf{r}}_j$ from the position of charge q_j to point A. The actual direction of the electric field $\mathbf{E}_j$ due to charge q_j is, however, determined by the product $q_j\,\hat{\mathbf{r}}_j$, and the sign of the charge must be taken into account. For example, the direction of $\mathbf{E}_3$ is $+\mathbf{i}$ because the negative sign of charge q_3 multiplied by $(-\mathbf{i})$ gives a direction $(+\mathbf{i})$. Numerical evaluation of Eq. (23–12) gives

$$\mathbf{E}_A = (9 \times 10^9\ \mathrm{N\cdot m^2/C^2}) \times \left[\frac{(2\times 10^{-6}\ \mathrm{C})}{(0.02\ \mathrm{m})^2}(\mathbf{i}) + \frac{(3\times 10^{-6}\ \mathrm{C})}{(0.04\ \mathrm{m})^2}(-\mathbf{i}) + \frac{(-2\times 10^{-6}\ \mathrm{C})}{(0.10\ \mathrm{m})^2}(-\mathbf{i})\right]$$
$$= (3\times 10^7\ \mathrm{N/C})\mathbf{i}.$$

The net electric field at point A is in the $+x$-direction, or toward the right.

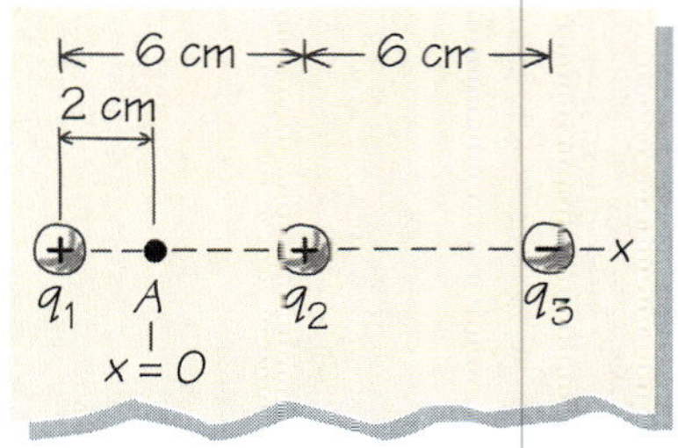

FIGURE 23–5 Example 23–2. The electric field at point A is due to three charges. The distance x is measured from point A.

Electric Dipoles and Their Electric Fields

An electric dipole

An **electric dipole** consists of two charges $+q$ and $-q$, of equal magnitude but opposite sign, that are separated by a distance L (Fig. 23–6a). The field of one charge decreases as $1/r^2$. If the two opposite charges (q_1 and $-q_2$) do not add to zero, their field would have the form $(q_1 - q_2)/r^2$ for $r >> L$. If equal and opposite charges were to sit precisely on top of one another, the two $1/r^2$ contributions to the field would cancel to a zero electric field. But because the two equal but opposite charges of a dipole are not quite on top of one another, we shall see that the resulting field decreases as $1/r^3$. The electric field depends only on the product qL, which is called the **electric dipole moment** of the neutral pair $(+q, -q)$; it is denoted by the letter p. We make $p = qL$ a vector by defining $\mathbf{L}$ to be directed from $-q$ to $+q$ (Fig. 23–6a). Thus, the electric dipole moment $\mathbf{p}$ is

$$\mathbf{p} = q\mathbf{L}. \tag{23–13}$$

The vector $\mathbf{p}$ points from the negative charge to the positive charge.

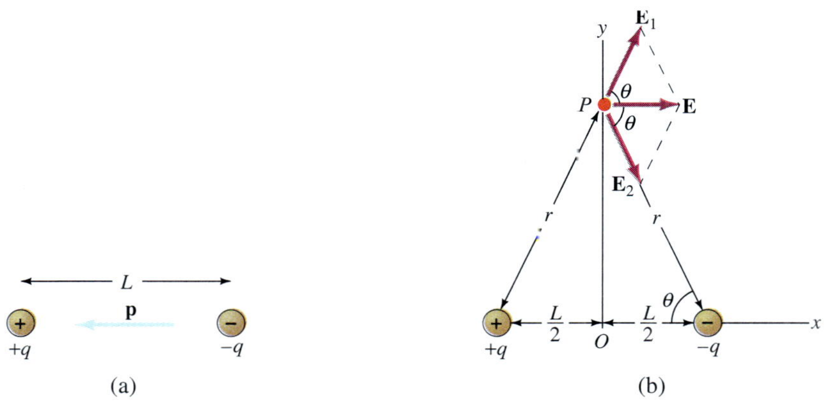

FIGURE 23–6 (a) An electric dipole consists of equal but opposite charges separated by a distance L. The electric dipole moment $\mathbf{p}$ is directed from the negative charge to the positive charge (b) The net field at point P, $\mathbf{E}$, acts only along the direction from $+q$ to $-q$.

EXAMPLE 23–3 Find the electric field of the electric dipole shown in Fig. 23–6b at a point P that is a large distance r ($r >> L$) from each charge. P lies along the perpendicular axis that bisects the line between the two charges.

Solution: The x- and y-axes are shown in Fig. 23–6b, and point P has xy-coordinates $(0, y)$. The net electric field at P is given by $\mathbf{E} = \mathbf{E}_1 + \mathbf{E}_2$, where the field $\mathbf{E}_1$ is due to the charge $+q$ and the field $\mathbf{E}_2$ is due to the charge $-q$. The magnitudes of the two fields are the same, but $\mathbf{E}_1$ points away from $+q$, whereas $\mathbf{E}_2$ points toward $-q$. The y-components of $\mathbf{E}_1$ and $\mathbf{E}_2$ exactly cancel each other, and we are left with a net x-component toward the right that is twice the x-component of the field due to either charge:

$$\mathbf{E} = E_x\,\mathbf{i} = (E_{1x} + E_{2x})\mathbf{i} = 2E_{1x}\,\mathbf{i},$$

where

$$E_{1x} = \frac{q}{4\pi\epsilon_0 r^2}\cos\theta.$$

From Fig. 23–6b, we see that $\cos\theta$ is given by

$$\cos\theta = \frac{L/2}{r} = \frac{L}{2r}.$$

Thus the total electric field of the dipole along the perpendicular bisector is

$$\mathbf{E} = \left(\frac{2q}{4\pi\epsilon_0 r^2}\right)\left(\frac{L}{2r}\right)\mathbf{i} = \frac{qL}{4\pi\epsilon_0 r^3}\,\mathbf{i}. \tag{23–14}$$

Equation (23–14) is the correct result along the perpendicular bisector even when the distance from the charge pair is not large. The electric field decreases with r as $1/r^3$. The field at points *along the bisecting axis* is given by

$$\mathbf{E} = -\frac{\mathbf{p}}{4\pi\epsilon_0 r^3}, \tag{23–15}$$

where we have used Eq. (23–13) for the electric dipole moment, $\mathbf{p}$. If $r >> L$, then $r \simeq y$ and

$$\text{along the bisecting axis:}\quad \mathbf{E} \simeq -\frac{\mathbf{p}}{4\pi\epsilon_0 y^3}. \tag{23–16}$$

The electric field from a dipole is *not* antiparallel to the dipole moment everywhere in space, although that is the case along the bisecting axis [Eqs. (23–15) and (23–16)].

The Importance of Electric Dipoles

Induced and permanent electric dipoles

In Example 23–3, we found that the electric dipole field along a bisecting axis does not depend on either q or L alone, but rather on their product. This is true for the dipole field at any point in space. Only the product $p = qL$ can be determined from the field of an electric dipole—not L or q separately.

Electric dipoles are of great interest because they occur so often in nature. An electric field is produced even though the total charge of a dipole is zero. External fields frequently induce charge separations in electrically neutral molecules and materials, leading to an excess of positive (or negative) charge on one side (and the other) and hence to an **induced electric dipole moment** (Fig. 23–7). There are also examples in nature of charge configurations with **permanent electric dipole moments** (dipole moments that are not induced by external fields). Many molecules that have an extended structure, with negatively charged electrons distributed preferentially in certain regions, have permanent electric dipole moments. Water is an example (Fig. 23–8). In cases such as common salt (NaCl) and hydrochloric acid (HCl), electrons cluster preferen-

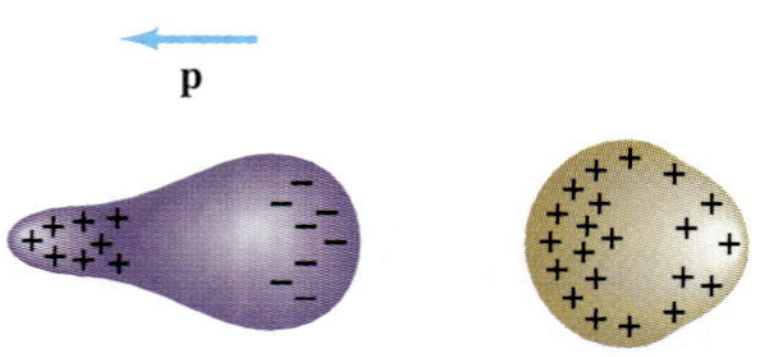

FIGURE 23–7 A nearby charge can induce a polarized charge, and hence an electric dipole, on a neutral object.

tially around one atom, giving that atom a negative charge. The other atom is left with a positive charge. In such molecules, there is always a permanent electric dipole moment. On the molecular level, where the effects of electric dipole fields have great physical importance, permanent dipole moments are always much larger than induced dipole moments. For example, $p \simeq 6 \times 10^{-30}$ C·m for a water molecule, whereas a hydrogen atom in the rather strong field $E = 3 \times 10^6$ N/C acquires an induced dipole moment of $p \simeq 3 \times 10^{-34}$ C·m.

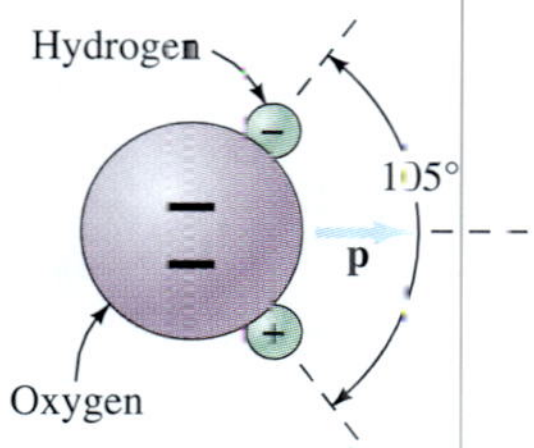

FIGURE 23–8 The water molecule, H_2O, is a permanent electric dipole. Both hydrogen electrons are shared with the oxygen atom, creating a strong electric bond that holds the molecule together (by what is called covalent bonding).

23–2 ELECTRIC FIELD LINES

The electric field due to a charge distribution and the force experienced by charged particles in the field can be visualized in terms of **electric field lines**. Their use was introduced by Michael Faraday around the middle of the nineteenth century—even before the concept of the electric field was clearly understood. Faraday used the phrase "lines of force."

We have already seen that we can map out the electric field by moving a test charge around in space. The field is easily expressed in algebraic form and is the best tool for obtaining algebraic or numerical results about electric forces. The field, however, is awkward to use in a visual sense. It is not easy to draw for each point of a region of space a vector whose varying length and direction represents the field. Electric field lines are continuous lines in space; they are more suited to visual representation than the field itself, and carry as much information as the field does.

Electric field lines

Electric field lines are smooth directional lines in space that are determined by the electric field, according to two simple rules:

1. Electric field lines are drawn so that the tangent to the field line at each point specifies the direction of the electric field **E** at that point. This rule relates the *direction* of the electric field lines to the direction of the electric field.

2. The *density* in space of electric field lines around a particular point is proportional to the strength of the electric field at that point.

PROBLEM SOLVING

Techniques to help you draw electric field lines

Properties of Electric Field Lines

Let's draw the electric field lines of a positive point charge q. We know that the electric field points radially away from a positive charge at every location in space. The field has the same magnitude all around a sphere centered on the charge, and this magnitude decreases with the distance r from the charge as $1/r^2$. Electric field lines are radial, pointing outward from the charge, and are distributed uniformly about the charge (Fig. 23–9a). We can use Fig. 23–9 to illustrate their properties.

Property One. The second electric field line rule, which states how electric field lines reflect the strength of the associated electric field, requires some explanation. The electric field does not change its direction abruptly as we move through a region of charge-free space. Thus, in a small region, electric field lines are very nearly parallel to one another. In this small region, we can take a small area that is oriented perpendicular to the nearly parallel field lines. The density of electric field lines is then the number of electric field lines that cross this small area, divided by that area. Note that the density is the number of lines *per unit area.*

Property Two. How do we set the number of electric field lines and the density of electric field lines? For convenience, we can *choose* to draw N field lines that originate at a given charge q; N is any number. Then the number of lines that leave the other charges is determined. For example, the number of field lines leaving a positive charge $q/2$ is $N/2$ and, more generally, the number of field lines leaving a positive charge q_i is $N_i = (q_i/q)N$. Now we can use the rule about the density of lines to show that *lines can start or terminate only on charges, never in empty space.* Suppose that N lines

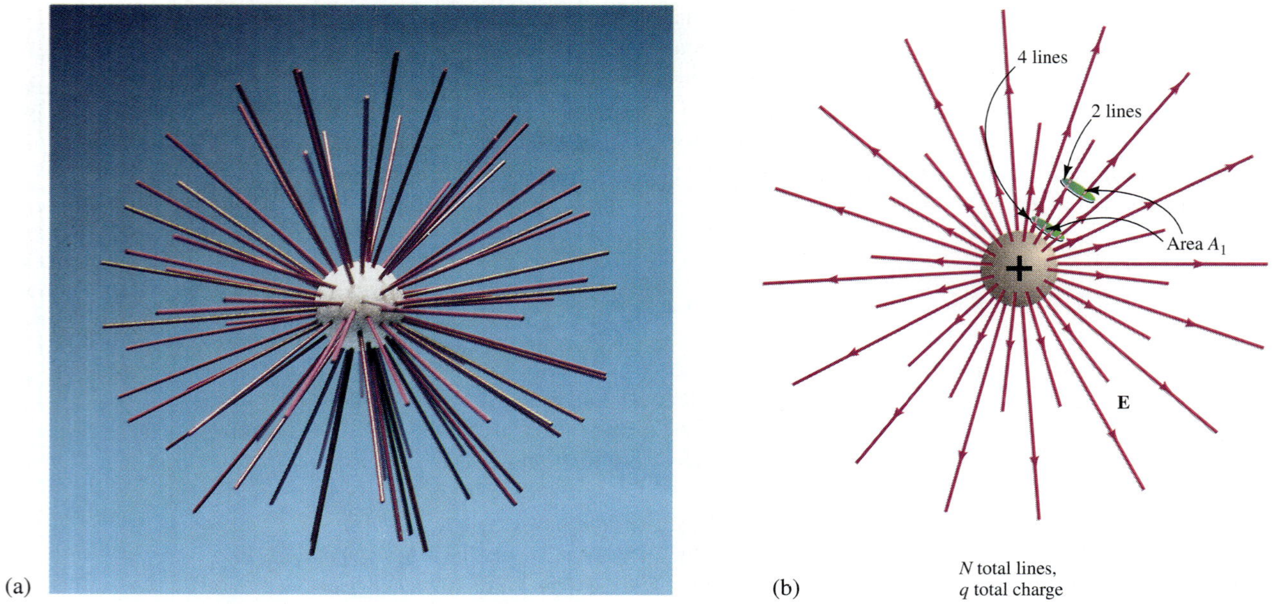

FIGURE 23–9 (a) Representation of the radial electric field lines from a point charge. (b) Fewer field lines pass through the same-size area farther from the charge. Note that field lines extend out to infinity.

originate at the point charge q in Fig. 23–9b, and that lines are neither created nor destroyed. If we consider the number of field lines that pass through an area the size of a dime, we see from the figure that many more field lines would pass through this area when the area is close to the point source than when the same area is far away. *If no new electric field lines are created as we move away from the charge*, then the density of lines at a radius R from the charge will be N divided by the area of the surface perpendicular to the lines. This surface is a sphere of radius R centered on the charge, and the density of lines is $N/4\pi R^2$. We know that the density of lines is proportional to the strength of the field, and that the field falls off as $1/r^2$. Therefore, *the connection between the strength of the field and the density of the electric field lines is automatic if electric field lines are neither created nor destroyed in regions where there is no charge.* We have demonstrated this only for a single point charge but, with the help of the superposition principle, we can show that this is true in general. Note that the electric field lines of a point charge go off to infinity. This will be true for any localized collection of charges with a net charge. At distances that are large compared with the dimensions of the region that contains such a collection, the net charge appears to be localized at a point, and electric field lines will be distributed evenly over a distant sphere that surrounds the net charge.

Property Three. Electric field lines originate on, and run outward from, positive charges. They run toward, and terminate on, negative charges. This reflects the fact that electric charges are the sources of electric fields, which point away from positive charges and toward negative charges.

Property Four. *No two field lines ever cross.* They cannot cross because the electric field has a definite magnitude and direction at any point in space. If two or more electric field lines were to cross at some point, then the direction of the electric field at that point would be ambiguous.

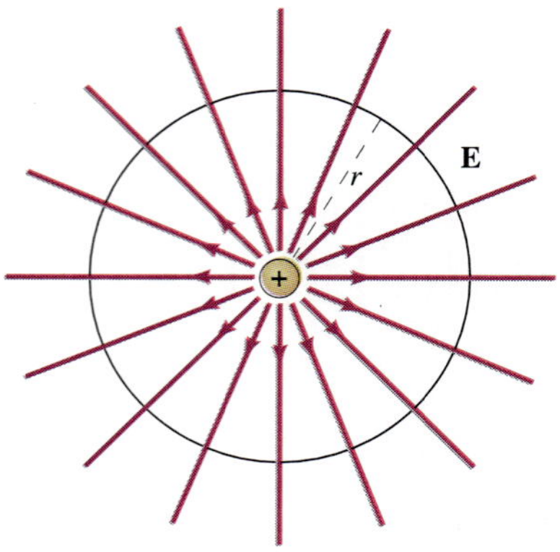

FIGURE 23–10 Electric field lines due to a point charge $+q$. Note the number of field lines that cross the circle (sphere) at radius r.

Symmetry is often a useful guide to drawing electric field lines. A point charge looks the same when viewed from any direction. It has spherical symmetry, and the field lines follow the only direction that respects this symmetry—namely, they are radial. Similarly, if we are dealing with a long line of charge, there is a symmetry around the line, and the field lines must project radially outward from the line; they would be perpendicular to a cylinder that surrounds the line.

We often draw field lines on a flat page. For example, you might draw Fig. 23–10 as the representation of the field lines for an isolated charge. *Such a drawing should not be used carelessly for determining the field strength.* You cannot simply count the field lines that cross a particular line. Figure 23–10 shows a circle of radius r centered on a positive charge. The number of lines that cross this circle is fixed at N, so the density of lines that cross the circle is N divided by the circumference of the circle, or $N/2\pi r$. Yet we know that the field decreases as $1/r^2$, not as $1/r$. The density of lines, which determines the magnitude of the electric field, is a density per unit *area*, not per unit length. Nevertheless, these planar drawings of electric field lines remain useful for visualizing the field and its effect on other charges.

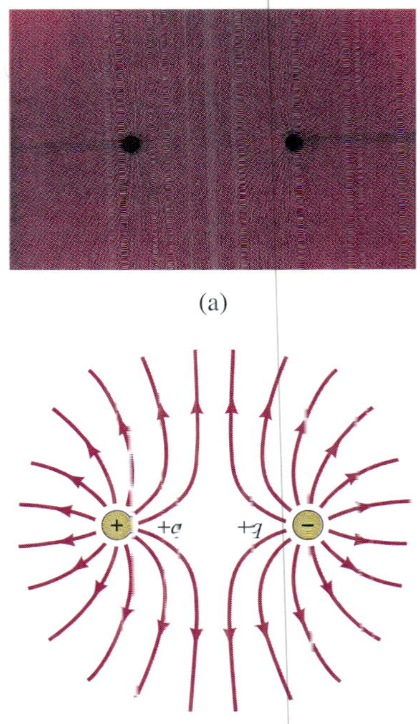

FIGURE 23–11 (a) The electric field lines due to two point charges $+q$, shown by threads in oil. (b) Schematic diagram of the field lines, which go off to infinity and appear to repel each other.

Some Examples

The easiest way to demonstrate the usefulness of electric field lines is to look at several examples beyond that of the isolated point charge. Figures 23–11a and 23–11b show the electric field lines on a plane that passes through two positive charges of equal magnitude. The field lines all extend to infinity, because there are no negative charges on which the lines can terminate. The field lines that approach each other between the two positive charges appear to repel each other because two field lines cannot cross. If we were to place a positive charge q' in the region shown in Fig. 23–11b, the field lines would show us the direction of the force on the charge (and likewise the acceleration). Once we have the field lines, it is easy to see the direction of force a given charge would have in the electric field. We emphasize that, although the electric field itself has physical meaning, electric field lines are simply an aid to picturing the electric field and how a charge would react when placed in that field.

Figures 23–12a and 23–12b depict the field lines of an electric dipole. The charges have equal magnitude, $\pm q$, so an equal number of field lines are attached to them, and every field line that originates on $+q$ terminates at $-q$. Near each charge the field lines are purely radial, but they must deviate from the radial direction in order to reach the other charge. Notice that the field lines in Fig. 23–12b are consistent with the field $\mathbf{E}$ determined in Example 23–3 (compare Figs. 23–12b and 23–6).

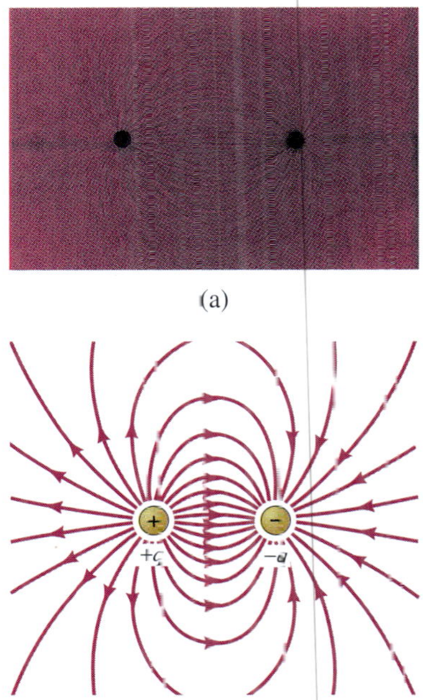

FIGURE 23–12 (a) The electric field lines due to point charges $+q$ and $-q$, a dipole, as indicated by threads in oil. (b) Schematic diagram of the field lines, all of which begin on $+q$ and end on $-q$; those that appear to be broken actually continue far from the charge.

EXAMPLE 23–4 Draw the electric field lines for a system of two charges, $+2q$ and $-q$.

Solution: The positive charge has twice the charge of the negative one, and we decide arbitrarily to show 24 lines that originate from $+2q$. Accordingly, only 12 lines will terminate at $-q$ (Fig. 23–13a). We draw these lines in two dimensions. Near the charges, we draw the field lines as radial. Twelve of the lines that originate from $+2q$ must terminate on $-q$. None of the field lines may cross, so we take the 12 lines of $+2q$ nearest to $-q$ to be the lines that terminate at $-q$ (Fig. 23–13b). What happens to the remaining 12 lines that emerge from $+2q$? Although they will initially curve toward $-q$, we realize that, very far away, they will be pointing radially outward from a net charge $+q$ $(+2q - q = +q)$. The fact that 12 lines remain is consistent with our original choice of 24 lines for the line density (Fig. 23–13c).

23-3 THE FIELD OF A CONTINUOUS DISTRIBUTION

We have thus far concentrated on electric fields due to point charges or collections of point charges. But *continuous* distributions of charge also produce fields, and such distributions are very important in practice. We will consider charges that are distributed *uniformly* throughout a region in space, whether a line, a surface, or a volume. We will also emphasize distributions where there is symmetry. For charge distributions that are not uniform or not symmetrical, the problem of determining the resulting electric field can be more complex.

We have already set up a general framework for calculating electric fields due to line, surface, and volume distributions (Chapter 22). Consider the calculation of the

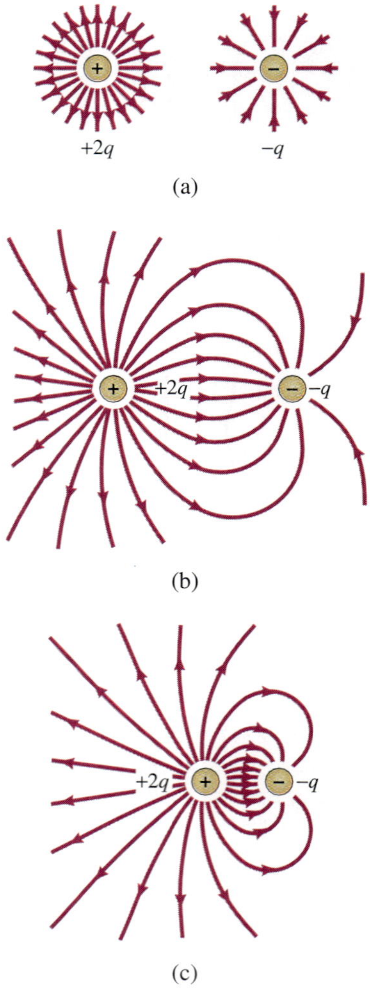

FIGURE 23–13 Example 23–4. (a) The electric field lines close to the $+2q$ and $-q$ point charges are those of a point charge. (b) Half the electric field lines that emerge from $+2q$ end up on $-q$. (c) Far from the point charges, the electric field lines are those of a point charge $+q$.

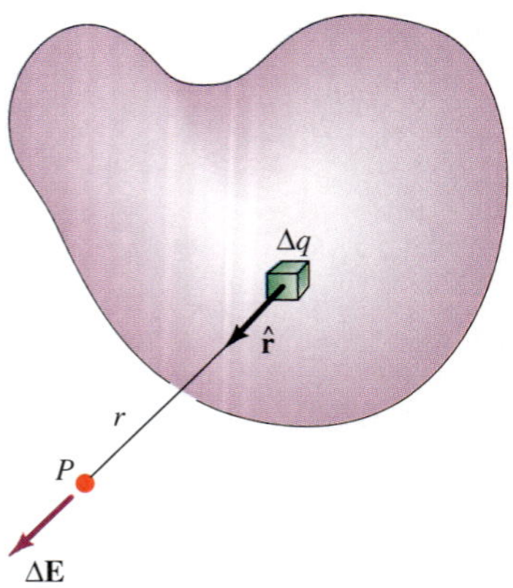

FIGURE 23–14 To find the electric field due to a continuous charge distribution, add all the electric fields $\Delta\mathbf{E}$ due to the charge elements Δq.

electric field at point P due to the charge distribution shown in Fig. 23–14. We divide the charge distribution into tiny elements, each of charge Δq. We first find the electric field $\Delta\mathbf{E}$ at an external point P that is due to a tiny charge element Δq, whose distance from P is r:

$$\Delta\mathbf{E} = \frac{\Delta q}{4\pi\epsilon_0 r^2}\hat{\mathbf{r}}. \tag{23–17}$$

Here, $\hat{\mathbf{r}}$ is the unit vector pointing away from the charge element. Superposition applies, and the total electric field at P is found by summing the infinitesimal fields $\Delta\mathbf{E}$:

$$\mathbf{E} = \sum \Delta\mathbf{E}. \tag{23–18}$$

In differential notation, Eq. (23–17) becomes

$$d\mathbf{E} = \frac{dq}{4\pi\epsilon_0 r^2}\hat{\mathbf{r}}. \tag{23–19}$$

The total electric field is found by integrating over the entire charge distribution:

$$\mathbf{E} = \sum_{\substack{\text{lim} \\ \Delta q \to 0}} \Delta\mathbf{E} = \int d\mathbf{E} = \frac{1}{4\pi\epsilon_0}\int \frac{dq}{r^2}\hat{\mathbf{r}}. \tag{23–20}$$

The formal expression of Eq. (23–20) is identical to the one we found for the force on a point charge q due to a charge distribution, Eq. (22–11), divided by q. This is exactly how we define the electric field.

The work of Chapter 22 tells us how to find the electric field due to a charge distribution along a line in terms of the linear charge density λ, the electric field due to a charge spread over a surface in terms of the surface charge density σ, and the electric field due to a charge distribution over a volume in terms of a volume charge density ρ (Fig. 23–15). In each case, the electric field is simply the expression for the force on the charge q divided by q; that is, Eqs. (22–13), (22–15), and (22–17) divided by q, respectively.

Constant Charge Densities

An important simplification for calculating the electric field due to a charge distribution occurs when the charge distribution, whether one-, two-, or three-dimensional, is *uniform*. This means that the charge densities are constants that can be removed from beneath the integrals expressing the electric field. Moreover, we can express the charge density in terms of the total charge Q contained in the distribution and the size of the distribution. In particular, if charge Q is distributed uniformly along a line segment of length L on a line, the linear charge density λ is

$$\lambda \equiv \frac{Q}{L}. \tag{23–21}$$

Dividing Eq. (22–13) by the test charge q, we have

$$\mathbf{E} = \frac{\lambda}{4\pi\epsilon_0}\int \hat{\mathbf{r}}\,\frac{dx}{r^2}. \tag{23–22}$$

Here, r is the distance to the point P at which the field is to be calculated, while $\hat{\mathbf{r}}$ is a unit vector from the segment within the sum to point P. For a charge Q distributed uniformly on a surface of area A, the surface charge density σ is

$$\sigma \equiv \frac{Q}{A}. \tag{23–23}$$

Then

$$\mathbf{E} = \frac{\sigma}{4\pi\epsilon_0} \int_{\text{Surface}} \hat{\mathbf{r}} \frac{dS}{r^2}. \tag{23–24}$$

Finally, if charge Q is distributed uniformly throughout a volume V, the volume charge density ρ is

$$\rho \equiv \frac{Q}{V}. \tag{23–25}$$

and the electric field is

$$\mathbf{E} = \frac{\rho}{4\pi\epsilon_0} \int_{\text{Volume}} \hat{\mathbf{r}} \frac{dV}{r^2}. \tag{23–26}$$

Examples 23–5 and 23–6 illustrate how integrations for the electric field of a charge distribution are performed in practice.

EXAMPLE 23–5 A straight insulating rod of length $2L$ carries a uniform linear charge density λ. Determine the electric field at point P, a distance R from the rod along the perpendicular bisector (Fig. 23–16). First find the field in the limit that the rod is much longer than $R (L >> R)$. Then find it for a distance very far from the rod $(R >> L)$.

Solution: Equation (23–22) applies because the charge distribution is linear. The origin is the midpoint of the rod, which we place along the y-axis. We use Eqs. (23–20) and (23–22), with $\hat{\mathbf{r}} = \cos\theta\,\mathbf{i} - \sin\theta\,\mathbf{j}$ (Fig. 23–16) to find that

$$\mathbf{E} = \frac{\lambda}{4\pi\epsilon_0} \int_{-L}^{L} \frac{dy}{r^2} (\cos\theta\,\mathbf{i} - \sin\theta\,\mathbf{j}). \tag{23–27}$$

The charge dq at a distance y *below* the x-axis gives rise to a field $d\mathbf{E}$ that is a mirror image of the field $d\mathbf{E}$ due to another charge dq' at a distance y *above* the axis. Thus we expect the net y-component of the field to vanish by symmetry. Here, we shall demonstrate this formally by performing the integration; normally, we would take advantage of the symmetry to reduce the mathematical calculation.

It is often true that the key to performing integrations such as that of Eq. (23–27) is to find the right variables. In this case, the simplest variable to use is the angle θ. Both y and $\mathbf{r}$ depend on θ, and we must change the integration variable from y to θ. We must find the dependence of both y and r on θ. We have

$$\tan\theta = \frac{y}{R} \tag{23–28}$$

and

$$\cos\theta = \frac{R}{r}. \tag{23–29}$$

From Eq. (23–28), we get

$$dy = R\,d(\tan\theta) = R\sec^2\theta\,d\theta = \frac{R}{\cos^2\theta}\,d\theta.$$

With Eq. (23–29), the combination dy/r^2 that appears in Eq. (23–27) is

$$\frac{dy}{r^2} = \frac{1}{r^2}\frac{R}{\cos^2\theta}\,d\theta = \frac{1}{r^2}\frac{R}{(R/r)^2}\,d\theta = \frac{1}{R}\,d\theta.$$

The factor $1/R$ is a constant and comes out of the integral, leaving

$$\mathbf{E} = \frac{\lambda}{4\pi\epsilon_0 R} \int_{-\theta_0}^{\theta_0} (\cos\theta\,\mathbf{i} - \sin\theta\,\mathbf{j})\,d\theta.$$

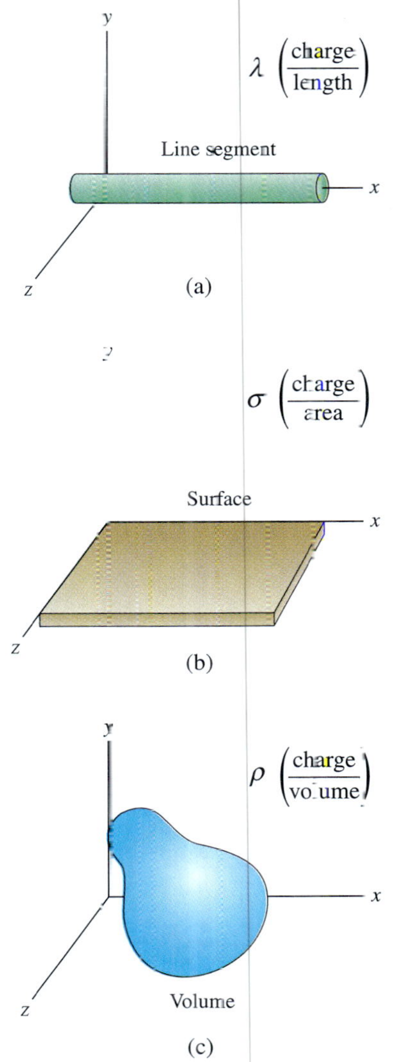

FIGURE 23–15 (a) One-dimensional, (b) two-dimensional, and (c) three-dimensional charge distribution. The charge density is labeled λ, σ, and ρ, respectively.

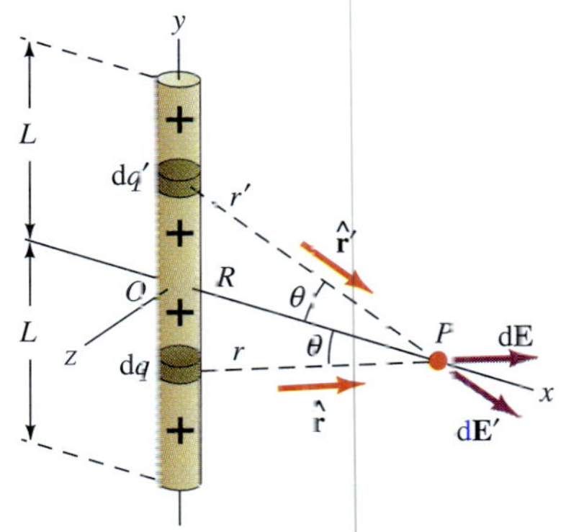

FIGURE 23–16 Example 23–5. The electric field due to a rod of length $2L$ that carries a uniform charge density λ, at a distance R from the rod. Note that the y-components of $d\mathbf{E}$ and $d\mathbf{E}'$ cancel each other.

The electric field near a long, uniformly charged straight rod

The limits $-\theta_0$ and θ_0 are the maximum values of θ, corresponding to the two ends of the line of charge. The integrals are elementary. The coefficient of **j**, which is the y-component of the field, is proportional to $\cos\theta_0 - \cos(-\theta_0) = 0$. Thus, as we expected, there is no y-component to the field. The coefficient of **i** is the x-component,

$$E_x = \frac{\lambda}{4\pi\epsilon_0 R}\int_{-\theta_0}^{\theta_0}\cos\theta\, d\theta = \frac{\lambda}{4\pi\epsilon_0 R}\sin\theta\Bigg|_{-\theta_0}^{\theta_0} = \frac{\lambda}{2\pi\epsilon_0 R}\sin\theta_0. \qquad (23\text{–}30)$$

We can use $\sin\theta_0 = L/\sqrt{L^2 + R^2}$, if desired.

For a rod with length $L >> R$, $\sin\theta_0 \simeq 1$, and the component E_x given by Eq. (23–30) becomes in this limit

$$\text{for } L >> R: \qquad E_x = \frac{\lambda}{2\pi\epsilon_0 R}. \qquad (23\text{–}31)$$

Equation (23–31) gives the electric field for an almost infinitely long rod (or for a point very close to a finite rod). The direction of the field is perpendicular to the rod. Notice that Eq. (23–31) states that the electric field varies as $1/R$ for an infinitely long rod, as opposed to the inverse-square dependence ($1/R^2$) for the point charge. The reason is that there is an infinite amount of charge in an infinite rod with a finite charge density. The summation over all the fields from charges in the rod—even the ones that are very distant from the point at which the field is measured—builds up a net field that decreases more slowly than the field of a finite charge distribution.

For the case of $R >> L$, $\sin\theta_0 \simeq L/R$, and in this limit Eq. (23–30) becomes

$$\text{for } R >> L: \qquad E_x = \frac{\lambda L}{2\pi\epsilon_0 R^2} = \frac{Q}{4\pi\epsilon_0 R^2}, \qquad (23\text{–}32)$$

where $Q = 2\lambda L$ is the total charge on the rod. In this case ($R >> L$), we have obtained the point-charge result because a rod of finite length looks like a point when it is viewed from large distances.

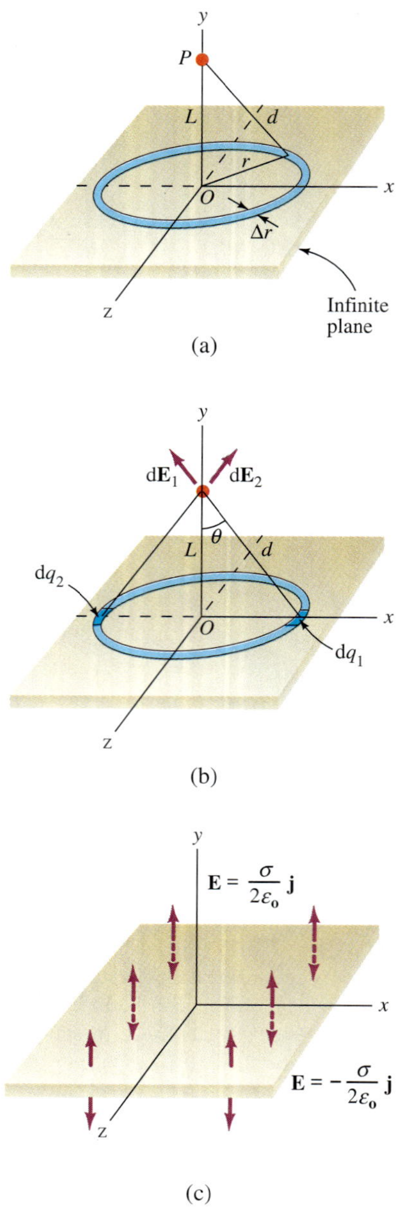

FIGURE 23–17 Example 23–6. (a) To find the electric field at a point P a distance L above an infinitely charged plane, break up the plane into concentric circles about the axis passing through P. (b) The field due to one such circle has no horizontal component at P because the horizontal contributions from opposite points on the circle cancel; moreover, the vertical component is easy to find because every point of the circle is equidistant from P. (c) The net field after summation from all circles is uniform and perpendicular to the plane.

EXAMPLE 23–6 Find the electric field at a distance L from an infinite plane sheet with a uniform surface charge density σ. (This example relates to capacitors, important circuit elements.)

Solution: Refer to Fig. 23–17a, where we establish the xy-plane in the plane sheet. We want to find the electric field at a point P, which is a distance L above the plane, and we choose the y-axis so that P is on it.

To integrate the effect of the entire plane, we break it up into pieces for which it is simple to find the field. One good way to break up the plane is with a series of thin concentric circular regions centered around a point below P (Fig. 23–17a). These circles have a width Δr that is so small that the area of the circular regions is $2\pi r\Delta r$. To see why this is a good choice, first note that the magnitude of the field at P from any given tiny segment of the circle is the same as from any other segment of the same size because every point on the circle is exactly the same distance d from P. If we denote the field from the entire ring as $\Delta\mathbf{E}$, then this field has magnitude

$$\Delta E = \frac{\Delta Q}{4\pi\epsilon_0 d^2} = \frac{\sigma(2\pi r\,\Delta r)}{4\pi\epsilon_0 d^2}.$$

Here, we have written the charge ΔQ in the ring as the surface charge density times the area of the ring.

Next let's think about the *direction* of the field from the circular region. Figure 23–17b shows that the horizontal components of $\Delta\mathbf{E}$ cancel from opposite segments, and that only the vertical, or y-, component remains. This component has magnitude

$$\Delta E_y = \Delta E\cos\theta = \Delta E\frac{L}{d} = \frac{\sigma(2\pi r\,\Delta r)}{4\pi\epsilon_0 d^2}\frac{L}{d} = \frac{\sigma L}{2\epsilon_0}\frac{r\,\Delta r}{(r^2 + L^2)^{3/2}}.$$

We have used the fact that $d = \sqrt{r^2 + L^2}$. In thinking about the direction of $\Delta\mathbf{E}$, symmetry has been an important guide.

Now we can see that a sum over rings of all different radii r from 0 to ∞ covers the entire plane. Thus we can find the field due to the entire plane by summing all the ΔE_z. This sum is a single integral over the radii r. We have

$$E_z = \sum \Delta E_z = \sum \frac{\sigma L}{2\epsilon_0} \frac{r\,\Delta r}{(r^2 + L^2)^{3/2}} = \frac{\sigma L}{2\epsilon_0} \int_0^\infty \frac{r\,dr}{(r^2 + L^2)^{3/2}}.$$

The integral can be looked up in a table of integrals:

$$\int_0^\infty \frac{x\,dx}{(x^2 + L^2)^{3/2}} = \frac{-1}{(x^2 + L^2)^{1/2}}\bigg|_{x=0}^{x=\infty} = \frac{1}{L}.$$

Thus the field due to the entire plane has magnitude $\sigma/2\epsilon_0$ and is oriented in the z-direction:

$$\text{for a uniformly charged plane:} \qquad \mathbf{E} = \frac{\sigma}{2\epsilon_0}\mathbf{j}. \qquad (23\text{–}33)$$

The electric field from a large, uniformly charged plane

The final result has the electric field everywhere perpendicular to the plane and *constant* in both magnitude and direction: The field **E** does not even depend on how far the point is from the plane (Fig. 23–17c). This is reasonable physically: If the plane is infinite and has a uniform charge distribution, it looks the same from everywhere.

In reality, we cannot have planes of infinite extent. For finite planes, the result above holds for distances much closer to the finite plane than the distance to the edge of the plane.

PROBLEM SOLVING

Reducing multiple integration to single integration

The preceding example illustrates how a formal two-dimensional integration can be reduced to a single integration. There are often several ways to approach this type of problem. For example, we could have divided our plane into narrow straight strips. The field due to a strip was found in Example 23–5, and we could have used that result. The effect of the entire plane is then found by summing over all the strips making it up. Once again, the integral would have been a single integral, and the integration variable would have been the differing distances of the strips from point P.

The Electric Field between Two Uniformly Charged Planes with Opposite Charge

Example 23–6 shows that the electric field for a positively charged plane of uniform surface charge density σ is uniform and directed away from the plane perpendicularly (Fig. 23–18a). If the plane were negatively charged, the field would be similar but would be directed *toward* the plane (Fig. 23–18b). What happens if we place the two planes, oppositely charged but with the same magnitude of charge

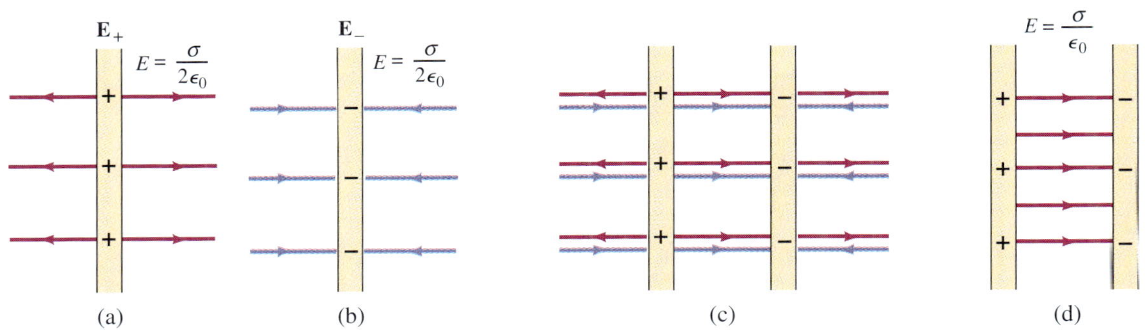

FIGURE 23–18 (a) The electric field due to a positively charged plane is directed away from the plane; (b) that due to a negatively charged plane is directed into the plane. (c) With two parallel planes carrying equal but opposite charge, the electric field cancels to zero outside the planes but is additive inside. (d) The field inside is σ/ϵ_0 and is directed from the positive plane to the negative plane.

density σ, parallel to each other? As shown in Fig. 23–18c, the fields outside the parallel planes will exactly cancel each other, but the fields between the planes are additive. The resulting field is shown in Fig. 23–18d. For two parallel, oppositely charged planes, the electric field is zero everywhere except between the planes, where the field has magnitude

$$\text{for parallel planes of opposite uniform charge:} \qquad E = \frac{\sigma}{\epsilon_0} \tag{23–34}$$

and is directed from the positively to the negatively charged plane (remember that the direction of the electric field is always the direction of the force on our positive test charge q_0).

23-4 MOTION OF A CHARGE IN A FIELD

We have been concerned with the construction of the electric field of a given collection of charges. Let's turn now to the force that charged particles will experience in an external electric field. Newton's second law becomes

$$\mathbf{F} = q\mathbf{E}_{\text{ext}} = m\mathbf{a}, \tag{23–35}$$

where a particle of mass m and charge q has an acceleration **a** due to a given external electric field $\mathbf{E}_{\text{ext}}$. We then solve Newton's second law as usual. Example 23–7 demonstrates the procedure.

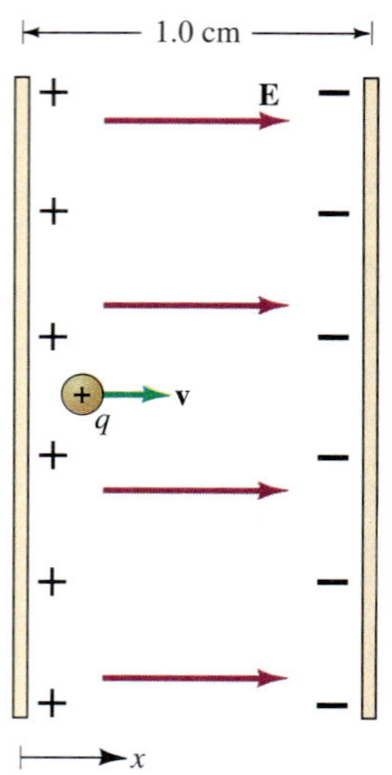

FIGURE 23–19 Example 23–7. A charge $+q$ moving between parallel plates.

EXAMPLE 23–7 Consider two oppositely charged parallel plates (Fig. 23–19). The magnitude of the surface charge density on each plate has a constant value of $\sigma = 1.0 \times 10^{-6}\ \text{C/m}^2$, and the plates are 1.0 cm apart. (a) If a proton is released from rest near the positively charged plate, with what speed will it strike the negatively charged plate? (b) What will the proton's transit time be?

Solution: (a) We first calculate the electric field and the acceleration of the proton; then we can use kinematic relations to determine the speed and transit time. Refer to the coordinate system in Fig. 23–19. The field **E** has only an x-component, $E_x = \sigma/\epsilon_0$, according to Eq. (23–34). From Eq. (23–35), the acceleration a_E due to the electric field is

$$a_E = a_x = \frac{qE_x}{m} = \frac{q\sigma}{m\epsilon_0}$$

$$= \frac{(1.6 \times 10^{-19}\ \text{C})(1.0 \times 10^{-6}\ \text{C/m}^2)}{(1.67 \times 10^{-27}\ \text{kg})(8.85 \times 10^{-12}\ \text{C}^2/\text{N}\cdot\text{m}^2)} = 1.08 \times 10^{13}\ \text{m/s}^2, \tag{23–36}$$

where we use the known charge q and mass m of the proton.

The problem is now one of one-dimensional kinematics with constant acceleration. From Section 2–5, with $a_E = a$, we have $v^2 - v_0^2 = 2ax$. But because the initial speed v_0 is zero,

$$v^2 = 2ax. \tag{23–37}$$

We insert the value of the acceleration from Eq. (23–36) and the distance traveled between the plates ($x = 1.0$ cm) to find that

$$v^2 = 2(1.08 \times 10^{13}\ \text{m/s}^2)(1.0 \times 10^{-2}\ \text{m}) = 2.2 \times 10^{11}\ \text{m}^2/\text{s}^2;$$

$$v = 4.7 \times 10^5\ \text{m/s}.$$

(b) Because the proton starts from rest, the transit time is found by dividing the final speed by the acceleration:

$$t = \frac{v}{a} = \frac{4.7 \times 10^5\ \text{m/s}}{1.08 \times 10^{13}\ \text{m/s}^2} = 4.3 \times 10^{-8}\ \text{s}.$$

The plates accelerate protons and thus represent a charged-particle accelerator.

Deflection of Moving Charged Particles

Let's consider what happens when we inject a charged particle into a region of uniform **E** between two plates. The particle has initial velocity $\mathbf{v}_0$ perpendicular to **E** (Fig. 23–20). For practical consideration, assume that the particle has negative charge (an electron, for example), so that in Fig. 23–20, it will be deflected up. From Eq. (23–35) the acceleration vector is

$$\mathbf{a} = a_x\,\mathbf{i} + a_y\,\mathbf{j} = \frac{qE}{m}\mathbf{j}. \tag{23–38}$$

Note that the x-component of the acceleration is zero. Because the initial velocity is only in the x-direction ($\mathbf{v}_0 = v_0\mathbf{i}$), the velocity vector becomes

$$\mathrm{v} = v_x\,\mathbf{i} + v_y\,\mathbf{j} = v_0\,\mathbf{i} + \frac{qE}{m}t\,\mathbf{j}. \tag{23–39}$$

The charged particle travels a horizontal length L_1 between the charged plates in the time T, determined by

$$x = v_0T = L_1; \tag{23–40}$$

$$T = \frac{L_1}{v_0}. \tag{23–41}$$

The particle's deflection in the y-direction is then

$$y = \frac{1}{2}a_y t^2 = \frac{1}{2}\frac{qE}{m}T^2 = \frac{1}{2}\frac{qE}{m}\frac{L_1^2}{v_0^2}. \tag{23–42}$$

The charged particle emerges from the plates at a position (x, y) given by Eqs. (23–40) and (23–42). The charged particle is then free from the influence of any force (ignoring gravity) and continues past the plates in a straight line at an angle θ from its initial direction:

$$\tan\theta = \frac{v_y}{v_x} = \frac{(qE/m)(L_1/v_0)}{v_0} = \frac{qEL_1}{mv_0^2}. \tag{23–43}$$

APPLICATION
The oscilloscope

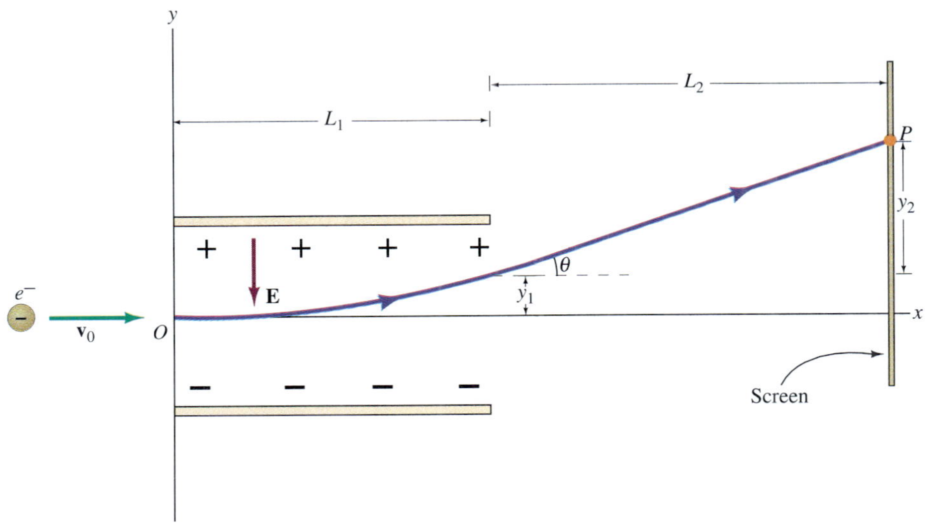

FIGURE 23–20 Example 23–8. An electron passing between vertical-deflection plates.

FIGURE 23–21 Schematic diagram of a cathode-ray tube, an element of oscilloscopes, televisions, and computer display terminals. Electrons are emitted from the cathode (point C), a hot filament; controlled by the grid (G); focused by a focusing anode (F); and then accelerated (A) by a high voltage while being formed into beams (collimated) through small apertures. Vertical-deflection plates (which are horizontal) deflect the beam vertically according to the voltage applied between the plates. Voltage is supplied also to the horizontal-deflection plates to sweep the beam regularly across the screen at a rate that can be varied. The electrons land on the screen and cause the spots on which they land to fluoresce.

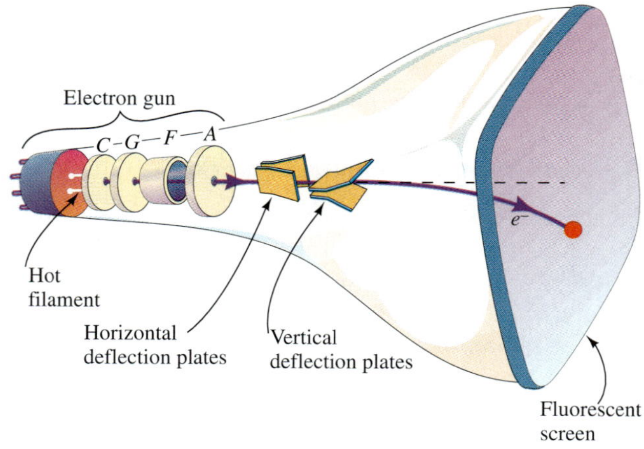

The possibility of guiding charged particles in this way can be demonstrated with oscilloscopes of older design (Fig. 23–21).

EXAMPLE 23–8 An electron moving horizontally at a speed $v_0 = 3 \times 10^6$ m/s enters the region between two horizontally oriented plates of length $L_1 = 3$ cm. A fluorescent screen is located $L_2 = 12$ cm past these plates. Find the electron's total vertical deflection on the screen from its initial direction if the electric field between the plates points downward with a magnitude of $E = 10^3$ N/C. There is no horizontal deflection.

Solution: Figure 23–20 illustrates the situation. The total vertical deflection of the electron is the deflection y_1, acquired in passing between the plates, as well as the additional deflection y_2, a result of the straight path after the electron leaves the plates. We use Eq. (23–42) to find the deflection y_1 of the electron while it is between the plates:

$$y_1 = \frac{1}{2}\frac{qE}{m}\frac{L_1^2}{v_0^2}.$$

The electron is deflected; after it leaves the region between the plates, it travels at an angle to its original direction, given by Eq. (23–43) as $\tan\theta = qEL_1/mv_0^2$. The y-deflection over the distance L_2 to the screen is then

$$y_2 = L_2 \tan\theta = \frac{qEL_1L_2}{mv_0^2}.$$

Finally, the total deflection is

$$y = y_1 + y_2 = \frac{1}{2}\frac{qEL_1^2}{mv_0^2} + \frac{qEL_1L_2}{mv_0^2} = \frac{qEL_1}{mv_0^2}\left(\frac{1}{2}L_1 + L_2\right).$$

From this solution, we see the important result that y is proportional to E. This is the feature that makes the oscilloscope so useful.

Numerical evaluation with a minus sign for E (**E** points downward), gives

$$y = \frac{(-1.6 \times 10^{-19}\text{ C})(-10^3\text{ N/C})(3 \times 10^{-2}\text{ m})}{(9.11 \times 10^{-31}\text{ kg})(3 \times 10^6\text{ m/s})^2} \times \left[\frac{1}{2}(3 \times 10^{-2}\text{ m}) + (12 \times 10^{-2}\text{ m})\right]$$

$$= 8.0 \times 10^{-2}\text{ m}.$$

Of this 8.0 cm, the deflection $y_1 = 0.9$ cm, and the deflection $y_2 = 7.1$ cm.

APPLICATION: ELECTROSTATIC PRECIPITATORS

Electric fields are central to these devices, which remove particles from the air (Fig. 23AB–1). They work in two steps: In the first step, air passes a series of thin charged wires alternating with thin metal strips. As we shall see in Chapter 24, the field near these wires is high enough to lead to ionization of the air molecules ("corona discharge"). The field draws the positive ions toward the metal strips and, as they pass through the air, they attach themselves to the particles to be removed. In the second step, the particles, which are now charged, are drawn on by air currents to regions between a series of metal plates of alternating charge. The fields set up between those plates draw the particles to them. They are deposited on the plates and thereby removed from the air. The clean air continues on.

The uses of precipitators are many: removing pollen and other allergens in the home; toxic aerosols such as paint spray in the factory; and bacteria from sterile environments. The most stringent requirements are in semiconductor manufacturing: The size of an active element on an integrated circuit is 1 micron or less, so the number of particles in the air larger than this size that could land on such a circuit and foul it must be reduced to a minimum. There precipitators limit such particles to less than one per cubic *meter* of air.

FIGURE 23AB–1 The high-voltage wire devices are placed outside this residential electrostatic precipitator unit.

23–5 THE ELECTRIC DIPOLE IN AN EXTERNAL ELECTRIC FIELD

In Section 23–1, we introduced the electric dipole, which has a total charge of zero but a positive and a negative center of charge separated by a distance L. Permanent electric dipoles (for example, polar molecules such as NaCl and H_2O) exist in nature. Let's discuss how permanent electric dipoles move in external electric fields.

Consider the permanent electric dipole discussed in Example 23–3. The electric dipole moment of the dipole has a vector character, and we assigned its direction as indicated in Fig. 23–7. The expression for the dipole moment **p** is given by Eq. (23–13). We place the electric dipole in a uniform external field (Fig. 23–22). The forces on $+q(F_+)$ and $-q(F_-)$ are

$$\mathbf{F}_+ = q\mathbf{E},$$

$$\mathbf{F}_- = -q\mathbf{E} = -\mathbf{F}_+.$$

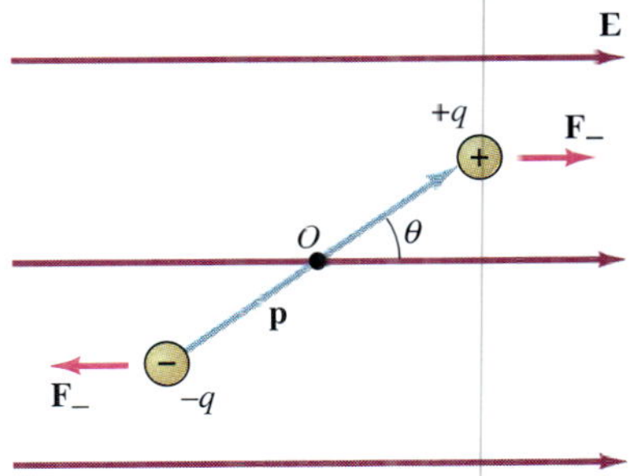

FIGURE 23–22 A dipole placed in a uniform external electric field experiences no net force but can experience a torque.

We notice that the two forces are equal and opposite and therefore cancel. *There is no net force on the dipole.*

However, there is a torque that tends to rotate the dipole. To calculate the torque and the corresponding rotation, we must choose a reference point, and it is convenient to choose this point to be the midpoint of the dipole, which is located at point O in Fig. 23–22. The actual motion will be independent of the choice of reference point. The torque, $\boldsymbol{\tau}$, about a point due to a force that acts on another point that is a displacement **r** away is given by Eq. (10–6):

Recall from Chapter 10 that the motion of a system on which torque acts is independent of the choice of origin.

$$\boldsymbol{\tau} = \mathbf{r} \times \mathbf{F}, \tag{23–44}$$

Recall that the right-hand rule determines the direction of a vector product.

where **r** is measured from point O. The resulting torque from the force on each charge is then clockwise, with magnitudes

$$\tau_+ = \left(\frac{L}{2}\right) qE \sin\theta,$$

$$\tau_- = \left(\frac{L}{2}\right) qE \sin\theta,$$

where the subscripts + and − refer to the charges. Because both τ_+ and τ_- are clockwise rotations, the total torque is also clockwise, with magnitude

The torque on a dipole in an electric field

$$\tau = \tau_+ + \tau_- = qLE \sin\theta. \qquad (23\text{–}45)$$

We can represent this expression for the torque on a dipole as the vector product of **p** and **E**:

$$\boldsymbol{\tau} = \mathbf{p} \times \mathbf{E}, \qquad (23\text{–}46)$$

which gives both the magnitude ($pE \sin\theta$) and the direction (into the page in Fig. 23–22) of the torque.

The maximum torque ($\tau = pE$) occurs when **p** and **E** are perpendicular ($\theta = \pi/2$). The torque is zero when **p** and **E** are parallel ($\theta = 0$) or antiparallel ($\theta = \pi$). The torque tends to rotate the electric dipole until **p** is parallel to **E**. The position $\theta = 0$ corresponds to a stable equilibrium, but the position $\theta = \pi$ is one of unstable equilibrium because a small deviation will cause the dipole to rotate toward $\theta = 0$.

Without a mechanism to dissipate the dipole's energy, the dipole will oscillate about $\theta = 0$ forever if it starts at some nonzero value of θ. As the dipole rotates toward $\theta = 0$, it gains kinetic energy and passes through $\theta = 0$ to the other side. The torque, however, then becomes counterclockwise, and the dipole slows down, stops, returns to $\theta = 0$, and passes through it again to the original side. Table 23–2 illustrates a time sequence for a rotating dipole in an electric field.

The Energy of a Dipole in an External Electric Field

Work must be done for an electric dipole to rotate in an external electric field. The electric field, for example, does positive work to rotate the dipole from $\theta = \pi/4$ to $\theta = 0$. An external agent would have to do positive work (and the electric field would do negative work) to rotate the dipole from $\theta = \pi/4$ to $\pi/2$. We learned in Section 10–6 that the work, W, done by the external agent while it exerts a torque τ on the system and moves it from angle θ_0 to θ is given by

$$W = \int \mathrm{d}W = \int_{\theta_0}^{\theta} \tau \, \mathrm{d}\theta.$$

Thus, for a dipole with dipole moment p,

$$W = \int_{\theta_0}^{\theta} pE \sin\theta \, \mathrm{d}\theta = pE(\cos\theta_0 - \cos\theta), \qquad (23\text{–}47)$$

where E is the external electric field. The work performed by the external agent is transformed into potential energy of the electric dipole, so the change in potential energy, $\Delta U = U - U_0$, is

$$U - U_0 = pE(\cos\theta_0 - \cos\theta). \qquad (23\text{–}48)$$

We are free to choose the constant U_0, and we choose it such that $U_0 = 0$ at $\theta_0 = \pi/2$. Thus the potential energy at angle θ is given by

The potential energy of a dipole in an external electric field

$$U = -pE\cos\theta. \qquad (23\text{–}49)$$

Notice that Eq. (23–49) is consistent with our choice for the zero of the potential energy, because $U_0 = -pE\cos\theta_0$, which is zero when $\theta_0 = \pi/2$.

Equation (23–49) can be written more compactly by using the scalar dot product of **p** and **E**:

TABLE 23–2 An Electric Dipole Rotating in a Uniform Electric Field

	Electric Field	Torque, τ	Angular Velocity, ω
Initial condition	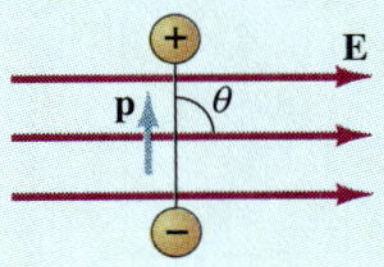	Maximum, into page	Zero
	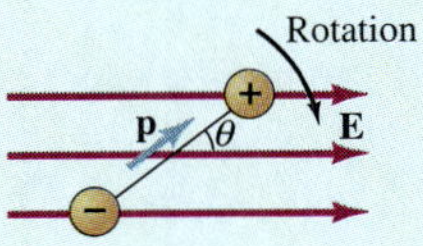	Decreasing, into page	Increasing, into page
	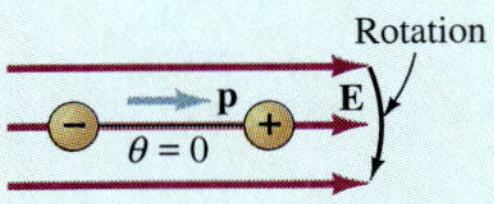	Zero	Maximum, into page
Time ↓	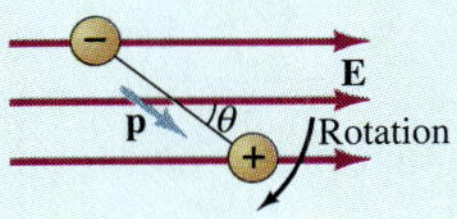	Changed direction, out of page, increasing	Decreasing, into page
	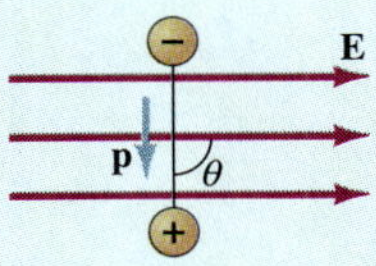	Maximum, out of page	Zero
	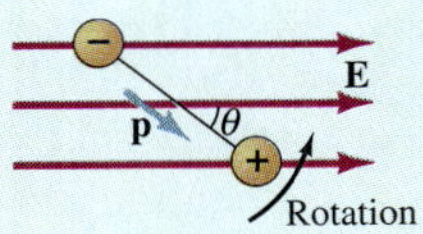	Decreasing, out of page	Changed direction, increasing, out of page

$$\text{for a dipole:} \qquad U = -\mathbf{p}\cdot\mathbf{E}. \tag{23–50}$$

Earlier, we discussed the stability of the equilibrium of the dipole in an external field. We can see directly from Eq. (23–50) that the orientation in which **p** is aligned with **E** is a point of stable equilibrium because U has a minimum there. In contrast, U has a maximum when **p** is antiparallel to **E**, and therefore this is a point of unstable equilibrium.

The Electric Dipole in a Nonuniform Electric Field

If a dipole is placed in a nonuniform external electric field, then, in addition to a torque, there may be a net force on the dipole. The resulting motion would be a combination of linear acceleration and rotation. The details of the motion depend critically on the particular electric field configuration.

The effect of a nonuniform electric field on an induced dipole explains the attraction of a neutral cork ball coated with conducting paint to a Teflon rod rubbed against fur (Fig. 23–23). The charged Teflon rod induces an electric dipole on the cork ball,

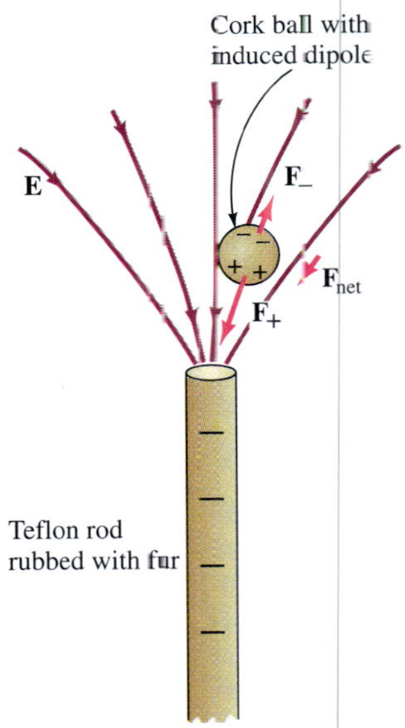

FIGURE 23–23 A dipole placed in a nonuniform external electric field can experience a net force. In this case the external electric field induces a dipole in the cork ball, which then experiences a force due to the electric field. This effect can be understood by using Coulomb's law.

which is in the nonuniform electric field of the rod. Another example of the action of a nonuniform electric field on an induced dipole is the attraction between small bits of paper and a comb that has just been charged by passing it through hair. The bits of paper have induced dipole moments and are attracted to the comb in its nonuniform field.

SUMMARY

Charge distributions set up electric fields in the space around them. The electric field vectors can be mapped out by moving a small, positive test charge q_0 around in this field. The field **E** is defined as the force **F** on this test charge, divided by q_0:

$$\mathbf{E} \equiv \lim_{q_0 \to 0} \frac{\mathbf{F}}{q_0}. \tag{23–2}$$

The electric field has units of N/C (or V/m). The force on a point charge q' in a given external electric field is

$$\mathbf{F} = q' \, \mathbf{E}_{\text{ext}}. \tag{23–7}$$

Electric field lines aid in the visualization of the direction and magnitude of the electric field produced by various charge configurations. Electric field lines begin on positive charges and end on negative charges, but are otherwise continuous. At a given point, an electric field line is tangent to the electric field at that point, and the density in area of the electric field lines is proportional to the strength of the field.

From Coulomb's law, the electric field due to a point charge q is

$$\mathbf{E} = \frac{q}{4\pi\epsilon_0 r^2}\hat{\mathbf{r}}. \tag{23–5}$$

Electric fields obey the principle of superposition:

$$\mathbf{E}_{\text{net}} = \mathbf{E}_1 + \mathbf{E}_2 + \mathbf{E}_3 + \cdots = \sum_i \mathbf{E}_i, \tag{23–10}$$

where $\mathbf{E}_i$ labels the field of the components that make up a charge distribution.

In its simplest form, an electric dipole consists of a positive charge q separated by a distance L from a negative charge $-q$. Such a configuration, or a configuration that is electrically neutral but has an imbalance of positive and negative charge from one side to another, occurs often in nature and produces an electric field. This field decreases with distance r as $1/r^3$ and, for the simple dipole, is proportional to the magnitude of the electric dipole moment **p**, given by

$$\mathbf{p} = q\mathbf{L}. \tag{23–13}$$

The direction of **L** (and of **p**) is from the negative charge to the positive charge. This direction determines the angular dependence of the electric dipole field.

The electric field due to a continuous charge distribution is

$$\mathbf{E} = \frac{1}{4\pi\epsilon_0}\int \frac{\mathrm{d}q}{r^2}\hat{\mathbf{r}}. \tag{23–20}$$

Here, r is the distance of a charge element $\mathrm{d}q$ from the point where the field is measured. To use this result, it is necessary to know how $\mathrm{d}q$ varies throughout space.

The electric field due to an infinitely long wire is radial and perpendicular to the wire. A charged plane, infinite in area, with a charge per unit area of σ has an electric field that is uniform and directed perpendicular to the plane, with

$$E = \frac{\sigma}{2\epsilon_0}. \tag{23–33}$$

In addition to producing an electric field, an electric dipole experiences a torque in a uniform external electric field:

$$\tau = \mathbf{p} \times \mathbf{E}. \tag{23–46}$$

The dipole in the external field has a potential energy of

$$U = -\mathbf{p} \cdot \mathbf{E}. \tag{23–50}$$

UNDERSTANDING KEY CONCEPTS

1. In older movies, you may see gasoline trucks dragging metal wires along the road? Why?
2. Why can't two electric field lines cross?
3. We have introduced the concept of an electric field. Why might it be useful to introduce an analogous gravitational field? In what ways would this field be like an electric field and in what ways would it be different?
4. An inflated rubber balloon is charged by rubbing it with fur. Explain what happens when that balloon is placed against (a) a metal wall; (b) an insulating wall.
5. Electric field lines originate from positive charges and terminate at negative charges, as exemplified by the field lines due to a dipole. Does this statement contradict the depiction of field lines due to a single positive point charge?
6. A pair of equal and opposite charges forms a dipole, and the electric field of a dipole is not zero. But if we were to look at a dipole from very far away, the two charges would appear to be on top of one another and to cancel; that is, we would see no charge, and hence we should see no field. How do you reconcile these statements?
7. Can the electric dipole induced on a spherical conducting ball cause the ball to rotate? How about the electric dipole induced on a long rod?
8. After you comb your hair, the comb can often attract small pieces of paper. The act of combing may induce an electric charge on the comb, but the combing does not itself affect the paper. What accounts for the attraction?
9. Explain how the water molecule, H_2O, acts as an electric dipole (see Fig. 23–8), given that there are *two* spatial regions (around the H atoms) with negative charge.
10. Explain why the electric-field-line technique would not be useful for a point charge if Coulomb's experiments had shown the electric force to decrease as $1/r$ or as $1/r^{2+\delta}$.
11. The internal motion of a liquid can be described by a velocity field, which is the velocity vector of the element of fluid at a given point. In what ways is this field like an electric field, and in what ways is it different?
12. Can you invent an arrangement of charges whose electric field would be radially directed into a point in some region of empty space? The correct answer has implications for the stability of charges placed in static electric fields.
13. Suppose that a small electric dipole ($+q$ and $-q$) is placed somewhere on a line that is perpendicular to, and bisects, a second (fixed) dipole ($+Q$ and $-Q$), as seen in Fig. 23–24. If the small dipole is free to pivot about its center, what will it do?

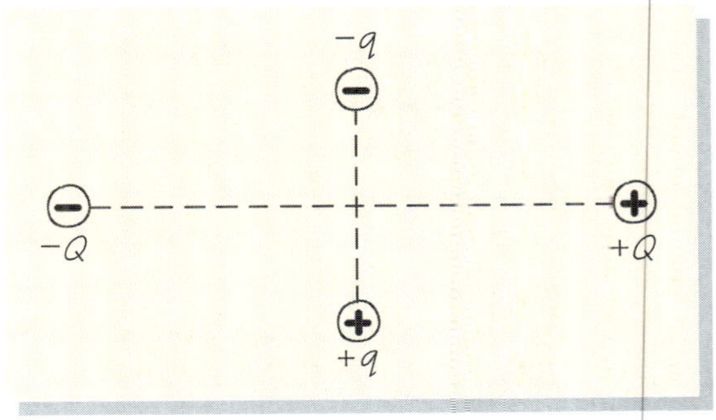

FIGURE 23–24 Question 13.

14. Consider a large number of identical dipoles centered in the xy-plane and pointing in the z-direction, distributed with uniform density. What is the electric field in the limit that the dipoles form a continuous distribution?
15. A large, flat, positively charged plate (of uniform charge density) is placed on the ground. A positively charged pellet, starting from rest, is released from above the plate. Ignore all air resistance. Qualitatively describe the motion of the pellet, according to the height from which it is dropped.
16. Suppose that a positively charged pellet is dropped from above onto the north pole of a large, positively charged sphere (of uniform charge density). Disregarding air resistance, and any small instabilities that would make the pellet move away from the vertical, describe the motion of the pellet.

PROBLEMS

23–1 Electric Field

1. (I) A 5-μC charge is located at $(x, y) = (5\text{ cm}, 0\text{ cm})$. Determine the electric field at (2 cm, 4 cm).
2. (I) Calculate the electric field at the origin due to the following distribution of charges: $+q$ at $(x, y) = (a, a)$, $+q$ at $(-a, a)$, $-q$ at $(-a, -a)$, and $-q$ at $(a, -a)$ (Fig. 23–25).
3. (I) Find the electric field due to the nucleus of an atom of gold ($Z = 79$) at a point 1 nm from the nucleus, which you can assume is a point. What is the force on an electron at that point?
4. (II) Charges of $+4\ \mu$C, $-3\ \mu$C, $+5\ \mu$C, and $-4\ \mu$C are located at the four corners of a square 6 cm on each side (Fig 23–26). Calculate the electric field at the center of the square

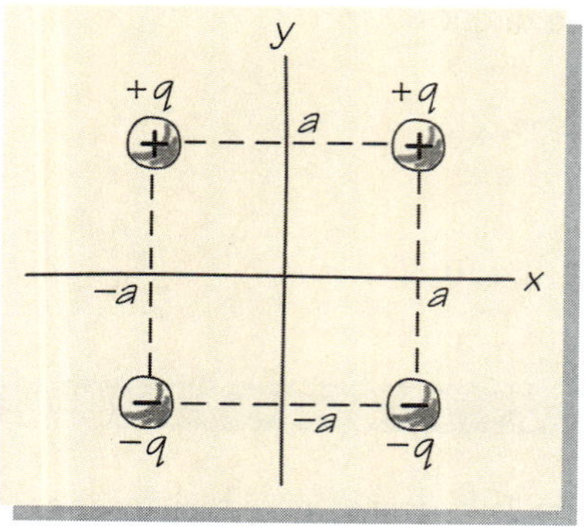

FIGURE 23–25 Problem 2.

[*Hint*: You may want to use a coordinate system with axes along the diagonals.]

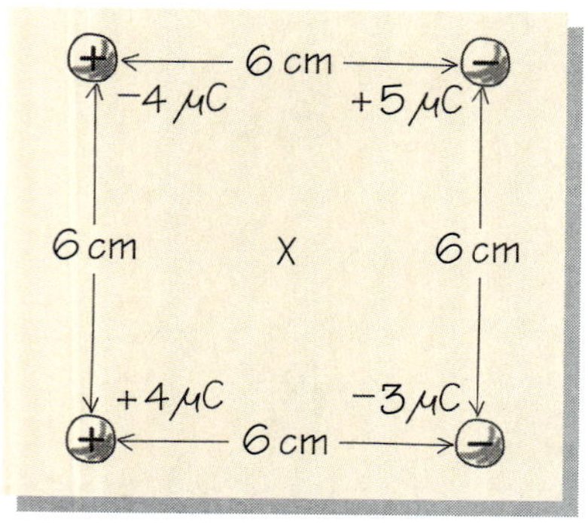

FIGURE 23–26 Problem 4.

5. (II) Five charges are located at five of the corners of a regular hexagon with sides of 10 cm, as shown on Fig. 23–27. Find the electric field at the sixth corner of the hexagon.

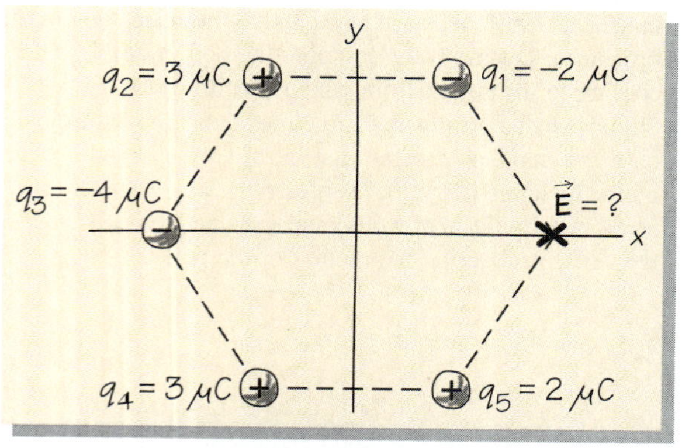

FIGURE 23–27 Problem 5.

6. (II) Charges $q_1 = 2\ \mu\text{C}$ and $q_2 = -3\ \mu\text{C}$ are a distance 20 cm apart. (a) Calculate the electric field of q_1 at the position of q_2; (b) the force acting on q_2, and (c) the total electric field at the midpoint between q_1 and q_2.

7. (II) A charge $-q$ is located at $y = -\ell/2$, and a second charge $+q$ is located at $y = +\ell/2$ (Fig. 23–28). (a) What is the electric field at the origin? (b) If the charge at $-\ell/2$ were instead $+q$, what would the electric field be at the origin? (c) For part (b), what would the electric field be in the entire xz-plane specified by $y = 0$?

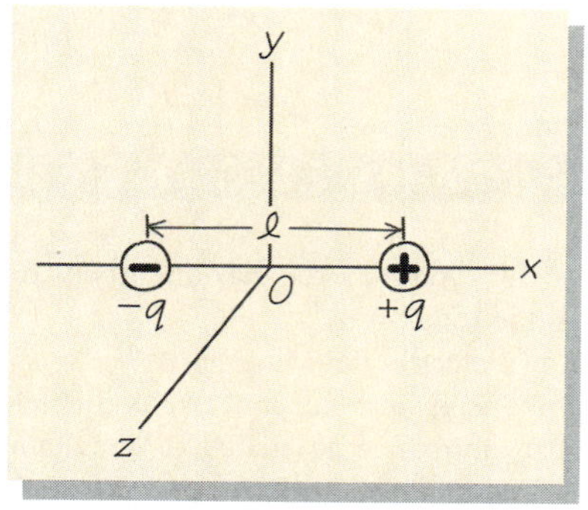

FIGURE 23–28 Problem 7.

8. (II) Identical charges Q are placed at $x = a$ and $x = -a$, respectively. (a) What is the electric field at $x = 0$? (b) Suppose that a positive test charge q_0 is placed at $x = 0$. Will it be in stable or unstable equilibrium? [*Hint*: Assume that the test charge is displaced a distance δ in a direction perpendicular to the x-axis. What will the net force on the test charge be at the new location?]

9. (II) Calculate the electric field *along the axis* of a dipole at a distance r from the center of the dipole shown in Fig. 23–6. Work out the field for $r >> L$.

10. (II) A succession of n alternating positive and negative charges q are placed along the x-axis, each a distance d from its neighbors. The arrangement is symmetrical about the y-axis, with the first $+q$ charge at $x = d/2$, the first $-q$ charge at $x = -d/2$, the second $-q$ charge at $3d/2$, the second charge $+q$ at $-\ 3d/2$, and so forth (Fig. 23–29). What is the field at a distant point $y = Y$ (where $Y >> nd$) on the y-axis?

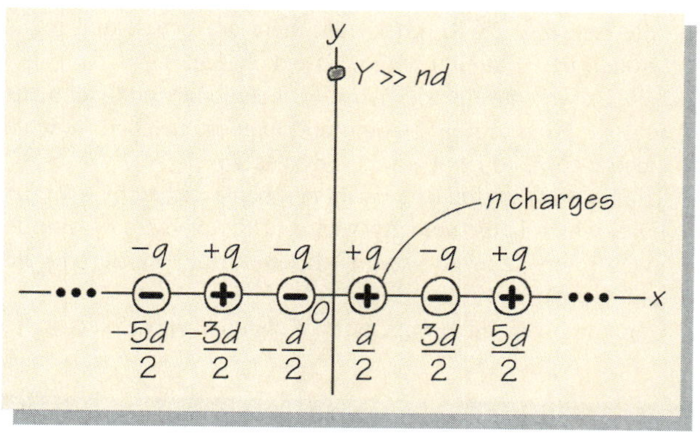

FIGURE 23–29 Problem 10.

11. (III) Suppose that the positive test charge in Problem 8 is constrained to move along the x-axis only. Will $x = 0$ be a stable equilibrium position? If it is, then the test charge should oscillate about $x = 0$ for small enough displacements. If that were the case, what would the frequency of oscillation be for a test charge of mass m? [*Hint*: Assume that the charge is displaced to a point $x = \delta$, where $\delta << a$, and calculate the magnitude and the direction of the electric field there. Use the approximation $1/(a + \delta)^2 = (1/a^2) - (2\delta/a^3) + \cdots$, valid for $\delta << a$.]

23–2 Electric Field Lines

12. (I) Draw the electric field lines due to charges of $+1\ \mu C$ and $+2\ \mu C$ located 10 cm apart.

13. (I) A pair of parallel plates have equal and opposite uniform charge distributions. The field is represented by parallel lines drawn with a density N per m^2. The charge density on the plates is tripled. How should the density of electric field lines be changed?

14. (I) A very thin rod of length L is placed along the x-axis. A charge Q is distributed uniformly on the rod. Sketch the electric field lines in the xy-plane.

15. (I) Consider the rod in Problem 14, and put another rod of the same length but with charge $-Q$ parallel to the first and some distance away in the xy-plane. Sketch the electric field lines in the xy-plane.

16. (II) Consider charges q placed along the x-axis at $x = na$, with $n = 0, \pm 1, \pm 2, \pm 3, \ldots$. Sketch the electric field lines.

17. (II) The field lines due to an electric dipole $\mathbf{p}$ are shown in Fig. 23–12b; by definition, the direction of $\mathbf{p}$ points from $-q$ to $+q$. Sketch the field lines for the combination of this dipole and (a) a dipole $-\mathbf{p}$ adjacent and parallel to the dipole $\mathbf{p}$; (b) a dipole $\mathbf{p}$ adjacent and parallel to the dipole $\mathbf{p}$; (c) a dipole $-\mathbf{p}$ on the axis of $\mathbf{p}$ some distance away past the $-q$ charge; (d) a dipole $\mathbf{p}$ on the axis of $\mathbf{p}$ some distance away past the $-q$ charge (Fig. 23–30).

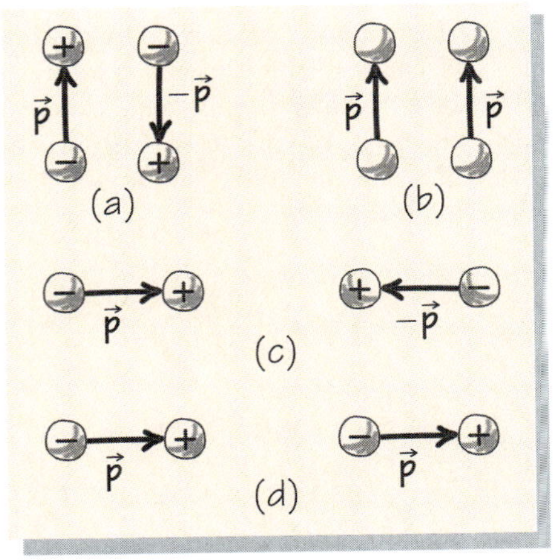

FIGURE 23–30 Problem 17.

18. (II) Charges q, q, and $-q$ form an equilateral triangle whose sides each have length 5 cm. Sketch the electric field lines. What is the magnitude of the electric field 0.01 cm from $-q$; what is it at a distance of 10 m from $-q$?

23–3 The Field of a Continuous Distribution

19. (I) Calculate the electric field due to an infinitely long, thin, uniformly charged rod with a charge density of 0.3 $\mu C/m$ at a distance 20 cm from the rod. Assume that the rod is aligned with the x-axis.

20. (I) A thin rod uniformly charged with a total charge of 2 μC and length 20 cm is placed along the z-axis, centered at the origin. Find the electric field at $(x, y, z) = (8\text{ cm}, 0\text{ cm}, 0\text{ cm})$ and $(0\text{ cm}, 8\text{ cm}, 0\text{ cm})$.

21. (I) Sketch the electric field lines between a point charge Q and a uniformly charged, flat square of area L^2 and total charge $-Q$. The point charge is located a distance L above the center of the plane.

22. (I) Consider two infinite plane sheets of insulator with uniform surface charge densities σ_1 and σ_2, respectively. The two sheets are parallel to each other and a distance L apart. What is the force on a point charge Q placed midway between the sheets?

23. (II) A negative charge is distributed uniformly on a long cylindrical shell. Sketch the field lines both inside and outside the shell. Do not include the ends of the cylinder.

24. (II) Consider positive charges distributed uniformly with a charge density λ on a circle of radius R. (a) Use symmetry arguments to deduce the direction of the electric field at a point in the plane of the circle but outside the circle. (b) What is the magnitude of the electric field at a distance L along the axis of the circle for $L >> R$?

25. (II) A rod with a uniform negative charge is bent into a semicircle. Make a rough sketch of the electric field lines in the plane of the rod.

26. (II) Two infinite plates with a uniform charge density of 8 $\mu C/m^2$ are placed along the yz-plane with one plate passing through $x = 3$ cm and the other through $x = -3$ cm. Determine the electric field at $(x, y, z) =$ (a) $(0\text{ cm}, 0\text{ cm}, 0\text{ cm})$; (b) $(5\text{ cm}, 0\text{ cm}, 0\text{ cm})$; (c) $(5\text{ cm}, 2\text{ cm}, 3\text{ cm})$.

27. (II) Two large, flat, vertically oriented plates are parallel to each other, a distance d apart. Both have the same uniform positive charge density σ. What is the electric field in the space around and between them?

28. (II) The axis of a hollow tube of radius R and length L is aligned with the x-axis; the tube's left-hand edge is at $x = 0$, as shown in Fig. 23–31. It carries a total charge q distributed uniformly along its surface. By integrating the result for a field due to a hoop of charge along the axis of the hoop (see Example 22–6), find the electric field along the x-axis due to the tube as a function of x.

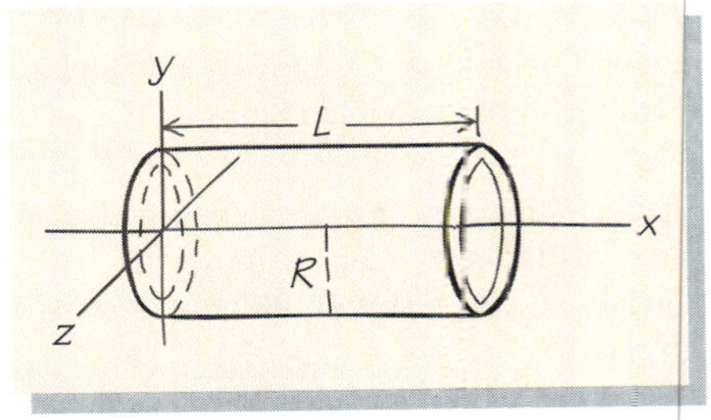

FIGURE 23–31 Problem 28.

29. (II) A thin, circular disk of radius R is oriented in the xy-plane with its center at the origin. A charge Q on the disk is distributed uniformly over the surface. (a) Find the electric field due to the disk at the point $z = z_0$ along z-axis. (b) Find the field in the limit $z_0 \to \infty$. (c) Find the field in the limit that $R \to \infty$. Are the limits of parts (b) and (c) the same?

30. (II) Consider a thin, uniformly charged rod 30 cm long that is bent into a semicircle. The total charge on the rod is 0.6 μC. What are the magnitude and direction of the electric field at the center of the semicircle?

31. (II) A rod 30 cm long is charged uniformly with a charge density of 15 μC/m. A charge of 3 μC is placed 30 cm from the midpoint of the rod along a line perpendicular to the rod. Calculate the electric field at a point halfway between the point charge and the center of the rod.

32. (III) A total charge Q is distributed uniformly over a rod of length L. The rod is aligned on the x-axis, with one end at the origin and the other at the point $x = L$ (Fig. 23–32). Calculate the electric field at a point $(0, D)$, and compare this result with the field at the point $(L/2, D)$.

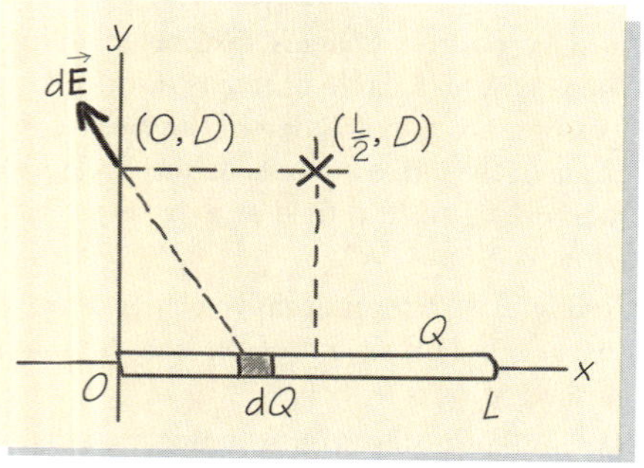

FIGURE 23–32 Problem 32.

33. (III) Consider a point at a height z_0 directly above the midpoint of a square with sides of length $2L$. The (nonconducting) square carries a uniform charge density σ. (a) Use the method of Example 23–6 to write an integral for the electric field at z_0. (b) How does the integral simplify in the limit $L \to \infty$? (c) $z_0 \to 0$?

23–4 *Motion of a Charge in a Field*

34. (I) An infinite plate carries a uniform charge density $\sigma = 3.56 \times 10^{-6}$ C/m^2. A pellet of mass 0.732 g is placed at rest 0.215 m from the plate. The pellet carries a negative charge $q = -1.14 \times 10^{-6}$ C. What is its speed when it reaches the plate? Ignore all forces except the electrostatic attraction.

35. (I) A small object, mass 120 mg, is observed to undergo an acceleration of magnitude 4.6 m/s^2 when it is placed in a constant electric field of magnitude 850 N/C. What is the charge on the object?

36. (I) A sheet of uniform charge density 4.2×10^{-9} C/m^2 is placed at $z = 0$ in the xy-plane. An electron with zero initial velocity is placed at $z = 0.50$ m. What will be its velocity after 2.4 ns?

37. (I) A large, flat plate with unknown, uniform charge density σ is placed on a horizontal tabletop. A cork ball of mass 0.83 g, carrying a charge 8.5×10^{-7} C, is placed at rest above the plate and remains at rest. What is σ?

38. (I) An alpha particle approaches a gold atom head on, stops, and turns around at a distance of 10^{-11} m from the nucleus. What is the electric field due to the gold nucleus at this point? Ignore the effects of the gold atom's orbiting electrons. What is the acceleration of the alpha particle when it is stopped? An alpha particle is a helium nucleus, composed of two protons and two neutrons.

39. (II) Consider an infinite wire with uniform charge density λ along the z-axis. A negatively charged particle moves in a circle in the xy-plane centered on the wire. Calculate the particle's speed, and show that the speed is independent of the radius of the circle. Ignore all forces except those due to the wire.

40. (II) A negative charge $-q$ is restricted to move in a plane in which there is a continuous line of positive charge and a charge density λ. The negative charge, of mass m, can pass the line of positive charge freely. What is the equation of motion for the negative charge?

41. (II) A positive charge q can travel in a circular orbit about a negatively charged line with uniform charge density λ. Show that the period of the orbit is proportional to the radius of the orbit. Compare this to the dependence of the period of a circular orbit on the radius of the orbit for a point charge that interacts with another point charge.

42. (II) A cork ball of mass 1.2 g is placed between two large horizontal plates. The bottom plate has a uniform charge density of $+1.4 \times 10^{-6}$ C/m^2, whereas the upper plate has a uniform charge density of -0.30×10^{-6} C/m^2. The cork ball, which carries an unknown charge, is placed between the plates and is observed to float motionlessly. What are the sign and magnitude of the charge on the ball?

43. (II) Consider the cathode-ray tube of Example 23–8. This time an electron enters the region between the vertical-deflection plates with a total speed of $v_0 = 5.0 \times 10^6$ m/s. The direction is such that the velocity has a vertical component $v_{0y} = +2.0 \times 10^5$ m/s. Find the total vertical deflection of the electron when it reaches the screen.

44. (II) A beam of electrons is accelerated by passing it through a region between two large charged parallel plates (Fig. 23–33). Calculate the charge density on the plates if the electrons accelerate from 10^6 m/s to 4×10^7 m/s between the plates. (This illustrates the role of electrodes G and A of the electron gun in Fig. 23–21.)

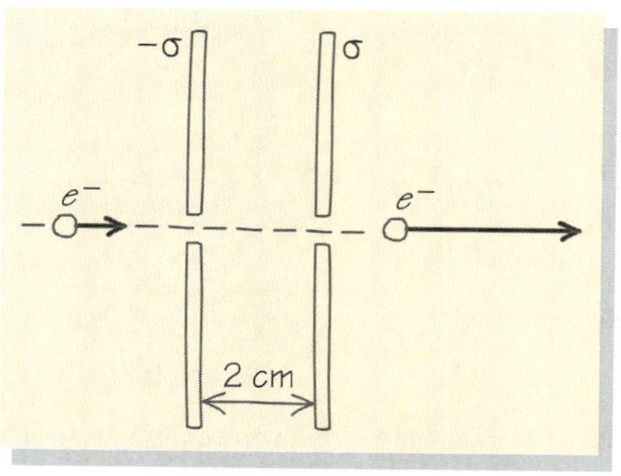

FIGURE 23–33 Problem 44.

45. (II) A cork ball of mass 5 g, carrying a charge of -2 μC, is suspended from a string 1 m long above a horizontal, uniformly charged plate of charge density 1 μC/m^2. The ball is displaced from the vertical by a small angle and allowed to swing. Show that the ball moves in simple harmonic motion, and calculate the angular frequency of that motion.

46. (III) A proton moves at speed $v = 5 \times 10^5$ m/s in the $+x$-direction and enters a certain region. An electric field in the region also is oriented in the $+x$-direction. The field's strength drops linearly with x: At the beginning of the region, $x = 0$ m, the field strength is 500 N/C; at $x = 3$ m, the field strength is zero. How much time does it take for the proton to traverse this region? [*Hint*: The equation of motion will be more familiar in terms of the variable $x' = x - 3$.]

23–5 The Electric Dipole in an External Electric Field

47. (I) An electric dipole consists of two opposite charges of magnitude 2 μC placed 10 cm apart (Fig. 23–34). The dipole is placed in a uniform electric field of 10 N/C along the x-axis, with the direction of **p** at an angle of $+45°$ from the x-axis in the xy-plane. Determine the torque on the dipole.

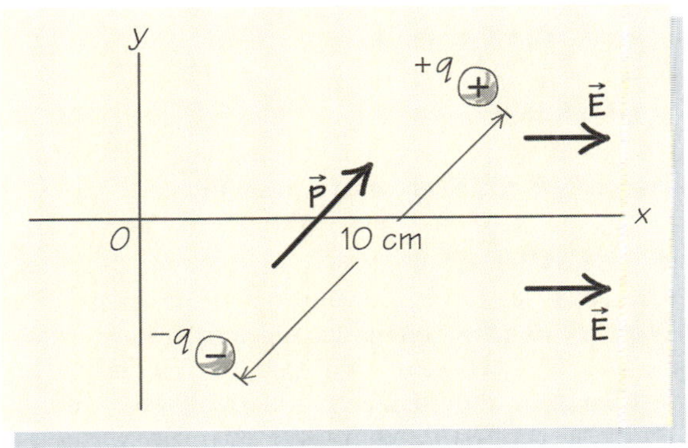

FIGURE 23–34 Problem 47.

48. (I) The magnitude of the two opposite charges that form an electric dipole is increased by a factor of 3 while the separation between the charges is doubled. What is the change in magnitude of the torque on the dipole in a uniform electric field?

49. (I) A water molecule has a permanent electric dipole moment of magnitude 6×10^{-30} C-m. Estimate the size of the electric field it produces at the position of a neighboring water molecule, which is 3×10^{-9} m away.

50. (II) Describe the motion of the dipole in Problem 47. How much work does the electric field do when the dipole moves from its initial position to alignment with the electric field?

51. (II) Two molecules with permanent electric dipole moments **p** are aligned (Fig. 23–35). Calculate the force between the molecules if they are separated by a distance that is large compared with the dimension of the dipoles. (A useful relation is $(1 + x)^{-2} \simeq 1 - 2x + 3x^2 + \cdots$ for small x.)

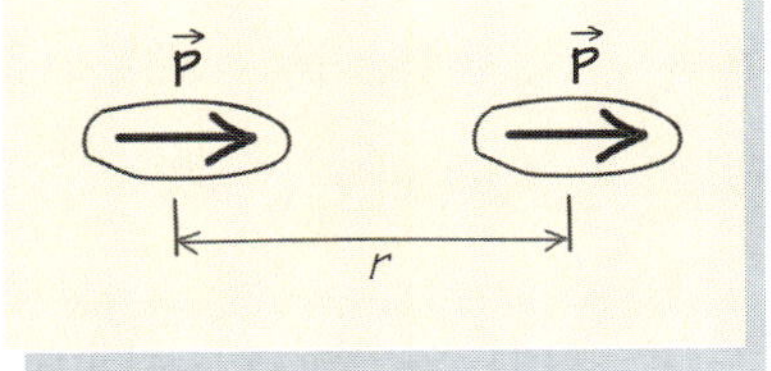

FIGURE 23–35 Problem 51.

52. (II) A molecule of lithium fluoride (LiF) has a permanent dipole moment. The molecule is placed in a uniform electric field of strength 10^4 N/C, and the difference between the maximum and minimum potential energies of the molecule in this field is 4.4×10^{-25} J. What is the electric dipole moment of the LiF molecule?

General Problems

53. (II) A point charge $-q$ is fixed at the center of a hollow spherical conductor of charge $+q$. Draw the electric field lines both inside and outside the sphere.

54. (II) A point charge $+q$ is fixed at the center of a hollow spherical conductor also of charge $+q$. Draw the electric field lines both inside and outside the sphere.

55. (II) Draw the electric field lines for a point charge $+q$ near an infinitely long, positively charged wire.

56. (II) A cork ball of radius 0.3 cm and a charge of $+2.0$ nC is covered with conducting paint. What is the electric field strength just outside the surface? A nickel nucleus, with a radius of 5×10^{-15} m, has a positive charge of $28e$. What is the electric field strength just outside the surface of the nucleus?

57. (II) Two infinitely long, uniformly charged rods, with charge densities of λ and $-\lambda$, respectively, are lined up parallel to each other and separated by a distance R. What are the magnitude and direction of the electric field due to the two rods at points that lie (a) on a line joining the two rods, and (b) along a perpendicular bisector of that line? Draw a figure to show the configuration, and use symmetry.

58. (II) What is the force per unit length that one of the two rods in Problem 57 exerts on the other?

59. (II) Two uniformly charged infinite plates with charge densities -5 μC/m^2 and 3 μC/m^2 are placed at right angles, the first one along the xz-plane, the second along the yz-plane. A test particle of mass 1 g and charge 1×10^{-7} C is placed a distance of 1 m from both planes; that is, its initial position is $(x, y, z) = (1$ m, 1 m, 0 m). What is the location of the test particle after a short time t (before it hits a plate)?

60. (II) Two infinite lines of charge density 3 μC/m are parallel to the z-axis. One line passes through $(x, y) = (2$ cm, 0 cm); the other, through $(x, y) = (-2$ cm, 0 cm) (Fig. 23–36). Find (a) the electric field at the origin; (b) the force on a 1-μC charge at the origin; (c) the force on a 2-μC charge located at (x, y, z) $= (6$ cm, -4 cm, 0 cm).

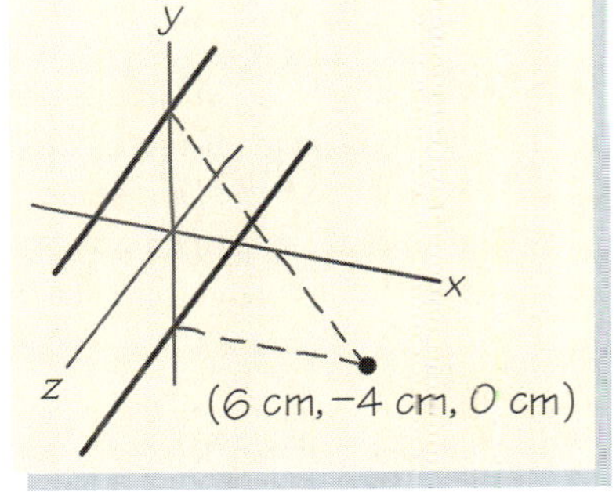

FIGURE 23–36 Problem 60.

61. (II) A proton with kinetic energy of 2×10^6 eV is fired perpendicular to the face of a large metal plate that has a uniform surface charge density of $\sigma = 8.0 \times 10^{-6}\ \text{C/m}^2$. (a) Calculate the magnitude and direction of the force on the proton. (b) How much work must the electric field do on the proton to bring it to rest? (c) From what distance should the proton be fired so that it stops right at the surface of the plate?

62. (II) The electric charge with the smallest magnitude that can be isolated is the charge on the electron or the proton. In 1909, Robert A. Millikan developed a classic method to measure this charge, known as the *oil drop experiment.* Millikan was able to place charges on tiny droplets of oil, which would fall at a given terminal velocity under the influence of gravity and air drag. By placing these droplets between parallel, horizontal charged plates, as in Fig. 23–37, the electric field between the plates produces a force on the charged droplet that is directed upward and can partly cancel the gravitational force. If the mass and size of the droplet are known, then, by seeing how fast droplets fall with and without the electric field, the charge can be measured.

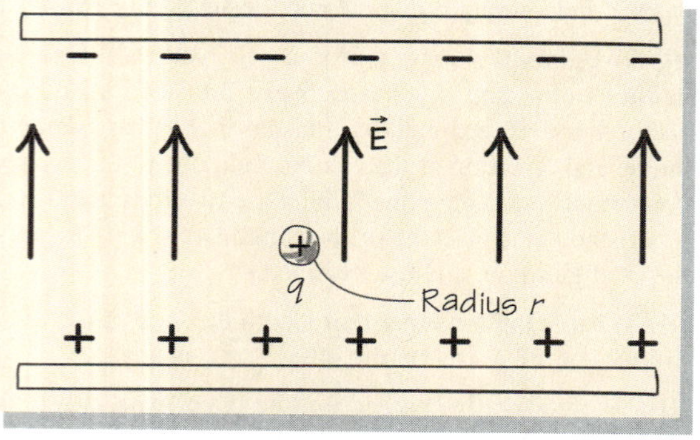

FIGURE 23–37 Problem 62.

The drag force on a droplet of radius r that falls at a steady speed v through air is also directed upward and is given by *Stokes's law*, $F_{\text{drag}} = 6\pi\eta r v$, where η is the viscosity of air. (a) Show from Newton's second law that the terminal velocity v_0 of the *uncharged* drop is $v_0 = \frac{2}{9}r^2\,\rho g/\eta$, where ρ is the density of the oil and g is the acceleration due to gravity. (b) Suppose that the charge on the drop, q, is positive and that the field is directed vertically upward, as in the figure, so that the electric force points up. Show by using Newton's second law that the charge is given by

$$q = \frac{18\pi(v_0 - v_1)}{E}\sqrt{\frac{v_0\eta^3}{2\rho g}},$$

where v_1 is the terminal velocity when the electric field E is imposed. (c) Take the minimum charge as 1.6×10^{-19} C, the oil's density as $0.85\ \text{g/cm}^3$, and the radius of the droplet as 2.0×10^{-4} cm. The droplet has the minimum charge. Find the value of E that will hold the droplet stationary between the plates.

63. (II) We will learn in Chapter 24 that the electric field near a conductor *must be perpendicular to the conducting surface.* Using this fact, draw the electric field lines for the following configurations: (a) a point charge $+q$ above an infinite, uncharged conducting plane; (b) a point charge $-q$ near an infinitely long, positively charged conducting wire; (c) a point charge $+q$ a distance $L/2$ above a charged conducting plane of area L^2 and charge $+q$.

64. (II) The field due to a line of uniform charge density λ varies with a radial distance r from the line as $1/r$. Suppose that a point charge q of mass m is placed at rest a distance R from the line, and that the force on the point charge due to the field of the line is attractive. Use dimensional analysis to calculate how the time it will take for the charge to drop to the charged line depends on λ, q, m, R, and ϵ_0.

65. (III) Consider the straight, nonuniformly charged rod of length L aligned along the x-axis, with the ends at $x = \pm L/2$, in Example 22–7. We showed there that the force on a charge q located at a point $x = R$ on the x-axis, to the right of the right-hand end of the rod, is

$$\mathbf{F} = \frac{q\lambda_0}{2\pi\epsilon_0 L}\left\{\ln\left[\frac{R-(L/2)}{R+(L/2)}\right] + R\left[\frac{1}{R-(L/2)} - \frac{1}{R+(L/2)}\right]\right\}\mathbf{i}.$$

Show that for $R >> L$, the force reduces to that of a dipole acting on q, $\mathbf{F} \simeq (q\lambda_0 L^2/12\pi\epsilon_0 R^3)\mathbf{i}$. What is the dipole moment? [*Hint*: Use the approximate forms $(1-x)^{-1} \simeq 1 + x + x^2 + x^3 + \cdots$ and $\ln(1+x) \simeq x - (x^2/2) + (x^3/3) - \cdots$, both appropriate for $x << 1$.]

66. (III) The field of an electric dipole decreases as $1/r^3$ when the distance of a given point to the dipole, r, is much larger than the separation between the charges. The only way to arrange two charges with a total charge of zero is to form a dipole. There are, however, many ways to arrange four charges with a total charge of zero in a compact pattern. An arrangement with an electric field that behaves at great distances as $1/r^4$ is an *electric quadrupole*. (a) For four charges aligned with alternating sign (such as $+ - - +$ so that the combination acts like dipoles of opposite orientation) along an axis, show that the field on the axis perpendicular to the line of charges decreases as $1/r^4$, where r is much larger than any separation distance within the quadrupole. [*Hint*: Use the approximation

$$(r^2 + \delta^2)^{-3/2} \simeq \frac{1}{r^3} - \frac{3\delta^2}{2r^5} + \cdots$$

(good when $\delta << r$) for each of the four charges; do not forget the sign of the charge in determining the field.] (b) For the arrangement shown in Fig. 23–38, sketch the field, using the field-line technique.

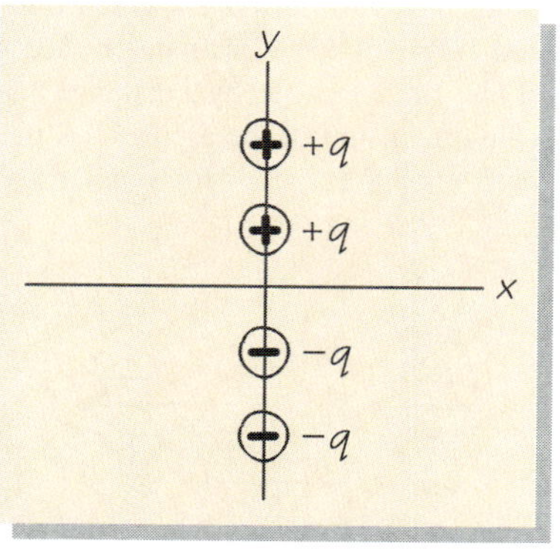

FIGURE 23–38 Problem 66.

CHAPTER

24

Sparks from a high-voltage electrostatic generator at the Boston Museum of Science do not harm the operator who is sitting inside a grounded Faraday cage.

Gauss' Law

Gauss' law is a different way of stating Coulomb's law; thus, it is one of the fundamental laws of electromagnetism.† Gauss' law is also more convenient than Coulomb's law in many situations. In Chapter 23, we learned the meaning of the electric field and how to use Coulomb's law to calculate the field due to a stationary distribution of charges. Gauss' law facilitates the calculation of electric fields when there is symmetry in the charge distribution, which is a common situation. Gauss' law also gives us insight into the behavior of conductors. To use Gauss' law, we must be able to calculate and use the electric flux—a quantity analogous to the flux we encountered in our study of fluid flow. In addition to learning how to use Gauss' law, we'll examine here the extent to which Gauss' law and Coulomb's law have been verified experimentally.

24–1 ELECTRIC FLUX

The electric flux is easiest to understand if we draw an analogy to the flow of water. Consider the water in a pipe. Let's assume that the water flows smoothly with a uniform velocity, magnitude v, in the horizontal, or x-, direction (Fig. 24–1). We could dip into the water a flat wire strainer in the form of a square of area $A = L^2$. Depending on the strainer's orientation, different volumes of water would pass through it in a given time (Fig. 24–2). If the strainer is oriented perpendicular to the flow, the volume of

Flux was first discussed when we studied fluid flow in Chapter 16.

†Karl Friedrich Gauss, a great mathematician of the nineteenth century, worked on celestial mechanics, electromagnetism, optics, and the theory of errors.

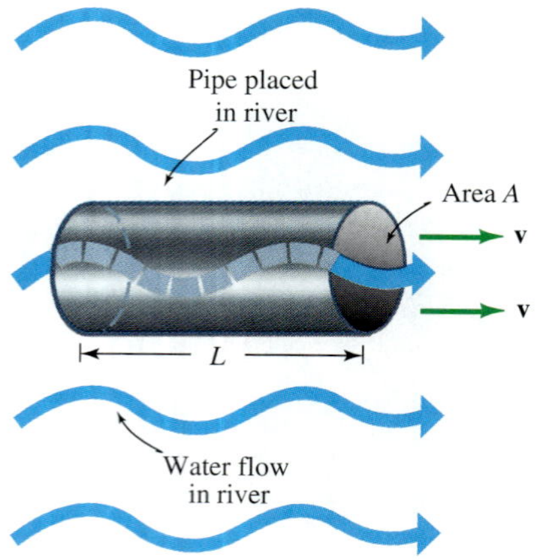

FIGURE 24–1 The rate at which water flows through this pipe can be measured.

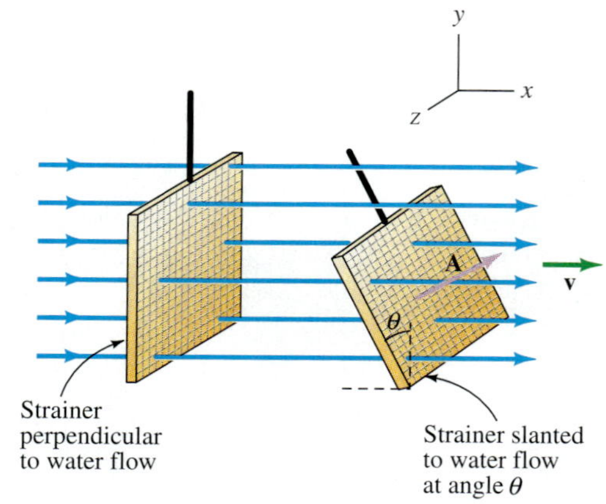

FIGURE 24–2 A jet of water moving in the x-direction passes through a square strainer. When the strainer is tilted, less water flows through it, by a factor cos θ. The vector **A** is perpendicular to the strainer.

water that flows through it in 1 s is given by Φ_w, where

$$\Phi_w = vL^2 = vA.$$

The volume of water that passes per unit time is the *flux*. Less water passes through the loop if it is tilted to make an angle θ with the vertical. This is because the vertical length that faces the stream is reduced from L to L cos θ. Thus the flux is reduced to

$$\Phi_w = vL^2 \cos\theta = vA \cos\theta.$$

We can find a more general form for the flux by assigning a direction to the area of the strainer. The area becomes a vector **A** with magnitude A and direction perpendicular to the strainer (Fig. 24–2). We can also remove the restriction that the velocity **v** is in the x-direction. Because $\mathbf{v}\cdot\mathbf{A} = vA\cos\theta$, where θ is now the angle that **A** makes with the water velocity, we can express the flux as

$$\Phi_w = \mathbf{v}\cdot\mathbf{A}.$$

If the surface through which the water passes is irregular, like a fish net or a butterfly net (Fig. 24–3) whose orientation is different in different places, then we can find the total flux by summing the fluxes through infinitesimal areas d**A**. Each area forms a tiny plane, and the direction of d**A** is perpendicular to its plane. We sum the infinitesimal fluxes $d\Phi_w = \mathbf{v}\cdot d\mathbf{A}$ by using integration. Thus the flux of water, the volume that passes through some nonplanar surface S per unit time, is

$$\Phi_w = \int_{\text{surface } S} d\Phi_w = \int_{\text{surface } S} \mathbf{v}\cdot d\mathbf{A}. \qquad (24\text{–}1)$$

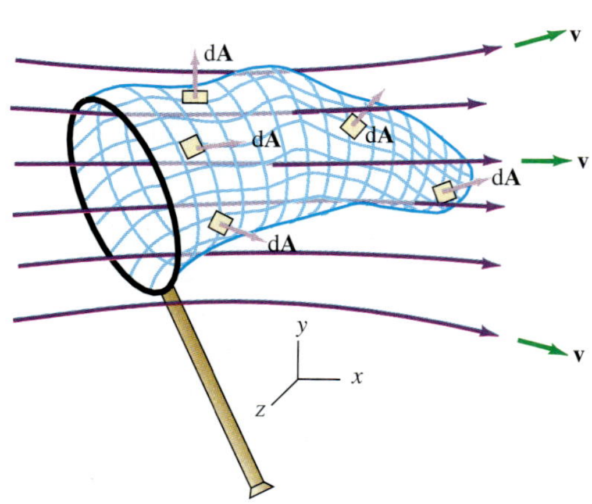

FIGURE 24–3 The surface through which the water flows can be irregular, with the orientation vector **A** changing from place to place. The velocity vector need not be constant, either.

Note that the velocity **v** may be different at each point over the surface. We can indicate that **v** varies with position by writing it as a function of y and z: $\mathbf{v}(y, z)$. For example, suppose that water flows in the $+x$-direction, but with a speed that depends on y and z. If the surface S through which the flux is calculated is in the yz-plane, then the differential cross-sectional area is $dA = dy\,dz$, the direction of $d\mathbf{A}$ is the x-direction (so that $\mathbf{v}\cdot d\mathbf{A} = v\,dA$), and the flux is

$$\Phi_w = \int_{\text{surface } S} v(y, z)\,dy\,dz. \tag{24–2}$$

There is a similarity between fluid flow and the electric field. This can be seen in a comparison of streamlines in a fluid (see Fig. 15–19) and electric field lines (Fig. 24–4). Let's now extend the notion of flux to the electric field. The **electric flux**, Φ (or Φ_E, when we need to distinguish it from some other kind of flux), is defined as

Definition of electric flux

$$\Phi \equiv \int_{\text{surface } S} \mathbf{E}\cdot d\mathbf{A}. \tag{24–3}$$

PROBLEM SOLVING

The electric flux through a surface is proportional to the number of field lines that pass through the surface.

Note that the "surface" that we use to calculate the flux is generally imaginary. No actual "net" has to form the surface. We shall be imagining many different surfaces for our convenience. The electric flux will prove to be an enormously useful quantity: We can use it to help find electric fields, and it appears in the formulation of the fundamental laws of electricity and magnetism.

There is an important difference between fluid flow and electric flux: Although water may actually move through a surface, electric fields do not "move." *No physical movement is involved in the electric flux.*

As we discussed in Chapter 23, the magnitude of the electric field is proportional to the density of electric field lines through an area perpendicular to the lines. Let N be the number of electric field lines that pass through a surface S, where the area perpendicular to **E** is $A_\perp$. For this simple calculation we suppose that E is constant over the surface area, so that $E \propto N/A_\perp$, and $N \propto EA_\perp = \Phi$. Because the flux defined in Eq. (24–3) is proportional to the electric field times the area through which it passes, that equation tells us that *the electric flux through a surface is proportional to the number of electric field lines that pass through that surface.*

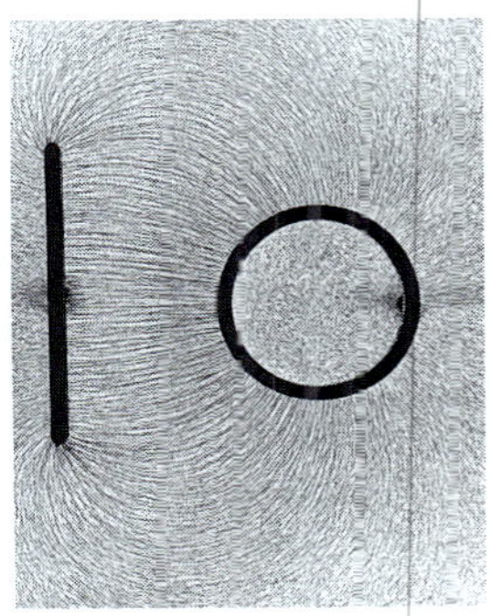

(a)

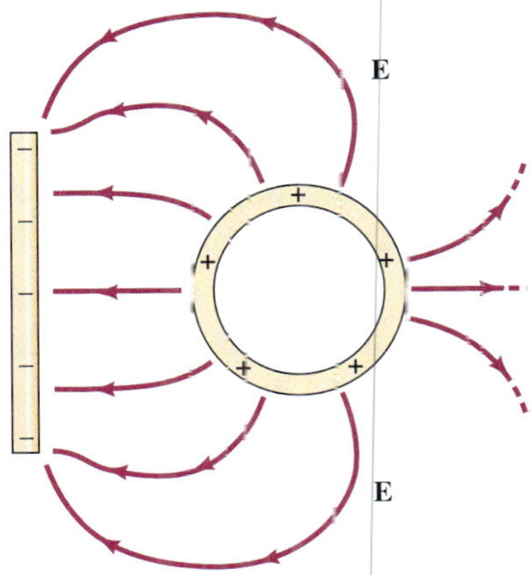

(b)

FIGURE 24–4 (a) The electric field lines due to a charged conducting cylinder close to an oppositely charged conducting plate, shown by threads in a shallow dish of oil. (b) The same situation, with the electric field lines drawn.

The Gaussian Surface

In order to use Gauss' law, we must determine the electric flux through a *closed* surface. Such surfaces—which we will imagine for our own convenience—may have the shape of a sphere or a cylinder or any other shape. We refer to such an imaginary closed surface as a **Gaussian surface**. Figure 24–5 shows a Gaussian surface together with electric field lines that pass into and out of the surface area. The electric flux passing through a closed surface has the same form as the flux through an open surface (such as the surface formed by our strainer), with only one refinement: We define the direction of an infinitesimal surface element $d\mathbf{A}$ of the Gaussian surface to be perpendicular to the surface and *pointing to the outside of the closed surface.* Thus, from Eq. (24–3), the electric flux through a closed surface is

$$\text{through a closed surface:}\quad \Phi = \oint_{\text{surface } S} d\Phi = \oint_{\text{surface } S} \mathbf{E}\cdot d\mathbf{A}, \tag{24–4}$$

where the circle on the integral indicates that we are integrating over a closed surface.

Figure 24–5 shows the direction of area elements $d\mathbf{A}$ at four different spots on the Gaussian surface. Notice that for $d\mathbf{A}_1$ and $d\mathbf{A}_2$, **E** and $d\mathbf{A}$ are oriented so that the product of $\mathbf{E}\cdot d\mathbf{A}$ is negative; for $d\mathbf{A}_3$ and $d\mathbf{A}_4$, the dot product is positive. The flux is negative for the part of the surface where the electric field lines enter the closed surface

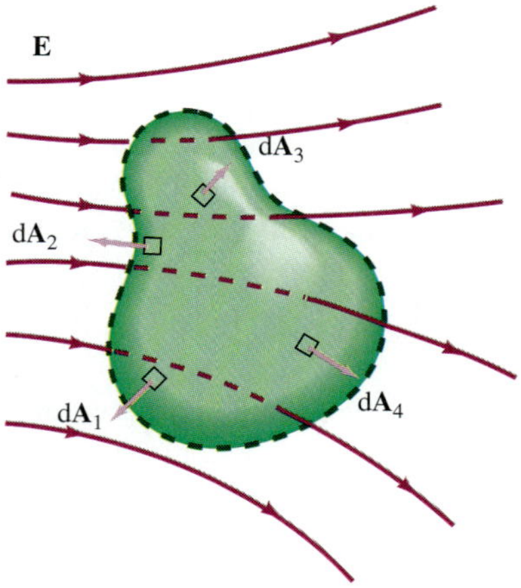

FIGURE 24–5 By convention, the directions of the areas d**A** are away from, and perpendicular to, the surface area. The electric field lines pierce the closed surface area. The total electric flux for this surface, which is dashed to remind you that it is imaginary, is zero.

FIGURE 24–6 A closed fish net in water is a mechanical analogue to a Gaussian surface in an electric field. Unless there is a source (a faucet) or sink (a drain) within the net, all the water that flows into the net flows out.

and positive where the electric field lines exit the surface. The total flux for the case shown in Fig. 24–5 is zero because all the field lines that enter the surface exit it. We can use simple physical reasoning for this case rather than performing the integration indicated by Eq. (24–3). In Chapter 23, we learned that electric field lines must originate from charges and must end on charges. If we have a Gaussian (closed) surface that surrounds no charges, then no electric field lines can originate or terminate inside that closed surface. The same number of field lines that enter must exit, so the integration in Eq. (24–4) must be zero. Thus, *if there is no charge inside a closed surface, the electric flux through the surface is zero.* Here again, the analogy to fluid flow is helpful. Suppose that our closed surface is formed by a fish net, and that net is placed in a river (Fig. 24–6). If there is no source (for example, a faucet) or sink (for example, a drain) inside the net, then all the water that flows into the net must flow out of it. The water flowing in the river in which the net is placed is analogous to the electric field in the region where the Gaussian surface is placed.

Gauss' Law Refers to Net Charge

We can again make use of the fluid analogy to demonstrate that the flux through a Gaussian surface is also zero if no *net* charge is enclosed by the surface. Consider a closed wire basket placed into a river. Two hoses lead into the interior of the basket. One hose pumps water in at a certain rate, while the other hose pumps water out at the same rate. The flowing river is analogous to an external electric field. The end of the inflow hose is analogous to a positive electric charge. The end of the outflow hose is analogous to a negative electric charge. Again, all the water that flows into the region enclosed by the basket flows out of it again, and the flux of water through the basket is zero. In just the same way, if a closed surface surrounds equal amounts of positive and negative charge, then the electric flux through that surface is zero. Figure 24–7 shows the electric field due to a dipole as described in Chapter 23. Imagine a series of Gaussian surfaces of any convenient shape placed wherever we choose. For example (Fig. 24–7a), if we place an imaginary Gaussian surface (surface 1) around charge $+q$ of a dipole, all the electric field lines exit the Gaussian surface, and the total electric flux is positive. If we place a second Gaussian surface (surface 2) around charge $-q$, all the electric field lines enter the Gaussian surface, and the electric flux is negative. Any Gaussian surface, such as surface 3, that surrounds *neither* charge has no net electric flux through it because the same number of electric field lines enter and exit such a surface. If the Gaussian surface surrounds *both* charges (Fig. 24–7b), then the num-

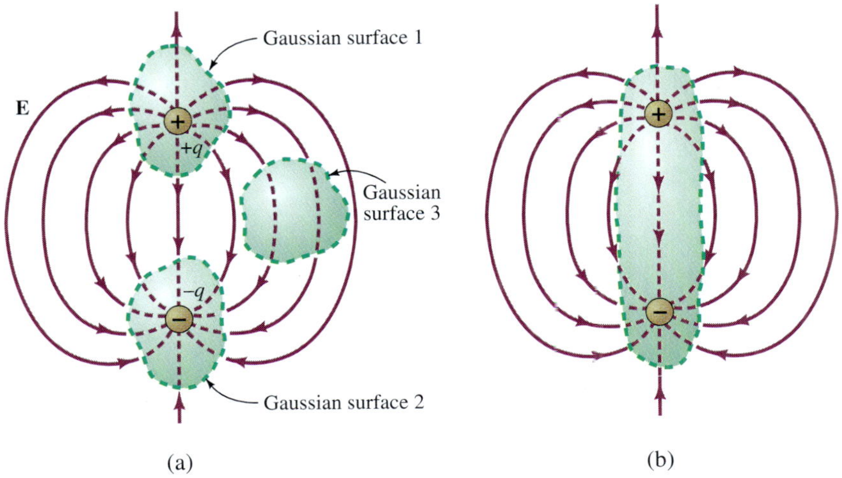

FIGURE 24–7 (a) Three Gaussian surfaces (imaginary and therefore dashed) in the electric field of a dipole. For surface 1, which surrounds the $+q$ charge, the electric flux is positive; for surface 2, which surrounds the $-q$ charge, the flux is negative; and for surface 3, which surrounds no charge, the flux is zero. (b) A Gaussian surface surrounding both charges. The flux through this surface is proportional to the net charge and is therefore zero.

ber of field lines that enter and exit the surface is again equal, and the total flux is zero. This observation is an important one. Our result applies to any Gaussian surface that surrounds a configuration of charges, as long as there is no *net* charge. We summarize:

The electric flux through a closed surface that encloses no net charge is zero.

The electric flux through a closed surface is zero if the surface encloses no net charge.

24–2 GAUSS' LAW

We have seen that the electric flux through a closed surface that encloses no net charge is zero. When the surface does enclose a net charge, however, the electric flux through the surface is not zero; in this case, *Gauss' law* expresses that flux in terms of the charge enclosed. Let's take a look at the flux through a Gaussian surface that encloses a point charge. Figure 24–8 shows an (imaginary) Gaussian sphere of radius R centered on a static point charge q. The centered sphere is chosen because the electric field has constant magnitude at a fixed distance from a charge, and it will be easy to find the flux through the sphere. We use Eq. (24–3) to find the electric flux that passes through the Gaussian surface. The electric field due to a point charge q was found in Eq. (23–5) to be

$$\mathbf{E} = \left(\frac{q}{4\pi\epsilon_0 r^2}\right)\hat{\mathbf{r}}.$$

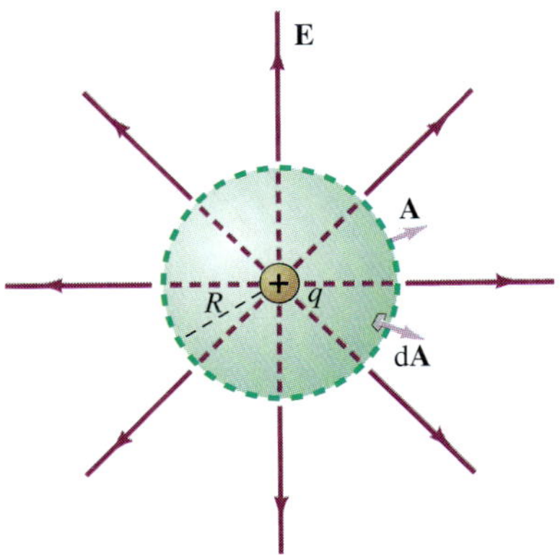

FIGURE 24–8 A simple choice for a Gaussian surface for a point charge q is a sphere of radius R.

The electric field points in the radial direction and is directed outward for positive q. Because the direction of the infinitesimal area $d\mathbf{A}$ for a small area on the sphere also points outward in the radial direction, the product $\mathbf{E} \cdot d\mathbf{A} = E\,dA$. Because the electric field has the constant value $q/4\pi\epsilon_0 R^2$ everywhere on our sphere, the infinitesimal electric flux through the infinitesimal area dA is

$$d\Phi = E\,dA = \frac{q}{4\pi\epsilon_0 R^2}\,dA.$$

We can now move the (constant) field E outside the integral sign in the integral that represents the total flux:

$$\Phi = \oint d\Phi = \oint \mathbf{E} \cdot d\mathbf{A} = \oint E\,dA = \oint \frac{q}{4\pi\epsilon_0 R^2}\,dA = \frac{q}{4\pi\epsilon_0 R^2}\oint dA.$$

The integral of the area element dA over the closed surface is the total area of the closed surface, $A = 4\pi R^2$. Thus,

$$\Phi = \frac{q}{4\pi\epsilon_0 R^2}A = \frac{q}{4\pi\epsilon_0 R^2}4\pi R^2 = \frac{q}{\epsilon_0}. \qquad (24\text{–}5)$$

This result is independent of the radius of our Gaussian sphere. The electric flux emanating from a point charge is q/ϵ_0.

We have considered a sphere centered on a charge and already see that the electric flux is independent of the sphere's radius. But we can extend this much further and show that the flux through *any* closed surface that surrounds the charge gives this same result! Such surfaces include Gaussian spheres off center from the point charge or, in fact, any regular or irregular surface around the charge (Figs. 24–9 and 24–10). In order to establish our result, recall that we have shown in Section 24–1 that the flux through a surface is proportional to the number of electric field lines that pass through that surface. Now, because electric field lines originate from and stop on charges, the number of electric field lines that pass through any surface that surrounds our single charge is the same as the number of electric field lines that pass through a sphere centered on the point charge. So the flux through any of these Gaussian surfaces is exactly the same. Equation (24–5) is thus established for any Gaussian surface S *as long as S surrounds the point charge* q:

The electric flux through any closed surface that encloses a point charge q is proportional to q.

$$\oint_{\text{closed surface } S} \mathbf{E} \cdot d\mathbf{A} = \frac{q}{\epsilon_0}. \qquad (24\text{–}6)$$

Gaussian surface 3 in Fig. 24–7a is one case in which the charge is outside the Gaussian surface. Here, as many lines leave as enter, and the charge gives no net flux through the surface.

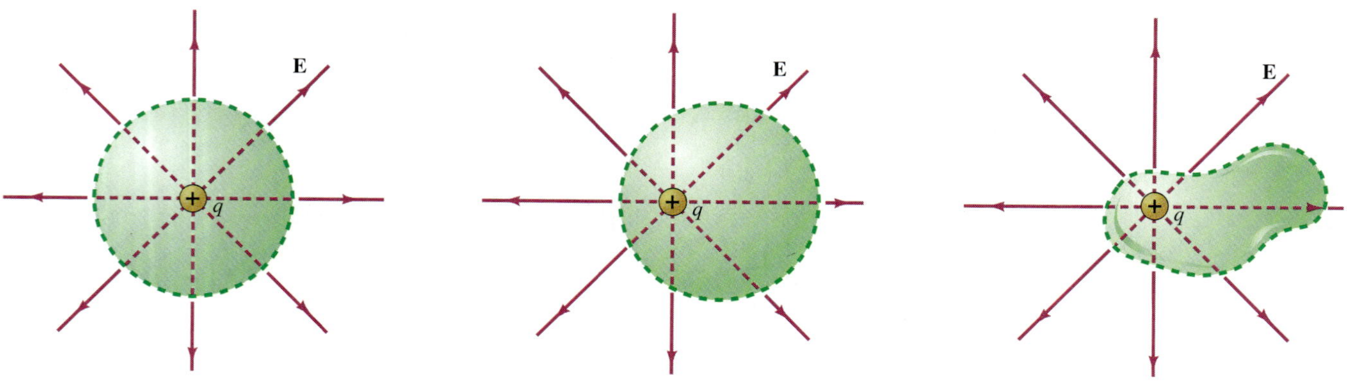

FIGURE 24–9 All the Gaussian surfaces shown give the same result for the electric flux. The same number of electric field lines pierce each surface.

We must generalize Eq. (24–6) for the case of multiple point charges and continuous charge distributions. We know that the *total or net charge* Q can be broken down into an assembly of point charges q_i. Further, from the superposition principle, we know that the total electric field $\mathbf{E}$ is the sum of the fields $\mathbf{E}_i$ due to point charges q_i. The total flux Φ through a Gaussian surface due to the net charge is then just the sum of the fluxes Φ_i due to the charges q_i:

$$\Phi = \sum_i \Phi_i = \oint_{\text{closed surface } S} \sum_i \mathbf{E}_i \cdot d\mathbf{A} = \frac{1}{\epsilon_0}\sum_i q_i = \frac{Q}{\epsilon_0}.$$

Our result is generally referred to as **Gauss' law**,

Gauss' law

$$\oint_{\text{closed surface } S} \mathbf{E} \cdot d\mathbf{A} = \frac{Q}{\epsilon_0}. \qquad (24\text{–}7)$$

The closed surface is *any* Gaussian surface that surrounds the *net* charge Q. The case in which the net charge is zero is included here—either because no charge whatsoever is enclosed by S or because there is an equal amount of positive and negative charge.

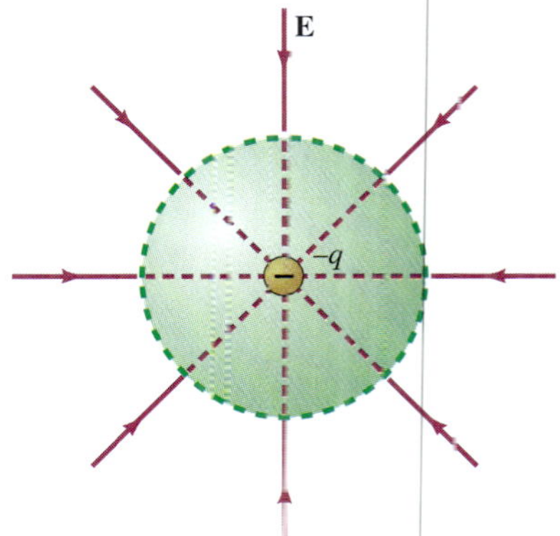

FIGURE 24–10 Similar to Fig. 24–9, but with negative charge. The flux is negative in this case.

Coulomb's Law and Gauss' Law

We have treated Gauss' law as a development of Coulomb's law by using the electric field of the point charge that was determined by using Coulomb's law. This procedure can be reversed, and we can derive Coulomb's law from Gauss' law. To do so, we center a Gaussian sphere on a point charge q (Fig. 24–8). The electric field of the charge, $\mathbf{E}$, is assumed to be unknown. Gauss' law tells us only that the electric flux integrated over the surface of the sphere is q/ϵ_0. This is insufficient to determine the field because the flux through any tiny surface element of the sphere depends on the value of the field in that region. We can, however, use a symmetry argument. All directions around a point charge should be equivalent. The only configuration of field around a charge that does not favor some particular direction is a radial field. The surface element $d\mathbf{A}$ of a Gaussian sphere is also radial. If we assume that $\mathbf{E}$ is parallel to $d\mathbf{A}$ at all locations—rather than antiparallel—we may at worst have made a sign error, and we can repair this later. Thus,

$$\mathbf{E} \cdot d\mathbf{A} = E\, dA.$$

Moreover, symmetry—or the assumption that there is no preferred direction—also implies that $\mathbf{E}$ will have the same magnitude everywhere on the centered sphere. We can then remove E from the integral that expresses the total flux through the sphere:

$$\oint \mathbf{E} \cdot d\mathbf{A} = \oint E\, dA = E \oint dA = EA = E(4\pi r^2) = \frac{q}{\epsilon_0},$$

where r is the radius of the Gaussian sphere. The last term in this equality is just Gauss' law. The equation can be solved for the magnitude of the electric field:

$$E = \frac{q}{4\pi\epsilon_0 r^2}.$$

This result is consistent with Eq. (23–5). Because E is positive, we correctly chose the direction of $\mathbf{E}$ to be radially outward for a positive charge. The symmetry of the situation tells us only that the electric field must be radial: either outward or inward. Gauss' law determines the orientation of $\mathbf{E}$ to be radially outward. Coulomb's law follows directly from the previous equation if we put another charge, q', in the electric field and use $\mathbf{F} = q'\mathbf{E}$.

In Examples 24–1 and 24–2, we use the fact that Gauss' law does not require us to use any particular surface. This is important because the flux through one surface may be much easier to calculate than the flux through another.

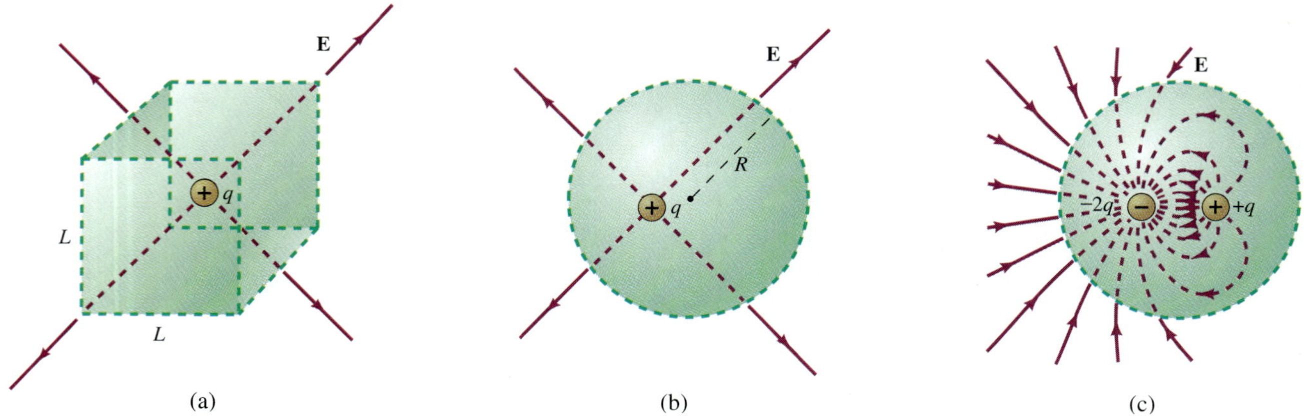

FIGURE 24–11 Example 24–1.

EXAMPLE 24–1 Find the electric flux through the Gaussian surfaces in Fig. 24–11: (a) a cube of sides L that surrounds the charge q; (b) a sphere of radius R that surrounds the charge q, which is off center; (c) a sphere of radius b that surrounds the charges $-2q$ and $+q$.

Solution: (a) We do not need to do the direct integration of the electric field over the cube. According to Gauss' law, the total electric flux is simply q/ϵ_0 because q is the net charge enclosed by the cubic Gaussian surface. The shape of the surface is immaterial, as is the position of the charge inside.

(b) It does not matter that the Gaussian sphere is off center. The total electric flux is still q/ϵ_0.

(c) We do not need to concern ourselves with the positions of the two charges within the cube. The total net charge Q enclosed by the Gaussian surface is $-2q + q = -q$, and the total electric flux through the Gaussian surface is $-q/\epsilon_0$.

EXAMPLE 24–2 Consider a point charge $q = 1$ mC placed at a corner of a cube of sides 10 cm in an electric field **E**. Determine the electric flux through each face of the cube.

Solution: The situation is sketched in Fig. 24–12. We can use symmetry to make this problem an easy one to solve. First, consider the three faces of the cube that the charge touches. For each of these faces, the product $\mathbf{E}\cdot\mathbf{A} = 0$ because the electric field lies along each of the three faces. The electric flux through each of the remaining three faces of the cube must be equal by symmetry.

What is the total electric flux through the cube? It would take seven other similarly placed cubes to surround the point charge q completely. Because each of the eight cubes is placed symmetrically about the charge, each of the cubes has an electric flux of $q/8\epsilon_0$. Therefore, each of the three faces of the cube that touch the charge must have an electric flux of $q/24\epsilon_0$. Note that the electric flux through each face is independent of the size of the cube.

Numerical evaluation gives

$$\Phi_{\text{face}} = \frac{q}{24\epsilon_0} = \frac{10^{-3}\,\text{C}}{24(8.85\times10^{-12}\,\text{C}^2/\text{N}\cdot\text{m}^2)} = 4\times10^6\,\text{N}\cdot\text{m}^2/\text{C}.$$

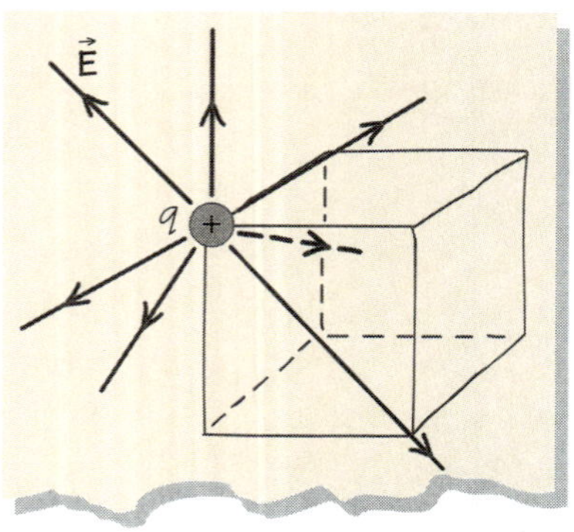

FIGURE 24–12 Example 24–2.

24–3 APPLICATIONS OF GAUSS' LAW

Gauss' law is a fundamental law in its own right. It is also a powerful tool for the determination of electric fields in situations where there is a high degree of symmetry. If there is enough symmetry so that the electric field is constant over a simple surface and can be removed from the integral that expresses the flux, then we can solve the equation that expresses Gauss' law for the field magnitude. Under these circumstances, we do not need to perform complex integrations. We shall study this technique using several examples that involve continuous charge distributions: the line of charge, the charged plane, the spherical shell, and the uniformly charged sphere. These cases were discussed in Chapter 23 and (for gravitation) in Chapter 12. We shall see that Gauss' law determines the fields briefly and simply. But the real power of solution by Gauss' law is revealed when we discuss conductors in Section 24–4. There, we find the fields in situations that are entirely new.

In Examples 24–3 through 24–6, we use these techniques together with Gauss' law, Eq. (24–7), to determine the field.

EXAMPLE 24–3 *Line of charge.* Determine the electric field due to an infinitely long, straight charged rod with positive, constant charge density λ.

Solution: Figure 24–13a illustrates the situation. We have oriented the rod along the z-axis. To find the appropriate Gaussian surface, we want to see what symmetry tells us about the direction and magnitude of the electric field lines. These lines must leave the positively charged rod and, to be symmetric, the electric field lines must extend away from the rod radially in the xy-plane (Fig. 24–13b). The electric field lines cannot have a component along the rod because there is no way to decide whether the field would be oriented in the $+z$- or $-z$-direction. The direction of the field can easily be identified by visualizing the force on a positive test charge placed outside the rod; the rod will repel it with a force $\mathbf{F} = q\mathbf{E}$. Moreover, again by symmetry, the magnitude of the field must be the same on every point of a circle centered on the rod. Thus the field magnitude can depend only on the radial distance from the rod. The Gaussian surface that takes advantage of the symmetry is a closed cylinder of radius r and height h, centered on the rod (Fig. 24–13c). This surface will enable us to find the field at a distance r from the rod.

We now want to find the flux through the cylinder. We can express this flux as

$$\Phi = \int_{\text{top surface}} \mathbf{E}\cdot d\mathbf{A} + \int_{\text{bottom surface}} \mathbf{E}\cdot d\mathbf{A} + \int_{\text{side surface}} \mathbf{E}\cdot d\mathbf{A}.$$

For the top and bottom, note that $\mathbf{E}$ is parallel to these surfaces, so the surface element $d\mathbf{A}$ is perpendicular to $\mathbf{E}$. Thus

$$\text{for the top and bottom surfaces:} \qquad \mathbf{E}\cdot d\mathbf{A} = 0.$$

The flux through the top and bottom is zero. For the curved side, the electric field is perpendicular to the surface, so

$$\text{for the side surface:} \qquad \mathbf{E}\cdot d\mathbf{A} = E\,dA.$$

This expression must be integrated over the side to find the flux through the side of the cylinder. But we have chosen the cylindrical surface so that the electric field has constant magnitude over the surface, and the field magnitude at a distance r from the wire can be removed from the integral. Thus

$$\Phi = \int_{\text{side surface}} \mathbf{E}\cdot d\mathbf{A} = E\int_{\text{side surface}} dA.$$

The remaining integral over the side is just the side area of a right cylinder of height h, namely, $2\pi rh$. We thus have, for the total flux through the cylinder,

$$\Phi = 2\pi rhE.$$

Now that the flux has been calculated, let's apply Gauss' law. The net charge inside the cylinder, q, is the charge on a portion of the rod of length h. This charge is just the charge density, λ, times the length, $q = \lambda h$. Equation (24–7), Gauss' law,

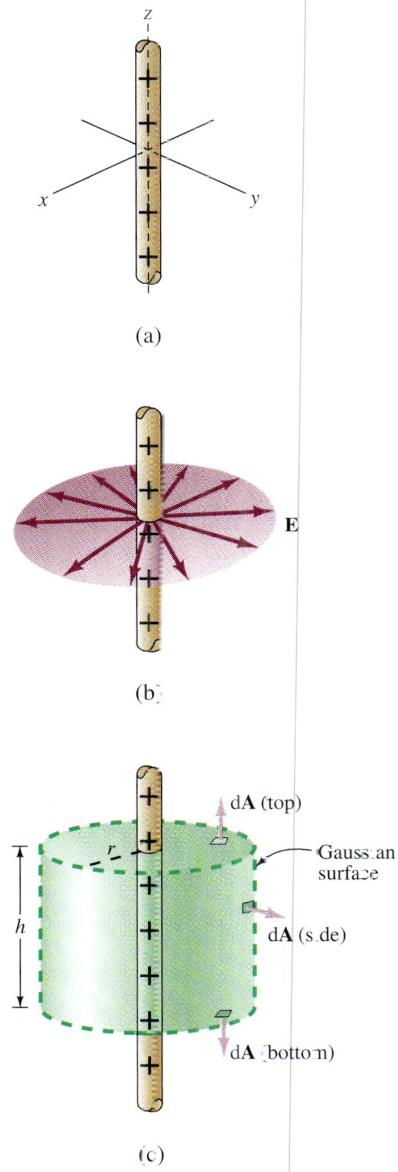

FIGURE 24–13 (a) Example 24–3. A line of charge is oriented along the z-axis. (b) By symmetry, the direction of the electric field $\mathbf{E}$ is radial in the xy-plane. (c) The best Gaussian surface to use to determine the electric field of a line charge is a cylinder. The directions of the areas $d\mathbf{A}$ for the various surfaces of the cylinder are shown.

PROBLEM-SOLVING TECHNIQUES

To use Gauss' law to find electric fields given a charge distribution, the following steps are helpful:

1. Make a sketch of the charge distribution. It will help you to recognize any appropriate symmetry.
2. Identify any spatial symmetry of the charge distribution and the electric field it produces. For example, a point charge has spherical symmetry because it looks the same from all around a sphere centered on it. The spherical symmetry of the point charge implies that the field must be radial.
3. Choose a Gaussian surface that is matched to the symmetry. This is the most important step in determining electric fields by Gauss' law. The experience we have gained so far in dealing with and visualizing electric fields is helpful here. Choose a Gaussian surface for which the field is either parallel to the surface ($d\Phi_E = 0$) or perpendicular to the surface ($d\Phi_E = E\,dA$) at various locations; further, choose the surface so that the field is constant over the part of the surface to which it is perpendicular. For example, the Gaussian surface best suited to a point charge is a sphere centered on the charge.
4. With surfaces chosen as in step 3, it should be possible to remove the electric field from inside the integral that expresses the flux. Then Gauss' law becomes an algebraic expression for the *magnitude* of the field.

now reads

$$2\pi rhE = \frac{q}{\epsilon_0} = \frac{\lambda h}{\epsilon_0}.$$

We can solve for E:

$$E = \frac{\lambda}{2\pi\epsilon_0 r}. \tag{24–8}$$

The arbitrary height h has canceled. In SI units, the charge density is in coulombs per meter. Thus $\epsilon_0 E$ has the units of coulombs per square meter, as it must.

Compare the ease with which we obtained Eq. (24–8) with the direct integration technique [Eq. (23–30)]. Why can't we use Gauss' law to find the field of a *finite* line of charge? Gauss' law continues to hold for *any* distribution of charge but, for a finite line of charge, the symmetry that allows us to determine the direction of **E** and remove it from the flux integration is not present. If the ends of the line are in view, they provide a guide to tell us where we are along the wire—for example, we can look to see that we are close to one end or the other. The symmetry along the wire is lost. This loss of symmetry has two consequences: First, the electric field will have a component along the wire; and second, the magnitude of the field will vary *along* the line.

EXAMPLE 24–4 *Spherical shell.* Determine the electric field both inside and outside a spherical shell of radius R that has a total charge Q distributed uniformly on its outer surface.

Solution: We show the configuration in Fig. 24–14a. First, we find the electric field outside the shell. From symmetry, the electric field must be directed radially outward for a positive charge Q and must have a constant magnitude everywhere at the distance r. We take a Gaussian surface that is a sphere of radius r centered on the shell. The product $\mathbf{E}\cdot d\mathbf{A} = E\,dA$ because **E** and $d\mathbf{A}$ are in the same direction. Gauss' law becomes

$$\frac{Q}{\epsilon_0} = \oint_{\text{surface}} \mathbf{E}\cdot d\mathbf{A} = E \oint_{\text{surface}} dA = E4\pi r^2. \tag{24–9}$$

Here, the total area of the Gaussian sphere is $4\pi r^2$, and the total charge enclosed by the Gaussian sphere is the charge Q on the surface of the spherical shell. From Eq.

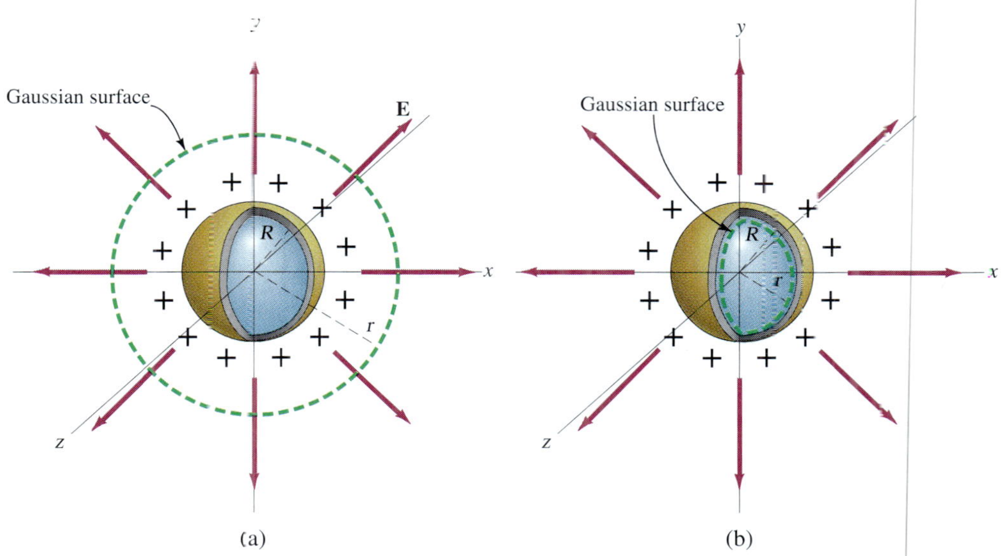

FIGURE 24–14 (a) Example 24–4. The best Gaussian surface to determine the electric field outside a uniformly charged spherical shell. The symmetry is spherical. (b) The best Gaussian surface to determine the electric field inside a uniformly charged spherical shell is a sphere inside the shell.

(24–9), the electric field outside the shell is

$$\text{outside a spherical shell, } r \geq R: \qquad E = \frac{Q}{4\pi\epsilon_0 r^2}. \qquad (24\text{–}10)$$

Thus the electric field is the same as the field of a point charge of the same total magnitude Q at the center of the spherical shell.

For a point *inside* the spherical shell, we again have spherical symmetry and so we draw another Gaussian sphere inside the shell (Fig. 24–14b). We find the electric flux in terms of any electric field as before. In this case, however, no charge is enclosed by the Gaussian sphere, so the left-hand side of Eq. (24–9) must be zero. Therefore, *the electric field inside a uniformly charged spherical shell must be zero*:

$$\text{inside a spherical shell, } r < R: \qquad E = 0. \qquad (24\text{–}11)$$

We noted in Chapter 12 that these same results hold for the force of gravity due to a spherical shell of matter. The mathematical problem is identical because the gravitational force has the same inverse-square form as the Coulomb force. We gave only the results in Chapter 12—without derivation—because the direct integration technique is fairly complicated. The Gauss' law derivation provided here is a very simple one. It is interesting to note that Newton delayed the publication of his theory of gravitation by some 20 years because of his lack of a simple proof of these results. If he had known Gauss' law, Newton would have saved a lot of time!

EXAMPLE 24–5 *Solid sphere*. Find the electric field outside and inside a solid, nonconducting sphere of radius R that contains a uniformly distributed total charge Q.

Solution: This charge distribution, shown in Fig. 24–15a, exhibits the same symmetry as does Fig. 24–14a: The electric field must be purely radial and can vary only with the distance r from the center of the sphere. It is best to take all Gaussian surfaces as spheres centered on the charged sphere, and the flux through any of these spheres will have the form

$$\Phi = \oint_{\text{sphere}} \mathbf{E}\cdot d\mathbf{A} = E \oint_{\text{sphere}} dA = E4\pi r^2.$$

Here, E is the field at a distance r from the center. In applying Gauss' law, we must be careful about the charge enclosed by the Gaussian surface.

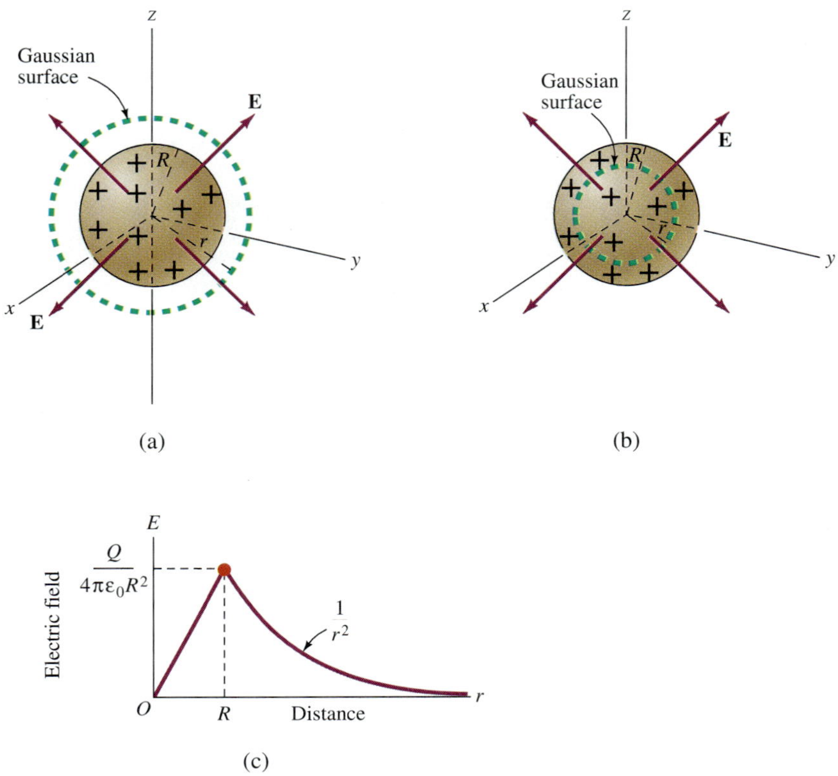

FIGURE 24–15 (a) Example 24–5. The best Gaussian surface to determine the electric field outside a uniformly charged, nonconducting sphere. The symmetry is spherical. (b) The best Gaussian surface to determine the electric field inside a uniformly charged, nonconducting sphere is a Gaussian sphere inside the solid sphere. Only the charge inside the Gaussian sphere contributes to the electric field at r. (c) The electric field due to a uniformly charged, nonconducting sphere as a function of the distance from the center of the sphere.

Let's first consider the field outside the solid sphere (Fig. 24–15a). The charge enclosed by a Gaussian sphere at $r > R$ is just Q and, as for a spherical shell,

$$\text{outside a solid sphere, } r > R: \qquad E = \frac{Q}{4\pi\epsilon_0 r^2}. \tag{24-12}$$

The situation is different inside the solid sphere, however. The charge Q' enclosed by our Gaussian sphere (Fig. 24–15b) is given by the charge density ρ times the volume $\frac{4}{3}\pi r^3$. The charge density is $\rho = Q/\text{total volume} = Q/(\frac{4}{3}\pi R^3)$. Thus

$$Q' = \rho\frac{4}{3}\pi r^3 = Q\frac{\frac{4}{3}\pi r^3}{\frac{4}{3}\pi R^3} = Q\frac{r^3}{R^3}.$$

From Gauss' law, the field at a radius $r < R$ is

$$E = \frac{Q'}{4\pi\epsilon_0 r^2} = Q\frac{r^3}{R^3}\frac{1}{4\pi\epsilon_0 r^2};$$

$$\text{inside a solid sphere, } r < R: \qquad E = \frac{Q}{4\pi\epsilon_0}\frac{r}{R^3}. \tag{24-13}$$

Inside, the charge enclosed increases with radius as r^3, whereas the area increases with radius as r^2; therefore, $E \propto Q/\text{area} \propto r$. The charge located outside r gives a net field that adds to zero at r. The electric field due to a solid sphere has the radial dependence displayed in Fig. 24–15c. As symmetry demands, the field is zero at the center of the sphere. The field increases linearly with r up to the radius of the sphere and then decreases inversely as the square of r. The fields in Eqs. (24–12) and (24–13) match at the point $r = R$.

Generally speaking, for a spherically symmetric charge distribution, the field at a radius r is that of a point charge at the center whose magnitude is the total charge within the sphere of radius r. We have seen that this is easy to prove by using Gauss' law. It holds not only for thin shells and solid spheres but indeed for *any* distribution of charge whose charge density varies only with the radius.

EXAMPLE 24–6 *Plane of charge.* Find the electric field outside an infinite, nonconducting plane of charge with uniform charge density σ.

Solution: We show a charged plane in Fig. 24–16. If the plane is positively charged, then, from symmetry, the electric field will be perpendicular to, and point away from, the plane. We can verify this fact by placing a test charge near the plane. The force on the test charge will be either directly away from or directly toward the plane. Symmetry also dictates that the electric field has a magnitude that depends at most on the perpendicular distance from the plane. Because the electric field is perpendicular to the plane, a good choice for the Gaussian surface is any right solid (such as a cylinder) with its top and bottom (area A) parallel to the charged plane (Fig. 24–16). Every facet of this Gaussian surface is either parallel or perpendicular to the electric field. The differential areas $d\mathbf{A}$ for the top and bottom of the Gaussian surface also point away from the charged plane, so the product $\mathbf{E}\cdot d\mathbf{A}$ for the three surfaces is

$$\text{for the top:}\quad \mathbf{E}\cdot d\mathbf{A} = E\,dA;$$

$$\text{for the bottom:}\quad \mathbf{E}\cdot d\mathbf{A} = E\,dA;$$

$$\text{for the side:}\quad \mathbf{E}\cdot d\mathbf{A} = 0.$$

The last equation follows because $d\mathbf{A}$ for the side points everywhere parallel to the plane, but $\mathbf{E}$ is everywhere perpendicular to the plane.

Equation (24–7) for Gauss' law now becomes

$$\frac{Q}{\epsilon_0} = \oint \mathbf{E}\cdot d\mathbf{A} = \int_{\text{top surface}} \mathbf{E}\cdot d\mathbf{A} + \int_{\text{bottom surface}} \mathbf{E}\cdot d\mathbf{A} = \int_{\text{side}} \mathbf{E}\cdot d\mathbf{A}\,|\,\pi = EA,$$

where we have used the fact that E is constant over the top and bottom area A of the Gaussian surface.

The total charge enclosed by the Gaussian surface is the charge on the plane within the surface. Because the charge density is σ and the area enclosed is A, we must have $Q = \sigma A$. The previous equation becomes

$$\frac{Q}{\epsilon_0} = \frac{\sigma A}{\epsilon_0} = 2EA;$$

$$E = \frac{\sigma}{2\epsilon_0}. \tag{24–14}$$

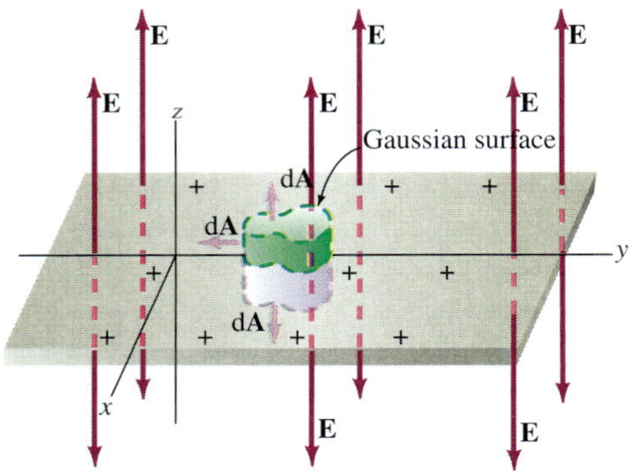

FIGURE 24–16 Example 24–6. A convenient Gaussian surface for a uniformly charged infinite plane can be any shape whose sides are perpendicular to the plane and whose top and bottom are parallel to the plane.

In SI units, σ is measured in coulombs per square meter. Thus the units of $\epsilon_0 E$ are C/m^2, which is a correct result. Equation (24–14) is the same result that we found with much more difficulty by direct integration in Chapter 23 [Eq. (23–33)]. Note that E is independent of the distance from the plane.

Equation (24–14) also expresses the electric field at a given point due to a *finite* charged plane as long as the distance from the point to the ends of the plane is much greater than the perpendicular distance of the point from the plane.

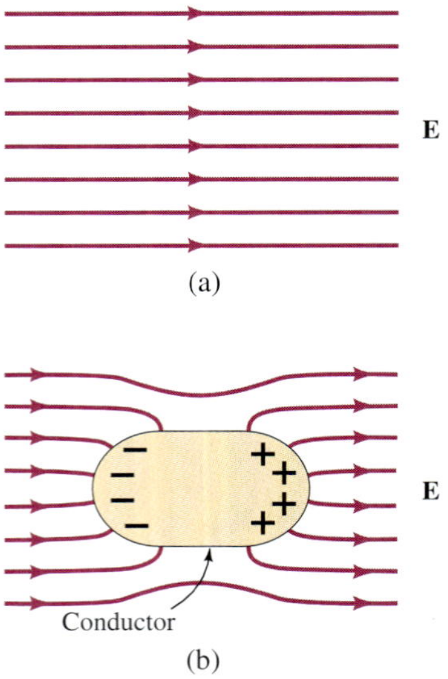

FIGURE 24–17 An uncharged conductor in an external electric field. (a) The electric field before a conductor is introduced. (b) Charges are induced on the surface of the conductor such that the electric field inside the conductor is zero. The induced charges modify the field outside the conductor, so the field no longer has its original form.

24–4 CONDUCTORS AND ELECTRIC FIELDS

A good conductor, such as silver, copper, or aluminum, has a large number of "free" electrons, which can move within the (electrically neutral) material. Any electric field that may appear inside the metal due to the presence of an external electric field will cause these electrons to move. In less than a microsecond, they rearrange themselves into a configuration that cancels the electric field inside the material. If any field whatsoever remained inside the material, it would cause the electrons of the conductor to move until they reached equilibrium. (We refer to electrostatic equilibrium.) *Conductors have no internal static electric field.*

There is no static electric field inside a conductor.

This property of conductors is illustrated in Fig. 24–17. A conductor is placed in a spatially constant and static external field that points to the right (Fig. 24–17a). Some electrons in the metal move to the left side of the conductor, which leaves a deficiency of electrons on the right side of the conductor. The arrangement of excess electrons on the left and a deficiency of electrons on the right forms a new, internal electric field that points to the left. This internal field will precisely cancel the external field, with the result that there is no net field within the conductor (Fig. 24–17b).

The fact that there are no static electric fields within conductors has implications for the behavior of conductors when charges are put on or near them, or when they are placed in external electric fields. This behavior is determined using Gauss' law. Let's consider what happens when an excess charge is added to a conductor. Figure 24–18 shows such a conductor as well as a Gaussian surface just inside the metal surface. If we apply Gauss' law to this surface, we find that, because there is no field, there is no flux, and hence there is no net charge inside the metal. Where is the excess charge? *In electrostatic equilibrium, all excess charge is on the outside surface of a conductor.*

Free charges move to the outside surfaces of conductors.

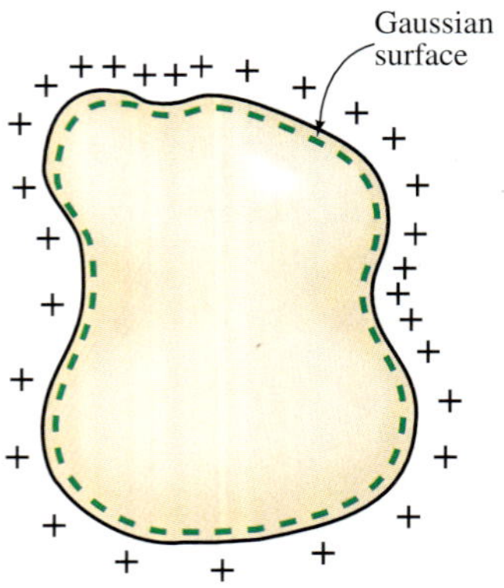

FIGURE 24–18 To find the electric field inside a conductor of arbitrary size and shape, choose a Gaussian surface just inside the surface, so that the closed surface encloses no charge.

Imagine a bubble within a conductor. The bubble is filled with a nonconducting medium (such as air) (Fig. 24–19a). Suppose that there is no excess charge within the bubble. It is only on the surface of the bubble—the interior surface of the conductor—that charge might accumulate. A Gaussian surface surrounding the bubble, but drawn within the conductor, has no electric flux through it because there is no static field within any conductor. Thus there is no net charge within that Gaussian surface. We have thereby shown that there can be no net charge on the interior surface of the conductor. *Any excess charge placed on a conductor, even if the conductor contains nonconducting bubbles, moves to the outside surface of the conductor*, provided there is no charge within the nonconducting bubbles.

We must modify our reasoning when there is charge within nonconducting bubbles in the conductor. Suppose that a nonconducting bubble inside a conductor contains a charge $+Q$ (Fig. 24–19b). Again, draw a Gaussian surface within the metal to surround the bubble. Because there is no field inside the metal, the net charge enclosed must be zero. In this case, a charge of $-Q$ will be induced on the *inner* surface of the metal, that is, on the bubble surface. This induced negative charge keeps the electric field zero *inside the conductor.*

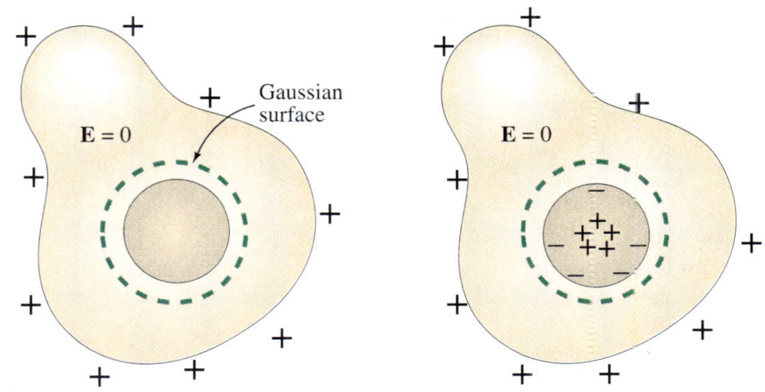

FIGURE 24–19 (a) A nonconducting, hollow space inside a conductor with no charge inside. All the charge must be on the outer surface of the conductor. (b) If we place a charge inside the hollow space, an induced charge will appear on the inside surface of the conductor, making the electric field zero inside the conducting material. A Gaussian surface drawn just outside the hollow space helps to show these results.

Electrostatic Fields near Conductors

We can draw two important conclusions about electrostatic fields around conductors from this discussion. First, *the electric field immediately outside a conductor must be perpendicular to the conductor's surface.* If there were a parallel component, then the charges resting on the surface would react and move in contradiction to our assumption of equilibrium. Second, by using Gauss' law, we can find the value of this perpendicular electric field near the surface in terms of the charge density on the surface. To do so, consider the conductor shown in Fig. 24–20, with a tiny Gaussian surface whose side is perpendicular and whose top is parallel to the conductor's surface. It is tiny because the surface charge density σ may vary over the conductor; we shall be referring to σ only at the point where the Gaussian surface is erected. The electric field is zero inside the metal surface, and it is parallel to the side of the Gaussian surface. Thus, the only contribution to the flux comes from the top. If the Gaussian surface is small enough, **E**, which is perpendicular to the top surface, can be regarded as constant over it, and

The electric field near a conductor is perpendicular to the conductor's surface.

$$\frac{Q}{\epsilon_0} = \oint \mathbf{E} \cdot d\mathbf{A} = EA,$$

where A is the area of the top of the Gaussian surface. The total charge Q enclosed by the Gaussian surface is σA, so the previous equation becomes

$$\frac{Q}{\epsilon_0} = \frac{\sigma A}{\epsilon_0} = EA.$$

The area cancels. The electric field just outside the surface is proportional to the local charge density σ:

$$\mathbf{E} = \frac{\sigma}{\epsilon_0}. \qquad (24\text{–}15)$$

The electric field just outside the surface of a conductor

This result holds only near the conductor's surface. Whether this result is useful or not depends on our knowing the charge density σ; the magnitude of the field will vary around the surface of the conductor as σ does. The electric field will always be perpendicular to the conductor near its surface (Fig. 24–21). We can check this result by considering a conductor that is a sphere of radius R and total charge Q. In this case, symmetry demands that the charge is spread evenly over the surface, and

$$\sigma = \frac{Q}{\text{area}} = \frac{Q}{4\pi R^2}.$$

For the field just outside the sphere, Eq. (24–15) would then give $E = Q/4\pi\epsilon_0 R^2$, which agrees with our earlier result, Eq. (24–12).

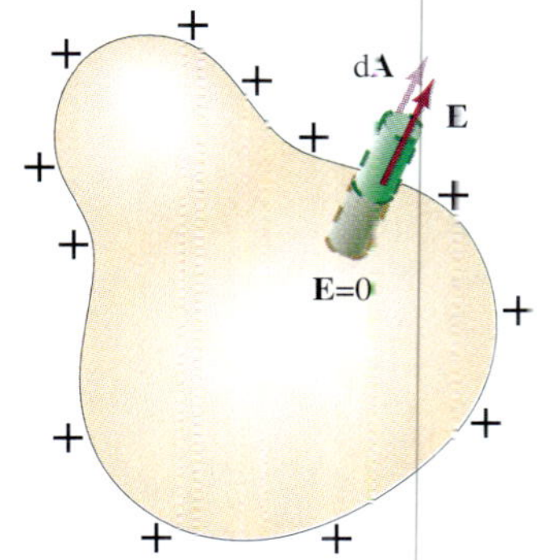

FIGURE 24–20 To find the electric field outside a conductor of arbitrary size, we choose a small right circular cylinder for the Gaussian surface. The only part of the cylinder through which there is a nonzero electric flux is the outside end of the cylinder.

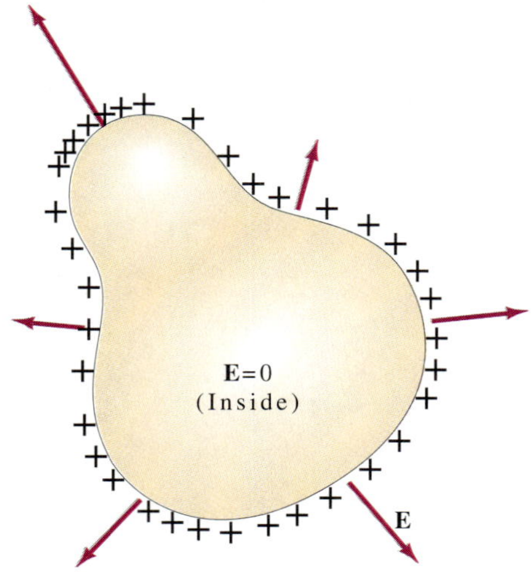

FIGURE 24–21 The electric field in and around a conductor in equilibrium. The electric field inside the conductor is zero, and, just outside the conducting surface, it must be perpendicular to the surface. The magnitude of the electric field varies according to the surface charge density σ, which may not be constant everywhere on the surface.

The field just outside a conductor ($E = \sigma/\epsilon_0$) is twice as large as the field of a nonconducting charged plane with the same charge density ($E = \sigma/2\epsilon_0$). The charge on a surface of area dA, σ dA, gives rise to a certain number of field lines. For a nonconducting plane, the field lines divide equally between the two sides of the plane. For a conducting plane, there are no field lines on the conducting side, so all the field lines must emerge on the open side.

Properties of electric fields in and around conductors

We can summarize what we have learned about conductors as follows:

1. The electrostatic field inside a conductor is zero.

2. The electrostatic field immediately outside a conductor is perpendicular to the surface and has the value σ/ϵ_0, where σ is the local surface charge density.

3. A conductor in electrostatic equilibrium—even one that contains nonconducting bubbles—can have charge only on its outer surface, as long as the bubbles possess no net charge.

We can add one more important result. Suppose that we have a bubble in a metal, with no charge in it. We now know that there is no field within the metal and, moreover, no net charge on the inner surface of the metal surrounding the bubble. Even for nonsymmetric situations, it can be shown that, as long as there is no charge within the cavity, *the electric field is zero everywhere within the bubble.*

24–5 ARE GAUSS' AND COULOMB'S LAWS CORRECT?

Gauss' law is equivalent to Coulomb's law only because Coulomb's law is an inverse-square law. Gauss' law is one of the cornerstones of our understanding of electricity and magnetism, and we must ask just how well it is known and test it as precisely as possible. The errors implicit in any measurements of Coulomb's law set limits on our knowledge of Gauss' law; these limits have been improved by various means right up to the present time. It is one of the characteristics of science to be eternally skeptical of yesterday's experiment. It is not so much that yesterday's experiment is wrong; rather, a more accurate experiment can be done with more modern apparatus. Let's review the precision with which the equations of electrostatics are known, and look in detail at one particularly sensitive technique for testing Gauss' law.

The earliest tests are associated with the discovery of Coulomb's law itself. Gauss' and Coulomb's law were discovered in a peculiar order. Joseph Priestley knew that

APPLICATION: SHIELDING FROM ELECTRIC FIELDS

The fact that there are no electric fields within charge-free cavities in metals has practical applications. Research laboratories often have enclosures known as *Faraday cages* formed by copper screens or sheets. These "shielded" rooms are necessary for taking careful electronic measurements that are unaffected by outside electrical interference (Fig. 24AB–1). The enclosure is simply a cavity within a metal—in this case the conducting material is simply the copper screens. As long as there is zero net charge inside of the enclosure, there is no electric field within it due to any external effects. (If there were a net charge inside, charge would be induced on the inside of the copper screens, forcing the electric field inside the copper to be zero, and there would be an electric field inside the enclosure.)

The consequences of these properties go beyond the laboratory. The interior of your car is a safe place in the event of nearby lightning, but, for the same reason, your car radio does not work as well when the radio is located within the "cage" formed by a metal bridge.

FIGURE 24AB–1 By Gauss' law, there is no static electric field in an empty cavity in a metal. Researchers take advantage of this fact by working within a metal cage, inside of which fields due to outside sources are minimized.

there is no gravitational field entirely within a spherically symmetric mass distribution. He made the inspired speculation that a similar behavior of the gravitational and electric force laws would explain why a charged cork ball placed inside a charged metal container is not attracted to the walls of the container. This effect had first been seen—if not understood—in 1755 by Benjamin Franklin, who reported it to Priestley. Priestley thus had experimental evidence of Coulomb's law and reported it in 1767, although he made no truly quantitative test.

Franz Aepinus read of Franklin's work and guessed at an inverse-square dependence. Aepinus published his idea in Latin in a Russian journal in 1759. John Robison somehow heard about this idea and, in 1769, carried out direct experimental tests of the distance dependence of the force between charges. Robison expressed the uncertainties in his result as a *deviation* from Coulomb's law; that is, he supposed that the force is not an inverse-square law but rather has the distance dependence

$$F \propto \frac{1}{r^{2 \pm \delta}}.$$

He then gave limits on the parameter δ, as in Table 24–1. When $\delta = 0$, the inverse-square law is exact. The smaller the limit on δ, the closer the law is known to be an inverse-square law. Unfortunately, Robison made his results known only at an obscure meeting, and did not publish his results until 1801, well after Coulomb's publication.

The next discovery of Coulomb's law was made in 1773 by the rather eccentric Henry Cavendish. Through his knowledge of gravitation, he knew that the presence or absence of charge on the inner surface of a closed conductor is a consequence of what we now call Gauss' law, and it is therefore a direct test of a $1/r^2$ force law. He placed one conducting sphere inside another and connected the two by a wire. After placing a charge on the apparatus, he disconnected the wire and looked for any charge that re-

TABLE 24–1 Experimental Measurements of Deviation from an Inverse-Square Force Law[†]: Force $\alpha\ 1/r^{2 \pm \delta}$

Investigators	*Date*	*Maximum* δ
Robison	1769	0.06
Cavendish	1773	0.02
Coulomb	1785	0.10
Maxwell	1873	5×10^{-5}
Plimpton and Lawton	1936	2×10^{-9}
Williams, Faller, and Hill	1971	3×10^{-16}

[†]For more information on this subject, see A. S. Goldhaber and M. M. Nieto, "The Mass of the Photon," *Scientific American*, p. 86, May 1976.

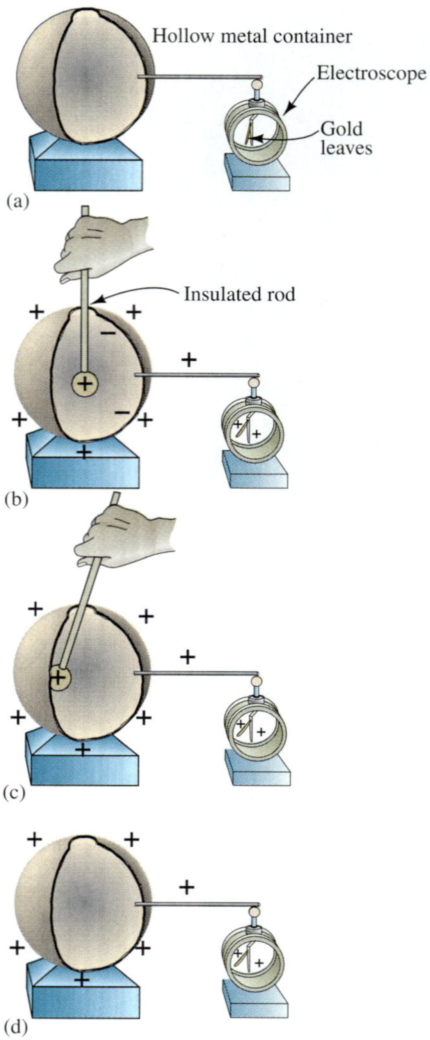

FIGURE 24–22 An electroscope is attached to the outside surface of a hollow conducting sphere to show the presence of charge. (a) No charge is present, and the gold leaf hangs down. (b) A charged metal ball on the end of an insulated rod is placed inside the sphere, and charge is induced. (c) If the metal ball touches the inside surface of the hollow conductor, all the charge passes to the outside surface. The electroscope's gold leaf indicates no change in the charge on the outside of the hollow conductor. (d) When the insulated metal ball is pulled outside, the charge remains on the outside of the hollow conductor, and the metal ball has no charge.

mained on the inner sphere. To the accuracy of his experiment, he found none. He was able to describe his result by saying that δ must be smaller than a certain value. Cavendish was even worse than Robison at publishing his results, which did not appear in print until more than 100 years later! Cavendish's experiment now goes under the general name of the *Faraday "ice-pail" experiment*, after Michael Faraday; it is the basis for many of the modern high-precision tests of Gauss' law.

It was not until 1785 that Charles Coulomb came into the picture, but he published his results promptly and now bears the credit for the force law. He tested the force law directly by using a torsion balance much like that used by Henry Cavendish in 1798 to test the gravitational force (see Chapter 12). As Table 24–1 shows, Coulomb's limit on δ was, in fact, worse than either Robison's or Cavendish's. Notable improvements in experiments similar to Cavendish's were made by James Clerk Maxwell in 1873, by Samuel J. Plimpton and Willard E. Lawton in 1936, and by Edwin R. Williams, James E. Faller, and Henry A. Hill in 1971. The degree of precision of these experiments, which are direct tests of Gauss' law, is astonishing.

A Null Experiment

Let's describe a simple version of the Faraday ice-pail experiment. We require an electroscope, which is the free charge detector introduced in Chapter 22. We also need a hollow metal container with a hole in the top, as in Fig. 24–22a. We introduce charge to the inside of the container with a small metal ball on the end of an insulating rod. The electroscope is attached to the outside of the container and thus indicates whether there is charge on the outside.

Next a positive charge, $+Q$, is placed on the metal ball, and the ball is inserted through the small hole into the hollow container without touching it (Fig. 24–22b). Gauss' law states that there is no net charge *inside* the nearly closed metal container; therefore, a charge of $-Q$ is induced on the inside surface of the container. (The hole can be made smaller and smaller until its presence does not matter.) Because the metal container is neutral, a charge of $+Q$ is induced on the outside, which the electroscope indicates. If the ball is moved around, there is no change whatsoever in the electroscope, which is consistent with Gauss' law. The metal ball is subsequently touched to the interior of the hollow container (Fig. 24–22c). If Gauss' law is correct, the charge on the ball neutralizes the $-Q$ charge induced on the inside surface, leaving the $+Q$ charge on the outside surface. The electroscope indicates this result because it does not change at all. When the metal ball is removed from the container, the container's outer surface remains charged (Fig. 24–22d). By touching the metal ball to another electroscope, we can verify that it carries no charge.

The description of this experiment shows why it is potentially so precise: If Gauss' law is correct, there is *no change* in the position of the gold leaf when the inner surface is touched. Equivalently, the Cavendish experiment tests for the *absence* of charge on the inner of two spheres. Experiments such as Coulomb's, which look for small changes in comparison with larger effects, are inherently less precise than experiments such as Cavendish's, which look for small changes in comparison with no effect. Experiments that test for small change versus no change are called **null experiments**. It is far easier to make a precision test of Gauss' law than of Coulomb's law because a null experiment can be done.

Coulomb's Law Holds over Small and Large Distances

This is not the end of the story, however. First, the experiments that we have listed in Table 24–1 test the laws only over a distance of about 1 m. Yet the laws of electrodynamics are supposed to hold in atomic systems and over galactic distances. Second, other evidence about the framework of the laws of physics suggests strongly that *a deviation from Coulomb's law of the form* $1/r^{2+\delta}$ *is not possible*. Instead, a way to characterize a deviation from Coulomb's law is with the *approximate* form

$$F \propto \frac{e^{-\mu r}}{r^2},$$

where e is the exponential constant 2.78... and μ is a constant. If Coulomb's law is correct, the parameter $\mu = 0$. We have seen the exponential form earlier; it is a function that decreases with r over a distance that depends on μ. The larger μ is, the faster the exponential decreases, and the larger the violation of Coulomb's law. Any violation is, we now know, more properly expressed by limits on μ. We can determine limits on μ, and hence tests of the accuracy of Coulomb's law, from the previously reported experiments. The experiment of Williams, Faller, and Hill, for example, implies that μ is smaller than $6 \times 10^{-8}\ \mathrm{m}^{-1}$. These limits can be extended by observing the space dependence of Earth's magnetic field and also of Jupiter's magnetic field, as measured by the spacecraft *Pioneer 10*. Although we have not yet studied magnetism, we can say that the limits on μ found thereby are indeed those associated with Gauss' law. In addition to being direct, the planetary measurements give values of μ that are smaller by an order of magnitude or more than those given by the laboratory experiments; they have the further advantage of testing Gauss' law out to large distances.

Finally, how well do we know Gauss' law at short distances? The colors of light given off by excited hydrogen atoms are very sensitive indicators of the Coulomb force at distances on the atomic scale, about 10^{-10} m. The accuracy with which Gauss' (and therefore Coulomb's) law is known is comparable to the accuracy of the experiments of Plimpton and Lawton (see Table 24–1); that is, to about one part in 1 billion. Even down to nuclear distances—about 10^{-15} m—experiments indicate consistency with the basic theory that leads to Coulomb's law.

SUMMARY

The electric flux due to the electric field **E** that intersects a surface S is

$$\Phi = \oint_{\text{surface } S} \mathbf{E} \cdot d\mathbf{A}. \tag{24–3}$$

Gauss' law relates the electric flux through a closed Gaussian surface—an imaginary closed surface—to the total charge enclosed by the surface, Q:

$$\oint_{\text{closed surface } S} \mathbf{E} \cdot d\mathbf{A} = \frac{Q}{\epsilon_0}. \tag{24–7}$$

Gauss' law is equivalent to Coulomb's law for static situations and, unlike Coulomb's law, it holds even when we consider nonstatic fields. It is thus one of the fundamental equations of electromagnetism.

Gauss' law is also a powerful tool for determining electric fields due to charge distributions with a high degree of symmetry. It can be used to derive in simple fashion the electric fields due to a straight-line charge or due to a conducting plane. For a general spherically symmetric charge distribution centered at the origin of a coordinate system, Gauss' law gives a simple derivation of the field at a distance r from the origin. If q is the total charge contained within a Gaussian sphere of radius r, then the electric field at r is the same as that of the field of a point charge q at the origin, $E = q/(4\pi\epsilon_0 r^2)$.

Conductors react in special ways to electric fields and to charges:

1. The electrostatic field inside a conductor is zero.
2. The electrostatic field immediately outside a conductor is perpendicular to the surface and has the value σ/ϵ_0, where σ is the local surface charge density (which is not necessarily constant).
3. If there are no nonconducting holes that contain charge, a conductor in electrostatic equilibrium can have charge only on its outer surface.

Gauss' law (and its equivalent, Coulomb's law) has been subjected to many experimental tests since the mid-eighteenth century. The inverse-square law dependence on distance has been verified to a precision that ranges from one part in 10^9 to one part in 10^{16} over distances between 10^{-10} m and 10^9 m.

UNDERSTANDING KEY CONCEPTS

1. A temperature field is defined when the temperature of every point of a region of space is specified. Is it possible to compute a flux associated with this field?
2. In the text we refer to the Faraday ice-pail experiment and discuss one version of it in detail. The discussion concerns a sphere with a hole cut in it, and we speak of the inside and outside of this open sphere (Fig. 24–22a). Yet an open sphere does not have a clear inside and outside because, unlike a closed, hollow sphere, it can be deformed continuously to a plane. Why is it possible to talk of the inside and outside of an open sphere, and why does the open sphere behave like a closed, hollow sphere (a bubble) in the experiment?
3. Use Gauss' law to show that electric field lines must be continuous and must originate from and end on charges.
4. Describe the way in which Gauss' law would fail if the field of a point charge were to decrease as $1/r$ rather than as $1/r^2$.
5. If a large, thin, flat plate is positively charged, the field extends in both directions from the plate and has a magnitude of $\sigma/2\epsilon_0$. If a second plate of equal but opposite charge is placed parallel to the first plate, the field around the first plate extends only toward the second plate and has a magnitude of σ/ϵ_0, where σ is exactly the same as before. How do you reconcile this second case with Gauss' law?
6. Consider an electric field $\mathbf{E}$ that is zero at every point on a closed surface S. Does this mean that there are no charges within this surface? Give an example for which there are charges inside a surface while $\mathbf{E} = 0$ on the surface.
7. Analyze Gauss' law as it applies to the flow of fluids. How would you formulate it? Suppose that in a certain region there are no sources of fluid, only sinks. What would you learn from Gauss' law? How does the possibility of evaporation of a fluid affect Gauss' law? What would be its counterpart in the electric Gauss' law?
8. What would Gauss' law look like for the gravitational field, which is defined by force/unit mass of a test body?
9. Charge is distributed uniformly on a circular wire that is surrounded by a torus (doughnut) for which the wire serves as an axis. Does the symmetry allow us to say anything about the electric field due to the charge on the circular wire?
10. A positive point charge and a negative point charge of equal magnitude are fixed on the surface of a conductor of arbitrary shape. What, if anything, can be said about the resulting electric field lines?
11. A region in space has a uniform electric field. What can we say about whether or not any charges are inside the region?
12. To derive the electric field of an infinitely long line of charge, we used a Gaussian surface in the form of a right cylinder centered on the line. Why does the use of such a surface not allow us to find the field of a line of charge of *finite* length?
13. Suppose that the electric field in some region is known to have only x- and y-components, and that the components depend only on x and y, not on z. What can you deduce about the charge distribution that gives rise to this field?
14. You have a probe that measures the electric field at any point in space. For a region in which you know independently that the charge density is constant, how can you use the probe to measure that charge density?

PROBLEMS

24–1 Electric Flux

1. (I) An infinitely large, nonconducting, thin plate carries a uniform charge density σ. (a) What is the electric flux through a circle of radius R placed parallel to the plane? (b) What is the flux through that circle if the plane of the circle is tilted at a 30° angle with respect to its original orientation (Fig. 24–23)?

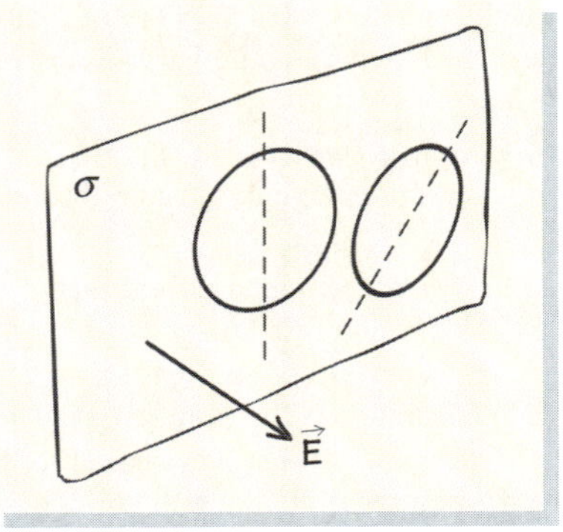

FIGURE 24–23 Problem 1.

2. (I) A region of space contains a constant electric field of magnitude 550 N/C. A wire frame forming a square 0.46 m on a side is placed in the region, oriented so that the perpendicular to the square makes an angle of 72° with the field. What is the magnitude of the electric flux through the frame?

3. (I) The electric field due to an infinitely long, straight line of charge with uniform charge density λ points straight away from the line and has magnitude $E = \lambda/2\pi\epsilon_0 r$, where r is the distance from the wire. Calculate the flux of this electric field through a right cylinder of height h and radius R, concentric with the charged line. Repeat the calculation for a cylinder of radius $2R$.

4. (I) The electric field in a certain region of space points in the z-direction and has magnitude $E = 5xz$, where x and z are measured from some origin. Calculate the flux of that field through a square perpendicular to the z-axis; the corners of the square are at $(x, y, z) = (-1, -1, 1)$, $(-1, 2, 1)$, $(2, 2, 1)$, and $(2, -1, 1)$. (All fields are measured in N/C, all distances in m.)

5. (I) An electric field has the components $E_x = 5x$, $E_y = -3y$, and $E_z = 4z$. Calculate the electric flux through the sides of a unit cube, whose corners are $(x, y, z) = (0, 0, 0)$, $(1, 0, 0)$, $(1, 1, 0)$, $(0, 1, 0)$, $(0, 0, 1)$, $(1, 0, 1)$, $(1, 1, 1)$, and $(0, 1, 1)$. (All fields are measured in N/C, all distances in m.)

6. (I) An electric field of 10 N/C points in the x-direction. A wire loop that is 1 cm^2 in area is placed perpendicular to the x-axis. (a) What is the electric flux through the loop? (b) If the loop is rotated about the y-axis so that the normal to the loop makes an angle of 30° with the x-axis, what is the flux through the loop now? (c) How does the flux change if the angle is increased to 330°?

7. (II) An electric field that is constant in direction is perpendicular to the plane of a circle of radius R. This electric field has a magnitude of $E_0(1 - r/R)$ at a distance r from the center of the circle. Calculate the electric flux through the plane of the circle.

8. (II) By direct calculation (that is, without using Gauss' law), find the flux of a constant electric field **E** through a hemispherical surface of radius R whose circular base is perpendicular to the direction of the field. Your result should be the same as the flux through the top surface of a cylinder whose circular base, of radius R, is oriented perpendicular to the field direction (Fig. 24–24). [*Hint*: The area of an infinitesimal strip at a latitude θ and a thickness $R d\theta$ is $2\pi R^2 \sin\theta\, d\theta$; θ varies from 0 at the North Pole to $\pi/2$ at the equator.]

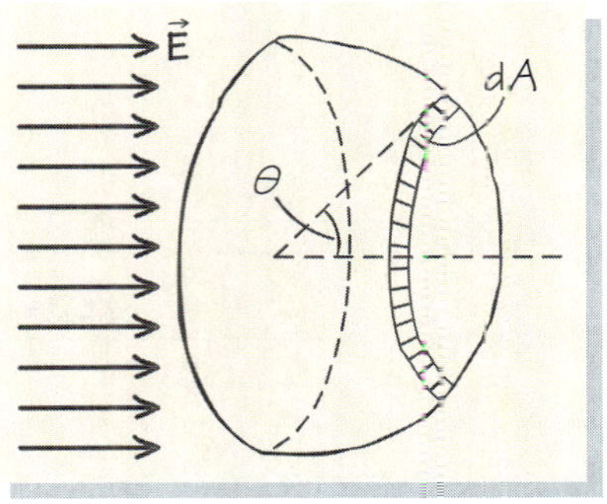

FIGURE 24–24 Problem 8.

9. (II) A point charge q is placed in the middle of a cylindrical surface of radius R, height $2h$. Find the electric flux through the surface by direct integration. [*Hint*: Use the angle θ in Fig. 24–25 as your variable of integration. Only the upper half of the cylinder is shown in the figure.]

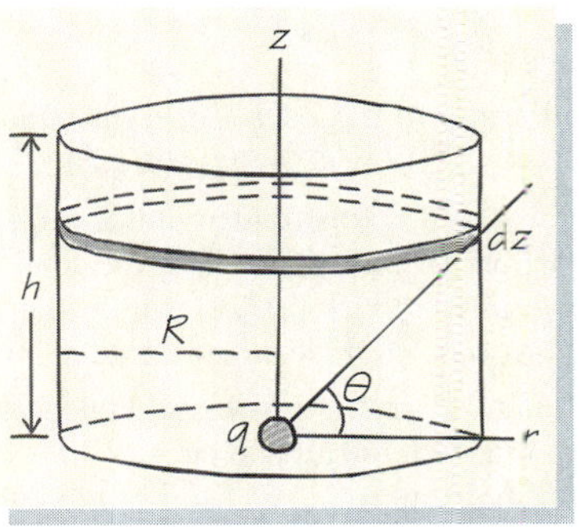

FIGURE 24–25 Problem 9.

10. (II) A charge q is placed just above the center of a horizontal circle of radius r, and a hemisphere of this radius is erected about the charge. Compute the electric flux through the closed surface that consists of the hemisphere and the planar circle (Fig. 24–26). Do not use Gauss' law.

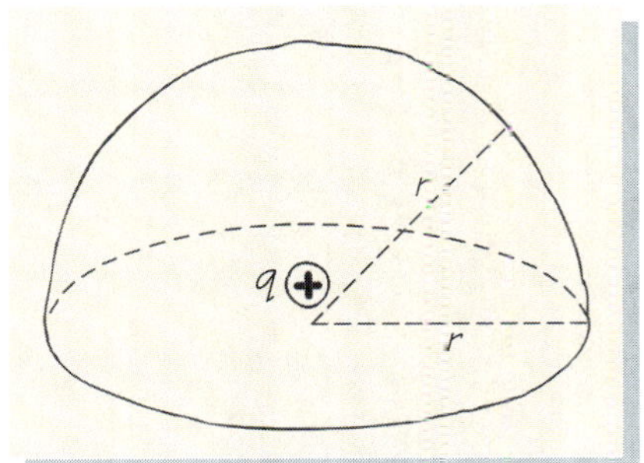

FIGURE 24–26 Problem 10.

11. (III) Consider an infinitesimal parallelepiped located at the point (x, y, z) with sides dx, dy, and dz along the x-, y-, and z-axes (Fig. 24–27). Show that the electric flux of the electric

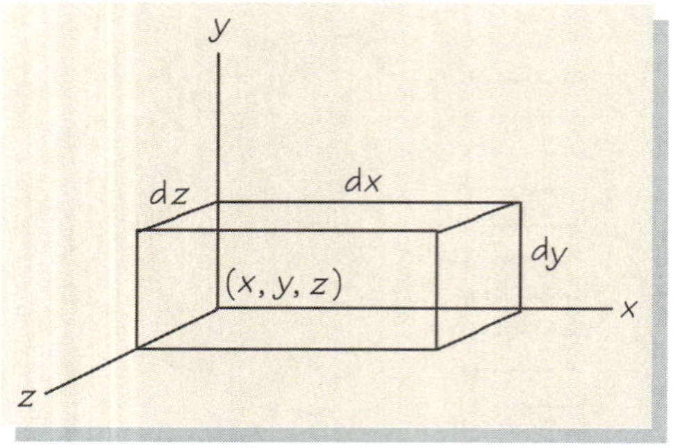

FIGURE 24–27 Problem 11.

field given by $\mathbf{E} = E_x\,\mathbf{i} + E_y\,\mathbf{j} + E_z\,\mathbf{k}$ through the surface that bounds this volume is given by

$$\Phi = \left(\frac{\partial E_x}{\partial x} + \frac{\partial E_y}{\partial y} + \frac{\partial E_z}{\partial z}\right) dx\,dy\,dz.$$

The quantity in parentheses (the coefficient of $dx\,dy\,dz$) is called the *divergence* of the vector field **E**.

24–2 Gauss' Law

12. (I) The flux through a closed surface surrounding a single charge is $-2.6 \times 10^3\ \mathrm{N\cdot m^2/C}$. What is the value of the charge?

13. (I) A charge of 10^{-3} C is distributed uniformly on the surface of a sphere of radius 1 cm. Calculate the total electric flux through a concentric sphere (a) just within the charged surface, and (b) just outside the charged surface.

14. (I) A 30-nC point charge is placed just inside the center of one face of an imaginary Gaussian cube. What is the flux that passes through the sum of all six faces of the cube?

15. (I) Consider the charge distribution shown in Fig. 24–28, where $q = 1\ \mu$C. Draw a spherical Gaussian surface centered on the origin with a radius of 5 cm. (a) What is the electric flux through the Gaussian sphere? (b) Do any of the electric field lines from the three charges pierce the Gaussian surface? Explain.

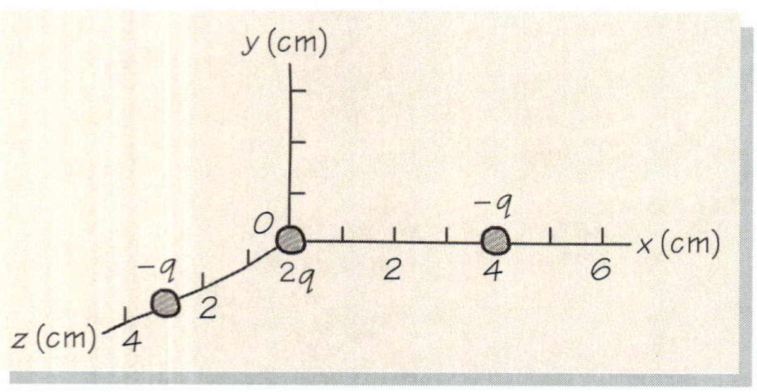

FIGURE 24–28 Problem 15.

16. (I) A charge of 1.2×10^{-4} C is placed inside a cylinder at the midpoint of the axis of the cylinder. The flux through one end of the cylinder is $4.5 \times 10^6\ \mathrm{N\cdot m^2/C}$. What is the flux through the curved part of the cylinder's surface?

17. (I) The net electric flux passing through a given closed surface is $-4 \times 10^2\ \mathrm{N\cdot m^2/C}$. What charge is contained inside the surface if that surface is (a) a sphere of radius 3 cm, (b) a cube of sides 3 cm, (c) a right circular cylinder of height 3 cm and radius 1 cm?

18. (II) A 0.6-μC charge is placed at the center of a cube of sides 10 cm. Determine the electric flux through each of the sides.

19. (II) A given region has an electric field that is a sum of two contributions: a field due to a charge $q = 5 \times 10^{-8}$ C at the origin, plus a uniform field of strength $E_0 = 3000$ N/C in the $-x$-direction. Calculate the flux through each side of a cube with sides of length 20 cm that are parallel to the x-, y-, and z-directions; the cube is centered at the origin.

20. (II) The *gravitational field* **g** due to a point mass M may be obtained by analogy with the electric field by writing an expression for the gravitational force on a test mass, and dividing by the magnitude of the test mass, m. Show that Gauss' law for the gravitational field reads $\Phi = \oint_{\text{surface}} \mathbf{g}\cdot d\mathbf{A} = -4\pi GM$, where G is the gravitational constant. Use this result to calculate the gravitational field at a distance r from the center of a sphere of radius R and uniform density for $r < R$ and for $r > R$.

24–3 Applications of Gauss' Law

21. (I) Calculate the electric field outside a long cylinder of finite radius R with a uniform (volume) charge density ρ spread throughout the volume of the cylinder.

22. (I) Use Gauss' law to show that the electric field outside a large, thin, nonconducting plate with uniform charge density σ is given by $E = \sigma/2\epsilon_0$.

23. (I) Charge is distributed on a long, straight rod with uniform density $\lambda = 6.5 \times 10^{-8}$ C/m. Compare the magnitude of the field 1 cm from the rod to the field 1 cm from a point charge $q = 6.5 \times 10^{-8}$ C.

24. (II) An infinitely long cylinder of radius R carries a uniform (volume) charge density ρ. Calculate the field everywhere inside the cylinder.

25. (II) On a clear day in Nebraska, the electric field just above the ground is 110 N/C and points toward the ground. Our planet Earth is a reasonable conductor and contains no electric field. How much net charge is contained on the surface of a 60-acre corn field (1 acre $\simeq$ 4000 m^2)?

26. (II) Two long, thin cylindrical shells of radii r_1 and r_2, respectively, are oriented coaxially (one cylinder is centered inside the other). The cylinders carry equal and opposite linear charge densities λ. Describe the resulting electric field inside the smaller cylinder, between the cylinders, and outside the larger cylinder (Fig. 24–29).

27. (II) A balloon of radius 15 cm carries a charge of 5×10^{-7} C distributed uniformly over its surface. What is the electric field at a distance of 50 cm from the center of the balloon? Suppose that the balloon shrinks to a radius of 10 cm but loses none of its charge. What is the electric field at a distance of 50 cm from the center?

28. (II) A thin, cylindrical copper shell of diameter 1.2 cm has along its axis a thin metal wire of diameter 2×10^{-2} cm. The wire and the shell carry equal and opposite charges of 2×10^{-8} C/cm, distributed uniformly. Calculate the electric field in the region between the wire and the cylinder, and the mag-

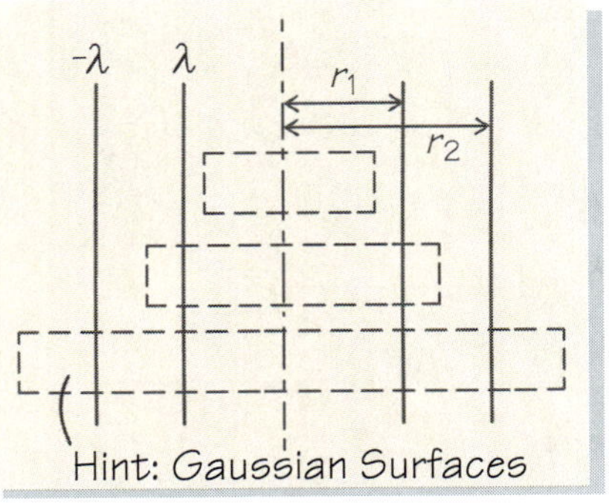

FIGURE 24–29 Problem 26.

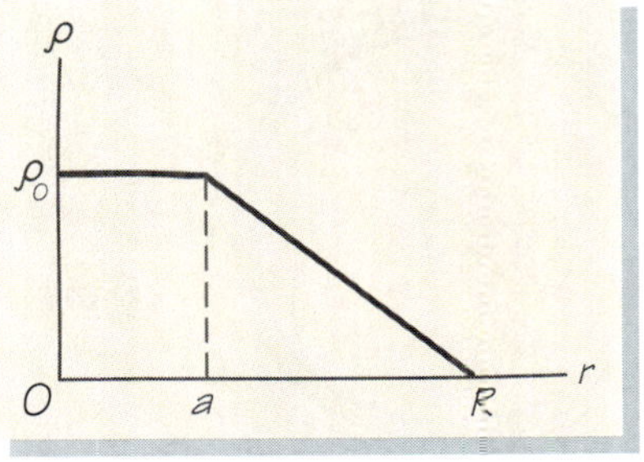

FIGURE 24–30 Problem 36.

nitude of the electric field at the surface of the wire and at the inner surface of the cylinder.

29. (II) A long, cylindrical shell of inner radius r_1 and outer radius r_2 carries a uniform volume charge density ρ. Find the electric field due to this distribution of charge everywhere in space.

30. (II) A Teflon rod of radius 0.60 cm and height 15.0 cm is being charged uniformly over its cylindrical surface. How much charge can the rod hold before the surrounding air breaks down electrically, which happens when the electric field in air is 2.0×10^6 N/C? Ignore the likelihood of breakdown at the sharp edges.

31. (II) A thick, nonconducting spherical shell with a total charge of Q distributed uniformly has an inner radius R_1 and an outer radius R_2. Calculate the resulting electric field in the three regions $r < R_1$, $R_1 < r < R_2$, and $r > R_2$.

32. (II) Two infinite-plane nonconducting, thin sheets with uniform surface charges of 5 μC/m^2 and -3 μC/m^2, respectively, are parallel to each other and 0.20 m apart. What are the electric fields between the sheets and outside them?

33. (II) Two infinite-plane sheets that are just like those of Problem 32 are placed at right angles to each other. What are the fields in the four regions into which space is divided by the planes?

34. (II) A slab of nonconducting material forms an infinite plane. The slab has a thickness t and carries a uniform positive charge density ρ. It is oriented parallel to the xy-plane, with its upper surface at $z = t/2$ and its lower surface at $z = -t/2$. Use Gauss' law to find the electric field both above and below the surface, as well as at an arbitrary value of z in the interior of the slab.

35. (II) Consider a solid sphere of radius 3 cm that carries a negative charge of 2 μC distributed uniformly. The sphere is placed concentrically in a spherical shell of radius 8 cm that has a positive charge of 5 μC distributed uniformly over it. Calculate the electric field as a function of radius r for $0 < r < 15$ cm.

36. (III) Charge is distributed throughout a sphere with the charge density given by $\rho = \rho_0$ for $r < a$, $\rho = \rho_0(r - R)/(a - R)$ for $a < r < R$, and $\rho = 0$ for $R < r$ (Fig. 24–30). Calculate the flux through the spherical surfaces at $r = a$, $r = R$, and $r = 10R$, and calculate the corresponding electric fields at these radii.

37. (III) Consider the charge distribution given in Problem 36. Plot the charge density, the flux through a concentric shell of radius r, and the electric field as a function of r. Use $R = 3a$.

24–4 Conductors and Electric Fields

38. (I) Two large, thin, metallic plates are placed parallel to each other, separated by 15 cm. The top plate carries a uniform charge density of 24 μC/m^2, while the bottom plate carries a uniform charge density of -38 μC/m^2. What is the electric field halfway between the plates?

39. (I) Two concentric metallic shells—conductors—have radii of R and $2R$, respectively. A charge q is placed on the inner shell, and a charge $-2q$ is placed on the outer shell. What are the electric fields in all of space due to the two shells?

40. (I) Two oppositely charged, parallel metal plates give rise to a field of 3×10^6 N/C between them. The plates are square and have dimensions 0.1 m $\times$ 0.1 m. How much charge must there be on each plate? Assume that the charge distribution and electric field are uniform, as if the plates were infinite in size. This will be a good approximation if the distance between the places is much smaller than 0.1 m.

41. (I) Charge is placed on a large spherical surface. What is the maximum surface charge density that avoids electrical breakdown in air ($E_{max} = 3 \times 10^6$ N/C)?

42. (I) A metal sphere of radius 25 cm is concentrically surrounded by a thin spherical metal shell whose inner radius is 35 cm. The electric flux through a concentric spherical Gaussian surface at a radius of 50 cm is 2.4×10^7 N·m^2/C, and that through a concentric spherical Gaussian surface at a radius of 30 cm is 1.2×10^7 N·m^2/C. What is the ratio of the charges on the inner and outer spheres?

43. (I) What is the ratio of the charge densities on the inner and outer spheres in Problem 42?

44. (I) A solid copper cube is placed in a constant electric field which points in the $+x$-direction. The faces of the cube are parallel to the x-, y-, and z-axes, and one corner is at the origin. Draw the field lines as they would be observed looking down on the cube toward the xy-plane. Show at least two electric field lines starting or stopping on each of the four sides of the cube perpendicular to the xy-plane.

45. (II) A metal sphere of radius a is surrounded by a metal shell of inner radius b and outer radius R. The flux through a spherical Gaussian surface located between a and b is Q/ϵ_0, and the flux through a spherical Gaussian surface just outside radius R is $2Q/\epsilon_0$ (Fig. 24–31). What are the total charges on the inner sphere and on the shell? Where are the charges located, and what are the charge densities?

46. (II) The electric field near Earth's surface on a given day is 100 N/C, pointing radially inward. If this were true everywhere

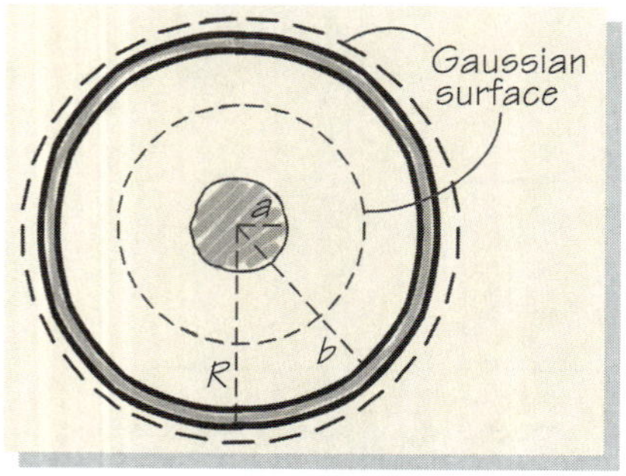

FIGURE 24–31 Problem 45.

on Earth's surface, what would the sign and magnitude of the total charge be on Earth? If Earth is treated as a conductor, where is the charge located? What is the charge density?

47. (II) A point charge q is placed a distance $L/2$ over the center of a conducting square plate of area L^2. (a) Draw the electric field lines on both sides of the plate, which has charge $-q$. (b) Repeat part (a) for a charge on the plate of $2q$.

48. (II) The center of a solid conducting sphere of radius 3 cm and charge 2 μC is placed 5 cm above and away from the center of a flat, horizontal conducting square plate of area 100 cm^2 and charge 1 μC. Draw the electric field lines.

General Problems

49. (II) Consider a cube of sides a located at the origin (Fig. 24–32). Suppose that an electric field is present and given by $\mathbf{E} = bx^2\,\mathbf{i}$, where b is a constant. Calculate the flux through each side of the cube, and use this to find the charge within the cube.

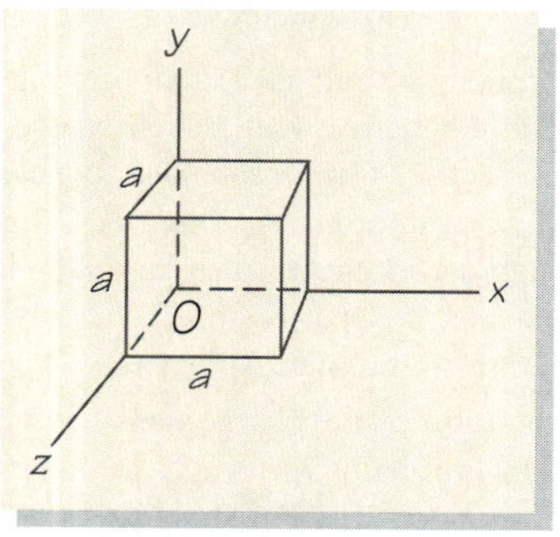

FIGURE 24–32 Problem 49.

50. (II) Consider a solid sphere of radius R with a charge Q distributed uniformly. Suppose that a point charge q of mass m, with a sign opposite that of Q, is free to move within the solid sphere. Charge q is placed at rest on the surface of the solid sphere and released. Describe the subsequent motion. In particular, what is the period of the motion, and what is the total energy of the point charge? [*Hint*: Recall the properties of the motion for which the force varies linearly with the distance from a fixed point and is a restoring force.]

51. (II) A constant electric field is inside a tube of square cross section with sides of length L and is parallel to the sides of the tube. A plane surface cuts the interior of the tube at an angle θ (Fig. 24–33). Show by explicit calculation that the flux

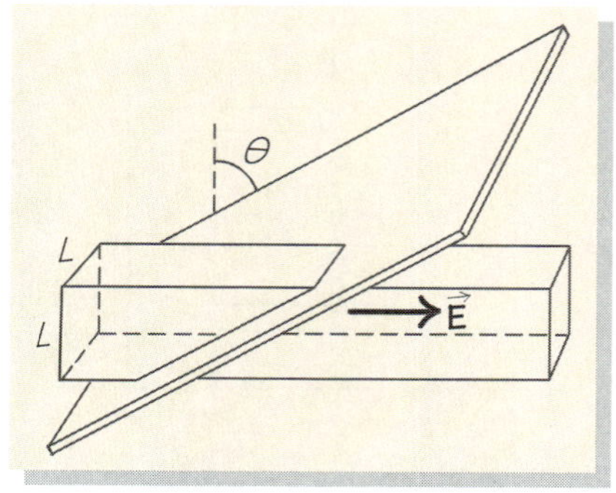

FIGURE 24–33 Problem 51.

through this surface is independent of the angle θ. How would you show this without explicitly calculating the flux through the surface?

52. (II) A conducting sphere of radius 0.30 m is centered at the origin of a coordinate system, as is a surrounding conducting shell of radius 0.80 m. The inner sphere has a charge density of 30 μC/m^2 over its surface, and the outer sphere has a uniform charge density half that large. (a) Find the electric field at a distance 0.50 m from the origin; (b) at a distance 0.70 m from the origin. (c) How would your answers to parts (a) and (b) change if the outer shell were not present? (d) What is the electric field at a distance 1.0 m from the origin?

53. (II) A constant electric field E that points in the $+z$-direction passes through an equilateral tetrahedron whose base is in the xy-plane and whose six edges have length L (Fig. 24–34). Calculate the total flux through the three upper sides of the tetrahedron.

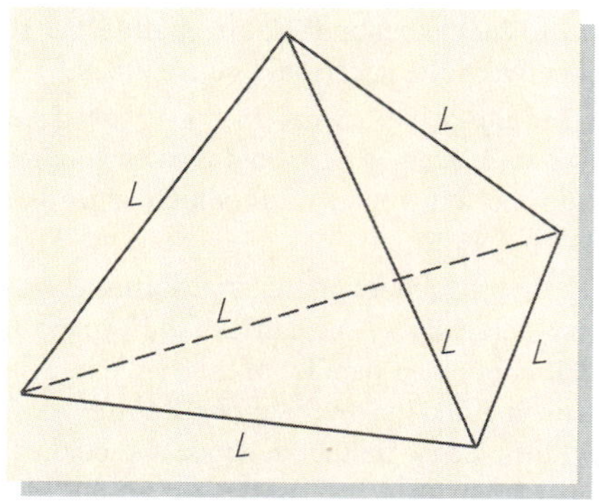

FIGURE 24–34 Problem 53.

54. (II) How should the charge density of a sphere of radius R vary with the distance from the center of the sphere to give a radial field of constant magnitude within the sphere? What happens at the origin, and why?

55. (II) A certain experiment requires an electric field that points symmetrically away from an axis and has a constant magnitude. Describe the charge distribution capable of creating such a field.

56. (II) A right solid conducting cylinder has a charge of -20 mC. Inside the cylinder a $+5$-mC charge rests at the center of a hollow spherical space (Fig. 24–35). (a) What is the charge on the surface of the hollow spherical space? (b) What is the charge on the outside surface of the cylinder?

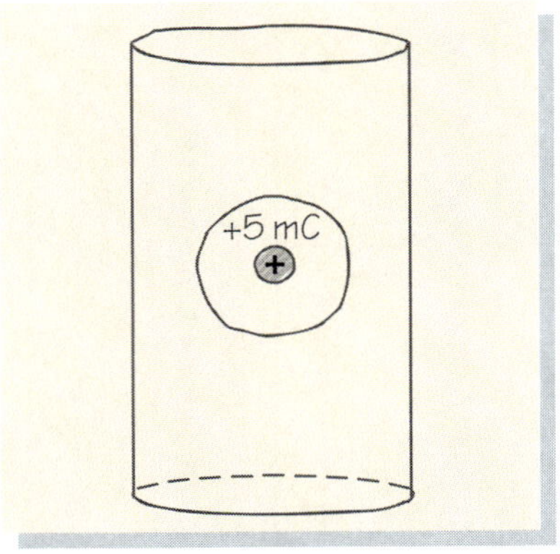

FIGURE 24–35 Problem 56.

57. (II) A conductor has a surface oriented in the yz-plane that marks the boundary of a region in which there is an electric field oriented in the $+x$-direction. The strength of this field increases linearly as x increases from $x = 0$ m to $x = 0.5$ m. At the beginning of the region, at $x = 0$ m, the field strength is 0; at $x = 0.5$ m, the field strength has increased to 3000 N/C. Describe the distribution in the x-direction of the charge that produces this field.

58. (III) A nonconducting sphere of radius R is charged uniformly with charge density ρ. Use Gauss' law to show that the electric field inside the sphere at a point P whose displacement vector from the sphere's center is **r** is given by $\mathbf{E} = (\rho/3\epsilon_0)\mathbf{r}$. A small sphere centered at the point whose displacement from the origin is **a** is cut out of the sphere (Fig. 24–36). Use the superposition principle to calculate the electric field inside the cavity. [*Hint*: The cavity can be created by inserting in the original sphere a sphere of opposite charge density, $-\rho$, and radius b, centered at **a**.]

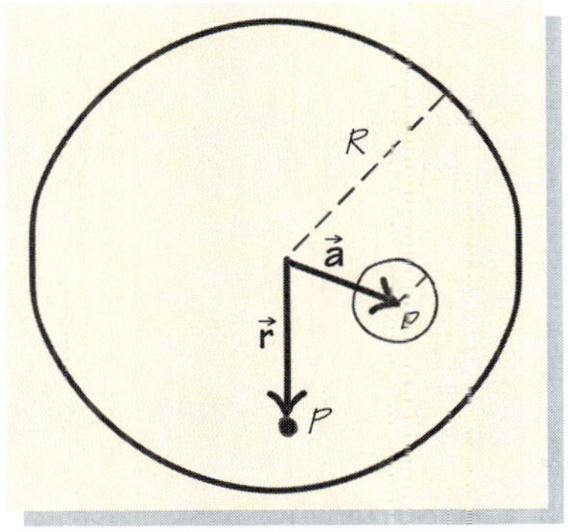

FIGURE 24–36 Problem 58.

59. (III) Use Gauss' law to show that a test charge in the electric field due to any given static charge distribution cannot be in stable equilibrium. [*Hint*: At an equilibrium point, the net electric field must be zero. What must the fields in the vicinity of that point be so that the equilibrium is stable?]

CHAPTER

25

The discharge of lightning bolts provides an impressive demonstration that there is energy in electric fields. A significant electric potential difference exists between Earth and the clouds or between different clouds when lightning forms.

Electric Potential

The electric, or Coulomb, force is a conservative force. Thus we expect a collection of charges to have a potential energy. This potential energy can become kinetic energy—just as the potential energy of a rock balanced on the edge of a cliff becomes kinetic energy when the rock falls. In this chapter, we describe electric potential energy. The concepts of conservative forces, work, and potential energy have already been developed in Chapters 6 and 7. Moreover, many of the results we develop here are similar to those for gravitation (Chapter 12) because the gravitational force and Coulomb's law have the same inverse-square form.

Electric forces concern the interaction of a charge distribution and a second charge. We found it useful to employ the electric field, which isolates the effect of the charge distribution alone, instead of the force. The force is the product of the second charge and the electric field. Similarly, the electrical potential energy is the energy of the charge distribution together with a second charge. In this chapter, we define the electric potential, measured in volts, which is a property of the charge distribution alone. The potential bears the same relation to the electric field as the potential energy bears to the force.

25-1 ELECTRIC POTENTIAL ENERGY

We have already learned that the concept of an energy of position, or a potential energy, is extremely useful. For example, we know that a mass m at a height h (much less than Earth's radius) above Earth's surface has a potential energy that can be written as $U(h) = mgh$. This helps us to determine the object's speed at any other height

The properties of conservative forces and potential energy were studied in Chapter 7.

if we know its speed at one height. *Any* conservative force has a potential energy associated with it. This potential energy is a function of position, and it can be converted to kinetic energy in accordance with the conservation of energy: The total energy is $E = K + U$, where K is the kinetic energy. Conservation of energy means that the change in E is zero, so $\Delta E = 0 = \Delta K + \Delta U$, or $\Delta K = -\Delta U$. Thus any change in U will be matched by an equal but opposite change in K.

The electric force on charge q_0 due to charge q, separated by a distance r, is

$$F = \frac{qq_0}{4\pi\epsilon_0}\frac{1}{r^2}\hat{\mathbf{r}}, \tag{25–1}$$

where $\hat{\mathbf{r}}$ is the unit vector that points radially outward from the position of q. This force bears a striking resemblance to the gravitational force between a mass m_0 and a mass m separated by a distance r,

$$\mathbf{F} = -Gmm_0\frac{1}{r^2}\hat{\mathbf{r}}. \tag{25–2}$$

(The gravitational force is always attractive, whereas the electric force is attractive or repulsive according to whether qq_0 is negative or positive.) Each force is conservative, so a potential energy is associated with each force. This potential energy takes the same form for both cases.

Only changes in potential have physical meaning.

Only *changes* in potential energy have meaning. From Eq. (7–9), we can express the change in potential energy of our system as the charge q_0 (or, in the case of gravitation, the mass m_0) moves from an initial point a at position $\mathbf{r}_a$ to a final point b at position $\mathbf{r}_b$ through the displacement $\mathbf{s}$ (Fig. 25–1) as

$$\Delta U = U_f - U_i = U(\mathbf{r}_b) - U(\mathbf{r}_a) = -\int_{\mathbf{r}_a}^{\mathbf{r}_b}\mathbf{F}\cdot d\mathbf{s}. \tag{25–3}$$

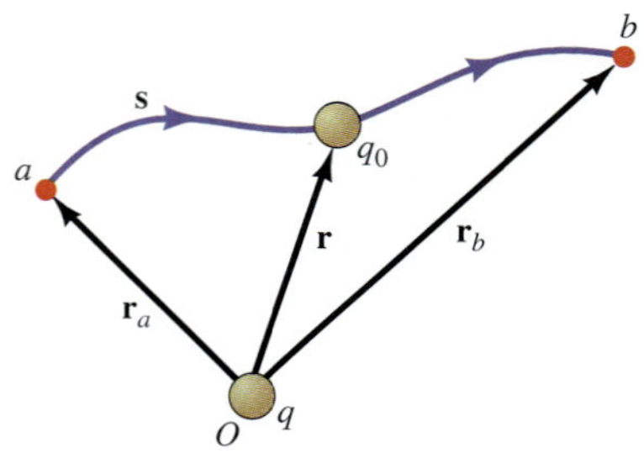

FIGURE 25–1 When a test charge q_0 moves from point a to point b in the presence of a charge q that is fixed in place, the potential energy of the system changes.

For conservative forces, the integral in this expression is a line integral whose value is *independent of the path of integration* between points a and b.

Let's now evaluate the change in electric potential energy for the point charge q at the origin and the point charge q_0 that moves from point a to point b. We start with the simplest situation (Fig. 25–2a), in which point a is on the same radius as point b. Then we take the path from a to b along the dashed line shown in Fig. 25–2a. Because the Coulomb force points outward along the radial direction we have chosen, we have for our path

$$\mathbf{F}\cdot d\mathbf{s} = F\,dr.$$

Then, from Eq. (25–3), the potential energy change when charge q_0 moves from a to b is

$$\Delta U = -\int_{r_a}^{r_b} F\,dr = -\int_{r_a}^{r_b}\frac{qq_0}{4\pi\epsilon_0 r^2}\,dr$$

$$= -\frac{qq_0}{4\pi\epsilon_0}\int_{r_a}^{r_b}\frac{dr}{r^2} = -\frac{qq_0}{4\pi\epsilon_0}\left(\frac{-1}{r}\right)\Bigg|_{r_a}^{r_b} = \frac{qq_0}{4\pi\epsilon_0}\left(\frac{1}{r_b} - \frac{1}{r_a}\right). \tag{25–4}$$

Equation (25–4) is the change in electrical potential energy when charge q_0 moves from any point a distance r_a from charge q to any other point a distance r_b from charge q.

What if charge q_0 moves between two points that do not lie on the same radius, as in Fig. 25–2b? In this case, we follow the dashed path shown. [Remember, the result of the integration in Eq. (25–3) is path independent.] For segment 1, which runs outward radially from a to a distance r_b from the origin, the result is identical to Eq. (25–4). For segment 2, which follows a circumference at a distance r_b from the origin, the integral is zero because the force is perpendicular to the path segment $d\mathbf{s}$ everywhere. The result for the change in potential energy is still given by Eq. (25–4).

Let's look at the physical content of Eq. (25–4). Suppose first that the charges move closer together ($r_a > r_b$). If the charges repel (qq_0 is positive), the change in potential energy is positive. This is like moving a mass *up* a mountain. If the charges attract (qq_0 is negative), the system loses potential energy when the charges move closer

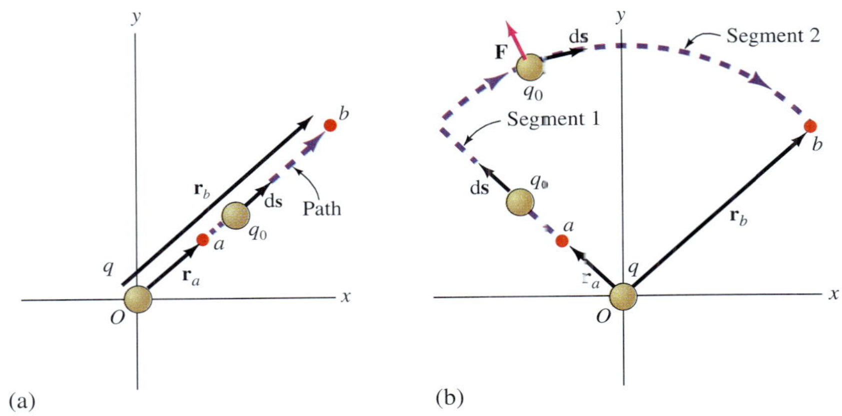

FIGURE 25–2 The change in potential energy of the system of two charges q and q_0 when the charge q_0 moves from point a to point b, in terms of a path-independent line integral. (a) Charge q_0 moves along a radius, and the path is taken directly along the radius. (b) The two points are not along the same radius. The path runs radially outward to the radius of point b, then follows the circumference at that radius.

together. This is like moving a mass *down* the mountain. As with any potential energy, electric potential energy can be converted into kinetic energy. If there are no additional forces acting, then like-sign charges slow down—or lose kinetic energy—when they move closer together. Similarly, charges of opposite sign speed up—or gain kinetic energy—when they move closer together. We draw a similar set of conclusions when the charges move farther apart ($r_a < r_b$). Charges that repel lose electric potential energy and, if there are no other forces, gain kinetic energy. Opposite charges (which attract) gain electric potential energy when they move farther apart and lose kinetic energy in the absence of other forces.

Equation (25–3) shows that the change in electric potential energy is given by the difference of two functions, $U(r_b)$ and $U(r_a)$. We know from Chapter 7 that only *changes* in potential energy have physical consequences. We can therefore choose zero of the potential energy function to be at whatever value of r we like. It is convenient and natural to choose zero potential energy to be at infinity. We can do this if we let $r_a \to \infty$ and let r_b take on a general value r in Eq. (25–4):

$$\Delta U = U(r) - U(r_a)\big|_{r_a \to \infty} \to \frac{qq_0}{4\pi\epsilon_0}\frac{1}{r}.$$

We then say that the potential energy of a charge q_0 a distance r from charge q is the difference in potential energy between that point and infinity. When we reverse the roles of q and q_0, the potential energy of q at a distance r from q_0 is again $qq_0/4\pi\epsilon_0 r$. We may thus say that the **electric potential energy** $U(r)$ for a system of two point charges q and q_0 separated by a distance r is

Electric potential energy between two point charges. Zero potential energy is chosen to be at infinity.

$$U(r) = \frac{qq_0}{4\pi\epsilon_0}\frac{1}{r}. \tag{25–5}$$

It is indeed true that $U(r) = 0$ in the limit $r \to \infty$. Thus the system has no potential energy when the two charges are infinitely far apart. Note that the potential energy of the two charges depends *only* on the distance r between them; it does not depend on any angle. Equation (25–5) has the same form as Eq. (12–9), calculated in Chapter 12 for the gravitational potential energy. As we know from Chapter 7, the physical significance of potential energy is that the value of potential energy when two charges have a finite separation is the work needed to bring the charges from infinity to that separation distance.

EXAMPLE 25–1 Nuclear fission is the breakup of nuclei. Fission occurs in heavy nuclei because of the Coulomb repulsion of the many positively charged protons in the nucleus. The greater the number of protons that are in a nucleus, the stronger is the repulsion. The attractive nuclear force is not quite strong enough to overcome the repulsion among all the protons in a heavy nucleus. In calculating the repulsive forces between the protons, we want to find the potential energy between the protons. Find the electrostatic potential energy between two of the 92 protons in a ^{236}U nucleus in which the protons are as close as they can be inside a uranium nucleus, which is about 2×10^{-15} m apart. The radius of a uranium nucleus is about 8×10^{-15} m.

Solution: We take zero potential energy to be at infinity and use Eq. (25–5) to calculate the electrostatic potential energy. The proton charge (see Table 22–1) is 1.6×10^{-19} C. Then,

$$U = \frac{(+e)^2}{4\pi\epsilon_0 r} = \frac{(9 \times 10^9\ \mathrm{N \cdot m^2/C^2})(1.6 \times 10^{-19}\ \mathrm{C})^2}{(2 \times 10^{-15}\ \mathrm{m})} = 1 \times 10^{-13}\ \mathrm{J}.$$

This is a typical energy value on the nuclear scale and is about 10^5 times larger than the energy that holds the proton and electron together in a hydrogen atom (see Example 25–3). When fission occurs, this potential energy is converted into kinetic energy of the various fragments. This kinetic energy is, in turn, the energy source of nuclear reactors.

25–2 ELECTRIC POTENTIAL

A point charge q is the source of an electric field **E** that exists in the surrounding space. The electric field affects any charge q_0 introduced into that space because there is a force **F** on q_0 given by $\mathbf{F} = q_0\mathbf{E}$. We saw in Section 25–1 that the introduction of a charge q_0 a distance r from q gives rise to the potential energy $U(r)$ of Eq. (25–5). If we write $U(r) = q_0V(r)$, we can make a statement analogous to the statement about the electric field: A charge q is the source of an **electric potential** (or just **potential**) $V(r)$, which affects any charge q_0 a distance r from q by creating potential energy $U(r) = q_0V(r)$. Strictly speaking, we should deal with a small *test charge* q_0, so that its presence does not disturb charge q or indeed any more general charge distribution that gives rise to the electric potential. The definition of electric potential, which is a *work per unit charge*, due to a charge distribution is then

Definition of electric potential

$$V(\mathbf{r}) \equiv \frac{U(\mathbf{r})}{q_0}, \tag{25–6}$$

where $U(\mathbf{r})$ is the potential energy of the test charge q_0 in the presence of the charge distribution. The potential $V(\mathbf{r})$ is independent of q_0, just as the electric field, defined by $\mathbf{E} \equiv \mathbf{F}/q_0$, is independent of the test charge:

Electric potential, like the electric field, is a property only of the charge or charges producing it, not of the test charge q_0.

The electric potential is a property only of the charge distribution that produces it.

The Electric Potential of a Point Charge

Let's calculate the electric potential of the simplest possible system: one point charge. Consider two point charges q and q_0 separated by a distance r. As Eq. (25–5) shows, the potential energy of the system is $U(r) = q_0q/4\pi\epsilon_0 r$. If we think of q_0 as a test charge, then $U/q_0 = q/4\pi\epsilon_0 r$. We have found *the electric potential of a point charge q at a point a distance r from the charge*:

The electric potential of a point charge

$$V(r) = \frac{q}{4\pi\epsilon_0 r}. \tag{25–7}$$

In Eq. (25–7), we have assumed that zero potential energy is at infinity and, as a consequence, *we have taken zero electric potential due to a charge q to be at infinity.* Just to emphasize this point, we might say that Eq. (25–7) is the potential of a single charge *with respect to infinity.*

The *electric potential difference* due to the charge q between the points a and b at locations $\mathbf{r}_a$ and $\mathbf{r}_b$ is given by (Fig. 25–3)

$$\Delta V = V_b - V_a = \frac{U_b - U_a}{q_0} = \frac{q}{4\pi\epsilon_0}\left(\frac{1}{r_b} - \frac{1}{r_a}\right). \qquad (25\text{–}8)$$

Here, we have abbreviated V as a function of r_a, or $V(r_a)$, as V_a, and so forth.

We can obtain another formulation of the electric potential difference by using Eqs. (25–3) and (25–8) and substituting $\mathbf{F} = q_0\mathbf{E}$:

$$\Delta V = \frac{U_b - U_a}{q_0} = -\int_{r_a}^{r_b} \mathbf{E}\cdot d\mathbf{s}. \qquad (25\text{–}9)$$

The electrical potential difference is an integral over the electric field.

Here, the electric potential difference is expressed as a *path-independent* integral over an electric field. There is no reference in Eq. (25–9) to the electric field of the point charge. Equation (25–3) is the potential energy change when a test charge q_0 moves from point a to point b in the field of *any* charge distribution. Thus Eq. (25–9) is a general expression for the electric potential difference between two points. Any charge distribution produces an electric field, and an electric potential is associated with any charge distribution. Electric potential is a useful concept in part because it is a scalar quantity. It is easier to deal with than the vector quantity that determines it—the electric field.

We recall that the change in the potential energy of a system is equal to the negative of the work done by the system in moving an object from point a to point b. Equivalently, $U_b - U_a$ is the work done by an external agent to move the object. These relations hold for the changes in electric potential energy when a test charge moves. Therefore we can interpret Eq. (25–9) to mean that

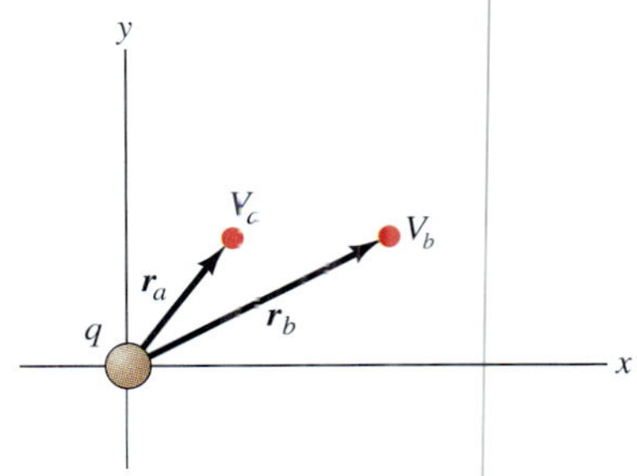

FIGURE 25–3 The potential at two different locations in space. Only the potential *difference* has physical meaning.

The electrical potential difference $V_b - V_a$ is the work per unit charge that must be done to move a test charge from point a to point b without changing its kinetic energy.

This work is performed by an external agent; for example, we may literally push the charge. If there is no external agent, then a change in potential, which corresponds to a change in potential energy of the test charge, must be accompanied by a corresponding change in the kinetic energy of the test charge.

If we know the electric potential $V(\mathbf{r})$ due to a charge distribution and we know the magnitude of a test charge q_0, then we also know the potential energy $U(\mathbf{r})$ when q_0 is placed at a point a displacement $\mathbf{r}$ away:

$$U(\mathbf{r}) = q_0 V(\mathbf{r}) \qquad (25\text{–}10)$$

In the absence of other forces, this equation tells us that a positive test charge q_0 in the presence of an electric potential *will move toward lower values of the potential* because the potential energy decreases in that way. The charge speeds up as it moves to lower potentials.

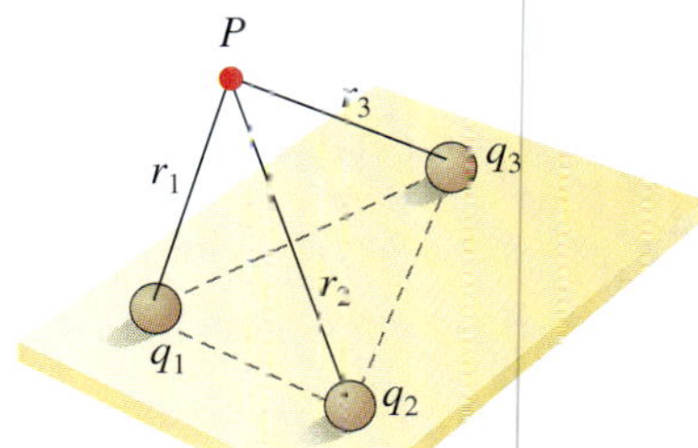

FIGURE 25–4 The superposition principle determines the potential at point P due to multiple charges. We simply add the potential due to each of the charges.

The Electric Potential of Different Charge Distributions

The electric field obeys the superposition principle. Therefore the electric potential of a system of charges can also be determined from the superposition principle. The superposition principle states that the electric field of a collection of charges is the sum of the electric fields of each charge. Thus *the electric potential at a point P due to n point charges $q_1, q_2, \ldots, q_n$* (Fig. 25–4 shows three charges) at distances $r_1, r_2, \ldots, r_n$ from point P is just

$$V_P = \frac{q_1}{4\pi\epsilon_0 r_1} + \frac{q_2}{4\pi\epsilon_0 r_2} + \cdots + \frac{q_n}{4\pi\epsilon_0 r_n}$$

or

Electric potential of a collection of point charges

$$V_P = \frac{1}{4\pi\epsilon_0}\sum_{i=1}^{n}\frac{q_i}{r_i}, \qquad (25\text{–}11)$$

where r_i is the distance from point charge q_i to point P. The electric potential due to a collection of charges is the scalar sum of the potentials due to single charges. This scalar sum is much easier to perform than the vector sum that expresses the electric field due to a collection of point charges.

The calculation of the electric potential due to a continuous charge distribution is also straightforward. We first find the electric potential dV at a point P due to a small charge dq (Fig. 25–5). Because electric potential is a scalar quantity, the addition of all the tiny potentials dV is given by scalar integration. Thus *the potential due to a continuous charge distribution* takes the symbolic form

Electric potential due to a continuous charge distribution

$$V = \int dV = \frac{1}{4\pi\epsilon_0}\int \frac{dq}{r}. \qquad (25\text{–}12)$$

The integration must be done over the entire charge distribution. In Section 25–5, we shall discuss techniques for its calculation for specific cases.

Units of Electric Potential

Units of electric potential

The dimension of electric potential is energy per charge; thus the SI unit is joules per coulomb (J/C). Because electric potential is used frequently, it has a separate name in the SI: the **volt** (V). It is named after Alessandro Volta, who did research at the beginning of the nineteenth century on the nature of electric energy:

$$\text{V} \equiv 1 \text{ J/C}. \qquad (25\text{–}13)$$

Note that electric potential has the dimensions of electric field times length, so the dimensions of electric field must be the dimensions of potential divided by length (V/m):

$$1 \text{ N/C} = 1 \text{ V/m}. \qquad (25\text{–}14)$$

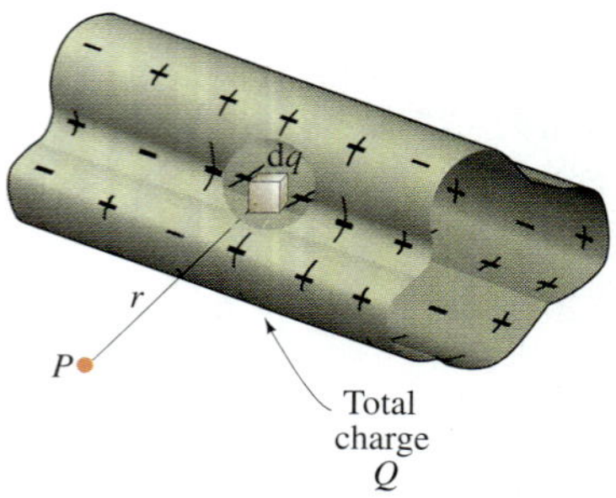

FIGURE 25–5 To find the potential at point P due to a continuous charge, integrate over the differential charges dq as if each dq were a point charge.

The Potential Energy in a System of Charges

Equation (25–10) gives the potential energy $U(r) = q_0V(r)$ of a test charge q_0 placed in the electric potential of a charge distribution. If the charge distribution is a collection of charges, then the electric potential V_P is given by Eq. (25–11), and the potential energy of the test charge is $U(r) = q_0V_P$. But it would be a mistake to call this the potential energy of the entire system of charges $q_0, q_1, q_2,\ldots, q_n$, because the product q_0V_P just represents the work that needs to be done to bring charge q_0 in from infinity. It does *not* take into account the work that must be done to bring the charges $q_1, q_2,\ldots, q_n$ in from infinity. To calculate the potential energy of a collection of three charges, for example, we assemble them one by one. To bring the first charge, q_1, in to the point P_1 requires no work by the external agent if the kinetic energy of the charge is unchanged. To bring the second charge, q_2, in from infinity to the point P_2 does require work because of the potential due to q_1. For our two charges, the work the external agent must do to bring q_2 in from infinity—the potential energy—is given by

$$U_{12} = q_2V_1 = \frac{q_1q_2}{4\pi\epsilon_0 r_{12}}, \qquad (25\text{–}15)$$

where r_{12} is the distance between charges q_1 and q_2.

What happens if we bring a third charge, q_3, in from infinity? We must calculate the additional work done by an external force to bring q_3 in. This work is given by the product of q_3 and electric potentials V_1 and V_2 due to q_1 and q_2 in place. Thus, the additional contribution to the potential energy of the system is

$$U_{13} + U_{23} = \frac{q_1q_3}{4\pi\epsilon_0 r_{13}} + \frac{q_2q_3}{4\pi\epsilon_0 r_{23}}, \qquad (25\text{–}16)$$

where r_{13} and r_{23} are the distances between q_3 and q_1, q_3 and q_2, respectively. The total potential energy U of the system is the sum of U_{12}, U_{13}, and U_{23}:

$$U = \frac{1}{4\pi\epsilon_0}\left(\frac{q_1q_2}{r_{12}} + \frac{q_1q_3}{r_{13}} + \frac{q_2q_3}{r_{23}}\right). \tag{25–17}$$

The potential energy of a system of charges

This can be generalized to any number of charges, and the resulting formula for the *electric potential energy of the system* is a simple generalization of Eq. (25–17):

$$U = \frac{1}{4\pi\epsilon_0}\sum_{i<j}\frac{q_iq_j}{r_{ij}}, \tag{25–18}$$

where r_{ij} is the distance between the locations of the charges q_i and q_j. The sum over i and j includes all charge pairs in the system, and the inequality $i < j$ avoids the counting of pairs more than once. We can eliminate that restriction by writing the equivalent expression

$$U = \frac{1}{2}\sum_{\substack{i,j\\ i\neq j}}\frac{q_iq_j}{4\pi\epsilon_0 r_{ij}}.$$

Now the sum is unrestricted, except that we omit the case $i = j$, which is not in the original sum, Eq. (25–18). Thus we can rewrite Eq. (25–18) as

$$U = \frac{1}{2}q_1\sum\frac{q_j}{4\pi\epsilon_0 r_{1j}} + \frac{1}{2}q_2\sum\frac{q_j}{4\pi\epsilon_0 r_{2j}} + \frac{1}{2}q_3\sum\frac{q_j}{4\pi\epsilon_0 r_{3j}} + \cdots$$

$$= \frac{1}{2}q_1V_1 + \frac{1}{2}q_2V_2 + \frac{1}{2}q_3V_3 + \cdots, \tag{25–19}$$

where V_1 is the electric potential due to all the other charges at the location of charge q_1, and so on. It should be stressed that the potential energy of q_1 in a given potential V_1 is still q_1V_1; this means that q_1V_1 can be converted into the kinetic energy of the particle that carries charge q_1. This potential energy must be distinguished from the potential energy of the *entire* charge configuration, Eq. (25–18) or (25–19). The potential energy of the entire charge configuration is the energy that would be made available if *all* the charges that appear in the problem were to escape to infinity.

In Examples 25–2 and 25–3, we illustrate calculation techniques for the electric potential energy and the electric potential when two or more point charges are involved.

EXAMPLE 25–2 In an experiment to investigate the effects of electricity, Benjamin Franklin might have placed two point charges, $q_1 = 2\ \mu\text{C}$ and $q_2 = -4\ \mu\text{C}$, at points P_1 and P_2, respectively (Fig. 25–6). (a) Find the electric potential at points a and b due to these two point charges. (b) Find the potential difference between points b and a. (c) How much energy would Franklin have had to supply to bring a third charge, of magnitude 3 μC, in from infinity to point b?

Solution: (a) We use Eq. (25–11) to determine the electric potential. Let's label the 2-μC charge, located at P_1, as q_1, and the -4-μC charge, located at P_2, as q_2. First, we find the potential at point a. The distance from point a to point P_1 is $r_{1a} = 2$ m, and the distance from point a to point P_2 is $r_{2a} = \sqrt{(2\text{ m})^2 + (3\text{ m})^2} = 4$ m (note that we only work with one significant figure here). The electric potential V_a at point a is then

$$V_a = \frac{1}{4\pi\epsilon_0}\left(\frac{q_1}{r_{1a}} + \frac{q_2}{r_{2a}}\right)$$

$$= (9\times10^9\ \text{N}\cdot\text{m}^2/\text{C}^2)\left(\frac{2\times10^{-6}\ \text{C}}{2\ \text{m}} + \frac{-4\times10^{-6}\ \text{C}}{4\ \text{m}}\right) = 0\ \text{V}.$$

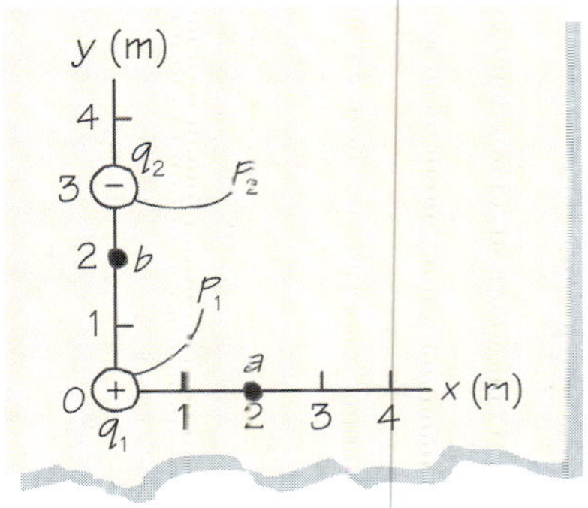

FIGURE 25–6 Example 25–2.

The units of potential are volts, a consequence of our use of SI units everywhere. In this case, the unit combination is $N \cdot m/C = J/C = V$. Such a check is always useful.

Next we find the electric potential at point b. The distance from charge q_1 to b is $r_{1b} = 2$ m; similarly, $r_{2b} = 1$ m. Therefore, the potential V_b is

$$V_b = \frac{1}{4\pi\epsilon_0}\left(\frac{q_1}{r_{1b}} + \frac{q_2}{r_{2b}}\right)$$

$$= (9 \times 10^9\ \mathrm{N \cdot m^2/C^2})\left(\frac{2 \times 10^{-6}\ \mathrm{C}}{2\ \mathrm{m}} + \frac{-4 \times 10^{-6}\ \mathrm{C}}{1\ \mathrm{m}}\right).$$

$$= -2.7 \times 10^4\ \mathrm{V} = -27\ \mathrm{kV}.$$

(b) The potential difference $V_b - V_a = -27\ \mathrm{kV} - 0\ \mathrm{kV} = -27\ \mathrm{kV}$. Thus the electric potential is higher at point a than at point b.

(c) The new charge acts like a test charge, $q_0 = 3\ \mu\mathrm{C}$. We now know the electric potential of the original system of two charges, so we use Eq. (25–10), $U_b = q_0 V_b$, to find the potential energy of the new charge at point b:

$$U_b = q_0 V_b = (3\ \mu\mathrm{C})(-27\ \mathrm{kV}) = (3 \times 10^{-6}\ \mathrm{C})(-27 \times 10^3\ \mathrm{V}) = -0.08\ \mathrm{J}.$$

The answer is in units of joules because we have used SI units throughout.

The work that Franklin would have done to bring q_3 in from infinity is equal to the change in the potential energy of the system, or -0.08 J. Does this make sense? The electric potential at point b is negative. The new charge is positive and will be attracted to the negative potential. Franklin would not have done positive work to bring the charge in to point b; on the contrary, he would have done negative work—just as we have calculated. He would have done positive work to bring the same charge back out to infinity. It is useful to think of what is happening physically rather than to rely solely on the signs in the equations. It is all too easy to make a sign error.

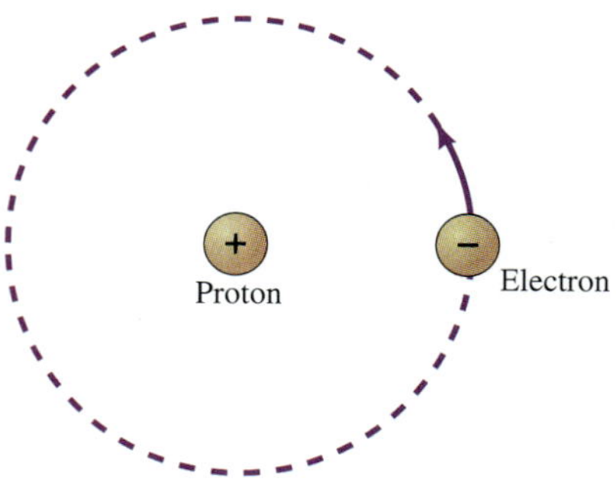

FIGURE 25–7 Example 25–3. A simplistic representation of an electron that orbits a proton in the hydrogen atom.

EXAMPLE 25–3 The hydrogen atom in its normal, unexcited configuration has an electron that revolves around a proton at a distance of 5.3×10^{-11} m (Fig. 25–7). At the position of the electron, what is the electric potential due to the proton? Determine the electrostatic potential energy between the two particles. This energy is relevant to understanding the chemical activity of atoms.

Solution: The electric potential V_P due to the proton can be found by using Eq. (25–7). We have

$$V_P = \frac{+e}{4\pi\epsilon_0 r} = \frac{(9 \times 10^9\ \mathrm{N \cdot m^2/C^2})(1.6 \times 10^{-19}\ \mathrm{C})}{5.3 \times 10^{-11}\ \mathrm{m}} = 27\ \mathrm{V}.$$

The electrostatic potential energy is found by using Eq. (25–15), and we simply multiply the potential by the charge of the electron (the moving particle):

$$U = (-e)V_P = (-1.6 \times 10^{-19}\ \mathrm{C})(27\ \mathrm{V}) = -4.3 \times 10^{-18}\ \mathrm{J}. \qquad (25\text{–}20)$$

The Electron-Volt

The electron-volt

We often determine energy by multiplying charge times voltage, as we have done in every example thus far in this chapter. Because the charge on an electron is needed so frequently, a useful unit of energy is that of the charge of an electron (or proton) times 1 V. We call this unit of energy an **electron-volt** (eV). An electron-volt is simply the energy an electron gains when it is accelerated though a potential difference of one volt. The electron-volt is not an SI unit. The relation between the electron-volt and the SI unit joule is

$$1\ \mathrm{eV} = (1.6 \times 10^{-19}\ \mathrm{C})(1\ \mathrm{V}) = 1.6 \times 10^{-19}\ \mathrm{J}.$$

This unit is especially valuable for calculations in atomic, nuclear, and particle physics. In Example 25–1, the electrostatic potential energy between the two close protons becomes 6×10^5 eV, or 0.6 MeV. In Example 25–3, the electrostatic potential energy between the proton and electron of the hydrogen atom becomes -27 eV. In chemistry, energy is typically on the scale of one electron-volt; in atomic physics, energy runs typically from one to as many as one thousand electron-volts; the scale in nuclear physics is normally 1 MeV (10^6 eV); and in particle physics, a typical scale is 1 GeV (10^9 eV).

EXAMPLE 25–4 Calculate the electric potential due to an electric dipole whose dipole moment has magnitude p at an arbitrary point P (Fig. 25–8).

Solution: A dipole consists of two pointlike charges, so Eq. (25–11) determines the potential, which will be zero at infinity. Let the distance from the $+q$ charge to point P be r, and the distance from the $-q$ charge to P be $r + \Delta r$. The point is also specified in Fig. 25–8 by the angle between **p** and the line between the $-q$ charge and P. Equation (25–11) gives

$$V = +\frac{+q}{4\pi\epsilon_0 r} + \frac{-q}{4\pi\epsilon_0(r + \Delta r)} = \frac{+q(r + \Delta r) - qr}{4\pi\epsilon_0 r(r + \Delta r)} = \frac{q}{4\pi\epsilon_0}\frac{\Delta r}{r(r + \Delta r)}. \quad (25\text{–}21)$$

If **p** is the dipole moment, the distance between the charges is $\ell = p/q$, and the distance Δr is

$$\Delta r = \ell \cos\theta = \frac{p\cos\theta}{q}. \quad (25\text{–}22)$$

When this result is substituted into Eq. (25–21), we find

$$V = \frac{qp\cos\theta}{4\pi\epsilon_0}\left[\frac{1}{r(qr + p\cos\theta)}\right]. \quad (25\text{–}23)$$

As we have mentioned, the dipole charge distribution occurs repeatedly in nature. Individual molecules that have dipole moments—H_2O is an example—are referred to as *polar molecules*. The exact electric dipole potential derived in Example 25–4 takes a simple approximate form far from the dipole, when $r >> \ell$. The condition $r >> \ell$ is equivalent to $qr >> q\ell = p$, and we can drop the second term in the denominator of Eq. (25–23). The result is

$$\text{for } r >> \ell: \qquad V = \frac{p\cos\theta}{4\pi\epsilon_0 r^2}, \quad (25\text{–}24)$$

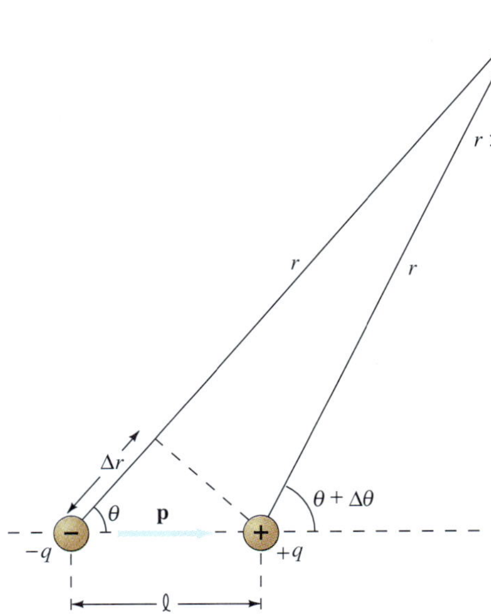

FIGURE 25–8 Example 25–4. Using geometry to find the potential at a point P for an electric dipole. The dipole moment $p = q\ell$.

where we now measure θ from anywhere between the two charges of the dipole. Note that the potential of the dipole for distant points decreases as $1/r^2$, as compared to the $1/r$ dependence for a point charge.

25–3 EQUIPOTENTIALS

Equipotentials defined

Regions where the electric potential of a charge distribution has constant values are called **equipotentials**. They are particularly interesting and worth investigating. Suppose that a system of charges produces a certain potential. The positions in space that have the same electric potential form surfaces in three dimensions and lines in two dimensions. We say that the places where the potential has a constant value form **equipotential surfaces** in three dimensions or **equipotential lines** in two dimensions. As an example, consider the equipotential surfaces formed by a point charge. The electric potential is proportional to $1/r$ and has a constant value at any fixed radial distance from the charge. Therefore, a sphere centered on the charge forms an equipotential surface (Fig. 25–9). Any other sphere centered on the charge forms a different equipotential because the potential varies according to the radius of the sphere.

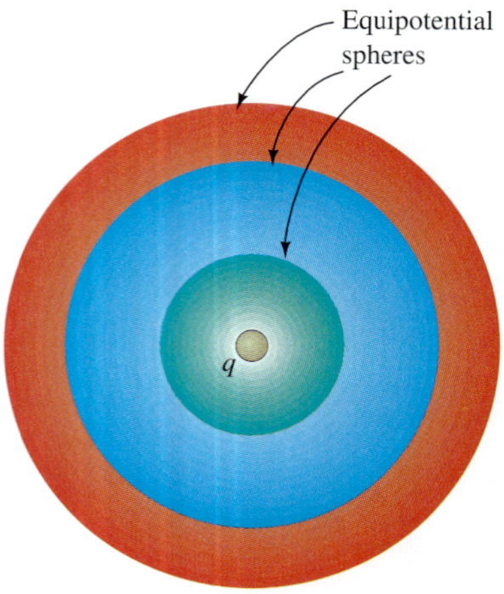

FIGURE 25–9 The equipotential surfaces for a point charge are spheres that surround the point charge.

Equipotentials are analogous to contour lines on a topographic map—lines for which the elevation from sea level is constant (Fig. 25–10). Because the gravitational potential energy of a mass depends only on the mass's elevation, the gravitational potential energy does not change when a mass moves along a contour line. Consequently, the force of gravity has no component along contour lines. Gravity acts in a direction perpendicular to a contour line; a ball that starts on a particular contour line will accelerate in a direction perpendicular to the line, or what we would call straight down the hill. What holds for contour lines holds for any equipotential surface or line: Any conservative force acts in a direction perpendicular to the equipotential because it can have no component along the equipotential.

Because the potential has exactly the same value over an equipotential, so does the potential energy of a test charge. No work is done when the test charge moves at constant speed on an equipotential surface or line. For the point charge discussed previously, the equipotentials are spheres centered on the charge (Fig. 25–9). A test charge can move freely about any one such surface without work being done by the electric field.

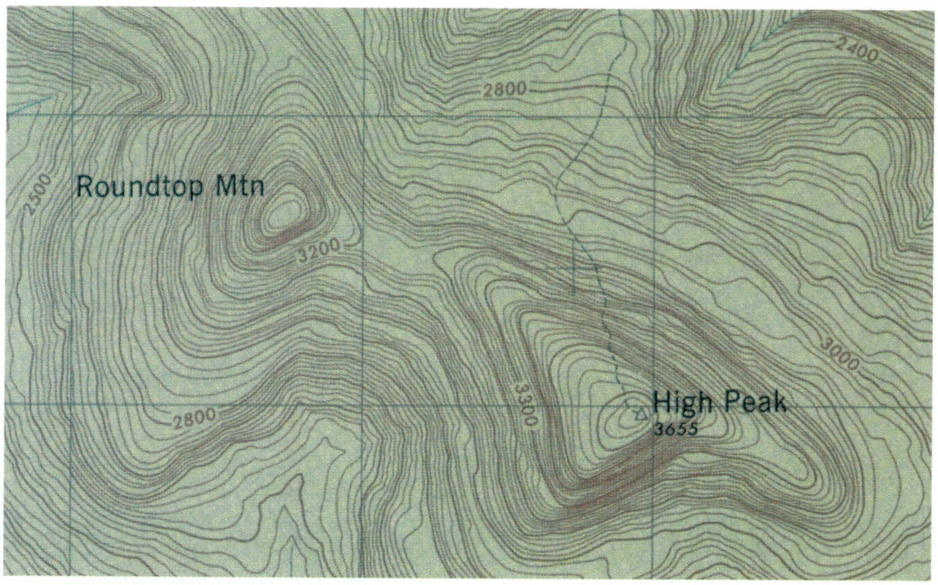

FIGURE 25–10 The contour lines on topographic maps are lines of constant elevation. These are also lines of constant gravitational potential energy. The force of gravity has no component *along* contour lines, only perpendicular to them. This map shows the contours of two peaks in the Catskill Mountains of New York.

Because no work is done by the electric force when a test charge moves on an equipotential, we can understand why the electric field cannot have a component along an equipotential surface. If it did, then that component of the electric field would do work to move a charge on the equipotential surface. This is not possible. Thus *the electric field must be everywhere perpendicular to the equipotential surface.* Furthermore, because all the charge on a conductor in equilibrium resides on the surface, a potential difference between two points on the surface would be quickly equalized by a flow of free charge, so *the surface of a conductor must be an equipotential.* In fact, the same reasoning shows that the entire conductor will then be at that same electric potential.

Electric field lines are perpendicular to the equipotential surfaces due to a system of charges.

Electric Field Lines from Equipotentials

PROBLEM SOLVING

Using or finding equipotential surfaces

The fact that the electric field and the equipotentials are everywhere perpendicular to each other is very helpful in finding equipotential surfaces if the field is known, and in finding the electric fields if the equipotentials are known. We can illustrate this for some charge configurations for which the fields are known. Consider the point charge. In Fig. 25–9, we showed spherical equipotential surfaces. In Fig. 25–11, we add the electric field lines, which extend outward radially for a positive charge. The equipotential surfaces are necessarily spheres (perpendicular to the radii/vectors that emerge from the origin).

Consider the electric field lines between two oppositely charged plates (Fig. 25–12). The equipotential surfaces are planes parallel to the charged plates. If the charged plates are conductors, then they must also be equipotential surfaces. Thus we have a series of n equipotential planes V_1, V_2, V_3,…, V_n between and including the two charged plates.

Let's next examine the equipotential surfaces due to an electric dipole, such as a polar molecule (Fig. 25–13). If we draw our equipotential surfaces everywhere per-

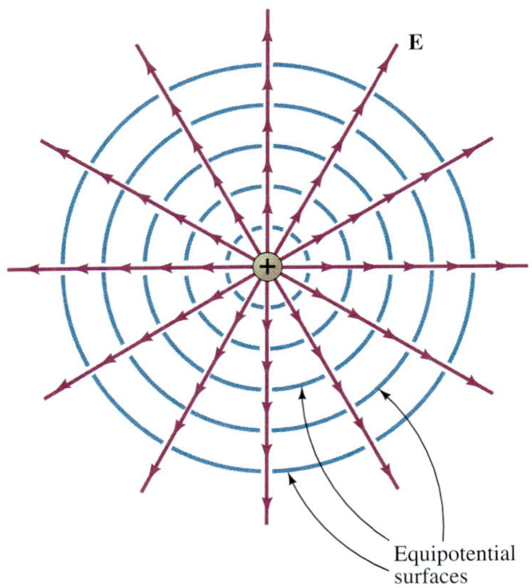

FIGURE 25–11 The electric field lines for a positive (negative) point charge extend outward (inward) radially, perpendicular to the equipotentials.

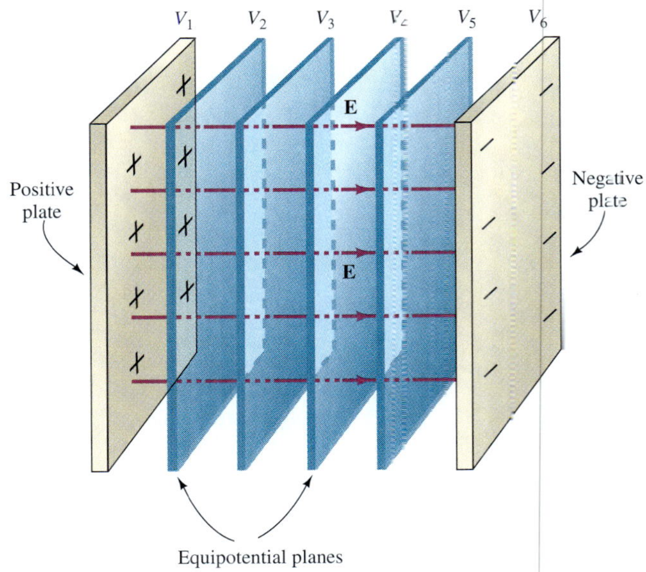

FIGURE 25–12 The electric field lines (burgundy) and the equipotentials (blue) for two oppositely charged parallel plates.

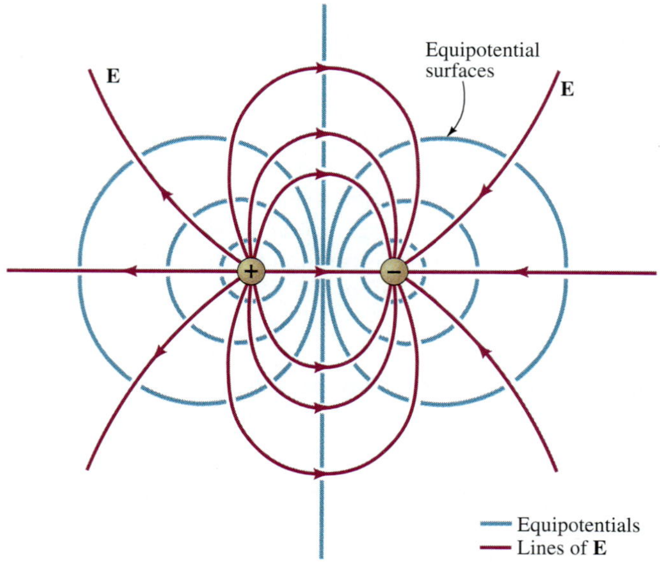

FIGURE 25–13 The electric field lines and equipotentials for an electric dipole.

pendicular to the electric field lines, we will arrive at the blue lines shown in Fig. 25–13. Even without using the electric dipole potential derived in Example 25–4, we have already determined visually what the equipotential surfaces must look like.

25–4 DETERMINING FIELDS FROM POTENTIALS

As we have already seen, if we know the electric field, **E**, Eq. (25–9) determines the potential difference $V_b - V_a$ between any two points a and b:

$$V_b - V_a = \int_{\mathbf{r}_a}^{\mathbf{r}_b} dV = -\int_{\mathbf{r}_a}^{\mathbf{r}_b} \mathbf{E} \cdot d\mathbf{s}$$

Because electrostatic forces are conservative, the potential difference is independent of the path taken between a and b in the line integral, and we can choose this path for convenience.

In this section, we find that we can reverse this procedure and calculate the electric field if the potential is known. Such a calculation is completely analogous to finding the force between objects if their potential energy is known as a function of position. Consider Fig. 25–14, which shows a set of electric field lines and two closely spaced equipotential surfaces. These equipotential surfaces are, by construction, perpendicular to the field lines. If the spacing between the equipotentials is very small, then so is the potential difference between them, which we write as dV. From Eq. (25–9), if the distance between the initial and final points a and b is infinitesimally small, then we are no longer integrating over a finite path in the integral $\int_{\mathbf{r}_a}^{\mathbf{r}_b} \mathbf{E} \cdot d\mathbf{s}$. The integral sign can be dropped, and we have

$$dV = -\mathbf{E} \cdot d\mathbf{s}. \tag{25–25}$$

It is simplest to take our infinitesimal path as in Fig. 25–14, with $d\mathbf{s}$ perpendicular to the two equipotential surfaces. As we have argued already, the electric field also points in that direction, so Eq. (25–25) reads

$$dV = -E\, ds.$$

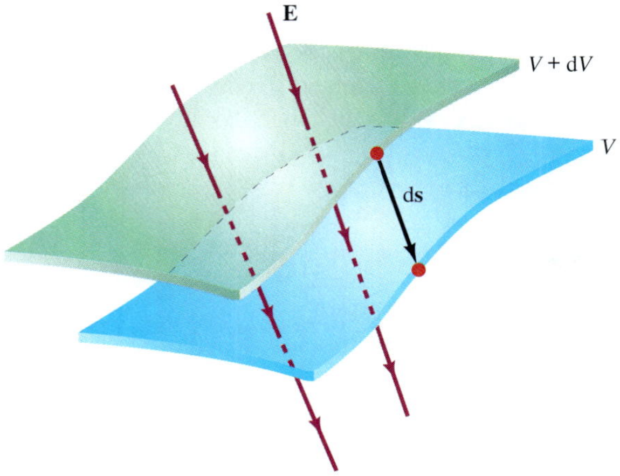

FIGURE 25–14 Two equipotentials differ by dV. The displacement **ds** along the direction of **E** between the equipotentials is perpendicular to them.

Equivalently,

$$E = -\frac{\mathrm{d}V}{\mathrm{d}s}. \qquad (25\text{–}26)$$

This equation gives the magnitude of the electric field in terms of the rate of change of V in a direction perpendicular to the equipotential at that point. The direction of the field is along that perpendicular direction, oriented to decreasing values of the potential. In other words, for closely spaced equipotentials:

The field points along the shortest direction from one equipotential to the next.

When equipotentials form concentric spheres, as they do for a point charge, the perpendicular direction lies along a radius. Therefore the electric field points in the radial direction, with magnitude

$$E = -\frac{\mathrm{d}V}{\mathrm{d}r}.$$

The same expression holds for equipotentials that form concentric cylinders but, in this case, the variable r is the distance to the cylindrical axis.

How the Potential Determines the Field in Cartesian Coordinates

We can look at this from a different point of view by supposing that an arbitrary displacement vector **ds** is decomposed into Cartesian coordinates:

$$\mathbf{ds} = \mathrm{d}x\,\mathbf{i} + \mathrm{d}y\,\mathbf{j} + \mathrm{d}z\,\mathbf{k}.$$

Here **i**, **j**, and **k** are the unit vectors in the x-, y- and z-directions, respectively. Then the scalar product in Eq. (25–25) takes the form

$$\mathrm{d}V = -\mathbf{E}\cdot\mathbf{ds} = -E_x\,\mathrm{d}x - E_y\,\mathrm{d}y - E_z\,\mathrm{d}z, \qquad (25\text{–}27)$$

where we have separated the field **E** into Cartesian components. In general, the potential depends on all three space coordinates, $V = V(x, y, z)$. The change in V in going from an initial position $\mathbf{r} = x\mathbf{i} + y\mathbf{j} + z\mathbf{k}$ to a new position $\mathbf{r} + \mathbf{ds} = (x + \mathrm{d}x)\mathbf{i} + (y + \mathrm{d}y)\mathbf{j} + (z + \mathrm{d}z)\mathbf{k}$ is

$$\mathrm{d}V = \frac{\partial V}{\partial x}\,\mathrm{d}x + \frac{\partial V}{\partial y}\,\mathrm{d}y + \frac{\partial V}{\partial z}\,\mathrm{d}z. \qquad (25\text{–}28)$$

Note the use of the partial derivatives here; this is necessary because V depends on all three Cartesian coordinates. Recall that partial derivatives are very simple to use: The partial derivative with respect to x means that y and z are held fixed while the ordinary derivative with respect to x is taken. To illustrate, if $V = xz^2$, then $\partial V/\partial x = z^2$, $\partial V/\partial y = 0$, and $\partial V/\partial z = 2xz$.

We can equate the coefficients of dx, dy, and dz in Eqs. (25–27) and (25–28):

$$E_x = -\frac{\partial V}{\partial x}, \quad E_y = -\frac{\partial V}{\partial y}, \quad E_z = -\frac{\partial V}{\partial z}.$$

The electric field in terms of derivatives of the electric potential

Equivalently, *the electric field vector is given in terms of derivatives of the electric potential* by

$$\mathbf{E} = -\frac{\partial V}{\partial x}\mathbf{i} - \frac{\partial V}{\partial y}\mathbf{j} - \frac{\partial V}{\partial z}\mathbf{k}. \qquad (25\text{–}29)$$

Equation (25–29) gives the Cartesian components of the electric field in terms of the potential. We have found a way to express a particular vector, the electric field, in terms of the derivatives of a scalar, the electric potential. The derivative operation in Eq. (25–29) produces an electric field vector that points in the direction of the greatest decrease in the potential. This direction is perpendicular to the equipotential surfaces.

EXAMPLE 25–5 Use the electric potential of a point charge q to find its electric field.

Solution: We know the electric potential and are asked to find the electric field. In this case, the potential is a function only of the radial distance from the charge, $V = q/4\pi\epsilon_0 r$. The equipotential surfaces are therefore spheres at a constant distance from the charge. According to our discussion of Eq. (25–26), the electric field is therefore directed outward radially—perpendicular to the equipotentials, in the direction of decreasing potential—and has magnitude

$$E = -\frac{\mathrm{d}V}{\mathrm{d}r} = -\frac{q}{4\pi\epsilon_0}\frac{\mathrm{d}}{\mathrm{d}r}\left(\frac{1}{r}\right) = \frac{q}{4\pi\epsilon_0 r^2}.$$

We are merely reproducing what we already know here, but the technique is useful in contexts where we do not already know the answer!

EXAMPLE 25–6 The potential with respect to infinity due to a certain charge distribution can be written as the function of position $V = Axy^2 - Byz$, where A and B are constants. Find the associated electric field.

Solution: Here, again, we are given the electric potential and must find the electric field. We require a simple application of Eq. (25–29). First we find the partial derivatives:

$$\frac{\partial V}{\partial x} = Ay^2;$$

$$\frac{\partial V}{\partial y} = 2Axy - Bz;$$

$$\frac{\partial V}{\partial z} = -By.$$

The electric field is therefore

$$\mathbf{E} = -Ay^2\mathbf{i} - (2Axy - Bz)\mathbf{j} + By\mathbf{k}.$$

Let's conclude this section with a comment on the electric dipole. Equation (25–24) gives the dipole potential, which is proportional to the factor $\cos\theta$ in Fig. 25–8.

On the bisecting axis, $\theta = 90°$, and, because $\cos 90° = 0$, the electric potential will be zero there. However, this does not mean that the electric field will be zero on the bisecting axis. The electric field is determined by *derivatives* of the potential at a particular r or θ. What counts in determining the field is how fast the potential is changing, not whether it is zero at some point. We have $\partial V/\partial\theta \propto \sin\theta$. Along the bisecting axis, where $\theta = 90°$, the electric field is a maximum.

25–5 THE POTENTIALS OF CHARGE DISTRIBUTIONS

PROBLEM SOLVING
Techniques for calculating electric potentials

We rarely deal with the electric field and potential of a single point charge. More often, we have collections of charges spread over regions of space, as when a charge spreads over the surface of a metal, or when the field of a complicated ionic molecule determines its chemical or biological behavior. We must therefore be able to find the potentials of continuous charge distributions. These charge distributions may not be simple ones, and we must develop strategies for calculating the corresponding electric potentials. In this section, we first summarize the underlying techniques, then illustrate them with a series of examples.

The qualitative shapes of equipotential surfaces due to a charge distribution are most easily found by graphical techniques. For quantitative calculations, we have learned two different ways to determine the electric potential of a charge distribution:

1. If the electric field is known, then Eq. (25–9) can be used to determine the potential:

$$\Delta V = V_b - V_a = -\int_{\mathbf{r}_a}^{\mathbf{r}_b} \mathbf{E}\cdot d\mathbf{s}.$$

2. If the electric field is not known, we may calculate the potential directly by using one of various forms:

for one point charge, Eq. (25–7): $$V = \frac{q}{4\pi\epsilon_0 r};$$

for many point charges, Eq. (25–11): $$V = \frac{1}{4\pi\epsilon_0}\sum_i \frac{q_i}{r_i};$$

for a continuous charge distribution, Eq. (25–12): $$V = \frac{1}{4\pi\epsilon_0}\int \frac{dq}{r}.$$

In a direct calculation of electric potential, we must decide the location of zero potential. In fact, the convention that zero potential is at infinity is already implicit in Eqs. (25–7), (25–11), and (25–12). This is almost always the most convenient choice for a charge distribution that does not extend all the way to infinity. If a potential *difference* is calculated directly, no decision need be made about the zero level.

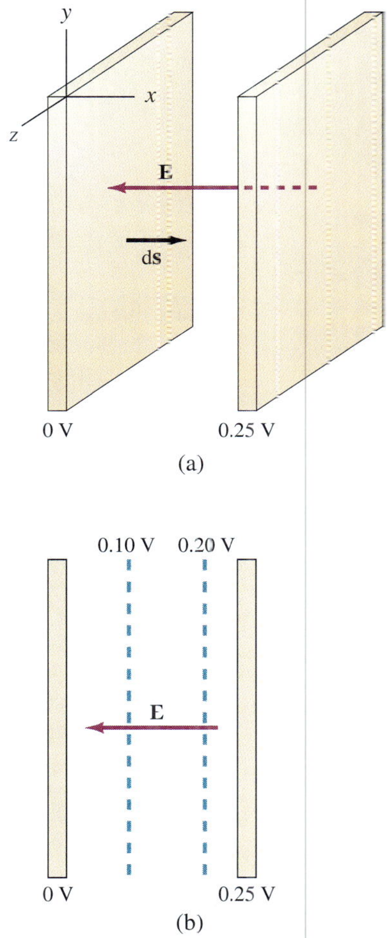

FIGURE 25–15 (a) Two parallel plates, viewed from the side, have a potential difference of 0.25 V. A differential displacement **ds** is indicated. (b) Example 25–7. Equipotentials for $V = 0.1$ V and 0.2 V. The plates are themselves equipotentials. The electric field points from right to left when the potential of the left-hand plate is lower than that of the right-hand plate.

Parallel Plates

Let's first look at the relation between electric field and potential for two parallel conducting plates (a *parallel-plate capacitor*—we shall revisit this in Chapter 26), each brought to different potentials (Fig. 25–15a). We suppose that the plates are close enough together or large enough that we can ignore the distortions of the field near the edges. This case is therefore an approximation to the central regions of real parallel-plate capacitors. In the figure, the left-hand plate is at a lower potential than the right-hand plate. The electric field between parallel plates is known to be constant, and it runs from regions of higher potential to lower potential—from right to left in this case. In Eq. (25–9), we take the path to be a straight line from the left to the right plate, so that **E** is antiparallel to **ds**. Let the direction from left to right define the x-axis, with the left plate at $x = 0$. Then the potential difference between the plates is given in terms of the field E between the plates and the separation ℓ of the plates by

$$\Delta V = V_{\text{right}} - V_{\text{left}} = -\int_{\text{left}}^{\text{right}} \mathbf{E} \cdot d\mathbf{s} = +\int_0^\ell E\,dx = E\int_0^\ell dx = E\ell.$$

The electric field between parallel plates

In turn, the electric field between parallel plates has magnitude

$$E = \frac{\Delta V}{\ell}. \tag{25–30}$$

Equation (25–30) is an important practical result. It states:

> **The constant electric field between parallel conducting plates is the potential difference between the plates divided by the distance between the plates.**

This result gives the electric field between two parallel-plane conducting plates whose potential difference is ΔV. We can also use it to find the equipotential surfaces associated with *any* constant field. These surfaces are planes perpendicular to the field, and the potential difference for a plane a distance ℓ from a reference equipotential changes *linearly* with ℓ, $\Delta V = E\ell$. Note the sign: The potential decreases along the direction in which the electric field points.

EXAMPLE 25–7 Two parallel metal plates have area $A = 225\ \text{cm}^2$ and are separated by $\ell = 0.50$ cm. They have a potential difference of 0.25 V (Fig. 25–15a). Find the numerical value of the electric field. What are the charge density and total charge on each plate? Draw the equipotential surfaces at 0.10 V and 0.20 V.

Solution: Equation (25–30) directly applies here. We know the potential difference ΔV between the plates as well as their separation. Thus the magnitude of the electric field is

$$E = \frac{\Delta V}{\ell} = \frac{0.25\ \text{V}}{0.0050\ \text{m}} = 5.0 \times 10^1\ \text{V/m},$$

and it points from right to left (Fig. 25–15b).

We know from Chapter 23 that the electric field between the plates is σ/ϵ_0:

$$E = 50\ \text{V/m} = \frac{\sigma}{\epsilon_0};$$

$$\sigma = 50\epsilon_0\ \text{V/m} = (50\ \text{V/m})(8.85 \times 10^{-12}\ \text{C}^2/\text{N}\cdot\text{m}^2) = 4.4 \times 10^{-10}\ \text{C/m}^2.$$

The answer must be in coulombs per square meter because we used SI units consistently. We know the area of the plates, so we can calculate the total charge on each plate to be

$$Q = \sigma A = (4.4 \times 10^{-10}\ \text{C/m}^2)(225\ \text{cm}^2)(10^{-4}\ \text{m}^2/\text{cm}^2) = 1.0 \times 10^{-15}\ \text{C}.$$

The electric field is constant between the parallel plates. Thus, from Eq. (25–30), we find that, at a distance d from the left plate, the potential differs by an amount $\Delta V = Ed$ from its value at the left plate, $d = 0$. The equipotential surface for 0.10 V is then

$$d = \frac{\Delta V}{E} = \frac{0.10\ \text{V}}{50\ \text{V/m}} = 0.20\ \text{cm}$$

from the left plate. For 0.20 V, we determine a distance of 0.40 cm (Fig. 25–15b).

EXAMPLE 25–8 *The Charged Ring.* Find the electric potential due to a uniformly charged ring of radius R and total charge Q at a point P on the axis of the ring.

Solution: We use the geometry shown in Fig. 25–16 and find the electric potential at point P a distance x along the axis from the center of the ring. This problem is a straightforward application of Eq. (25–12). We set up a differential charge dq along the ring that is a constant distance $r = \sqrt{R^2 + x^2}$ from point P. Because r is constant, we can bring it outside the integral of Eq. (25–12) to obtain

$$V = \frac{1}{4\pi\epsilon_0 r}\int dq = \frac{Q}{4\pi\epsilon_0 r} = \frac{Q}{4\pi\epsilon_0\sqrt{R^2 + x^2}}. \qquad (25\text{–}31)$$

We could also have found the electric field along the axis by applying the derivative operations of Eq. (25–29) (see Problem 41). This method is easier than the direct integration technique presented in Chapter 23 for the electric field.

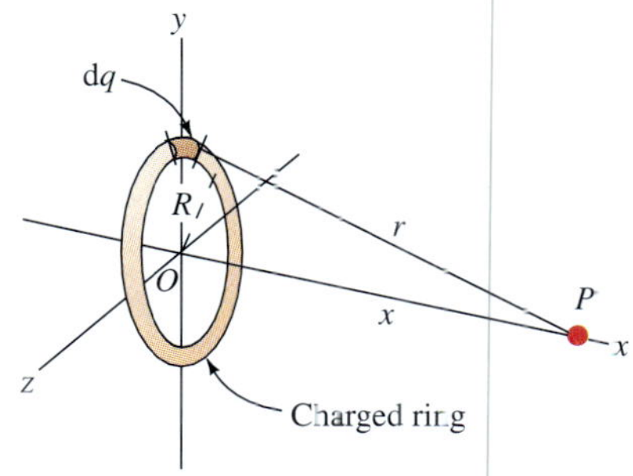

FIGURE 25–16 Example 25–8. Geometry to find the potential at a point P on the axis of a charged ring of radius R by using a differential charge dq.

EXAMPLE 25–9 *The Charged Disk.* Find the electric field due to a thin, flat, uniformly charged disk of radius R and total charge Q at a point P along its axis by first calculating the electric potential at this point.

Solution: The situation is shown in Fig. 25–17. In order to find the electric field, we first find the electric potential. We know the potential due to a ring from Example 25–8, so we divide the disk into a series of concentric rings with the intent of using the superposition principle. In Fig. 25–17, we divide the disk into a series of rings of radius r and width dr. The constant charge density over the disk is $\sigma = Q/\pi R^2$, and the total charge contained in a ring of differential area dA is

$$dq = \sigma\, dA = \sigma 2\pi r\, dr.$$

We use Eq. (25–31) for the potential due to the ring, and then integrate over all the rings to determine the total potential of the disk:

$$dV = \frac{dq}{4\pi\epsilon_0\sqrt{r^2 + x^2}} = \frac{2\pi\sigma}{4\pi\epsilon_0}\frac{r\, dr}{\sqrt{r^2 + x^2}};$$

$$V = \frac{\sigma}{2\epsilon_0}\int_0^R \frac{r\, dr}{\sqrt{r^2 + x^2}} = \frac{\sigma}{2\epsilon_0}\sqrt{r^2 + x^2}\bigg|_0^R = \frac{\sigma}{2\epsilon_0}\left(\sqrt{R^2 + x^2} - x\right)$$

$$= \frac{Q}{2\pi\epsilon_0 R^2}\left(\sqrt{R^2 + x^2} - x\right). \qquad (25\text{–}32)$$

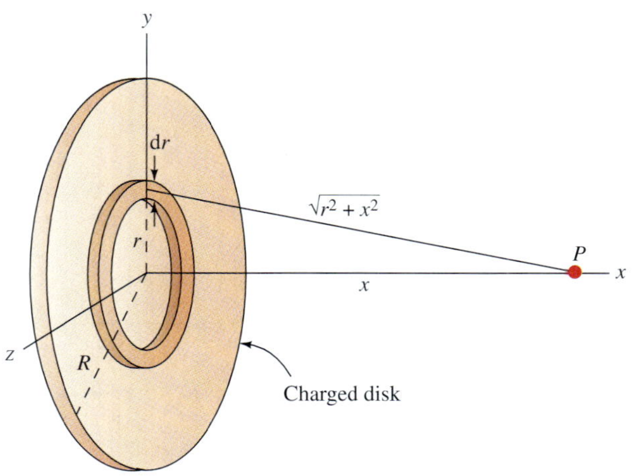

FIGURE 25–17 Example 25–9. Geometry to find the potential at a point P on the axis of a charged disk of radius R. The potential due to the ring of radius r and width dr is first found and then integrated.

Although the integration is a more difficult one than that of Example 25–8, it required no vectorial manipulation.

Because the potential depends only on x, the electric field has just an x-component, $\mathbf{E} = E_x\mathbf{i}$. The component E_x is given by

$$E_x = -\frac{\partial V}{\partial x} = -\frac{Q}{2\pi\epsilon_0 R^2}\left(\frac{x}{\sqrt{R^2 + x^2}} - 1\right).$$

That $\mathbf{E}$ has only an x-component should not come as a surprise: By symmetry, only the x-component of the field receives contributions that do not cancel out. As a check, this result can be shown in the limit that $x >> R$ to reduce correctly to the point charge limit $Q/4\pi\epsilon_0 x^2$.

EXAMPLE 25–10 *The Charged Line*. Find the electric potential as a function of the radial distance R from an infinite charged line of uniform charge density λ.

Solution: We have previously found the electric field for an infinite charged line and can use Eq. (25–9) to find the potential from the electric field. Equation (23–32) gives us the electric field, which has only a radial component. We integrate along a radial direction, so that $\mathbf{ds} = \mathbf{dr}$ (Fig. 25–18a). Equation (25–9) becomes

$$\Delta V = -\int E_r\,\mathrm{d}r = -\frac{\lambda}{2\pi\epsilon_0}\int\frac{\mathrm{d}r}{r}.$$

The potential difference depends, of course, on the end points of the integration. Let zero potential be at $r = a$, so that

$$\Delta V = V_R - V_a \equiv V = -\frac{\lambda}{2\pi\epsilon_0}\int_a^R\frac{\mathrm{d}r}{r} = -\frac{\lambda}{2\pi\epsilon_0}\ln r\Big|_a^R;$$

$$V = -\frac{\lambda}{2\pi\epsilon_0}\ln\frac{R}{a}. \qquad (25\text{–}33)$$

Note that it is not possible to set zero potential at infinity in this case because the logarithm is infinite at $a = \infty$. Physically, this is because the line itself reaches to in-

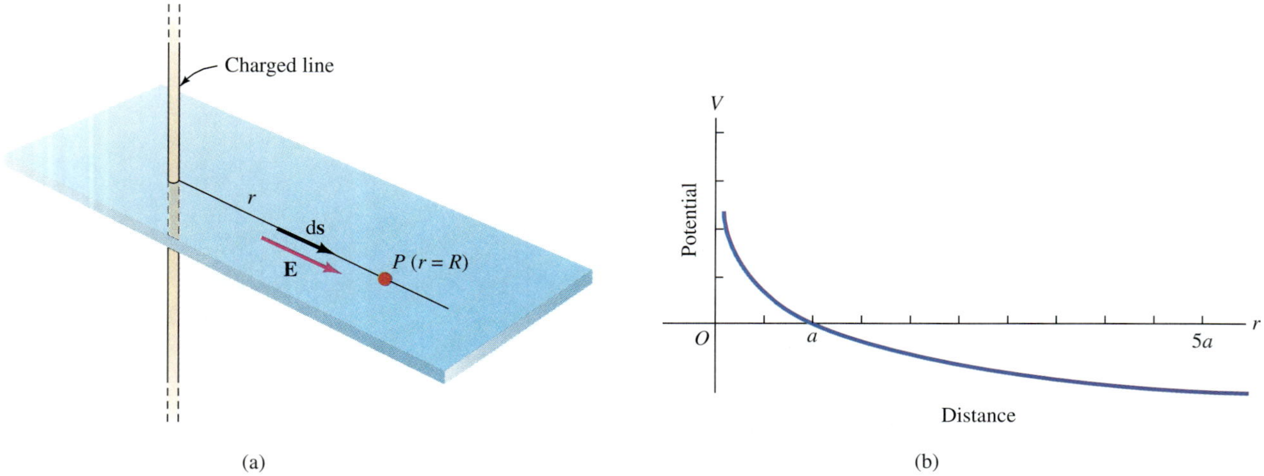

FIGURE 25–18 Example 25–10. (a) An infinite charged line has a radial electric field. To find the potential at the point P, we consider a displacement $\mathbf{ds}$ in the direction of the electric field $\mathbf{E}$ and use the known expression for the electric field. (b) The resulting potential, defined to be zero at $r = a$, goes to positive infinity at $r = 0$ and continues to negative infinity for large r.

finity; we can never get far away from it. We graph the potential of Eq. (25–33) in Fig. 25–18b, where we have assumed that the charge of the line is positive.

EXAMPLE 25–11 *The Charged Spherical Shell.* Find the potential for a uniformly charged spherical shell of total charge Q and radius R at positions both inside and outside the shell. Set zero potential at infinity.

Solution: We already know the electric field of the spherical shell from Example 24–4 and can use Eq. (25–9) to determine the electric potential. As a second fixed point, we choose $r = \infty$ and determine the potential difference, $\Delta V = V(r) - V(\infty)$. From Example 24–4, we have Eqs. (24–10) and (24–11):

$$\text{outside a spherical shell, } r > R: \qquad E = \frac{Q}{4\pi\epsilon_0 r^2};$$

$$\text{inside a spherical shell, } r < R: \qquad E = 0.$$

The electric field is purely radial, and Eq. (25–9) gives the potential difference, so we integrate along a radius from infinity to an arbitrary radial distance r. The differential element $d\mathbf{r}$ points out from the origin, so $\mathbf{E} \cdot \mathbf{ds} = E_r\,\mathrm{d}r = E\,\mathrm{d}r$. For a point outside the spherical shell,

$$\Delta V = V(r) - V(\infty) = -\int_{\infty}^{r} E\,\mathrm{d}r = -\frac{Q}{4\pi\epsilon_0}\int_{\infty}^{r}\frac{\mathrm{d}r}{r^2} = \frac{Q}{4\pi\epsilon_0}\left(\frac{1}{r} - \frac{1}{\infty}\right) = \frac{Q}{4\pi\epsilon_0 r}.$$

If we choose $V(\infty) = 0$, then the potential $V(r)$ equals ΔV:

$$\text{outside the shell, } r > R: \qquad V = \frac{Q}{4\pi\epsilon_0 r}. \tag{25–34}$$

The difference between the potential for a position inside the shell and the potential at infinity is

$$\Delta V = -\left(\int_{\infty}^{R} E_{\text{outside}}\,\mathrm{d}r + \int_{R}^{r} E_{\text{inside}}\,\mathrm{d}r\right).$$

Because $E_{\text{inside}} = 0$, the second integral drops out, and the integration is similar to the previous one with zero potential again at infinity,

$$\text{inside the shell, } r \leq R: \qquad V = \frac{Q}{4\pi\epsilon_0 r} = \text{a constant.} \tag{25–35}$$

Even though the electric field is zero inside the shell, the potential is not if we define it to be zero at infinity. We plot the electric field for the spherical shell in Fig. 25–19a and the potential in Fig. 25–19b.

We have now calculated the electric field and potential for several different distributions of charge. We summarize these results in Table 25–1.

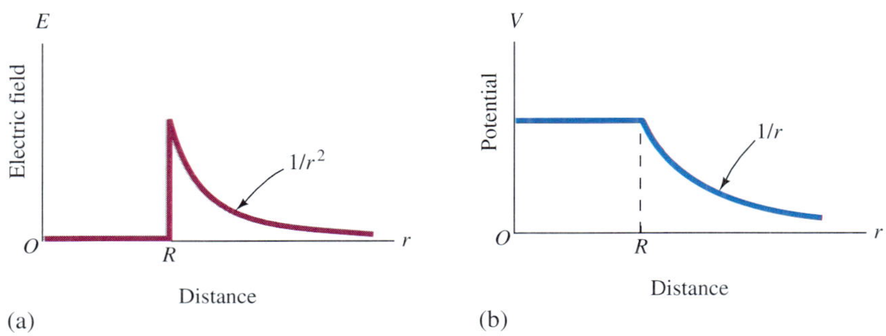

FIGURE 25–19 Example 25–11. (a) The electric field and (b) electric potential for a spherical shell of radius R. Even though the electric field is zero inside the shell, the potential has a constant value equal to that on the shell's surface.

TABLE 25–1 Electric fields and potentials for various charge configurations

Charge Configuration	*Magnitude of Electric Field*	*Electric Potential*	*Location of Zero Potential*
Point charge	$\frac{q}{4\pi\epsilon_0 r^2}$	$\frac{q}{4\pi\epsilon_0 r}$	∞
Infinite line of uniform charge density λ	$\frac{\lambda}{2\pi\epsilon_0 r}$	$-\frac{\lambda}{2\pi\epsilon_0}\ln\frac{r}{a}$	$r = a$
Parallel, oppositely charged plates of uniform charge density σ, separation d	$\frac{\sigma}{\epsilon_0}$	$\Delta V = -Ed = -\frac{\sigma d}{\epsilon_0}$	Anywhere
Charged disk of radius R, along axis at distance x	$\frac{Q}{2\pi\epsilon_0}\left(\frac{\sqrt{R^2+x^2}-x}{\sqrt{R^2+x^2}}\right)$	$\frac{Q}{2\pi\epsilon_0 R^2}(\sqrt{R^2+x^2}-x)$	∞
Charged spherical shell of radius R	$r \geq R$: $\frac{Q}{4\pi\epsilon_0 r^2}$	$r > R$: $\frac{Q}{4\pi\epsilon_0 r}$	∞
	$r < R$: 0	$r \leq R$: $\frac{Q}{4\pi\epsilon_0 R}$	∞
Electric dipole	Along bisecting axis only, far away: $\frac{p}{4\pi\epsilon_0 r^3}$	Everywhere, far away: $\frac{p\cos\theta}{4\pi\epsilon_0 r^2}$	∞
Charged ring of radius R, along axis	$\frac{Qx}{4\pi\epsilon_0 (R^2+x^2)^{3/2}}$	$\frac{Q}{4\pi\epsilon_0\sqrt{R^2+x^2}}$	∞
Uniformly charged nonconducting solid sphere of radius R	$r \geq R$: $\frac{Q}{4\pi\epsilon_0 r^2}$	$r \geq R$: $\frac{Q}{4\pi\epsilon_0 r}$	∞
	$r < R$: $\frac{Qr}{4\pi\epsilon_0 R^3}$	$r < R$: $\frac{Q}{8\pi\epsilon_0}\left(3-\frac{r^2}{R^2}\right)$	∞

25–6 POTENTIALS AND FIELDS NEAR CONDUCTORS

The most important cases of continuous charge distributions probably occur on metals; we shall give several practical examples in Section 25–7. These distributions are rarely uniform because charges are free to move on and within metals. Nevertheless, we can learn a surprising amount about the electric potentials near metals.

We have already learned that, in electrostatics, the electric field inside a conducting material must be zero, that the net charge on a conductor must lie on its outside surface, and that the electric field just outside the conductor's surface must be normal to the surface. Because there is no component of the electric field along the conducting surface, we have also noted that the conducting surface must itself be an equipotential. Because the electric field everywhere inside a conducting material is zero, the potential inside must have the same value as it has at the surface. We have already seen an effect of this nature in Example 25–11, the charged spherical shell. The potential at the surface and everywhere inside the shell has a constant value (that is, it is an equipotential).

We can also say something about the electric potential outside a conductor—whether the conductor is charged or not. If it is charged, the fact that the electric field is perpendicular to the surface means that the equipotentials near the surface will be parallel to the surface. This will be true even if the conductor is uncharged. Consider, for example, the electric field shown in Fig. 25–20a due to two parallel plates (which are not shown). If we place an uncharged conductor of arbitrary size in this electric field, the field around the conductor will be greatly modified (Fig. 25–20b). Charge

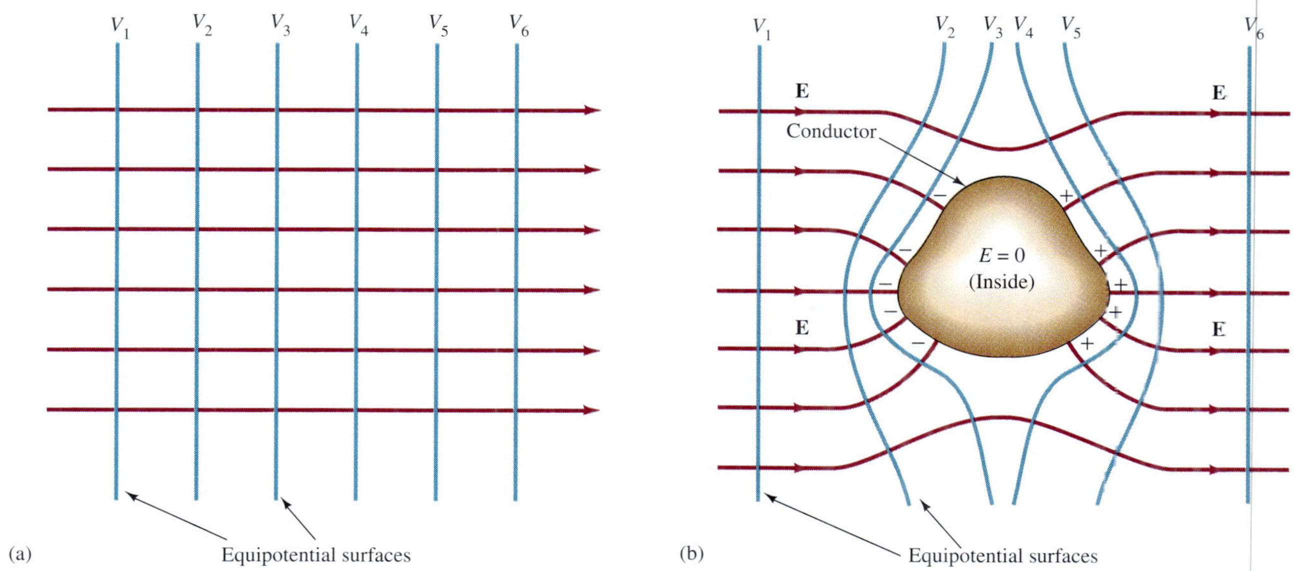

FIGURE 25–20 (a) A uniform electric field before an uncharged conductor is placed in the field. (b) Afterward, the electric field is changed dramatically, with no electric field inside the conductor. Induced charges, which make the electric field inside the conductor zero, appear on the outside surface of the conductor. These charges affect the electric field outside the conductor.

will be induced on the outside of the conductor in equilibrium, thereby forcing the electric field to be normal to the conducting surface and, again, *the equipotential surfaces near a conductor of arbitrary shape must be parallel to the conductor's surface.*

Near a conductor, the equipotential surfaces are parallel to the conductor's surface.

In Chapter 24, we found the magnitude of the electric field near the surface of a conductor in terms of the charge density at that point. The charge density, however, can vary for an irregular surface. Here we shall see how the concept of potential allows us to say more about the charge density and hence the fields near irregularly shaped charged conductors.

The Role of Sharp Points on Conducting Surfaces

Consider the irregular conductor shown in Fig. 25–21. We can characterize these two sides by inscribing spheres in the ends and measuring their respective radii. The left region is more sharply curved than the right region, and $r_1 < r_2$. We model this conductor with a two-step process. First, consider the two spherical conductors shown in Fig. 25–22a. The sizes of these spheres match the two ends of our irregular conductor of Fig. 25–21: The charges q and q' are placed on the two spheres. The electric potentials at the spheres are, respectively,

$$V_1 = \frac{q}{4\pi\epsilon_0 r_1}$$

and

$$V_2 = \frac{q'}{4\pi\epsilon_0 r_2}.$$

Now connect the two spheres by a long conducting wire (Fig. 25–22b). The entire system will come to the same potential after charge flows rapidly between the two spheres. This potential is

$$V = \frac{q_1}{4\pi\epsilon_0 r_1} = \frac{q_2}{4\pi\epsilon_0 r_2},$$

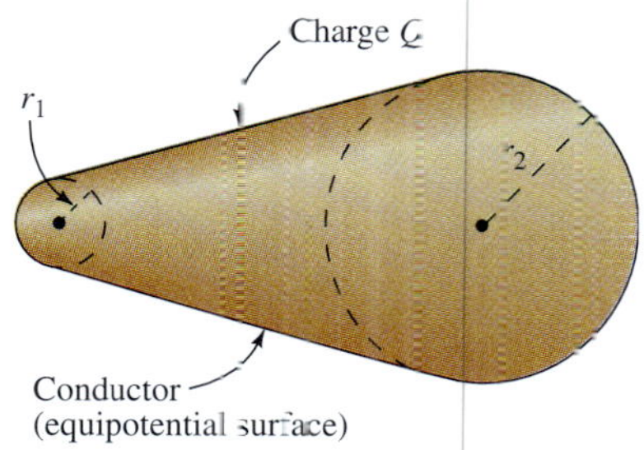

FIGURE 25–21 A conductor of odd shape is approximated by spheres of radii r_1 and r_2 at its ends

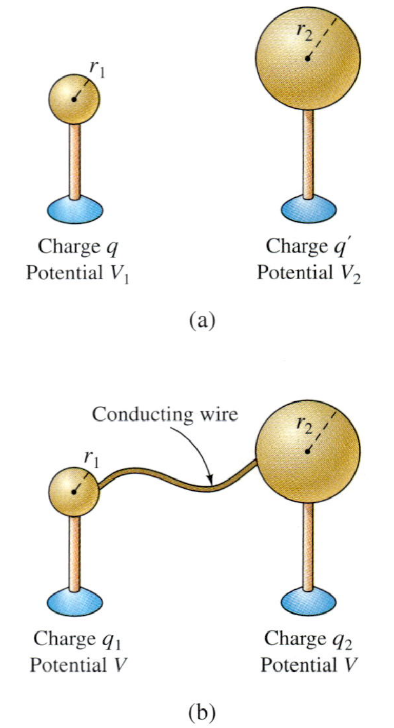

FIGURE 25–22 (a) Two conductors initially are at different potentials dependent on their respective charges. (b) If the conductors are connected by a wire, charge must flow to make the potentials equal everywhere.

The electric field near a conductor is larger near regions of sharp curvature.

APPLICATION

Lightning rods

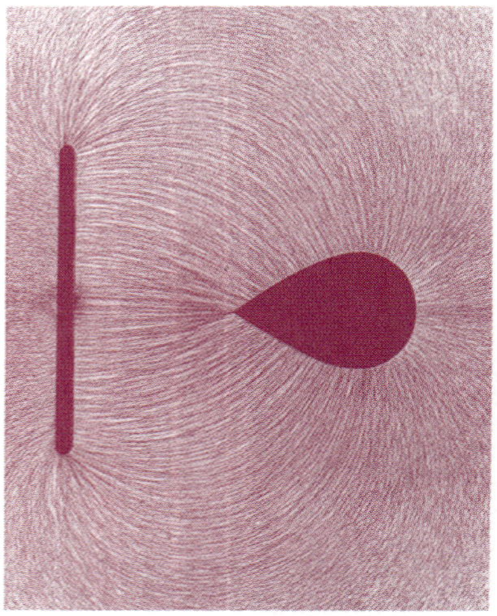

FIGURE 25–23 The electric field near small radii of curvature can be quite large, as seen for the point of the charged object.

where q_1 and q_2 are the equilibrium charges on the two spheres. The charges and radii are related by

$$\frac{q_1}{r_1} = \frac{q_2}{r_2}.$$

Because we have used only the fact that the entire system is at a single potential, this calculation will apply also to our irregular conductor of Fig. 25–21. The connected spheres form a model for the relative size of the electric fields at the two ends of the conductor. Note that q and q' no longer appear; all that remains is the charge-conservation requirement that $q + q' = q_1 + q_2$.

The surface charge density σ on a sphere is determined by the charge on the sphere and the surface area:

$$\sigma = \frac{q}{\pi r^2}.$$

For our two spheres, therefore, the equation $q_1/r_1 = q_2/r_2$ is

$$\frac{\sigma_1 \pi r_1^2}{r_1} = \frac{\sigma_2 \pi r_2^2}{r_2};$$

$$\sigma_1 r_1 = \sigma_2 r_2. \qquad (25\text{–}36)$$

The electric field E_i just outside each conducting sphere is equal to σ_i/ϵ_0, so that we can replace σ_i by $\epsilon_0 E_i$. Thus $\epsilon_0 E_1 r_1 = \epsilon_0 E_2 r_2$, or

$$E_1 r_1 = E_2 r_2;$$

$$\frac{E_1}{E_2} = \frac{r_2}{r_1}. \qquad (25\text{–}37)$$

The electric fields are inversely related to the radii. *For a small radius, the surface charge density and the corresponding electric field are large.*

The effect just described is important in conductors with sharp points (Fig. 25–23). Even if the conductor is at a low electric potential, particular regions of the surface with small radii of curvature can have large electric fields nearby. When an electric field is strong enough to overcome the attraction between electrons in metal and the forces that contain them in the metal, *corona discharge* occurs. The electrons move into the air and are sufficiently accelerated by the fields so that, through collisions, they can increase the internal energy of the atoms of the air. These atoms can then lose energy by emitting light, which is observed as a greenish glow. The electrons are ripped loose from the molecules, which are said to be *ionized*. In air, this occurs for fields on the order of 3×10^6 V/m. Sailors long ago saw these glows at the pointed tops of their masts and spars and dubbed the phenomenon *St. Elmo's fire*. This phenomenon is more generally called *dielectric breakdown* (see Fig. 25–24).

The positively charged ionized molecules are accelerated by the large electric fields and move away from the region of breakdown. Thus after dielectric breakdown, the air effectively becomes a conductor that carries away the excess charge. This lowers the electric potential around the original conductor. In thunderstorms, there are large potential differences between the ground and clouds because of a buildup of charges in the clouds. Lightning rods cause a local dielectric breakdown and lower the potential difference between the rod and the clouds. The rods do not *attract* the lightning strikes—by lowering the potential difference, they actually prevent lightning from striking in the vicinity.†

†Conversely, if the field between the cloud and the ground is too large to be dispersed by slow discharge, then the fact that corona discharge occurs off a sharp point leads to the formation of a "leader." The leader is formed by a set of ions of the proper sign that stream toward the cloud, making a cascade as they do so. In that case, lightning does indeed strike at the sharp point formed by the lightning rod rather than other areas of a building.

For spherical conductors, the potential and electric field on the surface are related by $V = RE$. Dielectric breakdown can thus occur with a relatively low potential on a conductor, if the conductor has sharp points. The maximum potential that can be put on a metal without air ionization is $V_{max} = R(3 \times 10^6\ \mathrm{V/m})$. For a needle point with a radius of curvature $R = 0.1$ mm, the maximum potential is only 300 V; but, for a domed structure of $R = 3$ m, the maximum potential is about 10^7 V.

Sparking. When an electric field in air at standard temperature and pressure reaches approximately 3×10^6 V/m, any electrons that are present are accelerated to sufficiently high speeds so that when they collide with atoms of the air, electrons are knocked completely out of those atoms. Those electrons, in turn, can be accelerated, freeing still other electrons. This process is enhanced if a metal is nearby because electrons pulled from the metal are then added to the mix. The number of electrons cascades rapidly, and we have a catastrophic buildup of moving electrons that carry charge and cancel the field. This catastrophic charge transfer is known as a spark (Fig. 25–24). Lightning is a familiar example (see chapter-opening photo and Fig. 25–25).

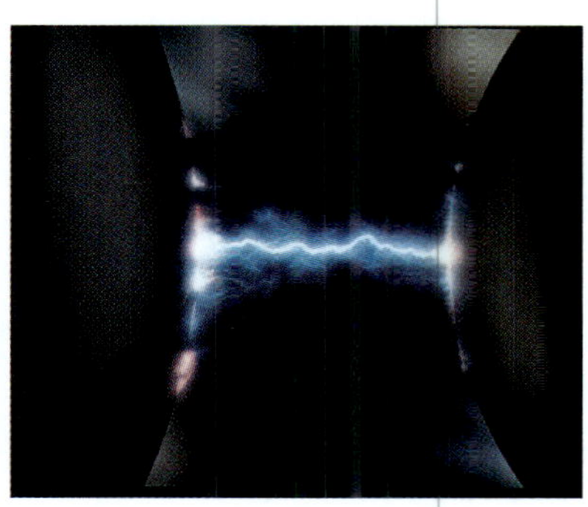

FIGURE 25–24 The electric field in the space between the metal plates is large enough to cause electrical breakdown in air: a spark.

FIGURE 25–25 Lightning is a spectacular example of the effects of dielectric breakdown under the influence of large electric fields

25–7 ELECTRIC POTENTIALS IN TECHNOLOGY

Even though electrostatics represents only a small portion of our study of electromagnetism, it has important applications. We briefly mention a few of them here to indicate how the understanding of basic principles can be put to good practical use.

The Van de Graaff Accelerator

If we place a charge anywhere in a conductor, the charge will move to the outside surface; the field inside the conductor will be zero. Robert Van de Graaff took advantage of this concept in 1931 to build an *accelerator*: an apparatus that produces highly energetic charged particles. Such particles are useful for microscopic probes of matter and as cancer treatments. Van de Graaff used a device similar in concept to the apparatus shown schematically in Fig. 25–26a. An insulated belt (or chain) continuously brings charge to the inside of a hollow conductor, which then moves to the outside surface of the conductor. The electric potential on the spherical conducting surface increases as charge flows to its surface ($V = q/4\pi\epsilon_0 R$). An *ion source* produces charged atoms whose sign is such as to be repelled from the region of high potential and thus accelerated. Such devices are called **Van de Graaff accelerators** or **Van de Graaff generators** (Fig. 25–26b).

The Field-Ion Microscope

The phenomenon that large electric fields occur at sharp points on a conductor is carried to its extreme in **field-ion microscopy**. The high electric fields involved in this technique allow us to produce images of individual atoms in the crystalline structure of the sharp point, or tip, of a metal. A fine tip of the crystalline material is prepared, commonly by dipping a mechanically formed tip in an electrolyte, a substance that dissolves atoms off the end. Depending on the particular metal and the crystal being prepared, these tips are as small as 200 nm across. On the 200-nm scale, such a tip looks smooth, but it is still very rough at the atomic level. The tip is then introduced into a vacuum, and a large positive potential of several kilovolts is applied (Fig. 25–27). The end of the tip is smoothed off even further as atomically sharp points of atoms of the metal itself are driven off in the form of positive ions by the large fields. The smoothing process leaves a tip like those in Figs. 25–28a and 25–28b. Oranges stacked in layers into a semipyramidal shape provide a familiar example of the possible structure of a tip.

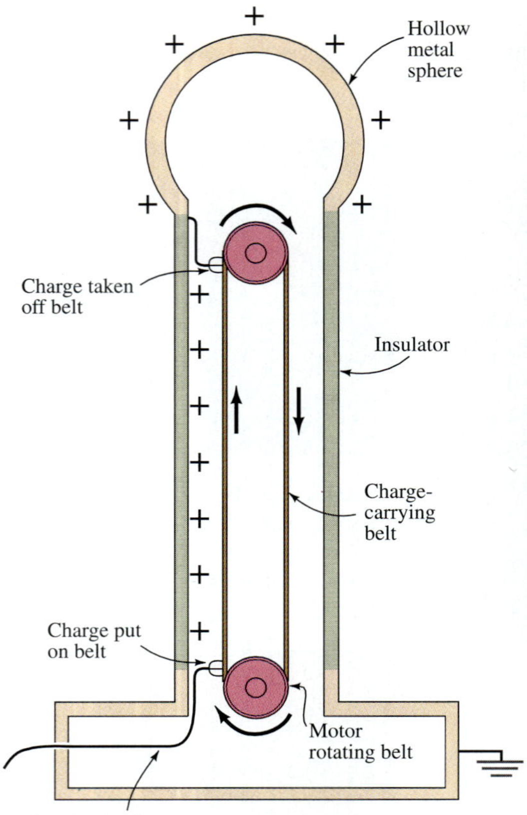

(a)

(b)

FIGURE 25–26 (a) Schematic diagram of a simple Van de Graaff accelerator. Charge is sprayed on the rotating belt at the bottom and taken off at the top. The charge goes to the outside surface of the conductor, and the potential continues to build up to high values. The symbol in the bottom right indicates that the base of the accelerator has been grounded. (b) The children touching this Van de Graaff generator are brought to a high electric potential. The individual hairs behave like the leaves of an electroscope.

In the next stage, a dilute gas such as helium or neon—called the *imaging* gas—is introduced into the chamber that contains the tip. The positive potential on the tip is increased, again to several kilovolts, until the gas atoms just begin to be ionized. This happens only where the field is largest: right above individual tip atoms. The field is strong enough to ionize the gas. The gas ions are then driven away from the tip, following the electric field lines out from the tip atoms to a grounded screen, where the impinging gas ions leave a visible trace. The image formed corresponds to the posi-

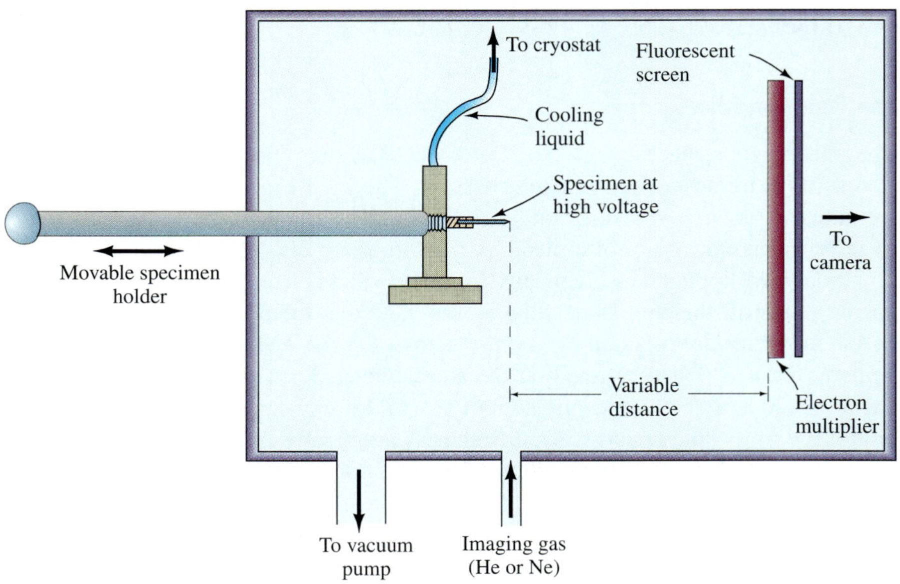

FIGURE 25–27 Schematic diagram of a field-ion microscope. The cryostat maintains a steady, low temperature in the chamber.

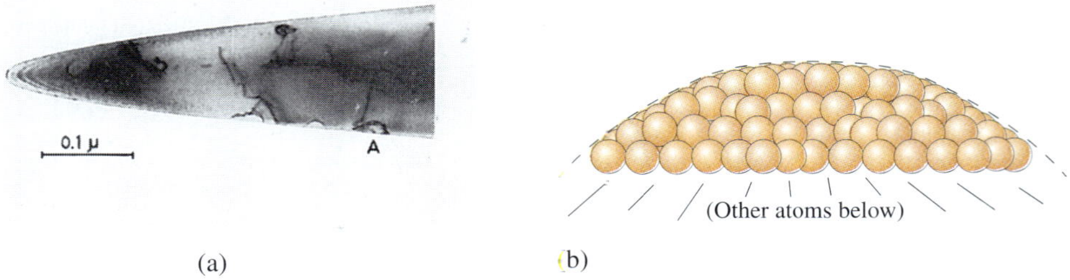

FIGURE 25–28 (a) Magnified iron tip. (b) The spheres represent individual atoms in a field-ion tip.

tion of the individual atoms of the tip, which thus become visible in a picture such as Fig. 25–29. Field-ion microscopy is useful for observing crystal structures and the effects of impurities and defects in crystals. Even some noncrystalline materials can be investigated this way.

APPLICATION

Field-ion microscopy

FIGURE 25–29 Field-ion micrograph of an iridium tip. The features are formed by individual atoms.

Xerography

Photocopying machines take advantage of electrostatics in several steps of **xerography**, or photoreproduction. The process begins with a positively charged plate coated with photoconducting material—something that is a good conductor in light but not in the dark—such as selenium (Fig. 25–30a). Light reflected from the original to be copied passes through a lens onto the charged plate, where the dark areas remain charged, but charges flow to the plate underneath at the areas that receive light (Fig. 25–30b). The resulting image of the dark areas is represented by the remaining charges. Negatively charged toner (a black powder) is added to the positively charged plate, leaving the original dark areas with black toner on the plate (Fig. 25–30c). In the next step, paper that has also been positively charged is placed over the plate and attracts the black, negatively charged toner (Fig. 25–30d). Heat is used to fuse the toner (and thus the image) to the paper.

Electric Potentials and Quantum Engineering

Electrons are trapped within a metal by electrostatic attractions to their parent ions. These forces can be represented in classical physics by a potential barrier that the electrons cannot cross. Figure 25–31a is an energy diagram of the potential energy of an electron as well as the electron's (constant) total energy. Classically, the electron cannot enter the region where its total energy is less than its potential energy. As we de-

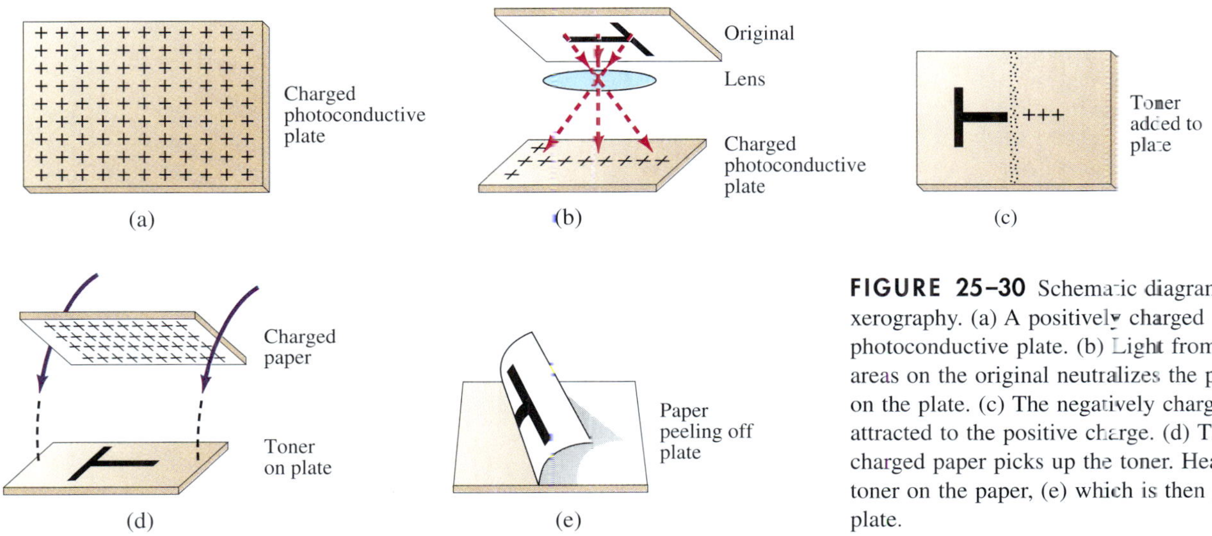

FIGURE 25–30 Schematic diagram of xerography. (a) A positively charged photoconductive plate. (b) Light from the white areas on the original neutralizes the positive charges on the plate. (c) The negatively charged toner is attracted to the positive charge. (d) The positively charged paper picks up the toner. Heat seals the toner on the paper, (e) which is then peeled off the plate.

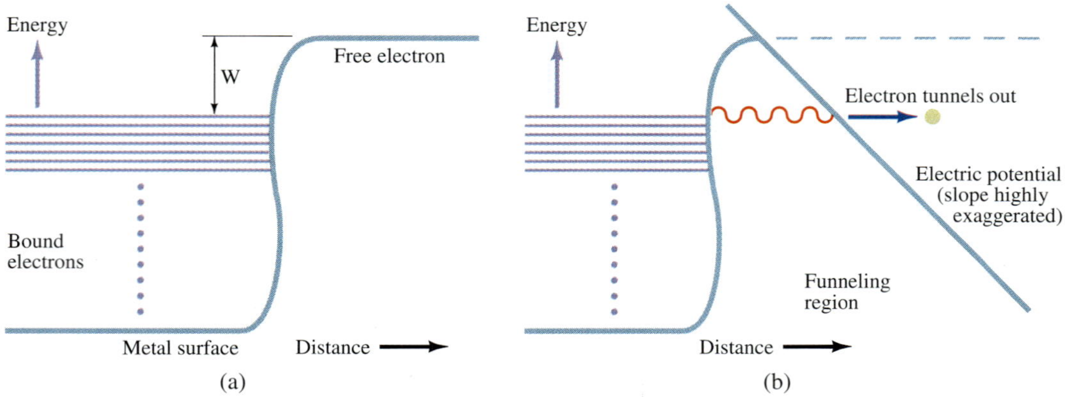

FIGURE 25–31 (a) Electrons in a metal take on a range of energies up to a highest level. This level is still negative compared to the energy of a free electron. It takes an energy W—typically 4 to 5 eV—to sufficiently raise the electron energy to allow the electron to leave the metal. The potential energy curve is drawn in blue. (b) When an external potential is applied to the metal, that potential plus the existing potential yields an effective potential barrier. Electrons can tunnel through this barrier.

scribed in Chapter 7, however, the electron has some very small probability of "tunneling" through the barrier due to quantum effects.

Some positively charged object brought near the metal surface pulls on the electrons. The positively charged object has an electric potential with respect to the metal, and an electron has a potential energy due to the external object (Fig. 25–31b). When the potential energy due to the external object is added to the original potential energy that holds the electron within the metal, the barrier is, in effect, reduced. Even if the external potential is too weak to lower the maximum potential energy below the electron's total energy and allow the electrons to escape classically, the fact that it lowers the barrier makes it easier for the electrons to tunnel through the barrier. This "potential-assisted tunneling" is utilized in the *scanning tunneling microscope* (Fig. 25–32a). A weak positive potential is placed on an ultrafine tungsten needle. The needle scans the surface of a sample and provides the necessary potential to help electrons escape the sample by tunneling. These electrons are attracted to the needle and form a current through it whose magnitude depends on the distance between the needle and the surface (Fig. 25–32b). This effect is used in two ways:

APPLICATION

Scanning tunneling microscopy

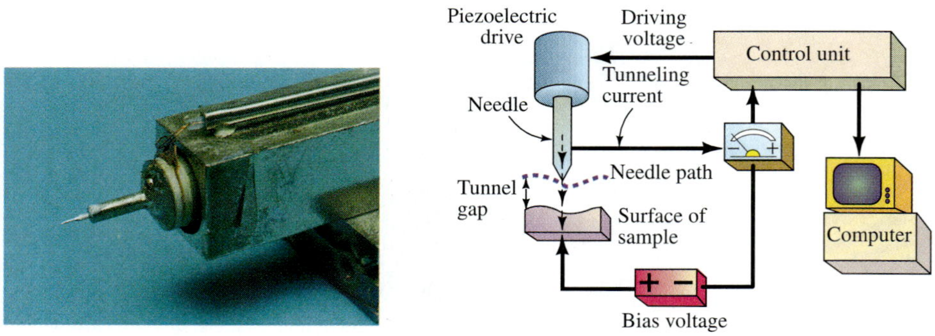

(a) (b)

FIGURE 25–32 (a) Close-up view of the needle of a scanning tunneling microscope. (b) Schematic diagram of a scanning tunneling microscope. The fine-tipped needle in the scanning head comes to within 1 nm of the sample; this distance is the tunnel gap. The tunneling current across this gap holds the gap distance constant as the tip scans the sample surface and thereby provides a map of that surface. The base voltage leads to the tunneling current, which can be used to form the driving voltage. The driving voltage moves the tip through piezoelectricity, a phenomenon in which a voltage is applied to and thereby moves a crystal.

APPLICATION: FIELD-EFFECT TRANSISTORS

The field-effect transistor lies behind almost all battery-controlled electronics: cellular phones, wristwatches, laptop computers, and so forth. It is a switch whose operation can be most simply described in terms of potentials. Figure 25AB–1 shows a side view of a strip of semiconductor—a material within which fields can be maintained yet charges can still move—containing charge carriers. The two ends, one called the *source* and the other the *drain*, have a given potential difference $V_s - V_d$. Held above the middle of the strip is what is referred to as the *gate*, here a cylinder, viewed end on; the potential V_g on the cylinder can be independently controlled (Fig. 25AB–2).

There is a constant electric field E_{sd} across the strip due to the source and drain, and a radial electric field E_g that drops off radially due to the gate. Near the source, $\mathbf{E}_g$ has a left-pointing component. This component is larger nearer the surface, so it is relatively more difficult for a charge closer to the surface to move from source to drain than for one farther within the strip. This is easily seen by plotting the corresponding potentials for different depths (Fig. 25AB–3): Charges near the surface have a potential hill to climb to get to the other side, whereas charges at greater depth can go entirely "downhill." By increasing the potential on the cylinder, the depth at which the downhill ride starts increases and thus the amount of charge that moves from source to drain decreases. The potential on the gate thus controls the amount of charge that moves across and can act as a switch.

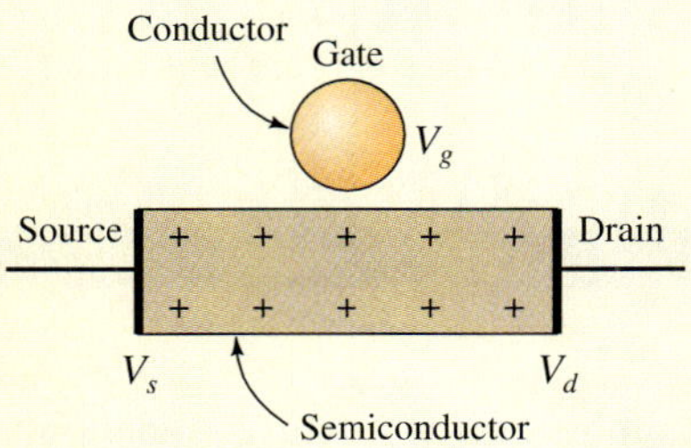

FIGURE 25AB–1 By maintaining a potential difference between the source and drain of a semiconducting strip, charges, here donated with a plus sign, move across the semiconductor.

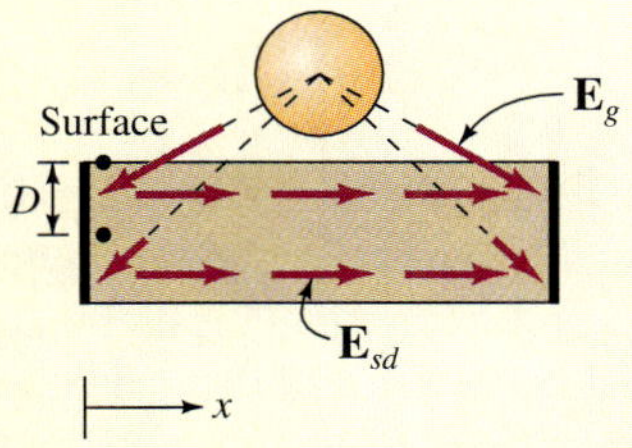

FIGURE 25AB–2 A gate with a potential on it is held above the semiconducting strip; the field it sets up tends to cancel the field due to the source-drain potential difference near the source, but tends to reinforce that field near the drain.

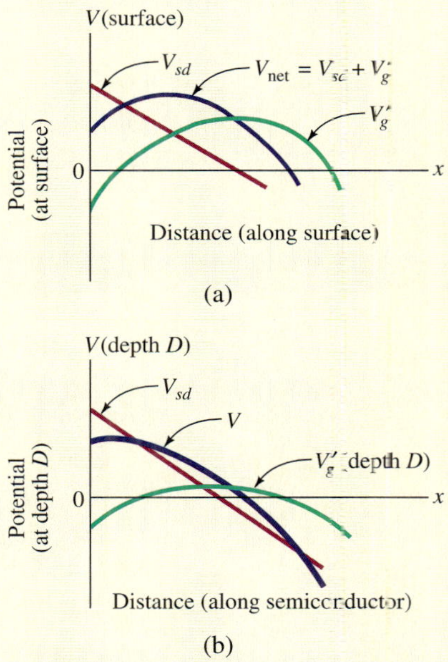

FIGURE 25AB–3 Due to the cancellation-reinforcement mechanism described in Fig. 25AB–2, the net potential when the gate is turned on has a barrier that inhibits the motion of charge from source to drain. The effect depends on the distance from the gate; that is, on the depth into the semiconducting strip. (a) For shallow depths, the barrier is quite pronounced. (b) For greater depths, the barrier is small. The result is that the gate potential acts as a sensitive switch for current flowing from source to drain.

1. A feedback mechanism that constantly repositions the needle can be set up so that the current is constant. The distance between the needle tip and the sample's surface is therefore constant. The repositioning can be measured, and the topography of the surface is thereby mapped (Fig. 25–33).

2. The potential on the needle can exert a slight pull on whole atoms. Just as the nonuniform field of a charged comb induces a dipole moment of neutral pieces of paper and attracts them, the needle tip induces a dipole moment on the atoms and attracts them out of the sample material. In this way, atoms can be moved *one at a time* to new positions (Fig. 25–34). This effect may allow the construction of new molecules and ultrasmall logic circuits (switching circuits in computers).

The combination of electrostatics and quantum mechanics is rapidly becoming a tool in what is aptly called *quantum engineering.*

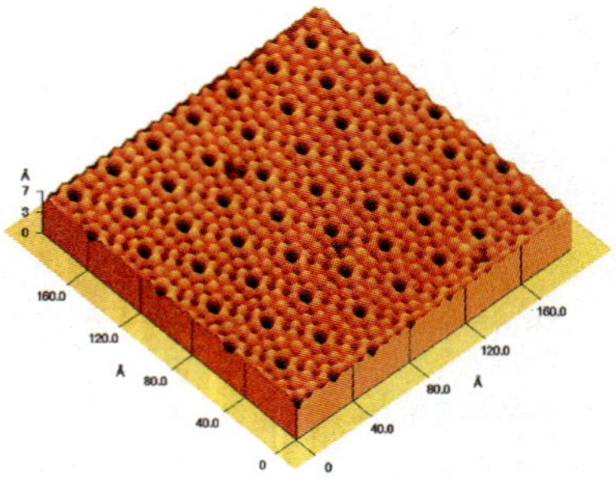

FIGURE 25–33 Scanning tunneling microscope image of the surface of a silicon crystal. Resolution down to the atomic level—the dark spots are individual atoms—has made this microscope a standard tool of materials scientists.

FIGURE 25–34 Individual xenon atoms (whose size is on the order of tenths of a nanometer) have been moved to line up in a row. The vertical scale of the figure is enhanced by a factor of ten for a more dramatic effect. The xenon atoms are actually spherically symmetric.

SUMMARY

The Coulomb force is conservative, so a potential energy—electric potential energy—is associated with it. If a test charge q_0 moves from point a to point b in the presence of a point charge q at the origin, the change in potential energy is given by

$$\Delta U = \frac{qq_0}{4\pi\epsilon_0}\left(\frac{1}{r_b} - \frac{1}{r_a}\right). \tag{25–4}$$

The electric potential difference due to any charge distribution between points a and b is defined as the change in potential energy divided by the magnitude of a test charge q_0:

$$\Delta V = V_b - V_a = \frac{U_b - U_a}{q_0} = -\int_{r_b}^{r_a} \mathbf{E} \cdot \mathrm{d}\mathbf{s}. \tag{25–9}$$

Here, $\mathbf{E}$ is the electric field due to the charge distribution. The integral in Eq. (25–9) is independent of the path between the end points. The potential difference $V_b - V_a$ is the work done per unit charge by an external agent in moving a test charge from point a to point b with no change in kinetic energy. The potential is independent of the test charge.

The electric potential can be determined by the following methods, in addition to graphical methods:

1. If the electric field is known, then Eq. (25–9) may be used.

2. If the electric field is not known, it is generally easier to calculate the potential directly by using one of these forms:

for one point charge: $$V = \frac{q}{4\pi\epsilon_0 r}; \tag{25–7}$$

for many point charges: $$V = \frac{1}{4\pi\epsilon_0}\sum_i \frac{q_i}{r_i}; \tag{25–11}$$

for a continuous charge distribution: $$V = \frac{1}{4\pi\epsilon_0}\int \frac{\mathrm{d}q}{r}. \tag{25–12}$$

In each of these cases, zero potential is chosen to be at infinity.

The SI unit of electric potential is the volt (V); 1 V = 1 J/C. A useful unit of energy for atomic and subatomic systems is the electron-volt (eV); 1 eV = 1.6×10^{-19} J.

The electric field can be determined if the potential is known in terms of derivatives of the potential:

$$\mathbf{E} = -\frac{\partial V}{\partial x}\mathbf{i} - \frac{\partial V}{\partial y}\mathbf{j} - \frac{\partial V}{\partial z}\mathbf{k}. \tag{25–29}$$

The electric field between two parallel plates is constant and is given by the potential difference divided by the distance between the plates:

$$E = \frac{\Delta V}{\ell}. \tag{25–30}$$

The electric field and potential for several charge configurations are given in Table 25–1.

Equipotential surfaces are surfaces at a fixed potential. The electric field is perpendicular to equipotentials. The surfaces of conductors form equipotentials, and the potentials inside conductors in equilibrium are everywhere the same as the potential on the surface. Electric fields just outside conductors are inversely proportional to the radius of curvature, so there are high electric fields near sharp points on conductors even if the conductors are at low potentials.

Applications of electrostatics include the Van de Graaff accelerator, the field-ion microscope, xerography, and the control of tunneling phenomena.

UNDERSTANDING KEY CONCEPTS

1. How many joules are in 1 V·C?
2. An infinite plane is uniformly charged with positive charge of density σ. How would you use a known negative test charge to measure σ?
3. How would you create an electric field inside the hollow space of a spherical metal shell that is constant in magnitude and direction in a small region of the interior space?
4. In good weather, the electric field in the lower atmosphere is approximately 100 V/m, pointing downward. What happens when a 3-m metal rod is planted in the ground?
5. When an electric field moves a charge by doing work on it, what is the source of the energy to do the work? Where did this energy come from originally?
6. In describing the potential difference as the work per unit charge to move a test charge, we added the phrase "without changing the kinetic energy" (see the boldface statement on p. 665). Why is this important?
7. Will a conductor always be an equipotential? If not, under what circumstances will that occur?
8. Using Eq. (25–29), explain why changing the location of zero potential does not affect the value of the electric field.
9. A small Van de Graaff generator can be used as a lecture demonstration device. If a person touches the dome, his or her hair stands up (see Fig. 25–26b). Explain why. Why should the person stand on an insulated mat during this demonstration?
10. Earth is typically defined to be at zero potential with respect to infinity. Does this mean that Earth can have no net charge? If Earth does have a net charge, can it still be at zero potential?
11. If we know the electric potential at a certain point do we also know the electric field? What can we know about the electric field if we know the electric potential at two points arbitrarily close to one another?
12. Is the electric potential energy of a system of point charges independent of the order in which the system is assembled?
13. Why are there so many curved surfaces on the Van de Graaff generator?
14. How do we really know that electric forces are conservative?
15. In the potential associated with a point charge, we chose zero potential to be infinitely far from the charge. What would change in our predictions about electric charges if we had chosen the potential to be zero at $r = 10^{-10}$ m from the charge?
16. If we start with point charges, for each of which zero potential is at infinity, is it possible for a superposition of charges to have zero potential other than at infinity?
17. The potential of a configuration of point charges is zero at certain points. Does this mean that the force on a test charge is zero at these points?
18. Is it possible to arrange charges so that the potential is zero over a small but finite region?

PROBLEMS

25–1 Electric Potential Energy

1. (I) Two protons are separated by a nuclear diameter of 5×10^{-15} m. What is their mutual electrostatic energy?
2. (I) What is the electrical potential energy between the nucleus of a uranium atom (92 protons) and a single electron located 10^{-12} m from the nucleus?
3. (I) A charge of 7.0×10^{-7} C is fixed at the origin of a coordinate system. A charge of 3.0×10^{-6} C is placed on a raisin

of mass 0.30 g. The raisin is then brought from far away to a point 20 cm from the origin. What is the electric potential energy of the system?

4. (I) Suppose that the raisin of Problem 3 is released from rest from its position 20 cm from the origin. If no other forces act on the raisin, where will it move? What will its final kinetic energy be?

5. (I) A 3-μC charge is brought in from infinity and fixed at the origin of a coordinate system. (a) How much work is done? (b) A second charge, of 5 μC, is brought in from infinity and placed 10 cm away from the first charge. How much work does the electric field of the first charge do when the second charge is brought in? (c) How much work does the external agent do to bring the second charge in if that charge moves with unchanging kinetic energy?

6. (I) Charges $q_1 = 6.0 \times 10^{-5}$ C and $q_2 = -4.0 \times 10^{-4}$ C are placed at rest 0.50 m apart. How much work must be done by an outside agent to move these charges slowly and steadily until they are 0.40 m apart?

7. (II) A positive charge of magnitude 3.0×10^{-6} C is placed 5.0 cm above the origin of a coordinate system, and a negative charge of the same magnitude is placed 5.0 cm below the origin, both on the z-axis. What is the potential energy of a positive charge of magnitude 0.20×10^{-6} C placed at the position $(x, y, z) =$ (30 cm, 0 cm, 50 cm)? at (30 cm, 0 cm, 0 cm)?

8. (II) Repeat the calculation of Problem 7 for the case that (a) both charges on the z-axis are positive and the third charge is negative; (b) the signs and magnitudes of all charges are the same.

9. (II) A charge of 1.5 μC is placed at the point $x = 12$, $y = 25$, $z = 0$ (all distances given in centimeters). Calculate the work done in bringing a charge of –3 μC from $x = 12$, $y = 60$, $z = 50$ to the point $x = 12$, $y = 50$, $z = -25$, assuming that the charge is moved at a steady speed.

10. (II) Use potential energy arguments to show that charges of the same sign cannot form a system with a closed circular orbit.

25–2 Electric Potential

11. (I) Two equal charges of −4 mC are placed along the y-axis at −3 mm and 4 mm, respectively. Where is the electric potential zero?

12. (I) Two charges are placed along the x-axis; 5 μC at 10 cm, and −3 μC at 25 cm. Find those points along the x-axis where the potential is zero.

13. (I) The electrostatic potential of a single unknown charge 2.5 mm from that charge is 0.12 V. (The value of the potential at infinity is zero.) What is the value of the charge?

14. (I) A proton moves from point A to point B under the sole influence of an electric field, losing speed as it does so from $v_A = 3 \times 10^4$ m/s to $v_B = 3 \times 10^3$ m/s. What is the potential difference between the two points?

15. (I) An external force steadily moves a point charge of $+3 \times 10^{-7}$ C from a negatively charged to a positively charged plate. The plates are large and parallel, and the negatively charged plate is at a potential of +3 kV, whereas the positively charged plate is at a potential of +17 kV. How much work does the external force do?

16. (I) Consider two very long coaxial cylinders that carry opposite charges. The interior cylinder, negatively charged, is at a potential of +3 kV, whereas the exterior cylinder, positively charged, is at a potential of +17 kV. An external force steadily moves a point charge of 3×10^{-7} C from the negatively to the positively charged cylinder (Fig. 25–35). How much work does the external force do?

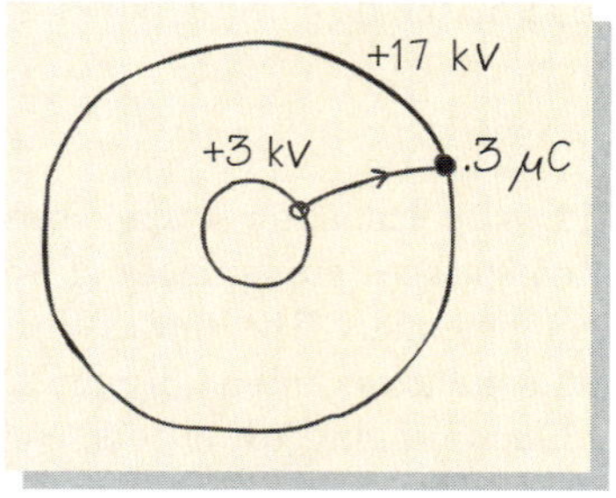

FIGURE 25–35 Problem 16.

17. (II) Three charges are at rest on the z-axis—$q_1 = 2$ mC at $z = 0$ m, $q_2 = 0.5$ mC at $z = 1$ m, and $q_3 = -1.5$ mC at $z = -0.5$ m. What is the potential energy of this system?

18. (II) Charges $+q$, $-q$, $+q$, and $-q$ are placed on successive corners of a square in the xy-plane. Plot all the locations in the xy-plane where the potential is zero.

19. (II) Consider two charges of 24×10^{-2} μC and -10×10^{-2} μC, respectively, at opposite ends of a diameter of a circle of radius 25 cm (Fig. 25–36). (a) What is the potential on a point of the circle that is 30 cm from the positive charge? (b) How much work is required to bring a charge of −0.2 μC from infinity to that point on the circle?

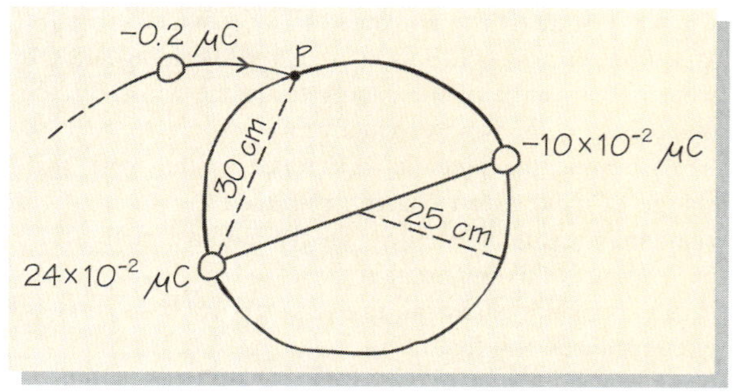

FIGURE 25–36 Problem 19.

20. (II) The origin of a coordinate system is at the intersection point of the perpendicular bisectors of the sides of an equilateral triangle of sides 10 cm. Calculate the potential at the origin due to three identical charges of 0.8 μC placed at the corners of the triangle.

21. (II) Consider a square of sides 14 cm. Charges are placed on the corners of the square as follows: 2 μC at (0 cm, 0 cm); −3 μC at (0 cm, 14 cm); 5 μC at (14 cm, 14 cm); +3 μC at (14 cm, 0 cm). What is the potential at the point (30 cm, 30 cm)?

22. (II) A 5-μC charge is fixed at $(x, y) = (15\text{ mm}, 20\text{ mm})$, a -3-μC charge is fixed at (15 mm, 30 mm), and a -2-μC charge is fixed at (25 mm, 20 mm) (Fig. 25–37). What is the potential energy of the system? Does the order in which the charges are brought in from infinity matter?

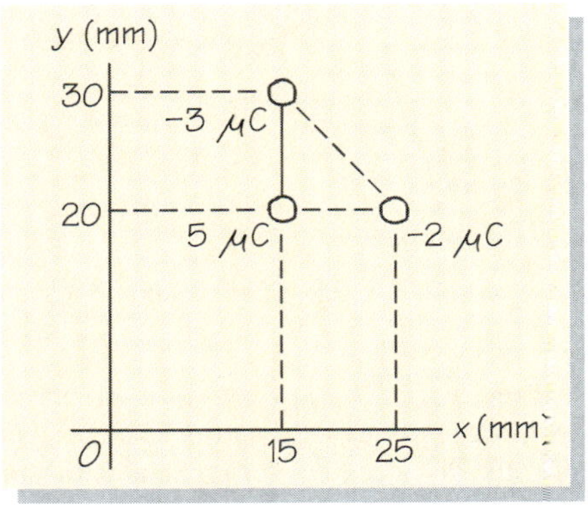

FIGURE 25–37 Problem 22.

23. (II) Calculate the electric potential at the origin due to the three charges considered in Problem 22.

24. (II) Two parallel conducting plates are brought to a potential difference of 3000 V, and a small pellet of mass 2 mg carrying a charge of 10^{-7} C accelerates from rest at one plate. With what speed will it reach the other plate?

25. (II) Figure 25–38 shows the cross section of a very large insulating slab that is uniformly charged to a charge density of 10^{-5} C/m^3. The thickness of the slab is 2 cm. (a) Determine the electric field of the charge on the slab at points *A*, *B*, and *C*. (b) Calculate the potential at points *B* and *C*, assuming that it is zero at *A*. (c) Plot the electric field and the potential as a function of distance from the center of the slab.

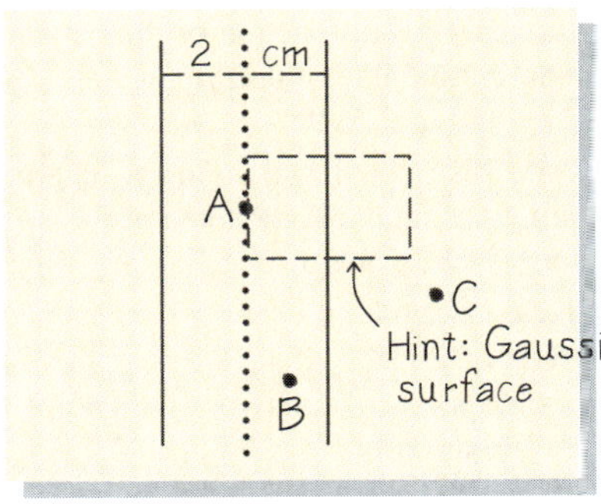

FIGURE 25–38 Problem 25.

26. (II) A charge Q is distributed uniformly over the surface of a spherical shell of radius R. How much work is required to move these charges to a shell with half the radius? The charges are again distributed uniformly.

27. (III) Calculate the potential inside and outside a sphere of radius R and charge Q, in which the charge is distributed uniformly throughout. [*Hint*: The additive constant for the potential inside the charged sphere must be chosen so that the two potentials, inside and outside, agree at $r = R$.]

25–3 Equipotentials

28. (I) Draw the equipotential surfaces for (a) a thin disk charged uniformly over its area and (b) a charged ring.

29. (I) Draw four equipotential surfaces for the charges shown in Fig. 25–39.

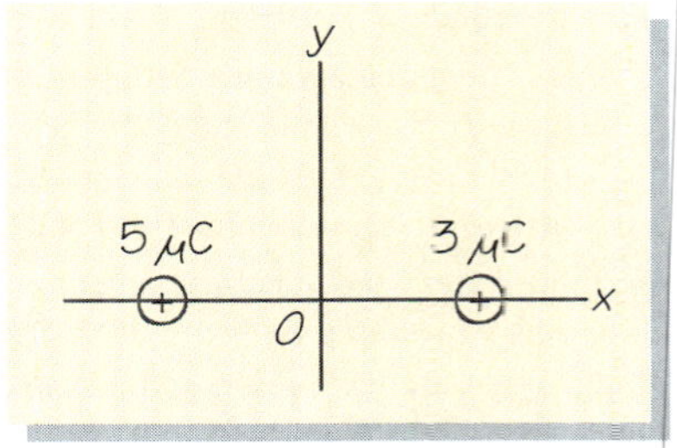

FIGURE 25–39 Problem 29.

30. (I) Sketch the equipotential surfaces for the charges shown in Fig. 25–40. Assume that the rod is an insulator.

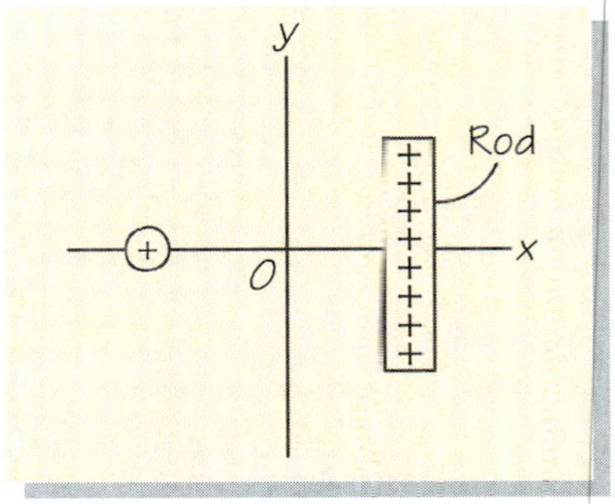

FIGURE 25–40 Problem 30.

31. (I) Sketch the electric fields and the equipotentials for the charge distribution shown in Fig. 25–41. Assume that the rod (of infinite length) is an insulator.

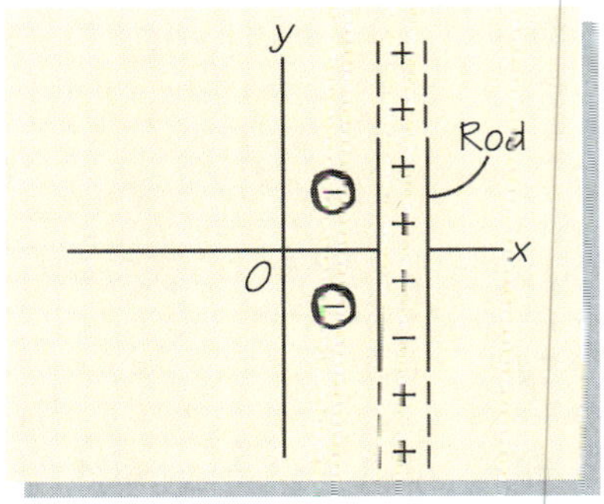

FIGURE 25–41 Problem 31.

32. (I) Three metallic spheres are placed on the corners of an equilateral triangle. The radii of the spheres are 1/3 of the side of the triangle. All the spheres are at the same potential V_0. Sketch the equipotentials for this system.

33. (II) A uniformly charged metal rod is placed parallel to an infinite, uncharged metal plate. Sketch the equipotentials in a plane perpendicular to the plate and to the rod, and in a plane perpendicular to the plate but parallel to the rod.

34. (II) Sketch the equipotentials in the xy-plane due to an infinite number of identical point charges q that lie on a line and are separated by a distance a, so that the coordinates of the point charges are $x_n = na$ and $y_n = 0$, where $n = 0, \pm 1, \pm 2, \pm 3, \ldots$.

35. (II) Two charges of equal magnitude but opposite sign are separated by a distance L. Sketch the equipotentials. What equipotential surfaces will have a potential of zero when the separate potentials for the two charges are chosen to be zero at infinity?

36. (II) Two infinite plates, each charged uniformly with charge density σ, are placed at right angles to each other and are almost touching. What are the equipotential surfaces? What are the equipotential surfaces if one of the plates has charge density $-\sigma$?

25–4 Determining Fields from Potentials

37. (I) The electric potential of a charge distribution within some region of space is $V(x, y, z) = Q/4\pi\epsilon_0 x$. Find the electric field in this region.

38. (I) Find the electric field of a charge distribution if the electric potential of the distribution is $V = Ax^2y^2 + Byz^2 + C$, where A, B, and C are constants.

39. (I) In a certain region of space, the electric potential due to a charge distribution varies only with x, changing according to $V = a_0 + a_1 x$ where $a_0 = 12.7$ V, $a_1 = -6.68$ V/m, and x is in meters. Find the electric field, magnitude, and direction in this region.

40. (II) Starting from the solution in Example 25–8 of the potential due to a uniformly charged ring, use the derivative operations in Eq. (25–29) to find the electric field along the axis of the ring.

41. (II) Find the electric field far away along the bisecting axis of an electric dipole from the potential given in Eq. (25–24).

42. (II) Consider charge distributed in an infinitely long cylinder of radius R whose axis forms the z-axis. The charge distribution depends only on the distance r from the z-axis. The potential is given for $r < R$ by $V(r) = (Q/2\pi\epsilon_0)\,[A(r/R) + B(r/R)^2 + C]$, where A, B, and C are constants. What is the electric field within the rod? What is the value of C if the potential is defined to be zero on the cylinder's surface?

43. (II) The potential $V(r)$ of a spherically symmetric charge distribution is given by $V(r) = (Q/4\pi\epsilon_0 R)\,[-2 + 3(r/R)^2]$ for $r < R$ and $V(r) = Q/4\pi\epsilon_0 r$ for $r > R$ (Fig. 25–42). Calculate the electric field.

44. (II) Use the results of Problem 43 and Gauss' law applied to Gaussian surfaces at various radii to calculate the charge distribution that gives rise to the potential given in that problem.

45. (III) The potential in the xy-plane due to a certain charge distribution is given by

$$V(x, y) = \frac{Q}{4\pi\epsilon_0 L} \times \left[\arctan\left(\frac{y}{x - a_0}\right) - 2\arctan\left(\frac{y}{x}\right) + \arctan\left(\frac{y}{x + a_0}\right)\right],$$

where L and a_0 are constant lengths. Show that the electric field at distances $x \gg a_0$, $y \gg a_0$ is proportional to $a_0{}^2$, and find its dependence on x and y. Express your answer in terms of r, the distance to the origin, and θ, the angle that the line from the origin to the point (x, y) makes with the x-axis.

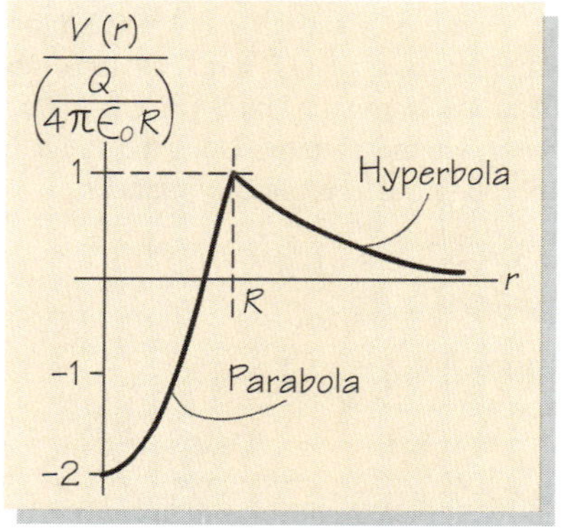

FIGURE 25–42 Problem 43.

25–5 The Potentials of Charge Distributions

46. (I) Two large, metal, parallel plates have a potential difference of 8 kV, and the electric field between them has magnitude 3×10^5 V/m. What is the separation distance between the plates?

47. (I) The voltage along the axis of a uniformly charged ring of radius 10 cm is 5 V at a point 15 cm from the center of the ring. How much charge is on the ring?

48. (I) In fair weather, there is a constant electric field near Earth's surface whose magnitude is roughly 100 V/m, directed downward. (a) Find the potential associated with this field. (b) What is the most convenient point to choose for zero potential? (c) How does the potential energy of a test charge near Earth compare in form with the potential energy of gravity? (d) How much negative charge would have to be placed on a person of mass 50 kg to have the electric force balance the force of gravity?

49. (II) Find the potential as a function of the perpendicular distance R from an infinite line of uniform charge density by using Gauss' law and Eq. (25–9).

50. (II) Charges are distributed with uniform charge density λ along a semicircle of radius R, centered at the origin of a coordinate system. What is the potential at the origin?

51. (II) A rod that is 20 cm long is given a uniformly distributed charge of 2 μC (Fig. 25–43). Calculate the potential at a point P, which is a distance of 10 cm from the end of the rod, assuming that $V = 0$ at infinity.

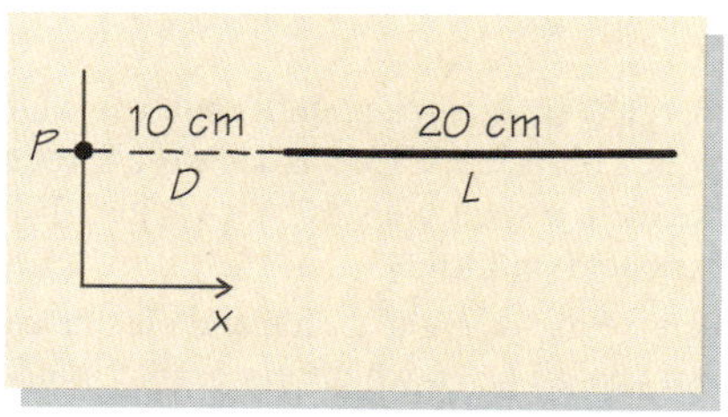

FIGURE 25–43 Problem 51.

52. (III) Find an expression for the electric potential at all points due to a rod of length L and uniform charge density λ, using Eq. (25–12). The rod is oriented along the z-axis, with its center at the origin (Fig. 25–44). Show that at distances much greater than L from the rod, the potential reduces to that of a point charge $Q = \lambda L$ at the origin.

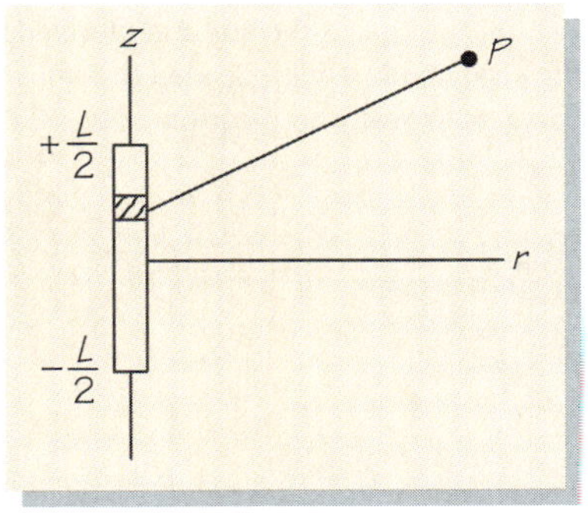

FIGURE 25–44 Problem 52.

53. (III) A charge $3q_0$ is placed on the x-axis at the point $x = x_0$ (where x_0 is positive), and a second charge, $-q_0$, is placed on the x-axis at the point $x = -x_0/2$. (a) What is the potential on the x-axis of this distribution of charges? Assume that zero potential for a point charge is at infinity. (b) Show that your result for part (a) can be approximated for large x by a term proportional to $1/x$, plus a term proportional to $1/x^2$, plus higher powers of $1/x$. (c) Show that the expansion of part (b) is that of a point charge at the origin, plus an electric dipole oriented along the x-axis and centered at the origin, plus other terms. Find the strength of the point charge as well as the dipole moment of the electric dipole. (d) How large must x be so that the approximation of a point charge plus a dipole comes within 1 percent of the exact answer? [*Hint*: Use the approximation $(1 + z)^k = 1 + kz + \frac{1}{2}k(k - 1)z^2 + \cdots$ for $z \leq 1$.]

25–6 *Potentials and Fields Near Conductors*

54. (I) A thin disk of radius 18 cm carries a total charge of 2.5×10^{-6} C spread evenly over its surface. What is the minimum work required to bring a charge $q = 0.2 \times 10^{-6}$ C at rest from infinity to a distance of 58 cm from the disk along its axis?

55. (I) A thin ring of radius 24 cm carries a uniformly distributed charge of 3.5×10^{-7} C. A negative charge $q = -8.5 \times 10^{-8}$ C is placed on the axis of the ring 28 cm from the plane of the ring (Fig. 25–45). How much work must an external agent do to move the charge slowly and steadily to a distance 85 cm away, also on the axis?

56. (I) An electric field of 2.8×10^6 V/m is sufficiently large to cause sparking in air. Find the highest potential to which a spherical conductor of radius 3 cm can be raised before breakdown occurs in the air surrounding it. Assume that zero potential is taken at infinity.

57. (I) Consider two charged metallic spheres. The spheres have radii r_1 and r_2, and carry charge q_1 and q_2, respectively. What is the amount of charge that flows through a wire that is brought in and connected to the two spheres?

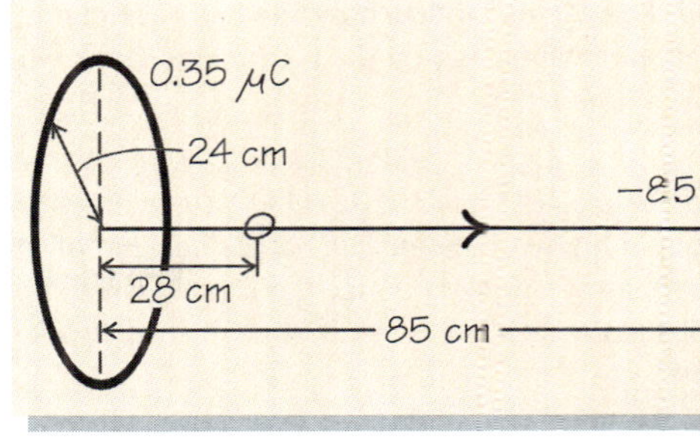

FIGURE 25–45 Problem 55.

58. (II) The same charges are placed on two identical drops of mercury. The drops are isolated and take perfectly spherical shapes, and the electric potential at the surface of each drop is 900 V. The drops coalesce into a larger drop with a net charge double that of either smaller charge. What is the potential at the surface of this larger charge?

59. (II) Two conducting spheres of different sizes are connected by a thin conducting wire. The radius of the larger sphere is three times that of the smaller sphere. If a total charge Q is placed on this apparatus, what fraction of Q sits on each sphere?

60. (II) Concentric metal shells, perfect conductors, have radii R and $1.5R$, respectively. A charge q is placed on the inner shell, and a charge $-3q$ is placed on the outer shell. (a) What are the electric fields in all space due to the two shells? (b) What is the potential difference between the two shells? (c) If a thin, perfectly conducting wire now joins the two shells, how does the charge redistribute itself?

61. (II) Two spherical conductors of radii 20 mm and 100 mm are connected by a thin wire and carry charges q_1 and q_2, respectively. If the wire is cut and the centers of the spheres are 250 mm apart, there is a repulsive force of 3.5 N between them. Use this information to calculate (a) q_1 and q_2 and (b) the electric fields at the surfaces of the conductors when they are connected by the wire.

62. (II) A balloon of radius 30 cm is sprayed with a metallic coating so that the surface is conducting. A charge of 3×10^{-6} C is placed on the surface. (a) What is the potential on the balloon's surface? (b) Suppose that some air is let out of the balloon, so that its radius shrinks to 21 cm. What is the new potential on the balloon's surface? (c) What happens to the energy associated with the change in potential energy?

25–7 *Electric Potentials in Technology*

63. (I) A proton is accelerated from rest in a Van de Graaff accelerator through a potential of 5.5×10^6 V. (a) What energy does the proton have—in electron-volts and joules? (b) What is the proton's final speed?

64. (I) A small Van de Graaff generator is used to demonstrate the effects of high potential. The device has a radius of 11 cm and stands in air. What is its maximum potential, and how much charge does the dome hold?

65. (II) Early Van de Graaff accelerators were built to operate in air without high-pressure gases. (a) How much voltage could an accelerator with a domed surface of radius 1.3 m produce, if air breaks down at 3×10^6 V/m? (b) How much kinetic energy could the protons produced by such an accelerator have? (c) What is the total charge on the accelerator dome when the maximum field is attained?

General Problems

66. (I) We have a high-voltage power supply capable of producing 5000 V, and we want to ionize the air molecules between parallel plates. What plate separation will give us electrical breakdown?

67. (I) The potential at a point x due to a thin, flat, uniformly charged disk of radius R and total charge Q at a point on the disk's axis a distance x away from the disk is given in Eq. (25–32). Consider two such identical disks, one in the yz-plane at $x = -a$, and another in the yz-plane at $x = +a$, both centered on the x-axis (Fig. 25–46). What is the potential at an arbitrary point on the x-axis between the two disks?

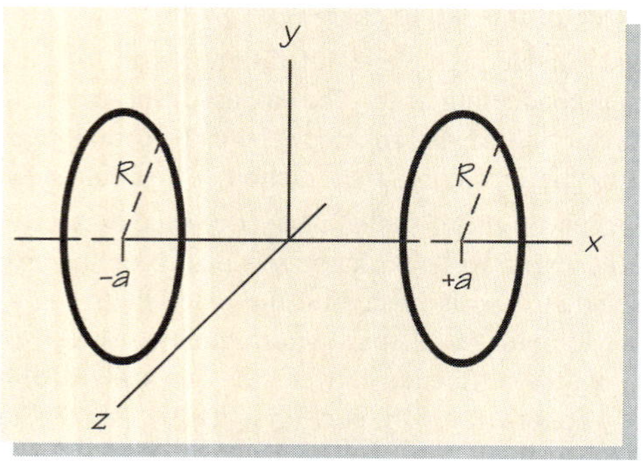

FIGURE 25–46 Problem 67.

68. (I) A metallic ring of radius R, carrying charge Q, has an associated electric potential at a distance x from the center of the ring along its axis given in Table 25–1. What is the potential at some point x due to two rings, one carrying charge Q and the other $-Q$, both in the yz-plane, centered on the x-axis, one located at $x = a$, the other at $x = -a$?

69. (II) What is the electric field at a point x on the axis due to the two charged rings described in the previous problem for $x >> a$? [*Hint*: Use the approximation $(x^2 + 2ax + b^2)^{-1/2} \simeq x^{-1}(1 - a/x + (3a^2 - b^2)/2x^2)$.]

70. (II) Write an expression for the total energy of two point charges—one positive and of magnitude Q, fixed at the origin; the other negative and of magnitude q and mass m, located at a point a distance r from the origin. Suppose the charge q, instead of being stationary, moves in a circular orbit of radius r around the charge Q. Assuming that the Coulomb attraction is responsible for the centripetal acceleration, calculate the energy. Why is the angular velocity of this motion constant?

71. (II) A nonconducting sphere of radius R carries a charge $+Q$ distributed uniformly throughout its volume. What is the potential energy of a point charge $-q$ a distance r ($r < R$) from the center of the sphere? Show that if there is a hole drilled through the sphere so that the point charge can move through it, then the point charge oscillates as though it were attached to a spring. Find the spring constant.

72. (II) Three electrons are located along the x-axis at positions –2 nm, 0 nm, and 2 nm, respectively. How much energy was required to move each of the electrons in turn from infinity? Does the order in which they were moved matter?

73. (II) A salt crystal consists of an array of positive Na and negative Cl ions, both carrying an elementary charge of magnitude e. Assume that a small "seed" crystal consists of four ions, forming a square of side 0.25 nm (Fig. 25–47). Find the electric force acting on one of the sodium atoms due to the other atoms of the seed and the work needed to remove this ion from the seed. Give your result in electron-volts.

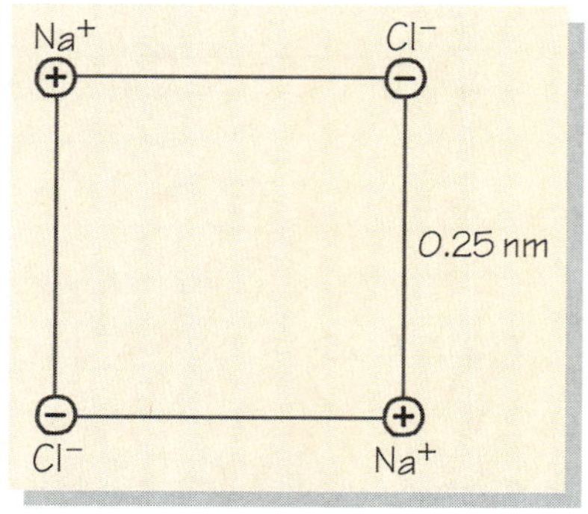

FIGURE 25–47 Problem 73.

74. (II) Calculate the potential at the point $P(x, y)$ due to the dipole in Fig. 25–48, which consists of a charge $+q$ placed at $(0, a)$ and a charge $-q$ placed at $(0, -a)$, and use this potential to calculate the electric field at point P.

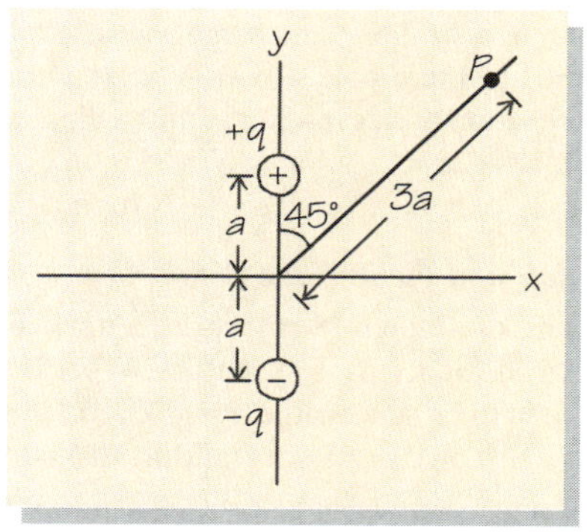

FIGURE 25–48 Problem 74.

75. (II) Find the electric field along the axis of a uniformly charged ring of radius R and total charge Q by taking the appropriate derivatives of the potential found in Example 25–8. Set up the problem by using the direct integration techniques presented in Chapter 23. Compare the difficulties of the two ways of calculating the electric field.

76. (II) A positron (charge $+e$ and same mass as that of the electron) approaches a proton (charge $+e$) head-on. As a result of the repulsion, the positron turns around a distance $r_0 = 1.0 \times 10^{-10}$ m from the proton. What is the kinetic energy of the positron when it is very far from the proton? You may assume that the proton motion can be neglected.

77. (II) An electron is moving in the field of a helium nucleus (atomic number $Z = 2$). What is the change in the electron's potential energy when it moves from a circular orbit of radius 3×10^{-10} m to one of radius 2×10^{-10} m? What is the change in kinetic energy? in the total energy of the electron? Energy conservation is not violated in this process because it can be carried away by radiation that is emitted during the change of orbits.

78. (II) An electric dipole fixed in space consists of a charge $+q$ at the point $x = -0.2$ m and a charge $-q$ at the point $x = +0.2$ m, where $q = 5\ \mu$C. A test charge $q_0 = 3\ \mu$C is steadily moved from the point $x = +0.6$ m to the point $x = -0.4$ m by following a semicircular path of radius 0.5 m that takes the test charge through the y-axis (Fig. 25–49). How much work is required to move the test charge?

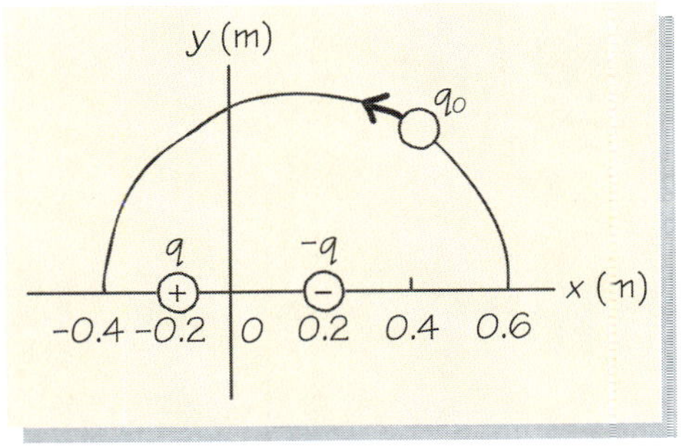

FIGURE 25–49 Problem 78.

79. (II) Two identical cork balls of charge 2.0 μC are suspended from the same point by thin threads 0.80 m long. (a) Calculate the mass of the cork balls if the threads each make a 30° angle with the vertical. (b) Calculate the potential energy of the system of two balls due to the presence of charges and to the presence of gravity as a function of the angle θ the threads make with the vertical. Choose zero gravitational potential energy to correspond to $\theta = 0$.

80. (II) A large, square plane with sides of length L, parallel to the yz-plane and located at x_1, has charge density σ_1. A similar plane, located at x_2, has charge density σ_2. How much work must be done to bring the second plane to within a distance a of the first one? Neglect end effects; that is, calculate the fields as though the planes were infinite.

81. (III) An infinitely long cylinder of radius R is filled with uniform charge density ρ. Calculate the potential inside and outside the cylinder.

82. (III) The inner radius of a spherical dielectric shell is 6 cm, and the outer radius is 15 cm. The shell carries a charge of 0.5×10^{-6} C, distributed uniformly. Sketch the shape of the potential for all values of r, the distance from the center of the shell, and evaluate it at the center and at the inner and outer radii

83. (III) A solid sphere of radius R has uniform charge density ρ. Calculate the total potential energy by calculating the energy required to bring a spherical shell of thickness dr and charge density ρ from infinity to a distance r from the sphere's center in the potential due to a uniformly charged sphere of radius r.

CHAPTER

26

The interior of the target chamber of the NOVA laser at Lawrence Livermore National Laboratory. This project aims to produce controlled nuclear fusion by means of depositing large amounts of energy at the target; the energy can be delivered quickly because it is stored in large capacitor banks.

Capacitors and Dielectrics

In electrostatics, any charged conducting object is characterized by an electric potential that is constant everywhere on and within that object. The potential difference of two charged conductors can lead to an acceleration of test charges, and thus the system stores energy. A capacitor is a device of this type; it stores energy because it stores charge. The relation between the amount of charge a capacitor stores and the potential difference of its components depends on the geometry of the capacitor. The relation between stored charge and potential difference is also affected by the insulating (nonconducting) material—called dielectric material—placed between the charged components of the capacitor. In this chapter, we study the role capacitors play in circuits, as well as how geometry and dielectric materials affect the properties of capacitors. We also consider the microscopic structure of dielectrics and thereby extend our fundamental knowledge of the behavior of matter.

26–1 CAPACITANCE

The definition and properties of a capacitor

A pair of conductors, whether separated by empty space or by a nonconducting material, forms a **capacitor**. Capacitors store separated charges. In their most common and useful form, capacitors are made of two conductors with equal but opposite charges Q. There is a potential difference V between the conductors.[†] We showed in Chapter 25 that this potential difference is linearly dependent on the charge; that is, $V \propto Q$. Thus, if we double the charge, we double the potential difference between the two conduc-

[†]We will sometimes use the terms "potential," "voltage drop," or "voltage" rather than "potential difference."

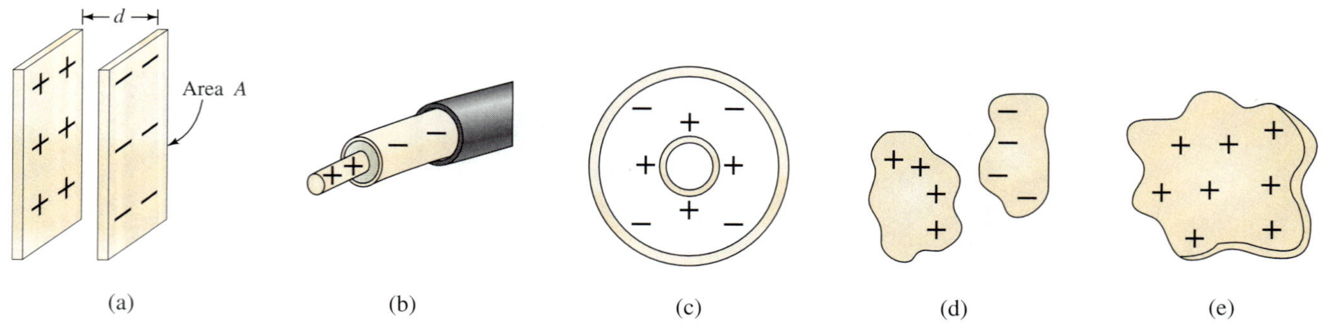

FIGURE 26–1 Various kinds of capacitors: (a) parallel-plate; (b) coaxial cable; (c) spherical (two hollow conducting spheres); (d) conductors of arbitrary size; (e) isolated conductor infinitely far from second conductor.

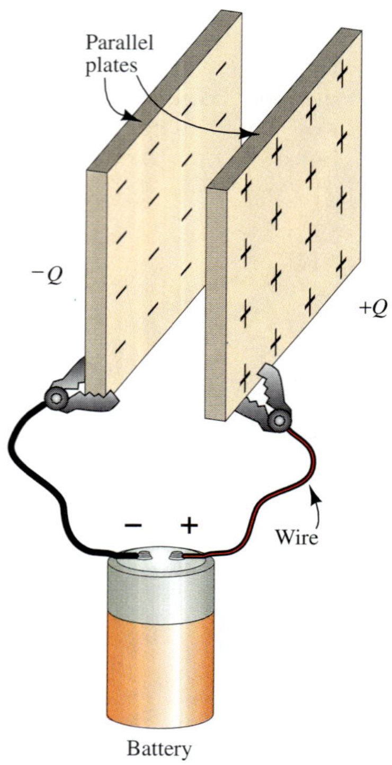

FIGURE 26–2 We can place charge $+Q$ on one conductor and charge $-Q$ on another conductor by using a battery.

tors. In other words, the ratio of Q to V between two conductors is constant. The constant ratio Q/V depends on the shape and arrangements of the two conductors of a capacitor—that is, on geometry—and on the material between the conductors. Figures 26–1a to 26–1e illustrate different configurations of conductors that can act as capacitors. In some cases, a capacitor is formed when one conductor with charge Q induces a corresponding charge $-Q$ on an adjacent conductor. The second conductor may be very far away, even at infinity (Fig. 26–1e).

Capacitors are important for several reasons. Different forms of capacitors can hold different amounts of charge for a given potential difference or can maintain different potential differences for a given amount of charge. With the appropriate capacitor, we can control the storage and delivery of charge. We can similarly use capacitors to control potential differences. Almost any device with an electronic circuit contains capacitors. Because they involve a potential difference, capacitors store energy as well as charge. A lightning strike is the spectacular discharge of a large capacitor formed by the system of a cloud and Earth. Capacitors are particularly useful for short-term storage of charge and energy. A camera photoflash contains a capacitor that stores energy and then discharges it when the flash is fired. Another important use of capacitors is the slow but smooth delivery of energy when capacitors are coupled with other circuit elements. Emergency backup systems for computers use capacitors in this way.

Let's place equal but opposite charges, $+Q$ and $-Q$, on two conductors that form a capacitor. This is easily done by touching the two conductors with wire leads from the $+$ and $-$ terminals of a battery (Fig. 26–2). The amount of charge accumulated depends on the shape of the conductors and on their relative positions. We call Q the *charge on the capacitor* even though the *net* charge on the oppositely charged pair of elements is zero.

We have argued that the charge on a capacitor is proportional to the potential difference. The proportionality constant is called the **capacitance**, C, determined by the relation

$$Q = CV. \tag{26–1}$$

In other words, the capacitance of the capacitor is defined as the ratio of the potential difference, V, that results when charges $\pm Q$ are placed on the two conductors:

The definition of capacitance

$$C \equiv \frac{Q}{V}. \tag{26–2}$$

When there is charge on a capacitor, we say it is *charged*; when a capacitor *discharges*, as when a flashbulb fires, it can deliver its stored energy rapidly.

The SI unit of capacitance

C is always taken to be positive; that is, Eq. (26–2) should contain absolute values. The unit of capacitance is coulombs per volt (C/V), but capacitance occurs so frequently that it has been given its own SI unit, the **farad** (F), in honor of Michael Faraday:

$$1\ \text{F} = \frac{1\ \text{C}}{1\ \text{V}}. \tag{26–3}$$

In practice, the farad is inconveniently large and, for practical use, units of μF, nF, and pF (often called "puffs") are much more common.

Calculating Capacitance

The capacitance of a capacitor can be calculated easily if the geometry is simple enough. The most basic capacitor consists of two parallel conducting plates of area A, separated by a distance d, with charges $+Q$ and $-Q$ respectively, distributed uniformly over the plates (Fig. 26–3). If the dimensions of the plates are large compared with d, then the electric field between the plates is to a very good approximation constant. Neglecting (small) edge effects, we found previously that $E = \sigma/\epsilon_0$, where σ is the charge density Q/A, ϵ_0 is the permittivity of free space, and the potential difference between the plates is $V = Ed$. We combine these results to find that

$$V = Ed = \frac{\sigma}{\epsilon_0} d = \frac{Q}{A}\frac{d}{\epsilon_0} = Q\frac{d}{\epsilon_0 A}.$$

Thus $Q/V = \epsilon_0 A/d$, and the capacitance $C \equiv Q/V$ of a parallel-plate capacitor is

$$C = \frac{Q}{V} = \frac{\epsilon_0 A}{d}. \tag{26–4}$$

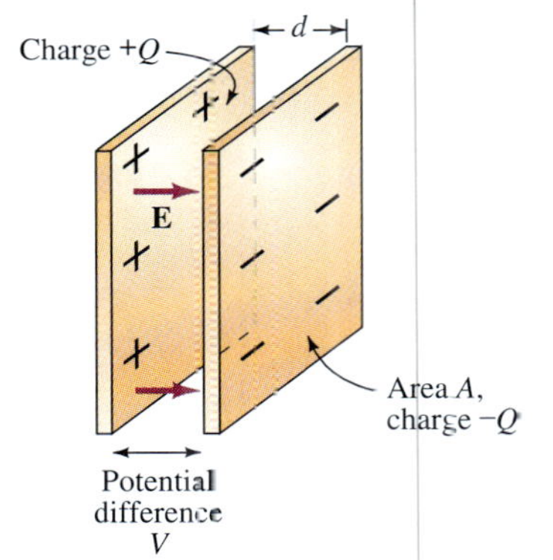

FIGURE 26–3 Two parallel plates with equal and opposite charges make up the most basic capacitor.

The fields near the edge are called *fringe fields* (Fig. 26–4). Their effect is small if the linear dimensions of the plates are much larger than the separation. Fringe fields do not affect the linear relationship between the charge and the potential, but do modify the simple form of Eq. (26–4).

Equation (26–4) gives us a second commonly used unit for the permittivity of free space, ϵ_0, namely farads per meter (F/m):

$$\epsilon_0 = 8.85 \times 10^{-12}\ \mathrm{C^2/N \cdot m^2} = 8.85 \times 10^{-12}\ \mathrm{F/m} = 8.85\ \mathrm{pF/m}.$$

Either unit is consistent with the SI.

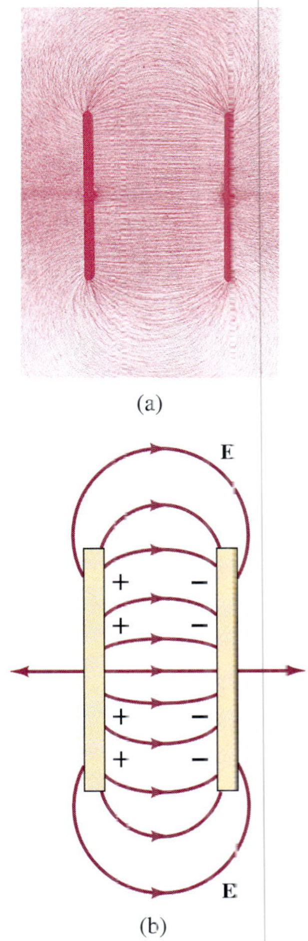

FIGURE 26–4 (a) The electric field lines due to a charged parallel-plate capacitor, shown by threads in oil. (b) The electric field between the two conducting plates of a parallel-plate capacitor.

EXAMPLE 26–1 (a) Calculate the capacitance C of parallel plates of area $A = 100\ \mathrm{cm^2}$ separated by a distance $d = 1$ cm. (b) Find the area of a parallel-plate capacitor with plate separation of 1 cm and a capacitance of 1 F.

Solution: We know both A (100 cm^2) and d (1 cm), so we can use Eq. (26–4) to determine the unknown capacitance. We find that

$$C = \frac{\epsilon_0 A}{d} = \frac{(8.85\ \mathrm{pF/m})(10^{-2}\ \mathrm{m^2})}{10^{-2}\ \mathrm{m}} = 8.85\ \mathrm{pF}.$$

The plate area is rather large, yet the capacitance is only 8.85 pF.

(b) In this case, we find A given C (1 F) and d (1 cm). From Eq. (26–4), we find

$$A = \frac{dC}{\epsilon_0} = \frac{(1.0 \times 10^{-2}\ \mathrm{m})(1.0\ \mathrm{F})}{8.85 \times 10^{-12}\ \mathrm{F/m}} = 0.11 \times 10^{10}\ \mathrm{m^2}.$$

This represents a square with sides of length $0.33 \times 10^5\ \mathrm{m} = 33$ km! Most common practical capacitors have capacitances much smaller than 1 F.

EXAMPLE 26–2 *The Coaxial Cable.* A *coaxial cable* transmits information, as in cable television systems (Fig. 26–5a). It is made of a solid (or stranded) cylindrical conducting wire of radius a surrounded by a coaxial conducting sheath of radius b (Fig. 26–5b). Find the capacitance per unit length of a coaxial cable, assuming that there is a vacuum between the central wire and the sheath.

Solution: We want to use the geometry shown in Fig. 26–5b and find the potential difference for a given charge. The ratio of these quantities determines the ca-

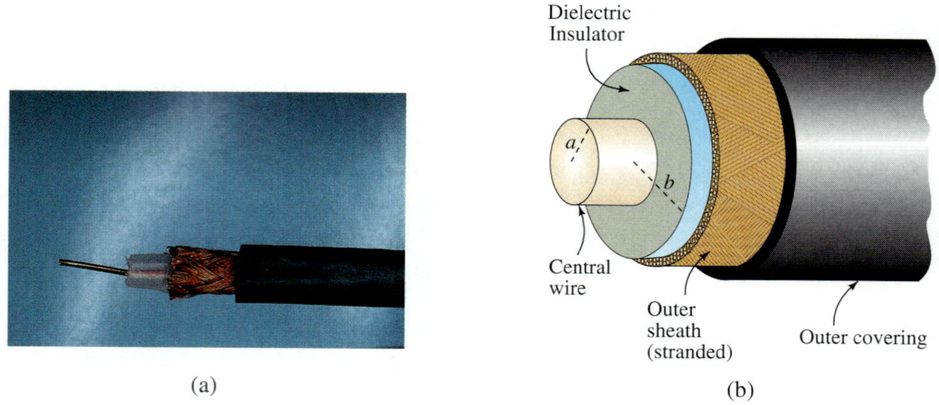

FIGURE 26–5 Example 26–2. (a) A coaxial cable. (b) Schematic diagram of a coaxial cable with an inner solid wire of radius a and an outer sheath of radius b. An insulator is placed between the two conductors.

pacitance. For an infinitely long coaxial cable, however, we can specify only the charge *per unit length*, so we must similarly calculate the capacitance per unit length.

To find the capacitance per unit length, we set a charge per unit length $+\lambda$ on the outer (and $-\lambda$ on the inner) conductor of the cable and calculate the resulting potential between the conductors. This calculation was already done in Example 25–10. There, we saw how Gauss' law gives the electric field and hence the potential outside a wire of finite radius. That example applies directly to the potential in the space *between* the conductors of the coaxial cable. From Example 25–10, we have

$$V_b - V_a = \frac{\lambda}{2\pi\epsilon_0} \ln \frac{b}{a}.$$

Dividing the charge per unit length λ by the potential difference $V = V_b - V_a$ gives the capacitance per unit length:

The capacitance of a coaxial cable

$$\frac{C}{\text{length}} = \frac{\lambda}{V_b - V_a} = \frac{2\pi\epsilon_0}{\ln(b/a)}. \qquad (26\text{–}5)$$

Equivalently, the capacitance of a length L of the cable is L times this value.

An isolated conductor can have a capacitance because, when such a conductor is charged, the charge must be brought in from infinity. In effect, the second conductor is at infinity. Example 26–3 illustrates how to calculate the capacitance of an isolated conductor.

EXAMPLE 26–3 *The Isolated Sphere.* What is the capacitance of an isolated conducting sphere of radius R? Calculate Earth's capacitance.

Solution: We find the potential that results from placing a charge Q on the sphere, and the ratio of these quantities is, by definition, the capacitance. In this case, the potential difference involved is just the potential of the sphere if zero potential is at infinity. We found in Example 25–11 that the potential for a conducting sphere is

$$V = \frac{Q}{4\pi\epsilon_0 R}.$$

The capacitance of the isolated sphere is Q divided by V:

$$\text{for an isolated sphere:} \qquad C = \frac{Q}{V} = 4\pi\epsilon_0 R. \qquad (26\text{–}6)$$

The capacitance of an isolated sphere

The dependence of the capacitance on R in Eq. (26–6) could have been obtained from dimensional analysis. The permittivity, ϵ_0, has units F/m and so must be multiplied by a length to give an acceptable capacitance. The length in this problem is the radius of the sphere.

Earth's capacitance is determined by setting R equal to Earth's radius, $R_E = 6.38 \times 10^6$ m. Therefore

$$C = 4\pi\epsilon_0 R = 4(3.14)(8.85 \times 10^{-12}\ \text{F/m})(6.38 \times 10^6\ \text{m}) = 7.10 \times 10^{-4}\ \text{F}.$$

The farad is indeed a large unit.

The technique for determining the capacitance is always the same. We assume that the conductors have a charge $\pm Q$, then we find the potential difference V between the conductors due to this charge. The ratio Q/V gives the capacitance.

PROBLEM SOLVING

A technique for determining capacitance

26–2 ENERGY IN CAPACITORS

There is an electric field between the two conductors of a charged capacitor, and this field can accelerate a test charge. Thus, a charged capacitor is capable of doing work and must contain energy. We can determine the energy contained in a charged capacitor by seeing how much work is required to charge it. Start by taking a positive charge dq from one neutral conductor and moving it to the other neutral conductor. The second conductor then has charge $+dq$, and the first conductor has charge $-dq$. Assume now that we have continued to move charge and that the conductors are already charged to a potential difference V with charge q (Fig. 26–6). V is given in terms of q by $V = q/C$. To move an additional charge dq from the negatively charged conductor to the positively charged one, we must do work

$$dW = V\,dq = \frac{q}{C}\,dq.$$

As represented in Fig. 26–7, the total work done when we start with zero charge and end up with charges $\pm Q$ is obtained by integrating the above expression from $q = 0$ to $q = Q$:

$$W = \int dW = \int_0^Q \frac{q}{C}\,dq = \frac{1}{C}\int_0^Q q\,dq = \frac{Q^2}{2C}. \qquad (26\text{–}7)$$

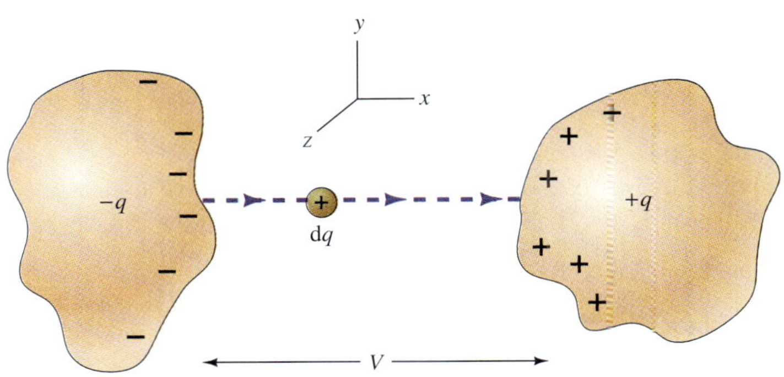

FIGURE 26–6 When an infinitesimal charge dq is moved between the two conductors of a capacitor charged to a potential difference V, work must be done.

Note that we have not restricted the derivation to parallel-plate capacitors: This is a general result for all capacitors.

Forms for the energy contained in a charged capacitor

The work done in charging a capacitor is stored in the capacitor as potential energy capable of doing work. It is this potential energy that is able to move a test charge placed between the conductors or to cause a flashbulb to flash (Fig. 26–8). The potential energy of a charged capacitor is

$$U = \frac{Q^2}{2C}. \tag{26–8}$$

Because $Q = CV$, this result is equivalent to

$$U = \frac{CV^2}{2}. \tag{26–9}$$

The first form is used when the charge is known; the second when the potential is known. Another equivalent form that is useful when the charge and voltage are known is

$$U = \frac{QV}{2}. \tag{26–10}$$

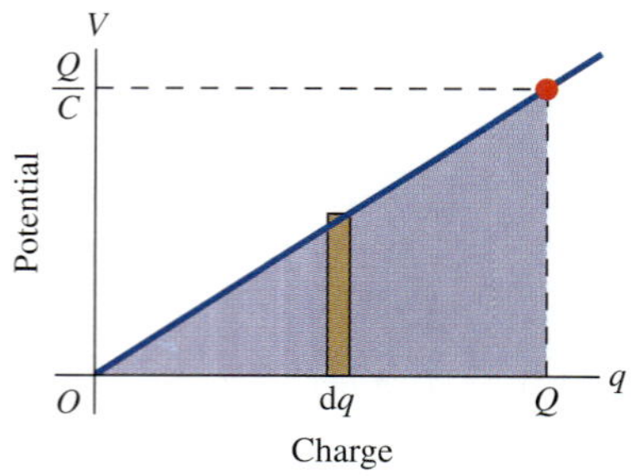

FIGURE 26–7 The amount of work $dW = V\,dq$ that is done to move a charge element dq from one capacitor plate to the other increases as the charge on the plates—and hence the potential V across them—builds up. The total work is found by adding the work elements together. This work is the area under a curve of potential difference versus charge on the plates: the integral of $V\,dq$.

As we know from Chapter 25, the electric potential energy associated with the movement of a charge Q through a fixed potential V is $U = QV$. This expression differs by a factor of 2 from the capacitor energy $QV/2$, Eq. (26–10). The reason for this difference is that, as a capacitor is charged, the potential is steadily increasing so, in effect, the average potential as the charging takes place is $V/2$.

Batteries versus Capacitors

A *battery* is a chemical device for the storage of energy. Whereas the potential of a capacitor decreases as the capacitor delivers its charge, a battery is able to maintain a fixed potential between two points (terminals) as it delivers charge. If we want to raise a capacitor to a certain potential, a battery is appropriate because it can hold the desired potential even as it delivers charge to the capacitor.

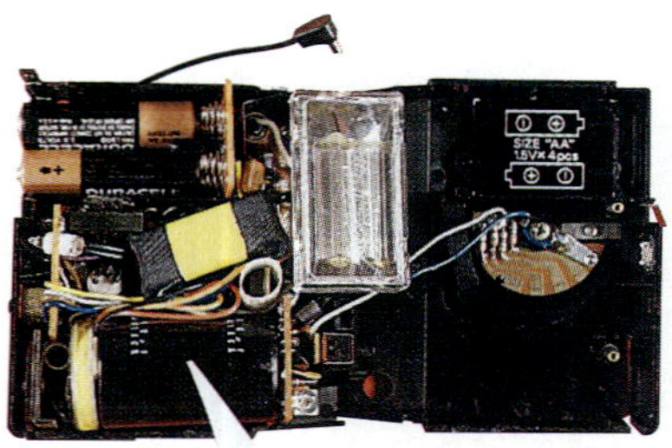

FIGURE 26–8 The energy stored in a capacitor is released in the forms of visible light and heat when the charge on the capacitor passes through the photoflash of a camera. Here we show a typical charging/discharging circuit for such a device.

EXAMPLE 26–4 A 12-V car battery is used to charge a 100-μF capacitor. (a) How much energy is stored in the capacitor? (b) Compare this energy with the energy stored in the battery itself, if the battery is capable of delivering a total charge of $Q = 3.6 \times 10^5$ C at the given voltage. (This is the charge that can be delivered by a battery rated at 100 ampere-hours, which is a standard unit for charge.)

Solution: (a) When a battery charges a capacitor, the voltage difference between the capacitor plates is the voltage rating of the battery. We therefore want to find the energy stored in the known capacitor when it is at 12 V, given by Eq. (26–9):

$$U = \frac{CV^2}{2} = \frac{(100 \times 10^{-6}\ \text{F})(12\ \text{V})^2}{2} = 7.2 \times 10^{-3}\ \text{J}.$$

The answer is in joules because we have used SI units throughout.

(b) The electric potential energy associated with the movement of a charge Q through a fixed potential V is $U = QV$. Thus the potential energy in the battery is

$$U = QV = (3.6 \times 10^5\ \text{C})(12\ \text{V}) = 4.3 \times 10^6\ \text{J}.$$

The battery contains a factor of 6×10^8 more energy than is stored in the capacitor!

We saw in Example 26–4 that a car battery has a potential energy of around 10^6 J—far more energy than that of a practical capacitor of 100 μF charged to a moderate potential. A potential of approximately 10^{10} V would be required to store the equivalent amount of energy in this capacitor. The largest available commercial capacitors have capacitance on the order of 1 F. Such capacitors, however, can be taken to a potential of only several volts, whereas it would require a potential of around 1000 V for a 1-F capacitor to contain as much energy as that contained in the car battery. A battery's energy is stored in chemical bonds rather than in the macroscopic separation of charge. Batteries provide a practical way to store large amounts of energy for long periods, but they are not a practical way to deliver the energy quickly. Conversely, we have noted that the possibility of quick energy delivery is one advantage of a capacitor.

26–3 ENERGY IN ELECTRIC FIELDS

The electric field is of fundamental physical significance. In this section, we develop the concept that the energy of a capacitor is contained in the electric field itself. Because the two conductors of a capacitor are charged, electric field lines point from the positively charged conductor to the negatively charged one. It is this electric field that causes the acceleration of a test charge placed between capacitor plates.

Let's relate the expression for the energy in a capacitor to the strength of the electric field in that capacitor. The parallel-plate capacitor is convenient for this purpose because both the capacitance and the field are known (Fig. 26–4). Equation (26–4) gives the capacitance for this case, $C = \epsilon_0 A/d$, where A is the area of the plates and d is their separation. The field has constant strength E, and the potential difference between the plates is $V = Ed$. Thus Eq. (26–9) gives the energy

The energy density in an electric field

$$U = \frac{CV^2}{2} = \frac{\epsilon_0 A}{2d}(Ed)^2 = \frac{\epsilon_0 E^2}{2}(Ad). \qquad (26\text{–}11)$$

We have written Eq. (26–11) so that the volume of the space between the plates, Ad, stands out. This is the volume that contains the electric field and, because the field is constant, the coefficient of the volume in Eq. (26–11) is the **energy density**, u, or energy per unit volume:

$$u \equiv \frac{U}{\text{volume}} = \frac{\epsilon_0 E^2}{2}. \qquad (26\text{–}12)$$

The energy of a capacitor is located where the electric field is in space (Fig. 26–9a). We could now imagine forgetting about the plates, leaving only the field (Fig. 26–9b). In fact, *Eq. (26–12) is a general expression for the local energy density in free space even for a variable electric field.* Wherever there is an electric field—even one that varies throughout space—the energy density, or energy per unit volume, at a particular location in space is found by squaring the electric field there and multiplying by $\epsilon_0/2$. Later, we shall see that we must modify the coefficient $\epsilon_0/2$ when dielectrics are present, but the energy density is always proportional to the square of the field.

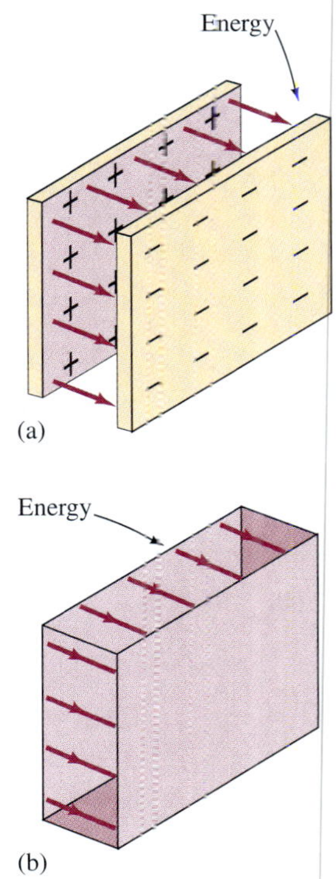

FIGURE 26–9 Electric fields have energy whether they are (a) inside a capacitor or (b) in free space.

EXAMPLE 26–5 (a) Determine the energy density due to an isolated, charged spherical conductor of radius R at each point in space as a function of the distance r from the sphere's center. (b) Use this energy density to compute the system's total energy. (c) Compare this total energy to the work done in charging the sphere.

Solution: (a) Equation (26–12) expresses the energy density in terms of the electric field. Equation (23–5) gives the electric field at a radius r outside the charged sphere. This field is radial and has magnitude

$$E = \frac{Q}{4\pi\epsilon_0 r^2}.$$

Inside the conducting sphere, the field is zero. The energy density is then

$$u = \frac{\epsilon_0 E^2}{2} = \frac{Q^2}{32\pi^2\epsilon_0 r^4}$$

outside the sphere and zero inside.

(b) The total energy of the electric field is the integral over space of the energy density. Because $u = 0$ inside the sphere, we need integrate only over radii $r > R$. Because of the spherical symmetry, it is useful to regard the volume as a series of concentric spherical shells. The volume $\mathrm{d}V$ of a thin shell of thickness $\mathrm{d}r$ is the product of $\mathrm{d}r$ and the surface area of the shell, $4\pi r^2$. Then the volume element $\mathrm{d}V = 4\pi r^2\,\mathrm{d}r$, and

$$U = \int_{\text{Volume}} u\,\mathrm{d}V = \frac{Q^2}{32\pi^2\epsilon_0}\int_R^\infty \frac{4\pi r^2}{r^4}\,\mathrm{d}r = \frac{Q^2}{8\pi\epsilon_0}\int_R^\infty \frac{1}{r^2}\mathrm{d}r$$

$$= -\frac{Q^2}{8\pi\epsilon_0}\,\frac{1}{r}\bigg|_R^\infty = \frac{Q^2}{8\pi\epsilon_0 R}.$$

(c) To calculate the work required to charge the sphere by bringing charge in from infinity, suppose that the sphere already has charge q and is at a potential of $V = q/4\pi\epsilon_0 R$. The additional work to bring charge $\mathrm{d}q$ in is $\mathrm{d}W = V\,\mathrm{d}q$, and the total work is

$$W = \int \mathrm{d}W = \int V\,\mathrm{d}q = \int_0^Q \frac{q}{4\pi\epsilon_0 R}\,\mathrm{d}q = \frac{1}{4\pi\epsilon_0 R}\int_0^Q q\,\mathrm{d}q = \frac{Q^2}{8\pi\epsilon_0 R}.$$

The work done to bring the charge in from infinity is indeed equal to the total energy of the system found in part (b). This is a good check on the validity of our calculations.

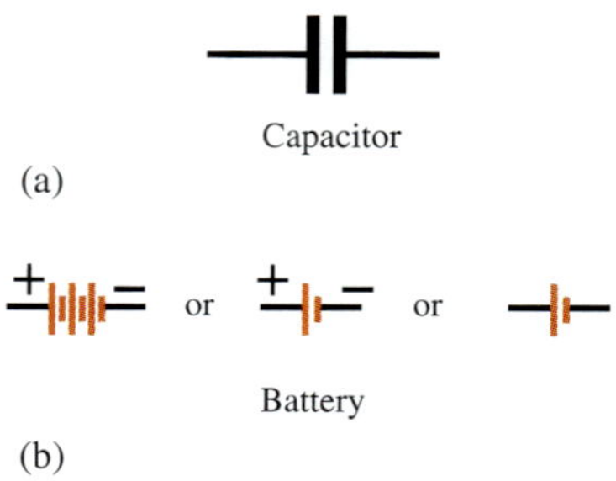

FIGURE 26–10 The symbols used to indicate (a) capacitors and (b) batteries in electric circuits.

26–4 CAPACITORS IN PARALLEL AND IN SERIES

Parallel and series combinations of circuit elements

Now let's begin our discussion of *electric circuits*. We have already mentioned two *circuit elements*: capacitors and batteries. We use the universal symbols shown in Fig. 26–10 for batteries and capacitors in *circuit diagrams*, schematic drawings that illustrate the connection of various circuit elements. The lines connecting circuit elements in these diagrams are assumed to be perfectly conducting wires, which means that they form equipotentials. We can use circuit diagrams to single out some interesting capacitor combinations. In a **parallel** combination, capacitors are connected as in Fig. 26–11a; in a **series** combination, capacitors are connected as in Fig. 26–11b. We want

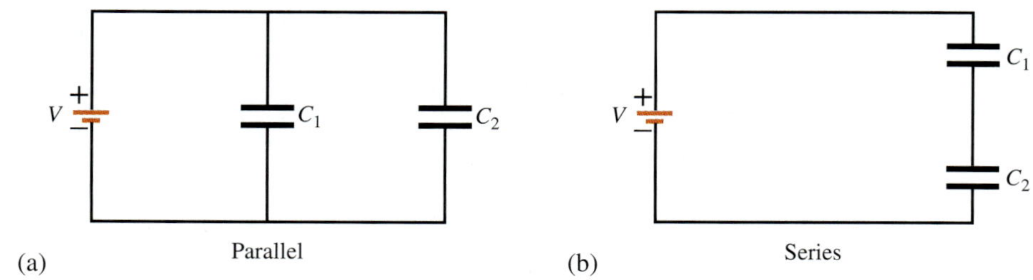

FIGURE 26–11 (a) Circuit with two capacitors connected in parallel. (b) Circuit with two capacitors connected in series.

to find the capacitance of the single *equivalent capacitor* that can replace these combinations without changing the potential across them; the use of equivalent capacitances greatly simplifies the treatment of circuits.

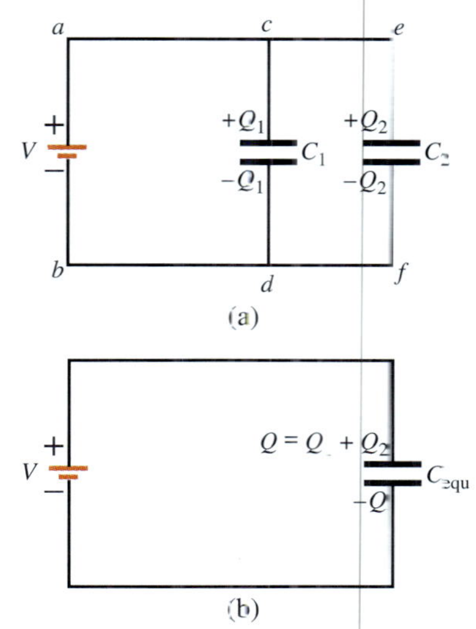

FIGURE 26–12 (a) A battery is used to place the same voltage V across the two capacitors connected in parallel. The charge on each capacitor depends on their individual capacitances. The two capacitors can be replaced (b) by an equivalent capacitor C_{equ}.

The equivalent capacitance for parallel connection

Parallel Connection

The battery in Fig. 26–12a maintains a potential V across the points a and b. Because the connecting wires are perfect conductors, the line ace is an equipotential, as is the line bdf. The potential difference from points c to d and from e to f must therefore be V also. *Capacitors connected in parallel have the same potential between their conductors.*

What is the equivalent capacitance, C_{equ}, of the single capacitor—defined by Fig. 26–12b—that replaces the parallel combination of capacitors C_1 and C_2? The same potential difference V and total charge Q must be maintained on the equivalent capacitor as are on the parallel combination of C_1 and C_2. The charges on C_1 and C_2 are related to the voltage across each capacitor:

$$Q_1 = C_1 V \quad \text{and} \quad Q_2 = C_2 V.$$

The charge for the equivalent capacitor is $Q = C_{equ}V$, and the total charge produced by the battery for the circuit in Fig. 26–12a is $Q_1 + Q_2$, so

$$Q = Q_1 + Q_2 = C_1V + C_2V = (C_1 + C_2)V = C_{equ}V.$$

The last equality shows that

$$C_{equ} = C_1 + C_2. \tag{26–13}$$

With n capacitors connected in parallel, we can similarly show that the equivalent capacitance is

$$C_{equ} = C_1 + C_2 + \cdots + C_n. \tag{26–14}$$

When capacitors are arranged in parallel, the total capacitance is larger than any of the individual capacitances.

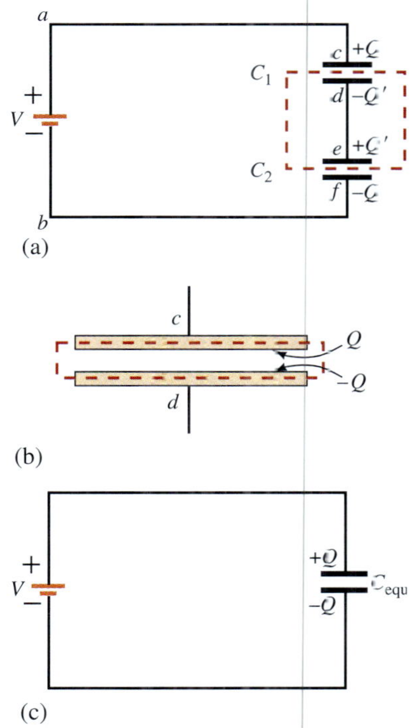

FIGURE 26–13 (a) Two capacitors connected in series must have identical charges $\pm Q$ but can have different voltages. The net charge within the dashed region is zero. (b) By drawing a Gaussian surface we establish that $Q' = Q$. (c) The two capacitors can be replaced by an equivalent capacitor C_{equ}.

Series Connection

Now let's determine the single capacitor equivalent to capacitors connected in series (Fig. 26–13a). Because the battery maintains a fixed potential V, charge $+Q$ appears at c and charge $-Q$ at f. The positive charge at c induces a charge $-Q'$ at d, and because the isolated piece of metal enclosed by the dashed line in Fig. 26–13a is neutral, $+Q'$ also appears at e. We now show that $Q' = Q$ by drawing a Gaussian surface like that shown in Fig. 26–13b. Because there is no field within the metal plates and because there is no electric flux through the short side portions, the flux through the surface is zero, the net charge enclosed must be zero, and $Q' = Q$. Capacitors C_1 and C_2 thus have identical charges Q. *Capacitors connected in series have identical charges.*

The single equivalent capacitor C_{equ} in the circuit of Fig. 26–13c must carry the identical charges $+Q$ and $-Q$ and have the same potential difference V across it as there are between c and f in the circuit of Fig. 26–13a. Capacitor C_1 has potential $V_1 = Q/C_1$; similarly, capacitor C_2 has potential $V_2 = Q/C_2$. The total potential $V = V_1 + V_2$, so we have

$$V = V_1 + V_2 = \frac{Q}{C_1} + \frac{Q}{C_2} = Q\left(\frac{1}{C_1} + \frac{1}{C_2}\right) = \frac{Q}{C_{equ}}.$$

The equivalent capacitance for series connection

The last equality shows that the value of the equivalent capacitance is determined by

$$\frac{1}{C_{equ}} = \frac{1}{C_1} + \frac{1}{C_2}; \tag{26–15}$$

$$C_{equ} = \frac{C_1 C_2}{C_1 + C_2}. \tag{26–16}$$

With n capacitors connected in series, the equivalent capacitance is given by

$$\frac{1}{C_{equ}} = \frac{1}{C_1} + \frac{1}{C_2} + \cdots + \frac{1}{C_n}. \tag{26–17}$$

When capacitors are arranged in series, the total capacitance is given by the reciprocal of the expression in Eq. (26–17), and this capacitance is less than any of the individual capacitances.

EXAMPLE 26–6 Determine the equivalent capacitance for the capacitors in the circuit shown in Fig. 26–14a.

Solution: In this circuit, the capacitors are combined both in series and in parallel, and we want to find the single equivalent capacitance for the whole combination. Our first step is to combine the two capacitors in parallel into one capacitor of value C'_{equ} (Fig. 26–14b); we then combine the three remaining capacitors in series into one final equivalent capacitor (Fig. 26–14c). We use Eq. (26–14) to combine the parallel capacitors:

$$C'_{equ} = 10\ \mu F + 6\ \mu F = 16\ \mu F.$$

Having reduced the capacitor arrangement to the one shown in Fig. 26–14b, we combine the series capacitors by using Eq. (26–17), as in Fig. 26–14c:

$$\frac{1}{C_{equ}} = \frac{1}{5\ \mu F} + \frac{1}{16\ \mu F} + \frac{1}{2\ \mu F} = \frac{61}{80}\,(\mu F)^{-1};$$

$$C_{equ} = 1.3\ \mu F.$$

26–5 DIELECTRICS

Many materials, such as paper, plastics, and glass, do not conduct electricity easily; we referred to them earlier as *insulators*. Nevertheless, they modify the external electric fields in which they are placed. In this context, we call these materials **dielectrics**. We shall see that a dielectric placed in a capacitor allows larger charges on the capacitor

FIGURE 26–14 (a) Example 26–6. Four capacitors of this electric circuit can be combined into one. (b) The two parallel capacitors are combined, giving three capacitors in series, (c) which are then combined into the equivalent capacitor C_{equ}.

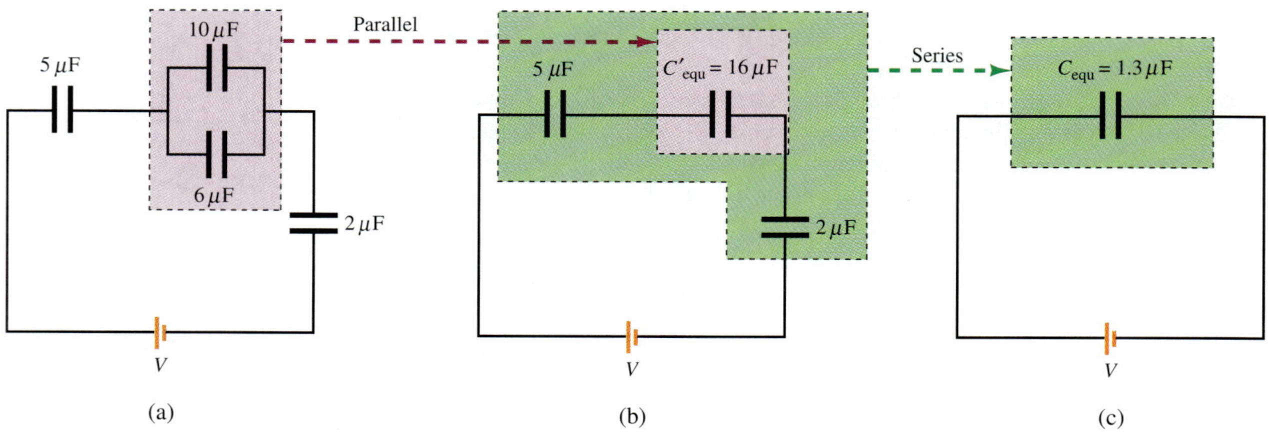

APPLICATION: THE MARX GENERATOR

A Marx generator ingeniously employs capacitors in *both* series and parallel to provide high voltages. It consists of N capacitors all connected in parallel by weakly conducting wires—in practice, resistors or inductors (Fig. 26AB–1). A power source with potential difference V is also connected by weakly conducting wires as shown; it slowly charges the capacitors, so that the potential difference across the capacitors would asymptotically rise to V. However, a spark gap is installed between the positive terminal of each capacitor and the negative terminal of its neighbor, making a diagonal ladder of spark gaps. At a certain potential difference that is slightly less than V, the air across these spark gaps ionizes, forming conducting channels. The capacitors are now connected in series. (The weakly conducting parallel connections are unimportant during the breakdown.) The potential difference between the positive end of the first capacitor and the negative end of the last capacitor is therefore N times V; for example, 20 capacitors and a 50 KV power supply will at this moment generate a 1 MV potential difference.

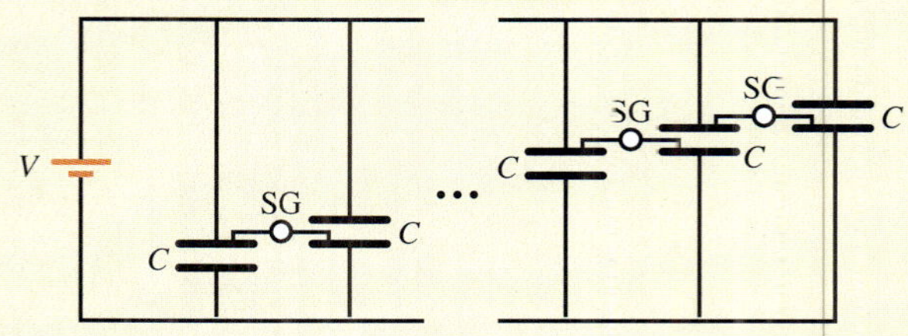

FIGURE 26AB–1 N identical capacitors are connected in parallel and charged by a source of emf to a potential nearly equal to V. A series of spark gaps, here denoted SG, connect the positive plate of each capacitor to the negative plate of its neighbor. Just below V, these gaps break down, effectively putting the capacitors in series and generating a voltage NV across the array.

The very high potential is used together with the energy stored in the capacitors for a variety of purposes. For example, ore can be fractured along crystal boundaries in order to more easily extract given compounds that concentrate there. The generator can also be combined with an X-ray tube to give short bursts of very penetrating (high-energy) X-rays. Such X-rays can be used to view the interior structures of operating machinery.

for a given voltage. A solid dielectric placed between a capacitor's two conductors also lends strength and mechanical stability to the capacitor. Finally, a dielectric can reduce the possibility of sparking across the plates of a capacitor. In this section, we take an experimental point of view about dielectrics; in the next section, we explore the microscopic origin of this behavior.

APPLICATION

Uses of dielectrics

Michael Faraday is generally given credit for performing the first experiments that showed that *the capacitance increases when insulating materials are placed between the two conductors of a capacitor*. If C_0 is the capacitance of a given capacitor in a vacuum (or in air), then the capacitance, C, of the same capacitor with dielectric between its conductors is larger than C_0 by a factor called the **dielectric constant**, κ. We have

The presence of a dielectric in a capacitor increases the capacitance.

$$C = \kappa C_0; \tag{26–18}$$

Definition of the dielectric constant

$$\kappa \equiv \frac{C}{C_0}. \tag{26–19}$$

The dielectric constant, which is larger than unity for all materials, depends on the material as well as on external conditions such as temperature. Depending on material and conditions, κ can run from only slightly larger than one—κ for air under normal conditions is 1.0005—to as large as several hundred. Table 26–1 gives a representative set of values of κ, but the temperature dependence of many of these values is so strong that the values must be used with care.

Experimental Evidence for the Behavior of Dielectrics

To simulate the experiment that led Faraday to his conclusions about dielectrics, let's first use a battery to charge a parallel-plate capacitor in air to a potential V_0 and a charge $Q_0 = C_0V_0$ (Fig. 26–15a). Here, the subscripts refer to the quantity in air—for

TABLE 26–1 Dielectric Properties of Materials[†]

Material	*Dielectric Constant,* κ	*Dielectric Strength,* E_{max} *(10^6 V/m)*
Vacuum	1.0	
Air	1.00054	3
Paraffin	2.0–2.5	10
Teflon	2.1	60
Polystyrene	2.5	24
Lucite	2.8	20
Mylar	3.1	
Plexiglass	3.4	40
Nylon	3.5	14
Paper	3.7	16
Fused quartz	3.75–4.1	
Pyrex	4–6	14
Bakelite	4.9	24
Neoprene rubber	6.7	12
Silicon	12	
Germanium	16	
Water	80	
Strontium titanate	332	8

[†] Values for some materials depend strongly on temperature and the frequency of oscillating fields.

example, C_0 is the capacitance when there is air between the plates. We disconnect the battery and use a voltmeter to measure the voltage, or potential difference (Fig. 26–15b). We then slide a dielectric, such as plexiglass, between the plates (Fig. 26–15c). *The voltage is reduced.* We call the reduction factor κ:

$$V = \frac{V_0}{\kappa}.$$

Because plexiglass is an insulator, the charge Q_0 on the capacitor plates cannot change, yet the voltage does change. The capacitance must therefore change from the original value C_0 to a new value C when the plexiglass is inserted:

$$C = \frac{Q_0}{V} = \frac{Q_0}{V_0/\kappa} = \kappa\frac{Q_0}{V_0} = \kappa C_0.$$

This verifies Eq. (26–18).

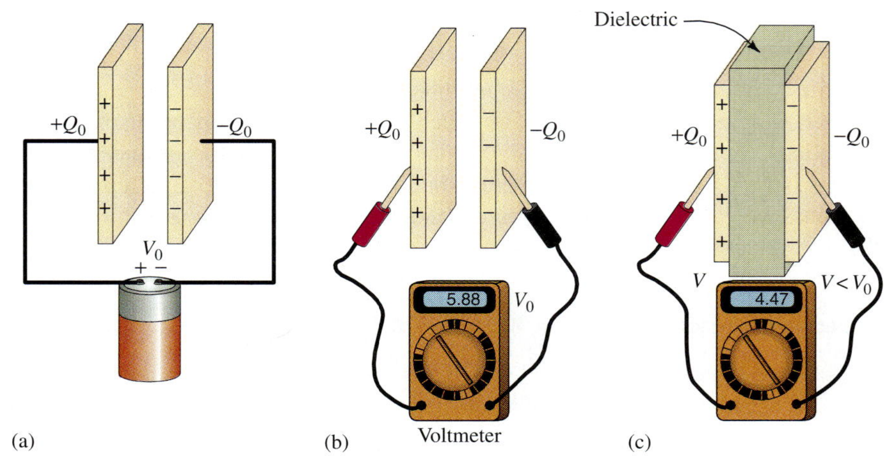

FIGURE 26–15 (a) A battery charges a capacitor to charge Q_0 and potential V_0. (b) If we take the battery away and measure the voltage with a voltmeter, we measure V_0 for the voltage. (c) If a dielectric is inserted into the capacitor, the voltage drops to $V < V_0$.

We can do one further experiment. Let's leave the battery connected to the capacitor after it is charged in air (Fig. 26–16a). The potential continues to be the battery voltage V_0 after we insert the plexiglass (Fig. 26–16b). However, we observe that *the charge on the conducting plates increases* by a factor of κ ($Q = \kappa Q_0$). Our experimental result remains in agreement with Eq. (26–18):

$$C = \frac{Q}{V_0} = \frac{\kappa Q_0}{V_0} = \kappa C_0.$$

Let's reinterpret these results in terms of permittivity. If we take a parallel-plate capacitor, for which $C_0 = \epsilon_0 A/d$, we have

$$C = \kappa C_0 = \frac{\kappa \epsilon_0 A}{d}. \qquad (26\text{–}20)$$

If the charge is held fixed, the new voltage is

$$V = \frac{V_0}{\kappa} = \frac{Q_0}{\kappa C_0} = \frac{\sigma_0 A}{\kappa C_0} = \frac{\sigma_0 d}{\kappa \epsilon_0}. \qquad (26\text{–}21)$$

When there is a dielectric between the plates of a capacitor, the electric field is decreased.

The new electric field between the plates is reduced in magnitude to

$$E = \frac{V}{d} = \frac{\sigma_0}{\kappa \epsilon_0} = \frac{E_0}{\kappa}, \qquad (26\text{–}22)$$

where $E < E_0$. Inspection of Eqs. (26–20) through (26–22) shows that they can be summarized *by replacing the permittivity of free space*, ϵ_0, *by a new permittivity*, ϵ, which depends on the dielectric used and on external conditions:

$$\epsilon = \kappa \epsilon_0. \qquad (26\text{–}23)$$

The effect of a dielectric is that ϵ_0 is replaced by ϵ in all formulas of electrostatics.

Although we have shown this simple rule only for the parallel-plate case, the substitution of ϵ for ϵ_0 when a dielectric is involved applies to all geometries and to all equations in which the permittivity appears, such as in the expressions for field strength, potential, and energy density.

Dielectrics in Capacitors

We have spoken of the important possibility of using a dielectric to increase the charge storage in a capacitor. Dielectrics also play a role in *voltage breakdown* across the plates of a capacitor. If the electric field in (or voltage across) a material becomes too great, electrons are pulled away from their atoms and cascade across the material, dis-

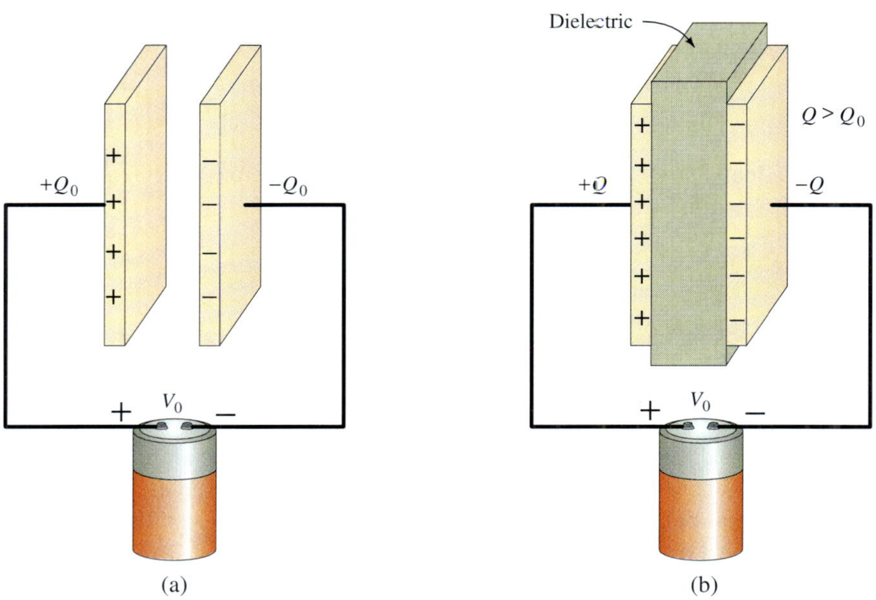

FIGURE 26–16 (a) Again, a battery charges the capacitor to charge Q_0 and potential V_0, but (b) this time we leave the battery connected when we insert the dielectric. The potential must remain at V_0, but the new charge is $Q > Q_0$.

charging the capacitor. Such events will damage dielectrics. Each dielectric has a *dielectric strength*, E_{max}, which is the maximum electric field a dielectric will support without breakdown. Table 26–1 contains some representative values. The dielectric strengths of commercial capacitors are indicated with a maximum allowable voltage.

EXAMPLE 26–7 A parallel-plate capacitor has area $A = 20.0\ \text{cm}^2$ and a plate separation $d = 4.0$ mm. (a) Find the capacitance in air and the maximum voltage and charge the capacitor can hold. (b) A Teflon sheet is slid between the plates, filling the entire volume. Find the new capacitance and maximum charge. (c) Before the insertion of the Teflon, the plates are set to a voltage of 24 V by a battery that is then disconnected. What are the energies in the capacitor before and after the Teflon is inserted? Was work done in inserting the Teflon?

Solution: (a) Equation (26–4) gives the capacitance in air, which we denote here by C:

$$C = \frac{\epsilon_0 A}{d} = \frac{(8.85 \times 10^{-12}\ \text{F/m})(2.00 \times 10^{-3}\ \text{m}^2)}{(4.0 \times 10^{-3}\ \text{m})} = 4.4 \times 10^{-12}\ \text{F} = 4.4\ \text{pF}.$$

The maximum charge depends on the maximum voltage. From Table 26–1, the dielectric strength of air is 3×10^6 V/m, so

$$V_{max} = E_{max}d = (3 \times 10^6\ \text{V/m})(4.0 \times 10^{-3}\ \text{m}) \simeq 10^4\ \text{V} = 10\ \text{kV}.$$

In turn,

$$Q_{max} = CV_{max} = (4.4 \times 10^{-12}\ \text{F})(10^4\ \text{V}) \simeq 5 \times 10^{-8}\ \text{C}.$$

(b) From Table 26–1, the dielectric constant of Teflon is 2.1. If we denote new values by primes, we have

$$C' = \kappa C = (2.1)(4.4\ \text{pF}) = 9.2\ \text{pF}.$$

Table 26–1 gives the dielectric strength of Teflon to be 6.0×10^7 V/m, so

$$V'_{max} = E'_{max}d = (6.0 \times 10^7\ \text{V/m})(4.0 \times 10^{-3}\ \text{m}) = 2.4 \times 10^5\ \text{V}$$

and

$$Q'_{max} = C'V'_{max} = (9.2 \times 10^{-12}\ \text{F})(2.4 \times 10^5\ \text{V}) = 2.2\ \mu\text{C}.$$

Both the maximum voltage and maximum charge are greatly increased after the Teflon is inserted.

(c) Equation (26–9) determines the energy in a charged capacitor. Before the Teflon is inserted, the energy is

$$U = \frac{CV^2}{2} = \frac{(4.4 \times 10^{-12}\ \text{F})(24\ \text{V})^2}{2} = 1.3 \times 10^{-9}\ \text{J}.$$

After the Teflon is inserted, C increases by the factor κ, whereas V decreases by the same factor. The product CV^2 thus *decreases* by a factor of κ:

$$U' = \frac{C'V'^2}{2} = \frac{(\kappa C)(V/\kappa)^2}{2} = \frac{U}{\kappa} = \frac{1.3 \times 10^{-9}\ \text{J}}{2.1} = 6 \times 10^{-10}\ \text{J}.$$

Because the potential energy has decreased, the capacitor does positive work as the Teflon is inserted.

When the material in the space between the plates of a capacitor is replaced by a dielectric of higher dielectric constant, the energy decreases (see Example 26–7). The capacitor therefore does positive work as the new dielectric is inserted, and so there must be a force that pulls in the dielectric. With sensitive instruments, the tug on the dielectric can be measured.

EXAMPLE 26–8 Suppose that the Teflon sheet inserted between the capacitor plates in Example 26–7 is only 2 mm thick and fills only half the volume (Fig. 26–17). Before the Teflon is inserted, the disconnected capacitor carries a charge of 1 nC. Find the electric field everywhere inside, and find the new capacitance.

Solution: Although the Teflon sheet is on the right in Fig. 26–17, this choice is immaterial. By Gauss' law, the electric field strength in air remains

$$E_0 = \frac{\sigma}{\epsilon_0} = \frac{Q}{\epsilon_0 A} = \frac{10^{-9}\ \text{C}}{(8.85 \times 10^{-12}\ \text{F/m})(2 \times 10^{-3}\ \text{m}^2)} \simeq 6 \times 10^4\ \text{V/m}.$$

The field inside the Teflon must be reduced by the factor κ:

$$E_{\text{Tef}} = \frac{E_0}{\kappa} = \frac{6 \times 10^4\ \text{V/m}}{2.1} \simeq 3 \times 10^4\ \text{V/m}.$$

To find the total voltage drop across the plates (which we need to find the capacitance), we calculate the integral

$$V = \int_0^d E\, dx = \int_0^{2\ \text{mm}} E_0\, dx + \int_{2\ \text{mm}}^{4\ \text{mm}} E_{\text{Tef}}\, dx$$

$$= E_0 \int_0^{2\ \text{mm}} dx + E_{\text{Tef}} \int_{2\ \text{mm}}^{4\ \text{mm}} dx = E_0(2\ \text{mm}) + E_{\text{Tef}}\,(2\ \text{mm})$$

$$= (5.6 \times 10^4\ \text{V/m})(2 \times 10^{-3}\ \text{m}) + (2.7 \times 10^4\ \text{V/m})(2 \times 10^{-3}\ \text{m}) \simeq 170\ \text{V}.$$

Notice that this calculation is independent of the precise location of the Teflon sheet.

Finally, the capacitance is, by definition, $C = Q/V$:

$$C = \frac{10^{-9}\ \text{C}}{170\ \text{V}} = 6\ \text{pF}.$$

This value is intermediate to the capacitances of the system empty or filled with Teflon. This capacitor is equivalent to two capacitors in series—one empty and of width $d/2$; the other filled with Teflon and of width $d/2$.

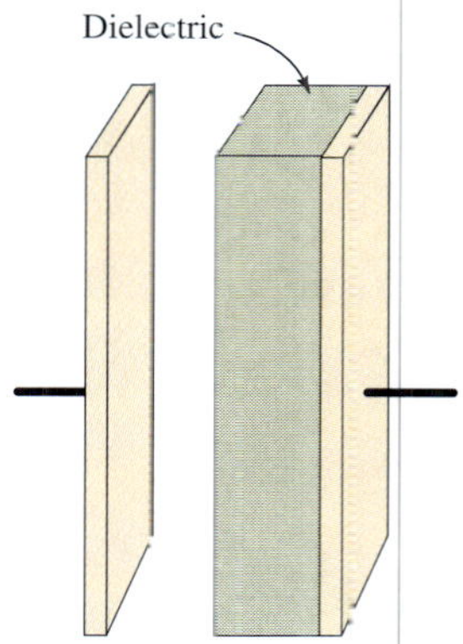

FIGURE 26–17 Example 26–8. A dielectric inserted in a parallel-plate capacitor fills only half the volume.

EXAMPLE 26–9 Consider the coaxial cable of Example 26–2. There are equal but opposite charges per unit length λ on the two elements of the cable, which therefore acts as a capacitor. A plug of insulating material of dielectric constant κ is inserted between the wire and the sheath to a depth x from the end (Fig. 26–18). What is the change in potential energy of the charged cable? What electric force acts on the plug as it is inserted?

Solution: In Example 26–2, we identified the potential difference between the conductors in the coaxial cable as

$$V_b - V_a = V = \frac{\lambda}{2\pi\epsilon_0} \ln\frac{b}{a}.$$

The potential energy per unit length is half the product of the voltage difference and the charge per unit length:

$$\frac{U}{\text{unit length}} = \frac{1}{2}\lambda V = \frac{\lambda^2}{4\pi\epsilon_0} \ln\frac{b}{a}$$

[see Eq. (25–19)]. In the region in which the dielectric is located, ϵ_0 is replaced by $\epsilon = \kappa\epsilon_0$. Thus the change in potential energy per unit length when the dielectric plug is inserted is

$$\frac{\Delta U}{\text{unit length}} = \frac{\lambda^2}{4\pi}\left(\frac{1}{\epsilon_0} - \frac{1}{\epsilon}\right) \ln\frac{b}{a}.$$

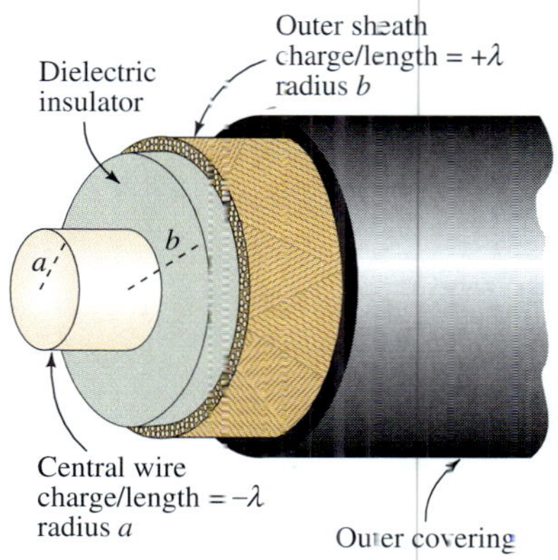

FIGURE 26–18 Example 26–9.

If the plug penetrates to a depth x, the total change in potential energy is

$$\Delta U = \frac{x\lambda^2}{4\pi}\left(\frac{1}{\epsilon_0} - \frac{1}{\epsilon}\right)\ln\frac{b}{a}.$$

Because $\kappa > 1$, $\epsilon > \epsilon_0$ and $\Delta U < 0$. The energy decreases when the plug is inserted, so we expect the plug to be pulled into the space between the conductors. The force exerted on the dielectric as it moves into the cable is obtained from

$$F = -\frac{dU}{dx} = \frac{\lambda^2}{4\pi}\left(\frac{1}{\epsilon_0} - \frac{1}{\epsilon}\right)\ln\frac{b}{a}.$$

F is positive because $\epsilon > \epsilon_0$. The plug is indeed pulled into position.

Large Capacitors

APPLICATION

How large commercial capacitors are constructed

Two types of capacitors comprise the bulk of modern capacitors. They achieve large capacitance in a small space by different strategies. In *multilayer ceramic capacitors*, metal sheets separated by ceramic insulators with dielectric constants as high as 20,000 are folded into a compact form. The dielectric constant is so high that capacitances on the order of thousands of microfarads are achieved. *Electrolytic capacitors* have capacitances of roughly the same size in even smaller volumes. In them, the dielectric—a nonconducting metal oxide—is deposited in a thin layer on a sheet of metal. The second conductor is a conducting paste or liquid that adheres well to the metal oxide. The dielectric layer between the conductors can be made quite thin—as thin as 10^{-8} m. Moreover, by etching the metal before the dielectric layer is deposited, a series of sharp valleys is created in the metal, greatly increasing its surface area. If we recall that the capacitance of parallel plates is inversely proportional to the distance between the plates and proportional to the area of the plates, we see that electrolytic capacitors can have large capacitances.

The technology now exists for making capacitors with capacitances on the order of farads. Let's look at what is required for this. With a suitable dielectric, a 6-V potential difference across a gap of as little as 0.25 μm is possible without breakdown. A 1-F capacitor for this situation must therefore have an area $A = (0.25 \times 10^{-6}\ \text{m})(1\ \text{F})/(8.85 \times 10^{-12}\ \text{F}\cdot\text{m}^{-1}) = 2.8 \times 10^4\ \text{m}^2$; this corresponds to a square more than 100 m on a side! To make such capacitors, sheets of mylar, a dielectric, are tightly folded. The mylar is then immersed in a conducting liquid, which plays the role of the other plate.

26–6 MICROSCOPIC DESCRIPTION OF DIELECTRICS

We refer to molecules with permanent electric dipole moments, such as H_2O, as *polar* molecules. In the absence of an external electric field, the directions of the dipole moments of polar molecules in a material are randomly distributed (Fig. 26–19a). However, when the material is placed in an external electric field, as in Fig. 26–19b, the dipoles are subject to a torque and tend to align themselves with the field. Thermal agitation disturbs the individual alignments, leaving only an average alignment. It will be more pronounced for stronger electric fields and lower temperatures. There is a wide range of values of the external field for which the average degree of alignment grows *linearly* with the external electric field.

Nonpolar molecules are those without a permanent dipole moment. In the absence of an external electric field, their charge distributions are symmetric; that is, no particular direction is picked out (Fig. 26–19c). As we discussed in Chapter 23, when these molecules are placed in an external electric field, they acquire an induced dipole moment that is fully aligned with the field (Fig. 26–19d). The magnitude of this di-

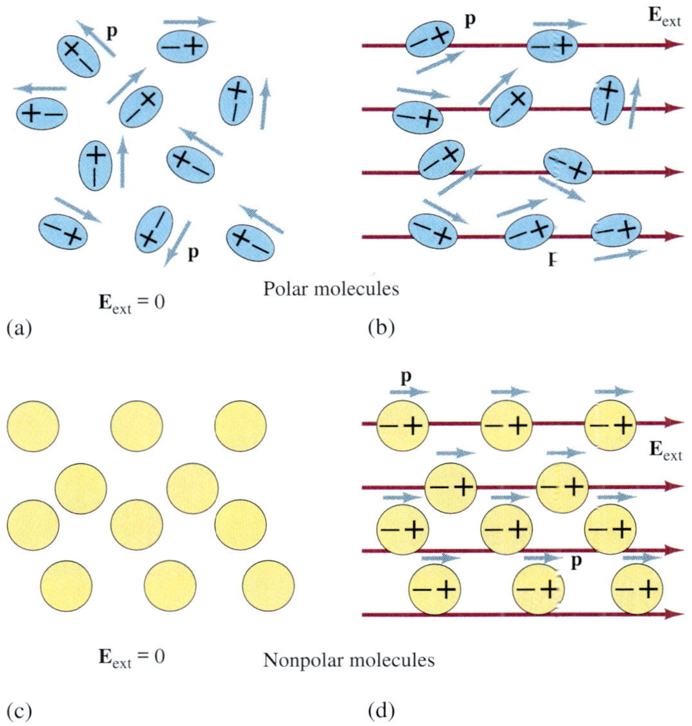

FIGURE 26–19 (a) For polar molecules not in an external electric field, the dipole moments, **p**, are randomly oriented. (b) In an external electric field, the dipole moments align themselves with the field. (c) For nonpolar molecules not in an electric field, there is no indication of a charge distribution. (d) In an external electric field, nonpolar molecules obtain an induced dipole moment aligned with the electric field.

pole moment increases as the external field increases, and again there is an important range of values of the external electric field for which the dipole moment grows *linearly* with the field.

If an insulating slab made of either polar or nonpolar molecules is placed in a charged capacitor, the effect on the capacitor is as shown in Fig. 26–20a. There is no free charge to move large distances in an insulator. However, the dipole moments, either permanent or induced, become aligned with the electric field. The inside of the dielectric remains electrically neutral, but the charge distribution is *polarized* so that *induced charge* appears on the two outside surfaces of the slab. We denote the induced surface charge density as σ_{ind}. The two dielectric surfaces have equal but opposite charge densities. The induced surface charge density is proportional either to the degree to which the permanent dipole moments of polar molecules are aligned, or to the magnitude of the induced dipole moments of nonpolar molecules. There will thus be an important range of values of the external field for which σ_{ind} *is proportional to the external field.*

We can now sort out the electric fields that appear in dielectrics. We can distinguish three fields. The external field $\mathbf{E}_0$ would be present whether the dielectric were inserted or not. An *induced electric field* $\mathbf{E}_{ind}$ is produced by the induced surface charge, which is due to the external field. Because of the way the induced charge forms, $\mathbf{E}_{ind}$ is opposite to $\mathbf{E}_0$. Finally, the net electric field inside the dielectric, **E**, is

$$\mathbf{E} = \mathbf{E}_0 + \mathbf{E}_{ind}. \qquad (26\text{–}24)$$

The direction of **E** is indicated in Fig. 26–20b. From Eq. (26–22), we can see that the resultant field **E** will be proportional to $\mathbf{E}_0$ only if $\mathbf{E}_{ind}$ (which we know is proportional to σ_{ind}) is proportional to $\mathbf{E}_0$. This proportionality can be expressed with the constant κ:

$$\mathbf{E} = \frac{\mathbf{E}_0}{\kappa}.$$

The constant κ is, in fact, the dielectric constant, as we can see from Eq. (26–22).

Let's translate this argument in terms of charge densities. Suppose that the original electric field $\mathbf{E}_0$ is produced by a surface charge density σ on the capacitor plates.

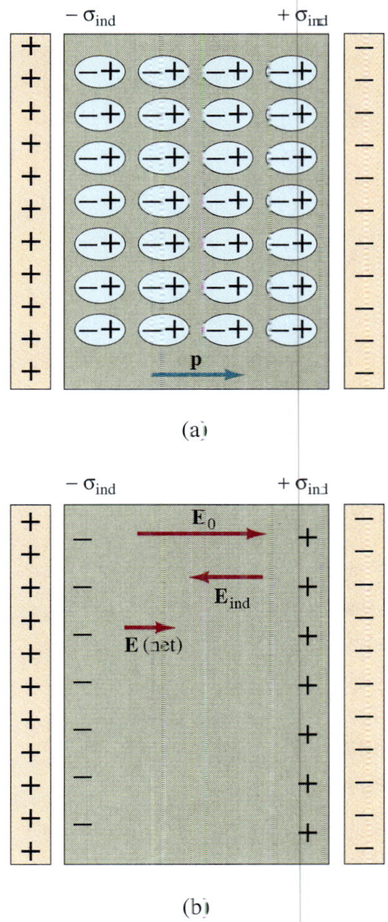

FIGURE 26–20 (a) When a dielectric is inserted in a capacitor, induced charges appear on the surface. (b) The induced charges cause an induced electric field $\mathbf{E}_{ind}$ opposite to the external electric field $\mathbf{E}_0$ caused by the free charge on the capacitor plates. The net effect is a reduced electric field $\mathbf{E} = \mathbf{E}_0 - \mathbf{E}_{ind}$ within the dielectric. In the region between the conducting plates where there is no dielectric, the net electric field remains $\mathbf{E}_0$.

We refer to σ as the density of *free charge*. We know that the magnitude of $\mathbf{E}_0$ is $E_0 = \sigma/\epsilon_0$ and that $\mathbf{E}_{\text{ind}}$ has magnitude $E_{\text{ind}} = \sigma_{\text{ind}}/\epsilon_0$ and points to the left if $\mathbf{E}_0$ points to the right. From Eq. (26–22), we have $E = \sigma/\kappa\epsilon_0$, so Eq. (26–24) becomes

$$\frac{\sigma}{\kappa\epsilon_0} = \frac{\sigma}{\epsilon_0} - \frac{\sigma_{\text{ind}}}{\epsilon_0}. \tag{26–25}$$

We cancel ϵ_0 and solve for σ_{ind}:

$$\sigma_{\text{ind}} = \sigma\left(1 - \frac{1}{\kappa}\right). \tag{26–26}$$

Because $\kappa > 1$ for all dielectrics, the induced charge is always less than the free charge. This is evident in our microscopic model; if E_{ind} were to exceed E_0, we would actually reverse the field in the material.

Gauss' Law and Dielectrics

We have mentioned that the presence of a dielectric causes ϵ_0 to be replaced by ϵ. How does this result affect Gauss' law? For the Gaussian surface drawn in Fig. 26–21, Gauss' law in its original form gives

$$\int_{\text{closed surface}} \mathbf{E}\cdot d\mathbf{A} = \frac{Q_{\text{encl}}}{\epsilon_0} = \frac{Q - Q_{\text{ind}}}{\epsilon_0}. \tag{26–27}$$

The enclosed charge is the free charge minus the induced charge. From Eq. (26–25),

$$\frac{Q}{\kappa} = Q - Q_{\text{ind}}.$$

Gauss' law in the presence of a dielectric

Thus we have an alternative form of Gauss' law when dielectrics are present:

$$\int_{\text{closed surface}} \mathbf{E}\cdot d\mathbf{A} = \frac{Q}{\kappa\epsilon_0}. \tag{26–28}$$

The constant ϵ_0 is replaced by $\epsilon = \kappa\epsilon_0$ *provided Q is the free charge.* This is a general result even though Fig. 26–21 refers to a parallel-plate capacitor. If the dielectric is not uniform throughout, κ will not have the same value throughout. In this case, κ should be brought under the integral, and Gauss' law reads

$$\int_{\text{closed surface}} \kappa\mathbf{E}\cdot d\mathbf{A} = \frac{Q}{\epsilon_0}. \tag{26–29}$$

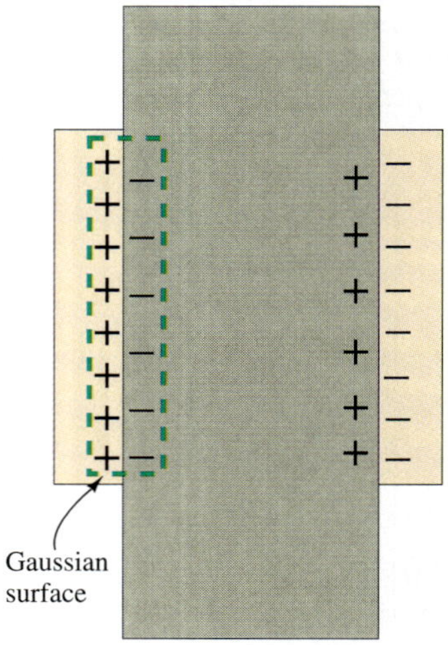

FIGURE 26–21 A dielectric fills the entire volume of a capacitor. A Gaussian surface surrounding the interface region between the dielectric and each plate surrounds both free and induced charges. The total charge enclosed by the Gaussian surface when Gauss' law is applied includes both charges.

Consequences of the Microscopic Model of Dielectrics

Our model of induced charges explains the experimental behavior of the dielectrics we have described to this point. This model is also the basis for a variety of testable experimental predictions:

1. There are two classes of dielectrics: those made of either nonpolar or polar molecules. The dipole moments of induced dipoles are generally much smaller than those of permanent dipoles, so the value of κ for nonpolar dielectrics should be much closer to one than that for polar dielectrics.

2. Due to decreasing disruptive thermal effects, polar dielectrics should line up more easily—have larger values of κ—at lower temperatures. Kinetic theory (see Chapter 19) shows that the dielectric constant takes the more precise form

$$\kappa = 1 + \frac{\text{a constant}}{T}. \tag{26–30}$$

This temperature dependence holds rather well. Equation (26–30) is called *Curie's law.* Nonpolar dielectrics should not obey such a law.

3. The polarization of solid substances with a permanent dipole moment can change if the planes of their lattice structure are stressed by being twisted or pressed. Under such stress, the internal electric fields change, and the changing fields produce an electrical signal. This phenomenon, known as *piezoelectricity*, is the principle behind the operation of some microphones, electronic pilot lights, and strain gauges.

APPLICATION

Piezoelectricity

SUMMARY

Capacitors are devices for storing electric charge and energy and typically consist of two conductors with equal and opposite charges Q and potential difference V. Capacitance is defined as

$$C \equiv \frac{Q}{V}. \tag{26–2}$$

The capacitance of a parallel-plate capacitor in air is given by

$$C = \frac{\epsilon_0 A}{d}, \tag{26–4}$$

where A is the plate area and d is the plate separation. The SI unit of capacitance is the farad: 1 F = 1 C/V.

The potential energy of a capacitor can be written as

$$U = \frac{Q^2}{2C} = \frac{CV^2}{2} = \frac{QV}{2}. \tag{26–8, 26–9, 26–10}$$

The energy density, or energy per unit volume, of an electric field is

$$u = \frac{\epsilon_0 E^2}{2} \tag{26–12}$$

Capacitors connected in parallel can be replaced by an equivalent capacitor with capacitance

$$C_{\text{equ}} = C_1 + C_2 + \cdots + C_n. \tag{26–14}$$

Capacitors connected in series can be replaced by an equivalent capacitor according to the relation

$$\frac{1}{C_{\text{equ}}} = \frac{1}{C_1} + \frac{1}{C_2} + \cdots + \frac{1}{C_n}. \tag{26–17}$$

Dielectrics are insulators with a characteristic property called the dielectric constant, κ; $\kappa > 1$. When a dielectric fills the space between the two conducting plates of a capacitor, the value of the capacitance is increased:

$$C = \kappa C_0, \tag{26–18}$$

where C_0 is the value of the capacitor with a vacuum (or air) between its conductors. Our previous results can be modified for the presence of dielectrics by replacing the permittivity of free space, ϵ_0, by the permittivity ϵ given by

$$\epsilon = \kappa \epsilon_0. \tag{26–23}$$

Each insulator also has a characteristic property called the dielectric strength, which gives the approximate maximum electric field that the insulating material can withstand before it breaks down and ionizes.

The behavior of capacitors can be understood by considering the molecular structure of matter. Polar and nonpolar molecules of a dielectric become aligned with the external electric field, reducing the effects of that field. An alternative form of Gauss' law when dielectrics are present in a capacitor is

$$\int_{\text{closed surface}} \mathbf{E} \cdot d\mathbf{A} = \frac{Q}{\kappa\epsilon_0}. \qquad (26\text{–}28)$$

where Q is the free charge.

UNDERSTANDING KEY CONCEPTS

1. There are two common ways to write SI units of permittivity. Does the fact that there is more than one way present a problem?
2. You have two parallel plates, a battery, a voltmeter, and a piece of unknown plastic. Devise a method to determine the dielectric constant of the plastic.
3. What argument can you give to show that the electric field of a parallel-plate capacitor cannot drop abruptly to zero as we pass outside the region between the plates? Recall the fact that the voltage drop around any closed path must be zero.
4. What is the meaning of a capacitor with zero capacitance?
5. If the radius of the inner wire of the coaxial cable in Example 26–2 approaches zero, the capacitance per unit length of the coaxial cable also approaches zero. What is the physical significance of that?
6. It is not possible to break up every combination of capacitors into a sequence of parallel and series capacitors. Find an example of a combination that cannot be decomposed in this way.
7. From our discussion of the physical nature of dielectrics, can you imagine a physical system in which the dielectric constant is less than one?
8. The plates of a charged parallel-plate capacitor are disconnected from the charging battery and are pushed together. What happens to the potential difference, the capacitance, and the stored energy?
9. The plates of a parallel-plate capacitor, still connected to a battery with potential difference V, are pushed together. What happens to the charge on the plates, the capacitance, and the stored energy?
10. You are given a thin, metal sheet of area A. You can make it into a spherical shell, roll it into two concentric cylinders, or cut it to make a parallel-plate capacitor. Which arrangement would give the largest possible capacitance?
11. What happens if you short out (connect with a conductor) the two plates of a large, charged capacitor? Could this be dangerous?
12. For finite parallel plates there is a fringe field (see Fig. 26–4). What effect would you expect this phenomenon to have on the capacitance of a parallel-plate capacitor?
13. Is it possible for a pair of nonconductors carrying equal but opposite charges to act as a capacitor? In what ways would such an arrangement differ from, or be similar to, the capacitors treated in this chapter?
14. Why is it a good idea to short out (connect with a conductor) the plates of a large capacitor when the capacitor is not in use?
15. Would you expect the term "dielectric strength" to have meaning for a vacuum?
16. Air, particularly on humid days, can cause charge leakage. Why, then, can capacitors with air between their plates hold charge in a way that is useful for circuits?

PROBLEMS

26–1 Capacitance

1. (I) (a) What is the capacitance of two square metal plates, each 50 cm^2 in area, separated by 1 mm? (b) What is the radius of a conducting sphere with the same capacitance?
2. (I) A coaxial cable has an inner wire of radius 2 mm and an outer sheath of radius 4 mm. What is the capacitance of a kilometer of the cable?
3. (I) At different times, a 4-μF capacitor has a charge of (a) 4 μC, (b) 10 μC, and (c) 1 mC. What is the voltage across the capacitor in each case?
4. (I) How much charge can be stored on the plates of a 1-μF capacitor if the plates are attached to a battery that can give a potential difference of (a) 2 V? (b) 12 V?
5. (I) You must design a capacitor to store 2×10^{-6} C of charge, but you have only a 3-kV power supply and two metal plates of area 250 cm^2 each. What limits do you put on the separation between the plates?
6. (I) What is the capacitance of a piece of coaxial cable 25 cm long for which the radius of the inner conductor is 0.50 mm and the radius of the outer conducting sheath is 1.5 mm?
7. (II) Calculate the capacitance of two concentric spherical conductors of radii r and R, respectively. Discuss the limits of (a) finite r, $R \to \infty$; (b) $(R - r) << r$.
8. (II) Two concentric conducting spheres have radii of 3.0 cm and 15 cm, respectively, and an equal but opposite charge of 1.4×10^{-7} C. What is the potential difference between them? [*Hint*: Use the results of Problem 7.]
9. (II) A parallel-plate capacitor of area 0.040 m^2 carries a charge $q = 4.0 \times 10^{-8}$ C. The potential across the plates increases with time t according to the equation $V = 50.0$ mV + (0.10

mV/s)t, as a result of a time-dependent increase of the separation between the plates. Find the function of time that describes the separation.

10. (II) A parallel-plate capacitor has square plates 15 cm on a side, separated by 2 mm. The capacitor is charged to 156 V, then disconnected from the charging power supply. What is the charge density on the plates? the total charge on each plate?

11. (II) The capacitance of a variable capacitor used in a radio varies from 0.2 μF to 0.01 μF. The capacitor is charged to a potential difference of 300 V at maximum capacitance and then isolated. At minimum capacitance, what is the voltage?

26–2 Energy in Capacitors

12. (I) A thundercloud has a charge of 900 C and a potential of 90 MV with respect to the ground 1 km below it. (a) What is the capacitance of the system? (b) How much energy is stored in the thundercloud system?

13. (I) A capacitor holds a charge of 0.068 C at a potential of 2900 V. How much energy was required to charge the capacitor?

14. (I) A small capacitor in a computer with a capacitance of 0.1 pF has 6 V across it. How much energy is contained within the electric field of the capacitor?

15. (I) A fully charged flash attachment for a camera has electrical energy of 27 J. The potential across it is 300 V. What is the capacitance of the flash attachment?

16. (I) How much energy is stored on a metal sphere of radius 35 cm when a charge of 3.0×10^{-5} C is placed on it?

17. (II) A coaxial cable with an inner wire of diameter 3 mm and an outer sheath wire of diameter 8 mm has a potential of 1 kV between the wires. (a) What is the capacitance of 10 m of the cable? (b) How much energy is stored in the 10-m piece of cable? in a 1-km piece?

18. (II) Two concentric conducting spheres of radii 15.0 cm and 35 cm, respectively, are given equal but opposite charges of 3.2×10^{-7} C. How much energy is stored in the system?

19. (II) A capacitor consists of two parallel plates, each of area A. It is charged using a battery of potential V_0, which is then disconnected. (a) How much does the energy of the capacitor change if the separation of the plates is changed from d_0 to d_1? (b) How much work is done by the external force used to move the plate? (c) Suppose that the plates of the capacitor remain connected to the battery as they are moved. How much does the energy stored in the capacitor change under these conditions? (d) Is this change related to the work done by the force moving the plate?

26–3 Energy in Electric Fields

20. (I) The electric field in a large thunderstorm is 125,000 V/m. How much energy is contained in 1 m^3? in 1 km^3?

21. (I) The electric field due to an infinitely long, uniformly charged wire is calculated in Example 23–5. What is the electrical energy density in the space around the wire as a function of the distance from the wire?

22. (I) A Van de Graaff accelerator with a spherical dome of radius 2.0 m has a potential of 300,000 V in air. Assume that the accelerator is, in effect, a charged sphere. How much energy is stored in its electric field?

23. (I) Approximately how much energy is stored in a cube of sides 5 cm that is 1.0 m from a point charge of magnitude 5×10^{-4} C?

24. (I) The energy density in the space between the plates of a parallel plate capacitor is 10^{-6} J/m^3. What is the voltage between the plates if the separation of the plates is 1 cm?

25. (II) An isolated metal sphere of radius 18 cm is at potential 8300 V. What is the charge on the sphere? What is the energy density of the electric field outside the sphere? Integrate this to obtain the total energy in the electric field.

26. (II) A metal sphere of radius 0.10 m carries a charge of 8.5×10^{-6} C. How much energy is contained in a spherical region of radius 25 cm that is concentric with the sphere?

27. (II) A Geiger–Muller tube is a device used to detect ionizing particles (radioactive products). It is a cylindrical capacitor with the outer metal cylinder at zero potential and the central wire at about 500 V (Fig. 26–22). (a) Calculate the capacitance of the tube if its length is 15 cm, the radius of the outer cylinder is 2 cm, and that of the central wire is 0.02 cm. (b) When an ionizing particle enters the detector, it creates free electrons and ions, the gas breaks down, and the capacitor discharges. How much energy is needed to recharge the Geiger–Muller tube?

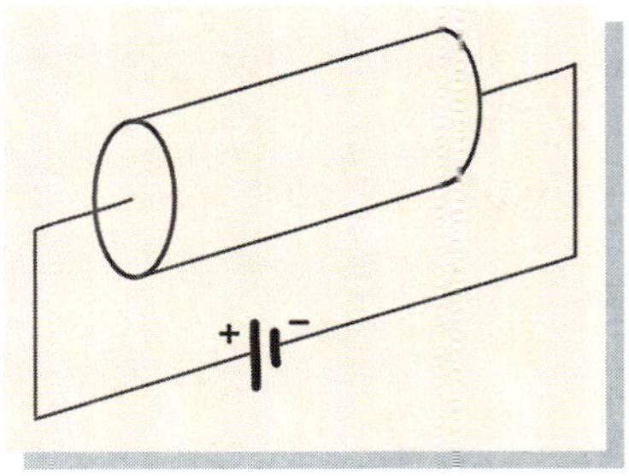

FIGURE 26–22 Problem 27.

28. (III) A nonconducting sphere of radius 0.10 m carries a uniformly distributed charge of 8.5×10^{-6} C. How much energy is contained in a spherical region of radius 25 cm that is concentric with the sphere?

29. (III) The plates of a parallel-plate capacitor are 400 cm^2 in area and 0.5 cm apart. The potential difference between the plates is 1500 V. (a) What is the field between the plates? (b) the charge on each plate? (c) the force exerted by the field on one of the plates? (d) Suppose that the plates are pulled apart so that the separation increases by 20 percent. What is the change in the stored energy? Is this consistent with the answer to part (c)? [If not, you probably answered (c) incorrectly.]

30. (III) Assume that an electron consists of a sphere of radius R with its charge distributed uniformly on the surface. (a) What is the electric field outside of the radius R? (b) What is the total electrostatic energy stored in the electric field? (c) Assume that all the energy of part (b) is solely responsible for the rest energy of the electron. (Rest energy is the energy associated with an object's mass, according to the theory of special relativity, even if the object is at rest. It takes the form mc^2, where in this case m is the electron's mass, 0.9×10^{-30} kg, and c is the speed of light, 3×10^8 m/s.) What must the radius R of the electron be?

26–4 Capacitors in Parallel and in Series

31. (I) When two capacitors are connected in parallel, the resulting combination has capacitance 6.5 μF. When the same two capacitors are connected in series, the resulting combination has capacitance 1.4 μF. What are the capacitances of the two capacitors?

32. (II) Find the capacitance of the parallel-plate system in Fig. 26–23. Can this system be represented by two pairs of parallel plates of half the total area connected in series or in parallel?

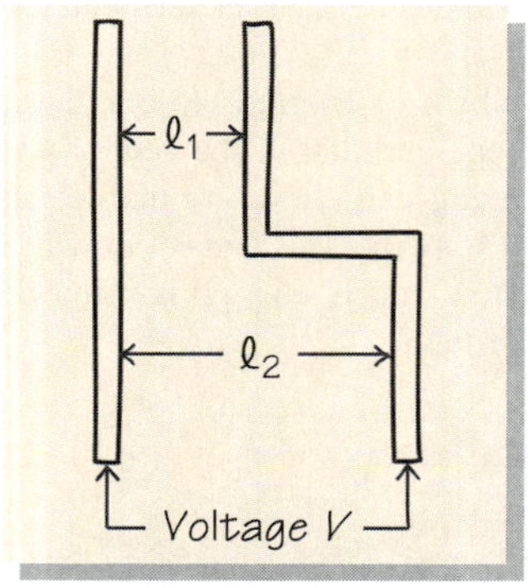

FIGURE 26–23 Problem 32.

33. (II) Find the equivalent capacitance of the circuit shown in Fig. 26–24.

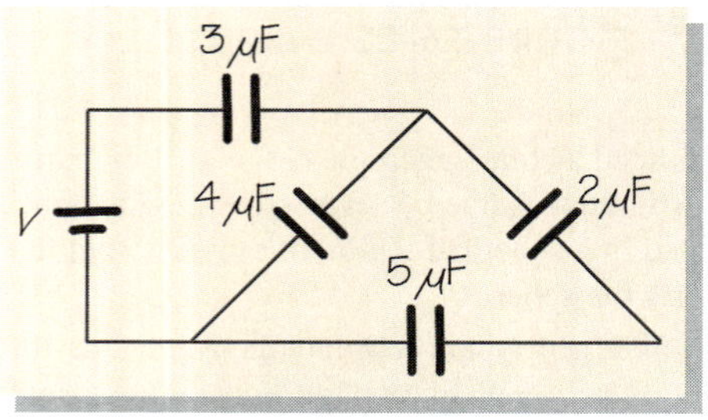

FIGURE 26–24 Problem 33.

34. (II) Two large, thin metal plates of area A and thickness d, carrying charges Q and $-Q$, respectively, are placed a distance D apart (Fig. 26–25). Suppose that an uncharged, thin metal plate of the same area and thickness is placed between them, such that the distance between the uncharged plate and the positively charged plate is x. What is the capacitance of the combined system as a function of x?

35. (II) What is the capacitance of the two concentric, spherical conductors of radii 3.0 mm and 12 mm, respectively, connected as shown in Fig. 26–26a? Suppose that the conductors are connected as shown in Fig. 26–26b. What is the capacitance now?

36. (II) Find the equivalent capacitance of the circuit shown in Fig. 26–27. The capacitance of each capacitor is 5 μF.

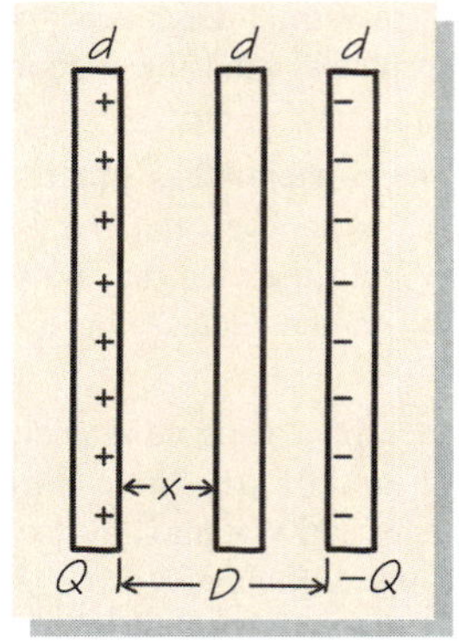

FIGURE 26–25 Problem 34.

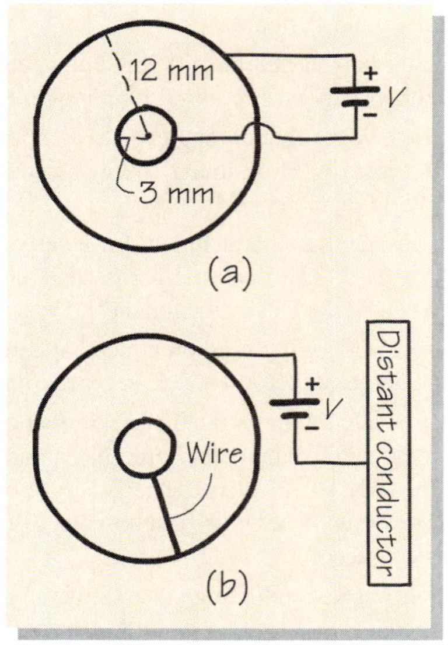

FIGURE 26–26 Problem 35.

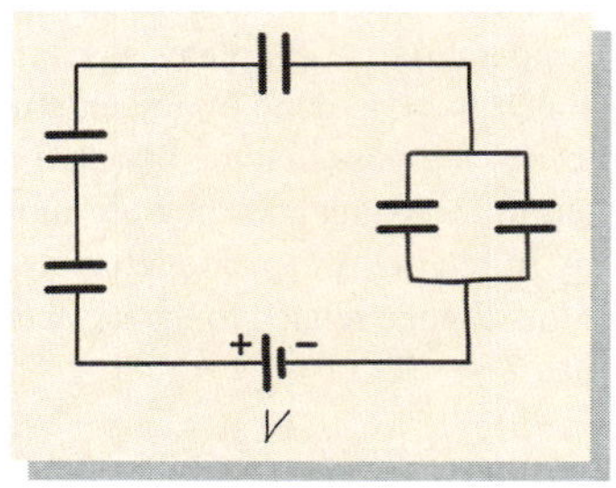

FIGURE 26–27 Problem 36.

37. (II) (a) Find the equivalent capacitance of the combination of capacitors shown in Fig. 26–28. (b) Assume that the potential difference between b and a is 300 V and find the charge on each of the capacitors in the figure.

38. (II) Figure 26–29 illustrates a set of five capacitors connected together across the points a and b. What is the value of a single capacitor that could replace this system and collect the same total charge for a given voltage drop V_{ab}?

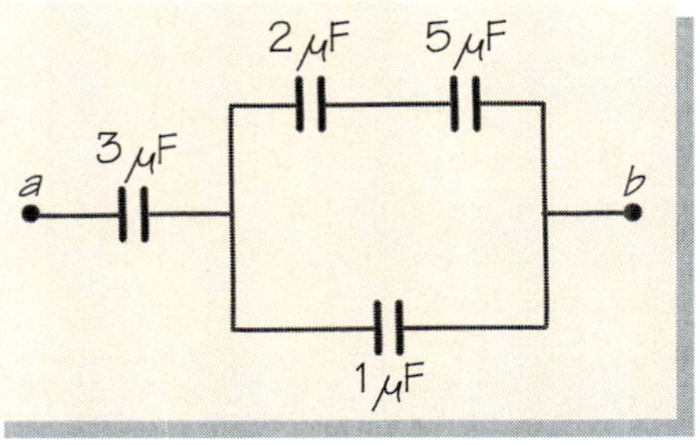

FIGURE 26–28 Problem 37.

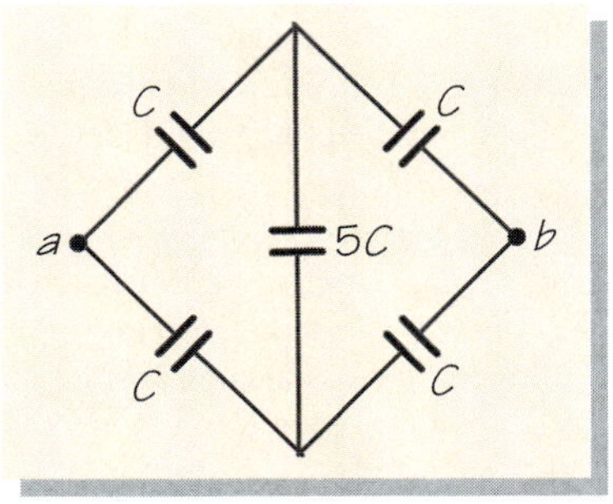

FIGURE 26–29 Problem 38.

39. (II) Figure 26–30 shows a network of identical capacitors. What is the equivalent capacitance between the points a and b? What is it between a and c? between b and d?

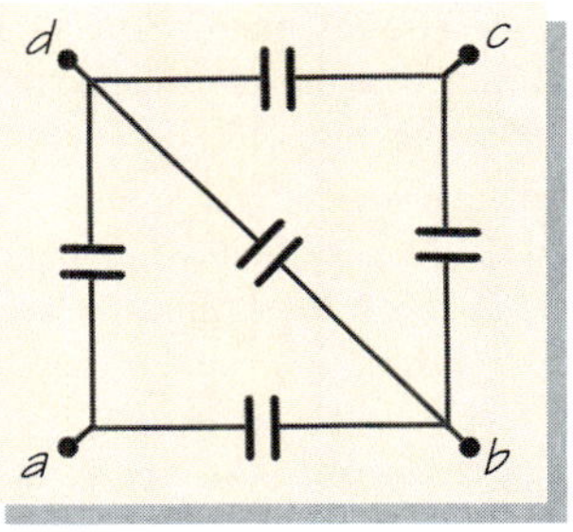

FIGURE 26–30 Problem 39.

40. (II) Capacitor C_1 has a capacitance of 10 μF; capacitor C_2 has a capacitance of 4 μF. A charge of $q = 25$ μC is placed on C_1, whereas C_2 is brought to a potential difference between its plates of 3 V. (a) What is the total energy stored in the two capacitors? (b) The negatively charged plate of C_1 is connected to the positively charged plate of C_2. What will change in the system, if anything? Neglect the fringe fields at the ends of the capacitors.

41. (II) Consider the capacitors of Problem 40, with C_2 modified so that it holds a charge of 25 μC at a potential difference of 5 V between the plates. (a) How is the capacitance of C_2 modified? (b) What is the charge on the capacitor equivalent to the whole system when the negatively charged plate of C_1 is connected to the positively charged plate of C_2?

42. (II) You have four capacitors whose capacitances are 2 μF, 3 μF, 4 μF, and 5 μF, respectively. Describe a circuit with an equivalent capacitance smaller than the 5-μF capacitor by 0.032 μF.

26–5 Dielectrics

43. (I) Consider a parallel-plate capacitor in which the space between the plates is filled with Teflon. With the charge held fixed, the Teflon is replaced by plexiglass. If the voltage across the capacitor was 600 V in the first case, what is it after the change?

44. (I) You have a piece of plastic whose dielectric constant you want to measure with two parallel plates, a 9-V battery, and a voltmeter. You charge the plates with the battery and then disconnect them. After you slide the plastic into the full volume between the plates, the voltmeter indicates a voltage drop from 9 V to 3.6 V. What is the dielectric constant?

45. (I) A 12-V automobile battery can store 4×10^6 J of energy. Find the area of a parallel-plate capacitor that can store the same amount of energy, if the separation between the plates is 1 mm and a dielectric with dielectric constant $\kappa = 3$ is between the plates.

46. (II) Repeat Problem 17 for polystyrene placed between the wires of the coaxial cable.

47. (II) Calculate the change in capacitance of an isolated sphere that becomes embedded in a dielectric with dielectric constant κ. If the capacitance change is due to a charge induced on the surface of the dielectric, what is the ratio of the induced charge density to that of the original surface charge density?

48. (II) Two large, parallel metal plates have a potential difference of 20 kV, and the electric field between them has a magnitude of 1.2×10^6 V/m. A material with a dielectric constant of 2.8 is inserted between the plates, with the plate separation adjusted so that the capacitance is unchanged. Calculate the new plate separation.

49. (II) A parallel-plate capacitor carrying charge q_0 is modified by the insertion of a dielectric with $\kappa = 1.8$ between the plates. As a consequence, the energy stored in the capacitor triples. What will the charge be after the dielectric is inserted?

50. (II) A coaxial cable has an inside wire of radius 0.6 mm and an outside metal sheath of radius 2.5 mm. The intermediate region is filled with a material of dielectric constant 2.6. (a) What is the capacitance of such a cable 100 m long? (b) If the potential difference between the inner and outer conductors is 500 V, what is the charge on the inner conductor, and how much energy is stored in 100 m of cable?

51. (II) A dielectric slab of thickness d and dielectric constant κ is inserted in the middle of a parallel-plate capacitor of plate separation D. What is the new capacitance of the capacitor, given that the area of each plate is A?

52. (II) A parallel-plate capacitor of dimensions 20 cm × 28 cm and separation distance 1.6 cm contains a dielectric slab of thickness 0.6 cm and dielectric constant 1.8. The potential difference between the plates is 600 V (Fig. 26–31). What are the electric fields in the empty space and inside the dielectric?

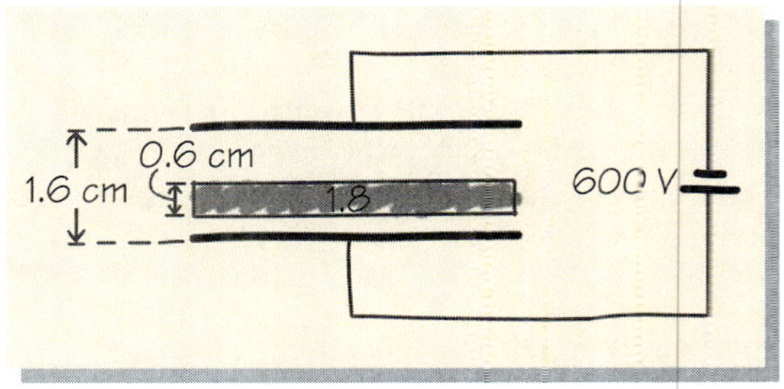

FIGURE 26–31 Problem 52.

53. (II) A parallel-plate capacitor of area 10 cm^2 and plate separation 5 mm holds how much free charge if the voltage between its plates is 300 V, and the following materials are inserted between its plates: air, paper, neoprene, Bakelite, and strontium titanate? (Use Table 26–1.)

54. (II) Two parallel plates of area 100 cm^2 with plexiglass inserted between them break down when a voltage of 10 kV is applied to the plates. How much charge will the plates hold when the plexiglass is removed? (Use Table 26–1.)

55. (II) A capacitor consists of two concentric spherical shells of radii r_1 and r_2, respectively. Calculate the capacitance if the space between the shells is filled with a dielectric of dielectric constant κ. If the capacitor starts out with air between the shells and carries a charge Q, and if the space is then filled with the dielectric, what is the change in energy?

56. (II) A parallel-plate capacitor has area $L \times L$ and separation $D << L$. One-half the space between the plates is filled with a dielectric for which $\kappa = \kappa_0$, and the other half with a dielectric for which $\kappa = \kappa_1$ (Fig. 26–32). Find the capacitance of this capacitor.

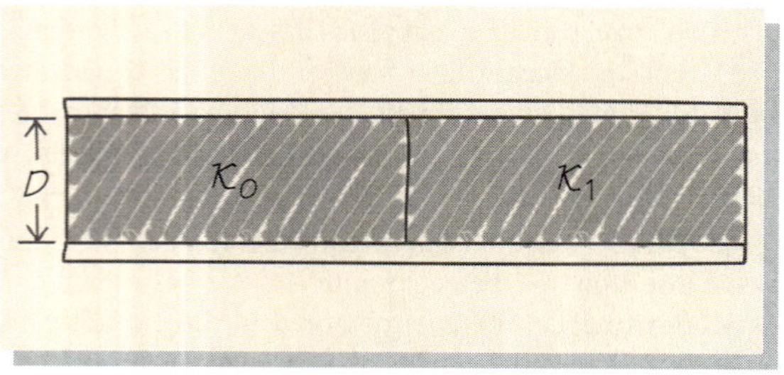

FIGURE 26–32 Problem 56.

57. (II) A capacitor consists of 10 plates attached alternately to a positive and negative terminal. The plates are 6.0 cm $\times$ 8.0 cm in size and are 1.2 mm apart. What is the capacitance? Suppose that the region between the plates is filled with material of dielectric constant 2.8. What will the capacitance be?

26–6 Microscopic Description of Dielectrics

58. (I) By measuring the capacitance and voltage of a capacitor containing a dielectric with dielectric constant 3.1, the free charge on the capacitor is measured to be 2.9 μC. What is the induced charge?

59. (II) Use Gauss' law and Eq. (26–24) to show, from Fig. 26–20, that $E_{ind} = \sigma_{ind}/\epsilon_0$.

60. (II) A charge Q is placed on a parallel-plate capacitor of area $L \times L$ and plate separation d. The capacitor is then filled with a dielectric of dielectric constant κ. If $L = 0.15$ m, $d = 3$ mm, $Q = 0.3\ \mu$C, and $\kappa = 2.5$, what is the surface charge induced on the dielectric? What is the magnitude of the electric field in the dielectric? How much energy is stored in this capacitor?

61. (II) A capacitor filled with a polar dielectric is used as a temperature sensor. Its capacitance is 3.2 μF at 23°C and 2.65 μF at 87°C. What is the capacitance at 48°C?

General Problems

62. (I) An uncharged metal plate is inserted midway between the plates of a parallel-plate capacitor carrying charges Q and $-Q$ on the plates. Will the plate be sucked in or will it have to be pushed in? Give a simple explanation for your result.

63. (II) *Estimate* how much charge you pick up when you walk across a carpet on a dry winter day. [*Hint*: View yourself as a good conductor, spherical in shape, and notice how close your hand has to come to a doorknob before the inevitable spark occurs. Use Table 26–1.]

64. (II) You have 300 cm^2 of aluminum plate (which you can cut into pieces) and a 500-cm^2 sheet of 5-mm thick Bakelite (which you can also cut). Neither material can be sliced into thinner sheets or rolled, and the minimum separation between any aluminum plates you cut is 5 mm. You have a power supply of a single voltage, 1200 V. (a) Design a system that will hold the maximum amount of charge. What charge and energy can this system hold? (b) Design a system that has the maximum electric field, and find this field. Is this the same system as part (a)?

65. (II) Calculate the energy of a composite capacitor that consists of N identical capacitors of capacitance C_1 that are connected (a) in series; (b) in parallel. In parts (a) and (b), the total potential difference across the composite capacitor is V. (c) Assume that the total charge is Q, and repeat the calculation.

66. (II) A capacitor consists of two flat metal plates of area 0.016 m^2 and plate separation of $d = 5.0$ cm. A flat metal plate of the same area and of thickness 1.0 cm is inserted midway between the plates of the capacitor, leaving two spaces of thickness 2.0 cm each. (a) Find the new capacitance. (b) If the original capacitor has charge Q, what is the surface charge density induced on the intermediate plate? (c) Suppose that the original charge on the external plates remains the same. How does the energy of the new system compare to the energy of the system without the inserted plate? (d) Compare the capacitor with the metal inserted to the same capacitor with a dielectric of the same dimensions inserted.

67. (II) A parallel-plate capacitor has an area of $L \times L$ and a plate separation of $D << L$. It is filled with a nonuniform dielectric whose dielectric constant varies linearly across the capacitor (Fig. 26–33). At $x = 0$, $\kappa = \kappa_0$, and at $x = L$, $\kappa = \kappa_1$. We can express κ as a function of x: $\kappa = \kappa_0 + [(\kappa_1 - \kappa_0)x/L]$. Treat the capacitor plates as broken into a set of capacitors connected in parallel with plates that are strips of width dx, and calculate the capacitance.

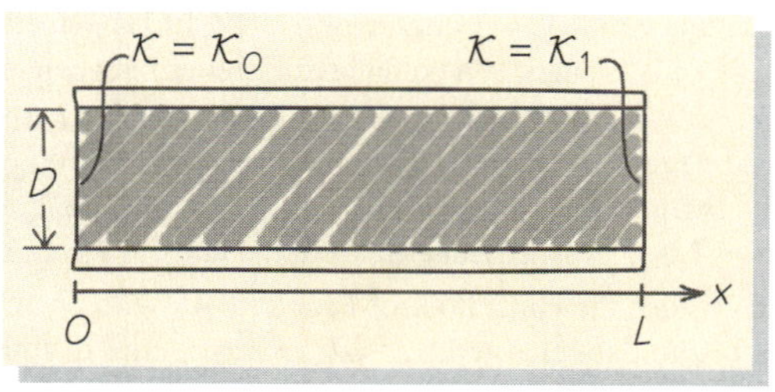

FIGURE 26–33 Problem 67.

68. (II) A thunderstorm is a fairly complicated phenomenon in terms of the distribution of charges, but we can estimate that there is a voltage drop of as much as 10^8 V between Earth and the bottom of a thundercloud, and the charges involved may run into the hundreds of coulombs. Estimate the capacitance of the Earth–cloud system and the energy contained in the space between the cloud and Earth.

69. (II) Consider the arrangement of the four initially uncharged capacitors shown in Fig. 26–34. Capacitors A, B, C, and D have capacitances 5.4 μF, 4.3 μF, 3.2 μF, and 2.1 μF, respectively. Suppose that a battery applies a potential difference of 3000 V across the circuit, which is then disconnected from the battery. What is the potential difference across each capacitor?

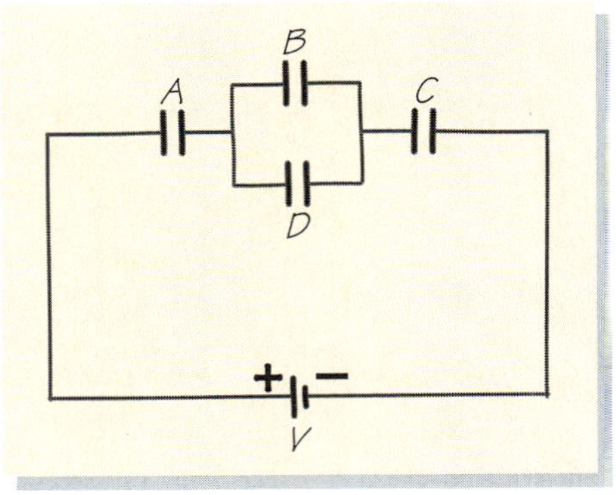

FIGURE 26–34 Problem 69.

70. (II) Consider a parallel-plate capacitor of plate area 1.0 m^2 and plate separation 4.0 mm. (a) Assume that the maximum electric field strength (before breakdown) in air is 3.0×10^6 V/m. What are the capacitance and the charge stored at the maximum voltage? (b) Suppose that the capacitor is immersed in oil of dielectric constant $\kappa = 2.4$, and the maximum charge that can be stored is a factor of 10 larger than that without the oil. What is the maximum field strength the oil can maintain?

71. (II) A parallel-plate capacitor has a capacitance of 3.0 μF. The plates are charged to 1500 V. What is the energy stored in the capacitor? How much work is required to insert a dielectric of $\kappa = 2.8$ between the plates? Assume that the capacitor is disconnected from the voltage source before the dielectric is inserted.

72. (II) A dielectric of dielectric constant κ is inserted a distance x into a parallel-plate capacitor with square plates of area A and plate separation d. What is the capacitance as a function of x? Calculate the amount of energy stored in the capacitor for a potential difference V.

73. (II) A parallel-plate capacitor has an area of $L \times L$ and a plate separation of $D << L$. It is filled with a nonuniform dielectric whose dielectric constant varies linearly from one plate to another (Fig. 26–35). At the bottom plate, the dielectric constant is κ_0; at the upper plate, it is κ_1. If y is the distance measured up from the bottom plate to the top plate, then $\kappa = \kappa_0 + [(\kappa_1 - \kappa_0)y/D]$. Treat the capacitor as a set of capacitors connected in series, and calculate the capacitance.

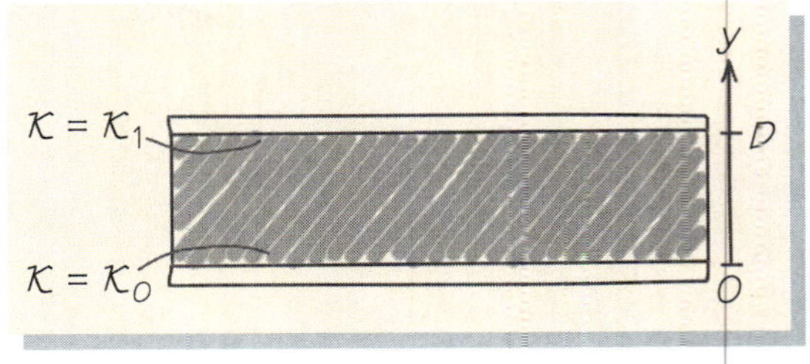

FIGURE 26–35 Problem 73.

74. (II) Three capacitors of strengths 3 μF, 5 μF, and 6 μF, respectively, can be connected in various ways between two points. What arrangement gives the smallest equivalent capacitance, and what arrangement gives the largest capacitance?

75. (II) Show that when capacitors are arranged in series, the total capacitance is less than any of the individual capacitances.

76. (III) Two identical capacitors of capacitance C are connected in series across a total potential V. A dielectric slab of dielectric constant κ can fill one of the two capacitors and is slowly inserted into that capacitor (Fig. 26–36). Compute the changes in the total electric energy of the two capacitors, in the charge on each capacitor, and in the potential drop across each capacitor. Account for any energy change by a corresponding change in energy in some other part of the system.

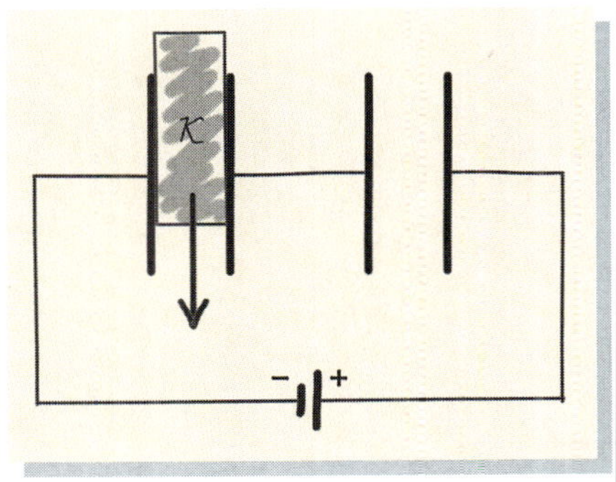

FIGURE 26–36 Problem 76.

CHAPTER

27

An electric arc welder can be used to attach two conducting objects. Currents are so large that the metals will melt and fuse together.

Currents in Materials

Electric current describes the motion of charges. The motion of charged particles in free space—for example, the electron beam of a television tube—is familiar to us from Chapter 23. More frequently, however, we encounter electric currents within the materials that make up circuits. The motion of charges within a material is determined by characteristics of the material. The effect of a material's structure is like the effect of drag: On average, the charges move at constant terminal speed. Because of the draglike forces, we must expend energy to make charges pass through materials, and we produce thermal energy. To describe this picture at the level of bulk materials, we introduce resistance, resistivity, and conductivity, which are characteristics of the materials. A fundamental understanding of resistance requires the ideas of quantum physics. Quantum physics also explains the differences among conductors, insulators, semiconductors, and superconductors.

27–1 ELECTRIC CURRENT

Electric current (or just **current**) is defined as the total charge that passes through a given cross-sectional area per unit time. In Fig. 27–1, we have drawn the charge that passes through a wire. Recall that *charge is conserved* (see Chapter 22). Even when charges flow through a region of empty space, the conservation of charge allows us to follow the flow systematically. Here, we concentrate on the general notions of current—whether that current describes the motion of charges within free space or within conducting materials.

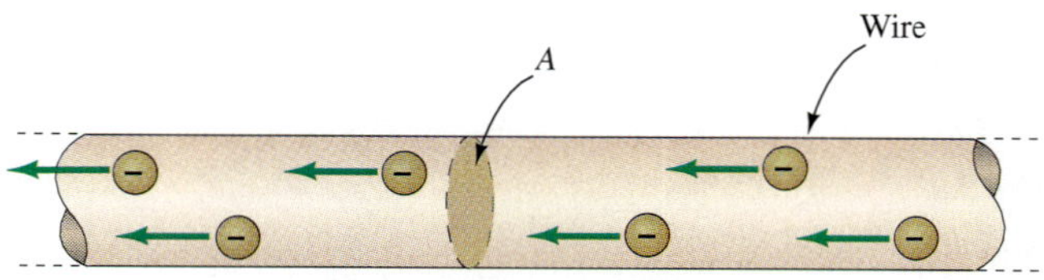

FIGURE 27–1 Charges move in a cross section of wire.

If ΔQ is the amount of charge passing through an area in a time interval Δt, then the *average current*, I_{av}, is defined as

$$I_{av} \equiv \frac{\Delta Q}{\Delta t}. \tag{27–1}$$

Definition of electric current

If the current changes with time, we define the *instantaneous* current, I, by taking the limit $\Delta t \to 0$, so that the current is the instantaneous rate at which charge passes through an area:

$$I \equiv \frac{dQ}{dt}. \tag{27–2}$$

Units of Current

The unit of current is the coulomb per second; this unit is also called the *ampere* (A), however, after André Marie Ampère, who performed pioneering work in electricity and magnetism early in the nineteenth century. The ampere will be defined more precisely in Chapter 29, but that definition will be equivalent to the simple relation

The definition of the ampere

$$1\ \text{A} = 1\ \text{C/s}.$$

We often use the term "amp" to refer to the ampere. Because the ampere is a rather large unit, current is expressed also in milliamps (mA), or 10^{-3} A; microamps (μA), or 10^{-6} A; or even nanoamps (nA), or 10^{-9} A.

Currents occur over a wide range of values (Table 27–1). Currents that have a harmonic time dependence are called alternating currents (AC), a phenomenon we shall study in more detail in Chapter 34. For such currents, the values in Table 27–1 represent the average magnitude of the oscillating current.

Current is a scalar quantity, but it has a sign associated with it. It is useful to indicate the sign of the current by a directional arrow. By convention, we associate the direction of the arrow with the flow of positive charges—even though it is actually the

TABLE 27–1 Values of Various Currents

Situation	*Current (A)*
Advanced computer technology chips	10^{-12} to 10^{-6}
Electron beam of a TV set	10^{-3}
Proton beam of the Fermilab accelerator	3×10^{-3}
Current that is dangerous when it passes through the human body	10^{-2} to 10^{-1}
Flashlight bulb	0.3
Household light bulb	1
Automobile starter	200
Peak current in a lightning strike	10^{4}
Maximum current carried by a superconducting niobium wire of 1 cm^2 cross section	10^{7}

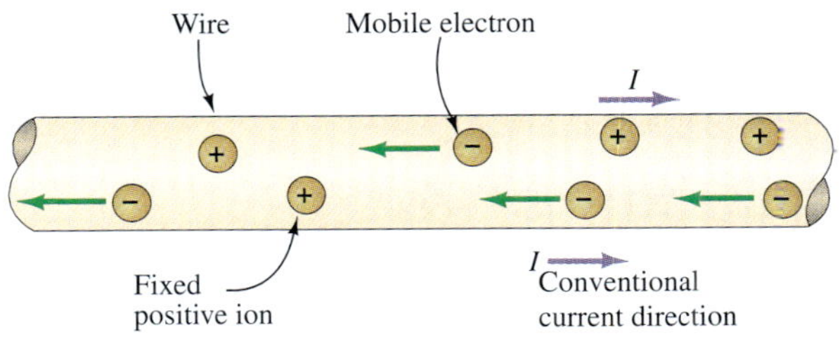

FIGURE 27–2 By convention, the current direction is the direction in which positive charge effectively moves, even though in conductors it is the negative charge, electrons, that actually moves.

negative charges that move in metals (Fig. 27–2). The positive charges—the atomic ions left behind by the electrons—are fixed in an ordered crystal lattice (see Chapter 21). This arbitrary convention for current direction causes no real problem; a flow of positive charge to the right and a flow of the same amount of negative charge to the left represents the same current. By simply measuring the current, it is *not possible* to determine the sign of the charges that move (the *charge carriers*). By convention, then,

Currents flow in the direction that positive charges would flow.

Currents are depicted as if the positive charges are moving, even though, in conductors, the carriers (the electrons) are negatively charged.

EXAMPLE 27–1 An accelerator used for research on the treatment of tumors ejects protons at the rate of 2.0×10^{13} protons/s. What is the current carried by this beam of protons?

Solution: The current is the charge per unit time carried by the beam. This current is

$$\begin{aligned} I &= (\text{number of protons/s})(\text{charge per proton}) \\ &= (2.0 \times 10^{13}\ \text{protons/s})(1.6 \times 10^{-19}\ \text{C/proton}) \\ &= 3.2 \times 10^{-6}\ \text{C/s} = 3.2 \times 10^{-6}\ \text{A}. \end{aligned}$$

Current Density

We must often deal with the *details* of charge motion, not just an overall movement of charge. Then, we must work with **current density**, **J**, which is the rate of charge flow per unit area through an infinitesimal area. Note that the flow rate can vary from one point to another and, to define the current density, we must take into account the local magnitude and direction of the charge flow. Unlike current, which is a scalar, current density is a *vector*, with units of amperes per square meter. The direction of **J** is defined to be the direction of the net flow of positive charges at the particular infinitesimal element of area.

What is the relation between current density and current? We determine this relation in a wire by dividing the finite area A through which the charge flows into infinitesimal areas d**A** (Fig. 27–3). This procedure is analogous to one we followed in treat-

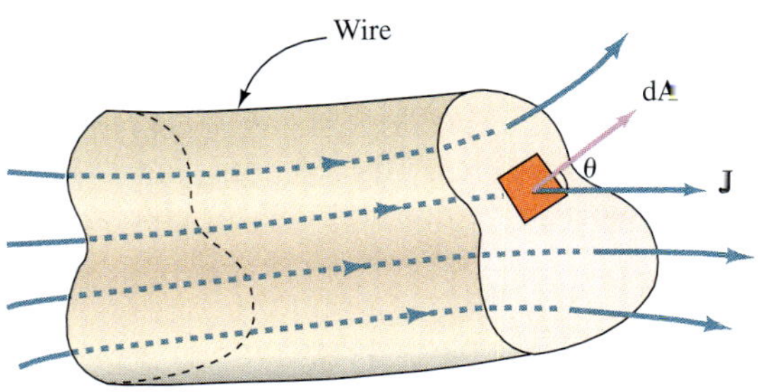

FIGURE 27–3 The area of a finite wire is divided up into differential areas d**A** with the current density, **J**, defined at every point. The direction of d**A** is normal to the differential area.

ing fluid flow (in Chapter 16) or electric flux (in Chapter 24). The differential current dI flowing through d**A** is

$$dI = \mathbf{J} \cdot d\mathbf{A} = J\,dA \cos\theta, \tag{27–3}$$

where θ is the angle between **J** and the area element d**A**. From Eq. (27–3), we see that dI is a maximum when **J** and d**A** are parallel and dI is zero when **J** is perpendicular to d**A**. The total current passing through the area A is a sum over the differential currents dI:

$$I = \int_{\text{surface } A} \mathbf{J} \cdot d\mathbf{A}. \tag{27–4}$$

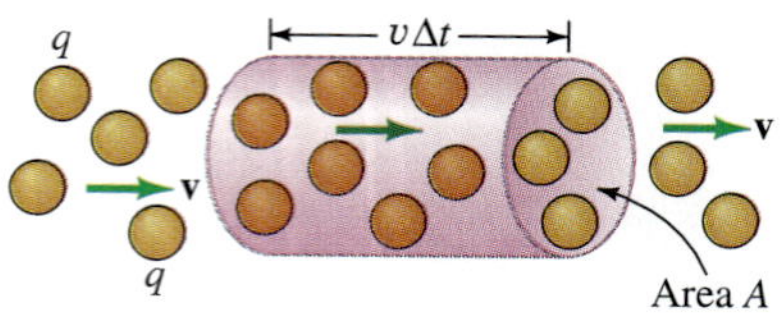

FIGURE 27–4 A collection of particles (each with charge q) with number density n_q all move to the right with velocity **v**. The total charge passing through an area A in time Δt is $\Delta Q = n_q qvA\,\Delta t$.

Current Density of Moving Charges

Let's find the current density of a group of moving charges. Suppose that we have a collection of particles with charge q. In some small region, the number of these charged particles per unit volume—the *number density*—is n_q. Suppose also that these particles all move with velocity **v**. Then, in a time interval Δt, the amount of charge passing through a given area A perpendicular to **v** is ΔQ, which is the charge contained in the volume $A(v\,\Delta t)$ swept out by the moving charges (Fig. 27–4):

$$\Delta Q = \left(\frac{\text{charge}}{\text{volume}}\right)(\text{volume}) = (n_q q)(Av\,\Delta t) = n_q qvA\,\Delta t. \tag{27–5}$$

We have used the fact that the charge per unit volume is the number density of the charge carriers times the charge per particle. Thus the current is given by

$$I = \frac{\Delta Q}{\Delta t} = n_q qvA. \tag{27–6}$$

Finally, the current density is I divided by A in the limit of *small* A, or $J = I/A$. The direction of **J** is specified by the direction of **v**:

$$\mathbf{J} = n_q q\mathbf{v}. \tag{27–7}$$

The classification of materials according to how well they carry charge was made in Chapter 22.

27–2 CURRENTS IN MATERIALS

We have defined *conductors* as materials through which charge moves easily, *insulators* as materials through which charge does not move easily, *semiconductors* as materials intermediate to conductors and insulators, and *superconductors* as materials that under certain circumstances—in particular, at low temperatures—carry charge with no inhibition whatsoever. How materials carry charge is of obvious importance. A simple model (to be developed further in Section 27–4) explains that metals are good conductors because they contain electrons that behave as though they were free. A free electron in a metal experiences a force $\mathbf{F} = -e\mathbf{E}$ and thus experiences an acceleration in a direction opposite to an electric field **E**.[†] Such electrons undergo frequent collisions with the positive ions that form the crystal lattice of the metal whether the field is present or not (Fig. 27–5). When there is no field, the electrons do not, *on average*, move in any particular direction. Their motion is random, like the motion of air molecules. When the electric field is present, there is a *net* movement of electrons in the direction of the electric force they experience. The collisions, in effect, give rise to a drag force on the flow of electrons. As with the fall of a parachute, drag acts to settle the motion to a steady flow in the direction of the force. The electrons move with a constant terminal velocity called the **drift velocity**, $\mathbf{v}_d$. Equation (27–6) gives the re-

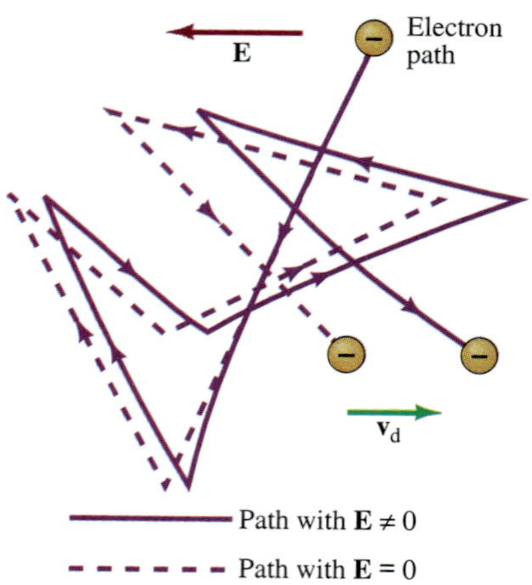

FIGURE 27–5 An electron collides frequently with the ions and impurities in a metal and scatters randomly. In an electric field, the electron picks up a small component of velocity opposite the field. The differences in the paths are exaggerated. The electron's path in an electric field is slightly parabolic.

[†]While in electrostatics, metals contain no electric fields, this is not electrostatic: Charges are moving continuously.

lation between drift speed and current. In the special case that n_q equals n_e, the density of free electrons in the metal, Eq. (27–6) gives

$$I = n_e q v_d A, \tag{27–8}$$

where A is the cross-sectional area of a metal wire.

We can solve Eq. (27–8) to find the drift speed in terms of the current:

$$v_d = \frac{I}{n_e q A}. \tag{27–9}$$

Equations (27–8) and (27–9) are the desired relations between current and drift speed. Remember that the direction of the electron's drift velocity is opposite to the defined direction of the current density because of the positive charge-carrier convention.

In Fig. 27–6, we show the relationships among the external electric field, **E**, the current, I, the current density, **J**, the electron's drift velocity, $\mathbf{v}_d$, and the movement of electrons. For the case of the wire, $J = I/A$; from Eq. (27–9), we have

$$\mathbf{J} = n_e q \mathbf{v}_d. \tag{27–10}$$

Note that **J** is indeed opposite to the direction of $\mathbf{v}_d$ because of the negative sign of the charge q ($-e$).

EXAMPLE 27–2 Estimate the drift speed, v_d, for electrons in a copper wire of diameter $d = 1$ mm that carries a current of 100 mA. Copper has about one free electron per atom available to carry charge and has a mass density of 8.92 g/cm³ and a molecular weight of 63.5 g/mol.

Solution: The situation is similar to that shown in Fig. 27–6. We use Eq. (27–9) to calculate the drift speed. We are given the current, I, and can find $A = \pi r^2$, where the wire's radius is $r = d/2$. However, we must find n_e from the given information about copper. Because copper has about one free electron per atom, the density of free electrons, n_e, is identical to the density of copper atoms, n. The atomic density is derived from the mass density of copper, $\rho_{Cu} = 8.92$ g/cm³; the number of atoms per mole, N_A; and the molar weight of copper, $M = 63.5$ g/mol:

$$n_e = n = \frac{N_A \rho_{Cu}}{M} = \frac{(6.02 \times 10^{23}\ \text{atoms/mol})(8.92\ \text{g/cm}^3)}{63.5\ \text{g/mol}} \frac{(1\ \text{electron})}{\text{atom}}$$

$$= 8.5 \times 10^{22}\ \text{electrons/cm}^3.$$

If we assume that the current and drift speed are constant across the wire, Eq. (27–9) gives

$$v_d = \frac{I}{n_e q A} = \frac{100 \times 10^{-3}\ \text{A}}{(8.5 \times 10^{22}\ \text{electrons/cm}^3)(1.6 \times 10^{-19}\ \text{C/electron})\pi(0.05\ \text{cm})^2}$$

$$= 9.4 \times 10^{-4}\ \text{cm/s} = 9.4 \times 10^{-6}\ \text{m/s}.$$

The drift velocity is so slow—only about 0.001 cm/s—that you might wonder how a measurable current can even flow. What happens when we switch on a household circuit? Certainly we do not have to wait for hours for the electrons to drift several feet. When the switch is thrown, the electric field that influences the electrons to move in the wire is set up throughout the wire at speeds approaching the speed of light. The

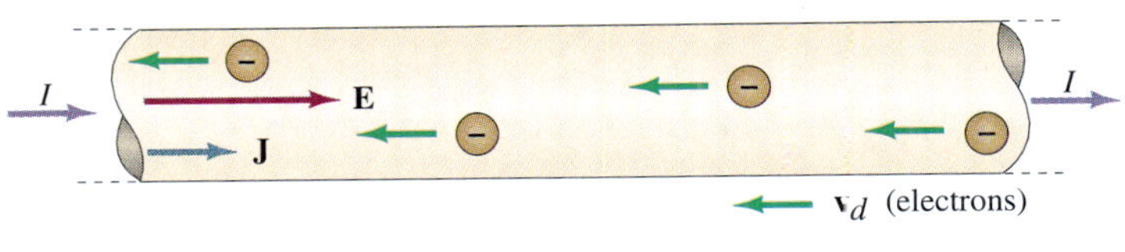

FIGURE 27–6 Electrons drift in the direction opposite that of the current, I, current density, **J**, and electric field, **E**.

free electrons are spread throughout the wire, and they all start moving at once in response to this field—those nearest the switch as well as those nearest the electrical appliance. A similar effect occurs in fluid flow. If you want to move a sprinkler while you water the lawn, you turn the water off, move the sprinkler, and then turn the water back on. Because the hose is already full of water, the sprinkler starts immediately: The force of the water at the faucet end is quickly transmitted all along the hose, and the water at the sprinkler end of the hose flows from the sprinkler almost the moment the faucet is opened.

Current and the Conservation of Charge

Current, like charge, is conserved.

How does the conservation of charge affect currents in materials? The conservation of mass implies that, in a steady state, the rate at which fluid enters a system of pipes is the rate at which it leaves the system. Similarly, the conservation of charge leads to the principle of the *conservation of current*. For steady flow, in which currents do not change with time, the total current that enters some section of wire is the total current that leaves that section. Thus the current is the same everywhere along a wire, even if the wire changes in area (Fig. 27–7a). Because the current is the same everywhere along a wire of varying area, the current density is inversely proportional to the area. In other words, if a wire has cross-sectional areas A_1 and A_2 at two points along it, then the conservation of current implies that $J_1A_1 = J_2A_2$, where J_1 and J_2 are the current densities at these respective regions; these current densities are assumed to be constant across the wire's cross section. The density of charge carriers and their charge are fixed in a given metal. Thus, by Eq. (27–10), if the current density is inversely proportional to the area of the wire, so is the drift speed (Fig. 27–7b). A familiar analogue to this situation occurs when you are driving and approach a region of the road where three lanes narrow into one. Motion in the three-lane section is painfully slow, but once you reach the one-lane section, your speed can increase.

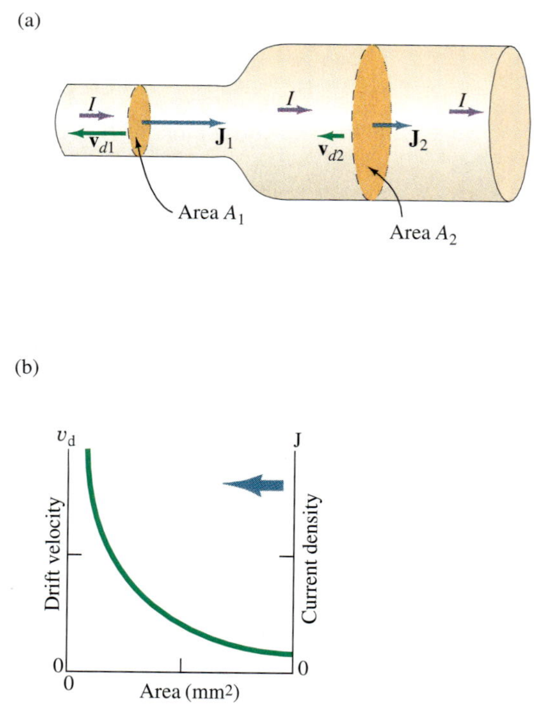

FIGURE 27–7 (a) A steady current is the same in all parts of a wire, even if the area of the wire varies. This means that the current density and drift speed will vary with the area: Both are larger when the wire cross section is smaller. (b) More precisely, the drift speed and current density are inversely proportional to the wire area.

APPLICATION: USAGE INDICATORS

A device that measures the total amount of current—and thus the number of hours of operation—is installed in many electronic or electrical products in order to control warranty, maintenance, and replacement (Fig. 27AB–1). Take a thin-walled capillary tube containing at each end a mercury "thread," one short and one long. Between the threads is a dilute solution of mercury nitrate, a conducting fluid; this fluid makes a short (about 1 mm) transparent break between the threads and acts as a "pointer" on an hour scale placed next to the tube. Two iron wires are inserted into the mercury threads at the ends of the tube and sealed to the glass capillary to form a hermetic system. The wires form the connections to the device.

This device is connected to a product in such a way that product use sends a steady electrical current through the device—the normal range being around 50 microamps to one milliamp. The current travels in along an iron wire, through the long mercury thread, through the solution, and into the short thread, exiting through the second wire. In the wires and the mercury, the charge carriers are conduction electrons; however, in the solution, the mercury nitrate is ionized and the charge carriers are mercury ions (positive) and nitrate ions (negative). When the mercury ions reach the mercury at the end of the short thread, they deposit on it and add to its volume. Similarly, when the nitrate ions reach the end of the long thread, they combine with the mercury to form mercury nitrate once more. The net effect is to dissolve mercury from the long thread and deposit it on the short thread. The gap and therefore the "pointer" moves along the tube, indicating the time that this current has been flowing. These devices can easily be reset by reversing the current flow and using a much larger current to shorten the reset time.

FIGURE 27AB–1 This usage indicator is only a few centimeters in size.

27–3 RESISTANCE

We have seen that a current flows when an electric field is applied to a conductor. We can consider the potential difference V due to the electric field as the source of the motion. The amount of current that flows through a material for a given potential difference across that material depends on the material's properties and geometry.

The definition of resistance

The **electrical resistance**, R, of a piece of material is a measure of how easily charge flows within that material. The electrical resistance is defined to be the ratio of the voltage (potential difference) across the material to the current that flows through it:

$$R \equiv \frac{V}{I}. \qquad (27\text{–}11)$$

Units of resistance

The units of resistance are volts per ampere, but a separate SI unit called the **ohm** (Ω) has been defined as the resistance through which a current of 1 A flows when a potential difference of 1 V is applied:

$$1\ \Omega = 1\,\frac{\text{V}}{\text{A}}.$$

Ohm's law applies whenever the resistance of a material is a constant.

Georg Simon Ohm was the first to study the resistance of different materials systematically. In 1826, he published his experimental result that, for many materials including most metals, *the resistance is constant over a wide range of potential differences*. This statement is called *Ohm's law*. It is not really a law at all but an empirical statement about the behavior of materials. When the resistance of a material is constant over a range of potential differences, we say that the material is *ohmic*. We shall

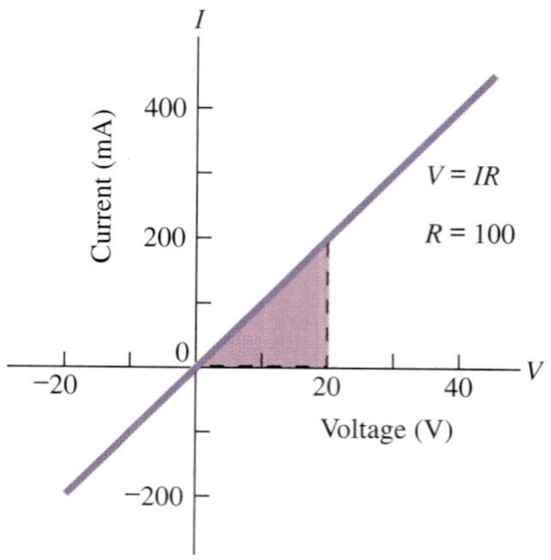

FIGURE 27–8 For ohmic materials, Ohm's law states that the ratio V/I is a constant.

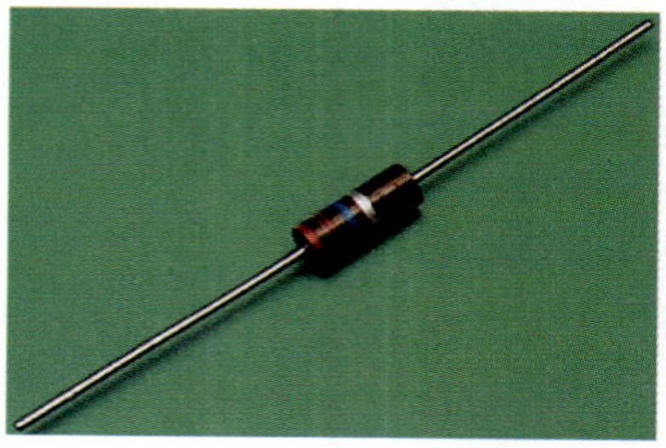

FIGURE 27–9 Resistors are color coded to indicate the value of their resistance.

continue the traditional practice of referring to this linear relation between voltage and current for these materials as a "law" and writing it as

$$V = IR. \tag{27–12}$$

where R is independent of V. *The resistance, R, is understood to be independent of V here.* Figure 27–8 illustrates the consequence of a constant R.

Resistors

A piece of ohmic material of significant resistance—a **resistor**—is the most mundane of the elements that make up an electric circuit (Fig. 27–9). ("Significant" depends on the application but, in practice, a resistance of a few ohms is small.) A resistor of given resistance R with a given potential V between its terminals allows the flow of a current $I = V/R$. Resistors are represented in circuit diagrams by zigzag lines, -\/\/\/-; they are connected to each other and to other elements, such as capacitors, by conducting wires that are generally assumed to have negligible resistance. Many resistors in electric circuits are given in kilohms (kΩ or 10^3 Ω) or megohms (MΩ or 10^6 Ω).

There are many *nonohmic materials*: materials for which the voltage and current do not obey the linear relation of Ohm's law. Figure 27–10a shows current versus voltage curves (ideal and typical) for a **diode**. A diode is a device that transmits current easily when the voltage is positive, but prevents charge flow (that is, it has a very high resistance) when the voltage is negative. Diodes are used in many electric devices (Fig. 27–10b). For example, they may be used to allow a battery to be charged but to prevent it from discharging, without the need for a switch.

Resistivity and Conductivity

The resistance of a conducting wire of a given material can vary with the wire's shape. Let's again consider a uniform wire. We may think of the resistance to a flow of charge in a conductor as the result of collisions of the moving charge carriers (electrons) with lattice atoms. When the length of the wire is doubled, the number of collisions doubles. Thus the *resistance of a wire is proportional to its length L.* Conversely, if the cross-sectional area of a wire is doubled, then twice as much current will flow through it, just as twice as much water will flow out of a bathtub with two identical drains as will flow out of a tub with only one drain. As long as the potential remains constant, a doubling of the current implies a halving of the resistance. Therefore, the *resistance of a wire of a given material is inversely proportional to its cross-sectional area A.*

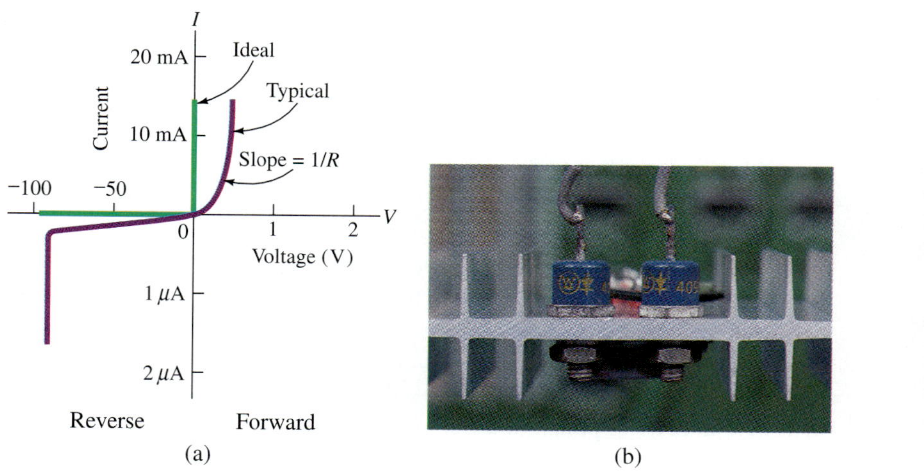

FIGURE 27–10 (a) Plot of current versus voltage for an ideal diode (green) and a typical real diode (purple). Note that, for the ideal diode, there is no current when the voltage is negative: The diode allows current to flow in only one direction. (b) Two (blue) diodes used to convert **AC** voltage to **DC** are bolted to an aluminum plate to dissipate heat energy (heat sink).

We combine these two results to define the **resistivity** of a material, ρ, by the relation

$$\rho \equiv R\frac{A}{L}. \tag{27–13}$$

The definition of resistivity

With this definition, and with the dependence of R on L and A that we have already established, ρ does not depend on the dimensions of the conductor, but only on the type of material. The units of resistivity are ohm-meters ($\Omega \cdot$ m); characteristic values for a variety of materials are given in Table 27–2. Equation (27–13) is typically rewritten as

Resistivity is a property only of the type of material, whereas resistance depends on both the type of material and its shape.

$$R = \rho\frac{L}{A}. \tag{27–14}$$

The reciprocal of the resistivity is the **conductivity**, σ:

The definition of conductivity

$$\sigma \equiv \frac{1}{\rho}. \tag{27–15}$$

Typical values of conductivity are in Table 27–2. The resistivities and conductivities of the materials shown in Table 27–2 vary over many orders of magnitude. The conductivity of a metal like aluminum (Fig. 27–11) is a factor of 10^{21} higher than that of a good insulator, such as Teflon.

We can write Eq. (27–12) in terms of resistivity and conductivity:

$$V = IR = \rho\frac{L}{A}I = \rho L\frac{I}{A};$$

$$\frac{V}{L} = \rho\frac{I}{A}.$$

The relation between electric field and current density

V/L is the magnitude E of the electric field applied to the material, and I/A is the magnitude of current density, J. Because charges move in the direction of the electric field, this translates into the vector relation

$$\mathbf{E} = \rho\mathbf{J}. \tag{27–16}$$

TABLE 27–2 Resistivities, Conductivities, and Temperature Coefficients (at 20°C)

Material	*Resistivity, ρ ($\Omega \cdot m$)*	*Conductivity, σ $(\Omega \cdot m)^{-1}$*	*Temperature Coefficient, α $(°C)^{-1}$*
Conductors			
Elements			
Aluminum	2.82×10^{-8}	3.55×10^{7}	0.0039
Silver	1.59×10^{-8}	6.29×10^{7}	0.0038
Copper	1.72×10^{-8}	5.81×10^{7}	0.0039
Iron	10.0×10^{-8}	1.0×10^{7}	0.0050
Tungsten	5.6×10^{-8}	1.8×10^{7}	0.0045
Platinum	10.6×10^{-8}	1.0×10^{7}	0.0039
Alloys			
Nichrome	100×10^{-8}	0.1×10^{7}	0.0004
Manganin	44×10^{-8}	0.23×10^{7}	0.00001
Brass	7×10^{-8}	1.4×10^{7}	0.002
Semiconductors			
Carbon (graphite)	3.5×10^{-5}	2.9×10^{4}	−0.0005
Germanium (pure)	0.46	2.2	−0.048
Silicon (pure)	640	1.6×10^{-3}	−0.075
Insulators			
Glass	10^{10} to 10^{14}	10^{-14} to 10^{-10}	
Neoprene rubber	10^{9}	10^{-9}	
Teflon	10^{14}	10^{-14}	

FIGURE 27–11 The three large black wires coming into the residential electrical service box connect the household to the electric utility service. One wire connects to ground, the remaining two connect to circuit breakers from which current goes to the house. Safety dictates that, while the exterior wires are aluminum, copper is used in the house.

Equivalently, from the definition of conductivity, Eq. (27–15), we have

$$\mathbf{J} = \sigma\mathbf{E}. \tag{27–17}$$

Equations (27–16) and (27–17) are general results—not limited to ohmic materials, for which ρ and σ do not vary with V or $\mathbf{E}$.

EXAMPLE 27–3 Determine the current density, resistance, and electric field for the copper wire of Example 27–2 if the wire is 10 m long.

Solution: According to Example 27–2, we know that the wire carries a 100-mA current. The wire has a diameter $d = 1$ mm, so its cross section is $A = \pi(\frac{1}{2}d)^2$. Thus we can compute the unknown current density from the definition of J:

$$J = \frac{I}{A} = \frac{100 \times 10^{-3}\ \text{A}}{\pi(0.5 \times 10^{-3}\ \text{m})^2} = 1.3 \times 10^5\ \text{A/m}^2.$$

If we take the resistivity of copper from Table 27–2, we can use Eq. (27–14) to determine the resistance:

$$R = \rho\frac{L}{A} = \frac{(1.72 \times 10^{-8}\ \Omega \cdot \text{m})(10\ \text{m})}{\pi(0.5 \times 10^{-3}\ \text{m})^2} \simeq 0.2\ \Omega.$$

Finally, given the resistivity, we can determine the electric field from Eq. (27–16):

$$E = \rho J = (1.72 \times 10^{-8}\ \Omega \cdot \text{m})(1.3 \times 10^5\ \text{A/m}^2) = 2.2 \times 10^{-3}\ \text{V/m}.$$

Note that both the current density and electric field are independent of the length of the wire. The voltage required to produce both, however, is dependent on the wire's length.

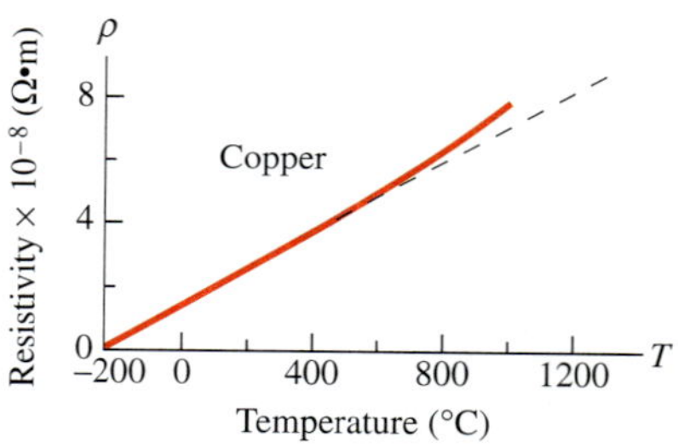

FIGURE 27–12 The resistivity of copper as a function of temperature.

The Temperature Dependence of Resistivity

Resistivities of some materials have a strong temperature dependence; that of copper is shown in Fig. 27–12. We can represent the temperature dependence with the following linear approximation, which is sufficiently accurate for most purposes:

$$\rho \simeq \rho_0[1 + \alpha(T - T_0)]. \tag{27–18}$$

The parameter α is the *temperature coefficient of resistivity*, and ρ_0 is the resistivity at the reference temperature T_0, normally 20°C. Values of ρ, σ, and α are given in Table 27–2 for $T_0 = 20$°C. Resistivities for most metals increase with temperature, as in Fig. 27–12; we'll discuss this further in the next section.

EXAMPLE 27–4 Calculate the resistance of a coil of platinum wire of diameter 0.5 mm and length 20 m at 20°C. Also determine the resistance at 1000°C.

Solution: We must find the resistance rather than the resistivity; however, the relation between these two quantities is given by Eq. (27–14). We can thus find the resistance at 20°C from the resistivity at 20°C in Table 27–2:

$$R_{20°\text{C}} = \rho\frac{L}{A} = (10.6 \times 10^{-8}\ \Omega \cdot \text{m})\frac{20\ \text{m}}{\pi\left[\frac{1}{2}(0.5 \times 10^{-3}\ \text{m})\right]^2} = 11\ \Omega.$$

To find the resistance at 1000°C, we combine Eq. (27–14), $R = \rho L/A$, with Eq. (27–18) to produce an equation that gives resistance as a function of temperature for conducting wires:

$$R = R_0[1 + \alpha(T - T_0)]. \tag{27–19}$$

From Table 27–2, we get the temperature coefficient for platinum and find that

$$R_{1000°\text{C}} = (11\ \Omega)[1 + (0.0039°\text{C}^{-1})(1000°\text{C} - 20°\text{C})] = 52\ \Omega,$$

where we have used $T_0 = 20$°C. The resistance is a factor of nearly 5 greater at the higher temperature.

*27–4 FREE-ELECTRON MODEL OF RESISTIVITY

A correct understanding of resistivity requires quantum mechanics. Nevertheless, there is a simple classical model of resistivity that can help us understand Ohm's law. It was first proposed in 1900 by Paul Drude and is known as the **free-electron model**, or the **Drude model**. Although the model has fundamental deficiencies, its study is worthwhile for two reasons: First, the model allows us to focus on the concept of resistivity. Second, the model provides us with an example of how model-building in the physical sciences proceeds, and how we can judge the success or failure of a model.

Let's start with the idea that solids contain "free" electrons, which can move within the material and carry charge. The density of free electrons, n_e, depends on the material and is responsible for the differences among conductors, insulators, and semiconductors (which we shall discuss in Section 27–5). In metals, the number of loosely attached electrons per atom (these are the electrons that behave as though they were free) lies in the range 1.0 to 1.3, but it can be as large as 3.5, as in aluminum.

Crystal lattices of solids are discussed in Chapter 21.

The model postulates that free electrons form a gas of independent particles at temperature T. When a current is produced, the electrons are accelerated by an applied electric field, but collisions with the atoms or ions that form the crystal lattice of the solid slow them down. In other words, drag forces act on the electrons in an average sense. The simplest drag force is proportional to the electrons' speed. If we assume this form, then Newton's second law for the component of electron motion that is parallel to the applied field is

$$ma = -eE - (\text{a constant})v,$$

where m is the mass of an electron. The constant must have dimensions of mass/time, and we write it as m/τ, where τ is a quantity with dimensions of time. It is reasonable to equate τ with the average time between collisions because it is the collisions that impede the electrons' motion. The acceleration drops to zero when the speed reaches the drift speed, v_d. When the acceleration is zero, $ma = -eE - (m/\tau)v_d = 0$, or

$$v_d = -\frac{eE\tau}{m}. \tag{27–20}$$

The minus sign indicates that the direction of the drift velocity is opposite to that of the electric field, as is appropriate for negative charge carriers. When this expression is inserted into Eq. (27–10) for the current density, we obtain

$$J = n_e(-e)v_d = \frac{n_e e^2 \tau}{m} E. \tag{27–21}$$

Comparison to Eq. (27–17) yields

$$\rho = \frac{n_e e^2 \tau}{m} \tag{27–22}$$

for the conductivity. Equivalently, the resistivity, $\rho = 1/\sigma$, is predicted to be

$$\rho = \frac{m}{n_e e^2 \tau}. \tag{27–23}$$

The quantities e and m are independent of the type of material. The average time between collisions may be expressed in terms of the *mean free path* λ and the average speed v_{av} of the electrons in the free-electron "gas" by using Eq. (19–51):

$$\tau = \frac{\lambda}{v_{av}}.$$

For normal electric fields, none of the quantities in Eq. (27–23) depend on E, and thus the resistivity (or conductivity) is constant. It was with this argument that the original atomic foundation of Ohm's law was laid down by Drude (and independently by Hendrik Lorentz) in 1900.

EXAMPLE 27–5 What is the free-electron model's prediction for the collision time of current-carrying electrons in copper, given that the resistivity of copper is $1.7 \times 10^{-8}\ \Omega \cdot \text{m}$? You may use the parameters of Example 27–2.

Solution: The connection between microscopic parameters such as resistivity and collision time in the free-electron model is given in Eq. (27–23). In Example 27–2, we estimated that the number density of current-carrying electrons is $n_e = 8.5 \times 10^{22}$ electrons/cm^3 $= 8.5 \times 10^{28}$ electrons/m^3. From Eq. (27–23),

$$\tau = \frac{m}{n_e e^2 \rho} = \frac{0.91 \times 10^{-30}\ \text{kg}}{(8.5 \times 10^{28}\ \text{electrons/m}^3)(1.6 \times 10^{-19}\ \text{C})^2(1.7 \times 10^{-8}\ \Omega \cdot \text{m})}$$

$$= 2.4 \times 10^{-14}\ \text{s}.$$

The Failure of the Free-Electron Model

The free-electron model helps us understand charge conduction in materials. If we take a closer look at the model's predictions by comparing them with experimental results, however, we find significant discrepancies. In particular:

1. The measured random thermal speed of electrons in metals is *more than a factor of 10* higher than the model predicts for copper at room temperature.
2. We know from Chapter 19 [see Eq. (19–33)] that the rms speed of the particles of a gas is proportional to $\sqrt{T}$, where T is the temperature. However, experimental values for the mean speed of conduction electrons are essentially independent of temperature.
3. The mean free path—the average distance an electron travels between collisions [see Eq. (19–15)]—should be independent of temperature according to the model. Experimentally, though, this quantity is much larger than expected and has a T^{-1} dependence.

Although the free-electron model is *qualitatively* correct in many aspects, it cannot be taken too literally. A correct model of electrical conduction requires the use of quantum mechanics. Conduction electrons do not act as a classical gas of noninteracting electrons; rather, they obey a velocity-distribution law based on quantum physics, and the movement of electrons depends on these quantum ideas. Quantum physics requires us to treat electrons as though they were waves scattering from the lattice structure of the material. Quantum physics predicts that *there would be no resistance to electron flow in a fixed, perfectly ordered crystal with no impurities.* Finite conductivities occur due to both the effects of impurities and the thermal vibrations of lattice atoms at finite temperatures. At high temperatures, resistivity to electron flow is caused primarily by thermal vibrations. At low temperatures, resistivity is due to electrons being scattered by impurities. There is ample evidence that the quantum physics ideas are correct. Indeed, *all* the properties described are correctly explained with these ideas.

*27–5 MATERIALS AND CONDUCTIVITY

Materials differ over an enormous range in their ability to conduct electricity. A good conductor might have a resistivity of $10^{-8}\ \Omega \cdot \text{m}$; a good insulator, about $10^{14}\ \Omega \cdot \text{m}$. The resistivity of semiconductors ranges from 10^3 to $10^{-5}\ \Omega \cdot \text{m}$ and depends sensitively on temperature. Superconductors have no measurable resistance at all below certain temperatures! A proper quantitative explanation of the resistivity of all materials requires quantum physics. In this section, we use a minimal amount of quantum mechanics to describe the critical properties that distinguish conductors, insulators, semiconductors, and superconductors.

In classical physics, the energy of an electron in a metal can take on any value; we say that the energy values form a *continuum*. In contrast, a quantum description of electrons in metals shows that the possible energy values of electrons confined to a metal are *quantized*; that is, the possible energies have discrete values. In other words, an electron cannot have *any* energy value, much as the frequencies of standing waves on a string cannot have any value, just a set of discrete values. In a sample of material whose size is large compared with atomic sizes (10^{-10} m), these energy values are so close together that they appear to be continuous, just as the separate dots in a newspaper photograph are not distinguishable from a large distance. Figure 27–13, an energy diagram, illustrates the allowed energy levels. It is important to keep in mind that this diagram illustrates only the *possible* energy levels. We do not necessarily have electrons in each energy level.

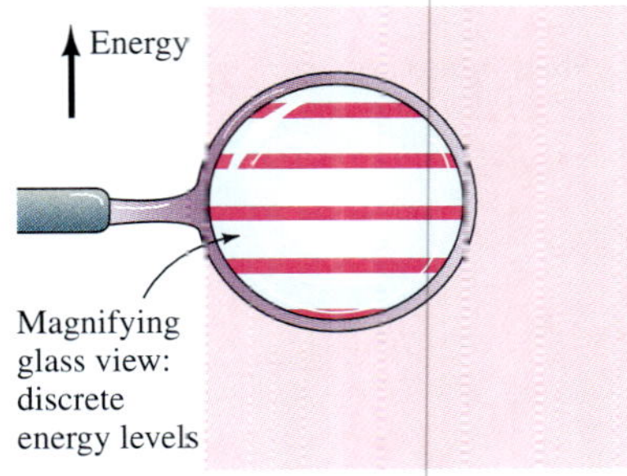

FIGURE 27–13 Energy diagram that shows the possible energy levels of an electron in a solid; it takes no account of the crystalline structure. Classical physics predicts a continuum of possible energies, but quantum mechanics shows that the possible levels are actually discrete but so closely spaced that they are hard to distinguish.

When a set of atoms forms a regular background lattice, the possible energy values of the electrons are modified still further. The allowed energies of an electron are still discrete but, instead of a tiny separation between neighboring levels, there are *energy gaps*, which are large regions of energy forbidden to the electron. The regions where the energy levels are close together are called *allowed bands* of energy levels (Fig. 27–14). The gaps are quite sizable on the scale of atomic physics—of the magnitude of electron-volts. Again, specifying the bands does not by itself specify which levels have electrons (whether the energy levels are *occupied* or not). The bands specify only the possible values of electron energies.

According to quantum physics, *there are at most two electrons in any one energy level*. This property, proposed by Wolfgang Pauli in 1925 and called the *Pauli exclusion principle*, has no counterpart in classical physics, and it plays a crucial role in determining the properties of materials. Let's consider a solid with many "free" electrons. In an equilibrium state of that material, these electrons fill the lowest energy levels available in the allowed bands—up to two in each level. When all the electrons are placed in the lowest possible energy state, we are left with two possible situations. In the first, the highest level to be filled is some intermediate level within a band; in the second, the electrons fill one or more bands completely. This description assumes that the material is at a sufficiently low temperature that the electrons cannot "jump" to higher energy levels due to thermal effects.

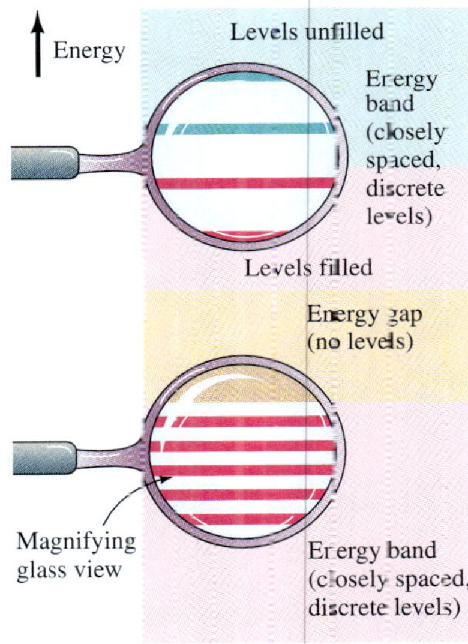

FIGURE 27–14 Energy diagram that shows the possible energy levels of an electron within a material made of a regular lattice of atoms. In contrast to the possibilities of Fig. 27–13, the electron energies are restricted to lie within allowed bands, and there is a large energy gap where no electrons are allowed. Even within the allowed bands, the possible electron energies are closely spaced discrete levels, as the magnified view shows. In the pink regions the electron energy levels are filled; in the blue regions, electron levels are present but are unfilled.

Suppose that we now add some energy to the free electrons—by imposing an electric field, for example. The electrons in the lower energy levels cannot accept that energy because they cannot move into an already-occupied higher energy level. The only electrons that can accept energy are those that lie in the top levels, and then only if there are nearby unoccupied levels into which they can move. Materials with a partly filled band are *conductors*. When the top layer of their electrons moves freely into the empty energy levels immediately above, there is a current. The electrons that jump from a lower level to a higher level are said to be *excited*. The energy-band structure for conductors is shown in Fig. 27–15a. *Conductors are characterized by having a highest-energy band with levels only partly occupied.*

If the highest-energy electrons of a material fill a band completely, then a small electric field will not give these electrons enough energy to jump the large energy gap to the bottom of the next (empty) band. We then have an *insulator* (Fig. 27–15b). An example of a good insulator is diamond (a form of carbon), whose energy gap is 6 eV.

In semiconductors, the highest-energy electrons fill a band (the *valence band*) at $T = 0$, as in insulators. Unlike insulators, semiconductors have a small energy gap between that band and the next, the *conduction band* (Fig. 27–16a). Because the energy gap is so small, a modest electric field (or finite temperature) will allow some electrons to jump the gap and thereby conduct electricity (Fig. 27–16b). Thus, there is a minimum electric field under the influence of which a material changes from insulator to conductor. Silicon and germanium have energy gaps of 1.1 eV and 0.7 eV, re-

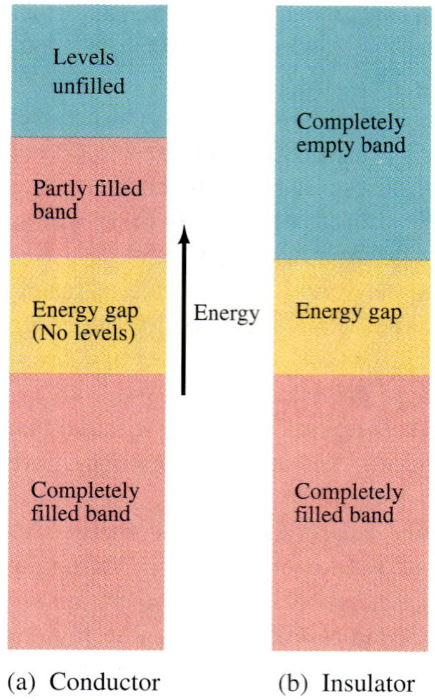

FIGURE 27–15 (a) Conductors have electrons in partly filled bands, whereas (b) insulators have an energy gap between a completely filled band and the next completely empty band. The pink and blue regions indicate where the allowed energy levels are filled and unfilled, respectively. Within each of the allowed bands, the possible energy levels form a set of closely spaced discrete levels.

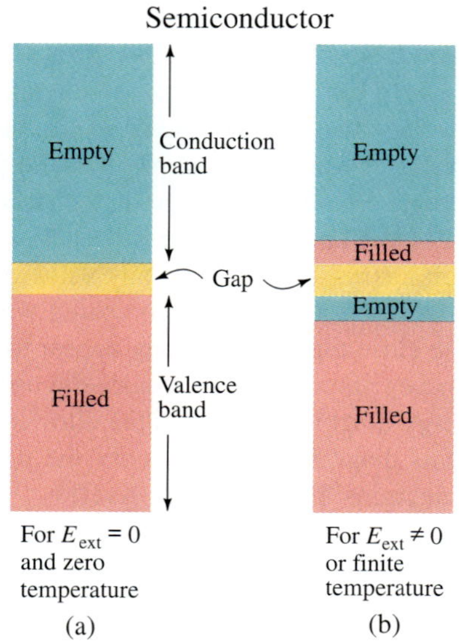

FIGURE 27–16 (a) For zero temperature and no external electric field, semiconductors have only a small energy gap between a completely filled band and the next highest, completely empty band. (b) A modest electric field $\mathbf{E}_{ext}$ or finite temperatures are enough to give some of the electrons sufficient energy to jump the energy gap, leaving holes in the valence band and conduction electrons in the previously empty conduction band.

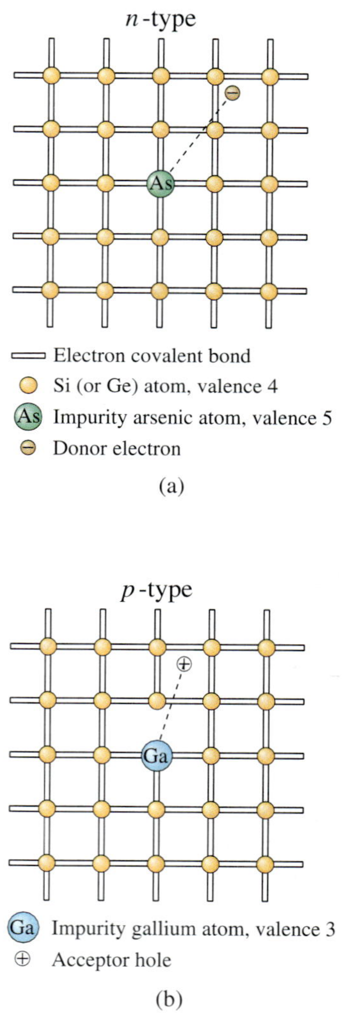

FIGURE 27–17 An (a) *n*-type and a (b) *p*-type semiconductor are created by doping the original lattice with atoms that have, respectively, more and less valence electrons than the atoms of the original lattice have.

spectively, and are semiconductors. For semiconductors, an increase in temperature will give a fraction of the electrons enough thermal energy to jump the gap. For an ordinary conductor, a rise in temperature *increases* the resistivity because the atoms, which are obstacles to electron flow, vibrate more vigorously. A temperature increase in a semiconductor allows more electrons into the empty band and thus *lowers* the resistivity.

When an electron in the valence band of a semiconductor crosses the energy gap and conducts electricity, it leaves behind what is known as a **hole**. Other electrons in the valence band near the top of the stack of energy levels can move into this hole, leaving behind their own holes, into which still other electrons can move, and so forth. The hole behaves like a positive charge that conducts electricity on its own as a positive charge carrier. An electron excited from the valence band to the conduction band is thus doubly effective at conducting electricity in semiconductors.

One of the major advances in materials technology has been our ability to produce new semiconductors. Semiconductor materials that are compounds, such as gallium arsenide, are called *hybrid* semiconductors, as opposed to *intrinsic* elemental semiconductors, such as silicon and germanium. Other special semiconductors are made by introducing impurities, small amounts of different elements, into the lattice. For example, an atom in the chemical group of phosphorus, arsenic, and antimony can replace one of the silicon atoms in a lattice without affecting the lattice itself too much. However, each of these impurity atoms has one more electron in its valence level than does a silicon atom; this extra electron, for which there is no room in the valence band, takes a place in the conduction band and can conduct electricity (Fig. 27–17a). A semiconductor with impurities of this sort is called an n-type semiconductor, and the extra electrons are called *donor* electrons. The semiconducting material, silicon in this case, is said to be *doped* by the impurity atoms.

Atoms of elements in the same chemical group as boron, aluminum, and gallium have one less valence electron than does silicon. If, as in Fig. 27–17b, such an atom is added to a lattice of silicon as an impurity, there is one less electron than is needed to form a bond that holds the lattice together. This electron must be provided by the electrons of the valence band of the lattice material, and holes are created in this band. These holes act as positive charge carriers. The impurity atoms are called *acceptors*, and a semiconductor with impurities of this sort is called a p-type semiconductor.

Many electronic devices, such as the diode mentioned earlier, depend heavily on the properties of semiconductors. Probably the best known of these devices are transistors, which can amplify electronic signals.

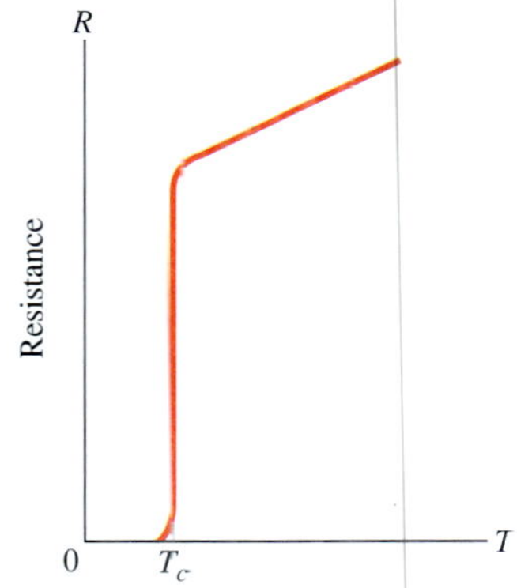

(a)

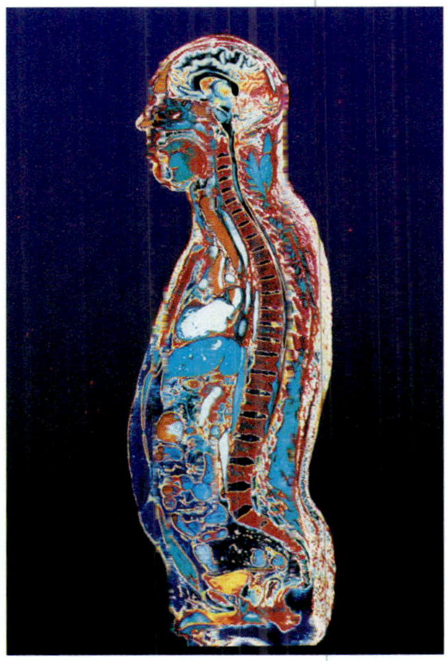
(b)

FIGURE 27–18 (a) For superconductors the resistance drops to zero at the critical temperature T_c. (b) An MRI image of a human made using a superconducting magnet.

Superconductors

In 1911, H. Kammerlingh Onnes found that mercury loses *all* its resistance abruptly at a *critical temperature* T_c of 4.1 K (Fig. 27–18a). This state of affairs persists at temperatures below T_c. When a material attains zero resistance at some critical temperature, it is called a *superconductor*. Detailed measurements on a superconducting ring in which a current had been induced showed that there was no observable decrease in the current after a year. From the measurements, it was possible to deduce that, if there were any resistive decrease of the current, it had to occur over a period of at least 10^9 years!

The prospect of having an *electric current that lasts forever* is an enticing one. It implies, among other things, the cheap transmission of electricity. The phenomenon of superconductivity cannot be understood as an extension of ordinary conductivity. The abruptness with which resistance disappears completely suggests that an ordinary conductor makes a transition to a totally different state of matter at T_c, much as liquid water turns into a crystal (ice) at 273 K. In 1957, John Bardeen, Leon Cooper, and Robert Schrieffer satisfactorily explained the *superconducting phase* with quantum physics in what is now known as the BCS theory.

Until 1986, the materials with the highest known values of T_c became superconducting at 23 K. Helium is liquid at such temperatures and is thus used for cooling superconductors. However, liquid helium is so expensive that superconducting devices have been limited to fairly specialized applications, such as magnets for particle accelerators (see Chapter 30) or nuclear magnetic resonance imaging machines in hospitals (see Chapter 32) (Fig. 27–18b). In 1986, however, K. Alex Muller and J. George Bednorz discovered a new class of materials for which T_c is much higher; superconductors have been discovered with a T_c above 120 K ($-153°$C). This discovery has great technological implications because such materials can be cooled relatively cheaply with nitrogen (which is liquid at 77 K). It seems likely that these materials will be used extensively in small devices, such as switches in supercomputers. The outlook for the use of the new high-temperature superconductors is undoubtedly a bright one, and research and development in the field of superconductors is sure to be extensive.

APPLICATION

Uses of superconductors

27–6 ELECTRIC POWER

Electric energy is sent to our homes and workplaces and supplies much of the energy used in our society. Efficient delivery of this energy is of paramount importance. In this section, we shall look at the ways in which resistance affects the delivery of electric energy.

We have compared electrical resistance to mechanical drag. When there is drag in mechanical motion, mechanical energy is converted to thermal energy. The second law of thermodynamics (Chapter 20) shows that some of this thermal energy is irretriev-

Power is discussed in Chapter 6.

ably lost in the sense that it cannot all be converted to mechanical work. Similarly, some electric energy is lost due to resistance. Just as mechanical friction generates heat, the passage of a current through a resistor generates heat. Sometimes it is the thermal energy that we want to use, as in the heating element of an electric stove. But it cannot all be converted to useful mechanical work.

To calculate the energy lost per unit time (the power lost) when a charge flows in a material, consider a small charge dq that moves through a potential difference V. The change in the potential energy of the charge (dU) is equal to the work done (dW) by the electric force due to the potential difference, and is given by $dU = V\,dq$. It follows that the power, the rate at which energy is expended by the force that pushes the charge, is

$$P = \frac{dW}{dt} = V\frac{dq}{dt}. \tag{27–24}$$

The power lost in resistance

Because the current $I = dq/dt$, the electric power lost, which is the power that must be delivered to move I through the potential V, is

$$P = VI. \tag{27–25}$$

This result is a general one, independent of the type of material—in particular, whether the material is ohmic or nonohmic—and of the nature of the charge movement. Power has SI units of watts (W), with 1 W = 1 J/s. By using Eq. (27–25), we have another unit for power:

$$1\ \text{W} = 1\ \text{V}\cdot\text{A}. \tag{27–26}$$

For ohmic materials, $V = IR$, where R is a constant. Thus the power expenditure for ohmic materials is

Alternative forms for the power lost in resistance

$$P = VI = V\left(\frac{V}{R}\right) = \frac{V^2}{R}. \tag{27–27}$$

Equivalently, we can use $V = IR$ in Eq. (27–27) to find that

$$P = VI = (IR)I = I^2R. \tag{27–28}$$

Whether we use Eq. (27–27) or Eq. (27–28) depends on what is known in a particular application. The power lost (rate of energy loss) in a resistor appears in the form of thermal energy and is variously called *ohmic heating, Joule heating*, and *I^2R loss*.

FIGURE 27–19 Example 27–6.

EXAMPLE 27–6 Nichrome is an alloy of nickel, chromium, and iron often used as a heating element in electrical devices. A nichrome wire (1.0 m in length) is crisscrossed along the bottom of a toaster oven (Fig. 27–19) that can carry a maximum current of 16 A when there is a 120-V potential difference from one end of this wire to the other.† If the resistivity of nichrome is $1.0 \times 10^{-6}\ \Omega\cdot\text{m}$, what is the radius of the wire? What power does the toaster use?

Solution: In this example, a wire of an unknown resistance carries a known current when a known voltage difference is applied. We can therefore solve for the resistance, R. We then find the area of the wire from the known value of the resistance, the length of the wire, and the resistivity of the material. The resistance of the nichrome wire is determined by setting $I = 16$ A when $V = 120$ V. From the definition of resistance,

$$R = \frac{V}{I} = \frac{120\ \text{V}}{16\ \text{A}} = 7.5\ \Omega.$$

†Real household electricity involves an oscillating voltage difference and an oscillating (or alternating) current. Ignore these effects here and in Example 27–7.

We now use Eq. (27–13), which relates the dimensions of the wire and the resistivity to the resistance. This equation is solved for the cross-sectional area A of the wire:

$$A = \frac{\rho L}{R} = \frac{(1.0 \times 10^{-6}\ \Omega \cdot \text{m})(1.0\ \text{m})}{7.5\ \Omega} = 1.3 \times 10^{-7}\ \text{m}^2.$$

The radius r of the wire is determined from $A = \pi r^2$ to be 0.20 mm.

The power the toaster consumes is determined by Eq. (27–25) from V and I:

$$P = VI = (120\ \text{V})(16\ \text{A}) = 1900\ \text{W}.$$

Electric power is also used in a context other than the one on which we have concentrated here. When electric energy is delivered to a home, the energy delivered per unit time is also called the electric power. This power is not always the power lost in resistance. We pay the electric company by the amount of energy we purchase from them. The energy unit in the electric power industry is the kilowatt-hour (kWh), which is not an SI unit:

$$1\ \text{kWh} = (1\ \text{kW})\left(\frac{1000\ \text{W}}{1\ \text{kW}}\right)\left(\frac{1\ \text{J/s}}{\text{W}}\right)(1\ \text{h})\left(\frac{3600\ \text{s}}{\text{h}}\right) = 3.6 \times 10^6\ \text{J}. \quad (27\text{–}29)$$

EXAMPLE 27–7 A 100-W bulb is left on in an outdoor storage room to keep paint from freezing. The 100-W rating refers to the power dissipated in the bulb's filament, which is a simple resistor. If electricity costs 8 cents/kWh, about how much does it cost to burn the light bulb for 3 months during winter?

Solution: We are given the power used by the bulb and the length of time over which the power is dissipated. From this information, we can find the energy used. We can find the total cost because we are given the price rate of electric energy. The total number of hours during the 3-month time $\simeq$ (3 months)(30 d/month)(24 h/d) = 2160 h. The amount of energy used is then the power multiplied by the length of time over which it is dissipated:

$$\text{energy} = (\text{power})(\text{time}) = (100\ \text{W})(2160\ \text{h}) \simeq 220\ \text{kWh}.$$

Thus

$$\text{cost} = (220\ \text{kWh})\left(\frac{\$0.08}{1\ \text{kWh}}\right) = \$17.60.$$

Although the primary purpose of a light bulb is to produce light, most of the electric energy it dissipates is converted into heat, not light.

EXAMPLE 27–8 When a wire is heated to a temperature T, it radiates energy in the form of electromagnetic waves. The power emitted per unit area of the wire is given by

$$P = 5.4 \times 10^{-8} T^4\ \text{W/(K}^4 \cdot \text{m}^2).$$

Consider a bus-stop shelter in the state of Minnesota, which is heated by a 1-m-long metallic coil that is 1 mm in diameter. The temperature of the coil is 2000 K. Assuming that the resistivity of the coil material at that temperature is given by $5 \times 10^{-7}\ \Omega \cdot \text{m}$, and that the efficiency of the conversion of electrical energy to radiant energy is 20 percent, estimate the current that must flow through the wire to maintain its temperature.

Solution: A wire of length L and diameter D has a surface area of πDL. The total power radiated, P_{rad}, is the area times the power radiated per unit area. With $T = 2000$ K,

$$P_{\text{rad}} = (5.4 \times 10^{-8}\ \text{W/(K}^4 \cdot \text{m}^2))(2 \times 10^3\ \text{K})^4(\pi DL)$$
$$= (8.6 \times 10^5\ \text{W/m}^2)\pi(10^{-3}\ \text{m})(1\ \text{m}) \simeq 3 \times 10^3\ \text{W}.$$

Next, we must find the electric energy delivered to the wire, $P = I^2R$. To do this, we must first find the resistance of the wire at 2000 K. We can find the resistance given the resistivity:

$$R = \rho\frac{L}{A} = \rho\frac{L}{\pi(D/2)^2} = (5 \times 10^{-7}\ \Omega \cdot \mathrm{m})\frac{1\ \mathrm{m}}{\pi(0.5 \times 10^{-3}\ \mathrm{m})^2} = 0.64\ \Omega.$$

To maintain the wire at the given temperature, we recall that only 20 percent of the electric power delivered to the wire is converted to radiated energy. In other words, $P_{\mathrm{rad}} = (0.20)I^2R$. We solve this equation for I:

$$I = \sqrt{P_{\mathrm{rad}}/(0.20\ R)} = \sqrt{(3 \times 10^3\ \mathrm{W})/[(0.20)(0.64\ \Omega)]} \simeq 150\ \mathrm{A}.$$

Resistors used in circuits are characterized not only by their resistance, but also by a power rating. This power rating states the maximum power that the resistor can dissipate without being damaged due to overheating. The power rating is measured in watts. According to Eq. (27–28), which states that the power dissipated in a resistor is $P = I^2R$, we can deduce the maximum allowed current from the power rating. One class of relatively inexpensive resistors, composition and carbon film resistors, is limited to about 2 W; a second, more expensive type known as wire-wound resistors have a power rating up to 50 W.

SUMMARY

Electric current is the rate at which charge passes. The instantaneous current is given by

$$I \equiv \frac{\mathrm{d}Q}{\mathrm{d}t}. \tag{27–2}$$

The unit of current is the ampere (A), 1 C/s. Currents are depicted as though the positive charges are moving, but it is actually the (negative) electrons that are mobile.

The current density, $\mathbf{J}$, is a vector quantity that gives the current that passes through an area per unit time. The current is related to the current density by

$$I = \int_{\text{surface } A} \mathbf{J} \cdot \mathrm{d}\mathbf{A}. \tag{27–4}$$

The free-electron model of conduction is useful as a qualitative description of current in a solid. The average, or drift, speed of the electrons that pass through the material is

$$v_d = \frac{I}{n_e qA}, \tag{27–9}$$

where n_e is the density of free electrons and q is the charge of an electron.

Electrical resistance, R, is the ratio of voltage to current

$$R \equiv \frac{V}{I}. \tag{27–11}$$

Many conducting metals show a linear relationship between voltage and current. The resistance is then constant over a wide range of voltages. This relation is called Ohm's law, $V = IR$.

Resistivity, ρ, is the quantity that distinguishes the part of the resistance that is intrinsic to each particular type of material. For wires of length L and area A, we have

$$R = \rho\frac{L}{A}. \tag{27–14}$$

The inverse of the resistivity is the conductivity, σ, which expresses how well a type of material conducts current:

$$\sigma \equiv \frac{1}{\rho}. \qquad (27\text{–}15)$$

Both ρ and σ depend on temperature.

The electric field and current density are related by

$$\mathbf{E} = \rho\mathbf{J} \qquad (27\text{–}16)$$

and by

$$\mathbf{J} = \sigma\mathbf{E}. \qquad (27\text{–}17)$$

The correct explanation of electrical conduction in metals depends on how electrons fill allowed energy bands and on the energy gap between these bands. Materials that conduct current easily have electrons in partly filled bands. Semiconducting materials, such as silicon and germanium, can be doped by impurity atoms to increase the density of charge carriers. The explanation of superconductivity requires both quantum mechanics and the presence of a new phase of matter in which electrons collectively transport electric current.

When a current moves through a potential difference, electric power, P, is dissipated (or produced), given by

$$P = VI. \qquad (27\text{–}25)$$

For resistive materials, the power is also given by

$$P = \frac{V^2}{R} = I^2R. \qquad (27\text{–}27,\ 27\text{–}28)$$

UNDERSTANDING KEY CONCEPTS

1. Consider the electron beam in a cathode-ray tube. The velocity of the electrons in the beam changes as the electrons are accelerated. Is the current the same everywhere in the beam?
2. How does the free-electron model for electrical resistance account for power dissipation? Does our microscopic picture agree with the voltage/current result?
3. The same current passes through two similar wires of unequal areas. Which wire will get hotter, and why?
4. The same current passes through two wires of the same area. One of the wires is made of aluminum, whereas the other is made of brass. Which wire will get hotter, and why?
5. What factors determine the differences in drift velocity of electrons in wires if the dimensions and current are the same?
6. If the movement of charges in a wire is similar to the flow of water in a hose, why, when a new hose is hooked up to a faucet, do we have to wait for a while until the water comes out, but when we hook a new wire up to a circuit, we do not have to wait for charge to come out the other end when the switch is turned on?
7. According to the discussion of Section 27–4, the resistivity in the free-electron model should vary with the square root of the temperature and thus should be zero at $T = 0$. Is this reasonable? How would you interpret this result?
8. We know that the resistivity of a metal is temperature dependent, and so therefore is the resistance of a wire. In Chapter 21, we saw that the dimensions of a piece of metal—such as a wire—change when the wire is heated. Does this provide an additional reason to change the resistance of a wire as it undergoes Joule heating? Would you expect the effect to be large?
9. When you throw a switch and charge flows in a household wire, does the wire become charged?
10. Suppose that we orient a wire between the plates of a charged capacitor so that there is an electric field along the cross section of the wire. Will the resistance of the wire change because all the charge-carrying electrons crowd to one side of the wire, thus effectively reducing the wire's cross section?
11. Gauss' law states that free charge in a conductor moves to the surface of the conductor. Does this mean that the current flowing through a wire is actually on the wire's surface?
12. The resistivity of most metals is on the order of $10^{-8}\ \Omega \cdot \text{m}$. Discuss why this might be so in terms of the result given by Eq. (27–23).
13. What is likely to happen when a current is so large that the power dissipation in a resistor through which the charge flows exceeds the resistor's power rating? What mechanism is responsible for such a disaster scenario?
14. What considerations would you have to take into account in designing a light bulb filament with emphasis on the length of filament versus the diameter of the filament?
15. The resistance of a wire is proportional to the length of the wire. Think of a wire as consisting of a series of shorter wires placed together in series (Fig. 27–20). Can you use this infor-

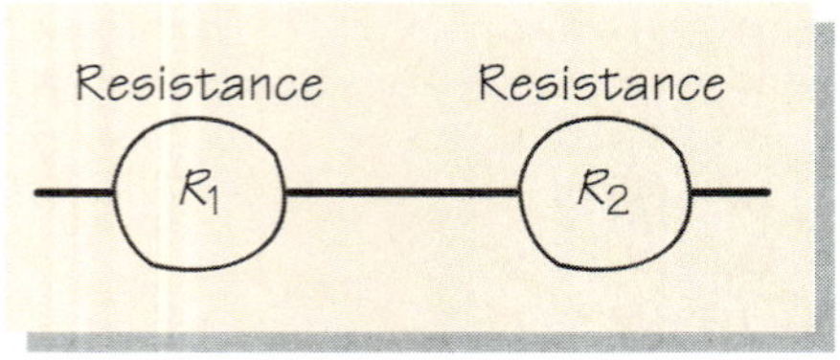

FIGURE 27–20 Question 15.

mation to predict the effective resistance of two different resistors of resistance R_1 and R_2 when they are placed in series in a circuit?

16. The resistance of a wire is inversely proportional to the area of the wire. By thinking of a wire as consisting of a number of thinner wires placed together side-by-side, can you use this information to predict the effective resistance of two different light bulbs of resistance R_1 and R_2 when these are placed in parallel in a circuit (Fig. 27–21)?

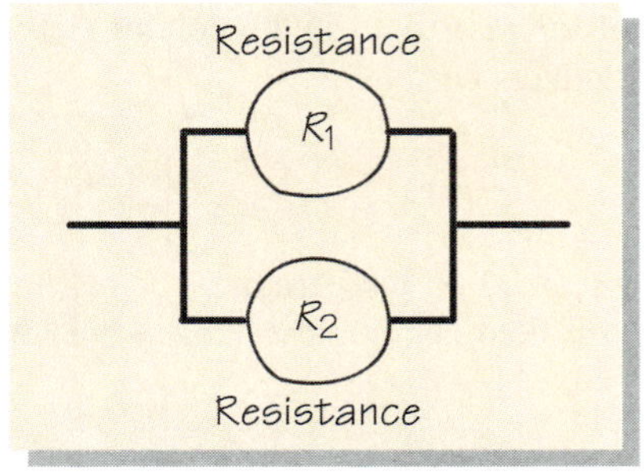

FIGURE 27–21 Question 16.

17. In high wattage light bulbs, why is the filament often curled up quite a bit?

PROBLEMS

27–1 Electric Current

1. (I) A wire of diameter 2.2 mm carries a current of 0.46 A. What is the average current density? How much charge crosses a fixed point in the wire per second?

2. (I) There is a 0.2-A current in a wire of cross-sectional area 6.8×10^{-6} m^2. The drift speed of the electrons that carry the current is 0.32×10^{-5} m/s. Find the density of current-carrying electrons.

3. (I) A jumper cable used to start a car carries a current of 100 A and has a cross-sectional area of 36 mm^2 and a length of 2 m. The free-electron density in the cable is 8.5×10^{22} electrons/cm^3. How long does it take a free electron to pass from one end of the cable to the other?

4. (I) Three straight wires of area 0.03 mm^2, 0.3 mm^2, and 3 mm^2, respectively, are aligned along the x-axis. They carry current densities along the x-axis of magnitude 2×10^5 A/m^2, 7×10^4 A/m^2, and 9×10^4 A/m^2, respectively. Find the current in each wire.

5. (I) A wire of radius 1.6 mm carries a current of 0.092 A. How many electrons cross a fixed point in the wire in 1 s?

6. (I) Charge carriers in a semiconductor have a number density $n_q = 3.5 \times 10^{24}$ carriers/m^3. Each carrier has a charge whose magnitude is that of an electron's charge. If the current density is 7.2×10^2 A/m^2, what is the speed of the charge carriers?

7. (I) The density of charge-carrying electrons in copper is 8.5×10^{28} electrons/m^3. If a current of 1.2 A flows in a wire 1.8 mm in radius, what is the speed of the electrons? How does that speed change in a second wire, of diameter 2.4 mm, connected end-to-end with the first wire?

8. (I) An electron accelerator in which electrons travel at a speed of 2.8×10^8 m/s produces a beam of electrons that carries a current of 5.0 mA. The effective area occupied by the beam is 0.50 cm^2. What is the density of electrons in the beam? Ignore all relativistic effects.

9. (II) In the National Synchrotron Light Source X-ray device at Brookhaven National Laboratory, there is an electron beam with an average current of 200 mA. The electrons have a kinetic energy of 2.5 GeV and a speed extremely close to the speed of light. How many electrons pass a given point in the accelerator per hour? How many electrons are contained in a 1-m length of the beam? Ignore all relativistic effects (a poor approximation, in this case).

10. (II) A cube of material is placed with one corner at the origin of a coordinate system; its sides, 1 cm long, are parallel to the three axes. The current density is $A\mathbf{i} + B\mathbf{j} + C\mathbf{k}$ throughout the cube. The units of A, B, and C are mA/cm^2. What are the currents along the x-axis, y-axis, and z-axis?

11. (II) The current density in a cylindrical wire of radius R is $J = J_0(1 - r^2/R^2)$, parallel to the axis of the wire (Fig. 27–22). Calculate the total current across a section perpendicular to the axis.

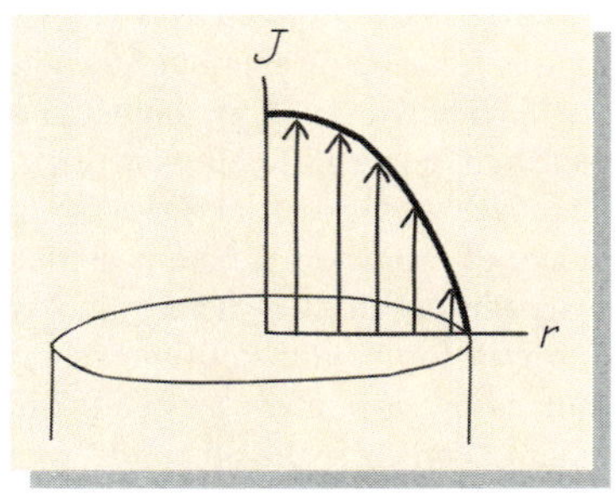

FIGURE 27–22 Problem 11.

12. (II) In a plasma containing equal densities n of electrons and (positive) ions, the ions move to the right. Their speed is a factor of 10^{-3} smaller than the speed with which the electrons move to the left. What is the (net) current density? Give its direction and magnitude.

13. (II) An aqueous solution contains 0.1 mol/L of NaCl. The NaCl is dissolved in the form of Na^+ and Cl^- ions. Calculate the velocities of the Na^+ and Cl^- ions, respectively, if there is a measured total current density of 40 A/m^2. Assume that the velocity of the Cl^- ions is about 50 percent greater than that of the Na^+ ions.

27–2 Currents in Materials

14. (I) Calculate the drift speed of electrons in the conduction cables of an automobile starter cable, which is made of copper and has a diameter of 4 mm, if you suppose that the cable carries 100 A. How would this speed change if the diameter of the wire were doubled? [*Hint*: Useful data are contained in Example 27–2.]

15. (I) A single charged elementary particle ($q = 1.6 \times 10^{-19}$ C) travels with a speed very close to that of light in a circular accelerator of diameter 5 km. What is the current represented by the particle, taking into account multiple traversals of the charge past a given point?

16. (I) How many particles like that described in Problem 15 must be present at a given time to give rise to a current of 30 mA?

17. (II) An aluminum wire of area 50 mm^2 placed along the x-axis passes 10,000 C in 1 h. Assume that there is one free electron for each aluminum atom. Determine the current, current density, and drift speed. The mass density of aluminum is 2.7 g/cm^3.

18. (II) Gold has one electron per atom available to carry charge. Given that the mass density of gold is 19.3×10^3 kg/m^3 and that its molecular weight is 197 g/mol, calculate the drift speed of the electrons in a gold wire that carries 0.1 A and has a circular cross section 0.3 mm in radius.

19. (II) Two parallel metal wires of diameter 0.2 cm and a charge-carrier density $n_e = 7 \times 10^{22}$ electrons/cm^3 carry a current of 3 A each. The wires join and then split into three identical but separate wires, each with a radius one-half that of the original wire (Fig. 27–23). All the wires are made of the same material. What are the drift speeds in both the larger and smaller wires? Can you explain the difference in speeds in terms of the speeds of water flow in pipes?

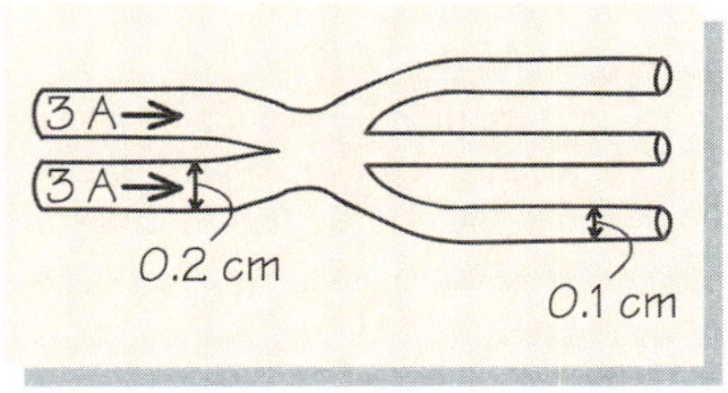

FIGURE 27–23 Problem 19.

20. (II) The charge carriers in a certain wire of circular cross section and radius R have a drift speed down the wire that is not constant across the wire. Instead, the drift speed rises linearly from zero at the circumference ($r = R$) to v_0 at the center ($r = 0$). Compare the total current carried by this wire with the current carried by a wire of the same radius, same density of charge carriers, and a constant drift speed of $v_0/2$.

21. (II) A thin copper wire carrying a current I is welded to the center of a circular copper plate capping a copper tube (Fig 27–24). The radius of the tube is R, the thickness of its wall and the top plate is d, and $d << R$. What is the current density in the tube and in the top plate?

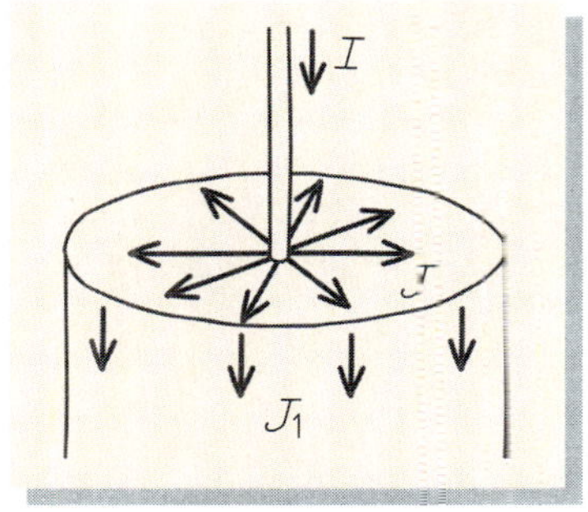

FIGURE 27–24 Problem 21.

22. (III) Charges q move longitudinally down a rod of circular cross section and radius R. The density of the charge carriers, n, decreases as a function of the radial distance r from the center of the rod according to $n = n_0 - n'r$. The speed v of the charge carriers varies with r according to $v = v_0 - v'r^2$, where n_0, n', v_0, and v' are constants. Calculate the current that passes through the rod.

27–3 Resistance

23. (I) You have two solid cylinders of the same material. Piece 2 has half the length and half the diameter of piece 1. How do the resistances of the two pieces differ?

24. (I) The conductivity of silver is 1.5 times that of gold. What is the ratio of the diameter of a silver wire to that of a gold wire of the same length if both wires are designed to have the same resistance?

25. (I) An underground wire made of aluminum is 528 m long and has an area of 0.12 cm^2. (a) What is its resistance? (b) What is the radius of a copper wire of the same length and resistance?

26. (I) An old house is wired with AWG #18 copper wire, which has a diameter of 0.0403 in. (a) What is the wire's resistance per 1000 ft? (b) One circuit consists of only one wire behind walls and has a resistance of 2.3 Ω. How long is this wire?

27. (I) The resistivity of copper is $1.72 \times 10^{-8}\ \Omega \cdot$ m. What is the resistance of a section of gauge #10 wire (diameter 0.2588 cm) that is 10 m long?

28. (I) A carbon rod used in a welding machine is 5.0 mm in diameter and 20.0 cm in length. What is its resistance and how much current will pass through it if the welding machine puts a voltage of 380 V across it?

29. (I) Cables used to jump an automobile can get hot if used for more than a few seconds. Calculate the resistance at 20°C of a 2-m-long copper cable of cross-sectional area 36 mm^2. By how much does the resistance increase as the temperature rises from 20°C to 100°C?

30. (I) In the text, we refer to a power line with a total resistance of 10 Ω. Suppose that the power line is made of copper with a resistivity of $1.72 \times 10^{-8}\ \Omega \cdot$ m, and is 200 km long. What is the radius of the wire?

31. (II) A current passes through a tungsten wire in an appliance. If you assume that there is a fixed potential drop from one end of the wire to the other, what is the fractional change in the power consumed as the temperature of the tungsten wire changes from 800°C to 1200°C? Ignore any effects due to the change in the wire's length by thermal expansion.

32. (II) An electrician tests for a short circuit by putting a potential difference of 1.5 V across two neighboring parallel wires that would be independent of each other if there were no short. A current of 0.14 A then flows in the wires. The wires consist of material with a resistivity of $1.7 \times 10^{-8}\ \Omega \cdot \text{m}$ and have a diameter of 0.24 mm (Fig. 27–25). Given that the short effectively makes the wires act like a single wire, how far away is the short?

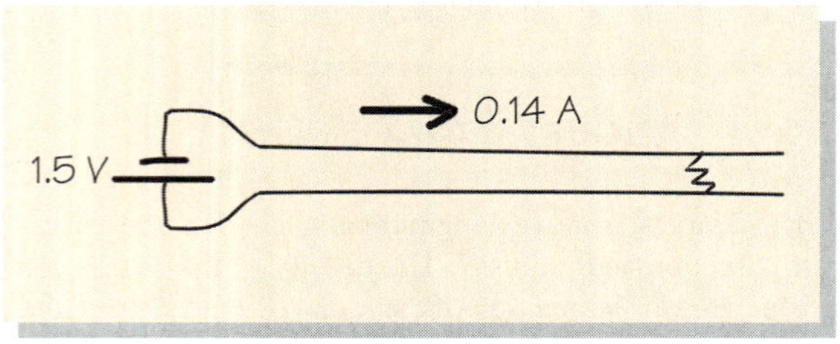

FIGURE 27–25 Problem 32.

33. (II) A nichrome wire of diameter 0.5 mm and length 50 cm is connected to a 50-V battery. What current passes through the wire at room temperature (25°C) and after the wire heats up to 400°C?

34. (II) The change in the resistance of a thin platinum wire can be used to measure temperature. Suppose that a constant 10 mA current passes through a platinum wire and that the potential drop measured at room temperature is 8.23 mV. What is the temperature of the wire when the potential drop is 9.11 mV?

35. (II) An aluminum wire of length L and a copper wire of length $5L$ have precisely the same resistance. Given that the resistivity of aluminum and copper are $2.8 \times 10^{-8}\ \Omega \cdot \text{m}$ and $1.7 \times 10^{-8}\ \Omega \cdot \text{m}$, respectively, what is the ratio of the radii of the two wires?

36. (II) You have a 100-m-long wire of area 0.5 mm² with a thin coating of insulation, but you cannot identify the type of material that makes up the wire. You have a 12.0-V battery and a device to measure current. When the battery is placed across the two ends of the wire, you measure a current of 1.07 A. What is the wire material? (Use Table 27–2.)

37. (II) A coil used to produce a magnetic field is made of copper wire of area 1.5 mm² wound many times around a spool of diameter 20 cm. The resistance of the wire is 1.35 Ω. We must know the number of turns of wire to know the magnetic field. How many turns of wire are there on the spool?

38. (II) You wish to double a current that flows through a wire of fixed length, but you can increase the voltage that drives the current by only a factor of 1.5. You have other wires made of the same material but of different radii. What is the smallest factor by which the radius of a replacement wire should differ from the radius of the original wire?

39. (II) Aluminum has a density of $2.7 \times 10^3\ \text{kg/m}^3$. What is the resistance of an aluminum wire 0.12 cm in diameter and 80 m long? What is the mass of the wire? What is the mass of a copper wire, of density $8.9 \times 10^3\ \text{kg/m}^3$, with the same length and same total resistance?

40. (II) What are the length and the radius of a copper wire (of circular cross section) whose resistance is 2 Ω and whose mass is 1.5 kg?

41. (II) How much silver (of density $10.5 \times 10^3\ \text{kg/m}^3$) would be needed to make a wire 1 km long, with a resistance of 5 Ω?

42. (II) A copper pipe has an inside diameter of 2.00 cm and an outside diameter of 2.20 cm. What length of copper pipe will have a resistance of 1.0 Ω?

43. (II) A copper resistor has the shape of a cylindrical shell. What is the resistance of this resistor if its length is 1 m, its inner radius is 0.1 cm, and its outer radius is 0.2 cm? What is the radius of a solid wire of circular cross section with the same length and the same resistance? Compare the masses of the two resistors.

44. (II) A *zener diode*, named for Clarence Zener, has the I–V curve shown in Fig. 27–26. Sketch the resistance of the diode versus both current and voltage. What is special about the critical voltage V_c?

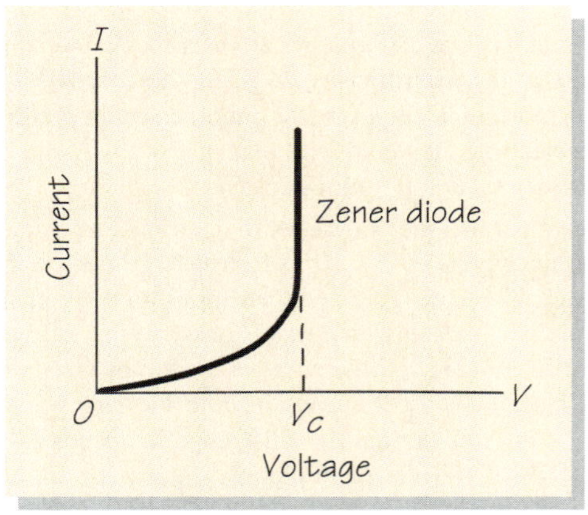

FIGURE 27–26 Problem 44.

*27–4 *Free-Electron Model of Resistivity*

45. (I) Using the average time between collisions as calculated in Example 27–5, determine the drift speed of charge carriers for a material in which the electric field is 2.0×10^{-3} V/m.

46. (I) Assuming the collision time from Example 27–5 and an average speed of 1.56×10^6 m/s, estimate the mean free path for an electron in copper.

47. (I) Recall Eq. (19–46), which relates the collision cross section to the mean free path of a particle. Use that result together with the results of Problem 46 to estimate the collision cross section of an electron with an ion in a copper lattice.

48. (II) In Problem 20, we described a wire of radius R within which the drift speed of charge carriers varies with the distance from the center of the wire as $v_d = v_0[1 - (r/R)]$. Supposing that this wire is made of ohmic material, describe how the resistivity must vary with r to produce this drift-speed profile.

*27–5 *Materials and Conductivity*

49. (II) If you treat electrons as a gas of independent particles, at what temperatures would an average electron have sufficient energy to cross the energy gap for silicon (1.1 eV), germanium (0.7 ev), and carbon (6 eV)?

27–6 *Electric Power*[†]

50. (I) What is the resistance of a 40-W headlight used on a 12-V car battery?

51. (I) What is the maximum voltage that can be applied to a 1000-Ω resistor rated at 1.5 W?

52. (I) Your little sister leaves a 100-W light bulb burning for an unnecessary hour. Assuming that electric power costs 10 cents per kilowatt-hour, what is the cost of her inaction?

53. (I) A graduate student in engineering has a collection of 100-Ω resistors with different power ratings of 1/8, 1/4, 1/2, 1, and 2 W. What is the maximum current that the student should use in each resistor?

54. (I) What is the maximum allowable current for (a) a 100-Ω, 5-W resistor? (b) a 1-kΩ, 2-W resistor?

55. (I) An electrostatic accelerator has a maximum attainable voltage of 8×10^6 V. If a current of 100 μA is produced by accelerating charges in this potential difference, what is the nominal power required to operate the accelerator? Assume that this power can be converted to the accelerator voltage with 100 percent efficiency.

56. (I) Consider a resistor of resistance R. If the maximum allowed power dissipation is P, what is the maximum allowed operating voltage?

57. (I) An electric heater draws a current of 10 A from a 120-V circuit. What is the cost per hour of operating the heater if electrical energy costs 7 cents per kWh?

58. (I) A 2-m-long copper cable with cross-sectional area 26 mm^2 draws 100 A when it is used to jump start an automobile. It takes 20 s to start the engine. Given the resistance is 9×10^{-4} Ω, calculate how much energy is dissipated in the cable during this operation.

59. (II) A heater uses nichrome wiring ($\rho = 10^{-6}\ \Omega \cdot \text{m}$) and generates 1250 W when connected across a 110-V line. How long must the wire be if its cross-sectional area is 0.2×10^{-6} m^2?

60. (II) Consider the terminals of a 12-V battery connected by a copper wire. How long must the wire be if its cross-sectional area is 3×10^{-5} m^2 and if the power dissipated is 1.2 kW?

61. (II) Buildings have circuit breakers, devices that switch the current off when it exceeds a critical value, to protect the electrical system from damage. One circuit for a building's lights has a 15-A breaker. (a) What is the maximum power that can be delivered by a 110-V line to this circuit? (b) How many light bulbs, each requiring 75 W, can this circuit handle?

62. (II) A 6-V battery is connected to two metal wires dipped in a pot of water (Fig. 27–27). A current of 50 mA flows for 18 h. How much energy is taken out of the battery during that time?

†When we refer to household electricity applications, assume in each case that their currents and voltage differences are of the simple, constant type discussed in this chapter.

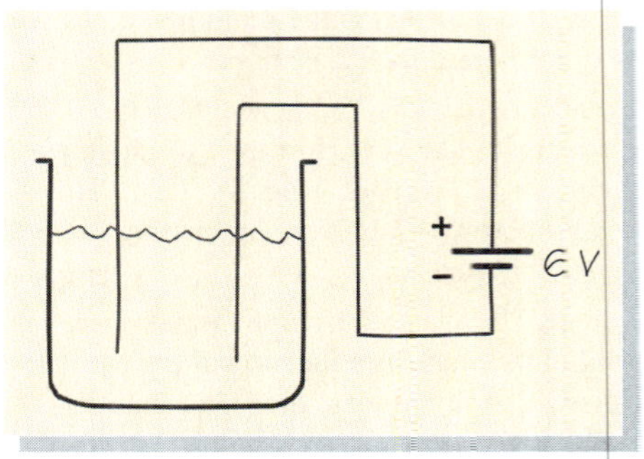

FIGURE 27–27 Problem 62.

63. (II) A 500-W electric heater is designed to operate on a line of 115 V. As the result of a brownout (a partial interruption of electrical power) the line voltage drops to 105 V. Assuming that the heating unit has a fixed resistance, what is the power of the heater now?

General Problems[†]

64. (II) An electric hot plate is used to boil water. The current drawn by the hot plate is 10 A. Make a rough estimate of the voltage and resistance, based on how long it takes to make a pot of tea.

65. (II) One month's electricity bill for an apartment is $25.33, and the cost of electricity is 8 cents/kWh. All appliances used in this apartment work at 120 V. How many electrons passed through the apartment's electrical meter that month?

66. (II) A wire of resistance r is drawn—pulled like taffy—to double its length. Assuming a constant voltage and a fixed volume, by how much does the power dissipation change?

67. (II) A Van de Graaff accelerator delivers 4-MeV protons at a current of 5 μA through a target onto a piece of tungsten that serves to stop the proton beam. (a) How many protons stop in the tungsten in 1 h? (b) How much energy is delivered to the tungsten in 1 h? (c) What is the power of the proton beam?

68. (II) A piece of brass is machined into a long, tapering cylinder. Its radius is expressed by $r = r_0 + \alpha x$, where α is a constant and x is measured from the narrow end of the tapering cylinder and runs from 0 to L (Fig. 27–28). Find an expression for the resistance of this piece.

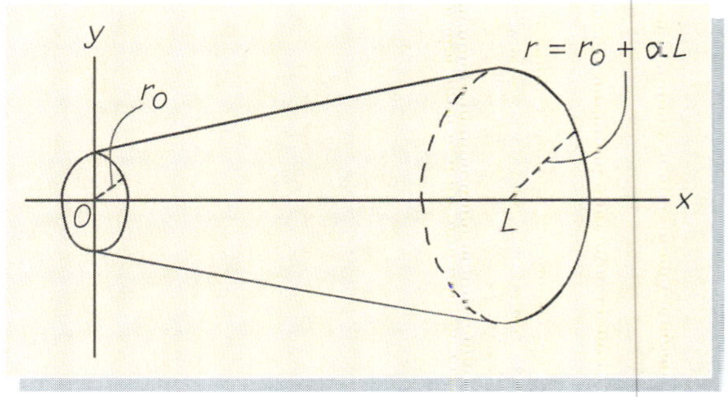

FIGURE 27–28 Problem 68.

69. (II) A bus bar (a conducting bar meant to carry a good deal of current) made of copper, of resistivity $1.72 \times 10^{-8}\ \Omega \cdot \text{m}$, is meant to carry 100 A over a distance of 0.25 m at a temperature of 300°C. What is the minimum cross section of the bus bar if no more than 0.2 W of power is to be dissipated?

70. (II) A generator delivers 30 A at a voltage of 12 V. What power does the generator deliver? How long would it take to raise the temperature of 10^{-4} m^3 of water by 75°C? How long would it take to boil away 0.1 L of water, starting at 25°C?

71. (II) Figure 27–10 shows the I–V curve of a typical semiconductor diode. Use the data from the figure to sketch the power dissipated in the diode as a function of the current. What is the power dissipated in the ideal diode, as shown in the figure?

72. (II) A potential is set up from one end of a copper wire to the other; as a result, current flows. The copper is thermally isolated to some extent. As the charge flows, the wire heats up, causing the resistivity to increase. Suppose that, during a short time period, the temperature of the wire as a function of time t is given by $T = T_0 + kt^2$. (a) Describe the current in the wire during this period. (b) What is the power dissipated by the wire as a function of time? (c) From the change with time of the dissipated power, will the wire continue to heat up and, perhaps, melt?

73. (II) The density of charge-carrying electrons in copper is 8.5×10^{28} electrons/m^3, its resistivity is $1.7 \times 10^{-8}\ \Omega \cdot$ m, and the drift speed in a copper wire is 1.2×10^{-5} m/s. The wire has a diameter of 1 mm and a length of 3 m. At what rate must thermal energy be carried off by a cooling medium if the wire is to maintain its temperature?

74. (II) A single layer of 200 turns of closely spaced wire of radius $r_1 = 0.6$ mm is wound in a coil of diameter $D_1 = 5$ cm (Fig. 27–29). A second coil of the same length but of diameter $D_2 = 8$ cm is composed of a single layer of closely spaced wire of radius $r_2 = 0.4$ mm. The wires are made of the same material. Find the ratio of the resistances of the two coils.

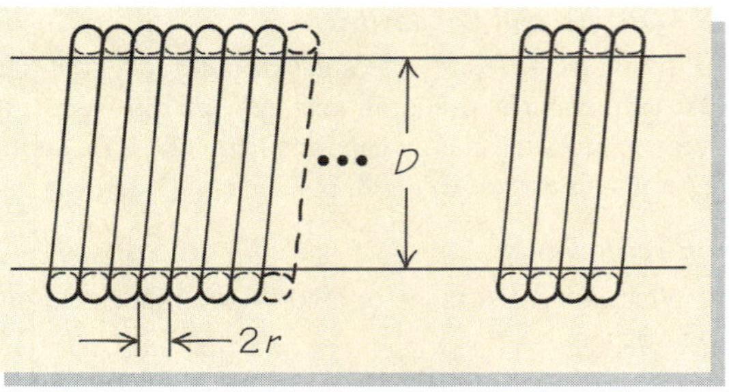

FIGURE 27–29 Problem 74.

75. (III) A thin wire of length L and cross-sectional area A oriented in the x-direction is made of an ohmic material whose resistivity varies along the wire according to the empirical law $\rho = \rho_0 e^{-x/L}$. (a) Describe how the field within the wire varies with position if the end at $x = 0$ is at a potential V_0 greater than the end at $x = L$. (b) How does the potential vary as you move along the wire? (c) What is the total resistance of the wire?

76. (III) If all the energy lost from Joule heating stays in a wire, and the temperature increases as a result, the resistivity will increase according to Eq. (27–18). The current will therefore change as a function of time, the Joule heating will change, and so forth. If the wire material has a constant heat capacity, the rate of energy loss in the wire will be proportional to the rate of change of temperature. Assuming that the potential stays constant, set up a differential equation that describes the rate of temperature change. If this equation is solved, how can the current be found as a function of time?

CHAPTER

28

The Honda Dream, a solar-powered car. The solar cells act as batteries, with energy from the Sun, and drive the electric circuits that make the car run.

Direct-Current Circuits

We have seen how charges move under the influence of potential differences, and how resistors and capacitors influence the flow of current and the movement of charge. When resistors, capacitors, and batteries are connected together by conducting wires, they form electric circuits. We can understand the flow of currents in circuits by applying just two simple physical principles: the conservation of current and the conservation of energy. In this chapter, we learn to apply these principles systematically to the analysis of circuits. We also discuss the common instruments that measure and monitor the current and voltage of electric circuits. The flow of energy to and from circuit elements provides an important theme for this chapter and leads us to the concept of time-varying currents and voltages.

28–1 EMF

The sources of electric energy that cause charges to move in electric circuits have historically been called sources of **electromotive force**. They are actually sources of energy, not of force; to avoid the word force, which is somewhat misleading here, we use the abbreviation **emf** instead. When we think of sources of emf, we usually consider batteries, but there is a wide variety of sources of electric energy made by humans. A battery converts chemical energy to an emf; a solar cell converts the energy in sunlight to an emf; a thermocouple produces an emf as a result of a difference in temperature; a large commercial electric power plant may burn coal, gas, or nuclear fuel, or use falling water, to drive a generator that produces an emf (Fig. 28–1).

In this chapter, we use the term "battery" to refer to any source of emf. We shall restrict ourselves to batteries for which the emf is *constant* with time. Up to Section 28–5, we shall also restrict our attention to phenomena such as current flows or potential differences that are similarly constant in time—we refer to *equilibrium*, or *steady-*

state, behavior. We shall also use the term **direct-current**, or **DC**, behavior, for such situations.

(a) (b) (c)

FIGURE 28–1 Various sources of emf that produce electrical energy include (a) solar panels (photovoltaic), (b) a thermocouple, and (c) a dynamic microphone.

Circuits

When batteries, resistors, capacitors, or other circuit elements (to be introduced later) are all connected by idealized resistanceless wires, they form a **circuit**. For example, when a switch is closed and a battery sends current through the filament of a light bulb, a circuit has been formed. Figure 28–2 illustrates a simple circuit by using the conventions for resistors, ideal wires (wires with no resistance), and batteries; the light bulb is a simple resistor. At this stage, we deal with steady currents, and we consider only circuits without capacitors.

Problems involving circuits typically require us to relate the currents and potential differences in them. We may want to know, for example, the potential drop across a capacitor or the current that passes through a resistor when there is a particular emf in the circuit.

The Role of Batteries

When a battery is part of a simple circuit such as the one in Fig. 28–2, a current flows from the battery terminal at the higher potential—the one marked positive. How much current flows depends on the rest of the circuit. In Fig. 28–2, the remainder of the circuit consists of a single resistor; we refer to its resistance, R, as the **load** resistance. There is current flow here because negative charge carriers (electrons) are attracted to the positive terminal. Because the current is defined as moving in a direction opposite to that of the electrons, it may be helpful to imagine the positive charges flowing to the negative terminal (or the terminal at the lower potential). The battery has a potential difference across its terminals called the **terminal voltage**.

Inside a chemical battery, a chemical process carries the positive charges back to the positive terminal. The battery can be thought of as a device that expends energy to pump charges, just as a water pump expends energy to pump water uphill to a tank with a higher gravitational potential energy. It is the internal pumping action of the battery that gives a precise definition of the emf. Suppose that it takes work $\mathrm{d}W$ to move a charge $\mathrm{d}q$ from the negative to the positive terminal. Then the emf of the battery is defined to be

The definition of emf

$$\mathcal{E} \equiv \frac{\mathrm{d}W}{\mathrm{d}q}. \tag{28–1}$$

The SI unit of emf is the volt, or joules per coulomb. The word *voltage* is sometimes used loosely to describe the emf, $\mathcal{E}$, but voltage more properly refers to the potential difference or terminal voltage across the emf terminals, which may be different from $\mathcal{E}$.

When a battery sets charges into motion, driving a current from the positive (higher-potential) terminal around the circuit to the negative (lower-potential) terminal, we say that the battery *discharges*. In discharging, the battery is expending its chemical energy. If a current is driven from the negative to the positive terminal, a process that can be accomplished in conjunction with a larger battery, the smaller battery is said to be *charging*. The battery may or may not be capable of restoring energy when it charges.

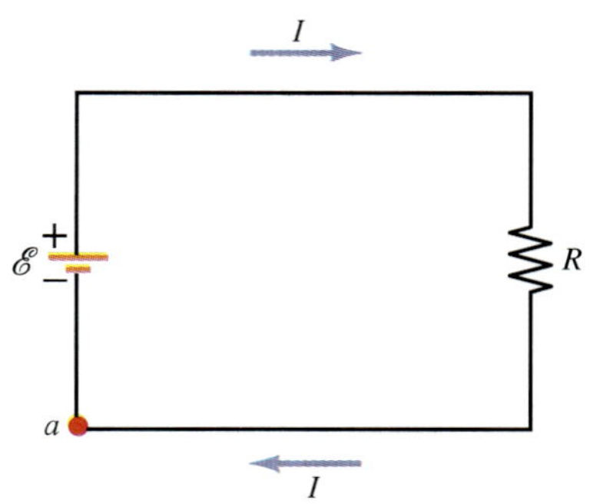

FIGURE 28–2 A simple circuit with a source of emf, $\mathcal{E}$, and a resistor, R.

Suppose that the potential difference across the battery terminals in Fig. 28–2 is $\mathcal{E}$. According to our reasoning, a current will flow around the circuit. To find this current, we can use the fact that the electric potential is associated with a conservative force. Therefore the net work done by this force in sending a charge around a closed loop is zero. In turn, the total potential drop involved in any round trip that starts from any point on a closed loop must be zero. Let's make such a round trip that starts at point a of Fig. 28–2 and follow the current around the circuit. There is no change in potential as we pass through the ideal (resistanceless) wire. In crossing the battery from the negative to the positive terminal, the potential *increases* by $\mathscr{E}$. When we cross the ohmic resistance, the potential *decreases* by an amount IR [see Eq. (27–12)]. The potential drop implies a decrease in the potential energy of the charges. This potential energy is converted into thermal energy in the resistor. The net potential change as we travel once around the circuit is zero, so

$$\mathscr{E} - IR = 0 \tag{28–2}$$

This equation determines the current, I:

$$I = \frac{\mathscr{E}}{R}. \tag{28–3}$$

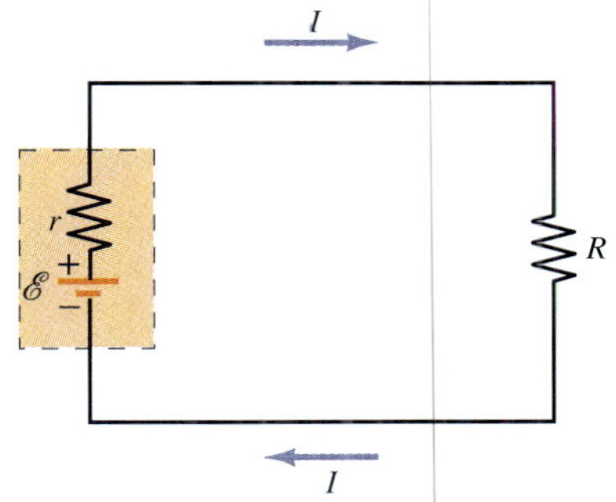

FIGURE 28–3 A source of emf also contains an internal resistance r. The shaded region includes both.

Internal Resistance

There is some energy loss in the operation of an actual—as opposed to an ideal—source of emf. For example, a car battery heats up to some extent when it discharges, a result of resistive heating. Thus a real battery contains an *internal resistance* r in addition to maintaining an emf. This resistance is sometimes shown separately from the emf (Fig. 28–3). If we calculate the net potential change around the circuit as before, we find that Eq. (28–2) becomes

$$\mathscr{E} - Ir - IR = 0. \tag{28–4}$$

Due to internal resistance, the potential across the terminals of a battery is different from its emf.

Because of internal resistance, *the potential difference across the battery terminals is no longer just* $\mathscr{E}$; it is given instead by

$$\text{with internal resistance:} \qquad V = \mathscr{E} - Ir. \tag{28–5}$$

This potential difference is a function of the current. Depending on the direction of current flow, the voltage across the terminals of a battery can be greater or less than the battery's emf. A second modification that results from internal resistance is that the current depends on it. From Eq. (28–4), the current in our circuit is

$$I = \frac{\mathscr{E}}{r + R}. \tag{28–6}$$

Compare this to Eq. (28–3).

PROBLEM SOLVING

What small internal resistance means

We can see from Eq. (28–5) that, for small internal resistance, the potential across the terminals is approximately the same as the emf. This is equally true if there is no current at all, so a reading of the potential across the terminals is also a reading of the emf when the external, or load, resistance in the circuit is very large. It is desirable that the internal resistance be small, but it is useful to remember that when we say "small," we mean small in comparison with external resistances, which vary depending on the application. The typical internal resistance of a car battery is less than 0.01 Ω, but it may be as large as 0.1 Ω for a flashlight battery. In many situations, the internal resistance is so small that it can be ignored in electric circuit analyses. Ordinary batteries run down with age not because their emf decreases, but because their internal resistance increases, which means that the current they can supply decreases.

EXAMPLE 28–1 One of two different resistors with respective resistances $R_1 = 5.00\ \Omega$ and $R_2 = 10.00\ \Omega$ can be placed into the circuit shown in Fig. 28–3. The emf, $\mathscr{E}$, and internal resistance, r, of the battery are unknown. When only R_1 is inserted, the current is $I_1 = 0.291$ A; when only R_2 is inserted, the current is $I_2 = 0.147$ A. Find $\mathscr{E}$ and r.

Solution: There are two unknowns here, $\mathscr{E}$ and r, and two different expressions for the current by which to determine them. Equation (28–4) used twice for the two different currents and resistances gives

$$\mathscr{E} - I_1 r - I_1 R_1 = 0,$$

$$\mathscr{E} - I_2 r - I_2 R_2 = 0$$

Multiplying the first equation by I_2 and the second by I_1 and then subtracting to solve for $\mathscr{E}$, we find that

$$\mathscr{E} = \frac{I_1 I_2}{I_1 - I_2}(R_2 - R_1).$$

If we insert this into either of the first two equations, we can solve for r:

$$r = \frac{I_2 R_2 - I_1 R_1}{I_1 - I_2}.$$

Numerical evaluation gives

$$\mathscr{E} = \frac{(0.291\ \text{A})(0.147\ \text{A})}{0.291\ \text{A} - 0.147\ \text{A}}(10.00\ \Omega - 5.00\ \Omega) = 1.48\ \text{V}$$

and

$$r = \frac{(0.147\ \text{A})(10.00\ \Omega) - (0.291\ \text{A})(5.00\ \Omega)}{0.291\ \text{A} - 0.147\ \text{A}} = 0.10\ \Omega.$$

Electric Power and Batteries

A source of emf (or electric energy) is also a source of *electric power*. The power is the rate at which the source delivers energy. From Eq. (27–30), the power of the source is the potential drop across the source times the current that passes through. For a source of emf, $\mathscr{E}$, we have

$$P = \mathscr{E} I. \tag{28–7}$$

Let's see how this works for the circuit of Fig. 28–3. Equation (28–6) tells us that $\mathscr{E}$ is given by $I(r + R)$, where I is the current in the circuit and r and R are the internal and load resistances, respectively. If we use this relation for $\mathscr{E}$ in Eq. (28–7), we find that

$$P = I^2 R + I^2 r. \tag{28–8}$$

Energy conservation dictates this result. The electric power of the source of emf is balanced by the sum of the power dissipated in both the internal and load resistances.

28–2 KIRCHHOFF'S LOOP RULE

A **single-loop circuit** is a circuit with a single path for the current. The simple circuit discussed in Section 28–1 (Fig. 28–3) is an example of a single-loop circuit. Let's repeat the exercise of going around the loop and examining the potential change at every step; to do so, let's redraw the circuit in Fig. 28–4a with various points indicated. The potential is graphed in Fig. 28–4b, where we follow the circuit along the current direction, with zero potential chosen arbitrarily at point a. The region inside the battery is normally not accessible. Because we are not interested in its details, we draw the internal resistance, r, in Fig. 28–4a as though it were separate from the emf; in Fig. 28–4b, we draw the rise of the emf as gradual. There is no potential change in an ideal conducting wire. There is some very small resistance in a real wire, but it can usually be ignored.

The net potential change in traversing the complete circuit is zero simply due to the conservation of energy. If we had followed the circuit in the opposite direction—against the current—the changes would all be of the opposite sign, but the end result would remain: the potential change in a complete circuit is zero. In the context of circuits, this simple law is given a special name, **Kirchhoff's loop rule**:[†]

[†]This rule is named after the nineteenth-century physicist Gustav Kirchhoff.

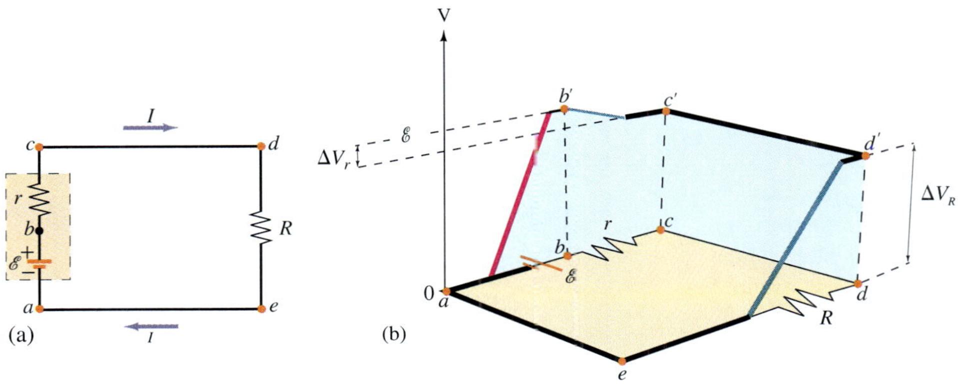

FIGURE 28–4 (a) A single-loop circuit showing the emf (ℰ), internal resistance (r), and load resistor (R). (b) The potential differences for the points labeled in part (a).

The sum of the potential changes around a closed path is zero, or

$$\sum_{\text{closed path}} \Delta V = 0. \tag{28–9}$$

Kirchhoff's loop rule

The loop rule is applicable to any closed path in any electric circuit. When a circuit is laid out in a diagram, many closed paths may be possible, and as we shall see in several examples, the loop rule applied to these many loops is an important tool for finding the desired circuit parameters. In Fig. 28–4a, there is only one closed loop.

In applying the loop rule, we must know the potential differences across various parts of a circuit. Therefore it is useful to summarize what we have learned here and in previous chapters about the potential changes ΔV across individual circuit elements. Here, ΔV is the change in potential when moving from an initial point to a final point; for example, Fig. 28–5, which is a summary figure for these rules, shows that $\Delta V = V_{ab} = V_b - V_a$.

Rules for potential differences across circuit elements

1. In going from the negative to the positive terminal of a battery with emf ℰ, the potential change is positive, $\Delta V = +\mathcal{E}$. In going from the positive to the negative terminal, the potential change is negative, $\Delta V = -\mathcal{E}$.
2. In moving across a resistance R *along* the direction of the current, I, the potential change is negative, $\Delta V = -IR$. The sign is opposite, $\Delta V = +IR$, in moving *against* the direction of the current.
3. In moving from the negatively to the positively charged plate of a capacitor of capacitance C and charge Q, the potential change is positive, $\Delta V = +Q/C$. The potential change is negative, $\Delta V = -Q/C$, when we move from the positively charged plate to the negatively charged plate.

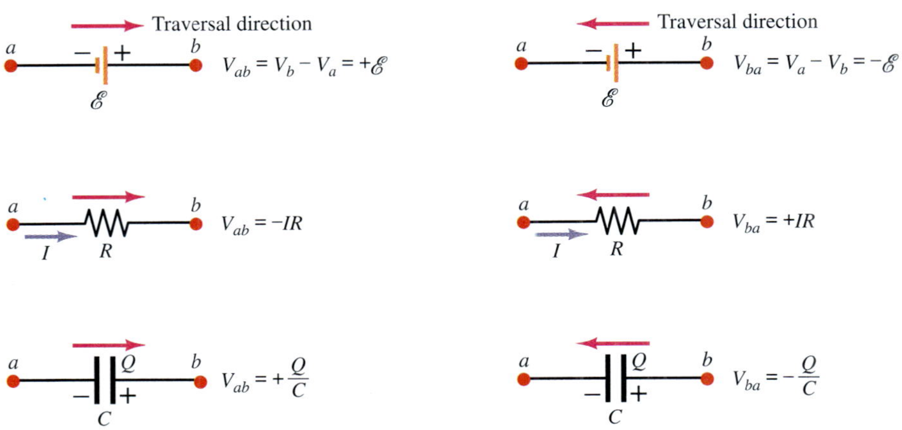

FIGURE 28–5 Rules for potential differences across circuit elements.

APPLICATION: CAPACITORS AS POWER SOURCES FOR LAPTOP COMPUTERS

In Chapter 26, we described the construction of capacitors with large capacitance—on the order of 1 F—that could be used with potential drops of up to 6 V. Such devices will make excellent energy storage devices for laptop computers. It has been traditional to use rechargeable batteries, but these have certain disadvantages: They can supply only a few hours of current before they require a time-consuming recharging. Moreover, they deteriorate with each recharge, and can survive only about 200 recharges—typically this is a lifetime somewhat less than a year. By comparison, a 1-F capacitor has an unlimited lifetime and can be recharged in less than 30 s with a current of 200 mA. But how many hours of use would a capacitor give?

A battery holds a steady potential difference, so that a constant current is produced if it is placed across an ordinary resistor. In order to use the capacitor as the battery is used, it is necessary to find a way to produce a constant current from it. In ordinary use, the potential across the plates of a capacitor decreases as it discharges, and the current produced will decrease if it is placed in series with a resistor (see Section 28–5). The development of new semiconductor devices allows us to circumvent this problem. Take a look at Fig. 27–10, which shows the I–V curve of a nonohmic material. There is a place on that curve where it is nearly flat. In that region of the curve, the current is nearly independent of the potential. A semiconductor device designed to behave this way will produce an I–V curve like that shown in Fig. 28AB–1. If this device, which we label as R, is placed in series with the charged capacitor, as in Fig. 28AB–2, the current that will be drawn is steady as the capacitor discharges and the potential across its plates decreases.

A steady current of the order of 2 μA is required to run a laptop. The time it takes the capacitor to discharge is then obtained from

$$I = \frac{dQ}{dt} = C\,\frac{dV}{dt},$$

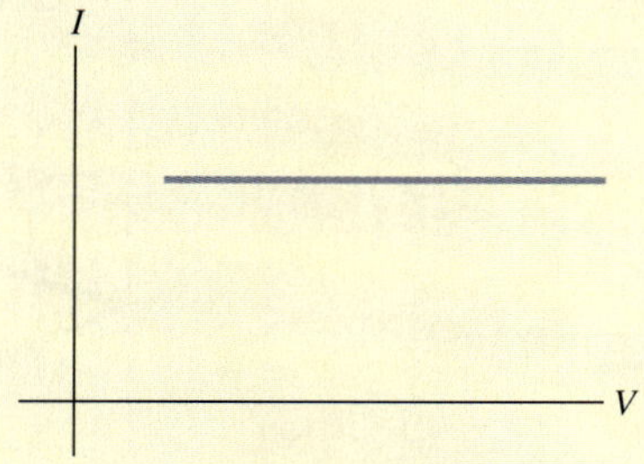

FIGURE 28AB–1 It is the flat part of the I–V curve that is useful for running a computer.

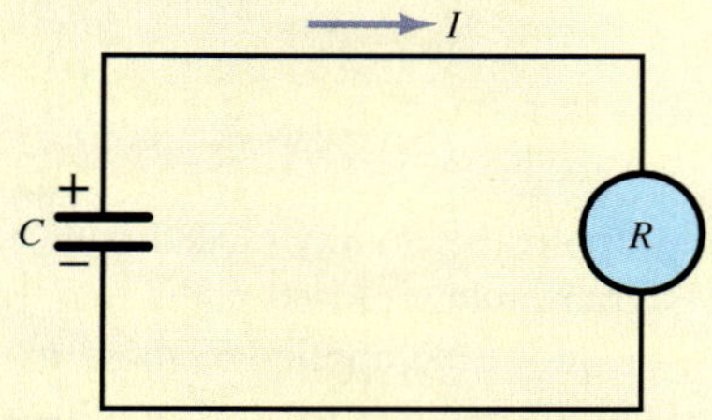

FIGURE 28AB–2 A material with the I–V characteristic of Fig. 28AB–1 allows a capacitor to power a computer.

where V is the potential drop across the capacitor. With both I and C constant, the solution to this differential equation for V is simple: V decreases linearly with time as the capacitor discharges:

$$\frac{V_f - V_i}{\Delta t} = \frac{I}{C},$$

where Δt is the time for the potential across the capacitor to drop from an initial value of 6 V to some final value. This final value is determined by the fact that the memory of the computer will no longer function; a reasonable value is 2.5 V. With $V_f - V_i = 3.5$ V, $I = 2\ \mu$A, and $C = 1$ F, we find $\Delta t = 20$ days. This is a great improvement over the time a battery can run a computer.

No actual current ever flows across a capacitor. When the current is changing with time, a capacitor acts *as if* a current flows across it. In steady-state operation, even this is not possible. In this case, the capacitor acts as an open switch—a place where there is a gap in a wire across which no current flows.

Resistors Connected in Series

Let's now apply Kirchhoff's loop rule to the circuit shown in Fig. 28–6a, which consists of an emf and three resistors. (The internal resistance of $\mathscr{E}$ will be ignored.) We start at point a and move toward the battery in the (clockwise) direction of the assumed current. (We could just as well follow the circuit in the opposite direction.) Kirchhoff's loop rule, Eq. (28–9), gives

$$\sum \Delta V = \mathscr{E} - IR_1 - IR_2 - IR_3 = 0.$$

The solution for the current is

$$I = \frac{\mathscr{E}}{R_1 + R_2 + R_3}. \qquad (28\text{–}10)$$

We compare the circuit of Fig. 28–6a with a second circuit, shown in Fig. 28–6b. The three resistors in Fig. 28–6a are said to be connected *in series*. We want to find a single equivalent resistor R_{eq} that, when it replaces the series combination, allows the same current to flow in the circuit. (Recall the combination of capacitors connected in series and replaced by a single equivalent capacitor from Chapter 26). The current in Fig. 28–6b is simply

$$I = \frac{\mathscr{E}}{R_{eq}}.$$

Comparison of the two preceding equations shows that the resistance equivalent to a set of n individual resistors connected *in series* is found by addition (Fig. 28–6c):

for n resistors connected in series: $$R_{eq} = R_1 + R_2 + R_3 + \cdots + R_n. \qquad (28\text{–}11)$$

The resistance equivalent to resistors connected in series

The analysis of single-loop circuits can be very simple. For example, let's reconsider the circuit of Fig. 28–4a, which includes a resistor R, an emf $\mathscr{E}$, and an internal resistance r. The two resistors are connected in series, and the equivalent series resistance is $r + R$. The current is the emf divided by the total resistance, $I = \mathscr{E}/(r + R)$, which is the result of Eq. (28–6).

EXAMPLE 28–2 Consider the circuit shown in Fig. 28–6a, where the internal resistance of the emf is small enough to ignore. Find the current in the loop if $\mathscr{E} = 12$ V, $R_1 = 2.0\ \Omega$, and $R_2 = R_3 = 6.0\ \Omega$.

Solution: Starting at point a, we use Kirchhoff's loop rule and traverse the closed loop in a clockwise manner. We have

$$\sum \Delta V = +\mathscr{E} - IR_1 - IR_2 - IR_3 = 0,$$

where I is the (unknown) current. Solving this equation for the current gives

$$I = \frac{\mathscr{E}}{R_1 + R_2 + R_3} = \frac{12\ \text{V}}{2.0\ \Omega + 6.0\ \Omega + 6.0\ \Omega} = \frac{12\ \text{V}}{14\ \Omega} = 0.86\ \text{A}.$$

We could realize (Figs. 28–6b, c) that the three resistors are connected in series and, using Eq. (28–11), add their resistances to obtain the equivalent resistance: $2.0\ \Omega + 6.0\ \Omega + 6.0\ \Omega = 14\ \Omega$. The current is as before: $(12\ \text{V})/(14\ \Omega) = 0.86$ A.

EXAMPLE 28–3 Find the current for the two-battery circuit shown in Figs. 28–7a and b. The values of the emfs and resistances are $\mathscr{E}_1 = 6$ V, $r_1 = 0.4\ \Omega$, $R_3 = 3\ \Omega$, $r_2 = 0.1\ \Omega$, $\mathscr{E}_2 = 12$ V, and $R_4 = 10\ \Omega$.

Solution: We assume that the current flows in the direction shown, and we proceed counterclockwise around the circuit from point a (just to show that the results that follow from Kirchhoff's loop rule depend neither on the direction followed nor on the assumed direction of the current). The loop rule gives

$$\sum \Delta V = +Ir_1 - \mathscr{E}_1 + IR_4 + \mathscr{E}_2 + Ir_2 + IR_3 = 0;$$

$$I = \frac{\mathscr{E}_1 - \mathscr{E}_2}{r_1 + R_4 + r_2 + R_3} = \frac{6\ \text{V} - 12\ \text{V}}{0.4\ \Omega + 10\ \Omega + 0.1\ \Omega + 3\ \Omega} = \frac{-6\ \text{V}}{13.5\ \Omega} \simeq -0.4\ \text{A}.$$

The minus sign indicates that we assumed the wrong direction for the current. The actual direction taken by the current is opposite to that in Fig. 28–7. Because

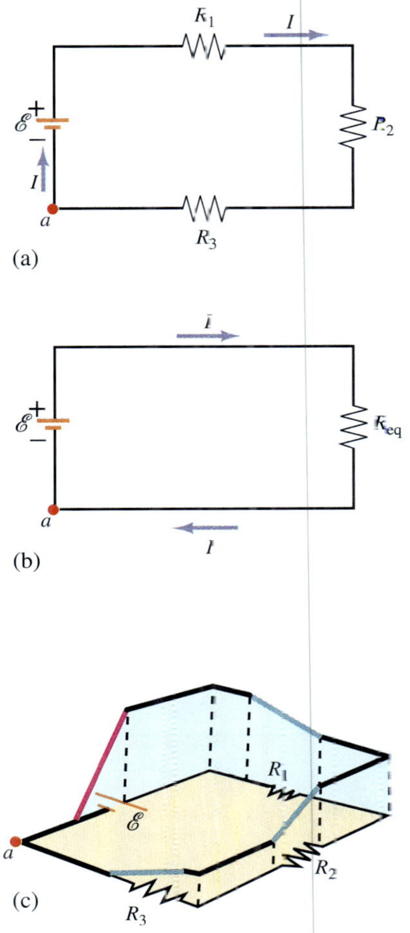

FIGURE 28–6 (a) The circuit with three resistors connected in series is equivalent to (b) the circuit with one resistor of value R_{eq}. (c) This way of looking at the circuit makes it clear that the equivalent resistance is the sum of the three resistances.

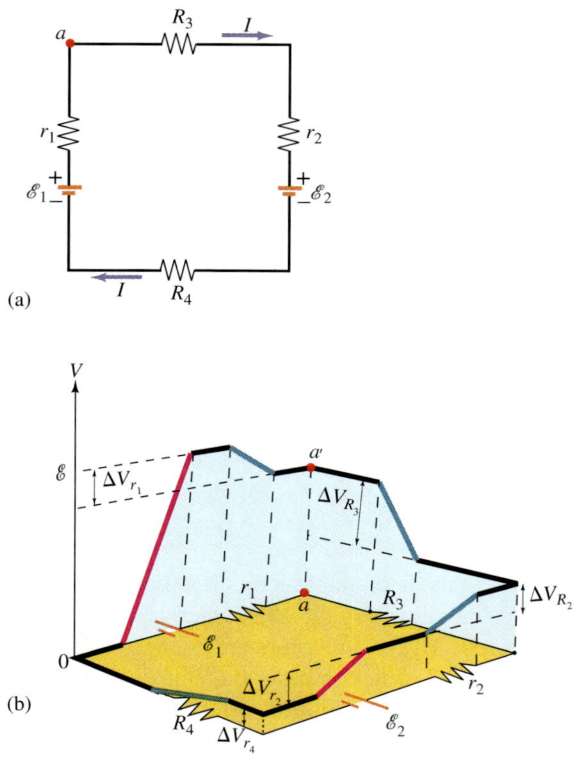

FIGURE 28–7 Example 28–3.

the emf $\mathscr{E}_2$ is larger than the emf $\mathscr{E}_1$, we could have realized from physical reasoning that the current will run counterclockwise. It is not normal for two emfs to oppose each other in single-loop circuits, except when one battery is charging another.

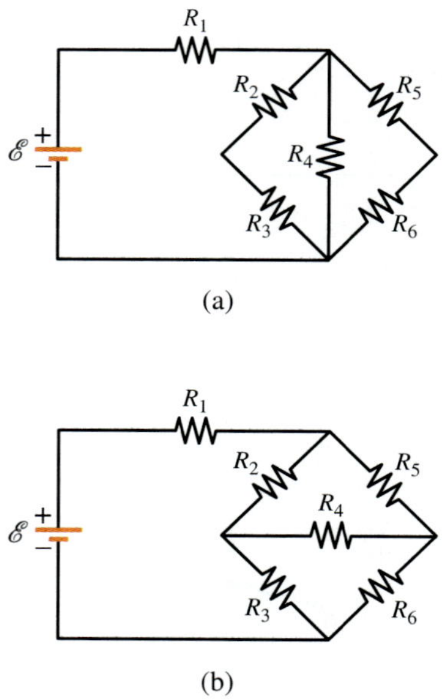

FIGURE 28–8 Two multi-loop circuits.

28–3 KIRCHHOFF'S JUNCTION RULE

Not all circuits are as simple as those we have discussed so far. Consider the two circuits shown in Fig. 28–8. These are examples of **multi-loop circuits**, circuits that have **junctions**, or *nodes*, which are places in the circuit where at least three lines (or wires) meet. Currents entering a junction divide their flow into the various branches of the junction. The circuit in Fig. 28–8a has two four-line junctions, and the circuit in Fig. 28–8b has four three-line junctions.

In steady-state operation, the current moving along a wire in an electric circuit is constant. If it were not, charge would build up at some point and change the electric field—in disagreement with our assumption of a steady state. This *conservation of current* also holds at a circuit junction where three or more wires come together. We know that charge is conserved, so, at any given time, the rate at which charge enters a junction is equal to the rate at which charge leaves the junction. **Kirchhoff's junction rule** (also known as Kirchhoff's first law) states that **the sum of the currents that enter a junction equals the sum of the currents that leave the junction**. We can state this another way: If we interpret a current that leaves a junction as the *negative* of a current of the same magnitude that enters the junction, then

Kirchhoff's junction rule

The algebraic sum of the currents that enter a junction equals zero, or

$$\sum I_{\text{in}} = 0. \tag{28–12}$$

In using Equation (28–12), we must remember that a positive current is literally entering a junction, whereas a negative current is literally leaving the junction. Equation (28–12) is true for each junction in a multi-loop circuit. These equations are additional tools that allow us to solve for currents, voltage differences, etc. The signs of the solution for currents tell us the directions: Positive signs correspond to real incoming currents; negative signs correspond to real outgoing currents.

Several times we have used water flow as a useful mechanical analogue to current flow. The conservation of current at junctions is analogous to the conservation of water flux at the junction of two rivers: The total water flow into the meeting point must equal the total water flow out.

Solving for the Behavior of Multi-Loop Circuits

Let's take a look at how the junction rule and loop rule help us solve for the unknown values in multi-loop circuits. Consider the circuit in Fig. 28–9a. Current flows between every junction. Thus we have currents I_1, I_2, and I_3 in the three separate *legs*, or *branches*, *afed*, *da*, and *dcba*, respectively (Fig 28–9b). (A leg is any line of the circuit that starts and finishes on a junction but does not itself include a junction. For example, the three possible lines that connect *a* and *d* in Fig. 28–9b are all legs.) We can postulate the direction of the three currents arbitrarily, but we shall see that the algebraic equations for the currents will determine them for us. If these equations give a negative value for any of the currents, that current actually travels in the opposite direction than the one we postulated. The junction rule can be applied at junctions *a* and *d* to obtain

$$\text{for junction } a: \quad -I_1 + I_2 + I_3 = 0; \tag{28–13a}$$

$$\text{for junction } d: \quad I_1 - I_2 - I_3 = 0. \tag{28–13b}$$

Notice that currents drawn as leaving the junction have minus signs, while currents that are drawn as arrows entering the junction have plus signs. Notice also that these two equations are not independent of each other; in fact, they are identical! This sometimes happens when the junction rule is applied; however, it is best to go ahead and write an equation for each junction, then check whether or not they are independent.

We have used the junction rule; how do we use the loop rule? We must start by counting the number of *independent* loops. Let's first state a rule for counting the number of such loops, and then see how it arises. For planar circuits (circuits that can be drawn in a plane with no two wires crossing each other), the number of independent loops is just the number of enclosed areas through which you could poke a pencil. According to this idea, the circuit in Fig. 28–9a has two *independent* loops. But we can draw three possible loops, those indicated by the circled integers 1, 2, and 3 of Fig. 28–9a, where 3 is around the perimeter of the circuit. Only two of these three loops can be independent. We can choose *any* two of the three possible loops as our

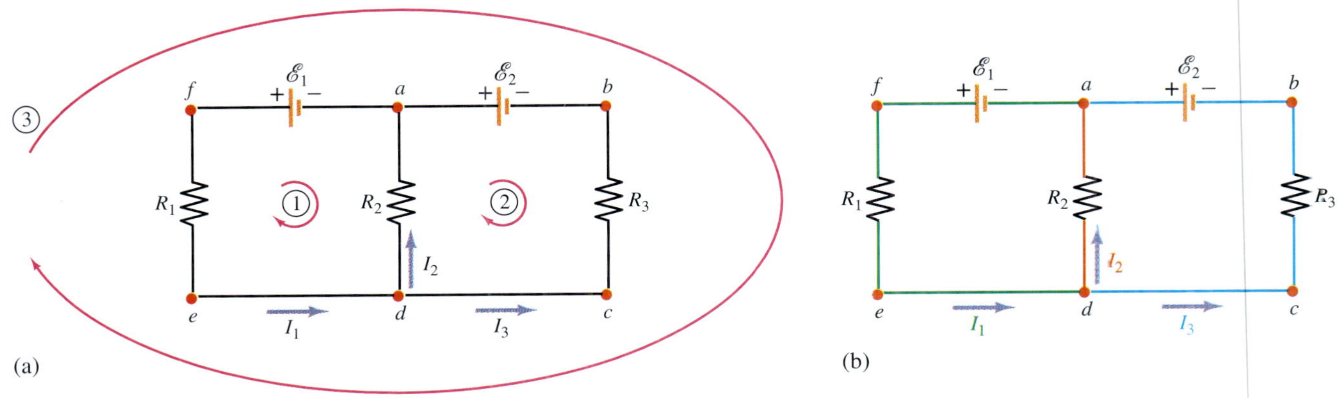

FIGURE 28–9 (a) A multi-loop circuit with two loops and two junctions. (b) The three legs of the circuit are shown in different colors.

independent loops. Before we choose, though, let's apply the loop rule for all three loops and verify that only two are independent. Point a is part of each of the three loops, so let's begin at point a for convenience and apply the loop rule of Section 28–2. The internal resistances for the two emfs are neglected. We traverse each loop in the direction of the loop arrow:

$$\text{for loop 1:} \qquad +I_2R_2 + I_1R_1 - \mathscr{E}_1 = 0; \tag{28–14a}$$

$$\text{for loop 2:} \qquad -\mathscr{E}_2 + I_3R_3 - I_2R_2 = 0; \tag{28–14b}$$

$$\text{for loop 3:} \qquad -\mathscr{E}_2 + I_3R_3 + I_1R_1 - \mathscr{E}_1 = 0. \tag{28–14c}$$

As we suspected, these three loop-rule equations are not independent; the sum of the first two produces the third.

If we now take one equation from Eqs. (28–13), the junction rule, and the first two equations from Eqs. (28–14), the loop rule, *we have a total of three equations to solve for the three unknowns* I_1, I_2, and I_3. If we insert values for resistances and emfs, we can solve the three linear equations for the currents. This example illustrates features common to many circuit problems. We can summarize the important features with a set of problem-solving techniques.

Resistors Connected in Parallel

Just as we were able to connect capacitors in parallel, we say that the three resistors in the circuit shown in Fig. 28–10a are connected *in parallel*. We want to determine whether or not a single equivalent resistor could replace these without affecting the current I supplied by the battery, as in Fig. 28–10b. From the single loop of Fig. 28–10b, we see that $R_{eq} = \mathscr{E}I$, so we can determine R_{eq} if we can find I. To do so, we turn to the analysis of the three-loop, four-junction circuit of Fig. 28–10a. This figure is labeled with the currents that we will need to find the equivalent resistance. No current is labeled in the legs from point a to point b and from point d to point c. We do not need to label these currents because application of the junction rule at junctions b and d indicates that the current from a to b is $I_{ab} = I_2 + I_3$, and the current from d to c is $I_{dc} = I_2 + I_3$. Therefore, $I_{ab} = I_{dc} = I_2 + I_3$. If we apply the junction rule for a, we find

$$I - I_1 - I_{ab} = I - I_1 - I_2 - I_3 = 0. \tag{28–15}$$

The same algebraic equation would follow for application of the junction rule at junction c, so the junctions at a and c are not independent of each other. We have thus applied all the junction rules (step three of the problem-solving techniques).

Application of the loop rule for the three loops indicated in Fig. 28–10a gives the following:

$$\text{for loop 1:} \quad \mathscr{E} - I_1R_1 = 0;$$

$$\text{for loop 2:} \quad -I_2R_2 + I_1R_1 = 0;$$

$$\text{for loop 3:} \quad -I_3R_3 + I_2R_2 = 0.$$

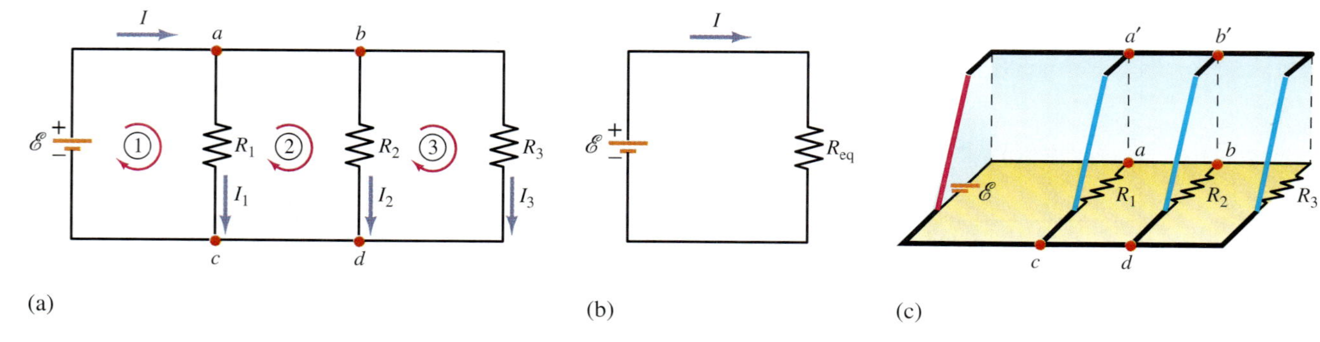

FIGURE 28–10 (a) The circuit with three resistors connected in parallel is equivalent to (b) the circuit with one resistor of value R_{eq}. (c) This way of viewing the circuit makes it clear that there is the same voltage drop across each resistor.

PROBLEM-SOLVING TECHNIQUES

In problems associated with multi-loop circuits, we must find unknown circuit parameters (such as resistance or current) when other parameters are given. To solve these problems, the following procedure may be helpful.

1. Draw a diagram with sources of emf, resistors, capacitors, and so forth clearly labeled. List the known and unknown parameters.
2. Assign a separate current for each leg of the circuit, and indicate that current on the diagram. The direction of current flow may not be immediately obvious; any direction can be assumed for the current, and the final algebraic solution will determine the correct direction. If the solution for a current turns out to be negative, your initial guess for the current direction was incorrect.
3. Apply the junction rule for the currents at each junction. Currents that you have chosen to draw as incoming have plus signs; those you have chosen to draw as outgoing have minus signs. If the circuit has N' junctions, then $N' - 1$ of the equations relating currents at the junctions will be independent.
4. Identify the number of loops N by counting the number of different ways that a pencil can poke through the circuit—a simple procedure for planar circuits. Indicate N loops on the diagram (for example, the loops labeled 1 and 2 on Fig. 28–9a).
5. Apply the loop rule to each of these loops.
6. Check to see that the number of linear equations from steps 3 and 5 matches the number of unknowns.
7. Solve these equations for the unknowns—whether they are currents or other parameters of the circuit. It is usually best to solve these equations *algebraically* and substitute numerical values later. Any special cases or simple limits can easily be checked this way.

These three equations can be written as

$$\mathcal{E} = I_1R_1 = I_2R_2 = I_3R_3.$$

Each of the three resistors has the same voltage, $V = \mathcal{E}$, across them (Fig. 28–10c), so the currents are

$$I_1 = \frac{\mathcal{E}}{R_1}, \quad I_2 = \frac{\mathcal{E}}{R_2}, \quad \text{and} \quad I_3 = \frac{\mathcal{E}}{R_3}.$$

The current from Eq. (28–15) becomes

$$I = \frac{\mathcal{E}}{R_1} + \frac{\mathcal{E}}{R_2} + \frac{\mathcal{E}}{R_3} = \mathcal{E}\left(\frac{1}{R_1} + \frac{1}{R_2} + \frac{1}{R_3}\right).$$

The circuit in Fig. 28–10a can be replaced by an equivalent circuit of the same emf, $\mathcal{E}$, and current, I, and with a single resistor with value R_{eq} (Fig. 28–10b). The current for this circuit has the value

$$I = \frac{\mathcal{E}}{R_{eq}}.$$

Comparison of the two previous equations shows that the equivalent resistance of n resistors connected in parallel is

The resistance equivalent to resistors connected in parallel

$$\text{for } n \text{ resistors connected in parallel:} \quad \frac{1}{R_{eq}} = \frac{1}{R_1} + \frac{1}{R_2} + \frac{1}{R_3} + \cdots + \frac{1}{R_n}. \qquad (28\text{–}16)$$

Note that the resistance of the equivalent resistor is less than that of *any* of the individual resistors: $R_{eq} < R_i$.

EXAMPLE 28–4 Reduce the circuit of Fig. 28–8a to a single-loop circuit with a battery, and find the equivalent resistance.

Solution: Resistances R_2 and R_3 are connected in series and can be replaced by

$$R_7 = R_2 + R_3.$$

Similarly, resistances R_5 and R_6 can be replaced by

$$R_8 = R_5 + R_6.$$

We replace the circuit diagram in Fig. 28–8a with the one in Fig. 28–11a. The resistances R_7, R_4, and R_8 are connected in parallel and can be replaced by R_9, given as in Fig. 28–11b by

$$\frac{1}{R_9} = \frac{1}{R_7} + \frac{1}{R_4} + \frac{1}{R_8}.$$

Finally, as in Fig. 28–11c, the series resistances R_1 and R_9 can be combined to give R_{10},

$$R_{10} = R_1 + R_9.$$

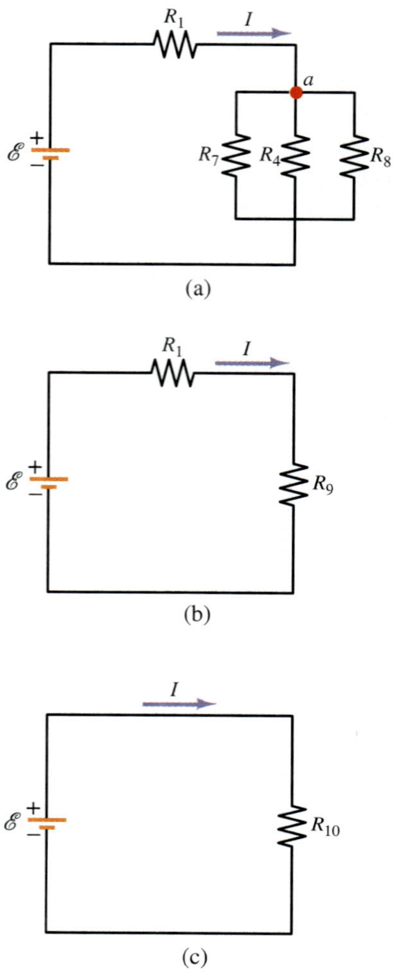

FIGURE 28–11 Example 28–4. A succession of circuits equivalent to the circuit of Fig. 28–8a.

EXAMPLE 28–5 Find the currents for the circuit of Fig. 28–9, given that $\mathscr{E}_1 = 6.00$ V, $\mathscr{E}_2 = 12.0$ V, $R_1 = 100.0\ \Omega$, $R_2 = 10.0\ \Omega$, and $R_3 = 80.0\ \Omega$.

Solution: Let's follow the problem-solving techniques. The circuit diagram is already given, and all parameters are labeled on Fig. 28–9. In this case, the unknowns are I_1, I_2, and I_3. Moreover, we have already written the junction and loop equations for this circuit: Eqs. (28–13) and (28–14), respectively. According to step 3, there is a single independent junction, and we choose Eq. (28–13b). According to step 4, there are two independent loops, and we choose Eqs. (28–14a) and (28–14b) for loops 1 and 2. We have three equations to solve for the three unknowns.

To solve, we substitute $I_1 = I_2 + I_3$ from the junction equation into the loop equations:

$$I_2R_2 + (I_2 + I_3)R_1 - \mathscr{E}_1 = 0; \qquad (28\text{–}17a)$$

$$-\mathscr{E}_2 + I_3R_3 - I_2R_2 = 0. \qquad (28\text{–}17b)$$

Solve Eq. (28–17b) for I_3:

$$I_3 = \frac{I_2R_2 + \mathscr{E}_2}{R_3}. \qquad (28\text{–}18)$$

When this result is inserted into Eq. (28–17a), we find an equation for I_2:

$$I_2\,(R_1 + R_2) + (I_2\,R_1 + \mathscr{E}_2)\frac{R_1}{R_3} - \mathscr{E}_1 = 0.$$

Solving for I_2, we find

$$I_2 = \frac{R_3\mathscr{E}_1 - R_1\mathscr{E}_2}{R_1R_2 + R_1R_3 + R_2R_3}. \qquad (28\text{–}19)$$

If we substitute this result into Eq. (28–18), we get a final expression for I_3:

$$I_3 = \frac{R_2(\mathscr{E}_1 + \mathscr{E}_2) + R_1\mathscr{E}_2}{R_1R_2 + R_1R_3 + R_2R_3}. \qquad (28\text{–}20)$$

Finally, I_1 is the sum of I_2 and I_3:

$$I_1 = \frac{R_2(\mathscr{E}_1 + \mathscr{E}_2) + R_3\mathscr{E}_1}{R_1R_2 + R_1R_3 + R_2R_3}. \qquad (28\text{–}21)$$

Equations (28–19), (28–20), and (28–21) constitute the desired algebraic solution. Is there a limit in which we can check them? If we take R_2 to be large, we would expect so much resistance in this leg that current I_2 should drop to zero. Indeed, in this limit $I_1 \to (\mathscr{E}_1 + \mathscr{E}_2)/(R_1 + R_3)$, I_3 tends to the same result, and $I_2 \to 0$. This is precisely what we would expect for the circuit if the segment containing R_2 were eliminated.

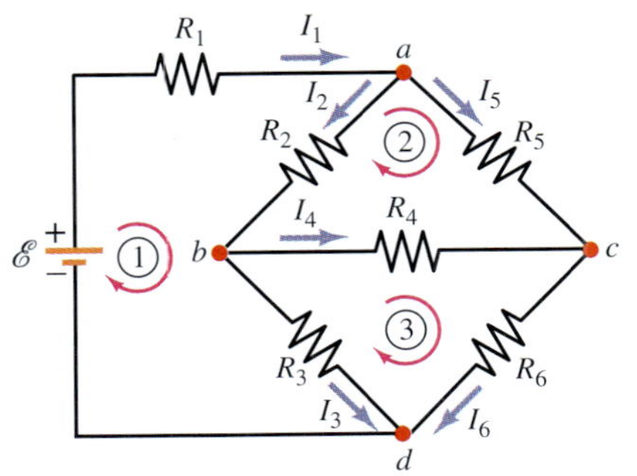

FIGURE 28–12 Example 28–6. The circuit of Fig. 28–8b is labeled with loops and junctions to be solved by circuit analysis.

Straightforward numerical substitution of the known circuit parameters into Eqs. (28–19), (28–20), and (28–21) gives $I_1 = 68$ mA, $I_3 = 141$ mA, and $I_2 = -73$ mA. Note that I_2 is negative, meaning that it flows in a direction opposite to the one we assumed in Fig. 28–9.

EXAMPLE 28–6 Assume that the emfs and resistors are known for the circuit in Fig. 28–8b. Express the linear equations that can be solved to find all currents.

Solution: Because the currents are not indicated in the circuit diagram, we redraw it with the currents in each separate segment indicated and the junctions labeled (Fig. 28–12). We have drawn three independent loops in the circuit. There are six currents, so we need six linear equations. There are four junctions, but, when the junction rule is applied, one will give a relation that is not independent of the others. Thus we may choose any three junctions to apply the junction rules, giving three equations. The three loop equations provide the necessary remaining three equations.

The six equations are

$$\text{for junction } a: \qquad I_1 - I_2 - I_5 = 0;$$

$$\text{for junction } b: \qquad I_2 - I_3 - I_4 = 0;$$

$$\text{for junction } c: \qquad I_4 + I_5 - I_6 = 0;$$

$$\text{for loop 1:} \qquad \mathscr{E} - I_1R_1 - I_2R_2 - I_3R_3 = 0;$$

$$\text{for loop 2:} \qquad -I_5R_5 + I_4R_4 + I_2R_2 = 0;$$

$$\text{for loop 3:} \qquad -I_6R_6 + I_3R_3 - I_4R_4 = 0.$$

With sufficient patience, we can solve these six equations for the six unknown currents.

EXAMPLE 28–7 Find the steady-state currents I_1 and I_2 in the circuit drawn in Fig. 28–13. Also find the resistance of resistor R_3 that will give a steady-state current $I_3 = 50$ mA. Finally, determine the potential drop across the capacitor. The values of the known elements are $\mathscr{E} = 6$ V, $R_1 = 100\ \Omega$, $R_2 = 80\ \Omega$, and $C = 2\ \mu$F.

Solution: This problem is unusual in two respects. There is a capacitor, and we are asking for the values of two currents, I_1 and I_2, and a resistor R_3, rather than three currents. There are two loops in the circuit and two junctions, but only one junction is independent. There will thus be three equations. As we have already remarked, no steady current can pass through a capacitor, which therefore acts as an open switch. Current I_2 will be zero, so in effect this is a simple single-loop circuit. We no longer need to worry about the equation for loop 2.

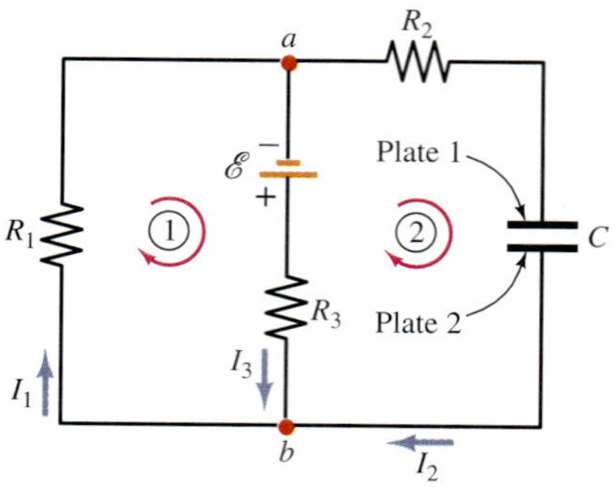

FIGURE 28–13 Example 28–7. Notice that, in steady-state operation, the capacitor acts as an open switch despite the fact that it has voltage across its plates.

The junction equation is particularly simple. Either junction gives $I_1 = I_2 + I_3$, and because $I_2 = 0$, $I_1 = I_3 = 50$ mA. The resistance R_3 is found from the loop equation

$$\mathcal{E} - I_3R_3 - I_1R_1 = 0;$$

$$R_3 = \frac{\mathcal{E} - I_1R_1}{I_3} = \frac{\mathcal{E} - I_3R_1}{I_3} = \frac{6\text{ V} - (50\text{ mA})(100\ \Omega)}{50\text{ mA}} = 20\ \Omega.$$

To find the potential drop V_C across the capacitor, note that the loop equation for loop 2 is valid even if there is no current in the loop. Because this equation expresses potential changes around this loop, it will give us V_C. Starting at point a and following the loop as indicated in Fig. 28–13,

$$-I_2R_2 + V_C + I_3R_3 - \mathcal{E} = 0.$$

The first term does not contribute because $I_2 = 0$, so

$$V_C = \mathcal{E} - I_3R_3 = 6\text{ V} - (50\text{ mA})(20\ \Omega) = 5\text{ V}.$$

Note the sign of the potential difference between the two plates. The potential increases when we go from plate 1 to plate 2, meaning that plate 2 is at a higher potential than plate 1. Positive charges have accumulated on plate 2.

28–4 MEASURING INSTRUMENTS

We have referred to the currents and voltages in various circuit elements, but we have not yet explained how these quantities are measured. A wide variety of instruments exists for this purpose: **Ammeters** measure current, **voltmeters** measure voltage, and **ohmmeters** measure resistance. These devices are often combined into one instrument called a **multimeter** or **VOM** (*v*olt-*o*hm-*m*illiammeter) (Fig. 28–14). We shall focus our attention on ammeters and voltmeters. Whether quantities such as current, voltage, or resistance are measured with analog (the continuous movement of the hands of a watch is an example) or digital (numeric) displays, and whatever the detailed mechanism of these measurements, a general principle must be respected: The measuring devices should not distort the operation of the circuit being measured. We shall emphasize this principle in the following discussion.

For most applications, analog devices have been supplanted by digital devices, which are usually less expensive and more accurate. However, analog devices (dials)—the gasoline gauge of your automobile, for example—remain superior for the visual recognition of trends in measured parameters. It will be simplest for us to illustrate the most important features of ammeters and voltmeters with analog devices.

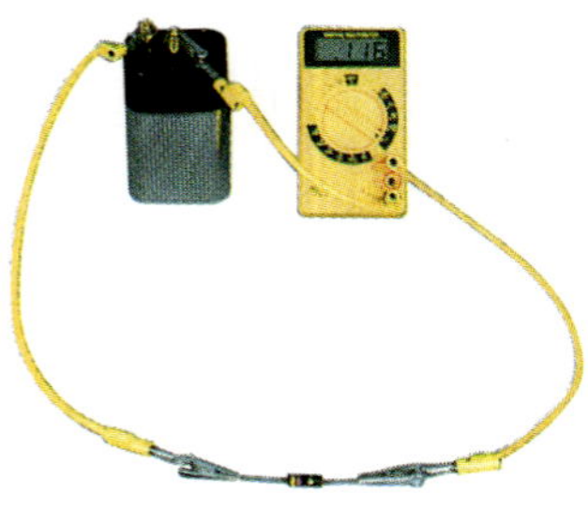

FIGURE 28–14 This digital multimeter is useful for the measurement of circuit properties.

How to Construct Analog Measuring Devices

Analog versions of ammeters and voltmeters typically utilize a *galvanometer* (to be studied in Chapter 30), which relies on magnetic effects. A galvanometer, indicated by a circled "*G*" in circuit diagrams, consists of a coil of wire that rotates in the magnetic field produced whenever a current passes through the wire. A needle is deflected by an amount proportional to the current that passes through the coil. The amount of deflection, properly calibrated, is the analog measurement.

A measuring device should not disturb the measured circuit.

Ammeters. Ammeters measure currents in circuit wires. That a galvanometer reacts to current does not mean that it is itself a good ammeter. Application of the general principle that an ammeter should not disturb the current being measured shows that *a good ammeter should have a resistance that is small compared to other resistances in the circuit*. This is necessary because the ammeter is placed in series in the circuit seg-

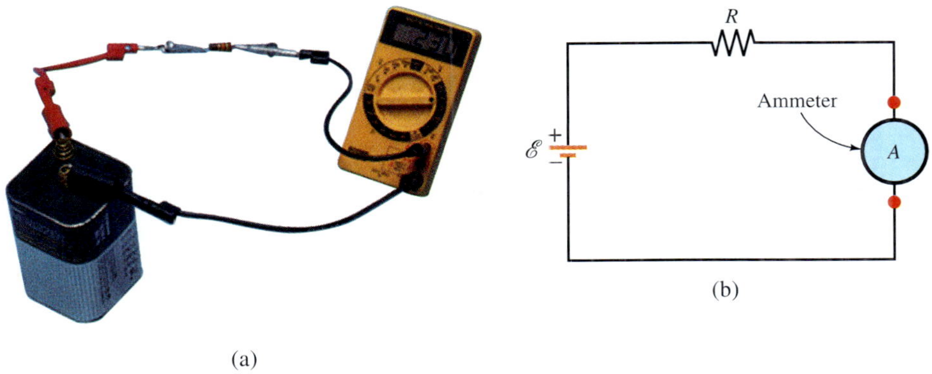

FIGURE 28–15 (a) An ammeter is placed in series where the current is to be measured. The resistance of an ammeter should be small so as not to change the current. (b) Schematic diagram of the circuit.

ment through which the current to be measured passes. Only if the ammeter has a small resistance will it not affect the current. Figure 28–15a shows a circuit being probed by an ammeter, indicated by a circled "A" in Fig. 28–15b. Suppose that the resistance of the ammeter is R_A and that the series resistance is R. Before the ammeter is inserted, the current is

$$I = \frac{\mathscr{E}}{R}.$$

After the ammeter is inserted, the current is

$$I = \frac{\mathscr{E}}{R + R_A}. \tag{28–22}$$

The current will be the same with or without the ammeter attached only if $R_A << R$.

An ammeter should have a small resistance.

Voltmeters. Voltmeters measure potential differences across circuit elements. They do so by being placed in parallel with those elements. Figure 28–16a shows a voltmeter—indicated by the circled "V" in Fig. 28–16b—used to measure the potential across a resistor. The figure makes it evident why the voltmeter, which measures the potential difference *across* a circuit element, is placed in parallel with that element. If we apply the general principle that a voltmeter should not disturb the potential difference being measured, this will show that *a good voltmeter should have a large resistance.* If the internal resistance of the voltmeter is R_V, the combination of voltmeter and resistance in Fig. 28–16b forms a parallel resistance circuit with equivalent resistance

$$\frac{1}{R_{eq}} = \frac{1}{R_V} + \frac{1}{R}.$$

When $R_V >> R$, then $R_{eq} \simeq R$, and none of the parameters of the original circuit are affected.

A galvanometer connected in series with a resistor of large resistance can serve as a voltmeter. If the current I passing through the galvanometer is known, then the potential drop across the voltmeter is approximately IR_V. For example, take a galvanometer that can measure a maximum current of 100 μA. This is equivalent to a measurement of 10 V if the internal resistance is set at

$$R_V = \frac{10\ \text{V}}{100 \times 10^{-6}\ \text{A}} = 10^5\ \Omega.$$

An analog voltmeter is then a galvanometer with a series resistor of resistance R_V (Fig. 28–17). In our example, we used a 10^5-Ω resistor for a full-scale reading of 10 V. If we had wanted to measure a full-scale voltage of 1000 V, we would have taken $R_V = 10^7\ \Omega$.

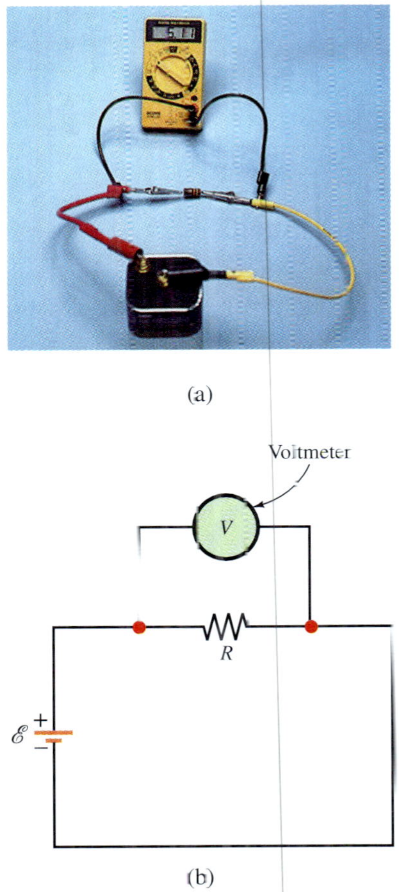

FIGURE 28–16 (a) A voltmeter is placed in parallel across the circuit element whose potential drop is to be measured. The resistance of a voltmeter should be large so as not to change the circuit. (b) Schematic diagram of the circuit.

A voltmeter should have a large resistance.

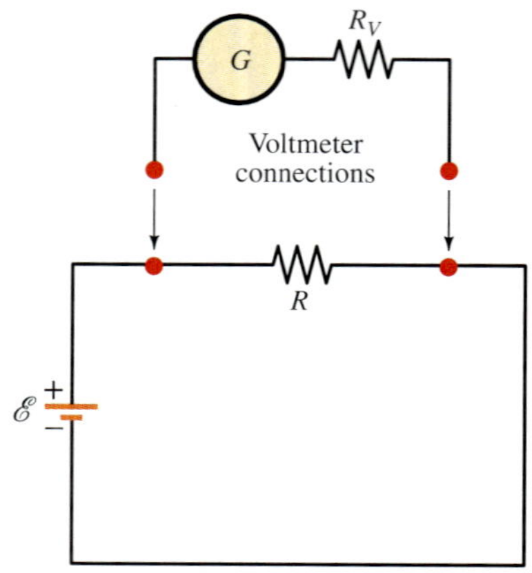

FIGURE 28–17 A galvanometer with a large series resistance can serve as a voltmeter.

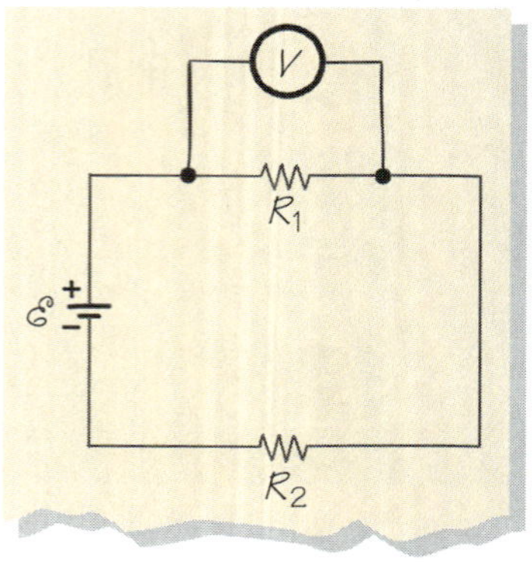

FIGURE 28–18 Example 28–8.

APPLICATION

Switching devices

EXAMPLE 28–8 A voltmeter with an internal resistance of $10^5\ \Omega$ is used to measure the voltage across resistor R_1 in the circuit of Fig. 28–18. Compare the potential drop with and without the voltmeter for $\mathscr{E} = 6$ V, $R_1 = 10$ kΩ, and $R_2 = 5$ kΩ. This describes the error in measurement caused by the voltmeter itself.

Solution: Without the voltmeter, the current flowing through the circuit is $\mathscr{E}/(R_1 + R_2)$, or (6 V)/(10 kΩ + 5 kΩ) = (6 V)/(15,000 Ω) = 0.4 mA. The voltage V_1 across R_1 is then

$$V_1 = IR_1 = (4 \times 10^{-4}\ \text{A})(10^4\ \Omega) = 4\ \text{V}.$$

When the voltmeter is connected, resistance R_1 is replaced by the equivalent resistance R_1', given by

$$\frac{1}{R_1'} = \frac{1}{R_1} + \frac{1}{R_V} = \frac{1}{10^4\ \Omega} + \frac{1}{10^5\ \Omega} = 1.1 \times 10^{-4}\ \Omega^{-1};$$

$$R_1' \simeq 9 \times 10^3\ \Omega.$$

The change in R_1 affects the current in the circuit and hence the voltage drop across it. This drop is just the voltage drop of the equivalent resistance R_1':

$$V_1' = IR_1' = \frac{\mathscr{E}R_1'}{R_1' + R_2} = \frac{(6\ \text{V})(9 \times 10^3\ \Omega)}{(9 \times 10^3\ \Omega) + (5 \times 10^3\ \Omega)} \simeq 3.9\ \text{V}.$$

The difference between 3.9 V and 4 V is 0.1 V, equivalent to a 3-percent error due to the voltmeter. The larger the voltmeter's resistance, the smaller the error so introduced.

In Fig. 28–19, we summarize the use of voltmeters and ammeters in a typical circuit.

28–5 RC CIRCUITS

RC circuits are circuits that contain both resistors and capacitors. They are interesting because their currents and potentials exhibit time-varying behavior. For steady-state currents, the capacitor acts as an open switch. Even for circuits containing time-*independent* sources of emf, we introduce time dependence in a circuit every time we open or close a switch. Circuits with capacitors, such as the circuit in Fig. 28–20, have time-dependent effects that are useful for the control of motors, machinery, or computers.

We first observed the effect of a capacitor in an electric circuit in Example 28–7, where we concentrated on steady-state operation with a fully charged capacitor. Now we want to examine the more complex, transient behavior that occurs when the capacitor is being charged and discharged. Consider the circuit shown in Fig. 28–20, with an initially uncharged capacitor. When the switch is closed (to position a) at $t = 0$, current begins to flow from the positive terminal of the battery, and positive charge begins to collect on plate 1 of the capacitor while an equal amount of negative charge collects on plate 2. Current flows everywhere in the circuit *except* through the plates of the capacitor. Immediately after the switch is closed, the current has its maximum value, but the charge that builds up on the capacitor plates opposes further charge flow, and the current decreases. When the potential across the capacitor plates equals the emf and equilibrium is reached, the current falls to zero. This occurs when the charge on the capacitor plates, Q_0, is such that $\mathscr{E} = Q_0/C$.

After equilibrium has been reached and the current has become zero, we change the switch to position b and take the battery out of the circuit. The circuit now consists only of the charged capacitor and the resistor. Current flows through the circuit

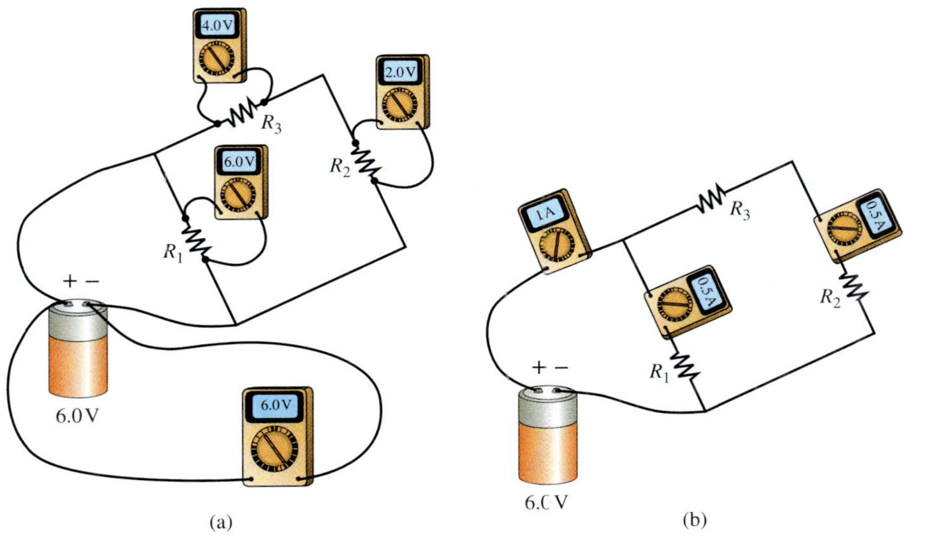

FIGURE 28–19 (a) Measuring voltage. (b) Measuring current.

from plate 1 of the capacitor to plate 2. The rate of flow is limited by the resistor. At first the current is high, but it decreases as the capacitor *discharges* through the resistor. Eventually the capacitor discharges completely and the current again falls to zero at equilibrium.

We first apply Kirchhoff's loop rule to the circuit of Fig. 28–20 for the switch at position a, when the capacitor is being charged. The loop rule gives

$$\mathscr{E} - IR - \frac{Q}{C} = 0. \tag{28–23}$$

In this equation, neither the current nor the charge on the capacitor is constant while the capacitor charges. Because $I = dQ/dt$, we can rewrite Eq. (28–23) as

$$\mathscr{E} - R\frac{dQ}{dt} - \frac{Q}{C} = 0. \tag{28–24}$$

The single variable in this equation is the charge Q. The differential equation (28–24) is straightforward to solve; let's omit the mathematical complexities and present its solution:

$$Q = C\mathscr{E}(1 - e^{-t/RC}). \tag{28–25}$$

By differentiating Eq. (28–25) with respect to time and substituting into Eq. (28–24), we can see that it is a solution (see Problem 56). More importantly, does it agree physically with what we expect? According to Eq. (28–25), the charge on the capacitor is zero at $t = 0$ and builds smoothly to $C\mathscr{E}$ at $t = \infty$, in agreement with our earlier discussion.

We can find the current in the circuit by differentiating Eq. (28–25) with respect to time:

$$I = \frac{dQ}{dt} = C\mathscr{E}\left(\frac{1}{RC}e^{-t/RC}\right) = \frac{\mathscr{E}}{R}e^{-t/RC}. \tag{28–26}$$

The sign of the current is positive, so we chose the correct current direction (clockwise). The maximum value of the current is $\mathscr{E}/R$ at $t = 0$, and the current is zero at $t = \infty$, which also agrees with our earlier discussion. Just after the switch is closed, the emf of the battery is $\mathscr{E} = IR$, and no potential drops across the capacitor because it is uncharged. As the capacitor charges, the current drops *exponentially* to zero.

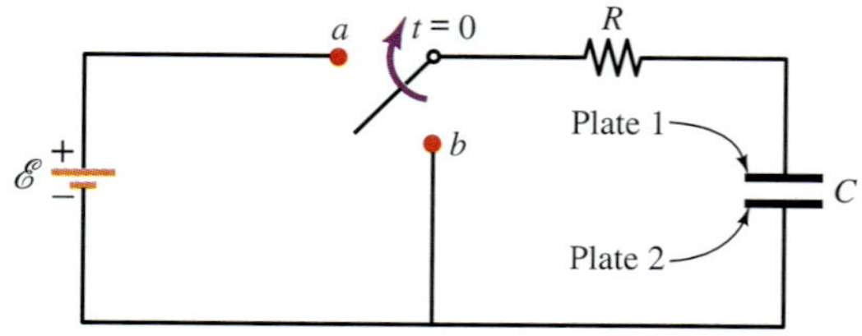

FIGURE 28–20 A circuit used to charge and discharge a capacitor through a resistor. When the switch is closed at a, the capacitor is charged by the source of emf, whereas the capacitor discharges through R when the switch is thrown to b.

Equations (28–25) and (28–26) show that the time dependence of both charge and current is determined by the product RC, which is called the **time constant**. It has units of time; with R and C in SI units, RC will be in seconds. The time constant determines how fast a capacitor charges and discharges. The smaller the value of RC, the more quickly the exponentials in the equations for Q and I fall; similarly, the larger the value of RC, the more slowly the exponentials change. Figure 28–21a shows the current in the circuit, and Fig. 28–21b shows the charge on the capacitor as a function of time while the capacitor is being charged (Fig. 28–21c). After a time RC, the current has dropped to $e^{-1} \simeq 0.37$ times its original value. After this same amount of time, the capacitor is $(1 - e^{-1}) \simeq 63$ percent fully charged. It is 86 percent charged at time $2RC$ and 95 percent charged at time $3RC$.

Let's return to the circuit of Fig. 28–20. Suppose that the switch has been in position a for a long time, the capacitor is fully charged, and there is no current. At time $t = 0$, we throw the switch to position b. Only the discharging capacitor and the resistor are now in the circuit (Fig. 28–22). The positive charge is on plate 1, and we assume as before that the current is clockwise. The loop rule now gives

$$-IR - \frac{Q}{C} = 0. \qquad (28\text{–}27)$$

Using $I = dQ/dt$, we have

$$R\frac{dQ}{dt} + \frac{Q}{C} = 0. \qquad (28\text{–}28)$$

This differential equation is solved by the function

$$Q = Q_0 e^{-t/RC}, \qquad (28\text{–}29)$$

where Q_0 is the initial charge on the capacitor when the switch is changed, $Q_0 = C\mathscr{E}$. Equation (28–29) may be substituted into Eq. (28–28) to verify that it is a solution

FIGURE 28–21 The time response of (a) the current I and (b) the charge Q across a capacitor as the capacitor is charged. The characteristic time response of the exponential behavior is RC. The value 0.37 in the graph of current is the factor e^{-1}; the value 0.63 in the graph of charge is the factor $(1 - e^{-1})$. (c) This oscilloscope screen shows the exponential current drop on a charging capacitor.

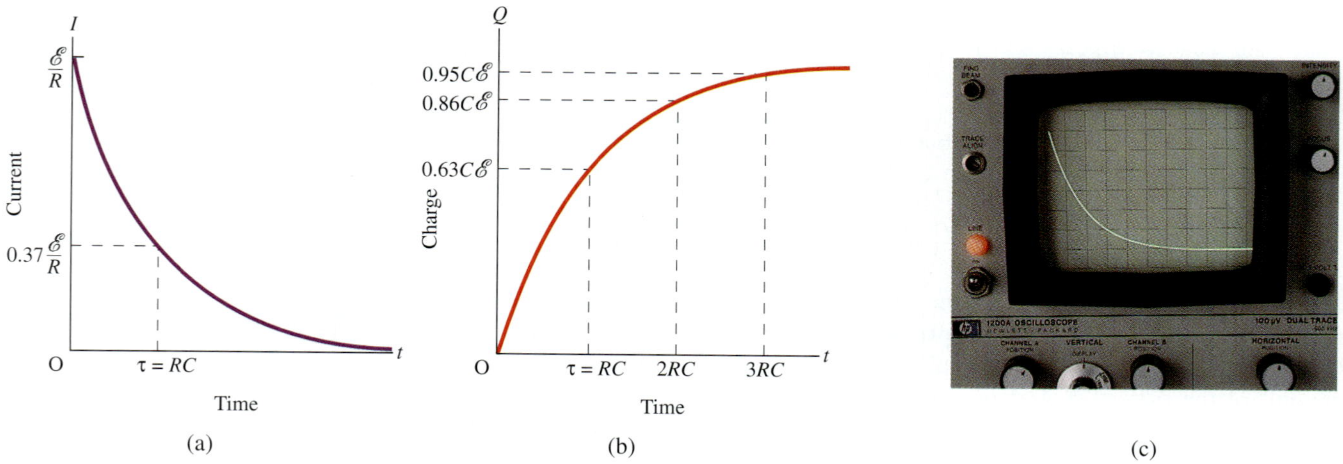

(see Problem 57). The charge on the capacitor decreases exponentially with the time constant RC, and, after a long time, there will be no charge on the capacitor.

We find the current by differentiating Eq. (28–29):

$$I = \frac{dQ}{dt} = -\frac{Q_0}{RC}e^{-t/RC}. \quad (28\text{–}30)$$

The current in this case is negative, indicating that the actual current is counterclockwise, opposite in direction to the current we assumed when we drew the diagram. It is again a maximum at $t = 0$, when the magnitude of the current is $Q_0/RC = \mathscr{E}/R$. After a long time, the current is again zero.

The behavior of the charge and current for the capacitor that discharges through a resistor is qualitatively what we expected from our earlier discussion. The magnitude of the current for this case is just as shown in Fig. 28–21a for the charging capacitor. The charge on the capacitor is plotted as a function of time in Fig. 28–23. Again, the factor 0.37 is e^{-1}.

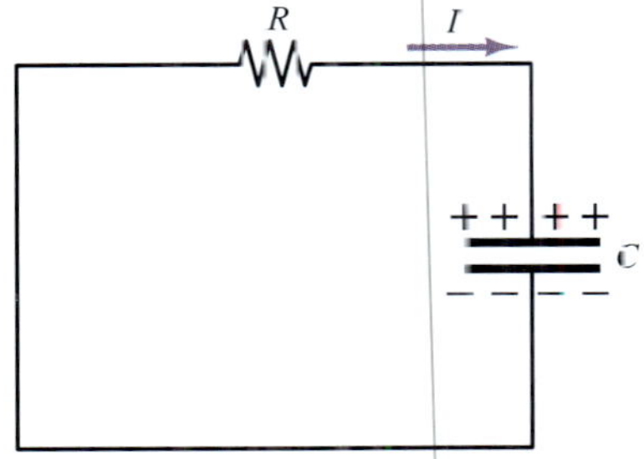

FIGURE 28–22 The circuit of Fig. 28–20 after the switch has been thrown to position b.

Energy in *RC* Circuits

Let's now examine the role of energy in the case of a charging capacitor. From the definition of potential, the amount of work done by the battery emf during the charging process is $\mathscr{E}$ times the total charge processed by the battery. This charge is the final charge $C\mathscr{E}$ on the capacitor plates after a long period of time. Thus, the work done by the battery, W_{bat}, is

The time scale of the time dependence in RC circuits is given by the product RC.

$$W_{bat} = \mathscr{E}(C\mathscr{E}) = C\mathscr{E}^2. \quad (28\text{–}31)$$

Where is the energy that matches this work? In part, the energy is stored in the capacitor. We know from Eq. (26–9) that the total energy stored by a capacitor is $CV^2/2$. The voltage V in this case is $\mathscr{E}$, so the energy stored by the capacitor, E_{cap}, is

$$E_{cap} = \frac{C\mathscr{E}^2}{2}. \quad (28\text{–}32)$$

Where has the other half of the work done by the battery gone? The only other circuit element is the resistor, and the other half of the work has gone into Joule heating of that resistor. From Eq. (27–30), we know that the power loss in the resistor is $P = I^2R$. We can integrate the power over time to find the energy loss in the resistor, E_{res}:

$$E_{res} = \int_0^\infty I^2R\,dt = \frac{\mathscr{E}^2}{R}\int_0^\infty e^{-2t/RC}\,dt$$

$$= \frac{C\mathscr{E}^2}{2}\left(-e^{-2t/RC}\right)\Big|_0^\infty = \frac{C\mathscr{E}^2}{2}. \quad (28\text{–}33)$$

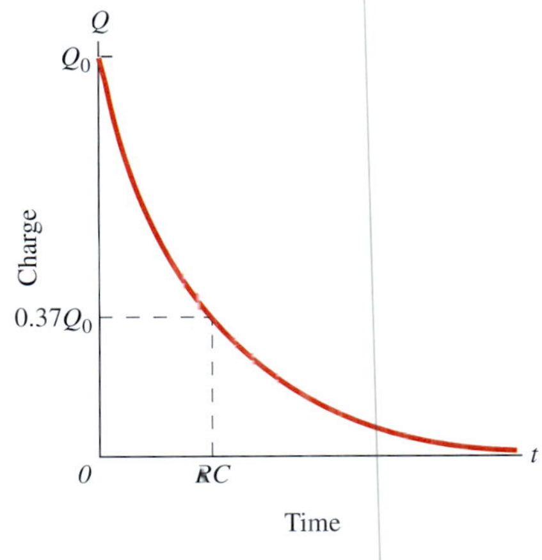

FIGURE 28–23 The capacitor charge as the capacitor of Fig. 28–22 discharges through the resistor as a function of time. The characteristic time response of the exponential behavior is again RC.

The thermal energy loss in the resistor indeed accounts for the other half of the work done by the battery. This 50-percent split of energy between the resistor and the capacitor is *independent* of $\mathscr{E}$, R, and C. For the case of the discharging capacitor, all the energy stored in the capacitor dissipates as heat in the resistor. A spectacular demonstration of the energy content is shown in Fig. 28–24.

EXAMPLE 28–9 The charging circuit shown in Fig. 28–20 (with a switch thrown to position a at $t = 0$) has the circuit elements $\mathscr{E} = 12$ V, $R = 100.0\ \Omega$, and $C = 10.0\ \mu$F. (a) Find the time constant, the final charge on the capacitor, and the work done by the battery. (b) How long does it take for the capacitor to be charged to 99.9 percent of its final charge?

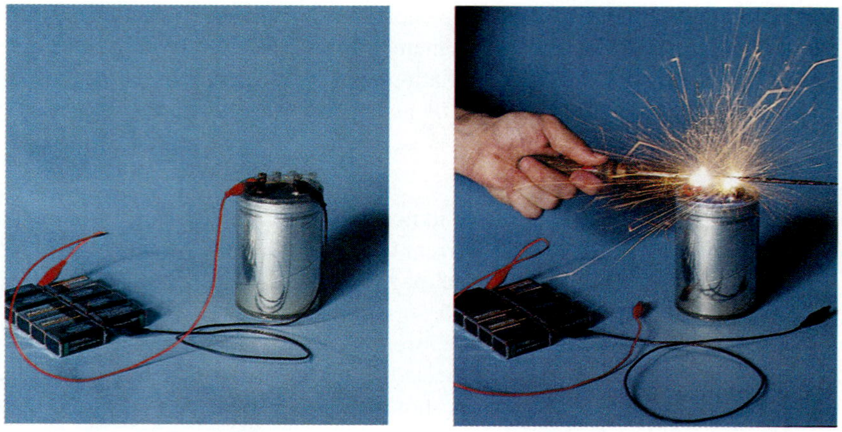

(a) (b)

FIGURE 28–24 When this capacitor is charged, as in part (a), it is capable of releasing its stored energy with the spectacular effects shown in part (b) when the capacitor connections are then shorted. The 14 mF capacitor is charged by placing 73 V across its terminals. Once charged, it stores 37 J of energy.

Solution: (a) The time constant is RC:

$$RC = (100.0\ \Omega)(10.0 \times 10^{-6}\ \text{F}) = 1.00 \times 10^{-3}\ \text{s} = 1.00\ \text{ms}.$$

The final charge on the capacitor, Q_f, according to Eq. (28–25) in the limit of large t, is $C\mathscr{E}$:

$$Q_f = C\mathscr{E} = (10.0 \times 10^{-6}\ \text{F})(12\ \text{V}) = 1.2 \times 10^{-4}\ \text{C}.$$

According to Eq. (28–31), the work done by the battery, W_{bat}, is

$$W_{\text{bat}} = C\mathscr{E}^2 = (10 \times 10^{-6}\ \text{F})(12\ \text{V})^2 = 1.4 \times 10^{-3}\ \text{J}.$$

(b) We use Eq. (28–25) to find how long it takes the capacitor to become 99.9 percent charged:

$$Q = 0.999Q_f = Q_f(1 - e^{-t/RC}).$$

We eliminate the maximum charge, Q_f, from this equation and rearrange the equation to obtain

$$e^{-t/RC} = 1 - 0.999 = 0.001.$$

Taking the natural logarithm of both sides gives

$$-\frac{t}{RC} = -6.91.$$

The time to reach 99.9 percent of its final charge is thus

$$t = 6.91RC = (6.91)(1.00\ \text{ms}) = 6.91\ \text{ms}.$$

SUMMARY

Sources of emf (electromotive force), $\mathscr{E}$, such as chemical batteries, are sources of electric energy. The emf is defined by the amount of work it can do to move charge:

$$\mathscr{E} \equiv \frac{dW}{dq}. \tag{28–1}$$

Batteries cause charges to move in circuits. The simplest circuits to analyze are direct-current circuits, in which no circuit parameters change with time.

Analysis of single- or multi-loop circuits is accomplished by the use of Kirchhoff's two rules. Kirchhoff's loop rule is "circuit language" for the conservative nature of electric force. It states that the sum of the potential changes around a closed path of a circuit is zero:

$$\sum_{\text{closed path}} \Delta V = 0. \tag{28–9}$$

We can specify the potential change across batteries, resistors, and capacitors. Any source of emf has an internal resistance r that may be large enough to require consideration.

Kirchhoff's junction rule is "circuit language" for the conservation of electric current. It states that the sum of the currents that enter a junction equals the sum of the currents that leave the junction. If we interpret an outgoing current as an ingoing current with a minus sign in front of it, then the junction rule takes the simple form:

$$\sum I_{in} = 0. \tag{28–12}$$

A total of n resistors connected in series can be combined into an equivalent resistance by the relation

$$R_{eq} = R_1 + R_2 + R_3 + \cdots + R_n. \tag{28–11}$$

A total of n resistors connected in parallel can also be combined into an equivalent resistance by the relation

$$\frac{1}{R_{eq}} = \frac{1}{R_1} + \frac{1}{R_2} + \frac{1}{R_3} + \cdots + \frac{1}{R_n}. \tag{28–16}$$

Ammeters, voltmeters, and ohmmeters measure current, voltage, and resistance, respectively. Ammeters must have a small internal resistance so that they do not affect the circuit leg in which the current is measured. Conversely, voltmeters need a large internal resistance because they are used in parallel with the circuit element being measured, and they too should not affect the circuit being measured.

A circuit exhibits time-varying behavior when a capacitor is being charged with a source of emf or when the capacitor is being discharged. For a simple circuit with an emf $\mathscr{E}$, a resistor R, and a capacitor C, an initially uncharged capacitor has charge

$$Q = C\mathscr{E}(1 - e^{-t/RC}) \tag{28–25}$$

and current

$$I = \frac{\mathscr{E}}{R} e^{-t/RC}. \tag{28–26}$$

When the source of emf is disconnected from the circuit and the capacitor is allowed to discharge through the resistor, the charge decreases exponentially:

$$Q = Q_0 e^{-t/RC}. \tag{28–29}$$

The time constant RC determines the time dependence of exponential increase or decrease of the charge and the current during the charging and discharging of a capacitor.

UNDERSTANDING KEY CONCEPTS

1. Why is it dangerous to be in a bathtub when an electrical appliance is standing on the edge of the tub?
2. An inexpensive voltmeter measures the voltage of a flashlight battery as 0.9 V, whereas a high-quality digital voltmeter measures 1.5 V. What might cause this difference?
3. Show that it is irrelevant whether, in drawing a circuit diagram, the internal resistance of a source of emf is placed before or after the emf itself.
4. Why does the potential drop linearly along the length of a resistor that is a uniform cylinder?
5. In Chapter 27, we showed that the resistance of a piece of material is proportional to its length and inversely proportional to its cross-sectional area. Explain how this result is equivalent to the rules for series and parallel resistances.
6. What sense does it make to draw a circuit diagram with resistanceless wires when real wires always have some resistance?
7. By taking a special combination of batteries of constant emf, resistors, and capacitors, is it possible to construct a circuit in which the emf around a closed loop is not zero?
8. Is it possible to break up any combination of resistors into a sequence of parallel and series resistances? If not, give an example of a combination that cannot be so decomposed.
9. A flash unit on a camera discharges by means of an RC circuit. Find out (say, from a local camera store) what the R and C values are, and obtain the RC time.
10. It would appear superficially that, in Section 28–5, if we had chosen the current in the wrong direction, we would have ended up with the equation $IR - (Q/C) = 0$ rather than Eq.

(28–7) to describe the current in an RC circuit. This would be disastrous because the solution of our new equation would be a rising exponential, $e^{+t/RC}$, rather than a falling exponential, and this expression would grow without limit. What is wrong with this reasoning?

11. When you reverse the polarities of all batteries in a circuit, the magnitudes of all currents stay the same. Why?

12. Suppose that you connect the terminals of two batteries of different emfs + to + and − to −. What do you expect to happen?

13. It might appear that the only effect of the internal resistance of a battery in a circuit is to change the battery's emf from $\mathscr{E}$ to $\mathscr{E}' = \mathscr{E} - Ir$, where $\mathscr{E}'$ acts as though there were no internal resistance. Could this be true? If not, why not?

14. In Example 28–4, we showed that the circuit of Fig. 28–8a can be reduced to a single-loop circuit by using an equivalent resistance. Is this true of the circuit of Fig. 28–8b?

15. Two reckless teenagers hang by both hands on a wire that can be connected to a constant voltage source. One teenager hangs from a position in which there is a resistor between his hands; the other one does not. The resistance of the resistor is much larger than that of the wire, which is very small. When the wire is connected to the battery, how do the teenagers experience that fact?

16. We have three identical light bulbs available to us. Two of them are connected in series to a battery. What is the relative brightness of the two bulbs? If the third bulb is connected in parallel with the second of the above bulbs, what are the relative brightnesses of the three bulbs?

PROBLEMS

28–1 EMF

1. (I) A 12-V car battery is rated at 80 A, meaning that it will send 80 A through a wire connected to its terminals. What is its internal resistance?

2. (I) The Magellan spacecraft that studied Venus in 1990 used two solar panels capable of producing 1200 W. If the solar array was capable of producing a total of 40 A, what was the terminal voltage of the device?

3. (I) A battery with an emf of 3.00 V sends a current of 1.99 A when it is connected in series with a 1.50-Ω resistor. What is the internal resistance of the battery?

4. (I) Nickel–cadmium batteries used in space flight are rated at 30 A·h (they can put out 30 A for 1 hr) and 30 V. How much energy do the batteries contain?

5. (I) A defibrillator used by emergency medical staff to restart an accident victim's heart has an internal resistance of 20 Ω, and a power supply that produces an average 5000 V over a short period of time. If the body resistance between the two electrodes is 230 Ω, how much current passes through the body to restart the heart?

6. (II) A flashlight battery with an internal resistance of 0.25 Ω produces a 320-mA current through a 18.5-Ω resistor. What is the emf of the battery? What is the terminal voltage of the battery in this usage?

7. (II) A certain automobile battery has an emf of 12 V. When it produces a current of 100 A, the terminal voltage reads 9.0 V. Calculate the internal resistance of the battery. What is the power dissipated in the battery when it produces this current?

8. (II) The resistance of the starter of a car, including the cables, is 0.14 Ω. On a frigid winter morning, cranking the engine for 10 s reduces the terminal voltage of the car's 12-V battery to 7 V. The car does not start. How much heat (in kilocalories) is produced in the battery during this time? Does it improve your chances of starting the car on a second try?

9. (II) The internal resistance of a battery whose emf is 12 V varies with the current according to the equation $r = (\alpha + \beta I)$, where $\alpha = 0.15\ \Omega$ and $\beta = 0.018\ \Omega/\text{A}$. Find the terminal voltages and the power dissipated in the battery when $I = 1.0$ A and $I = 10$ A.

28–2 Kirchhoff's Loop Rule

10. (I) Consider the circuits shown in Fig. 28–25. The light bulbs in all circuits are identical and the batteries are the same in the two circuits. Before working with these circuits, you are asked to make some predictions: (a) What is the brightness of the bulbs in circuit II relative to each other and to the bulb in circuit I? (b) If one of the bulbs is removed in circuit II, how will the brightness of the other bulb be affected? Does it matter which bulb is removed?

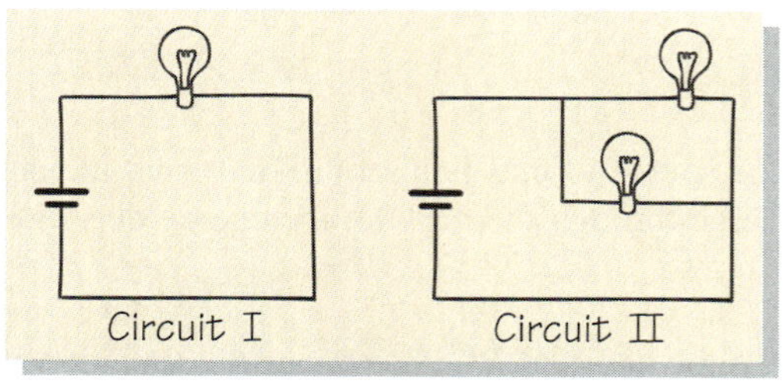

FIGURE 28–25 Problem 10.

11. (I) For what value of R_2 in the circuit of Fig. 28–26 is the voltage across the points a and b zero? For what value is the current in the circuit zero?

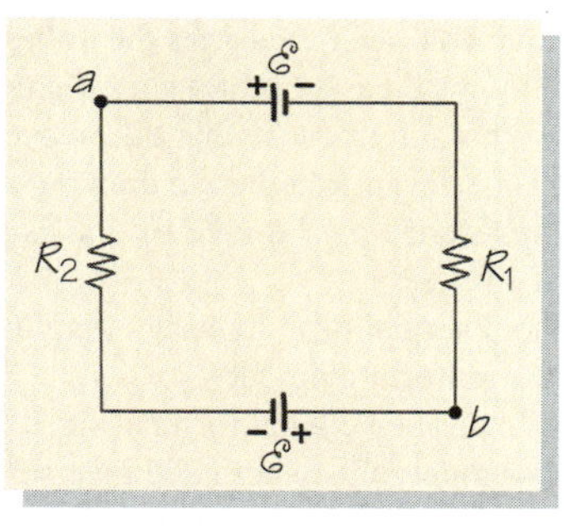

FIGURE 28–26 Problem 11.

12. (I) Five resistors of 32 Ω each are connected in series. If the potential difference between the ends of this set of resistors is 24 V, what current flows through the resistors? What is the power expended in the circuit?

13. (I) Two 60-Ω resistors are placed in series across two terminals whose potential difference is 120 V. What is the total power dissipated?

14. (I) The power dissipated in a resistor through which a current I passes is P_0. What is the power dissipated if the same current passes through three such resistors connected in series?

15. (II) A portion of a larger circuit is shown in Fig. 28–27. The potential drop between points b and a, known as V_{ba}, is $V_{ba} =$ 2 V; $V_{cb} = 3.5$ V, $V_{cd} = 2$ V, and $V_{df} = -0.5$ V Find the potential differences V_{gf}, V_{ag}, and V_{ca}.

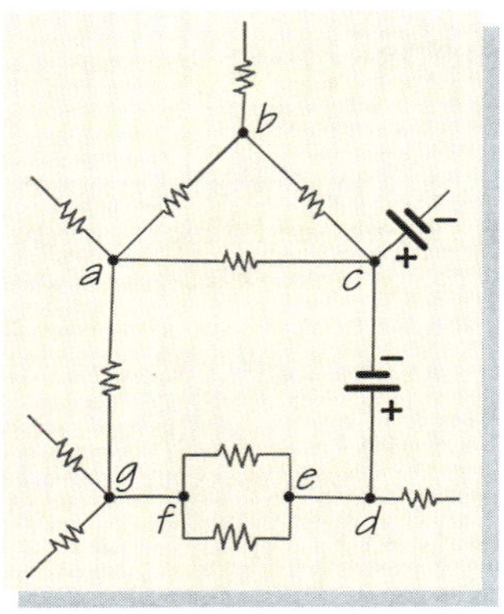

FIGURE 28–27 Problem 15.

16. (II) A generator (a "battery" that uses mechanical rather than chemical energy) of emf 110 V and internal resistance 0.80 Ω is used to charge a series of 32 batteries, each with emf 2.2 V and internal resistance 0.03 Ω. A series resistor is used to limit the charging current. (a) What is the terminal voltage of the generator? (b) the terminal voltage of the bank of batteries? (c) What series resistance must be included to allow a charging current of 15 A? (d) What is the power dissipated in all the resistors?

17. (II) A flashlight consists of two 1.5-V batteries connected in series to a bulb with resistance 10 Ω. (a) What is the power delivered to the bulb? (b) Batteries run down when they acquire an (internal) resistance. How large is the additional resistance if the power delivered to the bulb has decreased by one-third of its initial value?

28–3 Kirchhoff's Junction Rule

18. (I) Two resistors are placed in parallel. One of the resistors has twice the resistance of the other; the lesser of the two resistances is 150 Ω. What is the resistance of the parallel combination?

19. (I) In Fig. 28–28, the currents $I_1 = 2$ A, $I_2 = 0.5$ A, $I_3 = -3$ A, $I_4 = -0.5I_6$, and $I_5 = -I_6$. Find the unknown currents I_4, I_5, and I_6.

20. (I) Five identical light bulbs are placed in a circuit as shown in Fig. 28–29. What is the brightness of bulb #3 relative to bulb #1?

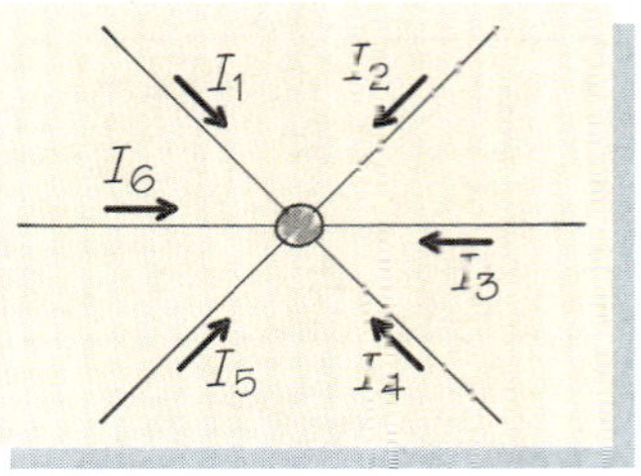

FIGURE 28–28 Problem 19.

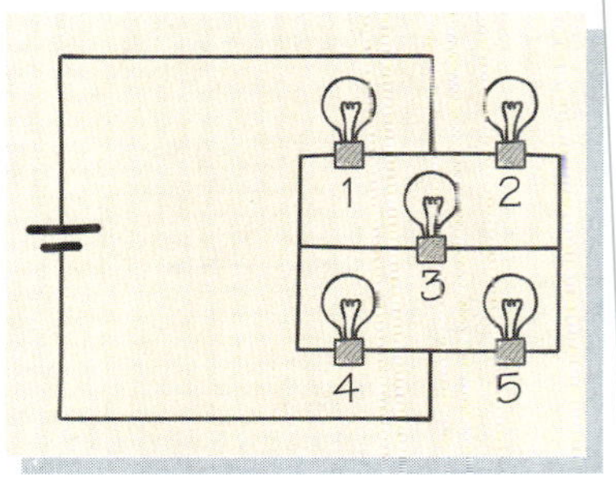

FIGURE 28–29 Problem 20.

21. (I) Find the equivalent resistance of the circuit shown in Fig 28–30.

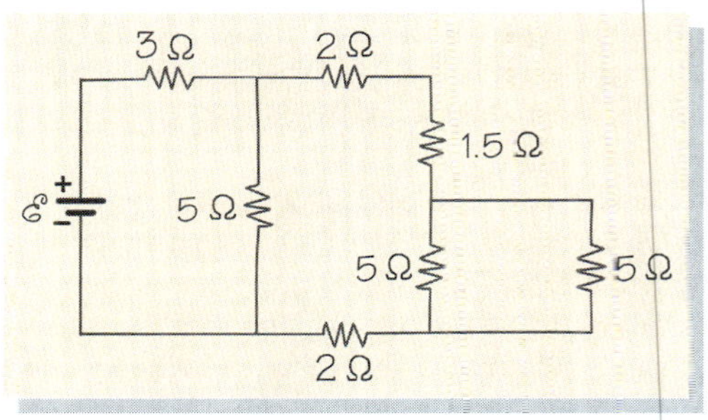

FIGURE 28–30 Problem 21.

22. (II) Find the current that passes through the 4-Ω resistor in the circuit shown in Fig. 28–31.

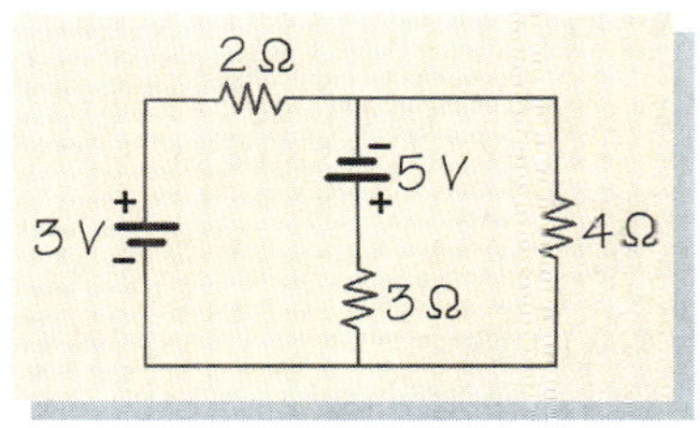

FIGURE 28–31 Problem 22.

23. (II) Find the current that passes through each of the resistors in the circuit shown in Fig. 28– 32.

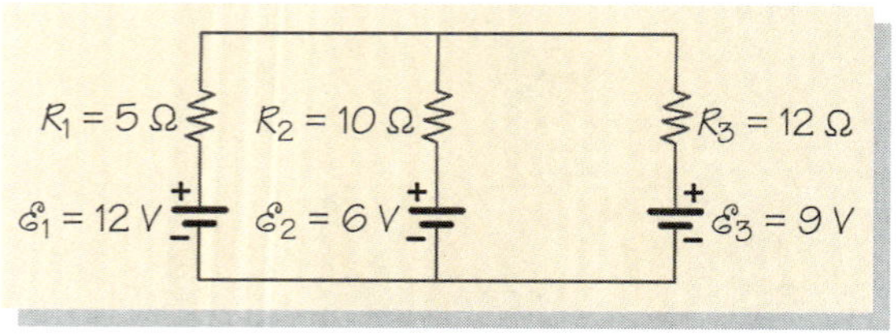

FIGURE 28–32 Problem 23.

24. (II) Can the resistors of the circuit in Fig. 28–33 be reduced to a single equivalent circuit by application of the rules for circuits with connections in parallel and in series? Solve for the currents through the three resistors.

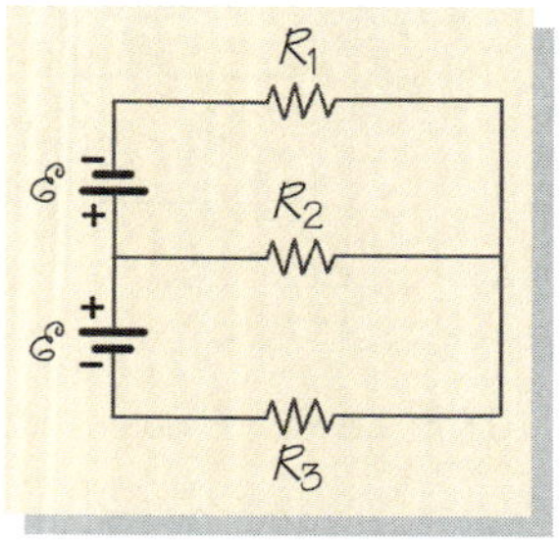

FIGURE 28–33 Problem 24.

25. (II) Three resistors connected in parallel have resistances of 250 Ω, 420 Ω, and 510 Ω, respectively. The total current passing through the set is 0.020 A. What is the potential difference across the set, and what are the currents in each of the resistors?

26. (II) The voltage at an electrical outlet is a constant 120 V. You have ten identical light bulbs whose maximum power consumption is 5 W (Fig. 28–34). (a) What is the resistance in each bulb if the power consumption is 50 W when the bulbs are connected in series? (b) If the ten bulbs are connected in parallel, an additional resistor is needed so that the bulbs do not burn out. The resistor is connected in series with the total set of light bulbs. What is the value of the resistance? What is the power loss in the resistor? (These are bulbs designed for use in a car.)

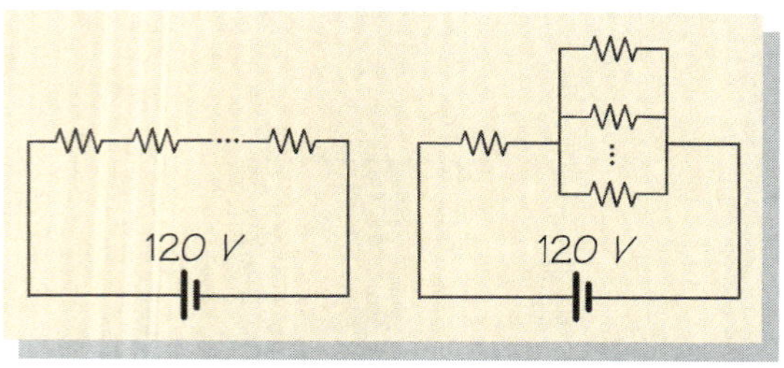

FIGURE 28–34 Problem 26.

27. (II) N identical batteries, with emf $\mathscr{E}$ and internal resistance r, are connected in parallel across a resistance R. Obtain the value for the current, and compare its value with that obtained if the batteries are connected in series.

28. (II) Figure 28–35 shows an example of a *voltage divider*, a device that allows a reduced voltage to be obtained. Calculate the potential difference across the line CD in terms of the potential difference across the line AB.

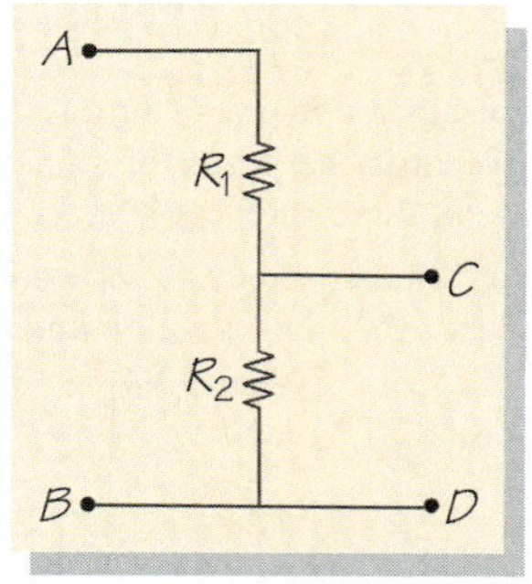

FIGURE 28–35 Problem 28.

29. (II) The circuit shown in Fig. 28–36 is an example of a loaded voltage-divider circuit (see Problem 28). By varying the values of R_1 and R_2, different values of V_L can be obtained; R_L represents the load. Let $R_1 = R_2 = 3.3$ kΩ and $\mathscr{E} = 10$ V. For load resistances of 20 kΩ, 200 kΩ, and 2 MΩ, how much different is V_L than 5 V?

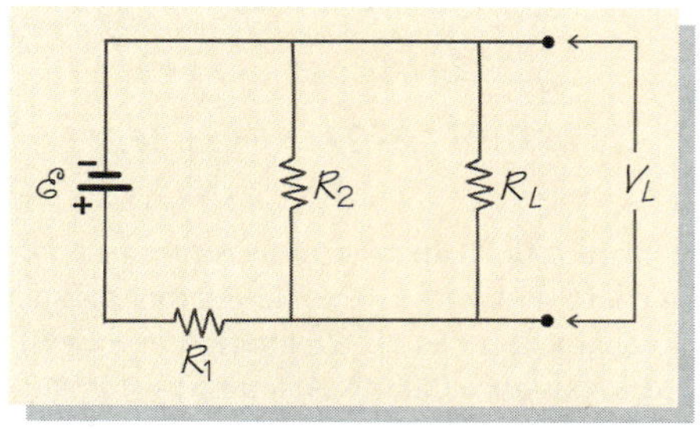

FIGURE 28–36 Problem 29.

30. (II) Consider the circuit shown in Fig. 28–10a. If $\mathscr{E} = 1.5$ V, and if all three resistors have identical resistances, find the resistance that ensures that the current I_3 will be 10 mA.

31. (II) How many independent junctions are there in the circuit shown in Fig. 28–37? To verify your answer, solve for all currents.

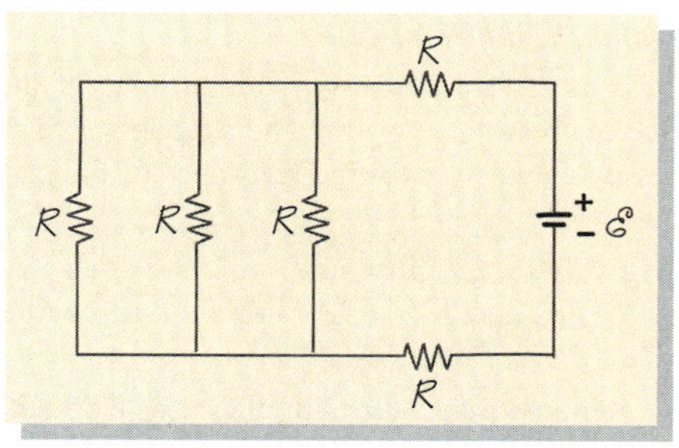

FIGURE 28–37 Problem 31.

32. (II) Replace the network of resistors in Fig. 28–38 by a single equivalent resistor. Can the combination of resistors be reduced to a single resistor by successive application of the rules for parallel and series resistors?

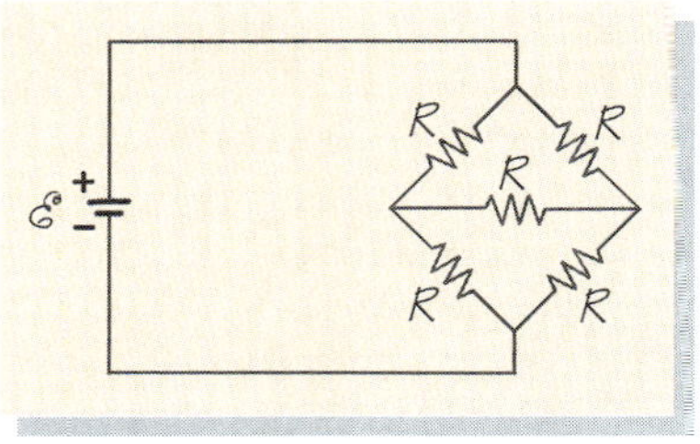

FIGURE 28–38 Problem 32.

33. (II) Two batteries with emf $\mathscr{E}_1 = 6$ V and $\mathscr{E}_2 = 9$ V, respectively, are connected to resistors with the resistances as marked in Fig. 28–39. (a) Calculate the power dissipated in the 50-Ω resistor. (b) Assume that the terminals on the 6-V battery are reversed, and repeat your calculation.

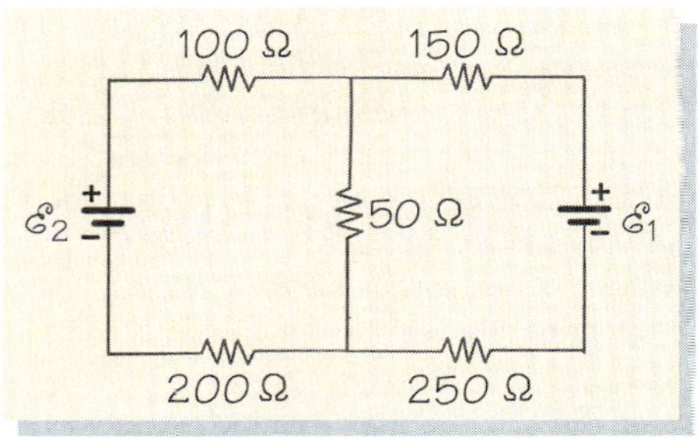

FIGURE 28–39 Problem 33.

34. (II) Consider the circuit shown in Fig. 28–40. Calculate the current and the power dissipated in the 4-Ω resistor as a function of the unknown resistance R_x.

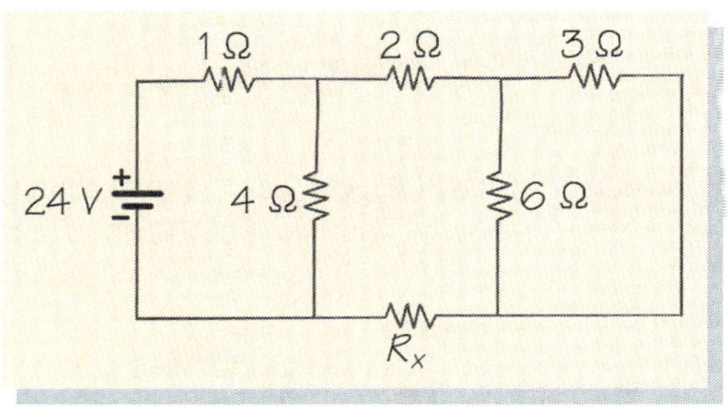

FIGURE 28–40 Problem 34.

35. (II) Consider the circuit shown in Fig. 28–41. Calculate the current through the 50-Ω resistor (a) by calculating the equivalent resistance for the circuit, and (b) by using Kirchhoff's rules.

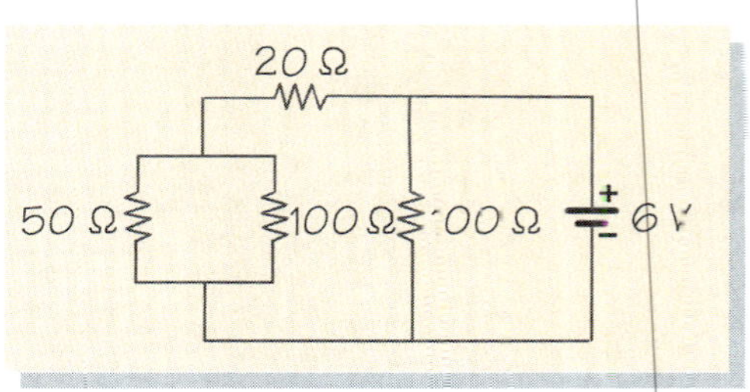

FIGURE 28–41 Problem 35.

36. (II) The known elements of the circuit in Fig. 28–42 are indicated. Find the value of R_3 that will give a current I_3 of 0.1 A with the indicated sign. Is there a value of R_3 that will give a current I_3 of the same magnitude but of opposite sign? If so, what is it?

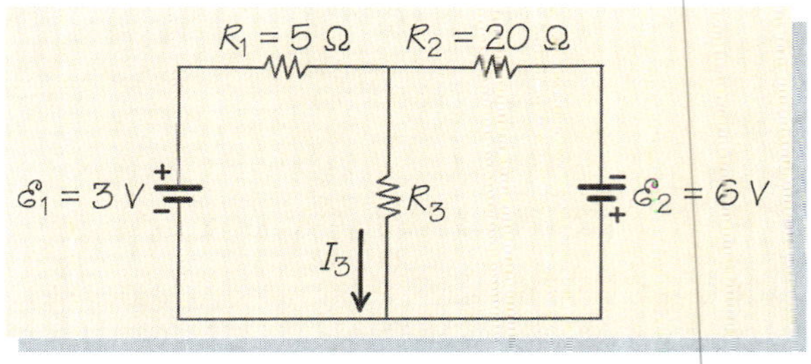

FIGURE 28–42 Problem 36.

37. (II) Two batteries are connected in parallel as in Fig 28–43 and supply current to a load resistor of 5 Ω. One of the batteries is freshly charged, with an emf of $\mathscr{E}_1 = 12$ V and an internal resistance of $r_1 = 0.1$ Ω. The other one is almost dead, with an emf of $\mathscr{E}_2 = 10$ V and an internal resistance of $r_2 = 10$ Ω. What is the current through the load resistor? How much of this current is supplied by each of the batteries?

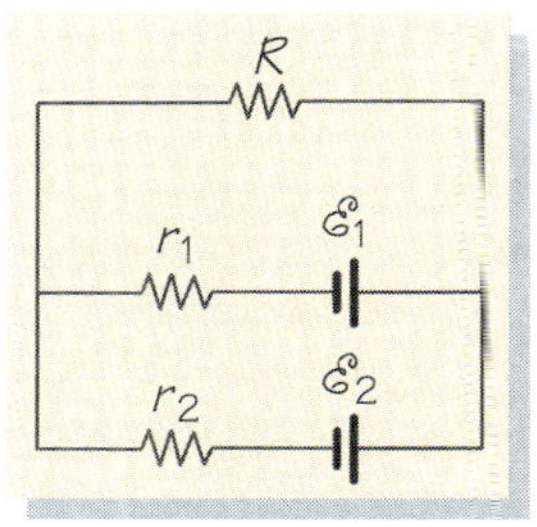

FIGURE 28–43 Problem 37.

38. (II) Points a and b are connected by the system of resistors shown in Fig. 28–44. A battery of 12 V and negligible internal resistance is connected across points a and b. (a) What is the equivalent resistance between points a and b? (b) the potential difference across the 75-Ω resistor? (c) the current flowing through the 33-Ω resistor?

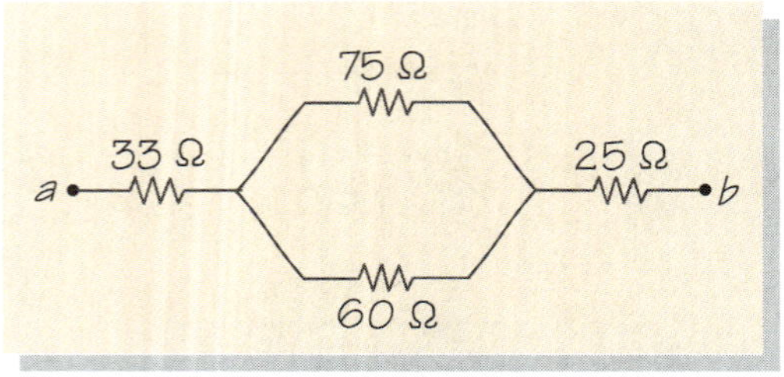

FIGURE 28–44 Problem 38.

39. (III) A cube consisting of identical wires, each of resistance R, is put across a line with voltage V (Fig. 28–45). What is the equivalent resistance of the cube? What is the current in each of the wires?

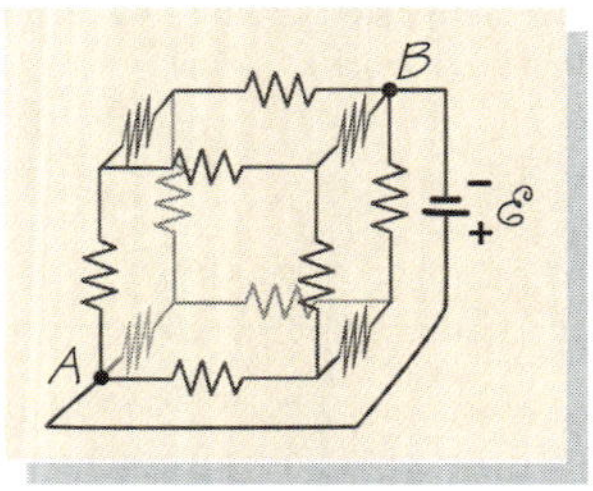

FIGURE 28–45 Problem 39.

40. (III) Consider a tetrahedron whose sides consist of identical wires, each with resistance 1 Ω (Fig. 28–46). Suppose that this arrangement is attached at two of its corners to a generator with potential 4 V. What is the power generated in each of the wires?

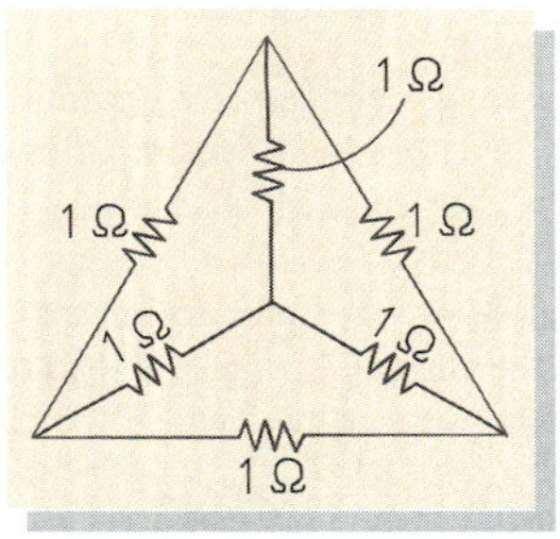

FIGURE 28–46 Problem 40.

41. (III) Figure 28–47 shows a ladder of resistors with n rungs. (a) Find the equivalent resistance between points P_1 and P_2 for $n = 1$; (b) for $n = 2$; (c) for $n = 3$; (d) for the limit $n \to \infty$ [*Hint*: Write an expression for R_n (the equivalent resistance of a ladder of n rungs) in terms of R_{n-1} (the equivalent resistance of a ladder of $n - 1$ rungs) and R, and use that equation in the limit $n \to \infty$.]

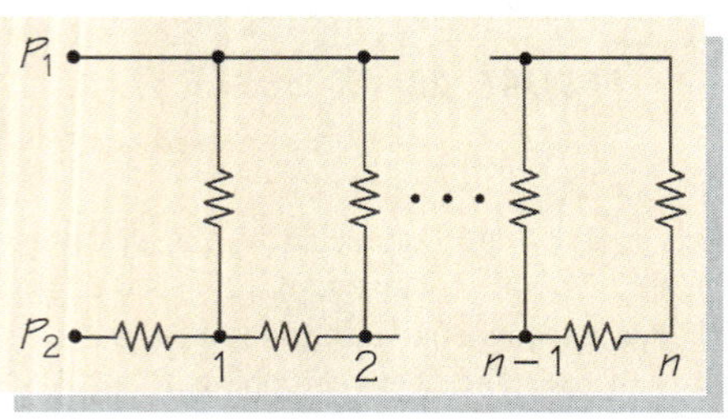

FIGURE 28–47 Problem 41.

28–4 Measuring Instruments

42. (I) A voltmeter with an internal resistance of 10 kΩ measures the voltage of a D-cell flashlight battery (of nominal voltage 1.5 V) as 1.24 V. What is the internal resistance of the worn-out battery?

43. (I) Currents produced with a 12-V source of emf and a range of resistances from 10 Ω to 1000 Ω are to be measured to an accuracy of at worst 0.1-percent with an ammeter. How small must the resistance of the ammeter be?

44. (I) A voltmeter is to be used to measure the voltage across a range of resistances from 100 Ω to 10,000 Ω. What is the minimum value of the internal resistance of the voltmeter such that a measurement can be carried out to 0.1-percent accuracy?

45. (I) A certain voltmeter has an internal resistance of 10^5 Ω. The voltmeter is used to measure the potential drop across resistors of (a) 10 Ω, (b) 10^5 Ω, and (c) 100 MΩ. In each case, what is the equivalent resistance of the voltmeter and the resistor across which it is placed? (The voltmeter is a suitable one if this equivalent resistance is as close as possible to the resistance of the resistor across which the potential drop is measured.)

46. (II) An ammeter that can measure a maximum current of 0.10 mA has an internal resistance of 3×10^{-4} Ω. What series resistance will convert it to a 0-to-3-V voltmeter?

47. (II) Suppose that the current to be measured by an ammeter is so large that a galvanometer deflected by the current would be pinned at its maximum reading. This problem can be resolved by the use of a *shunt resistor* (Fig. 28–48). Show that with the shunt resistor (resistance R_s) present, the current I is given in terms of a reduced current I_G flowing through the galvanometer by the formula $I = I_G[1 + (R_G/R_s)]$, where R_G is the resistance of the galvanometer. Thus, a reading of the reduced current I_G allows us to determine the current I.

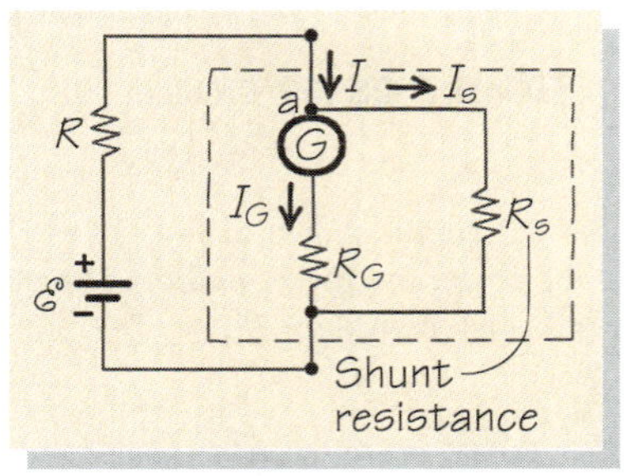

FIGURE 28–48 Problem 47.

48. (II) The output of the voltage-divider network shown in Fig. 28–49 is to be measured with two voltmeters of internal resistances 500 kΩ and 100 MΩ, respectively. What voltage will each indicate?

49. (II) A *Wheatstone bridge* is a device that measures resistances. In the circuit shown in Fig. 28–50, R is an unknown resistance. The resistances R_1, R_2, and R_3 are variable. A galvanometer, G, can be used to determine when the potential difference between B and C is zero, given that the battery is connected between A and D. The variable resistances are varied until there is no current in the galvanometer when the circuit is closed at the switch, S. Obtain an expression for R in terms of R_1, R_2, and R_3.

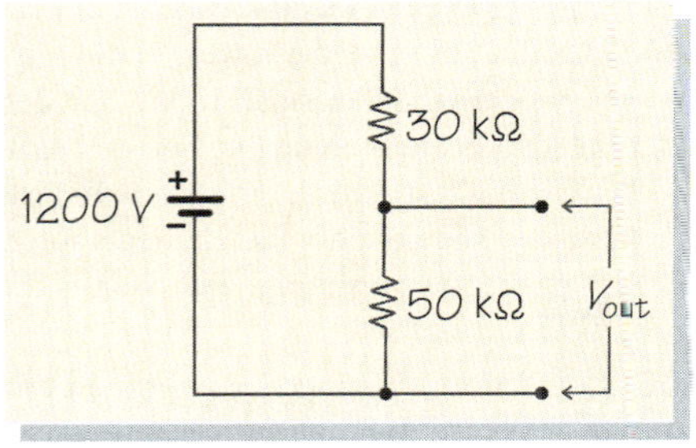

FIGURE 28–49 Problem 48.

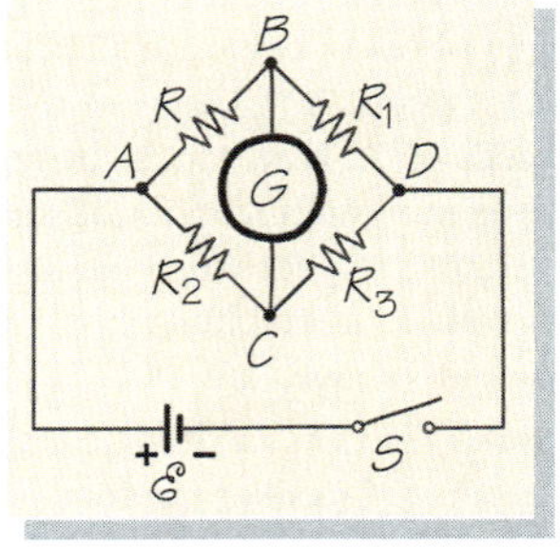

FIGURE 28–50 Problem 49.

50. (II) The circuit shown in Fig. 28–51 is used to measure the resistance R_x. Draw the circuit including internal resistances. V and I are the voltage and current measured, respectively. Find an exact expression for R_x in terms of the internal resistances of the voltmeter and ammeter. Under what conditions is $R_x = V/I$?

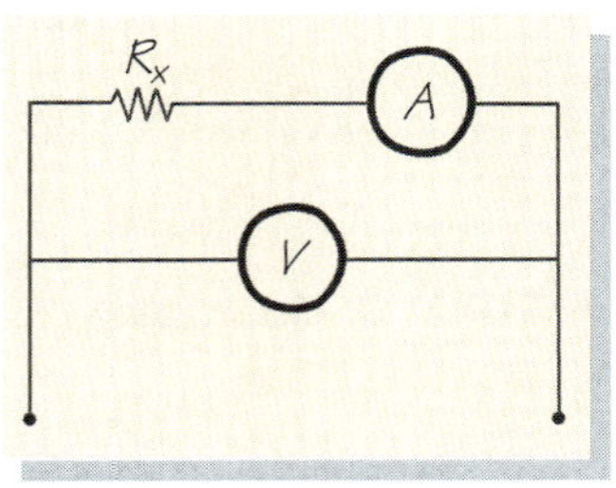

FIGURE 28–51 Problem 50.

51. (II) Repeat Problem 50 for the circuit shown in Fig. 28–52.

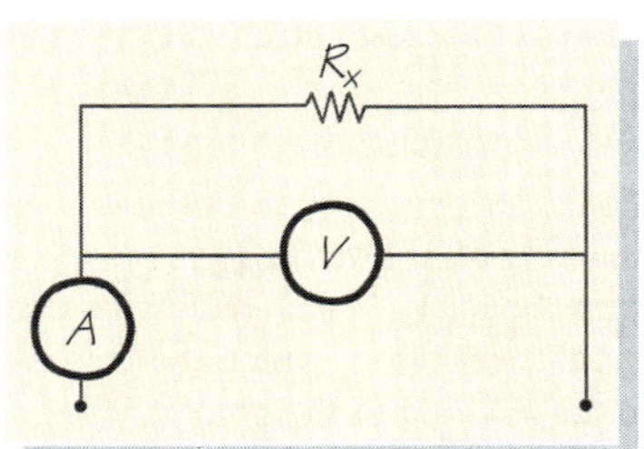

FIGURE 28–52 Problem 51.

28–5 RC Circuits

52. (I) A flashbulb in an RC circuit discharges with a time constant of 10^{-3} s. If the capacitor has a capacitance of 10 μF, what is the resistance in the RC circuit?

53. (I) A flashbulb mechanism operating through an RC circuit has a capacitor charged with a time constant of 2.0 s. If the resistance in the RC circuit is 10^5 Ω, what is the capacitance of the charging mechanism?

54. (I) Show that the product RC has units of seconds. Find the time constants for the following values of R and C: 5 MΩ, 30 μF; 8 kΩ, 3 μF; 20 Ω, 50 pF.

55. (I) The flash attachment of a camera has a capacitance of 600 μF. The flash-time—the characteristic time for the discharge of the capacitor—is 1/500 s. What is the resistance in the RC circuit?

56. (II) Show by direct substitution that Eq. (28–25) is a solution for the differential equation (28–24).

57. (II) Show by direct substitution that Eq. (28–29) is a solution for the differential equation (28–28).

58. (II) A resistor of resistance 5×10^6 Ω and a capacitor of capacitance 120 μF are connected in series to a 800-V power supply. Calculate (a) the time constant and (b) the current at a time when the charge on the capacitor has acquired 90 percent of its maximum value.

59. (II) Calculate the current in the battery as a function of time for the circuit shown in Fig. 28–53 if the switch S is closed at time $t = 0$.

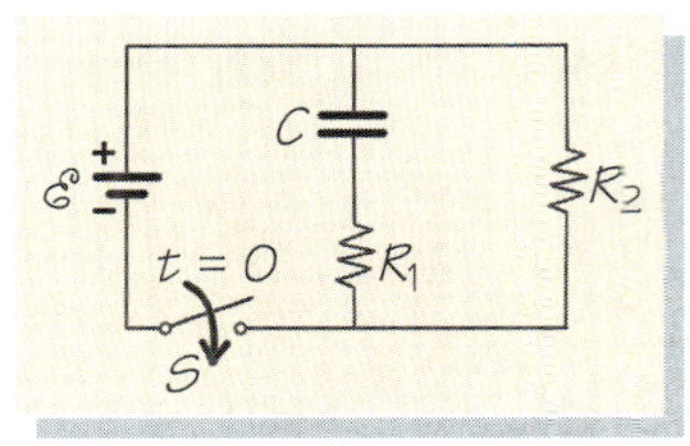

FIGURE 28–53 Problem 59.

60. (II) The circuit of Fig. 28–20 has ℰ = 200 V, R = 350 kΩ, and C = 20 μF. What are the voltage across the resistor and the charge on the capacitor 4 s after the switch is closed to position a?

61. (II) You have two capacitors of capacitance 5 μF and three resistors, one of resistance 250 Ω and the remaining two of resistance 300 Ω. Find the connection between these elements that will make a circuit whose time constant is 1 ms.

62. (II) Show that the time constant of a parallel-plate capacitor filled with a dielectric with a finite resistivity is independent of the area and separation of the plates.

63. (II) Polycarbonate, a so-called polar polymer, is a material with a dielectric constant $\kappa = 3.2$. It has a resistivity $\rho = 2 \times 10^{14}$ Ω·m. Suppose that it is used to fill the space in a parallel-plate capacitor of area 0.03 m^2 and plate separation 0.50 mm. A charge of $Q = 2$ μC is placed on the plates of the isolated capacitor. How long does it take for 70 percent of the charge to leak away?

64. (III) A capacitor C_1 is charged to Q_0 and connected to an uncharged capacitor C_2 via the resistor R (Fig. 28–54). Find the charge on each capacitor as a function of time, assuming that the switch is closed at $t = 0$.

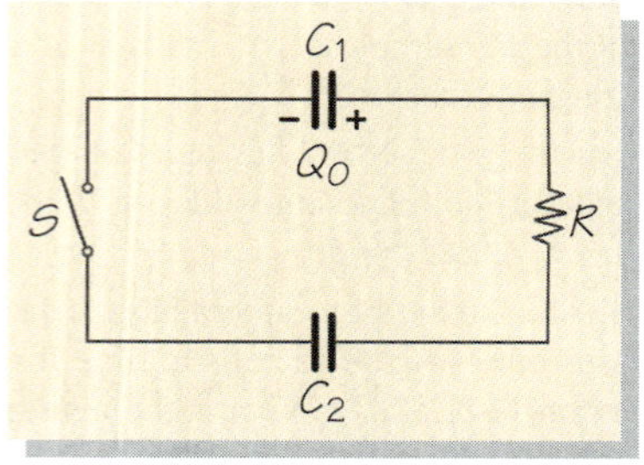

FIGURE 28–54 Problem 64.

General Problems

65. (I) Imagine that a household circuit uses direct current. Compare the current drawn from the main supply at 120 V if three household appliances of resistances 50 W, 60 W, and 20 W, respectively, are connected in parallel. (The appliances would not work in series because each appliance requires a voltage drop of 120 V. Actual household circuits use alternating currents; see Chapter 34.)

66. (II) Consider the circuit shown in Fig. 28–55, in which a 12-V battery is used to charge a 6-V battery. The resistance in the circuit is 20 Ω. Calculate (a) the current in the circuit; (b) the rate at which the energy of the smaller battery increases; (c) the total rate of energy dissipation in the resistor.

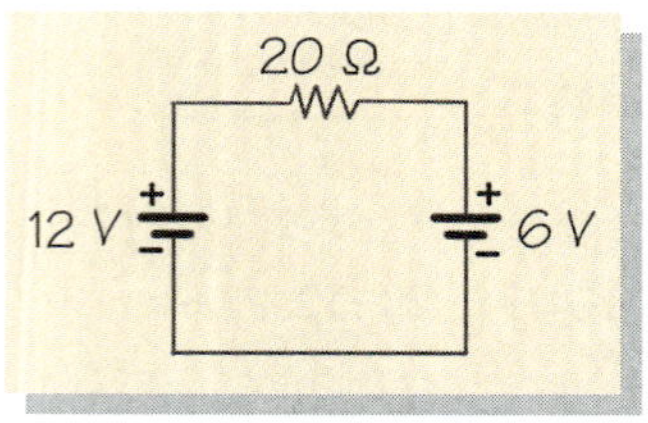

FIGURE 28–55 Problem 66.

67. (II) A student has a wide range of resistors, all rated for 5 W. How can a student combine identical resistors to obtain an effective resistance of 100 Ω rated for 30 W?

68. (II) If a battery of fixed emf and internal resistance r is connected to an external resistor of resistance R, show that the maximum power delivered to the external resistor occurs when $R = r$.

69. (II) The battery considered in Problem 9 is connected to a load resistor R. Calculate and plot the current in the circuit as a function of R in the range $R = 0$ (short circuit) to $R = 5\ \Omega$. Plot the ratio of the power delivered to the load to that dissipated in the internal resistance of the battery.

70. (II) When separately connected across a line with voltage V, two resistors generate power P_1 and P_2, respectively. What is the power generated when the two resistors are connected in series? in parallel?

71. (II) A student picks up an electric furnace at a yard sale. She discovers later that, even at the lowest setting, it delivers too much power to be used in her small room. She determines that the heater has two heating elements, one delivering 1 kW, the other 2 kW. The highest setting turns on both heaters simultaneously. How can she connect the elements to get a new setting lower than the lowest setting, and what is the power output for the new setting?

72. (II) A resistor R forms a single loop with an arrangement of two batteries of emf $\mathscr{E}$ and internal resistance r. The batteries are arranged (a) in series and (b) in parallel. Find the current through the resistance in both cases. Which arrangement gives the larger current for large R? for small R?

73. (II) An 800-W kitchen mixer, a 600-W vacuum cleaner, and a chandelier with 10 60-W bulbs are all plugged into the same outlet in a 120-V circuit. A *fuse* acts as a switch that opens if the current exceeds 15 A. How much current does each device draw? What is the minimum number of screwed-in bulbs that will blow the circuit? Do not worry about the oscillations of the current and voltage in real household circuits, but assume that all currents and voltages are DC.

74. (II) Consider three resistors of 10 Ω each. Each resistor can dissipate 5 W at most. Consider the four possible ways of arranging the three resistors, and calculate the maximum power that can be dissipated in each of the ways.

75. (II) Find the currents in each leg of the circuit shown in Fig. 28–56.

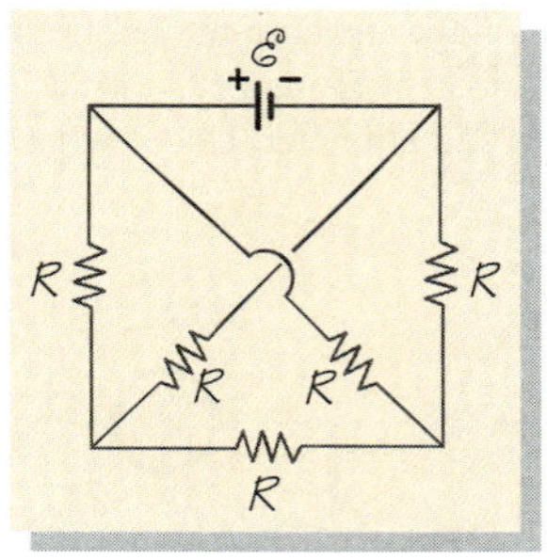

FIGURE 28–56 Problem 75.

76. (II) By using Table 27–2, compare the current density, electric field strength, and power loss in two cylindrical wires of the same length and same radius, one made of aluminum and the other made of copper. The wires are connected (a) in series and (b) in parallel. (c) If wires of a given length and radius were constructed from all the materials listed in Table 27–2, and if the same current were passed through each, which wire would have the largest current density, the weakest electric field, and the least power loss? Assume throughout that the temperature dependence is unimportant.

77. (II) A simple *potentiometer* circuit used to measure unknown voltages accurately is shown in Fig. 28–57. Here V_s is the known source voltage, V_x is the unknown voltage, and the resistor is a variable one from which the values R_1 and R_2 can be read from the position of the pointer. These resistances are varied until the current in the ammeter is zero. Show that the unknown voltage then has the value $V_x = V_s R_2/(R_1 + R_2)$.

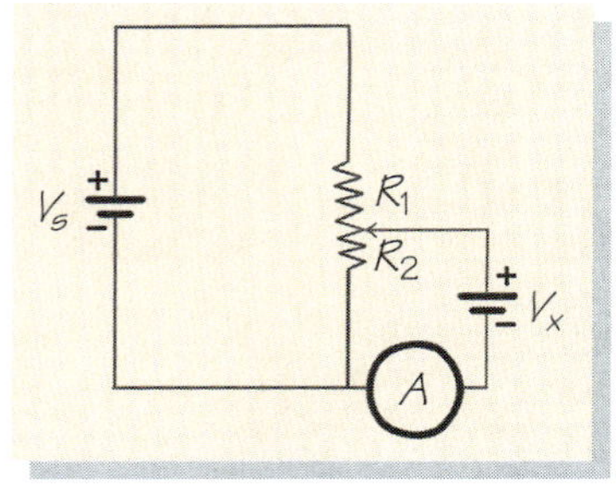

FIGURE 28–57 Problem 77.

78. (II) Two resistors and two capacitors are connected in series to a battery as shown on Fig. 28–58. Calculate the potential at B relative to that at A: (a) shortly after the closing of the switch and (b) a long time after the closing of the switch. (c) How fast does the circuit reach a steady state? (Give a time scale.)

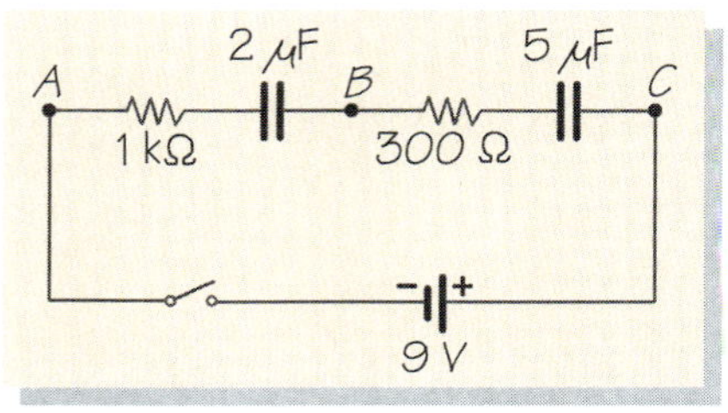

FIGURE 28–58 Problem 78.

79. (II) To avoid sparks accompanying the opening of a high-current circuit breaker, its terminals are connected to a large capacitor, as in Fig. 28–59. (a) How fast does the current decrease in the circuit shown in the figure? (Give a time scale.) (b) What is the charge on the plates of the capacitor a long time after the switch is opened?

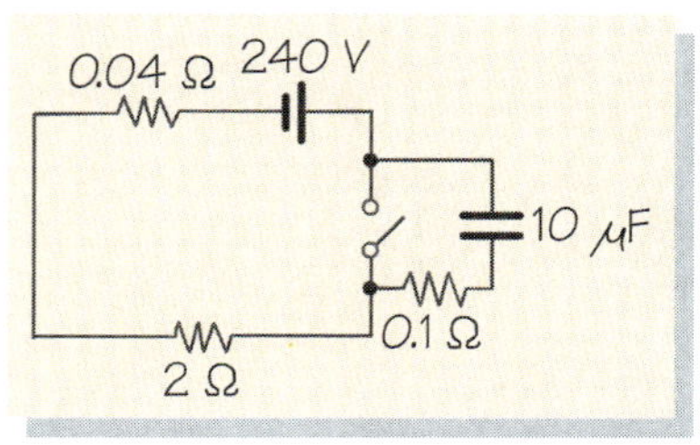

FIGURE 28–59 Problem 79.

80. (II) The circuit shown in Fig. 28–60 has been established for a long time. (a) What is the charge on the capacitor? Indicate which plate carries the positive charge and which one carries the negative charge. (b) Calculate the current flowing through the 35-Ω resistor.

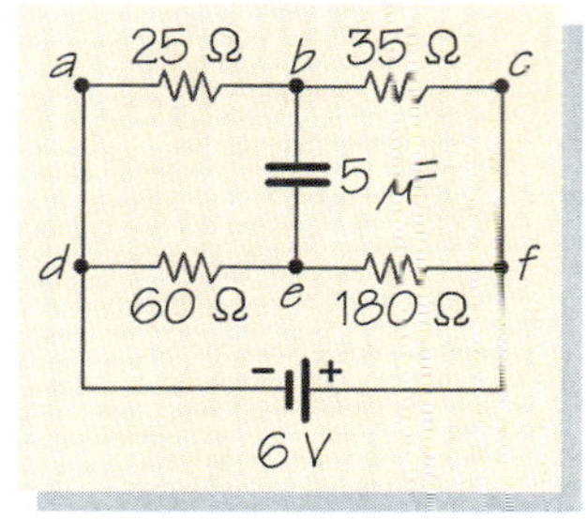

FIGURE 28–60 Problem 80.

81. (II) A single-loop circuit contains a battery of emf V_0 and negligible internal resistance and, connected in series with the battery, two circular plates of radius r separated by a distance $d << r$. The space between the plates is filled with a material of conductivity σ and dielectric constant 1. (a) What is the electric field between the two plates? (b) What is the current flowing in the circuit?

82. (III) Consider the infinite network of resistors shown in Fig. 28–61a. Calculate the resistance, R^*, of the network by noting that with an infinite set of resistors, adding one more rung to the ladder does not change the resistance. Thus the network may be broken up as shown in Fig. 28–61b.

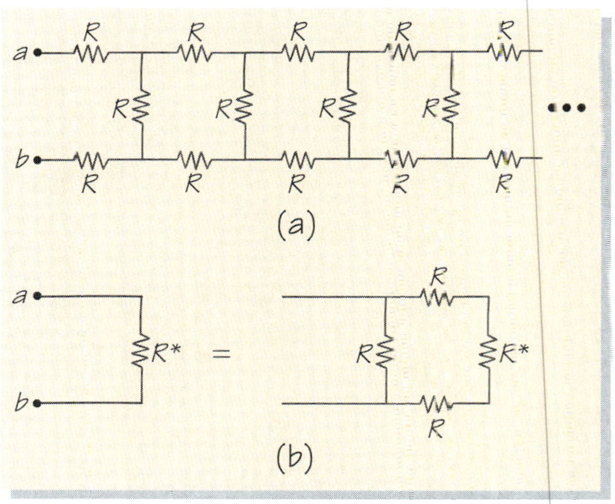

FIGURE 28–61 Problem 82.

CHAPTER

29

The aurora is a spectacular manifestation of Earth's magnetic field. Charged particles from the Sun that arrive within the influence of that field strike the atmosphere and produce light.

The Effects of Magnetic Fields

Sailors have used navigational compasses made from lodestone—the natural mineral magnetite—or treated iron for at least 800 years, and possibly much longer. In 1600, William Gilbert, an English physician who systematically studied electrical and magnetic phenomena, suggested that the compass behaves as it does because Earth is itself a giant lodestone, or magnet. Indeed, our use of the words "magnetic north and south poles" in connection with bar magnets comes from Gilbert's association of magnetism with Earth's geographic North and South Poles. Thus the compass works through magnetic forces.

Magnetism was not associated with electricity until 1820, when André Ampère used his own experiments and those of Hans Christian Oersted to show that magnetic effects arise when electric charges move. In fact, electrical and magnetic phenomena are both aspects of the interactions of electrically charged objects. In the 1820s, Michael Faraday uncovered the connection between electricity and magnetism, but it was James Clerk Maxwell who, in the late 1860s, made the ultimate synthesis of electricity and magnetism. This synthesis is described entirely by Maxwell's equations, which, together with Newton's work, the ideas of thermodynamics, and Einstein's special theory of relativity, summarize virtually all of classical physics. Our understanding of light and other electromagnetic waves rests on Maxwell's great achievement. In Chapters 29 to 35, we shall study magnetic phenomena, their connection to electrical phenomena, their practical applications, and other remarkable consequences of Maxwell's equations.

We shall discover the laws of magnetic forces by describing experiments with magnets and with electric currents. We'll also learn how magnetic forces are associated with magnetic fields—just as electric forces are associated with electric fields. In this chapter, we concentrate on the effects of magnetic fields on test objects. Just how magnetic fields are generated is the subject of the following chapter.

29–1 MAGNETS AND MAGNETIC FIELDS

When two bar **magnets** are brought close to each other, the forces between them—**magnetic forces**—become evident. (These forces are of a type we have not yet encountered.) Bar magnets have an orientation, or an axis. In some positions, they attract each other while in other positions, they repel; in still other positions, they exert torques on each other. We arbitrarily label the end of a magnet that is attracted to a point very near Earth's geographic South Pole as the *south pole*, S, and the other end of the magnet is called the *north pole*, N. If we experiment with two bar magnets labeled in this way, we find that the N end of one attracts the S end of the other, whereas the two N ends repel each other, as do the two S ends (Fig. 29–1). (We might note that the fact that the south pole of a bar magnet is attracted toward Earth's South Pole means that Earth's South Pole actually behaves like the north pole of a magnet! Similarly, Earth's North Pole behaves like the south pole of a magnet.) Based on our experience with charges, we might be tempted to conclude that a bar magnet contains "magnetic charges" (or *magnetic monopoles*) at each end and, further, that we could somehow extract them. Experiment shows that this is not possible; When you break a bar magnet in two, you end up with two bar magnets, not two separate magnetic charges.

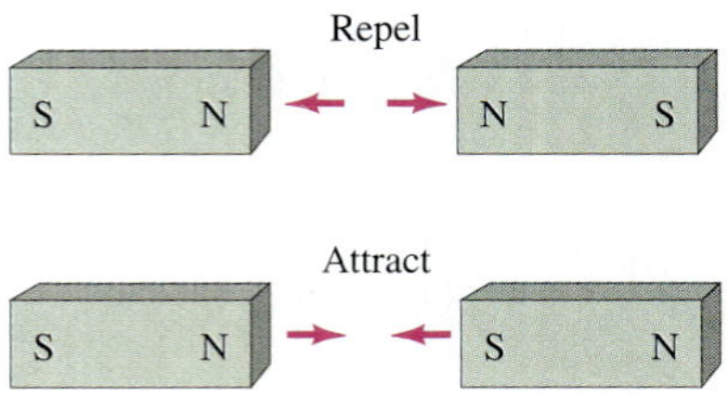

FIGURE 29–1 Bar magnets exert magnetic forces on each other.

Iron and a few other materials have a peculiar property: If we place such a material near a lodestone (a "natural" magnet), that material becomes a new magnet. Something about the very presence of a nearby magnet causes iron to become magnetic. Whatever the underlying cause, we can turn tiny shavings of iron (iron filings) into tiny magnets that can be used as test probes of magnetic forces.

Magnetic Fields

Magnets and electric currents set up magnetic fields.

If we scatter iron filings on a sheet of plastic above a bar magnet as in Fig. 29–2, the filings become aligned in certain directions—the magnetic forces line them up—and they clump more densely in certain regions, such as near the poles. For different magnets, the scattered filings have different densities and alignments. Figure 29–2 shows the distribution of iron filings around a straight bar magnet, a horseshoe magnet, and—surprise!—a wire that carries current. This last example is a clue that magnetic forces are associated with moving charges as well as with magnets. The magnetic force acts at a distance, just like gravitational and electric forces. Just as an object with mass sets up a gravitational field and a charged object sets up an electric field, a magnet or an electric current sets up a **magnetic field** throughout space. We denote this field with the symbol **B**. The bar magnet, horseshoe magnet, and current-carrying wire each set up a characteristic magnetic field. Magnetic forces have a marked directional character, so the magnetic field, like the electric field, is a vector.

The forces on iron filings can be used to map out magnetic fields.

How do we find the magnetic field due to a magnet or a current? We were able to map out electric fields due to charges by measuring the force on a small electric test charge. In the same way, iron filings can be used to map out the magnetic field due to a magnet or current by seeing how the filings respond in the neighborhood of the magnet or current. The analogy is not perfect; a test charge is accelerated due to an electric field, while a single iron filing is rotated. (In this respect, a filing in a magnetic field behaves more like an electric dipole in an electric field.) Iron filings map out the magnetic field, **B**, as follows. **B** is oriented along the alignment direction of the filings, and the magnitude of **B** is proportional to the density of the filings. Just as the electric field can be visualized with electric field lines, the magnetic field can be visualized

The magnetic field runs out of the north pole and into the south pole of a magnet.

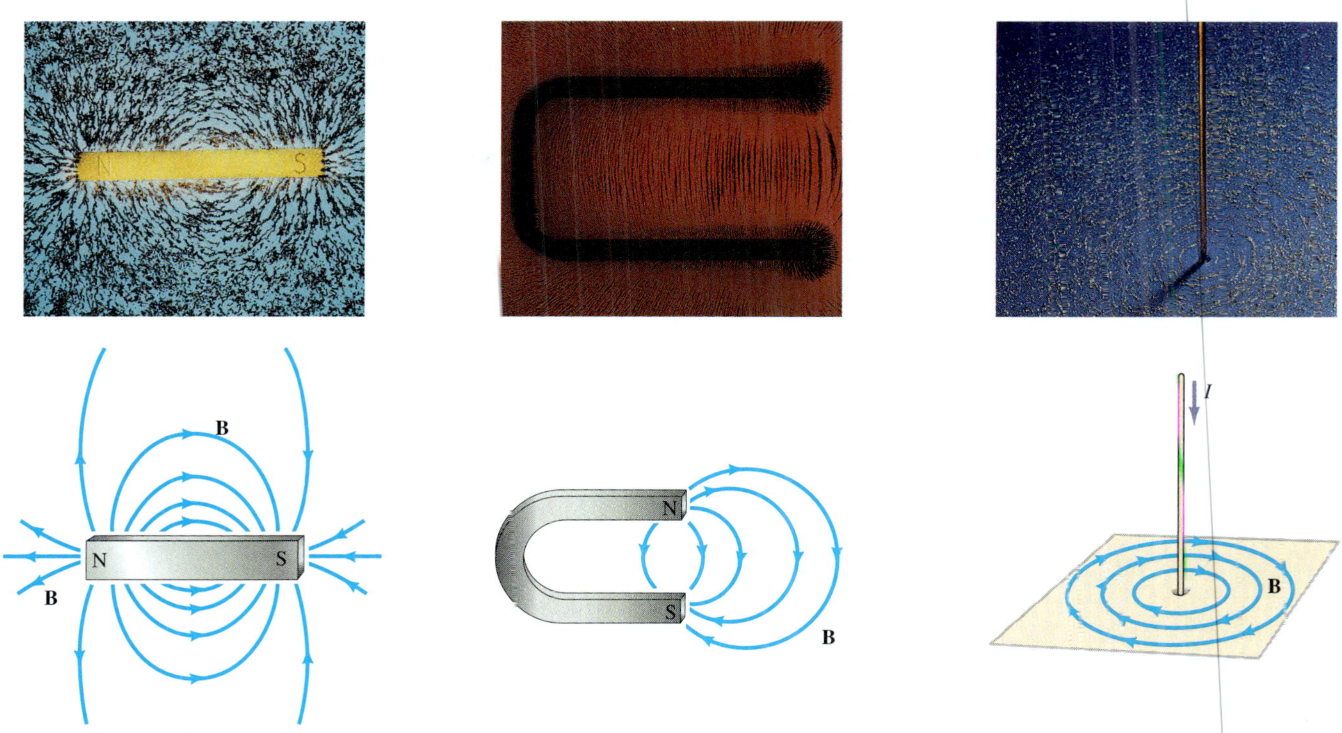

(a) (b) (c)

FIGURE 29–2 Iron filings map the magnetic field for (a) a straight bar magnet, (b) a horseshoe magnet, and (c) a current-carrying wire.

with magnetic field lines—continuous lines that run parallel to the direction of the field at every point and whose density (the number of lines per unit area) is proportional to the strength of the field. As an example of this mapping process, take a look at Fig. 29–2: The iron filings align themselves between the poles of magnets and, therefore, the magnetic field lines associated with the magnet run from pole to pole. We take the direction of the magnetic field of a magnet to run *from the N pole to the S pole*, just as we assign the electric field to run from positive electric charges to negative charges. Notice, however, that the magnetic field around a current-carrying wire has no magnetic pole. Once we have mapped out a magnetic field in this way, we can further investigate its effects and find the force laws associated with magnetism.

(a)

(b)

FIGURE 29–3 A compass needle can be used to determine the direction of a magnetic field, such as that of a bar magnet. (a) The N pole of the bar magnet is not attracted to the N end of the needle, but (b) the S pole of the bar magnet is attracted to the N end of the needle.

29–2 MAGNETIC FORCE ON AN ELECTRIC CHARGE

Experiments show that electric charges as well as bar magnets experience forces in the presence of magnetic fields; that is, they are accelerated by those fields. This phenomenon is easy to demonstrate (Fig. 29–3). We can use a bar magnet to deflect the electron beam of an oscilloscope. When a bar magnet is placed in different orientations near the beam, the beam deflects in various ways. The deflection allows us to measure the magnetic forces on the beam.

Consider a bar magnet with its N pole oriented so that the magnetic field is in the $+y$-direction (Fig. 29–4). The magnitude of the magnetic field at the position of a moving electric test charge q can be varied by altering the distance of the magnet from the charge. (If the actual experiment is done with an oscilloscope, keep in mind that the test charge, a moving electron, has a negative charge.) We observe the following phenomena, stated here under the assumption that the test charge is *positive*:

1. If q moves at speed v in the $+z$-direction, then q is deflected in the $-x$-direction (Fig. 29–4a). Furthermore, the larger v is, the stronger is the force $\mathbf{F}$. Detailed measurements show that the magnitude of $\mathbf{F}$ due to the magnetic field is proportional to v.
2. If q moves in the $+x$-direction, $\mathbf{F}$ is in the $+z$-direction, again proportional to v (Fig. 29–4b).
3. If q moves in the y-direction (+ or −), there is no change in the charge's direction or speed; that is, there is no force (Fig. 29–4c).
4. If q moves at speed v in an arbitrary direction, $\mathbf{F}$ is proportional to the velocity component perpendicular to the magnetic field, $v_\perp$, and perpendicular to the directions of both $\mathbf{v}_\perp$ and $\mathbf{B}$. This result summarizes points 1. through 3. In particular, if the charge is at rest, so that $v = 0$, there is no force (Fig. 29–4d).
5. $\mathbf{F}$ is proportional to the magnitude of $\mathbf{B}$.
6. $\mathbf{F}$ is proportional to the sign and magnitude of q. We can use an oscilloscope to study the effect of a negative charge, namely, that of an electron (Figs. 29–4e and 29–4f).

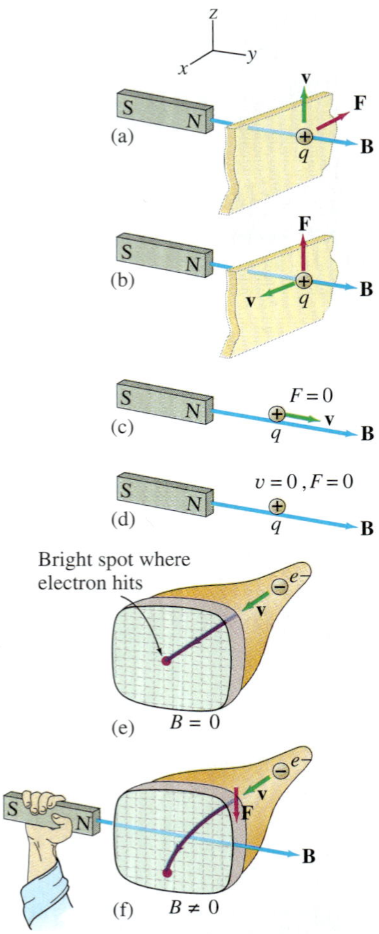

FIGURE 29–4 Experiments on a moving charge in a magnetic field. (a) through (d) The charge is positive. (e) An oscilloscope can be used to measure the effects of a magnetic field on a negative charge, namely the electron in the cathode-ray tube. (f) In the orientation shown, the electron is deflected down, not up [as was the case for the positive charge in part (b)].

The crucial feature of this seemingly complicated collection of results is the dependence on $\mathbf{v}$. The three results (1 to 3) are contained within result 4. In results 1 and 2, the initial velocity is purely in the z- or x-directions and is therefore perpendicular to $\mathbf{B}$. In result 3, there is no $\mathbf{v}_\perp$ and also no force. Figure 29–4 indicates that the force is perpendicular to both $\mathbf{v}$ and $\mathbf{B}$.

To summarize, the magnetic force $\mathbf{F}$ is proportional in magnitude to q, $v_\perp$, and B, and the direction of $\mathbf{F}$ is perpendicular to both $\mathbf{B}$ and $\mathbf{v}$ and depends on the sign of q. A direction perpendicular to both $\mathbf{v}$ and $\mathbf{B}$ can be represented by the *vector product*, which was discussed extensively in connection with torques and rotational motion. Recall that a vector $\mathbf{c} = \mathbf{a} \times \mathbf{b}$ (the vector product of $\mathbf{a}$ and $\mathbf{b}$) has magnitude $ab \sin \theta$, where θ is the angle between vectors $\mathbf{a}$ and $\mathbf{b}$ and is always taken to be less than 180° (Fig. 29–5). Vector $\mathbf{c}$ is perpendicular to both $\mathbf{a}$ and $\mathbf{b}$, in a direction determined by the right-hand rule. Thus our experiments have determined that the magnetic force on a test charge q moving with velocity $\mathbf{v}$ in a magnetic field $\mathbf{B}$ is given by

$$\mathbf{F}_B = q\,\mathbf{v} \times \mathbf{B}. \tag{29–1}$$

This important result is the **magnetic force law**. From this point on, we drop the subscript B unless we need to distinguish the magnetic force from some other force. If θ is the angle between vectors $\mathbf{v}$ and $\mathbf{B}$, the magnitude of $\mathbf{F}$ is given by

$$F = qvB \sin \theta = qv_\perp B. \tag{29–2}$$

Figure 29–6 shows how $\mathbf{F}$ is perpendicular to the plane formed by $\mathbf{v}$ and $\mathbf{B}$ according to a right-hand rule. Comparison with the observations in Fig. 29–4 shows that Eq. (29–1) is completely consistent with our experimental results. Recall that the vector product of two *parallel* vectors is zero; this explains why there is no magnetic force on a charge that moves along the axis of a bar magnet and no magnetic force associated with the component of $\mathbf{v}$ parallel to $\mathbf{B}$.

The vector product is summarized in Chapter 10.

The magnetic force law

Equation (29–1) shows that the dimensions of $\mathbf{B}$ are quite different from those of the electric field, $\mathbf{E}$. The SI unit of magnetic field is called the *tesla* (T), in honor of Nikola Tesla, who made important contributions to the technology of electrical energy generation. In terms of previously defined SI units,

The unit of magnetic field

$$1\ \text{T} = 1\frac{\text{kg}}{\text{C} \cdot \text{s}}. \tag{29–3}$$

Another (non-SI) unit for the magnetic field that is in common use is the *gauss* (G); 10^4 G = 1 T.

Table 29–1 contains some representative values of magnetic fields.

TABLE 29–1 Some Magnetic Fields

Location or Source	*Magnitude (T)*
Interstellar space	10^{-10}
Near Earth's surface	5×10^{-5}
Refrigerator magnet that holds notes	10^{-2}
Bar magnet near poles	10^{-2}–10^{-1}
Near surface of Sun	10^{-2}
Large scientific magnets	2–4
Largest steady-state magnet	30
Largest pulsed field in laboratory	500–1000
Near surface of pulsar	10^{8}
Near surface of atomic nucleus	10^{12}

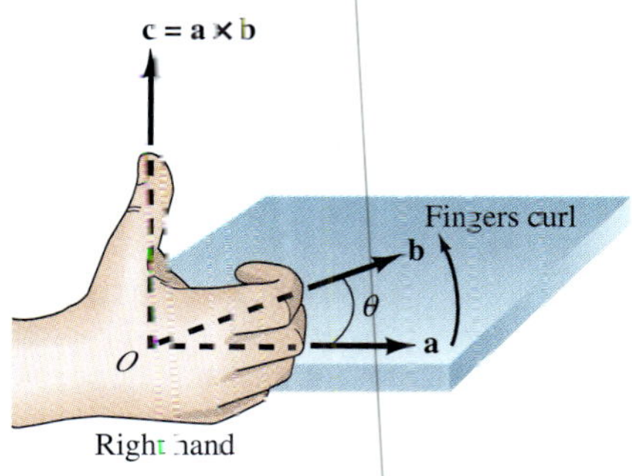

FIGURE 29–5 The vector product $\mathbf{c} = \mathbf{a} \times \mathbf{b}$ and the right-hand rule that determines the direction of the vector product. See Chapter 10, Fig. 10B1–1.

EXAMPLE 29–1 The undisturbed electron beam of an oscilloscope moves along the x-direction (Fig. 29–7). The south pole of a bar magnet approaches the cathode-ray tube from above and deflects the beam. The magnitude of the magnetic field from the magnet is 0.05 T in the vicinity of the beam, and the speed of the electrons in the beam is 2×10^5 m/s. What is the magnitude of the magnetic force on the electrons? What is the direction of this force; that is, which way is the beam deflected?

Solution: This exercise is a straightforward application of the magnetic force law, Eq. (29–1). We want to find the unknown force given the charge q, its velocity $\mathbf{v}$, and the magnitude of the magnetic field. The magnetic field is directed toward the south pole, so, as Fig. 29–7 indicates, the field of the bar magnet, $\mathbf{B}$, is in the $+y$-direction. The velocity $\mathbf{v}$ is perpendicular to $\mathbf{B}$, and the magnetic force is

$$\mathbf{F} = q\,\mathbf{v} \times \mathbf{B} = q(v\,\mathbf{i} \times B\,\mathbf{j}) = qvB(\mathbf{i} \times \mathbf{j}) = qvB\,\mathbf{k}.$$

Here, $\mathbf{i}$, $\mathbf{j}$, and $\mathbf{k}$ are the unit vectors in the x-, y-, and z-directions, respectively. With numerical values,

$$\mathbf{F} = (-1.6 \times 10^{-19}\ \text{C})(2 \times 10^{5}\ \text{m/s})(5 \times 10^{-2}\ \text{T})\mathbf{k} = (-1.6 \times 10^{-15}\ \text{N})\mathbf{k}.$$

Using the right-hand rule, we can verify that the vector product $\mathbf{v} \times \mathbf{B}$ is in the $+z$-direction. The electron charge, however, is negative, so the force on the electron is in the $-z$-direction, and the beam is deflected in this direction, as shown in Fig. 29–7.

The Lorentz Force

Further experimentation shows that charges react independently to electric and magnetic fields. Thus if an electric field is present in addition to a magnetic field, it produces an additional force $\mathbf{F} = q\mathbf{E}$ on the charge. The net force is then

$$\mathbf{F} = q[\mathbf{E} + (\mathbf{v} \times \mathbf{B})]. \tag{29–4}$$

The Lorentz force law

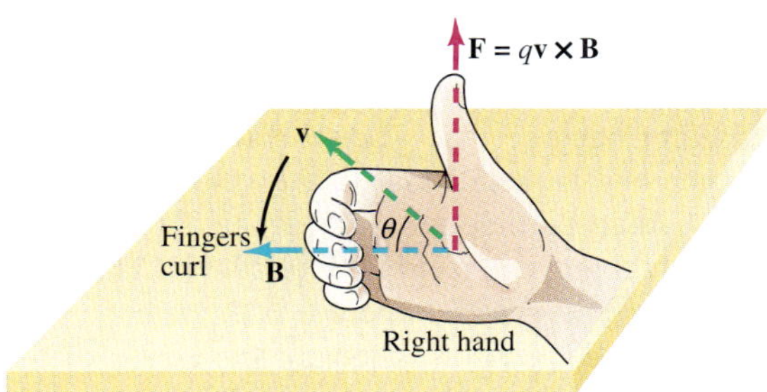

FIGURE 29–6 The right-hand rule for the magnetic force law $\mathbf{F} = q\,\mathbf{v} \times \mathbf{B}$.

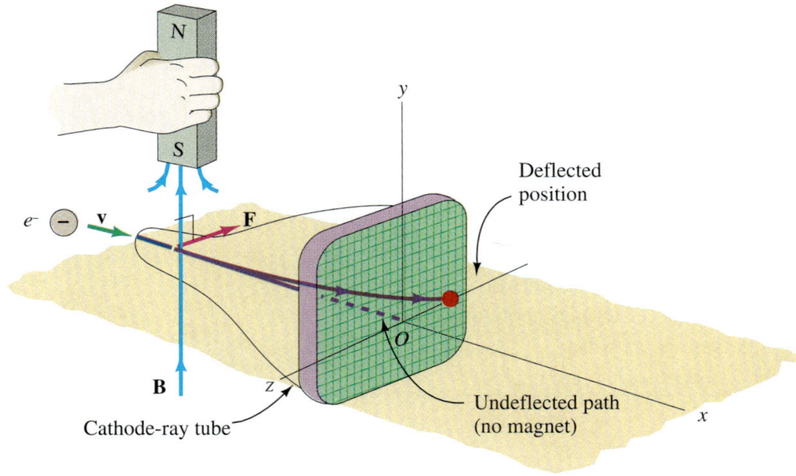

FIGURE 29–7 Example 29–1. The magnetic field points toward the south pole of the bar magnet and is therefore oriented in the $+y$-direction. The spot at which the beam reaches the screen is deflected in the $-z$-direction.

This equation is known as the **Lorentz force law**, named after the late-nineteenth-century physicist Hendrik A. Lorentz, who influenced the development of many areas of classical physics.

(a) ⊙ **B** out

(b) ⊗ **B** in

FIGURE 29–8 Conventions to indicate that a vector is (a) out of the page or (b) into the page. A useful mnemonic is to think of the circle with a dot as the point of the arrow coming toward you, and the circle with an x as the feathered tail of the arrow moving away from you.

A Notation for Vectors Perpendicular to the Page

Because the three-dimensional aspect of magnetic forces is so important, it is useful to have a notation for vectors oriented perpendicular to the page. Figures 29–8a and 29–8b show a vector coming out of and going into the page, respectively. We shall often use this convention when we illustrate magnetic fields.

PROBLEM SOLVING

Recognizing whether a vector is oriented into or out of the page

29–3 MAGNETIC FORCE ON A CHARGE: APPLICATIONS

Magnetic forces on charged particles have important implications that range from the functioning of electronic devices to phenomena in astrophysics and plasma physics. In this chapter, we assume that the magnetic fields are independent of time: We are therefore dealing with **magnetostatics**.

Circular Motion in a Constant Magnetic Field

The kinematics associated with acceleration perpendicular to the velocity were covered in Chapter 3.

The magnetic force law—Eq. (29–1)—states that only the velocity in the plane perpendicular to **B** contributes to the expression for the force. *The component of the velocity of a charged particle parallel to the magnetic field is not affected by the field*, and it is therefore unchanging in the absence of any other forces (such as electric forces). In addition, Eq. (29–1) states that *the force on the charge, and hence the charge's acceleration, is perpendicular to* ***B*** *and thus acts only in the plane perpendicular to* ***B***.

To explore the consequences of these observations more closely, consider a magnetic field **B** that is uniform in some region of space and a test charge q that enters this region with a velocity **v** perpendicular to the field (Fig. 29–9). What is the consequent motion of the charge? According to Eq. (29–1), the force will be perpendicular to **v** and have magnitude $F_B = qvB$. We saw in Chapter 3 that, when the acceleration (and hence the force) is constant in magnitude and perpendicular to the velocity, there is *circular motion at constant speed*. A charged particle moving perpendicularly to a constant, spatially uniform magnetic field will move in a circle (Fig. 29–9a). The magnitude of the acceleration for circular motion is $a = v^2/R$, where R is the radius of the particle's circular path; the direction of the acceleration is toward the center of the circular path. By Newton's second law, the force must have magnitude $ma = mv^2/R$.

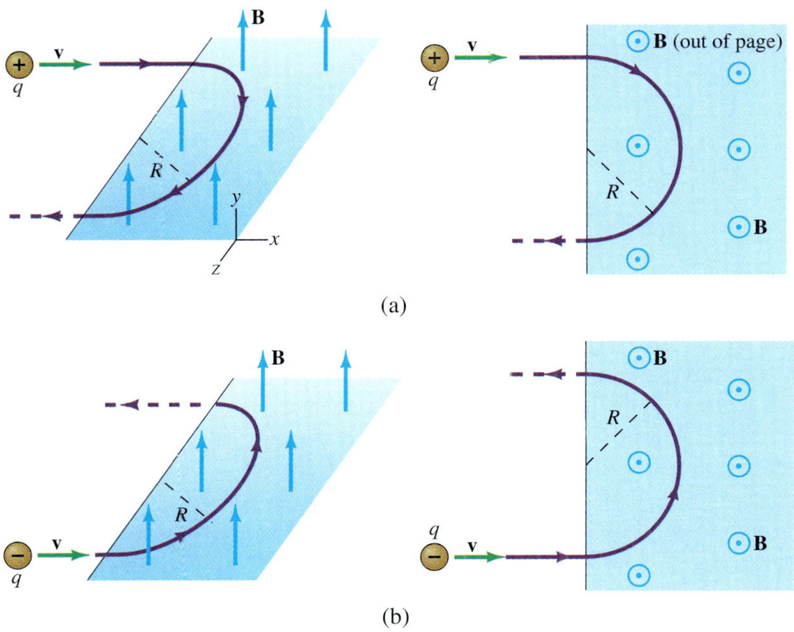

FIGURE 29–9 (a) A positively charged particle moves perpendicularly to a constant magnetic field, **B**, shown in both an oblique perspective and a view from above. The charged particle traces a circular path in the plane perpendicular to **B**, which is directed out of the plane of the page. (b) The direction of the curvature is opposite for a negatively charged particle.

In our case, the force responsible for the acceleration has magnitude F_B, and Newton's second law ($F_B = ma$) becomes (Fig. 29–10)

$$F_B = qvB = \frac{mv^2}{R}.$$

We solve for R:

$$R = \frac{mv}{qB}. \qquad (29\text{–}5)$$

R is proportional to the product of m and v—that is, to the momentum of the moving particle, $p = mv$—and is inversely proportional to the magnitudes of the charge q and of the field, **B**. Equation (29–5) holds *only* when the velocity is perpendicular to **B**. Whether the motion is clockwise or counterclockwise depends on the direction of **v** and on the sign of the charge, according to the right-hand rule. Figure 29–9b depicts the motion for a test charge of opposite sign to the test charge in Fig. 29–9a. The larger B is, the larger is the magnetic force, and the "tighter" is the curved path, which corresponds to a smaller radius of curvature (the radius of the fragment of a circle along which the charge moves at a given moment). The smaller the magnetic field is, the smaller is the force, and the larger is R. If the magnetic field varies in strength from place to place, then so will the radius of curvature of the path (Fig. 29–11).

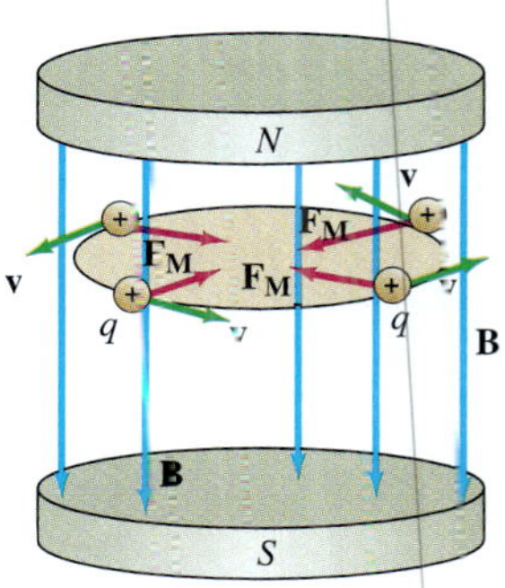

FIGURE 29–10 Forces on a charged particle that is moving perpendicularly to a uniform magnetic field.

The radius of the circle traced out by a charged particle that is moving perpendicularly to a constant magnetic field

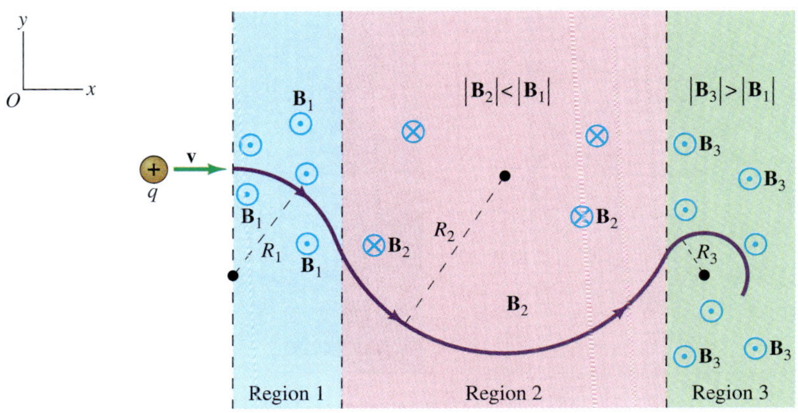

FIGURE 29–11 A particle of positive charge q moves with **v** in the x-direction. In region 1, the magnetic field $\mathbf{B}_1$ is of medium magnitude and oriented in the $+z$-direction. In region 2, $\mathbf{B}_2$ is small in magnitude and oriented in the $-z$-direction. In region 3, $\mathbf{B}_3$ is of large magnitude and oriented in the $+z$-direction.

The circular motion has a period $T = 2\pi R/v$, or, from Eq. (29–5),

$$T = \frac{2\pi}{v}\frac{mv}{qB} = \frac{2\pi m}{qB}. \tag{29–6}$$

Equivalently, the frequency $f = 1/T$ is

$$f = \frac{qB}{2\pi m}. \tag{29–7}$$

FIGURE 29–12 A cyclotron at the Lawrence Berkeley Laboratory. This photo was taken shortly before this cyclotron first operated in 1939. The ion source is to the right, while the region of magnetic field is to the left.

This frequency is called the **cyclotron frequency**. Notice that the period and frequency are *independent* of the speed. A slow particle traces out a tight circle in the same time that a fast particle traces out a large circle. The constancy of the cyclotron frequency is a guiding principle of a device called the *cyclotron* (Fig. 29–12; see Problem 31).

Equation (29–5), which specifies the radius of the circular path of a charged particle, has found application in many particle-detection devices—for example, in the *bubble chamber*. When charged particles produced in high-energy collisions speed through liquid hydrogen, they leave tracks that consist of very tiny bubbles, like a jet leaving a vapor trail in the atmosphere (Fig. 29–13). The momentum of these particles can be obtained by measuring the radius of curvature of their tracks when an external magnetic field is imposed. As we know from Chapter 8, information about momentum is helpful in deciphering collisions.

FIGURE 29–13 The tracks left by charged particles moving in a magnetic field.

Finally, when there is a component $v_{\parallel}$ of velocity $\mathbf{v}$ lies along $\mathbf{B}$, that component of the velocity does not change. The particle advances along $\mathbf{v}_{\parallel}$ while it moves in a circle in the plane formed by $\mathbf{v}_{\perp}$ (Fig. 29–14a). The resulting trajectory forms a *spiral* (or *helix*) with its axis along $\mathbf{B}$ (Fig. 29–14b). The circular motion in the plane perpendicular to $\mathbf{B}$ has a radius given by

$$R = \frac{mv_{\perp}}{qB}. \tag{29–8}$$

EXAMPLE 29–2 A particle of unknown charge q and unknown mass m moves at speed $v = 4.8 \times 10^6$ m/s in the $+x$-direction into a region of constant magnetic field (Fig. 29–15). The field has magnitude $B = 0.5$ T and is oriented in the $+y$-

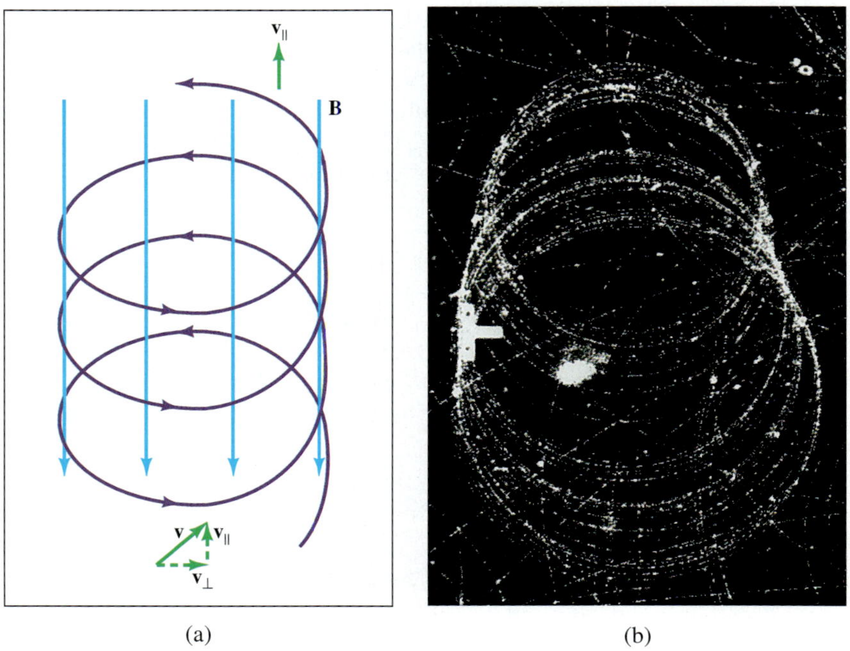

FIGURE 29–14 (a) A charged particle follows a helical path in a region where the magnetic field is constant. (b) An electron in a cloud chamber produced this 10-m-long spiral track. The electron's path begins at the bottom. The helix becomes more tightly wound about halfway up because the electron loses energy by radiation while moving in the helical path.

direction. The particle is deflected in the $-z$-direction and traces out a fragment of a circle of radius $R = 0.1$ m. What is the sign of the particle's charge, and what is the ratio q/m?

Solution: We are given the velocity **v** and the magnetic field, **B**, and can use the magnetic force on the particle to find the unknown charge and mass. To determine the sign of the charge, we need only relate the sign of the force, which acts initially in the $-z$-direction, to the sign of the charge. The force is given by Eq. (29–1), and the vector product $\mathbf{v} \times \mathbf{B}$ points, by the right-hand rule, initially in the $-z$-direction. In order for the force $q\,\mathbf{v} \times \mathbf{B}$ to point that way, the charge q must be positive.

The magnitude of the ratio q/m is determined by Eq. (29–5), $R = mv/qB$:

$$\left|\frac{q}{m}\right| = \frac{v}{BR} = \frac{4.8 \times 10^6\ \text{m/s}}{(0.5\ \text{T})(0.1\ \text{m})} = 9.6 \times 10^7\ \text{C/kg}.$$

This particle is a proton. To see why, let's assume that the unknown charge is that of an electron, $|q| = 1.6 \times 10^{-19}$ C. Then $m = |q|/(9.6 \times 10^7\ \text{C/kg}) = (1.6 \times 10^{-19}\ \text{C})/(9.6 \times 10^7\ \text{C/kg}) = 1.7 \times 10^{-27}$ kg, which is just the mass of a proton. Note, however, that the experiment described in this example can measure only the charge-to-mass ratio, not the charge or the mass alone.

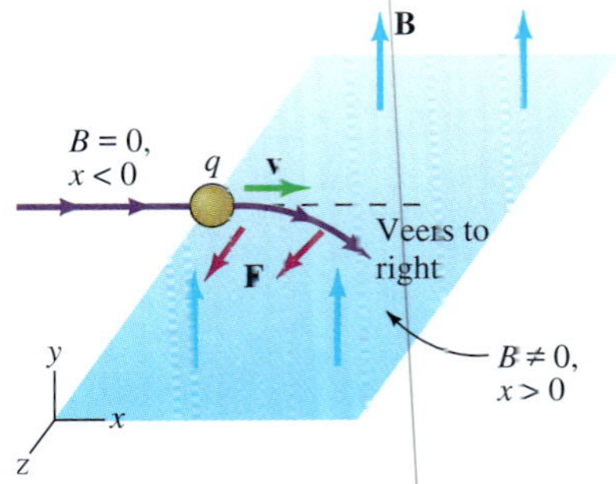

FIGURE 29–15 Example 29–2. If the speed of a particle of unknown charge and mass that moves in a region of constant magnetic field is known, then a measurement of the radius of curvature of the particle's path gives the charge-to-mass ratio of the particle.

Energy of a Charged Particle in a Static Magnetic Field

In the situations we have described, the speed of the charged particle never changes—the velocity component parallel to **B** is unaffected, and the perpendicular velocity component undergoes uniform circular motion in which its direction changes but its magnitude does not. Because the kinetic energy is $K = \frac{1}{2}mv^2$, it is generally true that *the kinetic energy of a charged particle in a static magnetic field is constant.* The work–energy theorem states that the work done by a force equals the change in kinetic energy; equivalently, *a static magnetic field does no work on a charge.*

A charged particle moving in a static magnetic field has a constant kinetic energy.

Velocity Selectors

A particular arrangement of electric and magnetic fields makes a **velocity selector**. Consider a region with uniform, mutually perpendicular **E** and **B** fields (Fig. 29–16). (We say that such fields are *crossed.*) A particle of mass m, (positive) charge q, and velocity **v** is directed perpendicularly to both **E** and **B** when it enters this region. We shall show that there is a certain value of v for which the particle traverses the region undeflected. At a speed other than v, the same particle *is* deflected; thus, in a beam of particles with a variety of speeds, only those particles with a certain speed pass through undeflected.

Both **E** and **B** fields are present, so we must use the Lorentz force law [Eq. (29–4)] and compute the contributions to the force from both fields. Referring to Fig. 29–16, the electric force is

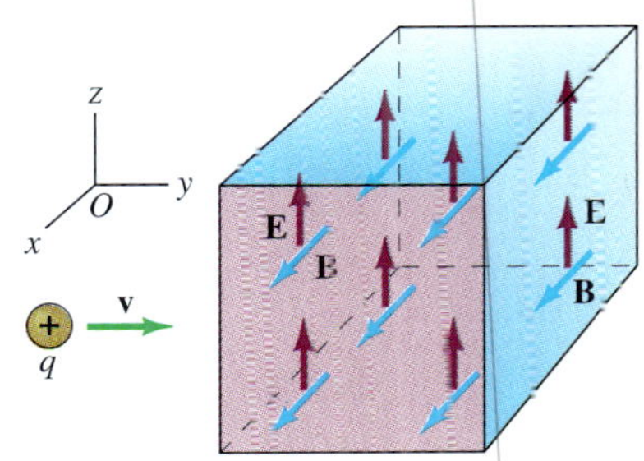

FIGURE 29–16 A charged particle enters a region of crossed electric and magnetic fields. If the speed of the particle is $v = E/B$, the particle will cross the region undeflected.

$$\mathbf{F}_E = qE\,\mathbf{k}.$$

By the right-hand rule, the magnetic force $q\,\mathbf{v} \times \mathbf{B}$ is

$$\mathbf{F}_B = -qvB\,\mathbf{k}.$$

The electric and magnetic forces will cancel if their magnitudes are equal because they point in opposite directions; in this case, the particle will travel undeflected. This cancellation occurs for $qvB = qE$, so the speed of a charged particle that passes through the crossed fields undeflected is

$$v = \frac{E}{B}. \quad (29\text{–}9)$$

If we reverse the sign of charge q, the electric force points in the $-z$-direction while the magnetic force points in the $+z$-direction; the forces still cancel when v is given by Eq. (29–9).

The speed of a charged particle that moves undeflected perpendicular to crossed magnetic and electric fields

EXAMPLE 29–3 A hole is cut in the wall of a container of a plasma—matter that has been broken into positively charged ions and electrons—creating a beam of particles of various charges. It is necessary to choose particles from the beam that have a speed of 3.2×10^6 m/s to do materials testing. The engineer designing the speed selector uses crossed **E** and **B** fields, and his **B** field comes from a magnet with field strength $B = 6.5 \times 10^{-4}$ T. What must he choose as the magnitude of the crossed electric field? Will all the particles so chosen have the same momentum and energy?

Solution: If we wish to design a crossed-field velocity selector for choosing a speed v, and if the magnetic field has magnitude B, then the electric field must have magnitude given by Eq. (29–9),

$$E = vB = (3.2 \times 10^6 \text{ m/s})(6.5 \times 10^{-4} \text{ T}) = 2.1 \times 10^3 \text{ V/m}.$$

Notice that the magnetic field is smaller than those usually associated with permanent magnets; unless the engineer is careful, Earth's magnetic field will spoil his design. A larger magnetic field, however, will entail a larger electric field.

The speed chosen is independent of the charge and the mass of the particle; thus, for example, electrons will be chosen in the same way as positive ions of a variety of charges. Both momenta and energy will take on a variety of values. This fact would make the particles so chosen not so useful to test materials, since their effect on the materials depends more on their momentum and energy than on their speed.

The Charge-to-Mass Ratio of the Electron. Sir Joseph John Thomson, the discoverer of the electron, performed a series of wide-ranging experiments whose results were crucial to our understanding of the electrical nature of matter. He used a velocity selector in 1897 as an important component of his experiment to measure the charge-to-mass ratio of the electron (Fig. 29–17). He first accelerated electrons in an electric field—not the electric field of the velocity selector—by passing them through an electric potential V. The work thereby done on the electrons is qV. The electrons gained a speed v determined by $mv^2/2 = qV$, so

$$v = \sqrt{\frac{2qV}{m}}. \tag{29–10}$$

The electrons accelerated in this way continued into a region of crossed electric and magnetic fields. Thomson adjusted the magnitudes of these fields until the electrons passed through the apparatus undeflected. When we combine Eqs. (29–9) and (29–10), we find that

$$v = \frac{E}{B} = \sqrt{\frac{2qV}{m}}.$$

FIGURE 29–17 (a) J. J. Thomson at work in his laboratory. (b) Schematic diagram of Thomson's apparatus for measuring the charge-to-mass ratio of the electron. The magnetic field is directed into the plane of the page.

(a)

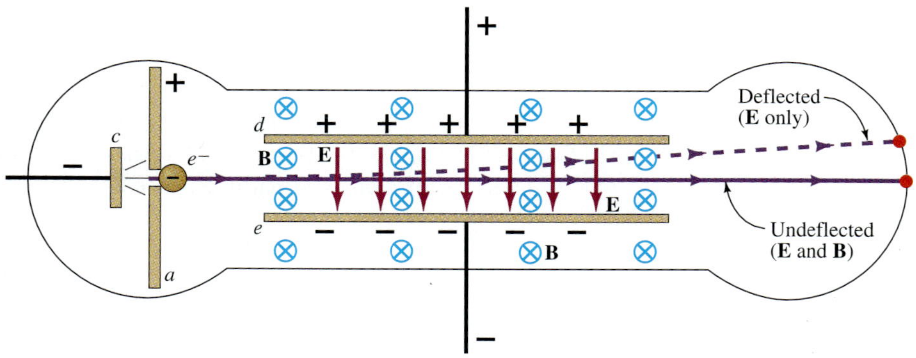

(b)

When both sides of this equation are squared, we can solve for q/m:

$$\frac{q}{m} = \frac{E^2}{2VB^2}. \tag{29–11}$$

The electrons in Fig. 29–17 were accelerated between plates c and a and moved on to the region between plates d and e where he could adjust the magnetic field until electrons were undeflected; he could then use Eq. (29–11) to measure q/m. More refined experiments based on Thomson's scheme have led to a value for q/m of 1.759×10^{11} C/kg.

Magnetic Fields in Outer Space

Magnetic fields exist in outer space. Throughout our galaxy, the magnetic field strength is in the range of 10^{-10} T. Charged particles (*cosmic rays*) are generated and accelerated by various stellar processes. If their momentum is less than a certain critical value p_c, they drift in gigantic circles within the galaxy due to the magnetic forces on them.† Cosmic rays with a momentum greater than p_c move on a circle with a radius of curvature greater than the galaxy's radius, and they therefore escape the galaxy. To estimate p_c for a cosmic ray whose charge is the electron charge e, we use the observation that the radius of the galaxy is about 5×10^{21} m. From Eq. (29–5), the critical momentum has magnitude

$$p_c = eBR = (1.6 \times 10^{-19}\ \text{C})(10^{-10}\ \text{T})(5 \times 10^{21}\ \text{m}) = 8 \times 10^{-8}\ \text{kg} \cdot \text{m/s}.$$

For a particle such as an electron or a proton, this momentum is enormously large. For comparison, an electron in the beam of a television picture tube typically has a momentum of 10^{-22} kg · m/s, whereas protons in the world's largest proton accelerator (at Fermilab) attain momenta of 5×10^{-16} kg · m/s. Because cosmic rays with a momentum greater than p_c leave the galaxy, we should expect to detect more cosmic rays that strike Earth with momenta lower than p_c than with momenta greater than p_c. Experimental observations of particles arriving from outer space help us to estimate the value of the interstellar magnetic field.

Van Allen Belts. Earth has a magnetic field like that of a huge bar magnet, directed from the geographic South Pole to the geographic North Pole. Charged particles, many streaming outward from the Sun in what is called the *solar wind*, arrive in Earth's vicinity. Collisions in Earth's vicinity slow them, and they subsequently are trapped and travel in spiral paths around Earth's magnetic field lines. Figure 29–18a illustrates how magnetic field lines become more dense, or "pinch," as they approach a pole. Detailed analysis of the force on a moving charge shows that when the lines "pinch," the helical path of the particles becomes flatter and eventually turns back on itself. A particle spiraling along the field lines from the South Pole toward the North Pole turns around near the North Pole and spirals back southward (Fig. 29–18b). There are, in effect, **magnetic mirrors** near the poles. Particles trapped in this way bounce back and forth between the poles and accumulate into regions called **Van Allen belts**. There are two such belts: one containing protons at a mean height of 3000 km above Earth's surface, and the other containing electrons at an elevation of about 15,000 km. James Van Allen discovered these belts in 1958 by using data from the Explorer I satellite. One consequence of the belts is the *aurora*, a phenomenon that occurs wherever the charged particles enter the atmosphere and excite the atoms in the air. The auroras are more noticeable near the poles, where the magnetic field lines dip toward Earth, carrying the Van Allen belts with them.

†We use momentum rather than speed because a calculation of the speed here gives a critical speed greater than the speed of light. This indicates that special relativity is necessary, and special relativity suggests that momentum should be used here.

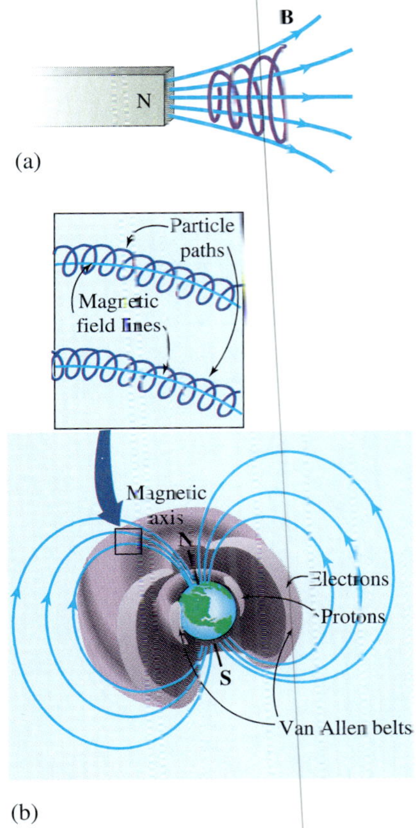

FIGURE 29–18 (a) Charged particles spiraling around the varying magnetic field near the pole of a magnet. The pitch, or inclination of the helix decreases to zero as particles near the poles, where the field lines are denser, and the particles reverse the direction of their helical path. (b) When this is applied to particles spiraling around Earth's magnetic field, the result is that the particles are trapped, bouncing back and forth between the polar regions. The trapped particles form the two Van Allen belts, one for electrons and one for protons.

EXAMPLE 29–4 Assume that a proton of speed 1.5×10^7 m/s approaches Earth at an angle of 40° to Earth's magnetic field lines and is captured in the lower Van Allen belt (at a mean altitude of 3000 km) without a change in speed. If the mean strength of the field at this altitude is 10^{-5} T, find the cyclotron frequency and the radius of curvature for the proton's motion.

Solution: The cyclotron frequency, from Eq. (29–7), is

$$f = \frac{qB}{2\pi m} = \frac{(1.6 \times 10^{-19}\ \text{C})(10^{-5}\ \text{T})}{2\pi(1.67 \times 10^{-27}\ \text{kg})} = 150\ \text{Hz}.$$

The proton moves in a spiral whose radius of curvature depends on the component of velocity perpendicular to the magnetic field. This component has magnitude $v_\perp = v \sin 40° = (1.5 \times 10^7\ \text{m/s})(0.64) \simeq 10^7$ m/s. Using Eq. (29–8), we then find that

$$R = \frac{mv_\perp}{qB} = \frac{(1.67 \times 10^{-27}\ \text{kg})(10^7\ \text{m/s})}{(1.67 \times 10^{-19}\ \text{C})(10^{-5}\ \text{T})}$$

$$= 10^4\ \text{m} = 10\ \text{km}.$$

The radius of curvature is much less than the altitude of the Van Allen belt.

29–4 MAGNETIC FORCES ON CURRENTS

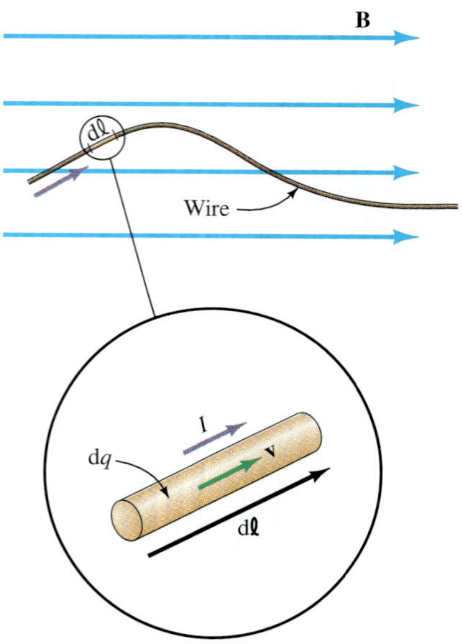

FIGURE 29–19 A wire carrying a current in a magnetic field. We isolate a segment $d\ell$ of the wire that contains the moving charge dq.

We have learned that there may be a force on moving charges in a magnetic field. Because electric currents in wires consist of moving charges, we expect that a magnetic field will exert a force on the charges in a current-carrying wire, and thus on the wire itself (Fig. 29–19). Experiment bears out this expectation.

A wire contains moving charges throughout, and the magnetic field may vary significantly along its length. The total force on a current-carrying wire is the vector sum of the magnetic forces on all of the moving charges within it. To find the total force, we first determine the force on a small segment of a current-carrying wire. We subsequently sum (integrate) the infinitesimal force on each segment.

Magnetic Forces on Infinitesimal Wires with Currents

Let's denote the small segment of a thin current-carrying wire by $d\boldsymbol{\ell}$: It has both an infinitesimal magnitude $d\ell$ and a direction along the instantaneous current carried by the wire at segment $d\boldsymbol{\ell}$. If the moving charge dq contained in a segment of wire $d\boldsymbol{\ell}$ has velocity $\mathbf{v}$ along the wire (Fig. 29–19), its displacement d in time dt is $d\boldsymbol{\ell} = \mathbf{v}\, dt$, so

$$\mathbf{v} = \frac{d\boldsymbol{\ell}}{dt}. \tag{29–12}$$

Because the current, I, is dq/dt by definition, the amount of charge within the segment is

$$dq = I\, dt. \tag{29–13}$$

Note that the magnetic field will be uniform over the length of the segment if the segment is small enough. With Eqs. (29–12) and (29–13) we can calculate the magnetic force $d\mathbf{F}$ that acts on our charge element dq and hence on the wire element:

$$d\mathbf{F} = dq\, \mathbf{v} \times \mathbf{B} = (I\, dt)\left(\frac{d\boldsymbol{\ell}}{dt} \times \mathbf{B}\right).$$

We cancel the factor dt to find the infinitesimal force on a wire element $d\boldsymbol{\ell}$ carrying current I in a magnetic field $\mathbf{B}$:

The magnetic force on a segment of current-carrying wire

$$d\mathbf{F} = I\, d\boldsymbol{\ell} \times \mathbf{B}. \tag{29–14}$$

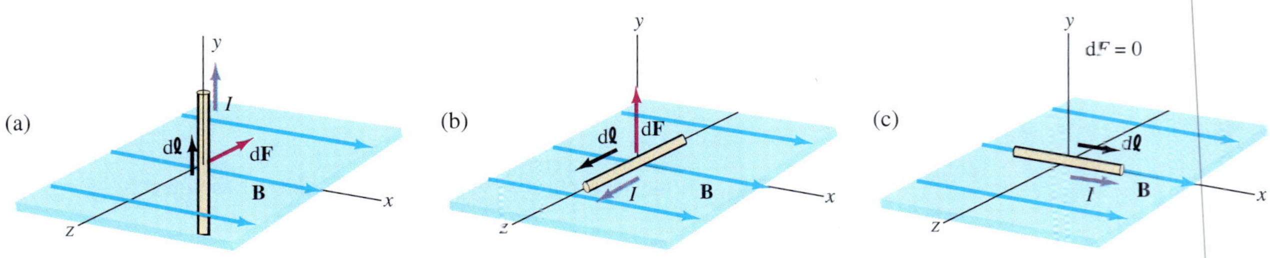

FIGURE 29–20 A wire segment, aligned with the (a) y-axis, (b) z-axis, and (c) x-axis, in a magnetic field, with the infinitesimal forces **dF** that act on the segment.

Note that the current is the same everywhere along the wire because current is conserved. The magnitude of the magnetic force $d\mathbf{F}$ is given by

$$dF = I\,d\ell\,B\sin\theta, \qquad (29\text{–}15)$$

where θ is the angle between the direction of the wire segment (the current's direction) and the direction of the magnetic field. As for the direction of the force, Fig. 29–20 shows three different placements of a segment of current-carrying wire in a uniform magnetic field that points in the $+x$-direction. In each case, the direction of the force on the wire segment is given by the right-hand rule and Eq. (29–14). In Fig. 29–20a, $d\ell$ points in the $+y$-direction, so that the force points in the $-z$-direction. In Fig. 29–20b, $d\ell$ points in the $+z$-direction, so the force points in the $+y$-direction. In Fig. 29–20c, $d\ell$ points in the $+x$-direction, parallel to **B**, so there is no force.

Magnetic Forces on Finite Wires with Currents

The *net* force **F** on a *finite* section of wire is the vectorial sum of the forces on the various infinitesimal segments that make up the wire. We find the net force by integrating **dF** [Eq. (29–14)] over the total length of the wire. Because I is the same everywhere along the wire, the summation of Eq. (29–14) has the form

$$F_B = I\int (d\ell \times \mathbf{B}). \qquad (29\text{–}16)$$

Whether the integral is easy to perform or not depends on the particular situation. A straight wire within a constant magnetic field represents an important special case. Let's suppose first that the wire, with length L and current I, is oriented perpendicular to the field (Fig. 29–21a). To perform the integration of Eq. (29–16), notice that each segment $d\ell$ of the wire points in the y-direction, $d\ell = d\ell\,\mathbf{j}$, and **B** is constant at each segment, $\mathbf{B} = B\,\mathbf{i}$. The infinitesimal force on each segment is therefore identical:

$$d\mathbf{F} = I[(d\ell\,\mathbf{j}) \times (B\,\mathbf{i})] = I\,d\ell B(\mathbf{j} \times \mathbf{i}) = I\,d\ell B(-\mathbf{k}),$$

which is a vector that points in the $-z$-direction. We can easily check the direction with the right-hand rule. The net force **F** is an integral over **dF**:

$$\mathbf{F} = \int d\mathbf{F} = \int d\ell B(-\mathbf{k}) = IB(-\mathbf{k})\int_0^L d\ell = -IBL\,\mathbf{k}. \qquad (29\text{–}17)$$

The net force points in the $-z$-direction (here, into the paper), and its magnitude is IBL.

Next, suppose that the wire makes an angle θ to the field (Fig. 29–21b). The only change we must make is in the expression for the magnitude of the infinitesimal force element: There is an additional factor $\sin\theta$ in dF [Eq. (29–15)]. The direction of the force remains in the $-z$-direction by the right-hand rule. Because θ is constant, it does not enter into the integration, which thus remains as in Eq. (29–17). The result is identical to Eq. (29–17) with an additional factor of $\sin\theta$:

$$\mathbf{F} = IBL\sin\theta\,\mathbf{k}. \qquad (29\text{–}18)$$

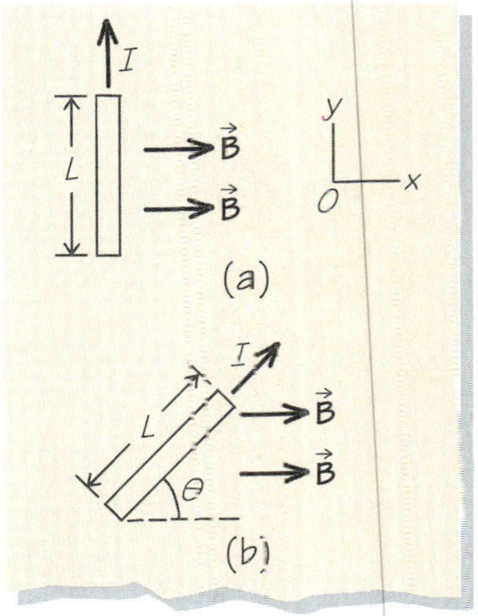

FIGURE 29–21 (a) A wire segment is oriented perpendicularly to a magnetic field. (b) The same wire segment as in part (a), but at an angle.

Equations (29–17) and (29–18) are useful and important results. They can be combined into one vectorial equation that gives the force on a thin, straight wire of length L in a magnetic field, namely,

$$\mathbf{F} = I\,\mathbf{L} \times \mathbf{B}, \tag{29–19}$$

where the vector $\mathbf{L}$ is oriented along the wire in the direction of the current.

EXAMPLE 29–5 A 12-cm long straight segment of wire carrying a current of 7.2 A is maneuvered entirely within a region known to contain a constant magnetic field until the force on the wire has a maximum magnitude of 0.37 N. Find the magnitude of the magnetic field.

Solution: We must recognize that, as the wire is moved around, the force on it takes on a maximum value when the wire is oriented perpendicular to the magnetic field. At this orientation, the force has magnitude given by Eq. (29–17). We solve that equation—actually its magnitude—for the magnetic field magnitude:

$$B = \frac{F}{IL} = \frac{0.37\ \text{N}}{(7.2\ \text{A})(12 \times 10^{-2}\ \text{m})} = 0.43\ \text{T}.$$

In fact, the measurement of forces on wires represents an important tool for the accurate measurement of magnetic fields.

The expression for the magnetic force on an isolated moving charge has led us directly to the expression for the magnetic force on a current-carrying wire. Historically, the order of discovery was just the reverse: Oersted, François Arago, and Ampère, who performed the first experiments on magnetic forces, observed those forces on current-carrying wires. Their results then led to an understanding of magnetic forces on moving charges.

29–5 MAGNETIC FORCE ON CURRENT LOOPS

Magnetic fields exert forces on all kinds of current-carrying wires, including those of closed loops. As we shall see, a uniform magnetic field actually exerts only a torque on a current loop. This phenomenon provides the torque that runs direct-current electric motors and the galvanometer, the device cited in Chapter 28 for use in ammeters and voltmeters.

Consider a stiff rectangular loop of wire carrying current I (Fig. 29–22a). The magnetic field is oriented in the $+x$-direction. The sides of the wire loop are denoted 1, 2, 3, and 4; sides 1 and 3 have length a, and sides 2 and 4 have length b. Figure 29–22b is a side view of the apparatus along the $+y$-direction. The direction perpendicular to the plane of the loop (the direction of the thumb when the fingers of the right hand follow the current direction) makes an angle ψ with the magnetic field.

We can calculate the force on each leg of the loop by using Eq. (29–19). The angle θ between the direction of the magnetic field and the direction of the current is shown in Fig. 29–22b for leg 1; legs 2 and 4 are perpendicular to the magnetic field, into and out of the page, respectively. Therefore the force on each leg is

$$F_1 = IaB \sin\theta, \text{ in the } -y\text{-direction}; \tag{29–20a}$$

$$F_2 = IbB, \text{ in the } -z\text{-direction}; \tag{29–20b}$$

$$F_3 = IaB \sin\theta, \text{ in the } +y\text{-direction}; \tag{29–20c}$$

$$F_4 = IbB, \text{ in the } +z\text{-direction}. \tag{29–20d}$$

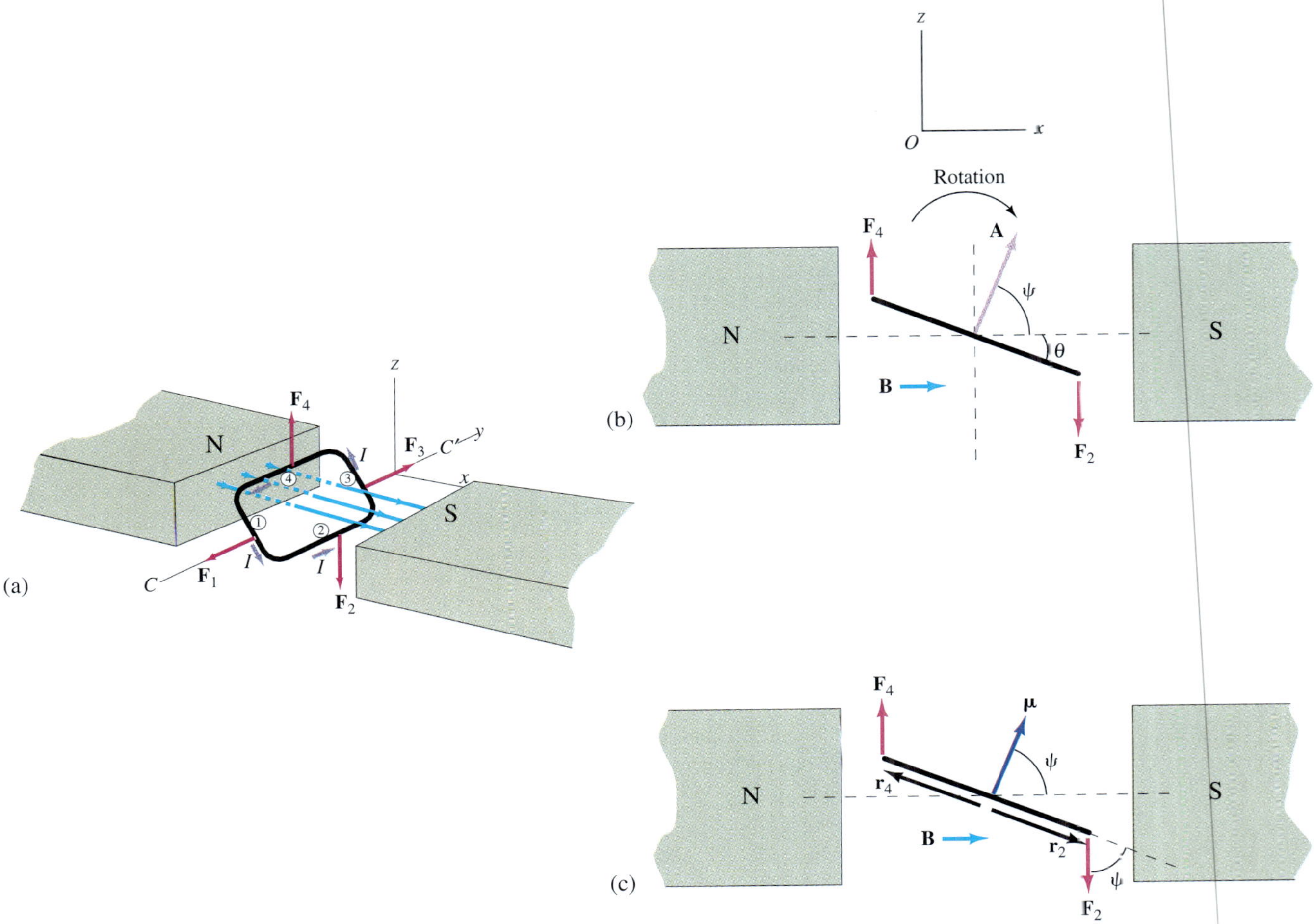

FIGURE 29–22 (a) A stiff, rectangular loop of wire is placed in a constant magnetic field. (b) A side view, looking along the y-axis, of the loop. (c) Geometry of the loop that allows us to calculate the torque on the loop. The torque tends to align the vector $\boldsymbol{\mu}$ with the magnetic field, $\mathbf{B}$.

These forces are indicated in Figs. 29–22a and 29–22b. Forces $\mathbf{F}_1$ and $\mathbf{F}_3$ are equal and opposite, as are forces $\mathbf{F}_2$ and $\mathbf{F}_4$, so there is no net force. However, there is an important difference between these two sets of forces: $\mathbf{F}_1$ and $\mathbf{F}_3$ act along the same axis (CC' in Fig. 29–22a) and exert no torque on the loop. $\mathbf{F}_2$ and $\mathbf{F}_4$ act along different axes, as emphasized in Fig. 29–22b, and therefore produce a torque that causes the wire loop to rotate clockwise in the magnetic field. When the wire has rotated into the yz-plane (when $\theta = 90°$ in Figure 29–22b), $\mathbf{F}_2$ and $\mathbf{F}_4$ act along the same axis, and there is no torque. When the loop is in the xy-plane ($\theta = 0°$), the torque is a maximum. Finally, when θ changes sign, so does the torque, and the loop will tend to rotate counterclockwise.

We can find the torque about the central axis CC' in Fig. 29–22a by using the results of Chapter 10. From Eq. (10–6), the net torque about this axis, $\boldsymbol{\tau}$, is

$$\tau = (\mathbf{r}_2 \times \mathbf{F}_2) + (\mathbf{r}_4 \times \mathbf{F}_4),$$

where $\mathbf{r}_2$ and $\mathbf{r}_4$ are the perpendicular vectors from axis CC' to legs 2 and 4, respectively (Fig. 29–22c). Both $\mathbf{r}_2$ and $\mathbf{r}_4$ have magnitude $a/2$. Figure 29–22c shows that ψ is the angle between $\mathbf{r}_2$ and $\mathbf{F}_2$ and between $\mathbf{r}_4$ and $\mathbf{F}_4$. The torque has magnitude

$$\begin{aligned}\tau &= r_2F_2 \sin\psi + r_4F_4 \sin\psi \\ &= \frac{a}{2}(IbB)\sin\psi + \frac{a}{2}(IbB)\sin\psi = iabB\sin\psi. \qquad (29\text{–}21)\end{aligned}$$

Here, we have used the values for F_2 and F_4 given by Eq. (29–20). According to the right-hand rule, both terms in the equation for τ point in the $+y$-direction, so the net torque is in this direction.

The torque on a current loop in a constant magnetic field

The torque on a current loop in a magnetic field as given by Eq. (29–21) can be summarized and generalized. First, the factor ab is the area A of the current loop. This result generalizes to *any* planar loop of area A—whatever its shape—by the calculus technique of decomposing a planar loop of any shape into tiny rectangles like the loop we have discussed. Next, the vectorial nature of the torque is handled neatly by defining a vector $\boldsymbol{\mu}$ perpendicular to the plane of the loop. Two vectors are perpendicular to any plane; which of them do we choose for $\boldsymbol{\mu}$? We choose to define $\boldsymbol{\mu}$ with a right-hand rule: Curl the fingers of the right hand in the direction of the current around the loop, and the right thumb gives the direction of $\boldsymbol{\mu}$ (Fig. 29–23). Try it for Fig. 29–22c, where we have indicated $\boldsymbol{\mu}$. The angle between $\boldsymbol{\mu}$ and $\mathbf{B}$ is ψ. We have thus shown that the direction and magnitude of the *torque on a current loop* are given by

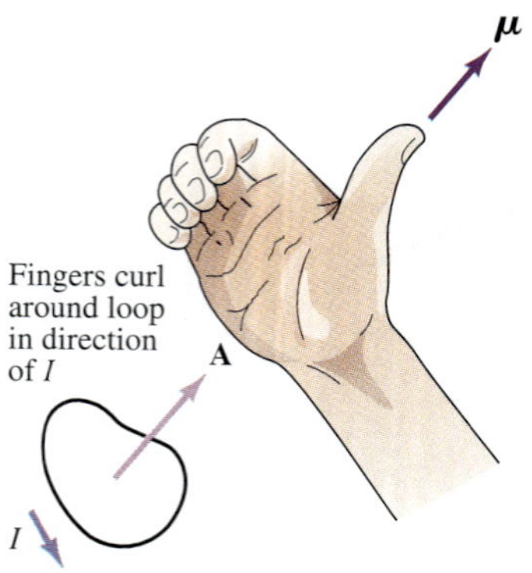

FIGURE 29–23 A right-hand rule indicates the direction of the magnetic dipole moment, $\boldsymbol{\mu}$, of a current loop.

$$\boldsymbol{\tau} = \boldsymbol{\mu} \times \mathbf{B}, \tag{29–22}$$

provided that the magnitude of $\boldsymbol{\mu}$ is taken to be

$$\mu = IA. \tag{29–23}$$

From these equations, the magnitude of the torque is

$$\tau = \mu B \sin \psi,$$

exactly as in Eq. (29–21), and the direction of the torque is the $+y$-direction. A torque in this direction acts to align $\boldsymbol{\mu}$ and $\mathbf{B}$. It is generally true that *the torque tends to rotate a current loop or coil in such a way that* $\boldsymbol{\mu}$ *and* $\mathbf{B}$ *become aligned.*

PROBLEM SOLVING

Magnetic dipoles tend to align their moments with an external magnet field.

The torque on a current loop in a uniform magnetic field is completely analogous to the torque on an electric dipole (a pair of equal and opposite electric charges) in a uniform electric field, discussed in Chapter 23. The electric dipole was described by the *electric dipole moment* $\mathbf{p}$, a vector aligned with the two charges and equal in magnitude to the charge times the distance of charge separation. In terms of the response to an external magnetic field, *the current loop is in all measurable respects a* **magnetic dipole**.† We therefore call $\boldsymbol{\mu}$, which plays a role analogous to $\mathbf{p}$, the **magnetic dipole moment** of the loop.

One more generalization is possible. Suppose that, instead of one turn of wire, our loop consists of N turns (each surrounding the same plane area). We might now refer to a *coil* rather than a loop. *Each* turn of the coil experiences the forces we have described, and the torque is multiplied by N. This factor is included with the other factors intrinsic to the coil, so the magnetic dipole moment will now take into account N:

$$\mu = INA. \tag{29–24}$$

The magnetic dipole moment of a coil of current

We have seen another system that aligns itself with external magnetic fields: the iron filings we used to make our preliminary definition of magnetic fields. These iron filings behave like little bar magnets that are also rotated by magnetic fields. Bar magnets react to fields just like current loops do. As we shall see in Chapter 30 when we find the magnetic fields *produced* by magnetic dipoles, bar magnets are themselves magnetic dipoles. This is so because, at a microscopic level, metals contain circulating currents, formed by the electrons that orbit atomic nuclei. We shall study this behavior in more detail in Chapter 32.

†One of the measurable properties—the defining property—of a dipole, magnetic or electric, is the characteristic field that the dipole itself *produces*. We deal with the calculation of the magnetic dipole field in Chapter 30.

Magnetic Charges

Our procedure in studying magnetic fields has been to probe the effects of these fields with test objects such as bar magnets or tiny magnetic dipoles. This is similar to the procedure we followed in probing the effects of electric fields, but there is one very fundamental difference. For the case of electric fields, we are able to use test objects with single electric charges, or electric monopoles. With single charges, it is quite straightforward to learn the force laws for electric fields. If we had used electric dipoles as probes, the corresponding force laws would have appeared to be much more complicated, and our job would have been much more difficult. Yet this is just what we have done for magnetism! Why did we not use magnetic *charges*, or magnetic monopoles, as probes? The answer is that, although many experiments have been performed to find magnetic monopoles, no one has ever observed a magnetic monopole unambiguously; it may be that such objects simply do not exist.

Motors and Galvanometers

With the apparatus shown in Fig. 29–22a, as soon as the wire loop rotates past the position in which it is aligned with the yz-plane, the torque on it changes sign and becomes counterclockwise. In fact, the torque changes direction when ψ goes through 0° or 180°. However, if we can make the *current* switch directions every time the loop passes $\psi = 0°$ or 180°, then the torque will continuously produce a clockwise rotation. Such a device is known as a *split-ring commutator* (Fig. 29–24): The loop continues to rotate under the influence of a torque whose sign does not change, and a type of *electric motor* has been created.

APPLICATION
Commutators and electric motors

We can use the fact that a magnetic field exerts a torque on a current loop to make a device that measures currents. In fact, we have already used such a device in Chapter 28: a *galvanometer*. For example, we can attach a spring to a loop to balance the torque due to a known magnetic field (Fig. 29–25a). The amount the spring stretches is a measure of the torque of the loop and hence of the current that passes through the loop.

FIGURE 29–24 (a) A current-carrying loop aligned in a magnetic field is fitted with a split-ring commutator. (b) Schematic diagram of a split-ring commutator. The torque on the loop serves to turn the loop and makes a motor.

(a)

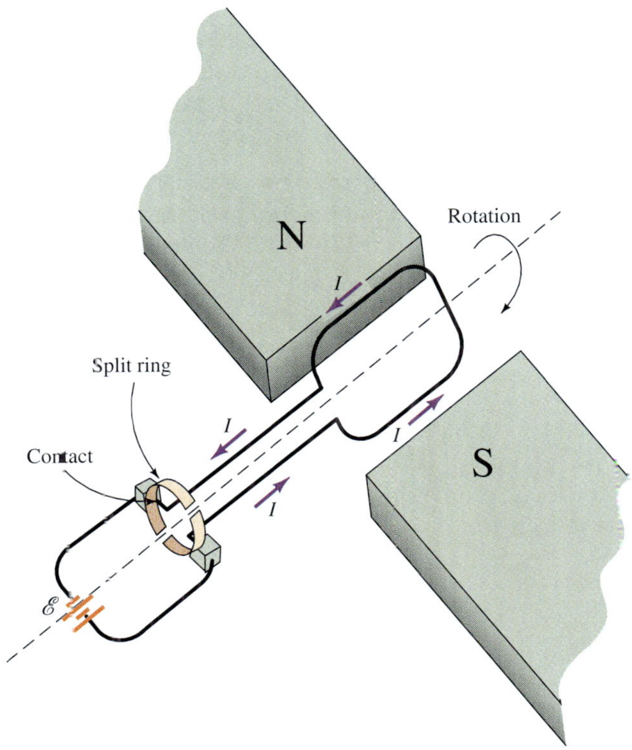

(b)

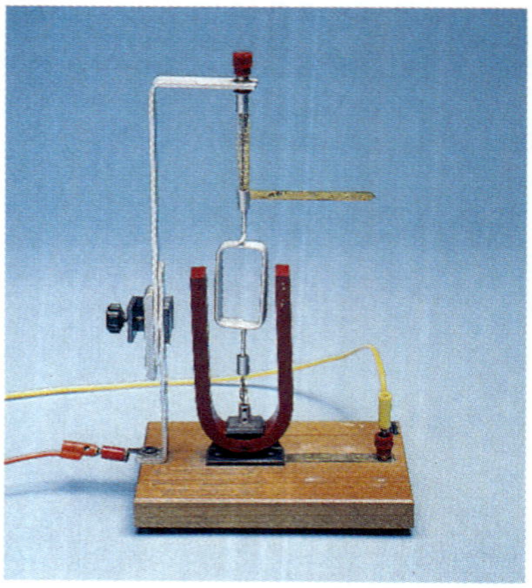

(a)

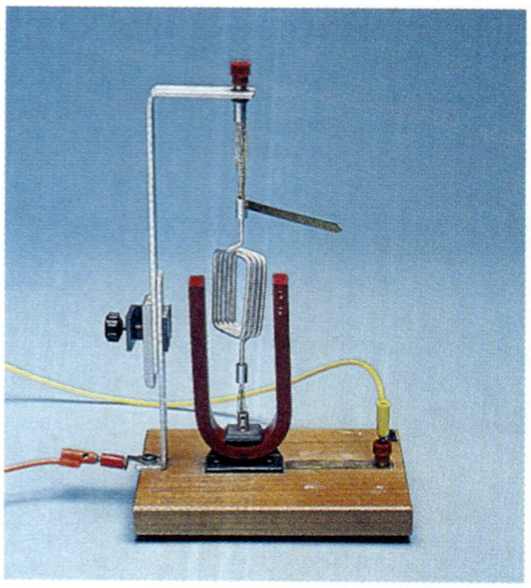

(b)

FIGURE 29–25 A galvanometer measures the current in a loop. Here, we picture a schematic version. (a) Without current in the coil, there is no effect on it when it is placed in a magnetic field. (b) The coil rotates within the field when a current runs through it. A torsion spring balances the torque from the magnetic forces on the loop. The amount by which the coil rotates against the torsion spring indicates the current.

The energy of a current loop in a constant magnetic field. The lowest value of potential energy occurs when the magnetic dipole moment and the magnetic field are aligned.

Energy and the Torque on Loops

When a magnetic field rotates a current loop, the field does work. For a constant field, the only variable in the work is the angle of rotation, ψ. We know from Chapter 7 that, when the force (or the torque) depends only on position, the concept of potential energy is useful because we can apply the principle of the conservation of energy. Thus we can associate a potential energy $U(\psi)$ with an oriented loop in a magnetic field, where ψ is the angle between $\boldsymbol{\mu}$ and $\mathbf{B}$. As always, only *changes* in the potential energy have physical consequences. The change in potential energy in rotating the coil from some initial angle ψ to a final angle of 90° is given by the negative of the work done by the magnetic field in moving the coil through these angles:

$$U(\psi) - U(90°) = -\int_{\psi}^{90°} \tau\, d\psi' = -\int_{\psi}^{90°} (\mu B \sin \psi')\, d\psi'$$

$$= -\mu B \int_{\psi}^{90°} \sin \psi'\, d\psi' = \mu B \cos 90° - \mu B \cos \psi.$$

The cosine of 90° is zero, so

$$U(\psi) - U(90°) = -\mu B \cos \psi. \tag{29–25}$$

Zero U can be chosen for convenience. It is customary to choose U to be zero at $\psi =$ 90°; that is, when $\boldsymbol{\mu}$ is perpendicular to $\mathbf{B}$. Thus we set $U(90°) = 0$ in Eq. (29–25), giving us *the potential energy of a current loop with a given magnetic dipole moment* $\boldsymbol{\mu}$ *in a constant magnetic field* $\mathbf{B}$:

$$U(\psi) = -\boldsymbol{\mu} \cdot \mathbf{B}. \tag{29–26}$$

The potential energy has a *minimum* when $\boldsymbol{\mu}$ is aligned along $\mathbf{B}$ (that is, when $\psi =$ 0°). Thus the orientation in which $\boldsymbol{\mu}$ is aligned with $\mathbf{B}$ is a stable equilibrium point. This agrees with our earlier result that the torque tends to rotate the loop to line up $\boldsymbol{\mu}$ and $\mathbf{B}$.

EXAMPLE 29–6 A current loop in a constant magnetic field is initially aligned so that $\boldsymbol{\mu}$ points in a direction slightly different from that of $\mathbf{B}$, and the loop is then released. There is no mechanism such as friction for energy loss; that is, no damping. Find the subsequent motion of the loop.

Solution: We start with the loop oriented slightly away from the stable equilibrium point; in other words, with small but nonzero ψ, where ψ is the angle between $\boldsymbol{\mu}$ and $\mathbf{B}$. To describe the behavior of the loop, we express $U(\psi)$ for small values of ψ. Using the small-angle approximation $\cos \psi \simeq 1 - (\psi^2/2)$ from Appendix IV–9, we find

$$U(\psi) = -\mu B \cos \psi \simeq -\mu B + \frac{\mu B \psi^2}{2}. \tag{29–27}$$

The term $-\mu B$ is a constant and therefore has no bearing on the loop's motion. The term $\mu B \psi^2/2$ is characteristic of a familiar physical system, the harmonic oscillator. This result is not surprising because we know that small motions about almost *any* stable equilibrium are harmonic. The harmonic oscillator of Chapter 13 has potential energy $kx^2/2$, a form associated with harmonic motion in the variable x about the equilibrium point $x = 0$. Here, ψ plays the role of x, whereas μB substitutes for the spring constant k. The loop in the magnetic field acts like a physical pendulum, so we should also substitute for the mass m in the harmonic oscillator the rotational inertia of the loop, I_M, about the rotation axis. (We use I_M to avoid any possible confusion with the current.) The motion will then be

$$\psi = \psi_m \sin(\omega t + \phi)$$

[see Eq. (13–1)]. Here, ψ_m and ϕ are the (given) amplitude and phase, respectively, and the angular frequency is

$$\omega = \sqrt{\frac{\mu B}{I_M}}.$$

We have harmonic motion in ψ about the stable equilibrium point $\psi = 0$, where $\boldsymbol{\mu}$ is aligned with **B**. There is an interchange between the potential energy, Eq. (29–27), and the kinetic energy term associated with the motion of the loop: When the potential energy is large, at the maximum value of ψ, the kinetic energy is zero; when the potential energy is zero, at $\psi = 0$, the kinetic energy is a maximum (see Problem 60).

We have already mentioned that bar magnets are influenced by magnetic fields in just the same way as are current loops. This suggests, correctly, that an undamped compass needle behaves much like the loop in this example because the needle is actually a bar magnet.

29–6 THE HALL EFFECT

The direction of a current does *not* itself determine the sign of the charge carriers in that current because a current to the right can be produced by the movement either of positive charges to the right or of negative charges to the left. The *Hall effect* allows us to find this sign. Consider a metal strip of length L on which an electric potential is applied from one end to the other, so that a current flows in the strip. The strip is placed in a uniform magnetic field that is perpendicular to the strip (Fig. 29–26). Let's now qualitatively describe the potential difference between points a and b.

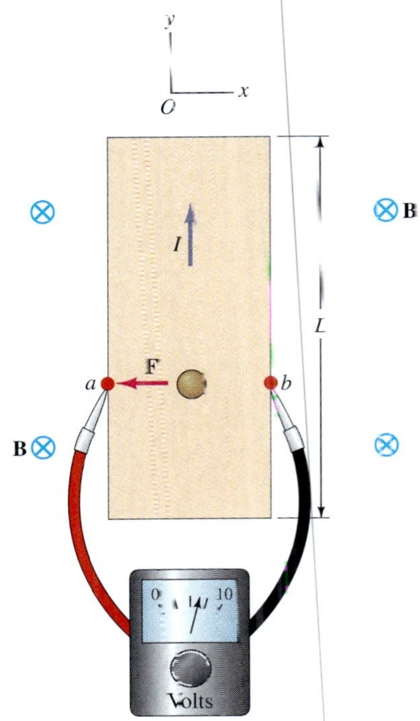

FIGURE 29–26 A conducting strip perpendicular to a constant magnetic field develops a potential called the Hall voltage between points a and b when the strip carries a current I.

The Hall effect allows us to determine the sign of the charge carriers in a metal.

Equation (29–19) gives the total force on the strip, $\mathbf{F} = I\,\mathbf{L} \times \mathbf{B}$. This force, by the right-hand rule, is directed in the $-x$-direction—it acts to the *left* in Fig. 29–26. By using the equivalent force law $\mathbf{F} = q\,\mathbf{v} \times \mathbf{B}$, we can show that the force on the charge carriers acts to the left, whatever the sign of the charge carriers. If the moving charges are positive (ions) and the current flows in the $+y$-direction, then the velocity of the charges is also in the $+y$-direction. According to the right-hand rule, $\mathbf{F} = q\,\mathbf{v} \times \mathbf{B}$ is then directed toward point a in Fig. 29–26. If the moving charges are negative (electrons), however, then the velocity of the charges is in the $-y$-direction. The vector product $\mathbf{v} \times \mathbf{B}$ is directed to the right, but q is negative, and $q\,\mathbf{v} \times \mathbf{B}$ again points to the left, moving these negative charges toward a. Either way, there is a buildup of the charge carriers at the left side of the strip. This buildup cannot continue indefinitely: Once enough charge carriers have moved to the left, they will supply a repulsive Coulomb force against the movement of other charge carriers there. An equilibrium is established once an electric potential is set up between points a and b that prevents further leftward drift of charge carriers. Charges then move up the strip as they would if there were no magnetic field. In fact, the charge separation leads to an electric field between points a and b. We therefore have crossed **E** and **B** fields, and the charges travel undeflected up the wire, just as in the velocity selector discussed in Section 29–3.

The sign of the potential difference between points a and b determines the sign of the charge carriers. This phenomenon is known as the **Hall effect**. If the charge carriers are negative, negative charges build up on the left side of the metal strip, and point a is at a lower potential than point b. Conversely, if the carriers are positive, positive charges build up on the left side of the strip, and point a is at a higher potential than point b. The first measurement of the sign of this *Hall potential* was performed by the physicist Edwin H. Hall in 1879. His measurement proved that the carriers of current in metals are negatively charged.

EXAMPLE 29–7 The apparatus of Fig. 29–26 that demonstrates the Hall effect sits in a magnetic field of 2.0 T. The width of the strip is 1.0 cm, and a voltage of magnitude 7.2 μV is measured across the strip. What is the speed v of the charge carriers in the strip?

APPLICATION: NONCONTACT DC SENSORS

Sensors are devices that measure the state of a system and can be used to send signals so that a response can be made to changes in the system. As such, they are key design elements in our technology. The Hall Effect has many industrial applications in the design of noncontact sensors. Such sensors are necessary when we must avoid material abrasion in a sensor involving physical contact, such as might occur in a button switch that is frequently used. The principle behind such sensors is that a change in the magnetic field in a Hall probe—the conducting strip across which we measure the Hall potential—gives rise to a corresponding change in the Hall potential. A change in this potential is a measurable effect. The advantage of a Hall Effect sensor is that it is also sensitive to very slow changes. This is not true for devices that rely on Faraday's law (see Chapter 31), which are sensitive only to relatively rapid changes in potentials. Here are some examples:

Computer keyboards that require constant and heavy use, such as in secretarial or textbook authors' offices, are designed using Hall Effect sensors. A small permanent magnet is attached to the bottom of each key on the keyboard. Underneath, there is a Hall probe. In modern design, such probes consist of a thin layer of conductor deposited on a rigid substrate that is often made of sapphire. The change in the position of the magnet relative to the conducting layer can, because of the $1/r^3$ dependence of the field, change the magnetic field at the probe by an order of magnitude. Such changes lead to changes in voltages which are then communicated to the computer, signaling that a given key has been pushed.

The Hall Effect is also used in the design of Rotation Position Sensing (RPS) devices. When the phase of a rotating shaft is to be measured—especially for slowly rotating systems—a permanent magnet is again attached to the shaft, and its distance from the Hall probe is measured by the resulting Hall current. This has uses in the design of antilock brakes. Permanent magnets on the four wheel shafts can pick up changes in the relative angular velocity of the rotating wheels, and thus signal the need for differential application of brakes.

Solution: We know the magnitude of the magnetic field and, from the voltage, we can find the magnitude of the crossed electric field. As we have already seen, the electric field set up across the strip is just such that the velocity-selector condition [Eq. (29–9)] holds, so the unknown speed is given by

$$v = \frac{E}{B}.$$

The electric field that is set up across the strip is determined from the electric potential, V, by

$$E = \frac{V}{d}$$

[recall Eq. (25–30)], where d is the width of the strip. Thus the charge-carrier's speed is

$$v = \frac{V}{Bd} = \frac{7.2 \times 10^{-6}\ \text{V}}{(2.0\ \text{T})(1.0 \times 10^{-2}\ \text{m})} = 3.6 \times 10^{-4}\ \text{m/s}.$$

This is a measurement of an electron's drift speed.

SUMMARY

Magnets, moving electric charges, and electric currents all experience magnetic forces. These forces can be described in terms of a magnetic field, **B**, whose spatial dependence can be mapped out with iron filings or by observing its effect on a moving electric test charge or a test current element. In terms of this field, the magnetic force on an electric charge q depends on the charge's velocity according to the magnetic force law,

$$\mathbf{F}_B = q\,\mathbf{v} \times \mathbf{B}. \tag{29–1}$$

The SI unit of magnetic field is the tesla, T: 1 T = 1 kg/(C · s). When both magnetic and electric fields are present, the Lorentz force law holds:

$$\mathbf{F} = q[\mathbf{E} + (\mathbf{v} \times \mathbf{B})]. \tag{29–4}$$

In a static magnetic field, the component of a charged particle's velocity parallel to the field is unaffected by that field. The magnitude of the force on the particle due to the field is proportional to the component of the velocity perpendicular to the field, and the direction of the force is perpendicular to this component of the velocity and to the field itself. It follows that the kinetic energy of a charged particle in a magnetic field is unchanging. When the field is constant, a charged particle traveling perpendicular to the field moves in a circle of radius

$$R = \frac{mv}{qB}. \tag{29–5}$$

The frequency of the particle's circular motion is the cyclotron frequency,

$$f = \frac{qB}{2\pi m}, \tag{29–7}$$

which is independent of the particle's velocity. The general path followed by a moving charge is a spiral around the magnetic field lines. Van Allen belts are regions near Earth where charged particles accumulate in spiraling paths under the influence of Earth's magnetic field.

When a charged particle has a particular velocity perpendicular to constant crossed electric and magnetic fields, the electric and magnetic forces cancel, and the particle passes through the fields undeflected. The magnitude of this special velocity is

$$v = \frac{E}{B}. \tag{29–9}$$

With the help of a velocity selector, an apparatus based in part on this phenomenon, the charge-to-mass ratio of the electron can be measured.

The infinitesimal magnetic force on an infinitesimal length of thin wire $d\boldsymbol{\ell}$ that carries a current I in the presence of a constant magnetic field is

$$d\mathbf{F} = I\, d\boldsymbol{\ell} \times \mathbf{B}. \tag{29–14}$$

To find the net force on a wire of finite length in a magnetic field, Eq. (29–14) is integrated. For example, the force on a straight wire of length L in a uniform magnetic field is given by Eq. (29–19), $\mathbf{F} = I\,\mathbf{L} \times \mathbf{B}$. A second example concerns a wire that carries a current and is formed into a loop (or coil) of N turns; the area of the face of the loop is A. When it is placed in a constant magnetic field, such a loop experiences a torque

$$\boldsymbol{\tau} = \boldsymbol{\mu} \times \mathbf{B}. \tag{29–22}$$

The loop reacts as a magnetic dipole with magnetic dipole moment μ. For a coil of N turns, μ has magnitude

$$\mu = INA \tag{29–24}$$

and direction perpendicular to the face of the coil, oriented by a right-hand rule on the current. The torque tends to rotate the loop so that $\boldsymbol{\mu}$ and $\mathbf{B}$ become aligned. The potential energy of the loop in a constant magnetic field can be expressed as

$$U(\psi) = -\boldsymbol{\mu} \cdot \mathbf{B}, \tag{29–26}$$

where ψ is the angle between $\boldsymbol{\mu}$ and $\mathbf{B}$.

The Hall effect exploits the equivalence between the force on a moving charge and the force on a current-carrying wire. This effect proves that the current carriers in metals are negatively charged.

UNDERSTANDING KEY CONCEPTS

1. A wire carrying a current is electrically neutral, yet a magnetic field acts on it. Why?
2. Explain how you might define and measure a magnetic field if magnetic monopoles existed.
3. An electron beam in an oscilloscope is deflected to the right on the screen. Could this be caused by an electric field *or* by a magnetic field? Explain how you could distinguish these possibilities.
4. An electron beam makes a spot in the center of the screen of a cathode-ray tube. A bar magnet is brought in from the left side (as seen from the front of the tube), with the S pole nearest to the beam. Which way will the spot move? Suppose that the N end of the bar magnet is brought near the beam from above. Which way will the spot move?
5. Much of the description of magnetic forces depends on the use of a right-hand rule. Does the magnetic force depend fundamentally on the fact that we have chosen the right rather than the left hand?
6. If you have just used a velocity selector for electrons and you wish to use it to choose positrons with the same speed, do you have to change any settings on the selector? Positrons are like electrons, but positively charged.
7. Induced charges give rise to electric forces even between electrically neutral objects. How do we know that the forces between bar magnets are not induced electric forces?
8. Imagine that an electrically neutral wire carrying a current moves in the presence of an external magnetic field. Do you expect that there will be an additional force on the wire due to the movement?
9. You have a fixed length of wire and want to use it to make a magnetic dipole with the largest possible magnetic dipole moment. Into what shape should you wind it? Are you better off making a single loop or N loops?
10. A small bar magnet forms a magnetic dipole; a current-carrying wire in the shape of a small loop also forms a magnetic dipole. If that is the case, the current loop should give rise to a magnetic field. Use this analogy to sketch the magnetic field lines that would be generated by such a current loop.
11. Consider two small circular current loops. Suppose the two loops are placed flat on a table close to each other (but not touching) and the two currents both flow in a counterclockwise direction. Will the two loops attract or repel? What happens if the directions of the currents are opposite?
12. Consider two small circular current loops. Suppose one loop is placed above the other (but not touching), with their areas oriented similarly. If the currents flowing in the two loops are both in the same direction, will the loops attract or repel?
13. Earth's magnetic south pole is near the geographic North Pole. Why would the geographic North Pole have been called a magnetic south pole?
14. Suppose that the coil of a direct-current electric motor consists of many turns rather than one turn of wire that carries a current I. Does the coil rotate faster than a single loop would? Does the split-ring commutator still work?
15. You have a large pail of water, a bar magnet with its N and S ends unmarked, a straight pin, and a cork. How could you make a compass? One of the things you need to know to construct this compass is how to distinguish north from south; you are allowed to watch the Sun to help with this part of the question.
16. Do magnetic north poles repel positive electric charges?
17. A classmate tells you that 1 T is 1 N/A · m. Is your classmate correct?

PROBLEMS

29–1 Magnets and Magnetic Fields

1. (II) Sketch the magnetic fields for the arrangements of bar magnets shown in Fig. 29–27.

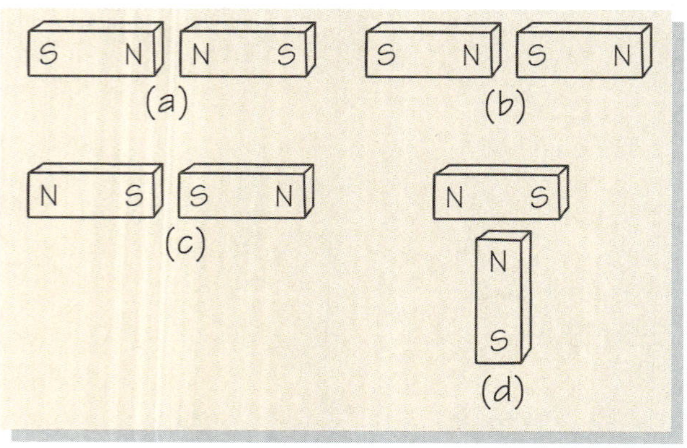

FIGURE 29–27 Problem 1.

2. (II) Consider the magnetic field generated by a current-carrying wire, as depicted in Fig. 29–2. Assuming that a reversal in the direction of the current also reverses the direction of the magnetic field, sketch the magnetic field due to two wires that are parallel to each other, and whose currents flow in the same direction. You may assume that the magnetic fields add vectorially, just like electric fields. Repeat the sketch for the situation in which the currents flow in opposite directions. Your sketches should show the field lines as seen by someone looking along the wires.

29–2 Magnetic Force on an Electric Charge

3. (I) An electron moving in the $-y$-direction enters a region of constant magnetic field and is observed to deflect to the $-x$-direction. What is the direction of the magnetic field?
4. (I) A proton with velocity $\mathbf{v} = (1.0 \times 10^6\ \text{m/s})\mathbf{i} + (2.0 \times 10^6\ \text{m/s})\mathbf{j} - (5.0 \times 10^5\ \text{m/s})\mathbf{k}$ moves through a magnetic field $\mathbf{B} = (0.2\ \text{T})\mathbf{i} - (0.3\ \text{T})\mathbf{j} + (0.4\ \text{T})\mathbf{k}$. Calculate the force on the proton.
5. (I) A proton of energy 100 keV moving in the $+x$-direction enters a region of uniform magnetic field perpendicular to the x-axis. Upon entry into that region, the proton experiences an acceleration of $3 \times 10^{12}\ \text{m/s}^2$ in the $+y$-direction. What are the magnitude and direction of the magnetic field?

6. (I) A pith ball charged to $+1\ \mu$C falls vertically at the equator. At this location, the magnitude of Earth's magnetic field is 0.5 gauss (0.5×10^{-4} T) and the field points to the north. When the ball falls to a speed of 5 cm/s, what is the magnetic force (magnitude and direction) on it?

7. (I) A proton moving with speed v enters a narrow (1.0 cm wide) region of magnetic field perpendicular to the direction of the proton's motion. As a result, the proton acquires a small component of speed perpendicular to its original direction of motion. This speed is much less than the original speed, and it is measured to be 3.3×10^5 m/s. What is the strength of the magnetic field?

8. (II) A cork ball carrying charge q has a mass of 0.6 g and is set in straight-line motion perpendicular to a uniform magnetic field of 0.03 T. What is the value of q if its direction of motion changes by 0.01° in 1.0 s?

9. (II) (a) A rapidly moving charged particle of charge e, mass m, and speed v passes through a region of magnetic field **B**, which points in a direction perpendicular to the motion. The particle spends a time interval Δt in the region. Estimate the angle θ through which it will be deflected during Δt, assuming that θ is small. (b) The particle is a proton, with $m = 1.7 \times 10^{-27}$ kg, and $e = 1.6 \times 10^{-19}$ C, and the speed is 1.4×10^7 m/s. The size of the magnetized region is 0.1 m across. How large must B be to give rise to a deflection of 0.1 rad?

10. (II) In an oversimplified model of Earth's magnetic field, the field is parallel to the rotation axis, has a constant magnitude of 10^{-4} T up to a height of 100 km, and then quickly drops to zero. A cosmic-ray particle with charge 1.6×10^{-19} C and mass 9.5×10^{-26} kg moves at a speed of 10^8 m/s directly toward the equatorial region from above. (a) In what direction is the particle deflected? (b) Estimate how much it will be deflected from the point of impact it would have if it were uncharged. (In fact, this is not a realistic example. Cosmic-ray particles as massive as this have greater charge.)

11. (II) Electrons travel at a speed of 6×10^7 m/s in a television tube. The electrons are affected by Earth's magnetic field. The tube, which is 0.4 m long, is located at a region where the magnetic field has a vertical component of 18 μT and a horizontal component of 24 μT (Fig. 29–28). If the initial direction of the electron beam is in the same direction as the horizontal component of Earth's magnetic field, in which direction, and by how much, is the electron beam deflected?

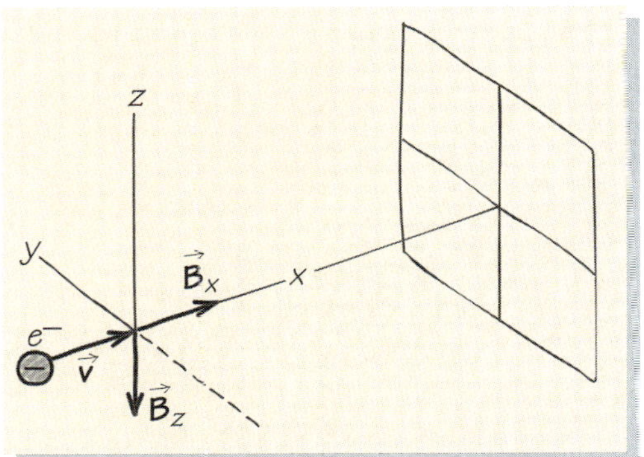

FIGURE 29–28 Problem 11.

29–3 *Magnetic Force on a Charge: Applications*

12. (I) A proton is sent into a region of constant magnetic field, oriented perpendicular to the proton's path. There the proton travels at a speed of 5×10^6 m/s in a circular path of radius 180 cm. What is the magnitude of the magnetic field?

13. (I) (a) Suppose that electrons from an electron gun with a voltage of 1600 V are injected into a region of constant magnetic field perpendicular to the electrons' velocity. What magnetic field will give the electrons a radius of curvature of 6 cm? (b) A magnetic field of what magnitude is necessary to give an alpha particle (charge $q = 2e$ and mass $m_\alpha = 7360$ times the mass of an electron) with a kinetic energy of 1200 eV a path with radius of curvature of 20 cm?

14. (I) Show that the radius of curvature of a proton moving at a velocity of 25 km/s in a magnetic field of 10^{-10} T is small compared to interplanetary distances. The protons therefore spiral around interplanetary magnetic field lines; we say that the protons are "tied to the magnetic field lines" in cosmic magnetic fields.

15. (I) The magnetic field at the surface of a neutron star has magnitude 3×10^7 T. What is the radius of the circular orbit of an electron that moves there at 0.1 percent of the speed of light? What is the magnitude of the magnetic force on the electron?

16. (I) With what frequency will deuterons, which have the same charge as protons but twice the mass, circulate in a cyclotron with a magnetic field of 1.2 T?

17. (I) If we want to triple the cyclotron frequency associated with a proton accelerator from an initial value of 6.1 MHz, what quantity must we change, and from what initial value to what final value?

18. (I) In a certain region, the average radius of curvature of the trajectory of electrons trapped in the Van Allen belt is 300 m and the average electron energy is 100 keV. What is the value of Earth's magnetic field in this region?

19. (I) Electrons of speed 10^6 m/s and protons of speed 10^4 m/s perpendicularly enter a region of constant magnetic field 10^{-5} T above Earth. What are the radii of their orbits? Why is the proton's radius greater? If the proton's speed were the same as that of the electron (10^6 m/s), what would be the radius of its orbit?

20. (II) The proposal for the Superconducting Supercollider envisaged that circulating protons would move in a ring of radius 13.8 km by means of magnetic fields of magnitude 5 T. What is the magnitude of the momentum of a proton that moves in this way? For protons with this momentum, the energy is given to excellent accuracy by the formula $E = pc$, where c is the speed of light, about 3×10^8 m/s. Calculate the energy of the proton in megaelectron-volts; 1 MeV = 1.6×10^{-13} J.

21. (II) Assume that the electrons in a television picture tube have an energy of 10 keV and move perpendicularly to Earth's magnetic field (see Table 29–1). (a) Calculate the final velocity (vector) of an electron when it hits the screen if the horizontal distance the electron travels is 40 cm. (b) What is the deflection (distance) of the electron perpendicular to its original direction?

22. (II) Earth acts as a giant magnet whose field lines are like those of a bar magnet, running from the magnetic north pole to the magnetic south pole. Thus the magnetic field at the equator is approximately constant, of magnitude 5×10^{-5} T, and runs from the geographic South Pole to the geographic North

Pole (Fig. 29–29). If we ignore air resistance and the gravitational force, a charged object could orbit Earth at the equator as a result of the magnetic force if it has just the right velocity. Suppose that such an object has a charge of -1 mC and a mass of 1.0 g. (a) What would its velocity have to be for it to travel in such an orbit? (b) Suppose that the gravitational force acts on this object as well. What is the ratio of the gravitational force to the magnetic force?

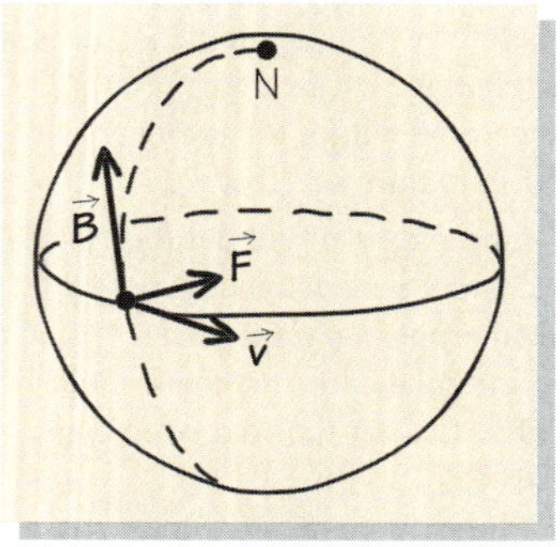

FIGURE 29–29 Problem 22.

23. (II) In Section 29–3, we calculated the critical momentum for an electron to stay within the galaxy. (a) Given that the energy of a high-energy particle is related to its momentum by $E = pc$ (see Problem 20), what is the energy of an electron with the critical momentum? What are the critical momentum and energy for (b) an alpha particle (charge $2e$ and mass four times the mass of a proton)? (c) an ion of uranium, with charge e and mass 240 times the proton mass?

24. (II) A proton moves horizontally perpendicular to a constant magnetic field oriented so as to deflect the proton upward. The magnitude of the field is 1 mT. What is the speed of the proton so that the magnetic force just cancels the gravitational force on the proton, leaving it in horizontal flight? This problem illustrates how very weak the gravitational force is compared to electromagnetic forces.

25. (II) An electron moves at a speed $v = 3 \times 10^5$ m/s in a region of constant magnetic field of magnitude 0.12 T. The direction of the electron when it enters this region is at 40° to the field, and the electron follows a helical path. When you look along the direction of the magnetic field, the path is a projected circle. How far has the electron traveled along the direction of **B** when one projected circle has been completed?

26. (II) An electron enters a bubble chamber that contains a constant magnetic field of strength 0.10 T and follows a helical path. The spacing between the turns of the path is 3.0 mm, as is the radius of the circular part of the path. Find the components of the velocity parallel and perpendicular to the field.

27. (II) A vacuum tube contains two axial cylinders (Fig. 29–30). The potential difference between these cylinders is 500 V. Some electrons are released from the inner cylinder and are accelerated by the electric field toward the outer cylinder; a small current thereby travels through the tube. Suppose that a uniform magnetic field is set up parallel to the axis of the tube. This curves the trajectories of the electrons and, at a critical magnetic field, the electrons will no longer reach the outer cylinder, and the current ceases to flow. What is the kinetic energy of the electrons that hit the outer cylinder at a magnetic field just below the critical value?

28. (II) A proton and an alpha particle, which has twice the charge and four times the mass of the proton, are each accelerated through the same potential difference and enter a region of constant magnetic field perpendicular to their paths. (a) What is the ratio of the radii of their orbits? (b) What is the ratio of the frequencies of their orbits?

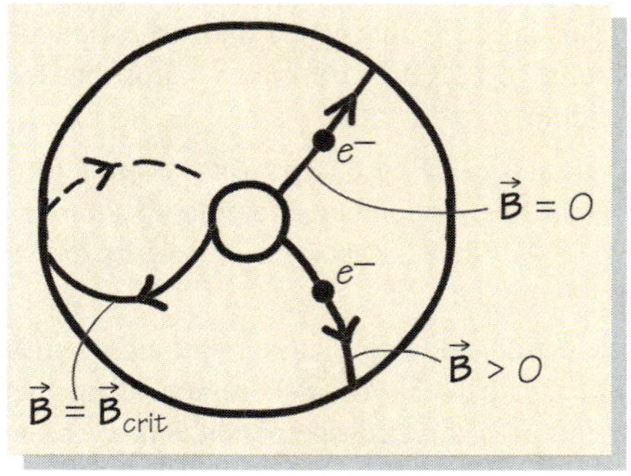

FIGURE 29–30 Problem 27.

29. (II) An electron is injected at $t = 0$ s with velocity $\mathbf{v}_0 = (2 \times 10^6\ \text{m/s})\mathbf{i}$ into a region with parallel electric and magnetic fields $\mathbf{E} = (1500\ \text{V/m})\mathbf{j}$ and $\mathbf{B} = (-0.2\ \text{T})\mathbf{j}$, respectively. Calculate the subsequent motion.

30. (II) You want to be able to tune a velocity selector such that you have the capacity to select electrons that have been accelerated from rest by a potential that runs from 1200 V to 12,000 V. If the magnetic field, B, is fixed at 0.1 T, what range of electric field strengths must be available? If the electric field strength were fixed at 100 V/cm, what range of magnetic field strengths must be available?

31. (II) Figure 29–31 is a schematic diagram of a cyclotron. A charged particle starts out at the central point and, for a given magnetic field perpendicular to the plane of motion, follows a circular path. The cyclotron takes advantage of the fact that the time for the particle to execute a half-circle is independent of the particle's velocity. An alternating voltage is applied across the gap between the two "dees" (the semicircular regions), so that, when the particle crosses the gap, the voltage acts to accelerate it. When the particle gets to the gap again after having completed a half-circle, the voltage has changed sign, and the particle is once again accelerated. The frequency of the oscillating voltage must match the cyclotron frequency. In this way, the particle is always accelerated, completing ever bigger circles in the same time, until the beam is extracted at the maxi-

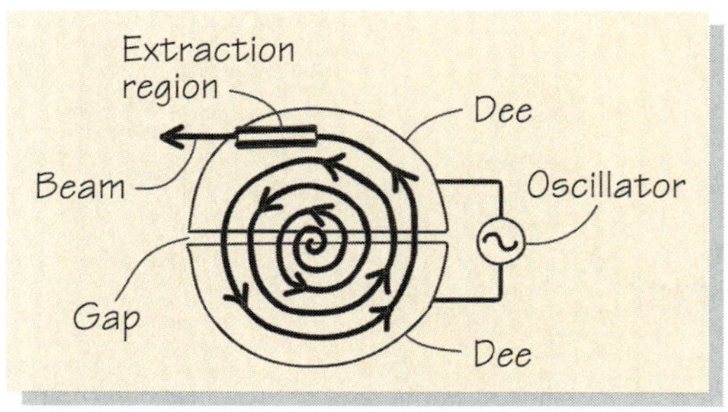

FIGURE 29–31 Problem 31.

mum radius. (a) If the magnetic field has strength 1 T and the circulating particle is a proton, $q = +e$ and $m = 1.7 \times 10^{-27}$ kg, what is the cyclotron frequency? (b) What is the maximum velocity of the proton for a maximum radius of 50 cm? (c) the corresponding maximum kinetic energy? (d) If the maximum voltage across the gap is 50 kV, how many full circles does the proton make before it reaches its maximum energy? (e) How much time does the proton spend in the accelerator?

32. (II) A cyclotron used for accelerating protons has a magnetic field of magnitude 1.7 T. The circular region in which the magnetic field exists has a radius of 40 cm. (a) What is the cyclotron frequency? (b) What is the largest kinetic energy that a proton accelerated in this machine can have? (c) Repeat parts (a) and (b) for a doubly ionized helium nucleus $^4He^{2+}$, with four times the mass of a proton and twice the charge.

33. (II) The particle accelerator at Fermilab, the Fermi National Accelerator Laboratory in Batavia, Illinois, can accelerate protons to relativistic speeds. The accelerator is circular and holds the protons in circular paths by increasing the strength of a magnetic field perpendicular to this path as the protons' momentum increases. (The momentum increases because the protons pass repeatedly through regions of electric potential.) The radius of the main Fermilab accelerator is 6.2 km, and the magnets are capable of maintaining magnetic field strengths between 1 T and 4.5 T. Given that the magnitude of a proton's electric charge is 1.6×10^{-19} C, what range of momenta can be accommodated in this accelerator? Because protons of such momenta are highly relativistic, their energies are given by the approximate relativistic formula $E = pc$. What range of energies can be reached at Fermilab? What would the speed of a baseball, mass 0.5 kg, be if it had the energy of the most energetic protons at Fermilab? (For the baseball, use the normal nonrelativistic formulas that relate energy and speed.)

34. (II) A proton, with charge q_p and mass m_p, is accelerated through an electric potential V. The proton then enters a region of constant magnetic field **B** oriented perpendicular to its path. In this region, the proton's path is circular with radius of curvature R_p. Another particle with the same charge as the proton but with mass m_x follows under the same conditions. Its radius of curvature in the magnetic field, R_x, is 1.4 times as large as R_p. What is the ratio of m_x to m_p?

The device we have described is a type of *mass spectrometer*, which can be used to identify a material by the masses of that material's constituent molecules (Fig. 29–32). Sometimes, instead of a simple electrostatic potential as in our example, a velocity selector of crossed **E** and **B** fields is used to select particles of a given speed.

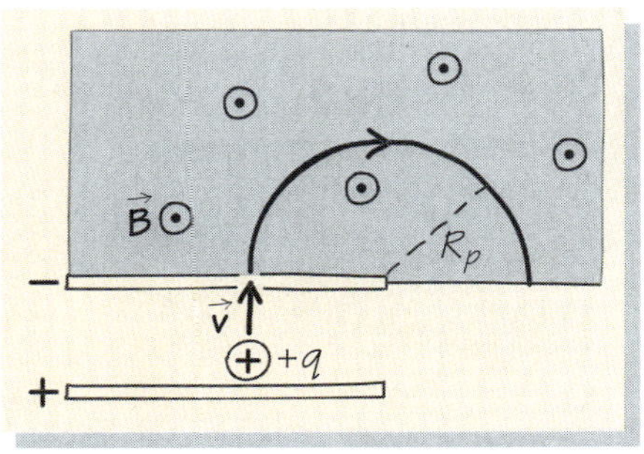

FIGURE 29–32 Problem 34.

35. (II) The apparatus shown in Fig. 29–33 is designed to measure the energy of alpha particles emitted by a radioactive source. (Alpha particles have a mass roughly four times the proton mass and a charge that is twice the proton charge.) The source is placed at the entrance of a channel that forms a quarter of a circle. A uniform magnetic field is applied perpendicular to the plane of the channel. Alpha particles with a specific velocity will make their way through the channel and be detected at the exit. All others will strike the walls and be lost. What is the range of values of B necessary to analyze alpha particles whose energies range up to 6 MeV?

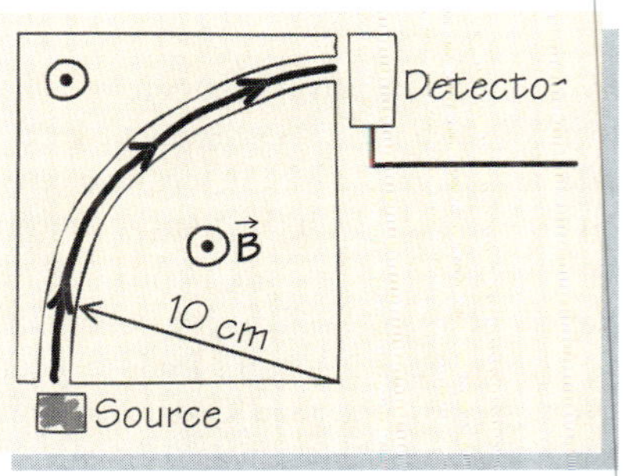

FIGURE 29–33 Problem 35.

36. (II) A 50-MeV proton, moving in the x-direction, enters a region in which there is a magnetic field. The proton experiences an acceleration of 10^{12} m/s² in the y-direction. What can you say about the magnetic field?

37. (III) A particle of mass m and charge $-q$ moves in a circular orbit of radius R about a fixed charge Q. The angular frequency for the orbit is given by

$$\omega_0^2 = \frac{qQ}{4\pi\epsilon_0 mR^3}.$$

A uniform magnetic field of magnitude B in a direction perpendicular to the plane of the orbit is turned on. As a result, the angular frequency is changed to $\omega_0 + d\omega$. Assuming that B is sufficiently small so that products of B and $d\omega$ can be neglected, calculate $d\omega$.

38. (III) You have an apparatus that can form an electric field of 2000 N/C and a magnetic field of 0.3 T. You want to build a velocity selector to select electrons of speed 2×10^4 m/s. (a) Draw the orientation of your apparatus, showing **E**, **B**, and **v**. (b) What are the minimum and maximum values of v that you can select? [*Hint*: Set the apparatus up so that **v** and **B** are not perpendicular to each other (Fig. 29–34).]

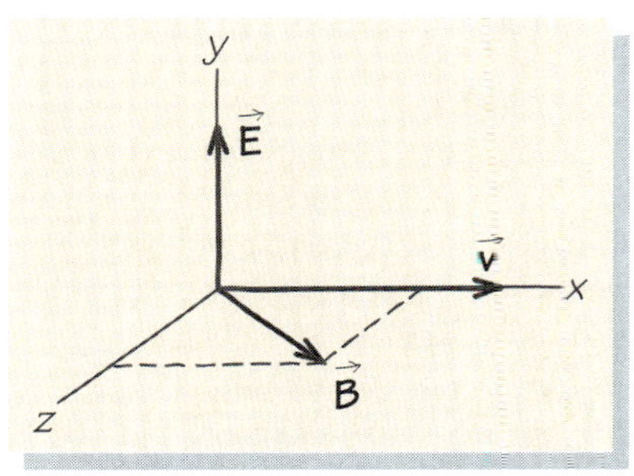

FIGURE 29–34 Problem 38.

29–4 Magnetic Forces on Currents

39. (I) A straight wire segment is placed in a region known to contain a constant magnetic field of unknown strength. A current of 6 A runs through the wire, which can be turned in various directions until the force per unit length acting on it takes on a maximum value of 0.18 N/m. What is the value of the magnetic field?

40. (I) A long wire carries a current of 8 A. A bar magnet is brought near the wire so that the charge carriers, of speed 3×10^{-4} cm/s, experience a magnetic field of 0.7 T perpendicular to their direction of motion. Calculate the force (a) on each moving charge carrier (electron) and (b) on a 1-m length of the wire.

41. (I) A thin, straight wire carries a current of 10 mA and makes an angle of 60° with a constant magnetic field of magnitude 10^{-6} T. The portion of the wire in this field has a length of 10 cm. Calculate the force, both direction and magnitude, on this segment of the wire.

42. (I) The length of a vertical lightning conductor from roof to ground is 20 m. A lightning stroke leads to a current of 10^4 A flowing through the conductor. Given that Earth's magnetic field is horizontal and of magnitude 0.5×10^{-4} T at the location of the building, what is the force on the conductor during the period the current flows?

43. (I) A straight wire is placed in a uniform magnetic field of magnitude 0.010 T. The direction of the field makes an angle of 30° with that of the wire, which carries a current of 10 A. What is the force on a 1.0-m segment of the wire?

44. (II) A current I flows through a circular wire loop of radius R that lies in the xy-plane (Fig. 29–35). Consider a constant magnetic field of magnitude B that points in the x-direction. Calculate the force on an element of the loop formed by an angle $d\theta$, located at an angle θ from the $+x$-axis.

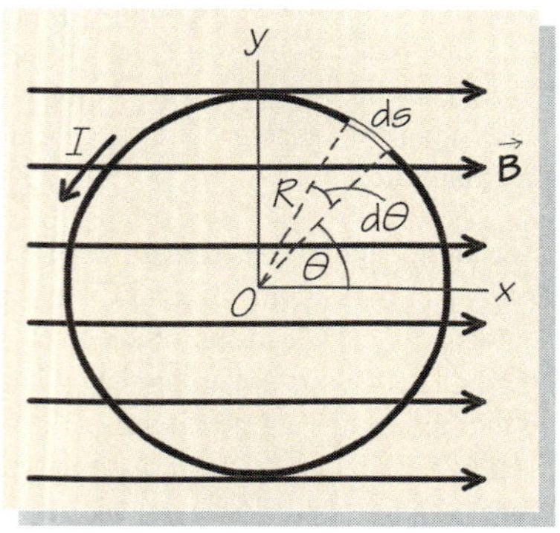

FIGURE 29–35 Problem 44.

45. (II) A wire of length L is suspended from two springs of spring constant k attached to a current source (Fig. 29–36). A magnetic field B, in a horizontal direction perpendicular to the wire (out of the page), is turned on; a current I then flows in the wire, which moves to a new equilibrium position. Which way will the wire move, and by how much?

46. (II) In a physics lecture demonstration, a thick copper wire of length 0.2 m and mass 10 g is attached to two thin wires and suspended so that it is horizontal (Fig. 29–37). A 0.05 T magnetic field pointing in the vertical direction is turned on. What angle will the supporting wires make with the vertical if a current of 5 A flows through the wire?

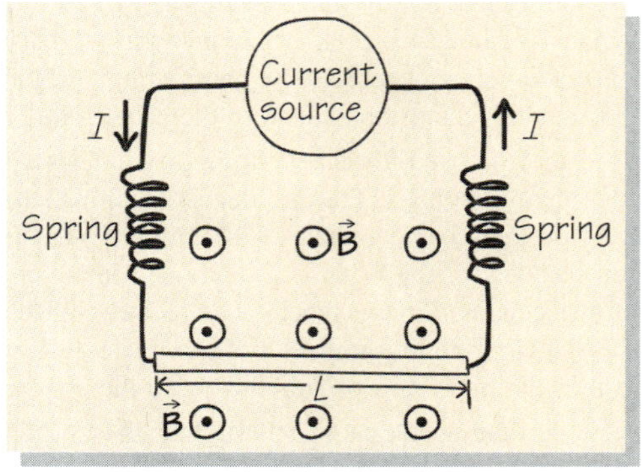

FIGURE 29–36 Problem 45.

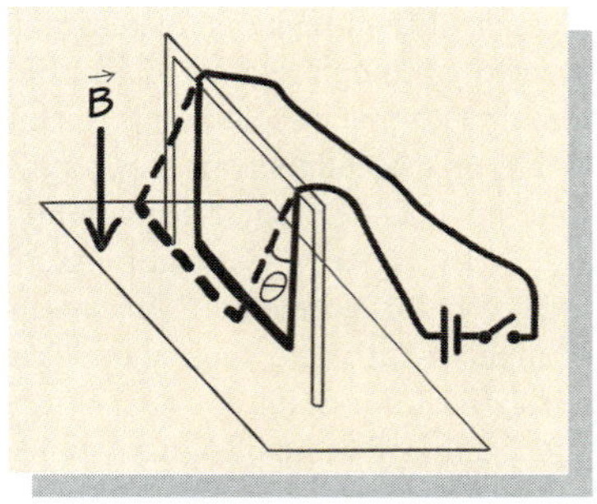

FIGURE 29–37 Problem 46.

47. (II) Figure 29–38 shows a possible device for measuring magnetic fields. A loop carrying a current I is dipped into a region of magnetic field. The loop is suspended from a spring of spring constant k that stretches if the magnetic field points in a certain direction. Here the loop has width $\ell = 1.2$ cm, $I = 100$ mA, the spring stretches 0.6 cm, $k = 5 \times 10^{-2}$ N/m, and the magnetic field is uniform. What is the magnitude of the field? How could such a device be used, or modified, to measure fields that are not uniform?

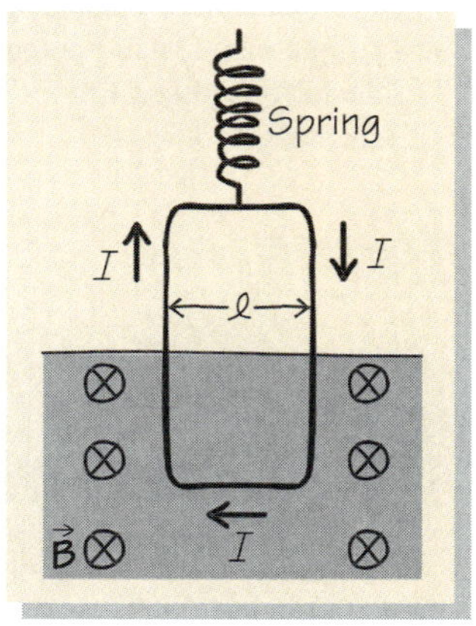

FIGURE 29–38 Problem 47.

29–5 *Magnetic Force on Current Loops*

48. (I) A wire coil of area 20 cm^2 with 180 turns experiences a maximum torque of 2×10^{-2} N · m when placed in a magnetic field of 0.3 T. What is the current through the coil?

49. (I) A rectangular wire loop of height 5 cm and width 3 cm consists of 60 turns and carries a current of 1.2 A. What are the magnitude and direction of the magnetic dipole moment? If a uniform magnetic field of 0.5 T is applied to the loop, and the field's direction makes an angle of 26° with the normal to the current loop, what is the torque (magnitude and direction) that acts on the loop?

50. (I) A circular coil of diameter 5.0 cm, consisting of 500 turns of wire, carries a current of 80 mA. How much work must be done to flip the coil through 180° when it is placed in a uniform magnetic field of 0.15 T? The field makes an initial angle of 50° with the direction of the coil's dipole moment.

51. (I) A wire forms a circular coil of N turns and radius R and carries a current I. The coil's magnetic dipole moment is initially aligned with a fixed external magnetic field, **B**. How much work must be done by an external torque to rotate the coil through an angle θ?

52. (I) A current loop of area 3.0 cm^2, carrying a current of 5.0 A, is placed in a uniform magnetic field of 0.25 T such that the normal to the loop is perpendicular to the direction of the magnetic field. There is a torque, and the loop changes direction. Because of friction in the bearings, it settles to the minimum energy orientation. How much energy was dissipated in the process?

53. (I) An atom can have a magnetic dipole moment of 10^{-23} J/T. Such an atom is placed in a magnetic field of 10 T. What is the range of potential energies involved?

54. (II) A wire carrying a current I splits into two channels of resistance R_1 and R_2, respectively, forming a circuit. The wire enters the space between the two poles of a magnet with a uniform magnetic field that runs from one pole piece to the other (Fig. 29–39). The circuit forms a loop; the field lies in the plane of the loop. What is the torque on the circuit about the wire axis, given that the wires are a distance d apart and that the length of the split is L?

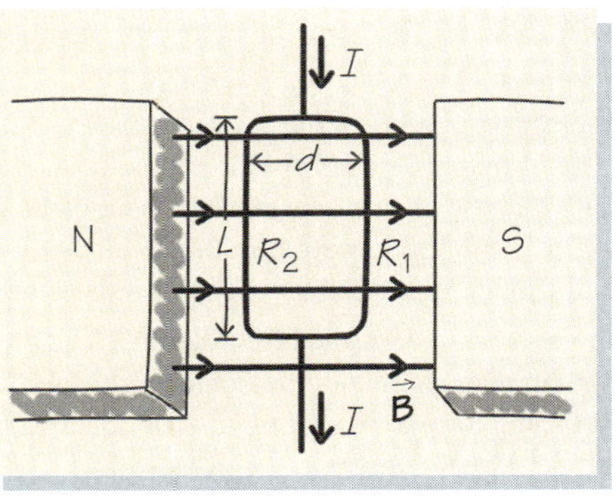

FIGURE 29–39 Problem 54.

55. (II) A circular wire coil of area 6 cm^2 has 50 turns. When the coil is placed in a magnetic field of 0.2 T, the maximum torque is 3×10^{-5} N · m. (a) What is the current in the coil? (b) What work is required to rotate the coil 180° in the magnetic field? Does the work depend on the initial angle?

56. (II) An electric motor consists of a current-carrying wire loop in a constant magnetic field **B** (Fig. 29–40). The field produces a torque that tends to rotate the loop so that the loop's magnetic dipole moment, $\boldsymbol{\mu}$, and **B** become aligned. When that happens, a split-ring commutator reverses the current direction, so that $\boldsymbol{\mu}$ changes its orientation by 180°, and the torque acts to continue the rotation. Suppose that $\boldsymbol{\mu}$ and **B** start out almost antiparallel. Plot the magnitude of the torque as a function of the angle between $\boldsymbol{\mu}$ and **B**, as this angle runs from −180° to 0°. At 0° the commutator reverses the current. Plot the torque through another half turn. What is the average value of the torque through a full turn if the current in the motor is 2.2 A, the magnitude of **B** is 0.10 T, and the area of the loop is 80 cm^2?

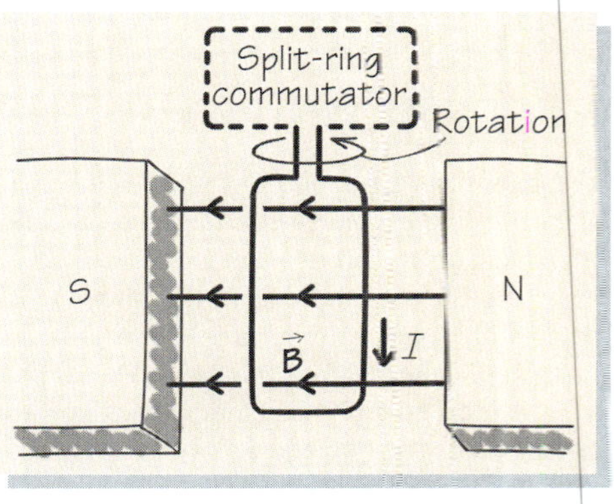

FIGURE 29–40 Problem 56.

57. (II) An electron, of charge $q = -1.6 \times 10^{-19}$ C, has a "size" of about 3×10^{-15} m, called its classical radius. The magnetic dipole moment of the electron is roughly 10^{-23} A · m^2. (a) Suppose that this magnetic moment were due to the entire charge q orbiting at the classical radius. What would the speed of the charge be to generate this magnetic moment? (b) Suppose that the electron's magnetic moment were perpendicular to a magnetic field of magnitude 1 T. What is the torque on the electron?

58. (II) (a) Calculate the magnetic dipole moment of a single atom, based on the following model: One electron travels at speed 2.2×10^6 m/s in a circular orbit of diameter 10^{-10} m. (b) The individual atomic magnetic dipoles of magnetic materials (such as iron) are preferentially lined up to point in the same direction. If a fraction f of the dipoles are so aligned along the long axis (with the rest oriented randomly so that their magnetic dipole moments add vectorially to zero), what is the net magnetic dipole moment of a piece of such material 1 cm^2 in area and 10 cm long? (The material may be viewed as an array of cubes, each of which contains one atom and is 10^{-10} m on a side.) (c) What is the torque experienced by the piece of material in part (b) in a field of 10^{-3} T when the magnetic field is directed at right angles to the long axis of the material?

59. (III) The current loop shown in Fig. 29–41 lies in the xy-plane and consists of a straight segment α of length $2R$ in the x-direction and a semicircular segment β, which has a radius of curvature R. There is a constant magnetic field of strength B

into the page. (a) Compute the magnetic force on segment α. (b) Find the magnetic force on segment β. You may wish to use symmetry arguments to simplify your task. (c) Add the results of parts (a) and (b) to find the net force on the loop. (d) How could you generalize your results to a loop of any shape in the xy-plane?

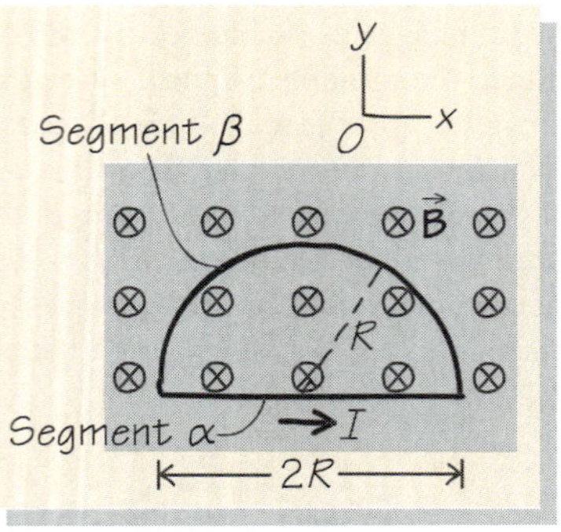

FIGURE 29–41 Problem 59.

60. (III) We showed in Example 29–5 that, when a current loop with magnetic dipole moment $\boldsymbol{\mu}$ is displaced slightly from perfect alignment of $\boldsymbol{\mu}$ and a magnetic field **B**, the rotational motion of the current loop due to the torque of the field is harmonic, with angular frequency $\omega = \sqrt{\mu B / I_M}$. I_M is the rotational inertia about the rotation axis. Calculate the kinetic energy, K, associated with this motion, and show that the sum of potential energy and kinetic energy is a constant.

61. (III) A coil carrying current $I = 50$ mA has moment of inertia $I_M = 7.5 \times 10^{-7}$ kg $\cdot$ m^2 about a rotational axis and an area of 6.0×10^{-4} m^2. The coil is placed in a magnetic field of magnitude 0.6 T, displaced 5° from alignment between its magnetic dipole moment, $\boldsymbol{\mu}$, and the field, and released from rest. Describe the subsequent motion. What is the maximum angular speed of the coil in that motion?

29–6 The Hall Effect

62. (II) Suppose that the strip of metal used in the apparatus that demonstrates the Hall effect has a cross section of width w and depth d_0. (The width is the space across which the Hall voltage ΔV is measured.) Show that the density n of charge carriers with charge e is independent of the width and is given by $n = IB/(d_0 e\, \Delta V)$. Knowing the density of carriers, find an expression for the drift speed as measured by a Hall apparatus.

63. (II) The probe that demonstrates the Hall effect is used to measure the density of charge carriers in an unknown sample of metal. A sample of the material 1.5 mm thick is placed in a magnetic field of 1.2 T. When a current of 1.8 A passes through the material, a Hall voltage of 6.2 μV is measured. What is the density of charge carriers?

64. (II) A Hall-effect probe can be used to measure the magnitude of a magnetic field. A researcher has lost her instruction booklet and forgotten the calibration procedure. However, when she places the Hall probe inside a known magnetic field of 1500 G, she measures a Hall voltage of 127 mV. What is the field of a magnet with a Hall voltage of 664 mV?

General Problems

65. (II) The wire coil of a galvanometer has an area of 2 cm^2 and 500 turns. The coil is placed in a magnetic field of magnitude 0.18 T and oriented so that its plane is initially parallel to the field. The restoring torque of the galvanometer spring is proportional to the angular deflection, with a proportionality constant of 10^{-8} N $\cdot$ m/° (see Example 29–5). What current corresponds to a deflection of 70°?

66. (II) The masses of atomic ions of known charge can be precisely measured by finding the time an atom takes to complete a circular trajectory in a known magnetic field. With a magnetic field of magnitude 1 T and an apparatus capable of measuring times to an accuracy of 10^{-9} s, how accurately can the mass of an ion with charge $+e$ be measured in 1 rev? If the mass is to be measured to an accuracy of 10^{-30} kg, how many revolutions must be measured?

67. (II) When an electron orbits a proton, the smallest circular orbit is one with a radius of about 0.5×10^{-10} m, the Bohr radius. The proton's electric field must have what magnitude to make the electron follow this orbit? Compare the magnitude of the magnetic field that would be required to make an electron move in a circle of the same radius at the speed it would have if it were orbiting a single proton.

68. (II) A massive charge Q is fixed at the origin of a coordinate system. A magnetic field **B** points in the $+z$-direction. A light particle of charge q and mass m moves in a circular orbit of radius r about the origin. For what value of **B** (as a function of r) is such motion possible, if Q and q have the same sign and if the angular momentum of the motion is a fixed constant L?

69. (II) Consider a parallel-plate capacitor with charge density $\pm 8.0 \times 10^{-7}$ C/m^2 on the two plates and an electric field that points in the $+z$-direction. What magnetic field is necessary to provide a velocity selector for 60-keV deuterons that move in the $+y$-direction? A deuteron has a mass of 3.2×10^{-27} kg and a charge of 1.6×10^{-19} C; 1 keV = 1.6×10^{-16} J.

70. (II) A narrow beam of particles of mass m and charge q travels in free space at speeds between v_1 and v_2. It enters a region of length L with a constant magnetic field that points perpendicular to the beam direction and parallel to the boundary between the field-free region and the region with the field. There, it follows a circular path with radius of curvature R until it exits that region. Show that the beam widens when it emerges from the region with the field, and calculate the spread in terms of a range of angles.

71. (II) An electron moving in the xy-plane is subject to forces due to a constant magnetic field **B** that points in the $+z$-direction. Assuming that the electron loses 10 percent of its energy after 20 turns, as a consequence of frictional forces, what will the fractional change in the radius of the orbit be after 20 turns?

72. (II) For the motion described in Problem 71, (a) what will the fractional angular-momentum change be during the 20 turns? (b) What is the torque exerted by the frictional forces in terms of the initial kinetic energy?

73. (II) Electrons are injected into a region with a constant magnetic field **B** by an electron gun with known voltage V. The electrons move in a plane perpendicular to **B** and follow an arc of radius R. Determine the charge-to-mass ratio e/m for the electrons in terms of the given parameters.

74. (II) Particles with mass $M_A = A(1.6 \times 10^{-27}$ kg) and charge $q = 1.6 \times 10^{-19}$ C are accelerated by a potential difference of

2×10^4 V and directed perpendicularly into a region of uniform magnetic field of strength 0.2 T. The region with the field is 30 cm deep. Calculate the angular deflection of the particles, θ, as a function of A.

75. (II) N electrons move at speed v in a circular orbit of radius R. (a) What is the angular momentum of the system of electrons? (b) the magnetic dipole moment associated with the current loop? (c) the ratio of the quantities in parts (a) and (b)?

76. (II) A rectangular wire loop of width a and height b is connected to a current source that, when turned on, gives rise to a current I in the wire. The loop is suspended in a uniform magnetic field **B** that points in a vertical direction (Fig. 29–42), and it would hang vertically if there were no current. We assume that the wire is massless, but two masses m are suspended at the lower corners. What is the angle θ at which the loop is in equilibrium? Calculate this in two ways: by using torques, and by expressing the potential energy as a function of θ and minimizing it. What happens if the direction of the current is reversed?

77. (III) Suppose that an experimental apparatus can have both electric and magnetic fields constant in magnitude and direction. In this apparatus a proton moving at a speed of 5.0×10^4 cm/s in the $+z$-direction does not accelerate, whereas a proton moving at a speed of 8.0×10^4 cm/s with no x-component at an angle of 42° with respect to the z-axis experiences an initial acceleration of magnitude 3.5×10^8 m/s^2 in the $-x$-direction. A proton moving in the xy-plane has a circular orbit. Find the values of **E** and **B** in the apparatus.

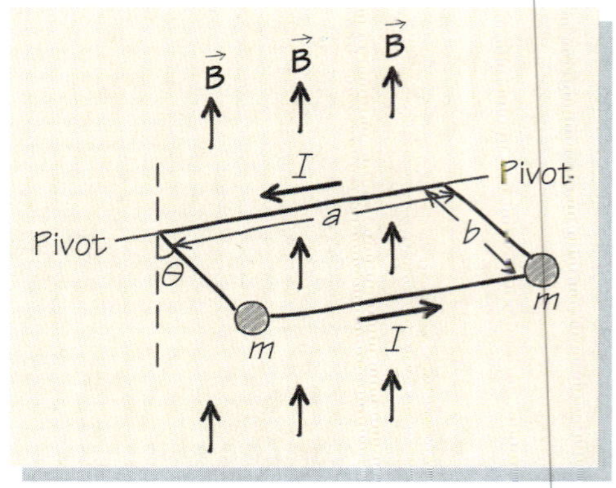

FIGURE 29–42 Problem 76.

CHAPTER

30

In the tokamak, an experimental device for the study of nuclear fusion-generated power, magnetic fields are used to contain a gas of positive ions. These fields, and the windings of wire that produce them, are topologically complex.

The Production and Properties of Magnetic Fields

We have seen that magnetic fields influence moving charges and current-carrying wires. But how are magnetic fields produced? We already know that electric charges exert forces on each other, and that electric charges produce electric fields. In this chapter, we shall see that moving charges and currents exert magnetic forces on each other, so that magnetic fields are *produced* by moving charges or, equivalently, by currents. We describe and explore the ways in which these magnetic fields are produced. We shall learn about Ampère's law (including Maxwell's generalization) and the Biot–Savart law, both of which describe the magnetic fields produced by moving charges or currents. We shall also begin an exploration of the intimate relation between electric and magnetic fields that will eventually lead to Maxwell's equations and to an understanding of the phenomenon of light.

30–1 AMPÈRE'S LAW

During the winter of 1819 to 1820, Hans Christian Oersted discovered that electric currents influence compass needles (Fig. 30–1). Until this discovery, there was only a suspicion of a connection between electricity and magnetism. Oersted, as well as André-Marie Ampère, soon showed that *current-carrying wires exert forces on each other.*

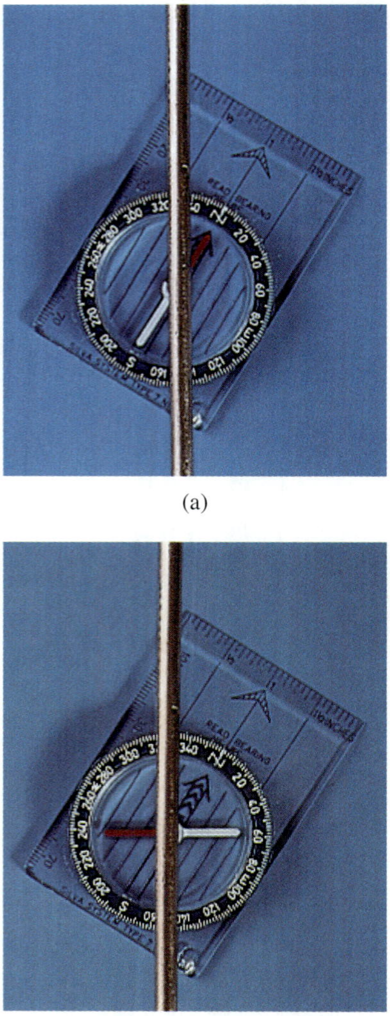

(a)

(b)

FIGURE 30–1 (a) The compass needle continues to point north when there is no current in the wire, but (b) when current flows through the vertical wire, the needle reacts to the magnetic field produced by the current.

Because such wires are everywhere electrically neutral, these forces are not electric. We recall from Chapter 29 that a current-carrying wire aligns iron filings on a plane perpendicular to the wire in a circular pattern (Fig. 29–2c). This suggests that a current-carrying wire is a source of a magnetic field. The force acting on one current-carrying wire in the presence of another is actually due to the magnetic field generated by the second wire. In this section, we shall find an expression for this magnetic field.

The Magnetic Field of a Straight Wire

In order to find the magnetic field produced by a single straight wire, we observe the force between two parallel current-carrying wires. We then interpret this force by saying that one wire is the source of a magnetic field, and that the force on the other wire is due to this field. Let's take two wires with parallel straight segments of length L that are labeled 1 and 2, respectively. The forces between the wires weaken so rapidly as the separation between the wires, d, increases that we need worry only about forces between segments closest to one another. When the currents are parallel, the wires attract each other (Figs. 30–2a, b); when the currents are antiparallel, they repel (Figs. 30–2c, d).

Suppose now that the wire segments are long, $L >> d$. In Fig. 30–3a, the force on wire 2, which is directed to the left, is due to the magnetic field of wire 1. Equation (29–19) describes the force on a segment of current-carrying wire. Using this equation, the force on wire 2 is of the form

$$\mathbf{F}_2 = I_2\,\mathbf{L}_2 \times \mathbf{B}_1, \tag{30–1}$$

provided only that the field $\mathbf{B}_1$ due to wire 1 is the same all along wire 2. This assumption is justified for long wires. Here, the vector $\mathbf{L}_2$, of magnitude L, is oriented

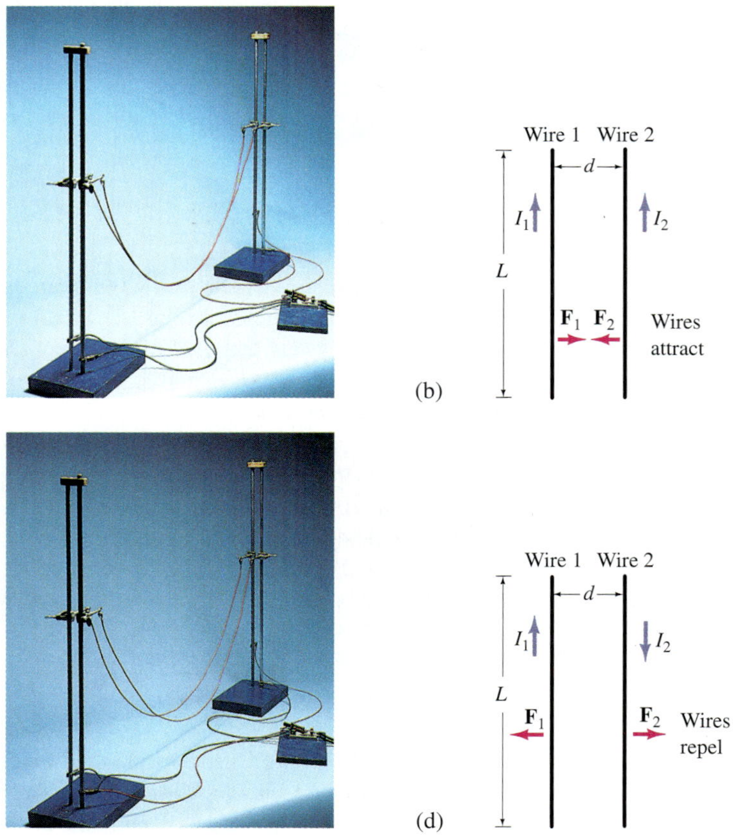

(a) (b) (c) (d)

FIGURE 30–2 Two parallel wires that carry currents exert forces on one another; these forces are larger when the wires are closer. (a) and (b) The currents are parallel, and the forces are attractive. (c) and (d) The currents are antiparallel, and the forces are repulsive.

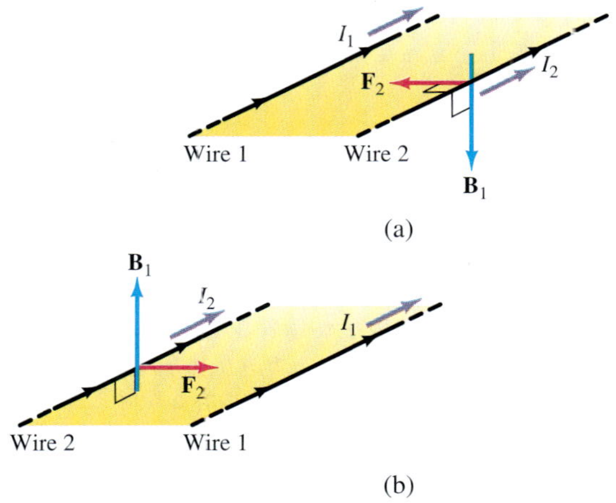

FIGURE 30–3 Determining the direction of the magnetic field due to wire 1. Currents I_1 and I_2 are parallel to each other. (a) According to the right-hand rule, $\mathbf{B}_1$ due to wire 1 is directed down when wire 2 is to the right of wire 1. (b) $\mathbf{B}_1$ due to wire 1 is directed up when wire 2 is to the left of wire 1.

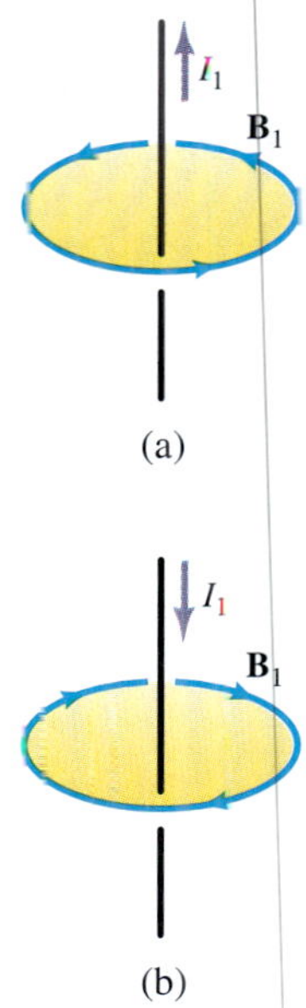

FIGURE 30–4 (a) The magnetic field due to wire 1, $\mathbf{B}_1$, traces out a circle around the wire in the direction shown. (b) If the current in wire 1 were reversed, the orientation of $\mathbf{B}_1$ would change as determined by a right-hand rule.

along the direction of I_2. As we move wire 2 around wire 1—always holding them parallel and at the same separation—we find that the force is always attractive and always has the same magnitude. Equation (30–1) and this observation about the force on wire 2 imply that *the magnetic field* $\mathbf{B}_1$ *due to wire 1 traces out a circle around wire 1*. Equation (30–1) shows that force $\mathbf{F}_2$ is insensitive to any component of $\mathbf{B}_1$ that is *parallel* to the wires because the vector product of two parallel vectors is zero. As for the component of $\mathbf{B}_1$ perpendicular to wire 2, application of a right-hand rule in Eq. (30–1) shows that $\mathbf{B}_1$ must be directed down when wire 2 is in its original position to the right of wire 1 (Fig. 30–3a). If wire 2 is moved to the left of wire 1, however, field $\mathbf{B}_1$ at wire 2 will be directed up because the two wires continue to attract each other (Fig. 30–3b). By using this argument for other positions, we find that, as we suspected, the magnetic field $\mathbf{B}_1$ due to wire 1 traces out a circle around wire 1 (Fig. 30–4a; see Fig. 29–2c). Another important result: If we reverse the direction of the current in wire 1, then the force on wire 2 also reverses. A repeat of the experiments just described would show that the magnetic field lines again trace circles around wire 1, but *in the opposite direction* (Fig. 30–4b).

We can demonstrate with one further experiment that there is no component of $\mathbf{B}_1$ oriented parallel to wire 1. If we wrap wire 2 in a circle around wire 1, as in Fig. 30–5, a component of $\mathbf{B}_1$ parallel to wire 1 would cause a force to be exerted on wire 2. As Eq. (30–1) indicates, such forces would tend to expand or contract the circle traced by wire 2, and this effect would be measurable. In this configuration, wire 2 experiences no forces. The magnetic field $\mathbf{B}_1$ about wire 1 indeed traces out circles around the wire.

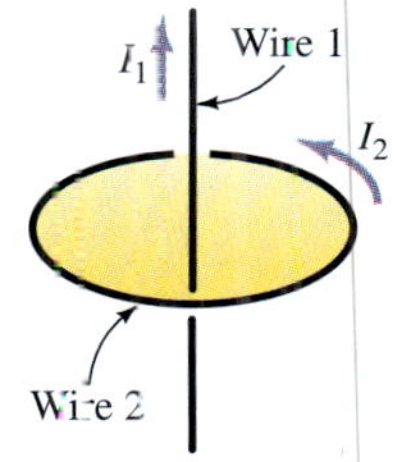

FIGURE 30–5 If wire 2 traced a circle around wire 1, it would react to any components of $\mathbf{B}_1$ due to wire 1 that are parallel to wire 1. No such forces are found.

We can summarize all this by saying that the direction of **B** is determined by a right-hand rule (Fig. 30–6):

If the thumb of the right hand is oriented along the direction of current flow in a wire, the fingers curl in the direction of the magnetic field.

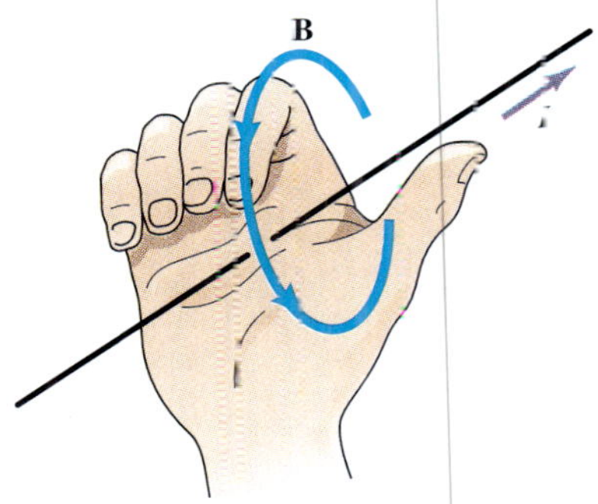

FIGURE 30–6 A right-hand rule determines the direction of the magnetic field around a current-carrying wire.

Having found the direction of the magnetic field produced by the current, how do we find its magnitude? We find this quantity by measuring the magnitude of the force between the wires. Such measurements show that the magnitude of the force between two parallel, straight segments of wire is

$$F = \frac{CI_1I_2L}{d},$$

where I_1 and I_2 are the currents in wires 1 and 2, respectively, d is the separation between the wire segments, and L is their length. The proportionality constant C depends on how we define the units of current. Conversely, if we use a *defined* proportionality constant, the force between two current-carrying wires determines the units of current. In the SI, we choose this latter alternative: C is defined according to

$$F = \frac{\mu_0 I_1 I_2 L}{2\pi d}, \tag{30–2}$$

where the constant μ_0, called the **permeability of free space**, is defined to be

$$\mu_0 \equiv 4\pi \times 10^{-7}\ \mathrm{T \cdot m/A}. \tag{30–3}$$

The definition of the ampere

With this definition of μ_0, 1 A is defined as the current that travels in two long, parallel wires of length L that are 1 m apart, such that the attractive force between them is $(2 \times 10^{-7}\ \mathrm{N/m})\ L$. Is this result consistent with other definitions already given? In Chapter 22, we defined the coulomb as the charge on two pointlike objects such that there is a certain force between them. We provisionally defined 1 A as 1 C/s in Chapter 27. The definition of the coulomb in terms of a force between charges depends on another defined constant, ϵ_0, in exactly the same way that the definition of the ampere depends on μ_0. Thus, for our relations to be consistent, *the definitions of μ_0 and ϵ_0 must be consistent*. If both μ_0 and ϵ_0 are defined, the same must be true for their product, which is given by

$$\mu_0 \epsilon_0 \equiv c^{-2} = (2.99792458 \times 10^8\ \mathrm{m/s})^{-2}. \tag{30–4}$$

The constant c is precisely the speed of light! We shall see in Chapter 35 why this is so.

Comparison between Eqs. (30–1) and (30–2) shows that a long, straight wire that carries a current I gives rise to a magnetic field whose magnitude at a distance r from the wire is

The magnetic field produced by a long, straight wire

$$B = \frac{\mu_0 I}{2\pi r}. \tag{30–5}$$

Ampère's Law

We can find a more universal form for the magnetic field produced by a current by expressing Eq. (30–5) differently. Imagine a line integral over the magnetic field, **B**, that follows a circular path of radius r around a wire. The integration path thus follows the direction of **B**. This path, labeled C, is shown in Fig. 30–7. The sign $\oint$ indicates that the path of the line integral goes all the way around the circle, or is closed. For the path chosen, **B** and the infinitesimal distance element d**s** of the integral are parallel, so $\mathbf{B} \cdot d\mathbf{s} = B\, ds$. Because B is a constant when the distance r from the wire is constant,

$$\oint \mathbf{B} \cdot d\mathbf{s} = B \oint ds = B(2\pi r). \tag{30–6}$$

The factor $2\pi r$ is the length of the path, which is the circumference of the circle of radius r. If we use Eq. (30–5), we find that

$$\oint \mathbf{B} \cdot d\mathbf{s} = \frac{\mu_0 I}{2\pi r} 2\pi r = \mu_0 I. \tag{30–7}$$

Equation (30–7) includes a right-hand-rule convention in which path C must be in the direction of the fingers of the right hand when the thumb is oriented along I.

A current passes through the path described. Let's now consider a path through which *no* current passes. In particular, consider path C' shown in Fig. 30–8. Path C' is broken into the segment from a to b, the nearly full circle C_2, the segment from c to d, and the nearly full circle C_1. We wish to compute $\oint_{C'} \mathbf{B} \cdot d\mathbf{s}$. The total contribution

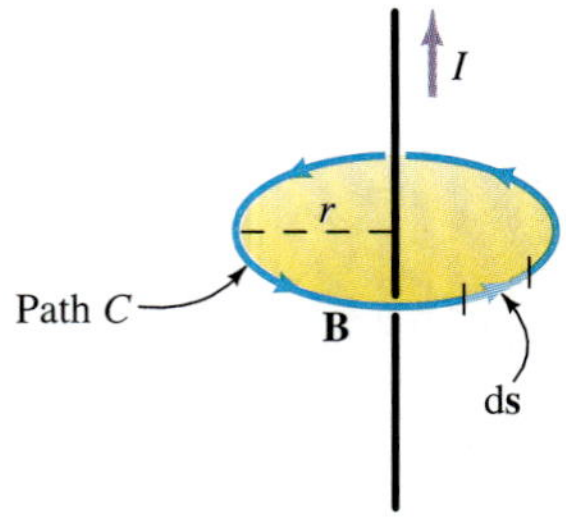

FIGURE 30–7 Path C circles a current-carrying wire at a constant distance r from the wire and follows the direction of the magnetic field, **B**, around the wire.

of the two paths from a to b and c to d is zero, because **B** is perpendicular to the path there. Thus

$$\oint_{C'} \mathbf{B}\cdot d\mathbf{s} = \int_{C1} \mathbf{B}\cdot d\mathbf{s} + \int_{C2} \mathbf{B}\cdot d\mathbf{s}$$
$$= -B_1(2\pi r_1) + B_2(2\pi r_2), \qquad (30\text{–}8)$$

where B_1 is the magnitude of the magnetic field at a distance r_1 from the wire and B_2 is the magnitude of the field at a distance r_2. The first term is negative because **B** is oriented opposite to the path direction on segment C_1. From Eq. (30–5), we see that the two contributions to Eq. (30–8) cancel:

$$\oint_{C'} \mathbf{B}\cdot d\mathbf{s} = -\frac{\mu_0 I}{2\pi r_1}(2\pi r_1) + \frac{\mu_0 I}{2\pi r_2}(2\pi r_2)$$
$$= -\mu_0 I + \mu_0 I = 0. \qquad (30\text{–}9)$$

What is the difference between Eq. (30–7) and Eq. (30–9)? Earlier, the path C enclosed current I, whereas we see from Fig. 30–8 that path C' encloses no current. We have taken the first steps toward the generalization that follows. Let the quantity I_{enclosed} be the total current enclosed by *any closed path*. Then

$$\oint \mathbf{B}\cdot d\mathbf{s} = \mu_0 I_{\text{enclosed}}, \qquad (30\text{–}10)$$

where the integral is taken around that closed path. Equation (30–10), which was formulated by Ampère in the 1820s during his extensive work on magnetism, is known as **Ampère's law**. The direction of the loop integral must be specified: If the fingers of the right hand curl in the same sense as the integral path, the thumb points in the direction a positive current takes in passing through the loop. The total current may thus include both positive and negative contributions. Remember that *the path does not have to be circular, just closed.*

The generalization we have made includes an experimental result that is worth pointing out: *The magnetic fields produced by different currents superpose*, just as the electric fields of different charges add according to the superposition principle.

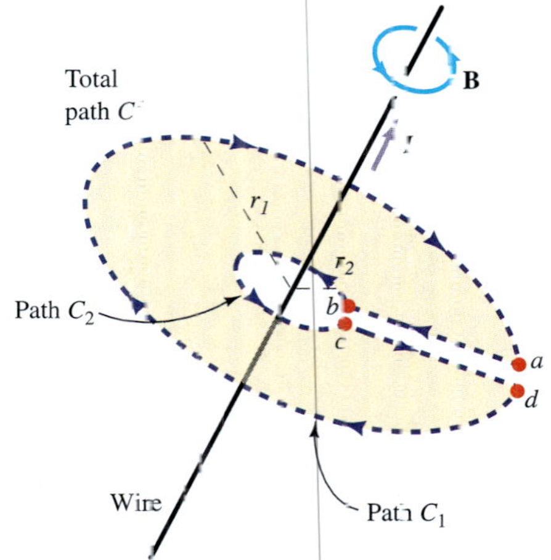

FIGURE 30–8 Path C' consists of a clockwise circle C_1 of radius r_1, a leg from a to b that moves in to a distance r_2 from the wire, a counterclockwise circle C_2 of radius r_2, and a leg from c to d that moves out to r_1. The magnetic field follows a counterclockwise circle.

Ampère's law. The magnetic field produced by a current obeys this law.

Using Ampère's Law

If there is some symmetry that suggests that the integral over a particular path is simple, then Ampère's law [Eq. (30–10)] can be used to *find* the magnetic field, in analogy with the way we use Gauss' law to find electric fields. In the case of Gauss' law, the integral is taken over a closed surface, and **E** is related to the electric charge enclosed. In the case of Ampère's law, the integral taken is along a closed path, and **B** is related to the electric current enclosed by the path.

EXAMPLE 30–1 The current, I, within a wire whose cross section has radius R is known to be distributed uniformly over that cross section. What is the magnetic field as a function of the distance r from the wire's axis outside the wire and what is it within the wire?

Solution: The wire looks the same as we move around it (it has cylindrical symmetry). Therefore we expect any magnetic field not to vary with the angle around the wire, but to be a function only of the radial distance r from the central axis. If we apply Ampère's law—Eq. (30–10)—for a circular path of radius r centered on the middle of the wire, **B** will be the same all along this path; we can use information about the current enclosed by the path to determine the field as a function of r. The amount of current enclosed depends on whether the path lies outside or inside the wire.

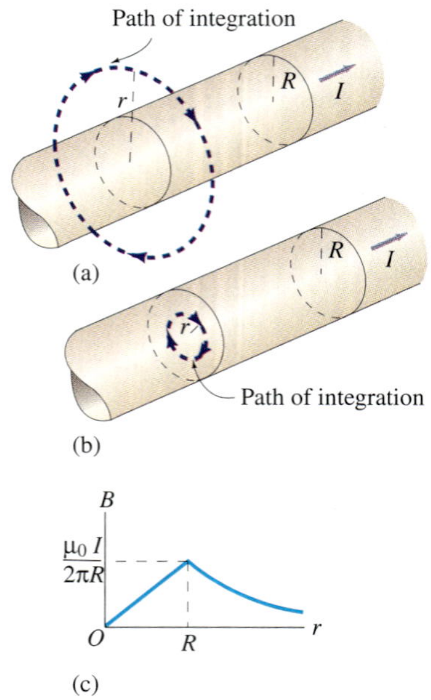

FIGURE 30–9 (a) Example 30–1. A circular path of radius r is used to determine the magnetic field outside a wire that carries a current I. (b) A similar path inside the wire. (c) The magnitude of the magnetic field versus r.

Figure 30–9a shows the path *outside* the wire that will determine the field outside. By the right-hand rule, **B** is oriented in the direction of the path, so $\mathbf{B}\cdot d\mathbf{s} = B\,ds$. The magnetic field magnitude is constant over the chosen path and thus comes out of the integral, leaving just the circumference of the path. The current enclosed is the *total* current carried by the wire. Thus Ampère's law becomes

$$\oint \mathbf{B}\cdot d\mathbf{s} = \oint B\,ds = B\oint ds = B(2\pi r) = \mu_0 I.$$

We can solve for B to find that

$$B = \frac{\mu_0 I}{2\pi r}.$$

Note that the magnetic field outside the wire is independent of the size of the wire, just as the electric field outside a spherically symmetric charge distribution is independent of the size of the distribution.

We continue to use symmetry to find the field inside the wire, but this time we take our circular path *inside* the wire (Fig. 30–9b). The current enclosed by the path is I times the ratio of the area of the circle of radius r to the area of the wire:

$$I_{\text{enclosed}} = I\frac{\pi r^2}{\pi R^2}.$$

As before, Ampère's law gives

$$\oint \mathbf{B}\cdot d\mathbf{s} = B(2\pi r) = \mu_0 I\left(\frac{\pi r^2}{\pi R^2}\right).$$

If we solve for B, we find that

$$B = \frac{\mu_0 I}{2\pi R^2}\,r.$$

By analogy with Gauss' law for electricity, any current outside a circle of radius r makes no contribution to the magnetic field at radius r. Inside the wire, the magnetic field decreases linearly to zero as r approaches zero. As a check, we see that the results for outside and inside the wire agree at $r = R$. Figure 30–9c is a graph of the magnitude of the magnetic field.

30–2 GAUSS' LAW FOR MAGNETISM

Gauss' law for electricity expresses an important relation obeyed by the flux of electric field; that is, by an integral over an area of the electric field through that area. Gauss' law relates the electric charge contained within a closed surface to the flux through that closed surface. In this section, we shall look at an analogous expression for magnetic fields. Static electric fields begin and end on electric charges. *Unlike electric field lines, magnetic field lines form closed curves.* If magnetic charges analogous to electric charges existed, then magnetic field lines would originate and terminate on magnetic charges—just as electric field lines originate and terminate on electric charges. Despite much experimental effort, however, *magnetic charges have never been discovered.* Thus, there is nothing on which magnetic field lines can begin or end.

Magnetic charges have never been discovered. Consequently, magnetic field lines are always closed.

In the absence of magnetic charges, a relation like Gauss' law for electricity holds for magnetism, but with the electric charge replaced by zero. Let's define the **magnetic flux**, Φ_B, for a magnetic field **B** over a surface S, open or closed, by

The magnetic flux

$$\Phi_B(S) \equiv \int_{\text{surface } S} \mathbf{B}\cdot d\mathbf{A}. \qquad (30\text{–}11)$$

APPLICATION: CLIP-ON AMMETERS

It is not always easy to insert an ammeter directly into a circuit. We may want to measure the current carried by a bundle of many wires, as in a telephone cable; we may want to measure the current coming out of a large power station; or, we may need to measure the current in a water pipe (local differences in potential cause such currents, which lead to electrolytic corrosion where different metals meet—for example, a brass fitting on a copper pipe).

A clip-on ammeter (Fig. 30AB–1) utilizes Ampère's law by sensing the magnetic field that surrounds a current-carrying conductor such as a cable. The heart of such a device is in the mechanism by which the magnetic field is measured. In its simplest form, it is a permanent dipole magnet pointer mounted on a center point bearing like a magnetic compass, but with a return hairspring that aligns the dipole in the direction of the clip, and therefore along the axis of the cable. When a current is present in the cable, the magnetic field generated orthogonal to the cable deflects the pointer. The deflection angle depends on the torque induced by the magnetic field—and hence the current—and on the opposing torque due to the hairspring. Such ammeters can be purchased in automobile parts stores; it is often necessary to measure the current through a starter motor because such motors become partially short-circuited by oil damage to the insulation. In more sophisticated versions of the clip-on ammeter, the magnetic field sensor is a Hall effect sensor (see the Application Box in Chapter 29), and, in the most accurate instruments, a so-called flux gate is used.

FIGURE 30AB–1 This clip-on ammeter uses Faraday's law to measure the current through the wire it surrounds, here 5.0 A. The jaws of the ammeter can be opened to allow its positioning.

Then **Gauss' law for magnetism** is

$$\text{for a closed surface:} \qquad \Phi_B = \int_{\text{closed surface}} \mathbf{B} \cdot d\mathbf{A} = 0. \qquad (30\text{–}12)$$

Gauss' law for magnetism

As for the electric flux, infinitesimal surface elements $d\mathbf{A}$ are perpendicular to the surface and, for a closed surface, are oriented outward. Another way of stating Gauss' law for magnetism is that the number of magnetic field lines that enter a closed surface, minus the number that leave the surface, is zero (Fig. 30–10). Any magnetic field line entering a closed surface must leave it somewhere because there are no magnetic charges on which magnetic field lines can begin or end.

The SI unit for magnetic flux is the unit of magnetic field times area; that is, tesla times square meters ($\text{T}\cdot\text{m}^2$). This unit occurs often enough to be given its own name in SI, the **weber** (Wb), after Wilhelm Eduard Weber:

$$1\ \text{Wb} \equiv 1\ \text{T}\cdot\text{m}^2. \qquad (30\text{–}13)$$

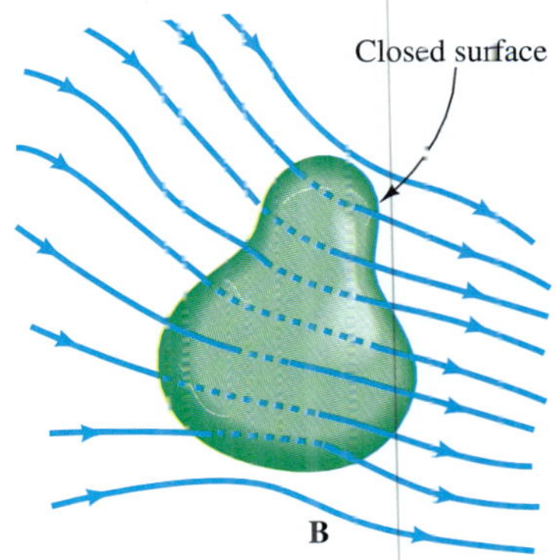

FIGURE 30–10 Magnetic field lines are everywhere parallel to the magnetic field; their density measures the field's strength. There are no magnetic charges, so magnetic field lines do not end, and the magnetic flux through a closed surface is zero. This is Gauss' law for magnetism.

The Field Lines of a Bar Magnet

When we drew the magnetic field lines for a bar magnet in Chapter 29, it may have seemed natural to think of the field lines as starting on the north pole and ending on the south pole. In light of what we have just learned, we now realize that magnetic field lines never start or stop—they are *continuous*. Therefore, we must reconsider our view of the magnetic field lines for a bar magnet. In fact, the field lines do not start or stop at the poles *but pass through the bar magnet*. The field lines that run from the north pole to the south pole outside the magnet return within the magnet to form closed

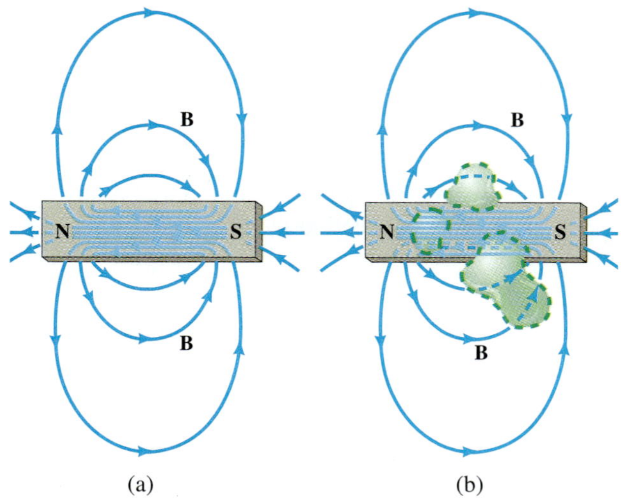

FIGURE 30–11 (a) The magnetic field lines of a bar magnet are continuous, forming closed loops. They continue within the magnet, running from the south pole to the north pole. (b) The continuity of the magnetic field lines follows from Gauss' law for magnetism: The same number of magnetic field lines that enter any closed surface will leave it.

loops (Fig. 30–11a). This view is consistent with Gauss' law for magnetism, Eq. (30–12), because, for any closed surface that can be drawn in and around a bar magnet, the same number of field lines enter the surface as leave it. Three surfaces are shown in Fig. 30–11b. The form of the magnetic field lines outside a bar magnet is just like the electric field lines outside an electric dipole, which consists of equal and opposite electric charges. However, if we look *between* the charges of the electric dipole and compare the electric field there to the magnetic field within a bar magnet, there is a crucial difference. As Fig. 30–12 shows, the fields are in different directions, and hence the fluxes are of different signs in the two cases.

Using Gauss' Law

Gauss' law for magnetism is useful for limiting the forms a magnetic field may take. As an example, let's use Gauss' law for magnetism to show that the magnetic field around a straight current-carrying wire can have no radial component. We need a suitable closed (imaginary) surface to construct about the wire in order to exploit any symmetries. For a straight wire, cylindrical symmetry is appropriate. Our closed surface will therefore be a right cylinder of radius R and length L whose central axis lies on the wire (Fig. 30–13). The wire looks the same from any point on the surface of the

FIGURE 30–12 Although electric field lines outside an electric dipole resemble magnetic field lines outside a bar magnet, the fields within are quite different. The electric field lines begin and end on the electric charges, whereas the magnetic field lines are continuous.

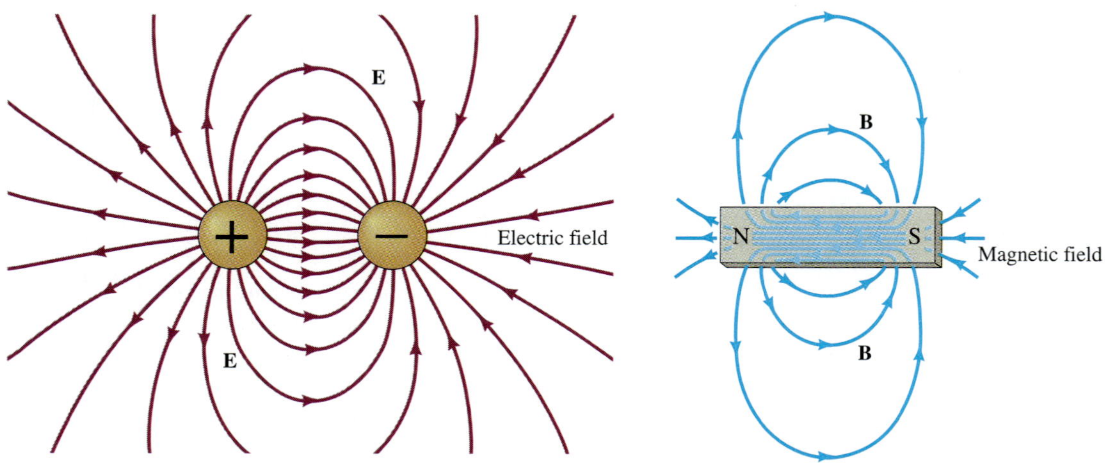

cylinder, so the magnetic field cannot depend on the angle around the axis of the cylinder. Thus, if there were a radial component B_r at some fixed radial distance, it would have to be the same all around the wire.

To find the net magnetic flux through the closed cylinder, Φ_B, we must consider contributions from its ends and sides. Only a longitudinal component of **B** (along the wire) would contribute to the flux at the ends. But the contribution from one end must cancel the contribution from the other end—if the longitudinal component enters the surface at one end, it must leave the surface at the other end. The flux through the ends is therefore zero.

The contribution to the magnetic flux from the sides is due to the radial component of the field, B_r. We have for the net flux

$$\Phi_B = \int_{\text{end surface}} \mathbf{B}\cdot d\mathbf{A} + \int_{\text{side surface}} \mathbf{B}\cdot d\mathbf{A} = \int_{\text{side surface}} \mathbf{B}\cdot d\mathbf{A} = B_r(2\pi RL),$$

where $2\pi RL$ is the area of the cylinder's sides. This must equal zero by Gauss' law and, because the area of the cylinder sides is not zero, B_r must be zero. We have shown by Gauss' law that there can be no radial component of the magnetic field.

We shall see in later chapters that the magnetic flux plays a central role in other fundamental laws of electromagnetism. It is therefore important to know how to calculate the magnetic flux for both closed and open surfaces. Example 30–2 is an exercise of this type.

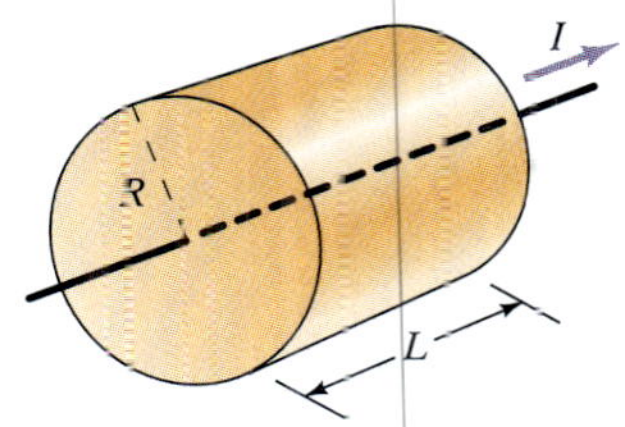

FIGURE 30–13 A Gaussian surface that exploits the symmetry of a long, straight wire.

EXAMPLE 30–2 The region between the poles of a tabletop electromagnet contains a constant magnetic field, $B = 0.0030$ T, oriented in the $+x$-direction. A square wire loop of sides $L = 1.0$ cm is oriented at a 30° angle to the field (Fig. 30–14a). Find the magnetic flux through the loop.

Solution: To find the flux through a given surface, we must specify an area element d**A** of the surface, find its scalar product with the magnetic field, and integrate over the entire surface. Figure 30–14b shows that the surface element d**A**, which is perpendicular to the wire loop, makes an angle $\theta = 60°$ with **B**. Thus $\mathbf{B}\cdot d\mathbf{A} = B$

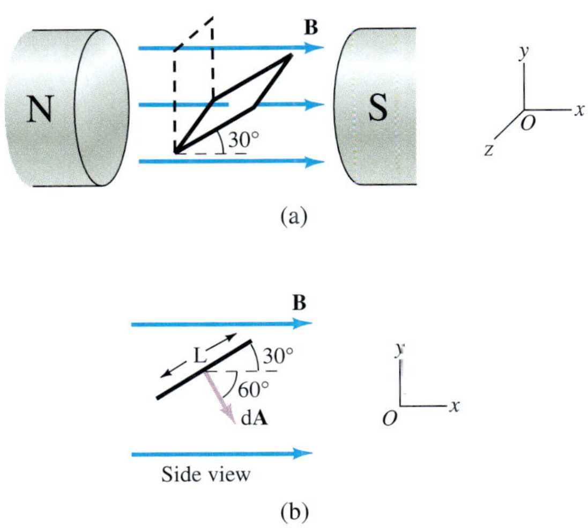

FIGURE 30–14 (a) Example 30–2. A tabletop electromagnet for which **B** is oriented in the $+x$-direction. (b) The surface element d**A** of the square wire loop is oriented perpendicular to the surface.

$\cos\theta\, dA$. Because B and θ are constants, they can be removed from the area integral in Eq. (30–11) for the magnetic flux:

$$\Phi_B(S) = \int_{\text{surface } S} \mathbf{B}\cdot d\mathbf{A} = \int_{\text{surface } S} B\cos\theta\, dA = B\cos\theta \int_{\text{surface } S} dA = B\cos\theta\, L^2.$$

The numerical value is

$$\Phi_B(S) = (0.0030\text{ T})(\cos 60°)(1.0 \times 10^{-2}\text{ m})^2 = 1.5 \times 10^{-7}\text{ Wb}.$$

30–3 SOLENOIDS

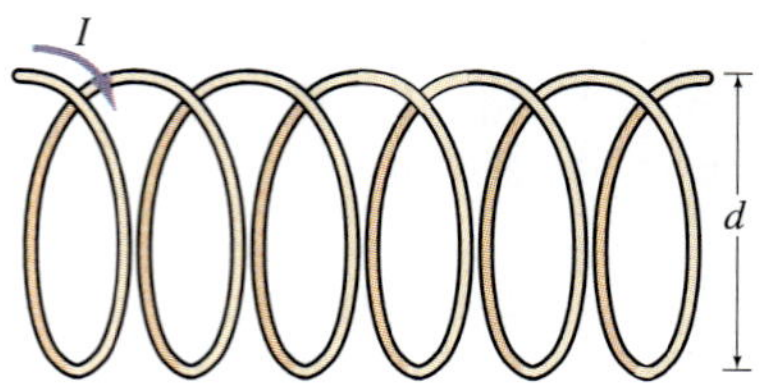

FIGURE 30–15 An ideal solenoid is an infinitely long cylinder made from a uniformly wound coil carrying current I (view shown is exaggerated).

A **solenoid** is an electrical element that generates magnetic fields. It is analogous to a parallel-plate capacitor: A solenoid generates a constant magnetic field in a region of space just as a parallel-plate capacitor sets up a constant electric field. An ideal solenoid consists of a coil of wire wound uniformly into an infinitely long cylinder, as in Fig. 30–15. In this figure, we have exaggerated the spacing between the wires, which normally are closely wound. The diameter of the cylinder is d, the current carried is I, and the wires are wound so that there are n turns per unit length, where the length is measured along the axis of the solenoid.

Let's sketch what the magnetic field of a solenoid might look like. Figure 30–16a is a cross-sectional view of several loops of a solenoid, which are again spaced more widely than they normally would be. Very near the wires, the magnetic fields form circles around the wires because the field approximates that of a single straight wire. Figure 30–16a shows that, between adjacent turns of the wire, the fields from adjacent wires tend to *cancel* each other. Inside the solenoid, these fields *add together* to form a large component that points to the right along the axis of the solenoid. Every loop of wire contributes constructively to make this interior field along the axis a strong one. The scenario is somewhat different outside the cylinder. A region at the top gets constructive contributions from the circular fields due to the wires at the top, making a field that points to the left at the top of Fig. 30–16a and to the right at the bottom of that figure. The circular fields from the wires at the bottom contribute also and make a field outside the cylinder that points to the right at the top and to the left at the bottom. This field tends to cancel the contribution from the wires at the top. As the wires are placed closer to each other, this effect increases (Fig. 30–16b).

The emerging qualitative model of the field lines is that the fields from the different loops of the coil reinforce inside the cylinder and make a net field that is parallel to the cylinder axis and whose direction is determined by a right-hand rule: If the fingers curl in the direction of the current, the thumb shows the direction of the magnetic field (Fig. 30–17). Outside the cylinder, the field points primarily in the opposite

FIGURE 30–16 (a) One section of a three-turn solenoid, showing the superposed magnetic fields (view shown is exaggerated). (b) As the winding density increases, the magnetic field takes on a simple form.

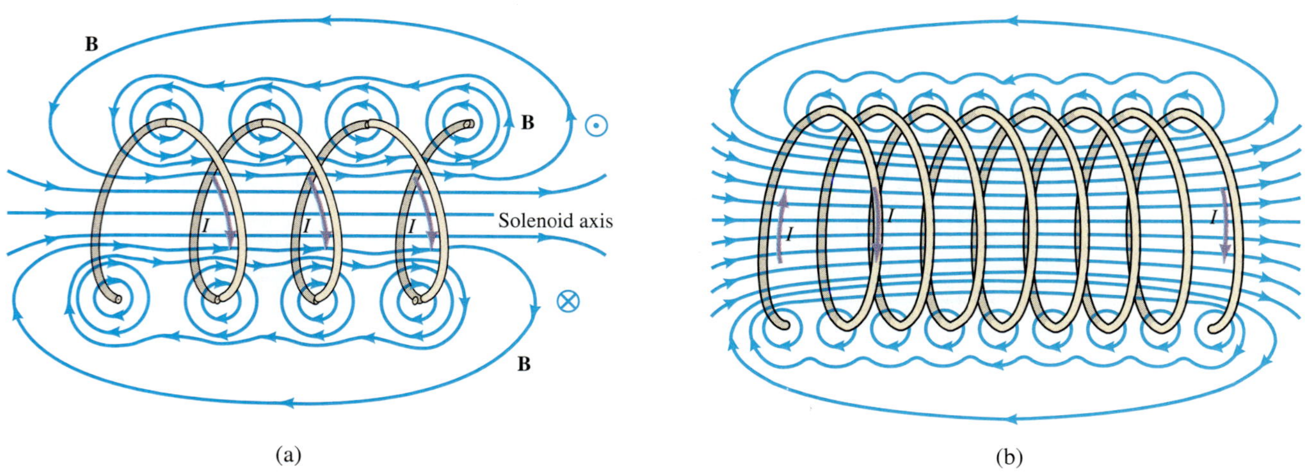

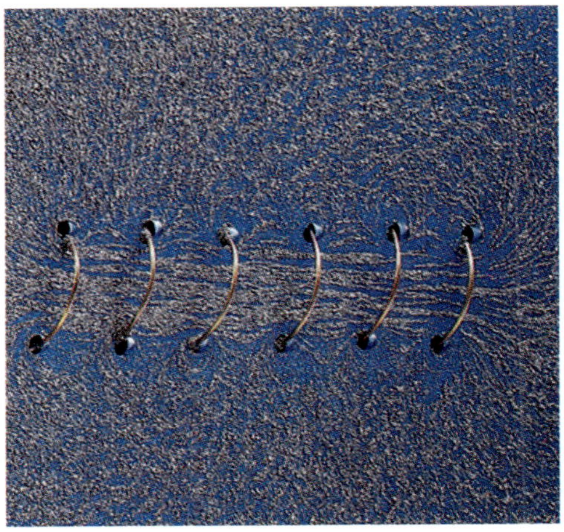

(a)

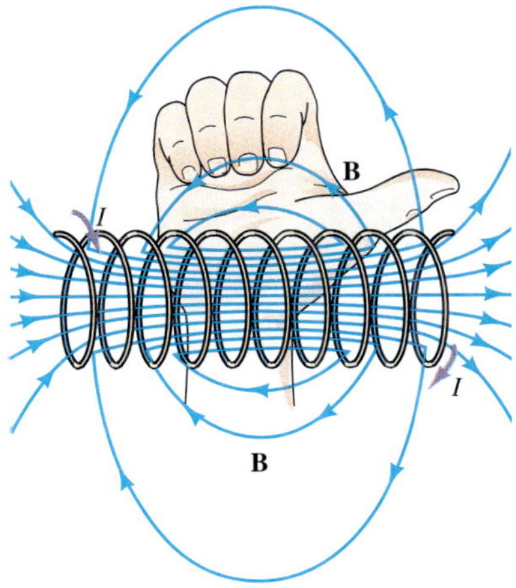

(b)

FIGURE 30–17 (a) Magnetic field lines of a solenoid, as shown by iron filings that align with the field. (b) A right-hand rule gives the direction of the magnetic field within a solenoid.

direction and is much weaker. Another way to see that the field is weaker outside the cylinder is to recall that magnetic field lines close. Because these lines cannot cross each other (why not?), the lines are squeezed together in the limited space inside the cylinder but spread throughout space on the outside. *Even though the magnetic field is not exactly zero outside a real solenoid, it is a good approximation to take the field there to be insignificant.*

Using Ampère's Law to Find the Magnetic Field in a Solenoid

Now that we understand qualitatively that a long solenoid has a large magnetic field inside—parallel to the solenoid axis—and a weak field outside, we can apply Ampère's law to calculate quantitatively the magnetic field inside the solenoid. Figure 30–18 shows a solenoid that carries a current I, and a closed *imaginary* loop consisting of four legs in a rectangle of length ℓ and height w on which to apply Ampère's law. The wire of the solenoid passes N times from above through the imaginary loop. The path about the imaginary loop is taken to be clockwise, so the net current into the imaginary loop, NI, is positive by the right-hand rule. We now calculate the line integral on the left-hand side of Eq. (30–10). There is only a very small contribution from leg 2 (point b to point c), because the field outside is insignificant. There is no contribution from leg 1 (point a to point b) or from leg 3 (point c to point d) for two reasons. First, the field outside is insignificant, and the field inside is parallel to the cylinder axis and hence perpendicular to the path. Second, any contributions from these two legs would cancel each other because they are in opposite directions. From point d to point a (leg 4), the field is parallel to the path. Along this portion of the path, the field has a constant unknown value B. The contribution to the integral is $B\ell$, and thus Ampère's law gives

$$\oint \mathbf{B}\cdot d\mathbf{s} = B\ell = \mu_0 I_{\text{enclosed}} = \mu_0 NI. \tag{30–14}$$

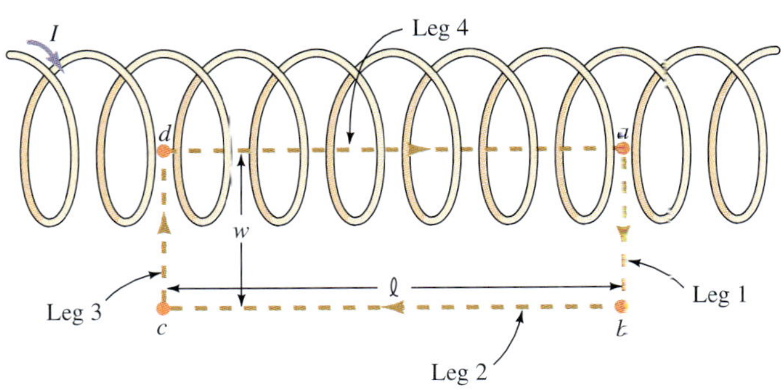

FIGURE 30–18 An imaginary rectangular loop is drawn half inside and half outside a solenoid. This loop provides a path for the application of Ampère's law.

We can eliminate the explicit dependence on ℓ by noting that the number of times the solenoid wire passes through the imaginary loop, N, is the length ℓ times the number of solenoid turns per unit length, n: $N = n\ell$. The quantity n is the *turn density* of the solenoid. We have

$$B\ell = \mu_0 n \ell I,$$

The magnetic field within a solenoid

and the *interior magnetic field of a long solenoid* has magnitude

$$B = \mu_0 n I. \tag{30–15}$$

Note that Eq. (30–15) contains no reference to the distance from the axis on the inside of the loop. Our derivation is completely independent of how close the path in Fig. 30–18 comes to the solenoid axis, and any choice of this distance would give the same field. *The magnetic field inside a long solenoid is uniform throughout the interior.* The magnetic field depends linearly on the current. The magnetic field produced when a current runs through a solenoid can be used as a switch (Fig. 30–19).

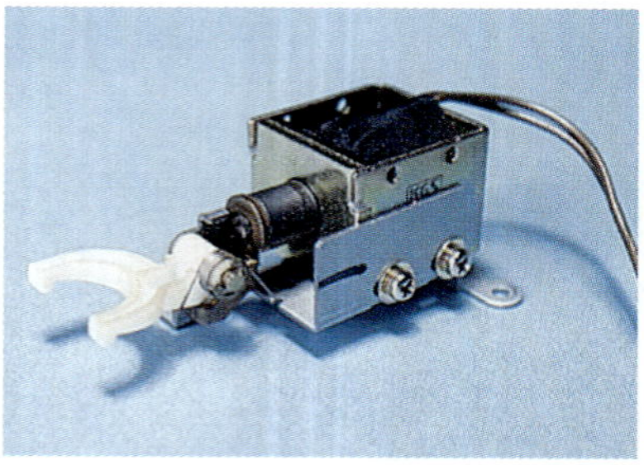

FIGURE 30–19 This solenoid switch works because an iron core retracts when a current passes through the solenoid, pulling against a spring and causing the white holder to move.

EXAMPLE 30–3 A solenoid consists of wires—each of diameter $d = 0.6$ mm—that can carry a maximum current of $I = 0.03$ A; the wires are tightly wound in a single layer. (a) What is the maximum magnitude of the field inside the solenoid? (b) Assume that the maximum current a wire can carry is proportional to the area of the wire, and that the wire's diameter is the only variable under consideration. What should the wire diameter be in order to double the magnetic field inside?

Solution: (a) From Eq. (30–15), we can find the magnitude of the unknown magnetic field inside the solenoid given the current, I, and the turn density, n. We know I, but we must calculate n. If the wire has diameter d, then we have one turn every length d, and $n = 1/d$. Thus

$$B = \mu_0 n I = \frac{\mu_0 I}{d} = \frac{(4\pi \times 10^{-7}\ \text{T}\cdot\text{m/A})(0.03\ \text{A})}{0.6 \times 10^{-3}\ \text{m}} = 0.6\ \text{T}.$$

PROBLEM SOLVING

How to find the magnetic field within a long solenoid

(b) If B is to double, then the product nI must double according to Eq. (30–15). Let's see how these factors depend on the diameter d. Because the solenoid is tightly wound, the turn density $n \propto 1/d$. The maximum current is proportional to the wire area, which, in turn, is proportional to the wire diameter squared. Thus $I \propto d^2$. Combining our results, $nI \propto (1/d)d^2 \propto d$. If the field is to be doubled, the diameter of the wire used in the solenoid should also be doubled.

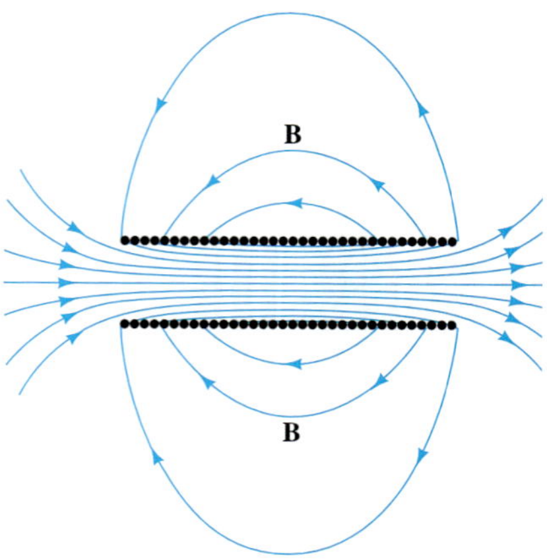

FIGURE 30–20 The magnetic field of a solenoid of finite length. (After E. M. Purcell, *Berkeley Physics Course: Electricity and Magnetism*, McGraw-Hill, 1990, p. 229.)

The technique of using Ampère's law to calculate the magnetic field inside a long, cylindrical solenoid is directly applicable to noncylindrical geometry with exactly the same results. Equation (30–15) holds *even if the cross section of the winding is not circular*. We require only that the solenoid is long and that the cross-sectional area is constant.

The results we have found for the ideal solenoid hold rather well even for a solenoid of finite length. Figure 30–20 shows the magnetic field lines of a solenoid in a plane that cuts through the center of the solenoid. The solenoid's length is four times its diameter.

The exterior field of the solenoid of finite length illustrated in Fig. 30–20 looks just like the magnetic field of a bar magnet, Fig. 29–2a. Does this mean that the field of a bar magnet has the same physical origin as that of a solenoid? The answer is yes. A bar magnet is made of aligned, atom-sized current loops. Ampère suggested that permanent magnets are associated with internal currents. In fact, the interior field of a bar magnet is also the same as the uniform interior field of a solenoid. The origin of magnetism in matter is discussed further in Chapter 32.

A Toroidal Solenoid

A real solenoid has finite length and therefore its magnetic field has some end effects. These end effects can be eliminated by making the solenoid into a doughnut shape, called a *torus* (Fig. 30–21). This shape does introduce some variation in the magnetic field within the solenoid. The radius of the coil is r_0, and the overall radius of the torus—the distance from the center to the circular axial line—is R. Symmetry implies that the magnetic field runs within the coil parallel to its walls. The same arguments that we gave for the straight solenoid imply that this field acts in the direction of the thumb when the fingers of the right hand are curled in the direction of the current. Ampère's law can then be used to find the magnitude of **B**. Take a path within the coil whose distance from the center is R' (Fig. 30–21). The magnitude of the field is the same all along the path, so

$$\oint \mathbf{B}\cdot d\mathbf{s} = B(2\pi R') = \mu_0 NI.$$

Here, N is the total number of loops that form the coil. Thus the magnitude of the field at a distance R' is

$$B = \frac{\mu_0 NI}{2\pi R'}. \qquad (30\text{–}16)$$

Although there are no end effects, the field is not constant across the cross-sectional area of the torus. The field at one value of R' is different from the field at a different value of R'. If the coil's radius r_0 is much less than the overall radius R of the torus, the possible values of R', from $R - r_0$ to $R + r_0$, do not vary much. The magnetic field within the torus will not vary very much either.

30–4 THE BIOT–SAVART LAW

Ampère's law is a general one. However, its usefulness as a tool for calculating magnetic fields depends on symmetry in the system of currents that create the magnetic field. Here we find a direct expression for the magnetic field produced by a current. This expression, the *Biot–Savart law*, can be applied even when there is no symmetry. There is a simple analogue to this procedure in electrostatics. When there is symmetry in a charge distribution, Gauss' law provides a powerful tool for finding the electric field. When there is no symmetry, we can always find the net electric field by using the superposition of the electric fields of point charges (as determined by Coulomb's law). In analogy, the Biot–Savart law gives us the magnetic field due to an infinitesimal distribution of current segments. We then use the superposition principle to determine the magnetic field of a finite arrangement of currents.

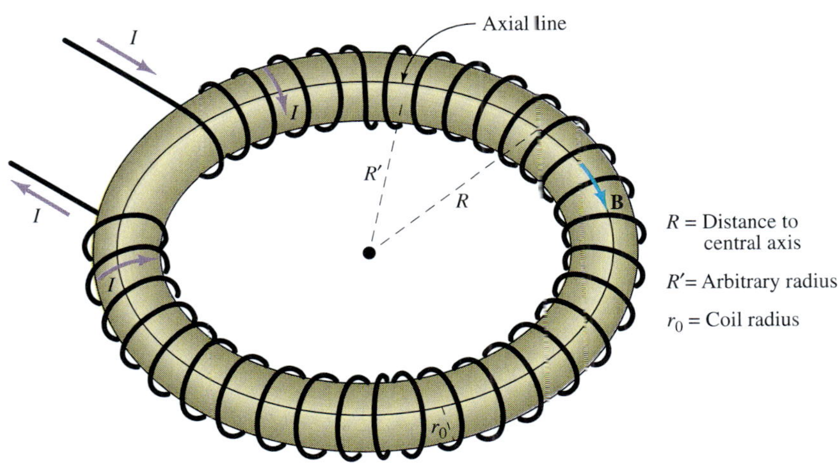

FIGURE 30–21 A torus wrapped with a wire that carries a current I has a magnetic field inside, which we can calculate by using Ampère's law. The overall radius of the torus is R, whereas the radius of the coil is r_0. The distance R' is not equal to R.

Let's start with a result we already know. Ampère's law applied outside a long, straight wire that carries a current I shows that the magnitude of the magnetic field at a radial distance r from the wire is

$$B = \frac{\mu_0 I}{2\pi r},$$

Eq. (30–5). The field lines form circles around the wire with the direction given by the right-hand rule. We expect this field to be the sum of the contributions of all the infinitesimal current elements $I\,d\ell$ that make up the wire. To find the form of the individual contributions, we make the observations that follow.

The $1/r$ dependence of the magnetic field resembles the $1/r$ dependence of the electric field due to a long, charged rod of constant linear charge density λ, as given in Eq. (23–31):

$$E = \frac{1}{2\pi\epsilon_0}\frac{\lambda}{r}.$$

This result was obtained by integrating the component of the electric field perpendicular to the wire due to the charge in an infinitesimal length $d\ell$ of the charged rod, from Eq. (23–27):

$$dE_\perp = \frac{1}{4\pi\epsilon_0}\frac{\lambda\,d\ell}{r^2}\cos\phi.$$

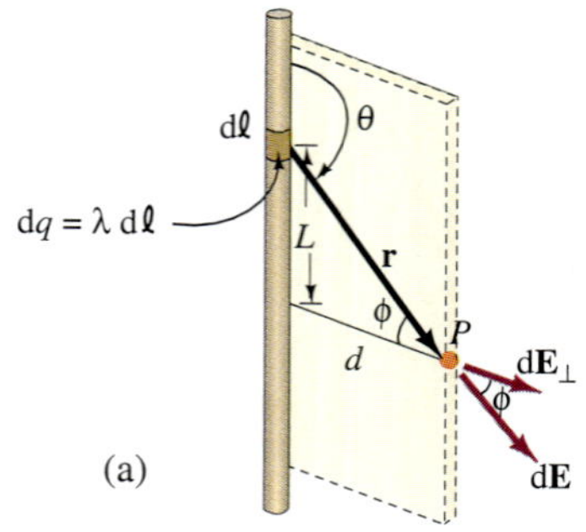

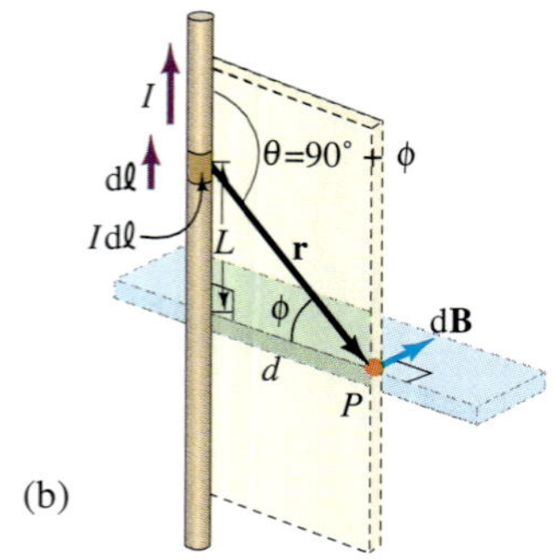

FIGURE 30–22 (a) A rod that carries a net charge density λ can be broken up into segments that contribute an electric field at any point. Integration over the contributions from the segments gives the net electric field. (b) A wire carrying current I similarly has a magnetic field that can be calculated by integrating the contributions of segments of the thread. For a segment of length $d\ell$, the vector $d\boldsymbol{\ell}$ is oriented along the segment in the direction of I.

Here, the particular element of the charged rod is a distance $r = \sqrt{L^2 + d^2}$ from the point where the field is measured, and ϕ is as shown in Fig. 30–22a. The factor $(1/4\pi\epsilon_0)(\lambda\,d\ell)/r^2$ is just the electric field strength for a point charge $dq = \lambda\,d\ell$. The second factor, $\cos\phi$, is present only because we are looking at the perpendicular component. The resemblance between Eqs. (23–31) and (30–5), in which $1/\epsilon_0$ is replaced by μ_0 and λ is replaced by I, suggests that we try a similar procedure for finding the magnetic field that results from a length $d\ell$ of a wire that carries a current I. The result has the same form as that for the electric field of the charged wire, Eq. (23–27), with the replacements just specified:

$$dB = \frac{\mu_0}{4\pi}\frac{I\,d\ell}{r^2}\cos\phi. \qquad (30\text{–}17)$$

The electric field is directed radially away from the charge element. The magnetic field still forms circles; at point P in Fig. 30–22b, **B** is directed as shown, as required by the right-hand rule. The direction of the current is indicated by making the infinitesimal length $d\ell$ a vector $d\boldsymbol{\ell}$ whose direction is along I. Note that **B** is perpendicular to both $d\boldsymbol{\ell}$ and **r**. When we speak of a vector perpendicular to two other vectors, we are reminded of the vector product. A short exercise in trigonometry shows that if $\theta = \phi + 90°$ (as in Fig. 30–22b), then $\cos\phi = \cos(\theta - 90°) = \cos\theta\cos 90° + \sin\theta\sin 90° = \sin\theta$. Equation (30–17) becomes

$$dB = \frac{\mu_0}{4\pi}\frac{I\,d\ell\sin\theta}{r^2} = \frac{\mu_0}{4\pi}I\frac{d\ell\, r\sin\theta}{r^3} \qquad (30\text{–}18)$$

in the direction perpendicular to $d\boldsymbol{\ell}$ and **r**. The presence of a $\sin\theta$ factor confirms our guess that a vector product is involved. We have found *the magnetic field* **dB** *produced by a segment of wire* $d\boldsymbol{\ell}$ *that carries a current I at a displacement* **r** *from the segment:*

The Biot–Savart law

$$d\mathbf{B} = \frac{\mu_0}{4\pi}\frac{I\,d\boldsymbol{\ell}\times\mathbf{r}}{r^3}. \qquad (30\text{–}19)$$

The vector product $d\boldsymbol{\ell}\times\mathbf{r}$ has a magnitude $d\ell\, r\sin\theta$ and the proper direction for this situation is into the page. Notice that there is no ambiguity as to the vector $d\boldsymbol{\ell}$. If the segment is short enough, it may be treated as a straight line. Equation (30–19) is the

Biot–Savart law, named after the two physicists who first formulated it, Jean-Baptiste Biot and Félix Savart. This law is analogous to Coulomb's law in electricity. It even has the same overall distance dependence of $1/r^2$—note that the magnitude of r appears in the numerator of Eq. (30–19). The angular factors are quite different, however.

Using the Biot–Savart Law

We can use the Biot–Savart law to find the magnetic field due to nonsymmetric current distributions. Thus it is used in the same way Coulomb's law is used in electricity. The net magnetic field is found by integrating over $d\mathbf{B}$:

$$\mathbf{B} = \int d\mathbf{B} = \frac{\mu_0}{4\pi}\int \frac{I\,d\boldsymbol{\ell} \times \mathbf{r}}{r^3}. \tag{30–20}$$

Equation (30–20) is often too complicated for practical use. In practice, magnetic fields due to nonsymmetric current distributions are measured experimentally, although sophisticated computer programs are now available to design the current windings needed to produce a desired field. Such programs superpose infinitesimal field contributions $d\mathbf{B}$, each given by the Biot–Savart law.

EXAMPLE 30–4 A straight segment of wire of length L carries a current I. Use the Biot–Savart law to find the magnetic field in the plane perpendicular to the wire and passing through the midpoint of the wire segment.

Solution: Orient the wire along the x-axis, as in Fig. 30–23, with its center positioned at the origin. We then wish to find the field in the yz-plane. At a given distance D from the wire, the wire looks the same from anywhere in this plane. Thus we can calculate the field anywhere around the wire and choose the point $y = D$, $z = 0$. The right-hand rule shows that the quantity $d\boldsymbol{\ell} \times \mathbf{r}$ points out of the plane where $d\boldsymbol{\ell}$ and $\mathbf{r}$ are as shown in the figure. The field forms circles around the wire, as before. Let's now concentrate on finding the magnitude at point D.

From Eq. (30–18), the magnetic field from the segment $d\boldsymbol{\ell}$ has magnitude

$$dB = \frac{\mu_0 I}{4\pi}\,dx\,\frac{\sin(\pi - \theta)}{r^2} = \frac{\mu_0 I}{4\pi}\,dx\,\frac{\sin\theta}{r^2} = \frac{\mu_0 I}{4\pi}\,dx\,\frac{\cos\phi}{r^2},$$

where we have replaced $d\ell$ by dx and used the angle ϕ defined in Fig. 30–23. To find the net magnetic field, we sum over the contributions of segments from $x = -L/2$ to $x = L/2$:

$$B = \frac{\mu_0 I}{4\pi}\int_{-L/2}^{L/2} \frac{\cos\phi}{r^2}\,dx.$$

Both ϕ and r depend on x. The integral is computed most simply if we use trigonometric variables; we therefore change variables from x to ϕ. We find

$$\frac{x}{D} = \tan\phi;$$

$$dx = D\,d(\tan\phi) = D\sec^2\phi\,d\phi = \frac{D}{\cos^2\phi}\,d\phi.$$

In addition, $r = D/\cos\phi$, so the combination that appears in the integral becomes

$$\frac{\cos\phi}{r^2}\,dx = \cos\phi\,\frac{1}{(D/\cos\phi)^2}\,\frac{D}{\cos^2\phi}\,d\phi = \frac{1}{D}\cos\phi\,d\phi.$$

Thus

$$B = \frac{\mu_0 I}{4\pi D}\int_{-\phi_0}^{+\phi_0} \cos\phi\,d\phi,$$

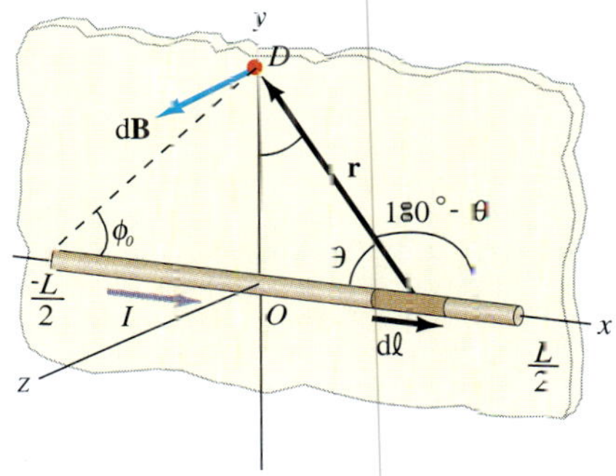

FIGURE 30–23 Example 30–4. A straight segment of wire of length L, carrying a current I, is oriented on the x-axis and centered at the origin.

where $+\phi_0$ is the limit of integration—the largest value taken on by ϕ_0. The integral of the cosine is the sine, and

$$B = \frac{\mu_0 I}{4\pi D}[\sin\phi_0 - \sin(-\phi_0)] = \frac{\mu_0 I}{2\pi D}\sin\phi_0.$$

We can reexpress this result in terms of L and D by using

$$\sin\phi_0 = \frac{L/2}{\sqrt{(L/2)^2 + D^2}}$$

to find that

$$B = \frac{\mu_0 I}{4\pi}\frac{L}{D\sqrt{(L/2)^2 + D^2}}. \tag{30–21}$$

The form given in Eq. (30–21) shows that the field depends not only on the distance D from the wire but also on the relative magnitudes of D and L. We can check this result by taking the limit that L is large; we should then recover Eq. (30–5). In the limit $L >> D$, the square root in Eq. (30–21) becomes $L/2$, and

$$B \to \frac{\mu_0 I}{4\pi}\frac{L}{D(L/2)} = \frac{\mu_0 I}{2\pi D},$$

which is the correct result as given by Ampère's law.

How numerically significant is it that the wire of Example 30–4 has finite length? Suppose that the current is 1 A, and that the wire segment is 0.1 m long. Then, 1 cm from the wire, Eq. (30–21) gives a field of 1.96×10^{-5} T, whereas the magnetic field of the infinitely long wire carrying the same current [Eq. (30–5)] is 2.00×10^{-5} T. In this case, we are a distance from the wire of one-tenth the length of the wire and the field is only 2 percent less than the infinite length result. But let's now move away from the wire so that we are one wire's length away: $L = 0.1$ m, and $D = 0.1$ m. In this case, the field from the wire has a value some 22 times smaller than the field of the wire of infinite length.

Perhaps you wonder why we could not have used Ampère's law directly in Example 30–4. The reason is that there are contributions from the current-carrying wires that bring the current into our finite segment of length L and take it out (not shown in Fig. 30–23). These wires destroy the cylindrical symmetry. It is only such symmetry that makes Ampère's law useful. Ampère's law is still valid, of course, but it is not useful here. The Biot–Savart law can *always* be used. When the segment becomes very long—so that the contributions from other segments of wire are small—then Ampère's law is again useful. As we have seen, the Biot–Savart law and Ampère's law then give the same result.

PROBLEM SOLVING

When to use Ampère's law and when to use the Biot–Savart law

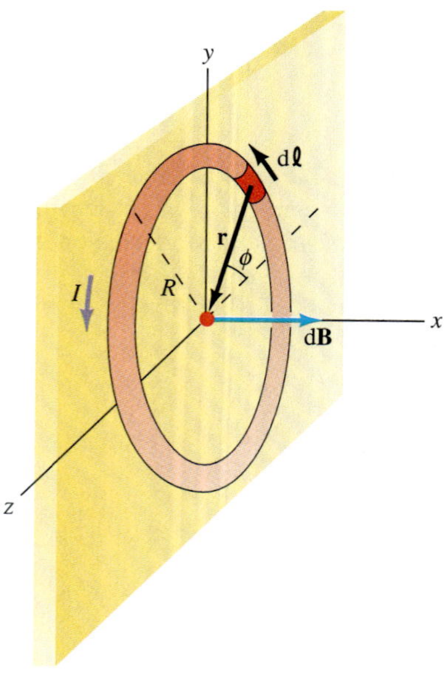

FIGURE 30–24 Example 30–5. An integration is required to find the net magnetic field at the center of the loop.

EXAMPLE 30–5 A wire forms a circular loop of radius $R = 12$ cm. A current $I = 8.0$ A flows counterclockwise in the wire (Fig. 30–24). Find the magnetic field at the center.

Solution: There is not enough symmetry here to allow us to draw a path along which the magnetic field is constant, so we cannot use Ampère's law. We must use the Biot–Savart law and integrate over the contributions $\mathbf{dB}$ of the different elements of the wire to find the unknown total field $\mathbf{B}$. Vector $\mathbf{r}$ runs from the current element $\mathbf{d\ell}$ to the center of the circle, the point where we want to find $\mathbf{B}$. The quantity $\mathbf{d\ell} \times \mathbf{r}$ for the current element shown in Fig. 30–24 is directed along the x-axis, and this will be true for all the current elements that make up the loop. The net magnetic field at the center is thus directed out of the page.

PROBLEM-SOLVING TECHNIQUES

In the examples of this chapter and Chapter 29, we have studied two aspects of problems on static magnetic fields. We may want to find the magnetic fields produced by a given time-independent set of currents or we may want to find the magnetic forces on currents or on moving charges. Two sets of key laws contain all that is generally necessary to approach such problems, and you should understand the symbols in these formulas and what the laws mean. First, we have the laws that determine the magnetic field due to moving charges or currents, which can be written in the two forms

Ampère's law: $$\oint \mathbf{B}\cdot d\mathbf{s} = \mu_0 I_{\text{enclosed}}, \quad (30\text{–}10)$$

Biot–Savart law: $$d\mathbf{B} = \frac{\mu_0}{4\pi}\frac{I\,d\boldsymbol{\ell}\times\mathbf{r}}{r^3}. \quad (30\text{–}19)$$

Second, we have the laws that express the force on a moving charge or a current due to a given magnetic field, namely,

$$\mathbf{F}_B = q\,\mathbf{v}\times\mathbf{B}, \quad (29\text{–}1)$$

$$d\mathbf{F} = I\,d\boldsymbol{\ell}\times\mathbf{B}. \quad (29\text{–}14)$$

Each set of laws involves a right-hand rule, and you should know how to use it. For Ampère's law, if the thumb of the right hand follows the current, the fingers curl in the direction of the integration path. For the force laws and the Biot–Savart law, the right-hand rule for a vector product applies.

Based on these laws, we can suggest a list of habits to develop. Many of these are the very same habits that are useful for solving *any* problem in physics.

1. Draw a figure that indicates the physical situation with the quantities known; include directions if appropriate.

2. Write down what is known and what is to be determined. Are you dealing with moving charges or with currents?

3. What physical principles connect the unknown quantities to the known ones?

4. If the problem concerns a force, do you have sufficient information to determine the force directly from the force laws? If not, you may need to compute a magnetic field or integrate an infinitesimal force.

5. In a situation with enough symmetry (for example, for long, straight wires), we can use Ampère's law to calculate a magnetic field. When applicable, Ampère's law will usually give the answer more easily than the Biot–Savart law.

6. If the system is not sufficiently symmetric, the Biot–Savart law is always available. In using it, be sure that the infinitesimal element $d\boldsymbol{\ell}$ and the position vector $\mathbf{r}$ are identified properly. A partial symmetry may rule out one or more directions for the magnetic field. If only one direction is indicated by symmetry, then the other components will cancel in the calculation of the integral over $d\boldsymbol{\ell}$, and you must only integrate for the component desired.

7. Checks of dimensions and units are always appropriate.

8. Substitute numbers only at the last stage; any checks you can find of limits or special cases are always helpful.

The magnitude of the field due to the element shown is given according to Eq. (30–18) by

$$dB = \frac{\mu_0}{4\pi}\frac{I\,d\ell}{R^2}.$$

There is no sine factor here because $d\boldsymbol{\ell}$ is perpendicular to $\mathbf{r}$. The integral of $d\ell$ around the circle is the circumference $2\pi R$, so the net field at the center has magnitude

$$B = \int dB = \int \frac{\mu_0}{4\pi}\frac{I\,d\ell}{R^2} = \frac{\mu_0 I}{4\pi}\frac{1}{R^2}\int d\ell = \frac{\mu_0 I}{4\pi}\frac{2\pi R}{R^2}$$

$$= \frac{\mu_0 I}{2R}. \quad (30\text{–}22)$$

The numerical value of this field is

$$B = \frac{(4\pi\times10^{-7}\,\text{T}\cdot\text{m/A})(8.0\,\text{A})}{2(0.12\,\text{m})} = 4.2\times10^{-5}\,\text{T}.$$

This value is about that of Earth's magnetic field at Earth's surface.

The orientation of the magnetic field we have calculated is consistent with a right-hand rule in which the fingers of the right hand curl with the current and the thumb indicates the direction of the field. This orientation rule is the same one that we found for the magnetic field of a solenoid; indeed, the loop is nothing more than a compressed solenoid.

EXAMPLE 30–6 Reconsider the circular wire loop of Example 30–5, drawn in Fig. 30–24. Find the magnitude of the magnetic field all along the axis of the loop, oriented as in Fig. 30–25. What is the limit of your result (a) at the center and (b) at large distances along the axis?

Solution: We want to find the magnetic field all along the axis of the loop, not just at the center as in Example 30–5. As we stated there, Ampère's law is not useful here. We must first use the Biot–Savart law to find the magnetic field on the axis from a particular current element, and then sum over the contributions of the elements. The loop is now aligned so that its axis is along the x-axis and its center is at the origin (Fig. 30–25). Consider a point P a distance d from the center. We have chosen an element $\mathbf{d}\boldsymbol{\ell}$ where the loop passes through the $+y$-axis. Its contribution $\mathbf{dB}$ is perpendicular to both $\mathbf{d}\boldsymbol{\ell}$ and $\mathbf{r}$, so it has *both* a component along the loop axis and a $+y$-component. If we had chosen an element on the opposite side of the loop, where the loop cuts the $-y$-axis, we would have found a $\mathbf{dB}$ with a component along the loop axis in the same direction as the contribution from the first loop element, and also a component in the $-y$-direction. The y-component of $\mathbf{dB}$ from the first element cancels the y-component from the second element. This will be true for all pairs of elements around the loop, so we must calculate only the component of $\mathbf{dB}$ along the loop axis, which is the x-component. This understanding of the role played by symmetry is important, and a moment or two spent looking for such symmetries is time well spent.

From Fig. 30–25, we see that the vectors $\mathbf{d}\boldsymbol{\ell}$ and $\mathbf{r}$ are perpendicular, so the cross product $\mathrm{d}\boldsymbol{\ell} \times \mathbf{r} = (\mathrm{d}\ell)(r)$. Then

$$\mathrm{d}B_x = \frac{\mu_0}{4\pi}\frac{I\,\mathrm{d}\ell}{r^2}\cos(\gamma).$$

The infinitesimal length $\mathrm{d}\ell = R\,\mathrm{d}\phi$, the $\cos(\gamma) = \sin(90°-\gamma) = R/r$, and the magnitude of $\mathbf{r}$ is $r = \sqrt{R^2 + d^2}$. Collecting terms, we have

$$\mathrm{d}B_x = \frac{\mu_0}{4\pi}\frac{IR^2\,\mathrm{d}\phi}{(R^2 + d^2)^{3/2}}.$$

The net field is

$$B_x = \int \mathrm{d}B_x = \frac{\mu_0 IR^2}{4\pi(R^2 + d^2)^{3/2}}\int_0^{2\pi}\mathrm{d}\phi = \frac{R^2}{(R^2 + d^2)^{3/2}}. \qquad (30\text{–}23)$$

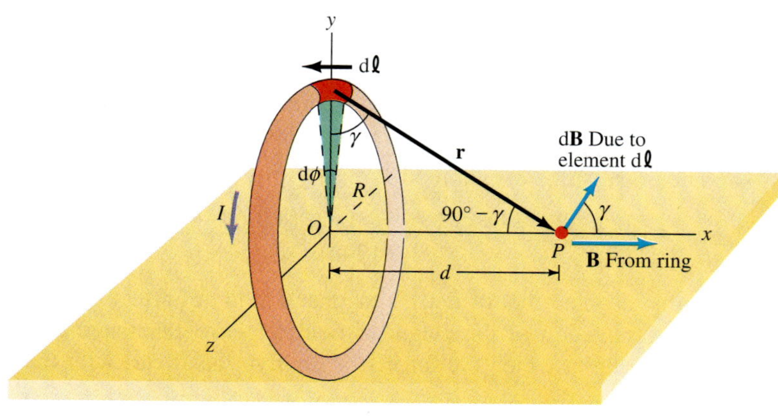

FIGURE 30–25 Example 30–6. The net magnetic field along the loop axis due to a current loop carrying a current I is oriented along the axis, according to a right-hand rule. The contribution from an infinitesimal element $\mathbf{d}\boldsymbol{\ell}$ has components along the axis and in other directions as well.

As our symmetry argument states, the magnetic field points along the $+x$-axis.

(a) For the center of the loop, Eq. (30–23) correctly reduces to Eq. (30–22) when d is zero.

(b) At large distances, $d >> R$, the axial magnetic field in Eq. (30–23) reduces to

$$B = \frac{\mu_0 I}{2}\frac{R^2}{d^3} = \frac{\mu_0}{2\pi}\frac{I\pi R^2}{d^3}. \tag{30–24}$$

Magnetic Dipoles

Equation (30–24) is similar to our expression for the electric field along the axis of an electric dipole, Eq. (23–14). Indeed, careful calculation of the magnetic field over all space due to the current-carrying loop of Examples 30–5 and 30–6 shows that the magnetic field lines, as in Fig. 30–26, are just like the electric field lines of the electric dipole, Fig. 23–12. The current loop forms a *magnetic dipole*, and the strength of the magnetic dipole is characterized by the *magnetic dipole moment*, μ. (See Chapter 29, where we studied the *response* of such a loop to an external field.) In place of the electric dipole moment p for the electric field of an electric dipole, we find the quantity μ,† which is defined by

The magnetic dipole moment. Magnetic dipoles are formed from current loops.

$$\text{for a circular current loop:} \qquad \mu \equiv I\,\pi R^2. \tag{30–25}$$

The magnetic field is shown in Fig. 30–26. At a distance d far from the loop, the magnitude of the magnetic field along the axis is

$$\text{along the axis:} \qquad B = \frac{\mu_0}{2\pi}\frac{\mu}{d^3}. \tag{30–26}$$

In fact, for *any* closed loop of area A that carries a current I, there is an analogous result: The magnetic field decreases as $1/d^3$ far from the loop. The strength of the field is proportional to a dipole moment μ that is equal to the product of the current in the loop and the area of the loop:

$$\mu = IA. \tag{30–27}$$

The direction of the field is determined if we form a vector $\boldsymbol{\mu}$ from μ. For any plane loop, $\boldsymbol{\mu}$ is perpendicular to the plane of the loop according to a right-hand rule: Curl the fingers of the right hand around the direction of current flow in the loop, and the

†Not to be confused with the permeability of free space, μ_0

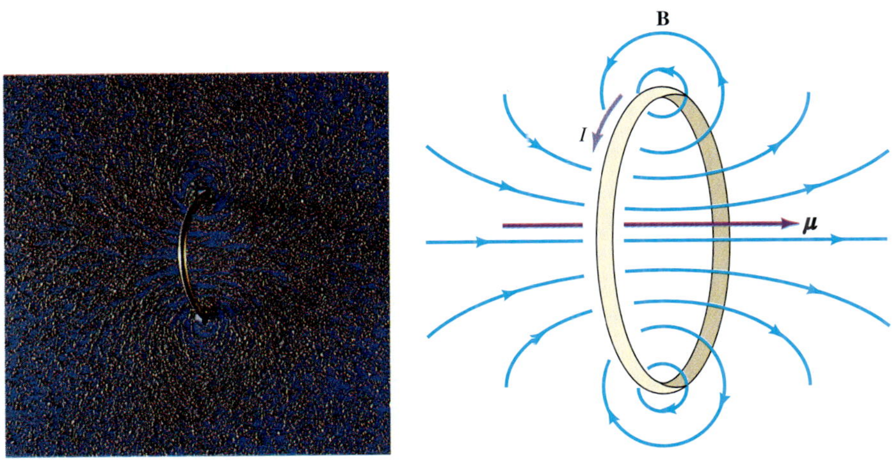

(a) (b)

FIGURE 30–26 (a) Magnetic field lines for a circular loop of current, as shown by iron filings. (b) The field for such a loop is a magnetic dipole field.

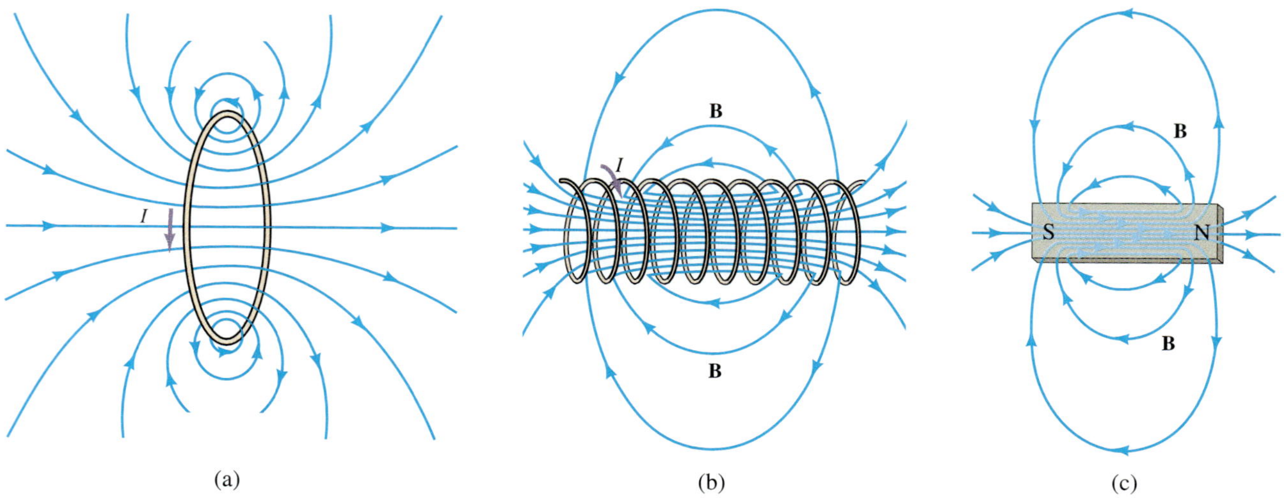

FIGURE 30–27 (a) The magnetic field lines of a circular loop of current have the same form as the magnetic field lines of (b) a solenoid and of (c) a bar magnet.

thumb points in the direction of $\boldsymbol{\mu}$. Note that magnetic charges do not exist, so the magnetic dipole *cannot* be formed from a pair of equal but opposite magnetic charges, but only from a closed loop of current. The magnetic field of a current loop (Fig. 30–27a) has the same form as the fields of a solenoid and of a bar magnet (Figs. 30–27b, c). Indeed, both solenoids and bar magnets act as magnetic dipoles.

In Section 29–5, we saw that the magnetic dipole responds to an external field: The field $\mathbf{B}_{ext}$ exerts a torque $\tau = \boldsymbol{\mu} \times \mathbf{B}_{ext}$ on the loop that tends to line up the vector $\boldsymbol{\mu}$ with $\mathbf{B}_{ext}$. As a comparison of Eqs. (29–22) and (30–27) shows, the same vector $\boldsymbol{\mu}$ determines both the magnetic field of the loop and the reaction of the loop to an external magnetic field. A magnetic dipole, like an electric dipole, both produces a field and responds to a field.

30–5 THE MAXWELL DISPLACEMENT CURRENT

There is a logical flaw in Ampère's law when there is time dependence in the current. In 1865, James Clerk Maxwell modified the law to remove this flaw. This modification was crucial to the completely unified theory of electricity and magnetism that will be discussed in Chapter 35.

Ampère's law is applied with an integration over some closed path. The right-hand side of Ampère's law, Eq. (30–10), contains what we called the current enclosed by a path. By "the current enclosed by a path," we mean the rate of charge flow through a surface whose boundary is the closed path. Such a surface can be chosen in many different ways (Fig. 30–28), but *when the current is continuous, the current that crosses any one of these surfaces must be the same as the current that crosses any other*. The freedom in how the surface is chosen therefore presents no problem. There is a situation, however, in which the freedom to choose the surface presents a difficulty, and that situation arises when the current deposits charge on the plates of a capacitor. Figure 30–29 shows two surfaces with the same loop as their boundary. A current I crosses surface 1 in the positive sense, while no current crosses surface 2. Yet there is no unambiguous way to distinguish the two surfaces in the application of Ampère's law.

Maxwell noted that even if no current passes through surface 2, there is nevertheless a distinguishing feature for this surface; namely, *there is a changing electric flux through it*. As the charge builds up on the plates of the capacitor, so does the re-

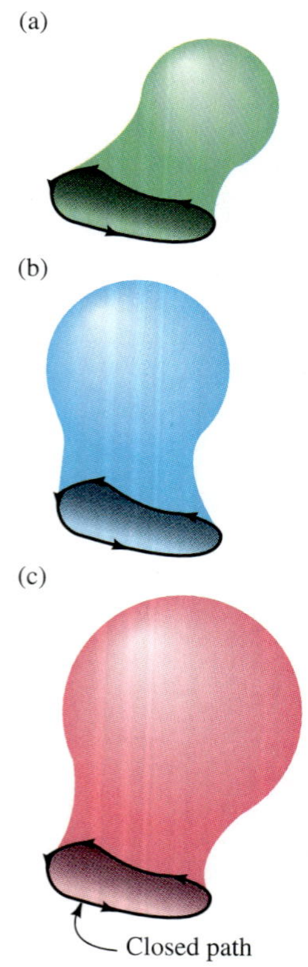

FIGURE 30–28 A closed path defines an infinite number of surfaces.

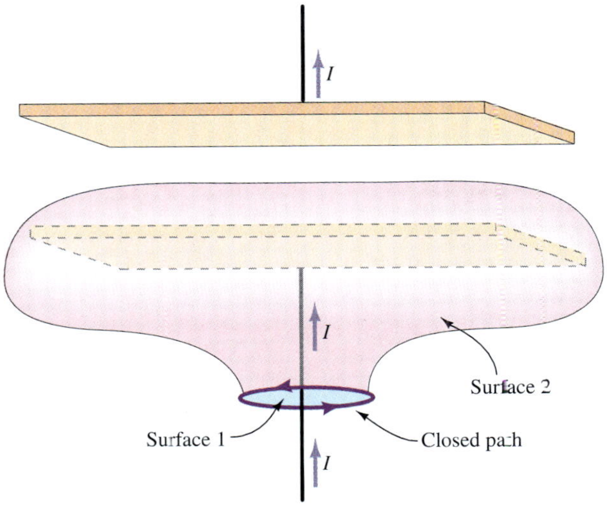

FIGURE 30–29 Two surfaces bounded by the same closed path. A current passes through surface 1 but not through surface 2.

sulting electric field, and hence so does the electric flux in the space between the plates.

Suppose that we have a capacitor with large, planar plates of area A (Fig. 30–30). As the current I accumulates on one of the plates, it deposits a charge $+q$. We can then use the results of Chapters 23 and 26: The field **E** of such a capacitor is uniform between the plates and small outside the region between the plates. The field points from the plate with positive charge (where the charge accumulates) to the plate with negative charge. The electric flux associated with a surface that passes between the plates, Φ_E, is then just EA. The magnitude of the electric field is given by $E = (1/\epsilon_0)(q/A)$. Because $\Phi_E = EA$ in this case, this result is equivalent to

$$\epsilon_0 \Phi_E = q. \tag{30–28}$$

If we take a time derivative, we find the relation

$$\epsilon_0 \frac{d\Phi_E}{dt} = \frac{dq}{dt} = I. \tag{30–29}$$

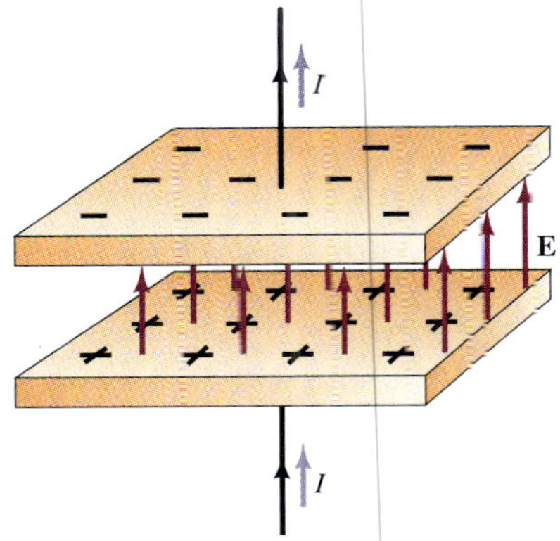

FIGURE 30–30 The electric field between two parallel plates that carry charges $+q$ and $-q$, respectively, is uniform inside and small outside the region between the plates.

Equation (30–29) implies that whatever the value of the current that passes through the wire that leads to the capacitor, that current equals the quantity $\epsilon_0\, d\Phi_E/dt$ between the plates. Therefore if we replace I in Ampère's law by the *sum* of the two terms in Eq. (30–29),

$$I + \epsilon_0 \frac{d\Phi_E}{dt},$$

Ampère's law would be satisfied for *any* surface we could draw for the path of Fig. 30–29. For surface 1, only the term I in this sum applies; for surface 2, only the changing flux term applies. The second term, $\epsilon_0\, d\Phi_E/dt$, is written in a way that does not refer explicitly to the plane geometry. Indeed, Maxwell was able to show that if the sum of these two terms is used, any surface gives the same answer in Ampère's law. Maxwell called the changing flux term the **displacement current**, I_d:

The displacement current

$$I_d \equiv \epsilon_0 \frac{d\Phi_E}{dt}. \tag{30–30}$$

Note that the displacement current is present only when there are changing currents; a truly steady current cannot involve changing flux. But any current that enters a capacitor *must* be changing. For example, the current decreases as the capacitor plates build up charge; when the plates are charged to saturation, the current stops.

The sum of the ordinary current and the displacement current through any closed loop is unchanging.

Even though the current is not continuous when capacitors are present, *the sum of the ordinary current and the displacement current is continuous.*

Maxwell's generalized form of Ampère's law is accordingly

The generalized Ampère's law. It applies even when the currents change with time.

$$\oint \mathbf{B} \cdot d\mathbf{s} = \mu_0(I + Id) = \mu_0 I + \mu_0 \epsilon_0 \frac{d\Phi_E}{dt}. \tag{30–31}$$

Here, the sum $I + I_d$ is calculated with reference to *any* surface that spans the closed path defining the line integral of the magnetic field.

That a changing electric flux produces a magnetic field has great importance for electromagnetic waves, as we shall see in Chapter 35. Example 30–7 shows, however, that it has very little practical effect otherwise.

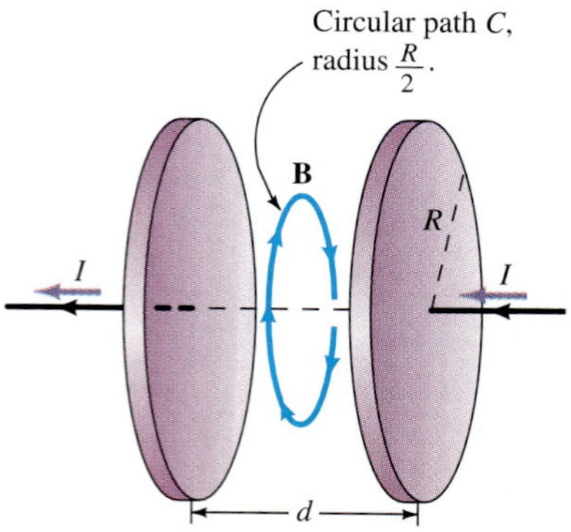

FIGURE 30–31 Example 30–7. Symmetry requires that the value of the magnetic field is the same everywhere on path C.

EXAMPLE 30–7 The planar circular plates of a capacitor are being charged. At a given moment, the charge is being built up at the rate of 1 C/s. The plates have a radius $R = 0.1$ m and a separation $d = 1$ cm. Calculate the magnetic field due to the displacement current midway between the plates at a radius equal to half the plate radius.

Solution: Because the plates are circular, symmetry requires that the value of the magnetic field is the same everywhere on path C, a circular path centered on the plates' axes and of radius $R/2$ (Fig. 30–31). The line integral in Ampère's law is taken in the sense drawn. Because **B** has constant magnitude on this path and points along the path, we may remove it from the integral. The remaining integral is the circumference of the path, $2\pi(R/2)$:

$$\oint \mathbf{B} \cdot d\mathbf{s} = B \oint ds = B\left(2\pi \frac{R}{2}\right).$$

In order to calculate the displacement current, we note that, in terms of the charge q on the plates, the electric field is constant and has magnitude given by

$$E = \frac{1}{\epsilon_0} \frac{q}{\pi R^2}.$$

We must now calculate the electric flux through the area bounded by path C (and not the *total* electric flux in the capacitor). The flux through path C is E times the area $\pi(R/2)^2$:

$$\Phi_E = \frac{1}{\epsilon_0} \frac{q}{\pi R^2} \pi \left(\frac{R}{2}\right)^2 = \frac{1}{4\epsilon_0} q.$$

Thus the displacement current is

$$I_d = \epsilon_0 \frac{d\Phi_E}{dt} = \epsilon_0 \left(\frac{1}{4\epsilon_0} \frac{dq}{dt}\right) = \frac{1}{4} \frac{dq}{dt}.$$

We can find the magnitude B by using Eq. (30–31):

$$B\, 2\pi \frac{R}{2} = \mu_0 I_d = \frac{\mu_0}{4} \frac{dq}{dt};$$

then

$$B = \frac{\mu_0}{4\pi} \frac{1}{R} \frac{dq}{dt}.$$

Numerically, $dq/dt = 1$ C/s $= 1$ A and $R = 0.1$ m, so

$$B = \frac{4\pi \times 10^{-7}\ \mathrm{N/A^2}}{4\pi} \frac{1}{0.1\ \mathrm{m}} (1\ \mathrm{A}) = 10^{-6}\ \mathrm{T}.$$

This is indeed a small field; for comparison, recall that Earth's magnetic field at Earth's surface is around 10^{-4} T.

SUMMARY

Magnetic fields are produced by electric currents and, equivalently, by moving charges. The magnetic field lines about a long, straight wire that carries a constant current form circles around the wire in the plane perpendicular to the wire. The direction of the field lines in these circles is determined when the thumb of the right hand points along the direction of the current flow: The fingers then curl in the direction of the magnetic field. The magnitude of the field at a radial distance r from the wire is

$$B = \frac{\mu_0 I}{2\pi r}. \tag{30–5}$$

The defined constant $\mu_0 = 4\pi \times 10^{-7}\ \mathrm{T \cdot m/A}$ is the permeability of free space.

The magnetic fields produced by unchanging currents obey Ampère's law:

$$\oint \mathbf{B} \cdot d\mathbf{s} = \mu_0 I_{\text{enclosed}}. \tag{30–10}$$

Here, the line integral follows any closed path through which the current I_{enclosed} passes. A second law obeyed by the magnetic field springs from the absence of magnetic equivalents to the electric charge. Because there are no magnetic charges on which magnetic field lines begin or end, magnetic field lines must close on themselves. This fact is expressed by Gauss' law for magnetism:

$$\text{for a closed surface:} \qquad \Phi_B = \int_{\text{closed surface}} \mathbf{B} \cdot d\mathbf{A} = 0. \tag{30–12}$$

This law states that the magnetic flux, Φ_B, through any closed surface is zero; equivalently, the number of magnetic field lines that enter a closed surface is the same as the number of lines that leave the surface.

Ampère's law is an important practical tool for determining magnetic fields when there is enough symmetry to allow a path choice in which the integral simplifies, as in the determination of the interior field of a long solenoid. A solenoid is a wire wound uniformly into a coil to form a tube. When current flows, a magnetic field is produced within the tube, has constant magnitude, and is aligned with the tube axis. The magnitude of the interior field is

$$B = \mu_0 n i, \tag{30–15}$$

where n is the number of windings of wire per unit length of the solenoid. Because its interior magnetic field is constant, a solenoid is to magnetism what a capacitor is to electricity.

When there is not enough symmetry to allow Ampère's law to be used to determine the magnetic field produced by a given configuration of currents, the Biot–Savart law can be used instead. According to this law the magnetic field $d\mathbf{B}$ produced by a segment of wire $d\boldsymbol{\ell}$ that carries a current I at a displacement $\mathbf{r}$ from the segment is given by

$$d\mathbf{B} = \frac{\mu_0}{4\pi}\frac{I\, d\boldsymbol{\ell} \times \mathbf{r}}{r^3}. \tag{30–19}$$

The magnetic field from an infinitesimal segment can be integrated to find the net magnetic field due to a finite segment of wire.

Application of the Biot–Savart law shows that the magnetic field due to a ring of current, or the exterior field of a solenoid of finite length, is a magnetic dipole field. The form of this magnetic dipole field is the same as that of a bar magnet or, equivalently, has the same form as the exterior electric field produced by an electric dipole. The current loop forms a magnetic dipole, characterized by a magnetic dipole moment, $\boldsymbol{\mu}$, which is aligned perpendicular to the surface of the loop according to a right-hand rule. Its magnitude is

$$\mu = IA, \tag{30–27}$$

where A is the loop area.

If currents are not constant, as when wires are interrupted by the presence of charging capacitor plates, then one surface that spans a closed path might not cross the wire that another surface might cross, and the concept of the current enclosed by a path becomes ambiguous. This ambiguity is remedied by Maxwell's modification of Ampère's law to

$$\oint \mathbf{B}\cdot d\mathbf{s} = \mu_0(I + I_d) = \mu_0 I + \mu_0\epsilon_0 \frac{d\Phi_E}{dt}. \qquad (30\text{–}31)$$

The quantity I_d, proportional to the rate of change of electric flux, is known as the Maxwell displacement current. The surface through which the sum of I and I_d passes is any surface that spans the closed integration path.

UNDERSTANDING KEY CONCEPTS

1. Suppose you move a compass needle near a straight wire that carries a current. Describe how the compass needle reacts when the compass is moved slowly in a circle centered on the wire and perpendicular to it.
2. In the definition of the ampere, does the length of the two parallel wires matter?
3. A wire connected to a battery is placed in the yoke of a tabletop electromagnet when a switch is open. When the switch is closed, the wire may take a big jump upward or it may not, according to which side of the battery terminals the wires are attached. Why?
4. Suppose the torus of Figure 30–21 is replaced by a tube that has an almost rectangular cross section and whose shape is irregular (e.g., an extended ellipse). Current-carrying wire is wrapped closely around the tube. What can you say about the magnetic field?
5. In the definition of the ampere, must we worry about the Coulomb forces between the charges in the two wires?
6. Why is the Biot–Savart law written in differential form? Explain why it cannot be written as in Eq. (30–19) but without the differential signs.
7. Why is it preferable to define current in terms of the force between two long, parallel wires rather than in terms of the rate at which charge passes a point?
8. Is it possible to arrange a set of electric currents and produce a magnetic field that, at large distances from the apparatus responsible, is everywhere directed radially away from the apparatus? Feel free to choose your apparatus, and give either a proof that it is impossible or a description of the apparatus.
9. Suppose that the space between the plates of the capacitor discussed in Section 30–6 is not empty but is filled with a dielectric. How would the treatment in that section, and the determination of the displacement current, change?
10. Suppose that magnetic charges were discovered. What are some practical consequences?
11. What are the SI units of the ratio E/B, where E is an electric field and B is a magnetic field?
12. When two bar magnets are placed side by side, they will (a) attract or (b) repel if the adjacent poles are (a) opposite or (b) the same. If you draw magnetic field lines for the combination of two magnets in both cases, the net magnetic field between the magnets will tend to (a) cancel or (b) be doubled. What conclusions can you draw?

PROBLEMS

30–1 Ampère's Law

1. (I) Sketch the magnetic field lines due to two current-carrying wires that are parallel to each other for the cases that (a) the currents move parallel to each other, and (b) the currents move in opposite directions.
2. (I) Consider an array of parallel, current-carrying wires arranged so that they all lie in a plane and are separated by equal distances. Sketch the magnetic field due to this array. Assume that all the currents point in the same direction.
3. (I) What is the force per meter between two long, parallel wires, each carrying 50,000 A but in opposite directions, if the two wires are 30 cm apart? Such currents are normal in the electrolytic production of aluminum.
4. (I) A long, straight wire carries a current of 250 A. What is the magnetic field at a distance of 5.0 m from the wire?
5. (I) A lightning conductor carries a current of 5×10^4 A for a short period. During that time, what is the magnitude of the magnetic force per unit length exerted on a parallel wire 4 m away in which a current of 100 μA flows?
6. (I) Two 2-cm-long parallel wires in a handheld calculator are 4 mm apart. The currents are parallel and have values 0.5 μA and 2.2 μA, respectively. What is the force between the wires due to the currents?
7. (I) Find the dimensions of μ_0 and ϵ_0, and use your expressions to show that the product $\mu_0\epsilon_0$ has dimensions of $(1/\text{speed})^2$. Find the value of that speed in SI units.
8. (I) A coaxial cable consists of a central wire that carries current I to the right and a tube centered on the central wire that carries the same current to the left. Find the magnetic field outside the cable.
9. (II) (a) In a thick, straight wire carrying a current that is uniform through its cross section, where is the magnetic field the

greatest? (b) If the radius of the wire is R and the current is I, what is the value of the maximum magnetic field? (c) What is the minimum magnetic field, and where does this occur? Consider regions both inside and outside the wire. (d) Plot the magnetic field as a function of the distance from the center of the wire.

10. (II) The current-carrying capacity of superconducting wires is limited by the fact that superconductivity breaks down if a large magnetic field is present. Estimate the largest current that can be transported via a NbTi wire, 0.5 mm in diameter, if the critical (break-down) magnetic field is 10 T. [*Hint*: What is the wire's own magnetic field?]

11. (II) Plot the curves of constant magnetic field in the xy-plane for values B_0, $2B_0$, $3B_0$, and $4B_0$ of the field about a straight wire that carries current along the z-axis. B_0 is some field value that you can choose. These curves are the intersections with the xy-plane of the surfaces of constant field.

12. (II) A very thin, infinitely long metal sheet lies in the xy-plane, between $x = -w$ and $x = w$. A current of density h A/m flows in the x-direction (Fig. 30–32). What are the magnitude and direction of the magnetic field at a distance $z << w$ above and below the sheet? Neglect end effects.

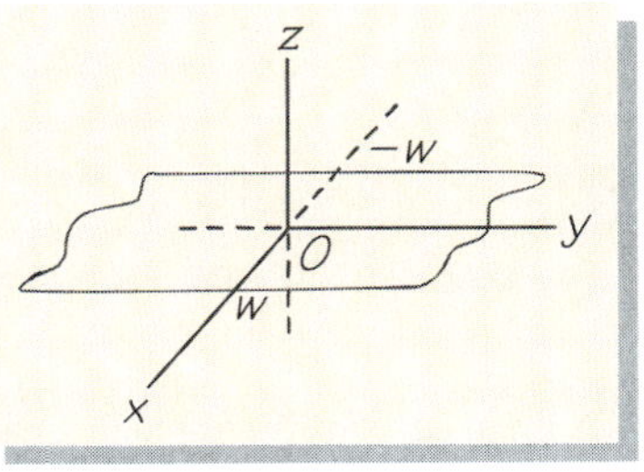

FIGURE 30–32 Problem 12.

13. (II) Current is carried from a battery to a device by a copper "ribbon" 1 in wide and 1/32 in thick. What is the magnetic field over the surface of the ribbon, if the current it carries is 120 A?

14. (II) Consider two parallel metal sheets, such as the sheet of Problem 12, with currents flowing in opposite directions. What are the magnetic fields between and outside the sheets? What is the situation when the currents are parallel rather than antiparallel?

15. (II) Consider a wire that passes through the origin along the z-axis and carries a current I (Fig. 30–33). (a) Calculate the x- and y-components of the magnetic field at a point whose coordinates are $(x, y, 0)$. (b) Use this result to obtain the magnetic field due to two wires that are parallel to the z-axis, cross the xy-plane at $(a, 0)$ and $(-a, 0)$, and carry current I in the $+z$-direction. (c) What are the fields when the currents are in opposite directions?

16. (II) A uniform current with current density h A/m flows parallel to the z-axis on a cylindrical metal sheath, where the radius of the cylinder is R. What is the magnetic field outside the sheath? inside the sheath?

17. (II) Current flows up the inner cylinder of a coaxial cable and returns on the outside cylinder. The radius of the inner cylinder is 0.1 cm, and the radius of the thin outer cylindrical shell is 0.5 cm. Calculate the magnetic field on the cylindrical surface midway between the inner and outer surfaces, given that the current is 10 A. Ignore end effects.

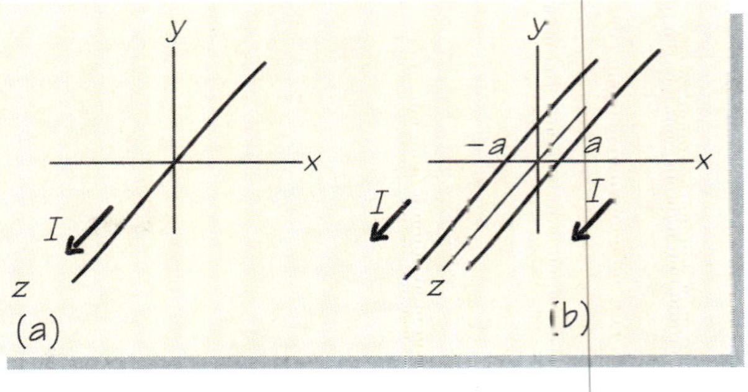

FIGURE 30–33 Problem 15.

18. (II) Two long, parallel wires carrying a current I in the same direction each have a mass density λ. The wires are initially a distance D apart and are then released. Write a differential equation for the distance between the wires that describes the relative motion of the wires. Ignore all forces other than the magnetic force.

19. (III) An electron beam contains electrons that move along the $+x$-axis at $0.020c$. The beam enters a region of length 1.0 m and runs parallel to a wire that carries a 0.20-A current in the $+x$-direction. The beam is 10.0 cm from the wire. (a) Specify the direction in which the beam is deflected, if at all. (b) Find the deflection of the beam as it passes through the 1-m region by calculating the impulse it receives during its brief passage through that region. (c) After it has passed the wire, does the beam have the same energy it had when it entered the region that contains the wire?

30–2 *Gauss' Law for Magnetism*

20. (I) Figure 30–34 shows the magnetic field lines that emerge from one pole of a bar magnet; these lines resemble the electric field lines that emerge from one end of an electric dipole. Sketch the magnetic field lines in the region of the pole inside the magnet. For comparison, also sketch the electric field lines in the central region of an electric dipole. What would the magnetic lines look like if the N and S poles of a magnet represented magnetic monopoles, which would be point sources of magnetic field?

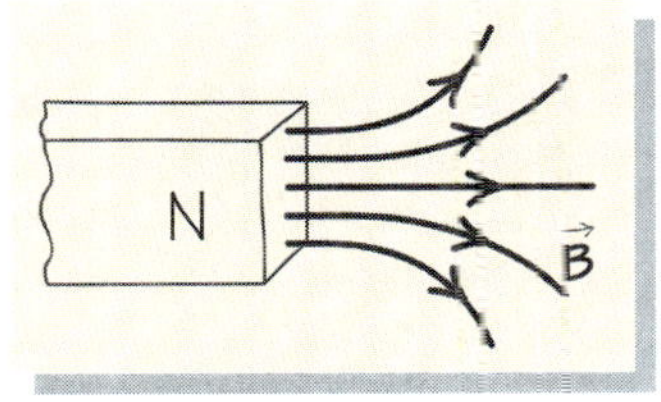

FIGURE 30–34 Problem 20.

21. (II) A long, current-carrying wire is oriented vertically; next to it is drawn a square whose area lies in the same plane as the wire (Fig. 30–35). Using the distances indicated, find the magnetic flux through the square.

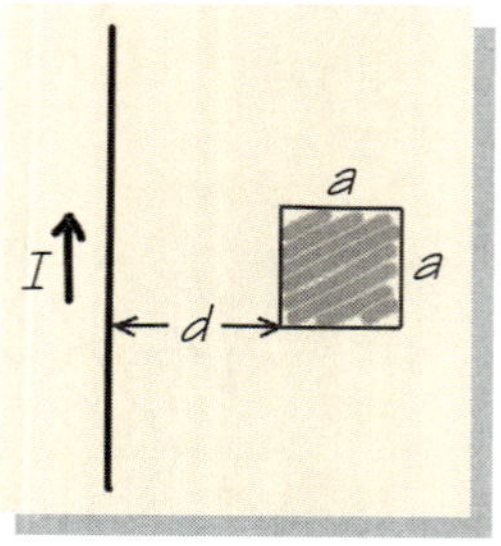

FIGURE 30–35 Problem 21.

22. (II) Using Gauss' law for magnetism, show that a magnetic field with only an x-component must be constant as x varies.

23. (II) Show that Gauss' law is satisfied for the magnetic field due to a straight wire that carries a current I in the $+z$-direction for a volume that represents a portion of a cylindrical shell of height h, extending from a radius r to a radius R and formed by an angle θ (Fig. 30–36).

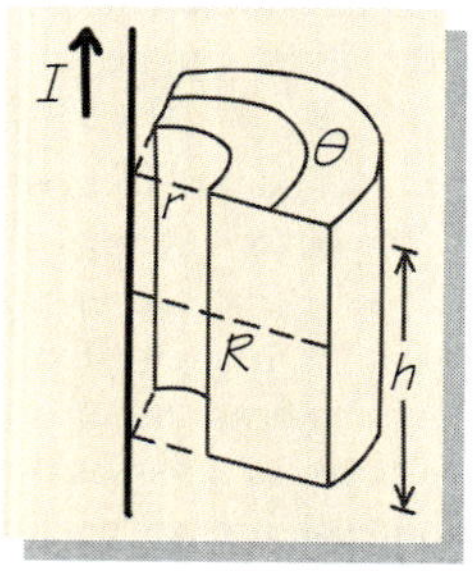

FIGURE 30–36 Problem 23.

24. (III) Apply Gauss' law for magnetism to a parallelepiped of dimensions a, b, and c, one of whose corners is located at point (x, y, z), as in Fig. 30–37. Assume that the dimensions in the x-, y-, and z-directions (a, b, and c) are small enough so that $B(x + a, y, z) = B(x, y, z) + a\, \partial B/\partial x$, and so on. Show that Gauss' law leads to the condition $(\partial B_x/\partial x) + (\partial B_y/\partial y) + (\partial B_z/\partial z) = 0$ in this limit.

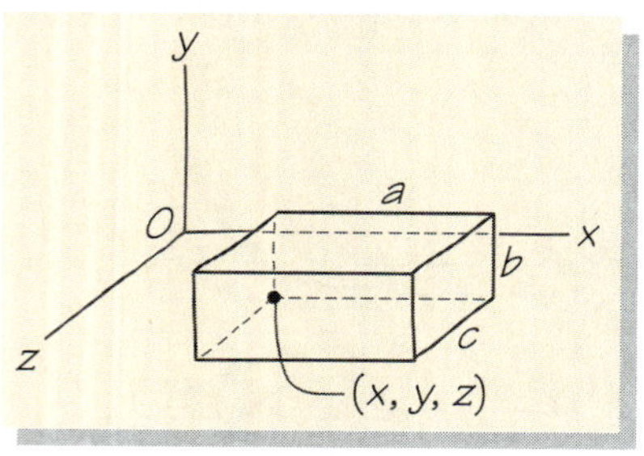

FIGURE 30–37 Problem 24.

30–3 Solenoids

25. (I) A solenoid of diameter 5 cm has a length of 25 cm and 320 turns of wire. What is the magnetic field at the center of the solenoid when the current in the coil is 3 A?

26. (I) A long, superconducting solenoid is wound with fine niobium–tin wire so that there are 5×10^4 turns/m. If a power supply produces 65 A, what is the magnetic field inside the solenoid?

27. (I) You are told that a toroidal solenoid, carrying a current of 0.36 A through 2500 turns, produces a magnetic field of 5.1×10^{-4} T at its central axis. What can you conclude about the radius of the circle made by that axis?

28. (I) A wire is wound around a torus with outer radius of 30 cm and inner radius of 27 cm. There are 3000 turns in all and the wire carries a current of 10 A. What is the range of the magnetic field inside the torus? What percentage change is there from the center to the outside?

29. (I) The magnetic field inside a cylindrical solenoid of area 4 cm^2 is 0.15 T along the axis of the solenoid. What is the magnetic flux through a disk of radius 3 cm placed perpendicular to the solenoid axis (Fig. 30–38)?

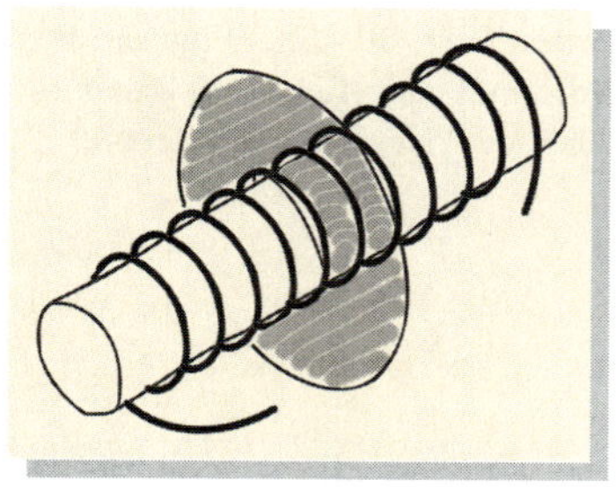

FIGURE 30–38 Problem 29.

30. (I) Show that the magnetic flux through an ideal cylindrical solenoid of radius R is given by the formula $\Phi_B = \mu_0 n I \pi R^2$, where n is the turn density.

31. (II) Consider a toroidal solenoid with a square cross section, each side of which has length L. The inner wall of the torus forms a cylinder of radius R. The torus is wound evenly with N loops of wire, and a current I flows through the wire. What is the total magnetic flux through the torus?

32. (II) A toroidal solenoid—similar to the one considered in Problem 31—has a square cross section of side length 3 cm and an inner radius of $R = 12$ cm. It is wound with 200 turns of 0.3 mm-diameter copper wire. The wire is connected to a 3-V battery with negligible internal resistance. (a) Calculate the largest and smallest magnetic field across the cross section of the toroid. (b) Calculate the magnetic flux through the torus. (c) Do you need to cool the solenoid?

30–4 The Biot–Savart Law

33. (I) Two long wires are placed along the y- and z-axes, respectively. They carry the same current I in the positive directions. Calculate the magnetic field along the x-axis.

34. (I) A single loop of wire forms a rectangle whose sides have lengths 2.5 cm and 3.7 cm. The wire carries a current of 170 mA. What is the magnetic dipole moment of the loop?

35. (II) An infinitely long L-shaped wire is placed so that a current I flows in along the y-axis toward the origin, then out from the origin along the x-axis. What is the magnetic field at a point on the z-axis at a height H above the origin?

36. (II) Consider a straight segment of wire of length L that carries a current I. Use the Biot–Savart law to find the magnetic field along the axis of the wire, beyond the wire itself, due to this segment.

37. (II) A differential length $\mathbf{dL}$ of wire carrying a current of 2 A is positioned at the origin of a coordinate system and points in the $+x$-direction. Find the magnetic field due to this wire segment at the following (x, y, z) positions, given in centimeters: (a) (0, 0, 3), (b) (0, 6, 0), (c) (3, 0, 0), and (d) (6, 0, 6). Give both the magnitude and direction of the magnetic field.

38. (II) Consider a thin dielectric ring 3 cm in diameter that rotates around a stem perpendicular to the plane of the ring and through its center at the rate of 200 rev/s. Assume that the ring is charged uniformly and carries a total charge of 2×10^{-5} C. What is the magnetic field produced at the center of the ring by the rotating charge?

39. (II) Repeat the calculation of Problem 38 for a solid disk 5 cm in diameter, with the same total charge.

40. (II) A current loop consists of a square with sides of length L. A current I circulates counterclockwise around the loop. Find the direction and magnitude of the magnetic field at the center of the square. Compare this to the field at the center of a circular loop of diameter L that carries the same current.

41. (II) Consider the wire shown in Fig. 30–39. Calculate the magnetic field at point P, the center of the half-circle of radius R around which the wire turns, as a function of R and the current I carried by the wire.

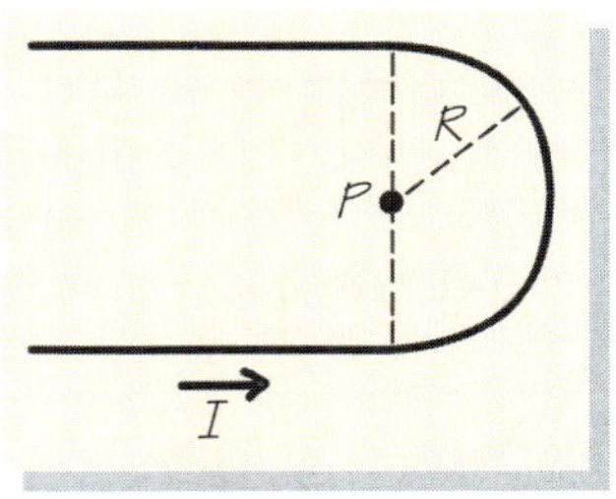

FIGURE 30–39 Problem 41.

42. (II) A very long wire is aligned along the $+x$- and $-y$-axes, making a right angle at the origin. A current I travels in the $-y$-direction and continues in the $+x$-direction. What is the magnetic field at the point (x, y), where both x and y are positive?

43. (II) Consider the wire shown in Fig. 30–40 with the inner and outer radii of the semicircle given as 5 cm and 8 cm, respectively. Given that the current in the wire is 12 A, what is the magnetic field at point P, the center of the semicircles?

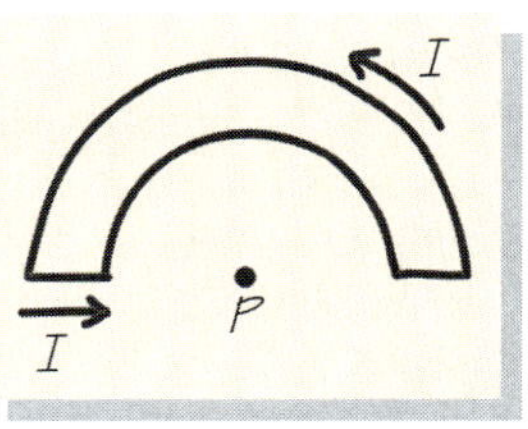

FIGURE 30–40 Problem 43.

44. (II) Calculate the magnetic field at the center of a wire square that consists of 200 loops and has sides of length 5 cm and carries a current of 0.1 A.

45. (II) A charge q moves at instantaneous speed v when it crosses the axis of a ring of current with a magnetic dipole moment μ. At that instant, q is located a distance d from the center of the ring in the direction of the dipole moment vector and is moving perpendicular to the axis (Fig. 30–41). What is the resulting instantaneous motion of the charge? Find the instantaneous radius of curvature of its motion.

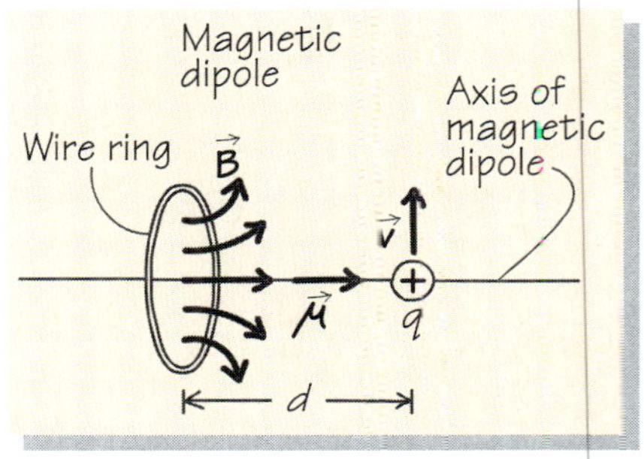

FIGURE 30–41 Problem 45.

46. (II) A circular current loop of radius R produces a magnetic field. At what distance along the axis of the loop does the field have magnitude 0.9 times the magnitude at the center of the loop? At what distance is the magnitude of the field reduced to 1/10 the value at the center? Give your answer in units of R.

47. (II) Find the magnetic field at point P in Fig. 30–42 if a current of 8 A flows in the infinitely long wire; the radius R of the semicircle is 1.2 cm.

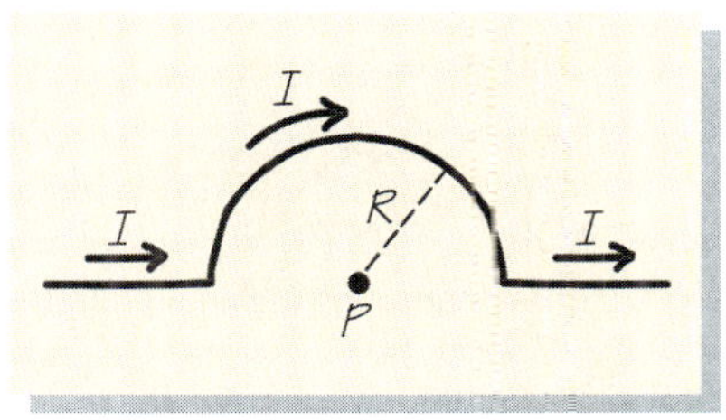

FIGURE 30–42 Problem 47.

48. (III) A segment of wire forms a straight line of length L and carries a current I. Find the magnetic field due to the wire segment in the plane perpendicular to it and passing through one end.

49. (III) By integration, find the magnetic dipole moment of a spherical shell of radius R that carries a total charge Q, distributed uniformly, if the shell rotates with angular velocity ω oriented along the z-axis

50. (III) Consider a long, thin-walled metal pipe that carries a total current I distributed evenly along the walls of the pipe. A simple application of Ampère's law indicates that the magnetic field inside the pipe is zero. Show by a simple geometric argument that the same result follows from the Biot–Savart law.

30–5 *The Maxwell Displacement Current*

51. (I) Consider the *RC* circuit shown in Fig. 30–43. Switch *S* is closed at time $t = 0$. Calculate the displacement current in the capacitor.

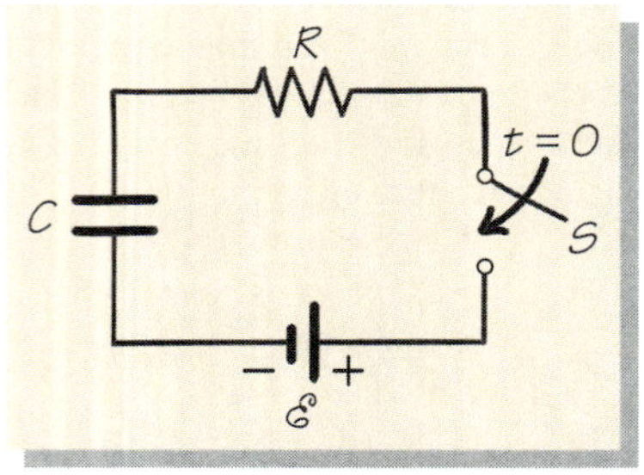

FIGURE 30–43 Problem 51.

52. (I) Consider the *RC* circuit shown in Fig. 30–43; this time the switch is closed. At some time, the switch is opened. What is the displacement current?

53. (I) A parallel-plate capacitor is being charged at a rate of $I = 0.2$ A. The plates have an area of 0.25 m^2 and are separated by 1.0 cm (Fig. 30–44). What is the value of $\int \mathbf{B} \cdot d\boldsymbol{\ell}$ for a closed path midway between the plates and covering an area of 5.0×10^{-2} m^2?

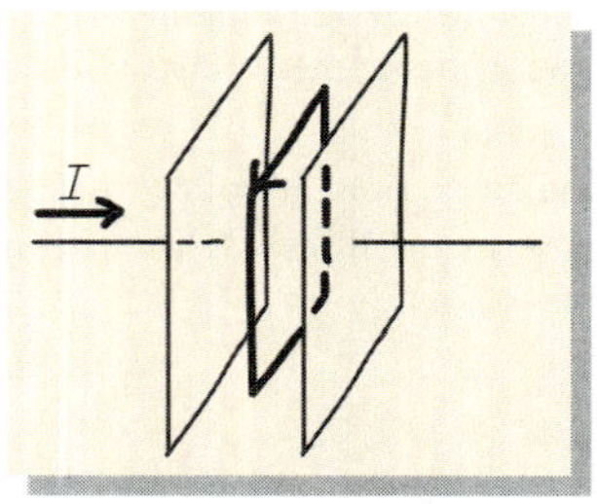

FIGURE 30–44 Problem 53.

54. (II) A 5.0-μA current starts flowing in a circuit with a 2.0×10^{-7} F capacitor of area 300 cm^2 at $t = 0$ s. (a) How fast is the voltage across the capacitor plates changing at $t = 0$ s? (b) Use the result of (a) to calculate *explicitly* $d\Phi_E/dt$ and the displacement current at $t = 0$ s.

55. (II) An alternating voltage of the form $V = V_0 \cos(\omega t)$ is connected across a capacitor *C*. What is the displacement current in the capacitor?

56. (II) A voltage of the form $V = V_0 \cos(\omega t)$, with $\omega = 2 \times 10^4$ rad/s and $V_0 = 0.1$ V, is applied across the plates of a 5-nF capacitor; the plates are 1.5 cm apart. (a) What is the maximum rate of change in electric field between the plates? (b) the maximum value of current leading to the capacitor?

57. (II) A conducting sphere of radius *R* initially has a uniform surface-charge density σ_0. Beginning at $t = 0$ this charge is drained off over a period t_0 such that $\sigma = \sigma_0 [1 - (t/t_0)]$. Find the displacement current at the surface of the sphere as a function of time. Compare the displacement current to the current carried off by the wire.

General Problems

58. (II) Calculate the force per unit area between two metal sheets that carry identical currents in the same direction. The sheets carry a current of linear density *h* A/m as in Problem 12.

59. (II) Equal but opposite currents *I* travel in the inner and outer wires of a coaxial cable. As a function of the distance from the central axis, find the magnetic field (a) inside the inner wire; (b) in the region between the wires; (c) in the outer (tubular) wire; (d) outside the outer wire.

60. (II) A hydrogen atom may be described as consisting of an electron that moves in a circular orbit around a proton. The force that gives rise to the motion is the Coulomb attraction between the proton and the electron, which have charges $\pm e$, respectively, where $e = 1.6 \times 10^{-19}$ C. The motion is further constrained by the requirement that the angular momentum has the value $nh/2\pi$, where *n* is an integer and $h = 6.63 \times 10^{-34}$ J·s, Planck's constant. Calculate the magnitude and direction of the magnetic field at the location of the proton. What is the magnetic moment of the current loop?

61. (II) Find the force between the long, straight wire and the rectangular wire loop shown in Fig. 30–45 for currents $I_1 = 10$ A and $I_2 = 5$ A.

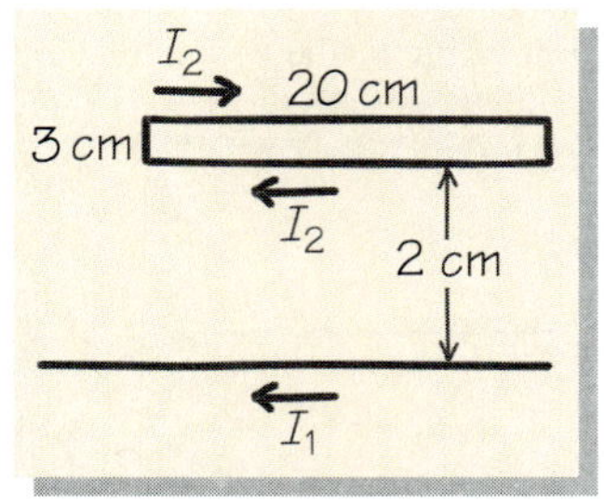

FIGURE 30–45 Problem 61.

62. (II) The mechanical integrity of solenoids may present a problem that has to be anticipated in technical design. To illustrate the forces that can be present, calculate the force between two neighboring turns of a superconducting solenoid. The radius of the torus is 4 cm, the diameter of the wire is 0.3 mm, and the current in the solenoid is 60 A.

63. (II) Consider two parallel wires spaced a distance $d = 1$ cm apart, which each carry a current $I = 1$ A. (a) Compare the magnetic force between these wires to the electric force they would exert on each other if the current carriers (electrons) were not neutralized by a background of positive charges. Use 10^{21} per cm as the linear density of charge carriers in the wire. (b) What excess of electrons per unit length over the positive background would make the electric force equal the magnetic force between the wires? (c) What fraction of the total number of charge carriers is the excess calculated in part (b)?

64. (II) A long wire carries a current I_1. A segment of a second wire, which carries a current I_2, is oriented radially away from the first wire. The segment has length *L*, and its closest end is a distance *d* from the first wire. Calculate the torque, direction and magnitude, on the wire segment about the axis defined by the long wire.

65. (II) In Example 30–6, we found the magnetic field due to a circular wire of radius R, carrying a current I, at a point a distance d away from the center of the ring but along the axis to be

$$B = \frac{\mu_0 I}{2} \frac{R^2}{(R^2 + d^2)^{3/2}}.$$

A pair of such coils placed coaxially a distance R apart makes up a *Helmholtz coil*, for which the magnetic field everywhere inside is fairly constant (Fig. 30–46). (a) Determine the magnetic field on the axis as a function of x, with $x = 0$ marking the location of the left-hand coil. Evaluate the field at $x = 0$, $x = R/4$, and $x = R/2$. (b) Show that $dB_x/dx = 0$ and $d^2B_x/dx^2 = 0$ at $x = R/2$.

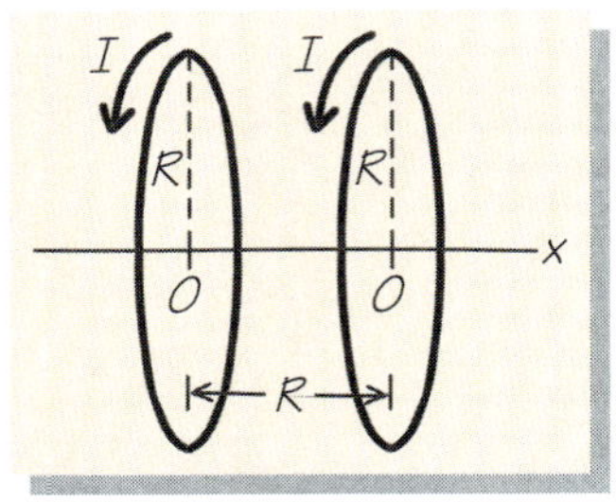

FIGURE 30–46 Problem 65.

66. (II) A sensitive experiment has to be performed in zero magnetic field. To achieve this, a Helmholtz coil of radius $R = 20$ cm, with 200 turns is used to compensate for Earth's magnetic field of 5×10^{-5} T. (A Helmholtz coil is a current loop that sets up a magnetic field to cancel an external field, such as that due to Earth.) (a) What should be the current in the coil? (b) *Estimate* the residual field if the sensitive component of the equipment is confined to a thin cylindrical volume 10 cm long?

67. (III) A demonstration apparatus consists of a large glass bulb containing a small electron gun. The bulb is filled with rarified inert gas, which makes the trajectory of the electron visible. By placing the equipment in the magnetic field of a Helmholtz coil, the experiment described in Example 29–2 can be performed, and the ratio e/m for an electron can be determined. Design the equipment and select appropriate parameters for this demonstration.

68. (III) A certain electric current distribution produces a magnetic field of the form $\mathbf{B} = \beta(y\,\mathbf{i} - x\,\mathbf{j})$ near the origin of a coordinate system. Find the current distribution responsible.

CHAPTER

31

Michael Faraday delivering one of his famous public lectures in 1856. These lectures earned him great popular success.

Faraday's Law

Our treatment of the magnetic field, its sources, and its effects on moving charges and currents may have given you some hints of the intimate connection between electricity and magnetism. In this chapter, the connection becomes explicit in a new physical law, Faraday's law. This law describes how changes in magnetic fields, or changes in circuits that have magnetic fields nearby, produce electric fields. Faraday's law has far-reaching technological applications. It lies behind our entire system of electrical power generation and plays a role in most of the electronic devices we use.

31–1 FARADAY AND MAGNETIC INDUCTION

A great experimentalist recognizes the significance of an odd or unexpected measurement. He or she realizes that a small effect is not always experimental error. He or she pursues the effect systematically, checks its reality, and considers its ramifications from as many points of view as possible.

Michael Faraday exhibited all these qualities in the discovery of what we now know as Faraday's law. Faraday's law states that *changing magnetic fields generate electric fields*. More precisely, the *change* of the magnetic flux through any surface bounded by a closed line causes an emf around that line. This emf can induce a current in a wire. Faraday referred to an **induced emf** and called the current produced by a changing magnetic flux an **induced current**; he called the general phenomenon **magnetic induction**. Within several days of his observations of the first small effects, he completed a series of experiments that revealed essentially all the aspects of magnetic induction.

Faraday's 1831 discovery was not an accident. At the time, it was known that an electric current produces a magnetic field. Ampère's law formalized this; a current passing through the wires of a solenoid produces a magnetic field inside the solenoid. In a period of enthusiastic experimentation on electricity and magnetism, it was natural to ask if magnetic fields could themselves make currents flow. Faraday had pursued this sort of question for several years—an entry dated 1822 in his notebook sets the goal "convert magnetism into electricity." In 1831, he carried out the experiment shown schematically in Fig. 31–1. The battery sends a current through the coil on the left side of an iron ring, which acts as a solenoid. The galvanometer is used to indicate any current in the coil on the right side of the ring. The only unfamiliar element is the iron ring, which does two things: It carries the magnetic field set up by the left-hand coil within the torus, and hence through the right-hand coil (recall that magnetic field lines are closed), and—as we shall discuss in Chapter 32—it *magnifies* the size of the field set up in the left-hand coil. We say that the iron ring *links* the two coils. To Faraday's disappointment, he observed no effect on the galvanometer when a *steady* current passed through the left-hand coil. Faraday's intuition served him well, however, when he noticed a very small twitch of the galvanometer *as the switch that controlled the flow of current in the left-hand coil was opened or closed.*

One of the first things Faraday did was to eliminate the possibility that the battery was itself important to the effect. In a second experiment, illustrated in Fig. 31–2, two bar magnets make a vee shape. A large magnetic flux passes through an iron rod when it touches the ends of the two magnets as shown. This rod is surrounded by a coil attached to a galvanometer. The galvanometer deflects—indicating a current in the coil—either when the rod is brought into contact with the ends of the two bar magnets or when the rod is pulled away. The crucial phenomenon leading to the current is that the magnetic flux through the coil *changes.*

Faraday's discovery was greeted with great excitement because the possibility of converting mechanical energy to electric energy thereby became a reality. Indeed, electricity generation worldwide is based on Faraday's results, so we might hazard a guess that Faraday's discovery has had a greater effect on the material welfare of humans than has any other discovery before or since.

31–2 FARADAY'S LAW OF INDUCTION

Let's illustrate some aspects of Faraday's law of induction. Figure 31–3a shows a bar magnet in the vicinity of a wire loop. There is an induced current in the wire loop only when there is movement, or, more precisely, when the magnetic flux through the loop changes. A faster movement—a more rapidly changing flux—results in a larger current than does a slower movement. Figure 31–3b shows a switch closing in the circuit of one loop; a current momentarily appears in the second loop. Figure 31–3c shows a current-carrying loop in the vicinity of a second loop; a current appears in the second loop when the distance between the loops changes. Figure 31–3d shows two similar loops; a current appears in the second loop when the orientation of the two loops change. Note that it is the change not just of the magnetic field but of the magnetic flux that induces a current. For example, if the bottom coil of Fig. 31–3c is squeezed, *simply*

The magnetic flux and its properties were first discussed in Chapter 30.

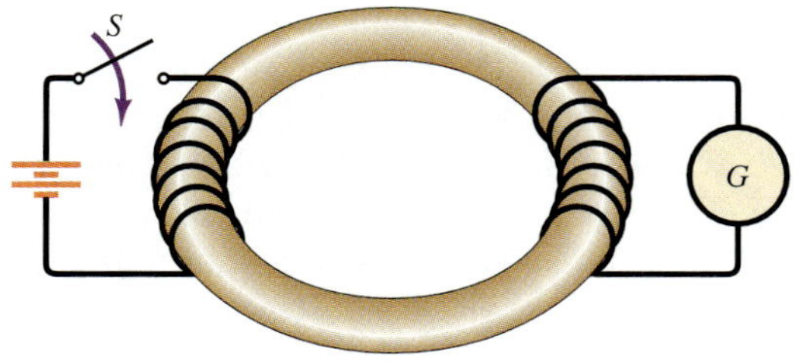

FIGURE 31–1 Faraday's ring. A changing magnetic flux in the iron ring induces a current in the galvanometer coil on the right; the changing flux is due to the opening or closing of a switch connected to the battery of the coil on the left.

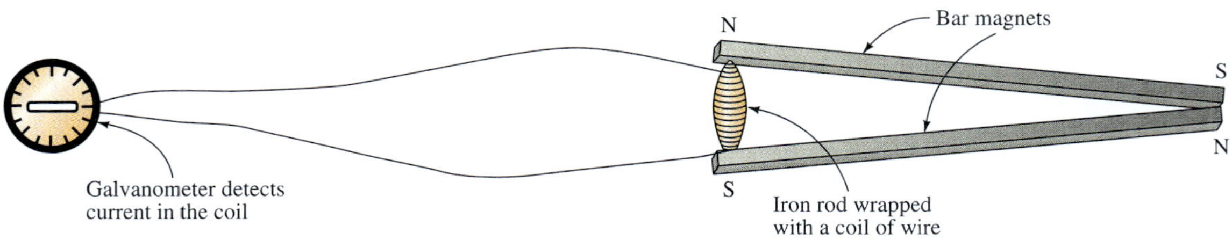

FIGURE 31–2 A figure from one of Faraday's lectures. The movement described in the text creates a changing magnetic flux and induces a current even when no battery is present. The figure is from Faraday's book *Experimental Researches in Electricity* in 1839; the labeling is our own.

changing its area, the magnetic flux through it changes even though the magnetic field due to the upper coil does not change, and an emf is induced in the bottom coil.

Faraday's law summarizes these observations: the negative of the *time rate of change* of the magnetic flux through a surface, Φ_B, equals an emf around the closed loop that bounds the surface. We know from Chapter 28 that an emf, $\mathscr{E}$, is an electric potential change; that is, a line integral of an electric field. In this case we are interested in the line integral around a *closed* loop:

$$\mathscr{E} = \oint \mathbf{E} \cdot d\mathbf{s}. \tag{31–1}$$

Faraday's law of induction

The precise statement of **Faraday's law of induction** is

$$\mathscr{E} = \oint \mathbf{E} \cdot d\mathbf{s} = -\frac{d\Phi_B}{dt}. \tag{31–2}$$

PROBLEM SOLVING

A right-hand rule for Faraday's law

Here, Φ_B is the magnetic flux through the surface S that spans the loop:

$$\Phi_B = \int_{\text{surface } S} \mathbf{B} \cdot d\mathbf{A}. \tag{30–11}$$

The loop around which the emf is defined, Eq. (31–1), must bound the surface through which the flux is calculated, and the orientation of that surface is determined by the direction of the loop integral and a right-hand rule. This right-hand rule works as follows: If the fingers of the right hand curl in the direction of the loop, the thumb indicates the direction of the surface for calculating the flux—the direction the surface element $d\mathbf{A}$ takes (Fig. 31–4).

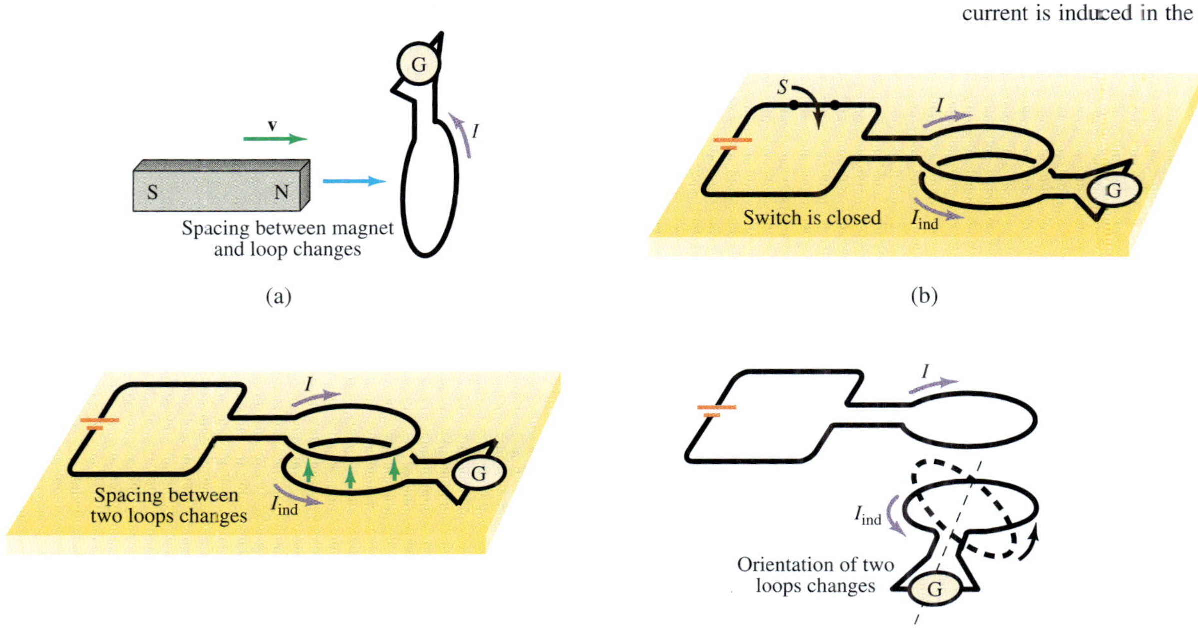

FIGURE 31–3 Ways to make a magnetic flux through a loop change and thereby induce a current I_{ind}. (a) The distance between a wire loop and a bar magnet changes, and a current is induced in the loop. (b) A switch is closed to start a current in one loop, and a current is induced in a second, nearby loop. (c) The distance between a current-carrying loop and a second loop changes, and a current is induced in the second loop. (d) A second loop rotates toward the current-carrying loop, and a current is induced in the second loop.

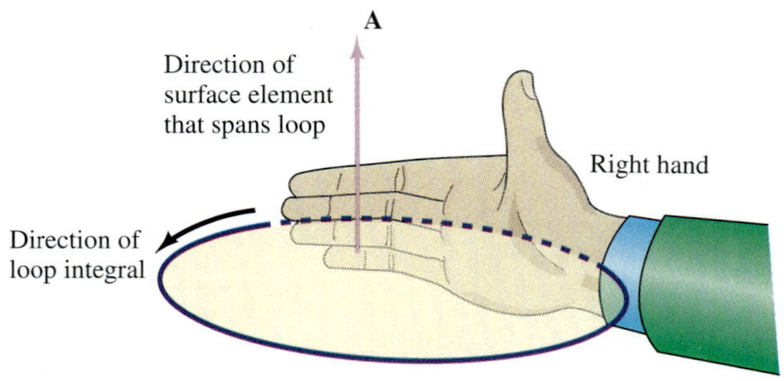

FIGURE 31–4 When the direction around a loop is given, the orientation of the surface that spans the loop is specified by a right-hand rule. Here we show a single vector **A** for the entire surface, which is flat. For a curved surface, the directions of infinitesimal areas d**A** vary from point to point.

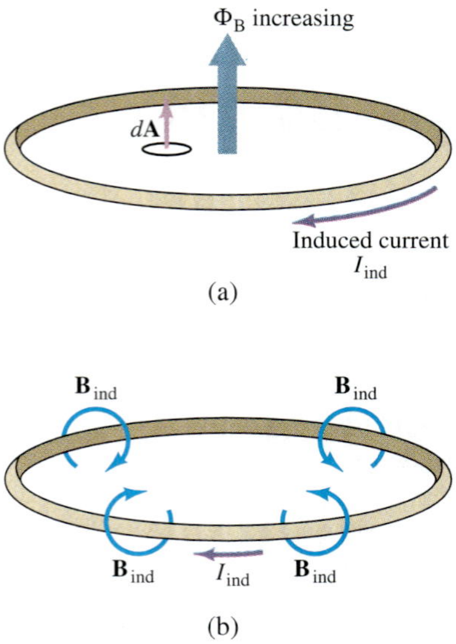

FIGURE 31–5 (a) A magnetic flux change induces a current. (b) The induced current produces its own magnetic field; that field tends to keep down the flux change that induced the current in the first place.

Lenz's Law and the Direction of Induced Current

The minus sign in Eq. (31–2) is so critical that it deserves a special discussion. Let's look at a loop of wire and suppose the magnetic flux *increases* through it in the sense shown in Fig. 31–5a. (This could happen in a number of ways; for example, we could move a bar magnet with the proper orientation toward it.) When we say that the flux is increasing, we mean that its time derivative is positive, so the right-hand side of Faraday's law is negative. The induced emf is *negative*. When we apply the right-hand rule to Fig. 31–5a, we see that the positive direction is counterclockwise; thus the negative sign means that the induced emf is *clockwise* in that figure. The resulting induced current will similarly be clockwise.

Now, we know that currents produce magnetic fields and the induced current is no exception. By using the right-hand rule, we can see that the magnetic field produced by the induced current is directed down through the loop (Fig. 31–5b). This field tends to *decrease* the magnetic flux through the loop. If we recall that the original flux change that induced the current in the first place was positive, we see that, in effect, the induced current has acted to oppose the flux change that caused it. Further analysis shows that the induced current always tends to keep the flux from changing. This way of thinking about Faraday's law is due to Heinrich Emil Lenz, and it is called **Lenz's law**:

Lenz's law

> **Induced currents produce magnetic fields that tend to oppose the flux changes that induce those currents.**

Lenz's law is a useful technique for determining the *direction* of an induced current.

PROBLEM SOLVING

Using Lenz's law

EXAMPLE 31–1 The north pole of a bar magnet is thrust toward the face of a fixed metal ring (Fig. 31–6). Use Lenz's law to determine the direction of any induced current in the ring.

Solution: As the north pole of the magnet approaches the ring, the magnetic field lines near the ring become denser. The magnetic flux through the surface of the ring, which is perpendicular to the magnet, *increases*. Lenz's law states that the induced current will *oppose* the change of magnetic flux that passes through the ring.

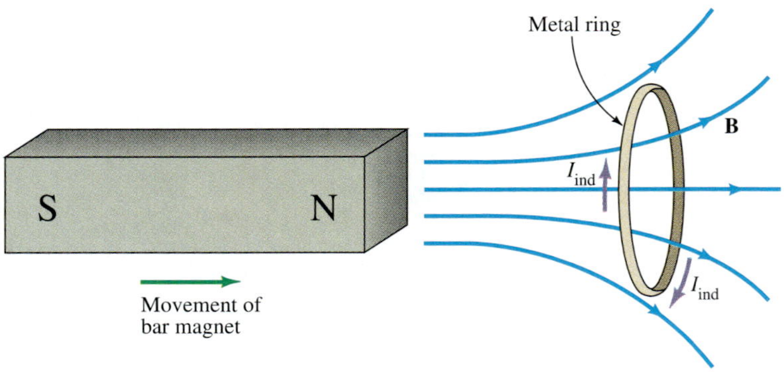

FIGURE 31–6 Example 31–1. When the bar magnet approaches the ring, a current is induced in the ring.

The induced current must therefore produce a field that serves to decrease the magnetic flux (and magnetic field). The induced magnetic field will be directed to the left. If we use the right-hand rule, the current in the ring that will produce this field is oriented clockwise as seen by an observer looking toward the north pole of the magnet.

EXAMPLE 31–2 A constant magnetic field has only a y-component B_0 in the region $x < 0$, and is zero for $x > 0$ (Fig. 31–7). A square metal loop with sides of length L is oriented in the xz-plane and pulled through the field with steady velocity $\mathbf{v} = v\mathbf{i}$. The total resistance of the loop is R. Find the induced current in the wire as a function of time, assuming that the front edge of the square crosses the line $x = 0$ at $t = 0$. Evaluate your result for $B_0 = 1.0$ T, $L = 0.10$ m, $R = 0.065\ \Omega$, and $v = 10.0$ cm/s.

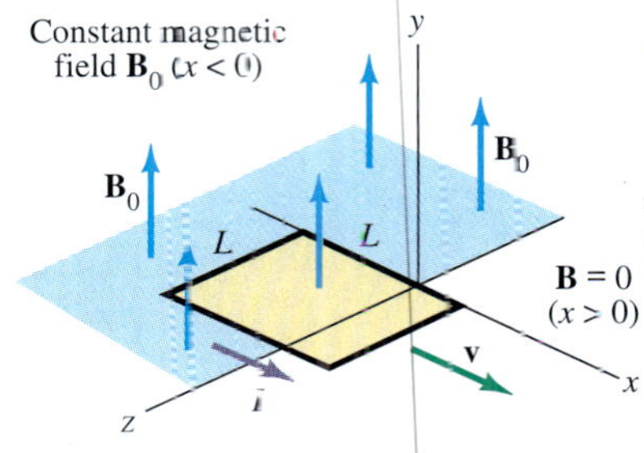

FIGURE 31–7 Example 31–2.

Solution: As the wire loop passes out of the region $x < 0$, the magnetic flux upward through the loop decreases, and an emf is induced in the loop, given by Faraday's law, Eq. (31–2). We'll calculate the flux through the planar surface of the loop. The loop integral as seen from above is taken to be counterclockwise. With this orientation for the surface element, the infinitesimal surface elements $d\mathbf{A}$ that make up the integral are oriented upward.

For $t < 0$, the flux through the loop takes on the constant value because the same magnetic field passes through the entire loop:

$$\Phi_B = \int_{\text{surface } S} \mathbf{B} \cdot d\mathbf{A} = B_0 \int_{\text{surface } S} dA = B_0 L^2.$$

This is unchanging with time for $t < 0$, so there is no induced emf and no current.

In the time period $t = 0$ to $t = L/v$, the loop is in the process of leaving the region of the magnetic field. Thus the flux changes. Only that part of the loop for which $x = L - vt$ remains in the region of the field. In other words, only an area $(L - vt)L$ remains in the field, and

$$\Phi_B = B_0(L - vt)L.$$

This flux is not constant, and for this time period

$$\frac{d\Phi_B}{dt} = -B_0 vL.$$

The emf counterclockwise around the loop equals the *negative* of this value: $\mathscr{E} = +B_0 vL$. A counterclockwise current is induced in the loop between $t = 0$ and $t = L/v$:

$$I = \frac{\mathscr{E}}{R} = \frac{B_0 vL}{R}. \tag{31–3}$$

In this time period, the numerical value of the induced current is

$$I = \frac{(1.0\ \text{T})(10.0 \times 10^{-2}\ \text{m/s})(0.10\ \text{m})}{0.065\ \Omega} = 0.15\ \text{A}.$$

For $t > L/v$, the loop has moved out of the region of constant field, so the flux takes on a constant value (zero), and there is neither an induced emf nor a current.

The Surface Used in Faraday's Law

The magnetic flux takes on the same value for any surface bounded by a given loop.

In our statement of Faraday's law, we did not specify the surface formed by the loop in question. We shall now show that *any surface bounded by the loop in question is suitable because the flux is the same through any such surface.* Let's recall two features of magnetic flux. Because magnetic charges do not exist, *all magnetic field lines are continuous; they neither begin nor end on charges.* Moreover, *the magnetic flux through a surface is proportional to the net number of field lines that pass through a surface.*

Let's now examine the consequences of these properties of magnetic field lines. Consider, two surfaces, S_1 and S_2, bounded by a loop (Fig. 31–8a). Because the lines are continuous, the number of lines that pass through the two surfaces must be the same; hence the flux through the two surfaces must be the same. Some lines may not pass through S_1, such as line 1 in Fig. 31–8b, but if a line passes into some third surface, S_3, and not into surface S_1, then the line must also pass out of surface S_3 and therefore does not contribute to the net flux through S_3.

We have come to the important conclusion that *the magnetic flux through one surface bounded by a closed loop is the same as the magnetic flux through any other surface bounded by the same loop. Both surfaces must be oriented by the right-hand rule.* This is a very helpful result because it shows that we can choose for convenience the surface through which to calculate the magnetic flux.

PROBLEM SOLVING

Choose the bounding surface for Faraday's law as the one for which the flux is easiest to calculate.

A good problem-solving technique is to find a surface over which the flux is easily calculated when it is necessary to compute the flux, as it is in Faraday's law. Example 31–3 illustrates this technique.

EXAMPLE 31–3 Suppose that a certain region has a magnetic field with the constant value $\mathbf{B} = B_0\mathbf{j}$, where $\mathbf{j}$ is the unit vector in the y-direction. Find the magnetic flux upward through the hemisphere of radius R shown in Fig. 31–9.

Solution: We can find the flux through the hemisphere by finding the flux through any other surface bounded by the same loop. The boundary of the hemisphere is a circle of radius R in the xz-plane, and the simplest surface with which to calculate the flux is that planar circle in the xz-plane. For the circle, $\mathbf{B}$ is parallel to $d\mathbf{A}$. Thus the sign of the flux is positive if B_0 is positive. Mathematically, the equality of the flux through the hemisphere and through the planar circle can be written as

$$\Phi_B\big|_{\text{hs}} = \int_{\text{pc}} \mathbf{B}\cdot d\mathbf{A} = B_0 \int_{\text{pc}} dA = B_0\pi R^2.$$

Although the direct calculation of the flux through the hemisphere is difficult, the calculation becomes trivial when the flat surface is used.

EXAMPLE 31–4 A loop of wire and a straight wire each lie on a tabletop (Fig. 31–10a). The straight wire, which is connected to a battery and carries a current in the direction shown, is moved toward the loop. What is the direction of the induced current, if any, in the loop?

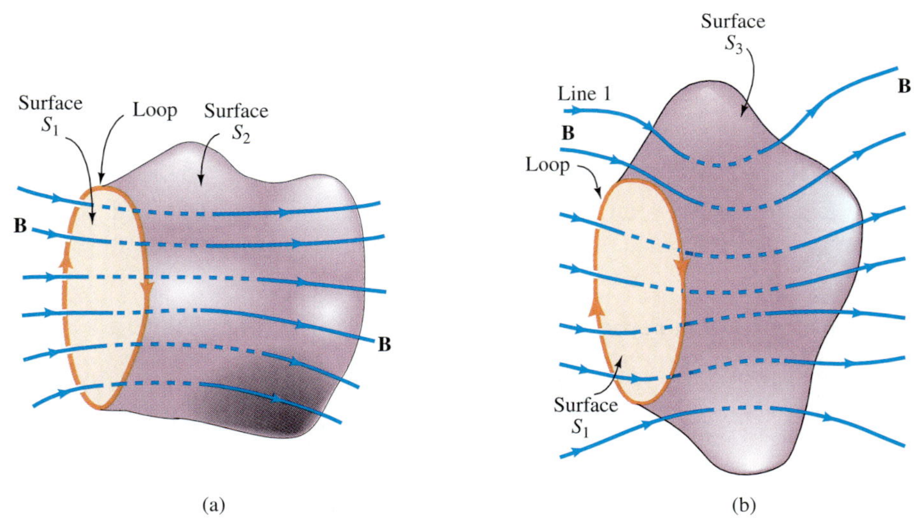

FIGURE 31–8 (a) The same net number of magnetic field lines pass through any two surfaces S_1 and S_2 bounded by a closed loop. (b) Surfaces S_1 and S_3 are both bounded by a loop. If magnetic field line 1 passes through surface S_3 but not through surface S_1, it must pass through surface S_3 again and hence does not contribute to the net magnetic flux through surface S_3.

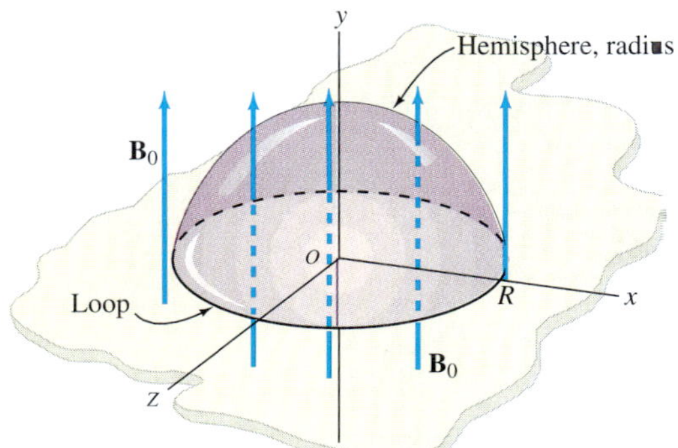

FIGURE 31–9 Example 31–3. The vector $\mathbf{B}_0$ is constant and vertical.

Solution: First, does the movement of the straight wire lead to an induced emf in the loop? In Fig. 31–10b, we have drawn some of the magnetic field lines for the constant current in the straight wire. This field is stronger near the wire, so as the wire moves toward the loop, the flux down through the loop into the table increases. There is indeed an induced emf in the loop. What is its direction (which will also be the direction of the induced current)? We can easily answer a question about the sign of the induced emf by using Lenz's law. If we look down on the table, the current is induced in the counterclockwise direction, then the magnetic field associated with this current will come up through the center of the loop, creating a flux that opposes the flux change from the movement of the straight wire (Fig. 31–10c). This, then, is the direction of the induced current.

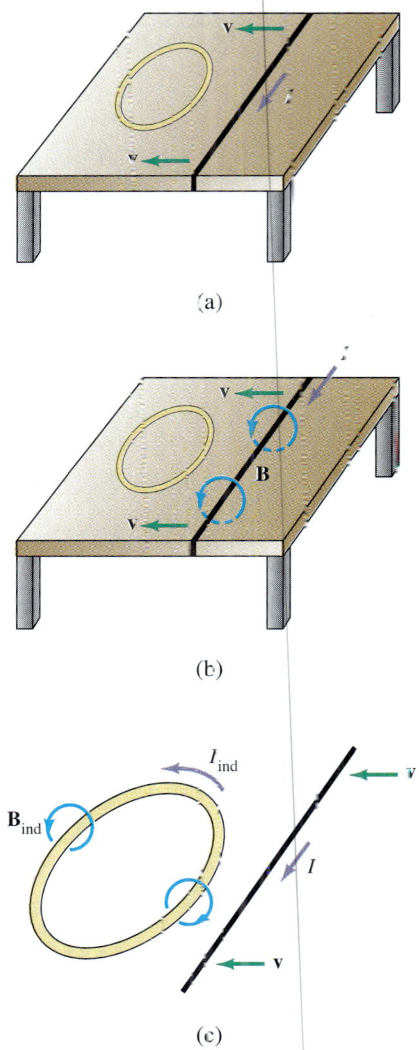

FIGURE 31–10 (a) Example 31–4. (b) The field associated with the current in the wire. (c) The current induced in the loop produces a magnetic field of its own.

EXAMPLE 31–5 A closed loop is constructed of a fixed wire shaped as a square-ended U and a conducting crossbar free to move in the x-direction, all in the xz-plane (Fig. 31–11a). The square base of the U-shaped section is at $x = 0$. A constant magnetic field is oriented in the y-direction, $\mathbf{B} = B_0\mathbf{j}$. (a) The movable crossbar is pulled at a constant speed v to the right, starting from $x = 0$ when $t = 0$. (Thus its position at time t is $x = vt$.) If the resistance of the loop varies with the total length L according to $R = \alpha L$, with α a constant coefficient, what is the direction and value of the current in the loop as a function of time? (b) All segments of the loop are copper wire of radius 0.25 cm, the length $D = 7.0$ cm, the speed $v = 28$ cm/s, and $B_0 = 0.18$ T. What is the current magnitude at $t = 2.0$ s?

Solution: (a) The area of the loop increases. Therefore, the flux through the loop also increases, and Faraday's law of induction applies. Orienting the surface element upward, along $\mathbf{B}$, the flux through the loop is

$$\Phi_B = \int_{\text{surface}} \mathbf{B}\cdot d\mathbf{A} = \int_{\text{surface}} B_0\, dA = B_0 \int_{\text{surface}} dA = B_0 A.$$

The area formed by the loop is $A = (vt)D$, so that

$$\Phi_B = B_0 A = B_0 Dvt;$$

$$\frac{d\Phi_B}{dt} = B_0 Dv.$$

The magnetic flux increases with time.

Lenz's law states that the induced current viewed from above will be in the clockwise direction because the magnetic field due to the induced current must be directed down in order to decrease the flux (Fig. 31–11b). This is in accord with Faraday's law, which, in this case, reads

$$\oint \mathbf{E}\cdot d\mathbf{s} = \mathscr{E} = -\frac{d\Phi_B}{dt} = -B_0 Dv.$$

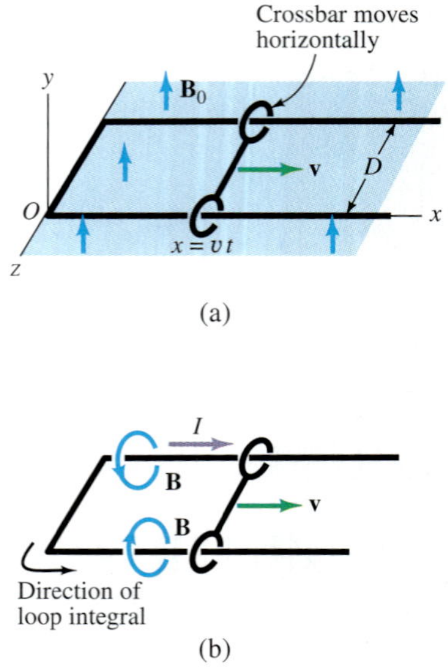

FIGURE 31–11 (a) Example 31–5. (b) Lenz's law shows that the current induced in the circuit that contains the moving crossbar will be in the direction shown.

The loop integral $\oint \mathbf{E} \cdot d\mathbf{s}$ is negative, corresponding to a clockwise induced emf. To relate the emf to the induced current, we must find the total resistance of the loop, R. We have

$$R = \alpha L = \alpha(2D + 2vt) = 2\alpha(D + vt).$$

The induced current is

$$I = I_{\text{ind}} = \frac{\mathcal{E}}{R} = -\frac{B_0 D v}{2\alpha(D + vt)}.$$

You should be able to check the dimensions of this current.

(b) Here, we apply the result of part (a) once we find α. To do so, we note that the resistivity of copper is $\rho = 1.72 \times 10^{-8}\ \Omega \cdot \text{m}$ and that, according to Eq. (27–14), the resistance of a wire of length L and cross-section A is $R = \rho(L/A)$. By comparing the form $R = \alpha L$, we see that

$$\alpha = \frac{\rho}{A} = \frac{(1.72 \times 10^{-8}\ \Omega \cdot \text{m})}{\pi(0.25 \times 10^{-2}\ \text{m})^2} = 8.8 \times 10^{-4}\ \Omega/\text{m}.$$

At this point, all quantities in the equation for the current are known and, at $t = 2$ s, the magnitude of the induced current is

$$I_{\text{ind}} = \frac{B_0 D v}{2\alpha(D + vt)} = \frac{(0.18\ \text{T})(0.070\ \text{m})(0.28\ \text{m/s})}{2(8.8 \times 10^{-4}\ \Omega/\text{m})[(0.070\ \text{m}) + (0.28\ \text{m/s})(2.0\ \text{s})]} = 3.2\ \text{A}.$$

Some Comments

A Changing Flux Does Not Necessarily Mean a Changing Magnetic Field. Example 31–5 illustrates several noteworthy physical features about Faraday's law, among them that *the magnetic flux can change not only because the magnetic field changes with time, but also because the area of the loop through which the flux is calculated may change with time*. In Example 31–5, the magnetic field is constant, yet there is nevertheless an induced emf.

Induced Electric Fields Are Nonconservative. We note that induced fields differ fundamentally from the electric fields we have previously encountered. In our earlier work, electric fields were always associated with *conservative* forces. The work done by those fields in moving a charge around a closed loop is always zero:

$$\text{conservative:} \qquad \mathcal{E} = \oint \mathbf{E} \cdot d\mathbf{s} = 0.$$

This is precisely what is *not* true for the fields that result from Faraday's law; the emf about a closed loop is specified by the changing flux:

$$\text{nonconservative:} \qquad \mathcal{E} = \oint \mathbf{E} \cdot d\mathbf{s} = -\frac{d\Phi_B}{dt}.$$

The induced electric field cannot be described by a potential that is a function of space.

31–3 MOTIONAL EMF

We have seen that an emf can be induced in a conductor that moves in a magnetic field. We call this emf a **motional emf**. Consider a conducting rod that moves with constant speed in a magnetic field (Fig. 31–12a). We shall show that Faraday's law leads to an accumulation of positive charges at the bottom of the rod and negative charges at the top, resulting in an emf along the rod. We shall see that the same effect may be viewed as a consequence of the Lorentz force law. In Fig. 31–12a, the rod has length L and

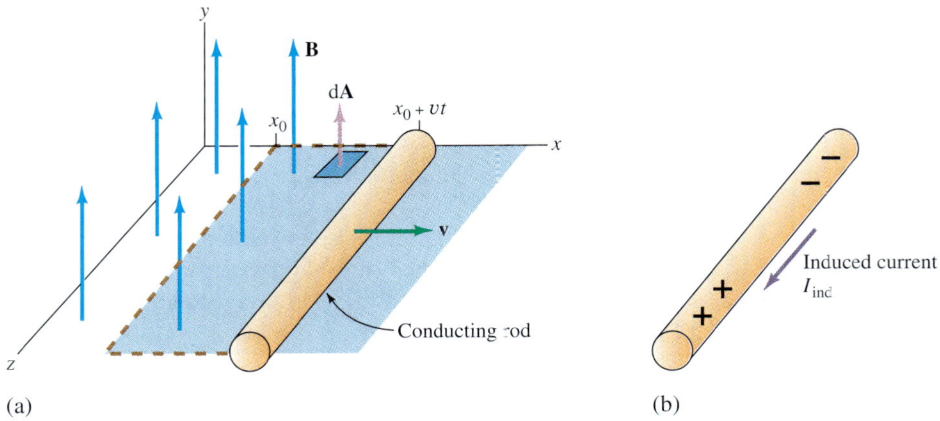

FIGURE 31–12 (a) A conducting rod of length L moves with constant speed v in the $+x$-direction through a constant magnetic field directed in the y-direction. The rod is oriented in the z-direction. The dashed lines represent the imaginary closure of a loop, to which we apply Faraday's law. (b) The induced emf in the rod can be computed by Faraday's law.

moves at speed v through the field, with axes chosen so that the position of the rod along the x-axis is $x = vt$. How does Faraday's law apply here? The answer is that Faraday's law applies to *any* loop, whether it is a real conducting loop, an imaginary loop in space, or a loop that is part real conductor and part imaginary—the type we shall now consider. Let's imagine attaching the ends of the moving rod to a *fixed* imaginary line, indicated in Fig. 31–12a by a dashed line. The rod and line form a closed loop. The loop formed by the imaginary line and the rod is situated in the xz-plane with its area elements $d\mathbf{A}$ oriented in the $+y$-direction and with its area equal to $L(vt - x_0)$. The flux through the closed loop is therefore given by

$$\Phi_B = B_0L(vt - x_0),$$

and the rate of change of this flux is

$$\frac{d\Phi_B}{dt} = B_0Lv. \tag{31–4}$$

This rate of change is independent of the position of any part of the imaginary closure of the loop. The emf taken in the counterclockwise direction—the direction that must be taken if we orient the flux in the $+y$-direction—is then

$$\mathscr{E} = -B_0Lv. \tag{31–5}$$

If the dashed line were conducting wire, this emf would drive a current clockwise (as seen from above) around the circuit. Because there is no actual closed circuit, current flows only until sufficient positive charges accumulate toward the bottom end of the rod to set up a canceling emf (Fig. 31–12b).

Motional emf can be viewed as the result of the magnetic force on moving charges.

The resulting *motional emf* is also simple to understand in terms of the Lorentz force law. Each charge carrier in the rod is moving in a magnetic field and therefore feels a force that equals $q\,\mathbf{v} \times \mathbf{B}$. This force acts in the $+z$-direction for positive charges. The force per unit charge, or the electric field that produces this force, has magnitude vB_0. Because this force is constant along the entire length of the rod, the potential difference from one end of the rod to the other is $\Delta V = EL = B_0vL$. This potential is just the emf that acts in the rod, Eq. (31–5). The sign in that equation dictates that the higher potential is at the lower end, in agreement with our Lorentz force analysis. Again, charges will move until a canceling electric field is set up within the conductor and equilibrium is achieved.

We conclude that we can view motional emf as due to the Lorentz force law or to Faraday's law of induction. We verify this with a second look at Example 31–2, in which a square loop moves out of a region of constant magnetic field into a region with no field. We redraw the situation in Figure 31–13. According to Faraday's law, there is no current in the loop for $t < 0$, when the loop is entirely within the constant field. A Lorentz force analysis shows that the Lorentz force has no effect when the loop is entirely within the field region. The force on positive charges in leg a is di-

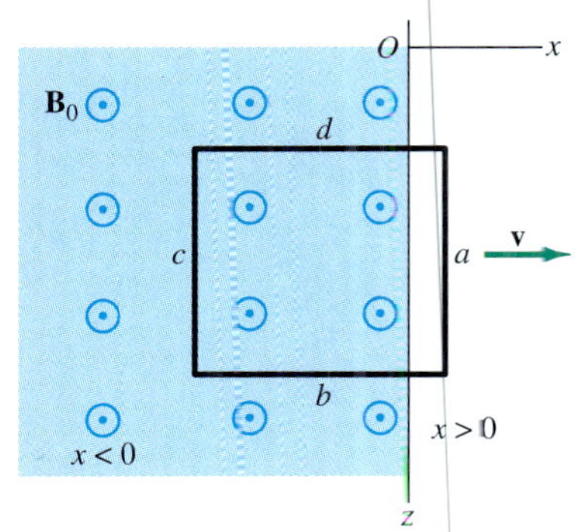

FIGURE 31–13 An analysis of the moving loop of Example 31–2 in terms of the Lorentz force law.

rected downward, as is the force on the positive charges in leg c, and no charges can move around the loop. The same conclusion holds for legs b and d. Once leg a passes out of the field region, however, there is a net force due to the charges in leg c, and this tends to push the positive charges in the counterclockwise direction, as we deduced in Example 31–2. The emf will be B_0vL, just as we found in the example. Finally, there are no magnetic forces once the loop has left the field region.

EXAMPLE 31–6 A rod of length L lying in the xy-plane pivots with constant angular velocity ω counterclockwise about the origin (Fig. 31–14). A constant magnetic field of magnitude B_0 is oriented in the z-direction. Find the motional emf in the rod by applying Faraday's law of induction.

Solution: We complete a loop by drawing the dashed lines shown in Fig. 31–14. The area element will be oriented upward, so the loop integral representing the emf must be taken to be counterclockwise. Here $\theta = \omega t$, and the area of the loop is the area of the segment swept out by the rod through the angle θ:

$$\text{area} = \frac{1}{2}\theta L^2 = \frac{1}{2}\omega t L^2.$$

The magnetic flux through the loop is B_0 times the area; the rate of change of the flux is

$$\frac{d\Phi_B}{dt} = \frac{d}{dt}\left(\frac{1}{2}B_0\omega L^2 t\right) = \frac{1}{2}B_0\omega L^2.$$

The motional emf is then

$$\mathscr{E} = -\frac{1}{2}B_0\omega L^2.$$

The sign indicates that the emf drives positive charges radially out from the origin. We can verify this direction using the right-hand rule to indicate the direction of the Lorentz force vector $\mathbf{v} \times \mathbf{B}$.

Eddy Currents

We have talked about induced emfs in the motion of wires and rods. But we can also move large pieces of metal in a magnetic field. Our treatment of motional emf suggests that here, too, currents that are distributed throughout the conductor are induced. These currents are known as **eddy currents**. Figure 31–15 shows the eddy currents set up in a flat, vertically oriented plate dropping through a region with a horizontal magnetic field.

APPLICATION

Reducing undesirable induced eddy currents

These eddy currents are dissipated in Joule heating through the resistivity of the metal plate. This heating can be a significant advantage in certain applications; when it is undesirable, it can be reduced by eliminating paths for the current flow. This is done

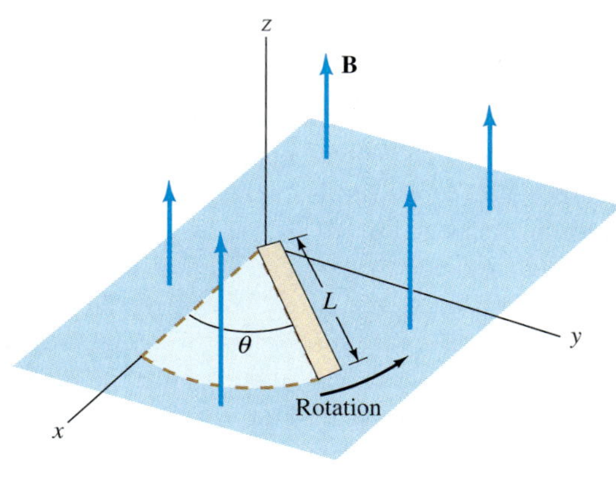

FIGURE 31–14 Example 31–6. The dashed lines represent the imaginary closure of a loop, to which we apply Faraday's law.

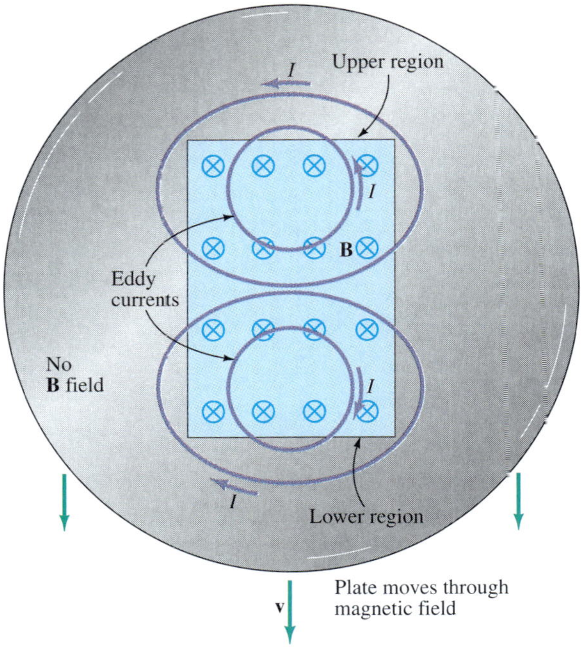

FIGURE 31–15 As the circular metal plate drops through a small region of constant magnetic field directed into the plate, eddy currents are induced in the plate. The direction of these currents is given by Lenz's law.

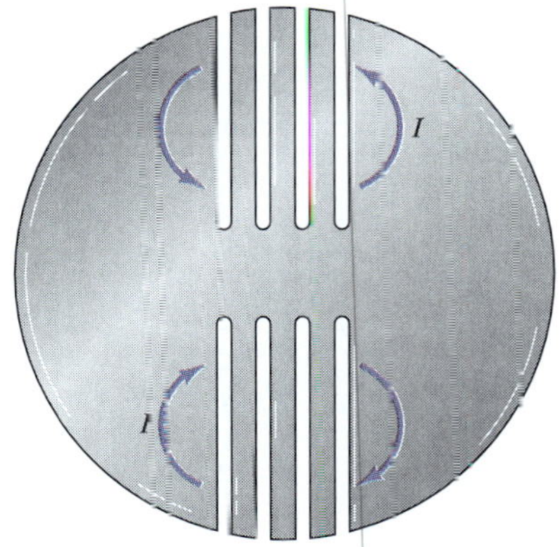

FIGURE 31–16 To inhibit the development of eddy currents in the moving metal plate, slots can be cut in the plate.

either by cutting slots in the metal plate (Fig. 31–16) or by laminating the metal with an insulator. Lamination, for example, may be used in the iron core of an electromagnet, which consists of a wound coil of wire—a solenoid—surrounding an iron core (Fig. 31–17). (The core increases the magnetic field in the solenoid—see Chapter 32.) With lamination, currents will still be induced in the iron sheets of the core. But because lamination limits the area of the iron sheets of the core, it also limits the eddy currents.

Eddy currents are not always a disadvantage, as we shall see in Section 31–4.

31–4 FORCES AND ENERGY IN MOTIONAL EMF

Magnetic forces on induced currents inhibit motion.

The presence of induced currents due to changing magnetic fluxes has one further implication: We already know that ordinary currents experience forces in magnetic fields, and this must be true for the induced currents. Thus, wires in which currents are induced will experience forces. Moreover, *the magnetic force on the induced current always inhibits the motion that produces the motional emf.* This is a consequence of the sign implied by Lenz's law.

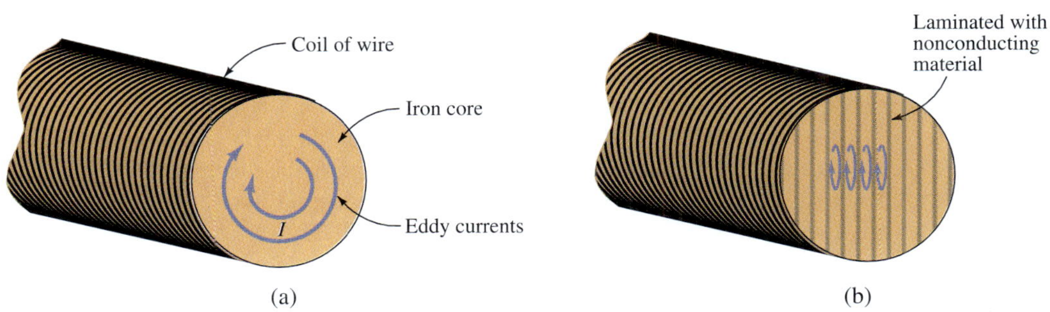

FIGURE 31–17 (a) Eddy currents are set up in the iron core of an electromagnet if the current in the surrounding solenoid is not steady. (b) To inhibit these currents, the core can be constructed of iron sheets laminated with sheets of nonconducting material.

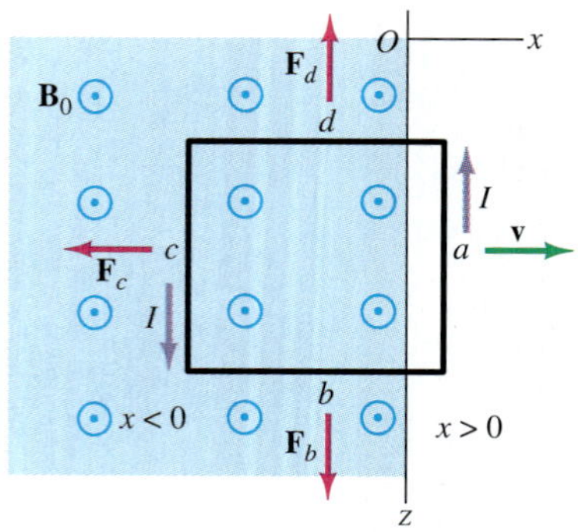

FIGURE 31–18 The magnetic forces on the loop of Example 31–2 act to slow down the loop.

In the moving loop of Example 31–2, an induced current, I, appears as the loop leaves the region of magnetic field. The force on the wire is given by Eq. (29–16),

$$\mathbf{F}_B = I \int (d\boldsymbol{\ell} \times \mathbf{B}), \tag{31–6}$$

where $d\boldsymbol{\ell}$ describes a length element of the wire and $\mathbf{B}$ describes the magnetic field at that wire element. Let's once again draw the loop—this time including the forces on each leg (Fig. 31–18). There is no force on those legs or portions of legs that are out of the magnetic field. Using the right-hand rule, the force on the portion of leg b that is located in the field is directed down, and it is canceled by the force on the portion of leg d located in the field. The only contribution to the net force comes from leg c. Here, application of Eq. (31–6) gives a force

$$F_c = ILB_0 \quad \text{to the left.} \tag{31–7}$$

Recall from Example 31–2, Eq. (31–3), that the magnitude of the current is $I = B_0 vL/R$, where R is the total resistance of the loop and v is the speed with which the loop moves. The force on the loop is thus

$$F_c = \frac{vL^2B_0^2}{R}. \tag{31–8}$$

This force acts to slow down the loop. Energy must be supplied to the loop at a certain rate to keep it moving with the constant velocity $\mathbf{v}$. In other words, power must be expended by an external force $\mathbf{F}$ to keep the loop moving. The external force will have the magnitude F_c, and it will be directed to the right, so the power expended is

$$P = \mathbf{F}\cdot\mathbf{v} = F_c v = \frac{v^2L^2B_0^2}{R}. \tag{31–9}$$

Let's compare the power expended by the external force to the power lost in the resistor. The power loss, or energy loss per unit time, is

$$P = I^2R = \left(\frac{B_0 vL}{R}\right)^2 R = \frac{v^2L^2B_0^2}{R}. \tag{31–10}$$

Equations (31–9) and (31–10) are the same. *The power loss due to the current flow through the resistor is matched by the power required to keep the loop moving.* The principle of conservation of energy suggests that this result was a foregone conclusion.

Magnetic Drag

Using magnetic drag forces

The force on the loop as expressed by Eq. (31–8) is proportional to the speed at which the loop moves. This is generally true for the forces on induced currents due to motional emf because their origin is the Lorentz force, which is also proportional to the speed. We have previously referred to such forces as *drag forces*, and the motion of conductors in magnetic fields is analogous to motion in a viscous medium. This can be quite useful. Eddy currents in a piece of metal that moves through a magnetic field act as brakes. Brakes of this type have practical applications that range from use in large electric motors to damping in a sensitive mass balance whose oscillations are a disadvantage. The metal plate falling through a magnetic field (see Section 31–3) and pictured in Fig. 31–15 illustrates magnetic drag. The drag force leads to a terminal velocity.

We can carry this to the extreme by supposing that the metal plate is a superconductor: a material with no resistance and with the property that the magnetic field does not penetrate its interior. If we drop such a piece of material into the field region, the piece is more than slowed down by its entry into the field; it is repelled and bounces back up. This bounce is perfectly elastic because there is no Joule energy loss in a material with no resistance. We have magnetic levitation (Fig. 31–19); one application is in magnetically levitated trains (see Fig. 2–28).

EXAMPLE 31–7 A square loop of wire has sides of length 5.0 cm. This loop falls at speed v under the influence of gravity through a region with a constant magnetic field of magnitude 15 T—only large scientific magnets can attain such a high field—into a region with no magnetic field (Fig. 31–20). The field is oriented perpendicular to the loop—into the page. The loop is constrained to remain vertically oriented. The total resistance of the loop is 1.0 Ω and its mass is 150 g. (a) Find the terminal velocity of the loop as it passes the boundary between the two fields. (b) Calculate the total energy lost to Joule heating in the loop during this period. As an approximation, assume that the loop moves at its terminal velocity when it enters the magnetic field region and that this velocity remains constant during the loop's passage.

Solution: (a) Whether the loop enters or leaves the region with the magnetic field, the drag force is oriented upward and is given by Eq. (31–8), $F = vL^2B_0^2/R$. The force of gravity is directed downward with magnitude mg. The terminal velocity v_t is reached when the drag force equals the force of gravity:

$$\frac{v_tL^2B_0^2}{R} = mg;$$

$$v_t = \frac{mgR}{L^2B_0^2}. \qquad (31\text{–}11)$$

Numerically,

$$v_t = \frac{(0.15\ \text{kg})(9.8\ \text{m/s}^2)(1.0\ \Omega)}{(5.0 \times 10^{-2}\ \text{m})^2(15\ \text{T})^2} = 2.6\ \text{m/s}.$$

(b) As long as the loop moves at its terminal velocity, the total energy lost to Joule heating will be a constant given by the product of power and the time the loop spends in the transition region. The power dissipated in Joule heating is, according to Eq. (31–10),

$$P = I^2R = \frac{v_t^2L^2B_0^2}{R}.$$

The time spent in transition is $t = L/v_t$, so the energy loss is

$$\Delta E = Pt = \frac{v_t^2L^2B_0^2}{R}\left(\frac{L}{v_t}\right) = \frac{v_tL^3B_0^2}{R}.$$

Substituting for v_t from Eq. (31–11), we find that

$$\Delta E = \frac{mgR}{L^2B_0^2}\,\frac{L^3B_0^2}{R} = mgL.$$

This is just the change in gravitational potential energy. All of this energy goes into Joule heating because the loop moves through this change without a corresponding change in kinetic energy. We find

$$\Delta E = (0.15\ \text{kg})(9.8\ \text{m/s}^2)(0.050\ \text{m}) = 0.074\ \text{J}.$$

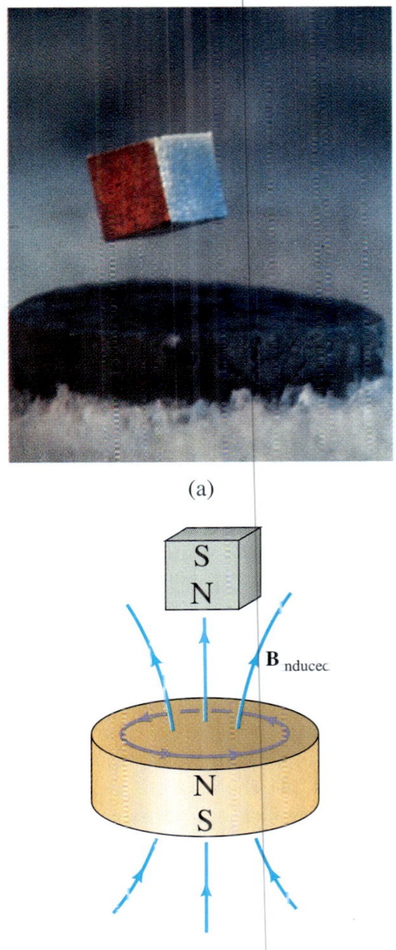

FIGURE 31–19 (a) A levitated magnet. A bar magnet moves toward the superconducting material, inducing persistent currents in the superconductor. (b) The magnetic forces between the superconductor and the magnet are repulsive and sufficiently strong to support the magnet's weight.

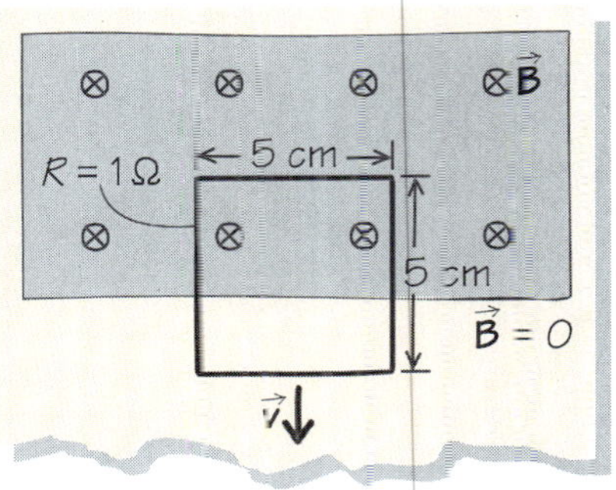

FIGURE 31–20 Example 31–7.

Forces and Lenz's Law

Lenz's law gives us a second way to think about the forces on induced currents. The magnetic field of a current loop or solenoid is the same as that of a bar magnet. We have already seen that both magnets and current loops produce magnetic dipole fields. Suppose that the north pole of a bar magnet moves toward a conducting ring that initially has no current. The magnetic flux through the ring increases and an emf is induced in the ring, which causes a current to flow (Fig. 31–21a). The induced current, in turn, produces a magnetic field whose flux tends to cancel the increase of flux due to the moving bar magnet. The direction of this induced magnetic field is thus opposed

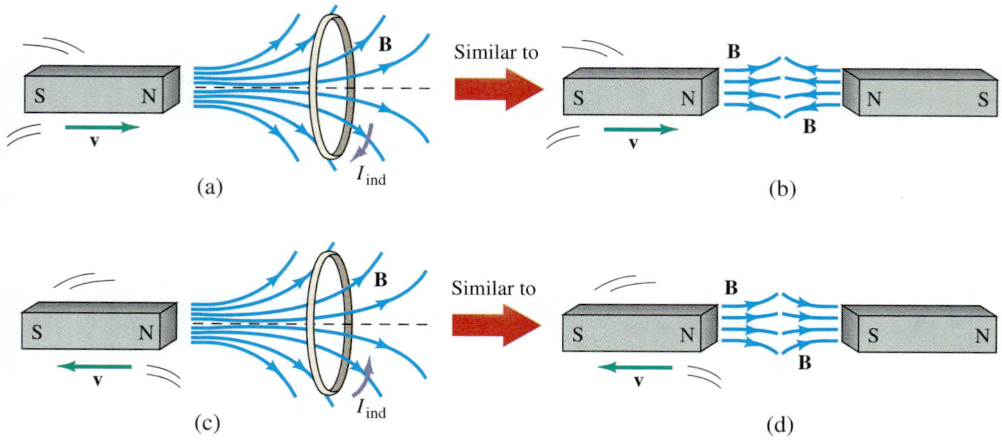

FIGURE 31–21 (a) The magnetic flux through a conducting ring is increasing because the north pole of a bar magnet moves toward it. The ring is repelled. (b) The direction of the induced current in the ring gives the ring the field of a bar magnet with its north pole to the left, and two north poles repel. (c) The bar magnet is pulled away. In this case, a current is induced in the ring in the opposite direction. (d) The magnetic field of the ring is then that of a bar magnet with its south pole to the left, and it is attracted to the receding bar magnet.

to the field of the bar magnet. We can think of the induced magnetic field as the magnetic field of a second bar magnet, with a field as shown in Fig. 31–21b. The situation is one of two north poles meeting, and we know that two north poles repel each other.

If we now pull the original bar magnet away from the ring, the induced current in the ring points in the opposite direction (Fig. 31–21c). The induced magnetic field changes direction, and the ring's magnetic field is like that of a bar magnet with its south pole adjacent to the north pole of the original bar magnet (Fig. 31–21d). We know that opposite poles attract each other, so the magnet pulls the ring along with it.

31–5 TIME-VARYING MAGNETIC FIELDS

The magnetic flux through a loop can change in a variety of ways.

1. The loop can move or rotate in the presence of a magnetic field that is not changing with time.

2. The source of the magnetic field can move, moving the field along with it, such as when a bar magnet moves.

3. The source of the magnetic field and hence the field itself can have explicit time dependence, such as when the current through a solenoid is made to change. In cases (2) and (3), it is not enough to invoke the Lorentz force, as is possible in case (1). Yet there is no way to tell whether the loop moved or the field changed. While *motional* emf can be interpreted in terms of the Lorentz force law, the fact that we can induce an emf with a time-varying magnetic field is a truly new aspect of Faraday's law.

Finding Induced Electric Fields

Faraday's law shows that if a magnetic field changes with time, then an electric field is induced in space such that Eq. (31–2) is satisfied. *If* there is sufficient symmetry in a situation, then it is possible to calculate the induced electric field in a way similar to the way in which we used Ampère's law to determine a magnetic field.

EXAMPLE 31–8 The two circular pole faces of an electromagnet, both of radius $R = 0.5$ m, are oriented horizontally with the north pole underneath (Fig. 31–22a). The electromagnet produces a field that is uniform throughout the volume between the faces. The field is increased linearly from 0.1 T to 1.1 T over a period of 10 s. Describe the electric field that results in the region between the poles.

Solution: The unknown electric field is induced by the changing magnetic flux. There is cylindrical symmetry between the pole faces, so the induced electric field can vary only with the distance r from the central axis of the pole faces; it cannot

APPLICATION: THE AUTOMOBILE SPEEDOMETER

This device consists of a shallow open-ended cup made from a conducting material—usually aluminum. A shaft supported on bearings is attached on an axis to the outside of the cup. A bar magnet attached orthogonally to a second shaft nearly spans the cylindrical interior wall of the cup across a diameter. This bar magnet is connected by a flexible cable to a part of the automobile that turns at an angular velocity proportional to the rotational speed of the wheels—for example, one of the axles of a front wheel. At any instant in time, the magnetic field from the magnet poles—and hence a magnetic flux—passes through the cup at the two areas opposite the magnet's poles. As the magnet rotates, the flux changes across these two areas, inducing eddy currents in these areas. In addition, forces are induced such that the cup tends to follow the magnet poles in order to hold the flux changes down; in other words, there is a torque on the cup. Now the size of the induced currents and thus the size of the induced torque depends on the speed at which the magnet poles pass. Left to itself, the cup would rotate with the magnet; however, a spiral hairspring provides an opposite torque on the cup shaft proportional to the angle of displacement. The shaft is therefore rotated by an amount proportional to the torque, and therefore proportional to the automobile's speed. All that is necessary to complete the picture is a pointer attached to the shaft moving over a dashboard speed scale.

vary with the angle around this axis. We must find the magnetic flux through a horizontal circle of radius r centered on the axis of the pole faces. The magnetic field strength, $\mathbf{B}$, is constant over the area of the circle and runs upward from the north pole, so the flux upward through the circle is

$$\Phi_B = \pi r^2 B.$$

As described in the problem statement, the magnetic field has a linear time dependence of the form

$$B = B_0 + \alpha t.$$

The rate of change of flux is therefore

$$\frac{d\Phi_B}{dt} = \pi r^2 \frac{dB}{dt} = \pi r^2 \frac{d}{dt}(B_0 + \alpha t) = \pi r^2 \alpha.$$

Faraday's law, Eq. (31–2), requires the rate of change of Φ_B to be the negative of the integral around the circle of the induced electric field:

$$\oint \mathbf{E}\cdot d\mathbf{s} = -\frac{d\Phi_B}{dt} = -\pi r^2 \alpha. \qquad (31\text{–}12)$$

The direction of the loop integration is counterclockwise as we look down the north pole because the flux is oriented upward (Fig 31–22b). With the minus sign in Eq. (31–12), the induced electric field points clockwise. Symmetry demands that the electric field have the same magnitude all around the circle, so the electric field is given by

$$\oint \mathbf{E}\cdot d\mathbf{s} = E2\pi r = -\pi r^2 \alpha.$$

Thus, the magnitude of $\mathbf{E}$ is

$$E = \frac{\pi r^2 \alpha}{2\pi r} = \frac{1}{2}\alpha r.$$

The induced electric field increases as we go out from the center.

Numerically, the coefficient α is found by knowing that B increases from 0.1 T to 1.1 T in 10 s, so the rate of increase is $\alpha = (1.1\ \text{T} - 0.1\ \text{T})/10\ \text{s} = 0.1\ \text{T/s}$. The induced electric field is then

$$E = \frac{1}{2}(0.1\ \text{T/s})r = (0.05\ \text{T/s})r,$$

which increases from 0 N/C at $r = 0$ m to a maximum of $E_{max} = (0.05\ \text{T/s})(0.5\ \text{m}) = 2.5 \times 10^{-2}$ N/C at $r = 0.5$ m.

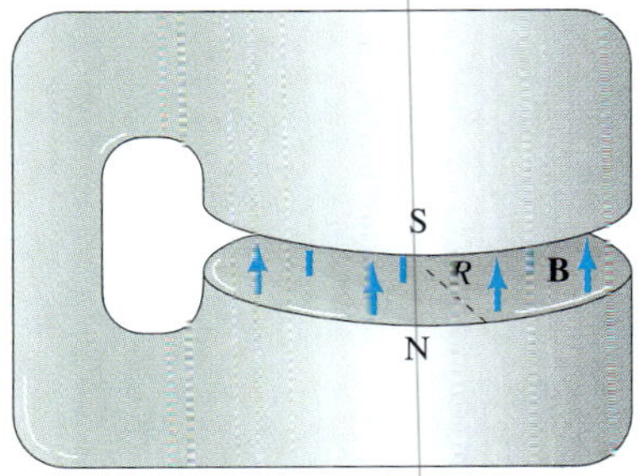

(a)

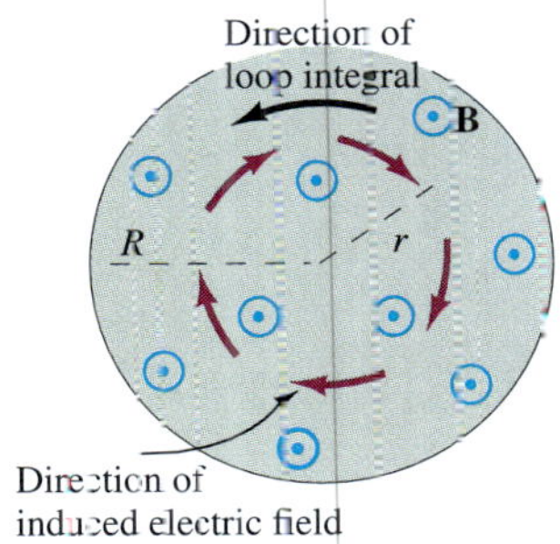

(b)

FIGURE 31–22 (a) Example 31–8. The current windings that produce the field are not shown. (b) A view straight down at the north pole. A changing magnetic field between the pole faces of an electromagnet induces an electric field. Symmetry allows us to specify the electric field, not just its integral around an arbitrary loop.

Is a Magnetic Field Needed Where a Current is Induced?

One aspect of Faraday's law is not very obvious. Suppose that charged particles such as protons are moving, but they are so far away from the region of the pole faces of a magnet that the magnetic field in their location is very small. Would the protons still accelerate when the magnetic field is changed? Is an electric field induced even in regions with no magnetic field whatsoever, as in the space outside a toroidal solenoid? Experiment confirms that the answers to our questions are affirmative. According to Faraday's law, all that counts is the change of the flux through the loop in question, regardless of the loop's location. A *changing* magnetic flux may induce electric fields in regions far from where the magnetic field is large.

31–6 GENERATORS AND MOTORS

The generation of electric energy in our society is based largely on the Faraday induction law. The conversion of mechanical energy (for example, the rotating blades of a steam turbine) to electric current is accomplished with the **alternating-current (AC) generator**.

Imagine a coil of N turns that makes a circle of area A. The coil is placed in a constant magnetic field, $\mathbf{B}$, and rotated at angular speed ω around an axis perpendicular to the field (Fig. 31–23a). The ends of the wire that makes up the coil are brought to the exterior through some sort of sliding contact with a fixed wire. As the coil rotates in the magnetic field, the magnetic flux through it changes, and an emf is induced. According to Fig. 31–23b, the magnetic flux through the loop is $\Phi_B = \mathbf{A}\cdot\mathbf{B} = AB\cos\theta$. We imagine starting the rotation at $t = 0$, so $\theta = \omega t$. Then the time derivative of the magnetic flux is

$$\frac{d\Phi_B}{dt} = AB\frac{d}{dt}\cos(\omega t) = -AB\omega\sin(\omega t). \tag{31–13}$$

There are N turns of wire, so the total emf induced across the two ends of the coil is

$$\mathscr{E} = -N\frac{d\Phi_B}{dt} = NAB\omega\sin(\omega t). \tag{31–14}$$

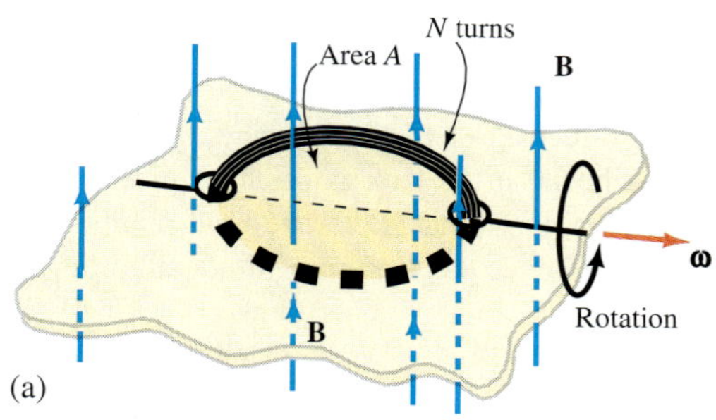

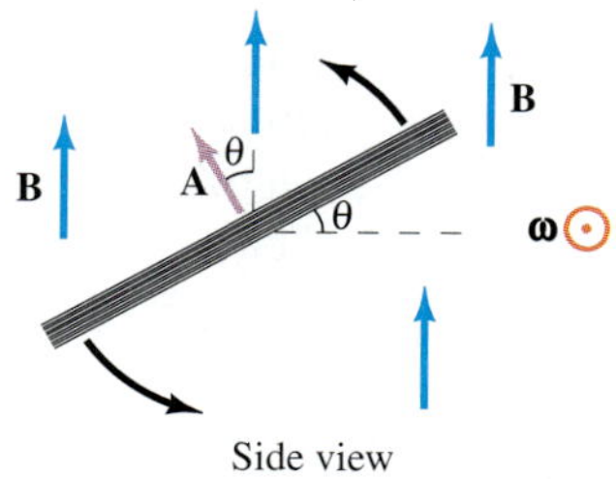

FIGURE 31–23 (a) An external force rotates a coil with angular velocity $\boldsymbol{\omega}$ in a magnetic field. An emf with sinusoidal time dependence of angular frequency ω is induced in the coil. (b) Side view of the process, looking at $\boldsymbol{\omega}$.

This arrangement makes up our generator, which is denoted by a circle enclosing a sine wave (Fig. 31–24).

If the wire of the generator coil is connected as a series element of a circuit with a resistance R, then a current is generated in the circuit:

$$I = \frac{\mathscr{E}}{R} = \frac{NAB\omega}{R}\sin(\omega t). \tag{31–15}$$

This *alternating current* oscillates in sign and has a maximum magnitude of $NAB\omega/R$.

The power delivered to this circuit, P, is given by the product of the emf and the current:

$$P = \mathscr{E}I = INAB\omega\sin(\omega t). \tag{31–16}$$

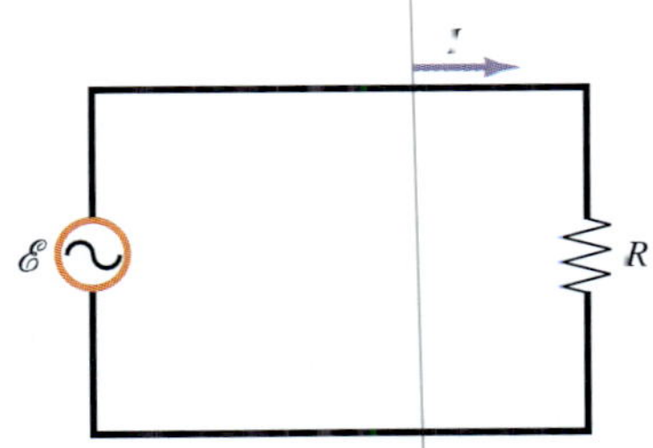

FIGURE 31–24 When the sinusoidally varying emf that results from rotating a coil in a constant magnetic field (as in Fig. 31–23a) is part of a circuit, a sinusoidal current with the same angular frequency ω results.

The mechanical force that rotates the loop must be the source of this power. We know that a loop that carries a current forms a magnetic dipole; we also know that a magnetic dipole experiences a torque that tends to align it with the direction of the magnetic field. Thus the force that rotates the coil must do work against this torque. Let's compute the rate at which this work is done. The torque on a dipole of magnetic dipole moment $\boldsymbol{\mu}$ in a field $\mathbf{B}$ has magnitude

$$\tau = |\boldsymbol{\mu} \times \mathbf{B}| = \mu B\sin\theta,$$

where we refer to Fig. 31–23b for θ and recall that the magnetic dipole moment is perpendicular to the current loop. The mechanical power P_{mech}, or work per unit time, that must be expended by the force that rotates the loop against this torque is

$$P_{\text{mech}} = \tau\omega = \mu B\omega\sin\theta.$$

The magnetic dipole moment of a current loop with N turns is INA [Eq. (29–24)]. Thus

$$P_{\text{mech}} = INAB\omega\sin\theta. \tag{31–17}$$

This result is the same as that given by Eq. (31–16). As expected, the electric power is accounted for entirely by the mechanical power expended.

The explicit time dependence of the power is found by taking the product of the current, Eq. (31–15), and the emf, Eq. (31–14) or, more simply, by evaluating the product V^2/R. We then find

$$P = \frac{(NAB\omega)^2}{R}\sin^2(\omega t). \tag{31–18}$$

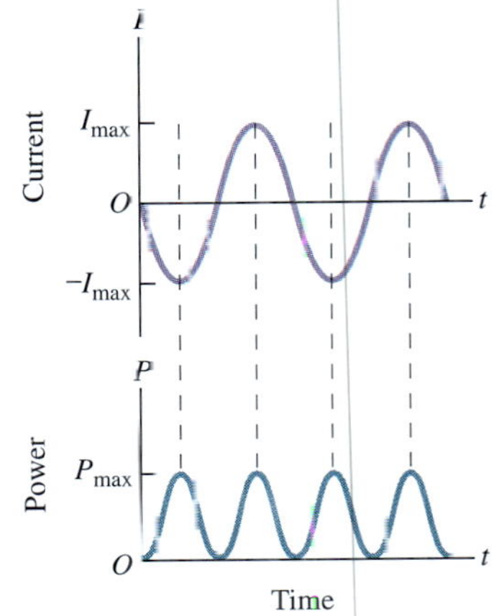

FIGURE 31–25 The power dissipated in the circuit of Fig. 31–24 is always positive, unlike the current, which alternates in sign. The maximum current is $I_{\max} = NAB\omega/R$, while the maximum power is $P_{\max} = I_{\max}^2 R$.

This quantity is always positive, as opposed to the emf or the current, both of which alternate in sign. The distinction is illustrated in Fig. 31–25, which shows a plot of the current and the power in the circuit of Fig. 31–24 as a function of time.

We may want to consider the *time average* of the power. To do so, note that the sine-squared function oscillates between 0 and 1 and averages over one period to 1/2. The average power dissipated in this circuit is then

$$P_{\text{av}} = \frac{1}{2}\frac{(NAB\omega)^2}{R} = \frac{1}{2}\frac{V_{\max}^2}{R}. \tag{31–19}$$

APPLICATION

AC motors

Electric power is generated by using mechanical energy to produce electric current (Fig. 31–26); the current is transmitted in AC form. If we run a generator in reverse, we can convert electric energy into mechanical energy: We have a *motor*. We already saw an example of one such device in Chapter 29 when we discussed how, if a direct current runs through a loop with a split-ring commutator, there is a torque on the loop—always in the same direction—and the loop turns. By employing an alternating current, we can dispense with the commutator because the current automatically reverses direction after one-half cycle. The design of efficient motors is obviously of great technological importance.

*31–7 THE FRAME DEPENDENCE OF FIELDS

FIGURE 31–26 A hydroelectric generator, which produces electric power by using the mechanical power of water to turn turbines in a magnetic field.

Let's now describe and resolve the problem of the consistency of the laws of electricity and magnetism with the principle of Galilean invariance—the relativity principle. According to this principle, there should be no way to tell which of two inertial frames is moving and which is not, if they are moving with respect to one another. The laws of motion should be the same in all inertial frames, so they do not provide a way to tell which frame is moving. Thus an observer O and an observer O′, who is moving with constant velocity **u** relative to observer O, should express the laws of motion in exactly the same way. In particular, Newton's second law,

$$\frac{d\mathbf{p}}{dt} = \mathbf{F},$$

with $\mathbf{p} = m\mathbf{v}$, must look the same to observer O, who sees a particle move with velocity **v**, and to observer O′, who sees the same particle move with velocity $\mathbf{v}' = \mathbf{v} - \mathbf{u}$. This will be the case if **u** is constant (as it must be for inertial frames) and if the force on the particle is the same to both observers.

If the particle is charged and subject to a Lorentz force, there is an apparent difficulty because the Lorentz force depends on the velocity of the particle. The problem would be only an apparent one if both observers saw the same Lorentz force. Let's determine what is necessary for the Lorentz force to be the same for both observers. Observer O sees the Lorentz force [Eq. (29–4)]

$$\mathbf{F} = q[\mathbf{E} + (\mathbf{v} \times \mathbf{B})]. \tag{31–20}$$

Observer O′ sees the force

$$\mathbf{F}' = q\{\mathbf{E}' + [(\mathbf{v} - \mathbf{u}) \times \mathbf{B}']\}. \tag{31–21}$$

Here, we have allowed for the possibility that the fields seen by observer O (**E** and **B**) are different from those seen by observer O′ (**E′** and **B′**). This possibility makes it feasible to reconcile the forces. The forces seen by the two observers will be the same *provided that* the fields they see are related by

$$\mathbf{E}' = \mathbf{E} + (\mathbf{u} \times \mathbf{B}) \tag{31–22}$$

and

$$\mathbf{B}' = \mathbf{B}. \tag{31–23}$$

Electric and magnetic fields mix according to an observer who is moving between inertial frames.

Eqs. (31–22) and (31–23) are the correct solution to the problem of reconciling the forces when speeds are not large compared to the speed of light. *Under the transformation from one inertial frame to another, electric and magnetic fields get mixed up* (*transform among themselves*) in a very special way.

This discussion shows how stationary and moving observers view the source and effect of motional emf. Suppose that an observer O is in a reference frame where there is a magnetic field but no electric field. If a conducting rod moves through this field, there is a motional emf. Its source, according to observer O, is the magnetic force on the conducting electrons in the rod (Section 31–3). Observer O′, moving with the rod, sees the rod at rest and also sees the same constant magnetic field seen by observer O, from Eq. (31–23). Therefore observer O′ sees no magnetic force. However, observer O′ also sees an electric field **E′** [Eq. (31–22)], and this electric field has a magnitude that makes the conducting electrons in the rod accelerate in just the way that observer O sees them accelerate. Each observer attributes the observed effects to different combinations of fields.

The question of the consistency of physical law for observers moving with respect to one another is a very important one; it led Albert Einstein to formulate the theory of special relativity.

SUMMARY

When the magnetic flux Φ_B through an open surface changes with time, an emf, $\mathscr{E}$, is induced around the line that bounds the surface. Faraday's law states the relation:

$$\mathscr{E} = \oint \mathbf{E} \cdot d\mathbf{s} = -\frac{d\Phi_B}{dt}. \tag{31–2}$$

Here, the line integral is over the bounding line; that line forms a closed loop. When the loop is a physical object capable of carrying current, such as a loop of wire, then the induced emf results in an induced current. In this case, the minus sign in Faraday's law can be interpreted in more physical terms as Lenz's law: Induced currents produce magnetic fields that tend to cancel the flux changes that induce them.

The flux change to which Faraday's law refers can occur either because the magnetic field changes with time or because the area or orientation of the surface through which the flux is calculated changes with time. In the latter case, the induced emf is called a motional emf, and it can be derived directly from application of the Lorentz force law. Application of Lenz's law to motional emfs shows that the induced current must lead to magnetic forces that inhibit the motion of the object in which the emf is induced. The power loss due to resistive flow of induced currents is matched by the power required to keep the conductor moving. When an induced emf occurs because magnetic fields change with time, Faraday's law is a new physical principle.

When a changing magnetic flux passes through a conducting solid, Faraday's law manifests itself by the induction of eddy currents in the material.

The AC generator, the foundation of electrical power generation, is an application of Faraday's law. When a coil rotates in a magnetic field, an emf is induced in the coil. The mechanical energy of the rotation is thus transformed into electric energy in the form of a current in circuits connected to the coil.

UNDERSTANDING KEY CONCEPTS

1. A spherical surface is placed in a changing magnetic field. Will there be an induced electric field along the equator?
2. Must there be a real conducting loop in a region with a changing magnetic flux in order for an electric field to be induced?
3. Electric leads (the wires that run from one part of the apparatus used for experiment to another) for sensitive experiments are almost never separated but are close together or even twisted around one another. Explain why this might be done.
4. Can the magnetic field change over some region without a change in the magnetic flux through a surface in the region? If so, give as many examples as you can.
5. Each part of Fig. 31–27 shows a current being induced in a conducting loop by a changing magnetic flux through the loop. In each case, is the direction of the induced current correct as shown? (a) A magnet approaches a loop; (b) a current-carrying conducting loop approaches a loop at rest; (c) a switch is closed in the first loop, causing a current to flow; (d) a switch is closed in the straight wire, causing a current to flow. Here the wire and loop are both on a flat table.
6. A conducting hoop is rolled in a straight line at a constant speed in an east–west direction in the northern hemisphere through Earth's magnetic field. Will a current flow in the hoop? If so, in what direction will it circulate?
7. A rectangular loop is moving across a uniform magnetic field such that the induced emf is zero. What can you say about the angle between the normal to the surface of the loop and the direction of the magnetic field during the motion?
8. A sheet of metal is placed between the pole pieces of a permanent magnet, perpendicular to the direction of the field lines. Does it take positive work to pull the sheet of metal out? If so, why?
9. When a bar magnet is moved toward a current loop, a current is induced in the loop. How, if at all, will physically measur-

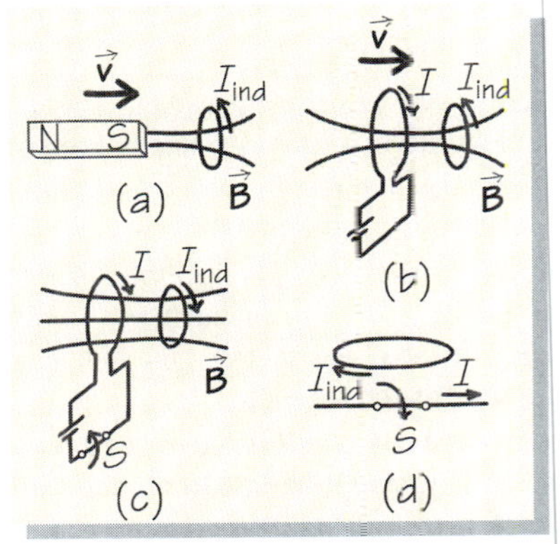

FIGURE 31–27 Question 5.

able quantities change if the loop is moved toward the magnet rather than vice versa?

10. If a flat plate hung by a cord and oriented parallel to the pole faces of a magnet—the faces are in a vertical plane, parallel to each other—moves as a pendulum bob through the pole faces, it slows down. If the magnet is sufficiently strong, the plate comes to rest. Why? How could this phenomenon be prevented?

11. When a gas is heated sufficiently, its atoms ionize into electrons and positive ions. The material forms a plasma. The flow of such a material may be viewed as a superposition of equal and opposite currents. If the plasma is forced to flow in a channel perpendicular to a magnetic field, an electric potential builds up across the channel. The device based on this phenomenon is the *magnetohydrodynamic* (*MHD*) *generator*. Given the magnetic field strength and the potential, we calculate the velocity of the plasma. The charge and density of the charge carriers in the plasma do not enter into the result. Why not? [*Hint*: Review the Hall Effect.]

12. What happens when a bar magnet is dropped down a long, vertical copper tube?

13. In a demonstration, an aluminum ring is placed around a projection of an iron core wound with a wire and connected to a battery (Fig. 31–28). The ring jumps when the circuit is closed. Why? What happens if a gap is cut in the ring?

14. A cylindrical piece of iron is inserted inside a solenoid to increase the magnetic field. A voltage varying harmonically with time is placed across the solenoid leads. A copper ring is slipped down over the solenoid, so that the solenoid passes through the ring, and is held there. (a) Explain why the copper ring becomes hot even though nothing touches it. (b) What is the source of the thermal energy? (c) Explain how energy is conserved in this case.

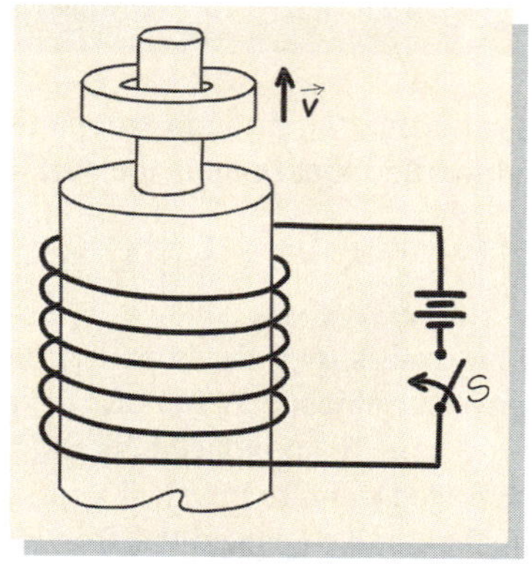

FIGURE 31–28 Question 13.

15. In Question 14, the solenoid axis is vertical. It is possible to find a particular ring of copper that, when slipped over the solenoid and placed horizontally, remains suspended in space around the solenoid. (a) Why does this work? (b) What are the criteria for selecting the particular piece of copper?

16. Discuss how a bicycle light generator utilizing the Faraday effect might work.

PROBLEMS

31–2 Faraday's Law of Induction

1. (I) You are given a horseshoe magnet, a coil of wire, and a flashlight bulb. How would you get the bulb to light up? How would you make the light brighter?

2. (I) Explain in words why the induced currents are as shown in each case in Figure 31–4. Describe the induced magnetic field.

3. (I) A loop of wire of area 12 cm^2 is placed between the pole pieces of an electromagnet, at right angles to the direction of the magnetic field lines. What is the emf generated around the loop if the magnetic field is changed at a uniform rate from 1.5 T to 2.0 T in 5.7 s? Assume that the magnetic field is uniform across the area of the loop.

4. (I) Suppose that the wire in Problem 3 has a resistance of 0.12 Ω. How much power will be lost to ohmic heating while the magnetic field increases?

5. (I) A magnetic field that changes with time but is uniform in space is directed along the x-axis. A conducting ring of diameter 7 cm and resistance 1.5×10^{-3} Ω is placed in the yz-plane. If the current in the ring is 2 A, how fast is the magnetic field changing?

6. (II) The magnetic field in a region is uniform. It varies with time as shown on Fig. 31–29. Plot the current through a ring that has an area of 14 cm^2, a resistance of 0.02 Ω, and whose plane is perpendicular to the magnetic field.

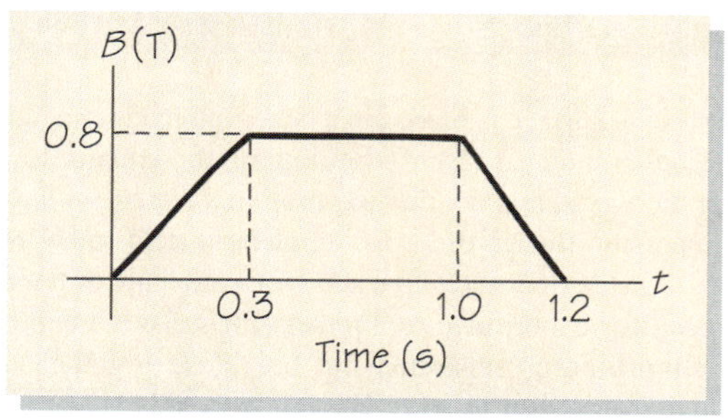

FIGURE 31–29 Problem 6.

7. (II) A square wire loop of dimensions $L \times L$ oriented in the xz-plane enters a region where the magnetic field is first oriented in the $+y$- and then in the $-y$-direction (Fig. 31–30). The width of each region is L. The loop moves at speed v in the $+x$-direction. Find the emf, sign and magnitude, induced in the loop as it enters and passes through the regions with the magnetic field.

8. (II) A long, straight wire oriented in the z-direction carries a current of 0.50 A. A square loop with sides of length 1.0 cm is in the xz-plane with its nearest edge 30 cm from the wire (Fig. 31–31). In a time of 0.10 s, the square loop moves uniformly 10 cm closer to the wire. What is the emf induced in the loop while it is moving? Ignore the variation in the wire's magnetic field *across* the loop.

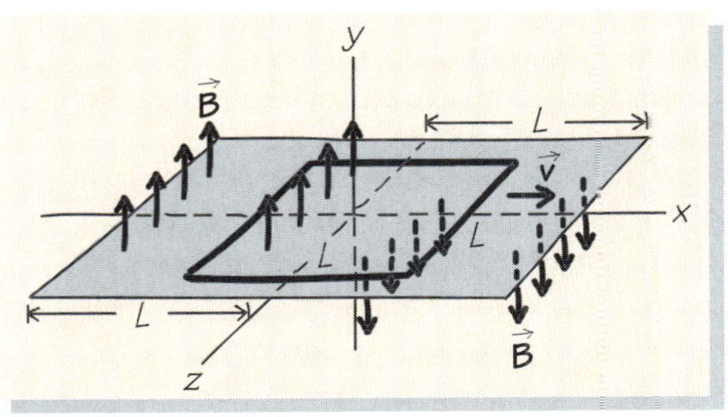

FIGURE 31–30 Problem 7.

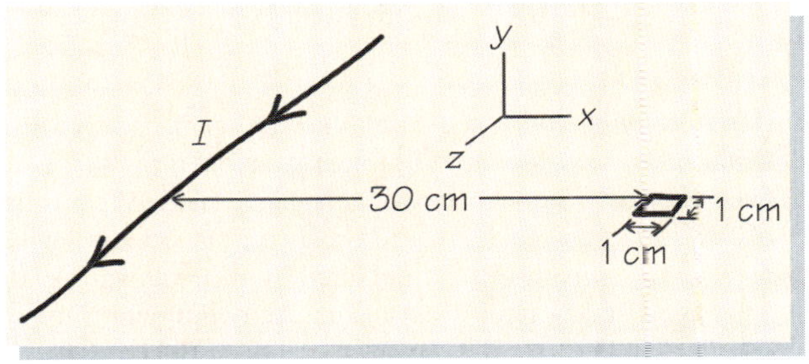

FIGURE 31–31 Problem 8.

9. (II) What is the peak emf produced by a 100-turn square coil 8 cm on each side, rotating on a diagonal axis with a frequency of 15 Hz in a magnetic field of 0.30 T perpendicular to the axis?

10. (II) A coil with 125 turns, a radius of 2.0 cm, and a resistance of 3.0 Ω is rotating about a diameter in a uniform magnetic field of 0.50 T. How fast must it rotate to produce a maximum current of 6.0 A in the coil?

11. (II) There is a constant magnetic field $\mathbf{B} = B_0(\mathbf{i} + \mathbf{j} + \mathbf{k})$ in the region $x > 0$, $y > 0$, $z > 0$. A square loop of dimensions $L \times L$ whose sides are parallel to the x- and y-axes moves with constant velocity $\mathbf{v} = v_0(\mathbf{i} + \mathbf{j})$ in the xy-plane such that its center moves along the line $x = y$. Calculate the emf induced in the loop, given that its leading corner passes the origin at the time $t = 0$.

12. (II) A vertical loop rotates with angular velocity $\boldsymbol{\omega}$ as shown in Fig. 31–32. At time $t = 0$, it is aligned perpendicular to a constant magnetic field oriented in the x-direction. Use Lenz's law to find the direction of the emf induced in the loop at $t = 0$, $t = T/4$, $t = T/2$, and $t = 3T/4$, where T is the rotation period of the loop.

13. (II) A closed loop is constructed of a fixed wire shaped as a squared-off U and a crossbar free to move in the x-direction, all in the xz-plane. The square base of the U-shaped segment is at $x = 0$. A magnetic field oriented in the y-direction varies with x according to $\mathbf{B} = Cx\mathbf{j}$; it is zero at $x = 0$. The situation and the relevant dimensions are as in Fig. 31–33. Suppose that the movable crossbar is pulled at a constant speed v to the

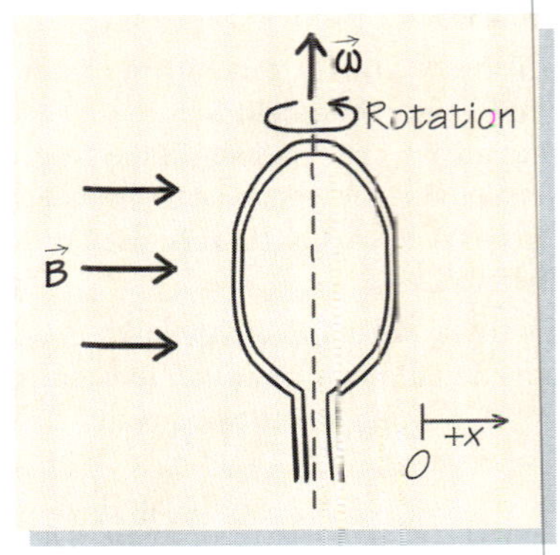

FIGURE 31–32 Problem 12.

right, starting at $x = 0$ when $t = 0$. Its position at any time is $x = vt$. If the resistance of the loop varies with the total length L according to $R = \alpha L$, what is the current in the loop as a function of time? Compare your answer with Example 31–5, and explain any differences.

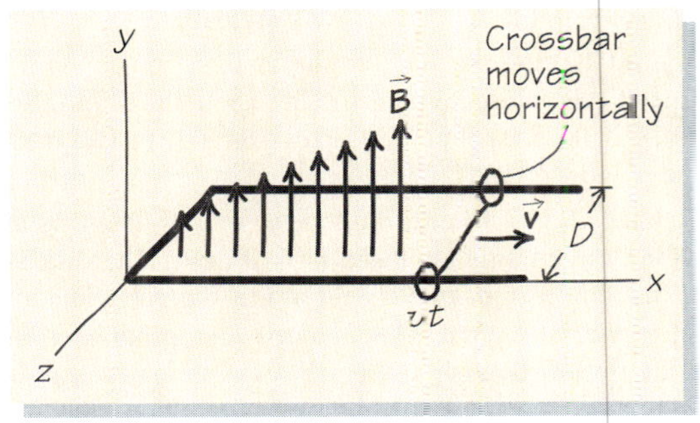

FIGURE 31–33 Problem 13.

14. (II) Suppose the magnetic field in Example 31–5 is a constant field oriented in the z-direction, $\mathbf{B} = B_0\mathbf{k}$. Find the induced current as a function of time.

15. (II) Suppose the magnetic field in Example 31–5 varies linearly with z and is oriented in the y-direction, $\mathbf{B} = Cz\mathbf{j}$. Find the induced current as a function of time.

16. (II) A metal ring is constructed so as to expand or contract freely. In a region with a constant magnetic field $\mathbf{B}_0$ oriented perpendicular to it, the ring expands, with its radius growing linearly with time as $r = r_0(1 + \alpha t)$. As the ring expands and grows thinner, its resistance *per unit length* changes according to the empirical rule $R = R_0(1 + \beta r)$. Find the current induced in the ring as a function of time. Specify the direction as well as the magnitude of the current.

17. (II) A circular loop of area A rotates with angular frequency ω about its vertical diameter. The rotating loop is placed in a horizontal constant magnetic field B. What is the emf induced in the loop?

18. (II) Work Example 31–3 by direct computation of the magnetic flux through the hemispherical surface.

31–3 Motional EMF

19. (I) The spacecraft *Voyager I* is moving through interstellar space, where the magnetic field is 2×10^{-10} T. Assume that *Voyager I* has an antenna 5 m long. If the spacecraft moves so that the antenna rod is perpendicular to the magnetic field when *Voyager I* has a speed of 8×10^3 m/s, what is the emf induced across the antenna?

20. (I) A 747 is flying due north at 900 km/h in a location where Earth's magnetic field consists of an upward vertical component of 2×10^{-5} T and a northward component of 3×10^{-5} T. If the wingtip-to-wingtip length of a 747 is 35 m, find the emf induced across the wings. If the airplane were flying due east instead of due north, how would your answer change?

21. (I) A metal rod is pulled through a magnetic field perpendicular to it with a velocity perpendicular to both the rod and the magnetic field as in Fig. 31–12a. The rod has length 0.25 m, its speed is 1.7 cm/s, and the magnetic field has magnitude 0.069 T. What is the magnitude of the potential difference, if any, from one end of the rod to the other?

22. (I) A metal rod 30 cm long falls to the ground from a height of 20 m. It stays horizontal and oriented in an E–W direction throughout the fall. If we assume that, in the region where the metal falls, Earth's magnetic field is 4×10^{-4} T and points in the N–S direction, then at what rate does the potential difference between the ends of the rods increase?

23. (I) A rod 10 cm long lying in the xy-plane pivots with angular speed 100 rad/s counterclockwise about the origin (See Fig. 31–14). If the measured emf across the rod is 100 mV, what is the magnetic field?

24. (II) A metal disk 18 cm in diameter rotates about its axis of symmetry at an angular speed of 620 rad/s. The disk is situated in a uniform magnetic field of 0.2 T perpendicular to the plane of the disk. What is the induced voltage between the axis and the rim of the disk?

25. (II) A metal bar of length 0.7 m is moved to the right at a speed of 5 m/s. The bar makes an angle of 60° with respect to its direction of motion. It is passing through a region of uniform magnetic field of magnitude 5×10^{-3} T oriented perpendicular to the plane swept out by the bar (out of the page in Fig. 31–34). What is the potential difference between the two ends of the bar as it moves through the magnetic field?

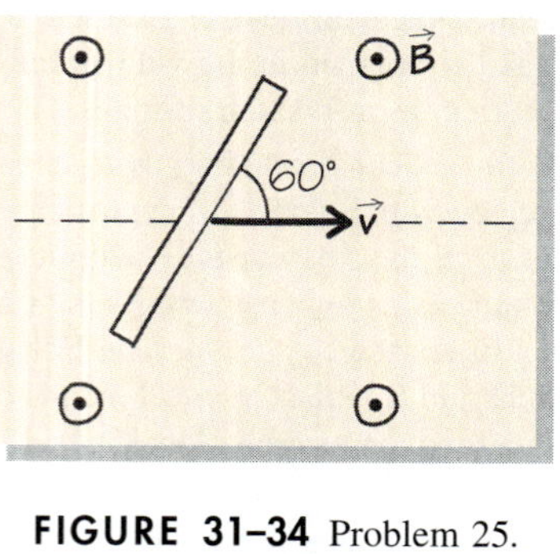

FIGURE 31–34 Problem 25.

26. (II) A circular metal plate moves as a pendulum bob between the poles of a tabletop electromagnet. The plate is oriented so that it is parallel to the faces of the magnet. Describe qualitatively the eddy currents induced in the plate as it moves.

27. (II) A rod of length L moves at constant speed v into the region between the poles of a horseshoe magnet, where there is a constant magnetic field perpendicular to the rod in a circular region (Fig. 31–35). $L = 2R$, the radius of the circular region. What is the emf induced in the rod as a function of time?

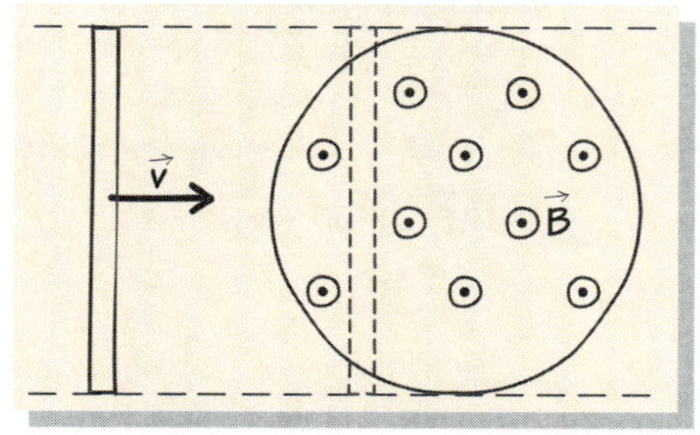

FIGURE 31–35 Problem 27.

28. (II) A *rotating coil* is a common device for measuring magnetic fields. Consider a coil of area A and N turns that is rotated at angular frequency ω in a magnetic field. The position of the coil is adjusted so as to produce a maximum induced current I_{max}, which can be measured by using an appropriate ammeter. R is the total resistance of the coil circuit. Find the relationship between the unknown magnetic field and I_{max}.

29. (II) If the rotating coil of Problem 28 is used with the split-ring commutator described in Chapter 29, DC current can be measured with a sensitive galvanometer. (a) Sketch the current as a function of time for several periods, where the period T is given by $2\pi/\omega$. (b) Calculate the average of this rectified current. (c) Find the relationship between the unknown magnetic field and the average measured DC current. [*Hint*: The average of an oscillating function with period T is $(1/T)\int_0^T f(t)\,dt$.]

30. (II) A long, straight wire carries a current of 16 A. A thin metal rod 25 cm long is oriented perpendicular to the wire and moves with a speed of 4.2 m/s in a direction parallel to the wire. What are the size and direction of the emf induced in the rod if the nearest point of the rod is 5 cm away from the wire, and if the rod moves in a direction opposite to the current?

31–4 Forces and Energy in Motional EMF

31. (I) A loop of metal, total resistance $R = 25\ \Omega$, moves with speed 35 m/s through a region of magnetic field such that at time $t = 0$ the rate of change of magnetic flux through the loop is given by $17\ \mathrm{T \cdot m^2/s}$. The magnetic field in the vicinity of the loop has instantaneous value 0.16 T at $t = 0$. Assume the shape of the loop is such that the resulting net force on the loop is due entirely to a straight section of the loop, 12 cm in length, that is perpendicular both to the magnetic field and to the direction of the loop's motion. What is the drag force on the loop at $t = 0$, and what is the instantaneous power at $t = 0$ expended by the force that must be used to keep the loop moving with constant velocity?

32. (II) A square wire loop of dimensions $L \times L$ lies in a plane perpendicular to a constant magnetic field. The field exists only in a certain region, with a sharp boundary (Fig. 31–36). The sides of the loop make a 45° angle with this boundary, and an external force moves the loop at a speed v out of the region of constant field. How much power must be supplied by the external force as a function of time?

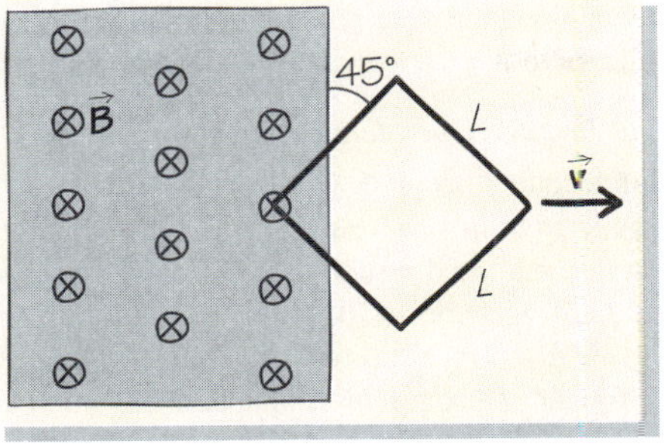

FIGURE 31–36 Problem 32.

33. (II) A conducting bar slides frictionlessly on two parallel horizontal rails 30 cm apart. The bar and rails form a closed circuit with a resistor of resistance 0.05 Ω, assumed to be constant throughout the motion. The circuit is placed in a uniform vertical magnetic field of 0.28 T perpendicular to the circuit's plane. The bar is pulled at a constant speed of 60 cm/s along the rails. (a) What is the magnitude of the force required to pull the bar? (b) What is the rate of Joule heating in the resistor?

34. (II) A long, straight wire carries a constant current I_0. A square loop with sides of length L and two sides parallel to the wire is pulled away at uniform speed v in a direction perpendicular to the wire. The nearest side of the loop is initially a distance D from the wire; the resistance of the loop is R. (a) Calculate the force necessary to pull the loop. (b) At what rate is work being done by the force? (c) How does your answer to part (b) compare with the Joule heating in the loop?

35. (II) In Example 31–7, what happens if the initial speed of the loop is (a) less than v_t, and (b) greater than v_t?

36. (II) When eddy currents are induced in a piece of metal moving through a magnetic field, drag forces that are proportional to the velocity of the metal act on it. Consider a thin metal disk rotating in a plane between the poles of a magnet. Show that the equilibrium angular velocity of the disk is proportional to the torque on the disk. Note that in household electricity meters, this arrangement—with the torque on the disk proportional to the power consumed—allows us to correlate the number of turns of the disk with the power consumption.

31–5 Time-Varying Magnetic Fields

37. (I) Consider a length of wire looped back on itself in a magnetic field B. The shape of the loop is not given, but it lies in a plane. The wire has negligible resistance, but there is a resistor R at one end. A cross bar, which rests on opposite legs of the loop, is pulled along so that the flux enclosed by the wire and crossbar varies with time as $\Phi_B(t)$. What is the instantaneous power needed to pull the crossbar?

38. (I) A long, straight wire carries a current $I = I_0 \cos(800\pi t)$, where t is time. Two sides of a fixed rectangular loop are 20 cm long and are parallel to the wire; the other sides are 1 cm long. The nearest long side is 3 cm from the wire. What is I_0 if the maximum emf induced in the loop is 1.5 μV? (Ignore the small variation of the magnetic field across the loop.)

39. (I) A loop of wire is placed between the poles of a large electromagnet. The loop is oriented so that the vector that characterizes the orientation of its planar surface runs from one pole to the other. When the magnet is turned on, its magnetic field builds up according to the formula $B = B_0(1 - e^{-at})$. We may assume that the magnetic field has the same value all across the surface of its poles and runs from one pole to another. (This formula gives $B = 0$ at $t = 0$ and $B = B_0$ at $t = \infty$.) What is the magnitude of the emf around the loop?

40. (I) A metal ring of diameter 10 cm is left between the pole tips of an electromagnet, such that the plane of the loop is perpendicular to the magnetic field. The electromagnet is turned on and reaches its full magnetic field of 1.0 T in 2.0 s at a linear rate. If the ring has resistance 0.08 Ω, how much current passes through the ring during the 2.0 s?

41. (II) A solenoid of radius r wound with n turns per unit length carries a current given by $I = I_0 \cos(\omega t)$, where t is the time. What are the magnitude and direction of the induced electric field just outside the solenoid?

42. (II) A very long cylindrical solenoid of radius r made from n turns of wire per unit length carries a current with the time dependence $I = I_0 e^{-t/t_0}$. Coaxial with and surrounding the solenoid are two turns of wire that make a circular loop slightly larger than the circular cross section of the solenoid (Fig. 31–37). The loop with two turns is far from the ends of the solenoid and has a resistance R. Find the current in the loop with two turns, I', as a function of time.

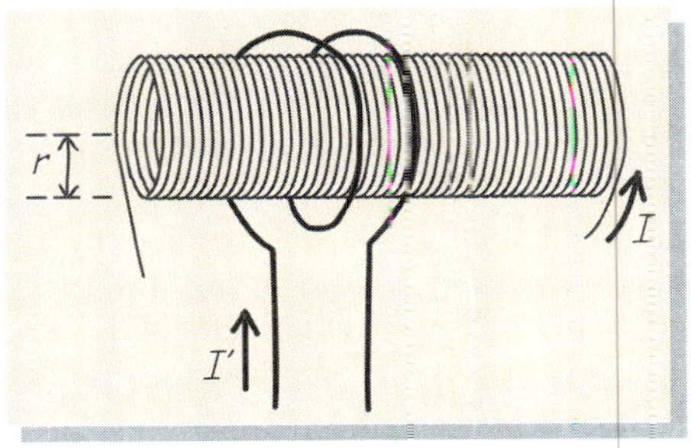

FIGURE 31–37 Problem 42.

43. (II) The uniform magnetic field of the electromagnet of Example 31–8, with circular pole faces of radius $R_0 = 0.08$ m, decreases linearly from 1.5 T to 0.7 T in 25 ms. What is the emf induced around the path drawn in Fig. 31–38 that consists of quarter arcs at radial distances $R_0/4$ and $R_0/2$, connected by radial lines? The path is clockwise.

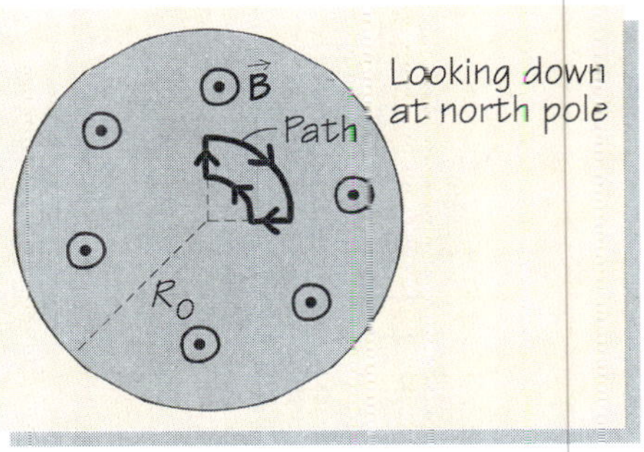

FIGURE 31–38 Problem 43.

44. (II) A long solenoid of radius R and n turns per unit length carries an alternating current $I = I_0 \sin(\omega t)$ (Fig. 31–39). What are the electric fields induced within the solenoid at a distance $R/2$ and outside the solenoid at a distance $2R$? [*Hint*: Apply Faraday's law to the two paths shown, and use symmetry.]

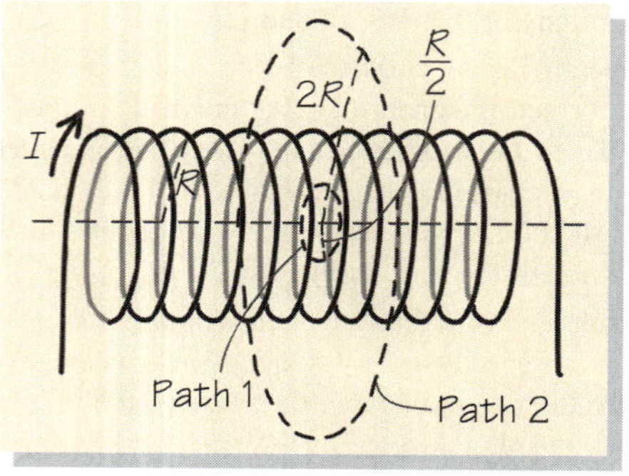

FIGURE 31–39 Problem 44.

31–6 Generators and Motors

45. (I) A coil of area 6.0 cm^2 with 180 turns of wire is connected to a resistor of resistance 3 Ω. It is rotated by hand at a frequency of 0.6 rev/s in a magnetic field of 0.40 T. (a) What is the maximum amount of current produced? (b) the average power produced?

46. (II) You have 25 m of wire, a constant magnetic field of 0.15 T, and a device that can rotate a coil at a fixed frequency of 90 Hz. What size circular coil will produce an AC emf of maximum voltage 120 V?

47. (II) The headlight of a bicycle is powered by a small generator that is driven by a wheel of the bicycle. The generator contains two coils fixed at the sides of the generator and connected in series with appropriate polarity (Fig. 31–40). Each coil consists of 70 turns and has an area of 8 cm^2. A small permanent magnet is rotated in front of the coils, so that the magnitude of the magnetic field in the coils varies between 0.1 T and zero. At what speed of the bicycle will the maximum emf be 6.4 V, given that the radius of the friction wheel is 1 cm?

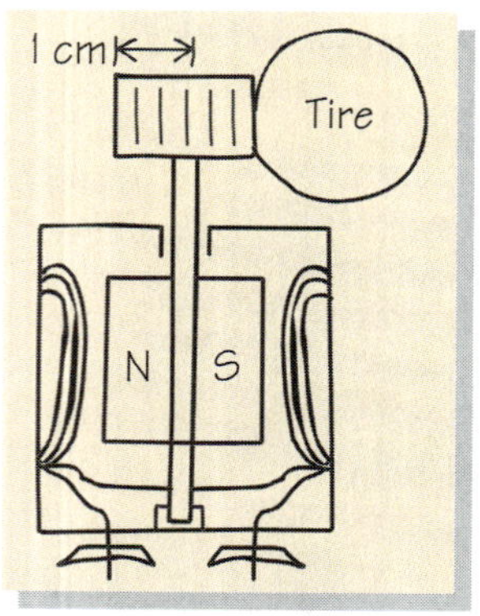

FIGURE 31–40 Problem 47.

48. (II) A bicycle wheel of radius $R = 33$ cm rotates at angular speed 38 rad/s in a plane perpendicular to a constant magnetic field of magnitude 0.14 T. What is the emf generated between the center of the wheel and its rim? When one end of a wire is attached to the center and the other end to a circular track in contact with the rim, a direct current is generated in the wire. Such a device is called a *homopolar generator*.

*31–7 The Frame Dependence of Fields

49. (II) Suppose that observer O sees an electric field $\mathbf{E} = E\mathbf{i}$ and a magnetic field $\mathbf{B} = B\mathbf{k}$. In what direction and at what (constant) speed u should a second observer move so as to see no electric field whatsoever? Use the nonrelativistic relation Eq. (31–22). If $E = 10^3$ V/m, for what range of values of B is the nonrelativistic approximation appropriate?

General Problems

50. (II) A 30-cm-long wire of square cross section with a mass of 25 g and a resistance of 0.05 Ω slides without friction down parallel conducting rails of negligible resistance (Fig. 31–41). The rails are connected to each other at the bottom by a resistanceless rail parallel to the wire so that the wire and rail form a closed rectangular conducting loop. The plane of the rails makes an angle of 35° with the horizontal, and a uniform vertical magnetic field of 0.18 T, pointing upward, exists throughout the region. What is the steady speed of the wire?

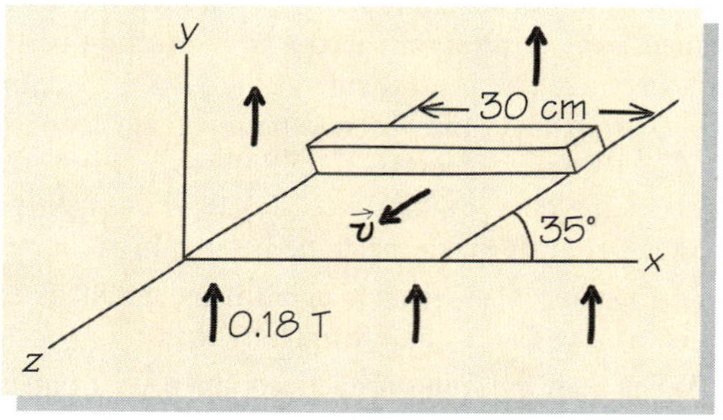

FIGURE 31–41 Problem 50.

51. (II) A straight wire carries a current $I = 150$ A near a rod that moves across two conducting wires (Fig. 31–42). The resistor has $R = 0.20$ Ω, and the rod moves at speed 45 cm/s. (a) What is the emf induced in the rod? (b) What is the current in the circuit? (c) How much work is done to move the rod 100 cm to the right? What force does this work?

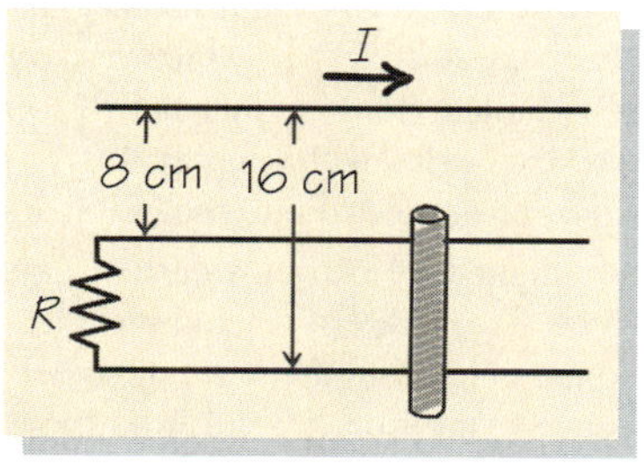

FIGURE 31–42 Problem 51.

52. (II) A conducting crossbar bracketing two vertical conducting wires slides down the wires. The wires are connected with a resistor R to form a closed circuit (Fig. 31–43a). (a) If there is a horizontal magnetic field B perpendicular to the plane of the loop, how fast does the bar fall after the initial accelerating period? (b) A battery is added to the circuit (Fig. 31–43b). What polarity and emf of the battery are needed to lift the bar with the same velocity?

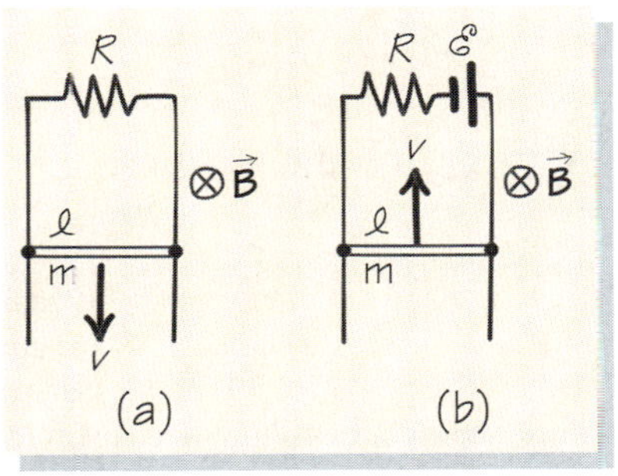

FIGURE 31–43 Problem 52.

53. (II) A large, circular coil of N turns and radius R carries a steady current I and is rotated at a constant angular speed ω about a horizontal diameter. At the center of this coil is a small, fixed, horizontal circular ring of radius r. (a) What is the emf induced in the small ring? (b) What is the angle between the plane of the coil and that of the ring when this emf is a maximum?

54. (II) Consider a 1.50-V battery attached to two conducting, frictionless rails 0.200 m apart. There is a magnetic field **B** of magnitude 0.400 T perpendicular to the rails, and a conducting bar can slide over the rails perpendicular to them as well as to the field (Fig. 31–44). The bar is placed on the rails, starts from rest, and accelerates. (a) What is the direction of its motion? (b) the direction of the emf induced? (c) Given that the total resistance of the closed circuit is 0.30 Ω, calculate the current in the bar when its speed is 12.0 m/s.

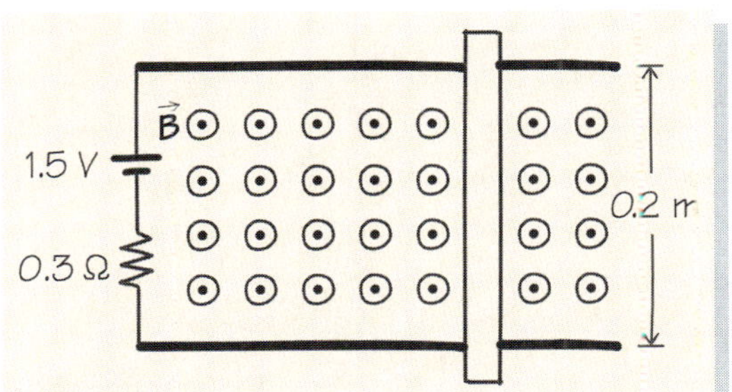

FIGURE 31–44 Problem 54.

55. (II) A wire carrying a current I is oriented in a horizontal direction. To its side, a wire loop is oriented so that it and the straight wire lie in the same horizontal plane. The straight wire is moved toward the loop. If a current is induced in the loop, what is its direction, and what is the direction of the force on the loop?

56. (II) If the plasma in a magnetohydrodynamic generator (see Question 11) is forced to flow in a channel perpendicular to a magnetic field, an electric potential builds up between points a and b, which are 1 m apart (Fig. 31–45). If the magnetic field has a strength 2.5 T, what must the speed of the plasma be in order that the potential be 1000 V?

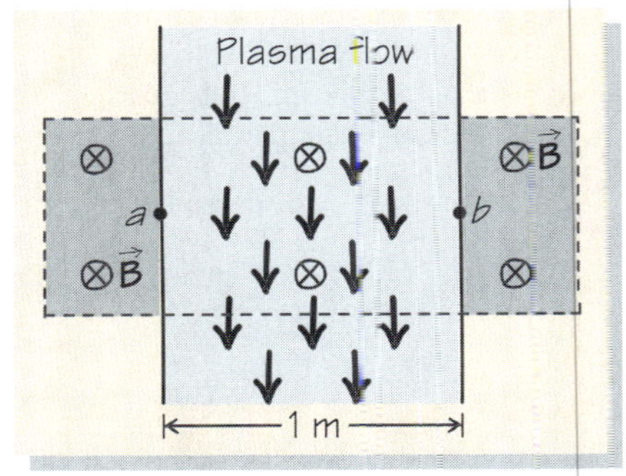

FIGURE 31–45 Problem 56.

57. (II) A coil with 200 turns, a diameter of 8.0 cm, and a resistance of 5.6 Ω is placed perpendicular to a uniform magnetic field of 1.4 T. The magnetic field suddenly reverses direction. What is the total charge that passes through the coil?

58. (II) A constant magnetic field of 1.5 T is directed along the x-axis. A wire coil of 50 turns and area 2.0 cm^2 is placed in the yz-plane. The coil of wire, called a *flip coil*, is then turned over (in other words, rotated by 180°). (a) If the total charge that passes through the coil when it is flipped is 0.024 C, what is the resistance of the coil circuit? (b) The same flip coil is used to measure an unknown magnetic field. The coil is flipped in several directions until it attains its maximum charge of 0.011 C, when the coil is flipped with its face in the xy-plane. What is the magnitude of the magnetic field? (c) What is the direction of the magnetic field in part (b)?

59. (II) A circular ring of area 100 cm^2 is connected to a 15-μF capacitor. The circuit has a resistance of 2 Ω. A uniform time-dependent magnetic field of magnitude $B = (0.03\ \text{T/s})t$ is perpendicular to the ring (Fig. 31–46). Calculate the current in the ring and the charge on the capacitor. Give the direction of the current and the polarity of the charge.

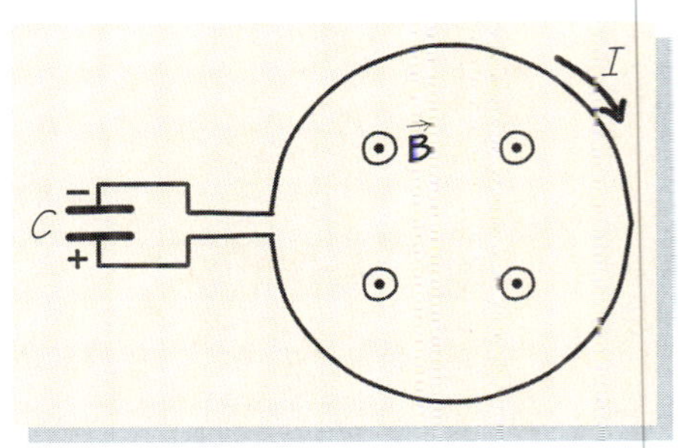

FIGURE 31–46 Problem 59.

60. (II) A current $I = I_0 \cos(\omega t)$ passes through a solenoid of area 10 cm^2 and 10^5 turns/m. The frequency is 60 Hz, and $I_0 = 10$ A. A small coil—a sense coil—is used to sense the changing

flux. This sense coil has an area of 20 cm^2 with 10 turns and is placed across the solenoid so that the face of the coil is perpendicular to the solenoid axis; the two coils are concentric. (a) What is the emf induced in the sense coil? (b) If the resistance of the sense coil circuit is 5 Ω, what is the current?

61. (II) A wire that is bent into a semicircle is rotated with angular velocity ω about the diameter shown in Fig. 31–47. The bent wire and its supports are placed in a uniform magnetic field perpendicular to the plane of the supports. What is the emf induced in the circuit shown? If the resistance of the closed loop is R, what is the average power dissipated?

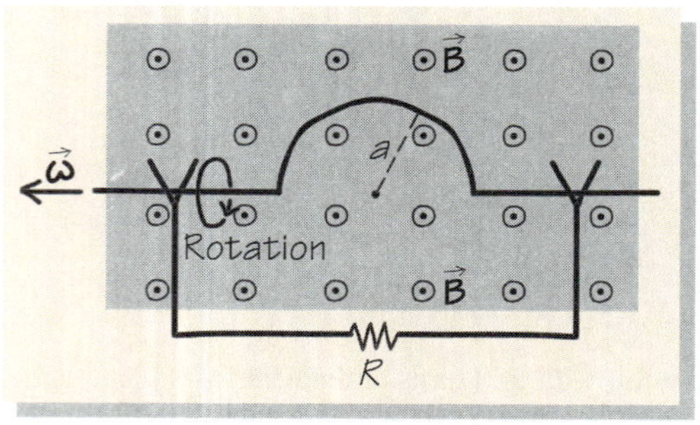

FIGURE 31–47 Problem 61.

62. (II) A pendulum consists of a metal bar suspended from two thin wires attached to a fixed conducting bar (Fig. 31–48). The resistance of this closed circuit is 2 Ω. The pendulum is placed in a vertical magnetic field of magnitude 0.03 T. The pendulum is displaced by a small angle from the equilibrium position and allowed to oscillate. What is the ratio of the power dissipated to the energy of the oscillator?

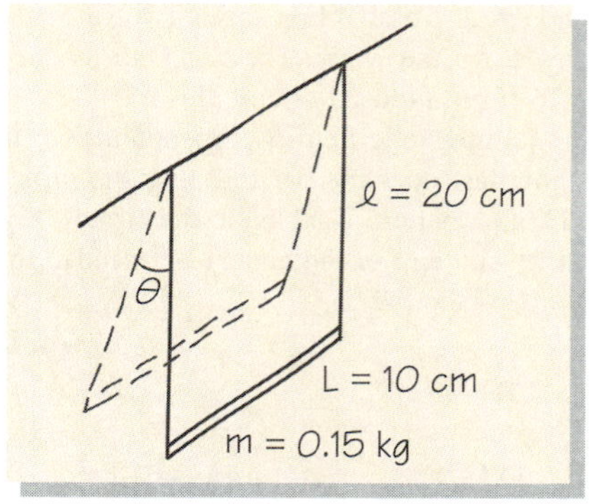

FIGURE 31–48 Problem 62.

63. (III) An electron follows a circular path of radius $R = 1$ m while traveling in a plane perpendicular to a spatially constant magnetic field of magnitude 10^{-6} T. As viewed along the magnetic field lines, the electron follows a counterclockwise path. (a) What is the speed of the electron? (b) Assuming that the motion of the electron is nonrelativistic, what is the energy, E, of the electron? (c) The magnitude of the magnetic field is reduced smoothly by a certain percentage during an interval Δt. Show that the fractional energy change of the electron, $\Delta E/E$, is independent of the radius of the electron's orbit as well as of the electron's speed. (This effect is the basis for low-energy operation of a particle accelerator called the *betatron*.) (d) If the magnetic field is reduced in time $\Delta t = 5$ s by 10 percent, estimate $\Delta E/E$.

CHAPTER

32

Large magnets have a ferromagnetic core. Here, such a magnet is used to lift scrap metal in a junkyard.

Magnetism and Matter

Why is it possible to make a slab of soft iron act like a bar magnet at some times but not at other times? Why can't the same be said for a piece of aluminum? How do we explain the fact that magnets pick up needles but not pieces of paper? The answers to these questions lie in understanding the magnetic properties of matter. This understanding is necessary for the construction of computer memories, electric motors, generators, transformers, particle accelerators, and medical scanners. Just as the dielectric properties of materials depend on the polarizability of atoms and molecules, the magnetic properties of materials depend on the magnetic properties of atoms and molecules. We shall describe the mechanism responsible for ferromagnetism, which is a characteristic of permanent magnets. We shall explore the magnetic properties of superconductors and describe the atomic origin of nuclear magnetic resonance, an important tool in materials science and in medicine.

32–1 THE MAGNETIC PROPERTIES OF BULK MATTER

Let's measure the magnetic field $\mathbf{B}_0$ produced by, and near one end of, a current-carrying solenoid (Fig. 32–1a).[†] How does the presence of materials affect this field? Insert a wooden or copper core into the solenoid and repeat the measurement. Only very sensitive instruments would reveal that the field has been changed by one part in 10^6.

[†] $\mathbf{B}_0$ represents any field, not necessarily constant, that is present before we introduce materials.

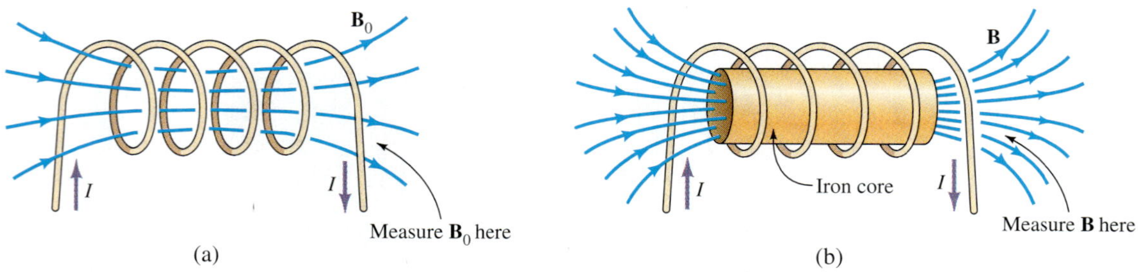

FIGURE 32–1 (a) A solenoid that carries a current has a magnetic field. (b) When the cylindrical volume of the solenoid is filled with an iron core, the magnetic field may be greatly amplified.

Ferromagnetism

Now replace that core by a soft iron core (Fig. 32–1b). This time, the measured field is increased by a factor many times that of the original field. Furthermore, if we then remove the iron core, the iron will act like a bar magnet, even if it showed no such properties before insertion. This last observation shows that the solenoid somehow magnetizes the soft iron core, but does not magnetize the wooden or copper cores. Materials that can form permanent magnets are said to be **ferromagnetic**, whereas materials that display magnetic properties only in the presence of external magnetic fields are *nonferromagnetic*. Ferromagnetic effects are much larger than nonferromagnetic effects.

The magnetic behavior of bulk materials—gases, liquids, and solids—is characterized by their **magnetization**, **M**, which we define as the *magnetic dipole moment per unit volume*. The field outside material with a net magnetization is that of a magnetic dipole, the same as that of a bar magnet. A magnetic moment has dimensions of current times area, so magnetization has dimensions of current times area divided by volume, or current per length. The SI units of **M** are amperes per meter (A/m).

The magnetic field of a solenoid was given in Chapter 30.

The net magnetic field of a solenoid S_1 with an iron core is the vector sum of contributions from the original magnetic field of the solenoid, $\mathbf{B}_0$, and the contribution of the magnetization of the core. The core may itself be viewed as a second solenoid, S_2, because any magnetic dipole has a magnetic field like that of an equivalent solenoid. Suppose that S_2 can be treated as a solenoid of area A, length L, and N loops that carry current I. From Eq. (30–15), the magnetic field of S_2 is

$$B_2 = \frac{\mu_0 NI}{L} = \frac{\mu_0 NIA}{LA} = \mu_0 \frac{m_2}{V},$$

where m_2 is the magnetic moment of S_2, $m_2 = NIA$,[†] and μ_0 is the permeability of free space. The factor m_2/V is the magnetic moment per unit volume—the magnetization, **M**. Thus *the contribution of the core to the total magnetic field is* $\mu_0\mathbf{M}$, and the total field of the solenoid with the core is

The net magnetic field in a material with magnetization **M**

$$\mathbf{B} = \mathbf{B}_0 + \mu_0\mathbf{M}. \tag{32–1}$$

Of course, an actual solenoid is not required: Eq. (32–1) applies whatever the source of the original field $\mathbf{B}_0$ in which the bulk material is placed. This situation is similar to that of dielectrics, where an electric field within a material has contributions from an original electric field due to free charges and from an internal, induced distribution of charge. We saw in Chapter 26 that we can separate the effect of these types of charges through the introduction of a dielectric constant and a generalized permittivity. Similarly, we want to separate the effects of the internal magnetization from the magnetic fields due to ordinary currents. We refer to such currents as *free*, or *real*, currents. The effects of free currents are isolated by defining the **magnetic intensity**, **H**, a quantity for which the effect of the magnetization of the material is subtracted out:

The definition of magnetic intensity

$$\mathbf{H} \equiv \frac{\mathbf{B}}{\mu_0} - \mathbf{M}. \tag{32–2}$$

[†]To avoid confusion with the magnetic permeability, μ, to be introduced shortly, we use the notation m for the magnetic dipole moment throughout this chapter. Don't confuse it with a mass!

The dimensions of **H** are those of **M**, *not* of **B**. By replacing **B** in Eq. (32–2) with $\mathbf{B}_0 + \mu_0\mathbf{M}$ according to Eq. (32–1), we find

$$\mathbf{B}_0 = \mu_0\mathbf{H}. \tag{32–3}$$

Equation (32–3) shows that *the magnetic intensity measures the magnetic field due to free currents*. Another form for the relation among **B**, **H**, and **M** is found by combining Eq. (32–1) and (32–3):

The magnetic intensity isolates the effects of free currents.

$$\mathbf{B} = \mu_0\mathbf{H} + \mu_0\mathbf{M}. \tag{32–4}$$

Nonferromagnetic materials have no magnetization unless they are in the presence of an external magnetic field $\mathbf{B}_0$ that induces magnetization. We know by experiment that, over a large range of conditions, *the magnitude of the magnetization of nonferromagnetic materials varies linearly with the external magnetic field*. The direction of the magnetization is more complicated: The original field $\mathbf{B}_0$ and the field $\mu_0\mathbf{M}$ due to the magnetization are parallel for one class of materials but antiparallel for a second class. For nonferromagnetic materials, **M** depends linearly on $\mathbf{B}_0$, and hence on **H**. The **magnetic susceptibility**, χ_m, is defined as the coefficient of the linear relation between the magnetization and the magnetic intensity:

The definition of magnetic susceptibility

$$\mathbf{M} \equiv \chi_m\mathbf{H}. \tag{32–5}$$

If the susceptibility of a material is positive, its magnetization is aligned along the external field; if the susceptibility is negative, the magnetization is aligned opposite to the external field. Because **M** and **H** have the same dimensions, χ_m is dimensionless. Table 32–1 gives a range of susceptibilities found in nature.

With our definition of χ_m, we can express the relation between the magnetic field in a material and the magnetic intensity. From Eq. (32–1),

$$\begin{aligned}\mathbf{B} &= \mathbf{B}_0 + \mu_0\mathbf{M} = \mu_0\mathbf{H} + \mu_0\chi_m\mathbf{H} \\ &= \mu_0(1 + \chi_m)\mathbf{H}.\end{aligned} \tag{32–6}$$

We define the coefficient of **H** in this equation as the **permeability**, μ, of the material:

$$\mu \equiv \mu_0(1 + \chi_m). \tag{32–7}$$

TABLE 32–1 Some Magnetic Susceptibilities (at 20°C unless indicated otherwise)

Material	*Susceptibility,* χ_m
Diamagnetic	
Water	$-9.$[illegible] $\times 10^{-6}$
Copper	-9.6×10^{-6}
Silver	-2.4×10^{-5}
Carbon (diamond form)	-2.2×10^{-5}
Bismuth	-1.7×10^{-4}
Paramagnetic	
Sodium	7.2×10^{-6}
Cupric oxide	2.6×10^{-4}
Aluminum	2.2×10^{-5}
Liquid oxygen (90 K)	3.5×10^{-3}
Ferromagnetic	
Iron (annealed)	5.5×10^{3}
Permalloy (55% Fe, 45% Ni)	2.5×10^{4}
Mu-metal (77% Ni, 16% Fe, 5% Cu, 2% Cr)	1×10^{5}

The relation between the total magnetic field in a material and the magnetic intensity, which is a measure of the effect of free currents, is then

$$\mathbf{B} = \mu\mathbf{H}. \tag{32–8}$$

Just as the electric permittivity, ϵ, replaces the permittivity of free space, ϵ_0, in expressions for electric fields in materials if the charge is free charge, so μ replaces μ_0 when the current is the free current in materials. From Eq. (32–3), we see that when there is a magnetic field in a vacuum, that field is related to the intensity by a relation like that of Eq. (32–8), but with μ_0 appearing in the place of μ; thus μ_0 is the permeability of the vacuum. As we can deduce from Table 32–1, μ is very close to μ_0 for nonferromagnetic materials.

The various quantities we have defined are all useful in characterizing the bulk magnetic behavior of materials. In Table 32–2, we summarize these quantities and their relations.

EXAMPLE 32–1 A straight solenoid of diameter 5 cm and length 25 cm is wrapped with 200 turns of wire that carries a current of 5 A. The solenoid is filled with a material of magnetic susceptibility $\chi_m = 10^{-5}$. Find (a) the magnetic intensity within the solenoid and (b) the magnetic field within the solenoid. (c) By what factor is the magnetic field changed due to the presence of the material?

Solution: (a) The magnetic intensity is associated with the free currents of the solenoid. It is found from Eq. (32–3) to have magnitude

$$H = \frac{B_0}{\mu_0} = \frac{\mu_0 nI}{\mu_0} = nI,$$

where we have used Eq. (30–15) for the interior field of a solenoid. Here, n is the number of turns per unit length:

$$n = \frac{200 \text{ turns}}{0.25 \text{ m}} = 800 \text{ turns/m}.$$

Thus

$$H = (800 \text{ turns/m})(5 \text{ A}) = 4000 \text{ A/m}.$$

TABLE 32–2 Magnetic Bulk Properties and their Relations

Symbol	*Property*
$\mathbf{B}_0$	Applied magnetic field, produced independently of type of material by a nearby magnet or currents
$\mathbf{H}$	Magnetic intensity, proportional to the applied magnetic field
$\mathbf{M}$	Magnetization, the magnetic dipole moment per unit volume of a material
$\mathbf{B}$	Net magnetic field, the sum of the applied magnetic field and a term proportional to the magnetization
μ_0	Permeability of free space
χ_m	Susceptibility of a material
μ	Permeability of a material, $\mu = \mu_0(1 + \chi_m)$

Some Relations

	$\mathbf{B}_0$	$\mathbf{H}$	$\mathbf{M}$	$\mathbf{B}$
$\mathbf{B}_0 =$	—	$\mu_0\mathbf{H}$	Not used	Not used
$\mathbf{H} =$	$\mathbf{B}_0/\mu_0$	—	$\mathbf{M}/\chi_m$	$\mathbf{B}/\mu$
$\mathbf{M} =$	$\chi_m\mathbf{B}_0/\mu_0$	$\chi_m\mathbf{H}$	—	Not used
$\mathbf{B} =$	$(1 + \chi_m)\mathbf{B}_0$	$\mu_0(1 + \chi_m)\mathbf{H}$	$\dfrac{\mu_0(1 + \chi_m)}{\chi_m}\mathbf{M}$	—

(b) The total magnetic field, **B**, includes the effect of the field due to the material that fills the solenoid. B can be found from, for example, Eqs. (32–8) and (32–7):

$$B = \mu H = \mu_0(1 + \chi_m)H$$
$$= (4\pi \times 10^{-7}\ \text{T}\cdot\text{m/A})(1 + 10^{-5})(4000\ \text{A/m}) = 5 \times 10^{-3}\ \text{T}.$$

(c) The factor by which the field changes is

$$\frac{\Delta B}{B} = \frac{\mu_0(1 + \chi_m)H - \mu_0 H}{\mu_0 H} = \chi_m.$$

The Magnetic Properties of Materials

Diamagnetism

Table 32–1 reveals that materials break down into three broad classes. The first class—with large positive susceptibilities—is composed of ferromagnetic materials. As we have already described, these substances can form permanent magnets.

Substances with very small negative susceptibilities are called **diamagnetic** materials. In such materials, the magnetization direction is *opposite* to the direction of the inducing field. The magnetic field inside such materials is *reduced* from its value outside the material. If a diamagnetic material is placed near the north pole of a magnet, the magnetization produces a field that points toward the pole (Fig. 32–2). The diamagnetic material acts as though it has a north pole adjacent to the external north pole: The diamagnetic material is *repelled* by the magnet. The behavior of diamagnetic substances is similar to that of dielectrics, for which polarization effects tend to cancel the electric field associated with free charges.

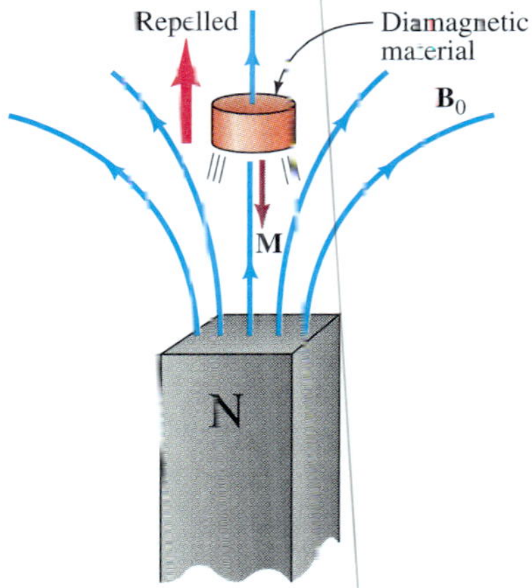

FIGURE 32–2 Diamagnetic substances are repelled by one pole of a nearby bar magnet.

Paramagnetism

Substances with small positive susceptibilities are called **paramagnetic** materials. For them, the external magnetic field aligns the atomic magnetic dipole moments parallel to itself. The magnetization points in the same direction as the field of an external magnet (Fig. 32–3), and it is as if the paramagnetic substance has a south pole oriented toward the magnet's north pole: The piece of paramagnetic material is *attracted* to the magnet. Ferromagnetism, diamagnetism, and paramagnetism will be discussed further in Sections 32–3 to 32–5, respectively.

Table 32–3 gives the forces acting on some samples near a sizable (3.0 T) electromagnet. Only the relative scale of the numbers matters in the table, since we are providing no details about the geometry of the magnet. The force is expressed in units of the weight of each sample. The − sign reflects repulsion (diamagnetism); the + sign reflects attraction (paramagnetism and ferromagnetism). Note the differences in the sizes of these forces for the different classes of materials.

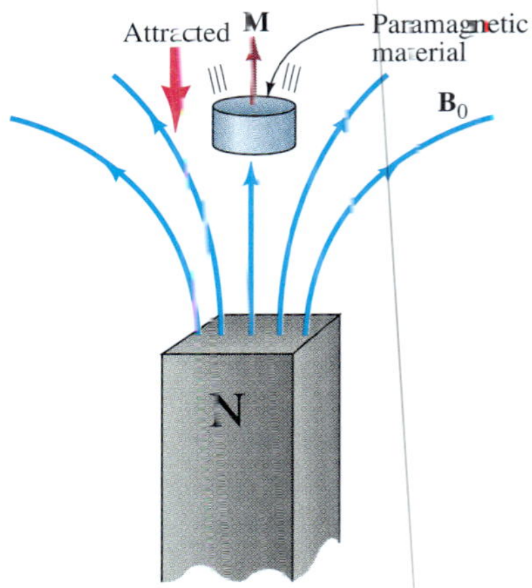

FIGURE 32–3 Paramagnetic substances are attracted to one pole of a nearby bar magnet.

32–2 ATOMS AS MAGNETS

The magnetic field of a solenoid is the same as that of a bar magnet. How can two such apparently dissimilar systems have the same magnetic field? The answer lies in the magnetic properties of atoms. To understand these properties qualitatively, we start

TABLE 32–3 Magnetic Forces on Materials

Material	*Material Class*	*Force (in units of sample weight)*
Copper (pure)	Diamagnetic	-1.3×10^{-3}
Lead	Diamagnetic	-19×10^{-3}
Graphite	Diamagnetic	-56×10^{-3}
Sodium	Paramagnetic	$+10.2 \times 10^{-3}$
Copper chloride	Paramagnetic	$+143 \times 10^{-3}$
Iron	Ferromagnetic	$+204$
Magnetite	Ferromagnetic	$+61$

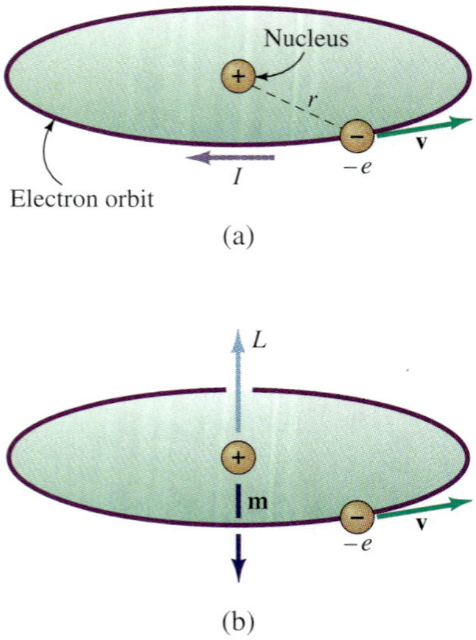

FIGURE 32–4 (a) An electron in a circular orbit around a nucleus. Over numerous orbits, the effect is the same as a continuous current ring about the nucleus. (b) The circulating electron forms a magnetic dipole with magnetic moment **m** ($\mathbf{m}_{\text{orbital}}$) oriented downward, opposite to the angular momentum, **L**.

with a classical planetary model for atoms and add some necessary quantum mechanical features. The orbiting electrons in atoms form ring currents like those of a solenoid and the atoms then act as magnetic dipoles. Thus we expect atoms to have magnetic dipole moments. The magnetic moments in a large assembly of atoms point in random directions, so that the net magnetic moment of a macroscopic material is generally zero. The exception occurs in ferromagnetic materials, in which the atomic magnetic moments tend to line up with their neighbors. It is possible to destroy "permanent" magnetic effects in such a material by heating it or by pounding it with a hammer. Such actions tend to randomize the collective atomic alignments.

The magnetic properties of individual atoms are affected by a tendency of their electrons to pair off such that the magnetic moments due to their orbits point in opposite directions. The net result is that a pair of electrons has no magnetic moment. In effect, for each electron with a clockwise orbit around the nucleus, there is an electron with the same orbit in a counterclockwise sense. This reasoning suggests that atoms with an even number of electrons will have no magnetic dipole moment, whereas only a last unpaired electron matters in atoms with an odd number of electrons.†

The Magnetic Dipole Moment of Atoms

Suppose that a single electron of charge $-e$ moves at speed v in a circular orbit of radius r around a heavy nucleus—as in Fig. 32–4a. The orbital period is $T = 2\pi r/v$. Because electric current is charge per unit time, the current around the nearly stationary nucleus is

$$I = \frac{-e}{T} = -\frac{ev}{2\pi r}. \tag{32–9}$$

The minus sign indicates that the current is in a direction opposite to that of the motion of the electron. The current loop has an area of πr^2; following Eq. (30–28), the magnitude of the *orbital magnetic dipole moment* is therefore given by

$$m_{\text{orbital}} = I\pi r^2 = \frac{ev}{2\pi r}\pi r^2 = \frac{1}{2}evr. \tag{32–10}$$

The direction of the magnetic moment vector **m** is determined by a right-hand rule (Fig. 32–4b).

EXAMPLE 32–2 Estimate an atomic orbital magnetic moment by taking the radius of the orbit to be of roughly atomic size, 10^{-10} m, and the kinetic energy to be a typical atomic energy of 1 eV. Compare this to the magnetic moment of a macroscopic loop of area 1 cm^2 that carries a current of 1 mA.

Solution: The unknown magnetic moment is determined from Eq. (32–10). We are given the radius r of the orbit and can deduce the speed v of the electron in its orbit from the given energy. If we convert to SI units, our energy estimate of 1 eV is 1.6×10^{-19} J. From this energy, we find v from

$$\tfrac{1}{2}m_e v^2 = 1.6 \times 10^{-19}\ \text{J}.$$

The electron mass is, to the same accuracy as the orbit radius, $m_e \simeq 10^{-30}$ kg, so

$$v = \sqrt{\frac{2(1.6 \times 10^{-19}\ \text{J})}{m_e}} \simeq \sqrt{\frac{3.2 \times 10^{-19}\ \text{J}}{10^{-30}\ \text{kg}}} = 5 \times 10^5\ \text{m/s}.$$

Equation (32–10) then gives the magnetic moment

$$m_{\text{orbital}} = \frac{1}{2}(1.6 \times 10^{-19}\ \text{C})(5 \times 10^5\ \text{m/s})(10^{-10}\ \text{m}) \simeq 4 \times 10^{-24}\ \text{A}\cdot\text{m}^2.$$

†Depending on the element, there may be more than one unpaired electron. Iron is an example.

For comparison, the magnetic moment of the macroscopic loop is

$$m = (10^{-3}\ \text{A})(10^{-4}\ \text{m}^2) = 10^{-7}\ \text{A}\cdot\text{m}^2,$$

which is some 2×10^{16} times larger than the magnetic moment of the single atom.

We can think of a magnetic dipole moment as due to a circulating charge, with the magnetic moment proportional to the angular momentum of the circulating charge. It is useful to express magnetic moments in terms of angular momentum because angular momentum is a fundamental physical quantity. For an electron in an atom, the relation is as follows: Eq. (32–10) can be written in the form

The magnetic moment of a single-electron atom

$$m_{\text{orbital}} = \frac{1}{2}evr = \frac{e}{2m_e}m_e vr = \frac{e}{2m_e}L, \qquad (32\text{–}11)$$

where m_e is the electron's mass and $L = m_e vr$ is the angular momentum of the electron in its circular orbit. If we include the vectorial properties of both the angular momentum and the magnetic moment, then Eq. (32–11) becomes

$$\mathbf{m}_{\text{orbital}} \equiv g_L \mathbf{L}. \qquad (32\text{–}12)$$

The coefficient g_L connecting the magnetic moment and the angular momentum is known as the **gyromagnetic ratio**. For the orbital motion, we have just seen that

$$g_L = -\frac{e}{2m_e}. \qquad (32\text{–}13)$$

The minus sign is present because $\mathbf{m}_{\text{orbital}}$ and $\mathbf{L}$ point in opposite directions (Fig. 32–4b).

According to the quantum mechanical quantization rules for circular orbits (Section 10–5), the magnitude of L is $\ell\hbar$, where $\hbar \equiv h/2\pi$, h is Planck's constant, and ℓ is an integer. Thus the vector $\mathbf{m}_{\text{orbital}}$ has magnitude

The magnetic moment of single-electron atoms is quantized.

An electron acts as a tiny magnet.

$$m_{\text{orbital}} = \left(\frac{e}{2m_e}\hbar\right)\ell \equiv m_B\ell, \qquad \ell = 0,1,2\ldots; \qquad (32\text{–}14)$$

m_B is called the **Bohr magneton**, after Niels Bohr, one of the founders of quantum mechanics. Its value is

$$m_B = \frac{e}{2m_e}\hbar = 9.27 \times 10^{-24}\ \text{A}\cdot\text{m}^2. \qquad (32\text{–}15)$$

In addition to a magnetic moment associated with orbital motion, electrons also carry an *internal* magnetic moment that cannot be identified with any real current. *The electron itself behaves as a tiny magnet!* Therefore, to the orbital magnetic moment, we must add the contribution of the electron's *intrinsic magnetic moment,* $m_{\text{intrinsic}}$. The value of $m_{\text{intrinsic}}$ turns out to be approximately m_B, to an accuracy of 0.1%.

Bulk Effects Are Due to the Alignment of Atomic Magnetic Dipoles

The magnetic field of an individual atom is tiny compared to the magnetic field of a bar magnet. But there are so many atoms that, if all the atomic magnetic moments were perfectly aligned in a material, we would have a very large effect. In fact, the alignment of the atomic magnetic moments need be only very slight to produce noticeable bulk effects (Example 32–3).

EXAMPLE 32–3 Consider 1 mol of atoms with individual magnetic moments $m_0 = 10^{-23}\ \text{A}\cdot\text{m}^2$. Assume that the magnetic moments can point only in the $+z$- and $-z$-directions with a fraction f pointing "up" and $1 - f$ pointing "down." What value of f gives the same magnetic moment as a 1-cm^2 wire loop that carries a current of 10 mA?

Solution: The magnetic moment of the specified loop is given by

$$IA = (10^{-2}\text{ A})(10^{-4}\text{ m}^2) = 10^{-6}\text{ A}\cdot\text{m}^2.$$

Now we find the net magnetic moment of the sample for a given f. Where the magnetic moment of an atom points "up," the magnetic moment is $+m_0$; where it points "down," the magnetic moment is $-m_0$. The fraction up is f, and the fraction down is $1 - f$. Thus the net magnetic moment of the atoms is

$$m = N_A m_0[f - (1 - f)] = N_A m_0(2f - 1).$$

where N_A is Avogadro's number. We equate this net magnetic moment to that of the current loop to find that

$$N_A m_0(2f - 1) = IA.$$

We solve for f to find that

$$f = \frac{1}{2}\left(1 + \frac{IA}{N_A m_0}\right) = \frac{1}{2} + \frac{m}{2N_A m_0}$$
$$= \frac{1}{2} + \frac{10^{-6}\text{ A}\cdot\text{m}^2}{2(6\times10^{23}\text{ atoms})(10^{-23}\text{ A}\cdot\text{m}^2/\text{atom})} = \frac{1}{2} + (8\times10^{-8}).$$

A random distribution would have $f = \frac{1}{2}$, and *a departure from complete randomness of one part in 10 million gives rise to macroscopic effects.*

If only a small deviation from randomness leads to significant bulk effects, as in Example 32–3, why aren't most materials magnetic? The reason is that in a sample of 10^{24} atoms, statistical fluctuations away from an average magnetic moment of zero are expected to lead, on average, to an excess of only 10^{12} atoms that point in a particular direction. (In statistics, $\sqrt{N}$ is a typical fluctuation from the mean when N objects or events are involved.) Thus, the typical value for the fraction f of atoms that point in a particular direction is $10^{12}/10^{24} = 10^{-12}$, and this leads to an infinitesimally small net magnetic moment. Our argument must be reexamined when atoms pack closely to form a solid or a liquid. In that case, forces between the atoms may cause neighboring atoms to line up with each other and lead to significant magnetic moments in large regions of the material. When this happens, permanent magnets form.

The Connection Between Microscopic and Macroscopic Quantities

A piece of material will have significant magnetic properties—be a magnet—if the directions of the magnetic dipole of its many component atoms or molecules are not completely random. In that case, the vector sum of the atomic magnetic moments will not be zero. If we divide the vector sum of the magnetic moments by the number of atoms, we get a *net* magnetic moment $\mathbf{m}_0$ per constituent (atom or molecule). The magnetization is then

$$\mathbf{M} = n\mathbf{m}_0, \tag{32–16}$$

where n is the number of constituents per unit volume. Once we have determined $\mathbf{M}$, we can determine the other bulk magnetic properties from the discussion in Section 32–1.

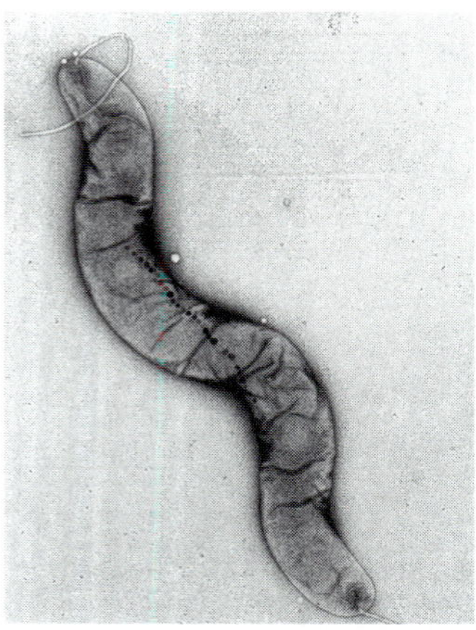

FIGURE 32–5 Some biological systems make use of ferromagnetic materials. This bacterium uses ferrites, here visible as dark spots, to navigate in Earth's magnetic field.

32–3 FERROMAGNETISM

Ferromagnetic materials, which include the elements iron, cobalt, nickel, gadolinium, and dysprosium, together with their alloys, can have large permanent magnetizations (Fig. 32–5). The direction and size of the magnetization can be set by an external magnetic field.

In ferromagnetic materials, the intrinsic magnetic dipole moments of the electrons in atoms align themselves in large numbers and lead to large magnetic effects. There is no classical mechanism that can align the intrinsic magnetic moments sufficiently strongly and the explanation is purely quantum mechanical. In 1928, Werner Heisenberg, one of the creators of the quantum theory, suggested that, in certain materials, as a consequence of the exclusion principle (see Chapter 27), electrons with parallel intrinsic magnetic moments arrange themselves in orbits that tend to maximize the distance between them. This reduces the potential energy of Coulomb repulsion between them and thus makes a state with parallel magnetic moments a state of lower energy. Thus there is a preference for the intrinsic magnetic moments of electrons to line up parallel with one another.

FIGURE 32–6 Photomicrograph of magnetic domains in a sample of iron with 3 percent silicon. A strong net magnetic field is associated with each domain. Domains with different orientations appear in different colors.

The intrinsic magnetic moments of the unpaired electrons of different atoms—each iron atom, for example, has two such electrons—do not ordinarily become aligned *throughout* a piece of ferromagnetic material. Rather, the alignment takes place between adjacent atoms in regions called *magnetic domains*, which may contain 10^{17} to 10^{21} atoms and occupy a volume on the order of 10^{-12} to 10^{-8} m^3 (volumes from 0.1 mm to 1 mm on a side). The magnetic field within these domains is quite large, but the material may be made of thousands of such domains, each with a magnetization aligned differently. Thus, without some special mechanism, the magnetization of the entire material will average to zero. A sample of magnetic material may look something like Fig. 32–6, which clearly shows the boundaries between domains, called *domain walls*. Figure 32–7a is a schematic diagram of the domains with their individual magnetic moments.

Ferromagnetism is due to the large-scale alignment of the magnetic moments of electrons.

The special mechanism that can align the magnetizations of different domains is provided by an applied magnetic field. If a field $\mathbf{B}_0$ is applied to a piece of ferromagnetic material, two things can happen to transform the material into a permanent magnet. First, the size of domains with their magnetic moments already aligned with $\mathbf{B}_0$ may enlarge at the expense of neighboring domains. Second, the magnetic moments of some of the domains may rotate to the direction of $\mathbf{B}_0$ through an overall realignment of their constituents (Fig. 32–7b). (Remember that a state with a magnetic moment aligned along $\mathbf{B}_0$ is a state of lower energy.)

The process we have described can be understood by a simple analogy. Imagine a large marching band whose members face in random directions (Fig. 32–8a). The band leader orders them by loudspeaker to face the same direction but fails to say *which* direction. Perhaps influenced by the random choices of a few band members (*A*, *B*, or *C* in Fig. 32–8b), the band members align themselves into three separate regions of alignment. An analogous situation applies in ferromagnets, where the arrows in Fig. 32–8 represent the intrinsic dipoles. Before an external field is applied, the dipoles are aligned over differently oriented domains because of the quantum effects we have mentioned. How can we get alignment over larger distances? The band members all align themselves in the same direction only when that precise direction is given over the loudspeaker. In a ferromagnetic material, the instructions for a precise direction are provided by the external field. Just as the band members will remain aligned even when the loudspeaker is turned off, the atomic magnetic dipole moments remain aligned even when the external field is removed—the magnetization remains.

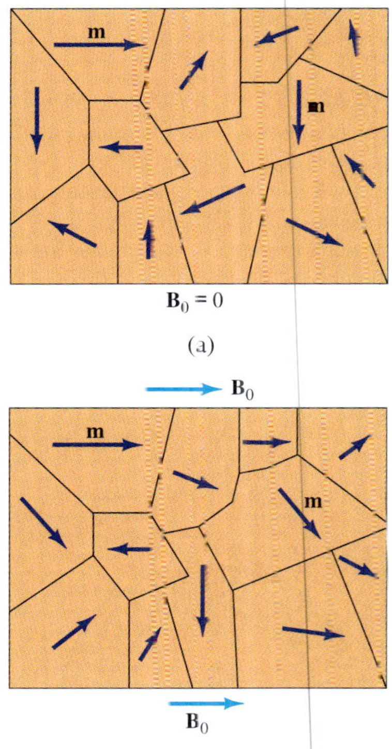

FIGURE 32–7 (a) Domain formation in ferromagnetic materials in the absence of an external magnetic field. The arrows indicate the magnetic moments of individual domains. (b) The presence of an external magnetic field influences the domains, making some larger and realigning others.

When a ferromagnet is heated, the increased movement of the atoms leads to a randomization of their orientation and thus to a decrease in the alignment. At the *Curie temperature*, T_c (after Pierre Curie), the randomization is complete, and the material is no longer a ferromagnet. The value of T_c varies from material to material; in iron, $T_c = 1043$ K, in gadolinium, $T_c = 292$ K. Below T_c, ferromagnetism appears, just as water forms the ordered lattice we know as ice below 273 K. When a ferromagnet cools below T_c, it does not automatically become a permanent magnet for the same reason that when a lake freezes, it does not form one huge ice crystal. The transition to ferromagnetic behavior, like freezing, takes place in domains, as described by our marching-band analogy.

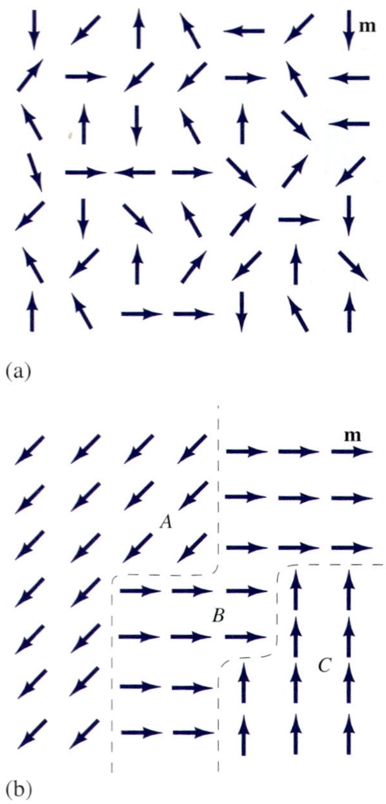

FIGURE 32–8 (a) The members of a marching band (or atomic magnetic moments in a ferromagnet) are oriented randomly. (b) The band members (or the atomic magnetic moments) influence one another in aligning themselves. Unless there is some external guide, the alignment will occur in small regions called domains.

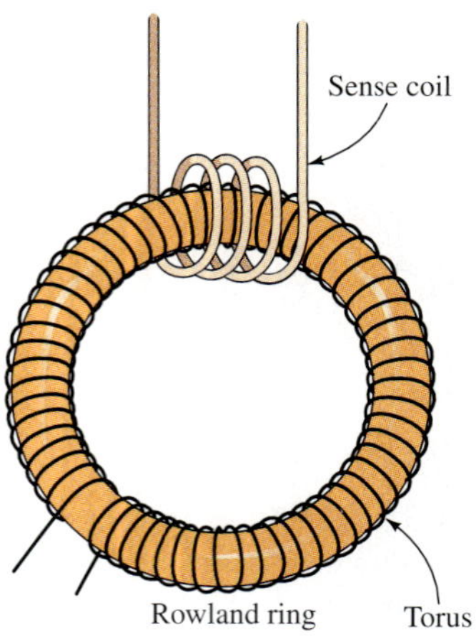

FIGURE 32–9 A Rowland ring is a wound core (solenoid) of material that may be used to measure the relation between **B** and **H** of that material. A sense coil measures changes in **B**.

EXAMPLE 32–4 Estimate the maximum possible magnetization in a single domain of iron.

Solution: We'll assume for this estimate that the atomic magnetic moments are aligned perfectly in a single domain with maximum magnetization. We first find the atomic magnetic moment of iron. The value of the intrinsic magnetic moment of an electron is given by Eq. (32–15), $m_{\text{intrinsic}} = m_B = 9.3 \times 10^{-24}\ \text{A}\cdot\text{m}^2$. The maximum possible magnetization comes when $m_{\text{intrinsic}}$ of the two unpaired electrons in an atom of iron are aligned with each other and with all the atoms' unpaired electrons—again, an assumption we make for the purposes of this estimate. The number density n of unpaired electrons in iron is

$$n = \left(\frac{2\ \text{unpaired electrons}}{1\ \text{atom}}\right)\left(\frac{6.02 \times 10^{23}\ \text{atoms}}{1\ \text{mol}}\right)\left(\frac{1\ \text{mol}}{56\ \text{g}}\right)\left(\frac{7.8\ \text{g}}{1\ \text{cm}^3}\right)\left(\frac{10^6\ \text{cm}^3}{1\ \text{m}^3}\right)$$
$$= 1.7 \times 10^{29}\ \text{unpaired electrons/m}^3.$$

We multiply by m_B for each unpaired electron to find a total magnetization of

$$M_{\max} = nm_B = (1.7 \times 10^{29}\ \text{unpaired electrons/m}^3)(9.3 \times 10^{-24}\ \text{A}\cdot\text{m}^2)$$
$$\simeq 1.6 \times 10^6\ \text{A/m}.$$

This result can be compared to the experimental $M_{\max}$ of annealed (tempered) iron, 1.7×10^4 A/m. The difference of a factor of 100 means that our assumptions were too strong and that the domains are never perfectly aligned.

Hysteresis

The relation between the magnetic field, B, and the magnetic intensity, H, is more complicated in ferromagnets than in other materials. In order to measure the relation between B and H in a ferromagnetic material, that material is demagnetized by heating. It is cooled, shaped into a ring, and wound with a wire that carries a current I. This experimental arrangement is called a *Rowland ring*. Without the ferromagnetic material, the magnetic field inside the ring, or toroidal solenoid, has the nearly constant value

$$B_0 = \mu_0 H = \mu_0 nI, \tag{32–17}$$

provided that the torus is "thin." Here n is the number of windings per unit length. As we know, when the ferromagnetic material is inserted into the torus, the magnetic field increases tremendously to a new value, B. We measure B by using a sense coil outside the torus (Fig. 32–9). The sense coil measures an induced emf proportional to the time rate of change of the magnetic field. As we raise the current in the toroidal coil at a given rate, we know the magnetic intensity $H = nI$, and the sense coil measures B. Figure 32–10 shows one example of a measured relation between H and B; a plot of H versus B is called a *magnetization curve*. Knowing H and B, we can determine the magnetization from the relation $B = \mu_0 H + \mu_0 M$.

For ferromagnetic materials, we observe a magnetization curve like that shown in Fig. 32–10; this curve is known as a **hysteresis loop**, and it indicates the phenomenon of **hysteresis**. The presence of hysteresis demonstrates an irreversibility in the magnetization process. When the current I in the solenoid is slightly changed and then changed back again, the original magnetization is generally not recovered. For example, if we start on curve c in Fig. 32–10 at a value of 10^{-4} T for $\mu_0 H$, B in the ferromagnetic material is negative. If $\mu_0 H$ is increased to 3×10^{-4} T and then brought back to 10^{-4} T, B is now positive, following curve b. Hysteresis results from the fact that the magnetic domains do not return to their original zero-external-field status when the current decreases. They "remember" the rise in field and do not automatically revert to their original alignments.

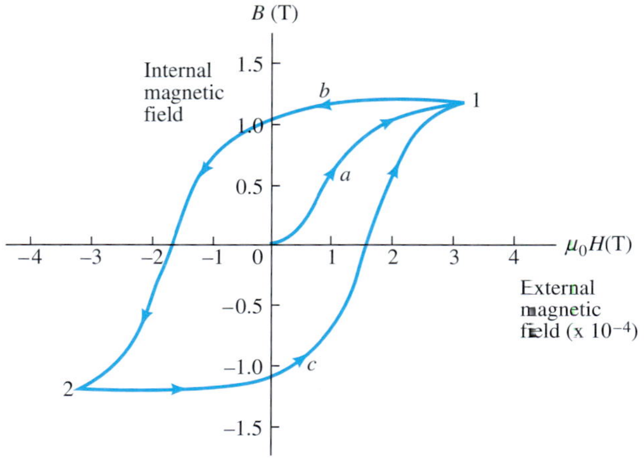

FIGURE 32–10 A magnetization curve illustrates the phenomenon of hysteresis in ferromagnetic materials. The material starts at the origin with zero magnetization. When a magnetic intensity H is applied, the material responds by becoming magnetic and is magnetic even when H is again zero.

APPLICATION

Uses of magnetically soft and magnetically hard materials

Some materials have narrow hysteresis loops, meaning that the alignment of the domains follows the external field rather closely (Fig. 32–11a). This type of curve holds for materials that are considered to be *magnetically soft*, such as iron. Such materials are often used in transformer cores. (Transformers are devices that transform AC currents—or voltages—from one value to another.) Other materials have broad hysteresis loops, meaning that their domains respond only to large external fields (Fig. 32–11b). Such materials, including carbon and tungsten, are said to be *magnetically hard*. They are difficult to magnetize but, once magnetized, they make good permanent magnets because they are equally difficult to demagnetize. Magnetically hard materials are especially important for making computer memories, magnetic tapes, or floppy disks because such materials are stable against changes due to nearby magnetic fields (Fig. 32–12). Generally speaking, the question of what materials are magnetically hard and what materials are magnetically soft is a complicated one.

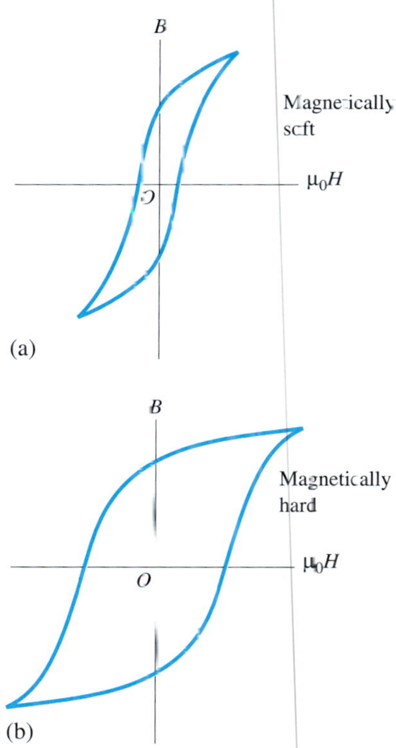

FIGURE 32–11 Hysteresis loops for materials that are (a) magnetically soft and (b) magnetically hard.

*32–4 DIAMAGNETISM

In diamagnetic materials, the induced magnetic field caused by $\mathbf{B}_0$, the applied field, is *opposite* to $\mathbf{B}_0$. Thus, the resulting net magnetic field is *less* than $\mathbf{B}_0$. Diamagnetism, as distinguished from ferromagnetism and paramagnetism, occurs in materials whose atoms have no permanent magnetic dipole moments, either orbital or intrinsic. A classical model helps us to understand the phenomenon qualitatively. Let's consider two electrons with identical orbits, except that the motion in one is counterclockwise and in the other clockwise (Fig. 32–13a). With no external magnetic field, the orbital magnetic moments of the two electrons cancel ($\mathbf{m}_1 + \mathbf{m}_2 = 0$), and there is no magnetization.

Let's suppose that there is an applied magnetic field, $\mathbf{B}_0$, perpendicular to the orbits of the electrons (Fig. 32–13b). For the electron on the left of the figure, the flux

FIGURE 32–12 (a) The magnetic hard disk of a computer. Magnetic heads on the disk "write" and "read" data in the form of digital signals. (b) A closeup of a magnetic hard disk.

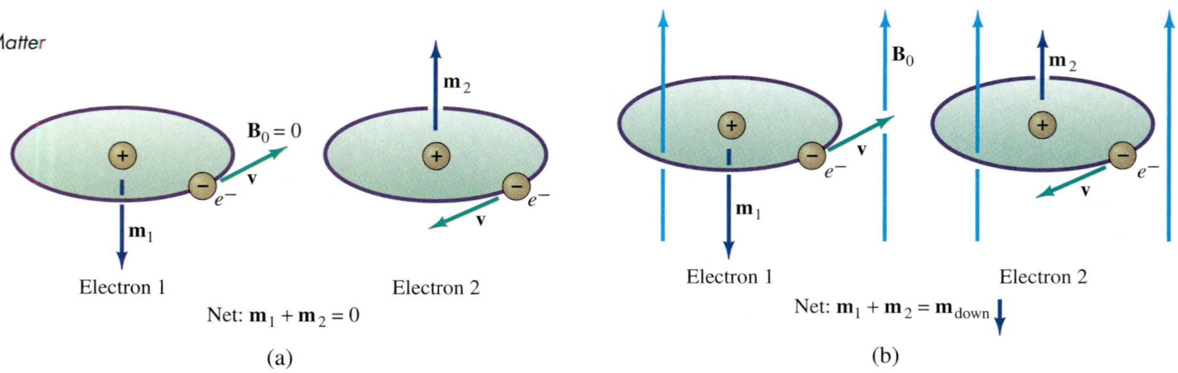

FIGURE 32–13 (a) A two-electron atom with a net orbital magnetic moment of zero in the absence of an applied magnetic field. (b) In the presence of an applied magnetic field, the orbital magnetic moment associated with each electron is changed, and there is a net magnetic moment that points down, opposite to the external field.

through its orbit increases as the external field increases; by Lenz's law, the electron responds to counter the increasing flux. Accordingly, the negatively charged electron speeds up, increasing its angular momentum as well as the magnitude of its orbital magnetic moment $\mathbf{m}_1$, which points downward. When the external field levels off, angular momentum conservation ensures that the new value of $\mathbf{m}_1$ persists. The electron on the right must slow down to oppose the increase in flux through its orbit, so the magnitude of its magnetic moment $\mathbf{m}_2$ is reduced. The result is that $\mathbf{m}_1 + \mathbf{m}_2$ now has a net value that points downward, and a magnetic field is produced that opposes the increasing external field. This is the origin of the negative magnetic susceptibility. The classical model must be revised for a proper quantum mechanical treatment of the atom because, in quantum mechanics, the angular momentum is *quantized*. It cannot change only slightly when the field changes slightly. Nevertheless, a correct treatment of a large collection of atoms reproduces the effect of the classical discussion.

When it is applied quantitatively, the simple qualitative model just described leads to reasonable estimates for the size of diamagnetic effects. The model correctly implies that *diamagnetism is present in all materials*, but it is masked for materials whose atoms have permanent magnetic moments.

*32–5 PARAMAGNETISM

Paramagnetism occurs in materials whose molecules have permanent magnetic dipole moments due to the intrinsic magnetic moments of unpaired electrons. In the absence of an external magnetic field, these dipoles are randomly oriented due to thermal motion, and the net magnetization of the materials is zero. Recall from Chapter 29 that the energy of a magnetic dipole moment $\mathbf{m}$ in a magnetic field $\mathbf{B}$ is, by Eq. (29–24), $U = -\mathbf{m} \cdot \mathbf{B}$. The lowest energy occurs when $\mathbf{m}$ and $\mathbf{B}$ are *parallel*. Thus an external magnetic field $\mathbf{B}$ tends to align the atomic magnetic moments along $\mathbf{B}$ and produces a positive magnetic susceptibility.

Two effects determine the extent to which the permanent magnetic dipoles become aligned. The first is the external field, which encourages alignment, and the second is the thermal motion, which randomizes the alignment. The relative importance of these two factors is measured by the relative size of the magnetic energy factor mB and the thermal energy factor kT, where T is temperature. If T is so large that $kT >> mB$, the average alignment over a large number of electrons will be weak. Conversely, if T is so low that $kT << mB$, the average alignment will be strong. For intermediate temperatures, the average alignment is proportional to the ratio of these energies, $m_{av} =$ (a constant)$(mB)/(kT)$. At room temperature, the intrinsic magnetic moments of most paramagnetic materials are only very slightly aligned, but large bulk effects come from very small alignments—as we saw in Example 32–3. In 1895, Pierre Curie observed the linear relation that we now call *Curie's law*:

$$\mathbf{M} = C\frac{\mathbf{B}}{T}, \tag{32–18}$$

where C is *Curie's constant*. This law is often expressed in terms of magnetic susceptibility, defined according to Eq. (32–5) as $\mathbf{M} = \chi_m \mathbf{H}$. If we anticipate that the susceptibility will be small, as it is for paramagnetic materials, then we can replace $\mathbf{B}$ in Eq. (32–18) by $\mu_0 \mathbf{H}$:

$$\mathbf{M} = C\frac{\mu_0 \mathbf{H}}{T},$$

or

$$\chi_m = \frac{\mu_0 C}{T}. \tag{32–19}$$

C is material dependent. The susceptibility is positive, which is characteristic of paramagnetism.

The temperature dependence in Eq. (32–18) is the same as that of the very similar phenomenon for dielectrics [see Eq. (26–30)]. There, too, this dependence is called *Curie's law*. We expect the law to fail at sufficiently low temperatures and/or large field. If the intrinsic magnetic moments are aligned perfectly with the field, they cannot be still further aligned to produce still higher magnetization. This *saturation* phenomenon is well known. A more quantitative version of our arguments can be used to predict—successfully—the size of the paramagnetic susceptibility.

Unlike diamagnetism, paramagnetism is not a universal phenomenon because relatively few materials have molecules with unpaired electrons (Fig. 32–14). When it is present, paramagnetism is normally a larger effect than diamagnetism. However, diamagnetism dominates at sufficiently high temperatures.

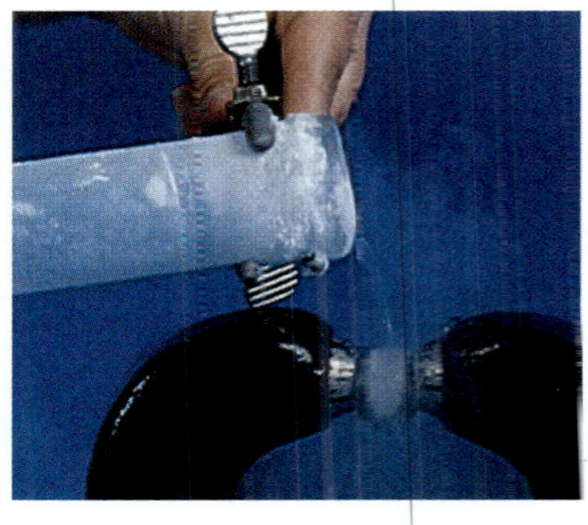

FIGURE 32–14 Oxygen is paramagnetic and is therefore attracted by the poles of a magnet. Here, liquid oxygen poured between two poles is held in place by the forces between it and the permanent magnet.

*32–6 MAGNETISM AND SUPERCONDUCTIVITY

Superconductors have magnetic properties just as extraordinary as their electric properties. The same collective quantum physical mechanism that makes any electric field inside a superconductor exactly zero *also makes the magnetic field inside zero*. A superconductor acts as a perfect diamagnet in the sense that currents are induced that precisely cancel any magnetic field inside. Alternatively, we say that the magnetic field lines are *expelled* from the superconductor—a phenomenon known as the *Meissner effect*. In Type I superconductors, the field is expelled entirely (Fig. 32–15). In Type II superconductors, the field is isolated in nonsuperconducting filamentary structures

The magnetic field within a superconducting material is zero.

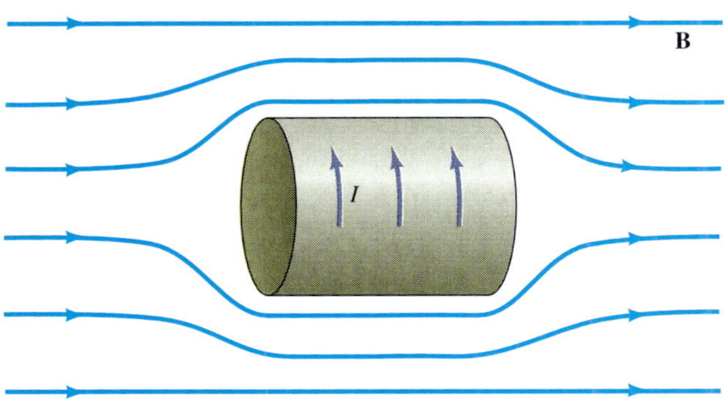

FIGURE 32–15 A Type I superconductor expels magnetic field from its interior by acting as a perfect diamagnet: Surface currents that just cancel the applied field inside are established.

APPLICATION: AN OXYGEN-IN-HYDROGEN MONITOR

The largest electrical power generators are cooled by blowing hydrogen through them. There are two reasons for this: First, the thermal conductivity of hydrogen is much greater than that of air due to the higher rms velocity of the atoms, and second, the viscous forces are very much lower because the hydrogen atoms have lower momenta than do atoms of air. (This latter point is important because the presence of viscous forces transfers heat energy to the coolant, thus defeating its purpose.) With hydrogen as a coolant, the energy density in the generator—both in the current in the windings and in the magnetic field—can be increased considerably.

Unfortunately, hydrogen is very dangerous when oxygen is present: A mixture of hydrogen with anywhere from 1 percent to 99 percent oxygen is explosively inflammable. If hydrogen is to be used in an area containing ignition sources such as sparks, the oxygen content must be maintained below 1 percent. A monitor for this condition takes advantage of the fact that oxygen—unlike the other components of air and unlike hydrogen—is paramagnetic (positive susceptibility). The instrument consists of three tubes fastened together to form a capital H. The cooling gas is allowed to flow equally through the two vertical legs of the H. By symmetry, there should be no flow of gas along the crossbar of the H. If, however, an inhomogeneous magnetic field acts across the crossbar, then there will be a force on the oxygen molecules in the crossbar, moving them to the region of higher field. This is the same mechanism that attracts liquid oxygen to the poles of a magnet (Fig. 32–14). The field will draw in oxygen molecules from the gas stream at the low field junction with the crossbar and insert them into the gas stream at the high field junction. This flow can be detected by placing a *thermistor*—a resistor with a highly temperature-sensitive resistance—at the center of the crossbar. Current is passed through this resistance to supply a constant amount of power, which heats it above the temperature of the hydrogen cooling gas. If there is no oxygen flow, the cooling of this thermistor is less than it would be if oxygen is flowing and thus carrying off thermal energy from the thermistor. Therefore, the measurement of the current in the thermistor is equivalent to a measurement of the oxygen flow in the crossbar. The reliability of this very stable device is far better than that of a mass spectrometer. It can easily measure a 0.1 percent contamination of oxygen in the hydrogen cooling gas.

within the material (Fig. 32–16). When such filaments are present, the resistivity of the material is no longer exactly zero. Currents circulate on the surfaces of these filaments, shielding the rest of the material from the magnetic field. Quantum physics sets a minimum amount of magnetic flux within each filament.

The expulsion of magnetic field from the interior of a superconductor translates into a statement about its magnetic susceptibility. Equation (32–6) expresses the internal field in terms of the magnetic intensity, **H**, and this internal field must be zero:

$$\mathbf{B} = \mu_0(1 + \chi_m)\mathbf{H} = 0.$$

The magnetic intensity is due to the free currents and is not zero. Thus $1 + \chi_m = 0$, or

$$\chi_m = -1.$$

The fact that there is no magnetic field inside a superconductor shows that there can be no currents inside. Imagine, as in Fig. 32–16, that some internal region of a superconductor carries current. Then we can draw a loop around this region and apply Ampère's law, which states that if there were a current through the loop, there would be a magnetic field; but this is not possible. We conclude that *all current carried by superconductors must be carried on their surfaces* (any boundary between superconducting and nonsuperconducting phases). For example, current can be carried on the walls of the filaments in Fig. 32–16.

APPLICATION
An important use of superconductivity

In the presence of a sufficiently large magnetic field (a *critical field*), a material in a superconducting phase jumps back to the normal (nonsuperconducting) phase even if the temperature is held fixed. This could be a serious problem because large electromagnets are ideally made from superconductors; these magnets do not undergo Joule heating in spite of the large currents in them. However, the large mag-

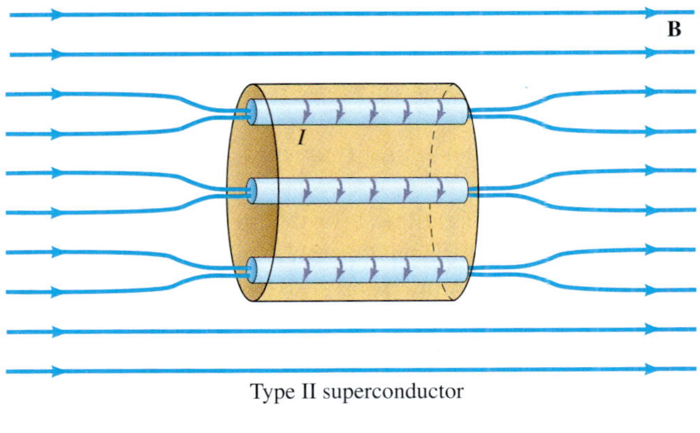

FIGURE 32–16 In Type II superconductors, the magnetic field is confined to filamentary structures. Inside the filaments, the material is not in its superconducting phase.

netic field may itself destroy the superconductivity. Type II superconductors channel the magnetic field into filaments, providing a way to make superconductors with much higher critical fields. These materials are used for the construction of superconducting magnets.

*32–7 NUCLEAR MAGNETIC RESONANCE

Atomic nuclei consist of protons and neutrons, which are far more massive than electrons. Their large masses mean that the protons and neutrons in atoms are very nearly fixed, and we can ignore their orbital motions. However, like electrons, protons and neutrons have intrinsic magnetic dipole moments. As is implied in Section 32–2, these magnetic moments are some 2000 times smaller than that of the electron because protons and neutrons are some 2000 times more massive than electrons. These moments arise from an intrinsic angular momentum called the *spin* and labeled **S**. Unlike the orbital angular momentum, which, according to quantum mechanics, can have a magnitude that is only an integer multiple of $\hbar$ (Planck's constant, h, divided by 2π), the electron or proton spin has magnitude $\frac{1}{2}\hbar$. If the proton magnetic moment is labeled $\mathbf{m}_p$, then for the orbital angular momentum [see Eq. (32–12)], there is a gyromagnetic ratio for the spin, g_S, defined by

$$\mathbf{m}_p \equiv g_S \mathbf{S}. \tag{32–20}$$

Let's consider the torque on, say, a proton due to an external magnetic field **B**. This torque is given by $\tau = \mathbf{m}_p \times \mathbf{B}$. Now, the torque on the proton is the rate of change of its internal angular momentum, $d\mathbf{S}/dt$, so with $\mathbf{S} = \mathbf{m}_p/g_S$, we have

$$\frac{1}{g_S}\frac{d\mathbf{m}_p}{dt} = \mathbf{m}_p \times \mathbf{B}. \tag{32–21}$$

This equation gives the rate of change of the magnetic moment of a proton in a magnetic field. The magnitude of $\mathbf{m}_p$ cannot change, but its direction can, and Eq. (32–21) describes the *precessional* motion of $\mathbf{m}_p$ about the direction of **B** (Fig. 32–17). This precession, called *Larmor precession*, is analogous to the precession of a spinning top under the influence of gravity (see Chapter 10). In Larmor precession, the direction of $\mathbf{m}_p$ traces a cone around the direction of **B**, as in Fig. 32–17. Moreover, quantum mechanics tells us that only two such cones are allowed: one with spin "up" and one with spin "down." In Problem 35, we describe how to calculate the angular speed of precession, which is

$$\omega_0 = g_S B. \tag{32–22}$$

Finally, let's recall one other feature of this motion: There is a potential energy associated with a magnetic moment **m** in an external field, given by $U = -\mathbf{m}\cdot\mathbf{B}$, and

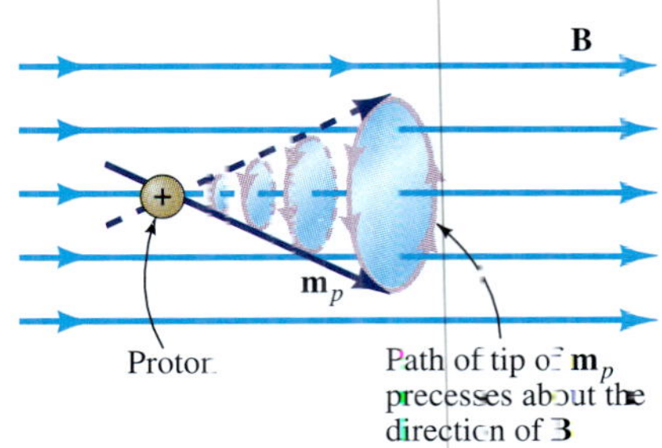

FIGURE 32–17 The magnetic moment of a proton, $\mathbf{m}_p$, precesses about the direction of an external magnetic field.

the motion with the "up" cone has lower energy than does the motion with the "down" cone.

Now we add a new ingredient in the form of an additional, oscillating magnetic field. (Such a field occurs in electromagnetic waves; see Chapter 35.) This field would not normally have much effect, but when the angular frequency ω of the oscillating field *exactly matches* the angular frequency ω_0 of the precession (a condition known as *resonance* and described in Chapter 13), the direction of the magnetic moment can flip. In this case, the oscillating magnetic field supplies *just the precise amount of energy*—$2\mathbf{m}_p \cdot \mathbf{B}$—necessary to flip the spin of the proton from up to down, or absorbs this amount of energy to flip the spin from down to up.

APPLICATION

Magnetic Resonance Imaging

This effect is called **nuclear magnetic resonance** (NMR). When the spin flips due to a transfer of energy between the oscillating field and the proton, there is a detectable signal. Thus, by tuning the frequency of the oscillating magnetic field, we can measure the frequency $\omega_0 = g_S B$ with very high precision.

If the external magnetic field is known, the NMR method may be used to measure the gyromagnetic ratio, g_S. It is in this way that the gyromagnetic ratios for protons and neutrons are known to contain the coefficients 2.79 and -1.91, respectively, in addition to classical factors. These coefficients suggest that the structures of protons and neutrons are more elaborate than that of electrons. NMR measurements can also be extended to nuclei, where they are used to study nuclear structure and the forces that give rise to it. For example, the deuteron nucleus—which may be described as a bound state of a proton and neutron structured in such a way that the intrinsic spins, and therefore the intrinsic magnetic moments, are parallel to each other—is expected to have a magnetic moment that is the sum of the moments of the proton and the neutron. NMR measurements give a slightly smaller result than this sum. It is possible to conclude from this that the deuteron is slightly cigar-shaped, and this gives us information about the forces that bind it.

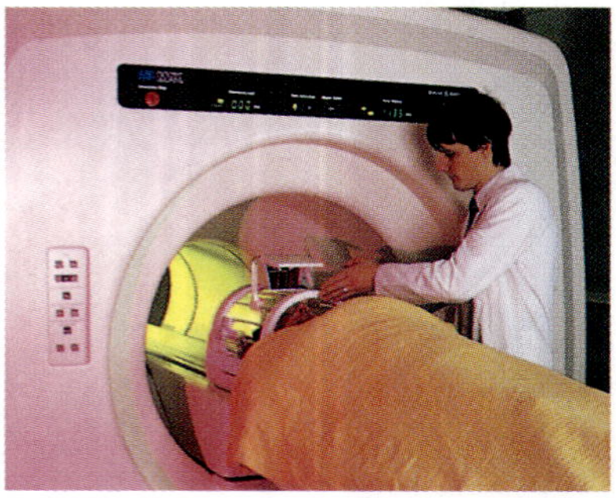

(a)

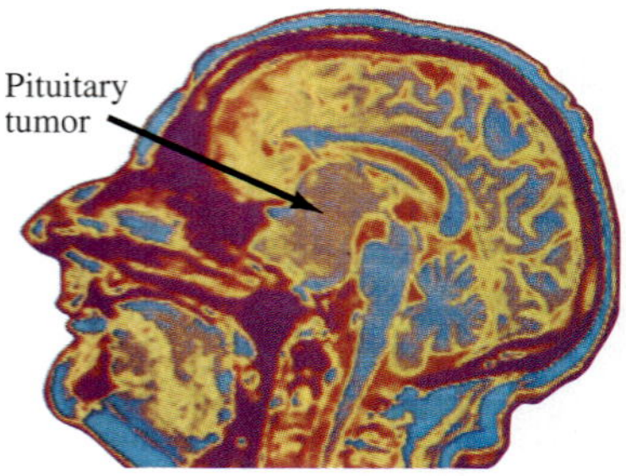

(b)

FIGURE 32–18 (a) This machine is used to produce cross-sectional images of patients by magnetic resonance imaging. At its "heart" is a superconducting magnet. (b) A magnetic resonance image of the head that indicates a pituitary tumor.

Magnetic Resonance Imaging

NMR has important applications in the study of materials and in medical diagnostics, where the procedure is called *magnetic resonance imaging* (MRI). (The word "nuclear" was dropped due to patients' fears that nuclear radiation was being used. In fact, MRI is regarded as an especially safe procedure.) The magnetic field, B, associated with materials contains a contribution from the electrons as well as from the nuclei within the material. Thus an NMR measurement gives us information about the atoms and molecules to which the nuclei belong. In the case of medical diagnostics, MRI works primarily on hydrogen atoms, whose nuclei contain single protons. MRI locates concentrations of hydrogen atoms in patients. Fat, which has a high concentration of hydrogen, can be distinguished from muscle, which has a much lower hydrogen concentration. Tumors can be distinguished from nerve tissue, and bones, which have little hydrogen, are hardly seen at all (Figs. 32–18a and b).

SUMMARY

The magnetic properties of bulk matter are summarized in the magnetization, $\mathbf{M}$, the magnetic dipole moment per unit volume. In the presence of an external **magnetic field** $\mathbf{B}_0$, there is a field in a material given by

$$\mathbf{B} = \mathbf{B}_0 + \mu_0 \mathbf{M}. \qquad (32\text{–}1)$$

The effect of free (real) currents (as opposed to the induced atomic effects) is contained in the magnetic intensity, $\mathbf{H} = \mathbf{B}_0/\mu_0$:

$$\mathbf{H} \equiv \frac{\mathbf{B}}{\mu_0} - \mathbf{M}. \qquad (32\text{–}2)$$

The magnetic susceptibility, χ_m, describes the response of a material to a magnetic field of external origin:

$$\mathbf{M} \equiv \chi_m \mathbf{H}. \quad (32\text{–}5)$$

In terms of χ_m, the net magnetic field is given by

$$\mathbf{B} = \mu_0(1 + \chi_m)\mathbf{H} = \mu\mathbf{H}. \quad (32\text{–}6), (32\text{–}8)$$

Here, μ is the permeability of the material:

$$\mu \equiv \mu_0(1 + \chi_m). \quad (32\text{–}7)$$

Magnetism in matter is due ultimately to the magnetism of its atomic constituents, and particularly to the unpaired electrons of atoms An orbiting electron produces an atomic orbital magnetic moment

$$\mathbf{m}_{\text{orbital}} \equiv g_L \mathbf{L}, \quad (32\text{–}12)$$

where g_L is the gyromagnetic ratio. Quantum mechanics implies that these magnetic moments take the value

$$m_{\text{orbital}} = \left(\frac{e}{2m_e}\hbar\right)\ell \equiv m_B \ell, \quad (32\text{–}14)$$

where the factor m_B is the Bohr magneton and ℓ is an integer. In addition, electrons have intrinsic magnetic moments equal in magnitude to m_B. Even a very slight alignment of atomic magnetic moments leads to large magnetic effects in bulk matter.

Ferromagnetic materials have large permeabilities. The atomic dipole moments are lined up in small regions called domains due to forces of quantum mechanical origin. The imposition of an external field leads to the dipole moments of the domains lining up together and produces permanent magnets. The fact that a ferromagnetic material "remembers" the orientation of the external field that magnetizes it leads to the phenomenon of hysteresis, in which the magnetization curve depends on how the magnetization was produced.

Diamagnetic materials have small negative susceptibilities that are ultimately due to Faraday's law. Diamagnetism is always present but may be masked by other effects. Paramagnetic materials have small positive susceptibilities due to the intrinsic magnetic moments of unpaired electrons, which find it energetically favorable to line up with an external field. Paramagnetism is strongly temperature dependent. Superconductors expel magnetic field from their interiors.

In nuclear magnetic resonance (NMR), the intrinsic magnetic moments of nuclei and nuclear constituents precess about an applied magnetic field. This precession is detected by the response of a material to the imposition of an electromagnetic wave of just the right frequency—a frequency ultimately characteristic of the material involved.

UNDERSTANDING KEY CONCEPTS

1. When we calculated the magnetic dipole moment associated with orbital motion, why was it reasonable to think of the nucleus as stationary and the electron as circulating around it?
2. When an electron orbits the nucleus in a planetary model, the system forms an electric dipole. Why does this electric dipole not produce a measurable electric dipole field around the atom?
3. Under what circumstances will Gauss' law for the magnetic field also hold for the magnetic intensity?
4. In a Rowland ring measurement of the magnetic field inside a piece of magnetic material, is it helpful to wrap the sense coil around the material many times?
5. Does iron exhibit diamagnetic properties? How could you determine them?

6. Aluminum is separated in junk yards by using large magnets. How is this possible?
7. Should the magnetic latch on a refrigerator door be made from magnetically hard or soft material?
8. Why should computer floppy disks not be made from magnetically soft material?
9. Why does diamagnetism dominate over paramagnetism at sufficiently high temperatures?
10. Explain how a permanent bar magnet attracts an unmagnetized iron needle.
11. You are given two identical iron rods—one magnetized, the other not. How can you determine which is the magnet, without using a third magnet (for example, Earth)?
12. Suppose that an electron in a circular orbit around a nucleus is placed in an external magnetic field. Will the angular momentum of the electron change if the field is aligned perpendicular to the plane of motion? parallel to the plane of motion?
13. Is it possible to arrange for a classical current loop to have a magnetic moment but no angular momentum? Assume first that you have both positive and negative charge carriers to work with, and then that you have only negative ones.
14. What is the value of **H** in an isolated permanent magnet?
15. It takes an external field to establish a macroscopic magnetization inside a permanent magnet cooled below its Curie temperature. What could have done this for lodestones, which are permanent magnets found in nature?
16. In a uniform magnetic field, a magnetic dipole experiences no net force, only a torque. How do two bar magnets repel or attract each other?

PROBLEMS

32–1 The Magnetic Properties of Bulk Matter

1. (I) A cylindrical rod of palladium (magnetic susceptibility $\chi m = 8 \times 10^{-4}$), of radius 1 cm and length 5 cm, is placed in and aligned with a uniform magnetic field of 1.0 T. What is the magnetic dipole moment of the rod?
2. (I) A thin, toroidal coil of total length 85 cm is wound with 920 turns of wire. A current of 2.4 A flows through the wire. What is the magnitude of **B** inside the torus if the core consists of a ferromagnetic material of magnetic susceptibility $\chi_m = 2.8 \times 10^3$? What is the magnitude of **H**?
3. (I) The coil of a solenoid wound with a turn density of 3400 turns/m is filled with a material of unknown magnetic susceptibility χ_m. When the wire carries 0.450 A, the magnetic field within is 1.907×10^{-4} T. What is χ_m?
4. (I) A solenoid magnet wound with a turn density of 1000 turns/m, with permalloy inserted inside the windings, has a magnetic field of 5.0 T inside. How much current flows in the windings?
5. (I) A permalloy magnet is 5 cm in diameter, 30 cm long, and has magnetic intensity **H** = 30 A/m at its pole. How many turns/m must an empty solenoid of the same dimensions have to give rise to the same intensity if it carries a current of 6 A?
6. (I) What is the magnetic moment of the equivalent solenoid in Problem 5?
7. (II) In a vacuum, a solenoid with a current I has a magnetic field B_0. (a) If silver is placed inside the solenoid, what is the change in the magnetic field? (b) What happens if cupric oxide is placed inside the solenoid?
8. (II) A 1.0-cm^3 cube of copper is placed between the poles of a magnet with a magnetic field of 6.0 T. What is the induced magnetization in the copper?
9. (II) A long solenoid filled with ferromagnetic material of permeability $\mu = 1320\mu_0$ is wound with wire so that there are 15 turns per cm. What current must flow through the wire to produce a magnetic field of 1.6 T within the solenoid?

32–2 Atoms as Magnets

10. (I) Suppose that 1 mol of atoms in a material have individual magnetic moments of 2.3×10^{-23} A·m^2. In the absence of any alignment, the magnetic moments form an *average* angle of 90° with some external axis. By how much does the average angle differ from 90° if the material has the same magnetic moment as a 2-cm^2 loop of wire that carries a current of 1 A? (The magnetic moment of the loop is aligned with the external axis.) Assume that the components of the atomic magnetic moments add algebraically.
11. (I) The atomic number of iron is 26, its atomic weight is 55.8, and its density is 7.87 g/cm^3. (a) How many electrons are there in 10 cm^3 of iron? (b) Suppose that each electron has the magnetic moment estimated in Example 32–2 (5×10^{-24} A·m^2), and that the magnetic moments "up" and "down" make up the fractions $(1/2)(1 + 2 \times 10^{-7})$ and $1/2(1 - 2 \times 10^{-7})$, respectively. What is the magnetization of the iron?
12. (II) Consider an electron in a circular orbit around a single proton (a hydrogen nucleus) whose total energy is −13.5 eV. Find the value of the orbital magnetic moment.
13. (II) The electron has a *classical radius* given by $r_0 \equiv e^2/4\pi\epsilon_0 m_e c^2 = 2.8 \times 10^{-15}$ m. This quantity is suggested by dimensional analysis: The particular combination of classical quantities is the only one that can be formed with dimensions of length. Use Eq. (32–11), with m_{orbital} equal to the Bohr magneton, m_B, to show that any charge at the distance of the classical radius will be moving faster than the speed of light. (Assume that all the charge is concentrated at a belt of radius r_0.) Treatment of the magnetic moment of an electron as a classical quantity leads to trouble!
14. (II) The current I in a circular loop of radius R is due to the flow of free electrons in the loop. Show that the gyromagnetic ratio of the loop is independent of I, R, and the density of atoms.

32–3 Ferromagnetism

15. (I) A torus is wound with 800 turns/m of wire. A current of 5 A runs through the wire. If the core of the torus is iron, the internal magnetic field in the core is 1.8 T. What is the magnetization? What is the value of μ/μ_0 for the iron core?
16. (I) An electromagnet with a ferromagnetic core, $\chi_m = 15{,}000$, produces a maximum magnetic field of 4.5 T. What is the maximum current carried by the coil if the turn density of the coils is 4 turns/cm?

17. (I) A long, tightly wound solenoid contains a magnetic field of magnitude $B = 2.4 \times 10^{-3}$ T. An iron core, with susceptibility χ_m, is inserted so that it fills the space inside. What is the new value of B?

18. (I) We may view the iron core in Problem 17 as equivalent to another solenoid that is concentric with the outer one and essentially coincident with it. How could you view the situation if the iron core were quite a bit shorter than the length of the solenoid? if the radius of the iron core were smaller than that of the solenoid? Draw magnetic field lines for these cases.

19. (II) A current of 0.5 A flows through a solenoid with 400 turns/m. An iron bar, with $\mu/\mu_0 = 640$, is placed along the solenoid axis (Fig. 32–19). (a) What is the magnetic field inside the iron bar? (b) outside the iron bar, but still within the solenoid?

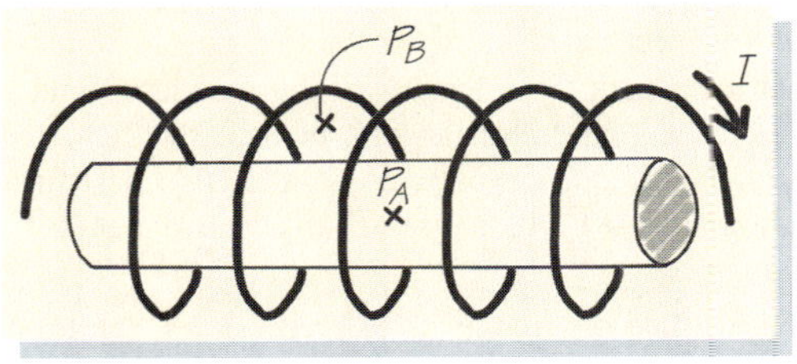

FIGURE 32–19 Problem 19.

20. (II) A disk-shaped permanent magnet has a thickness of 0.5 cm and a diameter of 2.0 cm. It is magnetized perpendicular to the plane of the disk, with the magnetic field on the axis near its north pole of magnitude 0.3 T. What is the current carried by a 100-turn coil of the same dimensions that gives this same value of the magnetic field on the axis (Fig. 32–20)?

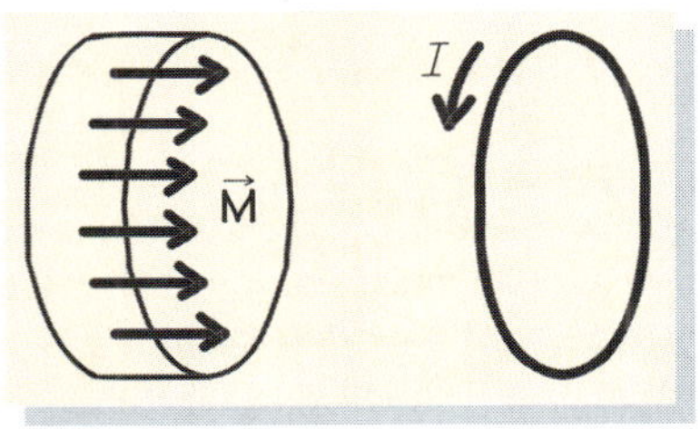

FIGURE 32–20 Problem 20.

21. (II) A current of 0.16 A is carried by a 50-turn coil 5.5 cm in diameter and 1.0 cm in length. Suppose a piece of iron with susceptibility 4.8×10^3 is placed inside the coil. What is the magnetic field inside the iron? What is the magnetic intensity there?

22. (II) A torus with a central radius of 25 cm and a tube radius of 2 cm is filled with iron of permeability $2800\mu_0$ (Fig. 32–21). There are 1200 turns around the torus. How much current must flow in the winding coil to produce a magnetic field of 1.5 T inside the torus? Treat the torus as having a constant magnetic field equal to the field at the central radius of the torus.

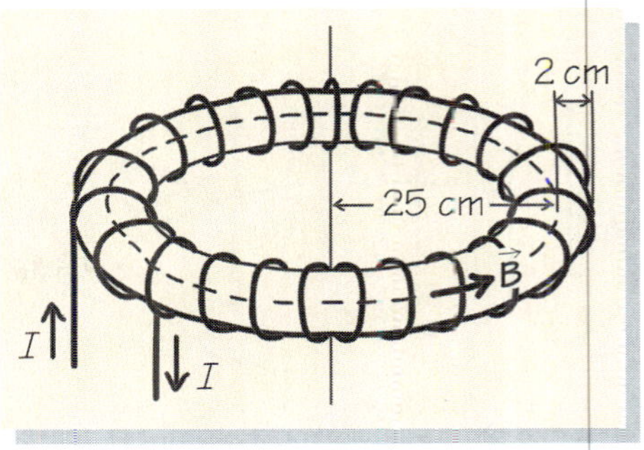

FIGURE 32–21 Problem 22.

23. (II) Soft iron (susceptibility = 6000) is used as the core of a transformer (see Chapter 34 for a discussion of this device). If large distortions of current are to be avoided, the transformer must be designed in such a way that the proportionality between **B** and **H** applies even for the strongest field. Estimate the maximum allowable value of H, knowing that the magnetic moment of individual iron atoms is 2.2 m_B.

24. (II) A Rowland ring measures the charge Q that passes through a sense coil by integrating the current in the sense coil over time (see Fig. 32–9). Both the sense coil and primary coil are wrapped tightly around a material. These coils have an area A. The number of turns in the sense coil is N. The emf induced in the sense coil by a changing magnetic flux in the material (due to a switch that passes current through the primary coil when the switch closes) is $\mathscr{E}$, and the sense coil has resistance R. Obtain a relation between the change in magnetic field, ΔB, and the charge Q. [*Hint*: Remember that $I = dQ/dt$.]

25. (II) A sense coil with a resistance of 0.1 Ω is wrapped tightly in 40 turns around a magnetic material of area 0.02 m^2. When a switch is closed in the primary coil, a charge of 5 mC flows through the sense coil (Fig. 32–22). If the magnetic field was initially zero, what is the new magnetic field in the material? (See Problem 24.)

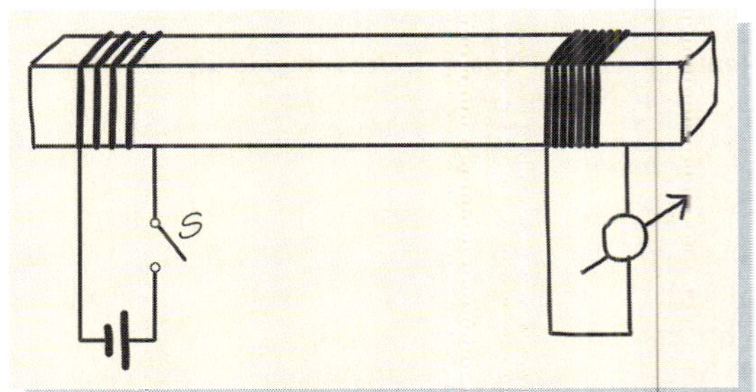

FIGURE 32–22 Problem 25.

*32–4 *Diamagnetism*

26. (III) An electron under the influence of some central force moves at speed v_i in a counterclockwise circular orbit of radius R. A uniform magnetic field **B** perpendicular to the plane of the orbit is turned on (Fig. 32–23). Suppose that the magnitude of the field changes at a given rate dB/dt. (a) What are the magnitude and direction of the electric field induced at

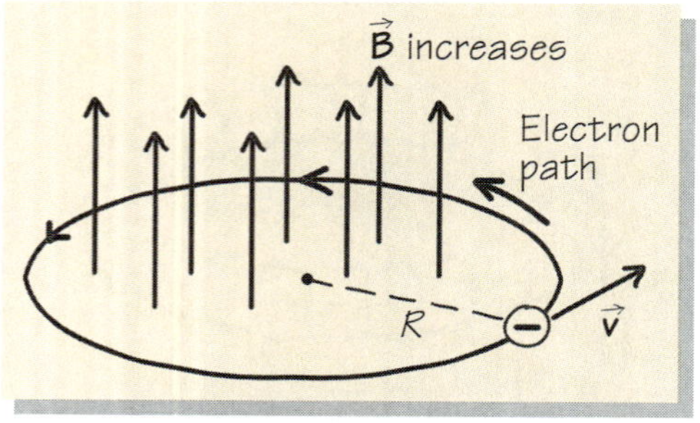

FIGURE 32–23 Problem 26.

the radius of the electron orbit? (b) The tangential force on the electron due to the induced electric field increases the electron's speed. Find the value of dv/dt. (c) Assuming that the initial orbital speed was v_i, find the final speed v_f as the magnitude of the magnetic field steadily increases from zero to a final value B_f by integrating dv/dt with respect to time. (d) Using your result for the change in speed, find the change in orbital angular momentum. (e) Use Eq. (32–11) to relate a change in the orbital magnetic moment to the change in the angular momentum.

27. (III) Refer to Fig. 32–23, but this time assume that the electron circulates clockwise rather than counterclockwise at speed v_i. By applying the same sequence of steps, show that the change in the magnetic moment of the electron's orbit is opposite the direction of change in the external field, just as in the case in which the electron circulates counterclockwise.

28. (III) Refer to Problems 26 and 27. Suppose that there are now two electrons moving at speed v_i in circular orbits of radius R, one clockwise and one counterclockwise. (a) What is the net orbital magnetic moment when the external field is zero? (b) after the external field has reached $\mathbf{B}_f$? (c) Show that the magnetic susceptibility for this system is $\chi_m = -(\mu_0 e^2 R^2/4m_e)\rho_e$, where ρ_e is the electron density.

29. (III) Using the techniques of Problems 26 through 28, estimate the magnetic susceptibility of copper, which has 29 electrons per atom. Assume that all the electrons move in orbits of the same radius, and that 14 move clockwise while 15 move counterclockwise. You will need to calculate the number density of electrons in copper.

*32–5 *Paramagnetism*

30. (II) The temperature of a sample of $FeCl_3$ (the ferric ions have a magnetic moment) inside a magnetic field is held constant as the field is increased. Sketch the induced magnetic moment as the magnetic field is increased.

31. (II) A long, straight conducting wire is embedded within an insulating paramagnetic material of magnetic susceptibility 2.6×10^{-4} at 300 K and carries a current of 10 mA. Find the value of the magnetic intensity as a function of the distance from the wire, as well as the magnetic field. What is the change in the magnetic field when the temperature is lowered to 86 K?

*32–7 *Nuclear Magnetic Resonance*

32. (I) Find the magnetic moment of the neutron, given that its gyromagnetic ratio is $-3.82e/2m_n$.

33. (II) Assume that it is possible to align perfectly the magnetic moments of protons in 1 mol of hydrogen gas at standard temperature and pressure. What are the magnetization and magnetic field inside the gas?

34. (II) In ^{17}O (oxygen with 17 nucleons in its nucleus) the nuclear magnetic moment is $-9.54 \times 10^{-27} A \cdot m^2$. The atomic electrons are lined up in such a way that they make no contribution to the magnetic moment of the atom. Suppose it were possible to align the oxygen atoms such that 50.5 percent pointed in one direction and 49.5 percent pointed in the opposite direction. What would be the magnetization of 1 mol of ^{17}O gas under those conditions at standard temperature and pressure?

35. (II) Express the equation $d\mathbf{m}/dt = g_S \mathbf{m} \times \mathbf{B}$ relevant to NMR in component form for the case that $\mathbf{B} = B\mathbf{k}$ and $\mathbf{m} = m_x \mathbf{i} + m_y \mathbf{j} + m_z \mathbf{k}$. (a) Show that m_z is a constant; (b) that $m_x^2 + m_y^2 + m_z^2$ is a constant; and (c) that $m_x = m_1 \cos(\omega t)$ and $m_y = -m_1 \sin(\omega t)$ satisfy the equation of motion, where ω is *the angular frequency of precession.*

36. (II) Archaeological objects are often located by detecting their minute influence on Earth's magnetic field. The equipment used for such measurements is the *proton magnetometer*, which measures the intensity of the magnetic field by measuring the angular frequency of protons in that field. Determine the angular frequency of protons in Earth's field at a typical location ($B = 50\ \mu T$) and the change in frequency caused by a change $\Delta B = 1$ nT.

37. (II) Calculate the frequency of precession f (see Problem 36) for a proton's magnetic moment in a field of 10^{-1} T. This frequency is in the so-called rf (radio-frequency) range.

38. (II) Gauss' law for magnetism,

$$\int_{\text{closed surface}} \mathbf{B} \cdot d\mathbf{A} = 0,$$

remains unchanged when materials are present because materials do not give rise to magnetic monopoles. Use this law to show that the magnetic field **B** does not change at the interface of two materials if the interface is perpendicular to the direction of the field. [*Hint*: Recall the method used to derive the electric field due to a surface distribution of charges.]

39. (II) If magnetic matter is present, Ampere's law changes to

$$\oint (\mathbf{B}/\mu) \cdot d\mathbf{s} = \oint \mathbf{H} \cdot d\mathbf{s} = I_{\text{enclosed}}.$$

Use this equation to find the magnetic field in a narrow gap cut through the core of a toroidal coil (Fig. 32–24). (The purpose of such an arrangement is to allow samples to be placed within the gap.) The average radius of the torus is 30 cm, $\mu = 1200\mu_0$ for the core material, and the coil consists of 500 turns carrying a current of 8 A. Calculate B for gap widths of 1 mm and 3 cm.

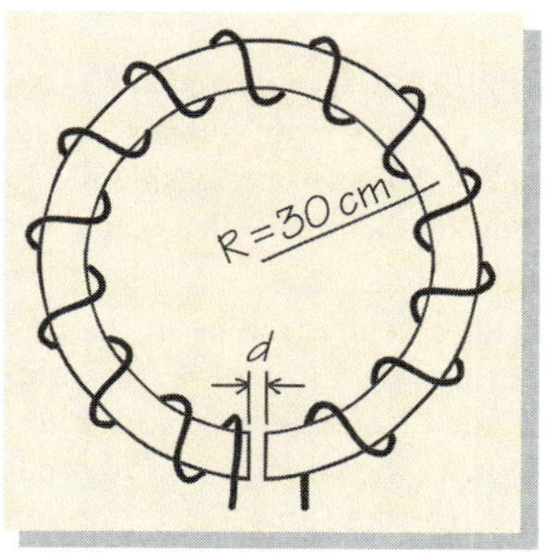

FIGURE 32–24 Problem 39.

General Problems

40. (II) Estimate the diamagnetic susceptibility of the diamond form of carbon, using the formula derived in Problem 28. Take the density of diamond to be 3.5 g/cm^3; the atomic weight, 12; the atomic number, 6; and the atomic radius, 0.75×10^{-10} m. Assume that all the electrons circulate at this radius. Your estimate should be rather good. Compare this result to that found in Problem 29.

41. (II) Large magnets typically consist of wound toruses of ferromagnetic material. There is a gap in the torus that forms a space between pole faces. Show that the magnetic field across the pole faces is the same as the magnetic field inside the ferromagnet by applying Gauss' law for magnetism to a closed surface partly in and partly out of one of the pole faces. (Gauss' law for magnetism holds independent of the presence of materials. Its validity rests on the fact that there are no magnetic monopoles, and materials introduce no such animals.)

42. (II) Two parallel conducting strips are each 6.0×10^{-4} m thick and 10 cm wide and are separated by a distance of 5.0×10^{-3} m. The space between the strips is filled with a ferromagnetic material whose permeability is $500\mu_0$. Each strip carries a uniform current of 2.0 A, in opposite directions. Find the value of the magnetic field, and magnetic intensity in the space between the strips.

43. (II) Earth's magnetic field is close to that of a dipole, and the strength of the field at the magnetic north pole is about 0.6×10^{-4} T. Calculate Earth's magnetic moment. If this magnetic moment is due to a magnetized iron core whose radius is half Earth's radius, what is the magnetization of the core (Fig. 32–25)? If the magnetic moment were due to a circulating belt of current at the radius of the core, what would be the magnitude of this current (Fig. 32–25)?

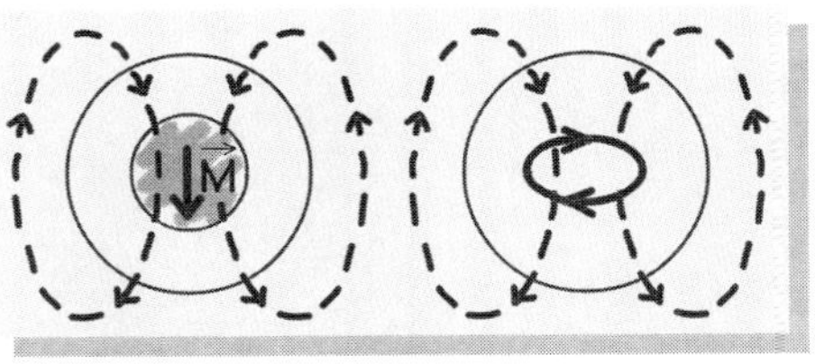

FIGURE 32–25 Problem 43.

44. (II) A torus of central radius 6 cm and tube radius 0.5 cm is filled with silver. It is wound with 200 turns of wire and has a 1-A current. The magnetic susceptibility of silver is -2.4×10^{-5}. Determine (a) the magnetic intensity, **H**; (b) the magnetic field, **B**; (c) the magnetization, **M**. (d) Repeat parts (a) through (c) for a torus filled with nickel instead of silver. The susceptibility of nickel is 95.

45. (II) In one of two simple classical models of the electron spin, the charge circulates at the classical electron radius r_0 (see Problem 13). In the other, the total electron charge is spread uniformly over a disk whose radius is the classical radius. Calculate the ratio of magnetic moments for the case in which the overall charge occurs entirely at the classical radius versus the case in which the charge is spread over the disk.

46. (II) The neutron has an internal spin **S** of magnitude $\hbar/2$ and a magnetic moment related to the spin by the gyromagnetic ratio, g_S, as in Eq. (32–20). The gyromagnetic ratio is $g_S = -3.82(e/2m_n)$. Suppose that a neutron consists of a heavy, positively charged particle of mass M and magnetic moment $e\hbar/2M$, with a lighter, negatively charged particle of mass m but no intrinsic magnetic moment orbiting the heavier particle with orbital angular momentum $\hbar$. What would mass m have to be to explain the observed magnetic moment of the neutron? (For simplicity, ignore the motion of the heavier particle about the center of mass.)

47. (III) In our discussion of kinetic theory from Chapter 19, we noted that, according to Boltzmann, the number of systems with a given energy E in a collection of systems in equilibrium at temperature T is given by $Ce^{-E/kT}$. Here, C is a constant determined by the requirement that, when all the systems are summed, we find the same total number of systems that we started with. For a collection of N magnetic dipoles at rest in an external magnetic field, we have $N(T) = Ce^{-(-\mathbf{m}\cdot\mathbf{B})/kT} = Ce^{(mB\cos\theta)/kT}$, where θ is the angle between the direction of the dipole and that of the external magnetic field. C is determined by the requirement that the total number of systems N is $N = C\int_0^{\pi} 2\pi \sin\theta \, d\theta \, e^{(mB\cos\theta)/kT}$. (a) Calculate C. (b) Calculate the average value of $\cos\theta$. (c) Plot $\langle\cos\theta\rangle$ as a function of mB/kT.

CHAPTER

33

The electrical arc shown here does not form when a switch is closed: It forms only when the switch is opened because Faraday's law acts to maintain an existing current. The arc is a manifestation of the existing current jumping an air gap.

Inductance and Circuit Oscillations

Capacitors are circuit elements that store energy in an electric field. Energy can also be stored in a magnetic field, using inductors as the circuit elements. Their operation is based on Faraday's law, which describes the effects of changing magnetic fields. Inductors are active only when currents change. For this reason, inductors are crucial to the control of time-dependence in circuits. Electric circuits containing inductors, capacitors, and resistors are analogous to damped harmonic oscillators, and all the features of such mechanical systems are also seen in electric circuits.

33–1 INDUCTANCE AND INDUCTORS

When a circuit contains a changing electric current, that current produces a changing magnetic field. According to Faraday's law, an additional emf may thereby be induced in the circuit. Such effects occur frequently because so many applications of circuits involve time dependence (from computers to televisions to the electrical systems of automobiles).

Any electric circuit in which the current has some time dependence serves to illustrate the general principle. Figure 33–1a shows a circuit with a switch that closes at $t = 0$. When the current increases from zero (Fig. 33–1b), the magnetic field around the wire also increases. As the magnetic field grows, the magnetic flux (directed into the page) through the area enclosed by the loop increases. According to Faraday's law,

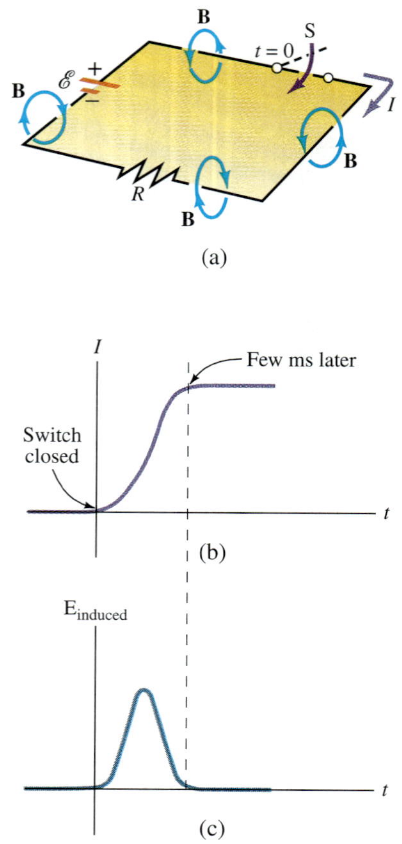

FIGURE 33–1 (a) When the current in a circuit changes, the flux through the circuit changes. (b) The current in the circuit increases as a function of time. (c) The induced emf is proportional to the derivative of the current according to Faraday's law.

as formulated by Lenz, an emf is induced in the loop and opposes this increase in flux (Fig. 33–1c). The induced emf therefore *opposes* the emf of the battery and slows down the flow of current. The principle illustrated here is a simple one: Lenz's formulation of Faraday's law tells us that induced currents always oppose any change in magnetic flux. The result is that *changing currents in circuits lead to induction effects that act to reduce the rate of change of those currents.*

In addition to the single-loop effect just illustrated, there is another possible effect associated with the changing flux through surfaces other than those defined by the current loop. If a second circuit loop is in the general vicinity of the first, the changing magnetic field due to the first circuit can change the magnetic flux through the second circuit, and a current will be induced in the second circuit. This effect, in turn, produces a changing flux that can affect the first circuit, and so forth. In this case, the two circuits are said to be **linked**. When the first loop induces an emf in itself, we say that there is a **self-inductance**, or **inductance** for short. When the first loop induces a current or emf in a second loop, we say that there is a **mutual inductance** between the two loops. The same physical principle lies behind both self-inductance and mutual inductance: Faraday's law.

Self-Inductance. When a wire carries a current I, a magnetic field is set up whose strength is proportional to I. Thus the magnetic flux appearing through a loop of that wire is also proportional to I. The proportionality constant is defined to be the inductance, which we write as L. L depends on the particular surface involved; that is, on the geometry of the loop around which the emf is induced. For a single loop through which a current I flows, the inductance is defined by expressing the magnetic flux, Φ_B, that passes through a surface bounded by the loop as

The definition of inductance

$$\Phi_B \equiv LI. \tag{33–1}$$

According to Faraday's law, the emf induced in this loop, $\mathcal{E}$, is the rate of change of the flux through the loop:

The emf associated with an inductance

$$\mathcal{E} = -\frac{d\Phi_B}{dt} = -L\frac{dI}{dt}\ . \tag{33–2}$$

The addition of an induced emf term proportional to the rate of change of a current will have a marked effect on how charges flow through the circuit.

Mutual Inductance. Let's now consider the two adjacent circuits shown in Fig. 33–2. If a current I_1 flows in loop 1 and a current I_2 flows in loop 2, there is a magnetic flux $\Phi_B(1)$ through the area of loop 1 given by

$$\Phi_B(1) = L_1I_1 + M_{12}I_2. \tag{33–3}$$

The first term is due to the current flowing in loop 1, and the constant of proportionality L_1 is the self-inductance of loop 1. The second term is due to the current flowing in loop 2 (Fig. 33-3a), and the constant M_{12} is the mutual inductance of loop 1 due to loop 2. Equation (33–3) defines the mutual inductance. Both L_1 and M_{12} are positive

FIGURE 33–2 (a) The flux through a circuit or circuit element may be due to its own current or to the current carried by an adjacent circuit or circuit element. (b) The two coils mounted on the iron core demonstrate mutual induction.

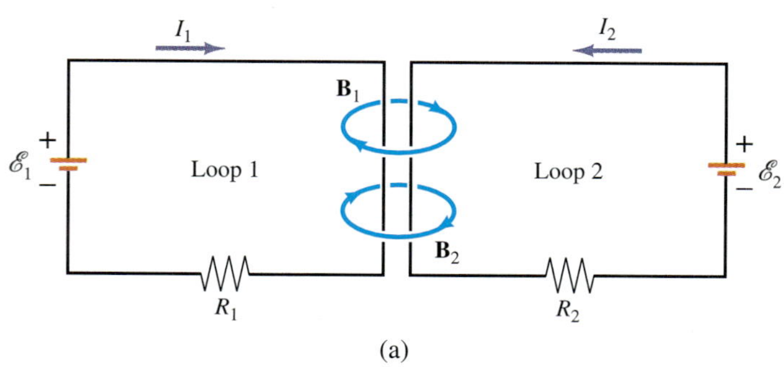

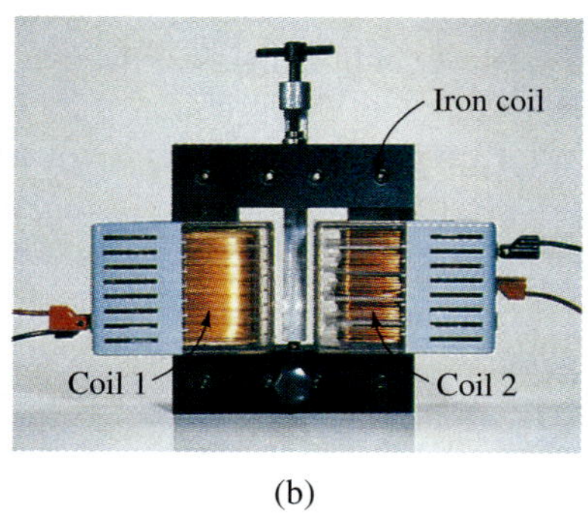

by definition. They depend *only* on geometry and on the medium in which the circuit is embedded. The inductances do not depend on the currents themselves.

The emf associated with a mutual inductance

The term *mutual* implies a degree of symmetry between the two loops. The magnetic flux through loop 2 has a term proportional to its own current and also a term proportional to the current in loop 1 (Fig. 33–3b):

$$\Phi_B(2) = L_2 I_2 + M_{21} I_1. \tag{33–4}$$

The second term involves what might appear to be a new constant, M_{21}. Although it is not obvious, *the mutual inductances are equal*: $M_{12} = M_{21}$. We will not provide the proof here. It is customary to drop the subscripts and write $M = M_{12} = M_{21}$, the mutual inductance of two loops.

Faraday's law gives the emf induced in loop 2 due to the change in current of loop 1:

$$\mathscr{E}_{21} = -M\frac{dI_1}{dt}. \tag{33–5}$$

A similar expression gives the emf induced in loop 1 due to the current in loop 2.

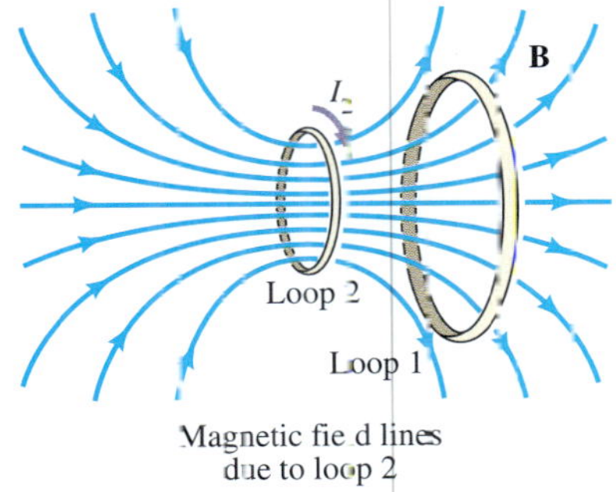

(a)

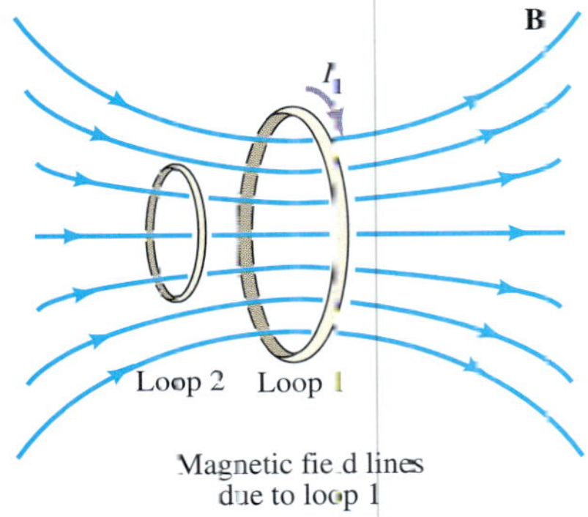

(b)

FIGURE 33–3 Mutual inductance. (a) There is a magnetic flux through loop 1 due to the magnetic field from the current I_2 in loop 2. (b) There is a magnetic flux through loop 2 due to the current I_1 in loop 1.

The inductances L and M have SI units of magnetic flux divided by current, or webers per ampere (Wb/A). Inductance is given its own unit in the SI, the **henry** (H) (Fig. 33–4):

The SI unit of inductance

$$1\ \text{H} = 1\ \text{Wb/A} = 1\ \text{T}\cdot\text{m}^2/\text{A}. \tag{33–6}$$

An inductance of 1 H is large but not unrealizable. For example, a cylindrical solenoid of area 10 cm^2, length 20 cm, and a winding density of 10 turns/cm has an inductance of 0.25 mH.

Elements within circuits with a significant self-inductance provide another source of emf to be taken into account when the loop rule is used for potential changes around a circuit. Such elements (usually in the form of solenoids) are as useful as capacitors and resistors and are typically introduced into circuits deliberately. These elements are called **inductors** (Fig. 33–5). They are represented in circuit diagrams by the symbol ꝏꝏ. We add this effect to our list of rules for application of Kirchoff's loop rule to circuits:

In moving across an inductor of inductance L *along* the presumed direction of the current, I, the potential change is $\Delta V = -L\,dI/dt$. The potential change is $\Delta V = +L\,dI/dt$

FIGURE 33–4 Joseph Henry, as depicted in a stained-glass window in the First Presbyterian Church of Albany, New York, the site of Henry's baptism. Henry investigated many effects of induction at about the same time as did Faraday.

APPLICATION

Inductors

when we trace the circuit *against* the presumed direction of the current. Note: The actual sign of this potential drop depends on the sign of the rate of change of the current; the sign of the potential drop will come out when we solve the circuit equations.

Mutual inductance is often too small to be a factor in the loop rule. The role of mutual inductance in linked circuits is a special one with an important application in *transformers*, devices to be described in Chapter 34 that are used to change the magnitude of time-varying voltages.

Finding the Inductance

In order to use the loop rule with inductors present, we must be able to calculate or measure the self-inductance or mutual inductance. Like capacitance, inductance can easily be calculated for only a few simple, but important, geometries. The most important of these is the solenoid. Consider the ideal solenoid of Fig. 33–6, which has length ℓ and radius R. For $\ell >> R$, the magnetic field within the solenoid is longitudinal and constant and is given by Eq. (30–15):

PROBLEM SOLVING

Rules for potential differences across inductors

$$B = \mu_0 n I,$$

where μ_0 is the permeability of free space, n is the number of turns per unit length of solenoid, and I is the current the solenoid carries. The magnetic flux through one turn of the solenoid is the field B times the cross-sectional area A, $\Phi_B = BA = \mu_0 AnI$. The total magnetic flux is this value times the *total* number of turns $N = n\ell$:

$$\Phi_B = \mu_0 A n^2 \ell I. \qquad (33\text{–}7)$$

By comparison with Eq. (33–1), the self-inductance is the coefficient of the current:

The inductance of an ideal solenoid

$$\text{for an ideal solenoid:} \qquad L = \mu_0 A \ell n^2. \qquad (33\text{–}8)$$

FIGURE 33–5 A collection of inductors.

EXAMPLE 33–1 During a short time period, the current in a cylindrical coil of length 10 cm, radius 0.5 cm, and 1000 turns of wire in a single layer is increased at the steady rate of 10^3 A/s. Find the emf induced during this period.

Solution: We are given dI/dt; the induced emf is, according to Eq. (33–2),

$$\mathscr{E} = -L\frac{dI}{dt}.$$

We must next determine the inductance of the solenoid, given by Eq. (33–8). The density of turns is $n = (1000 \text{ turns})/(0.1 \text{ m}) = 10^4$ turns/m. The area of the solenoid is given by $A = \pi r^2$, so

$$L = \mu_0 A \ell n^2 = (4\pi \times 10^{-7}\ \text{T}\cdot\text{m/A})[\pi(0.005\ \text{m})^2](0.1\ \text{m})(10^4\ \text{m}^{-1})^2 = 10^{-3}\ \text{H}.$$

The rate of change of current is given as 10^3 A/s, and

$$\mathscr{E} = -(10^{-3}\ \text{H})(10^3\ \text{A/s}) = -1\ \text{V}.$$

As we go around the circuit in the direction of the current, the induced emf will be negative, -1 V. In many electric circuits, 1 V is a significant emf.

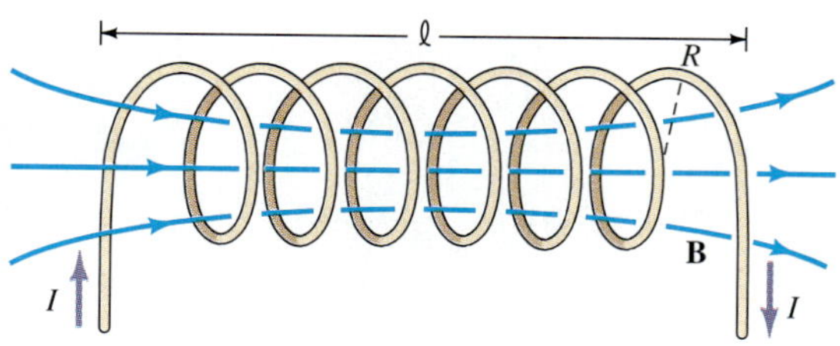

FIGURE 33–6 A solenoid of length ℓ, radius R, and turn density n carries a current I. Only the magnetic field **B** inside is shown; the outside field is zero in the limit where the solenoid is infinitely long.

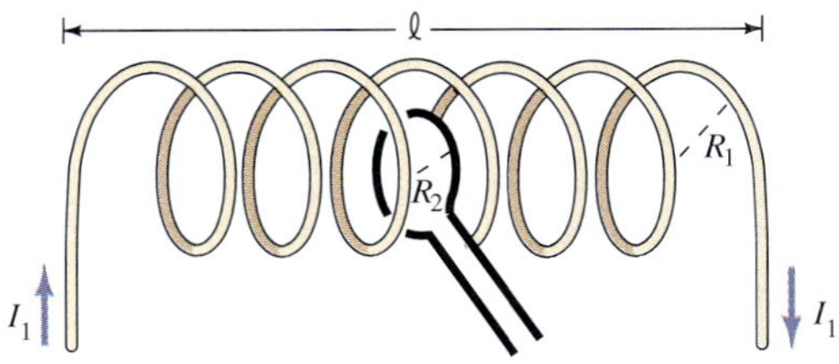

FIGURE 33–7 The area of the single, small loop is oriented perpendicular to the axis of the solenoid.

As an example of a calculable mutual inductance, consider a solenoid (of length ℓ, radius R_1, winding density n_1, and current I_1) that contains within it a single loop of radius R_2 whose area is oriented perpendicular to the axis of the solenoid (Fig. 33–7). The magnetic field of the solenoid is given by Eq. (30–15), $B = \mu_0 n_1 I_1$. The magnetic flux that passes through the single loop is

$$\Phi_B = BA_2 = B\pi R_2^2 = \mu_0 \pi R_2^2 n_1 I_1.$$

By definition the mutual inductance, M, is the coefficient of I_1:

$$M = \mu_0 \pi R_2^2 n_1. \tag{33–9}$$

If the single loop is replaced with a second solenoid with a total number of turns N_2, then the total flux that links that second solenoid contains a factor N_2, and M must be increased by this same factor:

$$M = \mu_0 \pi R_2^2 n_1 N_2. \tag{33–10}$$

EXAMPLE 33–2 Consider the single loop and the solenoid shown in Fig. 33–7. Suppose the loop carries a current I_2 that is a function of time. Find the emf in the solenoid induced by current I_2.

Solution: To apply Faraday's law, we must find the magnetic flux, Φ_B, due to the loop that links the solenoid. This flux is the mutual inductance of the loop and solenoid times current I_2. The mutuality of the inductance allows us to use the result of Eq. (33–9) for the inductance:

$$\phi_B = MI_2 = \mu_0 \pi R_2^2 n_1 I_2.$$

The emf in the solenoid is the negative time derivative of this flux:

$$\mathscr{E} = -\frac{d\Phi_B}{dt} = -\mu_0 \pi R_2^2 n_1 \frac{dI_2}{dt}.$$

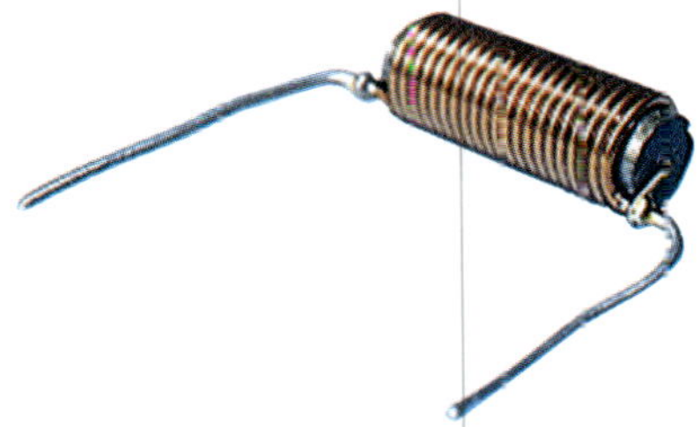

FIGURE 33–8 The inclusion of an iron core in this solenoid (an actual inductor) increases the magnetic field produced by a given coil current.

The Effects of Magnetic Materials on Inductance

In Chapter 32, we considered modifications due to materials with magnetic properties. We showed how we can write expressions that include the free (or real) current only if we replace the permeability of free space, μ_0, with the permeability of the material, μ. The permeability is given by Eq. (32–7),

$$\mu = \mu_0(1 + \chi_m).$$

Here, χ_m is the magnetic susceptibility of the material—negative and small for diamagnets, positive and small for paramagnets, and positive and large for ferromagnets. If a solenoid were filled with a magnetic material, its self-inductance would change due to the replacement of μ_0 by μ in Eq. (33–8). For ferromagnetic materials, self-inductances may be thereby increased manyfold (Fig. 33–8). In Fig. 33–9, we illustrate an application of inductors that uses the magnetic properties of materials.

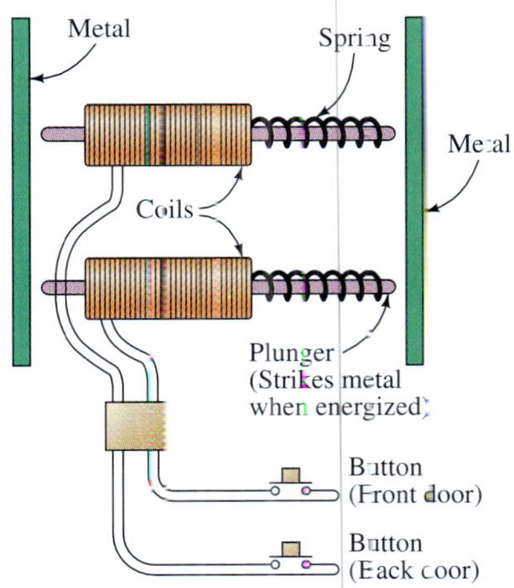

FIGURE 33–9 The doorbell is a simple application of inductance. The touch of a doorbell induces a current in the rods within the solenoid, which then are thrust from the solenoid according to Lenz's law. They thereby strike a chime.

33–2 ENERGY IN INDUCTORS

An inductor plays a role for the magnetic field analogous to that of a capacitor for the electric field: It is a device for storing energy in the magnetic field. The work–energy theorem establishes that there is energy in an inductor. Because any emf induced in the inductor opposes the change in current, work must be done by an external source, such as a battery, to cause a current to pass through an inductor. Just how much work is done is a measure of the energy stored in the inductor. To calculate this energy, we proceed in a manner analogous to that for the capacitor in Chapter 26: Calculate the work performed by some external emf, the work required to pass a current through the inductor.

We derive the general expression for the rate $\mathrm{d}W/\mathrm{d}t$ (the power) at which an external emf, $\mathscr{E}_{\text{ext}}$, does work when a current I flows from Eqs. (27–26) and (27–27):

$$\frac{\mathrm{d}W}{\mathrm{d}t} = I\mathscr{E}_{\text{ext}}.$$

If we have only the external emf and an inductor, the external emf must be equal but opposite to the induced emf in the inductor, given by Eq. (33–2). Thus

$$\frac{\mathrm{d}W}{\mathrm{d}t} = +LI\frac{\mathrm{d}I}{\mathrm{d}t}. \tag{33–11}$$

If the current is increasing, the power is positive, meaning that the external source must do positive work in supplying energy to the inductor; the internal energy U_L in the inductor is increasing. If the current is decreasing, the power is negative, meaning that the external source takes energy from the inductor; the inductor's internal energy is decreasing. The net change ΔU_L in the total magnetic energy of the inductor as the current changes from a value I_1 to a value I_2 between the times t_1 and t_2 can be found by integrating the work done by the external source as the current changes. We integrate Eq. (33–11) for $\mathrm{d}W/\mathrm{d}t$ from an initial time t_1 to a later time t_2:

$$\Delta U_L = \int_{t_1}^{t_2} \frac{\mathrm{d}W}{\mathrm{d}t}\mathrm{d}t = \int_{t_1}^{t_2} LI\frac{\mathrm{d}I}{\mathrm{d}t}\mathrm{d}t = L\int_{I_1}^{I_2} I\,\mathrm{d}I$$
$$= \frac{1}{2}LI_2^2 - \frac{1}{2}LI_1^2. \tag{33–12}$$

The energy contained in an inductor

In particular, if the inductor carries a current I, then the increase in energy as the current increases from zero, which we refer to simply as the energy of the inductor, is

$$U_L = \frac{1}{2}LI^2. \tag{33–13}$$

Equation (33–13) should be compared to the expression for the energy U_C contained in a capacitor of capacitance C that carries charge Q, from Eq. (26–8):

$$U_C = \frac{1}{2}\frac{Q^2}{C}.$$

Note that the equation for the energy of a capacitor contains the factor $1/C$, whereas the expression for the energy of an inductor contains the factor L.

EXAMPLE 33–3 A solenoid is designed to store $U_L = 0.10$ J of energy when it carries a current I of 450 mA. The solenoid has a cross-sectional area A of 5.0 cm^2 and a length ℓ of 0.20 m. How many turns of wire must the solenoid have?

Solution: We want to find the number of turns N, given the current, the length and area of the solenoid, and the energy it stores. To find N, we must express the energy in terms of it; for that, we need the inductance. The self-inductance of a solenoid is given by Eq. (33–8). This inductance can be written in terms of the total number of turns N rather than the turn density n by using $N = n\ell$, where ℓ is the length of the solenoid. Thus

$$L = \frac{\mu_0 AN^2}{\ell}.$$

The expression for the energy, Eq. (33–13), then reads

$$U_L = \frac{1}{2}\frac{\mu_0 A N^2}{\ell} I^2.$$

We can solve for N:

$$N = \frac{1}{I}\sqrt{\frac{2U_L\ell}{\mu_0 A}} = \frac{1}{4.5 \times 10^{-1}\text{ A}}\sqrt{\frac{2(0.10\text{ J})(0.20\text{ m})}{(4\pi \times 10^{-7}\text{ N/A}^2)(5.0 \times 10^{-4}\text{ m}^2)}}$$

$$= 1.8 \times 10^4 \text{ turns}.$$

Note that an inductor has an energy given by Eq. (33–13) even if the current is steady. We have argued that the origin of the effects of inductance is Faraday's law, which involves changes in current. How can we reconcile these two statements? The energy of an inductor carrying steady current arises from the original buildup of current—even if it occurred in the distant past. It is indeed *changes* in current that are at the origin of changes in magnetic energy.

33–3 ENERGY IN MAGNETIC FIELDS

In Chapter 26, we demonstrated that the electric energy associated with a capacitor is located in the electric field within the capacitor. Similarly, the energy of an inductor is located in its magnetic field. The ideal solenoid presents us with a tool to find the energy density in a magnetic field because the magnetic field within a solenoid is uniform.

The inductance of an ideal solenoid of area A and length ℓ is given by Eq. (33–8), so, from the expression for the total energy of an inductor [Eq. (33–13)], we find

$$U_L = \frac{1}{2}LI^2 = \frac{1}{2}\mu_0 A\ell n^2 I^2. \tag{33–14}$$

We also know that the magnetic field in the solenoid is proportional to the current. Equation (30–15) gives the precise connection, $B = \mu_0 nI$. If we substitute for I in terms of B in Eq. (33–14), we get

$$U_L = \frac{1}{2}\frac{B^2}{\mu_0}A\ell. \tag{33–15}$$

The volume within the solenoid is $A\ell$. Because the magnetic field is uniform within the solenoid, we may identify the **energy density**, u_B, the energy per unit volume of the magnetic field, as

The energy density of a magnetic field

$$u_B = \frac{1}{2}\frac{B^2}{\mu_0}. \tag{33–16}$$

This result generalizes to the case of a nonuniform magnetic field, no matter how it is produced. It should be compared to our expression for the energy density of an electric field, Eq. (26–12):

$$u_E = \frac{1}{2}\epsilon_0 E^2,$$

a result derived in a similar way. It is important to realize that *energy is located within the electric and magnetic fields themselves.*

When both magnetic and electric fields are present, the energy density is the sum of both magnetic and electric energy densities:

$$u = u_B + u_E = \frac{1}{2}\left(\frac{B^2}{\mu_0} + \epsilon_0 E^2\right). \tag{33–17}$$

EXAMPLE 33–4 A large electromagnet produces a magnetic field of 1 T. Compare the energy density associated with this field to that of the largest electric field in air, about 10^6 V/m.

Solution: Equations (33–16) and (26–12) express the energy density in magnetic and electric fields, respectively. From them, the ratio of the magnetic and electric energy densities is

$$\frac{u_B}{u_E} = \frac{\dfrac{1}{2}\dfrac{B^2}{\mu_0}}{\dfrac{1}{2}\epsilon_0 E^2} = \frac{1}{\mu_0\epsilon_0}\frac{B^2}{E^2}.$$

Note that, as in Eq. (30–4), the factor $1/\mu_0\epsilon_0$ has dimensions of speed squared, and this speed is the speed of light, c. In our case, the magnitudes of both the magnetic and electric fields are given, and

$$\frac{u_B}{u_E} = \frac{1}{(4\pi \times 10^{-7}\ \text{N/A})(8.85 \times 10^{-12}\ \text{F/m})}\frac{(1\ \text{T})^2}{(10^6\ \text{V/m})^2} = 9 \times 10^4.$$

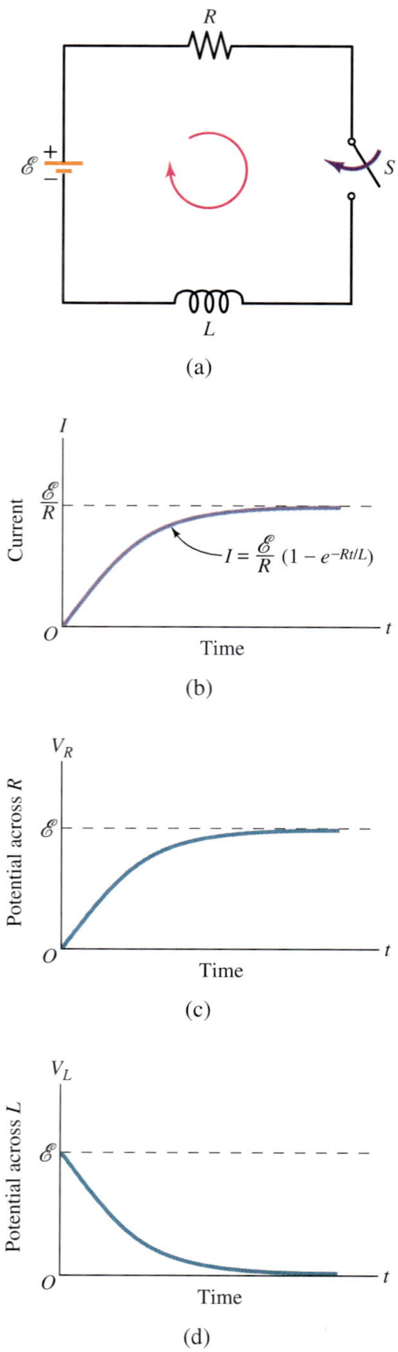

FIGURE 33–10 (a) An *RL* circuit, including a source of constant emf. (b) If the switch is closed at $t = 0$, the current rises from zero to a steady-state value only after a period of time determined by the ratio L/R. (c) The potential across R changes with time. (d) The potential across L changes with time.

33–4 *RL* CIRCUITS

We expect the presence of inductors in circuits to lead to new time-dependent phenomena because the potential drop across an inductor depends on how rapidly the current passing through it changes. For example, if we attempt to stop a current already flowing, say by opening a switch, then an emf is induced that attempts to keep the current flowing. Another illustration involves a circuit with a source of emf $\mathscr{E}$, a resistor of resistance R, and an inductor of inductance L (Fig. 33–10). We call such a circuit an **RL circuit**. A switch allows us to control the initial conditions. When we close the switch, the inductor opposes the changing current by Lenz's law. As a result, the current cannot jump suddenly but must build up over time. If we apply the loop rule to the circuit in the direction of the pink arrow in Fig. 33–10a, we can see this quantitatively. We have

$$\mathscr{E} - IR - L\frac{dI}{dt} = 0. \tag{33–18}$$

To find the solution to this differential equation for I, compare Eq. (33–18) with Eq. (28–24), which comes from applying the loop rule to the *RC* circuit shown in Fig. 33–11:

$$\mathscr{E} - \frac{Q}{C} - R\frac{dQ}{dt} = 0.$$

Here, Q is the charge on the capacitor. The first differential equation determines the current in *RL* circuits, and has exactly the same *form* as the equation for the charge in *RC* circuits. The similarities between the *RC* and *RL* circuits are summarized in Table 33–1, and the solution of the differential equation for current, Eq. (33–18), is the same as the solution of Eq. (28–24) for charge if we make the substitutions indicated in the table. In particular, if we replace *RC* with L/R, the current in the *RL* circuit will have the time dependence $e^{-t/(L/R)}$. The *RL* circuit is said to have a *time constant* L/R. Note that, as for the *RC* circuit, the time dependence is transient. Large values of the time constant (large L and/or small R) mean that the transient behavior is slow to disappear; small values of the time constant (small L and/or large R) mean that the transient behavior quickly disappears.

For complete solutions of Eq. (33–18) for the current, including the initial conditions, we can directly apply the results of Chapter 28 for the solution of the differential equa-

APPLICATION: THE AUTOMOBILE SPARK COIL

The spark that is fired by the spark plug within the cylinder of a car's engine comes from the energy stored in an iron core inductor in series with a capacitor and the car's 12 V battery. The capacitor is initially short-circuited—either by means of a mechanical switch or an electronic switch using a transistor (Fig. 33AB–1a). The current through the inductor increases with a time constant L/R to the asymptotic value V/R, where V is the battery potential and R is the inductor coil's resistance. When a spark is required, the switch is thrown; the circuit becomes a resonant one with an initial current and no initial potential drop across the capacitor (Fig. 33AB–1b). This circuit has a current with damped harmonic behavior. The current decreases until, at the end of the first quarter period of the harmonic cycle, there is no current; it is at this point in the cycle that the rate of change of the current, dI/dt, is a maximum, with a potential difference of some 300 V across the inductor. Now a new element enters the picture. Wrapped around the iron core inductor is a secondary coil. The magnetic flux from the first coil passes through this second coil and, because the rate of change of that flux depends on the rate at which the current in the first coil changes, there is a Faraday emf across the second coil. This emf is a maximum at the one-quarter stage of the oscillation. By adjusting the windings, it is possible for this emf to be a large one—typically 5–10 kV; this is large enough to make a spark when the second coil is connected appropriately to the spark plug. In effect, almost all the energy—some 3–6 mJ—stored in the first coil when the capacitor is short-circuited out is transferred to the spark. The loss in the spark is great enough so the oscillations damp out quickly, and the system is ready to restart its cycle.

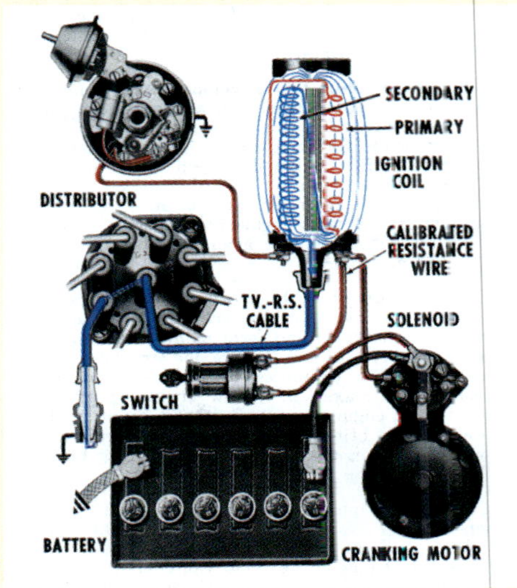

FIGURE 33AB–1 Schematic wiring diagram of an ignition system with the primary circuit in red and the secondary circuit in blue. The secondary voltage can be as high as 25 kV, but usually about 5 to 10 kV is applied to the spark plug.

tion for the RC circuit. As in Chapter 28, it is important to understand *physically* how different initial conditions will affect the solution. As a guiding principle, keep in mind that

> **The current in an inductor never changes instantaneously, but after the current settles down to a constant value, the inductor plays no role in the circuit.**

To take an example, suppose that the switch in Fig. 33–10a is closed at $t = 0$. Before $t = 0$, there is no current. After a long time, the current settles down to the constant value $I = \mathcal{E}/R$ because, after the transient behavior disappears, the inductor plays no role in the circuit. The function with exponential time dependence that satisfies these limits is

$$I = \frac{\mathcal{E}}{R}\left[1 - e^{-t/(L/R)}\right] = \frac{\mathcal{E}}{R}\left(1 - e^{-Rt/L}\right). \tag{33–19}$$

TABLE 33–1 Analogy Between *RC* and *RL* Circuits

	RC Circuit Parameter	*RL Circuit Parameter*
Variable	Q	I
Coefficient of variable	$1/C$	R
Coefficient of $\frac{d}{dt}$ (variable)	R	L
Time constant	RC	L/R

PROBLEM SOLVING

Using the correspondence between RC and RL circuits

The current in an inductor cannot change instantaneously. When current is constant, an inductor plays no role in a circuit.

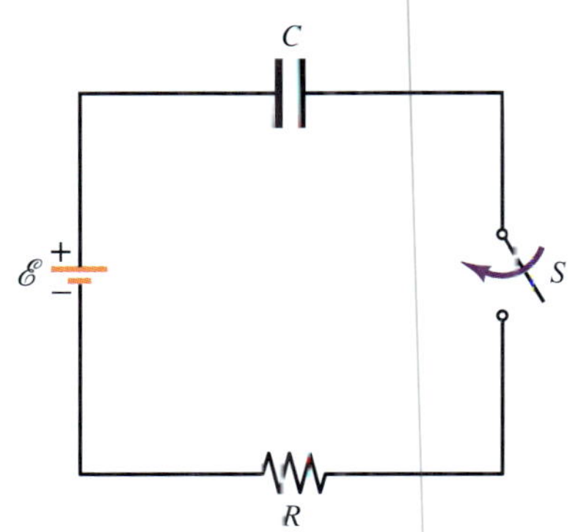

FIGURE 33–11 An RC circuit is analogous to the RL circuit of Fig. 33–10a.

Figure 33–10b is a graph of current versus time for this case. At $t = 0$, the exponential term is unity and $I = 0$, consistent with the principle that the current cannot change instantaneously. As $t \to \infty$, the exponential term drops out and I approaches $\mathscr{E}/R$—as though the inductor were not present at all. Similarly, we could, for example, find the potentials across the inductor or the capacitor (Figs. 33–10c, d). Substitute Eq. (33–19) into Eq. (33–18) to verify that it is indeed a solution (see Problem 45).

33–5 OSCILLATIONS IN CIRCUITS

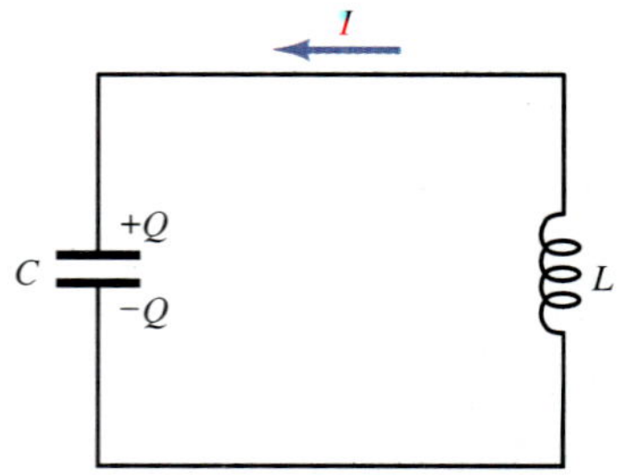

FIGURE 33–12 An *LC* circuit, consisting simply of an inductor and a capacitor in series.

A circuit with both inductance and capacitance exhibits oscillatory behavior. Such a circuit is called an ***LC* circuit**, and it turns out to be analogous to a mass on the end of a spring—a familiar mechanical system. The equivalence can be seen immediately if we write the loop rule for a circuit of this type (Fig. 33–12). Following the clockwise direction, we have

$$+\frac{Q}{C} + L\frac{dI}{dt} = 0. \tag{33–20}$$

This is not yet a differential equation for a single variable because both the current and the capacitor charge appear. However, if we recall that the current I is dQ/dt, we do indeed find a differential equation for the charge:

$$+\frac{Q}{C} + L\frac{d^2Q}{dt^2} = 0. \tag{33–21}$$

An *LC* circuit is analogous to a mass on an ideal spring.

To find the solution, let's recall that the equation of motion for a mass m on the end of an ideal spring of spring constant k is $-kx = ma$; because acceleration is the second derivative of displacement,

$$kx + m\frac{d^2x}{dt^2} = 0. \tag{33–22}$$

PROBLEM SOLVING

Helpful correspondences between mechanical systems and circuits

This equation for x has the same form as the equation for Q, and so the solutions for these variables have the same form. The mechanical motion is harmonic, so the same is true for the charge on the capacitor. We must make the appropriate change of variable names, and this can most easily be done by noting the correspondences summarized in Table 33–2. Using these equivalences, the angular frequency of the mechanical system is $\omega = \sqrt{k/m}$, so from the table the angular frequency of the *LC* circuit is

$$\omega = \frac{1}{\sqrt{LC}}. \tag{33–23}$$

The general form of the charge on the capacitor is then

$$Q = Q_1 \sin(\omega t) + Q_2 \cos(\omega t), \tag{33–24}$$

or, equivalently,

$$Q = Q_0 \cos(\omega t + \phi). \tag{33–25}$$

TABLE 33–2 Analogy Between *LC* Circuit and Mass on a Spring

	Mass on Spring Parameter	*LC Circuit Parameter*
Variable	x	Q
Coefficient of variable	k	$(1/C)$
Coefficient of $\frac{d^2}{dt^2}$ (variable)	m	L
Natural frequency	$\sqrt{k/m}$	$1/\sqrt{LC}$

The constants, either Q_1 and Q_2 or Q_0 and ϕ, are determined by the initial conditions. For example, if we know that the capacitor has a given total charge at time $t = 0$ and that the current at $t = 0$ is also zero (because a switch allowing current to flow in the LC circuit is closed at that point), then we can determine the two unknown constants.

Once we know the charge on the capacitor, we also know the current flowing through the inductor; we need only apply the relation $I = \mathrm{d}Q/\mathrm{d}t$. For example, Eq. (33–25) gives

$$I = \frac{\mathrm{d}}{\mathrm{d}t}[Q_0 \cos(\omega t + \phi)] = -Q_0\omega \sin(\omega t + \phi). \quad (33\text{–}26)$$

Thus the current, like the charge, is oscillatory with the same frequency $\omega = 1/\sqrt{LC}$.

What is happening here is that positive charge is flowing off a plate of the capacitor, through the inductor, and building up on the other side of the capacitor, only to flow back again and repeat the motion; the inductor prevents this from happening instantaneously. We shall see in Section 33–7 that this can also be interpreted in terms of an energy flow.

EXAMPLE 33–5 The capacitor charge in the circuit of Fig. 33–12 has the form $Q = Q_0 \cos(\omega t)$. (a) Find the voltage across the inductor. (b) If $L = 12\ \mu\mathrm{H}$ and $C = 0.80\ \mu\mathrm{F}$, what is the period for the voltage in part (a)?

Solution: (a) We are given the capacitance and inductance and can compute any changes in the current from the specified changes in the charge on the capacitor plates. In terms of these known quantities, the unknown voltage across the inductor is

$$V_L = -L\frac{\mathrm{d}I}{\mathrm{d}t} = -L\frac{\mathrm{d}^2Q}{\mathrm{d}t^2}.$$

We have

$$\frac{\mathrm{d}^2Q}{\mathrm{d}t^2} = Q_0\frac{\mathrm{d}}{\mathrm{d}t}\left[\frac{\mathrm{d}}{\mathrm{d}t}\cos(\omega t)\right] = Q_0\frac{\mathrm{d}}{\mathrm{d}t}[-\omega \sin(\omega t)] = -\omega Q_0[\omega \cos(\omega t)]$$
$$= -\omega^2 Q_0 \cos(\omega t) = -\omega^2 Q.$$

Thus

$$V_L = \omega^2 LQ = \omega^2 LQ_0 \cos(\omega t).$$

Just as in simple harmonic motion, where the acceleration of the mass on the end of a spring is proportional to the mass's displacement, in LC circuits the voltage across the inductor is proportional to the charge on the capacitor.

(b) The unknown period of the voltage across the inductor is identical to the period for the charge on the capacitor:

$$T = \frac{2\pi}{\omega}.$$

With $\omega = 1/\sqrt{LC}$, we have

$$T = 2\pi\sqrt{LC} = 2\pi\sqrt{(12 \times 10^{-6}\ \mathrm{H})(0.80 \times 10^{-6}\ \mathrm{F})} = 1.9 \times 10^{-5}\ \mathrm{s}.$$

33–6 DAMPED OSCILLATIONS IN CIRCUITS

When resistance, inductance, and capacitance are all present in a circuit, as in Fig. 33–13, we have an ***RLC* circuit**. We assume that a current is present in the circuit and apply the loop rule by following the direction of the current:

$$-L\frac{\mathrm{d}I}{\mathrm{d}t} - IR - \frac{Q}{C} = 0. \quad (33\text{–}27)$$

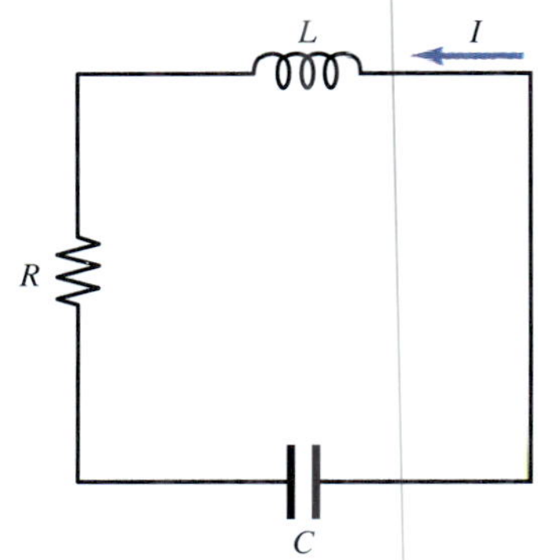

FIGURE 33–13 A prototype RLC circuit.

Because $I = \frac{dQ}{dt}$, Eq. (33–27) can also be written as

$$-L\frac{d^2Q}{dt^2} - R\frac{dQ}{dt} - \frac{Q}{C} = 0. \tag{33–28}$$

Damped harmonic oscillators are discussed in Section 13–7.

This equation is a differential equation for the charge Q; with the appropriate initial conditions, it determines the charge on the capacitor, including its time dependence. The current in the circuit is then found by differentiation of the charge.

Equation (33–28) has an analogue in mechanics problems that involve masses on springs in the presence of drag: To Newton's second law for a mass on an ideal spring (Section 33–5), add a term for a drag force. The drag force $-bv$ is proportional to the velocity or first derivative of x. When we put the acceleration term on the same side of the equation as the forces, Newton's second law for the damped harmonic oscillator becomes

An *RLC* circuit is analogous to a damped harmonic oscillator.

$$-m\frac{d^2x}{dt^2} - b\frac{dx}{dt} - kx = 0. \tag{33–29}$$

This equation describes the motion of a mass at the end of a spring immersed in a fluid that gives rise to a drag force. This is a physical system about which we have some intuition. The mathematical equivalence of Eqs. (33–28) and (33–29), described in detail in Table 33–3, is immensely helpful in understanding the *RLC* circuit.

PROBLEM SOLVING

More helpful correspondences between mechanical systems and circuits

Damping modulates the harmonic behavior of a mass on an ideal spring by changing the period slightly and imposing an envelope on the motion, in the form of a falling exponential. Thus the charge on the capacitor of an *RLC* circuit will also have harmonic time dependence within an envelope that falls exponentially with time. Current is the time derivative of charge and, because the derivatives of sines, cosines, and exponentials are cosines, sines, and exponentials, the current will also be harmonic in time within an exponential envelope.

We can again follow the techniques of Chapter 13—as applied to the damped harmonic oscillator—to find a solution to Eq. (33–27). Using our solution to the damped oscillator, Eq. (13–54), along with the equivalences in Table 33–3, we find

$$Q = Q_0 e^{-\alpha t}\cos(\omega' t + \phi). \tag{33–30}$$

The constants α and ω' are determined either by substitution back into the original differential equation, Eq. (33–28), or simply by using the correspondences with the results of Chapter 13. They are

$$\alpha = \frac{R}{2L} \tag{33–31}$$

and

$$\omega'^2 = \frac{1}{LC} - \frac{R^2}{4L^2} = \frac{1}{LC} - \alpha^2 = \omega^2 - \alpha^2. \tag{33–32}$$

The quantity α determines the rate of exponential damping and has dimensions of 1/time. The constants Q_0 and ϕ are determined from the initial conditions.

TABLE 33–3 Analogy Between *RLC* Circuits and Damped Harmonic Motion

	RLC Circuit Parameter	*Damped Harmonic Motion Parameter*
Variable	Q	x
Coefficient of variable	$1/C$	k
Coefficient of $\frac{d}{dt}$ (variable)	R	b
Coefficient of $\frac{d^2}{dt^2}$ (variable)	L	m

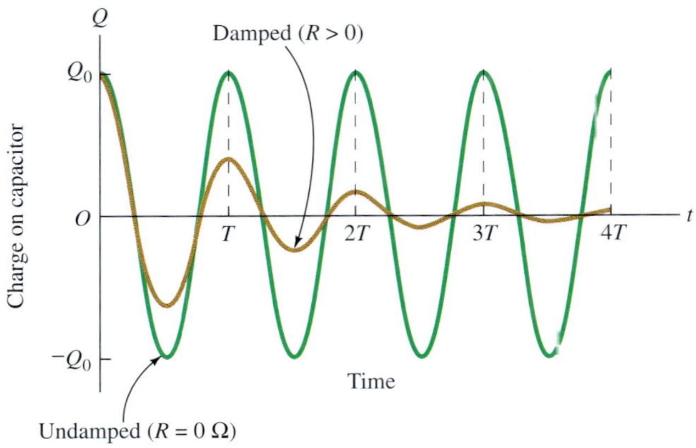

FIGURE 33–14 Comparison of an *RLC* circuit with and without damping: the charge on the capacitor versus time. The damped case has a very slightly larger period. *T* is the period of the undamped oscillator.

The exponential damping constant α depends only on L and R, as in *RL* circuits. The damping factor α for the full *RLC* circuit is *half* the size of the damping factor in *RL* circuits. The resistive element R is the crucial element in damping because energy is dissipated within it. The damping factor $e^{-\alpha t}$ forms a decreasing envelope for the harmonic behavior. In addition, the angular frequency ω' of the harmonic behavior differs from the angular frequency ω of the undamped circuit (an *LC* circuit) by the α-dependence of ω'. If the damping constant α is *small* compared to the undamped angular frequency $\omega = 1/\sqrt{LC}$, then ω' is only slightly less than ω. In Fig. 33–14, we plot the behavior of the capacitor charge for the case $L = 1$ H, $C = 1$ F, and $R = 0.3$ Ω, comparing it to the previous case where we had set $R = 0\ \Omega$. The period in the damped case is only very slightly larger than the period for the undamped case.

Figure 33–15 shows what happens to the charge if R is increased to 2 Ω and to 4 Ω. What is the explanation for this behavior? Equation (33–32) shows that when R is increased to a critical value R_c, ω'^2 decreases to zero. For $R = R_c$, we have

$$0 = \frac{1}{LC} - \frac{R_c^2}{4L^2},$$

which has the solution

$$R_c = 2\sqrt{\frac{L}{C}}. \tag{33–33}$$

When ω'^2 is zero, there is no more oscillation; as in Chapter 13, we refer to this case as *critical damping*. The value $R = 2\ \Omega$ in Fig. 33–15 represents this case: When $L =$ 1 H and $C = 1$ F, we have $R_c = 2\ \Omega$. For values of R that are larger than the critical value, there is *overdamping*. This kind of motion has its mechanical analogue in the motion of a mass at the end of a spring when the mass is moving in a jar of thick molasses. An example of this behavior illustrated in Fig. 33–15 is the case $R = 4\ \Omega$ (a value greater than R_c).

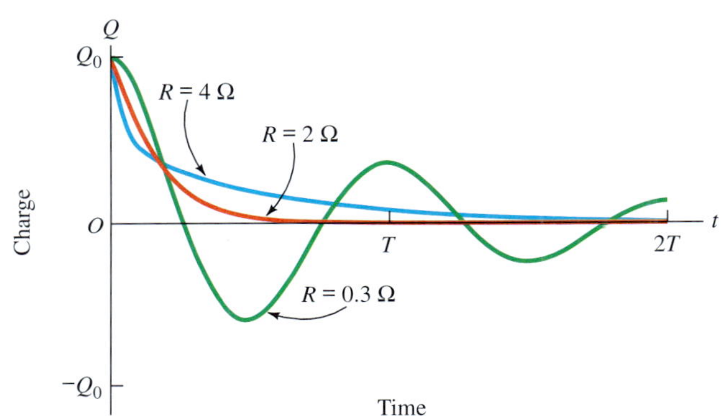

FIGURE 33–15 The *RLC* circuit of Fig. 33–13 for various values of R. The critical value R_c is $R = 2\ \Omega$. There is no oscillatory behavior, only damping, when $R > R_c$.

EXAMPLE 33–6 A resistor of 0.052 Ω is introduced by means of a switch into an LC circuit that is undergoing oscillatory behavior. The values of the inductance and the capacitance are 75 mH and 16 μF, respectively. How much time passes between $t = 0$, when the switch is thrown, and the moment when the amplitude of the oscillations has decreased to one-half of its original value (prior to $t = 0$)? How many oscillations does the circuit undergo during this time?

Solution: The relevant factor when damping is added is

$$\alpha = \frac{R}{2L} = \frac{(0.052\ \Omega)}{2(75 \times 10^{-3}\ \mathrm{H})} = 0.35\ \mathrm{s}^{-1}.$$

By comparison, the angular frequency of the oscillations of the (undamped) LC circuit is

$$\omega = 1/\sqrt{LC} = 1/\sqrt{(75 \times 10^{-3}\ \mathrm{H})(16 \times 10^{-6}\ \mathrm{F})} = 9.1 \times 10^{2}\ \mathrm{rad/s}.$$

The factor ω is much larger than α; thus, the modified angular frequency of the full RLC circuit differs only very slightly from the undamped angular frequency—the resistor is a small one. With a small value of α, the envelope that characterizes the decay of the oscillations changes slowly compared to the oscillation time. It is then meaningful to speak of the decay of the amplitude itself, according to

$$\text{amplitude} = A_0 e^{-\alpha t}.$$

Here, the amplitude could be the amplitude of the charge on the capacitor or of the current in the circuit or even of the potential across any element—all such quantities decay with the same exponential time factor. We want to find the time Δt over which the amplitude drops to one-half its initial value,

$$A_0\, e^{-\alpha \Delta t} = A_0/2.$$

A_0 drops out of this equation, and what remains determines Δt. By taking the natural logarithm of both sides, we find that

$$-\alpha \Delta t = \ln(1/2) = -\ln(2) = -0.69.$$

Thus

$$\Delta t = \frac{0.69}{\alpha} = \frac{0.69}{0.35\ \mathrm{s}^{-1}} = 2.0\ \mathrm{s}.$$

If $T = 2\pi/\omega$ is the period of the oscillator, then, during the time Δt, the circuit undergoes

$$\frac{\Delta t}{T} = \frac{\Delta t}{2\pi/\omega} = \frac{\omega\, \Delta t}{2\pi} = \frac{(9.1 \times 10^{2}\ \mathrm{rad/s})(2.0\ \mathrm{s})}{2\pi} \simeq 290\ \text{oscillations}.$$

33–7 ENERGY IN *RLC* CIRCUITS

Energy is an important consideration in the harmonic oscillator, and we expect it to be similarly important in RLC circuits. Let's reconsider instances with and without damping.

No Resistance. We set $R = 0$: no damping. Suppose that the initial conditions are such that the charge on the capacitor is given by

$$Q = Q_0 \cos(\omega t). \tag{33–34}$$

The current flowing through the circuit is then

$$I = \frac{\mathrm{d}Q}{\mathrm{d}t} = -\omega Q_0 \sin(\omega t). \tag{33–35}$$

One full period of both of these functions is plotted in Fig. 33–16. The energy contained in a capacitor is given by Eq. (26–8), $U_C = Q^2/2C$, or

$$U_C = \frac{Q_0^2}{2C}\cos^2(\omega t). \tag{33–36}$$

Equation (33–14) gives the magnetic energy in an inductor, $U_L = \frac{1}{2}LI^2$:

$$U_L = \frac{1}{2}L\omega^2 Q_0^2 \sin^2(\omega t) = \frac{Q_0^2}{2C}\sin^2(\omega t). \tag{33–37}$$

We have used $\omega = 1/\sqrt{LC}$. The functions U_C and U_L are plotted in Fig. 33–17. The energy in each element is never negative, but one rises to a maximum as the other falls to zero. The *total* energy in the inductor and capacitor is the sum of these two terms and is *constant*:

$$U = \frac{Q_0^2}{2C}[\cos^2(\omega t) + \sin^2(\omega t)] = \frac{Q_0^2}{2C}. \tag{33–38}$$

The two circuit elements swap the constant total energy back and forth harmonically, just as in the mechanical oscillator—for which the constant total energy is made up of a back-and-forth exchange of the potential energy of the spring and the kinetic energy of the attached mass. In Figs. 33–18a to 33–18e, we have drawn the progression over a full period T, starting with no current and a fully charged capacitor at $t = 0$. Without a resistor to dissipate energy, the oscillation will continue forever.

Resistance Is Introduced. We can easily understand the role of resistance in terms of energy. The power, P, is voltage times current and, for a resistor P, is

$$P_R = IV_R = I^2R. \tag{33–39}$$

This power is proportional to the current squared and is always positive. Energy is *always* lost to Joule heating in a resistor, regardless of the sign of the current. This

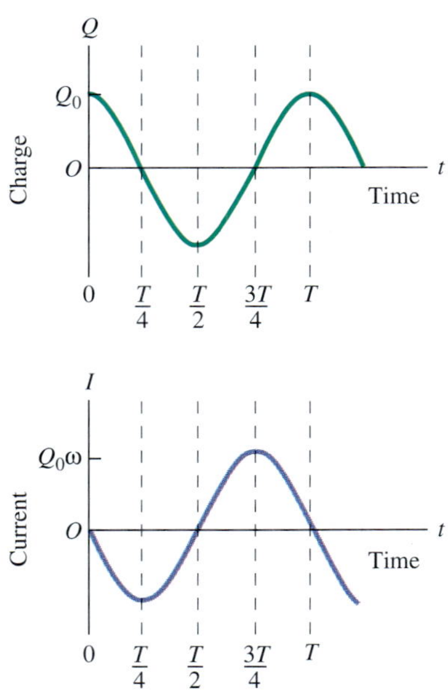

FIGURE 33–16 The capacitor charge and the current in a circuit that contains only inductance and capacitance are harmonic functions of time.

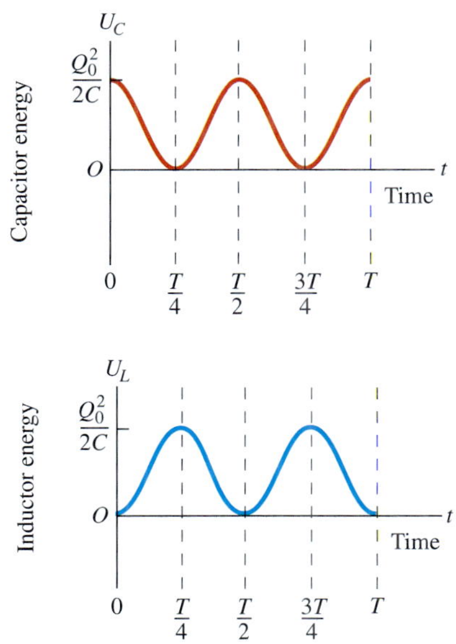

FIGURE 33–17 The energies of capacitor and inductor for charge and current of Fig. 33–13, but with $R = 0$. These oscillate out of phase.

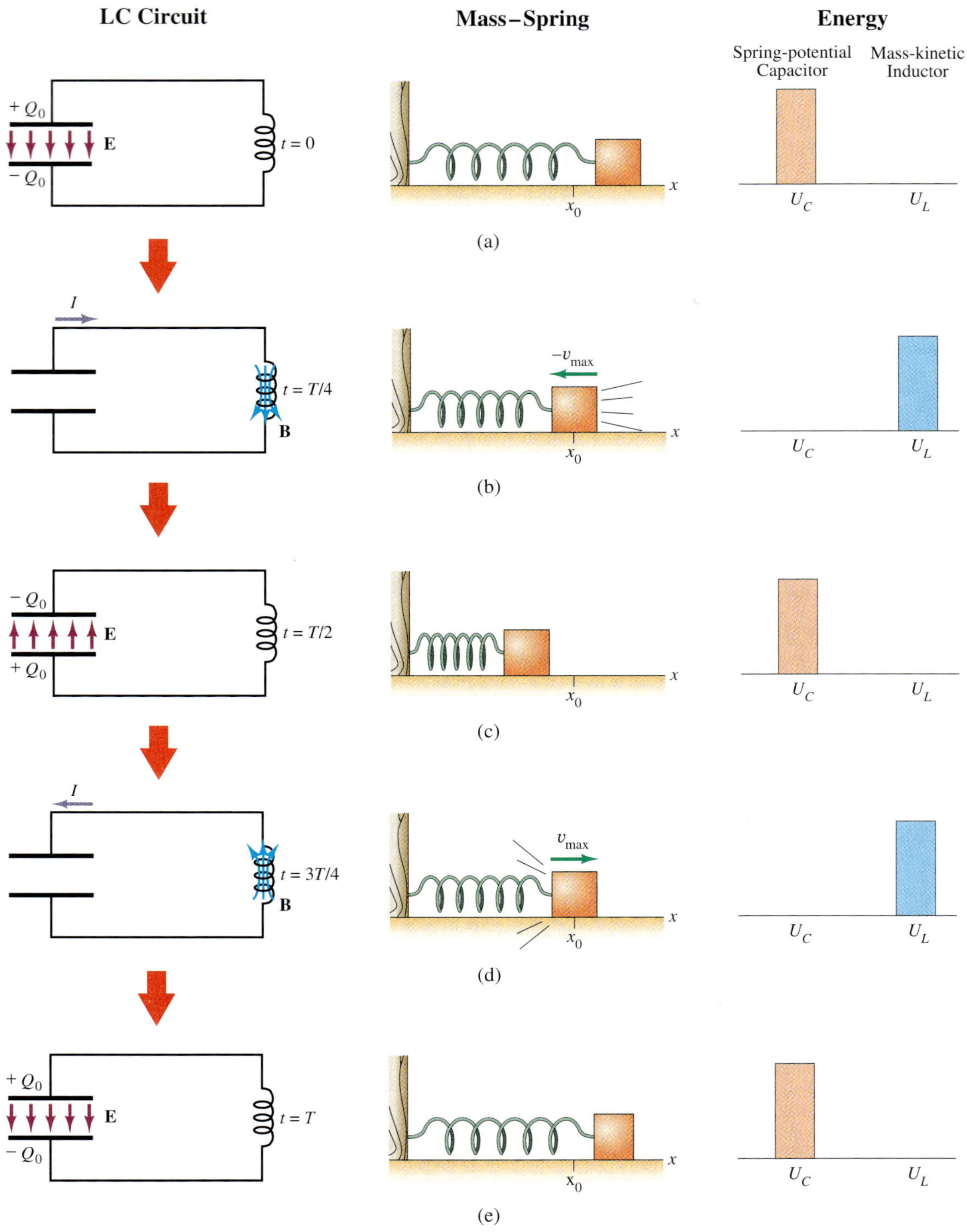

FIGURE 33–18 (a)–(e) Schematic diagrams of how the electric and magnetic fields, and therefore the energies, of Fig. 33–17 are realized within the capacitor and inductor: a time sequence. We also show the analogous mechanical system at corresponding times, as well as the balance of energy in the capacitor and inductor for those times.

is the origin of the exponential damping in *RLC* circuits. The power loss in resistors should be contrasted to the equivalent expressions for inductors ($P_L = IV_L$) or capacitors ($P_C = IV_C$). In each case, the rate of energy expenditure can be positive or negative, according to the situation [see Eq. (33–11)]. Unlike the resistor, these elements sometimes take energy from the other circuit elements and sometimes they give it back.

SUMMARY

An inductor is a circuit element that behaves as a current-carrying loop, such as a solenoid. It has an inductance, L, defined by the ratio between the magnetic flux and the current that passes through it:

$$\Phi_B \equiv LI. \tag{33–1}$$

By Faraday's law, the emf induced in this circuit element is

$$\mathcal{E} = -\frac{d\Phi_B}{dt} = -L\frac{dI}{dt}. \tag{33–2}$$

Faraday's law also shows that when there are adjacent loops in a circuit (or pair of circuits), the changing current in one loop induces an emf in the adjacent loop. In this case, the geometrical factor is the mutual inductance, M, which measures both the emf induced in loop 1 due to the current in loop 2 and the emf induced in loop 2 due to the change in current in loop 1,

$$\mathcal{E}_{21} = -M\frac{dI_1}{dt}. \tag{33–5}$$

Inductance is measured in henries (H) in the SI. The emf in an inductor is one more term to add to the loop rule.

A simple, calculable inductance is that of an ideal solenoid:

$$\text{for an ideal solenoid:} \qquad L = \mu_0 A\ell n^2. \tag{33–8}$$

Here, A is the area, ℓ is the length of the solenoid, and n is the number of turns per unit length.

The energy carried in an inductor is given by

$$U_L = \frac{1}{2}LI^2. \tag{33–13}$$

Just as the energy of a capacitor is carried by the electric field in the capacitor, the energy of an inductor is in the magnetic field. The energy density, or energy per unit volume, carried by a magnetic field is found by comparing the known field within a solenoid with the energy carried by the solenoid, and is given by

$$u_B = \frac{1}{2}\frac{B^2}{\mu_0} \tag{33–16}$$

The combination of inductance, capacitance, and resistance in circuits with and without batteries leads to interesting time dependence for currents and charges. The current in the inductor never changes instantaneously. When an inductor is placed in a circuit with a battery and a resistor, the loop rule produces an equation for the current characterized by transient exponential behavior, with time constant L/R. Such a circuit is called an LC circuit. When a capacitor is added to the circuit, the loop rule produces an equation for the charge on the capacitor that has the same form as the equation for the position of a mass on the end of a spring, with damping proportional to the speed of the mass. The electric circuit, called an RLC circuit, has damped oscillations. In particular, if the resistor is eliminated, the resulting circuit supports harmonic oscillations with a continual exchange of energy between the capacitor and the inductor. The angular frequency of these free oscillations is

$$\omega = \frac{1}{LC}. \tag{33–23}$$

UNDERSTANDING KEY CONCEPTS

1. Two electric circuits are placed near one another. Each circuit has self-inductance. Must there be a mutual inductance?
2. Does it take more work to cause current to flow through a coil of wire than through the same wire when it is straight?
3. Consider two circular coils that are in a variety of configurations (Fig. 33–19). Assuming that the separation between the coils is roughly the same for the various configurations, can you order the mutual inductances from largest to smallest?

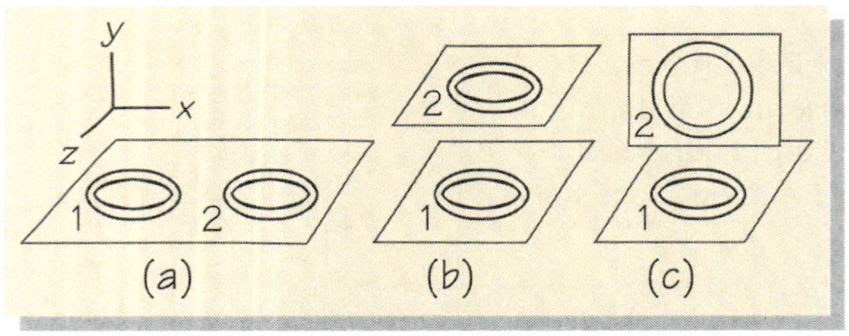

FIGURE 33–19 Question 3.

4. Why might Faraday's law cause the lights in a house to dim when an electrical appliance that uses a lot of energy, such as an electric clothes dryer, is turned on?
5. Is it possible to calculate the mutual inductance between a straight wire and a wire loop?
6. Why do you sometimes see a spark at a light switch when the switch is turned off? Is there a spark when the switch is turned on? Why or why not?
7. Describe how you could measure inductance with a battery of known emf, a known resistor, a voltmeter, and a timer.
8. A light bulb is placed in series with a resistor and in parallel with a coil of large inductance and negligible resistance. When a switch that connects a battery to this circuit is closed, the light bulb flashes before glowing dimly. When the switch is opened, the bulb flashes again before going out. Explain.
9. In the oscillations of an LC circuit, the energy is transferred from the electric field around the capacitor to the magnetic field around the inductor. How does the energy get from one place to the other?
10. Given your knowledge of the largest and smallest practical sizes of capacitors and inductors, what would you estimate is the electronic oscillator with the smallest frequency possible? the largest?
11. A solenoid has magnetic flux outside as well as inside, because magnetic field lines must close on themselves. Does this mean that there is magnetic energy outside the solenoid as well as inside it?
12. In Section 33–3, the magnetic energy calculation for a solenoid used the inductance of a portion of the solenoid, and that was translated into the energy density expression $B^2/2\mu_0$. The magnetic field must come around the outside of the solenoid because all magnetic field lines are closed. Why does the above calculation give the correct answer for the energy density?
13. We have made an analogy between a damped harmonic oscillator and an RLC circuit. What mechanical quantities are analogous to the energies $LI^2/2$ and $Q^2/2C$ of the RLC circuit?
14. How would you go about finding a generalization of Eq. (33–13) when two circuits with different currents are placed in such close proximity that their mutual inductance plays a role?
15. The magnetic energy density $B^2/2\mu_0$ has the dimensions of pressure and may be viewed as a magnetic pressure. Use this interpretation to justify the attraction/repulsion of two parallel wires that carry currents in the same/opposite directions.

PROBLEMS

33–1 Inductance and Inductors

1. (I) A wire loop has an inductance of 2 mH when a current of 30 mA passes through the circuit. What is the value of the magnetic flux that passes through the loop?
2. (I) What is the self-inductance per unit volume of a solenoid?
3. (I) Calculate the mutual inductance of a solenoid 25 cm long of radius 1.8 cm with 600 turns, and a single loop of radius 3.0 cm centered on the solenoid, with its area perpendicular to the axis of the solenoid.
4. (I) An investigator passes through a solenoid a current that changes at the rate of 25 A/s and measures an induced emf of 0.14 V. The length and diameter of the solenoid are 18 cm and 2.5 cm, respectively. What is the number of turns?
5. (I) The emf induced in an isolated circuit when the current in the circuit is changing by 10 A/s is 0.3 V. What is the self-inductance?
6. (II) An electrical engineer needs an inductor capable of producing an emf of 45 mV. The current source available produces current of the form $I = I_0 \cos(\omega t)$, with $I_0 = 1.0$ A and $\omega = 3.8 \times 10^2$ rad/s (of frequency 60 Hz). What size inductor should be used?
7. (II) A current of 1 A flows through a circuit placed in isolation. A magnetic flux of 0.010 $\mathrm{T \cdot m^2}$ passes through the circuit area. When this circuit is placed near another circuit with a current flow of 2 A, the magnetic flux through the first circuit increases to 0.012 $\mathrm{T \cdot m^2}$. (a) What is the mutual inductance of the two circuits? (b) How much magnetic flux passes through the second circuit, whose self-inductance is 1 mH?
8. (II) What is the self-inductance of the single inductor equivalent to two inductors of values L_1 and L_2, respectively, placed in series? Neglect the mutual inductance.
9. (II) A solenoid of length L consists of two coils tightly placed on top of each other. One coil has N_1 turns, the other N_2, and the area of the coils is A. Calculate the self-inductances of the coils (a) if only one of the two coils is used; (b) if the two coils are connected in series with their windings going in the same direction; (c) if the two coils are connected in series with their

windings going in the opposite direction. (d) Calculate the mutual inductance of the two coils.

10. (II) Consider two inductors with inductances L_1 and L_2, respectively, connected in parallel. What is the value of the single equivalent inductance that could replace the two inductances, assuming that the mutual inductance can be neglected?

11. (II) A solenoid of length L and area A contains two windings—one tightly placed on top of the other—with N_1 and N_2 turns, respectively. What happens if the two windings are connected in parallel and the composite coil is included in a circuit with a variable current?

12. (II) Equation (33–8) was derived for an ideal cylindrical solenoid. Show that this result holds also for a solenoid of any shape cross section, provided that the length is large compared to any cross-sectional measure.

13. (II) Consider the cylindrical solenoid and ring illustrated in Fig. 33–20. The solenoid, of diameter $d_1 = 2.0$ cm, has length $\ell = 20$ cm and 120 turns of wire. The ring of wire inside, with diameter $d_2 = 1.5$ cm and area perpendicular to the solenoid's axis, is connected by two wires to a single resistor of resistance $R = 33\ \Omega$. The current I_1 is in the form of a pulse that starts to rise linearly at $t = 0$ s. The current reaches a maximum of 30 A at $t = 0.30$ s, then starts to descend linearly; when the current reaches 0 A at $t = 0.60$ s, it ceases to flow. Find the current I_2 induced in the ring as a function of time.

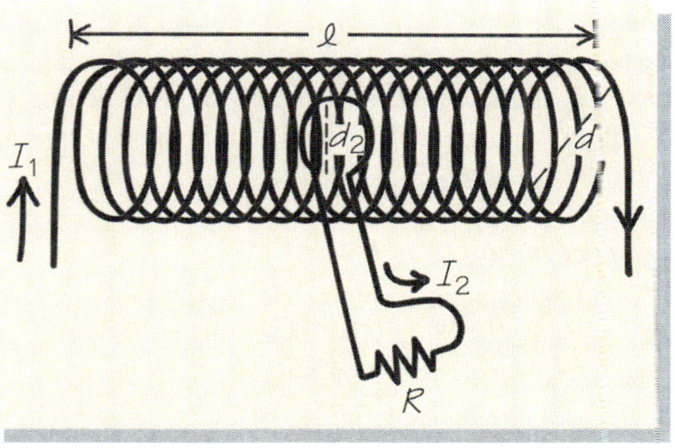

FIGURE 33–20 Problem 13.

14. (II) The ring contained within the cylindrical solenoid of Example 33–2 is replaced by a second cylindrical solenoid, of length ℓ_2, radius R_2, and turn density n_2. Calculate the mutual inductance of this system.

15. (II) The current in an inductor has the periodic triangular form plotted in Fig. 33–21, with an amplitude of 0.50 A and a period of $T = 0.45$ s. What is the voltage across the inductor as a function of time if $L = 2.3 \times 10^{-3}$ H? Either express your answer algebraically or plot it.

16. (II) A coaxial cable has a central conducting wire of radius r_0 surrounded by a conducting tube of radius r_1. The space in between is filled with a material of magnetic permeability μ. Show that if the wire has length ℓ, the self-inductance is $L = (\mu\ell/2\pi)\ln(r_1/r_0)$. [*Hint*: You must calculate the flux in the region between the cylinders.]

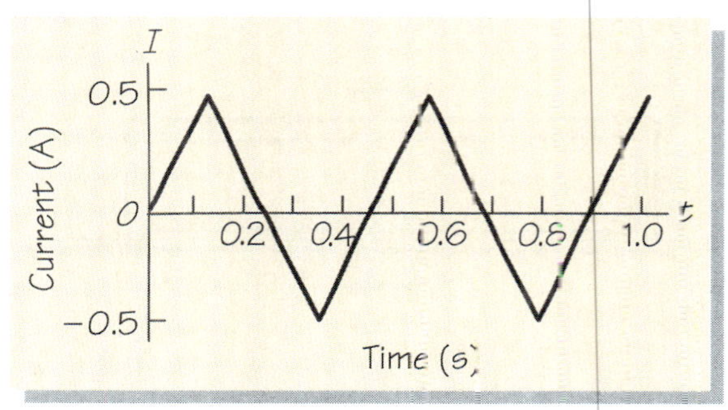

FIGURE 33–21 Problem 15

17. (II) Consider two identical solenoids placed end to end, with the windings going in the same direction. Prove that the total self-inductance of the combined system is $2(L + M)$, where L is the self-inductance of either solenoid and M is the mutual inductance.

18. (II) A torus of rectangular cross section with width w, height h, and inner radius R is wound with N turns of wire (Fig. 33–22). What is the self-inductance of the torus? Use the approximation $\ln(1 + x) = x$, valid for $x << 1$, to discuss the case where $R >> w$ and its relation to the self-inductance of a solenoid.

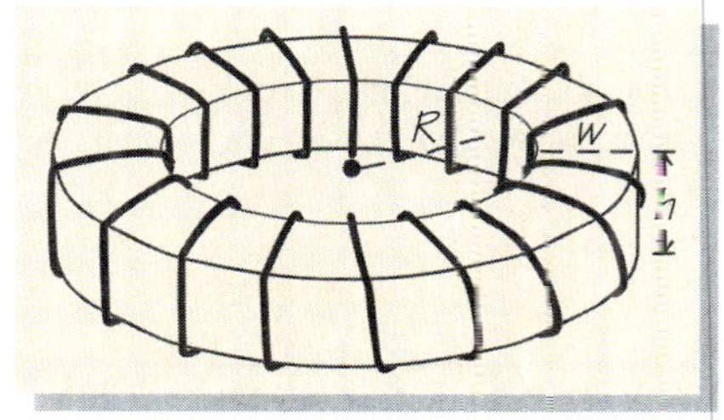

FIGURE 33–22 Problem 18.

19. (II) Consider a torus of square cross section. The radius of the torus (distance from the symmetry axis to the center of the square) is 35 cm; the sides of the square are 5.0 cm. The torus is wound with 1650 turns of wire. (a) What is the self-inductance of the torus? (b) What is the self-inductance if the core of the torus is made of soft iron, with $\mu = 4200\mu_0$?

20. (II) Consider a toroidal coil wound around an empty core whose self-inductance is 1.8 mH. The current in the coil changes uniformly by 4.5 A in 1.0 s. (a) What is the induced emf? (b) If the hollow center of the torus is filled with an iron core, with $\mu = 3400\mu_0$, what is the induced emf?

21. (II) Calculate the inductance of an elongated rectangular circuit, such as a length of a two-wire ribbon cable (Fig. 33–23). The length of the circuit is L and its width is $a << L$. If the radius of the wire is much less than a, can it be neglected?

22. (II) Figure 33–24 shows a straight wire that carries a current I and a square loop of wire with one side oriented parallel to the straight wire a distance d away. The square has sides of length a. Calculate the mutual inductance of this system. [*Hint*: The magnetic field due to the straight wire through a slice of the square of width dx parallel to the wire is constant, so the flux through this slice is easily calculable. Integrate to find the total flux through the square.]

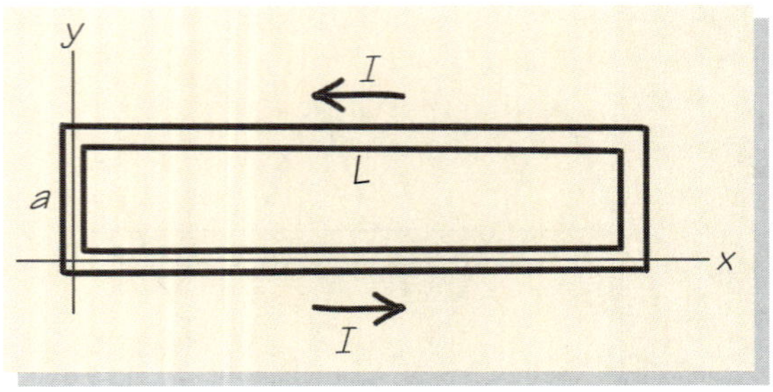

FIGURE 33–23 Problem 21.

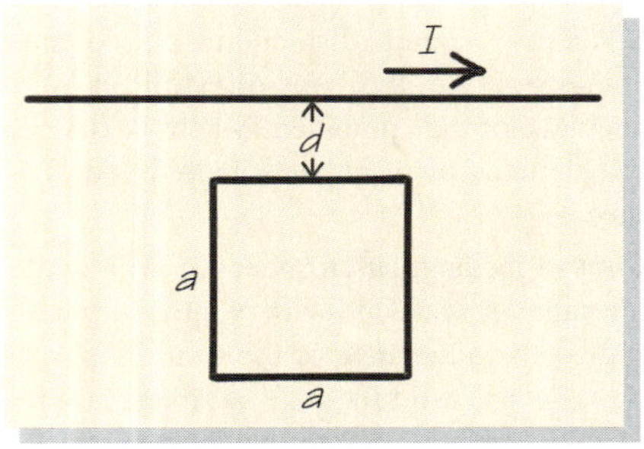

FIGURE 33–24 Problem 22.

23. (III) Calculate the mutual inductance of the two elongated rectangular circuits shown on Fig. 33–25. Assume that $L >> a, b$.

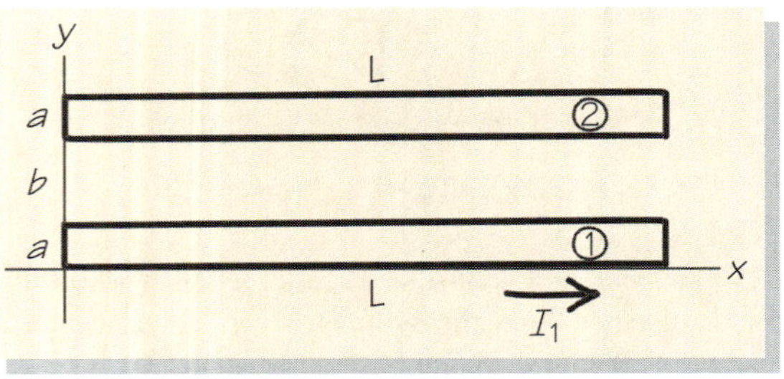

FIGURE 33–25 Problem 23.

33–2 *Energy in Inductors*

24. (I) A cylindrical solenoid of radius 2.0 cm is wound with a turn density of 1200 turns/m. It carries a current of 1.5 A. How much energy is stored per meter length of the solenoid?

25. (I) Consider an inductor with $L = 16$ H and an internal resistance of 0.1 Ω. We wish to use this inductor to store 0.1 MJ of energy. What is the rate at which energy is lost to Joule heating in this system? It is not practical to store large amounts of energy in large inductors unless the wire is superconducting.

26. (I) An inductor with $L = 0.49$ mH has an internal resistance small enough to be ignored. How much work would a battery have to do to increase the current through the inductor from 65 mA to 140 mA?

27. (I) A capacitor with $C = 0.02\ \mu$F has a charge of 15 μC. What is the equivalent steady current that should be carried by an inductor of $L = 20\ \mu$H if the inductor is to store the same amount of energy?

28. (I) A doorbell circuit contains a solenoid with 1000 turns of wire, a cross section of 10 cm^2, and a length of 10 cm. When the doorbell button is pushed, 0.2 A passes through the circuit. How much energy is contained within the solenoid at this time?

29. (II) A current with time dependence $I = I_0 e^{-\alpha t}$ passes through an inductor with $L = 2$ mH; $I_0 = 4.0$ A and $\alpha = 0.02$ s^{-1}. Compute the power expended in the inductor as a function of time.

30. (II) The voltage across an inductor with $L = 5.0$ mH is fixed at 12 V. The current is increased (a) from 0.0 A to 0.10 A, (b) from 0.10 A to 0.20 A, and (c) from 0.20 A to 0.30 A. What average power must be supplied from an external source in each step?

31. (II) Consider an inductor with $L = 1$ H and a capacitor with $C = 1$ F. (a) Compare the energy contained in the inductor when a current of 10 A flows through it with the energy in the capacitor if the charge is the amount of charge contained in the 10-A current, flowing for 1 s. (b) Repeat part (a) for a current of 1 mA.

32. (II) An electrical engineer constructs a cylindrical solenoid of area 8 cm^2 and length 25 cm from 150 m of thin wire. The wire will handle a maximum current of 50 mA. (a) What is the inductance of the solenoid? (b) How much energy can the inductor store?

33. (II) The inductance of a small superconducting solenoid is 8 H. The current is gradually increased from 0 A to 40 A. (a) How much energy is stored in the solenoid? (b) When the current reaches 40 A, the solenoid "quenches"; i.e., the wire of the solenoid loses its superconductive property because of the large magnetic field. The current decreases to zero rapidly, and the magnetic energy is dissipated in the liquid helium coolant. Given that the latent heat of vaporization of helium is 2.7×10^3 J/L, how much of the liquid helium coolant is evaporated during the quench?

33–3 *Energy in Magnetic Fields*

34. (I) The magnetic field in interstellar space has an approximate magnitude of 10^{-10} T. How much magnetic field energy is there in a spherical region around the Sun of radius 4 ly?

35. (I) The two circular pole pieces of a magnet are 63 cm in diameter and 21 cm apart. The magnetic field between them is 0.10 T. What is the magnetic energy stored in the field?

36. (I) An ideal cylindrical solenoid carrying a current of 55 mA has a winding density of 85 turns/cm. If the core is filled with iron, $\chi_m = 5500$, what is the energy density contained in the magnetic field within?

37. (II) A straight wire carries a current $I = 20$ A. Find the energy density in the surrounding magnetic field as a function of the distance r from the wire. At what distance from the wire does the energy density equal that of a parallel-plate capacitor with a charge of 10^{-7} C and a capacitance of 6.3×10^{-9} F, if the separation between the plates is 1.5 mm?

38. (II) (a) What is the energy density of the magnetic field outside a straight wire of radius a that carries a current I? (b) What is the total energy per unit length, due to that magnetic field, that is contained in a cylinder of radius R ($R > a$) centered about the wire?

39. (II) Consider a torus of radius R, wound with n turns per unit length of a wire that carries a current I. The cross section of

the torus forms a square with sides of length b; $b << R$. We know that the magnetic field inside the torus has the nearly constant value $B = \mu_0 nI$. Use this result and the two expressions related to the magnetic energy [Eqs. (33–13) and (33–15)] to show that, for this torus, $L = 2\pi\mu_0 n^2 Rb^2$.

40. (II) (a) What is the magnetic field energy density inside a straight wire of radius a that carries current I uniformly over its area? (b) What is the total magnetic field energy per unit length inside the wire?

41. (II) A coaxial cable consists of a wire 0.15 cm in diameter with a return path for the current in the shape of a very thin cylindrical conductor of diameter 0.8 cm. A current of 0.25 A flows through the cable. Calculate the magnetic energy per unit length of cable within the inner wire. [*Hint*: It will help to do Problem 40 first.]

33–4 RL Circuits

42. (I) You wish to make a circuit in which a resistor and an inductor are connected in series to a battery such that when the switch is closed, the current builds up to within 3% of its steady-state value in 10^{-4} s. You have a series of inductors with inductances ranging from 0.01 H to 0.1 H. You must acquire a set of resistors with what range of resistance?

43. (I) Show that the time constant L/R that characterizes RL circuits has dimensions of time.

44. (II) Consider the RL circuit of Fig. 33–10a; the switch is closed at time $t = 0$. For the circuit elements, $\mathscr{E} = 6$ V, $R = 3.3$ kΩ, and $L = 2.5$ mH. Using Eq. (33–19), find how much charge flows in the circuit during the first (a) 1 μs, (b) 1 ms, (c) 1 s.

45. (II) Show by direct substitution that Eq. (33–19) is a solution of Eq. (33–18).

46. (II) Consider a circuit in which a resistor and an inductor are connected in series to a battery. When the battery is suddenly shorted out so that only a closed RL circuit remains, the original current, V/R, decays to zero. Calculate the form of the current as a function of time by solving the equation $L\,dI/dt + RI = 0$—with a guess that the solution might be of the form $e^{-\alpha t}$.

47. (II) Consider the situation discussed in Problem 46. Calculate the total energy dissipated in the resistor from the time when the switch that shorts out the battery is thrown to the time $t = \infty$. Show that this is the energy stored in the inductor just before the switch is closed.

33–5 Oscillations in Circuits

48. (I) An electric oscillator consists of a parallel-plate capacitor and a long, cylindrical solenoid. If the resonant frequency of the oscillator is ω_0, what is the frequency of a similar oscillator in which both the capacitance and inductance are reduced by a factor of 2?

49. (I) You have an inductor with an inductance of 40 mH. Using it, you want to make a circuit with oscillations of frequency 20 Hz. What capacitor do you ask your roommate to pick up at the corner electronics store?

50. (II) Advanced electronic techniques utilize microscopic structures. Consider a single-turn solenoid in which the radius and the length of the solenoid are both of the order 10^{-6} m, and a parallel plate capacitor in which the plate separation and the radius of the plates are also of the 10^{-6} m. *Estimate* the order of magnitude of the frequency of oscillation of such a microscopic LC circuit.

51. (II) Two electric oscillators are made of exactly the same materials, but all the linear dimensions of the second circuit are ten times larger than the dimensions of the first circuit. Obtain the relation between (a) the undamped frequencies, (b) the damping factors, and (c) the damped frequencies of the two oscillators.

52. (II) An open circuit consists of a capacitor C and an inductor L connected in series. A charge q is placed on the capacitor, and the circuit is closed at time $t = 0$ by means of a switch. Find the maximum value of the current, as well as the times for which this maximum value occurs.

33–6 Damped Oscillations in Circuits

53. (I) An RLC circuit is composed of a resistor $R = 0.883$ Ω, an inductor $L = 1.75$ H, and a capacitor $C = 133$ pF, all arranged in series. What is the angular frequency of current oscillations in this circuit?

54. (I) An RLC circuit has $R = 1.5$ Ω, $L = 5$ mH, and $C = 25$ μF. (a) Find the damping factor and angular frequency. (b) If the resistance is variable, what value of R will give critical damping?

55. (I) Consider a series RLC circuit for which the initial capacitor charge is Q_0. If R is chosen such that there is critical damping, what is the instantaneous power consumption in the resistor? [*Hint*: Try the formula in Eq. (33–30).]

56. (II) Consider an RLC circuit at critical damping, with $L = 170$ mH. What is the value of R if the current decays by 10 percent in 0.02 s?

57. (II) Show that Eqs. (33–30) through (33–32) solve Eq. (33–27).

58. (II) Suppose that the values of R, L, and C in a series RLC circuit are such that $\omega'^2 < 0$. Assuming that the solution for the charge on the capacitor takes the form $Q = Q_1 e^{-\alpha_1 t} + Q_2 e^{-\alpha_2 t}$, find the values of α_1 and α_2.

59. (II) Consider the basic RLC circuit. By making an appropriate approximation of Eq. (33–32), show that when α is small compared to $\omega = 1/\sqrt{LC}$, the modified angular frequency ω' of the damped RLC circuit is $\omega' \simeq \omega - R^2\sqrt{C/L}/8L$. Find a similar relation for the periods of the undamped and slightly damped cases.

33–7 Energy in RLC Circuits

60. (II) Calculate the energy in an LC circuit, assuming that the initial conditions are such that the charge on the capacitor is $Q = Q_0 \cos(\omega t + \delta)$. Show that the energy is constant.

61. (II) A circuit consists of a capacitor of capacitance $C = 20$ nF connected in series with an inductor of inductance $L = 2 \times 10^{-5}$ H. If a charge of 30 nC is put on the capacitor, there is an oscillation in the circuit. (a) What is the maximum current that moves through this circuit? (b) Find the maximum energy within the inductor. (c) What is the ratio of the maximum energy in the inductor to the maximum energy in the capacitor?

62. (II) An LC circuit consists of a 4-mH inductor and a 200-μF capacitor. If the maximum energy stored in the circuit is 10^{-4} J, what are the maximum charge on the capacitor and the maximum current in the circuit? What are the minimum values?

63. (III) The 3-mF capacitor of an RLC circuit is initially charged to 30 μC. The 1.5-mH inductor has a very small resistance. At a particular instant, after 100 oscillations, the current through the inductor is zero while the capacitor is still charged to 5 μC. (a) What is the resistance of the circuit? (b) What are the energies of the circuit before and after the 100 oscillations? (c) Why are the two values of the energy in part (b) different? Where has the energy gone?

64. (III) Consider an *RLC* circuit. The energy is given by $E = LI^2/2 + Q^2/(2C)$. Show that the rate of change of this energy is equal to the power loss in the resistor (the ohmic heating power).

General Problems

65. (II) By considering the definition of inductance, show that, if the voltage V across an inductor changes with time, the total current passing through the inductor in that time is given by $I = (1/L)\int V\, dt$.

66. (II) Suppose that a square wave of voltage, as plotted in Fig. 33–26, is set across an inductor with $L = 0.01$ H. Use the result of Problem 65 to plot the current as a function of time.

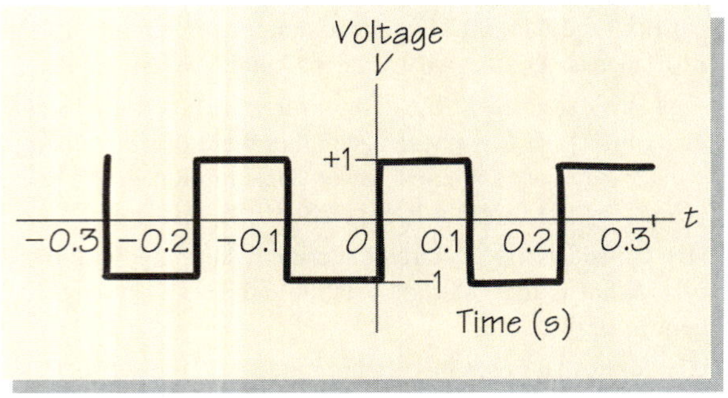

FIGURE 33–26 Problem 66.

67. (II) The switch in the circuit shown in Fig. 33–27 has been closed for a long time. (a) What is the current in each leg of the circuit? (b) When the switch is opened, the current in the inductor drops by a factor of 2 in 8 μs. What is the value of the inductance? (c) What is the current passing in each leg at 12 μs?

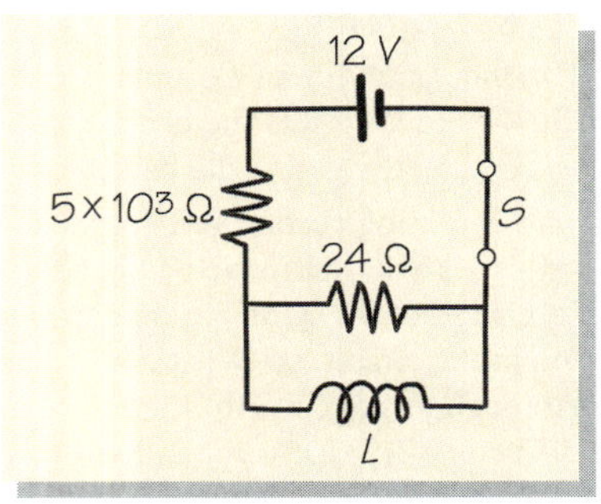

FIGURE 33–27 Problem 67.

68. (II) As a way of preventing arc formation between the terminals of a switch, a capacitor is connected to the two terminals (Fig. 33–28). What is the minimum capacitance of the capacitor if no voltage larger than 200 V is to be allowed in the circuit. [*Hint*: Assume that very little power can be dissipated during the time that the capacitor charges.]

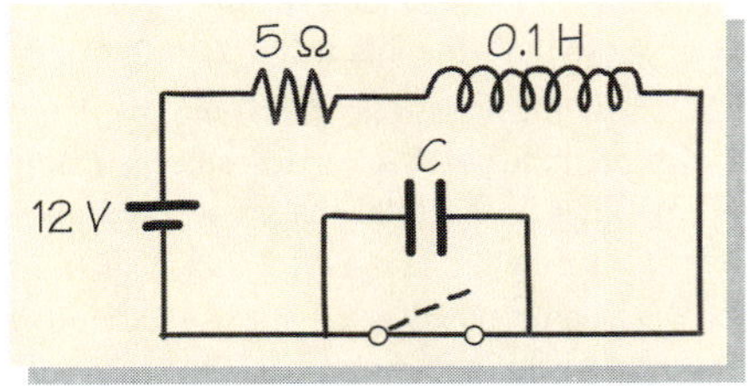

FIGURE 33–28 Problem 68.

69. (II) A coaxial cable has an inner, solid wire of radius r_1 and an outer, hollow wire of radius r_2. A current I flows through the inner wire and returns through the outer wire. Assuming that the cable is infinitely long, find the magnetic field energy per unit length. Include any field energy inside the inner wire and outside the outer wire.

70. (II) Molybdenum is paramagnetic, with a magnetic susceptibility of 1.2×10^{-4} at 300 K, which is about one-half its value at 20 K. Suppose that the self-inductance of a solenoid filled with molybdenum is $L = 0.1$ mH at 300 K. What is the fractional change in self-inductance between 300 K and 20 K?

71. (II) Consider a cavity uniformly filled with oscillating electric and magnetic fields. (a) Show that the ratio of the amplitude of these fields, E_0 and B_0, respectively, has the dimensions of $[\text{velocity}]^{-1}$. (b) For what value of this ratio is the magnetic energy density equal to the electric energy density?

72. (II) What are the currents in the three resistors of Fig. 33–29 immediately after the switch is closed? after a long time?

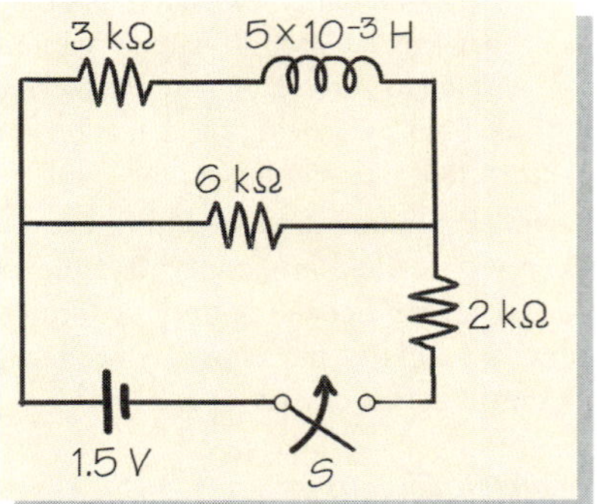

FIGURE 33–29 Problem 72.

73. (II) Two solenoids are wound on a common soft iron core (Fig. 33–30). Solenoid S_1 is connected in series to a battery and a variable resistor. Starting with the resistor set at *A* (low resistance), the sliding contact is moved to *B* (large resistance) and back to *A* again. Sketch the voltage *V* across solenoid S_2 while this is happening.

74. (II) The two identical coils in the circuit of Fig. 33–31 are placed close to each other, and their mutual inductance is 3 mH. Suppose that the switch has been closed for a long time and is then opened at $t = 0$. Calculate the current in the circuit at $t = 10$ ms.

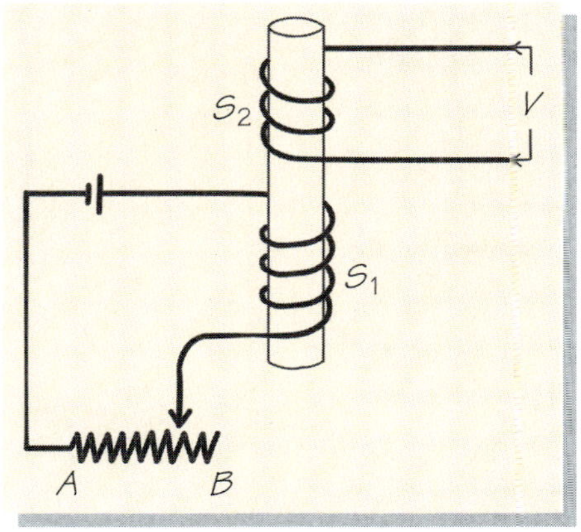

FIGURE 33–30 Problem 73.

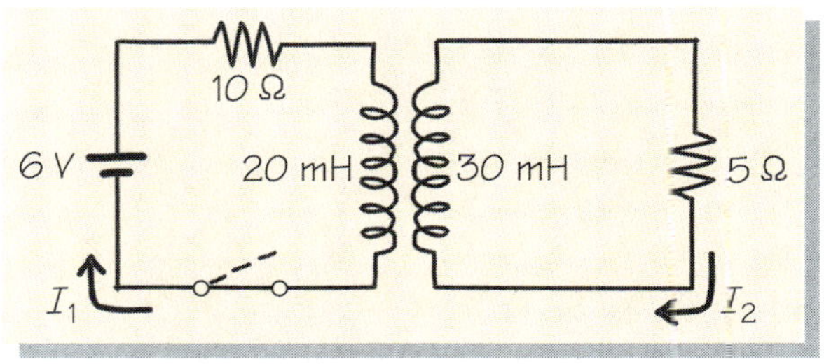

FIGURE 33–31 Problem 74.

75. (II) A ferromagnetic torus is part of a device to be used in a region where the magnetic permeability has the constant value $\mu = 2500\mu_0$. The torus has a circular cross section of 4 cm^2. Over its total length of approximately 35 cm, the torus is wrapped with 220 turns of wire. Immediately surrounding this winding is a secondary winding of 40 turns of (insulated) wire. What is the mutual inductance of the two windings? What is the role of the iron core, if any, in determining this mutual inductance?

76. (II) A torus of inner radius r_i and outer radius r_0 has a square cross section (Fig. 33–32). It is wound with N turns of wire that carries a current I. (a) Use Ampère's law to find the magnetic field inside the torus. (b) Calculate the magnetic energy density within the torus. (c) Integrate the magnetic energy density to find the total magnetic energy within the torus. (d) Use the formula $U_L = \frac{1}{2}LI^2$, Eq. (33–13), to compute the self-inductance of this torus.

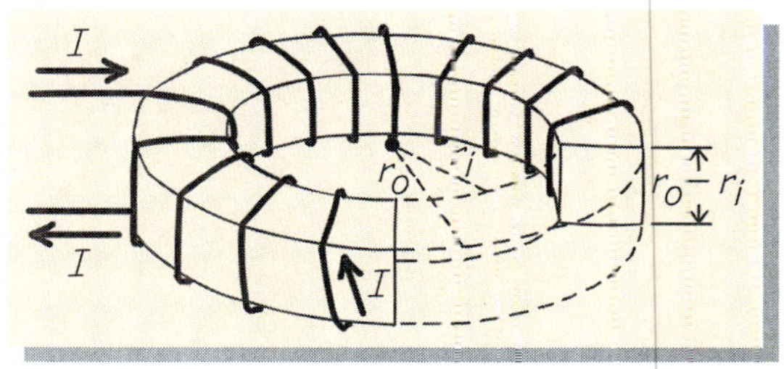

FIGURE 33–32 Problem 76.

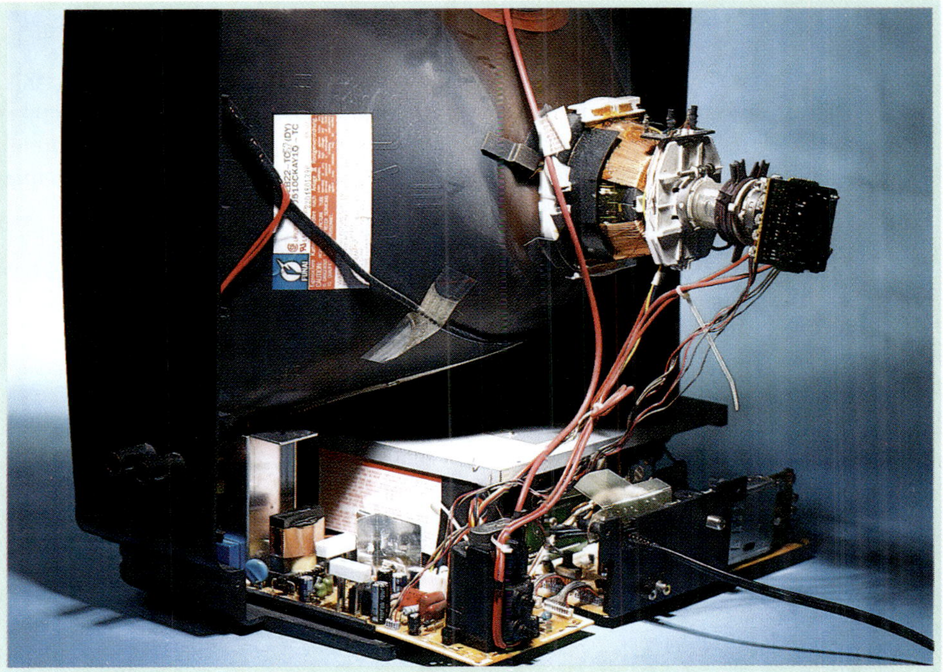

CHAPTER

34

Starting from an input alternating current with a voltage amplitude of 120 V, the transformer at the base of this television—the black box visible at the bottom—produces a voltage with an amplitude of 15,000 V. This potential is used to accelerate the electrons within the picture tube. The many circuits within the television use alternating currents in a variety of ways.

Alternating Currents

In Chapter 31, we learned that Faraday's law specifies how a changing magnetic flux induces an emf. This law implies that when a coil rotates in the presence of a magnet, an emf is induced that varies sinusoidally with time. The induced emf produces an alternating current (AC), which is a source of AC power. AC generators based on this principle convert the mechanical energy of falling water or the pressure of hot steam into electric currents. Such generators are the starting point for the delivery of electric power. Here, we shall see how we can vary the maximum voltage of a harmonically oscillating emf; this is an important element in the delivery of electric power. By including resistors, inductors, and capacitors in circuits with AC sources of emf, currents and voltages with new types of time-dependent behavior become possible. Such circuits exhibit the resonance phenomenon.

34–1 TRANSFORMERS

We described an alternating-current generator in Chapter 31 that was based on Faraday's law. Recall that an alternating current is characterized by harmonic (sine and cosine) time dependence, as are the other variables of the circuit, such as voltage.† We would like to be able to vary the maximum AC voltage because high or low voltages are useful in differing circumstances. For example, it is more economical to transport electric energy at high voltage, but high voltages are dangerous and in-

†"AC" stands for any kind of current or voltage that varies harmonically in time.

efficient in small appliances. Let's suppose that an AC generator produces an emf of the general form

$$\mathscr{E} = V_0 \sin(\omega t) \tag{34–1}$$

[see Eq. (31–14)]. The factor V_0 is the *voltage amplitude* of the source of emf, and this is the quantity we want to vary. In this section, we shall describe a device that can take an AC emf as an input and produce another AC emf with a *different* voltage amplitude. This device is called a **transformer**, and it can be constructed by using the principle of mutual inductance (Fig. 34–1).

APPLICATION
Transformers

Mutual inductance is discussed in Chapter 33.

Let's consider two fully linked ideal solenoids. We can arrange this in either of the two ways shown in Fig. 34–2. The solenoids have equal cross-sectional areas A but a different total number of turns: N_1 and N_2. Across the first coil (the *primary coil*) is an AC emf $\mathscr{E}_1$ with an amplitude V_1, as in Eq. (34–1):

$$\mathscr{E}_1 = V_1 \sin(\omega t). \tag{34–2}$$

We shall find the emf $\mathscr{E}_2$ across the second coil (the *secondary coil*) and show that $\mathscr{E}_2$ does indeed have an amplitude different from that of $\mathscr{E}_1$. Figure 34–3 shows the standard representation of this situation. Because $\mathscr{E}_1$ is time dependent, the current through coil 1 changes, and there is a changing magnetic flux through it. Faraday's law then implies that the total current I_1 in coil 1 [see Eq. (33–2)] is determined by the equation

$$\mathscr{E}_1 = -L\frac{dI_1}{dt}, \tag{34–3}$$

where L is the self-inductance of coil 1. We are not interested so much in the value of I_1 as we are in the fact that, at the same time, an emf $\mathscr{E}_2$ is induced across coil 2. This emf is induced because the changing current in coil 1 produces a changing magnetic flux through coil 2. By definition, $\mathscr{E}_2$ depends on the mutual inductance, M:

$$\mathscr{E}_2 = -M\frac{dI_1}{dt}. \tag{34–4}$$

FIGURE 34–1 A variac transformer produces an output AC with a continuum of amplitudes for a given input AC.

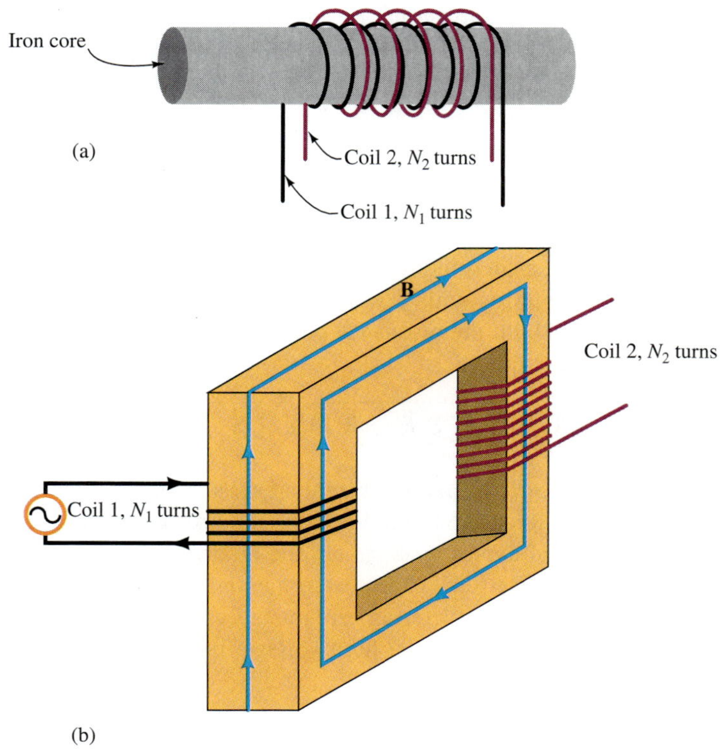

FIGURE 34–2 Two methods for creating fully linked coils: (a) one coil tightly wound over the other; (b) two coils wrapped around a common core of ferromagnetic material, which has the property of keeping the magnetic field lines within it.

If we substitute for dI_1/dt from Eq. (34–3), we find

$$\mathscr{E}_2 = M\frac{\mathscr{E}_1}{L};$$

$$\frac{\mathscr{E}_2}{\mathscr{E}_1} = \frac{M}{L}. \tag{34–5}$$

Equation (34–5) contains important information: The ratio M/L is a constant, and therefore $\mathscr{E}_2$ *has the same harmonic time dependence as* $\mathscr{E}_1$. If the angular frequency of the current in the primary coil is ω, as in Eq. (34–1), so is that of the current induced in the secondary coil.

Let's now use the fact that the two coils are fully linked ideal solenoids. In this case, we know both M and L. The mutual inductance of the two coils is a special case of the mutual inductance of a solenoid and a ring, Eq. (33–9); from that equation, we have

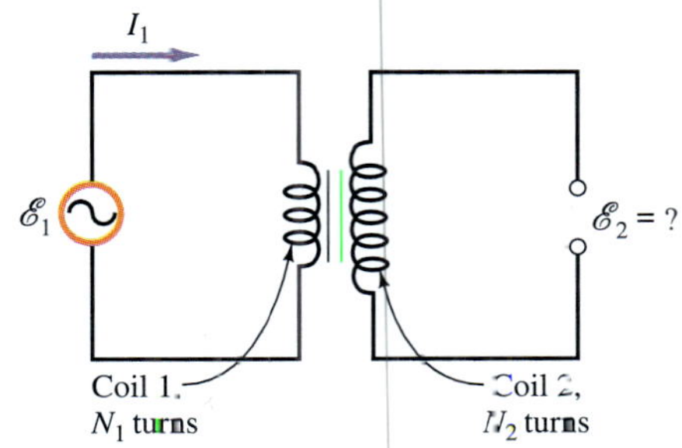

FIGURE 34–3 Schematic circuit-diagram symbol for two fully linked coils. When one coil is in the same circuit with an AC source of emf, this circuit acts as a transformer.

$$M = \mu_0 A\frac{N_1}{\ell_1}N_2. \tag{34–6}$$

From Eq. (33–8), the self-inductance of coil 1 is

$$L = \mu_0 A\frac{N_1^2}{\ell_1}. \tag{34–7}$$

Substituting these results into Eq. (34–5), we have

$$\frac{\mathscr{E}_2}{\mathscr{E}_1} = \frac{\mu_0 AN_1N_2}{\ell_1}\frac{\ell_1}{\mu_0 AN_1^2};$$

The ratio of the emfs in the two coils of a transformer is equal to the ratio of the number of turns in the coils.

$$\frac{\mathscr{E}_2}{\mathscr{E}_1} = \frac{N_2}{N_1}. \tag{34–8}$$

Because the time dependence of the AC is identical in $\mathscr{E}_1$ and $\mathscr{E}_2$, Eq. (34–8) relates the voltage *amplitudes* V_1 and V_2 (the coefficients of the sinusoidal time dependence of the emfs) in the two coils:

$$\frac{V_2}{V_1} = \frac{N_2}{N_1}. \tag{34–9}$$

The transformer is a tool for manipulating these voltage amplitudes. When $N_2 > N_1$, the transformer is a *step-up transformer*, and the voltage amplitude in the secondary coil is greater than that in the primary coil. When $N_2 < N_1$, the transformer is a *step-down transformer*; the voltage amplitude in the secondary coil is smaller than that in the primary coil. Note that the terms primary and secondary do not imply any fundamental distinction between the two coils.

How are the currents in the two coils related? Energy must be conserved when the transformer operates. If the transformer is constructed efficiently—resistance is reduced to a minimum and there is no loss to Joule heating—then the rate of energy flow, given by the product $I\mathscr{E}$, must be equal in the two transformer coils. We find

$$I_1\mathscr{E}_1 = I_2\mathscr{E}_2.$$

We substitute in Eq. (34–8) to find that

The ratio of the currents in the two coils of a transformer is equal to the inverse of the ratio of the number of turns in the coils.

$$\frac{I_1}{I_2} = \frac{N_2}{N_1} \tag{34–10}$$

In other words, the current amplitude in the secondary coil of a step-up transformer is decreased, while it is increased in the secondary coil of a step-down transformer.

(a)

(b)

FIGURE 34–4 (a) A receiving substation that transforms high voltages from a generating station to lower voltages that are used locally. (b) Power transmission lines carry AC at such high voltages that great care must be taken to keep them isolated electrically from their surroundings. If you stand under one of these lines, you can hear the electrical breakdown in the air that surrounds the wires.

EXAMPLE 34–1 A step-down transformer has 5000 turns in the primary coil, which handles an AC current with voltage amplitude $V_1 = 20{,}000$ V, and 220 turns in the secondary coil. If the current amplitude desired in the secondary coil is 100 A, what is the maximum power that must be delivered by the primary coil to the secondary coil?

Solution: We must find the power P_2 delivered to the secondary coil. Power is related to the unknown voltage amplitude V_2 and to the known current I_2 by the relation $P_2 = I_2V_2$. We can find V_2 from V_1 and the number of coil turns. V_2 is determined by Eq. (34–9):

$$V_2 = \frac{N_2}{N_1}V_1 = \frac{220}{5000}(20{,}000\ \text{V}) = 880\ \text{V}.$$

If the maximum current carried by the secondary coil is 100 A, the maximum power that it carries is $P_2 = (100\ \text{A})(880\ \text{V}) = 88$ kW. If the transformer is constructed efficiently, this is equal to the maximum power that can be carried in the primary coil.

Power Transmission

Low voltages are more practical for local use because they are more easily insulated against breakdown than are high voltages. Conversely, it is far more efficient to transmit electric energy at *high* voltages from a generating plant (Fig. 34–4a) to the places where it is to be used locally (Fig. 34–4b). Transformers allow us to reconcile the different voltage requirements of long-distance transmission and local use.

Let's demonstrate that it is more efficient to transmit electric power at high voltages than at low voltages, whether AC or DC. A power transmission line delivers energy for local use at a certain rate, P, and at a certain voltage, V. This means that a current passes through the transmission line given by $P = IV$. The transmission line will itself have a certain resistance, R, and the power dissipated in the line will be

$$P_{\text{lost}} = I^2R = \frac{P^2R}{V^2}. \tag{34–11}$$

A measure of the efficiency of transmission is the ratio of the power delivered to the power lost in Joule heating. According to Eq. (34–11), this ratio is

$$\frac{P}{P_{\text{lost}}} = \frac{V^2}{PR}. \tag{34–12}$$

This ratio increases rapidly as V increases. To take a realistic example, a transmission line delivering power $P = 1$ MW might have a total resistance of 10 Ω. If the power were delivered at 110 V, the ratio in Eq. (34–12) would be intolerably low: $(110\ \text{V})^2/(1\ \text{MW})(10\ \Omega) = 1.2 \times 10^{-3}$. In this case, 99.9 percent of your electricity bill would be for power lost in the wires! If, however, the power were delivered at 500,000 V, the ratio would be $(500{,}000\ \text{V})^2/(1\ \text{MW})(10\ \Omega) = 2.5 \times 10^4$, and most of the electric energy produced would be delivered.

34–2 SINGLE ELEMENTS IN AC CIRCUITS

Let's examine the effects of placing an AC source of emf, for which $\mathcal{E} = V_0 \sin(\omega t)$, in circuits that contain only single elements of resistance, capacitance, and inductance.

Resistive Circuit

We begin with the resistive circuit shown in Fig. 34–5a. The loop rule for the potential change around the circuit is

$$V_0 \sin(\omega t) - IR = 0. \tag{34–13}$$

The voltage across the resistor is then $V_R = IR = V_0 \sin(\omega t)$, and the current I through the resistor is

$$I = \frac{V_R}{R} = \frac{V_0 \sin(\omega t)}{R}. \qquad (34\text{–}14)$$

The current through the resistor and the voltage across the resistor have the same sinusoidal time dependence (Fig. 34–5b). We say that they are *in phase*. The amplitude of I is $I_{max} = V_0/R$.

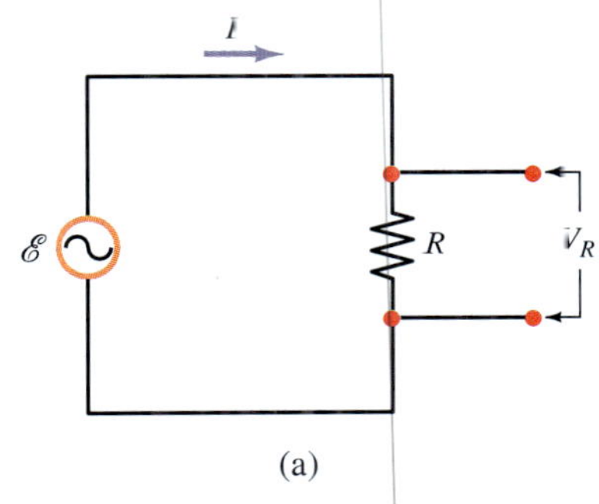

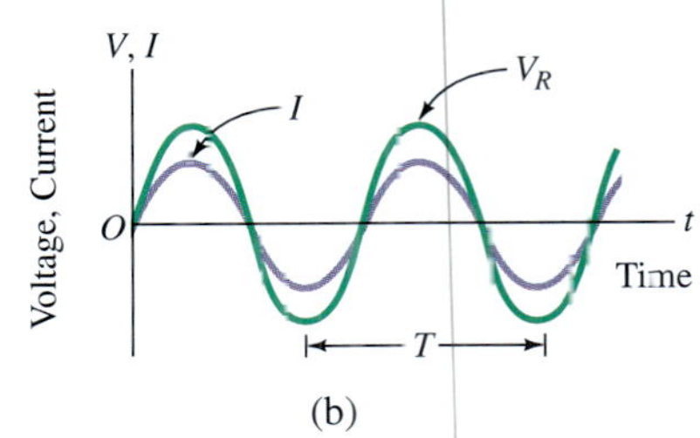

FIGURE 34–5 (a) A resistor connected in series with an AC source of emf. (b) The voltage across the resistor and the current through it are in phase.

Capacitive Circuit

We now place the AC emf across a pure capacitance in the circuit (Fig. 34–6a). We apply the loop rule and calculate the current in the circuit and the voltage across the capacitor. The loop rule gives

$$V_0 \sin(\omega t) - \frac{Q}{C} = 0. \qquad (34\text{–}15)$$

The voltage across the capacitor is simply the driving emf $V_C = V_0 \sin(\omega t)$. To find the current, we first find the charge Q from Eq. (34–15):

$$Q = CV_0 \sin(\omega t). \qquad (34\text{–}16)$$

The current I is then the time derivative of the charge:

$$I = \frac{dQ}{dt} = \omega C V_0 \cos(\omega t). \qquad (34\text{–}17)$$

With the identity $\sin[\theta + (\pi/2)] = \cos\theta$, Eq. (34–17) becomes

$$I = \omega C V_0 \sin\left(\omega t + \frac{\pi}{2}\right). \qquad (34\text{–}18)$$

The maximum value of the current in the circuit is $\omega C V_0$. If we relate the maximum voltage across the capacitor to the maximum current in the circuit, we find that

$$I_{max} = \omega C V_0. \qquad (34\text{–}19)$$

We can compare this equation to a similar equation for the resistive circuit, for which the corresponding relation was $I_{max} = V_0/R$. The *effective resistance* for a capacitive circuit is called the **capacitive reactance**, X_C, defined by

$$X_C \equiv \frac{1}{\omega C}. \qquad (34\text{–}20)$$

Equation (34–19) now takes the form

$$I_{max} = \frac{V_0}{X_C}. \qquad (34\text{–}21)$$

The capacitive reactance has units of ohms.

In Fig. 34–6b, we plot the current I and the voltage V_C versus time. Note that the voltage across the capacitor is zero at time $t = 0$, but the current in the circuit is at a maximum. The phase of the current differs by $\pi/2$ rad (90°) from that of the voltage. We say that the current *leads* the voltage by a phase $\pi/2$.

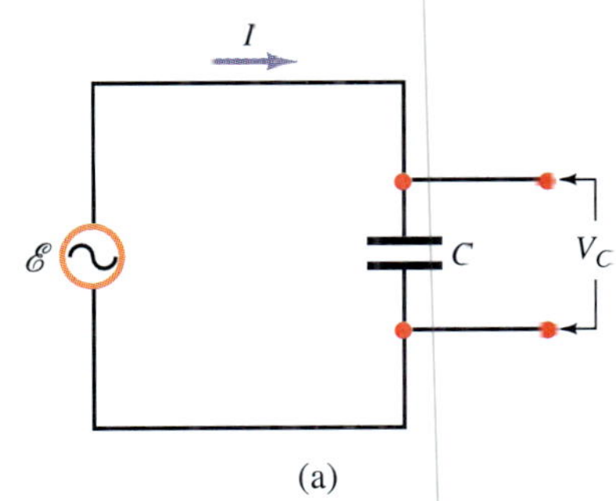

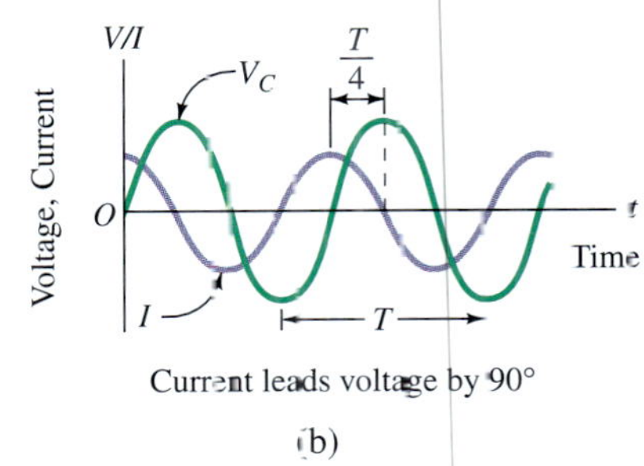

FIGURE 34–6 (a) A capacitor connected in series with an AC source of emf. (b) The current in the circuit leads the voltage across the capacitor by 90°.

EXAMPLE 34–2 The circuit shown in Fig. 34–6a has an emf given by $\mathcal{E} = V_0 \sin(\omega t)$, with $V_0 = 6.0$ V, and a capacitance $C = 1.0\ \mu$F. (a) What are the peak currents for frequencies of exactly 60 Hz and 6 MHz? (b) What are the currents I and voltages V_C at time 2.0 ms for the 60-Hz frequency?

Solution: (a) The unknown peak current is found from Eq. (34–19) and depends on the angular frequency. The angular frequencies ω for the two cases are found from $\omega = 2\pi f$ and are given by 2π (60 Hz) $= 2\pi$ (60 s^{-1}) $=$ 377 rad/s and 2π ($6 \times 10^6\ s^{-1}$) $= 3.77 \times 10^7$ rad/s, respectively. Thus, from Eq. (34–19),

$$\text{for 60 Hz:} \qquad I_{max} = (377 \text{ rad/s})(1.0 \times 10^{-6} \text{ F})(6 \text{ V}) = 2.3 \text{ mA};$$

$$\text{for 6 MHz:} \qquad I_{max} = (3.77 \times 10^{7} \text{ rad/s})(1.0 \times 10^{-6} \text{ F})(6 \text{ V}) = 230 \text{ A}.$$

The higher frequency makes a significant difference in the maximum current. For the higher frequency, 6 MHz, the capacitive reactive is so small ($X_C = 1/\omega C = 0.027\ \Omega$) that the circuit has almost no resistance to current flow.

(b) We want to specify the time in the full expressions for current and voltage, Eqs. (34–18) and $V_0 \sin(\omega t)$, respectively. We can substitute Eq. (34–19) into Eq. (34–18). For $f = 60$ Hz, we have

$$I = I_0 \sin(\omega t + \phi)$$

and

$$V = V_0 \sin(\omega t),$$

where $I_0 = 2.3$ mA, $\phi = \pi/2$ rad, $V_0 = 6.0$ V, and $\omega = 377$ rad/s. For $t = 2.0$ ms, the current is

$$I = (2.3 \text{ mA}) \sin\left[(377 \text{ rad/s})(0.0020 \text{ s}) + \frac{\pi}{2} \text{ rad}\right] = 1.7 \text{ mA},$$

and the voltage is

$$V_C = (6.0 \text{ V}) \sin[(377 \text{ rad/s})(0.0020 \text{ s})] = 4.1 \text{ V}.$$

By $t = 2.0$ ms, the current is coming down from its peak toward zero, while the voltage is rising toward its peak. At $t = 2.0$ ms, they are both about 70 percent of their peak values.

Inductive Circuit

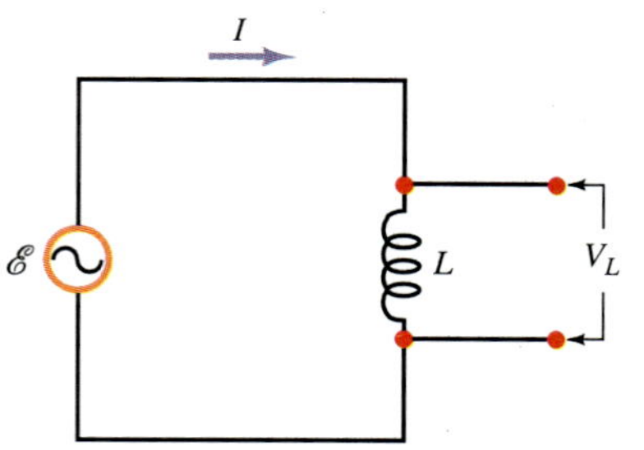

(a)

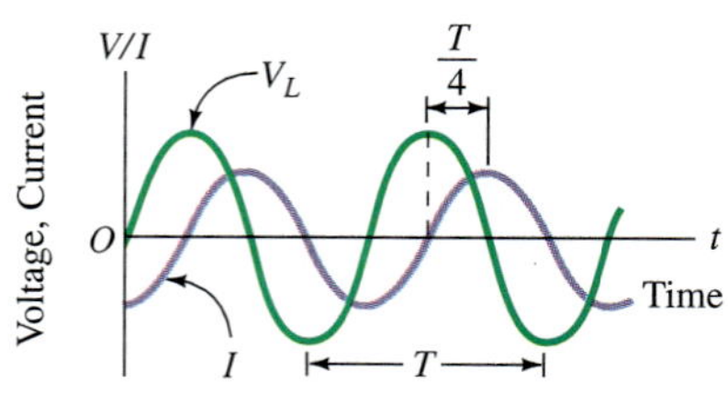

Current lags voltage by 90°

(b)

FIGURE 34–7 (a) An inductor connected in series with an AC source of emf. (b) The current in the circuit lags the voltage across the inductor by 90°.

Now let's replace the capacitor with an inductor in the circuit that has an AC emf (Fig. 34–7a). Repeating the previous procedure, we first apply the loop rule to the potentials around the circuit:

$$V_0 \sin(\omega t) - L\frac{dI}{dt} = 0. \tag{34–22}$$

The voltage drop across the inductor must be the emf $V_L = V_0 \sin(\omega t)$. To find the current through the inductor, we must solve Eq. (34–22) for the current I. We rewrite the equation as

$$dI = \frac{V_0}{L} \sin(\omega t)\, dt.$$

We find the current from the integral of this equation:

$$I = -\frac{V_0}{\omega L} \cos(\omega t).$$

We use the trigonometric identity $\cos\theta = -\sin[\theta - (\pi/2)]$ (Appendix IV–4) to rewrite this equation as

$$I = \frac{V_0}{\omega L} \sin\left[\omega t - \left(\frac{\pi}{2}\right)\right]. \tag{34–23}$$

The maximum current through the inductor is

$$I_{max} = \frac{V_0}{\omega L}. \tag{34–24}$$

If we compare this equation with the similar one from the purely resistive circuit, $I_{max} = V_0/R$, we see that the effective resistance for an inductive circuit is ωL. We call this the **inductive reactance**, defined by

$$X_L \equiv \omega L. \tag{34–25}$$

The inductive reactance has units of ohms.

For an inductive circuit, the effective resistance to current flow *increases* at higher frequencies. This is physically reasonable because inductors react to *oppose* any change in the current flow through them. A higher frequency means that the voltage, and therefore the current, is changing more rapidly. A current is induced to oppose the change.

We plot the current and voltage of the inductor versus time in Fig. 34–7b. As for the capacitive circuit, one sinusoidal curve is displaced from the other by a quarter cycle, although the role of the current and voltage curves is reversed in the two cases. This time the current *lags* the voltage.

EXAMPLE 34–3 Use the parameters of Example 34–2, but replace the capacitor with an inductor of inductance $L = 1.00$ mH. Calculate the inductive reactances.

Solution: We found in Example 34–2 that the angular frequencies are 377 rad/s and 3.77×10^7 rad/s, respectively. The inductive reactances are determined from Eq. (34–25):

for 60 Hz: $X_L = \omega L = (377 \text{ rad/s})(1.00 \times 10^{-3} \text{ H}) = 0.377\ \Omega;$

for 6 MHz: $X_L = \omega L = (3.77 \times 10^7 \text{ rad/s})(1.00 \times 10^{-3} \text{ H})$
$= 3.77 \times 10^4\ \Omega.$

As expected, the resistance to current flow increases dramatically for the higher frequency—the opposite behavior from that of the capacitive circuit.

Some Mathematical Devices

Two techniques simplify the treatment of circuits with time dependence. The first involves **phasors**, which make it easier to follow phases. The second, **complex analysis**, is a powerful tool that simplifies all aspects of the problem.

Phasors. A phasor is a vector that can be associated with any harmonic function

$$f(t) = f_0 \sin(\omega t + \phi). \tag{34–26}$$

The phasor lies in the xy-plane with its tail fixed at the origin. The length of the phasor is the function amplitude, f_0. The time dependence is described by a counterclockwise rotation of the phasor with angular speed ω. The function $f(t)$ is the instantaneous projection of the phasor on the y-axis (Fig. 34–8). Thus, for example, the function $V(t) = V_0 \sin(\omega t)$ has a phasor that starts at $t = 0$ aligned with the positive x-axis. As time increases and the phasor rotates counterclockwise, the y-component of the phasor increases until it reaches a maximum when $\omega t = \pi/2$. The y-component then decreases as the phasor moves into the second quadrant.

PROBLEM SOLVING
Phasors

By comparing the phasors of different harmonic functions that appear in an AC problem, we learn about the relative phases of these functions; that is, which function leads or which function lags. To see how this works, let's apply it to the purely inductive AC circuit of Fig. 34–7. The input voltage—and hence the voltage V_L across the inductor—has the form $V_0 \sin(\omega t)$, whereas the current I_L takes the form $I_{max} \sin(\omega t - 90°)$ according to Eq. (34–23). The phasors for these two quantities (Fig. 34–9a) rotate in the counterclockwise sense as time advances; the phasor for the voltage is always ahead of the phasor for the current. In this diagram, the idea that the voltage *leads* the current for the inductive circuit is vivid and easy to understand. We can easily perform the same exercise for the purely capacitive and purely resistive circuits; the phasor diagrams are in Fig. 34–9b and 34–9c, respectively. Again, we see the phase

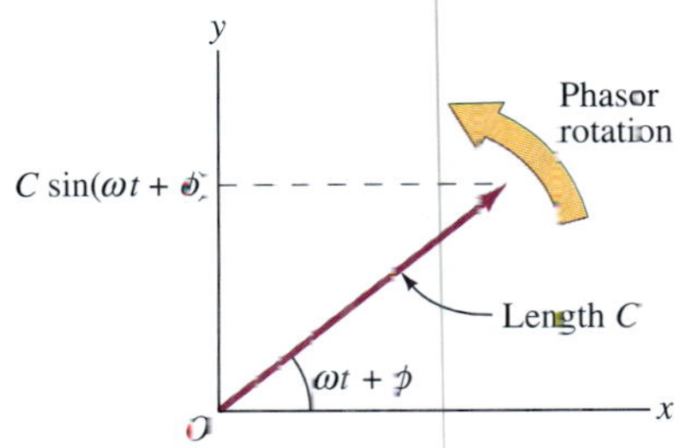

FIGURE 34–8 The projection of the phasor on the y-axis gives the value of the associated harmonic function.

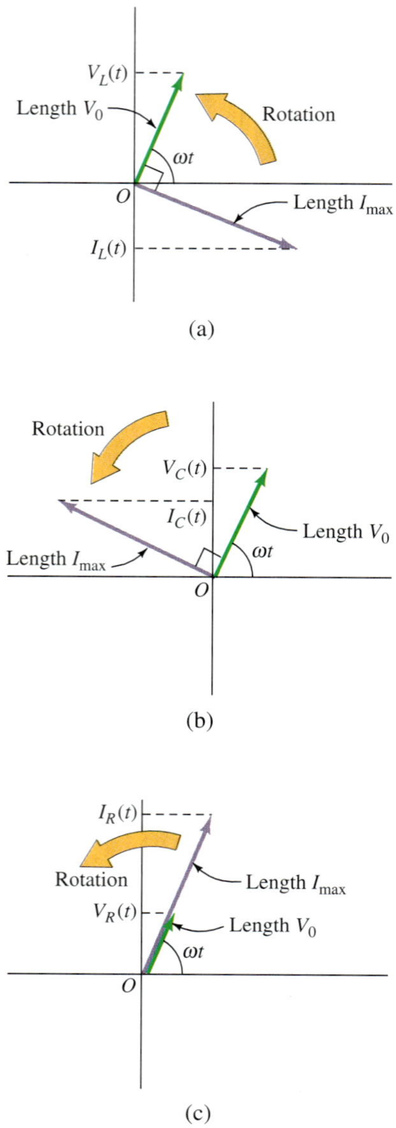

FIGURE 34–9 Phasors associated with the voltage across and current through (a) the inductor; (b) the capacitor; and (c) the resistor of an *RLC* circuit.

PROBLEM SOLVING

Complex analysis

The driven harmonic oscillator is discussed in Chapter 13.

relationship directly: The current in the purely capacitive AC circuit leads the voltage across the capacitor, and the corresponding current and voltage are in phase in the purely resistive circuit.

When elements are combined in certain AC circuits, the phasors combine as vectors. This is one way of treating the series *RLC* circuit to be discussed in Section 34–3.

***Complex Analysis.** Complex analysis is a powerful technique. Among other things, it is the working electrical engineer's basic tool for the analysis of circuits with time dependence. It depends on two ideas. The first is a *complex variable* z that can represent two physical quantities through the two variables x and y. This is accomplished through the use of $i \equiv \sqrt{-1}$; z is then given the form $z = x + iy$. x and y are extracted from z by taking the real and imaginary part of z—just the coefficients of i^0 and i^1, respectively. (Higher powers of i, which may appear through algebraic manipulation, reduce; for example, $i^2 = -1$.) In the case of circuits, the two physical quantities of interest are the amplitude and the phase. The second idea upon which it depends is one of the most remarkable relations in mathematics:

$$e^{i\vartheta} = \cos\vartheta + i\sin\vartheta.$$

(This result follows from the series expansions of sines, cosines, and the exponential.) Using this identity, we can write an input voltage like that of Eq. (34–1) as

$$V_0 \sin(\omega t) = \text{Im}(V_0 e^{i\omega t}),$$

where Im(z) is the imaginary part of z. We then represent all oscillating functions—currents, or potentials across individual elements—as exponentials rather than as sines or cosines, including possible phases. At the end, we take the appropriate real or imaginary part.

What is the advantage of this procedure? The answer is simple: Upon differentiation, exponentials remain exponentials. Therefore, the differential equations that describe the circuit behavior contain overall powers of exponentials that ultimately cancel. These differential equations reduce easily to algebraic equations—a much simpler prospect to deal with. Problems 83 to 87 provide some sampling of complex analysis.

34–3 AC IN SERIES *RLC* CIRCUITS

In mechanics, a *driven harmonic oscillator* is a device in which a harmonic external force acts on (for example) a mass fixed to a spring. If the driving force has been acting for some time, the mass has no choice but to move with the angular frequency ω of the force, even though the mass is attached to the spring and undergoes some damping. This system illustrates the important physical phenomenon of *resonance*, characterized by a large amplitude when the driving frequency ω is near the natural frequency ω_0.[†]

In Chapter 33, we noted the similarities between the mass–spring system and series *RLC* circuits without a driving term. If we add an AC source of emf to a series *RLC* circuit, the analogy between the mechanical system and the circuit continues to hold. In particular, *such a circuit exhibits resonant behavior.* Figure 34–10 illustrates

[†]When no confusion is possible, we use the term "frequency" rather than "angular frequency" for ω.

our driven circuit, and applying the loop rule for the potential changes around the circuit gives

$$V_0 \sin(\omega t) - L\frac{dI}{dt} - \frac{Q}{C} - IR = 0. \tag{34–27}$$

Because $I = dQ/dt$, we can reexpress this result in terms of the single variable Q, the charge on the capacitor:

$$V_0 \sin(\omega t) - L\frac{d^2Q}{dt^2} - R\frac{dQ}{dt} - \frac{Q}{C} = 0. \tag{34–28}$$

The unknown quantity in this differential equation is Q. Once we solve for Q, differentiation with respect to time will give the current. We can then find the voltage drops across the various circuit elements.

Equation (34–28) can be compared to Eq. (13–58), which expresses Newton's second law for a driven harmonic oscillator with damping (Fig. 34–11). It is rewritten here in the form

$$F_0 \sin(\omega t) - m\frac{d^2x}{dt^2} - b\frac{dx}{dt} - kx = 0, \tag{34–29}$$

where the first term is the driving force, with amplitude F_0. This differential equation for the position x as a function of time has a long-time solution in which *the position of the mass oscillates with the angular frequency of the driving force* (see Chapter 13). This solution is given by Eq. (13–59); it is convenient to shift a phase of that solution and write it as

$$x = -A\cos(\omega t - \phi). \tag{34–30}$$

As we described in Chapter 13, not only is the frequency determined, but *the amplitude A and phase ϕ are also determined*. We find their values by direct substitution into the differential equation, with the results

$$A = \frac{F_0}{\sqrt{m^2(\omega^2 - \omega_0^2)^2 + b^2\omega^2}} \tag{34–31}$$

and

$$\tan\phi = \frac{1}{b}\left(\omega m - \frac{k}{\omega}\right). \tag{34–32}$$

Here, ω_0 is the natural frequency of the oscillator, given by $\omega_0 = \sqrt{k/m}$.

The force and the position are both harmonic with the same frequency, but they are out of phase. For example, the function $\sin(\omega t)$—proportional to the force—rises

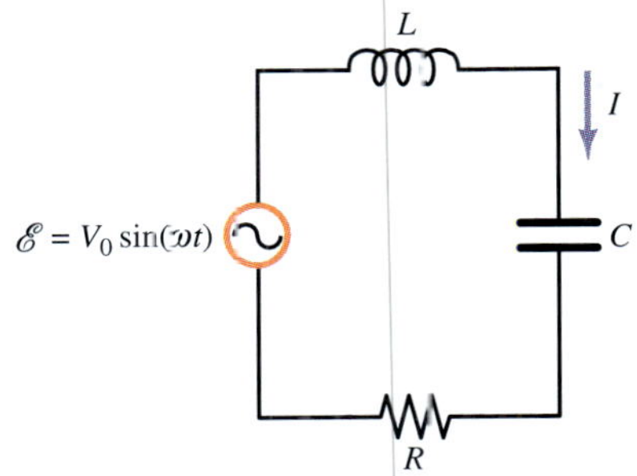

FIGURE 34–10 An *RLC* circuit is driven by an AC emf.

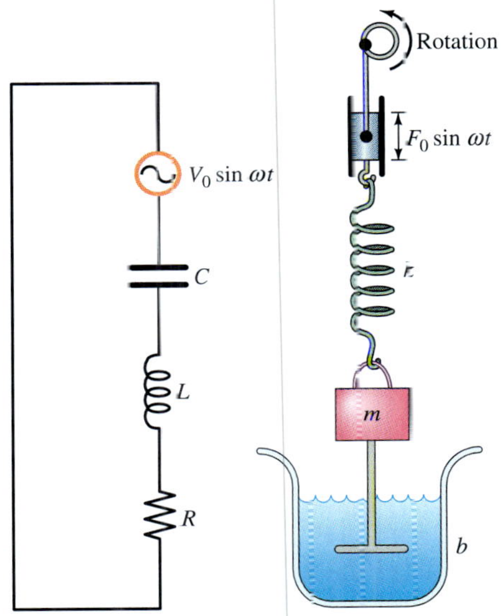

FIGURE 34–11 The *RLC* circuit and a damped mass-spring system are analogous.

TABLE 34–1 Analogy Between Driven *RLC* Circuits and Driven Spring Motion

	Circuit	*Mass–spring*
Variable	Charge Q	Position x
Coefficient of variable	$\frac{1}{C}$	k
Coefficient of $\frac{d(\text{variable})}{dt}$	R	b
Coefficient of $\frac{d^2(\text{variable})}{dt^2}$	L	m
Driving term	$V_0 \sin(\omega t)$	$F_0 \sin(\omega t)$
Natural frequency	$\frac{1}{\sqrt{LC}}$	$\sqrt{\frac{k}{m}}$

from zero at $t = 0$, whereas the function $\cos(\omega t - \phi)$—proportional to the position—rises from zero when its argument $\omega t - \phi$ is $-90° = -\pi/2$ rad, which occurs when $t = (\phi - \pi/2)/\omega$.

In order to solve Eq. (34–27) for the charge on the capacitor, we need only make the formal substitutions of Table 34–1, relating the parameters of the harmonic oscillator to the circuit parameters. The solution for the charge is thus

$$Q = -Q_{\text{max}} \cos(\omega t - \phi), \tag{34–33}$$

where

$$Q_{\text{max}} = \frac{V_0}{\sqrt{L^2(\omega^2 - \omega_0^2)^2 + R^2\omega^2}} \tag{34–34}$$

and

$$\tan \phi = \frac{1}{R}\left(\omega L - \frac{1}{\omega C}\right). \tag{34–35}$$

In this case, the natural frequency ω_0 of the circuit is the frequency of the pure LC circuit (no damping term, $R = 0$) from Eq. (33–23):

$$\omega_0 = \frac{1}{\sqrt{LC}}. \tag{34–36}$$

We can easily find the current in the circuit once we have the charge on the capacitor; we simply use $I = \text{d}Q/\text{d}t$. Differentiation of Eq. (34–33) gives

$$I = I_{\text{max}} \sin(\omega t - \phi), \tag{34–37}$$

where

$$I_{\text{max}} = Q_{\text{max}}\omega = \frac{\omega V_0}{\sqrt{L^2(\omega^2 - \omega_0^2)^2 + R^2\omega^2}}. \tag{34–38}$$

All the results given here reduce to the cases we treated in Section 34–2, in which only one circuit element is present at a time. We need only replace the values of L, R, or $1/C$ by zero, as appropriate (keeping in mind that $\omega_0 = 1/\sqrt{LC}$).

Impedance

We have already defined the reactances $X_L = \omega L$ and $X_C = 1/\omega C$ [Eqs. (34–20) and (34–25)]. They enter immediately into the phase that appears in the driven RLC circuit according to

$$\tan \phi = \frac{1}{R}(X_L - X_C). \tag{34–39}$$

Impedance

The reactances play the role of an effective resistance for the single element circuits. As we shall see, the effective resistance of the RLC circuit is the **impedance**, Z, defined by

$$Z \equiv \sqrt{\left(\omega L - \frac{1}{\omega C}\right)^2 + R^2} = \sqrt{(X_L - X_C)^2 + R^2}. \tag{34–40}$$

The impedance has units of ohms. Note that a cable with impedance generally has negligible resistance so, unlike resistance, impedance is *independent of length*.

To see how the impedance plays the role of a resistance, let's express the current in terms of it. It is a matter of a little algebra (see Problem 36) to show that, in terms of these quantities, Eq. (34–34) becomes $Q_{\text{max}} = V_0/\omega Z$ and hence, with $I_{\text{max}} = \omega Q_{\text{max}}$, we have $I_{\text{max}} = V_0/Z$. In other words,

$$I = I_{\text{max}} \sin(\omega t - \phi) = \frac{V_0 \sin(\omega t - \phi)}{Z}. \tag{34–41}$$

The current takes the form of an AC emf divided by the impedance. This equation is analogous to the DC equation $I = V/R$. *Impedance thus plays the role of resistance in an AC circuit.*

In contrast to the resistance, the impedance depends on the frequency; we can understand this on physical grounds. Inductance opposes a change in current, and larger values of angular frequency mean more rapid changes in the current. However, inductance has no effect when static potentials, corresponding to $\omega \to 0$, are involved. These properties are reflected in the frequency dependence of $X_L = \omega L$. A capacitor has just the opposite properties: No constant current can pass through a capacitor, but the capacitor has little effect when the current changes so rapidly that little charge can accumulate. These properties are reflected in the frequency dependence of $X_C = 1/\omega C$.

Recall from Section 28–5 that a capacitor does not allow a steady current to pass; from Section 33–4 an inductor does not allow a very rapidly changing current to pass.

EXAMPLE 34–4 The series *RLC* circuit in Fig. 34–10 is driven with an AC source of emf of the form $\mathscr{E} = V_0 \sin(\omega t)$, where V_0 is exactly 110 V, the frequency f is exactly 60 Hz, and $\omega = 2\pi f$. If $R = 20.0\ \Omega$, $L = 5.00 \times 10^{-2}$ H, and $C = 50.0\ \mu$F, find the potential drops across the inductor at $t = 0$ and at the first time—$t = t_1$—after $t = 0$ that $\mathscr{E}$ reaches a maximum.

Solution: We want to find the potential drop across the inductor, and it is given by $V_L = -L\, dI/dt$. The current I in this circuit is given by Eq. (34–41), so

$$V_L = -L\frac{d}{dt}\left[\frac{V_0}{Z}\sin(\omega t - \phi)\right] = -\frac{LV_0\omega}{Z}\cos(\omega t - \phi).$$

At the moment $t = 0$, V_L takes the form

$$V_L = -\frac{LV_0\omega}{Z}\cos(-\phi) = -\frac{LV_0\omega}{Z}\cos\phi.$$

The emf reaches a maximum at time t_1 when $\omega t_1 = \pi/2$, or

$$t_1 = \frac{\pi}{2\omega} = \frac{\pi}{4\pi f} = \frac{1}{4f}.$$

At $t = t_1$ we have

$$V_L = -\frac{LV_0\omega}{Z}\cos\left(\frac{\omega}{4f} - \phi\right) = -\frac{LV_0\omega}{Z}\cos\left(\frac{\pi}{2} - \phi\right) = -\frac{LV_0\omega}{Z}\sin\phi.$$

To evaluate these results, we have $\omega = 2\pi(60\ \text{Hz}) = 377$ rad/s. Then

$$X_L = \omega L = (377\ \text{rad/s})(5.00 \times 10^{-2}\ \text{H}) = 18.9\ \Omega,$$

$$X_C = \frac{1}{\omega C} = \frac{1}{(377\ \text{rad/s})(5.00 \times 10^{-5}\ \text{F})} = 53.0\ \Omega.$$

In turn, we have

$$Z = \sqrt{(X_L - X_C)^2 + R^2} = \sqrt{(18.9\ \Omega - 53.0\ \Omega)^2 + (20.0\ \Omega)^2} = 39.6\ \Omega$$

and

$$\tan\phi = \frac{1}{R}(X_L - X_C) = \frac{1}{20.0\ \Omega}(18.9\ \Omega - 53.0\ \Omega) = -1.71, \text{ or } \phi = -59.6°.$$

Thus, at $t = 0$, we have

$$V_L = -\frac{LV_0\omega}{Z}\cos\phi$$

$$= -\frac{(5.00 \times 10^{-2}\ \text{H})(110\ \text{V})(377\ \text{rad/s})}{39.6\ \Omega}\cos(-59.6^0) = -26.5\ \text{V}.$$

At $t = t_1$,

$$V_L = -\frac{LV_0\omega}{Z}\sin\phi$$
$$= -\frac{(5.00\times10^{-2}\text{ H})(110\text{ V})(377\text{ rad/s})}{39.6\ \Omega}\sin(-59.6^0) = 45.2\text{ V}.$$

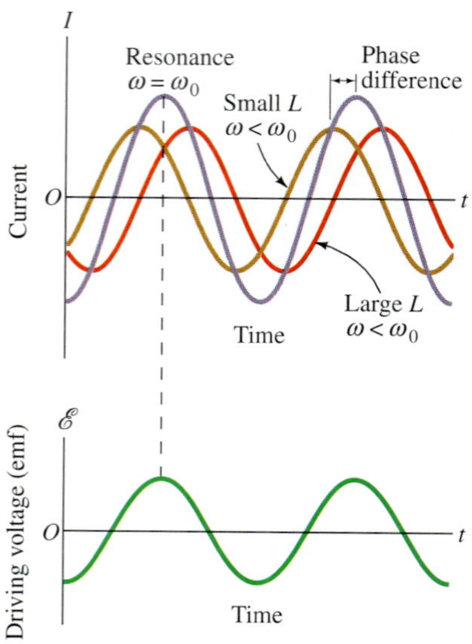

FIGURE 34–12 Given the driving voltage shown, the current through an *RLC* circuit varies harmonically with the frequency of the driving voltage but differs from it in amplitude and phase in a way that depends on whether or not the driving frequency, ω, equals the undamped natural frequency, ω_0. At resonance ($\omega = \omega_0$), the current amplitude is a maximum and the phase is that of the driving voltage. In the plot of current versus time, we have controlled the natural frequency by varying *L*.

Resonance in Driven *RLC* Circuits

The current amplitude exhibits the phenomenon of resonance—as do the amplitude of the charge on a capacitor and the voltage amplitudes across any of the elements of an *RLC* circuit. The amplitudes are inversely proportional to the impedance. When Z is a minimum, these amplitudes are *peaked*, or *resonate*. This occurs when the driving frequency ω is near the undamped natural frequency ω_0 (Fig. 34–12). The amount and sharpness of the peaking depends on the damping factor, which is the resistance R. Without R, the amplitudes would be infinite when the driving frequency ω equals ω_0. However, there is always some resistance in real circuits. Figure 13–21 illustrates the resonance peak for several values of the damping parameter b of a mechanical oscillator. Similar plots apply to *RLC* circuits. The smaller the damping—the smaller the resistance—the larger and sharper the resonance peak is. The maximum in this peak occurs when $X_L = X_C$,

$$\omega L = \frac{1}{\omega C}.$$

This condition implies that

$$\omega_{\max}^2 = \omega_0^2 = \frac{1}{LC}. \tag{34–42}$$

At this value of the angular frequency, $Z = R$. The current amplitude at this value of the driving frequency is given simply by

$$\text{at resonance:}\qquad I_{\max} = \frac{V_0}{R}. \tag{34–43}$$

Both the value of $I_{\max}$ at resonance *and the sharpness of the resonance peak* increase as R decreases. It is useful to note one last feature: At resonance, the tangent of the phase [Eq. (34–35)], and hence the phase itself, is zero.

APPLICATION

Resonance in receiver tuning

Resonance phenomena have many applications in circuits. The most familiar concern radio and television reception (Fig. 34–13). A receiver can be tuned by changing either an inductance or a capacitance, thus changing the resonant frequency. The receiver then picks up broadcast signals at that frequency.

34–4 POWER IN AC CIRCUITS

Energy flows in an *RLC* circuit. This energy is lost only in the resistance of such a circuit. Energy held by either the capacitor or the inductor is temporarily stored for some parts of the harmonic cycle but is not lost. Thus the power dissipated in the circuit is

$$P = I^2R = \frac{[V_0\sin(\omega t - \phi)]^2}{Z^2}R. \tag{34–44}$$

This power is always positive, but it oscillates between zero and a maximum of V_0^2R/Z^2. For engineering purposes, it is more important to know the *average* power dissipated over time. Let's indicate time-averaged quantities with angle brackets, $\langle\ \rangle$. The average of a sine (or cosine) squared over one cycle is one-half:

$$\langle[\sin(\omega t - \phi)]^2\rangle = \frac{1}{2},$$

so

$$\langle P\rangle = \frac{V_0^2 R}{Z^2}\langle[\sin(\omega t - \phi)]^2\rangle = \frac{1}{2}\frac{V_0^2 R}{Z^2}. \tag{34–45}$$

If we substitute Eq. (34–40) for the impedance, we find an explicit form for the average power:

$$\langle P\rangle = \frac{1}{2}\frac{V_0^2 R}{[\omega L - (1/\omega C)]^2 + R^2} = \frac{1}{2}\frac{V_0^2 R\omega^2}{L^2(\omega^2 - \omega_0^2)^2 + \omega^2 R^2}. \tag{34–46}$$

The resonant behavior of AC circuits is evident in this result for the power dissipated. As the driving frequency ω increases through ω_0, the power dissipated has the typical peaked behavior of resonance. The dissipated power is a *maximum* at resonance, when the driving angular frequency ω equals the angular frequency $\omega_0 = \sqrt{1/LC}$. At resonance, the power reduces to

$$\langle P\rangle_{res} = \frac{1}{2}\frac{V_0^2}{R}. \tag{34–47}$$

The current displays the same resonant behavior as does the power. Equation (34–41) shows that the current oscillates with time. It is useful to characterize the current (and other harmonically varying quantities in AC) with an *rms* (*root mean square*) value. The rms value, x_{rms}, of any quantity x is defined as the square root of the time average of the square of that quantity:

$$x_{rms} \equiv \sqrt{\langle x^2\rangle}.$$

In particular, if x varies harmonically—if $x = x_0\cos(\omega t - \phi)$—we can use the fact that the time average of the cosine squared is one-half to show that

$$x_{rms} = \frac{x_0}{\sqrt{2}}. \tag{34–48}$$

When we apply this concept to the AC current, we see from Eq. (34–41) that

$$I_{rms} = \frac{V_0}{(\sqrt{2})Z} = \sqrt{\frac{V_0^2\omega^2}{2[L^2(\omega^2 - \omega_0^2)^2 + \omega^2 R^2]}}. \tag{34–49}$$

Note from Eq. (34–46) that I_{rms} and the average power, $\langle P\rangle$, obey the same power–current relation as we found for DC quantities, namely

$$\langle P\rangle = I_{rms}^2 R. \tag{34–50}$$

Figure 34–14 plots I_{rms}^2 as a function of the driving angular frequency ω for three values of resistance. The sharpness in the peak of the average power (or of I_{rms}^2) versus ω is characterized by the *width* of the peak or, more precisely, the *total width at half-maximum* $\Delta\omega$. This is commonly called the **bandwidth** in the context of AC. To calculate the bandwidth, we find the angular frequencies at which the power drops to half the peak value and take the difference between these angular frequencies. This calculation shows that, for small values of resistance, the bandwidth is given by

$$\Delta\omega = \frac{R}{L}. \tag{34–51}$$

(a)

(b)

FIGURE 34–13 Receiver tuners can be constructed by (a) rotating the overlapping areas of the plates to change the capacitance or by (b) moving an iron core in and out of a solenoid to vary the inductance.

The concept of the root mean square was first introduced in Chapter 19.

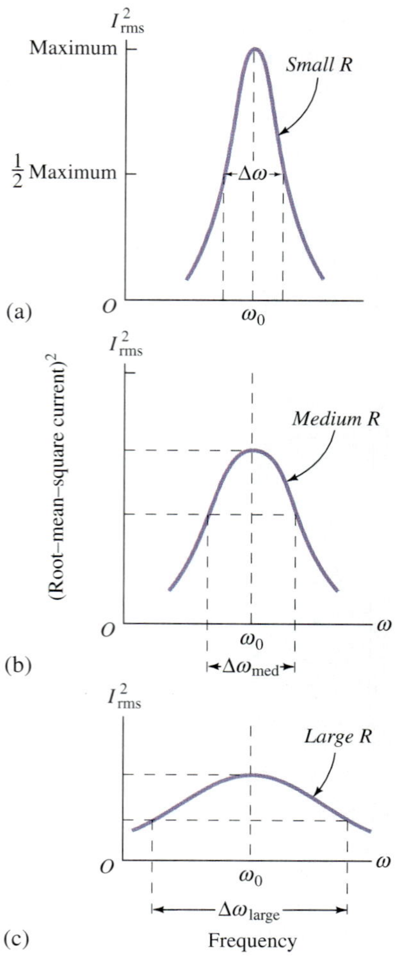

FIGURE 34–14 The rms current squared in a series *RLC* circuit with an AC source of emf such as the circuit of Fig. 34–10. (a) There is a resonance phenomenon when the driving frequency ω matches the natural frequency ω_0 of the circuit. The bandwidth $\Delta\omega$ measures the width of the rms current squared. (b, c) As R increases, the peak in the current as a function of ω broadens.

The smaller the resistance and the larger the inductance, the smaller the bandwidth. We can understand the importance of a small bandwidth by thinking about a radio or television receiver whose tuning circuit depends on the resonance phenomenon. If the resonance is sharp, the receiver will more effectively pick out only the desired frequency over others nearby (Fig. 34–14a). Conversely, if the resonance is broad, the circuit will respond to frequencies in the AC signal far from the desired frequency (Fig. 34–14b, c).

Another measure of the sharpness is the ratio $\omega_0/\Delta\omega$. This quantity is defined as the *quality factor*, or *Q-factor*,

$$Q \equiv \frac{\omega_0}{\Delta\omega} = \frac{\omega_0 L}{R}. \tag{34–52}$$

This factor is often used by electrical engineers to represent the sharpness of a resonant circuit.

EXAMPLE 34–5 Two FM radio stations broadcast at the same strength from the same nearby distance, one at a frequency of 91.3 MHz and the other at 91.1 MHz. You like the former very much but do not care for the latter. You want to construct a simple series *RLC* circuit to act as a receiver that is unique to your favored station, given an inductor with an inductance L of exactly 1 μH and adjustable resistance and capacitance. To limit the power received from the unwanted station to 1 percent of the power received from the desired station, what values would you choose for R and C?

Solution: There are two problems here: One, to make a circuit with resonance at $\omega_0 = 2\pi f = 2\pi(91.3\text{ MHz}) = 5.74 \times 10^8$ Hz; two, to make the resonant peak sharp enough to limit the power from the station broadcasting at $\omega_1 = 2\pi(91.1$ MHz$) = 5.72 \times 10^8$ Hz. The resonant frequency is determined from L and C alone. With a known value of L, C is determined: $\omega_0^2 = 1/LC$, or

$$C = \frac{1}{\omega_0^2 L} = \frac{1}{(5.74 \times 10^8\text{ Hz})^2(1\ \mu\text{H})} = 3.04 \times 10^{-12}\text{ F}.$$

The sharpness requirement determines R. The power delivered by the signal at resonance is given by Eq. (34–47), whereas the power delivered off resonance is given by Eq. (34–46). The two stations have the same strength, so it is appropriate to use the same value of V_0. Thus

$$\frac{\langle P\rangle_{\omega 1}}{\langle P\rangle_{res}} = 0.01$$

$$= \left[\frac{1}{2}\left(\frac{V_0^2}{R}\right)\right]^{-1}\frac{1}{2}\,\frac{V_0^2 R\omega_1^2}{L^2(\omega_1^2 - \omega_0^2)^2 + \omega_1^2 R^2} = \frac{R^2\omega_1^2}{L^2(\omega_1^2 - \omega_0^2)^2 + \omega_1^2 R^2}.$$

So

$$[L^2(\omega_1^2 - \omega_0^2)^2 + \omega_1^2 R^2](0.01) = R^2\omega_1^2.$$

To a good approximation, we can ignore the $\omega_1^2 R^2$ term on the left. We then take the square root of both sides:

$$L(\omega_0^2 - \omega_1^2)(0.1) = L(\omega_0 - \omega_1)(\omega_0 + \omega_1)(0.1) = R\omega_1.$$

The factor $(\omega_0 + \omega_1)$ is, to a good approximation, equal to $2\omega_1$, so we have

$$2\omega_1 L(\omega_0 - \omega_1)(0.1) \simeq R\omega_1.$$

Thus

$$R = 2L(\omega_0 - \omega_1)(0.1) = 2L(2\pi)(f_0 - f_1)(0.1)$$
$$= 2(10^{-6}\text{ H})(6.28)[(5.74 \times 10^8\text{ Hz}) - (5.72 \times 10^8\text{ Hz})](0.1) = 2.51\ \Omega.$$

APPLICATION: PHASE OSCILLATIONS IN POWER GRIDS: IS AC OR DC PREFERABLE?

The major advantage of AC in a power grid is that transformers can be used to step the potential up or down. There is, however, a price to be paid. In an AC generator, a "prime mover"—flowing water, burning diesel fuel, or a nuclear reactor—supplies the energy to rotate a coil within a magnetic field. When such generators are connected together on a grid, the phase of the AC of each generator must be the same—otherwise there would be a time-dependent difference in potentials from generator to generator and currents would flow where they are not wanted. This means that the angular position of the coils on different generators must be the same. Controlling these positions is a difficult task; it ordinarily involves an overshoot of the desired angle and subsequent oscillatory behavior.

If, for some reason, one generator suddenly drops out of a network of coupled generators, the retarding Faraday torque on the remaining generators increases abruptly, because they must each supply more power to the grid. Their angular velocities decrease, and the prime mover must supply additional torque. During this recovery period, the angular position of each generator will be affected and will undergo oscillations. If all the generators were identical, supplied with power from an identical prime mover, and identically controlled, these positional (phase) oscillations would all be synchronous and would cause no problem.

Unfortunately, differences between generators means that the phase response of a power station to a step input of electrical power demand is likely to differ for each station. During the recovery period, there are large flows of electrical energy into and out of a generator from the other generators on the net, further complicating the phase stability problem. The effect of these flows may be to overload the generator and take it off the net, providing yet another abrupt power demand transient. Worse, once one or two stations have left the net, the remainder may not have the capacity to supply the net demand, and the entire grid may break down.

A large region such as the eastern seaboard is then without power (Fig. 34AB–1). The switches on the consuming appliances are still on; given that no single station can meet the load, how is it possible to restart? The net has to be disconnected from the load, the generating stations must restart one by one, and phase synchronization must be painfully reestablished. Once the net is reestablished, it is then connected to the load—the consumers—district by district. Each connection produces a load transient, and it is necessary to wait until stable operation is reached before adding the next district. Recovery is therefore slow.

Today, with the development of high-power semiconductor devices that can freely convert between AC and DC, we have a real choice, and AC power networks are being modified so that the coupling between generators is DC. Rotational position synchronization between generators will no longer be required.

FIGURE 34AB–1 Policemen holding flashlights lead commuters through a subway tunnel in New York City on November 9, 1965 after a massive power failure affected eight northeast states and Canada.

The Power Factor

The power in AC circuits is commonly given in a form other than that given in Eq. (34–46). We find this form with the aid of the trigonometric identity

$$\cos^2 \phi = \frac{1}{(\tan^2 \phi) - 1}.$$

If we now use Eq. (34–39) for $\tan \phi$, we find that

$$\cos^2 \phi = \frac{1}{[(1/R)(X_L - X_C)]^2 + 1} = \frac{R^2}{(X_L - X_C)^2 + R^2} = \frac{R^2}{Z^2};$$

$$\cos \phi = \frac{R}{Z}. \qquad (34\text{–}53)$$

Then, using Eqs. (34–49) and (34–51), we see that Eq. (34–50) becomes

$$\langle P \rangle = I_{\text{rms}}^2 R = I_{\text{rms}}^2 Z \cos \phi. \qquad (34\text{–}54)$$

The term $\cos \phi$ in Eq. (34–54) is called the *power factor*. For a circuit without resistance, it is zero, whereas for a pure resistance it is a maximum, with a value of one.

34–5 SOME APPLICATIONS

Most electronic circuits in use today involve elements beyond those we have studied here. These elements may perform amplifying functions, as in transistors, or have resistance that depends on the direction of current flow, as it does in diodes. Modern circuits are typically constructed in integrated form with many thousands of elements included together from the start, and perform rather general functions. Nevertheless, several principles of such devices beyond those we have already discussed can be understood with a small addition to the elements we have in place.

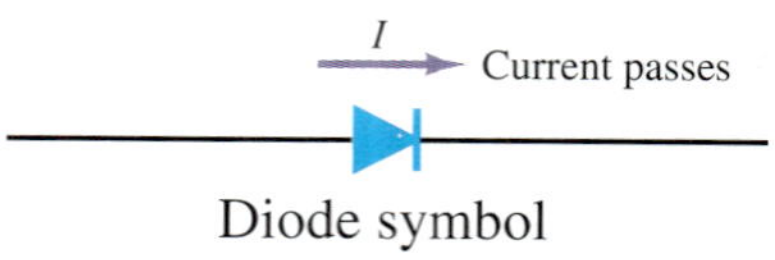

FIGURE 34–15 The symbol for a diode in a circuit diagram. Current can flow only in the direction shown.

Diodes and Rectifiers

Many sources of electric power produce AC voltage. However, many applications of power use require DC voltage. For example, the alternator of an automobile produces AC, but the car battery requires DC to be charged. We need a simple way to change from AC to DC voltage. The process by which this is accomplished is called *rectification*, and the tool used is the *diode*. A **diode** is a semiconductor device with a high resistance to current that flows in one direction, but a low resistance to current that flows in the other direction—the direction of the arrow in the diode symbol (Fig. 34–15). In effect, the diode allows current flow only in the direction of the arrow.

FIGURE 34–16 This DC power supply turns AC current into DC current by means of a rectifier.

The diode can be used to construct a **rectifier**, a circuit element that changes AC into DC (Fig. 34–16). Let's consider the circuit shown in Fig. 34–17a. The voltage across the load resistor can be negative or positive. When a diode is placed in the circuit, however, the negative voltages are blocked, leaving only positive voltages across the load resistor (Fig. 34–17b). Such a circuit is called a *half-wave rectifier*. This circuit can suffice as an approximation to a source of DC voltage although the voltage between points a and b, V_{ab}, is certainly not smooth and constant.

The foregoing situation can be improved considerably by the circuit shown in Fig. 34–17c, called a *full-wave rectifier*. Although this circuit appears to be more complex, it is quite simple: It is arranged such that *the voltage V_{ab} is always positive* even though the input voltage oscillates from positive to negative. When the emf produces positive voltage, positive current flows clockwise and passes through the path *cabd* in the direction of the rectifier arrows. The voltage V_{ab} is positive. When the emf produces negative voltage, *positive current flows counterclockwise* and the path of the current is *dabc*. In this case also, the voltage V_{ab} is *positive*. Note that now the voltage V_{ab} is positive for all half-cycles, and the rms voltage is much higher than it is for the half-wave rectifier. The use of *filters* allows the voltage peaks to be smoothed, producing a more nearly constant voltage.

FIGURE 34–17 (a) The AC voltage across a resistor. (b) A half-wave rectifier is applied. (c) A full-wave rectifier is applied.

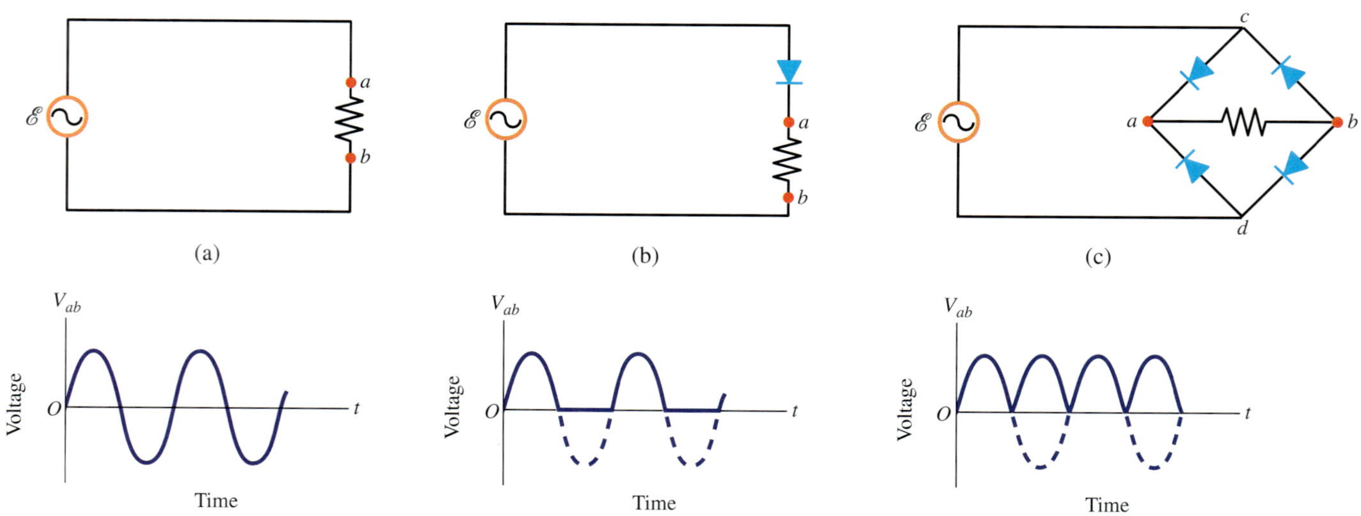

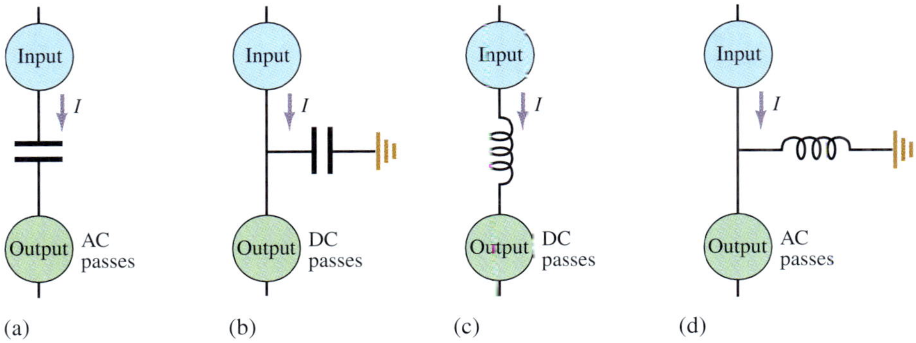

FIGURE 34–18 AC and DC filters formed from capacitors or inductors. With a capacitor: (a) AC current passes through to the output side. (b) DC current passes to the output side. With an inductor: (c) DC passes through to the output side. (d) AC passes through to the output side, but DC does not.

Filters

A **filter** is a device that takes an input signal from one part of a circuit that may be a mixture of AC and DC and passes only the AC or only the DC signal to a different part of the circuit. Our discussion of the capacitive and inductive reactances shows that either a capacitor or an inductor can act as such a filter. Consider Fig. 34–18, in which the current I from the input is a mixture of DC and AC: for example,

$$I = I_0 + I_1 \sin(\omega t).$$

Here I_0 and I_1 are constants. A constant current cannot pass across a capacitor, whereas the impedance of the capacitor goes to zero if ω becomes large. Thus for the capacitor in Fig. 34–18a, only the AC part of the current passes to the other side. For the capacitor in Fig. 34–18b, AC passes through the capacitor to ground, and the DC passes through to the output side of the circuit. An inductor works in just the opposite way: DC passes through without impedance ($X_L \to 0$ as $\omega \to 0$, and $\omega = 0$ corresponds to DC), whereas the impedance is large for AC with large ω. Thus for the inductor in Fig. 34–18c, DC passes through to the output; for the inductor in Fig. 34–18d, AC passes through.

EXAMPLE 34–6 Consider the circuit shown in Fig. 34–19, where $C = 1.0$ μF, $R = 0.20\ \Omega$, $V_0 = 0.10$ V, and $V_1 = 0.25$ V. What is the value of ω for which the voltage amplitude across the resistor is 50 percent of the value of the maximum voltage of the generator?

Solution: The generator produces a combination of DC and AC. We can apply the superposition principle by calculating the result from applying the loop rule that corresponds to the DC and AC terms separately, and then add the voltage drops. For the AC term, the current in the circuit is, from Eq. (34–41),

$$I_{AC} = \frac{V_1 \sin(\omega t - \phi)}{Z} = \frac{V_1}{\sqrt{1/\omega C)^2 + R^2}} \sin(\omega t - \phi).$$

The voltage drop across the resistor from the AC is $I_{AC}R$ and therefore has the amplitude

$$\frac{V_1 R}{\sqrt{(1/\omega C)^2 + R^2}}.$$

The capacitor acts as a perfect filter for the constant term in the input voltage because no constant current can pass. There is thus no voltage drop across the resistor associated with the V_0 term.

The maximum value of the input voltage is $V_0 + V_1$, and the ratio of the voltage amplitude across the resistor to the maximum input voltage is

$$\frac{V_1 R/\sqrt{(1/\omega C)^2 + R^2}}{V_0 + V_1}.$$

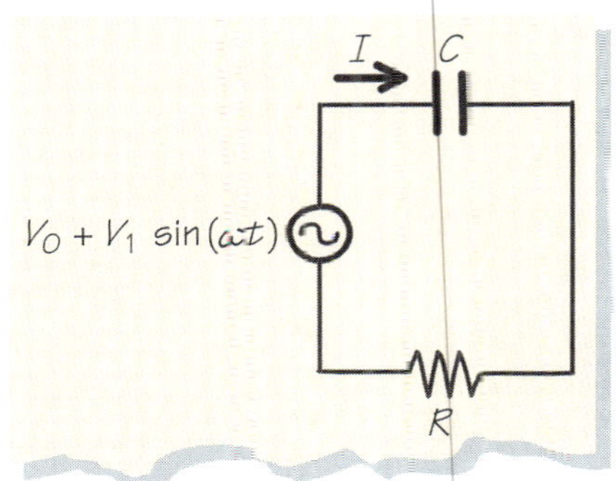

FIGURE 34–19 Example 34–6.

We want this factor to equal 50 percent. We set it to 0.50 and solve for ω:

$$\omega = \frac{1}{RC\sqrt{\dfrac{V_1^2}{(V_0 + V_1)^2(0.50)^2} - 1}}$$

$$= \frac{1}{(0.20\ \Omega)(1.0 \times 10^{-6}\text{F})\sqrt{\dfrac{(0.25\ \text{V})^2}{(0.10\ \text{V} + 0.25\ \text{V})^2}\dfrac{1}{(0.50)^2} - 1}} = 4.9\ \text{MHz}.$$

Application of a filter to the rectified voltage shown in Fig. 34–17b, for example, will smooth out the peaks and valleys in the curve of voltage versus time. If the RC time constant of the filter is much larger than the period of the rectified voltage, the resulting voltage is a much better approximation to DC voltage (Fig. 34–20). Figure 34–21 shows how filters can modify a signal that is a mix of many different frequencies.

Impedance Matching

Another aspect of AC of great practical importance concerns **impedance matching**, which refers, as in our discussion of filters, to the *connection* between different parts of a circuit. Figure 34–22a shows such a situation, in which some combination of circuit elements makes up circuit 1, connected at points a and b to circuit 2. The two circuits have impedances Z_1 and Z_2, respectively. We are not concerned here with the origin of currents in these circuits as much as we are with our ability to deliver power from circuit 1 to circuit 2. We therefore assume that the origin of these currents is within circuit 1 and break that circuit down as in Fig. 34–22b. The primary question is, if Z_1 is fixed, what are the requirements for Z_2 so that the power delivered to circuit 2 is a maximum? If, for example, a stereo amplifier is connected to a loudspeaker, what should the loudspeaker's impedance be in order that maximum power is delivered to it?

The answer is found by computing the average power $\langle P \rangle$ to circuit 2, which, from Eq. (34–50), is $I_{\text{rms}}^2 R_2$. The current in the loop of Fig. 34–22b is given by

$$I_{\text{rms}} = \frac{\mathscr{E}_{\text{rms}}}{Z_{\text{total}}}. \tag{34–55}$$

Here, $\mathscr{E}_{\text{rms}}$ is the rms value of the generator, whose maximum voltage, or amplitude, is V_0. If the generator produces a sinusoidal emf of the form of Eq. (34–1), then Eq. (34–48) shows that $\mathscr{E}_{\text{rms}} = V_0/\sqrt{2}$. The total impedance Z_{total} is found by separately adding the capacitive reactances, inductive reactances, and resistances, a result that follows from our knowledge of how series combinations of C, L, and R add (see Problem 42):

$$Z_{\text{total}} = \sqrt{[(X_{L_1} + X_{L_2}) - (X_{C_1} + X_{C_2})]^2 + (R_1 + R_2)^2}. \tag{34–56}$$

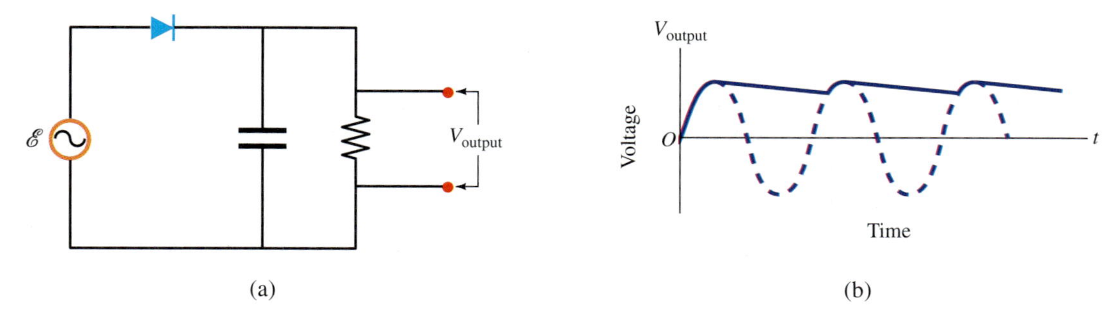

FIGURE 34–20 (a) An RC circuit acts as a filter for rectified AC voltage. Such a filter can produce a voltage that is nearly DC. (b) The slow decrease of the nearly constant-voltage segments is governed by the time constant of the RC circuit.

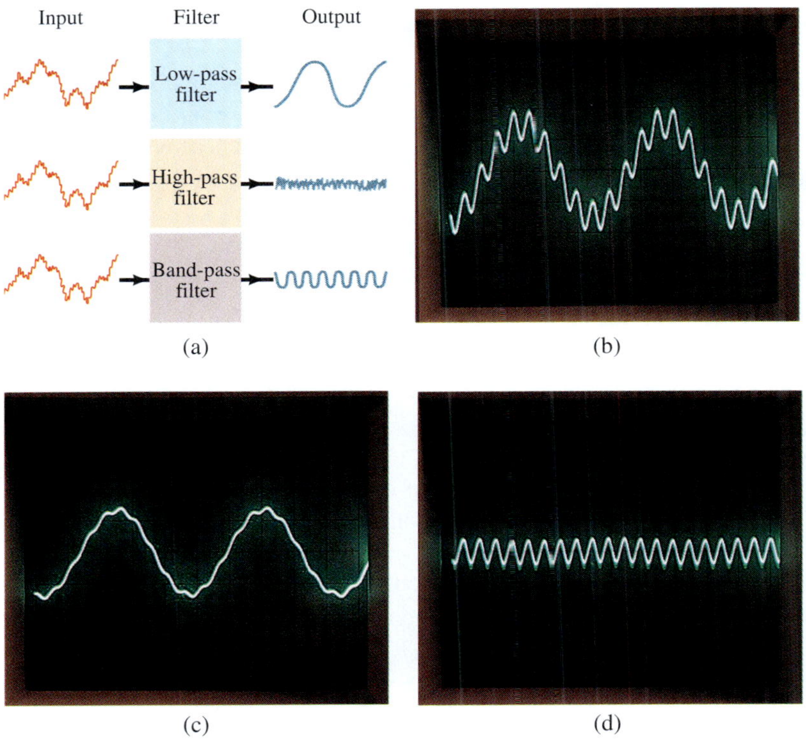

FIGURE 34–21 (a) A low-pass filter allows the low frequencies contained in an input signal to pass. (b) An input voltage containing both low and high frequencies. (c) The input voltage in part (b) has been sent through a low-pass filter, which eliminates the high frequencies. (d) The input voltage in part (b) has been sent through a high-pass filter, which eliminates the low frequencies.

Thus the average power delivered to circuit 2 is

$$\langle P \rangle = \frac{\mathcal{E}_{\rm rms}^2 R_2}{Z_{\rm total}^2} = \frac{\mathcal{E}_{\rm rms}^2 R_2}{[(X_{L_1} + X_{L_2}) - (X_{C_1} + X_{C_2})]^2 + (R_1 + R_2)^2}. \quad (34\text{–}57)$$

That there is a value of the parameters of Z_2 that maximizes this power is clear: If Z_2 is too small, the factor R_2 will also be small and $\langle P \rangle$ will be small; if Z_2 is too large, it will dominate the denominator of Eq. (34–57), and $\langle P \rangle$ will again be small. An intermediate value of the parameters of Z_2 will give a maximum value of $\langle P \rangle$. Two independent parameters are involved here: the resistance R_2 and the total reactance term for circuit 2, $X_{L_2} - X_{C_2}$. Formally, we find the value of the parameters that maximize $\langle P \rangle$ by taking the derivative of $\langle P \rangle$ with respect to these quantities and setting it equal to zero. From this exercise the power is maximized when

$$R_2 = R_1 \quad \text{and} \quad X_{L_2} - X_{C_2} = -(X_{L_1} - X_{C_1}). \quad (34\text{–}58)$$

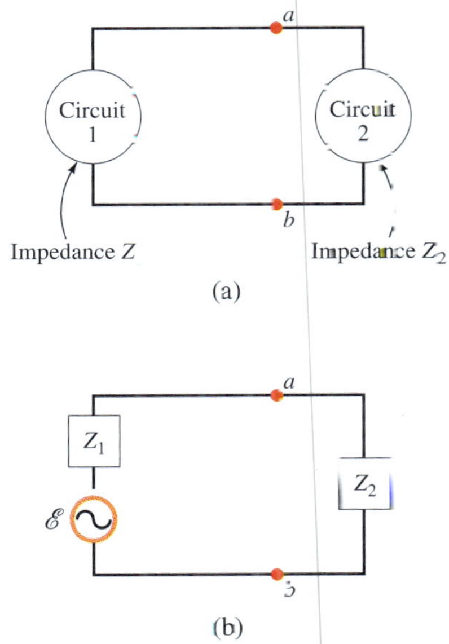

FIGURE 34–22 (a) Circuit diagram to illustrate impedance matching. (b) Circuit 1 of part (a) is broken down into a source of emf $\mathcal{E}$ and an impedance Z_1. Circuit 2 is assumed to include only an impedance Z_2.

The second condition—that the reactance term of Z_2 is equal but opposite to that of Z_1—follows because it means that the reactance terms in the denominator of Eq. (34–57) cancel, thus maximizing $\langle P \rangle$ whatever the value of the resistances. The first condition—that the resistances are equal—is perhaps less intuitive but nevertheless follows directly from the requirement that the derivative of $\langle P \rangle$ is zero (see Problem 66). When the conditions of Eq. (34–58) are met, the impedances are said to be matched.

Impedance matching is desirable when you wish to deliver maximum power to one part of a circuit. It is worth noting that we do not always wish to deliver maximum power. A voltmeter, for example, should have an impedance *mis*match because we want it to draw as little current as possible.

The impedances of two parts of a circuit should be matched when you want to deliver maximum power from one part of the circuit to the other.

The subject of circuit analysis is highly developed. We have been able to do no more than describe its principles, and this chapter will not have taught you to fix, much less design, TVs or computers. But the principles we have described here apply to all electric circuits.

SUMMARY

The presence of AC sources of emf in circuits with resistors, inductors, and capacitors introduces a variety of new possibilities. Transformers allow us to vary the voltage amplitude of AC emfs. The relation between the emfs and the numbers of turns of the primary and secondary coils of a transformer is

$$\frac{\mathscr{E}_2}{\mathscr{E}_1} = \frac{N_2}{N_1}. \tag{34–8}$$

Conservation of energy implies that the currents carried by the respective coils are related inversely:

$$\frac{I_1}{I_2} = \frac{N_2}{N_1}. \tag{34–10}$$

A series RLC circuit with an AC source of emf of frequency ω behaves like a damped harmonic oscillator driven by a harmonically varying force. Solutions for currents, voltages, and charges in such circuits can be found by using the solutions already developed for the driven harmonic oscillator. For such circuits the impedance, Z, is a quantity that plays the role of a resistance. The impedance is frequency dependent:

$$Z \equiv \sqrt{\left(\omega L - \frac{1}{\omega C}\right)^2 + R^2} = \sqrt{(X_L - X_C)^2 + R^2}. \tag{34–40}$$

where X_C is the capacitive reactance and X_L is the inductive reactance. The current in the driven circuit is then

$$I = I_{\max} \sin(\omega t - \phi) = \frac{V_0 \sin(\omega t - \phi)}{Z}, \tag{34–41}$$

where V_0 is the amplitude of the driving emf and ω is its frequency. The phase ϕ is given by

$$\tan \phi = \frac{1}{R}\left(\omega L - \frac{1}{\omega C}\right). \tag{34–35}$$

Such circuits exhibit resonant behavior when the driving frequency is near the natural frequency $\omega_0 = \sqrt{1/LC}$. This type of behavior is most clearly seen in the power dissipated in driven RLC circuits. Averaged over time, the power lost is

$$\langle P \rangle = I_{\text{rms}}^2 R = I_{\text{rms}}^2 Z \cos \phi. \tag{34–54}$$

where $I_{\text{rms}} = V_0/\sqrt{2}Z$ is the rms current. Near resonance, the power dissipated is a maximum, and the width of the peak of average power versus driving frequency has a width at half-maximum of

$$\Delta\omega = \frac{R}{L}, \tag{34–51}$$

a result that holds as long as the resistance is not too large.

If we add diodes—devices that allow current to pass in one direction only—to our arsenal of circuit elements, we can construct a variety of electronic devices, including rectifiers, which produce a positive (or negative) emf from an AC source, and filters, which take a mixed AC and DC signal and pass predominantly either the constant (DC) part or the variable (AC) part of a given frequency. Impedance matching refers to constraints that describe how different parts of a circuit can be connected together with minimal power loss.

UNDERSTANDING KEY CONCEPTS

1. Why is the material used to make the core of transformers so important?
2. What are some applications for step-up and step-down transformers?
3. Without R, the current amplitude in a series RLC circuit would be infinite when the driving frequency ω equaled ω_0, but this possibility could never happen, because there is always some resistance in real circuits. How do you reconcile this statement with the existence of superconductors?
4. To find the rms current [Eq. (34–49)], we square the current, then take the time average, and then take the square root. Why do we not simply take the time average of the current?
5. In discussing AC current, we found it useful to describe its root mean square; why didn't we bother with this in our discussion of power?
6. In Example 34–6, we took the input emf to be a mixture of DC and AC, then treated the effect of the AC and DC parts separately and added the two parts. How is this procedure justified?
7. A capacitor and a lamp are connected in series with an AC generator of constant voltage but variable frequency (Fig. 34–23). Which of the following three statements is true? The lamp will (a) not light, because the capacitor is connected in series with the lamp; (b) burn brightest when the frequency is high; (c) burn with the same brightness for all frequencies.

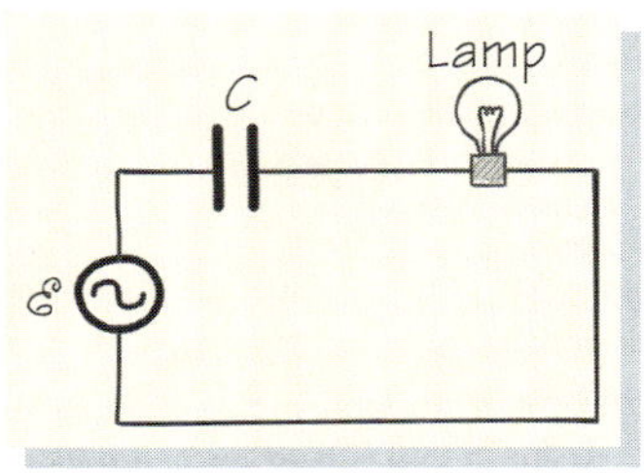

FIGURE 34–23 Question 7.

8. A capacitor, lamp, and resistor are connected to an AC generator of constant voltage but variable frequency (Fig. 34–24). Which of the following statements is true? The lamp will (a) not burn, because the capacitor shorts out the lamp; (b) burn brightest when the frequency is low; (c) burn brightest when the frequency is high; (d) burn with the same brightness for all frequencies.

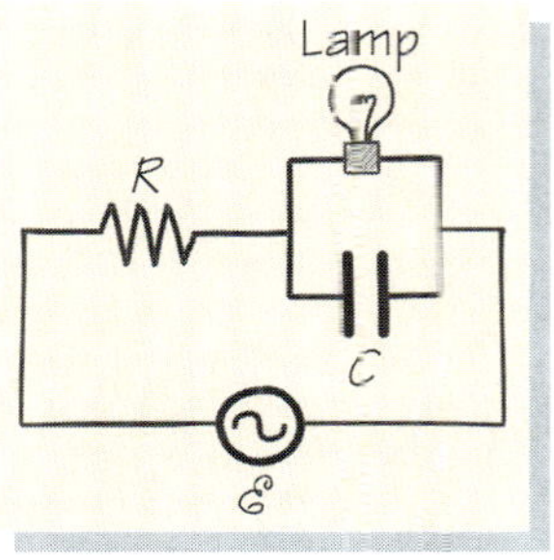

FIGURE 34–24 Question 8.

9. A particular appliance or household circuit is rated for a maximum current. Why is that, and why must currents not exceed that maximum?
10. The primary purpose of an electric heater is to produce heat. Why would such a heater require a 220-V socket rather than a 120-V socket?
11. Two circuits have impedances Z_1 and Z_2, respectively. When these circuits are placed in series with one another, why is the total impedance not given by $Z = Z_1 + Z_2$?
12. The reactance X_C is infinite when the input voltage is DC. Does that mean that the impedance is not defined for this situation?
13. If a capacitor has a large impedance for DC and an inductor has a large impedance for AC, how can a series LC circuit pass any current?
14. Television antenna wires normally have negligible resistance and an impedance of 75 Ω. Why is it important to use antenna wires with the same impedance throughout?
15. Some appliances that operate off a 120-V line yet draw in excess of 20 A of current have different plugs than regular 120-V household devices have. Why are these plugs different? What might happen if they were not different?
16. Do all television sets have transformers? How might transformers be used in a television set?
17. If electricity is transported in power transmission lines at 200 kV to 500 kV, does the power have to be generated at these voltages? Why or why not?

PROBLEMS

34–1 Transformers

1. (I) A transformer has 100 turns in the primary coil and 1500 turns in the secondary coil. If the amplitude of the AC voltage in the primary coil is 600 V, what is the voltage amplitude in the secondary coil?
2. (I) Suppose that electric power costs 15 cents/kWH. Consider a transmission line that delivers 1 MW of power and has a resistance of 10 Ω. Calculate the dollars lost annually due to the transmission line if the power is delivered at (a) 500,000 V and (b) 440 V.
3. (I) Many electrical devices, such as doorbells or buzzers, operate on 12 V AC. A small transformer used to produce this voltage has a primary coil of 550 turns and takes an input of 110 V AC. How many turns must the secondary coil have?
4. (I) The primary coil of a step-down transformer is connected to house current, 115 V at 60 Hz. If the secondary coil of the transformer delivers a current with an amplitude of 2.0 A at 24 V, what is the current drawn by the primary coil? Ignore losses in the transformer.

5. (I) A transformer whose output voltage can be varied is used to obtain AC power from a 120-V, 10-A supply. The secondary coil consists of 1200 turns of wire. The variable transformer works by connecting different numbers of turns of wire on the secondary coil. The secondary voltage can thereby be regulated. When all 1200 turns act as the secondary coil, the output voltage has an amplitude of 120 V. How many turns of wire should be used to obtain 45 V (Fig. 34–25)? How much current will flow in this case?

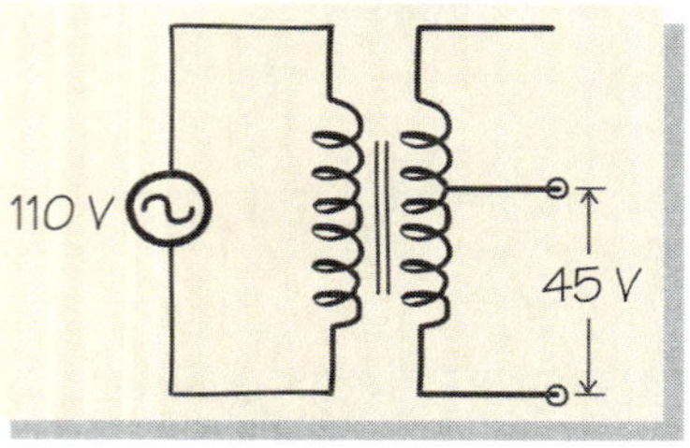

FIGURE 34–25 Problem 5.

6. (I) A transformer has one coil with an inductance of 47 mH, area 75 cm^2, and length 20 cm. It is fully linked with another coil having the same area and length, but not the same number of turns. Their mutual inductance is 23.5 mH. How many turns does each coil have?

7. (II) Figure 34–26 shows an ideal transformer with 220 V on the primary coil supplying power to a resistor of resistance R. If the resistor dissipates 88 W, what is the current in the primary coil?

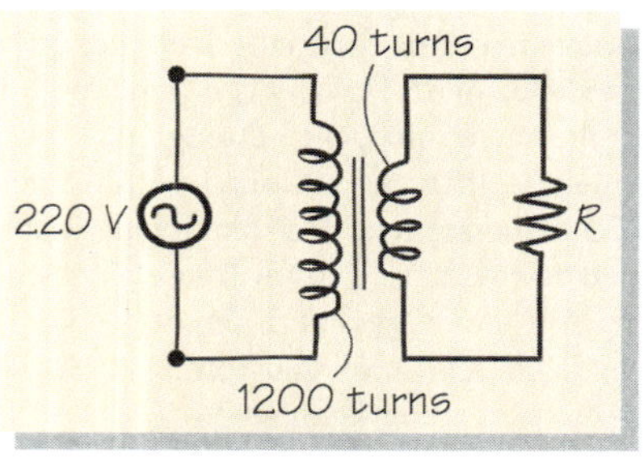

FIGURE 34–26 Problem 7.

8. (II) A step-down transformer has a turn ratio (N_1/N_2) of 5:1. (a) If the primary coil is connected across a 220-V oscillating-voltage generator, what voltage appears across the secondary coil? (b) Assuming that there are no power losses in the transformer, what current would have to flow through the primary coil so that a 40-Ω resistor placed across the secondary coil draws all the power of the circuit? (c) What resistance connected across the 220-V voltage generator would draw the same total power?

9. (II) The transformer shown in Fig. 34–27 has two secondary windings; one supplies 220 V, the other, 11 V. The input voltage at the primary coil is 110 V. If the 220-V secondary coil has 1000 turns, how many turns does the 11-V secondary coil have?

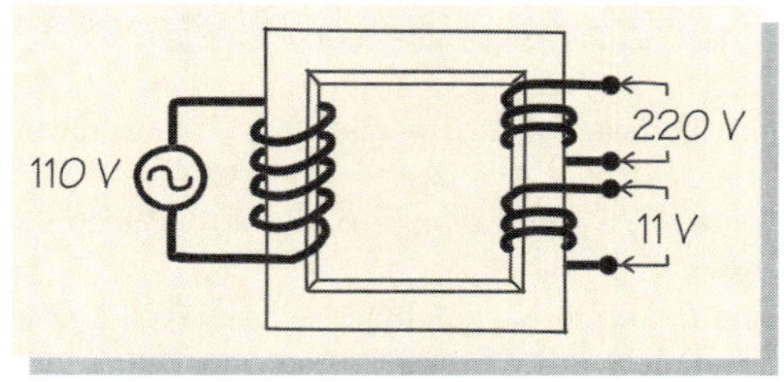

FIGURE 34–27 Problem 9.

10. (II) Suppose that a transformer consists of two separate windings of wire on the same core. The core material has a magnetic permeability μ. How does the ratio of the emfs in the two coils depend on μ?

34–2 Single Elements in AC Circuits

11. (I) A 12-μF capacitor is used in series with an AC generator. Measurement of the current shows that the capacitive reactance is 1.0 Ω. What is the input frequency?

12. (I) A 120-Ω resistor is connected across a power supply that produces a voltage of the form $V_0 \sin(\omega t)$, where $f = \omega/2\pi =$ 60 Hz and $V_0 = 163$ V. What is the current passing through the resistor?

13. (I) An alternating current of maximum value 2 A in a solenoid of self-inductance $L = 15$ mH induces an emf of maximum value 330 V. What is the angular frequency of the alternating current?

14. (I) An AC power supply with frequency 60 Hz is connected to a capacitor of capacitance $C = 40\ \mu$F. The maximum instantaneous current that passes through the circuit is 2.26 A. What is the maximum voltage?

15. (I) A current flowing through a circuit that contains only a capacitor and an AC power supply has the form $I_0 \cos[2\pi ft - (\pi/6)]$, where $I_0 = 2.45$ A and $f = 180$ Hz. If the maximum voltage supplied by the generator is 95 V, what is the capacitance?

16. (I) An AC power supply operating at a frequency of 2000 Hz is connected across an inductor. The maximum voltage of the source is 5 V, and the maximum current is 0.3 A. What is the inductive reactance? What is the inductance of the circuit?

17. (I) An AC circuit contains an inductor of 0.3 H and capacitor of 2 μC in series. The circuit is driven with an AC source of emf with an angular frequency range of 300–1000 rad/s. What are the maximum values of the capacitive and inductive reactances?

18. (II) A current $I = I_0 \sin[(\omega t - \pi/3]$ flows in a circuit for which $I_0 = 2.3$ A and $\omega = 2\pi(60\text{ Hz})$. (a) At what times does the peak current flow? (b) If the current flows through an inductance of 0.25 H, what is the peak voltage on the inductor? At what times does this peak voltage occur?

19. (II) The average of the square of the voltage in an inductive circuit (a circuit with no capacitors and no resistors) driven by an AC emf is $(30\text{ V})^2$, and the average of the square of the current is $(2\text{ A})^2$. What is the inductive reactance? If the inductance is 25 mH, what is the frequency of the alternating current?

20. (II) The voltage across an inductor takes the form $V(t) = (3\text{ V})\sin[(500\text{ s}^{-1})t] + (3\text{ V})\sin[(1500\text{ s}^{-1})t]$. Determine the current through the inductor if $L = 5$ mH.

21. (III) (a) Draw the phasor for the function $D \cos(\omega t + \phi)$ on the graph that contains the phasor for the function $C \sin(\omega t + \phi)$. Which phasor is more advanced in phase—that is, points in a direction corresponding to a larger angle, as measured from the $+x$-direction? (b) What is the angle between the two phasors on the plot you drew for part (a)? Which phasor leads? (c) Repeat the exercise for the function $f(t) = A \cos(\omega t) + B \sin(\omega t)$.

34–3 AC in Series RLC Circuits

22. (I) Consider an LC circuit driven by an AC source of emf. It differs from the one shown in Fig. 34–10 in that the inductor and the capacitor are in parallel, while there is no resistor. Given that the input voltage is $V_0 \sin\omega t$, determine the form of the current through the inductor without using Kirchhoff's rules.

23. (I) Consider a radio circuit with a fixed inductance of 14 μH. What is the value of the tunable capacitance for the reception of a 42-m radio wave?

24. (I) What is the range needed for a variable capacitor to be combined with a 0.15-mH coil so that a tuned circuit could be formed to cover the range of broadcast-band frequencies from 540 kHz to 1600 kHz?

25. (I) The driving frequency in a driven RLC circuit is at the resonant frequency. The maximum current carried by the circuit is found to be insufficient for the desired application. By what factor should the resistance be changed to double the maximum current?

26. (I) An AC generator with a voltage amplitude of 50 V and a frequency of 750 Hz is built to drive a circuit meant to be resonant. The resistance of the circuit is 0.5 Ω, and the inductance is 5 mH. What must the value of the capacitance be?

27. (I) A series RLC circuit of frequency 60 Hz has a maximum current of 100 mA. What is the maximum charge on the capacitor? If the impedance is 40 Ω, what is the emf?

28. (II) A series RLC circuit has parameters $R = 50.0\ \Omega$, $L = 40.0$ mH, and $C = 5.00\ \mu$F. Find the capacitive reactance, inductive reactance, and impedance for the frequencies (a) 50 Hz, (b) 1000 Hz, and (c) 20,000 Hz.

29. (II) A series RLC circuit consists of a 1200.0-Hz AC emf with $V_0 = 80$ V; $R = 500\ \Omega$, $L = 92$ mH, and $C = 2\ \mu$F. Find X_C, X_L, Z, Q_{max}, ϕ, and I_{max}.

30. (II) You want to build an AC series circuit with the smallest possible impedance. You have a fixed frequency generator with an angular frequency of 1000 rad/s, and the following circuit elements available: two capacitors, of 1 μF and 100 μF; two inductors, of 10 mH and 25 mH; and two resistors, of 10^3 and 3×10^3 Ω. You may use only one of each type element. What is the lowest circuit impedance and which values of R, L, and C would you choose to make this circuit?

31. (II) Find the voltages across the capacitor and inductor in the AC circuit of Problem 29 at $t = 0.10000$ s if the emf is switched on at $t = 0$ s. All circuit elements initially have no charge or current.

32. (II) An unknown impedance Z is investigated with an oscilloscope. It is connected in series with a 2.0-Ω resistor, and connected to a 60-Hz AC power supply. The horizontal deflection plates of the oscilloscope are connected to the known resistor so that the horizontal deflection of the electron beam is proportional to the potential drop on the resistor. The potential drop on the impedance Z is measured by the vertical displacement of the electron beam (Fig. 34–28). (a) Sketch the shape of the figure on the oscilloscope's screen if Z consists of a coil with inductance $L = 5$ mH and resistance $r = 1.0\ \Omega$. (b) Repeat the calculation if Z is a 3-mF capacitor in series with a resistance $r = 1.0\ \Omega$. (c) How can you tell if Z is capacitive or inductive?

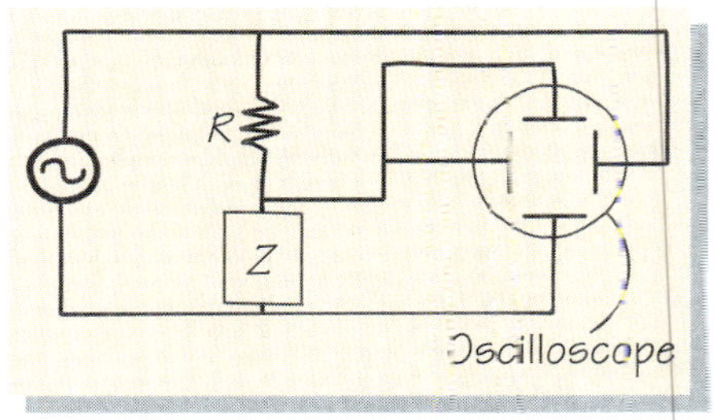

FIGURE 34–28 Problem 32.

33. (II) Given that the maximum voltage in the circuit shown in Fig. 34–29 is 110 V and the frequency of oscillation is 60 Hz, calculate the maximum current and the maximum potential drops across the resistor, capacitor, and inductor.

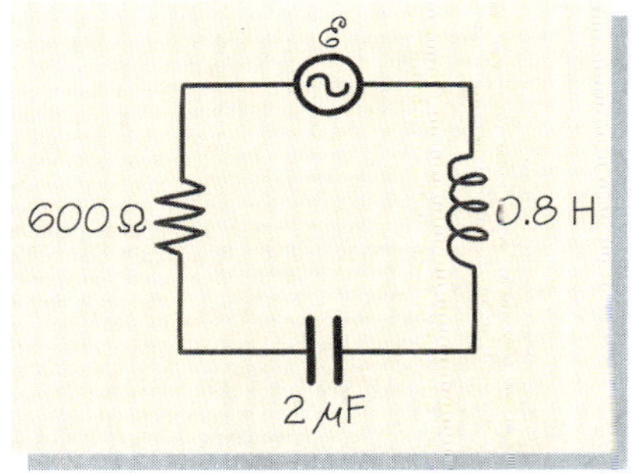

FIGURE 34–29 Problem 33.

34. (II) What is the resonant angular frequency ω_0 of the circuit in Problem 33? Suppose that the voltage generator has a variable angular frequency ω. For what values of ω will the current have half the value it has at resonance?

35. (II) An AC circuit consists of a parallel-plate capacitor and a long, cylindrical solenoid. Suppose that all the dimensions of the apparatus, including the wire sizes, are scaled down by a factor of 2. (Note that the turn density doubles.) How would the resonant frequency of the circuit change? Assume that there are changes in resistance.

36. (II) Show that Eq. (34–33) satisfies Eq (34–28) by direct substitution. Determine the maximum charge Q_{max} on the capacitor in terms of the impedance.

37. (II) Sketch the current and voltage for the following AC series circuits: (a) a pure capacitive circuit, (b) a pure inductive circuit, (c) an RL circuit, (d) an RC circuit, and (e) an LC circuit.

38. (II) A resistor draws 5 A when connected to a 12-V, 60-Hz line. A capacitor of what capacitance, when connected in series with the resistor, will drop the current to 2 A? What are the voltage drops across the capacitor and the resistor?

39. (II) A 16-μF capacitor is connected in series with a coil whose resistance is 30 Ω and whose inductance can be varied. The circuit is connected across a 12-V, 60-Hz generator. What is the potential difference across the capacitor and across the inductor–resistor combination when the frequency is the resonant frequency?

40. (II) Suppose that the maximum voltages across the resistor, capacitor, and inductor of a series RLC circuit driven by an AC generator of frequency f are identical. If the resistor has a resistance R, find the values of C and L in terms of R and f.

41. (II) A series RLC circuit contains a 70-nF capacitor and a 0.2-Ω resistor. If the circuit is resonant at a frequency of 180,000 Hz, what is the inductance?

42. (II) Consider an RLC circuit in which two resistors, R_1 and R_2, are connected in series, as are two capacitors, C_1 and C_2, and two inductors, L_1 and L_2. Show that the resulting total impedance is of the form

$$Z_{\text{total}} = \sqrt{[(X_{L_1} + X_{L_2}) - (X_{C_1} + X_{C_2})]^2 + (R_1 + R_2)^2}.$$

43. (III) There is an AC source of emf in a single-loop circuit that produces a potential drop in the form $V(t) = V_0 \sin(\omega t)$, while the current in the circuit takes the form $I(t) = I_0 \sin(\omega t - \phi)$. Make a phasor diagram for the current and potential drop across each element if the circuit contains (a) a resistor and a capacitor, and (b) a resistor and an inductor.

34–4 Power in AC Circuits

44. (I) What is the average power dissipated in the resistor for the circuit in Problem 12?

45. (I) An AC power supply with a frequency of 75 Hz dissipates energy at a rate of 150 W in a 12-Ω resistor. If the current at time 0 s is 5.0 A, what is the current at time 0.04 s?

46. (I) Consider an AC voltage of the form $V_0 \sin(\omega t)$ connected to a capacitor of capacitance C. Calculate the instantaneous power VI delivered by the source of emf, and find the average power dissipated in the circuit. You should have been able to obtain the answer to the second part of this question without doing any calculations. Why is that?

47. (I) Write down expressions for the average power in an RLC circuit in the two limits (a) ω very large, and (b) ω very small. Can you explain why the power goes to zero in the second case?

48. (I) A portable electric heater operating on AC voltage of amplitude 110 V is rated at a power of 1.5 kW. (a) What is the resistance of the heater? (b) Find the rms current. (c) Find the maximum current.

49. (I) What are the power factors for (a) pure capacitive circuits, (b) pure inductive circuits, and (c) pure resistive circuits?

50. (II) Show that, on average, no power is dissipated in a purely inductive circuit (a circuit with neither capacitors nor resistors).

51. (II) What are the power factors for (a) RL circuits, (b) RC circuits, and (c) LC circuits?

52. (II) An AC source of emf operating at a frequency of 85 Hz produces an rms voltage of 150 V. Find the voltage amplitude. The source of emf is connected in series with an impedance of $Z = 70\ \Omega$. Find the rms current and the current amplitude.

53. (II) When a coil draws 200 W from a $V_{\text{rms}} = 110$-V, 60-Hz line, the power factor is 0.6. If the same coil with a capacitor added in series is to draw the same power from a $V_{\text{rms}} = 220$-V, 60-Hz line, what must the capacitance be? If the aim were to maintain the same power factor rather than the same rms power, how would your answer change?

54. (II) A machine shop uses 120 A from a 220-V, 60-Hz line. However, due to the primarily inductive load—motors—the voltage and current are out of phase by 40°, wasting a lot of heat in the cables. A large capacitor connected parallel to the machines can solve this problem (Fig. 34–30). (a) How large should the capacitance be? (b) What will be the current in the main cable with the capacitor attached? (c) What is the total power of the machines in the shop?

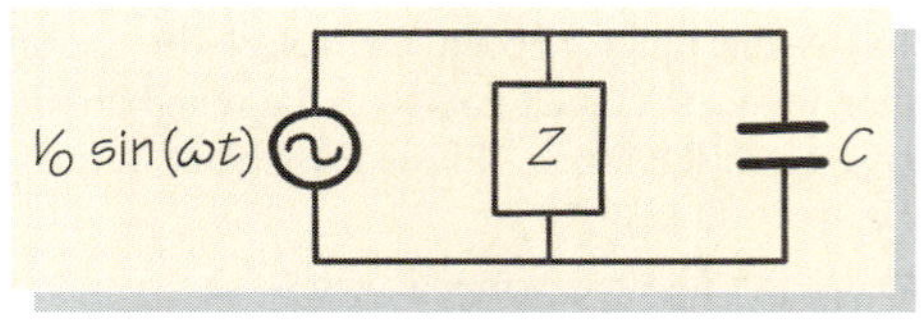

FIGURE 34–30 Problem 54.

55. (II) An electric motor consumes 5 kW of power at 220 V (voltage amplitude) with a power factor of 0.80. This motor is to be run at the end of a power transmission line with a total resistance of 2.5 Ω. What voltage and power must be supplied at the input end of the transmission line?

56. (II) An AC transmission line transfers energy to a device with a power factor of 0.7 at the rate of $\langle P \rangle = 3$ kW and a voltage of 220 V. If the transmission line has a resistance of 3.5 Ω, how much energy is lost to Joule heating in the transmission line?

57. (II) A 220-V generator has a current-carrying capacity of 80 A. What is the maximum rate at which energy can be taken from this generator by an impedance with a power factor of 0.55? for a power factor of 0.95?

58. (II) A 200-Ω resistor and a 15-μF capacitor are connected in series to a 110-V, 60-Hz power supply. Calculate the current, power, and power factor. How will these numbers change if an inductance of 0.24 H is connected in series with this circuit?

59. (II) House current, which has an rms voltage of 110 V and a frequency of 60 Hz, drives a resistor of a variable resistance set at $R = 50\ \Omega$, a capacitor of fixed capacitance $C = 20\ \mu$F, and an inductor of variable inductance, connected in series. (a) What is the power absorbed by the circuit if $L = 10$ mH? (b) What would the power drawn be if the resistance were halved without changing the setting of the inductance? (c) What is the maximum power drawn in part (b)?

60. (II) For a driven series RLC circuit, show that

$$\frac{R}{Z} = \frac{1}{\sqrt{1 + Q^2\left(\frac{\omega}{\omega_0} - \frac{\omega_0}{\omega}\right)^2}}.$$

Here, Q is the quality factor, or Q-factor, defined by $Q \equiv \omega_0 L/R$. This equation shows the resonant characteristic of the power loss when $\omega = \omega_0$. For large values of Q, the resonance is very sharp. For small values of Q, the resonance is broad.

61. (II) For Problem 60 plot R/Z for values of ω/ω_0 from 0.4 to 2.5 and values of Q of 1, 10, and 100. Use a computer program and graphics output, if available.

62. (II) For a driven series RLC circuit, show that Q is related to $\Delta\omega$ by the relation

$$\frac{\Delta\omega}{\omega} = \frac{1}{Q}\frac{\omega_0}{\omega} \simeq \frac{1}{Q}.$$

34–5 Some Applications

63. (II) Consider the circuit treated in Example 34–6 and drawn in Fig. 34–19. Take $C = 5$ nF and $R = 120\ \Omega$, but assume now that the input emf has the purely sinusoidal form $V_1 \sin(\omega t)$, where $V_1 = 0.20$ V. Calculate the potential across the capacitor for (a) $f = 100$ Hz, (b) $f = 10^5$ Hz, and (c) $f = 10$ MHz.

64. (II) Design a high-pass RC filter that will remove voltages with frequencies lower than 8 kHz.

65. (II) Design a high-pass RL filter for filtering out signals with frequencies lower than 8 kHz.

66. (II) The first condition for impedance matching is that the resistances are equal [Eq. (34–58)]. Show that this is true by starting with Eq. (34–57) in the case that the reactance terms are equal and opposite. Take a derivative of the resulting average power with respect to R_2, set it equal to zero, and show that this gives the equal resistance condition.

67. (II) A diode, through which current can flow only when the emf is positive, acts as a filter for an AC generator of angular frequency ω. The current has maximum magnitude I_0. Find its average and rms values.

68. (III) An RC filter circuit like that shown in Fig. 34–19 is called a *high-pass* filter circuit when the voltage output is taken across the resistor. Plot the ratio V_{out}/V_{in} as a function of frequency. Why does such a circuit block signals of low frequency but allow high-frequency signals to pass?

69. (III) An RC filter circuit like that shown in Fig. 34–19 is called a *low-pass* filter circuit when the voltage output is taken across the capacitor. Plot the ratio V_{out}/V_{in} as a function of frequency. Why does such a circuit block high-frequency signals but allow low-frequency signals to pass?

70. (III) Consider the LC filter of Fig. 34–31 with the emf $V_0 \sin(\omega t)$. Assume that $X_L >> X_C$ (or $\omega >> \omega_0$). (a) Show that $V_{out} = (X_C/X_L)V_0$. (b) Show that the circuit of Fig. 34–31 is generally effective in reducing the AC components, but not the DC components, of emf.

71. (III) Given that the driving voltage of the RLC circuit shown in Fig. 34–32 is $V = V_0 \cos(\omega t)$, calculate the currents in the three elements. Is there a resonant frequency? [*Hint*: Write down the circuit equations, and substitute the trial solution $I = I_0 \cos(\omega t + \phi)$].

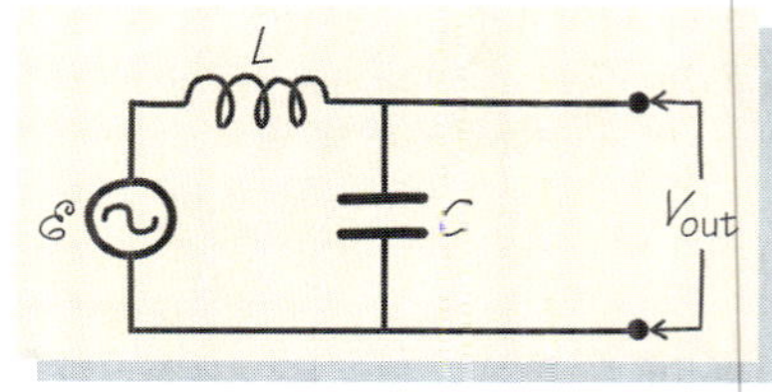

FIGURE 34–31 Problem 70.

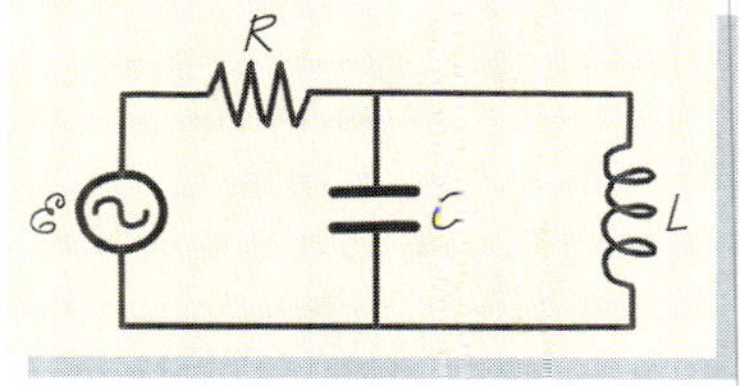

FIGURE 34–32 Problem 71.

General Problems

72. (II) Consider the circuit shown in Fig. 34–33. The emf has an amplitude of $V_0 = 5$ V and a frequency of 8000 Hz; $L = 30\ \mu$H, $C_1 = 6\ \mu$F, and $C_2 = 14\ \mu$F. Find (a) the maximum current; (b) the resonant frequency.

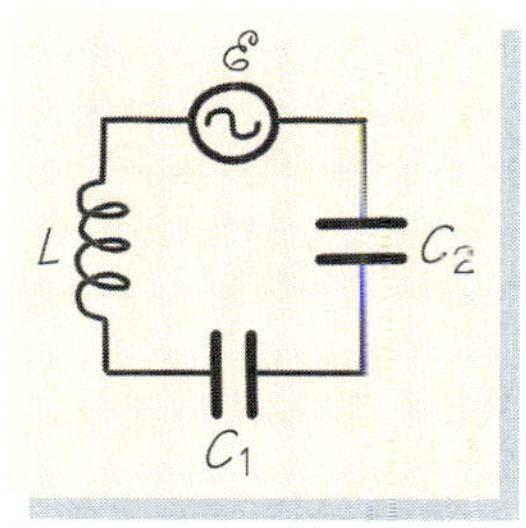

FIGURE 34–33 Problem 72.

73. (II) Calculate (a) the maximum instantaneous voltage across each capacitor; and (b) the maximum instantaneous voltage across the inductor for the circuit shown in Fig. 34–33. Use the parameters specified in Problem 72.

74. (II) An amplifier with an equivalent impedance of 15,000 Ω is to be connected to an 8-Ω speaker through a transformer (Fig. 34–34). What should the turn ratio of the transformer be?

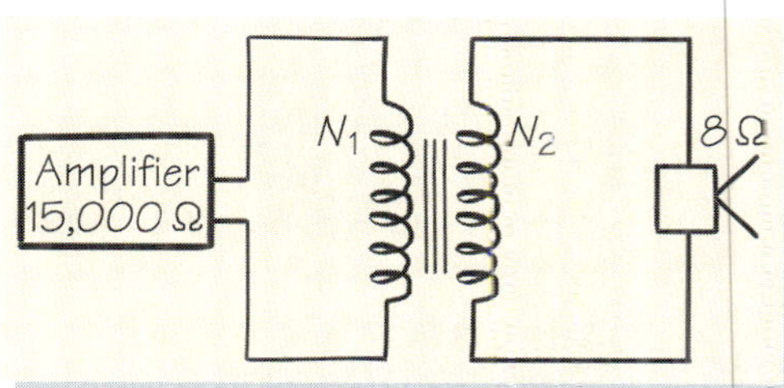

FIGURE 34–34 Problem 74.

75. (II) The impedance Z_1 in Fig. 34–35 can be regarded as a pure resistance $R_1 = 15\ \Omega$, whereas the impedance Z_2 is associated with a series resistance $R_2 = 8\ \Omega$ and a capacitance $C = 2\ \mu$F. If $f = 3000$ Hz and $V_0 = 3$ V, what is the power dissipated in Z_2?

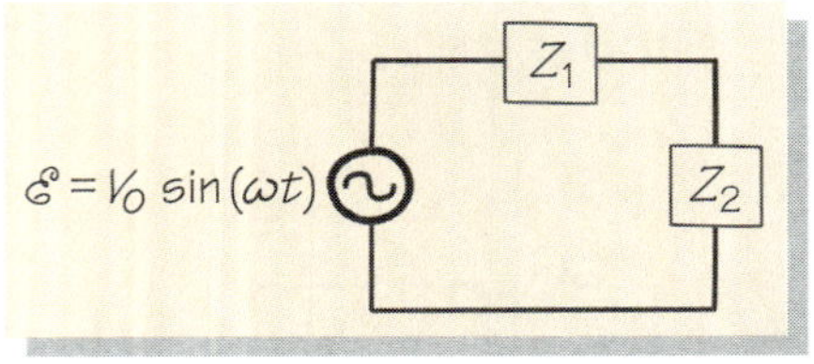

FIGURE 34–35 Problem 75.

76. (II) Consider the circuit shown in Fig. 34–36. The emf has an amplitude of $V_0 = 12$ V and a frequency of 400 Hz; $L = 10$ mH, $C_1 = 20\ \mu$F, and $C_2 = 30\ \mu$F. Find (a) the maximum current in each leg; and (b) the resonant frequency.

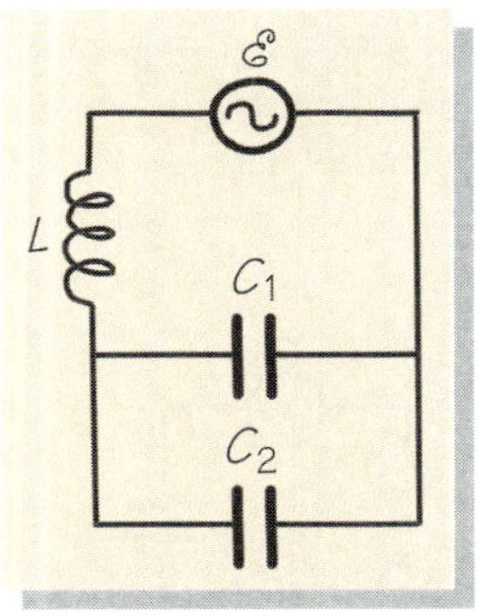

FIGURE 34–36 Problem 76.

77. (II) Calculate (a) the maximum instantaneous voltage across each capacitor; and (b) the maximum instantaneous voltage across the inductor for the circuit shown in Fig. 34–36. Use the parameters specified in Problem 76.

78. (II) Write down the two equations that specify the currents I_1 and I_2 in the two loops of the circuit shown in Fig. 34–37.

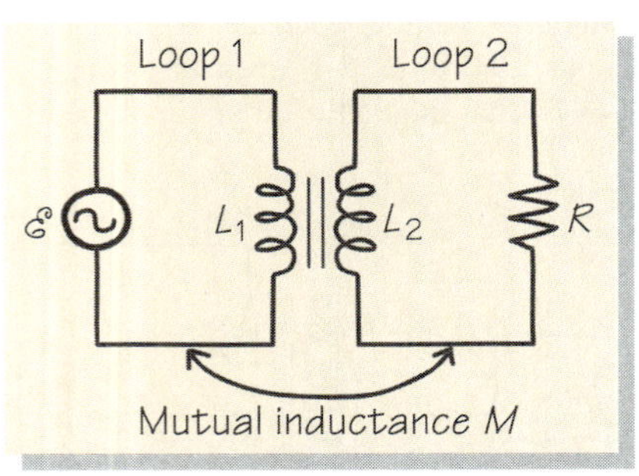

FIGURE 34–37 Problem 78.

79. (II) A series RLC circuit is to be designed to have a resonant frequency of 18 MHz, and the curve of power versus frequency f is to have a full width of 4.0 kHz. If the only capacitor available has a capacitance of 33 pF, what must R and L be?

80. (II) A resistor with $R = 2\ \Omega$ draws a current from a wall plug; a capacitor is connected in parallel with this resistor. The current source has an amplitude of 110 V and a frequency of 60 Hz, and the reactance of the capacitor is 8 Ω at this frequency. What is the current drawn by the parallel combination?

81. (II) A 15-μF capacitor connected in series with a resistor of variable resistance R is connected to a $V_{rms} = 110$-V, 60-Hz AC supply. Plot the variation of the rms current with R, and calculate the value of R for which the power delivered is maximum.

82. (II) An AC circuit supplies $V_{rms} = 220$ V at 60 Hz to a 10-Ω resistor, a 35-μF capacitor, and an inductor of variable self-inductance in the 30 mH to 300 mH range, all in series. The capacitor is rated to stand a maximum voltage of 1200 V. (a) What is the largest current possible that does no damage to the capacitor? (b) To what value can the self-inductance be safely set?

Problems with Complex Variables

In the following set of problems, we suppose there is an AC source of emf that produces a potential drop of the form $V_0 \sin(\omega t)$. In the complex variable technique, we work instead with the complex form $V_C(t) = V_0 e^{i\omega t}$, with the instruction that the original potential drop is given by the imaginary part of this form, $\text{Im}[V_C(t)]$. All quantities with oscillating time dependence are given this treatment. Note: the subscript "C" does not refer to a capacitor, it rather indicates a complex quantity.

83. (III) Consider a capacitive circuit, for which the loop rule takes the form $V_C(t) = Q_C(t)/C$. (a) Calculate $Q_C(t)$, and use it to calculate the complex current $I_C(t)$. (b) Show that the current $I(t)$ calculated according to the prescription at the head of these problems is identical to the one obtained in Eq. (34–17).

84. (III) Consider an inductive circuit, for which the loop rule takes the form $V_C(t) = L\, dI_C(t)/dt$. (a) Calculate $I_C(t)$ in two ways: (i) by direct integration of the equation; (ii) by noting that no matter how often the function $e^{i\omega t}$ is differentiated with respect to time, the time dependence remains $e^{i\omega t}$. (b) Show that the current $I(t)$ calculated according to the prescription at the head of these problems is the same as the one obtained in Eq. (34–23).

85. (III) In complex analysis, the loop rule for the series RLC circuit is $V_C(t) = L(dI_C(t)/dt) + Q_C(t)/C + RI_C(t)$. (a) Make use of the fact that the time dependence of $\exp(i\omega t)$ remains the same no matter how often it is differentiated with respect to time in order to calculate Q_{0C}, defined by $Q_C(t) = -iQ_{0C}e^{i\omega t}$. (b) Use the formula obtained for Q_{0C} to construct the complex quantities $Q_C(t)$ and $I_C(t)$. (c) Use the imaginary part prescription described at the head of these problems to calculate $Q(t)$ and $I(t)$.

86. (III) Consider the RLC circuit of Problem 85. Define the complex impedance Z_C to be $V_C(t)/I_C(t)$, and use the results of Problem 85 to show that Z_C is independent of time, with an absolute magnitude given by Eq. (34–40).

87. (III) Consider an RLC circuit without a driving emf, so that the loop rule reads $L(dI_C(t)/dt) + Q_C(t)/C + RI_C(t) = 0$. Solve this equation by using the substitution $Q_C(t) = Q_{0C}e^{i\omega t}$. Use the equation to obtain a value for ω. Show that your solution leads to the results of Eq. (33–27).

CHAPTER

35

A radar image, taken from the orbiting Space Shuttle, of a volcano in Russia. Radar is just one region of possible wavelengths in electromagnetic radiations; visible light corresponds to another such region.

Maxwell's Equations and Electromagnetic Waves

Faraday's law shows that electricity and magnetism are fundamentally connected; James Clerk Maxwell's introduction of the displacement current enhances this connection. We have now encountered a complete, consistent set of fundamental laws for electricity and magnetism—collectively known as Maxwell's equations. The individual experiments that led to their discovery never hinted at the rich implications that Maxwell's equations have when they are used *together*. The most dramatic prediction of Maxwell's equations is the existence of electromagnetic waves that propagate through empty space at a predictable speed—the speed of light. Light itself is such a wave. In this chapter, we shall discuss the orientation and relationship of the electric and magnetic fields contained in electromagnetic waves; the energy and momentum carried by these waves; and a phenomenon associated with the orientation of the fields called polarization.

35–1 MAXWELL'S EQUATIONS

Let's first list and then comment on **Maxwell's equations**, which fully describe the behavior of electric and magnetic fields in the presence of electric charges and currents.

Maxwell's equations

I. Gauss' law for electric fields

$$\int_{\text{closed surface}} \mathbf{E} \cdot d\mathbf{A} = \frac{Q}{\epsilon_0}. \qquad (35\text{–}1)$$

II. Gauss' law for magnetic fields

$$\int_{\text{closed surface}} \mathbf{B}\cdot d\mathbf{A} = 0. \quad (35\text{–}2)$$

III. Generalized Ampère's law

$$\oint \mathbf{B}\cdot d\mathbf{s} = \mu_0 I + \mu_0\epsilon_0 \frac{d}{dt}\int_{\text{surface}} \mathbf{E}\cdot d\mathbf{A}. \quad (35\text{–}3)$$

IV. Faraday's law

$$\oint \mathbf{E}\cdot d\mathbf{s} = -\frac{d}{dt}\int_{\text{surface}} \mathbf{B}\cdot d\mathbf{A}. \quad (35\text{–}4)$$

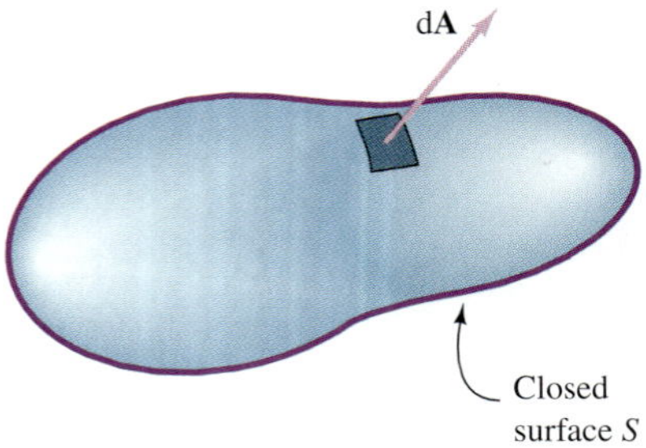

FIGURE 35–1 An infinitesimal surface element d**A** on the closed surface *S*.

I. Gauss' law, which is equivalent to Coulomb's law in static situations, relates the electric flux through a closed surface (the surface can be imaginary) to the charge enclosed [see Eq. (24–7)]. The surface element dA is normal to the surface S and is directed outward with magnitude dA (Fig. 35–1). The charge Q is the total charge contained within the closed surface. The factor ϵ_0 (the permittivity of free space) is associated with our choice of units. Gauss' law holds even for time-dependent electric fields.

II. Magnetic monopoles—which would be the magnetic analogues of electric charge—have never been discovered. Their nonexistence leads to Gauss' law for magnetic fields [see Eq. (30–12)]. This equation holds even for time-dependent magnetic fields.

III. Ampère's law describes the relation between a magnetic field and the current that gives rise to that field. The left-hand side of this equation is the expression for the integral of the magnetic field's tangential component along an arbitrary closed loop C (Fig. 35–2). The right-hand side has two contributions: One is the total current flowing through any surface S bounded by the closed loop C; the other is the rate of change of the electric field flux through such a surface, the displacement current contribution. As we described in Chapter 30 [see Eq. (30–31)], Maxwell was responsible for introducing the displacement current. The presence of the parameter $\mu 0$ (the permeability of free space) is a consequence of SI units.

IV. Faraday's law describes the induced electric field generated by a changing magnetic flux [see Eq. (31–2)]. The left-hand side is the integral of the tangential component of the induced electric field around an arbitrary closed loop C. The right-hand side measures the rate of change of the magnetic flux through any surface S bounded by C, just as in Fig. 35–2. Equation (35–4), as well as Eq. (35–3), implies a sign convention given by a right-hand rule. The minus sign is very important: It represents the fact that the induced electric field, were it to act on charges, would give rise to an induced current that opposes the change in the magnetic flux (Lenz's law).

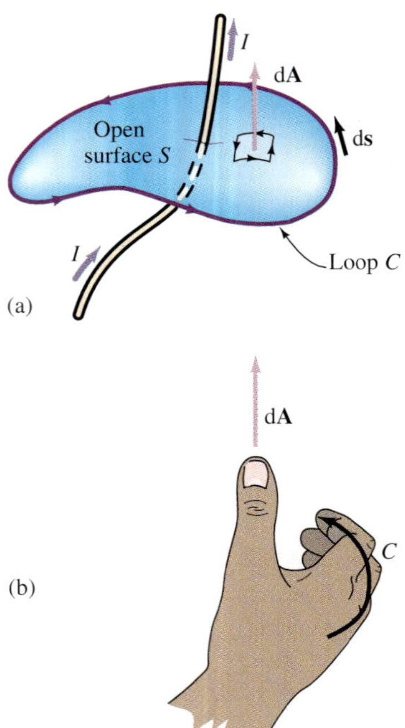

FIGURE 35–2 (a) A surface *S* bounded by the closed loop *C*. A current *I* passes through the surface. If the integration along *C* proceeds counterclockwise, (b) then the direction of the surface elements d**A** that make up *S* is given by a right-hand rule.

Maxwell's equations display a symmetry between electric and magnetic fields. This symmetry is not perfect because magnetic monopoles apparently do not exist. Faraday's law contains no term like the $\mu_0 I$ term in Ampère's law because there is no free magnetic charge to form a magnetic current. In a vacuum, where there are no electric charges, the symmetry is perfect.

In the presence of matter, Maxwell's equations are simply modified. For most types of materials, we can simply replace ϵ_0 by $\epsilon = \kappa\epsilon_0$, where κ is the dielectric constant. Except for ferromagnetic materials, the additional rule that the permeability of the vacuum (μ_0) is to be replaced by the material's permeability (μ) does not affect matters much because μ is very close to μ_0.

35–2 PROPAGATION OF ELECTROMAGNETIC FIELDS

A glance at Eqs. (35–3) and (35–4) shows that when the electric and magnetic fields are time dependent, they influence each other; they are said to be *coupled*. As a consequence of this coupling, electric and magnetic fields can transport energy (and momentum) over much larger distances than might be suggested by the $1/r^2$ falloff of the electric field in Coulomb's law or by the $1/r^2$ falloff of the magnetic field in the Biot–Savart law. The coupled fields produce traveling waves called **electromagnetic waves**. These waves are all around us: Radio and television, microwaves, visible light, and X-rays are examples.

The mechanical construction shown in Fig. 35–3 illustrates what we mean by the coupling of electric fields and magnetic fields. The red rope can support waves in a vertical plane, while the blue rope can support waves in a horizontal plane. A set of threads tie the motion of one rope to the motion of the other; thus, when there is a wave motion in the vertical rope, there must be a wave motion in the horizontal rope. The threads *couple* the two ropes. While there are no ropes and no threads in the case of electric and magnetic fields, Maxwell's equations imply a coupling. If there are oscillations in the electric field, there will be oscillations in the magnetic field, and vice versa.

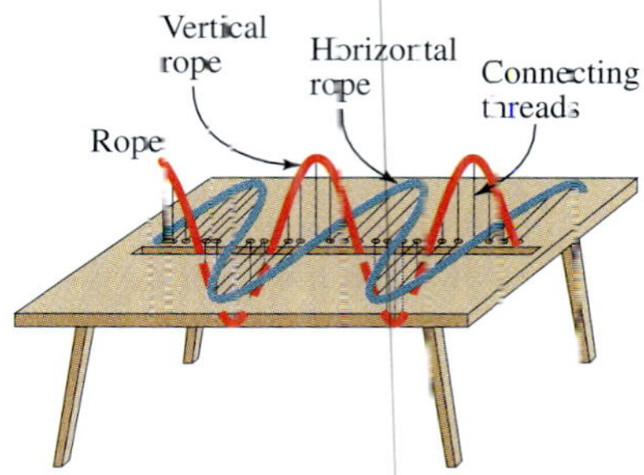

FIGURE 35–3 A mechanical system illustrating coupled waves. When the ropes are tied together with threads, one rope's wave motion is transverse in the vertical direction and produces a wave motion of the other rope, which is transverse in the horizontal direction, and vice versa.

The following argument conveys the physical mechanism by which the fields couple and electromagnetic waves propagate. Consider a straight wire that is aligned with the x-axis and carries a current I (Fig. 35–4a). A magnetic field forms rings around the wire; if the current changes, so does the magnetic field. Specifically, let's take the current to be increasing. The magnetic field increases, as does the magnetic flux through an area A_1 in the xz-plane. According to Faraday's law, Eq. (35–4), a changing magnetic flux induces an emf around the boundary of this area. This emf is associated with the induced electric field shown in Fig. 35–4b. Lenz's law determines the direction of the field.

Let's consider now the top edge of area A_1. Along that edge, the electric field has been induced in the $-x$-direction. This induced electric field changes because it is due to a changing magnetic field; in our example, the electric field is increasing. Now, according to the generalized Ampère's law, Eq. (35–3), we do not need flowing charges to induce a magnetic field. *A changing electric field also produces a magnetic field by giving rise to a displacement current.* The displacement current in this case is along the direction of the changing electric field, which is the direction of the original current. The displacement current is at higher values of z, however, than the original current. At this point, we can see how the propagation works: The displacement current produces a secondary magnetic field $\mathbf{B}'$ at still higher values of z (Fig. 35–4b). In the xz-plane, $\mathbf{B}'$ is perpendicular to that plane; that is, it points in either the $+y$- or $-y$-direction. Because the displacement current varies with time, $\mathbf{B}'$ is also changing. Therefore this secondary magnetic field produces an induced emf aligned in the x-direction at still higher values of z, and the process repeats itself to higher and higher z values.

Let's look at some qualitative conclusions we can draw from this discussion.

1. With the changing current *restricted to a line*, the fields propagate in a cylindrically symmetric way outward from the current line. (In our example, we chose the imaginary loop A_1 to lie in the xz-plane, and this led to a picture in which fields propagate perpendicular to the charge motion, in the z-direction.) The electric field was aligned parallel to the current, and the magnetic field was aligned perpendicular to both the electric field and to the direction of propagation. *This is a general feature of electromagnetic waves.*
2. The current that is the original source of the fields must change with time. A steady current would simply produce a static magnetic field. Equivalently, *charges that produce propagating electric and magnetic fields must be accelerating.* It is reasonable to expect that, if the motion of the charges is harmonic in time, then the electric and magnetic fields will also have a harmonic time dependence. In Section 35–3, we shall verify this expectation.

Only accelerating charges can produce propagating electromagnetic fields

35–3 ELECTROMAGNETIC WAVES

In this section, we shall see how the qualitative discussion at the end of Section 35–2 can be made quantitative. Using Maxwell's equations, we demonstrate that the electric and magnetic fields obey wave equations, and we find their speed. The electric fields

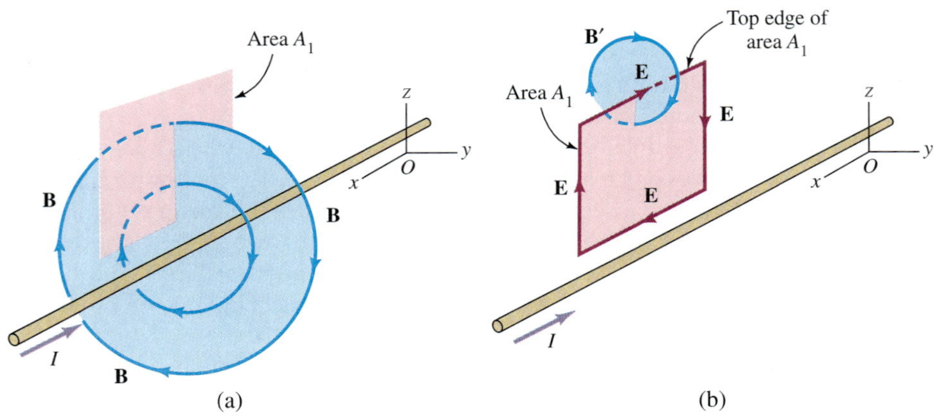

FIGURE 35–4 (a) As we know from Ampère's law, a current-carrying wire aligned in the x-direction has a magnetic field that forms circles in the yz-plane. (b) If the current in the wire changes with time, the magnetic field it produces changes with time, inducing a changing electric field, which in turn induces a changing secondary magnetic field, and so forth.

and magnetic fields forming these electromagnetic waves are in phase, and their magnitudes are closely related.

Instead of a current-carrying wire, we work with a *sheet of current*, which can be formed by a set of wires placed side by side and oriented in the xy-plane. The current is aligned with the x-axis (Fig. 35–5). With this configuration of currents, we shall see that the wave propagates only in the z-direction, perpendicular to the sheet. According to the mechanism discussed in Section 35–2, we expect that the electric fields will be oriented in the same direction as the current (parallel to the x-axis), while the magnetic fields will be oriented in the y-direction. These fields will depend on time in a way that mirrors the time dependence of the current. The electromagnetic fields form *plane waves*, which we recall from Chapters 14 and 15 refers to waves that advance along planar wave fronts—in this case, planes parallel to the xy-plane.

We discussed wave equations in Chapters 14 and 15.

To quantitatively understand how we can get fields that behave like this, we use two of Maxwell's equations—Faraday's law and the generalized Ampère's law—to derive an alternate set of equations for the fields. We actually derive these equations in the subsection "How to Get Maxwell's Equations as Differential Equations Leading to Waves" on pp. 943. Here we write them down:

$$-\frac{\partial B_y}{\partial z} = \mu_0\epsilon_0\frac{\partial E_x}{\partial t} \tag{35–5}$$

and

$$-\frac{\partial B_y}{\partial t} = \frac{\partial E_x}{\partial z}. \tag{35–6}$$

The field components B_y and E_x both depend on the value of z and on the time t. (Recall that partial derivatives appear whenever quantities such as fields depend on two or more variables. In taking a partial derivative with respect to one variable, the other variables are held fixed.)

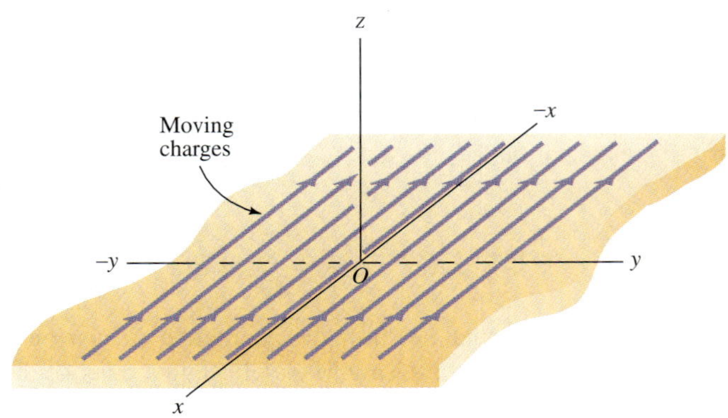

FIGURE 35–5 Current flows in a sheet along the $-x$-direction. It can be approximated by aligning wires side by side in the x-direction. If the current is oscillatory, charges move first in the $-x$-direction, then in the $+x$-direction.

Equations (35–5) and (35–6) lead directly to wave equations for the field components. These two equations couple the two fields. We can combine and simplify them by a straightforward procedure. The partial derivative of Eq. (35–5) with respect to time gives

$$-\frac{\partial^2 B_y}{\partial t\,\partial z} = \mu_0\epsilon_0 \frac{\partial^2 E_x}{\partial t^2}.$$

Similarly, the partial derivative of Eq. (35–6) with respect to z gives

$$-\frac{\partial^2 B_y}{\partial z\,\partial t} = \frac{\partial^2 E_x}{\partial z^2}.$$

Because the order of partial differentiation does not matter, the left-hand sides of these two equations are identical. We can therefore equate the right-hand sides:

$$\frac{\partial^2 E_x}{\partial z^2} = \mu_0\epsilon_0 \frac{\partial^2 E_x}{\partial t^2}. \tag{35–7}$$

This equation for E_x has the same form as an equation we have seen before [Eq. (14–25)]: *It is the wave equation*! A solution of this wave equation is a harmonic plane wave propagating in the $+z$-direction:

$$E_x = E_0 \cos(kz - \omega t + \phi), \tag{35–8}$$

where E_0 is an amplitude, k is a wave number and ω is an angular frequency. Direct substitution of this expression into the wave equation, Eq. (35–7), will verify that it is indeed a solution. The phase angle ϕ is included because we shall want to look at how the phase of the magnetic field, which also has an oscillating solution, is related to that of the electric field.

Some Properties of the Solution to the Wave Equation

We recall from our discussion in Chapter 14 of wave motion that Eq. (35–8) represents a wave of wavelength $\lambda = 2\pi/k$ and frequency $f = \omega/2\pi$. The propagation speed is $v = \lambda f = \omega/k$. This speed is found immediately from the wave equation itself, as comparison with the original form of the wave equation, Eq. (14–25), shows. We have

$$v^2 = \frac{1}{\mu_0\epsilon_0}. \tag{35–9}$$

When we use the numerical values for μ_0 and ϵ_0, we find

$$v^2 = \frac{1}{(1.257 \times 10^{-6}\ \text{T}\cdot\text{m/A})(8.854 \times 10^{-12}\ \text{C}^2/\text{N}\cdot\text{m}^2)}$$
$$= 8.999 \times 10^{16}\ \text{m}^2/\text{s}^2 = (3.00 \times 10^8\ \text{m/s})^2 = c^2,$$

where c is the speed of light. Thus

The speed of electromagnetic waves

$$c = \frac{1}{\sqrt{\mu_0\epsilon_0}}. \tag{35–10}$$

The Relation Between *E* and *B* in an Electromagnetic Wave

To see how E and B for an electromagnetic wave are related, we can start with Eqs. (35–5) and (35–6) and show that B_y also obeys a wave equation similar to that of E_x; namely,

$$\frac{\partial^2 B_y}{\partial z^2} = \mu_0\epsilon_0 \frac{\partial^2 B_y}{\partial t^2}.$$

Like the x-component of the electric field, the y-component of the magnetic field forms a wave that propagates at speed c in the z-direction. However, because Eqs. (35–5) and

(35–6) couple the fields, the waves of B_y do not propagate independently from those of E_x. If we have a wave solution for E_x, Eq. (35–8), then, from Eq. (35–5),

$$\frac{\partial B_y}{\partial z} = -\mu_0\epsilon_0\frac{\partial E_x}{\partial t} = -\mu_0\epsilon_0\frac{\partial}{\partial t}[(E_0\cos(kz - \omega t + \phi)]$$
$$= -\mu_0\epsilon_0\omega E_0\sin(kz - \omega t + \phi). \quad (35\text{–}11)$$

Equation (35–6) becomes

$$\frac{\partial B_y}{\partial t} = -\frac{\partial E_x}{\partial z} = -\frac{\partial}{\partial z}[(E_0\cos(kz - \omega t + \phi)] = kE_0\sin(kz - \omega t + \phi). \quad (35\text{–}12)$$

From these two expressions for the derivatives of B_y, it is easy to check that the following equation has the correct spatial and time dependence:

$$B_y = B_0\cos(kz - \omega t + \phi). \quad (35\text{–}13)$$

Relations between Amplitudes. The amplitude B_0 of the magnetic field wave is not independent of the amplitude E_0 of the electric field wave, as Example 35–1 shows.

EXAMPLE 35–1 Consider the electromagnetic traveling wave for which the electric and magnetic fields are given by Eqs. (35–8) and (35–13). Use the derivative relations we have found to show that the amplitudes are related by $E_0 = cB_0$.

Solution: Here we are proving a relation between the two coupled quantities E and B, so we must use the equations that couple them. Equation (35–11) is a coupling equation that relates a derivative of B_y to a derivative of E_x. This equation states that if E_x is given by Eq. (35–8), then

$$\frac{\partial B_y}{\partial z} = -\mu_0\epsilon_0\omega E_0\sin(kz - \omega t + \phi).$$

With B_y given by Eq. (35–13), we can compute the partial derivative

$$\frac{\partial B_y}{\partial z} = \frac{\partial}{\partial z}[B_0\cos(kz - \omega t + \phi)] = -kB_0\sin(kz - \omega t + \phi).$$

We equate these two results:

$$-kB_0\sin(kz - \omega t + \phi) = -\mu_0\epsilon_0\omega E_0\sin(kz - \omega t + \phi).$$

The sine factor cancels, and we are left with

$$B_0 = \frac{\mu_0\epsilon_0\omega}{k}E_0.$$

The factor $\omega/k = c$, whereas $\mu_0\epsilon_0 = 1/c^2$, so we are left with $B_0 = E_0/c$, the relation we needed to show.

The relation between electric and magnetic field amplitudes in an electromagnetic wave

The relation between the electric and magnetic field amplitudes in an electromagnetic wave is independent of the currents that set up the original wave. We have, in general,

$$E = cB, \quad (35\text{–}14)$$

where E and B are the amplitudes of the fields in an electromagnetic wave.

In an electromagnetic wave, the electric and magnetic fields and the direction of propagation are mutually perpendicular.

The Transversality of Electromagnetic Waves. The fields described by Eqs. (35–8) and (35–13) form traveling waves that propagate in the z-direction. Even though the fields are oriented in the x- and y-directions, these fields do not depend on x or y. Waves of this type have the same fields everywhere on a plane parallel to the xy-plane and are said to describe *plane waves*. The fact that they are plane waves has to do with the currents we used to set up the waves in the first place, and other forms are possible. In particular, there is nothing special about the x- and y-directions. If we had set up our currents to run in the y- rather than the x-direction, we would have found another set of solutions, with **E** in the y-direction and **B** in the x-direction. The wave propagation would still have been in the z-direction. It is generally true that *the electric field and the magnetic field are perpendicular to each other*, or

$$\mathbf{E}\cdot\mathbf{B} = 0. \qquad (35\text{–}15)$$

Moreover, an electromagnetic wave is *transverse* because the direction of the fields involved is perpendicular to the direction of wave propagation. *Neither the electric field nor the magnetic field has a component in the direction of propagation of the wave.*

The electric field and magnetic field in an electromagnetic wave are in phase.

The Electric Field and Magnetic Field Are in Phase. The phases that appear in the harmonic expressions for B_y and E_x in Eqs. (35–13) and (35–8), respectively, are exactly the same. When the electric field is a maximum, the magnetic field is also a maximum; when one is zero, the other is zero, and so forth. The fields oscillate together as shown in Fig. 35–6. The fields are *in phase*.

Figure 35–6 illustrates the important features of electromagnetic waves: The fields are in phase, transverse (perpendicular to the direction of propagation), and perpendicular to each other.

Electromagnetic Waves Are Real

From the numerical evaluation of v, Eq. (35–9), Maxwell recognized that the wave's speed is *the speed of light*. He immediately drew the conclusion that light (the subject of Chapter 36) is an electromagnetic wave. What is special to us about light is

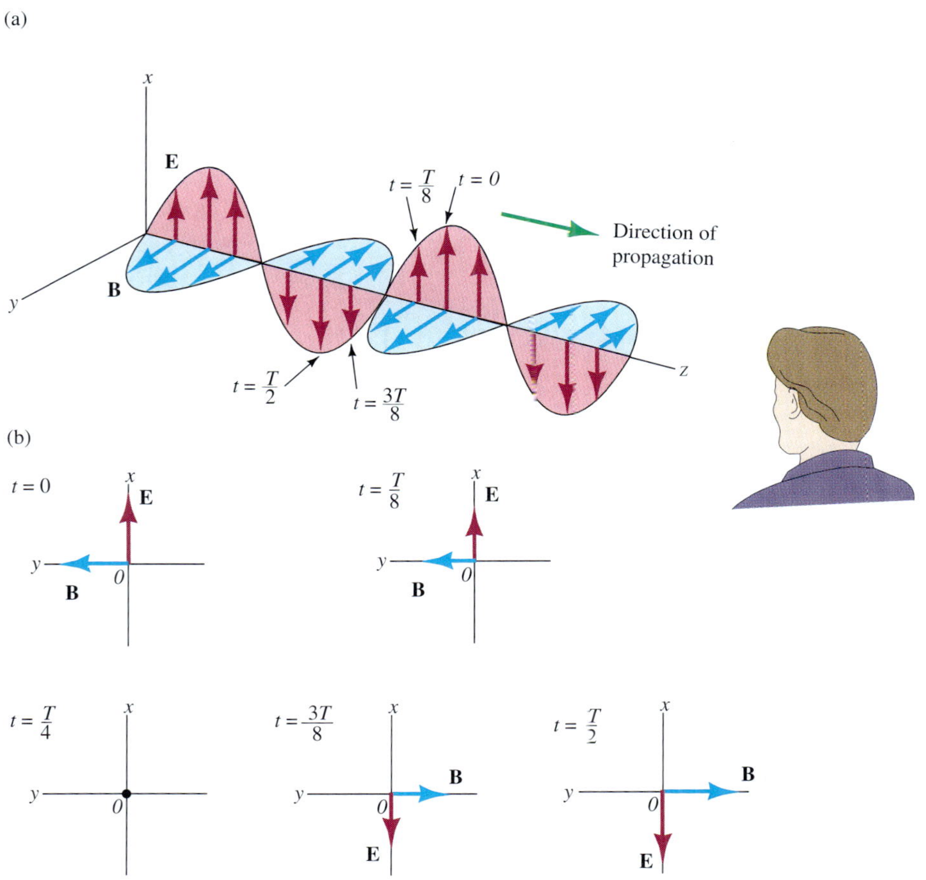

FIGURE 35–6 (a) A view at one particular time of the transverse electric and magnetic fields that propagate along the z-axis. (b) A view downward from the $+z$-direction of the electric and magnetic fields of an electromagnetic wave in an xy-plane over time.

that, through evolutionary adaptation, our eyes have become particularly good detectors of the range of wavelengths in which the Sun emits radiation most strongly and that pass easily through the atmosphere. We call this range of wavelengths the *visible spectrum*. Within the visible spectrum, we interpret different frequencies as colors. The shortest wavelengths of the visible spectrum are violet; the longest wavelengths are red.

Maxwell's treatment of the displacement current and his prediction of electromagnetic waves were published in 1864. A number of the leading physicists of his time found the notion of the displacement current difficult to accept, and it was more than 20 years before all resistance to the theory collapsed. It was not possible to confirm the existence of electromagnetic waves when Maxwell proposed them. That is because there was no technology to create AC currents of sufficiently high frequency to provide detectable radiation. In 1887, Heinrich Hertz devised the first direct test of Maxwell's waves. Hertz used the sparks that cross a "spark gap" when there is a high potential difference between the two points of the gap (Fig. 35–7a) The sparks have a rhythm associated with a back-and-forth motion of charge in the gap. To confirm that this oscillatory motion of charges produces radiation, Hertz took a wire bent into a circle with a (second) gap and placed it near the original spark gap (Fig. 35–7b). The electromagnetic wave that propagated in the space between the spark gap and the circular wire loop gave rise to sparks in the secondary gap, which thereby acted as a detecting antenna. Hertz also reflected waves from metallic surfaces, focused them with a concave metallic mirror, and found that they generally shared many of the properties of light that we shall study in Chapters 36 and 37. The frequencies of the electromagnetic waves studied by Hertz are quite different from the frequencies of the waves that form visible light. The wave equation for electromagnetic waves admits solutions for *any* frequency, and the collection of all frequencies is known as the **electromagnetic spectrum**. In the century since Hertz's work, the electromagnetic spectrum has been explored across an enormous range of frequencies (Fig. 35–8a). The term *electromagnetic radiation* refers to the entire electromagnetic spectrum, including visible light, ultraviolet radiation, infrared radiation, microwaves, radio waves, X-rays, and gamma rays. As the photographs in Fig. 35–8b illustrate, we have developed an elaborate technology to convert images made with many parts of the spectrum into something visible to our eyes.

Our discussion to this point has involved the formation of waves in a vacuum. Electromagnetic waves also propagate in matter. In (transparent) nonmetallic media,

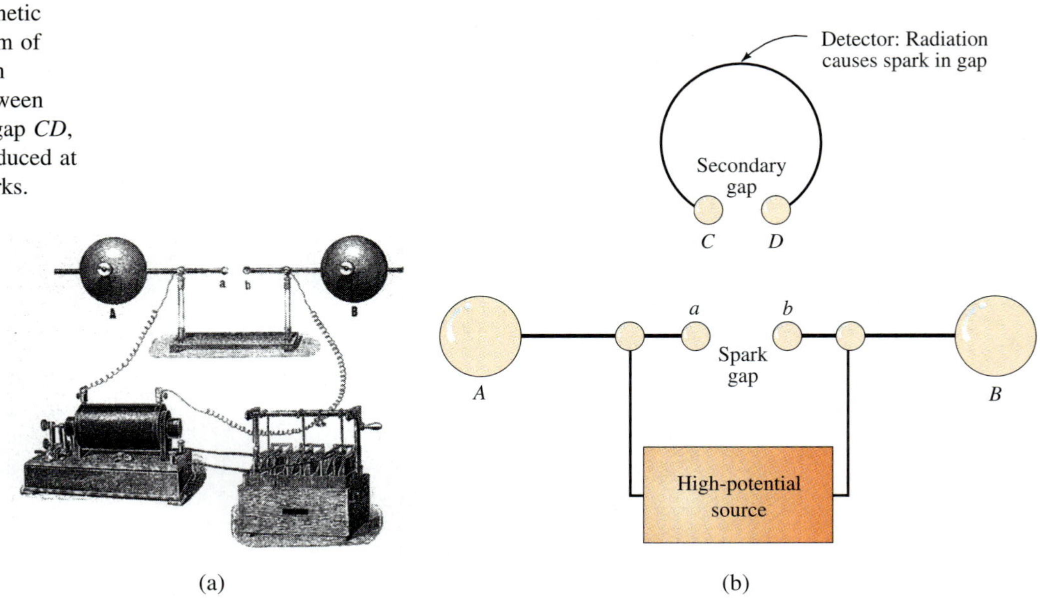

FIGURE 35–7 (a) Hertz's apparatus for the detection of electromagnetic radiation. (b) Schematic diagram of Hertz's apparatus. The radiation propagates from the region between the oscillating spark *ab* to the gap *CD*, which detects the radiation produced at gap *ab* by forming its own sparks.

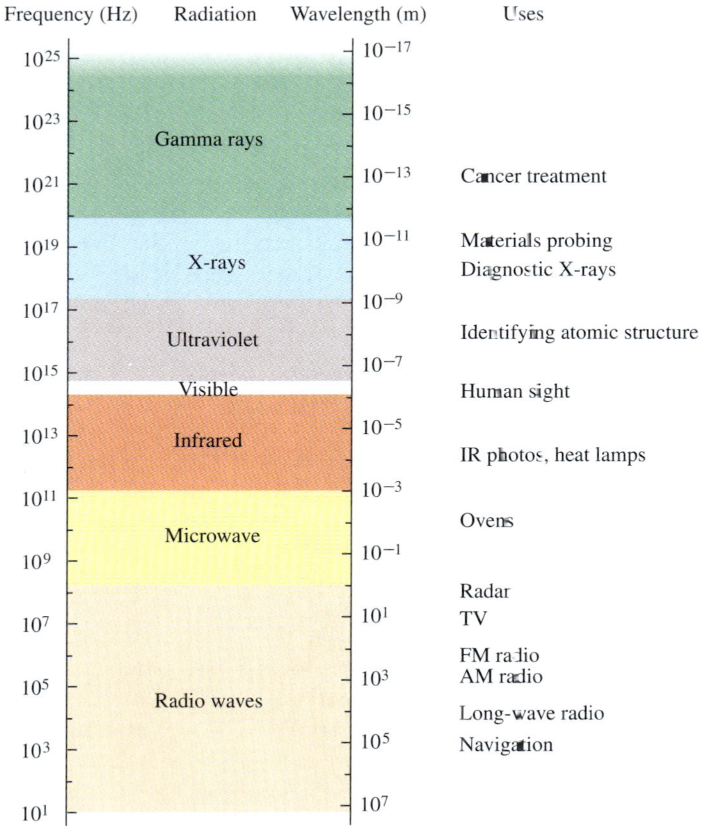

(a)

(b)

FIGURE 35–8 (a) The spectrum of electromagnetic radiation. (b) View from the Landsat 4 satellite of southeastern New England. Imaging of Earth by satellites is not necessarily limited to wavelengths in the visible range, but is limited to wavelengths for which Earth's atmosphere is transparent. The lighter shades in this image are regions that are relatively poor in vegetation—cities.

Maxwell's equations are modified only slightly: The speed of electromagnetic waves is reduced by a factor n according to

$$v = \sqrt{\frac{1}{\mu\epsilon}} \equiv \frac{c}{n}. \qquad (35\text{–}16)$$

The quantity n is the **index of refraction** of a given medium. In all except ferromagnetic materials (introduced in Chapter 32), μ is very close to μ_0 and $\epsilon = \kappa\epsilon_0$, where κ is the dielectric constant of the medium. Thus

The definition of the index of refraction

$$v = \sqrt{\frac{1}{\mu_0\epsilon_0\kappa}} = \frac{c}{\sqrt{\kappa}}; \qquad (35\text{–}17)$$

in other words, the index of refraction $n = \sqrt{\kappa}$. It should be noted that the dielectric constant can depend on the frequency of the electromagnetic wave. When the speed of the wave depends on the frequency, the medium is said to be *dispersive*.

*How to Get Maxwell's Equations as Differential Equations Leading to Waves

Starting from a set of accelerating charges and Maxwell's equations, let's derive the equations that lead us directly to electromagnetic waves. The particular set of charges that we use form currents in the xy-plane, oscillating back and forth in the x-direction, as in Fig. 35–5. As in the qualitative discussion of Section 35–2, we know that the moving charges will give rise to changing electric and magnetic fields. We concentrate on time-dependent fields that vary with z but not with x and y. This implies that, for a given z, the fields are the same out to infinity in the x- and y-directions. This cannot strictly be true in a physical situation; thus, we shall keep in the back of our minds that somewhere, for large enough values of x and y, the fields actually taper off to zero.

Let's draw an imaginary loop C in the yz-plane (at $x = 0$) that goes from $y = b$ to $y = -b$ at some value of z and returns from $y = -b$ to $y = b$ at $z + dz$ (Fig. 35–9). We are going to apply the generalized Ampère's law to the loop. Sides $y = \pm b$, going from z to $z + dz$, are very short. We shall ignore the contribution from the short sides because we can make these sides infinitesimally short. Moreover, our qualitative argument in Section 35–2 gives us no reason to believe that there is a field B_z. (This can be verified with the help of Gauss' law.) Application of the generalized Ampère's law, Eq. (35–3), now becomes easy. All we need to calculate for the line integral in Ampère's law are the contributions from the long (horizontal) sides of the loop. We have

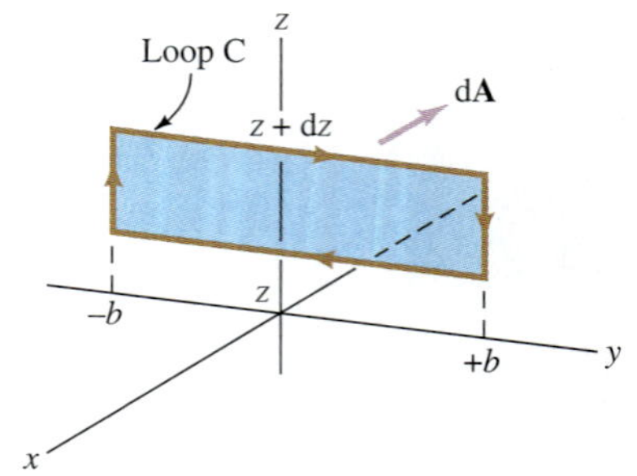

FIGURE 35–9 A loop used to derive a relationship between $\partial B_y/\partial z$ and $\partial E_x/\partial t$ from Ampère's law.

$$B_y(z + dz, t)(2b) - B_y(z, t)(2b) = \mu_0\epsilon_0 \frac{d}{dt} \int_{\text{loop area}} \mathbf{E}\cdot d\mathbf{A}. \tag{35–18}$$

From the definition of a derivative, the difference $B_y(z + dz, t) - B_y(z, t)$ is the rate of change of B_y with respect to z times dz, so

$$2b[B_y(z + dz, t) - B_y(z, t)] = 2b\left(\frac{\partial B_y}{\partial z} dz\right). \tag{35–19}$$

The partial derivative appears because we keep t constant in $B_y(z, t)$.

Now let's consider the right-hand side of Eq. (35–18). In using Ampère's law, a right-hand rule dictates that for loop C in the direction shown in Fig. 35–9, the surface element $d\mathbf{A}$ is oriented in the $-x$-direction, so $\mathbf{E}\cdot d\mathbf{A} = -E_x\, dA$. In addition, the area $A = 2b\, dz$ is infinitesimally small, so we can assume that E_x does not vary over the surface and we can remove it from the integral. Finally, the time derivative on the right-hand side of Eq. (35–18) acts only on E_x, because the surface is itself fixed. Thus

$$\mu_0\epsilon_0 \frac{d}{dt} \int_{\text{loop area}} \mathbf{E}\cdot d\mathbf{A} = -\mu_0\epsilon_0 \frac{\partial}{\partial t} E_x \int_{\text{loop area}} dA = -\mu_0\epsilon_0 \frac{\partial E_x}{\partial t} A$$

$$= -\mu_0\epsilon_0 \frac{\partial E_x}{\partial t} 2b\, dz. \tag{35–20}$$

We have used a partial derivative because z is a second variable that is held fixed. We now equate the two right-hand sides of Eqs. (35–19) and (35–20):

$$2b\left(\frac{\partial B_y}{\partial z} dz\right) = -\mu_0\epsilon_0 \frac{\partial E_x}{\partial t} 2b\, dz;$$

that is,

$$\frac{\partial B_y}{\partial z} = -\mu_0\epsilon_0 \frac{\partial E_x}{\partial t},$$

which is Eq. (35–5).

We next make use of Faraday's law, Eq. (35–4), the fourth of Maxwell's equations. We apply it to a loop C' that goes from $x = a$ to $x = -a$ at some value of z and returns from $x = -a$ to $x = a$ at $z + dz$ (Fig. 35–10). Then, a nearly identical derivation to the one that led us to Eq. (35–5) leads us to Eq. (35–6),

$$\frac{\partial B_y}{\partial t} = -\frac{\partial E_x}{\partial z}.$$

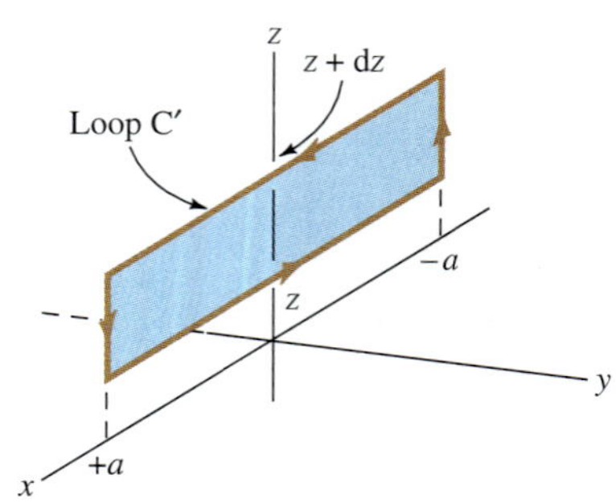

FIGURE 35–10 A loop used to derive a relationship between $\partial B_y/\partial t$ and $\partial E_x/\partial z$ from Faraday's law.

Equations (35–5) and (35–6) are the ones we used in Section 35–3 to find the wave equation for electromagnetic waves.

The Energy of Electromagnetic Waves

We can get sunburned because electromagnetic waves carry energy. We can find the energy causing that burn by using our earlier results on energy in electric and magnetic fields. The energy density, given in Eq. (33–17), is

$$u = \frac{1}{2}\left(\frac{B^2}{\mu_0} + \epsilon_0 E^2\right) = \frac{\epsilon_0}{2}\left(\frac{B^2}{\mu_0\epsilon_0} + E^2\right) = \frac{\epsilon_0}{2}(c^2B^2 + E^2). \quad (35\text{–}21)$$

Let's apply this result to an electromagnetic wave traveling in the z-direction. The fields are given by Eqs. (35–8) and (35–13):

$$E_y = E_0 \cos(kz - \omega t + \phi) \quad \text{and} \quad B_x = -B_0 \cos(kz - \omega t + \phi),$$

where $E_0 = cB_0$. For this wave, the energy density is

$$u = \frac{\epsilon_0}{2}(c^2B_0^2 + E_0^2)\cos^2(kz - \omega t + \phi). \quad (35\text{–}22)$$

In this expression, the two terms are the contributions of the magnetic and electric parts of the wave, respectively. Because $E_0 = cB_0$, *the energy contained in an electromagnetic wave is shared equally between the magnetic field and the electric field.* Equivalently, we could take the contribution of either the electric or the magnetic terms and multiply by 2 to find the total energy density in an electromagnetic wave:

The energy contained in the electric field of an electromagnetic wave is equal to the energy contained in the magnetic field.

$$u = \epsilon_0 E^2 = \frac{1}{\mu_0}B^2. \quad (35\text{–}23)$$

For practical purposes, the oscillations in electromagnetic waves are so rapid that we can simply consider the average of the energy density over one period, which we write as $\langle u \rangle$. The average of the cosine-squared factor in Eq. (35–22) over one period is one-half, so that

$$\langle u \rangle = \frac{\epsilon_0}{2}E_0^2 = \frac{1}{2\mu_0}B_0^2. \quad (35\text{–}24)$$

The Transport of Energy

The $\cos^2(kz - \omega t + \phi)$ time dependence and space dependence of the energy density in Eq. (35–22) shows that the energy in an electromagnetic wave is itself *transported* as a wave; it travels at speed $v = \omega/k = c$ in the z-direction. The amount of energy $d\mathscr{E}_t$ transported across a surface of area A perpendicular to the transport direction in a time interval dt is the energy contained in the volume of area A times the distance $c\,dt$ (Fig. 35–11); that is, the energy density u times this volume,

$$d\mathscr{E}_t = u(Ac\,dt).$$

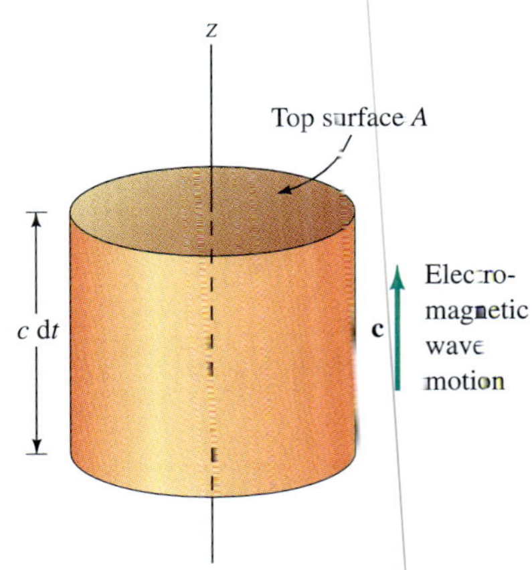

FIGURE 35–11 Electromagnetic energy contained in a volume $Ac\,dt$ is delivered in time dt to the area A.

Thus the rate of energy transport, or, equivalently, the power delivered by the electromagnetic wave, is

$$\frac{d\mathscr{E}_t}{dt} = cuA.$$

Finally, the power delivered per unit area to a surface perpendicular to the direction of propagation—the *energy flux*—is given by

The energy delivered by an electromagnetic wave per unit time and area

$$S = \frac{1}{A}\frac{d\mathscr{E}_t}{dt} = cu. \quad (35\text{–}25)$$

Definition of the Poynting vector, which gives the energy flux of electromagnetic fields

This flux has a direction associated with it and is a vector. The vector **S** that describes the energy flux is known as the **Poynting vector**. It is given by

$$\mathbf{S} \equiv \frac{1}{\mu_0}\mathbf{E} \times \mathbf{B}. \tag{35–26}$$

Let's check that the magnitude and direction of the vector **S** are indeed correct. Because the fields **E** and **B** are at right angles to each other, we notice that the magnitude of **S** is just EB/μ_0. This can be rewritten in terms of the electromagnetic energy density, $u = \epsilon_0 E^2$, as

See vector products in the Problem-Solving Techniques box in Chapter 10.

$$S = \frac{1}{\mu_0}EB = \frac{\epsilon_0}{\epsilon_0\mu_0}E\left(\frac{E}{c}\right) = \epsilon_0 c^2\frac{E^2}{c} = c\epsilon_0 E^2 = cu. \tag{35–27}$$

This is the same magnitude we found in Eq. (35–25). As for the direction, recall that the vector product of two vectors is perpendicular to both of them. From Fig. 35–6, we see that the direction of the vector product **E** × **B** is the $+z$-direction, the direction of wave propagation. More generally, the transversality of the electromagnetic wave will lead to a Poynting vector that lies along the direction of propagation.

The energy density, and hence the magnitude of the Poynting vector, varies with time. The *average* value of the magnitude of **S** over one cycle of the electromagnetic wave is called the **intensity**, I, of the radiation. Equation (35–27) allows us to relate the intensity to the amplitude E_0 of the electric field in the wave:

The definition of intensity

$$I = \langle S \rangle = c\epsilon_0\langle E^2 \rangle = \frac{1}{2}c\epsilon_0 E_0^2. \tag{35–28}$$

Note from Eq. (35–24) that the intensity is also related to the average energy density in the wave, $I = c\langle u \rangle$.

EXAMPLE 35–2 A characteristic number for the rate per unit area at which solar energy is delivered to a spot on Earth's surface is 1000 W/m^2. Use this number to estimate the amplitude of the electric and magnetic fields in the waves that deliver this energy.

Solution: Let's assume that all this energy comes in a single wave so that we can employ some of our results. The power delivered per unit area is the average of the energy flux—the intensity I, Eq. (35–28). From that equation, we solve for the amplitude, E_0, of the electric field:

$$E_0 = \sqrt{\frac{2I}{c\epsilon_0}} = \sqrt{\frac{2(1000\ \text{W/m}^2)}{(3 \times 10^8\ \text{m/s})(9 \times 10^{-12}\ \text{C}^2/\text{N}\cdot\text{m}^2)}} = 0.9 \times 10^3\ \text{V/m}.$$

Once we have the amplitude of the electric field, we can find the amplitude of the magnetic field by recalling that in an electromagnetic wave these amplitudes are related by

$$B_0 = \frac{E_0}{c} = \frac{0.9 \times 10^3\ \text{V/m}}{3 \times 10^8\ \text{m/s}} = 0.3 \times 10^{-5}\ \text{T}.$$

It is instructive to compare the field strength in sunlight (Example 35–2) with the field strength in, say, the waves emitted by a TV station 1 km from the station. For purposes of our comparison, let's suppose that the station broadcasts with $P = 50{,}000$ W and that all of this power is distributed equally in all directions. (This will not be true for real stations.) This intensity is P divided by the area of a sphere of radius r, $I = P/(4\pi r^2)$. We then find the electric field amplitude as we did in the example, with the result $E_0 \simeq 0.02$ V/m, some 45,000 times smaller than the field strength for solar energy.

Momentum in Electromagnetic Waves

An electromagnetic wave carries momentum as well as energy. To see this qualitatively, let's reconsider a plane wave that travels in the z-direction. We simplify matters by employing a wave with electric and magnetic fields along the x- and y-directions, respectively. When such a wave impinges on a particle with charge $+q$, the fields exert forces on the particle. Suppose that, at a given time, the oscillating electric field of the wave points in the $+x$-direction so there is a force qE in the $+x$-direction. The charge accelerates and moves with some velocity $\mathbf{v}$ in the $+x$-direction. If $\mathbf{E}$ points in the $+x$-direction, then $\mathbf{B}$ (which oscillates in phase with $\mathbf{E}$), must point in the $+y$-direction. The magnetic force $q\,\mathbf{v} \times \mathbf{B}$ on the charge acts in the $+z$-direction and pushes the charge in that direction. When the electric field later reverses sign, the electric force on the charge acts in the $-x$-direction, and the velocity picks up a component in the $-x$-direction. The magnetic field has also reversed sign, and the magnetic force *continues to act in the $+z$-direction*. All the forces in the x- and y-directions average to zero, but the force in the z-direction is always positive and there is a net force in the $+z$-direction. The charge has an increased momentum in the $+z$-direction; by momentum conservation, this momentum had to have been brought in by the electromagnetic wave.

The momentum density in electromagnetic waves

In this way, the **momentum density** of an electromagnetic wave—the amount of momentum carried by the wave per unit volume—can be shown to be $\mathbf{S}/c^2$. The magnitude of the momentum density is given by

$$\frac{S}{c^2} = \frac{u}{c}, \tag{35–29}$$

and the direction is that of $\mathbf{S}$.

Radiation Pressure. When electromagnetic waves are absorbed or reflected, they transfer momentum to the material on which they impinge. The rate at which momentum is transferred per unit area is a force exerted per unit area; that is, a pressure: Here we refer to **radiation pressure**. When an electromagnetic wave is absorbed, which happens when light falls on a black surface, all the momentum carried by the wave is transferred. The amount of radiation-produced momentum that falls in a perpendicular direction on a surface A in a time interval dt is given by the momentum density multiplied by the volume $A(c\,dt)$. Thus the momentum dp transferred is

$$dp = \left(\frac{S}{c^2}\right)(Ac\,dt) = \frac{S}{c}A\,dt.$$

The force per unit area (radiation pressure) is given by

$$\frac{F}{A} = \frac{1}{A}\frac{dp}{dt} = \frac{1}{A}\frac{S}{c}A = \frac{S}{c} = u, \tag{35–30}$$

where we have used Eq. (35–25). This expresses the radiation pressure when radiation is totally absorbed. When the electromagnetic wave is reflected, which happens when it falls on a shiny, metallic surface, then the momentum of the wave is reversed upon reflection. Thus the momentum density transferred to the metallic surface is double the previous result, $2u/c$ and the radiation pressure is $2u$.

EXAMPLE 35–3 Consider a 10^4-W searchlight that projects a cylindrical beam 0.6 m in diameter. What is the radiation pressure on a metallic mirror placed at right angles to the beam? Ignore the spreading of the beam.

Solution: The power delivered by the electromagnetic wave to a surface at right angles to the beam is given by

$$P = (\text{energy flux})(\text{area}) = SA = cuA,$$

where u is the energy density in the beam at the surface. The area of the beam is $A = \pi r^2 = \pi(0.3\text{ m})^2 = 0.28\text{ m}^2$. Thus

$$u = \frac{P}{Ac} = \frac{10^4\text{ J/s}}{(0.28\text{ m}^2)(3 \times 10^8\text{ m/s})} = 1.2 \times 10^{-4}\text{ J/m}^3.$$

In turn, the radiation pressure is

$$\frac{F}{A} = 2u = 2.4 \times 10^{-4}\ \text{N/m}^2.$$

The pressure is therefore on the order of 10^{-9} atm—a very tiny number.

EXAMPLE 35–4 The intensity (average energy flux) of solar radiation that falls on Earth is $1.4 \times 10^3\ \text{W/m}^2$. Compare the force exerted by solar radiation on a totally absorbing dust particle of diameter 10^{-6} m and mass density $3 \times 10^3\ \text{kg/m}^3$ with that due to the gravity of the Sun. The particle is taken to be at the Earth–Sun distance. The mass of the Sun is $M_{\text{Sun}} = 2 \times 10^{30}$ kg, and the distance between the Sun and Earth is $R = 1.5 \times 10^{11}$ m.

Solution: The intensity $I = uc$ (we assume time averages throughout) leads to a radiation pressure

$$\frac{F}{A} = u = \frac{I}{c} = \frac{1.4 \times 10^3\ \text{W/m}^2}{3 \times 10^8\ \text{m/s}} = 0.5 \times 10^{-5}\ \text{N/m}^2.$$

The area presented by the dust particle is $A = \pi r^2 = \pi(0.5 \times 10^{-6}\ \text{m})^2 = 0.8 \times 10^{-12}\ \text{m}^2$, so the force is

$$F = uA = (0.5 \times 10^{-5}\ \text{N/m}^2)(0.8 \times 10^{-12}\ \text{m}^2) = 0.4 \times 10^{-17}\ \text{N}.$$

The mass of the dust particle is

$$m = \rho V = \rho \frac{4}{3}\pi r^3 = (3 \times 10^3\ \text{kg/m}^3)\frac{4\pi}{3}(0.5 \times 10^{-6}\ \text{m})^3 = 1.6 \times 10^{-15}\ \text{kg}.$$

The force of gravity has magnitude

$$F_g = \frac{GmM_{\text{Sun}}}{R^2} = \frac{(6.67 \times 10^{-11}\ \text{N}\cdot\text{m}^2/\text{kg}^2)(1.57 \times 10^{-15}\ \text{kg})(2 \times 10^{30}\ \text{kg})}{(1.5 \times 10^{11}\ \text{m})^2}$$

$$= 0.9 \times 10^{-17}\ \text{N}.$$

Thus, the two forces are comparable, and the radiation pressure may keep the dust particles from falling into the Sun. This kind of dust grain is typical of those found in interplanetary space.

35–5 DIPOLE RADIATION

FIGURE 35–12 The use of electromagnetic waves of different frequencies has become a normal part of our technology. Three antennas are visible here; one sends and receives, while the other two receive.

We need accelerating charges to produce electromagnetic waves. We refer to systems in which accelerating charges initiate electromagnetic waves as *broadcasting antennas*; we refer to systems in which we detect the response of charges to the fields of an electromagnetic wave as *receiving antennas* (Fig. 35–12). Here, we shall describe one of the simplest systems that can act as an antenna, the *dipole antenna*. Radiation emitted with the characteristic pattern of this antenna is called **dipole radiation**.

A dipole antenna is formed by a charge that moves back and forth in harmonic motion along a line. Usually the oscillating charge is one of the pair of charges of a dipole. The second charge either is at rest or is oscillating as well (Fig. 35–13). Such an antenna is easy to construct with the help of an AC generator. When the dimensions of the antenna are small compared with the wavelength of the radiation, the current throughout the antenna is in phase. The resulting electric and magnetic fields are oriented as shown in Fig. 35–13. They form an outgoing electromagnetic wave whose frequency is that of the oscillating charges.

How the Intensity of Radiation from an Antenna Decreases with Distance

One of the most important characteristics of the electromagnetic waves radiated by an antenna is the rate at which the intensity decreases with increasing distance from the antenna. To understand this feature, it is not important that the antenna be a dipole antenna. Any antenna will do, including an antenna that radiates electromagnetic waves symmetrically in all directions, such as the Sun. From a distance, the electromagnetic waves emitted by the Sun appear to come from a point source, and we can use this fact to study the magnitudes of the electric and magnetic field strengths. As we learned in Section 35–4, the energy flux (the rate of flow of energy per unit area) is given by $S = cu = c\epsilon_0 E^2$. The total energy flow per unit time (the power) across any surface A is

$$P = \int_{\text{surface}} \mathbf{S} \cdot d\mathbf{A}.$$

If the magnitude of the electric field is independent of direction, as we would expect for a point source, then the *total* rate of energy flow across a sphere of radius R centered on the source is

The electric and magnetic fields in the electromagnetic wave emitted by a point source decrease as $1/R$.

$$P = c\epsilon_0 E^2(4\pi R^2). \tag{35–31}$$

But all the radiation emitted must eventually pass through any sphere that surrounds the source and has any radius, so P does not depend on R. From our expression for P, we see that this is possible only if the electric field decreases as $1/R$. The magnetic field must similarly fall off as $1/R$ because the magnetic and electric fields are proportional in an electromagnetic wave. Contrast this result with the typical $1/R^2$ behavior of static electric fields (discussed in Chapter 23).

We can express the result of Eq. (35–31) in terms of intensity. The quantity $c\epsilon_0 E^2$ is the magnitude of the Poynting vector, and its average value, which is defined as the intensity, I, is one-half this value [Eq. (35–28)]. Thus

$$P = 2I(4\pi R^2). \tag{35–32}$$

Because P is independent of R, the intensity of the electromagnetic wave from a point source decreases as $1/R^2$. Example 35–5 illustrates this important property.

EXAMPLE 35–5 A 100-W light bulb emits electromagnetic radiation equally in all directions. Assume that 10 percent of the 100 W is converted into radiation in the visible spectrum. What is the intensity of the visible radiation 1.5 m from the bulb?

Solution: We must find the intensity of visible radiation at a distance R from a source of radiation, given the total energy per unit time (power) radiated by the source into the visible spectrum. This total power is $P_0 = 10$ percent of 100 W $= 10$ W; it is spread uniformly across the surface of any sphere centered on the source. If R is the radius of such a sphere, from Eq. (35–32) $P_0 = 2I(4\pi R^2)$, where I is the intensity at radius R. Thus, with $R = 1.5$ m,

$$I = \frac{P_0}{8\pi R^2} = \frac{10\ \text{W}}{8\pi(1.5\ \text{m})^2} = 0.2\ \text{W/m}^2.$$

Compare this value to the 1400 W/m^2 in sunlight incident at the top of Earth's atmosphere, or to the 1000 W/m^2 of solar energy that reaches Earth's surface. About half this solar energy is in light in the visible part of the spectrum, whereas the light bulb emits most of its energy in the infrared region of the spectrum.

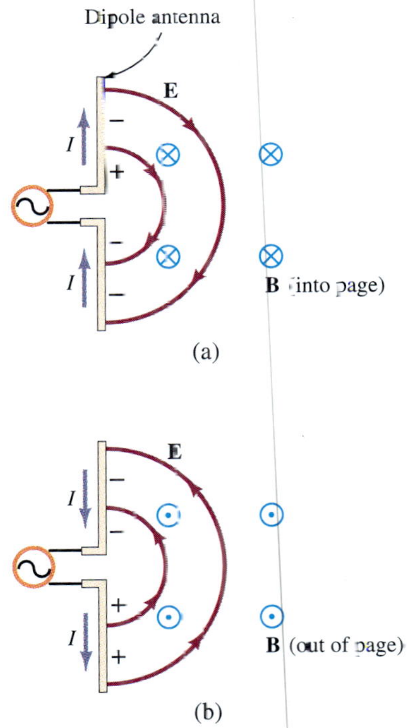

FIGURE 35–13 Pairs of equal but opposite charges move in simple harmonic motion along a line (vertical, here). These pairs of charges form a dipole antenna. In (a) and (b), the two sides of the antenna are oppositely charged, and the electric field directions are reversed.

In our discussion of static electric fields due to a point source, we used symmetry considerations to argue that the *electric field vector points in a radial direction*. This cannot be the case for electromagnetic waves emitted from a point source, however, because we showed that the fields for such waves are transverse to the direction in which they travel. When the waves come from a point source, they travel outward radially. The resolution to what appears to be a paradox is simple: *There are no truly pointlike sources of electromagnetic radiation*. The Sun is not a point, and it radiates because charges within it move and accelerate. The Sun is really a large collection of dipole antennas whose orientation is random.

The Angular Pattern of Dipole Radiation

The variation of the intensity of electromagnetic radiation with the angle of observation is an important property of radiation from an antenna. In our simple dipole antenna (Fig. 35–13), charges execute simple harmonic motion along the antenna direction (we shall call this the z-axis). The motion of the charge determines a preferred direction—along the z-axis. An observer looking along the z-axis would see no motion. An observer looking along a line perpendicular to the z-axis would see the full range of motion of the charges. An observer at an angle θ to the z-axis would see the charges move harmonically with an amplitude reduced from the full amplitude by a factor $\sin\theta$. The electric field that observer would see is thus proportional to $\sin\theta$. Because the intensity is proportional to the square of the electric field in the wave, the intensity of the radiation emitted by a dipole antenna along the direction of θ is proportional to $\sin^2\theta$:

$$S \propto \frac{\sin^2\theta}{R^2}. \tag{35–33}$$

Here, we have also included the $1/R^2$ factor that describes how the intensity varies with the distance R from the antenna. This intensity pattern describes the **angular distribution** of the power emitted by charges oscillating along a line (Fig. 35–14). Figure 35–15 shows the pattern of the electric field lines (plus magnetic

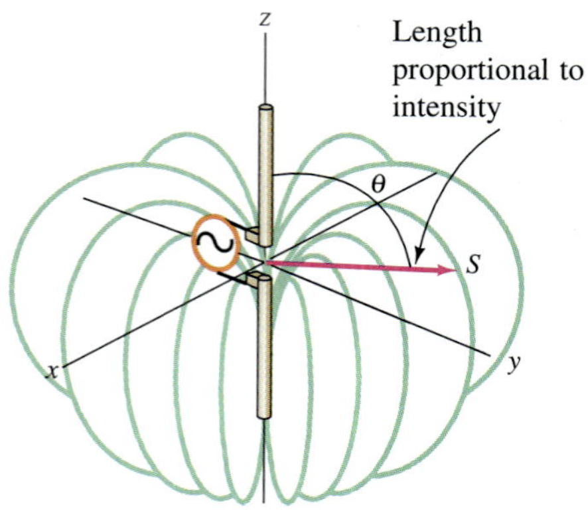

FIGURE 35–14 The intensity distribution S for a radiating dipole antenna. The curve illustrates the relative amount of power emitted as a function of the angle θ.

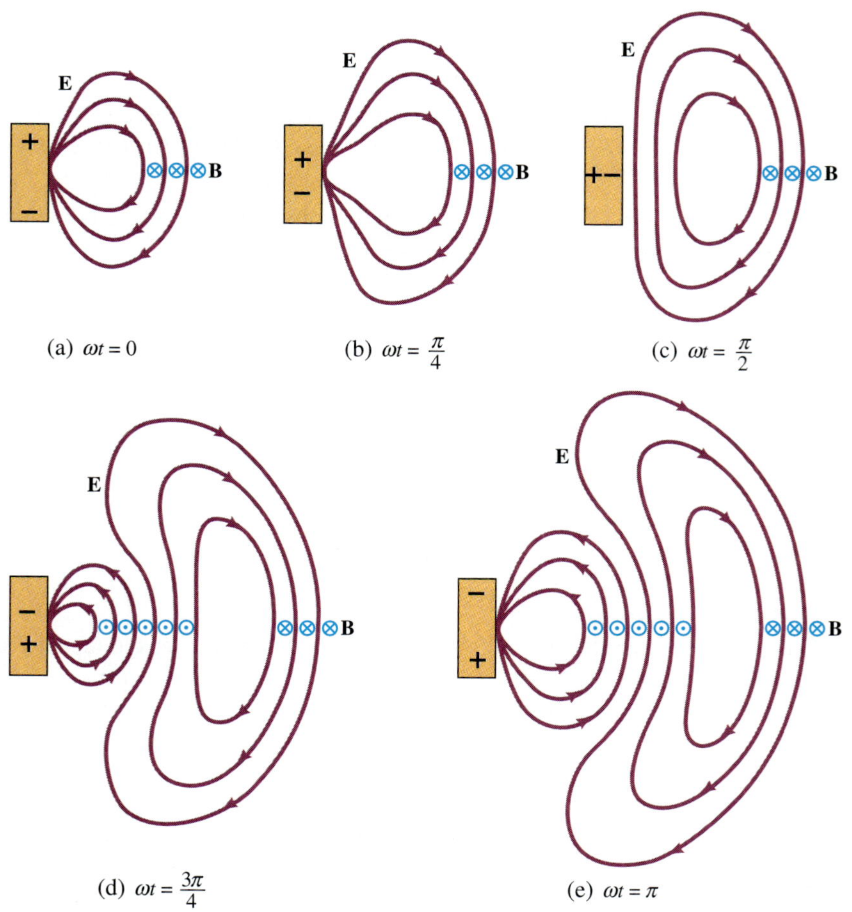

FIGURE 35–15 Electric and magnetic field lines produced by a radiating dipole antenna for a sequence of five times. (After P. Lorrain et al., *Electromagnetic Fields and Waves*, New York: W. H. Freeman, 1988.)

APPLICATION: TELEVISION ANTENNAS

Television signals generated by antennas at the station are often directly received by an antenna in the home. These signals are polarized with an electric field alignment determined by the sending antenna. In the early days of television—the late thirties—signals were transmitted with the electrical field in the vertical plane, and thus the sending and receiving dipoles were vertical. This had to be changed, and to the present day television signals are sent with the electrical field in the horizontal direction; that is, with horizontal polarization. Why was the change necessary? With a vertical dipole sending antenna the radiation pattern is isotropic about the vertical direction, whereas the horizontal dipole produces two radiation maxima orthogonal to the direction of the dipole. Thus an aircraft flying in the vicinity of a vertical antenna will reflect the signal equally all around the antenna, while for a horizontal dipole the reception of the reflected signal will be strong when the aircraft is in a direction orthogonal to the dipole, but will decrease for all other directions. The need to decrease unwanted aircraft reflections, which produce shimmering "ghosts" on the resultant picture, has led to the universal adoption of a horizontal electrical field as the polarization of both television and FM signals. In reality a multiple dipole array is used which greatly reduces the width of the radiation pattern; nevertheless it remains desirable for the reasons given above to have horizontally polarized radiation. There is one disadvantage to this choice: the receiving antenna must be aligned with its wires in the horizontal direction in order to pick up the horizontally polarized signal, and a horizontal dipole makes an excellent bird perch!

field lines) for the electric field radiated by such an antenna for a sequence of five times. Each time corresponds to one-eighth the oscillation period; one-half cycle is traced out.

35–6 POLARIZATION

A little experimentation with *polarizing* sunglasses at the seashore shows that a change in the orientation of the glasses' axis results in a change of the intensity of the light transmitted (Fig. 35–16). This occurs because the sunglasses are made of a material that is sensitive to the direction of the vector electric field. Light reflected from water or sand is **polarized**, meaning that its electric field is oriented in a particular way; the glasses "detect" the polarization of the electromagnetic wave (light).

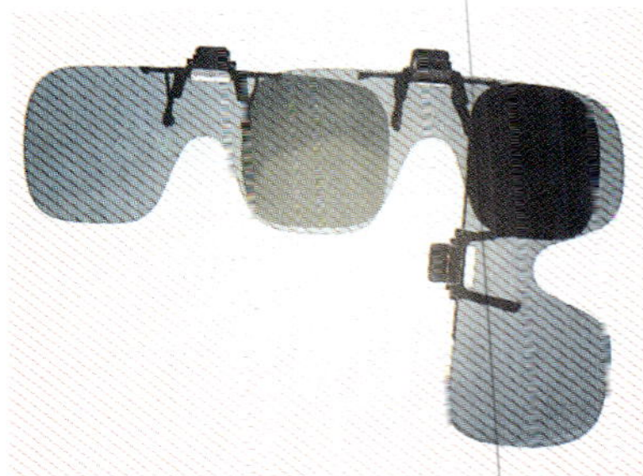

FIGURE 35–16 The extent to which polarizing materials, which have a kind of internal axis, pass light or other electromagnetic waves depends on how that axis is oriented relative to the orientation of the electric field vector of the light. Little light passes through the region where the glasses' axes are crossed—light passing through one pair cannot pass through the second pair.

To further study what polarization means, let's reconsider a charge that oscillates along the z-axis, as in Fig. 35–14. We found that, if we look along the x-direction, we would detect an electromagnetic plane wave that propagates along the x-direction, with an electric field aligned along the z-direction: $\mathbf{E} = E_z\mathbf{k}$, with $E_z = E_1 \cos(kx - \omega t)$. Here, we have set the phase $\phi = 0$. We say that the light is **linearly polarized** along the direction of the electric field vector when that vector has a definite orientation. Suppose a dipole antenna emits radiation with wavelength in the centimeter range. The **polarization** can be detected as follows: A current is induced in a receiving antenna and the rms current detected can be measured (Fig. 35–17). Place a metal grid (such as an oven rack) between the transmitter and the receiver. The diameter of the wires in the grid should be much less than 1 cm, and the grid spacing should be on the order of 1 cm or less. Then, the intensity of the radiation at the receiver depends on the orientation of the metal grid, and we say that the grid acts as an **analyzer** for the polarization.

Here is why the grid acts as an analyzer. The electrons in the grid wires are accelerated by the electric field of the wave along the field direction. When the wires in the grid are parallel to the electric field, the electrons in the grid wires can move in response to the field (Fig. 35–18a). Because they are set into motion, they *absorb large amounts of energy from the field*. This energy is lost in ohmic heating. The electric field of the radiation that passes through is reduced in magnitude because energy has been removed from the incident wave. In effect, the grid is opaque to the polarized ra-

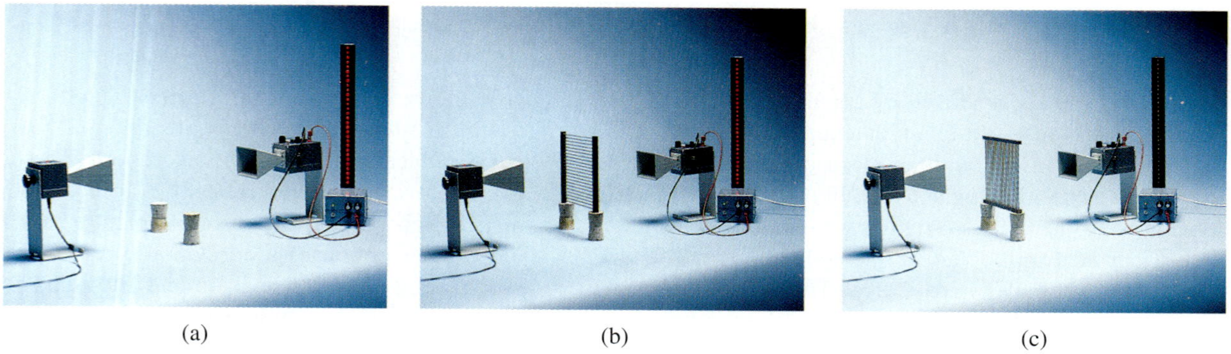

FIGURE 35–17 (a) A receiver and detector for determining the polarization of microwave radiation. The red lights indicate the presence of a signal. (b) A horizontal grid is placed between them, oriented so that the radiation passes. (c) The grid is now oriented vertically so that the radiation cannot pass. The grid's orientation reveals the polarization of the radiation, which is vertical.

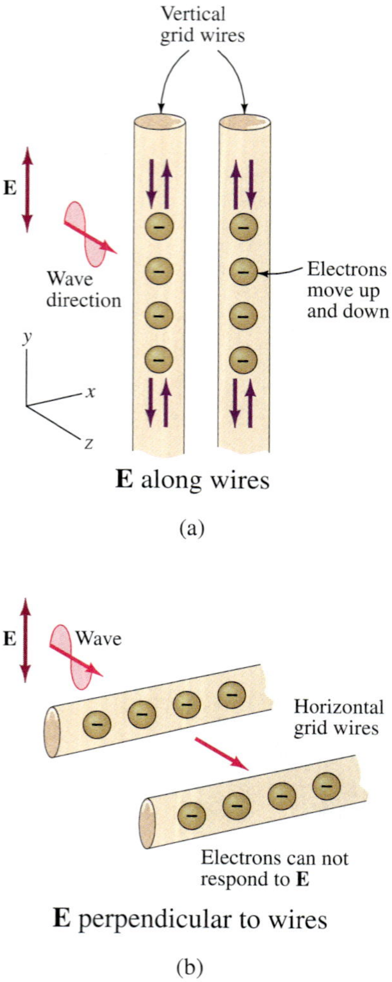

FIGURE 35–18 (a) When the grid wires are oriented in the direction of the electric field of an incoming wave, the electrons of the grid wires can respond and absorb energy from the wave. The transmitted wave is reduced in amplitude. (b) When the wires are perpendicular to the electric field of the wave, the electrons of the grid wires are constrained and cannot respond. Little energy is absorbed, and the wave passes through with little attenuation.

diation when it is oriented along the electric field vector. When the wires in the grid are perpendicular to the z-direction (Fig. 35–18b), the electrons in the metal are accelerated across the diameter of the wire. But, because the diameter is small, the electrons in the grid wires cannot respond fully and cannot absorb large amounts of energy from the incident wave. The energy remains in the transmitted wave. The grid acts as if it were transparent when it is oriented perpendicular to the polarization direction of the wave.

Certain materials, such as Polaroid, are analyzers for visible light. They are made of long molecules aligned parallel to each other. Electrons can easily move along the molecules but not across them and, because the molecular spacings are appropriate to the wavelengths of visible light, these materials behave like the microwave grid does.

A microwave grid or a piece of Polaroid is not simply an analyzer; it is also a **polarizer**: The microwave radiation that passes through the grid becomes polarized perpendicular to the grid wires. This is easily understood. Suppose that unpolarized microwave radiation approaches the grid. *Unpolarized radiation* is radiation that consists of a mixture of waves whose electric field vectors are as likely to point in any one direction as in another, as long as the direction is perpendicular to the direction of wave propagation. As we have seen, only those waves with the electric field oriented perpendicular to the grid can pass through, whereas the waves with the electric field parallel to the grid are absorbed. Thus the radiation that passes through the grid has become polarized perpendicular to the direction of the grid.

Malus's Law

When unpolarized radiation moving in the z-direction falls on a polarizer whose polarizing axis (the axis perpendicular to the "grid wires" within the polarizing material) makes an angle θ with the x-axis, for example, then only the component of any electric field along the polarizing axis will get through. What emerges is radiation that is linearly polarized along a line that makes an angle θ with the x-axis. We take the magnitude of the electric field that has passed through the polarizer to be E_0. The corresponding intensity is then

$$I_0 = \langle S \rangle = (\text{a constant})E_0^2. \tag{35–34}$$

Let's now place a second polarizer so that its axis lies along the x-axis (Fig. 35–19). The amplitude for the electric field in the wave incident on the polarizer is

$$\mathbf{E}_0 = E_0 \cos\theta\, \mathbf{i} + E_0 \sin\theta\, \mathbf{j}. \tag{35–35}$$

Only the component that is parallel to the axis of the second polarizer—the x-axis—passes through. Thus the field behind the second polarizer (which acts here as an analyzer) is given by $E_0 \cos\theta\,\mathbf{i}$. The intensity of the transmitted light is therefore

$$I = (\text{a constant})(E_0 \cos\theta)^2, \tag{35–36}$$

and the intensity of the light is reduced:

$$I = I_0 \cos^2\theta. \tag{35–37}$$

Equation (35–37) is known as **Malus's law**. In particular, when the axes of the polarizer and analyzer are perpendicular to each other ($\theta = \pi/2$), radiation is not transmitted.

One of the important consequences of Malus's law is that when unpolarized light passes through a plane polarizer, it has *half* its original intensity (see Example 35–6).

APPLICATION

Polarizing materials

Malus's law

EXAMPLE 35–6 Light passes through the glass plate of a transparency projector and emerges unpolarized with intensity I_0. (a) A Polaroid sheet is placed on the glass plate with its polarizing axis aligned with the 12 o'clock position. What are the polarization and intensity of the emerging light? (b) A second Polaroid sheet, with its polarizing axis along the 2 o'clock position, is placed over the first. Again find the polarization and intensity of the emerging light.

Solution: (a) We can set up the solution by supposing that the light wave propagates in the z-direction and that the 12 o'clock position is aligned along the $+x$-axis. The polarizer then passes light with its polarization in the x-direction, so the emerging light is polarized in the x-direction. To find the intensity, we must understand that unpolarized light consists of a series of wave bursts with an electric field aligned in different directions—always perpendicular to the propagation direction. If the projection of the incoming electric field on the x-axis is, for some particular burst, $E\cos\theta$, the intensity passed for that burst is $I = I_0\cos^2\theta$. We must now *average* this intensity over all θ. Because the average of the cosine squared is one-half, the average intensity passed is $I_0/2$.

(b) The 2 o'clock direction is at an angle $\theta = \frac{2}{12}(360°) = 60°$ to the 12 o'clock direction. The intensity of the light incident on the second sheet is $I_0/2$; thus, according to Malus's law, the intensity of the radiation that passes through both sheets of Polaroid is

$$I_f = \frac{I_0}{2}\cos^2(60°) = \frac{I_0}{2}\left(\frac{1}{4}\right) = \frac{I_0}{8}.$$

The emerging light is polarized along the 2 o'clock direction.

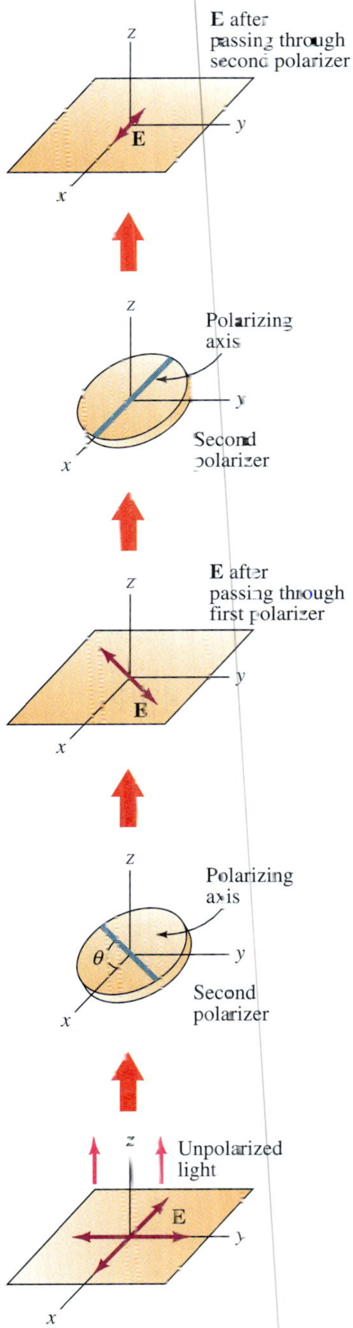

FIGURE 35–19 An unpolarized beam passes first through a polarizer whose axis makes an angle θ with the x-axis, and the beam is then polarized linearly in this direction. A second polarizer aligned with the x-axis allows only the component of the electric field aligned along the x-axis to pass.

How to Produce Polarized Radiation

We have seen two ways to produce polarized radiation: by accelerating charges in an oriented dipole antenna and by passing unpolarized radiation through a polarizer. Let's look at two more ways to polarize radiation.

Polarization by Scattering. If you look at the beam of an automobile headlight in a rainstorm or a snowstorm, there is a pronounced scattering of the light. The beam is much less visible from the side on a dry night (if there is not much dust in the air). Nevertheless, even in the absence of droplets or dust particles, light and other forms of electromagnetic radiation are scattered by air molecules. The scattering mechanism is the following: The oscillating electric field $\mathbf{E}$ of the incoming radiation sets in motion the electrons in the air molecules. The electrons act like oscillators subject to an

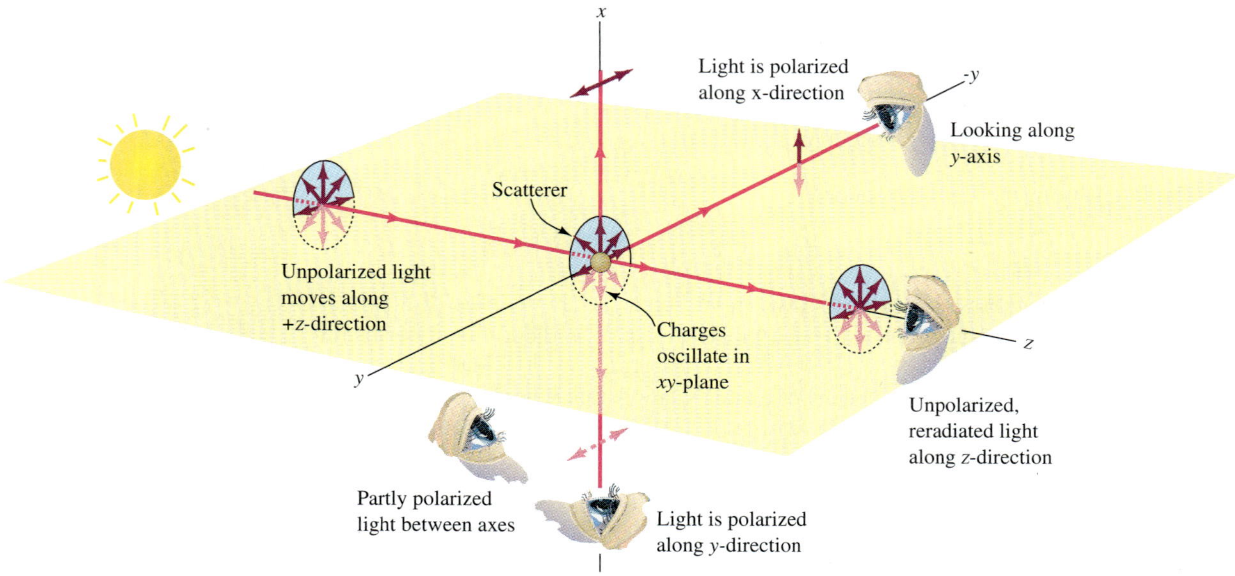

FIGURE 35–20 The polarization of radiation by scattering. An electromagnetic wave can propagate through a material because the wave's electric and magnetic fields cause electrons in the material to oscillate at the radiation frequency; these electrons in turn radiate new waves of the same frequency. The electric field of these new waves is aligned with the electron's motion. Here, unpolarized radiation is perpendicularly incident on the xy-plane in a gas; the electric fields of the radiation lie in that plane but are otherwise unrestricted. An observer along the z-axis sees the full range of motion of the electrons in that plane and hence sees unpolarized light. An observer at 90° to the original wave direction can see a side view of the plane from which the light is radiated and hence sees light fully polarized; the polarization direction is parallel to the plane's edge. At intermediate angles, the reradiated light is partly polarized.

external harmonic force and oscillate with the frequency of the incoming field. The electrons move in a plane perpendicular to the incident radiation and, if the incident wave is unpolarized, then there is no preferred direction to the electron motion as long as it occurs in the plane. An observer looking at an electron from a direction close to that of the incident radiation will see a radiated field that is unpolarized because there is no preferred direction. In contrast, an observer looking at the electron from a direction perpendicular to the direction of the incident radiation will see the electron moving in just one direction (and will not see the component of the motion toward or away from him or her). This observer thus sees 100 percent linearly polarized light (Fig. 35–20). The polarization is partial for angles between these directions. If you live where the atmosphere is clear, you can easily observe this by holding a piece of Polaroid and look toward (*but not at*) the Sun. The light intensity will change when the Polaroid is rotated, showing that the light scattered by the air molecules is polarized.

Polarization by Reflection. When unpolarized radiation is reflected from a surface such as water, the reflected light is partly polarized (Fig. 35–21a). When the angle of incidence is just right, the reflected light is fully polarized (Fig. 35–21b). This is for much the same reason that scattered light is polarized (Fig. 35–22). Unpolarized light incident at an angle θ_i (the *angle of incidence*) impinges on a surface. In general, we may decompose the electric field of the incident wave into two components—each perpendicular to the direction of propagation. One of these directions, the z-direction, is perpendicular to the surface of the page and parallel to the reflecting surface; we label the other the a-direction. When the wave arrives at the surface, its electric field accelerates electrons. These accelerated charges radiate and give rise to both the transmitted and the reflected wave.

Let's first discuss the radiation caused by the component of **E** in the z-direction, which is perpendicular to the plane of the paper. The electrons accelerated by that component of the incoming electric field move at right angles to the direction of the reflected wave. An observer looking back along the line of the reflected wave sees the full motion of these electrons. Thus there is strong reflection of this part of the incident wave. Next, let's consider the radiation induced by the component of the electric field of the incident wave in the a-direction. The electrons that absorb this incident radiation move parallel to the a-direction. An observer who looks along the line of the reflected wave sees a foreshortened motion of the electrons and thus only a limited amount of reflected radiation. Thus there is a preferential polarization direction for the reflected light. In the special case that the direction of the reflected wave is at right angles to the a'-direction (perpendicular to the direction of the transmitted radiation), there is no reflected radiation polarized along the a'-direction because, for this right angle, the motion of the absorbing and reradiating electrons along the a'-direction cannot be seen. *The reflected radiation is plane-polarized with an electric field in the z-direction, parallel to the plane of the reflecting surface.* For this special angle, the reflected and transmitted waves must be at 90° to one another.

The angle of incidence for which the reflected and transmitted (or refracted) rays are perpendicular to one another can be calculated when the rules for these rays (Snell's law) are developed, which we will do in Chapter 36. We give the result here: When the *angle of incidence* is the angle θ_B, known as **Brewster's angle**, the reflected ray is plane-polarized. This angle, which is the incident angle in Fig. 35–22, is given by

$$\tan\theta_B = \sqrt{\frac{\epsilon}{\epsilon_0}} = n. \tag{35–38}$$

As we have already noted, the effect is present but less dramatic for other angles. An analyzer whose polarizing axis is oriented in a direction perpendicular to the z-direction (the direction of polarization of the reflected wave) will absorb most of the reflected radiation. Thus Polaroid sunglasses, worn to cut down glare, need to have their polarizing axis aligned in a vertical direction.

(a)

(b)

FIGURE 35–21 Radiation polarized by reflection. (a) Here we see a shop window with confusing reflections. (b) The same scene, but with the camera lens fitted with a polarizing filter. The reflected light passing through the filter is greatly reduced.

*35–7 ELECTROMAGNETIC RADIATION AS PARTICLES

One of the most astonishing discoveries of the early part of the twentieth century is the discovery that *electromagnetic radiation consists of particles*. The research of Max Planck, of Albert Einstein, and of Arthur Compton established that what we call an electromagnetic wave consists of a large number of individual particles called **photons**. These particles are indivisible: It is not possible to have 0.3 photons, for example. For radiation characterized by a frequency $f = \omega/2\pi$, the energy carried by a single photon is

$$E = hf, \tag{35–39}$$

where $h = 6.63 \times 10^{-34}$ J·s is Planck's constant. A photon also carries momentum, given by

$$p = \frac{E}{c} = \frac{hf}{c} = \frac{h}{\lambda}. \tag{35–40}$$

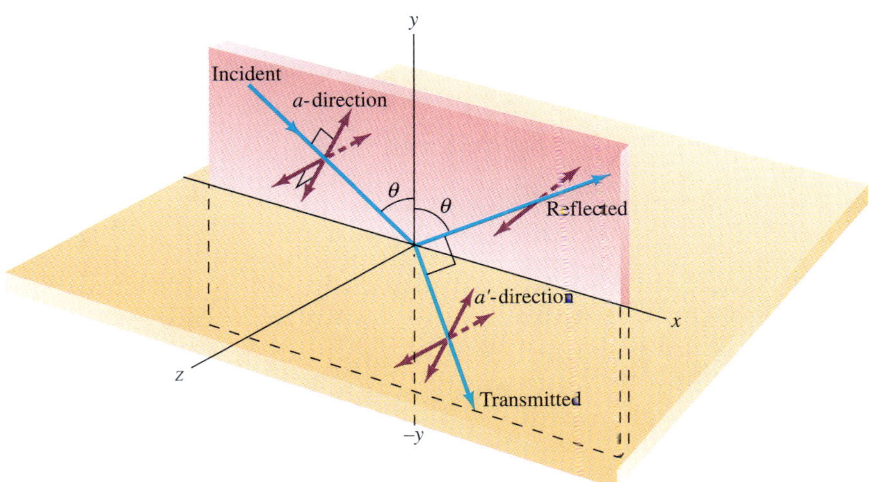

FIGURE 35–22 There is an angle of incidence θ_i for which the reflected wave is fully polarized. The electric field components are perpendicular to the rays, both in the marked plane and out of it.

The particle nature of electromagnetic radiation was established through Compton's experiments on the scattering of radiation by free electrons in carbon. Photons scattered preferentially through a given angle have energy that can be calculated by treating each photon as a relativistic billiard ball that collides elastically with each electron at rest. The momentum of the outgoing photon depends on the collision angle. Equation (35–40) then implies that the frequency of the scattered radiation also depends on the collision angle in a way that can easily be calculated.

That h is small explains why we think of light as a continuous phenomenon rather than a series of individual photons. Someone standing under Niagara Falls does not feel as if he or she is being bombarded by droplets of water! Today, our instruments routinely detect individual photons. Had the evolutionary history of the human eye been a little different—so that the eye could easily respond to a single photon—the notion of radiation as consisting of particles would have been obvious to everyone.

EXAMPLE 35–7 At what rate does a 60-W light bulb emit photons? For simplicity, assume that the light is emitted with a single wavelength of 590 nm.

Solution: In 1 s, the light bulb emits a total energy of 60 J. If there are N photons of frequency $f = c/\lambda = (3 \times 10^8\ \text{m/s})/(5.9 \times 10^{-7}\ \text{m}) = 0.51 \times 10^{15}$ Hz, then

$$N = \frac{E}{hf} = \frac{60\ \text{J}}{(6.6 \times 10^{-34}\ \text{J}\cdot\text{s})(0.51 \times 10^{15}\ \text{s}^{-1})} = 1.8 \times 10^{20}.$$

Thus, on the order of 10^{20} photons are emitted every second.

SUMMARY

Maxwell's equations—which comprise Gauss' laws for electric and magnetic fields, the generalized Ampère's law, and Faraday's law [Eqs. (35–1) to (35–4)]—imply that it is possible to have propagating electric and magnetic fields even in the absence of currents and charges. In the absence of free charges, the electric and magnetic fields obey the wave equation, which has the generic form

$$\frac{\partial^2 E_x}{\partial z^2} = \mu_0 \epsilon_0 \frac{\partial^2 E_x}{\partial t^2}. \tag{35–7}$$

In the case that $E_z = 0$ and $B_z = 0$, the waves propagate along the z-direction. Whatever the direction, the speed of propagation is given by

$$v^2 = \frac{1}{\mu_0 \epsilon_0}. \tag{35–9}$$

This speed is the speed of light, $v = c \simeq 3 \times 10^8$ m/s. In material media characterized by the dielectric constant κ, the speed of propagation is $c\sqrt{\kappa} = c/n$, where $n = \sqrt{\kappa}$ is the index of refraction. There are solutions of the wave equation (electromagnetic waves) in which the fields have the harmonic form

$$E_x = E_0 \cos(kz - \omega t + \phi) \tag{35–8}$$

and

$$B_y = B_0 \cos(kz - \omega t + \phi). \tag{35–13}$$

These waves propagate in the z-direction. More generally, electric and magnetic fields of waves that propagate in a given direction are transverse to that direction. The electric and magnetic field amplitudes are related by

$$E = cB, \tag{35–14}$$

and the fields are perpendicular to each other:

$$\mathbf{E}\cdot\mathbf{B} = 0. \tag{35–15}$$

Electromagnetic waves carry energy with energy density

$$u = \frac{1}{2}\left(\frac{B^2}{\mu_0} + \epsilon_0 E^2\right). \qquad (35\text{–}21)$$

This energy is carried in equal amounts by the electric and magnetic fields. Electromagnetic waves also carry momentum, with momentum density $\mathbf{S}/c^2$, where $\mathbf{S}$ is the Poynting vector, given by

$$\mathbf{S} = \frac{1}{\mu_0}\mathbf{E} \times \mathbf{B}. \qquad (35\text{–}26)$$

Thus radiation can transfer momentum; when a material absorbs radiation, there is a radiation pressure on the material, given by

$$\frac{S}{c} = u. \qquad (35\text{–}30)$$

Charged particles radiate when they are accelerated. For a charge q undergoing an acceleration along the z-direction, the energy flux is proportional to

$$S \propto \frac{\sin^2\theta}{r^2}, \qquad (35\text{–}33)$$

where θ is the angle with the z-axis and r is the distance from the charge. Radiation with a $\sin^2\theta$ angular dependence is called dipole radiation.

The polarization of an electromagnetic wave is the direction of the transverse electric field vector. It can be measured because polarizers transmit electromagnetic waves only along a particular polarization axis. Polarizers may be used to detect as well as to polarize electromagnetic waves. If a second polarizer is placed with its axis making an angle θ with the first one, then the electric field E of the transmitted wave is reduced in magnitude from the electric field E_0 of the incident wave according to $E = E_0 \cos\theta$. Thus the intensity I (the average of the energy flux) of the transmitted light is reduced from the incident intensity I_0 according to Malus's law:

$$I = I_0 \cos^2\theta \qquad (35\text{–}37)$$

Waves can be polarized by reflection. If light falls on a medium of dielectric constant κ at an angle θ_B (Brewster's angle), for which

$$\tan\theta_B = n \qquad (35\text{–}38)$$

(n is the material-dependent index of refraction), then the reflected light is polarized in a direction perpendicular to both the incoming direction and the reflected direction of the wave. Light can also be polarized by scattering.

UNDERSTANDING KEY CONCEPTS

1. Stable charged particles that move uniformly produce no electromagnetic waves. How does the conservation of energy suggest that this must be true?

2. Short-wave radio signals have wavelengths of several tens of meters. Such waves are particularly well reflected by Earth's ionosphere (an upper layer of the atmosphere that contains many free charges). Why would Earth's ionosphere reflect, rather than absorb, these waves?

3. The production and detection of polarization depends on the electric field vector. Can there be, in principle, a polarization associated with the magnetic field vector?

4. Can there be standing electromagnetic waves as well as traveling ones? Recall that mechanical standing waves on a string are possible when certain boundary conditions are satisfied, such as the ends of the string being fixed. How can we control the values of electric or magnetic fields on fixed boundaries?

5. Would the presence of magnetic monopoles analogous to electric charges change the nature of electromagnetic waves in free space?

6. A *solar sail* is a large surface on which the radiation pressure of the sun's radiation can act and thereby push the sail along.

Solar sails have been proposed for spaceships to travel throughout our solar system. What properties must such a sail have, and what difficulties do you see in the proposal?

7. Rockets are propelled forward when mass is ejected backward from them. Could a source of light (or other electromagnetic radiation) be used in place of the mass?

8. How can you tell whether light is plane polarized or not?

9. Can a sound wave in air be polarized?

10. Incident light is linearly polarized along the x-axis. We would like to rotate the direction of polarization so that it lies along the y-axis. Can this be done with one polarizer? Can it be done with two? What is the minimum reduction in intensity when two polarizers are used? Can there be even less intensity reduction with three polarizers?

11. In Hertz's test of the existence of electromagnetic waves, sparks appear in a secondary gap as the result of an AC current in a primary circuit. Hertz interpreted these sparks as due to the effect of electromagnetic waves and not to the effects of Faraday induction. What sort of checks did Hertz need to make in order to rule out Faraday induction?

12. Can you use the example and the arguments given in Section 35–2 to show that electromagnetic waves are transverse? Are there any pitfalls?

13. We showed that electromagnetic radiation carries momentum by thinking about its effect on a free charge. Consider its effect on an electric dipole to see whether it might carry angular momentum as well. Start by orienting the dipole with its axis along the direction of the electric field vector of the electromagnetic wave.

14. Consider the solar sail described in Question 6. Is it better to make a solar sail reflective (shiny) or absorbing (black)?

15. In the subsection on momentum in electromagnetic waves, we mentioned that an electromagnetic wave will accelerate a charged particle, giving it momentum at the same time. Does this mean the wave loses energy and momentum? In other words, if we sit behind a receiving antenna, do we pick up less radiation because of the presence of the antenna?

16. When electromagnetic radiation interacts with matter—for example, when light propagates in a crystal—it is always the electric and not the magnetic field that determines the behavior. Can you explain this, recalling the relation between the two fields in an electromagnetic wave?

PROBLEMS

35–1 Maxwell's Equations

1. (I) Verify the consistency of the dimensions of both sides of each of the four Maxwell's equations.

2. (II) Gauss' laws for electric fields and for magnetic fields differ due to the lack of magnetic charges. Assume that magnetic monopoles (magnetic charges) exist; denote them by the symbol M. Rewrite Gauss' law for magnetic fields, and give the SI units of M.

3. (II) Ampère's and Faraday's laws differ due to the lack of a currentlike term in Faraday's law. Assume that magnetic monopoles exist (call them M), and rewrite Faraday's law. Discuss the physical significance of any new terms added.

4. (II) A region is bounded by an imaginary closed surface. Cut the region into two subregions with an arbitrary surface. Show that if Maxwell's first and second equations (which involve the surface integrals) are valid for both subregions, they are also valid for the whole region. Show that the validity of the third and fourth equations (which involve line integrals) for the two parts of the surface of the regions created by the cut implies the validity of these equations for the full region.

35–3 Electromagnetic Waves

5. (I) If the electric field for a plane wave is given by $E_x = 0$, $E_y = E_0 \cos(kz + \omega t)$, and $E_z = 0$, what are $\mathbf{B}$ and the direction of propagation of the wave?

6. (I) Use dimensional analysis to show that $1/\sqrt{\mu_0\epsilon_0}$ has the dimensions of speed, $[LT^{-1}]$.

7. (I) An FM radio station announcer identifies the station as "Q94;" the number 94 stands for the frequency in some units. What is the wavelength and frequency of the waves emitted by the radio station?

8. (I) What is the relation between the amplitudes of the electric and magnetic fields in an electromagnetic wave propagating in a medium whose dielectric constant is κ? Assume the magnetic permeability of the medium is that of the vacuum.

9. (I) A superposition of electromagnetic waves traveling in the $+z$-direction and electromagnetic waves traveling in the $-z$-direction gives rise to standing waves. Check that a standing wave whose x-component of electric field has the form

$$E_0 \sin(kz)\cos(\omega t)$$

satisfies the wave equation [Eq. (35–7)].

10. (II) Find the approximate wavelength, wave number, frequency, and angular frequency for electromagnetic waves associated with (a) your favorite AM station; (b) your favorite FM station; (c) a microwave oven; (d) yellow light; (e) X-rays.

11. (II) Use Gauss' law to show that electromagnetic waves must be transverse. [*Hint*: Choose as your Gaussian surface a pillbox, with one of the plane surfaces chosen such that $\mathbf{E}$ or $\mathbf{B}$ vanish on it.]

12. (II) Starting from Eqs. (35–5) and (35–6), derive a wave equation for the y-component of the magnetic field. What is the speed of the resulting wave?

13. (II) Write the counterparts of Eqs. (35–5) and (35–6) for electromagnetic fields B_y and E_z that lie in the yz-plane and propagate in the x-direction. [*Hint*: Start with Figs. 35–9 and 35–10. Then relabel the axes according to $x \to y \to z \to x$.]

14. (II) A plane harmonic wave of electromagnetic radiation with wavelength λ is propagating in the $-x$-direction. The z-component of the electric field has magnitude E_0, and there is no y-component. (a) Write an expression for the electric field. (b) Use this expression and the result of Problem 13 to calculate the magnetic field. What vector components will this field have?

15. (II) A plane wave propagates along the direction in the xy-plane that makes an angle θ with the x-axis. Show that the electric field is given by $\mathbf{E}_0 \cos(kx \cos\theta + ky \sin\theta - \omega t + \phi)$. What directions can $\mathbf{E}_0$ have?

16. (II) A plane wave of wavelength 200 m propagates in the z-direction. The electric field points in the y-direction and has an amplitude of 0.2 V/m. Write an expression for the magnetic field, including its amplitude in SI units. Assume that the electric field is at its maximum at $z = 0$, $t = 0$.

17. (II) An electromagnetic wave of wavelength 600 nm propagates in the z-direction. The magnetic field points in the y-direction, and has a magnitude of 10^{-8} T. Write an expression for the electric field, including numerical values and units. Assume that the magnetic field is maximum at $z = 0$ m, $t = 0$ s.

18. (II) An electromagnetic traveling wave is generated at the left-hand end of a tube oriented in the z-direction; the wave travels in the $+z$-direction. At the ends of the tube, $z = 0$ and $z = L$, are highly reflective mirrors. The electric field of the incident wave is $\mathbf{E} = E_1 \cos(kz - \omega t)\, \mathbf{i}$, and the electric field of the wave reflected at $z = L$ is $\mathbf{E} = E_1 \cos(kz + \omega t + \phi)\, \mathbf{i}$. Show that the net electric field forms a standing wave and, by computing B_y, that the associated magnetic field B_y also has the form of a standing wave.

19. (II) Consider the standing electromagnetic wave in Problem 9. If the standing wave is confined to a region lying between $z = 0$ and $z = L$ by two metallic plates, what is the relation between the allowed wavelengths of the radiation and L? (Recall from Chapter 24 that the electric field along a conducting surface must vanish on that surface.)

20. (III) A pulse of electromagnetic radiation travels in the $-z$-direction. The electric field is oriented in the x-direction and is given by $\mathbf{E} = E_0 e^{-(z + ct)^2/a^2}\, \mathbf{i}$. What is the orientation of the magnetic field? Make a guess of the space–time dependence of the magnetic pulse, and use Eqs. (35–5) and (35–6) to find a form for $\mathbf{B}$ that satisfies Maxwell's equations.

35–4 Energy and Momentum Flow

21. (I) The intensity of an electromagnetic wave is 6×10^6 W/m^2. What is the amplitude of the magnetic field in this wave?

22. (I) A radio station emits a signal with a power of 30 kW. What are the values of the electric field and magnetic field at distances of 5 km and 100 km? Assume that the signal far from the antenna is transmitted with equal intensity in all directions. (Real radio stations cannot afford to transmit their energy in this way, and their antennas distribute energy with a high degree of directionality.)

23. (I) The electric field for a given electromagnetic wave has a peak value of 140 mV/m. What is the intensity of the wave?

24. (I) A laser emits a beam with an intensity of 10^{12} W/m^2 across an area of 1 mm^2. What force would the laser beam exert on a black (perfectly absorbing) object?

25. (I) A harmonic plane wave of wavelength 0.45 μm and an electric field amplitude of 3 V/m impinges on a totally reflecting surface of area 200 cm^2. What is the radiation pressure exerted by the wave?

26. (I) A plane electromagnetic wave with maximum electric field amplitude of 380 V/m is incident on a perfectly absorbing surface perpendicular to the direction of propagation. What is the rate of energy absorption per unit area of the surface?

27. (I) The rate at which the Sun emits energy in the form of radiation is 3.8×10^{26} W. (a) Calculate the magnitude of the Poynting vector at a distance of 1.5×10^{11} m from the Sun. (b) What is the radiation pressure exerted on a totally absorbing surface perpendicular to the direction of the radiation?

28. (II) (a) Sketch on the same graph $\sin x$ and $\cos x$ as a function of x. (b) On a separate graph sketch $\sin^2 x$ and $\cos^2 x$. Observe that the periodic functions $\sin^2 x$ and $\cos^2 x$ are identical to one another, except that one is displaced from the other by an interval $\pi/2$. (c) Use your sketch from part (b) to show that the area under the $\sin^2 x$ curve in the interval $0 \leq x \leq 2\pi$ is the same as the area under the $\cos^2 x$ curve in the same interval (d) Given the fact that $\sin^2 x + \cos^2 x = 1$, use the results obtained in parts (a)–(c) to show that the averages $\langle \sin^2 x \rangle$ and $\langle \cos^2 x \rangle$ are equal to each other and thus equal to 1/2.

29. (II) The magnetic field for a given electromagnetic wave has an rms value of 7×10^{-9} T. What is the intensity of the wave? How much energy is transported per minute through a 0.1-m^2 area?

30. (II) A typical lecture-demonstration laser of power 0.40 mW has a beam of diameter 1.2 mm. (a) What are the peak values of the electric and magnetic fields? (b) Suppose—as is in fact possible—that the beam is focused to a circular area with diameter of one wavelength. What is the peak value of the electric field, given that $\lambda = 633$ nm?

31. (II) Assume that a 75-W-light bulb emits light equally in all directions. What are the peak and rms values of the electric and magnetic fields at a distance of 0.5 m?

32. (II) A 100-W light bulb radiates uniformly in all directions, and 7 percent of this energy is emitted as electromagnetic radiation in the visible light range. What is the electromagnetic energy density of visible light at a distance of 80 cm from the bulb? What are the rms values of the corresponding electric and magnetic fields there?

33. (II) The total electromagnetic power emitted by the Sun is 3.8×10^{26} W. What is the radiation pressure exerted on a totally reflecting surface a distance $r = 10^{10}$ m from the Sun?

34. (II) What are the dimensions and SI units for the Poynting vector? Reduce your answer to the dimensions and units of mass, length, and time, then reexpress it in terms of watts and meters.

35. (II) Solar energy delivered to a horizontal surface in Washington, D.C., averaged over a full year is 160 W/m^2. Assuming that this radiation is fully absorbed on a particular square meter of ground, what is the approximate total momentum delivered to this area in 1 y? Compare this number to an estimate of the momentum absorbed by a baseball catcher in catching a single pitch.

36. (II) The radiation pressure of a beam of electromagnetic radiation is equal to atmospheric pressure. Calculate the intensity, energy density, and rms electric and magnetic fields of this beam. Assume that the beam is totally absorbed.

37. (II) Suppose that you want to use the radiation pressure from a beam of light to suspend a piece of paper in a horizontal position; the paper has an area of 50 cm^2 and a mass of 0.2 g (Fig. 35–23). Assume that there is no problem with balance, that the paper is dark and absorbs the beam fully, and that the

entire beam can be used to hold the paper against the pull of gravity. How many watts must the light produce? Given your answer, what do you suppose would happen to the paper?

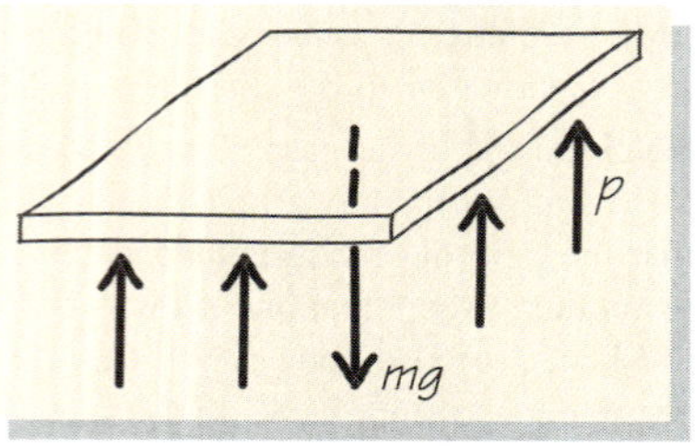

FIGURE 35–23 Problem 37.

38. (II) Tiny flakes of mica are kept aloft by a beam of light projected vertically upward. If the mass of a typical flake is 2.5×10^{-9} kg, and if on the average the area presented to the beam by a flake is $1.0\ \text{mm}^2$, what is the intensity of the beam? Assume that all of the light is reflected.

39. (II) A light beam with a given Poynting vector falls on a flat, fully reflecting surface at an angle of incidence θ (with respect to the vertical) (Fig. 35–24). What is the momentum transferred to the surface per unit area?

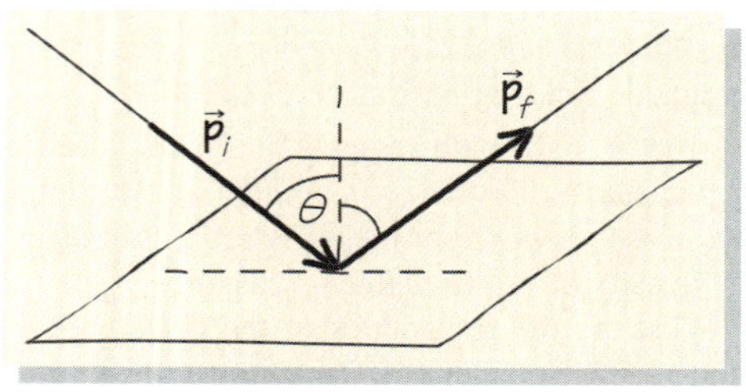

FIGURE 35–24 Problem 39.

40. (II) A laser delivers 10^2 J of energy in a pulse that lasts 10^{-8} s. What are the peak electric and magnetic fields for a laser beam of diameter 10^{-4} m?

41. (III) The short side of a thin, stiff rectangle 3.0 cm $\times$ 1.0 cm is attached to a vertical axis. Half of each side is painted black and is fully light absorbent; the other half is a shiny, reflecting metal (Fig. 35–25). The back of each half is different from the front. There is no friction at the axis. The apparatus is bathed in a well-collimated (nonspreading) beam of light whose Poynting vector has magnitude $0.5\ \text{kg/s}^3$ and that travels perpendicular to the vertical axis. Is there a net torque on the rectangle's surface? If so, what is its average value due to the light over a full, uniform rotation of the rectangle about the axis?

42. (III) The total power of a broadcasting dipole antenna is 20 MW. Calculate the intensity of its radiation at a distance of 1000 m, in the direction of the intensity maximum. Compare this to the intensity that would have been obtained if the intensity were distributed uniformly in every direction.

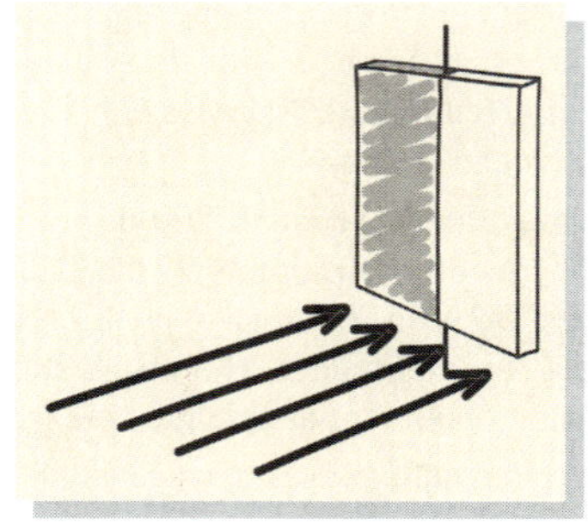

FIGURE 35–25 Problem 41.

35–5 Dipole Radiation

43. (I) Suppose that a vertical tower 120 m tall acts as a dipole antenna, with currents running back and forth along the tower to generate electromagnetic waves in a dipole pattern. If the wavelength of each electromagnetic wave is the height of the tower, what is the period of the current oscillation in the tower?

44. (I) A charge moves harmonically along a 1-m length in the z-direction, emitting dipole radiation. Two observers detect this radiation. Observer A is at a position that is 20 km from the charge and at an angle of 15° with respect to the z-axis, while observer B is at a position that is also 20 km from the charge, but at an angle of 73° with respect to the z-axis. What is the ratio of the intensity detected by the two observers?

45. (II) A broadcasting dipole antenna is oriented along the y-axis. For the geometry shown in Fig. 35–26, give the following information for a point P far away along the z-axis: (a) the direction of the electric field; (b) the direction of the magnetic field; (c) the direction of the Poynting vector. (d) Repeat parts (a) – (c) for the electromagnetic wave one-half cycle later.

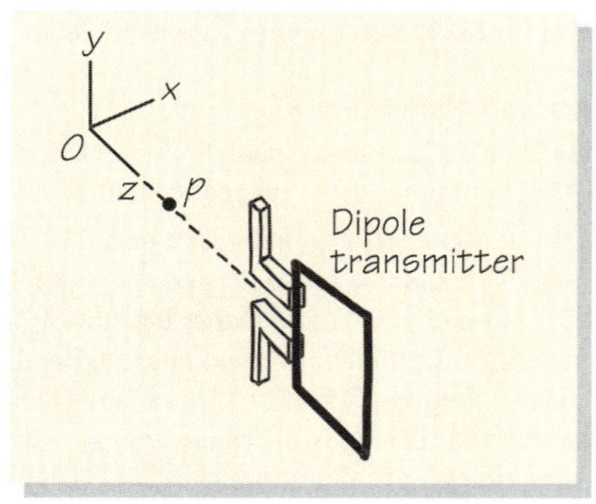

FIGURE 35–26 Problem 45.

46. (III) A cross-shaped antenna lies in the xy-plane, centered at the origin (Fig. 35–27). The charges oscillate with the same frequency within each arm of the cross. Find the Poynting vector along the z-axis as a function of z if charges moving in the $+x$-direction in the x-arm pass the origin at the same moment that (a) charges moving in the $+y$-direction in the y-arm pass the origin; (b) charges moving in the $-y$-direction in the y-arm pass the origin.

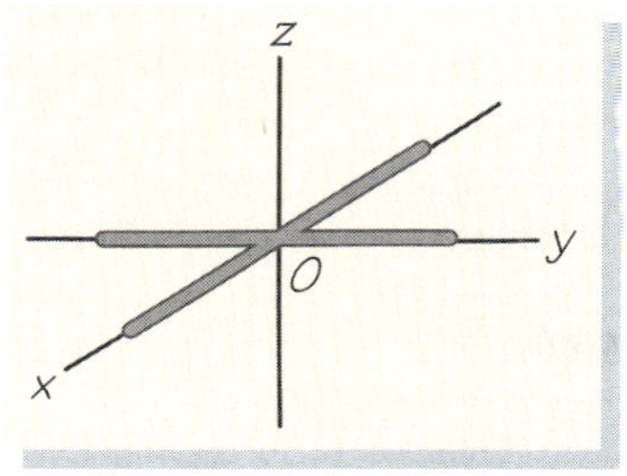

FIGURE 35–27 Problem 46.

35–6 *Polarization*

47. (I) At what angle should the axes of two ideal Polaroid sheets be placed to reduce the intensity of a given source of unpolarized light to (a) 7/10; (b) 3/10; (c) 3/20; (d) 1/20?

48. (I) The axes of four ideal Polaroid sheets are stacked, each at 30° with respect to the previous one. What fraction of initially unpolarized light passes through all four sheets?

49. (I) The Moon reflects light off a still pond at night. At what angle above the horizon is the polarization a maximum? The index of refraction of water is 1.33.

50. (I) Polarized light of intensity 10^6 W/m^2 is incident on a polaroid sheet placed perpendicular to the light beam with the polarizing axis of the sheet at an angle of 40° to the polarization vector of the light. What is the intensity of the beam after passing through the polarized sheet?

51. (I) The beam of Problem 50, after passing through the Polaroid sheet described in that problem, then passes through another Polaroid sheet, this one with its polarizing axis at an angle of 80° to the original polarization vector. What is the final intensity of the beam?

52. (II) A beam of light propagating in the z-direction is polarized in the y-direction. Two superposed Polaroid sheets are placed perpendicular to the beam. The polarization axis of one makes a 47° angle with respect to the y-direction, and the axis of the other makes a 63° angle with respect to the axis of the first sheet (Fig. 35–28). What is the intensity of the transmitted beam?

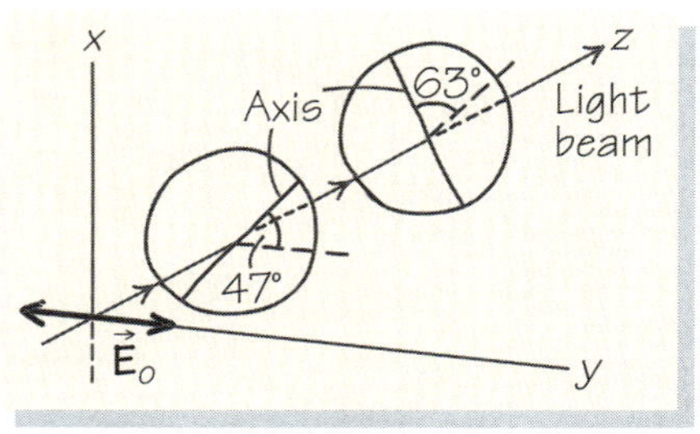

FIGURE 35–28 Problem 52.

53. (II) What fraction of initially unpolarized light passes through two Polaroid sheets placed at right angles to each other? What happens if a third sheet is placed between the two sheets, with its axis at an angle of 45° to the two?

54. (II) If light of intensity I_0 moving in the z-direction is polarized linearly in the x-direction, it will not pass through a piece of Polaroid that passes light polarized in the y-direction. Figure 35–29 shows a way in which this light can pass the y-direction analyzer if a second analyzer is used. If the lower analyzer makes an angle of θ with respect to the x-direction, what is the intensity of the light that passes the upper analyzer?

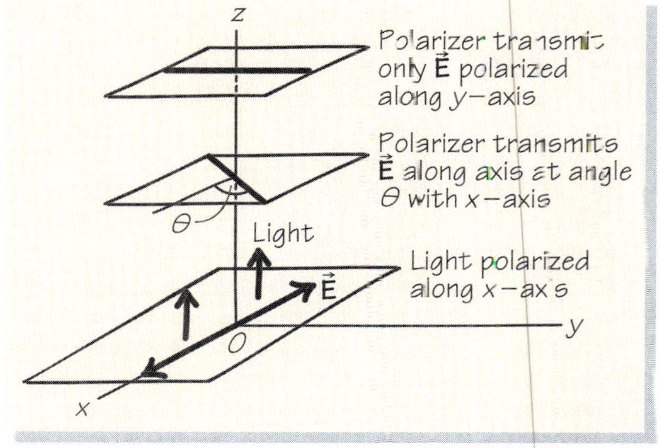

FIGURE 35–29 Problem 54.

55. (II) Unpolarized light of intensity I_0 passes through two pieces of Polaroid successively. (a) What is the intensity of the light after it passes through the first piece of Polaroid? (b) The second piece is rotated so that the intensity of the transmitted light goes to zero. What angle does the polarizing axis of the second piece make with that of the first piece? (c) A third piece of Polaroid is inserted between the two pieces in place. Calculate the intensity as a function of the angle θ that the axis of the third piece makes with the axis of the first piece (Fig. 35–30). (d) Show that the intensity of the transmitted light is no longer zero unless the axis of the third piece is parallel to that of either of the other pieces.

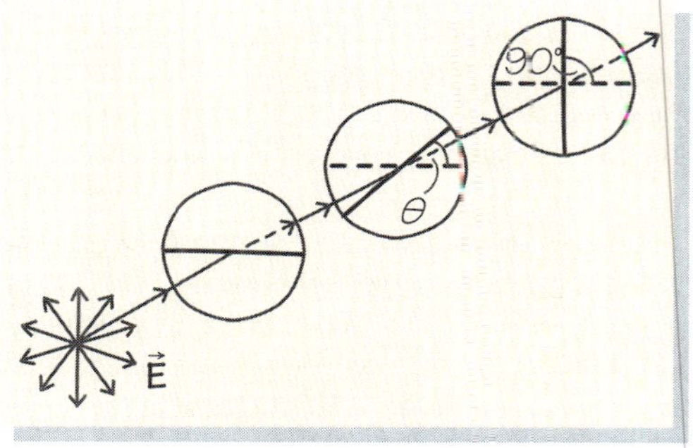

FIGURE 35–30 Problem 55.

*35–7 *Electromagnetic Radiation as Particles*

56. (II) Calculate the number of photons emitted by an FM radio station that broadcasts at a frequency of 10^3 Hz that are required to equal the energy contained in one photon of visible light at a wavelength of 600 nm.

57. (II) A scientist wishes to study the behavior of individual photons. To do that, she must decrease the intensity of her 1-mm^2 laser beam—the laser emits radiation with wavelength 630 nm—to a level at which there is no more than one photon in her apparatus at any given time. The path length of the light beam from source to detector is 2 m. What should be the intensity?

58. (II) The power generated by the Sun is 3.8×10^{26} W. Assuming that it is all emitted at an average wavelength of 550 nm, calculate the number of photons emitted per second.

General Problems

59. (II) Consider a solenoid of n turns/m with radius R. A current $I = I_0 \cos(\omega t)$ goes through the solenoid. (a) Calculate the magnetic field inside the solenoid. (b) Calculate the induced electric field inside the solenoid as a function of the distance r from the axis. (c) Calculate the Poynting vector, **S**. In particular, find its direction at different times during one cycle.

60. (II) The electric and magnetic fields of an electromagnetic wave act on a charge q. With what speed must the charge move so that the magnetic force on the charge is, at most, 10 percent of the electric force? If the electromagnetic wave is traveling in the z-direction and the electric field has only an x-component, what is the direction (or directions) of motion of q so that the magnitude of the magnetic force is greatest?

61. (II) Consider the solar sail described in Question 6. A solar sail can be aligned with its area perpendicular to a radial line from the Sun so that the sail is pushed straight outward. Show that in this configuration the force on the sail always has the same sign and is proportional to $1/r^2$, where r is the distance from the sail to the Sun. (Assume that only the radiation pressure and the gravitational force due to the Sun act on the sail.) This economical method of propulsion has been proposed for travel to the far reaches of the solar system when transit time is not an important factor.

62. (II) A solar sail (see Question 6) is to be designed such that, when it is aligned perpendicular to the Sun's rays and is 1.5×10^{11} m from the Sun the radiation pressure on it, P, just cancels the gravitational attraction of the Sun. The density of the material of the sail, which forms a sheet of constant thickness, is ρ. (a) Find P, given that the energy flux from the Sun is 1.4 kW/m^2 at the radius of Earth's orbit. (b) Express the sail's thickness in terms of ρ, P, the mass of the Sun, and the gravitational constant. If ρ is 2.0×10^3 kg/m^3, what is the thickness of the sail material? Your result is independent of the sail's area.

63. (II) Find an expression for the electric field of a plane electromagnetic wave with the following properties: (a) the frequency is 10^{14} Hz; (b) the wave travels in a medium of index of refraction 1.4; (c) the wave propagates along a line that lies in the xy-plane and makes a 30°-angle with the x-axis; (d) the wave is polarized along the z-axis; (e) the average value of the Poynting vector is 500 W/m^2.

64. (II) A swimming pool has underwater lights. What is Brewster's angle for reflection off the upper surface of water? The index of refraction of water is 1.33.

65. (II) The amount of solar energy reaching your body when you sunbathe on an ocean beach in summer is about 800 W/m^2. Assume that your body absorbs 40 percent of this incident radiation and that your exposed body area is 0.5 m^2. How much solar energy do you absorb in 1 h? Estimate how much perspiration must evaporate to dissipate this energy (see Chapter 18).

66. (II) A high-powered, pulsed laser used to confine plasma for nuclear fusion studies is rated at 10 MW. The laser beam is focused on an area of 1 mm^2. Calculate the intensity, peak electric and magnetic fields, and average energy density in this beam. Compare your results to Tables 23–1 and 29–1, which list some values for electric fields and magnetic fields, respectively, in other contexts.

67. (II) What is the number of photons/m^3 contained in a beam of electromagnetic radiation in a plane wave with a wavelength of 2 cm and an electric field amplitude 10 V/m?

68. (II) The solar energy flux at a distance $R_0 = 1.5 \times 10^{11}$ m from the Sun (the radius of Earth's orbit) is 1400 W/m^2. (a) What is the total energy flow from the Sun in watts? (b) Use your result to calculate the rate at which photons are emitted. Assume an average wavelength of 600 nm. (c) Using the result of part (b), find the number of photons/s that strike a 1 mm $\times$ 1 mm surface at a distance R_0. The surface is oriented perpendicular to the Sun.

69. (II) A laser emits N photons of frequency f. The beam strikes a mirror that is moving with speed v in the direction of propagation of the laser beam. Assuming that the kinetic energy of the mirror is much larger than that of the beam, use energy conservation and momentum conservation to find the frequency of the reflected beam. Treat the photon as a particle of energy hf and momentum hf/c.

70. (III) Consider a current I that flows through a cylindrical wire of length L, radius b, and resistance R (Fig. 35–31). The current flows uniformly across the cross section of the wire. Calculate the electric fields inside and on the surface of the wire. The current in the wire gives rise to a magnetic field, which you can calculate. Use these fields to find the direction and magnitude of the Poynting vector on the surface of the wire. Show that the rate of energy flow into the wire through its surface is IR^2, the power dissipated in ohmic heating.

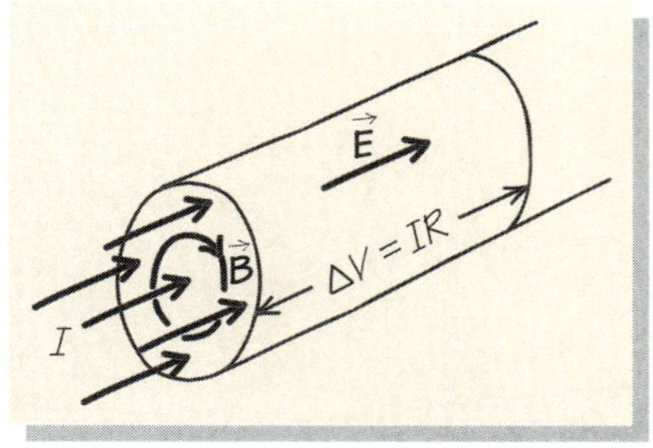

FIGURE 35–31 Problem 70.

71. (III) Consider a capacitor that consists of two circular metal plates of radius R a distance d apart (Fig. 35–32). R is so much larger than d that all fringe fields can be neglected. If the charge on the plates, Q, changes with time, then according to Ampère's law a magnetic field will be induced in the region between the plates. (a) What is the induced magnetic field? (b) Using the induced magnetic field and a calculation of the

electric field between the plates, find the Poynting vector. (c) Show that with this Poynting vector, the net energy flow into the capacitor is the rate of change of the capacitor energy $Q^2/2C$.

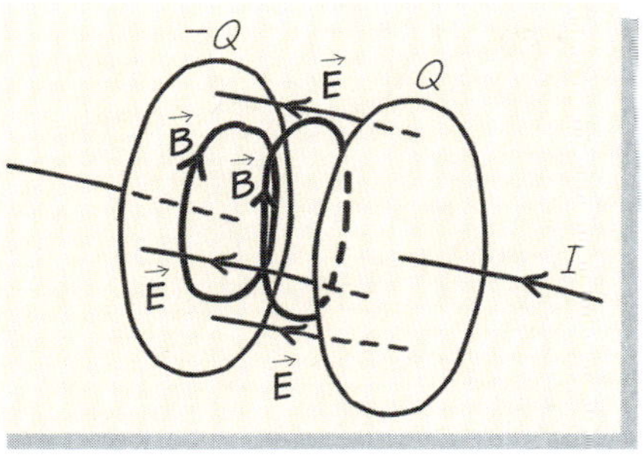

FIGURE 35–32 Problem 71.

72. (III) Consider a plane electromagnetic wave of frequency f that propagates in the z-direction in a cubic box whose sides are length L, with L much larger than the wavelength. The electric field of the radiation has the form $\mathbf{E} = E_0 \sin(kz - \omega t)\,\mathbf{i}$. Alternatively, we can say that the radiation consists of N photons, each propagating in the z-direction with energy hf, where h is Planck's constant. Use two alternative expressions for the energy of the radiation to express E_0 in terms of h, f, N, and L.

CHAPTER

36

Rainbows are a common occurrence at large waterfalls due to the mist in the air. This one is at spectacular Victoria Falls in Zimbabwe.

Light

You may think of light as something that travels in straight lines. You may have come to this conclusion from observing the rays that appear when light penetrates a dark and dusty room through a tiny opening. This property of light strongly suggests that light is composed of particles emitted by a source, and Isaac Newton, whose earliest work was on optics, finally supported that view. The phenomena he considered include many with which we have some everyday experience: the reflection of light from mirrors, refraction as light passes through glass lenses or water, and the dispersion of light, which leads to rainbows and the prismatic separation of the colors. The particle model provided such a good explanation of these observations that it is surprising that the idea that light consists of waves could have taken root in Newton's day. Yet Robert Hooke's idea that light is some type of oscillatory activity in an unidentified medium led Christian Huygens to propose a wave theory of light in 1687. In this chapter, we shall show that the wave theory of light can explain everything that the particle theory can.

By the early nineteenth century, it had become apparent that certain observations could not be explained by the particle theory; an explanation of these observations demanded that light behave like a wave. For example, when we look very closely, light does not cast *sharp* shadows; to some extent, light bends around corners. (Newton did not have the equipment to make this observation, and, in fact, he argued against the wave theory on the basis that light does *not* appear to bend around corners!) Under controlled conditions, we can also see that beams of light interfere with each other in just the same way that the waves discussed in Chapter 15 interfere with each other. Definitive experiments by Thomas Young in 1801 on the wave aspects of light eventually established the preeminence of the wave theory (Fig. 36–1). The phenomena associated with the wave aspect of light are the subject of Chapters 38 and 39. The prediction from Maxwell's equations that light is an electromagnetic wave would seem to have settled the question of whether light is a particle or a wave once and for all. In the twentieth century, however, we have had to revise our view once more, as new experimental evidence suggested that some aspects of light can be explained only if light sometimes behaves as particles. Now we are not forced to choose between a particle theory and a wave theory of light; a quantum mechanical explanation encompasses them both.

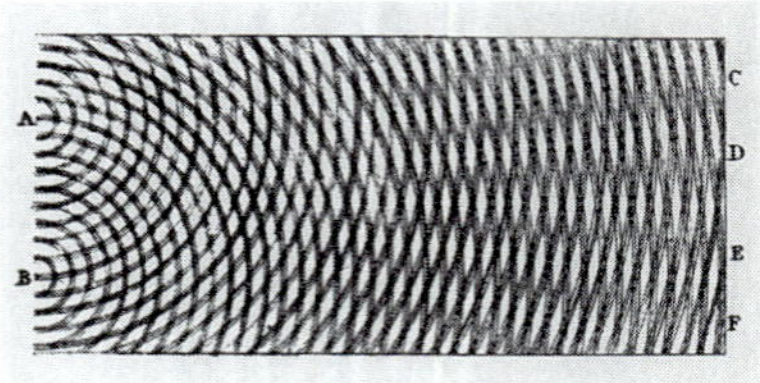

FIGURE 36–1 Young's view of the wave nature of light. In this sketch published in 1807 from his lectures, points A and B represent pinholes; points C and E represent spots where the waves reinforce each other, and points D and F, spots where they do not.

36–1 THE SPEED OF LIGHT

It requires some ingenuity to show that light travels at finite speed. Although Galileo had thought that the speed of light might be finite, his attempts to observe a time delay in the passage of light from one mountaintop to another were failures. Olaus Roemer observed the eclipses of the moons of Jupiter in 1675 and found that the timing of these phenomena could be explained only if a large, but finite, value were assumed for the speed of light. The solar system data available to Roemer at the time gave a value of 2×10^8 m/s for the speed of light—certainly a result of the correct order of magnitude. The first terrestrial measurements were made in 1849 by Hippolyte Fizeau, who used the ingenious device shown in Fig. 36–2. A light source is placed behind a toothed wheel that can be rotated at high speeds. The light passes through an inclined glass plate and then between

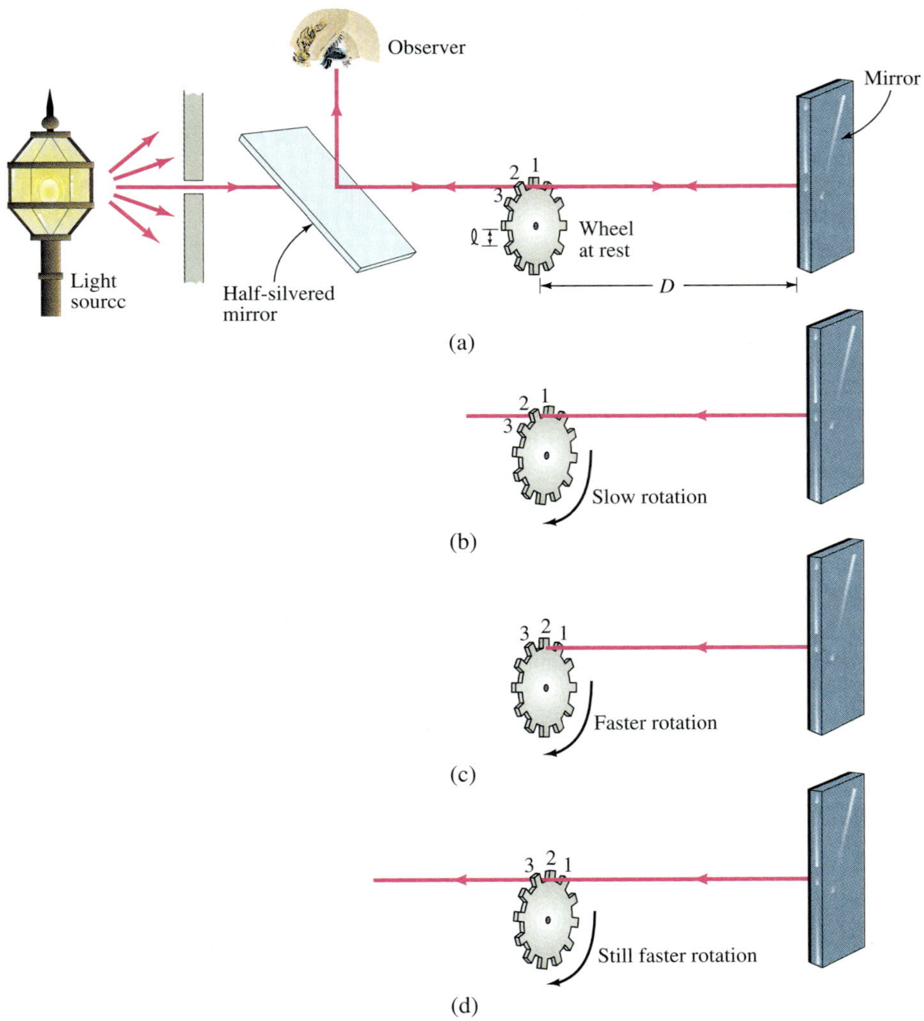

FIGURE 36–2 Fizeau's method for measuring the speed of light: (a) a sketch of the apparatus he used. (b) Incident and reflected light pass through the same gap. (c) When the wheel rotates faster, the reflected light does not manage to pass through the original or the next gap. (d) When the rotation speed is still higher, the reflected light passes through the next gap.

two teeth of the rotating wheel. It then travels to a mirror and is reflected straight back. If the speed of light were infinite, light would come straight back through the gap it had entered—between cog 1 and cog 2—independent of the rotational speed of the cogged wheel (Fig. 36–2b). Light traveling at a finite speed would also pass through the same gap if the rotational speed of the wheel were not too fast. But if the wheel is rotating faster and the speed of light has a finite value c, then, at a certain value of angular velocity, the wheel moves enough during the time the light travels to the mirror and is reflected back that the light will strike cog 2 (Fig. 36–2c). No light reaches the detector in this case. As the wheel rotates even faster, light once again reaches the detector, but this time it passes through the *next* gap in the wheel, the one between cogs 2 and 3 (Fig. 36–2d). If the wheel is moving at speed v when Fig. 36–2d first applies, then we can equate the time $2D/c$ for light to make a round trip from the wheel to the mirror and back with the time ℓ/v for the wheel to move a distance of one gap:

$$\frac{\ell}{v} = \frac{2D}{c}.$$

Here, ℓ is the cog spacing and D is the distance from the wheel to the mirror. If D is much larger than ℓ, then a value of v much smaller than c would suffice to measure c accurately.

The best measurement would be limited by our ability to measure time (currently we can measure time to one part in 10^{13}) and distance (currently to four parts in 10^9). Therefore, today we *define* the speed of light in a vacuum to be $c = 299{,}792{,}458$ m/s and use time along with the definition of c to measure distances. In this new system, the meter is not defined but measured: One meter is 1/299,792,458 times the distance traveled by light in 1 s. For practical purposes, you could use $c = 3.00 \times 10^8$ m/s.

TABLE 36–1
Indices of Refraction for Various Substances (at $\lambda = 600$ nm)

Material	*Index of Refraction, n*
Air (1 atm, 0°C)	1.00029
Carbon dioxide (1 atm, 0°C)	1.00045
Ice	1.31
Water (20°C)	1.33
Ethyl alcohol	1.36
Castor oil	1.48
Benzene	1.50
Fused quartz	1.46
Glass (crown)	1.52
Glass (flint)	1.66
Diamond	2.42

The Index of Refraction

Fizeau also found that the speed of light in transparent materials such as water or glass is *less* than the speed of light in empty space. We reserve the symbol c for the speed of light in empty space, and express the speed of light in a material as

$$v_m = \frac{c}{n}, \tag{36–1}$$

where n is the *index of refraction* of the material, a quantity introduced in Chapter 35. Table 36–1 lists indices of refraction for a variety of materials.

We should mention one other aspect of the speed of light in materials: *The index of refraction is a function of wavelength.* For example, violet light, which has a shorter wavelength than red light, travels more slowly in glass than does red light. We shall see that this property explains the separation of white light into the colors of a rainbow or a prism.

The index of refraction depends on the wavelength.

We saw in Chapter 35 [Eqs. (35–16) and (35–17)] that the index of refraction for a material with a dielectric constant κ is

$$n = \sqrt{\kappa}. \tag{36–2}$$

(Here, we have assumed that the relative magnetic permeability $\kappa_m \simeq 1$; this is a good approximation for substances that are transparent to light.) The variation of n with wavelength occurs because the dielectric constant can also vary with wavelength. We must therefore use κ at the appropriate wavelength and not use its static value in Eq. (36–2). Frequency, f, and wavelength, λ, are related by $f\lambda = v$, so, in a medium of index of refraction n, we find from Eq. (36–1) that

$$f\lambda = \frac{c}{n}. \tag{36–3}$$

Equation (36–3) shows that the product of f and λ is inversely proportional to n. Note that as radiation passes from one medium to another, *the frequency does not change.* This is easy to understand: Consider two observers on either side of an air–glass inter-

The frequency of a wave remains the same as the wave passes from one medium to another.

face. Each wave front that passes one observer must pass the other—otherwise wave fronts would pile up or disappear, neither of which can happen. As a consequence, the wavelength of light changes with the index of refraction in such a way that $c/f = n\lambda$ is constant; that is, f is the same for both media. Thus, when light passes between media 1 and 2,

$$n_1\lambda_1 = n_2\lambda_2. \tag{36–4}$$

36–2 DOES LIGHT PROPAGATE IN STRAIGHT LINES?

One goal for this chapter is to describe a wave model of light. A particle model easily accounts for what seems to be the most elementary features of light propagation; namely, the fact that light seems to travel in a straight line and casts a sharp shadow. How can a wave model explain these features, and do these features persist when we take a closer look?

In Chapter 35, we discussed electromagnetic waves that propagate along the $+z$-axis. The space dependence and time dependence of the electric or magnetic fields are described by a function such as $\cos(kz - \omega t)$. This function has a series of crests and troughs, and one peak occurs at $kz - \omega t = 0$, or

$$z = \frac{\omega t}{k} = ct.$$

Thus the peak propagates at speed c. We called this a plane wave because all points in the xy-plane defined by a fixed value of z have the same fields, whatever the x- or y-value (Fig. 36–3a). We may thus describe the planes for which the argument $kz - \omega t$ is constant as representing *wave fronts*. Figure 36–3b shows a sequence of wave fronts transverse to the direction of the electromagnetic (light) wave. It is useful to think of these wave fronts as representing a particular set of field values along the wave. For

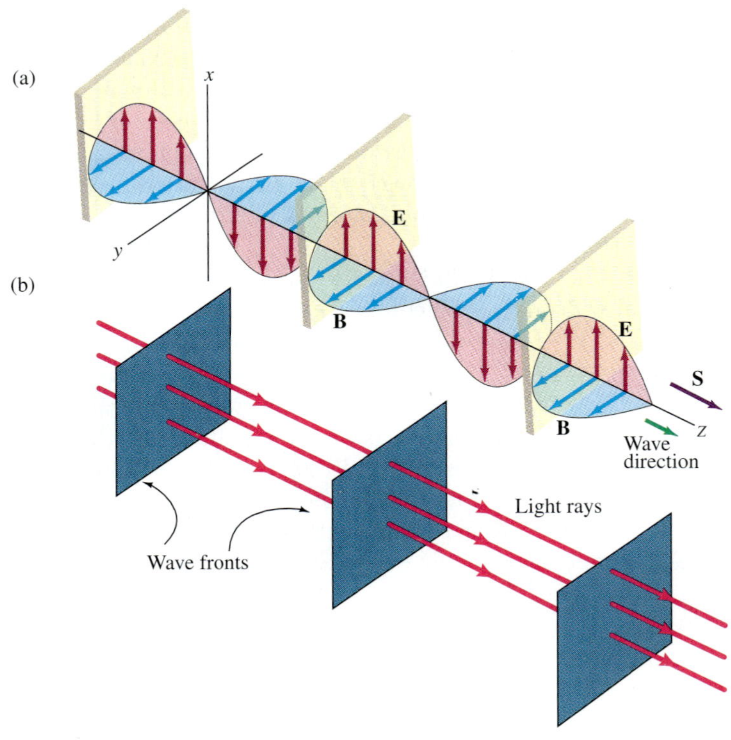

FIGURE 36–3 (a) The electric and magnetic fields of an electromagnetic wave propagating in the direction of the Poynting vector **S** (see Chapter 35). (b) Wave fronts are chosen arbitrarily at points where the electric and magnetic fields are maximal; the fronts could be chosen at the fields' zero points instead.

example, the wave fronts could represent the points where the electric field is a maximum or the points where the field is zero. (Fig. 36–3b arbitrarily sets the wave fronts at the points where both fields are maxima.) As long as there are no obstacles, the sequence of wave fronts moves at speed c along the original direction of propagation. In other words, the wave moves in a straight line—just as we observe light to move.

Huygens' Principle

We can understand these results from another point of view. Consider a wave front that propagates in the $+z$-direction (Fig. 36–4a). The location of the wave front after a time interval Δt is obtained by viewing every point on the original wave front as a source of light emitting a spherical pulse (or *wavelet*) of radiation (Fig. 36–4b). The radius of the sphere in empty space is $c\,\Delta t$, the distance the light travels in time Δt. In a medium in which the speed of light is c/n, the radius of the sphere is reduced by a factor of n to $(c/n)\,\Delta t$. In the limit that the separation between all the emission points is small, the envelope of all these tiny spheres, *taken in the direction of propagation of the initial wave front*, is the new wave front. In empty space, the wave fronts generated in this way remain planes parallel to the xy-plane. Thus, the straight-line propagation of wave fronts is assured. Christian Huygens developed this treatment of waves, which is called **Huygens' principle**. Huygens' principle can be justified from a detailed study of the behavior of waves in Maxwell's equations, although we shall not do so here.

Let's now see what Huygens' principle gives when a wave front approaches a slot in a wall (Fig. 36–5). When the wave front arrives at the wall, only the part of the wave at the slot can continue to propagate. This part of the wave front generates waves that travel through the slot, with the additional feature that the spherical wavelets emitted near the edges of the slot have no neighbors, and *a wave that spreads away from the slot edges is generated past the slot*. This spreading of the wave around the edges of the slot is known as **diffraction** (see Chapter 39). Huygens' principle suggests that the spreading is significant (in terms of the fraction of energy in the bent waves) only if the wavelength is about the same as, or larger than, the size of the slot. If the slot width is much larger than the wavelength, only a small fraction of the energy goes into the bent waves, and it is adequate to view the entire slot as a source of a plane wave front. Light has wavelengths around 5×10^{-7} m; therefore the slot must not be too much larger than this size for the effect to be significant. This explains why the diffraction of light was not observed in Newton's day.

For now, we are interested in a wave front propagating in a straight line. We can follow the front's direction by means of **rays**, lines perpendicular to wave fronts. Light

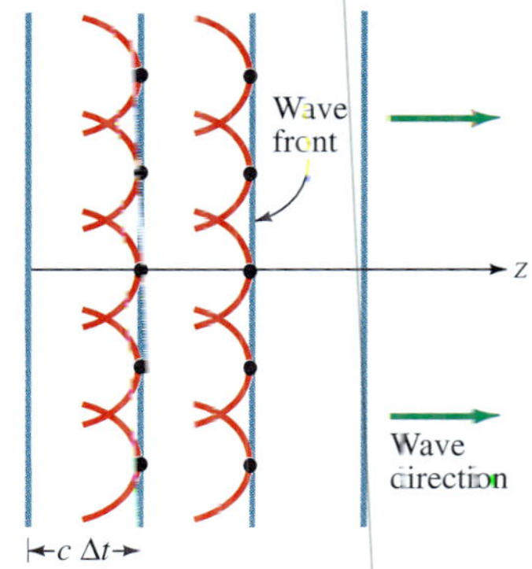

(a)

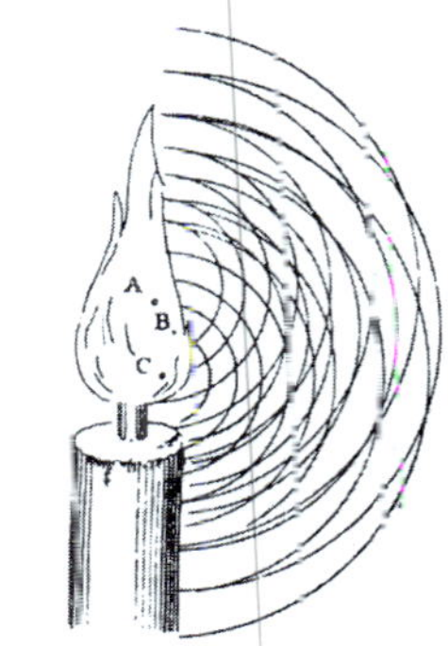

(b)

FIGURE 36–4 (a) Huygens construction of wave fronts. Wavelets emitted at each point along the wave front add to a new wave front and produce plane waves. (b) Huygens' illustration of wavelets, from his book *Traité de la Lumière* (1678).

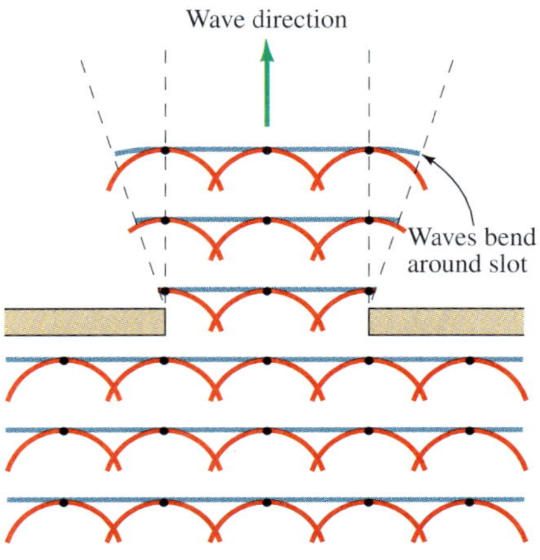

FIGURE 36–5 Huygens' construction of wave fronts that approach and pass through an open slot in a wall. Past the slot, the wave fronts bend around the slot edges.

FIGURE 36–6 A beam of light generated by a laser. The beam is visible because the air has scattered the light. This beam is used to make a reference for astronomical instruments that can adjust for atmospheric turbulence.

entering a darkened room through a pinhole or the beam of a searchlight provide vivid pictures of the propagation of light in the form of rays (Fig. 36–6). We normally do not *see* rays, but a ray can be made visible when light is scattered by dust particles, for example. The description of light based on the straight-line propagation of rays is called **geometric optics**.

36–3 REFLECTION AND REFRACTION

Reflection

A light ray **reflects**—it "bounces back"—when it strikes a smooth surface such as a mirror. The *incident ray* makes an angle θ with a line normal to the surface at the point of reflection. The *reflected ray* lies in the plane formed by the incident ray and the normal. The angle θ' that the reflected ray makes with the normal obeys the equation known as the **law of reflection**:

The law of reflection

$$\theta' = \theta \tag{36–5}$$

(Fig. 36–7). The consequences of this law are shown in Fig. 36–8a for the reflection of a set of parallel incident rays (a *bundle* of rays) from a flat mirror and in Fig. 36–8b for that from a smooth curved surface. For the curved surface, the angles of incidence and reflection are indeed equal, but the direction of the normal to the surface varies from point to point, and the reflected rays radiate in various directions.

EXAMPLE 36–1 A light ray is incident on two mirrors set at 90° with respect to each other. The incident ray is in the plane perpendicular to the two mirrors (Fig. 36–9). Show that after two successive reflections, the outgoing ray will travel along the direction of incidence no matter what angle the incident ray has with respect to the mirrors.

Solution: We see from Fig. 36–9 that angle ACB—the angle between two normals to surfaces perpendicular to each other—is 90°. Consequently, angle BCD is also 90°. Triangles ABC and DBC are congruent so that angle CDB—the angle marked ϕ—is equal to θ. Thus, the incoming and outgoing rays make the same angle with respect to line AD, and they are parallel.

Corner Reflectors

APPLICATION

The uses of corner reflectors

Example 36–1 illustrates the principle behind the construction of *corner reflectors* (Fig. 36–10). In a corner reflector, three mirrors are placed together at mutual 90° angles, like the interior corner of a room, or a set of such corners is placed adjacent to one an-

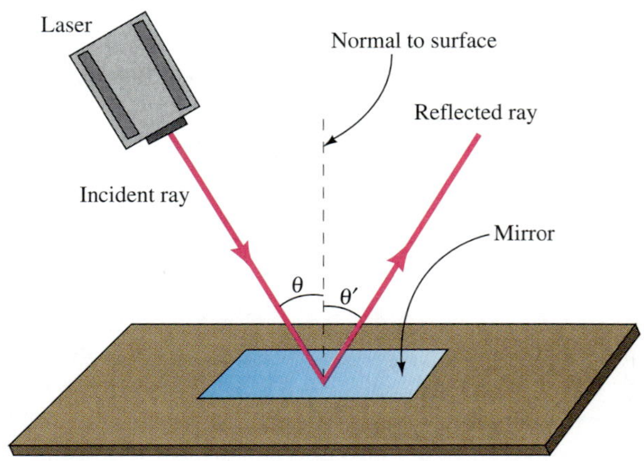

FIGURE 36–7 The angle of incidence θ equals the angle of reflection θ'.

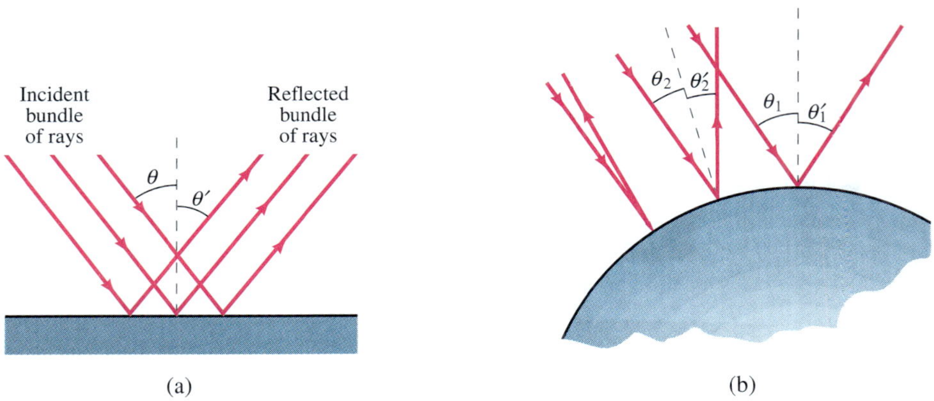

FIGURE 36–8 Reflection of rays from (a) flat and (b) curved surfaces.

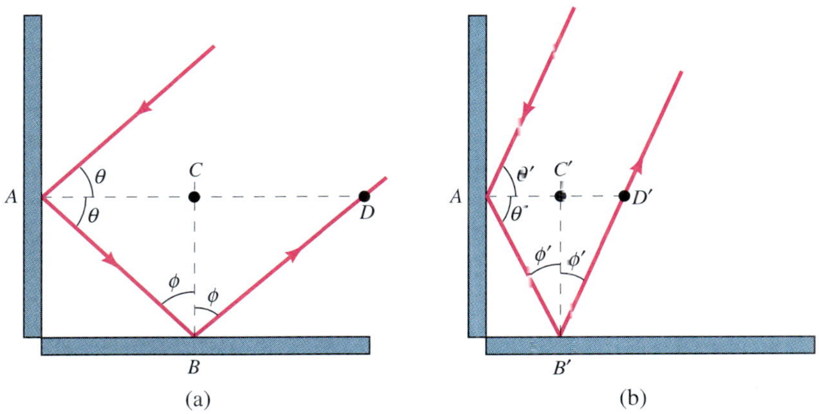

FIGURE 36–9 Example 36–1. (a) The incoming ray that strikes point A eventually emerges along BD. (b) The same for a ray incident at a different angle.

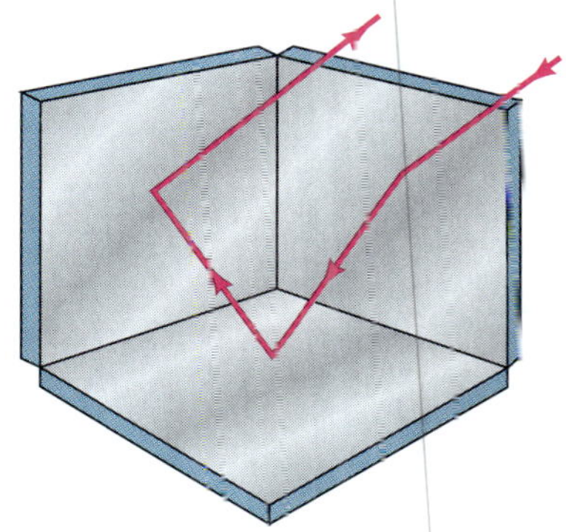

FIGURE 36–10 A corner reflector. A laser beam that has been sent into the device produces a reflected beam that is parallel to the incoming beam

other. Geometry of the kind used in Example 36–1 shows that a ray incident at any angle reflects out from a corner reflector along a ray exactly parallel to the incident ray. Corner reflectors are used routinely on highways because the light from a vehicle's headlight is automatically reflected back to that vehicle. Other applications stem from the fact that it is possible to measure time delays with great accuracy; a pulse from a laser directed at a corner reflector comes back, and the time delay gives a measurement of distance. In this way, Earth's slight movements across fault lines can be accurately surveyed. Similarly, Fig. 36–11 shows a corner reflector placed on the Moon that tells us the Moon's distance from the light source on Earth to within 15 cm!

FIGURE 36–11 This corner reflector was left on the Moon by the Apollo 14 astronauts. By shining a laser beam from Earth at the reflector and looking for the return light on Earth, the distance to the Moon can be accurately measured.

Refraction

When the light forming a ray moves from one medium to another—say, from air to water—the incident ray changes direction at the boundary between the media; the ray is said to undergo **refraction** (Fig. 36–12a). Let the index of refraction of the medium with the incident ray be n_1 and that of the medium with the *refracted ray* be n_2. The angles that the incident and refracted rays make with the line normal to the boundary between the media are θ_1 and θ_2, respectively (Fig. 36–12b). Then

$$n_1 \sin\theta_1 = n_2 \sin\theta_2. \tag{36–6}$$

This result, found by Willibrord Snell in 1621, is known as **Snell's law**. The index of refraction of air is very close to unity, so the angle of the refracted ray θ_2 at the interface for light that passes from air into a medium with index of refraction n is given by

$$\sin\theta_1 = n \sin\theta_2. \tag{36–7}$$

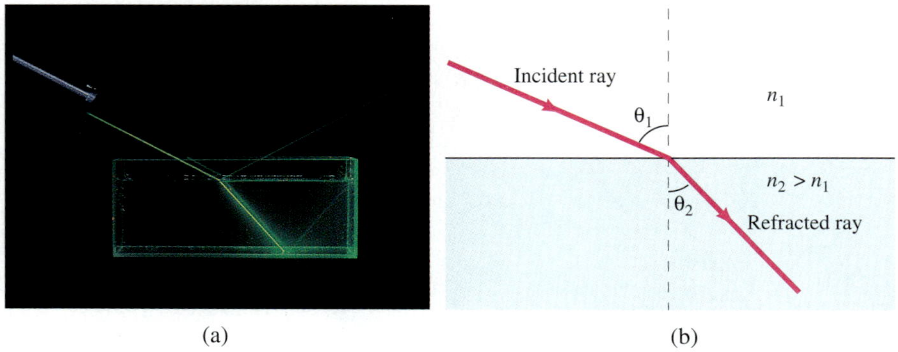

FIGURE 36–12 (a) A beam of light is refracted as it enters a tank of water. (b) Refraction from a medium with index of refraction n_1 into a medium with index of refraction n_2. In this case $n_2 > n_1$, and the refracted ray is bent toward the normal to the boundary surface. If n_2 had been less than n_1, the refracted ray would have bent away from the normal.

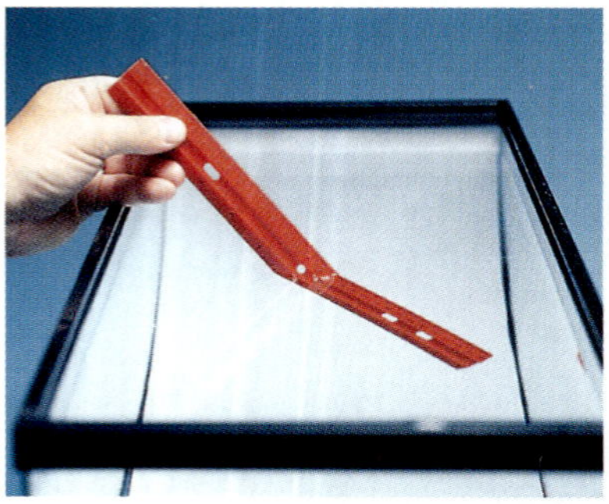

FIGURE 36–13 Water has a higher index of refraction than does air, so the immersed part of this ruler seems to bend away from the normal to the water surface.

Because n is generally larger than one, it follows that $\theta_2 < \theta_1$; that is, *the light is bent toward the normal to the boundary surface*. Equation (36–6) also shows that when light enters a medium with a lower index of refraction, such as when a ray of light travels from water to air, the ray is bent farther away from the line normal to the boundary (Fig. 36–13).

EXAMPLE 36–2 Consider light that is refracted by a prism shaped like an equilateral triangle (Fig. 36–14). The incident ray is parallel with the prism's base. What is the total deflection of the ray, given that the index of refraction of the prism material is 1.50?

Solution: The light passes through two surfaces, and we must apply Snell's law at each interface. Figure 36–14 shows that the angle the incident ray makes with the line normal to the first surface is $\theta = 30°$. The angle ϕ that the refracted ray makes with the line normal to the first surface is given by Eq. (36–7) as

$$\sin\theta = n\sin\phi.$$

When $n = 1.50$ and $\theta = 30°$, this expression gives $\phi = 19.5°$.

Figure 36–14 also shows that the refracted ray acts as an incident ray at the second surface. If the angle that the refracted ray makes with the line normal to the second surface is denoted by ψ, then $\phi + \psi + 120° = 180°$, or $\psi = 60° - \phi = 60° - 19.5° = 40.5°$. The angle θ' that the second refracted ray makes with the line normal to the second surface is given by

$$\sin\theta' = n\sin\psi = 1.50\sin(40.5°) = 0.97.$$

Thus, $\theta' = \arcsin(0.97) = 77°$ and we see from Fig. 36–14 that the angle the outgoing light ray makes with the base of the prism is $\theta' - \theta = 77° - 30° = 47°$.

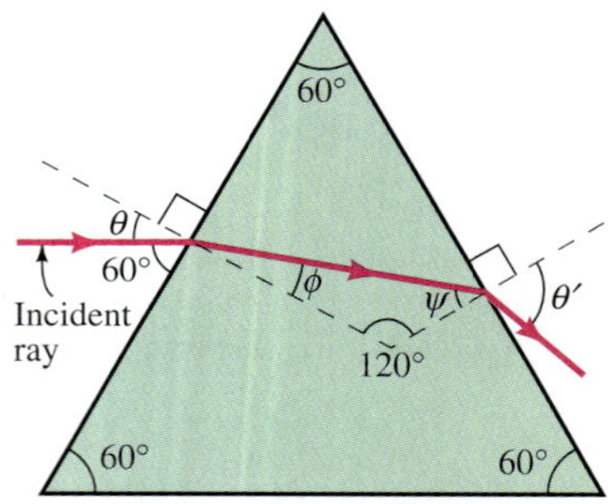

FIGURE 36–14 Example 36–2. Geometric construction of the path followed by a ray incident on a prism.

Refraction is responsible for some curious optical effects: You are perhaps familiar with the problem of spearing a fish (Fig. 36–15). The person with the spear assigns the source of the light ray from the fish to a point that lies on a straight line along the direction at which the ray enters the observer's eye. Thus the observer thinks the image I is the actual position of the fish.

FIGURE 36–15 The fish appears less deep than it really is because of refraction. This makes it hard for the fisherman to spear it from outside the water.

Energy in Reflection and Refraction

Refraction is generally accompanied by reflection. As we saw in Chapter 35, the incident ray carries electromagnetic energy. At the boundary between media, this energy is apportioned among the reflected and refracted rays such that the total energy is conserved. Maxwell's equations show that when light is perpendicularly incident on a surface that separates a medium of index of refraction n_1 from a medium

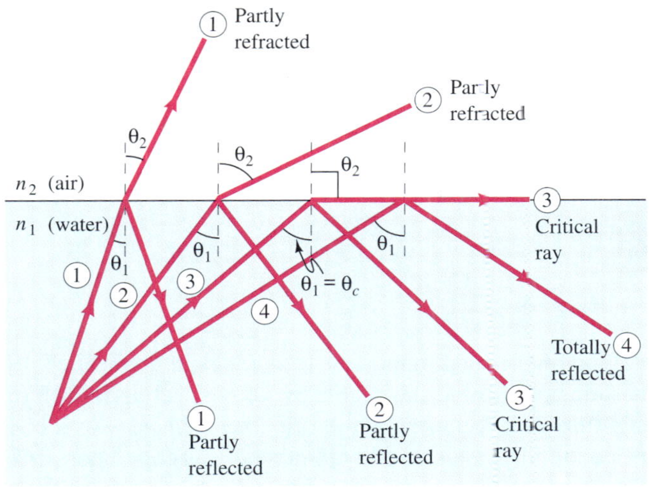

(a)

(b)

FIGURE 36–16 (a) Various rays traveling from a medium with a larger index of refraction (water) to a medium with a smaller index of refraction (air). When the incident angle is θ_c, there is total internal reflection. (b) Refraction and total internal reflection off the air–water interface in a water tank.

of index of refraction n_2, the intensity of the reflected light, I_r, is related to the incident intensity, I_0, by

$$\frac{I_r}{I_0} = \frac{(n_2 - n_1)^2}{(n_2 + n_1)^2}. \qquad (36\text{–}8)$$

For light perpendicularly incident from air ($n = 1.0$) into glass ($n = 1.5$), only 4 percent of the incident light is reflected. The intensity of the reflected light varies with the angle of incidence.

Total Internal Reflection

For some incident angles, all the incident energy is contained in the reflected ray. This situation, known as **total internal reflection**, can occur only when light travels from a medium with a larger index of refraction toward a medium with a smaller index of refraction, such as when light passes from water toward air. Simple geometry explains this phenomenon.

The critical angle for total internal reflection

Let's consider a light ray incident from a medium with an index of refraction n_1 to a medium with an index of refraction n_2; this time $n_1 > n_2$. Snell's law, Eq. (36–6), may be written in the form $\sin\theta_2 = (n_1/n_2)\sin\theta_1$. Because the factor n_1/n_2 is larger than unity, θ_2 reaches 90° before θ_1 does as θ_1 increases. Figure 36–16 shows what happens for various values of θ_1. When $\theta_2 = 90°$, the ray in medium 2 skims along the interface of the two media. This occurs when θ_1 reaches a critical angle θ_c such that $(n_1/n_2)\sin\theta_c = \sin 90° = 1$, or

$$\sin\theta_c = \frac{n_2}{n_1}. \qquad (36\text{–}9)$$

When θ_1 exceeds θ_c, there is no angle θ_2 that can satisfy Snell's law. The electromagnetic energy carried by the incident ray must go somewhere, and the ray is reflected. There is no diminution of the intensity of the reflected ray; the reflection is total.

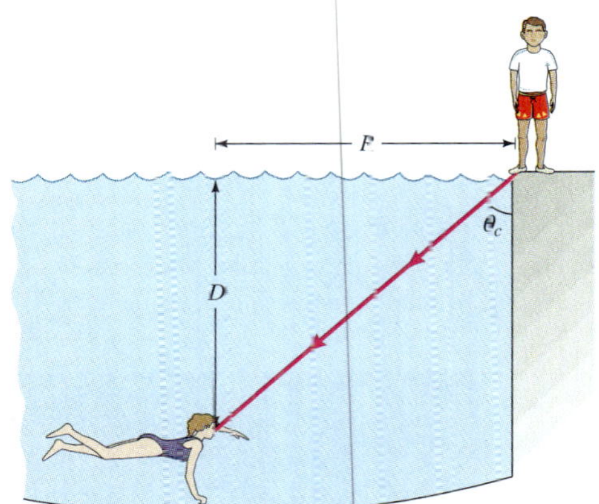

FIGURE 36–17 Example 36–3. When the swimmer's eyes are the deepest they can be in the pool and she can still see the child's feet, the ray between her eyes and his feet is at the critical angle.

EXAMPLE 36–3 A swimmer is in a deep pool and her eyes are a horizontal distance $R = 1.5$ m from the pool's edge (Fig. 36–17). How far below the surface are her eyes if she is just able to see the full height of a child who is standing on the edge of the pool? The index of refraction of water is 1.33.

Solution: We are interested in a ray from the child's feet to the swimmer's eyes. (If the swimmer can see the child's feet, the rest of the child will be visible.) When the swimmer's eyes are at the deepest point they can be in the pool and she can still see the feet, the reversed ray—the ray from the swimmer's eye to the base of the feet—must be at the critical angle θ_c (Fig. 36–17). We have, from Eq. (36–9),

$$\sin\theta_c = \frac{n_{\text{air}}}{n_{\text{water}}} = \frac{1}{n_{\text{water}}} = \frac{R}{\sqrt{R^2 + D^2}},$$

where D is the greatest possible depth of the swimmer's eyes. We can solve this equation for D:

$$D^2 = R^2(n_{\text{water}}^2 - 1) = (1.5\text{ m})^2(1.33^2 - 1) = 1.7\text{ m}^2;$$

$$D = 1.3\text{ m}.$$

Fiber optics represents one of the most important technological applications of total internal reflection. The principle behind this technique of conducting light from one place to another is straightforward: A transparent plastic fiber will serve as a conductor of light if any ray inside the fiber undergoes total internal reflection upon striking the side of the fiber (Fig. 36–18a). Figure 36–18b shows a ray in air ($n = 1$) entering a cylinder of diameter D at an angle θ_i with the axis of the cylinder. If n_f is the index of refraction of the fiber, then the angle that the ray makes with the axis inside the fiber is θ_f, where $\sin\theta_f = \sin\theta_i/n_f$. This ray will strike the wall of the cylinder at an angle $(90° - \theta_f)$ with the normal to the wall. There will be total internal reflection if $n_f\sin(90° - \theta_f) > 1$; that is, if $n_f\cos\theta_f > 1$. We have

$$n_f\ \cos\theta_f = n_f\sqrt{1 - \sin^2\theta_f} = n_f\sqrt{1 - \frac{\sin^2\theta_i}{n_f^2}} = \sqrt{n_f^2 - \sin^2\theta_i} > 1.$$

Because $\sin^2\theta_i \le 1$, we find

$$\sqrt{n_f^2 - \sin^2\theta_i} \ge \sqrt{n_f^2 - 1}.$$

Thus we automatically satisfy the condition for total internal reflection, $n_f\cos\theta_f > 1$, if

$$\sqrt{n_f^2 - 1} > 1. \qquad (36\text{–}10)$$

Because the largest value of $\sin\theta_i$ is 1 (the light first enters the cylinder from the end), Eq. (36–10) is a condition for internal reflection for *all* of the light that enters the fiber. Equation (36–10) is satisfied for any material with $n_f > \sqrt{2}$. A typical fiber has an index of refraction of 1.62, which is larger than the critical value. Note that once a ray is in the fiber, it remains inside *even if the fiber curves.*

*36–4 FERMAT'S PRINCIPLE

As an example of Huygens' principle in operation, let's consider the law of reflection [Eq. (36–5)]. Figure 36–19a shows a sequence of wave fronts as they approach a mirror. In Fig. 36–19b, point C_2 is the center of a reflected spherical wave, one of many along the mirror. An outgoing (reflected) wave front—here, the line tangential to point

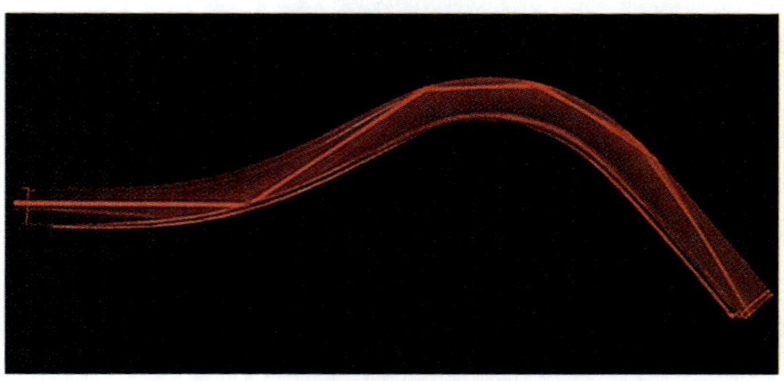

(a)

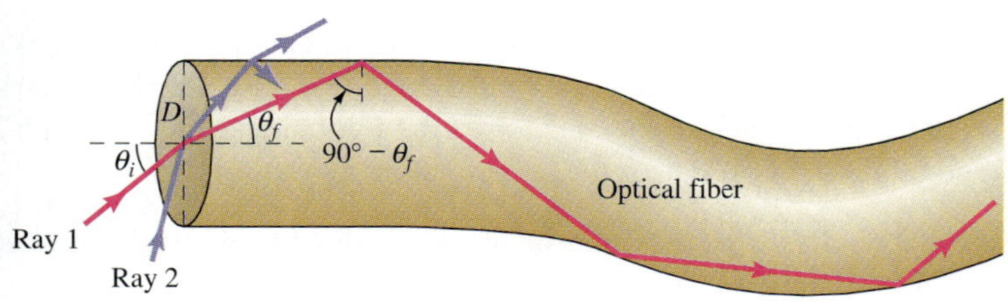

(b)

FIGURE 36–18 (a) Total internal reflection in an optical fiber. (b) Detailed construction of ray angles.

APPLICATION: FIBER OPTICS

Optical fibers, typically with diameters in the range of 50 μm (roughly the diameter of a human hair), now play an important role in technology (Fig. 36AB–1). The ideal situation outlined in the text is modified in a real fiber. The internal reflection is somewhat less than total if there are impurities such as moisture, dust, or oil on the surface because electromagnetic energy can leak across the thin "barrier" formed by the air layer between the fiber and the impurity. Light may be reflected thousands of times per meter, and it is therefore important to have no leakage of light. This problem is controlled largely by *cladding*—coating each fiber with a transparent covering whose index of refraction is lower than that of the fiber. In addition, the light intensity generally decreases as the ray propagates in a medium because the medium is not perfectly transparent. This effect is reduced by making the fiber from fused quartz, a highly transparent material, and purifying it to remove all traces of water. For devices such as *fiberscopes*, which are used for internal examinations of the human body, the distance covered by the light does not exceed several meters, and such refinements are not so critical. For communication by light pulses in optical fibers, much longer distances come into play. For the trans-Atlantic cable TAT-8, which can carry 40,000 conversations over two pairs of glass fibers simultaneously, it is necessary to boost the signal every 50 km with a repeater station. This is still much less expensive than systems of metal wire, which require boosting every kilometer. In the near future, much thinner fibers, 10 μm in diameter, will carry laser light; the number of boosting elements needed is reduced even further.

FIGURE 36AB–1 Light passes through a coiled fiber optic cable used for telephone lines.

D_2 of the semicircle centered on point C_2—forms. The distance the wave travels in time Δt is the same for incoming and outgoing waves, so a simple geometrical argument yields the result described by Eq. (36–5); namely, that the angle of reflection equals the angle of incidence. Figure 36–19c shows a later part of the sequence.

Snell's law may also be obtained by an application of Huygens' principle. The bending of the wave front is associated with the slowing down of the light waves in the medium. The bending can be visualized by analogy with the direction change of a wide column of soldiers who march at an angle toward a sidewalk, with orders given that as soon as a soldier steps on the sidewalk, he or she must walk slower without changing the distance between soldiers in each row. Rather than going through the derivation of Snell's law from the Huygens construction, let's demonstrate it from the principle enunciated by Pierre de Fermat in 1657. **Fermat's principle** states that

Fermat's principle

The path of a ray of light between two points is the path that minimizes the travel time.

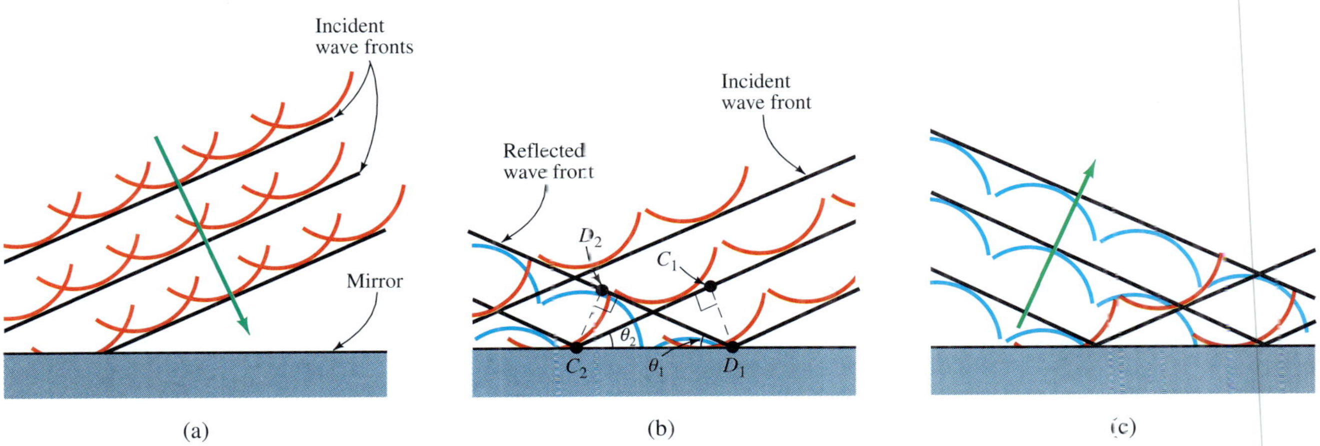

FIGURE 36–19 (a) Incident wave fronts approaching a plane mirror. (b) Later in the sequence, wave fronts reflect. (Note that fewer wavelets are shown, for clarity.) The fronts are generated by Huygens' construction. The relation $C_2D_2 = C_1D_1$ leads by geometrical reasoning to equal angles of incidence and reflection. (c) Even later, when most of the wave fronts are reflected.

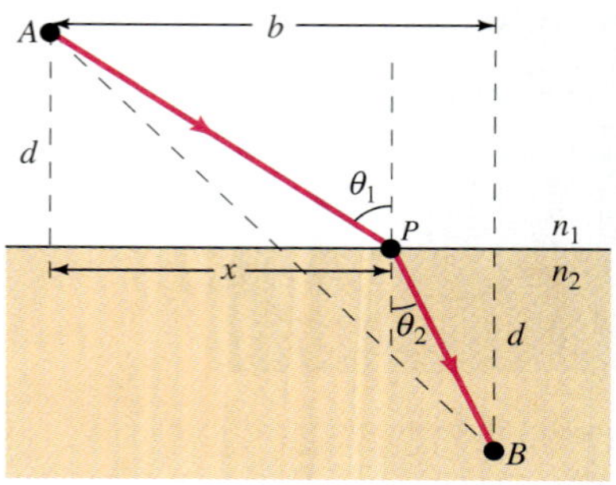

(a)

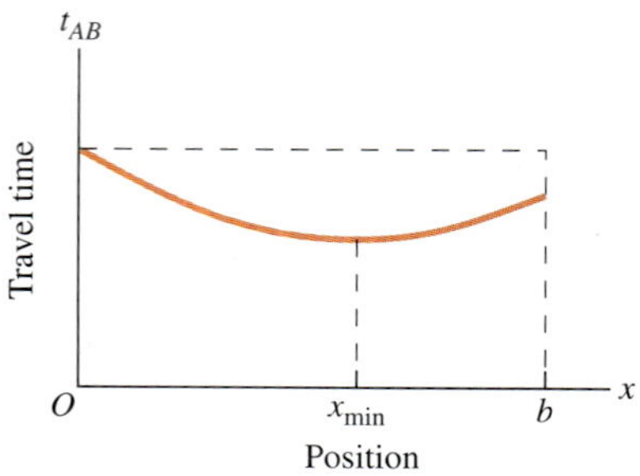

(b)

FIGURE 36–20 (a) Geometry for proving Snell's law by Fermat's principle. (b) The travel time t_{AB} for the ray as a function of x.

To derive Snell's law from Fermat's principle, let's consider a point A in medium 1 with index of refraction n_1, and a point B in medium 2 with index of refraction n_2 (Fig. 36–20a). We want to find the path between points A and B that takes a ray of light the least amount of time to travel. We choose A to be a distance d above the boundary and B a distance d below the boundary; we choose the horizontal distance between A and B to be b. The straight line connecting A to B crosses the boundary at a distance $b/2$ from the normal dropped from A onto the boundary but, because the indices of refraction are different, the ray's path crosses the boundary at a point P. Figure 36–20a shows that the distance from A to the intersection point P is $\sqrt{d^2 + x^2}$, and the distance from the intersection point P to B is $\sqrt{d^2 + (b - x)^2}$. The time for the ray to travel a distance D in a medium of index of refraction n is given by $t = D/v = D/(c/n) = nD/c$. Thus the total travel time is

$$t_{AB} = t_{AP} + t_{PB} = \frac{n_1\sqrt{d^2 + x^2} + n_2\sqrt{d^2 + (b - x)^2}}{c}. \qquad (36\text{–}11)$$

Figure 36–20b is a graph of t_{AB} as a function of x. The minimum travel time is obtained by finding the place at which the slope of t_{AB} as a function of x is flat; that is, the value of x at which

$$\frac{dt_{AB}}{dx} = 0.$$

This condition implies that

$$\frac{dt_{AB}}{dx} = \left(\frac{1}{c}\right)\left[\frac{n_1 x}{\sqrt{d^2 + x^2}} - \frac{n_2(b - x)}{\sqrt{d^2 + (b - x)^2}}\right] = 0. \qquad (36\text{–}12)$$

Now observe from Fig. 36–20a that

$$\frac{x}{\sqrt{d^2 + x^2}} = \sin\theta_1 \qquad (36\text{–}13a)$$

and that

$$\frac{b - x}{\sqrt{d^2 + (b - x)^2}} = \sin\theta_2, \qquad (36\text{–}13b)$$

where θ_1 and θ_2 are the angles the two rays make with respect to the normals in their respective media. Thus Eq. (36–12) may be rewritten as

$$n_1 \sin\theta_1 = n_2 \sin\theta_2, \qquad (36\text{–}14)$$

which is just Snell's law.

Both the straight-line propagation of light in a single medium and the law of reflection can also be derived from Fermat's principle. Fermat's principle follows directly from Maxwell's equations, although we cannot perform the derivation here. As Example 36–4 shows, principles such as Fermat's principle (more generally termed minimum principles) can apply in surprising circumstances.

EXAMPLE 36–4 A girl located at point B in Fig. 36–21a spots a ball at point A. Point A is in tall grass, where the girl can run at 1.1 m/s, and point B is in short grass, where the girl can run at 2.2 m/s. The whole area is flat. At what point x should she cross the boundary between the grasses so that she retrieves the ball as quickly as possible?

Solution: We let x be arbitrary for the moment and choose it so that the travel time for the child is minimized. Let v_1 be the speed in tall grass and v_2 be the speed in short grass. We have $v_2 = 2v_1$. For arbitrary x, the distance to be covered in tall grass is $L_1 = \sqrt{d^2 + x^2}$, and the time spent there is $t_1 = L_1/v_1$. The distance covered

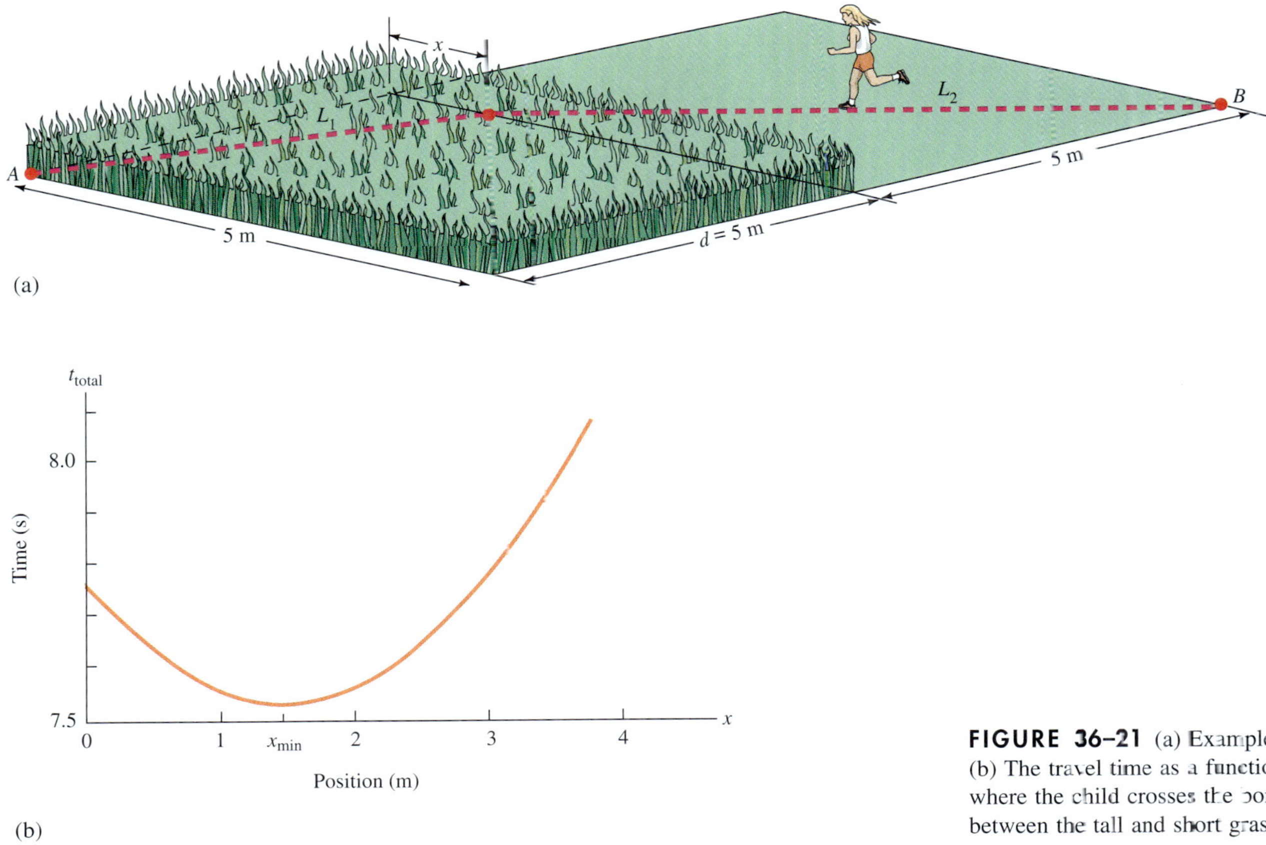

FIGURE 36–21 (a) Example 36–4. (b) The travel time as a function of where the child crosses the border between the tall and short grass.

in the short grass is $L_2 = \sqrt{d^2 + (d - x)^2}$, and the time spent there is $t_2 = L_2/v_2 = L_2/2v_1$. Thus the total time spent is

$$t_{total} = t_1 + t_2 = \frac{L_1}{v_1} + \frac{L_2}{2v_1} = \frac{\sqrt{d^2 + x^2}}{v_1} + \frac{\sqrt{d^2 + (d - x)^2}}{2v_1}.$$

We now want to minimize this time by varying x. This can be done by setting to zero the derivative of t_{total} with respect to x, but the algebra is fairly complicated. Alternatively, we can plot t_{total} as a function of x; in doing so we must use the numerical values of v_1, v_2, and d (Fig. 36–21b). There is indeed a minimum, at about $x = 1.5$ m. Compare this to the 2.5-m value x would take if the child were to run in a straight line.

The path followed looks like the path a light ray would take in traveling from a medium with a smaller value for the speed of light to a medium with a larger value for the speed of light. The two situations are in fact analogous!

36-5 DISPERSION

Let's now explore another property of the index of refraction, a property with some truly spectacular consequences: *In general, the index of refraction depends on the wavelength (or color) of the light being transmitted.* Table 36–2 shows how n varies with wavelength for glass—wavelengths near and including wavelengths of visible light. Different wavelengths are accordingly refracted to different degrees. In this way, white light (a mixture of different wavelengths) can be separated into its constituent colors of the rainbow. Newton himself studied light in this way. He observed the splitting of sunlight into the colors red, orange, yellow, green, blue, and violet, the colors that make

When the speed of light in a medium depends on the wavelength of that light, we have dispersion.

TABLE 36–2
Index of Refraction of Glass as a Function of Wavelength

Wavelength in Air (nm)	$\omega^2 = (2\pi c/\lambda)^2$ (rad²/s² × 10³¹)	n	*Color*
361	2.72	1.539	Near ultraviolet
434	1.89	1.528	Blue
486	1.50	1.523	Blue-green
589	1.02	1.517	Yellow
656	0.82	1.514	Orange
768	0.60	1.511	Red
1200	0.25	1.505	Infrared

(a)

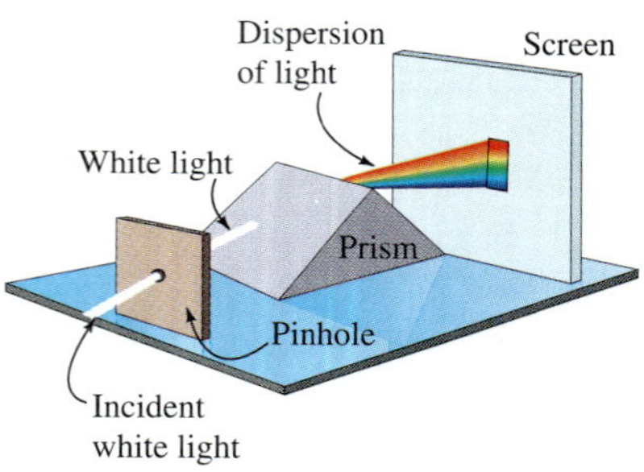

(b)

FIGURE 36–22 (a) Bright white light is collimated by a slit before being dispersed by a prism. (b) White light enters the prism, and light of different wavelengths follows different paths. The result is a beam separated by color.

up a rainbow (Fig. 36–22). When he projected each of the colored beams on a second prism, they refracted, but there was no further change in color. The dependence of refraction on the wavelength of light is called **dispersion**.

Rainbows and the Blue Sky

The colors of a rainbow result from dispersion in the scattering of light from individual water droplets in the air. When sunlight falls on a raindrop, light is reflected once before it leaves the drop. Many paths are possible; two are shown in Fig. 36–23a. The geometry is such that no ray can emerge after one reflection at an angle *steeper* than about 42°. Thus when the Sun is low and behind an observer on the ground, no raindrop high in the sky sends light back to the observer. Equivalently, when the Sun is behind the viewer, only drops that lie within a cone with an opening angle of about 42° reflect sunlight back to the observer's eye (Fig. 36–23b); moreover, *all* the drops in this cone reflect light to the observer. (We shall refer to a disk that fits into the cone because the depth of the cone is irrelevant.) One other feature of the disk is that light is reflected most strongly from raindrops at the edge, around 42°.

Dispersion has played no role in our discussion so far. The effect of dispersion is to make the angle of the outer radius of the disk slightly different for different colors. As Fig. 36–23c indicates, the disk for red light is larger than the disk for blue light. Because the intensity of the light in the disk is strongest at the edges, what we see is a red ring outside of a blue ring (with other colors placed accordingly). All the disks overlap inside the rainbow, giving white light. A *secondary rainbow* can be produced when there are two internal reflections within the raindrops (Fig. 36–24). The order of the colored disks produced by the raindrops will now be reversed, with red light at the bottom and blue light at the top of the secondary rainbow. Figure 36–25a illustrates how an observer sees rainbows and how the pattern of dispersion leads to the color inversion of a secondary rainbow compared with a primary rainbow. The light is brightest below the primary rainbow and above the secondary rainbow because the disks overlap; it is relatively darker between the two rainbows (Fig. 36–25b).

Dispersion is a widespread phenomenon. Like refraction, the scattering of light by matter has a frequency (or wavelength) dependence. It was shown by Lord Rayleigh in 1872 that the fraction of incident light scattered by air molecules varies as f^4 for light in the visible range. This explains the color of the setting Sun. As the Sun sets, the rays of light pass through more and more atmosphere, and an increasing number of the high-frequency (low-wavelength) components are scattered. The Sun's color changes from white to yellow to orange and finally to red as the higher frequencies are scattered away from the observer.

The frequency dependence of the amount of light scattered is also responsible for the fact that the sky looks blue. Blue light has a higher frequency than red; thus, the blue component of sunlight is scattered into our eyes by the atmosphere. Above the atmosphere, where there are no molecules to scatter the light, astronauts see a black sky.

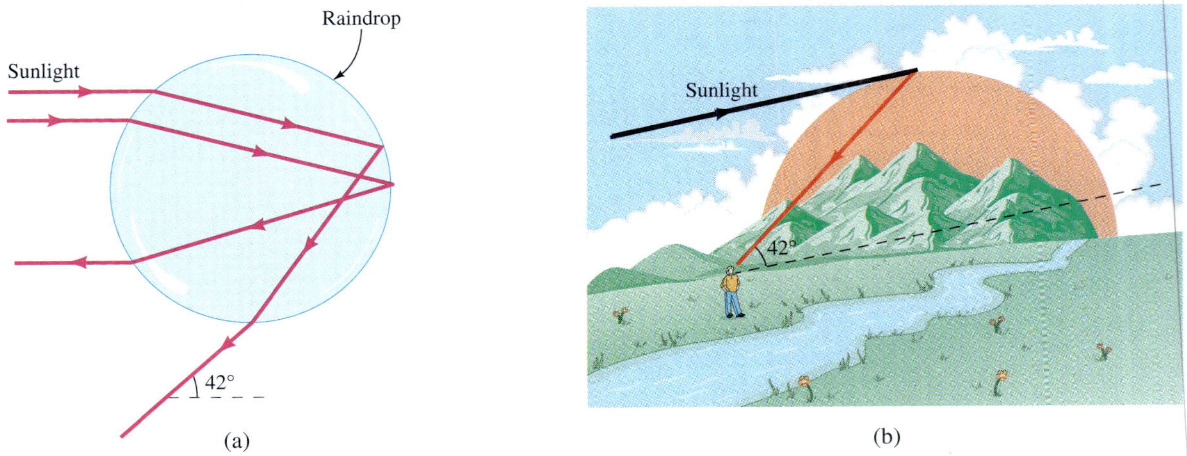

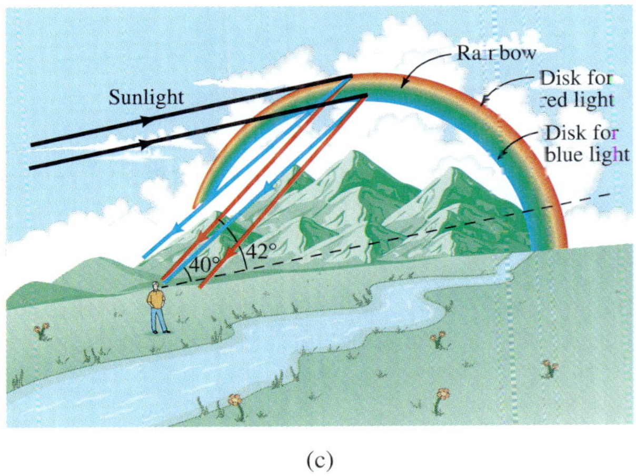

FIGURE 36–23 (a) When sunlight enters a raindrop from the horizontal and exits after one reflection, no light exits at an angle steeper than about 42°. (b) As a result, light comes back to an observer from all the raindrops that lie within a cone with an angle of about 42°, as seen by the observer. (c) Sunlight is a variety of colors. Due to dispersion, the disks that fit into the cones for different colors are of slightly different sizes, with red forming the largest cone and violet, the smallest. This figure exaggerates the effect.

If the sky is blue, why are clouds white? The f^4 law applies to the scattering of light by objects much smaller than the wavelength of the light. Thus it applies to the scattering of light by air molecules. Dust particles and water droplets that make up a cloud are comparable in size to, or larger than, the wavelength of light impinging on them and they act like mirrors. For surfaces larger than the wavelength, reflection rather than scattering occurs, and the geometrical laws of reflection hold for scattering off the droplets. But because the many droplets in a cloud form an irregular surface, we do not get images of the kind we are used to seeing in mirrors. Instead, we get *diffuse* reflection, and the clouds look white or gray (which is just a less intense white).

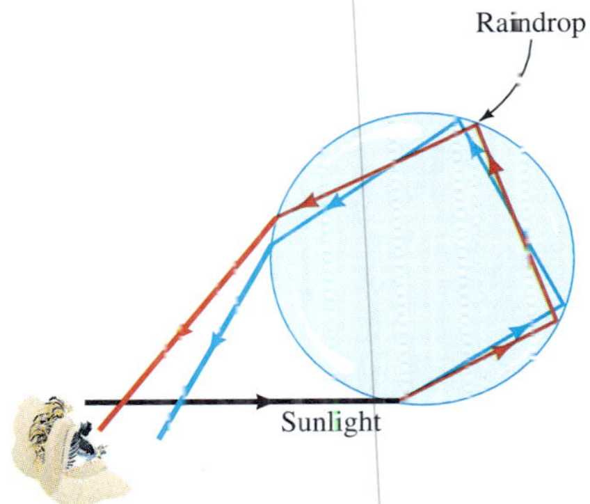

FIGURE 36–24 The light that reaches the eye from a secondary rainbow has undergone two internal reflections in a set of raindrops. Light of shorter wavelengths (blue light) emerges at a steeper angle than light of longer wavelengths (red light) does, in contrast to the light that undergoes only one internal reflection, which forms the primary rainbow.

The Atomic Theory of Dispersion

Dispersion occurs because of the atomic structure of dielectric media. The atoms in such media contain a positive charge at the center and negative charges, in the form of electrons, distributed on the outside over a region with linear dimensions of 0.1 nm. For our purposes, let's think of each atom as a single electron (of mass m) that oscillates about a positive ion as though the electron were bound to that ion by a spring. If the spring constant is k, then the angular frequency of oscillation (the natural frequency) is ω_0, given by $\omega_0{}^2 = k/m$. If no other forces are acting, the motion of the electron along the z-axis would then be of the form

$$z = A\cos(\omega_0 t). \qquad (36\text{–}15)$$

where A is the amplitude of motion.

The driven harmonic oscillator and the associated resonance phenomena are treated in Section 13–8.

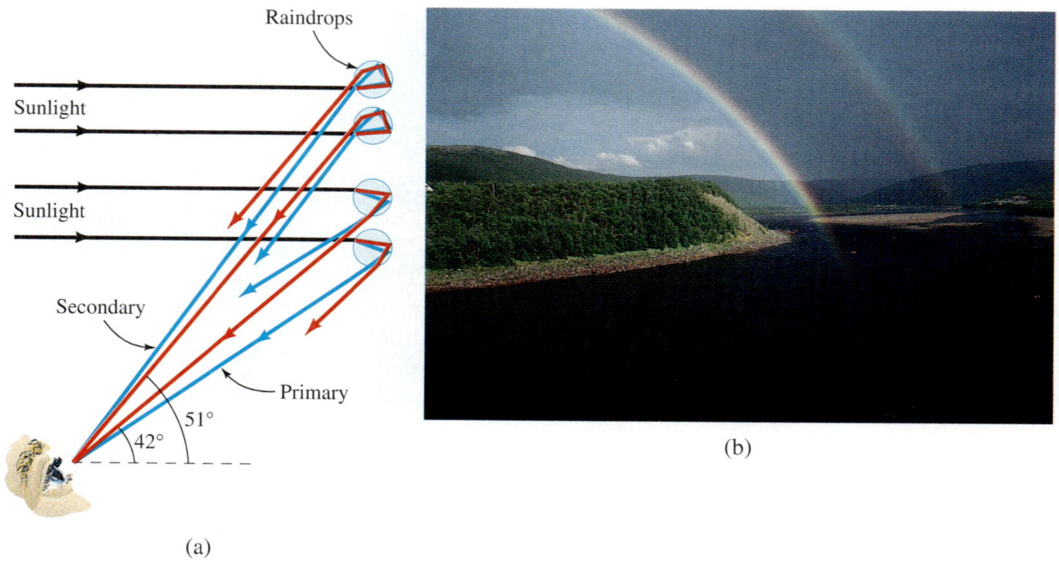

FIGURE 36–25 (a) When the eye sees a region of sky that contains raindrops illuminated by sunlight, the light reflected from individual drops forms a primary rainbow (one internal reflection within each drop) and a secondary rainbow (two internal reflections). (b) The brighter, primary rainbow is on the bottom. The order of colors is reversed in the two rainbows due to the extra reflection, which produces the fainter secondary rainbow. The disks overlap, so it is relatively brighter below the primary rainbow and above the secondary rainbow, but darker between them.

Now suppose that a plane electromagnetic wave oscillating with angular frequency ω is incident on an atom, with the electric field oriented in the z-direction. The electric force on an electron in the atom is then oscillatory with frequency ω. The situation is that of a driven harmonic oscillator. The motion of the electron is oscillatory with the driving frequency ω. The amplitude exhibits resonance, meaning that it becomes large when $\omega_0 \simeq \omega$. Thus, based on our knowledge of the driven harmonic oscillator, the motion is of the form

$$z \propto \frac{1}{\omega_0^2 - \omega^2}\cos(\omega t). \tag{36–16}$$

For materials such as water and glass, ω_0 is on the order of 5 to 6 times larger than the characteristic angular frequencies of visible light.

An accelerating charge (the electron) *radiates* electromagnetic energy, and the intensity of the radiation is proportional to the average of the acceleration squared. From Eq. (36–16), the average acceleration is proportional to

$$\frac{d^2z}{dt^2} \propto \frac{\omega^2}{\omega_0^2 - \omega^2}\langle\cos^2(\omega t)\rangle \propto \frac{\omega^2}{\omega_0^2}.$$

In the last step, we have used the fact that $\omega_0^2 >> \omega^2$ for visible light in these materials. The intensity, I, of the radiation is proportional to the acceleration squared. Thus I varies as ω^4 or, equivalently, as f^4. The wavelength, λ, is related to ω by $\lambda = c/f = 2\pi c/\omega$, so *the intensity of the radiation emitted by a charge set in oscillation by an external electric field is proportional to* $1/\lambda^4$, *where* λ *is the wavelength of the oscillating field.* This is the result first obtained by Lord Rayleigh. We used this result to explain why the sky is blue.

This discussion shows how the original electric field is modified by the addition of a radiated field as a result of the electron's motion; in fact, it is such a modification that is described by a dielectric constant κ, and hence an index of refraction n given by Eq. (36–2), $\kappa = n^2$. We accordingly expect n to depend on the frequency; that is, to exhibit dispersion. If we take a more detailed look at how the external field is modified in the presence of atomic electrons, we find

$$\frac{1}{n^2} = 1 - \frac{C}{\omega_0^2 - \omega^2}, \tag{36–17}$$

where C is a constant. This equation describes how the index of refraction varies with frequency. Because $\omega_0 >> \omega$ for visible light, *the index of refraction increases as the frequency of the light increases.* In fact, atoms and molecules have many resonant frequencies, so a more accurate version of Eq. (36–17) must contain several terms of the form $C_k/(\omega_{0k}^2 - \omega^2)$ added together.

SUMMARY

The properties of light are well understood in terms of the wave theory of light. The speed of propagation of light waves is $c = 3.00 \times 10^8$ m/s in a vacuum. In transparent media, the speed of propagation is c/n, where n is the index of refraction of the medium. In general, the index of refraction depends on the wavelength of the light.

The propagation of light can be described either in terms of wave fronts, which form an envelope of spherical wavelets built upon earlier wavelets (Huygens' principle), or in terms of rays, which are lines perpendicular to the wave fronts. Light rays travel in straight lines unless they meet boundaries. Upon reflection from a surface, the angle θ that the incident ray makes with the normal to the surface is equal to the angle θ' that the reflected ray makes with the surface (the law of reflection):

$$\theta' = \theta. \tag{36–5}$$

In the passage from a medium of index of refraction n_1 to a medium of index of refraction n_2, the incident angle θ_1 and the refracted angle θ_2 are related by Snell's law of refraction:

$$n_1 \sin \theta_1 = n_2 \sin \theta_2. \tag{36–6}$$

These results can be established by using the geometry of wave fronts. They can also be derived with the help of Fermat's principle, which states that the path taken by a light ray between two points is the path that takes the shortest time. One consequence of Snell's law is that total internal reflection occurs when light moving in a medium with index of refraction n_1 strikes a boundary of a medium with index of refraction n_2, where $n_1 > n_2$. This holds true provided that the angle of incidence is larger than a critical angle θ_c, given by

$$\sin \theta_c = \frac{n_2}{n_1}. \tag{36–9}$$

The dependence of the index of refraction on wavelength is called dispersion. Dispersion causes the different wavelengths in a beam of white light to refract through different angles. The colors of a prism, the rainbow, and the blue sky are all naturally occurring dispersion phenomena. Dispersion can be understood in terms of the atomic theory of matter.

UNDERSTANDING KEY CONCEPTS

1. If light travels only in straight lines, how does a light burning in one room give light in another room?
2. How difficult would it be to reflect light back to Earth from the Moon by using two perpendicular plane mirrors? Why does it help if there are three mutually perpendicular mirrors?
3. If fish could think, they might realize that the relative indices of refraction of water and air allow them to outwit fishermen. Why?
4. A person swimming underwater sees a lifeguard who is standing in the shallow part of the pool; the water comes up to the lifeguard's waist. In what way does the swimmer see the lifeguard's upper body distorted?
5. A fisherman standing up to his waist in a lake appears, to an observer outside the lake, to have shorter-than-normal legs. How will a fish in a horizontal position near the bottom of the lake appear to the observer?
6. A plane wave of radiation has an electric field of the form $\mathbf{E}_0 \cos(kz - \omega t)$ when it propagates in empty space. How do k and ω change when the plane wave enters a medium with index of refraction n?
7. As the Sun sets, its color changes from white to yellow to orange and finally to red. As the lowest part of the Sun sinks below the horizon, the Sun appears squashed, more egg-shaped than circular. Why?

8. A coin lies at the bottom of a pool of water. Starting from a point immediately above the coin, you observe the coin from the level of the surface. You then move your head horizontally away from the coin across the surface of the water. Is there a horizontal distance at which the coin is no longer visible?

9. Light from the sky refracts near the surface of hot sand, giving the impression that there is a bright surface that could be interpreted as water: a mirage (Fig. 36–26). The air near the surface of hot sand is hotter than the surrounding air. Does light travel faster or slower in hot air than in cold air?

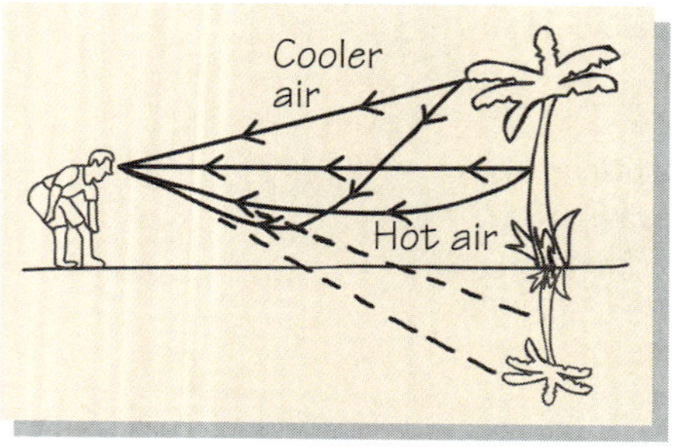

FIGURE 36–26 Question 9.

10. Mirages can occur when a layer of cold air lies closer to the surface. How would such an air layer affect the appearance of distant houses?

11. Why does the sky look black rather than blue, as it does from Earth, to astronauts in orbit?

12. What is the index of refraction of a vacuum?

13. For a moment, you are lying in the middle of a circular swimming pool—at the bottom of the pool—which is filled to a depth of 1 m with water and is surrounded by trees. A 2-m tall lifeguard is standing in the water about 3 m from you. What do things look like as you scan in all directions?

14. Laser light directed into the end of a glass rod comes out the other end with almost the original intensity. If another glass rod touches the side of the first rod, making a 30° angle with the lengthwise direction of the first rod, nothing happens. But if the point of contact is lubricated with glycerin, some of the original light beam is "stolen" by the second rod. Explain what happens.

15. White light is incident onto a pane of glass. Is there a dispersion of colors in the reflected light?

16. Stick a pin into the underside of a cylinder of cork, then float the cork in water. Even if you do not stick the pin in very far, you may not be able to see it from outside the water. Why not?

PROBLEMS

36–1 The Speed of Light

1. (I) What are the speeds of light in ice, ethyl alcohol, benzene, and diamond?

2. (I) The nearest star to our solar system (aside from the Sun) is Alpha Centauri, some 4.2 ly from Earth. How far is this in meters?

3. (I) A light wave of red light ($\lambda = 650$ nm) passes from air into water, where the index of refraction is 1.32. What are the wavelength and frequency of the light in water?

4. (I) Yellow light, whose frequency is 5×10^{14} Hz, impinges on glass, $n = 1.5$. What are the wavelengths of this light in a vacuum and in glass? What is the index of refraction of a material within which the wavelength of yellow light is one-half its value in a vacuum?

5. (II) Suppose that you have a version of Fizeau's apparatus in which the round-trip distance for the light beam is $2D = 1000$ m. The width of the opening between the teeth on the cogged wheel is 0.70 mm, and the center-to-center distance between these gaps is 1.5 mm. The wheel has a radius of 15.0 cm. What would the minimum rotational speed be, in revolutions per minute, so that light entering through the center of one gap would come out through the center of the next gap? Is such an apparatus realizable?

6. (II) Figure 36–27 shows an exaggerated view of the eclipsing of Io, the innermost moon of Jupiter, as seen from two different points on Earth's orbit around the Sun. If Earth were stationary at a point nearest Jupiter, N, a particular eclipse would begin at a precise time. When Earth is at point F, the eclipse starts somewhat later than expected because the light has to travel the additional distance of a diameter of the Earth–Sun orbit. The mean distance from Earth to the Sun is 1.50×10^{11} m. How much later will the eclipse be seen at point F compared with point N?

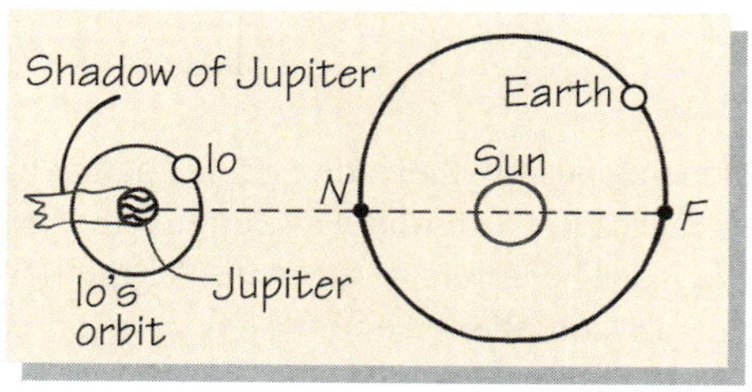

FIGURE 36–27 Problem 6.

7. (II) Personal computers can perform as many as 5×10^7 steps every second. This means that some leads connecting different parts of the computer may carry this many pulses per second. If you assume that the pulses travel at the speed of light, what is the distance between pulses? Does this result have implications for the design of these machines?

8. (II) Telephone connection between Europe and North America can be carried by cable or by the use of a geosynchronous communication satellite. Estimate the time it takes for a signal to travel 5000 km via cable, assuming the speed is close to the speed of light. How does this compare to the time required for the same signal to travel via satellite, 40,600 km from the center of Earth?

9. (II) The speed of light in a vacuum is defined to be 299,792,458 m/s. A lunar ranging experiment measures the time for a light pulse to reach the Moon and reflect back to Earth. Such experiments allow us to determine the distance between the Moon and Earth, which is approximately 3.84×10^8 m, to an accuracy of 15 cm. What is the smallest time interval that can be measured by the clock used to determine the time it takes for light to go to the lunar reflector and back?

10. (II) Galileo attempted to measure the speed of light with the help of lights and a clock on two adjacent mountains. In essence, a shutter over a light was opened on the first mountain, an observer on the second mountain saw that signal and returned a second signal, and the experimenter on the first mountain looked for a delay between the time the shutter was opened and the time the signal was returned. Use your knowledge of human reaction time to estimate the time measured by the first experimenter for the total round trip. How long would it actually take for light to travel back and forth between two mountaintops separated by 4 km? Your answers explain why Galileo's attempt did not work.

11. (II) Imagine an experiment similar to Fizeau's, with a cogged wheel of diameter 20 cm. A laser beam shines through one opening, travels 1500 m, and is reflected back. Given that the fastest rotation rate of the wheel is 1.2×10^5 rev/min, what should be the separation between adjacent cogs on the rim of the wheel?

36–3 Reflection and Refraction

12. (I) A fixed projector emits a narrow beam of light onto a plane mirror. At what angle with respect to the beam should you place the mirror in order to turn the beam by 90°?

13. (I) The critical angle for a particular material (used in air) is observed to be 38°. What is the material's index of refraction?

14. (I) A horizontal beam of light is reflected from a plane mirror that revolves about a vertical axis at a rate of 30 rev/min. The reflected beam sweeps across a screen that, at the point nearest the mirror, is 20 m away (Fig. 36–28). With what speed does the spot of light move across the screen at the point nearest the mirror?

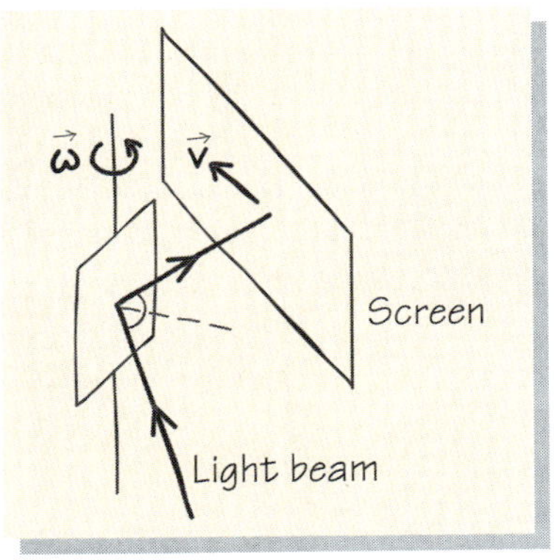

FIGURE 36–28 Problem 14.

15. (I) An intense light beam is incident at 45° to the surface of a clear lake. If the lake is 500 m deep and has a flat bottom, how far does the light beam travel before it hits bottom?

16. (I) A beam of light is sent from medium 1, index of refraction n_1, into a medium 2, index of refraction n_2; here, $n_2 > n_1$. We know that a refracted ray is bent toward the perpendicular to the boundary. Are there any incident angles for which the angle of the refracted ray is 90°? If not, what is the largest possible angle of refraction? Give a numerical value to your answer in the case of air-to-water: $n_1 = 1$ and $n_2 = 1.33$.

17. (I) A burglar stands in front of a department store window and directs his flashlight into the store. What fraction of the light is reflected at the window's surface, assuming that the index of refraction of the glass is 1.43? Ignore all reflections except for that at the outside interface between the glass and air.

18. (I) A swimmer is at the bottom of a large, shallow swimming pool. Through what angle must she move her eyes so that her direct gaze swings across the whole sky? Water's index of refraction is 1.33.

19. (I) The index of refraction of air is $1 + 2.93 \times 10^{-4}$. Assume that the atmosphere may be treated as a uniform medium of thickness 8.3 km, which covers Earth's surface; further, suppose a ray of light hits the top of the atmosphere parallel to the top of the atmosphere—grazing incidence. What is the angle that the refracted ray makes with the *horizontal*?

20. (II) What is the critical angle for total internal reflection in crown glass (used in air), for which $n = 1.52$? Show that it is possible to use a 45°–45°–90° triangular prism of crown glass (see Table 36–1) to make a perfect reflector of light (Fig. 36–29).

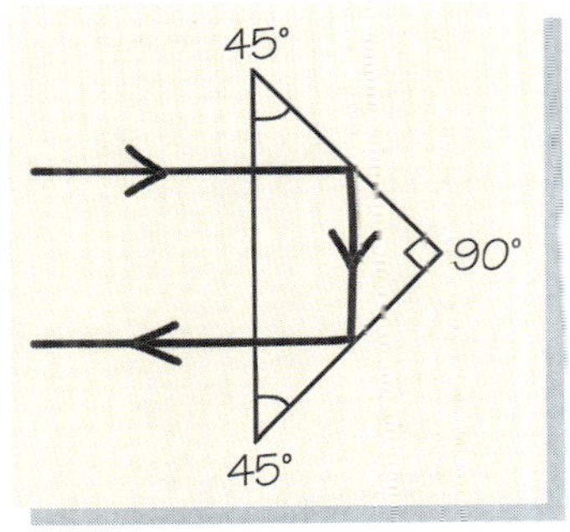

FIGURE 36–29 Problem 20.

21. (II) A thick glass plate ($n = 1.53$) lies on the bottom of a tank of water ($n = 1.33$). A light ray enters the water from air, making an angle of 72° with the normal to the surface. What angle does the ray make with the normal when the ray is in the water? What angle does it make with the normal when it is in the glass?

22. (II) Light in air enters a stack of three parallel plates with indices of refraction 1.50, 1.55, and 1.60, respectively. The incident beam makes a 60° angle with the normal to the plate surface. At what angle does the beam emerge into the air after passing through the stack?

23. (II) Light approaches a glass–air interface from the glass side ($n = 1.6$) at an angle θ_i. If $\theta_i > \theta_c$, total internal reflection occurs; if $\theta_i < \theta_c$, some light passes through the interface. Is it possible to modify this property of the interface by adding a stack of layers with carefully chosen indices of refraction?

24. (II) The composition of a glass block varies as a function of the distance x from the top surface (Fig. 36–30). As a consequence, the index of refraction increases as a function of x according to $n(x) = 1.6 - (0.2\ \text{cm}^2)/(x + 1\ \text{cm})^2$, with x in centimeters. A beam of light strikes the surface at an angle of incidence of 65° from the vertical. What will be the direction of the beam deep inside the block?

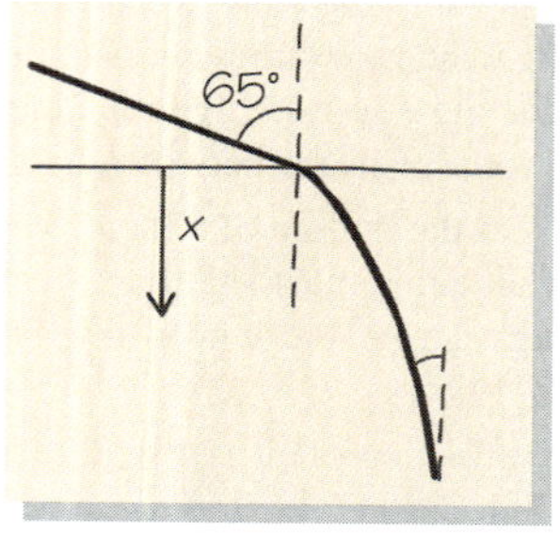

FIGURE 36–30 Problem 24.

25. (II) A glass sphere ($n = 1.6$) is centered at the origin of a coordinate system, with its equatorial plane defining the xy-plane. A beam of light enters the glass sphere at a latitude of 40°, parallel to x-axis in the xz-plane. Make a careful drawing to determine the angle at which the beam will strike the back of the sphere. Will there be total internal reflection?

26. (II) White light is refracted by the triangular prism shown in Fig. 36–31. A beam of light enters the prism along a path parallel to the prism base. The light is observed on a screen that is located 10 m from the prism and is perpendicular to the emerging rays. How far apart on the screen are the spots of blue light ($n = 1.528$) and red light ($n = 1.514$)?

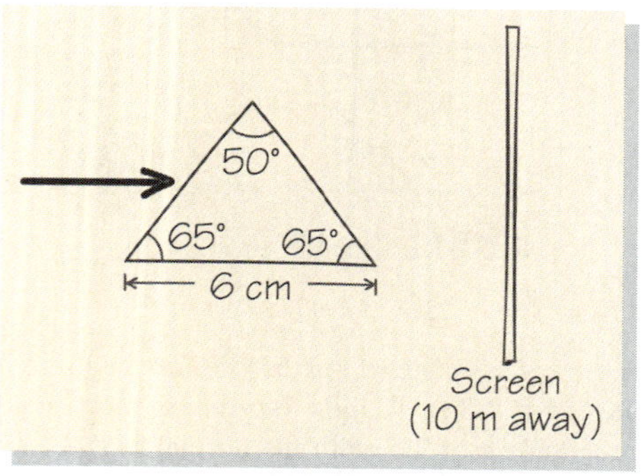

FIGURE 36–31 Problem 26.

27. (II) A very wide light beam strikes a white screen at 90° to the surface of the screen. An isosceles prism is placed in the way of the beam, as shown in Fig. 36–32. How will the screen be illuminated if the index of refraction of the glass of the prism is $n = 1.5$?

28. (II) A lifeguard 1.7 m tall stands in water 120 cm deep. From a vantage point at the bottom of the pool, a swimmer sees the lifeguard's head to be along a line at a 42° angle to the vertical. How far is the swimmer's eye from the lifeguard's feet? (For water, $n = 1.33$.)

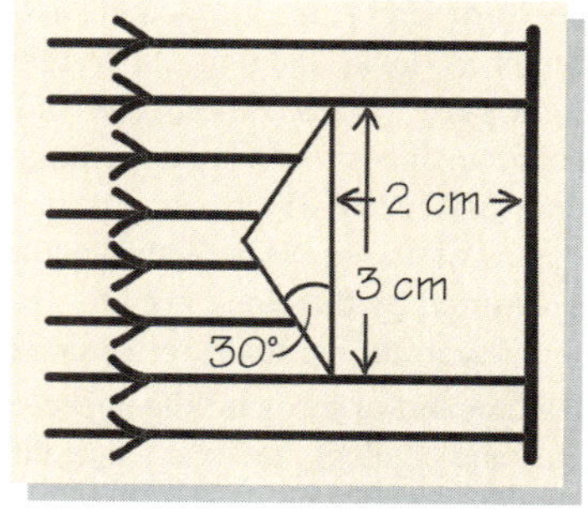

FIGURE 36–32 Problem 27.

29. (II) Suppose that you look at an aquarium with your eyes at the level of the water surface (Fig. 36–33). A duck swims on the surface of the water. When you look at the duck from the front, everything seems normal. However, when you look at the duck at an angle to the glass surface, the duck seems to be split in half, with the feet paddling ahead of the upper body. Explain this phenomenon. Suppose that both the duck and your eyes are at a distance of 1 m from the glass, and the line connecting them forms a 30° angle with the glass. Calculate the difference between the directions of the line of sight of the upper and lower halves of the duck.

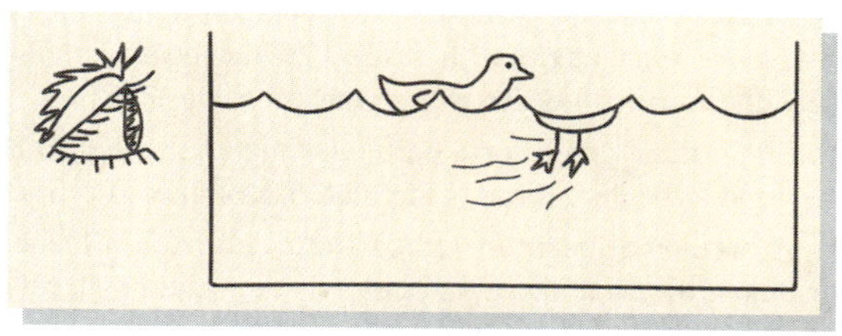

FIGURE 36–33 Problem 29.

30. (II) A narrow beam of light is incident at a 30° angle from the normal onto a glass pane 6 mm thick. Describe the position of the exit beam of light. What is its direction? Is it displaced from the incident beam? If so, by how much? (For the glass, $n = 1.60$.)

31. (II) Light is incident on an equilateral triangular prism ($n = 1.55$) at a 35° angle from the normal to one of the faces (Fig. 36–34). What is the exit angle?

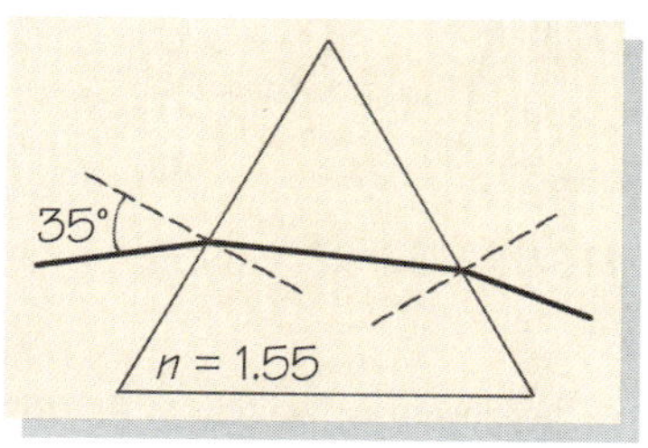

FIGURE 36–34 Problem 31.

32. (II) Consider a solid glass rod of length 30 cm and diameter 2 cm, with index of refraction 1.53. The ends of the rod are perpendicular to the lengthwise direction. (a) Light enters the center of the end of the rod from air. What is the maximum angle of incidence for which the light is totally reflected inside the rod? (b) Repeat part (a) for a similar rod totally immersed in water ($n = 1.33$).

33. (II) A ray of light impinges at a 60° angle of incidence on a glass pane of thickness 5 mm and index of refraction 1.54. The light is reflected by a mirror that touches the back of the pane (Fig. 36–35). By how much is the beam displaced compared with the return path it would have if the pane were absent?

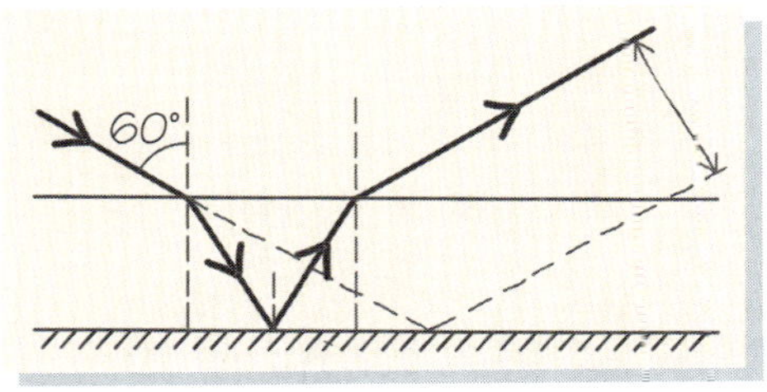

FIGURE 36–35 Problem 33.

34. (II) When a light beam is reflected by a conventional mirror, part of the light is reflected by the front surface of the glass pane, and part by the silvered back surface. What is the distance between the two reflected beams if the mirror is 2 mm thick, if it is made of glass with $n = 1.45$, and if the angle of incidence is 70°. (To avoid this double reflection, many optical instruments use mirrors with their front surfaces silvered.)

35. (II) At noon, a 2.0-m-long vertical stick casts a shadow 1.0 m long. If the same stick is placed in a flat-bottomed pool of water half the height of the stick (still at noon), how long is the shadow on the floor of the pool? (For water, $n = 1.33$.)

36. (II) You have three transparent liquids labeled 1, 2, and 3 that do not mix. When light is sent from liquid i to liquid j, there is an angle of incidence θ_i and an angle of refraction θ_j. Two separate experiments show the following: $1 \rightarrow 2$, $\theta_i = 22°$ and $\theta_j = 32°$; $2 \rightarrow 3$, $\theta_i = 35°$ and $\theta_j = 51°$. Find the ratios of the indices of refraction for each pair of liquids.

37. (II) Consider light that is perpendicularly incident on a triangular prism of the kind shown in Fig. 36–36. The index of refraction of the prism material is $n_1 = 1.814$. Suppose that the two reflecting sides are coated with a thin, uniform layer of a dielectric with index of refraction $n_2 = 1.380$. Will the glass–dielectric interface be totally reflecting? How large can n_2 be so that the interface is still totally reflecting?

38. (II) A prism has a cross section in the shape of an isosceles triangle with a base-to-height ratio of 1/3. A beam of light is incident upon the left side, parallel to the base. At what angle relative to the base will the beam leave the right side of the prism, which is made of glass with $n = 1.62$?

39. (III) Use Huygens' construction to prove Snell's law by working out the geometrical details in Fig. 36–37.

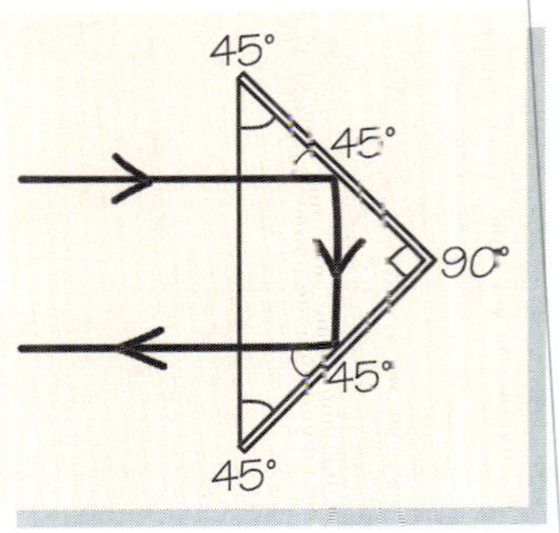

FIGURE 36–36 Problem 37.

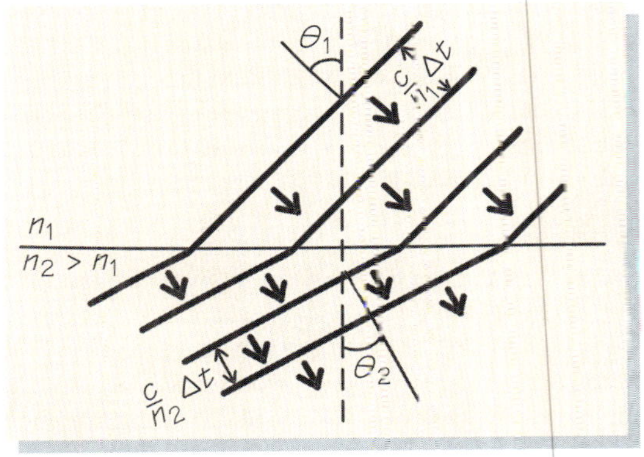

FIGURE 36–37 Problem 39.

*36–4 *Fermat's Principle*

40. (II) Use Fermat's principle to show that the critical angle for total internal reflection is given by $\sin\theta_c = 1/n$, where n is the index of refraction of the medium in which the light ray originates (Fig. 36–38). The outside medium is air.

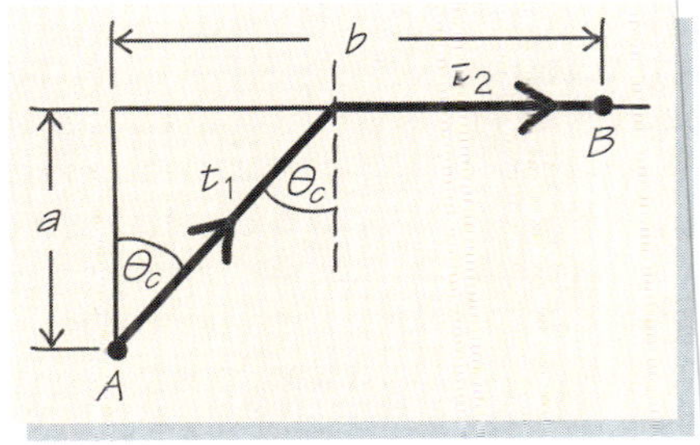

FIGURE 36–38 Problem 40.

41. (II) Show that the law of reflection follows from Fermat's principle.

42. (II) By using Fermat's principle, show that if two media have exactly the same index of refraction, then a beam of light travels in a straight line when it crosses the boundary between them.

43. (II) Use Fermat's principle to show that a beam of light that enters a plate of glass of uniform thickness emerges parallel to its initial direction (Fig. 36–39).

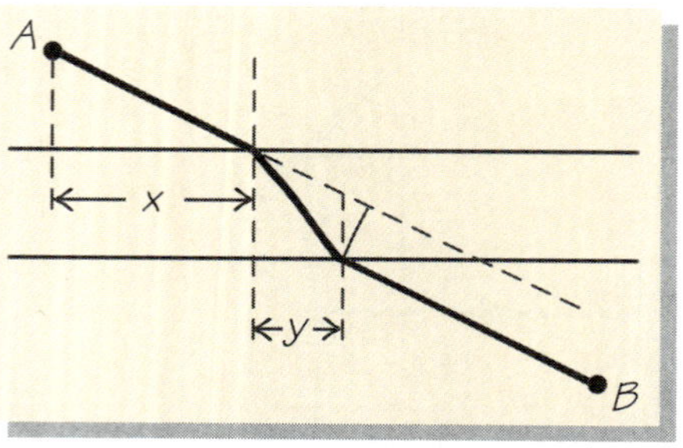

FIGURE 36–39 Problem 43.

44. (III) Calculate the parallel displacement of a beam of light that strikes a vertical slab of glass, with index of refraction n and thickness D, at an angle ϕ with the horizontal (use Fermat's principle). (In Problem 43, we showed that a ray of light that passes through a slab of glass emerges parallel to its initial direction. The ray is, however, displaced from its original line, and that displacement is what we want here.)

36–5 Dispersion

45. (I) At 0° C, the index of refraction in water of light of wavelength 397 nm (violet) is 1.3444, whereas it is 1.3319 for a wavelength of 656 nm (red). What is the difference in angles of refraction for rays refracting from water near the freezing point into air for these two wavelengths? The angle of incidence is exactly 30° in each case. Take $n = 1$ for air and ignore dispersion in this medium.

46. (I) By what percent does the speed of red light in a type of glass ($\lambda = 656$ nm, $n = 1.571$) exceed that of blue light in the same glass ($\lambda = 486$ nm, $n = 1.585$)?

47. (II) A beam of white light, whose frequencies are mixed with equal intensity, passes within a piece of glass and impinges on a boundary to the air at an angle of incidence θ. The index of refraction of the glass increases with increasing angular frequency according to the formula $n^2 = 1 + [C/(\omega_0^2 - \omega^2 - C)]$, where $C = 529 \times 10^{30}$ rad²/s² and $\omega_0^2 = 685 \times 10^{30}$ rad²/s². (a) What is the largest angular frequency that passes through the glass into the air? (b) At what angle of incidence should the light approach the boundary if we wish to allow only frequencies of $\omega = 3.2 \times 10^{15}$ rad/s (red light) and below to pass through to the air?

48. (II) Use the data in Problem 47 to calculate the critical angles for total internal reflection for five values of wavelengths in the range 430 nm to 770 nm. Plot your results.

49. (III) We wish to select a glass to construct a prism that can separate the yellow ($\lambda = 590$ nm) component of light from the blue-green ($\lambda = 490$ nm) component. The prism is to be a bar with the cross section of an equilateral triangle. If a ray of white light arrives parallel to the base of the prism, it must leave the prism with the two colors separated by at least 2°. What must the difference in indices of refraction be for the two colors? [*Hint*: Because the difference of angles is small, so is the difference of indices of refraction. Keep only leading terms in differences of angle and of index of refraction.]

General Problems

50. (I) Light of wavelength 660 nm enters a piece of glass with index of refraction 1.52. What are the wavelength and speed of that light in the glass?

51. (II) A pin is partly inserted perpendicularly into the flat surface of a cork with a 1.5-cm radius (Fig. 36–40). The cork, with the pin on the underside, is set afloat in a pool. A length of 1.2 cm of cork is under the water surface. Because of the effects of refraction, much of the pin is hidden from view from above the surface. What length of pin can be hidden in this way?

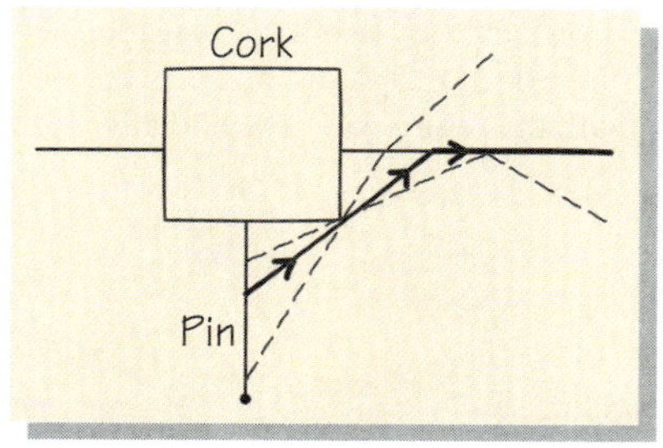

FIGURE 36–40 Problem 51.

52. (II) A beam of light is incident at an angle of 30° to the vertical on a horizontal glass plate of thickness 2.0 cm. The index of refraction of the glass is $n = 1.52$. The beam emerges on the other side. What is the perpendicular distance between the straight-line extrapolation of the incident ray and the ray refracted by the glass plate?

53. (III) Show that if an incident ray of white light that is parallel to the base of a prism in the shape of an isosceles triangle (apex angle 2ϕ) is separated into two components that exit the prism with an angular separation $\Delta\theta << 1$, then the difference in the indices of refraction for the two colors is proportional to $\Delta\theta$. Find the equation that expresses the relation between the differences in the indices of refraction and in $\Delta\theta$. [*Hint*: Consider the angle of emergence for a given n, and then find Δn as a function of $\Delta\theta$.]

54. (III) Sound can refract like light does. Suppose that a submarine lies flat 240 m below the water surface, and that there are three thermal layers of water (each 80 m deep) of different temperatures (Fig. 36–41). The speed of sound in water depends on temperature. In the bottom layer, the speed is 1.19 times that in the top layer; in the middle layer, the speed is 1.11 times that in the top layer. A detection device at surface level determines that sound from the submarine arrives at the surface at a 45° angle with the horizontal. What is the horizontal distance between the detector and the submarine?

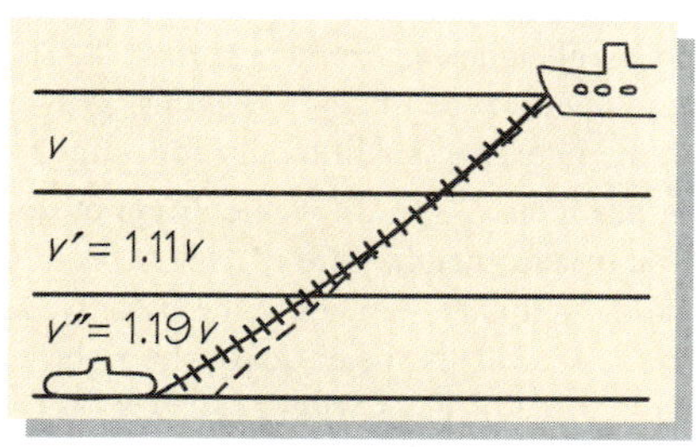

FIGURE 36–41 Problem 54.

55. (III) A ray of light is incident at an angle of incidence θ_i on one surface of a prism whose cross section is an isosceles triangle (apex angle 2ϕ). The light exits the prism at a total deflection angle Θ (Fig. 36–42). The prism has index of refraction n and is in a vacuum, which has an index of refraction of exactly 1. For what angle θ_i is the angle of deflection Θ a minimum?

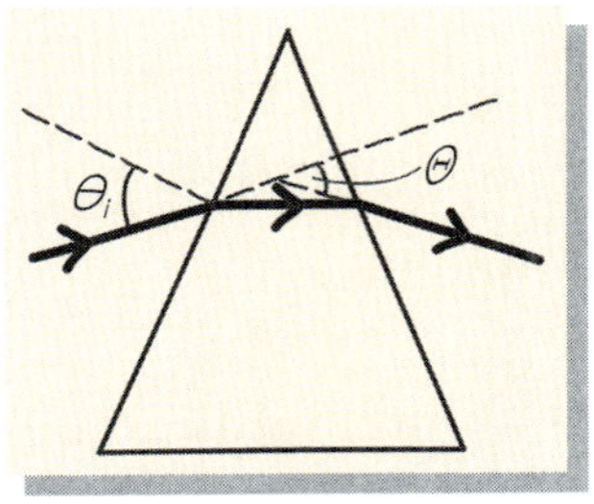

FIGURE 36–42 Problem 55.

56. (III) A ray of light incident from air onto a glass pane is partly reflected and partly refracted at the two surfaces of the pane (Fig. 36–43). The glass has an index of refraction n and a thickness d. Express in terms of n, d, and θ_i the displacement d' of the ray drawn, which enters the glass, reflects off the back surface, and exits.

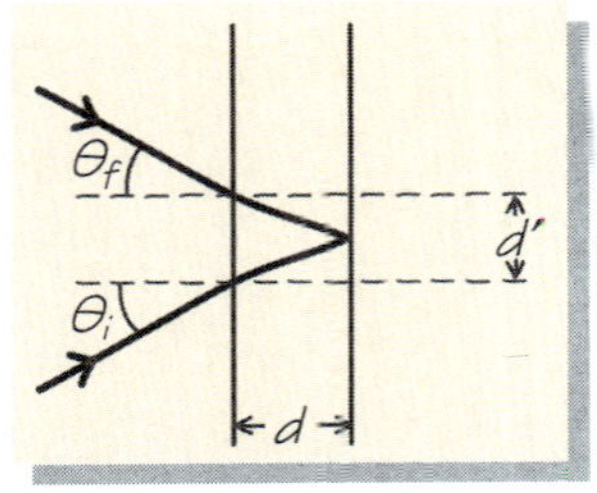

FIGURE 36–43 Problem 56.

57. (III) The first successful measurement of the speed of light, made by Olaus Roemer in 1675, was based on the following method. The mean orbital period of Io, a moon of Jupiter, is 42.5 h; however, that period is measured to be about 15 s less than this value when Earth in its orbit is approaching Jupiter, and about 15 s more when Earth is receding from Jupiter. (a) Given that Earth's orbital speed around the Sun is about 30 km/s, and that Earth is on a part of its orbit when it is moving toward Jupiter, how much closer will Earth have moved toward Jupiter during one orbit of Io? (b) Use the information given to estimate the speed of light.

CHAPTER

37

The image in this mirror depends on the position of the man standing before it and on the shape of the mirror. In this chapter we'll see why.

Mirrors and Lenses and Their Uses

Tools and instruments that can explore previously inaccessible domains often open new doors to understanding nature. Astronomy owes its progress to the invention of the telescope, and modern biology could not have been created without the microscope. In this chapter, we shall discuss the ideas that govern the construction of optical instruments. Their functioning is based on two very simple laws introduced in Chapter 36: The law of reflection explains the images that we see in mirrors and the functioning of the reflecting telescope; Snell's law explains the eye, cameras, magnifying glasses, refracting telescopes, and microscopes. This aspect of the study of light is called geometric optics because these two laws can be applied simply by tracing the geometrical paths of light rays.

37–1 IMAGES AND MIRRORS

The simplest reflecting surface is a flat (or plane) mirror. You see an image of yourself when you look at yourself in a mirror. What is this image, and how is it formed? Let's begin by considering rays going directly from a point source (S) to a person's eye in Fig. 37–1a and reflecting from a plane mirror according to the law of reflection that the angles of incidence and reflection are equal in Fig. 37–1b. We could, in fact, draw an infinite number of such rays as close to one another as we like. Rays that are near one another form *bundles*, as shown at number 2 in Fig. 37–1.

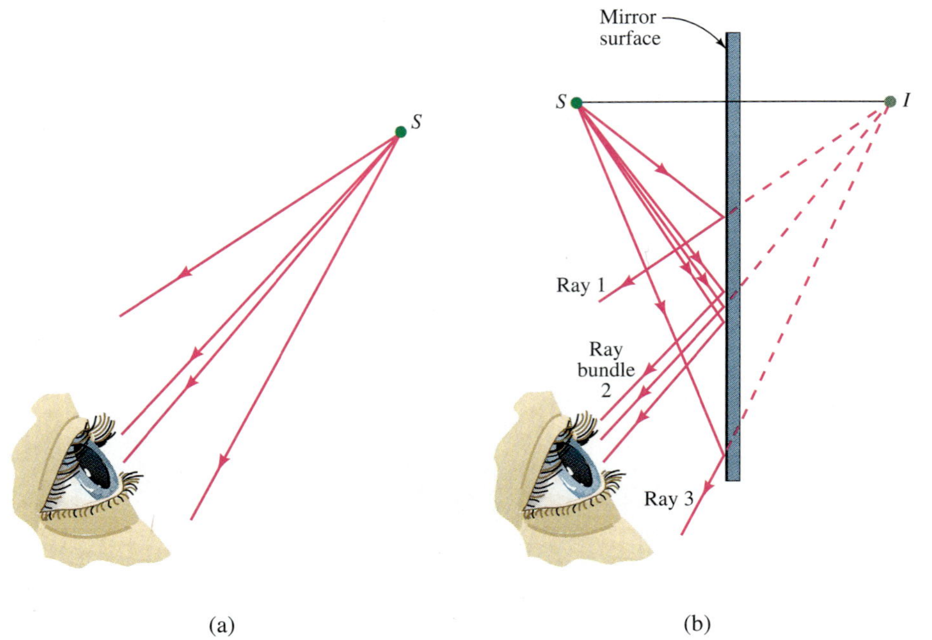

FIGURE 37–1 Rays leaving source point S (a) go to an eye and (b) reflect from a plane mirror before going to the eye. A bundle of such rays enters the eye, apparently from point I.

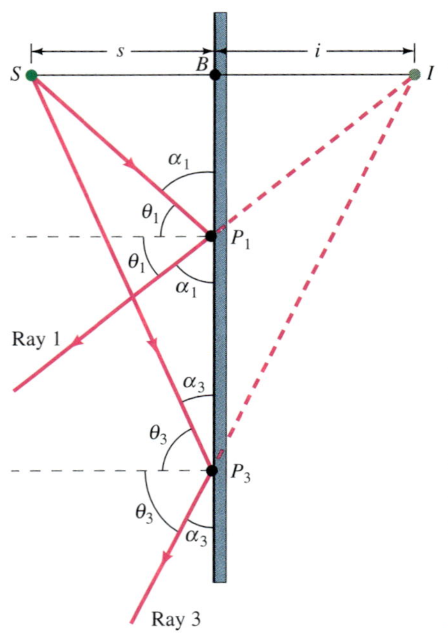

FIGURE 37–2 All the reflected rays from point S trace back to point I. The geometry implies that the perpendicular distance from the mirror to point S, s, equals the distance from the mirror to point I, i.

Simple geometry allows us to see that *all the reflected rays trace back to the same point I*. To see this, take a look at rays 1 and 3 in Fig. 37–2. We have indicated the equal angles of incidence and reflection θ_1 and θ_3 for these rays, respectively, as well as the angles α_1 and α_3. The angle formed by BP_1I is then equal to α_1. Thus, if point B is formed by dropping a perpendicular to the mirror from point S and if point I lies along the continuation of this line, triangles BIP_1 and BSP_1 are similar triangles. By the same method, so are triangles BIP_3 and BSP_3. Because both rays 1 and 3 emanate from the same point S, the distance BS forms the base of both triangles to the left of the mirror (the *object side*), and the distance BI forms the base of both triangles to the right of the mirror (the *image side*). The (imaginary) continuations of rays 1 and 3 to the image side meet at point I, as would the continuation of *any* reflected ray. An *image point* is in fact any point other than the object from which an unlimited number of rays emanate or appear to emanate when the rays are extended back in straight lines.

We have calculated the location of point I. Because BIP_1 and BSP_1 are similar triangles, the distances BS and BI are equal. How does the eye/brain "know" where to put I? Two eyes (or one eye that moves a little) sense a bundle of rays rather than a single ray. The eye/brain can measure their degree of divergence and is capable of extrapolating this diverging bundle back to point I.

The Image of an Extended Object

If our light source is extended, a second point on the source forms a second image point. Moreover, this second image point is as close to the first image point as the second source point is to the first source point. A set of nearby source points forms nearby image points (Fig. 37–3). The entire **object**, or *source* (we use the terms interchangeably), forms a set of matching image points, which together constitute an **image**. *An image is a set of contiguous points from which reflected rays emanate—or act as if they emanate—when the rays are extrapolated back in straight lines.* Figure 37–3 illustrates that the source and the image formed by our plane mirror have the same size. We say that a plane mirror does not produce a *magnified* image.

The definition of an image

No light rays actually pass through the image formed by a plane mirror; thus, we call it a **virtual** image. We shall see that we can form a **real** image with curved surfaces—a continuous set of points through which any number of rays of reflected light do pass.

Virtual and real images

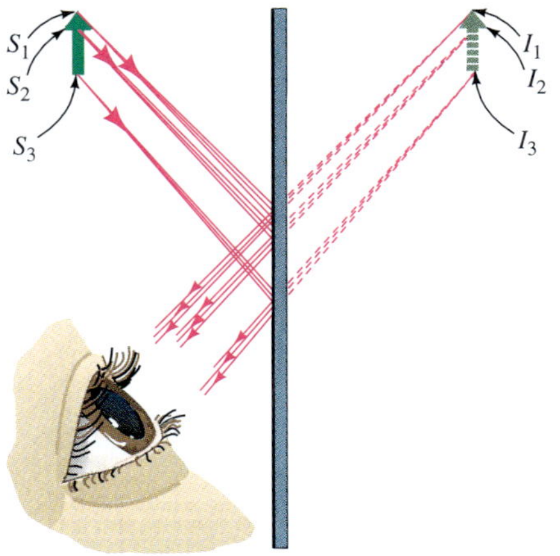

FIGURE 37–3 When the source is an extended one, there is an image point for every source point. This means that a bundle of rays will enter your eye from every image point no matter what your position before the mirror. Geometry can be used to locate the image $I_1I_2I_3$ of source $S_1S_2S_3$.

We have said that the image formed by a plane mirror is perfect. But it does have one striking peculiarity. If your right eye is blackened, your image, viewed as if you were meeting yourself on the street, has a left eye that is blackened. From Fig. 37–4, however, we can see that the *actual* reversal is a front-to-back reversal (the nose of the object points in the $+x$ direction in the figure, whereas the nose of the image points in the $-x$ direction), and this is what is referred to as a left-to-right reversal of the image.

Multiple Reflections

If the image produced by one reflecting surface is reversed left-to-right, the image produced by successive reflection from two mirrors is not. What do we mean by the image produced by successive reflection? We have seen that the image produced by a source comes from the reflections of the rays emitted by the source. But these reflected rays form a set of diverging rays, which is how the image is formed. The reflected rays can reflect a second time from another mirror, and we can think of the diverging set of rays incident on the second mirror as coming from a second source. In other words, *a first image can act as a source for a second image* (Fig. 37–5). It makes no difference whether the first image is virtual or real. This idea allows us to calculate the effects of optical systems with many elements, whether they are plane or curved mirrors or lenses. The corner reflector described in Chapter 36 provides an example.

PROBLEM SOLVING

An image can serve as the source for a subsequent reflection.

EXAMPLE 37–1 A ray of light is incident at angle θ on a plane mirror suspended vertically. If the mirror is rotated about a vertical axis through an angle α, by what angle ϕ is the reflected ray rotated?

Solution: Figure 37–6 is an overhead view of the situation. The incident ray initially makes an angle θ with the normal N to the mirror, as does the reflected ray, labeled I. A rotation of the mirror through an angle α moves the normal through an angle θ to a new normal position, N'. The new angle of incidence is $\theta + \alpha$, which

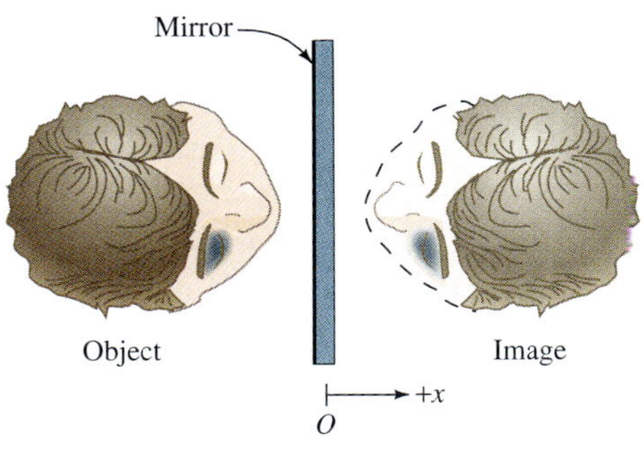

FIGURE 37–4 An image is reversed front to back. This means that if your right eye is black, your image has a black left eye.

FIGURE 37–5 Multiple reflections can be obtained with two plane mirrors. Where is the second mirror in this case?

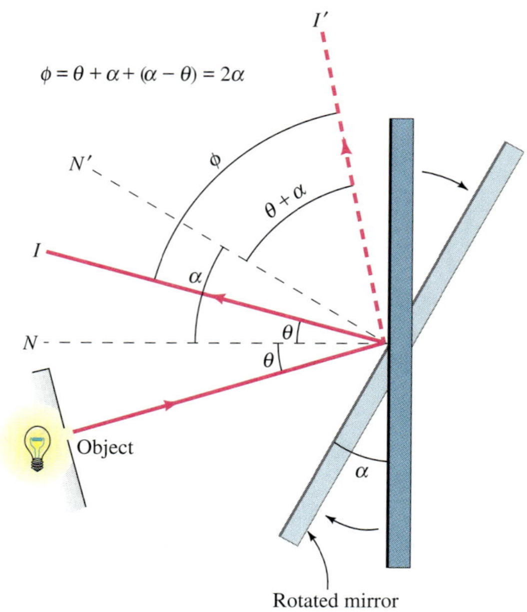

FIGURE 37–6 Example 37–1. Overhead view of a ray of light incident on a mirror that rotates about a vertical axis.

is also the new angle of reflection from N'. The angle between the incident ray and the new reflected ray I' increases from 2θ to $2(\theta + \alpha)$. Because the incident ray has not moved, the reflected ray is rotated by $\phi = 2\alpha$.

37–2 SPHERICAL MIRRORS

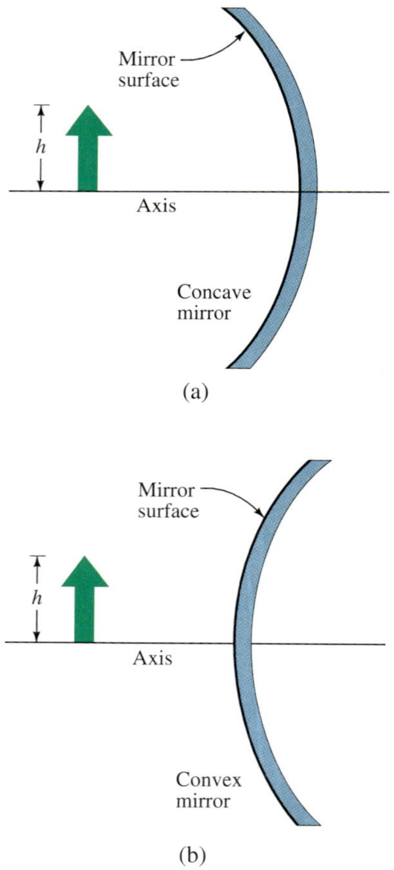

FIGURE 37–7 (a) Concave and (b) convex spherical mirrors. The object shown, an arrow, acts as an extended source.

We saw in Fig. 37–3 that plane mirrors produce images that are the same size as the object. We can construct mirrors that produce images of altered sizes, or real as well as virtual images, by using curved surfaces. Here, we shall study mirrors with surfaces forming a segment of a sphere. As a tool to analyze the effects of reflections from such a surface, we shall extend the *ray-tracing techniques* of Section 37–1, in which we followed rays or bundles of rays that reflect from a plane mirror. In particular, you will see that ray tracing is useful because there are some particularly simple and significant rays to follow.

We can think of our mirror as a segment of a sphere with a center somewhere in space. If the light source (object) is on the same side of the surface as this center, the mirror is *concave* (Fig. 37–7a); if the source is on the other side, the mirror is *convex* (Fig. 37–7b).

In order to apply our ray-tracing techniques, let's make some simplifying assumptions. Let's call the line perpendicular to the center point of the mirror the *axis* of the mirror. We shall study objects on or near the axis. The tips of the arrows shown in Fig. 37–7 are a distance h from the axis, and if h is small compared to the radii of curvature of the mirrors, we say that the object is *near the axis*. Finally, let's simplify the mathematics by considering only rays that are so close to being parallel to the axis that we can use small-angle approximations in studying the reflections. Such rays are said to be **paraxial**.

The Concave Mirror

Figure 37–8a illustrates a concave mirror. We consider rays from a very distant point source (to the left of the figure) on the axis CB. The source is so far away that *all the rays from it arrive practically parallel to the axis*. (We say that the source is *at infinity*.) Point C indicates the position of the center of the sphere (of radius R) of which

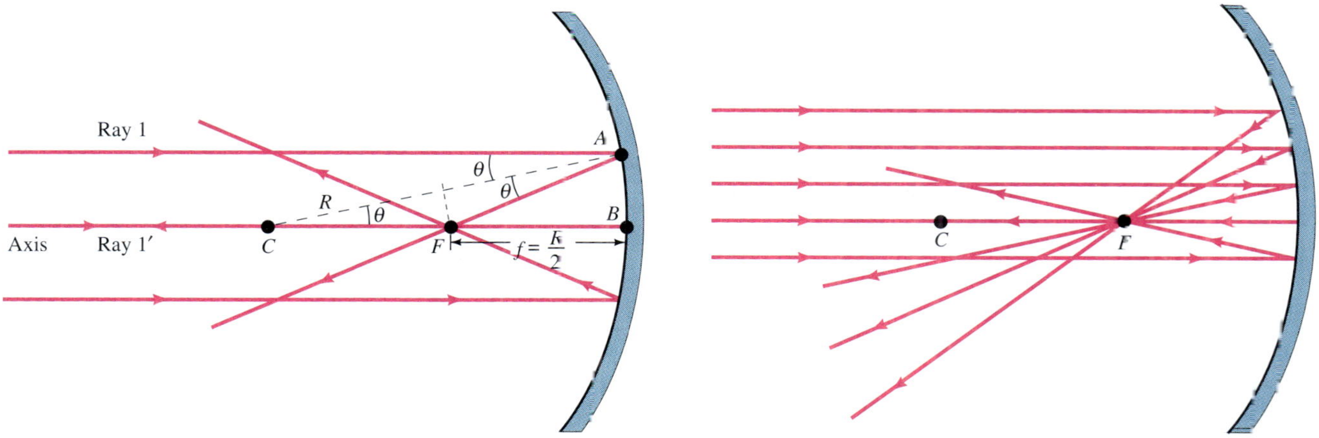

(a) (b)

the mirror is a segment. The position of C (called the *center of curvature*) is therefore a distance R from the mirror surface, and all lines from point C to the mirror are perpendicular to the mirror.

Location of the Focal Point. We now look at ray 1, which is reflected at point A in the direction AF in Fig. 37–8a. Angle θ is the angle of incidence (line CA is perpendicular to the mirror); hence, θ is also the angle of reflection. How far is point F from the mirror? Ray 1 arrives parallel to the axis, so the angle of incidence is equal to the angle ACB, which we have also labeled θ. Triangle ACF is therefore isosceles with a base of length R. Thus, by dropping a perpendicular from point F to the base AC, we see that the distance CF is $CF = (R/2)/\cos\theta$. For small θ, $\cos\theta \simeq 1$; hence $CF = R/2$, or $BF = R - CF = R/2$, *independent* of θ.[†] *All* the parallel rays near the axis reflect through point F, a distance $R/2$ from the mirror. An unlimited number of rays diverge from point F; hence it is the image point of the distant source point. Unlike the image points produced by a plane mirror, it is a real image point because rays actually cross there (Fig. 37–8b).

The point F at which the rays from a source point at infinity are brought together to form an image (are *focused*) is known as the **focal point** (or *focus*), and its distance f from the mirror is called the **focal length**. This term can be applied to any optical system that produces images—including plane mirrors, whose focal length is infinite. The focal length is the distance from the image point to the optical system—the mirror or lens or whatever—when the source point is at infinity. For concave mirrors, we have shown that

$$f = \frac{R}{2}. \tag{37–1}$$

The focal length of a mirror with radius of curvature R

If we were to reverse the directional arrow on the rays in Fig. 37–8a, we would see that a pointlike light source placed at the focus F will form an outgoing beam of light parallel to the axis of the mirror. Thus, spherical concave mirrors are useful for searchlights and flashlights.

The Image of an Extended Object. Let's take an extended object—as in Fig. 37–7—that is small compared to the radius of curvature of the mirror and close enough to the axis so that the rays are paraxial. In Fig. 37–9a, we label the object, which is *upright*, with the letters S and S'. As happened for the plane mirror and for the source point on the axis, entire bundles of rays from a given spot on the object pass through a corresponding spot in space after reflection, and thus an image I to I' of the entire object is formed. We want to determine the position and size of the image.

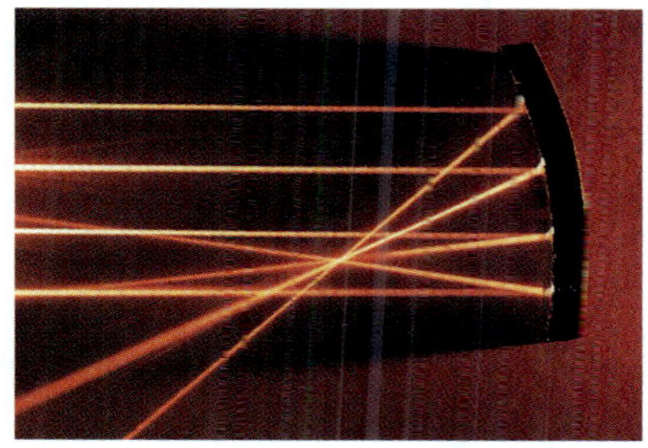

(c)

FIGURE 37–8 (a) Rays emitted by an object at infinity are all parallel to the axis. Ray 1 is reflected by the concave mirror surface and passes through the focal point, F. (b) To a good approximation, F is independent of θ; that is, any incoming ray parallel to the axis is reflected through it. Thus all rays from infinity cross the axis at F, which is therefore an image point. (c) Parallel rays of light reflected by a concave mirror.

[†]This result is accurate to 1 percent for angles θ less than about 10° (see Problem 35).

PROBLEM-SOLVING TECHNIQUES

Principal Rays

For optical systems such as mirrors and lenses, it is important to be able to find the size and location of the images, given a source (or object). Because all rays cross at the image point of a source point (or behave as though they do in the case of virtual images), we must only find the crossing points of a few rays from any point on the object, which we call *principal rays*. It is easy to trace the paths of these rays and see where they cross. Even if the optical system is such that a ray does not actually exist—for example, there may be a hole at the center of a mirror—we can pretend that it does because we simply use the principal rays as a tool to learn where the rest of the rays go.

We use here a convex mirror (Section 37–2), as in Fig. 37B1–1, and a converging lens (Section 37–4), as in Fig. 37B1–2. However, the technique applies equally well to concave mirrors (Section 37–2), diverging lenses (Section 37–4), and (under the conditions listed here and in the text) to single refracting surfaces (Section 37–3). These cases are all illustrated in the text.

We count four principal rays, numbered 1 through 4, from a given source point S:

1. Rays that enter the system parallel to the optical axis. By definition, these paraxial rays are reflected or refracted to the focal point F.

2. Rays that pass through (or are aligned as though they passed through) the focal point as they enter the system. These rays are just reversed versions of type 1 rays and thus are reflected or refracted to leave the system parallel to the axis.

3. Rays that pass through (or are aligned as though they pass through) the center of curvature C of the sphere from which a mirror or refracting surface is formed. These rays are perpendicularly incident on the surface and will be reflected or refracted along the line of arrival. Such rays have no analogue in a thin lens, which has two surfaces and two radii of curvature.

4. Rays that strike the center of the mirror surface. The reflected rays make the same angle with the axis as the incident rays do (except for sign). For the case of thin lenses, the ray drawn directly to the center of the lens passes through it in a straight line. There is no analogous rule for single refracting surfaces.

By drawing these principal rays from any given point S on a source, we can see where their reflections or refractions cross (or appear to cross) and learn the location of the image point I of source point S. When an optical system has more than one reflecting or refracting surface (an "element"), we can apply the simple rule that the image formed by one element can serve as a new object for the next element. In that case, *the principal rays must be redrawn for the new object* as they apply to the next optical element. We thereby locate the next image.

The origin of these rules is discussed more thoroughly in the text.

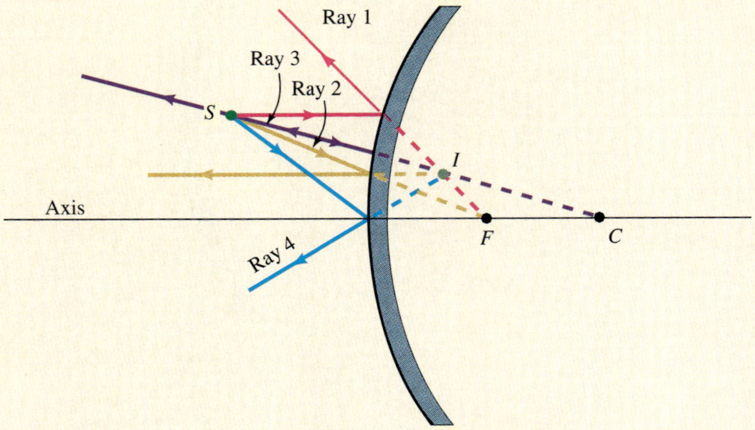

FIGURE 37B1–1 The four principal rays for reflection from a convex mirror.

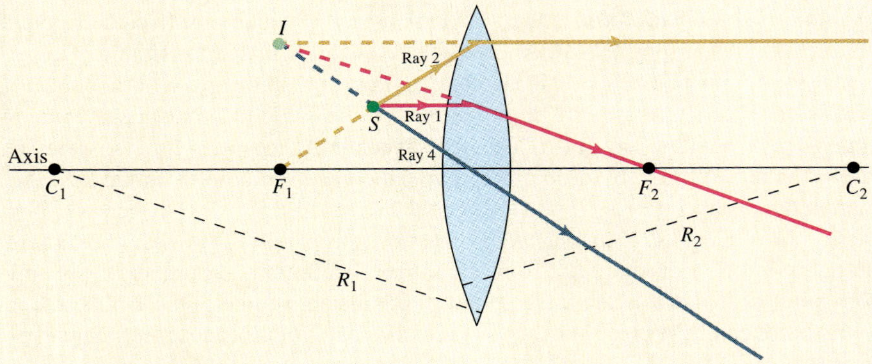

FIGURE 37B1–2 The three principal rays for refraction through a converging lens.

Principal rays for a concave mirror

To do so, we use the principal-ray technique described in detail in the Problem-Solving Techniques box. In Fig. 37–9b, the principal rays drawn to find point I of the image that corresponds to point S of the source are solid, and in Fig. 37–9c, those

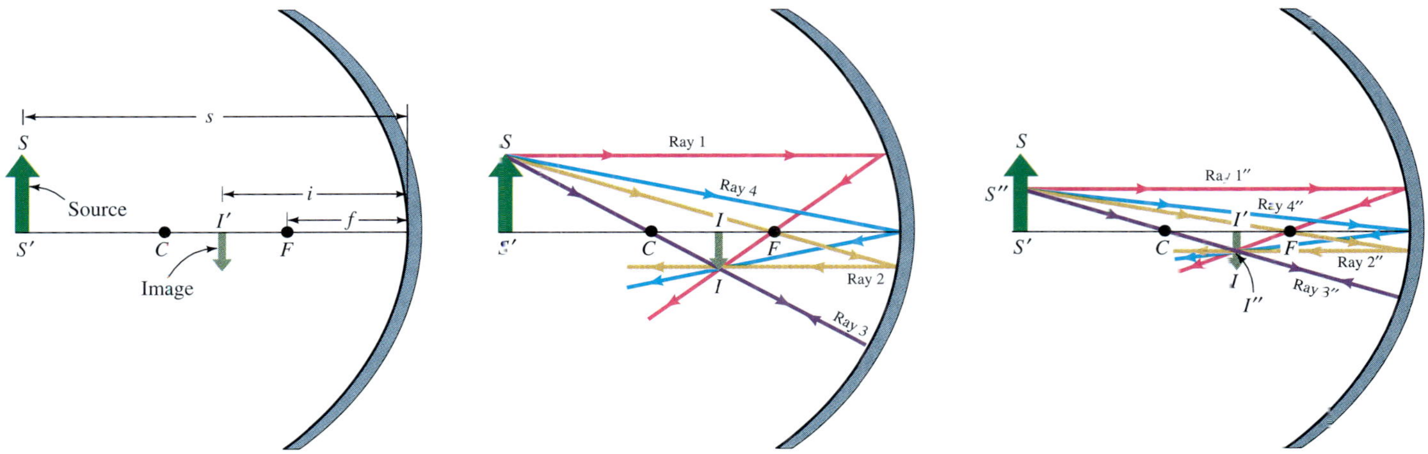

(a) (b) (c)

drawn to find the image point I'' that corresponds to source point S'' are dashed. These sets of rays are

1. ray 1, which approaches the mirror parallel to the axis and is reflected through the focus F (whose position we know);

2. ray 2, which passes through the focus and reflects off the mirror parallel to the axis;

3. ray 3, which passes through the center of curvature of the mirror and is reflected back in the direction from which it came;

4. ray 4, which goes to the center of the mirror surface and whose reflection makes the same angle with the axis as the incident angle.

The rays leaving source points S and S'' do indeed cross at the respective image points I and I''. All other points of the source have image points that can be constructed in the same way, and the image is thus constructed as in Figs. 37–9b and 37–9c. Note that both image point I' and source point S' lie along the optical axis. Our construction shows that a vertical source gives a vertical image. This is true for all optical systems that we study and allows us to compute the location of just one image point rather than all of them. For example, the entire image can be constructed if we find only image point I of the top of the source (that is, of source point S) and, in fact, any two of the four principal rays are sufficient to determine point I. For example, rays 1 and 2 are sufficient for locating point I in Fig. 37–9b.

The image in Fig. 37–9a, located by the procedure in Figs. 37–9b and 37–9c, is *real* (real light rays pass through points I and I' and those in between), in contrast to the virtual image produced by plane mirrors. The image of the object in Fig. 37–9 is *inverted* (*upside-down*) and *reduced in size*. As the source moves farther from the mirror, the image moves closer and closer to the focal point F and also becomes smaller. When the source moves closer to the mirror, the image becomes larger and moves away from the mirror.

Let's now consider the situation depicted in Fig. 37–10, with the source inside the focal point of the mirror. The principal rays from point S are drawn according to the method of the Problem-Solving Techniques box. Note that rays 2 and 3 only *behave* as though they pass through points F and C respectively. In this case, the reflected rays from source point S do not actually cross but are aligned as though they come from behind the mirror at image point I. In this case, the image is virtual. It is also *upright* and enlarged. These are useful features of the concave mirrors employed for cosmetic purposes and in dentistry. As the source moves closer to the mirror, so does the image, which becomes smaller; as the source moves closer to the focal point, the image moves farther and farther away, becoming hugely magnified.

(d)

FIGURE 37–9 (a) An extended object a distance s from a mirror forms an image a distance i from the mirror. (b) Ray tracing for the concave mirror, with the principal rays for source point S. (c) Principal rays for souce point S''. By repeating this exercise, we can build up the entire image. The principal rays are a guide; *any* ray from S that reflects will cross the image point I. (d) The object has produced an inverted and reduced image in a concave mirror. The object is outside the focal length of the mirror.

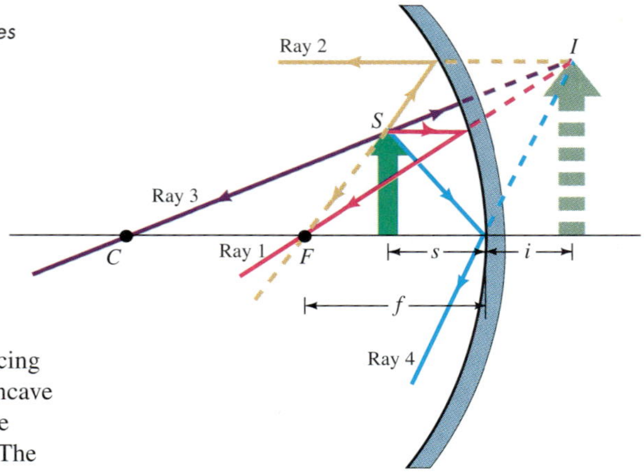

(a)

(b)

FIGURE 37–10 (a) Ray tracing with the principal rays for a concave mirror, for a source closer to the mirror than that of Fig. 37–9b. The image becomes virtual when the source moves inside focal point F. (b) An object is placed within the focal length of a concave mirror. This time the image it produces is upright and magnified.

The Convex Mirror

The same ray-tracing techniques we used for concave mirrors allow us to understand convex mirrors. Point C is the center of curvature of the sphere (of radius R) of which the convex mirror is a segment. All lines from point C to the mirror are perpendicular to the mirror.

Location of the Focal Point. We start by finding the focal point, the spot where rays from a point source at infinity (that is, a set of rays parallel to the axis) are focused. Figure 37–11 shows that the reflected rays diverge, so *the image is virtual*; the rays appear to originate at a common point at F behind the mirror. Of course, we must prove that there *is* a common point. To do so, we use ray 1 in Fig. 37–11. The line from its intersection with the mirror to point C is the normal to the mirror at that point. The angles marked θ are the angles of incidence and reflection for ray 1. By following the same trigonometric reasoning we used for the concave case, we can show that the distance f is again given by Eq. (37–1),

$$f = \frac{R}{2}$$

(see Problem 12). This result is *independent of* θ for small angles θ, so *all* the reflected rays appear to diverge from F. We have an image, and it is virtual. The focal point of a concave mirror is on the same side as the object, whereas the focal point of a convex mirror is on the side opposite the object.

FIGURE 37–11 (a) When a spherical mirror is convex, the focal point lies behind the mirror, as ray tracing shows. The reflected rays diverge, and their extensions all lead back to the focal point. (b) Parallel rays of light reflected by a convex mirror.

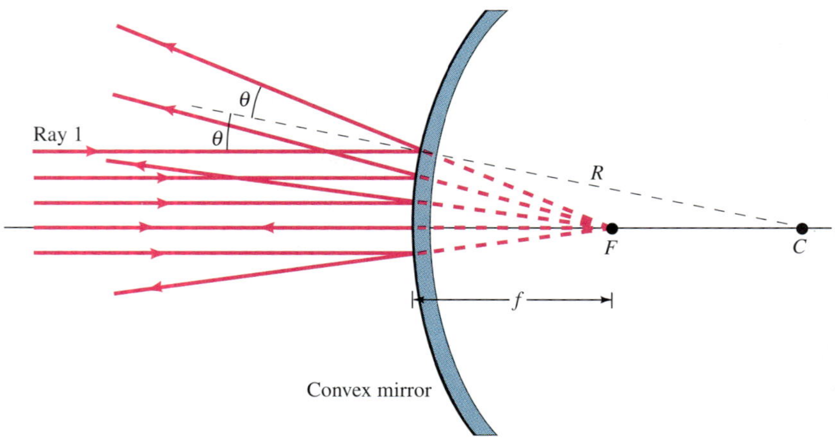

(a)

(b)

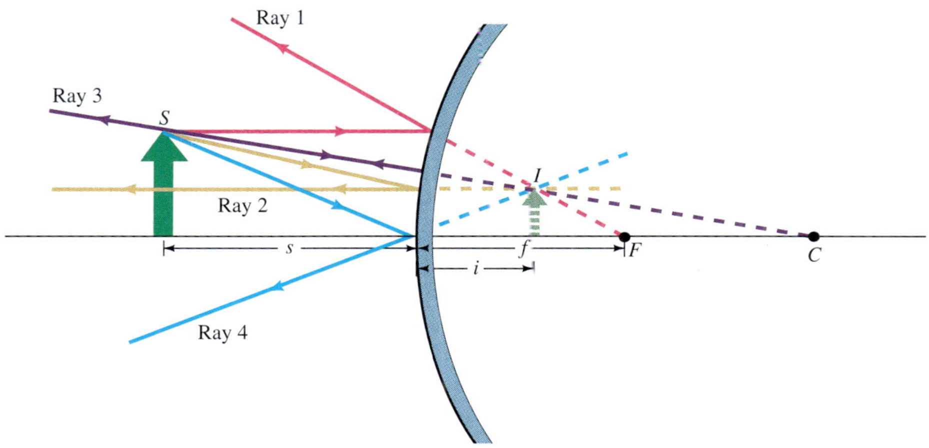

(a)

(b)

FIGURE 37–12 (a) Ray tracing describes the formation of a virtual image by a convex spherical mirror. (b) The image produced by this convex mirror is upright, reduced, and behind the mirror.

Principal rays for a convex mirror

The Image of an Extended Object. Let's now consider the object shown in Fig. 37–12. We can trace the four principal rays from source point S: ray 1—parallel to the optical axis, and whose reflection extends back along the line from the mirror to point F; ray 2—drawn as though it would pass through F, and whose reflection is parallel to the axis; ray 3—drawn as though it would pass through C, and whose reflection returns along the line of incidence; and ray 4—striking the center of the mirror surface, and whose reflection makes the same angle with the axis as the incident angle. A careful drawing shows that the reflected rays diverge from each other, but all four (indeed, *all* rays from S) appear to originate at point I. *Point I is the virtual image of point S.*

We can similarly find that there is a virtual image of the entire source; that image is upright and smaller than the source. When the source moves farther away, the image becomes smaller and remains upright, but there is no transition from virtual to real image, as there is in the concave case. These properties make convex spherical mirrors useful for rearview mirrors in motor vehicles. Figure 37–13 shows a more extreme case in which an entire sphere acts as a mirror.

The Relation between Source Distance and Image Distance

In Figs. 37–9a, 37–10, and 37–12 we have indicated the distance s from the mirror to the source, the distance i to the image, and the focal length f. Given f (or the radius of curvature $R = 2f$) and the source position, we can find the image position, its height, and whether or not it is inverted. Finding the quantitative relation among s, i, and f is generally a matter of straightforward geometry. The case of a concave spherical mirror is worked out in the subsection following this one. The relation is

$$\frac{1}{s} + \frac{1}{i} = \frac{1}{f}. \tag{37–2}$$

FIGURE 37–13 Reflection by a spherical mirror that is a large portion of a sphere, seen here in the etching *Hand with Reflecting Sphere* by Maurits C. Escher

The relation among source distance, image distance, and focal length for mirrors

With a set of conventions about signs, we shall see how *the same relation holds for the convex mirror*. Equation (37–2) is easily understood in two limits. When the object is far away ($s \to \infty$), then $1/s \to 0$ and $i = f$ (which is the definition of f). When the object is at the focus, $s = f$, then $1/i = 0$: The image is very far away. This result is reasonable because an object that is at the focal length and produces an image at infinity is simply the ray-reversal of an object that is at infinity and produces an image at the focus.

If the object is between the concave mirror and the focus, as in Fig. 37–10, then s is smaller than f, and Eq. (37–2) implies that i *must be negative*! We associate a neg-

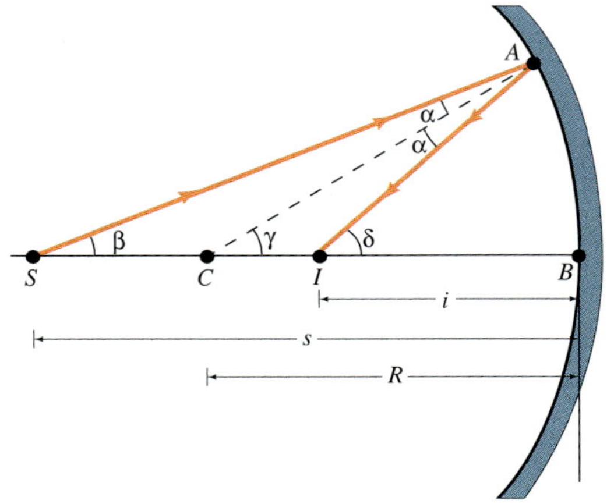

FIGURE 37–14 Geometric construction for deriving Eq. (37–2) for a spherical mirror.

ative i with the image on the far side of the mirror; that is, with a virtual image. Likewise, i is positive when the image is real. Equation (37–2) may be applied to a convex mirror if we follow the convention that *the focal length f is negative when the focus is on the "virtual image" side of the mirror*. This is equivalent to saying that if the mirror's center of curvature is on the back (nonreflecting) side of the mirror, f is negative. In the application of Eq. (37–2) to a convex mirror, s is always positive, and f is always negative. If s is positive (as it is when our optical system consists only of the convex mirror), then i will be negative (the image is virtual). Furthermore, the image in this case must be between the mirror and the focal point. That is because $1/s = (-1/|f|) + (1/|i|)$ is positive, so $1/|i| > 1/|f|$, and hence $|i| < |f|$. The rules for the sign of the object distance s will be discussed in Section 37–3.

How to Get Equation (37–2)

To arrive at Eq. (37–2), we consider two points on an optical axis—the light source (or object), S, and its image, I—and a concave spherical surface. We can easily see from Fig. 37–14 and from the fact that the sum of the internal angles of a triangle is π that the following relationships hold:

$$\gamma = \beta + \alpha; \tag{37–3}$$

$$\delta = \gamma + \alpha = \gamma + (\gamma - \beta) = 2\gamma - \beta. \tag{37–4}$$

In arriving at Eq. (37–4), we have used Eq. (37–3). The distances of Fig. 37–15 are related to the angles by the exact relation $AB = R\gamma$ and by the *approximate* (small-angle) equations $AB = i\delta = s\beta$. These relations allow Eq. (37–4) to be rewritten as

$$\frac{AB}{i} = \frac{2AB}{R} - \frac{AB}{s}.$$

We divide out the common factor AB and use the focal length $f = R/2$ for a spherical surface. We get

$$\frac{1}{i} + \frac{1}{s} = \frac{1}{f},$$

the relation we set out to establish: Eq. (37–2).

Magnification

Our geometric constructions show that an image may not be the same size as its source. Consider the convex mirror in Fig. 37–15. Ray 4 to the center point A of the mirror is useful because all the angles marked θ are the same, so triangles $AS'S$ and $AI'I$ are sim-

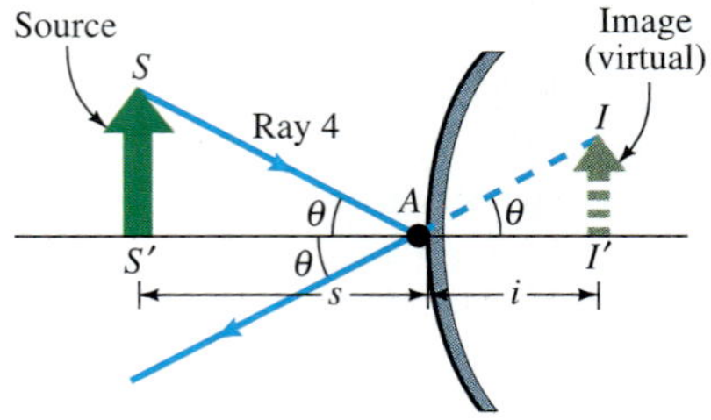

FIGURE 37–15 Geometry for the calculation of magnification.

ilar triangles. Thus, the magnitude of the **magnification**, M, defined as the ratio of the heights of the source and image, is

$$|M| \equiv \frac{|II'|}{|SS'|} = \frac{|i|}{|s|}. \tag{37–5}$$

We usually go a little further and specify whether the image is upright or inverted. This can be done in a very simple way, by writing

$$M = -\frac{i}{s}. \tag{37–6}$$

The magnification

If M is negative, the image is inverted; if M is positive, the image is upright. We can verify that this form works in explicit cases:

1. When the mirror is concave and the source is outside the focal point, the image is real (i is positive). By Eq. (37–6), M is then negative and the image should be inverted, as it is in Fig. 37–9a.

2. When the mirror is concave and the source is inside the focal point, the image is virtual (i is negative). By Eq. (37–6), M is then positive and the image should be upright, as it is in Fig. 37–10.

3. When the mirror is convex, the image is virtual (i is negative). By Eq. (37–6), M is then positive and the image should be upright, as it is in Fig. 37–12.

Equation (37–2) can be rewritten as

$$\frac{1}{i} = \frac{1}{f} - \frac{1}{s} = \frac{s-f}{fs}.$$

We can thereby find M in a form in which the image distance does not appear:

$$M = -\frac{i}{s} = -\frac{fs/(s-f)}{s} = \frac{f}{f-s}. \tag{37–7}$$

This form can be applied to both concave and convex mirrors if we recall that f is negative for convex mirrors. In the convex case, $f - s$ is always negative, so M is always positive; the image is always upright. Also, $|f - s| = |f| + |s|$ is always larger than $|f|$ for convex mirrors, so the image is always reduced in size.

EXAMPLE 37–2 A convex spherical mirror of radius of curvature R of magnitude 20.0 cm produces an upright image precisely one-quarter the size of an object, a candle. What is the separation distance between the object and its image?

Solution: We can find f from Eq. (37–1), $f = R/2 = -10.0$ cm. (The negative sign indicates that the mirror is convex.) Equation (37–7) then gives

$$s = f\left(1 - \frac{1}{M}\right).$$

In this case, we know that $M = \frac{1}{4}$ (it is positive because the image is upright), so

$$s = f\left(1 - \frac{1}{\frac{1}{4}}\right) = -3f = -3(-10.0\text{ cm}) = 30.0\text{ cm}.$$

The problem asks for the separation distance between the object and the image; hence we must also find i. It can be found in terms of s from Eq. (37–6):

$$i = -sM = -(30.0\text{ cm})\left(\frac{1}{4}\right) = -7.5\text{ cm}.$$

The minus sign is consistent with our knowledge that the image of a convex mirror is virtual (on the far side of the mirror). Then the object–image separation $s - i$ is

$$s - i = 30.0\text{ cm} - (-7.5\text{ cm}) = 37.5\text{ cm}.$$

The geometric construction in Fig. 37–16 confirms these calculations.

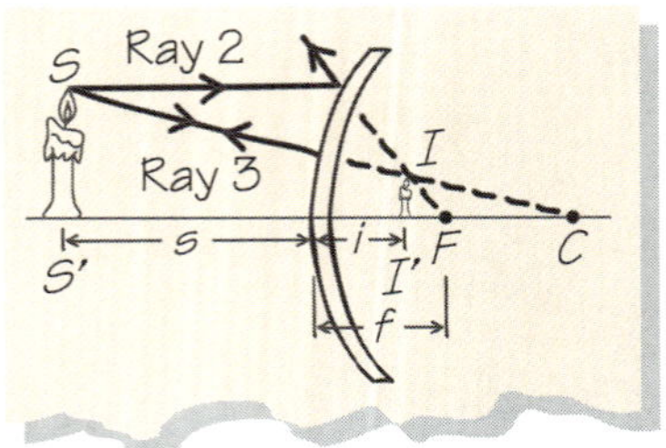

FIGURE 37–16 Example 37–2. Ray tracing to find the image.

The relation between source distance, image distance, and focal length [Eq. (37–2)], the expression for magnification [Eq. (37–6)], and the ray-tracing techniques are applicable to lenses as well as to mirrors. These three elements are sufficient for dealing with all the cases of interest to us.

The Plane Mirror as a Special Case of a Curved Surface

A plane mirror represents a concave or convex spherical mirror in the limit that $R \to \infty$, or, equivalently, $f \to \infty$. In such a case, Eq. (37–2) implies that $i = -s$. This is just our earlier result that the image is virtual and as far behind the mirror as the source is in front of it. We can calculate the magnification of a plane mirror from Eq. (37–7):

$$\text{for a plane mirror:} \qquad M = \frac{f}{f - s} \xrightarrow[f\to\infty]{} \frac{f}{f} = 1.$$

The magnification is one and the image is upright.

Ray Tracing

We have used ray-tracing techniques to learn about the images produced by certain types of mirrors—and we shall use the same techniques in Section 37–4 to learn about the images made by some simplified lenses. These techniques are the basis for the design of optical systems made from mirrors and lenses, especially the most sophisticated systems. We may want an optical system to produce a very sharp image over a very limited range of source distances; for example, orbital satellites never require focusing to short distances. Or we may want to sacrifice acuity in order for an optical system to operate in dim light. Real systems may have nonspherical mirrors, or thick, multielement lenses in which the elements move relative to one another, as in zoom lenses. Designers of optical systems use computer programs capable of tracing large numbers of rays in a system design, of previewing the quality and placement of the image, and of creating modified designs in which the desired optical properties are attained.

We conclude this section with a comment on signs. Do not bother to try to keep track of the signs of the various quantities we have discussed. Develop your ray-tracing techniques, and you will be able to derive the signs on your own. Table 37–1 gives the signs of all the quantities necessary for mirrors, refracting surfaces, and lenses. We recommend, however, that you do not rely on the table too heavily.

TABLE 37–1 Sign Conventions for Mirrors, Refracting Surfaces, and Lenses

In applying the information in this table, we must distinguish two "sides" to a reflecting or refracting surface:

Side *A*, the side from which light originates, and
Side *B*, the side to which light passes.

For mirrors, side *B* is identical to side *A*; for refracting surfaces and lenses, the two sides are opposite. Only the sign of the source position is determined by side *A*. All other quantities are determined by reference to side *B*.

Determined by Side A	
Source distance s	Positive if object is on side A (real object) Negative if object is on side opposite to side A (virtual object)
Determined by Side B	
Image distance i	Positive if image is on side B (real image) Negative if image is on side opposite to side B (virtual image)
Curvature R	Positive if center of curvature is on side B Negative if center of curvature is on side opposite to side B
Focal point	Positive if on side B Negative if on side opposite to side B

37–3 REFRACTION AT SPHERICAL SURFACES

Mirrors change the direction of rays of light and create real or virtual images of objects. The same objectives can be accomplished by using refraction rather than reflection. Where the law of equal angles of incidence and reflection is enough to determine the behavior of mirrors, Snell's law is enough to determine the behavior of lenses. As for mirrors, we want only to understand the qualitative behavior of lenses, not to design real ones, and we shall assume that the surfaces are spherical sections without too much curvature and consider only paraxial rays.

Let's first consider light rays that traverse from a medium with one index of refraction to another across a curved boundary. By repeatedly applying the rules we develop for a single boundary, we shall be able to understand lenses. In Chapter 36, we studied Snell's law, which describes the refraction of light that crosses the boundary between one medium with index of refraction n_1 and another medium with index of refraction n_2 (Fig. 36–9). The angles of incidence and refraction satisfy Snell's law, Eq. (36–6):

$$n_1 \sin \theta_1 = n_2 \sin \theta_2.$$

We want to apply this law to a boundary that is not flat but forms a segment of a sphere of radius of curvature R. Let's take a convex surface, one whose center of curvature—point C in Fig. 37–17—is in the region to which light passes. We choose $n_1 < n_2$, so that the light that passes from medium 1 to medium 2 bends toward the perpendicular to the surface. Nevertheless, the formulas we develop apply more generally.

The Focal Point of a Single Refracting Surface

As is true for a spherical mirror, a single refracting surface has a focal point F that we find by tracing rays that come from a very distant source, parallel to the axis. For the convex surface in Fig. 37–17, ray 1 bends toward the axis and crosses it at a point F. This point will be a focal point if all the incident rays that are parallel to the axis cross F—that the crossing point of a given ray is independent of the angle of incidence θ_1. For paraxial rays, the angle of incidence, θ_1, and that of refraction θ_2, are both small, so the relation $\sin\theta \simeq \theta$ is a good approximation. Thus Snell's law becomes

$$n_1\theta_1 \simeq n_2\theta_2. \qquad (37\text{–}8)$$

Simple geometry shows that $\phi_1 = \theta_1$, and therefore $\phi_2 = \theta_1 - \theta_2$. For small angles, the relation between BF and the arc length AB in Fig. 37–17 is given by

$$BF(\theta_1 - \theta_2) \simeq AB.$$

Because $AB = R\theta_1$, this result, along with Eq. (37–8), implies that

$$BF \simeq \frac{R\theta_1}{\theta_1 - \theta_2} \simeq \frac{Rn_2}{n_2 - n_1}.$$

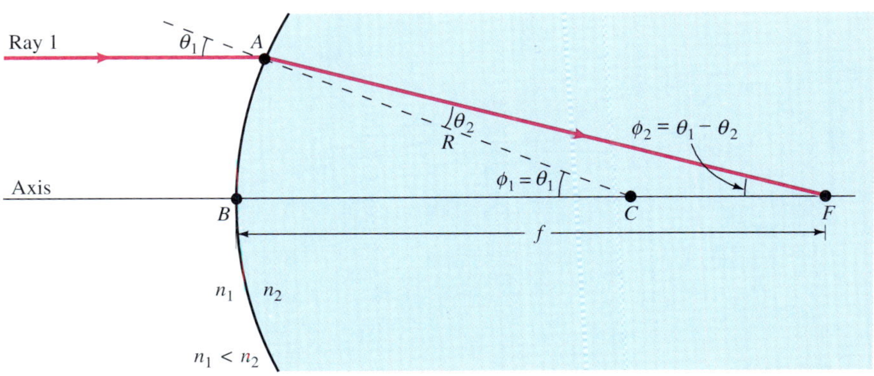

FIGURE 37–17 Ray tracing of a ray that enters a medium whose index of refraction is greater than that of the medium from which the ray came requires us to use Snell's law of refraction. Here we see refraction at a convex spherical surface.

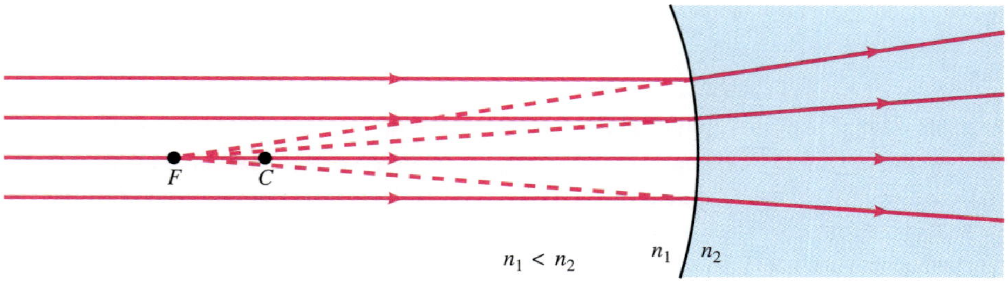

FIGURE 37–18 Refraction at a concave spherical surface.

This distance is indeed independent of θ_1 for small angles, so *all* parallel rays near the axis pass through point F, and F is the image of a point source at infinity. The focal length f is the distance BF:

$$f = \left(\frac{n_2}{n_2 - n_1}\right)R. \tag{37–9}$$

Note that the focal point for a single refracting surface can be farther from the surface than the center of curvature, as shown in Fig. 37–17. Whether it is or not depends on the relative sizes of n_1 and n_2. Although we have derived Eq. (37–9) for a convex surface, we can derive it for a concave surface just as easily. The center of curvature C of the concave surface is on the side from which the light is incident. We find exactly the same formula—except that the focal point is to the left of the surface, on the same side as C. We see in Fig. 37–18 that the focal point F for such a surface is virtual.

The Image of an Extended Object

PROBLEM SOLVING

For the principal rays, see the Problem-Solving Techniques box on p. 994.

Convex Surface. Consider next a vertical object that stands erect on the optical axis. We already know enough about the principal rays to proceed with ray tracing. For a single refracting surface, only two of the four principal rays from any source point, such as point S, are useful. In Fig. 37–19, ray 1 is incident parallel to the axis and refracts such that it crosses the axis at F; ray 3 forms the straight line through C. Ray 3 is perpendicular to the surface and hence does not deflect. The two refracted rays meet at point I. Again, these rays are only two of an unlimited number of rays that leave point S and pass through I. For example, we have drawn a "wild" ray in Fig. 37–19. We will not carry out the detailed geometry that shows that the wild ray passes through I. By drawing the principal rays for any point on the object, we can reconstruct the entire image, which is *real* and *inverted* for the distance drawn.

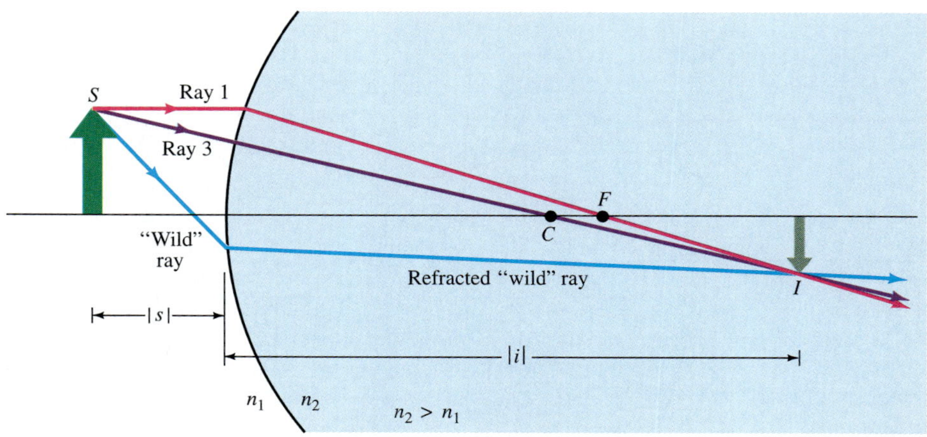

FIGURE 37–19 Ray tracing shows how a real image is formed by a convex spherical refracting surface.

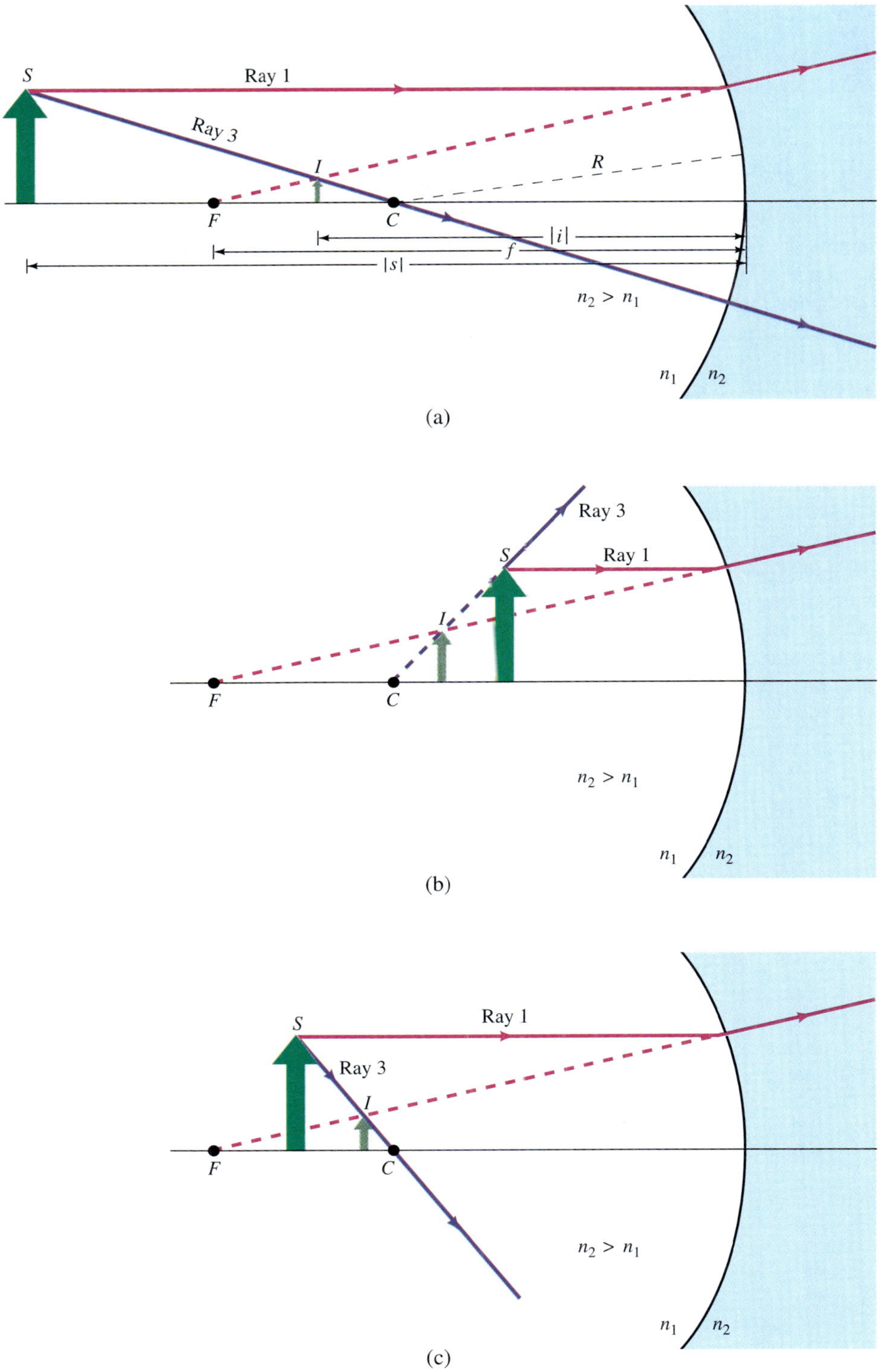

FIGURE 37–20 Ray tracing with principal rays for image formation for a concave spherical refracting surface: situations in which (a) $s > f$; (b) $R > s$; (c) $f > s > R$.

Concave Surface. In Fig. 37–20, we offer three possibilities that depend on whether the source distance s is larger than f, smaller than R, or smaller than f but larger than R. In each case, we use the same two principal rays from point S that we did for the convex case to locate the image of S at point I (Fig. 37–19). In Figs. 37–20b and 37–20c, ray 3 may not pass through the curved surface. We can extend the surface in order to pretend that ray 3 does pass through it because the principal rays are just tools for determining where all the rays that pass through the surface cross. In each case, the image is upright and virtual (the diverging rays appear to come from point I).

The Relation Between Source Distance and Image Distance

The relationship between the positions of a source and an image for a single (concave) refracting surface is derived in the following subsection. There, we find the result

$$\text{for a refracting surface:} \quad \frac{n_1}{s} + \frac{n_2}{i} = \frac{n_2 - n_1}{R}. \tag{37–10}$$

This equation is analogous to Eq. (37–2), which holds for a mirror. In deriving it, we assumed that the surface was convex; that is, that the center of curvature of the surface is on the side of the surface to which the light passes. Let's suppose that this corresponds to a positive value of R, as for the concave mirror. In addition, s was positive from the start. Whether i is positive or negative according to Eq. (37–10) is a numerical question; both are allowed. When i is positive, it is on the side of the surface to which light passes and the image is real, meaning that light does pass through it. When i is negative, it is on the side from which light is emitted and the image is virtual, meaning that light only *appears* to radiate from it when it is observed from medium 2. We can show that the result does not depend on taking $n_2 > n_1$ (see Problem 25), although a geometrical drawing would depend on it.

We can repeat this exercise with a concave spherical surface between the two media (see Problem 26). In this case, the center of the spherical surface is on the side of the light source. The important result is that *Eq. (37–8) continues to hold, but with a negative value of R*. Just as when the image is on the side to which light goes, i is positive and the image is real, R is positive when it is on that side. When the image is on the side from which light radiates, i is negative and the image is virtual. Similarly, R is negative when it is on that side. See Table 37–1 for a summary of the sign conventions.

How to Get Equation (37–10)

Let's now use Fig. 37–21 to arrive at Eq. (37–10), the relation between source distance, image distance, and focal length for a refracting surface with radius of curvature R, for a convex surface. We have taken $n_1 < n_2$, although the final result does not depend on this assumption. For small angles, the approximate form of Snell's law [Eq. (37–8)] holds. Simple geometry shows that $\theta_2 = \beta - \alpha$ and $\theta_1 = \beta + \gamma$, so

$$n_1(\beta + \gamma) = n_2(\beta - \alpha). \tag{37–11}$$

The distances of Fig. 37–22 obey the exact relation $AB = R\beta$ and the small-angle approximations $AB = s\gamma = i\alpha$. Thus, Eq. (37–11) becomes

$$n_1\left(\frac{AB}{R} + \frac{AB}{s}\right) = n_2\left(\frac{AB}{R} - \frac{AB}{i}\right).$$

With cancellation of the arc length AB, we obtain Eq. (37–10).

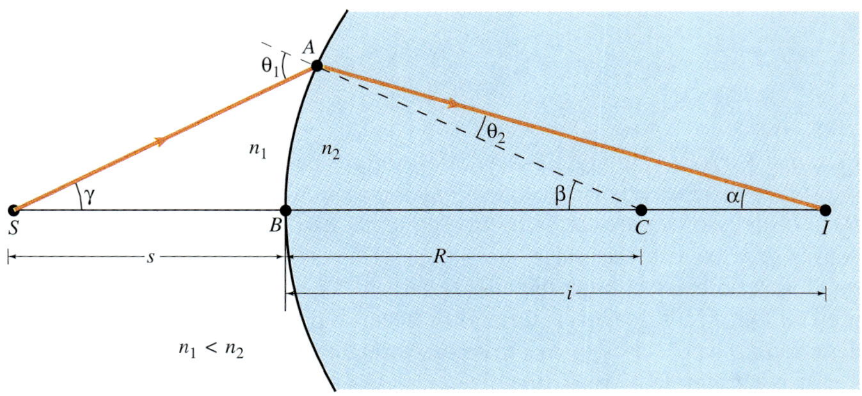

FIGURE 37–21 Geometric construction for deriving Eq. (37–10) for a spherical refracting surface.

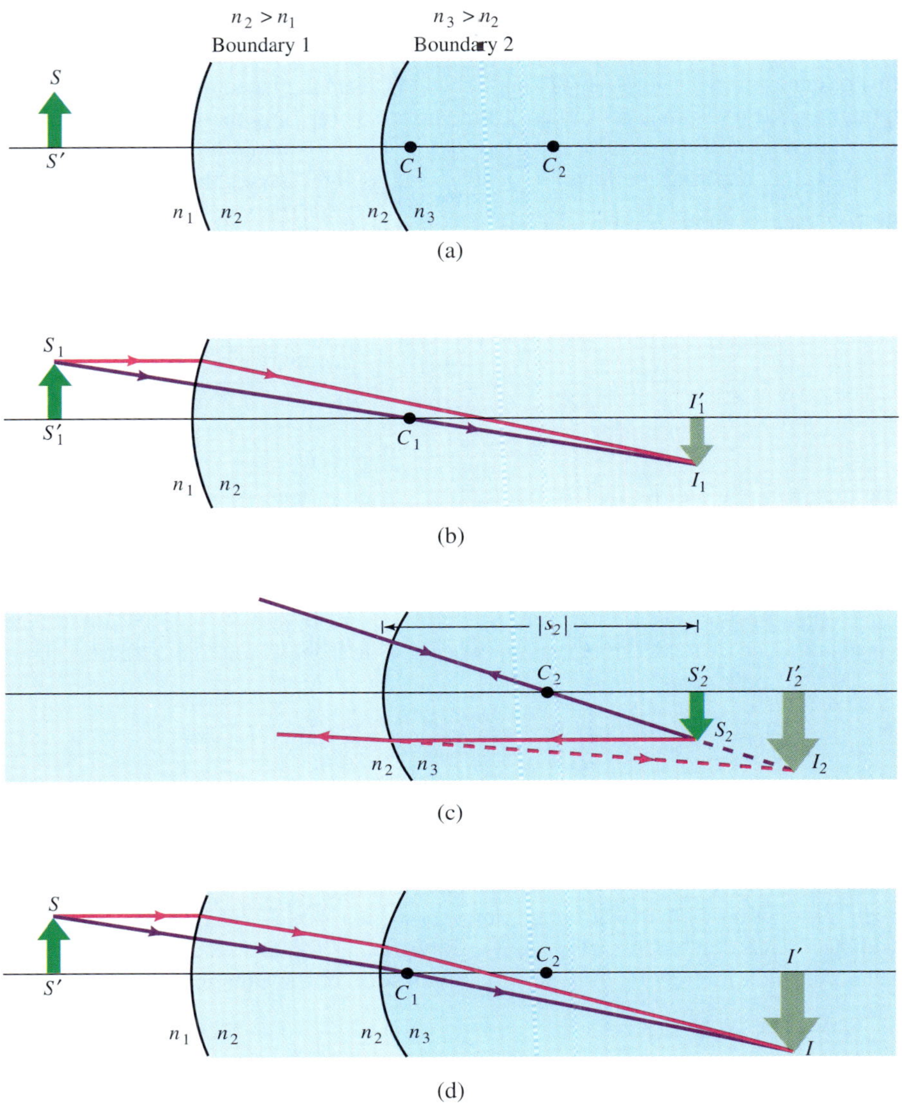

FIGURE 37–22 (a) Image construction when two refracting surfaces are involved. To simplify, (b) we can split the problem up by first finding the image point I_1 from boundary 1. (c) The resulting image serves as a virtual object; we use virtual source point S_2 for the interaction with boundary 2 to find the final image, at image point I. (d) The actual ray path.

The Sign of the Object Distance

We have established that a negative image distance and radius of curvature have meaning. Is it possible for s to take on a negative value? A positive s corresponds to a real object on the side from which the light radiates. For positive s, the light rays diverge from the object as they approach the boundary surface. Negative values of s correspond to the rays that *converge* as they approach the boundary, so their extrapolation would be on the side of the boundary to which the light passes. How is it possible to arrange this state of affairs? Certainly it cannot be done with a real object. But it is possible if the image produced by one surface acts as the source object for a second surface (Fig. 37–22a). The rays from the source will be refracted at both boundaries 1 and 2. We can break up the problem and find first the image point I_1 produced by boundary 1 (Fig. 37–22b). In actuality, the light never forms the image point I_1 because boundary 2 intervenes. However, the image becomes the virtual object for the light refracted at boundary 2 (Fig. 37–22c). According to our convention, the source distance s_2 is negative because the rays converge toward boundary 2, and the object is on the opposite side of the boundary from which the light radiates. The light comes from the left side of boundary 2, but the virtual object is on the right side. Equation (37–10) holds for the refraction at boundary 2, with a negative object distance s_2. The actual paths of the light rays through both boundaries are shown in Fig. 37–22d.

EXAMPLE 37–3 Consider a cylinder of glass 50 cm long, with $n = 1.6$, in air (Fig. 37–23). Surface 1 has a radius of curvature $R_1 = 0.20$ m; surface 2 has a radius of curvature $R_2 = 0.40$ m. A small object (a leaf) is placed perpendicular to the optical axis at a distance of 120 cm from surface 1. (a) Find the location of the object's image due to refraction at surface 1. (b) Let this image be the source object for surface 2, and find the location of *its* image as light passes through surface 2.

Solution: (a) We use Eq. (37–10) to calculate the distance i_1 of the image point I_1 from surface 1. We have $n_2 = 1.6$ and $n_1 = 1.0$ (air). The center of curvature is on the side to which light passes, so R_1 is positive. Finally, $s = +1.20$ m. We then have

$$\frac{n_1}{s} + \frac{n_2}{i} = \frac{n_2 - n_1}{R} = \frac{1.0}{1.20\text{ m}} + \frac{1.6}{i_1} = \frac{1.6 - 1.0}{0.20\text{ m}};$$

$$i_1 = +0.74\text{ m}.$$

The image is real and located 74 cm to the right of surface 1.

(b) We now use the first image as an object for surface 2. Because the surfaces are separated by 50 cm, this new object is 24 cm to the right of surface 2. The object is on the side to which light passes, so its distance from surface 1 is negative: $s_2 = -0.24$ m. For this second step, $n_1 = 1.6$, $n_2 = 1.0$, and $R_2 = -0.4$ m (surface 2 is concave, so its center of curvature is on the side from which light radiates). Thus Eq. (37–10) now gives

$$\frac{1.6}{-0.24\text{ m}} + \frac{1}{i_2} = \frac{1.0 - 1.6}{-0.4\text{ m}};$$

$$i_2 = +0.12\text{ m}.$$

This is positive, so the second image is real, or to the right of surface 2.

In Fig. 37–24, we have followed the principal rays 1 and 3 for surface 1 through to the production of the final image point, I_2, which is real and inverted. The intermediate image point, I_1, is also shown. Ray tracing for this example requires too much precision in drawing to be useful. Application of the equations is better in this case.

FIGURE 37–23 Example 37–3.

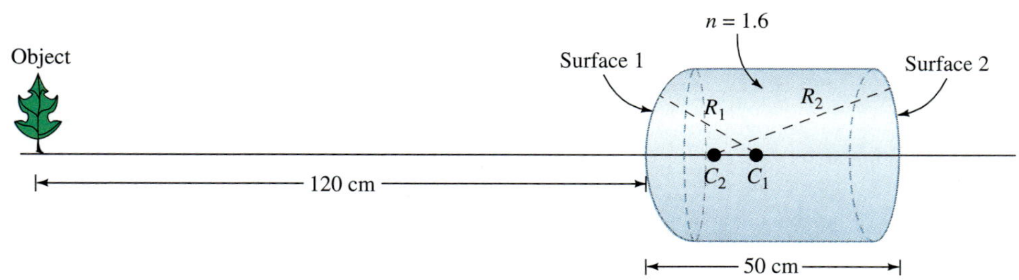

FIGURE 37–24 Two rays that form the final image in Example 37–3.

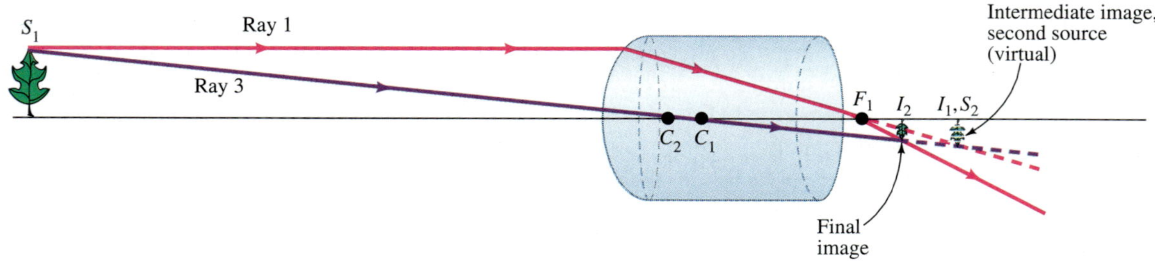

37-4 THIN LENSES

For our purposes, a **lens** consists of transparent material of refractive index n embedded in a material of refractive index n_1, normally air, for which $n_1 = 1$ (Fig. 37–25). We shall assume that $n > 1$ and $n_1 = 1$. We shall also assume that lenses are thin, so that the distance from the object and the image to the lens is independent of which lens surface is involved. This simplifies the treatment considerably. The two surface boundaries (1 and 2 in Fig. 37–26) are concave or convex spherical segments with respective radii of curvature R_1 and R_2. Whether these radii are positive or negative depends on whether the center of curvature is on the side to which light passes (positive R) or the side from which light radiates (negative R). For example, in Fig. 37–26a, R_1 is positive, whereas R_2 is negative.

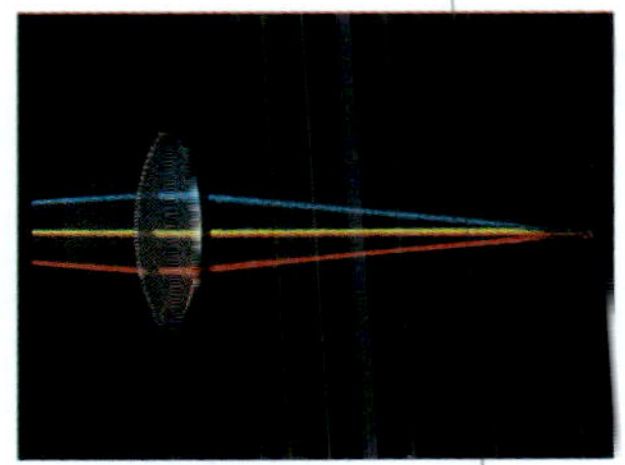

FIGURE 37–25 Light passing through a lens.

Let's suppose that a real object is a distance s_1 to the left of a thin lens. We can locate the final image and identify its features by using Eq. (37–10) twice in succession for image-making at a single surface, much as we did in Example 37–3. The image produced by the first surface serves as the object for the second surface. We do not have to worry about whether the various objects and images are real or virtual, upright or inverted, because the equation will automatically handle these questions. At surface 1, we have

$$\frac{1}{s_1} + \frac{n}{i_1} = \frac{n-1}{R_1}, \qquad (37\text{–}12a)$$

which we rewrite as

$$\frac{1}{i_1} = \frac{n-1}{nR_1} - \frac{1}{ns_1}. \qquad (37\text{–}12b)$$

Now, the image point I_1 produced by surface 1 serves as an object point S_2 for surface 2, producing a final image point at I_2. What is the sign of i_1? If i_1 is positive, the image is on the right of surface 1 and hence on the right of surface 2. This corresponds to an object distance s_2 for surface 2 that is negative. Similarly, if i_1 is negative, the image is to the left of both surfaces, corresponding to a positive object distance s_2 for surface 2. We must then reverse the sign of i_1 when we use it as the source distance s_2 for surface 2. Finally, note that in applying Eq. (37–10) a second time, $n_1 = n$ and $n_2 = 1$. Thus

$$\frac{n}{s_2} + \frac{1}{i_2} = \frac{1-n}{R_2};$$

$$-\frac{n}{i_1} + \frac{1}{i_2} = \frac{1-n}{R_2}.$$

When we substitute Eq. (37–12b), we find that

$$-n\left(\frac{n-1}{nR_1} - \frac{1}{ns_1}\right) + \frac{1}{i_2} = \frac{1-n}{R_2}.$$

If we now write $s_1 = s$ for the original object and $i_2 = i$ for the final image, we find (upon rearrangement)

for a thin lens in air: $$\frac{1}{s} + \frac{1}{i} = (n-1)\left(\frac{1}{R_1} - \frac{1}{R_2}\right). \qquad (37\text{–}13)$$

Relation between image and object distances for a thin lens

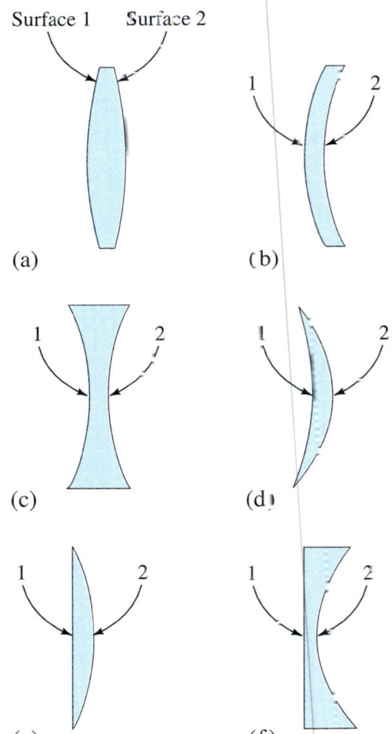

FIGURE 37–26 Six types of simple thin lenses with surfaces of different radii of curvature: (a) $R_1 > 0$, $R_2 < 0$; (b) $R_1 > 0$, $R_2 > 0$; (c) $R_1 < 0$, $R_2 > 0$; (d) $R_1 < 0$, $R_2 < 0$; (e) $R_1 = \infty$, $R_2 < 0$; (f) $R_1 = \infty$, $R_2 > 0$.

Equation (37–13), which applies *only* to thin lenses in air, is the *lens-maker's equation*. By Eq. (37–13), the image can be positive or negative; that is, real or virtual. The signs are summarized in Table 37–1, and ray tracing will alternatively allow you to understand the image.

Equation (37–13) can be used to find the focal point of a lens. In the limit where $s \to \infty$, $i = f$, the distance of the focal point from the lens:

The focal length of a thin lens

$$\frac{1}{f} = (n-1)\left(\frac{1}{R_1} - \frac{1}{R_2}\right). \qquad (37\text{–}14)$$

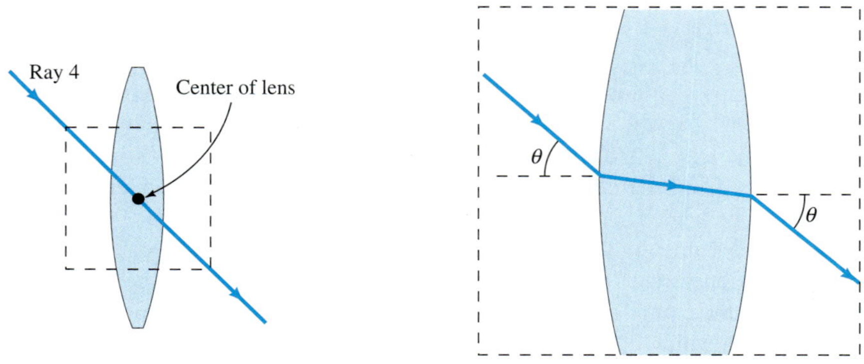

(a) (b)

FIGURE 37–27 (a) The ray to the center of a lens passes through without changing angle, because at its axis the lens is like a pane of glass. (b) An enlarged view of the same lens. If the lens is thin, the displacement of the ray is small.

Equation (37–2) applies to lenses as well as to mirrors.

If we substitute this result into Eq. (37–13), we get Eq. (37–2), which we originally derived for mirrors. The sign of f is determined by the signs of the radii of curvature, but we can say that f is positive if the image of a point source at infinity is on the side to which light passes (real image); f is negative if the image of the source at infinity is on the side from which light radiates (virtual image).

Principal ray 1 for a lens

Ray 1, light coming in parallel to the optical axis of the lens, crosses (or behaves as though it crosses) the axis at f. Note that there is a certain degree of symmetry in Eq. (37–13). When light arrives from the right of the lens rather than the left, R_1 and R_2 reverse their roles, and light from infinity is focused the same distance from the lens, but on the opposite side, from the first focal point. In turn, if light radiates (or behaves as though it does) from one of the two symmetric focal points of the lens, ray 2, the light emerges as a set of parallel rays.

Principal ray 2 for a lens

Principal ray 4 for a lens

One more ray is useful for understanding the behavior of lenses. If, as in Fig. 37–27a, ray 4 (remember, principal ray 3 is not applicable to lenses) is drawn to the center of the lens, it behaves as though it passes straight through. We can see this by looking at the enlarged section (Fig. 37–27b). To a good approximation, the two lens surfaces are in the middle of the lens, so the ray behaves like a ray that passes through a pane of glass. (There is a *small* displacement of the ray, but it drops to zero as the lens becomes thinner.)

PROBLEM SOLVING

For the principal rays, see the Problem-Solving Techniques box on p. 994.

We have thus found three principal rays that can be used to find the image. Let's look at an example. Figure 37–28 shows a lens that collects light from an object. With the principal rays, we can easily find image point I of object point S. The construction works for any point on the object. The image is real and inverted, in this case. In general, if a lens causes rays that pass through it to come together, it is called a *converging lens*, and if it causes rays that pass through it to spread out, it is a *diverging lens*. Converging lenses have positive focal lengths, whereas diverging lenses have negative focal lengths. Some simple ray tracing will show that a lens like that of Fig. 37–26a is a converging lens, and one like that of Fig. 37–26c is a diverging lens.

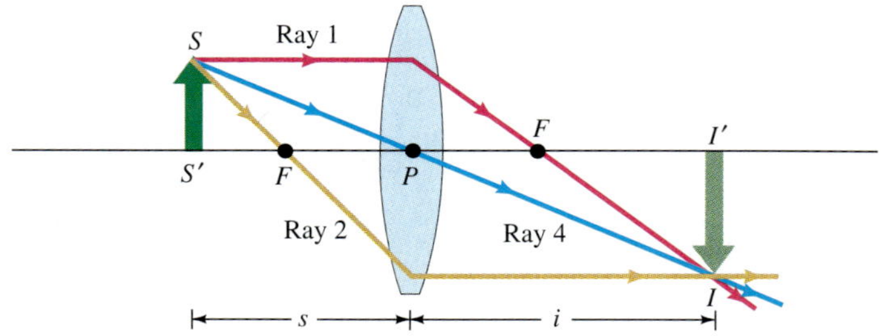

FIGURE 37–28 Ray tracing shows how a real image is formed with one type of thin lens. Point P marks the center of the lens.

Magnification

A thin lens produces a perfect image to the extent that the small-angle approximation is valid. Thus we can find the magnification by direct use of similar triangles. In Fig. 37–28, the magnification of the image has magnitude

$$M = \frac{II'}{SS'}.$$

From the geometry of the similar triangles $SS'P$ and $II'P$, we see that the magnitude of the magnification is $|M| = |i|/|s|$. Just as for mirrors, a systematic look at signs shows that we can decide with a single sign whether the image is upright or inverted:

$$M = -\frac{i}{s}.$$

This is Eq. (37–6)—the same form we found for mirrors. If M is positive, the image is upright; if it is negative, the image is inverted. From Eq. (37–2), we have the alternate form

$$M = \frac{f}{f - s},$$

which is Eq. (37–8), also used for mirrors.

EXAMPLE 37–4 A converging lens like that shown in Fig. 37–26a has surfaces with radii of curvature $R_1 = 80$ cm and $R_2 = 36$ cm. An emerald that is 2.0 cm tall is placed 15 cm to the left of the lens, for which $n = 1.63$. Where will the image be located, and what will its size be?

Solution: We first calculate the focal length from Eq. (37–14). The radius of curvature of the first surface is positive, $R_1 = 80$ cm, whereas the second surface has negative curvature, $R_2 = -36$ cm. Thus

$$\frac{1}{f} = (n-1)\left(\frac{1}{R_1} - \frac{1}{R_2}\right) = (1.63 - 1)\left(\frac{1}{80\text{ cm}} - \frac{1}{-36\text{ cm}}\right) = 0.025\text{ cm}^{-1}.$$

The object distance is positive, $s = 15$ cm, so Eq. (37–2) gives

$$\frac{1}{i} = \frac{1}{f} - \frac{1}{s} = 0.025\text{ cm}^{-1} - \frac{1}{15\text{ cm}} = -0.041\text{ cm}^{-1}.$$

Thus $i = -24$ cm. The minus sign indicates that the image is virtual and on the same side as the light source. The magnification is given by

$$M = -\frac{i}{s} = -\frac{-24\text{ cm}}{15\text{ cm}} = 1.6.$$

The positive value indicates that the image is upright. Figure 37–29 illustrates the ray paths.

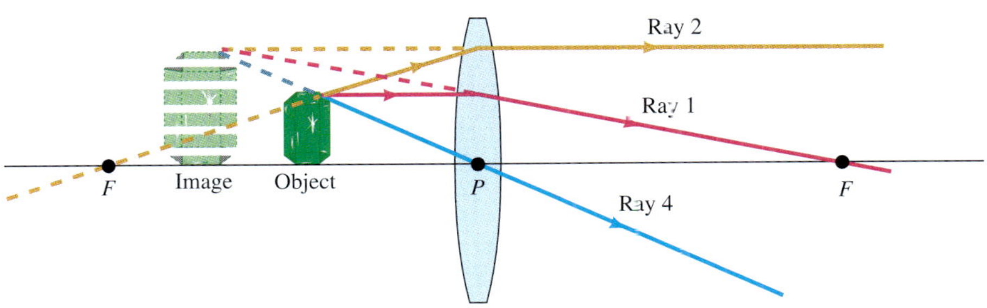

FIGURE 37–29 Example 37–4. When the object lies inside the focal point of the lens, ray tracing shows that the image formed is virtual.

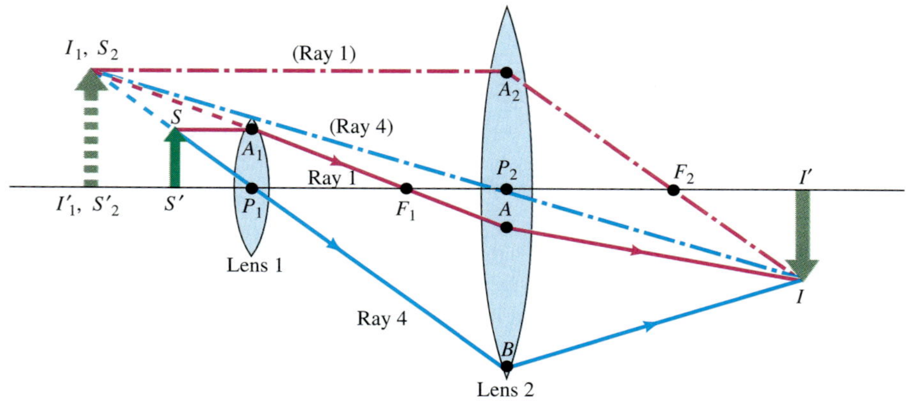

FIGURE 37–30 Ray tracing shows how two converging lenses produce a real magnified image.

We saw in Section 37–3 how the image produced by refraction at one surface acts as an object for the second surface. This principle extends to combinations of two or more lenses and lies at the heart of the design of complex optical instruments. Figure 37–30 gives an image construction for two thin converging lenses. The object SS' lies inside the focal length of lens 1 and thus gives rise to a virtual, enlarged image $I_1\ I'_1$. That image serves as an object $S_2S'_2$ for lens 2. Ray tracing uses the parallel ray $I_1A_2F_2I$ and I_1P_2I to determine the position of the real image, but the particular rays chosen really follow the path SA_1F_1AI and SP_1BI. This example shows that it is possible to obtain a magnified *real* image with two converging lenses, in conditions where it is not possible with one lens.

In Example 37–5, the object for the second lens is a negative distance from the lens.

EXAMPLE 37–5 Lens 1 in Fig. 37–31a is a converging lens with a focal length of 22 cm. An object is placed 32 cm to its left. Lens 2, which is a diverging lens with a focal length of 57 cm, lies 41 cm to the right of lens 1. Describe the position and other properties of the final image.

Solution: We simply apply Eq. (37–2) twice. The focal length of lens 1 (a converging lens) is $f_1 = +22$ cm, and the real object distance $s = s_1 = +32$ cm. If lens 2 were not present, the object would form an image $I_1I'_1$, which is determined by Eq. (37–2) to be at a position i_1 such that

$$\frac{1}{i_1} = \frac{1}{f_1} - \frac{1}{s} = \frac{1}{22\text{ cm}} - \frac{1}{32\text{ cm}}.$$

This gives $i_1 = +70$ cm, a real image (Fig. 37–31b). Because lens 2 is present, this image is not actually formed.

Lens 2 is a diverging lens. It takes parallel rays from infinity and makes them diverge, so the image from a source at infinity is virtual. The focal length is thus negative, $f_2 = -57$ cm. In addition, the source $S_2S'_2$ is the image $I_1I'_1$, which lies (70 cm − 41 cm) = 29 cm to the *right* of lens 2, that is, the side of lens 2 to which light passes. The source distance s_2 is thus negative, $s_2 = -29$ cm. The final image is then at distance i_2, determined by

$$\frac{1}{i_2} = \frac{1}{f_2} - \frac{1}{s_2} = \frac{1}{-57\text{ cm}} - \frac{1}{-29\text{ cm}} = \frac{1}{59\text{ cm}},$$

or $i_2 = +59$ cm. The final image is real, 59 cm to the right of lens 2.

The total magnification is found by applying the magnification formula twice:

$$M_{\text{tot}} = M_{\text{lens 1}}M_{\text{lens 2}} = -\frac{i_1}{s_1}\left(-\frac{i_2}{s_2}\right) = \frac{-(+70\text{ cm})}{32\text{ cm}}\left[\frac{-(+59\text{ cm})}{-29\text{ cm}}\right] = -4.5.$$

The final image is inverted (M_{tot} is negative) and 4.5 times the size of the object. *The sign is automatically taken care of in the product of the two magnifications.*

Our calculations are verified qualitatively by ray tracing (Fig. 37–31b).

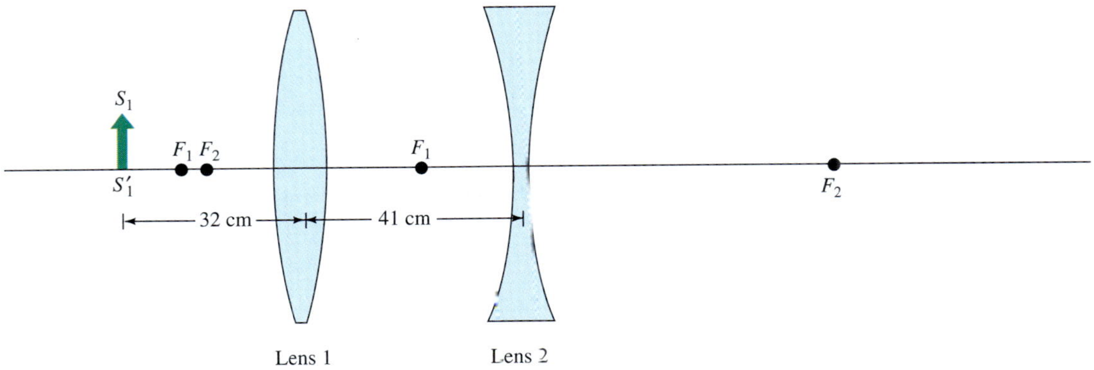

(a)

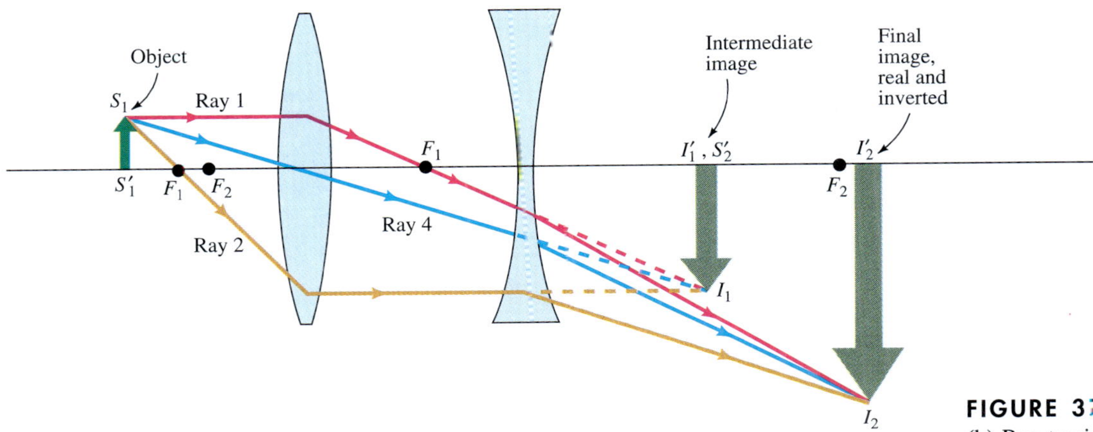

(b)

FIGURE 37–31 (a) Example 37–5. (b) Ray tracing confirms the calculations.

37–5 OPTICAL INSTRUMENTS

The Eye

The typical vertebrate eye—the basic structure of which is shown in Fig. 37–32—is a remarkable optical instrument. Light enters the eye proper through the *pupil*, the size of which can be changed by contraction or expansion of a membrane called the *iris* according to the intensity of the incident light. The light then passes through a convergent *crystalline lens* into a chamber filled with the *vitreous humor*, a fluid with index of refraction near that of water. The light is focused onto the back of the eye, the *retina*, which is covered with sensitive receptor cells. The stimulation of these cells by light produces a message that is sent to the brain along the *optic nerve*, and the brain reconstructs the image.

When a normal eye is relaxed, objects at infinity form an image precisely on the retina, a distance of about 1.7 cm from the lens. When objects are brought closer, the lens is compressed by surrounding muscles and becomes more convergent. The focal length is reduced, and the image continues to be focused on the retina. There is a limit to this power of *accommodation*. Objects closer than the *near point*, about 25 cm from the lens (or less for younger people), appear blurred. The near point tends to increase with age because the lens becomes unable to compress as far as it once did, and the image of a near object is beyond the location of the retina. Converging lenses correct this problem (Fig. 37–33a). In cases of nearsightedness, the image of an object at infinity is in front of the retina. A diverging lens will provide the necessary correction (Fig. 37–33b).

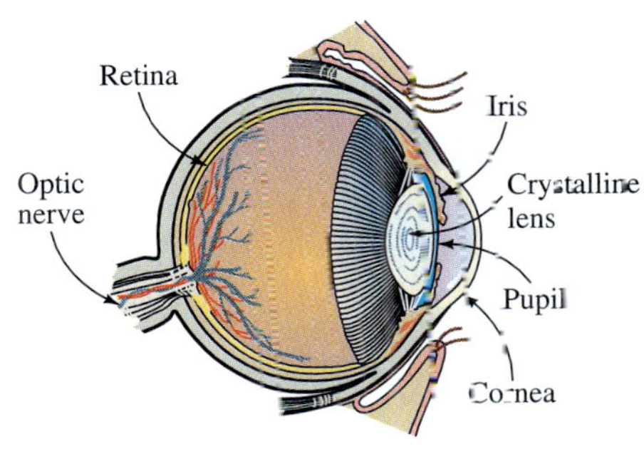

FIGURE 37–32 Schematic diagram of the human eye and some of its important features.

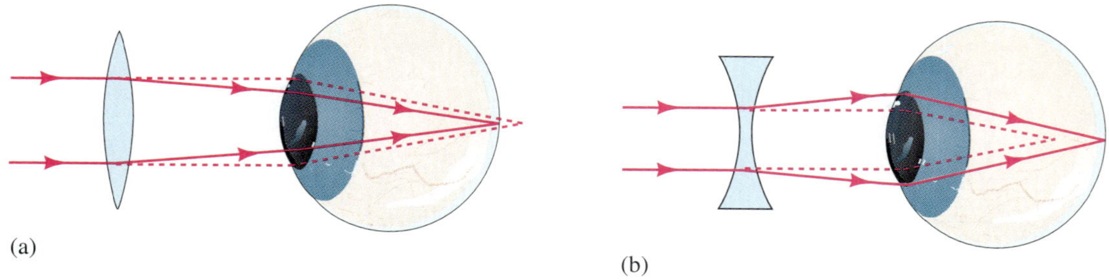

FIGURE 37–33 The dashed lines indicate the paths rays would take if no correcting lens were present. The solid lines mark the path of rays when a correcting lens is included. (a) A converging lens causes rays from an object, in this case at infinity, to focus closer to the lens of the eye. Such a lens corrects farsightedness by allowing the near point to be moved closer to the eye. (b) A diverging lens causes rays from an object, in this case at infinity, to focus farther from the lens of the eye. Such a lens corrects nearsightedness.

APPLICATION

The camera

The Camera

With one important exception, the camera is optically equivalent to an eye. There is a converging lens in front, and the *film*, which plays the role of the retina, is in back. There is an *aperture*, an opening equivalent to the pupil, and a *shutter*, which provides an approximation to an instantaneous image and avoids blurring of the picture (Fig. 37–34). The difference between the simple camera shown and the eye is that the focal length of the lens changes in the eye, whereas the focal length is fixed in a simple camera. Instead, the camera lens moves in and out (changing the image distance) to enable objects of different source distances to produce a focused image on the film.

Angular Magnification

For those optical instruments used for observing the world closely, *angular magnification* is a critical concept, and we shall discuss it before we cover some other instruments.

From Eq. (37–8), we see that the magnification of a lens or mirror is infinite when $s = f$. This is less important than it might appear to be because the image distance i also becomes infinite in that case. More important than the actual size of the image is the angle the image takes up in our field of vision. Given the limits of our own vision, *it is this angular coverage that determines how much detail we can see in an observed source.*

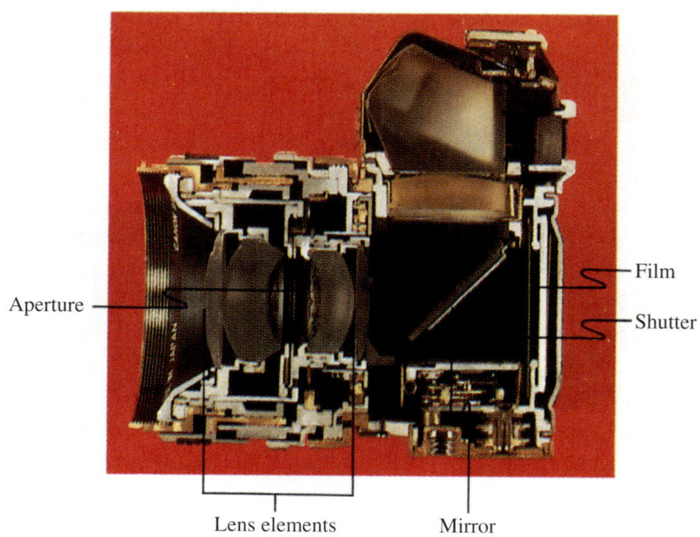

FIGURE 37–34 A cutaway view of the optical system of a camera.

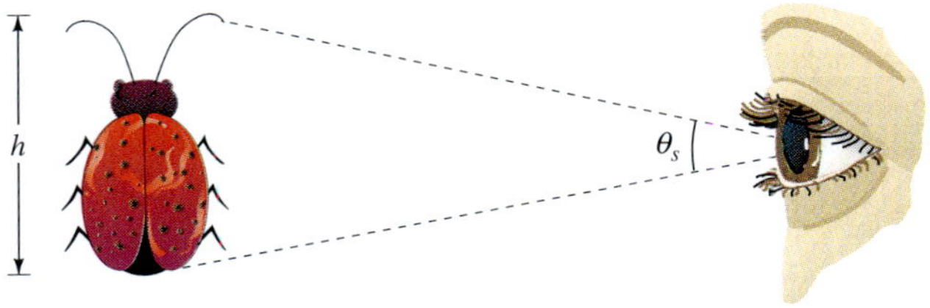

FIGURE 37–35 The angular size of an object, θ_s, is the relevant quantity for our ability to see detail in the object.

Imagine that you are a distance d from some object of height h (Fig. 37–35). For a source that does not cover an enormous part of your vision, the angular size θ_s of the source is

$$\theta_s \simeq \frac{h}{d}. \qquad (37\text{–}15)$$

For normal, unaided vision, this angular size can be maximized when the object is brought to the near point of vision, around $d = d_{min} = 25$ cm, and it is $\theta s \simeq h/(25\text{ cm})$ that is used as a reference for the angular magnification. Suppose now that we use an optical system to observe our source and that the image of the source as seen through the system has an angular size θ_i. Then the **angular magnification** of the system is

Angular magnification

$$M_\theta \equiv \frac{\theta_i}{\theta_s}. \qquad (37\text{–}16)$$

We do not bother with signs here and keep track only of the magnitudes of the angular sizes. If we know the angular magnification of two elements that are superposed in an optical system, then the net angular magnification is the product of the angular magnifications of each element.

The Simple Magnifier

A converging lens has a positive focal length. By Eq. (37–2),

$$\frac{1}{i} = \frac{1}{f} - \frac{1}{s}.$$

For a real object, i passes from positive (real image) to negative (virtual image) as the object moves toward the lens through the point $s = f$. At this point, i shifts to $-\infty$. A *simple magnifier* is a converging lens with the object placed near $s = f$ (Fig. 37–36). If the object size is h, the image size is, by definition, $h_i = Mh$, where $M = i/s$ is the magnitude of the magnification. The image size is infinite if i is infinite, but the *angular size of the image is finite*. When $s = f$, we have for the angular size

$$\theta_i = \frac{h_i}{i} = \frac{Mh}{i} = \frac{h}{s}\Big|_{s=f} = \frac{h}{f}. \qquad (37\text{–}17)$$

Note that we have no trouble seeing an image at infinity. At the near point, $d_{min} = 25$ cm, the angular size of our object is $\theta_{object} = h/d_{min}$. Thus the angular magnification of the magnifier is

$$M_\theta = \frac{\theta_i}{\theta_{object}} = \frac{h/f}{h/d_{min}} = \frac{d_{min}}{f}. \qquad (37\text{–}18)$$

If we choose a converging lens with a focal length of 2 cm, we get an angular magnification of (25 cm)/(2 cm) = 12.5.

The Microscope

The **compound microscope**, invented around 1590 by Zacharias Janssen, provides high angular magnification for nearby objects (Fig. 37–37a). A two-lens system illustrates the principle of the microscope (Fig. 37–37b). The *objective* is a converging lens with

APPLICATION
Microscopes

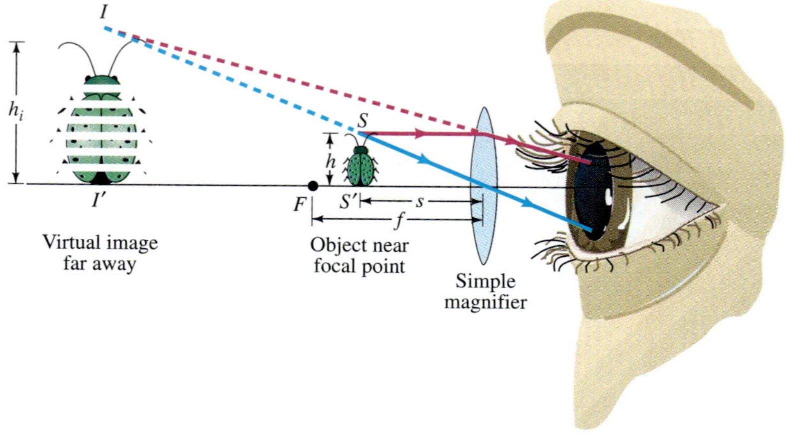

(a)

(b)

FIGURE 37–36 (a) The simple magnifier is a converging lens with an object placed near the focal point. The image is virtual and far away. (b) Such a magnifier in use.

a short focal length f_1, and the object to be viewed (of size h_0) is placed just outside the focal point of that lens. The image distance is given by $1/i = (1/f) - (1/s)$; i is large and positive for s just larger than f. (The image distance is roughly the distance between the two lenses, L.) This intermediate image has size $h_1 = Mh_0 \simeq Lh_0/f_1$. The second lens, the *eyepiece*, or *ocular*, acts as a simple magnifier: It is placed so that the intermediate image is at its focal length f_2 ($<< L$). Then the angular size θ_i of the final

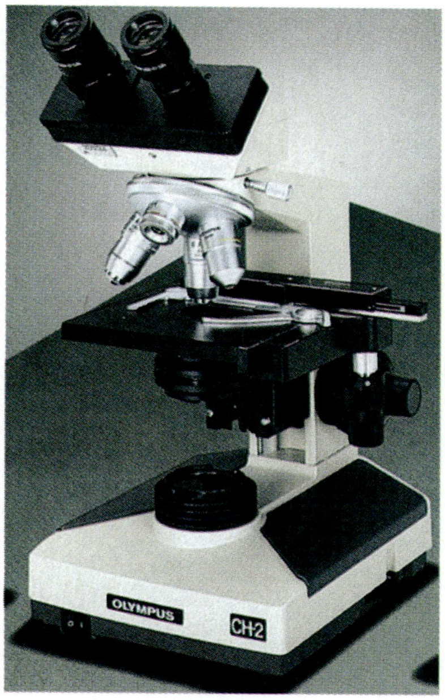

(a)

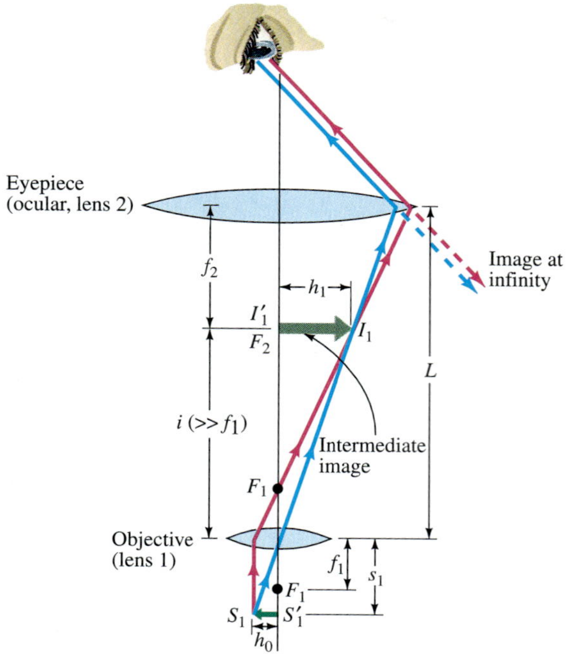

(b)

FIGURE 37–37 (a) The modern compound microscope, with binocular eyepieces and multiple objective lenses that can be rotated into position. (b) Schematic diagram of a compound microscope. The objective produces an image close to the focal point of the eyepiece.

image seen by an eye at the position of the eyepiece is given by Eq. (37–17) with $h_i \to h_1$ and $f \to f_2$:

$$\theta_i = \frac{h_1}{f_2} = \frac{Lh_0}{f_1 f_2}.$$

The net angular magnification is then given by Eq. (37–18):

$$M_\theta = \frac{\theta_i}{\theta_{\text{object}}} = \frac{Lh_0/f_1 f_2}{h_0/d_{\text{min}}} = \frac{Ld_{\text{min}}}{f_1 f_2}.$$

If we take typical values such as $L = 15$ cm, $f_1 \simeq 5$ mm, and $f_2 \simeq 2$ cm (d_{min} is always 25 cm), then the net angular magnification is 375.

The Telescope

APPLICATION
Telescopes

The **telescope** magnifies very distant objects. It was invented in Holland at the beginning of the seventeenth century and made an impact on astronomy soon thereafter. Galileo built his own telescope in 1609 (Fig. 37–38a).

The *refracting telescope* (a telescope with strictly refracting elements) resembles the microscope but, for the telescope, the original object is in effect at infinity (Fig. 37–38b). The first lens, the objective, creates an intermediate image very close to the focal point of that lens. If that point coincides with the focal point of the eyepiece, then the eyepiece again acts as a simple magnifier. The final image is magnified. Let's calculate the angular magnification for an object that has angular size θ_s. (The Moon, for example, has angular size of 1/2°. We can distinguish with the naked eye stars separated by about 1′ of arc [1/60 of 1°].) If the original object has size h_0, the objective produces an image of size $h_1 = Mh_0 = ih_0/s = i\theta_s = f_1\theta_s$. The final image then has an angular size given by Eq. (37–17) with $h \to h_1$ and $f \to f_2$, namely, $\theta_i = h_1/f_2 = \theta_s f_1/f_2$. In turn, the angular magnification is

$$M_\theta = \frac{\theta_i}{\theta_s} = \frac{f_1}{f_2}.$$

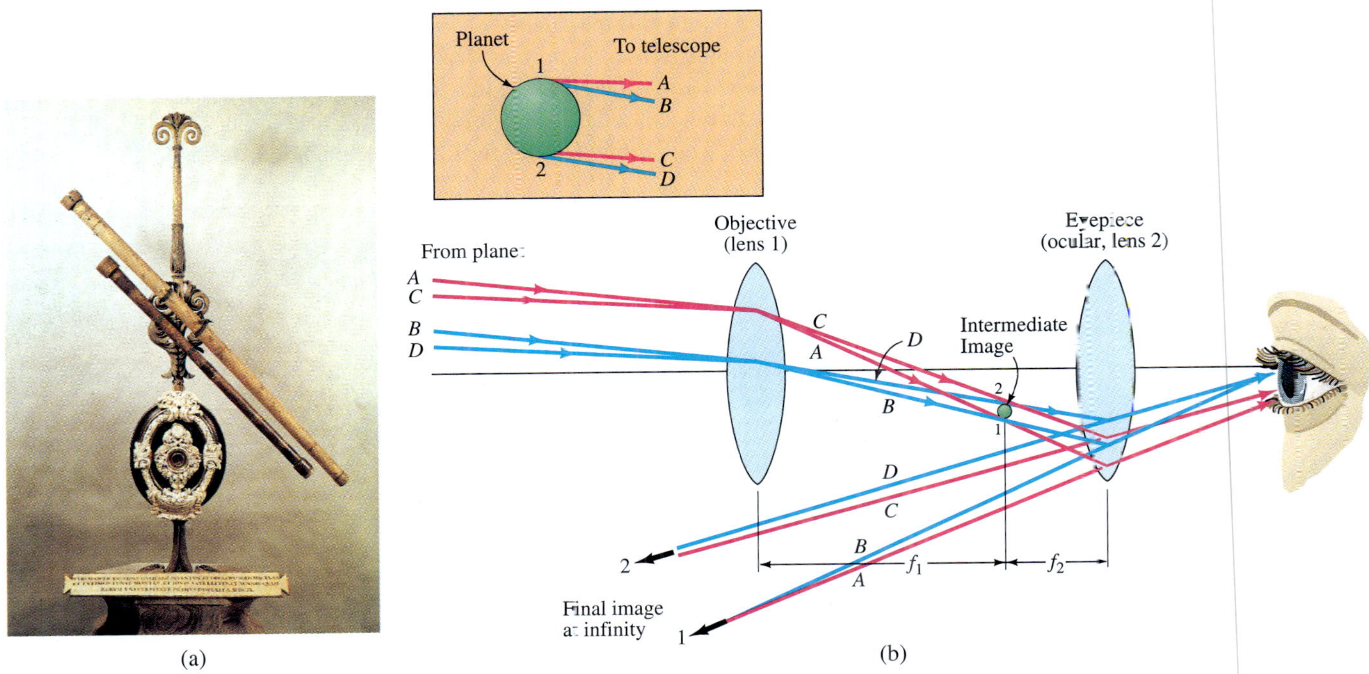

FIGURE 37–38 (a) Galileo's refracting telescope, used for viewing distant objects. (b) Schematic diagram of a refracting telescope.

Thus, in contrast to the microscope, the objective lens of a telescope should have as long a focal length f_1 as is practical.

The study of distant galaxies depends on an examination of the spectrum of the light they emit and of their energy output. The incident light from very distant objects is rather low in intensity, and more light is needed at the eyepiece in order to study spectra. To be most efficient at collecting light, the diameter of the optical system must be large. Large lenses are more difficult to construct than large mirrors, so most large telescopes are *reflecting telescopes* (Fig. 37–39a) rather than refracting telescopes. In a reflecting telescope, a mirror replaces the objective for the purpose of creating an intermediate image, which is then magnified by the eyepiece (Fig. 37–39b). Another advantage of such a telescope is that it has no chromatic aberration (see Section 37–6).

(a)

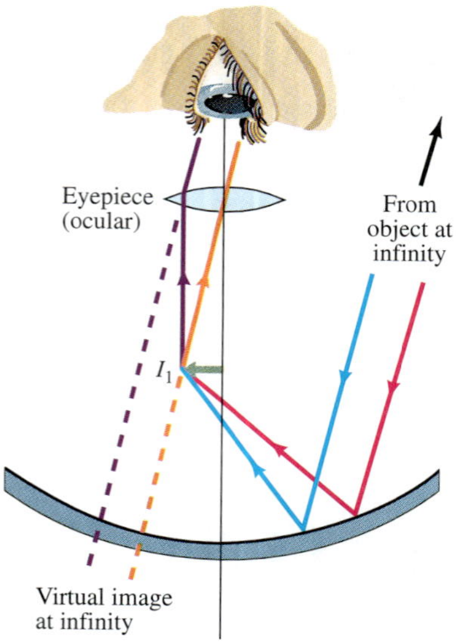

(b)

FIGURE 37–39 (a) A reflecting telescope, in which the objective of the refracting telescope is replaced by a concave mirror. (b) Schematic diagram of a reflecting telescope.

Geometric Optics without Curved Surfaces

Recent developments in manufacturing techniques hold the promise of being able to reproduce many of the features of optical systems without the need to mold or grind curved surfaces. These techniques allow the production of glasses whose indices of refraction vary with position in a controlled way. Light bends smoothly within such glasses. We can think of them as consisting of a series of infinitesimally thin layers of glass, each layer having an index of refraction slightly different from its neighbor. A light ray will bend by an infinitesimally small angle as it crosses the boundary between each layer. Figure 37–40 shows the path of a ray within material of this type. By tailoring the way the index of refraction varies with position, it would become possible, for example, to produce eyeglass lenses made of thin, flat plates of glass.

*37–6 ABERRATION

An accurate calculation would show that all rays that arrive at a spherical mirror or refracting surface from infinity cross in a small but finite region rather than at a sharp point. This is but one example of **aberration** (Fig. 37–41). Aberration should be distinguished from *distortion*, in which an image is not identical in form to the object, as in Fig. 37–13. For scientific purposes, the image of Fig. 37–13 is not necessarily a bad one because every ray from the object has its precise location in the image. Aberration concerns what we might call the quality of an image, not its geometric form.

We can distinguish two important types of aberrations in geometric optics. *Monochromatic aberrations* describe the fact that, in real optical systems, the rays from a given point on an object are not focused on a single image point (Fig. 37–42a). The correction for this type of aberration depends on the application. An optical system that

FIGURE 37–40 A sample of glass across which the index of refraction varies. As a result, light bends within the glass.

APPLICATION: MOSAIC TELESCOPES

The traditional reflecting telescope employs a single large mirror whose surface is ground with a large component of hand labor to a parabolic form—the shape that gives an aberration-free image. Unfortunately, there is a severe limit to the size of such a mirror, the 5-meter-diameter mirror in the Mount Palomar telescope representing one of the largest instruments. Why the limit? First, the massive mirror demands a very heavy and expensive telescope frame to support it, and, more importantly, the mirror distorts under its own weight as the elevation of the telescope is changed to view different parts of the sky.

The modern generation of large telescopes is made from an array—a mosaic—of spherical mirrors, arranged on a frame that positions the mirrors across a paraboloidal surface (Fig. 37AB–1). As any given mirror is relatively small, the error due to its sphericity is insignificant in the formation of the final image. The mirrors are cast as disks and machine-ground. The edges are then cut off to make them into hexagons, allowing the array to fit together neatly. Computer-controlled positioning devices are then used to mount the hexagons onto the main support frame. Each hexagonal mirror can therefore be orientated by computer command so that its contribution to the final image is as would be expected from a perfect parabolic mirror. Moreover, as the elevation of the telescope is changed or as thermal expansion due to ambient temperature changes distorts the support frame, the individual mirrors can be repositioned to maintain an aberration-free image.

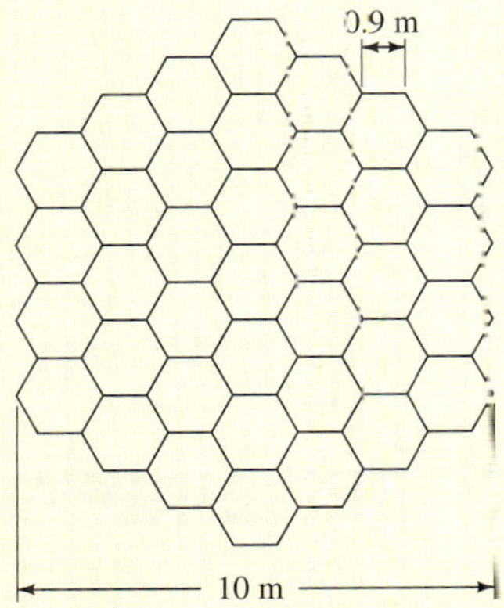

FIGURE 37AB–1 A large collecting mirror is made by assembling hexagonal elements.

collects images only from distant objects will have no aberration when a parabolic surface is used (Fig. 37–42b). Although such surfaces are difficult to construct from glass, a pool of mercury spinning about a vertical axis forms a parabolic surface, and such surfaces are employed in some modern telescopes. Alternatively, this type of aberration is minimized when the spherical section of the lens surface or mirror is small, although then the system collects less light.

Chromatic aberrations appear in refracting systems but not in mirrors. We treated dispersion in Chapter 36, in which the index of refraction of a material depends on the wavelength of the light. The optical path of a ray at one wavelength will differ from that of a ray at another wavelength (Fig. 37–43a). If a given point on an object is the source of a mixture of wavelengths (as is true of white light), then the image of this point is spread out according to the wavelength. A simple correction for chromatic aberration is to use filters that allow only a narrow band of wavelengths to pass. More commonly, several lenses are superposed (Fig. 37–43b). Different elements are designed to have canceling dispersion to minimize the net dispersion from successive passage.

FIGURE 37–41 This poor image is the result of spherical aberration, a type of monochromatic aberration.

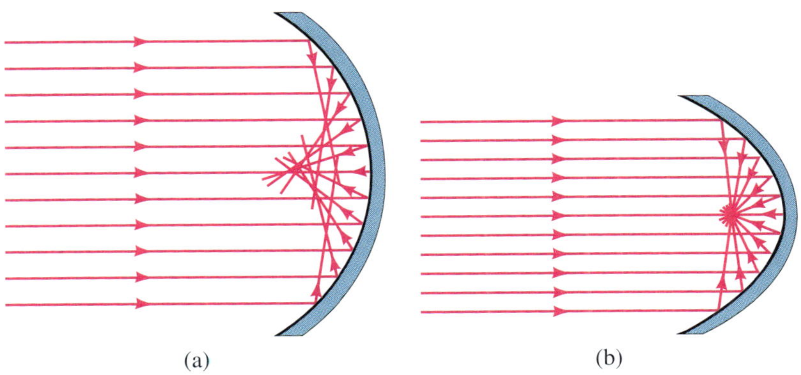

FIGURE 37–42 (a) In monochromatic aberrations, all the rays from infinity do not pass through the same point for a spherical mirror, so the focus is not sharp. (b) This type of aberration is eliminated by use of a parabolic mirror.

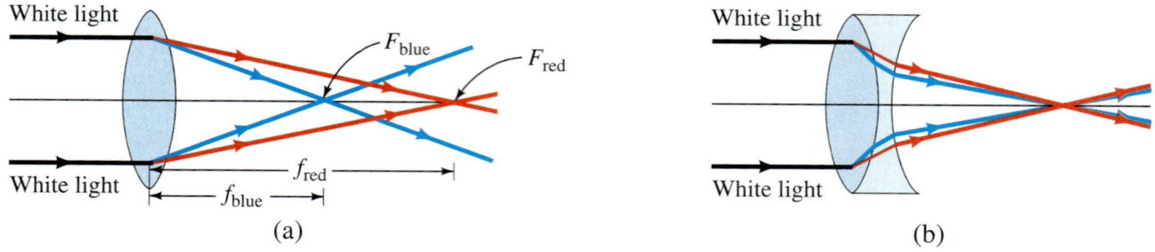

FIGURE 37–43 (a) In chromatic aberrations, the focal point of a converging lens may be different for different wavelengths. Here we show rays only for red light and blue light. (b) This type of aberration is eliminated by combining the lens with another lens with a different dispersion.

A good camera lens may consist of a dozen elements of different types of glass, with complicated geometric relations, which, in the case of zoom lenses, are variable. Unfortunately, even with all the corrections such lenses provide, the wave nature of light provides a fundamental and unavoidable limitation on the ability of optical systems to produce sharp images (see Chapters 38 and 39).

SUMMARY

Geometric optics is based on two basic laws of the behavior of light rays: In reflection, the angle of incidence on a reflecting surface is equal to the angle of reflection. In refraction, Snell's law, $n_1 \sin\theta_1 = n_2 \sin\theta_2$, holds, where θ_1 and θ_2 are the angles of incidence and refraction, respectively, for a ray incident from medium 1 on medium 2. Ray tracing is a technique that allows us to locate the image of a given source.

A spherical reflecting or refracting surface forms an image of an object that is a distance s from the surface. Bundles of light rays pass through an image point (for a real image) or are directed as though they all come from such a point (for a virtual image). The image point is a distance i from the surface. One limit of such a surface is the plane mirror, for which the distance of the virtual image from the mirror is given by

$$i = -s.$$

Parallel rays falling on a reflecting or refracting surface approximately converge to the focal point, a distance f from the element. For a mirror, $f = R/2$ [Eq. (37–1)], where R is the radius of curvature of the spherical section. The distance of the object and the distance of the image to the surface, and the focal length for a spherical mirror are related by

$$\frac{1}{s} + \frac{1}{i} = \frac{1}{f}. \quad (37\text{–}2)$$

Equation (37–2) applies to both convex and concave mirrors if proper account is taken of the signs of s, i, and R. The image size is magnified by a factor M, the magnification, times the object size, where

$$M = -\frac{i}{s} = \frac{f}{f - s}. \quad (37\text{–}6,\ 37\text{–}7)$$

For a positive M, the image is upright; for a negative M, the image is inverted.

For a spherical boundary between a medium of refractive index n_1 and a medium of refractive index n_2, with light incident from medium 1, Eq. (37–2) is replaced by

$$\frac{n_1}{s} + \frac{n_2}{i} = \frac{n_2 - n_1}{R}. \quad (37\text{–}10)$$

The same formula applies to both convex and concave spherical surfaces if proper account is taken of the signs of s, i, and R.

Thin lenses are understood by thinking of the image due to refraction at the surface nearest the object as the object for refraction at the second lens surface. For thin lenses in air, Eq. (37–2) applies, with

$$\frac{1}{f} = (n-1)\left(\frac{1}{R_1} - \frac{1}{R_2}\right). \tag{37–14}$$

Equations (37–2) and (37–14) combine into

$$\frac{1}{s} + \frac{1}{i} = (n-1)\left(\frac{1}{R_1} - \frac{1}{R_2}\right). \tag{37–13}$$

Equations (37–6) and (37–7) apply also to thin lenses.

Thin lenses can be used singly or in combination to make up optical instruments, including magnifiers, eyeglasses, cameras, microscopes, and telescopes. Angular magnification, which measures the ratio of the angular size of an object as seen through the instrument to the object's angular size as the naked eye sees it, is a fundamental consideration in these instruments. So is the quality of the image they produce.

UNDERSTANDING KEY CONCEPTS

1. Why is "ƎƆИAᒧUᙠMA" written on the front of an ambulance?
2. Consider a large room with walls covered with mirrors; at the center of the room is a candelabra with burning candles. Is the room brighter than a comparable room with black drapes in place of the mirrors?
3. The image of a distant candle is projected by a converging lens on a screen placed at the focal length of the lens. A piece of paper is taped over the lower half of the lens. Will only half of the image be seen?
4. Draw a right-handed coordinate system and its image in a plane mirror. Is the image a right-handed or a left-handed coordinate system? (In a right-handed system, the vector product **i** × **j** points along **k**.)
5. Would a dental mirror, the small mirror a dentist uses to examine your teeth, be concave, convex, plane, or sometimes one or another?
6. The side-view mirrors of some cars are labeled "Objects seen in this mirror may be closer than they appear." Is the mirror plane, convex, or concave?
7. For each of the simple lenses shown in Fig. 37–26, Eq. (37–6)—for magnification—shows that the size of the image of a ball placed at the focal point is infinite. Can you see by ray tracing why this must be?
8. Figure 37–13 shows the reflection made by a spherical surface. Parts of all four walls of the room, even the wall behind the sphere, are visible in the image. Why?
9. Does the focal length of a lens change when the lens is in water?
10. When a magnifying glass is lined up perpendicular to the line between it and the Sun, a hot spot forms on the side of the lens away from the Sun. What is the relation between the distance of this hot spot from the lens and the focal length of the glass? Why does the spot become hot?
11. A camera works by forming a real image on a film plate. Can a camera take a picture of a virtual image?
12. In William Golding's novel *The Lord of the Flies* (1954), some boys rediscover fire with the aid of the Sun shining through the eyeglasses of Piggy, a nearsighted boy. Has Golding made a mistake?
13. Are any principal rays useful for a point *on* the axis of an optical system?
14. When you have an eye exam, even one as simple as reading an eye chart, the examiner may dilate (open) the pupil by putting drops in your eye. Why is that useful?
15. We mentioned in Section 37–5 the possibility of making a flat eyeglass lens with material in which the index of refraction varies with position. Sketch the profile of the index of refraction for a converging lens constructed in this way.
16. Legend has it that Archimedes, acting as an advisor to the ruler of Syracuse, devised an optical system made of shields that could concentrate sunlight sufficiently well to set enemy boats on fire from a distance. How plausible is this legend?

PROBLEMS

37–1 *Images and Mirrors*

1. (I) Consider two mirrors at right angles to each other (Fig. 37–44). How many virtual images will a pointlike light source have?
2. (I) Consider two parallel mirrors that face each other, placed along the x-axis at $x = a$ and $x = -a$. Assume that a point source of light is placed at $x = x_0$ between the mirrors. What are the locations of the four images of the point source with the smallest values of image distance i?
3. (I) Two mirrors, each 2.0 m wide, are placed facing each other and parallel to each other and are separated by 10 cm. A ray

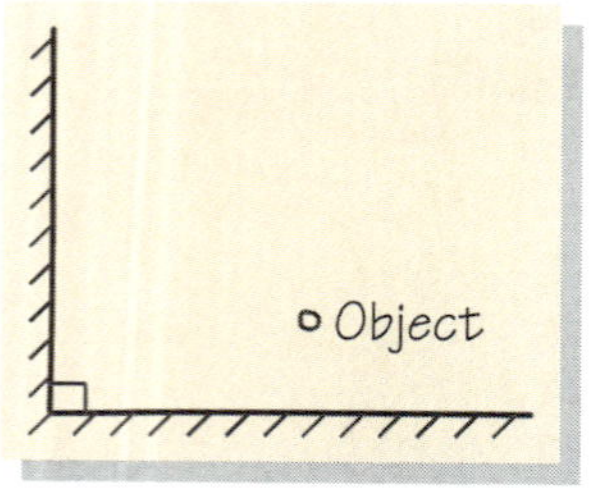

FIGURE 37–44 Problem 1.

of light enters the gap between them, grazes the edge of one mirror and strikes the other mirror at an angle of 30° with respect to the normal to the mirrors. At each reflection, the intensity of the light beam is attenuated by 5 percent. By how much is the beam attenuated when it finally leaves the space between the mirrors?

4. (II) A mirror is exactly half your height, and the top of the mirror is aligned with the top of your head. (a) If your eyes were at the top of your head, how close would you need to be to the mirror in order to be able to see your feet? (b) If your height is 165 cm and your eyes are 9 cm below the top of your head, what would have to be done with the mirror so that you could see both the top of your head and your feet?

5. (III) A kaleidoscope contains three plane mirrors forming a prism with an equilateral triangular base of side a (Fig. 37–45). Consider a small object placed on the axis of the kaleidoscope. Construct the position of the images formed by single as well as double reflection. How far are these images from the axis?

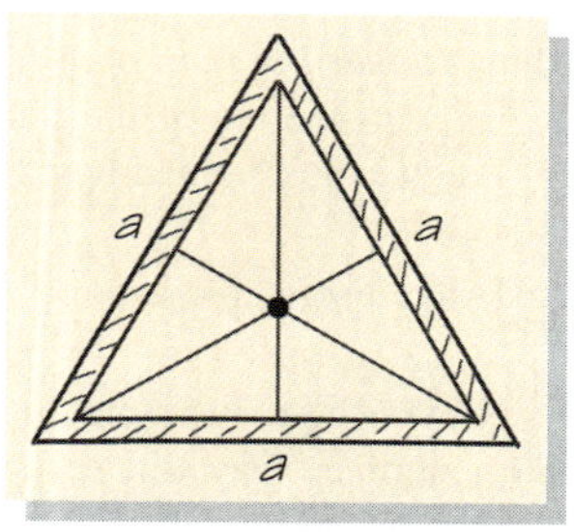

FIGURE 37–45 Problem 5.

6. (III) Suppose that two plane mirrors meet at an angle of 60° (Fig. 37–46). An object is placed between the mirrors, on the line that bisects this angle. Use graphical methods or trigonometric methods to locate all the images.

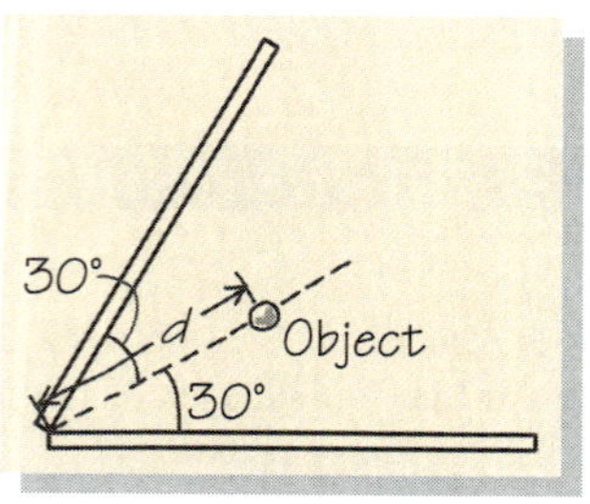

FIGURE 37–46 Problem 6.

37–2 Spherical Mirrors

7. (I) A dime 60 cm away from, and on the optical axis of, a concave spherical mirror produces an image 20 cm away from the mirror. If the dime is moved on the axis to 35 cm from the mirror, where will the image move? How large is the radius of the sphere of which the mirror is a section? Draw the system for the second case described.

8. (I) A paper clip is placed on axis 15 cm away from a convex mirror, part of a sphere of radius 60 cm. Where will the image be located, and what is the magnification? Make a sketch, including rays.

9. (I) An object of height 2 cm is placed 20 cm from a concave mirror. The real image is found to be 8 cm from the mirror. On which side of the mirror is the image, and how tall is it? Is it inverted?

10. (I) A concave mirror has a radius of curvature of 176 cm. What is the size of the image of an object 6.00 cm tall that is placed 133 cm from the mirror?

11. (I) A concave mirror is cut from a spherical surface of radius of curvature 2.0 m. A pencil 10 cm long is placed perpendicular to the axis of the mirror at a distance of 80 cm from the mirror. Where is the image and how large is it?

12. (II) By using the same reasoning that we used in the text for the case of the concave mirror, show that the reflection of ray 1 in Fig. 37–11 appears to originate at point F, independent of the angle θ. Your argument shows that, at point F, there is an image of a source point at infinity.

13. (II) Use ray tracing for parallel rays far from, as well as near to, the optical axis to show that parabolic mirrors more accurately focus parallel rays than spherical mirrors do.

37–3 Refraction at Spherical Surfaces

14. (I) A sphere of glass (n = 1.60) of radius 5.0 cm is immersed in water (n = 1.33). A small flower (at point B) is 2.0 cm outside the sphere (Fig. 37–47). What are the location and nature (real or virtual) of the flower's image made by refraction at the first surface?

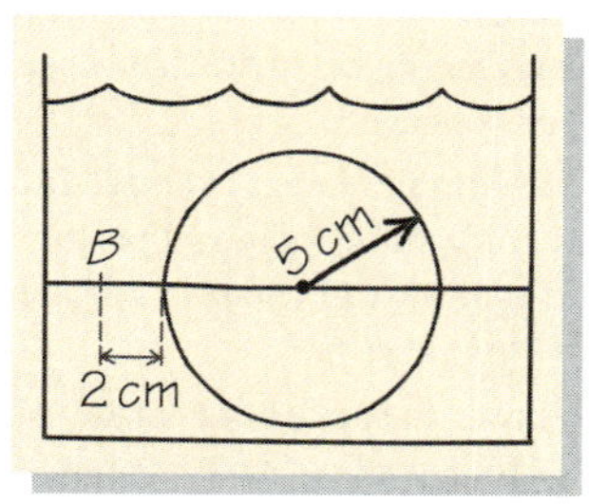

FIGURE 37–47 Problem 14.

15. (I) An object is placed 15 cm in air from the convex surface (radius of curvature 10 cm) of a very thick piece of glass (n = 1.5). Where is the image?

16. (I) The single refracting surface of a piece of glass in air has a radius of curvature R = 2.7 cm. A ray parallel to the axis of the curved piece of glass is bent toward the axis inside the piece of glass and crosses that axis at a point 6.5 cm into the glass. What is the index of refraction of the glass?

17. (I) A fish is located at a distance of 40 cm from the glass pane of an aquarium. How far from the glass does the fish appear to be located to an observer looking from the outside? (Use $n_{\text{water}} = 1.33$.)

18. (I) A small fish is cast into the center of a glass sphere of radius $R = 3$ cm and $n = 1.5$. Where will an observer see the fish? Where will the observer see a decorative background pattern painted on the back side of the sphere?

19. (II) A glass rod of refractive index $n = 1.6$ and diameter 1.6 cm has a hemispherical cap (Fig. 37–48). There is a fault in the glass 2.3 cm from the end. Can you see this fault if you look at the rod through the spherical cap? From about how far away should you look?

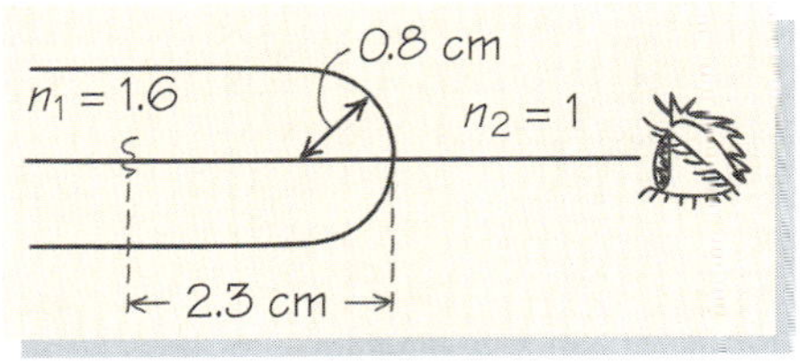

FIGURE 37–48 Problem 19.

20. (II) By applying Eq. (37–10), show that, if light is incident on a convex refracting surface with $n_2 > n_1$ (see Fig. 37–19), there is a critical distance s_c such that the image of an object closer than s_c will be virtual. Find s_c, and show by ray tracing that the virtual image when $s < s_c$ is upright and magnified.

21. (II) Consider the situation described in Problem 20. Use ray tracing to find the image when $s = s_c$, and when $s = 3s_c$.

22. (II) Consider a convex spherical boundary between two media with an upright object whose extreme point is at S, as in Fig. 37–19. Suppose that $n_2 < n_1$ rather than $n_2 > n_1$. Find the nature of the image (inverted or upright, virtual or real, reduced or magnified) by tracing rays from S. Is there a critical distance at which the nature of the image changes, as in Problem 20?

23. (II) A convex spherical boundary produces an image whose distance from the boundary surface is governed by Eq. (37–10). Suppose that $n_2 > n_1$. (a) Show that when an object is very far from the surface, the image is a distance $i = n_2 R/(n_2 - n_1)$ from the surface, and that the image is inverted, reduced, and real. (b) What is the distance s at which the image distance becomes infinite? (c) What is the position of the image for s just less than the critical value found in part (b)? Is it real? (d) As s continues to decrease, what happens to the position of the image?

24. (II) Consider a concave surface of radius of curvature R that separates two media with indices of refraction n_1 and n_2, where $n_2 > n_1$ (see Fig. 37–20). Find the distance s of an object for which the image, which is virtual, is superimposed on the object.

25. (III) In deriving Eq. (37–10), we took $n_1 < n_2$ in Fig. 37–21. Make a new drawing appropriate to $n_2 < n_1$ for a convex surface. Apply the same kind of reasoning, using small angles to show that the same algebraic formula applies whatever the relative sizes of n_2 and n_1.

26. (III) Show that Eq. (37–10) holds for a concave refracting surface with $n_2 > n_1$ by drawing a figure analogous to Fig. 37–21 and by making small-angle assumptions.

37–4 Thin Lenses

27. (I) The image of an object placed 24 cm away from a thin lens forms at a distance of 51 cm on the other side of the lens. (a) What is the focal length? (b) What type of lens is it? (c) Is the image real? upright? (d) What is the magnification?

28. (I) A double concave lens has radii of curvature of 1.5 cm and 2.3 cm. If the index of refraction of the lens is 1.68, what is the focal length?

29. (I) An apple is placed 15 cm in front of a diverging lens with a focal length of 22 cm. (a) Where is the image? (b) Is the image real? (c) upright? (d) What is the magnification?

30. (I) Find the condition that a single thin lens produces a real image, starting with a real source.

31. (II) We want to form an image of an insect magnified twofold by using a converging lens with a focal length of 25 cm. (a) Where should the object be placed for the image to be real? (b) Repeat part (a) for a virtual image.

32. (II) An object 3 cm high is placed on one side of a thin converging lens of focal length 25 cm. What are the location, size, and orientation of the image when the object is (a) 50 cm from the lens, (b) 30 cm from the lens, (c) 20 cm from the lens, (d) 5 cm from the lens?

33. (II) The two surfaces of a thin lens have radii of the same sign and magnitude. Show by ray tracing that the focal length of this lens is infinite. Is the image produced by this lens real or virtual?

34. (II) A thin converging lens forms an image of a distant mountain at a distance of 25 cm from the lens. (a) What is the focal length of the lens? (b) A pine cone is placed 100 cm from the lens. Describe the resulting image: its magnification and distance from the lens, and whether it is real or virtual, upright or inverted. (c) The lens glass has an index of refraction of 1.6. The lens is immersed in a clear liquid with an index of refraction of 1.4. What is its focal length in this medium?

35. (II) Consider the thin lenses shown in Figs. 37–26a through 37–26d. Suppose that in each case the magnitudes of the radii of curvature are $R_1 = 25$ cm and $R_2 = 60$ cm, and that $n = 1.55$. (a) Find the focal lengths for each of the four lenses, and use the sign of the focal lengths to obtain the locations of the image of a distant source. (b) In each case, is the image upright or inverted, real or virtual? (c) Calculate the magnification, M, from Eq. (37–6), and check that it is consistent with your results in part (b).

36. (II) An object is placed 25 cm to the right of each of the lenses of Problem 35. For each case, locate the image, state whether it is upright or inverted and real or virtual, and give the magnification.

37. (II) Repeat Problem 36 for an object placed 65 cm to the right of each of the lenses.

38. (II) Two thin lenses of focal length f_1 and f_2, respectively, are aligned along the same axis and placed very close together. Show that the focal length f of the combination is given by

$$\frac{1}{f} = \frac{1}{f_1} + \frac{1}{f_2}.$$

37–5 *Optical Instruments*

39. (I) The eyes of an elderly person have near points of 70 cm. What must the focal length of corrective lenses be in order for this person to read a book at a distance of 30 cm?

40. (I) A nearsighted person has near and far points of 14 cm and 34 cm, respectively. (The *far point* is the farthest point at which a person can see clearly.) (a) Determine the lens required for this person to be able to see clearly at infinity. (b) What does the lens correction of part (a) do to the near point? Can the person still easily read a book?

41. (I) You have a thin lens with $f = 9$ cm. If you want to see an insect magnified by a factor of 3, how close should you hold the glass to the insect? (Let the image be virtual.)

42. (I) What is the magnification of a telescope that has an objective lens with a focal length of 60 cm and an eyepiece with a focal length of 2.9 cm?

43. (I) You are trying to construct a compound microscope given two lenses with focal lengths $f_1 = 1.0$ cm and $f_2 = 4.0$ cm. How far apart should you place the lenses in order to obtain an angular magnification of 60?

44. (II) The two lenses of a telescope with magnification of 120× are separated by 70 cm. What are the focal lengths of the lenses?

45. (II) Calculate the angular magnification of the reflecting telescope shown in Fig. 37–39b.

46. (II) Galileo's original telescope had a convex objective and a concave eyepiece. The focal points of the two lenses coincided, as shown in Fig. 37–49. What is the angular magnification for a distant (but not infinitely far) object, and what is it for an infinitely far object?

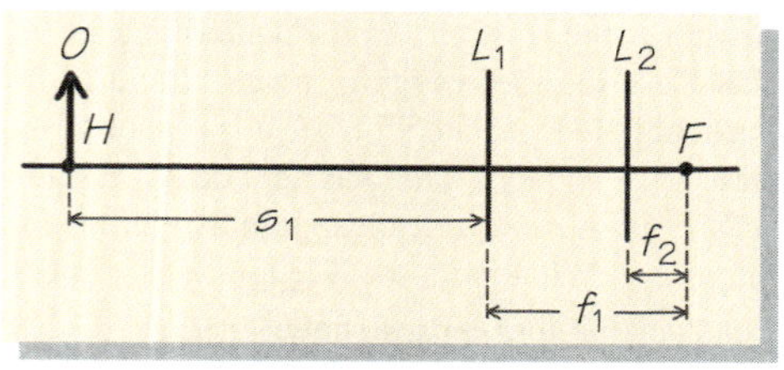

FIGURE 37–49 Problem 46.

*37–6 *Aberration*

47. (II) Consider a spherical mirror without making the paraxial approximation (Fig. 37–50). Show that when a ray parallel to the axis makes an angle θ with the radius R at the point of contact, then f, which in this case is the distance at which the ray crosses the axis, is given by

$$f = R\left(1 - \frac{1}{2\cos\theta}\right).$$

Show that for small angles, this formula reduces to $f = R/2$. Note that F is not the focal point here (there is no sharp focus), but only the point at which some particular ray crosses the axis.

48. (II) Use the result of Problem 47 to calculate the spread in values of f for a spherical mirror of radius 0.3 m and arc length 50 cm.

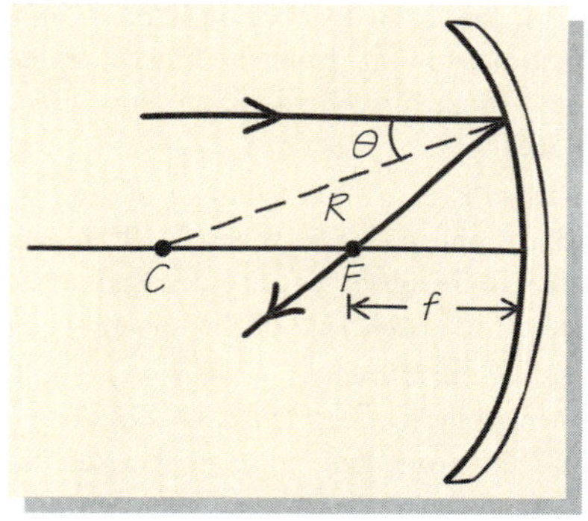

FIGURE 37–50 Problem 47.

49. (II) The index of refraction of optical glass used for a thin lens with $R_1 = +20.00$ cm and $R_2 = +28.75$ cm is $n = 1.48523$ for light of wavelength $\lambda = 587.6$ nm and $n = 1.48135$ for light of wavelength $\lambda = 768.2$ nm. What is the difference in the focal length for these two wavelengths?

General Problems

50. (II) Consider a circular concave mirror of focal length f and diameter d, where $f >> d$. This mirror's optical axis is aligned with the Sun. What is the area of the spot that contains the reflected rays as a function of the distance L from the mirror if $L < f$ (Fig. 37–51)? Sunlight has an intensity I as it arrives at the mirror. Find the intensity of the reflected rays as a function of L. Treat the Sun as a point source.

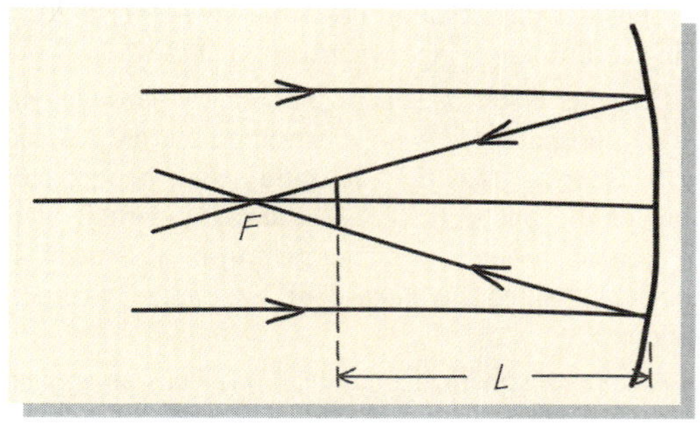

FIGURE 37–51 Problem 50.

51. (II) You are given a converging lens (Fig. 37–26a) with equal radii of curvature and a diverging lens (Fig. 37–26c) with the same radii of curvature as those of the converging lens. The lenses are made of material with $n = 1.50$, and the radii of curvature are all 35 cm. They are placed at opposite ends of a tube 15 cm long, and the nearer lens is 10 cm from an object (Fig. 37–52). What is the location of the image that results from the two refractions? Does it make a difference whether the converging or the diverging lens is closer to the object?

52. (II) Consider a 50-cm-long cylinder of glass in air, with $n = 1.6$, like the cylinder shown in Fig. 37–23. The two ends are shaped into sections of spheres; each has radius 20 cm. A small object is placed perpendicular to the optical axis at a distance of 15 cm from one of the spherical surfaces. (a) Find the location of the object's image due to refraction at surface 1. (b)

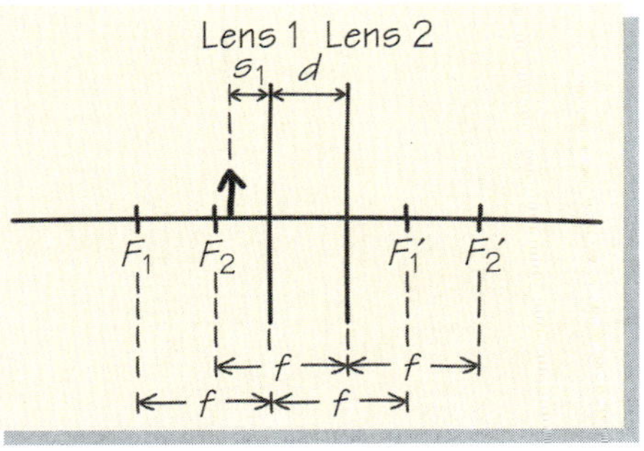

FIGURE 37–52 Problem 51.

Let this image be the object for surface 2, and find the location of *its* image as light passes through surface 2. (c) Use ray tracing to determine if the final image is upright or inverted.

53. (II) Two concave mirrors M_1 and M_2 face each other. They have respective radii of curvature of 32 cm and 14 cm and are separated by 50 cm. A lightbulb is placed on the optical axis 7 cm from M_1. (a) Where is the image of the bulb formed by M_1? Draw the system. (b) The image of the lightbulb formed by M_1 can in turn form an image as the result of reflections from M_2. Construct this second image by ray tracing, starting from the source lightbulb.

54. (II) A lens, made of glass with $n = 1.5$, has the configuration shown in Fig. 37–53, and a candle is placed 50 cm from surface 1. The lens cannot be thought of as thin; it has a thickness of 5 cm. (a) Where is the image made by surface 1? Is it inverted? What is the magnification? (b) By using the image made by surface 1 as an object for surface 2, find the final image's location relative to the candle as well as the magnification of the image. Is it inverted or upright?

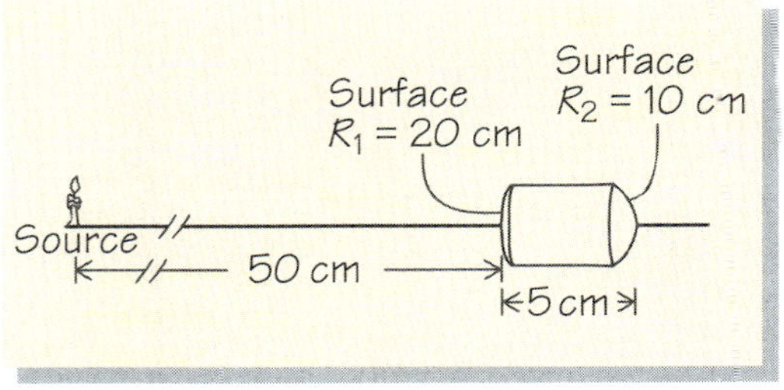

FIGURE 37–53 Problem 54.

55. (II) A thin lens with focal length f_1 is placed a distance d in front of a concave mirror with focal length f_2. What is the focal length of the combination?

56. (II) Consider the sphere of a glass with $n = 2$ in Fig. 37–54. Any incoming ray is parallel to an axis through the middle of the sphere and will be refracted, striking the rear surface of the sphere at the axis. Demonstrate that this holds true for paraxial rays. If the back surface is painted with a reflecting material, symmetry shows that the ray will come back out in the opposite direction. Tiny spheres of this type are used for highway reflectors.

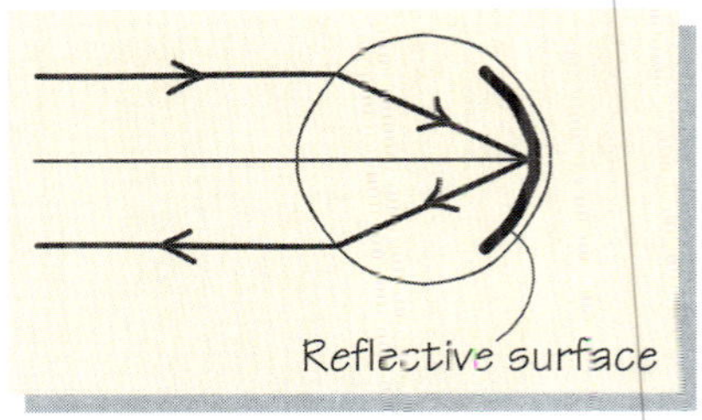

FIGURE 37–54 Problem 56.

57. (III) Rays of light strike a spherical glass surface parallel to the optical axis (Fig. 37–55). The incoming ray makes an angle θ with the normal to the surface. Show that the rays will cross the optical axis at a distance $d = R/(\sqrt{n^2 - \sin^2\theta} - \cos\theta)$ beyond the center of the sphere, where n is the index of refraction of the glass. To what does this expression reduce for small angles?

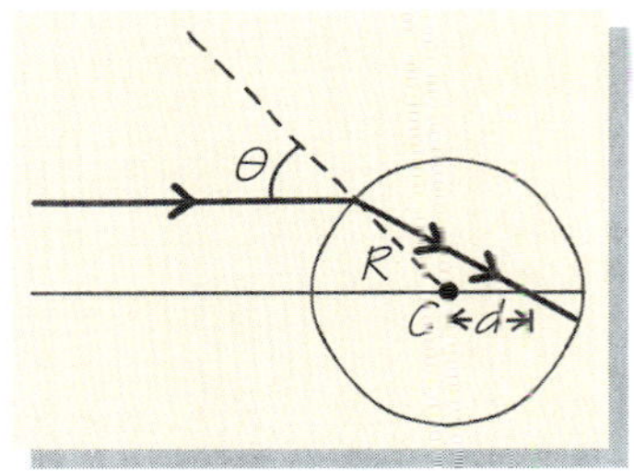

FIGURE 37–55 Problem 57.

58. (III) The index of refraction of a particular type of glass varies from 1.615 (for blue light) to 1.596 (for red light). Use the result of Problem 57 to calculate the color spread on the axis for light that strikes a hemispherical cap at the end of a glass rod, at an angle $\theta = 0.6$ rad. Take the radius of curvature of the sphere to be $R = 0.50$ cm. What is the spread for paraxial rays?

59. (III) An optical system contains a thin lens, $n = 1.4$, with positive curvature of radius $R_1 = 25$ cm for surface 1 and negative curvature of radius $R_2 = -25$ cm for surface 2. This lens collects light from the right side. Where to the left of the lens should a flat plate of thickness t of the same glass be placed, and how thick should it be, if you want the light that radiates from a distant object to be focused on a screen 35 cm to the left of the lens?

CHAPTER

38

The brilliant colors of the ruby-throated hummingbird feathers are due not to pigmentation, but to interference of the light reflected from them.

Interference

In Chapters 36 and 37, we emphasized the geometrical properties of light. We discussed reflection and refraction by treating light in terms of linear rays, and did not address the fact that light is a wave phenomenon. However, if we look more carefully at the behavior of light when obstacles or holes have dimensions comparable to the wavelength of the light, geometric optics is inadequate; the wave nature of light becomes important. Geometric optics cannot explain the colors observed in thin-walled soap bubbles or oil slicks. Similarly, if we look closely at shadows, we find that they are not completely sharp, in contradiction to the predictions of geometric optics. These phenomena are due to interference and diffraction, the subjects of this chapter and Chapter 39. *Physical optics*, which takes into account the wave nature of light, can explain a wider range of observations than can geometric optics.

38–1 YOUNG'S DOUBLE-SLIT EXPERIMENT

When two or more harmonic waves superpose, they interfere—whether they are water waves, waves on a string, or light waves. Interference between two light waves occurs because *the electric (or magnetic) fields of the two waves add vectorially*. The resulting electromagnetic wave is a wave with a new value of the electric field. Consider the superposition of two light waves from different sources at particular points in space that are propagating along the x-axis at a given time (Fig. 38–1). Where

The wave phenomena discussed in Chapters 14 and 15 apply to all waves, including those of light.

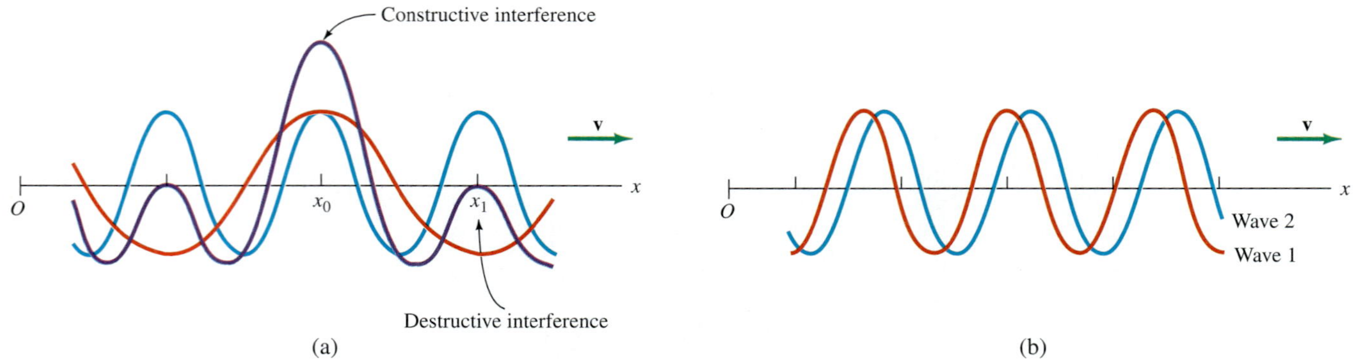

FIGURE 38–1 (a) Constructive interference between two waves occurs at point x_0 when the peaks coincide. Destructive interference between two waves occurs at point x_1 when their amplitudes cancel. (b) Two coherent waves have the same wavelength and a constant phase difference.

Waves can produce an interference pattern only if they are coherent.

the two waves add to a wave with a larger magnitude, we say that the waves interfere constructively, as they do at point x_0 in Fig. 38–1a. The two waves interfere destructively where they cancel one another, as they do at point x_1. If the two waves have similar wavelengths, they may interfere either constructively or destructively over a wide region of space.

For light waves from two sources to produce an interference pattern in space, there must be a definite relation between their respective wavelengths and phases *at their respective sources*. The waves must be *coherent*. The two light waves shown in Fig. 38–1b have exactly the same wavelength and a constant phase difference. These very long wave trains are the type of waves emitted by a laser, a source of coherent (and monochromatic) light, and it is easy to demonstrate in the classroom the interference pattern produced by laser light (see Chapter 39). If the waves emitted at one or both sources consisted of a mixture of waves of different wavelengths and phases, then there would be no interference pattern. An incandescent lightbulb and a candle flame produce light from many independent atomic sources at different times and places within the filament or flame. Such light is said to be *incoherent*.

Thomas Young was the first to observe interference phenomena in light, in 1802; however, he did not have a laser. How did he produce an interference pattern? Let's start with the incoherent light from a lightbulb. We might first pass it through a prism and choose a single color. We would then be dealing with monochromatic light, which contains only a narrow range of wavelengths. However, monochromatic light is still incoherent because it consists of many successive and overlapping bursts of different phases. The bursts of light from individual atomic sources within a lightbulb are a quantum phenomenon. These bursts last for a time on the order of $\tau \simeq 10^{-8}$ s, and the length of the resulting individual wave trains is therefore $c\tau$, or several meters. We can now produce coherent light at two sources by illuminating a single aperture (a slit or a hole), S, with our monochromatic source. The aperture must be so small that only one burst of light enters at a time. In this way, there are no phase differences among light from different bursts that may have entered different parts of the aperture. The single wave train of light that passes through forms a cylindrical wave that can illuminate two other apertures, S_1 and S_2 (Fig. 38–2). *These two apertures are two sources of coherent light.* If S_1 and S_2 are equidistant from S, the light from S travels the same distance to reach S_1 and S_2, and the light is in phase as it passes through the two apertures (Fig. 38–2a). If S_1 and S_2 are not equidistant from S, the light waves that pass through them are still coherent because they have a definite phase difference (Fig. 38–2b). Finally, although the interference pattern produced by a given burst lasts only for 10^{-8} s, the next burst, which has the same wavelength, produces exactly the same pattern as the first burst. The pattern is therefore a stable and observable one.

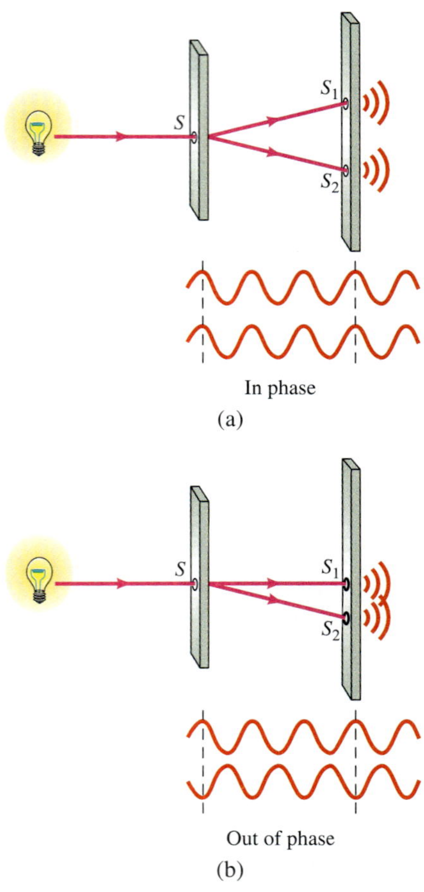

FIGURE 38–2 The light waves that pass through slits S_1 and S_2 are coherent. (a) These waves are also in phase, because the light travels the same path length from slit S to S_1 and S_2. (b) When the path lengths are different, there is a phase difference.

The Two-Source Interference Pattern

Let's review the spatial interference pattern produced when light from two sources of coherent waves interfere. Waves that pass through the rectangular, vertical slit S form cylindrical waves, and the same is true for the light that subsequently passes through slits S_1 and S_2 in Fig. 38–3a. Figure 38–3b is a view from above. If S_1 and S_2 are the same distance from S, then waves of the same frequency and phase emanate from S_1 and S_2 in the form of spreading circles from the source slits. The circles represent the crests (or troughs) of the spreading waves. Where the crests (or troughs) overlap, the waves interfere constructively. The pattern of these overlap points is apparent: As the waves progress, the positions of these points advance and form lines. There is constructive interference—wave motion with increased amplitude—*all along* these lines; therefore, the places where the lines intersect the screen are bright. There is destructive interference in between the regions of constructive interference, and the screen is dark. The result is a series of bright and dark areas on the screen.

The interference pattern of waves from two coherent sources was treated in Chapter 15.

Let's investigate this double-slit configuration more closely. Consider the geometry shown in Fig. 38–4. Along ray 1 and ray 2, the waves travel distances L_1 and L_2, respectively, to arrive at point P on the screen. Because the rays travel different distances, they may no longer be in phase at P, although they were in phase at the sources S_1 and S_2. Whether they are in phase or not depends on the *path-length difference* $\Delta L = L_2 - L_1$. The waves arrive in phase if ΔL is zero or if ΔL is an integral multiple of one wavelength. The electric fields can still interfere constructively even if the path lengths differ by many (integral) wavelengths. Similarly, if the peak of one wave arrives half a wavelength behind the peak of the other, the maximum electric field of one wave will occur at the same place in space as the minimum electric field of the other wave. The waves therefore cancel if ΔL is a multiple of one-half wavelength (180° out of phase). The interference is constructive when the waves add together and destructive when the waves cancel one another:

for constructive interference: $$\Delta L = n\lambda, \quad n = 0, \pm 1, \pm 2, \ldots; \quad (38\text{–}1a)$$

for destructive interference: $$\Delta L = \left(n + \frac{1}{2}\right)\lambda, \quad n = 0, \pm 1, \pm 2, \ldots. \quad (38\text{–}1b)$$

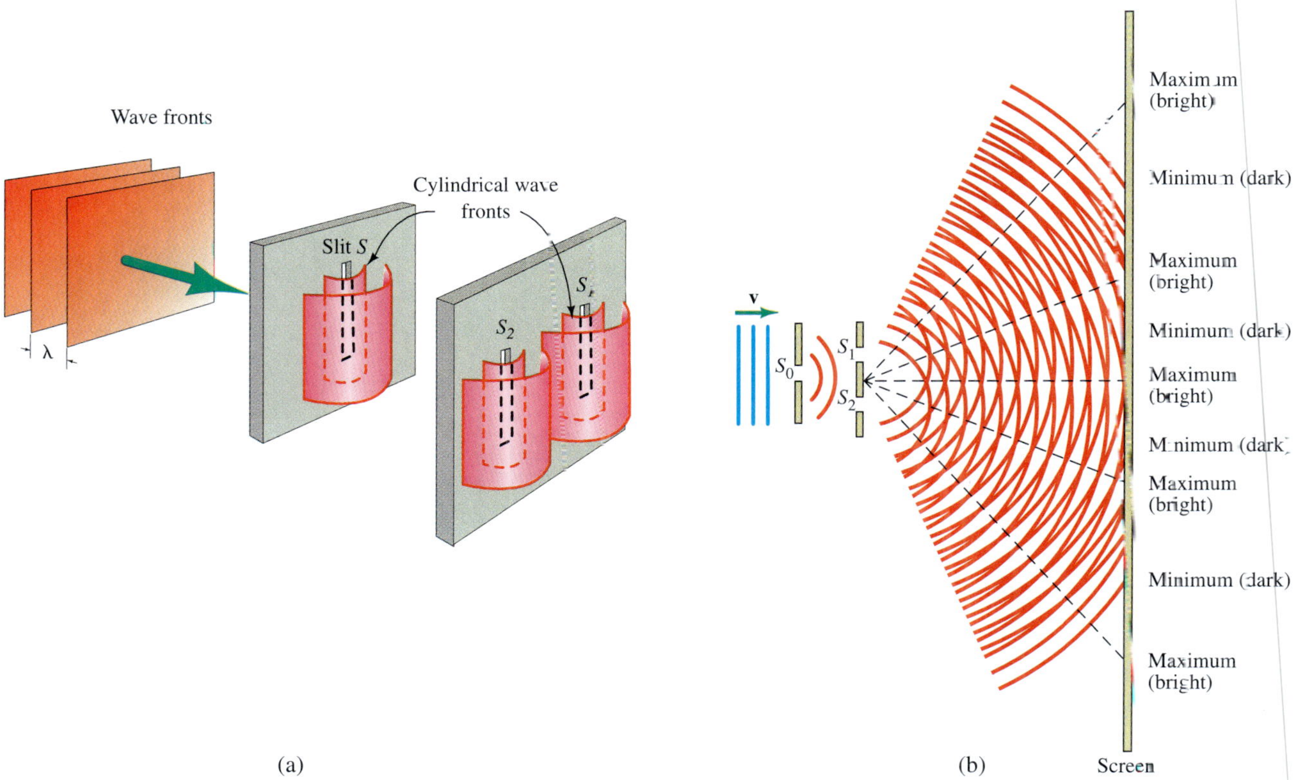

FIGURE 38–3 (a) Plane waves perpendicularly incident on thin slit S produce a series of cylindrical waves. When that series of waves reaches slits S_1 and S_2, the resulting cylindrical waves are coherent. (b) The view from above. Constructive interference occurs everywhere along the directions where the concentric circles, representing the crests of the spreading waves, overlap, because the waves are in phase along these directions. Alternating bright and dark places will be observed on a screen parallel to the walls.

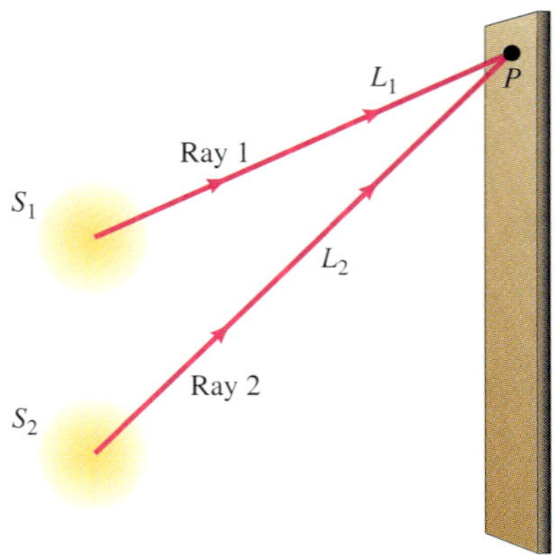

FIGURE 38–4 Light waves, indicated by rays 1 and 2, may not be in phase at point P despite being in phase at their sources, S_1 and S_2.

The two-source interference pattern

The resulting series of bright and dark lines on the screen is indicated in Fig. 38–3 and shown in Fig. 38–5.

The geometry shown in Fig. 38–6 determines conditions for constructive and destructive interference. We assume that the distance R to the screen is much greater than the distance d between the two slits. In this case, we must worry only about the single angle θ made by the line from the center point between the slits to P. From Fig. 38–6, we see that the angle formed by S_2S_1K is also θ. Thus

$$\Delta L = d \sin \theta. \tag{38–2}$$

According to Eqs. (38–1a) and (38–1b), maxima (bright regions) and minima (dark regions) thus occur on the screen for angles given by

$$\text{for constructive interference:} \quad \sin \theta = n\frac{\lambda}{d}, \qquad n = 0, \pm 1, \pm 2, \ldots; \tag{38–3a}$$

$$\text{for destructive interference:} \quad \sin \theta = \left(n + \frac{1}{2}\right)\frac{\lambda}{d}. \qquad n = 0, \pm 1, \pm 2, \ldots. \tag{38–3b}$$

The point that is aligned with the sources ($\theta = 0$) is a maximum ($n = 0$). Alternating minima and maxima lie on either side of this centerline. The value of n that labels the maxima is known as the *order*. The maxima on either side of the central maximum (the zeroth order) are first-order maxima ($n = \pm 1$). If θ is small, so that $\sin \theta \simeq \theta$, the maxima and minima are equally spaced in θ. This experiment shows conclusively that light behaves as a wave. Geometric optics cannot explain the result shown in Fig. 38–5.

EXAMPLE 38–1 In the double-slit experiment shown in Fig. 38–6, y is the distance from the center maximum along the screen. Find the positions of the maxima as a function of y. If the slit-to-screen distance $R = 3$ m, the slit separation $d = 0.2$ mm, and light comes from a helium–neon laser ($\lambda = 633$ nm) far away, determine y for the ninth-order maximum.

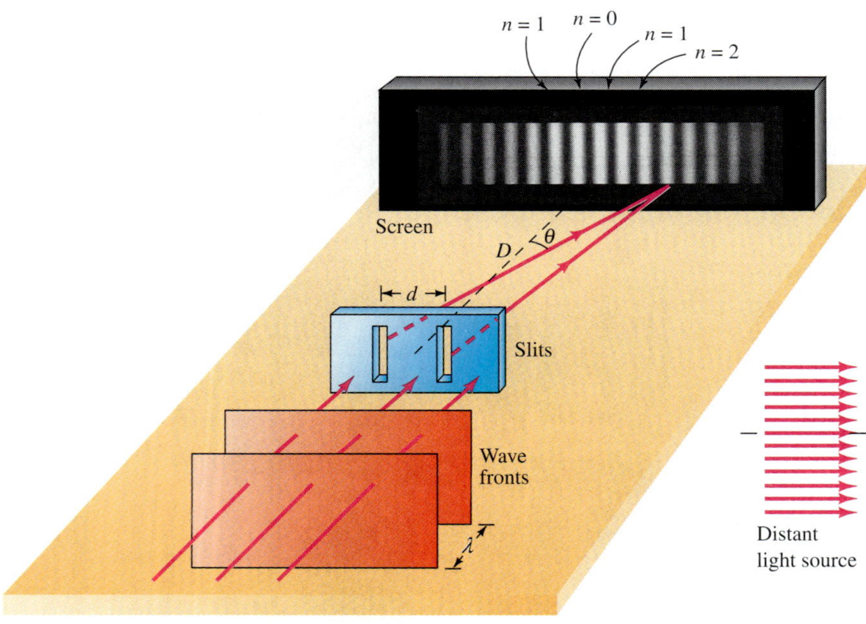

FIGURE 38–5 The interference pattern produced by double vertical slits is a series of alternating bright and dark vertical lines on a screen. The fall-off of the intensity toward the edge of the figure is a single-slit effect (see Chapter 39).

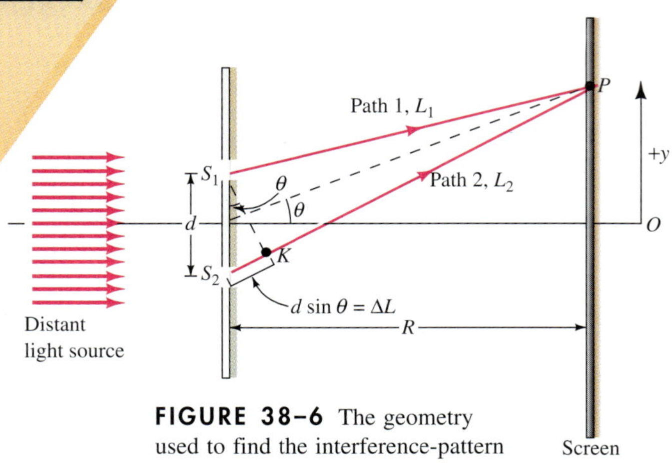

FIGURE 38–6 The geometry used to find the interference-pattern conditions for the light that reaches point P. The path-length difference $\Delta L = d \sin \theta$.

Solution: Equation (38–3a) gives the angles for which maxima occur. From Fig. 38–6, the distance y is given by

$$y = R \tan \theta.$$

If $R >> y$, $\tan \theta \simeq \sin \theta$, and we can insert the value of $\sin \theta$ from Eq. (38–3a) for the maxima. The maxima are then located at

$$y = \frac{n\lambda R}{d}. \tag{38–4}$$

For the ninth order, we have $n = 9$, and $y = 9\lambda R/d$. If we substitute the given values for λ, R, and d, we get for the ninth-order maximum

$$y = \frac{9(633 \times 10^{-9}\ \text{m})(3\ \text{m})}{0.2 \times 10^{-3}\ \text{m}} = 8.5\ \text{cm}.$$

The distance between each of the maxima is therefore about 1 cm.

Note that such measurements could be used to find λ: If we measure y, R, and d, we can solve Eq. (38–4) for λ.

The double-slit pattern exhibits a feature characteristic of all wave phenomena. The observed interference effects depend on the ratio λ/d; *as this ratio increases, the angular separation of the interference pattern increases*. For example, the separation Δy between maxima on the screen is, from Eq. (38–4), $\Delta y = R\lambda/d$. Thus, to see the interference pattern in a water ripple tank where the wavelength is measured in centimeters, the slit separations should be on the order of centimeters.

The angular separation of the pattern of maxima and minima in interference phenomena increases as λ/d increases.

38–2 INTENSITY IN THE DOUBLE-SLIT EXPERIMENT

The previous discussion relied on geometrical arguments to determine the angles for which maxima and minima can be obtained. We now turn our attention to the *intensity* of the light that reaches the screen. The intensity (or brightness, for light) measures the energy delivered by a wave per unit time per unit area.

Light intensity was discussed in Chapter 36.

The energy in a given mechanical wave or superposition of waves is proportional to the displacement squared. For light, the quantity that plays the role of displacement is the electric (or magnetic) field. The intensity of a light wave (the energy delivered by the wave per unit time per unit area) is the time average of the Poynting vector (introduced in Chapter 35), which is proportional to the product of electric and magnetic field vectors in the wave. Because the magnetic field is itself proportional to the electric field in an electromagnetic wave, the intensity is proportional to the electric field squared ($I \propto E^2$). In order to find the intensity of a collection of waves, we add the electric fields of all the waves and square the sum of the net field. For example, with two sources of equal intensity I_0, the maximum electric field is twice the electric field E_0 from each source, so

$$\frac{I_{max}}{I_0} = \frac{(E_0 + E_0)^2}{E_0^2} = 4,$$

where I_{max} is the maximum intensity. The maximum intensity $I_{max} = 4I_0$. Similarly, the minimum intensity occurs when the electric fields exactly cancel, and $I_{min} = 0$.

The simple argument just given, which is based on energy, is so useful that we shall develop it further. The intensity at any point P on the screen in Fig. 38–6 is proportional to the net Poynting vector, which is in turn proportional to the square of the net electric field. The net instantaneous electric field $\mathbf{E}_{net}$ at P is the sum of the instantaneous electric fields of the light waves emitted at the two sources: $\mathbf{E}_{net} = \mathbf{E}_1 + \mathbf{E}_2$. The net Poynting vector therefore has magnitude $S \propto (\mathbf{E}_1 + \mathbf{E}_2)^2 = E_1^2 + 2\mathbf{E}_1 \cdot \mathbf{E}_2 + E_2^2$. But light waves oscillate rapidly, and thus it is not the Poynting vector that is of interest; rather, it is the *time average* of the Poynting vector (that is, the intensity, I) at P. This is where the coherence of the light is important.

For incoherent sources, the intensities of the individual sources add.

If we denote time averages with triangle brackets, then

$$I_{\text{net}} \propto \langle E_1^2 \rangle + 2\langle \mathbf{E}_1 \cdot \mathbf{E}_2 \rangle + \langle E_2^2 \rangle. \tag{38–5}$$

For incoherent light, there is no correlation—no definite phase relation—between the electric fields from the two sources. One moment the sources have one relative phase, the next moment the relative phase is different, and *the term* $\langle \mathbf{E}_1 \cdot \mathbf{E}_2 \rangle$ *is zero*. Thus

$$I_{\text{incoh}} = I_1 + I_2. \tag{38–6}$$

The Intensity Pattern for Two Coherent Sources

For coherent waves, $\langle \mathbf{E}_1 \cdot \mathbf{E}_2 \rangle$ in Eq. (38–5) is not zero. If, at a given time, there is constructive interference at point P, where $\mathbf{E}_1 = \mathbf{E}_2$, the constructive interference will persist because the waves are coherent. Similarly, if there is destructive interference at a given time, where $\mathbf{E}_1 = -\mathbf{E}_2$, it also persists through later times. For destructive interference, $\langle \mathbf{E}_1 \cdot \mathbf{E}_2 \rangle \propto -I_1$, and Eq. (38–5) gives $I_{\text{net}} = I_1 - 2I_1 + I_1 = 0$.

Suppose then that the electric fields of the light waves from our coherent sources S_1 and S_2 at a single point P in space are identically oriented and have magnitudes

$$E_1 = E_0 \sin(\omega t), \tag{38–7a}$$

$$E_2 = E_0 \sin(\omega t + \phi). \tag{38–7b}$$

The phase difference ϕ for E_2 results from the path-length difference between the waves. If $\phi = 2\pi n$, where n is an integer, the fields are identical, and there is constructive interference. This phase difference of $2\pi n$ corresponds to a path-length difference of $\Delta L = n\lambda$. The ratio of ϕ to $2\pi n$ is the same as ΔL to $n\lambda$, so we have

$$\frac{\phi}{2\pi n} = \frac{\Delta L}{n\lambda};$$

$$\frac{\phi}{2\pi} = \frac{\Delta L}{\lambda}. \tag{38–8}$$

For the distant-screen geometry of Fig. 38–6, we can use Eq. (38–2) to transform Eq. (38–8) to

$$\phi = 2\pi \frac{\Delta L}{\lambda} = \frac{2\pi}{\lambda} d \sin \theta. \tag{38–9}$$

Now, the net electric field at P has magnitude

$$E_{\text{net}} = E_1 + E_2 = E_0[\sin(\omega t) + \sin(\omega t + \phi)].$$

If we apply the equation $\sin \theta_1 + \sin \theta_2 = 2 \cos[(\theta_1 - \theta_2)/2] \sin[(\theta_1 + \theta_2)/2]$(see Appendix IV–4), with $\theta_1 = \omega t$ and $\theta_2 = \omega t + \phi$, we find

$$E_{\text{net}} = 2E_0 \cos\left(\frac{\phi}{2}\right) \sin\left[\omega t + \left(\frac{\phi}{2}\right)\right]. \tag{38–10}$$

The Poynting vectors $\mathbf{S}_1$ and $\mathbf{S}_2$ of the light from the individual sources have magnitudes

$$S_1 \propto E_1^2 = E_0^2[\sin(\omega t)]^2 \quad \text{and} \quad S_2 \propto E_2^2 = E_0^2 [\sin(\omega t + \phi)]^2, \tag{38–11}$$

respectively, whereas the net Poynting vector at P has magnitude

$$S_{\text{net}} \propto E_{\text{net}}^2 = 4E_0^2 \cos^2\left(\frac{\phi}{2}\right) \sin^2\left[\omega t + \left(\frac{\phi}{2}\right)\right]. \tag{38–12}$$

To find the intensities (the time averages of the Poynting vectors), we need know only that the time average of $\sin^2(at + b) = \frac{1}{2}$. If we write the individual intensities as $I_0 \propto E_0^2/2$, then the net intensity from the two sources in terms of I_0 is

$$I_{\text{net}} = 4I_0 \cos^2\left(\frac{\phi}{2}\right). \tag{38–13}$$

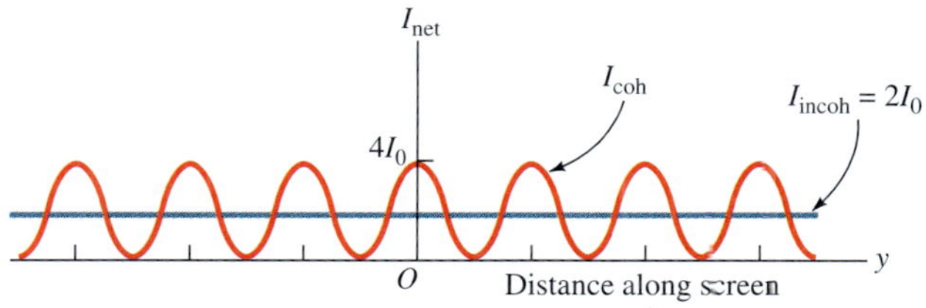

FIGURE 38–7 The net intensity of light from the double slit as a function of the distance from the center point on the screen ($y \simeq \sin\theta$). Compare the results for coherence and incoherence. The same amount of light energy reaches the screen in both cases but in the coherent case, it occurs in peaks and valleys. We have assumed that each slit is infinitely narrow; that is why the fall-off visible at the edges of Fig. 38–5 is not visible here.

When the phase ϕ in Eq. (38–7b) is related to the path-length difference by Eq. (38–8), then Eq. (38–13) for the intensity on the distant screen becomes

$$I_{net} = 4I_0 \cos^2\left(\frac{\pi d}{\lambda}\sin\theta\right). \tag{38–14}$$

The intensity of the two-source interference pattern

This is the expression for the intensity in Young's classic double-slit experiment.

The maxima and minima occur at the angles specified by Eq. (38–3). Figure 38–7 is a plot of the intensity at the screen as a function of $\sin\theta$. This figure also serves as a plot of the intensity as a function of the distance y from the center maximum along the screen: For small θ, $y \simeq R\sin\theta$, where R is the distance from the screen. If the apertures are narrow vertical slits, then the bright maxima on the screen are vertical lines called *fringes*.

We have also plotted in Fig. 38–7 the intensity $2I_0$ that would be present on the screen if the light sources were incoherent. The result for incoherence is constant and shows no interfering maxima and minima. However, *averaged over the entire screen*, the energy reaching the screen is exactly the same in the two cases, as required by the conservation of energy. To average the energy that reaches the screen in the case for coherence, we need only use the fact that the average of the cosine-squared factor in Eq. (38–14) is $\frac{1}{2}$, and $4I_0(\frac{1}{2}) = 2I_0$. The energy emitted at each source is the same whether the light from these sources is coherent or incoherent, and the total energy arriving at the screen must also be the same. The energy is spread evenly over the screen when the sources are incoherent, whereas it is distributed in peaks and valleys when the sources are coherent.

EXAMPLE 38–2 Suppose you live at point H, 20 km from a vertical radio dipole antenna that broadcasts at a frequency of 1100 kHz from point B (Fig. 38–8). How well your radio picks up the signal is a direct function of the intensity of the signal. A second antenna is constructed at point A, located $d = 100$ m from the first. The second antenna broadcasts a signal identical to the first one. Find the new intensity at your radio in terms of the old intensity Is your signal improved?

Solution: The situation is like the double-slit situation for light because there are two sources of coherent radiation. The only difference is that, in this case, the relevant wavelengths are much longer. Because your distance from the antennas is much greater than their separation distance, the geometrical approximations we used in discussing the double-slit experiment apply. These approximations tell us that the difference in distances between you and the two antennas is $R_A - R_B = d\cos\theta$ (Fig. 38–8). We must find the net electric field at point H. The electric fields, which are parallel to the antennas, are vertical (perpendicular to the page). Suppose that the first antenna (at point B) broadcasts a signal whose z-component at point H is $E_B = E_0\cos(\omega t)$ and that the original intensity is $I_0 = CE_0^2$, where C is some constant. The second antenna's electric field takes the same form except that there is a phase difference due to the fact that the antenna is a distance $R_A - R_B$ farther away:

$$E_A = E_0\cos(\omega t + \phi),$$

where $\phi = 2\pi(R_A - R_B)/\lambda = (2\pi d\cos\theta)/\lambda$. The net field at point H is then

$$E_{net} = E_A + E_B = E_0[\cos(\omega t) + \cos(\omega t + \phi)].$$

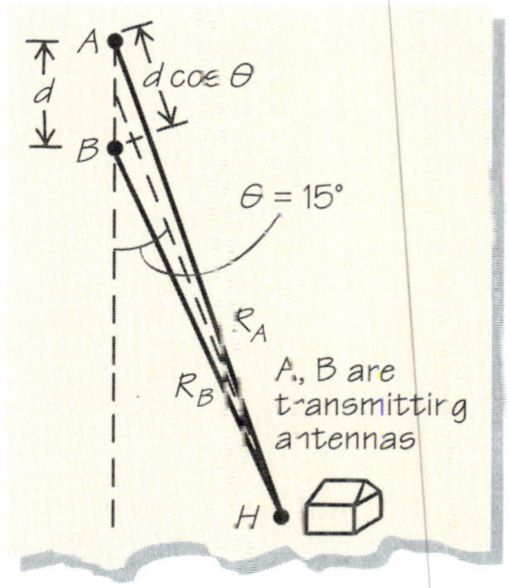

FIGURE 38–8 Example 38–2. R_A and R_B are the distances between your home and two antennas at points A and B, respectively.

This sum is similar to the one we needed to find Eq. (38–10), $E_{net} = 2E_0 \cos(\phi/2) \times \sin[\omega t + (\phi/2)]$. The net intensity is given similarly in Eq. (38–13),

$$I_{net} = CE_{net}^2 = 4CE_0^2 \cos^2\left(\frac{\phi}{2}\right) = 4I_0 \cos^2\left(\frac{\phi}{2}\right).$$

It remains to calculate the factor $\cos^2(\phi/2)$ and see by what factor the intensity is changed. We require that

$$\cos\left(\frac{\phi}{2}\right) = \cos\left(\frac{2\pi d \cos\theta}{2\lambda}\right) = \cos\left(\frac{\pi d \cos\theta}{\lambda}\right).$$

We have $d = 100$ m and $\theta = 15°$. The wavelength λ comes from the frequency $f = 1100$ kHz $= 1.1 \times 10^6$ Hz. We get $\lambda = c/f = (3.0 \times 10^8\ \text{m/s})/(1.1 \times 10^6\ \text{s}^{-1}) = 273$ m. Thus

$$\cos\left(\frac{\phi}{2}\right) = \cos\left[\frac{\pi(100\ \text{m})(\cos 15°)}{273\ \text{m}}\right] = \cos(1.11\ \text{rad}) = 0.44.$$

The net intensity $I_{net} = 4I_0 \cos^2(\phi/2)$ is a factor $4[\cos^2(\phi/2)] = 4(0.44)^2 = 0.79$ times the original intensity I_0. The signal you receive at point H has actually become weaker because there is partial destructive interference at your radio between the signals of the two antennas.

38–3 INTERFERENCE FROM REFLECTION

A ray may be partly reflected from and partly transmitted through a surface. If the transmitted ray is subsequently reflected from a second surface, then the two reflected rays may interfere with each other. The two reflected rays are automatically coherent because they originate from the same ray.

Interference Fringes from the Space Between Two Glass Plates

Let's look at two flat glass plates, each with one very flat surface. They have been placed together with a spacer at one side, so that the flat surfaces touch along one side and are separated by a small distance d at the other (Fig. 38–9). When these plates are illuminated from above with monochromatic light, a series of alternating light and dark bands is seen, starting with a black band along the side where the plates touch.

To approach this phenomenon more precisely, let's take the plates to have index of refraction n, the space between the plates to be air ($n = 1$), and the wavelength of the light to be λ. We can think of the light as consisting of rays perpendicularly inci-

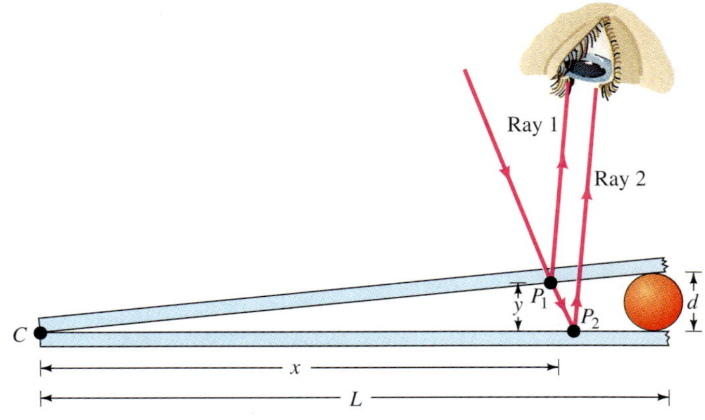

FIGURE 38–9 Two pieces of glass are placed together with a spacer (exaggerated in size) at one edge. The adjacent surfaces are flat on the scale of the wavelength of light.

dent on the plates. These rays are *perpendicular to the wave fronts*; when we speak of rays interfering, we are referring to the interference of the waves that form those fronts. In this situation, the rays can reflect and refract to form new rays in various ways, but one possibility leads to the observed interference here: An incident ray in Fig. 38–9 passes almost vertically through the top piece of glass and is partly reflected (ray 1) and partly refracted (ray 2) at point P_1, on the bottom surface of the top plate. Ray 2 then continues through the air gap and is reflected at point P_2, at the top of the bottom plate. To a good approximation, ray 2 travels a distance $2P_1P_2$ in air farther than ray 1 because the incident ray is almost vertical. Rays 1 and 2, exiting at the top, will interfere; the result is the series of bright and dark bands.

Now, as the points P_1 and P_2 approach the side where the plates touch (point C), the path-length difference distance $2P_1P_2$ decreases. As we get very close to point C, we might expect to find a *bright* region caused by the constructive interference of rays 1 and 2 traveling almost the same distance. Conversely, along the line where the plates touch, there is no gap between them, so there is no boundary; we might then expect a *dark* region at the touching edge! The second expectation is the correct one: There is a dark line at the edge where the plates touch. From the point of view of rays 1 and 2, the dark region there is an indication of destructive interference. *Destructive interference occurs because one of the rays undergoes a 180° phase change during reflection.* The light undergoes a phase change of 180° when it is reflected at P_2, whereas the light undergoes no such change when it is reflected at P_1.

In Chapter 15, we discussed a corresponding phenomenon in the reflection of one-dimensional waves on strings. Suppose that two strings of different densities are connected (Fig. 38–10). The connection point forms a boundary at which there may be reflection and transmission. We discussed experiments in Chapter 15 that showed that if the wave speed on the side from which the wave comes is greater than the wave speed on the far side of the boundary, then the reflected wave is inverted, corresponding to a phase shift of 180° if the wave is harmonic. We learned also that whether the pulse is inverted or not depends on certain boundary conditions—in this case, the densities of the strings.

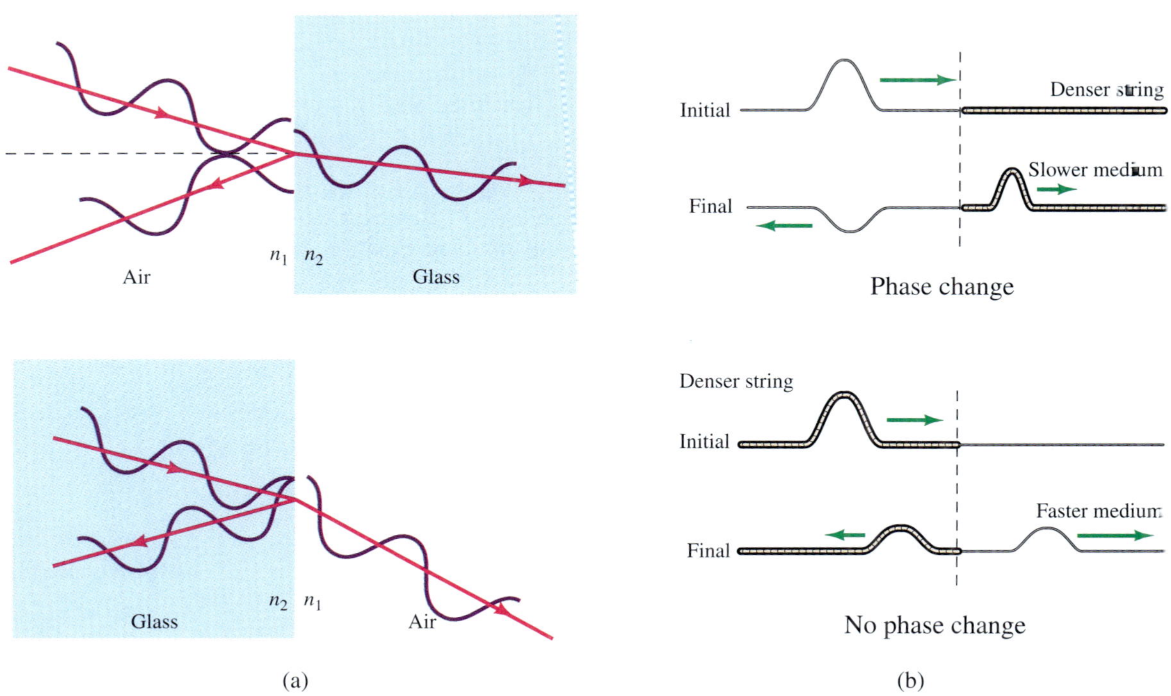

FIGURE 38–10 (a) The phase change of light upon reflection is similar to the inversion that occurs when a pulse that moves along a string meets a denser string. For light, a phase change of 180° occurs when the second medium has a higher index of refraction. (b) No phase change occurs when the string encounters a less dense string or, for light, when the second medium at the reflection boundary has a lower index of refraction.

Maxwell's equations determine what happens to the electric and magnetic fields of an electromagnetic wave at a boundary between two dielectric media through a set of boundary conditions. For most situations of physical interest to us, these boundary conditions depend on the relative speeds of the waves in the media. The electric field changes sign or does not change sign according to whether the wave speed in the medium on the far side of the boundary is greater than or less than the wave speed in the medium from which the wave comes. The boundary conditions imply no such change of sign for the magnetic field of the wave. When the electric field changes sign but the magnetic field does not, the result is a 180° phase change in the reflected electromagnetic wave. The speed of an electromagnetic wave is inversely proportional to the index of refraction: $v = c/n$. The result for the phase change can therefore be stated as:

How phases change when light waves reflect from boundaries between different media

The phase of an electromagnetic wave that moves from a medium of index of refraction n_1 toward a medium with index of refraction n_2 will change by 180° upon reflection when $n_2 > n_1$ and will not change when $n_2 < n_1$.

Now we can understand the interference pattern for the plates. The phase change of 180° occurs only for the reflection at P_2 because ray 2 goes from air ($n = 1$) to glass ($n > 1$). Because a phase shift of π rad (180°) corresponds to a shift of one-half wavelength, the condition for constructive interference becomes

The condition for constructive interference for an air space between plates

for constructive interference: $$\Delta L = 2P_1P_2 = \left(m + \frac{1}{2}\right)\lambda,\ m = 0, \pm 1, \pm 2, \ldots . \quad (38\text{–}15)$$

How the path-length difference $2P_1P_2$ varies with the distance from the touching edges is a geometrical question. The alternate bright and dark bands correspond to plate separation distances for which a monochromatic wave is in alternate constructive and destructive interference. In particular, the touching edge, where $\Delta L = 0$, has destructive interference; there is a dark band along that edge. This is consistent with a much simpler model of what happens at that edge: We have a film of air of zero thickness—that is, no film at all—and it is as though there is no boundary, hence no reflection, hence a dark region. The consistency of these two pictures gives us confidence that our discussion of the phase change is correct.

EXAMPLE 38–3 Two flat glass plates of length $L = 10$ cm touch at one end but are separated by a wire of diameter $d = 0.01$ mm at the other end (Fig. 38–9). Light shines almost perpendicularly on the glass and is reflected into the eye as shown. What is the distance x between the observed maxima if the incident (blue) light has $\lambda = 420$ nm?

Solution: Here, we have the conditions for constructive interference that we have described; we must translate those conditions into a band separation. Figure 38–9 gives us the necessary information. Our condition for constructive interference is

$$2y = \left(m + \frac{1}{2}\right)\lambda,\ m = 0, \pm 1, \pm 2, \ldots .$$

This condition can be translated into a condition for the x-position of the bands by using a geometrical relation between similar triangles in Fig. 38–9:

$$\frac{y}{d} = \frac{x}{L}.$$

We solve this equation for x and insert it into the equation for constructive interference:

$$x = \frac{L}{d}y = \frac{L}{d}\frac{1}{2}\lambda\left(m + \frac{1}{2}\right).$$

The difference in x from one maximum to another corresponds to a shift in m of 1, so

$$\Delta x = \frac{L}{d}\frac{\lambda}{2}\left\{\left(m + \frac{1}{2}\right) - \left[\left(m - 1\right) + \frac{1}{2}\right]\right\} = \frac{L}{d}\frac{\lambda}{2}$$

$$= \frac{(10 \times 10^{-2}\ \text{m})(420 \times 10^{-9}\ \text{m})}{(0.01 \times 10^{-3}\ \text{m})2} \simeq 2\ \text{mm}.$$

For glass plates 10 cm long, there will be about 50 bands of constructive interference.

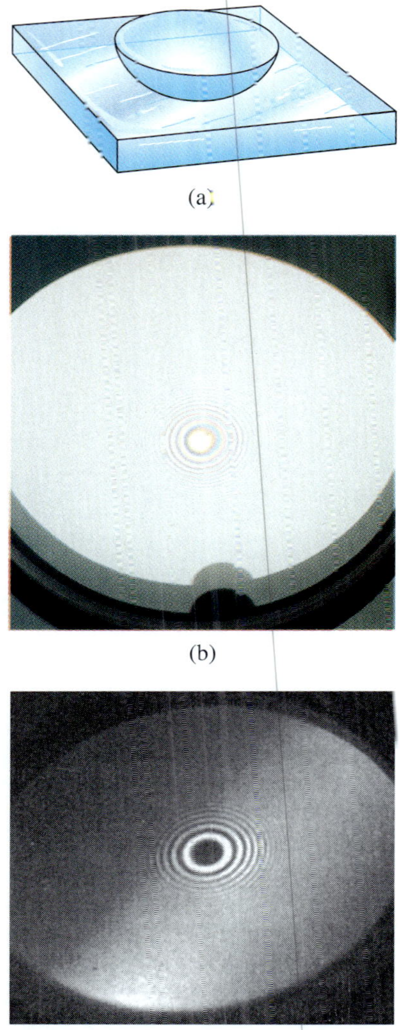

(a) (b) (c)

FIGURE 38–11 (a) A piece of glass whose bottom surface is curved rests on a flat glass surface. (b) The system is illuminated with white light from above, and a vertical view reveals colored rings called Newton's rings. (c) Concentric rings of alternating bright and dark appear when Newton's rings form from monochromatic light.

Newton's Rings

A variant on the previous example is known as *Newton's rings*, after Isaac Newton, who studied it in the 1700s along with contemporaries Robert Hooke and Robert Boyle. Here, a curved piece of glass is placed on a flat piece of glass and illuminated from above with white light (Fig. 38–11a). Observation from above reveals rings of color (Fig. 38–11b). If monochromatic rather than white light shines down on the glass, then a series of bright and dark concentric rings appear (Fig. 38–11c). Most significantly, the region around point C—the place where the glass pieces touch—is dark (Fig. 38–12). From our earlier discussion, we know that the region around C must be dark: We can think of this as a region without a reflecting boundary or the result of a phase change where ray 2 reflects from the second glass surface.

How do we explain the colors we observe when white light such as sunlight, which consists of all wavelengths, is incident on the glass? For a particular radial distance from point C, there may be only one wavelength in the visible range, say, for the color blue, for which there is *destructive* interference. At that radius, we see the color of sunlight with blue subtracted. If we look slightly farther away from C, where the distance P_1P_2 is greater, there will be destructive interference for a slightly larger wavelength, say, for green, and we see sunlight with green subtracted. These colors are not as vivid as those of the rainbow because they remain mixtures of different frequencies with some frequencies subtracted.

Testing for Flatness with Interference. The interference phenomena we have described in this section give us a practical method for determining how flat a given glass surface is. If the surface to be tested is placed on top of an *optical flat* (a glass with a surface known to be flat to a fraction of a wavelength of visible light), then, if the tested surface is indeed flat, no regions of constructive interference will appear (Fig. 38–13). In practice, a tiny spacer may be placed between the piece to be measured and the optical flat to open a wedge-shaped space between them. If the two pieces are so nearly flat and parallel that, without the spacer, *no* interference fringes are observed, the pieces may bond together and be virtually impossible to pull apart! Weeks of grinding and much money will have been lost.

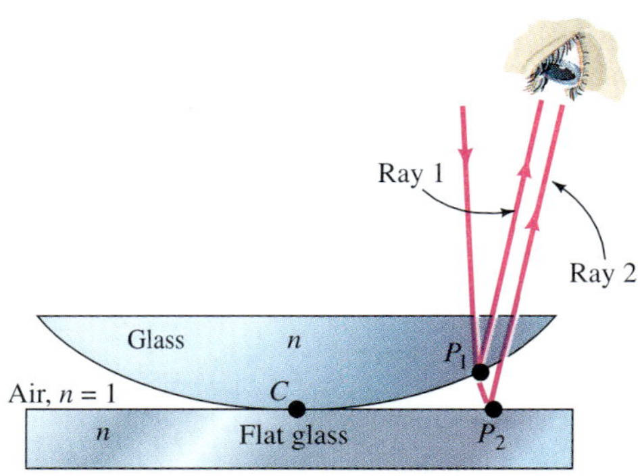

FIGURE 38–12 The geometry used to obtain the conditions for constructive interference in Newton's rings. The two rays moving toward the eye interfere after they reflect from different surfaces.

Mirrors are available commercially that are flat to better than 5 percent of one wavelength, or about 25 nm. They are used in lasers as well as in *interferometers*—devices that use the interference of light to measure distance down to a fraction of one wavelength (see Section 38–4).

APPLICATION
Optical flats

FIGURE 38–13 Fringes that occur when two nonflat pieces of glass are placed on top of one another. If each surface were perfectly flat and parallel, there would be no interference pattern. If the surfaces were flat but not parallel, the fringes would form straight lines. This is a good test of flatness for mirrors. The distortions are caused by surface irregularities.

Thin-Film Interference

The colors seen in soap bubbles and oil slicks are a manifestation of *thin-film interference*, which is another example of interference from reflection (Fig. 38–14). The interference occurs between the light reflected from the two surfaces of the thin film. Consider light ray 1 that is incident on the thin film in Fig. 38–15. Part of the light is reflected at boundary I and forms ray 2. Part of ray 1 is refracted at boundary I and then reflected at boundary II. This light wave is partly refracted again at boundary I before forming ray 3.

Because rays 2 and 3 both originate from ray 1 at point P_1, the conditions for constructive or destructive interference depend on the path-length difference $\Delta L = P_1P_2P_3 - P_1P_4$ as well as on any phase changes that may occur during the reflection. The rule discussed earlier for phase changes upon reflection indicates that ray 1 undergoes a phase change upon reflection at surface I but not at II. Thin films such as soap bubbles have varying thicknesses, so different wavelengths will interfere destructively on different parts of the bubble, and the colors that appear in the reflected light represent the original light minus the wavelength that interferes destructively. There is also an enhancement of those colors for which there is constructive interference. For an oil film floating on water, the oil may have an index of refraction between that of air and water, in which case there is a 180° phase change at *both* the air–oil surface and the oil–water surface. The phase changes due to reflection then cancel each other in the interference from light reflected from these two surfaces, and any phase difference is due only to a difference in the path length.

A new feature that does not arise in the case of Newton's rings appears with thin-film interference. The additional path length lies *within* the thin film of material. The phase change is then computed according to Eq. (38–9), but *the wavelength that appears is the wavelength within the material*, $\lambda_{\text{film}} = \lambda/n$, where n is the index of refraction of the material.

FIGURE 38–14 The colors of soap bubbles are due to thin-film interference of light.

EXAMPLE 38–4 The soap bubble shown in cross section in Fig. 38–15 has thickness t and index of refraction n. Light of wavelength λ in air falls vertically on the bubble and is reflected back. (a) Express the condition for constructive interference for the reflected light. (b) If $t = 400$ nm and $n = 1.3$, what color or colors will interfere constructively in the reflected light?

Solution: (a) When the incident light in Fig. 38–15 is vertical, distance $P_1P_4 = 0$. Light reflects at P_1 and P_2, undergoing a 180° phase change only at P_1, where the

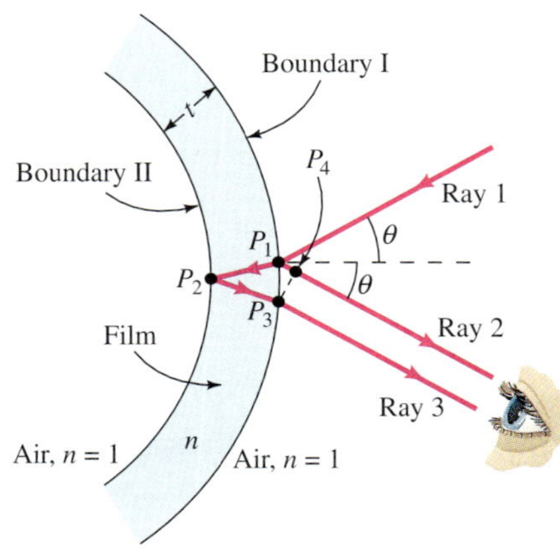

FIGURE 38–15 Geometry for thin-film interference. A light ray reflects from both the front and back boundaries of the film, and the reflected waves interfere.

index of refraction of the medium on the far side—the soap film—is greater than that of air. The path-length difference is $\Delta L = 2t$, and the phase difference $\phi_{\Delta L}$ between the two reflected waves due to ΔL is found from Eq. (38–8) in terms of the wavelength λ_n in the soap film:

$$\frac{\phi_{\Delta L}}{2\pi} = \frac{2t}{\lambda_n} = \frac{2tn}{\lambda}.$$

The overall phase difference between the two reflected waves is then

$$\phi = \pi + \phi_{\Delta L} = \pi + \frac{4\pi tn}{\lambda},$$

where π appears as an additive factor because of the 180° phase change at P_1. There is constructive interference when $\phi = 2\pi m$, where m is an integer. Thus $2\pi m = \pi + (4\pi tn/\lambda)$, or

$$\text{for constructive interference:} \qquad 4nt = (2m - 1)\lambda, \quad m = 1, 2, 3, \ldots. \qquad (38\text{–}16)$$

(b) When there is constructive interference for a given wavelength, the corresponding color is strong in the reflected light. Equation (38–16) gives the condition for constructive interference in this problem, and we need only apply it to find λ, given $t = 400$ nm and $n = 1.3$:

$$\lambda = \frac{4nt}{2m - 1} = \frac{4(1.3)(400\text{ nm})}{2m - 1} \simeq \frac{2100\text{ nm}}{2m - 1}.$$

For $m = 1$ through $m = 4$, these values of λ are

$$\lambda \simeq 2100\text{ nm},\ 700\text{ nm},\ 420\text{ nm},\ 300\text{ nm}.$$

Only the wavelengths 700 nm (red) and 420 nm (blue) are in the visible spectrum and these are the colors that interfere constructively in the reflected light. The color for the wavelength in between ($\lambda \simeq 560$ nm) interferes destructively and will be absent from the reflection.

EXAMPLE 38–5 A thin film of water, $n_2 = 1.33$, floats on cinnamon oil, which is denser than water and has index of refraction $n_3 = 1.65$. White light reflected at 45° has a maximum intensity for a wavelength around 600 nm. What is the minimum possible thickness of the film?

Solution: Figure 38–16 illustrates the paths of the two reflected rays that arrive in parallel at a distant point. *Both* reflected rays undergo a phase shift of 180°, and *no net phase difference is associated with reflection*. There is still a phase shift associated with different path lengths for the two rays—the phase shift ϕ_1 from the additional path P_1P_4 of ray 1,

$$\frac{\phi}{2\pi} = \frac{P_1P_4}{\lambda};$$

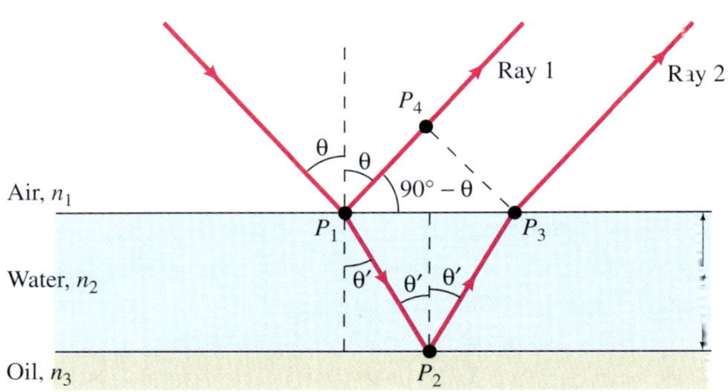

FIGURE 38–16 Example 38–5. Rays 1 and 2 arrive at a distant point where they interfere.

and the phase shift ϕ_2 for the additional path $P_1P_2 + P_2P_3$ of ray 2,

$$\frac{\phi}{2\pi} = \frac{P_1P_2 + P_2P_3}{\lambda_{H_2O}} = \frac{(P_1P_2 + P_2P_3)n_2}{\lambda}.$$

Note that the relevant wavelength for ray 1 is the wavelength in air ($n_1 = 1$), λ, whereas the relevant wavelength for ray 2 is $\lambda_{H_2O} = \lambda/n_2$. It is the difference $\phi_2 - \phi_1$ that enters into the net phase difference between the two waves. The net phase difference is then

$$\phi = \phi_2 - \phi_1 = (P_1P_2 + P_2P_3)\frac{2\pi n_2}{\lambda} - (P_1P_4)\frac{2\pi}{\lambda}.$$

The geometry necessary to calculate the path lengths is shown in Fig. 38–16. The incident ray enters at an angle θ perpendicular to the surface and is either reflected at angle θ or refracted into the water at angle θ'. We note that $P_1P_2 = P_2P_3 = t/(\cos\theta')$. Also, $P_1P_4 = P_1P_3\cos(90° - \theta) = (2t\tan\theta')\cos(90° - \theta)$. The phase difference is then

$$\phi = \left(\frac{2t}{\cos\theta'}\right)\frac{2\pi n_2}{\lambda} - (2t\tan\theta')\cos(90° - \theta)\frac{2\pi}{\lambda}.$$

We solve the equation for t, the film thickness:

$$t = \frac{\lambda\phi}{4\pi}\,\frac{1}{(n_2/\cos\theta') - \tan\theta'\cos(90° - \theta)}.$$

From Snell's law, $\sin\theta' = (\sin\theta)/n_2$, and with $\theta = 45°$, $\theta' = 32°$. The problem states that there is a maximum in the intensity (that is, constructive interference) for $\lambda = 600$ nm. For constructive interference, $\phi = m(2\pi)$, $m = 1, 2, 3, \ldots$. (Is $m = 0$ allowed here?) Thus

$$t = \frac{\lambda m(2\pi)}{4\pi}\,\frac{1}{(n_2/\cos\theta') - \tan\theta'\cos(90° - \theta)}$$

$$= \frac{600\text{ nm}}{2}m\,\frac{1}{(1.33/\cos 32°) - \tan 32°\cos(90° - 45°)} = (1065\text{ nm})m.$$

The minimum film thickness, for $m = 1$, is 1065 nm.

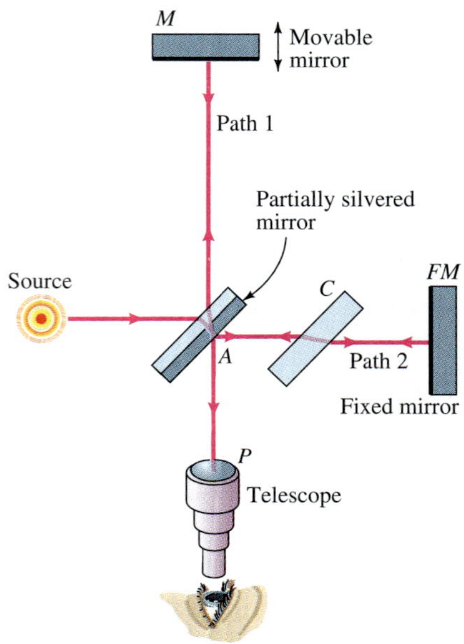

(a)

(b)

FIGURE 38–17 (a) Schematic diagram of a Michelson interferometer. Light is split by the partially silvered mirror. The resulting light travels two different paths before it returns to and interferes at point A and is subsequently observed through the telescope. (b) A modern Michelson interferometer.

* 38–4 INTERFEROMETERS

Optical interferometers are devices that utilize the interference between light waves to measure quantities such as wavelength, small path-length differences, wave speeds, and indices of refraction. Figure 38–17a is a schematic diagram of one type of optical interferometer called the **Michelson interferometer**. In this device, developed by Albert Michelson in the 1880s, a light source is split by a beam splitter—for example, a partially silvered mirror—into two coherent waves that may travel different distances or through different media before they rejoin and interfere (Fig. 38–17b).

Monochromatic light from the source in Fig. 38–17a is split at the mirror at point A. The two beams then travel along paths 1 and 2 before they rejoin at A. The recombined beam is formed from the superposition (and therefore interference) of the two beams that arrive at A. The element C (a *compensator*) is added to make sure that the two light waves travel through the same amount of glass. If the path lengths are exactly the same, the two light waves will constructively interfere. If mirrors M and FM are precisely perpendicular, the combined beam undergoes constructive interference and is bright. If the path lengths are not precisely the same because the mirrors are tilted slightly, the interference will produce alternating dark and bright lines, much like those discussed in Example 38–3, between two flat glass plates. The fringes will shift

APPLICATION: ANTIREFLECTIVE COATINGS

The lenses of some cameras, binoculars, and other optical devices are coated with a thin layer of material in order to reduce the intensity of the reflected light. Many optical devices have multiple lenses, and each lens typically reflects 4 percent of the energy of incident light. After several reflections, a significant amount of the intensity is lost, and the final image may be degraded by the light reflected back and forth between the surfaces. Antireflective coatings are also placed on solar cells (components of solar batteries) to increase the intensity of the transmitted light. Magnesium fluoride, MgF_2, is a material commonly used to coat glass lenses. A single coating can reduce the reflected energy by over a factor of 2. A coating material must have an index of refraction between that of air and the glass to be coated. For MgF_2, $n = 1.38$ for a wavelength of 550 nm.

If light is reflected from both surface I and surface II, as shown in Fig. 38AB–1, then there will be a phase change in both reflections. The overall reflection is minimized when the two reflected light waves *destructively interfere*. If the thickness of the coating is t and the wavelength of light in air is λ, destructive interference will occur when the difference in optical path lengths, $2t$, is equal to $\frac{1}{2}\lambda_n$, $\frac{3}{2}\lambda_n$, $\frac{5}{2}\lambda_n$, . . . , where λ_n is the wavelength within the coating. The condition of destructive interference is then

$$2t = \left(m + \frac{1}{2}\right)\lambda_n = \left(m + \frac{1}{2}\right)\frac{\lambda}{n}. \qquad \text{(AB–1)}$$

The coatings are applied as thinly as possible (that is, for $m = 0$), because this choice cuts down the reflection better. The thickness t for the antireflective coating then becomes

$$t = \frac{\lambda}{4n}. \qquad \text{(AB–2)}$$

A coating of *quarter-wave* thickness gives destructive interference for only one wavelength in the visible range of light. A 100-nm thickness of MgF_2 on a glass lens reduces reflectivity at 550 nm, which is in the middle of the visible range. The colors at the ends of the visible spectrum—red and violet—are still reflected, and the reflected light has a slight purplish hue. Reflected energies as low as 0.5 percent can be obtained with more sophisticated multiple-layer coatings. The phase condition that fixes the thickness of the coating is only part of the story. The destructive interference produced by the antireflective coating will be *total* only if the amplitudes of the two reflecting rays are equal. Because the amplitude of reflected light depends on the indices of refraction of the media involved, the condition for equal amplitudes becomes a condition on the index of refraction of the coating material (see Problem 48).

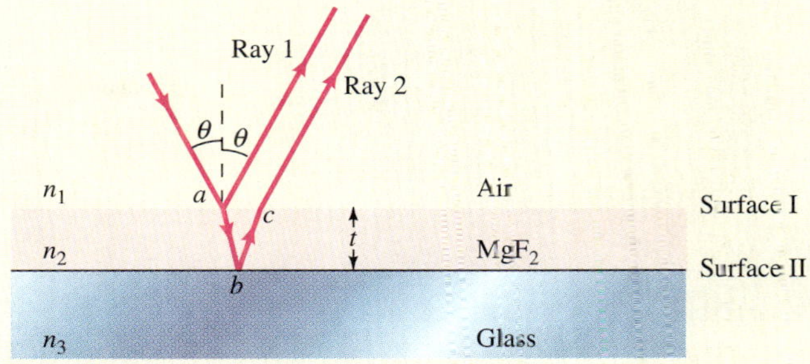

FIGURE 38AB–1 Thin coatings of materials such as MgF_2 can serve to reduce reflection by taking advantage of destructive interference.

Sometimes we want to *increase* the intensity of reflected light. Lasers require mirrors that reflect light strongly at their operating wavelengths. For this purpose, a single coating must have thickness $t = \lambda/2n$, so that there will be constructive interference between the two reflected light waves. Multiple coatings are even more effective in increasing the reflected intensity. It is possible to reflect more than 99 percent of the energy.

if the (screw-mounted) movable mirror is moved slightly. A movement of the movable mirror of only $\lambda/2$ will cause a shift from one fringe maximum to the adjacent one.

Unknown wavelengths of light can be determined by accurately measuring the movement of the movable mirror and counting the number of maxima that pass across the telescope eyepiece. If ΔL is the distance the mirror is moved and N is the number of maxima that pass across the eyepiece for this mirror movement, then

$$N = \frac{\Delta L}{\lambda/2} = \frac{2\Delta L}{\lambda}.$$

The wavelength is then

$$\lambda = \frac{2\Delta L}{N}. \qquad \text{(38–17)}$$

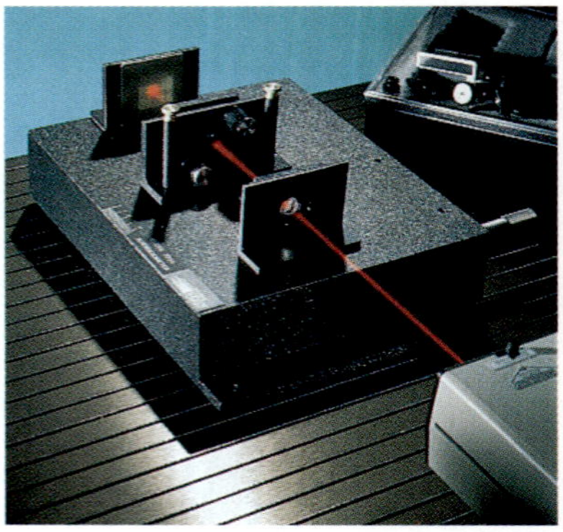

(a)

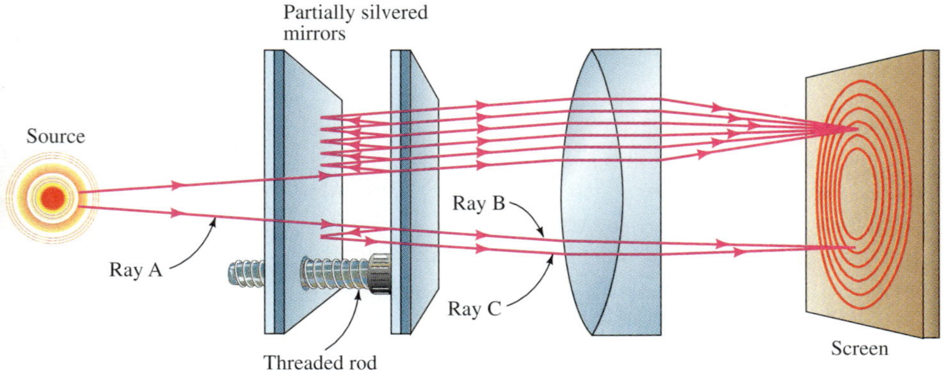

(b)

FIGURE 38–18 (a) A Fabry–Perot interferometer. (b) Schematic diagram of a Fabry–Perot interferometer. Some of the light is transmitted and some reflected at the partially silvered mirrors. Rays B and C may constructively or destructively interfere, depending on the extra path length of ray C. Multiple reflection is an important part of the operation of the Fabry–Perot interferometer and serves to make the maxima sharper and more easily locatable.

By performing measurements with a large number of maximum shifts, wavelengths can be accurately measured. If the wavelength is known, the same technique can be used to measure very small distances, in this case, ΔL.

The Fabry–Perot Interferometer

The most widely used interferometer is the *Fabry–Perot interferometer* (Fig. 38–18a), invented by Charles Fabry and Alfred Perot and illustrated schematically in Fig. 38–18b. It contains two end plates (partially silvered mirrors) that are precisely parallel and flat and are connected by a rod that allows the distance between the plates to be changed smoothly by a screw thread. An incident laser beam (ray A) is partly transmitted (ray B) and partly reflected; the reflected ray is in turn partly reflected and partly transmitted to make ray C, which interferes with ray B. The major improvement incorporated into the Fabry–Perot interferometer is that the plates are silvered such that multiple reflections are possible, and the interference is between the multiple rays formed by these multiple reflections. The multiple reflections reinforce the regions of constructive interference, making the maxima stronger. The maxima become easier to locate and the pattern becomes more distinct. It becomes easier to tell when these maxima have been shifted, so distances can be measured with more precision. If we compare Figs. 38–19a and 38–19b—interference patterns produced by a Michelson and by a Fabry–Perot interferometer, respectively—the greater sharpness of the Fabry–Perot pattern is evident.

APPLICATION

The Fabry–Perot interferometer is a powerful measurement device.

To measure distances with a Fabry–Perot interferometer, we begin with the plates at known positions and then count the number of interference maxima changes (fringes) as the plate separation varies by the desired distance. Because the location of maxima can be determined to great accuracy, the distance change can be measured to within an error of only a fraction of a wavelength of the laser light. The fringe counting is done automatically by electronic sensors. The number of fringes involved in distances of about 1 m is on the order of the number of wavelengths of visible light contained in 1 m, around 50 million.

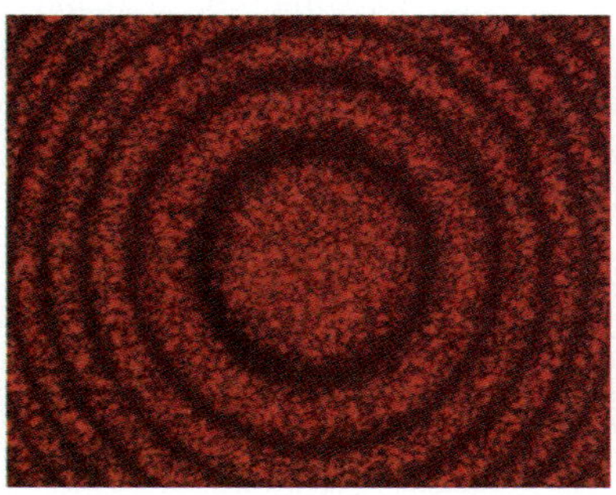

(a)

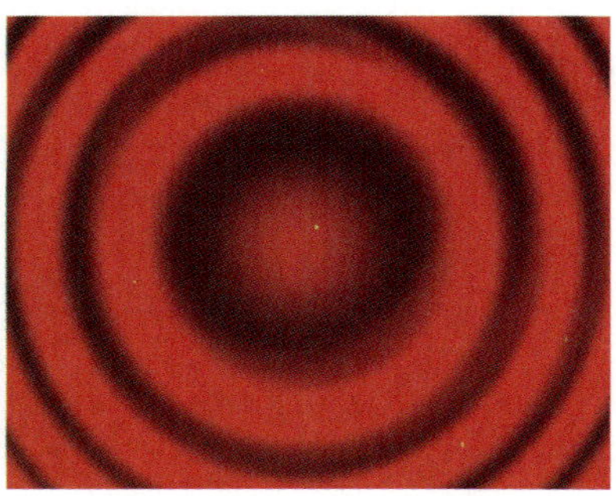

(b)

FIGURE 38–19 The interference patterns produced by (a) a Michelson interferometer and (b) a Fabry–Perot interferometer. The Fabry–Perot fringes are noticeably sharper.

SUMMARY

The wave theory of light explains experimental phenomena such as interference and diffraction that geometric optics cannot explain. Thomas Young was the first to substantiate the wave nature of light through his double-slit experiment, which produces interference maxima and minima. If λ is the wavelength and d is the distance between the (narrow) slits, there will be maxima and minima at angles θ on a distant screen given by

$$\text{for constructive interference:} \qquad \sin\theta = n\frac{\lambda}{d} \qquad n = 0, \pm 1, \pm 2, \ldots; \qquad (38\text{–}3a)$$

$$\text{for destructive interference:} \qquad \sin\theta = \left(n + \frac{1}{2}\right)\frac{\lambda}{d}, \qquad n = 0, \pm 1, \pm 2, \ldots . \qquad (38\text{–}3b)$$

The intensity pattern for the double slit is

$$I_{\text{net}} = 4I_0 \cos^2\left(\frac{\pi d}{\lambda}\sin\theta\right), \qquad (38\text{–}14)$$

where I_0 is the maximum intensity for a single slit and I_{net} is the total observed intensity.

Newton's rings appear when light shines vertically down upon a curved glass that rests on a flat piece of glass. The light reflecting from two surfaces interferes. For monochromatic light, alternating rings of bright and dark are observed. For sunlight, colors are observed. In both cases, the center is dark because the phase of an electromagnetic wave will change by 180° upon reflection when the wave moves from one medium to another of higher index of refraction. The condition for constructive interference in Newton's rings is

$$\Delta L = \left(n + \frac{1}{2}\right)\lambda, \qquad n = 0, \ \pm 1, \ \pm 2, \ldots, \qquad (38\text{–}15)$$

where ΔL is the difference in path length of the interfering rays.

Thin-film interference is responsible for the colors observed in bubbles and oil slicks. For light falling perpendicularly from air onto the surface of the film of thickness t, the condition for constructive interference is

$$4nt = (2m - 1)\lambda, \qquad m = 1, 2, 3, \ldots, \qquad (38\text{–}16)$$

where n is the index of refraction of the film. Light reflected from the two surfaces of the thin film interferes destructively for some wavelengths but not for others. If the incident light is white, the wavelength undergoing destructive interference is subtracted from the reflected light. This phenomenon is used in the application of thin coatings to lenses and other optical surfaces in order to reduce the intensity of reflected light.

Optical interferometers utilize the interference of light waves to measure distances with great precision. Two interferometers with practical applications are the Michelson interferometer and the Fabry–Perot interferometer.

UNDERSTANDING KEY CONCEPTS

1. Why is a laser a more practical source of coherent light than a bright lightbulb and a series of slits?

2. In Fig. 38–3b, the intersections represent places where constructive interference occurs on the crests of the advancing wave fronts. The troughs also reinforce each other between the circles. Do such reinforced troughs also represent a region of constructive interference? In the regions between the crests and troughs, where each wave is zero, the waves add to zero. Is this a region of constructive interference?

3. In Fig. 38–3b, the intersections represent sites where constructive interference occurs. Will these bright spots be seen in air, or is a screen required for us to see the alternating bright and dark spots?

4. We refer to lines, or fringes, when we use two elongated slits for interference. If we used two holes rather than two slits, would we still observe lines or would we observe something else?

5. Why does a single thin coating produce destructive interference in the reflection of light for only one wavelength in the visible region of light?

6. In discussing interference from thin films, we have mentioned viewing the reflected light from afar. Why? Is there no interference if the light is viewed from close to the film?

7. There is no light intensity and therefore no energy at an interference minimum from two sources of coherent light. Yet each of the two interfering sources alone would produce energy at that point. Why does this situation not violate the conservation of energy?

8. When there is a minimum in the intensity of light reflected from thin films, there is less energy in the reflected beam. Does this mean that energy conservation is violated? If not, what happens to the energy that would otherwise have been in the reflected light? It may be helpful to think about what the light that passes all the way through a Newton's rings apparatus must look like.

9. Is it possible to use two beams of light that travel in the z-direction—one polarized with its electric field vector aligned along the x-direction, and the other polarized with its electric field vector aligned along the y-direction—to make an interference pattern?

10. We estimated in Example 38–3 that almost 50 bands of constructive interference can occur along the glass plates. Can all these bands be observed? Will the color still appear blue?

11. In our discussion of Newton's rings, we did not consider light that reflects from the top surface of the curved piece of glass, nor that from the bottom surface of the bottom plate. Why not?

12. In Example 38–2, a second antenna has been constructed whose signal is as strong as that of the first, yet the signal you receive weakens. How is this consistent with the conservation of energy?

13. How is it possible to use interferometry to measure a small fraction of a wavelength when the distance between maxima represents a full-wavelength difference?

14. When you look into a good-quality camera lens, you will see a color tint (generally purple). What is the origin of the observed color?

15. It might appear that a given antireflective coating will work for any surface to which it is applied, provided that the thickness of the coating is given by Eq. (AB–2); however, this is not true. Why not? You can go back to the discussion of the energy in reflection (in Chapter 36) to find out why the index of refraction of the surface to be coated must be considered.

16. Why is it important that each slit be as narrow as possible in the double-slit experiment?

17. Antireflective coatings are always applied to the front surface—the side from which light comes—of an optical element, never to the back surface. Why?

PROBLEMS

38–1 Young's Double-Slit Experiment

1. (I) A coherent source of monochromatic light of unknown wavelength shines on double slits separated by 0.20 mm. Bright spots separated by 0.70 cm appear on a screen 3.0 m away. What is the wavelength of the light?

2. (I) Red light ($\lambda = 650$ nm) shines on a double slit with slit separation $d = 0.3$ mm. How far away from the central axis will the first minimum be on a screen 5 m from the double slit?

3. (I) A double-slit interference experiment is done in a ripple tank. The slits are 3.5 cm apart, and a viewing screen is 0.8 m from the slits. The wave speed of the ripples in water is 0.12 m/s, and the frequency of the vibrator producing the ripples is 12 Hz. How far from the centerline of the screen will the first maximum be found?

4. (I) The source for a double-slit experiment has a wavelength of 525 nm. The slits are a distance of 120 μm apart, and a screen is 40 cm away from the wall that contains the slits. How far from the center will the third maximum occur?

5. (I) Light of wavelength 590 nm falls on a wall with two slits 0.12 mm apart. A photographic plate is placed at a distance R from the wall. The $n = 3$ maximum appears 18 cm from the central maximum on the photographic plate. How far is the plate from the wall?

6. (I) Two tiny speakers are 80 cm apart. They broadcast a signal of frequency $f = 468$ Hz; the signals are in phase. A sensitive microphone is placed 1.0 m from the midpoint of the two speakers, on the line perpendicular to the line joining the two speakers. Is the intensity of sound a minimum or a maximum there? How far would the microphone have to be moved along an arc of radius 1.0 m centered on the midpoint in order to pick up a signal of maximum intensity?

7. (II) A double-slit experiment produces fringes on a distant screen. How does the linear separation between the bright maxima on the screen change when (a) the wavelength of the light doubles? (b) the separation between the slits doubles? (c) the distance between the slits and the screen doubles? (d) the intensity of the light doubles?

8. (II) A laser emitting light with $\lambda = 633$ nm shines on a double slit with a separation of 0.50 mm and produces interference fringes. If the maxima are separated by 0.80 cm, how far away is the screen on which the fringes are observed?

9. (II) In a double-slit experiment to determine an unknown wavelength of light, the measured total distance between 16 maxima (8 on each side of the central maximum) is 16.8 cm. The screen is located 3.45 m from the double slits, whose centers are 0.21 mm apart. What is the wavelength?

10. (II) Suppose that a double slit illuminates a distant screen. The light from sources S_1 and S_2 has come from a single monochromatic source S that is one-half wavelength closer to S_1 than to S_2. Use the geometry of Fig. 38–6 for the relation between the screen and the double slit to express the locations of maxima and minima. Is the point at $\theta = 0$ a maximum, a minimum, or neither?

11. (II) A double slit with variable separation d is superimposed on a single slit at right angles to the double slit, leading to a two-point source (Fig. 38–20). Then the double slit is rotated relative to its original direction by an angle ϕ. If light of wavelength λ shines through the system, determine the position of the interference maxima on a screen a distance R away. Do the fringes move inward or outward as ϕ increases?

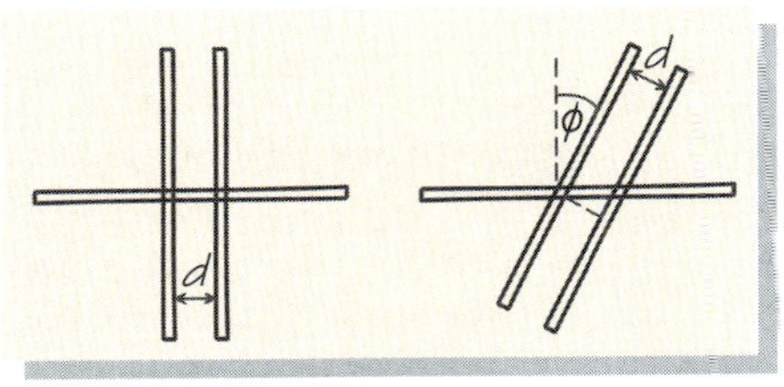

FIGURE 38–20 Problem 11.

12. (II) Two microwave sources are 20 cm apart. They radiate coherently with a frequency of 2.5×10^{10} Hz but with a phase difference α between the two sources. A microwave detector is moved along a line 3.0 m away from the sources. How far from the center ($\theta = 0$) will the first maximum occur, as α varies from 0 to 2π?

13. (II) Two sources radiate at almost identical frequencies f and $f + \Delta f$. How fast will the interference fringes on a screen a distance R away move, assuming that the sources are a distance d apart and that the velocity of wave propagation is v? Evaluate the result for two sources in a ripple tank ($f = 10$ Hz, $\Delta f = 10^{-6} f$, $R = 1$ m, $d = 5$ cm, $v = 0.15$ m/s). Do the same for an optical double-slit experiment ($f = 4.7 \times 10^{14}$ Hz, $\Delta f = 10^{-6} f$, $R = 1$ m, $d = 0.25$ mm, $v = 3 \times 10^8$ m/s). What can you conclude about the degree of coherence in the two cases?

14. (II) Light of two different wavelengths, λ_1 and λ_2, is incident on a double slit. On a distant screen, the twentieth maximum of λ_1 overlies the nineteenth minimum of λ_2. Show that the relative difference $(\lambda_1 - \lambda_2)/\lambda_1$ is small, and find a numerical value for this ratio.

15. (II) Consider two narrow slits illuminated from behind; the light impinges on a screen that is close rather than far (Fig. 38–21). Assume that the wavelength of the light, λ, is comparable to the separation d between the slits and to the screen distance R. The angle θ is measured from the point midway between the slits. (a) Show that the point corresponding to $\theta = 0$ continues to be a maximum of the pattern of light that reaches the screen. (b) At what angle θ is there a first maximum? (c) Show that your result for part (b) reduces to the distant-screen result for $R >> d$ (and $R >> \lambda$).

38–2 *Intensity in the Double-Slit Experiment*

16. (I) Green light of wavelength 620 nm shines on two slits separated by 0.2 mm. Find the intensity ratio I/I_0 at positions 0.4 mm and -0.7 mm from the central maximum on a screen 1 m from the slits.

17. (I) Two coherent light sources of the same wavelength, each with intensity of 10^3 W/m^2, interfere at a point at which the phase difference is 60°. What is the net intensity of light at this point?

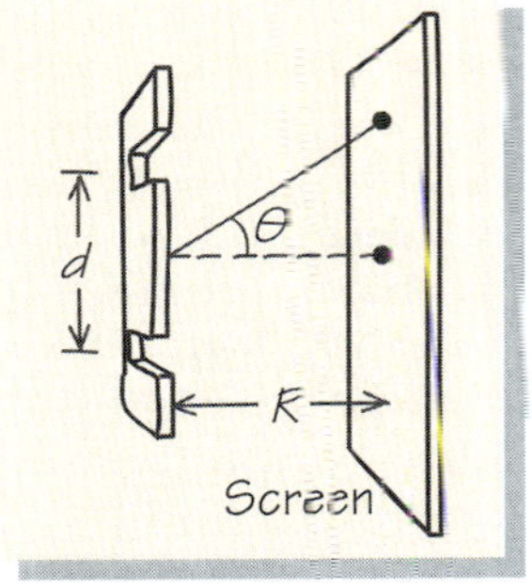

FIGURE 38–21 Problem 15.

18. (I) The net intensity from two coherent sources of equal strength is exactly twice the intensity due to either source at a certain point in space. What is the phase difference between the waves arriving from the two sources at that point?

19. (I) In a classic Young double-slit experiment, the net intensity is 1/4 of the individual source intensities at an angle of 27° off the central axis. There are no minima between this point and the central maximum. The source separation is 485 nm. What is the wavelength of the (monochromatic) light?

20. (I) The intensity at the central maximum of a double-slit diffraction pattern is I_{max}. If the wavelength of the light is 540 nm and the nearest location of maximum intensity is 10^{-3} rad from the central axis, what is the separation of the slits?

21. (II) Use the result of Young's classic double-slit experiment to find the average intensity on the screen by integrating the intensity over the surface of the screen. This result should be twice the average intensity from one slit alone and shows that energy conservation holds even when there is interference.

22. (II) Consider the double-slit arrangement shown in Fig. 38–22. The center of the screen C is a point of constructive interference. A container of thickness w, holding a liquid of refractive index n, is placed in the path of the ray from slit S_2 to C. Plot, qualitatively, the intensity of light at C as a function of w, assuming that the separation between the slits is d, and that the screen is a distance L from each slit.

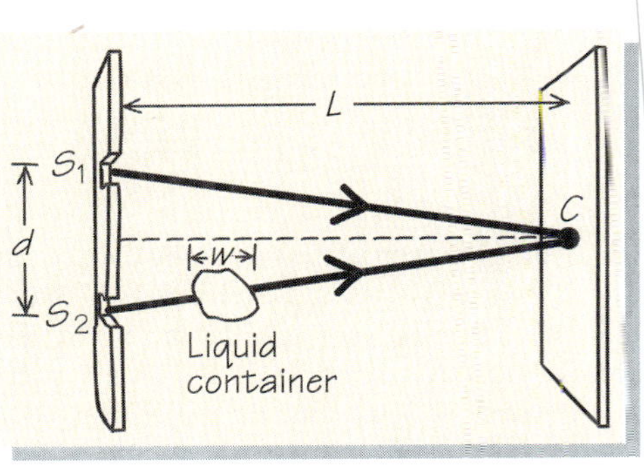

FIGURE 38–22 Problem 22.

23. (II) The *angular width* of a maximum of the intensity pattern due to double-slit interference is defined to be the angular separation $\Delta\theta$ of the points where the intensity is half its maximum value. Express the width of the central maximum in terms of the wavelength and the slit separation. Are the widths of all the maxima the same?

24. (II) A He–Ne laser, $\lambda = 633$ nm, shines on double slits separated by 0.5 mm. At what minimum angle θ is the intensity 50 percent of the maximum? If the screen is located 5 m away, what is the distance between the two angles on either side of the maximum for which this intensity occurs? This distance is the *full width at half maximum* of the central peak.

25. (II) Two point sources of radio waves, 12 m apart, radiate in phase with a frequency of 3.8×10^7 Hz. (a) If the average intensity of each single source is 5×10^{-4} W/m^2 at a certain distance, what is the direction in which the combined intensity is maximized? (b) What is the magnitude of the maximum intensity? (c) At what angle will the intensity have fallen to half its maximum value?

26. (II) Suppose that the two slits in a double-slit experiment are not exactly the same size, so the electric field from one of the slits at a particular point P on the screen is $E_1 \sin(\omega t)$, whereas the other one is $E_2 \sin(\omega t + \phi)$, where the phase ϕ is due to the path-length difference [compare Eq. (38–7)]. Show that the intensity at P is given by

$$\langle I_{\text{net}} \rangle = \langle I_1 \rangle + \langle I_2 \rangle + 2\sqrt{\langle I_1 \rangle \langle I_2 \rangle} \cos \phi,$$

where $\langle I_1 \rangle$ and $\langle I_2 \rangle$ are the intensities due to the light from the individual slits. [*Hint*: You will need $\langle \sin^2(\omega t) \rangle = \frac{1}{2}$; $\langle \sin(\omega t) \cos(\omega t) \rangle = 0$.]

27. (II) Point sources S and S' radiate with the same intensity and the same frequency, corresponding to a wavelength of 0.020 m (Fig. 38–23). They are 45° out of phase and 2.5 m apart. Plot the intensity as a function of distance along the x-axis for values of x much larger than the source separation.

FIGURE 38–23 Problem 27.

28. (III) Find an expression for net intensity for the situation in Problem 27 in terms of x, the wavelength (λ), and the source separation (d). Take into account that the electric field of the electromagnetic wave decreases as the inverse power of the distance between the source and the receiver, and that the distance between source S and any point on the axis is different from the distance between source S' and that point. [*Hint*: Use the results of Problem 26.]

38–3 Interference from Reflection

29. (I) Consider the two glass plates of Example 38–3 and the configuration of Fig. 38–11. The plates are 25 cm long. When light of wavelength 656 nm from hydrogen shines down perpendicularly to the glass, 102 interference fringes appear. How thick is the wire that separates the two glass plates at one end?

30. (I) Two rectangular pieces of glass are laid on top of one another on a plane surface. A thin strip of paper is inserted between them at one end, so that a wedge of air is formed. The plates are illuminated by perpendicularly incident light of wavelength 715 nm, and twelve interference fringes per centimeter-length of wedge appear. What is the angle of the wedge?

31. (I) For what thicknesses of a soap bubble ($n = 1.3$) that are less than 500 nm thick will blue light ($\lambda = 420$ nm) interfere constructively?

32. (I) A thin, uniform film of oil is spread on a glass plate. The oil has an index of refraction between that of air and of the glass. Write the condition for constructive interference for light of wavelength λ in air perpendicularly incident from air and reflecting back into the air from the air–oil–glass interface.

33. (II) A curved piece of glass in the form of the cap of a sphere of radius R is placed on a plane surface of glass. What will be the radius of the first dark Newton ring for light of wavelength λ? [*Hint*: It follows from plane geometry that, for a spherical cap, the radius of the circle bounding the cap, r, is related to the radius of the sphere R and h, the maximum thickness of the cap, by the relation $r^2 = h(2R - h)$, which can be approximated by $2Rh$ for $h << R$.]

34. (II) When two flat glass plates are placed on top of one another and a slip of paper is inserted between them at one edge, a thin wedge filled with air is produced between them. Interference bands form in reflection when monochromatic light falls vertically on the plates. Is the first band near the edge where the plates are in contact light or dark? Why?

35. (II) In a standard Newton's rings experiment, there is a dark spot where the convex surface touches the flat plate. Light of wavelength 500 nm is perpendicularly incident on the system. The convex lens is pulled slowly away from the flat plate until the minimum of the convex lens is 0.25 mm from the flat plate. A series of maxima and minima will appear at the center as the lens moves. How many maxima pass? Do the rings appear to move in to the center or away from the center?

36. (II) Consider the Newton's rings apparatus of Problem 35, with the minimum of the convex lens 0.25 mm from the flat plate. Water is poured into the space between the plates. Do the rings appear to move in to the center or away from the center, and how many maxima pass?

37. (II) A lens whose curved surface is part of a sphere of radius 5.0 m is placed over a flat glass plate and Newton's rings are observed. Determine the diameter of the fourth and seventh dark fringe for a wavelength of 520 nm.

38. (II) Light with a wavelength of 560 nm gives rise to a system of Newton's rings formed with a convex lens resting on a plane surface. The twentieth bright ring is at a radial distance of 0.98 cm. What is the thickness of the air film there, and what is the radius of curvature of the lens surface?

39. (II) The radius of curvature of a convex surface used for a Newton's rings apparatus is R (Fig. 38–24). Find the position x, measured from the point where the convex surface touches the flat surface, of the nth dark ring for light of wavelength λ perpendicularly incident from above.

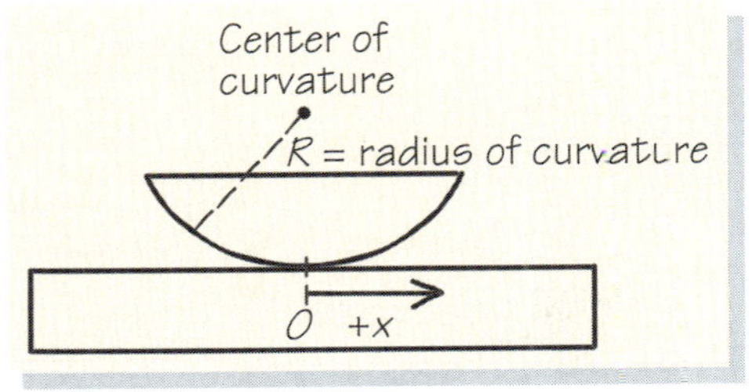

FIGURE 38–24 Problem 39.

40. (II) Constructive interference occurs when a soap bubble reflects light of wavelength 420 nm. What is the minimum thickness of the bubble if its index of refraction is 1.38?

41. (II) What minimum thickness of antireflective coating of MgF_2 is required to minimize the reflection of red light at 650 nm (the wavelength in air)? For MgF_2, $n = 1.38$.

42. (II) For light emitted by a He–Ne laser, $\lambda = 633$ nm, what *nonzero* minimum thickness of MgF_2 coating allows *maximum* reflectivity?

43. (II) The reflected light from an oil film floating on water shows constructive interference for light of wavelengths 434 nm and 682 nm incident along the normal. The index of refraction of the oil is $n = 1.51$. What is the film's minimum possible thickness?

44. (II) The reflected light from an oil film floating on water shows destructive interference for light of wavelengths 540 nm, 600 nm, and 675 nm incident along the normal. The index of refraction of the oil is $n = 1.60$. What is the film's minimum possible thickness?

45. (II) White light reflected at perpendicular incidence from a uniform soap film has an interference maximum at 666 nm and a minimum at 555 nm with no minima between 666 nm and 555 nm. If $n = 1.34$ for the film, what is the film thickness?

46. (II) A thin layer of CaF_2 ($n = 1.41$) is deposited onto glass, with $n = 1.6$. The layer is viewed in reflected light at 45° using a white light source (Fig. 38–25). For what layer thicknesses will red light of 660 nm show constructive interference? Are there possible thicknesses for which blue light of 440 nm interferes constructively as well?

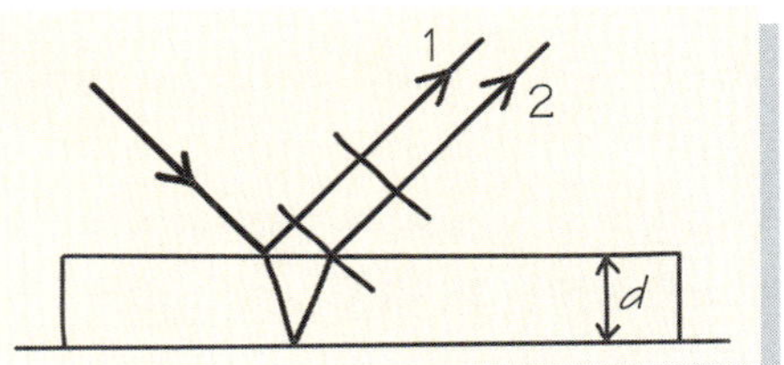

FIGURE 38–25 Problem 46.

47. (II) Light is perpendicularly incident on an oil film with $n = 1.2$, suspended in air. (a) If green light ($\lambda = 550$ nm) is reflected back most strongly, what is the minimum thickness of the film? (b) If n were increased, would the maximally reflected light have a longer or shorter λ? (c) If the film were suspended on the interface between water ($n = 1.33$) and air, what would be seen?

48. (II) We mentioned in Chapter 36 that when light is perpendicularly incident from a medium of index of refraction n_1 and refracts into a medium of index of refraction n_2, then the intensity of the reflected light, I_r, is related to the incident intensity, I_0, by $I_r/I_0 = (n_2 - n_1)^2/(n_2 + n_1)^2$. In order for a coating to eliminate reflections, it is not enough that the light reflecting from the two surfaces differs in phase by 180°; the interference is totally destructive only if the amplitudes are equal. Show that, when multiple reflections at interfaces are neglected, the destructive interference in light reflecting from coated glass in air is maximized when

$$(n_{air}/n_{coat}) = (n_{coat}/n_{glass}).$$

49. (II) You would like to eliminate the reflected light from a flat glass pane for perpendicularly incident light of wavelength 600 nm. If the index of refraction of the glass is 1.55 and you have a coating material with an index of refraction of 1.25, what minimum thickness of coating material will have the desired effect?

50. (II) Blue light, $\lambda = 480$ nm, is perpendicularly incident on a vertical soap film held in a plane. A sequence of bright horizontal bands appears in the reflected light (Figure 38–14). Note the film is illuminated with white, not monochromatic, light.) What is the rate of change with height of the thickness of the soap film if the horizontal bright bands are 0.3 cm apart? For the soap solution, $n = 1.36$.

51. (II) The material to be used for an antireflective coating has index of refraction of 1.25. How thick should the coating be to give the best result for $\lambda = 550$ nm and an angle of incidence of 30° with the normal?

*38–4 Interferometers

52. (I) Laser light with $\lambda = 633$ nm enters a Michelson interferometer. How many fringes will pass through the field of view if one of the mirrors is moved 0.10 mm?

53. (I) If the mirror of one arm of a Michelson interferometer is moved along the arm by 0.31 mm, 980 fringes traverse the field of view. What is the wavelength of the light used?

54. (II) A very sharp wedge of glass of index of refraction 1.52 is introduced perpendicularly in the path of one of the interfering beams of a Michelson interferometer illuminated by a narrow beam of light of wavelength 581 nm. This causes 1800 dark fringes to sweep across the field of view. Calculate the thickness of the glass wedge at the point where the beam passes through it.

55. (II) A scientist wants to measure the wavelength of yellow light ($\lambda \approx 590$ nm) to a precision of 0.1 percent. The minimum motion of the movable mirror is 0.03 mm. What is the minimum number of fringe shifts that must be counted?

56. (II) A laser with light of wavelength 582.5 nm is used to calibrate a Fabry–Perot interferometer. As the screw controlling the position of one end plate rotates exactly 100 turns, 714 fringes are counted. Calculate the wavelength of another light source that shifts only 593 fringes for 100 turns of the screw.

General Problems

57. (II) Two ordinary lightbulbs S_1 and S_2 are 1 m apart, each emitting light waves with intensity I, mainly at a wavelength of 550 nm. What is the pattern (not the absolute value) of intensity on a screen 100 m away?

58. (II) AM radio waves with a wavelength of 350 m travel 23 km to your home. Halfway between the transmitting tower and your home, but off to the side, is a tower that reflects radio waves; the reflected wave has no phase shift due to the reflection. How far off the direct line is the tower if destructive interference occurs between the direct waves and the reflected waves (Fig. 38–26)?

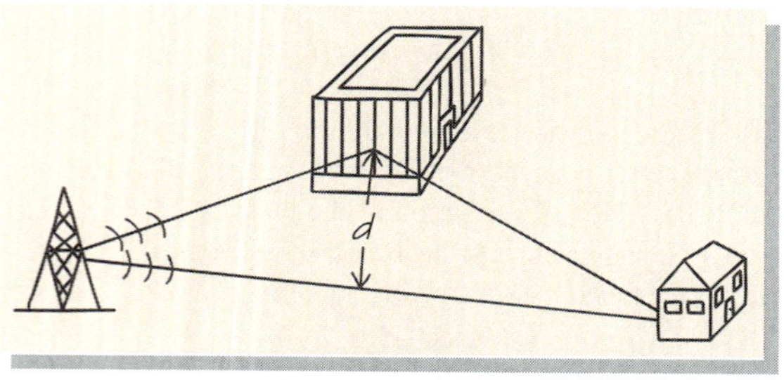

FIGURE 38–26 Problem 58.

59. (II) Two point sources of identical strength radiate in phase with the same frequency f. They are separated by a distance L. What is the energy density (discussed in Section 35–4) as a function of distance from one of the sources along the line that connects the sources in the central region, where the variation of the amplitudes with distance can be neglected (Fig. 38–27)?

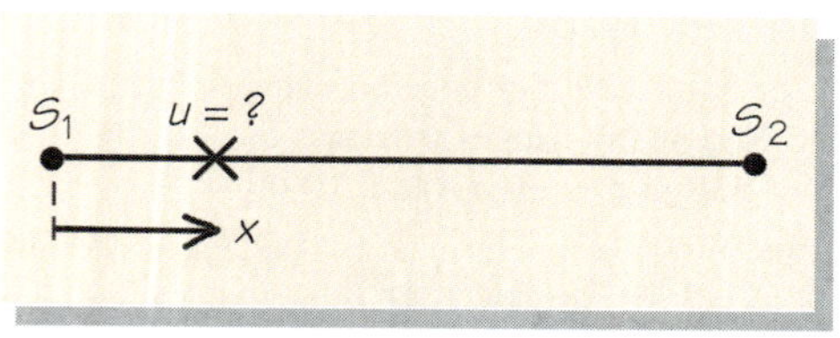

FIGURE 38–27 Problem 59.

60. (II) You want to hear a radio station that broadcasts at 97.9 MHz. You live on a direct line between the antenna and a large building that acts as a mirror for the radio waves broadcast by the station; you are exactly 100 m from the building (Fig. 38–28). Calculate the intensity of the signal you receive in terms of the intensity you would receive if the building were not present. Assume that the reflection from the building is total, with no phase shift, and ignore any decrease of the signal with distance.

61. (II) Coherent microwave radiation reflects from two identical obstacles (Fig. 38–29). Each obstacle, of size a, is much smaller than the wavelength of the radiation, λ, and much smaller than the obstacle separation d. Find an expression for the angles θ', defined in the figure, for which there are maxima on the distant screen.

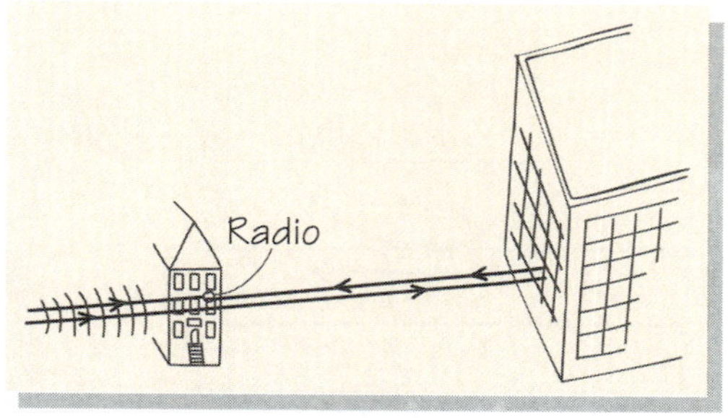

FIGURE 38–28 Problem 60.

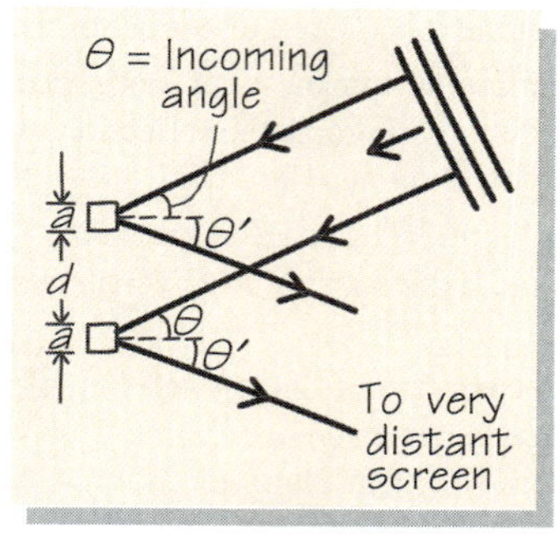

FIGURE 38–29 Problem 61.

62. (II) A radio wave undergoes a phase shift of 180° when it reflects from the calm surface of the ocean. A ship in a calm port nearing a shore station receives a 180-MHz signal from the station's antenna. This antenna is located 25 m above the sea surface, as is the ship's receiving antenna. The direct and reflected signals interfere, and a succession of maxima and minima are heard in the interfering signal at the ship (Fig. 38–30). How far is the ship from the station the first time the signal passes through a minimum? How fast is the ship moving if the time between this first minimum and the next one is 140 s?

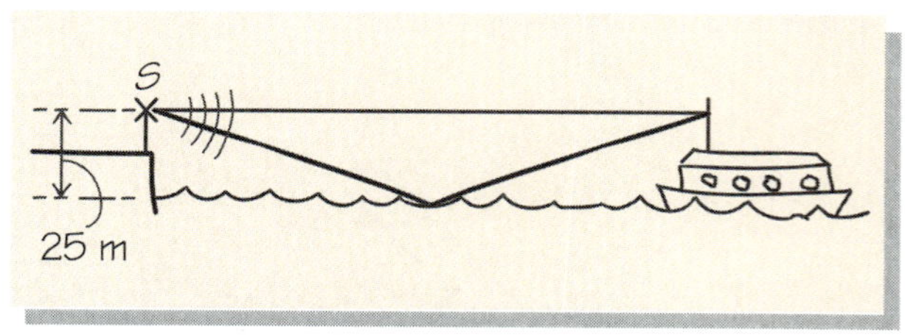

FIGURE 38–30 Problem 62.

63. (II) Sources S_1 and S_2 illuminate a distant screen; the distance to the screen is much larger than the separation between the sources. Each source emits light rays that are in phase at the sources, but the intensity I_1 of the light from S_1 is twice the intensity I_2 of the light from S_2. Give the ratio of the maximum to the minimum intensity of the light observed on the screen. (See Problem 26.)

64. (II) Figure 38–31 is an overhead view of two dipole radio antennas—vertical towers separated along the x-axis by a distance $\lambda/2$. This arrangement allows the radio signal to be beamed with greater intensities in some directions than in others, whereas either antenna alone would radiate its signal with the same intensity I_0 for any angle θ. (a) Find the intensity radiated by the antenna pair very far from the antennas as a function of θ, assuming that the signals of the two antennas are in phase. Describe the signal for all values of θ, from 0° to 360°. (b) The signal in the antennas is now 180° out of phase. How, if at all, does the distant intensity pattern change?

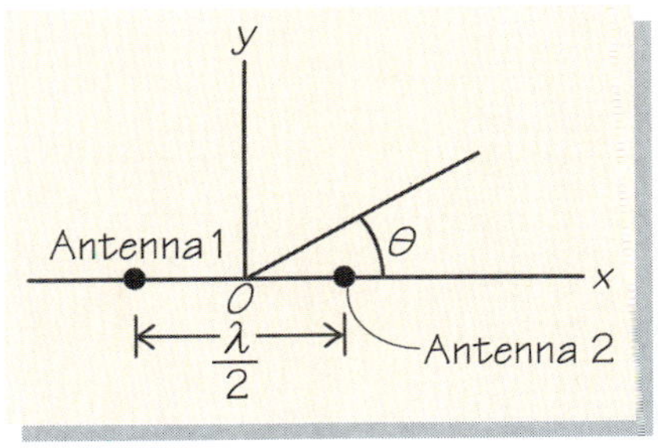

FIGURE 38–31 Problem 64.

65. (II) Consider the pair of dipole antennas in Problem 64. Suppose that the antennas are separated by one-quarter wavelength and that the signal in antenna 1 lags the signal in antenna 2 by 90° (one-quarter cycle). Show that the signal has a maximum in one direction, not two.

66. (II) An interference experiment uses a triplet slit, with slit separations d. What are the positions of the maxima on a screen a distance R away, if light of wavelength λ shines through the slits. What are the intensities of the maxima in terms of the intensity that each slit would produce alone?

67. (II) Four identical loudspeakers are placed at the corners of a square with sides of length $\lambda/\sqrt{2}$. The loudspeakers emit sound coherently with wavelength λ. A listener is situated very far from the square along one of the diagonals. If the intensity with just one loudspeaker on is I_0, what are the intensities when two, three, and four speakers are on? In a table, list all combinations and the resultant intensities.

68. (III) *Lloyd's mirror* is a mirror that reflects light at large angles of incidence from a point source to a screen (Fig. 38–32). As the angle of incidence nears 90° (a grazing angle), the source and image become close. (a) Will the direct light and reflected light interfere constructively or destructively in this limit? (b) Suppose that the (monochromatic) source is 20 cm from the center of the mirror and 0.50 mm above the plane of the mirror and that the screen is distant. What will the angular separation be between successive maxima of the interference pattern, where the angle is measured from the source point to the screen?

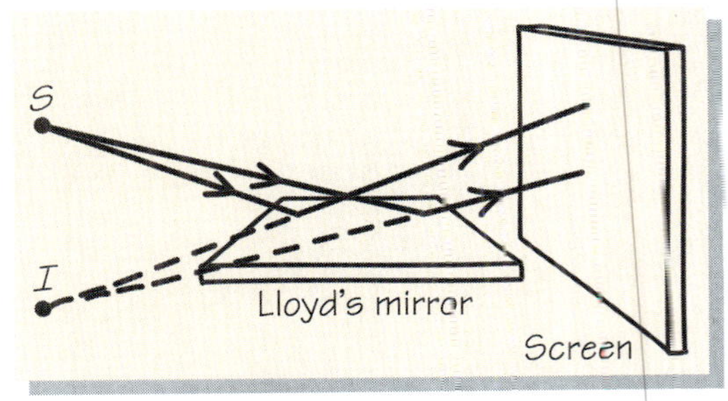

FIGURE 38–32 Problem 68.

69. (III) Light of wavelength λ is incident on a slab of glass of thickness h and index of refraction n resting on a mirror. The ray makes an angle θ with the vertical. The light reflected at the top of the glass surface and that reflected at the mirror will interfere. For what angles will the interference be totally constructive? [For angles even slightly away from grazing incidence (see Problem 68), light undergoes no phase shift upon reflection by a conductor such as a mirror.]

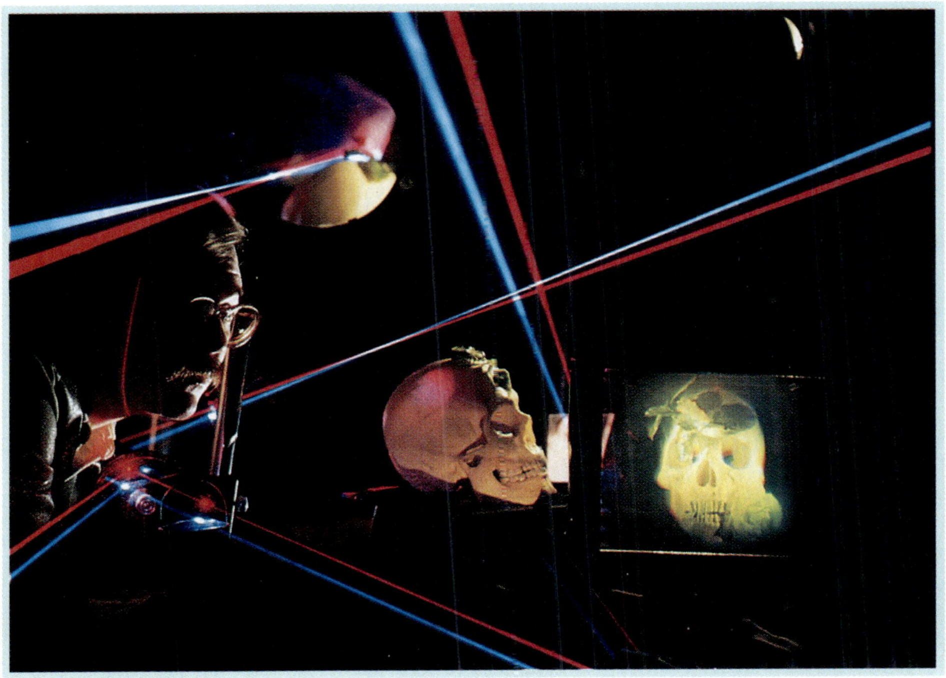

CHAPTER

39

A hologram of a skull in the process of production. Holograms depend crucially on the interference of coherent light, here supplied by laser beams.

Diffraction

We can clearly discern the wave nature of light only when there are obstacles or apertures with sizes comparable to the wavelength of light (several hundred nanometers), or when we look within comparable distances at the edges of shadows. In these cases, interference leads to effects that geometric optics cannot explain, such as the double-slit phenomena of Chapter 38. Diffraction is another manifestation of interference phenomena; this term can be used as an alternative to the term interference, but it usually refers specifically to the bending of waves around obstacles. We are already familiar with the circular spread of water waves passing through an opening narrower than their wavelength. This phenomenon is a diffractive effect. So is the pattern of maxima and minima that spreads across a screen in Young's double-slit experiment. The term diffraction also refers to interference between waves that emanate from a large number, or even a continuous set, of sources. Diffraction gratings have many slits or sources of coherent light and can be treated as a simple generalization of double slits. These gratings have important applications in the study of atomic systems and crystalline materials. We shall see that even a single, narrow slit produces characteristic and striking interference patterns that result from diffraction. Holography is a spectacular application of diffraction. Interference and diffraction effects are both consequences of the superposition of waves.

39–1 THE DIFFRACTION OF LIGHT

By the 1820s, serious attempts were underway to understand the consequences of the wavelike nature of light. Young's double-slit experiment had shown that there are clear interference effects associated with light. Although the bending of water waves around obstacles or around the edges of apertures is a familiar idea (Fig. 39–1), it is not part of our everyday experience for light. This is because *interference effects require coherent wave sources* and because *diffraction effects are typically most significant when the apertures or obstacles involved are comparable to the wavelength.* For light, the coherence condition is not realized in most situations, and wavelengths—several hundred nanometers—are tiny compared to the sizes of familiar objects.

FIGURE 39–1 The way water waves bend around and through obstacles is well known. Here, parallel waves pass through a hole, producing circular wave fronts on the far side.

In addition to his double-slit experiment, Young performed experiments that showed that light can pass around a single small obstacle, just as water waves can. Young firmly believed in the wave theory and publicly supported it despite considerable resistance in the scientific community. Augustin Fresnel, François Arago, and others produced further experiments and theories to establish even more firmly the reality of **diffraction phenomena** in light; in other words, situations in which light bends around obstacles by virtue of its wave properties. By 1821, Fresnel's research had progressed to the point where he was able to use a primitive version of an interferometer to make the first quantitative measurement of the wavelength of light.

Huygens' principle had been used much earlier to show how each point on a wave front can be treated as producing spherical waves that add by superposition to a new wave front. Fresnel treated various interference phenomena using the Huygens construction. He showed that even a single aperture creates its own diffraction pattern because waves passing through different parts of the aperture interfere with each other. Similarly, even a single obstacle creates a diffraction pattern because parts of the original plane wave have been blocked by the obstacle and no longer participate in the downstream regeneration of the wave.

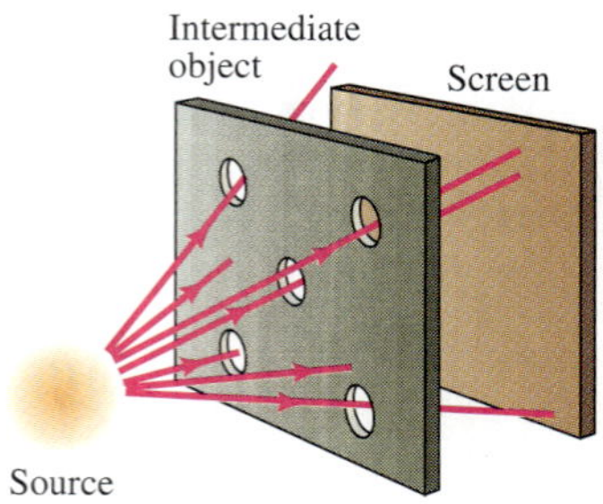

FIGURE 39–2 Diffraction effects can be observed by placing an intermediate object in the path of light that passes from a source to a viewing screen.

We can observe diffraction effects with a source of coherent light, an intermediate object in the form of an obstacle or a wall with holes, and a viewing screen (Fig. 39–2). Figure 39–3 is a photo of the diffraction pattern at the edge of the shadow of a razor blade. With a little ingenuity, the effects of diffraction are visible to the naked eye. For example, try looking at a distant mercury-vapor street lamp through the narrowest possible slit you can make between your fingers. Fresnel systematically investigated the interference patterns from apertures, edges, and small obstacles. If a source and viewing screen are each a finite distance from an intermediate object, the resulting diffraction pattern is known as *Fresnel diffraction*.

If both the source and the screen are far from the intermediate object, the mathematics is considerably simplified (Fig. 39–4). This special case is known as *Fraunhofer diffraction*, after Joseph von Fraunhofer. The Fraunhofer limit is easy to treat because the waves that originate at each position of the apertures or obstacles from the source and reach a given point on the screen are nearly parallel; this simplifies the calculation of path-length differences and phase differences. We treated the Fraunhofer case in Chapter 38 when we studied double-slit interference patterns. Starting with Section 39–2, we shall restrict ourselves to this case.

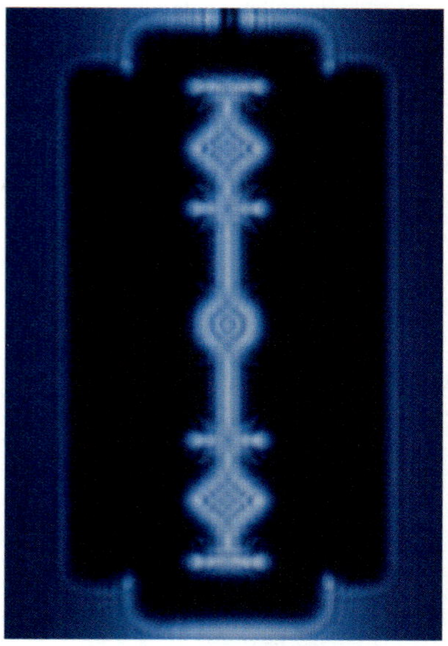

FIGURE 39–3 The diffraction of light around a razor blade.

Before considering in detail some of the cases mentioned here, though, let's first look at one rather spectacular demonstration of diffraction effects observed by Arago in 1818. Suppose that we place a perfectly round obstacle in the path of a point source of coherent light, as in Fig. 39–5a. Every point on the rim of the disk is equidistant from the source, and light falling on the rim is thus perfectly in phase. According to the Huygens construction, we can think of each of these points as a new source, and they are all in phase. All the rim points are equidistant from a point P on a screen, the point that lies on the symmetry axis between the disk and the source. Because the light reemitted from the rim arrives at P in phase, there is constructive interference at P, and hence a bright spot appears, known as the *Poisson spot*. This diffractive result is certainly inconsistent with geometric optics! Even an ordinary penny can be used to show this result (Fig. 39–5b), with the central maximum clearly visible.

39–2 DIFFRACTION GRATINGS

A simple way to generalize the double-slit interference experiment is to increase the number of narrow slits. There is a characteristic interference pattern *if the slits are regularly spaced*; a screen with such an arrangement is called a **diffraction grating**. Diffraction gratings are important for two reasons: First, the multiple slits allow more light through than do two slits, thus increasing the intensity; second, the interference maxima are much sharper than they are for two slits, allowing the wavelength of the light to be measured more precisely.

Although two slits were cut in an opaque screen in Young's double-slit experiment, the presence of transparent holes in an opaque screen is not necessary. We simply require an array of obstacles to serve as pointlike sources for the reradiation of spherical wavelets. Thus a diffraction grating can take many forms. For example, regular scratches or rulings can be inscribed on glass or metal plates. When light passes through a ruled glass plate, the scratches act as sources for the regeneration of spherical wavelets. This type of grating is called a *transmission grating*. When the scratches are made on metal plates, they act as regular point sources of reflected rather than transmitted light; such a grating is called a *reflection grating*. Even the marks on an ordinary ruler can be a reflection grating for laser light. What is important is that the light be scattered from regularly spaced centers. For all these gratings, the analysis of the diffraction pattern is similar; here, though, we shall concentrate on transmission gratings and draw them as though there are actual slits.

Fraunhofer made the first gratings from fine parallel wires. As we have stated, more slits for a fixed slit spacing lead to a sharper diffraction pattern and a grating more useful for the analysis of light. In the 1870s, Henry Rowland invented ruling machines capable of cutting gratings with thousands of grooves per centimeter (Fig. 39–6). This process revolutionized the use of diffraction gratings, and Rowland-type gratings were very important to the discoveries and progress of modern physics. *Spectroscopes*—optical devices used to measure the wavelength of light emitted by atoms after they have been excited—are typically based on diffraction gratings. The unique

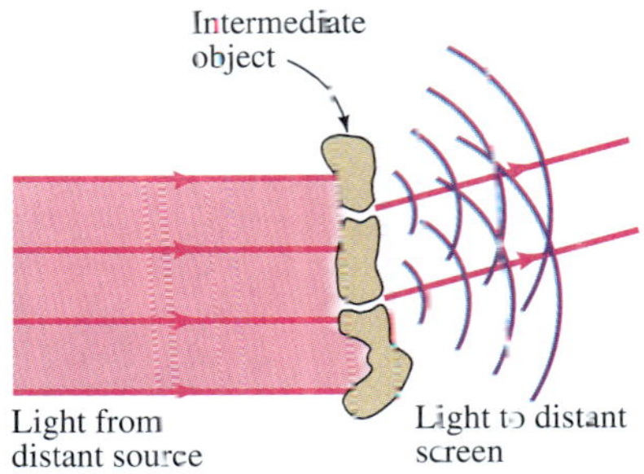

FIGURE 39–4 Parallel rays from a distant source are diffracted by an intermediate object and then viewed on a distant screen. When the screen is distant, we can treat the outgoing rays as parallel, and we have the case of Fraunhofer diffraction.

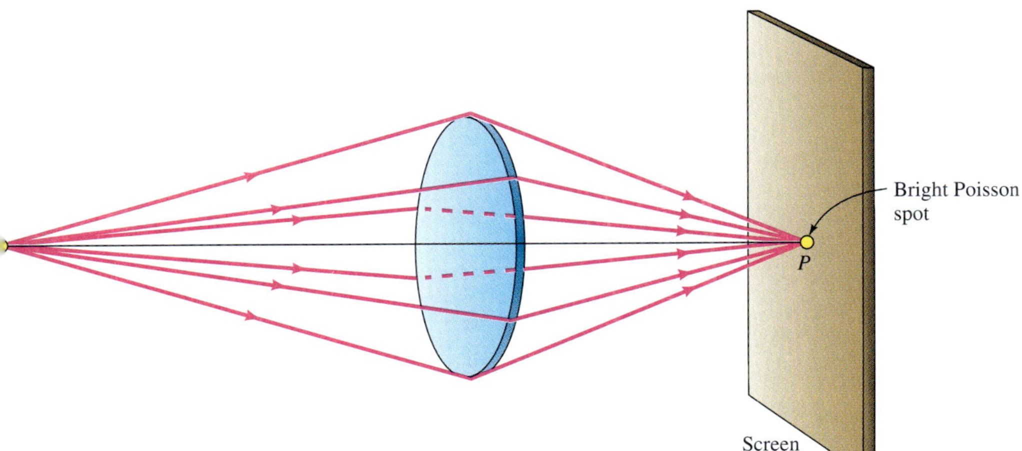

(a)

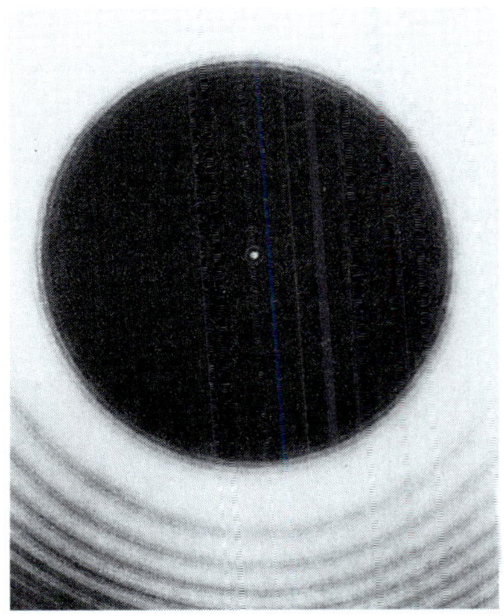

(b)

FIGURE 39–5 (a) A round, opaque object in the path of a point source of coherent light. Diffraction causes a bright spot (the Poisson spot) to be seen along the optical axis at point *P*. (b) The Poisson spot of a penny.

APPLICATION
The spectroscope

set of wavelengths (or frequencies) produced by excited states of each type of atom, ion, or molecule is called its *spectrum*. The measurement of these spectra led to quantum mechanics and an understanding of the structure of the atom. Diffraction gratings are still widely used in science and technology. Figure 39–7a is a schematic diagram of a spectroscope, and Fig. 39–7b shows the result of using a spectroscope to observe the light from a distant sodium-type street lamp. Arrays of antennas can act like diffraction gratings. Such antennas can both broadcast and receive, as in radio telescopes or ground-control approach antennas.

FIGURE 39–6 Henry Rowland's ruling machine for cutting gratings made the use of the patterns produced by gratings a standard tool for analyzing light.

Energy Conservation and Intensity

Let's turn now to the analysis of the Fraunhofer diffraction pattern. Figure 39–8 shows a few of the slits of a grating with N slits, separated from each other by a distance d; a monochromatic plane wave of wavelength λ approaches from a distant source. The plane wave arrives with the same phase at each slit, and spherical waves are emitted in phase at each slit. We look at wave propagation along the lines labeled W. Lines $W1$, $W2$, $W3$, and so on are oriented at an angle θ to the original wave-propagation direction. Note that we allow the screen to be so distant that these lines are approximately parallel even though they point toward a particular spot on the screen (or a lens focuses them on that spot).

If the wavelets are in phase along the front AA', defined by θ, then the light that eventually reaches the distant screen at this angle will also be in phase, and there will be constructive interference—a maximum. The condition under which the waves along AA' are in phase is that each path length differs from any other by integral multiples of the wavelength λ. The path-length difference between adjacent waves is $d \sin \theta$ (Fig. 39–8). Thus there are *principal maxima*, where the light from all the slits interferes constructively, at angles such that

The location of the principal maxima of a diffraction grating

$$\text{for principal maxima:} \quad d \sin \theta = m\lambda, \quad \text{where } m = 0, \pm 1, \pm 2, \ldots . \tag{39–1}$$

As is the case for the two-source pattern, the integer m specifies the *order* of the principal maxima. In this result, we see the universal diffraction phenomenon that *the pattern spreads in angle as the ratio λ/d increases*. Equation (39–1) is the same as

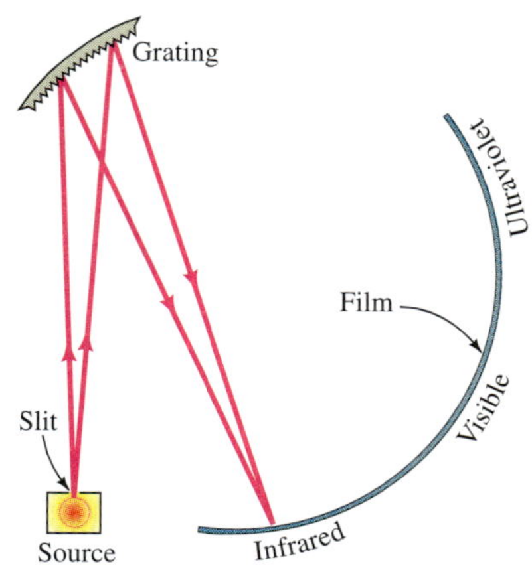

FIGURE 39–7 (a) Spectroscopes that capture an entire spectrum spread over a large sheet of film, as in this schematic diagram, are sometimes called spectrographs. The spectrograph depicted here forms part of a circle 7 m in radius, of which the film plate is a part, and employs a grating that is located along the circle. Light coming from a point on that circle is reflected back to the film. (b) The characteristic light spectrum produced by a sodium-type street lamp, as observed through a spectroscope.

(b)

Eq. (38–3a), which determines the maxima of the double-slit pattern. The intensity pattern on the screen, however, takes quite a different form.

Before performing the mathematical analysis for the intensity pattern, let's first do a qualitative analysis of intensity. If there were no interference whatsoever, the average intensity over the entire screen due to N slits would be NI_0, where I_0 is the average intensity for just one slit. Energy must be conserved whether or not there is interference, so the average intensity over the entire screen must be NI_0 even with interference. If the light intensity is zero in regions of destructive interference, there must be a higher light intensity in regions of constructive interference. What is the intensity at the principal maxima, where there is constructive interference? At such points, the electric or magnetic fields of the waves from all the slits add to N times the field from one slit. Because the intensity is proportional to the fields *squared*, the intensity at any maximum is N^2 times the intensity I_0 due to one slit:

The angular spread of the diffraction pattern increases when the ratio of the slit separation to the wavelength decreases.

$$I_{max} = N^2 I_0. \qquad (39\text{–}2)$$

The intensity of a maximum is proportional to the number of slits squared.

This result agrees with the calculation of the double-slit pattern, where we found that the heights of the maxima are $4I_0 = (2^2)I_0$. If the maximum intensity increases so markedly for N slits, then, in order to conserve energy, the space over which the maximum occurs must be much smaller. Suppose that the principal maxima have a *width* of $\Delta\theta$. (The width is the angular spread at the points where a principal maximum is half its maximum height of I_{max}.) To find $\Delta\theta$, we equate the intensity in a single maximum with the averaged intensity between maxima. Thus

$$I_{max}\,\Delta\theta \simeq NI_0 \times (\text{angular separation of successive maxima})$$
$$\simeq NI_0(\lambda/d).$$

We solve this equation for the width, and when we use Eq. (39–2), we find

$$\Delta\theta = \frac{NI_0(\lambda/d)}{I_{max}} = \frac{NI_0(\lambda/d)}{N^2 I_0} = \frac{1}{N}\frac{\lambda}{d}. \qquad (39\text{–}3)$$

The width of a maximum is inversely proportional to the number of slits.

To summarize, we have found that, *as the number of slits N increases, the height of the principal maxima increases as N^2. In addition, the width decreases—the principal maxima become sharper—as* $1/N$. This is the crucial feature that makes diffraction gratings so useful. A tall and narrow peak implies that the peak is easily seen even if the source is relatively weak, and that the peak can be located with great precision. Thus the wavelengths for which the principal maxima occur can be determined with great precision from Eq. (39–1).

Intensity Pattern

We can sum the electric fields at the screen associated with each of the slits in the N-slit pattern; in turn, we can find the intensity by squaring the time average of the net field. The result is in accordance with our energy argument:

$$\text{for multiple slits:}\qquad I = I_0\left[\frac{\sin(N\beta)}{\sin\beta}\right]^2. \qquad (39\text{–}4a)$$

The quantity β is the slit–to–slit phase difference. Using Fig. 39–8, we see that it is given by

$$2\beta = \frac{2\pi d\sin\theta}{\lambda}. \qquad (39\text{–}4b)$$

According to Eq. (39–1), $\sin\beta$ is zero at principal maxima. As $\sin\beta \to 0$, the ratio $\sin(N\beta)/\sin\beta$ approaches N. In this limit, Eq. (39–4) agrees with Eq. (39–2), which was much simpler to derive.

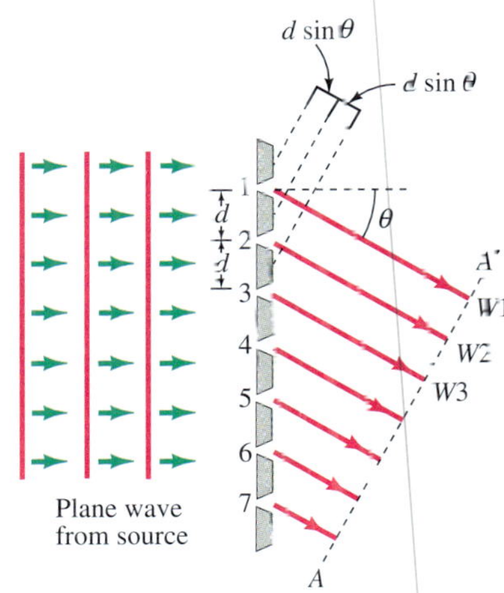

FIGURE 39–8 The geometry of a diffraction grating. Light passing through individual slits spreads in all directions. The interference of the light at a distant screen, along the directions indicated by $W1$, $W2$, $W3$, and so on, depends on the path-length differences $d\sin\theta$ between adjacent slits. The distant screen is not shown.

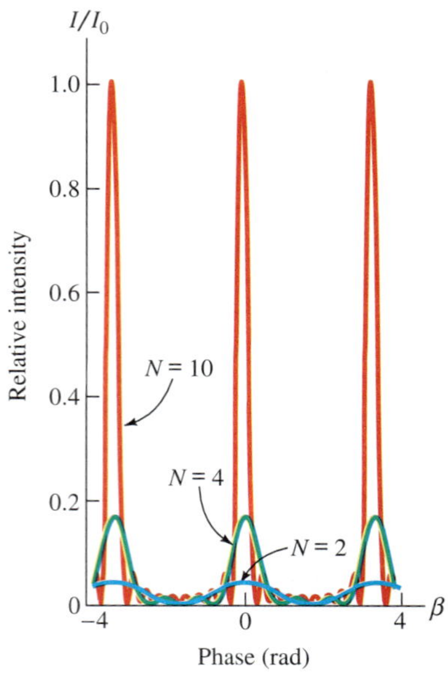

FIGURE 39–9 The ratio I/I_0 plotted against the phases $\beta = (\pi d \sin\theta)/\lambda$ for $N = 2$, 4, and 10 slits. The intensity pattern changes dramatically with N. The widths of the principal maxima decrease as $1/N$.

Equation (39–4a) is most easily understood by plotting it, as in Fig. 39–9 for $N =$ 2, 4, and 10: The ratio I/I_0 is plotted against β. The figure shows the principal maximum at $\beta = 0$ (corresponding to $\theta = 0$), the central spot on the screen, as well as the first principal maxima to the sides, corresponding to $m = \pm 1$ in Eq. (39–1). The widths of the principal maxima do indeed decrease as $1/N$. Note that there are also $N - 2$ small secondary maxima between each pair of principal maxima. They have intensities on the order of I_0 itself. For diffraction gratings in ordinary use, N is in the thousands, and the secondary maxima can safely be ignored. Physically, they occur because there is the possibility that the light from two or more of the non-neighboring slits may be in phase at a given angle, even though the light from *all* the slits is not in phase.

We can check our general result for N slits in the case $N = 2$. If we let $N = 2$ in Eq. (39–4) with $\sin(2\beta) = 2 \sin\beta \cos\beta$, we find

$$I = I_0\left[\frac{\sin(2\beta)}{\sin\beta}\right]^2 = I_0\frac{(2\sin\beta\cos\beta)^2}{(\sin\beta)^2} = 4I_0\cos^2\beta.$$

This is indeed the result for two slits, given in Eq. (38–13).

Resolution of Diffraction Gratings

Angular dispersion

Angular Dispersion. Diffraction gratings were formerly used to identify the characteristic wavelengths of elements. Now that these wavelengths are known, gratings are used primarily to *identify* elements, ions, and compounds through the characteristic light they emit or as a tool for understanding the structure of molecules. Because d, the distance between slits, is usually known, the angular location of a principal maximum ($m \neq 0$) gives λ according to Eq. (39–1). In this case, an important limitation is the ability to separate the spectral lines of nearly equal wavelengths λ_1 and λ_2. Two quantities determine the effectiveness of spectroscopic instruments. One is the **angular dispersion**, defined as $\Delta\theta/\Delta\lambda$, which measures the difference $\Delta\theta$ in the angles of the principal maxima due to two nearly equal wavelengths that differ by $\Delta\lambda$. We differentiate Eq. (39–1) to determine the angular dispersion:

$$d\,(\cos\theta)\,\Delta\theta = m\,\Delta\lambda;$$

$$\frac{\Delta\theta}{\Delta\lambda} = \frac{m}{d\cos\theta}. \tag{39–5}$$

The dispersion increases for higher orders, is inversely proportional to the distance between slits, and increases away from the central maximum.

Resolving power

Resolution. The angular dispersion alone does not tell us whether we can visually separate two similar wavelengths. This aspect of the effectiveness of a grating is characterized by the **resolving power** of the grating, defined by

$$R \equiv \frac{\lambda}{\Delta\lambda}. \tag{39–6}$$

Here, $\Delta\lambda$ is the smallest wavelength difference that can be observed with the grating. (The maxima of two wavelengths that are too close together lie so close to one another that they cannot be distinguished.) The larger the value of R, the better the grating can distinguish the relative wavelength difference of two closely spaced lines. It is the sharpness of the peaks that enables a grating to separate two closely spaced lines. By a detailed analysis of the peak widths of an N-slit system, it is possible to show that the resolving power of a grating is given by

The resolving power of a grating

$$R = mN. \tag{39–7}$$

The resolving power improves as the number of slits N increases and is better for larger orders; that is, for larger integers m.

EXAMPLE 39–1 Heated sodium provides an easily available source of light. It has a characteristic yellow–orange color and two intense wavelengths of 589.0 nm and 589.6 nm, called a *doublet*. (a) How many slits are required in a grating that resolves the doublet at the first-order maxima? (b) If the screen is exactly 4 m from a grating with exactly 2000 slits/cm, what are the screen positions of the two principal maxima of first order? Assume that the screen is far enough away so that the conditions of Fraunhofer diffraction apply.

Solution: The solution to this problem lies in our discussion of the resolving power and the position of principal maxima. We are given the two wavelengths and must look at the first-order maxima ($m = 1$).

(a) We first find the minimum resolving power needed to resolve the two closely spaced lines. The difference in wavelengths of the doublet is only $\Delta\lambda = 589.6\text{ nm} - 589.0\text{ nm} = 0.6\text{ nm}$. The resolving power needed to separate the doublet is given by Eq. (39–6),

$$R = \frac{\lambda}{\Delta\lambda} = \frac{589.0\text{ nm}}{0.6\text{ nm}} = 982.$$

Equation (39–7), with $m = 1$, gives $N = R = 982$. About 1000 slits are required.

(b) The angular positions of the first-order principal maxima are given by Eq. (39–1). To use it, we need the slit separation d, which is given by $d = 1/(2000\text{ slits/cm}) = 5.000 \times 10^{-6}\text{ m}$. Then Eq. (39–1), with $m = 1$, gives

$$\text{for } \lambda_1: \quad \sin\theta_1 = \frac{\lambda_1}{d} = \frac{589.0\text{ nm}}{5.000 \times 10^{-6}\text{ m}} = 0.1178;$$

$$\text{for } \lambda_2: \quad \sin\theta_2 = \frac{\lambda_2}{d} = \frac{589.6\text{ nm}}{5.000 \times 10^{-6}\text{ m}} = 0.1179.$$

The respective angles are $\theta_1 = 0.1181$ rad and $\theta_2 = 0.1182$ rad, and the distance from the centerline of the screen is $y = L\tan\theta$. These respective distances are

$$y_1 = (4\text{ m})\tan(0.1181\text{ rad}) = 0.4745\text{ m},$$

$$y_2 = (4\text{ m})\tan(0.1182\text{ rad}) = 0.4750\text{ m}.$$

The images are separated by only 0.5 mm, but the resolution will be sufficient for the two spectral lines to be distinguishable.

An Analogy with Interferometry

In Section 38–4, we discussed interferometers—instruments that rely on the interference between light rays that have undergone reflection. These instruments make precision measurements of distances by the observations of interference maxima and minima. We contrasted the Michelson interferometer, which utilizes the interference between two rays, with the Fabry–Perot interferometer, which utilizes multiple reflections. The Fabry–Perot interferometer can make much better measurements because the positions of maxima are much sharper. The two-slit and N-slit diffraction patterns are analogous to the Michelson and Fabry–Perot interferometers, respectively. The N-slit diffraction pattern utilizes the interference between multiple rays, and the positions of the maxima are much sharper than those of the double-slit pattern.

39–3 SINGLE-SLIT DIFFRACTION

Even coherent light passing through a *single* slit produces a diffraction pattern. This pattern forms because wavelets that have regenerated at different places across the single slit interfere with each other. If the width a of the single slit is comparable to or smaller than the wavelength of coherent light that passes through it, the diffraction pattern is a striking one. To explain the diffraction pattern, we treat the single slit as an infinitely large number of infinitesimal sources of wavelets. Thus the pattern produced by a single slit has much more in common with a diffraction grating than it does with a double slit.

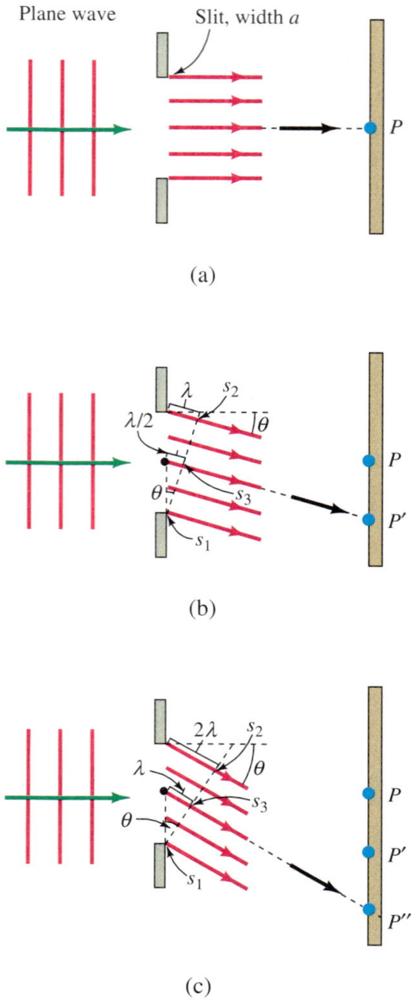

FIGURE 39–10 Monochromatic plane waves of light enter a narrow slit of width a (exaggerated here). We show three positions (P, P', and P'') on a distant screen. (a) We look along the incident direction at P. (b) We look along angle θ at P'. If the light from positions in the top half of the slit are out of phase with corresponding positions in the bottom half, destructive interference occurs at P'. (c) We look along a different angle θ at P''. Destructive interference can occur also at P'', as demonstrated by breaking the slit into halves and treating each half as we did for the direction toward point P'.

Let's consider a plane light wave of wavelength λ that moves toward an opaque barrier with a narrow rectangular slit of width $a(a > \lambda)$. A coherent wave arrives at the slit and regenerates spherical waves at each point across it. In Fig. 39–10a, we examine the light from the slit that proceeds along the initial direction toward the center point P of the distant screen. The regenerated wavelets are all in phase in this direction, and the central point of the screen is bright.

Along the direction that leads to point P' on the screen, with the angle given by $\sin\theta = \lambda/a$, there will be destructive interference under certain conditions (Fig. 39–10b). The path lengths from the line s_1s_2 to point P' are all the same because the screen is distant, so we must consider only the phase relations of the light waves at line s_1s_2. The wave emitted at the top of the slit has traveled a distance λ to point s_2, and the wave emitted from the midpoint of the slit has traveled a distance $\lambda/2$ to point s_3. Thus, along the line s_1s_2, the wave emitted at the top is *out of phase* with the wave emitted at the center of the slit (and similarly at point P'). Similarly, the wave emitted just below the top of the slit is out of phase with the wave emitted from just below the center. We can follow the points *in pairs* along the slit. For every point in the top half of the slit, there is a point in the bottom half, and the waves from the two points are precisely out of phase with each other. The result is destructive interference—a minimum (or dark spot)—on the screen at P' at the angle given by $\sin\theta = \lambda/a$.

Along the direction given by $\sin\theta = 2\lambda/a$, the wavelet emitted from the top of the slit travels a distance 2λ farther than the wavelet emitted from the bottom, and a distance λ farther than the wavelet emitted from the center point (Fig. 39–10c). We can think of the slit of width a as broken into two slits of width $a/2$. Along the direction chosen, we are in a situation similar to that in Fig. 39–10b. There is destructive interference at point P'' due to net destructive interference from both the top half and the bottom half of the slit. For example, for every wavelet emitted from a particular point in the top half of the slit, we can find a second wavelet emitted from another point in the top half that destructively interferes with the first wavelet, because their path-length difference is $\lambda/2$. The same happens for the bottom half. We will have another intensity minimum (dark spot) on the screen for angle $\sin\theta = 2\lambda/a$.

If we continue our analysis, we find that every time there is an additional path-length difference of λ between the top and bottom of the slit, destructive interference and a screen minimum result. Thus, we have

$$\text{for destructive interference:} \quad \sin\theta = \frac{m\lambda}{a}, \quad \text{where } m = \pm 1, \pm 2, \pm 3, \ldots . \qquad (39\text{–}8)$$

The value $m = 0$ is not part of this sequence of minima: For $m = 0$, $\sin\theta = m\lambda/a = 0$, and we have seen that this central point P must always be a maximum.

The interference pattern has the typical behavior of diffractive phenomena: Larger values of a/λ give a smaller angular spread in the interference pattern, and smaller values of a/λ give a larger angular spread. In the limit that $a >> \lambda$, the spread decreases so much that there is only a bright central spot on the screen. Do not confuse the angular spread on the screen with a projection of a slit width in geometric optics. The screen is very distant, and if geometric optics were to hold, the projection of the slit on the screen due to direct incident light would be a line with precisely the width of the slit. *The smaller the slit width compared to the wavelength, the larger the angular spread of the interference pattern.*

Approximately halfway between the successive minima, we have conditions for constructive interference and an intensity maximum (a bright region). Figure 39–11 shows the diffraction pattern of a single slit; most of the light is contained in the wide central maximum.

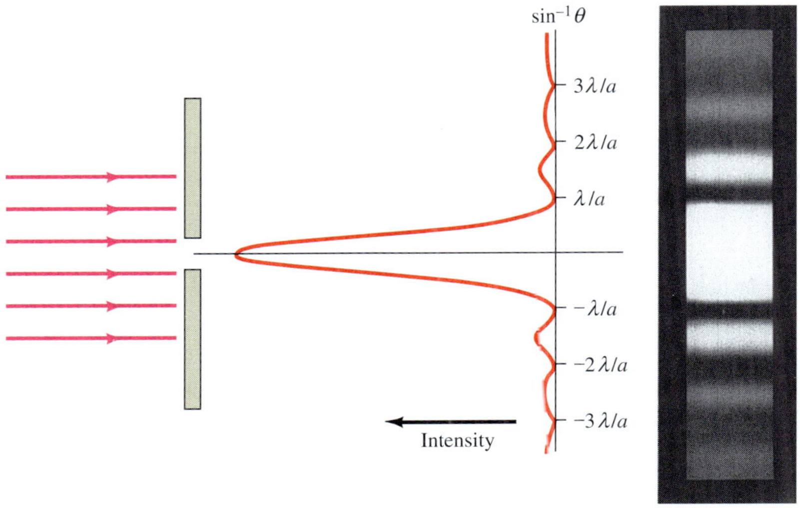

FIGURE 39–11 The interference pattern of single-slit diffraction and the relative intensities of such a pattern. Most of the light energy is in the bright central peak. The central peak is twice as wide as the secondary peaks.

EXAMPLE 39–2 Helium–neon laser light of wavelength 633 nm passes through a single slit of width 0.10 mm. The diffraction pattern is observed on a screen 3 m away—far enough so that the conditions for Fraunhofer diffraction apply. What is the distance between the two minima on either side of the central maximum?

Solution: The conditions for the single-slit pattern apply here; in particular, the slit width a is some 160 times larger than the wavelength. We can use Eq. (39–8) to find the angular positions of the minima that correspond to $m = 1$ and -1. These angles are, respectively,

$$\sin\theta = \frac{m\lambda}{a} = (\pm 1)\frac{633 \times 10^{-9}\text{ m}}{0.10 \times 10^{-3}\text{ m}} = \pm\,0.0063;$$

$$\theta = \pm 0.36^\circ.$$

At a distance of 3 m away, these minima will be (3 m) tan 0.36° = 1.9 cm from the central line. The total distance between them is thus 2(1.9 cm) = 3.8 cm.

EXAMPLE 39–3 A researcher embeds a single slit of width 5×10^{-4} m in a piece of glass of index of refraction n. He then shines laser light of wavelength 560 nm through one surface of the glass, which is diffracted through the slit and passes out another surface of the glass and on to a distant screen. The screen in which the slit is cut, the second surface of the glass, and the distant screen are all parallel to one another and perpendicular to the incoming laser beam (Fig. 39–12). He hopes to measure the index of refraction, n, of the glass by seeing how the position of the first diffraction minimum differs from the position of the same minimum when the slit is in air. Show that the researcher cannot measure n this way.

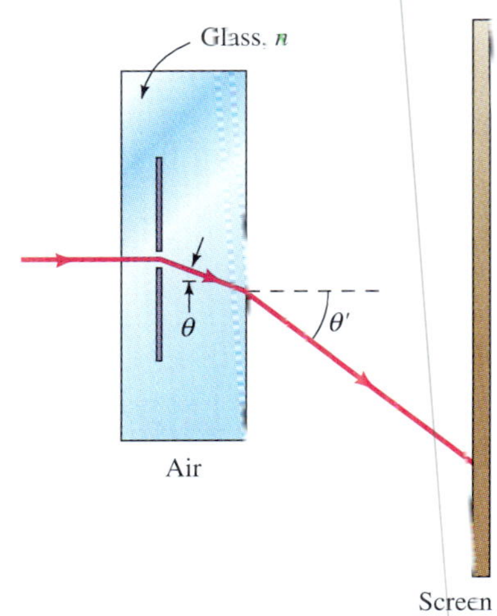

FIGURE 39–12 Example 39–3.

Solution: We want to show that the position of the first minimum on the screen is the same whether the slit is embedded in the glass or not. An important clue is that the ratio λ/a—where a is the slit width—is small so the diffraction is through small angles, and we can use small-angle approximations. With the slit in air, then, the first minimum occurs at an angle θ given by $\sin\theta \simeq \theta = \lambda/a$. With the slit embedded in the glass, the diffracted beam is refracted at the glass–air surface before arriving at the screen, as in Fig. 39–12. The angle of the first minimum within the glass, θ', is given by a similar formula, $\theta' = \lambda_g/a$. However, the wavelength within the glass is *not* equal to the wavelength in air. We have the general formula $\lambda f = v$, with v the wave speed; f cannot change to avoid a pileup of wave fronts at a surface, so λ must change if v does. The speed of light in the glass is c/n, hence λ_g is decreased to λ/n. The first diffraction minimum occurs at a decreased angle, $\theta' = \lambda/na = \theta/n$. The angle at which this line refracts is given by Snell's law for small angles, $\theta'' = n\theta' = n(\theta/n) = \theta$. Finally, because the glass piece is small and the screen is distant, the two equal angles θ'' and θ correspond to the same position on the screen.

The Intensity Pattern of Single-Slit Diffraction

The intensity pattern due to a single slit

We already did most of the work to find the detailed diffraction pattern of a single slit when we analyzed the N-slit grating. As our qualitative analysis shows, the crucial physical principle behind the diffraction pattern due to a single slit is that light from different parts of the slit interfere with each other. We can therefore formally divide the single slit into an infinite sequence of equally spaced and infinitely narrow pieces and analyze it as a grating. This work is done in the optional subsection that follows this one; the result is *the intensity pattern on a distant screen due to a single slit of width a*:

$$\text{for single slits:} \qquad I = I_{\max}\frac{\sin^2\alpha}{\alpha^2}, \tag{39–9}$$

where

$$\alpha = \frac{\pi a \sin\theta}{\lambda}. \tag{39–10}$$

Note that there is a central maximum at $\theta = 0$: At $\theta = 0$, Eq. (39–10) shows that $\alpha = 0$. Because the limit of $(\sin\alpha)/\alpha \to 1$ as $\alpha \to 0$, the intensity at this point on the screen is just $I_{\max}$.

Our qualitative discussion at the beginning of this section shows that angle α is simply the phase difference between the top and the middle of the slit. The intensity drops off rapidly as α increases. The intensity is zero for angles given by Eq. (39–8). We can now see that this occurs when α is an integral multiple of π:

$$\text{for minima:} \quad \alpha = n\pi = \frac{\pi a \sin\theta}{\lambda}, \quad \text{where } n = \pm 1, \pm 2, \pm 3, \ldots. \tag{39–11}$$

The intensities at the secondary maxima are estimated in Example 39–4.

EXAMPLE 39–4 Estimate the ratios of the intensities of the first and second maxima to the intensity of the central maximum for a single slit.

Solution: In order to find the exact positions of the maxima, we must take the derivative of the intensity with respect to α (or θ) and set that equal to zero (see Problem 25). However, we can make a rapid estimate by realizing that the maxima are approximately halfway between the minima:

$$\text{for maxima:} \qquad \alpha \simeq \left(n + \frac{1}{2}\right)\pi, \text{ where } n = 1, 2, 3, \ldots.$$

Let's denote the intensity at these secondary maxima by I_n. We use this approximation of α in Eq. (39–9) to get

$$\frac{I_n}{I_{\max}} = \left\{\frac{\sin[(n+\frac{1}{2})\pi]}{(n+\frac{1}{2})\pi}\right\}^2 = \frac{1}{(n+\frac{1}{2})^2\pi^2}.$$

[We have used $\sin^2(\pi/2) = \sin^2(3\pi/2) = \cdots = 1$.] With $n = 1$ and $n = 2$, we get

$$\frac{I_1}{I_0} = 0.045 \quad \text{and} \quad \frac{I_2}{I_0} = 0.016.$$

The intensities of the secondary maxima fall off rapidly. The central bright maximum is by far the most intense.

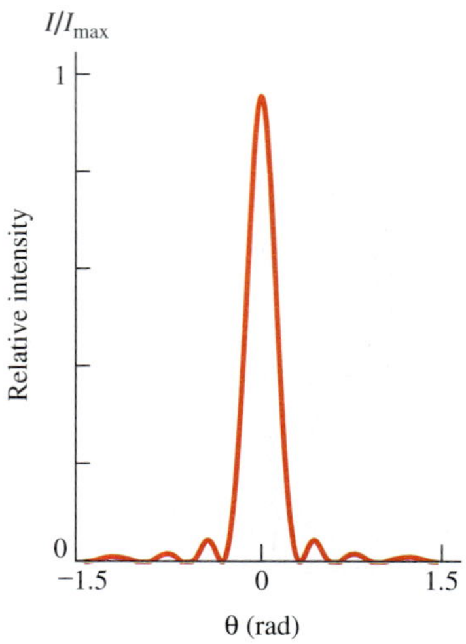

FIGURE 39–13 The intensity ratio $I/I_{\max}$ as a function of angle θ [from Eqs. (39–19) and (39–10)] for the slit width $a = 4\lambda$.

Figure 39–13 is a plot of the ratio of the intensity to its maximum value for the single slit as a function of θ. Note that the central maximum is *twice* as broad as the secondary maxima are—a feature that distinguishes the single-slit pattern from the double- or multiple-slit patterns. We have chosen $a = 4\lambda$ in this plot.

*How to Get the Single-Slit Intensity Pattern

Let's imagine breaking up a single slit of width a into N strips of width d (Fig. 39–14). (Each strip acts as a separate slit.) The entire slit is divided in this way, so

$$Nd = a.$$

In the limit that the number of strips $N \to \infty$, d must approach zero in order to keep a constant. Because each strip is infinitely narrow, we can treat these strips as the thin slits of a grating. We can then use the diffraction grating result of Eq. (39–4) directly to find that the intensity at the screen at angle θ is

$$I = \lim_{N\to\infty} I_0 \left[\frac{\sin(N\beta)}{\sin\beta}\right]^2.$$

This expression includes a slit width $d = a/N$ that approaches zero as $N \to \infty$. According to Eq. (39–4), $\beta = [\pi d \sin\theta]/\lambda = [\pi(a/N)\sin\theta]/\lambda = \alpha/N$, where the definition of α is contained in Eq. (39–10). Thus the intensity takes the form

$$I = \lim_{N\to\infty} I_0 \left[\frac{\sin\alpha}{\sin(\alpha/N)}\right]^2.$$

In the limit of large N, the factor $\sin(\alpha/N) \simeq \alpha/N$, and

$$I = N^2 I_0 \frac{\sin^2\alpha}{\alpha^2}.$$

The factor I_0 is the intensity due to one of the subslits of width d. We need only interpret the factor $N^2 I_0$ as the maximum possible intensity I_{max} of the single slit of width a to get Eq. (39–9).

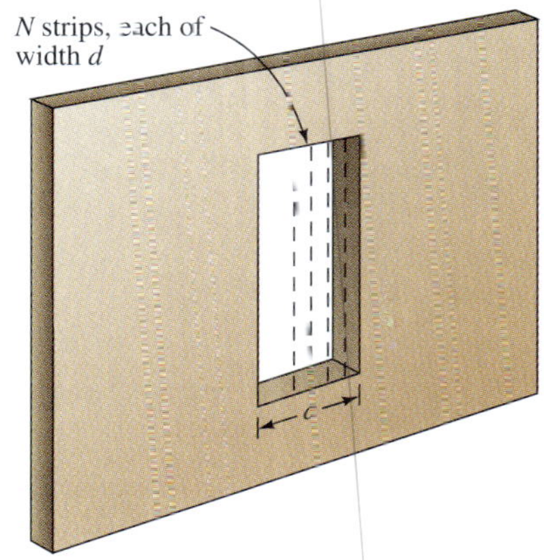

FIGURE 39–14 We can understand the intensity pattern of a single slit by supposing that the slit is composed of a large number of strips N and then use the result we derived for diffraction gratings.

39–4 DIFFRACTION AND RESOLUTION

We have already seen that the ability of a grating to resolve closely spaced lines is limited by the width of the principal maxima. Similarly, the fact that there is some spreading of light in a single aperture due to diffraction *intrinsically* limits the capacity of optical instruments to resolve objects. The resolution of instruments such as telescopes or microscopes is also limited by lens *aberration* (see Chapter 37). Better lens design decreases aberration, and there is no theoretical limit to such improvement. The limitation due to diffraction, however, is set by the aperture of the instrument and the wavelength of light.

Because most optical instruments rely on circular lenses or mirrors, we shall concentrate on circular apertures. Sir George Airy worked out the diffraction pattern from a distant point source that passes through a circular aperture (Fig. 39–15) in the 1830s. The bright central area—containing some 85 percent of the light intensity—is called an *Airy disk*. The rings outside the central area are the minima and secondary maxima of the diffraction pattern. The position of the first minimum occurs at an angle from the central axis given by

$$\theta_{min} = 1.22\frac{\lambda}{D}, \tag{39–12}$$

where D is the diameter of the aperture. This result can be compared to Eq. (39–8), which gives the angle of the first minimum for a slit of width $a \gg \lambda$ as $\theta_{min} = \lambda/a$.

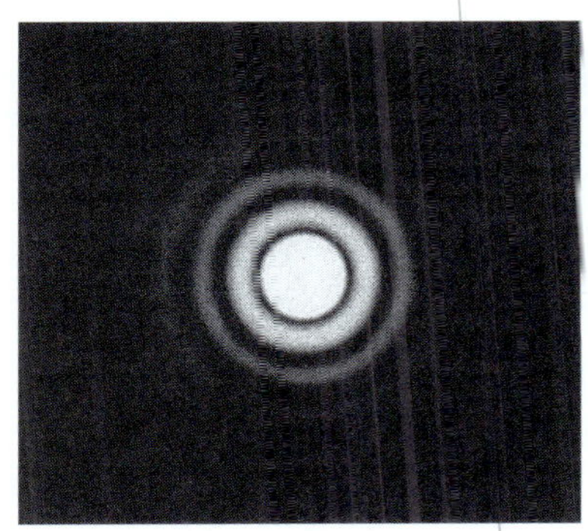

FIGURE 39–15 Diffraction pattern of light from a distant source after the light passes through a circular aperture. Some 85 percent of the light intensity is contained in the bright central maximum, called the Airy disk.

(Recall that sin $\theta \simeq \theta$ for small θ.) The factor 1.22 arises because the "width" of a circular aperture varies (in effect, $a \simeq D/1.22$). The precise factor 1.22 is actually irrelevant to us, and we prefer to drop it and give an approximate θ_{min} as

The approximate position of the first minimum of the diffraction pattern of an aperture

$$\theta_{min} \simeq \frac{\lambda}{D}. \tag{39–13}$$

Let's look at how the presence of a diffraction pattern limits our ability to make images. We may want to observe two closely spaced objects, such as a double star, or we may need to see detail in an X-ray taken of suspected stress fractures in piping for a power plant. The presence of the Airy disk means that even a very distant star does not produce a pointlike image but rather a disklike image with an angular spread described by Eq. (39–13). The image in a telescope of two stars that are so close together that their Airy disks overlap cannot be easily recognized as an image of two stars. In Fig. 39–16a, two objects have a sufficient angular separation θ that they are easily resolved. In Fig. 39–16b, the objects are just barely resolved. When they are even closer, as in Fig. 39–16c, the central diffraction peaks overlap too much to be resolved. It has become customary to describe the limiting case—shown in Fig. 39–16b—as the *Rayleigh criterion*:

The Rayleigh criterion

Two point sources are just resolved if the peak of the diffraction image of the first source overlies the first minimum of the diffraction image of the second source.

The Rayleigh criterion is satisfied when the angular separation of the objects is just θ_{min}, defined by Eq. (39–12). Equation (39–13) provides an alternative criterion. We emphasize that the limitation on the resolution of images due to the wave nature of light is fundamental.

This discussion refers to the angular separation of the image formed by an optical system. The methods of Chapter 37 allow us to show that this angle is also the angular separation of the objects. The angles are the same because the principal rays through the center of a lens are undeviated. When we use a microscope, for example, it is useful to know the spatial separation of two objects that can just be resolved. If the objects are placed at the focal point of the lens (Fig. 39–17), the minimum separation S_{min} for a lens of diameter D is given by

$$S_{min} = f\theta_{min} \simeq \frac{f\lambda}{D}. \tag{39–14}$$

This result can be applied to find the minimum separation of objects visible to the human eye.

EXAMPLE 39–5 Estimate the minimum separation between two objects such that the human eye can still perceive them as separate (the *minimum visible object separation*) if the objects are (a) at the near-point distance (25 cm) and (b) at a distance of 5 m. Take the pupil diameter to be 2.5 mm.

Solution: (a) We use the median range of visible wavelengths, or about 550 nm, in Eq. (39–14) to estimate the minimum visible object separation. For viewing an object at a given distance, the muscles of the eye adjust the focal length to that distance. Thus at 25 cm (the near point), we have

$$S_{min} = \frac{f\lambda}{D} = \frac{(0.25\ \text{m})(550 \times 10^{-9}\ \text{m})}{0.0025\ \text{m}} \simeq 0.055\ \text{mm}.$$

This is about the diameter of a thin thread or a human hair. It is also, roughly, the separation between the cells of the retina. In other words, the cells that receive light and send that message to the brain are no closer than the minimum separation that could ever be resolved, an admirable example of the economy of biological systems.

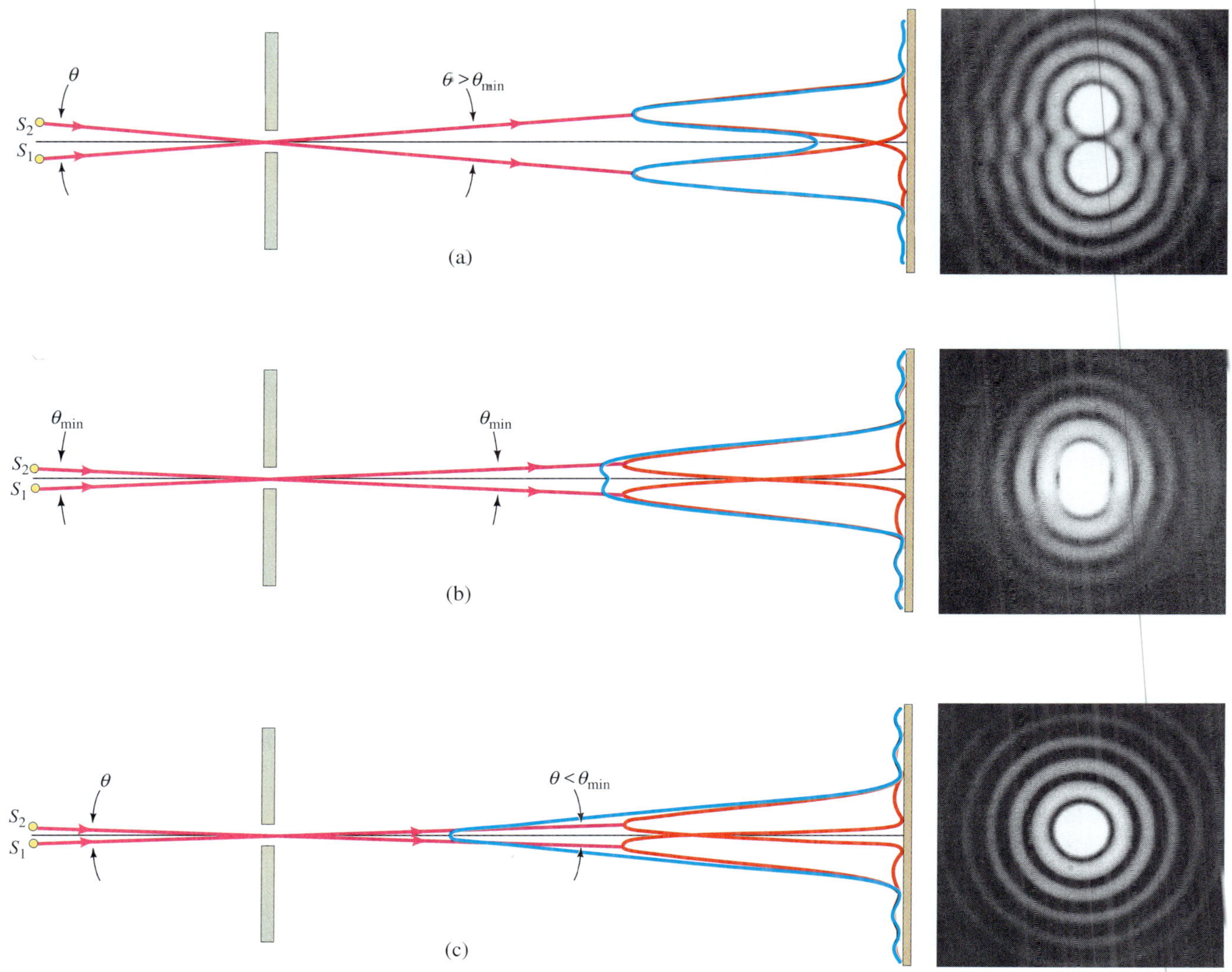

FIGURE 39–16 Light from two sources passes through an aperture (a) The objects are easily resolved. (b) The objects are just barely resolved. The angular separation is $\theta_{\min}$. (c) The objects are not resolved. The blue lines in each represent the sum of the two intensities.

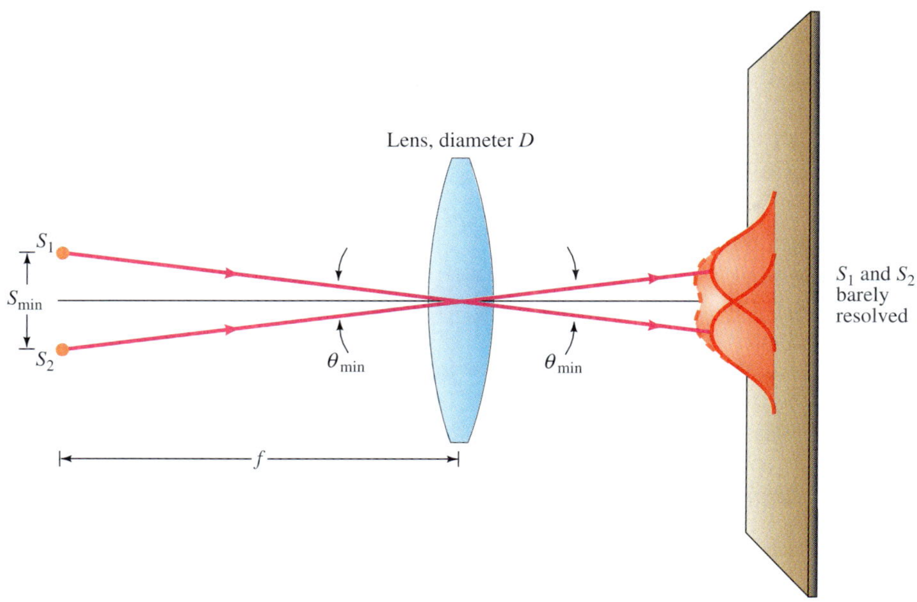

FIGURE 39–17 The minimum separation distance that is resolvable for a lens depends on $\theta_{\min}$ and the focal length of the lens.

(b) At 5 m, the minimum separation distance becomes

$$S_{\text{min}} = \frac{(5.0\text{ m})(550 \times 10^{-9}\text{ m})}{0.0025\text{ m}} \simeq 1\text{ mm}.$$

Thus, from a few meters away, we have an *intrinsic* inability to distinguish the millimeter markings on a meter stick.

FIGURE 39–18 The Keck telescope complex in Hawaii is one of the finest astronomical instruments in existence.

The Resolution of Telescopes

Diffraction effects can limit the effectiveness of telescopes. For the visible-light region of the electromagnetic spectrum, one of the world's largest telescopes is the 200-in-diameter Hale telescope on Mt. Palomar in southern California. Its diffraction limit alone implies an angular resolution of about 0.03″ of arc. Its real resolution is some 300 times worse than this due to the effects of atmospheric turbulence and ordinary aberration. The New Technology Telescope in Chile, another visible-light telescope, has resolved 0.36″ of arc. Radio telescopes such as that in Arecibo, Puerto Rico, operate in a wavelength range of roughly tens of centimeters. For a wavelength of 50 cm, the angular resolution of this apparatus, which has a diameter of 300 m, is about 7′ of arc.

EXAMPLE 39–6 The primary mirror on the Optical Telescope Assembly on the Hubble Space Telescope, which orbits 600 km above Earth, has a diameter of 2.4 m. (a) Calculate the minimum angular separation that it might resolve for visible light (about 550 nm). (b) Assume that the telescope is viewing Earth's surface. What is the separation of the most closely spaced objects that it might resolve? Ignore all atmospheric effects.

Solution: (a) This problem is a direct application of the resolution angle of Eq. (39–13):

$$\theta_{\text{min}} \simeq \frac{\lambda}{D} = \frac{(550 \times 10^{-9}\text{ m})}{2.4\text{ m}} = 2.3 \times 10^{-7}\text{ rad} = 0.06'' \text{ of arc.}$$

This resolution is superior to the best resolution obtained on Earth because of the effects of atmospheric turbulence.

(b) The separation distance corresponding to the angle found in part (a) from a distance of $L = 600$ km is

$$d_{\text{min}} = L\theta_{\text{min}} = (600\text{ km})(2.3 \times 10^{-7}\text{ rad}) = 0.14\text{ m} = 14\text{ cm}.$$

The 14-cm separation would be far from the real resolution. Atmospheric turbulence sets the true limit of such a satellite. Earth-observation satellites such as Landsat and Spot have resolution capabilities of a few meters, and classified spy satellites are reported to have resolutions of less than 1 m.

An optical device need not consist of a single aperture. The mirrors of many of the largest modern reflecting telescopes are separated into several pieces, like a mosaic, in part because a single large piece of glass is difficult to form and handle (Fig. 39–18). Such instruments diminish diffractive effects because the minimum angle that can be resolved is determined by the interference from the most widely separated pieces of the apparatus. If this maximum separation is D, then the minimum resolvable angle between objects is λ/D. (Note that D refers to a *transverse* separation; the distance between an eyepiece and an objective does not enter into diffractive effects.) An array of small electronically connected optical telescopes on the Moon spread over a region with a diameter of 10 km would have a resolution 100,000 times better than that of the best telescope on Earth. Were it not for Earth's atmosphere, such an instrument could pick up a newspaper headline on Earth!

*39–5 SLIT WIDTH AND GRATING PATTERNS

Our discussions of double- and multiple-slit diffraction patterns have treated each slit as a point source of light that emits a single spherical wave. The angular spread of the diffraction pattern depends on the parameter d/λ, where d is the slit separation. We have now seen that single slits of finite width have their own diffraction pattern. The angular spread of this pattern depends on a/λ, where a is the slit width. What effect does a finite slit width have on the multiple-slit pattern? For Fraunhofer diffraction, the overall intensity distribution is the product of the two intensity patterns. The pattern I_{mult}, corresponding to the double or multiple slit, is multiplied by the pattern I_{single}, corresponding to the single slit. The multiple-slit intensity is given by Eq. (39–4), whereas the single-slit intensity is given by Eq. (39–9). Thus their product is

$$I = I_{\text{mult}} I_{\text{single}} = I_{\text{max}} \left[\frac{\sin(N\beta)}{\sin \beta} \right]^2 \left(\frac{\sin \alpha}{\alpha} \right)^2, \tag{39–15}$$

where we recall that

$$\beta = \frac{\pi d \sin \theta}{\lambda} \quad \text{and} \quad \alpha = \frac{\pi a \sin \theta}{\lambda}.$$

In the equation for the intensity, we have combined the maximum intensity factors into a single maximum intensity I_{max}.

The fact that the intensity patterns are multiplied means that the broader pattern (usually due to the single slit) acts as an envelope for the narrower pattern. For example, suppose that $d = 3a$ for a double slit ($N = 2$). In this case, the individual slit pattern is much broader than the multiple-slit pattern. At the same time, let $a = 4\lambda$, so that the single-slit pattern is easily distinguishable. Figure 39–19 shows the single-slit pattern, double-slit pattern, and combined pattern. Note that certain maxima of the double-slit pattern are absent from the combined pattern because they fall where the minima of the single-slit diffraction pattern occur. These missing maxima are called *missing orders*. The locations of missing orders are independent of λ, as Problem 37 illustrates. A measurement of the pattern described here, but with $d = 10a$, is shown in Fig. 39–20.

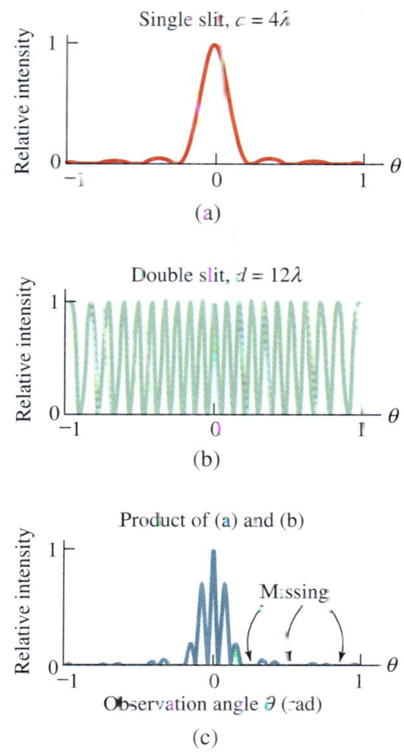

FIGURE 39–19 Intensity patterns as a function of observation angle θ for diffraction from multiple slits must include the effects of single-slit diffraction. For $a = 4\lambda$, (a) the intensity pattern for a single slit, (b) a double slit ($d = 12\lambda$), and (c) their product, which is the observed pattern. Missing orders occur; in this case, $d = 3a$, and the 3rd, 6th, 9th, . . . orders are missing from the overall pattern

FIGURE 39–20 Diffraction pattern for multiple slits where $d = 10a$. Note the missing orders.

*39–6 X-RAY DIFFRACTION

We have been emphasizing the use of gratings as a tool for the exploration of diffracted light. Light is just one form of electromagnetic radiation; here, let's take a look at the diffraction of X-rays. We shall see that crystalline solids form a natural kind of grating and that the diffraction of X-rays can be used to explore properties of solids. Diffraction gratings work because the apertures or obstacles serve as rescatterers. A powerful constructive interference occurs among rescattered light from *all* the apertures or obstacles because the sources of wavelets are in a *regular* pattern. The atoms in a crystalline solid serve nicely as a grating because they do indeed form a regular array of obstacles, even if that array is spread over three dimensions rather than two. Each atom in the array can serve as a rescatterer if the light can penetrate.

In 1895, Wilhelm Roentgen discovered that radiation was produced when he bombarded metal with high-energy *cathode rays* (now called electrons). This radiation was unlike any seen previously, and Roentgen called it *X-rays*. Shortly thereafter, he produced the first X-ray picture, a human hand (Fig. 39–21). We know now that X-rays are just electromagnetic radiation with wavelengths in the range of about 0.01 nm to 10 nm. This radiation is produced when atomic electrons change states within atoms, or when electrons are accelerated (or decelerated).

In the early 1900s, it was suspected that X-rays might be some form of electromagnetic radiation. A diffraction experiment reported in 1899 vaguely suggested that X-rays might have wavelengths of about 0.1 nm, much smaller than those of visible light. At the same time, some scientists suspected that solids might be made of atoms arranged in regular arrays. In 1912, Max von Laue had the idea of scattering X-rays from solids. If X-rays had about the same wavelength as the distance between the arrays of atoms ($\simeq 0.1$ nm), then diffraction effects would be significant. Von Laue was as interested in finding a tool for the precise measurement of the wavelengths of X-rays as he was in finding a tool for the exploration of crystals. He convinced two of his colleagues, Friedrich and Knipping, to perform an experiment, and the observation of X-ray diffraction soon followed. Von Laue's idea was a crucial step in the measurement of X-ray spectra and led to a revolution in our ability to study the nature of solids and the molecules that compose them. The precise knowledge that table salt, NaCl, has the three-dimensional structure shown in Fig. 39–22 is a consequence of X-ray diffraction experiments; virtually all of our knowledge of crystalline structure comes from such experiments. It was the use of X-ray diffraction on a crystallized form of DNA that led to the discovery of that molecule's double-helical structure (Fig. 39–23).

Because the rescattering centers (the atoms of a solid) are pointlike and three-dimensional rather than slitlike, the diffraction pattern of a crystalline solid consists of a regular array of spots rather than lines. The von Laue experiment on a crystalline solid, shown schematically in Fig. 39–24a, leads to a set of spots like those shown in Fig. 39–24b. Von Laue's idea was clarified almost immediately by W. L. Bragg in 1912, who proposed a simple and systematic way of showing just how the positions of the spots would be determined by the solid's crystalline structure. Bragg pointed out that, in any crystal, many sets of parallel planes (called *Bragg planes*) can be drawn that pass through the positions of the atoms, and that the planes of a set are separated by characteristic distances (*Bragg spacings*). Figure 39–25 shows where some of these planes cut a two-dimensional cross section of a cubic lattice similar to that formed by NaCl. One such plane is shown in a three-dimensional cutaway view of a crystal model in Fig. 39–26, and they are all throughout the crystal. The advantage of this approach

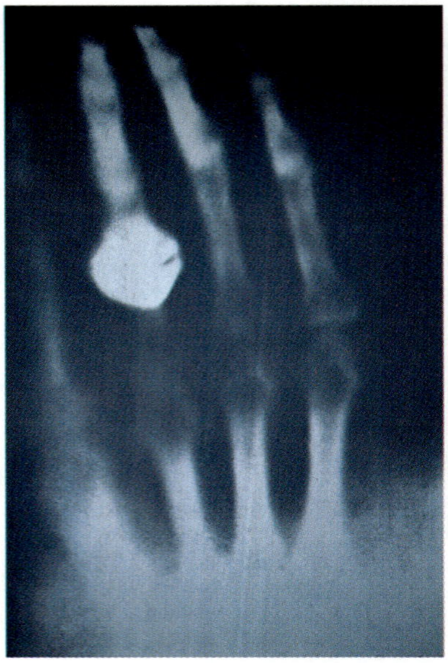

FIGURE 39–21 Roentgen's first X-ray photograph of a human—that of his wife's hand. X-rays have the well-known property of penetrating matter.

APPLICATION

Measuring the crystalline structure of solids

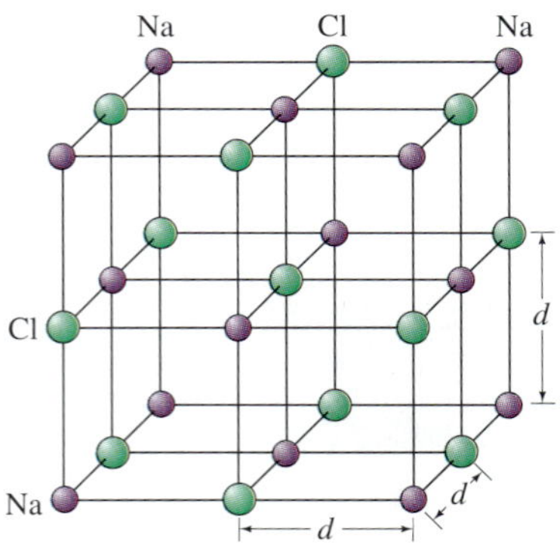

FIGURE 39–22 Crystals have three-dimensional structure with their atoms in regular arrays. This diagram shows one of the simplest, NaCl (table salt), which has a cubic structure.

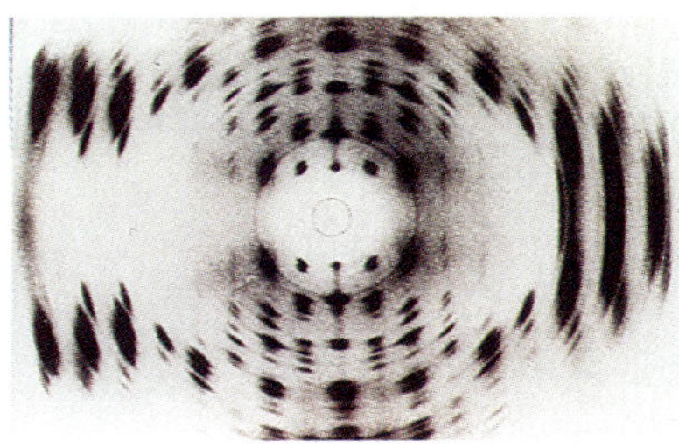

(a)

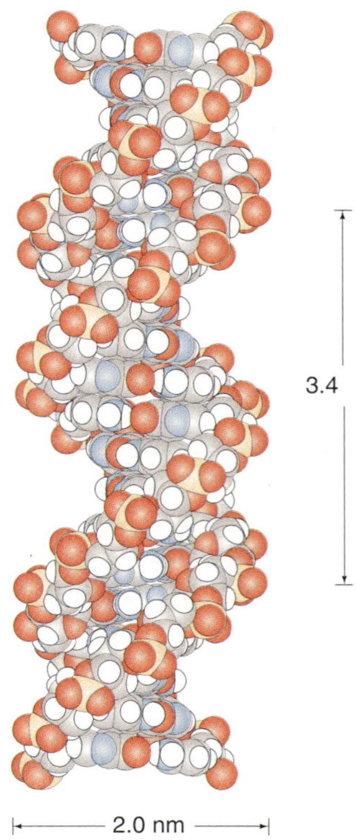

(b)

FIGURE 39–23 (a) Analysis of thousands of diffraction patterns produced by crystals of the large biological molecule deoxyribonucleic acid (DNA) showed that (b) the molecule has the shape of a double helix.

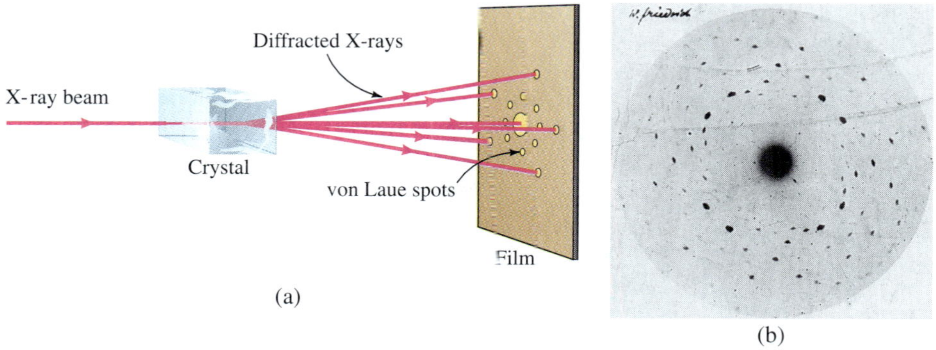

FIGURE 39–24 (a) Schematic diagram of the von Laue experiment for the diffraction of X-rays. (b) Von Laue spots in one of the first X-ray diffraction patterns. The large spot is undiffracted radiation.

is that we can think of each family of parallel planes as a slit-type diffraction grating for the X-rays. Figure 39–27 shows two rays scattered from two parallel planes within a crystal. These rays scatter from a given plane for which the reflection angle equals the incident angle because, at that reflection angle, the wavelets emitted by each atom within that plane add constructively. Now consider the interference between the scattered waves of *different* planes, which occurs because the X-rays penetrate the crystal. If the separation between the planes is d, then, from the geometry of Fig. 39–27, the difference in path lengths for the two lines is $2d \sin \theta$. Note that angle θ is measured from the plane surface rather than from the normal to the plane. Constructive interference for scattering from these two adjacent planes occurs when this path-length difference is an integer multiple of the wavelength. This relation is known as **Bragg's law**, or the **Bragg condition**:

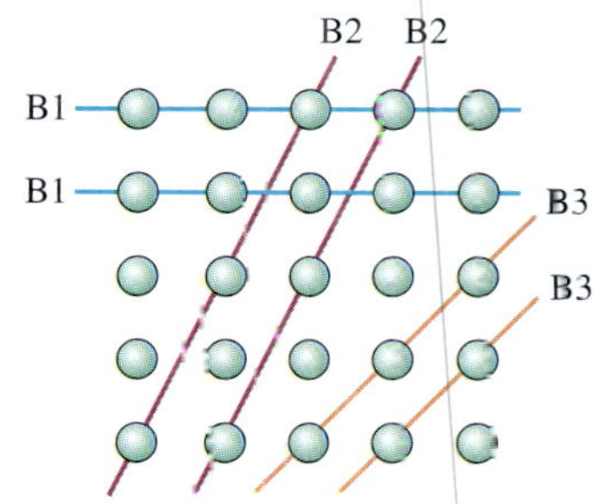

FIGURE 39–25 Many parallel Bragg planes can be drawn in a three-dimensional crystal.

$$\text{Bragg's law:} \quad 2d \sin \theta = n\lambda, \quad \text{where } n = 1, 2, 3, \ldots . \tag{39–16}$$

The angles for which there is constructive interference in scattering from crystalline arrays of atoms

Because the planes are equally spaced, the waves that scatter from the atoms in the entire set of planes in the direction specified by the Bragg condition *all* add constructively and, as in the case of a diffraction grating, the maximum is large and narrow.

Although this discussion is inadequate to explain the *intensities* of the spots, we may state generally that, if a particular family of planes contains more atoms than another does, the maxima those planes give are more intense. Intensity information is very important in determining the crystalline structure. More advanced mathematical methods are necessary for a complete model of the diffraction pattern, which is quite complicated because of the many possible planes and orders of scattering. This discussion describes the basis of the X-ray spectrometer (Fig. 39–28).

FIGURE 39–26 A cutaway model of crystal structure.

EXAMPLE 39–7 Figure 39–28 shows an X-ray tube, which produces a continuous distribution of wavelengths. If these wavelengths are scattered from a particular set of parallel planes of rock salt (NaCl) with a spacing $d = 0.282$ nm, what wavelengths will appear in the first and second orders at 25°?

Solution: In order to take advantage of Bragg's law, either the wavelength or the atomic-plane spacing must be known. In this case, the plane spacing is known,

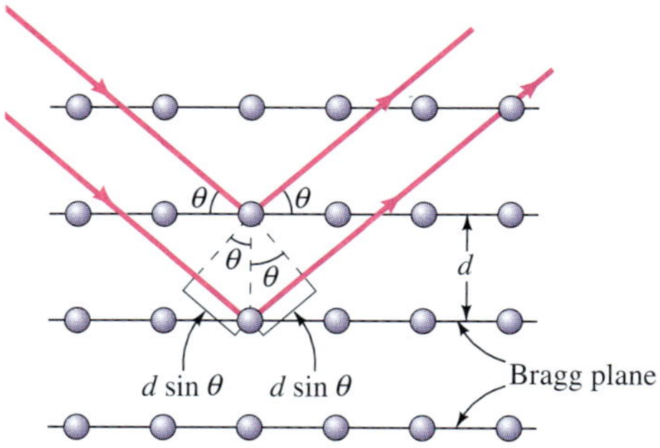

FIGURE 39–27 The geometry of X-ray diffraction between adjacent Bragg planes.

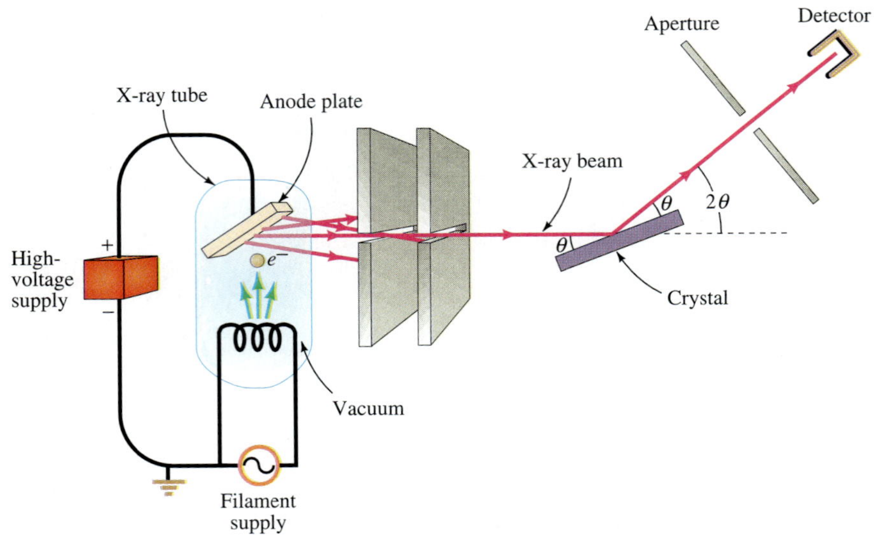

FIGURE 39–28 Schematic diagram of an X-ray spectrometer used to study properties of crystals. Electrons bombard the anode plate, producing X-rays that are collimated before being scattered by the crystal. A movable detector records X-ray intensity as a function of θ to determine where constructive interference occurs.

and the Bragg law can be used to identify unknown wavelengths or to select particular wavelengths to be used for further study. We use Eq. (39–16) to determine the wavelengths:

$$\lambda = \frac{2d \sin\theta}{n} = \frac{2(0.282\text{ nm})(\sin 25^\circ)}{n} = \frac{0.238\text{ nm}}{n}.$$

The wavelengths at 25° are 0.238 nm and 0.119 nm for the first ($n = 1$) and second ($n = 2$) orders, respectively. Note that if $\theta = 25^\circ$, the overall deflection from the original beam direction is $2\theta = 50^\circ$.

*39–7 HOLOGRAPHY

In 1947, Dennis Gabor proposed that interference effects between light emitted by an object (a source) and a second coherent beam can be recorded on film, which becomes a very special diffraction grating. When light is passed through this diffraction grating, it is diffracted and forms a fully three-dimensional image of the object, an image that can be viewed from different positions and angles, just like the original object. This process is **holography**, and the film on which the interference pattern is stored is a *hologram.*

In order to understand the principles, let's start with a distant point source that sends plane waves directly toward a piece of film (Fig. 39–29). At the same time, let's send a second beam—*coherent with the light from the source*—toward the film from an angle θ_r. This second beam is known as the *reference beam.* The reference beam interferes with the light from the source. Suppose that the wave in the reference beam interferes constructively with the source wave at point P_1. There will also be constructive interference at P_2, a distance d from P_1, if the wave path along line ℓ_2 differs from the path along line ℓ_1 by λ (or an integer m times λ; we consider only the case $m = 1$). When there is constructive interference at both P_1 and P_2, the relation between θ_r, the wavelength of the light, λ, and the separation d is

$$d \sin\theta_r = \lambda. \tag{39–17}$$

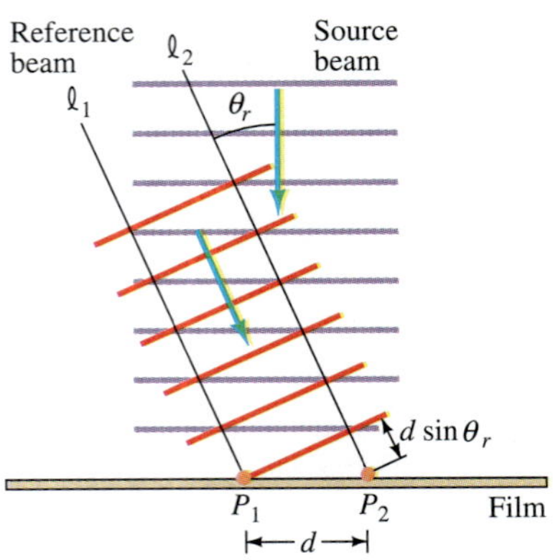

FIGURE 39–29 Coherent light from a distant source beam and from a reference beam interfere on film, producing a hologram.

On the slice of film shown in Fig. 39–29, constructive interference will occur at a series of equally spaced points, which will be recorded as dark spots on the film. (Recall that the film makes a negative.) These points are parts of continuous lines into or out of the page. The full interference pattern on the film thus consists of a set of curving lines that represent all the places where there is constructive interference. The film can record with shadings of gray places where the interference is not totally destructive.

Let's now turn to the question of how the image is viewed (or *reconstructed*). Suppose that we project a beam just like the reference beam, and at the same angle, onto the back of the film (Fig. 39–30). The dark areas on the film act as obstacles that rescatter the light. The direction indicated, *that of the original light from the source*, is a direction for which the diffracted light is a maximum as the geometry in Fig. 39–30 shows. Thus a viewer placed at point E will see light as though it comes from a distant point I, which we may think of as an image of the original source. Note that there is no requirement that the beam that produces the image be identical to the original reference beam, as long as it is coherent across the film. If the angle of the new beam is different, the only effect is to shift the angle of the viewed image.

Suppose now that the point source is closer to the film when the image is made. In this case, the spots of constructive interference will not be spaced equally across the screen (Fig. 39–31). At region A_1, the situation we have described is reproduced, but at region A_2, the points where there is constructive interference between the beams are different. When the exposed film is illuminated by a reference beam, a viewer at E_1 will see a plane wave along the direction from E_1 back to region A_1; that is, the observer will be looking back at the source from one angle. However, when the viewer is at E_2, the maximum of the diffraction pattern will indicate an image back along the direction from E_2 to A_2. *The viewer will be looking at the source from another angle.* There is a true three-dimensional image, which can be viewed from different angles.

When the object is more complicated than a point source, light arrives at any given point on the film from many points of the object. The interference pattern that this light makes with the reference beam is far more complicated and irregular, but it is never-

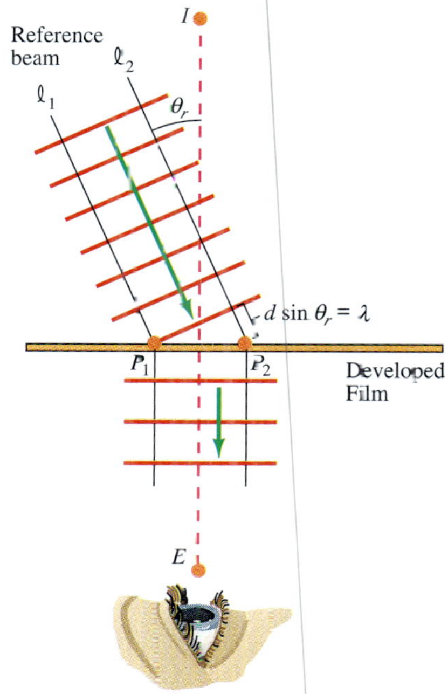

FIGURE 39–30 When a reference beam shines on holographic film, the image of the original object is reconstructed.

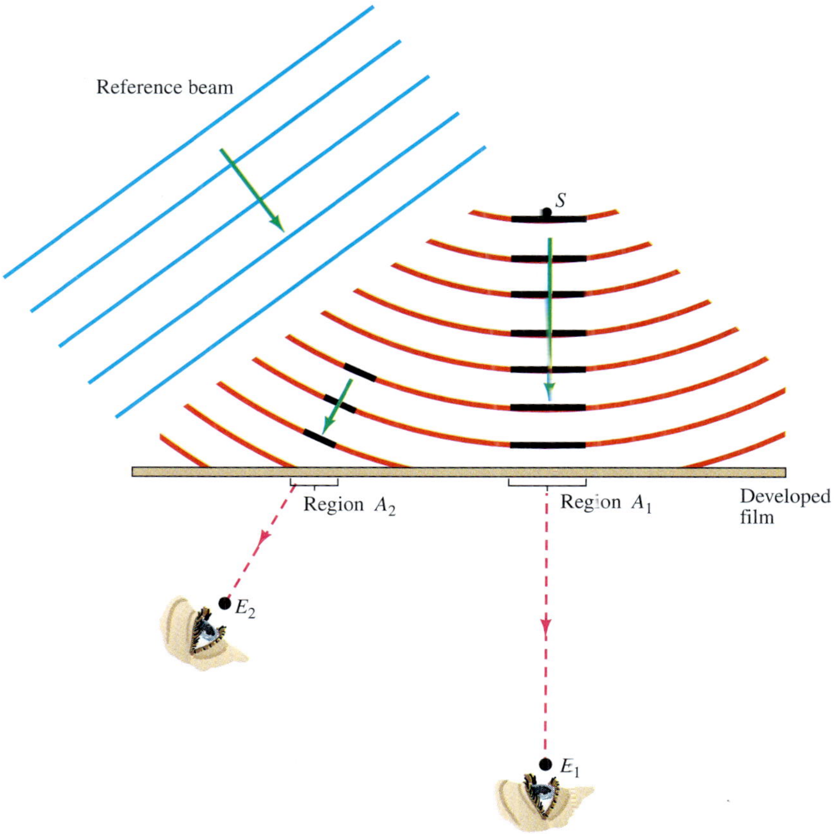

FIGURE 39–31 Holograms produce a three-dimensional image that allows the original object to be seen from different positions. A different image will be observed in each place.

APPLICATION: THE USES OF HOLOGRAPHY

The uses of holograms go beyond the simple beauty of the image. Holography has the potential to provide an extremely compact system of information storage. Because the light from every point on a printed page reaches every point of a hologram, every region of the film larger than several wavelengths across can reproduce the entire page, albeit with less detail. Moreover, successive holograms of successive pages can be made within thick photographic emulsions. If the exposure of each page is made with a reference beam oriented at a slightly different angle, then, by illuminating the resulting hologram with a light beam of that particular angle, only the corresponding page appears from a particular vantage point. All the paintings in a museum could be sequentially recorded in this way with great accuracy in a very small space indeed.

Another important use of holography involves making two holograms of the same object on the same film at successive times. If the object has moved slightly between the moments when the holograms are made, then the two images interfere with each other like the light from two surfaces of a soap film does. Figure 39AB–1 shows an interference hologram of a violin in motion under the influence of a vibrating string. This interference pattern, not to be confused with the interference that makes the hologram itself, reveals detail about the motion not otherwise visible. Similarly, density variations in air are visible as the interference between two successive images of the air made on the same hologram. In this way, the mechanisms by which a candle heats the air above it or an airplane produces shock waves can be studied.

FIGURE 39AB–1 Interference and diffraction were crucial in producing this image of a violin. The lines indicate the motion of the vibrating instrument and are associated with the interference of two holographic images.

theless unique to the object. Once this pattern is recorded, it serves as a diffraction grating for light from a reference beam to make a unique pattern that reproduces the light emitted by the original object, and from many angles. Although we have treated the film as a transmission grating, if the interference pattern can be recorded as scratches on a shiny surface, then it will act as a reflection grating, such as that in a hologram on a credit card.

How is the reference beam made coherent with the light from the source? We can take a laser beam and split it in two. One part illuminates the object while the other is routed to serve as a reference (Fig. 39–32). Figure 39–33 shows a close-up of a hologram itself, followed by some views of the image it produces.

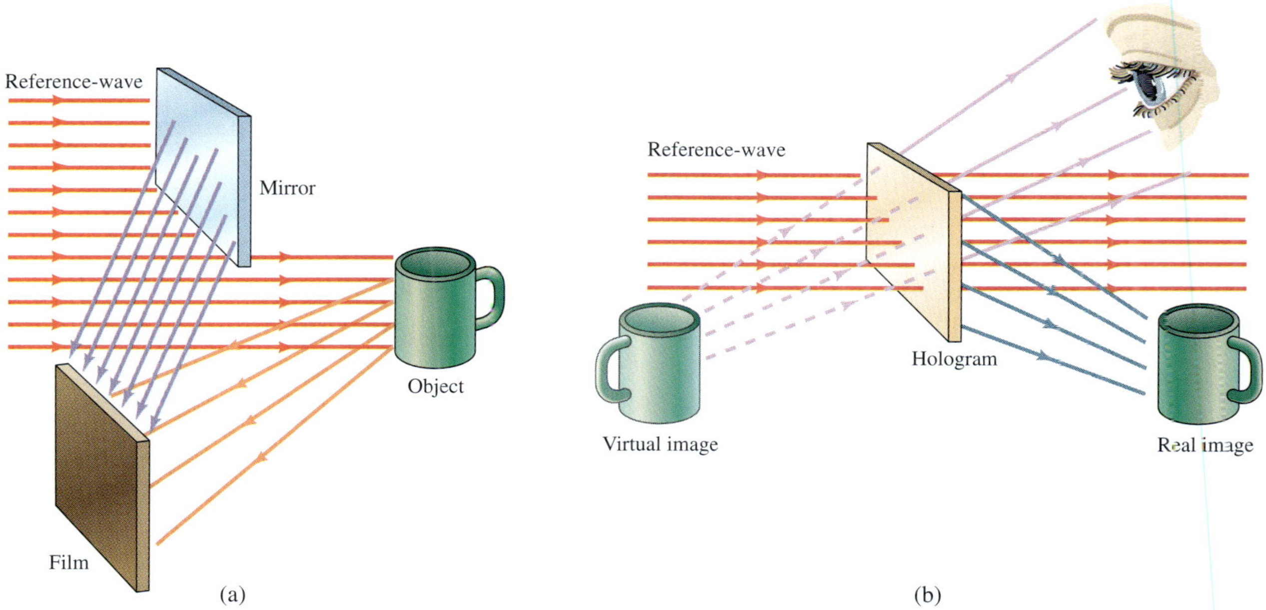

FIGURE 39–32 (a) Schematic diagram of the formation of a hologram. The reference beam and the light reflected from the object must be coherent so that they can interfere to make the hologram. (b) The holographic image is formed when the hologram acts as a diffraction grating.

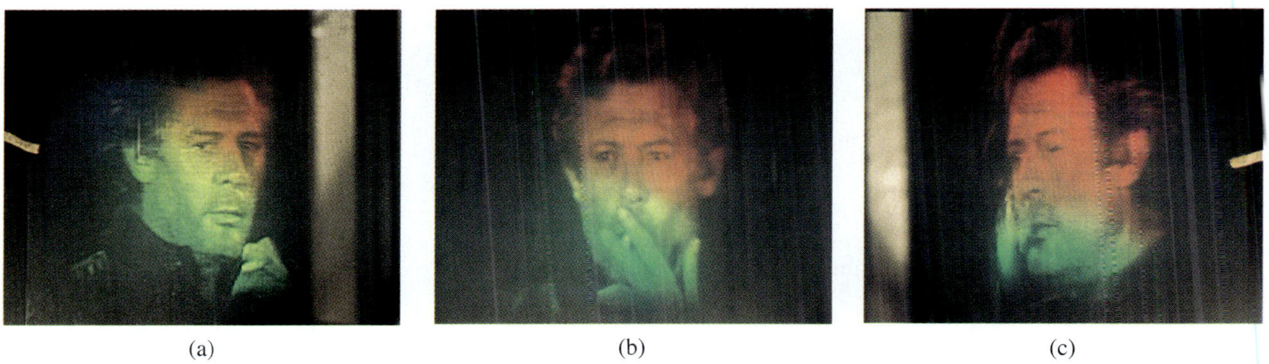

FIGURE 39–33 The three images (a), (b), and (c) are three holograms taken at successive times. Each can be viewed within a narrow angle; by moving from angle to angle a viewer in effect sees a sequence in time.

SUMMARY

Diffraction is a manifestation of interference among waves. Examples include the pattern produced by screens with evenly spaced multiple slits (diffraction gratings) and the patterns made by light that passes through single apertures or around obstacles.

If d is the distance between slits in a diffraction grating and θ is the angle of observation from the direction of incident light, principal maxima are observed for the condition

$$d \sin \theta = m\lambda, \quad \text{where } m = 0, \pm 1, \pm 2, \ldots. \tag{39–1}$$

Here, m is the order of the principal maxima. If the average intensity reaching the screen from any one slit is I_0, then the intensity of the light from the grating at the principal maxima is $N^2 I_0$, where N is the number of slits. In addition, the width of the principal maxima depends on N as $1/N$, so the diffraction peaks become sharper as N increases. This dependence on the parameters of the grating is contained in the expression for the intensity pattern:

$$I = I_0\left[\frac{\sin(N\beta)}{\sin\beta}\right]^2, \tag{39–4a}$$

where

$$\beta = \frac{\pi d \sin\theta}{\lambda}. \tag{39–4b}$$

Angular dispersion represents the change in observation angle θ as a function of a change in wavelength and is given by

$$\frac{\Delta\theta}{\Delta\lambda} = \frac{m}{d\cos\theta}. \tag{39–5}$$

The resolving power, R, is the ability of a grating to separate closely spaced lines:

$$R = \frac{\lambda}{\Delta\lambda} = mN. \tag{39–6, 39–7}$$

A single slit produces a diffraction pattern that can be derived by considering the slit to be composed of a large number of very thin slits. The criterion for destructive interference is

$$\sin\theta = \frac{m\lambda}{a}, \quad \text{where } m = \pm1, \pm2, \pm3, \ldots \tag{39–8}$$

and a is the width of the slit. The intensity pattern of a single slit is

$$I = I_{\max}\frac{\sin^2\alpha}{\alpha^2}, \tag{39–9}$$

where the angle α is given by

$$\alpha = \frac{\pi a \sin\theta}{\lambda}. \tag{39–10}$$

The minima are given in terms of α by

$$\alpha = n\pi, \quad \text{where } n = \pm1, \pm2, \pm3, \ldots. \tag{39–11}$$

Most of the light from the single slit is contained in the central peak; the secondary peaks are much less intense. The narrower the slit, the broader the diffraction pattern.

The Rayleigh criterion specifies that two point sources are just resolved if the peak of the diffraction image of the first source falls on the first minimum of the diffraction image of the second source. The minimum separation angle of two closely spaced sources obtained by a circular aperture of diameter D is approximated by

$$\theta_{\min} \simeq \frac{\lambda}{D}. \tag{39–13}$$

The minimum separation $S_{\min}$ of two closely spaced objects by a lens of diameter D is given by

$$S_{\min} = f\theta_{\min} \simeq \frac{f\lambda}{D}. \tag{39–14}$$

The practical limitation of Earth-based telescopes is due to air turbulence and not to diffraction limits.

Missing orders occur in the intensity patterns of diffraction spectra due to the overlapping effects of single slits and multiple slits.

X-rays are diffracted by the atom centers of regularly spaced Bragg planes—planes formed by the regular array of atoms in a crystal. The technique is important for many aspects of modern physics, including determining the structure of crystals and atomic composition. Bragg's law gives the observation angles θ (as measured from a plane surface in a crystal lattice) for which constructive interference is obtained from planes of spacing d:

$$2d \sin \theta = n\lambda, \quad \text{where } n = 1, 2, 3, \ldots. \tag{39–16}$$

Holography represents a special process by which three-dimensional images are captured. A hologram is a special diffraction grating formed by the interference of two coherent beams, one a reference beam and the other an object beam reflected from a three-dimensional object. The three-dimensional image of the object can be reconstructed by projecting a reference beam on developed holographic film.

UNDERSTANDING KEY CONCEPTS

1. Would the diffraction of water waves around the timbers of a pier be reduced by decreasing the diameter of the support poles? by increasing their diameter?
2. There are tentative plans to build telescopes for waves of various wavelengths, including visible light, on the Moon. What would the advantages of such facilities be?
3. Discuss how a Poisson spot might be obtained from a bowling ball. Would you want the source and screen to be close to or far away from the bowling ball? Explain.
4. Is it possible to obtain better resolution with a microscope with blue light than with red light? Why or why not?
5. Two waves are linearly polarized. The electric field of one wave is aligned with the x-axis and the other is aligned with the y-axis. In the absence of matter that might change the polarization, can these waves interfere with each other?
6. In a demonstration of diffraction peaks that involves the reflection of laser light from an ordinary ruler, does the light have to be at a glancing angle?
7. The spreading of light due to diffraction in an optical instrument is greater when the instrument uses red as opposed to blue light. Why?
8. A hologram contains information about an entire object, even in just a small portion of the film. Would you expect the image made by a small portion of the hologram to be as sharp as the image made by the entire hologram?
9. What are the differences between the interference patterns formed on a distant screen by coherent light that passes through a diffraction grating with thousands of rulings at a particular spacing and a double slit separated by the same spacing?
10. A lightbulb emits light with a spectrum characteristic of blackbody radiation. What pattern will this light produce when it is observed through a grating?
11. You are standing in the ocean, and a wave passes around you. Is this an example of diffraction?
12. How do the X-rays used in X-ray diffraction "know" that there is a given set of planes of atoms for which a diffraction pattern appears?
13. Does the fact that light bends around corners mean that, with a sensitive camera, you could read a newspaper from around a corner? (This is a serious question; try to estimate the amount of bending that the smallest obstacle would give for light, and how much information that light could contain.)
14. In the so-called 3-D movies introduced in the 1950s, a three-dimensional effect is achieved when different images are sent to each of your two eyes. How could you tell that they are not holographic images?

PROBLEMS

39–2 *Diffraction Gratings*

1. (I) Laser light is diffracted from a grating with 400 lines/cm. The central peak and the fourth peak are 10.34 cm apart on a screen 1.44 m away. The screen is perpendicular to the ray that makes the central peak. What is the wavelength of the light?
2. (I) A grating has a line density of 800/cm, and a screen perpendicular to the ray that makes the central peak of the diffraction pattern is 5 m from the grating. If light of two wavelengths, 620 nm and 635 nm, passes through the grating, what is the separation on the screen between the second-order maxima for the two wavelengths?
3. (I) A student finds a diffraction grating but does not know the spacing of the ruled lines. She shines light from a laser with $\lambda = 680$ nm through the grating and examines the maxima on a screen 265 cm away. If the distance between the tenth maxima on either side of the central peak is 14.3 cm, what is the rule spacing of the grating?
4. (II) A grating with 2×10^4 rulings spaced uniformly over 3 cm is illuminated at normal incidence by light of wavelength 530 nm. (a) What is the dispersion of the grating in the second order? (b) What is the smallest wavelength interval that can be resolved in the second order near $\lambda = 530$ nm?

5. (II) What is the resolving power of 3-cm-wide diffraction grating with 5000 lines/cm, for the first three orders? If light consisting of a series of discrete wavelengths around 420 nm is incident on the grating, what is the minimum wavelength separation that can be resolved in these three orders?

6. (II) Estimate the line spacing between two closely spaced lines near 580 nm if they are barely resolved in the fourth order by a grating with $N = 15{,}000$.

7. (II) A grating is to be inscribed on a 4-cm-wide glass plate so as to resolve two spectral lines with wavelengths 618.32 nm and 618.34 nm, respectively, in the first order. What is the minimum number of lines that must be ruled on the plate? What is the dispersion of the grating with this number of lines?

8. (II) A grating is made of five similar, uniformly spaced, narrow slits. For light of wavelength $\lambda = 633$ nm perpendicularly incident on the slits, the angular position of the first principal order is 0.18° to the normal. What is the slit separation? What is the angular position of the first principal order when the first and fifth slits are covered? when the second and fourth slits are covered?

9. (II) The resolving power of a certain grating for the first-order spectrum is 10^4. If the grating is 2 cm long, what angle separates the first- and second-order images for light with $\lambda = 580$ nm at normal incidence?

10. (II) White light shines on a diffraction grating with 4000 lines/cm. The diffracted light is observed on a screen 2 m away. Find the second- and third-order positions for blue light (440 nm), green light (560 nm), and red light (720 nm). Sketch a view of the screen.

11. (II) Visible light extends from wavelengths of 430 nm to 680 nm. If blackbody radiation, which contains all these wavelengths, is incident on a 5-cm-wide grating with 2500 slits/cm, what range of angles is covered for these wavelengths in the first-order maximum? in the second-order maximum?

12. (II) An atomic source emits two strong spectral lines, a red one of wavelength 615 nm and a blue one of wavelength 475 nm. The light falls on a diffraction grating with 5000 lines/cm that is 1.2 cm across, and passes to a screen 2 m away. On the screen, how far from the central maximum are the second-order maxima ($m = 2$) of the spectral lines? What is the width of these maxima?

13. (II) Light of wavelength λ is incident at an angle α to the normal of a transmission grating with spacing d between each slit (Fig. 39–34). At what angles β to the normal will diffraction maxima be located?

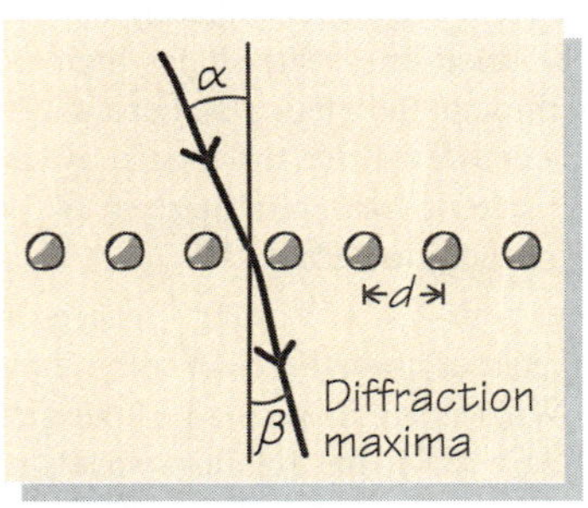

FIGURE 39–34 Problem 13.

39–3 *Single-Slit Diffraction*

14. (I) Light of wavelength $\lambda = 500.0$ nm falls on a slit of width $a = 0.50$ mm. At what angle θ from the normal to the wall in which the slit is cut does the second dark fringe occur?

15. (I) A single slit of width 2.8×10^{-5} m diffracts light of wavelength 495 nm to a screen. The distance between the minima on either side of the central maximum is 1.8 cm. How far away is the screen?

16. (I) A single slit diffracts laser light of wavelength 635 nm onto a screen 2.5 m away. The distance between the first-order maxima on either side of the central peak is 6.0 mm. How wide is the slit?

17. (II) Blue light ($\lambda = 470$ nm) passes through a slit 10 μm wide. What is the ratio between the maximum intensity of the central peak and the maximum intensity of the next adjacent peak? At what angle θ from the horizontal will the intensity of the central peak be half its maximum value? Would θ increase or decrease if red light ($\lambda = 670$ nm) were used instead?

18. (II) Plane waves of light of wavelength 560 nm are incident on a single slit of width 30 μm. A lens focuses the plane waves on a screen 60 cm away (Fig. 39–35). (a) What is the width of the central maximum on the screen? (b) What is the intensity ratio between the central maximum and the first-order maxima?

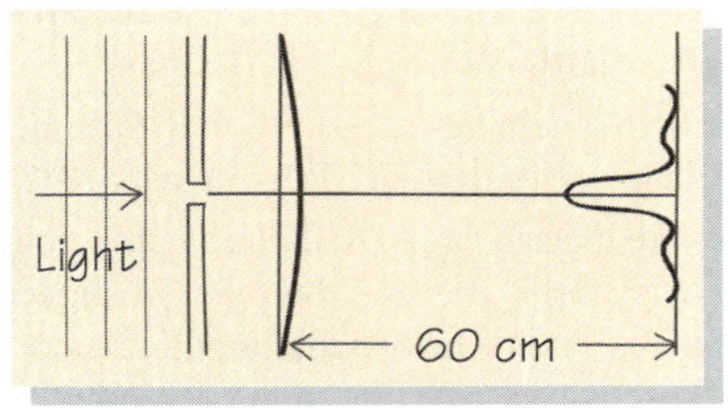

FIGURE 39–35 Problem 18.

19. (II) A single slit produces a diffraction pattern on a distant screen. Show that the separation distance between the two minima on either side of the central maximum is twice as large as the separation distance between all the other neighboring minima. Compare your result to the corresponding case for a double-slit pattern with very narrow slits.

20. (II) A diffraction pattern is formed by an adjustable slit. If the width of the slit is doubled, how do the following quantities change? (a) The distance of the first minima on the two sides of the central maximum; (b) the intensity at the central maximum; (c) the total power reaching the screen.

21. (II) The width of the central peak of a single-slit diffraction pattern can be characterized by the distance of the first-order minima on both sides of the maximum or the full width at half maximum, the latter defined by the points where the intensity decreases to 50 percent (Fig. 39–36). Compare the values of these widths.

22. (II) When light of wavelength 450 nm passes through a single slit of unknown width, the diffraction pattern displays a second maximum where the first minimum of light of an unknown wavelength had been observed to fall (Fig. 39–37). What is the unknown wavelength?

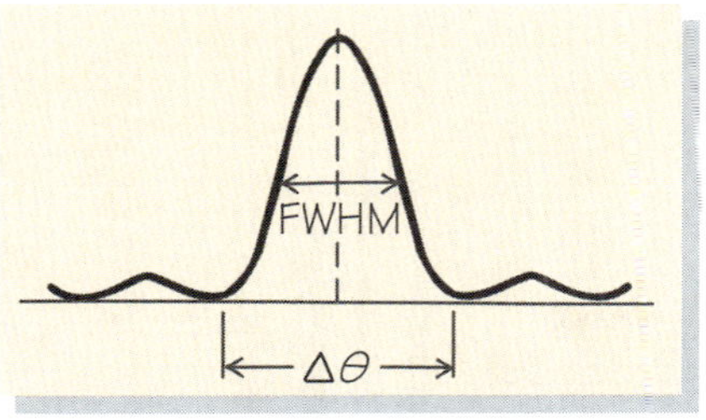

FIGURE 39–36 Problem 21.

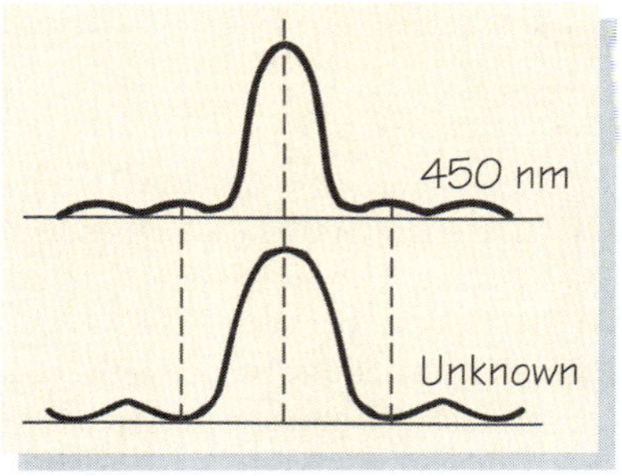

FIGURE 39–37 Problem 22.

23. (II) Light of wavelength λ arrives at a single slit of width a; the plane wave fronts arrive at the slit at an angle θ_i (Fig. 39–38). Find the angles θ for which minima appear on a very distant screen. Is there a "central maximum" in the direction defined by the incoming wave; that is, at $\theta = \theta_i$?

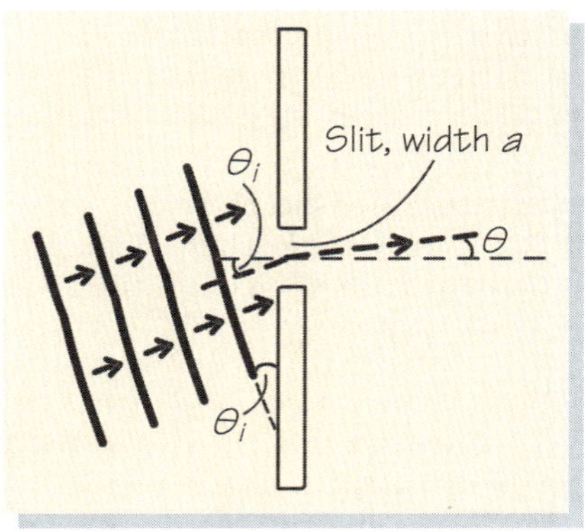

FIGURE 39–38 Problem 23.

24. (II) Suppose that light falls on a single slit at an angle ϕ with the normal to the wall that contains the slit (Fig. 39–39). Show that Eq. (39–9) still holds, but $\sin\theta$ must be replaced by $(\sin\theta + \sin\phi)$ in the expression for α [Eq. (39–10)].

25. (III) When we determined the position of the minima of the Fraunhofer diffraction pattern for a single slit, we argued that the maxima are located midway between the minima. To look at the accuracy of this assumption, (a) show that the maxima of the intensity pattern $(\sin^2\alpha)/\alpha^2$ are determined by the solutions of the transcendental equation $\alpha = \tan\alpha$. (b) Compare a numerical solution of this equation for the first and second maxima with angles that are midway between the first and second, and second and third, minima (you will need trigonometric tables). (c) By plotting the intersection points of $y = \tan\alpha$ and $y = \alpha$, show that the approximation improves as the order of the maximum increases.

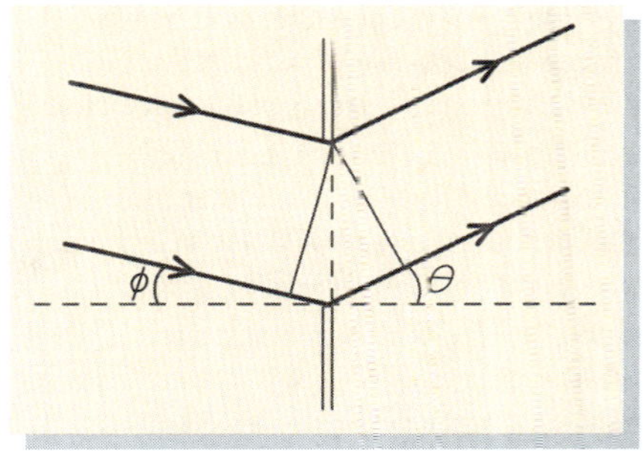

FIGURE 39–39 Problem 24.

39–4 Diffraction and Resolution

26. (I) A plane wave of microwave radiation, $\lambda = 1.5$ cm, passes through a circular aperture of diameter 5.0 cm. What is the angular position of the first minimum of the resulting Fraunhofer diffraction pattern?

27. (I) Astronauts leave two lunar rovers 5.00 km apart on the Moon. An Earth-based telescope of what minimum diameter is required to resolve laser beams ($\lambda = 650$ nm) emitted by the rovers toward the telescope? The rovers are 3.0 m long. A telescope of what diameter is required for the rovers themselves to be detected? Ignore air turbulence. (The Earth–Moon distance is 3.83×10^8 m.)

28. (I) An amateur astronomer uses a reflecting telescope of diameter 20 cm and focal length 200 cm to observe light of $\lambda \simeq 600$ nm from a star. (a) What minimum angular resolution can the astronomer obtain? (b) What is the diameter of the Airy disk? (c) What is the minimum separation distance of two objects on the Moon that the telescope can resolve?

29. (II) An astronaut in a satellite can barely resolve two point sources on Earth 220 km below. What is the separation distance between the sources, assuming ideal conditions, $\lambda = 480$ nm, and a pupil diameter of 1.5 mm?

30. (II) The two stars of a binary star system are just resolvable when observed by a telescope with a resolution of 3″ of arc and are 75 ly from Earth. Estimate their separation.

31. (II) A spy satellite is announced to be capable of distinguishing detail 10 in across. If the satellite orbits at a height of 220 mi, what must be the minimum size of the lens aperture, assuming maximum sensitivity at $\lambda = 525$ nm? Would it be better if the film (or other sensor) were sensitive to shorter or to longer wavelengths?

32. (II) What must be the lateral separation between two objects located 1 km from a camera that must resolve them? The camera lens's aperture is 5 mm in diameter, and the film is sensitive to light of wavelength 550 nm.

33. (II) The lens of a 35-mm camera is set in such a way that the image of a very distant object is ideally sharp on the film. The focal length of the objective is 50 mm. At what setting of the camera's aperture will the sharpest possible image of an object

5 m away form? [*Hint*: The image will be blurred both because it is off focus and because of diffraction. Find the aperture at which these two sources give equal contributions.]

34. (II) Use the Rayleigh criterion and make assumptions to estimate the distance at which the human eye should be able to resolve the headlines in a newspaper. Carry out an experiment to see how good your estimate is!

35. (II) The SR-71 Blackbird reconnaissance airplane could fly at over 70,000 ft. If the pilot's pupil has a diameter of 1.5 mm on a bright day, what is the distance between two objects on Earth that the pilot could just resolve from 60,000 ft? Take the wavelength of light to be 520 nm.

36. (II) The headlights of a car are 1.5 m apart. At night, the pupils of an oncoming driver have expanded to 4.8 mm. How close must the two cars approach before the headlights can be resolved? Take the wavelength of the light to be 550 nm.

*39–5 Slit Width and Grating Patterns

37. (II) Calculate the lowest missing order of a double-slit interference pattern if the separation of the two slits is three times their individual widths, $d = 3a$.

38. (II) The separation distance between two narrow slits is ten times the width of either slit. What is the intensity of the tenth interference maximum, taking the center as the first, when monochromatic light passes through the two slits and falls on a distant screen?

39. (II) Light of wavelength 690 nm from a ruby laser impinges on two slits 1.1 mm apart. Each slit is 0.20 mm wide. Find the intensity ratio I/I_0 on a screen 3.0 m away at the following distances from the central maximum: 0.050 mm, 0.50 mm, 1.5 mm, and 3.0 mm.

40. (II) Figure 39–40 shows the intensity as a function of diffraction angle (in radians) for a double slit with light of wavelength 600 nm. Estimate the separation of the slits as well as their widths.

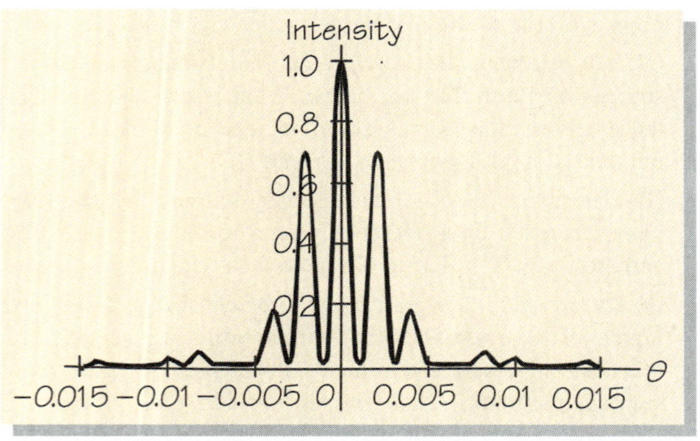

FIGURE 39–40 Problem 40.

41. (II) Figure 39–41 shows the intensity as a function of diffraction angle (in radians) for a multiple slit with light of wavelength 600 nm. Compare the pattern with double-slit pattern shown in Fig. 39–40 of the previous problem and determine the number of slits, their separation, and their width. [*Hint*: Look at Fig. 39–9.]

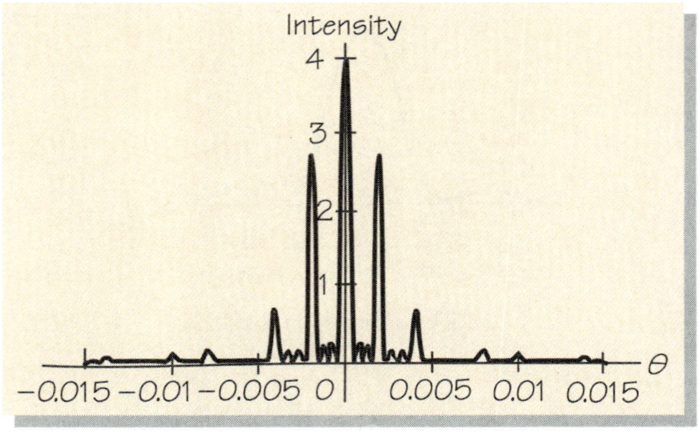

FIGURE 39–41 Problem 41.

42. (II) Light of wavelength 600 nm is perpendicularly incident on a diffraction grating. Two adjacent maxima occur at $\sin\theta = 0.30$ and $\sin\theta = 0.36$, respectively. The fourth order is missing. (a) What is the separation distance between adjacent slits? (b) What is the smallest possible individual slit width? (c) Name all orders that appear on the screen, consistent with the answers to parts (a) and (b).

43. (II) The centers of a double slit are separated by 1.2 mm; each slit is 0.4 mm wide. Are there missing orders? If so, at what angles are they missing on a distant screen if $\lambda = 589$ nm?

44. (II) The slit widths of a grating with 2500 slits/cm are one-third the slit spacing. What is the ratio of the intensities of the second-order and first-order principal maxima of the grating?

45. (II) Light of wavelength 625 nm is perpendicularly incident on a screen in which double slits of width $a = 0.25$ mm have been cut. The slits are a distance $d = 0.30$ mm apart. Find the first angle away from the central axis for which the intensity on a distant screen is exactly one-half the maximum intensity.

46. (II) A grating consists of slits of width a whose centers are separated by a distance d. Sketch the diffraction pattern for (a) $d >> a$ and (b) $d - a << a$ (the slits are wide compared to the strips between them).

*39–6 X-Ray Diffraction

47. (I) X-rays of wavelength 0.14 nm are aimed at an unknown crystal in a diffractometer. A first-order peak occurs at 38.2° (Fig. 39–42). What is the corresponding Bragg-plane spacing for the crystal?

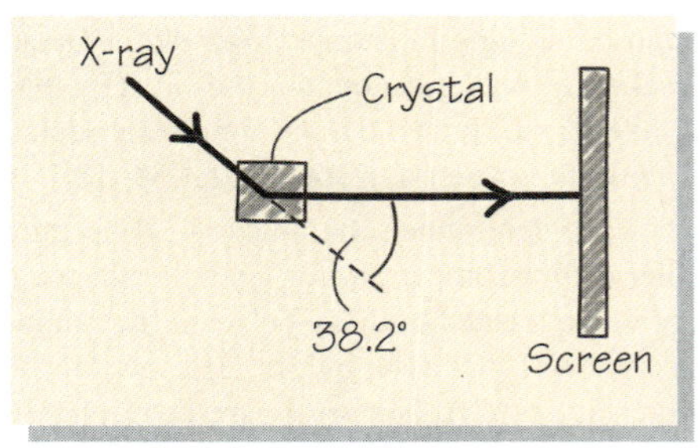

FIGURE 39–42 Problem 47.

48. (II) The distance between neighboring pairs of Bragg planes in calcite ($CaCO_3$) is 0.3 nm. At what angles to these planes will the first- and second-order diffraction peaks occur for X-rays of wavelength 0.12 nm?

49. (II) Consider a crystal consisting of identical cubes with atoms at the vertices. The spacing between adjacent atoms is 0.28 nm. X-rays of wavelength 0.14 nm scatter elastically from a set of planes parallel to the face of the cubes. At what angles will first-order Bragg diffraction be observed?

50. (II) Mica has a set of Bragg planes with spacing of 1.0 nm, whereas a set of planes in rock salt has a spacing of 0.28 nm. For an X-ray of wavelength 0.1 nm, which material produces a diffraction pattern with the greater angular separation? What is the difference in angular separation $\Delta\theta$ for each material for the Bragg planes above when the crystals are illuminated with X-rays of wavelengths 0.096 nm and 0.104 nm?

General Problems

51. (I) A ruby laser of wavelength 690 nm with a cross-sectional area of $10^{-3}\ \text{m}^2$ is aimed at the Moon, 3.84×10^8 m away. Estimate the minimum diameter of the light beam that reaches the Moon.

52. (I) A grating 5 cm long has 17,500 lines inscribed on it. A line of wavelength 550.000 nm is just resolved, in the third order, from a second line with a slightly longer wavelength. What is the wavelength of the second line?

53. (II) Radar is used to study the shapes of airplanes from as far away as 100 km. (a) Assuming that the distance scale determining a plane's shape (the size of the curves that distinguish one plane from another) is 1 m, what angular resolution is needed in the radar system? (b) Estimate the wavelength of the radar waves if the reflected radar signals are gathered in a dish of diameter 2.5 m.

54. (II) Deep-ocean waves move in linear fronts directly toward a harbor opening of width 50 m (Fig. 39–43). For what wavelength will there be a minimum within the harbor at an angle of 50° from the axial line of the opening?

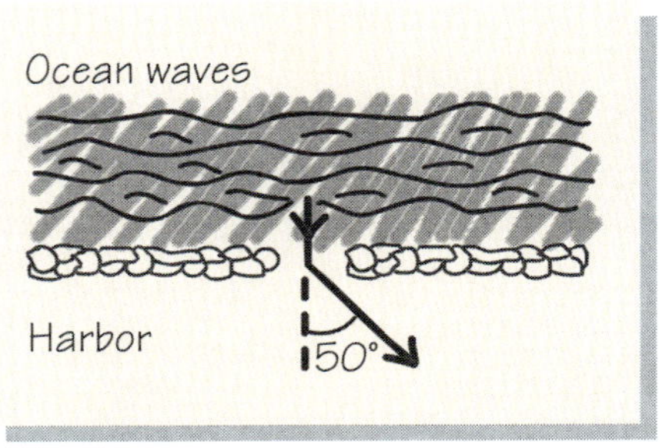

FIGURE 39–43 Problem 54.

55. (II) By varying the spacing between two vertical dipole antennas as well as the phase of the signal generated by each antenna, the antennas can give signals that are stronger in some directions than in others (see Chapter 38). Suppose that N antennas are lined up along the x-axis (Fig. 39–44). The total distance between the first and last antennas is λ, so the spacing between the antennas is $\lambda/(N-1)$. Any one antenna would radiate its signal with the same intensity I_0 for any angle θ. (a) Find the intensity radiated very far from the array by the system of antennas as a function of angle θ in terms of I_0, assuming that the signals of all the antennas are in phase. (b) Describe the signal for all values of θ from 0° to 360°.

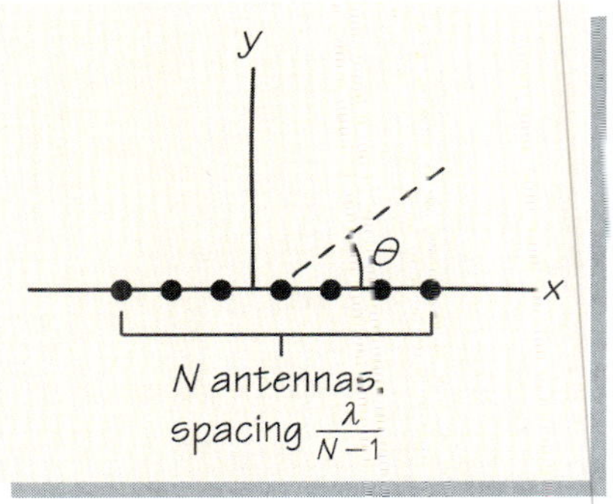

FIGURE 39–44 Problem 55.

56. (II) In Chapter 41, we shall see that an electron behaves like a wave whose wavelength, λ, is related to its momentum, p, by $\lambda = h/p$. Here, h is Planck's constant, $h = 6.63 \times 10^{-34}$ J · s. Electrons used in an electron microscope can be diffracted, and electron microscopes have a diffraction limit. If the energy of the electrons used in an electron microscope is 15 keV and if the aperture through which the electrons are channeled has a diameter of 0.02 mm, what, approximately, is the smallest angular separation the microscope can distinguish in an object?

57. (III) *Babinet's principle* is useful for the treatment of the diffraction of light by obstacles. It states that if light is incident on an opaque screen in which a hole (of any shape) is cut, then the diffraction pattern produced is the same (except at $\theta = 0$) as that obtained if the screen were removed and the hole were replaced by an obstacle. Use Babinet's principle to estimate the size of an opaque obstruction on a glass slide if a narrow laser beam (with $\lambda = 633$ nm) perpendicularly incident on the slide spreads to a spot of diameter 0.70 cm on a screen 2.5 m from the slide.

58. (III) What diffraction pattern is produced on a distant screen when light of wavelength λ is perpendicularly incident on a plane that contains N very thin hairs, each spaced a distance d apart from the next hair. [*Hint*: See Problem 57.]

59. (III) Electromagnetic radiation of frequency 1.25×10^{23} Hz is scattered by a nucleus of radius 3.2×10^{-15} m. The nucleus is totally radiation-absorbent and thus is a perfect obstacle. At what angle will the first diffraction minimum lie? [*Hint*: Use Babinet's principle (Problem 57).]

CHAPTER

40

Albert Einstein had a public fame few scientists have ever matched. He contributed in important ways to our understanding of space and time, of gravity and of quantum physics.

Special Relativity

There appears to be an important difference between electromagnetism and mechanics, at least superficially. The laws of mechanics look the same in all inertial frames—all reference frames that move with uniform velocity with respect to some standard inertial frame (for example, a frame at rest relative to distant stars). Electromagnetism, however, appears to violate this general law According to Maxwell's equations, electromagnetic waves propagate at the speed c, with no restrictions on the state of motion of the source or detector. This suggests the existence of an absolute frame for electromagnetism.

The special theory of relativity, proposed by Albert Einstein in 1905, extended to electromagnetism the principle that the fundamental laws of physics look the same in all inertial frames. This was accomplished not by altering Maxwell's equations but by modifying certain assumptions about our notions of space and time, assumptions that went unquestioned until 1905. Here, we explore these ideas of space and time as well as the physical consequences of Einstein's theory.

40-1 IS AN ETHER NECESSARY?

In the years following the discovery of Maxwell's equations, the absolute value of the speed of propagation of electromagnetic waves caused little concern. In an age of mechanical models, it was believed that electromagnetic waves need a medium to support them (just as sound waves need air). This presumed medium was thought to fill the universe and was called the **ether**. The ether was assumed to be at rest relative to the fixed stars. Maxwell's theory was assumed to give c for the speed of propagation of electromagnetic waves *relative to the ether's rest frame*, just as the speed of sound is given as 330 m/s relative to stationary air. This assumption was accepted even though the derivation of the speed of light in Maxwell's equations contains no reference to a frame. In a reference frame moving at speed u relative to the ether, the speed of light emitted by a source at rest relative to the ether would be $c + u$ if the frame were moving toward the source, and $c - u$ if the frame were moving away from the source. Earth represents such a moving frame because it travels at a speed of approximately 30 km/s relative to the fixed stars in its motion around the Sun. From the point of view of a frame fixed to Earth, the ether moves past at a speed of 30 km/s. Detecting the *ether wind*, though, was likely to be very difficult.

To see why, consider a standard speed-of-light measurement, with the light beam propagating along an axis that lies in the direction of the ether wind. If the distance from the source (and detector) to a mirror is L, then, in the absence of ether wind, the time for a single traversal of light from source to mirror and back is $t_0 = 2L/c$ (Fig. 40–1). With the ether wind blowing against the source, the speed of light that travels toward the mirror is $c - u$, and the speed of light that returns is $c + u$. Thus, the time for a single traversal is

$$t_1 = \frac{L}{c-u} + \frac{L}{c+u} = \frac{2cL}{c^2 - u^2}$$

$$= \frac{2L/c}{1 - (u^2/c^2)}. \qquad (40\text{–}1)$$

For $u = 30$ km/s and $c = 3 \times 10^5$ km/s, the factor $u^2/c^2 = 10^{-8}$. Thus, a time sensitivity of better than one part in 100 million would be required to detect such an ether wind.

The Michelson–Morley Experiment

The Michelson interferometer is described in Chapter 38.

In 1887, Albert A. Michelson and Edward W. Morley carried out a high-precision experiment to measure the possible effect of an ether wind. They used an interferometer designed by Michelson (Fig. 40–2). Let's suppose that the ether wind were aligned in the direction shown in Fig. 40–3a. The distance from the half-silvered mirror (a mirror that partly transmits and partly reflects light) to mirror M_1 in the direction aligned with the ether wind is L. The time that it takes for light to travel to mirror M_1 and back is t_1, given by Eq. (40–1). Let's next find t_2, the time it takes for the light to travel to mirror M_2 and back in the direction perpendicular to the presumed ether wind. Mirror M_2 is also a distance L away from the half-silvered mirror. (The assumption that the arms of the interferometer are equal in length is not essential, as we shall see later.) Because the second beam is perpendicular to the presumed ether wind, the light would have to travel a distance larger than $2L$; the beam would be blown "off course" by the ether wind—like the boat crossing the river in Section 3–6. As Fig. 40–3b shows, it travels a distance given by the hypotenuse of a triangle in which one leg has length L

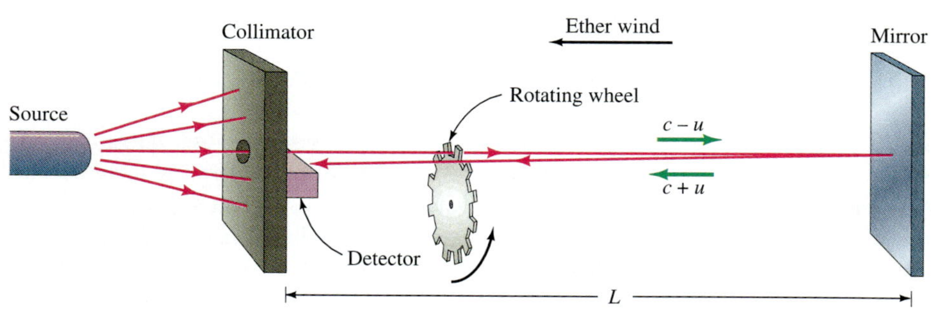

FIGURE 40–1 Schematic diagram of a light-speed measurement, with an ether wind blowing along one direction in which light travels. The ray speeds are different for the two directions of travel.

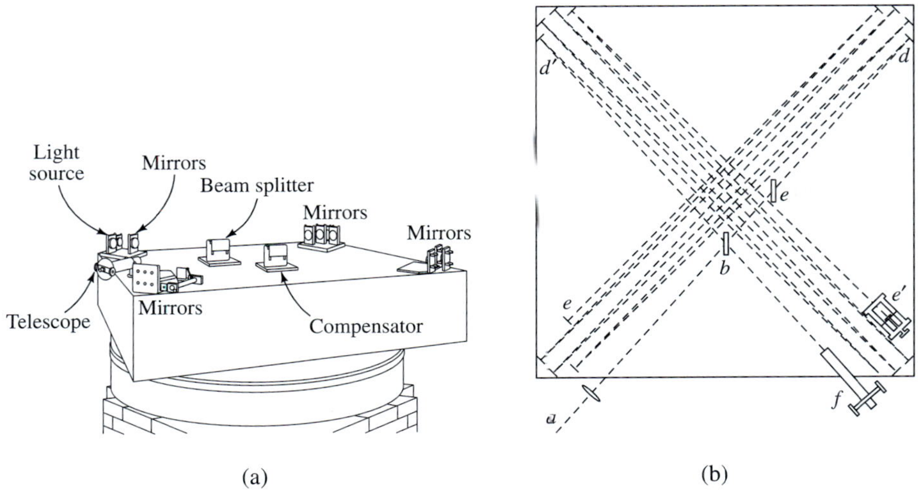

FIGURE 40–2 (a) Sketch of an interferometer built on a stone slab by Albert Michelson in 1887. (b) Schematic diagram of the workings of this interferometer. Follow the ray path from the source, point a, to the beam splitter, point b, and ultimately to the eyepiece, point f.

and the other leg is the transverse distance the beam is blown in time $t_2/2$; that is, $ut_2/2$. The distance is $\sqrt{L^2 + (ut_2/2)^2}$. Because the speed of light is c, we find

$$\sqrt{L^2 + \left(\frac{ut_2}{2}\right)^2} = \frac{ct_2}{2}.$$

We square both sides to find

$$L^2 + \left(\frac{u^2}{4}\right)t_2^2 = \left(\frac{c^2}{4}\right)t_2^2.$$

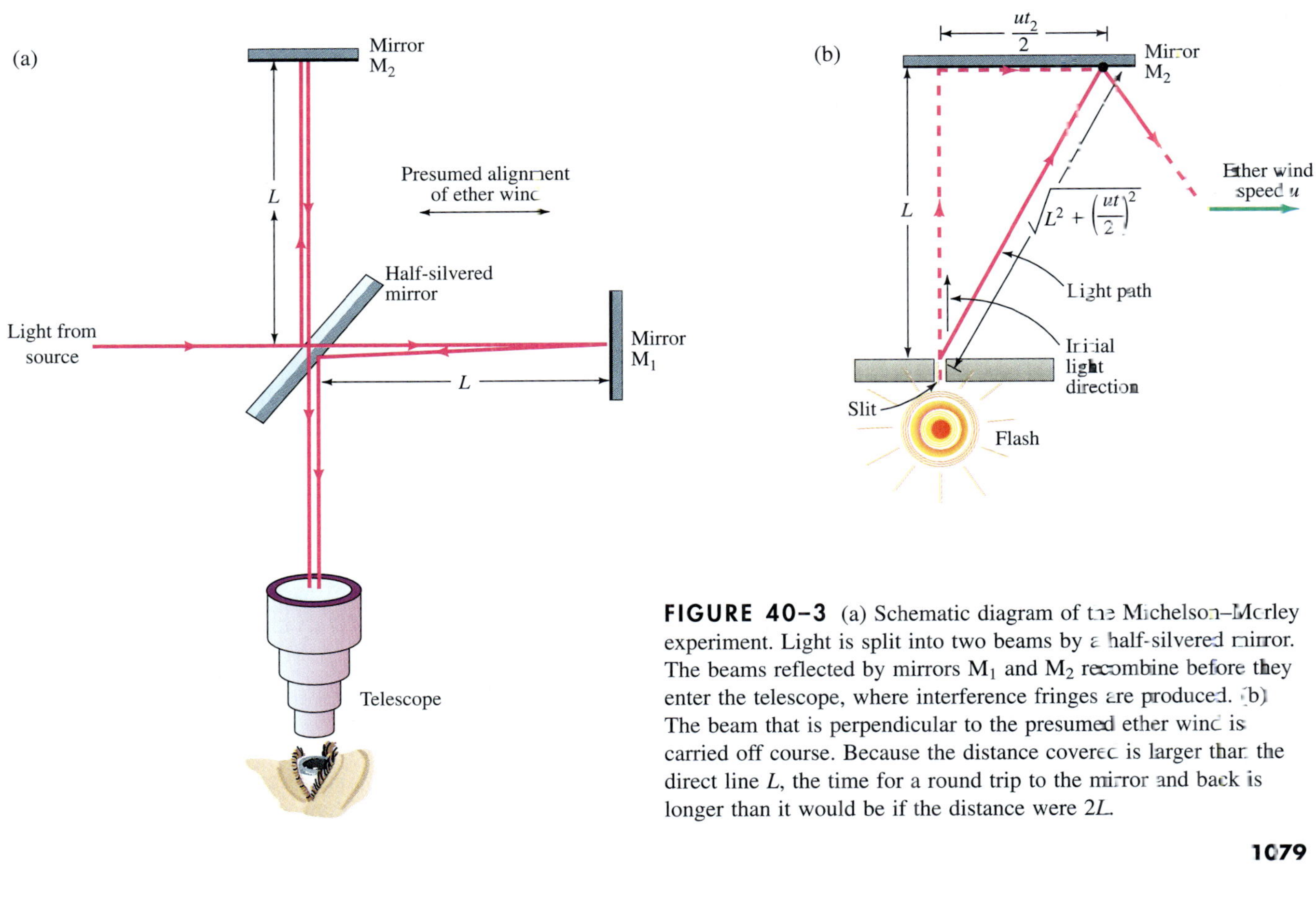

FIGURE 40–3 (a) Schematic diagram of the Michelson–Morley experiment. Light is split into two beams by a half-silvered mirror. The beams reflected by mirrors M_1 and M_2 recombine before they enter the telescope, where interference fringes are produced. (b) The beam that is perpendicular to the presumed ether wind is carried off course. Because the distance covered is larger than the direct line L, the time for a round trip to the mirror and back is longer than it would be if the distance were $2L$.

From this equation, it follows that

$$t_2^2 = \frac{4L^2}{c^2} - u^2;$$

$$t_2 = \frac{2L/c}{\sqrt{1-(u^2/c^2)}}.$$

We now use the fact that for small x, $1/(1-x) \simeq 1 + x$ and $1/\sqrt{1-x} \simeq 1 + (x/2)$, and apply these formulas for $x = u^2/c^2$. We find that the time difference between the arrivals of two parts of a wave pulse is

$$\Delta t \equiv t_1 - t_2 = \frac{2L}{c}\frac{u^2}{2c^2}. \tag{40–2}$$

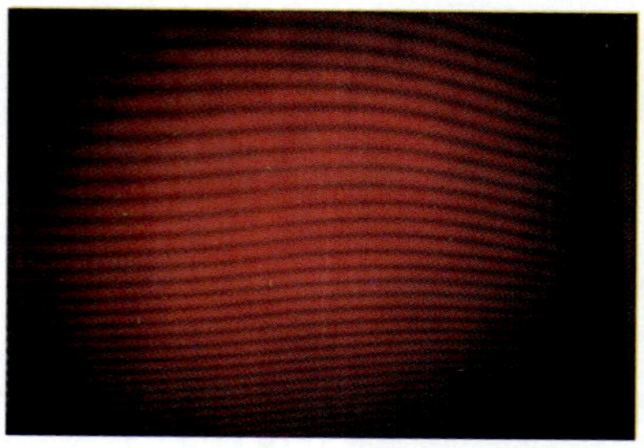

FIGURE 40–4 Interference fringes observed in a Michelson interferometer.

This corresponds to a path-length difference of $c\,\Delta t = Lu^2/c^2$. The two beams are combined upon their return from the two mirrors and, because they started out in phase, they will interfere according to the difference of their path lengths.

It is impossible to construct an apparatus in which the paths to the mirrors are exactly the same. In addition, because the mirrors are not exactly perpendicular to the beams, the path-length difference will vary slightly from one side of a mirror to another, and a view through the telescope yields a set of interference fringes (Fig. 40–4). Fortunately, any effects due to the apparatus itself can be accounted for by rotating the apparatus through 90°. The apparatus effects are unaltered, but the rotation effectively interchanges M_1 and M_2 and thus changes the path length to $-Lu^2/c^2$ (for unequal arm lengths, L is replaced by the average length). If there were an effect due to the ether, the fringe pattern would *shift* accordingly when the apparatus is rotated. The total path-length difference for the two orientations is $\Delta L = 2Lu^2/c^2$.

Result of the Michelson–Morley Experiment. A change in path length of $\Delta L = 2Lu^2/c^2$ leads to a shift of interference fringes of magnitude $\Delta L/\lambda = 2(L/\lambda)(u/c)^2$. Although the ratio u/c is very small, it is not an impossibly small number because L is so much larger than λ. The apparatus was capable of detecting a shift of as few as 0.04 fringes. If $(u/c)^2$ were as little as 10^{-8} (the value that would follow from Earth's movement around the Sun), the apparatus would give a shift of 0.4 fringes. The result of the experiment was that no shift was observed; that is, if there were any shift at all, it had to be less than 0.04 fringes. In other words, *there was no experimental evidence for the existence of an ether wind.*

More recent experiments performed with lasers show that the shift is less than 10^{-3} of the result that would be "expected" for Earth's movement through the ether. The Michelson–Morley experiment sharpened the difference between mechanics and electromagnetism. Maxwell's equations predict a definite speed of light, and all previous experience had suggested that such a speed must refer to a definite reference frame. This frame would have been the "preferred" frame of electromagnetism, yet the Michelson–Morley experiment showed that the preferred frame cannot be detected.

40–2 THE EINSTEIN POSTULATES

The postulates of special relativity

Albert Einstein was unaware of the Michelson–Morley experiment when he formulated the laws that explain its result. He conjectured that the laws of electricity and magnetism (electrodynamics), like those of mechanics, are the same in all inertial reference frames. The **special theory of relativity** is the result of this conjecture. Einstein himself referred to the following two postulates:

1. The laws of physics are the same in all inertial reference frames.
2. The speed of light in empty space is the same in all inertial frames.

Strictly speaking, the second postulate is part of the first because Maxwell's equations do not specify a frame when they predict the speed of light. As we have seen in Section

40–1, the second postulate might appear to be incompatible with the first. The combination of these apparently irreconcilable assertions led to the revolutionary insights into space and time that underlie the special theory of relativity.

The first postulate as applied to mechanics was introduced in Chapter 4, where we noted that the laws of mechanics are unchanged under the transformation

$$\mathbf{r}' = \mathbf{r} - \mathbf{u}t. \tag{40–3}$$

Figure 40–5 shows the two frames to which these equations apply. The primed variables are the coordinates in a coordinate system (or reference frame) F′. Unprimed variables are the coordinates in frame F. Frames F and F′ move relative to one another. Upon differentiation with respect to t, Eq. (40–3) leads to a transformation law for the velocities:

$$\mathbf{v}' = \mathbf{v} - \mathbf{u}. \tag{40–4}$$

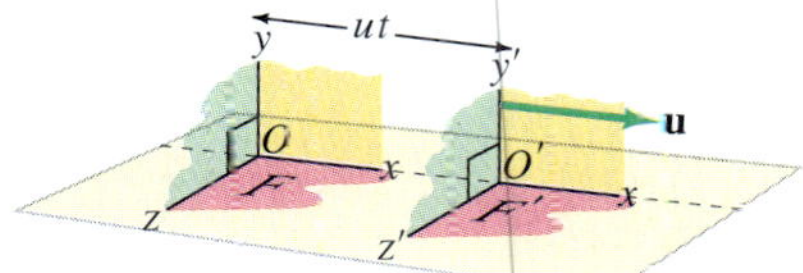

FIGURE 40–5 Reference frame F′ moves at velocity **u** with respect to frame F. An observer is in a particular frame when he or she measures the position and time of events in terms of the coordinate system and clocks of that frame.

The first postulate seemed to Newton to imply that time is the same in all frames. Although this may seem quite obvious to you, Newton recognized the need to make an assertion about the nature of time; in modern language, this statement is the trivial transformation law

$$t' = t. \tag{40–5}$$

This transformation law is a second one to be added to Eq. (40–3). Together, these equations form the *Galilean transformations*.

The Galilean transformation laws—Eqs. (40–3) and (40–5)—are incompatible with the second postulate. Suppose that we have a light source at rest at the origin of a frame F. When that source emits a flash of light in the form of a spherical wave that expands at the speed of light, then the location of the spherical wave front is given by

$$x^2 + y^2 + z^2 = \mathbf{r}^2 = c^2t^2 \tag{40–6}$$

(Fig. 40–6). Now let's consider how the light behaves according to an observer in a second frame, F′, that moves with respect to F, as in Fig. 40–5. Suppose that origin O' of F′ coincides with origin O of F at $t = t' = 0$, when the light flashes. The second postulate implies that *in* F′, *the pulse also forms a spherical wave front* because the speed of light is the same in F′ as it is in F, whatever the direction. Thus

$$x'^2 + y'^2 + z'^2 = \mathbf{r}'^2 = c^2t'^2. \tag{40–7}$$

If we set $t' = t$, Eq. (40–7) cannot be satisfied when the relation between **r** and **r**′, given by Eq. (40–3), is used because that would imply that $x'^2 + y'^2 + z'^2 = (x - ut)^2 + y^2 + z^2$. This conclusion shows the conflict between the Galilean transformation laws and the assertion of the absolute magnitude of the speed of light. It is clear that we must reexamine the very notions of space and time.

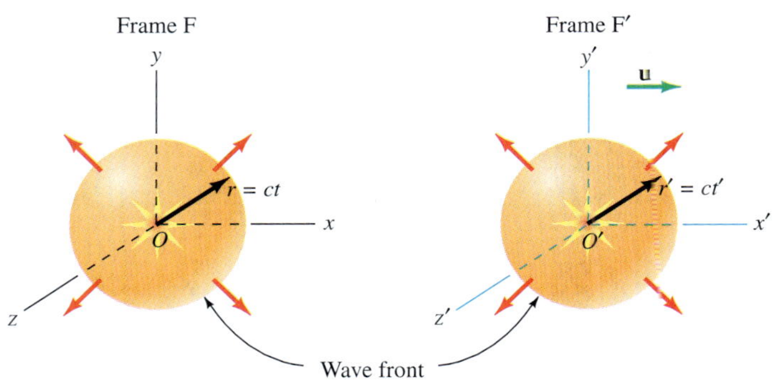

FIGURE 40–6 A spherical burst of light is emitted at the origin of frame F. To an observer in that frame, the light makes a spherical wave front centered at origin O. If origin O' of frame F′, moving with respect to frame F, is coincident with O when the burst occurs, then an observer at O′ moving with frame F′ will claim that the light makes a spherical wave front in frame F′ centered at O'.

40–3 SPACE, TIME, AND SIMULTANEITY

In studying the concepts of time and space, we must first define how we measure time and how we establish space coordinates. In Section 40–4, we shall construct a concrete example of a clock. Now, though, we require only that our clock be periodic: Time intervals between "ticks" are the same. We can put our clock and a light source at the origin of a coordinate system and then shine light along the x-axis. We place a mirror at some point x_1 and measure the time that it takes for the light to reach x_1 and return after reflection. If that time interval is two "ticks," we say that x_1 is one unit of length away from $x = 0$. Specifically, if the length of a tick is τ, then the distance to x_1 and back is $c\tau$. We now move the mirror farther away until the time for the light to reach there and back is four ticks. That point will be two length units ($2c\tau$) away from $x = 0$ along the x-axis. Proceeding in this way, we can in principle assign a coordinate (x, y, z) to every point in space.

To be able to discuss time at each point in the coordinate system, let's put a clock at every point for which x, y, and z are integer multiples of the unit of distance $c\tau$. We can synchronize all these clocks—set them all to the same time—as follows: At the origin at noon, a light signal is sent out to the point $x = 1$, $y = z = 0$. When the light ray arrives there, the clock operator at that point sets the clock to "one tick after noon," which corresponds to the point $x = 1$, $y = z = 0$. The operator at $x = 2$, $y = z = 0$ will set the clock there to "two ticks after noon" when the wave front reaches that point, and so on. In this way, all the times are synchronized and we have a reference frame in which space and time are well defined (Fig. 40–7). We have transmitted our signals in an unambiguous way because the speed of light is, by postulate (and by experiment!), independent of any motion. Transmission of signals by means of baseballs would mean that the velocity of the baseball would have to be measured, and this would involve us in complications having to do with the fact that to measure velocity, length and time must be unambiguously defined!

Although time intervals and distances within our given frame F (or another frame F′) are well defined by our setup, we must be careful about how we specify times and distances as seen from a moving frame. Let's reconsider frames F and F′ of Fig. 40–5. An observer at O' in frame F′ who wishes to see how fast a clock ticks in frame F must receive signals from the clock attached to O every time the clock in frame F ticks. Because O' is moving away from O in Fig. 40–5, the signals will be delayed.

Simultaneity

We define two events at different points in a given reference frame to be *simultaneous* when they occur at the same time in that frame. *This is what we mean when we say that two events are simultaneous in a given inertial frame*. The concept of simultaneity

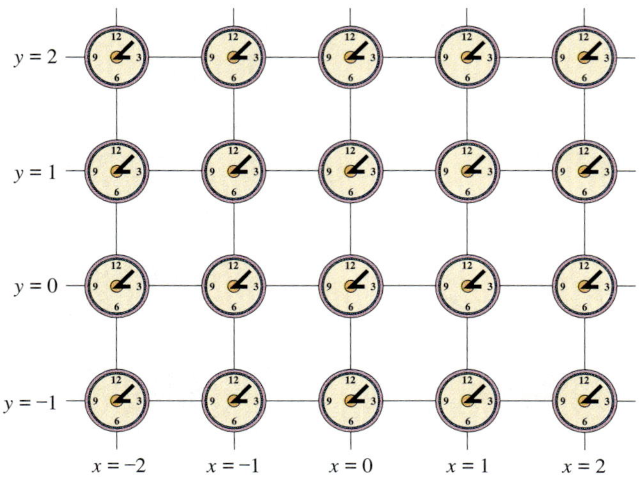

FIGURE 40–7 Clocks attached to lattice points that represent space coordinates separated by fixed distances. The location and the clock reading define the space–time coordinates of an event.

is crucial because it enters subtly into many kinds of measurements. As an example, suppose that an observer in the F′ frame wants to measure the length of a train at rest in the F frame (the observer sees the train move). To measure the length, the observer in F′ must take care that the locations x'_1 and x'_2, corresponding to the front and rear ends of the train, are marked off *at the same time*—simultaneously. Marking the position of the rear end of a moving train at midnight and the front end at 2 min past midnight and taking the difference between these two positions will not give a correct reading of the length of the train. Thus the notion of simultaneity enters into the length measurement. But as Einstein pointed out, our ordinary notion of simultaneity is strongly affected by the existence of a maximum speed for signals, the speed of light.

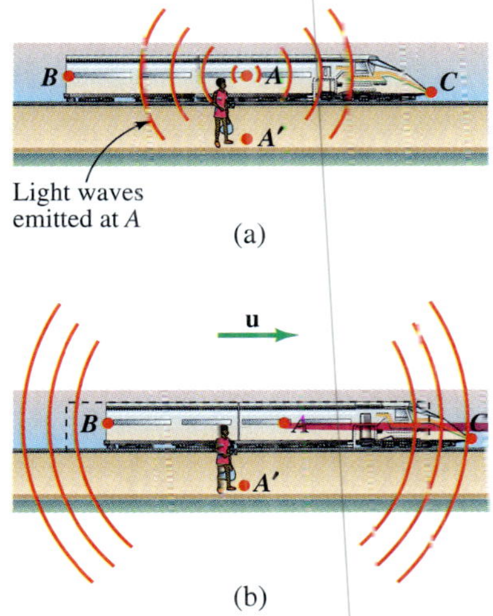

FIGURE 40–8 (a) Light emitted at point A, located midway between points B and C, reaches points B and C at the same time. (b) When B is moving toward A' and C away from A', the light reaches B before it reaches C.

Let's take a more careful look at the problem of measuring a train's length. Figure 40–8a shows a train of length L initially at rest with respect to a platform. L is measured while the train is at rest, so there is no difficulty in making the measurement; we merely lay meter sticks down and count them. The rear and front of the train are labeled B and C, respectively; two persons, A and A', are stationed at the exact midpoint of the train. Person A is *inside* the train and person A' is *outside* on the platform. For the moment, everything is measured with respect to a single inertial reference frame F. As we have already described, there is a set of synchronized clocks in this frame. We label all times in frame F as t. If A sends out a spherical light pulse at $t = 0$, light reaches B and C simultaneously, at $t = L/2c$.

Now suppose that the train is moving at uniform speed u (Fig 40–8b). The frame at rest with respect to the train is F, whereas F′ is a new frame at rest with respect to the platform. Frame F′ has its own system of clocks along the railroad tracks, and times in that frame are labeled t'. At the moment person A is adjacent to person A', person A fires a light pulse, and we can set the clocks to $t = 0 = t'$. From the point of view of frame F, all is as it was in the original situation: The light pulse reaches both points B and C at $t = L/2c$. *But this cannot be true from the point of view of frame* F′ *if the speed of light is the same for both frames*. Person A' sees point B approach even as the light pulse moves toward B at the (fixed) speed of light, so the light pulse reaches B at a time $t' = t'_B$ somewhat earlier. Similarly, the light pulse arrives at point C at a time $t' = t'_C$ somewhat later. In fact, if the train moves a distance ut'_B during the time point B moves toward the pulse, then the distance the beam moving toward B covers is $(L/2) - ut'_B$; according to Einstein's second postulate, this distance is ct'_B. Thus $(L/2) - ut'_B = ct'_B$, an equation that we can solve for t'_B:

$$t'_B = \frac{L/2}{c + u}. \qquad (40\text{–}8)$$

Similarly, according to an observer in F′, the light beam must travel an additional distance ut'_C to reach C. By the same argument,

$$t'_C = \frac{L/2}{c - u}. \qquad (40\text{–}9)$$

Time t'_B is different from time t'_C so *events that are simultaneous in* F *are not simultaneous in* F′! This rather counterintuitive idea that the concept of simultaneity is not absolute is the key to all relativity. Note that the time difference between t'_B and t'_C is very small for $u << c$; this fact explains the origin of our nonrelativistic intuition. If we lived in a world where ordinary speeds were comparable to the speed of light, we would have developed a different intuition.

Whether or not events are simultaneous depends on the frame in which they are measured.

40–4 TIME DILATION AND LENGTH CONTRACTION

The Einstein postulates, or the fact that events that are simultaneous in one reference frame are not simultaneous in a frame that is moving with respect to the first, has two dramatic consequences. These are the slowing down of moving clocks—time dilation—and the shortening of moving rods aligned with the direction of motion—length contraction. Let's examine both consequences.

Time Dilation

To clarify our discussion, consider a very simple clock (Fig. 40–9a).† It consists of a rod with a light bulb at one end and a mirror at the other end, a distance L apart. A mechanism is attached to the light bulb that makes the bulb flash whenever a previous flash returns after reflecting off the mirror. According to an observer at rest relative to the clock, the bulb flashes with a period $T = 2L/c$.

The clock behaves rather differently to a moving observer. Suppose that this observer is in an inertial frame, F′, that moves at speed u to the left (Fig. 40–9a). An observer in F′ will see the clock receding to the right (Fig. 40–9b). In F′, the light still travels at speed c to the mirror, but it now has farther to go: The mirror moves during the time that the light travels to it from the lightbulb. As in our discussion of the transverse light beam in the Michelson–Morley experiment, the time that it takes for the light to travel to the mirror and back is such that

$$\frac{cT'}{2} = \sqrt{L^2 + \left(\frac{uT'}{2}\right)^2};$$

that is, $T' = (2L/c)/\sqrt{1 - (u^2/c^2)}$, or

$$T' = \frac{T}{\sqrt{1 - (u^2/c^2)}}. \tag{40–10}$$

Time dilation

Time T' is greater than time T by a factor of $1/\sqrt{1 - (u^2/c^2)}$. The observer in frame F′ sees longer "ticks" for the clock; in other words, *the moving clock is slower by a factor* of $\sqrt{1 - (u^2/c^2)}$. This effect is known as **time dilation**: *Moving clocks run slower than clocks at rest do.* It is not that the clocks are physically altered; rather, time is different when seen from different inertial frames.

There are two important features of the time-dilation effect. First, it is a symmetric effect. If there were a clock at rest in frame F′ identical to the clock at rest in F, then the observer in F would see the clock in F′ running slow, just as the observer in F′ sees the clock in F running slow. Were it otherwise, the observers could use this asymmetry to decide who was moving and who was standing still, an explicit violation

†Here we follow N. D. Mermin in *Space and Time in Relativity*, New York: McGraw-Hill, 1968.

FIGURE 40–9 (a) Schematic diagram of a clock. The lightbulb at one end of the rod flashes whenever it receives a light flash reflected off the mirror a distance L away from the bulb. The time interval between flashes in the frame of the clock ("ticks") is $2L/c$. (b) The path that the light ray must take according to an observer for whom the clock is moving.

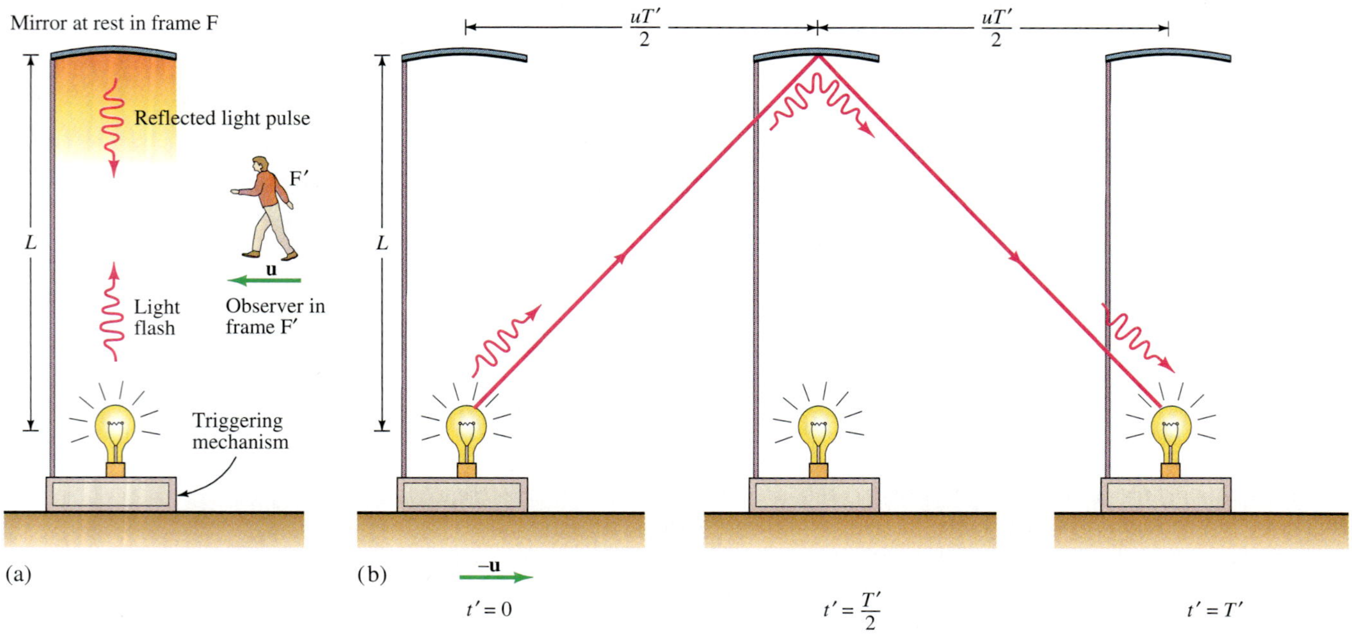

of the original premise that only relative motion has any meaning. Second, although the particular clock we have considered (the "light clock") is an odd one, it is universal in the sense that *every* clock imaginable must behave like it. Any additional clock in F can be synchronized with the light clock so that its ticks are directly and physically tied to the light clock's ticks. The ticks of a clock can take diverse forms from the periodic vibrations of an atomic system to the frequency of a light wave to the beating of a heart. *All* these clocks run slow according to an observer who is moving with respect to them.

Experimental Tests of Time Dilation. The time-dilation effect is real. We can produce experimental evidence with measurements of half-lives of radioactive nuclei or unstable particles in motion. For example, the unstable fundamental particle called the *muon* has a lifetime of 2.197×10^{-6} s when at rest. (In a large sample of muons, 63 percent will have decayed in 2.197×10^{-6} s.) The length of time it takes for 63 percent of a given large sample to decay can be regarded as the tick of a clock. Muons can be produced in a particle accelerator and will then travel at a speed u determined by the characteristics of the accelerator. It is found in the laboratory frame of the accelerator that 63 percent of the moving muons decay into electrons and neutrinos after a time $t = 2.197/\sqrt{1 - (u^2/c^2)}$ μs, *not* after a time 2.197 μs. In modern high-energy accelerators, it is possible to accelerate the muons to such a high velocity that the time-dilation factor $1/\sqrt{1 - (u^2/c^2)}$ can be as large as 10^6, and 63 percent of these muons decay in a period $10^6 \times (2.197\ \mu\text{s}) = 2.197$ s.

The time-dilation effect was also checked in a much more pedestrian way in 1972 (Fig. 40–10). A very accurate cesium clock was flown in a commercial airplane around the world and used to confirm time dilation to an accuracy of about 10 percent. The time-dilation effect has been confirmed so often that there is no doubt of its reality nor of the accuracy of Eq. (40–10).

FIGURE 40–10 A clock taken around the world on an airplane has been used to test time dilation.

EXAMPLE 40–1 Consider a clock taken on an airplane that travels at 1.0×10^3 km/h around the world along the equator. If the clock is synchronized with a stationary clock on departure, by how much will the two clocks differ after 1 round trip?

Solution: Equation (40–10) will provide us with the amount of time that each tick of the clock (here, 1 s) changes. We need to find the total number of seconds that it takes to fly around the world; this is obtained by dividing the distance traveled by the speed of travel. We take Earth's radius at the equator to be 6.38×10^3 km; the circumference is $2\pi r = 40.1 \times 10^3$ km. The speed of travel is 1.0×10^3 km/h, so the flight time is 40.1 h; that is, $(40.1\ \text{h})(3600\ \text{s/h}) = 1.44 \times 10^5$ s. The time-dilation factor is $\sqrt{1 - (u^2/c^2)}$: The clock records $\sqrt{1 - (u^2/c^2)}$ fewer seconds. With $u = (1.0 \times 10^3\ \text{km/h}) \times (10^3\ \text{m/km})/(3600\ \text{s/h}) = 280$ m/s, we can evaluate u/c. It is given by $u/c = (280\ \text{m/s})/(3.0 \times 10^8\ \text{m/s}) = 9.3 \times 10^{-7}$, and thus $u^2/c^2 = 8.6 \times 10^{-13}$. For such a small value of u^2/c^2, we can write

$$\sqrt{1 - \frac{u^2}{c^2}} \simeq 1 - \frac{u^2}{2c^2} = 1 - (4.3 \times 10^{-13}).$$

Thus N, the number of seconds lost, is

$$N = (4.3 \times 10^{-13})(1.44 \times 10^5\ \text{s}) = 6.2 \times 10^{-8}\ \text{s}.$$

To verify this result, we would need a clock that loses or gains no more than 6.2×10^{-8} s out of 1.44×10^5 s—a clock with an accuracy of about 1 s in 10^{12} s.

Here, we have neglected corrections due to Earth's rotation and to the presence of gravity. These factors must be included when actual measurements are compared with the predictions of special relativity.

The Twin Paradox. It should be stressed that all our considerations apply to clocks that move with uniform velocities. Insufficient attention to this restriction leads to the *twin paradox*. Consider identical twins on Earth. One of them takes off at high speed

v on a long journey. After traveling for a long time, that twin gently comes to rest and then retraces her steps. When the traveling twin returns to her starting point, the stay-at-home twin will observe that the traveling twin is considerably younger; they are no longer identical. The stay-at-home twin reasons that this is to be expected because, relative to herself, the traveling twin was moving with uniform velocity: The traveling twin's clock, metabolism, heart rate, and so on were slowed down by a factor of $\sqrt{1 - (v^2/c^2)}$. The deceleration and acceleration at the turning point are assumed to occur in such a short time that they do not affect this conclusion. The paradox appears if the traveling twin considers herself to be at rest while the stay-at-home twin was traveling at speed v in the opposite direction. The traveling twin would expect the stay-at-home twin to be younger. Surely both cannot be right!

There is no paradox: The traveling twin is not always in an inertial frame. She moves at uniform speed most of the time, but she does experience a deceleration and then an acceleration for the return. Thus she cannot make the same statements about the slowing down of clocks as her sister can. From the point of view of special relativity, only the stay-at-home twin, who is always in an inertial frame, can apply the theory to herself. In fact, a careful use of special relativity can show just how much younger the stay-at-home twin is at the end of the journey.

Length Contraction

The slowing down of moving clocks is accompanied by the contraction in length of moving objects along their direction of motion. We can begin by giving an argument that uses time dilation. Consider again the muon, an unstable particle with a lifetime of $\tau \simeq 2\ \mu s$. The lifetime is so small that even if a muon were moving near the speed of light, $c\tau$ would be much smaller than the atmosphere's height and, without relativity, muons produced at the top of the atmosphere would decay before they could reach the ground. But muons are in fact copiously produced in the upper atmosphere by cosmic rays and some muons reach the ground. Time dilation permits that. Suppose that, as seen from the ground, muons move at speed u. Because they are moving, their lifetime is increased to $\tau/\sqrt{1 - (u^2/c^2)}$. According to a ground-based observer, about half the muons will cover the distance L given by the speed times the increased lifetime:

$$L = \frac{u\tau}{\sqrt{1 - (u^2/c^2)}}. \tag{40–11}$$

This length is much greater than it would be if there were no time-dilation effect because the square-root factor approaches zero as u approaches c. The length could be larger than the height of the atmosphere; in this way, muons could reach the ground before they decay.

We now suppose that an "average" muon just reaches the ground before decaying, so L is the height of the atmosphere. Let's see how this looks to an observer who is moving with the muon—an observer who sees the muon at rest. He will measure the muon's lifetime to be its original value τ. This observer, however, will also detect that the muon reaches the ground if the Earth-based observer does; the collision with the ground is an event no observer could dispute. Thus, in time τ, he will see the whole atmosphere move past him at speed u. If our observer measures the atmosphere to have a height L', then the atmosphere will pass him in time L'/u. This must equal τ. Thus $L' = u\tau$, or, from Eq. (40–11),

$$L' = L\sqrt{1 - \frac{u^2}{c^2}}. \tag{40–12}$$

Length contraction

The observer moving with the muon measures the atmosphere to be thinner than an Earth-based observer does. To the moving observer, the atmospheric height, or any length in the direction of his motion, has undergone a **length contraction** by a factor of $\sqrt{1 - (u^2/c^2)}$.

Another way of seeing that there must be a length contraction along the direction of motion is to modify the clock constructed at the beginning of this section. We add an identical rod at right angles (transverse) to the original rod (Fig. 40–11). The mechanism is modified so that the bulb relights only when both reflected beams reach the lightbulb at the same time. This can be achieved by making the length of the rods identical. Each flash of the bulb is an event, and these events are observed from any inertial frame. Suppose now that the clock moves at speed u in the direction of the added rod with respect to an observer in frame F′. According to the observer in frame F′, the round-trip time for the light on the transverse rod is

$$T' = \frac{2L}{c}\frac{1}{\sqrt{1-(u^2/c^2)}}. \qquad (40\text{–}13)$$

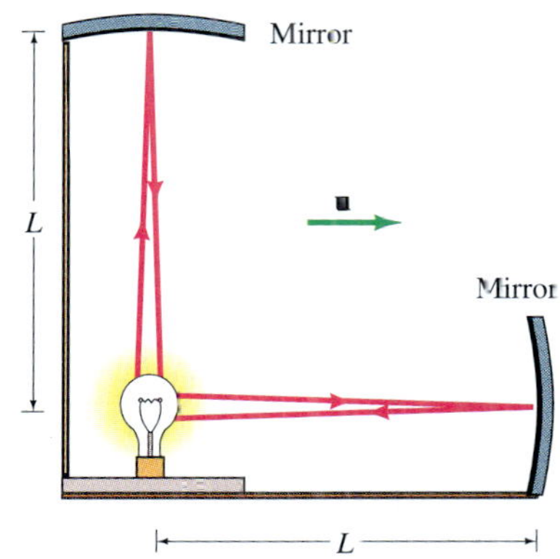

FIGURE 40–11 Schematic diagram of a two-armed clock used to exhibit length contraction. Note the similarity to the Michelson–Morley apparatus.

The round-trip time for the light that travels along the horizontal rod is the time t_1' it takes to get to the mirror added to the time t_2' it takes to return. The mirror is moving to the right, so the light on its outward trip has an extra distance ut_1' to travel. If, according to the observer in frame F′, the length of the rod is L' (the quantity we want to find), then

$$ct_1' = L' + ut_1'.$$

For the return trip, the bulb approaches the mirror at speed u, so the time t_2' to return is determined by

$$ct_2' = L' - ut_2'.$$

We solve these two equations for t_1' and t_2', respectively, and add:

$$t_1' + t_2' = \frac{L'}{c-u} + \frac{L'}{c+u} = \frac{2L'/c}{1-(u^2/c^2)}. \qquad (40\text{–}14)$$

This, however, must equal T', and a comparison of Eqs. (40–13) and (40–14) gives $L' = L\sqrt{1-(u^2/c^2)}$, which is the length contraction of Eq. (40–12).

Note that length contraction occurs only *along* the direction of motion. We can see that there is no change in directions transverse to the motion by the argument in Fig. 40–12. We made the unspoken assumption that there is no transverse length contraction in our derivation of time dilation; that derivation is now more solid.

EXAMPLE 40–2 The radius of our galaxy is 3×10^{20} m. (a) How fast would a spaceship have to travel to cross the entire galaxy in 300 yr as measured from within the spaceship? (b) How much time would elapse on Earth during the traversal?

Solution: (a) Our hypothetical traveler would be at rest within the spaceship and would see the galaxy approaching at some speed v. The galaxy is contracted along the direction of motion, and it is the contracted length that must be covered in 300 yr of "spaceship time" at speed v. If L is the diameter of the galaxy, then the contracted diameter, from Eq. (40–12), is

$$L' = L\sqrt{1-\frac{v^2}{c^2}}.$$

If the time measured in the spaceship is T, then the required speed is

$$v = \frac{L'}{T} = L\frac{\sqrt{1-(v^2/c^2)}}{T}.$$

Thus $v^2 = (L^2/T^2)[1-(v^2/c^2)]$, which we write as

$$x = \frac{L^2}{c^2T^2}(1-x),$$

where $x = v^2/c^2$. When we solve for x, we get

$$x = \frac{L^2}{L^2 + c^2T^2}.$$

We know that $L = 2(3 \times 10^{20}\text{ m}) = 6 \times 10^{20}\text{ m}$ and $T = (300\text{ yr})(3.15 \times 10^7\text{ s/yr}) = 9.5 \times 10^9\text{ s}$. This gives

$$x = \frac{(6 \times 10^{20}\text{ m})^2}{(6 \times 10^{20}\text{ m})^2 + (3 \times 10^8\text{ m/s})^2(9.5 \times 10^9\text{ s})^2}$$

$$= \frac{36 \times 10^{40}}{(36 \times 10^{40}) + (0.8 \times 10^{37})} \simeq \frac{1}{1 + (2 \times 10^{-5})} \simeq 1 - (2 \times 10^{-5}).$$

Thus

$$v/c = \sqrt{x} = \sqrt{1 - (2 \times 10^{-5})} \simeq 1 - 10^{-5} = 0.99999.$$

A spaceship must travel at a speed extremely close to that of light for it to traverse huge distances in a "reasonable" amount of time.

(b) As seen from Earth, the galaxy is not contracted and the spaceship moves at $0.99999c$. The time for the trip as seen from Earth is then

$$t_{\text{earth}} = \frac{L}{v} = \frac{6 \times 10^{20}\text{ m}}{(0.99999)(3 \times 10^8\text{ m/s})} \simeq 2 \times 10^{12}\text{ s} \simeq 64{,}000\text{ yr}.$$

The Earth-based observer will, however, see the spaceship's clock tick off only 300 yr as it travels from one end of the galaxy to the other.

The Doppler shift for sound was covered in Chapter 14.

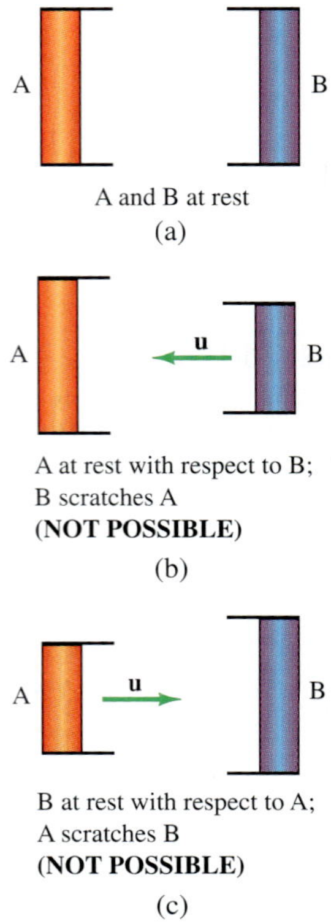

FIGURE 40–12 (a) Two rods, A and B, have the same length when they are at rest with respect to one another. (b) Now the rods approach each other. From the viewpoint of rod A, rod B might be shortened in a direction transverse to its direction of motion. Rod B could *simultaneously* scratch marks near the top and bottom of rod A as shown. (c) If the principle of relativity holds, then from the viewpoint of B, A would similarly be shortened. This shortening could be marked by scratches made by A onto B. But now bring the rods to rest and compare: A scratch at the 0.8-m mark of A, say, was made by the 1-m mark of B; a scratch at the 0.8-m mark on B was made by the 1-m mark of A. One "event," as recorded by the scratches, has been seen differently by two observers—an impossibility. The only possible resolution is that there can be no shortening in directions perpendicular to the motion.

40–5 THE RELATIVISTIC DOPPLER SHIFT

The Doppler shift for sound describes the changes in pitch of a train whistle as the train approaches, passes, and recedes from an observer. When a moving source that emits sound waves with frequency f travels toward an observer at rest relative to the air, the observed frequency f' is shifted from the source frequency according to

$$f' = \frac{f}{1 - (u/c)},$$

Eq. (14–50). We have changed the notation slightly, representing the source speed by u (instead of v_s) and the speed of sound by c (instead of v). If the source is at rest relative to the air and the observer is moving toward the source, then the frequency picked up by the observer is

$$f' = f\left(1 + \frac{u}{c}\right),$$

Eq. (14–54), with a similar change of notation. The frequencies f' are not the same in the two cases, so it is possible—by an accurate measurement of the frequency shift and a knowledge of the relative speed—to determine whether it is the source or the receiver who is moving relative to the medium (the air). The reason for the difference between the two shifts is that, for sound, there *is* a preferred frame; namely, the frame at rest relative to the air. The Doppler shift for electromagnetic radiation (including light) cannot distinguish between the two situations and must therefore have a different form.

To find the Doppler shift for light, let's consider a periodically flashing light that moves at speed u toward an observer (Fig. 40–13). The source is placed at the origin of frame F′. Suppose that one pulse of light is emitted for every time interval τ_0, so the frequency of emission is $f_0 = 1/\tau_0$ as seen by someone moving with the source. The stationary observer on the right sees frame F′ moving toward him at speed u. Let's calculate what he observes the frequency of the flashes to be. First, the flashing is time dilated; the moving clock runs slow, and has a period

$$\frac{\tau_0}{\sqrt{1 - (u^2/c^2)}}$$

according to the stationary observer. But there is a second effect. The time between two successive pulses is reduced because the flashing clock has moved toward the sta-

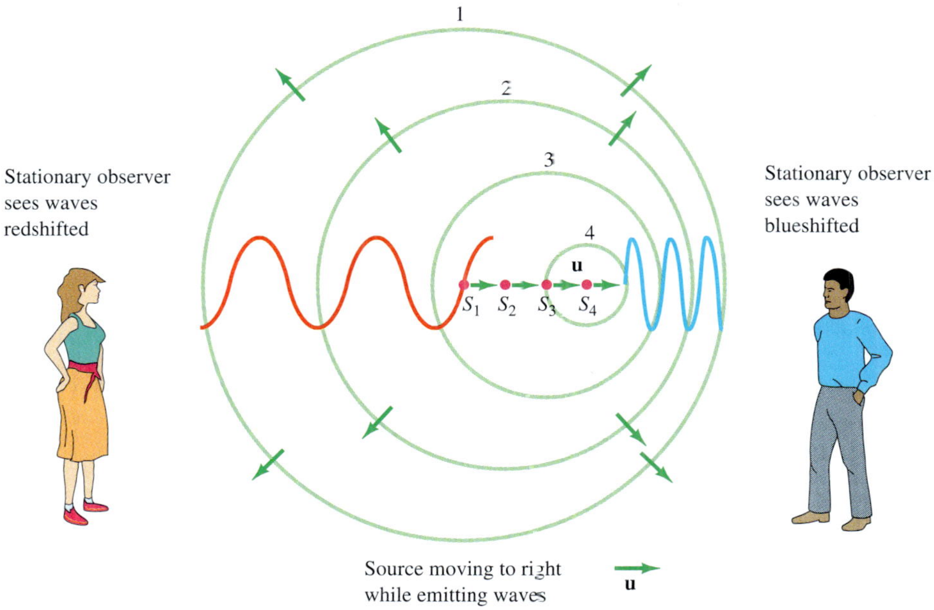

FIGURE 40–13 The Doppler effect associated with the relative movement of a light source and an observer. The observer toward whom the source moves sees the wave fronts crowded together (a decreased wavelength), and hence he sees a blueshift. The observer away from whom the source moves sees the wave fronts spread apart (an increased wavelength), and hence she sees a redshift. A time-dilation factor must be applied as well.

tionary observer between a first and a second pulse. If he—the stationary observer—says that the light is a distance L from him when the first pulse is sent, then he measures the light to be at a reduced distance

$$L - \frac{u\tau_0}{\sqrt{1-(u/c^2)}}$$

when the second pulse is sent. This reduces the time between pulses because the second pulse has a shorter distance to travel. If the first pulse arrives at a time $t_1 = L/c$, then the second pulse arrives at a time t_2 given by $\tau_0/\sqrt{1-(u/c)^2}$ plus the reduced distance divided by c:

$$t_2 = \frac{\tau_0}{\sqrt{1-(u/c)^2}} + \frac{1}{c}\left[L - \frac{u\tau_0}{\sqrt{1-(u/c)^2}}\right].$$

The stationary observer measures the period of the pulses to be $\tau = t_2 - t_1$; that is,

$$\tau = t_2 - t_1 = \frac{\tau_0}{\sqrt{1-(u/c)^2}} + \frac{1}{c}\left[L - \frac{u\tau_0}{\sqrt{1-(u/c)^2}}\right] - \frac{L}{c} = \frac{\tau_0}{\sqrt{1-(u/c)^2}}\left(1 - \frac{u}{c}\right).$$

If we use $1 - x^2 = (1-x)(1+x)$, we can write this result as

$$\tau = \tau_0\sqrt{\frac{1-u/c}{1+u/c}}. \qquad (40\text{–}15)$$

Finally, we can recognize that the frequency is the inverse of the period:

$$f_1 = f_0\sqrt{\frac{1+(u/c)}{1-(u/c)}}, \qquad (40\text{–}16a)$$

The Doppler shift for light

where f_0 is the frequency of the source in its rest frame and f_1 is the frequency observed from a frame that moves at speed u relative to the source. The two frames are moving toward each other at relative speed u, and the frequency is increased. Instead of the frequency, we can use $\lambda = c/f$ to express the wavelength λ_1 seen by the observer in terms of the wavelength λ_0 at the source:

$$\lambda_1 = \lambda_0\sqrt{\frac{1-(u/c)}{1+(u/c)}}. \qquad (40\text{–}16b)$$

We call this situation—in which the source moves toward the observer and the observed wavelength decreases—a *blueshift*. The speed u can be interpreted either as the speed of the source toward a stationary receiver or as the speed of the receiver toward a stationary source. According to the principles of special relativity, these two possibilities are not distinguishable. If the source is moving away from the observer, then we must change the sign of u in our results, and

$$f_1 = f_0\sqrt{\frac{1-(u/c)}{1+(u/c)}}. \tag{40–17a}$$

Equivalently, the observed wavelength is

$$\lambda_1 = \lambda_0\sqrt{\frac{1+(u/c)}{1-(u/c)}}. \tag{40–17b}$$

Thus, the frequency decreases (the wavelength increases) in this case. The visible spectrum is shifted toward the red colors, and Eqs. (40–17) are said to describe a relativistic *redshift*. This is the case for the stationary observer on the left in Fig. 40–13.

Cosmological Implications of the Doppler Shift for Light

Measurements of the Doppler shift of starlight have proven to be crucial in the evolution of modern astrophysics and cosmology. In one application, the Doppler shift is used to establish the velocities of stars or other radiating bodies. Radiation emitted by atoms and molecules is characterized by *spectral lines*: discrete or very narrow frequency bands of especially intense radiation. These spectral lines provide a signature for elements and compounds. If an entire sequence of spectral lines in starlight is observed to correspond to a sequence of laboratory-observed spectral lines all shifted by the same factor, then we know that the source of the starlight is the same as the laboratory source but moves at a velocity that can be calculated from equations such as Eq. (40–16) (Fig. 40–14).

One of the most interesting uses of the Doppler shift was made by the astronomer Edwin Hubble (Fig. 40–15). In the 1920s, 1930s, and 1940s he studied the spectral lines of a large number of stars in distant galaxies, and he estimated their distance from Earth by using the known characteristic brightnesses of these stars. Hubble discovered that the spectra of most of these distant stars are redshifted, which means that their

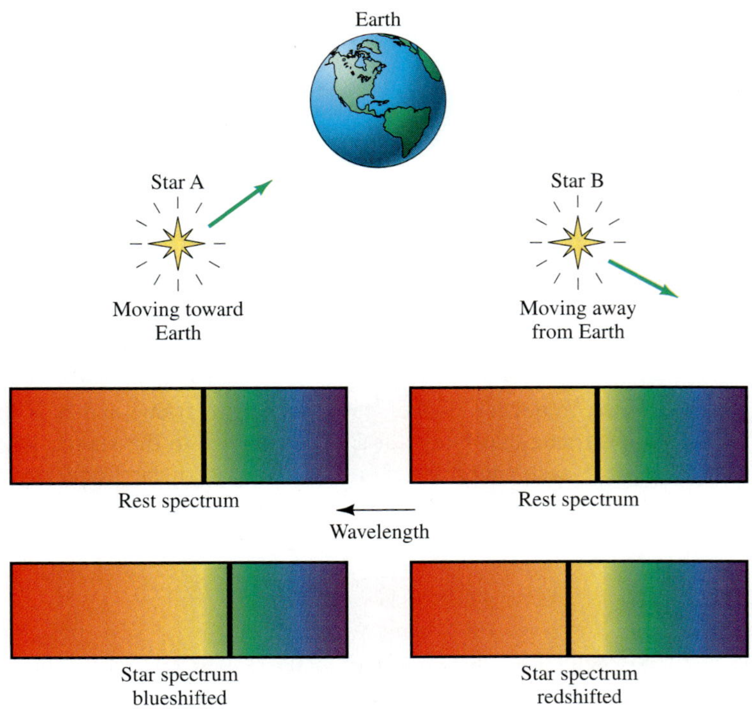

FIGURE 40–14 The spectral lines that can be attributed to a specific element are shifted if the source moves relative to the observer. The spectral lines emitted by Star A, moving toward Earth, are shifted to the blue, while those of Star B, moving away from Earth, are shifted to the red.

galaxies are receding from us. He found that *the recession velocity of the galaxies relative to our galaxy is proportional to their distance from Earth*. This result is known as **Hubble's law**, and it takes the mathematical form

Hubble's law

$$D = u/H. \tag{40–18}$$

Here, D is the distance to a galaxy, u is the recession speed relative to us, and H is the so-called *Hubble parameter*, measured to be $H \simeq 2.5 \times 10^{-18}\ \text{s}^{-1}$. The fact that the distance and speed are measured relative to Earth appears to give Earth a central position, which is deceptive. If all stars and galaxies are moving away from each other, then an observer located at any one of them would report the same effect. This is easily visualized if we consider a simple model of dots painted uniformly on a balloon (Fig. 40–16a). As the balloon is inflated, all the dots move farther away from each other, and each dot "sees" the others moving away from it (Fig. 40–16b). The model of galaxies that move away from each other is part of the cosmological theory of the *big bang*, which proposes that the universe started from a point and underwent a rapid expansion. In this theory, the age of the universe is on the order of H^{-1}, roughly 13 billion years.

Measurements of the Doppler shift, together with Hubble's law, allow us to calculate distances to galaxies. In the 1960s, astronomers working with radio telescopes discovered very powerful pointlike sources of radiation in which very large redshifts ($f_1/f_0 \simeq 0.3$) were observed. These sources are *quasars*, or *quasistellar objects*. Astronomers concluded that quasars emit huge amounts of energy, which left a puzzle as to a mechanism by which that much energy is produced. An explanation that is gaining acceptance is that enormous accelerations of matter caused by the presence of black holes lead to the large amount of radiation.

FIGURE 40–15 Edwin Hubble at Mt. Palomar Observatory in 1948.

EXAMPLE 40–3 Studies of a quasar show that a spectral line whose wavelength in the laboratory is 121 nm has a measured wavelength of 358 nm. With what speed is the quasar receding from Earth? Assuming that Hubble's law holds, what is the distance in light-years of the quasar from Earth?

Solution: Equation (40–17b) gives the Doppler-shift formula for wavelengths. From it, we find

$$\left(\frac{\lambda_1}{\lambda_0}\right)^2\left(1 - \frac{u}{c}\right) = 1 + \frac{u}{c}.$$

When we solve this equation for u/c, we find that

$$\frac{u}{c} = \frac{(\lambda_1/\lambda_0)^2 - 1}{(\lambda_1/\lambda_0)^2 + 1}.$$

The data give $\lambda_1/\lambda_0 = (358\ \text{nm})/(121\ \text{nm}) = 2.96$, so $u/c = 0.79$, and

$$u = (0.79)(3.00 \times 10^8\ \text{m/s}) = 2.38 \times 10^8\ \text{m/s}.$$

Application of Eq.(40–18) gives

$$D = \frac{u}{H} = \left(\frac{2.38 \times 10^8\ \text{m/s}}{2.5 \times 20^{-18}\ \text{s}^{-1}}\right) = 0.95 \times 10^{26}\ \text{m}.$$

Because $1\ \text{ly} = (3.15 \times 10^7\ \text{s})(3.00 \times 10^8\ \text{m/s}) = 0.95 \times 10^{16}\ \text{m}$, we obtain $D = 10^{10}$ ly. Such a distance is nearly at "the edge of the universe," and the light reaching Earth gives information about the quasar 10 billion yr ago!

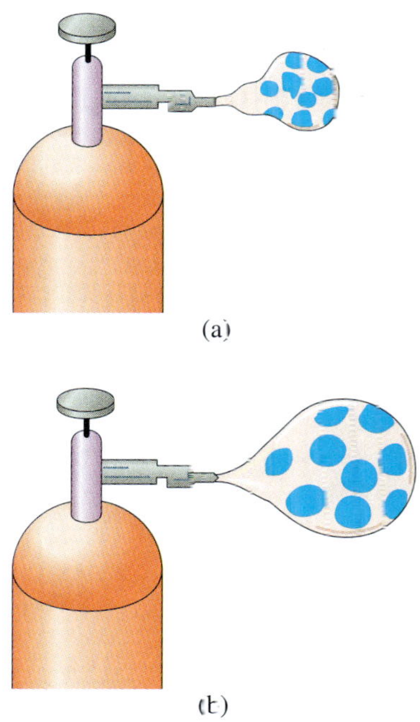

FIGURE 40–16 (a) Dots painted on the surface of a balloon represent an analogy to Hubble's expanding universe. (b) As the balloon expands, the dots move away from one another at a speed that depends on the distance between them.

The Relativistic Addition of Velocities

Suppose that observer A measures the velocity of an object as $\mathbf{v}_1$; in turn observer B measures observer A to move with velocity $\mathbf{u}$ with respect to him. According to the Galilean law of velocity addition [Eq. (40–4)], observer B will measure the object to move with velocity $\mathbf{v}_2 = \mathbf{v}_1 + \mathbf{u}$. As we shall soon see, this simple result cannot be consistent with special relativity.

We can use the Doppler shift for light to find the important relation that describes how velocities add. Suppose that a source emits light with the rest-frame frequency f_0 (Fig. 40–17). An observer moving away from the source at speed v_1 along the x-axis receives a redshifted frequency f_1. If that observer, O_1, immediately reradiates with frequency f_1 to another observer, O_2, who is moving away from O_1 in the same direction at speed v_2 (with respect to O_1), we would expect that the relationship between the received frequency f_2 and the original frequency f_0 would be the frequency f_0 shifted by the velocity V of the second observer relative to the source. In nonrelativistic mechanics, we would expect to find the correct shift with the Galilean form $V = v_1 + v_2$. We will now use this procedure to obtain the relativistic counterpart of the Galilean form for the addition of velocities. It should come as no surprise that the formula needs modification because, for $v_1 = v_2 = 0.8c$, for example, $v_1 + v_2$ is larger than c, which our expressions for time dilation and length contraction [Eqs. (40–10) and (40–12)] do not allow.

The frequency f_1 measured by observer O_1 is, according to Eq. (40–17a),

$$f_1 = f_0\sqrt{\frac{1-(v_1/c)}{1+(v_1/c)}}.$$

If that observer immediately reradiates the light toward observer O_2, moving away at speed v_2 relative to observer O_1, then the frequency seen by observer O_2 is

$$f_2 = f_1\sqrt{\frac{1-(v_2/c)}{1+(v_2/c)}}.$$

We may eliminate f_1 by expressing f_2 in terms of f_0 and the (as yet unknown) velocity V of observer O_2 relative to the source. We have

The relativistic addition of velocities

$$f_2 = f_0\sqrt{\frac{1-(v_1/c)}{1+(v_1/c)}}\sqrt{\frac{1-(v_2/c)}{1+(v_2/c)}},$$

which we rewrite in the form

$$f_2 = f_0\sqrt{\frac{1-(V/c)}{1+(V/c)}}.$$

When we square (f_2/f_0) in its two forms and equate the two, we obtain

$$\frac{1-(V/c)}{1+(V/c)} = \frac{[1-(v_1/c)][1-(v_2/c)]}{[1+(v_1/c)][1+(v_2/c)]}.$$

If we solve this equation for V, the result is the *law of addition of velocities*:

$$V = \frac{v_1 + v_2}{1 + (v_1 v_2/c^2)}. \quad (40\text{–}19)$$

FIGURE 40–17 A light source emits a frequency f_0, seen as frequency f_1 by an observer O_1 who moves at speed v_1 to the right. Observer O_1 re-emits the light with frequency f_1, which is seen by an observer O_2 who moves to the right at speed v_2 with respect to observer O_1. (Observer O_2 moves at speed V with respect to the original source.) Observer O_2 measures a light frequency f_2, consistent with the law of addition of velocities.

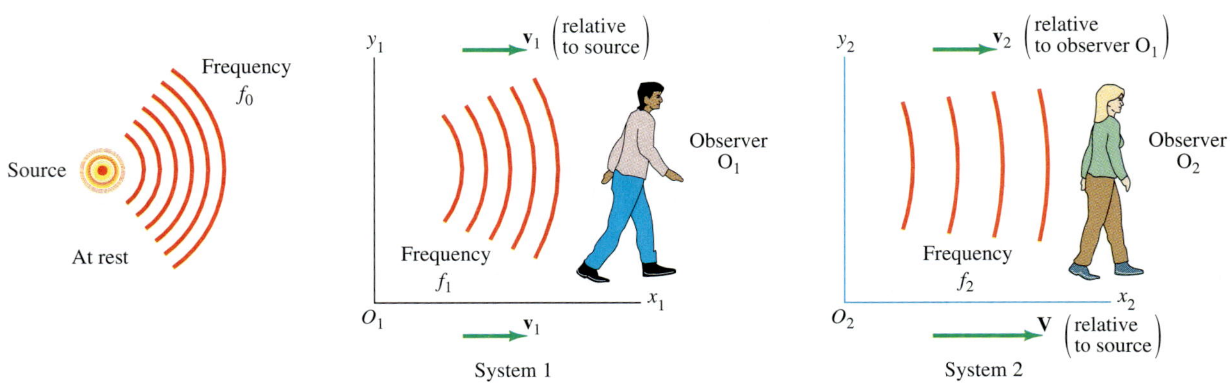

We see that in the nonrelativistic limit where v_1/c and v_2/c are both small, Eq. (40–19) reduces to the Galilean law of addition of velocities: $V \to v_1 + v_2$. But if $v_1 = v_2 = 0.5c$, then Eq. (40–19) gives $V = 0.8c$, not $1.0c$. If $v_1 = c$, we get $V = c$, *independent of* v_2, so the maximum speed is c, and it is never exceeded. Equation (40–19) is a beautiful illustration of how relativistic kinematics smoothly adjoins the nonrelativistic (Galilean) form while providing a very different general result.

EXAMPLE 40–4 A light source flashes with a frequency of 1.0×10^{15} Hz. The radiation is reflected by a mirror that moves at a speed of 100 km/s away from the source. As observed at the source, how different is the frequency of the reflected radiation from the original radiation?

Solution: We shall approach this problem in the same way that we approached the derivation of the law of addition of velocities. We imagine a hypothetical observer who travels with the mirror. This observer receives the (redshifted) radiation from the source and reradiates it toward the source with the frequency that observer saw the light to have. A second observer at the source sees this radiation redshifted once more, and this radiation is completely equivalent to what results from reflection in a receding mirror.

Let's first determine the frequency as observed at the mirror. The emitted frequency is f_0. Because the source is receding from the mirror at speed u, the frequency f' observed at the mirror is redshifted:

$$f' = f_0\sqrt{\frac{1-(u/c)}{1+(u/c)}}.$$

That is also the frequency of the radiation "emitted" by the mirror in its rest frame. The fact that it is a moving source that emits to a stationary observer is of no consequence because the same shift in frequency is obtained if the source is at rest and the observer is moving. The frequency f'' observed at the source is therefore further redshifted:

$$f'' = f'\sqrt{\frac{1-(u/c)}{1+(u/c)}}$$

Thus

$$f'' = f_0\left[\frac{1-(u/c)}{1+(u/c)}\right].$$

The reflected radiation differs from the original radiation by the frequency difference $f_0 - f''$. We have

$$f_0 - f'' = f_0\left[1 - \frac{1-(u/c)}{1+(u/c)}\right] = \frac{f_0}{1+(u/c)}\left[1 + \frac{u}{c} - \left(1 - \frac{u}{c}\right)\right] = \frac{2f_0(u/c)}{1+(u/c)}.$$

Here, $u/c = (10^5\ \text{m/s})/(3.0 \times 10^8\ \text{m/s}) = 0.30 \times 10^{-3}$ is very small, so we can drop it compared to 1 in the denominator. Thus

$$f_0 - f'' \simeq 2(1.0 \times 10^{15}\ \text{Hz})(0.30 \times 10^{-3}) = 6.0 \times 10^{11}\ \text{Hz},$$

only 6×10^{-4} times the original frequency.

Another way of obtaining this result is to think of the image of the source in the mirror as radiating directly to the observer at the source. If a mirror recedes at speed u, then the speed of recession of the image is given by Eq. (40–19), with $v_1 = v_2 = u$. When $V = 2u/[1 + (u^2/c^2)]$ is substituted into the normal redshift formula, we obtain the same result.

APPLICATION: SPEED DETECTORS

The speed detector used by police is a device that sends electromagnetic waves with frequency in the range of 10 to 30 kMHz ($1–3 \times 10^{10}$ Hz) toward a moving vehicle (Fig. 40AB–1). (Although this is called a radar, it is not a radar in the strict sense of the word because it provides no range information.) The vehicle is metal, so it must remain an equipotential; to cancel the incoming wave at its skin, it radiates an electromagnetic wave back toward the generating device. As the vehicle is both a moving absorber and a moving reradiator, there are two Doppler shifts of the radiation (Example 40–4). If the original radiation has angular frequency ω_1, then the returned signal—which is detected by the police device—has a slightly shifted angular frequency ω_2, with the amount of the shift determined by the speed of the vehicle. How then is the speed determined? There is no question of a direct measurement of ω_2—it is much too close to ω_1 for a direct comparison to be useful. Instead, a trick is used. The device multiplies the returning signal by the outgoing signal; that is, the function $\cos(\omega_1 t)\cos(\omega_2 t)$ is formed. This combination is equivalent by trigonometric identity to

$$\frac{1}{2}\{\cos[(\omega_1+\omega_2)t]+\cos[(\omega_1-\omega_2)t]\}.$$

The first term is another oscillation with frequency of tens of kMHz. The second term, however, is a "beat," with a frequency directly proportional to the difference of the two frequencies. This beat frequency turns out to be in the kHz region and is easily measurable. A simple frequency meter, calibrated in equivalent miles per hour, informs the police of the speed of the vehicle.

FIGURE 40AB–1 A policeman uses a radar gun to determine the speed of passing vehicles.

40–6 THE LORENTZ TRANSFORMATIONS

An observer in any frame F will describe an event by its location in space and time in that frame; that is, by its *space–time* coordinates. An observer in a second frame, F′, will describe the same event by its space–time coordinates in frame F′. Let frames F and F′ be inertial, moving at speed u along the x-axis with respect to each other. Let's take the origins and axes of frames F and F′ to coincide at time $t = t' = 0$. The origin of the moving frame F′ at a later time is given by $x = ut$ in frame F, whereas that origin in frame F′ is still given by $x' = 0$. An event, which may be an explosion, the collision of two particles, or the flash of a lightbulb, is described by the variables (x, t) in F and (x', t') in F′ (Fig. 40–18).

The assumption of a universal time would relate the times by Eq. (40–5),

$$\text{Galilean transformation:} \qquad t' = t,$$

and the position coordinates by Eq.(40–3),

$$\text{Galilean transformation:} \qquad x' = x - ut.$$

As an example of how these *transformation laws* apply, suppose that an explosion occurs at the origin of frame F ($x = 0$) at time t. It would be described in frame F′ as having occurred at $x' = x - ut = -ut$. This is physically sensible: The moving frame F′ will have left frame F behind. As we noted in Section 40–2, the Galilean transfor-

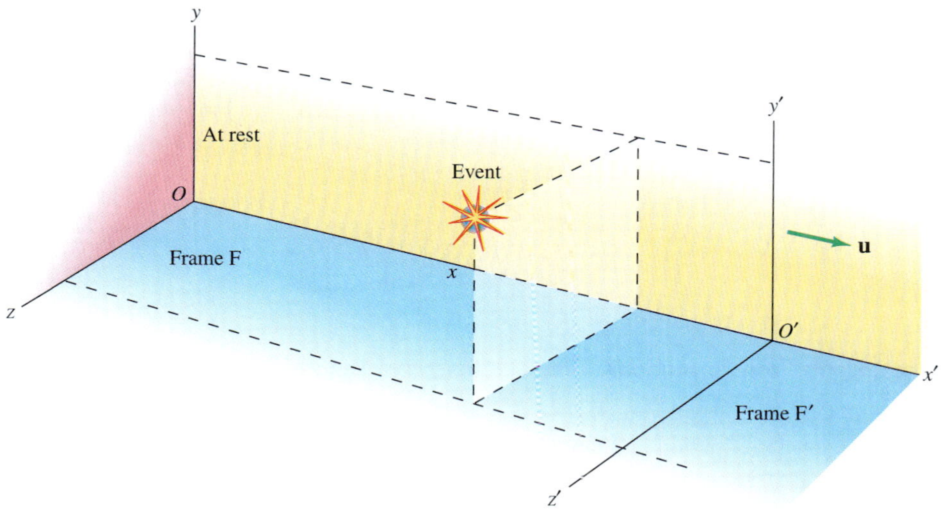

FIGURE 40–18 (a) The origins of the coordinate systems of frames F and F′ coincide at time $t = t' = 0$. (b) An event occurs some time later. An observer in frame F describes the space–time coordinates of the event differently than an observer in frame F′ would.

mation laws do not satisfy the condition that the speed of light is the same in all inertial frames. The correct transformation laws—known as the **Lorentz transformations**—were found by Hendrik A. Lorentz in 1890:†

Equations (40–20), (40–21), (40–23), and (40–24) are the Lorentz transformations.

Lorentz transformations:

$$x' = \gamma(x - ut); \tag{40–20}$$

$$t' = \gamma\left(t - \frac{ux}{c^2}\right), \tag{40–21}$$

with γ defined by

$$\gamma \equiv \frac{1}{\sqrt{1 - (u^2/c^2)}}. \tag{40–22}$$

To these equations, we should add that

$$y' = y \tag{40–23}$$

and

$$z' = z. \tag{40–24}$$

Equations (40–23) and (40–24) reflect the fact that the description of the y- and z-coordinates of an event is the same in frames F and F′.

We observe that when the relative speed of the two frames is small compared to c ($u/c << 1$), then $\gamma \simeq 1$ and the Lorentz transformations reduce to the Galilean transformations, Eqs. (40–3) and (40–4). This is why the necessity for the Lorentz transformations is not evident from ordinary mechanics. It can also be shown that, as in Eq. (40–6), the location of a wave front for a plane wave emitted from the joint origin of frames F and F′ at $t = t' = 0$ is given by $x^2 - c^2t^2 = 0 = x'^2 - c^2t'^2$ (see Problem 38). This result reflects the necessity that the speed of light be the same in both frames. In fact, the Lorentz transformations show more: *In general,*

$$x'^2 - c^2t'^2 = x^2 - c^2t^2. \tag{40–25}$$

Equation (40–25) states that the quantity $x^2 - c^2t^2$ is *invariant*, meaning that its value is the same in all inertial frames. There is a familiar three-dimensional analogue in geometry: For points on the surface of a sphere, $x^2 + y^2 + z^2$ is always the same.

An invariant quantity has the same value in all inertial frames.

†This date is well before 1905, when Einstein published his work. Einstein's key role was perhaps less in discovering new formulas than in formulating a conceptual whole.

We can obtain x and t in terms of x' and t' by solving the two simultaneous algebraic equations (40–20) and (40–21). We refer to the result as the inverse Lorentz transformations. Note that if frame F′ is moving at speed u along the x-axis away from frame F, then frame F may be viewed as moving at speed $-u$ along the x-axis away from frame F′. Thus, *the inverse transformation laws are found from the original transformation laws by the simple substitution of* $-u$ *for* u:

$$x = \gamma(x' + ut'); \tag{40–26}$$

$$t = \gamma\left(t' + \frac{ux'}{c^2}\right). \tag{40–27}$$

The Lorentz Transformations, Time Dilation, and Length Contraction

We may use the Lorentz transformations of Eqs. (40–20) and (40–21) to rederive the formulas for time dilation and length contraction. Let's consider a clock fixed in frame F at $x = 0$. It starts at time $t = t' = 0$, and its initial location is $x = x' = 0$. The starting time and place may be viewed as the first event. If the period of the clock is τ, then the clock's first "tick" will be the second event, which will occur at $x = 0$, $t = \tau$. With the help of the Lorentz transformations given in Eqs. (40–20) and (40–21), we find that

$$x' = -\gamma u\tau, \tag{40–28a}$$

$$t' = \gamma\tau. \tag{40–28b}$$

Equation (40–28b) tells us that an observer in frame F′ sees the interval between ticks of the moving clock to be longer by a factor of γ. An observer in that frame thus sees the clock as running slow. At the first tick, the clock is seen to be at $x' = -\gamma u\tau = -ut'$, the location of the origin of frame F as seen from frame F′.

We can find the formula for length contraction by considering a rod of length L at rest in frame F. The length of an object in its own rest frame is called its **proper length**. The coordinates of the rod are $x_1 = 0$ and $x_2 = L$. What is the length of the rod in frame F′? As we emphasized earlier, when an object is moving in a certain reference frame, a length measurement makes sense only when the coordinates of the two ends are located simultaneously. Thus we require that $t'_1 = t'_2$. It is convenient to choose both of these to be zero; that is, make the length measurement at time $t'_1 = t'_2 = 0$. Let's see what these times are in frame F. We have

$$t'_1 = \gamma\left(t_1 - \frac{ux_1}{c^2}\right).$$

Because $x_1 = 0$ and $t'_1 = 0$, we obtain

$$t_1 = 0.$$

We also get

$$t'_2 = \gamma\left(t_2 - \frac{ux_2}{c^2}\right).$$

With $t'_2 = 0$ and $x_2 = L$, we find that

$$t_2 = \frac{uL}{c^2}.$$

Thus the length measurements are made at different times in frame F. We again see that events that are simultaneous in one frame are not simultaneous in another. We may now find the coordinates of the two ends in frame F:

$$x'_1 = \gamma(x_1 - ut_1) = 0;$$

$$x'_2 = \gamma(x_2 - ut_2) = \gamma\left(L - \frac{u^2L}{c^2}\right) = L\gamma\left(1 - \frac{u^2}{c^2}\right) = \frac{L}{\gamma}. \tag{40–29}$$

This result is the length contraction as observed in frame F′.

EXAMPLE 40–5 As seen from Earth, spaceship A of proper length L is traveling east at speed v_1 and spaceship B of proper length $2L$ is traveling west at speed v_2. The pilot of spaceship A sets his clock to zero when the front of spaceship B passes him. (The spaceship pilots sit in the nose cone.) Use Lorentz transformations to calculate the time at which, according to the pilot of spaceship A, the tail of spaceship B passes him.

Solution: The Lorentz transformations provide the necessary information. To proceed, we must identify two events and describe them both in frame F, with coordinates (x, y, z), in which spaceship A is at rest, and frame F′, with coordinates (x', y', z'), in which spaceship B is at rest. We place the pilots at the origins of their respective reference frames. The Lorentz transformation formulas involve only relative speeds of frames F and F′. To calculate the relative speed u, we use the relativistic law of addition of velocities for objects moving toward each other. This law is given by Eq. (40–19),

$$u = \frac{v_1 + v_2}{1 + (v_1 v_2/c^2)}.$$

If the spaceships are aligned along their respective axes (x and x'), the front of spaceship A in its own rest frame is at $x = 0$, and its back is at $x = -L$ (Fig. 40–19a). Similarly, the front of spaceship B is at $x' = 0$, and its back is at $x' = 2L$. The time at which the two fronts just pass each other is the event that we may choose to take place at $t = t' = 0$. Our equations relating frames F and F′ are

$$x' = \gamma(x + ut)$$

and

$$t' = \gamma\left(t + \frac{ux}{c^2}\right).$$

Thus $x = 0$, $t = 0$ and $x' = 0$, $t' = 0$ are consistent.

The second event of interest to us is the alignment of the back end of spaceship B with the front end of spaceship A when $x' = 2L$ and $x = 0$ (Fig. 40–19b). Given x' and x, we can find t, the time for the event recorded in spaceship A's rest frame. When these values for x' and x are inserted into the equations that relate x' and t' to x and t, we get

$$2L = \gamma(0 + ut) = \gamma ut;$$

$$t = \frac{1}{u}\frac{2L}{\gamma}.$$

This is the time recorded by pilot A for the back of spaceship B to reach his position. This is a reasonable result: Pilot A sees the length of spaceship B contracted from $2L$ to $2L/\gamma$. Because spaceship B is moving at speed u relative to spaceship A, the time it takes to pass is its observed length $2L/\gamma$ divided by u.

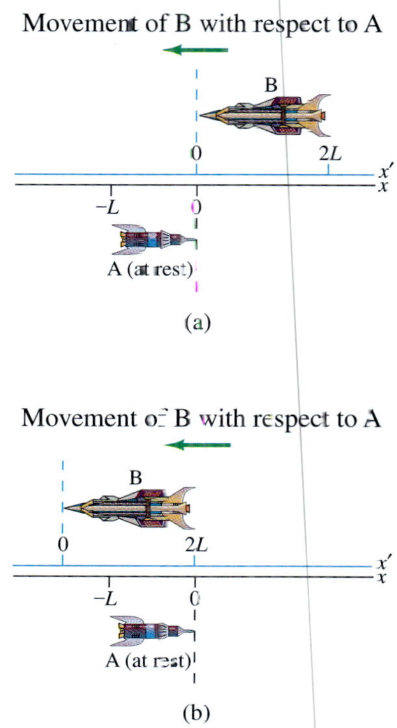

FIGURE 40–19 (a) Example 40–5. The event specified by the passing of the fronts of spaceships A and B. (b) The event specified by the passing of the rear of spaceship B by the front of spaceship A. Both cases are from the viewpoint of an observer in spaceship A, so spaceship B is shortened.

EXAMPLE 40–6 A train of proper length $2L = 500$ m approaches a tunnel of proper length $L = 250$ m. The train's speed u is such that $\gamma = 1/\sqrt{1 - (u^2/c^2)} = 2$. An observer at rest with respect to the tunnel measures the train's length to be contracted by a factor of 2 to 250 m and expects the whole train to fit into the tunnel. An observer on the train knows that the length of the train is 500 m, and the tunnel is contracted by a factor of 2 to 125 m. Thus the observer on the train argues that the train will not fit into the tunnel. Who is right?

Solution: To analyze this problem, we start with two frames: frame F, the rest frame of the tunnel, and frame F′, the rest frame of the train. An observer in frame F would assign the position of the left side of the tunnel as $x = 0$ and the right side of the tunnel as $x = L$ (Fig. 40–20a), while an observer in frame F′ would assign the

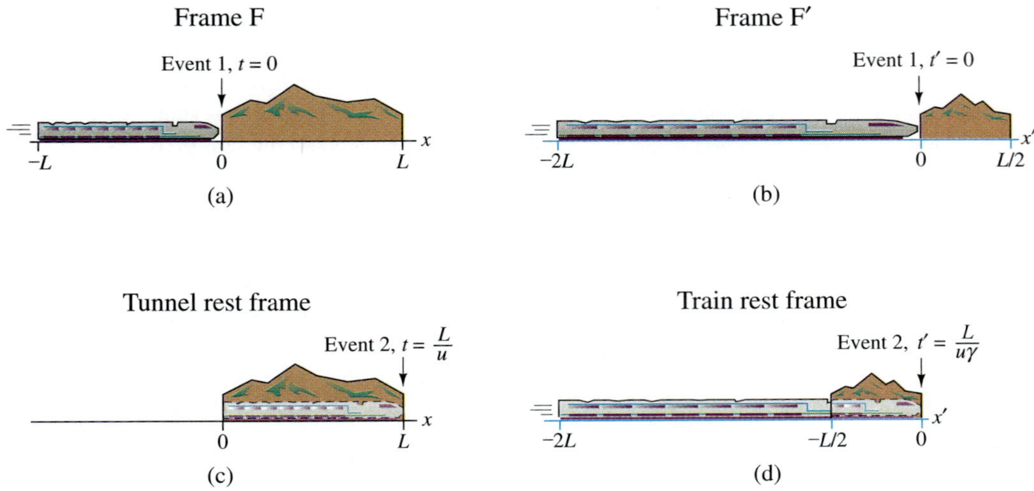

FIGURE 40–20 Example 40–6. Two events associated with the passage of a very fast train through a tunnel: the coincidence of the front of the train first with the left end of the tunnel, and second with the right end of the tunnel. Parts (a) and (c) show the events according to an observer in the rest frame of the tunnel; parts (b) and (d) show the events according to an observer in the rest frame of the train.

front of the train as $x' = 0$ and the rear of the train as $x' = -2L$ (Fig. 40–20b). We must first check that the train does indeed fit into the tunnel as measured in frame F. Then we must confirm that the train does not fit into the tunnel as measured in frame F′. In so doing, we can explain this paradox.

We identify two events, marked by arrows in Fig. 40–20, and describe them in both frames. Events must be specified by both position and time. The first event is the entry of the train into the tunnel. The clocks in the two frames are set so that at $t = t' = 0$, the front of the train ($x' = 0$) coincides with the left end of the tunnel ($x = 0$), as measured by an observer in frame F at time $t = 0$ (Fig. 40–20a) and by an observer in frame F′ at time $t' = 0$ (Fig. 40–20b). The second event is the alignment of the right end of the tunnel ($x = L$) and the front of the train ($x' = 0$), as measured by the observer in frame F (Fig. 40–20c) and the observer in frame F′ (Fig. 40–20d), respectively.

It follows from Eq. (40–20), $x' = \gamma(x - ut)$, that for the second event,

$$0 = \gamma(L - ut),$$

so $t = L/u$. This is the time of the second event as seen by the observer in frame F (Fig. 40–20c). Calculation of t' from the Lorentz transformation equation (40–21), $t' = \gamma[t - (ux/c^2)]$, yields

$$t' = \gamma\left(\frac{L}{u} - \frac{uL}{c^2}\right) = \frac{L}{u}\sqrt{1 - \frac{u^2}{c^2}} = \frac{L}{u\gamma} = \frac{t}{\gamma}.$$

This is the time of the second event as measured by the observer in frame F′ (Fig. 40–20d). Where is the rear end of the train at $t = L/u$ according to the observer in frame F? We have

$$x' = -2L = \gamma\left[x - u\left(\frac{L}{u}\right)\right] = \gamma(x - L).$$

We solve this for x and get $x = (-2L/\gamma) + L = 0$. Thus, at the time measured in frame F for the second event, the rear of the train is indeed at the tunnel entrance. As observed in frame F, the train fits into the tunnel. By this we mean that the observer in frame F sees that the front is at the exit of the tunnel and the rear is at the entrance of the tunnel *at the same time*, which is $t = L/u$.

Let's check that the observer in frame F′ does not see the train fit in the tunnel. That observer considers the train to be at rest and the tunnel to be moving toward the train. What is the location in frame F′ of the tunnel entrance at the time of the event specified by the alignment of the front of the train and the tunnel exit—the value of x' when $x = 0$ at time $t' = L/\gamma u$? Substituting these values into Eq. (40–26), we get the result that $x = 0$ implies that $x' = -ut' = -L/\gamma = -L/2$ (Fig. 40–20d). Thus, from the point of view of an observer in frame F′, only one-quarter of the train is in the tunnel when its front reaches the tunnel exit!

Let's make a last check in frame F′ to see the time t' at which the rear of the train is at the tunnel entrance: What is t' when $x' = -2L$ and $x = 0$? We have

$$t' = \gamma\left[\frac{L}{u} - (0)\left(\frac{u}{c^2}\right)\right] = \frac{L/u}{\sqrt{1 - (u^2/c^2)}} = \frac{\gamma L}{u}.$$

This time is different from the time $t' = L/u\gamma$ when the front of the train reaches the end of the tunnel. Thus, the observer on the train (in frame F′) measures the end of the train to pass the tunnel entrance at a later time. The observer in frame F′ will argue that the train's front passes the tunnel exit earlier than the rear passes the tunnel entrance. The differences between the interpretations of the two observers stem from their different notions of simultaneity. These differences allow both observers to be correct in their claims!

Lorentz Transformations of Electric and Magnetic Fields

Maxwell's equations predict the speed of light. Therefore, electric and magnetic fields play a distinct role in special relativity. In Chapter 31, we saw that, in order to maintain Galilean invariance (invariance under the transformation laws $\mathbf{r}' = \mathbf{r} - \mathbf{u}t$ and $t' = t$) of the Lorentz force equation, the electric and magnetic fields must mix among themselves when we observe them from different reference frames. In Eqs. (31–22) and (31–23), respectively, we derived Galilean transformation laws for these fields:

Galilean transformation laws for electromagnetic fields

$$\mathbf{E}' = \mathbf{E} + (\mathbf{u} \times \mathbf{B});$$

$$\mathbf{B}' = \mathbf{B}.$$

(As usual, the prime refers to the field as measured in frame F′.) The replacement of the Galilean transformation law for time and position by the Lorentz transformations means that the electromagnetic fields also transform differently. Here, we want to discuss briefly the physical origin of the transformation laws for these fields.

Electric fields result from the presence of electric charges, and magnetic fields result from the movement of electric charges. Let's now look at an example of how electric charge distributions are affected by Lorentz transformations. Suppose that an electrically neutral wire carries a current in the $+x$-direction in some frame F. The current consists of negative charges that move at some drift speed in the $-x$-direction against a background of stationary positive charges with the same spacing (Fig. 40–21a). This wire produces a magnetic field but no electric field.

Electric currents are described in Chapter 27.

Now let's consider the same wire as observed from a frame F′ that moves at the drift velocity $\mathbf{v}_d$ (Fig. 40–21b). An observer in frame F′ would see the electrons in the wire at rest and the positive charges (the ions) move in the $+x$-direction. We shall now

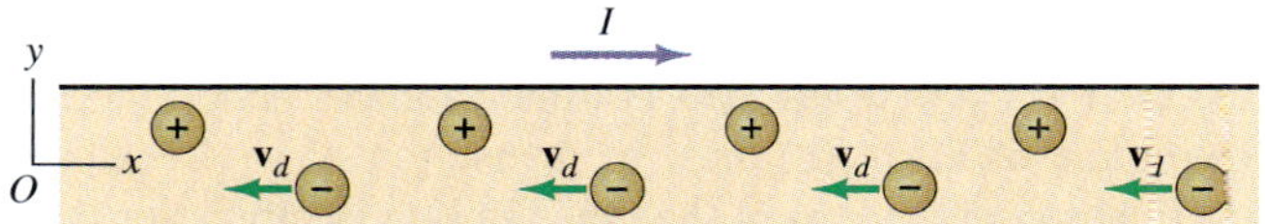

(a) Frame F: Positive charges (wire) at rest

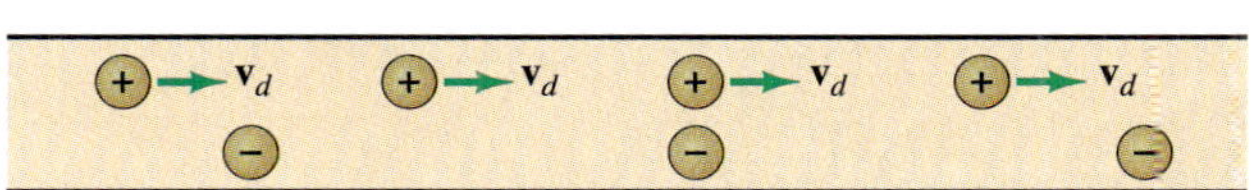

(b) Frame F′: Negative charges at rest

FIGURE 40–21 A current-carrying wire as observed from two frames. (a) The frame in which the wire is at rest. Here, the wire is electrically neutral. (b) The frame that moves with the negative charges. Here, the wire acquires a net charge density.

show that special relativity implies that *the wire is not electrically neutral in frame* F′. An observer in frame F would measure the electrons to have less space between them than the F′ observer does, due to Lorentz contraction. In other words, the F′ observer measures *more* space between the electrons than does the F observer. The F′ observer measures *less* space between the positive ions than does the F observer: To the F′ observer, the positive ions are moving. Thus if the F observer sees the same spacing between electrons as between positive ions, the F′ observer will not, and the wire will no longer be electrically neutral to the F′ observer. To the F′ observer, there is an electric field.

We have shown that, because of the way special relativity affects space, the presence of a magnetic field alone in one frame introduces an electric field in another. It is in this way that transformation laws between fields come about. A careful quantitative analysis of situations such as the one just described gives the full set of Lorentz transformations between electric and magnetic fields. These transformation laws, together with the Lorentz transformations for space and time, leave the physical consequences of Maxwell's equations invariant. Because the speed of light is one of the physical consequences of Maxwell's equations, that speed is the same in all frames. We have come full circle with a consistent description.

40–7 MOMENTUM AND ENERGY IN SPECIAL RELATIVITY

Momentum

Momentum and its properties are treated in Chapter 8.

The need to modify our notions of space and time suggests that the definitions of other kinematical quantities, which are based on measurements in space and time, also require modification. In nonrelativistic mechanics, the *momentum* of a particle that moves with velocity $\mathbf{v}$ is

$$\mathbf{p} = m\mathbf{v}.$$

We sometimes call the coefficient of $\mathbf{v}$ in this expression the *rest mass m*. In the absence of external forces, the sum of the momenta of interacting particles is constant; that is, the total momentum is conserved:

$$\sum \mathbf{p}_i = \mathbf{P} = \text{a constant}.$$

Momentum conservation has its origin in Newton's third law and is a principle that holds both nonrelativistically and relativistically.

Momentum in relativity is a quantity ascribed to moving particles and has the following properties: (a) In the absence of external forces, the sum of momenta of interacting particles is conserved, and (b) in the limit that $\mathbf{v} \to 0$, $\mathbf{p} \to m\mathbf{v}$. On purely dimensional grounds, we expect that

$$\mathbf{p} = mf(v)\mathbf{v},$$

where the function $f(v)$ must be 1 for $v = 0$ and $f(v)$ is dimensionless. The function $f(v)$ depends only on the magnitude of $\mathbf{v}$, so $f(v)$ must be a function of v^2. Because f is dimensionless, it must be a function of v^2/c^2 or of the now familiar combination $\gamma = 1/\sqrt{1 - (v^2/c^2)}$.

A somewhat lengthy analysis of collisions between equal-mass particles leads to the result that $f(v) = \gamma$, so

The relativistic momentum

$$\mathbf{p} = m\gamma\mathbf{v} = \frac{m\mathbf{v}}{\sqrt{1 - (v^2/c^2)}}. \tag{40–30}$$

For $(v/c) << 1$, this reduces to the familiar low-velocity result $\mathbf{p} = m\mathbf{v}$. Newton's second law now reads

$$\mathbf{F} = \frac{d\mathbf{p}}{dt} = m\frac{d}{dt}(\gamma\mathbf{u}). \qquad (40\text{–}31)$$

One consequence of the relativistic modification of the expression for momentum is that $\mathbf{F}$ and $d\mathbf{u}/dt$ no longer have to point in the same direction (see Problem 59).

Kinetic Energy

The relativistic kinetic energy

We may use Eq. (40–31) to derive the relativistic expression for the kinetic energy, K, from the work–energy theorem, Eq. (6–8). By the work–energy theorem, the work done to bring a particle of mass m from rest to speed v will be the kinetic energy of the particle. In the next subsection, we calculate the work and find that

$$K = mc^2(\gamma - 1) = mc^2\left(\frac{1}{\sqrt{1 - (v^2/c^2)}} - 1\right). \qquad (40\text{–}32)$$

For small v^2/c^2, $\sqrt{1 - (v^2/c^2)} \simeq 1 - (v^2/2c^2)$, so in the low-velocity limit, K reduces to the familiar $mv^2/2$ (as we pointed out in Section 6–6).

*How to Get the Relativistic Kinetic Energy

For simplicity, we derive the expression for the kinetic energy, K, by dealing in one dimension. We define K for a particle as the work done to bring the particle from rest (at $t = 0$) to a speed v (at time t). Thus

$$K = \int F\,dx = \int_0^t F\frac{dx}{dt}\,dt = \int_0^t Fv\,dt. \qquad (40\text{–}33)$$

We now use Eq. (40–31) for the force:

$$Fv = vm\frac{d}{dt}(\gamma v) = mv^2\frac{d\gamma}{dt} + m\gamma v\frac{dv}{dt}.$$

The first term contains

$$\frac{d\gamma}{dt} = \frac{d}{dt}\frac{1}{\sqrt{1 - (v^2/c^2)}} - \frac{1}{2}\left(1 - \frac{v^2}{c^2}\right)^{-3/2}\left(-\frac{2}{c^2}v\frac{dv}{dt}\right) = \left(\frac{\gamma^3}{c^2}\right)v\frac{dv}{dt}. \qquad (40\text{–}34)$$

Thus the integrand of Eq. (40–33) takes the form

$$\left[m\gamma^3\left(\frac{v^2}{c^2}\right) + m\gamma\right]v\frac{dv}{dt}.$$

The factor in square brackets in this expression is

$$m\gamma\left[\left(\frac{v^2}{c^2}\right)\gamma^2 + 1\right] = m\gamma\left[\frac{v^2/c^2}{1 - (v^2/c^2)} + 1\right] = m\gamma\left[\frac{1}{1 - (v^2/c^2)}\right] = m\gamma^3.$$

Thus the integrand is $m\gamma^3 v\dfrac{dv}{dt}$ and, as Eq. (40–34) shows, this is just $mc^2\dfrac{d\gamma}{dt}$. Thus

$$K = \int_0^t mc^2\frac{d\gamma}{dt}\,dt = mc^2\int_{\gamma'=1}^{\gamma'=\gamma} d\gamma' = mc^2(\gamma - 1) = mc^2\left[\frac{1}{\sqrt{1 - (v^2/c^2)} - 1}\right],$$

which is Eq. (40–32).

Energy Associated with Mass

Einstein pointed out that energy and mass are related. In order to demonstrate that inertia (mass) is associated with energy, consider a railroad car of mass M and length L standing on rails. Imagine a flashbulb attached to the left interior wall of the car (Fig.

40–22a). At a particular time, the bulb emits a burst of light toward the right wall of the railroad car. As we learned in Chapter 35, if the energy of the light pulse is E, then there is momentum associated with the pulse that has magnitude E/c. Momentum conservation implies that the railroad car must move with an equal and opposite momentum toward the left (Fig. 40–22b). Because the mass of the car is M, the car will move with a velocity such that

$$Mv = E/c.$$

The time t that the pulse spends between the walls is given by

$$ct = L - vt$$

because the right wall is now moving toward the light pulse. Thus

$$t = \frac{L}{c + v} = \frac{L}{c + (E/Mc)}.$$

The distance traveled by the car in that time is

$$D = vt = \frac{E}{Mc}\,\frac{L}{c + (E/Mc)} = \frac{EL}{E + Mc^2}. \qquad (40\text{–}35)$$

The car comes to a stop after it has moved a distance D because the light along with its momentum is absorbed by the right wall of the car.

If momentum is to be conserved, the center of mass of the railroad car with the flashing bulb must not move. Yet the car has moved to the left. From this, we must infer that the energy in the light is equivalent to a mass μ carried by the light flash and that this mass moves to the right when the car moves to the left such that the center of mass remains stationary. We can find μ by equating the position X of the overall center of mass before and after the flash event. Place the initial position of the left-hand side of the car at $x = 0$. The car can be treated as a point mass M, initially at $x = L/2$. Then, before the light is emitted,

$$X = \frac{(\mu)(0) + (M)(L/2)}{\mu + M} = \frac{M}{\mu + M}\,\frac{L}{2}.$$

After the light has been absorbed at the right-hand side, at position $x = L - D$,

$$X = \frac{(\mu)(L - D) + M[(L/2) - D]}{\mu + M}.$$

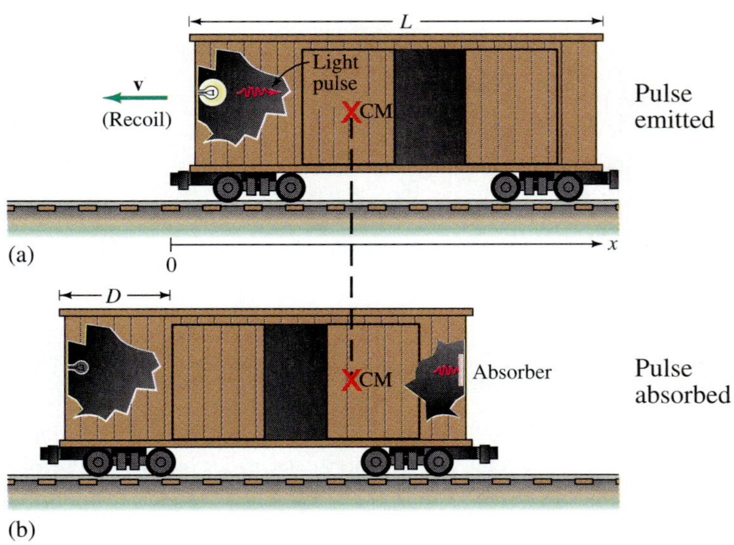

FIGURE 40–22 (a) A light source in a railroad car emits an electromagnetic pulse that carries energy E. The recoil momentum $p = E/c$ sets the car in motion. (The center of mass is exaggerated here.) (b) This motion ceases when the pulse is absorbed at the other end of the car.

We equate these and solve for μ to find that

$$\mu = \frac{MD}{L - D}.$$

From Eq. (40–35), $L - D = LMc^2/(E + Mc^2)$; when we substitute for $L - D$ as well as for D in the equation for μ, we get

$$\mu = \frac{MEL/(E + Mc^2)}{LMc^2/(E + Mc^2)} = \frac{E}{c^2}.$$

E is equivalent to a mass μ given by

$$E = \mu c^2. \tag{40–36}$$

The relation between mass and energy

This is a very important result because it shows that mass and energy are interchangeable concepts. We can reverse the reasoning here to say that if an object has a mass m, then it has a **rest energy** $E = mc^2$. Light has no mass in the usual sense, but any energy it has is equivalent to a mass E/c^2. The total energy, E, of any object is now the sum of the kinetic energy, the rest energy (or mass energy, E_{mass}), and the potential energy. For a particle on which no forces act, there is no potential energy, and

$$E = E_{mass} + K = \frac{mc^2}{\sqrt{1 - (v^2/c^2)}}. \tag{40–37}$$

It follows from this result and from Eq. (40–30) that

$$\mathbf{v} = \frac{c^2\mathbf{p}}{E}. \tag{40–38}$$

Velocity in terms of the relativistic momentum and energy

The reality of the relation $E_{mass} = mc^2$ has been tested innumerable times in a large variety of nuclear reactions. More dramatic confirmation of this law came with the discovery of *antimatter*. Quantum mechanics and relativity together show that, for each particle, there is a corresponding *antiparticle*. The antiparticle has a charge opposite that of the particle, and an antiparticle and a particle can annihilate each other to produce electromagnetic radiation (Fig. 40–23). Similarly, a particle and an antiparticle can be created out of radiation energy alone. The conversion of energy (in the form of radiation) into mass is thereby exhibited unambiguously.

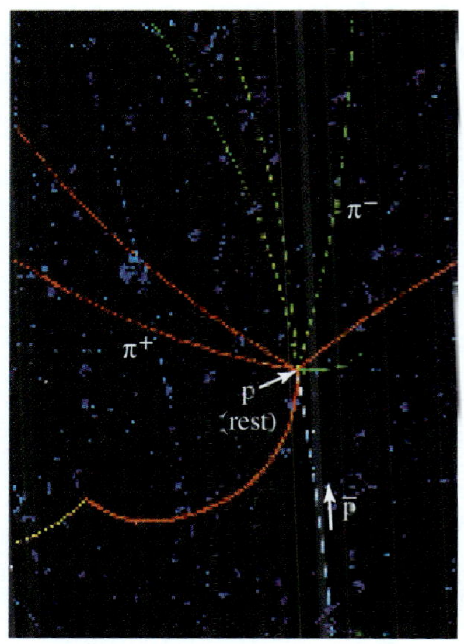

FIGURE 40–23 In this color-enhanced bubble-chamber photo, an incoming antiproton (light blue) strikes and annihilates with a proton at rest producing 4 positive pions (red) and 4 negative pions (green), which are antiparticles of the positive pions.

EXAMPLE 40–7 The nucleus ^{8}Be is an unstable isotope of beryllium. It decays into two alpha particles, $^8\text{Be} \rightarrow {}^4\text{He} + {}^4\text{He}$. The atomic masses (in atomic mass units u) are $M(^8\text{Be}) = 8.005305$ u and $M(^4\text{He}) = 4.002603$ u. Assume that a nucleus ^{8}Be decays while it is at rest. Find the kinetic energy of the helium nuclei (the alpha particles) in MeV.

Solution: Both energy and momentum must be conserved and these conservation laws provide the solution to the problem. The ^{8}Be is initially at rest, so the total momentum is always zero. The two helium nuclei must therefore have momenta equal in magnitude but opposite in direction, and hence the same kinetic energies. The initial energy is just the energy due to the mass of ^{8}Be. The final energy is the sum of the rest energies and the kinetic energies of the alpha particles. We equate the initial and final energies and get

$$E = M(^8\text{Be})c^2 = 2M(^4\text{He})c^2 + K_{total}.$$

The kinetic energy K_{total} is called the *Q value* in nuclear reactions. It is given by the difference between the initial and final rest energies:

$$\begin{aligned} K_{total} = Q &= [M(^8\text{Be}) - 2M(^4\text{He})]c^2 \\ &= [8.005305 \text{ u} - 2(4.002603 \text{ u})]c^2 = (0.000099 \text{ u})c^2. \end{aligned}$$

We make use of the conversion 1 u = 931.5 MeV/c^2 to obtain

$$K_{\text{total}} = (0.000099\ \text{u})\left(\frac{931.5\ \text{MeV}/c^2}{\text{u}}\right)c^2 = 0.092\ \text{MeV}.$$

Because we have shown above that momentum conservation implies that the kinetic energies of the two alpha particles are equal, each alpha carries kinetic energy of (0.092 MeV)/2 = 0.046 MeV.

EXAMPLE 40–8 A 1.0-kg meteorite of antimatter strikes Earth. (This is just a thought experiment; there is no solid evidence of antimatter in such large lumps.) How much energy is liberated in the annihilation process in which all the antimatter and an equal amount of matter are converted to radiant energy? Neglect the kinetic energy of the meteorite; $(v/c) << 1$.

Solution: The rest energy of mass M is Mc^2. The antimatter interacts with an equal amount of matter in the annihilation process, so the amount of energy liberated is $2Mc^2$:

$$E = 2\,Mc^2 = 2(1.0\ \text{kg})(3 \times 10^8\ \text{m/s})^2 = 1.8 \times 10^{17}\ \text{J}.$$

To get some idea of the significance of this number, the amount of (chemical) energy in 1 ton of TNT is 4.2×10^9 J. The meteorite explosion would generate the equivalent of 4×10^7 tons of TNT or about 40 hydrogen bombs. Antimatter–matter annihilation has been suggested as a source of fuel for manned planetary journeys: It is the most efficient fuel possible.

The Relation Between the Momentum and Energy of a Particle

There is an important relation between the momentum and energy of a moving particle. From Eq. (40–37), we have

$$E^2 = \frac{m^2c^4}{1 - (v^2/c^2)}.$$

Equation (40–30) gives us an expression for p^2c^2, which is a quantity with the same dimension as E^2:

$$p^2c^2 = \frac{m^2v^2c^2}{1 - (v^2/c^2)}.$$

The difference between these results is

$$E^2 - p^2c^2 = \frac{m^2c^4}{1 - (v^2/c^2)}\left(1 - \frac{v^2}{c^2}\right) = m^2c^4. \quad (40\text{–}39)$$

The relation of a particle's energy, momentum, and mass

This relation may also be written as

$$E = \sqrt{p^2c^2 + m^2c^4}. \quad (40\text{–}40)$$

We note that for *any* massless particle (a particle with $m = 0$),

$$E = pc. \quad (40\text{–}41)$$

Light obeys this relation and thus behaves as a massless particle. There are other particles that are massless, at least to the accuracy that measurements allow. The neutrino is such a particle. Because the velocity of a particle is measured by the ratio of p to E, as in Eq. (40–38), we see that *massless particles always move at the constant speed c*. There is no reference frame in which these particles can be brought to rest or, for that matter, to any speed other than c. We know this to be true for light, but it is true for any other massless particle as well.

We also see from Eq. (40–39) that the combination $E^2 - p^2c^2$ is *invariant*. Invariant quantities have the same value in every inertial frame, just as $c^2t^2 - x^2$ does, as shown in Eq. (40–25). This can be a valuable result in the analysis of relativistic collision phenomena.

*40–8 BEYOND SPECIAL RELATIVITY

The Equivalence Principle

Special relativity expresses the physical equivalence of all inertial reference frames. In noninertial, or accelerating, frames, another physical equivalence holds. It is expressed by the *equivalence principle*, formulated by Einstein in 1911:

The equivalence principle of general relativity

> **Provided that the observations take place in a small region of space and time, it is not possible by experiment to distinguish between an accelerating frame and an inertial frame in a suitably chosen gravitational potential.**

This is the principle behind *general relativity*, also known as Einstein's theory of gravitation. We described a number of consequences of the equivalence principle in Section 12–8. We list these again:

A. The Equality of Gravitational Mass and Inertial Mass. The *gravitational mass* m_g, the attribute of an object that appears in Newton's expression for the gravitational force,

$$F = -\frac{Gm_{1g}m_{2g}}{r^2},$$

and the *inertial mass* m_i, the attribute of an object that appears in the expression for the proportionality of force and acceleration,

$$F = m_i a,$$

must be equal. As a consequence, all objects fall at equal rates in a given gravitational field: If an object of mass m is subject to a gravitational force due to an object of mass M, the relation $F = ma$ reads

$$\frac{Gm_gM}{r^2} = m_i a;$$

if $m_i = m_g$, the acceleration does not depend on the mass of the object. The equality of the inertial mass and gravitational mass has been verified experimentally to an accuracy of one part in 10^{11}.

B. The Gravitational Deflection of Light. If a horizontal beam of light enters a pinhole in an elevator that is accelerating upward with an acceleration g, then, in the time t that it takes for the light to cross the elevator, the horizontal level of the pinhole will have moved upward by a distance $gt^2/2$. If the width of the elevator is L, the light will hit a spot Q that is a distance

$$\Delta = \frac{1}{2}g\left(\frac{L}{c}\right)^2$$

below the spot Q', which is horizontally across from the pinhole (Fig. 40–24). An observer within the elevator states that the light *falls* by that amount. According to the equivalence principle, no experiment can determine whether the elevator has accelerated upward or has remained at rest within a gravitational field. Therefore light must fall in a gravitational field. Note that Δ is the same distance that any other body would fall. The implications are twofold:

1. Everything that has energy falls downward with the same acceleration in a gravitational field.

2. Light is deflected as it passes stars, as we discussed in Chapter 12 (Fig. 40–25).

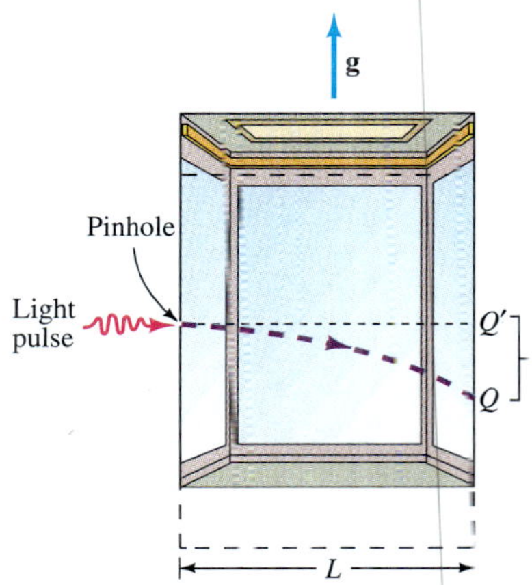

Light pulse enters elevator aimed at Q' but hits wall at Q as elevator rises

FIGURE 40–24 A pulse of light is directed into an elevator. If the elevator is at rest, the pulse would eventually arrive at point Q'. If the elevator accelerates upward with magnitude g as the light pulse crosses, an observer within the elevator would see the light pulse follow the parabolic path shown, eventually arriving at point Q on the opposite wall, a distance Δ below Q'.

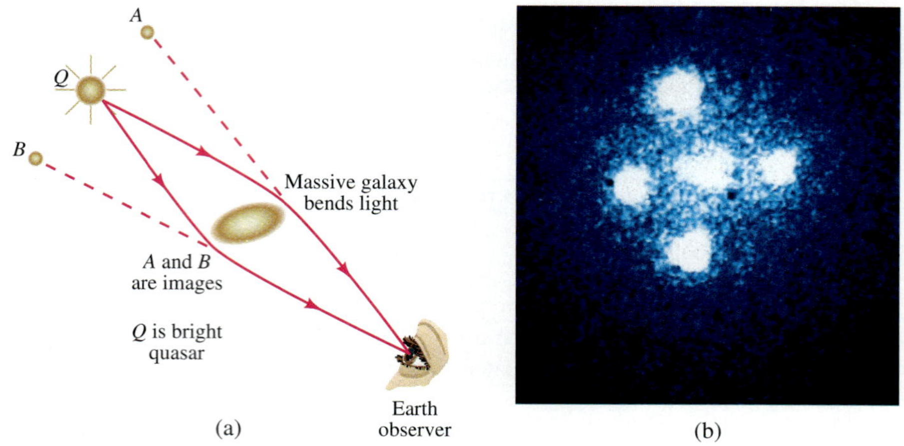

FIGURE 40–25 (a) The formation of an image by a gravitational lens. If a light ray "bends" toward mass, we should see such images in the sky. (b) A mass (a relatively nearby galaxy) has bent the light from a quasar behind it such that four distinct images of the quasar can be observed from Earth. The fuzzy spot in the center is the galaxy's core.

C. The Gravitational Redshift. When light "falls," it undergoes a frequency shift. Consider a source at a height x above the ground in the presence of local gravity, emitting radiation with frequency f (Fig. 40–26a). According to the equivalence principle, the physics should be described equally well by an observer who sees the system being accelerated upward with acceleration g in empty space. If both the source and the detector are at rest at the time of emission, then, according to the observer who sees the entire apparatus accelerating upward, in the time that the radiation has reached the detector ($t = x/c$), the detector will have acquired an upward velocity of magnitude $v = gt = gx/c$ (Fig. 40–26b). Thus the detector sees the radiation with a frequency f', Doppler shifted upward from the emitting frequency f. These are related by

$$\frac{f'}{f} = \sqrt{\frac{1 + (v/c)}{1 - (v/c)}} \simeq 1 + \frac{v}{c} = 1 + \frac{gx}{c^2}.$$

The factor gx may be regarded as a *gravitational potential*, ϕ: We express the potential energy of a mass m in the vicinity of Earth as $mgx = m\phi$. In this way, the effect

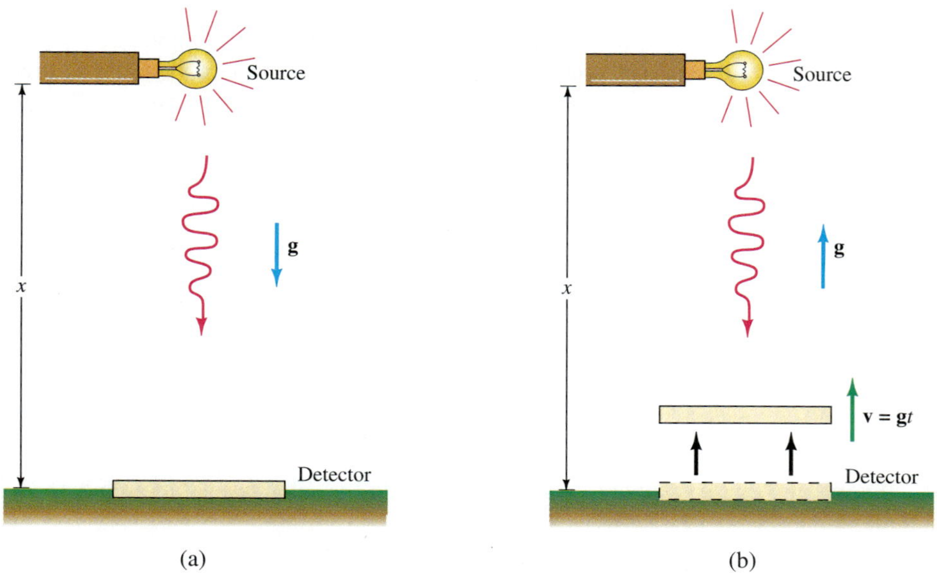

FIGURE 40–26 (a) An apparatus to measure the frequency of emitted light on Earth, in Earth's gravitational field; **g** is the (downward) acceleration due to gravity. (b) The measurement of a Doppler shift due to the acceleration of the detector (as part of the whole system) is indistinguishable from what is observed when the apparatus is in a gravitational potential; **g** is the (upward) acceleration of the system.

of Earth is isolated in the factor $\phi = gx$. Our result for the frequency ratio then takes the form

$$\frac{f'}{f} = 1 + \frac{\phi}{c^2}, \tag{40–42}$$

so

$$\frac{\Delta f}{f} \equiv \frac{f' - f}{f} = \frac{\phi}{c^2}. \tag{40–43}$$

Let's now apply our result to universal gravitation. The potential energy difference for a mass m and a star (of mass M and radius R) between the surface of the star and a point far from the star is $-GmM/R$. Thus we can assign to the star a corresponding gravitational potential of $-GM/R$. The frequency of light that is emitted from the star's surface and arrives at an Earth-based telescope is shifted by

$$\frac{\Delta f}{f} = -\frac{GM}{Rc^2}.$$

(We ignore the relatively insignificant "fall" down to Earth's surface.) The minus sign shows that the frequency is shifted downward. Thus the wavelength shifts upward, indicating a *gravitational redshift*. For the Sun, $M = 2 \times 10^{30}$ kg and $R = 7 \times 10^8$ m, so $\Delta f/f = 2.12 \times 10^{-6}$. Recent measurements of some characteristic spectral lines of sodium in the solar spectrum have confirmed this to an accuracy of 5 percent.

In a terrestrial measurement of the gravitational redshift carried out in 1960 by Robert Pound and Glen Rebka, light was "dropped" from a tower at Harvard University. This experiment measured a fractional shift of 3.3×10^{-15} to an accuracy of 1 percent.

When the gravitational potential is very large, the frequency shift $\Delta f/f$ becomes large. In the extreme case that $(\Delta f/f) = 1$, the frequency of the light is shifted to zero; that is, no light can be seen. A *black hole* has been formed. This occurs when $(GM/Rc^2) > 1$. When Einstein's full theory of gravitation is taken into account, it is for the modified condition $(GM/Rc^2) > \frac{1}{2}$ that a black hole is formed.

SUMMARY

The special theory of relativity is based on the postulates that

1. The laws of physics are the same in all inertial reference frames.
2. The speed of light in empty space is the same in all inertial frames.

The first postulate generalizes the notion of Galilean invariance that we applied to the laws of mechanics. The second postulate is justified by the result of Michelson and Morley, that there is no evidence for an absolute reference frame.

These postulates require a rethinking of the concepts of space and time. If two inertial frames F and F′ move at speed u with respect to each other, then

1. Two events that are simultaneous in frame F are not simultaneous in frame F′.
2. A time interval measured as T on a "clock" at rest in frame F is given by T' in frame F′:

$$\text{time dilation:} \quad T' = \frac{T}{\sqrt{1 - (u^2/c^2)}}. \tag{40–10}$$

3. A length of an object at rest in frame F measured as L in frame F has length L' in frame F′:

$$\text{length contraction:} \quad L' = L\sqrt{1 - \frac{u^2}{c^2}}. \tag{40–12}$$

4. A light source that radiates with frequency f_0 in frame F is observed to have frequency f_1 in frame F′:

$$f_1 = f_0\sqrt{\frac{1 - (u/c)}{1 + (u/c)}}\,. \tag{40–17a}$$

5. If two objects are moving at velocities v_1 and $-v_2$, respectively, with respect to an observer, then the velocity of one object as seen in the rest frame of the other is

$$\text{law of addition of velocities:} \qquad V = \frac{v_1 + v_2}{1 + (v_1 v_2/c^2)}\,. \tag{40–19}$$

The relativistic Doppler shift for light has important astronomical implications. The measurement of the redshift of the light from a galaxy, given by Eq. (40–17), leads to a value for the speed u with which that galaxy recedes from Earth. That speed is in turn related to the distance D of the galaxy by Hubble's law:

$$D = u/H; \tag{40–18}$$

here, H is the Hubble parameter, a number characteristic of the age of the universe.

The space–time coordinates of an event are described in reference frames F and F′ by (x, t) and (x', t'), respectively. These coordinates are related by the Lorentz transformations

$$x' = \gamma(x - ut) \tag{40–20}$$

and

$$t' = \gamma\left(t - \frac{ux}{c^2}\right), \tag{40–21}$$

where $\gamma \equiv 1/\sqrt{1 - (u^2/c^2)}$. An event can be specified by any two of (x, t, x', t'), and the other two coordinates can then be found from the Lorentz transformation laws, which must be supplemented by $y' = y$ and $z' = z$ when the relative motion is along the x-axis.

The relativistic momentum is defined by

$$\mathbf{p} = \frac{m\mathbf{v}}{\sqrt{1 - (v^2/c^2)}}\,. \tag{40–30}$$

The general relation between energy and mass μ is $E = \mu c^2$, Eq. (40–36). The total energy of a particle is

$$E = \frac{mc^2}{\sqrt{1 - (v^2/c^2)}} \tag{40–37}$$

and consists of a rest energy associated with the mass of the particle, $E_{\text{mass}} = mc^2$, plus the relativistic kinetic energy,

$$K = mc^2\left[\frac{1}{\sqrt{1 - (v^2/c^2)}} - 1\right]. \tag{40–32}$$

The velocity of a particle is given in terms of the momentum and energy according to

$$\mathbf{v} = \frac{c^2\,\mathbf{p}}{E}\,. \tag{40–38}$$

The energy and momentum are related by

$$E = \sqrt{p^2c^2 + m^2c^4}. \tag{40–40}$$

For massless particles, $E = pc$, Eq. (40–41).

Einstein extended his theory beyond inertial reference frames by proposing the equivalence principle, according to which

> **It is not possible by experiment to distinguish between an accelerated frame and an inertial frame in a suitably chosen gravitational potential, provided that the observations take place in a small region of space and time.**

This principle has important consequences:

1. Inertial and gravitational masses are equal, a result known to be accurate to one part in 10^{11}.
2. Light falls in a gravitational field.
3. A source radiating with frequency f will be observed in a gravitational potential ϕ at a frequency f' such that

$$\frac{f'}{f} = 1 + \frac{\phi}{c^2}. \tag{40–42}$$

UNDERSTANDING KEY CONCEPTS

1. Can there be such a thing as a perfectly rigid object?
2. Does the statement "moving clocks run slow" depend on the direction in which a clock is moving?
3. The rest mass of a proton is given as 937 MeV/c^2. How can a mass involve energy units?
4. As measured from Earth, what is the shortest possible travel time between Earth and Alpha Centauri, the second nearest star system to us, which is located 4.3 ly away? Why can it not be made any shorter?
5. Suppose that the Michelson–Morley experiment were carried out over one arbitrarily short time period, much less than 1 d, and showed no sign of movement through an ether. Is this result enough to rule out the presence of an ether?
6. According to one solution proposed to make the presence of an ether consistent with the results of the Michelson–Morley experiment, there is an *ether drag*: For some reason, Earth carries a bubble of the ether with it as it moves through space. Can you think of experimental consequences that could be used to rule out such an idea?
7. A golf ball is struck so well that it travels at a speed of $0.95c$ but is so durable that it is not deformed by the blow it receives. What shape does the golf ball have as it travels through the air, as measured by the golfer?
8. If a mirror recedes at speed $0.75c$ from a light source, does the image recede at speed $1.5c$ from the source?
9. By looking at a very distant quasar some years ago, astronomers found that the separation distances of certain peaks in brightness increase at a rate of 0.2 ms of arc/yr. The quasar is so far away that the separation speed translates into $v \simeq 8c$! Is this the death knell of the special theory of relativity?
10. A stick 2 m long travels toward a hole that is just short of 1 m in diameter at a speed such that $\gamma = 2$. Will the stick fall through the hole? How will things look to an observer in the rest frame of the stick?
11. Suppose an experimenter found that some particle, such as a neutrino, travels just a bit faster than the highest-frequency radiation that has been observed so far. Would we have to give up the special theory of relativity?
12. Folklore has it that, as a teenager, Einstein worried about what would happen if somebody were looking in a mirror while accelerating to a speed faster than that of light. What could Einstein have been worried about?
13. Suppose that current in a wire is carried by little green men who pass negative charges along a chain from person to person while standing on positive charges, such that the wire is electrically neutral. Would the changing mechanism for the current flow change the observed behavior of a charge q that lies outside the wire? How would this look to an observer who is moving with some velocity along the wire?
14. A closed box of little mass sits on a horizontal frictionless surface. The inside walls are perfect mirrors and reflect back all radiation. A laser is inside on the left-hand wall and projects a very short burst of light directly at the right-hand wall. What is observed from the outside, and why?
15. According to Hubble's law, a single spectral line characteristic of a single atom is redshifted by an amount proportional to the distance of the star from Earth. How do we know that the radiation of a particular color seen to come from a distant star is the redshifted radiation of a particular known spectral line?
16. The length-contraction experiment seems to imply that a meter stick accelerated to the speed of light would shrink to a point, and all the calibration markings on the meter stick would be lost. Is there something wrong with this reasoning?
17. Light falls in an accelerating elevator. But if an elevator moves upward at a constant velocity, a horizontal light beam would hit a spot *below* the horizontal projection on the opposite wall of the elevator. Does this mean that light falls in an elevator that moves at constant velocity?
18. When the supernova 1987a occurred, bursts of neutrinos—particles that are massless according to accelerator-based experiments and to within the accuracy of the experiments—arrived at detectors at various places on Earth's surface. These neutrinos are thought to have been emitted by the supernova all at once. How could differences in the arrival times of the neutrinos be used to test whether or not neutrinos have mass?

PROBLEMS

40–1 Is an Ether Necessary?

1. (I) An airplane flies at an air speed of 600 mi/h. It travels east from town A to town B and returns to A without stopping. In the absence of wind, the journey takes exactly 4 h. A town C is the same distance away from A and is located due north of A. Suppose that a wind with ground speed 60 mi/h is blowing

east to west. Calculate the times it takes for the plane to make journeys ABA and ACA.

2. (II) In one version of the Michelson–Morley experiment, light of wavelength 633 nm emitted by a He–Ne laser travels through a total path length of 6 m in each arm of the interferometer. To the accuracy of the apparatus, a shift of 1/20 of a fringe, no shift was seen. Estimate the greatest value possible for the speed of Earth through the ether.

40–2 *The Einstein Postulates*

3. (II) A small, powerful laser is placed on a turntable that rotates at 900 rev/s. The laser, whose beam makes a 25° angle with the horizontal, shines on clouds 70 km away. Calculate the speed with which the light spot on the clouds moves. Does this speed violate the limitation of the speed of light? Explain.

40–4 *Time Dilation and Length Contraction*

4. (I) A muon (a subatomic particle) moves at a speed of $0.4c$. How much slower does its "clock" tick than if it were at rest?

5. (I) Two twins wave goodbye to each other. One twin, an astronaut, travels to Mars. The trip takes 1 yr in each direction, and the average speed with respect to Earth is 20,000 km/h. What will the approximate time difference in the twins' clocks be when they are together again on Earth?

6. (I) Proxima Centauri, the star nearest our own, is some 4.2 ly away. (a) If a spaceship could travel at a speed of $0.80c$, how long would it take to reach the star according to the spaceship's pilot? (b) What would someone in the frame that moves along with the spaceship measure as the distance to Proxima Centauri?

7. (I) According to a passenger, how long will it take a spaceship moving at $0.999c$ to cross a galaxy with a diameter, as measured in the galaxy's rest frame, of 2.5×10^{19} m?

8. (I) The Space Shuttle orbits Earth at 16,300 mi/h in 111 min. How much time will an astronaut's atomic clock have lost during a total trip that takes 7 d?

9. (I) The diameter of our galaxy is about 10^5 ly, or 10^{21} m. Suppose that a proton moves at a speed such that $\sqrt{1 - (v^2/c^2)} \simeq 10^{-7}$. (Such speeds correspond to the most energetic cosmic rays known). How long does it take the proton to cross the galaxy in (a) the galaxy's rest frame? (b) the proton's rest frame?

10. (II) A meter stick is tilted to make an angle of 40° with the x-axis. How will an observer at rest in a frame F′ that moves at velocity $v = 0.95c$ in the $+x$-direction relative to the meter stick describe the stick?

11. (II) A student must complete a test in 1 h in the teacher's frame of reference F. The student puts on his rocket skates and soon is moving at a constant speed of $0.75c$ relative to the teacher. When 1 h has passed on the teacher's clock, how much time has passed on a clock that moves with the student, as measured by the teacher?

12. (II) As measured by an observer in an inertial frame, a small clock moving at a constant speed of $0.75c$ traverses a distance of 60 km. The moving clock records 10,000 ticks during the passage. How many ticks pass on an identical clock at rest relative to the observer?

13. (II) A spaceship of length 30 m travels at $0.6c$ past a satellite. Clocks in frame S′ of the spaceship and S of the satellite are synchronized within their respective frames of reference and are set to zero so that $t' = t = 0$ at the instant the front of the spaceship F passes point A on the satellite, located at $x' = x = 0$ (Fig. 40–27). At this time, a light flashes at F. (a) What is the length of the ship as measured by an observer on the satellite? (b) What time does the observer on the satellite read from her clock when the trailing edge B of the spaceship passes her? (c) When the light flash reaches B at the rear of the spaceship, what is the reading t_1' of a clock at B? (d) What is the reading t_1 on the clock on the satellite when, according to the observer on the satellite, the flash reaches B?

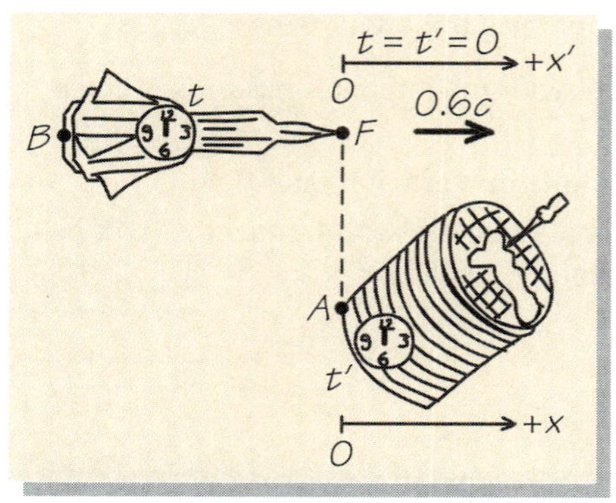

FIGURE 40–27 Problem 13.

14. (III) Jessica embarks on a cosmic journey at a speed of $(12/13)c$ relative to Earth. Before leaving, she tells her twin brother Tom, who stays on Earth, that she will travel outward for 26 yr of Earth time, then back for another 26 yr of Earth time. Tom will thus be 52 yr older when she returns. She promises to send a radio message on each of her birthdays (Fig. 40–28). According to an Earth-based clock, when will these messages reach Tom, and how much older than the age at which she leaves will Jessica be when she returns to Earth?

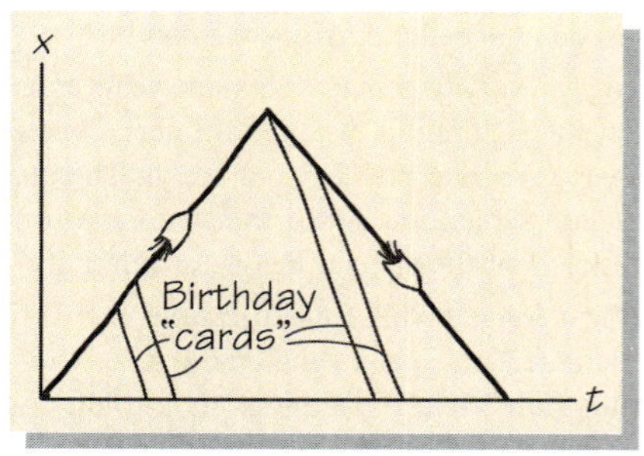

FIGURE 40–28 Problem 14.

15. (III) A relativistic sprinter running at speed v, near the speed of light, passes beneath a victory arch a height h above his eyes. Show that he will continue to see the arch, even though his eyes face forward, until he has run a distance $hv/[c\sqrt{1 - (v^2/c^2)}] = \gamma hv/c$ *beyond* the arch. [*Hint*: Work in the rest frame of the sprinter, and think of the top of the arch as emitting pulses of light, the last of which can be seen when it travels vertically downward toward the sprinter.]

40–5 The Relativistic Doppler Shift

16. (I) The sodium doublet refers to light waves emitted by sodium in a closely spaced pair of frequencies. The wavelengths of this doublet are at 589.0 nm and 589.6 nm. Suppose that the lower-wavelength member of this doublet is Doppler redshifted to a wavelength of 602.7 nm in the light emitted by a certain star. What happens to the wavelength of the second member of the doublet?

17. (I) A spaceship accelerates at a rate of 0.1 m/s^2 away from Earth. How long will it take (as measured in Earth's reference system) before a yellow beacon on Earth ($\lambda = 600$ nm) looks green ($\lambda = 500$ nm) to the crew of the spaceship?

18. (I) The wavelength of a spectral line in the laboratory is measured to be 108 nm. The same line is observed in light coming from a distant galaxy; in this observation, the wavelength is found to be 124 nm. What is the speed of motion of the galaxy relative to Earth?

19. (I) A particular spectral line measured in the emission of light by the star Alpha Centauri has wavelength $\lambda = 512\,311$ nm. That same line measured in the laboratory has wavelength $\lambda = 512.350$ nm. Determine the radial velocity of Alpha Centauri relative to Earth.

20. (II) A driver was caught running a red light. His defense is that he saw the light as green, as a result of the Doppler shift. He is arrested. What for? Estimate the seriousness of his transgression.

21. (II) Yellow light at 587.6 nm, characteristic of helium, is found to be redshifted as it is observed in a certain star; the wavelength is measured to be 611.7 nm. (a) How fast is the star receding from Earth? (b) Use Hubble's law to estimate the distance of the star from Earth.

22. (II) For a particular quasar, $(\lambda - \lambda_0)/\lambda_0 = 1.95$, where λ_0 is the wavelength of the radiation emitted as measured in the quasar's rest frame. What is the speed of the quasar relative to Earth, assuming that it is traveling in a radial direction away from Earth? How far away is the quasar according to Hubble's law?

23. (II) A source radiates light with a frequency of 2×10^{15} Hz. The signal is reflected by a mirror that is moving at speed 1 km/s away from the source. What is the shift of the frequency of the reflected radiation, as observed at the source?

24. (II) The equation $\lambda/\lambda_0 = \sqrt{(1 + \beta)/(1 - \beta)}$, where $\beta = v/c$ and v is the speed of a source that is moving away from an observer or of an observer who is moving away from the source, takes a simple form if v is small compared to c. Show that if $\lambda = \lambda_0(1 + x)$, then for small β, $x \simeq \beta$.

25. (III) During the journey described in Problem 14, Tom sends a radio message to Jessica on each of his birthdays adding to a total of 52 messages. With what interval in her rest frame does Jessica receive these messages during the outward part of the journey? During the return trip? Use this information to calculate how much Jessica ages during her trip according to an Earth-based clock.

26. (III) A source emits pulses with a frequency f_0. A spaceship moving at speed v_1 away from the source will receive a redshifted frequency f_1. Suppose that the spaceship immediately reemits the signals with the frequency f_1. A second spaceship, moving at speed v_2 relative to the first spaceship and in the same direction, will receive the signals with a redshifted frequency f_2. (a) Calculate f_1 and f_2. (b) If we were to eliminate the first spaceship, we could view f_2 as the redshifted frequency received by the second spaceship, which moves at some speed v relative to the source. Show that if both $v_1 << c$ and $v_2 << c$, then $v = v_1 + v_2$, as expected from the ordinary rules that govern relative motion. (c) Calculate v for arbitrary values of v_1 and v_2. This result is the relativistic law of addition of velocities, which differs from $v = v_1 + v_2$ when v_1 and v_2 are not very small compared with c.

40–6 The Lorentz Transformations

27. (I) Measurements of distant galaxies show that all galaxies are receding from one another at a speed proportional to their intergalactic distances. Suppose that we see galaxy 1 move away from us at a speed of $0.4c$ along the South Pole, and galaxy 2, equally far away, move away from us at the same speed along the North Pole (Fig. 40–29). What would an observer in galaxy 1 measure for the speed with which galaxy 2 moves away from him?

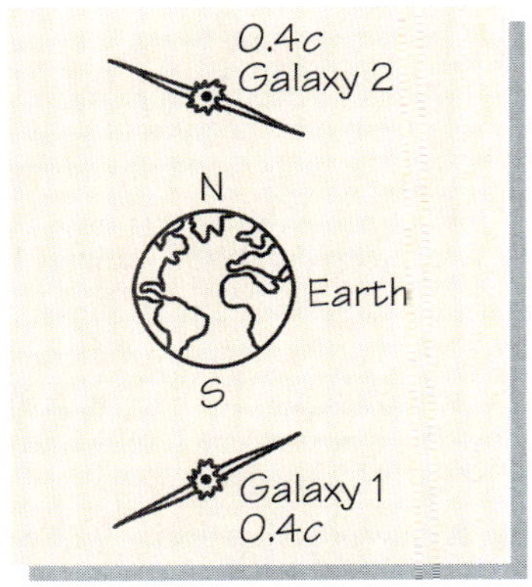

FIGURE 40–29 Problem 27.

28. (I) Events A and B are simultaneous in frame S and are 1 km apart on a line that defines the x-axis. A series of spaceships all pass at the same speed in the $+x$-direction, and they have synchronized their clocks so that together they make up a moving frame S′. They time events A and B to be separated by 10^{-6} s. What is the speed of the spaceships? How far apart in space do they measure the two events to be?

29. (II) Two friends decide to demonstrate the Lorentz contraction of a train. They sit at the two ends of the 100-m-long train, with their watches properly synchronized. At $t = 0$, each drops a small bag out of the window; these bags act as markers (Fig. 40–30). Later, they go back and measure the distance between the bags. (a) Does this distance represent the length of the train relative to a coordinate system fixed to the ground? (b) What is the distance between the bags if the speed of the train is $0.7c$? (Neglect the time it takes for the bag to reach the ground!)

30. (I) Let F and F′ represent two inertial frames moving at speed u along the x-axis with respect to one another. The origins and axes of these two frames coincide at time $t = t' = 0$. Suppose an observer in F and one in F′ agree that an event has occurred at the same time on their respective clocks, namely time $t = t' = t_0$. Is it possible that the event also occurred at the same place—that is, the same space coordinate—in each frame?

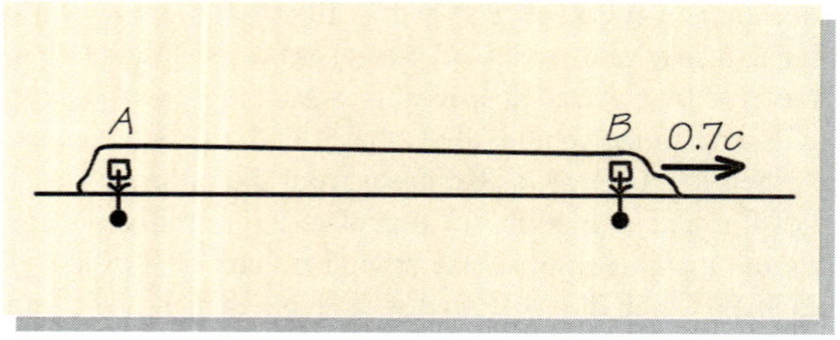

FIGURE 40–30 Problem 29.

31. (I) An unmanned spaceship launched with speed $0.8c$ explodes after 24 s as measured in its own rest frame. (a) Assuming that the launching coordinates are $x = t = 0$ in the spaceship coordinate system, what are the coordinates of the explosion point in the coordinate system of the spaceship? (b) If the launching coordinates in the rest frame of the launching pad are $x' = t' = 0$, what are the coordinates of the explosion point in the frame of the launching pad?

32. (II) In a given reference frame, event 1 occurs at time $t_1 = 0$ s and position $x_1 = 0$ m, while event 2 occurs at $t_2 = 1.5 \times 10^{-3}$ s and $x_2 = 4 \times 10^5$ m. Is there a second frame in which these events could be at the same position but different times? If so, specify its motion with respect to the first frame. If not, what is the frame in which the events have the least possible separation in distance? [*Hint*: Use invariants.]

33. (II) A new Klingon battleship races at a top speed of $0.2c$ away from the planet XG4T. The starship *Enterprise* follows at a speed of $0.25c$ relative to the Klingon ship. With what speed does the *Enterprise* appear to catch up with the Klingon ship according to an observer on the planet?

34. (II) Spaceship A moves with velocity $0.5c$ in the positive x-direction of a reference frame S. Spaceship B, moving in the same direction with a speed of $0.6c$, is 3×10^9 m behind (Fig. 40–31). At what times, in reference frame S and in the reference frame of spaceship A, will B catch up with A?

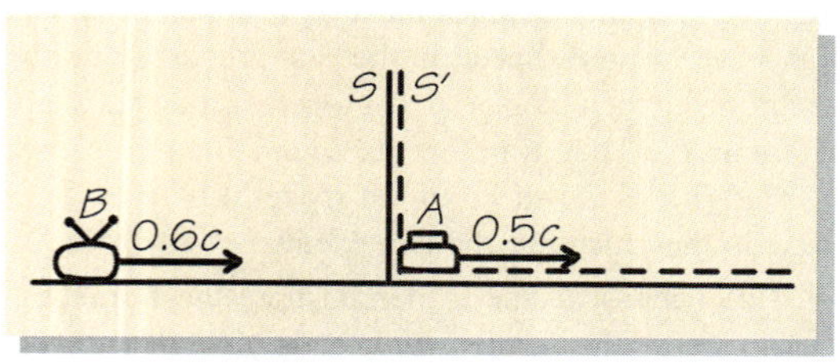

FIGURE 40–31 Problem 34.

35. (II) An observer in frame S measures two events to occur at the same point in space and separated by a time interval Δt. Show that in every other inertial frame, these events are separated by a larger time interval.

36. (II) A particle in frame F has velocity $v_x\mathbf{i} + v_y\mathbf{j}$. What is the velocity seen by an observer at rest in frame F′, a frame that moves at velocity $u\mathbf{i}$ relative to frame F?

37. (II) You shine light that moves at speed c/n through a medium of index of refraction n (Fig. 40–32). Suppose that the medium moves at speed u relative to you, parallel to the direction of the light. What is the speed of light in the medium as seen by you? [The result for $(u/c) << 1$ gives a result that is different from $(c/n) + u$, and it was first obtained by Augustin Fresnel in 1818. The measurements confirming the result were made by Hippolyte Fizeau in 1851.]

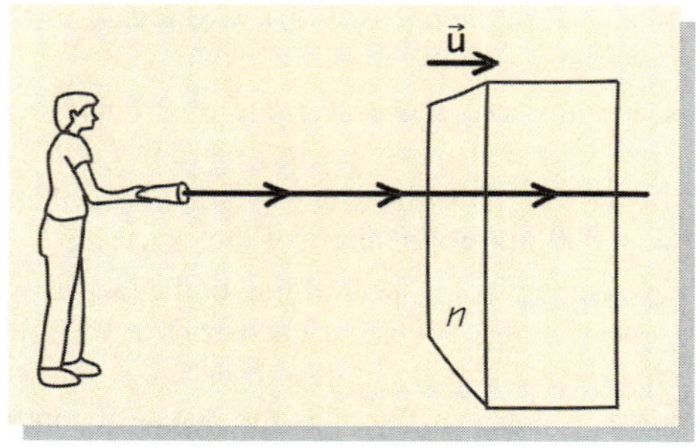

FIGURE 40–32 Problem 37.

38. (II) Use the Lorentz transformations to show that, as in Eq. (40–6), the location of a wave front for a plane wave emitted at $t = t' = 0$ from the joint origin of reference frames F and F′ is given by $x^2 - c^2t^2 = 0$ and by $x'^2 - c^2t'^2 = 0$.

39. (III) The electric field obeys the wave equation

$$\frac{\partial^2 E}{\partial t^2} - c^2 \frac{\partial^2 E}{\partial x^2} = 0$$

when electromagnetic waves propagate along the x-axis. What is the form of the equation for a wave seen in a Lorentz-transformed frame that moves at speed u along the x-axis? [*Hint*: Use the Lorentz transformation laws to obtain expressions for $\partial/\partial t$ and $\partial/\partial x$ in terms of $\partial/\partial t'$ and $\partial/\partial x'$.]

40–7 Momentum and Energy in Special Relativity

40. (I) A spaceship of mass 10^4 kg has a kinetic energy that is twice its rest energy. What is its total energy?

41. (I) The neutrino is a particle with zero rest mass. What is the momentum, in SI units, of a neutrino with energy 4 MeV?

42. (I) *Estimate* the mass lost when 1 million tons of TNT explodes. Assume that each chemical reaction between individual molecules involves 10 eV of energy.

43. (I) Humans generate energy at a rate of some 10^{13} W worldwide. (The United States, with less than 10 percent of the world's population, uses about 25 percent of the energy.) At what rate is mass being lost due to relativistic effects?

44. (I) A particle, the π^0, has a rest mass of 135 MeV/c^2. It decays at rest into two identical massless particles. What is the momentum of each of the two decay products of the π^0?

45. (II) Energy from the Sun reaches Earth (above the atmosphere) at a rate of about 1400 W/m^2. How fast is the sun losing mass due to energy radiation?

46. (II) What value of v/c must a particle of rest mass m have in order for its momentum to have magnitude $p = 10mv$?

47. (II) How much work has to be done on a proton to accelerate it (a) from rest to $0.01c$; (b) from $0.8c$ to $0.81c$; (c) from $0.9c$ to $0.91c$; (d) from $0.99c$ to c?

48. (II) An electron that is accelerated in the Stanford Linear Accelerator in California has a total energy of 50 GeV. How much of this is kinetic energy? What is the momentum of the electron? What is its speed?

49. (II) A proton accelerated at Fermi National Laboratory in Illinois has a momentum of 746 GeV/c. (a) What is the proton's velocity? (b) the proton's kinetic energy?

50. (II) An electron and its antiparticle of identical mass, the positron, annihilate each other and produce two photons (Fig. 40–33). Both the electron and the positron were initially at rest. What are the energy and momentum of each photon?

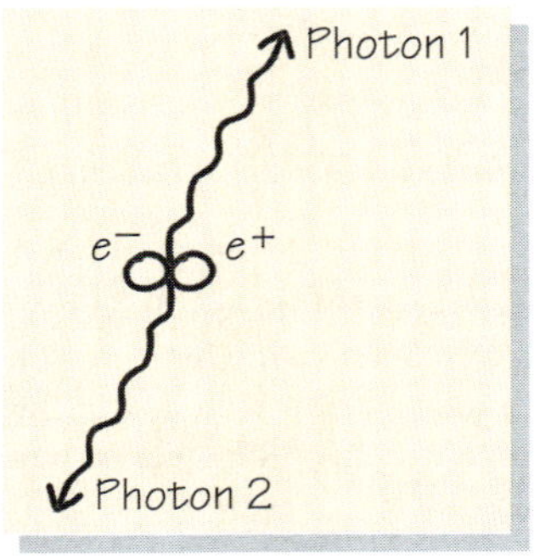

FIGURE 40–33 Problem 50.

51. (II) A photon is the quantum unit of light. It has an energy $E = hf$, where h is Planck's constant and f is the frequency of the light. Show that when a photon is absorbed by a free electron, without anything else occurring, energy and momentum cannot be conserved simultaneously.

52. (II) Show that energy and momentum conservation do not allow a high-energy photon to turn into a positron–electron pair with no other result. This process, called pair production, *can* occur if another object participates in the reaction (Fig. 40–34). What is the minimum energy required for the photon to give rise to an electron–positron pair, if the third object is very massive compared with the electron? (For the electron and positron $mc^2 = 511$ keV, and the third object, if it is a lead nucleus, has mc^2 about 4×10^5 times larger.)

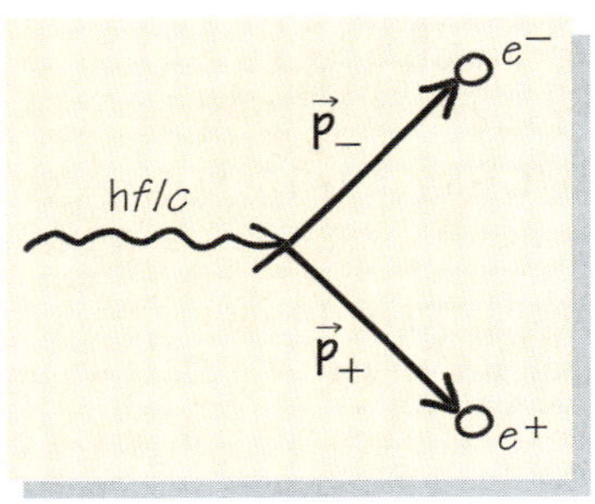

FIGURE 40–34 Problem 52.

53. (II) You analyze the track of a particle in a photographic plate placed in a magnetic field, and find that the total energy of the particle is 1130 MeV. The bending in the magnetic field gives information about momentum, and you learn that the particle's momentum is $p = 830$ MeV/c. What is the mass of the particle?

54. (II) In a generalization of Example 40–7, suppose that a ^{8}Be nucleus is moving in the x-direction with 10 MeV of kinetic energy when it decays into two alpha particles. Both alpha particles move off along the x-axis. What are their kinetic energies? [*Hint*: Note that $Q << M(^4\text{He})c^2$.]

55. (II) In the nuclear reaction $^{241}\text{Am} \rightarrow {}^4\text{He} + {}^{237}\text{Np}$, there is an energy release of 6 MeV. In the approximation in which the value of mc^2 for the Am nucleus is (241)(938 MeV), that of the He nucleus is 4(938 MeV) and that of the Np nucleus is (237)(938 MeV), calculate the value of pc (p = momentum) of the ^{4}He nucleus emitted from a ^{241}Am nucleus at rest (a) using nonrelativistic kinematics and ignoring the recoil of the ^{237}Np nucleus; (b) using nonrelativistic kinematics and taking the recoil motion of ^{237}Np nucleus into account; and (c) using relativistic kinematics and taking the recoil motion of the ^{237}Np nucleus into account.

56. (II) The lifetime of a particle called the neutral pion, π^0, is 0.9×10^{-16} s in the particle's rest frame. With what energy would one π^0 have to be produced so that its decay point is distinguishable from its production point in a photographic plate? Assume that a 1-mm separation is required for a measurement. The pion mass corresponds to $mc^2 = 135$ MeV.

57. (II) The decay products of a nucleus of mass M^* include another nucleus of mass M ($M < M^*$) and radiation. If the decaying nucleus is at rest, what is the kinetic energy of the remnant nucleus of mass M? [*Hint*: Use the fact that radiation of energy E carries momentum E/c].

58. (III) Experiments have shown that for the quantum of radiation (a photon), the energy and momentum are related by $E = pc$, corresponding to a particle with mass $m = 0$. Suppose that in the observation of a supernova 170,000 ly away, the first bursts of photons with an energy range of $E = 10$ eV to 10^4 eV arrive within 10^{-8} s of each other. What limits does this set on the mass of a photon? [*Hint*: Use the fact that mc^2 is small, so $E = pc + (m^2c^3/2p)$ is a good approximation.]

59. (III) Calculate an expression for the force $\mathbf{F}$ as defined by $m(d/dt)(\gamma\mathbf{u})$, and show that the force and the acceleration $d\mathbf{u}/dt$ do not necessarily point in the same direction.

*40–8 *Beyond Special Relativity*

60. (II) A neutron star has a mass of 3×10^{30} kg and a radius of 9 km. What is the gravitational redshift of radiation emitted with a frequency of 2×10^{19} Hz from the star's surface?

61. (III) A clock on a disk rotating with angular speed ω, when placed at a distance R from the center of the disk, experiences an acceleration toward the center of the disk. What gravitational potential will an observer at rest relative to the clock assume that he or she is in, by the equivalence principle? Will the clock be slow or fast relative to a clock at the center of the disk?

General Problems

62. (I) A photographer takes a flash-illuminated photo of a train car as the car passes moving to the left at 12 m/s. The flash goes off when the midpoint of the car is adjacent to the camera. Will two observers, one at each end of the 18-m-long train car, receive the flash simultaneously on their synchronized clocks? If not, which observer sees the flash earlier, and what is the difference in arrival time according to their synchronized clocks?

63. (II) Electrons and positrons (the antiparticles of electrons) of energy 35 GeV travel in opposite directions around a storage ring, a device in which the particles are held in circular orbits. What is the speed of each particle in the rest frame of the other?

64. (II) The Stanford Linear Accelerator accelerates electrons to a total energy of 50 GeV. How long does a meter stick at rest appear to a hypothetical observer at rest with respect to one such electron?

65. (II) An astronomer on Mars measures the optical spectrum of Earth. Averaged over a long period, she will see the spectral lines broadened by the Doppler effect due to Earth's rotation (Fig. 40–35). Calculate the width of the spectral line at 650 nm; i.e., the difference between the longest and shortest wavelength of a 650-nm line as seen by the astronomer, assuming that she is positioned in Earth's equatorial plane. Ignore the motion of Mars.

FIGURE 40–35 Problem 65.

66. (II) In 1990, an SR-71 Blackbird reconnaissance airplane on its way to retirement at the Smithsonian Air and Space Museum set several speed records. The plane averaged 2153 mi/h during the 2300-mi trip from Los Angeles to Washington D.C., and 2242 mi/h during the 311-mi trip from St. Louis to Cincinnati. What would have been the difference in time elapsed for the two record-setting segments between an atomic clock placed in the airplane and another atomic clock on the ground?

67. (II) A cosmic ray is approaching Earth from outer space. A hypothetical observer in a frame that moves with the cosmic ray measures Earth as a flattened ball whose thickness is 3/7 of its diameter (Fig. 40–36). (a) With what speed is the cosmic ray approaching Earth? (b) The cosmic ray is identified as a proton, with mass m and mass energy given by $mc^2 = 1$ GeV. What is the energy of the approaching proton, as seen from Earth?

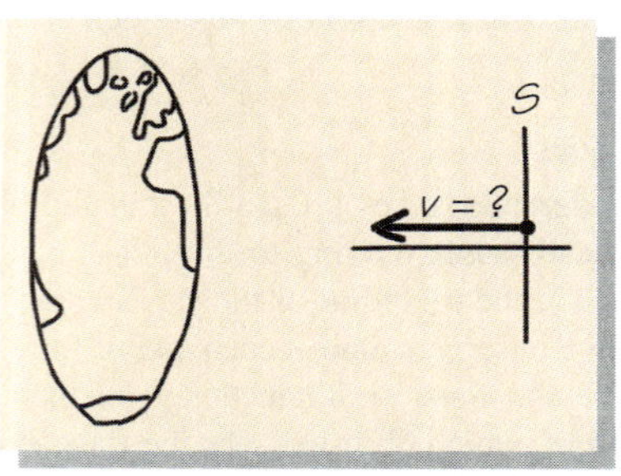

FIGURE 40–36 Problem 67.

68. (II) A new Klingon battleship has a proper length of 217 m and travels at speeds of $0.20c$ with respect to its home planet. The Klingons prepare to battle the *Enterprise*, which is moving at the same speed with respect to the same planet (Fig. 40–37). If the Klingons are heading straight at the *Enterprise*, what is the length of the Klingon ship as measured by Captain Kirk?

FIGURE 40–37 Problem 68.

69. (II) A particle of mass M is at rest. It decays into two identical particles, each of mass m, with $2m < M$. (a) If one of the two decay particles moves north with a momentum of magnitude p, what is the momentum of the other particle? (b) Use energy conservation to find p.

70. (II) In a quantum mechanical model, a proton and an antiproton annihilate each other and produce a pair of photons, light quanta whose frequency is related to their energy by the relation $E = hf$, where h is Planck's constant ($h \simeq 6.63 \times 10^{-34}$ J·s). The proton and the antiproton are nearly at rest when they annihilate. Find the frequencies of the emitted photons (Fig. 40–38). What are these frequencies if the proton and the antiproton are approaching each other in a head-on collision in which each particle has a kinetic energy of 500 MeV? For both protons and antiprotons, $mc^2 \simeq 938$ MeV.

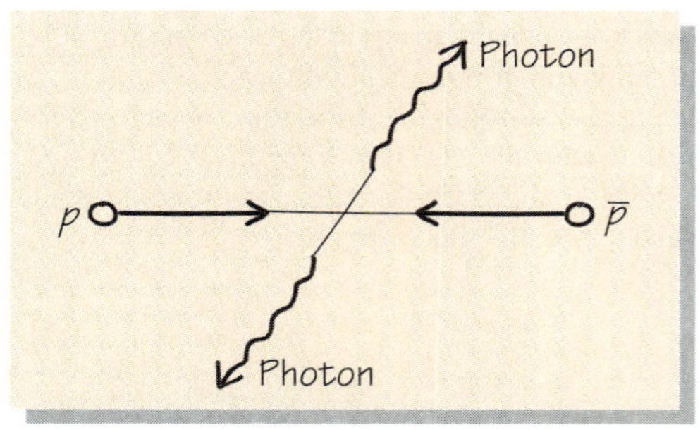

FIGURE 40–38 Problem 70.

71. (II) According to Chapter 19, the distribution function for the z-component of the velocity of a gas at temperature T is

$$G(v_z) \propto e^{-(mv_z^2/2kT)},$$

where k is Boltzmann's constant and m is the mass of one gas molecule. If a molecule at rest emits a spectral line, light at a characteristic frequency of f_0, then what is the distribution of frequencies of the light given off by the gas of such molecules when heated to temperature T? Assume that you are looking along the z-direction and take into account only motion in this direction. The effect described here is known as *Doppler broadening* of a spectral line. It is a tool for determining the temperature of stars and interstellar gases.

72. (II) Particles with energies as high as 10^{18} eV have been observed. Suppose that one of those particles collides with a photon of cosmic background radiation of wavelength $\lambda = 10^{-3}$ m. After a head-on collision, what will the final wavelength of the photon be? [*Hint*: These energies are so high that the particles can be treated as massless.]

73. (II) A charged pi meson (a particle of mass about 140 MeV/c^2) is ejected from a nuclear collision with a kinetic energy of 250 MeV. Pi mesons have a half-life (the time over which half of a given collection will decay radioactively) of about 1.5×10^{-8} s. Calculate (a) the pi meson's speed and (b) momentum. (c) How far will a collection of pi mesons travel before half of them decay?

74. (II) The *rapidity* V of a moving body is defined by $\tanh(V/c) = v/c$, where v is the body's relativistic speed. An observer in frame S′, moving at speed u in the $+x$-direction with respect to a frame S, measures a body to have speed v along the x-axis. Show that an observer in frame S measures the body to have rapidity W, given by $W = U + V$. Here, U is the rapidity of frame S′ with respect to frame S. The rapidity thus adds like a Galilean velocity.

75. (III) A proton moves with a momentum of magnitude p in the $+x$-direction. It strikes a second proton, which is at rest. Three protons and one antiproton result from the collision. The four particles in the final state remain together; that is, they have no motion relative to one another. Use energy and momentum conservations to find p. For both protons and antiprotons, $mc^2 \simeq 938$ MeV.

CHAPTER

41

Conferees in the 1927 Solvay Conference, including many of the founders of quantum mechanics. First row, left to right: I. Langmuir, M. Planck, M. Curie, H. A. Lorentz, A. Einstein, P. Langevin, C. E. Guye, C. T. R. Wilson, O. W. Richardson. Second row, left to right: P. Debye, M. Knudsen, W. L. Bragg, H. A. Kramers, P. A. M. Dirac, A. H. Compton, L. V. de Broglie, M. Born, N. Bohr. Standing, left to right: A. Piccard, E. Henriot, P. Ehrenfest, E. Herzen, T. De Donder, E. Schrödinger, E. Verschaffelt, W. Pauli, W. Heisenberg, R. H. Fowler, L. Brillouin.

Quantum Physics

Quantum mechanics represents a scientific revolution as deep and as important as that generated by Newton. The quantum physical phenomena described by quantum mechanics cannot be described by classical physics. Quantum mechanics describes the behavior of matter on a microscopic scale, far beyond what our senses can perceive. The results to be discussed in this chapter include the wave properties of matter, the particle nature of radiation, and the fact that position and momentum cannot be specified simultaneously with perfect precision. According to classical physics, given a definite set of initial conditions and knowledge of the forces acting, you are able to determine the behavior of a physical system unambiguously. In contrast, quantum mechanics only predicts the probabilities of physical events and implies that this is the only information we can have about these events. In this chapter, we describe some of the phenomena that classical physics cannot deal with and show how they are treated in quantum mechanics.

41–1 THE WAVE NATURE OF MATTER

One of the fundamental revelations of quantum mechanics is that *all matter exhibits wave properties*. Indeed, experiments have verified that particles such as electrons show typical wavelike properties, including interference. This possibility was conjectured by Louis de Broglie in 1924 (Fig. 41–1). It is possible to test the wave properties of electrons because quantum mechanics makes a clear statement about the wavelength associated with what is classically a particle. *When subject to experiments that test wave-*

like properties, particles with momentum p act like waves with a **de Broglie wavelength** λ, *given by*

$$\lambda = \frac{h}{p}. \tag{41–1}$$

Here, $h \simeq 6.626 \times 10^{-34}$ J·s is *Planck's constant*. Equivalently, the wave number $k = 2\pi/\lambda$ associated with a particle of momentum p is given by

$$k = \frac{2\pi}{\lambda} = \frac{2\pi p}{h} = \frac{p}{\hbar}. \tag{41–2}$$

Here, the commonly occurring combination $h/2\pi$ is written as $\hbar$.

$$\hbar \equiv \frac{h}{2\pi} = 1.05 \times 10^{-34}\ \text{J}\cdot\text{s}. \tag{41–3}$$

The fact that h is so small makes the wave character of matter evident only on a very small scale. A dust particle of mass 10^{-6} g traveling at a speed of only 10 m/s has a wavelength of

$$\lambda = \frac{h}{p} = \frac{h}{mv} = \frac{6.63 \times 10^{-34}\ \text{J}\cdot\text{s}}{(10^{-9}\ \text{kg})(10\ \text{m/s})} \simeq 6.6 \times 10^{-26}\ \text{m}.$$

This wavelength is too small to detect. Thus dust particles, baseballs, and airplanes do not reveal their wavelike aspects. On the atomic scale, however, things are quite different. Electrons ($m_e = 9.1 \times 10^{-31}$ kg) moving at a speed of 10^6 m/s, typical of electron speeds in atoms, have a wavelength

$$\lambda = \frac{6.63 \times 10^{-34}\ \text{J}\cdot\text{s}}{(9.1 \times 10^{-31}\ \text{kg})(10^6\ \text{m/s})} \simeq 0.7\ \text{nm}.$$

This wavelength is of the same magnitude as interatomic spacing in matter and is thus subject to tests by diffraction experiments like those described for X-rays.

In the quantum mechanical view, any particle is said to have a dual wave-particle nature. The situation is somewhat analogous to the duality of geometric and physical optics. Experiments involving the straight-line propagation of light or its reflection and refraction tell us nothing about the wave properties of light. Conversely, interference and diffraction form a second set of phenomena in which the wave properties of light emerge.

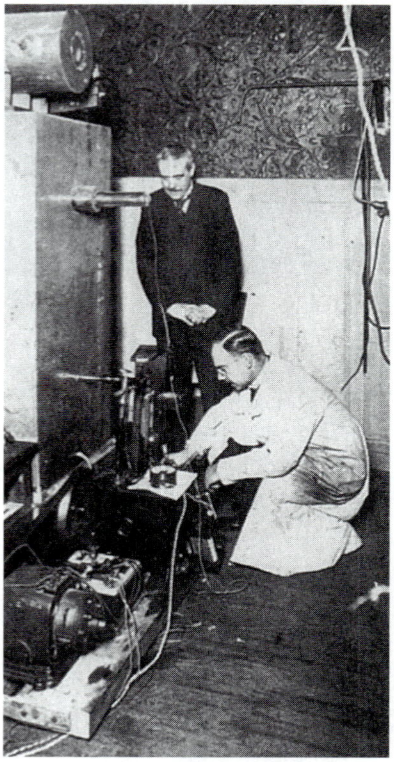

FIGURE 41–1 Louis de Broglie (standing).

The wavelength associated with any particle of momentum p

The wave number and wavelength of classical waves are described in Chapter 14.

EXAMPLE 41–1 What is the de Broglie wavelength of a neutron (mass $m = 1.6 \times 10^{-27}$ kg) with a speed $v = 1500$ m/s? (If v is taken as the rms speed of a gas of neutrons, the corresponding equilibrium temperature is around 35 K.)

Solution: The de Broglie wavelength is

$$\lambda = \frac{h}{mv} = \frac{6.63 \times 10^{-34}\ \text{J}\cdot\text{s}}{(1.6 \times 10^{-27}\ \text{kg})(1.5 \times 10^3\ \text{m/s})} = 0.28\ \text{nm}.$$

This value is comparable to the typical spacing between atoms in a crystal.

Experimental Evidence for the Wavelike Behavior of Matter

The experiments that confirmed de Broglie's conjecture were first carried out in 1927 by Clinton J. Davisson and Lester H. Germer, and independently by George Paget Thomson. They found that certain scattering directions are preferred when electrons are scattered by a crystal, as expected from constructive interference. We found the interference condition for waves, *Bragg's law*, in Chapter 39. When waves of wavelength λ are reflected from a succession of crystal planes separated by a distance d, there will be constructive interference for angles θ that satisfy Bragg's law:

$$k(2d)\sin\theta = 2\pi n, \tag{41–4}$$

where n is an integer and $k = 2\pi/\lambda$ is the wave number. From Eq. (41–4), we have

$$\lambda = \frac{2d}{n}\sin\theta. \qquad (41\text{–}5)$$

In the Davisson–Germer experiment, the spacing between the scattering planes in the crystal was determined by X-ray diffraction experiments to be $d = 0.091$ nm (Fig. 41–2). Davisson and Germer then scattered electrons with incident electron energy of 86.4×10^{-19} J (54 eV) and observed a diffraction maximum—wavelike behavior—at 65°. The kinetic energy of the electron corresponds to a momentum p:

$$p = \sqrt{2m_eE} = \sqrt{2(9.1 \times 10^{-31}\text{ kg})(86.4 \times 10^{-19}\text{ J})} = 39.7 \times 10^{-25}\text{ kg}\cdot\text{m/s}.$$

Now that we know the momentum, we can find the angle of constructive interference from Eq. (41–5),

$$\sin\theta = \frac{n\lambda}{2d} = \frac{nh}{2dp} = \frac{n(6.63 \times 10^{-34}\text{ J}\cdot\text{s})}{2(9.1 \times 10^{-11}\text{ m})(39.7 \times 10^{-25}\text{ kg}\cdot\text{m/s})} = 0.92n.$$

For $n = 1$, this yields $\theta = 66°$, which is in good agreement with the measured value.

EXAMPLE 41–2 At what angles do diffraction peaks occur for electrons of kinetic energy 120 eV incident on a crystal whose scattering planes are 0.12 nm apart?

Solution: We must apply the Bragg condition. To do so, we need the de Broglie wavelength, and hence the momentum, which we calculate from $E = p^2/2m = 120$ eV. We first convert the energy to joules: 120 eV = (120 eV)(1.6×10^{-19} J/eV) = 1.9×10^{-17} J. Then the wavelength is given by

$$\lambda = \frac{h}{p} = \frac{h}{\sqrt{2m_eE}} = \frac{6.63 \times 10^{-34}\text{ J}\cdot\text{s}}{\sqrt{2(9.1 \times 10^{-31}\text{kg})(1.9 \times 10^{-17}\text{ J})}}$$
$$= 1.1 \times 10^{-10}\text{ m} = 0.11\text{ nm}.$$

With a crystal plane separation of $d = 0.12$ nm, the angles for constructive interference are

$$\sin\theta = \frac{n\lambda}{2d} = \frac{n(0.11\text{ nm})}{2(0.12\text{ nm})} = 0.47n.$$

There will be diffraction peaks at $\theta = 28°$ ($n = 1$) and at 70° ($n = 2$).

Diffraction experiments have been performed with a variety of particles. In Example 41–1, we calculated the wavelength of neutrons that move at a certain speed; the wavelengths of these neutrons satisfy the conditions for substantial diffractive effects in scattering from crystals. The diffraction of neutrons by crystal surfaces is of practical importance in the study of those surfaces. Neutrons are ideal for such exper-

APPLICATION

Surface studies with neutrons

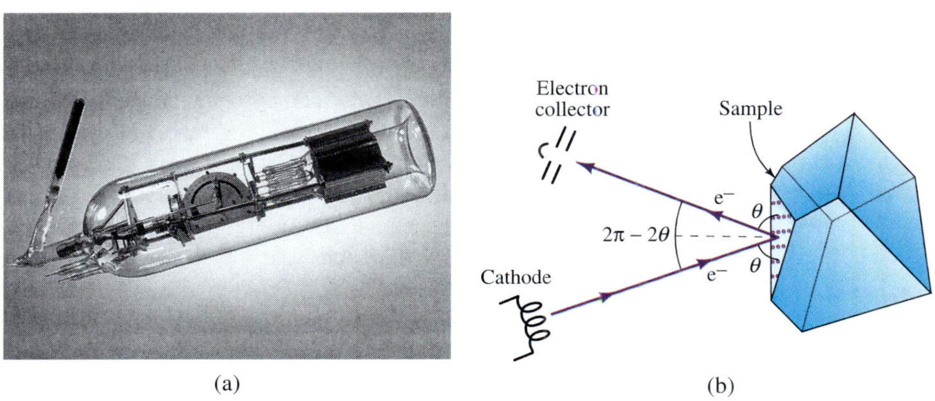

FIGURE 41–2 (a) The electron tube used in the Davisson–Germer electron diffraction experiment. (b) Experimental setup for the experiment. Electrons from a cathode strike a surface of a nickel crystal and are scattered to an electron collector.

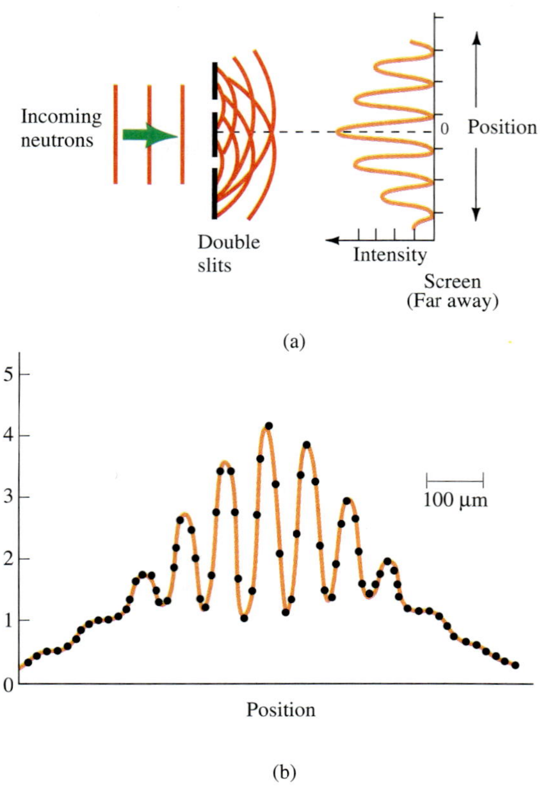

FIGURE 41–3 (a) Neutrons produce a diffraction pattern when they pass through double slits. (b) The measured data shown here involve double slits approximately 20 μm wide and separated by 104 μm (after A. Zeilinger et al., "Single- and Double-Slit Diffraction of Neutrons," *Reviews of Modern Physics*, **60**, p. 1067 [Oct. 1988]). The vertical axis for this data is proportional to the rate at which neutrons arrive at a screen.

iments because they can be slowed by collisions in hydrogenous materials such as paraffin, and slower neutrons have longer wavelengths. Figure 41–3 shows the results of a 1988 experiment in which neutrons of de Broglie wavelength 2 nm were incident on a screen with two slits approximately 100 μm apart.

The simplicity of the effect should not mask its extraordinary nature. In many respects, electrons and neutrons behave classically: They move according to Newton's second law, $F = ma$, when they are subject to forces. An ordinary water wave does not move as a classical particle. Yet electrons and neutrons deflected by regular structures produce interference patterns, just as a water wave would!

Potential energy barriers are discussed in Chapter 7; tunneling was introduced in Section 7–5.

Tunneling

Experiment shows that particles such as electrons or alpha particles (^{4}He nuclei) are able to pass through a potential energy barrier from one region in space to another. A classical particle with energy E_1 that starts within the region $r_1 < r < r_2$ in the energy diagram of Fig. 41–4 will always remain there. That is because the total energy of a particle in the region $r_2 < r < r_4$ would be less than the particle's potential energy, so its kinetic energy would be negative, which is impossible. However, quantum mechanics allows a real particle that starts in the region $r_1 < r < r_2$ to appear in the region $r > r_4$. An example of *tunneling*, or *barrier penetration*, is provided by nuclear fusion. An important application is the scanning tunneling microscope. Tunneling cannot be explained in terms of the particle aspects of matter. The wave properties of matter, however, do explain tunneling.

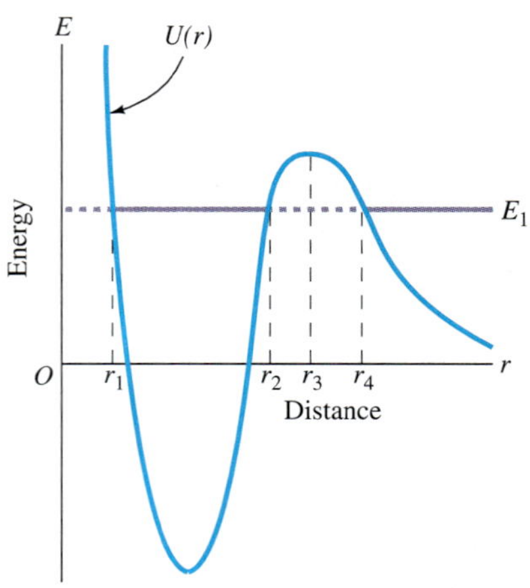

FIGURE 41–4 Energy diagram in which the potential energy forms a barrier for a classical particle with less energy than the potential energy maximum at distance r_3.

In order to see how wave phenomena can lead to tunneling, let's consider the phenomenon of total internal reflection in optics. When light traveling through glass reaches a glass–air surface at an angle that exceeds the critical angle, the reflection is total. However, the fields just outside the glass drop exponentially, not abruptly, to zero (Fig. 41–5a). If, then, the air forms just a thin strip that is no larger than a few wavelengths of the light between two regions of glass, an exponentially reduced field remains at the second interface. Starting from there, the fields can again propagate as sinusoidal waves in the second piece of glass, although reduced in amplitude (Fig. 41–5b). *The light waves have tunneled through the air gap* with a calculable intensity (Fig. 41–6).

In quantum mechanical tunneling, it is similarly possible to calculate the reduction of the intensity of quantum mechanical matter waves that tunnel through a potential barrier. This is a calculation of *the fraction of the number of particles that tunnel through the barrier*. The fraction is negligibly small under ordinary circumstances, which is why the phenomenon is not an intuitive one. The tunneling fraction depends primarily on the barrier width a and the difference $U - E$ of the barrier height and the energy. Approximately,

$$\text{fraction of particles getting through} \simeq e^{[-(2/\hbar)a\sqrt{2m(\langle U\rangle - E)}]}. \qquad (41\text{–}6)$$

In this equation $\langle U\rangle$ is the average height of the potential barrier. Barrier penetration is favored when the width of the barrier is small and when the energy E is not much below the barrier height.

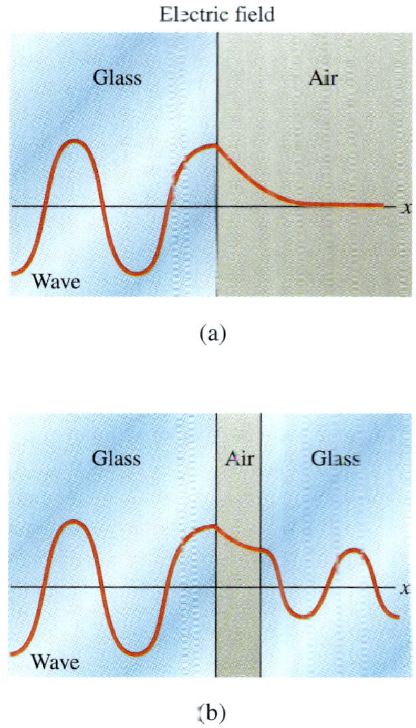

FIGURE 41–5 (a) Electric field at a glass–air interface for the case of total internal reflection. (b) Electric field at a narrow air gap between two pieces of glass, which shows the tunneling of an electromagnetic field through the air gap

41–2 THE HEISENBERG UNCERTAINTY RELATIONS

Electrons (and all particles) have wavelike characteristics, and the wave properties already discussed in Chapters 14 and 15 apply to them. The most important of these properties, stated by Werner Heisenberg, are known as the **Heisenberg uncertainty relations** (Fig. 41–7). The uncertainty relations take two forms. According to the *position–momentum uncertainty relation*,

We first saw an uncertainty relation in Chapter 8; in Chapter 15, we learned that the uncertainty relations are a general property of wave pulses.

The position–momentum uncertainty relation†

Any attempt to localize a particle within a distance Δx necessarily limits a simultaneous determination of the x-component of that particle's momentum to an uncertainty of Δp_x, where these uncertainties are related by

$$\Delta x\,\Delta p_x > \hbar. \qquad (41\text{–}7)$$

The second uncertainty relation is the *time–energy uncertainty relation*:

The time–energy uncertainty relation

If an energy measurement is to be carried out in a time Δt, then the accuracy ΔE with which the energy can be measured during this time interval is limited by the relationship

$$\Delta E\,\Delta t > \hbar. \qquad (41\text{–}8)$$

The small size of Planck's constant ($h \simeq 6 \times 10^{-34}$ J · s) guarantees that the uncertainty principle is important only on an atomic scale. For example, if we know the location of a dust particle to an accuracy of 10^{-6} m, then the uncertainty principle constrains our simultaneous knowledge of its momentum to an accuracy of 10^{-28} kg · m/s. This momentum uncertainty is so tiny that it is overwhelmed by other more mundane

†A more precise statement is $\Delta x\,\Delta p_x \geq \hbar/2$. We won't be requiring this precision.

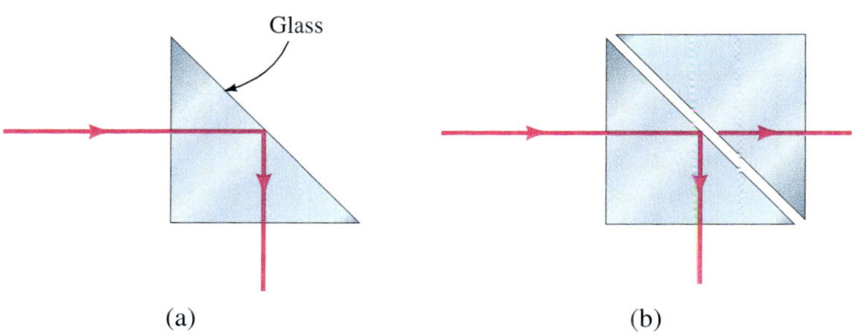

FIGURE 41–6 (a) Light incident from glass onto a glass–air interface that meets the proper geometric conditions will undergo total internal reflection. (b) If a second piece of glass is very close to the first piece, some light may cross the air gap.

FIGURE 41–7 Werner Heisenberg.

uncertainties. Thus the uncertainty principle has no practical role in the world of cars or dust particles.

When we deal with electrons in an atom however, the situation is quite different. The mass of an electron is about 10^{-30} kg, and its speed in an atom is in the range of 10^6 m/s. The momentum of an electron in an atom is then about 10^{-24} kg·m/s. The diameter of an atom is on the order of 10^{-10} m. If we try to pin down the location of an atomic electron to within 10 percent of the atom's size ($\Delta x \simeq 10^{-11}$ m), then the momentum becomes uncertain to about 10^{-23} kg·m/s, *10 times the value of the electron's momentum in its classical atomic orbit*. The momentum becomes so uncertain that we are not even sure that the electron will stay within the atom! The uncertainty relation is so important for atoms and nuclei that Newtonian momentum is a concept that must be used with care.

In the remainder of this section, we shall explore how the uncertainty relations resolve certain conflicts inherent in a dual wave–particle model. The uncertainty relations are also useful for making numerical estimates when quantum effects are important.

The Double-Slit Dilemma

Diffraction experiments carried out with electrons, protons, neutrons, and a variety of molecular beams verify that these particles possess the wave properties predicted by quantum theory. These results raise conceptual difficulties that are well illustrated by the following experiment. Consider a source of electrons. The electrons are emitted at some rate, so many per second. The stream of electrons impinges on a screen after they have passed through two slits in a wall (Fig. 41–8). If we think of electrons as particles, we expect that each electron will go through one slit or the other. In fact, if slit A is open for 5 min while slit B is closed, and then B is opened while A is closed for 5 min, then the electrons will arrive at the screen in two well-defined locations (Fig.

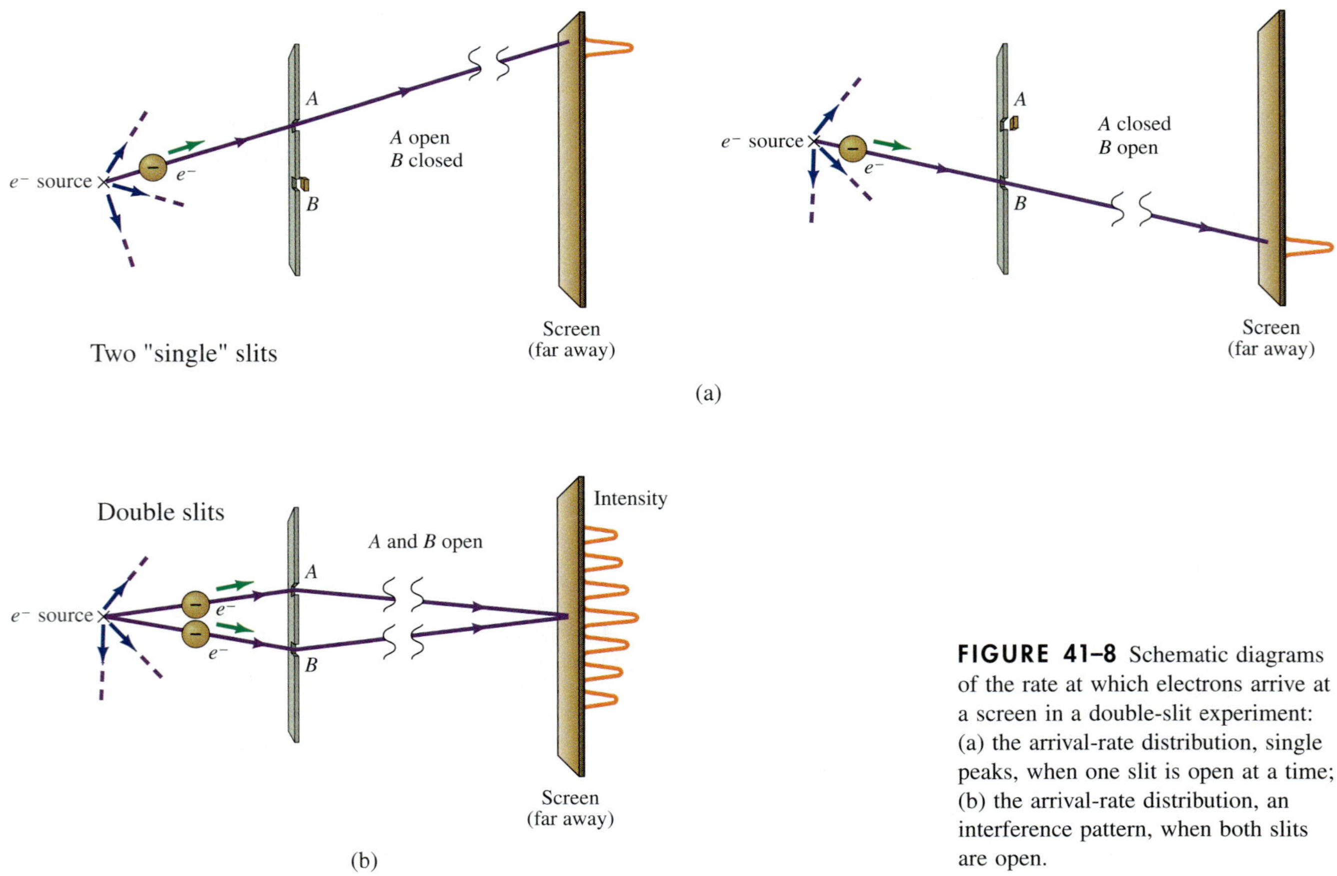

FIGURE 41–8 Schematic diagrams of the rate at which electrons arrive at a screen in a double-slit experiment: (a) the arrival-rate distribution, single peaks, when one slit is open at a time; (b) the arrival-rate distribution, an interference pattern, when both slits are open.

41–8a). But if the slits are open simultaneously for a total of 10 min, an interference pattern very much like that in Fig. 38–5 for light forms on the screen (Fig. 41–8b).

These two sets of results seem incompatible. Suppose that an observer could tell which slit the electron was about to pass through. The observer could then momentarily close the other slit. This process could be repeated for each electron. The system of slits, together with the alert observer, should (and does) give the same result as the opening of one slit at a time. *Somehow, changing the experiment so that we know which of the two slits the electron passes through destroys the interference pattern.* If quantum mechanics is to be a useful theory, it must be able to account for this peculiar state of affairs. The position–momentum uncertainty relation is the key to resolution of the conflict.

Resolution of the Double-Slit Dilemma

In our double-slit electron experiment, an observer must use a monitor of some kind to be able to tell which slit a given electron passes through. We can now show why, if a monitor enables us to determine the slit through which the electron passes, the interference pattern is destroyed. For this monitor to determine which slit an electron will pass through, it must be able to locate the electron's y-coordinate near the wall with the slits to an accuracy $\Delta y < d/2$, where d is the separation distance between the slits and y is taken to be the direction across the slits (Fig. 41–9). Any monitor must interact with the electrons to "see" where they are going. For example, the monitor may consist of a beam of light (like the beam of a searchlight) that reflects off the electron. Such a device will transfer momentum to the electron in a direction parallel to the screen (the y-direction in Fig. 41–9). If this momentum transfer is Δp_y, then the uncertainty relation states that

$$\Delta p_y > \frac{\hbar}{\Delta y} > \frac{2\hbar}{d}. \qquad (41\text{–}9)$$

This much momentum imparted to the electron is sufficient to wipe out the interference pattern. Recall that, for a slit separation d, the angles for constructive interference—interference maxima—are given by

$$d \sin\theta_n = n\lambda, \qquad (41\text{–}10)$$

where θ_n is the angle that the line leading to the nth maximum makes with the x-axis (Fig. 41–10). The distance between adjacent maxima on that screen is

$$\Delta y_{\max} = D\sin\theta_{n+1} - D\sin\theta_n = \frac{(n+1)D\lambda}{d} - \frac{nD\lambda}{d} = \frac{D\lambda}{d}, \qquad (41\text{–}11)$$

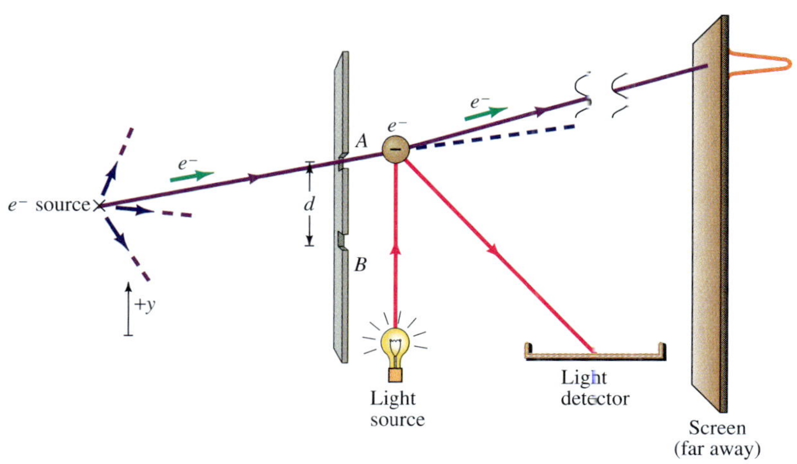

FIGURE 41–9 Schematic diagram of a monitor for a double-slit experiment designed to detect which slit an electron passes through. One mechanism: Shine light toward the slits and signal the passage of the electron by the pattern of reflected light.

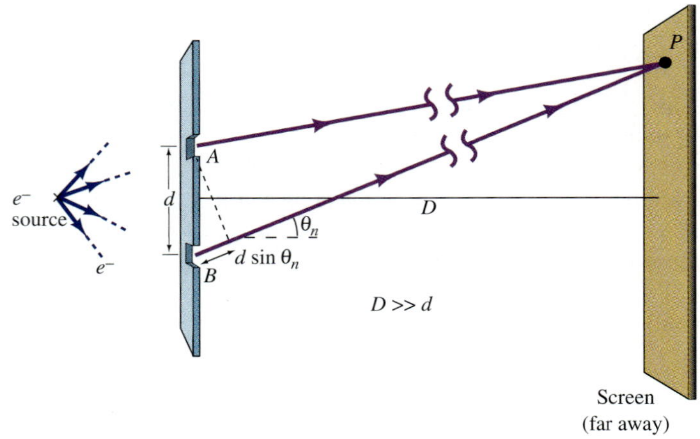

FIGURE 41–10 Geometry of path-length differences for a double-slit experiment.

where D is the distance to the viewing screen. Now suppose that the monitor gives the electron a sideways (y-direction) "kick." The electron thereby acquires an additional momentum Δp_y in the $+y$-direction. This changes the angle of deflection by $\Delta p_y/p$ and thus its location at the screen by $D\,\Delta p_y/p$. It follows from the uncertainty relation that the displacement at the screen is

$$D\frac{\Delta p_y}{p} > \frac{D(2\hbar/d)}{p} = \frac{2D/d}{k} = \frac{D\lambda}{\pi d}. \tag{41–12}$$

This displacement is comparable to the separation between the maxima, Eq. (41–11). Thus the interference pattern is wiped out, not just shifted.

What we have shown is that there is no paradox. A pure double-slit experiment and an experiment that includes a monitor to determine the electrons' paths are different experiments, and different patterns are predicted for them. A measurement that depends on the particle nature of an electron ("Which slit does it pass through?") must, at a minimum, disturb the system just enough to remove evidence of the wavelike nature of the electron. Quite generally, the uncertainty relation will remove any contradiction between the particle and wave aspects of a physical system. *Any attempt to determine whether an electron (or other physical system) is "really" particlelike, or "really" wavelike, disturbs the system so much that no determination can be made.*

The Uncertainty Relations and Numerical Estimates

The uncertainty relations may be used to estimate the smallest possible energy of a particle under the influence of a given force. This information determines the **ground-state** (or minimum) energies of atoms and molecules and is thus of great importance. Let's consider a particle with a potential energy $U(x)$. We choose a coordinate system such that the minimum of the potential energy is located at $x = 0$, and we change the potential energy by a constant—as we are free to do—so that $U(0) = 0$. Because the total energy of the particle is given by $E = (p^2/2m) + U(x)$, the energy is lowest when both the kinetic energy and the potential energy are lowest; classically this occurs for $p = 0$ and $x = 0$. With $U(0) = 0$, the lowest classical energy is $E = 0$. However, quantum mechanics does not permit a perfect localization in both p and x. If we suppose that the particle is at $x = 0$ with an uncertainty Δx, then we impose an uncertainty in the momentum p of magnitude larger than $\hbar/\Delta x$. This means that p^2 can only be known to an accuracy $(\Delta p)^2 > (\hbar/\Delta x)^2$. Because p^2 cannot be zero, the energy cannot be zero either. Thus the energy is a function of Δx. We can find the value of Δx for which the energy has its lowest value, *but that minimum energy value cannot be zero.*

To understand just how this works, let's take an example: a particle subject to the influence of a spring. The potential energy is $U(x) = m\omega^2x^2/2$, where ω is the angular frequency of the classical motion. The particle's energy is

$$E = \frac{p^2}{2m} + \frac{m\omega^2 x^2}{2}. \tag{41–13}$$

If the position of the particle is known only to an accuracy b ($\Delta x = b$) then the uncertainty in the momentum is $\Delta p > \hbar/b$. Thus the lowest value of the energy must obey the inequality

$$E > \frac{(\hbar/b)^2}{2m} + \frac{m\omega^2 b^2}{2}. \tag{41–14}$$

The right-hand side of Eq. (41–14) is plotted in Fig. 41–11. We see that it has a minimum as a function of b. Let's call the right-hand side of Eq. (41–14) $f(b)$ and find the minimum value of $f(b)$. We do so from the condition that the slope of $f(b)$ is zero at the minimum, $\mathrm{d}f/\mathrm{d}b = 0$:

$$\frac{\mathrm{d}f}{\mathrm{d}b} = -\frac{\hbar^2}{mb^3} + m\omega^2 b = 0.$$

When we solve this equation for b^2, we get $b^2 = \hbar/m\omega$. Substituting this value into the expression for $f(b)$, we obtain the minimum value of $f(b)$. According to Eq. (41–14), this is the minimum value of E:

$$E_{\text{min}} = \frac{\hbar^2}{2m(\hbar/m\omega)} + \frac{1}{2}m\omega^2\left(\frac{\hbar}{m\omega}\right) = \hbar\omega. \tag{41–15}$$

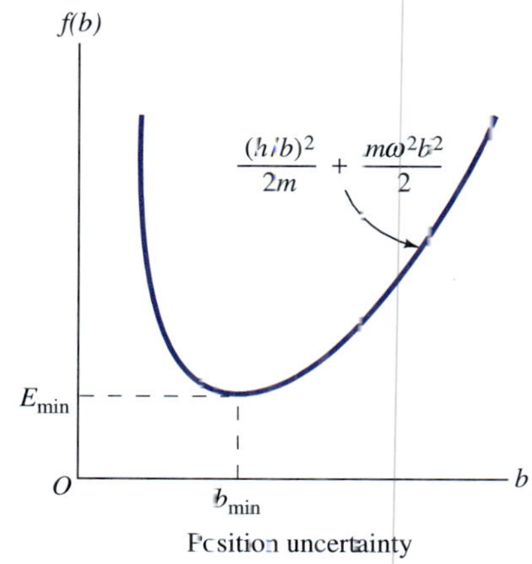

FIGURE 41–11 The right-hand side of Eq. (41–14), labeled $f(b)$, as a function of the position uncertainty b. The minimum energy is the minimum value of the curve $f(b)$.

This is an estimate of the quantum mechanical *zero-point energy*. A full quantum mechanical calculation yields $E_{\text{min}} = \hbar\omega/2$. The minimum energy is never zero; the particle on the end of the spring can never be brought completely to rest. This is certainly a nonclassical result!

The zero-point energy

EXAMPLE 41–3 Use the position–momentum uncertainty relation to estimate the lowest energy of a particle of mass m in a one-dimensional box of width L.

Solution: In this case Δx is specified from the beginning: If all that is known about the particle is that it is somewhere in the box, then $\Delta x = L$. In turn $\Delta p > \hbar/L$. Thus the particle momentum is undetermined to an accuracy Δp, and the lowest energy must satisfy $E_{\text{min}} = (\Delta p)^2/2m$, or

$$E_{\text{min}} = \hbar/(2mL^2).$$

EXAMPLE 41–4 Use the position–momentum uncertainty relation to estimate the lowest energy of an electron, mass m, moving in an attractive Coulomb potential of the form $U(r) = -e^2/4\pi\epsilon_0 r$. Express your result in electron-volts.

Solution: Zero electric potential is at $r = \infty$, and the total energy of the particle is given by

$$E = \frac{p^2}{2m} - \frac{e^2}{4\pi\epsilon_0 r}.$$

Classically, this describes a particle in the class of orbits described in Chapter 12, and the lowest energy is $-\infty$, corresponding to the particle at the center of force, $r = 0$. As in our previous discussion of the use of the position–momentum uncertainty relation, we express the minimum energy in terms of the uncertainty in radius, Δr. We do not need to worry about how Δr is related to Δx, Δy, and Δz because all the uncertainties are estimates. For simplicity, we let $\Delta r = R$. As the distance from the origin of an electron increases, the potential energy increases from $-\infty$. Thus, when the uncertainty in the position is of size R, the potential energy will be as large as $-e^2/4\pi\epsilon_0 R$. The uncertainty in momentum is larger than $\hbar/R$, and therefore the kinetic energy is larger than $(\hbar/R)^2/2m$. The lowest value of the energy obeys the inequality

$$E > \frac{\hbar^2}{2mR^2} - \frac{e^2}{4\pi\epsilon_0 R}.$$

We call the function on the right-hand side of this expression $f(R)$ and find its minimum value. When the function $f(R)$ is a minimum,

$$\frac{\mathrm{d}f}{\mathrm{d}R} = -\frac{\hbar^2}{mR^3} + \frac{e^2}{4\pi\epsilon_0 R^2} = 0.$$

This equation has the solution

$$R = \frac{4\pi\epsilon_0\hbar^2}{me^2}.$$

When this value of R is inserted into $f(R)$, we obtain the minimum value of f—that is, the minimum energy:

$$E_{\text{min}} = \frac{\hbar^2}{2m}\frac{(me^2)^2}{(4\pi\epsilon_0\hbar^2)^2} - \frac{e^2}{4\pi\epsilon_0}\frac{me^2}{4\pi\epsilon_0\hbar^2} = -\frac{1}{2}m\left(\frac{e^2}{4\pi\epsilon_0\hbar}\right)^2.$$

The minus sign of the energy indicates that the particle is bound to the attractive Coulomb potential. With $m = m_e = 0.9 \times 10^{-30}$ kg, $\hbar = 1.05 \times 10^{-34}$ J·s, and $e^2/4\pi\epsilon_0 = (1.6 \times 10^{-19}\ \text{C})^2(8.99 \times 10^9\ \text{m/F}) = 2.3 \times 10^{-28}\ \text{C}^2\cdot\text{m/F}$, the magnitude of E_{min} is 2.2×10^{-18} J. With 1 eV $= 1.6 \times 10^{-19}$ J, the energy in electron-volts is

$$E_{\text{min}} = -13.7\ \text{eV}.$$

In Example 41–4, we estimated the minimum energy—the ground-state energy—of the hydrogen atom by using the position–momentum uncertainty relation. We find a result that is extremely close to the correct value. In general, such estimates are not quite so accurate, but they can be counted on to give the correct dependence on parameters; here, $\hbar$, m, and $e^2/4\pi\epsilon_0$. The principal result is that quantum mechanics provides a minimum energy even for systems that have no such classical restrictions. The fact that there is a minimum energy accounts for the stability of all atoms.

Blackbody radiation was first discussed in Chapter 17.

41–3 THE PARTICLE NATURE OF RADIATION

FIGURE 41–12 Max Planck.

Blackbody Radiation

Quantum mechanics treats matter and radiation on the same footing. Just as particles exhibit wave properties, so do waves exhibit particle properties. Actually, the first inkling of quantum mechanics, as well as Planck's constant, appeared in 1900—preceding de Broglie's work by 24 years—when Max Planck was trying to understand the experimental data on *blackbody radiation* (Fig. 41–12). Electromagnetic radiation in a cavity in thermodynamic equilibrium at a temperature T has energy distributed in various frequencies, according to an energy density $u(f, T)$. The energy–density function has the following meaning: *The electromagnetic energy in a cavity of unit volume, for radiation with frequencies between f and f +* df, is $u(f, T)\,df$. Using a classical application of the equipartition of energy (see Chapter 19), Lord Rayleigh and James Jeans predicted that the energy density should have the form

$$u(f, T) = \frac{8\pi f^2}{c^3}kT.$$

This result agrees with experiment for low frequencies but disagrees badly for high frequencies. Planck found that by introducing the constant h (now called Planck's constant), he could fit the observed energy density over the full range of frequencies that had been measured with the formula given by Eq. (17–13),

$$u(f, T) = \frac{8\pi h}{c^3}\frac{f^3}{e^{hf/kT} - 1}.$$

Figure 17–18b shows the agreement between measured values of $u(f, T)$ and the Planck formula for a temperature of 2.7 K (see Section 17–5).

Planck's empirical fit could be derived only with a new assumption, formulated by Planck and more completely by Albert Einstein in 1905. The assumption was that *electromagnetic radiation consists of* ***quanta****, or identical, indivisible units, each carrying energy hf, where f is the frequency of the radiation.* In other words, the electromagnetic radiation comes in "bundles" of energy given by

$$E = hf. \tag{41–16}$$

The energy of a photon is related to its frequency by $E = hf$.

We shall soon discuss other experiments that show that quanta of radiation behave like particles. These particles have come to be called **photons**. Because the momentum and energy of any particle that travels at the speed of light are related by $p = E/c$, a photon of energy $E = hf$ carries momentum of magnitude

$$p = \frac{E}{c} = \frac{hf}{c}. \tag{41–17}$$

Note that for radiation,

$$\lambda = \frac{c}{f} = \frac{hc}{hf} = \frac{hc}{E} = \frac{h}{p}. \tag{41–18}$$

It was this formula that was later adopted for matter by de Broglie in his daring conjecture concerning the wave properties of matter [Eq. (41–1)].

EXAMPLE 41–5 An ordinary bright star easily visible to the naked eye emits radiation such that the intensity at Earth's surface is $I = 1.6 \times 10^{-9}\ \text{W/m}^2$ at a wavelength of 560 nm. Estimate the rate at which photons enter the night-adapted eye from such a star.

Solution: Here, we must convert the wavelength of the radiation into frequency and then into energy. The number of photons per square meter per second can then be found. The only estimate necessary is the area of the pupil for the night-adapted eye. We assume that the pupil is circular, with a diameter of 0.5 cm.

A wavelength λ corresponds to a frequency $f = c/\lambda$. The energy of each photon is thus $E = hf = hc/\lambda$. The intensity I is NE, where N is the number of photons striking the earth per square meter per second. Thus,

$$N = \frac{I}{E} = \frac{I\lambda}{hc} = \frac{(1.6 \times 10^{-9}\ \text{J/m}^2\cdot\text{s})(5.6 \times 10^{-9}\ \text{m})}{(6.64 \times 10^{-34}\ \text{J}\cdot\text{s})(3 \times 10^8\ \text{m/s})}$$
$$= 0.44 \times 10^{10}\ \text{photons/m}^2\cdot\text{s}.$$

The area of the pupil is estimated to be $A = \pi r^2 = \pi(2.5 \times 10^{-3}\ \text{m})^2 = 2 \times 10^{-5}\ \text{m}^2$, so the number of photons that enter the eye in 1 s is

$$NA = (0.44 \times 10^{10}\ \text{photons/m}^2\cdot\text{s})(2 \times 10^{-5}\ \text{m}^2) = 0.9 \times 10^5\ \text{photons/s}.$$

The human eye can register as few as several photons per second (see Problem 27).

The Photoelectric Effect

Further support for the quantum nature of radiation came from the work of Albert Einstein, who used it to explain the *photoelectric effect* in 1905. Heinrich Hertz discovered this effect in 1887 in experiments shown schematically in Fig. 41–13. The photoelectric effect can be summarized as follows:

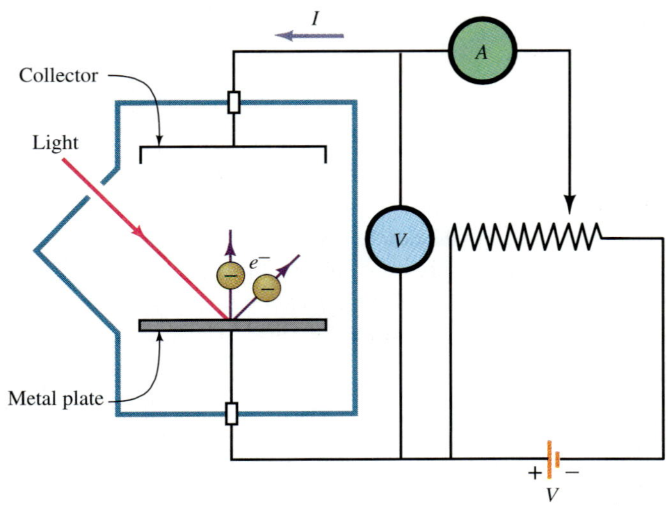

FIGURE 41–13 Schematic diagram of an experimental setup for measuring the photoelectric effect. Light strikes a metal plate in an evacuated chamber. The electron's current is measured by a collector, and the kinetic energy is determined by the grid voltage needed to slow the electrons down to rest.

1. When a polished metal plate is exposed to electromagnetic radiation, it may emit electrons. These electrons are sometimes termed *photoelectrons*.

2. Electrons will be emitted only if the frequency of the incident light exceeds a threshold value—that is, $f > f_0$. The value of f_0 may vary with the particular metal.

3. If the frequency is held constant, the magnitude of the emitted current of electrons is proportional to the intensity of the light source.

4. The maximum kinetic energy of the emitted electrons is independent of the intensity of the light source but varies linearly with the frequency of the incident light (Fig. 41–14).

5. Subsequent to 1905, experiments showed that, to an accuracy of 10^{-9} s, there is no measurable time delay between the arrival of the radiation and the appearance of the electron current.

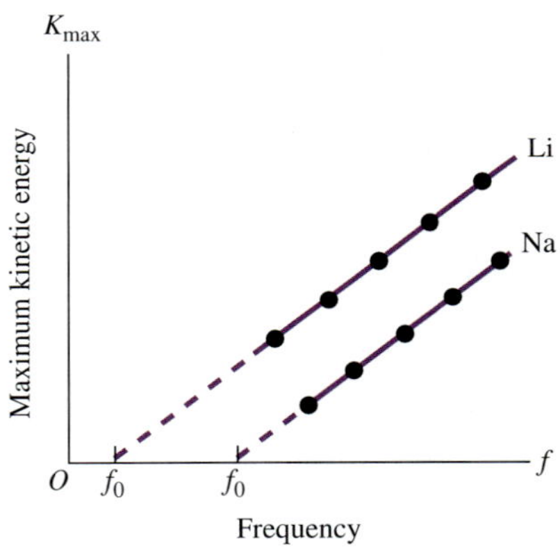

FIGURE 41–14 Data for the photoelectric effect, showing the maximum kinetic energy of the emitted electrons (of lithium, Li, and sodium, Na) as a function of light frequency. Note the linear relationship and the presence of a minimum frequency f_0.

The mere fact that electrons are emitted from metals subjected to electromagnetic radiation can be understood without invoking quantum ideas. Metals contain free electrons. Because electrons do not leak out of a metal freely, it is reasonable to expect that a minimum of energy must be deposited in the metal to liberate electrons. In classical electromagnetic theory, the energy delivered by radiation to the metal is proportional to the square of the electric field, **E**, that is, to the intensity of the incoming radiation. We would therefore expect the energy carried off by the electrons to be proportional to the intensity; for example, a doubling of the intensity would double the number of electrons emitted with a given kinetic energy. What is not comprehensible in the classical view is that electrons are emitted even when the incident radiation is of very low intensity, that the energy of the emitted electrons depends linearly on the radiation frequency, and that there is a frequency threshold. Classically, energy should be delivered for all radiation frequencies. In addition, we would expect that, with low-intensity radiation, the energy required to liberate a certain number of electrons would be collected over some time, and that there would be a time delay (that increases with decreasing intensity) before the electrons would appear.

Einstein explained these phenomena by postulating that electrons are emitted because individual electrons absorb photons. The photons correspond to radiation with frequency f and carry energy $E = hf$. If there is a minimum energy W required to liberate an electron, then no electrons will be emitted when hf is less than W. When hf exceeds W, the excess energy can go into kinetic energy of the emitted electrons:

$$\frac{1}{2}mv^2 = hf - W. \qquad (41\text{–}19)$$

The quantity W is a kind of potential energy that must be acquired before the electron can be liberated; it is called the *work function*. It is a characteristic of the particular

APPLICATION: NIGHT VISION VIEWERS

The devices that "see" at night depend on the photoelectric effect. They work by collecting individual photons, then using the photoelectric effect to amplify their presence. A lens system initially sends any collected light to a glass plate coated on the back side with a photoelectric material. The photoelectrons are accelerated through a potential difference of several hundred volts to a "channel plate" containing many fine holes typically 10 microns in diameter. The hole has an interior conducting surface, and there is an additional potential difference from one side of the channel plate to the other. When a photoelectron strikes a hole, it ionizes atoms at the point of impact. This releases several electrons, which are accelerated farther down the hole, striking the sides of the hole to make more electrons and so forth. Tens of thousands of electrons leave the hole as a result, and they are accelerated yet again to strike a fluorescent screen, where their effects are strong enough to be seen by the eye (Fig. 41AB–1). The constraint that the electron avalanche is held to the hole is what preserves the image quality. The image is produced in pixels that correspond one to each hole.

Noise keeps this device from being infinitely sensitive. Because the photons arriving on the photoelectric surface are discrete, their number fluctuates with time, and the smaller the rate of their arrival the larger the fluctuation at any given pixel. These fluctuations will manifest themselves as a fluctuating brightness termed "photon noise." To combat this, it is possible to integrate the arriving photons for longer periods by using a fluorescent screen in which the brightness builds up and decays slowly. This has its limits, though; if the integration time is too long, moving objects will become "smeared" across the viewing screen. Acceptable pictures can be obtained for 500 photons arriving at each pixel per integration time (0.25 seconds), a level easily reached for making observations by starlight.

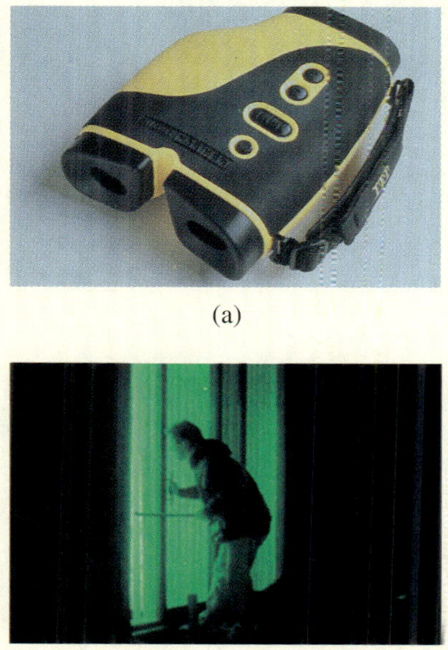

(a)

(b)

FIGURE 41AB–1 (a) A hand-held night vision viewer and (b) a view at night using this device.

metal that emits electrons. It takes one photon to liberate one electron. Therefore, the current of emitted electrons is proportional to the intensity of the radiation because the intensity is proportional to the number of photons in the electromagnetic wave. An electron absorbs a photon instantaneously, so the lack of time delay is also explained. The first accurate experiments on the photoelectric effect were carried out in 1916 by Robert Millikan, who was skeptical about Einstein's theory (Fig. 41–15). Millikan's experiments unequivocally confirmed Eq. (41–14) and thus the quantum explanation of the photoelectric effect.

FIGURE 41–15 Robert Millikan (left).

EXAMPLE 41–6 The largest wavelength of light that will induce a photoelectric effect in potassium is 564 nm. Calculate the work function for potassium in electron-volts.

Solution: The largest wavelength or, equivalently, the lowest frequency, to induce a photoelectric effect expresses the threshold frequency f_0, and we need f_0 to calculate the work function. The minimum frequency f_0 is given by $f_0 = c/\lambda_{\max}$. The work function is then given by

$$W = hf_0 = \frac{hc}{\lambda_{\max}} = \frac{(6.63 \times 10^{-34}\ \mathrm{J \cdot s})(3.00 \times 10^{8}\ \mathrm{m/s})}{5.64 \times 10^{-7}\ \mathrm{m}} = 3.53 \times 10^{-19}\ \mathrm{J}$$

$$= (3.53 \times 10^{-19}\ \mathrm{J}) \frac{1\ \mathrm{eV}}{1.60 \times 10^{-19}\ \mathrm{J}} = 2.20\ \mathrm{eV}.$$

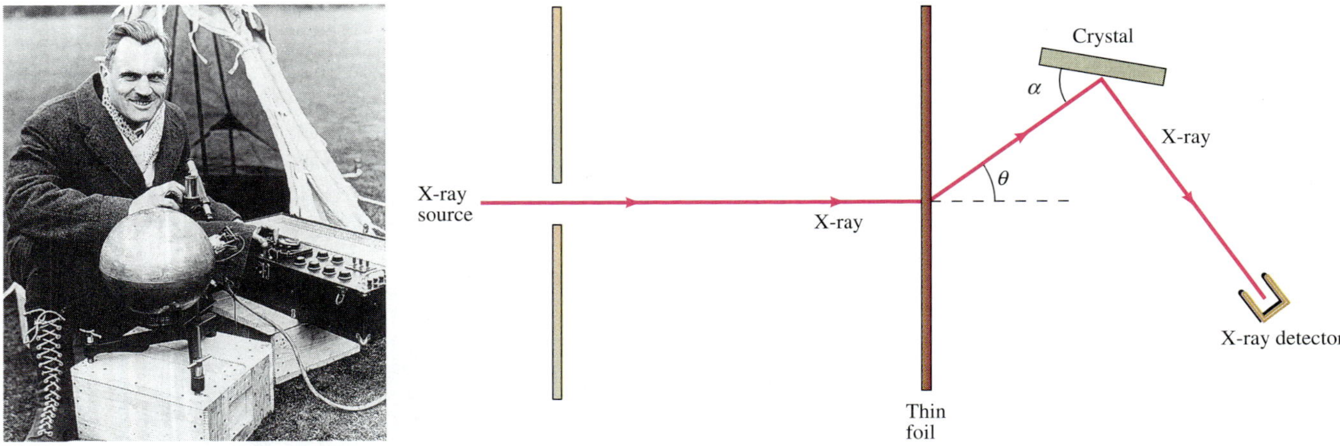

(a) (b)

FIGURE 41–16 (a) Arthur Compton. (b) Schematic diagram of a setup for Compton's experiment. The scattered X-rays are diffracted by a crystal, with the angle α used to determine the wavelength of the scattered radiation.

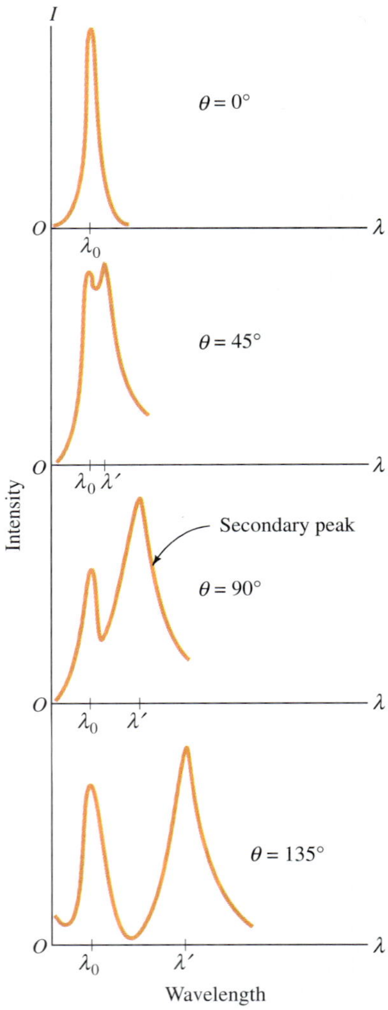

FIGURE 41–17 Experimental data for Compton's experiment. The secondary peak, due to X-ray scattering by free electrons, becomes more pronounced as the scattering angle increases.

The Compton Effect

Even more compelling evidence for the particle properties of photons came from experiments carried out by Arthur Compton in 1922 (Fig. 41–16a). Compton sent X-rays through thin metallic foils and discovered that the scattered X-rays emerge with one of two wavelengths (Fig. 41–16b). One component emerges with the same wavelength as the incident radiation. The other component emerges with a longer wavelength. This result is in contrast to the prediction of classical radiation theory, in which the electrons absorb radiation and reradiate it as dipole radiation without any change in wavelength. The experiments showed that for the second component, the wavelength varies with the scattering angle of the X-ray (Fig. 41–17). The dependence of the increased wavelength on angle fits the formula

$$\lambda' - \lambda = \frac{h}{mc}(1 - \cos\theta), \tag{41–20}$$

where λ is the wavelength of the incident X-rays and λ' that of the scattered X-rays and m is the electron mass. The presence of h indicates that quantum mechanics is involved, and the independence of the result on the metal used in the foils indicated to Compton that it had nothing to do with the metal's crystal structure.

Compton discovered that Eq. (41–20) can be derived by treating the photon as a particle of energy hf and momentum hf/c. The process is then a two-body collision, with a photon incident on a target electron at rest, complete with momentum conservation and energy conservation (Fig. 41–18). It is necessary to take the electron's energy in the relativistic form $E_e = \sqrt{p^2c^2 + m^2c^4}$, where p is the final electron momentum. The energy of the scattered photon differs from that of the incident photon, just as would be true in a collision of billiard balls. And because the energy of the scattered photon is different, so is its wavelength. The calculation shown in the following subsection shows that Equation (41–20) correctly emerges. The quantity h/mc, called the *Compton wavelength* of the electron, has the dimensions of length and magnitude 2.4×10^{-12} m.

The particle nature of radiation manifests itself in experiments designed to study that particle nature. Experiments that probe the wave character of radiation, such as interference experiments, confirm the wave character of radiation, even at extremely high frequencies.

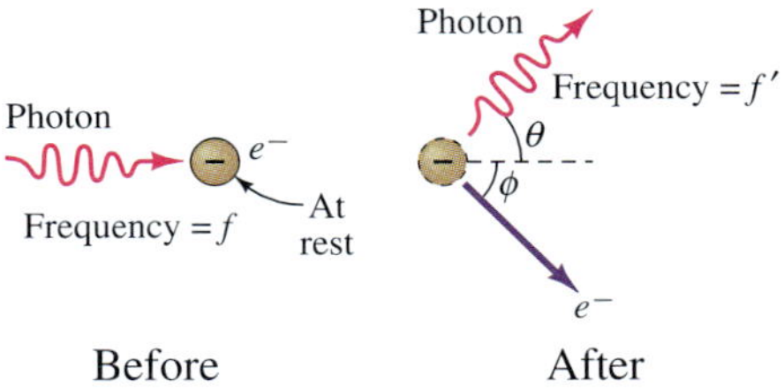

FIGURE 41–18 Kinematics for calculating the shift in frequency in the Compton effect. Light scatters from an electron as though light were a particle (photon). As in the ordinary collisions of particles, the photon's energy changes when it scatters, and by the laws of quantum mechanics, its frequency also changes to a new, calculable value.

How to Get Equation (41–20)

To derive Eq. (41–20), we formulate the equations of conservation of momentum and energy for the collision described in Fig. 41–18. Using energy and momentum labels of that figure, the momentum-conservation equations are

$$\text{in the } x\text{-direction:} \quad p = p' \cos\theta + p'_e \cos\phi.$$

$$\text{in the } y\text{-direction:} \quad 0 = p' \sin\theta - p'_e \sin\phi.$$

We rewrite these equations in the form

$$p - p'\cos\theta = p'_e \cos\phi,$$

$$p' \sin\theta = p'_e \sin\phi.$$

We can combine these equations by squaring both sides and adding them. With the relation $\sin^2 x + \cos^2 x = 1$ for both θ and ϕ, we find that

$$p^2 - 2pp'\cos\theta + p'^2 = p_e'^2. \tag{41–21}$$

The energy-conservation equation is

$$E + mc^2 = E' + E'_e,$$

where m is the electron mass. We substitute $E = pc$, $E' = p'c$, and $E'_e = \sqrt{p_e'^2 c^2 + m^2c^4}$ and rearrange the equation to the form

$$pc - p'c + mc^2 = \sqrt{p_e'^2 c^2 + m^2c^4}.$$

We now square both sides of this equation to eliminate the square root; the factor c^2 cancels:

$$p^2 + p'^2 + m^2c^2 - 2pp' + 2mpc - 2mp'c = p_e'^2 + m^2c^2. \tag{41–22}$$

Finally we subtract Eq.(41–22) from Eq. (41–21):

$$2pp'(1 - \cos\theta) - 2mpc + 2mp'c = 0.$$

We can rearrange this result as

$$\frac{1}{p'} - \frac{1}{p} = \frac{1}{mc}(1 - \cos\theta).$$

When we substitute $p = h/\lambda$ and $p' = h/\lambda'$, we get Eq. (41–20).

41–4 QUANTUM MECHANICS AND PROBABILITY

The phenomena we have discussed in this chapter are far from intuitive. This is perhaps most pronounced for an electron beam that passes through a double slit, the experiment discussed in Section 41–2. Let's reconsider this experiment. When both slits are open, an interference pattern is produced, whatever the rate at which electrons go through the double-slit system. Suppose that we significantly reduce the density of electrons in the beam. Then the interference pattern is built up by a one-by-one accumulation of electrons in regions of constructive interference, while no electrons accumulate in the regions of destructive interference. The arrival of any one electron is causally disconnected from the arrival of any other, so *each* electron must somehow carry information about the final interference pattern.

Quantum mechanics accounts for the experiment by describing the electron and the screen with the slits by means of a **wave function**, which contains all the information about the diffraction pattern. The wave function (strictly speaking, its square) gives only the *probability* that the electron will be found in one place or another on the screen. The probability is largest where the magnitude of the wave function is largest. The probability that an electron will land on the screen where the waves from the two slits interfere destructively is very small, whereas the probability for it landing where the interference is constructive is large. This implies that the outcome of the journey of any one electron is not determined; only the *probability* for a set of outcomes can be known.

An analogous situation arises when polarized light passes through an analyzer (see Chapter 35). If the polarizer makes a 45° angle with the polarization vector, then the intensity of the passed light is half that of the incident light. Classically, this is easy to understand. But the classical description no longer works if light is formed from photons. Imagine an incident light intensity that is so low that the photons arrive one at a time. How will a particular photon "decide" whether to pass the analyzer or not? The resolution of this problem is that, in quantum mechanics, a wave function describes the photon, the polarizer, and the analyzer, and with this wave function we can predict *only* that a given photon has a probability of $\frac{1}{2}$ of passing the analyzer. We cannot predict whether or not a given photon will pass.

Quantum mechanics predicts only probabilities.

To summarize: *Quantum mechanics is different from all other theories that we have studied so far in that it does not make predictions about the outcome of single events. It makes predictions only about the probabilities of different outcomes.*

A situation of this type arises in the de-excitation of an atomic electron from an excited state or in nuclear radioactivity. Radioactivity occurs when a nucleus decays to some state of lower energy by the emission, for example, of an electron, an alpha particle (^{4}He nucleus), or a photon. Quantum theory predicts (and all experiments confirm) that if we start with a certain number of radioactive nuclei, N_0, then, the number of nuclei left after a time t will be

$$N = N_0 e^{-t/\tau}. \tag{41–23}$$

The parameter τ has the dimensions of time and is called the *lifetime*, or *mean life*, of the radioactive nucleus. It can be calculated by using the machinery of quantum mechanics and is relatively easy to measure. After a time $t = \tau$, the number of nuclei that remain is $N_0 e^{-1}$, or 0.37 N_0. The value of τ is the same for a given nucleus, whether the nucleus has just been artificially produced in the laboratory or found in a rock 1 billion yr old. How does a given nucleus "know" when to decay? There is no evidence that the nucleus contains an internal clock that tells it when to decay (we say that there are no *hidden variables*). Because the quantum mechanical description of a single nucleus cannot contain information about what other nuclei are going to do, the only interpretation possible is that the *probability* that a nucleus will last a certain time t is part of the wave function of that nucleus, whereas a determination of precisely when the nucleus will decay is not.

It might appear that this is no different from the problem of life expectancy in a population. There is some probability that 100-yr-old people exist in a population, but actuarial tables make no predictions for an individual. There is, however, a difference: People do have internal clocks, and an examination of the habits and jobs of individuals can give us more information about their life expectancy. With sufficiently detailed medical information, we could at least *in principle* make a prediction about a given lifetime.

A characteristic of quantum systems is that a measurement has a well-defined effect on the system. We can illustrate this with a technique known as *radiometric dating* (Fig. 41–19). Consider a piece of wood discovered in an archeological dig. The method of radiometric dating depends on the fact that, in the atmosphere, the ratio of ^{14}C (whose nucleus has 6 protons and 8 neutrons) to ^{12}C (whose nucleus has 6 protons and 6 neutrons) is constant. Even though ^{14}C nuclei decay radioactively with a lifetime of 8270 yr, their number is replenished in the atmosphere through cosmic-ray bombardment of the stable nucleus ^{14}N. Once a tree dies and its wood stops taking in carbon from the atmosphere, the proportion of ^{14}C in the wood (relative to ^{12}C) steadily decreases according to the law given in Eq. (41–23). Suppose now that the proportion of ^{14}C atoms indicates that a sample of wood is 20,000 yr old. A measurement has been made. If the remaining (undecayed) ^{14}C atoms are now set aside, they will continue to decay in such a way that 37 percent will be left in 8270 yr. In other words, once an atom is measured not to have decayed, its clock, so to speak, has been reset to $t = 0$. This is quite different from the identification of 100-yr-old people. Once these people are identified as being alive, they are not in effect reborn!

These concepts are far removed from what we would call "common sense." It should be remembered, however, that common sense about the physical world is developed through observation, and there is no reason why the microscopic world should conform to the notions of what is sensible as developed from observation of the macroscopic world.

APPLICATION

Radiometric dating

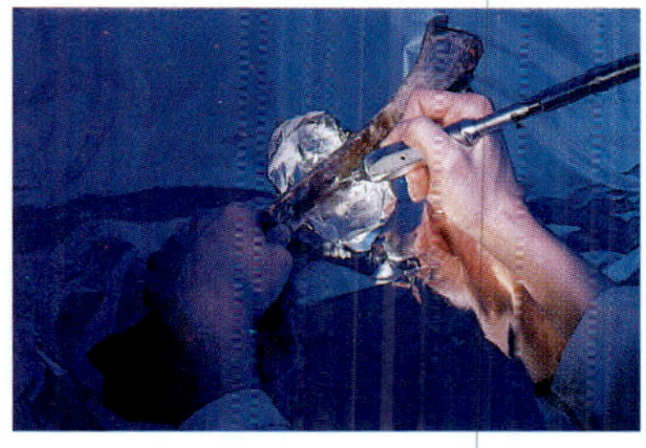

FIGURE 41–19 The researcher taking shavings from this reindeer bone will measure the ratio of C^{14} to C^{12} isotopes and thereby learn the time since the animal's death.

SUMMARY

The phenomena that comprise quantum physics are most important in microscopic systems such as atoms, molecules, and nuclei. These phenomena are rather unexpected to an intuition based on classical physics. Essentially, we learn that what we think of as particles behave in some respects like waves, and what we think of as waves (electromagnetic radiation, for example) behave in some respects like particles. Quantum mechanics provides us with a unified model of these phenomena. A "particle" of momentum p will have the properties of a wave of de Broglie wavelength

$$\lambda = \frac{h}{p}, \qquad (41\text{–}1)$$

where $h = 6.63 \times 10^{-34}\ \text{J}\cdot\text{s}$ is Planck's constant. The frequency associated with a particle can similarly be related to its energy. The wavelike properties of particles include interference, which has indeed been observed for particles such as electrons and neutrons in diffraction experiments. We also see evidence of the wavelike properties of matter in tunneling. The smallness of h is the reason the wave properties of matter are not evident on a macroscopic scale.

Just as classical particles have wavelike properties, electromagnetic "waves" have particlelike properties. Electromagnetic radiation of frequency f behaves as though it consists of particles (photons) with energy.

$$E = hf \qquad (41\text{–}16)$$

and momentum of magnitude

$$p = hf/c. \qquad (41\text{–}17)$$

Radiation exhibits its particlelike properties in the spectrum of blackbody radiation; in the photoelectric effect, in which electrons absorb incident photons and are ejected with specific energies from metals; and in the Compton effect, in which the wavelengths of photons scattered by electrons change.

There are apparent inconsistencies in the description of an "object" with both particle and wave attributes. Interference phenomena intrinsic to waves require the wave to be spread out in space and time, whereas a particle must have a definite, well-defined location. The inconsistencies are removed by the use of the Heisenberg uncertainty relations, which set limits on the use of classical variables such as position and momentum. Any attempt to specify the x-position with a precision Δx implies a limit with which the x-component of momentum can be simultaneously measured,

$$\Delta x \, \Delta p_x > \hbar, \tag{41–7}$$

where $\hbar = h/2\pi$. Similarly, there are limits on the precision of measurements of energy and time. An energy measurement is limited to a precision ΔE by the duration of time Δt that the measurement takes:

$$\Delta E \, \Delta t > \hbar. \tag{41–8}$$

These relations resolve potential inconsistencies between a simultaneous particle and wave description.

One of the consequences of the uncertainty relations is that a particle cannot be at rest at the minimum level of potential energy. The lowest energy of a quantum system (the ground-state energy) is always larger than what is expected by classical reasoning.

Quantum mechanics is special in that it cannot be used to predict the future behavior of a system, only the probability of a set of possible behaviors.

UNDERSTANDING KEY CONCEPTS

1. The shorter the wavelength of a photon, the more the photon behaves like a particle. Why?
2. The uncertainty relations provide a reason why the temperature $T = 0$ cannot be reached. What is that reason?
3. A lit cigarette can be seen at a distance of 500 m on a dark night. Outline how you would estimate the rate at which photons from the cigarette hit the retina of a night-adapted eye.
4. To probe very tiny regions of space (such as the inside of a proton) with electron beams, you need electron beams of very high energy. Why? Can you estimate the kind of energy needed to study a region of diameter d?
5. Given that electrons behave like waves, how is a Doppler shift described in terms of momentum?
6. Before the Planck formula was discovered. Rayleigh and Jeans had obtained the expression $u(f, T) = (8\pi f^2/c^3)kT$. How could we tell that something is wrong with this expression, even with no experimental data on the subject of blackbody radiation?
7. Suppose that half of a sample of radioactive nuclei has decayed in a given time T. How long will it be before half the remaining nuclei will have decayed?
8. You start with a large sample of radioactive nuclei; how much time will elapse before you can be sure that all the radioactivity has died away to zero?
9. The lifetime τ that measures the decay rate of a sample of radioactive particles is affected by the considerations of special relativity; that is, moving radioactive particles decay more slowly than stationary ones. How do the particles "know" that they are moving and that they should decay more slowly?
10. One electron is sent through a double-slit apparatus. In what sense, if any, can we say that there is an interference pattern on the screen?
11. Does the fact that all particles, however large, have wavelike properties mean that there is some probability that a baseball can tunnel through a catcher's mitt?
12. In discussing blackbody radiation, we spoke of a cavity. What does the cavity provide? Do we mean a real cavity in bulk material?
13. Do the uncertainty relations taken together imply that there are restrictions on the simultaneous measurement of position and time?
14. An electron microscope operates by the reflection of electrons, rather than by the reflection of light, from an object. Does the use of particles such as electrons eliminate the problems associated with diffraction through the viewing aperture of the microscope?
15. Does the fact that the speed of light is a definite, predictable quantity conflict with the uncertainty relations?
16. What increases the probability of tunneling the most: halving the energy difference between U and E or halving the barrier width?

PROBLEMS

41–1 The Wave Nature of Matter

1. (I) What is the de Broglie wavelength of an electron whose kinetic energy is (a) 5.0 eV? (b) 50 eV? (c) 50,000 eV? (d) 5.0×10^8 eV? (e) What size targets would you need in order to observe diffraction of electrons of each of these wavelengths?
2. (I) What is the kinetic energy of an electron whose de Broglie wavelength is that of visible red light, 600 nm?

3. (I) Ultracold neutrons can have speeds as low as 100 m/s. What is the de Broglie wavelength of such neutrons?

4. (I) What is the de Broglie wavelength of a proton (mass 1.7×10^{-27} kg) with kinetic energy (a) 1.0 MeV; (b) 10 MeV; (c) 300 MeV? Neglect relativistic effects.

5. (I) Consider a crystal with a planar spacing of 0.074 nm. (a) What energies would electrons need for you to be able to observe up to four interference maxima? (b) Repeat the problem for neutrons.

6. (I) The spacing between scattering planes in a crystal is 0.20 nm. What is the scattering angle from such a crystal with electrons of energy 40 eV for which a first maximum is observed?

7. (I) Express the de Broglie wavelength of a particle of mass m and kinetic energy E in terms of m and E. Suppose that the particle moves at a speed close to the speed of light. How is that relationship modified?

8. (I) In a neutron two-slit diffraction experiment, the slits are 100 μm apart. If the third diffraction maximum is located at an angle of 2×10^{-7} rad, what is the kinetic energy of the neutrons?

9. (II) Suppose that you want to carry out diffraction experiments with the protons of Problem 4. What spacing of scatterers would you need in each of the three cases of energy 1.0 MeV, 10 MeV, and 300 MeV?

10. (II) What is the de Broglie wavelength of a neutron whose kinetic energy is equal to the average kinetic energy of a gas of neutrons of temperature $T = 20$ K?

11. (II) Crystals are studied by means of electron and neutron diffraction as well as by X-ray diffraction. Recall that the typical interatomic distance in a crystal is 10^{-8} cm. Estimate the energy an electron must have to be useful for diffraction experiments on crystals. Repeat the exercise for neutrons.

12. (II) Although the working of an electron microscope does not depend on the wave nature of matter, the waves associated with electrons do set a limit on the resolving power of such instruments. (a) If the electrons in an electron microscope have a kinetic energy of 2.5×10^4 eV and the aperture of the microscope is 3.5×10^{-4} m, estimate the smallest angle that can be resolved. (b) How much energy would electrons need so that two objects separated by 5 nm could be resolved? Give your answer in electron-volts.

13. (II) A beam of electrons with energy 1.0 eV approaches a potential barrier with $\langle U \rangle = 2.0$ eV, whose width is 0.10 nm (Fig. 41–20). Estimate the fraction of electrons that tunnel through the barrier.

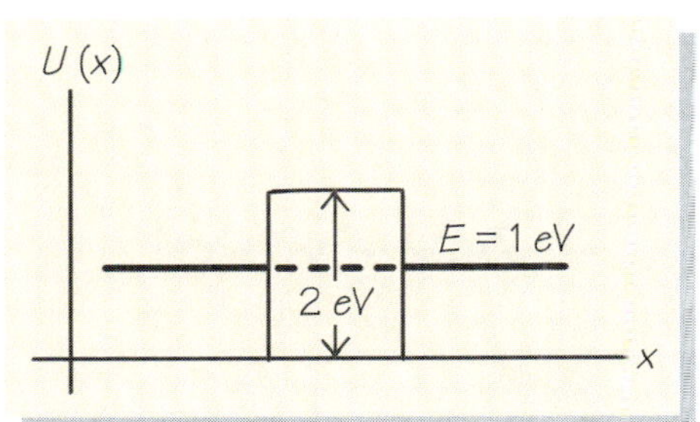

FIGURE 41–20 Problem 13.

14. (II) Suppose that the barrier in Problem 13 is 1.0 nm thick. What fraction of electrons tunnel through the barrier?

15. (II) A truck of mass 3000 kg rolls, with its transmission in neutral, at 2.0 m/s and approaches a smooth bump whose average height is 25 cm. and whose average width is 10 m (Fig. 41–21). Estimate the tunneling factor for the truck from Eq. (41–6).

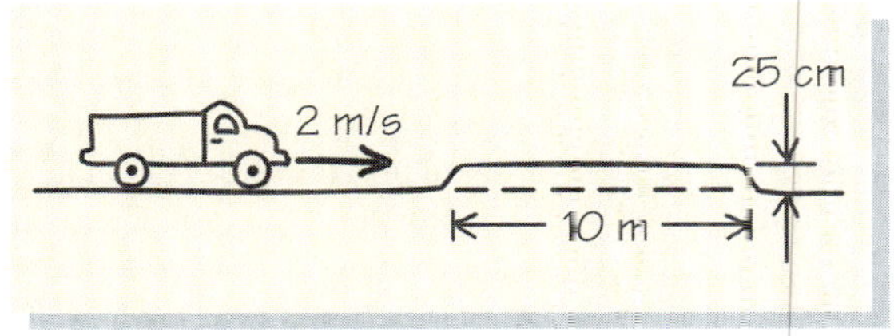

FIGURE 41–21 Problem 15.

16. (II) An electron in a semiconductor device is contained by a potential barrier of width 2.5 nm. The difference between the average potential energy and the energy of the electron is 1.4 eV. By what factor is the probability of tunneling changed if (a) the barrier width is halved? (b) the energy of the electron is increased by 1.2 eV? (c) the barrier width is halved and E is increased by 1.2 eV?

17. (III) The fraction F_n of particles of mass m and energy E that tunnel through a barrier whose potential energy is $U(x_n)$ and (minimal) thickness is Δx_n is given by Eq. (41–6).

$$F_n = e^{(-2/\hbar)\Delta x_n \sqrt{2m[U(x_n)-E]}}.$$

Suppose that a sequence of adjacent barriers at positions x_1, x_2, . . . have potential energy $U(x_1)$, $U(x_2)$, . . . and thickness Δx_1, Δx_2, . . . , respectively (Fig. 41–22). The fraction of particles that penetrate these barriers is given by the product of probabilities $F(x_1)$, $F(x_2)$, . . . that the particle can penetrate the individual barriers independently, namely $F = F(x_1)F(x_2) \ldots$. Show that in the limit that each barrier is infinitesimally thin, we get

$$e^{(-2/\hbar)\int dx \sqrt{2m[U(x_n)-E]}}.$$

for the fraction of particles that penetrate the barrier. Note that Eq. (41–6) is an approximation to this result.

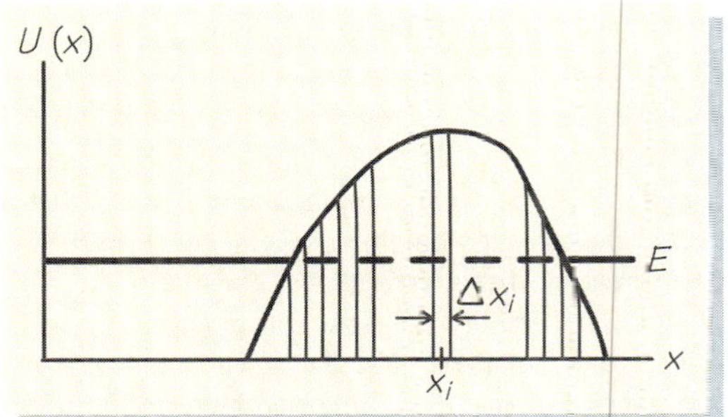

FIGURE 41–22 Problem 17

41–2 *The Heisenberg Uncertainty Relations*

18. (I) An electron is localized inside a cubic region of sides 0.10 nm. What is the uncertainity in its kinetic energy?

19. You are given a detector that can time the arrival of a particle to within an accuracy of 10^{-15} s. What is the accuracy of the energy determination possible with such a detector?

20. (I) The speed of an electron emitted from an atom is measured to a precision of ± 2.0 cm/s. What is the smallest uncertainty possible in the electron's position?

21. (I) Estimate the lowest energy of a neutron confined in a one-dimensional box the size of a uranium nucleus (diameter 1.5×10^{-14} m).

22. (I) One of the consequences of the uncertainty principle is that a measurement of the momentum of an atomic electron that is too precise will force such a great uncertainty in the electron's position that it is no longer sure that the electron remains in the atom. In other words, the measurement has knocked the electron out of the atom. Estimate that precision. Compare your estimate of Δp with the typical momentum of an atomic electron, which corresponds to a kinetic energy on the order of 27 eV.

23. (I) A proton in a tin nucleus is known to lie within a sphere whose diameter is about 1.2×10^{-14} m. What are the uncertainties in the momentum and kinetic energy of the proton?

24. (I) One measure of the uncertainty in the position of a particle in a simple harmonic oscillator is to set that uncertainty equal to the amplitude of the motion. In turn, we find the amplitude by writing the energy in terms of the classical amplitude $E = m^2\omega^2A^2/2$ and setting E equal to $n\hbar\omega$. Use the amplitude you find to *estimate* the uncertainty in the momentum.

25. (II) Monochromatic light of wavelength 720 nm passes through a fast shutter, which stays open for 1.0×10^{-9} s. What will the wavelength spread of the beam be after the light emerges through the shutter?

26. (II) The uncertainty in momentum of an electron with a kinetic energy of approximately 3 keV is 2 percent. What is the minimum uncertainty in its position?

27. (II) A completely free electron in empty space is measured to have a position within a sphere of radius $R = 1.0 \times 10^{-14}$ m, typical of an atomic nucleus. Within what radius can you say with assurance that the electron will be found after 1.0 s? Repeat the problem for an electron initially measured to lie within a sphere of radius $R = 1.0 \times 10^{-10}$ m, the radius of an atom.

28. (II) A radar gun measures the speed of a projectile, mass 7 g, to be 380 mi/h, to an accuracy of 1.5 percent. How well can the position of the projectile be determined in principle? Such accuracy cannot be achieved in practice.

29. (II) A wide beam of electrons of momentum p impinges on a slit of width a. Classically, the width of the beam of electrons passing through the slit and impinging on a screen that is a distance D from the slit would be a (Fig. 41–23). Even if we continue to treat electrons as particles, quantum effects operating through the uncertainty principle will modify this result. Using the uncertainty relations, find the spread of the beam on the screen. For what value of a will the width of the beam be a minimum?

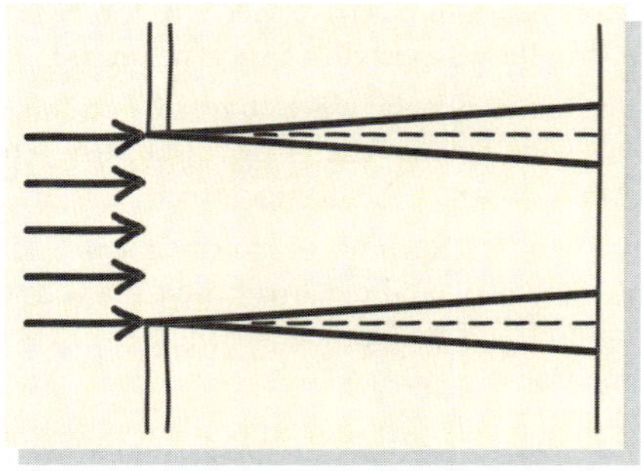

FIGURE 41–23 Problem 29.

30. (II) Consider a particle of mass m with potential energy of the form $U = -U_0(x/a)$ for $x < 0$ and $U = +U_0(x/a)$ for $x > 0$ (Fig. 41–24). Use the uncertainty relations to estimate the lowest energy the particle can have.

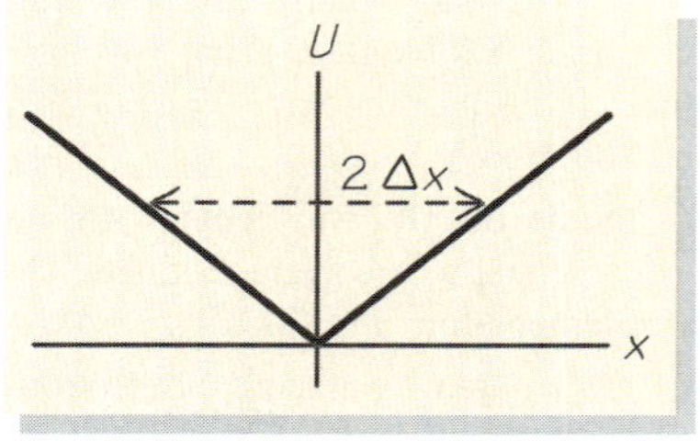

FIGURE 41–24 Problem 30.

31. (II) A neutron, of mass $m = 1.6 \times 10^{-27}$ kg, is localized inside a carbon nucleus, of radius 6.0 fm (1 fm $\equiv 10^{-15}$ m). Use the uncertainty relations to calculate a minimum (negative) potential energy for neutrons inside heavy nuclei.

32. (II) A hydrogen molecule can be modeled as a dumbbell in which the two masses—the hydrogen atoms—are connected by a spring whose unextended length is 0.074 nm (Fig. 41–25). The hydrogen molecule can vibrate along its axis with frequency 1.32×10^{14} Hz. (a) Calculate the minimum energy of this vibration. (b) A deuterium molecule D_2 differs from the picture we describe only in the masses of the atoms, which are twice that of hydrogen. Estimate the vibration frequency and zero point energy of the deuterium molecule.

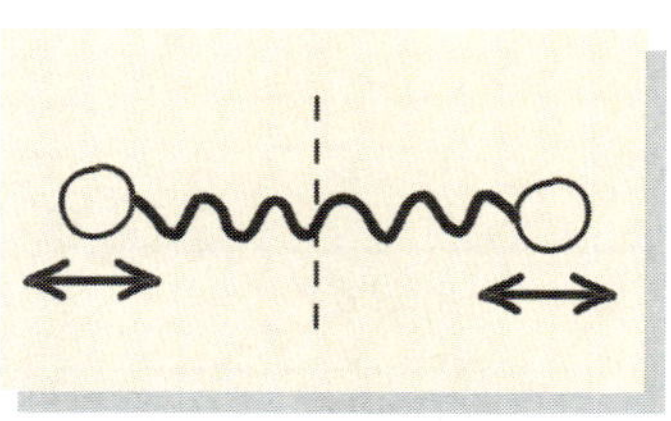

FIGURE 41–25 Problem 32.

41–3 *The Particle Nature of Radiation*

33. (I) For what kinetic energy is the de Broglie wavelength of an electron equal to its Compton wavelength? Express your answer in units of m_ec^2 in doing the calculation, and then use $m_ec^2 = 0.5$ MeV.

34. (I) What are the energy and momentum of a photon in He–Ne laser light of wavelength 632 nm?

35. (I) Find the energy of a photon for each of the following cases: (a) a microwave of wavelength 1.5 cm; (b) red light of wavelength 660 nm; (c) a radio wave of frequency 96 MHz; (d) an X-ray of wavelength 0.17 nm.

36. (I) Use the fact that the human eye can detect as few as 25 photons per second of red light to *estimate* the minimum power needed to trigger the optic nerve.

37. (I) Light of frequency 0.85×10^{15} Hz falls on a metal surface. If the maximum kinetic energy of the photoelectrons is 1.7 eV, what is the work function of the metal?

38. (I) The work function of sodium is 2.75 eV. What is the energy of electrons emitted in the photoelectric effect, if any, when light of wavelength 748 nm bombards sodium?

39. (I) Calculate the maximum wavelength for the initiation of a photoelectric current in the following elements: aluminum (work function $W = 4.28$ eV), cesium (2.14 eV), nickel (5.15 eV) and lead (4.25 eV).

40. (I) A metal has a work function of 1.6 eV. What is the maximum kinetic energy of a photoelectron if radiation of 450 nm wavelength falls on the surface?

41. (II) Estimate the energy of a photon of each of the following radiation types: (a) visible light; (b) X-rays; (c) microwaves; (d) television signals; (e) AM radio.

42. (II) Calculate the number of photons from an FM radio station broadcasting at a frequency of 10^8 Hz that is required to equal the energy contained in one photon of visible light at a wavelength of 550 nm.

43. (II) The power generated by the Sun is 4×10^{26} W. Assuming that it is emitted entirely at an average wavelength of 500 nm, calculate the number of photons emitted per second.

44. (II) The energy density of electromagnetic radiation in some region of space is 10^{-8} J/m^3. Assume that the radiation has a wavelength in the middle of the visible range, 550 nm. What is the photon density?

45. (II) Use the fact that the human eye can pick up as few as 5 photons/s in the visible range to estimate the intensity of the dimmest star that can be detected by a night-adapted eye. What is the ratio of this intensity to the intensity of noon sunlight, some 1400 W/m^2? This large intensity range means that the eye is indeed a very adaptable instrument.

46. (II) Show that the total energy density in a cavity filled with blackbody radiation at temperature T in degrees kelvin—that is, of energy density $u(f, T)$ given by the Planck radiation law, Eq. (17–14)—is $U(T) = aT^4$. This result is the *Stefan–Boltzmann law*. Calculate the value of the constant a, given the integral

$$\int_0^\infty \frac{x^3}{e^x - 1}\,dx = \frac{\pi^4}{15}.$$

47. (II) Use the Stefan–Boltzmann law [see Eq. (17–14)] to calculate the temperature of the Sun, given that the Sun has a radius of 0.7×10^9 m, it is 1.5×10^{11} m from Earth, and the rate at which the total solar radiation that reaches Earth is 1.36×10^3 J/m · s.

48. (II) The maximum energy of photoelectrons from aluminum is 2.3 eV for radiation of wavelength 200 nm and 0.90 eV for radiation of 216 nm. Use these data to calculate Planck's constant and the work function of aluminum.

49. (II) The threshold wavelength for the photoelectric effect in tungsten is 270 nm. Calculate the work function of tungsten, and calculate the maximum kinetic energy that a photoelectron can have when radiation of 120 nm falls on tungsten.

50. (II) A photon of energy 660×10^3 eV (660 keV) collides with an electron at rest. The photon is scattered through 70° (Fig. 41–26). What is its energy after the collision? What is the kinetic energy of the electron after the collision?

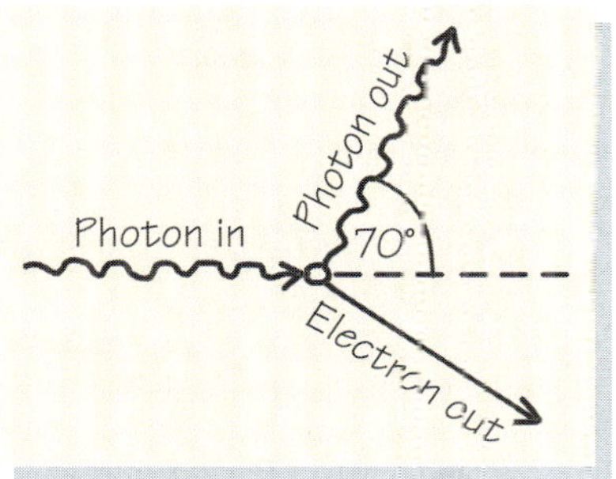

FIGURE 41–26 Problem 50.

51. (II) The wavelength of the incoming X-rays in a Compton scattering experiment is 7.078×10^{-2} nm, and the wavelength of the outgoing X-rays is 7.314×10^{-2} nm. At what angle was the scattered radiation measured?

52. (II) Consider a case of Compton scattering in which a photon collides with a free electron and scatters backward while it gives up half its energy to the electron (Fig. 41–27). (a) What are the frequency and energy of the incident photon? (b) What is the electron's velocity after the collision?

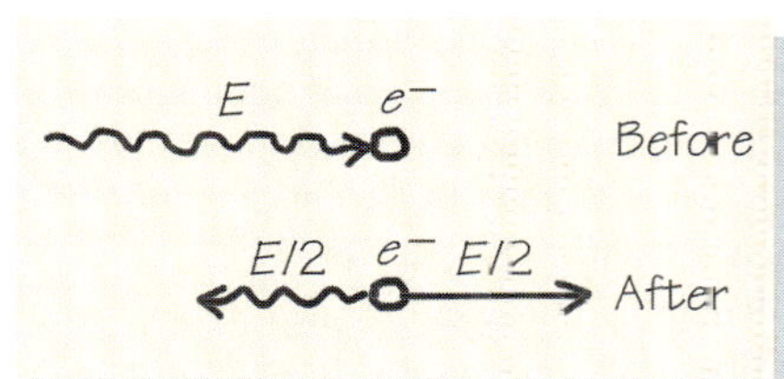

FIGURE 41–27 Problem 52.

53. (III) Consider a cavity that contains blackbody radiation at 6000°C. Calculate the energy density for radiation in the wavelength range 690 nm to 710 nm, and compare it with the energy density for radiation in the range 440 nm to 460 nm (Fig. 41–28). [*Hint*: To calculate energy density as a function of wavelength, use $u(f, T)\,df = u(f, T)\,d\lambda\,(df/d\lambda)$ and calculate the factor $(df/d\lambda)$. You must substitute $\lambda f = c$ in $u(f, T)$.]

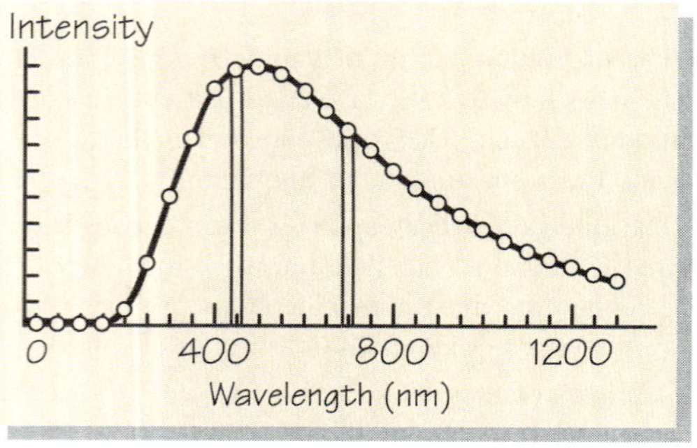

FIGURE 41–28 Problem 53.

54. (III) The Sun's radiation peaks at a wavelength of 500 nm. How much less is the radiation intensity at 400 nm and at 700 nm? Use the results of Problem 53.

55. (III) A copper crystal emits 8.3×10^{10} photoelectrons/m$^2 \cdot$ s. Given that the atomic weight of copper is $A = 64$ and the mass density of the crystal is 8900 kg/m^3, and assuming that the photoelectrons are supplied by the 5 layers of copper atoms nearest the surface of the crystal, how many atoms, on average, furnish one photoelectrons/s? Assume that the atoms form a cubic lattice.

41–4 Quantum Mechanics and Probability

56. (I) The number of atoms in an excited state whose mean lifetime for single-photon decay is τ is given by $N(t) = N(0)e^{-t/\tau}$. What is the number of photons emitted per second?

57. (I) A beam of light is sent along the $+z$-axis through a polarizer so that it is polarized in the direction of the $+x$-axis. This beam impringes on a second polarizer that makes a 55° angle with the x-axis. What is the probability that a photon will get through the second polarizer?

58. (II) The *half-life* of a set of radioactive nuclei is the time in which half of the nuclei decay. Express the half-life in terms of the lifetime, τ, that appears in Eq. (41–23).

59. (II) The half-life of ^{14}C is about 5730 yr (see Problem 58). Organisms accumulate this isotope from the atmosphere while they live but cease doing so upon dying. The skeleton of a mammoth is found to have a concentration of ^{14}C that is 15 percent of the atmospheric value (Fig. 41–29). When did the mammoth live? Assume that the concentration of atmospheric ^{14}C does not change.

60. (II) Volcanic eruptions can be dated by analyzing the potassium and argon contents of rocks from the eruption. Most rocks contain potassium (K), 0.012 percent of which is ^{40}K, a radioactive isotope that decays to ^{40}Ar (argon) with a half-life of 1.3×10^9 years. Since argon is an inert gas, it is likely that all of the argon in the rock originates from the decay of ^{40}K. Suppose that a test of a rock shows that the ratio of the number of non-decayed ^{40}K and ^{40}Ar isotopes is found to be 10:1. When did the eruption happen?

61. (II) Equation (41–23) may be written in the form $dN = -(N/\tau)dt$. Interpret this equation, including the sign. Does your interpretation support the assertion that the decay of any one radioactive nucleus is unaffected by the presence of others?

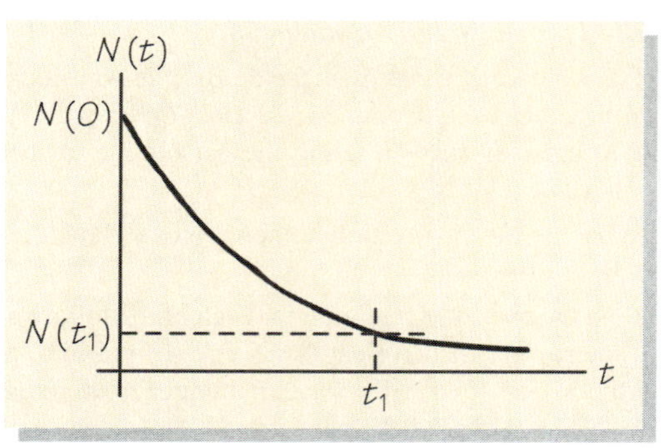

FIGURE 41–29 Problem 59.

62. (II) A sample of radioactive material undergoing 3.7×10^{10} disintegrations/s is said to have an activity of 1 *curie* (Ci), where the activity is the rate at which decays occur. The activity of 1 g of Ra (radium) is 1 Ci. Given that $A = 226$ for radium, estimate the lifetime of ^{226}Ra.

63. (II) What is the activity of 1 g of ^{232}Th (thorium), whose lifetime is 1.4×10^{10} yr? (See Problem 62.)

General Problems

64. (II) When high resolution is needed in biological microscopy, electron microscopes rather than light microscopes are used. Compare the resolution limit of an electron microscope that uses electrons of energy 30 keV with that of a light microscope.

65. (II) Find the average de Broglie wavelength of a gas of nitrogen molecules (each of mass 4.7×10^{-26} kg) at a temperature of 300 K. Assume that the gas is ideal and is at 1 atm of pressure. Calculate the interatomic spacing. What would the temperature have to be for the same amount of nitrogen gas in the same volume to have a de Broglie wavelength equal to the interatomic spacing?

66. (II) *Estimate* the rate at which photons in the visible range are emitted by a flashlight. Assume that 50 percent of the energy is emitted in the visible range and that the average wavelength in this range is 550 nm.

67. (II) It is consequence of the Planck radiation law that the energy density expressed as a function of wavelength and temperature, $U(\lambda, T)$, peaks at $\lambda = \lambda_{max}$, where $\lambda_{max} = (2.9 \times 10^{-3}\ \text{m} \cdot \text{K})/T$ and T is the absolute temperature (see Problem 53). (a) Use this result to calculate λ_{max} for the radiation emitted by the Sun, whose surface temperature is approximately 6000 K, and which, to a good approximation, radiates as a blackbody (Fig. 41–30). (b) Use this result to estimate how hot "red-hot" is.

68. (II) Show that, in the simple process of photon absorption by a free electron, without anything else occurring, energy and momentum cannot be conserved simultaneously.

69. (III) A tungsten block contains a microcrack of width 1 nm. What is the probability that an electron will tunnel from one side of the crack to the other, if the work function of tungsten is 4.6 eV?

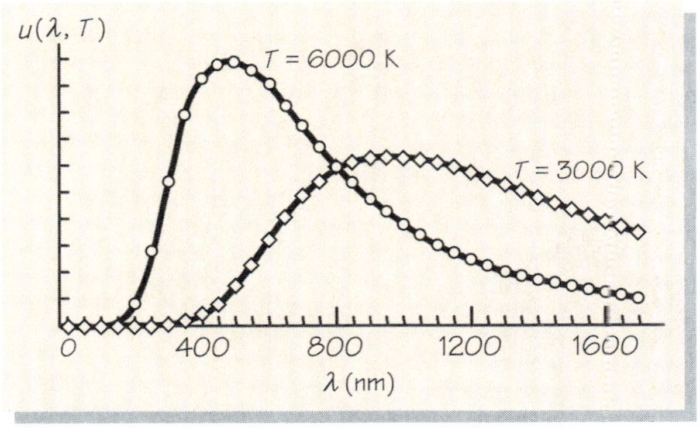

FIGURE 41–30 Problem 67.

70. (III) The abundances of ^{238}U and ^{235}U are assumed to have been equal when Earth formed. The mean life of ^{238}U is 0.6×10^{10} yr, and that of ^{235}U is 1.0×10^{9}. The present ratio of their abundances is $^{238}\mathrm{U}/^{235}\mathrm{U} = 140$. From these data and the assumption, estimate Earth's age.

71. (III) The de Broglie wavelength of a free electron of energy E may be written in the form $\lambda_0 = h/p = h/\sqrt{2m_e E}$. When the electron enters a region that has an electric potential $V(x)$, the electron acquires a potential energy $U(x) = -eV(x)$, and $E = (p^2/2m) + U(x)$. Write an expression for the de Broglie wavelength of the electron in this region, and show that the presence of the potential $V(x)$ means that the electron behaves as though it were moving in a region of refractive index $n(x) = \sqrt{[E - U(x)]/E}$.

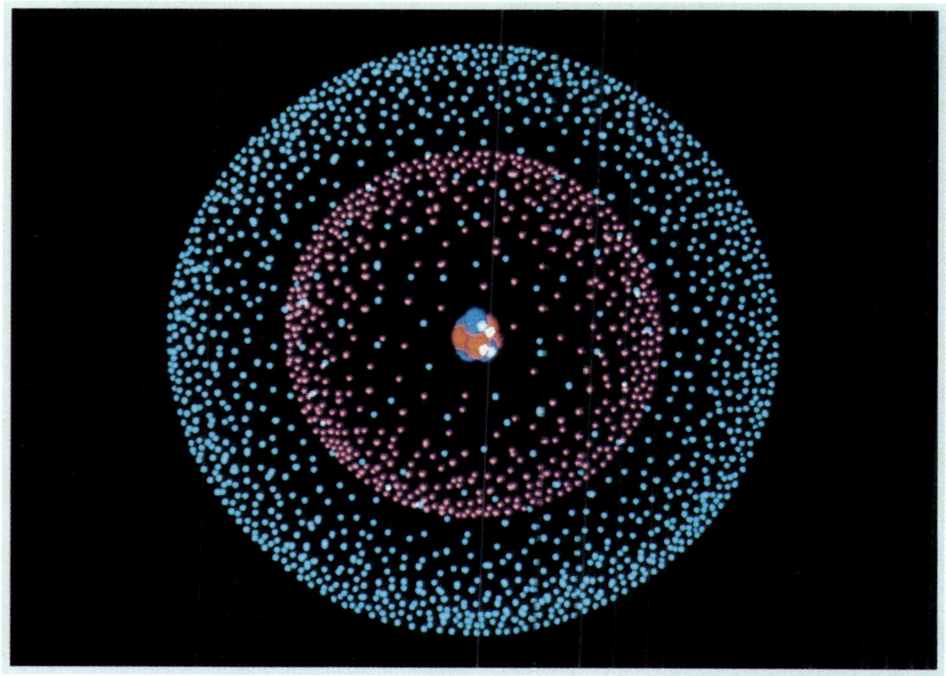

CHAPTER

42

This computer-generated graphic is a representation of a beryllium atom. The electrons "orbit" in the purple and blue regions. The size of the nucleus of the center is greatly exaggerated.

Quantization of Angular Momentum and of Energy Values

In classical physics, a particle or system of particles can have a continuous range of energies—whether it is free or confined in its movement by forces. Experiments have shown, however, that energy can take on only discrete values, a result that is most striking on an atomic scale. This fact can be explained in terms of quantum mechanics. The discreteness (or quantization) of energy values is closely associated with the quantization of angular momentum. The consequences of the quantizations of energy and angular momentum are important. When we couple the quantization principles with the so-called exclusion principle, we can explain many otherwise inexplicable features of microscopic systems, including the structure and stability of atoms and molecules. These ideas have important consequences even on a macroscopic level, where they can, for example, explain the band structure of metals.

42–1 ENERGY AND ANGULAR MOMENTUM ARE QUANTIZED

The energies and angular momenta of systems are quantized.

We have stressed throughout this book that matter consists of atoms. In a classical picture, an atom would be made of negatively charged electrons orbiting a positively charged nucleus (the nucleus approximately 10^5 times smaller than the atom), like a

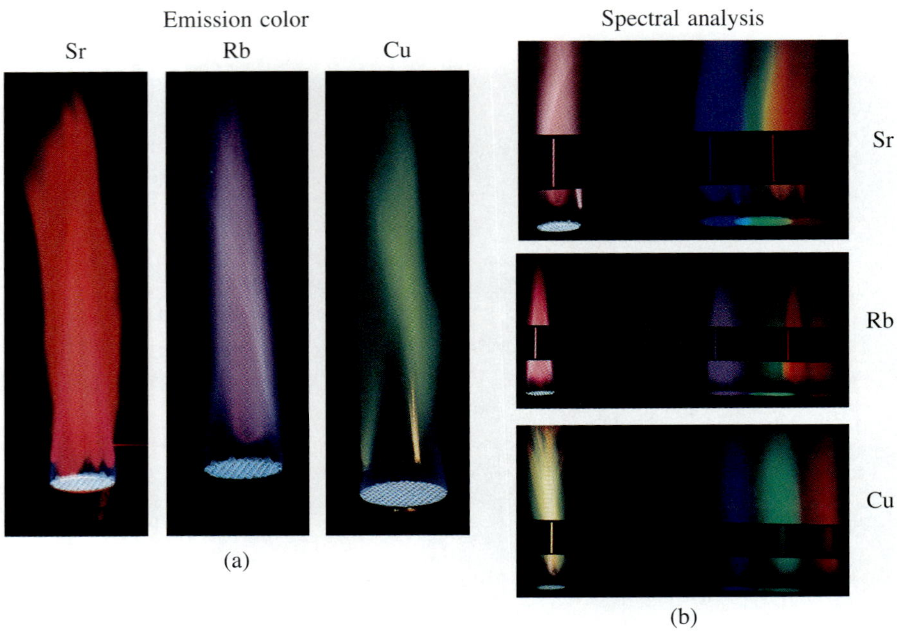

FIGURE 42–1 (a) The color emitted by atoms after they have been excited by heat is characteristic of the particular element they comprise: (i) strontium, (ii) rubidium, and (iii) copper. (b) When the color of each element in part (a) is analyzed spectrally, it is seen to contain discrete frequencies (lines).

miniature solar system; elliptical or circular electron orbits of any size would be allowed. Such variability in the forms of atoms is in sharp conflict with experiment. One hydrogen atom (^{1}H) is, *literally*, indistinguishable from another. We generally see only one form of hydrogen, one form of helium, one form of iron. Quantum mechanics explains this by a remarkable prediction: In contrast to classical systems, *quantum systems can have only certain—quantized—values of energy and of angular momentum.* Possible energy values are always separated by gaps. The allowed energy values have come to be called *energy levels*.

Virtually all the atoms in a bottle of helium at room temperature will be in their lowest energy level, the *ground state*. Ordinary thermal collisions between these helium atoms cannot supply enough energy to change very many helium atoms from the ground state to the next allowed state (an *excited state*). At sufficiently high temperatures, some fraction of the atoms may be "kicked" up into the next, higher state; these atoms are said to be *excited*. Excited atoms can "jump" back to the lowest energy level and emit light of *discrete frequencies*. The light carries an energy E that is the difference in energies between the excited state and the ground state. In this way, energy is conserved. The emitted light consists of photons with a frequency f determined by the frequency–energy relation of photons, $E = hf$. Thus atomic energy levels can be studied by looking at the discrete radiation frequencies emitted by atoms at high temperatures (Fig. 42–1).

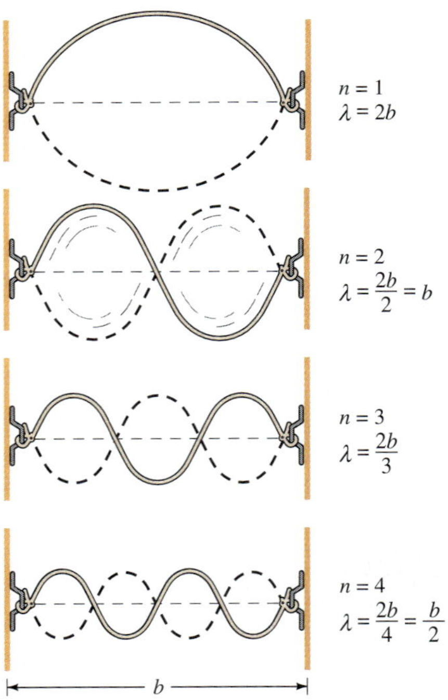

FIGURE 42–2 The wavelengths of standing waves on a string fixed at both ends are given by the length of the string divided by half-integers.

The Wave Nature of Matter and Energy Quantization

The wave properties of matter explain the energy quantization just described. Let's consider a particle confined to a one-dimensional box of length b. If the particle inside the box behaves like a standing wave on a string of length b that is fixed at its ends, then the wavelengths are constrained by the condition

$$\lambda = \frac{2b}{n}$$

[Eq. (14–8)], where $n = 1, 2, 3, \ldots$ (Fig. 42–2). The wavelength is related to the momentum, p, by the de Broglie relation, $\lambda = h/p$ (Eq. 41–1), so the particle's momentum has the form

$$p = \frac{nh}{2b}.$$

The energy of the particle in the box takes only the discrete values

$$E = \frac{n^2h^2}{8mb^2}. \qquad (42\text{–}1)$$

This expression gives us both the ground-state energy ($n = 1$) and the gaps between the allowed energies. Our more precise calculation improves the estimate we made in Chapter 41 with the help of the position–momentum uncertainty relation [Eq. (41–7)].

The Bohr Model of Hydrogen

The wavelike properties of matter allow us to understand the orbits of electrons in atoms. We start with a classical description of the hydrogen atom, consisting of one electron, of mass m_e and charge $-e$, and a nucleus that usually consists of a single proton, of mass $m_p >> m_e$ and charge $+e$. The classical configuration is that of the electron in circular (or elliptical) orbits with the proton at the center (or a focus), in analogy to planetary orbits. For a circular orbit of radius r, Newton's second law, $F = ma$, becomes

$$\frac{e^2}{4\pi\epsilon_0 r^2} = m_e a = \frac{m_e v^2}{r}. \qquad (42\text{–}2)$$

The left side of this equation is the Coulomb force. The total energy is

$$E = K + U = \frac{p^2}{2m_e} - \frac{e^2}{4\pi\epsilon_0 r}, \qquad (42\text{–}3)$$

where zero potential energy has been chosen to be at $r = \infty$. Newton's second law, Eq. (42–2), implies that

$$\frac{p^2}{2m_e} = \frac{1}{2}m_e v^2 = \frac{e^2}{8\pi\epsilon_0 r}.$$

When this result is inserted into Eq. (42–3), we find

$$E = \frac{e^2}{8\pi\epsilon_0 r} - \frac{e^2}{4\pi\epsilon_0 r} = -\frac{e^2}{8\pi\epsilon_0 r}. \qquad (42\text{–}4)$$

The energy is negative, as we would expect.

It will be useful to find the angular momentum **L** in the circular orbit. It has magnitude $L = m_e vr$. From Eq. (42–2), $v = e/\sqrt{4\pi\epsilon_0 m_e r}$, so $L = m_e vr = m_e(e/\sqrt{4\pi\epsilon_0 m_e r})r = e\sqrt{m_e r/4\pi\epsilon_0}$. We can solve for r to find the radius of the orbit for a given value of angular momentum:

$$r = \frac{L^2}{m_e e^2/4\pi\epsilon_0}. \qquad (42\text{–}5)$$

In the classical planetary model, the energy, the orbital radius, and the angular momentum can take on a continuum of values. But this model has a fatal flaw. A charge moving in a circular orbit has a constant acceleration, and we saw in Chapter 35 that an accelerating charge radiates energy. Consequently, an orbiting electron would steadily lose energy by radiation; as it did so, the radius would decrease [see Eq. (42–4)] until the proton swallowed up the electron. Detailed estimates (see Problem 19) show that this would happen in only 10^{-10} s!

The first treatment of the atom that incorporated an energy quantization condition was provided by Niels Bohr in 1913 (Fig. 42–3). Here are two of the assumptions Bohr made:

FIGURE 42–3 Niels Bohr.

1. The classical circular orbits, with variable energy and radius, are replaced by *stationary states*, so called because the energies of those states are fixed and hence electrons in these states cannot radiate classically. The energies of these stationary states take on only discrete values.

2. The energy values of stationary states are determined for circular orbits by imposing the condition that the angular momentum, L, is quantized in integer units of $\hbar$:

The Bohr quantization condition

$$L = n\hbar, \qquad \text{where } n = 1, 2, 3, \ldots . \tag{42–6}$$

The integer n is the *principal quantum number* for the orbit. Equation (42–6) is called the *Bohr quantization condition.*

To see how the Bohr quantization condition determines the energy values, we substitute L into Eq. (42–5); this determines the radius of a circular orbit in terms of L. We find that

The allowed radii of circular orbits for hydrogen

$$r_n = \frac{n^2\hbar^2}{m_e e^2/4\pi\epsilon_0} = n^2 a_0, \tag{42–7}$$

where the *Bohr radius* a_0 is the radius corresponding to $n = 1$:

$$a_0 = \frac{\hbar^2}{m_e e^2/4\pi\epsilon_0} = 0.53 \times 10^{-10}\ \text{m}. \tag{42–8}$$

The allowed radii are discrete, with an index n corresponding to the quantum number n for L. Once the radii can take on only discrete, fixed values, Eq. (42–4) shows that the energy also has discrete, fixed values. When the allowed values of r from Eq. (42–7) are inserted into Eq. (42–4) for the energy, we obtain the allowed energy values of a hydrogen atom in the Bohr model:

$$E_n = -\frac{e^2}{8\pi\epsilon_0 r_n} = \frac{e^2}{8\pi\epsilon_0}\frac{1}{[(n^2\hbar^2)/(m_e e^2/4\pi\epsilon_0)]}$$

$$= -\frac{m_e}{2n^2}\left(\frac{e^2}{4\pi\epsilon_0\hbar}\right)^2 = -\frac{21.8 \times 10^{-19}\ \text{J}}{n^2} = -\frac{13.6\ \text{eV}}{n^2}. \tag{42–9}$$

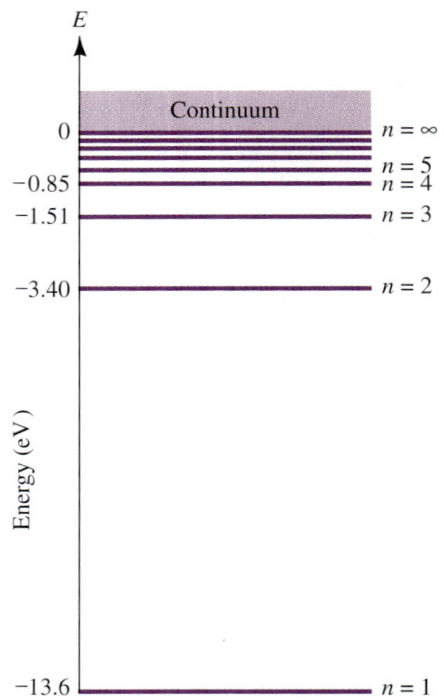

FIGURE 42–4 The energy levels in a hydrogen atom for circular orbits in the Bohr model, obtained from Eq. (42–9). The energy levels (here, not to scale) bunch up as the quantum number n increases. Above the ionization point $E = 0$, the electron and the proton are no longer bound together.

The ground-state energy is $E_1 = -13.6$ eV.

Figure 42–4 shows the energy values predicted by Eq. (42–9). The excited-state energies get closer and closer together as n increases, and the energy approaches zero as $n \to \infty$. Positive energies correspond to the situation in which the electron is not bound to the proton. There the energy can take on continuous values. The minimum energy required to remove an electron in the ground state from the atom is called the *ionization energy*. For hydrogen, this energy is (0 eV) − (−13.6 eV) = 13.6 eV.

We can think of the quantization of angular momentum as a result of a requirement that *the circumference of the orbit accommodates an integral number of de Broglie waves*. This condition is reminiscent of the frequency-fixing conditions for standing waves; it requires that, for an orbital radius labeled by the integer n,

$$\frac{2\pi r_n}{\lambda} = n. \tag{42–10}$$

Because $\lambda = h/p_n$ for that orbit, Eq. (42–10) takes the form

$$\frac{2\pi r_n}{h/p_n} = \frac{r_n p_n}{h/2\pi} = \frac{L}{\hbar} = n,$$

which is identical to Eq. (42–6).

Testing the Bohr Model

The Bohr model for hydrogen can be confirmed experimentally by the observation of the discrete light frequencies that hydrogen atoms emit after they have been excited. These frequencies are those of photons *emitted* when an electron "jumps" from one orbit to another one of lower energy (Fig. 42–5). Energy conservation determines the energies and hence the frequencies of these photons:

Atomic electrons can make transitions (jumps) from one allowed level with an initial energy E_i to another allowed level with a final energy E_f. When $E_i > E_f$, energy is released. The released energy can manifest itself in the appearance of a photon that carries off the excess energy $E_i - E_f$. Because a photon of energy E has frequency f given by $E = hf$, the frequency of the emitted photon† is determined by

The frequencies of light emitted by atoms

$$hf = E_i - E_f. \tag{42–11}$$

Conversely, an electron starting from some initial allowed energy can *absorb* a photon of the correct frequency and "jump" to a higher-energy (excited) state (Fig. 42–5). Only photons with frequency given by $hf = E_f - E_i$ are absorbed.

EXAMPLE 42–1 A certain laser emits light with a wavelength of 3391 nm. What is the energy difference, in electron-volts, between the two energy levels involved in producing this light?

Solution: The energy difference ΔE between the two energy levels is related to the wavelength of the emitted radiation, λ, by

$$\Delta E = hf = hc/\lambda.$$

Therefore

$$\Delta E = \frac{hc}{\lambda} = \frac{(6.63 \times 10^{-34}\ \text{J}\cdot\text{s})(3.0 \times 10^{8}\ \text{m/s})}{(3391 \times 10^{-9}\ \text{m})(1.6 \times 10^{-19}\ \text{J/eV})} = 0.37\ \text{eV}.$$

For hydrogen, the possible energies E_i and E_f are given by Eq. (42–9). We may rewrite Eq. (42–11) in terms of the wavelength, λ, rather than the frequency by using the relation $f = c/\lambda$. Then the wavelengths of the photons emitted when an electron jumps down from an excited state are restricted to the values

$$\frac{1}{\lambda} = \frac{E_i - E_f}{hc} = R_\infty\left(\frac{1}{n_1^2} - \frac{1}{n_2^2}\right), \tag{42–12}$$

where n_1 and n_2 are the quantum numbers of the initial and final energies, respectively, and $R_\infty \equiv (m_e/2hc)\,(e^2/4\pi\epsilon_0\hbar)^2 = 1.0974 \times 10^7\ \text{m}^{-1}$ is the *Rydberg constant*, named for Johannes Rydberg. The predicted wavelengths are in good agreement with the measured wavelengths of the spectral lines in hydrogen (Fig. 42–6). In fact, in 1890, Rydberg had already made a purely empirical fit to the frequencies of the emission spectrum of hydrogen with the formula in Eq. (42–12). Frequencies corresponding to many values of n_2 and n_1 ($n_1 < n_2$) could be closely approximated by this formula, and new frequencies were correctly predicted on its basis. Bohr was on the right track.

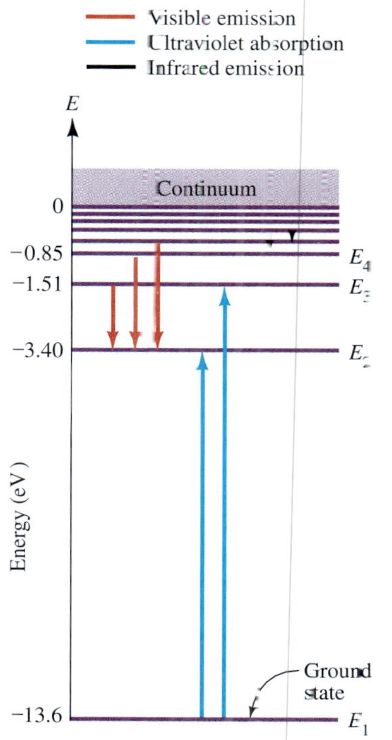

FIGURE 42–5 An atomic electron may jump down from one level to another with the emission of a photon or jump up with the absorption of a photon. The frequency of the photon is determined by the energy change.

EXAMPLE 42–2 Find the magnitude of the energy difference between the lowest energy of a hydrogen atom ($n = 1$) and the first excited state ($n = 2$). Estimate the temperature at which a substantial fraction of a gas of hydrogen atoms would occupy the first excited state.

Solution: To find the energy difference between the $n = 1$ and $n = 2$ levels of hydrogen, we must calculate only $E_2 - E_1$, where the energies are given by Eq. (42–9):

$$E_2 - E_1 = E_1\left[\frac{1}{(2)^2} - \frac{1}{(1)^2}\right] = (-13.6\ \text{eV})\left(\frac{1}{4} - 1\right) = 10.2\ \text{eV}.$$

To estimate the temperature T at which a substantial fraction of hydrogen atoms would occupy the first excited state, we find how much kinetic energy is needed per atom to arrive at 10.2 eV $= (1.6 \times 10^{-19}\ \text{J/eV})(10.2\ \text{eV}) \simeq 1.6 \times 10^{-18}$ J. If we set

†Energy conservation permits emission of two or more photons: the probability of this is small, however.

FIGURE 42–6 The emission spectrum of the hydrogen atom.

$\frac{3}{2}kT$ equal to this number, as the equipartition principle suggests, then $T \simeq \frac{2}{3}(1.6 \times 10^{-18}\ \text{J})/(1.38 \times 10^{-23}\ \text{J/K}) = 0.8 \times 10^5\ \text{K}$! This estimate dramatically confirms our earlier assertion that, under normal circumstances, matter will be in its ground state. We can nevertheless study the spectra of elements in the laboratory because a small fraction of atoms is excited, even at low temperatures.

The principles on which the Bohr model is based can be applied to any classical system. In Section 42–3 we apply them to the harmonic oscillator (which represents vibrations in molecules) and to rotating molecules. Example 42–3 presents another use of Bohr-model techniques.

EXAMPLE 42–3 Two massive quarks (subnuclear particles) attract each other with a constant force. Think of this situation as a single particle of mass M orbiting under the influence of a constant central force with potential energy $U = U_0(r/r_0)$. Here, r is the distance between the quarks, U_0 is a constant with the dimension of energy, and r_0 is a constant with the dimension of length. Find the quantized energies of circular orbits.

Solution: To use the Bohr rules, we proceed as we did for the hydrogen atom. We first express the energy of circular orbits in terms of angular momentum; we then quantize the angular momentum. The energy is

$$E = Mv^2/2 + U_0(r/r_0).$$

The force corresponding to our potential energy has magnitude $F = U_0/r_0$, so, for circular orbits, the dynamical equation $F = ma$ is

$$\frac{U_0}{r_0} = \frac{Mv^2}{r}.$$

The quantization of angular momentum takes the form $Mvr = n\hbar$, or $r = n\hbar/Mv$. Substituting into the dynamical equation, we get

$$\frac{U_0}{r_0} = \frac{Mv^2}{n\hbar/Mv} = \frac{M^2v^3}{n\hbar}.$$

Thus $v = (n\hbar U_0/M^2 r_0)^{1/3}$ and $r = n\hbar/Mv = (n\hbar)^{2/3}\,(r_0/MU_0)^{1/3}$. The possible energies are then

$$E_n = \frac{1}{2}M\left(\frac{n\hbar U_0}{M^2 r_0}\right)^{2/3} + \frac{U_0}{r_0}(n\hbar)^{2/3}(r_0/MU_0)^{1/3} = \frac{3}{2}\left(\frac{U_0^2\hbar^2}{Mr_0^2}\right)^{1/3} n^{2/3}.$$

Despite the successes of Bohr's quantization rules, which work well for single-electron atoms, the rules cannot be successfully applied to multi-electron atoms. Bohr's rules were rather artificially grafted onto classical laws, and there was no understanding of when an electron would decide to "jump" from one orbit to another, nor of where the electron was during its jumps. It was clear that what Bohr had done was provisional.

42–2 THE TRUE SPECTRUM OF HYDROGEN

Werner Heisenberg (in 1925) and Erwin Schrödinger (in 1926) generalized their extensive studies of the "old" quantum theory to make the leap to the correct formulation of quantum mechanics. The details are beyond the scope of this course; we simply quote some of the results of quantum mechanics without attempting to derive them. The complete hydrogen spectrum is one of the topics that must be treated in this way.

Heisenberg used the **Bohr correspondence principle** as a guide for his construction of a theory of quantum phenomena. This principle, which grew out of work on the original Bohr atom, states that quantum mechanical results should reduce to those that follow from a classical treatment when quantum numbers—the principal atomic quantum number n, for example—are large. Problem 20 illustrates how the correspondence principle can be applied.

A full quantum mechanical treatment applied to the hydrogen atom gives several noteworthy features:

1. The concept of orbits disappears completely. When an electron is bound to a nucleus, it is described by a *wave function* (see Chapter 41). The square of the wave function at a given point is the probability distribution function—in the sense described in Chapter 19—of finding the electron at that point. The electron can exist in one of a number of *states* characterized only by energy, angular momentum, and an orientation in space (the analogue of the tilt of a planetary orbit relative to some axis). These variables are quantized, specified by a set of *quantum numbers* taking on integral values. Once the quantum numbers are specified, the wave function (and hence all the properties of the atom) is determined.

2. The electron's angular momentum in the atom is quantized according to $L = \ell\hbar$, with ℓ taking on the values 0, 1, 2, ... (We shall soon see that there is a further restriction on ℓ.) Moreover, we can speak of that angular momentum only relative to some axis (Fig. 42–7). What we mean by this is the following: The component of angular momentum along the specified axis (which henceforth we shall call the z-axis) can take on only the values $m\hbar$ [as in Eq. (42–6)], with $m = \ell, \ell - 1, \ell - 2, \ldots 1, 0, -1, \ldots -(\ell - 1), -\ell$. In other words, the z-component L_z of the angular momentum vector is itself quantized. The square of the angular momentum is given by $\ell(\ell + 1)\hbar^2$, *not* by $\ell^2\hbar^2$. Figure 42–7 shows the angular momentum and its allowed projections for $\ell = 1$, $\ell = 2$, and $\ell = 3$. This behavior is totally at variance with classical mechanics, although for very large L (so that $\ell\hbar$ is some macroscopic number such as 1 g · cm²/s), the deviations from the classical description are very small.

In 1921, Otto Stern and Walter Gerlach measured the quantization of angular momentum directly by seeing how the atoms reacted in a magnetic field. Figure 42–8 uses a classical analogy to illustrate that the atoms will experience a force if the magnetic field is not uniform. A beam of atoms is passed through a region of thickness D in which there is a magnetic field $\mathbf{B}$ oriented in the z-direction (Fig. 42–9).The atoms are characterized by a magnetic dipole moment $\boldsymbol{\mu}$, and the potential energy of the atoms in the magnetic field is given by Eq. (29–26):

$$U_{\text{mag}} = -\boldsymbol{\mu} \cdot \mathbf{B} = -\mu_z B.$$

If $\mathbf{B}$ varies with z, then there is a force on the atom proportional to its magnetic dipole moment, in the z-direction:

$$F_z = -\frac{dU_{\text{mag}}}{dz} = \mu_z \frac{dB}{dz}. \tag{42–13}$$

Classically, μ_z, which is proportional to the angular momentum component of the atom in the z-direction [see Eq. (32–12)], is not quantized. From the classical point of view, therefore, the atoms would reach the far side of the apparatus distributed in the z-direction with a uniform spread. The maximum deflection angle θ characterizing this spread would be given approximately by

$$\theta \simeq \frac{\Delta p}{p} = \frac{\text{(force)(time in region containing the field)}}{mv}$$

$$= \mu_z \frac{dB}{dz}\frac{1}{mv}\frac{D}{v} = \frac{D}{mv^2}\mu_z\frac{dB}{dz}, \tag{42–14}$$

where $p = mv$ is the momentum of an atom of the beam.

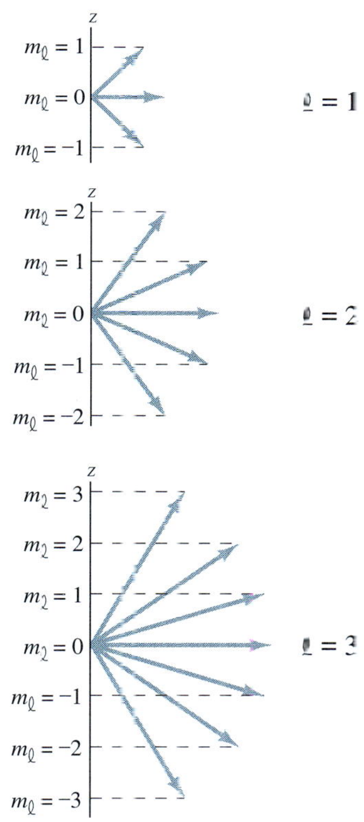

FIGURE 42–7 The directions of angular momentum vectors for $\ell = 1$, 2, and 3 allowed by the vector angular-momentum quantization conditions. The attempt to depict a quantum phenomenon with a classical model of restricted motion is successful only in that it indicates that not all spatial orientations are allowed.

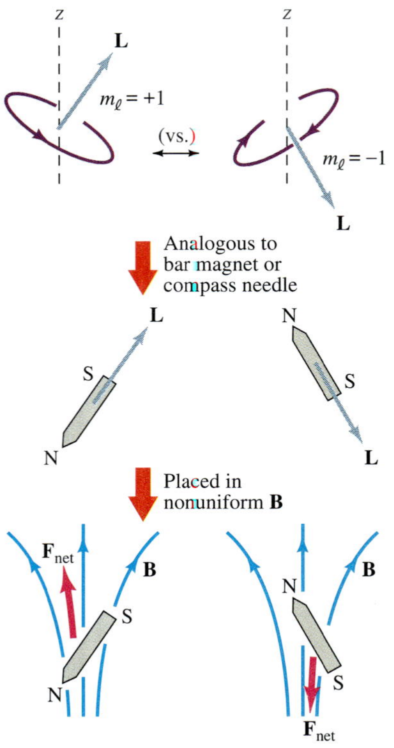

FIGURE 42–8 The atom with oriented angular momentum is analogous to an oriented bar magnet. The bar magnet will experience a net force whose direction depends on the bar magnet's orientation if it is placed in a nonuniform magnetic field.

Rather than seeing a uniform spread, Stern and Gerlach saw that their beams were split into components at specific angles. The number of components varied with the species of atom. For helium, there was only one (undeflected) component; a beam of oxygen atoms split into five components; and a beam of silver atoms split into two components. In all cases, the components were symmetrically distributed about the point of no deflection. These observations imply that the magnetic moments must be quantized, which implies that the angular momentum is quantized. The initial beam of a species of atom with angular momentum quantum number ℓ consists of an equal mixture of atoms in each of the allowed $2\ell + 1$ states that correspond to quantization of its z-component. Each of these $2\ell + 1$ states has a different magnetic dipole moment z-projection, proportional to $-\ell$, $-(\ell - 1)$, ... $(\ell - 1)$, ℓ. Thus the single component of helium corresponds to $\ell = 0$, and the five components of oxygen correspond to $\ell = 2$. We shall soon discuss the case of silver, for which two components are observed.

3. The energy values of the hydrogen atom are quantized according to

$$E_n = -\frac{13.6\ \text{eV}}{(n_r + \ell + 1)^2} \equiv -\frac{13.6\ \text{eV}}{n^2}. \qquad (42\text{–}15)$$

This is close to the prediction of the Bohr model, but in a generalized form. The original integer n of Bohr's model is replaced by $n_r + \ell + 1$, where $n_r = 0, 1, 2, 3, \ldots$ and ℓ is restricted to the integer values $0, 1, 2, 3, \ldots$. Note, however, that $n \geq \ell + 1$; in other words, $\ell \leq n - 1$ for a given n. In addition, a state with a given ℓ really refers to the collection of $2\ell + 1$ different states—all of which have the same angular momentum.

Equation (42–15) leads to a complex of energy levels that are a little more complicated than the naive Bohr model in that each energy level contains several states. Note that the lowest state has $n_r = 0$ and $\ell = 0$ $(n = 1)$, and it is unique. The next level, corresponding to $n = 2$, consists of 1 state for which $n_r = 1$ and $\ell = 0$ and $2\ell + 1 = 3$ states with $n_r = 0$ and $\ell = 1$; that is, there are 4 states for $n = 2$. The next level, corresponding to $n = 3$, consists of 1 state for which $n_r = 2$ and $\ell = 0$, 3 states with $n_r = 1$ and $\ell = 1$, and 5 states with $n_r = 0$ and $\ell = 2$—a total of 9 states. These results generalize in an obvious way: The total number of states labeled by n is n^2.

We have already noted that the conservations of energy and linear momentum continue to hold in quantum mechanics. The same is true of angular momentum conservation in the absence of external torques. This affects the state in which a given electron can end up when it makes a jump with the emission of a photon. Electromagnetic radiation carries angular momentum, and the angular momentum for a photon is to a good approximation $L_{photon} = \hbar$. Thus, the initial and final angular momenta of the electron

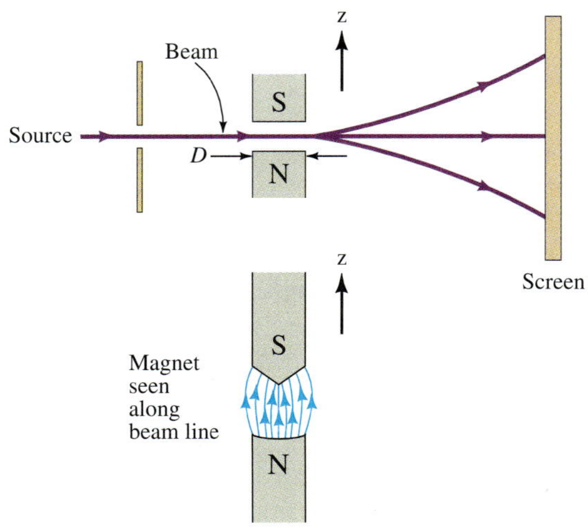

FIGURE 42–9 A Stern–Gerlach experiment. The magnetic field is not uniform. In this case, the beam is split into three components.

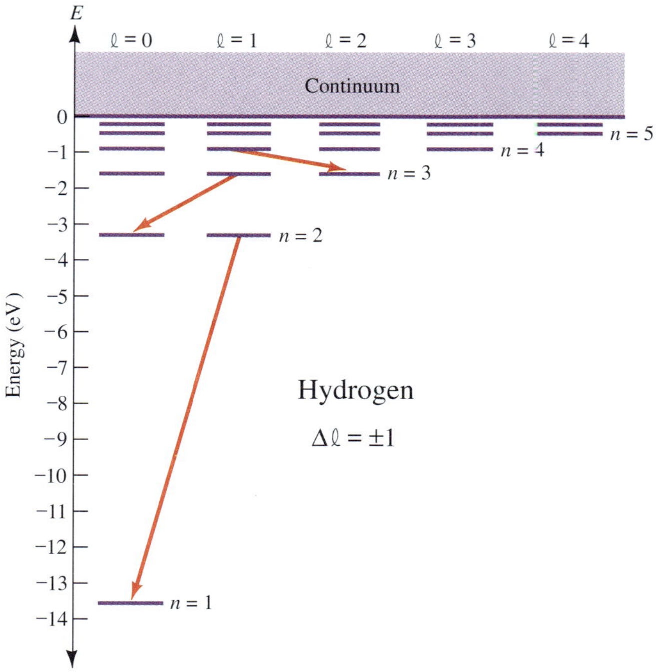

FIGURE 42–10 Some possible transitions of electrons in the quantum mechanical view of a hydrogen atom.

states must differ by one unit of $\hbar$. Some of the possible transitions are sketched in Fig. 42–10. All the results described here are in extremely good agreement with experiment.

EXAMPLE 42–4 The *Balmer series* is a series of spectral lines that correspond to atomic transitions in hydrogen that end with a principal quantum number $n = 2$. Sketch the allowed transitions that lead to this series, and compute the longest and shortest wavelengths in this series.

Solution: Figure 42–11 shows the transitions that end at $n = 2$, $\ell = 0$ (which start at $n = 3, 4, \ldots$ and $\ell = 1$); and also at $n = 2$, $\ell = 1$ (which start at $n = 3, 4, 5, \ldots$ and $\ell = 0, 2$). The frequencies of the transitions are given by the usual relation between the energy change of the emitting atom and the frequency of the emitted light:

$$f_2 = \frac{E_n - E_2}{2\pi\hbar} = -\frac{m_e}{2}\frac{1}{2\pi\hbar}\left(\frac{e^2}{4\pi\epsilon_0\hbar}\right)^2\left(\frac{1}{n^2} - \frac{1}{4}\right) = \frac{m_e}{16\pi\hbar^3}\left(\frac{e^2}{4\pi\epsilon_0}\right)^2\frac{n^2 - 4}{n^2}.$$

Hence

$$\lambda_2 = \frac{c}{f_2} = \frac{16\pi\hbar^3 c}{m_e(e^2/4\pi\epsilon_0)^2}\frac{n^2}{n^2 - 4} = 363 \times 10^{-9}\left(\frac{n^2}{n^2 - 4}\right)\text{m}.$$

The shortest wavelength, corresponding to $n = \infty$, is 363 nm; the longest wavelength, corresponding to $n = 3$, is 653 nm.

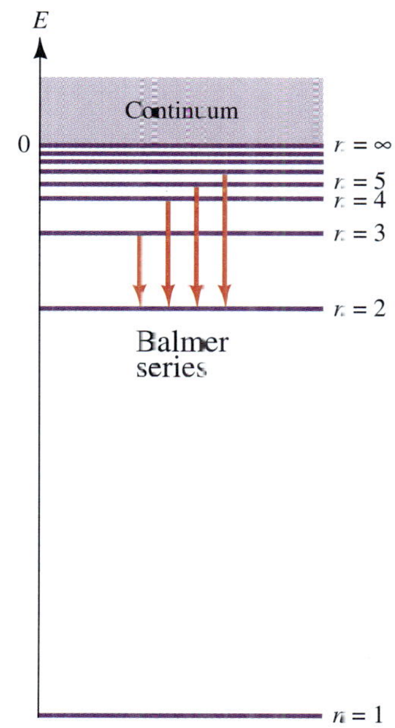

FIGURE 42–11 Example 42–4. This series of transitions is known as the Balmer series.

42–3 THE EXCLUSION PRINCIPLE AND ATOMIC STRUCTURE

The Spin of the Electron

When an atom with a magnetic dipole moment is placed in an external magnetic field, it has a potential energy $U_{mag} = -\boldsymbol{\mu} \cdot \mathbf{B}$ [Eq. (29–26)]. As we already saw in our discussion of the Stern–Gerlach experiment, each one of the $2\ell + 1$ orientations of an atom with angular-momentum quantum number ℓ has a magnetic dipole moment with a different z-component. When such an atom is placed in a magnetic field, the $2\ell + 1$ orientations no longer have the same energy. If hydrogen is placed in an external magnetic field, the energies of the first excited state with $\ell = 1$ are therefore slightly *split*

(Fig. 42–12). The frequencies of the radiation emitted (or absorbed) in a transition to or from one of the three (now-split) levels are accordingly not quite the same, and this is detectable by experiment.

The electron spin

States of a given ℓ that have a common energy in the absence of a magnetic field break up into $(2\ell + 1)$-member *multiplets* with slightly different energies in the presence of a magnetic field. Because $\ell = 0, 1, 2, 3, \ldots$, only odd-valued multiplets were expected. This turned out not to be the case: For some atoms—silver, for example—*doublets* appear; that is, there are two components, as we saw in our discussion of the Stern–Gerlach experiment. For a doublet, $2\ell + 1 = 2$, so $\ell = \frac{1}{2}$ for these atoms, and this was forbidden by the rules of quantum theory as they were understood in the early 1920s. In 1924, Wolfgang Pauli decided that the electron had to be described by one more quantum number, which could take only two values. A year later, George Uhlenbeck and Samuel Goudsmit proposed that the electron has an *intrinsic angular momentum*, or *spin*, $\hbar/2 \equiv s\hbar$. Whereas the angular momentum L discussed thus far is associated with the motion of an electron around a nucleus, the spin is an *internal* property of the electron.

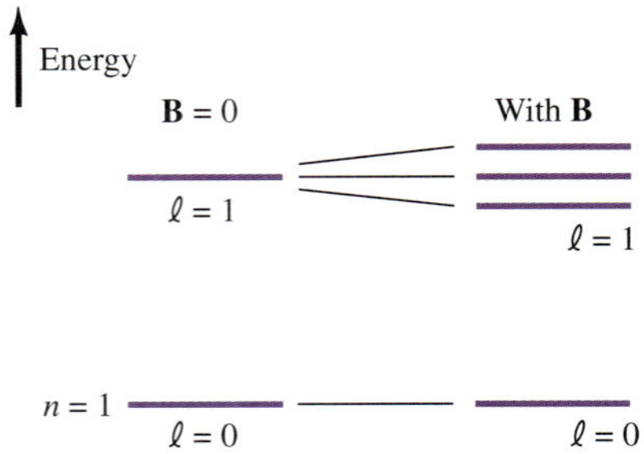

FIGURE 42–12 When a hydrogen atom is subject to an external magnetic field, the three states corresponding to $\ell = 1$, which all have the same energy in no magnetic field, have slightly different energies—the levels are split. The amount of splitting depends on the strength of the magnetic field. Note that the $\ell = 0$ state consists of only one level and is therefore not split.

The fact that electrons have an intrinsic angular momentum $s\hbar$, with $s = \frac{1}{2}$, means that $2s + 1 = 2$, and an electron can appear in two states. We call these states "up" and "down." In the absence of a magnetic field, the energy of an "up" electron is the same as that of a "down" electron in the hydrogen atom. However, when a magnetic field is present, the energies of these two states differ slightly. When the electrons jump from these two states, the frequencies of the photons emitted differ slightly. As a result of electron spin, the number of possible electron states that correspond to a given ℓ doubles from $2\ell + 1$ to $2(2\ell + 1)$. When $\ell = 0$, the number of states with the same energy is 2 (Fig. 42–13). For $\ell = 1$, there are $2 \times 3 = 6$ states, and so on. When observed closely in the presence of a magnetic field, the $\ell = 0$ state is always a doublet, the $\ell = 1$ state contains 6 levels, and so forth.

Multi-Electron Atoms and the Exclusion Principle

In multi-electron atoms each electron moves in the attractive Coulomb potential of the nucleus plus a repulsive potential due to the presence of the other $(Z - 1)$ electrons, where Z is the atomic number, the number of electrons in a neutral atom. The energy-level structure turns out to be qualitatively the same as that of the hydrogen atom. We still have an n-quantum number, which labels the total energy, and an ℓ-quantum number, which labels the angular momentum of the electron in that energy level. But it is no longer true that, for a given n, the $\ell = 0$, $\ell = 1, \ldots$ levels have the same energy. For a fixed value of n, the energy increases with ℓ (Fig. 42–14).

We might expect that in the ground state of any atom, all the electrons would be in the lowest energy level. The radiation spectrum from such an atom would be qualitatively the same as that of hydrogen. (The numbers would be different because the central charge is Ze and because there would still be effects of electron–electron repulsion.) Such a spectrum would bear little resemblance to the rich structure observed throughout the periodic table of the elements.

The Pauli exclusion principle

It was again Pauli who pointed out that a new ingredient was needed to understand the structure of multi-electron atoms ($Z \geq 2$). Pauli proposed his **exclusion principle**, according to which *each quantum state can accommodate only two electrons, one in the "up" state and one in the "down" state* (Fig. 42–15).

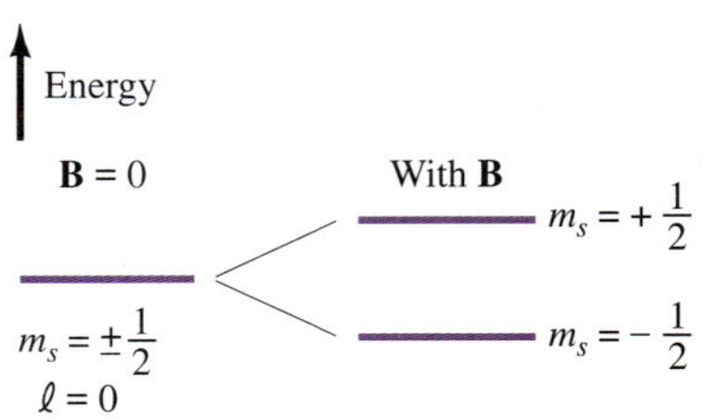

FIGURE 42–13 The existence of electron spin explains why some states are split into an even number of levels when a magnetic field is applied.

Let's examine what emerges when we start filling energy levels as per the exclusion principle (Fig. 42–16). Helium, $Z = 2$, has two electrons; both can fit into the $n = 1$, $\ell = 0$ state. There is no room for another electron in the lowest state. Helium is said to form a *closed shell*. Next let's consider lithium, $Z = 3$. Two electrons fit into the $n = 1$, $\ell = 0$ state, and the third electron has to go into the next lowest energy state, which is the $n = 2$, $\ell = 0$ state. The third electron is farther from the nucleus than the other two electrons are (remember that $r \propto n^2$), and the positive charge $+3e$ of that nucleus is partly screened by the negative charge of the two electrons in the $n = 1$ orbit. As a result, the third electron is less tightly bound to the nucleus and can therefore be

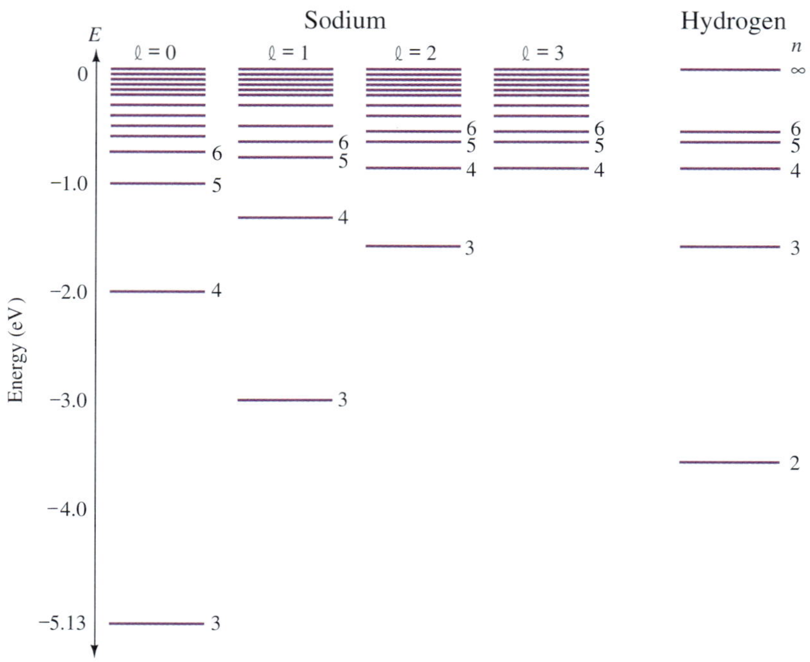

FIGURE 42–14 The schematic diagram of energy levels for an atom (here, sodium) is qualitatively similar to the hydrogen atom spectrum. The multiplets all have very nearly the same energy. For atoms with $Z > 2$, the energies for a given n value but for varying values of ℓ are no longer equal.

pulled more easily into the orbit of a nearby atom. Thus a lithium atom can bind with another atom to form a molecule and, like other atoms that have one electron outside a closed shell, is chemically very active.

For beryllium, $Z = 4$, we again fill a shell, the $n = 2$, $\ell = 0$ shell, and we expect beryllium to be less active than lithium. This is indeed the case. For $Z = 5$ through $Z = 10$, the $n = 2$, $\ell = 1$ levels are successively filled. The element $Z = 10$ is neon, which corresponds to another major closed shell, the $n = 2$ shell; it is an inert gas, one of a class of elements noted for their chemical inactivity. Fluorine, $Z = 9$, is one electron short of having a filled shell. Elements such as fluorine with a *hole* in a shell react particularly strongly with atoms such as lithium, which have one electron outside a filled shell. Fluorine has the lowest Z value of the *halogens* (atoms with a single hole in a shell), just as lithium has the lowest Z value of the *alkali metals* (atoms with one electron outside a filled shell).

All the details of the periodic table can be understood both qualitatively and quantitatively in a quantum mechanical description. Note that the existence of discrete energy levels, spin, and the exclusion principle are purely quantum mechanical phenomena. There is no classical hint of their existence.

FIGURE 42–15 Wolfgang Pauli.

EXAMPLE 42–5 An atom has $Z = 37$ electrons. What are the values of n and ℓ for the electron that is least tightly bound?

Solution: We list the possible levels for increasing values of n and ℓ, using the rule that for a given n, ℓ can take on values only up to $n - 1$, and then counting the number of electrons that fill each of the levels:

n	ℓ	*Number of Electrons*	*Cumulative Total of Electrons*
1	0	2	2
2	0	2	4
2	1	6	10
3	0	2	12
3	1	6	18
3	2	10	28
4	0	2	30
4	1	6	36
4	2	10	46

Thus the thirty-seventh electron is expected to lie in the $n = 4$, $\ell = 2$ shell.

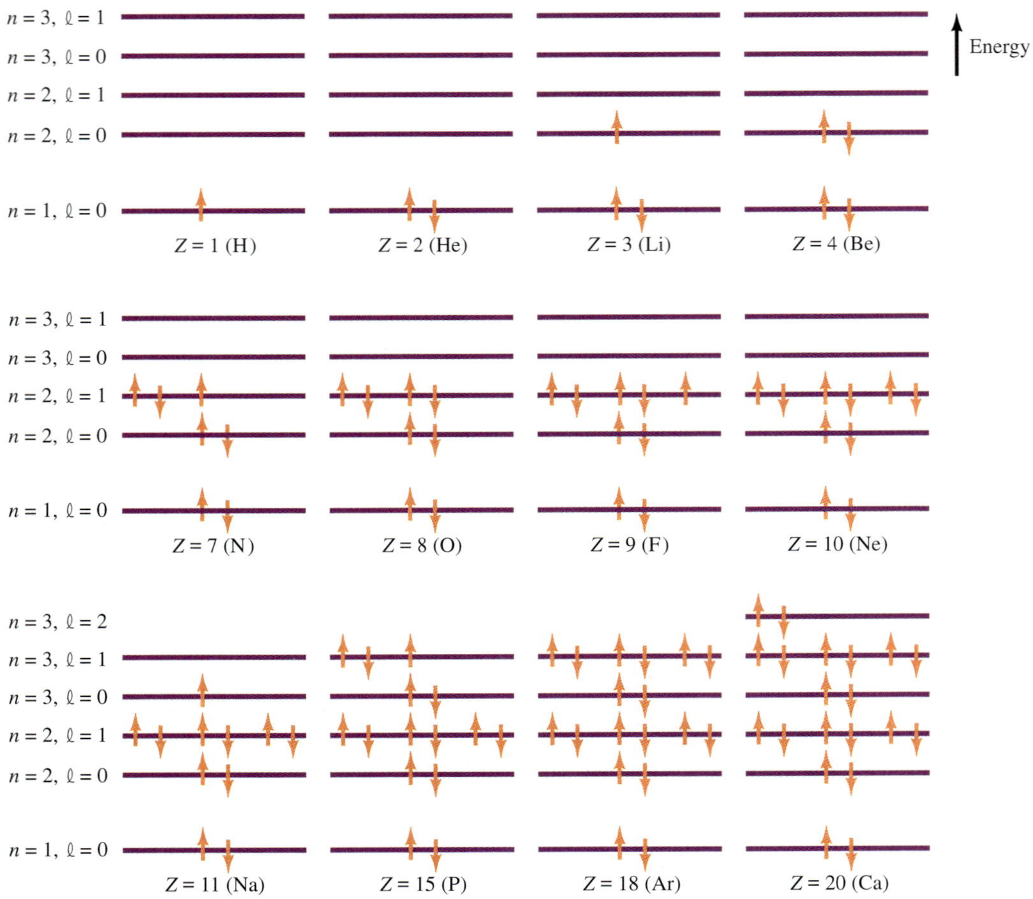

FIGURE 42–16 Pattern of electron energy-level occupation for elements from $Z = 1$ to $Z = 20$. The level splittings are not to scale.

In Example 42–5, we apply simple counting rules to determine the quantum numbers of the outermost electron in an atom with a certain number of electrons. As Z increases, these simple rules fail, because new effects enter. For example, the dynamics of the orbital motion leads to a tendency for an electron to align its spin with or against the orbital angular momentum. There is then a term in the energy that depends on the alignment of the spin and the orbital angular momentum, a so-called spin–orbit term. These effects may lead to slight differences in the order in which the levels are filled; deviations from our simple rules start at $Z = 19$.

Do All Particles Obey the Exclusion Principle?

Electrons have spin $\hbar/2$. Nuclear physics experiments have shown that protons and neutrons also have spin $\hbar/2$ and, according to a very general theorem, *all particles with spins* $\hbar/2$, $3\hbar/2$, $5\hbar/2$, . . . *obey the exclusion principle*. We call particles having half-integral spin **fermions**, named after Enrico Fermi. This has an important bearing on the structure of nuclei, which are made up of protons and neutrons. Nuclei have a shell structure analogous to that of atomic electrons. In nuclei, the average potential energy is the result of the mutual attraction of all the protons and neutrons by a *nuclear force*. This force is such that the energy levels tend to be spaced equally.

There are particles that do not obey the exclusion principle. Those particles with intrinsic spin of the form $s\hbar$, where $s = 0, 1, 2, \ldots$, behave differently from particles with spin $\hbar/2$. Unlike fermions, which are unable to share the same quantum state with one another, such particles "prefer" to be in the same state. Particles in this class, whose spin is an integer multiple of $\hbar$, are known as **bosons**, named after Satyendra Nath Bose. The photon is an example of a boson; it has intrinsic angular momentum $\hbar$.

Photons, like all particles with integer spin, show a preference for congregating in the same quantum state. We shall see some physical consequences of this congregation effect when we study lasers and liquid ^{4}He.

42–4 MOLECULAR STRUCTURE

The Formation of Molecules

Molecules are combinations of two or more atoms. We can calculate properties of molecules because even the lightest nucleus is some 2000 times more massive than an electron, so the nuclei in a molecule move much more slowly than their electrons. It is thus a good approximation to take the nuclei to be fixed. The electrons move rapidly around the nuclei and effectively create a smear of negative charge in which the nuclei are embedded.

Many of the important properties of simple molecules can be understood by focusing on a particular example. Let's consider the molecule H_2, which consists of two hydrogen atoms. We'll study the energy as a function of r, the separation between the two hydrogen nuclei. There are several contributions to the energy. First, there is the electrostatic repulsion between the positively charged nuclei. For the H_2 molecule, this contribution to the potential energy is $U(r) = e^2/4\pi\epsilon_0 r$.

We must also include the attraction of each of the two electrons to each nucleus and the electron–electron repulsion. Although a precise calculation requires quantum mechanics, we can estimate these contributions to the energy (the "electron contribution"). When r is large, the energy is lowest when one electron is close to one of the H nuclei and the other is close to the other H nucleus. In other words, for large r, the H_2 molecule looks like two independent hydrogen atoms. When $r \to 0$—the two H nuclei are directly on top of one another—the electrons "see" a nucleus of charge $2e$, which, for atomic purposes, is like a helium nucleus. Thus, the H_2 molecule looks like a helium atom for $r \to 0$.

Let's now put together some numerical values. The energy of a single hydrogen atom in its ground state is -13.6 eV, so at large r the total electron contribution for molecular hydrogen is twice this value, -27.2 eV. At small r, the electron contribution is that of a helium atom. This energy can be evaluated by noting that (*a*) each electron sees a nucleus of charge Ze, with $Z = 2$. The electron–nucleus energy is then $-Z^2(13.6 \text{ eV}) = -54.4$ eV for each electron, a total of -108.8 eV. (*b*) The two electrons are fairly close together, on average roughly half a Bohr radius. In that case the energy associated with the electron–electron repulsion is about double 13.6 eV, and positive. When we put (*a*) and (*b*) together, the electron contribution at small r is $(-108.8 \text{ eV}) + (27.2 \text{ eV}) \simeq -80$ eV. This is quite a good estimate; a correct quantum mechanical calculation of the binding energy of helium (and experiment!) gives the binding energy of helium as -78.5 eV.

To estimate the electron contribution for other values of r, we draw a smooth line between the small-r result (–78.5 eV) and the large-r result (-27.2 eV). Figure 42–17 shows the internuclear repulsion term, $e^2/4\pi\epsilon_0 r$; the electron contribution; and the sum of these two terms. *This sum has a minimum* at a nuclear separation of approximately 0.07 nm, and this is a stable equilibrium point. The experimental value of the separation of the H atoms in the H_2 molecule is very close to this value. This pattern holds for other stable diatomic molecules. Most simply stated, *the forces of attraction between the electrons and nuclei cancel the forces of repulsion between the electrons and between the nuclei at the position of stable equilibrium.*

Only certain combinations of atoms can form molecules. Qualitatively, molecules can form under the following circumstances:

1. The two atoms have *paired* electrons outside closed shells and the orbits of those electrons overlap. Here, "paired" means that one electron has spin up while the other has spin down. Because the electrons have opposite spin orientations, the Pauli exclusion principle does not prevent them from moving close to one another, into the region between the nuclei. In that region, the attraction of each electron to the "opposite" nucleus

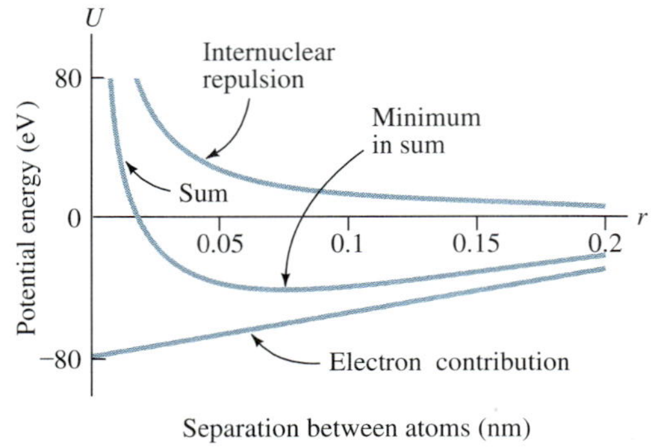

FIGURE 42–17 Energy diagram for the formation of the H_2 molecule. The upper curve is the potential energy associated with internuclear repulsion; the lower curve is an estimate of the potential energy associated with the electrons; the middle curve is the sum of these two terms. The sum has a minimum at $r \simeq 0.07$ nm, very close to the observed nuclear separation in the H_2 molecule.

more than compensates for their mutual repulsion and there is a net attraction. Paired electrons of this type form a *bond*: their interaction binds the two atoms together. The larger the number of bonds, the stronger the binding.

2. Only electrons that are not in closed shells can form bonds. An electron in a closed shell already has a second electron acting as its "partner" within its own shell. Thus only electrons in outer shells determine chemical properties.

3. An electron may not be in a closed shell but may nevertheless already be paired within its own, partly filled shell. Such electrons cannot pair up with an electron from another atom. More precisely, whether electrons pair up with partners from their own or from another atom is a matter of which configuration has the lower energy. This is not easily calculated, but chemists have developed a set of useful empirical rules for this question.

The H_2 molecule is an example of bonding between atoms that have unfilled shells. We say that such molecules are formed by *valence bonding* (Fig. 42–18a). *Ionic bonding* occurs when an atom with an electron outside a closed shell combines with an atom in which the outermost shell has one vacancy (a "hole") (Fig. 42–18b). As an example, consider sodium fluoride (NaF). Sodium has one electron outside a closed shell, and fluorine has a hole in its outer shell. Energy must be expended for the outermost electron of a sodium atom to be freed. This energy is the ionization energy of sodium. Its value is 5.1 eV, which is compensated by energy released if the free electron drops into the hole in a fluorine atom. This released energy is the negative of the energy it would take to remove that electron from a now-filled shell of fluorine, -3.5 eV. The net energy cost to move the electron from the sodium atom to the fluorine atom is $5.1\ \text{eV} - 3.5\ \text{eV} = 1.6\ \text{eV}$. However, we now have a positive (Na^+) ion and a negative (F^-) ion. These ions attract one another, and the energy associated with this attraction more than makes up for the 1.6 eV, so NaF has a lower energy than the separated Na and F atoms.

Van der Waals Forces

The conditions that we listed for the formation of molecules are not hard and fast. Under special circumstances, the inert gases—whose atoms have no electrons outside of closed shells—*do* form molecules. Ar_2 (diatomic argon) is an example. These mol-

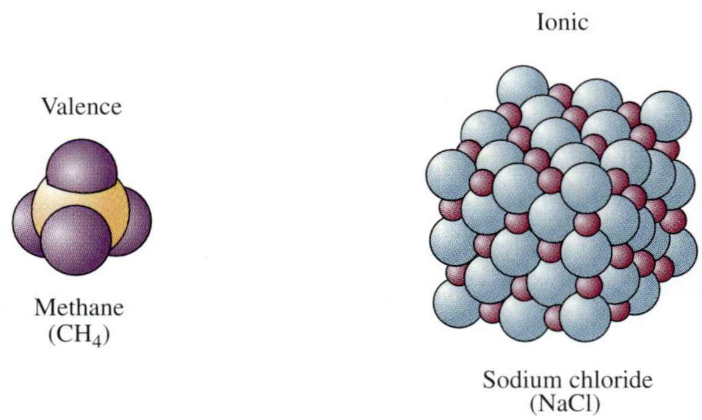

FIGURE 42–18 Methane is held together with valence bonding while a salt crystal is held together with ionic bonding.

APPLICATION: AUGER ELECTRON SPECTROSCOPY

It is important for organic and inorganic chemistry to determine the bonding state of a given atom in a molecule. For example, is an iron atom in a molecule in the divalent (ferrous) state or in the trivalent (ferric) state? In other words, how many outer orbital electrons does the iron atom share with its neighbors? Because of the influences of neighboring atoms and molecules—what we might term aggregating effects—the energies of the outer orbitals are not sharply defined. As a result, it is difficult to make direct use of the outer orbital energies to learn the bonding state. However, the outer orbits screen the inner part of the atom from aggregating effects, leaving the inner orbits as sharply defined levels.

Auger electron spectroscopy uses the *inner* levels to learn about the bonding state. In this process, photons or electrons with energies of as much as 50 KeV are projected to the material, sometimes ejecting the electrons in the inner orbitals. These holes persist in the atoms for only a short time before they are filled by an electron from a higher orbit. However there is a significant excess of energy when an electron from a higher orbit fills an inner orbital. One of the ways in which this is manifested is that the energy is transferred to another electron in the higher orbit. This electron is then ejected with a kinetic energy equal to the energy difference between the available energy from filling the inner orbital hole and its binding energy. The ejected electrons are known as *Auger electrons*. How does this reveal information about the chemical bonding? Quantum mechanically the "outer" electrons have a finite probability of being within the inner orbits! The outer electrons in effect screen the nucleus from the inner electrons to a certain degree, and therefore influence the energies of the inner levels. In turn, the energy of the emitted Auger electrons is affected. By accurately measuring the energy of the emitted Auger electrons, as in Fig. 42AB–1, the number of electrons resident in the outer shell can be determined, and therefore the bond structure.

As an illustration of the size of the effect, the difference of energy of Auger electrons between ferrous and ferric bonding for iron in a molecule is a few eV out of a total emitted electron energy of 4.5 KeV. Thus measurement of the electron energy to one part in 2000 can determine the valence state.

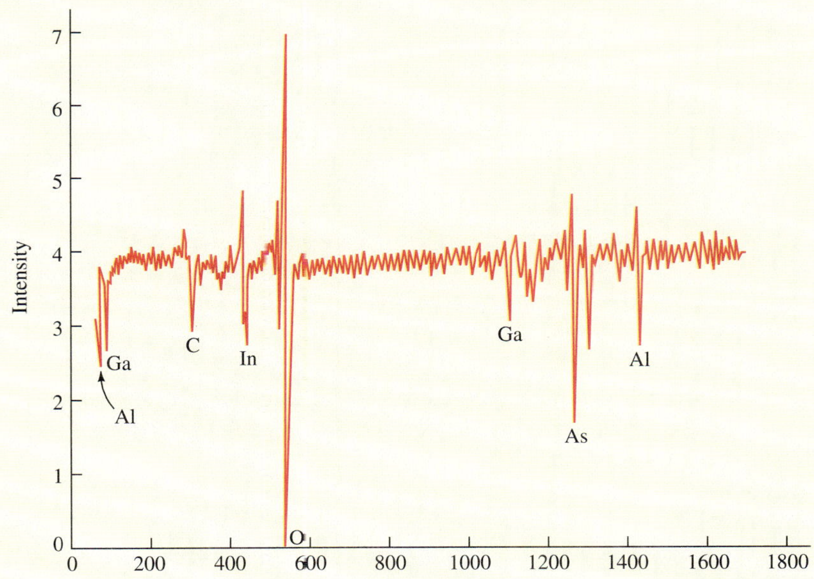

FIGURE 42AB-1 A spectrum of emitted Auger electrons reveals spikes characteristic of different atoms and of molecular bonding effects.

ecules form because there is an additional force between atoms, the *van der Waals force*. This force is due to electromagnetism but falls off much more rapidly than does the Coulomb force with distance. Despite the fact that the van der Waals force is rather weak, it is always attractive and thus can bind atoms. The van der Waals force arises for the same reason that a comb attracts uncharged bits of paper. When one atom approaches another, its charge distribution affects the other by giving rise to a small displacement of charges, such that an electric dipole structure is created. This electric dipole has an electric field, which interacts with the electric dipole of the first atom. The two dipoles attract with a force proportional to $1/r^5$ (see Problem 47).

Van der Waals forces, although very weak, are responsible for a number of additional phenomena, such as the adhesion between a liquid and the sides of its container, and the departure of a gas from ideal gas behavior (see Chapter 17).

Molecular Spectra

In molecules, nuclei separated by roughly 0.1 nm form nearly fixed centers of attraction for electrons. We may expect from the general ideas about quantum mechanics in Section 42–1 that molecules will have a series of electron energy levels separated by the kinds of energies that characterize atomic levels; that is, gaps on the order of 1 eV to 10 eV. Experiments, however, reveal a far richer spectrum—even for simple diatomic molecules (Fig. 42–19). The additional structure is due to quantized motions not allowed for the hydrogen atom: vibrational and rotational motion.

Vibrational Motion. For diatomic molecules, we saw that the potential energy curve has a minimum (Fig. 42–18). The minimum point is a point of stable equilibrium. Moreover, as we learned in Chapter 7, the potential energy curve can normally be approximated near the minimum by a parabolic curve; that is, by a harmonic oscillator potential. As a consequence, there will be energy levels associated with the vibrational motion of the nuclei in this harmonic oscillator potential. We show in Example 42–6 by using Bohr's techniques that these energy levels are given by $E_{vib} = n\hbar\omega$, where n is an integer and ω is a characteristic frequency given by $\sqrt{k/M}$. Here, k is the "spring constant" and M is the *reduced mass* of the two nuclei in vibrational motion, given by $M = M_1M_2/(M_1 + M_2)$, where M_1 and M_2 are the two nuclear masses.

EXAMPLE 42–6 Suppose that a mass m is attached to the end of a spring and moves in a circular orbit of radius r under the influence of the spring, which exerts an attractive centripetal force of magnitude $F = kr$ (Fig. 42–20). Use the condition of fitting an integral number of wavelengths into allowed circular orbits to determine the energy spectrum associated with these circular orbits.

Solution: We proceed exactly as we did for the hydrogen atom: We use $F = ma$ to obtain one relation between position r and speed v, and then impose the quantization condition to obtain another relation. For $F = kr$, the equation of motion $F = ma$ for circular motion is

$$kr = m\omega^2 r.$$

This relation can be solved for ω, yielding $\omega = \sqrt{k/m}$. The energy is the sum of the kinetic energy, K, and the potential energy, $U(r)$, of a three-dimensional harmonic oscillator:

$$E = K + U(r) = \frac{1}{2}mv^2 + \frac{1}{2}kr^2 = \frac{1}{2}m\omega^2r^2 + \frac{1}{2}m\omega^2r^2 = m\omega^2r^2.$$

The wavelength-fitting condition implies that

$$n\lambda = 2\pi r.$$

FIGURE 42–19 Comparison of the spectra of (a) molecular hydrogen and (b) atomic hydrogen. The far richer structure of the molecular spectrum is evident even in this case, where the molecule is as simple as can be imagined.

We now use $\lambda = h/p = h/mv = h/m\omega r$ in the equation $n\lambda = 2\pi r$ to obtain

$$\lambda = \frac{2\pi r}{n} = \frac{h}{m\omega r}.$$

It follows from this equation that

$$r^2 = \frac{nh}{2\pi m\omega}.$$

Substitution into the expression for the energy gives the desired allowed energies:

$$E_n = m\omega^2 r^2 = \frac{m\omega^2 nh}{2\pi m\omega} = \frac{nh\omega}{2\pi} = n\hbar\omega, \qquad \text{where } n = 1, 2, 3, \ldots. \quad (42\text{–}16)$$

Although we have found the energy levels that correspond to circular motion, we can always think of that motion as due to superimposed oscillatory motions (see Chapter 14). Thus we can think of our result as more generally representing the allowed energy levels of a three-dimensional harmonic oscillator. These levels are equally spaced in units of $\hbar\omega$.

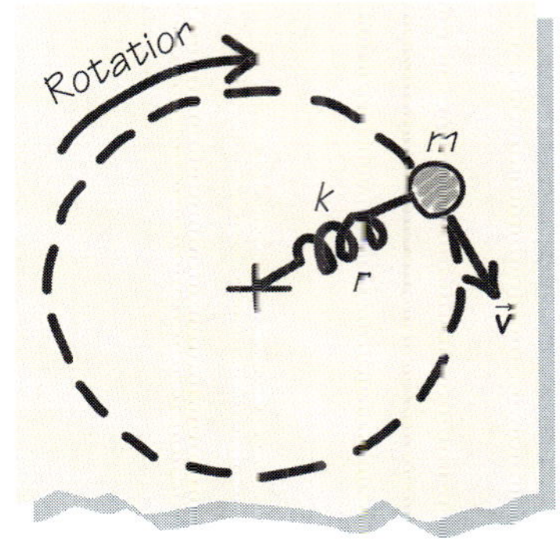

FIGURE 42–20 Example 42–6.

We have seen that the minimum in the potential energy (the "spring") is created by the electron cloud in which the two nuclei are embedded, and we might therefore expect that it is parameters such as the size of the atomic orbits that determine the value of k. We can *estimate* the value of k by dimensional analysis. The dimensions of any spring constant such as k are $[k] = [EL^{-2}]$, where $[E]$ is the dimension of energy, $[ML^2T^{-2}]$. To estimate k, we use this relation with typical atomic energies and distances, $E \simeq e^2/8\pi\epsilon_0 a_0$ and $L \simeq a_0$, where a_0 is the Bohr radius. Then

$$k \simeq \frac{e^2}{8\pi\epsilon_0 a_0^3} = \frac{e^2}{8\pi\epsilon_0}\left(\frac{e^2 m_e}{4\pi\epsilon_0\hbar^2}\right)^3 = \frac{1}{2}\frac{(m_e)^3}{\hbar^2}\left(\frac{e^2}{4\pi\epsilon_0\hbar}\right)^4. \quad (42\text{–}17)$$

Dimensional analysis cannot specify any additional numerical factors and we must add a little additional physical reasoning to improve our estimate. In particular, because molecules are somewhat larger than atoms, we could replace a_0 by a somewhat larger radius, say, $2a_0$. This leads to the estimate

$$k \simeq \frac{1}{16}\frac{(m_e)^3}{\hbar^2}\left(\frac{e^2}{4\pi\epsilon_0\hbar}\right)^4. \quad (42\text{–}18)$$

Once we know the effective spring constant, we can find the allowed energies of oscillation from Eq. (42–16):

$$E_{\text{vib}} = n\hbar\omega = n\hbar\sqrt{\frac{k}{M}} = \frac{n}{2}m_e\left(\frac{e^2}{4\pi\epsilon_0\hbar}\right)^2\sqrt{Z_1 Z_2\frac{m_e}{M}}, \quad (42\text{–}19)$$

where n is an integer and M is the reduced mass of the two nuclei.

We have written Eq. (42–19) in this way because it exposes the factor $(m_e/2)(e^2/4\pi\epsilon_0\hbar)^2$, which we recall from Eq. (42–9) is the magnitude of the ground-state energy of hydrogen, 13.6 eV. Let's call this factor E_0. Then

$$E_{\text{vib}} = nE_0\sqrt{Z_1 Z_2\frac{m_e}{M}}. \quad (42\text{–}20)$$

Numerically, the factor by which nE_0 is multiplied is on the order of 10^{-2}; that is, we estimate the vibrational energies are on the order of 10^{-1} eV.

Rotational Motion. Independent of vibrations, the nuclei of a diatomic molecule can rotate like a dumbbell of length r_0. Here r_0 is the separation for which the potential energy has a minimum. Our molecule will then have energy levels associated with these rotations. The classical energy E is given in terms of the angular momentum L by $E = L^2/2I$, and we find the quantized energies when we insist that L be quantized.

We must evaluate the rotational inertia. We are treating a diatomic molecule such as OH as two masses, M_1 and M_2, connected by a massless rigid rod of length r_0. The molecule rotates about an axis perpendicular to the rod and passing through the center of mass of the system. For this system the rotational inertia I is given by (see Chapter 9)

$$I = Mr_0^2,$$

where M is the reduced mass of the system. We now quantize L, $L^2 = \ell(\ell + 1)\hbar^2$. The rotational energy levels are then

$$E_{\text{rot}} = \frac{L^2}{2I} = \frac{\ell(\ell + 1)\hbar^2}{2I} \simeq \frac{\ell(\ell + 1)\hbar^2}{2Mr_0^2} = \ell(\ell + 1)\frac{m_e}{2}\left(\frac{e^2}{4\pi\epsilon_0\hbar}\right)^2 \frac{m_e}{4M}.$$

We have substituted $r_0 = 2a_0 = 8\pi\epsilon_0\hbar^2/e^2m_e$, as we did in the discussion of the vibrational energy levels. We again recognize the factor E_0, so

$$E_{\text{rot}} = \frac{\ell(\ell + 1)E_0 m_e}{4M}. \tag{42–21}$$

The energy levels are suppressed by an additional factor $\sqrt{m_e/M}$ from the factors of the vibrational levels and are on the order of 10^{-3} eV.

Let's summarize our results on molecular spectra. There are electronic levels on the order of electron-volts apart. Associated with each level is a series of vibrational levels, separated by roughly 10^{-2} of the electronic levels; associated with each of *these* levels is a series of rotational levels, with separations about 10^{-4} of the electronic levels (Fig. 42–21). The latter two sets of levels are described as vibrational bands and rotational bands. The study of rotational bands is important to chemistry because these are most easily excited. Molecular spectra involving transitions between the rotational levels consist of wavelengths on the order of 10^4 times atomic wavelengths and thus involve infrared rather than optical spectroscopy.

APPLICATION
Preventing electric breakdown

The complex structure of molecular energy levels is responsible for many of the subtleties of organic chemistry and of the chemical reactions that occur in biological systems. A technological application illustrates one consequence of the complexity of molecular structure. The high electric potentials that occur within transformers can cause the electrical breakdown of pockets of air in the transformer. In breakdown, high voltage ionizes air molecules. A fluorescent tube can be lit simply by holding it near a high-voltage supply (Fig. 42-22). The electrons liberated are accelerated by the large electric fields. In colliding with successive air molecules, they liberate more electrons, which are accelerated. An avalanche of electrons results, leading to currents that short out the transformer. An excellent method to prevent this type of breakdown is to introduce a gas that allows the liberated electrons to lose energy in ways that do not lead to further ionization. Sulfur hexafluoride (SF_6), which is one gas used in this way, has a very rich molecular structure, and there are many non-ionizing ways in which it can be excited by an electron colliding with it.

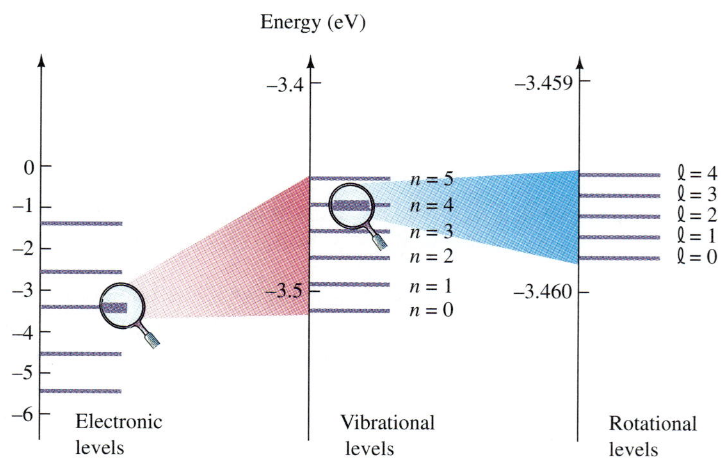

FIGURE 42–21 Molecular levels. The striking feature of these spectra is that the energy levels in each part are separated by energies 100 times finer than the energy-level separations of the previous part. The three parts represent electronic, vibrational, and rotational motions.

42–5 BAND THEORY

Many engineering applications of quantum phenomena arise in the area of semiconductor physics. In Section 27–5, we described the *band structure* in the energy levels of crystalline solids, in which there are regions of energy with no allowed energy levels. No electrons within a given solid can have energies in these forbidden regions; their energies are restricted to certain finite ranges of energy—the bands of allowed energies. We discussed how the presence of energy bands and the numbers of electrons that occupy the allowed energy levels determine why some materials are conductors whereas others are insulators or semiconductors. We are now in a better position to elaborate on why bands form, a discussion of what is known as *band theory*.

Let's start by recalling our discussion from Section 42–4 of the energy levels for two hydrogen atoms. We found that we can think about a helium atom as the result of moving two hydrogen atoms together. However, our naive expectation for the binding energy of helium—that the value of the ground-state energy of helium is a simple multiple of the binding energy of hydrogen—is incorrect. This is because the electrons of the two hydrogen atoms interact with each other. In other words, a new interaction added to an existing system will generally shift the energy levels of the existing system. Just how are the energy levels shifted? When the two hydrogen atoms are far apart, there are two electrons in a pair of identical levels of energy -13.6 eV, the ground states of the two separate hydrogen atoms. We say that the two electrons are in *degenerate* energy levels. (Actually, there are four degenerate levels, because each electron can be in a spin "up" or spin "down" state, and these states have the same energy values.) When the atoms are brought so close together that helium forms, we have seen that there is at least one level lower than the original -13.6 eV levels; namely, the true ground state of helium. The remaining three levels are pushed *upward* and correspond to excited states of helium. This is a general characteristic of energy levels:

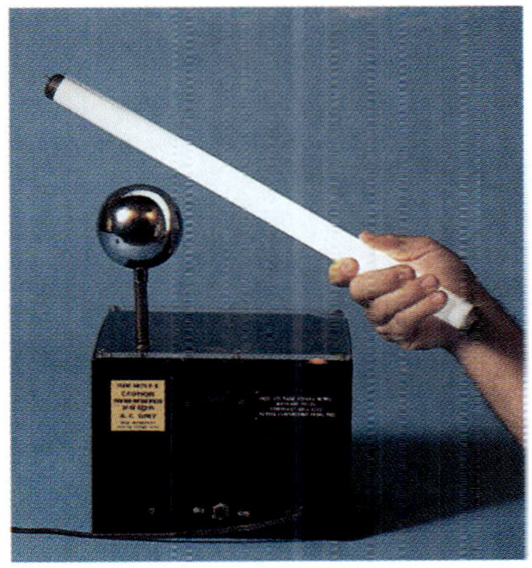

FIGURE 42–22 The high electric field produced by the power supply is enough to ionize the molecules inside the nearby fluorescent tube, which then produces light.

The splitting of degenerate energy levels

When there are degenerate energy levels and an additional interaction comes into play, the levels are split—some moving upward and some moving downward relative to their original positions.

The observation above plays an important role in band theory. Let's study N atoms of sodium, which have filled shells plus a single electron (the valence electron) in an $n = 3$, $\ell = 0$ state. There are $2N$ degenerate (equal-energy) states available to the valence electron when the atoms are far apart and there is no interaction between the atoms. (The factor of 2 is present because the electron can be in a spin "up" or "down" state, and these states have the same energy.) Now suppose that the N atoms form a crystal lattice with closely spaced lattice sites. Then the originally degenerate valence-electron states are split by new interactions between the atoms; N energy levels are pushed down and N are pushed up, each by a slightly different amount. The $2N$ energy levels then form a band of very narrowly spaced levels. The N valence electrons fill only half of this band; namely, the N lowest-lying energy levels (Fig. 42–23). As we discussed in Chapter 27, an electric field can easily lift some electrons at the top of the filled levels to the empty levels, where they can move without hindrance. Metallic sodium is thus a good conductor.

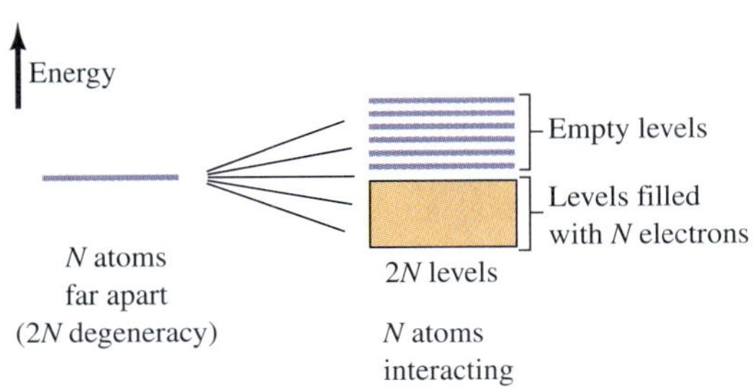

FIGURE 42–23 When many atoms are brought together to form a crystalline solid, the degenerate levels of N atoms split to form band structures, in which there are ranges of allowed electron energies and ranges (gaps) of forbidden energies. In sodium, the valence electrons fill one-half of a band.

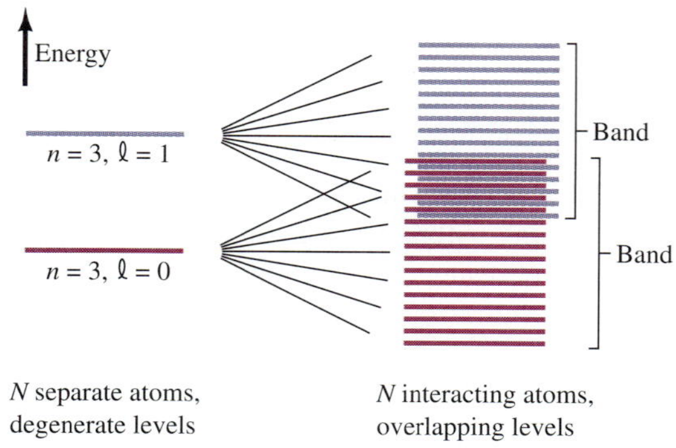

FIGURE 42–24 The origin of the band structure of N atoms of magnesium in a crystalline solid. The splitting of the degenerate levels with quantum number $n = 3$, $\ell = 0$ leads to a band that is completely filled with electrons. The degenerate levels $n = 3$, $\ell = 1$ also split, and the allowed band they form overlaps with the band formed from the $n = 3$, $\ell = 0$ levels.

What are the "new" interactions that occur when the atoms are assembled into a regular lattice, and that lead to the splitting of degenerate electron energy levels into a band structure? These are the interactions of electrons with ions other than the ones to which they were originally attached. In particular, if the ions were not present, the electrons would move freely in the "box" formed by the dimensions of the material. The electrons' wavelike aspect makes them behave analogously to waves on a string, namely as standing waves whose wavelengths must be integral multiples of the appropriate dimensions of the box. These waves are the waves formed by the wave function, and the electron is likely to be found where those waves have crests (or troughs). The crests and troughs are regularly spaced. At this point, we can imagine adding the ions at regularly spaced sites. Two modifications to our original depiction result: First, only those electrons whose wavelengths correspond to the ionic spacing can be supported in the material for reasons closely related to the Bragg scattering arguments we gave in Chapter 41. Second, half of the allowed waves of the electrons have peaks (or troughs) where the ions are found, and half have peaks (or troughs) exactly in between the ions. The energies of the half whose peaks are where the ions are found are lowered because these electrons are more likely to be near the ions to which they are attracted. The energies of the half whose peaks are in between the ion locations are raised because these electrons do not feel the force of attraction as strongly as they did when they were in an isolated atom.

For contrast, we can consider N atoms of magnesium. The Z value of magnesium is one higher than that of sodium. Magnesium has the same atomic structure as sodium, except that there are *two* electrons in the $n = 3$, $\ell = 0$ states. The band formed from the originally degenerate states when the magnesium atoms combine into a lattice is therefore completely filled. Magnesium could be an insulator. However, empty levels that originally came from the $n = 3$, $\ell = 1$ states, which were above the $n = 3$, $\ell = 0$ levels when the two sets of levels were each degenerate, fan out into a second band. For magnesium, it happens that the two bands overlap (Fig. 42–24). Thus there is room for electrons, under the influence of an electric field, to move into empty levels, and magnesium is a good conductor. There are other cases in which bands of this type do not overlap. Indeed, all the possibilities discussed in Chapter 27 can be realized.

SUMMARY

One of the important consequences of quantum mechanics is that energies of bound systems can take on only discrete values. In the Bohr model, the quantization of atomic energy levels follows from the restriction of angular momenta to integral multiples of $\hbar$:

$$L = n\hbar, \qquad \text{where } n = 1, 2, 3 \ldots . \tag{42–6}$$

With this condition the allowed energy values are

$$E_n = -\frac{m_e}{2n^2}\left(\frac{e^2}{4\pi\epsilon_0\hbar}\right)^2, \qquad \text{where } n = 1, 2, 3, \ldots, \tag{42–9}$$

in agreement with experiment. Energy conservation allows electrons to jump between levels with different n values while emitting photons of frequency given by

$$hf = E_i - E_f, \tag{42–11}$$

where E_i and E_f are the initial and final energies, respectively.

Quantum mechanics as developed by Heisenberg and Schrödinger shows that the structure of the possible energy levels is more complex than the Bohr model predicts. For each value of n, there are n^2 energy levels characterized by angular momentum $\ell\hbar$, where $\ell = 0, 1, 2, \ldots, (n-1)$, and $2\ell + 1$ spatial orientations are allowed for the vector angular momentum characterized by ℓ. However, the full structure of atoms cannot be understood until we add the fact that electrons carry an intrinsic angular momentum $\hbar/2$ called spin. Moreover, the Pauli exclusion principle shows that no more than two electrons can appear in any quantum state (corresponding to the $2s + 1$ states with $s = \frac{1}{2}$). With these additions, the complex structure of multi-electron atoms—as revealed in the periodic table of elements—can be explained.

Molecules form when the forces of attraction between the electrons and nuclei cancel the forces of repulsion between the electrons and between the nuclei at the position of stable equilibrium. This can occur if the atoms involved have electrons outside closed shells. Molecular spectra reveal energy levels of a molecule that are associated with vibrations of the atoms and rotations of the entire molecule. The electronic, vibrational, and rotational modes form a hierarchy.

Electrons in solids are attracted by the ions that form the crystal lattice. The effect of this attraction is to split previously degenerate levels, forming a range of very closely spaced levels called energy bands.

UNDERSTANDING KEY CONCEPTS

1. In the treatment of a particle confined to a box (in Section 42–1), why was the ground state characterized by $n = 1$ rather than $n = 0$?

2. Can we determine the atomic composition of distant objects by studying the wavelengths of their emitted photons?

3. What determines the shortest and longest wavelengths that a hydrogen atom can emit?

4. On the one hand, we say that electrons in atoms have discrete energies; on the other hand, we say that there is inherent uncertainty in our ability to measure energies. Is there a conflict here?

5. Because of the exclusion principle, only electrons with energy near the Fermi energy move under the influence of a field and create currents. If electrons did not obey the exclusion principle, how would conduction in metals differ?

6. In the Stern–Gerlach experiment, atoms of silver were observed to have only two components of angular momentum. Does that mean that all the electrons of a silver atom have angular-momentum quantum number ℓ equal to zero?

7. Is it true that all hydrogen atoms are indistinguishable from each other? How is it possible to distinguish between hydrogen with a nucleus consisting of a single proton (^{1}H), hydrogen with a nucleus consisting of a single proton and a single neutron (^{2}H, deuterium), and hydrogen with a nucleus consisting of a single proton and two neutrons (^{3}H, tritium)?

8. Why does an atom with a magnetic dipole moment need a magnetic field that varies spatially in order to deflect an atom? Would a constant magnetic field suffice?

9. The term *fine structure* is used to refer to closely spaced energy levels. Discuss the origin of such levels for some of the systems we described in this chapter.

10. Van de Graaff accelerators, which have terminals at very high voltages inside a pressurized tank, are used to accelerate nuclear particles to high energy for nuclear reactions. Would air or sulfur hexafluoride be better as an insulating gas in a van de Graaff accelerator?

11. Why does it take about twice as much energy to excite an electron from the $n = 1$, $\ell = 0$ state to the $n = 2$, $\ell = 1$ state for He^+ as it does for He (neutral helium)?

12. Would you expect the orbital radius of the lowest orbit in a helium atom to be less than, equal to, or greater than that in a hydrogen atom? Why?

13. In discussing the formation of molecules, we stated that the minimum of the net potential energy in the interaction of two hydrogen atoms is the position where the attractive forces between the electrons and the nuclei cancel the repulsive forces between each electron and between each nucleus at the position of stable equilibrium. How do we translate a statement about the potential energy minimum to a statement about the force?

14. Why do the arguments we made about the formation of the diatomic hydrogen molecule, H_2, not apply to the formation of a diatomic helium molecule, He_2, from two helium atoms?

15. Helium has two electrons. Both can be in spin "up" (or "down") states, or one can be up while the other is down. Taking into account the exclusion principle and the fact that angular-momentum states with quantum number $\ell = 1$ have a higher energy than angular-momentum states with quantum number $\ell = 0$, which spin arrangement will occur in the ground state?

16. Suppose that we add an electron to hydrogen. The second electron could be in the same orbit as the first (the spins would then have to point in opposite directions). What might prevent the existence of such a negatively charged atom? Would the existence of an atom consisting of one proton and three electrons be as likely, or unlikely?

PROBLEMS

42–1 *Energy and Angular Momentum Are Quantized*

1. (I) What wavelength of radiation is necessary to ionize hydrogen? [*Hint*: Recall that to ionize an atom, it is necessary to raise the energy of the electron to be emitted from its ground state to at least $E = 0$.]

2. (I) Consider double ionized lithium (lithium in which two electrons have been removed). How much energy is needed to remove the third electron?

3. (I) What are the energy and wavelength of the photon emitted when a hydrogen atom jumps from its second excited state ($n = 3$) to its ground state ($n = 1$)? from $n = 5$ to $n = 3$ (Fig. 42–25)?

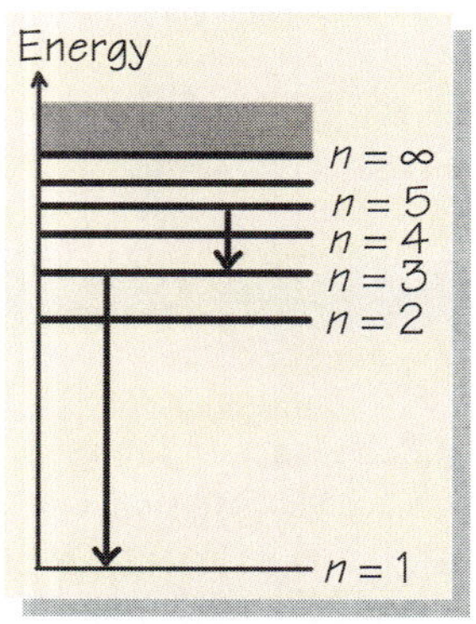

FIGURE 42–25 Problem 3.

4. (I) Bohr suggested an addition to the model named after him; specifically, that emitted radiation is very faint unless n changes by one unit. For what range of values of n will the radiation from singly ionized helium lie in the visible range; that is, with wavelengths in the range 400 nm to 700 nm (Fig. 42–26)?

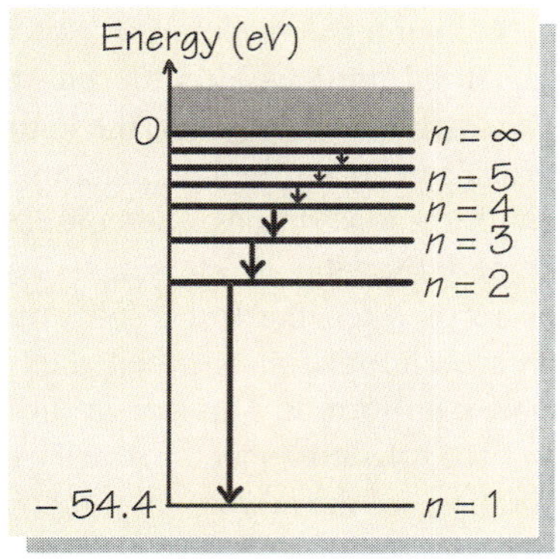

FIGURE 42–26 Problem 4.

5. (I) The negative muon (symbol μ^-) is a heavy version of the electron; it has a mass $m_\mu = 210\, m_e$. What is the ground state of the atom formed by a μ^- and a proton atom, assuming the Bohr rules apply?

6. (I) Consider the μ^--proton atom described in Problem 5. What is the energy difference between the first excited state and the ground state?

7. (II) The *Lyman series* is a series of spectral lines for hydrogen whose wavelengths correspond to Eq. (42–12) when $n_1 = 1$ (Fig. 42–27). (a) What are the quantum numbers of the states involved in the three transitions of the Lyman series with the longest wavelengths? (b) Calculate the wavelengths for the transitions of part (a). Are these wavelengths in the visible or ultraviolet regions?

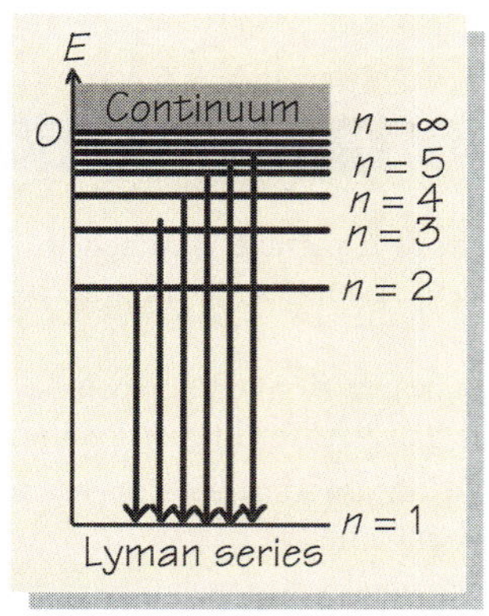

FIGURE 42–27 Problem 7.

8. (II) Singly ionized helium is a single-electron atom with $Z = 2$ and a nuclear mass four times as large as the mass of the hydrogen nucleus. Calculate the longest wavelength in the analogue of the Lyman series, those transitions that end in $n = 1$ (see Problem 7).

9. (II) What is the wavelength of the first line of the Lyman series (see Problem 7) in triply ionized beryllium ($Z = 4$)?

10. (II) The *Paschen series* is a series of spectral lines for hydrogen whose wavelengths correspond to Eq. (42–12) when

$n_1 = 3$ (Fig. 42–28). (a) What are the quantum numbers of the states involved in the three transitions of the Paschen series with the longest wavelengths? (b) Calculate the wavelengths for the transitions of part (a). Are these wavelengths in the visible or infrared regions?

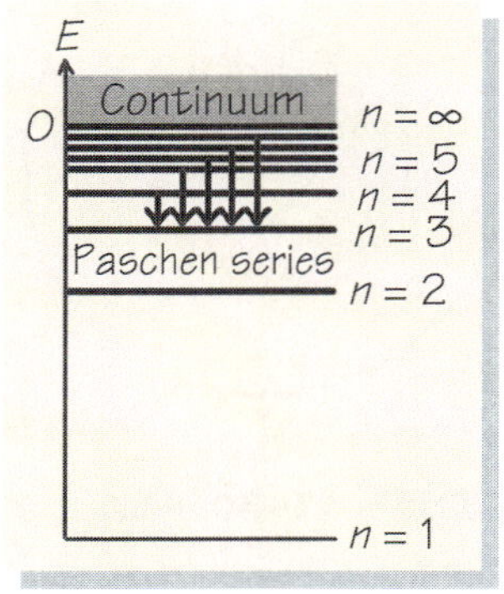

FIGURE 42–28 Problem 10.

11. (II) Calculate the wavelengths of the *Paschen* ($n_1 = 3$), *Brackett* ($n_1 = 4$), and *Pfund* ($n_1 = 5$) series transitions. (These series all refer to hydrogen.) Sketch the spectrum of atomic hydrogen for the 1000 nm to 5000 nm range of wavelengths. Calculate only wavelengths for values of $n_2 \leq 8$ in the n_1 to n_2 transitions.

12. (II) Derive a formula for the wavelengths of the transitions in single-electron ions whose nuclei have charge $+Ze$. For what range of Z values will radiation for $n = 2$ to $n = 1$ transitions be in the 30-nm range? in the 0.3-nm range?

13. (II) It is useful to introduce the *fine-structure constant* $\alpha \equiv e^2/4\pi\epsilon_0\hbar c$ in problems that involve atoms. (a) What are the dimensions and value of α? What is the value of $1/\alpha$, to the nearest integer? (This is a useful number to remember.) (b) Express in terms of α the energy of the nth level of the hydrogen atom, E_n. (c) Calculate in terms of α the speed of the electron in the lowest Bohr orbit of hydrogen.

14. (II) What are the orbital radius, speed, momentum, and energy of an electron in the $n = 4$ state of hydrogen? Assume a classical model to calculate the momentum and speed of the electron.

15. (II) The most strongly bound $n = 1$ electrons of a large atom are closer to the nucleus than the other electrons. As a consequence, their energy is primarily determined by their interaction with the nucleus, without much influence from the other electrons. Use this idea to *estimate* the energy required to remove an electron from the lowest energy state of copper ($Z = 29$) and tungsten ($Z = 74$) atoms. (Experiments by Moseley show that the nucleus is somewhat shielded by the other electrons, an effect that can be taken into account for $n = 1$ electrons by replacing Z of the nucleus by $Z - 1$).

16. (II) Consider the Rutherford planetary model of circular orbits. (a) Use the expression for the radius given in Eq. (42–5) to calculate the speed of an electron in terms of $e^2/4\pi\epsilon_0$, the angular momentum L and m_e. (b) Use the result of part (a) to calculate the electron's acceleration. (c) Calculate the period of the electron's orbital motion.

17. (II) According to classical electromagnetism, the power radiated by a particle of charge e that undergoes an acceleration a is

$$P = \frac{2}{3}\frac{e^2}{4\pi\epsilon_0 c^3}a^2.$$

Use this formula and the results of Problem 16 to calculate the energy radiated by an electron per unit time for a circular orbit of angular momentum L.

18. (II) (a) Use the results of Problem 17 to show that the fraction of energy radiated by an electron in a single period T is given by

$$\frac{\Delta E}{E} = \frac{8\pi}{3}\left(\frac{e^2}{4\pi\epsilon_0 Lc}\right)^3.$$

(b) Use your result to find the number of periods over which the electron would lose all its energy, and the time over which this would take place. Evaluate this result numerically for L when the orbital radius is 0.1 nm. (Your result should be on the order of 10^{-10} s.)

19. (II) Calculate the frequency of the radiation emitted by an electron in a hydrogen atom when it jumps from a level of quantum number $n + k$ to a level of quantum number n. Find a simple expression for it when $n >> k$. Express your result in terms of $L = n\hbar$ and compare it with the classical frequency $f_{c_1} = 1/T$, where T is the period calculated in Problem 16. Show that the two will be equal only for $k = 1$, so the requirement that the results of quantum mechanics coincide with classical results for very large values of n (the Bohr correspondence principle) means that only transitions with changes in n of value $\Delta n = 1$ are allowed.

20. (III) (a) Express the power ($P = \Delta E/\Delta t$) radiated by an accelerating electron (see Problem 17) in terms of $L = n\hbar$. The energy will be radiated in the form of a photon of energy hf, where f is the frequency that corresponds to a transition $\Delta n = 1$. (b) Find $\Delta t = \Delta E/P$ in terms of n. Use this formula to calculate a numerical value for Δt when $n = 1$. This time interval may be viewed as the time constant for the $n = 2$ state of the hydrogen atom to jump to the $n = 1$ (ground) state. (c) Compare your time constant with the period of an electron in the $n = 2$ orbit. From your comparison, could you argue that the $n = 2$ orbit is almost stable? (d) Roughly, Δt is the longest time you have available for measuring the energy of the $n = 2$ state. Use the time–energy uncertainty relation to estimate the spread in energy of the $n = 2$ state, and compare this spread with the transition energy $E_2 - E_1$.

21. (III) *Estimate* the deflection of a beam of electrons in a Stern–Gerlach experiment. Assume that the kinetic energy of the electrons are 100 eV each, and that they travel through a region of length 0.2 m, in which there is a magnetic field gradient of 1 T/cm.

42–2 The True Spectrum of Hydrogen

22. (I) What are the possible values for the quantum numbers ℓ and its corresponding m-value for an atomic electron in a state labeled with $n = 2$?

23. (II) A correction should be supplied to the formulas for the energy levels of hydrogenlike atoms: The electron orbits about the center of mass of the electron–proton system rather than about the proton itself (Fig. 42–29). This correction, known as the *reduced-mass effect*, is small because the proton, of mass m_p, is much more massive than the electron, of mass m_e. The result is that the energy levels of hydrogen should be corrected to

$$E_n = -\frac{m_e}{1 + (m_e/m_P)} \frac{1}{2n^2} \frac{e^2}{(4\pi\epsilon_0\hbar)^2} .$$

The energy levels for deuterium, an atom with a nucleus whose charge is that of the hydrogen nucleus but whose mass is about twice that of hydrogen, obey the same formula. Find the difference in the wavelengths of radiation emitted in the transition between the $n = 2$ and $n = 1$ states for the two atoms. It was the observation of this difference that led to the discovery of deuterium by Harold Urey in 1931.

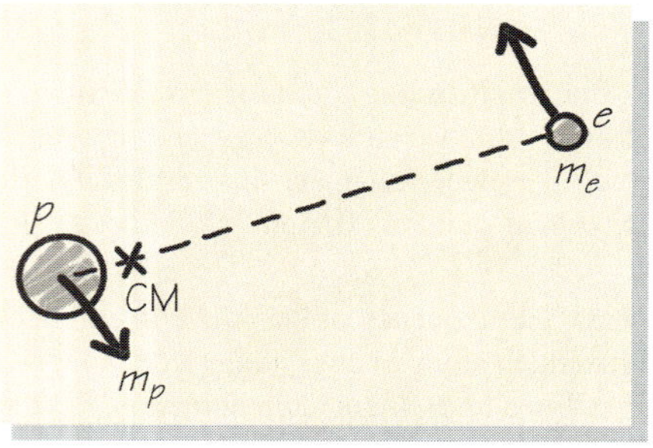

FIGURE 42–29 Problem 23.

24. (II) A heavy version of the electron, called the *muon*, μ, differs from the electron only in that its mass is $m_\mu \simeq 207m_e$. In the muonic hydrogen atom (a proton–muon atom), the reduced-mass effect of Problem 23 is much more important than in ordinary hydrogen. (a) What is the radius of the muonic hydrogen atom in its ground state? (b) Calculate the wavelength of radiation emitted in the transition of a muonic hydrogen atom from the $n = 2$ state to the $n = 1$ state.

25. (II) When an electron and a positron get close together at very low energies, they can form a bound state. This $e^- - e^+$ "positronium atom" lasts for a relatively long time before the electron and the positron annihilate. What are the energy levels of the positronium atom, assuming that the potential energy results from a pure Coulomb attraction? Which transitions are in the ultraviolet, the visible, and the infrared ranges?

42–3 The Exclusion Principle and Atomic Structure

26. (I) For what value of Z is the $n = 2$ level filled?

27. (I) Give the n and ℓ quantum numbers of the levels that are filled in the ground state of an atom of phosphorus, $Z = 15$.

28. (I) An atom has $Z = 32$ electrons. What are the n and ℓ values for the electron that is least tightly bound?

29. (I) What is the lightest element with a single electron in an $n = 3$ level?

30. (I) Sketch the pattern of electron energy-level occupation for the elements $Z = 12$ and 17 (see Fig. 42–16).

31. (II) In multi-electron atoms, the ordering of levels does not coincide with hydrogenlike atoms. In a particularly stable multi-electron atom, the following states are fully occupied: $n = 1, \ell = 0$; $n = 2, \ell = 0, 1$; $n = 3, \ell = 0, 1, 2$; $n = 4, \ell = 0, 1, 2$; $n = 5, \ell = 0, 1$. What is the Z value of this atom?

32. (II) An atom with $Z = 10$ has a closed shell; an atom with $Z = 11$ (sodium) may be viewed as a "nucleus" with a net charge of $+e$ and one electron on the outside (Fig. 42–30). In terms of this simplistic depiction, what would you expect sodium's ionization energy to be? [*Hint*: What levels are filled in the "nucleus"?]

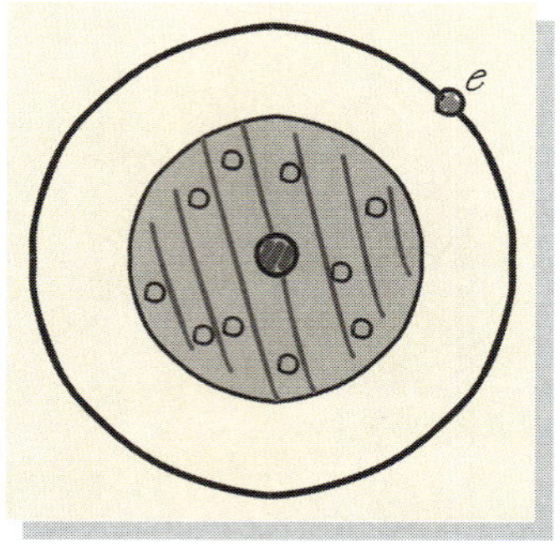

FIGURE 42–30 Problem 32.

33. (II) When an electron in an atomic state characterized by the quantum numbers (n, ℓ) is placed in a magnetic field, then the $(2\ell + 1)$ possible states no longer have the same energy $E_{n,\ell}$. The magnetic field *splits the degenerate levels* so the energy levels have the values $E_{n,\ell} + \kappa\ell B$, $E_{n,\ell} + \kappa(\ell - 1)B$, . . . , $E_{n,\ell} - \kappa(\ell - 1)B$, $E_{n,\ell} - \kappa\ell B$, where κ is a constant. (a) In the presence of a magnetic field, how many different spectral lines are there in the transition $(n = 2, \ell = 1) \to (n = 1, \ell = 0)$? (b) in the transition $(n = 3, \ell = 2) \to (n = 2, \ell = 1)$? [*Hint*: Different transitions in which the energy change ΔE is the same give a single spectral line of frequency $f = \Delta E/h$.]

34. (II) The potential energy U of a magnetic dipole with magnetic dipole moment μ in a magnetic field **B** is $U = -\boldsymbol{\mu}\cdot\mathbf{B}$. The magnetic dipole moment of an electron has magnitude $\mu = (e/m_e)S$. How much more (or less) energy does an electron with spin up have than an electron with spin down in the presence of an external magnetic field with magnitude $B = 0.8$ T, assuming that the field is parallel to the "up" direction?

35. (II) The halogens are a set of elements whose outermost shells of a given n are filled, except for one electron. Find the Z values of all the halogens for which $Z < 100$. [The approximation of successive filling of levels in the order $n = 1, 2, 3, \ldots, \ell = 0, 1, 2, \ldots, (n - 1)$ is incorrect for Z greater than about 20. Thus your numbers will not agree with the periodic table at the high end.]

36. (III) The binding energy of the $n = 1$ electron in an atom for which the nuclear electric charge is $+Ze$ is obtained by taking the hydrogen atom results and replacing e^2 by $(Ze)(e) = Ze^2$. The ionization energies of the least-tightly bound electrons for some atoms with Z values from 1 to 29 (that is, through the $n = 3$ levels) are as follows:

Z:	2	10	11	18	19	28	29
Ionization energy (eV):	24.6	21.6	5.1	15.8	4.3	7.6	7.7

Compare these values with those that you would obtain for the removal of an $n = 1$ electron. To what do you ascribe the huge discrepancy in most cases?

42–4 Molecular Structure

37. (I) What is the minimum (nonzero) rotational energy of the hydrogen molecule?

38. (I) Use the Boltzmann factor $e^{-E/kT}$ to estimate the temperature at which 25 percent of a gas of harmonic oscillators with angular frequency 10^{16} Hz will be excited from the ground state to the first excited state.

39. (I) Use the Boltzmann factor $e^{-E/kT}$ to estimate the temperature at which 25 percent of a gas of diatomic molecules in the ground state ($\ell = 0$) will be excited to the $\ell = 1$ rotational state, given that the rotational inertia of the molecule is $3 \times 10^{-47}\ \mathrm{kg \cdot m^2}$.

40. (II) What are the energies of the three lowest levels in the vibrational spectrum of the HCl molecule (containing ^{35}Cl)? Assuming that the allowed transitions correspond to $\Delta n = 1$, find the wavelengths of the allowed transitions between these levels.

41. (II) The wavelength of the $n = 1 \rightarrow n = 0$ transition in the vibrational spectrum of NaCl is 2.63×10^{-5} m. Use this to estimate the spring constant k in Eq. (42–18). Approximate the mass of the sodium and chlorine nuclei as 23 and 35 times the mass of a hydrogen nucleus, respectively, to compare your result with a calculation from Eq. (42–18). The discrepancy that you will find between the two numbers suggests that Eq. (42–18) is a very crude approximation to the spring constant.

42. (II) The $\ell = 1 \rightarrow \ell = 0$ rotational transition in the lowest electronic state of the KCl molecule has a wavelength of 43.4 mm. Estimate the rotational inertia of the KCl molecule and the equilibrium separation of the atoms. Approximate the mass of the K and Cl nuclei as 39 and 35 times the proton mass, respectively.

43. (II) The energy difference between the lowest state in the CN molecule and the first excited electronic state would give rise to a single spectral line at a wavelength near 387.4 nm if there were no rotations (or vibrations). However, the lowest and first excited states actually consist of a series of rotational states with superimposed energies $\ell(\ell + 1)\hbar^2/2I_0$ and $\ell(\ell + 1)\hbar^2/2I_1$, respectively (Fig. 42–31). Calculate I_0 and I_1 from the following data: Transition ($n = 1, \ell = 1) \leftrightarrow (n = 0, \ell = 0$) gives $\lambda = 387.4608$ nm; transition ($n = 1, \ell = 2) \leftrightarrow (n = 0, \ell = 1$) gives $\lambda = 387.3998$ nm; transition ($n = 1, \ell = 0) \leftrightarrow (n = 0, \ell = 1$) gives $\lambda = 387.5763$ nm. Why, physically, should I_0 and I_1 be different?

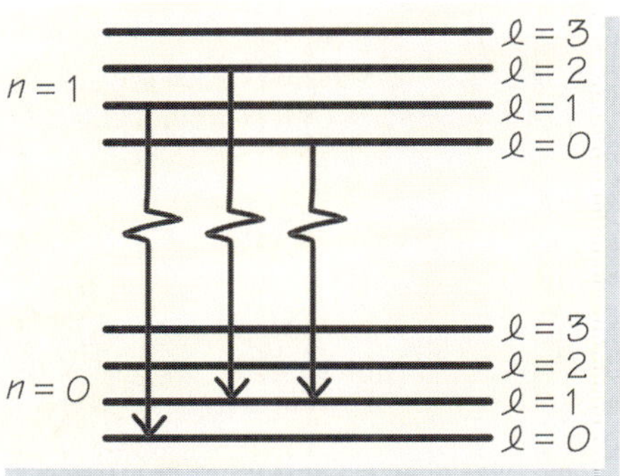

FIGURE 42–31 Problem 43.

44. (II) (a) Use the data of Problem 43 to find the internuclear separation between the C ($A = 12$) and N ($A = 14$) nuclei for the two electronic states $n = 1$ and $n = 0$. (b) Calculate the wavelengths for the transitions ($n = 1, \ell = 3) \rightarrow (n = 0, \ell = 2$). Compare your results with the measured value, $\lambda = 387.3369$ nm.

45. (II) Suppose that, when one atom is separated from another by a distance r, its charge distribution affects the other by giving rise to a small displacement of charges, with separation $d <<$ the atomic radius, and by creating an electric dipole. We know from Eq. (23–14) that the electric field due to such a dipole has magnitude $E \simeq ed/4\pi\epsilon_0 r^3$. (We have ignored angular factors such as sines or cosines.) The charge separation created in the second atom induces a dipole moment of magnitude αE. The coefficient α has the dimensions $4\pi\epsilon_0(\text{length})^3$, as can be seen from $E \simeq ed/4\pi\epsilon_0 r^3$, and this length turns out to be the charge separation d. Now the induced dipole interacts with the electric field E. Show that the resulting interaction leads to a potential energy in the interaction of the atoms of the general form

$$U(r) = -(\text{a constant})\,\frac{e^2}{4\pi\epsilon_0}\,\frac{d^5}{r^6}.$$

This potential energy is what leads to the van der Waals force between the atoms.

46. (III) The rotational motion of a diatomic molecule affects the equilibrium position of the nuclei. If R_0 is the separation for zero angular momentum and R is the average separation when there is rotation, then the rotational energy is $\ell(\ell + 1)\hbar^2/2MR^2$, where M is the reduced mass of the nuclei. In addition, for a given vibrational frequency ω, the vibrational potential energy in the presence of rotation is $\frac{1}{2}M\omega^2(R - R_0)^2$. Calculate the new equilibrium separation R by minimizing $E(R)$, the sum of the two new terms. Treat $R - R_0$ as small. How is the rotational inertia changed, and what effect does this have on the rotational spectrum of the molecule?

General Problems

47. (I) For what value of the angular momentum quantum number ℓ does the quantum mechanical angular momentum have the value $1.00\ \mathrm{kg \cdot m^2/s}$?

48. (II) Suppose that two electrons are in orbit around one proton (an H^- ion), both in an $n = 1$ level. By listing all the potential energy contributions, make a crude guess of how much energy it would take to ionize one of the electrons.

49. (II) An electron in an atom jumps from the first excited state to the ground state. The mean duration for the transition is 4.5×10^{-11} s. What is the uncertainty in the energy value of the first excited state? Give your answer in electron-volts and as a fraction of the energy of the state, which is 2.6 eV.

50. (II) By using the assumptions of the Bohr model, calculate the radius of the ground-state electron orbit for doubly ionized lithium, Li^{2+}.

51. (II) A marble of mass $m = 20$ g moves in a circular orbit near the bottom of a circular bowl (Fig. 42–32). The height of the sides of the bowl is given by $h = \alpha r^2$, where r is the radial distance from the bottom of the bowl. Given that $\alpha = 0.25\ \mathrm{cm^{-1}}$, find the separation between the successive allowed energies of the marble in the bowl. It is not surpris-

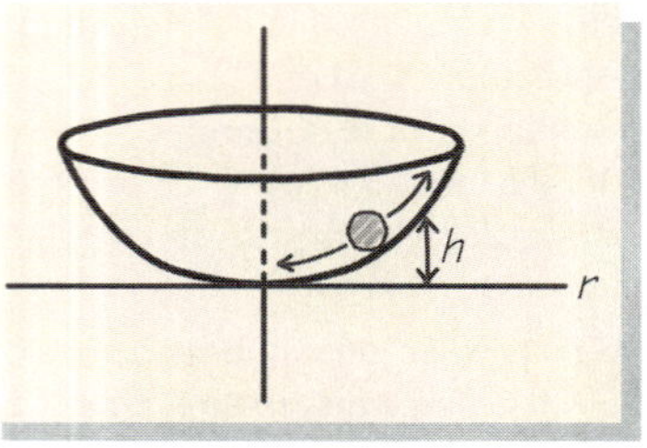

FIGURE 42–32 Problem 51.

ing that we have no intuitive feel for quantum mechanical phenomena.

52. (II) All integer values of the principal quantum number, n, even very large ones, are allowed in atoms. In practice, it is very hard to excite orbits that correspond to large n values in an atom unless the atom is totally isolated. Estimate the largest value of n that would be possible if you could make a gas of *atomic* hydrogen of density $\rho = 1.0 \times 10^{-10}$ g/cm^3. For practical purposes, we regard an interatomic spacing of at least three times the diameter of the large-n atom as total isolation.

53. (II) By finding the momentum of an electron in a circular orbit with orbital angular momentum L and by using your knowledge of the magnetic dipole moment due to a current loop (Fig. 42–33), show that this orbiting electron has a magnetic dipole moment

$$M = \frac{e\hbar}{2m_e}\frac{L}{\hbar}.$$

The quantity $e\hbar/2m_e$, known as the *Bohr magneton*, is the minimum quantized value for the possible magnetic dipole moments of electrons in their orbits.

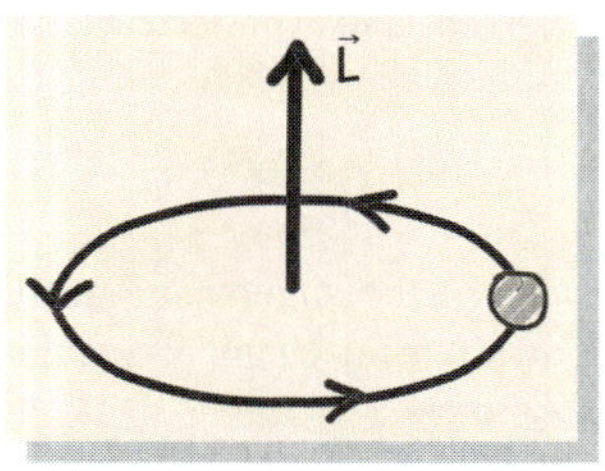

FIGURE 42–33 Problem 53.

54. (III) Repeat your calculation of the energy levels of the Bohr atom, but now assume that the potential energy is given by $U(r) = -(e^2/4\pi\epsilon_0 r) + (\sigma/r^2)$, where σ is a constant. Are your energy values larger or smaller than the corresponding values for the pure Coulomb case? Is your result physically reasonable? Is the effect of the added term more important for large n or for small n, and is that result plausible?

55. (III) According to the Heisenberg uncertainty relations, it is impossible to measure to high precision the position of an atomic electron without making its momentum highly uncertain. By finding the magnitudes of the momentum for the ground state and the first excited state of an electron in a hydrogen atom, calculate just how well the position of an electron in the ground state of hydrogen can be located before you can no longer be sure whether it is in the ground state or the first excited state. It is for the reasons outlined here that it makes little sense to think of an electron as following a classical orbit like those of the planets.

56. (III) Use the position–momentum uncertainty relation to prove that it is not possible to detect orbits in hydrogen, by the following argument: (i) a measurement of the nth orbit must be such that $\Delta x << r_{n+1} - r_n$. Calculate Δx. (ii) This gives rise to an uncertainty in the momentum of the electron in the orbit, and hence in the energy. Calculate the uncertainty in the energy. (iii) Show that the energy is larger than the energy difference between adjacent orbits, so an orbit cannot be "photographed."

57. (III) Consider a particle of mass m that moves in a circular orbit of radius r around a center of attraction that exerts a force on the particle. The particle's potential energy is $U = U_0(r/a)^4$, where U_0 is a constant with dimensions of energy, and a is a constant with dimensions of length. (a) Calculate the total energy (including kinetic energy) in terms of r by using the relation $mv^2/r = |F| = |dU/dr|$. (b) Use the Bohr quantization rules to obtain an expression for the quantized energy values of the particle.

CHAPTER

43

These laser beams are made visible by smoke in the air. Lasers operate as a result of quantum phenomena that are important on a macroscopic scale.

Quantum Effects in Large Systems of Fermions and Bosons

The particles that compose matter are of two types: Fermions have angular momentum $(\ell + \frac{1}{2})\hbar$, where $\ell = 0, 1, 2, \ldots$, and bosons have angular momentum $\ell\hbar$. Identical fermions obey the Pauli exclusion principle, and no more than two fermions are allowed to be in the same quantum state (the state specifies the energy of the particle). Identical bosons, however, obey a kind of "anti-exclusion principle" that translates into a tendency for them to congregate in the same, lowest state. The consequent behaviors of systems with many fermions or bosons are the subject of this chapter. We have already seen how the filling of the energy bands in materials is explained by the exclusion principle as it applies to electrons. Even with no dynamic forces present, identical fermions behave as though there were a repulsive force between them, just because they obey the exclusion principle. This "force" accounts for the incompressibility of matter. The congregating effects of identical bosons explain the laser and superfluidity. Under the right circumstances, identical fermions may in effect congregate in pairs, such that a large number of them behaves like a large system of bosons. This phenomenon lies behind the behavior of superconductors and leads to such properties as flux quantization and the Josephson effect—properties that go beyond the mere absence of resistivity in superconductors.

43–1 THE EXCLUSION PRINCIPLE IN BULK MATTER

Electrons in Metals and the Fermi Energy

The classical treatment of the electrical conductivity of metals (Section 27–4) starts with free electrons in metals. These electrons move under the influence of an externally imposed electric field. Resistance to current flow is due to the collisions between the electrons and ions, which leads to a retarding force and a kind of terminal speed for the electrons called the *drift speed*. The quantum mechanical description of conductivity is also based on the premise of free electrons in metals. The description of the motion of those electrons must be quantum mechanical, however, and this leads to some major differences with the classical estimates.

Picture the electrons as being confined to a one-dimensional box of macroscopic length L—several centimeters, for example. From Eq. (42–1), we get the energy levels for a single electron confined to a one-dimensional box:

$$E = \frac{\pi^2\hbar^2 n^2}{2m_e L^2}.$$

(Here, we have used $\hbar = h/2\pi$.) In a three-dimensional box, there are three such contributions, corresponding to motion in the x-, y-, and z-directions. The result is that

$$E = \frac{\pi^2\hbar^2(n_1^2 + n_2^2 + n_3^2)}{2m_e L^2}, \tag{43–1}$$

with each of the integers n_1, n_2, and n_3 allowed to take the values 1, 2, 3, ... These integers label a state, with the possibility that many states have the same energy. The lowest energy level is the one for which $n_1 = n_2 = n_3 = 1$, so that it is composed of a single unique state. The next energy level consists of three states: $(n_1, n_2, n_3) = (2, 1, 1)$, $(1, 2, 1)$, and $(1, 1, 2)$, and all three states have the same energy. Different states that have the same energy are degenerate; as the n value increases, the degree of degeneracy becomes high.

Suppose that we now start filling the levels described by Eq. (43–1) with electrons. According to the Pauli exclusion principle, each state can accommodate a maximum of two electrons. There are many electrons because, in a metal, one or more electrons per atom are free. *The lowest possible electron energy for a metal that contains N_e electrons corresponds to the states filled from the bottom—the lowest energy level—with two electrons per state.* The values of $n_1^2 + n_2^2 + n_3^2$ under consideration are very large. Suppose that an electron in the metal has energy 1 eV $= 1.6 \times 10^{-19}$ J—a value typical of the energy of electrons in conductors. Let's also take $L = 1$ cm, so that we are dealing with a small but definitely macroscopic piece of metal. It follows from Eq. (43–1) that

$$(n_1^2 + n_2^2 + n_3^2) = 2m_e\frac{EL^2}{\pi^2\hbar^2} \simeq \frac{2(0.9 \times 10^{-30}\,\text{kg})(1.6 \times 10^{-19}\,\text{J})(0.01\,\text{m})^2}{\pi^2(1.05 \times 10^{-34}\,\text{J}\cdot\text{s})^2}$$
$$\simeq 2.6 \times 10^{14}.$$

Thus we are dealing with values of n_1, n_2, and n_3 that are very large. If, for example, we let $n_1 = n_2 = n_3 = 10^8$, then the energy *difference* ΔE between the corresponding level and an adjacent level, with n_1 increased by 1 (but n_2 and n_3 unchanged), is determined by

$$\frac{\Delta E}{E} = \frac{(n_1 + 1)^2 - n_1^2}{n_1^2 + n_2^2 + n_3^2} \simeq \frac{2n_1}{n_1^2 + n_2^2 + n_3^2} \simeq 10^{-8}.$$

The energy levels are so close together that we say they *almost* form a continuum.

When an electron accelerates classically under the influence of an external field, its energy increases smoothly. In quantum mechanics, the electron energy must jump by a discrete amount. Here, the exclusion principle plays an important role: An electron cannot jump to a state of higher energy unless that state is free of electrons. Thus only electrons at the top of the filled levels can be accelerated by the electric field.

The Fermi energy

The highest-filled energy level is called the **Fermi energy**, E_F, after Enrico Fermi (Fig. 43–1). Fermi understood the importance of the exclusion principle for many-particle systems (Fig. 43–2). In general, the de Broglie wavelength of a particle roughly corresponds to the space that it occupies. Thus the closest that two electrons with the same energy and angular momentum can get to each other is about half a de Broglie wavelength. Any closer distance would effectively superimpose the electrons, a situation forbidden by the exclusion principle.

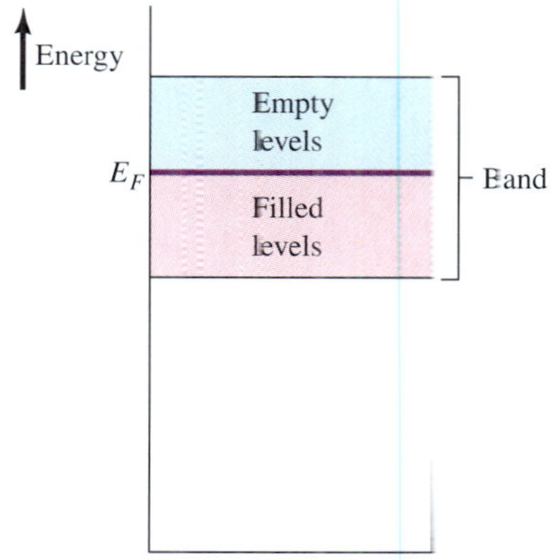

FIGURE 43–1 The energy levels in a crystalline solid form a band structure. In metals at $T = 0$, the levels in the topmost band containing electrons are filled only to the Fermi energy, E_F, of the material, leaving unoccupied levels above.

The Fermi energy can be calculated to within an accuracy of a few percent by taking the closest possible distance between two electrons to be one-half the de Broglie wavelength that corresponds to the *Fermi momentum* $p_F = \sqrt{2m_eE_F}$. If we denote this closest distance by d, then the total number of electrons N_e in a cubical box of sides L is

$$N_e = \left(\frac{L}{d}\right)^3; \tag{43–2}$$

$$d = \left(\frac{N_e}{L^3}\right)^{-1/3} = n_e^{-1/3}. \tag{43–3}$$

Here, n_e is the number density of free electrons in the metal. When we equate this closest distance to half the de Broglie wavelength at the Fermi energy, λ_F, we find that

$$d = \frac{\lambda_F}{2} = \frac{h}{2p_F} = \frac{h}{2\sqrt{2m_eE_F}} = \frac{\hbar\pi}{\sqrt{2m_eE_F}}. \tag{43–4}$$

We can combine Eqs. (43–3) and (43–4) and solve for E_F: $E_F = (\hbar^2/2m_e)(\pi^3 n_e)^{2/3}$. A more precise calculation of the number of electrons that can be accommodated with energy $E < E_F$, including degeneracy factors, leads to the replacement of the factor π^3 by $3\pi^2$, a very minor change:

$$E_F = \frac{\hbar^2}{2m_e}(3\pi^2 n_e)^{2/3}. \tag{43–5}$$

The magnitude of the Fermi energy depends on the density of free electrons. Copper, for example, has one free electron per atom. For copper, then, with an atomic weight of 63.5 g/mol and a mass density of 8.95 g/cm^3,

$$n_e = \left(\frac{1\text{ electron}}{\text{atom}}\right)\left(\frac{6.02\times10^{23}\text{ atoms}}{\text{mol}}\right)\left(\frac{1\text{ mol}}{63.5\text{ g}}\right)\left(\frac{8.95\text{ g}}{\text{cm}^3}\right)\left(\frac{10^6\text{ cm}^3}{\text{m}^3}\right)$$
$$= 8.5\times10^{28}\text{ electrons/m}^3.$$

Substituting this result into Eq. (43–5), we find $E_F = 1.12 \times 10^{-18}\text{ J} = 7.0\text{ eV}$.

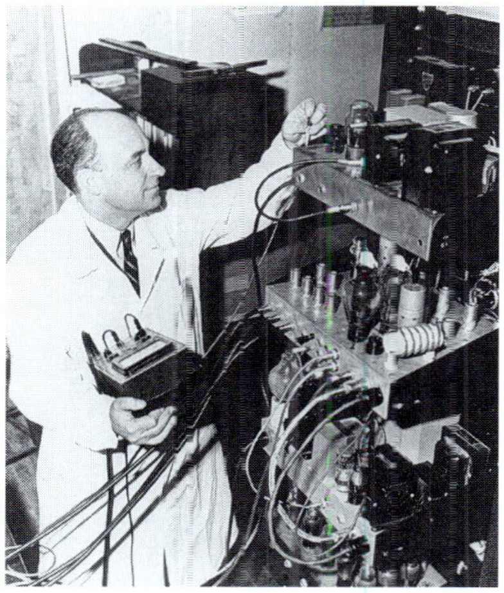

FIGURE 43–2 Enrico Fermi excelled as an experimentalist and as a theorist. Among his other accomplishments, he led the successful effort to build the first controlled nuclear fission reactor, completed in December 1942 in Chicago.

EXAMPLE 43–1 Calculate the Fermi energy for N *neutrinos* (massless particles of spin $\hbar/2$, symbolized as ν) confined to a cubic volume with sides of length L. As far as experiment can tell, neutrinos are massless, and thus obey the relationship $E = pc$ between energy and momentum.

Solution: Neutrinos are fermions with spin $\hbar/2$, so the exclusion principle still applies; that is, $\lambda_F = 2d$, where d is the spacing between neutrinos. The only difference between the treatment of neutrinos and our earlier treatment of electrons is that, for neutrinos, we use the massless-particle relation $E_F = p_Fc$ instead of $E_F = p_F^2/2m_e$. Thus

$$d = \frac{\lambda_F}{2} = \frac{h}{2p_F} = \frac{hc}{2E_F}.$$

Hence

$$E_F = \frac{hc}{2d} = \pi\hbar c n_\nu^{1/3} = \pi\hbar c\left(\frac{N}{L^3}\right)^{1/3}.$$

A more precise counting of states leads to an E_F value that is a factor of $\pi/3$ larger than our crude calculation.

With the picture we have sketched in mind, we can ask what happens when a weak electric field is applied to a metal. Only the electrons at the top of the filled levels can accelerate and move to states of higher energies because, as we have already mentioned, only they find empty levels to occupy (see Fig. 43–1). In the expression for electrical conductivity, Eq. (27–22), with the collision time given by Eq. (19–50), the appropriate quantity to use is the Fermi speed v_F, defined by $E_F = \frac{1}{2}m_e v_F^2$, rather than v_{rms} given by the kinetic theory of gases. Recall that the use of v_{rms} gave an incorrect $\sqrt{T}$ temperature dependence for the resistivity [see Eq. (27–25)]. The effect of temperature changes on the behavior of electrons with a given Fermi energy is negligible. A 100-K change in temperature changes the energy of an electron by $kT = (1.4 \times 10^{-23}\ \text{J/K})\ (100\ \text{K}) = 1.4 \times 10^{-21}\ \text{J} = (1.4 \times 10^{-21}\ \text{J})/(1.6 \times 10^{-19}\ \text{J/eV}) \simeq 0.9 \times 10^{-2}\ \text{eV}$. Compared with the Fermi energy of copper—7 eV—the factor kT is negligible. Thus the resistivity will be temperature independent for reasonable temperature changes.

The Incompressibility of Matter

The exclusion principle plays a crucial role in explaining the incompressibility of matter. A measure of this incompressibility is given by the bulk modulus, B, defined by Eq. (21–5),

$$B \equiv -\frac{p}{\Delta V/V}.$$

Here, p is a pressure change that brings about a fractional change $\Delta V/V$ in the volume of some sample of matter. Because an infinitesimal volume change dV is brought about by an infinitesimal pressure change, we rewrite this definition as

$$B = -V\frac{dp}{dV}. \tag{43–6}$$

Suppose now that the sample of material forms a cylinder of cross-sectional area A and that the pressure is applied to the ends. The work done in compressing the material along the cylinder's axis by an amount dL is $dW = -F\,dL$. Work dW is done, so energy $dE = dW$ is added to the sample. We have

$$dE = dW = -F\,dL = -\left(\frac{F}{A}\right)(A\,dL) = -p\,dV,$$

so the pressure p is the negative of the ratio of the energy change to the volume change:

$$p = -\frac{dE}{dV}. \tag{43–7}$$

When the volume of a metal changes, the number density changes, and so does the total energy. Let's calculate the bulk modulus under the assumption that the *only* resistance of a metal to compression is due to this energy change. The total energy of the free electrons in the material is equal to the number of electrons N_e multiplied by an average energy, which is a value somewhere between 0 and E_F. A calculation that we omit because of its length yields the result that the average energy factor is $\frac{3}{5}E_F$ (see Problem 51):

$$E \simeq \frac{3}{5}E_F N_e = \frac{3}{5}\frac{\hbar^2}{2m_e}\left(3\pi^2\frac{N_e}{V}\right)^{2/3} N_e.$$

Thus, from Eq. (43–7), the pressure is

$$p = -\frac{\mathrm{d}E}{\mathrm{d}V} = \frac{2}{5}\frac{\hbar^2}{2m_e}(3\pi^2)^{2/3}\left(\frac{N_e}{V}\right)^{5/3}. \tag{43–8}$$

According to Eq. (43–6), one further derivative of the pressure is necessary to find the bulk modulus:

$$B = -V\frac{\mathrm{d}p}{\mathrm{d}V} = -V\frac{\mathrm{d}}{\mathrm{d}V}\left[\frac{2}{5}\frac{\hbar^2}{2m_e}(3\pi^2)^{2/3}\left(\frac{N_e}{V}\right)^{5/3}\right]$$

$$= -V\left[\frac{2}{5}\frac{\hbar^2}{2m_e}(3\pi^2)^{2/3}(N_e)^{5/3}\right]\left(-\frac{5}{3}\right)\frac{1}{V^{8/3}}$$

$$= \frac{2}{3}\frac{\hbar^2}{2m_e}(3\pi^2)^{2/3}\left(\frac{N_e}{V}\right)^{5/3} = \frac{2\pi^{4/3}}{3^{1/3}}\frac{\hbar^2}{2m_e}n_e^{5/3}. \tag{43–9}$$

For copper, $n_e = 8.5 \times 10^{28}\ \mathrm{m}^{-3}$, so $B = 6.4 \times 10^{10}\ \mathrm{N/m^2}$. The experimental value of B for copper is $13.4 \times 10^{10}\ \mathrm{N/m^2}$. Given the uncertainties of our estimates, our neglect of interaction between electrons, and the Coulomb repulsion between the ions, the fact that our rough approximation is within a factor of 2 of the experimental result is impressive. The effective repulsion between electrons due to the exclusion principle plays a major role in the high degree of incompressibility of matter.

White Dwarfs and Neutron Stars

The exclusion principle also plays a critical part in the evolution of stars: It is responsible for preventing a star's collapse under the mutual gravitational attraction of its mass. Stars are generally formed from the clumping together of large clouds of gas through mutual gravitational attraction (mainly hydrogen, which forms most of the raw material of the universe). When the hydrogen atoms fall together, gravitational potential energy is converted into kinetic energy. The density and temperature increase. The hydrogen atoms are ionized and, as further compression occurs, the protons that form the hydrogen nuclei come close enough to undergo a variety of reactions. For example, through a sequence of reactions to be described in Chapter 45, four protons can combine to form a helium nucleus and two positrons (the antimatter of electrons). A great deal of energy is released in the process, which is responsible for the luminosity of many stars and represents a kind of thermonuclear burning. While the reactions occur, the temperature of the star remains high, and an equilibrium is established. Once the hydrogen fuel has burned to the point where the process can no longer occur, there is a substantial fraction of helium nuclei in the star. The gravitational pressure shrinks the size of the star further until the temperature is high enough that the helium nuclei begin to undergo reactions that produce still more energy. In a sequence of processes, heavier and heavier elements, up to iron, are produced. Iron does not take part in further thermonuclear reactions, and the reactions stop. However, the material continues its gravitational contraction.

We can estimate the gravitational pressure that acts to compress the matter. Suppose that the star contains N nucleons (protons and neutrons), each of mass M. (We can neglect the electron mass here.) A characteristic gravitational force at the star's surface has magnitude $G(NM)^2/R^2$, and thus a characteristic gravitational pressure (force per unit area) at the surface is

$$P_g = \frac{G(NM)^2/R^2}{4\pi R^2} = \frac{G(NM)^2}{4\pi(\frac{3}{4}V/\pi)^{4/3}} = 0.54\frac{G(NM)^2}{V^{4/3}},$$

where the volume $V = \frac{4}{3}\pi R^3$. Note that the gravitational pressure is directed *inward.* It actually varies through the star; a more rigorous calculation shows that the correct quantity for our calculation is three-fifths of the estimated value. Thus

$$p_g = 0.32\frac{G(NM)^2}{V^{4/3}}. \tag{43–10}$$

As V decreases, the pressure grows and, without some countering pressure, the star would collapse. The exclusion principle provides this countering pressure, called the *degeneracy pressure* of the electrons. This pressure, p_e, given by Eq. (43–8), points outward. (The subscript indicates that the degeneracy pressure is due to electrons in this case.) Note that the degeneracy pressure is largest when the mass of the fermion involved is smallest, which is why we use the degeneracy pressure of electrons rather than that of, say, protons. When the electron degeneracy pressure matches the gravitational pressure, equilibrium is reached. For less massive stars, the end product is a *white dwarf* (Fig. 43–3). Given that N_e = number of protons $\simeq N/2$, we can calculate the radius of the resulting star. We get a radius of some 7000 km for a star the mass of the Sun, 2×10^{30} kg (see Problem 54). This corresponds to a density of 7.5 metric tons/cm^3!

FIGURE 43–3 A white dwarf star is at the center of the gas cloud.

For a more massive star, p_g is larger and, to maintain equilibrium, the factor N_e/V in the degeneracy pressure [Eq. (43–8)] must increase. This means, however, that the Fermi energy of the electrons grows. When the Fermi energy of an electron is comparable to the electron's rest mass, we must treat the electron relativistically. In the extreme relativistic case, the result of Example 43–1 must be used. The total energy of the electron is again N_e multiplied by the average energy; this time, though, p_e is proportional to $(N_e/V)^{4/3}$. This implies that the factor V cancels from the equation $p_e = p_g$, which then cannot be satisfied by a judicious choice of V. The gravitational pressure always exceeds the electron degeneracy pressure when a star is more massive than about 1.4 solar masses (the *Chandrasekhar mass*) and gravitational collapse continues.

EXAMPLE 43–2 Show that if the electron's energy E in a star is so high that its rest mass can be ignored (see Example 43–1), then the gravitational pressure is on the order of magnitude of the degeneracy pressure for a star of one solar mass. (A star is electrically neutral and, to a good approximation, the number of neutrons equals the number of protons when the star is formed.)

Solution: The gravitational pressure is given by Eq. (43–10). We want to compare this to the degeneracy pressure, which can be calculated from the relation $p = -\mathrm{d}E/\mathrm{d}V$, Eq. (43–7). We again use $E \simeq \frac{3}{5}E_F N_e$, but we need the expression for the Fermi energy applicable to massless particles, $E_F = \pi\hbar c(N_e)^{1/3}V^{-1/3}$ (see Example 43–1). Thus,

$$p = -\frac{\mathrm{d}E}{\mathrm{d}V} = -\frac{\mathrm{d}}{\mathrm{d}V}\left[\frac{3}{5}E_F N_e\right] = -0.6N_e^{4/3}\pi\hbar c\frac{\mathrm{d}(V^{-1/3})}{\mathrm{d}V} = -0.2\pi\hbar c N_e^{4/3}\,V^{-4/3}.$$

We can relate the number of electrons to the mass of the star as follows. Electrical neutrality gives us $N_e = N_p$. The total mass of the star, $M_\odot$ (one solar mass), is the total number N of neutrons and protons times the mass M of a proton (or neutron), $M_\odot = NM$. Finally, the number of protons equals the number of neutrons, so $N_e \simeq N/2 = (NM)/2M = M_\odot/2M$. Thus, the ratio of the degeneracy pressure to the gravitational pressure is

$$\frac{p}{p_g} = \frac{0.2\pi\hbar c(0.5M_\odot/M)^{4/3}V^{-4/3}}{0.32G(NM)^2/V^{4/3}} = \frac{0.2\pi\hbar c(0.5M_\odot/M)^{4/3}}{0.32GM_\odot^{\,2}} \simeq 1.2.$$

This crude calculation does give the right order of magnitude. Chandrasekhar showed the more general result that if a star has a mass $sM_\odot$, then $p/p_g = 1.2s^{-2/3}$.

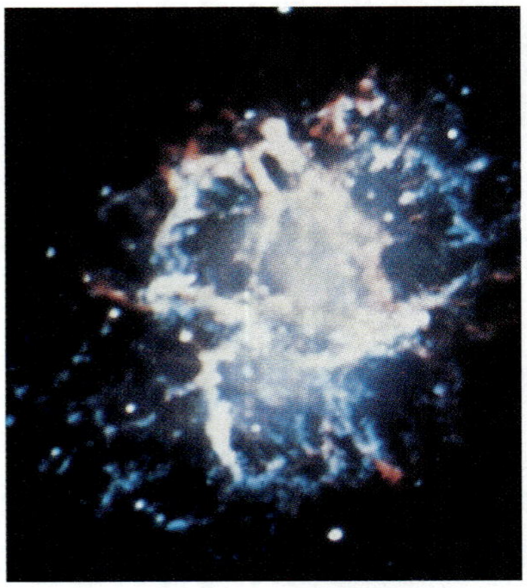

FIGURE 43–4 At the center of the crab nebula, shown here, is a neutron star, identifiable through its properties as a pulsar.

What happens when a star exceeds the Chandrasekhar mass and collapses can crudely be described by the statement that the gravitational pressure forces the electrons in the star to combine with protons into neutrons and neutrinos via the reaction

$$e^- + p \to n + \nu.$$

As we remarked in Example 43–1, neutrinos are apparently massless and, more important, they interact so weakly with matter that they immediately escape from the star. The end result is a star made of N neutrons only: a *neutron star* (Fig. 43–4).

Neutrons are fermions and hence also obey the exclusion principle. We have a new equilibrium condition, $p_g = p_n$, where p_n, the neutrons' degeneracy pressure, has the same form as that of p_e but replaces N_e with N and m_e with M. The relation

$$p_g = 0.32\frac{G(NM)^2}{V^{4/3}} = p_n = \frac{2}{5}\frac{\hbar^2}{2M}(3\pi^2)^{2/3}\left(\frac{N}{V}\right)^{5/3} \tag{43–11}$$

can be solved for V and thus for R. For a star of a few solar masses, numerical calculation yields R on the order of 10 km! The neutron star is as dense as a nucleus, and its compactness is responsible for the many properties that continue to intrigue astrophysicists. The existence of neutron stars was predicted in 1934 by Walter Baade and Fritz Zwicky. The pulsars first discovered by Anthony S. Hewish and collaborators in 1967 were identified by Tom Gold in 1968 as rapidly rotating neutron stars.

EXAMPLE 43–3 Calculate the radius of a neutron star whose mass is three solar masses.

Solution: When the gravitational pressure p_g and the degeneracy pressure p_n are equated, we find that

$$0.32\frac{G(3M_\odot)^2}{V^{4/3}} = \frac{2}{5}\frac{\hbar^2}{2M}(3\pi^2)^{2/3}\left(\frac{3M_\odot}{MV}\right)^{5/3},$$

where we have replaced N, the number of neutrons, by $3M_\odot/M$; $M_\odot$ is the solar mass, and M is the neutron mass. From this expression, we get

$$V^{1/3} = \left(\frac{4\pi}{3}\right)^{1/3}R = \frac{\frac{2}{5}(\hbar^2/2M)(3\pi^2)^{2/3}(3M_\odot/M)^{5/3}}{0.32G(3M_\odot)^2},$$

where R is the radius of the neutron star. Thus,

$$R = \left(\frac{3}{4\pi}\right)^{1/3}\frac{\frac{2}{5}(\hbar^2/2M)(3\pi^2)^{2/3}(3M_\odot/M)^{5/3}}{0.32G(3M_\odot)^2} = 1.0 \times 10^4 \text{ m}.$$

If the mass of a star is so large that even the pressure due to the Pauli exclusion principle applied to the neutrons proves inadequate to resist collapse, continued gravitational contraction occurs and a black hole forms. It is clear that the exclusion principle is a crucial ingredient in the understanding of matter, from ordinary terrestrial form to stellar cores.

43–2 LASERS AND THE BEHAVIOR OF BOSONS

The tendency of identical bosons to cluster into the same quantum mechanical state has its most direct application in the functioning of lasers. To understand lasers, we need some further detail about photon-producing transitions in atoms and molecules.

Transitions Between Energy Levels

The quantum of electromagnetic radiation—the photon—has a wavelength λ or a frequency f related to λ by $f = c/\lambda$. It also has a momentum vector $\mathbf{p}$ oriented in the direction of propagation of the photon. The magnitude of this momentum is given by Eq. (41–17):

$$p = \frac{hf}{c}.$$

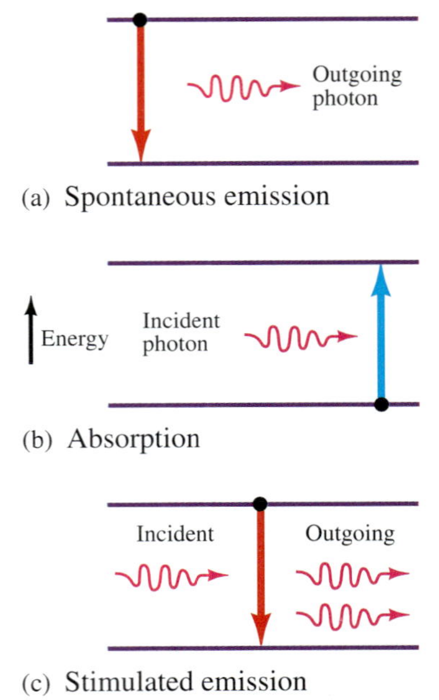

FIGURE 43–5 (a) In spontaneous emission, a photon is produced as an electron drops from an excited atomic state to a lower-lying level. (b) An atom can absorb a photon if the photon's frequency matches the energy difference between two atomic levels of the proper ℓ values. (c) In stimulated emission, the decay from an excited state occurs more readily in the presence of photons in the same state as the photon to be emitted.

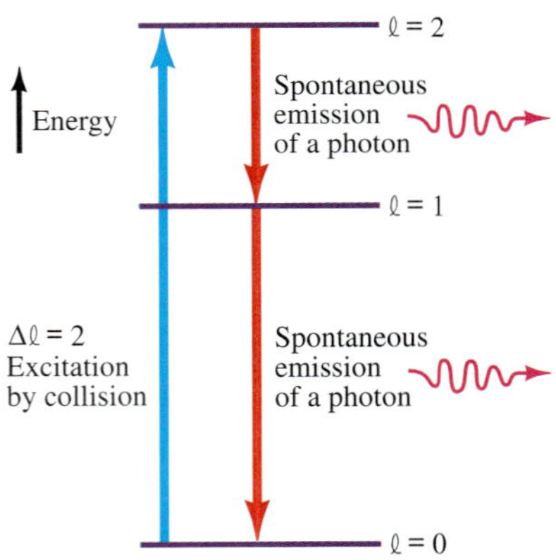

FIGURE 43–6 Excitation and deexcitation of atomic levels. An $\ell = 2$ level cannot rapidly decay by single photon emission to the ground state; a jump to an intermediate $\ell = 1$ level followed by a jump to the $\ell = 0$ level, with two photons emitted successively, is possible. Direct excitation from the $\ell = 0$ level to the $\ell = 2$ level is possible in collisions.

Photons are emitted (or radiated) when the electrons of atoms (or of molecules) undergo a transition ("jump") from a higher energy level to a lower energy level, as in Fig. 43–5a. Such photons are said to be *spontaneously emitted*. We have already mentioned that transitions in which photons are emitted are subject to the law of conservation of angular momentum; moreover, because photons act as though they have angular momentum quantum number $\ell = 1$ (with small mixtures of $\ell = 2$ and higher values), the most frequent transitions occur between atomic or molecular levels for which $|\ell_i - \ell_f| = 1$.

In order for an atom to radiate, it must first be excited into a higher energy level. To excite the atom, an atomic electron must be given the proper energy and angular momentum. One method of doing this is to bombard the atom with photons whose energy is the energy required for the transition. In this case, an atomic electron *absorbs* a photon and jumps into a higher energy state (Fig. 43–5b). If electrons in the ground state have $\ell = 0$, then (to high accuracy) the only states that can be reached in this way are those in which the electrons have $\ell = 1$. It is possible to reach other states—in this case, $\ell = 2, 3, \ldots$—by other mechanisms, such as collisions with other atoms. In fact, the discovery of states with $\ell > 1$ was possible only through these other excitation methods. How does an electron in a state with $\ell = 2$ fall back to the ground state? If an $\ell = 1$ energy level lies between the $\ell = 2$ state and the $\ell = 0$ ground state, the electron can jump first to the intermediate $\ell = 1$ state by emitting a photon, then jump to the $\ell = 0$ state by emitting a second photon (Fig. 43–6). What happens, though, if there is no intermediate state through which the electron can cascade downward? A jump accompanied by photon emission is still possible, but it is on the order of 10^4 times less probable. If the change in ℓ is 3 rather than 2, then a jump accompanied by photon emission is 10^8 times less probable. These probabilities can be illustrated by thinking of a given excited energy level as a tub filled with water. A large drain that empties it in a short time corresponds to the possibility of transitions in which the angular momentum changes by one unit of $\hbar$. There are two tiny holes that are 10^4 and 10^8 smaller in area than the drain hole. If the drain is open, we can ignore any leakage through the small holes, but if the drain is blocked (if there are no possible $\Delta\ell = 1$ transitions), leakage occurs through the next smallest hole. It takes much longer to empty the tub through the tiny hole. If there are no possible $\Delta\ell = 1$ transitions, then the energy level is called *metastable*, or nearly stable. Metastable states play a critical role in the operation of the laser.

We have thus far treated the possibility of *spontaneous transitions* in atoms. In 1917, Albert Einstein used thermodynamics together with the rudimentary quantum theory in existence at the time to predict the possibility of *stimulated transitions*. Consider an electron in a metastable state. It will stay in that state for a long time (long on the atomic scale, typically 10^{-8} s) before it decays with the emission of a single photon into the ground state. The energy of the photon will be $hf = E_i - E_f$, where E_i and E_f are the initial and final electron energies. If, however, photons of frequency f are present in the vicinity of the atom, then these photons will stimulate the transition, which occurs more rapidly as the number of stimulating photons grows. We say that there is *stimulated emission* (Fig. 43–5c). The more external photons, the more rapidly the stimulated emission occurs. There is another very important effect: *Because they are bosons, the photons emitted by stimulated emission will preferentially be in the same quantum state as the stimulating photons*; that is, they will have the same momentum, frequency, and phase. A *coherent* state of many photons is formed, a state in which the electromagnetic fields associated with the photons reinforce. This coherent state describes a single, intense, monochromatic plane wave, or beam. Moreover, because the photons have the same momentum (both magnitude and direction), the beam is extremely well collimated. The functioning of lasers relies on the existence of such stimulated transitions. In fact, *laser* is an acronym for "light amplification by stimulated emission of radiation."

Suppose that we have a collection of atoms in a cavity, each of which is in a metastable state. The presence of many photons of energy hf will stimulate a very rapid transition to the ground state, where f is the frequency corresponding to the transition to the ground state; otherwise, the atoms will remain in the metastable state for a long time. We may then picture the following sequence: After a long time, one atom decays to the ground state and emits a photon of energy hf. The photon is constrained to stay in the cavity by means of mirrors positioned to reflect the photon back into the cavity (Fig. 43–7a). Other atoms in the metastable state, stimulated by the first photon, undergo a slightly faster transition to the ground state and produce more photons in the same state. The ever-increasing number of the "right" photons—those that induce stimulated emission—rapidly produces an avalanche of decays. These photons form a well-collimated, coherent *laser beam* (Fig. 43–7b).

The preceding description represents only the simplest type of laser. It is possible to vary the mechanisms involved to produce lasers of different types. An example of a different mechanism is provided by the helium–neon laser, in which the working material is a mixture of helium and neon gas in a tube (Fig. 43–8a). This laser illustrates one way to provide a metastable energy level. Electrons in the helium atom are raised from the ground state ($n = 1$, $\ell = 0$) to a pair of metastable excited states—one 18.6 eV and the other 19.8 eV above the ground state—in which one of the electrons is in an $n = 2$, $\ell = 0$ level (Fig. 43–8b). These excitations cannot occur through absorption of a photon because $\Delta\ell = 0$ in the transition. Rather, the excitation occurs through collisions with free electrons introduced by an electrical discharge through the gas. The same reason that the state cannot be excited by photon absorption ensures that deexcitation will not occur through photon emission. When many helium atoms are put into these metastable states, we say that the states are highly populated.

The metastable helium levels are *not* the levels that emit photons and thereby produce the laser beam. Instead, the helium atoms in the metastable state collide with the neon atoms of the gas and transfer the energy of the excitation to the neon. Neon has two $\ell = 0$ levels at approximately the same energies as those of the metastable states in helium, so the energy is easily transferred. In the transfer, the helium goes back to its ground state, and a large population of neon atoms in the two excited $\ell = 0$ levels is formed. The neon atoms in the two $\ell = 0$ levels can now decay to lower-lying $\ell = 1$ states. The upper $\ell = 0$ state has a low-frequency transition with an infrared wavelength of 3392 nm. Decays to a lower-lying $\ell = 1$ level with radiation of wavelength 633 nm (red) are also possible. The lower of the two $\ell = 0$ levels decays to an $\ell = 1$ level by a laser transition in the infrared at wavelength 1153 nm. These neon atoms decay further by spontaneous (nonstimulated) photon emission to an $\ell = 0$ excited state, and the electrons in these states are deexcited by collisions with other atoms and with the cavity walls. The neon atoms in their ground state are then reexcited to the higher $\ell = 0$ states by collisions with the helium atoms in their metastable states. We call this continual reexcitation *optical pumping*.

The emitted photons are contained within the cavity, which often takes the form of a narrow tube, by mirror reflectors at each end of the tube. The photons bounce back and forth between the reflectors and stimulate transitions of the metastable atoms in the tube. Because the photons also have to be extracted to form a coherent beam, the mirrors at the ends of the tube are not perfect reflectors. Instead, one of the mirrors is partly transparent. The fraction of the radiation reflected back into the tube is over 99 percent, and thus the cascade of stimulated emission occurs rapidly. The photons in the tube emerge as a coherent, monochromatic, and well-collimated beam through the partly silvered mirror. The coherent state is so intense that even though only 1 percent of the photons emerge, the beam is still useful.

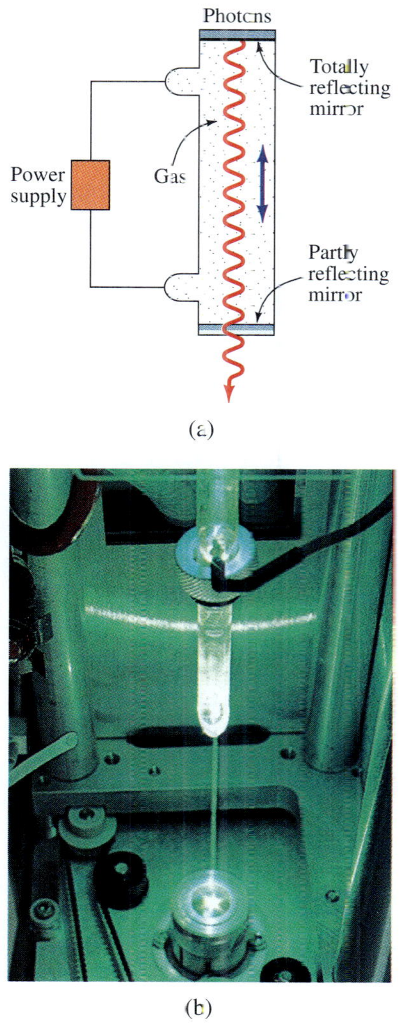

FIGURE 43–7 (a) Schematic diagram of a laser. The mirrors are present to contain photons and provide the conditions necessary for stimulated emission. (b) A krypton laser under test.

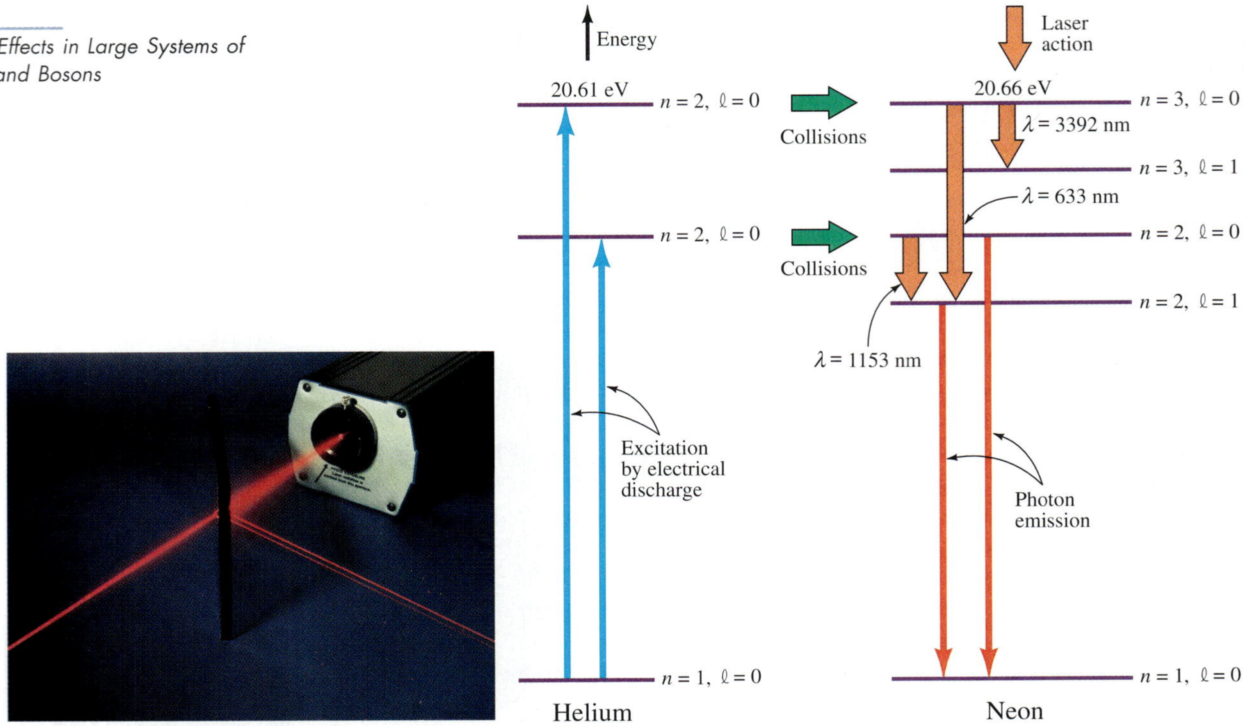

(a) (b)

FIGURE 43–8 (a) This beam from a helium-neon laser is partially transmitted and partially reflected. The two parts of the reflected beam are due to reflection from the upper and lower surfaces of the glass plate. (b) Energy levels in helium and neon relevant to the operation of the He–Ne laser. The wide amber arrows represent the transitions that produce the laser light; we say that these transitions are involved in "laser action" when photons of the proper frequency are present.

In summary, the construction of a laser requires:

1. A collection of atoms that contain a metastable energy level into which many atoms can be excited.

2. A mechanism for holding emitted photons in a cavity containing excited atoms so that there is a massive stimulated deexcitation of the excited atoms (a laser transition).

3. A mechanism for repopulating the excited level after the laser transition has taken place.

4. A way for the coherent laser beam to be extracted sufficiently well to be useful.

APPLICATION

Lasers in ranging, holography, isotope separation, and medicine

Some Uses of Lasers

We described in Chapter 36 how the ability to measure time intervals accurately allows us to use lasers to perform *ranging*, or a measurement of distance (see the subsection Corner Reflectors on p. 970). The coherence of the laser beam is crucial in *holography*, as we described in Chapter 39. Lasers are an important part of optical readers, as in CD players and checkout counters. Another application is isotope separation, the significance of which was discussed in Chapter 19 (see the Application box on p. 521). Laser separation of isotopes is a commercially important separation method, and it works as follows: The spectra of different isotopes of an element differ slightly because of tiny differences in their nuclear masses and magnetic dipole moments. Both properties affect the atomic energy levels. If atoms consisting of a mixture of isotopes are irradiated with a laser beam tuned very precisely to a transition frequency of only one of the isotopes, then the atoms of only those isotopes will be excited. For example, a suitably tuned laser beam can excite atoms of ^{235}U but not atoms of ^{238}U. A second laser beam with enough energy to ionize the already excited ^{235}U atoms but not ^{238}U atoms is then applied. The charged ions of ^{235}U can now be separated by electric fields from the un-ionized atoms of ^{238}U.

Lasers also have important applications in medicine. One of their first uses was to reattach detached retinas. The lens of the eye focuses a laser beam onto a small area of the retina, which fuses to the tissue from which it has become detached (Fig. 43–9). Because the energy is delivered in a short time, there is no need to immobilize the eye. Infections are less likely to occur than with surgical procedures. Lasers can also be used to cauterize internal wounds and stop bleeding. We can mention also the potential application of lasers for controlled thermonuclear fusion reactions (to be discussed further in Chapter 45). Laser beams focused on lightweight nuclei may be able to provide the energy necessary for them to fuse and produce additional energy. Figure 43–10 illustrates an apparatus designed for this job.

Finally, lasers hold promise as a way to manipulate microscopic systems as small as individual atoms.† The ability to do this will be of increasing importance in quantum engineering (see Chapter 44).

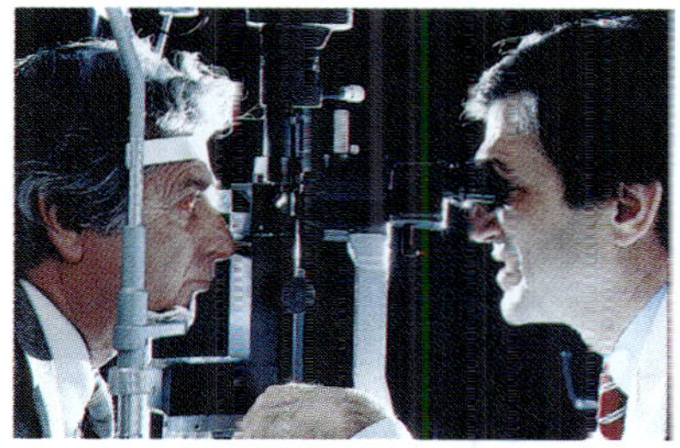

FIGURE 43–9 The reattachment of detached retinas by means of lasers is now a routine medical procedure.

43-3 SUPERCONDUCTIVITY

Cooper Pairs and the BCS Theory

In our discussion of fermions and the Fermi energy in metals, we ignored all dynamic interactions among the fermions. How can this make sense, given that electrons repel each other because of their electric charges? The exclusion principle explains why electron interactions are not very important in ordinary metals. Consider two electrons in different states. Their interaction would manifest itself in some change of state—the only way we would see that something had happened. However, the vast majority of electrons cannot change states because the states above and below their own state are already fully occupied, and the exclusion principle forbids further occupation. Interactions do lead to some modification of the predictions of the free-electron theory, and these modifications are well understood.

The properties and applications of superconductors and possible uses of high-temperature superconductivity were discussed in the optional Sections 27–6 and 32–6.

The phenomenon of *superconductivity* relies on the interactions between the electrons and the vibrating lattices of ions in a metal. For complicated reasons, the electron–ion interaction leads to a weak attraction between those electrons whose energies are close to the Fermi energy. Leon Cooper discovered this attraction, which leads to the formation of weakly bound pairs of electrons known as **Cooper pairs**. The two electrons that pair up may be as far apart as 100 nm (recall that the separation between ions in the lattice is in the range 0.1–0.4 nm), so the pairs do not form small, atomlike structures. Nevertheless, the motions of the two electrons in a pair are correlated and, in a given region, there are many Cooper pairs whose "bonds" are entangled. Because the pairs have lower energies than free electrons do (it takes some energy to separate the electrons that form a pair), an energy gap of width $2\Delta \simeq 10^{-3}$ eV develops about the Fermi energy. The gap width is small compared to the Fermi energy itself.

Cooper pairs

John Bardeen, Leon Cooper, and Robert Schrieffer constructed a theory in 1957—the so-called BCS theory—that explained satisfactorily all of the observations about superconductors known at the time. A pair has a total angular momentum of 0 or $\hbar$. Although—strictly speaking—a pair cannot be treated as a boson such as a helium atom, a pair shares some crucial properties with bosons. In particular, the BCS theory proposed that the Cooper pairs all condense into the same state when the material's temperature drops below a critical temperature $T_c \simeq \Delta$, the gap width. This condensation signals an abrupt change in the properties of the metal, a change somewhat analogous to the phase change from liquid to solid.

FIGURE 43–10 The intense laser beams of the OMEGA system at the University of Rochester focus simultaneously on a pellet in order to produce a state hot enough and dense enough to cause thermonuclear fusion.

Electromagnetic Properties of Superconductors

The most striking property of superconductors is that their resistivity is zero below T_c. This phenomenon is explained by the coherence of the superconducting state. In a normal metal, when a current starts and there is no potential difference to maintain it, the

†See S. Chu, "Laser Trapping of Neutral Particles," *Scientific American*, Feb. 1992, p. 70.

electrons scatter from the lattice and give up energy. The energy appears as ohmic heating and the current rapidly decays. In a superconductor, however, a large number of electron pairs move together in a coordinated way. For the superconductor to lose energy, *the entire coherent state would have to be broken up*, and this would require a large amount of energy to be supplied at once. The mechanisms of ordinary electrical resistance cannot work if a macroscopic number of electron–electron pairs (say, 10^{22}) must be slowed down *all at once*. An analogy is the difference between pushing your hand through liquid water (a "normal" metal) and pushing your hand against an icicle (a "superconductor"). In the first case, there are easy ways to lose just a little energy: You can form arbitrarily small waves and eddies. In the second case, you can lose a large amount of energy only all at once, when you break the icicle.

The magnetic properties of superconductors were first described in Chapter 32.

Another property, the *Meissner effect*, is that no magnetic field can penetrate a superconductor, and that a magnetic field already present is expelled when a metal is cooled below T_c (Fig. 43–11). The inability to establish a magnetic field within a superconductor is also a consequence of the coherence of the superconducting state. If you were to try to increase a magnetic field inside a superconductor, you would induce an emf within the material by Faraday's law. A current would be induced to oppose the change in magnetic flux. But, because all the Cooper pairs act together, even the tiniest change in magnetic flux can generate a current sufficiently large to cancel the inducing field entirely.

Flux quantization

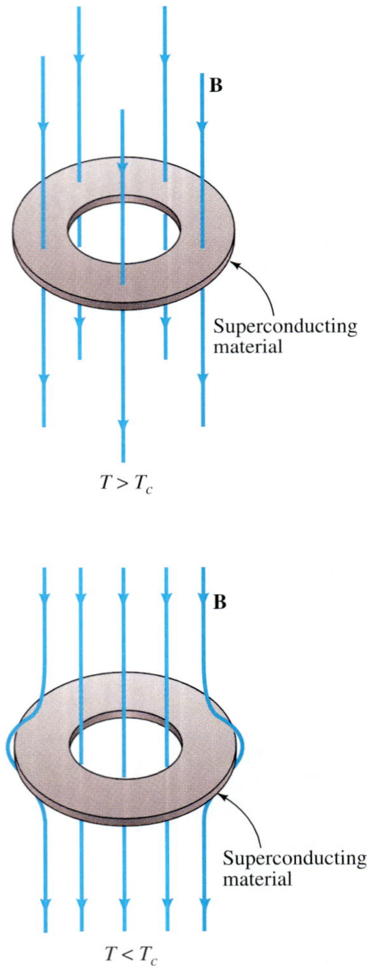

FIGURE 43–11 Magnetic field lines expelled by a ring of material that makes a transition to the superconducting state when the temperature drops below the critical temperature T_c. Flux is trapped by the hole.

Flux Quantization

Suppose that we place into a magnetic field a ring of superconducting material that is above its critical temperature. The ring lies in a plane perpendicular to the field direction. We now lower the temperature of the material below T_c so that the material becomes superconducting. The field is expelled, and all the field lines will be either outside the ring or confined within the ring.

The BCS theory can be used to show that the coherence of the superconducting state implies that *the magnetic flux through the ring is quantized in units of h/2e*:

$$\Phi_B = \frac{nh}{2e} \equiv n\Phi_0, \qquad \text{where } n = 1, 2, \ldots . \tag{43–12}$$

The quantity Φ_0 is the *magnetic flux quantum*, with value 2.07×10^{-15} Wb $= 2.07 \times 10^{-15}\ \mathrm{T \cdot m^2}$. The charge $2e$ that appears here is the charge of a Cooper pair. The prediction that flux is quantized has been successfully tested by experiment.

EXAMPLE 43–4 A single unit of the quantized flux Φ_0 is confined within a cylinder of copper wire of diameter 1.3×10^{-5} m. What is the magnitude of the magnetic field in the wire, assuming that it is uniform throughout the wire?

Solution: The unit of flux is $\Phi_0 = 2.07 \times 10^{-15}\ \mathrm{T \cdot m^2}$. If the magnetic field is uniform within a wire of radius R, its magnitude is given by

$$B = \frac{\Phi_0}{\pi R^2} = \frac{2.07 \times 10^{-15}\ \mathrm{T \cdot m^2}}{\pi (0.65 \times 10^{-5}\ \mathrm{m})^2} = 1.6 \times 10^{-5}\ \mathrm{T}.$$

This is only about a factor of 6 less than Earth's magnetic field and is therefore quite measurable.

Tunneling of Pairs and the Josephson Effects

Tunneling, which we studied in Chapters 7 and 41, can occur across junctions between normal metals, between normal metals and superconductors, and between superconductors. Here, "junction" refers to a very thin insulating strip between two pieces of metal. The insulating strip is formed from a material that can be evaporated and then deposited in thin, uniform layers. If the strip is thin enough (10 nm to 20 nm), an electron can tunnel across it from one metal to the other. Let's consider two normal met-

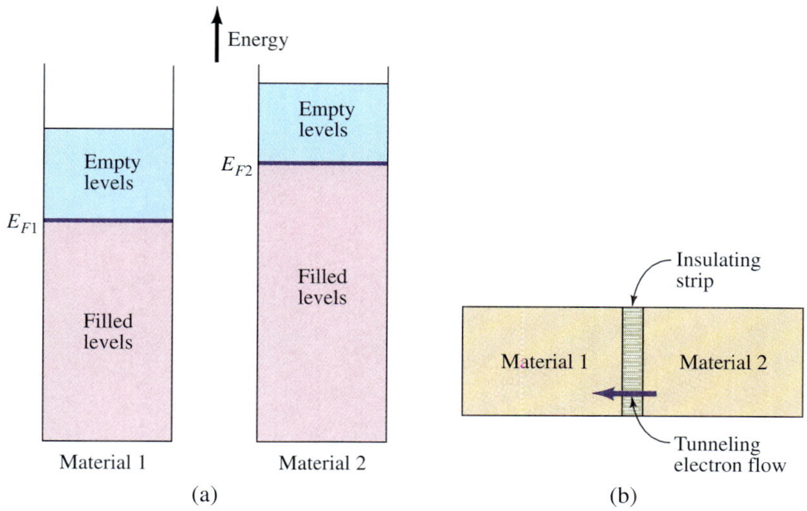

FIGURE 43–12 (a) Two metals with different Fermi energy levels. (b) The metals are joined with an insulating strip between them.

als labeled 1 and 2, with Fermi energies E_{F1} and E_{F2}, respectively. If $E_{F2} > E_{F1}$ (Fig. 43–12a), then, since they find empty space to occupy, electrons can tunnel through the barrier formed by the insulating strip, moving from metal 2 to metal 1 (Fig. 43–12b). If the two metals are identical, there will be no tunneling—at least at very low temperatures—because there are no empty states to occupy. However, if an external potential V_{ext} is imposed, the levels on one side of the junction will be lowered relative to those on the other side of the junction, and tunneling is again possible (Fig. 43–13).

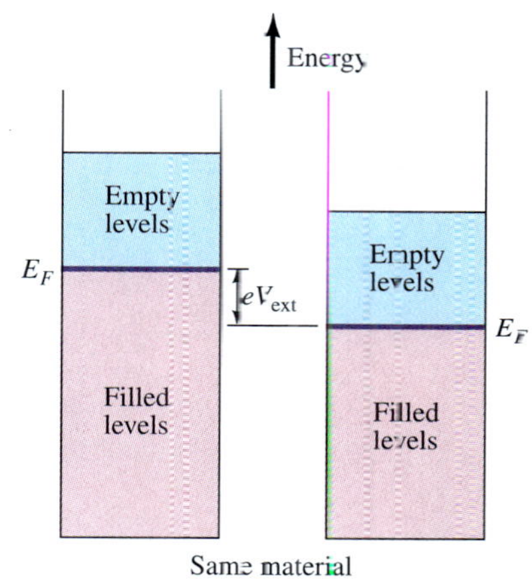

FIGURE 43–13 When an electric potential that lowers the levels on the right-hand side compared to those of the left-hand side is applied, a tunneling current flows from left to right across the insulating strip. In this case, the metals on both sides of the strip are the same.

Now, suppose that we have a junction between a metal in its normal state and a metal in a superconducting state (Fig. 43–14a) and that each piece has the same Fermi energy. With no external potential, no electrons flow. Suppose that an external potential V_{ext} that lowers the Fermi energy of the superconductor is applied. A current will flow between normal metals, but because of the energy gap in the superconductor, the external potential must exceed a minimum value before empty levels become available for tunneling into (Fig. 43–14b). This minimum value is given by $eV_{min} = \Delta$ and, by varying the external potential and observing the onset of tunneling, it is possible to measure the gap size. Figure 43–14c illustrates the relationship between current and potential in this case.

Let's now consider two superconductors separated by a thin insulating strip, an arrangement known as a *Josephson junction*. At first, there might appear to be little current across the barrier. If the probability of a single electron tunneling through is very small, then, because the probability of two independent electrons tunneling through is the product of the tunneling probabilities of the individual electrons, there would be very little tunneling for an electron pair. In 1962, Brian Josephson noted that if the barrier is less than 1 nm thick, the two superconductors form a single, coherent quantum system. As a result, what tunnels through the barrier is not two individual electrons, but a pair of electrons as a unit. This has two consequences (we shall prove neither here): First, *even in the absence of a potential difference between two identical superconductors, a tunneling current will flow across a Josephson junction.* This is called the *DC Josephson effect*. Second, *when there is a constant potential difference V across the barrier between two superconductors, the current that flows between them oscillates with angular frequency*

$$\omega = \frac{2e}{\hbar} V. \tag{43–13}$$

This phenomenon is called the *AC Josephson effect*. Because frequencies can be measured with great precision, the AC Josephson effect provides physicists and engineers with the most accurate way known of standardizing voltage measurements. Equivalently, it allows the most precise measurement possible of the fundamental ratio $e/\hbar$.

APPLICATION

High-precision voltage measurement

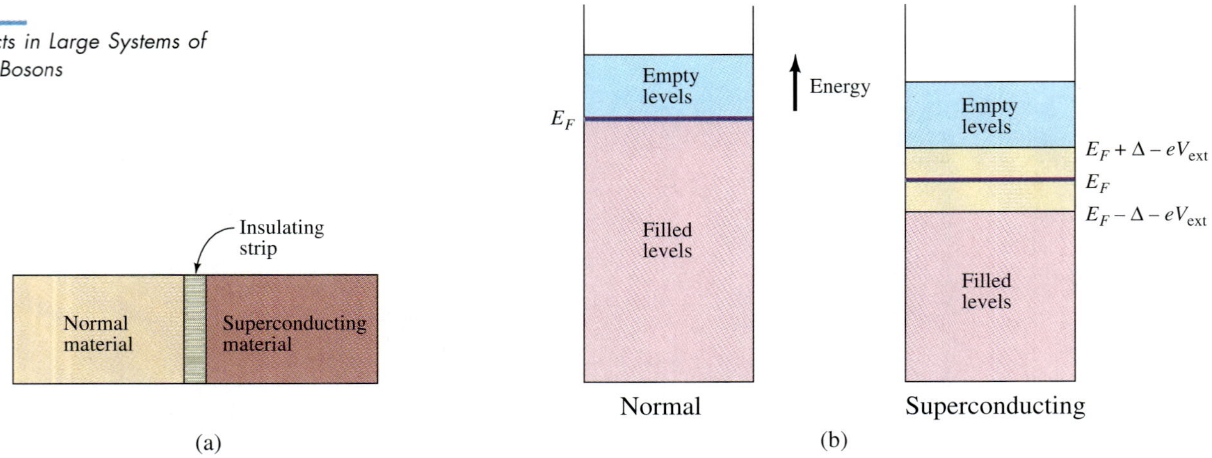

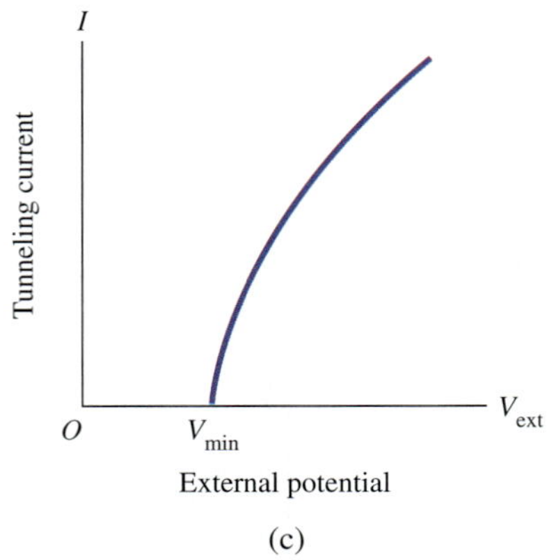

FIGURE 43–14 (a) A metal in its normal state separated by a thin insulating strip from the same metal in its superconducting state. (b) An energy diagram of this situation, including an external potential, V_{ext}, applied so that the energy levels in the superconducting metal are lowered with respect to those in the normal metal. Because an energy gap is present in the superconductor, the potential must be a certain minimum size before electrons from the normal metal can tunnel into the empty levels of the superconductor. This minimum value of V_{ext} is given by $eV_{min} = \Delta$. (c) The flow of tunneling current in the junction described in Fig. 43–14a as a function of V_{ext}. V_{ext} must have the minimum value V_{min} before current can flow.

Consider two Josephson junctions connected in parallel (Fig. 43–15). Like the interference pattern produced by an electron passing through a double slit, an interference pattern appears in the net current. The interference pattern is affected by a magnetic flux enclosed by the two arms of the pair of junctions. The flux plays the same role in the pair of Josephson junctions that the slit separation plays in a double-slit experiment. The dependence of the current on the enclosed magnetic flux Φ_B is given by

$$J = J_{max} \cos\left(\frac{e\Phi_B}{\hbar}\right). \tag{43–14}$$

APPLICATION

SQUIDs

This phenomenon allows us to measure magnetic flux precisely. The SQUID (Superconducting QUantum Interference Device), which is based on the phenomenon described, is widely used for high-precision measurements of very tiny magnetic fluxes

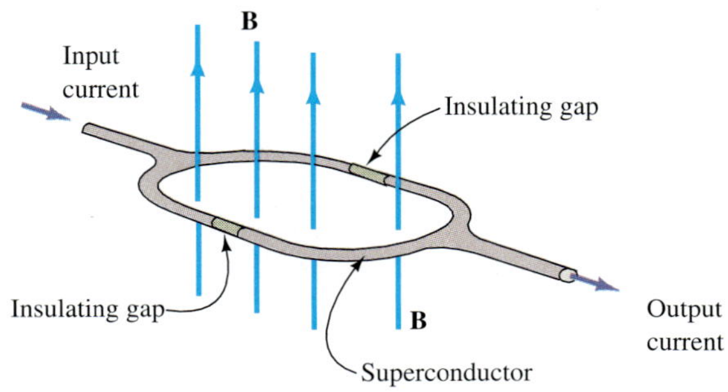

FIGURE 43–15 The current that passes through a pair of Josephson junctions enclosing a magnetic flux varies with that flux.

(Fig. 43–16). Magnetic fields as small as 10^{-13} T have been measured. With SQUIDs, an esoteric quantum mechanical phenomenon has become a major technological tool in medical diagnostics, where tiny variations in the electric currents generated by the heart or the brain can be measured by the magnetic fields they produce.

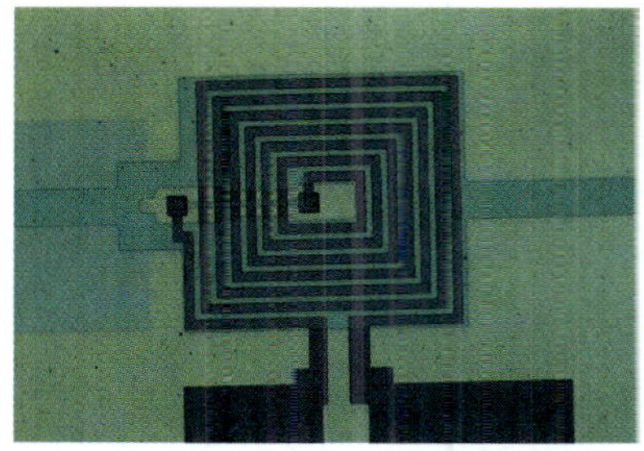

FIGURE 43–16 This SQUID (Superconducting QUantum Interference Device), manufactured from high-T_c superconducting material, is used as a high-sensitivity magnetometer, a device that measures magnetic fields.

43–4 SUPERFLUIDITY AND LIQUID HELIUM

Superfluidity

In addition to superconductivity, there is another technologically important low-temperature phenomenon that involves the condensation of bosons into a macroscopic coherent state. Helium atoms, ^{4}He, with a nucleus consisting of two protons and two neutrons, are bosons; their nucleus has an angular momentum of zero, and the two electrons of helium atoms have an angular momentum that is an integral multiple of $\hbar$. Helium condenses into an ordinary liquid at low temperatures; more remarkably, at a temperature of 2.17 K, there is another change of state. As the temperature decreases below this point, an increasing fraction of the liquid helium flows with no internal friction (viscosity). This property is called **superfluidity**. Superfluid helium can flow freely through the narrowest of channels. It creeps up the sides of a beaker containing it and flows to the outside (Fig. 43–17). It can be used to make the most nearly perfect (frictionless) heat engine—one that converts thermal energy to kinetic energy with no moving parts save that of the liquid helium itself. Finally, and most important technologically, it makes a nearly perfect heat conductor.

The fact that atoms of ^{4}He are bosons allows us to explain these properties. In an ordinary, nonsuperfluid, liquid, the excitations do not involve the entire fluid. Arbitrarily small amounts of energy can be lost, leading to viscosity at any speed. In superfluid helium, however, the helium atoms congregate in the same (coherent) ground state. When energy is added to the system, excitations of the entire fluid are produced. In particular, the lowest excitations are compressional, or sound, waves. (These waves are of the same type that form sound waves in solids.) Very little energy must be added to the system to produce these excitations. However, the energy of the excitation is not the whole story. Even though sound waves are the excitations of lowest energy, it is difficult to excite them mechanically, and this class of excitations is not useful for carrying off mechanical energy; that is, they produce no internal friction. To understand the significance of this remark, let's consider the process by which friction is produced.

FIGURE 43–17 The fountain effect of superfluid helium, which acts as though it has no viscosity.

Suppose that we drag a sphere of mass M through liquid helium. If the sphere gives up some energy and momentum to the liquid, it will create an excitation. Such excitations are discrete quantum mechanical states that correspond to motion of the entire system of liquid helium, and the excitations have a definite momentum $\mathbf{p}$ and energy E. Momentum conservation states that

$$M\mathbf{v}_i = \mathbf{p} + M\mathbf{v}_f,$$

where $\mathbf{v}_i$ is the initial velocity of the sphere and $\mathbf{v}_f$ is its velocity after it has given up some energy. Then $\mathbf{v}_f = \mathbf{v}_i - (\mathbf{p}/M)$. Similarly, energy conservation gives

$$\frac{1}{2}Mv_i^2 = E + \frac{1}{2}Mv_f^2.$$

We substitute $\mathbf{v}_f$ into the energy conservation equation:

$$\frac{1}{2}Mv_i^2 = E + \frac{1}{2}\left(Mv_i^2 - (2\mathbf{v}_i \cdot \mathbf{p}) + \frac{p^2}{M}\right).$$

If we drop the last term on the right because M is large (macroscopic), we can then solve for E:

$$E = \mathbf{v}_i \cdot \mathbf{p}. \tag{43–15}$$

APPLICATION: SUPERCONDUCTIVITY IN MRI

The magnetic solenoids used in medical magnetic resonance imaging (see Chapter 32) need an inside diameter large enough to allow a person to enter (Fig. 32–18). Unless the current flow in ordinary copper wire is less than 300 amperes per square centimeter of winding, it is impossible to remove the heat produced in the resistance. In order to obtain a typical MRI field strength of 4 T, an ordinary copper winding would then have to be a meter thick, and there would be a resulting power dissipation of several hundred kilowatts per meter of winding length. The cost of both the energy and the cooling is, as a consequence, impracticable.

Solenoids producing 4–5 T can be constructed using superconducting tin–niobium stranded wire as discussed in the chapter. Such a solenoid runs in the persistent current mode and, once energized, needs no more electrical power to maintain the magnetic field. The only power consumption is in the liquefier producing the liquid helium that maintains a superconducting state. Today, for each watt of heat evaporating liquid helium about one kilowatt of power is needed at the helium liquefier.

We can recall from Chapter 32 that MRI works by tuning a radio frequency to produce photons whose energies match the difference in energies of the proton spin "up" and spin "down" states. (The proton is the hydrogen nucleus, so that MRI is actually a detector of the hydrogen concentration, which in turn represents a soft tissue concentration.) A resonance phenomenon operates when the frequency matches the energy difference precisely. The energy difference in question is given by $2\mu_p B$, where μ_p is the proton magnetic moment and B is the field strength. Now if the solenoid produced a perfectly uniform magnetic field, then the energy difference between the two spin states would be the same throughout the patient's body; photons of just one frequency would excite all the hydrogen nuclei, and there would be no information as to just where the hydrogen concentration is high and where it is low. To fix this, small solenoids are placed so that the magnetic field varies slightly throughout the large solenoid's core, where the patient is. Then only the protons in one region of the patient are excited at a given radio frequency. By changing the radio frequency, protons in different regions of the patient are affected, and computer reconstruction then enables three-dimensional maps of the soft tissue density within the patient to be made.

As long as this relation can be satisfied, it will be possible for the sphere to lose kinetic energy by causing excitations in the liquid, and there will therefore be resistance to the motion of the sphere.

Let's see whether it is possible for sound waves to be excited in this way. For sound waves, the relation between energy and momentum is given by

$$E = v_s p,$$

where v_s is a constant, the velocity of a sound wave. Thus, for $v_i < v_s$, Eq. (43–15) cannot be satisfied, and *there is no way for the sphere to lose energy*. It therefore travels with no viscosity.

The creeping of liquid helium up the sides of a beaker can be understood in terms of the absence of viscosity together with the fact that the van der Waals forces between neighboring helium atoms and between the atoms of helium and of the beaker walls are stronger than the force of gravity on a helium atom. These forces then pull helium atoms up against gravity. The strange properties of liquid ^{4}He are another demonstration of quantum effects on a large scale.

The technological importance of these properties is closely linked to the importance of superconductivity. Superconducting systems such as the large arrays of superconducting magnets used in particle accelerators must be kept at temperatures below their critical temperature. The only practical superconductors that can be used for such arrays are those for which liquid helium is sufficiently cold to form a coolant. Here, the fact that the liquid ^{4}He is a nearly perfect heat conductor—its coefficient of thermal conductivity is hundreds of thousands of times greater than that of copper—is critical. The entire array of magnets, which, for the Fermilab accelerator, forms a ring 6 km long, can be cooled by a completely connected system of helium, and no "hot spots" can form. In particular, boiling cannot occur at any one location, and this offers crucial insurance against any of the magnet coils heating beyond its critical temperature.

SUMMARY

In large systems for which quantum mechanics governs the dynamics, the Pauli exclusion principle plays a crucial role if the system is composed of fermions. This principle states that only one fermion—for our purposes, a particle with spin $\hbar/2$—of a given type can occupy a particular state such as an energy level. In metals, this means that very closely spaced energy levels are filled with electrons up to the Fermi energy, E_F. The Fermi energy of electrons is

$$E_F = \frac{\hbar^2}{2m_e}(3\pi^2 n_e)^{2/3}, \tag{43–5}$$

where n_e is the electron number density in the metal. Only electrons near the top of the filled levels take part in dynamic processes such as current flow.

The exclusion principle is responsible for the incompressibility of matter. In ordinary matter the incompressibility can be estimated to within a factor of 2 by assuming that the electrons in the matter are free. The exclusion principle prevents stars that are not too massive from collapsing under the mutual gravitational attraction of their constituents.

A system composed of bosons—particles whose spins are integral multiples of $\hbar$, such as photons and ^{4}He atoms—manifests quantum mechanical effects quite different from those of fermionic systems. Identical bosons have an enhanced probability of aggregating in the same quantum state. As a consequence there can be stimulated emission of radiation (photons) into states already occupied by other photons. The operation of the laser is based on stimulated emission and on the bosonic properties of photons. To construct a laser, we need a way to populate an excited energy level in a collection of atoms; a way to build up many photons of the "proper" energy to provoke a massive stimulated transition from the excited level; and a way to extract the coherent laser beam formed by the stimulated emission.

Fermions can pair up to form bosons, and those bound pairs tend to congregate in the same quantum state. This is the mechanism behind superconductivity. Superconductors form when a collection of electron–electron pairs known as Cooper pairs occupy a single, coherent quantum state at low temperatures. Cooper pairs can tunnel, and this phenomenon has advanced our ability to observe very small magnetic fields. Superfluidity, characterized by flow without viscosity in superfluid ^{4}He, is another phenomenon that follows from the tendency of bosons to condense in the same quantum state at low temperatures.

UNDERSTANDING KEY CONCEPTS

1. How would the behavior of electrons in materials change if the Pauli exclusion principle were not applicable?

2. Most heavenly bodies rotate and have some angular momentum. Astrophysicists believe that pulsars are rapidly rotating neutron stars, some with periods as short as 0.005 s. Explain how such a small, massive object could rotate so fast.

3. Where does the fact that photons tend to congregate in the same quantum state play a role in our discussion of lasers?

4. Chemical compounds in metastable excited states can be formed in chemical reactions. Is this a suitable way to arrive at the conditions necessary for laser action?

5. The *Fermi temperature*, T_F, is defined by $E_F \equiv kT_F$. What does it mean if T_F in a given material is much larger or much smaller than the material's temperature?

6. Describe the steps in the evolution of a star that becomes a black hole.

7. A He–Ne laser emits light that is not coherent as well as coherent laser light. Why?

8. An electron on the Moon and an electron on Earth both have spin up and are in the ground state of hydrogen. Given that both electrons are in the same state, how is the Pauli exclusion principle bypassed?

9. Suppose that a crystal could be made from the atoms of the inert gases (by going to low temperatures, for example). Would such a crystal be likely to act as a metal?

10. The spins of the two electrons in the ground state of helium can be either parallel or antiparallel. The Pauli exclusion principle does not apply to antiparallel spins, but electrons with parallel spins must avoid each other. Would you expect the two possible spin arrangements to have the same energy? If not, which energy would be lower?

11. How could you tell if electrons in very distant galaxies obey the Pauli exclusion principle?

12. Given our discussion of stellar evolution, why is Jupiter not a star? (Assume that Jupiter is an aggregate of hydrogen atoms held together by gravity.)

13. An electron traveling through a collection of protons in a plasma is slowed down by collisions and captured into an orbit with a large principal quantum number, far from the proton it orbits. Given that radiative transitions are largely those for which $\Delta\ell = \pm 1$, would the electron end up in the ground state through one transition or through a sequence of transitions?

14. The filament of a lightbulb has many atoms that become thermally excited into the same excited states. Why doesn't the bulb act as a laser?

15. What difficulties might you encounter in an attempt to make an X-ray laser?

16. Use the argument as to why the flow of superconducting helium around an obstacle is nonviscous when the flow occurs at a speed less than that of sound to explain why electron waves propagate without energy loss through a rigid crystal lattice at $T = 0$. Why will the electrons experience resistance at finite temperatures?

17. In our treatment of the incompressibility of ordinary matter, we spoke about metals rather than a more general class of solids. Does our discussion apply to insulators? (See Chapter 27 for the difference between conductors and insulators.)

PROBLEMS

43–1 *The Exclusion Principle in Bulk Matter*

1. (I) What is the value of the Fermi energy of calcium, which has two valence electrons per atom? The gram-atomic weight of calcium is 40.1 g/mol, and its mass density is 2.30 g/cm^3.

2. (I) Find the Fermi temperature (see Question 5) of gold, for which the Fermi energy is 5.51 eV.

3. (I) Calculate the Fermi energies of sodium ($n_e = 2.65 \times 10^{28}$ m^{-3}), potassium ($n_e = 1.40 \times 10^{28}$ m^{-3}), and aluminum ($n_e = 18.1 \times 10^{28}$ m^{-3}). All these metals have a single valence electron per atom.

4. (I) Calculate the bulk moduli, B, of sodium, potassium, and aluminum, using the data given in Problem 3. (The experimental values of B for these materials are 0.64, 0.28, and 7.6, respectively, in units of 10^{10} N/m^2.)

5. (I) Zinc, a metal with two free electrons per atom, has a gram-atomic weight of 65.4 g/mol and a density of 7.13 g/cm^3. What is the density of electrons in zinc?

6. (I) The density of electrons in silver is 5.85×10^{28} electrons/m^3. Calculate the bulk modulus of silver. (The experimental value is 10.1×10^{10} N/m^2.)

7. (I) What values of $n_1(= n_2 = n_3)$ give an energy close to 1 eV for an electron in a cubical box whose sides are precisely 100 nm long? What is the actual energy of this state? of the state in which n_1 and n_2 are unchanged but n_3 is increased by 1?

8. (I) Consider a copper "nanoparticle," a cube of $10 \times 10 \times 10$ atoms (Fig. 43–18). Given that $n_e = 8.47 \times 10^{28}$ electrons/m^3, what is the energy difference between adjacent energy states in the neighborhood of the Fermi energy? Take the distance between neighboring atoms to be 0.25 nm.

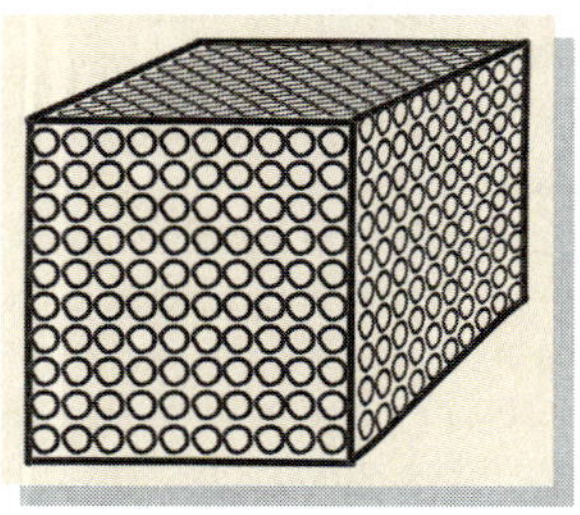

FIGURE 43–18 Problem 8.

9. (I) The Fermi energy of electrons in a metal, E_F, may be used to define the Fermi momentum, p_F, defined by $E_F \equiv p_F{}^2/(2m_e)$. Express the Fermi momentum in terms of n_e. Use your results to calculate the Fermi speed, v_F, the speed of an electron with the Fermi energy.

10. (I) What is the speed of an electron with the Fermi energy in magnesium, given that $n_e = 8.60 \times 10^{28}$ electrons/m^3?

11. (I) Estimate the gravitational pressure at the surface of the Sun.

12. (II) When the Fermi speed is $0.25c$, where c is the speed of light, it is more appropriate to use a relativistic connection between E_F and p_F rather than the nonrelativistic connection that we used in Problem 9. Estimate the value of n_e for which the relativistic connection is necessary.

13. (II) The conduction electrons of a metal have to be treated relativistically when the Fermi speed is about $0.1c$ or more. Because they are confined to a small volume, the inner-shell electrons may also reach high speeds when Z is large. (a) Derive the condition on the density of conduction electrons that allows a nonrelativistic treatment. (b) Derive the condition on Z for which the innermost electrons can be treated nonrelativistically.

14. (II) Five electrons are in an infinitely deep, one-dimensional potential energy well of width L. (a) What is the lowest possible energy value for this system? (b) What is the next highest energy value? In both parts, ignore Coulomb interactions between the electrons.

15. (II) Consider N identical, noninteracting fermions of mass m in an infinitely deep one-dimensional potential energy well of width L. Assume that $N >> 1$. Find (a) the Fermi energy, and (b) the total energy of this system of particles in its lowest energy state. (c) What is the smallest amount of energy that can be absorbed by the system; that is, what is the minimum excitation energy of the system from its ground state?

16. (II) Assume that the nucleus is composed of free neutrons and protons. What is the equation analogous to Eq. (43–1) for the case of a nucleus? Assume that the nucleus is a cubic box with sides L. L is such that the box volume is the same as a sphere

of radius $R = r_0 A^{1/3}$, where A is the total number of neutrons and protons and $r_0 \simeq 1.2 \times 10^{-15}$ m.

17. (II) Calculate the Fermi energy for the neutrons confined to the nucleus ^{96}Mo, which roughly forms a sphere of radius 5.5×10^{-15} m. (There are 54 neutrons in the ^{96}Mo nucleus.)

18. (II) The general definition of the pressure in a gas with internal energy U is $p = -(\partial U/\partial V)$, where V is the volume of gas. (This follows from the first law of thermodynamics and from the fact that the infinitesimal work done when a system changes volume by dV is $dW = p\ dV$.) Given the fact that for a gas of identical fermions at $T = 0$, $U = \frac{3}{5}NE_F$, show that at $T = 0$, $p = \frac{2}{3}U/V$ for the fermion gas. This result is the *same* as our result in Chapter 19, where we studied the atomic origin of pressure. (That result did not depend on the precise distribution, so it is not surprising that we recover it here.)

19. (II) Using the result of Problem 18, (a) calculate the pressure of an ideal gas of N identical fermions at $T = 0$ in terms of the volume, and (b) evaluate the pressure of the degenerate electron gas at $T = 0$ in a sample of sodium. The density of free electrons in sodium is 2.65×10^{22} electrons/cm^3.

20. (II) Calculate the ratio of the degeneracy pressure of electrons to that of protons for a star such as our Sun. Assume that the star is electrically neutral.

21. (II) Suppose that electrons are confined to a plane at a density of n_e electrons per unit area. Follow the steps that lead to Eq. (43–5) to calculate the Fermi energy for such a system.

22. (II) An iron nucleus consists of 26 protons and 30 neutrons in a sphere of radius 6×10^{-15} m. Assume that none of the particles interact with each other. Calculate the Fermi energies of the protons and the neutrons.

23. (II) (a) Calculate the radius of a neutron star as a function of its mass. (b) Find the Fermi energy and Fermi momentum of a neutron star in terms of its radius and its mass. (c) Estimate the mass for which it is necessary to use relativistic expressions in part (b). Express your results in solar masses.

24. (III) A proper calculation of the Fermi energy in a cubical box with sides L amounts to counting the number of electrons that fill all the states up to an energy E_F. There are two states for each set of positive integers $\{n_1, n_2, n_3\}$ that satisfy the condition $n_1^2 + n_2^2 + n_3^2 = 2(m_e E_F L^2/\pi^2\hbar^2) \equiv R^2$. In a "continuum approximation" to the problem of counting the number of states, appropriate to large values of n_1, n_2, and n_3, the number of states is twice the volume of an octant (one-eighth) of a sphere of radius R; that is, $2(\frac{1}{8})(4\pi/3)R^3$ (Fig. 43–19 is a two-dimensional version). Compare the answer in terms of the density with the approximate result that we gave in the text, $E_F \simeq (\hbar^2/2m_e)(\pi^3 n_e)^{2/3}$.

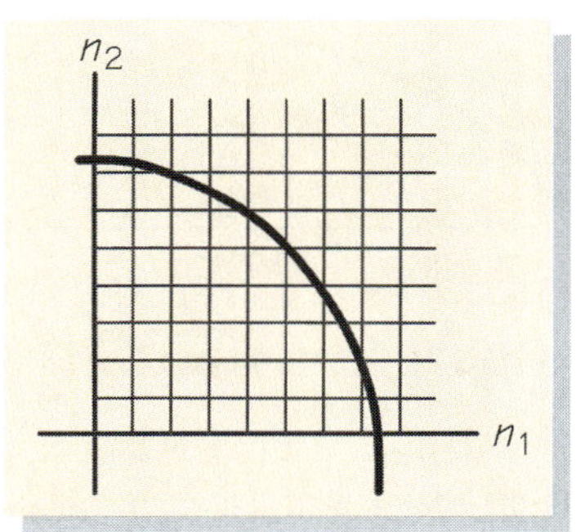

FIGURE 43–19 Problems 24, 25.

25. (III) By counting the number of states (that is, by using the techniques described in Problem 24), calculate the Fermi energy for a two-dimensional gas of free, noninteracting electrons (Fig. 43–19). Show that $d = \lambda_F/2$, an approximation we made in our text discussion, is not as good as it is for a three-dimensional gas.

43–2 *Lasers and the Behavior of Bosons*

26. (I) A laser emits radiation of wavelength 590 nm and delivers power of 25 mW. How many photons are emitted per second?

27. (I) The characteristic red light of a He–Ne gas laser (see Fig. 43–8a) is due to stimulated emission between neon levels at 20.66 eV and 18.70 eV. Calculate the wavelength and frequency of the photon emission.

28. (I) A He–Ne laser emits radiation with a power of 1/100 W at a wavelength of 632.8 nm. How many photons are emitted per second?

29. (I) A laser emits radiation at a wavelength of 555 nm. Photons are emitted at a rate of 8.5×10^{18} s^{-1}. What is the power of the laser?

30. (II) Free electrons in a magnetic field could form the basis of a tunable laser, in which the two levels involved in the transition are the "up" and "down" spin states of the electron in a magnetic field. How large a magnetic field is needed to operate the laser at 500 nm?

31. (II) Consider N identical bosons of mass m in an infinitely deep one-dimensional potential energy well of width L (Fig. 43–20) (a) What is the total energy of this system of particles in its lowest energy state? (b) What is the smallest amount of energy that can be absorbed by the system; that is, what is the minimum excitation energy of the system from its ground state?

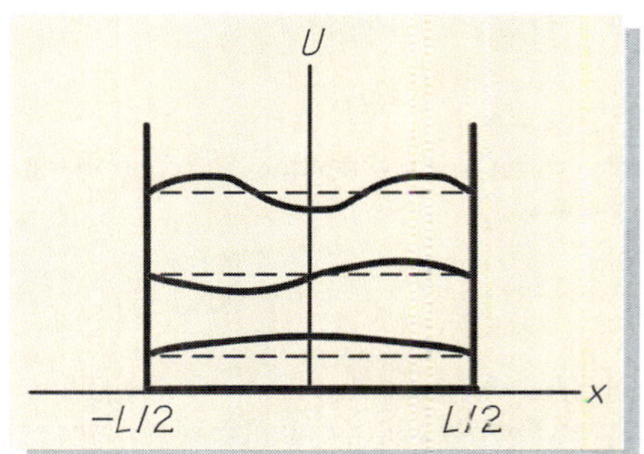

FIGURE 43–20 Problem 31.

32. (II) The collimation of a laser beam is limited by diffraction; that is, the angular beam spread $\Delta\theta$ is at least as large as $\Delta\theta \simeq \lambda/D$, where D is the diameter of the laser's aperture. What is the diameter of the beam spot projected on the Moon, 3.8×10^8 m from a laser of aperture diameter 0.75 cm emitting light of wavelength 550 nm?

33. (II) A laser beam projects light of wavelength 663 nm at the rate of 3×10^{17} photons/s. (a) What is the power of the laser? (b) What is the radiation pressure exerted by the beam on a shiny surface if the laser beam projects a circle 3.5 mm in diameter?

34. (III) In 1917, Einstein wrote a paper that describes the stimulated-emission process. We can summarize his results as follows: Suppose that an atom has two states. State 1 is the ground state; state 2 is an excited state. The atom is in an enclosure filled with blackbody radiation, which we recall from Chapter 17 has the energy-density function $u(f, T)$. The transition rate for excitation from state 1 to state 2, $\Gamma_{1\to2}$, is proportional to the number of atoms in state 1, N_1, times the density of photons present that have the proper frequency (such that the photons have the transition energy): $\Gamma_{1\to2} = N_1 u(f, T)B_{12}$, where B_{12} is a proportionality constant. The transition rate for deexcitation from state 2 to state 1, $\Gamma_{2\to1}$, has two terms: a stimulated-emission term proportional to the number of atoms in state 2 times the density of photons present with the proper frequency, $N_2 u(f, T)B_{21}$, and a spontaneous-emission term proportional to the number of atoms in state 2. Thus $\Gamma_{2\to1} = N_2[u(f, T)B_{21} + A]$, where A is a constant. (a) In thermodynamic equilibrium, the transition rate in the two directions must be equal. Use this fact, together with the Boltzmann relation for the populations, $(N_1/N_2) = e^{-E_1/kT}/e^{-E_2/kT}$, to show that

$$u(f, T) = \frac{A}{B_{12}e^{hf/kT} - B_{21}}\ .$$

(b) In the classical limit, $u(f, T) = 8\pi f^2 kT/c^3$. Show that the quantum mechanical expression for u can match this form for large T only if $B_{12} = B_{21}$, and find the value of the ratio A/B_{12} from the matching requirement.

35. (III) From Problem 34, the rate per atom at which atoms in equilibrium with radiation at temperature T emit photons of frequency f has the form

$$\frac{\Gamma_{2\to1}}{N_2} = C\left(1 + \frac{1}{e^{hf/kT} - 1}\right),$$

where C is a constant. (a) Use the hint at the end of this problem to show that the average number of photons of frequency f per unit volume, $\langle n\rangle$, is given by

$$\langle n\rangle = 1/(e^{-hf/kT} - 1).$$

(b) Use this result to show that the deexcitation transition rate per atom is given by

$$\frac{\Gamma_{2\to1}}{N_2} = A(1 + \langle n\rangle).$$

This establishes the result that the transition rate for deexcitation increases linearly with the number of photons of the proper frequency already present, as is required for laser operation. [*Hint*: The distribution function to find n photons of energy hf is given according to Planck's work on blackbody radiation as $P_n = e^{-nhf/kT}(1 - e^{-hf/kT})$. The sum representing the average number is $\sum P_n n$, and this sum can be performed by noticing that $\sum_n x^n$ is a geometric series, and that $\sum_n nx^n = x(d/dx)[\sum_n x^n]$.]

43–3 Superconductivity

36. (I) The critical temperature of a superconductor, T_c, varies with the isotopic mass of the element making up the superconductor, M, according to the relation $T_c\sqrt{M}$ = a constant. In lead, $T_c = 7.184$ K for the isotopic molar mass 207.7 g. What is the critical temperature for the isotopic molar mass 206.3 g?

37. (I) An unknown constant potential difference V is placed across a Josephson junction, and an alternating current of frequency 23.3 GHz is produced. What is the value of V?

38. (I) What is the frequency of the AC component of the current when a DC voltage of 1.38 V is placed across a Josephson junction? a DC voltage of 1.00 μV?

39. (II) Verify that the units of magnetic flux are consistent with Eq. (43–14).

40. (II) The current-carrying ability of superconductors is limited by the fact that superconductivity breaks down in a large magnetic field. How large a current can be carried by a niobium-tin wire of diameter 0.1 mm if the critical magnetic field is 1 T?

41. (II) When a current passes through a Josephson junction, the resulting AC current can be regarded as corresponding to photons whose frequency is that of the AC current. For a particular Josephson junction, if these photons have an energy of 5×10^{-6} eV, what is the DC potential difference across the junction?

42. (II) The loop of a SQUID surrounds an area of 1 cm^2. If the phase of the interference current can be determined to an accuracy of $\pi/2$, to what accuracy can a magnetic field be determined? What current produces a field like this 1 cm away?

43–4 Superfluidity and Liquid Helium

43. (II) According to one approach that takes into account the quantum mechanical "clustering" behavior of the bosons that make up a superfluid, the temperature at which the superfluid forms is given by $T_c = [N/(2.612)(0.886)C]^{2/3}$, where $C = 2\pi V(2mk/h^2)^{3/2}$, V is the volume in which the N bosons are confined, and m is the mass of the boson. Apply this result to liquid ^{4}He; take the density of the liquid to be 0.147 g/cm^3, and calculate T_c. (The experimental value of T_c for this system is 2.2 K.)

General Problems

44. (II) (a) The energy-distribution function of electrons at $T = 0$ is $n(E) = (V/2\pi^2)(2m/\hbar^2)^{3/2}\sqrt{E}$ (Fig. 43–21). Use this to calculate the speed-distribution function. (b) Calculate $\langle v^4\rangle$ by using the result of part (a).

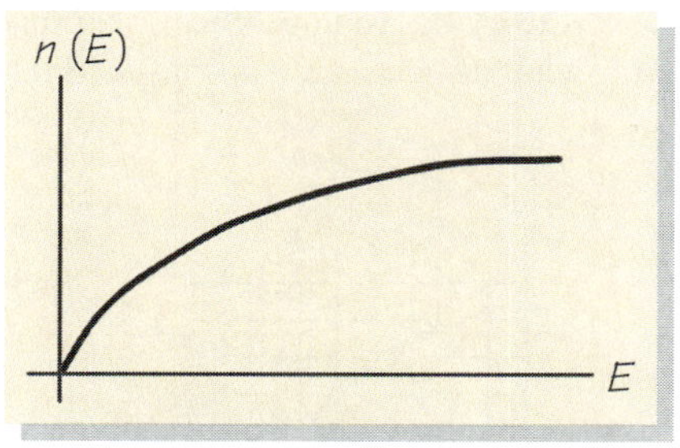

FIGURE 43–21 Problem 44.

45. (II) What is the radius of a star with a mass of 10^{-3} solar masses, assuming that the star has the density of the Sun? Compare your answer with the radius of Jupiter.

46. (II) Assume that the Fermi energy depends only on $\hbar$, the electron density (n_e), and the mass of the electron. Then use dimensional analysis to get the dependence.

47. (II) Assume that the Fermi energy depends only on $\hbar$, the electron density (n_e), and the speed of light (c). Then use dimensional analysis to get the dependence.

48. (II) You want to verify the isotopic effect described in Problem 36 with a set of data you collected. Show that if an isotope equation of the type $T_c M^\gamma$ = a constant were correct, the data on a plot of $\ln(T_c)$ versus $\ln(M)$ would be a straight line whose slope determines γ. What is the curve if $\gamma = \frac{1}{2}$?

49. (II) Show that the isotopic-effect equation $T_c\sqrt{M}$ described in Problem 36 implies that T_c is proportional to v, where v is the speed of sound in a material with isotopic mass M.

50. (II) (a) Using the methods in the text, show that in a gas of free electrons in a box with sides of length L at $T = 0$ K, the total number of electrons with energy less than E is

$$N = \left(E\,\frac{2m_e}{\hbar^2\pi^2}\,L^2\right)^{3/2}\frac{\pi}{3},$$

where $E < E_F$. (b) Use the results of part (a) to show that the number of electrons with energies between E and $E + dE$ is given by

$$n(E)\,dE = \frac{V}{2\pi^2}\left(\frac{2m}{\hbar^2}\right)^{3/2}\sqrt{E}\,dE,$$

where $n(E)$ is the energy-distribution function of the electrons. [*Hint*: The expression you want is given by $n(E)\,dE \equiv N(E + dE) - N(E)$.]

51. (II) Use the energy-distribution function of Problem 50 to show that the total energy of a gas of electrons in a box with sides of length L at $T = 0$ K is given by $E_{tot} = \frac{3}{5}NE_F$.

52. (III) The behavior of the atoms of a crystal at temperature T under the influence of the springlike forces between the atoms is analogous to the behavior of light in a cavity at temperature T (blackbody radiation). The behavior of the lattice can be described by *phonons*, collective oscillations of the lattice that can be treated as particles analogous to the photons in a hot cavity. The phonons are identical bosons that obey the quantum rules for such particles. There is one important difference between the lattice vibrations and blackbody radiation: The number of possible modes of a lattice made up of N atoms is limited to $3N$ (corresponding to the three-dimensional motion), and there is a maximum frequency f_{max}. By following the kind of reasoning Planck used in deriving his blackbody spectrum, it is possible to show that the total lattice vibration energy, U, of the crystal is given by

$$U = \frac{9NkT}{\Lambda^3}\int_0^\Lambda \frac{x^3}{e^x - 1}\,dx,$$

where $\Lambda \equiv T_D/T$; T_D, called the *Debye temperature*, is given by $T_D = hf_{max}/k$. (a) Show that in the limit of $T >> T_D$, the energy of the solid is the classical expression $3NkT$. (b) Show that in the limit $T << T_D$, $U \propto T^4$ (c) What is the temperature dependence of the specific heat at low temperatures? The curve formed by the expression for U matches well the experimental data over a large temperature range in many solids (Fig. 43–22). (After G. Burns, *Solid State Physics*, Academic Press, Inc., 1985, p. 369).

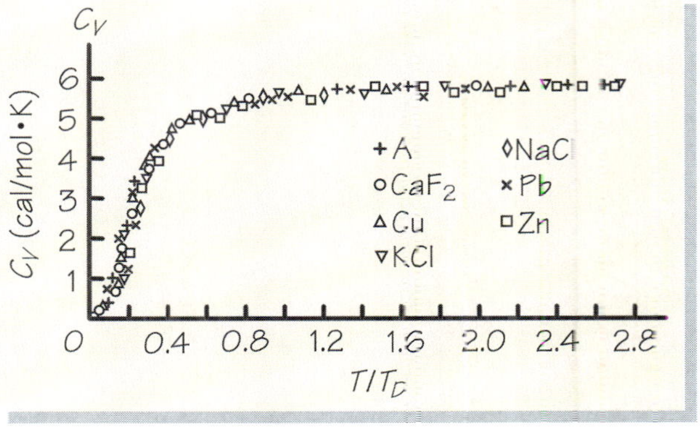

FIGURE 43–22 Problem 52.

53. (III) For stars whose mass is roughly one solar mass (2×10^{30} kg) and whose thermonuclear reactions have stopped because the fuel has been used up, the repulsive effects due to the Pauli exclusion principle *for electrons* are strong enough to keep the star from collapsing gravitationally. In that case, white dwarfs are formed. (a) Calculate the Fermi energy of the electrons in a white dwarf of mass M. (b) Calculate the radius of a white dwarf of mass M. If $M = M_\odot$ is one solar mass, what is the ratio of your result to the present radius of the Sun? [*Hint*: Follow our procedure for a neutron star. The gravitational pressure is the same as before, because only the protons and neutrons account for the mass. The number of electrons is the same as the number of protons, which is roughly half the total of protons plus neutrons. The radius of the Sun is 6.96×10^8 m.]

54. (III) When all the hydrogen in a star has burned, helium nuclei and electrons remain. Suppose that there are N electrons, and that the star is a sphere of radius R and mass $M \simeq (N/2)m_{He}$. (a) Use Eq. (43–8) to estimate the degeneracy pressure. (b) The gravitational force of the helium nuclei exerts an inward pressure, which is approximately $p_{grav} = E_{grav}/V \simeq (GM^2/R)/V$. Find p_{grav} in terms of R (c) The star will collapse until the two pressures cancel, when it has a radius R_f. Calculate R_f in terms of m_e, m_{He}, and N. (d) Evaluate R_f for a star of one solar mass. (The result will be surprisingly small.)

55. (III) Use the methods of Problem 24 to calculate the Fermi energy of a gas of electrons, assuming that the mass of electrons can be neglected, so that $E = pc$. Under what conditions will the star of Problem 54 *not* collapse?

56. (III) What is the bulk modulus of a nucleus that consists of N neutrons and Z protons and has a radius $R = r_0A^{1/3}$, where $r_0 \simeq 1.2 \times 10^{-15}$ m?

CHAPTER

44

A "quantum corral," produced by using a scanning tunneling microscope to place 48 individual iron atoms on a surface of copper. The ripples within are manifestations of the probability of finding an electron in the corral, and they show just the properties that quantum mechanics requires for a particle in a circular box.

Quantum Engineering

The close interplay of science and technology enriches both. At times, technology has led science, as in the case of thermal physics, which was stimulated by the need to understand the engines that propelled the Industrial Revolution. At other times, science has led technology, as with the development of the electric power industry out of the discoveries of Faraday and Maxwell. These sorts of processes have been frequent in the second half of the twentieth century. The discovery of quantum mechanics has led to an ever-better understanding of the detailed properties of materials; this understanding has been accompanied by rapid developments in the fabrication of materials required to satisfy the needs of technology. In this chapter, we begin with semiconductors and their applications. Then we make a rapid qualitative tour of more recent technological developments that rely on quantum physics. We conclude with a brief discussion of some of the devices whose potential for making major contributions to future technology has already been demonstrated.

44–1 SEMICONDUCTORS

Two major features govern the electrical conduction properties of solids: the Pauli exclusion principle, which is obeyed by all electrons (see Chapter 43), and the presence of a band structure, with gaps, that follows from the lattice of ions in the material (see Chapters 27 and 42). The Pauli exclusion principle implies that there are at most two electrons in each energy level. These levels are filled in order, starting from the lowest. The energy of the highest filled level is the *Fermi energy*, E_F. The band structure expresses the presence of energy gaps between allowed bands; there are no levels within

Conduction is discussed in Chapter 27.

TABLE 44–1 Gap Widths of Some Semiconductors†

Material	*Gap Width (in eV)*
Tin (gray form)	0.08
Tellurium	0.35
Lead sulfide	0.37
Germanium	0.67
Silicon	1.12
Gallium arsenide	1.43
Boron	1.5
Selenium	1.8

†At room temperature.

such gaps, so no electrons can have energies there. In conductors, levels are filled to the Fermi energy, but there are empty energy levels in the band containing the Fermi energy. In semiconductors and insulators, the *valence band*, which is the topmost band containing electrons, is filled right up to a gap. Materials with a totally filled valence band will, in general, act as insulators *unless the energy gap between the valence band and the next band is small.* In that case, it is possible to induce conductivity by thermally exciting some electrons across the gap. The next band above the valence band is therefore referred to as the *conduction band* (see Fig. 27–18a). Typical gap widths—the energy difference from the top to the bottom of the gap—at room temperature are listed in Table 44–1 for some semiconductors. In contrast, the gap width of the diamond form of carbon, which is regarded as an insulator, is 5.5 eV.

The Effects of Temperature

The behavior of semiconductors depends crucially on temperature. Therefore let's study the effects of finite temperatures. It is simplest to look first at finite temperature effects in conductors. At $T = 0$ K, we are certain to find an electron in every state up to an energy of E_F. If we suppose that the possible levels are so closely spaced as to represent a continuum, then, at $T = 0$ K, the probability $f(E)$ that the state with energy E is occupied by an electron (actually up to two electrons, due to spin) is

$$\text{for } E < E_F: \quad f(E) = 1,$$
$$\text{for } E > E_F: \quad f(E) = 0. \tag{44–1}$$

When $T > 0$ K, some of the electrons will be thermally excited into previously unoccupied levels, leaving vacant energy levels behind them. For temperatures that are not too large, this can happen only for electrons with energies near E_F because electrons with energies far below E_F are unlikely to get a sufficiently large thermal "kick." If electrons behaved according to classical rules, they would obey a Maxwell–Boltzmann distribution (see Chapter 19). But, because electrons obey the Pauli exclusion principle, their energy distribution is instead described by the **Fermi–Dirac distribution**:

The Fermi–Dirac distribution

$$f(E) = \frac{1}{1 + e^{(E - E_F)/kT}}. \tag{44–2}$$

You can easily check that we recover Eq. (44–1) from this distribution in the limit $T \to 0$. Figure 44–1 shows the distribution for the increasing temperatures $T = 0$, T_1, and T_2. Notice that for $(E - E_F) >> kT$, the exponential factor dominates in the denominator, and we get the approximation

$$f(E) \simeq e^{-(E - E_F)/kT}. \tag{44–3}$$

Equation (44–3) has the form of the Maxwell–Boltzmann distribution, so the distribution is classical when the energy difference $E - E_F$ is large compared to kT. What is a typical value of kT? A useful number to remember is that room temperature, $T =$ 300 K, corresponds to 1/40 eV.†

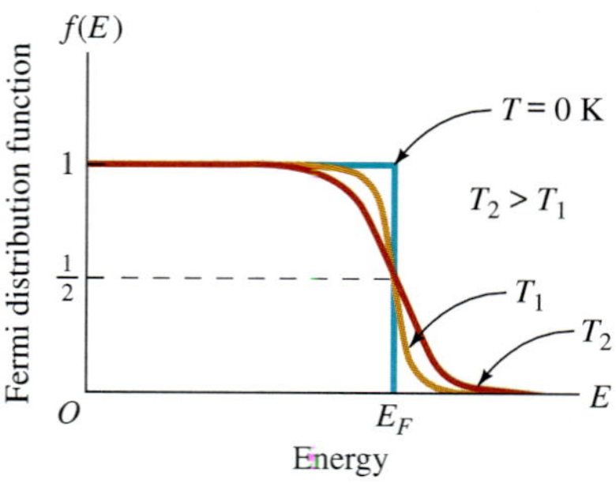

FIGURE 44–1 The Fermi–Dirac distribution, $f(E)$, which describes the probability of finding one of a set of identical fermions with an energy E. The distribution is sketched for three successively higher temperatures, $T = 0$, T_1, and T_2.

EXAMPLE 44–1 Assume that electrons in copper are distributed according to the Fermi–Dirac distribution. The Fermi energy in copper is $E_F = 7.04$ eV. (a) The *Fermi velocity* has magnitude v_F, the speed of an electron that moves with a kinetic energy equal to E_F. What is v_F for copper? (b) From the Fermi–Dirac distribution, $f(E_F) = \frac{1}{2}$. What is the value of $f(E)$ at a temperature of 300 K for an electron with a speed 1 percent higher than v_F? (c) Estimate the temperature at which the probability $f(E)$ is 10^{-10} when E is 1.2 eV above E_F.

†For $T = 300$ K, $kT = (1.38 \times 10^{-23}$ J/K$)(300$ K$)/(1.6 \times 10^{-19}$ J/eV$) = 2.6 \times 10^{-2}$ eV $= 1/40$ eV.

Solution: (a) We assume that the electrons are nonrelativistic, so $E_F = \frac{1}{2}m_e v_F^2$, or

$$v_F = \sqrt{\frac{2E_F}{m_e}} = \sqrt{\frac{2(7.04\text{ eV})(1.60 \times 10^{-19}\text{ J/eV})}{0.911 \times 10^{-30}\text{ kg}}} = 1.57 \times 10^6\text{ m/s}.$$

(b) When the speed of an electron increases by 1 percent, the energy—proportional to the speed squared—increases by approximately 2 percent. Thus, $E = (1.02)(7.04\text{ eV}) = 7.18\text{ eV}$. With $kT = 2.6 \times 10^{-2}$ eV at $T = 300$ K, $(E - E_F)/kT = 5.4$, and hence

$$f(E) = \frac{1}{1 + e^{5.4}} \simeq 4.6 \times 10^{-3}.$$

(c) We set $f(E)$ equal to 10^{-10} and solve for kT, given $E - E_F = 1.2$ eV. When $f(E) = 10^{-10}$, the exponential term in the denominator of $f(E)$ in Eq. (44–2) is much larger than 1. Thus,

$$f(E) \simeq e^{-(E - E_F)/kT} = 10^{-10},$$

and $-(1.2\text{ eV})/kT = \ln(10^{-10}) = -10\ \ln(10) = -23$. Thus, $kT = (1.2\text{ eV})/23 = 0.052$ eV. This is twice the value of kT at $T = 300$ K [see part (b)], so the desired estimate is 600 K.

Semiconductors, Electrons, and Holes

To apply our ideas about finite temperatures to semiconductors, we must reconsider the role of the energy gap. Suppose that the band structure is that of a semiconductor; that is, at $T = 0$, the valence band is filled, the conduction band is empty, and the gap energy $E_g \equiv E_c - E_v$ is relatively small. Here, E_v is the maximum energy of the valence band (the *edge of the valence band*) and E_c is the lowest energy of the conduction band (the *edge of the conduction band*) (Fig. 44–2).

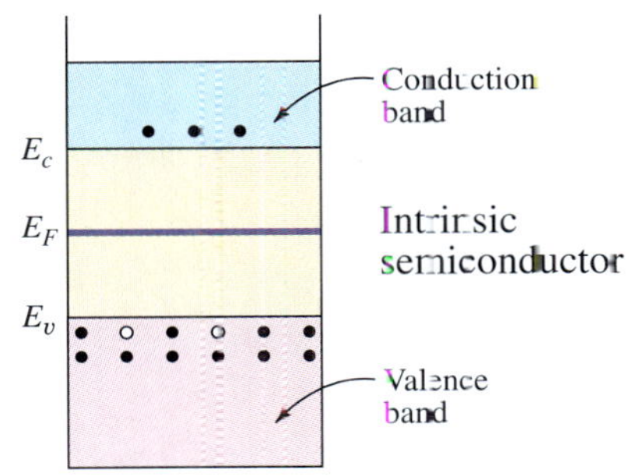

FIGURE 44–2 The band structure of an intrinsic semiconductor is that of a filled valence band, a relatively narrow band gap, and, at $T = 0$, an empty conduction band. The solid circles represent electrons, and the open circles represent holes. At $T = 0$, there are no electrons in the conduction band and no holes in the valence band.

But just what is Fermi energy for this situation? If we think of the Fermi energy as the energy above which there are no electrons (at zero temperature), then *any* energy value within the energy gap would serve, and that is not an answer to our question. Our procedure is as follows: We use Eq. (44–2) but replace the Fermi energy with an unknown quantity μ:

$$f(E) = \frac{1}{1 + e^{(E - \mu)/kT}}. \tag{44–4}$$

The quantity μ is a thermodynamic variable—the *chemical potential*—but the detailed properties of this variable need not concern us further. For the moment, μ is unknown, but we shall *determine* its value from a condition of the conservation of charge. *In the following discussion, we shall conform to established practice for semiconductor engineering by writing E_F rather than μ throughout.* When you see formulas for semiconductors with the parameter E_F, keep in mind that this is not the Fermi energy in the sense in which we first defined it.

When $T \neq 0$, some electrons go into the conduction band, and the valence band is correspondingly depleted of electrons (Fig. 44–3). At room temperature ($kT \simeq 1/40$ eV), $E_g >> kT$. We can then very roughly estimate that the relative number of electrons able to jump the gap in, say, silicon, for which $E_g \simeq 1$ eV, is $e^{-E_g/kT} \simeq e^{-25} \simeq 10^{-11}$. Thus only a very small fraction of electrons will reach the conduction band. This same reasoning would hold if, instead of E_g, we used the smaller quantity $E_c - E_F$—as long as $E_c - E_F$ were of the same order of magnitude as E_g (as it typically is). We would thus have a situation in which an energy distribution of the same form as Eq. (44–3) holds (although we have yet to determine E_F). Because the function in Eq. (44–3) decreases so rapidly as energy increases, the number of electrons per unit volume (the number density n) in the conduction band is well approximated by replacing E in Eq. (44–3) with the lowest energy in the conduction band, E_c:

$$n = N_c e^{-(E_c - E_F)/kT}. \tag{44–5}$$

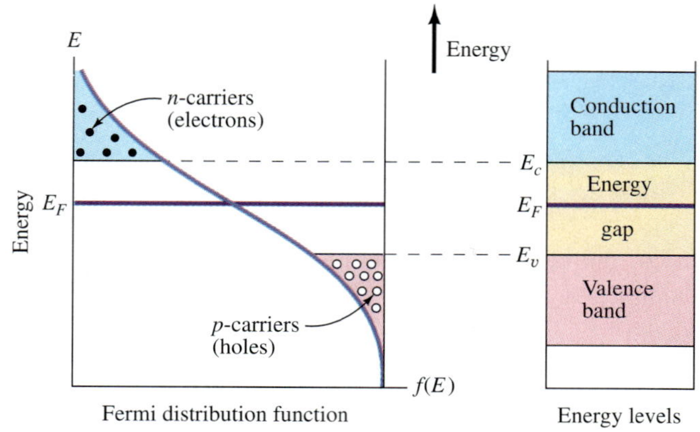

FIGURE 44–3 Electrons near the top of the valence band can jump the gap in a semiconductor at finite temperature. The Fermi–Dirac distribution describes the probability that an electron will have enough energy to be promoted to the conduction band. For every electron that can be promoted, a hole is left behind in the valence band.

The proportionality constant N_c will be treated later. The electrons in the conduction band are free to move when an electric field is applied. These electrons are called *n-carriers* (n for negative).

Electrons that are thermally "promoted" into the conduction band leave vacancies in the valence band. Only the electrons with energies near E_v are promoted, so the vacancies are located at those energies. Such a vacancy is called a **hole**. Holes can move: When a hole is filled by an electron, that leaves a hole in that electron's original position, which, in turn, is filled by an electron, and so on. If there is an electric field pointing in the $+x$-direction, then electrons will flow in the $-x$-direction, and thus the hole will flow in the $+x$-direction. (A helpful analogy might be an air bubble floating upward in a tank of water. The "vacancy" in the water, the air bubble, is constantly replaced by water falling down under the force of gravity; this water leaves its own vacancy, which thus continually moves upward.) The hole thus acts as a positively charged particle; we call such a hole a *p-carrier* (p for positive). The probability of finding a hole with energy E can be determined by the Fermi–Dirac distribution. If $f(E)$ is the probability of finding an electron with energy E, then the probability of *not* finding an electron with that energy is

$$1 - f(E) = 1 - \frac{1}{1 + e^{(E - E_F)/kT}} = \frac{e^{(E-E_F)/kT}}{1 + e^{(E-E_F)/kT}}$$

$$= \frac{e^{-(E_F-E)/kT}}{1 + e^{-(E_F-E)/kT}}. \qquad (44\text{–}6)$$

Here, E is the energy of the empty electron state (the hole), which is an energy in the valence band, so the factor $E_F - E$ is positive. For $E_F - E >> kT$, which is typically the case, the exponential in the denominator is very small compared with unity, and the right-hand side of Eq. (44–6) reduces to $e^{-(E_F-E)/kT}$. This exponential factor decreases so rapidly as the factor $E_F - E$ increases that, to a good approximation, we can replace $E_F - E$ by $E_F - E_v$. Thus the number density of holes is accurately given by

$$p = N_v e^{-(E_F-E_v)/kT}. \qquad (44\text{–}7)$$

We shall give the proportionality constant N_v later. Note that the product of the n-carrier concentration and the p-carrier concentration is given by

$$np = N_c e^{-(E_c-E_F)/kT} N_v e^{-(E_F-E_v)/kT} = N_c N_v e^{-(E_c-E_v)/kT} = N_c N_v e^{-E_g/kT}. \qquad (44\text{–}8)$$

Intrinsic semiconductors

When each hole in the valence band is created by the promotion of an electron into the conduction band, the number of p-carriers equals the number of n-carriers. Semiconductors for which this is the case are called *intrinsic semiconductors*. We can now use this condition to determine E_F. With a subscript i to indicate that we are dealing with an intrinsic semiconductor, Eq. (44–8) implies that

$$n_i = p_i = \sqrt{n_i p_i} = \sqrt{N_c N_v}\, e^{-E_g/2kT}. \qquad (44\text{–}9)$$

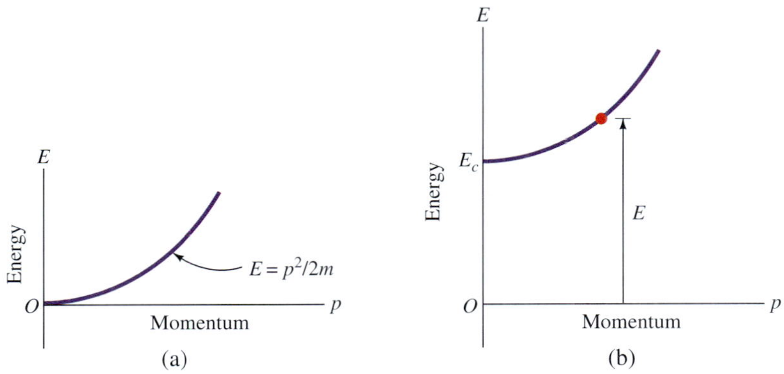

FIGURE 44–4 (a) Energy versus momentum for a free electron. (b) Energy versus momentum for an electron in the conduction band. The lowest possible energy for the electron is E_c.

Comparison of the exponential factor in this result with the exponential factors of Eqs. (44–5) and (44–7)—which is valid as long as N_c and N_v do not vary too rapidly with temperature—shows that

$$E_c - E_F = E_F - E_v = E_g/2. \tag{44–10}$$

In other words, E_F *(or more properly, the chemical potential) is at the halfway point of the gap*. This result justifies our expectation that $E_c - E_F$ and $E_F - E_v$ are much larger than kT if E_g is, too.

An additional effect is important to semiconductors. The energy of a free electron with momentum p is $E = p^2/2m_e$, as in Fig. 44–4a. Figure 44–4b shows the energy–momentum relation that would hold for the electrons in the conduction band if those electrons were free. Electron interaction with the lattice, however, changes this relationship. For small momentum, the principal effect is to change the steepness of the parabola, and this is equivalent to changing the electron mass to an *effective electron mass*, denoted by m^*. For semiconductors, m^* is typically an order of magnitude *smaller* than the true electron mass; for gallium arsenide (GaAs), $m^*/m_e \simeq 0.067$, and for indium antimonide (InSb), this ratio is only 0.015. Furthermore, the effective mass of n-carriers need not be the same as that of p-carriers; generally m^* is much smaller for n-carriers (Fig. 44–5). An understanding of why effective masses take on their values is beyond the scope of our discussion. Roughly speaking, the ratio m^*/m_e is small if E_g is small compared to the width of the conduction band.

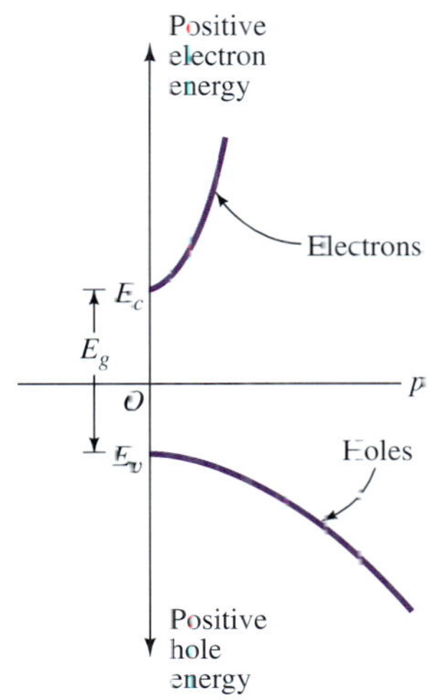

FIGURE 44–5 The energy–momentum relation for an electron in the conduction band (Fig. 44–4b) is modified because the electron behaves as though it has a modified mass m^*. Also shown is the energy–momentum relation for a hole, which in general has a different effective mass than does an electron.

Doping

Extrinsic semiconductors

In contrast to intrinsic semiconductors, in *extrinsic*, or *impurity*, *semiconductors*, additional n-carriers or p-carriers are supplied by the addition of impurity atoms to the crystal lattice. The addition of impurities (whether artificially or naturally) is known as *doping*. It works as follows: Suppose that we add some arsenic atoms to germanium. The crystal structure will remain unchanged (each arsenic ion replaces one germanium ion in the lattice), but the electronic structure will change. Germanium has four valence electrons, whereas arsenic has five; the arsenic atom makes one more electron available for conduction and is said to be a *donor impurity*. When n-carriers are added in this way, the doped material is said to be an *n-type semiconductor*.

If the additional electron were not bound at all to the arsenic ion, it could fill an energy level only in the conduction band and therefore could easily carry current. The energy such an electron would have is on the order of E_c. Although the additional electron is indeed bound to the arsenic ion, it is bound only very weakly. If the arsenic were not part of the semiconductor lattice, we might expect the electron to be bound to the arsenic ion with an energy like the Bohr energy [Eq. (42–9) with $n = 1$]:

$$E = -\frac{e^4 m_e}{2(4\pi\epsilon_0\hbar)^2} \simeq -13.6 \text{ eV}.$$

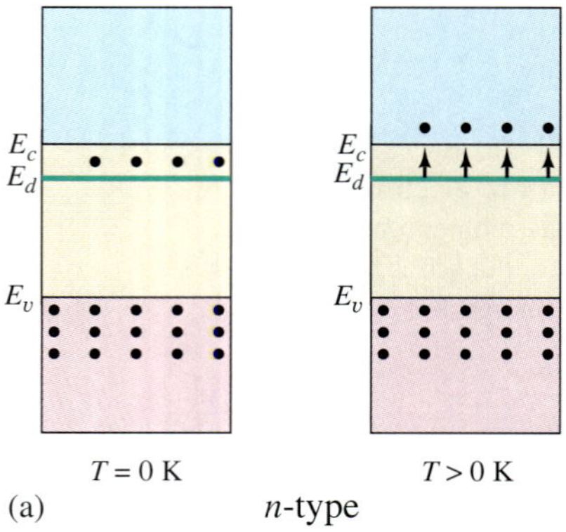

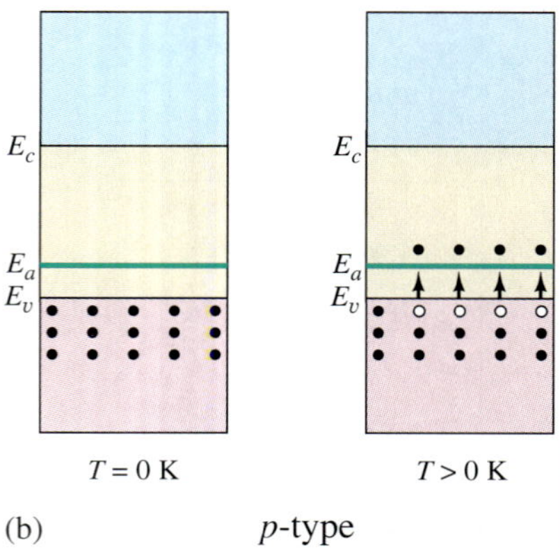

FIGURE 44–6 Energy-band structure for doped semiconductors. (a) In n-type semiconductors, E_F (labeled E_d) lies close to E_c. The levels of electrons (solid circles) from donor impurities lie above E_F and are easily excited at $T > 0$ into the conduction band. (b) In p-type semiconductors, E_F (labeled E_a) lies close to E_v. There are levels associated with the acceptor impurities that are quite close to E_v, so at finite temperatures, electrons from the valence band are easily promoted to these levels, leaving holes (open circles) in the valence band.

Within the semiconductor, this binding energy is weakened for two reasons. First, the mass of the electron is effectively reduced—typically by more than a factor of 10—from m_e to m^*. The radius of the electron's Bohr orbit is now so large that the electron is circulating within the larger environment of the semiconductor lattice. Second, the semiconductor medium has a dielectric constant κ (defined by $\epsilon = \kappa\epsilon_0$) that is typically more than 10 times larger than the value of 1 for a vacuum. These two factors change the binding energy to

$$E = -\frac{e^4 m^*}{2(4\pi\epsilon\hbar)^2}. \tag{44–11}$$

Our numerical values for m^* and ϵ show that the magnitude of this energy is less than $\frac{1}{1000}$ that of the typical Bohr energy. This small binding energy, on the order of $\frac{1}{100}$ eV or less, manifests itself in the presence of additional energy levels very close to the bottom of the conduction band (Fig. 44–6a). The amount by which these energies differ from E_c is on the order of kT or less at room temperature. Thus the additional electron, which is bound at $T = 0$, is very easily excited to the conduction band at room temperature. This may make the donor electrons far more important for conduction than the intrinsic electrons.

EXAMPLE 44–2 Calculate the ionization energy for a donor electron in doped indium antimonide, a semiconductor in which $\epsilon/\epsilon_0 = 17.9$ and $m^*/m_e = 0.015$.

Solution: The ionization energy E is the negative of the binding energy of the ground state of the donor atom [Eq. (44–11)], so

$$E = \frac{e^4 m^*}{2(4\pi\epsilon\hbar)^2} = (13.6\text{ eV})\left(\frac{m^*}{m_e}\right)\left(\frac{\epsilon_0}{\epsilon}\right)^2$$

$$= (13.6\text{ eV})(0.015)\left(\frac{1}{17.9}\right)^2 = 6.4 \times 10^{-4}\text{ eV}.$$

This much energy corresponds to a temperature of $T = E/k = 0.025$ K.

If an impurity has a deficit of valence electrons, a surplus of p-carriers is created because holes form when electrons from the intrinsic semiconductor atoms become strongly attached to the impurity atoms. Such impurities are known as *acceptor impurities*. Boron, for example, has three valence electrons and is an acceptor impurity for germanium. Semiconductors doped with acceptor impurities are known as *p-type semiconductors*. By the same sort of reasoning that we used for n-type semiconductors, p-type semiconductors have the level structure shown in Fig. 44–6b, with some additional levels close to E_v. The n-carriers from the valence band are easily promoted to the acceptor levels, leaving holes in the valence band that are very effective at carrying charge.

As in intrinsic semiconductors, the effective Fermi energy of doped semiconductors determines the probability of promoting electrons and holes to levels where they can conduct. The Fermi energy, E_F, is found by correctly relating the densities of n-carriers and p-carriers. Since the carrier densities depend on the difference between the energy and E_F, as in Eq. (44–3), we might expect that the Fermi energy is close to E_c (i.e., it increases) in n-type semiconductors and is close to E_v (i.e., it decreases) in p-type semiconductors. This is indeed the case. These results can be formally developed by starting with Eq. (44–8), which holds whether or not the semiconductor is doped. The constants N_c and N_v in Eq. (44–8) can be obtained by summing the number of electrons in the conduction band and the number of holes in the valence band, respectively:

$$N_c = 2\left(\frac{m_n^* kT}{2\pi\hbar^2}\right)^{3/2} \quad\text{and}\quad N_v = 2\left(\frac{m_p^* kT}{2\pi\hbar^2}\right)^{3/2}, \tag{44–12}$$

where m_n^* and m_p^* are the effective masses of the n-carriers and p-carriers, respectively. Then Eq. (44–8) reads

$$np = n_i^2 = N_c N_v e^{-(E_c - E_v)/kT} = N_c N_v e^{-E_g/kT}. \qquad (44\text{–}13)$$

We have rewritten Eq. (44–8) as Eq. (44–13) because we have given it a greatly extended reach. Equation (44–13) relates the n-carrier and p-carrier densities in a *doped* semiconductor to the intrinsic carrier densities, whereas in its original form it related the carrier densities in the undoped semiconductor.

Equation (44–13) has important consequences. For example, if we dope an intrinsic semiconductor with acceptor impurities to increase the density of holes, the density of electrons must decrease so that Eq. (44–13) continues to be satisfied. It is typical that the n-carriers dominate in an n-type extrinsic semiconductor, whereas p-carriers dominate in a p-type extrinsic semiconductor.

EXAMPLE 44–3 The effective masses in the semiconductor gallium arsenide are $m_n^*/m_e = 0.067$ and $m_p^*/m_e = 0.48$; the band gap is $E_g = 1.43$ eV. Find the intrinsic concentration of n-carriers or p-carriers in gallium arsenide at room temperature, where $kT = 0.026$ eV.

Solution: The concentration of intrinsic carriers in a semiconductor (doped or not) is given by Eq. (44–12). With the help of Eq. (44–13), we get

$$n_i = \sqrt{N_c N_v}\, e^{-E_g/2kT} = 2\left(\frac{m_e kT}{2\pi\hbar^2}\right)^{3/2}\left(\frac{m_n^* m_p^*}{m_e^2}\right)^{3/4} e^{-E_g/2kT}$$

$$= 2\left[\frac{(0.91 \times 10^{-30}\text{ kg})(1.38 \times 10^{-23}\text{ J/K})(300\text{ K})}{2\pi(1.05 \times 10^{-34}\text{ J}\cdot\text{s})^2}\right]^{3/2}$$

$$\times [(0.067)(0.48)]^{3/4} e^{-(1.43\text{ eV})/2(0.026\text{ eV})}$$

$$= 2.2 \times 10^{12}\text{ carriers/m}^3.$$

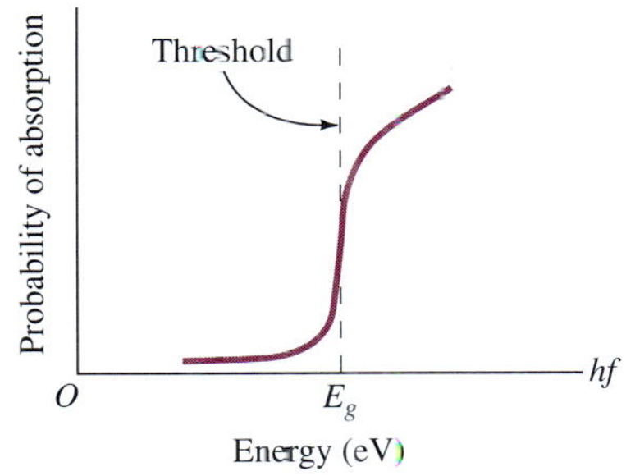

FIGURE 44–7 The probability of photon absorption in a semiconductor has a threshold when the photon energy equals the gap energy, because no energy levels are available for the absorbing electron at lower energies. Because a photon's energy is proportional to its frequency, the energy threshold can be expressed as a threshold in frequency.

Optical Effects in Semiconductors

When a photon strikes a semiconductor, there is a probability that it will be absorbed by the electron system. This process has a threshold as a function of the photon's frequency, f (Fig. 44–7). An electron in the material can absorb the photon only if the energy of the photon, hf, is greater than the gap energy; only at such energies are there empty levels available for electron occupation in the conduction band. Thus, some materials are transparent (a photon passes through without being absorbed) in certain frequency ranges but cease to be so for higher frequencies. An absorbing electron that is kicked up into the conduction band will lose energy by colliding with lattice atoms until its energy falls to near E_c; here, further energy loss is inhibited by the exclusion principle and/or the absence of energy levels below E_c. At that stage, the only way in which the electron can lose energy is to fall back into a hole in the valence band, a process called *recombination*. The electron–hole recombination can produce radiation whose frequency is such that $hf_g = E_g$ (Fig. 44–3). The radiation is called *photoluminescence*. The electrons can alternatively be initially excited into the conduction band by the bombardment of the material by high-energy electrons (*cathodoluminescence*) or by the action of an electric current in the material (*electroluminescence*).

When an electron–hole pair recombines rapidly, on the order of 10^{-8} s, the process is called *fluorescence*. A slower transition may arise when there are impurity levels in the gap, and the excited electron first undergoes a transition to those levels. If the impurity level has quantum numbers such that selection rules forbid the direct transition of an electron from it to the hole, the electron is said to be *trapped*. It will ultimately undergo a transition to a vacancy in the valence band (electron–hole recombination), but this may take as much as seconds or even minutes to occur. This process is termed *phosphorescence*, and the materials in which this occurs are called *phosphors*. The en-

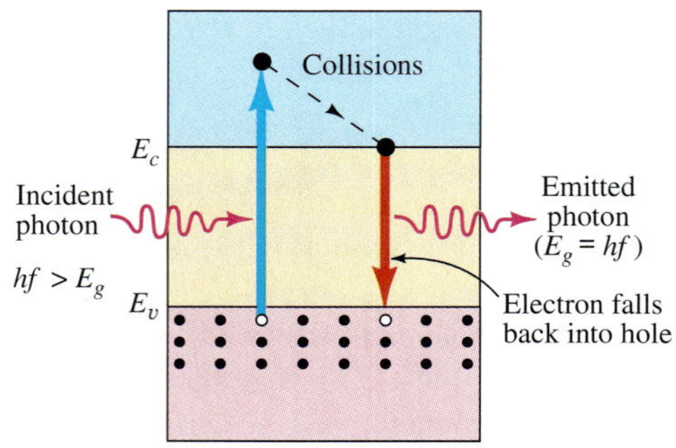

FIGURE 44–8 An electron excited to the conduction band by photon absorption loses energy by collision until its energy reaches E_c; at that point it recombines with a hole in the valence band, with photon emission.

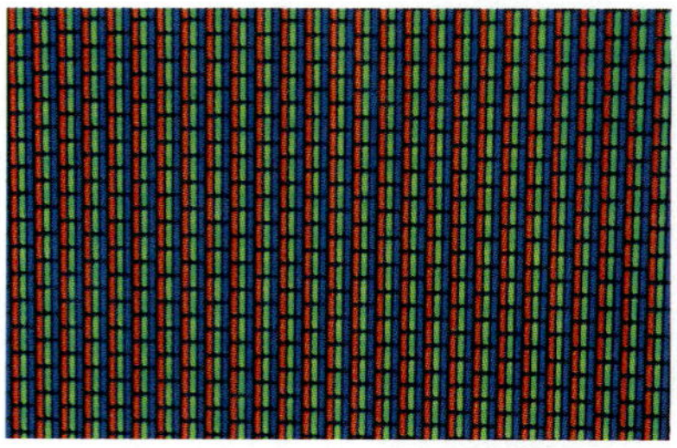

FIGURE 44–9 Excited red, green, and blue phosphors are visible in this photomicrograph of a color television screen.

ergy of the emitted photon is determined by the energy difference between the intermediate impurity level and the top of the valence band. We can thus choose the impurities such that radiation will be emitted in a variety of colors. In a color television tube, the screen is coated with phosphor dots that emit radiation in the primary colors (Fig. 44–9). The phosphors are excited by three incoming electron beams (cathodoluminescence), which are deflected to the proper spot on the screen by electric and magnetic fields determined by the incoming signal to the television. Each of the beams excites a different color.

APPLICATION
Monitor screens

A semiconductor exposed to radiation for which $hf > E_g$ has an increased number of n-carriers and p-carriers and thus an increased conductivity. Devices whose conductivity changes when they are exposed to light—*photoconductive* devices—are used for lights that automatically turn on at dusk and off at dawn, or for measuring light intensity, as in exposure meters for cameras (Fig. 44–10).

APPLICATION
Photoconductive devices

44–2 SEMICONDUCTOR STRUCTURES

The Fermi energies equalize across materials that are in electrical contact.

In applying our knowledge of energy distributions in single semiconductors to semiconductor structures (combinations of semiconductors in contact so that charge carriers can pass between them), one principle determines essentially all the important features: Like temperature, *the chemical potential—what we have agreed to call the Fermi energy for semiconductors—has the same value throughout a system in equilibrium.* This follows from the fact that in the absence of external potentials and in thermodynamic equilibrium, there is no net flow of charge or of energy across the boundary between two materials. The materials may be the same type of semiconductor doped differently, two different intrinsic semiconductors, or even a semiconductor and a metal. We prove this important result in the following subsection.

FIGURE 44–10 A light meter aids a photographer in finding the correct light settings for the camera. At the heart of the device is a material whose electrical response depends on the intensity of the light it receives.

How We Know that the Fermi Energy Is Constant Across a Boundary

Consider two materials, A and B, in contact so that charge carriers can pass between them; the materials are in thermal equilibrium. At thermal equilibrium, there is no net energy flow and no net charge flow across the boundary. The number of states in a unit volume per unit energy interval around the energy E that electrons can occupy in each material are $n_A(E)$ and $n_B(E)$, respectively—we refer to these quantities as the density of available states. (We consider the electrons alone because the behavior of the holes follows from the behavior of the electrons.) The conditions of no energy flow and no charge flow mean that the flow of electrons from A to B at any E value equals the flow of electrons from B to A at that same E value. The flow of electrons from A to B at energy E is proportional to the number of electrons present in A times the number of empty states present in B; that is, to the density of occupied states in A times the density of empty states in B. To express this mathematically, we need the densities of occupied and unoccupied states:

$$\text{density of occupied states} = (\text{density of available states})(\text{probability of occupation}) = n(E)f(E),$$

where $f(E)$ is the probability of occupation given by the Fermi–Dirac distribution, Eq. (44–2). Because the probability of a state being vacant is one minus the probability of that state being occupied, we also have

$$\text{density of unoccupied states} = (\text{density of available states})(\text{probability of vacancy}) = n(E)[1 - f(E)].$$

We apply these results to our situation:

$$(\text{flow of electrons from } A \text{ to } B \text{ at energy } E) \propto [n_A(E)f_A(E)]\{n_B(E)[1 - f_B(E)]\};$$

$$(\text{flow of electrons from } B \text{ to } A \text{ at energy } E) \propto [n_B(E)f_B(E)]\{n_A(E)[1 - f_A(E)]\}.$$

We set these equal to each other:

$$[n_A(E)f_A(E)]\{n_B(E)[1 - f_B(E)]\} = [n_B(E)f_B(E)]\{n_A(E)[1 - f_A(E)]\};$$

$$f_A(E) - [f_A(E)f_B(E)] = f_B(E) - [f_B(E)f_A(E)];$$

$$f_A(E) = f_B(E).$$

If we now compare this equation with Eq. (44–2), we see that it can be satisfied only if the parameters E_F in the two materials are equal.

The *p-n* Junction

When an n-type and a p-type semiconductor are brought into contact, a *p-n junction* is formed. Such a junction is of interest because it acts as a diode. We consider p-n junctions in which both semiconductors are made by doping the same intrinsic semiconductor. In Fig. 44–11a, we have used p_p and p_n to denote the hole densities and n_p and n_n to denote the electron densities in p-type and n-type semiconductors, respectively. The sketches are highly distorted because the ratio of the dominant carriers to the carriers of opposite sign on each side of the junction is typically on the order of 10^{11}; drawn to scale, the n_p and p_n levels would be indistinguishable from the horizontal axis. We shall henceforth refer to the regions of n-type and p-type semiconductors as the n-side and p-side, respectively.

The p-side has an excess of mobile holes, and the n-side has an excess of mobile electrons, not necessarily with the same density. Electrical neutrality is preserved on the p-side by the (immobile) negative acceptor ions in the crystal lattice and on the n-side by the (immobile) positive donor ions. When the two materials are put in contact, the mobile holes on the p-side tend to diffuse into the n-side, and the mobile electrons on the n-side tend to diffuse into the p-side. This mixing continues until the Fermi energies in the two materials reach the same level. This occurs by the following mechanism: Charges flow until the net positive charge remaining on the n-side (the right) and the net negative charge remaining on the p-side (the left) set up an electric field that stops the diffusion (Fig. 44–11b). The n-side will be at a higher potential, V_0, than the p-side. But electrons have a *negative* charge, so the energy of electrons on the right is *reduced* by eV_0.

Let's look at the p-n junction with the aid of energy diagrams. Figure 44–12a shows the energy levels of the semiconductors before they are brought together. The gap energies are the same because both semiconductors are made by doping the same intrinsic semiconductor. When the junction is formed, the conduction and valence bands will be distorted (Fig. 44–12b). The internal potential V_0 is what brings the Fermi en-

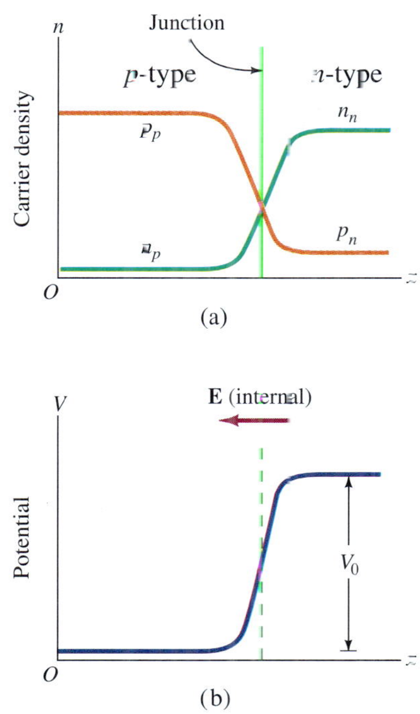

FIGURE 44–11 A p-n junction is formed by placing an n-type and a p-type semiconductor in contact. (a) The number densities of p-carriers (p_p and p_n, respectively), and the number densities of n-carriers (n_p and n_n, respectively) in p-type and n-type semiconductors. These densities are not drawn to scale. (b) As described in the text, the diffusion of carriers causes a potential difference to build up across the boundary. The direction of the resulting electric field is shown.

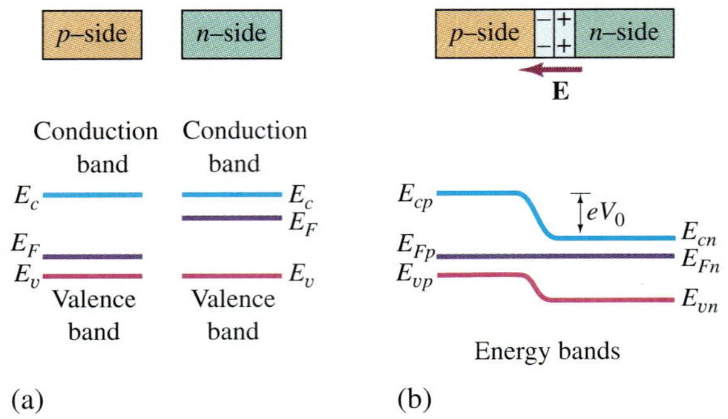

FIGURE 44–12 (a) The energy-level structure of n-type and p-type semiconductors before they are joined in a junction. (b) After the junction is formed and equilibrium is established, the energy levels adjust in such a way that the Fermi levels of the two materials equalize. This lowers the electron energy levels on the n-side and raises the electron energy levels on the p-side. (After B. G. Streetman, *Solid State Electronic Devices*, Prentice Hall, 1980, p. 141.)

ergies to the same level. From Figs. 44–8 and 44–9, we see this statement is equivalent to the relation

$$E_{Fn}^0 - E_{Fp}^0 = eV_0,$$

where E_{Fn}^0 and E_{Fp}^0 are the Fermi energies in the n-type and p-type semiconductors, respectively, *before* the materials are brought into contact. *After* the materials are brought into contact,

$$E_{cp} - E_{cn} = eV_0, \tag{44–14}$$

There is an internally generated potential difference across the boundary in a p-n junction.

where E_{cp} and E_{cn} are the edges of the conduction band in the p-type and n-type semiconductors, respectively. This relation follows directly from Fig. 44–12b. A similar expression holds for the edges of the valence bands.

Note that V_0, which is known as the *contact potential*, is *not* an externally imposed potential. It is a property of the junction itself.

Let's suppose that the potential jump across the boundary is abrupt and study the consequences. (In real materials, the potential change occurs across a finite width called the *transition region*, or *depletion region*, as in Fig. 44–13. However, we shall ignore effects due to the transition region.) The densities of n-carriers on the two sides are, according to Eq. (44–5),

$$n_n = N_c e^{-(E_{cn}-E_F)/kT} \quad \text{and} \quad n_p = N_c e^{-(E_{cp}-E_F)/kT}.$$

Thus

$$\frac{n_p}{n_n} = \frac{N_c e^{-(E_{cp}-E_F)/kT}}{N_c e^{-(E_{cn}-E_F)/kT}} = e^{-(E_{cp}-E_{cn})/kT}$$
$$= e^{-eV_0/kT}, \tag{44–15}$$

where we have used Eq. (44–14) in the last step. Similarly, the ratio of the equilibrium hole densities on the two sides is

$$\frac{p_n}{p_p} = e^{-eV_0/kT}. \tag{44–16}$$

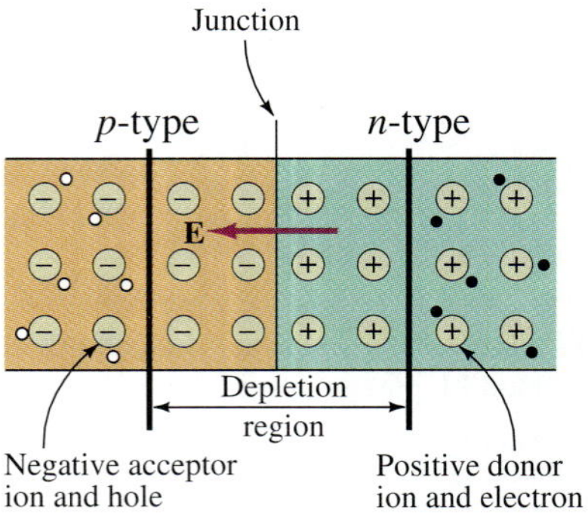

FIGURE 44–13 The potential in a p-n junction takes place over a region of finite width, called the depletion region. There is an internally generated electric field across the depletion region. (After Narciso Garcia and Arthur C. Damask, *Physics for Computer Science Students*, John Wiley & Sons, 1986, p. 457.)

The ratios in Eqs. (44–15) and (44–16) help us to understand just how the currents in the two directions balance. Let's consider the electrons—the flow of holes would work the same way. From Eq. (44–15), the concentration of electrons on the p-side is less than the concentration on the n-side by the factor $e^{-eV_0/kT}$, but these electrons have no potential barrier to climb to get to the n-side. There is a large density of electrons on the n-side; to get to the p-side, however, they need enough energy to climb a potential barrier of height eV_0—only a number reduced by the Maxwell–Boltzmann factor $e^{-eV_0/kT}$ will have that much energy. In this way, the flow of electrons in the two directions matches in equilibrium.

It follows from Eq. (44–14) that each of the products $p_n n_n$ and $n_p p_p$, which, from Eqs. (44–15) and (44–16) are equal in equilibrium, is also equal to n_i^2. Thus

$$p_n n_p = N_c N_v e^{-(E_g+eV_0)/kT} \quad \text{and} \quad p_p n_n = N_c N_v e^{-(E_g-eV_0)/kT}. \tag{44–17}$$

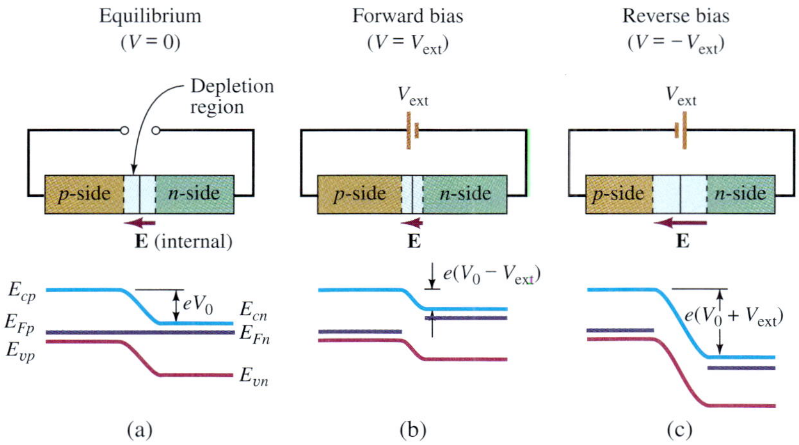

FIGURE 44–14 (a) In the unbiased p-n junction, the intrinsic contact potential leads to equilibrium. (b) In forward biasing of a p-n junction, the energy levels on the n-side are raised relative to those on the p-side. (c) In reverse biasing of the same junction, the energy levels on the n-side are lowered relative to those on the p-side. (After B. G. Streetman, *Solid State Electronic Devices*, Prentice Hall, 1980, p. 151.)

Biasing

Let's now apply an external potential, V_{ext}, across a p-n junction, which is then said to be *biased*. In this case, there can be a net flow of charge and energy, unlike the unbiased case of Fig. 44–14a. Let's first assume that the external potential is such that the energies on the p-side are brought closer to the energies on the n-side (Fig. 44–14b). This external potential, by which the magnitude of the potential difference between the n-side and the p-side is *reduced* from $|V_0|$ to $|V_0| - |V_{\text{ext}}|$, is described as a *forward bias* voltage. The electrons that diffuse from the n-side to the p-side have less of a potential barrier to overcome. The same is true for the holes that diffuse from the p-side to the n-side (the potential difference is the same, but the charges of the carriers have opposite signs so both have the same hill to climb). There is a net current from the p-side to the n-side, which rapidly increases as $|V_0| - |V_{\text{ext}}|$ decreases in magnitude. Note that there is also a small fixed current from the n-side to the p-side, associated with movement of electrons and holes without a potential barrier to overcome.

If the external potential is applied in such a way that the potential difference between the n-side and the p-side is *increased* from $|V_0|$ to $|V_0| + |V_{\text{ext}}|$, we have a *reverse bias* voltage (Fig. 44–14c). Both electrons from the n-side and holes from the p-side have an even harder time overcoming the additional barrier, and there is little current due to those charge carriers. However, the small fixed current from the n-side to the p-side is unaffected.

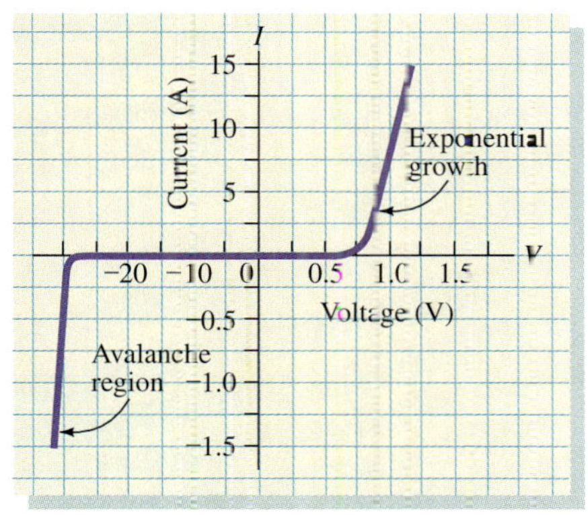

FIGURE 44–15 I-V characteristic of a p-n junction. The junction behaves as a diode under an external potential V_{ext}.

To make our argument more quantitative, let's suppose that a positive V_{ext} is a forward bias and a negative V_{ext} is a reverse bias. Consider first the current due to the motion of the holes, with positive current corresponding to positive charges moving to the right. The potential barrier $e(V_0 - V_{\text{ext}})$ inhibits positive carriers from the left (of density p_p) from flowing, and only the fraction $p_p e^{-e(V_0 - V_{\text{ext}})/kT}$ gets through. Positive carriers from the right do not encounter such a barrier, but their density is smaller by a factor $e^{-eV_0/kT}$, independent of V_{ext}. (The same is true separately for the n-carriers.) Take I_0, which is proportional to $e^{-eV_0/kT}$, to be the current of p-carriers from the right and n-carriers from the left. Without the external potential, I_0 would be canceled by the current of p-carriers from the left and n-carriers from the right, but those currents have now been modified by the factor $e^{eV_{\text{ext}}/kT}$. Thus the *net* current flowing to the right is given by

$$I_{\text{net}} = I_0(e^{eV_{\text{ext}}/kT} - 1). \qquad (44\text{–}18)$$

This current–voltage relation, called the *I-V characteristic* of the junction, is shown in Fig. 44–15.

The I-V characteristic shows that the p-n junction behaves like a *diode*, a device that allows current to pass in one direction but not in another. Diodes can be used to construct rectifiers (see Chapter 34). The p-n junction can also act as a *solar cell* or a *light-emitting diode* (see the Application box).

APPLICATION

Diodes

If a photon of light with frequency $f > E_g/h$ shines on a p-n junction, an electron from the valence band that absorbs that photon will be excited to a state in the conduction band. The electron now becomes an n-carrier and leaves behind a hole, a p-carrier. The intrinsic contact potential V_0 separates the electrons and the holes: The p-carriers on the p-side will move to the junction in the direction of the intrinsic electric field, while the n-carriers on the n-side will also move to the junction. The movement of these new carriers produces a potential difference opposite to V_0. It thus acts as a forward bias voltage across the junction, even in the absence of V_{ext}. A current is produced by this *photovoltaic effect*, and power can be delivered to an external circuit. All that is needed is a conductor attached to the two components of the junction. In silicon, photons of wavelength shorter than 1200 nm will give rise to a photovoltaic effect.

A solar cell, a device that produces electricity from sunlight, is a p-n junction constructed in such a way that the solar photons can be absorbed in the area closest to the built-in contact potential—near the junction (Fig. 44AB–1). To maximize that area, the sunlight is allowed to fall onto the top semiconductor, which is covered with an antireflective coating. The light penetrates the semiconductor and is absorbed in the junction region below. The materials are chosen with a band gap small enough to absorb even the long-wavelength component of solar light. It is also important to have a large absorption coefficient for the light. In order to conserve energy and momentum in crystalline silicon, an electron must give up some momentum and energy to the lattice, and this reduces by some two orders of magnitude the probability that a transition will occur. Thus, in crystalline silicon, the absorption is not very good. Amorphous silicon has 50 times the absorptive power of crystalline silicon, and copper indium diselenide is another factor of 10 more efficient for absorption because momentum in these materials can be conserved without the involvement of the ions.

Because of limitations on collecting light and on absorbing photons, there is a theoretical maximum of 28 percent for conversion of solar energy into electric energy in silicon. Practical silicon solar cells operate with efficiencies of around 10 percent. Perhaps the most serious barrier to large-scale utilization of solar cells for power generation is cost. In 1991, the cost per kilowatt-hour for energy from solar cells was 30 to 60 times more than the cost of energy produced from fossil fuels. With improvements in the technology of known materials, we expect that solar cells will become competitive with conventional power sources before the year 2000. Power generation by solar cells is already indispensable in satellites and for areas on Earth that are remote from other power sources (Fig. 44AB–2).

A light-emitting diode (or LED), which is a p-n junction acting as a signal light, is essentially a solar cell that operates in reverse. When a forward bias is applied across the junction, electrons flow from the n-side to the p-side and holes from the p-side to the n-side. When the electrons arrive at the p-side, they recombine with available holes and emit light in the process. The holes that arrive at the n-side recombine with available electrons and also emit light. LEDs are widely used as electronic signaling lights because they are long-lasting, use little energy, can be switched on and off rapidly, and are compact. LEDs are used also for digital readouts in some watch faces and calculator displays.

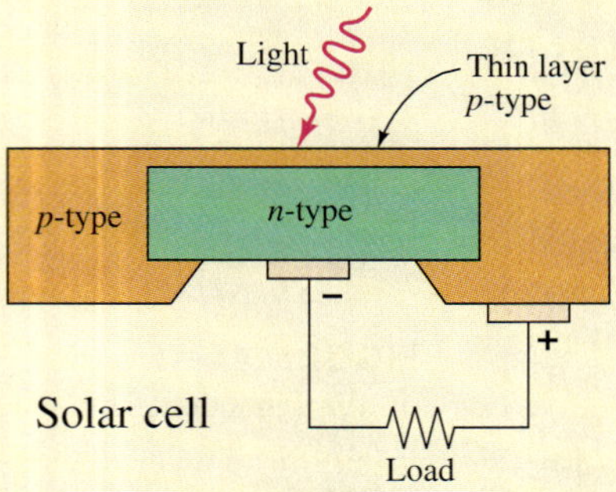

FIGURE 44AB–1 Schematic diagram of a solar cell, which generates current across the load (resistor) when light shines on the p-side of a p-n junction.

FIGURE 44AB–2 The high initial cost of solar cells is compensated when they operate in situations in which alternatives are few, as in the solar arrays of the Hubble Space Telescope.

EXAMPLE 44–4 Calculate the ratio of the currents through a p-n junction diode for positive voltages to those for negative voltages for voltages whose magnitude is given by $eV_{ext} = 0.1$ eV, 0.2 eV, and 0.3 eV at $T = 300$ K.

Solution: The current through a junction diode biased by a voltage is given by Eq. (44–18). Thus the ratio of the currents for positive to negative bias, R, is given in terms of $K \equiv eV_{ext}/kT$ by

$$R = \left| \frac{e^K - 1}{e^{-K} - 1} \right| = \frac{e^K - 1}{1 - e^{-K}}.$$

At $T = 300$ K, $kT = 2.6 \times 10^{-2}$ eV, so the parameter K has the values 3.85, 7.69, and 11.5 for $eV_{ext} = 0.1$ eV, 0.2 eV, and 0.3 eV, respectively. The respective values of R are then 48, 2.2×10^3, and 10.2×10^4. The rapid growth of these values as eV_{ext} increases illustrates the dramatic biasing power of relatively small potential differences.

The Bipolar Junction Transistor

The **transistor** is a three-terminal device that can be used for sensitive control of currents flowing through circuits, particularly for amplifying or switching currents. One of the first transistors, the *bipolar junction transistor* (BJT), started a revolution in electronics that is still going on. That transistor and the theory of its operation were developed in 1948 and 1949 by John Bardeen, Walter Brattain, and William Shockley (Fig. 44–16). Bipolar junction transistors can be described schematically as n-p-n or p-n-p three-terminal devices. In an n-p-n transistor, there are two junctions in which a narrow p-type *base* is sandwiched between two wide n-type semiconductors, called the *emitter* and *collector* (Fig. 44–17a). In a p-n-p transistor, a narrow n-type base is sandwiched between a p-type emitter and collector (Fig. 44–17b). Figure 44–18a shows the band structure for an n-p-n BJT where there is no external (bias) voltage.

FIGURE 44–16 "Bardeen's box" constructed in 1949, contains a circuit with transistors as elements. The instant turn-on of the circuit impressed audiences who had been used to waiting a long time for circuits containing tubes to warm up.

In Fig. 44–18b, we show how the band structure is altered when the emitter–base junction is forward biased and the collector–base junction is reverse biased, as can be arranged with the circuit of Fig. 44–19a. Because the base is so narrow, the transistor is more than just two n-p junctions back to back. Electrons that enter the base from the emitter can diffuse all the way across the base, and no electric field of the type encountered in the n-p junction described earlier will build up. Once electrons from the emitter reach the base–collector junction, they flow easily into and through the col-

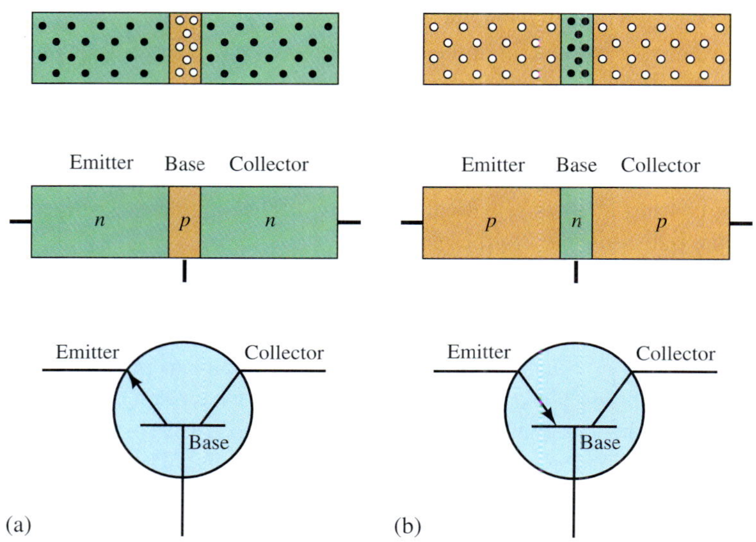

FIGURE 44–17 Diagrams of (a) n-p-n and (b) p-n-p bipolar junction transistors. The circuit-diagram symbols for these devices are shown at the bottom of the figures.

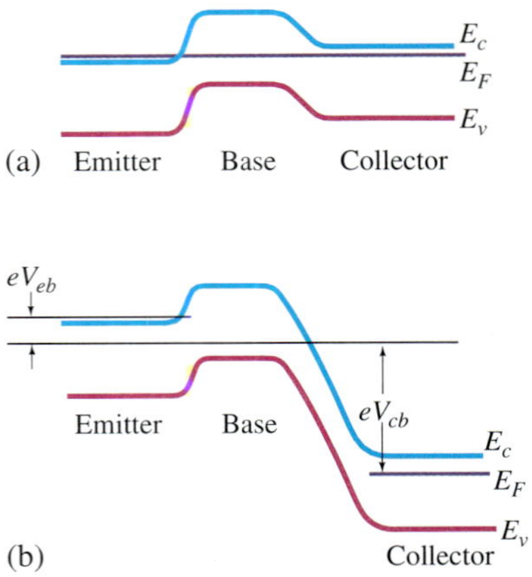

FIGURE 44–18 The band structure of an *n-p-n* bipolar junction transistor (a) in the absence of an external biasing voltage and (b) with forward biasing on the emitter–base junction and reverse biasing on the collector–base junction. (After M. Shur, *Physics of Semiconductor Devices*, Prentice Hall, 1990, p. 242.)

APPLICATION

The operation of a bipolar junction transistor

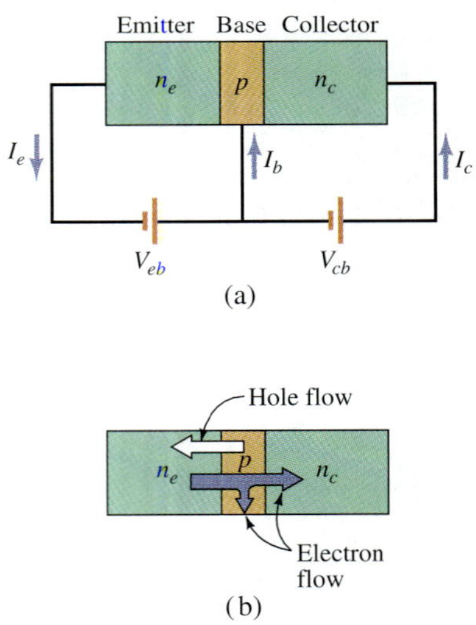

FIGURE 44–19 (a) The biasing of the *n-p-n* bipolar junction transistor of Fig. 44–18b is accomplished by a circuit such as this one. (b) Schematic diagram of the resulting flows of charged particles for this situation.

APPLICATION

Integrated circuits

lector (Fig. 44–19b). Note that the current flowing through the collector is controlled by the forward bias voltage across the emitter–base junction.

Let's define a *current transfer ratio*, α, by

$$\alpha \equiv \frac{I_c}{I_e}, \tag{44–19}$$

the ratio of the change, I_c, in the collector current with the change, I_e, in the emitter current. As can be seen in Fig. 44–19a, the emitter current is the sum of the collector current and base current (I_b):

$$I_e = I_c + I_b. \tag{44–20}$$

The *current gain*, β, is a measure of the change in collector current for a change, I_b, in base current, defined by

$$\beta \equiv \frac{I_c}{I_b}. \tag{44–21}$$

It follows from Eqs. (44–19) through (44–21) that

$$\frac{1}{\alpha} = \frac{I_e}{I_c} = \frac{I_c + I_b}{I_c} = 1 + \frac{1}{\beta}.$$

Thus

$$\beta = \frac{\alpha}{1 - \alpha}. \tag{44–22}$$

When α is close to unity, β can be large. For typical values for a transistor whose base is about 1 μm wide, forward bias is about 0.2 V, and backward bias is 5 to 10 V; α is in the range 0.97 to 0.99, corresponding, according to Eq. (44–21), to gains of from 32 to 99. Thus a small variation of the current into the base is *amplified* into a very large variation of the current into the collector. The transistor sensitively fulfills its amplification role by manipulating bias voltages.

The Field-Effect Transistor

The *field-effect transistor* is a semiconductor transistor structure widely used in modern electronics. A semiconductor plate called a *channel* is connected to two terminals called the *source* and the *drain*. A charge is induced in the channel by a metal plate known as a *gate*, which is separated from the semiconductor by insulating material. The channel and the gate together act as a capacitor. In the metal-oxide semiconductor field-effect transistor (MOSFET), the insulator separating the gate from the semiconducting channel is silicon dioxide (Fig. 44–20a). When a positive voltage is applied to the gate, a negative charge is induced in the channel; that is, *n*-carriers fill the channel. As the gate voltage increases, so does the concentration of *n*-carriers in the channel, and hence so does its conductance. If a drain-to-source voltage is applied, a current flows across the channel. This current can be changed by changing the gate voltage as just described. Thus, the field-effect transistor is a three-terminal device in which the current between two terminals is sensitively controlled by changing the voltage at the third terminal, the one connected to the gate.

MOSFETs form building blocks for today's electronics. By the techniques to be described in Section 44–3, thousands or even millions of tiny MOSFETs and other semiconductor structures making up complicated circuits are fabricated as one unit; we call these circuits *integrated circuits*. These circuits have enabled us to produce desktop computers many times more powerful than yesterday's mainframes. The layers of material that make up the devices in integrated circuits are as thin as 0.2 nm, and the area of a single MOSFET is as small as 5 μm $\times$ 5 μm (Fig. 44–20b). Typical currents in these circuits are on the order of 20 mA.

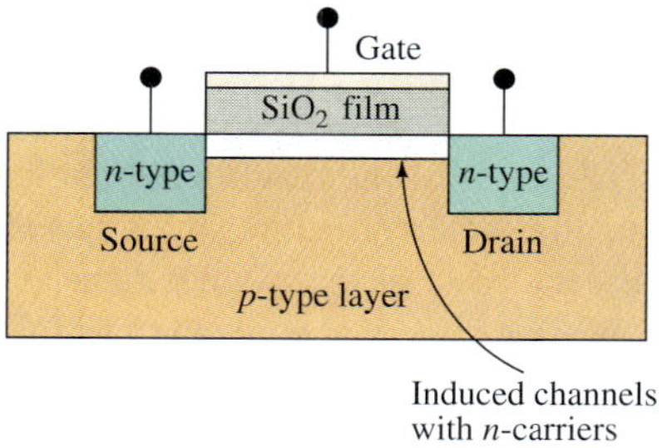

(a)

(b)

FIGURE 44–20 (a) Schematic diagram of a MOSFET, a type of field-effect transistor. (b) In 1992, IBM engineers produced this MOSFET, so tiny—its length is several hundred nanometers—that memory chips containing 4 gigabits of memory are made possible. (The largest memory chip available until then had a capacity of 16 megabits.) The wide yellow strip is the gate connection, while the dark patches on the side are the source and drain. This transistor is so small that many of its properties are quantum mechanical in nature.

FIGURE 44–21 The molecular-beam epitaxy machine shown here is used to construct microscopic structures molecular layer by molecular layer.

44–3 BAND-GAP ENGINEERING

Technological advances in the fabrication of semiconductor structures have created a new tool chest for the electronics engineer and computer designer. One advance involves the ability to place thin layers of one or several materials on a single underlying layer (a *substrate*) of another material with very similar crystal structure. (Such techniques are the basis of integrated-circuit construction.) This procedure, known as *epitaxy*, has reached its highest refinement with *molecular-beam epitaxy* (Fig. 44–21). Examples are provided in structures in which single crystals of gallium arsenide form a substrate and compounds of aluminum and gallium arsenide ($Ga_{1-x}Al_xAs$, which are alloys of variable content; x represents a fraction) form junctions with the substrate. Gallium, aluminum, and arsenic are evaporated in a vacuum chamber and projected as beams onto the gallium arsenide substrate. The beams form layers at a rate of about one per second on the substrate whose thickness can be controlled. The components of the deposited layers can be varied by exposing the substrate to different beams in different proportions. Impurities are easily added in desired proportions.

Lithography is another indispensable technology for integrated circuits. With this process, we can create detailed patterns on a crystal surface (Fig. 44–22). In scanning-electron-beam lithography, an opaque film on a transparent base is removed in a precise pattern by a moving electron beam. The resulting transparent features are as small as 20 nm. Light passes through this "negative" onto a semiconductor blank that has been coated with a polymer that disintegrates when light hits it. The coated semiconductor can then be etched by chemicals such as chlorine gas that act where the polymer has disintegrated. Alternatively, other materials can be deposited on the areas where the polymer has disintegrated, or doping ions can be introduced in those areas. A process such as this can be repeated to produce elaborate structures with many layers.

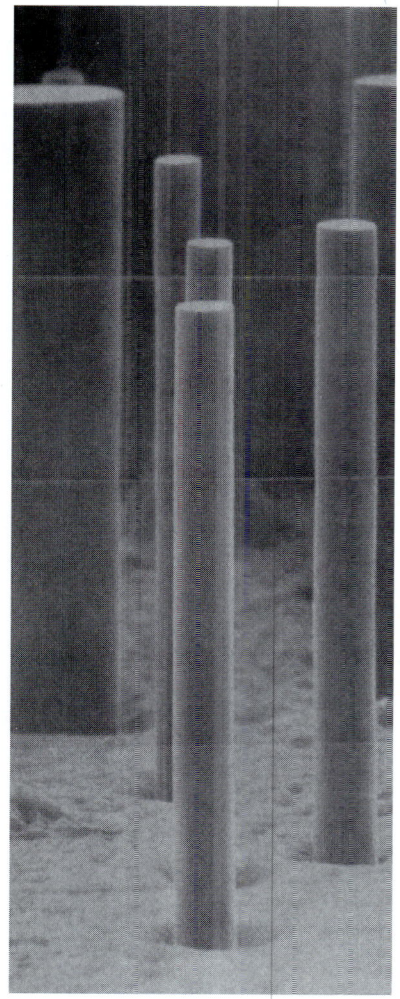

FIGURE 44–22 Some of the elaborate microscopic structures produced by lithography. They are less than a half micron high.

Heterojunctions

A *heterostructure* is a single crystal in which the occupancy of the ion lattice sites changes at an interface, or boundary. Such crystals are created by epitaxy. Examples are GeAs/GaAs and $Al_xGa_{1-x}As/GaAs$, where the slash indicates the interface. When we use two different semiconductors, we say that we have a *heterojunction*. Figure 44–23a shows the band-gap structure of two different semiconductors, each doped as an n-type material. The semiconductor on the right has a wider gap than the one on the left; in each case, the Fermi energy, E_F, lies closer to E_c, than to E_v. (The doping

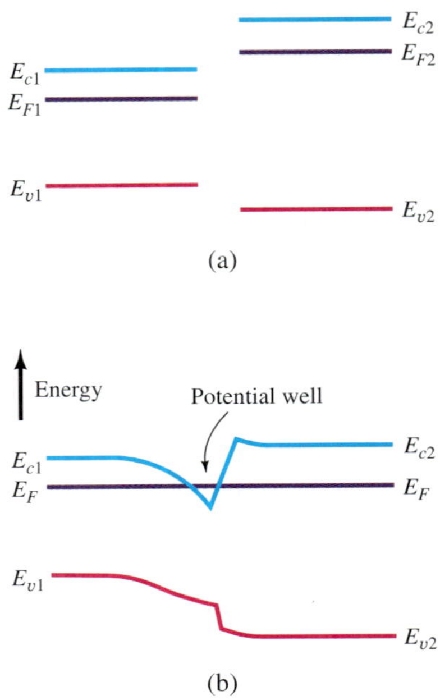

FIGURE 44–23 (a) Energy diagrams for two different semiconductors, each doped to n-type, before they are joined in a heterojunction. (b) When the semiconductors are joined to form a heterojunction, the edge of the conduction band forms a kink that can act as a potential energy well for electrons. There is no restriction to movement into or out of the page for an electron in this well.

in the semiconductors need not be uniform. In Fig. 44–23a, the donor impurities are located some distance away from what will become a junction.)

When the materials are joined to form a heterojunction, the flow of charges causes the Fermi energies to equalize, as in Fig. 44–23b. The form of the resulting potential difference depends on the materials at the junction. In a heterostructure, the edge of the conduction band (or, depending on the materials, the edge of the valence band) is distorted to contain the kink shown in Fig. 44–23b. We can understand this kink as follows. When the heterojunction is formed, the energies shift so as to equalize the Fermi energies of the two semiconductors. The gap separations on either side of the junction, however, must remain as they were before the junction was formed because those are intrinsic to the semiconductors. The only way to satisfy both constraints is to develop a kink such as the one shown in Fig. 44–23b.

As a result of the kink, *electrons in the conduction band at the interface find themselves in a potential energy well* for motion across the junction. The potential well is far from the donor ions, so if an electric field is applied in the direction perpendicular to the plane of Fig. 44–23b, electrons will flow along the channel formed by the well without finding many positive impurity ions in their path. These electrons can therefore move very rapidly, on the order of 1.8×10^5 m/s, some 1000 times faster than in MOSFETs. The drift speed of electrons is an important factor in devices that act as switches in high-speed computers.

APPLICATION

High-speed switches

Quantum Wells, Quantum Wires, and Quantum Dots

When two or more heterojunctions are put back to back, the peculiar shapes taken by the edge of the conduction band may form a region in which electrons have "walls" on two sides. Figure 44–24a shows a thin GaAs layer sandwiched between two AlGaAs layers, surrounded by GaAs that contains the donor ions; Fig. 44–24b is an energy diagram for this structure. The electrons in the GaAs layer move in a *quantum well*. Such potential wells are characterized by quantized energy levels. We can find the values of these energies by thinking of standing waves set up in the direction of the (small) distance a between the "walls" that confine the electron, just as atomic energies in the Bohr model follow from thinking about fitting standing waves into atomic orbits. The allowed energies of the quantum well are then given by

$$E_n = \frac{n^2\hbar^2\pi^2}{2m^*a^2}, \qquad (44\text{–}23)$$

where n is any integer. The confining distance a is typically 10 to 100 nm. Although a is large compared with atomic dimensions, the fact that m^* is small partly compensates for this effect. Only the energy of lateral motion is quantized: the motion in the direction perpendicular to the plane of the page for Fig. 44–24a is not quantized.

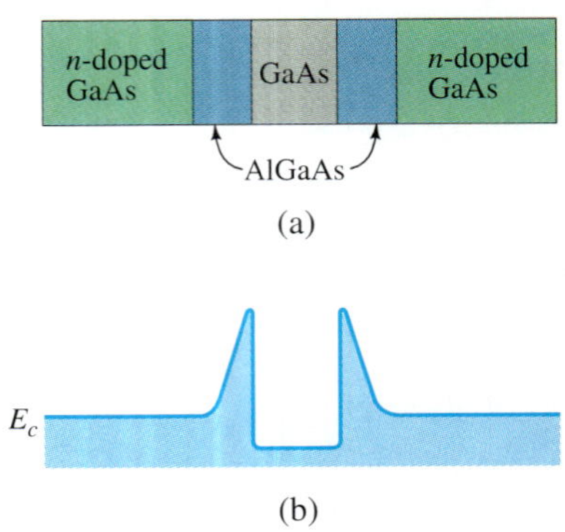

FIGURE 44–24 (a) A heterostructure within which (b) a potential energy well forms and confines electrons.

EXAMPLE 44–5 Calculate the energy difference between the ground state and the first excited state in a quantum well of width 20 nm, given that the effective mass is $m^* = 0.07m_e$. What is the wavelength of a photon emitted when an electron undergoes a transition between the two states?

Solution: We use Eq. (44–22) to calculate the energy difference between levels characterized by $n = 2$ (the first excited state) and $n = 1$ (the ground state). It follows that

$$\Delta E = E_2 - E_1 = \frac{(2^2 - 1^2)\hbar^2\pi^2}{2m^*a^2} = \frac{3\hbar^2\pi^2}{2m^*a^2} = \frac{3(1.05 \times 10^{-34}\,\text{J}\cdot\text{s})^2(3.14)^2}{2(0.07)(0.9 \times 10^{-30}\,\text{kg})(2 \times 10^{-8}\,\text{m})^2}$$

$$= 6.2 \times 10^{-21}\,\text{J} = \frac{6.2 \times 10^{-21}\,\text{J}}{1.6 \times 10^{-19}\,\text{J/eV}} = 3.9 \times 10^{-2}\,\text{eV}.$$

The frequency of a photon emitted in a transition is given by $f = \Delta E/2\pi\hbar$, and the wavelength is given by

$$\lambda = \frac{c}{f} = \frac{2\pi\hbar c}{\Delta E} = \frac{2(3.14)(1.05 \times 10^{-34}\ \text{J}\cdot\text{s})(3 \times 10^{8}\ \text{m/s})}{6.2 \times 10^{-21}\ \text{J}} = 3.2 \times 10^{-5}\ \text{m}.$$

A quantum well can act as a very sensitive switch. Figure 44–25 shows the *I-V* characteristic for a quantum well device. With zero voltage between points *A* and *B* (Fig. 44–25a), the quantum well presents a barrier to the *n*-carriers. There is no tunneling because the energy levels on both sides of the quantum well are occupied equally. When a voltage is applied, the *n*-carriers on the left are at a higher energy, and they can tunnel into empty levels. Thus, the current rises with *V*, as shown in Fig. 44–25b. When the bottom of the conduction band on the left is lifted high enough to match one of the low-lying quantized levels inside the well, there is a drastic increase in the flow of the current (Fig. 44–25c). A further voltage rise decreases the current because the bottom of the conduction band on the left no longer matches a well level (Fig. 44–25d). As the voltage increases further, the current decreases and then increases again as the second energy level is approached, as we see in Fig. 44–25e. The ability to pick the applied voltage that produces substantial tunneling makes for a sensitive switch.

The existence of discrete energy levels in the well can be confirmed by the selective absorption of laser light at frequencies that correspond to energy differences between the well levels, according to the relation $\Delta E = hf$. By making a long string of AlGaAs/GaAs/AlGaAs sandwiches, we can create a series of quantum wells that resemble a crystal lattice; the arrangement is called a *superlattice*. The energy-level structure of a superlattice is just like an ordinary band structure with an important difference: Crystal lattice spacings are on the order of 0.1 nm, whereas quantum well spacings are on the order of 10 to 100 nm. The superlattice then has band gaps in the range of millielectron-volts.

Quantum wires and *quantum dots* represent further steps in the fabrication of systems in which quantum behavior is dominant. They correspond to electron-confining "pipes" and "boxes," respectively. Electrons in a quantum dot, the equivalent of a tiny three-dimensional box, have widely separated energy levels (see Problem 34). Such structures are promising as very precise switching devices.

Semiconductor Lasers

Semiconducting materials are well suited to the construction of microscopic lasers. Such lasers have already found uses in compact-disk players, amplifiers in satellite receivers, and communications with fiber optics. An energy diagram for a heterostructure illustrates how these lasers may be constructed. The semiconductor within which the laser light will be produced (for example, GaAs) is sandwiched between two lay-

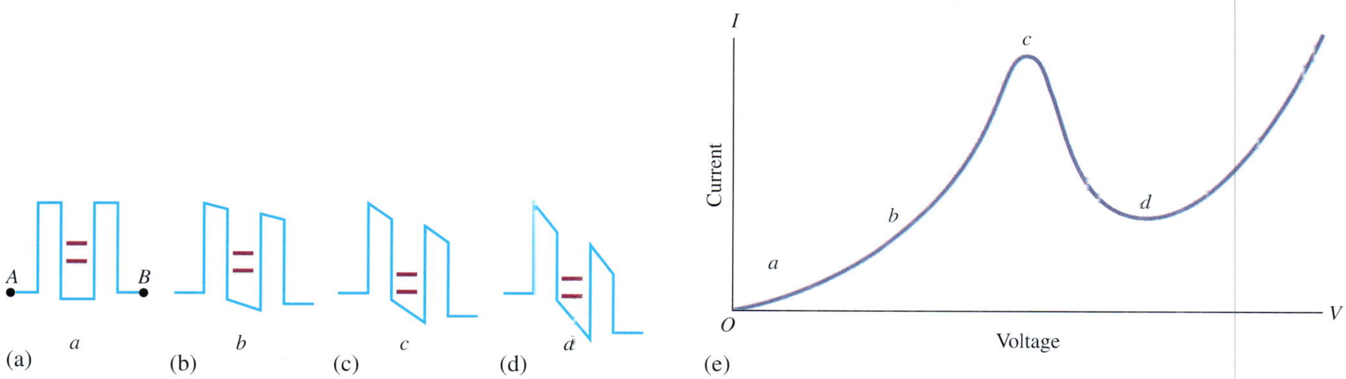

FIGURE 44–25 The quantum well of Fig. 44–24 subjected to an external potential. (a) through (d) As the potential is increased, there is an enhanced probability that an electron will tunnel from one side of the well to another when the location of allowed levels in the well matches an occupied level in the conduction band on the left. (e) *I-V* characteristic for the situation.

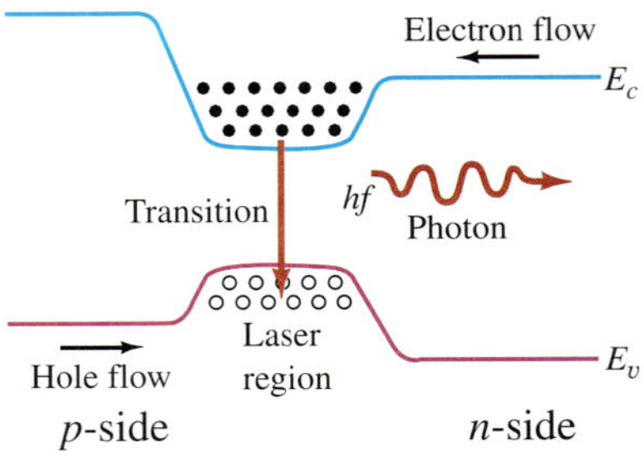

FIGURE 44–26 Energy diagram for a semiconductor laser. The laser light is produced in the recombination of n-carriers and p-carriers at the central region.

ers of appropriately doped AlGaAs. The right side of the energy diagram in Fig. 44–26 is doped to n-type and the left side is doped to p-type. A potential barrier keeps the electrons from flowing to the left near the bottom of the conduction band, and a counterpart barrier keeps the holes from flowing to the right near the top of the valence band. If a voltage is applied that causes more electrons to flow into the GaAs region and more holes to flow into the laser region, and the barrier prevents both n-carriers and p-carriers from flowing back, a population inversion is built up. Note that, unlike the lasers described in Chapter 43, the "pumping" is done automatically in these materials. The electrons and holes can recombine by the emission of photons, and these photons stimulate further, rapid recombination. This describes the process we treated in Chapter 43 for the production of laser light.

APPLICATION

Semiconductor microlasers

Figure 44–27 is an image of a set of *microlasers* that consist of a thin InGaAs quantum well, with stacks of AlAs, GaAs, and AlGaAs layers both above and below it. The laser action takes place in the quantum well and, because the well is narrow, the power required to make the laser work is very small. Many reflections of the light are required to get suitable amplification, and the large number of layers provide a certain amount of reflection at each interface. Even though the reflectivity at each interface is less than 1 percent, the large number of layers gives a total reflectivity back to the lasing material of 99 percent. Many lasers—millions of them!—each with cylindrical geometry, can be formed on a single substrate by molecular-beam epitaxy and subsequent etching. Such laser arrays are useful for optical communications in fiber optics (Fig. 44–28).

FIGURE 44–27 Scanning tunneling microscope (STM) image of a microlaser. The multilayer structure is evident.

44–4 SCANNING MICROSCOPY

Our ability to see an object is diffractively limited by the wavelength of the illuminating radiation. Thus, to resolve atoms, whose size is roughly 0.1 nm, the radiation must have a wavelength on the order of 0.1 nm or less. For electromagnetic radiation, this corresponds to a frequency of $f = c/\lambda = (3 \times 10^8 \text{ m/s})/(10^{-10} \text{ m}) = 3 \times 10^{18}$ Hz. A photon of this frequency has an energy $hf = (6.62 \times 10^{-34} \text{ J}\cdot\text{s})(3 \times 10^{18} \text{ s}^{-1}) = 2.0 \times 10^{-15} \text{ J} = (2.0 \times 10^{-15} \text{ J})/(1.6 \times 10^{-19} \text{ J/eV}) = 12$ keV. This energy is 1000 times larger than atomic ionization energy, and such photons are not readily available. The situation is improved somewhat when electrons are used, but the required electron energy is about 150 eV, which is still very large. J. A. O'Keefe pointed out in 1956 that it would be possible to bypass the diffraction limits with a new type of microscope. The basic idea is simple: Light passes through a tiny hole in a screen and illuminates the object to be viewed, which is on the other side of the hole. The transmitted light (or reflected light) is recorded as the object is moved back and forth across the hole. *The size of the hole determines the resolution rather than the wavelength of the light.* The use of *piezoelectric* materials has provided the technology for controlling the position of an object with the kind of precision needed to realize O'Keefe's idea (see Fig. 25–25). Piezoelectrics are ceramic materials that ex-

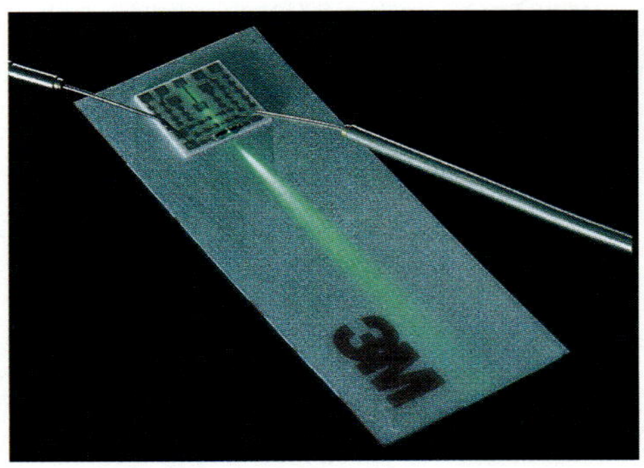

FIGURE 44–28 The blue-green light emitted by this semiconductor laser is clearly visible in this photograph.

pand or contract when an electric field is applied to them. The ability to move them with an accuracy of 10^{-5} nm (!) has made scanning microscopy into a superb tool for the study of surfaces.

Scanning Tunneling Microscopy

The scanning tunneling microscope (STM) is based on the quantum mechanical result that electrons can tunnel from one region to another through domains that are classically inaccessible. Figure 44–29a shows the energy levels in two metallic samples separated by a vacuum of width a. Electrons at the Fermi energy can cross over from one sample to the other in significant numbers only if (1) there are empty energy levels of the same energy, so that the exclusion principle does not prevent tunneling, and (2) the tunneling probability is sufficiently large. If an electric field is applied, thereby lowering the Fermi energy on one side as in Fig. 44–29b, then condition (1) is satisfied and tunneling can occur. An application of Eq. (41–6) shows that the rate at which electrons that cross such a barrier is

$$f = e^{-(2/\hbar)a\sqrt{2mW}}, \tag{44–24}$$

where m is the electron mass and W is the work function of the metal (the height of the potential barrier between the metals). The important feature of Eq. (44–24) is the extreme sensitivity of the tunneling fraction (that is, the current) to the magnitude of the separation.

Figure 44–30 shows how an STM is built. A tungsten tip acts as a "hole in the screen." When such a tip is poised above a surface at a potential different from that of the tip, then, according to Eq. (44–24), the current passing through the tip is a sensitive indicator of its distance to the surface. The resolution depends on the tip size. Tips that end in a *single* atom are made by heating a crude tungsten tip and applying a strong electric field to it. The field pulls atoms away layer by layer, leaving a single-atom tip. With such tips it is possible to resolve features 0.1 nm across: *We can see single atoms on a surface.*

There are two ways of scanning a surface. In one method, piezoelectric supports that hold the tip are arranged to sense the tunneling current and move the tip up or

STMs were first described in Chapter 7, and tunneling is treated in more detail in Chapter 41.

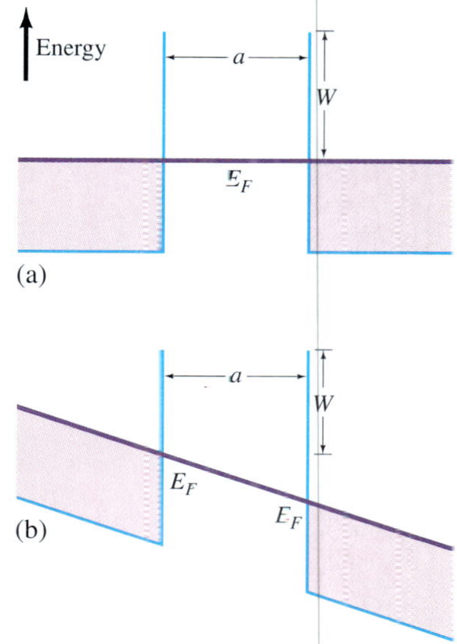

FIGURE 44–29 (a) When two metals are separated by a gap, there is no tunneling in equilibrium. (b) If an external potential is applied, the Fermi energy within one of the metals is lowered, leaving empty levels into which electrons from the side with the higher electron energies can tunnel. The height of the barrier to overcome is the metal's work function, W.

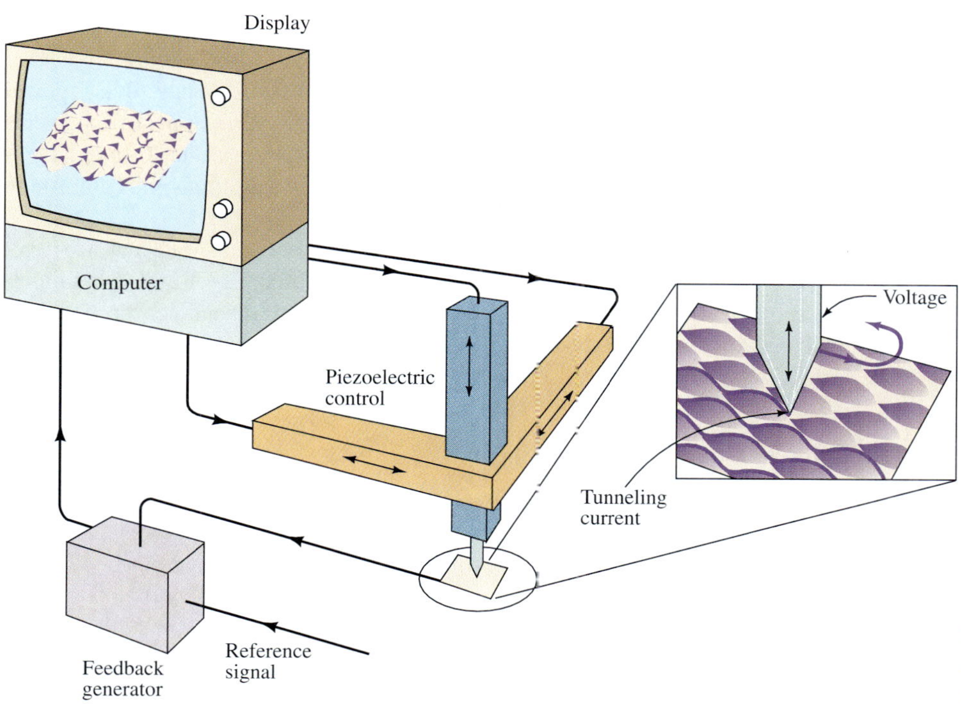

FIGURE 44–30 Schematic diagram of a scanning tunneling microscope.

down so as to maintain a constant current or, equivalently, a constant distance from the surface. The voltages applied to the piezoelectric supports during this process provide a record of the surface topography. A second way of scanning the surface is to move the tip horizontally across the surface and measure the dependence of the tunneling current at any given point on the voltage applied. STMs are especially effective because miniaturization of the apparatus helps in the reduction of "noise": The smaller the apparatus, the higher the frequency of the random thermal vibrations, and this noise is relatively easy to filter out. In addition, STMs work better in air than in a vacuum because the random bombardment by air molecules also tends to average out noise.

APPLICATION

The uses of scanning tunneling microscopy

STMs are used in a variety of surface studies. Electron charge distributions at surfaces can be measured, as can the way the distributions are affected by the deposition of very thin films on the surfaces. Another application involves the use of the STM tip to manipulate individual atoms or groups of atoms and move them around on the surface of the sample (Fig. 44–31; see also Fig. 25–28). There is potential for the creation of tailor-made molecular structures for specific tasks in microelectronic devices.

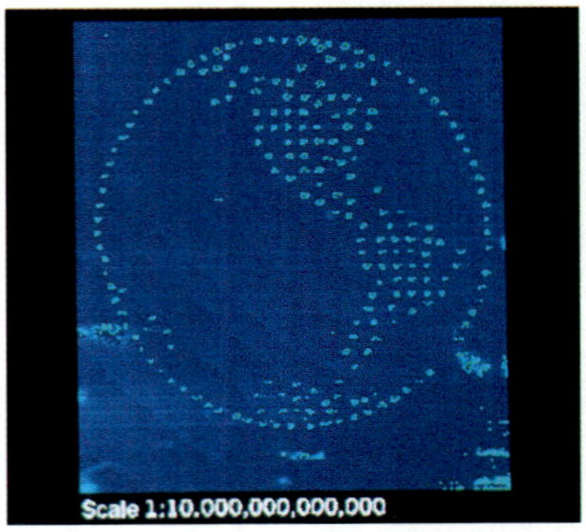

FIGURE 44–31 A map made of tiny dots that consist of just a few thousand atoms of gold. The map, about 1 micron across, was made by manipulating the gold dots with the tip of a scanning tunneling microscope.

Atomic Force Microscopy

APPLICATION

The uses of atomic force microscopy

To make images of insulating surfaces, we can use the *atomic force microscope* (AFM), which images a surface by measuring the effect of the force it exerts on a small cantilever (like a diving board) rather than by probing it with a current. This force is an atomic force, and the idea that it should be directly measurable sounds absurd. Nevertheless, if we calculate an interatomic spring constant $k = m\omega^2$, with $\omega = 10^{13}$ Hz (a typical atomic frequency) and $m = 10^{-25}$ kg, we find that $k = 10$ N/m. For comparison, the spring constant of a piece of aluminum foil 4 mm long and 1 mm wide acting as a cantilever is about 1 N/m. Because the atomic spring constant and the mechanical cantilever spring constant are comparable, atomic forces can be measured by sensing small displacements (0.1 nm) of such a cantilever.

In some AFMs, the cantilever is a piece of silicon or silicon oxide about 100 μm long and 1 μm thick. Such cantilevers have a spring constant in the range of 0.1 to 1.0 N/m. The sensor that detects the motion of the cantilever must detect displacements of less than 0.1 μm. One way of detecting this motion is to shine laser light on it and record the position of the reflected beam with a light detector. The light detector activates the piezoelectric base on which the sample is placed, and a current controls the base such that the distance between the sample and tip is held constant. The movement of the sample is translated into an image of the surface (Fig. 44–32).

The effect of the external environment on the AFM is quite negligible. Typical building vibration frequencies are about 10 to 20 Hz, whereas the resonant frequencies of the cantilever are about 10 to 100 kHz. Amplitude distortions due to external vibrations of frequency f are on the order of $(f/f_{\text{res}})^2$. Building vibrations thus contribute a correction of less than 10^{-3} nm to the displacement.

AFMs can be operated in a variety of environments: in air, in water, at low temperatures, or in a vacuum. They have been applied to the imaging of biological molecules such as amino acids, DNA, and proteins and have even imaged certain chemical processes *as they occur*. The AFM has also been used to study atomic-scale structure on the surfaces of graphite, mica, sodium chloride, and many other materials.

The Ultimate in Quantum Engineering

In this chapter, we have emphasized the interplay of physics and modern technology. We have touched on only a few topics—concentrating on those for which quantum physics is important. The ultimate aim of the kind of work we have described here is the production of devices in which single atoms or single electrons play the major role. For example, a computer memory that stores information by the presence or absence of an individual atom in a given surface site could conceivably store all the knowledge of the human race on a surface of area 1 cm^2! And transistors that operate by allow-

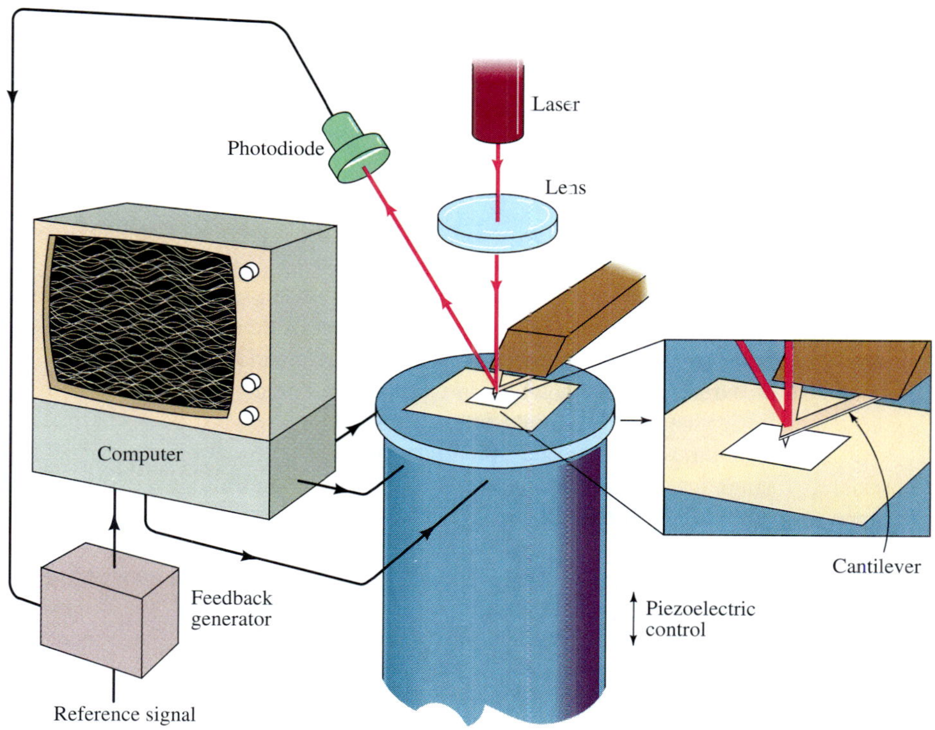

FIGURE 44–32 Schematic diagram of an atomic force microscope (AFM).

ing single electrons to tunnel could compose the smallest possible circuit. The realization of these ideas may not be impossibly distant. The link between physics and technology in the area of materials is so strong that the distinction between the two fields has all but disappeared.

SUMMARY

Because electrons obey the Pauli exclusion principle, their energies within crystalline solids are distributed according to the Fermi–Dirac distribution,

$$f(E) = \frac{1}{1 + e^{(E-E_F)/kT}}, \tag{44–2}$$

where E_F is the Fermi energy. Within semiconductors, whose band structure is that of a valence band separated from a conduction band by a small energy gap (with the Fermi energy in the middle of the gap), this distribution leads to strongly temperature-dependent electrical conduction. Both electrons promoted to the conduction band and the holes they leave behind can conduct electricity. Within semiconductors, electrons (and holes) have effective masses that can be quite different from the mass of free electrons. The conduction properties of semiconductors can also be affected by doping, the introduction of impurities that increase the number of negatively or positively charged carriers (n-carriers and p-carriers, respectively). Doping shifts the position of the Fermi energy but leaves the densities of the n-carriers and p-carriers linked in a way that depends on the behavior of the undoped semiconductor. Semiconductors may emit light when n-carriers and p-carriers recombine, leading to fluorescence and phosphorescence, and conductivity may depend on whether or not light strikes the material.

A pair of semiconductors placed in contact with one another, whether the same semiconductor doped differently or different semiconductors, forms a junction. Junctions exhibit a variety of behaviors and are the basis for much of the electronic technology in use today. The conservation of energy and of charge leads to the principle that the Fermi energy across a junction is constant. This principle dictates that an internally generated electric potential, the contact potential, is generated across the boundary of a p-n junction, a semiconductor structure consisting of a semiconductor doped with p-carriers in contact with the same semiconductor doped with n-carriers. The contact potential explains why p-n junctions act as diodes, devices for which current I_{net} in effect passes in one direction only when an external potential V_{ext} is applied:

$$I_{net} = I_0(e^{eV_{ext}/kT} - 1), \tag{44–18}$$

where I_0 is a current associated with the contact potential. A p-n junction can also act as a light-emitting diode or as a solar cell.

The transistor is a three-terminal semiconductor device in which small changes in potential across one pair of terminals can lead to large amplification of the current that passes through another pair of terminals. Such devices are therefore useful for the sensitive control of currents and of voltages in circuits. We studied in particular both the bipolar junction transistor and the field-effect transistor.

Modern techniques of material manipulation permit the elaboration of heterostructures, single-crystal structures built in a precisely controlled way from different materials. The electrical behavior of heterostructures depends intimately on quantum phenomena, even beyond the collective quantum effects that lead to any band structure. Microscopic semiconductor lasers and quantum wells are manifestations of this new technology whose limits are not yet known.

Scanning tunneling microscopes employ quantum tunneling to explore conductors, and atomic force microscopes allow us to explore nonconductors. Both instruments avoid resolution difficulties associated with diffraction and permit the study of matter at sizes around one Bohr radius, atomic size.

UNDERSTANDING KEY CONCEPTS

1. You have just calculated the number density of n-carriers in undoped silicon, a semiconductor, at room temperature. Is it necessary to make a second calculation to find the number density of holes?
2. When a hole moves from the p-side of a junction to the n-side, why does its energy increase?
3. Does the Fermi energy still lie in the middle of the energy gap when the effective masses of the n-carriers and p-carriers are different?
4. Is the mechanism that makes metals opaque to visible light but transparent to X-rays similar to the mechanism that makes semiconductors opaque to visible light but transparent to infrared light?
5. There are several methods by which electrons can be removed from a solid's surface. How can holes be removed?
6. What happens to the Fermi energy at low temperatures when a semiconductor is so heavily doped that the energy levels of the donor states overlap the conduction band?
7. How can the energy of a hole be raised as it travels from the p-side of a p-n junction to the n-side?
8. Why, physically, is it true that the smaller the effective mass of an n-carrier, the larger the drift velocity of the carrier?
9. We described why new states for electrons are available near the bottom of the conduction band in an n-type extrinsic semiconductor. Why are new states for holes available just above the top of the valence band in a p-type extrinsic semiconductor?
10. Is there a type of viscosity that acts on holes in a semiconductor and gives them a terminal velocity in an electric field, or do holes just accelerate under the influence of such a field? If there is a terminal velocity, is it the same as that of the electrons?
11. We referred to a current amplification in our discussion of transistors. Can this mean that the power is amplified? If so, where does the necessary energy come from?
12. The gap widths of the semiconductors in Table 44–1 are given at room temperature. How could a gap width depend on temperature?

PROBLEMS

44–1 Semiconductors

1. (I) In silicon, $E_g = 1.12$ eV, and the effective mass of the n-carriers is $m^* = 0.31m_e$, where m_e is the electron mass. Find the number densities of n-carriers at 100 K and at 300 K. Compare your result with the number density of "free" electrons in copper, 8.5×10^{22} electrons/cm^3.
2. (I) What is the density of electrons in the conduction band of silicon, for which $m_n^* = 0.31m_e$, at temperatures of (a) 77 K, (b) 293 K, and (c) 600 K? [*Hint*: see problem 1.]
3. (I) For sodium ($E_F = 3.23$ eV) at room temperature (300 K), calculate the occupation probability of an electron state of energy (a) 2 eV, (b) 3 eV, (c) 3.2 eV, (d) 3.26 eV, and (e) 4 eV.

4. (I) For copper ($E_F = 6.95$ eV), at what temperature is the probability that an energy state at 7 eV will be populated equal to 33 percent?

5. (I) Find the energy gap of an intrinsic semiconductor in which 1.02×10^{-30} of the available energy levels near the bottom of the conduction band are occupied at $T = 126$ K.

6. (I) The value of the Fermi–Dirac distribution function $f(E)$ for a particular energy value within a collection for electrons at a temperature $T = 34$ K is found to be equal to 0.4. What is the value of that function for the same energy value when $T = 54$ K?

7. (I) What is the probability of finding an electron at the bottom of the conduction band in GaAs, for which $E_g = 1.43$ eV, at $T = 300$ K?

8. (I) The Fermi temperature, T_F, is defined by E_F/k, where k is Boltzmann's constant. What is T_F for (a) silver ($E_F = 5.1$ eV), (b) copper ($E_F = 6.95$ eV), and (c) a white dwarf star ($E_F \approx 2 \times 10^5$ eV)?

9. (I) What is the maximum wavelength of light that will excite an electron from the valence band to the conduction band in (a) InSb ($E_g = 0.18$ eV), (b) GaSb ($E_g = 0.68$ eV), and (c) GaAs ($E_g = 1.43$ eV)?

10. (I) The following table gives the energy above the edge of the valence band, ΔE, for the energy levels of various acceptor impurities in silicon:

Impurity:	*Boron*	*Gallium*	*Indium*	*Nickel*	*Zinc*
ΔE (in eV):	0.045	0.065	0.16	0.22	0.31

Find the maximum wavelength of the radiation required to excite electrons from the top of the valence band to these acceptor levels.

11. (I) Calculate the intrinsic carrier concentration in silicon at 400 K, given that $m_n^* = 0.31m_e$ and $m_p^* = 0.56m_e$.

12. (I) At room temperature (300 K), undoped GaSb has an energy gap of 0.68 eV. (a) What is the probability that an electron occupies an energy state at the bottom of the conduction band? (b) What is the probability that there is a hole state at the top of the valence band?

13. (II) The effective masses of n- and p-carriers in germanium are given by $m_n^* = 0.55m_e$ and $m_p^* = 0.37m_e$, respectively. The concentration n_i of n-carriers at 300 K is 2.5×10^{19} m^{-3}. Use these data to calculate the gap width of germanium.

14. (II) The effective masses of n- and p-carriers in GaAs are given by $m_n^* = 0.067m_e$ and $m_p^* = 0.48m_e$, respectively. The gap width of GaAs is 1.43 eV. Use this information to find the intrinsic carrier density at 450 K.

15. (II) A sample of silicon is doped so that the donor density is $n = 2 \times 10^{23}$ m^{-3}. Find $E_c - E_F$ at $T = 300$ K, given that the effective n-carrier mass in silicon is $m_n^* = 0.31m_e$. Compare your result with E_g for silicon.

16. (II) A sample of silicon is doped so that the hole density is $p = 2 \times 10^{22}$ m^{-3}. Find $E_F - E_v$ at $T = 300$ K, given that the effective p-carrier mass in silicon is $m_p^* = 0.56m_e$.

17. (II) The intrinsic semiconductor InSb has a gap energy of 0.18 eV, and the effective masses of the electrons and holes are $0.015m_e$ and $0.39m_e$, respectively. Find the carrier densities at $T = 100$ K and 300 K.

18. (II) The *mobility*, μ, of a charge carrier in a material is defined as the drift speed of the carrier, v_d, divided by the magnitude of the electric field driving the current, $\mu \equiv v_d/E$. For germanium at 300 K, the mobilities of electrons and holes are about 0.45 $\text{m}^2/\text{V}\cdot\text{s}$ and 0.35 $\text{m}^2/\text{V}\cdot\text{s}$, respectively. (a) Use the definition of μ and the relation between v_d and the current to calculate the free-electron density, n, if the conductivity is 8.3×10^{-5} $(\Omega\cdot\text{m})^{-1}$ and $n >> p$. (b) Repeat part (a) for the hole density, p, assuming doping such that $p >> n$.

19. (II) Calculate the electrical conductivity at room temperature of germanium doped to n-type if the density of n-carriers and p-carriers are 2.5×10^{18} m^{-3} and 7.0×10^{13} m^{-3}, respectively. [*Hint*: Use the results of Problem 18.]

20. (II) The Hall Effect (see Chapter 29) is probed on an n-type semiconductor with carrier density n; the density of holes is negligible. Determine the magnitude and polarity of the Hall voltage if there is a current to the right with density j (Fig. 44–33). Repeat the derivation for a p-type semiconductor. Notice the difference in polarity.

FIGURE 44–33 Problem 20.

21. (II) Use the energy distribution for a collection of identical fermions at finite temperature T to (a) show that the distribution of speeds (the probability that a fermion has a speed between v and $v + dv$) is given by

$$n(v)dv = \frac{8\pi m^3}{h^2}\,\frac{v^2\,dv}{1 + e^{[(mv^2/2)-E_F]/kT}}.$$

(b) What is the distribution for $kT << E_F$? [*Hint*: See Problem 50 in Chapter 43.]

22. (II) The Fermi energy can be defined as the energy at which the probability of finding an electron is $\frac{1}{2}$. (a) Show that this is the case at any temperature. (b) Refine this idea by showing that if the probability that an electron has an energy E *above* E_F is P, then the probability that an electron has an energy E *below* E_F is $1 - P$.

23. (III) Using a computer, plot the Fermi distribution as a function of energy for a material with a Fermi energy of 11 eV at temperatures 20 K, 100 K, 300 K, and 1000 K.

44–2 Semiconductor Structures

24. (I) A silicon solar cell produces electric current when the frequency of the light that falls on it is above 2.68×10^{14} Hz. What is the width of the band gap?

25. (I) Plot the factor $e^{(eV_{ext})/kT} - 1$ [from Eq. (44–18)] as a function of temperature for a sample in which $eV_{ext} = 0.02$ eV over the range $100\text{ K} < T < 350$ K. Plot the same factor as a function of eV_{ext} (positive and negative) for $T = 300$ K.

26. (I) The ratio of carrier densities p_n/p_p for a *p-n* junction at 300 K is determined to be 0.90. What is the difference between Fermi energies for the *n*-type and *p*-type semiconductors?

27. (II) The Fermi energy of a heavily doped *n*-type semiconductor is close to E_c, whereas the Fermi energy of a heavily doped *p*-type semiconductor is close to E_v. (a) By using an energy diagram, show that when a *p-n* junction is formed from these semiconductors, a sufficient amount of reverse biasing will allow electrons to tunnel from the *p*-side to the *n*-side—a phenomenon known as the *Zener effect*. (b) The tunneling current is a sensitive function of the width of the barrier across which the electrons must tunnel. Show by means of energy diagrams that this width systematically decreases when the reverse bias increases, leading to a quickly building current. [*Hint*: Recall that the transition region has a finite width.] The Zener effect can be used in a *Zener diode*, in which a current flows when a reverse bias potential reaches a certain value.

28. (II) Consider a *p-n* junction formed from two samples of doped silicon. The two samples are as described in Problems 15 and 16 (Fig. 44–34). What is the value of eV_0 for this junction?

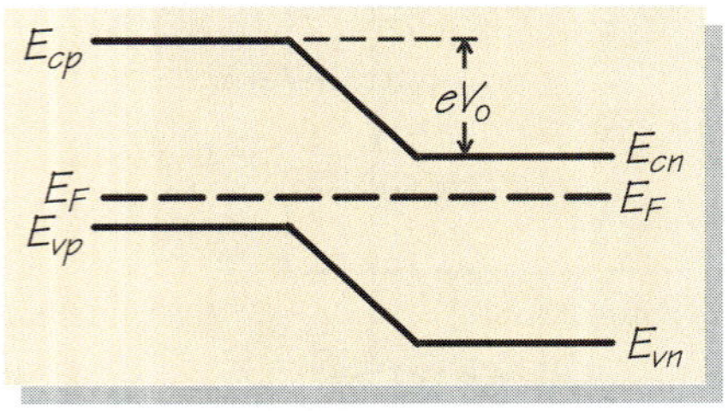

FIGURE 44–34 Problem 28.

29. (II) A germanium *p-n* junction has $n = 3.5 \times 10^{22}\ \mathrm{m^{-3}}$ on the *n*-side and $p = 2.8 \times 10^{23}\ \mathrm{m^{-3}}$ on the *p*-side. For germanium, $E_g = 0.67$ eV, $m_n^* = 0.55\ m_e$, and $m_p^* = 0.37\ m_e$. Find the locations of the Fermi energies before the junction is formed. Use your results to find eV_0 at (a) $T = 80$ K and (b) $T = 300$ K.

30. (II) Many LEDs give off red light. Given this fact, estimate the width of the band gap for the materials from which such LEDs are constructed.

31. (II) A current of 2.5 mA flows through an LED with a band-gap energy of 1.4 eV. Assume that each current-carrying electron drops into a hole, and a single photon is thereby emitted. (a) What is the power emitted in the light? (b) How many photons/s are emitted?

32. (II) Consider a heterostructure in which an intrinsic semiconductor with a narrow band gap is sandwiched between an intrinsic semiconductor with a larger energy gap. (a) Draw an energy diagram for this system. (b) What happens to any *n*-carriers in this system in the absence of external potentials?

44–3 Band-Gap Engineering

33. (II) A quantum well with a width of 9 nm is produced, and spectroscopic measurements show that the energy difference between the ground state and the first excited state within this well is 2.28×10^{-2} eV. What is the effective mass of electrons within the well?

34. (II) For a quantum dot, in which an electron is confined within three dimensions (Fig. 44–35), the allowed energy values have the form

$$E = \frac{\hbar^2\pi^2 n_1^2}{2m^* a_1^2} + \frac{\hbar^2\pi^2 n_2^2}{2m^* a_2^2} + \frac{\hbar^2\pi^2 n_3^2}{2m^* a_3^2},$$

where m^* is the effective mass of an electron; a_1, a_2, and a_3 are the three confining dimensions of the box, and the n_i values are integers. Suppose that a particular quantum dot has dimensions 15 nm × 40 nm × 40 nm and that $m^* = 0.55m_e$. (a) Calculate the energy gap between the first and second excited states. (b) What is the wavelength of the radiation emitted in the transition between these two states?

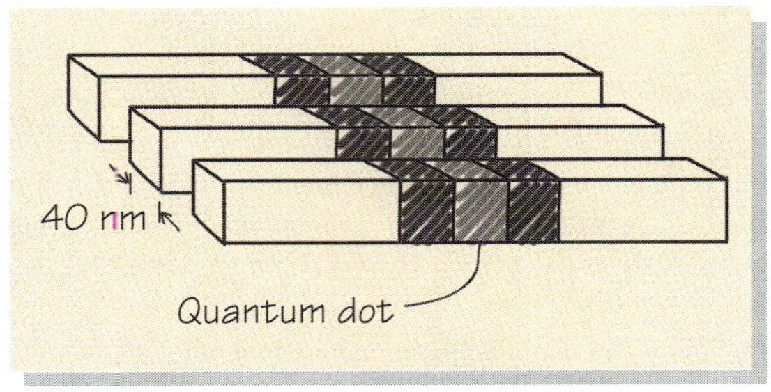

FIGURE 44–35 Problem 34.

44–4 Scanning Microscopy

35. (I) The current measured in a scanning tunneling microscope whose tip height is held fixed increases by a factor of 20 when the tip moves from point *A* to point *B* on the surface. How much closer to the tip is point *B* than point *A*? The work function of the sample is 2.3 eV.

36. (I) An atomic force microscope has a spring constant of 0.2 N/m. If a displacement as small as 0.1 nm must be detected, how much energy is needed to displace the cantilever?

37. (II) The cantilever of an atomic force microscope has a spring constant of 0.6 N/m, and a laser sensor detects that it has been deflected by 0.2 nm (Fig. 44–36). What is the ratio of the force acting on the cantilever to that of the force of gravity on a fly of mass 0.1 g?

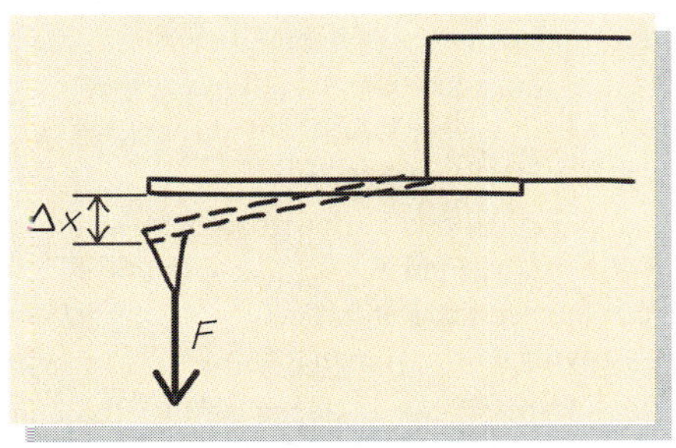

FIGURE 44–36 Problem 37.

General Problems

38. (I) An electric field of magnitude 10^6 V/m is applied in the $+x$-direction in a piece of intrinsic Si, for which the

mobilities of electrons and holes at room temperature are 0.14 $m^2/V \cdot s$ and 0.05 $m^2/V \cdot s$, respectively (see Problem 18). What are the drift velocities (direction and magnitudes) of electrons and holes at room temperature?

39. (I) A germanium sample is doped with 1.1×10^{23} atoms/m^3 of arsenic. Given that the intrinsic n-carrier density is $n_i = 7.0 \times 10^{16}$ carriers/m^3, what is the hole concentration at equilibrium (Fig. 44–37)?

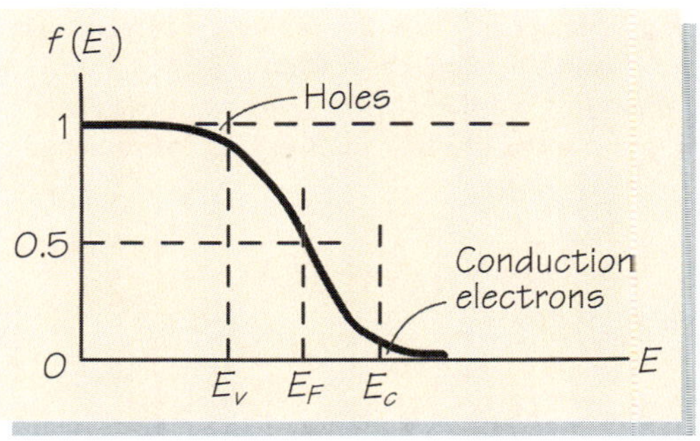

FIGURE 44–37 Problem 39.

40. (II) The effective electron mass in silicon is given by $m_n^* = 0.55m_e$. Calculate the value of N_c [from Eq. (44–12)] at 300 K, and use your result to find the location of E_F relative to the edge of the conduction band for the doped silicon sample of Problem 39.

41. (II) A donor electron moves in doped indium antimonide, for which $\epsilon/\epsilon_0 = 17.9$ and $m^* = 0.015m_e$. Find the radius of a circular atomic orbit of such an electron in terms of the Bohr radius a_0 (Fig. 44–38). The effective nuclear charge for such a loosely bound electron is $Z = 1$.

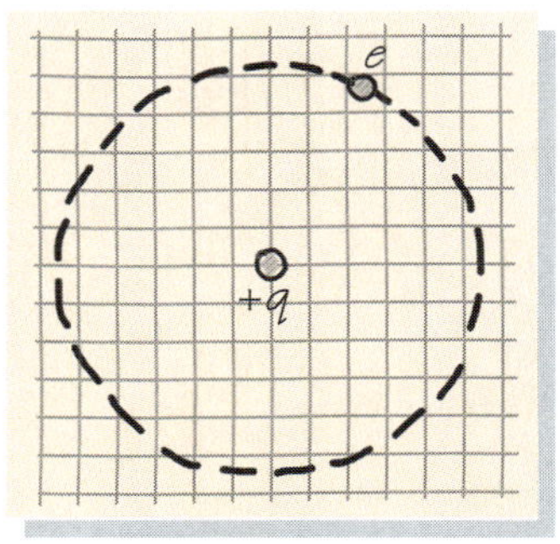

FIGURE 44–38 Problem 41.

42. (II) An electron in germanium moves about a positive center of charge $+e$ to form a hydrogen-like atom in its ground state. Calculate the wavelength of the photon needed to raise the electron to the first excited state. What is the change in the radius of the orbit? ($\kappa = 15.8$ and $m_n^* = 0.55m_e$ in germanium.)

43. (II) The impurity concentration in semiconductors must be low in order to keep the interaction between individual electrons or hole charge carriers negligible. Estimate the maximum concentration of As impurities in germanium that avoids an overlap between the ground state wave functions about adjacent impurities. Express your result in terms of a volume density (impurities/cm^3) as well as in terms of the ratio impurities/atoms. (For germanium $\kappa = 15.8$ and $m_n^* = 0.55$ m_e, $\rho = 5.32$ g/cm^3 and $A = 72.6$.)

44. (II) Repeat the calculation of Problem 43 for InSb with Sb impurities. (For InSb, $\kappa = 17.9$ and $m_n^* = 0.015m_e$, $\rho = 5.9$ g/cm^3 and $A_{\text{In}} = 114.8$, and $A_{\text{Sb}} = 121.8$.)

45. (II) A magnetic field of 1.80 T is applied to doped indium antimonide. What is the angular frequency of the donor electrons? [*Hint*: See Problem 41 for the parameters of this semiconductor, and Chapter 29 for a discussion of the cyclotron frequency.]

46. (II) The number of electrons with energy in the range E to $E + \Delta E$ contained in a volume V is given by $f(E)\,\Delta N(E)$, where $f(E)$ is the Fermi–Dirac distribution function and $\Delta N(E)$ is the number of states available to the electrons in the range E to $E + \Delta E$ (Fig. 44–39). Given that

$$\Delta N(E) = \frac{V}{2\pi^2}\left(\frac{2m_e}{\hbar^2}\right)^{3/2}\sqrt{E}\,\Delta E,$$

calculate the densities of electrons that have energies in the range 10.45 eV to 10.47 eV for gallium ($E_F = 10.40$ eV) at the temperatures 90 K and 300 K.

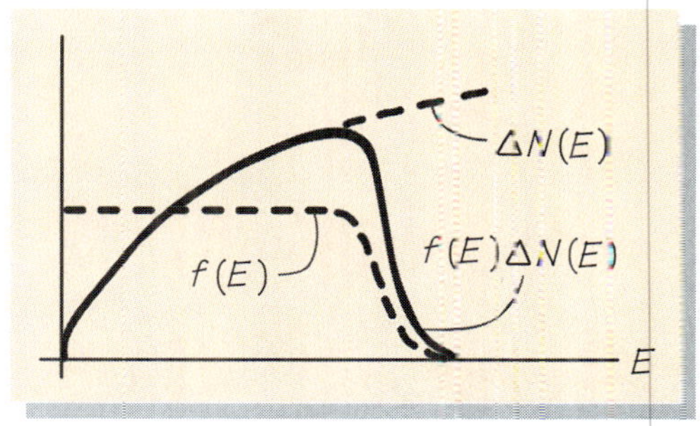

FIGURE 44–39 Problem 46.

47. (II) Repeat the calculation of Problem 46 for the energy interval between 12.65 eV and 12.67 eV.

48. (II) For a set of identical fermions at high temperature, the average occupation number of any given energy level $f(E) << 1$. Show that the Fermi–Dirac distribution approaches the Maxwell–Boltzmann distribution [Eq. (19–40)] in the limit of high temperatures. [*Hint*: The factor $e^{-E_F/kT}$ is determined from the normalization condition $\Sigma_i\, n_i = N$ where the index i labels the allowed energy levels.]

49. (II) The laser that reads a compact disk is a semiconductor laser of power in the milliwatt range with light of wavelength around 800 nm. (a) Estimate the rate at which photons are emitted. (b) If the compact disk turns beneath the (narrow) laser beam at several hundred revolutions per minute, how many photons will strike a region 0.1 mm in length of the disk?

50. (II) Consider a group of N atoms that obey Maxwell–Boltzmann statistics. The atoms have only two energy states, a ground state at energy E_0 and an excited state at energy E_1. The energy difference $\Delta E = E_1 - E_0 >> kT$. Show that the average energy of the atoms at temperature T is $E_0 - \Delta E e^{-\Delta E/kT}$.

51. (II) Monochromatic light shines on a thin film of silicon (E_g = 1.12 eV), and the electrical resistance is measured across the film. (a) What happens to the resistance as the wavelength of the light increases, starting with a wavelength in the ultraviolet range? (b) At what critical wavelength does a change in resistance occur? (c) Does the resistance increase or decrease at the critical wavelength?

52. (III) Use the expression for $\Delta N(E)$ given in Problem 46 to express (a) the total number of electrons in a volume V and (b) the average energy of an electron in a volume V at temperature T. (c) Find the limits of your results for parts (a) and (b) for $T \to 0$.

53. (III) A degenerate semiconductor is a semiconductor that is so heavily doped that the doping ions participate in the formation of their own band structure. The Fermi energy may lie within the original conduction band in a degenerate n-type semiconductor, and within the original valence band in a degenerate p-type semiconductor. When a p-n junction is formed from a semiconductor that is degenerate on both the n- and p-sides, a *tunnel diode* is formed. Under these circumstances tunneling is possible when there is an external voltage. Draw an energy diagram (a) for the equilibrium state (no external voltage across the junction); (b) for a reverse bias. How will the tunneling current vary as the reverse bias increases from zero? (c) Draw an energy diagram for a *small* forward bias. How will the tunneling current vary as the small forward bias increases from zero? (d) Show that when the forward bias increases beyond a critical value, the tunneling current begins to *decrease*. This phenomenon is called *negative resistance*.

CHAPTER

45

The nucleus of every atom is composed of neutrons and protons, whose interactions determine the properties of the nucleus. As we shall learn in Chapter 46, even neutrons and protons have structure.

Nuclear Physics

Once its existence became known, the nucleus, an object at the center of an atom that contains all the positive charge and almost all the atomic mass, became the subject of study. Early on, many of the scientists involved in this research were convinced that nuclear physics would have no practical applications. Of course, we now know that this is not true: Understanding the properties of the nucleus has had enormous consequences for technology and for society in general. Here, we shall describe the nuclear constituents, the static properties of nuclei, and the radioactive decays of certain nuclei. We shall see how nuclei react with one another. The practical applications of nuclear physics are widespread and include power generation, radiometric dating, microscopic crack and void testing, food conservation, and cancer treatment.

45–1 STATIC PROPERTIES OF NUCLEI

Nuclear Constituents

The *nucleus* is an object that sits at the center of the atom with a mass of more than 99.9 percent of the atom's total mass and a radius that is some 10^5 times smaller than the radius of the atom as a whole. The nuclear structure of the atom was seen in scattering experiments conducted by Ernest Rutherford in 1911. Even before Rutherford made his discovery, Joseph J. Thomson noted the presence of electrons in atoms. The number of electrons is equal to the *atomic number*, Z, a quantity defined by chemical properties. Another number used for atoms is its atomic weight, which is proportional to the mass, and it was known that the atomic number was less than half the atomic weight. In addition, Thomson had measured the charge-to-mass ratio of the electron; coupled with H. A. Wilson's and Robert Millikan's measurements of the electron charge, the electron's mass can be deduced to be much less than the atomic mass. Because the atom is neutral, it must contain a positive charge $+Ze$ in addition to the electron charge $-Ze$. What was *not* known was how the positive and negative charges were distributed within the atom or what carried the positive charge.

We described in Section 8–8 how we could learn of the existence of a steel pellet in a volume of cotton candy by firing BBs at the candy. Similar reasoning applies to the exploration of atoms by scattering. The atomic model favored at the turn of the twentieth century—proposed by Thomson and known as the Thomson model—had pointlike electrons embedded within a uniform distribution of positive charge. Rutherford undertook the study of atoms by scattering *alpha* (α) *particles* from atoms in a thin gold foil (Fig. 45–1). Alpha particles are produced when certain heavy elements decay, as we shall see in Section 45–3. Rutherford had analyzed the charge and mass of α particles and knew that they consist of doubly ionized helium atoms (charge $+2e$). An α particle is some 8000 times more massive than an electron and would scarcely be deflected by a collision with an electron. A smooth distribution of positive charge, such as the positively charged matrix of the Thomson model, would hardly deflect α particles that passed through the middle of the charge distribution because the effects of repulsion from each side of the distribution would tend to cancel.

Rutherford's experiment showed that far too many α particles scattered through large angles—a factor of 10^{10} for angles greater than 90°—than could be explained by the Thomson model. However, if the positive charge and the mass of the atom were concentrated in a central structure, the scattering angle could be large. We can estimate the typical scattering angle, θ, of an incident α particle deflected by its electrical interaction with a positively charged sphere of radius R as follows: $\theta \simeq \Delta p/p$, where $p = m_\alpha v$ is the momentum of the α particle, v is its speed, and Δp is the sideways momentum transfer that the α particle receives as it passes the spherical charge distribution. The typical deflection occurs when the α particle passes close to the edge of the positive charge distribution (Fig. 45–1b). In this case, we can approximate Δp by $F\,\Delta t$, where $F = (Ze)(2e)/4\pi\epsilon_0 R^2$ is the Coulomb force between the charge distribution and the α particle that reaches the edge of the distribution, and $\Delta t \simeq R/v$ is the time period over which the α particle is deflected. Thus

$$\theta \simeq \frac{F\,\Delta t}{p} = \frac{(Ze)(2e)}{4\pi\epsilon_0 R^2}\,\frac{R/v}{m_\alpha v} = \frac{2Ze^2}{4\pi\epsilon_0 m_\alpha v^2 R}. \qquad (45\text{–}1)$$

The factor $1/R$ in this result indicates that the smaller the radius of the nucleus, the larger the typical scattering angle will be. For R equal to the atomic radius (about 0.1 nm), the deflection is negligible. Rutherford's results could be explained with R on the order of 10^{-5} of the atomic radius.

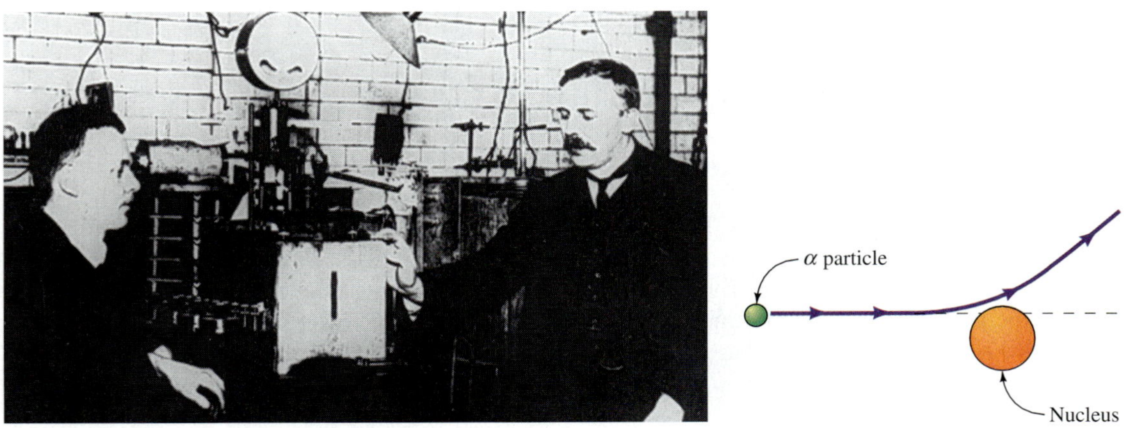

FIGURE 45–1 (a) Ernest Rutherford (here, on the right, with Hans Geiger) directed the scattering experiments that revealed the atom's nuclear structure. (b) Alpha-particle scattering from a nucleus. A typical deflection occurs when the α particle passes the edge of the nucleus.

EXAMPLE 45–1 An α particle with kinetic energy of 5 MeV heads directly toward a gold nucleus ($Z = 79$) at rest. The gold nucleus is much heavier than the α particle. Find how close the α particle comes to the gold nucleus before it turns around.

Solution: Coulomb repulsion between the gold nucleus and the α particle will cause the α particle to slow down, stop, and then turn around. Let's assume the gold nucleus is so heavy that it remains fixed and use the conservation of energy to solve this problem. The initial total energy of the system is $E = K + U$. We take the potential energy, U, to be zero when the colliding particles are far apart. E is then equal to K, or 5 MeV, and it remains that value throughout. When the α particle reaches the turnaround point, its kinetic energy is zero, and the energy is entirely potential energy. At this point, the separation is R (Fig. 45–2). The Coulomb potential energy of two objects with charges Z_1e and Z_2e is given by

$$U(r) = \frac{Z_1Z_2e^2}{4\pi\epsilon_0}\frac{1}{r}.$$

We apply the conservation of energy by equating the energies at $r = \infty$ and at $r = R$, setting $Z_1 = 2$ (the α-particle charge) and $Z_2 = 79$ (the gold nucleus charge):

$$K\bigg|_{r=\infty} + U\bigg|_{r=\infty} = K\bigg|_{r=R} + U\bigg|_{r=R};$$

$$K + 0 = 0 + \frac{(2)(79)e^2}{4\pi\epsilon_0}\frac{1}{R}.$$

We solve for R:

$$R = \frac{(2)(79)e^2}{4\pi\epsilon_0}\frac{1}{K} = (2)(79)(1.6 \times 10^{-19}\,\text{C})^2(9 \times 10^9\,\text{N}\cdot\text{m}^2\cdot\text{C}^{-2})$$

$$\times \frac{1}{5 \times 10^6\,\text{eV}}\,\frac{1\,\text{eV}}{1.6 \times 10^{-19}\,\text{J}} = 4.55 \times 10^{-14}\,\text{m}.$$

This distance is well inside even the innermost Bohr radius. You might recall from the Section 5–5 discussion on fundamental forces that nuclear forces are short range and the nuclear force will be negligible at this distance (see Section 45–2).

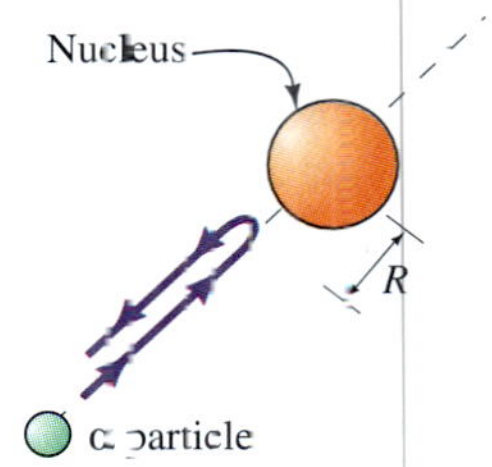

FIGURE 45–2 Example 45–1. A head-on collision in which an α particle is deflected straight back from a nucleus.

Scattering Distributions

Rutherford was able to make a more sophisticated calculation of what happens when a collimated beam of α particles is fired at a set of atoms constructed on the basis of the nuclear model. Some of the α particles follow a line directly toward a nucleus, and some follow a line off to the side (Fig. 45–3). For a given line, the force anywhere along that line is the Coulomb force, and the trajectory of the α particle can be calculated. If the beam of α particles is spread uniformly across a region with a collection of target nuclei, then the number of α particles that arrive at any distance off to the side of any nucleus can be calculated. Rutherford used this technique to predict a precise *distribution* of the number of α particles deflected at any given angle.†

Rutherford assumed that *all* the positive charge was at the center of the atom and was contained in a concentration much more massive than the projectile. Suppose that M is the target mass (of charge Z_2e) at rest and m is the projectile particle (of charge Z_1e) moving with kinetic energy K. Then if $M >> m$, the target particle will not recoil. We will consider only this case because of its simplicity. The probability that the particle will be scattered into a small solid angle $d\Omega$ at an angle θ is proportional to

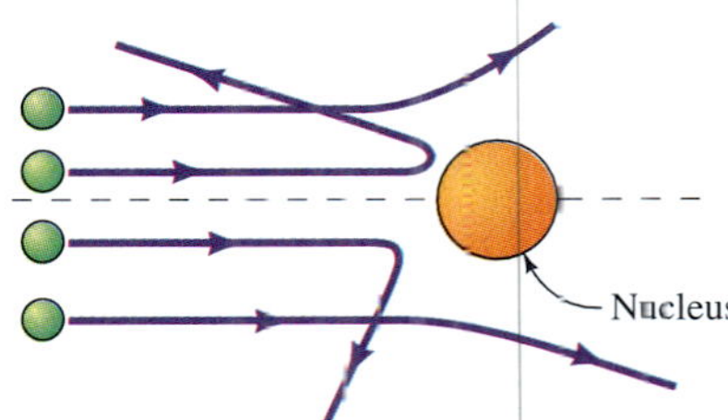

FIGURE 45–3 A beam of α particles sent into a sample of material will scatter from a nucleus at many different angles.

The collision cross section was described in Chapter 19.

†Rutherford could make only a classical calculation, but we now know that quantum physics dominates at these scales. By an extraordinary stroke of luck, the inverse-square law obeyed by the Coulomb force is the *only* force law for which a quantum mechanical calculation gives a distribution identical to that given by a classical calculation.

$d\Omega$ times the *collision cross section*, $\sigma(\theta)$. Rutherford showed that, for $M >> m$, this cross section is given by

$$\sigma(\theta) = \frac{(Z_1 Z_2 e^2)^2}{16(4\pi\epsilon_0)^2} \frac{1}{K^2} \frac{1}{\sin^4(\theta/2)}. \tag{45–2}$$

The critical features are the inverse-square dependence on K and the strong dependence on θ.

Rutherford's assistants Hans Geiger and Ernest Marsden painstakingly counted the scattering of α particles from a thin gold foil by observing the particles that arrived on a fluorescent screen (Fig. 45–4). These experiments verified Eq. (45–2) and hence also the existence of the nucleus and Rutherford's concept of the atom. Rutherford also noticed a deviation from the prediction of Eq. (45–2). When an α particle is on a path directly toward a nucleus, it backscatters at 180° (see Example 45–1). If the Z value of the target is small enough and the energy of the incident α particle is high enough (greater than 5 MeV in Rutherford's experiment), the α particle has enough energy to overcome the Coulomb force and penetrate into the nucleus. The force that the α particle experiences is thereby modified and Eq. (45–2) is no longer correct. The penetrating α particle experiences a nuclear force, and Rutherford's observation was the first time the nuclear force was seen in scattering.

Neutrons. The nucleus cannot be composed of protons alone: Almost all the mass of an atom is in its nucleus and the atomic weight is roughly twice the atomic number. Rutherford speculated in 1920 that a nucleus could consist of Z protons and $N = A - Z$ particles of about the same mass as the proton but with no electric charge—particles later called **neutrons**. Here, A is the *mass number*, or the number of protons plus neutrons in the nucleus. But neutral particles are hard to detect. When charged particles pass through matter, Coulomb forces leave a path of ionized atoms that is easy to observe by various detection techniques. When neutral particles such as photons or neutrons pass through matter, they do not ionize the matter very easily.

How, then, were neutrons observed? When atoms of boron or beryllium are bombarded with α particles, observation of the recoiling nuclei shows that neutral particles are produced. These neutral particles are not necessarily neutrons; they could, for example, be photons. In 1932, James Chadwick showed these particles to be neutrons by allowing them to pass into material that contains a good deal of hydrogen, such as paraffin. He observed that a rather energetic proton was occasionally produced (Fig. 45–5). He interpreted this as the result of a collision between the neutral component and a hydrogen nucleus, which consists of just a single proton. By measuring the momentum of the observed proton and by using momentum conservation, Chadwick verified that the neutral component could not be a photon. Instead, he concluded that the particle, which is now called the *neutron*, must have about the same mass as the proton. Thus, some 20 years after Rutherford's discovery of the nucleus, the nature of the nuclear components was firmly established.

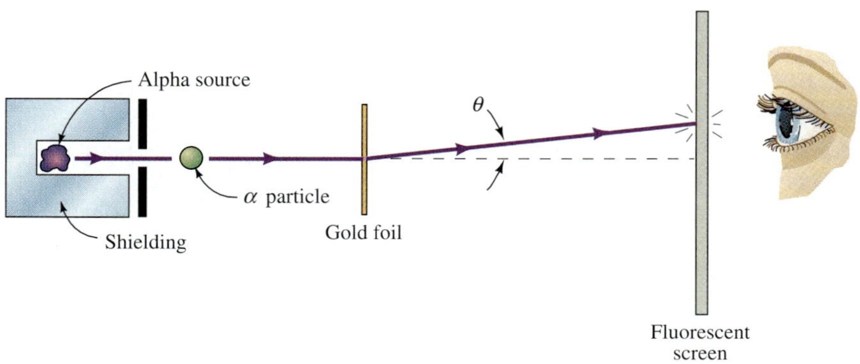

FIGURE 45–4 Schematic diagram of the experiment in which Geiger and Marsden observed the scattering of α particles from gold atoms in gold foil.

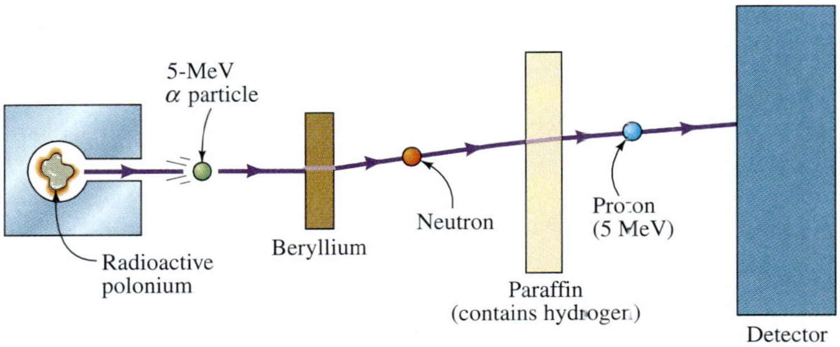

FIGURE 45–5 Schematic diagram of Chadwick's experiment to discover neutrons. Paraffin contains a high percentage of hydrogen and hence of nuclei consisting of single protons that act as targets for any incoming projectiles. A kinematic analysis shows that only neutrons scattering from hydrogen nuclei could have knocked forward the observed protons.

The proton and the neutron collectively are known as **nucleons** because they are both found in the nucleus and they share many of the same properties. An important difference between protons and neutrons is that, unlike protons, free neutrons are not stable, but decay with a lifetime $\tau \simeq 890$ s. Neutrons do *not*, however, decay within stable nuclei.

The lifetime of unstable particles is discussed in Chapter 41.

The properties of the nucleus depend on the force that holds it together, called the *nuclear*, or *strong*, force. We know that such a force must exist because, without it, the repulsive Coulomb force between protons would cause the nucleus to disintegrate.

Some Terminology

A given nucleus contains Z protons and N neutrons, constituting $A = Z + N$ nucleons.

Recall that the number of nucleons in a single nucleus is the mass number, A. This quantity very nearly determines the mass of the nucleus; the number of protons is the atomic number, Z; and the number of neutrons is $N = A - Z$. A **nuclide** is a nucleus of an element with a given Z and A and is written as

$$^{A}_{Z}X,$$

where X is the chemical symbol of the element. For example, the nuclide $^{4}_{2}$He is a helium nucleus ($Z = 2$), with 2 protons and 2 neutrons. The nuclide $^{3}_{2}$He also occurs naturally. This terminology is redundant because the chemical symbol alone specifies Z, so we normally write the simpler form ^{4}He. The proton is a hydrogen nucleus, $^{1}_{1}$H, but we shall more frequently use the standard symbol p for the hydrogen nucleus. As we have just seen for helium, the nucleus of a given chemical element has a particular Z value, but a range of N values are possible. For example, neon has three stable nuclides (^{20}Ne, ^{21}Ne, and ^{22}Ne) and several unstable ones. Nuclides with the same Z value are known as **isotopes** of the given element (see Chapter 19). The gram-atomic weight of a chemical element—the mass of 1 mol of that element—is actually an average over the different naturally occurring isotopes of that element weighted by the relative abundance of those isotopes.

Nuclear Masses and Binding Energies

The masses of nuclei are equal to the masses of the corresponding atoms to within less than 0.5 percent, and these masses can be determined at this accuracy by purely chemical methods. The mass of a nucleus, which we label $m(^{A}_{Z}X)$, can be determined more accurately, however, by first stripping its atom of one or more electrons, and then sending the nucleus through a mass spectrometer (see Chapter 29). This spectrometer may first use crossed electric and magnetic fields to find the speed of the ion, which has been accelerated in another electric field. The ion then passes through a region of fixed magnetic field; there, the radius of curvature of the ion's path accurately determines the momentum. Knowing the speed and the momentum, we can determine the mass. With such devices, the masses of ions can be determined to an accuracy of better than 1 part in 10^8. That atomic masses and not nuclear masses are obtained is not a handicap because the electron masses normally cancel in calculations with nuclei.

TABLE 45–1
Some Atomic Masses

Element	*Atomic Mass (u)*
$^{1}_{1}$H	1.007825
$^{2}_{1}$H	2.014102
$^{3}_{2}$H	3.016029
$^{4}_{2}$He	4.002603
$^{6}_{3}$Li	6.015121
$^{7}_{3}$Li	7.016003
$^{10}_{5}$B	10.012937
$^{12}_{6}$C	12.000000
$^{14}_{6}$C	14.003242
$^{14}_{7}$N	14.003074
$^{15}_{7}$N	15.000109
$^{16}_{8}$O	15.994915
$^{23}_{11}$Na	22.989768
$^{24}_{12}$Mg	23.985042
$^{27}_{13}$Al	26.981539
$^{35}_{17}$Cl	34.968853
$^{40}_{20}$Ca	39.962591
$^{55}_{25}$Mn	54.938047
$^{56}_{26}$Fe	55.934939
$^{60}_{28}$Ni	59.930788
$^{60}_{27}$Co	59.933820
$^{74}_{32}$Ge	73.921177
$^{90}_{40}$Zr	89.904703
$^{138}_{56}$Ba	137.905232
$^{181}_{73}$Ta	180.947992
$^{208}_{82}$Pb	207.976627
$^{209}_{82}$Pb	208.981065
$^{210}_{84}$Po	209.982848
$^{232}_{90}$Th	232.038051
$^{235}_{92}$U	235.043924
$^{238}_{92}$U	238.050785

In passing, let's note that nuclear abundances, the relative amounts of isotopes that occur in nature, can also be established from mass spectrometry. Ionized atoms from a naturally occurring sample of the element pass through the spectrometer. Because their masses are different, each isotope follows a different path, and the relative number that follow each path is a measure of their abundance.

A useful mass unit for nuclear and atomic physics is the *atomic mass unit* (u), which is defined as $\frac{1}{12}$ the mass of a ^{12}C atom. Such a mass standard is useful because the *relative* masses of atoms (and nuclei) can be measured with high precision, even if we can express with much less precision the atomic mass unit in terms of the kilogram:

$$1\ \text{u} = 1.66057 \times 10^{-27}\ \text{kg}. \qquad (45\text{–}3)$$

We can also express masses in eV/c^2, or MeV/c^2 by using the relativistic relation between mass and energy. We find

$$1\ \text{u} = 931.494\ \text{MeV}/c^2. \qquad (45\text{–}4)$$

In atomic mass units, the nucleon masses are

$$m_p = 1.00728\ \text{u} = 938.272\ \text{MeV}/c^2 = 1.67262 \times 10^{-27}\ \text{kg} \quad \text{and} \qquad (45\text{–}5a)$$

$$m_n = 1.0086\ \text{u} = 939.566\ \text{MeV}/c^2 = 1.67493 \times 10^{-27}\ \text{kg}, \qquad (45\text{–}5b)$$

whereas the mass of the electron is

$$m_e = 5.48578 \times 10^{-4}\ \text{u} = 0.510995\ \text{MeV}/c^2 = 0.910939 \times 10^{-30}\ \text{kg} \qquad (45\text{–}5c)$$

and the mass of the hydrogen atom is

$$M(^1\text{H}) = 1.007825\ \text{u} = 938.783\ \text{MeV}/c^2 = 1.67356 \times 10^{-27}\ \text{kg}. \qquad (45\text{–}5d)$$

Note in Eqs. (45–5) the standard notation of capital em (M) for atomic mass and lower case em (m) for nuclear and particle masses. Table 45–1 gives the values of some atomic masses in u. The corresponding nuclear mass can be found by subtracting Zm_e from the atomic mass.

Binding Energy. From Eqs. (45–5), the sum of the masses of 6 protons, 6 neutrons, and 12 electrons is 12.0989 u—about 1 percent greater than the mass of a ^{12}C atom, exactly 12 u. This discrepancy is due entirely to the nuclear mass: The mass of a carbon nucleus is less than the sum of the masses of 6 protons and 6 nucleons by this same amount. *In general, the mass of a stable nucleus is less than* $Zm_p + Nm_n$ due to the energies of binding the nucleons into the nucleus and to special relativity. To separate a nucleon (or collection of nucleons) from a stable nucleus takes energy.

The nuclear mass is about 1 percent less than the total masses of the nucleus's constituents.

The *binding energy* E_b of a nucleus is the energy released when the nucleus is formed from A independent constituent nucleons. To understand the relation between binding energy and nuclear mass, we must use mass–energy equivalence (see Chapter 40). Because energy must be added to a nucleus to transform it to free nucleons, the energy of the nucleus is less than the sum of the energies of free nucleons; moreover, by the equivalence of mass and energy, the mass of the nucleus is less than the sum of the masses of the constituents. These considerations hold for *any* bound system, but this effect is a small one for atoms and the other bound systems we have considered previously. For the nucleus, the attractive forces between nucleons are so strong that the binding energy is a significant fraction of the total mass. Thus, by the relativistic relation between mass and energy, *the mass of a nucleus is on the order of 1 percent less than the sum of its nucleon masses*. More precisely, the binding energy is c^2 times the difference between the atomic mass and the sum of the masses of Z hydrogen atoms and N neutrons:

$$E_b \equiv [ZM(^1\text{H}) + Nm_n - M(^A X)]c^2. \qquad (45\text{–}6)$$

This quantity is positive for stable nuclei. Table 45–2 gives the total binding energies of some nuclei. In the case of ^{12}C, we get $E_b = (12.0989\ \text{u} - 12.0\ \text{u})\ c^2(931.494\ \text{MeV/u}c^2) = 92.12$ MeV, from Eq. (45–6).

TABLE 45–2 Some Nuclear Binding Energies

Nucleus	*Total Binding Energy (MeV)*	*Binding Energy per Nucleon (MeV/nucleon)*
${}^{2}_{1}H$	2.23	1.1
${}^{4}_{2}H$	28.3	7.1
${}^{12}_{6}C$	92.1	7.7
${}^{16}_{8}O$	127.5	8.0
${}^{56}_{26}Fe$	492.3	8.8
${}^{63}_{29}Cu$	552.1	8.8
${}^{238}_{92}U$	1803	7.6

Table 45–2 also gives the *binding energy per nucleon*, E_b/A, in addition to the total binding energy. Once A is 12 or larger, the binding energy per nucleon is fairly constant, between 7 and 9 MeV/nucleon. We plot this quantity as a function of the mass number, A, in Fig. 45–6. Note the position on this curve of ${}^{4}He$, which has an exceptionally large E_b/A value and is therefore strongly bound, and the region around ${}^{56}Fe$, for which the curve exhibits a broad maximum. The fact that the binding energy per nucleon increases from zero for small nuclei and decreases as A increases for $A > 60$ tells us a lot about nuclear stability. We see that the total mass–energy per nucleon *decreases* (the total binding energy increases) when two small nuclei combine into an intermediate one (for instance, E_b/A for ${}^{2}H$ is 1.1 MeV, whereas E_b/A for ${}^{4}He$ is 7.1 MeV). Similarly, the total mass–energy per nucleon decreases when one large nucleus, such as ${}^{235}U$, for which $E_b/A \simeq 7.5$ MeV, forms two intermediate nuclei, for which E_b/A may be around 8.5 MeV, or when a large nucleus forms a remnant nucleus plus a ${}^{4}He$ nucleus (an α particle). Thus, decays of large nuclei into smaller nuclei are to be expected. Whether these decays actually occur or not, and the rate at which they occur, depend on details of the structure of the decaying nucleus. For example, a decay may occur only when there is tunneling through a potential barrier (see Chapter 41).

Other Properties

Size and Internal Distribution of Mass and Charge. Rutherford's experiments have been refined. When the energy of the bombarding particle is increased to the point at which the particle can penetrate the nucleus, we can study the *internal* structure of the nucleus by analyzing the distribution of the scattering. If the nucleus is bombarded with electrons, which interact *only* through the Coulomb force with the nucleus, we learn about the distribution of charge within the nucleus. If the nucleus is bombarded

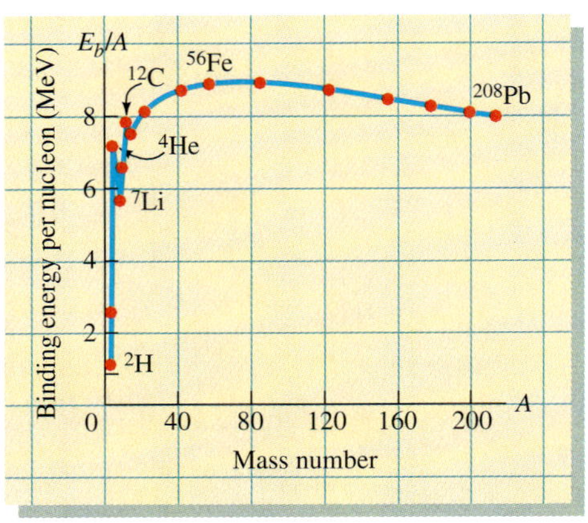

FIGURE 45–6 The binding energy per nucleon, E_b/A, as a function of the mass number A.

with α particles, protons, or neutrons, we learn instead about the distribution of nucleons within the nucleus. Both types of experiments give roughly the same picture of both the radius and the internal makeup of the nucleus; in other words, the protons and neutrons are distributed in much the same way within the nucleus. In particular, both the mass and the charge of a nucleus of mass number A are distributed *uniformly* out to a radius R_A given by

$$R_A = r_0 A^{1/3}, \tag{45–7}$$

where $r_0 \simeq 1.2 \times 10^{-15}$ m. The mass (and charge) density within this volume is very nearly the same for all nuclei. Beyond R_A, the mass and charge densities drop rapidly to zero. The fact that R_A is proportional to $A^{1/3}$ means that the volume, approximately equal to $\frac{4}{3}\pi R_A^3$, is proportional to A: volume $= \frac{4}{3}\pi R_A^3 = \frac{4}{3}\pi (r_0 A^{1/3})^3 = \frac{4}{3}\pi r_0^3 A$. In the classical model, the nucleus resembles a set of closely packed marbles (Fig. 45–7).

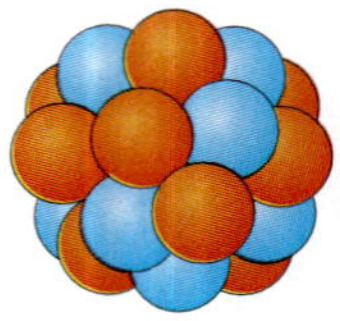

FIGURE 45–7 Nucleons are closely packed within nuclei. We have used red to indicate neutrons and blue to indicate protons.

Stability. Although nuclides with Z values up to $Z = 92$ (uranium) occur naturally, not all of these nuclides are stable. The nuclide $^{209}_{83}$Bi is the heaviest stable nucleus. Even though uranium is not stable, however, its longest-lived isotope, ^{238}U, has a half-life (see Chapter 41) of some 4 billion yr. At low Z values, nuclei tend to have the same number of neutrons and protons ($N \simeq Z$), but as Z increases, N tends to exceed Z. The nuclide ^{40}Ca is the heaviest stable nuclide for which $N = Z$. Figure 45–8 shows that the stable nuclides lie in a region called the *line of stability*. This curve has the rough form $N \simeq Z$ for small Z and $N \simeq 1.6Z$ for large Z.

The reason that the number of neutrons outnumbers the number of protons for large nuclei involves the short-range nature of the nuclear force and the long-range nature of the Coulomb force. Nucleons interact via the nuclear force mainly with their neighbors, while protons interact via the Coulomb force with all the other protons in the nucleus. The repulsive Coulomb interaction thus becomes more important for larger nuclei, and it is less energetically favorable to have as many protons as neutrons.

There are approximately 250 stable isotopes. When the number of stable isotopes is expressed in terms of whether Z and N are even or odd, a striking pattern emerges. About 60 percent of the stable isotopes have both Z and N even; about 20 percent have Z even, N odd; and 20 percent have Z odd, N even. Only 5 stable isotopes have both Z and N odd. The strong preference for nuclei to have both Z and N even can be understood by a pairing force of the kind that leads to Cooper pairs in superconductivity (see Chapter 43).

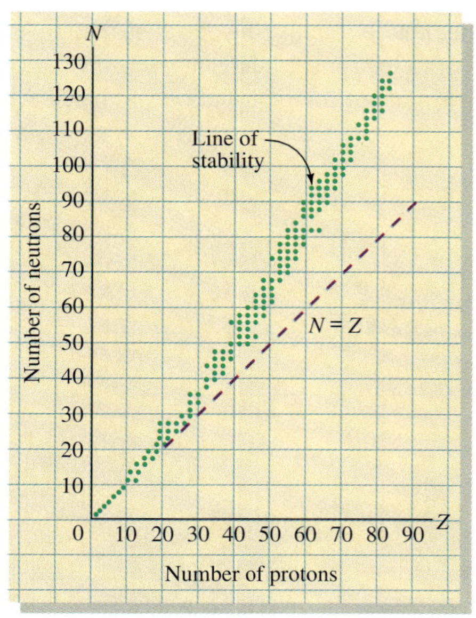

FIGURE 45–8 The stable nuclides plotted on a graph of neutron number, N, versus proton number, Z. Note that for heavier nuclides, N is larger relative to Z. The stable nuclides group along a curve called the line of stability.

Spins and Magnetic Dipole Moments. Nuclei have angular momentum, just as atoms do. This angular momentum is usually called the *spin* of the nucleus because, without knowing the internal structure of the nucleus, we can regard it as intrinsic. Nucleons have a spin of $\hbar/2$. As is true of electrons in atoms, the nucleons that compose the nucleus can have an orbital angular momentum. The total angular momentum of a nucleus results from adding the orbital angular momentum and the spins according to the rules of quantum mechanics, supplemented by the exclusion principle. (The exclusion principle states, for example, that two protons cannot move in the same orbit and both have spin up.)

Associated with the angular momentum of a nucleus is a magnetic dipole moment, and this magnetic moment provides the means of detecting the angular momentum of the nucleus. We described *nuclear magnetic resonance* in Chapter 32, in which the nuclear magnetic dipole moment is detected because of a resonance phenomenon. Another means of detecting the nuclear magnetic dipole moment is to note that a magnetic dipole produces a magnetic field and, because atomic electrons have both spin and orbital angular momenta, the energy of these electrons depends on this nuclear magnetic field. The small resulting difference in electron energy levels is referred to as *hyperfine splitting* of the atomic spectra, a splitting some thousand times smaller than the fine-structure splitting discussed in Chapter 42.

The orders of magnitude of nuclear magnetic moments are those of the magnetic dipole moments of the proton and neutron themselves. We would estimate these magnetic dipole moments in turn to be on the order of $e\hbar/2m_pc$ (a quantity known as the *nuclear magneton*) for the proton and zero for the (neutral) neutron, similar to the magnetic dipole moment of $e\hbar/2m_ec$ for the electron. (The composite nature of protons and neutrons—see Chapter 46—modifies this expectation a good deal.) Note that the nuclear magnetic moment is some 2000 times smaller than the electron magnetic moment because of the masses involved. That is why the hyperfine splitting of atomic spectra is some 1000 times smaller than the fine-structure splitting.

EXAMPLE 45–2 (a) From data on atomic masses, determine the total binding energy and binding energy per nucleon of ^{56}Fe. (b) How much energy does it take to remove the proton from ^{56}Fe that is least tightly bound?

Solution: (a) Table 45–1 and Eqs. (45–5) contain the masses necessary for us to be able to use Eq. (45–6) to determine the binding energy. When Eq. (45–6) is applied to ^{56}Fe, we find that

$$\begin{aligned} E_b(^{56}\mathrm{Fe}) &= [Z_{^{56}\mathrm{Fe}}M(^1\mathrm{H}) + N_{^{56}\mathrm{Fe}}m_n - M(^{56}\mathrm{Fe})]c^2 \\ &= [26M(^1\mathrm{H}) + 30m_n - M(^{56}\mathrm{Fe})]c^2 \\ &= [26(1.007825\ \mathrm{u}) + 30(1.008665\ \mathrm{u}) - 55.934939\ \mathrm{u}]c^2 \\ &= (0.52846\ \mathrm{u}c^2)(931.494\ \mathrm{MeV/u}c^2) = 492.3\ \mathrm{MeV}. \end{aligned}$$

The binding energy per nucleon is (492.3 MeV)/(56 nucleons) = (8.8 MeV)/nucleon.

(b) To determine how much energy binds the least-tightly bound proton, which we label S_p, we use a relation like Eq. (45–6) but instead compare the mass of an atom of ^{56}Fe to that of a hydrogen atom plus an atom of ^{55}Mn (the ^{55}Mn nucleus is what remains when a proton is removed from an ^{56}Fe nucleus):

$$\begin{aligned} S_p(^{56}\mathrm{Fe}) &= [M(^1\mathrm{H}) + M(^{55}\mathrm{Mn}) - M(^{56}\mathrm{Fe})]c^2 \\ &= [1.007825\ \mathrm{u} + 54.938046\ \mathrm{u} - 55.934939\ \mathrm{u}]c^2 \\ &= (0.0109\ \mathrm{u}c^2)(931.494\ \mathrm{MeV/u}c^2) = 10.2\ \mathrm{MeV}. \end{aligned} \qquad (45\text{–}8)$$

The energy required to remove just one nucleon from a nucleus is called the *separation energy*. The proton separation energy for ^{56}Fe is particularly large—10.2 MeV—because ^{56}Fe is a particularly strongly bound nucleus.

45–2 NUCLEAR FORCES AND NUCLEAR MODELS

Nuclear Forces

One of the major tasks of nuclear physics is to discover the nature of the nuclear force. It must be strong enough to overcome the electrical repulsion between protons. An energy of 5 to 10 MeV is required to free a nucleon from a nucleus. This value can be compared with the 13.6 eV of energy needed to ionize hydrogen or the 3 to 5 eV needed to free an electron from a metal (the work function). By this measure, the nuclear forces are about a million times stronger than the electric forces that bind atoms.

The nuclear force is a short-range force.

The fact that the nucleon separation energy is on the order of a few MeV even for nuclei as heavy as lead ($A \simeq 208$) implies that the number of "bonds" that attach a nucleon to a nucleus does not grow as the nucleus grows; we say that the nuclear force becomes *saturated* when a nucleon is surrounded by other nucleons. *The nuclear force has a short range comparable to the spacing between nucleons in a nucleus.* By the range, we mean the distance beyond which the nuclear force rapidly decreases. From our formula for the nuclear radius, Eq. (45–7), we can deduce that the spacing between nucleons is on the order of 1 **fermi** (fm), or 10^{-15} m (also known as a *femtometer*). The conclusion that the nuclear force has a short range is supported by scattering experiments, which also allow us to measure the range quantitatively.

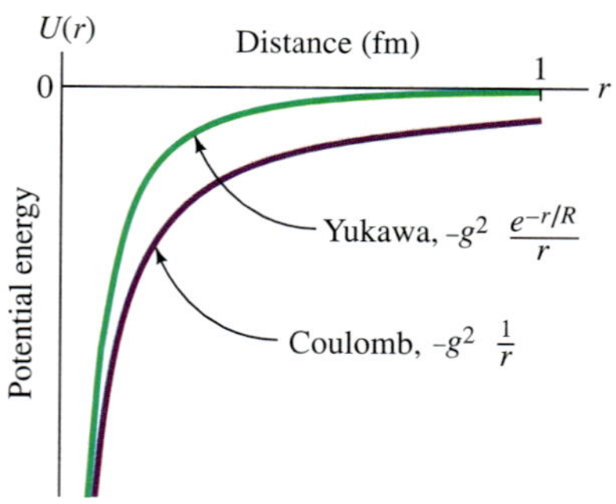

FIGURE 45–9 The Yukawa potential $U(r) = -g^2e^{-r/R}/r$, and the Coulomb form $-g^2/r$ for comparison. We have chosen $g^2 = 1$ and $R = 0.3$ fm.

We can summarize the quantitative behavior of nuclear interactions as follows. Except at distances much less than 1 fm, a reasonable representation of the nucleon-nucleon potential energy is

$$U(r) = -g^2\frac{e^{-r/R}}{r}, \tag{45–9}$$

where r is the distance between the two nucleons, and the parameter R, which we define as the range, is approximately 2×10^{-15} m. Figure 45–9 illustrates this potential energy as well as a potential energy of the same strength but with a Coulomb form, $U_c = -g^2/r$, for comparison; these two potential energies have the same behavior for small r. The potential energy of Eq. (45–9) is known as the *Yukawa potential*, named for its originator, Hideki Yukawa. It can be justified in the context of quantum mechanics (as we shall see in Chapter 46).

In our discussion, we have referred to the nucleon–nucleon force, not the proton–proton force, the proton–neutron force, or the neutron–neutron force. Numerous scattering experiments, as well as the study of nuclear binding energies for nuclei with the same A value but different Z values and N values, show that, basically, the only differences in the interactions of neutrons and protons are due to electromagnetism.

A full description of nuclear potentials takes us far beyond the Yukawa potential. First, there is a repulsion (a *core repulsion*) between the nucleons at separation distances less than about 1 fm. Second, nuclear forces have a substantial dependence on the orientation of the spin of the nucleons. Third, and perhaps most important, nuclear forces normally involve more than just two interacting particles. In atomic physics, this does not present much of a problem because the forces are relatively weak, and mathematical tools exist by which we can handle such forces when many particles are involved. In nuclear physics, the forces are *strong*, as measured by the parameter g^2 in Eq. (45–9). Analytic mathematical techniques for the systematic treatment of such forces do not exist. The problems posed by nuclear forces are too difficult to be solved from a fundamental theory. Despite this difficulty, scientists know much about nuclei. As in the case of materials, models and approximations that explain different, limited features of nuclei have been developed for describing the nucleus. We discuss two of these models, the shell model and the liquid-drop model.

The Shell Model

In Chapter 42, we learned that the energy levels of electrons in an atom form a shell structure under the influence of the Coulomb attraction of the nucleus. We have seen in this chapter that nuclear forces are quite different from Coulomb forces. Despite this fact, individual nucleons in a nucleus also appear to have well-defined energy levels. The description of the nucleus based on this idea is known as the **shell model** (or as an *independent-particle model*).

To examine what such a structure would look like, we assume that the force experienced by a given nucleon is a three-dimensional spring force. (This at least gives the correct orders of magnitude.) In three dimensions, the energy levels of a harmonic oscillator are a simple generalization of the result of Example 42–4, corresponding to the possibility of motion in three dimensions:

$$E = \hbar\omega\left(2n_r + \ell + \frac{3}{2}\right). \tag{45–10}$$

Here, the radial quantum number $n_r = 0, 1, 2, 3, \ldots$ and ℓ, the angular momentum quantum number, takes on the values $\ell = 0, 1, 2, 3, \ldots$. As usual, there are $2\ell + 1$ states with the same energy for each value of ℓ. We separately fill energy levels with protons and neutrons. The lowest-energy state of Z protons, for example, is obtained by filling successive levels with two (spin $\hbar/2$) protons each—that is, $2(2\ell + 1)$ protons could fit into the levels for a given ℓ value. Thus, the order of filling levels for protons is the following:

			Total protons
$E = \frac{3}{2}\hbar\omega;$	$n_r = 0, \ell = 0;$	2 protons	2
$E = \frac{5}{2}\hbar\omega;$	$n_r = 0, \ell = 1;$	6 protons	8
$E = \frac{7}{2}\hbar\omega;$	$n_r = 0, \ell = 2;$	10 protons	18
$E = \frac{7}{2}\hbar\omega;$	$n_r = 1, \ell = 0;$	2 protons	20
$E = \frac{9}{2}\hbar\omega;$	$n_r = 0, \ell = 3;$	14 protons	34
$E = \frac{9}{2}\hbar\omega;$	$n_r = 1, \ell = 1;$	6 protons	40

and so on; the same sequence holds for neutrons. Nuclei in which the states that correspond to given quantum numbers (shells) are filled (closed shells) are expected to be particularly stable. According to the list presented here, closed shells occur for $Z = 2$, $Z = 2 + 6 = 8$, $Z = 8 + (10 + 2) = 20$, $Z = 20 + (14 + 6) = 40$, and so on. Because the same list holds for neutrons, particularly stable nuclei ought to occur in this simple model for Z values or N values of 2, 8, 20, 40, 70, ... These values are called *magic numbers*.

The *experimental* values of Z for which the binding energy is particularly large are 2, 8, 20, 28, 50, 82, 126. These numbers do not represent a perfect match to the predicted magic numbers and indicate that the order of filling levels is different in reality than the simple harmonic oscillator potential would suggest. In 1949, Hans Jensen and Maria Goeppert-Mayer independently pointed out that the energy should contain terms not included earlier that involve both the spin of a single nucleon and its orbital angular momentum. This *spin–orbit coupling* is the same type of coupling that causes splitting of atomic levels (see Section 42–3). By including this type of coupling, splitting of levels occurs and the ordering of levels is upset. The observed magic numbers are thereby more correctly predicted, as is the complicated pattern of level splittings seen in nuclei.

The Liquid-Drop Model

The short range of nuclear forces and the small amount of space between the constituent nucleons suggested to Niels Bohr that a nucleus should behave in some ways like a continuous fluid, such as water. The model based on this idea, the **liquid-drop model** (also referred to as a *collective model*), contains no reference to the behavior of individual nucleons and in this sense is complementary to the shell model. We note two similarities between nuclear data and the behavior of incompressible fluids. First, the density of all nuclear matter is roughly constant, just as for an incompressible fluid. Second, just as the binding energy per nucleon, E_b/A, is roughly constant from nucleus to nucleus, the heat of vaporization per unit mass is constant for different size drops of an incompressible fluid. (The heat of vaporization is the energy required to separate a drop of fluid into its component molecules, and the total mass of a nucleus is proportional to A. Thus, E_b/A is analogous to the heat of vaporization per unit mass.)

Following this analogy, we can write an *empirical* formula for the mass of a nucleus, $M(^AX)$, by including terms of the type that would be included in computing the energy of a liquid drop, as follows:

1. A zeroth-order term that is just the sum of the rest masses of the nucleons:

$$ZM_H + Nm_n = ZM_H + (A - Z)m_n. \tag{45–11}$$

2. A term proportional to the volume of the nucleus (or $\propto A$, as we have shown), which takes into account the nearly constant binding energy per nucleon:

$$-a_V A, \tag{45–12}$$

where a_V is positive because a binding energy reduces the mass of the nucleus.

3. A term proportional to $A^{2/3}$ or, equivalently, to the surface area of the nucleus:

$$+a_S A^{2/3}. \tag{45–13}$$

A nucleon near the surface is not bound as strongly as one in the interior, so a_S takes into account a reduced binding energy and is positive. This term is the analogue of a surface-tension term in a liquid drop.

The remaining three terms take into account some simple observations specific to the nucleus:

4. A term for the Coulomb repulsion between all the protons in the nucleus. This term tends to increase the mass of the nucleus. If we suppose that the protons are distributed uniformly throughout a sphere of radius $R_A = r_0 A^{1/3}$ [Eq. (45–7)], then the energy required to assemble that charge is[†]

$$\frac{3}{5}\frac{Z^2e^2}{4\pi\epsilon_0 r_0 A^{1/3}} = 0.72\frac{Z^2}{A^{1/3}}\ \text{MeV} \tag{45–14}$$

(see Problem 59 of Chapter 12). The contribution of the energy to the mass is this quantity divided by c^2.

5. A term that has a minimum for $N = Z$ accounts for the tendency for the number of neutrons to equal the number of protons. If terms 1 through 4 were the only terms present, then the energy could be lowered by taking Z to zero; that is, nuclei would consist exclusively of neutrons. In fact, there are no nuclei with many more neutrons than protons, and nuclei are particularly stable when $N \simeq Z$. This term is written as

$$+a_A\frac{(A - 2Z)^2}{A}, \tag{45–15}$$

where a_A is positive.

6. Several terms that describe the tendency for nuclei with even numbers of protons and/or neutrons to be more deeply bound than nuclei with odd numbers of protons and/or neutrons. This tendency occurs because the spins of two nucleons in each "shell"

[†]In some instances, the calculated coefficient 0.72 of the factor $Z^2/A^{1/3}$ in Eq. (45–14) is replaced by the empirical parameter $a_C \simeq 0.60$. This is a very small difference in the contribution of the energy to the masses of nuclei. In Table 45–3, we use the empirical form.

are antiparallel. Such terms have the empirical form

$$\text{for } Z, N \text{ even:} \quad \Delta = -\frac{a_P}{\sqrt{A}} \tag{45–16a}$$

$$\text{for } Z \text{ even, } N \text{ odd; or } Z \text{ odd, } N \text{ even:} \quad \Delta = 0 \tag{45–16b}$$

$$\text{for } Z, N \text{ odd:} \quad \Delta = +\frac{a_P}{\sqrt{A}} \tag{45–16c}$$

The sum of all these terms is the **semiempirical mass formula**:

$$M_A = ZM_H + (A - Z)m_n - a_V A + a_S A^{2/3} + \frac{a_C Z^2}{A^{1/3}} + a_A \frac{(A - 2Z)^2}{A} + \Delta. \tag{45–17}$$

The parameters are determined empirically by fitting this relation with many different nuclear masses. One set of parameters is given in Table 45–3. For these values, the curve in Fig. 45–6 for $A = 20$ and above is in good agreement with Eq. (45–17). For very light nuclei, the notion of a nucleus that looks like a droplet with a volume and a surface loses meaning, and the mass formula does not work very well, but otherwise the fit is excellent.

A droplet can deform in shape. This changes the energy, because the surface area changes. In fact, droplets oscillate because of this effect, even when the fluid is incompressible. Surface forces tend to return the droplet to a spherical shape, but the Coulomb energy term favors a larger deformation. When we compare these two effects for heavy nuclei, we find that the liquid drop is unstable and breaks up into two smaller droplets. The liquid-drop model thus provides an explanation of the fission process.

TABLE 45–3
Parameters of the Semiempirical Mass Formula

Parameter	*Value (u)*
a_V	1.7×10^{-2}
a_S	1.8×10^{-2}
a_C	7.5×10^{-4}
a_A	2.5×10^{-2}
a_P	1.3×10^{-2}

EXAMPLE 45–3 (a) Use the semiempirical mass formula to calculate the atomic masses of ^{208}Pb and ^{209}Pb. (b) Calculate the neutron separation energy for ^{209}Pb.

Solution: (a) We use the semiempirical mass formula with the parameters of Table 45–3:

$$\begin{aligned} M(^{208}\text{Pb}) &= (82)(1.007825\text{ u}) + (208 - 82)(1.008665\text{ u}) - (1.7 \times 10^{-2}\text{ u})(208) \\ &\quad + (1.8 \times 10^{-2}\text{ u})(208)^{2/3} + (7.5 \times 10^{-4}\text{ u})\left[\frac{(82)^2}{(208)^{1/3}}\right] \\ &\quad + (2.5 \times 10^{-2}\text{ u})\left\{\frac{[208 - 2(82)]^2}{208}\right\} - \left(\frac{1.3 \times 10^{-2}\text{ u}}{\sqrt{208}}\right) \\ &= 207.921\text{ u;} \end{aligned}$$

$$\begin{aligned} M(^{209}\text{Pb}) &= (82)(1.007825\text{ u}) + (209 - 82)(1.008665\text{ u}) - (1.7 \times 10^{-2}\text{ u})(209) \\ &\quad + (1.8 \times 10^{-2}\text{ u})(209)^{2/3} + (7.5 \times 10^{-4}\text{ u})\left[\frac{(82)^2}{(209)^{1/3}}\right] \\ &\quad + (2.5 \times 10^{-2}\text{ u})\left\{\frac{[209 - 2(82)]^2}{209}\right\} + 0 \\ &= 208.915\text{ u.} \end{aligned}$$

(b) To find the neutron separation energy, we write an equation similar to Eq. (45–8):

$$\begin{aligned} S_n(^{209}\text{Pb}) &= [m_n + M(^{208}\text{Pb}) - M(^{209}\text{Pb})]c^2 \\ &= [1.008665\text{ u} + 207.912\text{ u} - 208.915\text{ u}]c^2 \\ &= (0.0057\text{ u}c^2)(931.494\text{ MeV/u}c^2) = 5.3\text{ MeV.} \end{aligned}$$

The experimental values of the masses are $M(^{208}Pb) = 207.97663$ u and $M(^{209}Pb) = 208.98107$ u, and the experimental value of the neutron separation energy is 3.9 MeV. The 5.3-MeV separation energy that we calculated from the semi-empirical mass equation does not take into account the shell-model magic numbers, for which nuclei are particularly strongly bound. The nuclide ^{208}Pb is an example of a "doubly magic" nucleus ($Z = 82$, $N = 126$), so it should be *very* strongly bound. The nuclide ^{209}Pb gives up its neutron rather easily in order to reach this special configuration. Therefore, the actual value, 3.9 MeV, is below our calculated value.

45–3 ENERGETICS OF NUCLEAR REACTIONS

By nuclear reactions, we generally mean the processes that occur when nuclei interact through collisions. All the possibilities for different types of collisions outlined in Chapter 8 occur. Two nuclei may have an elastic collision or various types of inelastic collisions. A nucleus may absorb energy and then decay by one of several modes, which we shall discuss in Section 45–4. Mass may be transferred from one nucleus to another by the exchange of nucleons. Colliding nuclei may coalesce—a process called **nuclear fusion**. The decay of a single unstable nucleus may be regarded as a nuclear reaction. Examples are α decay, in which a nucleus decays to a helium nucleus and another nucleus; and **nuclear fission**, in which the decay products include two nuclei of more or less equal size. In all these interactions, the conservation laws we have developed throughout this book apply, including the conservations of energy, momentum, angular momentum, and charge.

At the simplest level, collisions between nuclei allow us to measure kinematic quantities such as masses or angular momenta. A terminology special to nuclear reactions has been developed. Consider a nuclear collision in which both the initial and final states consist of two nuclei:

$$A + B \to C + D.$$

Energy conservation applies, but we must use relativistic relations. For any one nucleus (or particle) of mass M, this energy takes the form

$$E = K + Mc^2,$$

where K is the kinetic energy of the nucleus. The value of K depends on the frame of reference, but the conservation of energy, $E_A + E_B = E_C + E_D$, applies in any reference frame. If we suppose that nucleus B is the target nucleus, at rest in the laboratory frame of reference, then the conservation of energy takes the form

$$(K_A + M_Ac^2) + M_Bc^2 = (K_C + M_Cc^2) + (K_D + M_Dc^2) \tag{45–18}$$

in that reference frame. Depending on the masses of the nuclei involved, the reaction can be *exothermic* (having more kinetic energy in the final state than in the initial state) or *endothermic* (having less kinetic energy in the final state). A kinematic parameter that describes such properties is the *Q value* of the reaction:

$$Q \equiv K_C + K_D - K_A - K_B. \tag{45–19}$$

By including the possibility that nucleus B has kinetic energy, we have written a general relation. A positive Q value corresponds to an exothermic reaction, and a negative Q value corresponds to an endothermic reaction. By moving all the kinetic energy terms in Eq. (45–18) to one side of the equation and all the mass terms to the other, we can express the Q value in terms of atomic masses:

$$Q = (M_A + M_B - M_C - M_D)c^2. \tag{45–20}$$

By measuring kinetic energies in a reaction and comparing Eqs. (45–19) and (45–20), we can use the Q value for mass measurements that cannot otherwise be made. Note that although we have used laboratory frame quantities in Eq. (45–19), our original definition of Q, Eq. (45–20) shows that Q is actually an invariant.

The variety of measurements of different nuclear reactions—including their dependence on different variables of the collision process, such as the projectile energy and angle—gives us much of our knowledge of nuclear structure and nuclear forces. The most useful concept for collisions is the *collision cross section*, which, as we described in Chapter 19, measures the effective area taken up by the colliding nuclei. The cross section is a measure of the probability of interaction; the larger the cross section, the more probable the collision and hence the reaction. We can thus define a cross section *for each possible reaction* that measures the relative probability of each of these processes.

One of the important features of the cross section is the presence of *resonances*, which we define here as bumps in the curve of cross section versus collision energy of the same form as the resonance curve of Fig. 13–23. A bump in the cross section centered around E_0 of width ΔE implies that the nucleus is strongly affected by the incident projectile when the total energy is in the range $E_0 - \Delta E/2$ to $E_0 + \Delta E/2$. Here, the nucleus–projectile system behaves like a nucleus with an excited state around E_0, but spread over an energy range ΔE. In other words, *the presence of a resonance peak in a cross section signals the presence of an excited state at that energy*. Indeed, these excited states are often referred to as resonances. The lifetime τ of this excited state is related to the width by the Heisenberg uncertainty principle,

The width of a resonance is inversely proportional to its lifetime. See Chapters 13, 15, and 41.

$$\Delta E\,\tau \simeq \hbar. \tag{45–21}$$

Thus *the width in energy of the resonance is ħ divided by the lifetime of the resonance*. The shorter-lived the resonance is, the broader the resonance peak is. In nuclear physics, resonances may have widths ranging from about 0.1 eV, corresponding to a lifetime of 10^{-14} s, to as much as 1 MeV, corresponding to a lifetime of 10^{-21} s. In Fig. 45–10, we see resonance shapes in the probability (total cross section) for neutron absorption by ^{23}Na. Each peak corresponds to an excited state of the nuclide ^{24}Na, which then decays. In some interactions of nucleons and other particles that participate in the nuclear force, resonances occur with widths as large as 100 MeV. This corresponds to a lifetime of 10^{-23} s, about the time it takes for light to cross the nucleus!

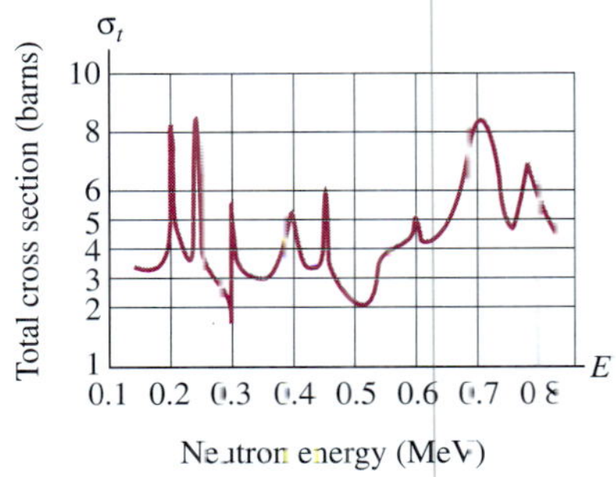

FIGURE 45–10 The total cross section for the absorption of neutrons by ^{23}Na. Energy is measured in the laboratory frame of reference. The peaks, also known as resonances, correspond to excited states of ^{24}Na. (After S. F. Mughabghab et al., *Neutron Cross Sections*, Academic Press, 1984.)

45–4 RADIOACTIVITY

Henri Becquerel discovered radioactivity in 1896, even before the existence of the nucleus had been established (Fig. 45–11). **Radioactivity**—the name was coined by Marie Curie—is the phenomenon of nuclear decay (Fig. 45–12). Nuclei can decay in a variety of *modes* or *channels*—for example, by emitting photons or α particles. Different physical laws may govern different decay modes, so each mode may have a different probability of occurring. The lifetime, τ, of a collection of radioactive nuclei is a constant with dimensions of time that determines the rate at which a set of such nuclei decay: If N_0 is the number of nuclei present at $t = 0$, then $N(t)$, the number of nuclei that remain after a time t, is given by Eq. (41–21), $N(t) = N_0 e^{-t/\tau}$. An alternate way to write this equation is to define the **decay constant**, λ, as $\lambda \equiv 1/\tau$, so

$$N(t) = N_0 e^{-\lambda t}. \tag{45–22}$$

The rate at which nuclei decay, the *decay rate*, is $-dN/dt$. From Eq. (45–22),

$$\frac{dN}{dt} = -\lambda N_0 e^{-\lambda t} = -\lambda N. \tag{45–23}$$

The decay rate λN, also called the *activity*, is high when the number of unstable nuclei is large and when the decay constant is large (the lifetime is small). We also employ the *half-life*, $t_{1/2}$, the time for half of a given sample of unstable nuclei to decay.

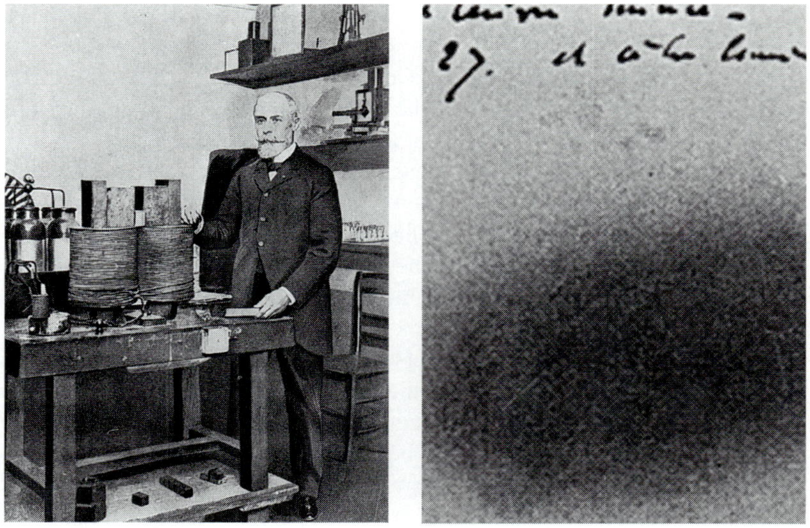

(a) (b)

FIGURE 45–11 (a) Henri Becquerel, who discovered radioactivity after he had left a photographic plate under uranium salts in a closed drawer for several days in 1896. (b) The blurred images that formed on Becquerel's photographic plate—the first evidence of radioactivity.

FIGURE 45–12 Marie Curie in her laboratory at the Sorbonne, Paris, circa 1908. Curie and her husband, Pierre Curie, performed important early work on radioactivity for which she earned two Nobel prizes.

The half-life is related to the lifetime by $t_{1/2} = 0.693\tau$ (see Problem 48 in Chapter 41). The SI unit of activity is the *becquerel* (Bq): 1 Bq = 1 decay/s.

If a nucleus can decay in two or more ways, then the total decay rate is a sum of decay rates for the various decay modes. Think of several holes in a leaky can: The total rate of flow from the can is the sum of the rates from each of the holes. The decay rate from the mode labeled by the subscript j (for example, j may label the α-decay mode) is proportional to the decay constant λ_j for that mode, so the total decay rate is proportional to the sum of all the decay constants. In other words, there is a total decay constant λ given by

$$\lambda = \sum_j \lambda_j. \tag{45–24}$$

If the decay in one mode, say, mode 1, is much more rapid than that in all the other modes, then λ_1 is much larger than the other decay constants, and the λ_1 term dominates the sum in Eq. (45–24). In that case, $\lambda \simeq \lambda_1$. It is as though one hole in our leaky can is much larger than the others. We say that the nucleus *decays predominantly through the channel* labeled by λ_1.

With the exception of fission reactions (to be discussed in Section 45–5), a typical radioactive decay involves a nucleus (the *parent nucleus*) that emits a particle and is thereby converted into a *daughter* nucleus. Three different types of decays can occur, labeled α, β, or γ according to the particles emitted.

EXAMPLE 45–4 A sample of uranium ore emits α radiation characteristic of ^{235}U at a rate of 9.3×10^5 decays/s. What mass of chemical uranium is present in the ore? The half-life of ^{235}U is 7.04×10^8 yr, and the abundance of naturally occurring ^{235}U in a sample of uranium is 0.72 percent.

Solution: We are given the decay activity and half-life for ^{235}U. From the half-life we can find the decay constant, λ; with λ and the activity we can find the number of ^{235}U nuclei. Then we can use the abundance of ^{235}U to find the total quantity of uranium. We have $\lambda = 1/\tau = 0.693/t_{1/2}$. From the relation for the activity, Eq. (45–23), we find the number of ^{235}U atoms to be

$$N = -\frac{1}{\lambda}\frac{dN}{dt} = -\frac{t_{1/2}}{0.693}\frac{dN}{dt}$$

$$= -\frac{(7.04 \times 10^8\ \text{yr})(3.16 \times 10^7\ \text{s/yr})}{0.693}(-9.3 \times 10^5\ \text{decays/s}) = 3.0 \times 10^{22}.$$

Given this number of ^{235}U atoms, we can use Avogadro's number to find the mass m of ^{235}U:

$$m = (3.0 \times 10^{22}\text{ atoms})\left(\frac{1\text{ mol}}{6.02 \times 10^{23}\text{ atoms}}\right)(235\text{ g/mol}) \simeq 12\text{ g}.$$

Because ^{235}U makes up 0.72 percent of naturally occurring uranium, the total amount of uranium is

$$(1/0.0072)(12\text{ g}) = 140(12\text{ g}) = 1.6\text{ kg}.$$

Alpha Decay

Decay in the α-**decay mode** is written as

$$^{A}_{Z}X \rightarrow {}^{A-4}_{Z-2}X' + {}^{4}_{2}\text{He}, \qquad (45–25)$$

where X and X' are the chemical symbols of the parent and daughter nuclei, respectively. Here, we have recognized that the emitted α particle is actually a ^{4}He nucleus. An example of this type of decay is $^{238}_{92}\text{U} \rightarrow {}^{234}_{90}\text{Th} + {}^{4}_{2}\text{He}$.

Alpha decay involves the quantum mechanical tunneling of an α particle through a potential barrier. For α decay to occur, the mass of the parent nucleus must be greater than the sum of the masses of an α particle and the daughter nucleus. The systematics of nuclear masses implies that α decay is increasingly more likely to occur as the size of a nucleus increases. Many nuclei above $Z = 83$ are unstable in this mode.

Tunneling is discussed in Chapter 41.

Nuclear forces are understood well enough to allow us to predict the probabilities of α decay. Because the rate at which tunneling occurs is extremely sensitive to the width and height of the potential barrier, the range of lifetimes of nuclei in the α-decay mode is huge, from 3×10^{-7} s to 1.4×10^{17} s!

EXAMPLE 45–5 ^{241}Am decays in the α-decay mode: $^{241}\text{Am} \rightarrow \alpha + {}^{237}\text{Np}$. Calculate the kinetic energy of the α particle if the ^{241}Am nucleus decays at rest and the atomic masses are $M(^{241}\text{Am}) = 241.05682$ u, $M(^{4}\text{He}) = 4.002603$ u, and $M(^{237}\text{Np}) = 237.04817$ u.

Solution: Both conservation of energy and conservation of momentum must be applied here. From the conservation of energy, we have

$$M(^{241}\text{Am}) = M(^{4}\text{He}) + M(^{237}\text{Np}) + K_f/c^2,$$

where K_f is the kinetic energy of the two-body final state. Thus

$$\begin{aligned} K_f &= [M(^{241}\text{Am}) - M(^{4}\text{He}) - M(^{237}\text{Np})]c^2 \\ &= (241.05682\text{ u} - 4.002603\text{ u} - 237.04817\text{ u})c^2 \\ &= 0.006047\text{ u}c^2 = (0.006047\text{ u}c^2)(931.474\text{ MeV/u}c^2) = 5.63\text{ MeV}. \end{aligned}$$

This kinetic energy is much less than the rest mass of either final nucleus, so the motion can be treated nonrelativistically. $K_f = K(^{4}\text{He}) + K(^{237}\text{Np})$, where $K = p^2/2M$ of the respective nuclei. The conservation of momentum applied to the final state gives $0 = p(^{4}\text{He}) + p(^{237}\text{Np})$, or $p(^{237}\text{Np}) = -p(^{4}\text{He})$. Thus

$$\begin{aligned} K(^{4}\text{He}) &= K_f - K(^{237}\text{Np}) = K_f - \frac{p(^{237}\text{Np})^2}{2M(^{237}\text{Np})} = K_f - \frac{[-p(^{4}\text{He})]^2}{2M(^{237}\text{Np})} \\ &= K_f - \frac{p(^{4}\text{He})^2}{2M(^{4}\text{He})}\,\frac{M(^{4}\text{He})}{M(^{237}\text{Np})} = K_f - K(^{4}\text{He})\frac{M(^{4}\text{He})}{M(^{237}\text{Np})}. \end{aligned}$$

We can solve this relation for $K(^{4}\text{He})$:

$$K(^{4}\text{He}) = \frac{K_f}{1 + \left[\dfrac{M(^{4}\text{He})}{M(^{237}\text{Np})}\right]} = \frac{5.63\text{ MeV}}{1 + (4.00\text{ u}/237\text{ u})} = 5.53\text{ MeV}.$$

The ^{237}Np nucleus is so much more massive than the α particle that it moves very slowly, and the α particle carries off most of the kinetic energy.

Beta Decay

In the **β-decay mode** of a nucleus, an electron and an *antineutrino* ($\overline{\nu}$) (or a positron and a neutrino) are emitted, leaving a daughter nucleus. The antineutrino is the antiparticle of the neutrino (ν), just as the positron is the antiparticle of the electron. The properties of the antineutrino and the neutrino are, for our purposes, identical.

The neutrino was introduced in Chapter 6.

Wolfgang Pauli postulated the existence of the neutrino in 1930 on the basis of experimental data on β decay. An important feature of the neutrino, which has zero charge and spin $\hbar/2$, is that it interacts so weakly with matter that any one neutrino will traverse *light-years* of ordinary matter before the probability of interacting with the matter is significant. It is therefore not surprising that the neutrino was not *independently* observed to initiate a collision process until 1956, 60 yr after β decay was first observed. As a result of the neutrino's weak interaction, what is actually observed in β decay is the electron and (sometimes) the daughter nucleus but never the neutrino. Finally, the neutrino has one other important property: Its mass, if any, is so small—at most several eV/c^2, about 10^{-5} times the electron's mass—as to be negligible.

When an electron is emitted in nuclear β decay, the daughter nucleus has a charge $+e$ more than that of the parent nucleus; that is, *Z increases by 1 in this β-decay process*. As we did in Eq. (45–25), we write this β-decay process symbolically as

$$^{A}_{Z}X \rightarrow {}_{Z+1}^{\;\;A}X' + e^- + \overline{\nu}. \tag{45–26}$$

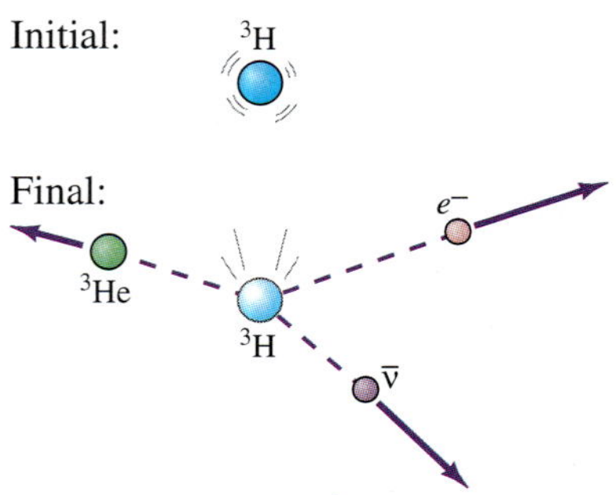

β decay of ^{3}H at rest
(three-body decay)
(a)

An example of this type of decay is $^3\text{H} \rightarrow {}^3\text{He} + e^- + \overline{\nu}$. In this form of β decay, a neutron has in effect been converted to a proton, an electron, and an antineutrino within the nucleus. Indeed, the most primitive β decay is that of the neutron itself:

$$n \rightarrow p + e^- + \overline{\nu}. \tag{45–27}$$

For *free* neutrons, the lifetime for this process is approximately 890 s. Thus neutrons by themselves are unstable. In contrast, neutrons within nuclei that do not undergo β decay are stable. The primary β-decay mode, Eq. (45–26), is possible only for the mass condition

$$M_P > M_D \tag{45–28}$$

(see Problem 32), where M_P and M_D are the atomic masses of the parent and daughter nuclei, respectively. For nuclei that undergo β decay, the kinetic energy of the electrons ranges up to 10 MeV, with 1 MeV being typical.

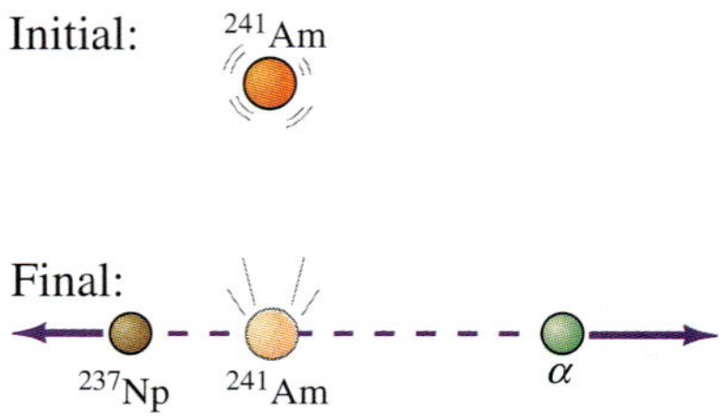

α decay of ^{241}Am at rest
(two-body decay)
(b)

FIGURE 45–13 (a) Beta decay is a three-body decay, as opposed to (b) α decay, a two-body decay. In each case the parent nucleus is at rest. In the three-body decay, a variety of directions and energies are possible for the decay products, subject to overall energy and momentum conservations. In the two-body decay, the decay products emerge back-to-back and with fixed energies.

If the neutrinos in β decay cannot be observed directly, how was it possible for Pauli to predict their existence and properties as early as 1930? The existence of the neutrino was inferred on the basis of several types of conservation laws. First, conservations of momentum and of energy have rather different consequences for three-body decay (one in which the parent nucleus decays to three particles; Fig. 45–13a) than for two-body decay (Fig. 45–13b). In two-body decay of a nucleus at rest, the energies of each of the two decay products are fixed: The conservations of momentum and of energy *uniquely* determine the magnitudes of the momentum and the energy of the decay products. This is not true in three-body decay, in which different configurations of the three bodies allow for a range of energies for all three bodies. Various configurations allow the electron's kinetic energy to range from zero up to a maximum value, as shown in Fig. 45–14. A similar plot for a two-body decay would show a single energy value at the energy maximum K_{max}.

Pauli surmised from the broad curve observed in Fig. 45–14 that three particles—including the otherwise invisible neutrino—must be present. Angular momentum conservation leads to the same conclusion. If the decay of the neutron were the two-body decay $\text{n} \rightarrow p + e^-$ rather than a three-body decay, we would have a discrepancy. Neutrons, like protons and electrons, have a spin of $\hbar/2$, but the rules of angular momentum addition do not allow two $\hbar/2$ spins to add to a total spin of $\hbar/2$. The addition of a third particle with a spin of $\hbar/2$ can resolve the discrepancy; in this way, Pauli was able to predict that the invisible, massless third particle, the neutrino, was a fermion.

Lifetimes of β decays vary from about 1 s to more than 10^{20} s. The energy dependence and low rate of β decay of some nuclei suggest that the β-decay process is not a quantum mechanical tunneling process but rather a primary manifestation of the *weak force* (one aspect of the electroweak force; see Chapter 5).

We can mention two other forms of β decay here—both of them consequences of the weak force. The first is *positron emission*, which takes the form

$$_{Z+1}^{A}X \rightarrow {}_{Z}^{A}X' + e^{+} + \nu. \tag{45–29}$$

This process is allowed under the mass condition

$$M_X > M_{X'} + 2m_e. \tag{45–30}$$

The second variant is *electron capture*, in which an electron from an atomic orbit is absorbed by the nucleus:

$$_{Z+1}^{A}X + e^{-} \rightarrow {}_{Z}^{A}X' + \nu. \tag{45–31}$$

Electron capture is possible under the mass condition

$$M_X > M_{X'}. \tag{45–32}$$

Note that the mass conditions for the primary decay mode and electron capture are identical. However, β decay tends to occur for neutron-rich nuclei, and electron capture for neutron-poor nuclei. Practically all the kinetic energy in electron capture goes to the neutrino because of the conservation of energy and linear momentum (see Example 45–5).

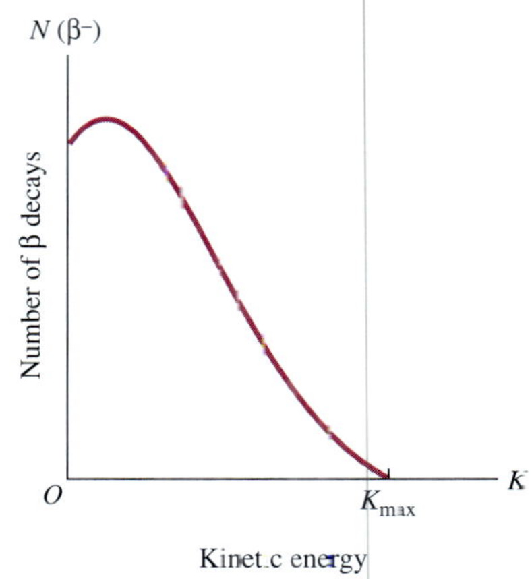

FIGURE 45–14 The energy of the electron for a large sample of β decays. The energy is not fixed, but rather spread, a consequence of the kinematics of three-body decay. The maximum possible electron energy, K_{max}, is the energy the electron would have if there were no neutrino in the decay.

Gamma Decay

A nucleus in an excited state can decay in the **γ-decay mode**: It emits photons as it decays to its ground state or to lower-lying excited states (Fig. 45–15), just as an atom in an excited electronic state can. Because the typical energy differences between nuclear excited states and the ground state are some 10^6 times greater than the corresponding energy differences in atoms, lighting a match is not enough to excite the nucleus, as it is to excite an atom. However, daughter nuclei are often produced in their excited states in α, β, or even γ decay, and these residual states can then decay by the γ-decay mode. Photons with energies as high as a few MeV were not at first recognized as electromagnetic radiation and were therefore given a special name, γ-rays; we speak of the γ-decay mode of the nucleus. The lifetimes of γ decay range from 10^{-17} s to as much as 10^8 s.

The electromagnetic forces responsible for the properties of the γ-decay modes are well known. Therefore we can use observations of lifetimes and the energies and angular distributions of the emitted photons as sensitive indicators of the quantum numbers and nature of the decaying excited states. In this way models (such as the shell model) and their predictions for the energies and quantum numbers of the excited states can be tested or developed.

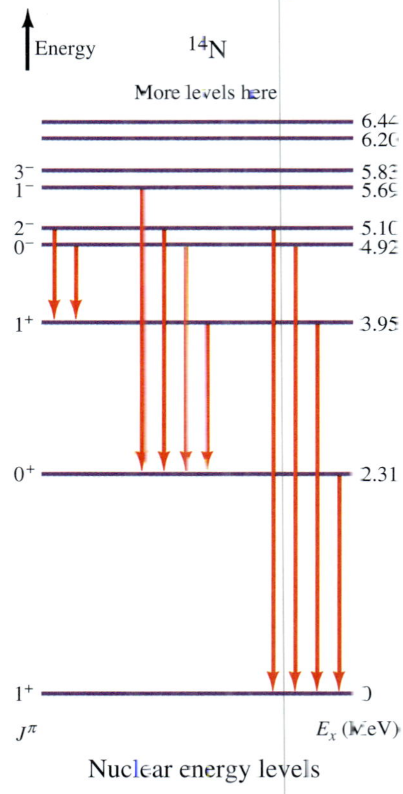

FIGURE 45–15 Some γ-ray decay transitions in ^{14}N. The notation J^{π} describes the respective quantum states labeled by quantum numbers not discussed here. E_x labels the energy of the states relative to the ground state. Not all possible transitions are shown. Many more levels exist above 6.44 MeV in ^{14}N.

Natural Radioactivity

It is often true that the daughter nucleus produced in a radioactive decay is itself unstable. The result is a *radioactive series*, in which a succession of nuclei decay in a cascade to some stable nucleus or nuclei. Along the steps of the cascade, some nuclei may have two competitive decay modes, such as an α-decay mode and a β-decay mode, in which case we say that there is *branching*. Figure 45–16 shows the uranium series, which starts with ^{238}U. It generally exists on the neutron-rich side of the line of stability (discussed in Section 45–1), also visible in the figure. Beta decays pull the branching line closer to the line of stability. Of the three possible decay modes, only α decay changes A, and by 4 units. Therefore, if a series starts with a parent whose A value

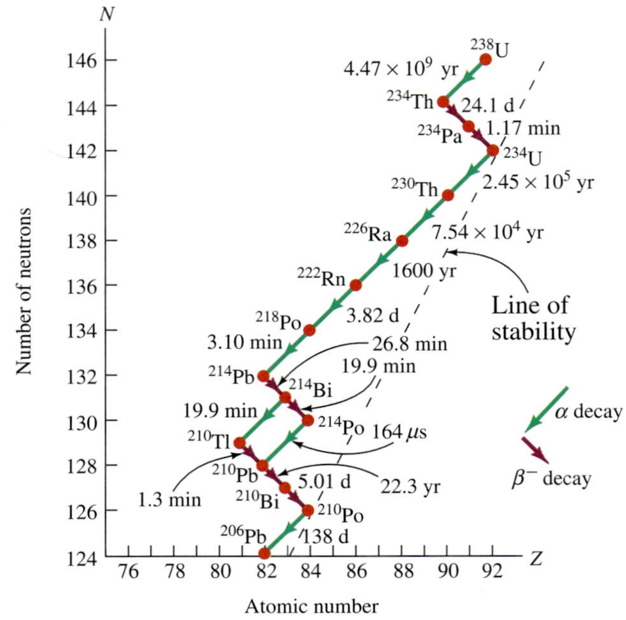

FIGURE 45–16 The decay series of ^{238}U plotted on a graph of N versus Z. Half-lives of each nuclide are also shown. Only one of the four different branchings of this series is shown.

has the form $A = 4n$, where n is an integer, then all the nuclides with $A = 4n - 4$, $4n - 8$, and so forth will be encountered until a stable nucleus is obtained. Similarly, if the starting nucleus has $A = 4n + 1$, then nuclei with $A = 4n - 3$, $4n - 7$, . . . are reached. A third series with $A = 4n + 2$ and a fourth with $A = 4n + 3$ are also possible, but the "fifth" series, with $A = 4n + 4$, is identical to the $A = 4n$ series. Thus *there are only four different radioactive series*. The series shown in Fig. 45–16 for $^{238}_{92}$U is the $A = 4n + 2$ series; a series starting with $^{232}_{90}$Th is the $A = 4n$ series.

The lifetimes of the original parent nuclei are 6.4×10^9 yr for $^{238}_{92}$U and 2.0×10^{10} yr for $^{232}_{90}$Th. The fact that these numbers are comparable to the age of the solar system explains why we can still find these nuclides in nature: If the lifetimes of the parent nuclei were shorter, all the parents would have decayed by now. Moreover, the fact that the daughter nuclei are continuously being replenished through decays of the parent nuclei explains why we find extremely short-lived nuclei in nature. For example, in the decay series starting with $^{238}_{92}$U, the sixth step produces $^{222}_{86}$Rn, and this nucleus undergoes α decay to $^{218}_{84}$Po with a half-life of 3.8 d, certainly much less than the age of the solar system. Yet $^{222}_{86}$Rn can be found by chemical means in uranium ore.

The end product of a radioactive series is a stable nucleus. The series starting with $^{238}_{92}$U includes 14 steps with 4 different branchings. In the series starting with $^{232}_{90}$Th, there are 10 steps with 1 branching. Despite the branchings, there is a unique end product in each case: the stable nuclides $^{206}_{82}$Pb and $^{208}_{82}$Pb, respectively.

Radioactivity and Life

Radioactivity has significant effects on biological systems. Some effects are desirable—medical diagnosis and treatment, for example—and some are just the opposite. Undesirable effects occur when one of the forms of radiation deposits energy in a normal living cell. That energy is typically absorbed when ionization occurs. Chemical bonds are broken, destroying needed substances in the cell. New ions may induce abnormal chemical reactions. We can distinguish three levels of damage: The chemical functioning of cellular processes may be so badly interrupted that the cell dies or is unable to reproduce; internal controls on the cell's behavior may be lost, so that the cell becomes cancerous; or the gene structure of the cell may be changed, resulting in abnormal offspring.

APPLICATION: A LEAD PAINT DETECTOR

Nuclear physics has been an important element in the environmental arena of technology for a long time. A recent example is a small, hand-held device, which, when placed close to a painted surface, will indicate whether the paint is lead-based. The device consists of a radioactive source that emits beta rays (electrons) with energies of several hundred kiloelectron-volts, surrounded by an annular silicon diode detector (Fig. 45AB–1). When this device is placed close to a surface, emitted electrons strike the surface coatings. Some of these electrons collide with and displace orbital electrons from the inner shells of the atoms in the coating. When these inner shell vacancies refill from the adjacent shells, photons (X-rays) of energies equal to the differences of shell energies are emitted. For light elements, these photons are only a few kiloelectron-volts in energy but, for lead, the characteristic photon energy is approximately 60 keV. Some of these photons travel into the silicon diode detector.

The detector consists of a large area silicon *p-n* diode (see Chapter 44) with a potential difference maintained across it in such a way that the electron and hole carriers are displaced away from the *p-n* junction toward the ohmic connections at the ends of the diode. The center body of the diode is therefore depleted of charge carriers. A photon entering this region can collide with and release an electron from the silicon. Because the binding energy is relatively small, the released primary electron has nearly the photon energy. This electron travels through the lattice, causing further ionization. The resulting cascade of electron–hole pairs diffuse to the ohmic contacts under the influence of the electric field in the depleted region. As the number of these pairs is proportional to the primary electron's energy, the total charge flow as the carriers are collected determines the energy of the photon. By counting pulses of charge from the detector that correspond to photons in the 50–70 keV energy range, the concentration of lead in the coating can be determined.

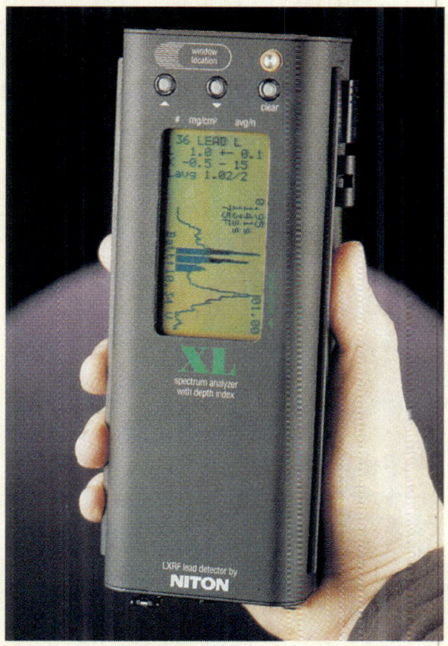

FIGURE 45AB–1 This lead paint detector is an important tool in environmental testing.

The danger of a given radioactive substance is associated with the activity, the energy of the radioactive decay products, and the amount of that energy deposited in the body. One measure of possible damage is the energy of the radiation absorbed per mass of the organism. The corresponding SI unit is the *gray* (Gy): 1 Gy ≡ 1 J/kg. A dose of several Gy over the entire human body may be lethal, and 10 Gy or more is lethal within a short period. Because damage depends on just how the energy is deposited, a more precise measure takes into account the type of radioactive decay products that produce a given dose.

Some cells, especially those that reproduce frequently, are more susceptible to radiation damage than others. Even a small dose can lead to long-term damage. In deciding whether a risk from radioactivity is worth a possible benefit, recall that we are constantly bombarded by cosmic rays, and that both the planet Earth and elements in our own bodies have some natural radioactivity. We receive a dose of some 10^{-3} Gy/yr from such sources. Exposure to radioactivity from human-made sources must be compared to this natural background. For example, an inhabitant of the United States receives on average less than half the natural dose in diagnostic X-rays, and the background radioactivity associated with nuclear power generation is only about 3×10^{-5} of this value. Nevertheless, a nuclear disaster would create a much greater risk from radiation exposure. Because we know little about the long-term effects of low-level radiation, we are conservative about the doses that are allowable; the recommended limit for whole-body exposure is 0.05 Gy/yr.

45-5 FISSION AND FUSION

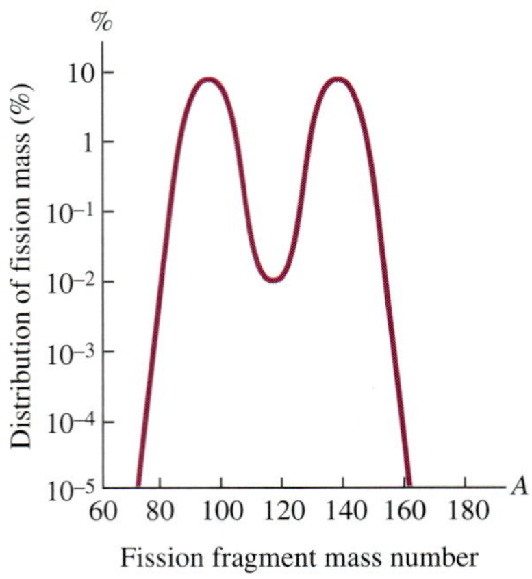

FIGURE 45–17 The relative probability of producing a fragment of atomic number A when a ^{236}U nucleus undergoes fission. When slow neutrons induce fission, a variety of daughter nuclides (fragments) are produced. Note that it is *less* likely that fragments with exactly half the size of the parent, in this case ^{236}U, are produced than that the fragments are of unequal size. (After W. E. Burcham, *Nuclear Physics*, McGraw-Hill, 1963.)

Fission

We have mentioned that α decay is a tunneling process. It is a type of **fission**, a process in which a nucleus breaks into two pieces of more or less equal size. Fission occurs in large nuclei because the energetics favor it. The liquid-drop model helps us to understand why fission occurs. When a large nucleus breaks into two fragments, the sum of the Coulomb repulsion terms from Eq. (45–14) for the two pieces is considerably less than that of the corresponding single term of the parent nucleus. The Coulomb term alone implies that the most energetically favorable fission process is one in which the parent nucleus breaks into two equal pieces. Other terms in the energy take into account detailed differences. Some nuclei are bound more strongly than others. In real fission processes, the two fragments are typically of different sizes (Fig. 45–17).

EXAMPLE 45–6 Find the difference in energies due to the Coulomb term alone between a $^{238}_{92}$U nucleus and two $^{119}_{46}$Pd (palladium) nuclei.

Solution: The difference in energies due to the Coulomb energy term, Eq. (45–14), for a parent nucleus with Z protons and A nucleons and the sum of the same terms for each of the two daughter nuclei with $Z/2$ protons and $A/2$ nucleons is

$$\Delta E = \frac{3}{5}\frac{Z^2e^2}{4\pi\epsilon_0 r_0 A^{1/3}} - 2\left[\frac{3}{5}\frac{(Z/2)^2e^2}{4\pi\epsilon_0 r_0 (A/2)^{1/3}}\right] = (1 - 2^{-2/3})\frac{3}{5}\frac{Z^2e^2}{4\pi\epsilon_0 r_0 A^{1/3}}.$$

We are interested in the case $Z = 92$ and $A = 238$. Numerical evaluation gives

$$\Delta E \simeq (1 - 2^{-2/3})(9.8 \times 10^8\ \text{eV}) = 3.6 \times 10^8\ \text{eV} = 360\ \text{MeV}.$$

This is a substantial amount of energy even for a nuclear reaction. But it is not the whole story, because other terms contribute to energy differences of this type. It does, however, give a reasonable order-of-magnitude estimate.

If the Coulomb energy destabilizes large nuclei, how can such nuclei exist? The presence of an energy proportional to the surface of a liquid drop provides a barrier against breakup. The energy term is positive and is minimized when the nucleus is spherical [Eq. (45–14)]. In Fig. 45–18, we show a sequence in which a large liquid-drop nucleus formed by neutron absorption in ^{235}U breaks into two fragments. As the fragments start to separate, the surface—and hence the surface energy—*increases* without the Coulomb energy changing very much. As the fragments separate further, the surface-energy term makes it energetically favorable for the separating fragments to "neck off." It is only when the nuclei have separated into two spheres that the surface energy no longer changes. The fact that the sum of surface and Coulomb energies initially increases as the fragments start to separate means that there is a potential barrier. This barrier is on the order of 5 MeV for nuclei such as uranium.

Because of the potential barrier, fission proceeds through quantum mechanical tunneling. This process is called *spontaneous fission*; it occurs at a slow rate. Another process, called *induced fission*, occurs when free neutrons are captured by heavy nuclei such as uranium. The resulting nucleus is an excited state, with the added neutron having a binding energy of about 5 MeV. This is about the height of the potential barrier against fission, and the new nucleus thus has enough energy to undergo fission easily. For ^{233}U and ^{235}U, the energy is carried *above* the potential barrier even for very slow neutrons, so fission is guaranteed. For ^{238}U, the neutron must also supply at least 1 MeV of kinetic energy; this will not happen if the neutrons are not sufficiently energetic. Induced fission is important in the possibility of using sustained fission for power production (we shall explore this possibility in Section 45–6).

Fusion

Heavy elements release energy during fission because the curve of binding energy per nucleon decreases as A increases for large A values (see Fig. 45–6). Conversely, this curve increases as A increases for *small* A values; light elements thereby release energy during **fusion**, the combining of small nuclei into larger nuclei. An example is the combining of free nucleons into nuclei: 2.23 MeV of energy is released when a free proton and a free neutron combine and form a deuteron, ^{2}H. More typically, the combining nuclei are each charged and, because the Coulomb force has long range whereas the nuclear force does not, each of the combining nuclei must have considerable energy in order to surmount the potential barrier due to the Coulomb force. Once the potential barrier is passed and the nuclei come close enough to fuse, the energy released by their fusion is much more than the total kinetic energy of the nuclei.

Fusion reactions are central to the "burning" processes of stars. These reactions occur in cycles of reactions or decays, the primary example of which is the *proton cycle*:

$$p + p \rightarrow {}^2H + e^+ + \nu, \quad \text{with 0.4 MeV of energy released;} \qquad (45\text{–}33a)$$

$$^2\mathrm{H} + p \rightarrow {}^3\mathrm{He} + \gamma, \quad \text{with 5.5 MeV of energy released;} \qquad (45\text{–}33b)$$

$$^3\mathrm{He} + {}^3\mathrm{He} \rightarrow {}^4\mathrm{He} + 2p + \gamma, \quad \text{with 13.0 MeV of energy released.} \qquad (45\text{–}33c)$$

The internal temperature of stars is high enough to give some colliding nuclei sufficient energy to overcome the Coulomb repulsion involved in all three of these reactions. The *net* effect of this cycle is to convert four protons into an α particle along with the emission of energy in the form of photons and a neutrino. Although the energy given to the neutrino is lost (in the sense that the neutrino does not interact further), the photons continue to heat the interior. This is the source of the energy that eventually arrives on Earth from our own star, the Sun.

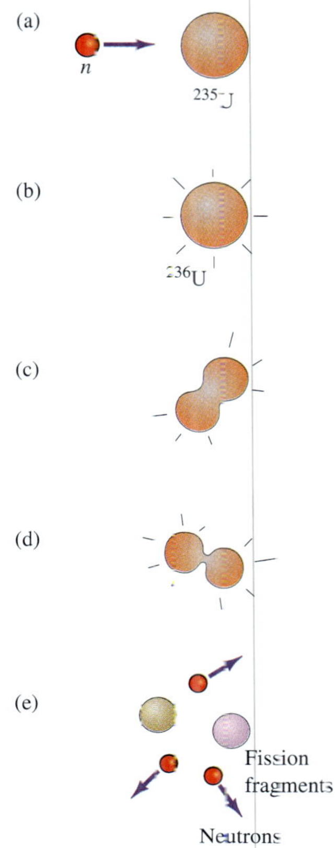

FIGURE 45–18 The fission process. (a), (b) The ^{236}U nucleus produced by neutron bombardment of ^{235}U is in unstable equilibrium, and (c)–(e) breaks into two daughter nuclei. Several free neutrons are also produced.

EXAMPLE 45–7 Calculate the total energy released each time a ^{4}He nucleus is produced in the proton cycle. To the energies described in Eqs. (45–33), add the energy released when each of the 2 positrons produced annihilates with 2 electrons (already present in the star), producing 4 photons and a kinetic energy equivalent to four times the electron mass—the positron and electron have the same mass, 0.51 MeV/c^2—namely, 2.0 MeV: $2e^+ + 2e^- \rightarrow 4\gamma + 2.0$ MeV.

Solution: For the third reaction of the cycle, Eq. (45–33c), two ^{3}He nuclei are present, so the reaction of Eq. (45–33b) must occur twice. Similarly, the reaction of Eq. (45–33a) must also occur twice. In effect, the chain of reactions that produces one ^{4}He nucleus is

$$\begin{aligned} 2(p+p) + p + p + 2e^- &\rightarrow 2({}^2\mathrm{H} + e^+ + \nu + 0.4\ \mathrm{MeV}) + p + p + 2e^- \\ &\rightarrow 2({}^3\mathrm{He} + \gamma + 5.5\ \mathrm{MeV}) + (2e^+ + 2\nu + 0.8\ \mathrm{MeV}) + 2e^- \\ &\rightarrow ({}^4\mathrm{He} + 2p + \gamma + 13.0\ \mathrm{MeV}) + (2\gamma + 11.0\ \mathrm{MeV}) \\ &\qquad + (2e^+ + 2\nu + 0.8\ \mathrm{MeV}) + 2e^- \\ &\rightarrow ({}^4\mathrm{He} + 2p + \gamma + 13.0\ \mathrm{MeV}) + (2\gamma + 11.0\ \mathrm{MeV}) + (2\nu + 0.8\ \mathrm{MeV}) \\ &\qquad + (4\gamma + 2.0\ \mathrm{MeV}) = {}^4\mathrm{He} + 2p + 7\gamma + 2\nu + 26.8\ \mathrm{MeV}. \end{aligned}$$

Notice that 2 protons are left; these protons can contribute anew to the proton cycle.

EXAMPLE 45–8 The total power output of the Sun, its *luminosity*, is about 3.9×10^{26} W. Assume that the Sun is made entirely of protons, that its luminosity remains constant, and that it "burns" protons via the proton cycle until all the protons have been converted into α particles. How long can the Sun burn in this way? You may use the results of Example 45–7.

Solution: From the results of Example 45–7, 4 protons are required to produce one α particle, so the energy released *per proton* is 26.8 MeV/4 = 6.7 MeV. We know the total mass of the Sun, 2.0×10^{30} kg, so the number of protons in the Sun is $N_p = (\text{mass of Sun})/m_p = (2.0 \times 10^{30}\ \text{kg})/(1.67 \times 10^{-27}\ \text{kg/proton}) = 1.2 \times 10^{57}$ protons. We can find the number of protons converted per second, N, from the Sun's total power:

$$N = \frac{3.9 \times 10^{26}\ \text{J/s}}{(6.7\ \text{MeV/proton})(1.602 \times 10^{-13}\ \text{J/MeV})} = 3.6 \times 10^{38}\ \text{protons/s}.$$

The time it would take for all of the Sun's protons to be burned can now be found by dividing the number of protons by their rate of use in the proton cycle:

$$\frac{1.2 \times 10^{57}\ \text{protons}}{3.6 \times 10^{38}\ \text{protons/s}} = 3.3 \times 10^{18}\ \text{s}.$$

This amounts to about 10^{11} yr, or 100 billion yr. The actual burning time for the Sun is more like 10 billion yr (half of which has passed) because only the innermost 10 percent of the Sun's mass is hot enough to burn through the proton cycle.

FIGURE 45–19 This jaw fragment has been radiometrically dated to have an age of several million years. Its close relation to the human jaw suggests that it may come from one of our earliest known ancestors.

45–6 APPLICATIONS OF NUCLEAR PHYSICS

Radiometric Dating

APPLICATION Measuring the ages of objects

In Chapter 41, we discussed how the radioactive nucleus ^{14}C can be used to date biological materials. The lifetime involved is 8268 yr, and that is the rough order of magnitude of ages that can be measured with this method. Other radioactive decays, with longer lifetimes, can be used to date longer geologic times, in the range 10^6 yr to 10^{10} yr (Fig. 45–19). For example, rocks form when liquid magma solidifies; after that point, chemically induced separation of elements is more difficult. As long as there is no lead in a sample of rock when that rock forms, any ^{206}Pb found in the rock must be the product of ^{238}U decays (for which $\tau = 4.5$ billion yr). Therefore the ratio of ^{238}U to ^{206}Pb in the rock allows us to find the rock's age. To use such a technique, we must have confidence that there was no lead in the liquid magma that formed the rock initially. We can similarly measure the percentages of radioactive intermediate products of the ^{238}U decay chain.

Radioisotopes

APPLICATION Uses of radioisotopes

Radioisotopes, unstable isotopes produced in nuclear reactions, have characteristic lifetimes for decay. Atoms with these nuclides behave chemically like stable isotopes of the same atom. Observation of the location of the decay of a radioisotope helps us to understand where a chemical or physical process has taken the radioisotope. The applications of radioisotopes include the use of ^{153}Gd, ^{67}Ga, ^{201}Pb, and ^{123}I in medical diagnosis—a cancerous tumor, for example, may concentrate one of these elements, and the observed location of α decays indicates the tumor site; ^{60}Co in cancer treatment—the nuclide can be placed in a tumor and, upon decaying, deposits large amounts of energy that kills cells there; and ^{85}Kr in leak detection—krypton can penetrate very small cracks, and decays indicate the location of those cracks.

Nuclear Power Generation

Here, we discuss two ways in which nuclei can be used for electric power generation. The first uses the fission process, and it is a proven, well-established technology. The second uses the fusion process, and it is not yet commercially feasible. In both cases, the energy released in the respective nuclear reactions is converted through collisions to thermal energy, which is then used to turn turbines and generate electric power. We shall not concern ourselves further with this aspect of the process.

Controlled Fission. It is possible to generate energy from fission because the proportion of mass in neutrons is larger for heavy nuclei such as those of uranium than for light nuclei. (In Example 45–6, we treated the decay of ^{238}U into two ^{119}Pd nuclei, but in fact the heaviest *stable* isotope of palladium has $A = 110$.) This fact, a consequence of the Coulomb repulsion of protons, implies that when a parent nucleus undergoes induced fission into two fragments, the daughter nuclei have relatively too many neutrons to be stable. They may undergo β decay, or, more typically, a few neutrons may be produced along with the two daughter fragments in fission (see Fig. 45–18e).

The induced fission process tends to occur more readily when the bombarding neutrons are slow. Although the neutrons that are *produced* in the fission process generally have too much kinetic energy to induce efficiently the fission of other uranium nuclei, their kinetic energy can be reduced by allowing them to collide within a *moderator*, a material whose atoms have light nuclei: The recoiling neutrons lose a substantial fraction of their kinetic energy in such collisions. Once this has happened, these neutrons can induce other fission processes. Of course, some neutrons do not do so, but in each fission reaction as many as 2 to 3 neutrons are produced, on average. If enough fissionable nuclei (^{233}U or ^{235}U, for example) are present, so that at least one released neutron induces another fission, the process can be sustained. We thereby have a *chain reaction* (Fig. 45–20), which takes place within a *reactor*.

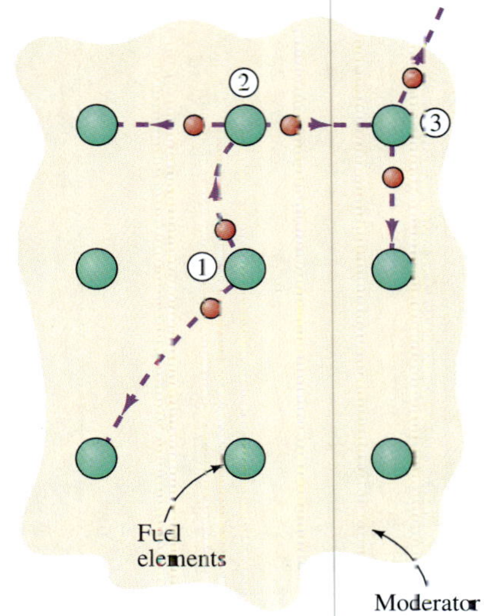

FIGURE 45–20 A chain reaction, in which the free neutrons produced in a fission process induce the fission of still other nuclei, indicated in the sequence ①, ②, ③. The moderator slows down the released neutrons and increases the likelihood that they will produce another fission.

By increasing or decreasing the relative amounts of either fissionable nuclei or moderating material in the form of "control rods," the reaction process can be either speeded up or slowed down and thus satisfactorily controlled. The problem of disposing of the waste products has proven to be more difficult to resolve. These waste products, which include the fission fragments themselves, are typically radioactive and have lifetimes that are long compared to a human lifespan. Because radioactivity can cause genetic damage and cancer, we must dispose of such waste material in such a way that it is isolated from humans and other living things for many generations.

APPLICATION

Power generation with nuclear fission

A Natural Reactor. There is evidence that sufficiently rich deposits of uranium occur in nature to have sustained a "natural" chain reaction. Such ore deposits must have not only a large concentration of uranium but also deuterium—which is contained in water—to act as a moderator. The ^{235}U-to-^{238}U ratio that exists today is not large enough for a natural reactor to operate. But because ^{235}U decays with a shorter lifetime than ^{238}U does, we know that ^{235}U was relatively more abundant in the past, and ore bodies could have sustained a chain reaction at earlier times. The remnants of a possible *natural reactor* that shut itself down after a certain amount of ^{235}U was burned have been discovered in Africa. The evidence includes a depleted supply of ^{235}U and a characteristic distribution of daughter nuclei descended from fission products.

Controlled Nuclear Fusion. Controlling the processes that power the Sun for direct power generation on Earth is an old idea. Deuterium, a hydrogen atom with a ^{2}H nucleus, which is abundant in sea water, could supply fuel for the fusion reaction, and the radioactive waste problem for nuclear fusion appears to be less acute than that for fission. But the technology of controlling fusion reactions has proved to be a difficult one to master. The essence of the problem is not simply to make nuclei collide with enough energy to overcome the Coulomb repulsion barrier and fuse, but to make such collisions occur in large numbers. A collection of nuclei could in principle be brought to energies high enough to fuse by heating them, but the temperatures required to do so correspond to energies on the nuclear scale rather than the atomic scale. We can estimate this energy to be $E \simeq e^2/4\pi\epsilon_0 R$, the Coulomb energy between two singly charged nuclei separated by a nuclear radius R, where R is on the order of 10^{-14} m. Upon substituting, we find that $E \simeq 150$ keV. This energy corresponds to a temperature of about

10^9 K, far higher than any ordinary container could withstand. Indeed, at such temperatures, atoms lose their electrons and matter breaks down to a fully ionized gas of electrons and nuclei, a *plasma*. Although in the Maxwell distribution there are enough particles with high speeds so that a temperature of some 10^8 K is still sufficient to produce reactions, this temperature is still much too high for conventional treatment.

Before we discuss current ideas of how to overcome the problem of handling hot plasma, let's discuss the reactions themselves. The most promising of these reactions is the so-called D–T reaction. The deuteron, D, is the ^{2}H nuclide, and the triton, T, is the ^{3}H nuclide. (The atom with a triton nucleus is tritium.) The D–T fusion reaction is

$$D + T \to {}^4\text{He} + n.$$

The Q value of this reaction is 17.6 MeV. A second reaction of possible interest is the D–D reaction,

$$D + D \to {}^3\text{He} + n,$$

in which a total of 4 MeV of kinetic energy is produced. In addition to the fact that more energy is produced in the D–T reaction, the cross section for the D–T reaction is almost 10 times higher than that of the D–D reaction when the energies of the colliding particles are in the range of 100 keV. These positive features largely override the negative features of tritium, which is highly radioactive and must be produced in other nuclear reactions. If the D–D reaction could be mastered, however, a virtually unlimited supply of energy would be available, because deuterium occurs so abundantly in water.

If the problem of producing and containing the hot plasma could be solved, the fusion reaction would be self-sustaining: The kinetic energy of the neutrons produced is more than enough to maintain the plasma at high temperatures even though a large fraction of that kinetic energy is used up in energy production. Two rather different approaches have been taken to address the problem of confinement of the hot plasma. The first is *inertial confinement*, a scheme in which tiny pellets of material that contains nuclei suitable for fusions are compressed by powerful laser beams or ion beams until the necessary temperatures and densities are reached (Fig. 45–21). In this scheme, there is no effort to hold the plasma confined for long periods; the beams are fired in pulses, and the pellets are replaced with each pulse. The principal difficulties of this approach are associated with understanding just how the pellets will behave when the energy-supplying beams strike them, with the construction and handling of the pellets, and with the guidance of the beams.

FIGURE 45–21 (a) The Particle-Beam Fusion Accelerator II at Sandia National Laboratories in Albuquerque, New Mexico, used for inertial-confinement experiments. (b) Schematic diagram of fusion by inertial confinement with laser beams or particle beams. D–T pellets are sent to a site where momentum from many laser or particle beams implodes them, bringing them to high enough temperatures and densities for significant amounts of fusion to occur. The neutrons from the fusion process produce tritium in the lithium shield, which also becomes hot. The tritium can be distilled out of the hot lithium and used to produce more fuel pellets. Deuterons are abundant in water. Electricity could be made from the thermal energy removed at the heat exchanger.

(a)

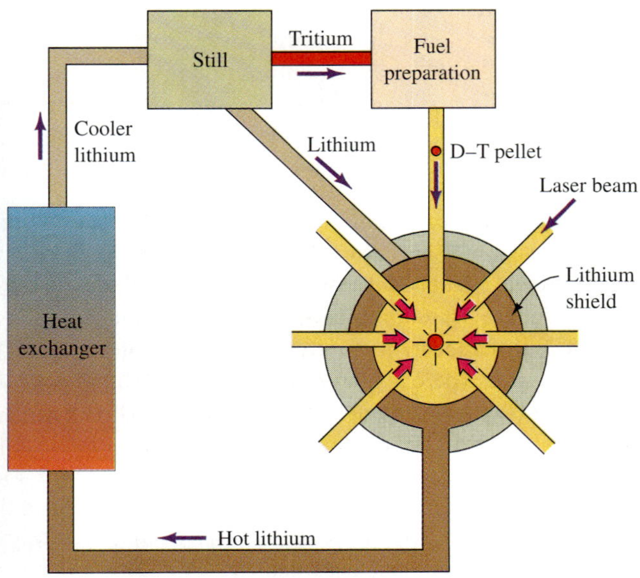

(b)

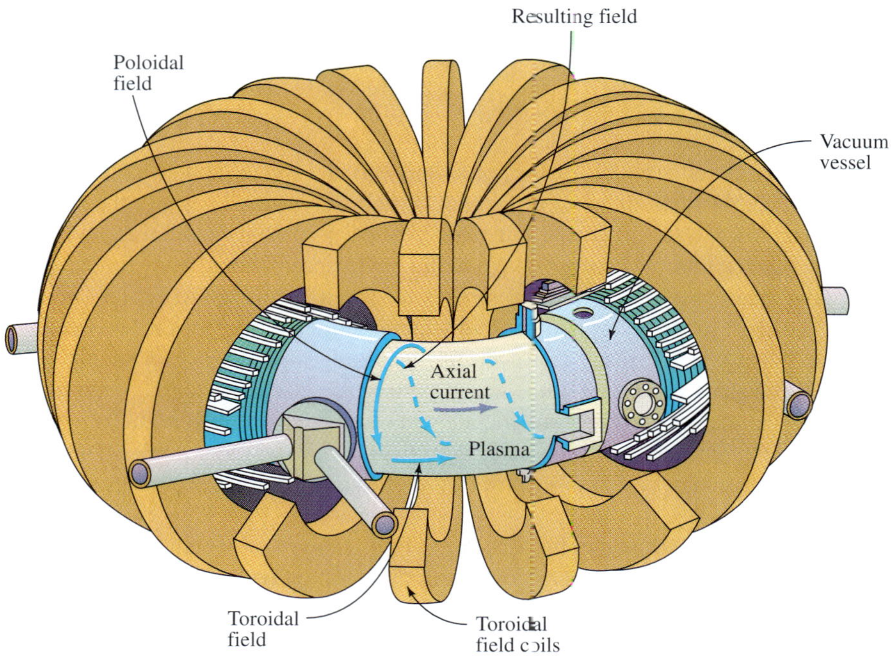

FIGURE 45–22 Schematic diagram of fusion by magnetic confinement, containment in the magnetic fields of a tokamak. The so-called poloidal field forms circles around the toroidal field. [After W. M. Stacey, Jr., "Fusion Reactor Development: A Review," *Advanced Nuclear Science and Technology*, **15**: 131 (1983).]

In the second scheme, *magnetic confinement*, the plasma, which is given additional thermal energy by the absorption of electromagnetic waves, is held in place by magnetic forces. The confinement can last for long periods. A difficulty faced in this approach is that magnetic fields do not affect the component of a charged particle's motion that is along the direction of the magnetic field lines, so the plasma tends to "squirt out" in one direction or another. To avoid this problem, complicated configurations of magnetic fields must be used. One field configuration employs magnetic mirrors like those that contain Earth's Van Allen belts (see Chapter 29). The *tokamak* is a device that attempts to hold the plasma with magnetic fields contained within a finite volume (Fig. 45–22; see the opening photograph of Chapter 30). (An example of a contained magnetic field is that of the toroidal solenoid.) In the early 1990s, large tokamaks came close to achieving self-sustaining reactions. Although both inertial confinement and magnetic confinement schemes show promise, we are still far from making fusion a commercial technology.

SUMMARY

The nucleus was discovered by means of scattering experiments that showed significant numbers of incident α particles scattered through large angles, which follows if the atom has a small, massive, positive core structure. The nucleus, held together by the nuclear force, is composed of Z protons and $N = A - Z$ neutrons, where A is the mass number. Nuclei with the same Z value but different N values are isotopes of a given atomic species. The masses of nuclei and of atoms are conveniently measured in atomic mass units, u, defined so that the mass of a ^{12}C atom is exactly 12 u. The binding energy, the energy released when a nucleus is formed from its A constituents (nucleons), is significant; a binding energy per nucleon of 7 to 9 MeV is typical of larger nuclei. This binding energy is so large that, due to the relativistic equivalence of mass and energy, the mass of a nucleus is on the order of 1 percent less than the sum of its constituent masses. To a good approximation, the constituents of the nucleus are closely packed into a sphere of radius

$$R_A = r_0 A^{1/3} \tag{45–7}$$

with $r_0 \simeq 1.2$ fm. The lifetimes of unstable nuclei, the spectrum of states, spins, and magnetic dipole moments all provide useful ways to understand nuclear behavior.

For distances greater than about 1 fm, the nuclear force between nucleons is well described in terms of the Yukawa potential,

$$U(r) = -g^2 \frac{e^{-r/R}}{r}, \tag{45–9}$$

where g^2 measures the strength of the force and R is the range, approximately 2 fm. Because we are not sure of the nuclear force at short distances, and because the strength of the nuclear force is so great that our mathematical tools for calculating its consequences are inadequate, we cannot make a full set of predictions from fundamental physical laws. However, a useful set of models has been developed to explain nuclear behavior. The shell model is an approach in which individual nucleons move in a central nuclear potential. The resulting spectrum, including the presence of magic numbers—Z and N values for which a nucleus is particularly strongly bound—can be explained by this model. The liquid-drop model treats the nucleus as a continuous medium and is able to explain behavior such as fission, the splitting of large nuclei. The semi-empirical mass formula, which is inspired by the liquid-drop model, successfully fits the masses of nuclei.

The strong binding of nuclei has important implications for whether nuclear reactions, including scattering and decay, can occur. Short-lived resonances, excited states of nuclei, are visible in cross sections for scattering, and their widths are inversely proportional to their lifetimes according to the Heisenberg uncertainty principle.

Nuclei can decay through three principal processes: β decay, α decay, and γ decay, depending on whether the decaying nucleus emits an electron, an α particle, or a γ-ray (a highly energetic photon), respectively. Collectively, these processes comprise the phenomenon of radioactivity. The rate of a single decay process is described by the equation

$$N(t) = N_0 e^{-\lambda t}. \tag{45–22}$$

Here, $N(t)$ is the number of undecayed nuclei that remain after time t, N_0 is the initial number at $t = 0$, and λ is the decay constant, which is different for each type of decay. Tunneling is an essential feature of α decay. Natural radioactivity is associated with decay chains, processes in which the products of radioactive decay are themselves radioactive and decay. The presence of such processes supplies us with useful radiometric dating techniques. Certain artificially produced radioactive nuclides have a variety of uses, including radioactive and medical applications.

The systematics of nuclear masses shows that fission, the breakup of large nuclei into smaller ones, and fusion, the combining of small nuclei into larger ones, are energetically favorable. When these processes occur, energy is released. However, both can occur only if a potential energy barrier is overcome. Fusion processes are the means by which the stars obtain energy. Fission induced by neutrons can occur in controlled chain reactions, in which a fissioning nucleus releases neutrons that can initiate other fission reactions. Such chain reactions are a commercially important source of energy.

UNDERSTANDING KEY CONCEPTS

1. Alpha particles have a charge of $+2e$ and a mass of roughly four nucleon masses. Describe experiments that would allow you to measure these quantities.
2. Neutron stars are stars that have collapsed under the influence of the gravitational force into what are essentially enormous nuclei composed mostly of neutrons. Why do these neutrons not decay?
3. In our discussion of why some nuclei undergo fission, we did not mention the effect of the energy term proportional to A in the liquid-drop model. Why not?
4. Protons and electrons each have intrinsic spins $\hbar/2$. Use the fact that the spin of ^{6}Li is $\hbar$ to show that the ^{6}Li nucleus cannot consist exclusively of protons and electrons.
5. Why is it easier to produce ions with a positive charge rather than with a negative charge from neutral atoms?
6. How can a mass spectrometer be used to determine the isotopic abundances of nuclides of an element such as oxygen?
7. Experiments to measure the inner structures of the nucleus typically use electrons as incident projectiles. Why might electrons be more useful than α particles for such experiments?

8. Of the calcium isotopes ^{39}Ca, ^{40}Ca, ^{41}Ca, and ^{42}Ca, which would you expect to have the smallest neutron separation energy, and why?
9. Of the nuclides ^{15}N, ^{16}O, and ^{17}F, which would you expect to have the largest proton separation energy, and why?
10. The mass conditions for β decay and electron capture are the same. Under what conditions is each one more likely to occur for a given nuclide?
11. Certain chemical elements tend to concentrate in bone marrow. Why should you avoid any such elements that are radioactive?
12. Why are light nuclei more efficient moderators for controlled nuclear fission than heavy nuclei are?
13. Neutrons that have been slowed down by a moderator are unable to induce rapid fission in ^{238}U nuclei. Why?
14. When we gave the (final) kinetic energies of the ^{4}He and neutron products in the D–T fusion reaction, we assumed that the collision takes place at rest. Yet we know there is a Coulomb energy potential barrier to overcome. Why was it possible for us to ignore the initial kinetic energy?
15. Most of the daughter nuclei that result from fission are radioactive. Why?
16. Why might you expect a neutron to penetrate farther into a sample of matter than an α particle of the same kinetic energy would?

PROBLEMS

45–1 Static Properties of Nuclei

1. (I) How many neutrons and how many protons do the following nuclides have: ^{9}Be, ^{13}C, ^{22}Na, ^{31}P, ^{57}Fe, ^{72}Ge, ^{107}Ag, ^{131}Cs, ^{208}Pb, ^{241}Am?
2. (I) What is the total binding energy of ^{6}Li? Its binding energy per nucleon?
3. (I) What is the Rutherford collision cross section of a 8.0-MeV α particle ($Z = 2$) on a gold nucleus ($Z = 79$) at a scattering angle of 60° (Fig. 45–23)? Express your answer in units of *barns* (1 barn $\equiv$ 10^{-24} cm^2).

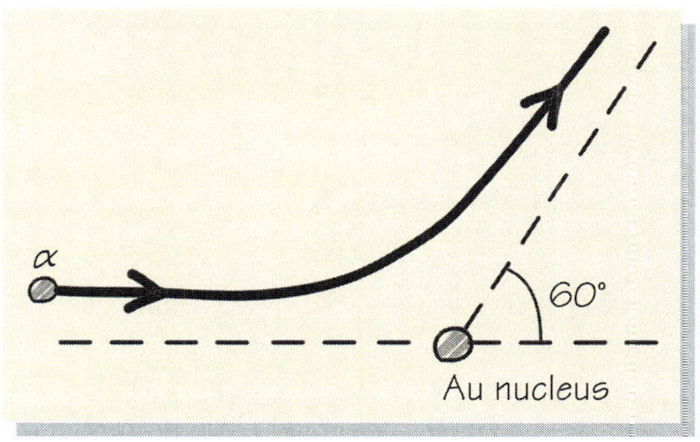

FIGURE 45–23 Problem 3.

4. (I) Consider the deflection of an α particle of kinetic energy 4 MeV by a tungsten atom, for which $Z = 74$. Estimate the typical angular deflection (a) for $R = 0.01$ nm; (b) for $R = 10$ fm (Fig. 45–24).
5. (II) An α particle with (nonrelativistic) speed v is approaching the center of a nucleus of radius R and charge $+Ze$. Calculate the distance of closest approach. Assume that the nucleus is much heavier than the α particle. For what value of v will the distance of closest approach equal R?
6. (II) Calculate the distance of closest approach that a 6.00-MeV α particle makes when it scatters head-on with an tungsten nucleus (see Fig. 45–2). Neglect recoil of the tungsten nucleus. Compare this with the sum of the nuclear radii for helium and tungsten.

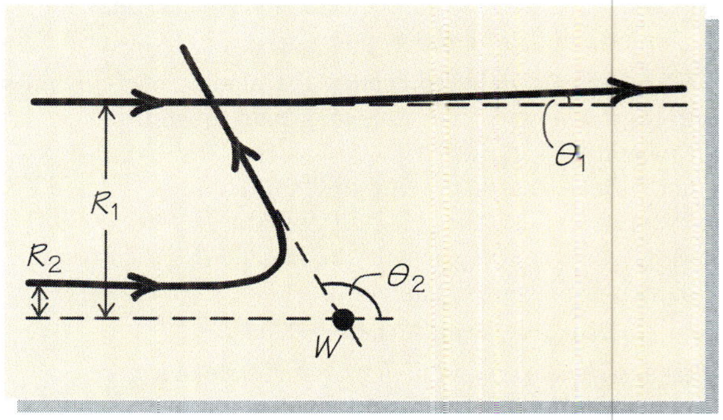

FIGURE 45–24 Problem 4.

7. (II) A proton with 6.0 MeV of energy scatters at a 60° angle from a gold atom. (a) How close did the proton come to the center of the nucleus? (b) To the surface of the gold nucleus? (Assume that the proton is pointlike.)
8. (II) The masses of the stable isotopes of calcium are 41.95863 u for ^{42}Ca, 42.95878 u for ^{43}Ca, and 43.95549 u for ^{44}Ca. Calculate the binding energies of each nuclide.
9. (II) Express the Rutherford collision cross section in terms of the momentum transfer, $\Delta\mathbf{p}$, of a scattered particle, which has an initial momentum $mv\mathbf{i}$ and a final momentum $(mv\cos\theta)\mathbf{i} + (mv\sin\theta)\mathbf{j}$ (Fig. 45–25).

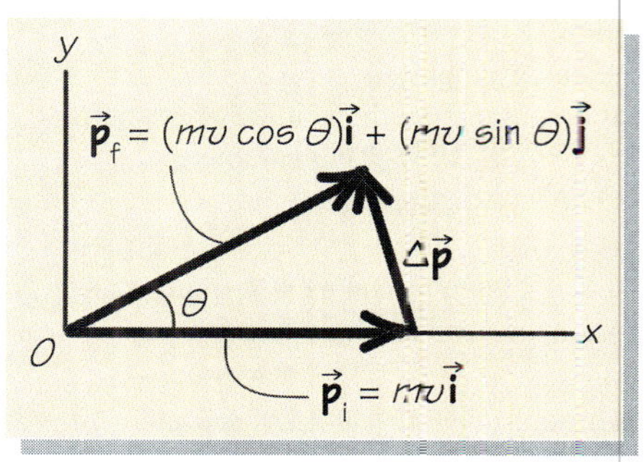

FIGURE 45–25 Problem 9.

10. (II) If we want to use an electron to probe distances to the size of a uranium nucleus, what energy should the electron have? What if we want to probe distances to the size of a nucleon, about 1 fm?

11. (II) Use the work–energy theorem to solve Example 45–1. Integrate the work over the distance from infinity to the turn-around distance R, the point at which the Coulomb force has done work equal to the initial kinetic energy of the α particle.

12. (III) *Mirror nuclei* are pairs of nuclei for which Z and N of each nucleus are opposites: For example, ${}^{23}_{11}\text{Na}$ and ${}^{23}_{12}\text{Mg}$ are mirror nuclei with $Z = 11$, $N = 12$ and $Z = 12$, $N = 11$, respectively. The idea that the *only* difference in the energies of these nuclei is due to a difference in electrostatic energy gives us another way to measure nuclear radii and in particular the constant r_0 in Eq. (45–7). If you solved Problem 69 in Chapter 25, you showed that when a charge Q is distributed uniformly throughout a spherical volume $\frac{4}{3}\pi R^3$, the electrostatic energy is $\frac{3}{5}Q^2/4\pi\epsilon_0 R$. Use this result to express the *difference* ΔE in the electrostatic energies of the pair of mirror nuclei above, and use that expression to find r_0, as defined by $R = r_0 A^{1/3}$, in terms of ΔE.

13. (III) Evaluate r_0 by comparing the binding energy and electrostatic energy of the following pairs of mirror nuclei (see Problem 45–12): ${}^{7}\text{Li}(7.016005)$ and ${}^{7}\text{Be}(7.016930)$; ${}^{9}\text{B}(9.01333)$ and ${}^{9}\text{Be}(9.012183)$; ${}^{11}\text{B}(11.009305)$ and ${}^{11}\text{C}(11.011433)$; ${}^{21}\text{Ne}(20.99385)$ and ${}^{21}\text{Na}(20.99765)$; ${}^{23}\text{Na}(22.9898)$ and ${}^{23}\text{Mg}(22.9941)$. The quantity in parenthesis is the mass of the particular nucleus in units of u.

45–2 Nuclear Forces and Nuclear Models

14. (I) (a) The energy of the first excited state of a nucleus of ${}^{70}\text{Ge}$ is 1.04 MeV. Calculate the frequency and wavelength of a photon needed to excite this state from germanium atoms at rest. (b) Do the same for the nuclide ${}^{181}\text{Ta}$, whose first excited state has an energy of 0.0062 MeV.

15. (I) Use the shell model to determine whether the following nuclides have a closed shell (magic number) for either neutrons or protons: ${}^{15}\text{N}$, ${}^{39}\text{K}$, ${}^{40}\text{Ca}$, ${}^{56}\text{Fe}$, ${}^{140}\text{Ce}$, ${}^{208}\text{Pb}$.

16. (I) Use the semiempirical mass formula to find the ratio of the surface contribution to the Coulomb contribution for the nucleus ${}^{6}\text{Li}$.

17. (II) Assume that the two protons inside a ${}^{4}\text{He}$ nucleus are approximately 4 fm apart. (a) Calculate the Coulomb force between them. (b) Use the binding energy of ${}^{4}\text{He}$ to *estimate* the nuclear force. Compare the two forces.

18. (II) Find the force that corresponds to the Yukawa potential energy, Eq. (45–9). Sketch it as a function of the separation distance; on the same plot, sketch an inverse-square force, assuming that that force corresponds to the same potential energy as Eq. (45–9) for small r.

19. (II) The energy required to assemble Z protons in a uniform distribution throughout the volume of a sphere of radius $r_0 A^{1/3}$ is $\frac{3}{5}(Z^2e^2/4\pi\epsilon_0 r_0 A^{1/3})$. The parameter r_0 is roughly 1.2×10^{-15} m. What fraction of the mass of a carbon nucleus does this energy represent?

20. (II) Use the semiempirical mass formula to determine the neutron, proton, and α-particle separation energies for the nuclide ${}^{56}\text{Fe}$.

21. (II) Use the semiempirical mass formula to find the Z value that minimizes the total mass $M(A, Z)$ for fixed A by solving $\partial M/\partial Z = 0$.

22. (III) Use the semiempirical mass formula to determine the relation between A and Z such that spontaneous fission into two identical fragments occurs; that is, for which $M(A,Z) - 2M(A/2,Z/2) > 0$.

45–3 Energetics of Nuclear Reactions

23. (II) The first nuclear reaction ever observed, by Ernest Rutherford, was the reaction $\alpha + {}^{14}\text{N} \rightarrow p + {}^{17}\text{O}$. If 5-MeV α particles are used to bombard ${}^{14}\text{N}$ at rest, determine the sum of the kinetic energies of the two outgoing particles from the atomic masses in Table 45–2 [$M({}^{17}\text{O}) = 16.999131$ u].

24. (II) The α particle Rutherford used to initiate the first observed nuclear reaction (see Problem 23) had to overcome the Coulomb barrier between the protons in ${}^{4}\text{He}$ and those in ${}^{14}\text{N}$. Assume that the α particle must come within 1 fm of a proton in the target nucleus for a nuclear reaction to occur. Calculate the minimum energy the α particle must have to initiate the reaction. Compare your result with the 5.26-MeV energy for the α particle in Rutherford's experiment. Was the reaction likely?

25. (II) One way to produce a beam of energetic neutrons is to bombard ${}^{3}\text{H}$ with accelerated ${}^{2}\text{H}$. The reaction ${}^{3}\text{H} + {}^{2}\text{H} \rightarrow {}^{4}\text{He} + n$ releases 17.6 MeV. What is the energy of a neutron produced by a 2 MeV ${}^{2}\text{H}$ beam, assuming that the ${}^{4}\text{He}$ nucleus moves in the direction of the ${}^{2}\text{H}$ beam (Fig. 45–26)?

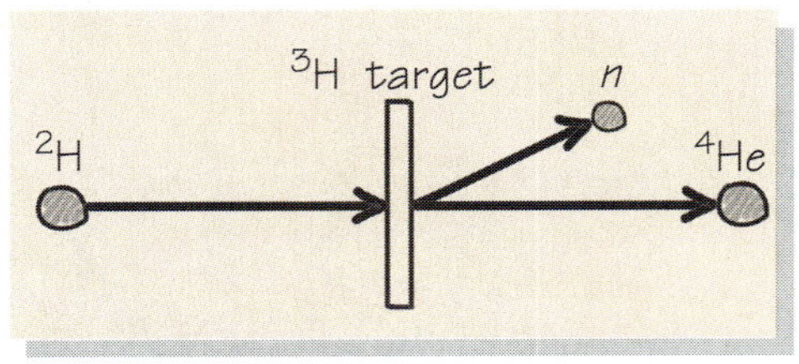

FIGURE 45–26 Problem 25.

26. (II) Nucleons are not static, but move within nuclei. The approximate energy of a nucleon inside a nucleus is about 20 MeV. Use this kinetic energy and the uncertainty principle to *estimate* the dimensions of a sphere that might contain the nucleon. Is this size approximately equal to that of a nucleus? Interpret your result.

27. (II) The nuclear reaction James Chadwick used in his experiment to identify the neutron was $\alpha + {}^{11}\text{B} \rightarrow n + {}^{14}\text{N}$. Assume that the α particle had energy 5.3 MeV and that the ${}^{11}\text{B}$ was at rest; $M({}^{11}\text{B}) = 11.009305$ u. If the kinetic energy of the ${}^{14}\text{N}$ is 0.8 MeV, what energy would the neutron have?

28. (II) Is the nuclear reaction ${}^{16}_{8}\text{O} + {}^{24}_{12}\text{Mg} \rightarrow {}^{40}_{20}\text{Ca}$ possible if the nuclei in the initial state are at rest? If so, how much energy is released? If not, how much extra energy is needed in the form of kinetic energy?

29. (II) The most frequently used laboratory neutron source works as follows: A source of ${}^{226}\text{Ra}$ emits 4.78 MeV α particles that are used to bombard beryllium. Neutrons are produced in the reaction ${}^{9}_{4}\text{Be} + {}^{4}_{2}\text{He} \rightarrow {}^{12}_{6}\text{C} + n$. The atomic mass of ${}^{9}_{4}\text{Be}$ is 9.012183. Find the other masses from Table 45–1 and use these to estimate the maximum kinetic energy of the neutrons.

30. (II) Which of the following reactions are not possible, and why not: (a) $\alpha + {}^{57}\text{Fe} \rightarrow n + {}^{60}\text{Ni}$; (b) $p + {}^{58}\text{Ni} \rightarrow {}^{3}\text{He} + {}^{57}\text{Co}$; (c) $n + {}^{157}\text{Gd} \rightarrow d + {}^{156}\text{Eu}$; (d) ${}^{241}\text{Am} \rightarrow \alpha + {}^{237}\text{Pu}$?

31. (III) A foil made of 20 mg of ^{58}Co (metallic cobalt) is exposed for 1 h to a beam of neutrons with a flux of 5×10^{13} neutrons/cm^2 s. If the *capture cross section* (the reaction ${}^{58}\text{Co} + n \rightarrow {}^{59}\text{Co}$) for such neutrons is 1700 barns (see Problem 3), how many ^{59}Co nuclei are formed? [*Hint*: Find the effective area taken up by the target cobalt atoms.]

45–4 Radioactivity

32. (I) Show that the primary β-decay process, Eq. (45–26), is possible only when the atomic mass of the parent nucleus is greater than the atomic mass of the daughter nucleus.

33. (II) Show that positron emission, Eq. (45–29), is possible only when $M_P > M_D + 2m_e$, where M_P is the atomic mass of the parent nucleus and M_D is the atomic mass of the daughter nucleus.

34. (I) Show that electron capture is possible only when the atomic mass of the parent nucleus is greater than the atomic mass of the daughter nucleus.

35. (I) The number of unstable nuclei $N(t)$ present at time t diminishes with time according to the differential equation

$$\frac{dN(t)}{dt} = -\frac{1}{\tau}N(t),$$

where τ is the lifetime. Show by direct substitution that the solution to this equation is indeed $N(t) = N_0e^{-t/\tau}$.

36. (I) What is the activity of a 1-kg sample of ^{235}U? The half-life of ^{235}U is 7.04×10^8 yr.

37. (II) The upper 0.3 m of soil contains on the average about 10^3 kg of uranium and 3×10^3 kg of thorium per square mile. Calculate the amount of radon gas formed as the product of the radiactive decay of these elements. Give your results in cm^3/yr at standard temperature and pressure. (Radon-220 is formed from thorium-232 with a half-life of 1.39×10^{10} yr, and radon-222 is formed from uranium-238 with a half-life of 4.51×10^9 yr. The half-life of intermediate nuclides is much shorter).

38. (II) The theory of α decay as a tunneling process allows us to predict an experimental relation between the half-life, $t_{1/2}$, and the kinetic energy, K, of α particles emitted during radioactivity. An approximate version of this prediction is

$$\ln t_{1/2} = \frac{Z(3.97\ \text{MeV}^{1/2})}{\sqrt{K}} - 123,$$

where K is in MeV. (a) Fill in the missing values in the following table:

Nucleus	*K(MeV)*	$t_{1/2}$
^{215}Ra	?	1.6 ms
^{223}Th	7.7	?
^{244}Pu	?	8×10^7 yr
^{248}Cm	5.16	?

(b) In order to perform an experiment, we need a source of α particles with a half-life of at least 5 days. What is the highest-energy source we are likely to find?

39. (II) By looking up the masses of the neutron, proton, and electron in Appendix II, find the maximum energy, in MeV, of an electron produced in the decay of a neutron.

40. (II) In a sample of $^{211}_{86}$Rn nuclei, 26 percent decay by α decay to $^{207}_{84}$Po and the rest decay by electron capture to $^{211}_{85}$At. What is the ratio of the lifetime for α decay, τ_α, to the lifetime for electron capture, τ_β?

41. (II) The half-life of ^{14}C is 5730 yr and the tissues of organisms accumulate this isotope from the atmosphere while the organisms are living. The skeleton of a mammoth is found to have 15 percent as much ^{14}C as the atmosphere has. When did the mammoth live? Assume that the concentration of atmospheric ^{14}C does not change.

42. (II) The age of marine sediments can be determined by their thorium-230 content. ^{230}Th is the decay product of ^{234}U and it decays to ^{220}Ra with a half-life of 75,200 yr. Measurements indicate that the ^{230}Th content of a sediment is 18 times larger at the surface than at a depth of 120 m. What is the deposition rate at that location in mm/g?

43. (II) ^{80}Br is one of a small number of nuclides that is able to decay by all three β-decay processes. Show that this is possible. $M(^{80}\text{Br}) = 79.918528$ u, $M(^{80}\text{Kr}) = 79.916376$ u, $M(^{80}\text{Se}) = 79.916521$ u.

44. (II) Suppose that the lifetime for γ decay of a nucleus at rest is so long that, for all practical purposes, the natural width of the spectral line can be ignored (that is, the γ energy takes on a single, fixed value). When the nucleus is part of a collection of atoms at finite temperature, it moves randomly, sometimes toward and sometimes away from a γ-ray detector, and the spectral line width is determined by the Doppler shift of the emitted γ-rays. This phenomenon is known as *Doppler broadening*. Estimate the width due to Doppler broadening for these atoms at a temperature $T = 800$ K.

45. (III) A radioactive nucleus a decays with a lifetime τ_1 into nucleus b, which decays with a lifetime τ_2 into nucleus c. The change in the number of nuclei a obeys the relation $dN_a = -(N_a/\tau_1)\,dt$ (see Problem 41–49). The number of nuclei b does not obey a simple equation of this type, because those nuclei are continuously being depleted through decays into nuclei c while they are replenished through decays of nuclei a. Show that the correct expression for the change in the number of nuclei b is $dN_b = -(N_b/\tau_2)\,dt + dN_a$. How will this expression generalize to a chain of decays $a \rightarrow b \rightarrow c \rightarrow d \ldots$ with lifetimes $\tau_1, \tau_2, \tau_3, \tau_4, \ldots$, respectively (Fig. 45–27)?

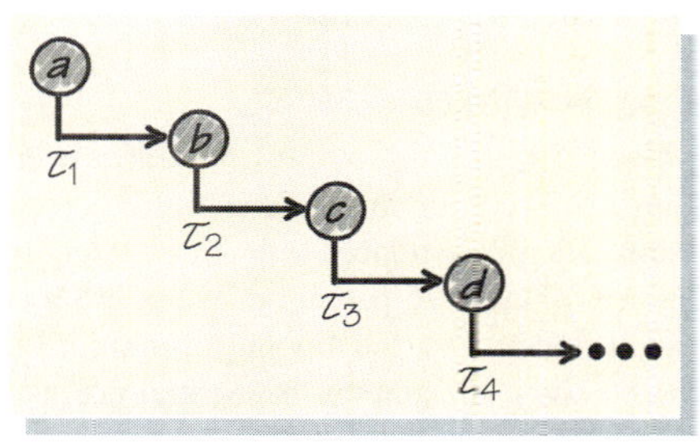

FIGURE 45–27 Problem 45.

46. (III) We have a chain of decays in a three-level system of nuclei a, b, c with $M_a > M_b > M_c$. Level a decays both to level b (with lifetime τ_{ab}) and level c (with lifetime τ_{ac}), while level

b decays to level c (with lifetime τ_{bc}). Thus the number of nuclei at level a at time t obeys the differential equation

$$\frac{dN_a(t)}{dt} = -\frac{1}{\tau_{ab}} N_a(t) - \frac{1}{\tau_{ac}} N_a(t) \equiv -\frac{1}{\tau_a} N_a(t).$$

(a) Show that the solution to this equation is $N_a(t) = N_0 e^{-t/\tau_a}$. (b) Show that the number of nuclei of level b at time t obeys the differential equation

$$\frac{dN_b(t)}{dt} = \frac{1}{\tau_{ab}} N_a(t) - \frac{1}{\tau_{bc}} N_b(t).$$

(c) Show by substitution that the number of level b nuclei is

$$N_b(t) = N_{0a} \frac{\tau_a \tau_{bc}/\tau_{ab}}{\tau_{bc} - \tau_a} (e^{-t/\tau_{bc}} - e^{-t/\tau_a}).$$

(d) Show that there is no need to write an equation for N_c once N_a and N_b are known as functions of time, because $N_c(t) = N_{0a} - N_a(t) - N_b(t)$ if we start with only a collection of N_{0a} nuclei at time $t = 0$.

47. (III) Show that when the rate of decay from nucleus a in Problem 46 is much smaller than the rate of decay from nucleus b to nucleus c, then after a time that is long compared to τ_{bc}, $N_b(t) \simeq (\tau_{bc}/\tau_{ab})N_a(t)$. (You may use the results derived in Problem 46.)

45–5 Fission and Fusion

48. (I) The energy of nuclear weapons is expressed in terms of megatons of TNT, where 1 megaton of TNT = 4.3×10^{15} J. What minimum mass of ^{235}U is required to produce a nuclear weapon of 100 kilotons? [*Hint*: Each fission of ^{235}U produces about 200 MeV of energy.]

49. (II) The temperature of the interior of stars is about 10^7 K. (a) Estimate the typical kinetic energy of the colliding protons that participate in the proton cycle. (b) Compare your answer to part (a) with an estimate of the kinetic energy that protons initially distant from one another must have to come within 1 fm of each other, a distance at which the strong forces can lead to fusion.

50. (II) Calculate the difference between the Coulomb energy for a parent nucleus (Z, A) and the sum of the Coulomb energies of two daughter nuclei (fZ, fA) and [$(1-f)Z$, $(1-f)A$], respectively, where f is some fraction between 0 and 1. Show that this energy difference is maximized when $f = \frac{1}{2}$; that is, from the Coulomb energy terms alone it is energetically favorable for fission to produce two equal-sized fragments.

45–6 Applications of Nuclear Physics

51. (II) A 440-MW (electrical output) nuclear power plant uses ^{235}U as its fuel. The uranium fuel contains 6 percent ^{235}U, and each fission reaction produces 200 MeV of energy. Assume that the power plant is 25 percent efficient in producing electric energy. (a) Calculate the amount of uranium fuel used in 1 yr. (b) Determine the amount of thermal energy released to the environment in that time. (c) How many fission events occur per second?

52. (II) One of the reactions in the proton cycle is Eq. (45–33b), $^2\text{H} + p \rightarrow {}^3\text{He} + \gamma$. Calculate the energy released in this reaction. Use the data given in Table 45–1, but take into account that atoms inside burning stars are completely stripped of their electrons.

General Problems

53. (I) The *carbon cycle*, a stellar burning process secondary to the proton cycle, consists of the following sequential series of nuclear reactions and decays:

$^{12}\text{C} + p \rightarrow {}^{13}\text{N} + \gamma$, with 1.9 MeV of energy released;
$^{13}\text{N} \rightarrow {}^{13}\text{C} + e^+ + \nu$, with 1.2 MeV of energy released;
$^{13}\text{C} + p \rightarrow {}^{14}\text{N} + \gamma$, with 7.6 MeV of energy released;
$^{14}\text{N} + p \rightarrow {}^{15}\text{O} + \gamma$, with 7.4 MeV of energy released;
$^{15}\text{O} \rightarrow {}^{15}\text{N} + e^+ + \nu$, with 1.7 MeV of energy released;
$^{15}\text{N} + p \rightarrow {}^{12}\text{C} + {}^4\text{He}$, with 5.0 MeV of energy released.

(a) What is the net effect of one cycle, and how much energy is produced per cycle? (b) Consider a star of mass 3×10^{30} kg, consisting mainly of hydrogen, with 0.1 percent of its mass in ^{12}C. The characteristic duration of a carbon cycle in the star is 5×10^6 yr. Estimate the energy produced each year by the carbon cycle, assuming that each carbon nucleus in the star acts as a catalyst for the cycle.

54. (II) (a) Use the semiempirical mass formula to determine an expression for the binding energy of a nucleus. (b) Apply your result to calculate the binding energy of ^{12}C.

55. (II) A fissioning ^{235}U nucleus produces two fission fragments, ^{139}Cs and ^{93}Rb (Fig. 45–28). Assume that the two fragments are spherical in shape and that the charge is uniformly distributed in both of them. What is the Coulomb potential energy of the system if the two fragments are just touching?

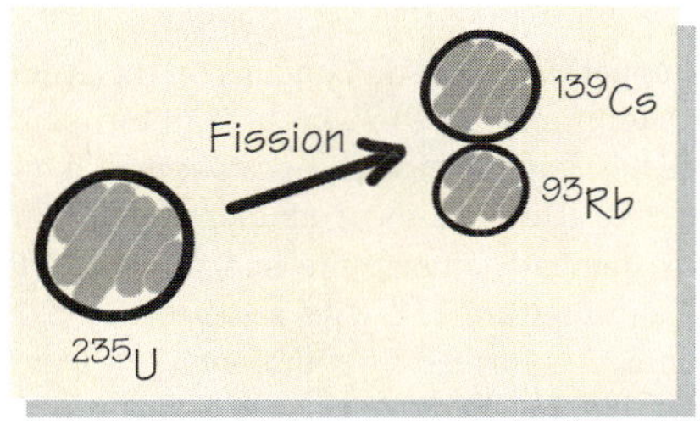

FIGURE 45–28 Problem 55.

56. (II) Use the semiempirical mass formula to find the neutron separation energy for the nuclides ^{16}O and ^{17}O. Explain the difference.

57. (II) Two protons within a star, each with energy 12 keV, collide head-on. What is the distance of their closest approach?

58. (II) A bone containing 20 g of carbon emits β particles from ^{14}C at the rate of 5.8 per second. At the time the ^{14}C is taken from the air and incorporated into bone, the activity of natural carbon due to its ^{14}C content is 15 Bq/mol. The lifetime of ^{14}C is 8268 yr. Estimate the bone's age.

59. (II) Two identical and neutral helium atoms approach each other. Because there is no Coulomb repulsion between them, they should be able to get very close together, close enough for the nuclei to touch. (a) What is wrong with this argument? (b) Given that the nuclear radius of helium is about 2 fm and that the center of mass of two approaching helium atoms is at rest, what energy must the helium atoms have for the nuclei to touch?

60. (II) A beam of neutrons of kinetic energy 1.5 GeV is produced through the bombardment of nuclei by protons. What distance does this beam travel before 5 percent of the neutrons have decayed (Fig. 45–29)? [*Hint*: The neutrons are relativistic, and time-dilation effects are important.]

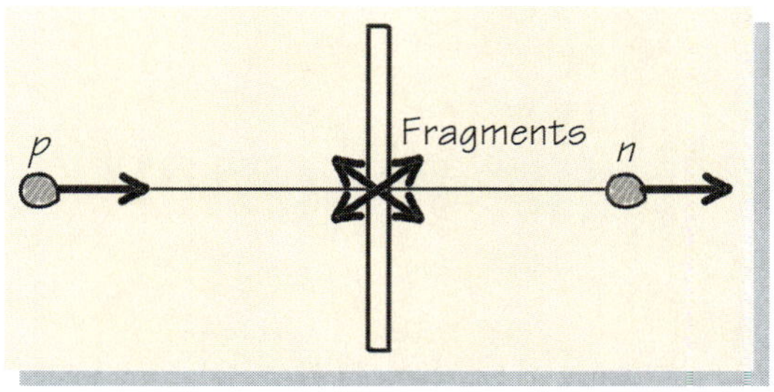

FIGURE 45–29 Problem 60.

61. (II) One of the early suggestions about the composition of the nucleus was that it contains A protons and $A - Z$ electrons. Use the uncertainty principle to estimate that if an electron is confined to a sphere with a radius of about 10^{-14} m, the size of a nucleus, then its momentum must be larger than about 10^2 MeV/c. Given that β rays emitted by the nucleus have energy on the order of 1 MeV, this estimate rules out the possibility that there are electrons in the nucleus.

62. (II) Calculate the overall kinetic energy produced in the carbon cycle (see Problem 53) after the two positrons produced in the cycle have annihilated against electrons.

63. (II) The energy of a collection of noninteracting fermions of mass m with number density n in a volume V is given by $E = V(\hbar^2\pi^3/10m)(3n/\pi)^{5/3}$. Use this expression to calculate the energy of the protons, E_p, in a nucleus in terms of A, Z, and r_0, where r_0 is the nuclear radius scale in Eq. (45–7). Repeat your calculation for the energy of the neutrons, E_n.

64. (III) Starting with the results of Problem 63, sum the proton and neutron energy terms, using the variables A and $\tau \equiv N - Z$. You will want to make the substitutions $Z = (A - \tau)/2$ and $N = (A + \tau)/2$. Approximate your result for $\tau << A$ by using $(1 + x)^p \simeq 1 + px + \frac{1}{2}p(p - 1)x^2$, an approximation good for $x << 1$. Show that this result has one term proportional to A and another term proportional to τ^2/A. Compare the coefficients of these two terms with the corresponding terms in the semiempirical mass formula. What does this tell you about the model of a nucleus as a collection of noninteracting protons and neutrons in a box?

CHAPTER

46

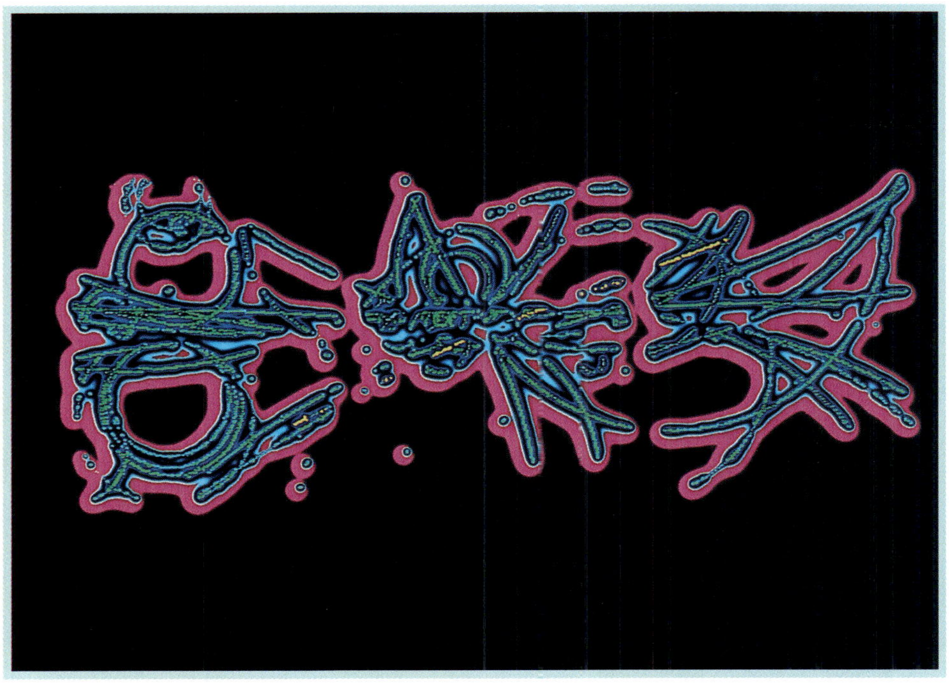

This enhanced image records the discovery of a fundamental carrier of the electroweak force, the Z particle, at CERN (Geneva) in 1983. The Z particle is produced by the collision of a proton–antiproton pair. Its tell-tale signature is an electron–positron pair (in gold) moving back-to-back away from the collision point. The remaining tracks represent other debris from the collision.

Particles and Cosmology

We believe that the same physical principles govern the interior of a proton and the far reaches of the universe. Because both realms are at the edges of our knowledge of the fundamental laws of physics, they act as proving grounds for our current ideas about those laws. But these realms converge in a more profound way. Particle physicists study the rules that govern the behavior of matter at the shortest distances we can examine. Astronomers attempt to see the behavior of the cosmos at the farthest distances. Because the speed of light is finite, when we look at a very distant object, we see the radiation that object emitted long ago: Looking at distant objects is equivalent to looking back in time. There is compelling evidence for the Big Bang theory, according to which the universe expanded from a tiny volume some 15 billion years ago. Thus our efforts to see back to the era when the universe began return us to the questions of the behavior of matter at short distances.

46–1 LEVELS OF MATTER

A "Russian doll" is a set of nested dolls; each but the smallest can be opened to reveal a smaller version of itself. The investigation of the structure of matter and of the forces that hold it together can be compared to the opening of a Russian doll. As the investigators of any given era have looked more closely at what they regarded as the fundamental constituents and interactions of matter, they have found—at smaller scales within that matter—other constituents interacting through other forces. The constituents and interactions at larger scales can be explained in terms of the physics at smaller scales. The first real step in "opening the Russian doll" was made by John Dalton in

1808. He knew that when chemical elements combine chemically, they do so in precise proportions, and he proposed that the simplest explanation for this phenomenon is that a variety of different types of atoms exist, one for each chemical element. The atoms of the chemical elements supply the building blocks for the construction of molecules. The interatomic forces that supply the "glue" for this construction are electrical in origin.

But atoms are not truly indivisible. We saw in Chapter 42 how an atom is constructed from electrons and a nucleus. The structure of the nucleus and of electrons represents the Russian doll within an atom. The atom is held together by electromagnetic forces between the electrons and the nuclei.

We have seen that nuclei have structure. They consist of nucleons—protons and neutrons. The nucleus is held together by the nuclear, or strong, force. This force differs from the electromagnetic force in that it has finite range. And the energy required to remove a neutron or proton from a given nucleus is so large that relativistic effects—those that relate energy to mass—are significant for the first time in our brief overview of the structure of the matter.

Hadrons

Do electrons and nucleons have structure? If so, what forces hold those structures together? As far as we know today, there is no evidence to suppose that electrons are composed of more fundamental constituents. This is not the case for nucleons and other strongly interacting particles, known collectively as **hadrons**. Scattering experiments carried out with particle accelerators have provided us with powerful evidence regarding their structure and the nature of the interactions involved.

Scattering

How do we determine the shape and size of an object in everyday life? We look at it. This means that we observe light (photons) bouncing off the object. Our eyes serve as a detector for observing a pattern of reflected light.

What prevents us from seeing atoms in this way? Light diffraction limits our ability to discern details of small objects, as we learned in Chapter 39. Resolution of details in an object of size d requires light of wavelength $\lambda \leq d$. Thus to study atoms in this way, we would need light of short wavelength. Because of the relation $\lambda = h/p$ [Eq. (41–1)], our light consists of photons with large momentum. The fact that matter has wavelike properties extends this limitation to the case in which we scatter massive projectiles—electrons, for example.

In Chapter 8, we discussed the connection between the momentum (or energy) of a projectile and the size of the target that can be studied. Let's review that relation, and use our knowledge of diffraction patterns this time. The characteristic angular size of the diffraction pattern made by a projectile of wavelength λ scattering from an object of size d is $\sin\theta \simeq \lambda/d$. The size of the momentum transfer in a scattering experiment is characteristically $\Delta p \simeq p \sin\theta$. Therefore

$$\Delta p \simeq p \sin\theta \simeq p\frac{\lambda}{d} = \frac{h}{d}, \tag{46–1}$$

where we have used the de Broglie momentum–wavelength relation, $p = h/\lambda$ in the last step. Thus an increasing momentum is needed to study ever smaller objects. In the laboratory frame (a moving projectile and a fixed target particle), much of the energy of the system is contained in the motion of the center of mass, which, because of momentum conservation, continues unaltered after the collision. This energy is unavailable for probing the target. *Thus, Eq. (46–1) actually applies to the center-of-mass momentum and momentum transfer.*

The Heisenberg uncertainty relations are discussed in Chapter 41.

The diffraction limit expressed in Eq. (46–1) is essentially contained in the Heisenberg uncertainty relations. Recall that the momentum–position uncertainty relation is $\Delta x\, \Delta p > \hbar$, Eq. (41–7), where $\hbar = h/2\pi$. If we interpret d as the position uncertainty Δx, then the connection is evident.

Let's look at some numerical aspects of Eq. (46–1). An atomic radius is approximately 10^{-10} m. The nucleus is roughly 10^5 times smaller, and we can take the size of a single proton to be characteristically 10^{-15} m. If we want to look to within a region of radius 1 percent of the size, we want $d \simeq 10^{-17}$ m. According to our relation, this requires us to have a momentum transfer of magnitude $\Delta p \simeq 6 \times 10^{-17}$ kg·m/s in our collider. This number corresponds to a large accelerator. The highest-energy accelerator currently operating is that of Fermilab, near Chicago, and it circulates protons of momenta ten times this size. In the Large Hadron Collider (LHC) scheduled to be built near Geneva, Switzerland through the first decade of the twenty-first century, the colliding protons will have 10 times this momentum. With the help of Eq. (46–1), we see that the distance to be probed at the LHC is on the order of $d \simeq 10^{-19}$ m.

EXAMPLE 46–1 The proton has a radius of approximately 1.2 fm. You want to use electrons as projectiles to study the structure of protons on a scale of 0.5 to 1.2 fm. Assuming that the maximum observable electron-scattering angle in the center-of-mass system is 30°, what center-of-mass momentum must the incident electrons have? Is it necessary to use relativistic kinematics?

Solution: Equation (46–1) gives us a relation between the incident momentum and the scale d that can be probed, namely,

$$p > \frac{h}{d \sin \theta}.$$

The largest value of p is determined by the smallest value of d, namely, $d = 0.5$ fm:

$$p \simeq \frac{6.63 \times 10^{-34}\ \mathrm{J \cdot s}}{(0.5\ \mathrm{fm})(\sin 30°)} = \frac{6.63 \times 10^{-34}\ \mathrm{J \cdot s}}{(0.5\ \mathrm{fm})(0.5)} \frac{1\ \mathrm{fm}}{10^{-15}\ \mathrm{m}} = 2.7 \times 10^{-18}\ \mathrm{kg \cdot m/s}.$$

Thus

$$pc = (2.7 \times 10^{-18}\ \mathrm{kg \cdot m/s})(3.0 \times 10^8\ \mathrm{m/s})$$
$$= (8.0 \times 10^{-10}\ \mathrm{J}) \frac{1\ \mathrm{eV}}{1.6 \times 10^{-19}\ \mathrm{J}} = 5.0 \times 10^9\ \mathrm{eV},$$

or 5.0 GeV. Because $m_e c^2 \simeq 0.5$ MeV for an electron, the value of pc required is some 10,000 times the rest mass, and the motion is highly relativistic.

Note that an electron with 5 GeV of energy in the center-of-mass reference frame is an electron with much higher energy in the laboratory reference frame, namely about 53 GeV, or 53,000 MeV (see Problem 8).

Features at Different Levels

The scattering experiments we have described are the tools by which we probe the different levels of the structure of matter. These levels have not turned out to be monotonous repetitions of one another—the same Russian doll does not emerge at each level. Two important features appear for the first time at the level of the structure of protons, the subnuclear level. Relativity dominates, and the constituents are confined. Let's look at these features in turn.

Relativity. We saw in Chapter 45 that even at the level of nuclear structure, the effects of relativity cannot be ignored. At the subnuclear level, relativity is crucial: The mass of a proton in its first excited state (a "proton resonance") is some 30 percent higher than that of the proton.

Similarly, the center-of-mass energies of the projectiles required to study proton structure are highly relativistic. A proton's radius is roughly 10^{-15} m. Suppose that we want to look at a region whose radius is 1/100 of this value. Then, with $\Delta r \simeq 10^{-17}$ m, we estimate a momentum transfer $\Delta p = \hbar/\Delta r$. If we use an electron as a projectile and assume that the momentum transfer is of the same order of magnitude as the initial momentum, the electron projectile must have a (relativistic) center-of-mass energy $E \simeq (\Delta p)c \simeq (\hbar/\Delta r)c$. With $\hbar c \simeq 200$ MeV·fm (see Appendix II), $E \simeq 20$ GeV. By comparison with the electron's mass, whose energy equivalent is about 0.5 MeV, our projectile is ultrarelativistic. An energy of 20 GeV is also much larger than the energy that corresponds to the proton's mass, on the order of 1 GeV.

Once the energies required in our "microscope" become so large, there is an immediate complication: Additional particles can be created from the incident energy (consistent with conservation laws, such as that of charge). In particular, we recognize the presence of antiparticles such as antiprotons, which are like their counterpart particles except that all the intrinsic quantum numbers—electric charge, for example—have the opposite sign. Particle–antiparticle pairs can always be created if the excess energy is sufficiently large, and the final state of a scattering experiment may contain many particles.

EXAMPLE 46–2 We wish to collide a pair of electrons head-on in a *symmetric collider* (an accelerator in which beams of equal energy meet head-on) to produce a proton-antiproton ($p\bar{p}$) pair. Given that the proton mass is 938 MeV/c^2, find the magnitude of the minimum momentum of either of the colliding electrons. The reaction takes the form $e^- + e^- \rightarrow e^- + e^- + p + \bar{p}$. Ignore the electron mass throughout.

Solution: We must satisfy conservation of both energy and momentum. The minimum momentum corresponds to a minimum energy; this is just enough energy to account for the rest mass of all the final particles, with nothing left over for kinetic energy. The outcome is a pair of electrons and a $p\bar{p}$ pair at rest in the laboratory reference frame, which is also the center-of-mass frame for a collider. At the same time, we automatically satisfy momentum conservation because the initial momentum of the pair of electrons is zero for a symmetric collider and, because all the final-state particles are at rest, this is the momentum of our minimum-energy final state as well.

Let the magnitude of the minimum incoming momentum of each of the two electrons be p_{min}. The total initial energy is then $2\sqrt{p_{\text{min}}{}^2c^2 + m_e^2c^4}$. This energy must equal the energy of a pair of electrons and a $p\bar{p}$ pair, all at rest:

$$2\sqrt{p_{\text{min}}{}^2c^2 + m_e^2c^4} = 2m_ec^2 + 2m_pc^2,$$

where we have used the fact that the mass of an antiproton is equal to that of a proton. We square both sides and subtract $m_e^2c^4$ to get

$$p_{\text{min}}{}^2c^2 = (m_ec^2 + m_pc^2)^2 - m_e^2c^4 = 2m_em_pc^4 + m_p^2c^4.$$

The proton mass is some 2000 times larger than the electron mass, so we can safely neglect the $2m_em_pc^4$ term, containing m_e. We are left with

$$p_{\text{min}} \simeq m_pc = \frac{m_pc^2}{c}.$$

Numerically, $p_{\text{min}} \simeq (938\ \text{MeV})/c = 0.938$ GeV/c. (Note the use of the convenient unit GeV/c for momentum.) Equivalently, the minimum total initial energy, E_{min}, is the energy corresponding to the rest mass of a $p\bar{p}$ pair:

$$E_{\text{min}} = 2m_pc^2 = 2[(0.938\ \text{GeV})/c^2]c^2 = 1.88\ \text{GeV}.$$

The minimum energy required for an electron incident on another electron at rest to produce a $p\bar{p}$ pair is much larger (see Problem 9).

Confinement. If we excite an atom, the energies of successive excitations increase until at some point the energy exceeds the binding energy of the atom. One or more electrons may escape from the atom, leaving behind a positive ion. With enough energy, all the electrons can be separated from the nucleus. Similarly, when we excite the nucleus, the energies of the excitations increase until the nucleus breaks apart into constituent daughter nuclei or nucleons. In this way, the constituents of the atom, or of a nucleus, can be isolated, or observed as free particles. *The constituents of the nucleons or other hadrons cannot, however, be separated and observed individually no matter how much energy is added.* This phenomenon is known as **confinement**. In particular, the excitations of the nucleons, whose masses are significantly greater than that of their ground-state counterparts, do not decay into the constituents of the nucleon. This does not mean that these excitations do not decay. They decay most readily into other (lower-lying) excitations or into the nucleons themselves and into other strongly interacting particles such as *mesons* (see Section 46–2). We shall briefly discuss the reasons for confinement in Section 46–3.

Confinement

46–2 NEW PARTICLES AND CONSERVATION LAWS

How the Nuclear Force Is Transmitted

In 1935, the goal of finding an ordinary potential energy for nuclear forces underwent a deep modification as a result of an insight by Hideki Yukawa. Yukawa proposed that, at the most fundamental level, forces between interacting particles result from the exchange of a special quantum that we shall call a "bonding" particle. One of two interacting particles emits a bonding particle, and the other interacting particle absorbs it. Yukawa was able to predict the mass of the bonding particle responsible for the nuclear force—he called that particle the *meson*—from knowledge of the range of nuclear forces.

The range of the nuclear force is described in Chapter 45.

The idea that a force is mediated by an exchange is not a complicated one; two people throwing a ball back and forth exert a (repulsive) force on one another through the exchange of the ball (Fig. 46–1). What is new here is that the exchanged object contains a substantial amount of energy. Take the mass of the meson to be m. After a meson is emitted, and before it is reabsorbed, the system of interacting particles has an energy deficit on the order of mc^2. Energy conservation can be consistent with such a deficit only if we cannot actually measure the energy to within a value of $\Delta E = mc^2$. The time–energy Heisenberg uncertainty relation, $\Delta E\, \Delta t > \hbar$, means that, if Δt is the time interval for making an energy measurement, then the energy is uncertain by at least $\Delta E \simeq \hbar/\Delta t$. Thus

$$\Delta t \simeq \frac{\hbar}{\Delta E} = \frac{\hbar}{mc^2}. \tag{46–2}$$

The principle of the conservation of energy is not violated in that we cannot *measure* the nonconservation of energy within this time period Δt. When a particle is emitted and reabsorbed in this way, we say that it exists in a *virtual* state and that the process is virtual.

Energy nonconservation in the exchange of quanta is unmeasurable because of an uncertainty relation

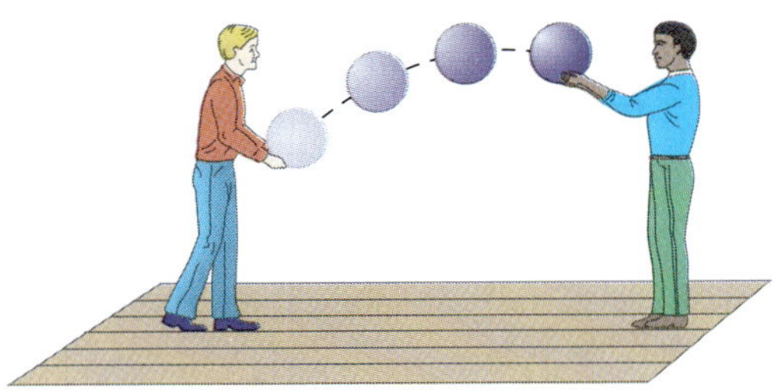

FIGURE 46–1 The exchange of a ball leads to effective forces between the two people.

Range. Let's see how far a virtual meson can travel, which is a measure of the distance over which the nuclear force extends. During the time Δt, the distance the meson can travel is $v\,\Delta t$ and, because v must be less than c, the greatest distance the particle can travel is $R = c\,\Delta t$. From Eq. (46–2),

$$R = c\,\Delta t \simeq \frac{c\hbar}{mc^2} = \frac{\hbar}{mc}. \tag{46–3}$$

We refer to this maximum distance as the *range* of the force; we have already described the range of nuclear forces in Chapter 45.

Yukawa's meson, the bonding particle of the nuclear force, is called the pi-meson, or *pion*, denoted by the symbol π. We can estimate its mass, m_π, if we take $R \simeq 1.4$ fm, an experimental value for the range of nuclear forces:

$$m_\pi = \frac{\hbar}{Rc} = \frac{1.05 \times 10^{-34}\ \mathrm{J \cdot s}}{(1.4 \times 10^{-15}\ \mathrm{m})(3.0 \times 10^{8}\ \mathrm{m/s})} = 2.5 \times 10^{-28}\ \mathrm{kg} = 0.15 m_p, \tag{46–4}$$

where m_p is the proton mass. Thus, $m_\pi c^2 \simeq (0.15)(938\ \mathrm{MeV}) = 140$ MeV. Yukawa also calculated the potential energy as a function of particle separation r for the force associated with meson exchange:

$$U(r) = g^2 \frac{e^{-m_\pi rc/\hbar}}{r}. \tag{46–5}$$

Here, g^2 is a measure of the strength of the force between the interacting particles, analogous to $e^2/4\pi\epsilon_0$ for the Coulomb force. Note that the Coulomb force fits Yukawa's framework perfectly well if it is regarded as the result of the exchange of a massless particle—the photon.

Several comments and qualifications about Yukawa's idea are worth making:

1. The potential energy of Eq. (46–5) is oversimplified; there may also be a complicated dependence on the spins of the interacting particles.

2. The meson must be a boson; that is, its intrinsic angular momentum (spin) must be an integral multiple of $\hbar$. That is because the particles that emit and absorb the meson—the nucleons—are fermions, and if angular momentum is to be conserved, they cannot themselves emit or absorb a particle with half-integral spin and remain fermions.

3. Electric charge conservation must be respected throughout the virtual process. Nuclear forces occur with nearly equal range between neutrons and neutrons, protons and protons, and neutrons and protons, and this means that Yukawa's meson occurs in the three charge states π^0, π^+, and π^-, all of which must have roughly the same mass.

Yukawa's idea is eminently testable: Pions, with predicted masses, should be produced in collisions of sufficient energy. In the late 1940s, accelerators with energies sufficient for this purpose were built, and all three pions, with masses close to the predicted values, were discovered.

EXAMPLE 46–3 A proton projectile collides with a stationary proton target. What minimum kinetic energy must the projectile have so that the final state contains two protons and a neutral pion?

Solution: In the laboratory reference frame, the incoming proton has momentum of magnitude p and energy E, related by $E = \sqrt{p^2c^2 + m_p^2c^4}$. The target proton has zero momentum and energy m_pc^2. The minimum energy in this frame is the energy that accounts for the final particle rest masses, with leftover kinetic energy only in the (constant) center-of-mass motion. Therefore, the three final particles—two protons and one π^0—have a total momentum p but no motion relative to each other. Thus the final state can be viewed as a single particle of momentum p and mass $2m_p + m_\pi$. Energy conservation gives the relation

$$E_i = E + m_pc^2 = \sqrt{p^2c^2 + m_p^2c^4} + m_pc^2 = E_f = \sqrt{p^2c^2 + (2m_p + m_\pi)^2c^4}.$$

We square both sides of this equation to find that

$$p^2c^2 + m_p^2c^4 + 2Em_pc^2 + m_p^2c^4 = p^2c^2 + (2m_p + m_\pi)^2c^4,$$

or, solving for E,

$$E = \frac{2m_p^2c^4 + 4m_pm_\pi c^4 - m_\pi^2c^4}{2m_pc^2}.$$

Thus the minimum kinetic energy needed is

$$K = E - m_pc^2 = 2m_\pi c^2 + \frac{m_\pi^2}{2m_p}c^2 = 2(140\text{ MeV}) + \frac{(140\text{ MeV})^2}{2(938\text{ MeV})} = 290\text{ MeV}.$$

An intensive program of experimentation produced pions in such large numbers that they could themselves be scattered from nuclear targets. One of the most striking discoveries of this program was the existence of many new particles, both bosons (mesons) and fermions. These particles are all hadrons because the ease with which they are produced suggests that they all interact through the nuclear force. The new hadrons differ by electric charge and by new quantum numbers. All these particles are as "elementary" as the nucleons and the pions. They had not been discovered earlier because they (and the pions) are all unstable, decaying with lifetimes between 10^{-8} s and as little as 10^{-23} s (the time it takes for light to cross two proton diameters!). The mesons can in principle be exchanged by any two hadronic fermions. These discoveries show that nuclear forces are far from simple. The underlying dynamics of the particles that interact through nuclear forces has been better understood by studying the forces between *constituents* of the nucleons.

Baryon Number

Relativity and quantum mechanics together predict that each particle should have an antiparticle with the opposite charge but the same mass and lifetime for decay. The positron (the antielectron) or the antiproton are easy to identify through their charges. But what is the antiparticle of the neutron, which is electrically neutral? The antineutron is identifiable through a new quantum number with some similarities to the electric charge. It is called the **baryon number**, B, and nucleons are each assigned the baryon number $B = 1$. The antineutron and the antiproton have the baryon number $B = -1$, and all the fermionic cousins of the nucleons discovered in scattering experiments have a nonzero baryon number. These particles—*all strongly interacting fermions*—collectively are referred to as **baryons**. The pion and all the other mesons have a baryon number $B = 0$.

Baryon number is a conserved and quantized quantum number.

Baryons

The existence of the baryon number, together with a conservation law for it, was postulated when the remarkable stability of matter was recognized. Charge conservation and angular momentum conservation would allow the decay

$$p \to e^+ + \pi^0. \qquad (46\text{–}6)$$

Unless this decay and others like it were somehow hindered, atoms would disappear. But experiment shows that *if the proton lifetime is not infinite, that lifetime is greater than 10^{30} yr*. How can we say this if the universe has existed for only some 10^{10} yr? We can measure such a lifetime by assembling 10^{30} protons and watching them carefully; if the lifetime were 10^{30} yr, we would expect one decay per year. An extraordinary experiment in which a tank filled with 8000 tons of water (about 10^{32} protons) was surrounded by electron counters showed that no decays like those in Eq. (46–6) occurred over a period of several years (Fig. 46–2). The tank was placed in a salt mine deep underground to avoid interference from high-energy particles coming from outer space. Because electrons and pions are assigned a baryon number $B = 0$ and because we postulate that *baryon number is conserved in any reaction*, the decay of Eq. (46–6) is forbidden.

FIGURE 46–2 The very pure water in this tank is surrounded by phototubes, detectors sensitive to the tiniest trace of light. Here the apparatus is being checked by a scientist with scuba equipment. The phototubes are activated when the lights are turned out; they are capable of capturing the tiny flash of light associated with a proton decay somewhere in the tank.

Although baryon number is conserved like electric charge, baryon number differs from electric charge in that there are no Coulomblike forces associated with baryon number. Another way in which baryon number differs from electric charge is that, under certain circumstances, the conservation of baryon number may be violated. Conditions in the early stages of the universe could have provided these circumstances.

The Quark Model

The large number of "elementary" particles discovered in scattering experiments is reminiscent of atoms and their excited states. The Bohr model of hydrogen allows us to explain all the states of hydrogen in terms of an electron, a proton, and the forces between them. Similarly, the *quark model* economically describes baryons and mesons in terms of a few constituents and their interactions. This model was proposed independently by Murray Gell-Mann and George Zweig in 1964.

In the quark model, neutrons, protons, and pions are composed of two fundamental constituents called **quarks**, each with spin $\hbar/2$: the *u-quark* ("up" quark) and the *d-quark* ("down" quark). Both types of quarks have an antiparticle, $\overline{u}$ and $\overline{d}$, respectively. The quantum numbers of these quarks are listed in Table 46–1. The electric charge of the quarks is a fraction of the electron charge. This does not mean that charge is no longer quantized, but that the true quantization is in units of $\frac{1}{3}e$. The fact that quarks and antiquarks have spin $\hbar/2$ means that the spin of an even number of quarks plus an-

TABLE 46–1 Properties of Quarks[†]

Quark	*Electric Charge*	*Baryon Number*	*Spin*
u	$+\frac{2}{3}e$	$+\frac{1}{3}$	$\frac{\hbar}{2}$
d	$-\frac{1}{3}e$	$+\frac{1}{3}$	$\frac{\hbar}{2}$
$\overline{u}$	$-\frac{2}{3}e$	$-\frac{1}{3}$	$\frac{\hbar}{2}$
$\overline{d}$	$+\frac{1}{3}e$	$-\frac{1}{3}$	$\frac{\hbar}{2}$

[†]Quarks actually come in three distinct sets, or *families*, totaling six quarks. These lead to the prediction of still more hadrons, most of them already observed.

tiquarks is an integer times $\hbar$, whereas the spin of an odd number of quarks plus antiquarks is a half-integer times $\hbar$.

The proton is a composite of three quarks, uud; the neutron has the structure udd. The antiproton is made of the combination $\bar{u}\bar{u}\bar{d}$, whereas the antineutron is $\bar{u}\bar{d}\bar{d}$. A combination of three quarks can have a variety of angular momenta, and these different combinations successfully explain all the baryons thus far discovered. For example, among the observed baryons is a quartet of states with similar masses, Δ^{++}, Δ^{+}, Δ^{0}, Δ^{-}, all with spin $3\hbar/2$ and baryon number $B = 1$. These are formed from the combinations uuu, uud, udd, and ddd, respectively. The quark model also accounts for mesons. Mesons have $B = 0$ and are bosons (their spin is an integral multiple of $\hbar$), so they must be formed from combinations of quarks and antiquarks. In particular, $\pi^{+} = (u\bar{d})$ and $\pi^{-} = (d\bar{u})$, whereas π^{0} is a combination of $u\bar{u}$ and $d\bar{d}$.

We mentioned in Section 46–1 that quarks are *confined* inside hadrons; they cannot be separated from a nucleon or other strongly interacting particle and exist independently. The evidence for the existence of quarks does, however, go beyond the fact that quarks in combination account for the observed hadrons. Complementary evidence is provided by scattering experiments. When electrons with high energy were first used to probe protons in a hydrogen target in the 1950s, the results quite clearly revealed a distribution of charge within the proton (that is, the proton is not pointlike). Figure 46–3 shows the observed distribution compared to the distribution a pointlike proton would give. There was wide expectation that further experiments would continue to verify that the proton's charge is distributed smoothly within it, much as was expected for the α-particle scattering from the atom before Rutherford's work proved otherwise. If the proton charge were distributed uniformly, electrons passing through the proton (and possibly breaking it apart) would be deflected very little. To the surprise of the physics community, experiments of this type performed in 1968 showed a very different behavior: Electrons were scattered in large numbers at large angles and in a pattern characteristic of a proton constructed of three pointlike particles, each carrying a fraction of the proton charge. These particles are the quarks that we used to describe the observed hadrons.

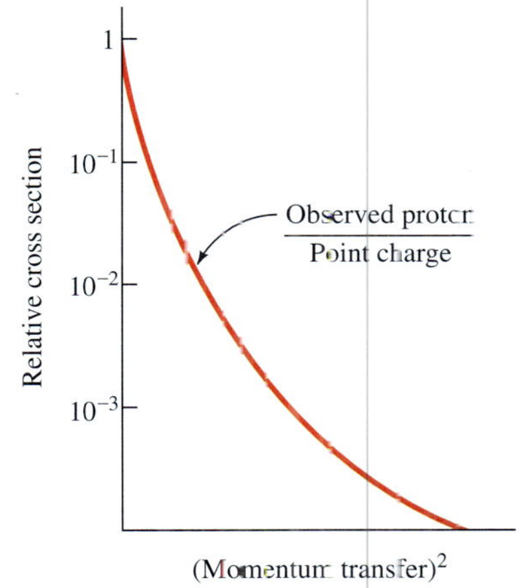

FIGURE 46–3 Elastic electron–proton scattering experiments show that the proton is not pointlike. The horizontal axis is the square of the momentum transfer to the target proton; we have already seen that this quantity measures how closely the target is probed. The vertical axis is the ratio of the measured cross section to the cross section that would follow if the proton were pointlike. This curve would be flat, not decreasing, if the proton were pointlike. We know from other scattering experiments that the corresponding curve for quarks is flat, to the accuracy with which we can measure it.

Lepton Number

Leptons

We have seen that a proper understanding of the forces that act in nature cannot be achieved until we have some knowledge of the constituents of matter. Even though they are not hadrons, both the electron and the neutrino (ν) are fundamental constituents of matter. These particles, and other particles that do *not* participate in the strong force, are known as **leptons**.

We recall from our discussion in Chapter 45 that the neutrino has no charge and little or no mass. But, along with the electron, it does carry a conserved **lepton number** analogous to the baryon number. Both the electron and the neutrino, each with spin $\hbar/2$, have the lepton number $+1$; their antiparticles e^{+} and $\bar{\nu}$ have the lepton number -1. Electrons interact through the forces of electromagnetism, whereas neutrinos do not. But both participate in the weak force associated with β decay. We shall see in Section 46–3 that these forces are closely related.

46–3 FUNDAMENTAL FORCES REVISITED

We introduced fundamental forces in Chapter 5 and are now in a position to add to that discussion. At present, physicists believe that three fundamental forces act in nature (Table 46–2): gravitation, the electroweak force, and the strong force. (The electroweak force is the result of a mid-1970s unification of electromagnetism and the weak force, responsible for β decay.) Each of the fundamental forces is thought to act through the exchange of one or more particles. In Table 46–2, we distinguish the "constituent" particles that interact by means of these forces and the exchange particles, or bonding

TABLE 46–2 Fundamental Forces

Force	*Particles on which Force Acts*	*Particles that Transmit Force*
Gravitation	Any particle with mass	Gravitons†
Electroweak force	Particles with electric charge or "weak" charge	Photons, weak bosons
Strong force	Quarks, gluons	Gluons

†Gravitons have never been observed in any sense, nor has the law of gravitation (general relativity) been satisfactorily integrated with quantum ideas.

particles, through which the force is transmitted. Note that the nuclear force is not listed; it is ultimately due to the strong force, the force between the quarks. Gravitation plays no practical role in our understanding of the behavior of matter at energies that are currently accessible. Indeed, the bonding particle for gravitation, the *graviton*, has never been observed. We shall discuss the other two forces in more detail.

Strong Force. The strong force involves the quarks, described earlier, and the **gluons**, a set of bonding particles with spin $\hbar$. The quarks carry a quantum number analogous to the electric charge that, like electric charge, is both quantized and conserved. This quantum number is called the *color*, but it should not be confused with the colors of the visible-light spectrum. The electric charge occurs in two different types; the color quantum number occurs in three different types. Gluons are emitted and absorbed by particles with a color quantum number (quarks), just as photons are emitted and absorbed by particles with electric charge. But there is a crucial difference between electromagnetism and the strong force: Photons do not *carry* electric charge, *but gluons carry the color quantum number and thus can themselves interact via the strong force through the exchange of other gluons*. This difference is crucial. The theory that describes these interactions is **quantum chromodynamics**.

The fact that gluons interact among themselves through the strong force leads to a force between quarks and gluons that *increases* with distance. (The restoring force due to a spring has this property, but the spring force is not a fundamental one, and a spring will distort and eventually break if it is stretched too far.) The farther apart the quarks and gluons are, the more strongly they attract one another. Any effort to pull quarks and gluons out of the nucleon by supplying energy is therefore bound to fail. *This is the basis of the confinement mechanism discussed earlier.*

Electroweak Force. At the energies typical of the everyday world, the two aspects of the electroweak force, electromagnetism and the "weak" force responsible for β decay, appear to be rather different. The photon is an exchange particle that transmits the electromagnetic aspect of the electroweak force. The weak force manifested in β decay appears to have a very different strength than the electromagnetic force, but this is not the case: The difference is that, in addition to photons, there is another set of force transmitters, the *weak bosons*, W^+, W^-, and Z^0. These bosons, whose superscripts indicate their electric charge, have very substantial masses, some 86 and 97 times the mass of a proton for the W (W^+ and W^- both) and Z^0, respectively. This fact implies that the weak force has not only a limited range (through Yukawa's reasoning), but also a certain energy dependence. If the energies of the particles that interact via the weak force are greater than the mass of the weak bosons, the strength of the "weak" interaction becomes comparable to that of electromagnetism. Thus the apparent difference of the electromagnetic and weak aspects of the electroweak force disappears at high energies.

Quarks as well as leptons can emit or absorb the carriers of the electroweak force and are subject to the conservation of electric charge, of baryon number, and of lepton number. Processes such as β decay, $n \rightarrow p + e^- + \overline{\nu}$, can be visualized as a two-step process in which a d-quark within a neutron is transformed to a u-quark:

$$d \rightarrow u + W^-, \quad W^- \rightarrow e^- + \bar{\nu}.$$

The unification of the electromagnetic and weak forces is an important step in our understanding of the fundamental forces. A number of predictions follow from the unification. For example, the dramatically increased rate at which reactions such as $e^- + p \rightarrow n + \nu$ occur at high energies is a consequence of the fact that the "weak" force becomes as strong as the electromagnetic force at high energies. To this point, all the predictions of the electroweak unification that have been tested have confirmed that this unification is correct. Further unification is a continuing goal of current research.

A Summary Chart

Figure 46–4 is a chart that summarizes much of the current state of knowledge of fundamental particles and forces.

46–4 TOOLS OF PARTICLE PHYSICS

Richard Feynman compared the process of understanding the behavior of elementary particles to understanding how a watch works by throwing two watches at each other and studying the array of gears and springs that come out. Actually, the situation is somewhat worse. We require high energies to see well at short distances, but high energies bring with them relativistic effects such as pair production (pairs of "watches and antiwatches" are produced). In addition, because of confinement, the gears and wheels cannot themselves be final products, just more watches, perhaps of a different design but made from those same gears and wheels. Many of the particles that result are unstable and decay so quickly that we can see only their decay products. Finally, the collisions being studied are relatively rare, so a high flux of colliding particles is necessary.

The problem of designing experiments to study the behavior of particles at short distances is twofold: First, we require particle accelerators that have the best possible combination of high energies and high beam currents. Second, we need detectors that enable us to analyze the results of collisions. In constructing accelerators and detectors, physicists have pushed technologies to their limits. The results have been scientifically and technologically rewarding.

Accelerators

The principle of modern particle accelerators is the repeated acceleration of a stable charged particle by an electric field. A magnetic field controls the path the beam of accelerated particles follows. Beyond these basic ideas, a variety of techniques are used in different accelerators.

Electrons or Protons. Electrons have revealed no substructure: They appear to be as elementary as quarks. In addition, it appears that electrons and quarks are *both* necessary in quantum mechanical models with calculable predictions. These facts suggest a possible unification between strongly interacting particles and particles that interact by means of the electroweak interactions. Thus there is as much interest in the observation of electron collisions as in the observation of proton collisions; accelerators exist that allow us to study pp, $p\bar{p}$, ep, and $e\bar{e}$ collisions. Proton collisions are of interest in the sense that they represent collisions involving quarks.

Fixed-Target Machines or Colliders. Until the 1960s, virtually all accelerators were *fixed-target machines*, in which an accelerated beam is incident on a target (Fig. 46–5).

FUNDAMENTAL PARTICLES AND INTERACTIONS

FERMIONS

matter constituents
spin = 1/2, 3/2, 5/2,...

Leptons spin = 1/2

Flavor	Mass GeV/c²	Electric charge
ν_e electron neutrino	$< 7\times10^{-9}$	0
e electron	0.000511	−1
ν_μ muon neutrino	< 0.0003	0
μ muon	0.106	−1
ν_τ tau neutrino	< 0.03	0
τ tau	1.7771	−1

Quarks spin = 1/2

Flavor	Approx. Mass GeV/c²	Electric charge
u up	0.005	2/3
d down	0.01	−1/3
c charm	1.5	2/3
s strange	0.2	−1/3
t top (initial evidence)	170	2/3
b bottom	4.7	−1/3

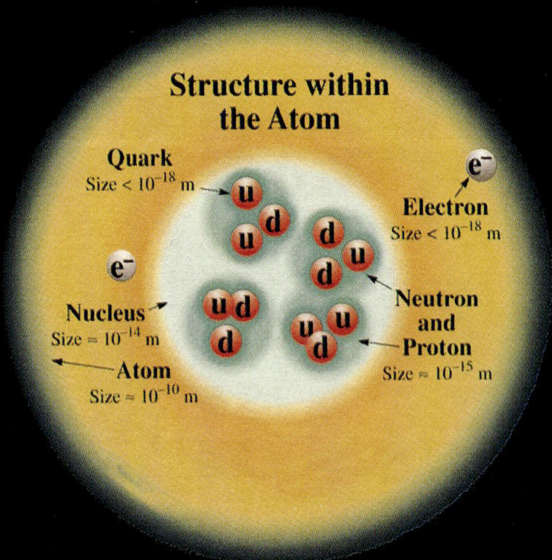

BOSONS

force carriers
spin = 0, 1, 2,...

Unified Electroweak spin = 1	Mass GeV/c²	Electric charge
γ photon	0	0
W^-	80.22	−1
W^+	80.22	+1
Z^0	91.187	0

Strong or color spin = 1	Mass GeV/c²	Electric charge
g gluon	0	0

PROPERTIES OF THE INTERACTIONS

Property \ Interaction	Gravitational	Weak (Electroweak)	Electromagnetic (Electroweak)	Strong: Fundamental	Strong: Residual
Acts on:	Mass – Energy	Flavor	Electric Charge	Color charge	See Residual Strong Interaction Note
Particles experiencing:	All	Quarks, Leptons	Electrically charged	Quarks, Gluons	Hadrons
Particles mediating:	Graviton (not yet observed)	W^+ W^- Z^0	γ	Gluons	Mesons
Strength for two u quarks at: 10^{-18} m (relative to electromagnetic)	10^{-41}	0.8	1	25	Not applicable to quarks
Strength for two u quarks at: 3×10^{-17} m (relative to electromagnetic)	10^{-41}	10^{-4}	1	60	
for two protons in nucleus	10^{-36}	10^{-7}	1	Not applicable to hadrons	20

Sample Fermionic Hadrons

Baryons qqq and Antibaryons $\bar{q}\bar{q}\bar{q}$

Symbol	Name	Quark content	Electric charge	Mass GeV/c²	Spin
p	proton	uud	1	0.938	1/2
$\bar{p}$	anti-proton	$\bar{u}\bar{u}\bar{d}$	−1	0.938	1/2
n	neutron	udd	0	0.940	1/2
Λ	lambda	uds	0	1.116	1/2
Ω^-	omega	sss	−1	1.672	3/2

Sample Bosonic Hadrons

Mesons $q\bar{q}$

Symbol	Name	Quark content	Electric charge	Mass GeV/c²	Spin
π^+	pion	$u\bar{d}$	+1	0.140	0
K^-	kaon	$s\bar{u}$	−1	0.494	0
ρ^+	rho	$u\bar{d}$	+1	0.770	1
D^+	D⁺	$c\bar{d}$	+1	1.869	0
η_c	eta-c	$c\bar{c}$	0	2.979	0

FIGURE 46–4 Electric charges of the particles listed here are given in units of the proton's charge, and spins are given in units of $\hbar$. There are two more neutrinos, two more leptons, and four more quarks than we describe in the text (the "top" quark was discovered as late as 1995); we do not now understand why these additional particles exist. There are many more fermionic hadrons and bosonic hadrons than are described here, all of which can be accounted for by quantum chromodynamics. More complete copies of this chart can be obtained from Science Kit & Boreal Laboratories, 777 East Park Drive, Tonawanda, NY 14150.

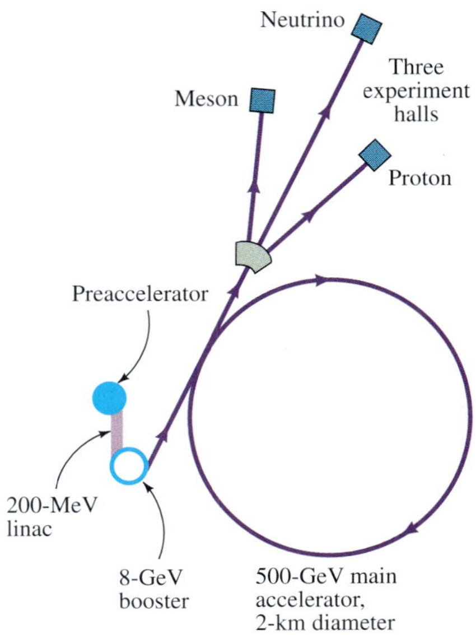

FIGURE 46–5 Schematic diagram of a fixed-target machine, in this case an early version of the accelerator at the Fermi National Accelerator Laboratory. Note the several stages of acceleration (beginning with the preaccelerator and a linear accelerator section, or linac) and the extensive target areas (experiment halls), where beams extracted from the ring of the main accelerator collide with fixed targets.

In a fixed-target machine, the beam is typically extracted and directed to a fixed target such as liquid hydrogen. The advantages of such an accelerator are that a huge number of target particles can be spread throughout any volume and that there is only one beam to manage. These machines have systematically been replaced by *colliders*, in which two moving beams of particles with equal but opposite momenta interact in head-on collisions with one another (Fig. 46–6). Table 46–3 lists the characteristics of some advanced colliders. We know that the momentum of the center of mass of a system of interacting particles is constant. When a moving particle collides with a stationary target, the center of mass has nonzero momentum, which is maintained after the collision. Thus a large fraction of the incident energy remains in the motion of the center of mass rather than going into the production of particles, the most interesting

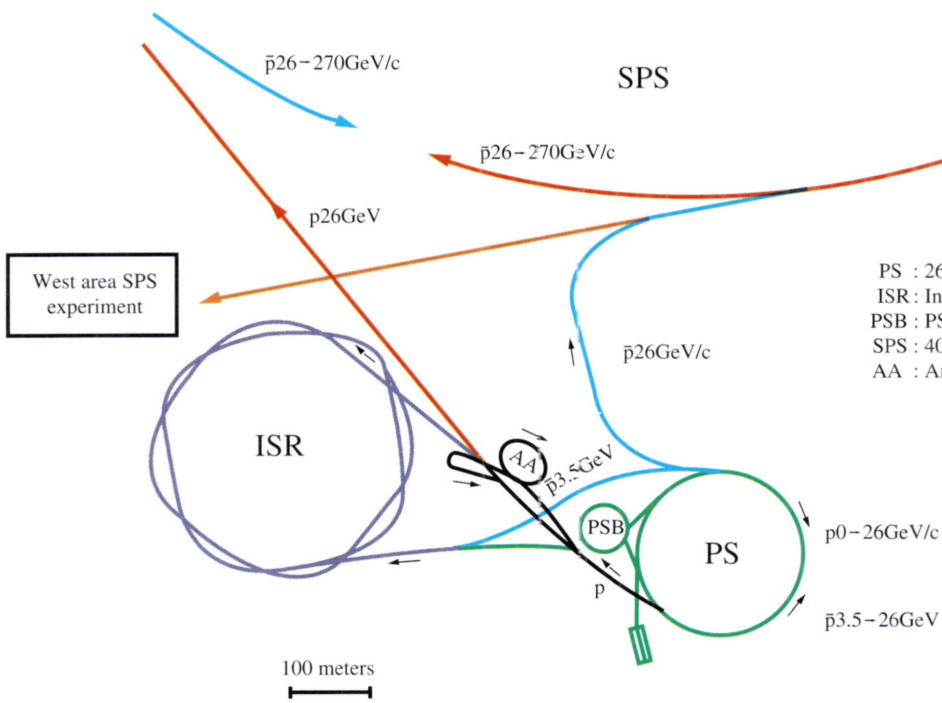

FIGURE 46–6 Schematic diagram of a collider, in this case the SPS near Geneva, Switzerland. The entire SPS ring is not shown because it has a circumference of some 27 km. Protons and antiprotons are injected into the SPS accelerator and travel in opposite directions before colliding.

TABLE 46–3 Characteristics of Some Colliders

Name	*Laboratory, Location*	*Particles*	*Maximum Beam Energy (GeV)*
Sp$\bar{p}$S (SPS)	European Laboratory for Particle Physics (CERN), Geneva, Switzerland	$p\bar{p}$	315
TEVATRON	Fermi National Accelerator Laboratory (FNAL), Batavia, Illinois	$p\bar{p}$	1000
HERA	Deutsches Elektronen–Synchrotron (DESY), Hamburg, Germany	ep	e: 26; p: 820
Large Electron–Positron Collider (LEP)	European Laboratory for Particle Physics (CERN), Geneva, Switzerland	$e\bar{e}$	60
Stanford Linear Collider (SLC)	Stanford Linear Accelerator Center (SLAC), Stanford, California	$e\bar{e}$	50

part of the reaction. In colliders, the system's center of mass is at rest relative to the accelerator, and all the energy goes into the production of particles. Given that the cost of an accelerator increases rapidly with the energy of the beam or beams it accelerates, a collider buys more center-of-mass energy per dollar.

EXAMPLE 46–4 The W^+ particle, one of the carriers of the electroweak force, has a mass of about 80 GeV/c^2. (a) For W^+ to be produced in the reaction $p + \bar{p} \rightarrow W^+ + \pi^-$, what would the minimum energy of the proton (and antiproton) have to be in a collider? (b) What would the minimum energy of the $\bar{p}$ have to be if the same reaction were to occur in a collision with a stationary proton?

Solution: Figure 46–7 shows the $p\bar{p}$ collision in the frames of reference appropriate to the collider and to the fixed-target machine, with the momenta and energies labeled. (a) Energy conservation in the center-of-mass reference frame gives $2E = E_W + E_\pi$. The minimum energy, $E = E_{\text{min}}$, corresponds to the W^+ and π^- both produced at rest, so $E_W = M_W c^2$ and $E_\pi = m_\pi c^2$. With $m_\pi << M_W$, we have $E_\pi << E_W$, so

$$E_{\text{min}} \simeq \frac{1}{2} M_W c^2 \simeq \frac{1}{2}(80 \text{ GeV}) = 40 \text{ GeV}.$$

(b) In the laboratory reference frame, the total momentum is k, the momentum of the initial $\bar{p}$. At the minimum energy, the W and π are at rest relative to each other

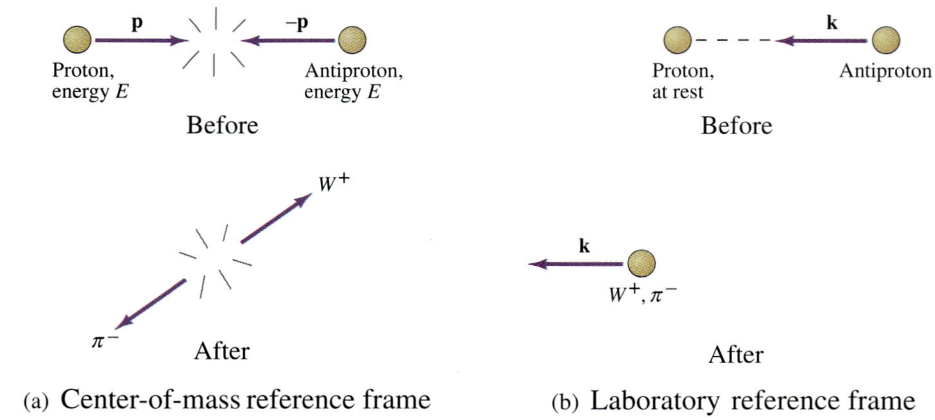

FIGURE 46–7 Example 46–4. (a) The reaction as seen in the center-of-mass reference frame. (b) The reaction as seen in the laboratory reference frame.

and act like a single particle of mass $M_W + m_\pi \simeq M_W$. Momentum conservation ensures that the momentum of the $W - \pi$ pair is k. Energy conservation then gives

$$E_{\text{lab, min}} + m_p c^2 = \sqrt{k^2c^2 + M_W^2 c^4},$$

where $E_{\text{lab, min}}$ is the minimum energy of the $\overline{p}$. We square both sides of this equation to get

$$E^2_{\text{lab, min}} + 2E_{\text{lab, min}} m_p c^2 + m_p^2 c^4 = k^2c^2 + M_W^2 c^4.$$

Finally, we use the fact that in the laboratory reference frame, the energy of the initial $\overline{p}$ is related to its momentum by $E_{\text{lab, min}} = \sqrt{k^2c^2 + m_p^2c^4}$. We square this relation and substitute for the term $E^2_{\text{lab, min}}$ in the preceding equation:

$$k^2c^2 + m_p^2c^4 + 2E_{\text{lab, min}} m_p c^2 + m_p^2 c^4 = k^2c^2 + M_W^2 c^4,$$

an equation we can solve for $E_{\text{lab, min}}$. With $m_p << M_W$, we get

$$E_{\text{lab, min}} = \frac{M_W^2 c^4}{2m_p c^2} \simeq \frac{(80\ \text{GeV})^2}{2(0.94\ \text{GeV})} = 3400\ \text{GeV}.$$

The huge difference between the answers to parts (a) and (b) illustrates the advantage of colliders. It is far cheaper to accelerate two beams to 40 GeV each than to accelerate one beam to 3400 GeV.

The difficulty in controlling the two beams of a collider is not as great as you might imagine. In Fig. 46–8, we can deduce that the same electric field that accelerates particles in one direction accelerates their antiparticles in the other direction, and the same magnetic field that forces charged particles to follow a counterclockwise path also forces their antiparticles to follow a clockwise path. Thus the same fields can be used for a beam of both electrons and positrons or protons and antiprotons. The major complication is that the two beams must be concentrated and steered so that they cross at isolated points where the collisions occur; at those points, elaborate detectors monitor the results of the collisions.

We have shown that colliders are energetically advantageous. A second advantage of a collider concerns the distribution of the many particles that are produced in a collision. In the center-of-mass reference frame, the particles produced are spread over all angles (Fig. 46–9), and this is the way the reaction appears in a collider. The fact that the center of mass moves in a fixed-target machine means that the particles produced are carried forward with it and hence concentrated in a narrow cone in the forward direction. As Fig. 46–10 shows, this concentration makes it hard to separate the products.

Each beam of a collider provides a target for the other beam that contains far fewer target particles than does a fixed-target machine Thus the rate at which collisions occur is less in a collider than in a fixed-target machine. This is not always a disadvantage because the reactions we observe at very high energies produce so much information that even the best computers cannot keep up with data that arrive too rapidly.

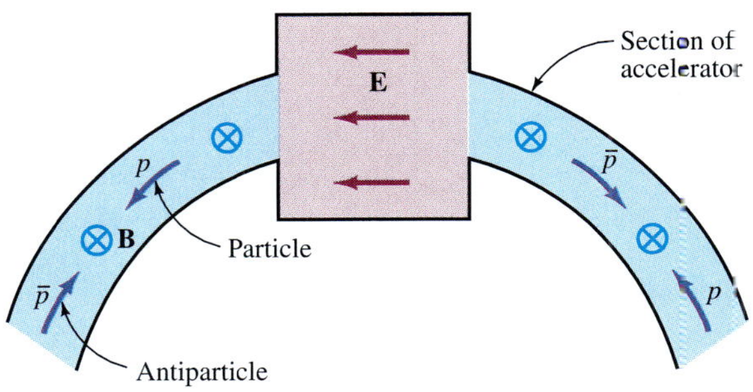

FIGURE 46–8 In a collider, the same magnetic fields guide both particle p and antiparticle $\overline{p}$, and the same electric fields accelerate both p and $\overline{p}$.

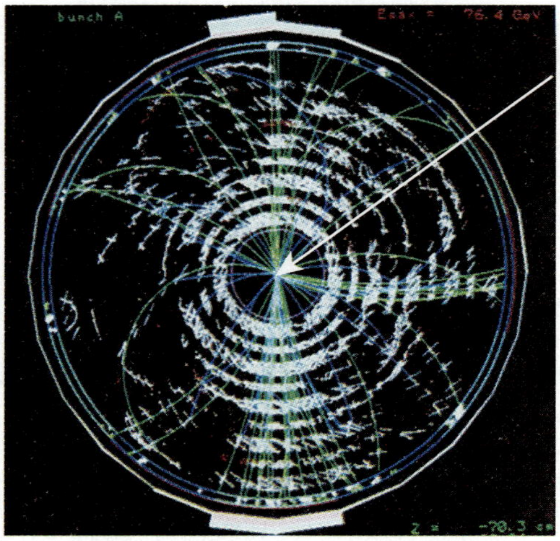

FIGURE 46–9 A collision at the TEVATRON collider at FNAL, as recorded in the Collider Detector Facility. The white and green lines are produced in the concentrically oriented cylindrical detectors and indicate the passage of particles. The view is an end-on one—that is, along the direction of the colliding beams. A substantial number of particles are produced at right angles to the beams.

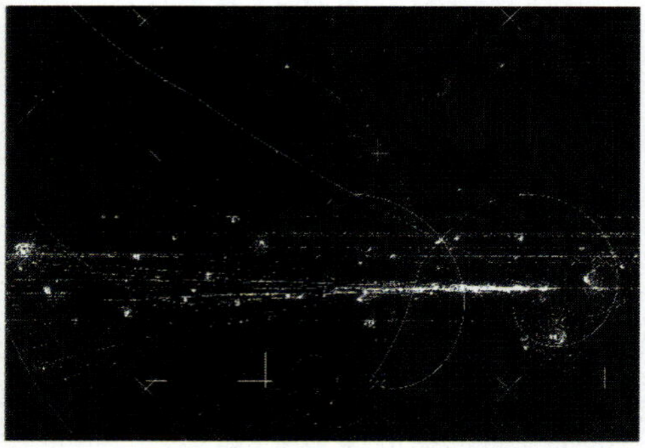

FIGURE 46–10 In this bubble chamber image, a proton with an energy of 300 GeV enters and strikes a stationary proton. Among the products of this collision are the 26 charged particles visible here.

But when rare reactions are to be studied, a collider may be at a real disadvantage compared to a fixed-target machine.

Circular Versus Linear Colliders. The problem with beams that are not very dense (compared to target densities in a fixed-target machine) is in part overcome in colliders that are circular, so that the beams cross and recross frequently. However, recall from Chapter 36 that energy is radiated in the form of electromagnetic waves (or photons) when charged particles accelerate. For the beams of an accelerator, this is undesirable in that we want the beam to retain as much energy as possible. This is unavoidable for the initial acceleration that gives the particle its energy. However, the circular motion in a circular collider means that the beam undergoes a centripetal acceleration that is proportional to the momentum squared. Thus the higher the energy of the beam, the greater the radiation of energy. Moreover, the particles' acceleration is inversely proportional to their mass, so lighter particles, such as electrons, radiate energy at a higher rate than do heavier particles, such as protons. LEP, a circular machine that accelerates electrons and positrons to 60 GeV each in a ring of radius 5 km (Table 46–3), is at about the limit for an electron collider in the form of a ring—at higher energies, the electron beam loses too much of its energy with each turn. The alternative is to construct a linear accelerator in which a new electron beam is repeatedly generated along a straight beam tube without following a circular trajectory. A linear collider—two linear machines with their beams meeting head-on—provides the best solution. The major disadvantage of such a machine, which has indeed been constructed on a demonstration basis (SLC in Table 46–3), is that the beams cross only once, so there is relatively little chance for collision. To overcome this difficulty, the beams are concentrated to only a few microns in size, a tiny fraction of the size of the beams of other accelerators; these beams must then cross with a remarkable precision. Such machines appear to be the only way to construct electron colliders with energies much greater than those of LEP.

Detectors

The apparatus that observe and analyze collisions in particle accelerators are called *detectors*. Detectors have undergone continual evolution since Joseph J. Thomson first detected the electron in 1897. Subnuclear particles can be detected because they leave a trace of energy as they pass through or decay within a detector. For example, a charged particle that passes through a *photographic emulsion*, a suspension used to coat photographic plates, film, and paper, leaves the same kind of trace that light does. Stacks of photographic emulsion formed the earliest detectors (Fig. 46–11)—they were in use until the 1950s—and allowed scientists to learn much about the charged particles that enter Earth's atmosphere (*cosmic rays*). The resolution of emulsions—their ability to pinpoint a particle track—is less than a micron (10^{-6} m). A similar detection technique uses the fact that a liquid near its boiling point forms bubbles of vapor along the path of a passing charged particle. In use until the 1980s, *bubble chambers* filled with liquid hydrogen utilized this technique (Fig. 46–12). Other, more modern types of detectors register the passage of a charged particle or of neutral particles, such as photons, by their characteristic effects on the electronic properties of the detector.

A detector may be sensitive to the total amount of energy a particle deposits as it is brought to a stop within the detector (a *calorimeter*), or the rate at which it loses energy as it passes through. In a supplementary technique, by measuring the amount of curvature in the path of a passing particle in a magnetic field of large, known magnitude within a detection region, the particle's momentum is determined. Knowing both the momentum and energy, we can calculate the mass, so the particle can be identified. Particle identification is important and not so simple at high energies.

Photographic emulsions record everything that passes through them. Techniques that use electronic *triggers* avoid this difficulty. For example, suppose that you are interested in viewing only those events in which a muon (see Chapter 40) of high energy is produced. Muons pass through matter more easily than do protons or other strongly interacting particles. Then it is possible to surround the interaction region with a great deal of matter—such as many tons of steel—and outside this matter to place a detector that registers electronically the passage of charged particles. This detector emits a signal that what has passed is likely to be a muon and thereby acts as a trigger so that the rest of the event will be registered. Modern detectors are huge and complex (Fig. 46–13). The electronic registration of the passage of dozens or even hundreds of particle tracks requires vast information storage facilities and powerful computers to extract the most useful information. The computers and analysis techniques that triggered detectors require have filtered down to more general use.

FIGURE 46–11 This image has been formed within a photographic emulsion. Where charged particles pass, a black line forms. At point A a reaction has occurred. One of the reaction products has either come to rest or decayed to neutral products at point B. Detailed track analysis can tell which.

FIGURE 46–12 Bubble chambers can record the passage of charged particles: A chain of bubbles forms around the particle tracks. The photo in part (a) is illustrated in part (b). In this false-color image, a π^- (green) collides with a stationary proton in the chamber and produces two neutral particles that travel invisibly through the chamber until they decay into charged particles. One neutral particle decays to a proton (red) and a π^- (green); the other neutral particle decays to a π^- (green) and a π^+ (yellow). The charged tracks curve because a magnetic field is present. Analysis of the decay tracks reveals properties of the neutral particles. The blue tracks were made by particles not involved in this interaction.

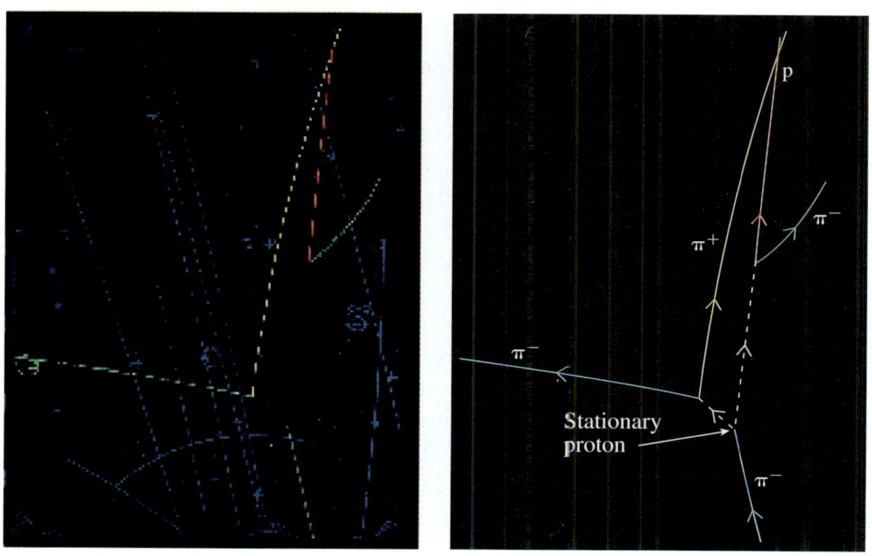

(a) (b)

FIGURE 46–13 This detector at Fermi National Accelerator Laboratory in Batavia, Illinois weighs some 4500 tons, including magnets and energy-measuring devices called calorimeters; it can be rolled into its proper position in the beamline.

46–5 THE EXPANDING UNIVERSE

Giordano Bruno was burned at the stake in 1600 in large part for claiming that the universe is infinite. Although passions have calmed somewhat, people have always had a lively interest in the nature of the cosmos as a whole (*cosmology*). Powerful instruments have allowed us to see so far back into the past that we now have, for the first time, concrete data about the very early stages of our universe.

Olbers's Paradox. *Olbers's paradox* was stated in 1823, when telescopic observations had made it clear that the universe contains many, many more stars than are visible with the naked eye. Heinrich Olbers argued that, if the universe were infinite and unchanging, then, regardless of the direction in which we looked, we would see a star and star light would cover the entire sky. We cannot dispute this argument by arguing that distance makes stars too faint to see. Suppose that N stars are contained in a thin spherical shell a distance R away from us. Then, because the area of a sphere increases as the radius-squared increases, there would be $4N$ stars in a shell a distance $2R$ away from us. The intensity of the light decreases as the distance-squared increases, so each star in the more distant shell is only one-fourth as intense as those in the nearer shell, but the total intensity is *independent* of the distance of the shell. The net intensity of an infinite number of concentric shells would be infinite. Yet—and this is what constitutes Olbers's paradox—the night sky is not bright. Before we can resolve Olbers's paradox, we must learn about more modern astronomical observations.

Hubble's Law

Before 1923, nebulae were thought to be clouds of gas within our own galaxy. (It was only a few years before this that Harlow Shapley had determined that what we now know to be our galaxy has a definite shape and that our Sun is far from the galaxy's center.) In 1923, Edwin Hubble identified the nebulae as galaxies like our own and thereby took a giant step toward the understanding of the universe. Hubble used the Mount Wilson telescope in southern California, the first telescope sufficiently powerful to distinguish individual stars in the nebulae. Hubble also measured the spectral lines of stars in the galaxies and showed that, on average, those lines are redshifted, indicating a movement away from us. Hubble's law (1929) relates the average speed u with which another galaxy retreats from our own to a distance D away from us [Eq. (40–18)]:

$$D = \frac{u}{H}.$$

H is the *Hubble parameter*, measured to be $H \simeq 2.5 \times 10^{-18}\ \mathrm{s}^{-1}$. The redshift gives the speed of recession directly, but how did Hubble know the distance of a given galaxy from our own? He used the cepheids, a class of powerful stars whose systematic waxing and waning make them easily identifiable. The intensity of these stars within our galaxy is characteristic, and Hubble supposed that any cepheid, even one in a distant galaxy, would emit light with that characteristic intensity. To assign a distance to a given cepheid, Hubble used the fact that the intensity at any distance decreases as the inverse square of the distance from the star.

There is also *independent* evidence that objects with large redshifts, such as quasars, are very distant: We see some of them through gravitational lenses (Chapter 12). This is possible only if these quasars are much farther than the galaxies that produce their images through the bending of the quasar light. The distances calculated from the lensing effect are consistent with those provided by Hubble's law.

The Big Bang Model

An immediate conclusion follows from the fact that other galaxies recede from ours at a speed proportional to their distance from us: *At some time in the distant past, all galaxies must have been very close together*. This fact defines what is conventionally called the Big Bang model of the history of the universe. Note that Hubble's law does not mean that Earth (or our galaxy) occupies a privileged position in the universe. In fact, it means just the opposite, as the following one-dimensional analogue shows. Figures 46–14a and 46–14b show a line of galaxies at times T and $2T$, respectively, after the moment when the Big Bang occurred, $t = 0$ Note that the galaxies are evenly spaced at time T; there is no preferred point in this arrangement. We have identified our galaxy (galaxy A) and supposed that Hubble's law holds. Therefore galaxy C, which is twice as far from us as galaxy B, recedes from us at twice the speed with which galaxy B recedes. Thus at time $2T$, galaxy C is still twice as far from us as galaxy B. The galaxies continue to be uniformly spaced, and no point along the line is preferred. Hubble's law can be contrasted with a law in which other galaxies retreat from ours at constant velocity. Even though the galaxies are evenly spaced at time T, at time $2T$ this is no longer the case, and galaxy A occupies a special point (Fig. 46–14c). In fact, Hubble's law is the unique law for galactic speeds by which no one galaxy occupies a special point within the cosmos. This can be confirmed by studying Figs. 46–14a and 46–14b from the point of view of galaxy B (see Problem 34). An inhabitant of galaxy B would agree that Hubble's law holds for him or her as well. A two-dimensional analogue of the situation described by Hubble's law is provided by dots painted on a spherical balloon (Fig. 40–16). When the balloon is blown up, *all* the dots recede from each other, with no dot occupying any special position, and a law like Hubble's law holds. If we lived on the surface of such a balloon, any voyage we took would eventually take us back home, and we would conclude that there is no "outside."

It has taken centuries of discussion for humankind to understand that it is not at the universe's physical center. There *is* no center. It is not as though we are now living in a space in which an explosion once occurred; such a description would imply a privileged point, the position of the explosion itself. Rather, all of space began as a small volume *with no outside*.

Hubble's law with a constant value of H can be used to find the time required for galaxies to separate by given amounts (see Problem 37). Unfortunately, if the law is used to find the time at which all matter was together, that time is infinite (see Problem 38)! Thus if there was an initial moment when the universe started to expand, Hubble's law with a constant value of H cannot hold all the way back to that time. We can, however, use the law to make an estimate: If we assume that the recession speed is constant, Eq. (40–18) states that the expansion time (the *Hubble time*) is just H^{-1}, some 13 billion years. Other information (for example, theoretical knowledge of stellar evo-

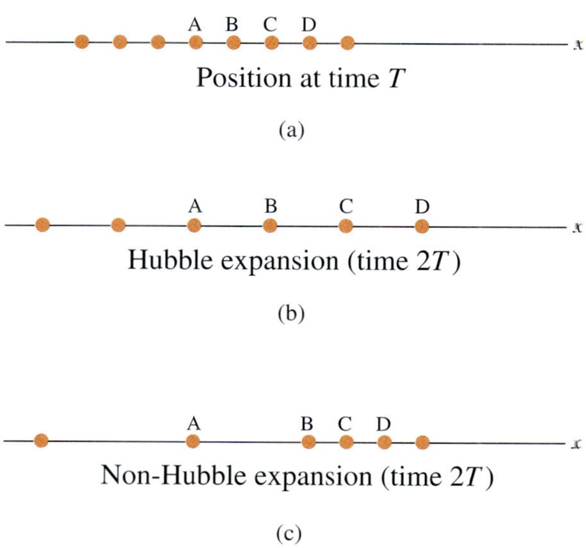

FIGURE 46–14 (a), (b) Two successive times in a one-dimensional version of Hubble's law. The points (galaxy positions, for example) remain equally spaced along the line. (a), (c) The same two successive times in a (non-Hubble) expansion law in which all the points recede from point A at the same speed, independent of their distance from A. Point A is a special point in the expansion law.

lution and experimental identification of the characteristics of the oldest stars) confirms this estimate to within a factor of 2.

We can cite three important pieces of evidence that support the idea that the entire visible universe has expanded away from a small volume. The first and most striking piece of evidence is the presence and character of the observed background blackbody radiation. According to the Big Bang model, this radiation is a relic of the universe at the tender age of 300,000 years. The second piece of evidence is that the relative amounts of light elements, such as hydrogen and helium, that are distributed throughout the visible universe are successfully predicted by the Big Bang model (see Section 46–6). The third piece of evidence is that stars and galaxies exhibit signs of having evolved over time; for example, there appears to have been a time when quasars were more abundant than they are now, and the optical emissions of older radio galaxies (galaxies that emit radio waves strongly) are different from those of younger radio galaxies.

We discussed background radiation and its discovery in Chapter 17.

The Big Bang model resolves Olbers's paradox. If the universe has existed for only a finite time, then any one star has been radiating for a finite time, and there has not been enough time for the radiation from all stars to have filled the universe and created a bright night sky.

What does it mean to say that the universe has a finite "size?" Or, to rephrase, what would happen if we tried to throw something beyond the limits of the universe? This troubling question can be answered in the context of general relativity, which allows space to have a curvature in the presence of matter. We can best describe the curvature of three-dimensional space by analogy with a two-dimensional space such as Earth's surface. If we were restricted to move *only* on Earth's surface, we would find that a long voyage would have no limits yet would still lead us back, after a long circumferential trip, to our starting point. The space on the surface of a sphere is closed—that is, finite—yet has no real boundary, even though the "size" of this surface is well defined. Earth's surface, as opposed to the surface of an infinite plane, has *curvature*. Whether or not our three-dimensional space has curvature is a question that can be answered, at least in principle, without making a round trip, which in the case of our universe would be a very long one indeed (see Question 1).

46–6 THE EARLY MOMENTS OF THE UNIVERSE

Special relativity tells us that energy cannot be distinguished from mass. When we move back in time to the early universe, the mass density of the universe increases, which means that the energy density increases. By the principles of thermodynamics, the temperature increases as well. Only one mathematical model, the *Friedmann model*, seems to describe the overall dynamic structure of an expanding universe in a way consistent with the formalism of general relativity—indeed, this model can be used to define the Big Bang theory. When Alexander Friedmann proposed his model in 1922, he had no idea that he was correctly predicting the expanding universe that Hubble observed. Friedmann was merely trying to find the simplest model predicted by Einstein's general relativity, and that model implies a dynamic, expanding universe.

The Friedmann model follows from two simple assumptions that are referred to as the *cosmological principle*: The universe looks the same in all directions (the universe is *isotropic*), and it looks the same from any observation point (the universe is *homogeneous*). Both assumptions are roughly consistent with observation. We say "roughly" because the presence of clumps of matter such as planets, stars, galaxies, and clusters of galaxies appear to be inconsistent with homogeneity. However, if we look from a sufficient distance, the distribution of matter is relatively uniform, as in Fig. 46–15. The Friedmann model has a definite prediction for temperature of the universe as a function of its size: For all but the earliest times in the history of the uni-

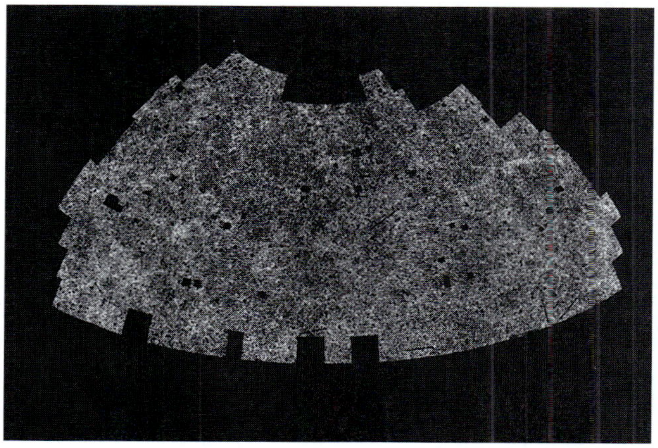

FIGURE 46–15 Structure is evident in this very large scale composite image of some one million galaxies. That structure repeats uniformly across the entire view.
On this very large scale, the universe is homogeneous—the "foamy" structure is the same all the way across the image.

verse, temperature is inversely proportional to "size," or *distance scale*, R:

$$T \propto \frac{1}{R}. \tag{46–7}$$

Thus, as we move forward in time and the universe expands (R increases), the temperature decreases. The Friedmann model also describes how R, and hence T, changes as a function of time.

Temperature and Energy Scales

The equipartition theorem of statistical mechanics states that the energy of an individual particle in a system at temperature T is on the order of kT. As the temperature increases, the increasing energy can bring us to the realm where atoms or nuclei can be broken up or where the collisions between nucleons are in the energy regime described in the first two sections of this chapter. Statistical techniques are sufficiently well developed to treat these complicated systems. At the high temperatures of the early universe, the interests of physicists studying the behavior of matter at the shortest distances and those studying cosmology converge. The further back in time we go, the higher the temperatures we encounter, from energies where the fundamental physics is well understood to energies where we are still groping for answers about the behavior of matter.

Equipartition is described in Chapter 19.

It is helpful to view the changes in the early universe in terms of a logarithmic time scale because we can "stretch" smaller and smaller fractions of a second into finite segments on the logarithmic scale. Such a scale never arrives at time zero because $t = 0$ is the point $-\infty$ on the logarithmic scale. In any case, we do not know how to go all the way to $t = 0$ because of our difficulty in reconciling quantum physics with general relativity, in which the presence of matter affects the nature of space and time. All that our knowledge really allows us to say is that, thus far, we have a satisfactory model of our dynamic universe back in time to when it occupied a very small but finite volume. To go further back in time is pure speculation. We shall now take a brief tour, starting with the earliest time our present understanding of the physics allows us to treat.†

A Time Chart of the Early Universe

Figure 46–16 illustrates the change in various parameters of the universe, in particular temperature, as a function of time. The temperatures in the chart are determined by

†For more information on this fascinating subject, among several accessible treatments we cite in particular Steven Weinberg, *The First Three Minutes: A Modern View of the Origin of the Universe*, Basic, 1988.

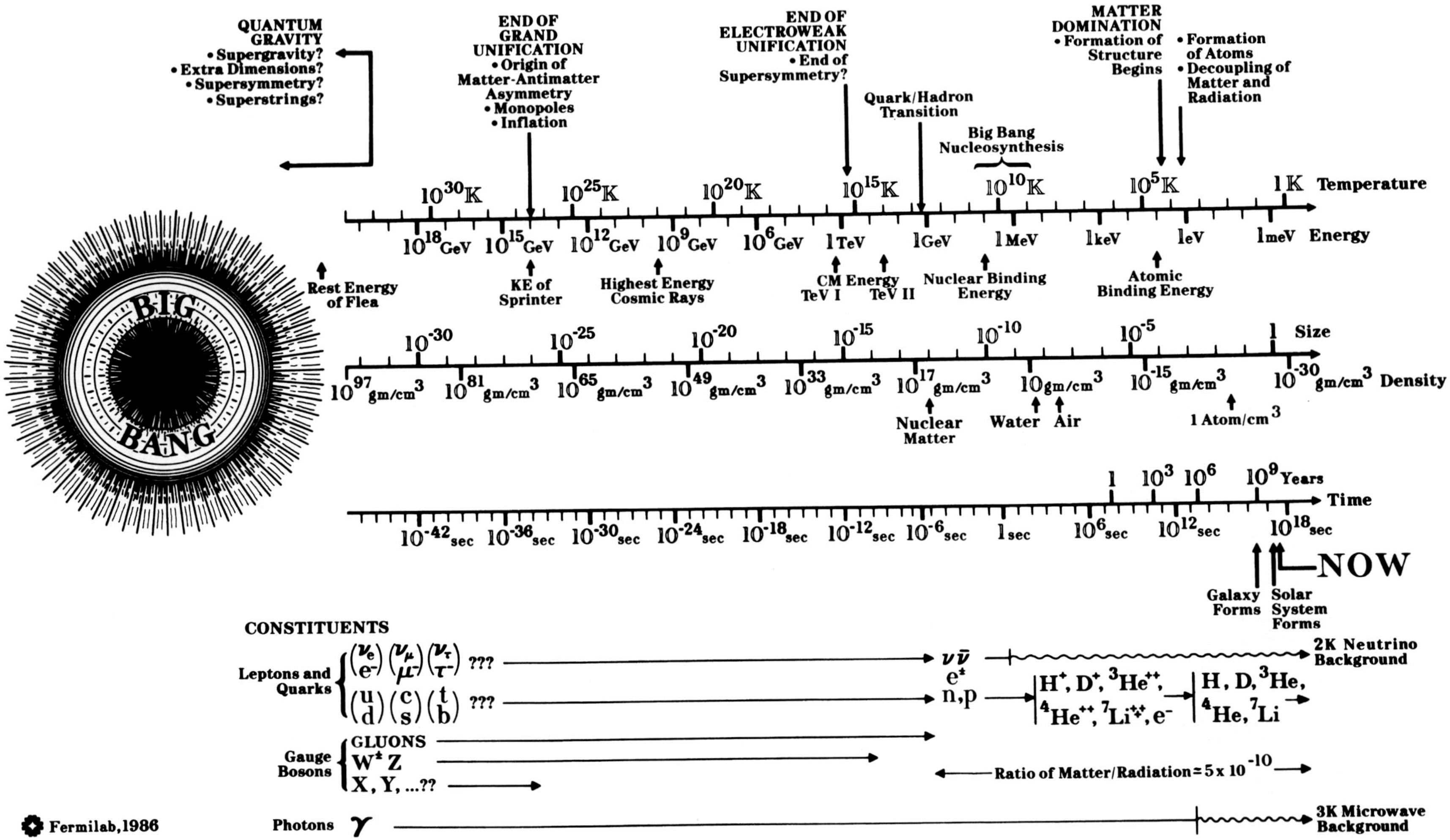

FIGURE 46–16 Stages in the early universe. One of the horizontal axes represents time increasing on a logarithmic scale. An increase in time means an increase in size and a decrease in temperature, characteristic energy, and density. In addition to signaling various "events" in the history of the universe, the chart indicates the changing composition of the universe.

working backward from the temperature of background radiation today. The stages labeled in this chart correspond to different "eras" in the development of the universe, and we shall comment on key aspects of these eras.

Time Period $t < 10^{-43}$ s This period can be called the era of quantum gravity. Physicists have not yet succeeded in combining general relativity with quantum mechanics, so this era is not understood in any fundamental way. Dimensional analysis suggests that this era must involve both quantum physics and gravitation: We can form a mass (or, equivalently, an energy) from Planck's constant, the gravitational constant, and the speed of light. This mass, known as the *Planck mass*, is given by

$$M_P \equiv \sqrt{\frac{\hbar c}{G}} \simeq 2.2 \times 10^{-8}\ \text{kg}. \tag{46–8}$$

(Keep in mind that although this mass may appear small, it is roughly equal to 1.3×10^{19} proton masses.) The energy equivalent to this mass is $M_P c^2 \simeq 1.2 \times 10^{19}$ GeV, corresponding to a temperature of 1.4×10^{32} K (see Problem 39). This energy is higher than any we shall ever achieve with an accelerator built with our current technology.

Time Period 10^{-43} s $< t < 10^{-12}$ s. This period also lies beyond what we can study with current accelerators, whose highest energy corresponds to $t \simeq 10^{-12}$ s. However, if our ideas about strong and electroweak interactions are correct, we can say that nucleons have not yet formed and that the universe is a soup of all the fundamental particles we know about. Particles and antiparticles, including quarks and antiquarks, are continuously being formed in pairs and annihilated.

One feature of this period may be associated with the apparent domination of matter over antimatter in the universe today. Although we have no direct evidence that distant galaxies are made of matter rather than antimatter (antimatter would form antiatoms with exactly the same emission spectra as atoms), there is much indirect evidence that the amount of antimatter in the universe is limited. For example, even though Earth is continually bombarded with cosmic rays, these particles from distant space are *particles*, not antiparticles. The collision of antimatter and matter would produce spectacular emissions of radiant energy. Yet if a hot early universe had equal amounts of matter and antimatter, which is a fair assumption, how could we have arrived at today's situation?

Experimental evidence from the decays of a particle called the K-meson indicates that physical processes are not quite symmetric under the reversal of time. This is not a statistical time's arrow, like that discussed in Chapter 20, but rather a real, if small, directionality of physical processes. To use the language of Chapter 20, a film of fundamental physical processes that is run in the forward direction would look slightly different than the same film run backward. Although we do not understand the origin of this effect, its implications have been explored. One of its consequences is that matter is preferentially created over antimatter, if only slightly, in the ensemble of possible reactions that could have occurred up to time $t = 10^{-12}$ s. For every 10^9 antiquarks created, roughly $10^9 + 1$ quarks are created. Later, when the universe had cooled to the point where the average energy of colliding objects in the cosmic soup was less than the mass of a particle (and of its antiparticle), then production of that particle–antiparticle pair no longer took place. Eventually, as the temperature dropped, no more particle–antiparticle production took place. The existing particles and antiparticles could then annihilate each other without being re-created in other collisions. (This occurred when the universe was about 1 s old.) At this point, the slight predominance of matter over antimatter was all-important, because the remaining bit of matter, the matter that exists around us today, had no antimatter to annihilate against, and it persisted.

Time Period 10^{-12} s $< t < 1$ s. The period from 10^{-12} s on covers a realm where we have investigated at least the basic interactions with Earth-based experiments. But we have no direct experiments on the behavior of large quantities of quarks and antiquarks. Thus our ability to trace the behavior of the universe is a combination of direct knowledge and deduction. A particularly significant moment in the evolution of the universe was that of the *quark–hadron transition*, when the energy density decreased to the point that quarks, antiquarks, and the gluons formed nucleons and other particles that participate in the nuclear force. This occurred at about 10^{-6} s, when the temperature was some 10^{13} K, corresponding to an energy of around 1 GeV. Of the strongly interacting particles, only the nucleons are stable on the time scale of 1 s, and even that fact must be tempered by our knowledge that neutrons decay due to the weak interaction. Even though up and down quarks occurred in equal numbers in the cosmic soup so the number of neutrons equaled the number of protons initially, the neutrons decayed such that, at around 1 s, there was one neutron for every five protons. Between 1 s and 100 s, the time at which some nuclei formed and held the neutrons permanently, many more neutrons decayed. At 100 s, neutrons and protons existed in about a 1:7 ratio.

Time Period 1 s $< t <$ 300,000 yr. This period is sometimes referred to as the *radiation era* because, during this time, the energy density of electromagnetic radiation was greater than that of matter. The Friedmann model shows that the distance scale was proportional to the square root of time for this period,

$$R \propto \sqrt{t}. \tag{46–9}$$

Both the energy density of radiation and the energy density of matter decreased as the universe expanded, but the energy density of radiation decreased more rapidly because the energy density in mass itself (as opposed to the kinetic energy of matter) was preserved. At roughly 300,000 yr, the two energy densities were equal. We may pick out a significant moment within this era, $t \simeq 100$ s, when the temperature had dropped to the point where neutrons and protons combined to form primarily the nuclei H^+ and $^4He^{2+}$, with some $^2H^+$, $^3He^{2+}$, and $^7Li^{3+}$. (Only these nuclei formed because they are so strongly bound that potential barriers exist against the formation of others.) All the positive charge in this set of nuclei was matched by an equal quantity of negative charge in the form of free electrons, which continuously absorbed and reradiated the electromagnetic radiation.

Time Period $t >$ 300,000 yr. The energy density of matter was greater than the energy density of electromagnetic radiation once the universe was older than about 300,000 yr, a period often called the *matter-dominated era*. Today, the energy density of matter is several thousand times greater than that of radiation. Since some 300,000 yr after the Big Bang, the expansion has followed the law

$$\mathrm{R} \propto t^{2/3}. \tag{46–10}$$

At about the moment that the universe became matter dominated, which was the moment when the temperature was some 3000 K, the temperature was no longer high enough to keep electrons and ions from combining into stable neutral atoms. The ratio of one neutron for every seven protons determined the number of nuclei formed at 100 s and hence the ratios of the stable elements, primarily hydrogen and helium. With the 1:7 ratio of neutrons to protons, the number of moles of hydrogen formed was 12 times the number of moles of helium formed, and hence the mass of hydrogen was roughly 3 times that of the helium (see Problem 43). All the stellar evolution that has occurred since that time has not significantly changed this difference (only 2 percent of the mass of today's universe is in elements heavier than helium), so this ratio of 3 in the mass abundance can be regarded as a prediction and has been confirmed by observation. This measurement is an important confirmation of the Big Bang theory.

All the free electrons are used to form stable atoms, and without free electrons, electromagnetic radiation no longer interacts as strongly with matter. The universe is said to have become *transparent* because radiation *decoupled* from matter. It is this radiation that we spoke of in Chapter 17 when we described the discovery of the 3-K background blackbody radiation. The wavelengths of this radiation increase as the universe expands; in fact, they are a measure of R. Blackbody radiation with larger wavelengths corresponds to blackbody radiation at a lower temperature. What was blackbody radiation at 3000 K at the moment of decoupling has become, through the expansion, radiation at 3 K. The fact that we observe this radiation, which is the oldest directly observable relic of the Big Bang, supports the ideas discussed here.

In the period that followed decoupling, stars formed, as did galaxies. The stellar evolution process, which involves series of nuclear reactions that occur in the hearts of stars, is on the whole well understood and confirmed by observation. According to this process, heavy elements, and hence the matter that makes up Earth and its inhabitants, are produced in supernovas, violent explosions of stars. We are thus part of a second generation of stars.

The Uniformity of the Universe

Matter distribution in the universe is granular, or clumped. Galaxies (clumps of stars) are grouped into clusters, and there are large regions with no visible matter (Fig. 46–17). The granular structure is thought to have developed from small, random fluctuations in a uniform energy density, forming small clumps of higher density. Such clumps act as seeds, gathering matter into them under the influence of the gravitational force. The process is much like the formation of clouds, whose droplets require "nuclei" of some kind. The initial density fluctuations that started the process must have begun at a time not too late in the evolution of the universe. The 3-K background radiation should contain some "memory" of these fluctuations. In 1992, the COBE satellite (see Chapter 17) discovered nonuniformities in the background radiation consistent with the observed granularity of the matter distribution.

The Future

Will the universe continue to expand forever? The answer to that question lies in a parameter of the Friedmann model. At large distances, the gravitational force controls the destiny of the universe. If there is more than a *critical density* of matter, the expansion would slow, stop, and reverse itself (resulting in a *closed* universe). An era of collapse would inevitably follow the expansion era. With less than the critical density, the expanding universe would continue to expand (an *open* universe). Between these two cases, there is a *critical* universe, with the critical density of matter (Fig. 46–18). Current observations can account directly for enough matter for only about 10 percent of the critical density. But indirect evidence—details about the rotation of galaxies, for example—indicates the presence of matter that we cannot observe directly (*dark matter*). On the cosmic scale, 10 percent is close enough to 100 percent to make it tempting to think that the actual density may equal the critical density! In that case, the universe

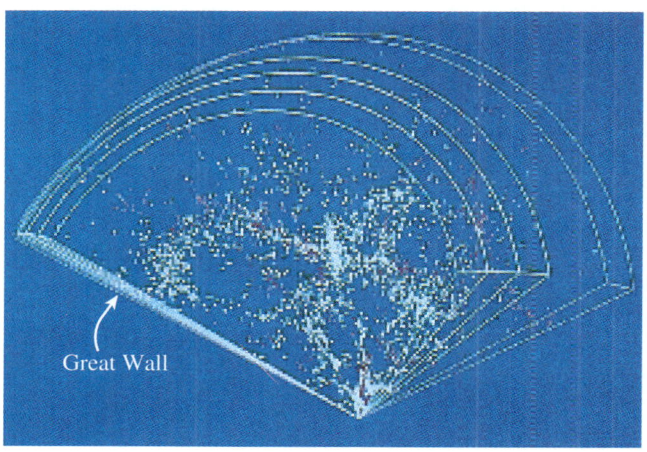

FIGURE 46–17 This composite view of galaxies, which covers a small region of the view in Fig. 46–15, reveals a well-developed structure of clumping and voids. One of these features has received a name—the "great wall."

would take forever to reach a point where it is neither expanding nor contracting. We shall surely learn much in trying to untangle the answer to our question.

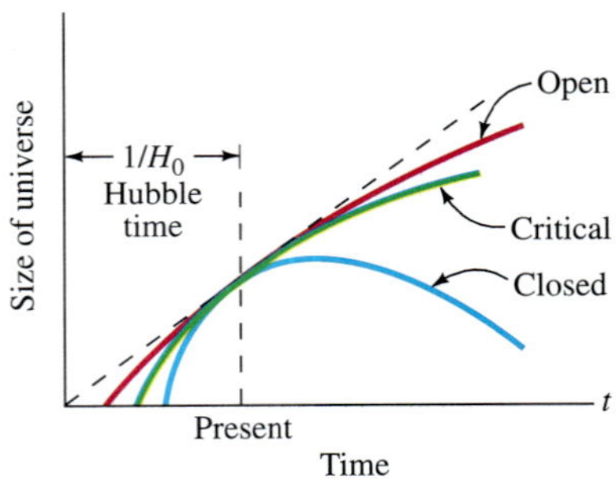

FIGURE 46–18 The evolution with time of open, closed, and critical universes. The vertical axis is some measure of the size of the universe.

46–7 A FEW LAST WORDS

In the introduction to this text, we spoke of macroscopic and microscopic realms. For centuries, engineers (or the inventors of technology) were limited to manipulating matter in the macroscopic realm on the basis of empirical knowledge. That matter was known to consist of atoms, and that the behavior of atoms was governed by quantum mechanics, initially had little impact on the way engineers worked. Scientists, by contrast, were interested from the start in the fundamental questions for their own time, those that appeared to be rather remote from practical application. The interest of engineers, with certain exceptions, came later and in unforeseeable ways. Thus the bulk of today's technology is built on quantum principles and on an understanding of the quantum nature of matter. It is unimaginable that our society could operate without the quantum-physics-based circuits that make computers or electronic communication possible or that a chemical engineer could construct new molecules without knowledge of their microscopic structure. While engineers continue to build bridges, those bridges are built on a solid understanding of the microscopic behavior of the materials that comprise the bridges and with the aid of computers that depend on the microscopic realm.

We are still learning how to apply quantum mechanical ideas to bulk materials. We are still in the process of evaluating how to make the best use of our scientific understanding of atoms and nuclei. And we have not yet seen how we can directly apply the knowledge we have gained in the areas of particle physics and cosmology. But we can say with confidence that if our society survives in anything like its present form, we shall use our new knowledge.

SUMMARY

At the microscopic scale, matter exhibits a series of levels, one within the other. Bulk matter is made of atoms and molecules (themselves made of atoms); atoms are built from electrons and nuclei; nuclei are constructed from nucleons (protons and neutrons); and nucleons and other particles that participate in the nuclear force are comprised of quarks. These different levels, which correspond to smaller and smaller distances, are probed by microscopes that must, because of the Heisenberg uncertainty relation between momentum and distance, use beams of higher and higher momenta. This fact implies that special relativity becomes important when we want to probe distances of nuclear size and less.

At the level of the structure of nucleons, a fundamentally new feature emerges: confinement. Quarks, unlike the constituents of matter at larger scales, are confined; that is, they have never been observed as isolated particles. It is believed that, by the nature of the strong force, they can never be isolated.

Three forces can be regarded as fundamental: the gravitational force, the electroweak force, and the strong force. The gravitational force, unlike the electroweak and strong forces, has not yet been successfully reconciled with the ideas of quantum mechanics. Particles exert these forces on one another through the exchange of still other particles. The electroweak force is a unified version of the electromagnetic force and the weak force, which nevertheless appear to be rather different from one another at low energies. The exchanged particles of the electroweak force are the photon, which is massless, and the W and Z bosons, whose masses are some 100 times that of the proton. The strong force is mediated by the exchange of the gluon. Because the gluon is itself subject to the strong force, this force increases with increasing distance, a fact that guarantees that not only quarks, but gluons as well, are confined.

Various conservation laws apply to the fundamental forces and manifest themselves in certain rules for how reactions can proceed. The conservation of electric charge, of baryon number, and of lepton number applies in strong and electroweak interactions.

The reactions that occur in high-energy collisions form the most important pieces of evidence for the deduction of the fundamental laws. Colliders, accelerators in which beams of electrons or protons collide head-on, allow us to produce reactions with the highest energies.

Experiments on high-energy reactions share a common ground with astronomical observations. Hubble's law describes an expanding universe that started from a tiny initial volume, at the so-called Big Bang. When the expansion started, the universe was hot and dense, conditions appropriate to the high-energy reactions where the fundamental forces act. As the universe expanded, the temperature dropped, and a succession of energy scales resulted. By applying their knowledge of the behavior of matter at these different energy scales, physicists have successfully accounted for the background blackbody radiation measured to be present as well as the observed amounts of hydrogen and helium in today's universe. The early evolution of the universe must have involved gravity in a form consistent with quantum ideas, as we can see from the Planck mass,

$$M_P = \sqrt{\frac{\hbar c}{G}}. \qquad (46\text{–}8)$$

The value of M_P is characteristic of very high energies, hence of very early times. Accounting for these earliest times is a formidable challenge for scientists.

UNDERSTANDING KEY CONCEPTS

1. Whether a surface has curvature can be determined from certain geometrical properties of that surface. The sum of the angles of a triangle drawn in a flat plane is 180°. Show that this is not the case for Earth's surface by summing the angles of a triangle that links the North Pole with any two points on the equator (Fig. 46–19).

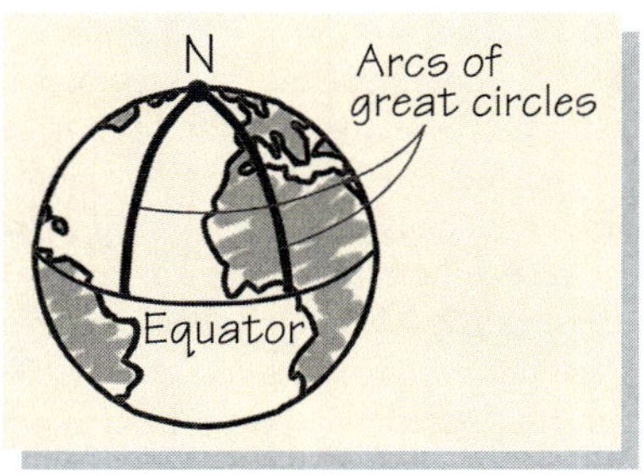

FIGURE 46–19 Question 1.

2. There are at least three known "families" of leptons. For example, in addition to the electron, there is a particle known as the muon (μ), which behaves in all respects like the electron except that it is more massive. By virtue of the extra mass, the muon is unstable. We also believe that the muon has a muon–lepton number, which, like the lepton number associated with the electron, is conserved. What experiment might tell you that the lepton numbers of the muon and the electron are conserved separately?

3. The lifetime of the proton is at least 10^{30} s, much greater than the age of the universe. Yet it is possible to create and destroy protons in particle reactions. How?

4. Why are electrons rather than protons used to study the charge distribution in the nucleus?

5. In a reaction between colliding high-energy electrons and positrons, why is the excess energy likely to create particles and antiparticles?

6. What fundamental forces most nearly account for the following phenomena: (a) walking; (b) the structure of gasoline molecules; (c) moonrise; (d) the source of sunlight; (e) the passage of sunlight from the Sun to Earth; (f) neutron stars (stars made mainly of neutrons packed to a density on the order of that of the nucleus)?

7. Can a particle be a lepton and hadron simultaneously? Meson and hadron? Hadron and baryon?

8. Suppose that electrons could not be removed from atoms, isolated, and studied independently. Would we still have experimental evidence of their existence?

9. Cosmic rays, consisting of protons and nuclei, arrive from outer space with energies that sometimes surpass those of the largest particle accelerators. What are the advantages and disadvantages of using such particles for beams for the study of high-energy reactions?

10. Earth orbits the Sun, the Sun moves within our galaxy, and our galaxy moves within its own galactic group. With all this movement, how could Hubble have discovered Hubble's law?

11. If the universe is closed, then a beam of light emitted from one point will eventually make a round trip and come back to that point, just as Magellan went all the way around the world without ever making a 180° turn. Does this mean that you would be able to see your own back if you looked through a sufficiently powerful telescope?

12. By energy and momentum conservations, an isolated electron cannot emit a photon and remain an electron. Why, then, can this happen in a virtual process?

13. Suppose that an accelerator can accelerate either electrons or protons to a given energy and that these projectiles are to collide with protons at rest. Which type of projectile should be chosen to obtain the greatest center-of-mass energy in the collision?

14. We know that the early universe was denser than it is now. What are some pieces of evidence that indicate that the early universe was hotter?

PROBLEMS

46–1 Levels of Matter

1. (I) Robert Hofstadter received a 1961 Nobel Prize for experiments in which 500-MeV electrons scattered from stationary nuclei. He found that the nucleus has a fairly constant density but, over a thickness of about 2.4 fm at its surface, the density drops to zero. With what spatial resolution could Hofstadter probe the nucleus with 500-MeV electrons?

2. (I) The Stanford Linear Accelerator accelerates electrons up to an energy of 52 GeV. These electrons can collide with stationary targets. (a) Determine the spatial accuracy with which such a probe can discern details of the nucleus. (b) In the 1960s, experiments performed there with 20-GeV electrons showed that the proton contains pointlike particles, the quarks. What is the best spatial resolution obtainable with such an electron beam?

3. (II) We want to use electrons as projectiles to study the structure of protons by means of the scattering reaction $e + p \rightarrow e + p$. The target proton is initially at rest in the laboratory; we measure those electrons that recoil directly backward. Suppose that all the motion is ultrarelativistic (you can ignore all rest masses except that of the particle at rest). Calculate the momentum transferred to the proton target (Fig. 46–20). If the energy of the incident electron is doubled, how much does the momentum transferred to the proton increase?

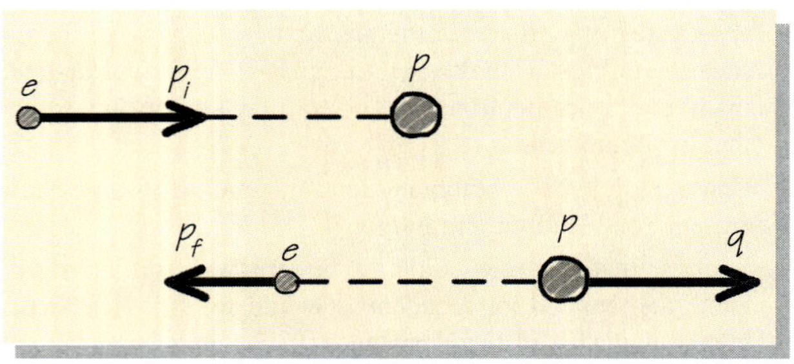

FIGURE 46–20 Problem 3.

4. (II) Problem 3 describes the ultrarelativistic collision of an electron with a stationary proton. Compare the final momenta obtained there with the momenta obtained in a similar situation with nonrelativistic kinematics.

5. (II) Consider the ultrarelativistic $e + p$ collision described in Problem 3, but now consider those events in which the electron is scattered at 90° to its initial direction (Fig. 46–21). What is the direction and magnitude of the proton's momentum after the collision?

6. (II) (a) Repeat Problem 3, but keep the rest mass of the proton, so that $E = \sqrt{(m_pc^2)^2 + (qc)^2}$ for a proton of momentum q. Continue to treat the ratio of the mass of the electron to that of the proton as zero. (b) Is there an initial electron energy for which the electron motion must be treated relativistically while the motion of the proton can be treated nonrelativistically?

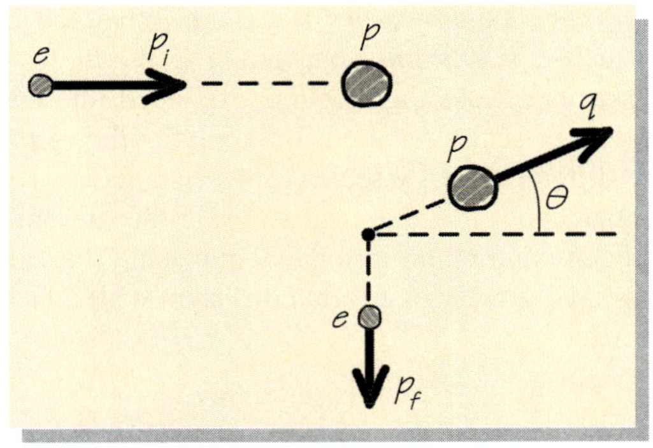

FIGURE 46–21 Problem 5.

7. (II) Equation (45–2) gives the collision cross section of a *pointlike* projectile of charge Z_1e, mass m, and kinetic energy K scattering from a *pointlike* target particle of charge Z_2e and mass M ($M >> m$) at rest, as a function of the scattering angle θ:

$$\sigma(\theta) = Z_1^2 Z_2^2 \left(\frac{e^2}{4\pi\epsilon_0}\right)^2 \frac{1}{16} \frac{1}{K^2 \sin^4(\theta/2)} .$$

Suppose that the initial and final momenta of the projectile are $\mathbf{p}_i$ and $\mathbf{p}_f$, respectively, with $|\mathbf{p}_i| = |\mathbf{p}_f| = p$. Then for large M the magnitude of the momentum transfer $\Delta \equiv \mathbf{p}_f - \mathbf{p}_i$ in the collision is given by $\Delta^2 = (\mathbf{p}_f - \mathbf{p}_i)^2 = 2p^2(1 - \cos\theta)$. (a) Express the cross section in terms of Δ^2, and plot it as a function of Δ^2. (b) Suppose that the target is not pointlike but spread out over a sphere of radius r_0. Give some qualitative arguments based on the position–momentum uncertainty relation of how the dependence of the cross section on Δ^2 would change.

8. (III) Consider a colliding electron and proton. Show that if in the center-of-mass reference frame the electron has energy 3000 MeV, then in the laboratory reference frame (in which the proton is at rest), the electron has an energy of about 20 GeV, or 20,000 MeV. [*Hint*: You may suppose that in the center-of-mass frame, the electron and proton collide and produce some object with zero total momentum, and with a mass M that can be calculated, given the electron's momentum. In the laboratory frame, that same object is created, and, by the conservation of momentum, moves with a momentum p_{lab} equal to that of the incoming electron. The energy of the object in the laboratory frame is then given by $E_{\text{object}} = \sqrt{(Mc^2)^2 + (p_{\text{lab}}c)^2}$. You may then use energy conservation,

$p_{\text{lab}}c + m_p c^2 = E_{\text{object}}$, and your calculation of the mass M to find $p_{\text{lab}}c$.]

9. (III) An electron collides with a second electron at rest. What is the minimum energy the incident electron must have to produce a proton–antiproton pair? The reaction takes the form $e + e \rightarrow e + e + (p\overline{p})$. Ignore the electron mass. (See the hint to Problem 8. Fig. 46–22 shows the reaction in the center-of-mass frame.)

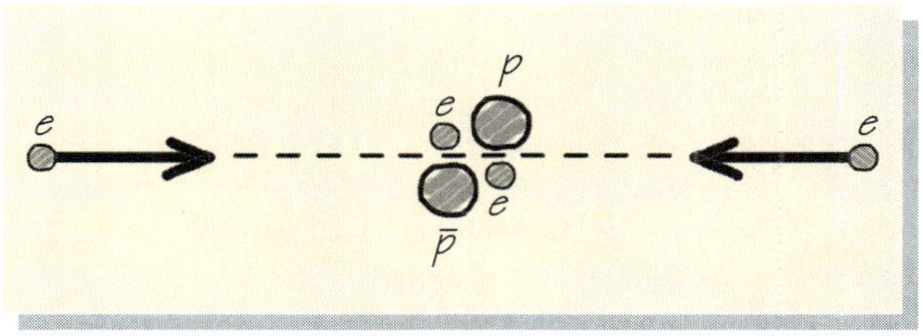

FIGURE 46–22 Problem 9.

46–2 New Particles and Conservation Laws

10. (I) Use Table 46–1 to find the electric charge, baryon number, and spin of a π^-, which consists of a $d\overline{u}$ quark combination.

11. (I) The K^+ meson (the *kaon*) decays by the process $K^+ \rightarrow \pi^+ + \pi^0$. Assume that in this decay the conservation laws of charge, baryon number, and lepton number are obeyed. What are the values of these quantum numbers for the K^+ meson?

12. (I) Show that the conservation laws of charge, baryon number, and lepton number are satisfied in neutron decay, $n \rightarrow p + e^- + \overline{\nu}$.

13. (I) The reaction $p + e^- \rightarrow \overline{p} + e^+$ does not occur in sensitive experiments designed to observe it. Why not?

14. (II) The π^- meson decays as $\pi^- \rightarrow \mu^- + \overline{\nu}$. What is the baryon number of the μ^-? What is its lepton number? Is it a boson or a fermion? What is its electric charge?

46–3 Fundamental Forces Revisited

15. (I) The W^+ meson has a mass of 81 GeV/c^2. Calculate the range of the force transmitted by the exchange of the W^+ meson.

16. (I) Describe the quark content of the antiproton.

17. (II) There exist mesons like the pion that can be emitted and absorbed by nucleons, but with mass some 5.5 times that of the pion. The exchange of these particles gives rise to a repulsive nucleon–nucleon force. (a) Estimate the range of this repulsive force. (b) Estimate at what energy, in the center-of-mass reference frame of a proton–proton collision, the repulsive force will begin to affect the scattering.

18. (II) You have 1 kg each of hydrogen (gram-atomic weight 1 g/mol) and uranium 235 (gram-atomic weight 235 g/mol). What are the baryon numbers of the two samples of material?

19. (II) The pion is the exchange boson for nuclear forces: It is exchanged between two protons or two neutrons and leads to the nuclear force. Given that the transition $n + \pi^0 \rightarrow n$ is possible (virtually), where π^0 is the uncharged pion, describe the minimum quark content for this particle.

20. (II) Positive pions (π^+) can be produced in the reaction $p + p \rightarrow \pi^+ + D$, where D stands for the deuteron, the ^{2}H nucleus. Use this information and the known quark content of protons (*uud*) and neutrons (*udd*) to get the quark content of these pions.

21. (II) The quark contents of some strongly interacting particles are (*uud*) for protons, (*udd*) for neutrons, $(u\overline{d})$ for π^+, $(d\overline{u})$ for π^-, and $(u\overline{u} - d\overline{d})/\sqrt{2}$ for π^0. Figure 46–23 is a sketch of the reaction $p + \pi^- \rightarrow n$ that uses this information. Here all lines represent particles moving to the right, with right-pointing arrows for quarks and left-pointing arrows for antiquarks. Moreover, quark lines never end in such a graph. Use this technique to draw graphs for the reactions (a) $p + \pi^- \rightarrow n + \pi^0$, (b) $p + \overline{p} \rightarrow \pi^+ + \pi^- + \pi^0$, and (c) $p + \pi^- \rightarrow p + n + \overline{p}$.

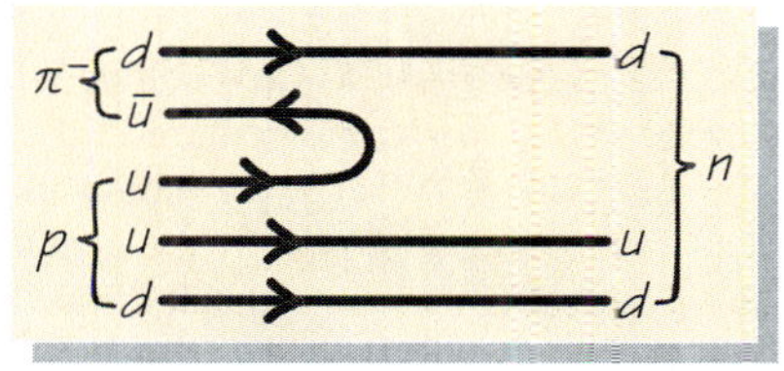

FIGURE 46–23 Problem 21.

22. (II) Combine the quark-line rules of Problem 21 with the additional graphical rules for emission of a W^- boson by a d-quark, emission of a W^+ boson by a u-quark, and $W^- \rightarrow e^- + \overline{\nu}$, as drawn in Fig. 46–24. Use these rules to give the graphical representation of the processes (a) $n \rightarrow p + e^- + \overline{\nu}$, (b) $\pi^+ \rightarrow \pi^0 + e^+ + \nu$, and (c) $\pi^- \rightarrow e^- + \overline{\nu}$, all of which proceed with a W boson present at an intermediate stage.

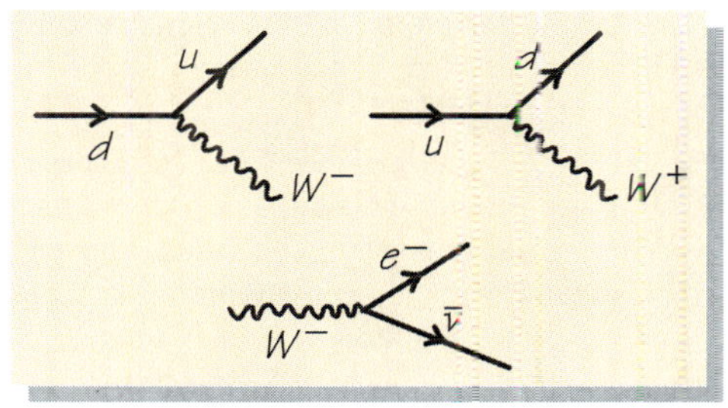

FIGURE 46–24 Problem 22.

46–4 Tools of Particle Physics

23. (I) According to some theories, a proton at rest can decay (with very long lifetime!) in the mode $p \rightarrow e^+ + \pi^0$. Use $m_p \simeq$ 940 MeV/c^2 and $m_\pi \simeq$ 135 MeV/c^2 to calculate the momentum, q, of the π^0 in the decay; neglect the electron mass (Fig. 46–25). Express your answer in terms of the combination qc, in units of MeV. Detectors looking for the decay would use this characteristic momentum as a signal.

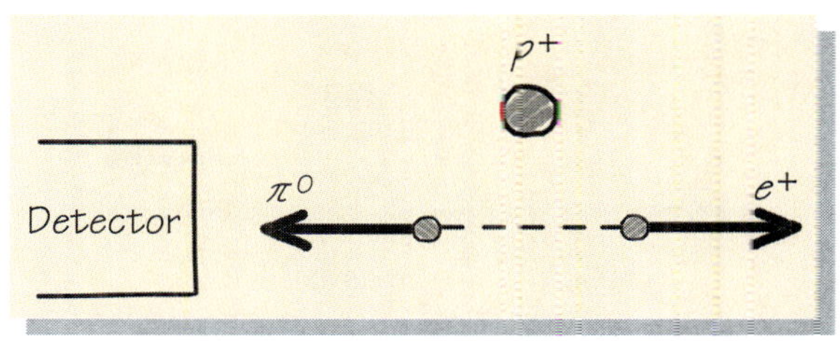

FIGURE 46–25 Problem 23.

24. (II) A proton–antiproton collider is actually a quark–antiquark collider, if we think of each quark in a proton (or each antiquark in an antiproton) as carrying one-third the momentum of the proton (or antiproton). What is the proton (and antiproton) energy in a symmetric $p\bar{p}$ collider that produces quark–antiquark collisions with the same total energy as the collisions produced by an electron–positron collider in which each electron carries a momentum of 11 GeV/c? Ignore the masses of the electron, positron, quark, and antiquark but not those of the proton and antiproton.

25. (II) Calculate the momentum in the center of mass of each incident proton for an accelerator in which 300-GeV protons are incident on stationary proton targets. (See the hint for Problem 8.)

26. (II) Antiprotons of energy 1000 GeV collide with protons of the same energy but opposite momentum in the TEVATRON collider (see Table 46–3). What is the momentum of antiprotons incident on stationary protons if the center-of-mass energy of the $p\bar{p}$ system is the same as that of the TEVATRON?

27. (II) An experimenter wants to study the reaction $e^- + e^+ \rightarrow Z^0$, where the Z^0, of mass 91.1 GeV/c^2, is the neutral intermediate boson of the electroweak interactions. The experimenter has a fixed-target machine that accelerates a beam of positrons that can be directed at atoms, which provide a stationary target of electrons. What is the minimum energy of the positron beam for which the reaction can occur?

28. (II) The *luminosity* of a colliding beam machine is the rate at which particles pass in one beam times the number of collisions per unit area of a second beam. The luminosity of the TEVATRON (see Table 46–3) is $10^{37}/\text{m}^2\cdot\text{s}$. If the total collision cross section for proton–proton interactions is 100 millibarns (1 barn $= 10^{-28}\ \text{m}^2$), how many interactions per second are there at the TEVATRON?

29. (II) The TEVATRON collider (see Table 46–3) has proton and antiproton beams that follow a circular path of circumference 6.3 km. What is the frequency with which a given proton in the proton beam crosses a given antiproton in the antiproton beam, as seen in the laboratory reference frame?

30. (II) A calorimeter whose active material is 100 cm^3 of water is used to measure the energy of protons. How large an increase of temperature is expected when 10^{12} protons of kinetic energy 1 GeV are stopped in the calorimeter?

31. (II) A proton–antiproton pair can be produced in the reaction $\gamma + p \rightarrow p + \bar{p} + p$ (Fig. 46–26). If the target proton is at rest, what is the minimum energy that the photon must have to produce this reaction?

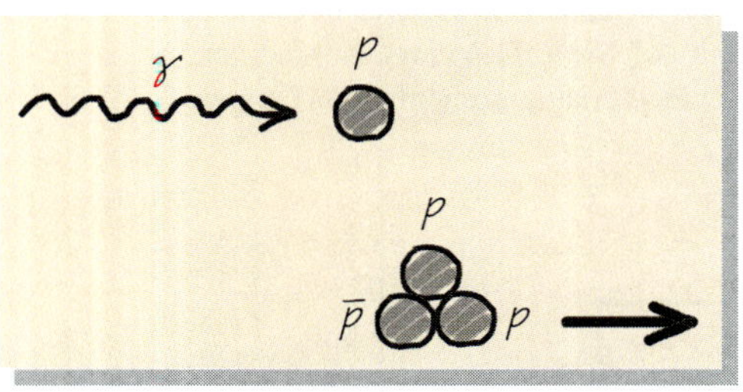

FIGURE 46–26 Problem 31.

32. (III) Table 46–3 shows one collider, HERA, that is asymmetric in that its two beams, one of electrons and the other of protons, are of different momenta. This collider is thus a kind of hybrid between symmetric colliders and fixed-target machines. In the following, neglect the baryon-number and lepton-number conservation laws that must be obeyed in proton–electron collisions. (a) What is the maximum number of neutral pions that can be produced in HERA collisions? (b) What momentum would beams of protons and electrons of equal momentum have in order to be able to produce the number of pions in part (a)? (c) What momentum would a beam of electrons incident on protons at rest have to produce the number of pions in part (a)?

33. (III) Electron accelerators that produce copious quantities of X-rays for research on atomic and condensed-matter physics employ electron beams of several GeV. The X-rays are produced when the electrons scatter from photons of lesser energy. (For example, the National Synchrotron Light Source at Brookhaven National Laboratory accelerates electrons to an energy of 2.5 GeV and produces photons with energies up to 310 MeV.) Consider laser photons (with wavelength of, say, 590 nm) that scatter head-on from 8-GeV electrons. Calculate the maximum energy of the recoil photon produced in such a (Compton scattering) reaction (Fig. 46–27). At what angle would this photon be produced? [*Hint*: You must keep the electron mass in the relativistic expression for the electron energy, E, even though $E >> m_e c^2$. You may, however, make the approximation $\sqrt{(pc)^2 + (m_e c^2)^2} \simeq pc + (m_e c^2)^2/2pc$ in this case.]

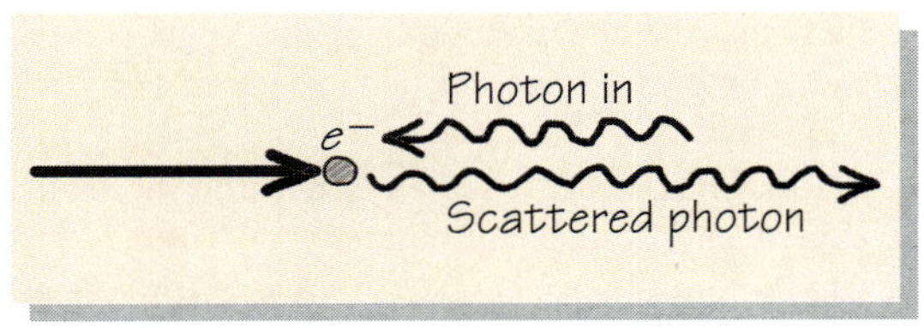

FIGURE 46–27 Problem 33.

46–5 The Expanding Universe

34. (II) Consider the one-dimensional line of galaxies in Figs. 46–14a and 46–14b. Show that an inhabitant of galaxy B would agree that Hubble's law holds, but with the distance being the distance from galaxy B. In this way you confirm that Hubble's law is the unique law consistent with no special central point.

35. (II) Assume that the size of the universe is determined by a distance scale factor R. This factor is increasing with time t at the rate $R = kt^n$, where k and n are constants. Hubble's law then becomes $dR/dt = HR$. Show that the Hubble parameter H is not constant, and find its time dependence.

36. (II) The universe is expanding, but the expansion may be slowing down. (a) If R varies with time as in Problem 35, find the deceleration of R. (b) The deceleration parameter, q, is defined by $q \equiv -R(d^2R/dt^2)/(dR/dt)^2$. Calculate the value of q.

37. (III) Use Hubble's law to find the time it would take for the distance from Earth to some galaxy to increase by 10 percent over what it is today. Assume that for the period over which you make your calculation, H is constant. [*Hint*: First find the time interval for the distance to change from x to $x + dx$, assuming that dx is an infinitesimal quantity. Then integrate your result.]

38. (III) By using the techniques of Problem 37, show that if Hubble's law with a constant value of H has held ever since the time when there was no separation between galaxies, it would have taken an infinite amount of time for the universe to have expanded to its present situation.

46–6 *The Early Moments of the Universe*

39. (I) Show that the Planck mass, 2.2×10^{-8} kg, is equivalent to an energy of 1.2×10^{19} GeV and to a temperature of 1.4×10^{32} K.

40. (I) In addition to the Planck mass, there is a distance referred to as the *Planck length*. Use dimensional analysis to form a length from the constants $\hbar$, G, and c.

41. (I) In addition to the Planck mass, there is a time referred to as the *Planck time*. Use dimensional analysis to form a time from the constants $\hbar$, G, and c.

42. (I) Use Fig. 46–16 to find the energy of a particle, the size of the universe, and the time when the temperature of the universe was 10^{25} K.

43. (II) Suppose that at the time when stable atoms formed, neutrons and protons were present in the ratio of 1:4. (a) Show that if hydrogen and helium formed with these nucleons (plus an appropriate number of electrons), then the number of moles of hydrogen formed was 6 times the number of moles of helium formed. (b) Show that the total mass of hydrogen formed is roughly 1.5 times that of helium formed then. (The actual evolution was such that the neutron to proton ratio was 1:7 when stable atoms formed.)

44. (III) Under conditions that apply to our universe, Einstein's theory of gravitation relates the density of matter to the Hubble parameter through the relation $\rho = 3H^2/8\pi G$. (a) Given the present value of the Hubble parameter, $H_0 = 2.5 \times 10^{-18}\ \mathrm{s}^{-1}$, calculate the present value of the density of matter, ρ_0. Assuming that matter consists (almost) entirely of hydrogen, with $M_H = 1.7 \times 10^{-27}$ kg, find the number density of hydrogen. (b) Show that the conservation of baryon number implies that the time dependence of the density is given by $\rho(t) = \rho_0[R_0/R(t)]^3$, where R_0 is the current distance scale of the universe and $R(t)$ is the distance scale at a general time t. (c) Combine the result of part (b) with the general relation between ρ and $H \equiv (1/R)\ dR/dt$ to show that $R(t) = kt^{2/3}$, and find k. [*Hint*: Substitute the general form $R(t) = kt^n$ into your equations.]

General Problems

45. (II) Below what temperature will stable carbon atoms form? He^+ ions? The ionization energies of H and C are 13.6 eV and 11.3 eV, respectively.

46. (II) What is the maximum number of antiprotons that can be created in a scattering reaction with 70-GeV protons incident on a stationary hydrogen target?

47. (II) Consider a fixed-target machine with protons of momentum 3.2×10^5 GeV/c. If the largest magnetic field we could use to guide the beam in a circular path has magnitude 9.5 T, what is the radius of such a machine?

48. (II) *Ĉerenkov counters* are detectors that measure the presence and speed of charged particles by observing the angle of the "wake" of light (Ĉerenkov radiation; see Fig. 14–24) produced by those particles when their speed through a medium is greater than the speed of light in that medium. What is the minimum energy of electrons that produce Ĉerenkov radiation in toluene ($n = 1.494$)?

49. (II) In a reaction in which 40-GeV electrons hit 40-GeV positrons head-on (as has been done at the Stanford Linear Accelerator Center), what is the maximum number of π mesons (and antimesons) that can be created for each colliding pair? The mass of a π meson is about 140 MeV/c^2.

50. (II) Find a relationship between the de Broglie wavelength and the kinetic energy, K, of ultrarelativistic particles (particles for which $K >> mc^2$).

51. (II) The muon, symbol μ, is for all practical purposes just like the electron, except that it is some 207 times more massive. The muon can be destroyed by a proton in the reaction $\mu^- + p \rightarrow n + \nu$. A muon sent into hydrogen gas slows and is finally captured in a hydrogenlike Bohr orbit. Once it is in the lowest orbit (radius R_0), it will be destroyed by the reaction above if it comes within a distance $L = 0.2$ fm of the proton (Fig. 46–28). The probability that a muon in the lowest orbit will come that close to the proton is estimated from quantum mechanics to be the ratio of the volume of a sphere of radius L to that of a sphere of radius R_0. Suppose that the lifetime for destruction from the lowest Bohr orbit in hydrogen is τ_c. Show that the lifetime for destruction from the ground state of a Bohrlike atom in which the muon orbits a nucleus with Z protons is τ_c/Z^4.

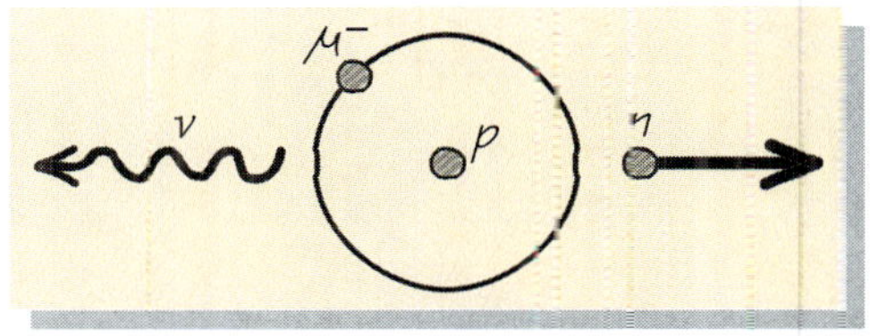

FIGURE 46–28 Problem 51.

52. (II) Consider muon destruction from the lowest state in a hydrogenlike Bohr atom (see Problem 51) What is the energy of the neutron that emerges from the reaction? If the process occurs when the muon is in orbit around a heavy nucleus, will the reaction look the same as in the hydrogenlike atom?

53. (II) Using energy conservation and momentum conservation, prove that an isolated electron cannot emit a photon and remain an electron. (See Question 12.)

54. (II) Estimate the repulsive force between protons in a uranium nucleus. Compare that force to the attraction between the uranium nucleus and the innermost electron in a uranium atom.

55. (II) Free neutrons have a lifetime of approximately 890 s. Given that the ratio of neutrons to protons was 1:5 when the universe was 1 s old, *estimate* the ratio at 1000 s. Take into account only the loss of neutrons through their decay.

56. (II) A pi meson (of mass approximately 140 MeV/c^2) is ejected from a nuclear collision with a kinetic energy of 350 MeV. Pi mesons at rest have a lifetime of about 2.6×10^{-8} s. Calculate (a) the speed of the pi meson, (b) its momentum, and (c) how far, on average, such a pi meson will travel before it decays.

57. (III) When an antiproton slows down by interacting with matter and approaches a proton, it can be captured and form a hydrogenlike atom called *protonium* through its Coulomb attraction to the proton. Find (a) the binding energy of the atom in electron-volts and (b) the radius of the atom in its ground state. (Do not forget the reduced-mass effect.)

58. (III) The probability that a proton and an antiproton in the ground state of protonium (see Problem 57) will come within a distance R_1 of each other is estimated from quantum mechanics to be the ratio of the volume of a sphere of radius R_1 to that of a sphere of radius $2R_0$, where R_0 is the ground-state radius of protonium. If the lifetime for annihilation of a proton and an antiproton is 10^{-22} s when they are within 0.2 fm of each other and is infinite beyond this distance, estimate how long the protonium atom lasts in its ground state.

APPENDIX I

The Système Internationale (SI) of Units

I–1 SOME SI BASE UNITS

Physical Quantity	Name of Unit	Symbol
length	meter	m
mass	kilogram	kg
time	second	s
electric current	ampere	A
thermodynamic temperature	kelvin	K
amount of substance	mole	mol

I–2 SOME SI DERIVED UNITS

Physical Quantity	Name of Unit	Symbol	SI Unit
frequency	hertz	Hz	s^{-1}
energy	joule	J	$kg \cdot m^2/s^2$
force	newton	N	$kg \cdot m/s^2$
pressure	pascal	Pa	$kg/m \cdot s^2$
power	watt	W	$kg \cdot m^2/s^3$
electric charge	coulomb	C	$A \cdot s$
electric potential	volt	V	$kg \cdot m^2/A \cdot s^3$
electric resistance	ohm	Ω	$kg \cdot m^2/A^2 \cdot s^3$
capacitance	farad	F	$A^2 \cdot s^4/kg \cdot m^2$
inductance	henry	H	$kg \cdot m^2/A^2 \cdot s^2$
magnetic flux	weber	Wb	$kg \cdot m^2/A \cdot s^2$
magnetic flux density	tesla	T	$kg/A \cdot s^2$

I–3 SI UNITS OF SOME OTHER PHYSICAL QUANTITIES

Physical Quantity	SI Unit
speed	m/s
acceleration	m/s^2
angular speed	rad/s
angular acceleration	rad/s^2
torque	$kg \cdot m^2/s^2$, or $N \cdot m$
heat flow	J, or $kg \cdot m^2/s^2$, or $N \cdot m$
entropy	J/K, or $kg \cdot m^2/K \cdot s^2$, or $N \cdot m/K$
thermal conductivity	$W/m \cdot K$

I–4 SOME CONVERSIONS OF NON-SI UNITS TO SI UNITS

Energy:
1 electron-volt (eV) = 1.6022×10^{-19} J
1 erg = 10^{-7} J
1 British thermal unit (BTU) = 1055 J
1 calorie (cal) = 4.186 J
1 kilowatt-hour (kWh) = 3.6×10^6 J

Mass:
1 gram (g) = 10^{-3} kg
1 atomic mass unit (u) = 931.5 MeV/c^2 = 1.661×10^{-27} kg
1 MeV/c^2 = 1.783×10^{-30} kg

Force:
1 dyne = 10^{-5} N
1 pound (lb or #) = 4.448 N

Length:
1 centimeter (cm) = 10^{-2} m
1 kilometer (km) = 10^3 m
1 fermi = 10^{-15} m
1 Angstrom (Å) = 10^{-10} m
1 inch (in or ″) = 0.0254 m
1 foot (ft) = 0.3048 m
1 mile (mi) = 1609.3 m
1 astronomical unit (AU) = 1.496×10^{11} m
1 light-year (ly) = 9.46×10^{15} m
1 parsec (ps) = 3.09×10^{16} m

Angle:
1 degree (°) = 1.745×10^{-2} rad
1 min (′) = 2.909×10^{-4} rad
1 second (″) = 4.848×10^{-6} rad

Volume:
1 liter (L) = 10^{-3} m^3

Power:
1 kilowatt (kW) = 10^3 W
1 horsepower (hp) = 745.7 W

Pressure:
1 bar = 10^5 Pa
1 atmosphere (atm) = 1.013×10^5 Pa
1 pound per square inch (lb/in^2) = 6.895×10^3 Pa

Time:
1 year (yr) = 3.156×10^7 s
1 day (d) = 8.640×10^4 s
1 hour (h) = 3600 s
1 minute (min) = 60 s

Speed:
1 mile per hour (mi/h) = 0.447 m/s

Magnetic field:
1 gauss = 10^{-4} T

APPENDIX II

Some Fundamental Physical Constants†

Constant	Symbol	Value	Error
speed of light in a vacuum	c	2.99792458×10^{8} m/s	exact
gravitational constant	G	6.67259×10^{-11} m³/kg · s²	128
Avogadro's number	N_A	6.02214×10^{23} mol^{-1}	0.6
universal gas constant	R	8.31451 J/mol · K	8.4
Boltzmann's constant	k	1.38066×10^{-23} J/K	8.5
elementary charge	e	1.60218×10^{-19} C	0.3
permittivity of free space	ϵ_0	$8.85418781762 \times 10^{-12}$ C²/N · m²	exact
	$1/4\pi\epsilon_0$	8.987552×10^{9} kg · m³ · s^{-2} · C^{-2}	
permeability of free space	μ_0	$4\pi \times 10^{-7}$ T · m/A	exact
electron mass	m_e	9.10939×10^{-31} kg	0.6
proton mass	m_p	1.67262×10^{-27} kg	0.6
neutron mass	m_n	1.67493×10^{-27} kg	0.6
Planck's constant	h	6.62608×10^{-34} J · s	0.6
$h/2\pi$	$\hbar$	1.05457×10^{-34} J · s	0.6
		$= 6.58212 \times 10^{-22}$ MeV · s	0.3
	$\hbar c$	197.327 Mev · fm	0.3
electron charge-to-mass ratio	$-e/m_e$	-1.75882×10^{11} C/kg	0.3
proton-electron mass ratio	m_p/m_e	1836.15	0.15
molar volume of ideal gas at STP		22414.1 cm³/mol	8.4
Bohr magneton	μ_B	9.27402×10^{-24} J/T	0.3
magnetic flux quantum	$\Phi_0 = h/2e$	2.067783×10^{-15} Wb	0.3
Bohr radius	a_0	0.529177×10^{-10} m	0.05
Rydberg constant	R_∞	1.09737×10^{7} m^{-1}	0.001

†From E. R. Cohen and B. N. Taylor, *The 1986 Adjustment of the Fundamental Constants,* Report of the CODATA Task Group on Fundamental Constants, CODATA Bulletin 63, Pergamon, Elmsford, N.Y. (1986).

We have given values of the measured constants to six significant figures, even though they may be known to greater accuracy. We quote the error, which expresses the uncertainty in the values of these constants, in parts per million. Defined constants have no error, and we give their full definition; they are indicated by the notation "exact" in the error column.

APPENDIX III

Other Physical Quantities

III–1.1 SOME ASTRONOMICAL CONSTANTS

Constant	Symbol	Value
standard gravity at Earth's surface	g	$9.80665\ m/s^2$
equatorial radius of Earth	R_e	6.374×10^6 m
mass of Earth	M_e	5.976×10^{24} kg
mass of Moon		7.350×10^{22} kg $= 0.0123\ M_e$
mean radius of Moon's orbit around Earth		3.844×10^8 m
mass of Sun	$M_\odot$	1.989×10^{30} kg
radius of Sun	$R_\odot$	6.96×10^8 m
mean radius of Earth's orbit around Sun	AU	1.496×10^{11} m
period of Earth's orbit around Sun	yr	3.156×10^7 s
diameter of our galaxy		7.5×10^{20} m
mass of our galaxy		2.7×10^{41} kg $= (1.4 \times 10^{11})\ M_\odot$
Hubble parameter	H	$2.5 \times 10^{-18}\ s^{-1}$

III–1.2 PLANETARY DATA

Planet	Diameter (in km)	Diameter Relative†	Relative Mass†	Average Density (in g/cm³)	Period of Rotation	Surface Gravity† (in g)	Escape Speed (in km/s)	Semimajor Axis (AU)	Period of Solar Orbit	Average Orbital Speed (in km/s)
Mercury	4,800	0.38	0.05	5.2	58 d 15 h	0.38	4.3	0.387	87.96 d	47.8
Venus	12,100	0.95	0.82	5.3	243 d 4 h	0.90	10.3	0.723	224.7 d	35.0
Earth	12,750	1.00	1.00	5.5	23 h 56 min	1.00	11.2	1.000	365.26 d	29.8
Mars	6,800	0.53	0.11	3.8	24 h 37 min	0.38	5.1	1.524	687.0 d or 1.88 yr	24.2
Jupiter	142,800	11.23	317.9	1.3	9 h 50 min	2.69	59.5	5.20	11.86 yr	13.1
Saturn	120,660	9.41	95.2	0.7	10 h 39 min	1.19	35.6	9.54	29.46 yr	9.7
Uranus	51,000	3.98	14.6	1.3	17 h	0.93	21.4	19.18	84.01 yr	6.8
Neptune	49,500	3.88	17.2	1.7	18 to 22 h	1.22	23.6	30.06	164.79 yr	5.4
Pluto	2,290	0.23	0.002	0.4	6 d 9 h 17 min	0.05	1.2	39.44	248 yr	4.7

†Relative to Earth.

III–2 ENERGY SUPPLY AND DEMAND†

†From the *Physics Vade Mecum,* Ed. Herbert L. Anderson, American Institute of Physics (New York, 1981); and U.S. Congress, Office of Technology Assessment, *Changing by Degrees: Steps to Reduce Greenhouse Gases,* OTA-O-482 (Washington, D.C.: U.S. Government Printing Office, February 1991).

III–2.1 FUEL RESOURCES (1980, ESTIMATED)

Resource	U.S. Resources	World Resources
coal (recoverable)	5×10^{21} J	2×10^{22} J
oil (not including oil shales)	10^{21} J	10^{22} J
natural gas	2×10^{21} J	10^{22} J
hydroelectric	10^{22} J/yr (North America)	6×10^{22} J/yr

III–2.2 ANNUAL USAGE OF RESOURCE (1989, PERCENTAGE OF TOTAL)

Resource	U.S. Usage (total = 9×10^{19} J)	World Usage (total = 4×10^{20} J)
coal	40	24
oil	23	35
natural gas	23	18
nuclear	7	5
hydroelectric	3	5
biomass	3	13

III–2.3 ENERGY CONTENT OF FUELS

Fuel	Energy Content (in J/kg)
bread	10×10^6
glucose ($C_6H_{12}O_6$)	16×10^6
white pine wood	20×10^6
methyl alcohol (CH_4O)	20×10^6
anthracite coal	32×10^6
domestic heating oil	45×10^6
propane (C_3H_8)	50×10^6
natural gas (96% CH_4)	51×10^6
fission of U^{235}	5.8×10^{11}
perfect mass-energy conversion	9×10^{16}

III–2.4 SOLAR ENERGY OUTPUT

total radiated power from the Sun	4×10^{26} W
power per unit area at the top of Earth's atmosphere	1.4 kW/m^2
average power per unit area delivered to an average horizontal surface in the United States in 1 yr	0.2 W/m^2

III–2.5 ENERGY CONSUMPTION IN TRANSPORTATION

Mode	Energy Consumption (J/passenger · km)
bicycle	5×10^4
foot travel	1.5×10^5
intercity bus	3×10^5
intercity train	9×10^5
automobile	1.5×10^6
747 jet airplane	2×10^6
snowmobile	6×10^6

III–2.6 ENERGY CONSUMPTION OF ELECTRICAL APPLIANCES

Appliance	Power (in W)	Energy Use Per Year (in kWh)
window air conditioner	1565	1390
clock	2	17
dishwasher	1200	363
window fan	200	170
hair dryer	380	14
iron	1000	144
microwave oven	1450	190
radio	71	86
refrigerator-freezer	615	1830
stove	12,200	1175
color television	200	440
vacuum cleaner	630	46
washing machine	512	103

APPENDIX IV

Mathematics

IV–1 SOME MATHEMATICAL CONSTANTS†

Constant	Value
π	3.14159
e (Euler's constant)	2.71828
$\sqrt{2}$	1.41421
$1/\sqrt{2}$	0.707107
$\ln(10)$	2.30259
$\ln(2)$	0.693147
1 rad	57.2958°
1°	0.0174533 rad

†To six significant figures.

IV–2 SOLUTION OF QUADRATIC EQUATIONS

Quadratic equation:

$$ax^2 + bx + c = 0$$

Two solutions:

$$x = \frac{-b \pm \sqrt{b^2 - 4ac}}{2a}$$

IV–3 BINOMIAL THEOREM

$$(x + y)^n = \sum_{k=0}^{n} \binom{n}{k} x^{n-k} y^k,$$

where

$$\binom{n}{k} = \frac{n!}{(n-k)!\,k!}.$$

The factorial $m! \equiv 1 \cdot 2 \cdot 3 \ldots \cdot m$; $0! \equiv 1$. Some particular cases of the binomial theorem:

(1) $(x \pm y)^2 = x^2 \pm 2xy + y^2$;
(2) $(x \pm y)^3 = x^3 \pm 3x^2y + 3xy^2 \pm y^3$;
(3) $(x \pm y)^4 = x^4 \pm 4x^3y + 6x^2y^2 \pm 4xy^3 + y^4$.

IV–4 TRIGONOMETRY

1. For a right triangle with sides a, b, and c (the hypotenuse), where the angle opposite side a is θ_a (Fig. A–1),

$$\text{sine of } \theta_a = \sin\theta_a \equiv \frac{a}{c};$$

$$\text{cosine of } \theta_a = \cos\theta_a \equiv \frac{b}{c};$$

$$\text{tangent of } \theta_a = \tan\theta_a = \frac{a}{b}.$$

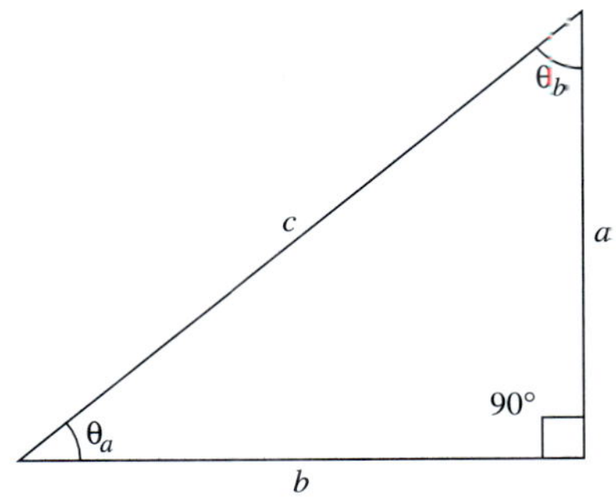

FIGURE A–1

2. The cosine function is even, $\cos(-x) = \cos x$; the sine function is odd, $\sin(-x) = -\sin x$.

3. (1) $\tan\theta = \dfrac{\sin\theta}{\cos\theta}$

(2) $\sec\theta = \dfrac{1}{\cos\theta}$

(3) $\operatorname{cosec}\theta = \dfrac{1}{\sin\theta}$

(4) $\cot\theta = \dfrac{1}{\tan\theta}$

4. (1) $\sin^2\theta + \cos^2\theta = 1$

(2) $\sec^2\theta - \tan^2\theta = 1$

(3) $\operatorname{cosec}^2\theta - \cot^2\theta = 1$

5. (1) $\sin(\theta_1 \pm \theta_2) = \sin\theta_1\cos\theta_2 \pm \cos\theta_1\sin\theta_2$

(2) $\cos(\theta_1 \pm \theta_2) = \cos\theta_1\cos\theta_2 \mp \sin\theta_1\sin\theta_2$

(3) $\sin\theta_1 \pm \sin\theta_2 = 2\sin\left(\dfrac{\theta_1 \pm \theta_2}{2}\right)\cos\left(\dfrac{\theta_1 \mp \theta_2}{2}\right)$

(4) $\cos\theta_1 + \cos\theta_2 = 2\cos\left(\dfrac{\theta_1 + \theta_2}{2}\right)\cos\left(\dfrac{\theta_1 - \theta_2}{2}\right)$

(5) $\cos\theta_1 - \cos\theta_2 = -2\sin\left(\dfrac{\theta_1 + \theta_2}{2}\right)\sin\left(\dfrac{\theta_1 - \theta_2}{2}\right)$

(6) $\tan(\theta_1 + \theta_2) = \dfrac{\tan\theta_1 + \tan\theta_2}{1 - (\tan\theta_1)(\tan\theta_2)}$

(7) $\cos\left(\theta \pm \dfrac{\pi}{2}\right) = \mp\sin\theta$

(8) $\sin\left(\theta \pm \dfrac{\pi}{2}\right) = \pm\cos\theta$

(9) $\sin\theta_1\sin\theta_2 = \dfrac{1}{2}[\cos(\theta_1 - \theta_2) - \cos(\theta_1 + \theta_2)]$

$$(10)\ \cos\theta_1 \cos\theta_2 = \frac{1}{2}[\cos(\theta_1 - \theta_2) + \cos(\theta_1 + \theta_2)]$$

$$(11)\ \sin\theta_1 \cos\theta_2 = \frac{1}{2}[\sin(\theta_1 - \theta_2) + \sin(\theta_1 + \theta_2)]$$

6. (1) $\sin(2\theta) = 2\sin\theta\cos\theta = \dfrac{2\tan\theta}{1 + \tan^2\theta}$

(2) $\cos(2\theta) = \cos^2\theta - \sin^2\theta = 2\cos^2\theta - 1 = 1 - 2\sin^2\theta$

(3) $\tan(2\theta) = \dfrac{2\tan\theta}{1 - \tan^2\theta}$

(4) $\sin\left(\dfrac{\theta}{2}\right) = \pm\sqrt{\dfrac{1 - \cos\theta}{2}}$

(5) $\cos\left(\dfrac{\theta}{2}\right) = \pm\sqrt{\dfrac{1 + \cos\theta}{2}}$

7. Expansions of Trigonometric Functions (θ in rad):

(1) $\sin\theta = \theta - \dfrac{\theta^3}{3!} + \dfrac{\theta^5}{5!} - \dfrac{\theta^7}{7!} + \cdots \qquad (\theta^2 < 1)$

(2) $\cos\theta = 1 - \dfrac{\theta^2}{2!} + \dfrac{\theta^4}{4!} - \dfrac{\theta^6}{6!} + \cdots \qquad (\theta^2 < 1)$

(3) $\tan\theta = \theta + \dfrac{1}{3}\theta^3 + \dfrac{2}{15}\theta^5 + \dfrac{17}{315}\theta^7 + \cdots \qquad \left(\theta^2 < \dfrac{\pi^2}{4}\right)$

IV–5 GEOMETRICAL FORMULAS

1. (circumference of a circle of radius r) = $2\pi r$

2. (area of a circle of radius r) = πr^2

3. (area of a sphere of radius r) = $4\pi r^2$

4. (volume of a sphere of radius r) = $\frac{4}{3}\pi r^3$

5. (area of a rectangle with sides of lengths L_1 and L_2) = $L_1 L_2$

6. For a right triangle with sides a, b, and c and angles θ_a and θ_b opposite the sides a and b, respectively (Fig. A–1):

(1) $a^2 + b^2 = c^2$ (the Pythagorean theorem)

(2) area = $\frac{1}{2}$(base)(height) = $\frac{1}{2}ab$

7. For a triangle with sides a, b, and c opposite the angles θ_a, θ_b, and θ_c, respectively (Fig. A–2):

(1) $\theta_a + \theta_b + \theta_c = 180° = \pi$ rad

(2) $a^2 = b^2 + c^2 - 2bc\cos\theta_a$

(3) $\dfrac{a}{\sin\theta_a} = \dfrac{b}{\sin\theta_b} = \dfrac{c}{\sin\theta_c}$

(4) $a = b\cos\theta_c + c\cos\theta_b$

(5) area $= \dfrac{1}{2}$(base)(height) $= \dfrac{1}{2}ab\sin\theta_c = \dfrac{1}{2}a^2\dfrac{\sin\theta_b \sin\theta_c}{\sin\theta_a}$

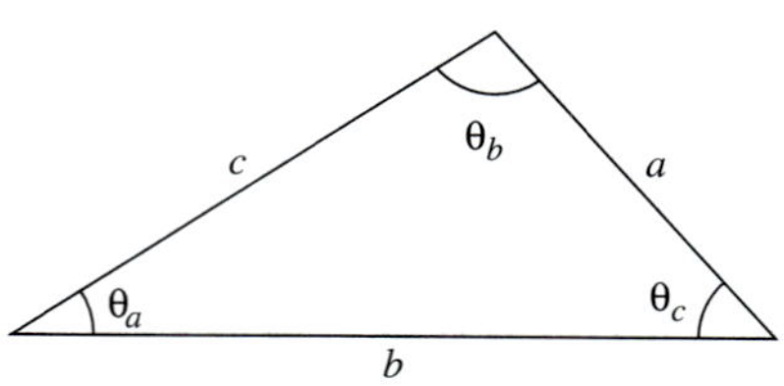

FIGURE A–2

8. (volume of a right cylinder of height h and radius r) = $\pi r^2 h$

IV–6 SOME PROPERTIES OF ALGEBRAIC FUNCTIONS

1. General properties

(1) $a^x a^y = a^{x+y}$

(2) $a^0 = 1$

(3) $(ab)^x = a^x b^x$

2. Properties of exponential of x, e^x:

(1) $e^{\ln(x)} = x$

(2) $e^{x_1}e^{x_2} = e^{x_1 + x_2}$

(3) $e^0 = 1$

(4) expansion: $e^x = 1 - x + \dfrac{x^2}{2!} - \dfrac{x^3}{3!} + \cdots$

3. Properties of the natural logarithm of x, $\ln(x)$:

(1) $\ln(e^x) = x$

(2) $\ln(x_1 x_2) = \ln(x_1) + \ln(x_2)$

(3) $\ln(x_1/x_2) = \ln(x_1) - \ln(x_2)$

(4) $\ln(1) = 0$

(5) expansion: $\ln(1 + x) = x - \dfrac{x^2}{2} + \dfrac{x^3}{3} - \dfrac{x^4}{4} + \cdots \ (x^2 < 1)$

IV–7 DERIVATIVES

In the following, b and p are constants and u and v are functions of x:

1. $\dfrac{db}{dx} = 0$

2. $\dfrac{d}{dx}(bu) = b\dfrac{du}{dx}$

3. $\dfrac{d}{dx}(u + v) = \dfrac{du}{dx} + \dfrac{dv}{dx}$

4. $\dfrac{d}{dx}(uv) = v\dfrac{du}{dx} + u\dfrac{dv}{dx}$

5. $\dfrac{dx^p}{dx} = px^{p-1}$

6. Chain rule: If u is a function of y and y is in turn a function of x, then $\dfrac{du}{dx} = \dfrac{du}{dy}\dfrac{dy}{dx}$

7. $\dfrac{d}{dx}(\sin x) = \cos x$

8. $\dfrac{d}{dx}(\cos x) = -\sin x$

9. $\dfrac{d}{dx}(\tan x) = \dfrac{1}{\cos^2 x}$

10. $\dfrac{d}{dx}(e^{bx}) = be^{bx}$

11. $\dfrac{d}{dx}\ln(x) = \dfrac{1}{x}$

IV–8 TAYLOR EXPANSION

If $f(x)$ is well-behaved near point $x = x_0$,

$$f(x) = f(x_0) + \left.\frac{df}{dx}\right|_{x=x_0}(x - x_0) + \left.\frac{d^2f}{dx^2}\right|_{x=x_0}\frac{(x - x_0)^2}{2!} + \cdots$$

IV–9 INTEGRALS

In the following, b and p are constants and u and v are functions of x:

1. $\int \frac{du}{dx}\,dx = u$

2. $\int_{x_1}^{x_2} \frac{du}{dx}\,dx = u(x_2) - u(x_1)$

3. $\int bu\,dx = b\int u\,dx$

4. $\int (u + v) = \int u\,dx + \int v\,dx$

5. $\int u\frac{dv}{dx}\,dx = uv - \int v\frac{du}{dx}\,dx$ (integration by parts)

6. If u is a function of y and y is in turn a function of x, then

$$\int u\,dy = \int u\frac{dy}{dx}\,dx$$

7. $\int x^p\,dx = \frac{x^{p+1}}{p+1}$ $(p \neq -1)$

8. $\int \frac{dx}{x} = \ln(|x|)$

9. $\int \sin x\,dx = -\cos x$

10. $\int \cos x\,dx = \sin x$

11. $\int e^{bx}\,dx = \frac{1}{b}e^{bx}$

12. $\int xe^{bx}\,dx = e^{bx}\left(\frac{x}{b} - \frac{1}{b^2}\right)$

13. Some definite integrals

(1) $\int_0^\infty x^n e^{-x}\,dx = n!$

(2) $\int_0^\pi \sin^2 x\,dx = \int_0^\pi \cos^2 x\,dx = \frac{\pi}{2}$

(3) $\int_0^\infty e^{-b^2x^2}\,dx = \frac{\sqrt{\pi}}{2b}$ $(b > 0)$

(4) $\int_0^\infty xe^{-x^2}\,dx = \frac{1}{2}$

(5) $\int_0^\infty x^2e^{-x^2}\,dx = \frac{\sqrt{\pi}}{4}$

(6) $$\int_0^\infty \frac{b}{b^2 + x^2}\,dx = \begin{cases} \frac{\pi}{2} & (b > 0) \\ 0 & (b = 0) \\ -\frac{\pi}{2} & (b < 0) \end{cases}$$

IV–10 SOME EXPANSIONS APPROPRIATE FOR $x^2 < 1$

1. The following expression is good for any n, positive or negative, integer or noninteger:

$$(1 + x)^n = 1 + nx + \frac{n(n-1)}{2!}x^2 + \frac{n(n-1)(n-2)}{3!}x^3 + \cdots$$

2. $$\sin x = x - \frac{x^3}{3!} + \frac{x^5}{5!} + \cdots$$

3. $$\cos x = 1 - \frac{x^2}{2!} + \frac{x^4}{4!} + \cdots$$

4. $$\tan x = x + \frac{x^3}{3} + \frac{2}{15}x^2 + \cdots$$

5. $$e^{ax} = 1 + ax + \frac{(ax)^2}{2!} + \frac{ax^3}{3!} + \cdots$$

IV–11 SOME MATHEMATICAL NOTATION

1.	$=$	is equal to
2.	$\simeq$	is approximately equal to
3.	$\propto$	is proportional to
4.	$\equiv$	is defined to be
5.	$\neq$	is unequal to
6.	$>$	is greater than
7.	$\geq$	is greater than or equal to
8.	$<$	is less than
9.	$\leq$	is less than or equal to
10.	Δx	the change in x
11.	$\lvert x \rvert$	the absolute value of x
12.	$O(N)$	on the order of the magnitude of N
13.	$\pm$	plus or minus
14.	$\mp$	minus or plus
15.	$\langle x \rangle$	average of x
16.	$\sum_{i=i_1}^{i_2} f_i$	the sum of all f_i over the integers i from a smallest integer i_1 to a largest integer i_2
17.	$\ln(x)$	natural logarithm of x
18.	$\log(x)$	logarithm to the base 10 of x
19.	$\int$	integral
20.	$\oint$	line integral around a loop

APPENDIX V

Periodic Table of the Elements

Atomic Mass → 1.00797
Atomic number → 1 H ← Symbol
Hydrogen ← Name

Metals

Nonmetals

1	2	3	4	5	6	7	8	9	10	11	12	13	14	15	16	17	18
1.00797 1 H Hydrogen																1.00797 1 H Hydrogen	4.0026 2 He Helium
6.939 3 Li Lithium	9.012 4 Be Beryllium											10.811 5 B Boron	12.01 6 C Carbon	14.01 7 N Nitrogen	15.999 8 O Oxygen	18.998 9 F Fluorine	20.182 10 Ne Neon
22.99 11 Na Sodium	24.31 12 Mg Magnesium											26.98 13 Al Aluminum	28.09 14 Si Silicon	30.97 15 P Phosphorus	32.06 16 S Sulfur	35.45 17 Cl Chlorine	39.95 18 Ar Argon
39.10 19 K Potassium	40.08 20 Ca Calcium	44.96 21 Sc Scandium	47.90 22 Ti Titanium	50.94 23 V Vanadium	52.00 24 Cr Chromium	54.94 25 Mn Manganese	55.85 26 Fe Iron	58.93 27 Co Cobalt	58.71 28 Ni Nickel	63.54 29 Cu Copper	65.37 30 Zn Zinc	69.72 31 Ga Gallium	72.59 32 Ge Germanium	74.92 33 As Arsenic	78.96 34 Se Selenium	79.91 35 Br Bromine	83.80 36 Kr Krypton
85.47 37 Rb Rubidium	87.62 38 Sr Strontium	88.91 39 Y Yttrium	91.22 40 Zr Zirconium	92.91 41 Nb Niobium	95.94 42 Mo Molybdenum	(98) 43 Tc Technetium	101.07 44 Ru Ruthenium	102.9 45 Rh Rhodium	106.4 46 Pd Palladium	107.9 47 Ag Silver	112.4 48 Cd Cadmium	114.82 49 In Indium	118.69 50 Sn Tin	121.75 51 Sb Antimony	127.6 52 Te Tellurium	126.90 53 I Iodine	131.30 54 Xe Xenon
132.91 55 Cs Cesium	137.34 56 Ba Barium	Lanthanides 57-71	178.49 72 Hf Hafnium	180.95 73 Ta Tantalum	183.85 74 W Tungsten	186.2 75 Re Rhenium	190.2 76 Os Osmium	192.2 77 Ir Iridium	195.09 78 Pt Platinum	196.97 79 Au Gold	200.59 80 Hg Mercury	204.37 81 Tl Thallium	207.19 82 Pb Lead	208.98 83 Bi Bismuth	(210) 84 Po Polonium	(210) 85 At Astatine	(222) 86 Rn Radon
(223) 87 Fr Francium	(226) 88 Ra Radium	Actinides 89-103															

Lanthanides	138.91 57 La Lanthanum	140.12 58 Ce Cerium	140.91 59 Pr Praseodymium	144.24 60 Nd Neodymium	(145) 61 Pm Promethium	150.4 62 Sm Samarium	151.96 63 Eu Europium	157.25 64 Gd Gadolinium	158.9 65 Tb Terbium	162.5 66 Dy Dysprosium	164.9 67 Ho Holmium	167.3 68 Er Erbium	168.9 69 Tm Thulium	173.0 70 Yb Ytterbium	175.0 71 Lu Lutetium
Actinides	(227) 89 Ac Actinium	232.04 90 Th Thorium	(231) 91 Pa Protactinium	238.03 92 U Uranium	(237) 93 Np Neptunium	(242) 94 Pu Plutonium	(243) 95 Am Americium	(247) 96 Cm Curium	(247) 97 Bk Berkelium	(251) 98 Cf Californium	(254) 99 Es Einsteinium	(253) 100 Fm Fermium	(256) 101 Md Mendelevium	(253) 102 No Nobelium	(257) 103 Lr Lawrencium

†Atomic masses given in parentheses refer to the most stable isotope of an unstable element.

APPENDIX VI

Significant Dates in the Development of Physics

History can rarely be stated as a simple series of dates, and the history of science is no exception. Throughout the text we have alluded to important discoveries in physics. The list below is a personal choice and should be thought of as a guide. It oversimplifies some of the history, including stories that are covered more thoroughly in the text (for example, Coulomb's law). Some of the dates are to be taken with a grain of salt, because discoveries are rarely made in a single identifiable moment. Our list includes some names (and discoveries) not mentioned in the text. Far more numerous are the names not listed, the names of those who built the experimental foundations, those who explored the false paths and cleared the way for those whose names we remember today, or those who verified the speculations that we call laws.

1583	Galileo	Pendulum motion
1600	Gilbert	Study of magnets
1602	Galileo	Early statement of Newton's first law
1602	Galileo	Laws of falling bodies
1609	Kepler	First two laws of planetary motion
1619	Kepler	Third law of planetary motion
1620	Snell	Law of refraction
1648	Pascal	Atmospheric pressure
1650	Grimaldi	Diffraction of light
1661	Boyle	Chemical elements
1669	Newton	Light dispersion in prisms
1678	Huygens	Wave propagation
1687	Newton	Laws of motion; universal gravitation
1760	Black	Calorimetry
1785	Coulomb	Coulomb's law
1789	Lavoisier	Conservation of mass
1798	Cavendish	Measurement of G
1800	Volta	Electric battery
1801	Young	Interference of light
1801	Dalton	Laws of chemical combination
~ 1802	Charles; Gay-Lussac	Ideal gases
1807	Dalton	Atomic theory
1812	Fourier	Decomposition of waves
1815	Fraunhofer	Discrete spectral lines
1819	Fresnel	Wave picture of light
1820	Oersted	Magnetic fields from currents
1820	Biot; Savart	Law of magnetic field produced by current
1824	Carnot	Second law of thermodynamics
1827	Ohm	Ohm's Law
1827	Ampère	Ampère's law
1831	Faraday; Henry	Induction
1842	Joule	Mechanical equivalent of heat
1847	Helmholtz	Conservation of energy
1849	Fizeau	Direct measurement of the speed of light
1865	Maxwell	The laws of electricity and magnetism; light waves
1877	Boltzmann; Gibbs	Statistical mechanics
1879	Stefan	Blackbody radiation
1885	Osmond	Crystalline structure of metals
1887	Hertz	Electromagnetic waves
1887	Michelson and Morley	Constancy of the speed of light

1896	Becquerel	Radioactivity
1897	Thomson	Charge-to-mass ratio of the electron
1900	Planck	Quanta in blackbody radiation
1903	Rutherford; Soddy	Isotopes
1905	Einstein	Special relativity; quanta in photoelectric effect
1908	Kammerlingh Onnes	Superfluidity
1911	Kammerlingh Onnes	Superconductivity
1911	Rutherford	Nuclear structure of atom
1911	Millikan	Quantization of charge
1912	von Laue	X-ray diffraction in crystals
1913	Bohr	Atomic structure
1916	Einstein	General relativity
1923	Hubble	Discovery of galaxies
1924	de Broglie	Wave nature of particles
1925	Pauli	Exclusion principle
1925	Heisenberg	Formulation of quantum mechanics
1925	Goudsmit and Uhlenbeck	Electron spin
1926	Davisson and Germer; Thomson	Diffraction of electrons by crystals
1926	Schrödinger	Alternate formulation of quantum mechanics
1926	Born	Probabilistic interpretation of quantum theory
1927	Heisenberg	Uncertainty relations
1929	Hubble	Hubble's law
1930	Dirac	Antiparticles
1932	Anderson	The positron
1932	Lawrence and Livingston	The cyclotron
1932	Chadwick	The neutron
1934	Yukawa	Nuclear forces and the pi meson
1948	Feynman; Schwinger; Tomonaga	Electromagnetism as a quantum theory
1954	Townes	The maser
1957	Lee and Yang	Nonconservation of parity
1957	Bardeen, Cooper, and Schrieffer	Theory of superconductivity
1962	Josephson	Josephson junction
1964	Gell-Mann; Zweig	Quarks
1964	Penzias and Wilson	Background radiation of the universe
1967–1970	Glashow; Salam; Weinberg	Unification of electromagnetic and weak forces

APPENDIX VII

Tables in the Text

APPENDIX VIII

Selected Text Boxes

PROBLEM-SOLVING TECHNIQUES

Chapter	Subject	Page
2	General problem-solving techniques	29
4	How to prepare a force diagram for a given body	97
5	Problems in which more than one mass appears	120
6	The scalar product	151
6	Integration	155
7	The energy diagram	183
8	Collisions	216
9	Rigid-body motion	250
10	The vector product	270
11	Suggestions for statics problems	304
16	Bernoulli's equation expresses conservation of energy for a fluid	448
22	We often need to calculate electric forces	597
24	To use Gauss' law to find electric fields	644
28	In problems associated with multi-loop circuits	757
30	Problems on static magnetic fields	823
37	Ray tracing	994

APPLICATION

Chapter	Title	Page
1	The Global Positioning System	6
2	Timed Lights	31
3	Achieving Controlled Free Fall	67
4	Safe Ejection Seats	90
5	Drag Forces in an Everyday Device	127
6	Jet Engines-Kinetic Energy Systems	148
7	Conversion of Energy in Electrical Power Generation	187
8	The Pile Driver	204
9	Angular Momentum Conservation in Drag Racing	254
10	Navigation through the Conservation of Angular Momentum	279
11	Construction Cranes	309
12	Mapping g	333
13	Quartz Watches	365
14	Soldering Circuit Boards	387
15	Silence by Sound Wave Cancellation	417
16	Reeded Instruments	451
17	Carbonated Beverages	471
18	Adiabatic Transformations in Diesel Engines	502
19	Isotope Separation	521
20	Mandating Efficiencies	545

APPLICATION

Answers to Odd-Numbered Problems

CHAPTER 1

1. 2.500×10^3 green jelly beans.
3. 10^7; 10^{14}.
5. 1.8×10^{10} atoms.
7. \$0.71/kg.
9. 32.2 ft/s^2.
11. 3.33 g/cm^3.
13. 34 mi/gal; 10 mi/gal.
15. 0.0402%; 8.49×10^{-6}%.
17. $V = 0.09 \pm 0.04$ m^3.
19. 2.5%.
21. $[MLT^{-1}]$.
23. $[ML^2T^{-1}]$.
25. (a) $[\alpha] = [L^{-1}]$, (b) $[A] = [L^4T^{-2}]$.
27. 8 tons.
29. 3.5×10^{-6} cm^2.
31. (a) 10^6 mechanics, (b) 10^6 mechanics, (c) 10^6 mechanics.
33. 10^{19} droplets.
35. 5×10^6 automobiles.
37. 16 cm.
39. 1.2×10^{57} hydrogen atoms; 9×10^{56} hydrogen atoms.
41. Her position can be described by the clockwise angle ϕ from the north-south line from the center to the starting point. Her direction of travel will be tangent to the circle with a magnitude of 2 m/s at an angle ϕ above the west direction.
43. $1.2\mathbf{i} + 8.2\mathbf{j}$ paces (8.3 paces, 82° north of east).
45. Catch: $0\mathbf{i} + 0\mathbf{j}$; first turn: $0\mathbf{i} + 0\mathbf{j} + 20\mathbf{j} = 0\mathbf{i} + 20\mathbf{j}$; second turn: $0\mathbf{i} + 20\mathbf{j} - 15\mathbf{i} = -15\mathbf{i} + 20\mathbf{j}$; third turn: $-15\mathbf{i} + 20\mathbf{j} + 10\mathbf{j} = -15\mathbf{i} + 30\mathbf{j}$; fourth turn: $-15\mathbf{i} + 30\mathbf{j} + 25\mathbf{i} = 10\mathbf{i} + 30\mathbf{j}$; fifth turn: $10\mathbf{i} + 30\mathbf{j} - 10\mathbf{i} = 30\mathbf{j}$; touchdown: $30\mathbf{j} + 65\mathbf{j} = 95\mathbf{j}$.
47. $V_x = +V \cos \alpha$.
49. $\mathbf{AB} = (7, 0) = 7$ at an angle of 0°; $\mathbf{BC} = (-7, 7) = 7\sqrt{2}$ at an angle of 135°; $\mathbf{CA} = (0, -7) = 7$ at an angle of 270°.
51. (a) $\mathbf{A} = -4\mathbf{i} + 2\mathbf{j}$, $\mathbf{B} = -\mathbf{i} + 4\mathbf{j}$, $\mathbf{C} = 2\mathbf{i} + 2\mathbf{j}$, $\mathbf{D} = 5\mathbf{i} - 3\mathbf{j}$, (b) $-11\mathbf{i} + 9\mathbf{j}$, $5\mathbf{j}$, 9.2.

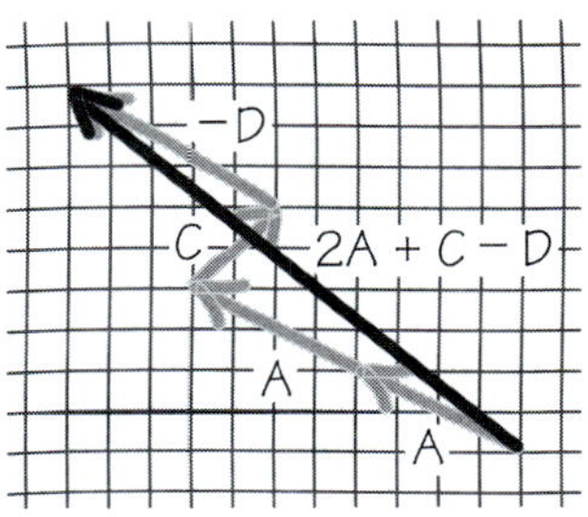

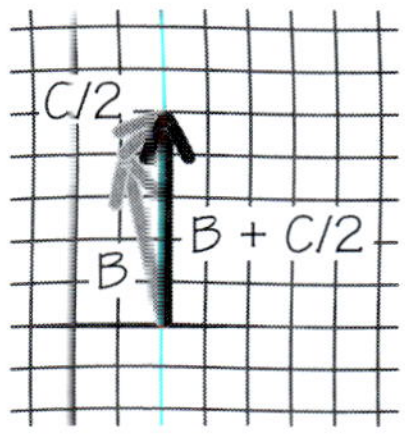

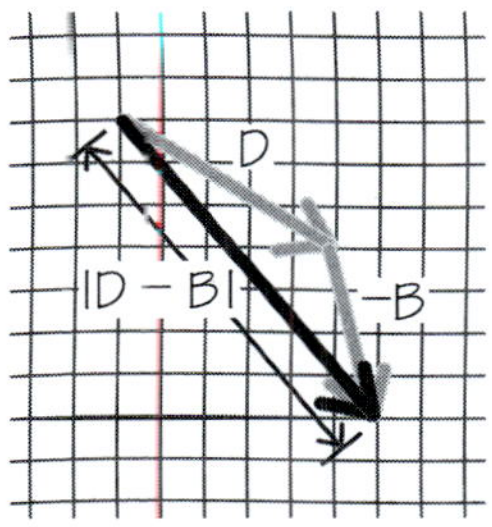

55. (a) $[M^{1/2}L^{3/2}T^{-1}]$, (b) $[ML^2T^{-2}]$.
57. (a) 10^{-6} metric tons/g, (b) 10^{-6} m^3/cm^3.
59. 1.3 s.
61. 3.0×10^{-26} kg/molecule.
63. 3×10^{-8} cm.
65. 2×10^{41} kg/galaxy; 1.2×10^{68} H atoms.
69. 10^{3} truck/day; 2×10^3 trucks/day.
71. 1.1×10^8 km; 29 yr.
73. 10^{44} molecules.
75. (a) $\mathbf{v} = -v \sin \theta\mathbf{i} + v \cos \theta\mathbf{j}$, or $\mathbf{v} = +v \sin \theta\mathbf{i} - v \cos \theta\mathbf{j}$
(b)

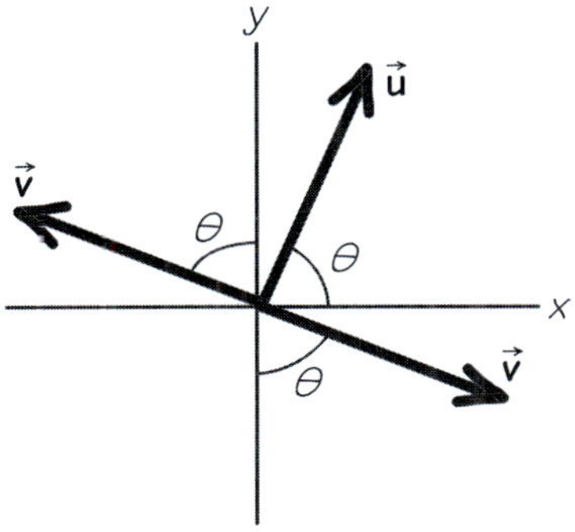

77. (a) $r \sin \theta \cos \phi$, (b) $r \cos \theta$, (c) $r \sin \theta \sin \phi$.
79. 50 cm.
81. $t_0 =$ (a constant) $\ell\, (\lambda/\tau)^{1/2}$.

CHAPTER 2

1. +21 cm; 21 cm from the origin in the positive direction.
3. 252 m; 0;

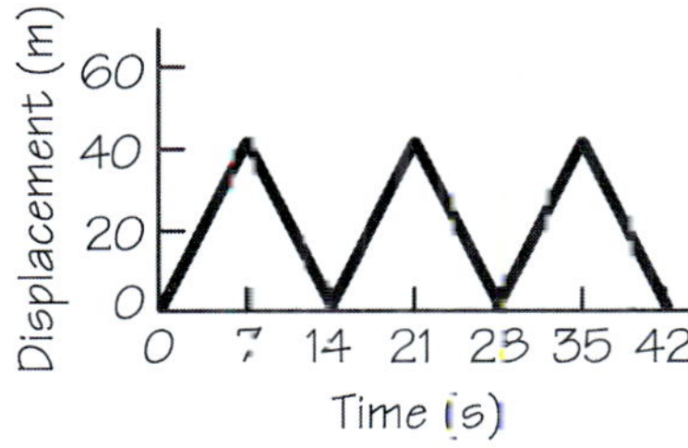

5. (a)

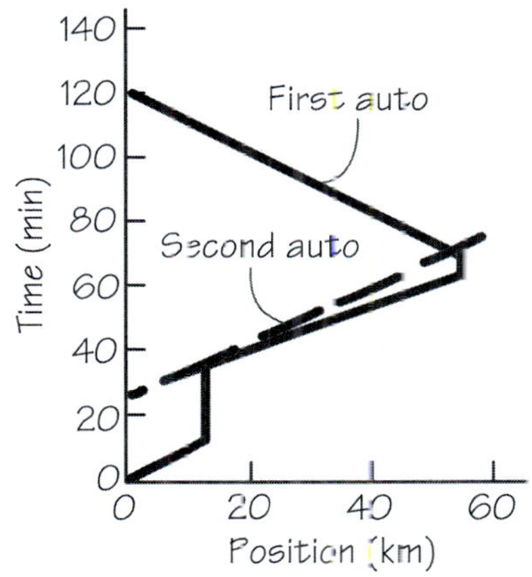

(b) 37 min and 15 km; 68 min and 53 km.
7.

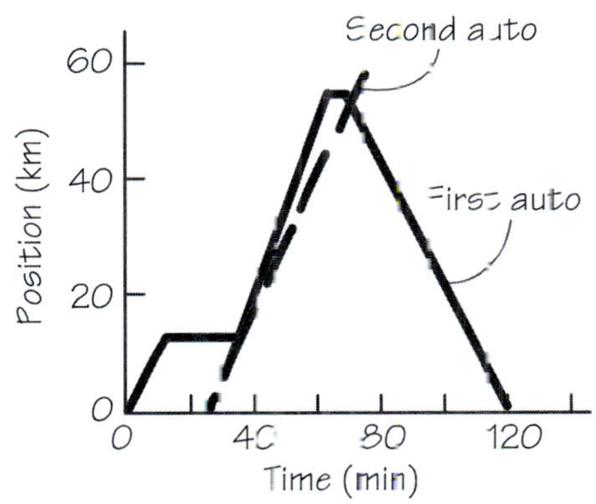

Figure continued at top of next page.

Figure continued from previous page.

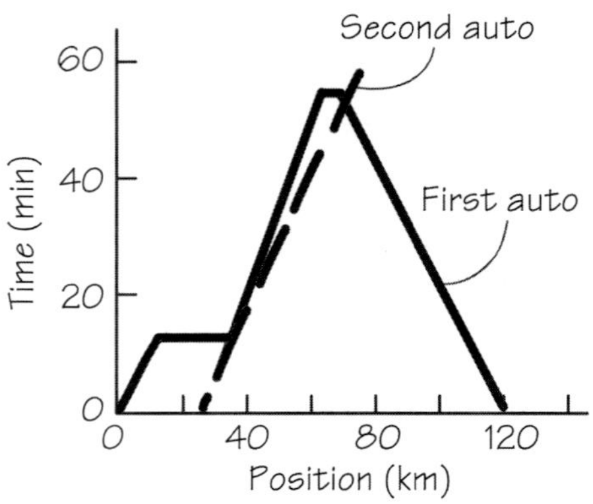

9. 6.4 m/s; 6.6 m/s; 6.3 m/s.

11.

t, s	v, m/s	x, m
0.0	0.00	0.00
0.5	0.75	0.19
1.5	1.75	1.44
2.5	8.75	6.69
3.5	21.75	21.94
4.5	39.75	52.69
5.5	62.75	103.94
6.5	90.75	180.69
7.5	122.75	287.44

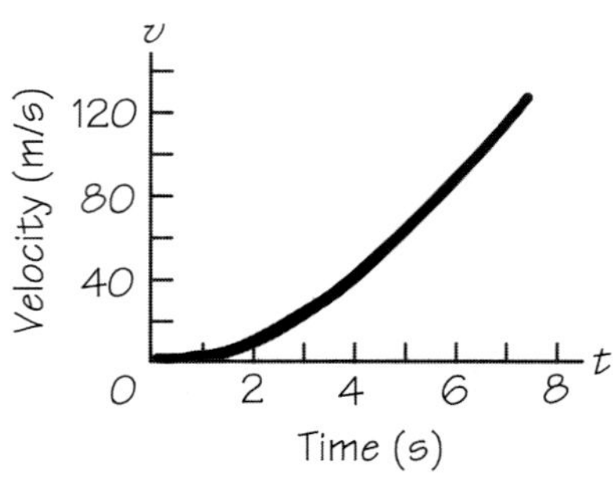

13.

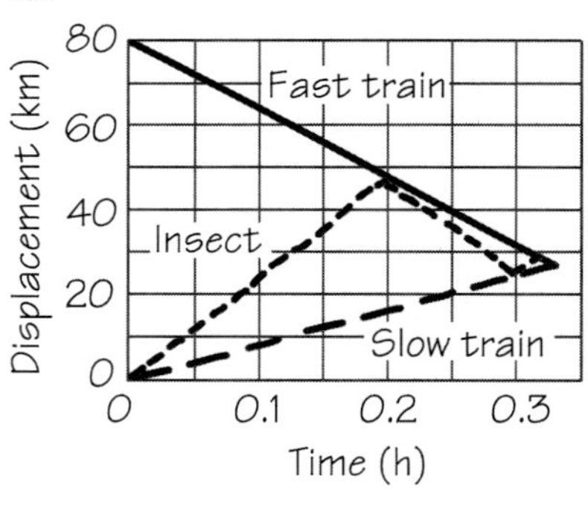

80 km.

15. $+1.9\mathbf{i}$ cm/s; $+2.5\mathbf{i}$ cm/s; $+6.0\mathbf{i}$ cm/s; $+3.0\mathbf{i}$ cm/s.

17. 30 s.

19. $0.30g$.

21. 1.6 m/s^2.

23. $(A\pi/12)\cos(\pi t/12)$; $-(A\pi^2/144)\sin(\pi t/12)$.

25. The particle never gets farther from the origin than A; it oscillates back and forth through the origin. The magnitude of the velocity is maximum at the origin and zero at $x = \pm A$. The magnitude of the acceleration is maximum at $x = \pm A$ and zero at the origin.

27.

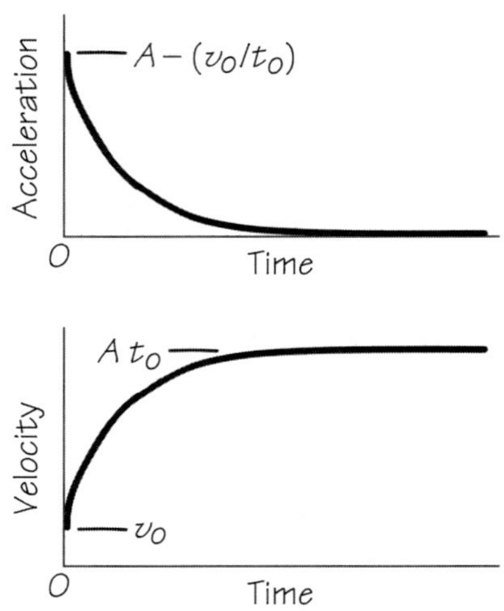

29. 2.02 ft/s^2 (0.61 m/s^2).

31. 2.25 m/s.

33. (a) $8.0 - 0.5t$, easterly with t in s, v in m/s, (b) 34 m to the east.

35. (a)

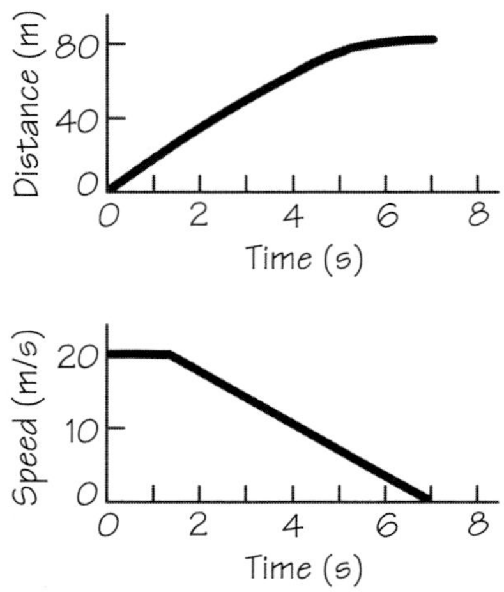

(b) -3.6 m/s^2.

37. 1.92 s.

39. (a) 2.96 m/s^2, (b) 97.2 s.

41. (a)

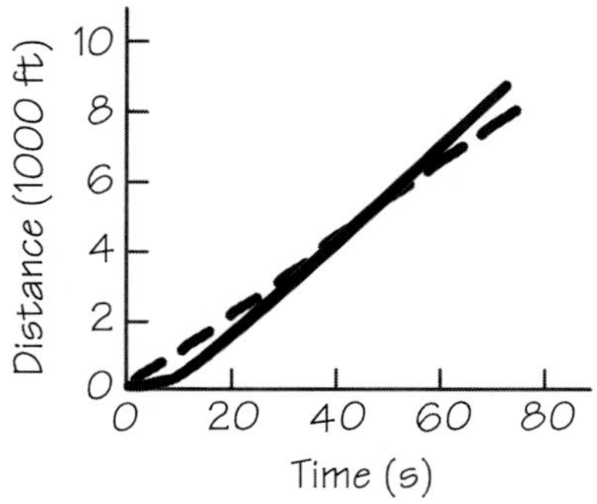

(b) 5940 ft (1.13 mi), (c) 54 s.

43. 3.65 s.

45. (a) 8.0×10^{-9} s, (b) 6.3×10^{14} m/s^2.

47. 4.0 m/s^2; 18 m.

49. 54 m.

51. 9.0×10^5 m/s^2; 6.7×10^{-4} s.

53. Not much time to say anything!

55. 1.1 s.

57. 1.26 s.

59. -22 m/s; 25 m; 20 m/s.

61. 28.4 m.

63. (a)

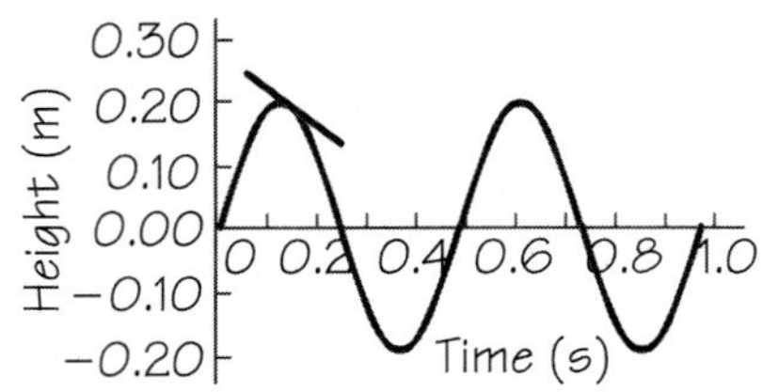

(b) -0.8 m/s, (c) 2.5 m/s,
(d)

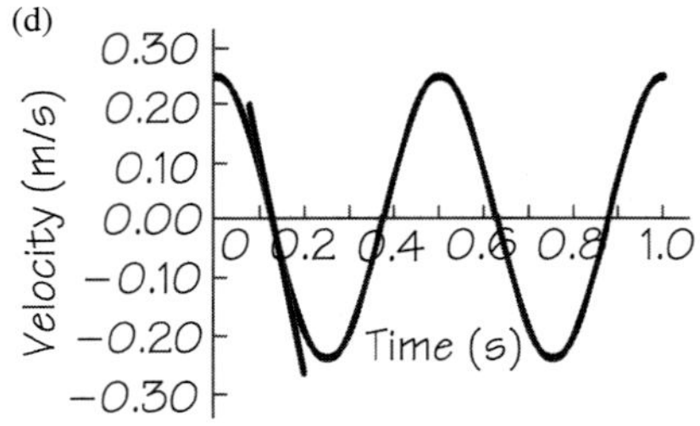

-30 m/s^2.

65. $(3v_f/\alpha)^{1/3}$.

67. 0.7 m; 3 m.

69. 12 m.

71. 1.6 m/s^2; 15 m.

73. 1.6 m; 0.08 s.

CHAPTER 3

1. $(15\mathbf{i} + 15\mathbf{j})$ km; $(30\mathbf{i} + 15\mathbf{j})$ km; $(30\mathbf{i} + 43\mathbf{j})$ km; 52 km, 55° N of E.

3. $\mathbf{r}_A = 0$; $\mathbf{r}_B = (25\text{ m})\mathbf{i}$; $\mathbf{r}_C = (25\text{ m})\mathbf{i} + (35\text{ m})\mathbf{j}$; $\mathbf{r}_D = (35\text{ m})\mathbf{j}$.

5.

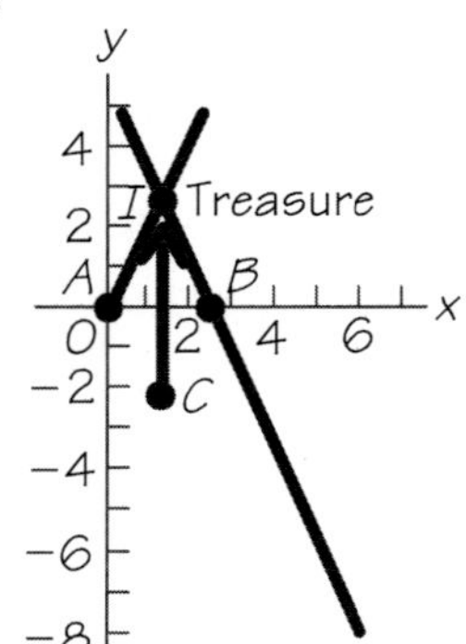

$(1.3\text{ km})\mathbf{i} + (2.7\text{ km})\mathbf{j}$; $(0.1\text{ km})\mathbf{i} + (4.9\text{ km})\mathbf{j}$.

7. $(3.60\text{ m/s})t\mathbf{i} + (6.00\text{ m/s})t\mathbf{j}$, $0 \le t \le 41.6$ s; $[-(141\text{ m}) + (7.00\text{ m/s})t]\mathbf{i} + (250\text{ m})\mathbf{j}$, $41.6\text{ s} \le t \le 77.3$ s.

9. $(2.0\text{ m})\mathbf{i} - (3.5\text{ m})\mathbf{j}$; 4.0 m; $-(4.0\text{ m})\mathbf{j}$; $d = 4.0$ m; $(4.0\text{ m})\mathbf{i}$; $d = 4.0$ m; $\theta(t) = -\pi t/T$.

11. 1.0 m/s^2, opposite to the direction of the velocities.

13. $(3.7\text{ m/s}^2)\mathbf{j} - (2.4\text{ m/s}^2)t\mathbf{k}$.

15. 30 km/h, 31.6° north of west; $(-4.3\mathbf{i} + 2.7\mathbf{j})$ km.

17. (a) $[(0.00225\text{ m}^2/\text{s}) - (0.0009\text{ m}^2/\text{s}^2)t]\mathbf{j}/[(0.0169\text{ m}^2) - (0.0045\text{ m}^2/\text{s})t - (0.0009\text{ m}^2/\text{s}^2)t^2]^{1/2}$; (b) $(0.017\text{ m/s})\mathbf{j}$; 0.

19. $\mathbf{v} = [-(4\pi/T)\sin(\pi t/T)\mathbf{i} - (4\pi/T)\cos(\pi t/T)\mathbf{j}]$ m/s; $\phi = (\pi/2) - (\pi t/T)$.
21.

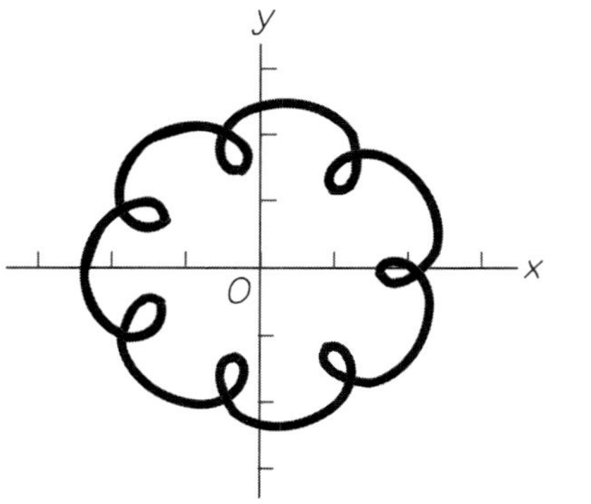

23. $[T^{-1}]$; $v_0 = 0$; $v \to -u$; $a_0 = -Bu$; $a \to 0$.
25. 12 m/s (horizontal); 7.4 m.
27. (a) 3.64 s, (b) 11.5 m, (c) -14.7 m/s (down); $+14.7$ m/s (up).
29. 5.2 m/s.
31. 187 m/s.
33. 40° below the horizontal.
37. 329 m.
39. (a) $t = 1.25$ s, (b) unsuccessful; 0.4 m below the bar.
41. 15° and 75°; 0° and 90°.
43. (a) 7.8×10^3 m/s, (b) 9.1 m/s^2 toward Earth's center.
45. 2.2 m/s^2.
47. 9.5 s.
49. 610 m.
51. (a) (0.43 m, 0.25 m), (b) $-79\mathbf{i}$ m/s^2, (c) $-79\mathbf{j}$ m/s^2.
53. $1.21R$.
55. 15.8 km/h, 18° south of east.
57. 50 km/h.
59. (a) 843 km/h, (b) 781 km/h, 26° W of S, (c) $-(1720\text{ km})\mathbf{i} - (3505\text{ km})\mathbf{j}$.
61. (a) 29.5 km/s, (b) 30.5 km/s, (c) 30.0 km/s.
63. (a) 36 m/s at 65°, (b) 62.3 m.
65. $(10\cos\theta + 6)t\mathbf{i} + (10\sin\theta)t\mathbf{j}$, with r in kilometers and t in hours; 127°; 68 s.
67. Hit the deck.
69. (a) 7.35 m/s, (b) 0; 2.45 m going up; 2.45 m coming down; 0, (c) 4.9 m.

CHAPTER 4

1. (a) Force of gravity (toward Earth), (b) force of gravity (down), normal force from the ice (up); and a small friction force from the ice (opposite to the motion), (c) essentially none.
3. 600 N in the $-y$-direction.
5. Yes.
7. (a) 3×10^{-4} N, (b) 3×10^{-4} N.
9. (a) The acceleration opposite to the motion is due to a retarding force. (b) The observer sees the car (initially at rest) move backward with increasing speed until it reaches v_0. She would say that this is due to a backward force from the wind, etc.
11.

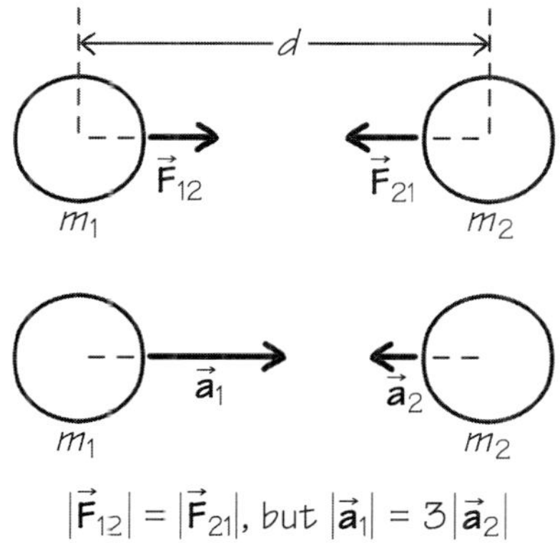

13. $(F/m)\mathbf{j}$; no change.
15. 6.5×10^2 N.
17. $T = F$; $0.433F$.
19. 78 N.
21. 1.6×10^{-21} m/s^2.
23. (a) The Earth, (b) the ice and Earth, (c) none.
25. (a) 0.30 m/s^2, (c) 3.6×10^4 N backward.
(b)

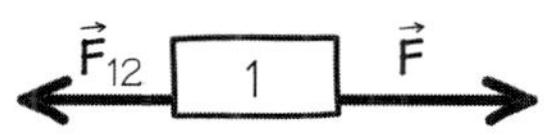

27. $(\mathbf{i} + \mathbf{j} + 2\mathbf{k})$ N.
29. Yes; yes; yes; 2 m/s^2 in the $-x$-direction; not real.
31. (a) Parallel to the window edge, (b) at an angle given by $\tan\theta = a'/g$, (c) parallel to the window edge.
33.

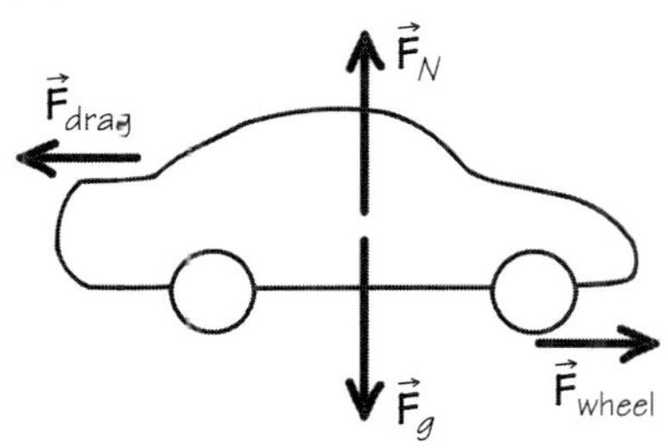

35.

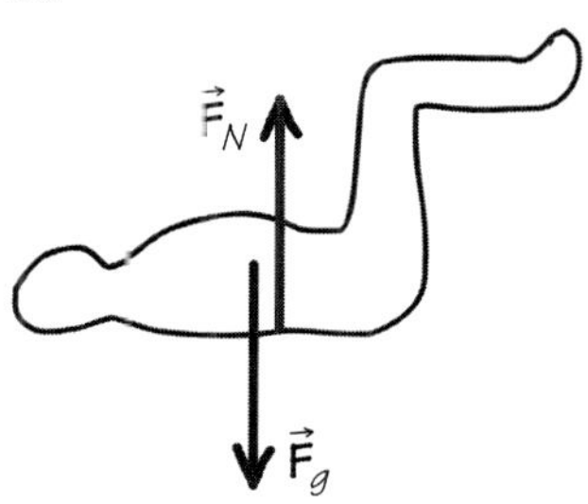

37.

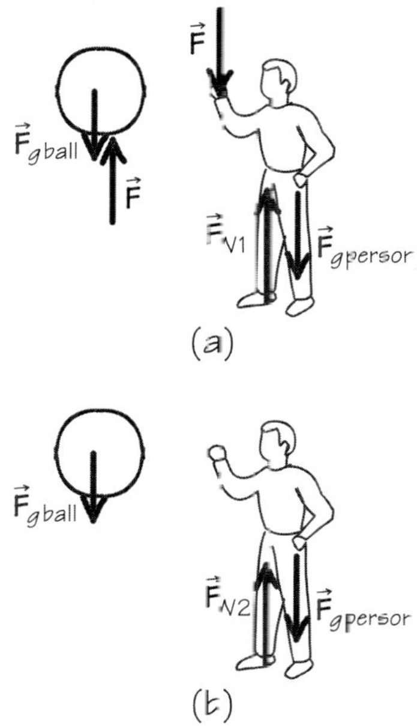

39.

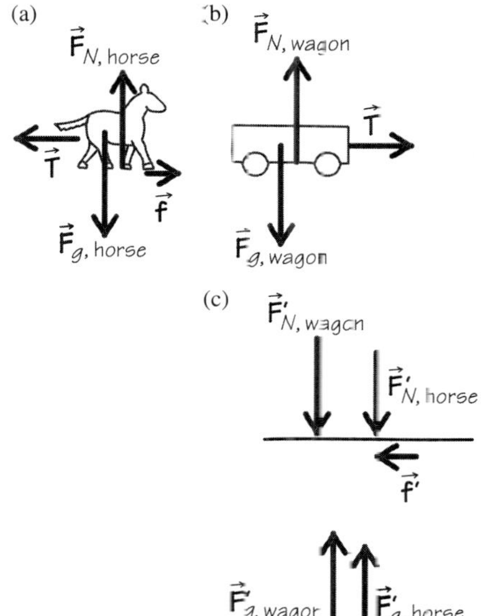

(d) a forward force from the Earth on the horse's hooves.
41.

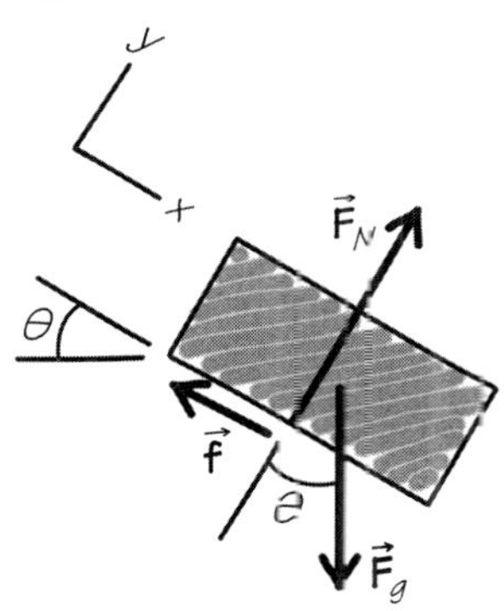

$\mathbf{F}_g = F_g\sin\theta\mathbf{i} - F_g\cos\theta\mathbf{j}$; $\mathbf{F}_N = F_N\mathbf{j}$; $\mathbf{f} = -f\mathbf{i}$.

43.

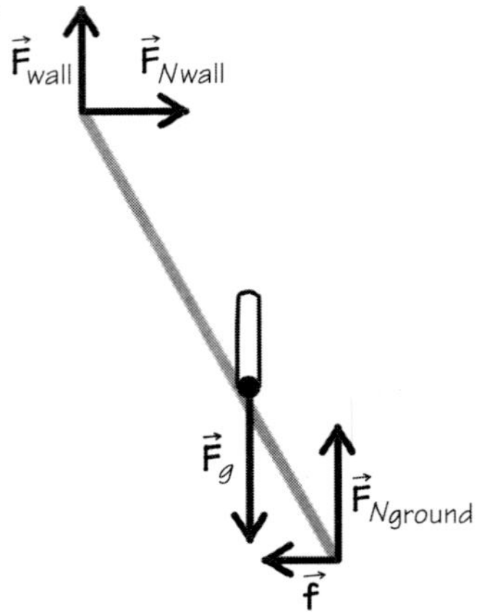

There is a net horizontal force from the wall.

45.

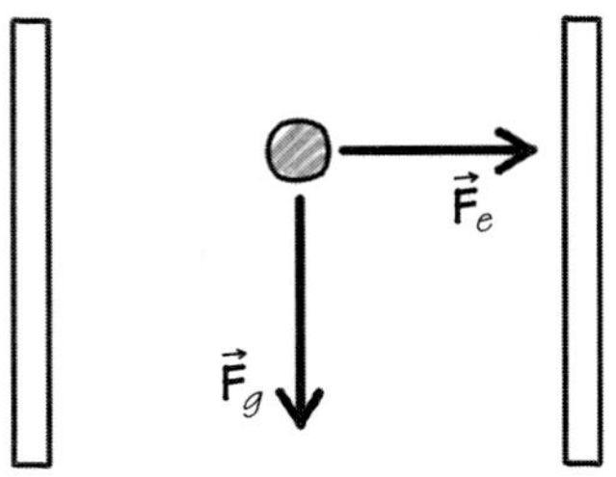

47. (a) $F_g = mg$ down; $F_N = mg$ up, (b) mg down; $2mg$ up, (c) mg down.
49. $(1/M)(F_1{}^2 + F_2{}^2)^{1/2}$; $\theta = \tan^{-1}(F_2/F_1)$.
51. $(9.0\ \text{N} \cdot \text{s}^{1/2})t^{-1/2}$.
53. (a) $c = \sqrt{k/m}$, (b) $A = 0$; $B = +v_0\sqrt{m/k}$.
55. (a)

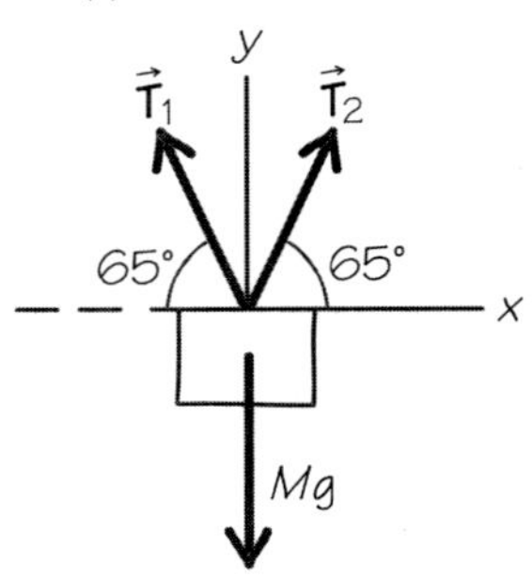

(b) 82.2 N up, (c) 1.55 m/s² up.

57.

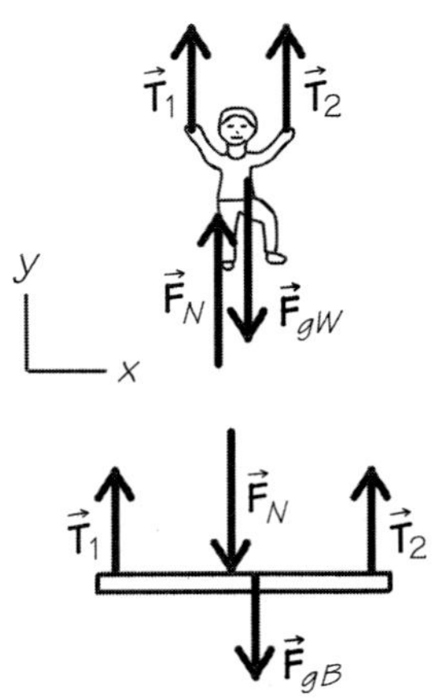

$T = (m_W g + m_B g)/4$; $a = 4T - (m_W g + m_B g)/(m_W + m_B)$.

59. (a) 2×10^4 N, (b) 0.20 m/s²; 1.2×10^4 N forward, (c) 8×10^3 N forward; 4×10^3 N backward; 0.20 m/s².
61. (a)

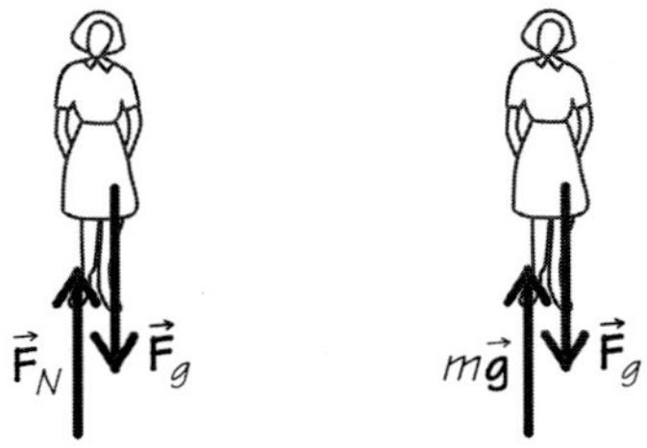

(b) same forces, but the upward force would be a fictitious force.
63. (a) $[A] = [ML^{-1}]$ with units of kg/m, (b) $(Av^2/m) - g$, (c) $v_t = \sqrt{mg/A}$.
65. (a)

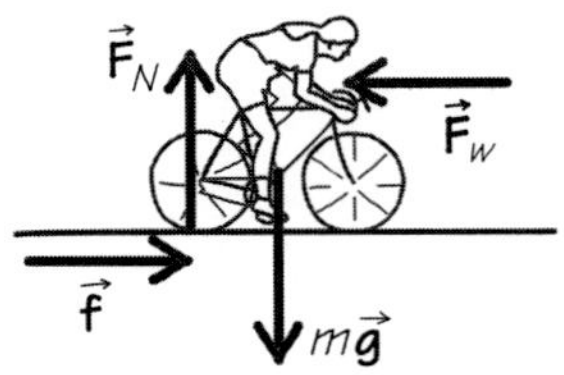

Force of gravity: mg (down); normal force of ground: F_N (up); friction force of ground: f (forward); wind resistance: F_w (backward), (b) 0, (c) 0; $v_w + v$; 0, (d) $v + v_w$; 0; 0.
67. $\phi = \theta$.

CHAPTER 5

1. 1.4×10^4 N.
3. 980 N.
5. 0.095 s.
7. (a) 24 N, (b) 7.8 N.
9. (a) 66 N, (b) 5.1 m/s² up.
11. (a) 2.5 m/s² forward, (b) 3.5×10^3 N forward.
13. $M_{max} = 9.2$ kg; $M_{min} = 4.5$ kg; 0.
15. 25.3 N; 50.6 N; 2.85 m/s² (up); 0.32 m/s² (up); −3.48 m/s² (down).
17. $a_1 = [(m_1m_2 + m_1m_3 - 4m_2m_3)/(m_1m_2 + m_1m_3 + 4m_2m_3)]g$;
$a_2 = [(m_1m_2 - 3m_1m_3 + 4m_2m_3)/(m_1m_2 + m_1m_3 + 4m_2m_3)]g$;
$a_3 = [(-3m_1m_2 + m_1m_3 + 4\,m_2m_3)/(m_1m_2 + m_1m_3 + 4m_2m_3)]g$;
$T_1 = [8m_1m_2m_3/(m_1m_2 + m_1m_3 + 4m_2m_3)]g$;
$T_2 = [4m_1m_2m_3/(m_1m_2 + m_1m_3 + 4m_2m_3)]g$;
$a_1 = [(m_1 - 2m_2)/(m_1 + 2m_2)]g$; $a_2 = a_3 = [(2m_2 - m_1)/(m_1 + 2m_2)]g$.
$T_1 = [4m_1m_2/(m_1 + 2m_2)]g$; $T_2 = [2m_1m_2/(m_1 + 2m_2)]g$.
19. 0.12.
21. 0.61.
23. 3.3×10^2 N.
25. 0.40.
27. (a) 52 m, (b) 3.9 s.
29. (a) 2.8 s, (b) 2.3 N, (c) 0.54.
31. (a) $370/(\cos\theta + 0.75\sin\theta)$ N, (b) 37°; 2.9×10^2 N.
33. 0.077.
35. (a) $a_1 = 0$; $a_2 = F/m_2$, (b) $a_1 = a_2 = F/(m_1 + m_2)$, (c) $\mathbf{F}$(lower on upper) $= \mu_k m_1 g\mathbf{i} + m_1 g\mathbf{j}$; (d) $a_1 = \mu_k g$; $a_2 = (F - \mu_k m_1 g)/m_2$.
37. 5.4×10^2 N.
39. 600 cm².
41. 0.46 m/s.
43. 0.67.
45. 1.3 m/s² toward the center.
47. 1.4×10^2 N toward the center.
49. $v_{max} = 32$ m/s; $v_{min} = 5.8$ m/s.
51. 0.59.
53. 1.0×10^3 m/s tangent to the orbit.
55. 4.43 m/s.
57. (a)

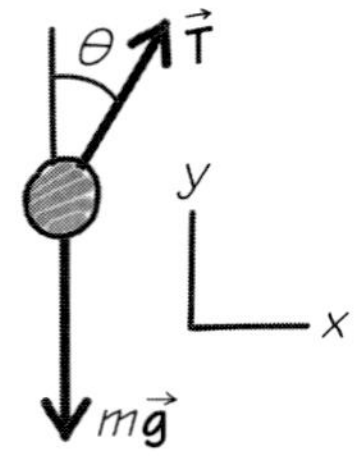

(b) 0.4 m/s.
59. 0.19 m.
61. 55 m/s.
63. (a) 40 N, (b) 35 N.
65. 1.6×10^{-12} kg/m³.
67. 1.2×10^2 N.
69. (a) 0.5 N, (b) 1.2 N.
71. (a) $\Delta x_2 = -\frac{1}{2}\Delta x_1$, (b) $a_1 = -1.78$ m/s² (down); $a_2 = +0.89$ m/s² (up), (c) 9.6 N.
73. $\omega = (g/\mu_s R)^{1/2}$.
75. $\omega = [g/(\ell\cos\theta)]^{1/2}$.
77. 26 m.
79. 11 m/s.

CHAPTER 6

1. (a) 39 J, (b) 1.0 m/s; 89 m/s, (c) 0.40 m/s.
3. (a) The force of gravity 98 N (down); the upward pull = 98 N; $F_{net} = 0$, (b) 0, (c) 98 J.
5. 1.0×10^4 J.
7. (a) 2.4×10^2 J, (b) friction force, (c) 0.
9. 730 J; −730 J.
11. −3.73 J.
13. 1.8×10^3 J.
15. 2.8×10^{11} J.
17. 0.127 J; 0.159 J; 0.175 J.
19. +3.
21. 0.89 m/s.
23. 2.57.
25. 3.9×10^3 J.
27. 6.1×10^3 J.
29.

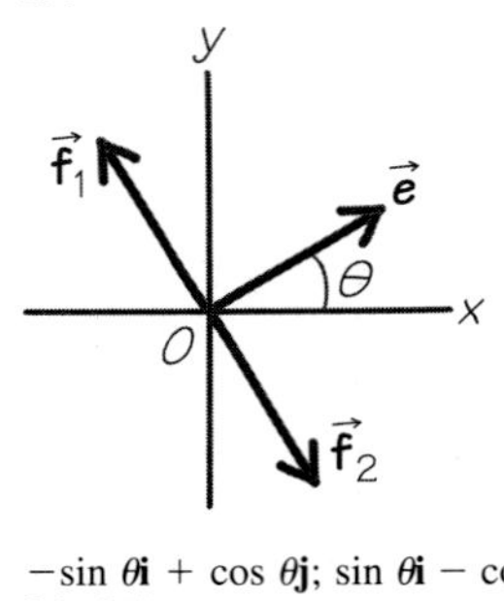

$-\sin\theta\mathbf{i} + \cos\theta\mathbf{j}$; $\sin\theta\mathbf{i} - \cos\theta\mathbf{j}$.
31. 3.4.
33. −24 J.
35. 0.067 J.
37. $2g_1 - 4g_2$.
39. 8.6 m/s.
41. 8.1×10^{-2} J.
43. 14 J.
45. 2.1×10^3 J.
47. −15 J; −5 J; conservative.
49. (a) 6.3×10^{-2} J, (b) 0, (c) 5.9×10^{-2} J, (d) -5.9×10^{-2} J.

51. (a) 4 J; 6 J; 6 J; 4 J, (b) +5 m, (c) conservative.
53. Constant forces are included; not a function of position only.
55. (a) $0.16C$ J; 0, (b) $0.32C$ J.
57. 22 bulbs.
59. 0.7 kW.
61. 52 kW.
63. 5.0 m/s; 1.3 m/s.
65. 4.7×10^7 W.
67. (a) 16 m/s, (b) 0.5 s, (c) 12 s.
69. 5.8×10^{-12} J; 5.8×10^{-9} J.
71. 0.62; 0.31.
73. −0.2 J.
75. 0.034.
77. (a) 2.2×10^7 J, (b) -2.2×10^{-7} J.
79. (a) mg = 26 N down; F_N = 22 N perpendicular to plane (up); f_k = 5.4 N parallel to plane (down), (b) −18 J; 0; −7.0 J, (c) 4.4 m/s.
81. 5.3 J.
83. (a) 0, (b) −0.20 J, (c) 9.8 J.
85. (b) $W_T = mgH(1 + \mu_k \cot\theta)$.
87. $mgL(\cos\theta_f - \cos\theta_i)$; $v = [2gL(1 - \cos\theta_i)]^{1/2}$.
89. 0.01 mK/x.

CHAPTER 7

1. (a) 196 J, (b) 29 J; 225 J.
3. 25 m.
5. 1.2×10^{-2} J.
7. (a) $-8x$ J, with x in m, (b) +18 J, (c) 4.1 m/s.
9.

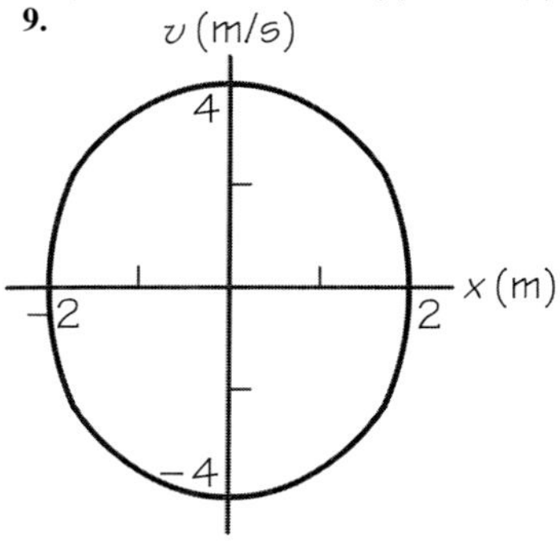

11. 0.541 m; 0.468 m.
13. 0.43 m; 0.48 m.
15. (a) 7.1 m/s, (b) −0.25 J, (c) 0.73.
17.

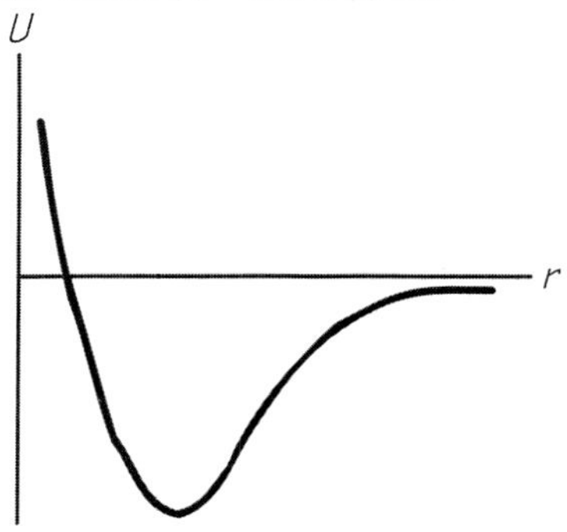

19. (a) to the right, (b) $v = (k/m)^{1/2}x$, (c) to the left; $v = -(k/m)^{1/2}x$.
21.

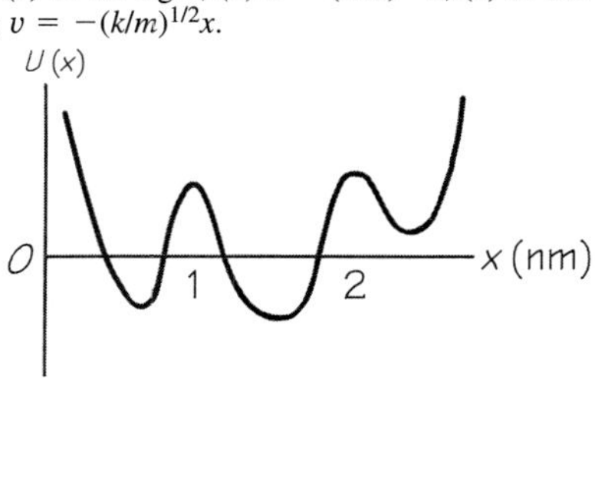

23. $x = 0$ is unstable; $x = \pm 0.24$ m is stable.
25. 503 m/s
27. 130 m/s, same.
29. (b) $\sqrt{2gh}$, (c) $2\sqrt{gh}$.
33. -5.0×10^4 J.
35. (a) 1.7 m/s, (b) 1.3 m/s, (c) 11.4 cm from point a.
37. (a) $U(x) = \frac{1}{2}k(\sqrt{h^2 + x^2} - L)^2$,
(b) $-\dfrac{k(\sqrt{h^2 + x^2} - L)x}{\sqrt{h^2 + x^2}}$.

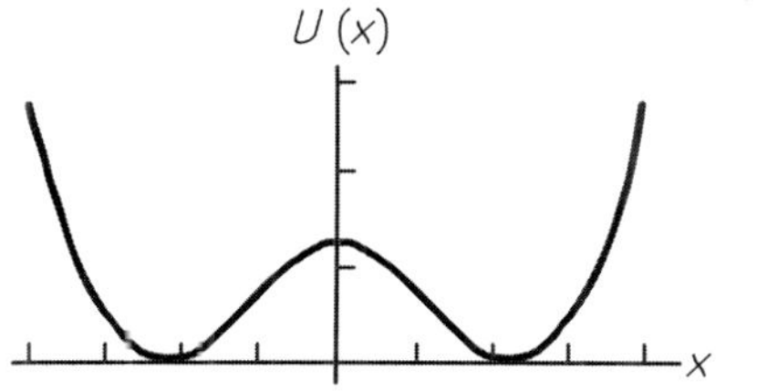

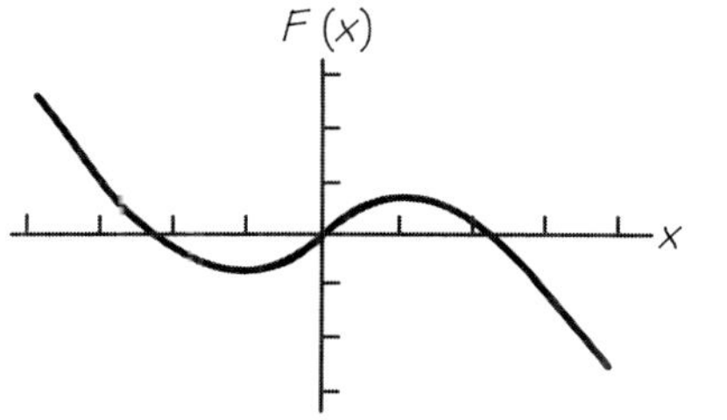

39. (a) GMm/r^2 toward M, (b) $GMm/2r$, (c) $-GMm/2r$.
43. -6.2×10^4 J.
45. 14.
47. 1.9×10^{11} s ($\simeq 6 \times 10^3$ yr).
49. 5.9×10^2 J.
51. (a) $3 + \dfrac{3x^2}{2} - \dfrac{0.2x^3}{3}$ J, with x in m,
(b)

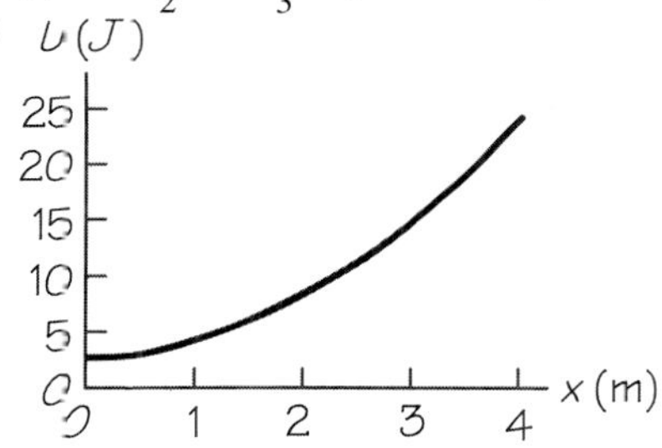

53. (a) conservative, (b) not conservative.
55. −2 J; −2 J; +2 J.
57. (a) 7.7 m/s; 14.7 m/s, (b) the skier leaves the surface.
59. (a) 4.4 m/s, (b) 3.7 m/s, (c) 5.9 N; 4.2 N.
61. (a) conservative force, (b) $U(r) = -C/r$, (c) 2.1×10^7 m/s.
63. (a) +1.13 J, (b) −0.083 J, (c) −0.30 J, (d) 0.70 m/s, (e) 8.8 cm, (f) 5.0 cm.
65. 1.1 m/s.
69. $\dfrac{\partial F_x}{\partial y} = \dfrac{\partial F_y}{\partial x}, \dfrac{\partial F_x}{\partial z} = \dfrac{\partial F_z}{\partial x}, \dfrac{\partial F_y}{\partial z} = \dfrac{\partial F_z}{\partial y}$.
71. $x = L[\sin\alpha + 2\sin\alpha\cos^2\alpha + 2(\cos^3\alpha - \cos^6\alpha)^{1/2}]$.

CHAPTER 8

1. (a) 1.2 kg · m/s; (b) 5.1 kg · m/s; (c) 7.0×10^2 kg · m/s; (d) 7.6×10^2 kg · m/s.
3. (a) 4.2×10^2 kg · m/s; (b) 6.0×10^6 kg · m/s; (c) 1.2×10^4 kg · m/s; (d) 3.3×10^{-22} kg · m/s; (e) 1.4×10^{-2} kg · m/s.
5. (a) −1.0 m/s; (b) 3.5 J.
7. (a) $(-1.9\mathbf{i} - 10.8\mathbf{j})$ kg · m/s; (b) $(-4.9\mathbf{i} - 6.8\mathbf{j})$ m/s; (c) $(-0.90\mathbf{i} - 2.9\mathbf{j} - 1.2\mathbf{k})$ m/s; (d) 24 J; 64 J; 19 J.
9. 40 m/s.
11. −12 kg · m/s, opposite to the original motion; 1.7×10^4 N, opposite to the original motion.
13. -1.2×10^3 kg · m/s, down; 1.5×10^4 N, up.
15. 2.2×10^2 kg · m/s, up.
17. 7.5×10^2 N, up.
19. (a) 5.7×10^2 N; (b) 1.4×10^2 J; (c) 0.011 s.
21. $0.60\sqrt{g(0.20 + 0.10N)}$ kg · m/s.
23. $V = -0.13$ m/s (opposite to the direction of the first cart).
25. 0.74 m/s.
27. $m_1/m_2 = 3$.
29. (a) $M/(m + M)$; (b) $v = [(m + M)/m]\sqrt{2gh}$.
31. (a) $v[m_2{}^2 + M(m_1 + m_2)]/M(m_2 + M)$ forward; (b) $v[m_1{}^2 + M(m_1 + m_2)]/M(m_1 + M)$ forward; (c) $v(m_1 + m_2)/M$ forward.
33. (a) −0.20 m/s; (b) 64%.
35. 2.59 m/s.
37. 15°.
39. (a) $\sqrt{2gh}$; (b) $\sqrt{2gh}$; (c) $[(m - 3M)/(m + M)]\sqrt{2gh}$, up; (d) $[(m - 3M)/(m + M)]^2h$; (e) $9h$.
41. $\mathbf{v}_{1f}$ = 1.2 m/s 60° below the $-x$-axis; not elastic.
43. (2.00 m/s)$\mathbf{i}$ + (1.00 m/s)$\mathbf{j}$; elastic.
45. 2.6 m/s.
47. 8.7 m/s, 20° above the horizontal.
49. The driver of the car was speeding!
51. 0.78 m from the heavier sphere.
53. X = 0.38 m; Y = −0.66 m.
55. 0.
57. X = 0; Y = 1.0 m from the bottom of the handle.
61. $4R/3\pi$ from the center of the arc along the bisector.
63. $a\dfrac{\ln[(L^2 + a^2)/a^2]}{2\tan^{-1}(L/a)}$.
65. -3.3×10^3 kg/s.
67. 1% of mass must be discarded.
69. 2.5×10^3 m/s.
71. (a) spit out seeds one at a time; (b) 6×10^{-3} m/s.
73. (a) 7.9 m/s (to the right); (b) 7.9 m/s (to the right); (c) −1.9 m/s (to the left); +1.1 m/s (to the right); (d) 2.1×10^4 N (to the right).
75. (a)

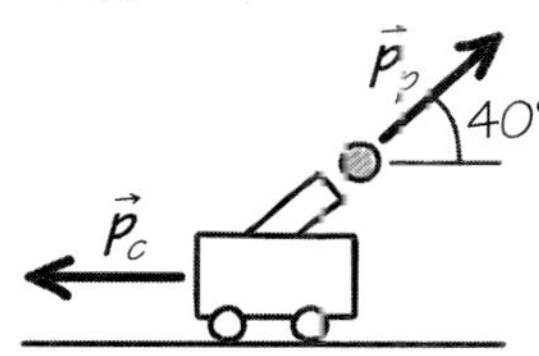

(b) 4.1 m/s recoil; (c) upward impulse provided by the ground.
77. (a) $[8M(M - m)/(M + m)^2]h$;
(b) $[(M - m)/(M + m)]^2h$.
79. $R/6$ from the center of the styrofoam sphere.
81. (a) 0.11 m/s; (b) 0; (c) −0.074 m; (d) 0.

CHAPTER 9

1. 0.68 rad/s².
3. −17 rad/s².
5. 7.3×10^{-5} rad/s from the South Pole to the North Pole; 2.0×10^{-7} rad/s perpendicular to the orbital plane.

7. (a) -2.62×10^2 rad/s^2, (b) 1.67×10^2 m.
9. (a) 0.10 rad/s^2, (b) 11 rev, (c) 2.1×10^2 m.
11. (a) $a = 1.5$ m/s^2; $x = x_0 + (0.75\text{ m/s}^2)t^2$, (b) 100 rad/s^2, (c) $\omega = (100\text{ rad/s}^2)t$.
13. 1.35×10^{-4} m.
17. 9.
19. 2.8×10^{-3} kg $\cdot$ m^2.
21. 1.3×10^2 J; 4.0×10^2 J.
23. (a) 0.021 kg $\cdot$ m^2, (b) 0.083 kg $\cdot$ m^2.
25. 9.8×10^{37} kg $\cdot$ m^2; $I_{\text{neutron star}} = 1.0 \times 10^{39}$ kg $\cdot$ m^2 $\simeq 10\, I_{\text{earth}}$.
27. $\frac{1}{2}M(R_1{}^2 + R_2{}^2)$.
29. $(3/10)(\tan^2 \alpha)MH^2$.
31. $(8/15)\pi[(\rho_1 - \rho_2)R_1{}^5 + \rho_2 R^5]$.
33. $(16/45)MR^2$.
35. 24 lb.
37. 98 kg.
39. $\frac{1}{2}L(F_{1y} + F_{2y})$ perpendicular to the rod.
41. (a) 2.3 N $\cdot$ m, (b) change 7-N force to 5.4 N.
43. 0.38 N $\cdot$ m.
45. (a) 2.13 kg $\cdot$ m^2, (b) 8.84 kg $\cdot$ m^2/s, (c) 0.886 rev, (d) 18.3 J.
47. 3.8 rad/s up; rotate in the direction of the original rotation of the wheel.
49. -2.46×10^{-5}.
51. 0.12.
53. 13.4 m; 17.9 m.
55. 0.050 m/s^2.
57. 2.3 rev.
59. 5.1×10^6 m.
61. $3v_0{}^2/4g \sin \theta$.
63. (a) 0.72 kg $\cdot$ m^2/s, (b) 3.8 J, (c) 7.6 J.
65. $3\mu MgR/(R + 2r)$.
67. $MgR^2/(R^2 + 2r^2)$.

CHAPTER 10

1. (a) 1.1×10^{10} kg $\cdot$ m^2/s down, (b) 7.9×10^9 kg $\cdot$ m^2/s down.
3. (a) 5×10^4 kg $\cdot$ m^2/s down, (b) 5×10^4 kg $\cdot$ m^2/s down.
5. 1.5 kg $\cdot$ m^2/s north.
7. $-7.4t^2\mathbf{k}$ kg $\cdot$ m^2/s (into the page).
9. $\frac{1}{2}mbvt^2\mathbf{i} - \frac{1}{2}mawt^2\mathbf{j} - \frac{1}{2}mavt^2\mathbf{k}$.
11. $\mathbf{r} \times m_2m_1/(m_1 + m_2)\, d\mathbf{r}/dt = \mathbf{r} \times \mu\, d\mathbf{r}/dt$.
13. (a) $6.4 \times 10^{-2}\mathbf{k}$ kg $\cdot$ m^2/s (along ω-direction), (b) same.
15. (a) $m\omega d^2$ along the axis of rotation, (b) $\frac{3}{4}m\omega d^2$ along the axis of rotation, (c) $\frac{1}{2}m\omega d^2$ along the axis of rotation.
17. $(17\mathbf{i} - 19\mathbf{j} + 14\mathbf{k})$ N $\cdot$ m.
19. $140\mathbf{k}$ N $\cdot$ m (perpendicular to table).
21. $MgR \sin(\omega t)$.
23. $mvd \sin(2\theta)$ up; 0.
25. $-7\mathbf{i} + 9\mathbf{j} + 10\mathbf{k}$.
27. 1.1 rad/s.
29. 30 m/s.
31. 0.103 rad (0.016 rev).
33. 5.3×10^2 N $\cdot$ m along the axis.
35. 409 rad/s.
37. 0.14 J.
39. $r_n = (n^2\hbar^2/mk)^{1/4}$, $n = 1, 2, \ldots$; $v_n = (n^2\hbar^2k/m^3)^{1/4}$, $n = 1, 2, \ldots$; $K_n = \sqrt{k/m}\, n\hbar/2$, $n = 1, 2, \ldots$.
41. No energy level 2.0 eV above the lowest state; excitation energies are 10.2 eV, 12.1 eV, 12.75 eV, ...
43. $\omega_p = Mg\ell/I\omega$ along the z-axis.
45. 2.8 rad/s.
47. 72 rad/s.
49. 0.62 rad/s; -33 J; friction.
51. (a) 3.8 m/s, (b) 0.78 s.
53. h.
55. (a) 19.4 m/s, (b) 0.127, (c) 0.132.
57. 26 kg; 52 rad/s; 12 kg.
59. 4.0×10^3 J.
61. (a) K/L, (b) L^2/MK, (c) $2\pi L^3/MK^2$, (d) MK^3/L^4.
63. (a) 2.4 m/s, (b) 2.4×10^2 rad/s.

CHAPTER 11

1. 40 kg.
3. 139 N down; 96 N down.
5. (2, 0, 1) m.
7. 1.7 m.
9. 0.8 m.
11. 14.2 N $\cdot$ m in xy-plane 27° from $-x$-axis $(\perp\mathbf{r})$.
13. $\frac{3}{4}L$; $11L/12$.
15. 0.57 Mg.
17. 1.2×10^2 N; $F_{NA} = 0$.
19. (a) 2.9 m.
21. 2.9×10^2 N 13.5° above the horizontal; 3.5×10^2 N down; 2.8×10^2 N.
23. 0.58 Mg outward.
25. 1.9×10^2 N right; 73 N left; 23 N left.
27. (a) $\frac{1}{2}mg\ell_2/\ell_1$ down; $\frac{1}{2}mg(\ell_2 - \ell_1)/\ell_1$ up, (b) seat will lose contact at B and turn clockwise; seat will turn counterclockwise, (c) 74 N; 25 N.
29. (a) 56 N, (b) 164 N; (164 N)$\mathbf{i}$ + (46 N)$\mathbf{j}$.
31. 353 kg.
33. (a) $F_{N1} = 0$; $F_{N3} = F_{N2} = \frac{1}{2}(M + m)g - (mg\, x/L)$; $F_{N4} = 2mgx/L$, (b) $F_{N4} = 0$; $F_{N1} = -2mgx/L$; $F_{N3} = F_{N2} = \frac{1}{2}(M + m)g + (mg\, x/L)$.
35. 1.33×10^4 N.
37. 69 N.
39. (a) $\frac{1}{2}Mg \cot \theta_0$, (b) $(3g \cos \theta_0)/2L$, (c) $\omega = \sqrt{(3g \sin \theta_0)/L}$.
41. 85 N.
43. 1.36×10^3 N.
45. (a) near legs: $Mg(\frac{1}{4} + \frac{1}{2}\mu_k)$ up; $\mu_k Mg(\frac{1}{4} + \frac{1}{2}\mu_k)$ to the left; far legs: $Mg(\frac{1}{4} - \frac{1}{2}\mu_k)$ up; $\mu_k Mg(\frac{1}{4} - \frac{1}{2}\mu_k)$ to the left, (b) 0.5.
47. $\tan(\frac{1}{2}\theta)$.
49. Top ball: mg; $0.577mg$; $1.155mg$; Bottom ball: mg; $1.155mg$; $2.000mg$; $0.577mg$.

CHAPTER 12

1. 2.975×10^{-19} s^2/m^3.
3. (a) $[ML^4T^{-2}]$, (b) $\sqrt{mh}$, (c) $2\pi\sqrt{m/h}$ = a constant.
5. (a) 1.9×10^{-6} N, (b) must be the same.
7. 4.2×10^{-10} N; $\simeq 10^{-7}\, W_{\text{fly}}$.
9. 6.65×10^6 m (330 km or 205 mi above the surface).
11. 1.51 h.
13. 9.56×10^6 m.
15. 4.9 m.
17. 0.25.
19. 6.18×10^5 m/s.
21. 123 m/s.
23. 1.6×10^{-3} rad/s.
25. 1.09×10^4 m/s.
27. (a) 1.85×10^8 m, (b) 7.99×10^8 m, (c) impossible.
29. 7.79 km/s tangent to the orbit; 1.52×10^{10} J; 2.56×10^{13} kg $\cdot$ m^2/s perpendicular to the orbit.
31. 1.74 h.
33. 1.49×10^{11} m.
35. 2.99 km/s.
37. (a)

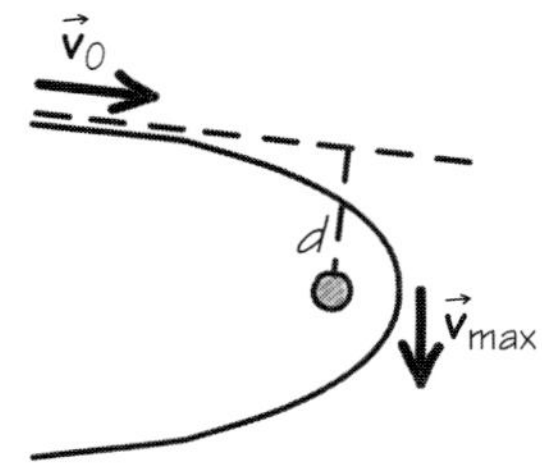

(b) 75.6 km/s perpendicular to the radius, (c) 4.76×10^7 km.
39. (a) 6.91×10^3 m/s, (b) 1.74×10^{13} kg $\cdot$ m^2/s, (c) 8.68×10^{12} kg $\cdot$ m^2/s, (d) the satellite crashes.
41. (a) 7.73×10^3 m/s, (b) 1.03×10^{14} kg $\cdot$ m^2/s, (c) -5.98×10^{10} J, (d) zero, (e) cannot be circular.
43. 0.996860; 0.996868.
45. 7.9×10^3 m/s.
47. 4.51×10^2 km from the Sun's center.
49. $4\pi\sqrt{R^3/GM}$.
51. 3.5×10^{-15} m.
53. 0.40 rad/s.
55. (a) 1.66×10^3 m, (b) 2.8 m/s, (c) speed of 1.76 m/s to orbit the asteroid.
57. (a) Circular orbits are supported, (b) $T^2/r^{n+1} = 4\pi^2(m/k)$ = a constant.
59. $\Delta U \simeq -4\pi fr$; $\Delta E = -2\pi fr$; $\Delta K \simeq 2\pi fr$.
61. (a) 1.92 m/s relative to himself, which is 1.97 m/s relative to the ship, (b) 1.9 h.
63. (a) possible, (b) 1.83×10^7 s, (c) unstable.

CHAPTER 13

1. $\pi/2$ rad.
3.

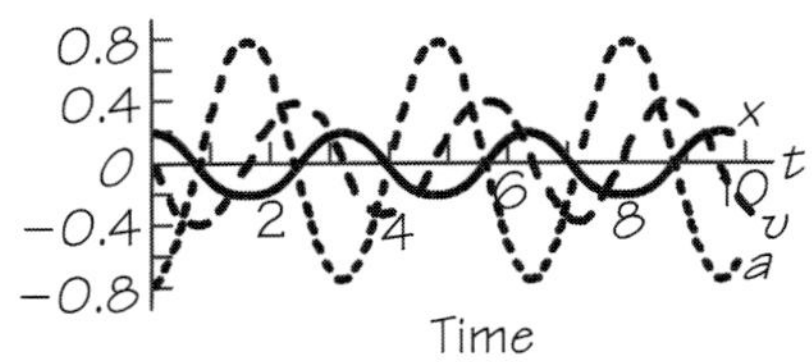

5. 4.09 s.
7. -0.038 m.
9. 7.0137×10^{-6} m.
11. $x = A \cos[(2.0\text{ rad/s})t - 1.17\text{ rad}]$, t in s.
13. (a) 3.97 s, (b) 1.26 m.
15. (a) 4.08 cm/s, (b) 6.41 cm/s; 23.6 cm/s^2.
17. (a) $x = (3.44\text{ cm}) \sin[(3.0\text{ rad/s})t + 2.08\text{ rad}]$, (b) 1.75 s, (c) 0, 0.71 s, 1.05 s, 1.75 s,
(d)

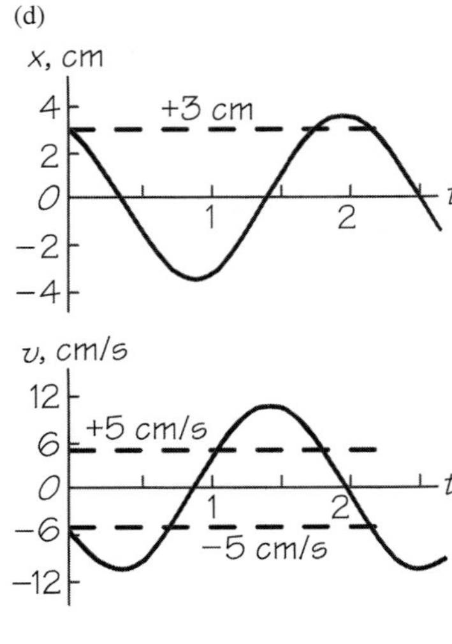

19. R.
21. $x = R\cos(\omega t + \delta)$, with $R = 1.5 \times 10^{11}$ m and $\omega = 1.99 \times 10^{-7}$ rad/s.
25. 5.1 kg.
27. 0.18 N/m.
29. 1.5×10^2 N.
31. 1.2 s.
33. $x = (0.01 \text{ m}) \sin[(71 \text{ rad/s})t + \pi/2]$.
35. 1.1 s.
37. 0.042 m/s.
39. $\omega_{\text{bottom}} = 6.1$ rad/s.
43. 1.67 m/s; 0.74 m; 0.27 J.
45. 9.79 m/s^2.
47. (a) 2.23 m, (b) 430 m.
49. (a) 1.7 s, (b) 0.05 m.
51. 0.76 s.
53. $2\pi\sqrt{\dfrac{2(L^2 - 3Ly + 3y^2)}{3(L - 2y)g}}$.
55. 0.81 Hz.
57. (a) $\frac{1}{2}MR^2 + M\ell^2$, (b) $-g\ell\ \theta = (\frac{1}{2}R^2 + \ell^2)\ d^2\theta/dt^2$, (c) $2\pi\sqrt{\dfrac{R^2 + 2\ell^2}{2g\ell}}$, (d) $T \to \infty$; no torque.
59. (a) 2.7 s, (b) 1.0×10^{-4} J, (c) 3.6×10^{-2} m/s.
63. 0.113; 8.4 min.
65. 0.095 kg/s.
67. $(2.0 + 0.00051)$ s.
69. $x_{3.0} = -5.40$ cm; $x_{4.8} = -1.48$ cm; $x_0 = -6.65$ cm.
73. 1.8 Hz.
75. (a) 2.2 N · s/m, (b) 0.25 s, (c) $\Delta\omega = 8$ rad/s; $Q = 3.1$.
77. 0.36 m.
79. $\sqrt{2}\ \omega$.
81. (a) $0.553H$, (b) $0.347H$.
83. (a) 1.2 s, (b) $\pi\sqrt{\dfrac{2m}{k}}$, (c) $\pi\sqrt{\dfrac{2m}{k}}$, (d) $2\pi\sqrt{\dfrac{m}{k}}$.
85. 9.0 s.
87. (a) $2A/e^2$, (b) $U = (-e^4/4A) + (e^4/4A)x^2$; $e^4/\sqrt{4A^3m}$.
89. $\dfrac{1}{2\pi}\sqrt{\dfrac{g}{2R}}$.
91. $\dfrac{1}{2\pi}\sqrt{2k/3m}$.

CHAPTER 14

1. 1.08 m; 0.54 m; 0.36 m;

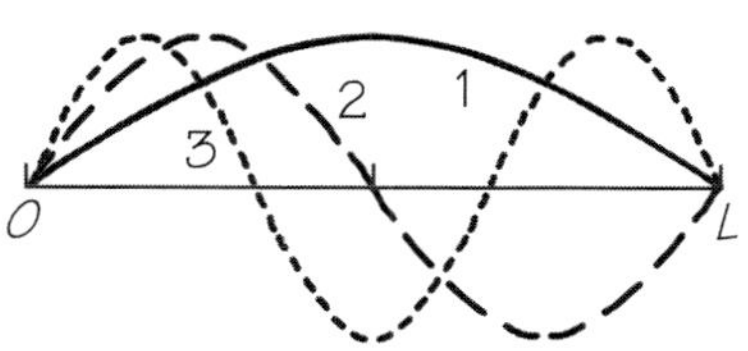

3. 13 N.
5. $m_2/m_1 = 1/9$.
7. 1.
9. 3.54 Hz, 22.2 rad/s, 0.28 s; 7.07 Hz, 44.4 rad/s, 0.14 s; 10.6 Hz, 66.6 rad/s, 0.094 s.
11. 8.49×10^2 N.
13.

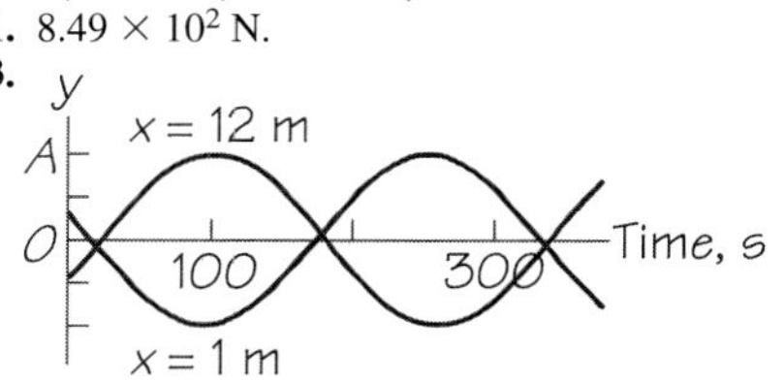

19. 0.3 kg · m/s.
21. 200 m/s.
23. 23 m^{-1}; 48 Hz.
25. 5.1×10^3 m/s; 3.5×10^3 m/s.
27. 25 km/h.
29. $v = \ell(k/m)^{1/2}$.
31. 2.1×10^3 W; 5.3×10^2 W.
33. 8.9×10^{-10} m.
35. $\mu v\omega^2 z_0{}^2 \cos^2(kx + \omega t)$.
37. $P = \mu v\left(\dfrac{\omega}{k}\right)^2\left[\dfrac{df}{d(x - vt)}\right]^2$.
39. 1.28 kg/m^3; 0.085 kg/m^3.
41. 89 dB; 65 dB; 39 dB.
43. 6.3×10^{-2} W.
45. 1.97×10^9 N/m^2.
47. 147 Hz.
49. (a) 0.874 m; 378 Hz, (b) 372 Hz.
51. 1.5×10^7 m/s.
53. 1.88×10^3 Hz.
55. $\dfrac{440 \text{ Hz}}{\sqrt{1 + 0.060t}}$.
57. 965 m/s (3.5×10^3 km/h).
59. $2.40/(2n - 1)$ m, where $n = 1, 2, 3, \ldots$.
61. 3.2×10^6.
63. 400 Hz; 1.08 m.
65. (a) 484 Hz; 543 Hz, (b) D; B.
67. 3.5×10^2 m/s.
69. $f_1 = \dfrac{1}{2L}\sqrt{\dfrac{T}{\mu} + \alpha\dfrac{\pi^2}{L^2}}, f_2 = \dfrac{2}{2L}\sqrt{\dfrac{T}{\mu} + 4\alpha\dfrac{\pi^2}{L^2}}$; the higher frequency.

CHAPTER 15

1. $A = \sqrt{2}$; $\tan\phi = 1$, or $\phi = \pi/4$.
3. $1.41z_0 \sin[kx - \omega t + (\pi/4)]$.
5. 3.6 cm.
7. $\delta = 2\pi/3$ and $-2\pi/3$.
9. $2z_0 \sin[kx + (\pi/4)] \cos[\omega t + (\pi/4)]$, which represents a standing wave; x(nodes) $= [n - (1/4)](\pi/k)$, where $n = 0, \pm1, \pm2, \ldots$.
11. (a) 11 Hz; 7.3 m/s; (b) $z_\ell = 3\sin(kx + \omega t)$; $z_r = 3\sin(kx - \omega t)$.
15. 6 m.
17. 506 Hz.
19. 422 Hz.
21. −1.6%.
23. $z = 2A\cos(\omega t - kL)$.
25. 0.59 cm.
27. (a) 37°; (b) 17.5°; (c) 37°.
31. (a) −1 cm, 0, +1 cm; (b) (6 cm) $\sin(\pi r - \pi vt)$, where $r = (y^2 + 4)^{1/2}$.
33. (d) Total energy is conserved.
35.

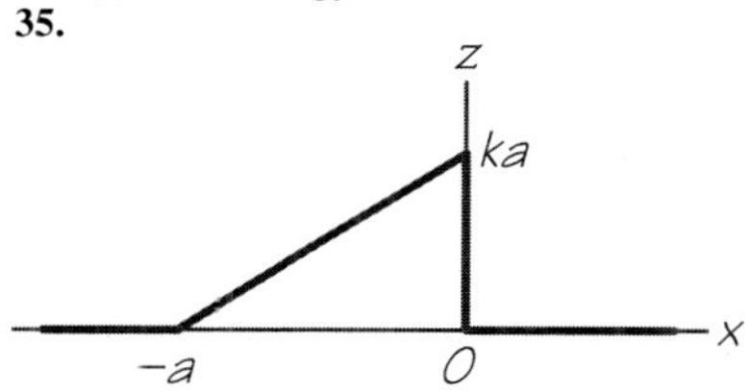

$z(x, t) = 0$ for $x < (vt - a)$ and $x > vt$; $z(x, t) = k(x - vt + a)$ for $(vt - a) < x < vt$.
37. $v_t = z_0\left[2(x - vt)\left(\dfrac{v}{\alpha^2}\right)\right]e^{-(x-vt)^2/\alpha^2}$;

$a_t = z_0\left[-\dfrac{2v^2}{\alpha^2} + 4(x - vt)^2\left(\dfrac{v^2}{\alpha^4}\right)\right]e^{-(x-vt)^2/\alpha^2}$;

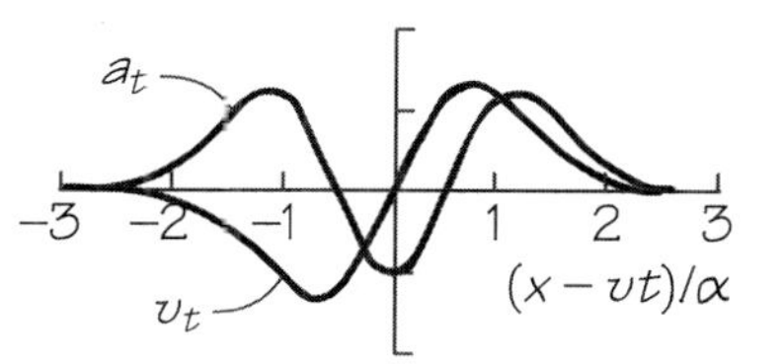

39.

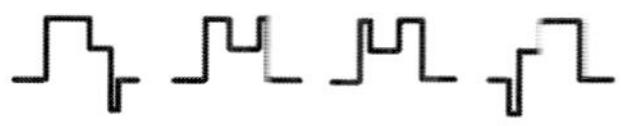

41. $-0.069A$; $1.07A$.
43. 6.6×10^{-3}.
45. $\dfrac{B}{A} = \dfrac{k_2 - k_1}{k_2 + k_1} = \dfrac{\sqrt{\mu_2} - \sqrt{\mu_1}}{\sqrt{\mu_2} + \sqrt{\mu_1}}$; $\dfrac{C}{A} = \dfrac{2k_2}{k_2 + k_1} = \dfrac{2\sqrt{\mu_1}}{\sqrt{\mu_2} + \sqrt{\mu_1}}$; $z_i = A\cos(k_1x - \omega t)$; $z_r = B\cos(k_1x + \omega t - \pi)$; $z_t = C\cos(k_2x - \omega t)$.
47. 1 MHz.
49. (a) 400 Hz; 0.4 m; 160 m/s; (b) 7.7 cm.
51. $\Delta T/T = 9.1 \times 10^{-3}$.
53. $69\cos(4\pi t)$ Hz.
55. 90 Hz; 101 Hz; 111 Hz.
57. 6.0×10^{-7} m; 3.9×10^{-7} m.
59. (a) and (b) are linear.
61. z(envelope) $= \sin[\Delta k(x/2) - \Delta\omega(t/2) + (\pi/4)]$; $f = (\omega_1 - \omega_2)/2\pi$; $\lambda = 2\pi/(k_1 - k_2)$.
63. (a) $y_1(x, t) = y_0\sin(k_1x + \omega_1t)$; $y_2(x, t) = y_0\sin(k_2x - \omega_2t)$, (b) $\omega_2 = k_2\omega_1/k_1$; (c) $\delta\omega = v\ \delta k$.

CHAPTER 16

1. 1.15×10^3 kg/m^3.
3. 50 L.
5. $\Delta d_1 = af/\sqrt{2}$; $\Delta d_2 = -af/\sqrt{2}$.
7. 2.1×10^5 N/m^2.
9. (a) 5.1×10^4 N; (b) 1.3×10^2 N (equivalent to lifting 13 kg).
11. 7.8×10^2 N/m^2; p will decrease.
13. 1.11×10^5 Pa; 2.02×10^5 Pa; 1.11×10^6 Pa; 1.01×10^3 Pa.
15. 1.02×10^5 Pa; 1.05×10^5 Pa.
17. $h_{\text{Venus}} = 0.838$ m; $h_{\text{Neptune}} = 0.6$[illegible]9 m.
19. 52 N; 2.2 mm.
21. Sink.
23. 8.4×10^2 kg/m^3.
25. 12.7 m^2.
27. 0.96×10^3 kg/m^3.
29. $\rho_1 < \rho_3 < \rho_2$; $D/H = (\rho_3 - \rho_1)/(\rho_2 - \rho_1)$.
31. 8.1×10^2.
33. 2.9 m/s; 0.23 kg/s.
35. 57 mi/h.
37. 0.81 cm.
39. 1.6×10^2 W (0.22 hp).
41. -4.7×10^2 Pa (gauge); 4.7×10^2 Pa.
43. 8.7×10^{-2} N; rise.
45. (a) 7.7 m/s; (b) 4.3 m.
47. (a) 1.9 m/s; (b) 2.16×10^5 Pa (2.14 atm).
49. $p_0 + \frac{1}{2}\rho v_0{}^2 + \rho gh_0 = p_1 + \frac{1}{2}\rho v_1{}^2 + \rho gh_0$; $p_0 + \frac{1}{2}\rho v_0{}^2 + \rho gh_0 = p_2 + \frac{1}{2}\rho v_2{}^2 + \rho gh_0$; $v_0D^2 = v_1d_1{}^2 + v_2d_2{}^2$.
51. $a = 1$, $b = 1$.
53. 5.4 cm.
55. 3.9 atm; less air.
57. (a) 27 m/s; (b) ≃200 (the rate of flow of the hose); (c) 1.9×10^{-3} m^2; (d) [illegible]$.65 \times 10^5$ Pa (gauge).
59. Yes.
61. (a) 6.0×10^2 kg/m^3; (b) $-0.39 \times 10^3\ \Delta$ N (up); (c) $0.39 \times 10^3\ \Delta$ N (down); (d) 1.44 Hz.
63. (a) 1.8×10^5 Pa; (b) 4.8×10^5 Pa; (c) 7.0×10^{-4} m/s; (d) 13 m/s; (e) 0.37 L/s.
65. $[2g(h_1 - h_0)]^{1/2}$.
67. 8.6 min.

CHAPTER 17

1. (b), (e), (f) are in thermal equilibrium.
3. The extensive variables are (a) and (d). The intensive variables are (b) and (c).

5. (a) $t_F = t_C = -40°$; (b) none; (c) $t_F = T = +575$.
7. 95.9°F to 108.5°F.
9. $\Delta p = 0.29$ Pa.
11. (a) 457°C; 854°F; (b) −196°C; −321°F; (c) 297 K; 24°C; (d) 2.2°C; 276 K.
13. $T = 1.25t_R + 273$; $t_F = 2.25t_R + 32$.
15. 672°R; 7.6°R; 10,800°R; 492°R; 430°R; 447°R.
17. 52°C (126°F).
19. 7.0 cm^3.
21. 1.4×10^3 cm^3.
23. 7.4×10^4 m^3.
25. 0.47 kg/m^3.
27. $p_2/p_1 = 0.562$.
29. 1.30×10^5 Pa; 0.28×10^5 Pa; 1.58×10^5 Pa.
31. 3.1×10^{-10} m.
33. (a) 6.06×10^4 Pa; 308 K; (b) 3.5×10^{-3} m^3; (c) 0.77 atm.
35. 22.4×10^{-3} m^3; 1.29 kg/m^3.
37. 6.07 kg/m^3.
39. 2.4×10^9 molecules.
41. (a) 71 atm; (b) 71 atm.
43. 40.5 cm^3; 142 cm^3

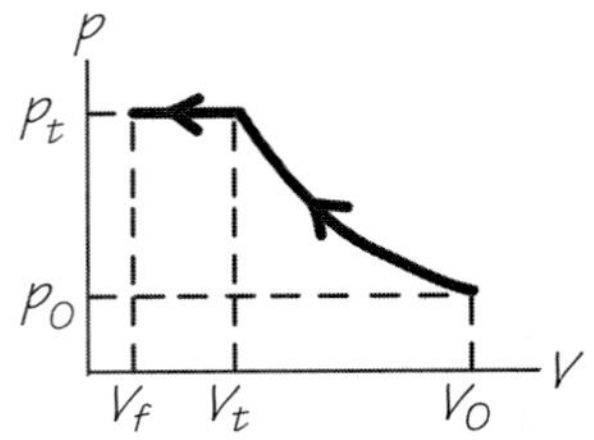

45. 1.35 kg/m^3; 0.74 m^3/kg.
47. (a) $p_2 = 10^5$ Pa, $V_2 = 600$ cm^3, $T_2 = 568$ K; (b) $p_3 = 1.30 \times 10^5$ Pa, $V_3 = 600$ cm^3, $T_3 = 762$ K; (c) $p_4 = 1.30 \times 10^5$ Pa, $V_4 = 300$ cm^3, $T_4 = 381$ K.
49. 4.0×10^{-2} m^3.
51. 2.56×10^3 J; 2.15×10^3 J; the energy of the gas is $\simeq 10^4$ the energy of the ball.
53. 373.15 K.
55. (b) 421 K (148°C).
57. $E_{3200°C} = 1.84 \times 10^4\ E_{25°C}$.
59. (a) 6.25×10^{10} Hz; (b) 5.83×10^{12} Hz; (c) 1.67×10^{13} Hz; (d) 6.25×10^{13} Hz.
61. 9.4×10^{-6} m.
63. 4.47×10^{26} W.
65. 2.05×10^{17} W; almost the same.
67. 5.67×10^{-8} W/m$^2 \cdot$ K^4.
69. 37 kg.
71. 0.14 m^3.
73. 26 m/s.
75. 0.754×10^{-23} J/K′.

CHAPTER 18

1. Reversible transformations: (c), (e), and (h); irreversible transformations: (a), (b), (d), (f), and (g)

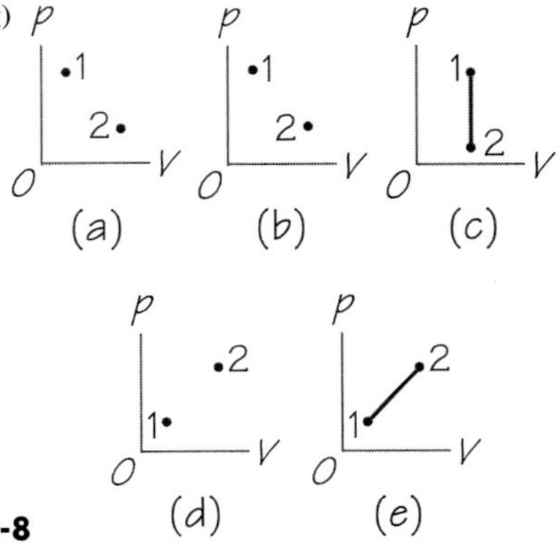

3. 5.0×10^2 cal/s.
5. $c'_p pV/R$.
7. (a) 262.8 g of water and 37.2 g of ice at 0°C; (b) 300 g of water at 7.6°C.
9. (a) 0.22 cal/g · K, 5.9 cal/mol · K; (b) 0.103 cal/g · K; 5.8 cal/mol · K; (c) 0.04 cal/g · K; 8 cal/mol · K.
11. 308 K.
13. 167 K.
15. 6.7°C.
17. 33 times.
19. 1 cal = 4.14 J.
21. (a) 11.9°C; (b) 10.1°C; (c) ΔT of the water is less than in part (a); (d) net negative temperature change.
23. 1.0×10^6 J.
25. 3.0×10^5 J.
27. 0.012 m^3; 0.0034 m^3; -1.6×10^3 J.
29. (a) $[ML^{-4}T^{-2}]$; (b) $(V_2 - V_1)[p_0 - \frac{1}{2}\beta(V_2 + V_1)]$; (c) same as part (b).
31. 4.1×10^3 J.
33. 633 cal.
35. (a) 2.7×10^4 cal; (b) -3.5×10^4 cal; liberated.
37. (a) $U_A = (c'_V/R)p_AV_A$; $U_B = (c'_V/R)p_BV_B$; $U_C = (c'_V/R)p_BV_A$; (b) $\frac{1}{2}(p_A - p_B)(V_B - V_A)$.
39. (a) 2.8×10^2 K; 6.1×10^2 K; (b) 1.2×10^3 J; (c) 2.3×10^2 J; (d) 1.4×10^3 J (into the gas).
41. −0.72 K (cooled).
43. 5.0×10^4 J.
47. 20°C.
49. (a) No; (b) no; (c) no.
51. $\Delta p_{\text{adiabatic}}/\Delta p_{\text{isothermal}} = 1.24$.
53. $C_V/C_p = 0.683$; $+1.43p_1V_1$; $+1.43p_1V_1$.
55. 61.7 cm^3; 151 cm^3.
59. (a)

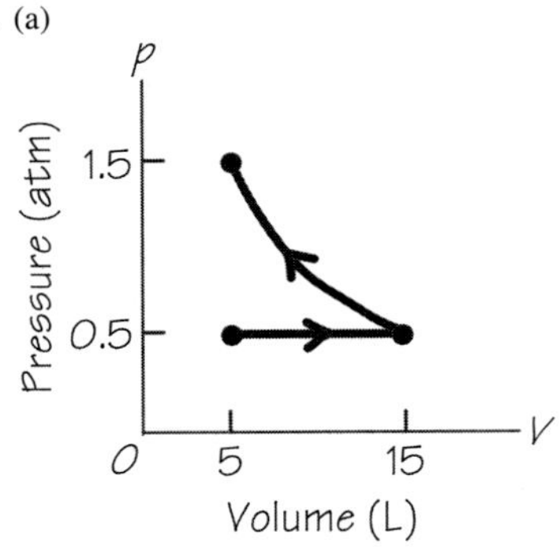

(b) 1.5 atm; (c) -3.2×10^2 J.
61. (a) $+7.1 \times 10^3$ J; (b) $+2.5 \times 10^3$ J.
63. $\text{slope}_{\text{adiabatic}}/\text{slope}_{\text{isothermal}} = \gamma$.
65. 3.85×10^3 J.
67. $dm/dt = 20 \times 10^3$ kg/s.
69. (a) 10 atm; (b) -5.7×10^3 J; (c) -5.7×10^3 J (from the gas).
71. 32°F.
73. $+2.48 \times 10^3$ J.
75. (a) 7.18×10^5 Pa (7.1 atm); (b) 5.9×10^{-2} m^3; (c) 2.0×10^4 J; (d) 137 K; (e) $+2.0 \times 10^4$ J; (f) $+1.2 \times 10^4$ J.

CHAPTER 19

1. (a) 1.0×10^{-4}; 5.8×10^{-2}; (b) 6.1×10^{-3}; 0.23; (c) 1.0×10^{-12}; 1.2×10^{-4}; (d) 6.1×10^{-11}; 4.9×10^{-4}.
3. 3.9×10^{24} components; 3.9×10^5 m^2.
5. 5.9 Pa.
7. (a) 3.8×10^3 J; (b) 304 K; (c) $\langle v^2 \rangle = 3.8 \times 10^5$ m^2/s^2; (d) 6.1×10^2 m/s.
9. 11 m/s.
11. 3×10^{-17} Pa.
13. (a) 290 K; (b) 2.90×10^5 K; (c) 3.6 J.
15. 6.1×10^{-21} J; 3.5×10^{-10} m/s.
17. 7.0×10^5 m/s.
19. 484 m/s; 517 m/s; 412 m/s; 1934 m/s.
21. (a) 1.2×10^4 m/s; (b) 1.9×10^3 m/s.
23. 4.9×10^{-3} m/s.
25. 7.2×10^{73} K.
27. (a) 61.8 mi/h; (b) 62.4 mi/h.
29. (a) 1/1296; (b) 1/324.
31. (a)

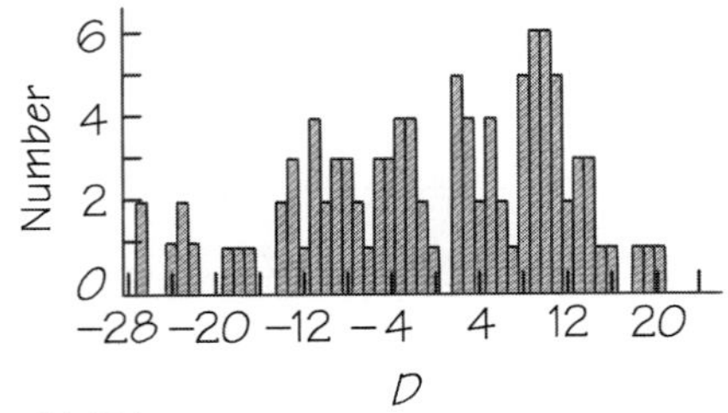

(b) 124.
33. (a) +6.4; (b) 474; (c) +20.
37. 0.984; 0.970; independent of the temperature.
39. $v_{\text{av}} = 2(2kT/\pi m)^{1/2}$.
41. (b) $\langle \mathbf{v} \rangle = \mathbf{u}$; $\langle \mathbf{v}^2 \rangle = v_{\text{rms},0}^{\ 2} + u^2$.
43. $s = 5$.
45. Monatomic gas; dumbbell shape; atoms not linear; linear arrangement.
51. 10^{-3} atm; 10^{-7} atm.
53. $\simeq 100$ m.
55. 5.6×10^{19} atoms; 5.6×10^{16} atoms; 30 Pa.
57. 0.63; no change.
59. (a) $v_{\text{esc}} = 4.1 \times 10^3$ m/s; (b) 1350 K; v_{rms} and T decrease; (c) similar effect; (d) lighter component will make up a smaller fraction of the atmosphere.
61. (a) 7.0×10^5 m; (b) The mean free path decreases by a factor of 1000 to 7.0×10^2 m.
63. 2.0×10^{-3} m/s.
65. $p = \frac{1}{3}n\langle E \rangle$.
67. 285 K (12°C); 2.8×10^4 Pa.
69. (a) 3.2×10^{11} atoms/m^3; (b) 5.6×10^4 s; (c) 7.1×10^7 m.
71. $\dfrac{4}{\sqrt{\pi}} \displaystyle\int_{\sqrt{3/2}\,v_{\text{escape}}/v_{\text{rms}}}^{\infty} u^2 e^{-u^2}\, du.$

CHAPTER 20

1. 50%.
3. (a) The possible outcomes are

1234	1235	1236	1245	1246
1256	1345	1346	1356	1456
2345	2346	2356	2456	3456

(b) 1/15; (c) 3/5.
5. 29%.
7. (a) 38%; (b) 0.38 cal.
9. (a) 325 K (52°C); (b) 9.1 kWh; 4.1 kWh.
11. Violation of the first law of thermodynamics.
13. 60.
15. Engine A is not reversible.
17. 127 W.
19. 13.7; 9.1.
21. $C_V(T_2 - T_3)$.
23. 5.5 W (0.007 hp).
25. The statement cannot be trusted.
27. $\simeq 1$ cent.
29. (a) $T_3 = p_LV_U/nR$; $T_4 = p_LV_L/nR$; (b) $-C_Vp_L(V_U - V_L)/nR$; (c) $Q_{IV} = +[1 + (C_V/nR)]p_L(V_U - V_L)$; (d) $nR(p_U - p_L)(V_U - V_L)/[C_V(p_UV_U - p_LV_L) + nRp_U(V_U - V_L)]$.

33. $1 - [(T_3 - T_4)/\gamma(T_2 - T_1)]$.
35. 7.9×10^{-2} J/K · min; not reversible.
37. 31°C; +10 J/K.
39. +8 J/K.
41. $\Delta S_{gas} = +44$ J/K; $\Delta S_{universe} = +44$ J/K.
43. 20 J/K.
45. The entropy of the gas decreases.
47. $\Delta T = 0$; +32.5 J/K.
49. (a)

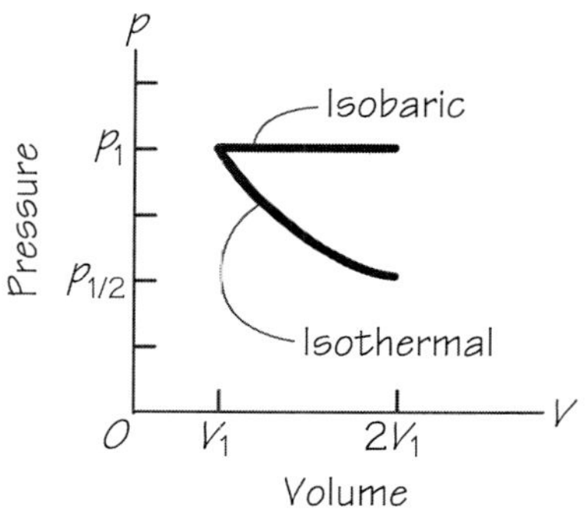

(b) p_1V_1; $0.693p_1V_1$; (c) $0.693nc'_p$; $0.693nR$.
51. Process $A \to B$ (isobaric): $Q = \frac{5}{2}nR(T_B - T_A)$, $W = nR(T_B - T_A)$, $\Delta U = \frac{3}{2}nR(T_B - T_A)$, $\Delta T = T_B - T_A$, $\Delta S = \frac{5}{2}nR\ln(T_B/T_A)$;
Process $B \to C$ (adiabatic): $Q = 0$, $\Delta U = \frac{3}{2}nR(T_C - T_B)$, $W = \frac{3}{2}nR(T_B - T_C)$, $\Delta T = T_C - T_B$, $\Delta S = 0$;
Process $C \to D$ (isothermal): $\Delta T = 0$, $\Delta U = 0$, $W = \frac{3}{2}nRT_C\ln(T_A^{2/3}T_C/T_B^{5/3})$, $Q = \frac{3}{2}nRT_C\ln(T_A^{2/3}T_C/T_B^{5/3})$, $\Delta S = \frac{3}{2}nR\ln(T_A^{2/3}T_C/T_B^{5/3})$;
Process $D \to A$ (constant volume): $Q = \frac{3}{2}nR(T_A - T_C)$, $W = 0$, $\Delta U = \frac{3}{2}nR(T_A - T_C)$, $\Delta T = T_A - T_C$, $\Delta S = \frac{3}{2}nR\ln(T_A/T_C)$.
53. $\Delta S_{gas} = Q/T < 0$ (decreases); $\Delta S_{reservoir} = -Q/T > 0$ (increases); no conflict, reversible process.
55. (a) $T = T_0$; (b) $+nR\ln(3)$; (c) $+nR\ln(3)$; (d) 0.
57. (a) +87 J; (b) 2.7 K; (c) +0.38 J/K.
59. (a) 2.25×10^{-2} m³; (b) 273 K; (c) −11.5 J/K.
61. (a) 3.07×10^{-2} m³; (b) 373 K; (c) 0.
63. (a) $Q_{h,engine} = 400$ J; $Q_{c,engine} = 300$ J; $Q_{c,refrigerator} = 300$ J; $Q_{h,refrigerator} = 400$ J; $Q_{h,net} = 0$; $Q_{c,net} = 0$; (b) $Q_{h,engine} = 500$ J; $Q_{c,engine} = 400$ J; $Q_{c,refrigerator} = 300$ J; $Q_{h,refrigerator} = 400$ J; $Q_{h,net} = -100$ J (from the reservoir); $Q_{c,net} = +100$ J (to the reservoir). (c) $Q_{h,engine} = 400$ J; $Q_{c,engine} = 300$ J; $Q_{c,refrigerator} = 240$ J; $Q_{h,refrigerator} = 340$ J; $Q_{h,net} = -60$ J (from the reservoir); $Q_{c,net} = +60$ J (to the reservoir).
65. $dQ_c/dt = 770$ MW; $dm/dt = 1.5 \times 10^4$ kg/s.
67. 0.30; 60% of the Carnot efficiency:
69. nc_pT.
71. (a) 437 K (164°C); 1020 K (747°C); (b) $Q_{abs}/m = 5.38 \times 10^6$ J/kg; $Q_{rej}/m = -3.73 \times 10^6$ J/kg; $Q_{net}/m = 1.65 \times 10^6$ J/kg (absorbed); (c) 1.65×10^6 J/kg.

CHAPTER 21

1. (a) 0.255 nm.
3. 0.524; 0.740.
5. 2.97 cm.
7. $(\Delta L/L)_c = 0.014$; 14 mm.
9. 2.5×10^2 N/m².
11. $\Delta y \simeq 10^{-13}$ m.
13. 0.12° below the horizontal.
15. 1.99928 cm.
19. 9.1×10^{-12} m²/N.
21. 7.0×10^4 MN/m².
23. 8.03×10^{11} dyne/cm².
25. Yes.
27. $Y = 3 \times 10^5$ MN/m²; $G = 8 \times 10^4$ MN/m²; $L = 1.3 \times 10^7$ m $= 1.3 \times 10^4$ km.
29. 0.5 mm.
31. $\alpha = 8.0 \times 10^{-4}$ K⁻¹.
33. 1.3×10^{-4} cm².
35. (a) steel; (b) 0.99971 Hz $\le f_{aluminum} \le$ 1.00029 Hz; 0.99979 Hz $\le f_{copper} \le$ 1.00021 Hz; 0.99987 Hz $\le f_{steel} \le$ 1.00013 Hz.
37. 2.1%
39. 7.6%.
41. 287 K (14°C).
43. 7.4 kW.
45. Pine: 144 Btu/h; Fiberglas: 62 Btu/h; $R_{eff} = 16$ ft² · h · °F/Btu.
47. (a) $\kappa_{eff} = \kappa_1\kappa_2(L_1 + L_2)/(\kappa_2L_1 + \kappa_1L_2)$; (b) $T_3 = (\kappa_1L_2T_1 + \kappa_2L_1T_2)/(\kappa_1L_2 + \kappa_2L_1)$.
53. $\frac{1}{L}\frac{dQ}{dt} = \frac{2\pi\kappa(T_2 - T_1)}{\ln[1 + (\alpha/R_1)]}$.
55. 9.6×10^2 W.
57. 0.90 W.
59. 2.0×10^5 N/m².

CHAPTER 22

1. 6.2×10^9 fewer electrons.
3. 4.82×10^4 C.
5. -2×10^{-10} C, 1.25×10^9 electrons; -1×10^{-10} C, 6.2×10^8 electrons.
7. 2.2×10^{-10}.
9. (a) 1.8×10^{51} electrons; (b) 3.5×10^{-39}.
11. (a) conserved; (b) not conserved; (c) conserved; (d) not conserved.
13. -8.5×10^{-24} C.
15. 46 N repulsion, 23 N attraction.
17. $q = 1.6 \times 10^{-19}$ C, 1 electron.
19. 3.5×10^{-10}, masses much larger.
21. (a) 3×10^9 esu in 1 C; (b) 4.8×10^{-10} esu.
23. $q_1/q = q_2/q = \frac{1}{2}$.
25. (a) 2.6×10^{-9} N toward the proton (centripetal); (b) $v = 9.2 \times 10^5$ m/s; (c) 4.9×10^{14} Hz; (d) 8.6 N/m.
27. $q = 1.4 \times 10^{-8}$ C.
29. $q_1 = +1.1 \times 10^{-7}$ C, $q_2 = -3.2 \times 10^{-7}$ C, $q_3 = +5.3 \times 10^{-7}$ C.
31. $r_f = 0.9$ m.
33. $x = 44.4$.
35. zero.
37. $F_+ = 39$ N 30° above the x-axis, the line joining the two "up" quarks, $F_- = 40$ N toward the center of the line joining the two "up" quarks.
39. (a) $\Sigma\mathbf{F} = 0$; (b) unstable; (c) stable.
41. (a) $(\sqrt{3})kq^2/2L^2$, 9.7° above the − x-axis; (b) $3.2kqQ/L^2$, 6.3° below the − x-axis.
43. $\mathbf{F} = (2kq\lambda/x_0)\mathbf{i}$.
45. $kQ\,\mathbf{i}/d(L + d)$.
47. 1.4 N away from the center.
49. $Q = 3.0 \times 10^{-7}$ C.
51. 0.52 N.
53. 5.7 N/m², same.
55. $\theta = 6.6°$.
57. (a) $v = (e^2/4\pi\epsilon_0 mR)^{1/2}$; (b) $L = (e^2mR/4\pi\epsilon_0)^{1/2}$; (c) $v = e^2/4\pi\epsilon_0 L$; (d) $R = 4\pi\epsilon_0 L^2/me^2$; (e) $\tau = 32\pi^3\epsilon_0^2L^3/me^4$; (f) $v = 2.2 \times 10^6$ m/s, $R = 5.3 \times 10^{-11}$ m, $\tau = 1.5 \times 10^{-16}$ s
59. (a) $F_{net} = kq^2\{(1/x^2) - [1/(\ell - x)^2]\}$ away from the closer charge, $x = \ell/2$ (b) $\mathbf{F}_{net} = -2kq^2\ell\,\Delta\mathbf{x}/[(\ell/2)^2 - (\Delta x)^2]^2$; (c) $f = (1/2\pi)(32kq^2/\ell^3m)^{1/2}$.
63. (a) $\mathbf{F} = kqQR\sum_{n=-\infty}^{\infty}\left\{\frac{1}{[(na)^2 - R^2]^{3/2}}\right\}\mathbf{j}$;
(b) $\mathbf{F} = \frac{k\lambda Q}{R}\int_{-\infty}^{\infty}\frac{du}{(u^2 + 1)^{3/2}}\mathbf{j}$.

CHAPTER 23

1. $1.8 \times 10^7(-0.60\mathbf{i} - 0.80\mathbf{j})$ N/C.
3. $(1.14 \times 10^{11}$ N/C$)\hat{\mathbf{r}}$, $-(1.82 \times 10^{-8}$ N$)\hat{\mathbf{r}}$ (toward the nucleus).
5. 3.19×10^6 N/C, 78° above the + x-axis.
7. (a) $-(1/4\pi\epsilon_0)(8q/\ell^2)\mathbf{j}$; (b) $\mathbf{E} = 0$; (c) $E = (1/4\pi\epsilon_0)\{2qd/[d^2 + (\ell/2)^2]^{3/2}\}$ away from the origin.
9. $\mathbf{E} = \frac{2\mathbf{p}}{4\pi\epsilon_0 r^3}\left\{\frac{1}{[1 + (L/2r)]^2[1 - (L/2r)]^2}\right\}$,
$\mathbf{E} = \frac{2\mathbf{p}}{4\pi\epsilon_0 r^3}$.
11. Stable, $(1/2\pi)(2q/\pi\epsilon_0 a^3m)^{1/2}$.
13. tripled.
15.

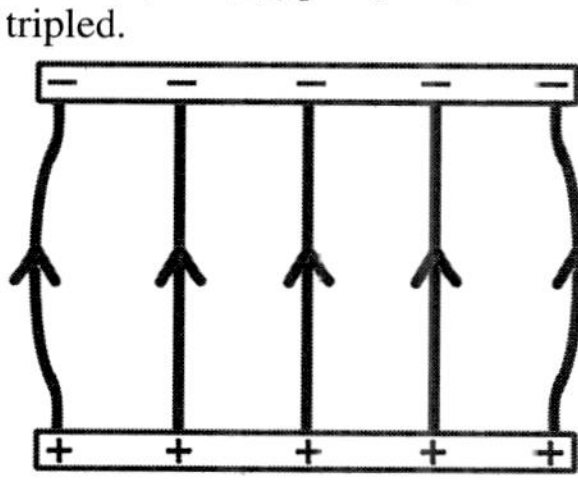

17.

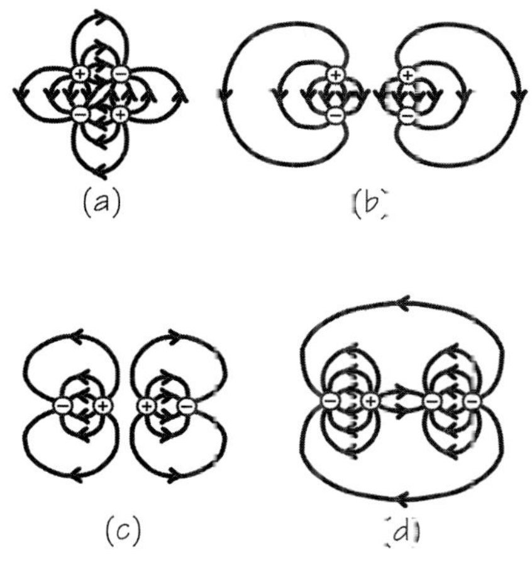

19. $\mathbf{E} = 2.7 \times 10^4$ N/C perpendicular to and away from the line
21.

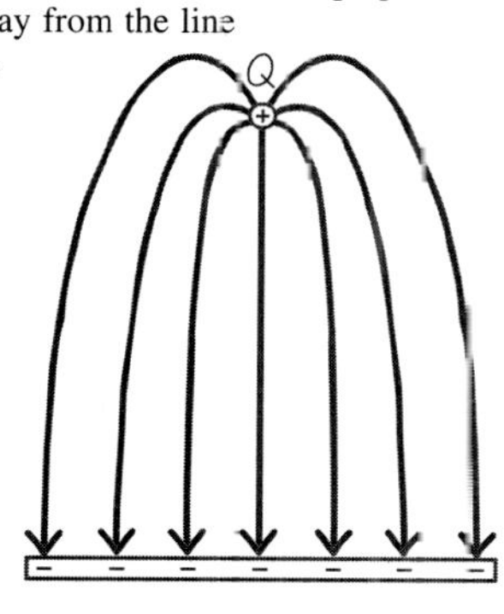

23.

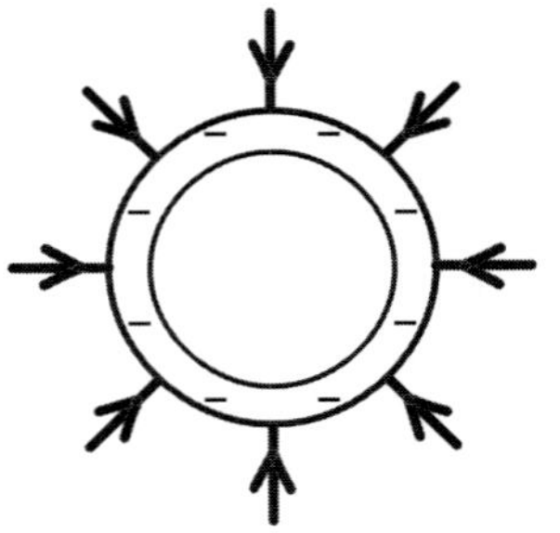

25.

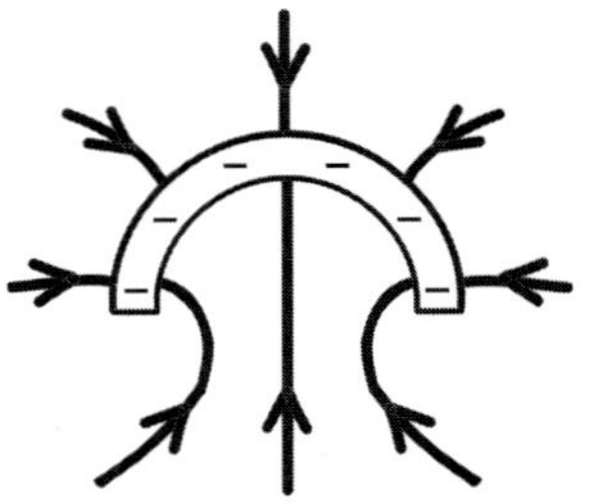

27. $E = (\sigma/2\epsilon_0) + (\sigma/2\epsilon_0) = \sigma/\epsilon_0$ perpendicular to the plates and away from them, $E = (\sigma/2\epsilon_0) - (\sigma/2\epsilon_0) = 0$.
29. (a) $\mathbf{E} = \dfrac{z_0 Q}{2\pi\epsilon_0 R^2}\left[\dfrac{1}{z_0} - \dfrac{1}{(R^2 + z_0^2)^{1/2}}\right]\mathbf{k}$;
(b) $Q/4\pi\epsilon_0 z^2\mathbf{k}$; (c) $Q/2\pi\epsilon_0 R^2\mathbf{k}$.
31. 7.3×10^4 N/C toward the point charge.
33. (a) $\dfrac{\sigma L z_0}{2\pi\epsilon_0}\mathbf{k}\displaystyle\int_{x=-L}^{x=L}\dfrac{dx}{(x^2 + z_0^2)\sqrt{L^2 + x^2 + z_0^2}}$;
(b) $(\sigma/2\epsilon_0)\mathbf{k}$; (c) $(\sigma/2\epsilon_0)\mathbf{k}$.
35. $q = 6.5 \times 10^{-7}$ C $= 0.65\ \mu$C.
37. $\sigma = 1.7 \times 10^{-7}$ C/m^2.
39. $v = (q\lambda/2\pi\epsilon_0 m)^{1/2}$.
43. 3.4 cm.
45. 5.7 s^{-1}.
47. $-(1.41 \times 10^{-6}$ N·m$)\mathbf{k}$.
49. 2×10^6 N/C.
51. $-6p^2/4\pi\epsilon_0 r^4$ (attraction).
53.

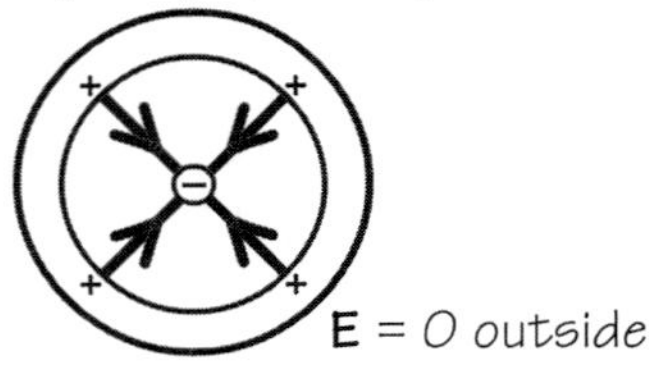

55.

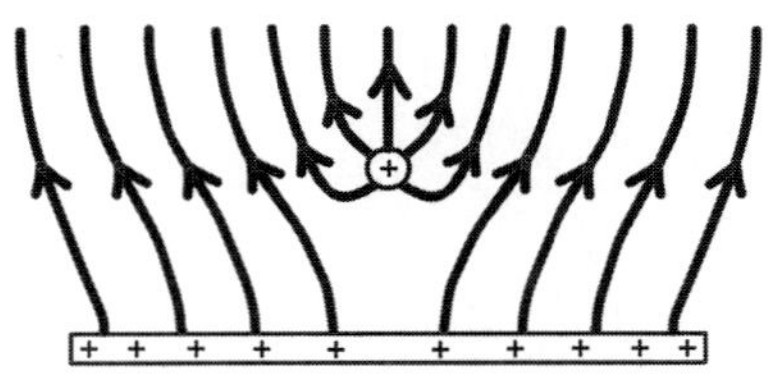

57. (a) $\{(\lambda R/2\pi\epsilon_0)/[y^2 - (R/2)^2]\}\mathbf{j}$;
(b) $-\{(\lambda R/2\pi\epsilon_0)/[x^2 + (R/2)^2]\}\mathbf{j}$.
59. $(1 + 8.5t^2)\mathbf{i} + (1 - 14t^2)\mathbf{j}$ m, with t in s.
61. (a) 7.2×10^{-14} N away from the plate; (b) -3.2×10^{-13} J; (c) 4.4 m.

63.

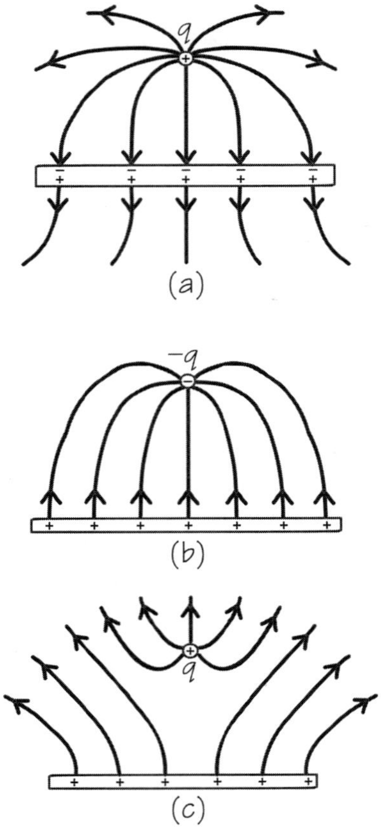

65. $p = \lambda_0 L^2/6$.

CHAPTER 24

1. (a) $\sigma\pi R^2/2\epsilon_0$; (b) $0.866\ \sigma\pi R^2/2\epsilon_0$.
3. $\lambda h/\epsilon_0$.
5. $+6$ N·m^2/C.
7. $\pi E_0 R^2/3$.
9. $\dfrac{q}{\epsilon_0}\dfrac{1}{\sqrt{1 + R^2/h^2}}$.
13. (a) zero; (b) 1.13×10^8 N·m^2/C.
15. (a) net flux is 0.
17. (a) $Q = -3.54 \times 10^{-9}$ C; (b) $Q = -3.54 \times 10^{-9}$ C; (c) $Q = -3.54 \times 10^{-9}$ C.
19. 9.4×10^2 N·m^2/C out of the sides parallel to the *xy*- or *yz*-planes, 10.6×10^2 N·m^2/C out of the side perpendicular to the $-x$-axis, 8.2×10^2 N·m^2/C out of the side perpendicular to the $+x$-axis.
21. $E = \rho R^2/2\epsilon_0 r$.
23. $E_{\text{rod}}/E_{\text{pt chge}} = 0.02 = 2\%$.
25. -2.3×10^{-4} C.
27. $(1.8 \times 10^4)\hat{\mathbf{r}}$ N/C, $(1.8 \times 10^4)\hat{\mathbf{r}}$ N/C.
29. $E = 0$ for $r < r_1$, $[\rho(r^2 - r_1^2)/2\epsilon_0 r]\hat{\mathbf{r}}$ for $r_1 < r < r_2$, $[\rho(r_2^2 - r_1^2)/2\epsilon_0 r]\hat{\mathbf{r}}$ for $r_2 < r$.
31. $E = 0$ for $r < R_1$, $[Q(r^3 - R_1^3)/4\pi\epsilon_0(R_2^3 - R_1^3)r^2]\hat{\mathbf{r}}$ for $R_1 < r < R_2$, $(Q/4\pi\epsilon_0 r^2)\hat{\mathbf{r}}$ for $R_2 < r$.
33. 1st quadrant: E at $-\theta$; 2nd quadrant: E at $180° + \theta$; 3rd quadrant: E at $180° - \theta$; 4th quadrant: E at θ, $E = 3.3 \times 10^5$ N/C, $\theta = 31°$.
35. $\mathbf{E} = (-6.7 \times 10^8)r\,\hat{\mathbf{r}}$ N/C with r in m, $r <$ 3 cm; $\mathbf{E} = [(-1.8 \times 10^4)/r^2]\hat{\mathbf{r}}$ N/C with r in m, 3 cm $< r <$ 8 cm; $\mathbf{E} = [(2.7 \times 10^4)/r^2]\hat{\mathbf{r}}$ N/C, with r in m, 8 cm $< r$.

37.

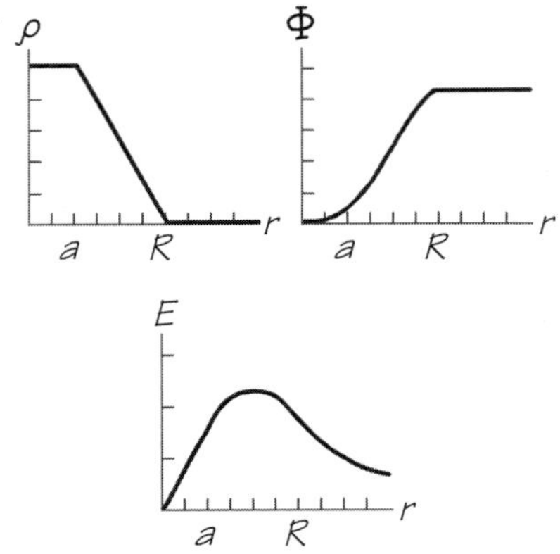

39. $r < R$: $\mathbf{E} = 0$; $R < r < 2R$: $\mathbf{E} = (q/4\pi\epsilon_0 r^2)\hat{\mathbf{r}}$; $2R < r$: $\mathbf{E} = -(q/4\pi\epsilon_0 r^2)\hat{\mathbf{r}}$.
41. 2.7×10^{-5} C/m^2.
43. $\sigma_{\text{sphere}}/\sigma_{\text{shell}} = 1.96$.
45. $\sigma_{\text{inner sphere}} = Q/4\pi a^2$, $\sigma_{\text{shell, inside}} = -Q/4\pi b^2$, $\sigma_{\text{shell, outside}} = +Q/2\pi R^2$.
47.

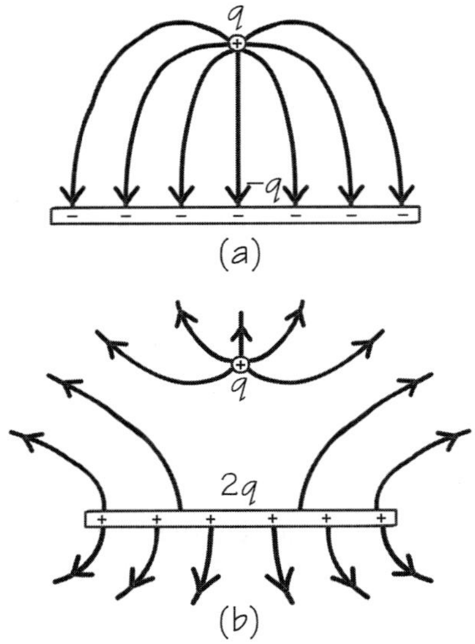

49. $\Phi_{x=0} = 0$, $\Phi_{x=a} = ba^4$, $\Phi_{y=0} = \Phi_{y=a} = \Phi_{z=0} = \Phi_{z=a} = 0$, $q = \epsilon_0 ba^4$.
53. $0.433EL^2$.
55. $\rho(r) = \epsilon_0 E/r$.
57. $\rho = +6000\epsilon_0$ C/m^3 (constant).

CHAPTER 25

1. 4.6×10^{-14} J.
3. 9.5×10^{-2} J.
5. (a) 0; (b) -1.35 J; (c) $+1.35$ J.
7. 1.36×10^{-3} J, 0.
9. -4.8×10^{-2} J.
11. $r = \infty$.
13. $q = 3.3 \times 10^{-14}$ C.
15. 4.2×10^{-3} J.
17. -5.0×10^4 J.
19. (a) $+5.0 \times 10^3$ V; (b) -1.0×10^{-3} J.
21. $+2.4 \times 10^5$ V.
23. $+4.3 \times 10^5$ V.
25. (a) $E_A = 0$, $E_B = (1.13 \times 10^6$ N/C · m$)x$, $x < 1$ cm, $E_C = 1.13 \times 10^4$ N/C, $x > 1$ cm; (b) $V_B - V_A = -(5.65 \times 10^5$ V/m$^3)x^2$, $x <$ 1 cm, $V_C = +5.65$ V $- (1.13 \times 10^4$ V/m$)x$, $x >$

1 cm; (c)

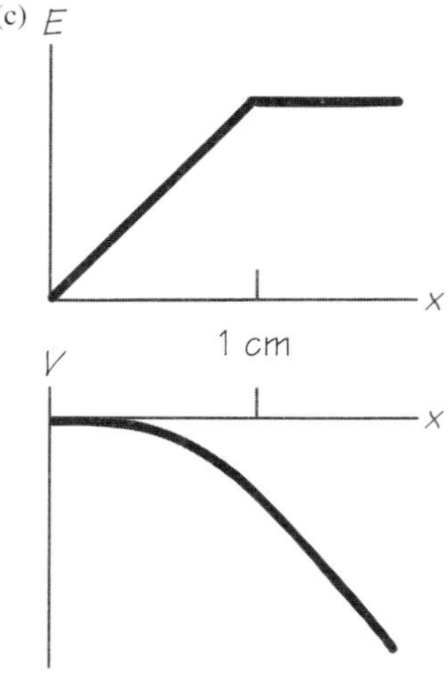

27. $V_{\text{outside}} = Q/4\pi\epsilon_0 r$, when $r > R$, $V_{\text{inside}} = (Q/8\pi\epsilon_0 R)[3 - (r/R)^2]$, when $r < R$.

29.

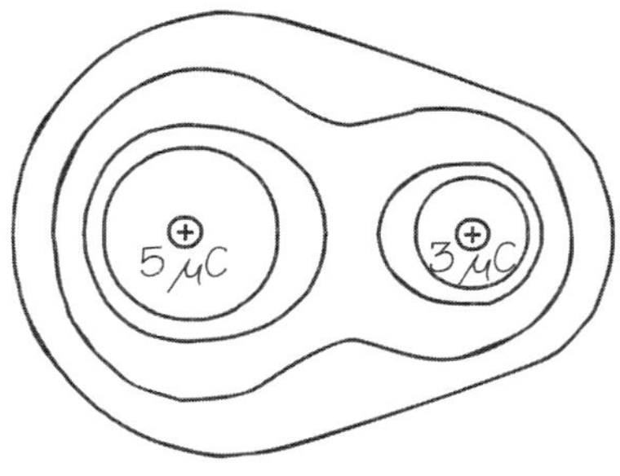

31.

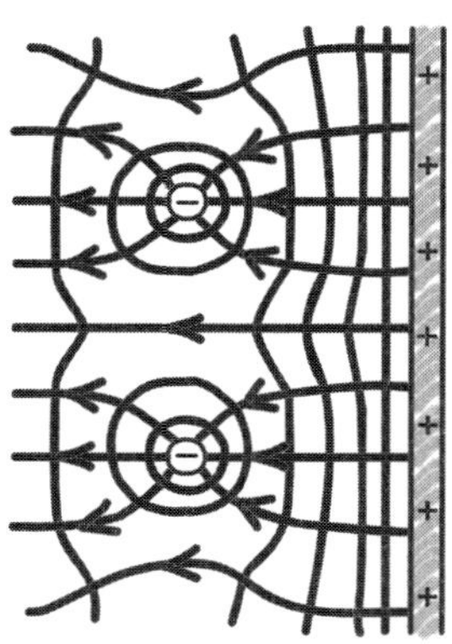

33.

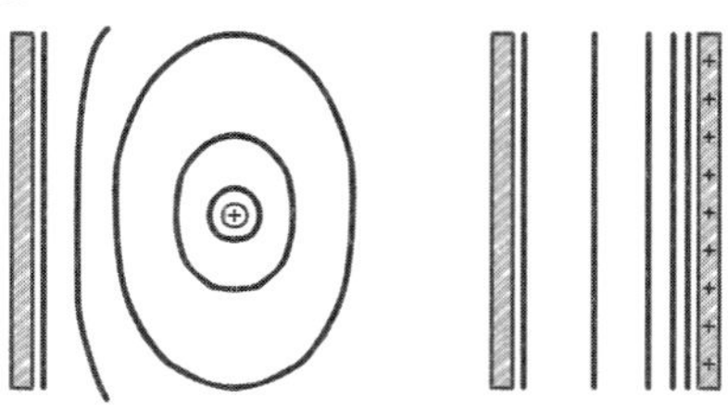

35.

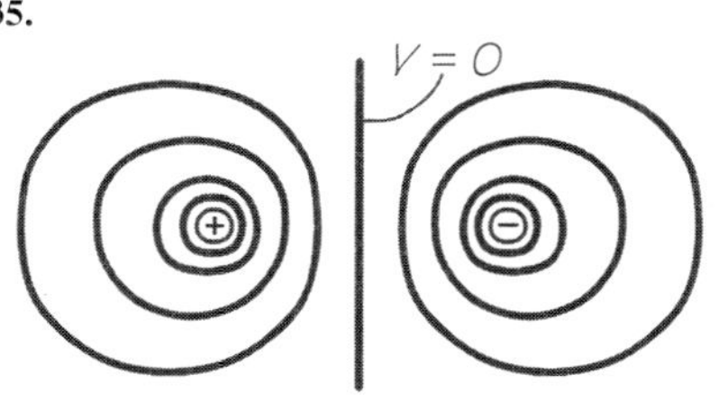

37. $\mathbf{E} = (Q/4\pi\epsilon_0 x^2)\mathbf{i}$.
39. $\mathbf{E} = (6.68\ \text{V/m})\mathbf{i}$.
41. $\mathbf{E} = -(p/4\pi\epsilon_0 r^3)\mathbf{i}$.
43. $\mathbf{E}_{r<R} = (Q/4\pi\epsilon_0)(6r/R^3)\hat{\mathbf{r}}$, $\mathbf{E}_{r>R} = (Q/4\pi\epsilon_0 r^2)\hat{\mathbf{r}}$.
45. $\mathbf{E} = \dfrac{2Qa_0^2}{4\pi\epsilon_0 L r^3}(-\sin\theta\mathbf{i} - 3\cos\theta\mathbf{j})$.
47. $Q = 1.0 \times 10^{-10}$ C.
49. $-(2\lambda/4\pi\epsilon_0)\ln(R)$ + (a constant).
51. 9.9×10^4 V.
53. (a) $V = \dfrac{q_0}{4\pi\epsilon_0}\left[\left(\dfrac{3}{x - x_0}\right) - \left(\dfrac{1}{x + \dfrac{x_0}{2}}\right)\right]$; (b) $V = \dfrac{q_0}{4\pi\epsilon_0}\left[\dfrac{2}{x} + \dfrac{7}{2}\dfrac{x_0}{x^2} + \dfrac{11}{4}\dfrac{x_0^2}{x^3} + \dfrac{25}{8}\dfrac{x_0^3}{x^4} - \cdots\right]$; (c) $q_{\text{net}} = 2q_0$, $p = 7q_0x_0/2$; (d) $|x| > 12.1x_0$.
55. $+4.2 \times 10^{-4}$ J.
57. $(q_1r_2 - q_2r_1)/(r_1 + r_2)$.
59. $Q_1 = 3Q/4$, $Q_2 = Q/4$.
61. (a) 2.2 μC, 11 μC; (b) 4.95×10^7 V/m, radial, 9.90×10^6 V/m, radial.
63. (a) $+5.5 \times 10^6$ eV, 8.8×10^{-13} J; (b) $v = 3.3 \times 10^7$ m/s.
65. (a) 3.9×10^6 V; (b) 3.9 MeV (6.2×10^{-13} J); (c) $Q = 5.6 \times 10^{-4}$ C.
67. $\dfrac{Q}{2\pi\epsilon_0 R^2}\left[\sqrt{R^2 + (a + x)^2} + \sqrt{R^2 + (a - x)^2} - 2a\right]$.
69. $Qa/2\pi\epsilon_0 x^2$.
71. $-(qQ/8\pi\epsilon_0 R)[3 - (r^2/R^2)]$, $k = qQ/4\pi\epsilon_0 R^3$.
73. 3.4×10^{-9} N toward the other Na^+, 1.2×10^{-18} J (7.4 eV).
75. $[Qx/4\pi\epsilon_0(R^2 + x^2)^{3/2}]\mathbf{i}$.
77. $\Delta U = -7.68 \times 10^{-19}$ J, $\Delta K = +3.84 \times 10^{-19}$ J, $\Delta E = -3.84 \times 10^{-19}$ J.
79. (a) $m = 9.9 \times 10^{-3}$ kg; (b) $(0.023/\sin\theta) + 0.078(1 - \cos\theta)$.
81. $-\dfrac{\rho r^2}{4\epsilon_0} - \dfrac{\rho R^2}{2\epsilon_0}\left[\ln\left(\dfrac{R}{a}\right) - \dfrac{1}{2}\right]$, $r < R$; $-\dfrac{\rho R^2}{2\epsilon_0}\ln\left(\dfrac{r}{a}\right)$, $r > R$.

CHAPTER 26

1. (a) 44 pF; (b) 40 cm.
3. (a) 1 V (b) 2.5 V; (c) 250 V.
5. 0.33 mm.
7. (a) $C \to 4\pi\epsilon_0 r$; (b) $C \to 4\pi\epsilon_0 r^2/(R - r) = \epsilon_0 A/d$.
9. $(4.43 \times 10^{-7}\ \text{m}) + (8.85 \times 10^{-10}\ \text{m/s})t$.
11. 6 kV.
13. 99 J.
15. 600 μF.
17. (a) 5.67×10^{-10} F; (b) 2.83×10^{-4} J, 2.83×10^{-2} J.
19. (a) $(\epsilon_0 AV_0^2/2d_0^2)(d - d_0)$; (b) $(\epsilon_0 AV_0^2/2d_0^2)(d - d_0)$; (c) $(\epsilon_0 AV_0^2/2dd_0)(d_0 - d)$; (d) energy has been stored in the battery.
21. $\lambda^2/8\pi^2\epsilon_0 r^2$.
23. 1.1×10^{-2} J.
25. 1.7×10^{-7} C, 7.0×10^{-4} J.
27. (a) 1.8 pF; (b) 2.3×10^{-7} J.
29. (a) 3.00×10^5 V/m; (b) 1.06×10^{-7} C; (c) 1.59×10^{-2} N; (d) 1.59×10^{-5} J.
31. 2.04 μF, 4.46 μF.
33. 1.93 μF.
35. 0.45 pF, 1.33 pF.
37. (a) 1.34 μF; (b) $Q_1 = 165$ μC, $Q_2 = Q_5 = 237$ μC, $Q_3 = 402$ μC.
39. $1.6C$, C, $2C$.
41. (a) 5.0 μF; (b) 25 μC.
43. 370 V.
45. $2.1 \times 10^{12}\ \text{m}^2$
47. $C - C_0 = (\kappa - 1)4\pi\epsilon_0 R$, $\sigma_{\text{ind}}/\sigma = (\kappa - 1)/\kappa$.
49. $q = 2.3q_0$.
51. $C = \kappa\epsilon_0 A/[(d + \kappa(D - d)]$
53. air: 5.3×10^{-10} C; paper; 2.0×10^{-9} C; neoprene: 3.6×10^{-9} C; Bakelite: 2.6×10^{-9} C; strontium titanate: 1.8×10^{-7} C.
55. $(1 - \kappa)Q^2/2C$ (a decrease).
57. 0.32 nF, 0.90 nF.
61. 2.96 μF.
63. $Q \simeq 10^{-6}$ C.
65. (a) $\frac{1}{2}C_1V^2/N$; (b) $\frac{1}{2}NC_1V^2$; (c) $U_{\text{series}} = \frac{1}{2}Q^2N/C_1$, $U_{\text{parallel}} = \frac{1}{2}Q^2/NC_1$.
67. $C = \frac{1}{2}(\kappa_0 + \kappa_1)(\epsilon_0 L^2/d)$.
69. 8.5×10^2 V, 7.2×10^2 V, 1.43×10^3 V.
71. 3.4 J, −2.2 J.
73. $C = (\kappa_1 - \kappa_0)[\epsilon_0 L^2/D\ln(\kappa_1/\kappa_0)]$.

CHAPTER 27

1. $1.2 \times 10^5\ \text{A/m}^2$, 0.46 C.
3. 1.0×10^4 s (2.8 h)
5. 5.8×10^{17} electrons.
7. 8.7×10^{-6} m/s, 2.0×10^{-5} m/s
9. 4.5×10^{21} electrons, 4.2×10^9 electrons.
11. $\pi J_0 R^2/2$.
13. $v_+ = 1.7 \times 10^{-6}$ m/s, $v_- = -2.5 \times 10^{-6}$ m/s.
15. 3.1×10^{-15} A.
17. 2.8 A, $5.6 \times 10^4\ \text{A/m}^2$, 5.8×10^{-6} m/s.
19. 8.5×10^{-5} m/s, 2.3×10^{-4} m/s.
21. $J_{\text{tube}} = I/2\pi Rd$ along the tube, $J_{\text{plate}} = I/2\pi rd$ radial.
23. $R_2 = 2R_1$.
25. (a) 1.2 Ω; (b) 0.15 cm.
27. 3.27×10^{-2} Ω.
29. 9.6×10^{-4} Ω, 3.0×10^{-4} Ω.
31. −0.27.
33. $I_{25} = 19.6$ A, $I_{400} = 17.1$ A.
35. $r_{\text{Cu}}/r_{\text{Al}} = 1.74$.
37. 188 turns.
39. 2.0 Ω, 0.24 kg, 0.48 kg.
41. 33.4 kg.
43. 1.82×10^{-3} Ω, $r = 0.17$ cm, masses are the same.
45. 8.4×10^{-6} m/s.
47. $\sigma = 2.2 \times 10^{-22}\ \text{m}^2$.
49. $T_{\text{Si}} = 8.5 \times 10^3$ K $T_{\text{Ge}} = 5.4 \times 10^3$ K, $T_{\text{C}} = 4.6 \times 10^4$ K.
51. 39 V.
53. 35 mA, 50 mA, 71 mA, 0.10 A, 0.14 A.
55. 800 W.
57. 8.4 ¢/h.
59. 1.9 m.
61. (a) 1.65 kW; (b) 22 bulbs.
63. 417 W.
65. 5.94×10^{25} electrons.
67. (a) 1.1×10^{17} protons; (b) 7.2×10^4 J; (c) 20 W.
69. $A = 4.5 \times 10^{-4}\ \text{m}^2$.

71.

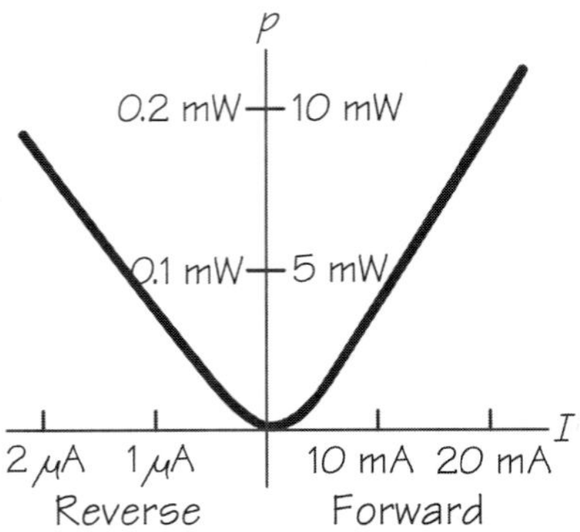

73. 1.1×10^{-3} W.
75. (a) $E_0 e^{-x/L}$; (b) $V_0(e^{-x/L} - e^{-1})/(1 - e^{-1})$; (c) $(\rho_0 L/A)(1 - e^{-1})$.

CHAPTER 28

1. 0.15 Ω.
3. 0.0075 Ω.
5. 20 A.
7. 0.030 Ω, 300 W.
9. 11.8 V, 0.17 W; 8.7 V, 33 W.
11. $R_2 = R_1$, $R_2 \to \infty$.
13. 120 W.
15. $V_{gf} = 0$, $V_{ag} = -4.0$ V, $V_{ca} = 5.5$ V.
17. (a) 0.90 W; (b) 1.0 Ω/battery.
19. $I_4 = +0.5$ A, $I_5 = +1.0$ A, $I_6 = -1.0$ A.
21. 6.1 Ω.
23. $I_1 = +0.45$ A, $I_2 = +0.38$ A, $I_3 = -0.068$ A.
25. $I_1 = 9.6$ mA, $I_2 = 5.7$ mA, $I_3 = 4.7$ mA.
27. $I_a = \mathscr{E}/[R + (r/N)]$, $I_b = \mathscr{E}[r + (R/N)]$.
29. 0.38 V, 0.04 V, 0.004 V.
31. One independent junction, $I_1 = 3\mathscr{E}/7R$, $I_2 = I_3 = I_4 = \mathscr{E}/7R$.
33. (a) 60.5 mW; (b) 6.8 mW.
35. (a) 0.075 A; (b) 0.075 A.
37. 2.35 A, 2.52 A, no current.
39. $2V/5R$, $V/5R$.
41. (a) $2R$; (b) $5R/3$; (c) $13R/8$; (d) $\frac{1}{2}(1 + \sqrt{5})R$.
43. 0.010 Ω.
45. (a) 10 Ω; (b) 5×10^4 Ω; (c) 10^5 Ω.
49. $R = R_1R_2/R_3$.
51. $R_x = (V/I)/[1 - (V/IR_V)]$, $R_V >> R_x$.
53. 20 μF.
55. 3.3 Ω.
59. $(\mathscr{E}/R_1)e^{-t/R_1C} + (\mathscr{E}/R_2)$.
61. 300-Ω resistors in parallel with each other and in series with the 250-Ω resistor and the two capacitors.
63. 6.8×10^3 s (1.9 h).
65. 0.417 A, 0.500 A, 0.167 A.
67. Six 600-Ω resistors in parallel, six 16.7-Ω resistors in series, two series 150-Ω resistors in parallel with two more sets of two series 150-Ω resistors, three series 66.7-Ω resistors in parallel with three series 66.7-Ω resistors.
69. $I = \{-(R + \alpha) + [(R + \alpha)^2 + 4\beta\mathscr{E}]^{1/2}\}/2\beta$.

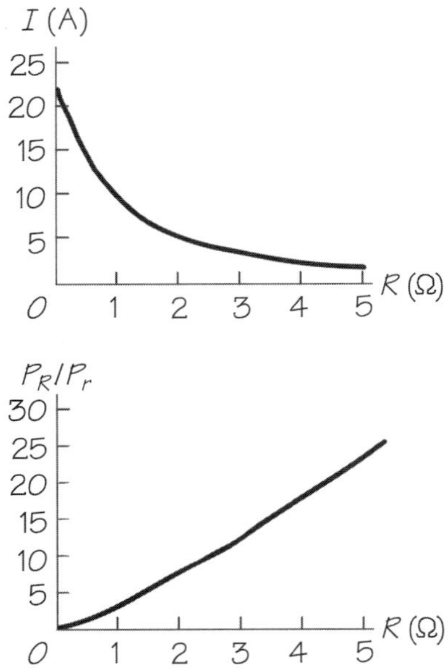

71. 667 W.
73. $I_{mixer} = 6.67$ A, $I_{vacuum} = 5.00$ A, $I_{chandelier} = 5.00$ A, 7 bulbs.
75. 0, $\mathscr{E}/2R$, $\mathscr{E}/R$.
79. (a) 21.4 μs; (b) 2.4 mC.
81. (a) $E = V_0/d$; (b) $\sigma\pi r^2V_0/d$.

CHAPTER 29

1.

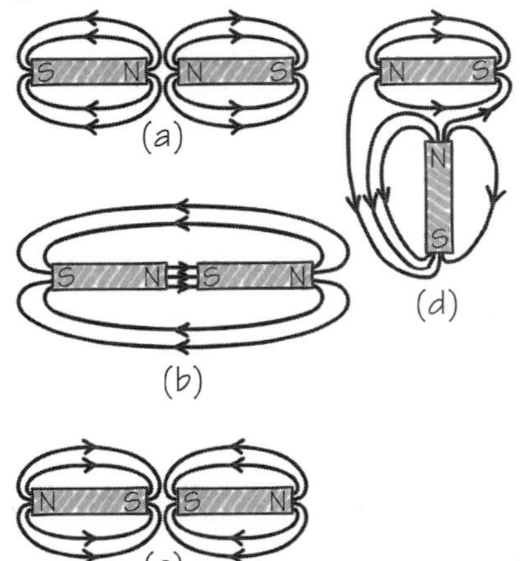

3. $-z$-direction.
5. $\mathbf{B} = -(7.2 \times 10^{-3}\ \text{T})\mathbf{k}$.
7. 34 μT.
9. (a) $\theta = (eB\,\Delta t)/m$; (b) 0.15 T.
11. −4.2 mm (horizontal).
13. (a) 2.25×10^{-3} T; (b) 2.5×10^{-2} T.
15. 5.7×10^{-14} m, 1.4×10^{-6} N.
17. Magnetic field, 0.40 T, 1.20 T.
19. $R_e = 0.57$ m, $R_p = 10.4$ m, 1.04 km.
21. (a) 5.9×10^7 m/s, 3.4° from the original direction; (b) 1.2 cm.
23. (a) 24 J (15×10^{19} eV); (b) 16×10^{-8} kg · m/s, 48 J; (c) 8×10^{-8} kg · m/s, 24 J.
25. 6.8×10^{-5} m.
27. 500 eV.
29. The circular motion is parallel to the xz-plane with radius 5.7×10^{-5} m. The motion in the y-direction has a constant acceleration of -2.6×10^{14} m/s^2.
31. (a) 1.5×10^7 Hz; (b) 4.8×10^7 m/s tangential; (c) 1.9×10^{-12} J (1.2×10^7 eV); (d) 120; (e) 8.0×10^{-6} s.
33. 9.9×10^{-16} kg · m/s $< p < 45 \times 10^{-16}$ kg · m/s, 1.9 TeV $< E <$ 8.4 TeV, 2.3×10^{-3} m/s.
35. 3.54 T.
37. $d\omega = qB/2m$.
39. 0.03 T.
41. 8.7×10^{-10} N perpendicular to the wire and to **B**.
43. 5.0×10^{-2} N perpendicular to the wire and to **B**.
45. $\Delta y = ILB/2k$.
47. 0.25 T.
49. 0.11 A · m^2 perpendicular to the loop, 0.024 N · m parallel to the loop and perpendicular to **B**.
51. $IN\pi R^2B(1 - \cos\theta)$.
53. $(-10^{-22}\ \text{J}) \leq U \leq (+10^{-22}\ \text{J})$.
55. (a) 5.0 mA; (b) $6 \times 10^{-5} \cos\theta_i$ J.
57. (a) 4×10^{10} m/s; (b) 10^{-23} N · m.
59. (a) $2IRB\,\mathbf{j}$; (b) $-2IRB\,\mathbf{j}$; (c) 0; (d) 0.
61. 0.43 rad/s.
63. 1.5×10^{27} carriers/m^3.
65. 0.11 mA.
67. 5.8×10^{11} N/C, 2.6×10^5 T.
69. 0.037 T.
71. $\Delta r/r = 0.05$ (decrease).
73. $e/m = 2V/B^2R^2$.
75. (a) $L = NmvR$ perpendicular to the orbit; (b) $\frac{1}{2}NevR$; (c) $2m/e$.
77. $\mathbf{E} = 0$, $\mathbf{B} = (6.8 \times 10^{-3}\ \text{T})\mathbf{k}$.

CHAPTER 30

1.

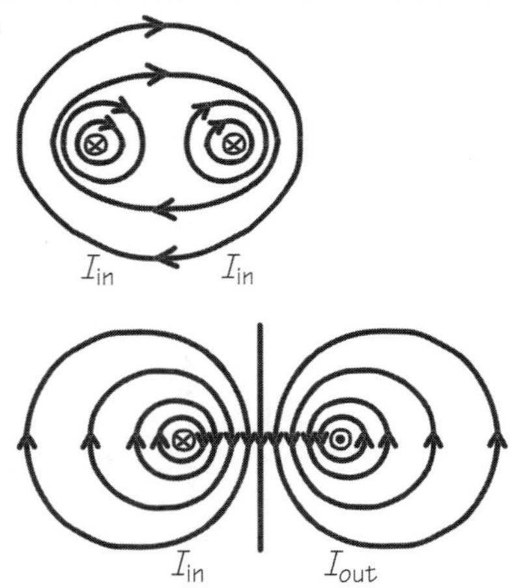

3. 1.7×10^3 N/m.
5. 0.25 μN/m attraction.
7. $[MLC^{-2}]$, $[C^2M^{-1}L^{-3}T^2]$, 3.00×10^8 m/s.
9. (a) $r = R$; (b) $B_{max} = \mu_0I/2\pi R$ at $r = R$; (c) $B_{min} = 0$ at $r = 0$ and $r = \infty$;

(d)

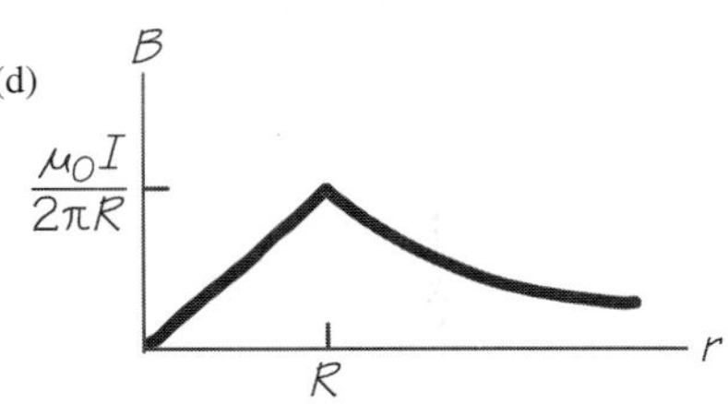

11.

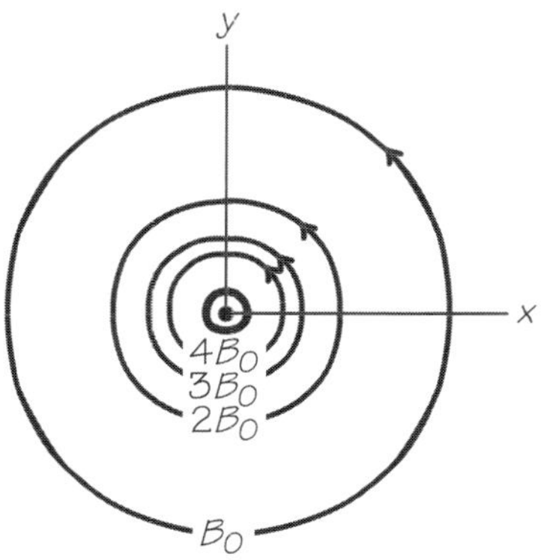

13. 3.0×10^{-3} T.

15. (a) $\frac{\mu_0 I}{2\pi(x^2 + y^2)}(-y\mathbf{i} + x\mathbf{j})$;

(b) $\frac{\mu_0 I}{2\pi}\left\{-\left[\frac{y}{(x-a)^2 + y^2} + \frac{y}{(x+a)^2 + y^2}\right]\mathbf{i} + \left[\frac{x-a}{(x-a)^2 + y^2} + \frac{x+a}{(x+a)^2 + y^2}\right]\mathbf{j}\right\}$;

(c) $\frac{\mu_0 I}{2\pi}\left\{\left[\frac{-y}{(x-a)^2 + y^2} + \frac{y}{(x+a)^2 + y^2}\right]\mathbf{i} + \left[\frac{x-a}{(x-a)^2 + y^2} - \frac{x+a}{(x+a)^2 + y^2}\right]\mathbf{j}\right\}$.

17. 6.7×10^{-4} T circular.

19. (a) away from the wire; (b) 5.9×10^{-3} m; (c) same energy.

21. $\frac{\mu_0 I a}{2\pi}\ln\left(\frac{a+d}{d}\right)$.

25. 4.8×10^{-3} T.

27. 0.35 m.

29. 6.0×10^{-5} Wb.

31. $\Phi_B = (\mu_0 NIL/2\pi)\ln[(R + L)/R]$.

33. $B = \sqrt{2}(\mu_0 I/2\pi x)$ in the yz-plane 45° below the y-axis.

35. $B = \sqrt{2}(\mu_0 I/4\pi H)$ in the xy-plane 45° from the $-x$-axis and the $-y$-axis.

37. (a) $-(2.2 \times 10^{-4}$ T/m$)dL\,\mathbf{j}$; (b) $(0.56 \times 10^{-4}$ T/m$)dL\,\mathbf{k}$; (c) 0; (d) $-(0.20 \times 10^{-4}$ T/m$)dL\,\mathbf{j}$.

39. 2.0×10^{-7} T along the axis.

41. $(5.14 \times 10^{-7}\ \text{T}\cdot\text{m/A})I/R$ out of the page.

43. 2.8×10^{-5} T into the page.

45. $R = 2\pi mvd^3/q\mu\mu_0$.

47. 2.1×10^{-4} T into the page.

49. $\boldsymbol{\mu} = \frac{1}{3}Q\omega R^2\mathbf{k}$.

51. $(\mathscr{E}/R)e^{-t/RC}$.

53. 5.0×10^{-8} T·m.

55. $-CV_0\omega\sin(\omega t)$.

57. $-4\pi R^2\sigma_0/t_0$, $-4\pi R^2\sigma_0/t_0$.

59. (a) $B = (\mu_0 I/2\pi R_1^2)r$ circular CCW, $r < R_1$; (b) $B = \mu_0 I/2\pi r$ circular CCW, $R_1 < r < R_2$; (c) $B = (\mu_0 I/2\pi r)(R_3^2 - r^2)/(R_3^2 - R_2^2)$ circular CCW, $R_2 < r < R_3$; (d) $B = 0$, $R_3 < r$.

61. $-6.0 \times 10^{-5}\,\mathbf{j}$ N (attraction).

63. (a) $F_B/F_E = 4.3 \times 10^{-26}$; (b) 2.1×10^8 electrons/cm; (c) 2.1×10^{-13}.

65. (a) $\frac{\mu_0 I}{2R}\left\{\frac{1}{[1 + (x/R)^2]^{3/2}} + \frac{1}{[2 - 2(x/R) + (x/R)^2]^{3/2}}\right\}$, $B(0) = 0.677\mu_0 I/R$, $B(R/4) = 0.713\mu_0 I/R$, $B(R/2) = 0.716\mu_0 I/R$.

67. $NI = 336$ A·turns.

CHAPTER 31

1. Rotate magnet around coil; increase the speed of rotation.

3. 1.05×10^{-4} V.

5. $d\mathbf{B}/dt = (0.78\ \text{T/s})\mathbf{i}$.

7. $-BLv$ (clockwise), $0 < t < (L/v)$; $+2BLv$ (counterclockwise), $(L/v) < t < (2L/v)$; $-BLv$ (clockwise); $(2L/v) < t < (3L/v)$.

9. 18 V.

11. $-2B_0 v_0^2 t$ $0 < t < L/v_0$.

13. $-CDv^2 t/2\alpha(D + vt)$ clockwise.

15. $-CD^2 v/4\alpha(D + vt)$ clockwise.

17. $BA\omega\sin(\omega t)$.

19. 8×10^{-5} V.

21. 2.9×10^{-4} V.

23. 0.20 T.

25. 1.5×10^{-2} V, bottom end at higher potential.

27. $-2Bv(2Rvt - v^2t^2)^{1/2}$, $0 < t < 2R/v$.

29. (a)

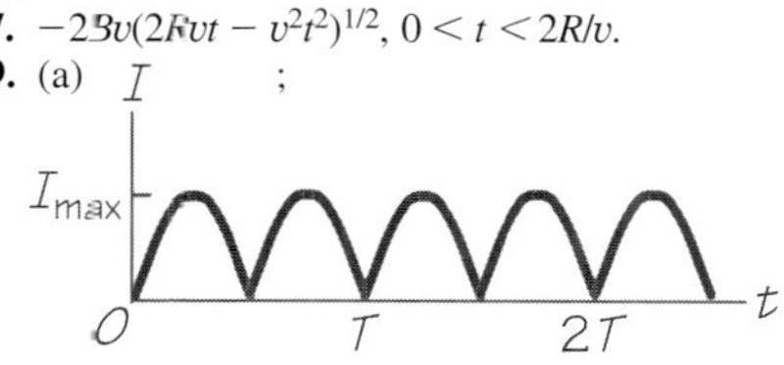

;

(b) $2NEA\omega/R\pi$; (c) $B = \pi I_{av}R/2NA\omega$.

31. 1.3×10^{-2} N, 0.46 W.

33. (a) 0.085 N; (b) 0.051 W.

35. (a) Speed increases until v_t is reached; (b) speed decreases until v_t is reached.

37. $(d\Phi_B/dt)^2/R$.

39. aAB_0e^{-at}.

41. $E = +(\mu_0 n I_0 \omega r/2)\sin(\omega t)$, circular.

43. -0.030 V (counterclockwise).

45. (a) 54 mA; (b) 4.4 mW.

47. 5.7 m/s.

49. Constant speed of E/B in the $-y$-direction, $B > 3 \times 10^{-5}$ T.

51. (a) -9.4×10^{-6} V (up); (b) 47 μA; (c) 9.7×10^{-10} J, external force.

53. (a) $(\pi r^2\mu_0 NI\omega/2R)\sin(\omega t)$; (b) 90°.

55. Counterclockwise, away from the wire.

57. 0.50 C.

59. $Q = (4.5 \times 10^{-9}\ \text{C})[1 - e^{-t/(3.0\times10^{-5}\text{s})}]$, with lower plate positive, $I = (1.5 \times 10^{-4}\ \text{A})e^{-t/(3.0\times10^{-5}\text{s})}$, clockwise.

61. $\mathscr{E} = (B\pi a^2\omega/2)\sin(\omega t)$, $P_{av} = B^2\pi^2a^4\omega^2/8R$.

63. (a) 1.8×10^5 m/s; (b) 1.4×10^{-20} J; (d) -10%.

CHAPTER 32

1. 1.0×10^{-2} A·m² in the direction of the magnetic field.

3. 992.

5. 5 turns/m.

7. (a) $-(2.4 \times 10^{-5})\mathbf{B}_0$; (b) $+(2.6 \times 10^{-4})\mathbf{B}_0$.

9. 0.64 A.

11. (a) 2.2×10^{25} electrons; (b) 2.2 A/m.

15. $M = 1.4 \times 10^6$ A/m, 3.6×10^2.

17. $(2.4 \times 10^{-3}\ \text{T})\chi_m$.

19. (a) 0.16 T; (b) 2.5×10^{-4} T.

21. 1.45×10^2 A/m, 0.88 T.

23. $H_{max} = 289$ A/m.

25. 6.3×10^{-4} T.

29. -2.2×10^{-4}.

31. $H = (1.6 \times 10^{-3}\ \text{A})/r$ circular, $B = (2.0 \times 10^{-9}\ \text{T}\cdot\text{m})/r$ circular, $\Delta B = (1.3 \times 10^{-12}\ \text{T}\cdot\text{m})/r$.

33. 0.38 A/m, 4.7×10^{-7} T.

37. 4.2×10^6 Hz.

39. 1.96 T, 0.16 T.

43. $m = 8 \times 10^{22}$ A·m², $M = 6 \times 10^2$ A/m, 2×10^9 A.

45. $m_{ring}/m_{disk} = 2$.

47. (a) $C = \left(\frac{NmB}{2\pi kT}\right)/(e^{mB/kT} - e^{-mB/kT})$;

(b) $\langle\cos\theta\rangle = \frac{e^{mB/kT} + e^{-mB/kT}}{e^{mB/kT} - e^{-mB/kT}} - \frac{kT}{mB}$.

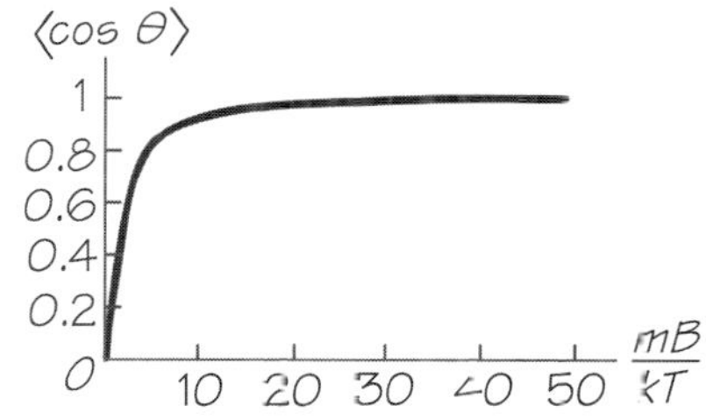

CHAPTER 33

1. 6.0×10^{-5} Wb.

3. 3.1 μH.

5. 30 mH.

7. (a) $M_{12} = 1.0$ mH; (b) 3×10^{-3} Wb.

9. (a) $\mu_0 AN_1^2/\ell$ or $\mu_0 AN_2^2/\ell$; (b) $\mu_0 A(N_1 + N_2)^2/\ell$; (c) $\mu_0 A(N_1 - N_2)^2/\ell$; (d) $\mu_0 AN_1N_2/\ell$.

13. $I = 0$ for $t < 0$, -0.40 μA for $0 < t < 0.30$ s, $+0.40$ μA for 0.30 s $< t <$ 0.50 s, $I = 0$ for $t > 0.60$ s.

15. $\mathscr{E} = -10$ mV for $0 < t < T/4$, $3T/4 < t < 5T/4, \ldots$; $\mathscr{E} = +10$ mV for $T/4 < t < 3T/4$, $5T/4 < t < 7T/4, \ldots$.

19. (a) 3.9×10^{-2} H; (b) 16 H.

21. $L = \frac{\mu_0\ell}{\pi}\left[\frac{1}{2} + \ln\left(\frac{a}{r}\right)\right]$. The radius of wire cannot be neglected.

23. $\frac{\mu_0 L}{2\pi}\ln\left[\frac{(a+b)^2}{(2a+b)b}\right]$.

25. 1.2 kW.

27. $I = 24$ A.

29. $0.64\,e^{-(0.04/s)t}$ mW.

31. (a) same; (b) same.

33. (a) 6.4×10^3 J; (b) $V = 2.4$ L

35. 2.6×10^2 J.

37. 11 cm.

41. 1.6×10^{-9} J/m.

47. $V^2L/2R$.

49. 1.6 mF.

51. (a) $\omega_1/10$; (b) $\alpha_1/100$; (c) $\omega_2'^2 = (\omega_1^2/100) - (\alpha_1/100)^2$.

53. $\omega' = 6.6 \times 10^4$ rad/s.

55. $2Q_0^2/(LC^3)^{1/2}e^{-2\alpha t}$.

59. $T + \frac{1}{4}\pi R^2(C^3/L)^{1/2}$.

61. (a) 47 mA; (b) 2.2×10^{-8} J; (c) 1.

63. (a) 4.0×10^{-3} Ω; (b) 1.5×10^{-7} J at $t = 0$ 4.2×10^{-9} J at $t = 100$ oscillations; (c) Joule heating.

67. (a) $I_2 = 0$, $I_1 = I_\varepsilon = I_L = 2.4$ mA (b) 0.28 mH; (c) $I_1 = I_\varepsilon = 0$, $I_L = I_2 = 0.85$ mA

69. $(\mu_0 I^2/4\pi)[(1/4) + \ln(r_2/r_1)]$.

71. (b) c.

73.

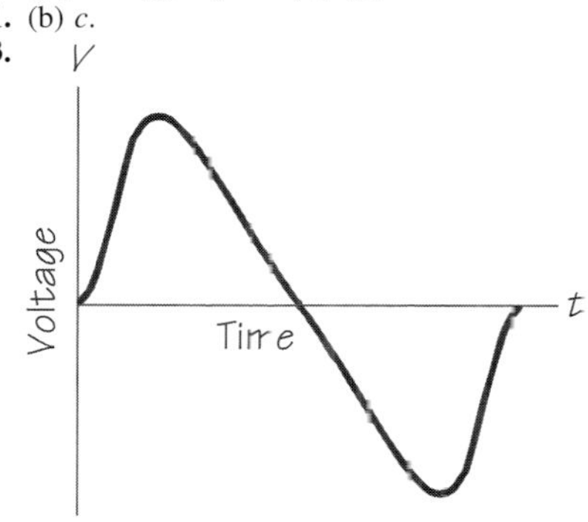

75. 3.2×10^{-2} H.

CHAPTER 34

1. 9000 V.
3. 60 turns.
5. 450 turns, 27 A.
7. 0.40 A.
9. 50 turns.
11. 13 kHz.
13. $\omega = 1.1 \times 10^4$ rad/s.
15. 22.8 μF.
17. $X_{C\max} = 1.7 \times 10^3\ \Omega$, $X_{L\max} = 300\ \Omega$.
19. 15 Ω, 95 Hz.
21. (a) **D** is more advanced in phase; (b) 90°, **D** leads; (c) $\tan\delta = A/B$.

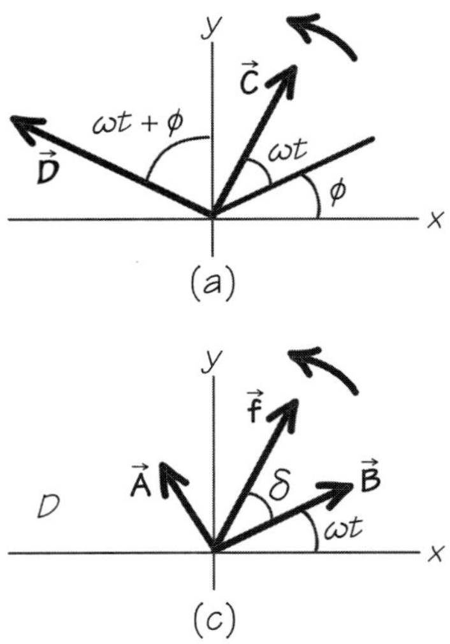

23. 35 pF.
25. $R_2 = \frac{1}{2}R_1$.
27. 2.7×10^{-4} C, 4.0 V.
29. $X_C = 66.3\ \Omega$, $X_L = 694\ \Omega$, $Z = 802\ \Omega$, $Q_{\max} = 13\ \mu$C, $\phi = +51.5°$, $I_{\max} = 0.10$ A.
31. $V_C = -4.05$ V, $V_L + 43.1$ V.
33. $I_{\max} = 92$ mA, $V_{R\max} = 55.5$ V, $V_{C\max} = 123$ V, $V_{L\max} = 27.9$ V.
35. Double.
37.

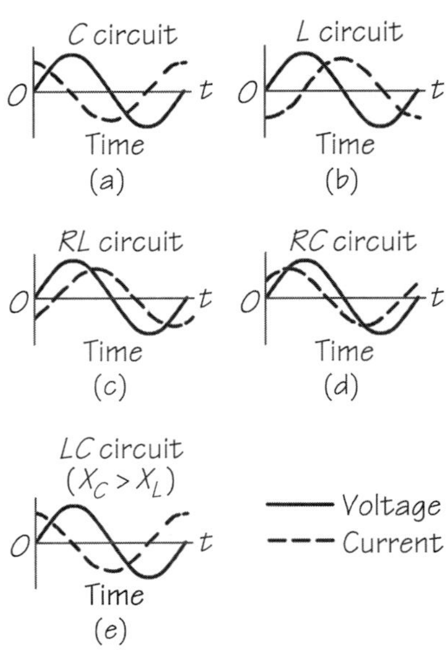

39. $V_{C\max} = 66$ V, $V_{RL\max} = 67$ V.
41. 11 μH.
43.

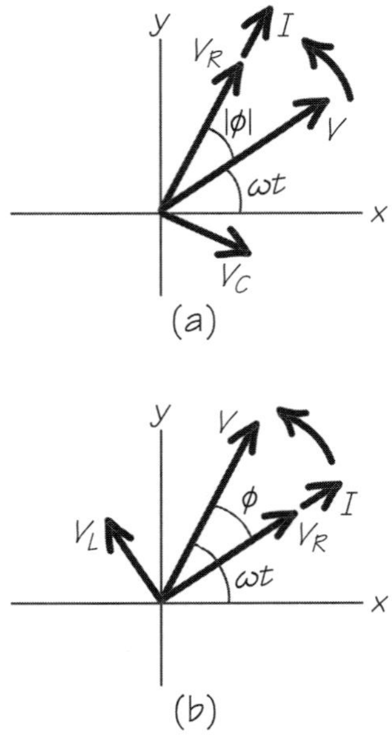

45. 5.0 A.
47. (a) $\langle P\rangle \to \frac{1}{2}V_0^2 R/\omega L \to 0$; (b) $\langle P\rangle \to \frac{1}{2}V_0^2 R\omega C \to 0$, no current through the capacitor.
49. (a) 0; (b) 0; (c) 1.
51. (a) $R/(X_L^2 + R^2)^{1/2}$; (b) $R/(X_C{}^2 + R^2)^{1/2}$; (c) 0.
53. 27 μF, 46 μF.
55. 345 V, 9.05 kW.
57. 9.7 kW, 16.7 kW.
59. (a) 31 W; (b) 17 W; (c) 34 W.
61.

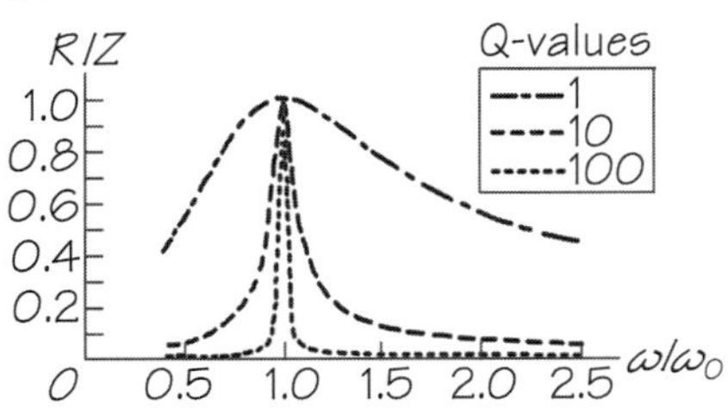

63. (a) 0.20 V; (b) 0.19 V; (c) 5.3×10^{-6} V.
65. $R = 2.5$ kΩ and $L = 50$ mH.
67. $\langle I\rangle = I_0/\pi$, $I_{\rm rms} = I_0/2$.
69. At low frequency, $Z \to X_C$; at high frequency, $Z \to R$.

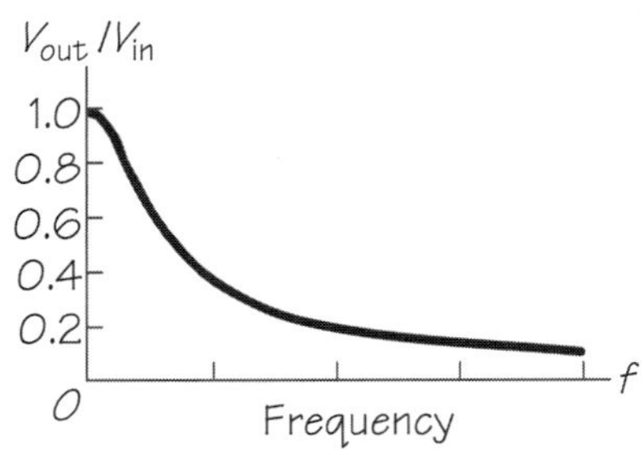

71. $I = (V_0/Z)\cos(\omega t + \phi)$, $I_C = (V_0/X_C)[(R/Z)\sin(\omega t + \phi) - \sin(\omega t)]$, $I_L = (V_0/X_L)[-(R/Z)\sin(\omega t + \phi) + \sin(\omega t)]$, there is a resonant frequency.
73. (a) 5.14 V, 2.20 V; (b) 2.34 V.
75. 29 mW.
77. (a) 5.7 V; (b) 17.8 V.
79. $L = 2.4\ \mu$H, $R = 0.060\ \Omega$.
81. $R = 177\ \Omega$.
83. (a) $Q_c(t) = CV_0 e^{i\omega t}$, $I_c(t) = i\omega C V_0 e^{i\omega t}$.
85. (a) $Q_{0c} = \dfrac{V_{0c}}{\omega\left[\mathbf{i}\left(L\omega - \dfrac{1}{\omega C}\right) + R\right]}$; (b) $Q_c(t) = -\mathbf{i}(V_0/\omega Z)e^{i(\omega t - \phi)}$, $I_c(t) = (V_0/Z)e^{i(\omega t - \phi)}$; (c) $Q = -(V_0/\omega Z)\cos(\omega t - \phi)$, $I = (V_0/Z)\sin(\omega t - \phi)$.

CHAPTER 35

3. $\oint \mathbf{E}\cdot d\mathbf{s} = \mu_0 \dfrac{dM}{dt} - \dfrac{d}{dt}\iint_S \mathbf{B}\cdot d\mathbf{A}$.
5. $\mathbf{B} = (E_0/c)\cos(kz + \omega t)\mathbf{i}$, traveling in the $-z$-direction.
7. 94×10^6 Hz, 3.2 m.
13. $+\dfrac{\partial B_y}{\partial x} = \mu_0\epsilon_0 \dfrac{\partial E_z}{\partial t}$ and $+\dfrac{\partial B_y}{\partial t} = \dfrac{\partial E_z}{\partial x}$.
15. Plane formed by the z-axis and the line $y = -(\tan\theta)x$.
17. (3 V/m) $\cos[(1.05 \times 10^7\ \text{m}^{-1})z - (3.15 \times 10^{15}\ \text{rad/s})t]\mathbf{i}$.
19. $\lambda = 2\pi/k = 2L/n$, $n = 1, 2, 3, \ldots$.
21. $B_0 = 2.2 \times 10^{-4}$ T.
23. 2.6×10^{-5} W/m^2.
25. 8.0×10^{-11} N/m^2.
27. (a) 1.3×10^3 W/m^2; (b) 4.5×10^{-6} N/m^2.
29. 1.2×10^{-2} W/m^2, 0.070 J.
31. $E_0 = 134$ V/m, $E_{\rm rms} = 95$ V/m, $B_0 = 4.47 \times 10^{-7}$ T, $B_{\rm rms} = 3.16 \times 10^{-7}$ T.
33. 2.0×10^{-3} N/m^2.
35. 17 kg·m/s; 3 kg·m/s, about 1/5 of the solar result.
37. $P = 5.9 \times 10^5$ W, vaporize.
39. $\Delta p/A = 2(S/c)\cos^2\theta\,\Delta t$.
41. There is a net torque, 3.1×10^{-16} N·m.
43. 4.0×10^{-7} s.
45. (a) $+y$-direction; (b) $-x$-direction; (c) $+z$-direction; (d) electric field in $-y$-direction, magnetic field in $+x$-direction, Poynting vector in $+z$-direction.
47. (a) no solution; (b) 39°; (c) 57°; (d) 72°.
49. 37°.
51. 3.5×10^5 W/m^2.
53. 0, 1/8.
55. (a) $(1/2)I_0$; (b) 90°; (c) $(1/8)I_0\sin^2(2\theta)$.
57. 4.7×10^{-5} W/m^2.
59. (a) $\mathbf{B} = \mu_0 n I_0 \cos(\omega t)\mathbf{k}$ (along the axis); (b) $\mathbf{E} = \frac{1}{2}\mu_0 n I_0 \omega r \sin(\omega t)$ (circular); (c) $\mathbf{S} = (1/4)\mu_0 n^2 I_0^2 \omega r \sin(2\omega t)\hat{\mathbf{r}}$; $0 < t < \frac{1}{4}T$: $\mathbf{S}$ is $+\hat{\mathbf{r}}$; $\frac{1}{4}T < t < \frac{1}{2}T$: $\mathbf{S}$ is $-\hat{\mathbf{r}}$; $\frac{1}{2}T < t < \frac{3}{4}T$: $\mathbf{S}$ is $+\hat{\mathbf{r}}$; $\frac{3}{4}T < t < T$: $\mathbf{S}$ is $-\hat{\mathbf{r}}$.
63. $\mathbf{E} = (5.19 \times 10^2\ \text{V/m})\cos[(2.54 \times 10^6\ \text{m}^{-1})x + (1.47 \times 10^6\ \text{m}^{-1})y - (2\pi \times 10^{14}\ \text{s}^{-1})t]\mathbf{k}$.
65. 5.8×10^5 J, 0.25 kg.
67. 4.5×10^{13} photons/m^3.
69. $f' = [(c - v)/(c + v)]f$.
71. (a) $\mathbf{B} = (\mu_0 r/2\pi R^2)dQ/dt$ circular, for $r < R$; (b) $\mathbf{S} = -(Qr/2\pi^2 R^4 \epsilon_0)(dQ/dt)\hat{\mathbf{r}}$.

CHAPTER 36

1. $v_{\rm ice} = 2.29 \times 10^8$ m/s, $v_{\text{ethyl alcohol}} = 2.21 \times 10^8$ m/s, $v_{\rm benzene} = 2.00 \times 10^8$ m/s, $v_{\rm diamond} = 1.24 \times 10^8$ m/s.
3. 492 nm, 4.62×10^{14} Hz.
5. 2.9×10^4 rev/min.
7. 6 m.
9. 1.0 ns.
11. 1.3 cm.

13. 1.62.
15. 590 m.
17. 0.031.
19. 1.4°.
21. $\theta_{\text{water}} = 45.7°$, $\theta_{\text{glass}} = 38.5°$.
23. Cannot be changed.
25. 7.4°, no total internal reflection.
27. Outside the "shadow" of the prism, direct illumination. For a distance of 0.57 cm from each edge, no illumination. For the next distance of 0.20 cm, illumination from one half. For the next distance of 0.73 cm to the center of the pattern, illumination from both halves.
29. 8.3°.
31. 74° from the normal.
33. 5.3 mm.
35. 0.86 m.
37. Prism will not be totally reflecting, $n_2 \leq 1.283$.
45. 0.48°.
47. (a) $\omega \leq 12.5 \times 10^{15}$ rad/s; (b) 27.7°.
49. 0.009.
51. 1.31 cm.
53. $\Delta n = [(\cos\theta_4)(\cos\theta_2)/\sin(2\phi)]\,\Delta\theta$, where $\cos\theta_2 = \cos\{\sin^{-1}[(\sin\phi)/n]\}$ and $\cos\theta_4 = \cos\{\sin^{-1}[n\sin(2\phi - \{\sin^{-1}[(\sin\phi)/n]\})]\}$.
55. $\theta_i = \sin^{-1}(n\sin\phi)$.
57. (a) 4.6×10^6 km; (b) 3.1×10^8 m/s.

CHAPTER 37

1. 3 images.
3. 82.5% of the beam has been dissipated.
5. $0.577a$, a.
7. $i_2 = 26.3$ cm, $R = 30$ cm.

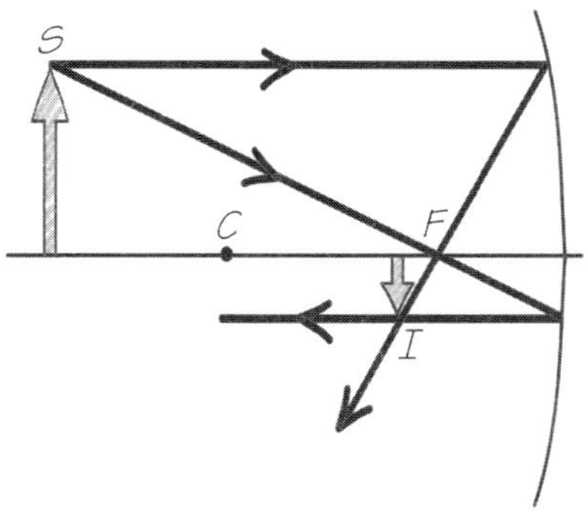

9. In front of mirror, 0.8 cm, inverted.
11. $i = -400$ cm (behind mirror), $+50$ cm.
15. $i = -90$ cm (in air in front of glass).
17. $i = -30$ cm (behind the glass).
19. $i = +18.4$ cm, 43 cm from the surface.
21.

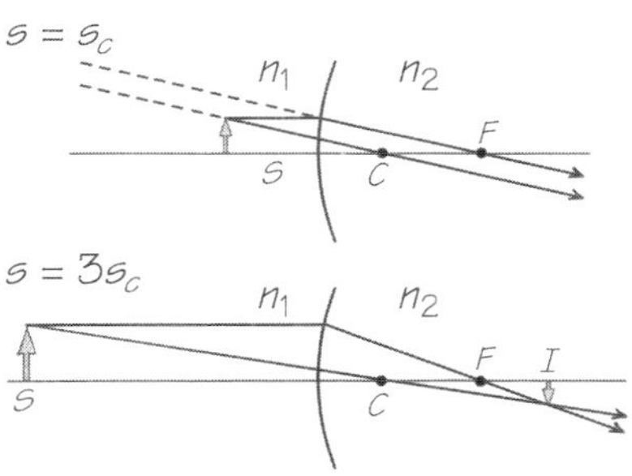

23. (b) $s = n_1R/(n_2 - n_1)$; (c) image is very far in front of the boundary: (d) image approaches the boundary.
27. (a) +16 cm; (b) converging; (c) inverted; (d) −2.1.
29. (a) 8.9 cm in front of the lens; (b) no; (c) yes; (d) +0.60.
31. (a) +38 cm; (b) +13 cm.
33. virtual image.

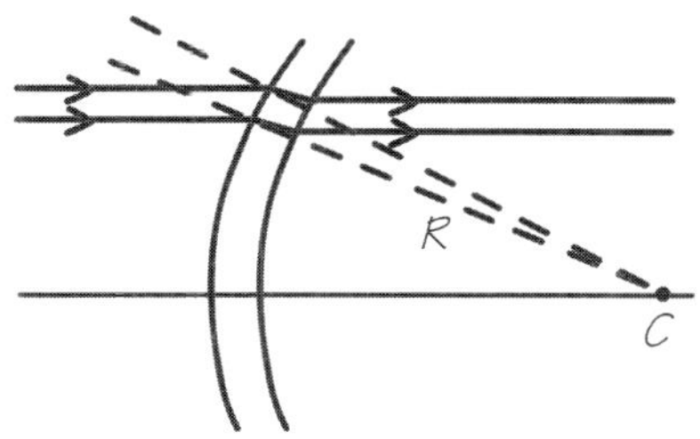

35. Lens *a*: (a) $f_a = 32.1$ cm; (b) inverted and real; (c) M −; Lens *b*: (a) $f_b = +77.9$ cm; (b) inverted and real; (c) M −; Lens *c*: (a) $f_c = -32.1$ cm; (b) upright and virtual; (c) M +; Lens *d*: (a) $f_d = -77.9$ cm; (b) upright and virtual; (c) M +.
37. Lens *a*: image is 63.4 cm to the left of the lens, inverted, real, with $M = -0.98$; Lens *b*: image is 393 cm to the right of the lens, upright, virtual with $M = +6.0$; Lens *c*: image is 21.5 cm to the right of the lens, upright, virtual, with $M = +0.33$ Lens *d*: image is 35.4 cm to the right of the lens, upright, virtual, with $M = +0.55$.
39. +53 cm.
41. +5 cm.
43. 9.6 cm.
45. f_1/f_2.
49. 1.09 cm.
51. Near *positive* lens: image is 15.8 cm in front of the negative lens, or 0.8 cm from the positive lens on the object side; near the *negative* lens, image is 65.4 cm in front of the positive lens, or 50.4 cm from the negative lens on the object side; the order of the lenses is important.
53. (a) first image is 12.4 cm behind M_1; (b) second image is 7.9 cm in front of M_2.

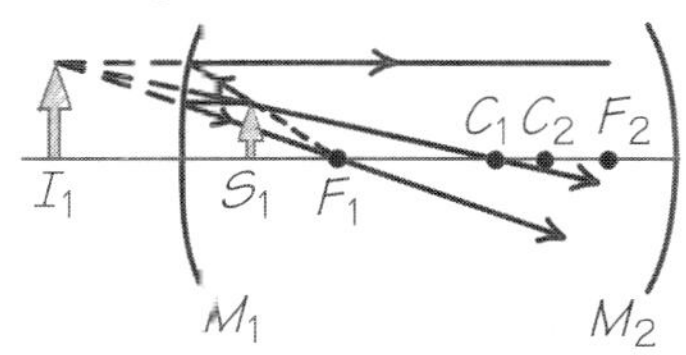

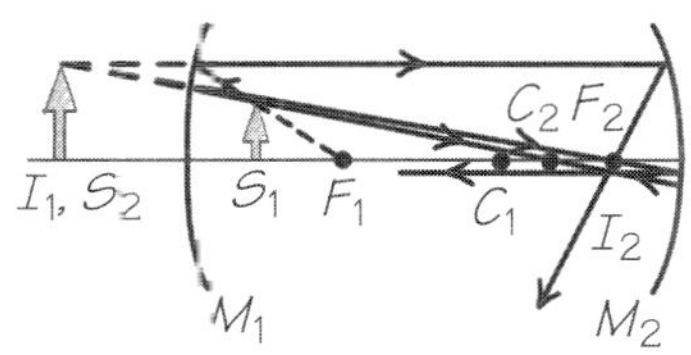

55. $f = [f_1(d^2 - f_1d - 2f_2d + f_1f_2)]/(d^2 - 2f_1d - 2f_2d + 2f_1f_2 + f_1^2)$.
57. $d \to R/(n-1)$.
59. The plate may be placed anywhere between the lens and the screen, with thickness 13.1 cm.

CHAPTER 38

1. 4.7×10^2 nm.
3. 0.24 m.
5. 12.2 m.
7. (a) doubles; (b) reduces by $\frac{1}{2}$; (c) doubles; (d) no change
9. 6.4×10^2 nm.
11. Inward.
13. $v_{\text{fringe,ripple}} = 3 \times 10^{-6}$ m/s, $v_{\text{fringe,optical}} = 1.2 \times 10^6$ m/s.
15. (b) $\theta = \tan^{-1}\left(\lambda\sqrt{\dfrac{1}{d^2 - \lambda^2} + \dfrac{1}{4P^2}}\right)$.
17. $3 \times 10^{+3}$ W/m².
19. 151°, 525 nm.
21. $I_{av} = 2I_0$.
23. $\Delta\theta = 2\sin^{-1}(\lambda/4d)$.
25. (a) along the perpendicular bisector; (b) 2×10^{-3} W/m²; (c) 9.5°.
27.

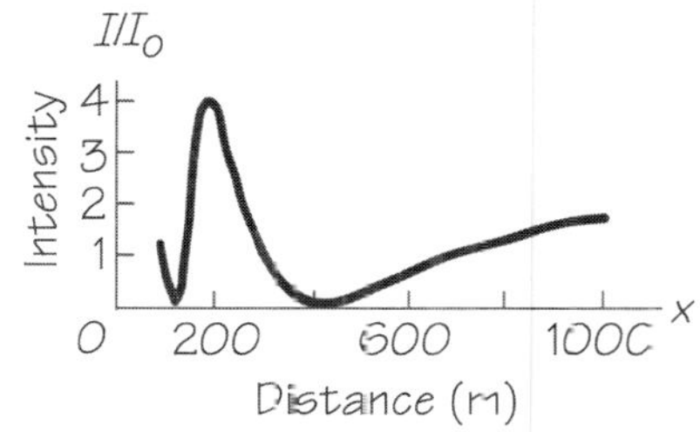

29. 33.6 μm.
31. 81 nm, 242 nm, 404 nm.
33. $r = (\lambda R)^{1/2}$.
35. 1000 maxima pass. rings move in to the center.
37. 6.4 mm, 8.5 mm.
39. $x = (n\lambda R)^{1/2}$.
41. 118 nm.
43. 790 nm.
45. 621 nm.
47. (a) 115 nm; (b) longer; (c) no reflection of green light.
49. 120 nm.
51. 120 nm.
53. 633 nm.
55. 100 fringes to 0.1 of a fringe.
57. Uniform.
59. $u \propto \sin^2[2\pi f(x - \frac{1}{2}L)/c]$.
61. $\sin\theta' = \sin\theta - (m\lambda/d)$, $m = 0, \pm1, \pm2, \ldots$.
63. 34.0.
67. Two: 0, or $4I_0$; three: I_0; four: 0.
69. $\theta = \sin^{-1}\{n^2 - [\lambda(m - \frac{1}{2})/2h]^2\}^{1/2}$, $m = 1, 2, 3, \ldots$.

CHAPTER 39

1. 450 nm.
3. 39.7 lines/cm.
5. 15,000, 30,000, 45,000; 0.028 nm, 0.014 nm, 0.0093 nm.
7. 3.1×10^4 lines, 8.83×10^{-4} rad/nm
9. $\Delta\theta = 18.6°$.
11. First order: 6.17° to 9.79°, second order: 12.42° to 19.88°.
13. $\sin\beta = \sin\alpha - (m\lambda/d)$, $m = 0, \pm1, \pm2, \ldots$.
15. 0.51 m.
17. $I_0/I_1 = 22.2$, 1.19°, increase θ.
21. $\Delta\theta_h/\Delta\theta_l = 0.443$.
23. $\sin\theta = \sin\theta_i - (m\lambda/a)$, where $m = \pm1, \pm2, \ldots$; there is a "central maximum."
25. (b) $\alpha_{\max 1} = 257.43°$, midway between 1st and 2nd minima = 270°; $\alpha_{\max 2} = 442.61°$, midway between 2nd and 3rd minima = 450°.
27. 5.0 cm, 83 m
29. 70 m.
31. 0.73 m, shorter wavelengths.
33. 2.4 mm.
35. 21 ft.
37. Third order.

39. I/I_0: 3.97, 1.79, 2.43, 0.26.
41. $a = 0.1$ mm, $d = 0.3$ mm.
43. $n = 3m = 3, 6, 9, \ldots$; 0.084°, 0.169°, 0.253°, … .
45. 0.0274°.
47. 0.21 nm.
49. 29.0°.
51. 15×10^3 m.
53. (a) 1.0×10^{-5} rad $= (5.7 \times 10^{-4})°$; (b) 25 μm.
55. (a) $I_0\{\sin[(N\pi\cos\theta)/(N-1)]/\sin[(\pi\cos\theta)/(N-1)]\}^2$; (b) slight decrease in I for θ close to 0° and 180° and a buildup to the maxima at 90° and 270°.
57. 0.45 mm.
59. 22°.

CHAPTER 40

1. 4.04 h, 4.02 h.
3. 3.6×10^8 m/s, no violation.
5. 0.011 s.
7. 3.7×10^9 s = 120 yr.
9. (a) 3.3×10^{12} s (10^5 yr); (b) 3.3×10^5 s ($\simeq$ 4 days).
11. 0.66 h.
13. (a) 24 m; (b) $+13.3 \times 10^{-8}$ s; (c) 10×10^{-8} s; (d) 5.0×10^{-8} s.
17. 5.4×10^8 s ($\simeq$ 17 yr).
19. 2.3×10^4 m/s toward the earth.
21. (a) 1.2×10^7 m/s; (b) 4.8×10^{24} m (5.1×10^8 ly).
23. $f_2 - f_0 = -1.3 \times 10^{10}$ Hz.
25. 5 yr, (1/5) yr, 20 yr.
27. $0.69c$.
29. (a) distance between the bags is not a measured length; (b) 140 m.
31. (a) $x = 0$, $t = 24$ s; (b) 9.6×10^9 m, 40 s.
33. $0.23c$.
37. $(c/n)[1 + (un/c)]/[1 + (u/cn)]$.
39. Same form.
41. 2.1×10^{-21} kg · m/s.
43. 1.1×10^{-4} kg/s (3.5×10^3 kg/yr).
45. 4.4×10^9 kg/s.
47. (a) 0.047 MeV; (b) 36 MeV; (c) 110 MeV; (d) not possible.
49. (a) $[1 - (7.9 \times 10^{-7})]c$; (b) 745 GeV.
53. 767 MeV/c^2.
55. (a) $p_{\text{He}}c = 212.19$ MeV; (b) $p_{\text{He}}c = 210.42$ MeV; (c) $p_{\text{He}}c = 210.50$ MeV.
57. $K_M = (M^* - M)^2c^2/2M^*$.
59. $\mathbf{F} = m\gamma(d\mathbf{u}/dt) + [m(\gamma^3u/c^2)\,du/dt]\mathbf{u}$.
61. $\phi = -\frac{1}{2}R^2\omega^2$, slow.
63. c.
65. 2×10^{-3} nm.
67. (a) $v = 0.90c$; (b) 2.33 GeV.
69. (a) p, south; (b) $p = \frac{1}{2}c(M^2 - 4m^2)^{1/2}$.
71. $G(f) \propto e^{-\{mc^2[(f-f_0)/f_0]^2/2kT\}}$.
73. (a) $0.933c$; (b) 3.6×10^2 MeV/c (1.9×10^{-19} kg · m/s); (c) 12 m.
75. 6.5 GeV/c (3.47×10^{-18} kg · m/s).

CHAPTER 41

1. (a) 0.55 nm; (b) 0.17 nm; (c) 5.5×10^{-3} nm; (d) 2.49×10^{-15} m; (e) same size as the wavelengths.
3. 4.0 nm.
5. (a) 1.8×10^{-16} J (1.1×10^3 eV); (b) 9.6×10^{-20} J (0.60 eV).
7. $\lambda = h/p = h/(2mE)^{1/2}$, $\lambda = hc/(E^2 + 2mc^2E)^{1/2}$.
9. 1.4×10^{-14} m, 4.6×10^{-15} m, 8.5×10^{-16} m.
11. 2.4×10^{-17} J (150 eV), 1.3×10^{-20} J (0.082 eV).
13. 0.36.
15. 0.
19. 10^{-19} J (0.6 eV).
21. 1.5×10^{-14} J (0.092 MeV).
23. 8.8×10^{-21} kg · m/s, 2.3×10^{-14} J (0.14 MeV).
25. 2.7×10^{-4} nm.
27. 3×10^8 m, 6×10^5 m.
29. $\Delta y = a + (hD/\pi pa)$, $a_{\min} = (hD/\pi p)^{1/2}$.
31. -0.14 MeV.
33. 0.2 MeV.
35. (a) 8.3×10^{-5} eV; (b) 1.9 eV; (c) 4.0×10^{-7} eV; (d) 7.3 keV.
37. $W = 1.8$ eV.
39. $\lambda_{\text{max, Al}} = 290$ nm, $\lambda_{\text{max, Cs}} = 579$ nm, $\lambda_{\text{max, Ni}} = 241$ nm, $\lambda_{\text{max, Pb}} = 292$ nm.
41. (a) 2.4 eV; (b) 12 keV; (c) 1.2×10^{-4} eV; (d) 1.2×10^{-7} eV; (e) 1.2×10^{-8} eV.
43. 1.0×10^{45} photons/s.
45. 6×10^{-14} W/m², $I/I_0 = 4 \times 10^{-17}$.
47. 5.8×10^3 K.
49. $W = 4.60$ eV (7.37×10^{-19} J), $K_{\max} = 5.75$ eV (9.21×10^{-19} J).
51. 88.4°.
53. 2.0×10^{-2} J/m³, 2.6×10^{-2} J/m³, 0.77.
55. 1.15×10^9 atoms/(photoelectron/s).
57. 0.33.
59. 1.57×10^4 yr ago, 13,700 B.C.
61. Does support the assertion.
63. 1.6×10^{-7} Ci.
65. $\lambda = 0.027$ nm, $d = 3.5$ nm, $T = 0.02$ K.
67. (a) 480 nm; (b) 4500 K.
69. 1.2%.
71. $\lambda = h/\{2m_e[E - U(x)]\}^{1/2}$.

CHAPTER 42

1. 91.2 nm.
3. 12.1 eV, 103 nm; 0.97 eV, 1280 nm.
5. -2.86×10^3 eV.
7. (a) $n = 2$, 3, and 4; (b) 122 nm, 103 nm, 97 nm, ultraviolet region.
9. 7.58 nm.
11. Paschen: 1876 nm, 1287 nm, 1094 nm, 1005 nm, 955 nm, … , 820 nm; Brackett: 4052 nm, 2626 nm, 2165 nm, 1945 nm, … , 1459 nm; Pfund: 7460 nm, 4654 nm, 3740 nm, … , 2280 nm.

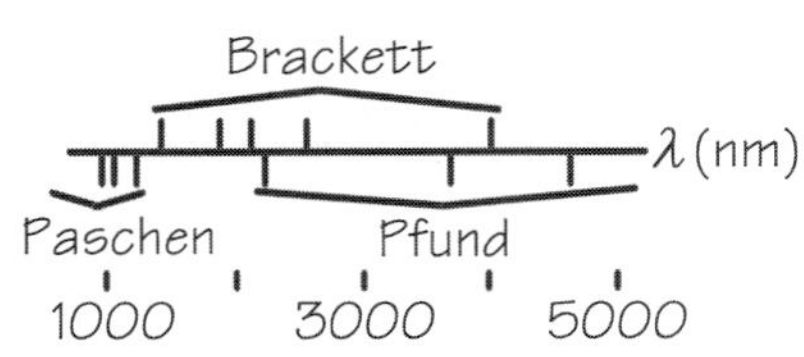

13. (a) dimensionless constant, $\alpha = 0.00731$, $1/\alpha = 137$; (b) $-m_ec^2\alpha^2/2n^2$; (c) $v_1 = \alpha c$.
15. $E_{\text{Cu}} = 10.7$ keV, $E_{\text{W}} = 72.5$ keV.
17. $\frac{2}{3}(e^2/4\pi\epsilon_0)^7m_e^2/c^3L^8$.
19. $f = (m_e/4\pi\hbar^3)(e^2/4\pi\epsilon_0)^2\{(1/n^2) - [1/(n+k)^2]\}$; $f \simeq (m_e/2\pi L^3)(e^2/4\pi\epsilon_0)^2k$, $n >> k$; $f(n >> k) = k(1/T)$.
21. $(3.3 \times 10^{-4})°$.
23. 0.033 nm.
25. $-(6.8\text{ eV})/n^2$; $n = 1$ series, all transitions in the UV; $n = 2$ series, all transitions in the IR; all other series, all transitions are in the IR.
27. $n = 1$, $\ell = 0$; $n = 2$, $\ell = 0$; $n = 2$, $\ell = 1$; $n = 3$; $\ell = 0$; $n = 3$, $\ell = 1$.
29. Na ($Z = 11$).
31. $Z = 54$ (Xe).
33. (a) 3 spectral lines; (b) 7 spectral lines.
35. $Z = 1$ (H), $Z = 9$ (F), $Z = 27$ (Co, should be Cl, $Z = 17$), $Z = 59$ (Pr, should be Br, $Z = 35$).
37. 7.4×10^{-3} eV.
39. 24 K.
41. 119 kg/s², greater than 97 kg/s² from Eq. (42–18).
43. $I_1 = 1.43 \times 10^{-46}$ kg · m², $I_0 = 1.48 \times 10^{-46}$ kg · m², potential has nonlinear terms.
47. $\ell = 9.48 \times 10^{33}$.
49. $\Delta E = 1.5 \times 10^{-5}$ eV, 5.6×10^{-6}.
51. 2.3×10^{-33} J (1.4×10^{-14} eV).
55. $\Delta x = 1 \times 10^{-10}$ m $\simeq 2a_0$.
57. (a) $E = 3U_0(r/a)^4$; (b) $E_n = 3(\hbar^2/4ma^2)^{2/3}U_0^{1/3}n^{4/3}$.

CHAPTER 43

1. 6.15 eV.
3. Na: 3.25 eV; K: 2.13 eV; Al: 11.7 eV.
5. 1.31×10^{29} electrons/m³.
7. $n_1 = n_2 = n_3 = 94$, 0.999 eV, $\Delta E = 7.1 \times 10^{-3}$ eV, 1.006 eV.
9. $p_F = \hbar(3\pi^2n_e)^{1/3}$, $v_F = (\hbar/m_e)(3\pi^2n_e)^{1/3}$.
11. 5.3×10^{13} N/m².
13. (a) $n_e \leq 5.9 \times 10^{32}$ electrons/m³; (b) $Z \leq 14$.
15. (a) $E_F = (\pi^2\hbar^2/8m_e)(N/L)^2$; (b) $E_{\text{total}} = \pi^2\hbar^2N^3/24m_eL^2$; (c) $\Delta E = (\pi^2\hbar^2/2m_eL^2)N$.
17. 36 MeV.
19. (a) $(\hbar^2/5m)(3\pi^2)^{2/3}(N/V)^{5/3}$; (b) 5.5×10^9 N/m².
21. $E_F = (\hbar^2\pi^2/2m_e)n_e$.
23. (a) $12/N_{\text{sm}}^{1/3}$ km, N_{sm} = number of solar masses; (b) $E_F = (\hbar^2/2m_n)(9\pi/4)^{2/3}(N_{\text{sm}}M_{\text{sun}}/m_n)^{2/3}/R^2 = 60N_{\text{sm}}^{4/3}$ MeV, $p_F = 335N_{\text{sm}}^{2/3}$ MeV/c; (c) $4.7M_{\text{sun}}$.
25. $E_F = (\hbar^2\pi/m_e)n_e$, 1.57.
27. 634 nm, 4.73×10^{14} Hz.
29. 3.0 W.
31. (a) $\pi^2\hbar^2N/2mL^2$; (b) $3\pi^2\hbar^2/2mL^2$.
33. (a) 0.090 W; (b) 6.2×10^{-5} N/m².
37. 48.3 μV.
41. 2.5 μV.
43. 3.20 K.
45. $R_{\text{star}} = 6.96 \times 10^7$ m.
47. $E_F = \hbar cn_e^{1/3}$.
53. (a) $E_F = 0.19N_{\text{sm}}^{4/3}$ MeV; (b) $R = (7.2 \times 10^3)/N_{\text{sm}}^{1/3}$ km, N_{sm} = number of solar masses, 0.01.
55. $M < 3 \times 10^{30}$ kg ($\simeq$ 1.5 solar masses).

CHAPTER 44

1. $5.5 \times 10^{-11}n$-carriers/cm³, 1.7×10^9n-carriers/cm³ $\simeq 10^{-14}$ that of copper.
3. (a) 1; (b) 0.99986; (c) 0.76; (d) 0.24; (e) 1.2×10^{-13}.

5. 1.5 eV.
7. 1.03×10^{-12}.
9. (a) 6.9 μm; (b) 1.8 μm; (c) 0.867 μm.
11. $9.3 \times 10^{11}/\text{cm}^3$.
13. $E_g = 0.65$ eV.
15. $E_c - E_F = 0.080$ eV.
17. $n_i = n_n = n_p = 3.0 \times 10^{18}\ \text{m}^{-3}$, $n_i = n_n = n_p = 1.6 \times 10^{22}\ \text{m}^{-3}$.
19. $0.18\ (\Omega \cdot \text{m})^{-1}$.
21. (b) $n(v)\,dv = \left(\dfrac{8\pi m_e^3}{h^3}\right) e^{-[(m_e v^2/2)-E_F]/kT} v^2\, dv$.
23.

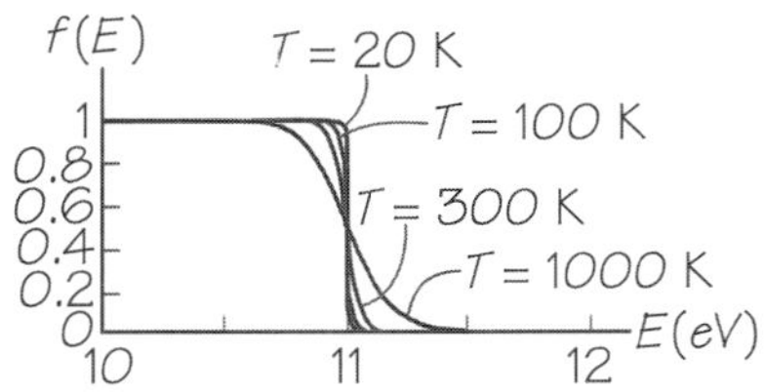

25.

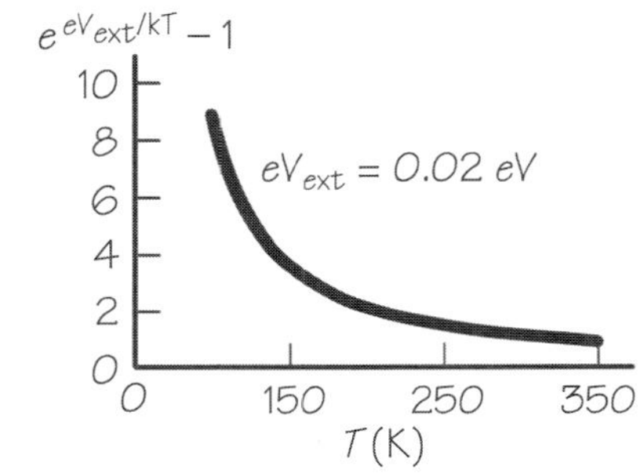

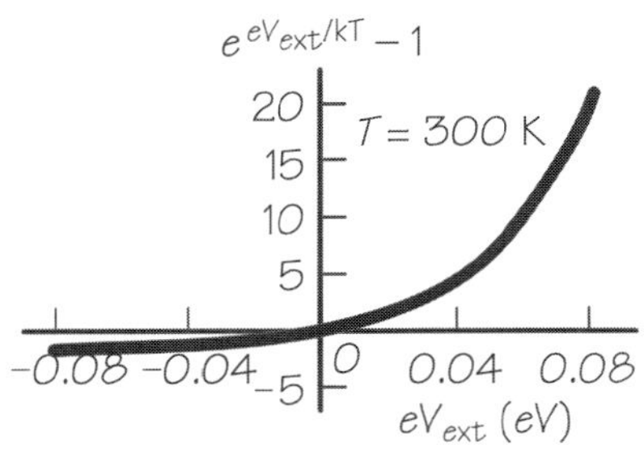

29. (a) 0.64 eV; (b) 0.45 eV.
31. (a) 3.5 mW; (b) 1.6×10^{16} photons/s
33. $m^* = 0.61\ m_e$.
35. B is 0.19 nm closer to the tip than A.
37. 1.2×10^{-7}.
39. $p = 4.5 \times 10^{10}$ carriers/m^3.
41. $(1.2 \times 10^3)a_0$.
43. 6.8×10^{19} impurities/cm^3, 1.5×10^{-3} (0.15 atomic %).
45. 2.1×10^{13} rad/s.
47. 10^{-100} electrons/m^3, 6.1×10^{-12} electrons/m^3.
49. (a) $\simeq 4 \times 10^{15}$ photons/s; (b) $\simeq 4 \times 10^{11}$ photons.
51. (a) Rapid increase in resistance; (b) 1.10×10^3 nm; (c) increases.
53. (b) tunneling current will increase; (c) tunneling current will increase.

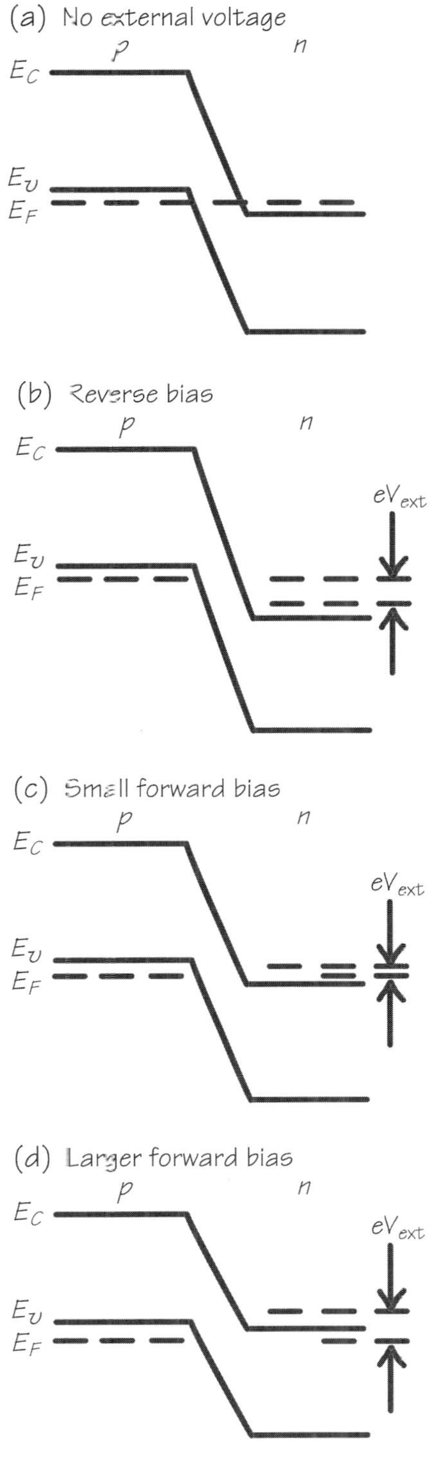

CHAPTER 45

1. ^{9}Be; $5n$, $4p$; ^{13}C: $7n$, $6p$; ^{22}Na: $11n$, $11p$; ^{31}P: $16n$, $15p$; ^{57}Fe: $31n$, $26p$; ^{72}Ge: $40n$, $32p$; ^{107}Ag: $60n$, $47p$; ^{131}Cs: $76n$, $55p$; ^{208}Pb: $126n$, $82p$; ^{241}Am: $146n$, $95p$.
3. 8.1 barns.
5. $r_{min} = 4Z(e^2/4\pi\epsilon_0)/mv^2, v = 2[Z(e^2/4\pi\epsilon_0)/mR]^{1/2}$.
7. (a) 18 fm; (b) 11 fm.
9. $\sigma = (Z_1 Z_2 e^2/4\pi\epsilon_0)^2[4m^2/(\Delta p)^4]$.
13. 3.67 fm, 3.50 fm, 2.16 fm, 1.85 fm, 1.75 fm.
15. ^{15}N: $N = 8$; ^{39}K: $N = 20$; ^{40}Ca: $Z = 20$, $N = 20$; ^{56}Fe: neither; ^{140}Ce: $N = 82$; ^{208}Pb: $Z = 82$, $N = 126$.
17. (a) $\simeq 14$ N; (b) $F_r \simeq 560$ N.
19. 1.0×10^{-3}.
21. $Z = A/(2.0 + 0.015A^{2/3})$.
23. $K_f = 3.80$ MeV.
25. $K_n = 18.3$ MeV.
27. $K_n = 4.7$ MeV.
29. $K_n = 10.49$ MeV.
31. 6.4×10^{16} nuclei formed.
37. $dV/dt = 0.031$ cm^3/yr.
39. 0.784 MeV.
41. $\simeq 13{,}700$ B.C.
45. $dN_a/dt = -N_a/\tau_1$; $dN_b/dt = -(N_b/\tau_2) + (N_a/\tau_1)$; $dN_c/dt = -(N_c/\tau_3) + (N_b/\tau_2)$; $dN_d/dt = -(N_d/\tau_4) + (N_c/\tau_3)$; … .
49. (a) 1.3 keV; (b) 0.7 MeV.
51. (a) 11.5 metric tons; (b) 4.2×10^{16} J; (c) 5.5×10^{19} fissions/s.
53. (a) 24.8 MeV released; (b) 7.5×10^{47} MeV $= 1.2 \times 10^{34}$ J.
55. 2.5×10^2 MeV.
57. $r_{min} = 60$ fm.
59. (a) repulsive Coulomb force between the nuclei; (b) $K = 0.7$ MeV for each helium atom.
63. $E_p = (\hbar^2/10m_p r_0^2)(3^7\pi^2 Z^5/16A^2)^{1/3}$, $E_n = (\hbar^2/10m_n r_0^2)[3^7\pi^2(A - Z)^5/16A^2]^{1/3}$.

CHAPTER 46

1. 2.5 fm.
3. $q = p_1 + \frac{1}{2}Mc$. $(4p_1 + Mc)/(2p_1 + Mc)$ increase.
5. $\theta \simeq 0$, and $q \simeq p$.
7. (a) $\sigma(\theta) = Z_1^2 Z_2^2(e^2/4\pi\epsilon_0)^2[4m^2/(\Delta^2)^2]$ (b) σ would be smaller than for a point charge.

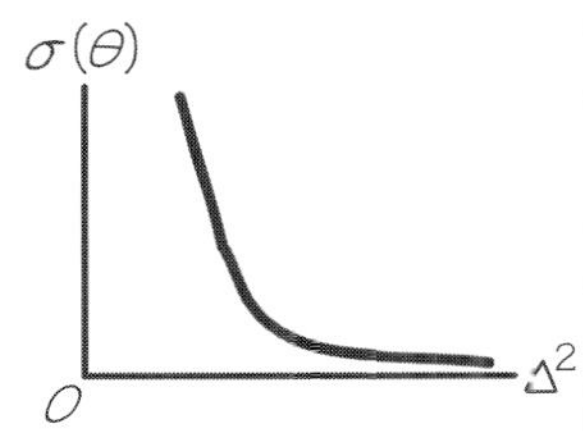

9. 3.44 TeV.
11. Charge: +1; Baryon number: 0; Lepton number: 0.
13. Reaction does not occur.
15. 2.5×10^{-3} fm.
17. (a) 0.26 fm; (b) 0.27 GeV.
19. $(u\bar{u})$ or $(d\bar{d})$ or a combination.
21.
(a)

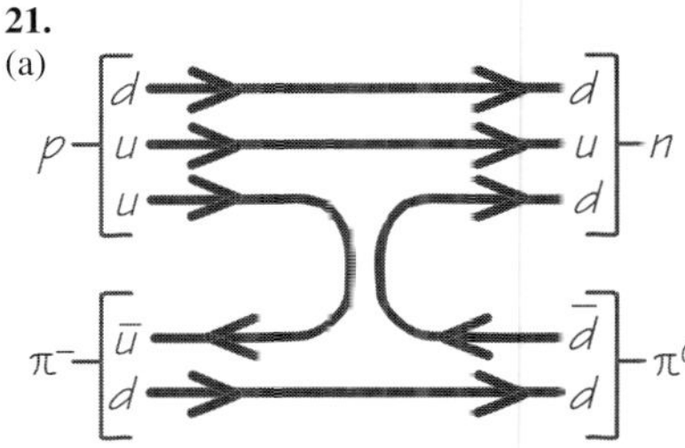

21.
(b)

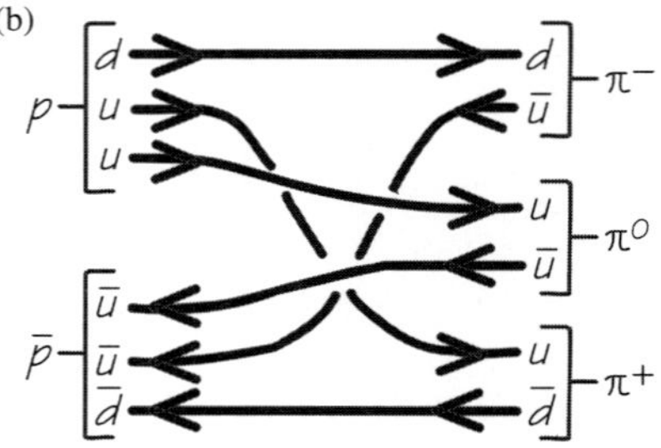

21.
(c)

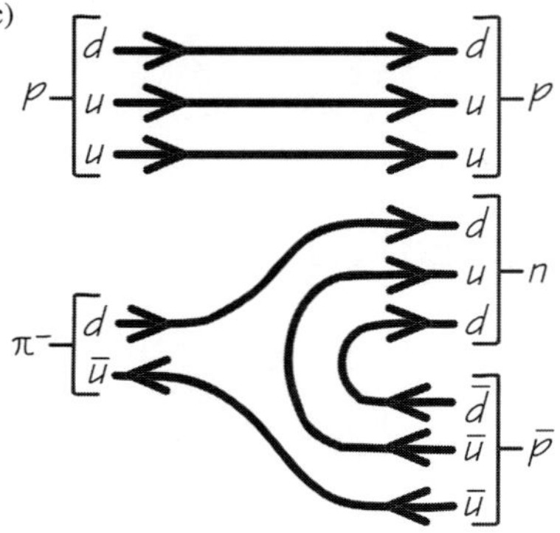

23. 460 MeV.
25. 11.9 GeV/c.
27. 8.1×10^9 MeV.
29. $9.5 \times 10^4\ s^{-1}$.
31. 3.75 GeV.
33. 1.6 GeV, 180°.
35. $H = n/t$.
37. 1.2×10^9 yr.
41. $t = (\hbar G/c^5)^{1/2}$.
45. 1.3×10^5 K, 6.3×10^5 K.
47. 1.1×10^2 km.
49. 570 mesons (285 pairs).
55. $\simeq$ 1:17.5.
57. (a) 1.25×10^4 eV; (b) 58 fm.

Index

Note: Italics indicate a definition or primary entry for multiple entries, where applicable.

Photo Credits

CHAPTER 1 **CO** David Madison Photography **1–1** Hale Observatories/Photo Researchers **1–2a** State Department of Public Safety **1–2b** Focus on Sports **1–2c** Patricia McDonough **1–3a** Tom Cogill **1–4** John G. Ross/Photo Researchers **1–5** National Christmas Tree Association **1–24** Photofest

CHAPTER 2 **CO** Lawrence Livermore **2–1** David Madison Photography **2–10** Robert Tringali/Sportschrome **2–13** Courtesy of S. T. Thornton **2–14a** Richard Megna/Fundamental Photographs **2–14b** Richard Megna/Fundamental Photographs **2–19** Richard Megna/Fundamental Photographs **AB–1** Bard Martin/The Image Bank **AB–2** Los Alamos National Laboratory

CHAPTER 3 **CO** The Stock Market **3–3** Ellis Herwig/Stock Boston **3–9** Richard Megna/Fundamental Photographs **3–10** Richard Megna/Fundamental Photographs **3–15** Richard Megna/Fundamental Photographs **3–16a** Tom Cogill **AB–1** NASA Headquarters

CHAPTER 4 **CO** Howard Dratch/The Image Works **4–1** Gregory K. Scott/Photo Researchers **4–4** Bernard Asset/Agence Vandystadt/Photo Researchers **4–8** Dan Guravich/Photo Researchers **4–9** Jerald Fish/Tony Stone Images **4–13a** Paul Silverman/Fundamental Photographs **4–13b** Paul Silverman/Fundamental Photographs **4–14a** Tom Cogill **4–14b** Tom Cogill **4–15** NASA Headquarters **4–24** John Dewaele/Stock Boston **4–25** NASA Headquarters

CHAPTER 5 **CO** NASA Headquarters **5–1** Martin Miller/Positive Images **5–11** Richard Megna/Fundamental Photographs **5–14** The Granger Collection **5–16a** Tom Cogill **5–17** David Madison Photography **5–19a** Paul Silverman/Fundamental Photographs **5–19b** Paul Silverman/Fundamental Photographs **5–25** Photofest **5–28** Richard Megna/Fundamental Photographs **AB–1** S. T. Thornton

CHAPTER 6 **CO** Holt Confer/Grant Heilman Photography **6–1a** Daemmrich/Stock Boston **6–1b** Daemmrich/Stock Boston **6–2** Movie Star News **6–5** Grant Heilman Photography **6–16** Rich Iwasak//Tony Stone Images **6–22** David Madison Photography **6–23** The Granger Collection **AB–1** Aram Gesar/The Image Bank **AB–1** Tom Cogill

CHAPTER 7 **CO** Monkmeyer/Van Etten **7–1** F.O.S. Inc. **7–2** Anthony A. Boccaccio/The Image Bank **7–3** Richard & Mary Magruder/Aristock, Inc **7–9** S. T. Thornton **7–12** Andy Levin/Photo Researchers **7–15** D.O.E./Science Source/Photo Researchers **AB–1** Creative Services/Virginia Power, Richmond, VA

CHAPTER 8 **CO** Otto Greule/Allsport **8–4** Joe Strunk/Loren Winters/NCSSM **8–10** Andy Hayt/Focus on Sports **8–24** Martha Swope/Time Picture Syndication **8–30** NASA Headquarters **AB–1** Manitowoc Engineering Company

CHAPTER 9 **CO** Harald Sund/The Image Bank **9–10** The Image Works **9–18** Tom Cogill **9–22** Richard Megna/Fundamental Photographs **9–30** Richard Megna/Fundamental Photographs **9–32** Thomas J. Cutitta/International Imaging **9–34** Richard Megna/Fundamental Photographs **AB–1** Alvis Upitis/The Image Bank

CHAPTER 10 **CO** Daniel Forster/Duomo Photography **10–10** Claudia Dhimitri/Picture Cube **10–20a** Gerard Lacz/NHPA **10–20b** (top) Focus on Sports **10–20b** (bottom) Blake/Sell/Reuters/Bettmann **10–21c** Tom Cogill **10–21d** Tom Cogill **10–26a** Thomas J. Cutitta/International Imaging **10–27** International Imaging **AB–1** Honeywell

CHAPTER 11 **CO** Patrick Piel/Gamma-Liaison **11–1** Addison Geary/Stock Boston **11–4** Richard Megna/Fundamental Photographs **11–4** Tom Cogill **11–5** Mary Ann Hemphill/Photobank **11–7a** Paul Silverman/Fundamental Photographs **11–7b** Paul Silverman/Fundamental Photographs **11–8** Rafael Macia **11–11** K. Bressert/Photobank **11–14a** Tom Cogill **AB–1** Simon Wilkinson/Stock Photos Unlimited

CHAPTER 12 **CO** NASA Headquarters **12–5** The Granger Collection **12–7b** S. T. Thornton **12–9a** NASA Headquarters **12–9b** NASA Headquarters **12–10b** National Optical Astronomy Observatories **12–12** NASA Headquarters **12–14** Charles M. Duke, Jr./NASA Headquarters **12–20a** John Elk III/Stock Boston **12–20b** John Elk III/Stock Boston **12–26** NRAO/AURA

CHAPTER 13 **CO** Paul Sakuma/AP/Wide World Photos **13–5** Chris Sorenson/The Stock Market **13–10** Richard Megna/Fundamental Photographs **13–14** Kuempel Chime Clock Works and Studio **13–21a** Farquharson, Neg. No. 4/University of Washington Libraries, Special Collections Division **13–21b** Farquharson, Neg. No. 4/University of Washington Libraries, Special Collections Division **13–22a** David Brownell/The Image Bank **AB–1** Swiss Time Promotions

CHAPTER 14 **CO** Arvind Garg/Photo Researchers **14–5** Paul Silverman/Fundamental Photographs **14–7** Used with permission of National Film Board of Canada **14–8** Dohrn/Science Photo Library/Photo Researchers **14–11** Susan Van Etten/Monkmeyer Press **14–14** Mitch Diamond/Photobank **14–21** Ben Rose/The Image Bank **14–27** J. Kim Vandiver **14–28** Energy Technology Visuals Collection **14–29** Jim Kahnweiler/Positive Images

CHAPTER 15 **CO** Dr. John S. Shelton **15–3** Runk/Schoenberger/Grant Heilman Photography **15–5** Thomas J. Cutitta/International Imaging **15–8** Dr. E. R. Degginger **15–12a** Thomas J. Cutitta/International Imaging **15–12b** Thomas J. Cutitta/International Imaging **15–21** Howard Sochurek/Medical Images **AB–1** Telex Communications

CHAPTER 16 **CO** Bob Woodward/The Stock Market **16–5** Michael Grecco/Stock Boston **16–6** Tom Cogill **16–9** Tom Cogill **16–13a** Bob Burch/Photobank **16–13b** Tom Cogill **16–14a** Tom Cogill **16–14b** Tom Cogill **16–14c** Tom Cogill **16–17** Kevork Djansezian/AP/Wide World Photos **16–18** Francoise Sauze/Science Photo Library/Photo Researchers **16–19** NASA Headquarters **16–20** PH Archives **16–21** Diane Schiumo/Fundamental Photographs **16–25a** Tom Cogill **16–25b** Tom Cogill **16–26** Cenco **16–28b** Tom Cogill **16–33a** Tony Freeman/Photo Edit **16–33b** NASA Headquarters **AB–1** Richard Megna/Fundamental Photographs

CHAPTER 17 **CO** NASA Headquarters **17–1** Glass/Monkmeyer Press **17–2a** Photo Researchers **17–2b** Tom Cogill **17–2c** Adam Hart-Davis/Science Photo Library/Photo Researchers **17–2d** Raytek **17–4a** Tom Cogill **17–4b** Tom Cogill **17–4c** Tom Cogill **17–12** A. G. Webster/AIP Emilio Segre Visual Archives **17–13** NASA/Science Source/Photo Researchers **17–14** AT&T Archives **17–16** NASA Headquarters **AB–1** S. T. Thornton

CHAPTER 18 **CO** Roger Ressmeyer/Starlight **18–4** Jeff Greenberg/Photo Researchers **18–6a** Grant Heilman/Grant Heilman Photography **18–6b** Hans Pfletschinger/Peter Arnold, Inc. **18–6c** Richard Megna/Fundamental Photographs **18–8** Courtesy of S. T. Thornton **18–9** Richard Megna/Fundamental Photographs **18–10** The Granger Collection **18–18** Homelite **AB–1** Cummings

CHAPTER 19 **CO** Ben Simmons/The Stock Market **19–1a** Tom Cogill **19–1b** Tom Cogill **19–5** Masterson/Photobank **19–14a** Paul Silverman//Fundamental Photographs **19–14b** Paul Silverman//Fundamental Photographs **19–14c** Paul Silverman//Fundamental Photographs

CHAPTER 20 **CO** Garry D. McMichael/Photo Researchers **20–8** Larry LeFever/Grant Heilman Photography **20–11** The Carrier Corporation **20–14a** Richard Megna/Fundamental Photographs **20–14b** AP/Wide World Photos **20–16** Richard Megna/Fundamental Photographs

CHAPTER 21 **CO** Tom Cogill **21–3** Herve Berthoule/Photo Researchers **21–4a** PH Archives **21–4b** PH Archives **21–5a** PH Archives **21–5b** PH Archives **21–8** G. Haling/Science Source/Photo Researchers **21–14a** Tom Cogill **21–14b** Tom Cogill **21–14c** Tom Cogill **21–15** Daemmrich/The Image Works **21–16** Tom Cogill

CHAPTER 22 **CO** The Granger Collection **22–1** S. T. Thornton **22–2a** Tom Cogill **22–4** The British Library **22–7a** Cenco **22–8** Brookhaven National Laboratory **AB–1** Grunnitus/Monkmeyer Press

CHAPTER 23 **CO** Tom Cogill **23–3a** Harold M. Waage **23–9a** Tom Cogill/S. T. Thornton **23–11a** Harold M. Waage **23–12a** Harold M. Waage **AB–1** Honeywell

CHAPTER 24 **CO** Peter Menzel/Stock Boston **24–4a** Harold M. Waage **24–6** Jeffery L. Rotoan/Peter Arnold, Inc. **AB–1** Tom Pantages

CHAPTER 25 **CO** Kent Wood/Photo Researchers, Inc. **25–23b** Harold M. Waage **25–24** Phil Jude/Science Photo Library/Photo Researchers, Inc. **25–25** Telegraph Colour Library/FPG International **25–26b** Lester V. Bergman & Associates, Inc. **25–28b** B. Loberg and H. Norden **25–29** Erwin Mueller/Science Photo Library/Photo Researchers, Inc. **25–32a** S. T. Thornton **25–33** Park Scientific Instruments **25–34** IBM Research

CHAPTER 26 **CO** Lawrence Livermore **26–2a** Tom Cogill/S. T. Thornton **26–4a** Harold M. Waage **26–5a** Archives/The Image Works **26–8** Leonard Lessin/Peter Arnold, Inc.

CHAPTER 27 **CO** Courtesy Lincoln Electric Co. **27–9** The Image Works **27–9b** Tom Pantages **27–11** S. T. Thornton **27–18b** Dan McCoy/Rainbow **27–19** Monkmeyer Press **AB–1** Tom Cogill

CHAPTER 28 **CO** Will Burgess Reuter/Bettman **28–1(1)** T. J. Florian/Rainbow **28–1(2), (3)** Tom Pantages **28–14** Thomas J. Cutitta/International Imaging Inc. **28–15a** Tom Cogill **28–16a** Tom Cogill **28–21c** International Imaging Inc. **28–24a, b** Tom Cogill

CHAPTER 29 **CO** Martin Marietta Corp. **29–2a** Richard Megna/Fundamental Photographs **29–2b** Werner H. Muller/Peter Arnold, Inc. **29–2c** Richard Megna/Fundamental Photographs **29–3a, b** International Imaging Inc. **29–12** Lawrence Berkeley Laboratory **29–13** Brookhaven National Laboratory/Science Photo Library/Photo Researchers, Inc. **29–14b** Lawrence Berkeley Laboratory **29–17b** The Granger Collection **29–25** Tom Cogill **29–26a, b** Tom Cogill

CHAPTER 30 **CO** Princeton University **30–1(1), (2)** Thomas J. Cutitta/International Imaging Inc. **30–2a, b, c** Tom Cogill **30–17a** Richard Megna/Fundamental Photographs **30–19** Tom Cogill/S. T. Thornton **30–26a** Richard Megna/Fundamental Photographs **AB–1** Tom Cogill/ S. T. Thornton

CHAPTER 31 **CO** Goodwin/Monkmeyer Press **31–20a** IBM Research/Peter Arnold, Inc. **31–29** H. Liverance/The Image Works **AB–1** John Broadhurst

CHAPTER 32 **CO** Wayne Decker/Fundamental Photographs **32–5c** Tony Buffa **32–6** Dr. John Unguris **32–12a** IBM **32–12b** IBM, Almaden Research Center **32–14** Donald Clegg **32–18a** Simon Fraser/Dept. of Neuroradiology/Photo Researchers, Inc. **32–18b** Mehau Kulyk/Science Photo Library/Photo Researchers, Inc.

CHAPTER 33 **CO** Thomas J. Cutitta/International Imaging, Inc. **33-2b** Cenco **33–4** Michael A. Gallitelli/Metroland Photo, Inc. **33–5** The Image Works **33–8** Tom Cogill/S. T. Thornton **AB–1** Delco Remy

CHAPTER 34 **CO** Tom Cogill/S. T. Thornton **34–1** Tom Cogill/S. T. Thornton **34–4a** Grant Heilman Photography **34–4b** John D. Cunningham/Visuals Unlimited **34–13a, b** Tom Pantages **34–16** Paul Silverman/Fundamental Photographs **34–21b, c, d** Tom Cogill/S. T. Thornton **AB–1** Bettman

CHAPTER 35 **CO** NASA Headquarters **35–7a** Ann Ronan Picture Library **35–8b** NASA Headquarters **35–12** Bruce Iverson **35–16** Tom Cogill **35–17a, b, c** Tom Cogill **35–21a, b** Leonard Lessin/Peter Arnold, Inc.

CHAPTER 36 **CO** Dieter Blum/Peter Arnold, Inc. **36–1** *Scientific American,* September 1968, page 55/PH Archives **36–4b** Courtesy of the Bancroft Library, University of California at Berkeley **36–6** Lawrence Livermore National Laboratory/Science Photo Library/Photo Researchers, Inc. **36–11b** NASA Headquarters **36–12** Tom Cogill **36–13** Tom Cogill **36–16b** Thomas J. Cutitta/International Imaging Inc. **36–18a** Cenco **36–22a** Grant Heilman Photography **36–25** Lucren Falco/Lucren Falco4 **AB–1** Bill Pierce/Rainbow

CHAPTER 37 **CO** Coco McCoy/Rainbow **37–5** Richard Megna/Fundamental Photographs **37–9c** Tom Cogill/S. T. Thornton **37–10d** Tom Cogill/S. T. Thornton **37–11b** Tom Cogill/ S. T. Thornton **37–12b** Tom Cogill/S. T. Thornton **37–13b** S. T. Thornton **37–14** M. C. Escher, "Hand with Reflecting Sphere," M. C. Escher Heirs, Cordon Art, Baarn, Holland **37–26** Richard Megna/Fundamental Photographs **37–37b** Richard Megna/Fundamental Photographs **37–38a** Olymbus America, Inc. **37–39** Tom Pantages **37–39a** Art Resource **37–40a** Science Photo Library/Photo Researchers, Inc. **37–41** Lightpath Technologies, Inc.

CHAPTER 38 **CO** Russell C. Hansen/Peter Arnold, Inc. **38–5** © by Springer-Verlag OHG, Berlin, 1962, from the *Atlas of Optical*/PH Archives **38–11b** Cenco **38–11c** Tom Pantages **38–13** Dr. E. R. Degginger **38–14** Paul Silverman/Fundamental Photographs **38–17b** Pasco Scientific **38–18a** Cenco **38–19a, b** Cenco

CHAPTER 39 **CO** Roger Ressmeyer/Corbis **39–1** PH Archives **39–3** Ken Kay/Fundamental Photographs **39–5a** P. M. Rinard/American Journal of Physics **39–5b** PH Archives **39–6** Johns Hopkins University **39–11** Springer-Verlag **39–15** Springer-Verlag **39–16a, b, c** PH Archives **39–18** Roger Ressmeyer/Corbis **39–20** Springer-Verlag **39–21** Photo Researchers, Inc. **39–23** Omikron/Photo Researchers, Inc. **39–24b** Burndy Library **39–26** International Imaging Inc. **39–33a, b, c** Yaov Levy/Phototake NYC

CHAPTER 40 **CO (left and right)** AP/Wide World Photos **40–4** Tom Pantages **40–11** AP/Wide World Photos **40–15** National Optical Astronomy Observatories **40–16** Bettman **40–24** Lawrence Berkeley Laboratory/Science Photo Library/Photo Researchers, Inc. **40–26b** NASA Headquarters **AB–1** David R. Frazier/Photo Researchers, Inc.

CHAPTER 41 **CO** Institute International de Physique Solvay/Courtesy AIP Niels Bohr Library **41–2a** AT&T Bell Laboratories **41–7** AIP Emilio Segre' Visual Archives **41–12** AIP Niels Bohr Library **41–15** AIP Niels Bohr Library **41–16a** AIP Niels Bohr Library **41–19** James King-Holmes/Science Photo Library/Photo Researchers, Inc. **AB–1a, 1b** Bear Advertising, Inc.

CHAPTER 42 **CO** Ken Eward/S. S./Photo Researchers, Inc. **42–1a(1), 1a(2), 1a(3)** Wolfgang Rueckner **42–1b(1), 1b(2), 1b(3)** Wolfgang Rueckner **42–7b** Dan McCoy/Rainbow **42–15** AIP Niels Bohr Library **42–22** Tom Cogill/S. T. Thornton

CHAPTER 43 **CO** Lawrence Livermore **43–2** AIP Niels Bohr Library **43–3** S. Heap/NASA Headquarters **43–8a** Richard Megna/Fundamental Photographs **43–9** Andy Levin/Photo Researchers, Inc. **43–10** Laboratory for Laser Energetics/University of Rochester **43–16** Conductus Inc. **43–17** St. Andrew's University

CHAPTER 44 **CO** IBM, Almaden Research Center **44–9** Jerome Wexler/Photo Researchers, Inc. **44–10** Minolta **44–16** University of Illinois **44–20b** IBM, Almaden Research Center **44–21** Texas Instruments **44–28** 3M **44–31** IBM **AB–2** NASA—Johnson Space Center

CHAPTER 45 **CO** Michael Gilbert/Science Photo Library/Photo Researchers, Inc. **45–1** AIP Emilio Segre' Visual Archives **45–11a** The Mansell Collection Limited **45–11b** AIP Niels Bohr Library **45–12** The Granger Collection **45–19** Tom D. White/Brill Atlanta/David L. Brill Photography **45–21** Walter Dickerman/Sandia National Laboratories **AB–1** Niton Corporation

CHAPTER 46 **CO** P. Loiez, Cern/Science Photo Library/Photo Researchers, Inc. **46–2** Joe Stancampiano/National Geographic Society **46–4** Science Kit and Boreal Laboratories **46–9** Fermilab Photo Dept./PH Archives **46–10** Fermilab Visual Media Services **46–11** C. Powell, P. Fowler & D. Perkins/Science Photo Library/Photo Researchers, Inc. **46–12a** Lawrence Berkeley Laboratory/Science Photo Library/Photo Researchers, Inc. **46–13** Fermilab Visual Media Services **46–15** Oxford Astrophysics **46–16** Fermilab Visual Media Services **46–17** Harvard-Smithsonian

TABLES IN THE TEXT

PROBLEM-SOLVING TECHNIQUES BOXES

SOME MATHEMATICAL CONSTANTS†

Constant	*Value*
π	3.14159
e (Euler's constant)	2.71828
$\sqrt{2}$	1.41421
$1/\sqrt{2}$	0.707107
ln(10)	2.30259
ln(2)	0.693147
1 rad	57.2958°
1°	0.0174533 rad

†To six significant figures.

THE GREEK ALPHABET

Alpha	A	α	Nu	N	ν
Beta	B	β	Xi	Ξ	ξ
Gamma	Γ	γ	Omicron	O	o
Delta	Δ	δ	Pi	Π	π
Epsilon	E	ϵ	Rho	P	ρ
Zeta	Z	ζ	Sigma	Σ	σ
Eta	H	η	Tau	T	τ
Theta	Θ	θ	Upsilon	Υ	υ
Iota	I	ι	Phi	Φ	ϕ
Kappa	K	κ	Chi	X	χ
Lambda	Λ	λ	Psi	Ψ	ψ
Mu	M	μ	Omega	Ω	ω

SOME USEFUL CONSTANTS

Constant	*Value*
Acceleration of gravity	$g = 9.81$ m/s^2
Density of air (STP)	1.29 kg/m^3
Specific heat of air at constant pressure	1.01×10^3 J/kg·K
Specific heat of air at constant volume	0.72×10^3 J/kg·K
Density of water (STP)	10^3 kg/m^3
Density of ice (STP)	0.917 kg/m^3
Speed of sound in air (STP)	331 m/s
Average range of audible frequencies	20 Hz–16,000 Hz
Index of refraction of water	1.33
Range of wavelengths for visible light	380 nm–750 nm
Resistivities of typical conductors	10^{-8}–10^{-6} Ω·m
Resistivities of typical insulators	10^{9}–10^{14} Ω·m